Pearson New International Edition

Campbell Biology
Concepts & Connections
J. Reece M. Taylor E. Simon J. Dickey
Seventh Edition

Pearson Education Limited
Edinburgh Gate
Harlow
Essex CM20 2JE
England and Associated Companies throughout the world

Visit us on the World Wide Web at: www.pearsoned.co.uk

PEARSON

ISBN 10: 1-292-02635-9
ISBN 13: 978-1-292-02635-0

British Library Cataloguing-in-Publication Data
A catalogue record for this book is available from the British Library

Printed in the United States of America

Table of Contents

Glossary

A

A site One of two of a ribosome's binding sites for tRNA during translation. The A site holds the tRNA that carries the next amino acid in the polypeptide chain. (A stands for aminoacyl tRNA.)
abiotic factor (ā′-bī-ot′-ik) A nonliving component of an ecosystem, such as air, water, or temperature.
abiotic reservoir The part of an ecosystem where a chemical, such as carbon or nitrogen, accumulates or is stockpiled outside of living organisms.
ABO blood groups Genetically determined classes of human blood that are based on the presence or absence of carbohydrates A and B on the surface of red blood cells. The ABO blood group phenotypes, also called blood types, are A, B, AB, and O.
abscisic acid (ABA) (ab-sis′-ik) A plant hormone that inhibits cell division, promotes dormancy, and interacts with gibberellins in regulating seed germination.
absorption The uptake of small nutrient molecules by an organism's own body; the third main stage of food processing, following digestion.
acetyl CoA (a-sē′til-kō′a′) (acetyl coenzyme A) The entry compound for the citric acid cycle in cellular respiration; formed from a fragment of pyruvate attached to a coenzyme.
acetylcholine (a-sē′-til-kō′lēn) A nitrogen-containing neurotransmitter. Among other effects, it slows the heart rate and makes skeletal muscles contract.
achondroplasia (uh-kon′-druh-plā′-zhuh) A form of human dwarfism caused by a single dominant allele; the homozygous condition is lethal.
acid A substance that increases the hydrogen ion (H^+) concentration in a solution.
acid precipitation Rain, snow, or fog that is more acidic than pH 5.2.
acrosome (ak′-ruh-som) A membrane-enclosed sac at the tip of a sperm. The acrosome contains enzymes that help the sperm penetrate an egg.
actin A globular protein that links into chains, two of which twist helically around each other, forming microfilaments in muscle cells.
action potential A change in membrane voltage that transmits a nerve signal along an axon.
activation energy The amount of energy that reactants must absorb before a chemical reaction will start.
activator A protein that switches on a gene or group of genes.
active immunity Immunity conferred by recovering from an infectious disease or by receiving a vaccine.
active site The part of an enzyme molecule where a substrate molecule attaches (by means of weak chemical bonds); typically, a pocket or groove on the enzyme's surface.
active transport The movement of a substance across a biological membrane against its concentration gradient, aided by specific transport proteins and requiring an input of energy (often as ATP).
adaptation An inherited characteristic that enhances an organism's ability to survive and reproduce in a particular environment.
adaptive immunity A vertebrate-specific defense that is activated only after exposure to an antigen and is mediated by lymphocytes. It exhibits specificity, memory, and self–nonself recognition. Also called acquired immunity.
adaptive radiation Period of evolutionary change in which groups of organisms form many new species whose adaptations allow them to fill new or vacant ecological roles in their communities.
adenine (A) (ad′-uh-nēn) A double-ring nitrogenous base found in DNA and RNA.
adhesion The attraction between different kinds of molecules.
adipose tissue A type of connective tissue whose cells contain fat.
adrenal cortex (uh-drē′-nul) The outer portion of an adrenal gland, controlled by ACTH from the anterior pituitary; secretes hormones called glucocorticoids and mineralocorticoids.
adrenal gland One of a pair of endocrine glands, located atop each kidney in mammals, composed of an outer cortex and a central medulla.
adrenal medulla (uh-drē′-nul muh-dul′-uh) The central portion of an adrenal gland, controlled by nerve signals; secretes the fight-or-flight hormones epinephrine and norepinephrine.
adrenocorticotropic hormone (ACTH) (uh-drē′-nō-cōr′-ti-kō-trop′-ik) A protein hormone secreted by the anterior pituitary that stimulates the adrenal cortex to secrete corticosteroids.
adult stem cell A cell present in adult tissues that generates replacements for nondividing differentiated cells. Adult stem cells are capable of differentiating into multiple cell types, but they are not as developmentally flexible as embryonic stem cells.
age structure The relative number of individuals of each age in a population.
agonistic behavior (a′-gō-nis′-tik) Confrontational behavior involving a contest waged by threats, displays, or actual combat that settles disputes over limited resources, such as food or mates.
AIDS (acquired immunodeficiency syndrome) The late stages of HIV infection, characterized by a reduced number of T cells and the appearance of characteristic opportunistic infections.
alcohol fermentation Glycolysis followed by the reduction of pyruvate to ethyl alcohol, regenerating NAD^+ and releasing carbon dioxide.
alga (al′-guh) (plural, **algae**) A protist that produces its food by photosynthesis.
alimentary canal (al′-uh-men′-tuh-rē) A complete digestive tract consisting of a tube running between a mouth and an anus.
allantois (al′-an-tō′-is) In animals, an extraembryonic membrane that develops from the yolk sac. The allantois helps dispose of the embryo's nitrogenous wastes and forms part of the umbilical cord in mammals.
allele (uh-lē′-ul) An alternative version of a gene.
allergen (al′-er-jen) An antigen that causes an allergy.
allergy A disorder of the immune system caused by an abnormally high sensitivity to an antigen. Symptoms are triggered by histamines released from mast cells.
allopatric speciation The formation of new species in populations that are geographically isolated from one another.
alternation of generations A life cycle in which there is both a multicellular diploid form, the sporophyte, and a multicellular haploid form, the gametophyte; a characteristic of plants and multicellular green algae.
alternative RNA splicing A type of regulation at the RNA-processing level in which different mRNA molecules are produced from the same primary transcript, depending on which RNA segments are treated as exons and which as introns.
altruism (al′-trū-iz-um) Behavior that reduces an individual's fitness while increasing the fitness of another individual.
alveolus (al-vē′-oh-lus) (plural, **alveoli**) One of the dead-end air sacs within the mammalian lung where gas exchange occurs.
Alzheimer's disease (AD) An age-related dementia (mental deterioration) characterized by confusion, memory loss, and other symptoms.
amino acid (uh-mēn′-ō) An organic molecule containing a carboxyl group and an amino group; serves as the monomer of proteins.
amino-acid-derived hormone A regulatory chemical consisting of a protein, peptide (a short polypeptide), or amine (modified amino acids). Amino-acid-derived hormones are water soluble (but not lipid soluble).
amino group A chemical group consisting of a nitrogen atom bonded to two hydrogen atoms.

ammonia NH_3; A small and very toxic nitrogenous waste produced by metabolism.
amniocentesis (am′-nē-ō-sen-tē′-sis) A technique for diagnosing genetic defects while a fetus is in the uterus. A sample of amniotic fluid, obtained by a needle inserted into the uterus, is analyzed for telltale chemicals and defective fetal cells.
amnion (am′-nē-on) In vertebrate animals, the extraembryonic membrane that encloses the fluid-filled amniotic sac containing the embryo.
amniote Member of a clade of tetrapods that have an amniotic egg containing specialized membranes that protect the embryo. Amniotes include mammals and birds and other reptiles.
amniotic egg (am′-nē-ot′-ik) A shelled egg in which an embryo develops within a fluid-filled amniotic sac and is nourished by yolk. Produced by reptiles (including birds) and egg-laying mammals, the amniotic egg enables them to complete their life cycles on dry land.
amoeba (uh-mē′-buh) A general term for a protist that moves and feeds by means of pseudopodia.
amoebocyte (uh-mē′-buh-sīt) An amoeba-like cell that moves by pseudopodia and is found in most animals; depending on the species, may digest and distribute food, dispose of wastes, form skeletal fibers, fight infections, and change into other cell types.
amoebozoan A member of a clade of protists in the supergroup Unikonta that includes amoebas and slime molds and is characterized by lobe-shaped pseudopodia.
amphibian Member of a clade of tetrapods that includes frogs, toads, salamanders, and cecilians.
amygdala (uh-mig′-duh-la) An integrative center of the cerebrum; functionally, the part of the limbic system that seems central in recognizing the emotional content of facial expressions and laying down emotional memories.
anabolic steroid (an′-uh-bol′-ik ster′-oyd) A synthetic variant of the male hormone testosterone that mimics some of its effects.
analogy The similarity between two species that is due to convergent evolution rather than to descent from a common ancestor with the same trait.
anaphase The fourth stage of mitosis, beginning when sister chromatids separate from each other and ending when a complete set of daughter chromosomes arrives at each of the two poles of the cell.
anatomy The study of the structures of an organism.
anchorage dependence The requirement that to divide, a cell must be attached to a solid surface.
androgen (an′-drō-jen) A steroid sex hormone secreted by the gonads that promotes the development and maintenance of the male reproductive system and male body features.
anemia (uh-nē′-me-ah) A condition in which an abnormally low amount of hemoglobin or a low number of red blood cells results in the body cells receiving too little oxygen.
angiosperm (an′-jē-ō-sperm) A flowering plant, which forms seeds inside a protective chamber called an ovary.
annelid (uh-nel′-id) A segmented worm. Annelids include earthworms, polychaetes, and leeches.
annual A plant that completes its life cycle in a single year or growing season.
antagonistic hormones Two hormones that have opposite effects.
anterior Pertaining to the front, or head, of a bilaterally symmetric animal.
anterior pituitary (puh-tū′-uh-tār-ē) An endocrine gland, adjacent to the hypothalamus and the posterior pituitary, that synthesizes several hormones, including some that control the activity of other endocrine glands.
anther A sac located at the tip of a flower's stamen; contains male sporangia in which meiosis occurs to produce spores that form the male gametophytes, or pollen grains.
anthropoid (an′-thruh-poyd) A member of a primate group made up of the apes (gibbons, orangutans, gorillas, chimpanzees, bonobos, and humans) and monkeys.
antibody (an′-tih-bod′-ē) A protein dissolved in blood plasma that attaches to a specific kind of antigen and helps counter its effects.
anticodon (an′-tī-kō′-don) On a tRNA molecule, a specific sequence of three nucleotides that is complementary to a codon triplet on mRNA.
antidiuretic hormone (ADH) (an′-tē-dī′-yū-ret′-ik) A hormone made by the hypothalamus and secreted by the posterior pituitary that promotes water retention by the kidneys.
antigen (an′-tuh-jen) A foreign (nonself) molecule that elicits an adaptive immune response.
antigen receptor A transmembrane version of an antibody molecule that B cells and T cells use to recognize specific antigens; also called a membrane antibody.
antigen-binding site A region of the antibody molecule responsible for the antibody's recognition and binding function.
antigenic determinant A region on the surface of an antigen molecule to which an antibody binds.
antigen-presenting cell (APC) One of a family of white blood cells (for example, a macrophage) that ingests a foreign substance or a microbe and attaches antigenic portions of the ingested material to its own surface, thereby displaying the antigens to a helper T cell.
antihistamine (an′-tē-his′-tuh-mēn) A drug that interferes with the action of histamine, providing temporary relief from an allergic reaction.
anus The opening through which undigested materials are expelled.
aorta (ā-or′-tuh) A large artery that conveys blood directly from the left ventricle of the heart to other arteries.
aphotic zone (ā-fō′-tik) The region of an aquatic ecosystem beneath the photic zone, where light does not penetrate enough for photosynthesis to take place.
apical dominance (ā′-pik-ul) In a plant, the hormonal inhibition of axillary buds by a terminal bud.
apical meristem (ā′-pik-ul mer′-uh-stem) Plant tissue made up of undifferentiated cells located at the tip of a plant root or in the terminal or axillary bud of a shoot. Apical meristems enable roots and shoots to grow in length.
apoptosis (ā-puh-tō′-sus) The timely and tidy suicide of cells; also called programmed cell death.
appendicular skeleton (ap′-en-dik′-yū-ler) Components of the skeletal system that support the fins of a fish or the arms and legs of a land vertebrate; in land vertebrates, the cartilage and bones of the shoulder girdle, pelvic girdle, forelimbs, and hind limbs. *See also* axial skeleton.
appendix (uh-pen′-dix) A small, finger-like extension of the vertebrate cecum; contains a mass of white blood cells that contribute to immunity.
aquaporin A transport protein in the plasma membrane of some plant or animal cells that facilitates the diffusion of water across the membrane (osmosis).
aqueous humor (ā′-kwē-us hyū′-mer) Plasma-like liquid in the space between the lens and the cornea in the vertebrate eye; helps maintain the shape of the eye, supplies nutrients and oxygen to its tissues, and disposes of its wastes.
aqueous solution (ā′-kwā-us) A solution in which water is the solvent.
arachnid A member of a major arthropod group (chelicerates) that includes spiders, scorpions, ticks, and mites.
Archaea (ar′-kē-uh) One of two prokaryotic domains of life, the other being Bacteria.
Archaeplastida One of five supergroups proposed in a current hypothesis of the evolutionary history of eukaryotes. The other four supergroups are Chromalveolata, Rhizaria, Excavata, and Unikonta.
arteriole (ar-ter′-ē-ōl) A vessel that conveys blood between an artery and a capillary bed.
artery A vessel that carries blood away from the heart to other parts of the body.
arthropod (ar′-thrō-pod) A member of the most diverse phylum in the animal kingdom. Arthropods include the horseshoe crab, arachnids (for example, spiders, ticks, scorpions, and mites), crustaceans (for example, crayfish, lobsters, crabs, and barnacles), millipedes, centipedes, and insects. Arthropods are characterized by a chitinous exoskeleton, molting, jointed appendages, and a body formed of distinct groups of segments.

artificial selection The selective breeding of domesticated plants and animals to promote the occurrence of desirable traits.
ascomycete Member of a group of fungi characterized by saclike structures called asci that produce spores in sexual reproduction.
asexual reproduction The creation of genetically identical offspring by a single parent, without the participation of sperm and egg.
assisted reproductive technology Procedure that involves surgically removing eggs from a woman's ovaries, fertilizing them, and then returning them to the woman's body. *See also in vitro* fertilization.
association areas Sites of higher mental activities, making up most of the cerebral cortex.
associative learning Learning that a particular stimulus or response is linked to a reward or punishment; includes classical conditioning and trial-and-error learning.
astigmatism (uh-stig′-muh-tizm) Blurred vision caused by a misshapen lens or cornea.
atherosclerosis (ath′-uh-rō′-skluh-rō′-sis) A cardiovascular disease in which fatty deposits called plaques develop on the inner walls of the arteries, narrowing their inner diameters.
atom The smallest unit of matter that retains the properties of an element.
atomic mass The total mass of an atom; also called atomic weight. Given as a whole number, the atomic mass approximately equals the mass number.
atomic number The number of protons in each atom of a particular element.
ATP Adenosine triphosphate, the main energy source for cells.
ATP synthase A cluster of several membrane proteins that function in chemiosmosis with adjacent electron transport chains, using the energy of a hydrogen ion concentration gradient to make ATP.
atrium (ā′-trē-um) (plural, **atria**) A heart chamber that receives blood from the veins.
auditory canal Part of the vertebrate outer ear that channels sound waves from the pinna or outer body surface to the eardrum.
autoimmune disease An immunological disorder in which the immune system attacks the body's own molecules.
autonomic nervous system (ot′-ō-nom′-ik) The component of the vertebrate peripheral nervous system that regulates the internal environment; made up of sympathetic and parasympathetic subdivisions. Most actions of the autonomic nervous system are involuntary.
autosome A chromosome not directly involved in determining the sex of an organism; in mammals, for example, any chromosome other than X or Y.
autotroph (ot′-ō-trōf) An organism that makes its own food (often by photosynthesis), thereby sustaining itself without eating other organisms or their molecules. Plants, algae, and numerous bacteria are autotrophs.
auxin (ok′-sin) A plant hormone (indoleacetic acid or a related compound) whose chief effect is to promote seedling elongation.
AV (atrioventricular) node A region of specialized heart muscle tissue between the left and right atria where electrical impulses are delayed for about 0.1 second before spreading to both ventricles and causing them to contract.
axial skeleton (ak′-sē-ul) Components of the skeletal system that support the central trunk of the body: the skull, backbone, and rib cage in a vertebrate. *See also* appendicular skeleton.
axillary bud (ak′-sil-ār-ē) An embryonic shoot present in the angle formed by a leaf and stem.
axon (ak′-son) A neuron extension that conducts signals to another neuron or to an effector cell. A neuron has one long axon.

B

B cell A type of lymphocyte that matures in the bone marrow and later produces antibodies. B cells are responsible for the humoral immune response.
bacillus (buh-sil′-us) (plural, **bacilli**) A rod-shaped prokaryotic cell.
Bacteria One of two prokaryotic domains of life, the other being Archaea.
bacteriophage (bak-tēr′-ē-ō-fāj) A virus that infects bacteria; also called a phage.
balancing selection Natural selection that maintains stable frequencies of two or more phenotypic forms in a population.
ball-and-socket joint A joint that allows rotation and movement in several planes. Examples in humans are the hip and shoulder joints.
bark All the tissues external to the vascular cambium in a plant that is growing in thickness. Bark is made up of secondary phloem, cork cambium, and cork.
Barr body A dense body formed from a deactivated X chromosome found in the nuclei of female mammalian cells.
basal metabolic rate (BMR) The number of kilocalories a resting animal requires to fuel its essential body processes for a given time.
basal nuclei (bā′-sul nū′-klē-ī) Clusters of nerve cell bodies located deep within the cerebrum that are important in motor coordination.
base A substance that decreases the hydrogen ion (H^+) concentration in a solution.
basidiomycete (buh-sid′-ē-ō-mī′sēt) Member of a group of fungi characterized by club-shaped, spore-producing structures called basidia.
basilar membrane The floor of the middle canal of the inner ear.
behavior Individually, an action carried out by the muscles or glands under control of the nervous system in response to a stimulus; collectively, the sum of an animal's responses to external and internal stimuli.
behavioral ecology The study of behavior in an evolutionary context.
benign tumor An abnormal mass of cells that remains at its original site in the body.
benthic realm A seafloor, or the bottom of a freshwater lake, pond, river, or stream.
biennial A plant that completes its life cycle in two years.
bilateral symmetry An arrangement of body parts such that an organism can be divided equally by a single cut passing longitudinally through it. A bilaterally symmetric organism has mirror-image right and left sides.
bilaterian Member of the clade Bilateria, animals exhibiting bilateral symmetry.
bile A mixture of substances that is produced by the liver and stored in the gallbladder. Bile emulsifies fats and aids in their digestion.
binary fission A means of asexual reproduction in which a parent organism, often a single cell, divides into two genetically identical individuals of about equal size.
binomial A two-part, latinized name of a species; for example, *Homo sapiens*.
biodiversity hot spot A small geographic area with an exceptional concentration of endangered and threatened species, especially endemic species (those found nowhere else).
biofilm A surface-coating colony of prokaryotes that engage in metabolic cooperation.
biogenic amine A neurotransmitter derived from an amino acid. Examples of biogenic amines are epinephrine and dopamine.
biogeochemical cycle Any of the various chemical circuits that involve both biotic and abiotic components of an ecosystem.
biogeography The study of the past and present distribution of organisms.
biological clock An internal timekeeper that controls an organism's biological rhythms, marking time with or without environmental cues but often requiring signals from the environment to remain tuned to an appropriate period. *See also* circadian rhythm.
biological control The intentional release of a natural enemy to attack a pest population.
biological magnification The accumulation of harmful chemicals that are retained in the living tissues of consumers in food chains.
biological species concept Definition of a species as a group of populations whose members have the potential to interbreed in nature and produce viable, fertile offspring, but do not produce viable, fertile offspring with members of other such populations.
biomass The amount, or mass, of organic material in an ecosystem.
biome (bī′-ōm) A major type of ecological association that occupies a broad geographic region of land or water and is characterized by organisms adapted to the particular environment.

bioremediation The use of living organisms to detoxify and restore polluted and degraded ecosystems.
biosphere The entire portion of Earth inhabited by life; the sum of all the planet's ecosystems.
biotechnology The manipulation of living organisms or their components to make useful products.
biotic factor (bī-o'-tik) A living component of a biological community; an organism, or a factor pertaining to one or more organisms.
bipolar disorder Depressive mental illness characterized by extreme mood swings; also called manic-depressive disorder.
birds Members of a clade of reptiles that have feathers and adaptations for flight.
birth control pill A chemical contraceptive that contains synthetic estrogen and/or progesterone (or a synthetic progesterone-like hormone called progestin) and prevents the release of eggs.
bivalve A member of a group of molluscs that includes clams, mussels, scallops, and oysters.
blastocoel (blas'-tuh-sēl) In a developing animal, a central, fluid-filled cavity in a blastula.
blastocyst (blas'-tō-sist) A mammalian embryo (equivalent to an amphibian blastula) made up of a hollow ball of cells that results from cleavage and that implants in the mother's endometrium.
blastula (blas'-tyū-luh) An embryonic stage that marks the end of cleavage during animal development; a hollow ball of cells in many species.
blood A type of connective tissue with a fluid matrix called plasma in which red blood cells, white blood cells, and platelets are suspended.
blood pressure The force that blood exerts against the walls of blood vessels.
blood-brain barrier A system of capillaries in the brain that restricts passage of most substances into the brain, thereby preventing large fluctuations in the brain's environment.
body cavity A fluid-containing space between the digestive tract and the body wall.
bolus A lubricated ball of chewed food.
bone A type of connective tissue consisting of living cells held in a rigid matrix of collagen fibers embedded in calcium salts.
bottleneck effect Genetic drift resulting from a drastic reduction in population size. Typically, the surviving population is no longer genetically representative of the original population.
Bowman's capsule A cup-shaped swelling at the receiving end of a nephron in the vertebrate kidney; collects the filtrate from the blood.
brain The master control center of the nervous system, involved in regulating and controlling body activity and interpreting information from the senses transmitted through the nervous system.
brainstem A functional unit of the vertebrate brain, composed of the midbrain, the medulla oblongata, and the pons; serves mainly as a sensory filter, selecting which information reaches higher brain centers.
breathing Ventilation of the lungs through alternating inhalation and exhalation.
breathing control center The part of the medulla in the brain that directs the activity of organs involved in breathing.
bronchiole (bron'-kē-ōl) A fine branch of the bronchi that transports air to alveoli.
bronchus (bron'-kus) (plural, **bronchi**) One of a pair of breathing tubes that branch from the trachea into the lungs.
brown alga One of a group of marine, multicellular, autotrophic protists belonging to the supergroup Chromalveolata; the most common and largest type of seaweed. Brown algae include the kelps.
bryophyte (brī'-uh-fīt) A plant that lacks xylem and phloem; a seedless nonvascular plant. Bryophytes include mosses, liverworts, and hornworts.
budding A means of asexual reproduction whereby a new individual develops from an outgrowth of a parent. The new individual eventually splits off and lives independently.
buffer A chemical substance that resists changes in pH by accepting hydrogen ions from or donating hydrogen ions to solutions.
bulbourethral gland (bul'-bō-yū-rē'-thrul) One of a pair of glands near the base of the penis in the human male that secrete a clear alkaline mucus.
bulk feeder An animal that eats relatively large pieces of food.

C

C_3 plant A plant that uses the Calvin cycle for the initial steps that incorporate CO_2 into organic material, forming a three-carbon compound as the first stable intermediate.
C_4 plant A plant that prefaces the Calvin cycle with reactions that incorporate CO_2 into four-carbon compounds, the end product of which supplies CO_2 for the Calvin cycle.
calcitonin (kal'-sih-tōn'-in) A peptide hormone secreted by the thyroid gland that lowers the blood calcium level.
Calvin cycle The second of two stages of photosynthesis; a cyclic series of chemical reactions that occur in the stroma of a chloroplast, using the carbon in CO_2 and the ATP and NADPH produced by the light reactions to make the energy-rich sugar molecule G3P.
CAM plant A plant that uses an adaptation for photosynthesis in arid conditions in which carbon dioxide entering open stomata during the night is converted to organic acids, which release CO_2 for the Calvin cycle during the day, when stomata are closed.
cancer A disease characterized by the presence of malignant tumors (rapidly growing and spreading masses of abnormal body cells) in the body.
capillary (kap'-il-er-ē) A microscopic blood vessel that conveys blood between an arteriole and a venule; enables the exchange of nutrients and dissolved gases between the blood and interstitial fluid.
capillary bed A network of capillaries in a tissue or organ.
capsid The protein shell that encloses a viral genome.
carbohydrate (kar'-bō-hī'-drāt) Member of the class of biological molecules consisting of single-monomer sugars (monosaccharides), two-monomer sugars (disaccharides), and polymers (polysaccharides).
carbon fixation The incorporation of carbon from atmospheric CO_2 into the carbon in organic compounds. During photosynthesis in a C_3 plant, carbon is fixed into a three-carbon sugar as it enters the Calvin cycle. In C_4 and CAM plants, carbon is fixed into a four-carbon sugar.
carbon skeleton The chain of carbon atoms that forms the structural backbone of an organic molecule.
carbonyl group (kar'-buh-nēl') A chemical group consisting of a carbon atom linked by a double bond to an oxygen atom.
carboxyl group (kar'-bok-sil) A chemical group consisting of a carbon atom double-bonded to an oxygen atom and also bonded to a hydroxyl group.
carcinogen (kar-sin'-uh-jin) A cancer-causing agent, either high-energy radiation (such as X-rays or UV light) or a chemical.
carcinoma (kar'-sih-nō'-muh) Cancer that originates in the coverings of the body, such as skin or the lining of the intestinal tract.
cardiac cycle (kar'-dē-ak) The alternating contractions and relaxations of the heart.
cardiac muscle A type of striated muscle that forms the contractile wall of the heart.
cardiac output The volume of blood pumped per minute by each ventricle of the heart.
cardiovascular disease (kar'-dē-ō-vas'-kyū-ler) Disorders of the heart and blood vessels.
cardiovascular system A closed circulatory system with a heart and a branching network of arteries, capillaries, and veins.
carnivore An animal that mainly eats other animals.
carpel (kar'-pul) The female part of a flower, consisting of a stalk with an ovary at the base and a stigma, which traps pollen, at the tip.
carrier An individual who is heterozygous for a recessively inherited disorder and who therefore does not show symptoms of that disorder but who may pass on the recessive allele to offspring.
carrying capacity In a population, the number of individuals that an environment can sustain.

cartilage (kar′-ti-lij) A flexible connective tissue consisting of living cells and collagenous fibers embedded in a rubbery matrix.

Casparian strip (kas-par′-ē-un) A waxy barrier in the walls of endodermal cells in a plant root that prevents water and ions from entering the xylem without crossing one or more cell membranes.

cation exchange A process in which positively charged minerals are made available to a plant when hydrogen ions in the soil displace mineral ions from the clay particles.

cecum (sē′-kum) (plural, **ceca**) A blind outpocket at the beginning of the large intestine.

cell A basic unit of living matter separated from its environment by a plasma membrane; the fundamental structural unit of life.

cell body The part of a cell, such as a neuron, that houses the nucleus.

cell cycle An ordered sequence of events (including interphase and the mitotic phase) that extends from the time a eukaryotic cell is first formed from a dividing parent cell until its own division into two cells.

cell cycle control system A cyclically operating set of proteins that triggers and coordinates events in the eukaryotic cell cycle.

cell division The reproduction of a cell through duplication of the genome and division of the cytoplasm.

cell plate A double membrane across the midline of a dividing plant cell, between which the new cell wall forms during cytokinesis.

cell theory The theory that all living things are composed of cells and that all cells come from other cells.

cell wall A protective layer external to the plasma membrane in plant cells, bacteria, fungi, and some protists; protects the cell and helps maintain its shape.

cell-mediated immune response The type of specific immunity brought about by T cells. The cell-mediated immune response fights body cells infected with pathogens. *See also* humoral immune response.

cellular metabolism (muh-tab′-uh-lizm) All the chemical activities of a cell.

cellular respiration The aerobic harvesting of energy from food molecules; the energy-releasing chemical breakdown of food molecules, such as glucose, and the storage of potential energy in a form that cells can use to perform work; involves glycolysis, the citric acid cycle, and oxidative phosphorylation (the electron transport chain and chemiosmosis).

cellular slime mold A type of protist that has unicellular amoeboid cells and aggregated reproductive bodies in its life cycle; a member of the amoebozoan clade.

cellulose (sel′-yū-lōs) A structural polysaccharide of plant cell walls composed of glucose monomers. Cellulose molecules are linked into cable-like fibrils.

centipede A carnivorous terrestrial arthropod that has one pair of long legs for each of its numerous body segments, with the front pair modified as poison claws.

central canal The narrow cavity in the center of the spinal cord that is continuous with the fluid-filled ventricles of the brain.

central nervous system (CNS) The integration and command center of the nervous system; the brain and, in vertebrates, the spinal cord.

central vacuole In a plant cell, a large membranous sac with diverse roles in growth and the storage of chemicals and wastes.

centralization The presence of a central nervous system (CNS) distinct from a peripheral nervous system.

centriole (sen′trē-ōl) A structure in an animal cell composed of cylinders of microtubule triplets arranged in a 9 + 0 pattern. An animal usually has a centrosome with a pair of centrioles involved in cell division.

centromere (sen′-trō-mēr) The region of a duplicated chromosome where two sister chromatids are joined (often appearing as a narrow "waist") and where spindle microtubules attach during mitosis and meiosis. The centromere divides at the onset of anaphase during mitosis and anaphase II during meiosis.

centrosome (sen′-trō-sōm) Material in the cytoplasm of a eukaryotic cell that gives rise to microtubules; important in mitosis and meiosis; also called the microtubule-organizing center.

cephalization (sef′-uh-luh-zā′-shun) An evolutionary trend toward concentration of the nervous system at the head end.

cephalopod A member of a group of molluscs that includes squids, cuttlefish, octopuses, and nautiluses.

cerebellum (sār′-ruh-bel′-um) Part of the vertebrate hindbrain; mainly a planning center that interacts closely with the cerebrum in coordinating body movement.

cerebral cortex (suh-rē′-brul kor′-teks) A folded sheet of gray matter forming the surface of the cerebrum. In humans, it contains integrating centers for higher brain functions such as reasoning, speech, language, and imagination.

cerebral hemisphere The right or left half of the vertebrate cerebrum.

cerebrospinal fluid (suh-rē′-brō-spī′-nul) Blood-derived fluid that surrounds, nourishes, and cushions the brain and spinal cord.

cerebrum (suh-rē′-brum) The largest, most sophisticated, and most dominant part of the vertebrate forebrain, made up of right and left cerebral hemispheres.

cervix (ser′-viks) The neck of the uterus, which opens into the vagina.

chaparral (shap′-uh-ral′) A biome dominated by spiny evergreen shrubs adapted to periodic drought and fires; found where cold ocean currents circulate offshore, creating mild, rainy winters and long, hot, dry summers.

character A heritable feature that varies among individuals within a population, such as flower color in pea plants or eye color in humans.

chelicerate (kē-lih-suh′-rāte) A lineage of arthropods that includes horseshoe crabs, scorpions, ticks, and spiders.

chemical bond An attraction between two atoms resulting from a sharing of outer-shell electrons or the presence of opposite charges on the atoms. The bonded atoms gain complete outer electron shells.

chemical cycling The use and reuse of a chemical element, such as carbon, within an ecosystem.

chemical energy Energy available in molecules for release in a chemical reaction; a form of potential energy.

chemical reaction The making and breaking of chemical bonds, leading to changes in the composition of matter.

chemiosmosis (kem′-ē-oz-mō′-sis) Energy-coupling mechanism that uses the energy of hydrogen ion (H^+) gradients across membranes to drive cellular work, such as the phosphorylation of ADP; powers most ATP synthesis in cells.

chemoautotroph An organism that obtains both energy and carbon from inorganic chemicals. A chemoautotroph makes its own organic compounds from CO_2 without using light energy.

chemoheterotroph An organism that obtains both energy and carbon from organic compounds.

chemoreceptor (kē′-mō-rē-sep′-ter) A sensory receptor that detects chemical changes within the body or a specific kind of molecule in the external environment.

chiasma (kī-az′-muh) (plural, **chiasmata**) The microscopically visible site where crossing over has occurred between chromatids of homologous chromosomes during prophase I of meiosis.

chitin (kī-tin) A structural polysaccharide found in many fungal cell walls and in the exoskeletons of arthropods.

chlamydia A member of a group of bacteria that live inside eukaryotic host cells. Chlamydias include human pathogens that cause blindness and nongonococcal urethritis, a common sexually transmitted disease.

chlorophyll A green pigment located within the chloroplasts of plants, algae, and certain prokaryotes. Chlorophyll *a* can participate directly in the light reactions, which convert solar energy to chemical energy.

chloroplast (klō′-rō-plast) An organelle found in plants and photosynthetic protists that absorbs sunlight and uses it to drive the synthesis of organic molecules (sugars) from carbon dioxide and water.

choanocyte (kō-an′-uh-sīt) A flagellated feeding cell found in sponges. Also called a collar cell, it has a collar-like ring that traps food particles around the base of its flagellum.

cholesterol (kō-les′-tuh-rol) A steroid that is an important component of animal cell membranes and that acts as a precursor molecule for the synthesis of other steroids, such as hormones.
chondrichthyan (kon-drik′-thē-an) Cartilaginous fish; member of a clade of jawed vertebrates with skeletons made mostly of cartilage, such as sharks and rays.
chorion (kō′r-ē-on) In animals, the outermost extraembryonic membrane, which becomes the mammalian embryo's part of the placenta.
chorionic villi (kōr′-ē-on′-ik vil′-us) Outgrowths of the chorion, containing embryonic blood vessels. As part of the placenta, chorionic villi absorb nutrients and oxygen from, and pass wastes into, the mother's bloodstream.
chorionic villus sampling (CVS) A technique for diagnosing genetic defects while the fetus is in an early development stage within the uterus. A small sample of the fetal portion of the placenta is removed and analyzed.
choroid (kōr′-oyd) A thin, pigmented layer in the vertebrate eye, surrounded by the sclera. The iris is part of the choroid.
Chromalveolata One of five supergroups proposed in a current hypothesis of the evolutionary history of eukaryotes. The other four supergroups are Rhizaria, Excavata,Unikonta, and Archaeplastida.
chromatin (krō′-muh-tin) The combination of DNA and proteins that constitutes eukaryotic chromosomes; often used to refer to the diffuse, very extended form taken by chromosomes when a cell is not dividing.
chromosome (krō′-muh-sōm) A threadlike, gene-carrying structure found in the nucleus of a eukaryotic cell and most visible during mitosis and meiosis; also, the main gene-carrying structure of a prokaryotic cell. A chromosomes consists of one very long piece of chromatin, a combination of DNA and protein.
chromosome theory of inheritance A basic principle in biology stating that genes are located on chromosomes and that the behavior of chromosomes during meiosis accounts for inheritance patterns.
chyme (kīm) The mixture of partially digested food and digestive juices formed in the stomach.
chytrid (kī-trid) Member of a group of fungi that are mostly aquatic and have flagellated spores. They probably represent the most primitive fungal lineage.
ciliate (sil′-ē-it) A type of protist that moves and feeds by means of cilia. Ciliates belong to the supergroup Chromalveolata.
cilium (plural, **cilia**) A short cellular appendage specialized for locomotion, formed from a core of nine outer doublet microtubules and two single microtubules (the 9 + 2 pattern) covered by the cell's plasma membrane.
circadian rhythm (ser-kā′-dē-un) In an organism, a biological cycle of about 24 hours that is controlled by a biological clock, usually under the influence of environmental cues; a pattern of activity that is repeated daily. *See also* biological clock.
circulatory system The organ system that transports materials such as nutrients, O_2, and hormones to body cells and transports CO_2 and other wastes from body cells.
citric acid cycle The chemical cycle that completes the metabolic breakdown of glucose molecules begun in glycolysis by oxidizing acetyl CoA (derived from pyruvate) to carbon dioxide. The cycle occurs in the matrix of mitochondria and supplies most of the NADH molecules that carry energy to the electron transport chains. Together with pyruvate oxidation, the second major stage of cellular respiration.
clade A group of species that includes an ancestral species and all its descendants.
cladistics (kluh-dis′-tiks) An approach to systematics in which common descent is the primary criterion used to classify organisms by placing them into groups called clades.
class In Linnaean classification, the taxonomic category above order.
cleavage (klē′-vij) (1) Cytokinesis in animal cells and in some protists, characterized by pinching in of the plasma membrane. (2) In animal development, the first major phase of embryonic development, in which rapid cell divisions without cell growth transforms the animal zygote into a ball of cells.
cleavage furrow The first sign of cytokinesis during cell division in an animal cell; a shallow groove in the cell surface near the old metaphase plate.
clitoris An organ in the female that engorges with blood and becomes erect during sexual arousal.
clonal selection (klōn′-ul) The production of a lineage of genetically identical cells that recognize and attack the specific antigen that stimulated their proliferation. Clonal selection is the mechanism that underlies the immune system's specificity and memory of antigens.
clone As a verb, to produce genetically identical copies of a cell, organism, or DNA molecule. As a noun, the collection of cells, organisms, or molecules resulting from cloning; colloquially, a single organism that is genetically identical to another because it arose from the cloning of a somatic cell.
closed circulatory system A circulatory system in which blood is confined to vessels and is kept separate from the interstitial fluid.
club fungus *See* basidiomycete.
clumped dispersion pattern A pattern in which the individuals of a population are aggregated in patches.
cnidarian (nī-dār′-ē-un) An animal characterized by cnidocytes, radial symmetry, a gastrovascular cavity, and a polyp and medusa body form. Cnidarians include the hydras, jellies, sea anemones, corals, and related animals.
cnidocyte (nī′-duh-sīt) A specialized cell for which the phylum Cnidaria is named; consists of a capsule containing a fine coiled thread, which, when discharged, functions in defense and prey capture.
coccus (kok′-us) (plural, **cocci**) A spherical prokaryotic cell.
cochlea (kok′-lē-uh) A coiled tube in the inner ear of birds and mammals that contains the hearing organ, the organ of Corti.
codominant Inheritance pattern in which a heterozygote expresses the distinct trait of both alleles.
codon (kō′-don) A three-nucleotide sequence in mRNA that specifies a particular amino acid or polypeptide termination signal; the basic unit of the genetic code.
coelom (sē′-lom) A body cavity completely lined with mesoderm.
coenzyme An organic molecule serving as a cofactor. Most vitamins function as coenzymes in important metabolic reactions.
coevolution Evolutionary change in which adaptations in one species act as a selective force on a second species, inducing adaptations that in turn act as a selective force on the first species; mutual influence on the evolution of two different interacting species.
cofactor A nonprotein molecule or ion that is required for the proper functioning of an enzyme. *See also* coenzyme.
cognition The process carried out by an animal's nervous system that includes perceiving, storing, integrating, and using the information obtained by the animal's sensory receptors.
cognitive map A representation, within an animal's nervous system, of spatial relations among objects in the animal's environment.
cohesion (kō-hē′-zhun) The sticking together of molecules of the same kind, often by hydrogen bonds.
collecting duct A tube in the vertebrate kidney that concentrates urine while conveying it to the renal pelvis.
collenchyma cell (kō-len′-kim-uh) In plants, a cell with a thick primary wall and no secondary wall, functioning mainly in supporting growing parts.
colon (kō′-lun) Large intestine; the portion of the vertebrate alimentary canal between the small intestine and the anus; functions mainly in water absorption and the formation of feces.
communication Animal behavior including transmission of, reception of, and response to signals.
community An assemblage of all the organisms living together and potentially interacting in a particular area.
companion cell In a plant, a cell connected to a sieve-tube element whose nucleus and ribosomes provide proteins for the sieve-tube element.
competitive inhibitor A substance that reduces the activity of an enzyme by binding to the enzyme's active site in place of the

substrate. A competitive inhibitor's structure mimics that of the enzyme's substrate.

complement system A family of innate defensive blood proteins that cooperate with other components of the vertebrate defense system to protect against microbes; can enhance phagocytosis, directly lyse pathogens, and amplify the inflammatory response.

complementary DNA (cDNA) A DNA molecule made *in vitro* using mRNA as a template and the enzyme reverse transcriptase. A cDNA molecule therefore corresponds to a gene but lacks the introns present in the DNA of the genome.

complete digestive tract A digestive tube with two openings, a mouth and an anus.

complete dominance A type of inheritance in which the phenotypes of the heterozygote and dominant homozygote are indistinguishable.

complete metamorphosis (met′-uh-mōr′-fuh-sis) A type of development in certain insects in which development from larva to adult is achieved by multiple molts that are followed by a pupal stage. While encased in its pupa, the body rebuilds from clusters of embryonic cells that have been held in reserve. The adult emerges from the pupa.

compost Decomposing organic material that can be used to add nutrients to soil.

compound A substance containing two or more elements in a fixed ratio. For example, table salt (NaCl) consists of one atom of the element sodium (Na) for every atom of chlorine (Cl).

compound eye The photoreceptor in many invertebrates; made up of many tiny light detectors, each of which detects light from a tiny portion of the field of view.

concentration gradient A region along which the density of a chemical substance increases or decreases. Cells often maintain concentration gradients of ions across their membranes. When a gradient exists, substances tend to move from where they are more concentrated to where they are less concentrated.

conception The fertilization of the egg by a sperm cell in humans.

cone (1) In vertebrates, a photoreceptor cell in the retina stimulated by bright light and enabling color vision. (2) In conifers, a reproductive structure bearing pollen or ovules.

coniferous forest A biome characterized by conifers, cone-bearing evergreen trees.

conjugation The union (mating) of two bacterial cells or protist cells and the transfer of DNA between the two cells.

conjunctiva A thin mucous membrane that lines the inner surface of vertebrate eyelids.

connective tissue Animal tissue that functions mainly to bind and support other tissues, having a sparse population of cells scattered through an extracellular matrix, which they produce.

conservation biology A goal-oriented science that endeavors to sustain biological diversity.

continental shelf The submerged part of a continent.

contraception The deliberate prevention of pregnancy.

controlled experiment An experiment in which an experimental group is compared with a control group that varies only in the factor being tested.

convergent evolution The evolution of similar features in different evolutionary lineages, which can result from living in very similar environments.

copulation Sexual intercourse, usually necessary for internal fertilization to occur.

cork The outermost protective layer of a plant's bark, produced by the cork cambium.

cork cambium Meristematic tissue that produces cork cells during secondary growth of a plant.

cornea (kor′-nē-uh) The transparent frontal portion of the sclera, which admits light into the vertebrate eye.

corpus callosum (kor′-pus kuh-lō′-sum) The thick band of nerve fibers that connect the right and left cerebral hemispheres in placental mammals, enabling the hemispheres to process information together.

corpus luteum (kor′-pus lū′-tē-um) A small body of endocrine tissue that develops from an ovarian follicle after ovulation and secretes progesterone and estrogen during pregnancy.

cortex In plants, the ground tissue system of a root, made up mostly of parenchyma cells, which store food and absorb minerals that have passed through the epidermis.

corticosteroid A hormone synthesized and secreted by the adrenal cortex. The corticosteroids include the mineralocorticoids and glucocorticoids.

cotyledon (kot′-uh-lē′-don) The first leaf that appears on an embryo of a flowering plant; a seed leaf. Monocot embryos have one cotyledon; dicot embryos have two.

countercurrent exchange The transfer of a substance or heat between two fluids flowing in opposite directions.

countercurrent heat exchange A circulatory adaptation in which parallel blood vessels convey warm and cold blood in opposite directions, maximizing heat transfer to the cold blood.

covalent bond (ko-vā′-lent) A strong chemical bond in which two atoms share one or more pairs of outer-shell electrons.

craniate A chordate with a head.

crista (kris′tuh) (plural, **cristae**) An infolding of the inner mitochondrial membrane.

crop A pouch-like organ in a digestive tract where food is softened and may be stored temporarily.

cross A mating of two sexually reproducing individuals; often used to describe a genetics experiment involving a controlled mating (a "genetic cross").

cross-fertilization The fusion of sperm and egg derived from two different individuals.

crossing over The exchange of segments between chromatids of homologous chromosomes during synapsis in prophase I of meiosis; also, the exchange of segments between DNA molecules in prokaryotes.

crustacean A member of a major arthropod group that includes lobsters, crayfish, crabs, shrimps, and barnacles.

cuticle (kyū′-tuh-kul) (1) In animals, a tough, nonliving outer layer of the skin. (2) In plants, a waxy coating on the surface of stems and leaves that helps retain water.

cyanobacteria (sī-an′-ō-bak-tēr′-ē-uh) Photoautotrophic prokaryotes with plantlike, oxygen-generating photosynthesis.

cystic fibrosis (sis′-tik fī-brō′-sis) A genetic disease that occurs in people with two copies of a certain recessive allele; characterized by an excessive secretion of mucus and vulnerability to infection; fatal if untreated.

cytokinesis (sī′-tō-kuh-nē-sis) The division of the cytoplasm to form two separate daughter cells. Cytokinesis usually occurs in conjunction with telophase of mitosis. Mitosis and cytokinesis make up the mitotic (M) phase of the cell cycle.

cytokinin (sī′-tō-kī′-nin) One of a family of plant hormones that promotes cell division, retards aging in flowers and fruits, and may interact antagonistically with auxins in regulating plant growth and development.

cytoplasm (sī′-tō-plaz′-um) The contents of a eukaryotic cell between the plasma membrane and the nucleus; consists of a semifluid medium and organelles; can also refer to the interior of a prokaryotic cell.

cytosine (C) (sī′-tuh-sin) A single-ring nitrogenous base found in DNA and RNA.

cytoskeleton A network of protein fibers in the cytoplasm of a eukaryotic cell; includes microfilaments, intermediate filaments, and microtubules.

cytotoxic T cell (sī′-tō-tok′-sik) A type of lymphocyte that attacks body cells infected with pathogens.

D

decomposer Prokaryotes and fungi that secrete enzymes that digest nutrients from organic material and convert them to inorganic forms.

decomposition The breakdown of organic materials into inorganic ones.

deductive reasoning A type of logic in which specific results are predicted from a general premise.
dehydration reaction (dē-hī-drā′-shun) A chemical reaction in which two molecules become covalently bonded to each other with the removal of a water molecule.
deletion The loss of one or more nucleotides from a gene by mutation; the loss of a fragment of a chromosome.
demographic transition A shift from zero population growth in which birth rates and death rates are high to zero population growth characterized by low birth and death rates.
denaturation (dē-nā′-chur-ā′-shun) A process in which a protein unravels, losing its specific structure and hence function; can be caused by changes in pH or salt concentration or by high temperature; also refers to the separation of the two strands of the DNA double helix, caused by similar factors.
dendrite (den′-drīt) A neuron fiber that conveys signals from its tip inward, toward the rest of the neuron. A neuron typically has many short dendrites.
density-dependent factor A population-limiting factor whose intensity is linked to population density. For example, there may be a decline in birth rates or a rise in death rates in response to an increase in the number of individuals living in a designated area.
density-dependent inhibition The ceasing of cell division that occurs when cells touch one another.
density-independent factor A population-limiting factor whose intensity is unrelated to population density.
deoxyribonucleic acid (DNA) (dē-ok′-sē-rī′-bō-nū-klā′-ik) A double-stranded helical nucleic acid molecule consisting of nucleotide monomers with deoxyribose sugar and the nitrogenous bases adenine (A), cytosine (C), guanine (G), and thymine (T). Capable of replicating, DNA is an organism's genetic material. *See also* gene.
dermal tissue system The outer protective covering of plants.
desert A biome characterized by organisms adapted to sparse rainfall (less than 30 cm per year) and rapid evaporation.
desertification The conversion of semi-arid regions to desert.
determinate growth Termination of growth after reaching a certain size, as in most animals. *See also* indeterminate growth.
detritivore (duh-trī′-tuh-vor) An organism that consumes organic wastes and dead organisms.
detritus (duh-trī′-tus) Dead organic matter.
deuterostome (dū-ter′-ō-stōm) A mode of animal development in which the opening formed during gastrulation becomes the anus. Animals with the deuterostome pattern of development include the echinoderms and the chordates.
diabetes mellitus (dī′-uh-bē′-tis me-lī′-tis) A human hormonal disease in which body cells cannot absorb enough glucose from the blood and become energy starved; body fats and proteins are then consumed for their energy. Type 1 (insulin-dependent) diabetes results when the pancreas does not produce insulin; type 2 (non-insulin-dependent) diabetes results when body cells fail to respond to insulin.
dialysis (dī-al′-uh-sis) Separation and disposal of metabolic wastes from the blood by mechanical means; an artificial method of performing the functions of the kidneys that can be life sustaining in the event of kidney failure.
diaphragm (dī′-uh-fram) The sheet of muscle separating the chest cavity from the abdominal cavity in mammals. Its contraction expands the chest cavity, and its relaxation reduces it.
diastole (dȳ′-as′-tō-lē) The stage of the heart cycle in which the heart muscle is relaxed, allowing the chambers to fill with blood. *See also* systole.
diatom (dī′-uh-tom) A unicellular, autotrophic protist that belongs to the supergroup Chromalveolata. Diatoms possess a unique glassy cell wall containing silica.
dicot (dī′-kot) A term traditionally used to refer to flowering plants that have two embryonic seed leaves, or cotyledons.
differentiation The specialization in the structure and function of cells that occurs during the development of an organism; results from selective activation and deactivation of the cells' genes.
diffusion The spontaneous movement of a substance down its concentration gradient from where it is more concentrated to where it is less concentrated.
digestion The mechanical and chemical breakdown of food into molecules small enough for the body to absorb; the second stage of food processing in animals.
digestive system The organ system involved in ingestion and digestion of food, absorption of nutrients, and elimination of wastes.
dihybrid cross (dī′-hī′-brid) An experimental mating of individuals differing in two characters.
dinoflagellate (dī′-nō-flaj′-uh-let) A member of a group of protists belonging to the supergroup Chromalveolata. Dinoflagellates are common components of marine and freshwater phytoplankton.
diploid In an organism that reproduces sexually, a cell containing two homologous sets of chromosomes, one set inherited from each parent; a $2n$ cell.
directional selection Natural selection in which individuals at one end of the phenotypic range survive and reproduce more successfully than do other individuals.
disaccharide (dī-sak′-uh-rīd) A sugar molecule consisting of two monosaccharides linked by a dehydration reaction.
dispersion pattern The manner in which individuals in a population are spaced within their area. Three types of dispersion patterns are clumped (individuals are aggregated in patches), uniform (individuals are evenly distributed), and random (unpredictable distribution).
disruptive selection Natural selection in which individuals on both extremes of a phenotypic range are favored over intermediate phenotypes.
distal tubule In the vertebrate kidney, the portion of a nephron that helps refine filtrate and empties it into a collecting duct.
disturbance In ecology, a force that changes a biological community and usually removes organisms from it.
DNA *See* deoxyribonucleic acid (DNA).
DNA ligase (lī′-gās) An enzyme, essential for DNA replication, that catalyzes the covalent bonding of adjacent DNA polynucleotide strands. DNA ligase is used in genetic engineering to paste a specific piece of DNA containing a gene of interest into a bacterial plasmid or other vector.
DNA microarray A glass slide carrying thousands of different kinds of single-stranded DNA fragments arranged in an array (grid). A DNA microarray is used to detect and measure the expression of thousands of genes at one time. Tiny amounts of a large number of single-stranded DNA fragments representing different genes are fixed to the glass slide. These fragments, ideally representing all the genes of an organism, are tested for hybridization with various samples of cDNA molecules.
DNA polymerase (puh-lim′-er-ās) A large molecular complex that assembles DNA nucleotides into polynucleotides using a preexisting strand of DNA as a template.
DNA profiling A procedure that analyzes DNA samples to determine if they came from the same individual.
DNA technology Methods used to study and/or manipulate DNA, including recombinant DNA technology.
doldrums (dol′-drums) An area of calm or very light winds near the equator, caused by rising warm air.
domain A taxonomic category above the kingdom level. The three domains of life are Archaea, Bacteria, and Eukarya.
dominance hierarchy The ranking of individuals within a group, based on social interactions and usually maintained by agonistic behavior.
dominant allele The allele that determines the phenotype of a gene when the individual is heterozygous for that gene.
dorsal Pertaining to the back of a bilaterally symmetric animal.
dorsal, hollow nerve cord One of the four hallmarks of chordates, a tube that forms on the dorsal side of the body, above the notochord.
double circulation A circulatory system with separate pulmonary and systemic circuits, in which blood passes through the heart

after completing each circuit; ensures vigorous blood flow to all organs.
double fertilization In flowering plants, the formation of both a zygote and a cell with a triploid nucleus, which develops into the endosperm.
double helix The form of native DNA, referring to its two adjacent polynucleotide strands interwound into a spiral shape.
Down syndrome A human genetic disorder resulting from the presence of an extra chromosome 21; characterized by heart and respiratory defects and varying degrees of mental retardation.
Duchenne muscular dystrophy (duh-shen′ dis′-truh-fē) A human genetic disease caused by a sex-linked recessive allele; characterized by progressive weakening and a loss of muscle tissue.
duodenum (dū-ō-dē′-num) The first portion of the vertebrate small intestine after the stomach, where chyme from the stomach mixes with bile and digestive enzymes.
duplication Repetition of part of a chromosome resulting from fusion with a fragment from a homologous chromosome; can result from an error in meiosis or from mutagenesis.

E

eardrum A sheet of connective tissue separating the outer ear from the middle ear that vibrates when stimulated by sound waves and passes the waves to the middle ear.
echinoderm (uh-kī′-nō-derm) Member of a phylum of slow-moving or sessile marine animals characterized by a rough or spiny skin, a water vascular system, an endoskeleton, and radial symmetry in adults. Echinoderms include sea stars, sea urchins, and sand dollars.
ecological footprint An estimate of the amount of land required to provide the raw materials an individual or nation consumes, including food, fuel, water, housing, and waste disposal.
ecological niche (nich) The role of a species in its community; the sum total of a species' use of the biotic and abiotic resources of its environment.
ecological species concept A definition of species in terms of ecological niche, the sum of how members of the species interact with the nonliving and living parts of their environment.
ecological succession The process of biological community change resulting from disturbance; transition in the species composition of a biological community, often following a flood, fire, or volcanic eruption. *See also* primary succession; secondary succession.
ecology The scientific study of how organisms interact with their environment.
ecosystem (ē′-kō-sis-tem) All the organisms in a given area, along with the nonliving (abiotic) factors with which they interact; a biological community and its physical environment.
ecotourism Travel to natural areas for tourism and recreation.
ectoderm (ek′-tō-derm) The outer layer of three embryonic cell layers in a gastrula. The ectoderm forms the skin of the gastrula and gives rise to the epidermis and nervous system in the adult.
ectopic pregnancy (ek-top′-ik) The implantation and development of an embryo outside the uterus.
ectotherm (ek′-tō-therm) An animal that warms itself mainly by absorbing heat from its surroundings. Examples include most amphibians, lizards, and invertebrates.
ectothermic Referring to organisms that do not produce enough metabolic heat to have much effect on body temperature. *See also* ectotherm.
effector cell (1) A muscle cell or gland cell that performs the body's response to stimuli, responding to signals from the brain or other processing center of the nervous system. (2) A lymphocyte that has undergone clonal selection and is capable of mediating an acquired immune response.
egg A female gamete.
ejaculation (ih-jak′-yū-lā′-shun) Expulsion of semen from the penis.
ejaculatory duct The short section of the ejaculatory route in mammals formed by the convergence of the vas deferens and a duct from the seminal vesicle. The ejaculatory duct transports sperm from the vas deferens to the urethra.
electromagnetic receptor A sensory receptor that detects energy of different wavelengths, such as electricity, magnetism, and light.
electromagnetic spectrum The entire spectrum of radiation ranging in wavelength from less than a nanometer to more than a kilometer.
electron A subatomic particle with a single negative electrical charge. One or more electrons move around the nucleus of an atom.
electron microscope (EM) A microscope that uses magnets to focus an electron beam through, or onto the surface of, a specimen. An electron microscope achieves a hundredfold greater resolution than a light microscope.
electron shell An energy level representing the distance of an electron from the nucleus of an atom.
electron transport chain A series of electron carrier molecules that shuttle electrons during the redox reactions that release energy used to make ATP; located in the inner membrane of mitochondria, the thylakoid membranes of chloroplasts, and the plasma membranes of prokaryotes.
electronegativity The attraction of a given atom for the electrons of a covalent bond.
element A substance that cannot be broken down to other substances by chemical means.
elimination The passing of undigested material out of the digestive compartment; the fourth and final stage of food processing in animals.
embryo (em′-brē-ō) A developing stage of a multicellular organism. In humans, the stage in the development of offspring from the first division of the zygote until body structures begin to appear, about the 9th week of gestation.
embryo sac The female gametophyte contained in the ovule of a flowering plant.
embryonic stem cell (ES cell) Cell in the early animal embryo that differentiates during development to give rise to all the different kinds of specialized cells in the body.
embryophyte Another name for land plants, recognizing that land plants share the common derived trait of multicellular, dependent embryos.
emergent properties New properties that arise with each step upward in the hierarchy of life, owing to the arrangement and interactions of parts as complexity increases.
emerging virus A virus that has appeared suddenly or has recently come to the attention of medical scientists.
endemic species A species whose distribution is limited to a specific geographic area.
endergonic reaction (en′-der-gon′-ik) An energy-requiring chemical reaction, which yields products with more potential energy than the reactants. The amount of energy stored in the products equals the difference between the potential energy in the reactants and that in the products.
endocrine gland (en′-dō-krin) A ductless gland that synthesizes hormone molecules and secretes them directly into the bloodstream.
endocrine system (en′-dō-krin) The organ system consisting of ductless glands that secrete hormones and the molecular receptors on or in target cells that respond to the hormones. The endocrine system cooperates with the nervous system in regulating body functions and maintaining homeostasis.
endocytosis (en′-dō-sī-tō′-sis) Cellular uptake of molecules or particles via formation of new vesicles from the plasma membrane.
endoderm (en′-dō-derm) The innermost of three embryonic cell layers in a gastrula; forms the archenteron in the gastrula and gives rise to the innermost linings of the digestive tract and other hollow organs in the adult.
endodermis The innermost layer (a one-cell-thick cylinder) of the cortex of a plant root; forms a selective barrier determining which substances pass from the cortex into the vascular tissue.
endomembrane system A network of membranes inside and around a eukaryotic cell, related either through direct physical contact or by the transfer of membranous vesicles.

endometrium (en′-dō-mē′-trē-um) The inner lining of the uterus in mammals, richly supplied with blood vessels that provide the maternal part of the placenta and nourish the developing embryo.
endoplasmic reticulum (ER) An extensive membranous network in a eukaryotic cell, continuous with the outer nuclear membrane and composed of ribosome-studded (rough) and ribosome-free (smooth) regions. *See also* rough ER; smooth ER.
endorphin (en-dōr′-fin) A pain-inhibiting hormone produced by the brain and anterior pituitary; also serves as a neurotransmitter.
endoskeleton A hard skeleton located within the soft tissues of an animal; includes spicules of sponges, the hard plates of echinoderms, and the cartilage and bony skeletons of vertebrates.
endosperm In flowering plants, a nutrient-rich mass formed by the union of a sperm cell with two polar nuclei during double fertilization; provides nourishment to the developing embryo in the seed.
endospore A thick-coated, protective cell produced within a bacterial cell. Endospore becomes dormant and is able to survive harsh environmental conditions.
endosymbiont theory (en′-dō-sim′-bī-ont) A theory that mitochondria and chloroplasts originated as prokaryotic cells engulfed by an ancestral eukaryotic cell. The engulfed cell and its host cell then evolved into a single organism.
endotherm An animal that derives most of its body heat from its own metabolism. Examples include most mammals and birds.
endothermic Referring to animals that use heat generated by metabolism to maintain a warm, steady body temperature. *See also* endotherm.
endotoxin A poisonous component of the outer membrane of gram-negative bacteria that is released only when the bacteria die.
energy The capacity to cause change, especially to perform work.
energy coupling In cellular metabolism, the use of energy released from an exergonic reaction to drive an endergonic reaction.
energy flow The passage of energy through the components of an ecosystem.
enhancer A eukaryotic DNA sequence that helps stimulate the transcription of a gene at some distance from it. An enhancer functions by means of a transcription factor called an activator, which binds to it and then to the rest of the transcription apparatus. *See also* silencer.
enteric division Part of the autonomic nervous system consisting of complex networks of neurons in the digestive tract, pancreas, and gallbladder.
entropy (en′-truh-pē) A measure of disorder. One form of disorder is heat, which is random molecular motion.
enzyme (en′-zīm) A macromolecule, usually a protein, that serves as a biological catalyst, changing the rate of a chemical reaction without being consumed by the reaction.
epidermis (ep′-uh-der′-mis) (1) In animals, one or more living layers of cells forming the protective covering, or outer skin. (2) In plants, the tissue system forming the protective outer covering of leaves, young stems, and young roots.
epididymis (ep′-uh-did′-uh-mus) A long coiled tube into which sperm pass from the testis and are stored until mature and ejaculated.
epigenetic inheritance The inheritance of traits transmitted by mechanisms not directly involving the nucleotide sequence of a genome, such as the chemical modification of histone proteins or DNA bases.
epinephrine (ep′-uh-nef′-rin) An amine hormone (also called adrenaline) secreted by the adrenal medulla that prepares body organs for action (fight or flight); also serves as a neurotransmitter.
epithelial tissue (ep′-uh-thē′-lē-ul) A sheet of tightly packed cells lining organs, body cavities, and external surfaces; also called epithelium.
erythrocyte (eh-rith′-rō-sȳ′t) A blood cell containing hemoglobin, which transports oxygen; also called a red blood cell.
erythropoietin (EPO) (eh-rith′rō-poy′uh-tin) A hormone that stimulates the production of erythrocytes. It is secreted by the kidney when tissues of the body do not receive enough oxygen.
esophagus (eh-sof′-uh-gus) A muscular tube that conducts food by peristalsis, usually from the pharynx to the stomach.
essential amino acid An amino acid that an animal cannot synthesize itself and must obtain from food. Eight amino acids are essential for the human adult.
essential element In plants, a chemical element required for the plant to complete its life cycle (to grow from a seed and produce another generation of seeds).
essential fatty acid An unsaturated fatty acid that an animal needs but cannot make.
essential nutrient A substance that an organism must absorb in preassembled form because it cannot synthesize it from any other material. In humans, there are essential vitamins, minerals, amino acids, and fatty acids.
estrogen (es′-trō-jen) One of several chemically similar steroid hormones secreted by the gonads; maintains the female reproductive system and promotes the development of female body features.
estuary (es′-chū-ār-ē) The area where a freshwater stream or river merges with the ocean.
ethylene A gas that functions as a hormone in plants, triggering aging responses such as fruit ripening and leaf drop.
eudicot (yū-dē-kot) Member of a group that consists of the vast majority of flowering plants that have two embryonic seed leaves, or cotyledons.
Eukarya (yū-kar′-ē-uh) Domain of life that includes all eukaryotes.
eukaryotic cell (yū-kar-ē-ot′-ik) A type of cell that has a membrane-enclosed nucleus and other membrane-enclosed organelles. All organisms except bacteria and archaea are composed of eukaryotic cells.
eumetazoan (yū-met-uh-zō′-un) Member of the clade of "true animals," the animals with true tissues (all animals except sponges).
Eustachian tube (yū-stā′-shun) An air passage between the middle ear and throat of vertebrates that equalizes air pressure on either side of the eardrum.
eutherian (yū-thēr′-ē-un) Placental mammal; mammal whose young complete their embryonic development within the uterus, joined to the mother by the placenta.
evaporative cooling The process in which the surface of an object becomes cooler during evaporation.
"evo-devo" Evolutionary developmental biology; the field of biology that combines evolutionary biology with developmental biology.
evolution Descent with modification; the idea that living species are descendants of ancestral species that were different from present-day ones; also, the genetic changes in a population from generation to generation.
evolutionary tree A branching diagram that reflects a hypothesis about evolutionary relationships among groups of organisms.
Excavata One of five supergroups proposed in a current hypothesis of the evolutionary history of eukaryotes. The other four supergroups are Chromalveolata, Rhizaria, Unikonta, and Archaeplastida.
excretion (ek-skrē′-shun) The disposal of nitrogen-containing metabolic wastes.
exergonic reaction (ek′-ser-gon′-ik) An energy-releasing chemical reaction in which the reactants contain more potential energy than the products. The reaction releases an amount of energy equal to the difference in potential energy between the reactants and the products.
exocytosis (ek′-sō-sī-tō′-sis) The movement of materials out of the cytoplasm of a cell by the fusion of vesicles with the plasma membrane.
exoskeleton A hard external skeleton that protects an animal and provides points of attachment for muscles.
exotoxin A poisonous protein secreted by certain bacteria.
exponential growth model A mathematical description of idealized, unregulated population growth.
external fertilization The fusion of gametes that parents have discharged into the environment.
extinction The irrevocable loss of a species.
extirpation The loss of a single population of a species.
extracellular matrix (ECM) The meshwork surrounding animal cells; consists of glycoproteins and polysaccharides.

extraembryonic membranes Four membranes (the yolk sac, amnion, chorion, and allantois) that form a life-support system for the developing embryo of a reptile, bird, or mammal.
extreme halophile A microorganism that lives in a highly saline environment, such as the Great Salt Lake or the Dead Sea.
extreme thermophile A microorganism that thrives in a hot environment (often 60–80°C).
eyecup The simplest type of photoreceptor, a cluster of photoreceptor cells shaded by a cuplike cluster of pigmented cells; detects light intensity and direction.

F

F factor A piece of DNA that can exist as a bacterial plasmid. The F factor carries genes for making sex pili and other structures needed for conjugation, as well as a site where DNA replication can start. F stands for fertility.
F_1 generation The offspring of two parental (P generation) individuals; F_1 stands for first filial.
F_2 generation The offspring of the F_1 generation; F_2 stands for second filial.
facilitated diffusion The passage of a substance through a specific transport protein across a biological membrane down its concentration gradient.
family In Linnaean classification, the taxonomic category above genus.
farsightedness An inability to focus on close objects; occurs when the eyeball is shorter than normal and the focal point of the lens is behind the retina; also called hyperopia.
fat A lipid composed of three fatty acids linked to one glycerol molecule; a triglyceride. Most fats function as energy-storage molecules.
feces The wastes of the digestive tract.
feedback inhibition A method of metabolic control in which a product of a metabolic pathway acts as an inhibitor of an enzyme within that pathway.
fertilization The union of the nucleus of a sperm cell with the nucleus of an egg cell, producing a zygote.
fertilizer A compound given to plants to promote their growth.
fetus (fē′-tus) A developing human from the 9th week of gestation until birth. The fetus has all the major structures of an adult.
fiber (1) In animals, an elongate, supportive thread in the matrix of connective tissue; an extension of a neuron; a muscle cell. (2) In plants, a long, slender sclerenchyma cell that usually occurs in a bundle.
fibrin (fȳ′-brin) The activated form of the blood-clotting protein fibrinogen, which aggregates into threads that form the fabric of a blood clot.
fibrinogen (fȳ′-brin′-uh-jen) The plasma protein that is activated to form a clot when a blood vessel is injured.
fibrous connective tissue A dense tissue with large numbers of collagenous fibers organized into parallel bundles. This is the dominant tissue in tendons and ligaments.
filtrate Fluid extracted by the excretory system from the blood or body cavity. The excretory system produces urine from the filtrate after removing valuable solutes from it and concentrating it.
filtration In the vertebrate kidney, the extraction of water and small solutes, including metabolic wastes, from the blood by the nephrons.
fimbria (plural, **fimbriae**) One of the short, hairlike projections on some prokaryotic cells that help attach the cells to their substrate or to other cells.
first law of thermodynamics The principle of conservation of energy. Energy can be transferred and transformed, but it cannot be created or destroyed.
fission A means of asexual reproduction whereby a parent separates into two or more genetically identical individuals of about equal size.
fixed action pattern (FAP) A genetically programmed, virtually unchangeable behavioral sequence performed in response to a certain stimulus.
flagellum (fluh-jel′-um) (plural, **flagella**) A long cellular appendage specialized for locomotion. The flagella of prokaryotes and eukaryotes differ in both structure and function. Like cilia, eukaryotic flagella have a 9 + 2 arrangement of microtubules covered by the cell's plasma membrane.
flatworm A member of the phylum Platyhelminthes.
fluid feeder An animal that lives by sucking nutrient-rich fluids from another living organism.
fluid mosaic A description of membrane structure, depicting a cellular membrane as a mosaic of diverse protein molecules embedded in a fluid bilayer of phospholipid molecules.
fluke One of a group of parasitic flatworms.
follicle (fol′-uh-kul) A cluster of cells that surround, protect, and nourish a developing egg cell in the ovary. Follicles secrete the hormone estrogen.
food chain A sequence of food transfers from producers through one to four levels of consumers in an ecosystem.
food web A network of interconnecting food chains.
foot In an invertebrate animal, a structure used for locomotion or attachment, such as the muscular organ extending from the ventral side of a mollusc.
foraging Behavior used in recognizing, searching for, capturing, and consuming food.
foraminiferan A protist that moves and feeds by means of threadlike pseudopodia and has porous shells composed of calcium carbonate.
forebrain One of three ancestral and embryonic regions of the vertebrate brain; develops into the thalamus, hypothalamus, and cerebrum.
forensics The scientific analysis of evidence for crime scene and other legal proceedings. Also referred to as forensic science.
fossil A preserved remnant or impression of an organism that lived in the past.
fossil fuel An energy-containing deposit of organic material formed from the remains of ancient organisms.
fossil record The chronicle of evolution over millions of years of geologic time engraved in the order in which fossils appear in rock strata.
founder effect Genetic drift that occurs when a few individuals become isolated from a larger population and form a new population whose gene pool is not reflective of that of the original population.
fovea (fō′-vē-uh) An eye's center of focus and the place on the retina where photoreceptors are highly concentrated.
fragmentation A means of asexual reproduction whereby a single parent breaks into parts that regenerate into whole new individuals.
free-living flatworm A nonparasitic flatworm.
frequency-dependent selection Selection in which the fitness of a phenotype depends on how common the phenotype is in a population.
fruit A ripened, thickened ovary of a flower, which protects developing seeds and aids in their dispersal.
functional group A specific configuration of atoms commonly attached to the carbon skeletons of organic molecules and involved in chemical reactions.
Fungi (fun′-ji) The kingdom that contains the fungi.

G

gallbladder An organ that stores bile and releases it as needed into the small intestine.
gametangium (gam′-uh-tan′-jē-um) (plural, **gametangia**) A reproductive organ that houses and protects the gametes of a plant.
gamete (gam′-ēt) A sex cell; a haploid egg or sperm. The union of two gametes of opposite sex (fertilization) produces a zygote.
gametogenesis The creation of gametes within the gonads.
gametophyte (guh-mē′-tō-fīt) The multicellular haploid form in the life cycle of organisms undergoing alternation of generations; mitotically produces haploid gametes that unite and grow into the sporophyte generation.
ganglion (gang′-glē-un) (plural, **ganglia**) A cluster of neuron cell bodies in a peripheral nervous system.

gas exchange The exchange of O_2 and CO_2 between an organism and its environment.
gastric juice The collection of fluids (mucus, enzymes, and acid) secreted by the stomach.
gastrin A digestive hormone that stimulates the secretion of gastric juice.
gastropod A member of the largest group of molluscs, including snails and slugs.
gastrovascular cavity A central compartment with a single opening, the mouth; functions in both digestion and nutrient distribution and may also function in circulation, body support, waste disposal, and gas exchange.
gastrula (gas′-trū-luh) The embryonic stage resulting from gastrulation in animal development. Most animals have a gastrula made up of three layers of cells: ectoderm, endoderm, and mesoderm.
gastrulation (gas′-trū-lā′-shun) The second major phase of embryonic development, which transforms the blastula into a gastrula. Gastrulation adds more cells to the embryo and sorts the cells into distinct cell layers.
gel electrophoresis (jel′ ē-lek′-trō-fōr-ē′-sis) A technique for separating and purifying macromolecules, either DNAs or proteins. A mixture of the macromolecules is placed on a gel between a positively charged electrode and a negatively charged one. Negative charges on the molecules are attracted to the positive electrode, and the molecules migrate toward that electrode. The molecules separate in the gel according to their rates of migration, which is mostly determined by their size: Smaller molecules generally move faster through the gel, while larger molecules generally move more slowly.
gene A discrete unit of hereditary information consisting of a specific nucleotide sequence in DNA (or RNA, in some viruses). Most of the genes of a eukaryote are located in its chromosomal DNA; a few are carried by the DNA of mitochondria and chloroplasts.
gene cloning The production of multiple copies of a gene.
gene expression The process whereby genetic information flows from genes to proteins; the flow of genetic information from the genotype to the phenotype.
gene flow The transfer of alleles from one population to another as a result of the movement of individuals or their gametes.
gene pool All the alleles for all the genes in a population.
gene regulation The turning on and off of genes within a cell in response to environmental stimuli or other factors (such as developmental stage).
gene therapy A treatment for a disease in which the patient's defective gene is supplemented or altered.
genetic code The set of rules that dictates the correspondence between RNA codons in an mRNA molecule and amino acids in protein.
genetic drift A change in the gene pool of a population due to chance. Effects of genetic drift are most pronounced in small populations.
genetic engineering The direct manipulation of genes for practical purposes.
genetic recombination The production, by crossing over and/or independent assortment of chromosomes during meiosis, of offspring with allele combinations different from those in the parents. The term may also be used more specifically to mean the production by crossing over of eukaryotic or prokaryotic chromosomes with gene combinations different from those in the original chromosomes.
genetically modified (GM) organism An organism that has acquired one or more genes by artificial means. If the gene is from another species, the organism is also known as a transgenic organism.
genetics The scientific study of heredity. Modern genetics began with the work of Gregor Mendel in the 19th century.
genital herpes A sexually transmitted disease caused by the herpes simplex virus type 2.
genomic library (juh-nō′-mik) A set of DNA fragments representing an organism's entire genome. Each segment is usually carried by a plasmid or phage.
genomics The study of whole sets of genes and their interactions.
genotype (jē′-nō-tīp) The genetic makeup of an organism.
genus (jē′-nus) (plural, **genera**) In classification, the taxonomic category above species; the first part of a species' binomial; for example, *Homo*.
geologic record A time scale established by geologists that divides Earth's history into three eons—Archaean, Proterozoic, and Phanerozoic—and further subdivides it into eras, periods, and epochs.
germinate To start developing or growing.
gestation (jes-tā′-shun) Pregnancy; the state of carrying developing young within the female reproductive tract.
gibberellin (jib′-uh-rel′-in) One of a family of plant hormones that triggers the germination of seeds and interacts with auxins in regulating growth and fruit development.
gill An extension of the body surface of an aquatic animal, specialized for gas exchange and/or suspension feeding.
gizzard A pouch-like organ in a digestive tract where food is mechanically ground.
glans The rounded, highly sensitive head of the clitoris in females and penis in males.
glia A network of supporting cells that is essential for the structural integrity and normal functioning of the nervous system.
global climate change Increase in temperature and change in weather patterns all around the planet, due mostly to increasing atmospheric CO_2 levels from the burning of fossil fuels. The increase in temperature, called global warming, is a major aspect of global climate change.
glomeromycete Member of a group of fungi characterized by a distinct branching form of mycorrhizae (symbiotic relationships with plant roots) called arbuscules.
glomerulus (glō-mer′-ū-lus) (plural, **glomeruli**) In the vertebrate kidney, the part of a nephron consisting of the capillaries that are surrounded by Bowman's capsule; together, a glomerulus and Bowman's capsule produce the filtrate from the blood.
glucagon (glū′-kuh-gon) A peptide hormone, secreted by the islets of Langerhaus in the pancreas, that raises the level of glucose in the blood. It is antagonistic with insulin.
glucocorticoid (glū′-kuh-kor′-tih-koyd) A corticosteroid hormone secreted by the adrenal cortex that increases the blood glucose level and helps maintain the body's response to long-term stress.
glycogen (glī′-kō-jen) An extensively branched glucose storage polysaccharide found in liver and muscle cells; the animal equivalent of starch.
glycolysis (glī-kol′-uh-sis) The multistep chemical breakdown of a molecule of glucose into two molecules of pyruvate; the first stage of cellular respiration in all organisms; occurs in the cytoplasmic fluid.
glycoprotein (glī′-kō-prō′-tēn) A protein with one of more short chains of sugars attached to it.
goiter An enlargement of the thyroid gland resulting from a dietary iodine deficiency.
Golgi apparatus (gol′-jē) An organelle in eukaryotic cells consisting of stacks of membranous sacs that modify, store, and ship products of the endoplasmic reticulum.
gonad A sex organ in an animal; an ovary or testis.
Gram stain Microbiological technique to identify the cell wall composition of bacteria. Results categorize bacteria as gram-positive or gram-negative.
gram-positive bacteria Diverse group of bacteria with a cell wall that is structurally less complex and contains more peptidoglycan than that of gram-negative bacteria. Gram-positive bacteria are usually less toxic than gram-negative bacteria.
granum (gran′-um) (plural, **grana**) A stack of membrane-bounded thylakoids in a chloroplast. Grana are the sites where light energy is trapped by chlorophyll and converted to chemical energy during the light reactions of photosynthesis.
gravitropism (grav′-uh-trō′-pizm) A plant's directional growth in response to gravity.
gray matter Regions within the central nervous system composed mainly of nerve cell bodies and dendrites.

green alga A member of a group of photosynthetic protists that includes chlorophytes and charophyceans, the closest living relatives of land plants. Green algae include unicellular, colonial, and multicellular species and belong to the supergroup Archaeplastida.
greenhouse effect The warming of Earth due to the atmospheric accumulation of CO_2 and certain other gases, which absorb infrared radiation and reradiate some of it back toward Earth.
ground tissue system A tissue of mostly parenchyma cells that makes up the bulk of a young plant and is continuous throughout its body. The ground tissue system fills the space between the epidermis and the vascular tissue system.
growth factor A protein secreted by certain body cells that stimulates other cells to divide.
growth hormone (GH) A protein hormone secreted by the anterior pituitary that promotes development and growth and stimulates metabolism.
guanine (G) (gwa′-nēn) A double-ring nitrogenous base found in DNA and RNA.
guard cell A specialized epidermal cell in plants that regulates the size of a stoma, allowing gas exchange between the surrounding air and the photosynthetic cells in the leaf.
gymnosperm (jim′-nō-sperm) A naked-seed plant. Its seed is said to be naked because it is not enclosed in an ovary.

H

habitat A place where an organism lives; the environment in which an organism lives.
habituation Learning not to respond to a repeated stimulus that conveys little or no information.
hair cell A type of mechanoreceptor that detects sound waves and other forms of movement in air or water.
haploid In the life cycle of an organism that reproduces sexually, a cell containing a single set of chromosomes; an *n* cell.
Hardy-Weinberg principle The principle that frequencies of alleles and genotypes in a population remain constant from generation to generation, provided that only Mendelian segregation and recombination of alleles are at work.
heart A muscular pump that propels a circulatory fluid (blood) through vessels to the body.
heart attack The damage or death of cardiac muscle cells and the resulting failure of the heart to deliver enough blood to the body.
heart murmur A hissing sound that most often results from blood squirting backward through a leaky valve in the heart.
heart rate The frequency of heart contraction, usually expressed in number of beats per minute.
heartwood In the center of trees, the darkened, older layers of secondary xylem made up of cells that no longer transport water and are clogged with resins. *See also* sapwood.
heat Thermal energy; the amount of energy associated with the movement of the atoms and molecules in a body of matter. Heat is energy in its most random form.
helper T cell A type of lymphocyte that helps activate other types of T cells and may help stimulate B cells to produce antibodies.
hemoglobin (hē′-mō-glō-bin) An iron-containing protein in red blood cells that reversibly binds O_2.
hemophilia (hē′-mō-fil′-ē-uh) A human genetic disease caused by a sex-linked recessive allele; characterized by excessive bleeding following injury.
hepatic portal vein A blood vessel that conveys nutrient-laden blood from capillaries surrounding the intestine directly to the liver.
herbivore An animal that mainly eats plants or algae. *See also* carnivore; omnivore.
herbivory Consumption of plant parts or algae by an animal.
heredity The transmission of traits (inherited features) from one generation to the next.
hermaphroditism (her-maf′-rō-dī-tizm) A condition in which an individual has both female and male gonads and functions as both a male and female in sexual reproduction by producing both sperm and eggs.
heterokaryotic stage (het′-er-ō-ker-ē-ot′-ik) A fungal life cycle stage that contains two genetically different haploid nuclei in the same cell.
heterotroph (het′-er-ō-trōf) An organism that cannot make its own organic food molecules and must obtain them by consuming other organisms or their organic products; a consumer or a decomposer in a food chain.
heterozygote advantage Greater reproductive success of heterozygous individuals compared to homozygotes; tends to preserve variation in gene pools.
heterozygous (het′-er-ō-zī′-gus) Having two different alleles for a given gene.
high-density lipoprotein (HDL) A cholesterol-carrying particle in the blood, made up of thousands of cholesterol molecules and other lipids bound to a protein. HDL scavenges excess cholesterol.
hindbrain One of three ancestral and embryonic regions of the vertebrate brain; develops into the medulla oblongata, pons, and cerebellum.
hinge joint A joint that allows movement in only one plane. In humans, examples include the elbow and knee.
hippocampus (hip′-uh-kam′-pus) An integrative center of the cerebrum; functionally, the part of the limbic system that plays a central role in the formation of memories and their recall.
histamine (his′-tuh-mēn) A chemical alarm signal released by injured cells of vertebrates that causes blood vessels to dilate during an inflammatory response.
histone (his′-tōn) A small protein molecule associated with DNA and important in DNA packing in the eukaryotic chromosome. Eukaryotic chromatin consists of roughly equal parts of DNA and histone protein.
HIV (human immunodeficiency virus) The retrovirus that attacks the human immune system and causes AIDS.
homeobox (hō′-mē-ō-boks′) A 180-nucleotide sequence within a homeotic gene and some other developmental genes.
homeostasis (hō′-mē-ō-stā′-sis) The steady state of body functioning; a state of equilibrium characterized by a dynamic interplay between outside forces that tend to change an organism's internal environment and the internal control mechanisms that oppose such changes.
homeotic gene (hō′-mē-ot′-ik) A master control gene that determines the identity of a body structure of a developing organism, presumably by controlling the developmental fate of groups of cells.
hominin (hah′-mi-nid) Member of a species on the human branch of the evolutionary tree; a species more closely related to humans than to chimpanzees.
homologous chromosomes (hō-mol′-uh-gus) The two chromosomes that make up a matched pair in a diploid cell. Homologous chromosomes are of the same length, centromere position, and staining pattern and possess genes for the same characteristics at corresponding loci. One homologous chromosome is inherited from the organism's father, the other from the mother.
homologous structures Structures in different species that are similar because of common ancestry.
homology Similarity in characteristics resulting from a shared ancestry.
homozygous (hō′-mō-zī′-gus) Having two identical alleles for a given gene.
horizontal gene transfer The transfer of genes from one genome to another through mechanisms such as transposable elements, plasmid exchange, viral activity, and perhaps fusions of different organisms.
hormone (1) In animals, a regulatory chemical that travels in the blood from its production site, usually an endocrine gland, to other sites, where target cells respond to the regulatory signal. (2) In plants, a chemical that is produced in one part of the plant and travels to another part, where it acts on target cells to change their functioning.
horseshoe crab A bottom-dwelling marine chelicerate, a member of the phylum Arthropoda.
human chorionic gonadotropin (hCG) (kōr′-ē-on′-ik gon′-uh-dō-trō′-pin) A hormone secreted by the chorion that maintains the

production of estrogen and progesterone by the corpus luteum of the ovary during the first few months of pregnancy. hCG secreted in the urine is the target of many home pregnancy tests.

Human Genome Project (hGP) An international collaborative effort to map and sequence the DNA of the entire human genome. The project was begun in 1990 and completed in 2004.

humoral immune response The type of specific immunity brought about by antibody-producing B cells. The humoral immune response fights bacteria and viruses in body fluids. *See also* cell-mediated immune response.

humus (hyū'-mus) Decomposing organic material found in topsoil.

Huntington's disease A human genetic disease caused by a single dominant allele; characterized by uncontrollable body movements and degeneration of the nervous system; usually fatal 10 to 20 years after the onset of symptoms.

hybrid An offspring of parents of two different species or of two different varieties of one species; an offspring of two parents that differ in one or more inherited traits; an individual that is heterozygous for one or more pairs of genes.

hybrid zone A geographic region in which members of different species meet and mate, producing at least some hybrid offspring.

hydrocarbon An organic compound composed only of the elements carbon and hydrogen.

hydrogen bond A type of weak chemical bond formed when the partially positive hydrogen atom participating in a polar covalent bond in one molecule is attracted to the partially negative atom participating in a polar covalent bond in another molecule (or in another region of the same molecule).

hydrolysis (hī-drol'-uh-sis) A chemical reaction that breaks bonds between two molecules by the addition of water; process by which polymers are broken down and an essential part of digestion.

hydrophilic (hī'-drō-fil'-ik) "Water-loving"; pertaining to polar or charged molecules (or parts of molecules) that are soluble in water.

hydrophobic (hī'-drō-fō'-bik) "Water-fearing"; pertaining to nonpolar molecules (or parts of molecules) that do not dissolve in water.

hydrostatic skeleton A skeletal system composed of fluid held under pressure in a closed body compartment; the main skeleton of most cnidarians, flatworms, nematodes, and annelids.

hydroxyl group (hī-drok'-sil) A chemical group consisting of an oxygen atom bonded to a hydrogen atom.

hypertension A disorder in which blood pressure remains abnormally high.

hypertonic Referring to a solution that, when surrounding a cell, will cause the cell to lose water.

hypha (hī'-fuh) (plural, **hyphae**) One of many filaments making up the body of a fungus.

hypoglycemia (hī'-pō-glī-sē'-mē-uh) An abnormally low level of glucose in the blood that results when the pancreas secretes too much insulin into the blood.

hypothalamus (hī-pō-thal'-uh-mus) The master control center of the endocrine system, located in the ventral portion of the vertebrate forebrain. The hypothalamus functions in maintaining homeostasis, especially in coordinating the endocrine and nervous systems; secretes hormones of the posterior pituitary and releasing hormones that regulate the anterior pituitary.

hypothesis (hī-poth'-uh-sis) (plural, **hypotheses**) A testable explanation for a set of observations based on the available data and guided by inductive reasoning.

hypotonic Referring to a solution that, when surrounding a cell, will cause the cell to take up water.

I

immune system An animal body's system of defenses against agents that cause disease.

immunodeficiency disease An immunological disorder in which the immune system lacks one or more components, making the body susceptible to infectious agents that would ordinarily not be pathogenic.

imperfect fungus A fungus with no known sexual stage.

impotence The inability to maintain an erection; also called erectile dysfunction.

imprinting Learning that is limited to a specific critical period in an animal's life and that is generally irreversible.

***in vitro* fertilization (IVF)** (vē'-tro) Uniting sperm and egg in a laboratory container, followed by the placement of a resulting early embryo in the mother's uterus.

inbreeding Mating between close blood relatives.

inclusive fitness An individual's success at perpetuating its genes by producing its own offspring and by helping close relatives to produce offspring.

incomplete dominance A type of inheritance in which the phenotype of a heterozygote (*Aa*) is intermediate between the phenotypes of the two types of homozygotes (*AA* and *aa*).

incomplete metamorphosis A type of development in certain insects in which development from larva to adult is achieved by multiple molts, but without forming a pupa.

indeterminate growth Growth that continues throughout life, as in most plants. *See also* determinate growth.

induced fit The change in shape of the active site of an enzyme, caused by entry of the substrate, so that it binds more snugly to the substrate.

induction During embryonic development, the influence of one group of cells on an adjacent group of cells.

inductive reasoning A type of logic in which generalizations are based on a large number of specific observations.

inferior vena cava (vē'-nuh kā'-vuh) A large vein that returns oxygen-poor blood to the heart from the lower, or posterior, part of the body. *See also* superior vena cava.

infertility The inability to conceive after one year of regular, unprotected intercourse.

inflammatory response An innate body defense in vertebrates caused by a release of histamine and other chemical alarm signals that trigger increased blood flow, a local increase in white blood cells, and fluid leakage from the blood. The resulting inflammatory response includes redness, heat, and swelling in the affected tissues.

ingestion The act of eating; the first main stage of food processing in animals.

ingroup In a cladistic study of evolutionary relationships, the group of taxa whose evolutionary relationships are being determined. *See also* outgroup.

inhibiting hormone A kind of hormone released from the hypothalamus that prompts the anterior pituitary to stop secreting hormone.

innate behavior Behavior that is under strong genetic control and is performed in virtually the same way by all members of a species.

innate immunity The kind of immunity that is present in an animal before exposure to pathogens and is effective from birth. Innate immune defenses include barriers, phagocytic cells, antimicrobial proteins, the inflammatory response, and natural killer cells.

inner ear One of three main regions of the vertebrate ear; includes the cochlea, organ of Corti, and semicircular canals.

insulin A protein hormone, secreted by the islets of Langerhans in the pancreas, that lowers the level of glucose in the blood. It is antagonistic with glucagon.

integration The analysis and interpretation of sensory signals within neural processing centers of the central nervous system.

integrins A transmembrane protein that interconnects the extracellular matrix and the cytoskeleton.

integumentary system (in-teg'-yū-ment-ter-ē) The organ system consisting of the skin and its derivatives, such as hair and nails in mammals. The integumentary system helps protect the body from drying out, mechanical injury, and infection.

interferon (in'-ter-fer'-on) An innate defensive protein produced by virus-infected vertebrate cells and capable of helping other cells resist viruses.

intermediate One of the compounds that form between the initial reactant and the final product in a metabolic pathway, such as between glucose and pyruvate in glycolysis.

intermediate filament An intermediate-sized protein fiber that is one of the three main kinds of fibers making up the cytoskeleton of eukaryotic cells. Intermediate filaments are ropelike, made of fibrous proteins.
internal fertilization Reproduction in which sperm are typically deposited in or near the female reproductive tract and fertilization occurs within the tract.
interneuron (in′-ter-nūr′-on) A nerve cell, located entirely within the central nervous system, that integrates sensory signals and relays signals to other interneurons and to motor neurons.
internode The portion of a plant stem between two nodes.
interphase The period in the eukaryotic cell cycle when the cell is not actually dividing. Interphase constitutes the majority of the time spent in the cell cycle. *See also* mitotic phase (M phase).
interspecific competition Competition between individuals or populations of two or more species requiring a limited resource.
interspecific interactions Relationships between individuals of different species in a community.
interstitial fluid (in′-ter-stish′-ul) An aqueous solution that surrounds body cells and through which materials pass back and forth between the blood and the body tissues.
intertidal zone (in′-ter-tīd′-ul) A shallow zone where the waters of an estuary or ocean meet land.
intestine The region of a digestive tract located between the gizzard or stomach and the anus and where chemical digestion and nutrient absorption usually occur.
intraspecific competition Competition between members of a population for a limited resource.
invasive species A non-native species that spreads beyond its original point of introduction and causes environmental or economic damage.
inversion A change in a chromosome resulting from reattachment of a chromosome fragment to the original chromosome, but in a reverse direction. Mutagens and errors during meiosis can cause inversions.
invertebrate An animal that lacks a backbone.
ion (ī-on) An atom or group of atoms that has gained or lost one or more electrons, thus acquiring a charge.
ionic bond (ī-on′-ik) A chemical bond resulting from the attraction between oppositely charged ions.
iris The colored part of the vertebrate eye, formed by the anterior portion of the choroid.
isomers (ī′-sō-mers) Organic compounds with the same molecular formula but different structures and, therefore, different properties.
isotonic (ī-sō-ton′-ik) Referring to a solution that, when surrounding a cell, has no effect on the passage of water into or out of the cell.
isotope (ī′-sō-tōp) One of several atomic forms of an element, each with the same number of protons but a different number of neutrons.

K

karyotype (kār′-ē-ō-tīp) A display of micrographs of the metaphase chromosomes of a cell, arranged by size and centromere position. Karyotypes may be used to identify certain chromosomal abnormalities.
kelp Large, multicellular brown algae that form undersea "forests."
keystone species A species that is not usually abundant in a community yet exerts strong control on community structure by the nature of its ecological role, or niche.
kilocalorie (kcal) A quantity of heat equal to 1,000 calories. Used to measure the energy content of food, it is usually called a "Calorie."
kin selection The natural selection that favors altruistic behavior by enhancing reproductive success of relatives.
kinesis (kuh-nē′-sis) (plural, **kineses**) Random movement in response to a stimulus.
kinetic energy (kuh-net′-ik) The energy of motion; the energy of a mass of matter that is moving. Moving matter does work by imparting motion to other matter.
kingdom In classification, the broad taxonomic category above phylum.
***K*-selection** The concept that in certain (*K*-selected) populations, life history is centered around producing relatively few offspring that have a good chance of survival.

L

labia majora (lā′-bē-uh muh-jor′-uh) A pair of outer thickened folds of skin that protect the female genital region.
labia minora (lā′-bē-uh mi-nor′-uh) A pair of inner folds of skin, bordering and protecting the female genital region.
labor The series of events that expel the infant from the uterus.
lactic acid fermentation Glycolysis followed by the reduction of pyruvate to lactate, regenerating NAD^+.
lancelet One of a group of small, bladelike, invertebrate chordates.
landscape Several different ecosystems linked by exchanges of energy, materials, and organisms.
landscape ecology The application of ecological principles to the study of the structure and dynamics of a collection of ecosystems; the scientific study of the biodiversity of interacting ecosystems.
large intestine *See* colon.
larva (lar′-vuh) (plural, **larvae**) A free-living, sexually immature form in some animal life cycles that may differ from the adult in morphology, nutrition, and habitat.
larynx (lār′-inks) The upper portion of the respiratory tract containing the vocal cords; also called the voice box.
lateral line system A row of sensory organs along each side of a fish's body that is sensitive to changes in water pressure. It enables a fish to detect minor vibrations in the water.
lateral meristem Plant tissue made up of undifferentiated cells that enable roots and shoots of woody plants to thicken. The vascular cambium and cork cambium are lateral meristems.
lateralization The phenomenon in which the two hemispheres of the brain become specialized for different functions during infant and child brain development.
law of independent assortment A general rule in inheritance (originally formulated by Gregor Mendel) that when gametes form during meiosis, each pair of alleles for a particular characteristic segregate independently of other pairs; also known as Mendel's second law of inheritance.
law of segregation A general rule in inheritance (originally formulated by Gregor Mendel) that individuals have two alleles for each gene and that when gametes form by meiosis, the two alleles separate, each resulting gamete ending up with only one allele of each gene; also known as Mendel's first law of inheritance.
leaf The main site of photosynthesis in a plant; typically consists of a flattened blade and a stalk (petiole) that joins the leaf to the stem.
learning Modification of behavior as a result of specific experiences.
leech A member of one of the three large groups of annelids, known for its bloodsucking ability. *See* annelid.
lens The structure in an eye that focuses light rays onto the retina.
leukemia (lū-kē′-mē-ah) A type of cancer of the blood-forming tissues, characterized by an excessive production of white blood cells and an abnormally high number of them in the blood; cancer of the bone marrow cells that produce leukocytes.
leukocyte (lū′-kō-sȳ′t) A blood cell that functions in fighting infections; also called a white blood cell.
lichen (lī′-ken) A close association between a fungus and an alga or between a fungus and a cyanobacterium, some of which are known to be beneficial to both partners.
life cycle The entire sequence of stages in the life of an organism, from the adults of one generation to the adults of the next.
life history The series of events from birth through reproduction to death.
life table A listing of survivals and deaths in a population in a particular time period and predictions of how long, on average, an individual of a given age will live.
ligament A type of fibrous connective tissue that joins bones together at joints.
light microscope (LM) An optical instrument with lenses that refract (bend) visible light to magnify images and project them into a viewer's eye or onto photographic film.
light reactions The first of two stages in photosynthesis; the steps in which solar energy is absorbed and converted to chemical energy

in the form of ATP and NADPH. The light reactions power the sugar-producing Calvin cycle but produce no sugar themselves.

lignin A chemical that hardens the cell walls of plants.

limbic system (lim′-bik) A functional unit of several integrating and relay centers located deep in the human forebrain; interacts with the cerebral cortex in creating emotions and storing memories.

limiting factor An environmental factor that restricts population growth.

linkage map A listing of the relative locations of genes along a chromosome, as determined by recombination frequencies.

linked genes Genes located near each other on the same chromosome that tend to be inherited together.

lipid An organic compound consisting mainly of carbon and hydrogen atoms linked by nonpolar convalent bonds, making the compound mostly hydrophobic. Lipids include fats, phospholipids, and steroids and are insoluble in water.

liver The largest organ in the vertebrate body. The liver performs diverse functions, such as producing bile, preparing nitrogenous wastes for disposal, and detoxifying poisonous chemicals in the blood.

lobe-fin A bony fish with strong, muscular bonus supported by bones.

locomotion Active movement from place to place.

locus (plural, **loci**) The particular site where a gene is found on a chromosome. Homologous chromosomes have corresponding gene loci.

logistic growth model A mathematical description of idealized population growth that is restricted by limiting factors.

long-day plant A plant that flowers in late spring or early summer, when day length is long. Long-day plants actually flower in response to short nights.

long-term memory The ability to hold, associate, and recall information over one's lifetime.

loop of Henle (hen′-lē) In the vertebrate kidney, the portion of a nephron that helps concentrate the filtrate while conveying it between a proximal tubule and a distal tubule.

loose connective tissue The most widespread connective tissue in the vertebrate body. It binds epithelia to underlying tissues and functions as packing material, holding organs in place.

low-density lipoprotein (LDL) A cholesterol-carrying particle in the blood, made up of thousands of cholesterol molecules and other lipids bound to a protein. An LDL particle transports cholesterol from the liver for incorporation into cell membranes.

lung An infolded respiratory surface of terrestrial vertebrates that connects to the atmosphere by narrow tubes.

lymph A colorless fluid, derived from interstitial fluid, that circulates in the lymphatic sytem.

lymph node An organ of the immune system located along a lymph vessel. Lymph nodes filter lymph and contain cells that attack viruses and bacteria.

lymphatic system (lim-fat′-ik) The vertebrate organ system through which lymph circulates; includes lymph vessels, lymph nodes, and the spleen. The lymphatic system helps remove toxins and pathogens from the blood and interstitial fluid and returns fluid and solutes from the interstitial fluid to the circulatory system.

lymphocyte (lim′-fuh-sīt) A type of white blood cell that is chiefly responsible for the acquired immune response and is found mostly in the lymphatic system. *See* B cell; T cell.

lymphoma (lim-fō ′-muh) Cancer of the tissues that form white blood cells.

lysogenic cycle (lī′-sō-jen′-ik) A type of bacteriophage replication cycle in which the viral genome is incorporated into the bacterial host chromosome as a prophage. New phages are not produced, and the host cell is not killed or lysed unless the viral genome leaves the host chromosome.

lysosome (lī-sō-sōm) A digestive organelle in eukaryotic cells; contains hydrolytic enzymes that digest engulfed food or damaged organelles.

lytic cycle (lit′-ik) A type of viral replication cycle resulting in the release of new viruses by lysis (breaking open) of the host cell.

M

macroevolution Evolutionary change above the species level, encompassing the origin of new taxonomic groups, adaptive radiation, and mass extinction.

macromolecule A giant molecule formed by the joining of smaller molecules, usually by a dehydration reaction: a protein, carbohydrate, or nucleic acid.

macronutrient A chemical substance that an organism must obtain in relatively large amounts. *See also* micronutrient.

macrophage (mak′-rō-fāj) A large, amoeboid, phagocytic white blood cell that functions in innate immunity by destroying microbes and in acquired immunity as an antigen-presenting cell.

major depression Depressive mental illness characterized by a low mood most of the time.

major histocompatibility complex (MHC) molecule *See* self protein.

malignant tumor An abnormal tissue mass that can spread into neighboring tissue and to other parts of the body; a cancerous tumor.

malnutrition A failure to obtain adequate nutrition.

mammal Member of a clade of amniotes that possess mammary glands and hair.

mantle In a mollusc, the outgrowth of the body surface that drapes over the animal. The mantle produces the shell and forms the mantle cavity.

marsupial (mar-sū′-pē-ul) A pouched mammal, such as a kangaroo, opossum, or koala. Marsupials give birth to embryonic offspring that complete development while housed in a pouch and attached to nipples on the mother's abdomen.

mass number The sum of the number of protons and neutrons in an atom's nucleus.

matter Anything that occupies space and has mass.

mechanoreceptor (mek′-uh-nō-ri-sep′-ter) A sensory receptor that detects changes in the environment associated with pressure, touch, stretch, motion, or sound.

medulla oblongata (meh-duh′-luh ob′-long-got′-uh) Part of the vertebrate hindbrain, continuous with the spinal cord; passes data between the spinal cord and forebrain and controls autonomic, homeostatic functions, including breathing, heart rate, swallowing, and digestion.

medusa (med-ū′-suh) (plural, **medusae**) One of two types of cnidarian body forms; an umbrella-like body form.

meiosis (mī-ō′-sis) In a sexually reproducing organism, the division of a single diploid nucleus into four haploid daughter nuclei. Meiosis and cytokinesis produce haploid gametes from diploid cells in the reproductive organs of the parents.

membrane potential The charge difference between a cell's cytoplasm and extracellular fluid due to the differential distribution of ions.

memory The ability to store and retrieve information. *See also* long-term memory; short-term memory.

memory cell A clone of long-lived lymphocytes formed during the primary immune response. Memory cells remain in a lymph nodes until activated by exposure to the same antigen that triggered its formation. When activated, a memory cell forms a large clone that mounts the secondary immune response.

meninges (muh-nin′-jēz) Layers of connective tissue that enwrap and protect the brain and spinal cord.

menstrual cycle (men′-strū-ul) The hormonally synchronized cyclic buildup and breakdown of the endometrium of some primates, including humans.

menstruation (men′-strū-ā′-shun) Uterine bleeding resulting from shedding of the endometrium during a menstrual cycle.

meristem (mer′-eh-stem) Plant tissue consisting of undifferentiated cells that divide and generate new cells and tissues.

mesoderm (mez′-ō-derm) The middle layer of the three embryonic cell layers in a gastrula. The mesoderm gives rise to muscles, bones, the dermis of the skin, and most other organs in the adult.

mesophyll (mes′-ō-fil) The green tissue in the interior of a leaf; a leaf's ground tissue system; the main site of photosynthesis.

messenger RNA (mRNA) The type of ribonucleic acid that encodes genetic information from DNA and conveys it to ribosomes, where the information is translated into amino acid sequences.
metabolic pathway A series of chemical reactions that either builds a complex molecule or breaks down a complex molecule into simpler compounds.
metabolic rate The total amount of energy an animal uses in a unit of time.
metabolism The totality of an organism's chemical reactions.
metamorphosis (met′-uh-mōr′-fuh-sis) The transformation of a larva into an adult. *See* complete metamorphosis; incomplete metamorphosis.
metaphase (met′-eh-fāz) The third stage of mitosis, during which all the cell's duplicated chromosomes are lined up at an imaginary plane equidistant between the poles of the mitotic spindle.
metastasis (muh-tas′-tuh-sis) The spread of cancer cells beyond their original site.
methanogen (meth-an′-ō-jen) An archaean that produces methane as a metabolic waste product.
methyl group A chemical group consisting of a carbon atom bonded to three hydrogen atoms.
microevolution A change in a population's gene pool over generations.
microfilament The thinnest of the three main kinds of protein fibers making up the cytoskeleton of a eukaryotic cell; a solid, helical rod composed of the globular protein actin.
micrograph A photograph taken through a microscope.
micronutrient An element that an organism needs in very small amounts and that functions as a component or cofactor of enzymes. *See also* macronutrient.
microRNA (miRNA) A small, single-stranded RNA molecule that associates with one or more proteins in a complex that can degrade or prevent translation of an mRNA with a complementary sequence.
microtubule The thickest of the three main kinds of fibers making up the cytoskeleton of a eukaryotic cell; a hollow tube made of globular proteins called tubulins; found in cilia and flagella.
microvillus (plural, **microvilli**) One of many microscopic projections on the epithelial cells in the lumen of the small intestine. Microvilli increase the surface area of the small intestine.
midbrain One of three ancestral and embryonic regions of the vertebrate brain; develops into sensory integrating and relay centers that send sensory information to the cerebrum.
middle ear One of three main regions of the vertebrate ear; a chamber containing three small bones (the hammer, anvil, and stirrup) that convey vibrations from the eardrum to the oval window.
migration The regular back-and-forth movement of animals between two geographic areas at particular times of the year.
millipede A terrestrial arthropod that has two pairs of short legs for each of its numerous body segments and that eats decaying plant matter.
mineral In nutrition, a simple inorganic nutrient that an organism requires in small amounts for proper body functioning.
mineralocorticoid (min′-er-uh-lō-kort′-uh-koyd) A corticosteroid hormone secreted by the adrenal cortex that helps maintain salt and water homeostasis and may increase blood pressure in response to long-term stress.
missense mutation A change in the nucleotide sequence of a gene that alters the amino acid sequence of the resulting polypeptide. In a missense mutation, a codon is changed from encoding one amino acid to encoding a different amino acid.
mitochondrial matrix The compartment of the mitochondrion enclosed by the inner membrane and containing enzymes and substrates for the citric acid cycle.
mitochondrion (mī′-tō-kon′-drē-on) (plural, **mitochondria**) An organelle in eukaryotic cells where cellular respiration occurs. Enclosed by two membranes, it is where most of the cell's ATP is made.
mitosis (mī′-tō-sis) The division of a single nucleus into two genetically identical nuclei. Mitosis and cytokinesis make up the mitotic (M) phase of the cell cycle.
mitotic phase (M phase) The part of the cell cycle when the nucleus divides (via mitosis), its chromosomes are distributed to the daughter nuclei, and the cytoplasm divides (via cytokinesis), producing two daughter cells.
mitotic spindle A football-shaped structure formed of microtubules and associated proteins that is involved in the movement of chromosomes during mitosis and meiosis.
mixotroph A protist that is capable of both autotrophy and heterotrophy.
mold A rapidly growing fungus that reproduces asexually by producing spores.
molecular biology The study of the molecular basis of genes and gene expression; molecular genetics.
molecular clock Evolutionary timing method based on the observation that at least some regions of genomes evolve at constant rates.
molecular systematics A scientific discipline that uses nucleic acids or other molecules in different species to infer evolutionary relationships.
molecule Two or more atoms held together by covalent bonds.
mollusc (mol′-lusk) A soft-bodied animal characterized by a muscular foot, mantle, mantle cavity, and visceral mass. Molluscs include gastropods (snails and slugs), bivalves (clams, oysters, and scallops), and cephalopods (squids and octopuses).
molting The process of shedding an old exoskeleton or cuticle and secreting a new, larger one.
monoclonal antibody (mAb) (mon′-ō-klōn′-ul) An antibody secreted by a clone of cells and therefore specific for the one antigen that triggered the development of the clone.
monocot (mon′-ō-kot) A flowering plant whose embryos have a single seed leaf, or cotyledon.
monogamous Referring to a type of relationship in which one male mates with just one female, and both parents care for the children.
monohybrid cross An experimental mating of individuals differing in a single character.
monomer (mon′-uh-mer) The subunit that serves as a building block of a polymer.
monophyletic (mon′-ō-fī-let′-ik) Pertaining to a group of taxa that consists of a common ancestor and all its descendants, equivalent to a clade.
monosaccharide (mon′-ō-sak′-uh-rīd) The simplest carbohydrate; a simple sugar with a molecular formula that is generally some multiple of CH_2O. Monosaccharides are the monomers of disaccharides and polysaccharides.
monotreme (mon′-uh-trēm) An egg-laying mammal, such as the duck-billed platypus.
morning after pill (MAP) A birth control pill taken within three days of unprotected intercourse to prevent fertilization or implantation.
morphological species concept A definition of species in terms of measurable anatomical criteria.
motor neuron A nerve cell that conveys command signals from the central nervous system to effector cells, such as muscle cells or gland cells.
motor output The conduction of signals from a processing center in the central nervous system to effector cells.
motor system The component of the vertebrate peripheral nervous system that carries signals to and from skeletal muscles, mainly in response to external stimuli. Most actions of the motor system are voluntary.
motor unit A motor neuron and all the muscle fibers it controls.
mouth An opening through which food is taken into an animal's body, synonymous with oral cavity.
movement corridor A series of small clumps or a narrow strip of quality habitat (usable by organisms) that connects otherwise isolated patches of quality habitat.
multipotent stem cell An unspecialized cell that can divide to produce one identical daughter cell and a more specialized daughter cell, which undergoes differentiation. A multipotent stem cell can form multiple types of cells, but not all types of cells in the body.
muscle fiber Muscle cell.

muscle tissue Tissue consisting of long muscle cells that can contract, either on its own or when stimulated by nerve impulses; the most abundant tissue in a typical animal. *See* skeletal muscle; cardiac muscle; smooth muscle.

muscular system The organ system that includes all the skeletal muscles in the body. (Cardiac muscle and smooth muscle are components of other organ systems.)

mutagen (myū′-tuh-jen) A chemical or physical agent that interacts with DNA and causes a mutation.

mutagenesis (myū′-tuh-jen′-uh-sis) The creation of a change in the nucleotide sequence of an organism's DNA.

mutation A change in the nucleotide sequence of an organism's DNA; the ultimate source of genetic diversity. A mutation also can occur in the DNA or RNA of a virus.

mutualism An interspecific relationship in which both partners benefit.

mycelium (mī-sē′-lē-um) (plural, **mycelia**) The densely branched network of hyphae in a fungus.

mycorrhiza (mī′-kō-rī′-zuh) (plural, **mycorrhizae**) A close association of plant roots and fungi that is beneficial to both partners.

mycosis A general term for a fungal infection.

myelin sheath (mī′-uh-lin) A series of cells, each wound around, and thus insulating, the axon of a nerve cell in vertebrates. Each pair of cells in the sheath is separated by a space called a node of Ranvier.

myofibril (mī′-ō-fī′-bril) A contractile strand in a muscle cell (fiber), made up of many sarcomeres. Longitudinal bundles of myofibrils make up a muscle fiber.

myosin A type of protein filament that interacts with actin filaments to cause cell contraction.

N

NAD^+ Nicotinamide adenine dinucleotide; a coenzyme that can accept electrons during the redox reactions of cellular metabolism. The plus sign indicates that the molecule is oxidized and ready to pick up hydrogens; the reduced, hydrogen (electron)-carrying form is NADH.

$NADP^+$ Nicotinamide adenine dinucleotide phosphate, an electron acceptor that, as NADPH, temporarily stores energized electrons produced during the light reactions.

natural family planning A form of contraception that relies on refraining from sexual intercourse when conception is most likely to occur; also called the rhythm method.

natural killer (NK) cell A cell type that provides an innate immune response by attacking cancer cells and infected body cells, especially those harboring viruses.

natural selection A process in which individuals with certain inherited traits are more likely to survive and reproduce than are individuals that do not have those traits.

nearsightedness An inability to focus on distant objects; occurs when the eyeball is longer than normal and the lens focuses distant objects in front of the retina; also called myopia.

negative feedback A primary mechanism of homeostasis, whereby a change in a physiological variable triggers a response that counteracts the initial change. Negative feedback is a common control mechanism in which a chemical reaction, metabolic pathway, or hormone-secreting gland is inhibited by the products of the reaction, pathway, or gland. As the concentration of the products builds up, the product molecules themselves inhibit the process that produced them.

negative pressure breathing A breathing system in which air is pulled into the lungs.

nematode (nem′-uh-tōd) A roundworm, characterized by a pseudocoelom, a cylindrical, wormlike body form, and a tough cuticle that is molted to permit growth.

nephron The tubular excretory unit and associated blood vessels of the vertebrate kidney; extracts filtrate from the blood and refines it into urine. The nephron is the functional unit of the urinary system.

nerve A cable-like bundle of neurons tightly wrapped in connective tissue.

nerve cord An elongated bundle of neurons, usually extending longitudinally from the brain or anterior ganglia. One or more nerve cords and the brain make up the central nervous system in many animals.

nerve net A weblike system of interconnected neurons, characteristic of radially symmetric animals such as a hydra.

nervous system The organ system that forms a communication and coordination network between all parts of an animal's body.

nervous tissue Tissue made up of neurons and supportive cells.

neural tube (nyūr′-ul) An embryonic cylinder that develops from the ectoderm after gastrulation and gives rise to the brain and spinal cord.

neuron (nyūr′-on) A nerve cell; the fundamental structural and functional unit of the nervous system, specialized for carrying signals from one location in the body to another.

neurosecretory cell A nerve cell that synthesizes hormones and secretes them into the blood and also conducts nerve signals.

neurotransmitter A chemical messenger that carries information from a transmitting neuron to a receiving cell, either another neuron or an effector cell.

neutron A subatomic particle having no electrical charge, found in the nucleus of an atom.

neutrophil (nyū′-truh-fil) An innate, defensive, phagocytic white blood cell that can engulf bacteria and viruses in infected tissue.

nitrogen fixation The conversion of atmospheric nitrogen (N_2) to nitrogen compounds (NH_4^+, NO_3^-) that plants can absorb and use.

node The point of attachment of a leaf on a stem.

node of Ranvier (ron′-vē-ā) An unmyelinated region on a myelinated axon of a nerve cell, where nerve signals are regenerated.

noncompetitive inhibitor A substance that reduces the activity of an enzyme without entering an active site. By binding elsewhere on the enzyme, a noncompetitive inhibitor changes the shape of the enzyme so that the active site no longer effectively catalyzes the conversion of substrate to product.

nondisjunction An accident of meiosis or mitosis in which a pair of homologous chromosomes or a pair of sister chromatids fail to separate at anaphase.

nonpolar covalent bond A covalent bond in which electrons are shared equally between two atoms of similar electronegativity.

nonself molecule A foreign antigen; a protein or other macromolecule that is not part of an organism's body. *See also* self protein.

nonsense mutation A change in the nucleotide sequence of a gene that converts an amino-acid-encoding codon to a stop codon. A nonsense mutation results in a shortened polypeptide.

norepinephrine (nor′-ep-uh-nef′-rin) An amine hormone (also called noradrenaline) secreted by the adrenal medulla that prepares body organs for action (fight or flight); also serves as a neurotransmitter.

notochord (nō′-tuh-kord) A flexible, cartilage-like, longitudinal rod located between the digestive tract and nerve cord in chordate animals; present only in embryos in many species.

nuclear envelope A double membrane that encloses the nucleus, perforated with pores that regulate traffic with the cytoplasm.

nuclear transplantation A technique in which the nucleus of one cell is placed into another cell that already has a nucleus or in which the nucleus has been previously destroyed.

nucleic acid (nū-klā′-ik) A polymer consisting of many nucleotide monomers; serves as a blueprint for proteins and, through the actions of proteins, for all cellular structures and activities. The two types of nucleic acids are DNA and RNA.

nucleic acid probe In DNA technology, a radioactively or fluorescently labeled single-stranded nucleic acid molecule used to find a specific gene or other nucleotide sequence within a mass of DNA. The probe hydrogen-bonds to the complementary sequence in the targeted DNA.

nucleoid (nū′-klē-oyd) A dense region of DNA in a prokaryotic cell.

nucleolus (nū-klē′-ō-lus) A structure within the nucleus where ribosomal RNA is made and assembled with proteins imported from the cytoplasm to make ribosomal subunits.
nucleosome (nū′-klē-ō-sōm) The bead-like unit of DNA packing in a eukaryotic cell; consists of DNA wound around a protein core made up of eight histone molecules.
nucleotide (nū′-klē-ō-tīd) A building block of nucleic acids, consisting of a five-carbon sugar covalently bonded to a nitrogenous base and one or more phosphate groups.
nucleus (plural, **nuclei**) (1) An atom's central core, containing protons and neutrons. (2) The genetic control center of a eukaryotic cell.

O

obesity The excessive accumulation of fat in the body.
ocean acidification Decreasing pH of ocean waters due to absorption of excess atmospheric CO_2 from the burning of fossil fuels.
ocean current One of the river-like flow patterns in the oceans.
omnivore An animal that eats animals as well as plants or algae.
oncogene (on′-kō-jēn) A cancer-causing gene; usually contributes to malignancy by abnormally enhancing the amount or activity of a growth factor made by the cell.
oogenesis (ō′-uh-jen′-uh-sis) The development of mature egg cells.
open circulatory system A circulatory system in which blood is pumped through open-ended vessels and bathes the tissues and organs directly. In an animal with an open circulatory system, blood and interstitial fluid are one and the same.
operator In prokaryotic DNA, a sequence of nucleotides near the start of an operon to which an active repressor protein can attach. The binding of a repressor prevents RNA polymerase from attaching to the promoter and transcribing the genes of the operon. The operator sequence thereby acts as a "genetic switch" that can turn all the genes in an operon on or off as a single functional unit.
operculum (ō-per′-kyuh-lum) (plural, **opercula**) A protective flap on each side of a fish's head that covers a chamber housing the gills. Movement of the operculum increases the flow of oxygen-bearing water over the gills.
operon (op′-er-on) A unit of genetic regulation common in prokaryotes; a cluster of genes with related functions, along with the promoter and operator that control their transcription.
opportunistic infection An infection that can be controlled by a normally functioning immune system but that causes illness in a person with an immunodeficiency.
opposable thumb An arrangement of the fingers such that the thumb can touch the ventral surface of the fingertips of all four fingers.
optimal foraging theory The basis for analyzing behavior as a compromise between feeding costs and feeding benefits.
oral cavity The mouth of an animal.
oral contraceptive *See* birth control pill.
order In Linnaean classification, the taxonomic category above family.
organ A specialized structure composed of several different types of tissues that together perform specific funstions.
organ of Corti (kor′-tē) The hearing organ in birds and mammals, located within the cochlea.
organ system A group of organs that work together in performing vital body functions.
organelle (ōr-guh-nel′) A membrane-enclosed structure with a specialized function within a cell.
organic compound A chemical compound containing the element carbon and usually the element hydrogen.
organism An individual living thing, such as a bacterium, fungus, protist, plant, or animal.
orgasm A series of rhythmic, involuntary contractions of the reproductive structures.
osmoconformer (oz′-mō-con-form′-er) An organism whose body fluids have a solute concentration equal to that of its surroundings. Osmoconformers do not have a net gain or loss of water by osmosis. Examples include most marine invertebrates.
osmoregulation The homeostatic maintenance of solute concentrations and the balance of water gain and loss.
osmoregulator An organism whose body fluids have a solute concentration different from that of its environment and that must use energy in controlling water loss or gain. Examples include most land-dwelling and freshwater animals.
osmosis (oz-mō′-sis) The diffusion of water across a selectively permeable membrane.
osteoporosis (os′-tē-ō-puh-rō′-sis) A skeletal disorder characterized by thinning, porous, and easily broken bones.
outer ear One of three main regions of the ear in reptiles (including birds) and mammals; made up of the auditory canal and, in many birds and mammals, the pinna.
outgroup In a cladistic study, a taxon or group of taxa known to have diverged before the lineage that contains the group of species being studied. *See also* ingroup.
ovarian cycle (ō-vār′-ē-un) Hormonally synchronized cyclic events in the mammalian ovary, culminating in ovulation.
ovary (1) In animals, the female gonad, which produces egg cells and reproductive hormones. (2) In flowering plants, the basal portion of a carpel in which the egg-containing ovules develop.
oviduct (ō′-vuh-dukt) The tube that conveys egg cells away from an ovary; also called a fallopian tube. In humans, the oviduct is the normal site of fertilization.
ovulation (ah′-vyū-lā′-shun) The release of an egg cell from an ovarian follicle.
ovule (ō-vyūl) In plants, a structure that develops within the ovary of a seed plant and contains the female gametophyte.
oxidation The loss of electrons from a substance involved in a redox reaction; always accompanies reduction.
oxidative phosphorylation (fos′-fōr-uh-lā′-shun) The production of ATP using energy derived from the redox reactions of an electron transport chain; the third major stage of cellular respiration.
ozone layer The layer of ozone (O_3) in the upper atmosphere that protects life on Earth from the harmful ultraviolet rays in sunlight.

P

P generation The parent individuals from which offspring are derived in studies of inheritance; P stands for parental.
P site One of two of a ribosome's binding sites for tRNA during translation. The P site holds the tRNA carrying the growing polypeptide chain. (P stands for peptidyl tRNA.)
paedomorphosis (pē′-duh-mōr′-fuh-sis) The retention in an adult of juvenile features of its evolutionary ancestors.
pain receptor A sensory receptor that detects pain.
paleoanthropology (pā′-lē-ō-an′-thruh-pol′-uh-jē) The study of human origins and evolution.
paleontologist (pa′-lē-on-tol′-uh-jist) A scientist who studies fossils.
pancreas (pan′-krē-us) A gland with dual functions: The digestive portion secretes digestive enzymes and an alkaline solution into the small intestine via a duct. The endocrine portion secretes the hormones insulin and glucagon into the blood.
Pangaea (pan-jē′-uh) The supercontinent that formed near the end of the Paleozoic era, when plate movements brought all the landmasses of Earth together.
parasite Organism that derives its nutrition from a living host, which is harmed by the interaction.
parasympathetic division The component of the autonomic nervous system that generally promotes body activities that gain and conserve energy, such as digestion and reduced heart rate. *See also* sympathetic division.
parathyroid gland (pār′-uh-thī′-royd) One of four endocrine glands that are embedded in the surface of the thyroid gland and that secrete parathyroid hormone.
parathyroid hormone (PTH) A peptide hormone secreted by the parathyroid glands that raises blood calcium level.
parenchyma cell (puh-ren′-kim-uh) In plants, a relatively unspecialized cell with a thin primary wall and no secondary wall; functions in photosynthesis, food storage, and aerobic respiration and may differentiate into other cell types.

Parkinson's disease A motor disorder caused by a progressive brain disease and characterized by difficulty in initiating movements, slowness of movement, and rigidity.
parsimony (par'-suh-mō'-nē) In scientific studies, the search for the least complex explanation for an observed phenomenon.
partial pressure The pressure exerted by a particular gas in a mixture of gases; a measure of the relative amount of a gas.
passive immunity Temporary immunity obtained by acquiring ready-made antibodies. Passive immunity lasts only a few weeks or months.
passive transport The diffusion of a substance across a biological membrane, with no expenditure of energy.
pathogen An agent, such as a virus, bacteria, or fungus, that causes disease.
pattern formation During embryonic development, the emergence of a body form with specialized organs and tissues in the right places.
PCR *See* polymerase chain reaction (PCR).
pedigree A family genetic tree representing the occurrence of heritable traits in parents and offspring across a number of generations. A pedigree can be used to determine genotypes of matings that have already occurred.
pelagic realm (puh-laj'-ik) The region of an ocean occupied by seawater.
penis The copulatory structure of male mammals.
peptide bond The covalent bond between two amino acid units in a polypeptide, formed by a dehydration reaction.
peptidoglycan (pep'-tid-ō-glī'-kan) A polymer of complex sugars cross-linked by short polypeptides; a material unique to bacterial cell walls.
per capita rate of increase The average contribution of each individual in a population to population growth.
perennial (puh-ren'-ē-ul) A plant that lives for many years.
peripheral nervous system (PNS) The network of nerves and ganglia carrying signals into and out of the central nervous system.
peristalsis (per'-uh-stal'-sis) Rhythmic waves of contraction of smooth muscles. Peristalsis propels food through a digestive tract and also enables many animals, such as earthworms, to crawl.
permafrost Continuously frozen ground found in the tundra.
peroxisome An organelle containing enzymes that transfer hydrogen atoms from various substrates to oxygen, producing and then degrading hydrogen peroxide.
petal A modified leaf of a flowering plant. Petals are the often colorful parts of a flower that advertise it to pollinators.
pH scale A measure of the relative acidity of a solution, ranging in value from 0 (most acidic) to 14 (most basic). The letters pH stand for potential hydrogen and refer to the concentration of hydrogen ions (H^+).
phage (fāj) *See* bacteriophage.
phagocyte (fag'-ō-sȳ't) A white blood cell (for example, a neutrophil or monocyte) that engulfs bacteria, foreign proteins, and the remains of dead body cells.
phagocytosis (fag'-ō-sī-tō'-sis) Cellular "eating"; a type of endocytosis in which a cell engulfs macromolecules, other cells, or particles into its cytoplasm.
pharyngeal slit (fā-rin'-jē-ul) A gill structure in the pharynx; found in chordate embryos and some adult chordates.
pharynx (fār'-inks) The organ in a digestive tract that receives food from the oral cavity; in terrestrial vertebrates, the throat region where the air and food passages cross.
phenotype (fē'-nō-tīp) The expressed traits of an organism.
phenotypic plasticity An individual's ability to change phenotype in response to local environmental conditions.
phloem (flō'-um) The portion of a plant's vascular tissue system that transports sugars and other organic nutrients from leaves or storage tissues to other parts of the plant.
phloem sap The solution of sugars, other nutrients, and hormones conveyed throughout a plant via phloem tissue.
phosphate group (fos'-fāt) A chemical group consisting of a phosphorus atom bonded to four oxygen atoms.
phospholipid (fos'-fō-lip'-id) A lipid made up of glycerol joined to two fatty acids and a phosphate group, giving the molecule two nonpolar hydrophobic tails and a polar hydrophilic head. Phospholipids form bilayers that function as biological membranes.
phosphorylation (fos'-fōr-uh-lā'-shun) The transfer of a phosphate group, usually from ATP, to a molecule. Nearly all cellular work depends on ATP energizing other molecules by phosphorylation.
photic zone (fō'-tik) The region of an aquatic ecosystem into which light penetrates and where photosynthesis occurs.
photoautotroph An organism that obtains energy from sunlight and carbon from CO_2 by photosynthesis.
photoheterotroph An organism that obtains energy from sunlight and carbon from organic sources.
photon (fō'-ton) A fixed quantity of light energy. The shorter the wavelength of light, the greater the energy of a photon.
photoperiod The relative lengths of day and night; an environmental stimulus that plants use to detect the time of year.
photophosphorylation (fō'-tō-fos'-fōr-uh-lā'-shun) The production of ATP by chemiosmosis during the light reactions of photosynthesis.
photopsin (fō-top'-sin) One of a family of visual pigments in the cones of the vertebrate eye that absorb bright, colored light.
photoreceptor A type of electromagnetic sensory receptor that detects light.
photorespiration In a plant cell, a metabolic pathway that consumes oxygen, releases CO_2, and decreases photosynthetic output. Photorespiration generally occurs on hot, dry days, when stomata close, O_2 accumulates in the leaf, and rubisco fixes O_2 rather than CO_2. Photorespiration produces no sugar molecules or ATP.
photosynthesis (fō'-tō-sin'-thuh-sis) The process by which plants, autotrophic protists, and some bacteria use light energy to make sugars and other organic food molecules from carbon dioxide and water.
photosystem A light-capturing unit of a chloroplast's thylakoid membrane, consisting of a reaction-center complex surrounded by numerous light-harvesting complexes.
phototropism (fō'-tō-trō'-pizm) The growth of a plant shoot toward light (positive phototropism) or away from light (negative phototropism).
phylogenetic species concept A definition of species as the smallest group of individuals that shares a common ancestor, forming one branch on the tree of life.
phylogenetic tree (fī'-lō-juh-net'-ik) A branching diagram that represents a hypothesis about the evolutionary history of a group of organisms.
phylogeny (fī-loj'-uh-nē) The evolutionary history of a species or group of related species.
phylum (fī'-lum) (plural, **phyla**) In Linnaean classification, the taxonomic category above class.
physiology (fi'-zē-ol'-uh-ji) The study of the functions of an organism's structures.
phytochrome (fī'-tuh-krōm) A plant protein that has a light-absorbing component.
phytoplankton (fī'-tō-plank'-ton) Algae and photosynthetic bacteria that drift passively in aquatic environments.
pineal gland (pin'-ē-ul) An outgrowth of the vertebrate brain that secretes the hormone melatonin, which coordinates daily and seasonal body activities such as the sleep/wake circadian rhythm with environmental light conditions.
pinna (pin'-uh) The flap-like part of the outer ear, projecting from the body surface of many birds and mammals; collects sound waves and channels them to the auditory canal.
pinocytosis (pē'-nō-sī-tō'-sis) Cellular "drinking"; a type of endocytosis in which the cell takes fluid and dissolved solutes into small membranous vesicles.
pistil Part of the reproductive organ of an angiosperm, a single carpel or a group of fused carpels.
pith Part of the ground tissue system of a dicot plant. Pith fills the center of a stem and may store food.

pituitary gland An endocrine gland at the base of the hypothalamus; consists of a posterior lobe, which stores and releases two hormones produced by the hypothalamus, and an anterior lobe, which produces and secretes many hormones that regulate diverse body functions.
pivot joint A joint that allows precise rotations in multiple planes. An example in humans is the joint that rotates the forearm at the elbow.
placenta (pluh-sen′-tuh) In most mammals, the organ that provides nutrients and oxygen to the embryo and helps dispose of its metabolic wastes; formed of the embryo's chorion and the mother's endometrial blood vessels.
placental mammal (pluh-sen′-tul) Mammal whose young complete their embryonic development in the uterus, nourished via the mother's blood vessels in the placenta; also called a eutherian.
plasma The liquid matrix of the blood in which the blood cells are suspended.
plasma cell An antibody-secreting B cell.
plasma membrane The membrane at the boundary of every cell that acts as a selective barrier to the passage of ions and molecules into and out of the cell; consists of a phospholipid bilayer with embedded proteins.
plasmid A small ring of independently replicating DNA separate from the main chromosome(s). Plasmids are found in prokaryotes and yeasts.
plasmodesma (plaz′-mō-dez′-muh) (plural, **plasmodesmata**) An open channel in a plant cell wall through which strands of cytoplasm connect from adjacent cells.
plasmodial slime mold (plaz-mō′-dē-ul) A type of protist that has amoeboid cells, flagellated cells, and an amoeboid plasmodial feeding stage in its life cycle.
plasmodium (1) A single mass of cytoplasm containing many nuclei. (2) The amoeboid feeding stage in the life cycle of a plasmodial slime mold.
plate tectonics (tek-tän′-iks) The theory that the continents are part of great plates of Earth's crust that float on the hot, underlying portion of the mantle. Movements in the mantle cause the continents to move slowly over time.
platelet A pinched-off cytoplasmic fragment of a bone marrow cell. Platelets circulate in the blood and are important in blood clotting.
pleiotropy (plī′-uh-trō-pē) The control of more than one phenotypic characteristic by a single gene.
polar covalent bond A covalent bond between atoms that differ in electronegativity. The shared electrons are pulled closer to the more electronegative atom, making it slightly negative and the other atom slightly positive.
polar ice A terrestrial biome that includes regions of extremely cold temperature and low precipitation located at high latitudes north of the arctic tundra and in Antarctica.
polar molecule A molecule containing polar covalent bonds and having an unequal distribution of charges.
pollen grain The structure that will produce the sperm in seed plants; the male gametophyte.
pollination In seed plants, the delivery by wind or animals of pollen from the pollen-producing parts of a plant to the stigma of a carpel.
polychaete (pol′-ē-kēt) A member of the largest group of annelids. *See* annelid.
polygamous Referring to a type of relationship in which an individual of one sex mates with several of the other.
polygenic inheritance (pol′-ē-jen′-ik) The additive effects of two or more gene loci on a single phenotypic characteristic.
polymer (pol′-uh-mer) A large molecule consisting of many identical or similar monomers linked together by covalent bonds.
polymerase chain reaction (PCR) (puh-lim′-uh-rās) A technique used to obtain many copies of a DNA molecule or a specific part of a DNA molecule. In the procedure, the starting DNA is mixed with a heat-resistant DNA polymerase, DNA nucleotides, and a few other ingredients. Specific nucleotide primers flanking the region to be copied ensure that it, and not other regions of the DNA, is replicated during the PCR procedure.
polynucleotide (pol′-ē-nū′-klē-ō-tīd) A polymer made up of many nucleotide monomers covalently bonded together.
polyp (pol′-ip) One of two types of cnidarian body forms; a columnar, hydra-like body.
polypeptide A polymer (chain) of amino acids linked by peptide bonds.
polyploid An organism that has more than two complete sets of chromosomes as a result of an accident of cell division.
polysaccharide (pol′-ē-sak′-uh-rīd) A carbohydrate polymer of many monosaccharides (sugars) linked by dehydration reactions.
pons (pahnz) Part of the vertebrate hindbrain that functions with the medulla oblongata in passing data between the spinal cord and forebrain and in controlling autonomic, homeostatic functions.
population A group of individuals belonging to one species and living in the same geographic area.
population density The number of individuals of a species per unit area or volume.
population ecology The study of how members of a population interact with their environment, focusing on factors that influence population density and growth.
population momentum In a population in which $r = 0$, the continuation of population growth as girls in the prereproductive age group reach their reproductive years.
positive feedback A type of control in which a change triggers mechanisms that amplify that change.
post-anal tail A tail posterior to the anus; found in chordate embryos and most adult chordates.
posterior Pertaining to the rear, or tail, of a bilaterally symmetric animal.
posterior pituitary An extension of the hypothalamus composed of nervous tissue that secretes hormones made in the hypothalamus; a temporary storage site for hypothalamic hormones.
postzygotic barrier A reproductive barrier that prevents hybrid zygotes produced by two different species from developing into viable, fertile adults. Includes reduced hybrid viability, reduced hybrid fertility, and hybrid breakdown.
potential energy The energy that matter possesses because of its location or arrangement. Water behind a dam possesses potential energy, and so do chemical bonds.
predation An interaction between species in which one species, the predator, eats the other, the prey.
prepuce (prē′-pyūs) A fold of skin covering the head of the clitoris or penis.
pressure flow mechanism The method by which phloem sap is transported through a plant from a sugar source, where sugars are produced, to a sugar sink, where sugars are used.
prevailing winds Winds that result from the combined effects of Earth's rotation and the rising and falling of air masses.
prezygotic barrier A reproductive barrier that impedes mating between species or hinders fertilization if mating between two species is attempted. Includes temporal, habitat, behavioral, mechanical, and gametic isolation.
primary consumer In the trophic structure of an ecosystem, an organism that eats plants or algae.
primary growth Growth in the length of a plant root or shoot, produced by an apical meristem.
primary immune response The initial immune response to an antigen, which appears after a lag of several days.
primary oocyte (ō′-uh-sīt) A diploid cell, in prophase I of meiosis, that can be hormonally triggered to develop into an egg.
primary phloem *See* phloem.
primary production The amount of solar energy converted to chemical energy (in organic compounds) by autotrophs in an ecosystem during a given time period.
primary spermatocyte (sper-mat′-eh-sīt′) A diploid cell in the testis that undergoes meiosis I.
primary structure The first level of protein structure; the specific sequence of amino acids making up a polypeptide chain.

primary succession A type of ecological succession in which a biological community arises in an area without soil. *See also* secondary succession.
primary xylem *See* xylem.
primers Short, artificially created, single-stranded DNA molecules that bind to each end of a target sequence during a PCR procedure.
prion An infectious form of protein that may multiply by converting related proteins to more prions. Prions cause several related diseases in different animals, including scrapie in sheep and mad cow disease.
problem solving Applying past experiences to overcome obstacles in novel situations.
producer An organism that makes organic food molecules from CO_2, H_2O, and other inorganic raw materials: a plant, alga, or autotrophic prokaryote.
product An ending material in a chemical reaction.
progestin (prō-jes′-tin) One of a family of steroid hormones, including progesterone, produced by the mammalian ovary. Progestins prepare the uterus for pregnancy.
programmed cell death The timely and tidy suicide (and disposal of the remains) of certain cells, triggered by certain genes; an essential process in normal development; also called apoptosis.
prokaryotic cell (prō-kār′-ē-ot′-ik) A type of cell lacking a membrane-enclosed nucleus and other membrane-enclosed organelles; found only in the domains Bacteria and Archaea.
prolactin (PRL) (prō-lak′-tin) A protein hormone secreted by the anterior pituitary that stimulates human mammary glands to produce and release milk and produces other responses in different animals.
prometaphase The second stage of mitosis, during which the nuclear envelope fragments and the spindle microtubules attach to the kinetochores of the sister chromatids.
promiscuous Referring to a type of relationship in which mating occurs with no strong pair-bonds or lasting relationships.
promoter A specific nucleotide sequence in DNA located near the start of a gene that is the binding site for RNA polymerase and the place where transcription begins.
prophage (prō′-fāj) Phage DNA that has inserted by genetic recombination into the DNA of a bacterial chromosome.
prophase The first stage of mitosis, during which the chromatin condenses to form structures (sister chromatids) visible with a light microscope and the mitotic spindle begins to form, but the nucleus is still intact.
prostate gland (pros′-tāt) A gland in human males that secretes a thin fluid that nourishes the sperm.
protein A functional biological molecule consisting of one or more polypeptides folded into a specific three-dimensional structure.
proteobacteria A clade of gram-negative bacteria that encompasses enormous diversity, including all four modes of nutrition.
proteomics The study of whole sets of proteins and their interactions.
protist A member of the kingdom Protista. Most protists are unicellular, though some are colonial or multicellular.
proton A subatomic particle with a single positive electrical charge, found in the nucleus of an atom.
proto-oncogene (prō′-tō-on′-kō-jēn) A normal gene that, through mutation, can be converted to a cancer-causing gene.
protostome A mode of animal development in which the opening formed during gastrulation becomes the mouth. Animals with the protostome pattern of development include the flatworms, molluscs, annelids, nematodes, and arthropods.
protozoan (prō′-tō-zō′-un) (plural, **protozoans**) A protist that lives primarily by ingesting food; a heterotrophic, "animal-like" protist.
proximal tubule In the vertebrate kidney, the portion of a nephron immediately downstream from Bowman's capsule that conveys and helps refine filtrate.
proximate cause In animal behavior, a condition in an animal's internal or external environment that is the immediate reason or mechanism for a behavior.
proximate question In animal behavior, a question that concerns the immediate reason for a behavior.
pseudocoelom (sū′-dō-sē′-lōm) A body cavity that is not lined with mesoderm and is in direct contact with the wall of the digestive tract.
pseudopodium (sū′-dō-pō′-dē-um) (plural, **pseudopodia**) A temporary extension of an amoeboid cell. Pseudopodia function in moving cells and engulfing food.
pulmonary artery A large blood vessel that conveys blood from the heart to a lung.
pulmonary circuit The branch of the circulatory system that supplies the lungs. *See also* systemic circuit.
pulmonary vein A blood vessel that conveys blood from a lung to the heart.
pulse The rhythmic stretching of the arteries caused by the pressure of blood during contraction of ventricles in systole.
punctuated equilibria In the fossil record, long periods of apparent stasis, in which a species undergoes little or no morphological change, interrupted by relatively brief periods of sudden change.
Punnett square A diagram used in the study of inheritance to show the results of random fertilization.
pupil The opening in the iris that admits light into the interior of the vertebrate eye. Muscles in the iris regulate the pupil's size.

Q

quaternary consumer (kwot′-er-ner-ē) An organism that eats tertiary consumers.
quaternary structure The fourth level of protein structure; the shape resulting from the association of two or more polypeptide subunits.

R

R plasmid A bacterial plasmid that carries genes for enzymes that destroy particular antibiotics, thus making the bacterium resistant to the antibiotics.
radial symmetry An arrangement of the body parts of an organism like pieces of a pie around an imaginary central axis. Any slice passing longitudinally through a radially symmetric organism's central axis divides it into mirror-image halves.
radioactive isotope An isotope whose nucleus decays spontaneously, giving off particles and energy.
radiolarian A protist that moves and feeds by means of threadlike pseudopodia and has a mineralized support structure composed of silica.
radiometric dating A method for determining the absolute ages of fossils and rocks, based on the half-life of radioactive isotopes.
radula (rad′-yū-luh) A toothed, rasping organ used to scrape up or shred food; found in many molluscs.
random dispersion pattern A pattern in which the individuals of a population are spaced in an unpredictable way.
ray-finned fish Bony fish; member of a clade of jawed vertebrates having fins supported by thin, flexible skeletal rays.
reabsorption In the vertebrate kidney, the reclaiming of water and valuable solutes from the filtrate.
reactant A starting material in a chemical reaction.
reading frame The way a cell's mRNA-translating machinery groups the mRNA nucleotides into codons.
receptor potential The electrical signal produced by sensory transduction.
receptor-mediated endocytosis (en′-dō-sī-tō′-sis) The movement of specific molecules into a cell by the inward budding of membranous vesicles, which contain proteins with receptor sites specific to the molecules being taken in.
recessive allele An allele that has no noticeable effect on the phenotype of a gene when the individual is heterozygous for that gene.
recombinant DNA A DNA molecule carrying nucleotide sequences derived from two or more sources.
recombination frequency With respect to two given genes, the number of recombinant progeny from a mating divided by the total number of progeny. Recombinant progeny carry

combinations of alleles different from those in either of the parents as a result of crossing over during meiosis.
Recommended Dietary Allowance (RDA) A recommendation for daily nutrient intake established by a national scientific panel.
rectum The terminal portion of the large intestine where the feces are stored until they are eliminated.
red alga A member of a group of marine, mostly multicellular, autotrophic protists, which includes the reef-building coralline algae. Red algae belong to the supergroup Archaeplastida.
red blood cell *See* erythrocyte.
red bone marrow A specialized tissue that is found in the cavities at the ends of bones and that produces blood cells.
red-green colorblindness A category of common, sex-linked human disorders involving several genes on the X chromosome; characterized by a malfunction of light-sensitive cells in the eyes; affects mostly males but also homozygous females.
redox reaction Short for **red**uction-**ox**idation reaction; a chemical reaction in which electrons are lost from one substance (oxidation) and added to another (reduction).
reduction The gain of electrons by a substance involved in a redox reaction; always accompanies oxidation.
reflex An automatic reaction to a stimulus, mediated by the spinal cord or lower brain.
regeneration The regrowth of body parts from pieces of an organism.
regulatory gene A gene that codes for a protein, such as a repressor, that controls the transcription of another gene or group of genes.
relative fitness The contribution an individual makes to the gene pool of the next generation, relative to the contributions of other individuals in the population.
releasing hormone A kind of hormone secreted by the hypothalamus that promotes the release of hormones from the anterior pituitary.
renal cortex The outer portion of the vertebrate kidney, above the renal medulla.
renal medulla The inner portion of the vertebrate kidney, beneath the renal cortex.
repetitive DNA Nucleotide sequences that are present in many copies in the DNA of a genome. The repeated sequences may be long or short and may be located next to each other (tandomly) or dispersed in the DNA.
repressor A protein that blocks the transcription of a gene or operon.
reproduction The creation of new individuals from existing ones.
reproductive cloning Using a somatic cell from a multicellular organism to make one or more genetically identical individuals.
reproductive cycle A recurring sequence of events that produces eggs, makes them available for fertilization, and prepares the female body for pregnancy.
reproductive isolation The existence of biological factors (barriers) that impede members of two species from producing viable, fertile hybrids.
reproductive system The organ system responsible for reproduction.
reptile Member of the clade of amniotes that includes snakes, lizards, turtles, crocodilians, and birds, along with a number of extinct groups, such as dinosaurs.
respiratory system The organ system that functions in exchanging gases with the environment. It supplies the blood with O_2 and disposes of CO_2.
resting potential The voltage across the plasma membrane of a resting neuron. The resting potential in a vertebrate neuron is typically around -70 millivolts, with the inside of the cell negatively charged relative to the outside.
restoration ecology The use of ecological principles to develop ways to return degraded ecosystems to conditions as similar as possible to their natural, predegraded state.
restriction enzyme A bacterial enzyme that cuts up foreign DNA (at specific DNA sequences called *restriction sites*), thus protecting bacteria against intruding DNA from phages and other organisms. Restriction enzymes are used in DNA technology to cut DNA molecules in reproducible ways. The pieces of cut DNA are called restriction fragments.
restriction fragment length polymorphism (RFLP) (rif′-lip) Variation in the length of a restriction fragment. RFLPs are produced when homologous DNA sequences containing SNPs are cut up with restriction enzymes.
restriction fragments Molecules of DNA produced from a longer DNA molecule cut up by a restriction enzyme. Restriction fragments are used in genome mapping and other applications.
restriction site A specific sequence on a DNA strand that is recognized as a "cut site" by a restriction enzyme.
retina (ret′-uh-nuh) The light-sensitive layer in an eye, made up of photoreceptor cells and sensory neurons.
retrovirus An RNA virus that reproduces by means of a DNA molecule. It reverse-transcribes its RNA into DNA, inserts the DNA into a cellular chromosome, and then transcribes more copies of the RNA from the viral DNA. HIV and a number of cancer-causing viruses are retroviruses.
reverse transcriptase (tran-skrip′-tās) An enzyme used by retroviruses that catalyzes the synthesis of DNA on an RNA template.
RFLP *See* restriction fragment length polymorphism (RFLP).
Rhizaria One of five supergroups proposed in a current hypothesis of the evolutionary history of eukaryotes. The other four supergroups are Chromalveolata, Excavata,Unikonta, and Archaeplastida.
rhizome (rī′-zōm) A horizontal stem that grows below the ground.
rhodopsin (ro-dop′-sin) A visual pigment that is located in the rods of the vertebrate eye and that absorbs dim light.
rhythm method A form of contraception that relies on refraining from sexual intercourse when conception is most likely to occur; also called natural family planning.
ribonucleic acid (RNA) (rī-bō-nū-klā′-ik) A type of nucleic acid consisting of nucleotide monomers with a ribose sugar and the nitrogenous bases adenine (A), cytosine (C), guanine (G), and uracil (U); usually single-stranded; functions in protein synthesis, gene regulation, and as the genome of some viruses.
ribosomal RNA (rRNA) (rī′-buh-sōm′-ul) The type of ribonucleic acid that, together with proteins, makes up ribosomes; the most abundant type of RNA in most cells.
ribosome (rī′-buh-sōm) A cell structure consisting of RNA and protein organized into two subunits and functioning as the site of protein synthesis in the cytoplasm. In eukaryotic cells, the ribosomal subunits are constructed in the nucleolus.
ribozyme (rī′-bō-zīm) An RNA molecule that functions as an enzyme.
RNA interference (RNAi) A biotechnology technique used to silence the expression of specific genes. Synthetic RNA molecules with sequences that correspond to particular genes trigger the breakdown of the gene's mRNA.
RNA polymerase (puh-lim′-uh-rās) A large molecular complex that links together the growing chain of RNA nucleotides during transcription, using a DNA strand as a template.
RNA splicing The removal of introns and joining of exons in eukaryotic RNA, forming an mRNA molecule with a continuous coding sequence; occurs before mRNA leaves the nucleus.
rod A photoreceptor cell in the vertebrate retina enabling vision in dim light.
root cap A cone of cells at the tip of a plant root that protects the root's apical meristem.
root hair An outgrowth of an epidermal cell on a root, which increases the root's absorptive surface area.
root pressure The upward push of xylem sap in a vascular plant, caused by the active pumping of minerals into the xylem by root cells.
root system All of a plant's roots, which anchor it in the soil, absorb and transport minerals and water, and store food.
rough endoplasmic reticulum (reh-tik′-yuh-lum) That portion of the endoplasmic reticulum with ribosomes attached that make membrane proteins and secretory proteins.
***r*-selection** The concept that in certain (*r*-selected) populations, a high reproductive rate is the chief determinant of life history.
rule of addition A rule stating that the probability that an event can occur in two or more alternative ways is the sum of the separate probabilities of the different ways.

rule of multiplication A rule stating that the probability of a compound event is the product of the separate probabilities of the independent events.

ruminant (rū′-min-ent) An animal, such as a cow or sheep, with multiple stomach compartments housing microorganisms that can digest cellulose.

S

SA (sinoatrial) node (sȳ′-nō′-ā′-trē-ul) The pacemaker of the heart, located in the wall of the right atrium, that sets the rate and timing at which all cardiac muscle cells contract.

sac fungus *See* ascomycete.

salivary glands Glands associated with the oral cavity that secrete substances to lubricate food and begin the process of chemical digestion.

salt A compound resulting from the formation of ionic bonds; also called an ionic compound.

sapwood Light-colored, water-conducting secondary xylem in a tree. *See also* heartwood.

sarcoma (sar-kō′-muh) Cancer of the supportive tissues, such as bone, cartilage, and muscle.

sarcomere (sar′-kō-mēr) The fundamental unit of muscle contraction, composed of thin filaments of actin and thick filaments of myosin; in electron micrographs, the region between two narrow, dark lines, called Z lines, in a myofibril.

saturated fatty acid A fatty acid in which all carbons in the hydrocarbon tail are connected by single bonds and the maximum number of hydrogen atoms are attached to the carbon skeleton. Saturated fats and fatty acids solidify at room temperature.

savanna A biome dominated by grasses and scattered trees.

scanning electron microscope (SEM) A microscope that uses an electron beam to study the fine details of cell surfaces or other specimens.

scavenger An animal that feeds on the carcasses of dead animals.

schizophrenia Severe mental disturbance characterized by psychotic episodes in which patients have a distorted perception of reality.

sclera (sklār′-uh) A layer of connective tissue forming the outer surface of the vertebrate eye. The cornea is the frontal part of the sclera.

sclereid (sklār′-ē-id) In plants, a very hard sclerenchyma cell found in nutshells and seed coats.

sclerenchyma cell (skluh-ren′-kē-muh) In plants, a supportive cell with rigid secondary walls hardened with lignin.

scrotum A pouch of skin outside the abdomen that houses a testis and functions in cooling sperm, keeping them viable.

search image The mechanism that enables an animal to find a particular kind of food efficiently.

second law of thermodynamics The principle stating that every energy conversion reduces the order of the universe, increasing its entropy. Ordered forms of energy are at least partly converted to heat.

secondary consumer An organism that eats primary consumers.

secondary endosymbiosis A theory that explains the evolution of protist diversity as the product of a symbiotic association that arose when an autotrophic eukaryotic protist (a product of primary endosymbiosis) was engulfed by a heterotrophic eukaryotic protist.

secondary growth An increase in a plant's diameter, involving cell division in the vascular cambium and cork cambium.

secondary immune response The immune response elicited when an animal encounters the same antigen at some later time. The secondary immune response is more rapid, of greater magnitude, and of longer duration than the primary immune response.

secondary oocyte (ō′-uh-sīt′) A haploid cell resulting from meiosis I in oogenesis, which will become an egg after meiosis II.

secondary phloem *See* phloem.

secondary spermatocyte (sper-mat′-uh-sīt′) A haploid cell that results from meiosis I in spermatogenesis and becomes a sperm cell after meiosis II.

secondary structure The second level of protein structure; the regular local patterns of coils or folds of a polypeptide chain.

secondary succession A type of ecological succession that occurs where a disturbance has destroyed an existing biological community but left the soil intact. *See also* primary succession.

secondary xylem *See* xylem.

secretion (1) The discharge of molecules synthesized by a cell. (2) In the vertebrate kidney, the discharge of wastes from the blood into the filtrate from the nephron tubules.

seed A plant embryo packaged with a food supply within a protective covering.

seed coat A tough outer covering of a seed, formed from the outer coat (integuments) of an ovule. In a flowering plant, the seed coat encloses and protects the embryo and endosperm.

seedless vascular plants The informal collective name for lycophytes (club mosses and their relatives) and pterophytes (ferns and their relatives).

segmentation Subdivision along the length of an animal body into a series of repeated parts called segments; allows for greater flexibility and mobility.

selective permeability (per′-mē-uh-bil′-uh-tē) A property of biological membranes that allows some substances to cross more easily than others and blocks the passage of other substances altogether.

self protein A protein on the surface of an antigen-presenting cell that can hold a foreign antigen and display it to helper T cells. Each individual has a unique set of self proteins that serve as molecular markers for the body. Lymphocytes do not attack self proteins unless the proteins are displaying foreign antigens; therefore, self proteins mark normal body cells as off-limits to the immune system. The technical name for self proteins is *major histocompatibility complex (MHC) proteins. See also* nonself molecule.

self-fertilize A form of reproduction that involves fusion of sperm and egg produced by the same individual organism.

semen (sē′-mun) The sperm-containing fluid that is ejaculated by the male during orgasm.

semicircular canals Fluid-filled channels in the inner ear that detect changes in the head's rate of rotation or angular movement.

semiconservative model Type of DNA replication in which the replicated double helix consists of one old strand, derived from the old molecule, and one newly made strand.

seminal vesicle (sem′-uh-nul ves′-uh-kul) A gland in males that secretes a thick fluid that contains fructose, which provides most of the sperm's energy.

seminiferous tubule (sem′-uh-nif′-uh-rus) A coiled sperm-producing tube in a testis.

sensitive period A limited phase in an individual animal's development when learning of particular behaviors can take place.

sensory adaptation The tendency of sensory neurons to become less sensitive when they are stimulated repeatedly. For example, a prominent smell becomes unnoticeable over time.

sensory input The conduction of signals from sensory receptors to processing centers in the central nervous system.

sensory neuron A nerve cell that receives information from sensory receptors and conveys signals into the central nervous system.

sensory receptor A specialized cell or neuron that detects stimuli and sends information to the central nervous system.

sensory transduction The conversion of a stimulus signal to an electrical signal by a sensory receptor.

sepal (sē′-pul) A modified leaf of a flowering plant. A whorl of sepals encloses and protects the flower bud before it opens.

sessile An organism that is anchored to its substrate.

sex chromosome A chromosome that determines whether an individual is male or female.

sex-linked gene A gene located on a sex chromosome. In humans, the vast majority of sex-linked genes are located on the X chromosome.

sexual dimorphism (dī-mōr′-fizm) Marked differences between the secondary sex characteristics of males and females.

sexual reproduction The creation of genetically unique offspring by the fusion of two haploid sex cells (gametes), forming a diploid zygote.

sexual selection A form of natural selection in which individuals with certain inherited traits are more likely than other individuals to obtain mates.
sexually transmitted disease (STD) A contagious disease spread by sexual contact.
shared ancestral characters A character shared by members of a particular clade that originated in an ancestor that is not a member of that clade.
shared derived characters An evolutionary novelty that is unique to a particular clade.
shoot system All of a plant's stems, leaves, and reproductive structures.
short tandem repeat (STR) A series of short DNA sequences that are repeated many times in a row in the genome.
short-day plant A plant that flowers in late summer, fall, or winter, when day length is short. Short-day plants actually flower in response to long nights.
short-term memory The ability to hold information, anticipations, or goals for a time and then release them if they become irrelevant.
sickle-cell disease A genetic condition caused by a mutation in the gene for hemoglobin. The mutation causes the protein to crystallize, which deforms red blood cells into a curved shape. Such blood cells produce a cascade of symptoms that can be life-threatening.
sieve plate An end wall in a sieve-tube element that facilitates the flow of phloem sap.
sieve-tube element A food-conducting cell in a plant; also called a sieve-tube member. Chains of sieve-tube elements make up phloem tissue.
signal In behavioral ecology, a stimulus transmitted by one animal to another animal.
signal transduction pathway In cell biology, a series of molecular changes that converts a signal on a target cell's surface to a specific response inside the cell.
silencer A eukaryotic DNA sequence that functions to inhibit the start of gene transcription; may act analogously to an enhancer by binding a repressor. *See also* enhancer.
silent mutation A mutation in a gene that changes a codon to one that encodes for the same amino acid as the original codon. The amino acid sequence of the resulting polypeptide is thus unchanged.
single circulation A circulatory system with a single pump and circuit, in which blood passes from the sites of gas exchange to the rest of the body before returning to the heart.
single nucleotide polymorphism (SNP) A one-nucleotide variation in DNA sequence found within the genomes of at least 1% of a population.
single-lens eye The camera-like eye found in some jellies, polychaetes, spiders, many molluscs, and vertebrates.
sister chromatid (krō′-muh-tid) One of the two identical parts of a duplicated chromosome in a eukaryotic cell. Prior to mitosis, sister chromatids remain attached to each another at the centromere.
skeletal muscle A type of striated muscle attached to the skeleton; generally responsible for voluntary movements of the body.
skeletal system The organ system that provides body support and protects body organs, such as the brain, heart, and lungs.
small intestine The longest section of the alimentary canal. It is the principal site of the enzymatic hydrolysis of food macromolecules and the absorption of nutrients.
smooth endoplasmic reticulum That portion of the endoplasmic reticulum that lacks ribosomes.
smooth muscle A type of muscle lacking striations; responsible for involuntary body activities.
SNP *See* single nucleotide polymorphism (SNP).
social behavior Any kind of interaction between two or more animals, usually of the same species.
social learning Modification of behavior through the observation of other individuals.
sociobiology The study of the evolutionary basis of social behavior.
sodium-potassium (Na-K) pump A membrane protein that transports sodium ions out of, and potassium ions into, a cell against their concentration gradients. The process is powered by ATP.
solute (sol′-yūt) A substance that is dissolved in a solution.
solution A liquid that is a homogeneous mixture of two or more substances.
solvent The dissolving agent of a solution. Water is the most versatile solvent known.
somatic cell (sō-mat′-ik) Any cell in a multicellular organism except a sperm or egg cell or a cell that develops into a sperm or egg.
spatial learning Modification of behavior based on experience of the spatial structure of the environment.
speciation The evolution of a new species.
species diversity The variety of species that make up a community. Species diversity includes both species richness (the total number of different species) and the relative abundance of the different species in the community.
sperm A male gamete.
spermatogenesis (sper-mat′-ō-jen′-uh-sis) The formation of sperm cells.
spermicide A sperm-killing chemical (cream, jelly, or foam) that works with a barrier device as a method of contraception.
sphincter (sfink′-ter) A ringlike band of muscle fibers that regulates passage between some compartments of the alimentary canal.
spinal cord A jellylike bundle of nerve fibers that runs lengthwise inside the spine in vertebrates and integrates simple responses to certain stimuli.
spirochete A member of a group of helical bacteria that spiral through the environment by means of rotating, internal filaments.
sponge An aquatic animal characterized by a highly porous body.
sporangium (spuh-ranj′-ē-um′) (plural, **sporangia**) A structure in fungi and plants in which meiosis occurs and haploid spores develop.
spore (1) In plants and algae, a haploid cell that can develop into a multicellular individual without fusing with another cell. (2) In prokaryotes, protists, and fungi, any of a variety of thick-walled life cycle stages capable of surviving unfavorable environmental conditions.
sporophyte (spōr′-uh-fīt) The multicellular diploid form in the life cycle of organisms undergoing alternation of generations; results from a union of gametes and meiotically produces haploid spores that grow into the gametophyte generation.
stabilizing selection Natural selection that favors intermediate variants by acting against extreme phenotypes.
stamen (stā′-men) A pollen-producing male reproductive part of a flower, consisting of a filament and an anther.
starch A storage polysaccharide in plants; a polymer of glucose.
start codon (kō′-don) On mRNA, the specific three-nucleotide sequence (AUG) to which an initiator tRNA molecule binds, starting translation of genetic information.
stem The part of a plant's shoot system that supports the leaves and reproductive structures.
stem cell An unspecialized cell that can divide to produce an identical daughter cell and a more specialized daughter cell, which undergoes differentiation.
steroid (ster′-oyd) A type of lipid whose carbon skeleton is in the form of four fused rings with various chemical groups attached. Examples are cholesterol, testosterone, and estrogen.
steroid hormone A lipid made from cholesterol that acts as a regulatory chemical, entering a target cell and activating the transcription of specific genes.
stigma (stig′-muh) (plural, **stigmata**) The sticky tip of a flower's carpel, which traps pollen grains.
stimulus (plural, **stimuli**) (1) In the context of a nervous system, any factor that causes a nerve signal to be generated. (2) In behavioral biology, an environmental cue that triggers a specific response.
stoma (stō′-muh) (plural, **stomata**) A pore surrounded by guard cells in the epidermis of a leaf. When stomata are open, CO_2 enters a leaf, and water and O_2 exit. A plant conserves water when its stomata are closed.
stomach An organ in a digestive tract that stores food and performs preliminary steps of digestion.
stop codon In mRNA, one of three triplets (UAG, UAA, UGA) that signal gene translation to stop.

STR *See* short tandem repeat (STR).
STR analysis Short tandem repeat analysis; a method of DNA profiling that compares the lengths of short tandem repeats (STRs) selected from specific sites within the genome.
strata (singular, **stratum**) Rock layers formed when new layers of sediment cover older ones and compress them.
stretch receptor A type of mechanoreceptor sensitive to changes in muscle length; detects the position of body parts.
stroke The death of nervous tissue in the brain, usually resulting from rupture or blockage of arteries in the head.
stroma (strō′-muh) The dense fluid within the chloroplast that surrounds the thylakoid membrane and is involved in the synthesis of organic molecules from carbon dioxide and water. Sugars are made in the stroma by the enzymes of the Calvin cycle.
stromatolite (strō-mat′-uh-līt) Layered rock that results from the activities of prokaryotes that bind thin films of sediment together.
substrate (1) A specific substance (reactant) on which an enzyme acts. Each enzyme recognizes only the specific substrate or substrates of the reaction it catalyzes. (2) A surface in or on which an organism lives.
substrate feeder An organism that lives in or on its food source, eating its way through the food.
substrate-level phosphorylation The formation of ATP by an enzyme directly transferring a phosphate group to ADP from an organic molecule (for example, one of the intermediates in glycolysis or the citric acid cycle).
sugar sink A plant organ that is a net consumer or storer of sugar. Growing roots, shoot tips, stems, and fruits are sugar sinks supplied by phloem.
sugar source A plant organ in which sugar is being produced by either photosynthesis or the breakdown of starch. Mature leaves are the primary sugar sources of plants.
sugar-phosphate backbone In a polynucleotide (DNA or RNA strand), the alternating chain of sugar and phosphate to which nitrogenous bases are attached.
superior vena cava (vē′-nuh kā′-vuh) A large vein that returns oxygen-poor blood to the heart from the upper body and head. *See also* inferior vena cava.
surface tension A measure of how difficult it is to stretch or break the surface of a liquid. Water has a high surface tension because of the hydrogen bonding of surface molecules.
surfactant A substance secreted by alveoli that decreases surface tension in the fluid that coats the alveoli.
survivorship curve A plot of the number of members of a cohort that are still alive at each age; one way to represent age-specific mortality.
suspension feeder An aquatic animal that sifts small food particles from the water; sometimes called a filter feeder.
sustainability The goal of developing, managing, and conserving Earth's resources in ways that meet the needs of people today without compromising the ability of future generations to meet theirs.
sustainable agriculture Long-term productive farming methods that are environmentally safe.
sustainable resource management Management of a natural resource so as not to damage the resource.
swim bladder A gas-filled internal sac that helps bony fishes maintain buoyancy.
symbiosis (sim′-bē-ō-sis) A physically close association between organisms of two or more species.
sympathetic division A set of neurons in the autonomic nervous system that generally prepares the body for energy-consuming activities, such as fleeing or fighting. *See also* parasympathetic division.
sympatric speciation The formation of new species in populations that live in the same geographic area.
synapse (sin′-aps) A junction between two neurons, or between a neuron and an effector cell. Electrical or chemical signals are relayed from one cell to another at a synapse.
synaptic cleft (sin-ap′-tik) In a chemical synapse, a narrow gap separating the synaptic terminal of a transmitting neuron from a receiving neuron or an effector cell.
synaptic terminal The tip of a transmitting neuron's axon, where signals are sent to another neuron or to an effector cell.
synaptic vesicle A membrane-enclosed sac containing neurotransmitter molecules at the tip of the presynaptic axon.
systematics A scientific discipline focused on classifying organisms and determining their evolutionary relationships.
systemic acquired resistance A defensive response in plants infected with a pathogenic microbe; helps protect healthy tissue from the microbe.
systemic circuit The branch of the circulatory system that supplies oxygen-rich blood to, and carries oxygen-poor blood away from, organs and tissues in the body. *See also* pulmonary circuit.
systems biology An approach to studying biology that aims to model the dynamic behavior of whole biological systems based on a study of the interactions among the system's parts.
systole (sis′-tō-lē) The contraction stage of the heart cycle, when the heart chambers actively pump blood. *See also* diastole.

T

T cell A type of lymphocyte that matures in the thymus and is responsible for the cell-mediated immune response. T cells are also involved in humoral immunity.
taiga The northern coniferous forest, characterized by long, snowy winters and short, wet summers, extending across North America and Eurasia to the southern border of the arctic tundra; also found just below alpine tundra on mountainsides in temperate zones.
tapeworm A parasitic flatworm characterized by the absence of a digestive tract.
target cell A cell that responds to a regulatory signal, such as a hormone.
taxis (tak′-sis) (plural, **taxes**) Virtually automatic orientation toward or away from a stimulus.
taxon A named taxonomic unit at any given level of classification.
taxonomy The scientific discipline concerned with naming and classifying the diverse forms of life.
technology The application of scientific knowledge for a specific purpose, often involving industry or commerce but also including uses in basic research.
telomere (tel′-uh-mēr) The repetitive DNA at each end of a eukaryotic chromosome.
telophase The fifth and final stage of mitosis, during which daughter nuclei form at the two poles of a cell. Telophase usually occurs together with cytokinesis.
temperate broadleaf forest A biome located throughout midlatitude regions, where there is sufficient moisture to support the growth of large, broadleaf deciduous trees.
temperate grassland A grassland region maintained by seasonal drought, occasional fires, and grazing by large mammals.
temperate rain forest Coniferous forests of coastal North America (from Alaska to Oregon) supported by warm, moist air from the Pacific Ocean.
temperate zones Latitudes between the tropics and the Arctic Circle in the north and the Antarctic Circle in the south; regions with milder climates than the tropics or polar regions.
temperature A measure of the intensity of heat in degrees, reflecting the average kinetic energy or speed of molecules.
tendon Fibrous connective tissue connecting a muscle to a bone.
tendril A modified leaf used by some plants to climb around a fixed structure.
terminal bud Embryonic tissue at the tip of a shoot, made up of developing leaves and a compact series of nodes and internodes.
terminator A special sequence of nucleotides in DNA that marks the end of a gene. It signals RNA polymerase to release the newly made RNA molecule and then to depart from the gene.
territory An area that one or more individuals defend and from which other members of the same species are usually excluded.

tertiary consumer (ter′-shē-ār-ē) An organism that eats secondary consumers.
tertiary structure The third level of protein structure; the overall three-dimensional shape of a polypeptide due to interactions of the R groups of the amino acids making up the chain.
testcross The mating between an individual of unknown genotype for a particular characteristic and an individual that is homozygous recessive for that same characteristic. The testcross can be used to determine the unknown genotype (homozygous dominant versus heterozygous).
testicle A testis and scrotum together.
testis (plural, **testes**) The male gonad in an animal. The testis produces sperm and, in many species, reproductive hormones.
testosterone (tes-tos′-tuh-rōn) An androgen hormone that stimulates an embryo to develop into a male and promotes male body features.
tetrad A paired set of homologous chromosomes, each composed of two sister chromatids. Tetrads form during prophase I of meiosis, when crossing over may occur.
tetrapod A vertebrate with two pairs of limbs. Tetrapods include mammals, amphibians, and birds and other reptiles.
thalamus (thal′-uh-mus) An integrating and relay center of the vertebrate forebrain; sorts and relays selected information to specific areas in the cerebral cortex.
theory A widely accepted explanatory idea that is broader in scope than a hypothesis, generates new hypotheses, and is supported by a large body of evidence.
therapeutic cloning The cloning of human cells by nuclear transplantation for therapeutic purposes, such as the generation of embryonic stem cells. *See* nuclear transplantation; reproductive cloning.
thermodynamics The study of energy transformation that occurs in a collection of matter. *See* first law of thermodynamics; second law of thermodynamics.
thermoreceptor A sensory receptor that detects heat or cold.
thermoregulation The homeostatic maintenance of internal temperature within a range that allows cells to function efficiently.
thick filament The thicker of the two protein filaments in muscle fibers, consisting of staggered arrays of myosin molecules.
thigmotropism (thig′-mō-trō′-pizm) A plant's directional growth movement in response to touch.
thin filament The thinner of the two protein filaments in muscle fibers, consisting of two strands of actin and two strands of regulatory protein coiled around each other.
three-domain system A system of taxonomic classification based on three basic groups: Bacteria, Archaea, and Eukarya.
threshold The minimum change in a membrane's voltage that must occur to generate an action potential (nerve signal).
thylakoid (thī′-luh-koyd) A flattened membranous sac inside a chloroplast. Thylakoid membranes contain chlorophyll and the molecular complexes of the light reactions of photosynthesis. A stack of thylakoids is called a granum.
thymine (T) (thī′-min) A single-ring nitrogenous base found in DNA.
thymus gland (thī′-mus) An endocrine gland in the neck region of mammals that is active in establishing the immune system; secretes several hormones that promote the development and differentiation of T cells.
thyroid gland (thī′-royd) An endocrine gland located in the neck that secretes thyroxine (T_4), triiodothyronine (T_3), and calcitonin.
thyroid-stimulating hormone (TSH) A protein hormone secreted by the anterior pituitary that stimulates the thyroid gland to secrete its hormones.
thyroxine (T_4) (thī-rok′-sin) An amine hormone secreted by the thyroid gland that stimulates metabolism in virtually all body tissues. Each molecule of this hormone contains four atoms of iodine.
Ti plasmid A bacterial plasmid that induces tumors in plant cells that the bacterium infects. Ti plasmids are often used as vectors to introduce new genes into plant cells. Ti stands for "tumor inducing."
tissue An integrated group of cells with a common function, structure, or both.
tissue system One or more tissues organized into a functional unit within a plant or animal.
tonicity The ability of a solution surrounding a cell to cause that cell to gain or lose water.
topsoil The uppermost soil layer, consisting of a mixture of particles derived from rock, living organisms, and humus.
trace element An element that is essential for life but required in extremely minute amounts.
trachea (trā′-kē-uh) (plural, **tracheae**) The windpipe; the portion of the respiratory tube that passes from the larynx to the two bronchi.
tracheal system A system of branched, air-filled tubes in insects that extends throughout the body and carries oxygen directly to cells.
tracheid (trā′-kē-id) A tapered, porous, water-conducting and supportive cell in plants. Chains of tracheids or vessel elements make up the water-conducting, supportive tubes in xylem.
trade winds The movement of air in the tropics (those regions that lie between 23.5° north latitude and 23.5° south latitude).
trait A variant of a character found within a population, such as purple or white flowers in pea plants.
trans fat An unsaturated fat, formed artificially during hydrogenation of vegetable oils, which is linked to health risks.
transcription The synthesis of RNA on a DNA template.
transcription factor In the eukaryotic cell, a protein that functions in initiating or regulating transcription. Transcription factors bind to DNA or to other proteins that bind to DNA.
transduction (1) The transfer of bacterial genes from one bacterial cell to another by a phage. (2) *See* sensory transduction. (3) *See* signal transduction pathway.
transfer RNA (tRNA) A type of ribonucleic acid that functions as an interpreter in translation. Each tRNA molecule has a specific anticodon, picks up a specific amino acid, and conveys the amino acid to the appropriate codon on mRNA.
transformation The incorporation of new genes into a cell from DNA that the cell takes up from the surrounding environment.
transgenic organism An organism that contains genes from another species.
translation The synthesis of a polypeptide using the genetic information encoded in an mRNA molecule. There is a change of "language" from nucleotides to amino acids.
translocation (1) During protein synthesis, the movement of a tRNA molecule carrying a growing polypeptide chain from the A site to the P site on a ribosome. (The mRNA travels with it.) (2) A change in a chromosome resulting from a chromosomal fragment attaching to a nonhomologous chromosome; can occur as a result of an error in meiosis or from mutagenesis.
transmission electron microscope (TEM) A microscope that uses an electron beam to study the internal structure of thinly sectioned specimens.
transpiration The evaporative loss of water from a plant.
transpiration-cohesion-tension mechanism A transport mechanism that drives the upward movement of water in plants. Transpiration exerts a pull that is relayed downward along a string of molecules held together by cohesion and helped upward by adhesion.
transport vesicle A small membranous sac in a eukaryotic cell's cytoplasm carrying molecules produced by the cell. The vesicle buds from the endoplasmic reticulum or Golgi and eventually fuses with another organelle or the plasma membrane, releasing its contents.
transposable element A transposable genetic element, or "jumping gene"; a segment of DNA that can move from one site to another within a cell and serve as an agent of genetic change.
TRH (TSH-releasing hormone) A peptide hormone that triggers the release of TSH (thyroid-stimulating hormone), which in turn stimulates the thyroid gland.
trial-and-error learning Learning to associate a particular behavior with a positive or negative effect.
triiodothyronine (T_3) (trī′-ī-ō-dō-thī′-rō-nīn) An amine hormone secreted by the thyroid gland that stimulates metabolism in

virtually all body tissues. Each molecule of this hormone contains four atoms of iodine.
trimester In human development, one of three 3-month-long periods of pregnancy.
triplet code A set of three-nucleotide-long "words" that specify the amino acids for polypeptide chains. *See* genetic code.
trisomy 21 *See* Down syndrome.
trophoblast (trōf′-ō-blast) In mammalian development, the outer portion of a blastocyst. Cells of the trophoblast secrete enzymes that enable the blastocyst to implant in the endometrium of the mother's uterus.
tropical forest A terrestrial biome characterized by high levels of precipitation and warm temperatures year-round.
tropics Latitudes between 23.5° north and south.
tropism (trō′-pizm) A growth response that makes a plant grow toward or away from a stimulus.
true coelom A body cavity completely lined with tissue derived from mesoderm.
true-breeding Referring to organisms for which sexual reproduction produces offspring with inherited traits identical to those of the parents. The organisms are homozygous for the characteristics under consideration.
tubal ligation A means of sterilization in which a segment of each of a woman's two oviducts (fallopian tubes) is removed. The ends of the tubes are then tied closed to prevent eggs from reaching the uterus (commonly referred to as having the "tubes tied").
tuber An enlargement at the end of a rhizome in which food is stored.
tumor An abnormal mass of rapidly growing cells that forms within otherwise normal tissue.
tumor-suppressor gene A gene whose product inhibits cell division, thereby preventing uncontrolled cell growth. A mutation that deactivates a tumor-suppressor gene may lead to cancer.
tundra A biome at the northernmost limits of plant growth and at high altitudes, characterized by dwarf woody shrubs, grasses, mosses, and lichens.
tunicate One of a group of invertebrate chordates, also known as sea squirts.

U

ultimate cause In animal behavior, the evolutionary reason for a behavior.
ultimate question In animal behavior, a question that addresses the evolutionary basis for behavior.
ultrasound imaging A technique for examining a fetus in the uterus. High-frequency sound waves echoing off the fetus are used to produce an image of the fetus.
uniform dispersion pattern A pattern in which the individuals of a population are evenly distributed over an area.
Unikonta One of five supergroups proposed in a current hypothesis of the evolutionary history of eukaryotes. The other four supergroups are Chromalveolata, Rhizaria, Excavata, and Archaeplastida.
unsaturated fatty acid A fatty acid that has one or more double bonds between carbons in the hydrocarbon tail and thus lacks the maximum number of hydrogen atoms. Unsaturated fats and fatty acids do not solidify at room temperature.
uracil (U) (yū′-ruh-sil) A single-ring nitrogenous base found in RNA.
urea (yū-rē′-ah) A soluble form of nitrogenous waste excreted by mammals and most adult amphibians.
ureter (yū-rē′-ter or yū′-reh-ter) A duct that conveys urine from the kidney to the urinary bladder.
urethra (yū-rē′-thruh) A duct that conveys urine from the urinary bladder to the outside. In the male, the urethra also conveys semen out of the body during ejaculation.
uric acid (yū′-rik) An insoluble precipitate of nitrogenous waste excreted by land snails, insects, birds, and some reptiles.
urinary bladder The pouch where urine is stored prior to elimination.
urinary system The organ system that forms and excretes urine while regulating the amount of water and ions in the body fluids.
urine Concentrated filtrate produced by the kidneys and excreted by the bladder.
uterus (yū′-ter-us) In the reproductive system of a mammalian female, the organ where the development of young occurs; the womb.

V

vaccination (vak′-suh-nā′-shun) A procedure that presents the immune system with a harmless variant or derivative of a pathogen, thereby stimulating the immune system to mount a long-term defense against the pathogen.
vaccine (vak-sēn′) A harmless variant or derivative of a pathogen used to stimulate a host organism's immune system to mount a long-term defense against the pathogen.
vacuole (vak′-ū-ōl) A membrane-enclosed sac that is part of the endomembrane system of a eukaryotic cell and has diverse functions.
vagina (vuh-jī′-nuh) Part of the female reproductive system between the uterus and the outside opening; the birth canal in mammals; also accommodates the male's penis and receives sperm during copulation.
vas deferens (vas def′-er-enz) (plural, **vasa deferentia**) Part of the male reproductive system that conveys sperm away from the testis; the sperm duct; in humans, the tube that conveys sperm between the epididymis and the common duct that leads to the urethra.
vascular bundle (vas′-kyū-ler) A strand of vascular tissues (both xylem and phloem) in a plant stem.
vascular cambium (vas′-kyū-ler kam′-bē-um) During secondary growth of a plant, the cylinder of meristematic cells, surrounding the xylem and pith, that produces secondary xylem and phloem.
vascular cylinder The central cylinder of vascular tissue in a plant root.
vascular plant A plant with xylem and phloem, including club mosses, ferns, gymnosperms, and angiosperms.
vascular tissue Plant tissue consisting of cells joined into tubes that transport water and nutrients throughout the plant body.
vascular tissue system A transport system formed by xylem and phloem throughout the plant. Xylem transports water and minerals, while phloem transports sugars and other organic nutrients.
vasectomy (vuh-sek′-tuh-mē) Surgical removal of a section of the two sperm ducts (vasa deferentia) to prevent sperm from reaching the urethra; a means of sterilization in males.
vector In molecular biology, a piece of DNA, usually a plasmid or a viral genome, that is used to move genes from one cell to another.
vein (1) In animals, a vessel that returns blood to the heart. (2) In plants, a vascular bundle in a leaf, composed of xylem and phloem.
ventilation A mechanism that provides for the flow of air or water over a respiratory surface.
ventral Pertaining to the underside, or bottom, of a bilaterally symmetric animal.
ventricle (ven′-truh-kul) (1) A heart chamber that pumps blood out of the heart. (2) A space in the vertebrate brain filled with cerebrospinal fluid.
venule (ven′-yūl) A vessel that conveys blood between a capillary bed and a vein.
vertebra (ver′-tuh-bruh) (plural, **vertebrae**) One of a series of segmented skeletal units that enclose the nerve cord, making up the backbone of a vertebrate animal.
vertebral column Backbone, composed of a series of segmented units called vertebrae.
vertebrate (ver′-tuh-brāt) A chordate animal with a backbone. Vertebrates include lampreys, chondrichthyans, ray-finned fishes, lobe-finned fishes, amphibians, reptiles (including birds), and mammals.
vesicle (ves′-i-kul) A sac made of membrane in the cytoplasm of a eukaryotic cell.
vessel element A short, open-ended, water-conducting and supportive cell in plants. Chains of vessel elements or tracheids make up the water-conducting, supportive tubes in xylem.
vestigial structure A feature of an organism that is a historical remnant of a structure that served a function in the organism's ancestors.

villus (vil′-us) (plural, **villi**) (1) A finger-like projection of the inner surface of the small intestine. (2) A finger-like projection of the chorion of the mammalian placenta. Large numbers of villi increase the surface areas of these organs.
viroid (vī′-royd) A plant pathogen composed of molecules of naked, circular RNA several hundred nucleotides long.
virus A microscopic particle capable of infecting cells of living organisms and inserting its genetic material. Viruses are generally not considered to be alive because they do not display all of the characteristics associated with life.
visceral mass (vis′-uh-rul) One of the three main parts of a mollusc, containing most of the internal organs.
visual acuity The ability of the eyes to distinguish fine detail. Normal visual acuity in humans is usually reported as "20/20 vision."
vital capacity The maximum volume of air that a mammal can inhale and exhale with each breath.
vitamin An organic nutrient that an organism requires in very small quantities. Many vitamins serve as coenzymes or parts of coenzymes.
vitreous humor (vit′-rē-us hyū′-mer) A jellylike substance filling the space behind the lens in the vertebrate eye; helps maintain the shape of the eye.
vocal cord One of a pair of bands of elastic tissue in the larynx. Air rushing past the tensed vocal cords makes them vibrate, producing sounds.
vulva The collective term for the external female genitalia.

W

water mold A fungus-like protist in the supergroup Chromalveolata.
water vascular system In echinoderms, a radially arranged system of water-filled canals that branch into extensions called tube feet. The system provides movement and circulates water, facilitating gas exchange and waste disposal.
wavelength The distance between crests of adjacent waves, such as those of the electromagnetic spectrum.
westerlies Winds that blow from west to east.
wetland An ecosystem intermediate between an aquatic ecosystem and a terrestrial ecosystem. Wetland soil is saturated with water permanently or periodically.
white blood cell *See* leukocyte.
white matter Regions within the central nervous system composed mainly of axons, with their whitish myelin sheaths.
whole-genome shotgun method A method for determining the DNA sequence of an entire genome. After a genome is cut into small fragments, each fragment is sequenced and then placed in the proper order.
wild-type trait The version of a character that most commonly occurs in nature.
wood Secondary xylem of a plant. *See also* heartwood; sapwood.
wood ray A column of parenchyma cells that radiates from the center of a log and transports water to its outer living tissues.

X

X chromosome inactivation In female mammals, the inactivation of one X chromosome in each somatic cell.
xylem (zī′-lum) The nonliving portion of a plant's vascular system that provides support and conveys xylem sap from the roots to the rest of the plant. Xylem is made up of vessel elements and/or tracheids, water-conducting cells. Primary xylem is derived from the procambium. Secondary xylem is derived from the vascular cambium in plants exhibiting secondary growth.
xylem sap The solution of inorganic nutrients conveyed in xylem tissue from a plant's roots to its shoots.

Y

yeast A single-celled fungus that inhabits liquid or moist habitats and reproduces asexually by simple cell division or by the pinching of small buds off a parent cell.
yellow bone marrow A tissue found within the central cavities of long bones, consisting mostly of stored fat.
yolk sac An extraembryonic membrane that develops from the endoderm. The yolk sac produces the embryo's first blood cells and germ cells and gives rise to the allantois.

Z

zoned reserve An extensive region of land that includes one or more areas that are undisturbed by humans. The undisturbed areas are surrounded by lands that have been altered by human activity.
zooplankton (zō′-ō-plank′-tun) Animals that drift in aquatic environments.
zygomycete (zī′-guh-mī-sēt) Member of a group of fungi characterized by a sturdy structure called a zygosporangium, in which meiosis produces haploid spores.
zygote (zī′-gōt) The diploid fertilized egg, which results from the union of a sperm cell nucleus and an egg cell nucleus.
zygote fungus *See* zygomycete.

Biology: *Exploring Life*

Biology: *Exploring Life*

BIG IDEAS

Themes in the Study of Biology
(1–4)

Common themes help to organize the study of life.

Image State Media Partners Limited

Evolution, the Core Theme of Biology
(5–7)

Evolution accounts for the unity and diversity of life and the evolutionary adaptations of organisms to their environment.

Neg./Transparency no. 326668. Courtesy Dept. of Library Services, American Museum of Natural History

The Process of Science
(8–9)

In studying nature, scientists make observations, form hypotheses, and test predictions with experiments.

Suzanne L. Collins/Science Source

Biology and Everyday Life
(10–11)

Learning about biology helps us understand many issues involving science, technology, and society.

National Geographic

Cyril Ruoso/Photolibrary

This young lemur spends much of its time on its mother's back. But it is also groomed and cared for by other females of the troop, a group of about 15 members led by a dominant female. About the size of large cats, these ring-tailed lemurs (scientific name, *Lemur catta*) are noted for their distinctive tails, dark eye patches and muzzle, and, as you can see on the cover of this book, entrancing eyes. They are also highly vocal primates, with 28 distinct calls, such as "predator alert," "stay out of my territory," and purrs of contentment.

About 33 species of lemurs live on the island of Madagascar, and before humans arrived 2,000 years ago, there were even more. The size of Texas, Madagascar is located about 400 km (240 miles) off the southeast coast of Africa. It is home to many plants and animals found nowhere else in the world. Its geographic history helps explain its remarkable biodiversity.

Madagascar was once part of a supercontinent that began breaking apart about 150 million years ago—before the primates (lemurs, monkeys, apes, and humans) had evolved. Apparently, 60 million years ago, ancestral lemurs floated on logs or vegetation from Africa to the island. Their new home was relatively free of predators and competitors and offered many different habitats, from tropical forests to deserts to highlands with spiny shrubs. Over millions of years, lemurs diversified on this isolated island.

The wonderful assortment of lemurs on Madagascar is the result of evolution, the process that has transformed life on Earth from its earliest beginnings to the diversity of organisms living today. In this chapter, we begin our exploration of biology—the scientific study of life, its evolution, and its amazing diversity.

Themes in the Study of Biology

1 All forms of life share common properties

Defining **biology** as the scientific study of life raises the obvious question: What is *life*? How would you describe what distinguishes living things from nonliving things? Even a small child realizes that a bug or a flower is alive, while a rock or water is not. They, like all of us, recognize life mainly by what living things do. **Figure 1** highlights seven of the properties and processes that we associate with life.

(1) *Order*. This close-up of a sunflower illustrates the highly ordered structure that typifies life. Living cells are the basis of this complex organization.

(2) *Reproduction*. Organisms reproduce their own kind. Here an emperor penguin protects its baby.

(3) *Growth and development*. Inherited information in the form of DNA controls the pattern of growth and development of all organisms, including this hatching crocodile.

(4) *Energy processing*. When this bear eats its catch, it will use the chemical energy stored in the fish to power its own activities and chemical reactions.

(5) *Response to the environment*. All organisms respond to environmental stimuli. This Venus flytrap closed its trap rapidly in response to the stimulus of a damselfly landing on it.

(6) *Regulation*. Many types of mechanisms regulate an organism's internal environment, keeping it within limits that sustain life. Pictured here is a typical lemur behavior with a regulatory function—"sunbathing"—which helps raise the animal's body temperature on cold mornings.

(7) *Evolutionary adaptation*. The leaflike appearance of this katydid camouflages it in its environment. Such adaptations evolve over many generations as individuals with traits best suited to their environment have greater reproductive success and pass their traits to offspring.

Figure 1 reminds us that the living world is wondrously varied. How do biologists make sense of this diversity and complexity, and how can you? Indeed, biology is a subject of enormous scope that gets bigger every year. One of the ways to help you organize all this information is to connect what you learn to a set of themes that you will encounter throughout your study of life. The next few modules introduce several of these themes: novel properties emerging at each level of biological organization, the cell as the fundamental unit of life, the correlation of structure and function, and the exchange of matter and energy as organisms interact with the environment. We then focus on the core theme of biology—evolution, the theme that makes sense of both the unity and diversity of life. And in the final two sections of the chapter, we look at the process of science and the relationship of biology to our everyday lives.

Let's begin our journey with a tour through the levels of the biological hierarchy.

? How would you define life?

● Life can be defined by a set of common properties such as those described in this module.

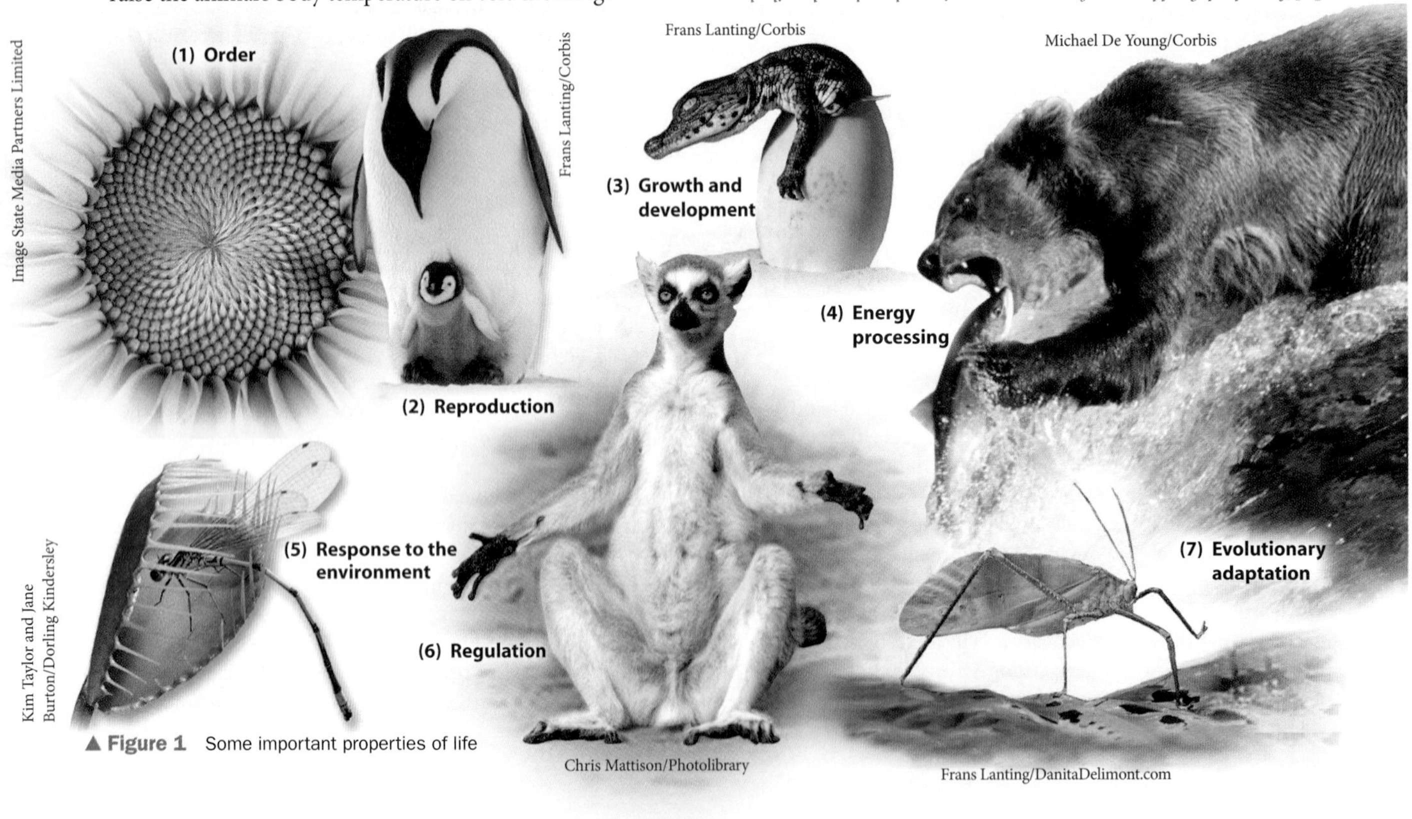

▲ **Figure 1** Some important properties of life

2 In life's hierarchy of organization, new properties emerge at each level

As **Figure 2** illustrates, the study of life extends from the global scale of the biosphere to the microscopic scale of molecules. At the upper left we take a distant view of the **biosphere**, all of the environments on Earth that support life. These include most regions of land, bodies of water, and the lower atmosphere.

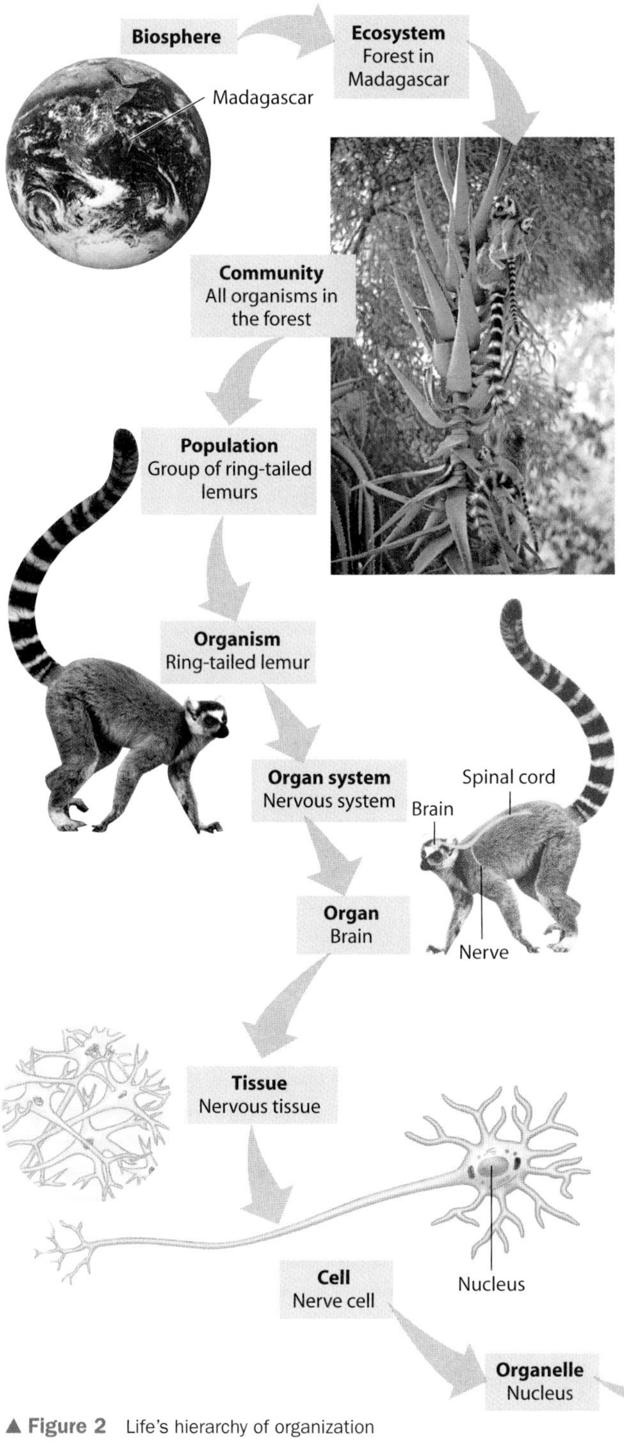

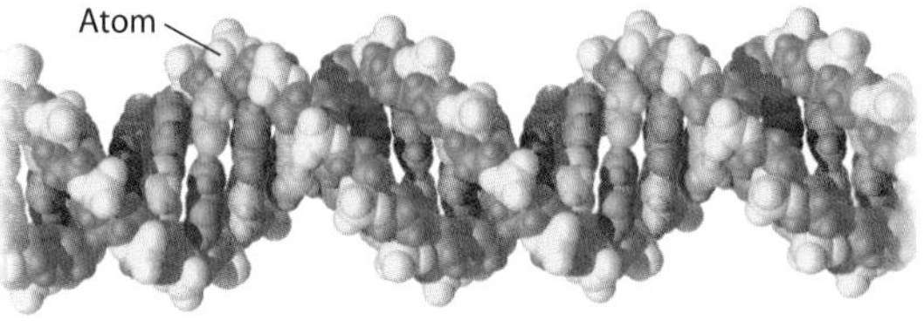

▲ **Figure 2** Life's hierarchy of organization

NASA/Johnson Space Center; Alain Mafart-Renodier/Photolibrary; Martin Harvey/Photolibrary

A closer look at one of these environments brings us to the level of an **ecosystem**, which consists of all the organisms living in a particular area, as well as the physical components with which the organisms interact, such as air, soil, water, and sunlight.

The entire array of organisms in an ecosystem is called a **community**. The community in this forest ecosystem in Madagascar includes the lemurs and the agave plant they are eating, as well as birds, snakes, and catlike carnivores called civets; a huge diversity of insects; many kinds of trees and other plants; fungi; and enormous numbers of microscopic protists and bacteria. Each unique form of life is called a species.

A **population** includes all the individuals of a particular species living in an area, such as all the ring-tailed lemurs in the forest community. Next in the hierarchy is the **organism**, an individual living thing.

Within a complex organism such as a lemur, life's hierarchy continues to unfold. An **organ system**, such as the circulatory system or nervous system, consists of several organs that cooperate in a specific function. For instance, the organs of the nervous system are the brain, the spinal cord, and the nerves. A lemur's nervous system controls its actions, such as climbing trees.

An **organ** is made up of several different **tissues**, each made up of a group of similar cells that perform a specific function. A **cell** is the fundamental unit of life. In the nerve cell shown here, you can see several organelles, such as the nucleus. An **organelle** is a membrane-enclosed structure that performs a specific function in a cell.

Finally, we reach the level of molecules in the hierarchy. A **molecule** is a cluster of small chemical units called atoms held together by chemical bonds. Our example in Figure 2 is a computer graphic of a section of DNA (deoxyribonucleic acid)—the molecule of inheritance.

Now let's work our way in the opposite direction in Figure 2, moving up life's hierarchy from molecules to the biosphere. It takes many molecules to build organelles, numerous organelles to make a cell, many cells to make a tissue, and so on. At each new level, there are novel properties that arise, properties that were not present at the preceding level. For example, life emerges at the level of the cell—a test tube full of organelles is not alive. Such **emergent properties** represent an important theme of biology. The familiar saying that "the whole is greater than the sum of its parts" captures this idea. The emergent properties of each level result from the specific arrangement and interactions of its parts.

? Which of these levels of biological organization includes all others in the list: cell, molecule, organ, tissue?

● Organ

3 Cells are the structural and functional units of life

The cell has a special place in the hierarchy of biological organization. It is the level at which the properties of life emerge—the lowest level of structure that can perform all activities required for life. A cell can regulate its internal environment, take in and use energy, respond to its environment, and develop and maintain its complex organization. The ability of cells to give rise to new cells is the basis for all reproduction and for the growth and repair of multicellular organisms.

All organisms are composed of cells. They occur singly as a great variety of unicellular (single-celled) organisms, such as amoebas and most bacteria. And cells are the subunits that make up multicellular organisms, such as lemurs and trees. Your body consists of trillions of cells of many different kinds.

All cells share many characteristics. For example, every cell is enclosed by a membrane that regulates the passage of materials between the cell and its surroundings. And every cell uses DNA as its genetic information. There are two basic types of cells. **Prokaryotic cells** were the first to evolve and were Earth's sole inhabitants for about the first 1.5 billion years of life on Earth. Fossil evidence indicates that **eukaryotic cells** evolved about 2.1 billion years ago.

Figure 3 shows these two types of cells as artificially colored photographs taken with an electron microscope. A prokaryotic cell is much simpler and usually much smaller than a eukaryotic cell. The cells of the microorganisms we call bacteria are prokaryotic. Plants, animals, fungi, and protists are all composed of eukaryotic cells. As you can see in Figure 3, a eukaryotic cell is subdivided by membranes into many functional compartments, called organelles. These include a nucleus, which houses the cell's DNA.

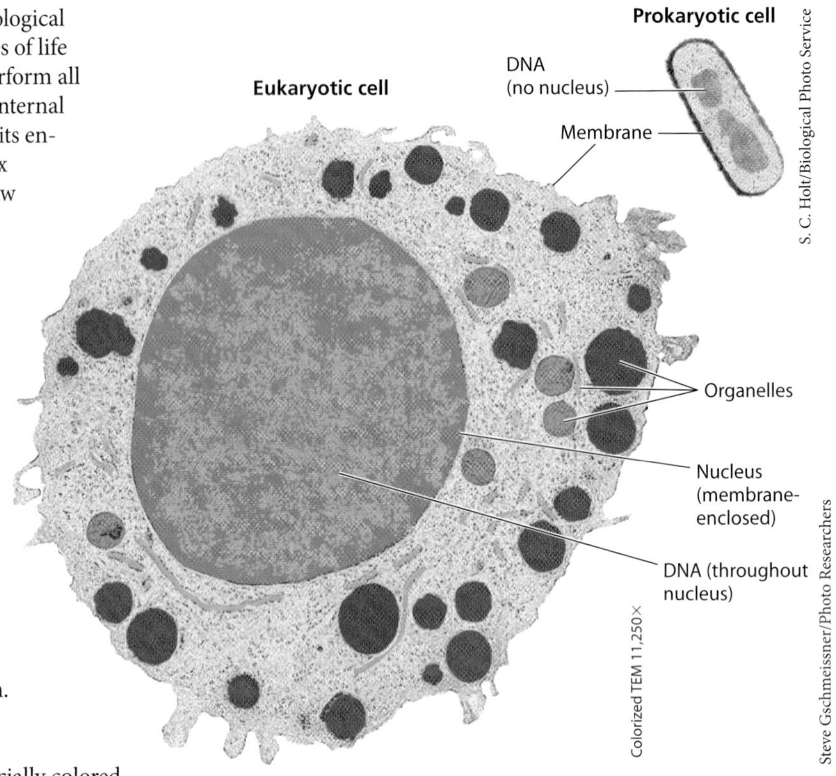

▲ **Figure 3** Contrasting the size and complexity of prokaryotic and eukaryotic cells (shown here approximately 11,250 times their real size)

The properties of life emerge from the ordered arrangement and interactions of the structures of a cell. Such a combination of components forms a more complex organization that we can call a *system*. Cells are examples of biological systems, as are organisms and ecosystems. Systems and their emergent properties are not unique to life. Consider a box of bicycle parts. When all of the individual parts are properly assembled, the result is a mechanical system you can use for exercise or transportation.

The emergent properties of life, however, are particularly challenging to study because of the unrivaled complexity of biological systems. At the cutting edge of large-scale research today is an approach called **systems biology**. The goal of systems biology is to construct models for the dynamic behavior of whole systems based on studying the interactions among the parts. Biological systems can range from the functioning of the biosphere to the molecular machinery of an organelle.

Cells illustrate another theme of biology: the correlation of structure and function. Experience shows you that form generally fits function. A screwdriver tightens or loosens screws, a hammer pounds nails. Because of their form, these tools can't do each other's jobs. Applied to biology, this theme of form fitting function is a guide to the structure of life at all its organizational levels. For example, the long extension of the nerve cell shown in Figure 2 enables it to transmit impulses across long distances in the body. Often, analyzing a biological structure gives us clues about what it does and how it works.

The activities of organisms are all based on cells. For example, your every thought is based on the actions of nerve cells, and your movements depend on muscle cells. Even a global process such as the cycling of carbon is the result of cellular activities, including the photosynthesis of plant cells and the cellular respiration of nearly all cells, a process that uses oxygen to break down sugar for energy and releases carbon dioxide. In the next module, we explore these processes and how they relate to the theme of organisms interacting with their environments.

? Why are cells considered the basic units of life?

● They are the lowest level in the hierarchy of biological organization at which the properties of life emerge.

4 Organisms interact with their environment, exchanging matter and energy

An organism interacts with its environment, which includes other organisms as well as physical factors. **Figure 4** is a simplified diagram of such interactions taking place in a forest ecosystem in Madagascar. Plants are the *producers* that provide the food for a typical ecosystem. A tree, for example, absorbs water (H_2O) and minerals from the soil through its roots, and its leaves take in carbon dioxide (CO_2) from the air. In photosynthesis, a tree's leaves use energy from sunlight to convert CO_2 and H_2O to sugar and oxygen (O_2). The leaves release O_2 to the air, and the roots help form soil by breaking up rocks. Thus, both organism and environment are affected by the interactions between them.

The *consumers* of the ecosystem eat plants and other animals. The lemur in Figure 4 eats the leaves and fruits of the tamarind tree. To release the energy in food, animals (as well as plants and most other organisms) take in O_2 from the air and release CO_2. An animal's wastes return other chemicals to the environment.

Another vital part of the ecosystem includes the small animals, fungi, and bacteria in the soil that decompose wastes and the remains of dead organisms. These *decomposers* act as recyclers, changing complex matter into simpler mineral nutrients that plants can absorb and use.

The dynamics of ecosystems include two major processes—the recycling of chemical nutrients and the flow of energy. These processes are illustrated in Figure 4. The most basic chemicals necessary for life—carbon dioxide, oxygen, water, and various minerals—cycle within an ecosystem from the air and soil to plants, to animals and decomposers, and back to the air and soil (blue arrows in the figure).

By contrast, an ecosystem gains and loses energy constantly. Energy flows into the ecosystem when plants and other photosynthesizers absorb light energy from the sun (yellow arrow) and convert it to the chemical energy of sugars and other complex molecules. Chemical energy (orange arrow) is then passed through a series of consumers and, eventually, decomposers, powering each organism in turn. In the process of these energy conversions between and within organisms, some energy is converted to heat, which is then lost from the system (red arrow). In contrast to chemical nutrients, which recycle within an ecosystem, energy flows through an ecosystem, entering as light and exiting as heat.

In this first section, we have touched on several themes of biology, from emergent properties in the biological hierarchy of organization, to cells as the structural and functional units of life, to the exchange of matter and energy as organisms interact with their environment. In the next section, we begin our exploration of evolution, the core theme of biology.

? Explain how the photosynthesis of plants functions in both cycling of chemical nutrients and the flow of energy in an ecosystem.

Photosynthesis uses light to convert carbon dioxide and water to energy-rich food, making it the pathway by which both chemical nutrients and energy become available to most organisms.

Dave Stamboulis/age fotostock

▲ **Figure 4** The cycling of nutrients and flow of energy in an ecosystem

Evolution, the Core Theme of Biology

5 The unity of life is based on DNA and a common genetic code

All cells have DNA, and the continuity of life depends on this universal genetic material. DNA is the chemical substance of **genes**, the units of inheritance that transmit information from parents to offspring. Genes, which are grouped into very long DNA molecules called chromosomes, also control all the activities of a cell. The molecular structure of DNA accounts for these functions. Let us explain: Each DNA molecule is made up of two long chains coiled together into what is called a double helix. The chains are made up of four kinds of chemical building blocks. **Figure 5** illustrates these four building blocks, called nucleotides, with different colors and letter abbreviations of their names. The right side of the figure shows a short section of a DNA double helix.

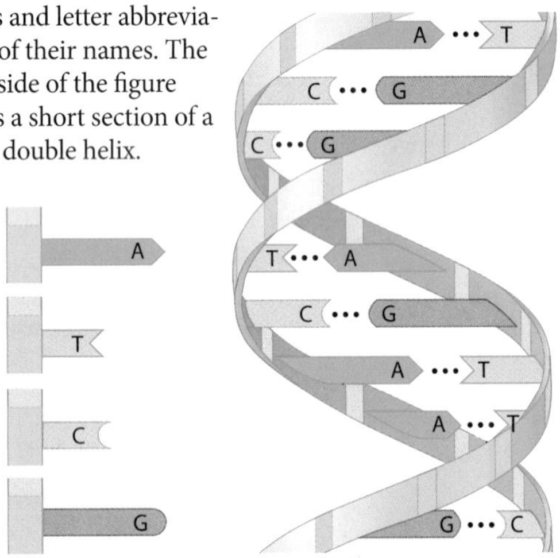

▲ **Figure 5** The four building blocks of DNA (left); part of a DNA double helix (right)

The way DNA encodes a cell's information is analogous to the way we arrange letters of the alphabet into precise sequences with specific meanings. The word *rat*, for example, conjures up an image of a rodent; *tar* and *art*, which contain the same letters, mean very different things. We can think of the four building blocks as the alphabet of inheritance. Specific sequential arrangements of these four chemical letters encode precise information in genes, which are typically hundreds or thousands of "letters" long.

The DNA of genes provides the blueprints for making proteins, and proteins serve as the tools that actually build and maintain the cell and carry out its activities. A bacterial gene may direct the cell to "Make a yellow pigment." A particular human gene may mean "Make the hormone insulin." All forms of life use essentially the same genetic code to translate the information stored in DNA into proteins. This makes it possible to engineer cells to produce proteins normally found only in some other organism. Thus, bacteria can be used to produce insulin for the treatment of diabetes by inserting a gene for human insulin into bacterial cells.

The diversity of life arises from differences in DNA sequences—in other words, from variations on the common theme of storing genetic information in DNA. Bacteria and humans are different because they have different genes. But both sets of instructions are written in the same language.

In the next module, we see how biologists attempt to organize the diversity of life.

? What is the chemical basis for all of life's kinship?

DNA as the genetic material

6 The diversity of life can be arranged into three domains

We can think of biology's enormous scope as having two dimensions. The "vertical" dimension, which we examined in Module 2, is the size scale that stretches from molecules to the biosphere. But biology also has a "horizontal" dimension, spanning across the great diversity of organisms existing now and over the long history of life on Earth.

Grouping Species Diversity is a hallmark of life. Biologists have so far identified and named about 1.8 million species, and thousands more are identified each year. Estimates of the total number of species range from 10 million to over 100 million. Whatever the actual number, biologists face a major challenge in attempting to make sense of this enormous variety of life.

There seems to be a human tendency to group diverse items according to similarities. We may speak of bears or butterflies, though we recognize that each group includes many different species. We may even sort groups into broader categories, such as mammals and insects. Taxonomy, the branch of biology that names and classifies species, arranges species into a hierarchy of broader and broader groups, from genus, family, order, class, and phylum, to kingdom.

The Three Domains of Life Until the 1990s, most biologists used a taxonomic scheme that divided all of life into five kingdoms. But new methods for assessing evolutionary relationships, such as comparison of DNA sequences, have led to an ongoing reevaluation of the number and boundaries of kingdoms. As that debate continues, however, there is consensus that life can be organized into three higher levels called **domains**. **Figure 6**, on the facing page, shows representatives of the three domains: Bacteria, Archaea, and Eukarya.

Domains **Bacteria** and **Archaea** both consist of prokaryotes, organisms with prokaryotic cells. Most prokaryotes are single-celled and microscopic. The photos of the prokaryotes in Figure 6 were made with an electron microscope, and the number along the side indicates the magnification of the image. Bacteria and archaea were once combined in a single kingdom. But much evidence indicates that they represent two very distinct branches of life, each of which includes multiple kingdoms.

Bacteria are the most diverse and widespread prokaryotes. In the photo of bacteria in Figure 6, each of the rod-shaped structures is a bacterial cell.

Many of the prokaryotes known as archaea live in Earth's extreme environments, such as salty lakes and boiling hot springs. Each round structure in the photo of archaea in Figure 6 is an archaeal cell.

All the eukaryotes, organisms with eukaryotic cells, are grouped in domain **Eukarya**. As you learned in Module 3, eukaryotic cells have a nucleus and other internal structures called organelles.

Protists are a diverse collection of mostly single-celled organisms and some relatively simple multicellular relatives. Pictured in Figure 6 is an assortment of protists in a drop of pond water. Although protists were once placed in a single kingdom, it is now clear that they do not form a single natural group of species. Biologists are currently debating how to split the protists into groups that accurately reflect their evolutionary relationships.

The three remaining groups within Eukarya contain multicellular eukaryotes. These kingdoms are distinguished partly by their modes of nutrition. Kingdom Plantae consists of plants, which produce their own food by photosynthesis. The representative of kingdom Plantae in Figure 6 is a tropical bromeliad, a plant native to the Americas.

Kingdom Fungi, represented by the mushrooms in Figure 6, is a diverse group, whose members mostly decompose the remains of dead organisms and organic wastes and absorb the nutrients into their cells.

Animals obtain food by ingestion, which means they eat other organisms. Representing kingdom Animalia, the sloth in Figure 6 resides in the trees of Central and South American rain forests. There are actually members of two other groups in the sloth photo. The sloth is clinging to a tree (kingdom Plantae), and the greenish tinge in the animal's hair is a luxuriant growth of photosynthetic prokaryotes (domain Bacteria). This photograph exemplifies a theme reflected in our book's title: connections between living things. The sloth depends on trees for food and shelter; the tree uses nutrients from the decomposition of the sloth's feces; the prokaryotes gain access to the sunlight necessary for photosynthesis by living on the sloth; and the sloth is camouflaged from predators by its green coat.

The diversity of life and its interconnectedness are evident almost everywhere. Earlier we looked at life's unity in its shared properties, two basic types of cell structure, and common genetic code. And now we have briefly surveyed its diversity. In the next module, we explore how evolution explains both the unity and the diversity of life.

? To which of the three domains of life do we belong?

Eukarya

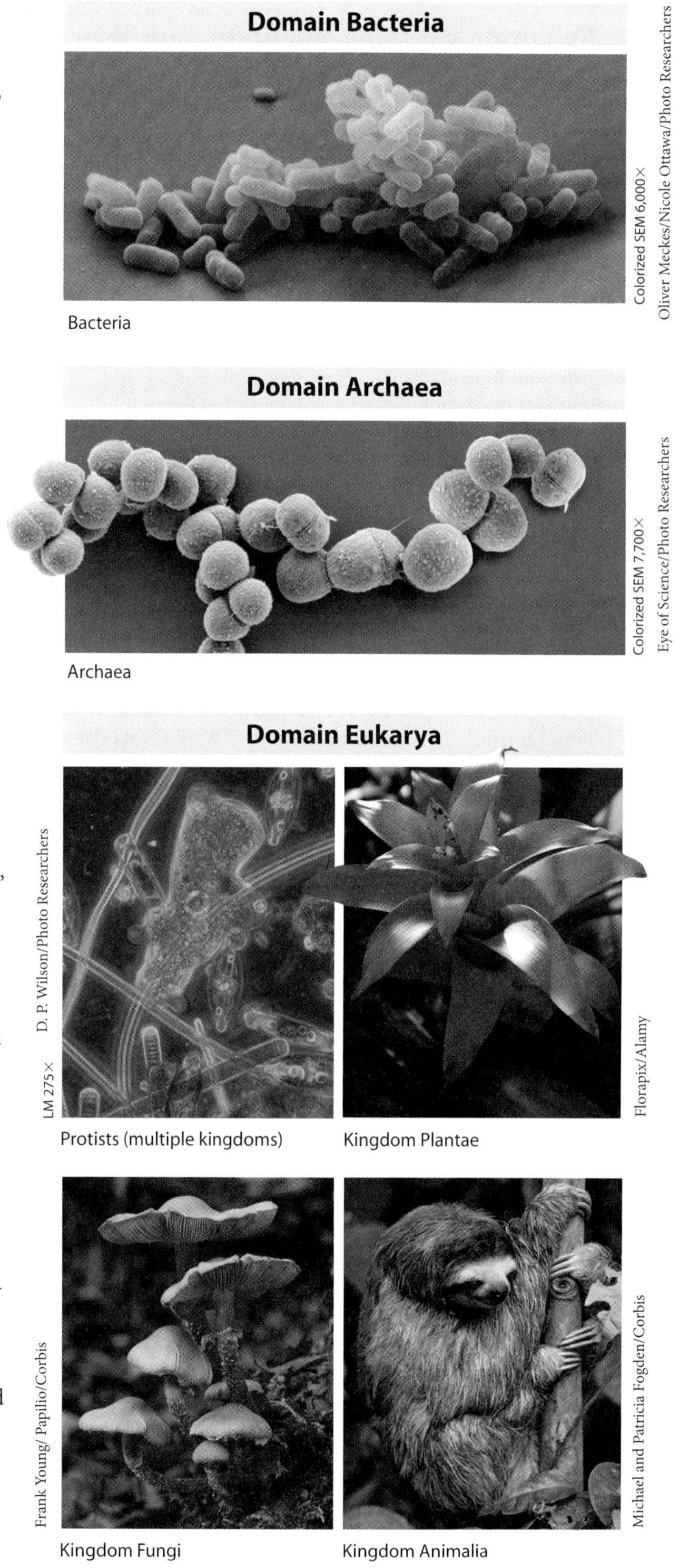

▲ **Figure 6** The three domains of life

7 Evolution explains the unity and diversity of life

The history of life, as documented by fossils, is a saga of a changing Earth billions of years old, inhabited by an evolving cast of living forms (**Figure 7A**). And yet, there is relatedness among these diverse forms, and patterns of ancestry can be traced through the fossil record and other evidence. Evolution accounts for life's dual nature of kinship and diversity.

In November 1859, the English naturalist Charles Darwin (**Figure 7B**) published one of the most important and influential books ever written. Entitled *On the Origin of Species by Means of Natural Selection*, Darwin's book was an immediate bestseller and soon made his name almost synonymous with the concept of evolution. Darwin stands out in history with people like Newton and Einstein, scientists who synthesized comprehensive theories with great explanatory power.

The Origin of Species articulated two main points. First, Darwin presented a large amount of evidence to support the idea of **evolution**—that species living today are descendants of ancestral species. Darwin called his evolutionary theory "descent with modification." It was an insightful phrase, as it captured both the unity of life (descent from a common ancestor) and the diversity of life (modification as species diverged from their ancestors).

Darwin's second point was to propose a mechanism for evolution, which he called **natural selection**. Darwin synthesized this idea from observations that by themselves were neither profound nor original. Others had the pieces of the puzzle, but Darwin saw how they fit together. He started with the following two observations: (1) Individuals in a population vary in their traits, many of which are passed on from parents to offspring. (2) A population can produce far more offspring than the environment can support. From these two observations, Darwin inferred that those individuals with heritable traits best suited to the environment are more likely to survive and reproduce than are less well-suited individuals. As a result of this unequal reproductive success over many generations, a higher and higher proportion of individuals will have the advantageous traits. The result of natural selection is evolutionary adaptation, the accumulation of favorable traits in a population over time.

Figure 7C uses a simple example to show how natural selection works. ❶ An imaginary beetle population has colonized an area where the soil has been blackened by a recent brush fire. Initially, the population varies extensively in the inherited coloration of individuals, from very light gray to charcoal. ❷ A bird eats the beetles it sees most easily, the light-colored ones. This selective predation reduces the number of light-colored beetles and favors the survival and reproductive success of the darker beetles. ❸ The surviving beetles reproduce. After several generations, the population is quite different from the original one. As a result of natural selection, the frequencies of the darker-colored beetles in the population have increased.

Darwin realized that numerous small changes in populations caused by natural selection could eventually lead to major alterations of species. He proposed that new species could evolve as a result of the gradual accumulation of changes over long periods of time.

We see the exquisite results of natural selection in every kind of organism. Each species has its own set of evolutionary adaptations that have evolved over time. Consider the two very

▲ **Figure 7B** Charles Darwin in 1859

Neg./Transparency no. 326668. Courtesy Dept. of Library Services, American Museum of Natural History

Seelevel.com

▲ **Figure 7A** Fossil of *Dimetrodon*. This 3-m-long carnivore was more closely related to mammals than to reptiles.

❶ Population with varied inherited traits

❷ Elimination of individuals with certain traits

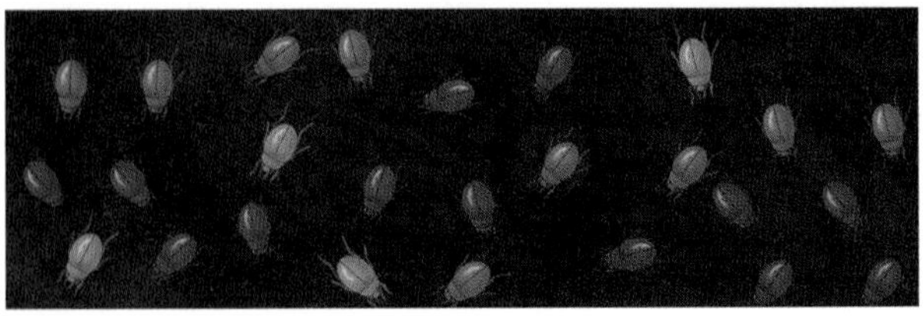

❸ Reproduction of survivors

▲ **Figure 7C** An example of natural selection in action

Photoshot Holdings Ltd/Alamy

Fred Felleman/Stone/Getty Images

▲ **Figure 7D** Examples of adaptations to different environments

different mammals shown in **Figure 7D**. The ground pangolin, found in southern and eastern Africa, has a tough body armor of overlapping scales, protecting it from most predators. The pangolin uses its unusually long tongue to prod ants out of their nests. The killer whale is a mammal adapted for life at sea. It breathes air through nostrils on the top of its head and communicates with its companions by emitting clicking sounds that carry in water. Killer whales use sound echoes to detect schools of fish or other prey. The pangolin's armor and the killer whale's echolocating ability arose over many, many generations as individuals with heritable traits that made them better adapted to the environment had greater reproductive success. Evolution—descent with modification—explains both how these two mammals are related and how they differ. Evolution is the core theme that makes sense of everything we know and learn about life.

? How does natural selection adapt a population of organisms to its environment?

On average, those individuals with heritable traits best suited to the local environment produce the greatest number of offspring. This unequal reproductive success increases the frequency of those traits in the population.

The Process of Science

8 Scientific inquiry is used to ask and answer questions about nature

The word *science* is derived from a Latin verb meaning "to know." Science is a way of knowing—an approach to understanding the natural world. It stems from our curiosity about ourselves and the world around us. And it involves the process of inquiry—a search for information, explanations, and answers to specific questions. Scientific inquiry involves making observations, forming hypotheses, and testing predictions.

Recorded observations and measurements are the data of science. Some data are *quantitative*, such as numerical measurements. Other data may be descriptive, or *qualitative*. For example, primatologist Alison Jolly has spent over 40 years making observations of lemur behavior during field research in Madagascar, amassing data that is mostly qualitative (**Figure 8**).

Collecting and analyzing observations can lead to conclusions based on a type of logic called **inductive reasoning**. This kind of reasoning derives generalizations from a large number of specific observations. "All organisms are made of cells" is an inductive conclusion based on the discovery of cells in every biological specimen observed over two centuries of time. Careful observations and the inductive conclusions they lead to are fundamental to understanding nature.

Observations often stimulate us to seek natural causes and explanations. Such inquiry usually involves the forming and testing of hypotheses. A **hypothesis** is a proposed explanation for a set of observations. A good hypothesis leads to predictions that scientists can test by recording additional observations or by designing experiments.

Deduction is the type of logic used to come up with ways to test hypotheses. In **deductive reasoning**, the logic flows from general premises to the specific results we should expect if the premises are true. If all organisms are made of cells (premise 1), and humans are organisms (premise 2), then humans are composed of cells (deduction). This deduction is a prediction that can be tested by examining human tissues.

Colin Radford

▲ **Figure 8** Alison Jolly with her research subjects, ring-tailed lemurs

Theories in Science

How is a theory different from a hypothesis? A scientific **theory** is much broader in scope than a hypothesis. It is usually general enough to generate many new, specific hypotheses that can then be tested. And a theory is supported by a large and usually growing body of evidence. Theories that become widely adopted (such as the theory of evolution) explain a great diversity of observations and are supported by a vast accumulation of evidence.

? Contrast inductive reasoning with deductive reasoning.

Inductive reasoning derives a generalization from many observations; deductive reasoning predicts specific outcomes from a general premise.

9 Scientists form and test hypotheses and share their results

Let's explore the elements of scientific inquiry with two case studies, one from everyday life and one from a research project.

A Case Study from Everyday Life We all use hypotheses in solving everyday problems. Let's say, for example, that your flashlight fails during a campout. That's an observation. The question is obvious: Why doesn't the flashlight work? Two reasonable hypotheses based on past experience are that either the batteries in the flashlight are dead or the bulb is burned out. Each of these hypotheses leads to predictions you can test with experiments or further observations. For example, the dead-battery hypothesis predicts that replacing the batteries with new ones will fix the problem. **Figure 9A** diagrams this campground inquiry.

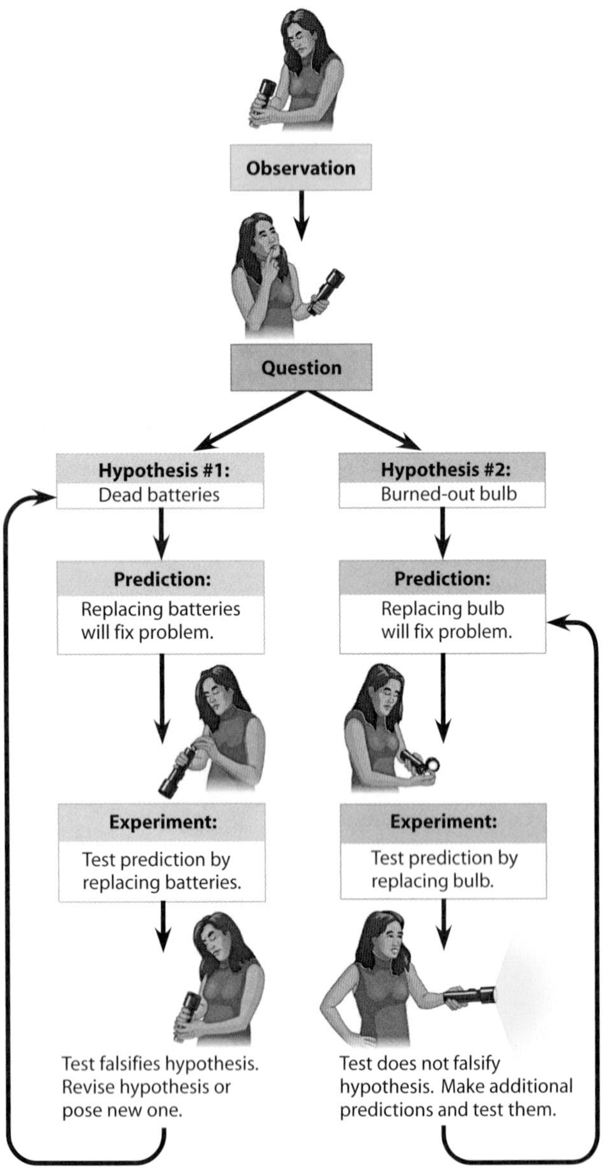

▲ **Figure 9A** An example of hypothesis-based science

The flashlight example illustrates two important points. First, a hypothesis must be *testable*—there must be some way to check its validity. Second, a hypothesis must be *falsifiable*—there must be some observation or experiment that could show that it is not true. As shown on the left in Figure 9A, the hypothesis that dead batteries are the sole cause of the problem was falsified by replacing the batteries with new ones. As shown on the right, the burned-out-bulb hypothesis is the more likely explanation. Notice that testing supports a hypothesis not by proving that it is correct but by not eliminating it through falsification. Perhaps the bulb was simply loose and the new bulb was inserted correctly. Testing cannot *prove* a hypothesis beyond a shadow of doubt, because it is impossible to exhaust all alternative hypotheses. A hypothesis gains credibility by surviving multiple attempts to falsify it, while alternative hypotheses are eliminated by testing.

A Case Study from Science To learn more about how science works, let's examine some actual scientific research.

The story begins with a set of observations and generalizations. Many poisonous animals are brightly colored, often with distinctive patterns. This so-called warning coloration apparently says "dangerous species" to potential predators. But there are also mimics. These imposters resemble poisonous species but are actually harmless. A question that follows from these observations is: What is the function of mimicry? A reasonable hypothesis is that mimicry is an evolutionary adaptation that reduces the harmless animal's risk of being eaten.

In 2001, biologists David and Karin Pfennig, along with William Harcombe, one of their undergraduate students, designed an elegant set of field experiments to test the hypothesis that mimics benefit because predators confuse them with the harmful species. A venomous snake called the eastern coral snake has warning coloration: bold, alternating rings of red, yellow, and black (**Figure 9B**, on the next page). (A *venomous* species delivers its poison by stinging, stabbing, or biting.) Predators rarely attack these snakes. The predators do not learn this avoidance behavior by trial and error; a first encounter with a coral snake would usually be deadly. Natural selection has apparently increased the frequency of predators that inherit an instinctive avoidance of the coral snake's coloration.

The nonvenomous scarlet king snake mimics the ringed coloration of the coral snake (**Figure 9C**). Both types of snakes live in North and South Carolina, but king snakes are also found in regions that have no coral snakes.

The geographic distribution of these snakes made it possible for the researchers to test a key prediction of the mimicry hypothesis: Mimicry should help protect king snakes from predators, but only in regions where coral snakes also live. Avoiding snakes with warning coloration is an adaptation of predator populations that evolved in areas where coral snakes are present. Therefore, predators adapted to the warning coloration of coral snakes will attack king snakes less frequently than will predators in areas where coral snakes are absent.

Steve Kaufman/Photolibrary

Suzanne L. Collins/Science Source

▲ Figure 9B Eastern coral snake (venomous)

▲ Figure 9C Scarlet king snake (nonvenomous)

To test this prediction, Harcombe made hundreds of artificial snakes out of wire covered with a claylike substance called plasticine. He made two versions of fake snakes: an *experimental group* with the color pattern of king snakes and a *control group* of plain brown snakes as a basis of comparison.

The researchers placed equal numbers of the two types of artificial snakes in field sites throughout North and South Carolina, including the region where coral snakes are absent. After four weeks, they retrieved the snakes and recorded how many had been attacked by looking for bite or claw marks. The most common predators were foxes, coyotes, and raccoons, but black bears also attacked some of the snakes (**Figure 9D**).

The data fit the key prediction of the mimicry hypothesis. The artificial king snakes were attacked less frequently than the artificial brown snakes only in field sites within the geographic range of the venomous coral snakes. The bar graph in **Figure 9E** summarizes the results.

This case study is an example of a **controlled experiment**, one that is designed to compare an experimental group (the artificial king snakes, in this case) with a control group (the artificial brown snakes). Ideally, the experimental and control groups differ only in the one factor the experiment is designed to test—in our example, the effect of the snakes' coloration on the behavior of predators. Without the control group, the researchers would not have been able to rule out other variables, such as the number of predators in the different test areas. The experimental design left coloration as the only factor that could account for the low predation rate on the artificial king snakes placed within the range of coral snakes.

The Culture of Science Science is a social activity, with most scientists working in teams, which often include graduate and undergraduate students. Scientists share information through publications, seminars, meetings, and personal communication. The Internet has added a new medium for this exchange of ideas and data. Scientists build on what has been learned from earlier research and often check each other's claims by attempting to confirm observations or repeat experiments.

Science seeks natural causes for natural phenomena. Thus, the scope of science is limited to the study of structures and processes that we can directly observe and measure. Science can neither support nor falsify hypotheses about supernatural forces or explanations, for such questions are outside the bounds of science.

The process of science is necessarily repetitive: In testing a hypothesis, researchers may make observations that call for rejection of the hypothesis or at least revision and further testing. This process allows biologists to circle closer and closer to their best estimation of how nature works. As in all quests, science includes elements of challenge, adventure, and luck, along with careful planning, reasoning, creativity, cooperation, competition, patience, and persistence.

? Why is it difficult to draw a conclusion from an experiment that does not include a control group?

Without a control group, you don't know if the experimental outcome is due to the variable you are trying to test or to some other variable.

David Pfennig

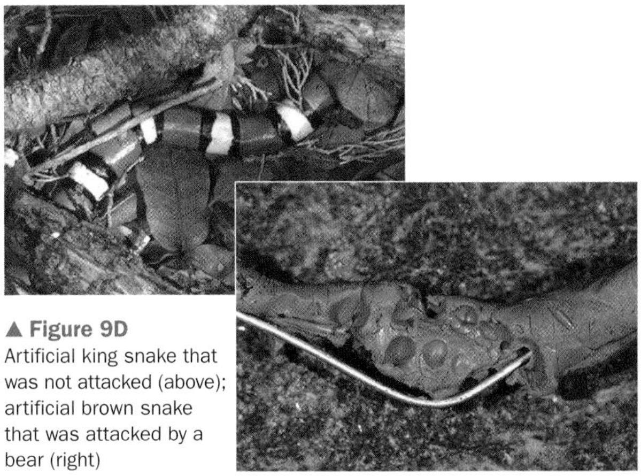

▲ Figure 9D Artificial king snake that was not attacked (above); artificial brown snake that was attacked by a bear (right)

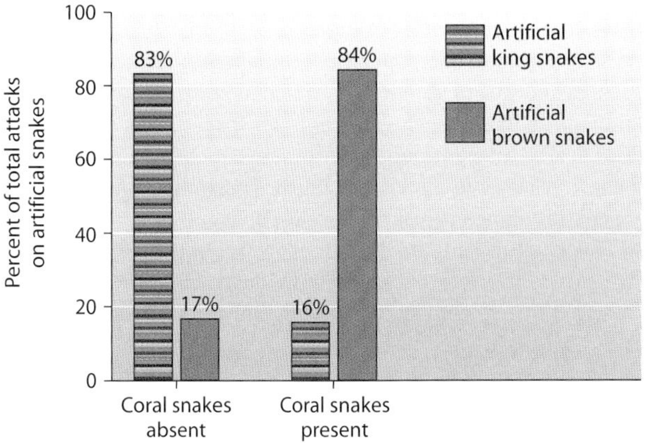

Data in graph based on D. W. Pfennig et al. 2001. Frequency-dependent Batesian mimicry. *Nature* 410: 323

▲ Figure 9E Results of mimicry experiment

Biology and Everyday Life

CONNECTION 10 Biology, technology, and society are connected in important ways

Many issues facing society are related to biology (**Figure 10**). Most of these issues also involve our expanding technology. Science and technology are interdependent, but their basic goals differ. The goal of science is to understand natural phenomena. In contrast, the goal of **technology** is to apply scientific knowledge for some specific purpose. Scientists often speak of "discoveries," while engineers more often speak of "inventions." The beneficiaries of those inventions also include scientists, who use new technology in their research. And scientific discoveries often lead to the development of new technologies.

Technology depends less on the curiosity that drives basic science than on the needs and wants of people and on the social environment of the times. Debates about technology center more on "should we do it" than "can we do it." Should insurance companies have access to individuals' DNA information? Should we permit research with embryonic stem cells?

Technology has improved our standard of living in many ways, but not without adverse consequences. Technology that keeps people healthier has enabled Earth's population to grow 10-fold in the past three centuries and to more than double to 6.8 billion in just the past 40 years. The environmental effects of this growth can be devastating. Global climate change, toxic wastes, deforestation, nuclear accidents, and extinction of species are just some of the repercussions of more and more people wielding more and more technology. Science can help us identify such problems and provide insight into what course of action may prevent further damage. But solutions to these problems have as much to do with politics, economics, and cultural values as with science and technology. Now that science and technology have become such powerful aspects of society, every citizen has a responsibility to develop a reasonable amount of scientific literacy. The crucial science-technology-society relationship is a theme that adds to the significance of any biology course.

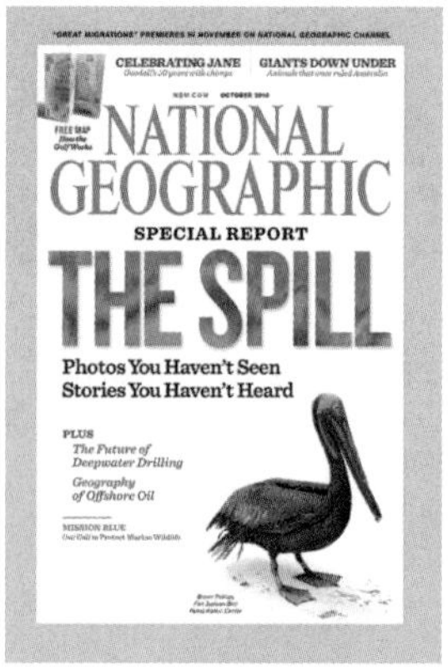

National Geographic

▲ **Figure 10** Biology and technology in the news

? How do science and technology interact?

New scientific discoveries may lead to new technologies; new technologies may increase the ability of scientists to search for new knowledge.

EVOLUTION CONNECTION 11 Evolution is connected to our everyday lives

Evolution is the core theme of biology. To emphasize the centrality of evolution to biology, we include an Evolution Connection module in each chapter in this book. But is evolution connected to your everyday life? And if so, in what ways?

Biologists now recognize that differences in DNA among individuals, populations, and species reflect the patterns of evolutionary change. The new technology of automatic DNA-sequencing machines has enabled scientists to determine the order of the billions of DNA bases in the human genome and in the genomes of other species. Comparisons of those sequences allow us to identify genes shared across many species, study the actions of such genes in other species, and, in some cases, search for new medical treatments. Identifying beneficial genes in relatives of our crop plants has permitted the breeding or genetic engineering of enhanced crops.

The recognition that DNA differs between people has led to the use of DNA tests to identify individuals. DNA profiling is now used to help convict or exonerate the accused, determine paternity, and identify remains.

Evolution teaches us that the environment is a powerful selective force for traits that best adapt populations to their environment. We are major agents of environmental change when we take drugs to combat infection or grow crops in pesticide-dependent monocultures or alter Earth's habitats. We have seen the effects of such environmental changes in antibiotic-resistant bacteria, pesticide-resistant pests, endangered species, and increasing rates of extinction.

How can evolutionary theory help? It can help us be more judicious in our use of antibiotics and pesticides and help us develop strategies for conservation efforts. It can help us create flu vaccines and HIV drugs by tracking the rapid evolution of these viruses. It can identify new sources of drugs. For example, by tracing the evolutionary history of the endangered Pacific Yew tree, once the only source of the cancer drug Taxol, scientists have discovered similar compounds in more common trees.

We hope you develop an appreciation for evolution and biology and help you apply your new knowledge to evaluating issues ranging from your personal health to the well-being of the whole world. Biology offers us a deeper understanding of ourselves and our planet and a chance to more fully appreciate life in all of its diversity.

? How might an understanding of evolution contribute to the development of new drugs?

As one example, we can find organisms that share our genes and similar cellular processes and test the actions of potential drugs in these organisms.

REVIEW

For Practice Quizzes, BioFlix, MP3 Tutors, and Activities, go to www.masteringbiology.com.

Reviewing the Concepts

Themes in the Study of Biology (1–4)

1 All forms of life share common properties. Biology is the scientific study of life. Properties of life include order, reproduction, growth and development, energy processing, response to the environment, regulation, and evolutionary adaptation.

2 In life's hierarchy of organization, new properties emerge at each level. Biological organization unfolds as follows: biosphere > ecosystem > community > population > organism > organ system > organ > tissue > cell > organelle > molecule. Emergent properties result from the interactions among component parts.

3 Cells are the structural and functional units of life. Eukaryotic cells contain membrane-enclosed organelles, including a nucleus containing DNA. Prokaryotic cells are smaller and lack such organelles. Structure is related to function at all levels of biological organization. Systems biology models the complex interactions of biological systems, such as the molecular interactions within a cell.

4 Organisms interact with their environment, exchanging matter and energy. Ecosystems are characterized by the cycling of chemical nutrients from the atmosphere and soil through producers, consumers, decomposers, and back to the environment. Energy flows one way through an ecosystem—entering as sunlight, converted to chemical energy by producers, passed on to consumers, and exiting as heat.

Dave Stamboulis/age fotostock

Evolution, the Core Theme of Biology (5–7)

5 The unity of life is based on DNA and a common genetic code. DNA is responsible for heredity and for programming the activities of a cell. A species' genes are coded in the sequences of the four building blocks making up DNA's double helix.

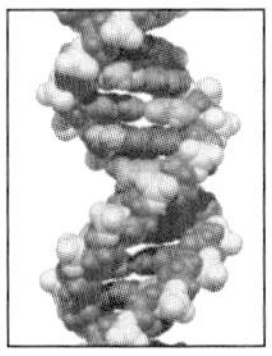

6 The diversity of life can be arranged into three domains. Taxonomy names species and classifies them into a system of broader groups. Domains Bacteria and Archaea consist of prokaryotes. The eukaryotic domain, Eukarya, includes various protists and the kingdoms Fungi, Plantae, and Animalia.

7 Evolution explains the unity and diversity of life. Darwin synthesized the theory of evolution by natural selection.

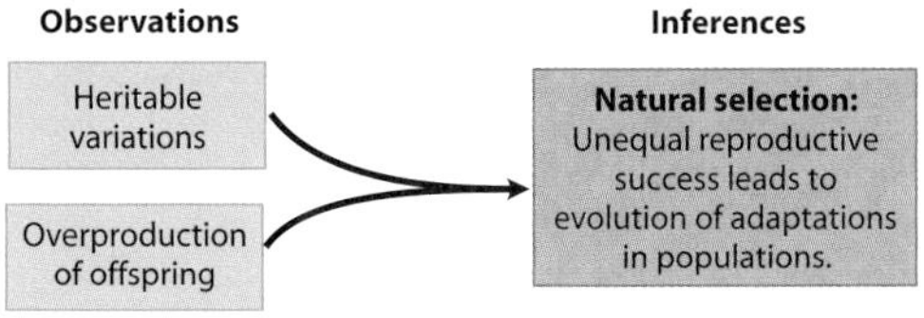

The Process of Science (8–9)

8 Scientific inquiry is used to ask and answer questions about nature. Scientists use inductive reasoning to draw general conclusions from many observations. They form hypotheses and use deductive reasoning to make predictions. Data may be qualitative or quantitative. A scientific theory is broad in scope, generates new hypotheses, and is supported by a large body of evidence.

9 Scientists form and test hypotheses and share their results. Predictions can be tested with experiments, and results can either falsify or support the hypothesis. In a controlled experiment, the use of control and experimental groups helps to demonstrate the effect of a single variable. Science is a social process: scientists share information and review each other's results.

Biology and Everyday Life (10–11)

10 Biology, technology, and society are connected in important ways. Technological advances stem from scientific research, and research benefits from new technologies.

11 Evolution is connected to our everyday lives. Evolutionary theory is useful in medicine, agriculture, forensics, and conservation. Human-caused environmental changes are powerful selective forces that affect the evolution of many species.

Connecting the Concepts

1. Biology can be described as having both a vertical scale and a horizontal scale. Explain what that means.
2. Complete the following map organizing some of biology's major concepts.

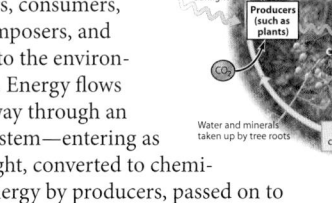

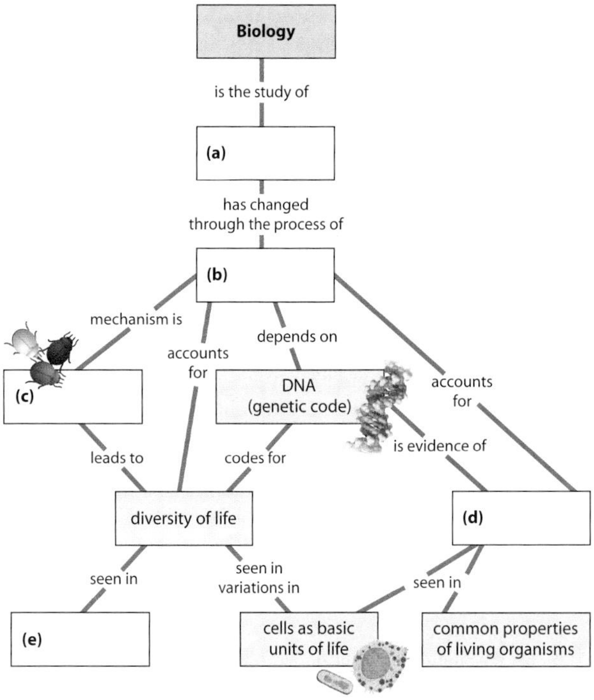

Testing Your Knowledge

Multiple Choice

3. Which of the following best describes the logic of the scientific process?
   a. If I generate a testable hypothesis, tests and observations will support it.
   b. If my prediction is correct, it will lead to a testable hypothesis.
   c. If my observations are accurate, they will not falsify my hypothesis.
   d. If my hypothesis is correct, I can make predictions and my results will not falsify my hypothesis.
   e. If my predictions are good and my tests are right, they will prove my hypothesis.
4. Single-celled amoebas and bacteria are grouped into different domains because
   a. amoebas eat bacteria.
   b. bacteria are not made of cells.
   c. bacterial cells lack a membrane-enclosed nucleus.
   d. bacteria decompose amoebas.
   e. amoebas are motile; bacteria are not.
5. A biologist studying interactions among the protists in an ecosystem could *not* be working at which level in life's hierarchy? (*Choose carefully and explain your answer.*)
   a. the population level
   b. the molecular level
   c. the community level
   d. the organism level
   e. the organ level
6. Which of the following questions is outside the realm of science?
   a. Which organisms play the most important role in energy input to a forest?
   b. What percentage of music majors take a biology course?
   c. What is the physical nature of the universe?
   d. What is the influence of the supernatural on current events?
   e. What is the dominance hierarchy in a troop of ring-tailed lemurs?
7. Which of the following statements best distinguishes hypotheses from theories in science?
   a. Theories are hypotheses that have been proved.
   b. Hypotheses are tentative guesses; theories are correct answers to questions about nature.
   c. Hypotheses usually are narrow in scope; theories have broad explanatory power.
   d. Hypotheses and theories are different terms for essentially the same thing in science.
   e. Theories cannot be falsified; hypotheses can be falsified.
8. Which of the following best demonstrates the unity among all living organisms?
   a. descent with modification
   b. related DNA sequences and common genetic code
   c. emergent properties
   d. natural selection
   e. the three domains
9. The core idea that makes sense of all of biology is
   a. the process of science.
   b. the correlation of function with structure.
   c. systems biology.
   d. evolution.
   e. the emergence of life at the level of the cell.

Describing, Comparing, and Explaining

10. In an ecosystem, how is the movement of energy similar to that of chemical nutrients, and how is it different?
11. Explain the role of heritable variations in Darwin's theory of natural selection.
12. Explain what is meant by this statement: The scientific process is not a rigid method.
13. Contrast technology with science. Give an example of each to illustrate the difference.
14. Explain what is meant by this statement: Natural selection is an editing mechanism rather than a creative process.

Applying the Concepts

15. The graph below shows the results of an experiment in which mice learned to run through a maze.

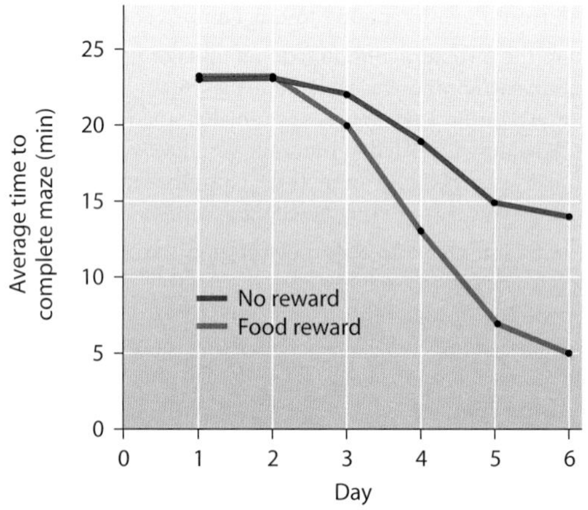

   a. State the hypothesis and prediction that you think this experiment tested.
   b. Which was the control group and which the experimental? Why was a control group needed?
   c. List some variables that must have been controlled so as not to affect the results.
   d. Do the data support the hypothesis? Explain.
16. In an experiment similar to the mimicry experiment described in Module 9, a researcher counted more predator attacks on artificial king snakes in areas with coral snakes than in areas outside the range of coral snakes. From those numbers, the researcher concluded that the mimicry hypothesis is false. Do you think this conclusion is justified? Why or why not?
17. The fruits of wild species of tomato are tiny compared to the giant beefsteak tomatoes available today. This difference in fruit size is almost entirely due to the larger number of cells in the domesticated fruits. Plant biologists have recently discovered genes that are responsible for controlling cell division in tomatoes. Why would such a discovery be important to producers of other kinds of fruits and vegetables? To the study of human development and disease? To our basic understanding of biology?
18. The news media and popular magazines frequently report stories that are connected to biology. In the next 24 hours, record the ones you hear or read about in three different sources and briefly describe the biological connections in each story.

Review Answers

1. The vertical scale of biology refers to the hierarchy of biological organization: from molecules to organelles, cells, tissues, organs, organ systems, organisms, populations, communities, ecosystems, and the biosphere. At each level, emergent properties arise from the interaction and organization of component parts. The horizontal scale of biology refers to the incredible diversity of living organisms, past and present, including the 1.8 million species that have been named so far. Biologists divide these species into three domains—Bacteria, Archaea, and Eukarya—and organize them into kingdoms and other groups that attempt to reflect evolutionary relationships.
2. a. life; b. evolution; c. natural selection; d. unity of life; e. three domains (or numerous kingdoms; 1.8 million species)
3. d 4. c 5. e (You may have been tempted to choose b, the molecular level. However, protists may have chemical communication or interactions with other protists. No protists, however, have organs.) 6. d 7. c 8. b 9. d
10. Both energy and chemical nutrients are passed through an ecosystem from producers to consumers to decomposers. But energy enters an ecosystem as sunlight and leaves as heat. Chemical nutrients are recycled from the soil or atmosphere through plants, consumers, and decomposers and returned to the air, soil, and water.
11. Darwin described how natural selection operates in populations whose individuals have varied traits that are inherited. When natural selection favors the reproductive success of certain individuals in a population more than others, the proportions of heritable variations change over the generations, gradually adapting a population to its environment.
12. In pursuit of answers to questions about nature, a scientist uses a logical thought process involving these key elements: observations about natural phenomena, questions derived from observations, hypotheses posed as tentative explanations of observations, logical predictions of the outcome of tests if the hypotheses are correct, and actual tests of hypotheses. Scientific research is not a rigid method because a scientist must adapt these processes to the set of conditions particular to each study. Intuition, chance, and luck are also part of science.
13. Technology is the application of scientific knowledge. For example, the use of solar power to run a calculator or heat a home is an application of our knowledge, derived by the scientific process, of the nature of light as a type of energy and how light energy can be converted to other forms of energy. Another example is the use of DNA to insert new genes into crop plants. This process, often called genetic engineering, stems from decades of scientific research on the structure and function of DNA from many kinds of organisms.
14. Natural selection screens (edits) heritable variations by favoring the reproductive success of some individuals over others. It can only select from the variations that are present in the population; it does not create new genes or variations.
15. a. Hypothesis: Giving rewards to mice will improve their learning. Prediction: If mice are rewarded with food, they will learn to run a maze faster.
    b. The control group was the mice that were not rewarded. Without them, it would be impossible to know if the mice that were rewarded decreased their time running the maze only because of practice.
    c. Both groups of mice should not have run the maze before and should be about the same age. Both experiments should be run at the same time of day and under the same conditions.
    d. Yes, the results fail to falsify the hypothesis because data show that the rewarded mice began to run the maze faster by day 3 and improved their performance (ran faster than the control mice) each day thereafter.
16. The researcher needed to compare the number of attacks on artificial king snakes with attacks on artificial brown snakes. It may be that there were simply more predators in the coral snake areas or that the predators were hungrier than the predators in the other areas. The experiment needed a control and proper data analysis.
17. If these cell division control genes are involved in producing the larger tomato, they may have similar effects if transferred to other fruits or vegetables. Cancer is a result of uncontrolled cell division. One could see if there are similarities between the tomato genes and any human genes that could be related to human development or disease. The control of cell division is a fundamental process in growth, repair, and asexual reproduction—all important topics in biology.
18. Virtually any news report or magazine contains stories that are mainly about biology or at least have biological connections. How about biological connections in advertisements?

The Chemical Basis of Life

The Chemical Basis of Life

BIG IDEAS

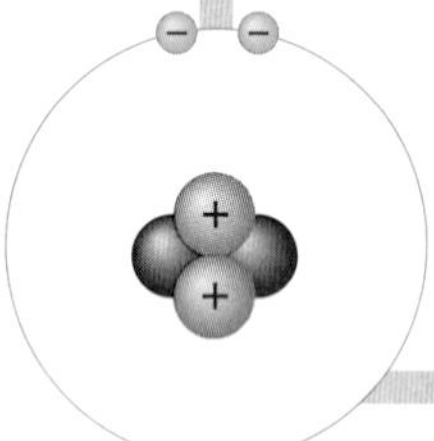

Elements, Atoms, and Compounds
(1–4)

Living organisms are made of atoms of certain elements, mostly combined into compounds.

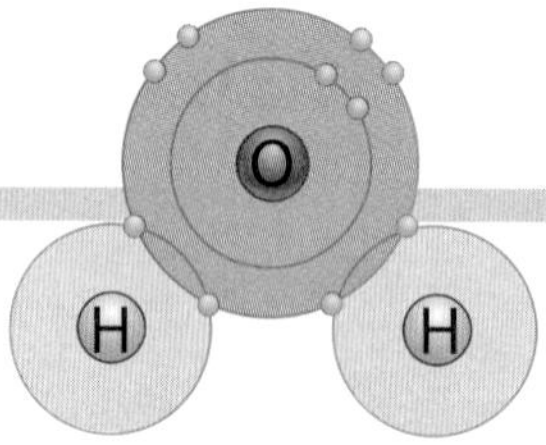

Chemical Bonds
(5–9)

The structure of an atom determines what types of bonds it can form with other atoms.

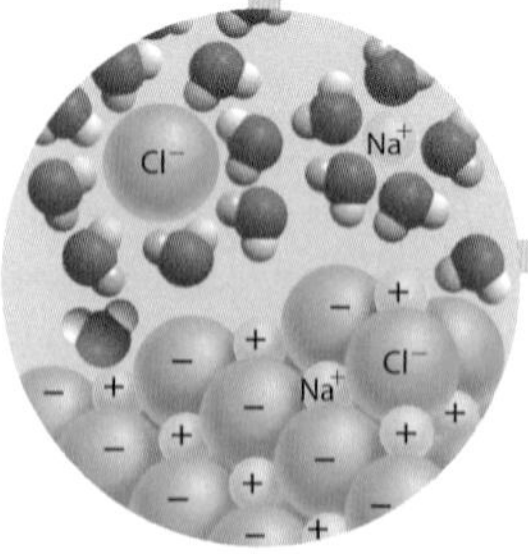

Water's Life-Supporting Properties
(10–16)

The unique properties of water derive from the polarity of water molecules.

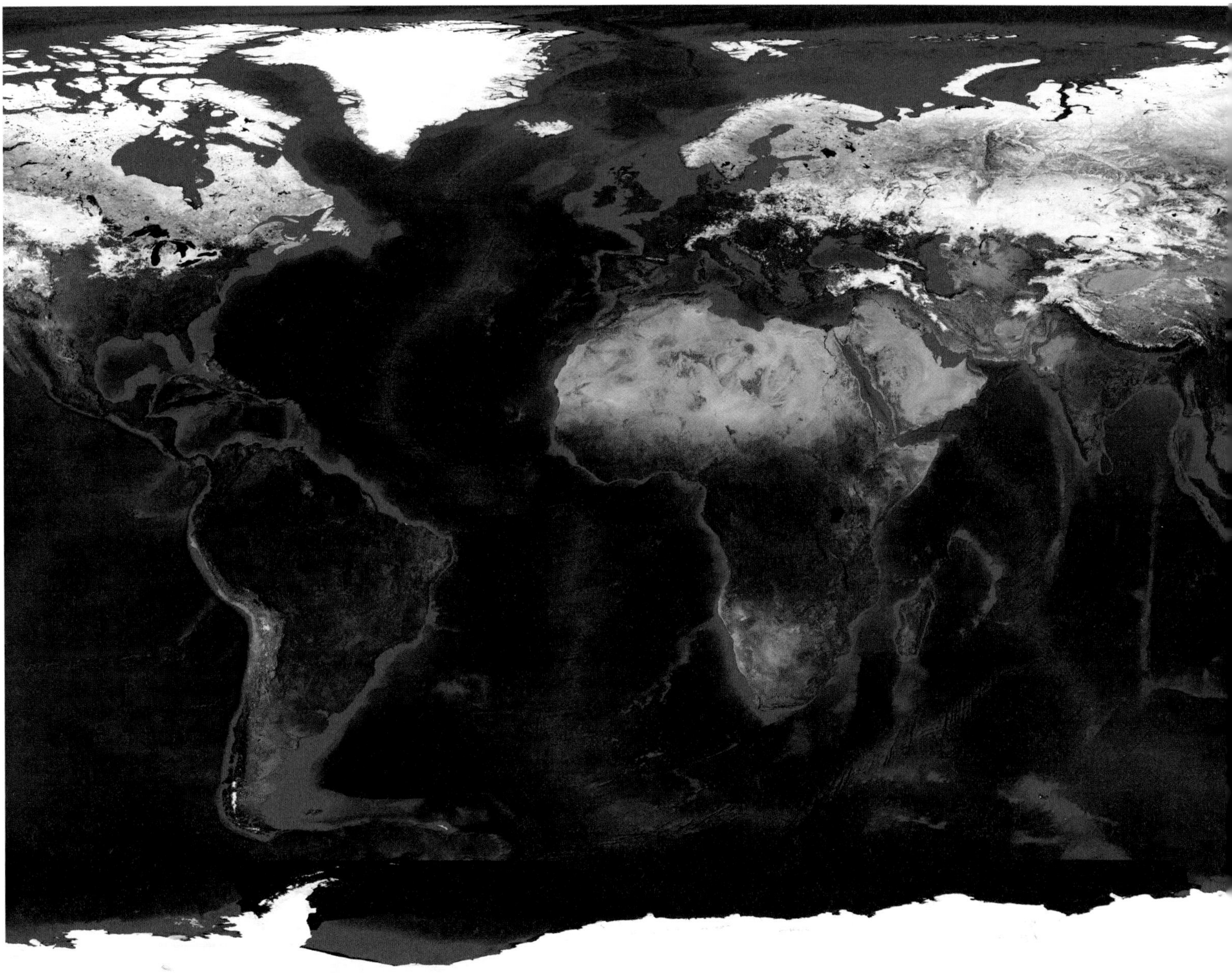

NASA

Biology is about life, but it is also about chemistry. Chemicals are the very stuff making up our bodies, those of other organisms, and the physical environment around us. Your cells, tissues, and organs are all made of chemical compounds. And the activities of your body—from reading these words to running a race—are all based on chemical reactions. So understanding and appreciating life begins with a basic understanding of chemistry: the elements that make up living matter and the ways in which the atoms of those elements interact.

Life's chemistry is tied to water. The chemical reactions of your body take place in cells whose water content ranges from 70% to 95%. As you can see in this satellite photograph of our planet, three-quarters of Earth's surface is covered by water. Life began in water and evolved there for 3 billion years before spreading onto land. And modern life, even land-dwelling life, is still dependent on water. All living organisms require water more than any other substance. What properties of the simple water molecule make it so indispensable to life on Earth? You'll find out in this chapter.

This chapter will also make connections to a variety of themes. One of these themes is the organization of life into a hierarchy of structural levels, with new properties emerging at each successive level. Emergent properties are apparent even at the lowest levels of biological organization—the ordering of atoms into molecules and the interactions of those molecules. The intricate structures and complex functions of all living organisms arise from these interactions. We begin our story of biology with the basic concepts of chemistry that will apply throughout our study of life.

Elements, Atoms, and Compounds

1 Organisms are composed of elements, in combinations called compounds

Living organisms and everything around them are composed of **matter**, which is defined as anything that occupies space and has mass. (In everyday language, we can think of mass as an object's weight.) Matter is found on Earth in three physical states: solid, liquid, and gas.

Types of matter as diverse as rocks, water, air, and biology students are all composed of chemical elements. An **element** is a substance that cannot be broken down to other substances by ordinary chemical means. Today, chemists recognize 92 elements that occur in nature; gold, copper, carbon, and oxygen are some examples. Chemists have also made a few dozen synthetic elements. Each element has a symbol, the first letter or two of its English, Latin, or German name. For instance, the symbol for sodium, Na, is from the Latin word *natrium*; the symbol O comes from the English word *oxygen*.

A **compound** is a substance consisting of two or more different elements combined in a fixed ratio. Compounds are much more common than pure elements. In fact, few elements exist in a pure state in nature.

Many compounds consist of only two elements; for instance, table salt (sodium chloride, NaCl) has equal parts of the elements sodium (Na) and chlorine (Cl). Pure sodium is a metal and pure chlorine is a poisonous gas. Chemically combined, however, they form an edible compound **(Figure 1)**. The elements hydrogen (H) and oxygen (O) exist as gases. Combined in a ratio of 2:1, they form the most abundant compound on Earth—water (H_2O). These are simple examples of organized matter having emergent properties: A compound has characteristics different from those of its elements.

Most of the compounds in living organisms contain at least three or four elements. Sugar, for example, is formed of carbon (C), hydrogen, and oxygen. Proteins are compounds containing carbon, hydrogen, oxygen, nitrogen (N), and a small amount of sulfur (S). Different arrangements of the atoms of these elements give rise to the unique properties of each compound.

About 25 elements are essential to life. As you can see in **Table 1**, four of these—oxygen, carbon, hydrogen, and nitrogen—make up about 96% of the weight of the human body, as well as that of most other living organisms. These four elements are the main ingredients of biological molecules such as proteins, sugars, and fats. Calcium (Ca), phosphorus (P), potassium (K), sulfur, sodium, chlorine, and magnesium (Mg) account for most of the remaining 4% of the human body. These elements are involved in such important functions as bone formation (calcium and phosphorus) and nerve signaling (potassium, sodium, calcium, and chlorine).

The **trace elements** listed at the bottom of the table are essential, but only in minute quantities. We explore the importance of trace elements to your health next.

TABLE 1 | ELEMENTS IN THE HUMAN BODY

Element	Symbol	Percentage of Body Weight (Including Water)	
Oxygen	O	65.0%	96.3%
Carbon	C	18.5%	
Hydrogen	H	9.5%	
Nitrogen	N	3.3%	
Calcium	Ca	1.5%	3.7%
Phosphorus	P	1.0%	
Potassium	K	0.4%	
Sulfur	S	0.3%	
Sodium	Na	0.2%	
Chlorine	Cl	0.2%	
Magnesium	Mg	0.1%	

Trace elements, less than 0.01% of human body weight: Boron (B), chromium (Cr), cobalt (Co), copper (Cu), fluorine (F), iodine (I), iron (Fe), manganese (Mn), molybdenum (Mo), selenium (Se), silicon (Si), tin (Sn), vanadium (V), zinc (Zn)

? **Explain how table salt illustrates the theme of emergent properties.**

The elements that make up the edible crystals of table salt, sodium and chlorine, are in pure form a metal and a poisonous gas.

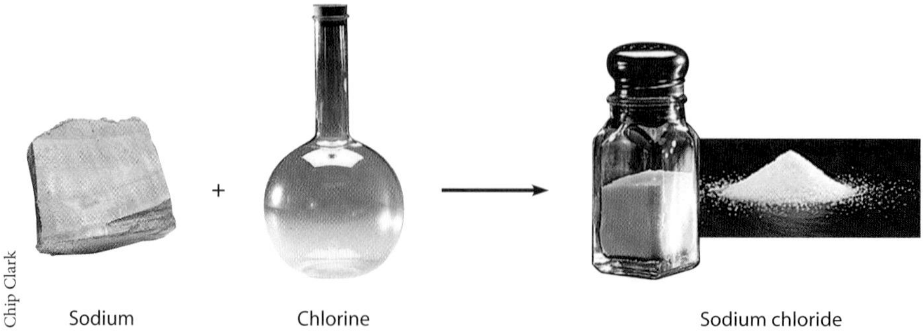

▲ **Figure 1** The emergent properties of the edible compound sodium chloride

CONNECTION 2 Trace elements are common additives to food and water

Some trace elements, such as iron (Fe), are needed by all forms of life. Iron makes up only about 0.004% of your body weight but is vital for energy processing and for transporting oxygen in your blood. Other trace elements, such as iodine (I), are required only by certain species. You need to ingest only a tiny amount of iodine each day, about 0.15 milligram (mg). Iodine is an essential ingredient of a hormone produced by the thyroid gland, which is located in the neck. An iodine deficiency in the diet causes the thyroid gland to grow to abnormal size, a condition called goiter (**Figure 2A**). Iodine deficiency is also linked to mental retardation. Adding iodine to table salt has reduced the incidence of iodine deficiency in many countries. Unfortunately, iodized salt is not available everywhere, and an estimated 2 billion people worldwide have insufficient iodine intake. Seafood, kelp, strawberries, and dark, leafy greens are good natural sources. Thus, deficiencies are often found in inland regions, especially in areas where the soil is lacking in iodine. Although most common in developing nations, iodine deficiencies may also result from excessive consumption of highly processed foods (which often use non-iodized salt) and low-salt diets intended to lower the risk of cardiovascular disease.

Alison Wright/Photo Researchers

▲ **Figure 2A** Goiter, a symptom of iodine deficiency, in a Burmese woman

Iodine is just one example of a trace element added to food or water to improve health. For more than 50 years, the American Dental Association has supported fluoridation of community drinking water as a public health measure. Fluoride is a form of fluorine (F), an element in Earth's crust that is found in small amounts in all water sources. In many areas, fluoride is added during the municipal water treatment process to raise levels to a concentration that can reduce tooth decay. If you mostly drink bottled water, your fluoride intake may be reduced, although some bottled water now contains added fluoride. Fluoride is also frequently added to dental products, such as toothpaste and mouthwash (**Figure 2B**).

Chemicals are added to food to help preserve it, make it more nutritious, or simply make it look better. Look at the nutrition facts label from the side of the cereal box in **Figure 2C** to see a familiar example of how foods are fortified with mineral elements. Iron, for example, is a trace element commonly added to foods. (You can actually see that iron has been added to a fortified cereal by crushing the cereal and then stirring a magnet through it.) Also note that the nutrition facts label lists numerous vitamins that are added to improve the nutritional value of the cereal. For instance, the cereal in this example supplies 10% of the recommended daily value for vitamin A. Vitamins consist of more than one element and are examples of compounds.

In the next module, we explore the structure of an atom and how this structure determines the chemical properties of elements.

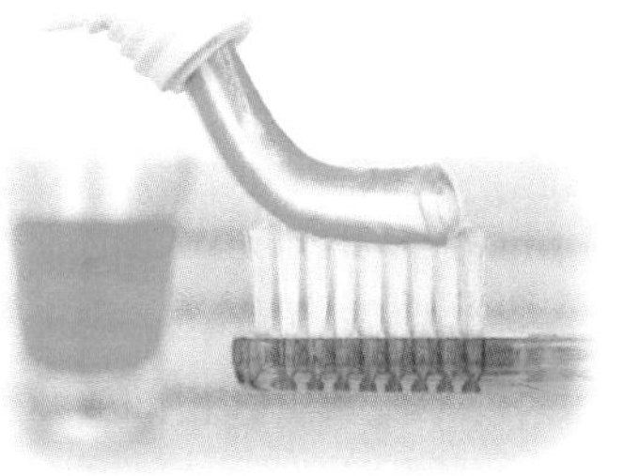

◀ **Figure 2B** Mouthwash and toothpaste with added fluoride

Anton Prado/Alamy

? **In addition to iron, what other trace elements are found in the cereal in Figure 2C? Does Total provide the "total" amount needed of these elements?**

● Zinc and copper: Total provides 100% of the zinc but only 4% of the copper needed in a day.

Kristin Piljay

Nutrition Facts
Serving Size ¾ cup (30g)
Servings Per Container about 17

Amount Per Serving	Whole Grain Total	with ½ cup skim milk
Calories	100	140
Calories from Fat	5	10
	% Daily Value**	
Total Fat 0.5g*	1%	1%
Saturated Fat 0g	0%	0%
Trans Fat 0g		
Polyunsaturated Fat 0g		
Monounsaturated Fat 0g		
Cholesterol 0mg	0%	1%
Sodium 190mg	8%	11%
Potassium 90mg	3%	8%
Total Carbohydrate 23g	8%	10%
Dietary Fiber 3g	10%	10%
Sugars 5g		
Other Carbohydrate 15g		
Protein 2g		
Vitamin A	10%	15%
Vitamin C	100%	100%
Calcium	100%	110%
Iron	100%	100%
Vitamin D	10%	25%
Vitamin E	100%	100%
Thiamin	100%	100%
Riboflavin	100%	110%
Niacin	100%	100%
Vitamin B_6	100%	100%
Folic Acid	100%	100%
Vitamin B_{12}	100%	110%
Pantothenic Acid	100%	100%
Phosphorus	8%	20%
Magnesium	6%	10%
Zinc	100%	100%
Copper	4%	4%

*Amount in cereal. A serving of cereal plus skim milk provides 1g total fat, less than 5mg cholesterol, 260mg sodium, 290mg potassium, 29g total carbohydrate (11g sugars) and 7g protein.
**Percent Daily Values are based on a 2,000 calorie diet. Your daily values may be higher or lower depending on your calorie needs:

	Calories	2,000	2,500
Total Fat	Less than	65g	80g
Sat Fat	Less than	20g	25g
Cholesterol	Less than	300mg	300mg
Sodium	Less than	2,400mg	2,400mg
Potassium		3,500mg	3,500mg
Total Carbohydrate		300g	375g
Dietary Fiber		25g	30g

▲ **Figure 2C** Nutrition facts from a fortified cereal

3 Atoms consist of protons, neutrons, and electrons

Each element consists of one kind of atom, which is different from the atoms of other elements. An **atom**, named from a Greek word meaning "indivisible," is the smallest unit of matter that still retains the properties of an element. Atoms are so small that it would take about a million of them to stretch across the period printed at the end of this sentence.

Subatomic Particles Physicists have split the atom into more than a hundred types of subatomic particles. However, only three kinds of particles are relevant here. A **proton** is a subatomic particle with a single positive electrical charge (+). An **electron** is a subatomic particle with a single negative charge (−). A **neutron**, as its name implies, is electrically neutral (has no charge).

Figure 3A shows two very simple models of an atom of the element helium (He), the "lighter-than-air" gas that makes balloons rise. Notice that two protons (⊕) and two neutrons (●) are tightly packed in the atom's central core, or **nucleus**. Two electrons (⊖) move around the nucleus at nearly the speed of light. The attraction between the negatively charged electrons and the positively charged protons holds the electrons near the nucleus. The left-hand model shows the number of electrons in the atom. The right-hand model, slightly more realistic, shows a spherical cloud of negative charge created by the rapidly moving electrons. Neither model is drawn to scale. In real atoms, the electrons are very much smaller than the protons and neutrons, and the electron cloud is much bigger compared to the nucleus. Imagine that this atom was the size of a baseball stadium: The nucleus would be the size of a fly in center field, and the electrons would be like two tiny gnats buzzing around the stadium.

Atomic Number and Atomic Mass All atoms of a particular element have the same unique number of protons. This number is the element's **atomic number**. Thus, an atom of helium, with 2 protons, has an atomic number of 2. Carbon, with 6 protons, has an atomic number of 6 (**Figure 3B**). Note that in these atoms, the atomic number is also the number of electrons. Unless otherwise indicated, an atom has an equal number of protons and electrons, and thus its net electrical charge is 0 (zero).

An atom's **mass number** is the sum of the number of protons and neutrons in its nucleus. For helium, the mass number is 4; for carbon, it is 12 (Figures 3A and 3B). The mass of a proton and the mass of a neutron are almost identical and are expressed in a unit of measurement called the dalton. Protons and neutrons each have masses close to 1 dalton. An electron has only about 1/2,000 the mass of a proton, so it contributes very little to an atom's mass. Thus, an atom's **atomic mass** (or weight) is approximately equal to its mass number—the sum of its protons and neutrons.

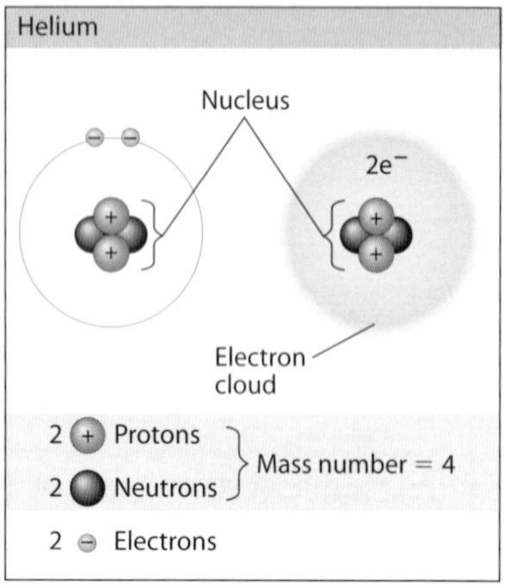

▲ **Figure 3A** Two models of a helium atom

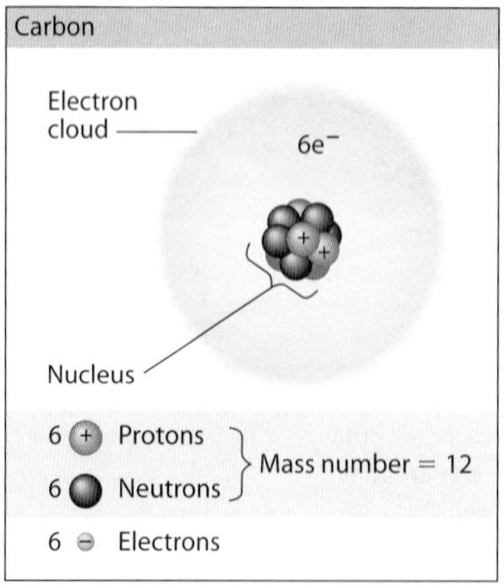

▲ **Figure 3B** Model of a carbon atom

TABLE 3 ISOTOPES OF CARBON

	Carbon-12		Carbon-13		Carbon-14	
Protons	6	Mass number 12	6	Mass number 13	6	Mass number 14
Neutrons	6		7		8	
Electrons	6		6		6	

Isotopes All atoms of an element have the same atomic number, but some atoms of that element may differ in mass number. The different **isotopes** of an element have the same number of protons and behave identically in chemical reactions, but they have different numbers of neutrons. **Table 3** shows the numbers of subatomic particles in the three isotopes of carbon. Carbon-12 (also written ^{12}C), with 6 neutrons, accounts for about 99% of the carbon in nature. Most of the remaining 1% consists of carbon-13 (^{13}C), with 7 neutrons. A third isotope, carbon-14 (^{14}C), with 8 neutrons, occurs in minute quantities. Notice that all three isotopes have 6 protons—otherwise, they would not be carbon.

Both ^{12}C and ^{13}C are stable isotopes, meaning their nuclei remain intact more or less forever. The isotope ^{14}C, on the other hand, is unstable, or radioactive. A **radioactive isotope** is one in which the nucleus decays spontaneously, giving off particles and energy. Radiation from decaying isotopes can damage cellular molecules and thus can pose serious risks to living organisms. But radioactive isotopes can be helpful, as in their use in dating fossils. They are also used in biological research and medicine, as we see next.

? **A nitrogen atom has 7 protons, and its most common isotope has 7 neutrons. A radioactive isotope of nitrogen has 9 neutrons. What is the atomic number and mass number of this radioactive nitrogen?**

Atomic number = 7; mass number = 16

CONNECTION

4 Radioactive isotopes can help or harm us

Living cells cannot distinguish between isotopes of the same element. Consequently, organisms take up and use compounds containing radioactive isotopes in the usual way. Because radioactivity is easily detected by instruments, radioactive isotopes are useful as tracers—biological spies, in effect—for monitoring the fate of atoms in living organisms.

Basic Research Biologists often use radioactive tracers to follow molecules as they undergo chemical changes in an organism. For example, researchers have used carbon dioxide (CO_2) containing the radioactive isotope ^{14}C to study photosynthesis. Using sunlight to power the conversion, plants take in CO_2 from the air and use it to make sugar molecules. Radioactively labeled CO_2 has enabled researchers to trace the sequence of molecules made by plants in the chemical route from CO_2 to sugar.

Medical Diagnosis and Treatment Radioactive isotopes may also be used to tag chemicals that accumulate in specific areas of the body, such as phosphorus in bones. After injection of such a tracer, a special camera produces an image of where the radiation collects. In most diagnostic uses, the patient receives only a tiny amount of an isotope.

Sometimes radioactive isotopes are used for treatment. As you learned in Module 2, the body uses iodine to make a thyroid hormone. Because radioactive iodine accumulates in the thyroid, it can be used to kill cancer cells there.

Substances that the body metabolizes such as glucose or oxygen, may also be labeled with a radioactive isotope. **Figure 4A** shows a patient being examined by a PET (positron-emission tomography) scanner, which can produce three-dimensional images of areas of the body with high metabolic activity. PET is useful for diagnosing certain heart disorders and cancers and for basic research on the brain.

The early detection of Alzheimer's disease may be a new use for such techniques. This devastating illness gradually destroys a person's memory and ability to think. As the disease progresses, the brain becomes riddled with deposits (plaques) of a protein called beta-amyloid. Researchers have identified a protein molecule called PIB that binds to beta-amyloid. PIB contains a radioactive isotope that can be detected on a PET scan. **Figure 4B** shows PET images of the brains of a healthy person (left) and a person with Alzheimer's (right) injected with PIB. Notice that the brain of the Alzheimer's patient has high levels of PIB (red and yellow areas), whereas the unaffected person's brain has lower levels (blue). New therapies are focused on limiting the production of beta-amyloid or clearing it from the brain. A diagnostic test using PIB would allow researchers to monitor the effectiveness of new drugs in people living with the disease.

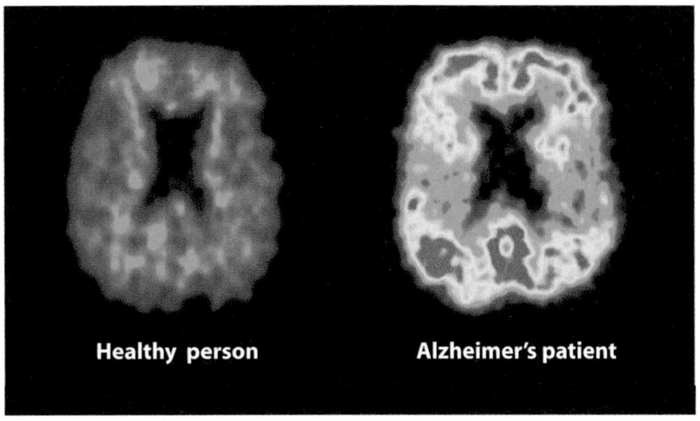

W. E. Klunk and C. A. Mathis, University of Pittsburgh

▲ **Figure 4B** PET images of brains of a healthy person (left) and a person with Alzheimer's disease (right). Red and yellow colors indicate high levels of PIB bound to beta-amyloid plaques.

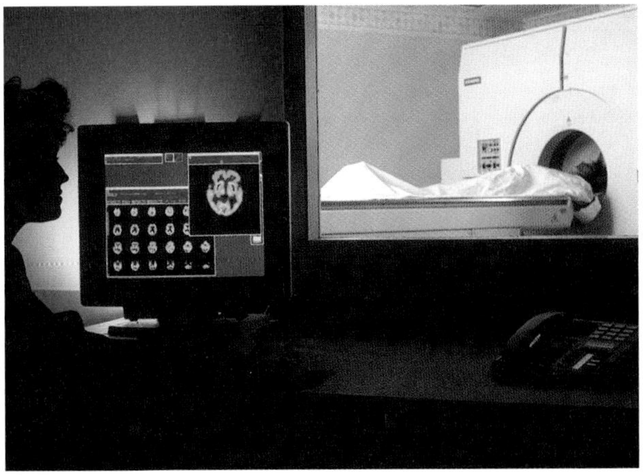

Will & Deni McIntyre/Photo Researchers

▲ **Figure 4A** Technician monitoring the output of a PET scanner

Dangers Although radioactive isotopes have many beneficial uses, uncontrolled exposure to them can harm living organisms by damaging molecules, especially DNA. The particles and energy thrown off by radioactive atoms can break chemical bonds and also cause abnormal bonds to form. The explosion of a nuclear reactor at Chernobyl, Ukraine, in 1986 released large amounts of radioactive isotopes into the environment, which drifted over large areas of Russia, Belarus, and Europe. A few dozen people died from acute radiation poisoning, and over 100,000 people were evacuated from the immediate area. Increased rates of thyroid cancer in children exposed to the radiation have been reported, and many thousands may be at increased risk of future cancers.

Natural sources of radiation can also pose a threat. Radon, a radioactive gas, may be a cause of lung cancer. Radon can contaminate buildings in regions where underlying rocks naturally contain uranium, a radioactive element. Homeowners can buy a radon detector or hire a company to test their home to ensure that radon levels are safe. If levels are found to be unsafe, technology exists to remove radon from homes.

? Why are radioactive isotopes useful as tracers in research on the chemistry of life?

Organisms incorporate radioactive isotopes of an element into their molecules, and researchers can use special scanning devices to detect the presence of these isotopes in biological pathways or locations in the body.

Chemical Bonds

5 The distribution of electrons determines an atom's chemical properties

To understand how atoms interact with each other—the main subject of this section—we need to explore atomic structure further. Of the three subatomic particles—protons, neutrons, and electrons—only electrons are directly involved in the chemical activity of an atom. Electrons vary in the amount of energy they possess. The farther an electron is from the positively charged nucleus, the greater its energy. Electrons move around the nucleus only at certain energy levels, called **electron shells**. Depending on an element's atomic number, an atom may have one, two, or more electron shells surrounding the nucleus.

Figure 5 is an abbreviated version of the periodic table of the elements. It shows the distribution of electrons for the first 18 elements, arranged in rows according to the number of electron shells (one, two, or three). Within each shell, electrons travel in different *orbitals*, which are discrete volumes of space in which electrons are most likely to be found.

Each orbital can hold a maximum of 2 electrons. The first electron shell has only one orbital and can hold only 2 electrons. Thus, hydrogen and helium are the only elements in the first row. For the second and third rows, the outer shell has four orbitals and can hold up to 8 electrons (four pairs). Note that the number of electrons increases by one as you read from left to right across each row in the table and that the electrons don't pair up until all orbitals have at least one electron.

It is the number of electrons in the outermost shell, called the valence shell, that determines the chemical properties of an atom. Atoms whose outer shells are not full (have unpaired electrons) tend to interact with other atoms—that is, to participate in chemical reactions.

Look at the electron shells of the atoms of the four elements that are the main components of biological molecules (highlighted in green in Figure 5). Because their outer shells are incomplete all these atoms react readily with other atoms. The hydrogen atom has only 1 electron in its single electron shell, which can accommodate 2 electrons. Atoms of carbon, nitrogen, and oxygen also are reactive because their valence shells, which can hold 8 electrons, are also incomplete. In contrast, the helium atom has a first-level shell that is full with 2 electrons. Neon and argon also have full outer electron shells. As a result, these elements are chemically inert (unreactive).

How do chemical interactions between atoms enable them to fill their outer electron shells? When two atoms with incomplete outer shells react, each atom will share, donate, or receive electrons, so that both partners end up with completed outer shells. These interactions usually result in atoms staying close together, held by attractions known as **chemical bonds.** In the next two modules, we look at two important types of chemical bonds.

? How many electrons and electron shells does a sodium atom have? How many electrons are in its valence shell?

11 electrons; 3 electron shells; 1 electron in the outer shell

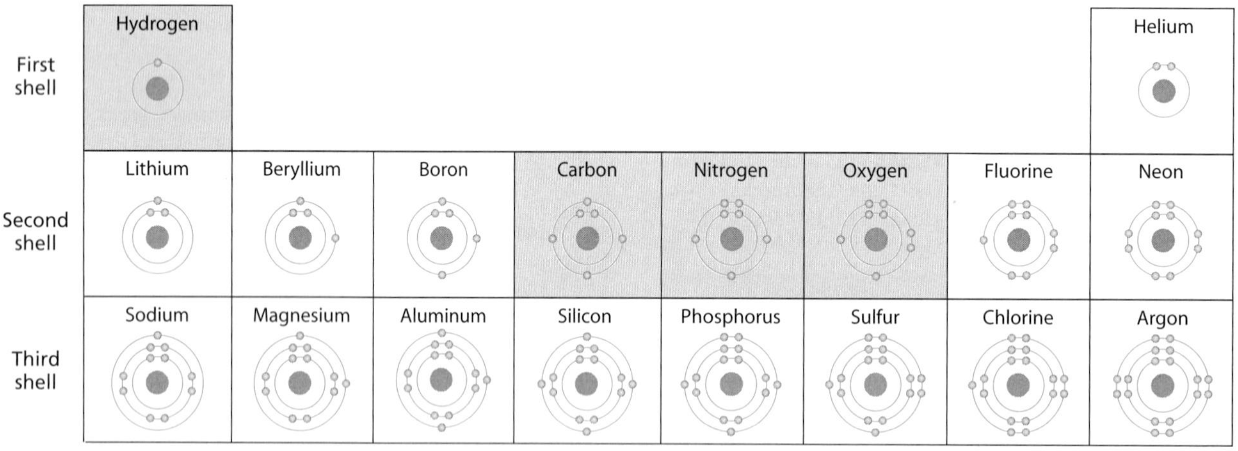

▲ **Figure 5** The electron distribution diagrams of the first 18 elements in the periodic table

6 Covalent bonds join atoms into molecules through electron sharing

The strongest kind of chemical bond is the **covalent bond**, in which two atoms *share* one or more pairs of outer-shell electrons. Two or more atoms held together by covalent bonds form a **molecule**. For example, a covalent bond connects two hydrogen atoms in a molecule of the gas H_2.

Table 6, on the next page, shows four ways to represent this molecule. The symbol H_2, called the molecular formula, tells you that a hydrogen molecule consists of two atoms of hydrogen. The electron distribution diagram shows that the atoms share two electrons; as a result, both atoms fill their outer (and only)

TABLE 6 ALTERNATIVE WAYS TO REPRESENT FOUR COMMON MOLECULES

Molecular Formula	Electron Distribution Diagram	Structural Formula	Space-Filling Model
H_2 Hydrogen	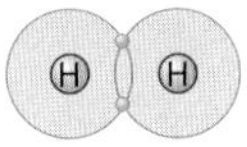	H—H Single bond	
O_2 Oxygen	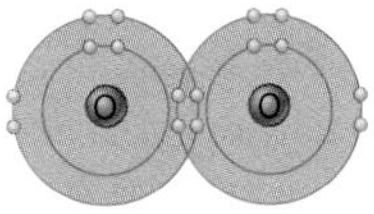	O=O Double bond	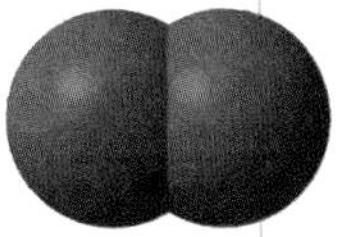
H_2O Water	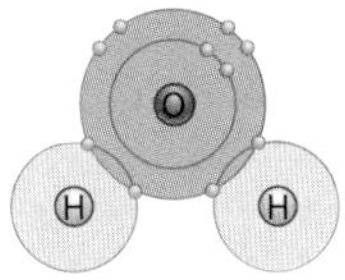	O H H	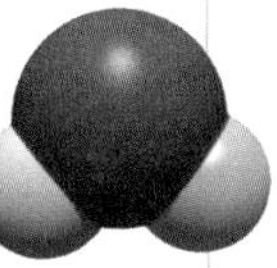
CH_4 Methane	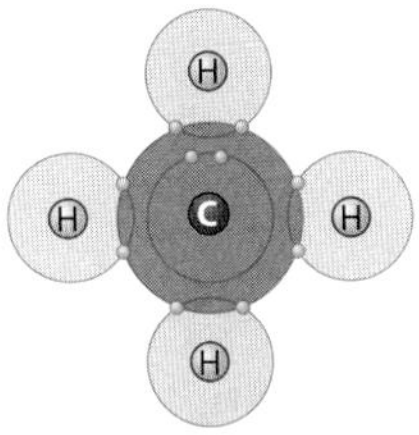	H H—C—H H	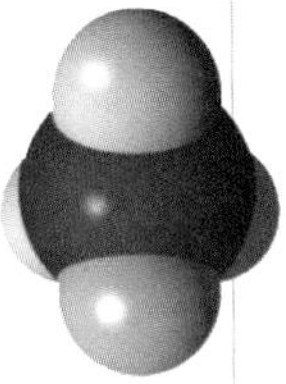

shells. The third column shows a structural formula. The line between the hydrogen atoms represents the single covalent bond formed by the sharing of a pair of electrons (1 electron from each atom). A space-filling model, shown in the fourth column, uses color-coded balls to symbolize atoms and comes closest to showing a molecule's shape.

How many covalent bonds can an atom form? It depends on the number of additional electrons needed to fill its outer, or valence, shell. This number is called the *valence*, or bonding capacity, of an atom. Looking back at the electron distribution diagrams in Figure 5, we see that H can form one bond; O can form two; N, three; and C, four.

In an oxygen molecule (O_2), shown next in Table 6, the two oxygen atoms share two pairs of electrons, forming a double bond. A double bond is indicated by a pair of lines.

H_2 and O_2 are molecules composed of only one element. The third example in the table is a compound. Water (H_2O) is a molecule in which two hydrogen atoms are joined to oxygen by single bonds. And as shown at the bottom of Table 6, it takes four hydrogen atoms, each with a valence of 1, to satisfy carbon's valence of 4. This compound, methane (CH_4), is a major component of natural gas.

Atoms in a molecule are in a constant tug-of-war for the shared electrons of their covalent bonds. An atom's attraction for shared electrons is called its **electronegativity**. The more electronegative an atom, the more strongly it pulls shared electrons toward its nucleus. In molecules of only one element, such as O_2 and H_2, the two identical atoms exert an equal pull on the electrons. The bonds in such molecules are said to be **nonpolar covalent bonds** because the electrons are shared equally between the atoms. Compounds such as methane also have nonpolar bonds, because the atoms of carbon and hydrogen are not substantially different in electronegativity.

In contrast to O_2, H_2, and CH_4, water is composed of atoms with different electronegativities. Oxygen is one of the most electronegative of the elements. (Nitrogen is also highly electronegative.) As indicated by the arrows in **Figure 6**, oxygen attracts the shared electrons in H_2O much more strongly than does hydrogen, so that the electrons spend more time near the oxygen atom than near the hydrogen atoms. This unequal sharing of electrons produces a **polar covalent bond**. In a polar covalent bond, the pulling of shared, negatively charged electrons closer to the more electronegative atom makes that atom partially negative and the other atom partially positive. Thus, in H_2O, the oxygen atom actually has a slight negative charge and each hydrogen atom a slight positive charge. Because of its polar covalent bonds and the wide V shape of the molecule, water is a **polar molecule**—that is, it has an unequal distribution of charges. It is slightly negative at the oxygen end of the molecule (point of the V) and slightly positive at each of the two hydrogen ends.

In some cases, two atoms are so unequal in their attraction for electrons that the more electronegative atom strips an electron completely away from its partner, as we see next.

? What is chemically nonsensical about this structure?

H—C=C—H

Each carbon atom has only three covalent bonds instead of the four required by its valence.

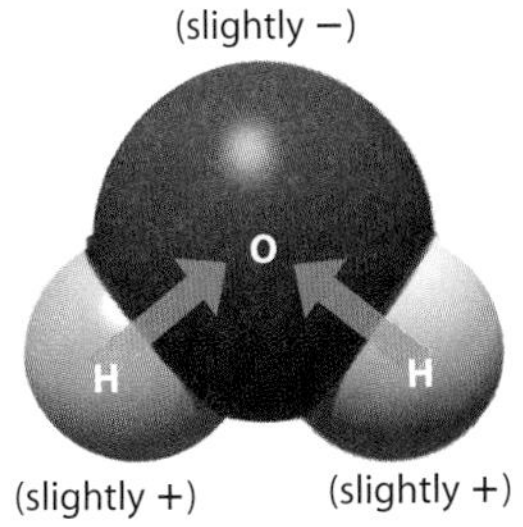

▲ **Figure 6** A water molecule, with polar covalent bonds

7 Ionic bonds are attractions between ions of opposite charge

Table salt is an example of how the transfer of electrons can bond atoms together. **Figure 7A** shows how a sodium atom and a chlorine atom can form the compound sodium chloride (NaCl). Notice that sodium has only 1 electron in its outer shell, whereas chlorine has 7. When these atoms interact, the sodium atom donates its single outer electron to chlorine. Sodium now has only two shells, the outer shell having a full set of 8 electrons. When chlorine accepts sodium's electron, its own outer shell is now full with 8 electrons.

Remember that electrons are negatively charged particles. The transfer of an electron between the two atoms moves one unit of negative charge from one atom to the other. As you can see on the right in Figure 7A, sodium, with 11 protons but now only 10 electrons, has a net electrical charge of 1+. Chlorine, having gained an extra electron, now has 18 electrons but only 17 protons, giving it a net electrical charge of 1−. In each case, an atom has become what is called an ion. An **ion** is an atom or molecule with an electrical charge resulting from a gain or loss of one or more electrons. (As shown in the figure, the ion formed from chlorine is called a chloride ion.) Two ions with opposite charges attract each other. When the attraction holds them together, it is called an **ionic bond**. The resulting compound, in this case NaCl, is electrically neutral.

Sodium chloride is a familiar type of **salt**, a synonym for an ionic compound. Salts often exist as crystals in nature. **Figure 7B** shows the ions Na^+ and Cl^- in a crystal of sodium chloride. An NaCl crystal can be of any size (there is no fixed number of ions), but sodium and chloride ions are always present in a 1:1 ratio. The ratio of ions can differ in the various kinds of salts.

The environment affects the strength of ionic bonds. In a dry salt crystal, the bonds are so strong that it takes a hammer and chisel to break enough of them to crack the crystal. If the same salt crystal is placed in water, however, the ionic bonds break when the ions interact with water molecules and the salt dissolves, as we'll discuss in Module 13. Most drugs are manufactured as salts because they are quite stable when dry but can dissolve easily in water.

? Explain what holds together the atoms in a crystal of table salt (NaCl).

Opposite charges attract. The positively charged sodium ions (Na^+) and the negatively charged chloride ions (Cl^-) are held together by ionic bonds, attractions between oppositely charged ions.

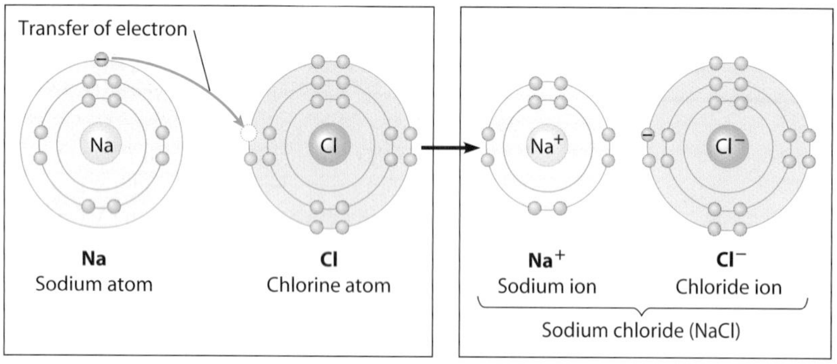

▲ **Figure 7A** Formation of an ionic bond, producing sodium chloride

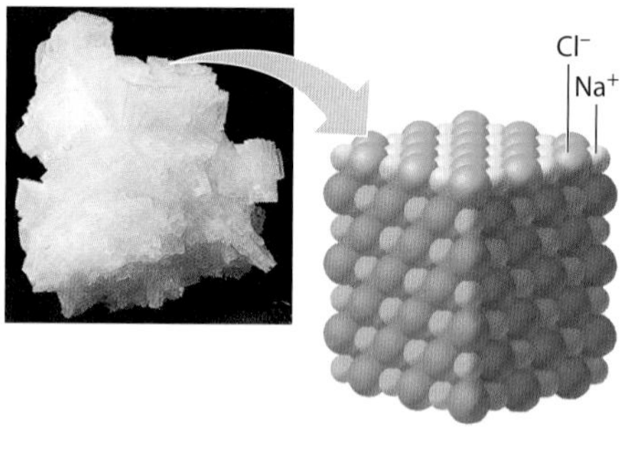

▲ **Figure 7B** A crystal of sodium chloride

8 Hydrogen bonds are weak bonds important in the chemistry of life

In living organisms, most of the strong chemical bonds are covalent, linking atoms to form a cell's molecules. But crucial to the functioning of a cell are weaker bonds within and between molecules, such as the ionic bonds we just discussed. Most large molecules are held in their three-dimensional shape by weak bonds. In addition, molecules in a cell may be held together briefly by weak bonds, respond to one another in some way, and then separate.

As you saw in Module 6, a hydrogen atom that has formed a polar covalent bond with an electronegative atom (such as oxygen or nitrogen) has a partial positive charge. This partial positive charge allows it to be attracted to a nearby partially negative atom (often an oxygen or nitrogen) of another molecule. These weak but important bonds are best illustrated with water molecules, as shown in **Figure 8**. The charged regions on each water molecule are electrically attracted to oppositely charged regions on neighboring molecules. Because the positively charged region in this special type of bond is always a hydrogen atom, the bond is called a **hydrogen bond**. As Figure 8 shows, the negative (oxygen) pole of a water molecule can form hydrogen bonds (dotted lines) to two hydrogen atoms. And each hydrogen atom of "*a water molecule*"

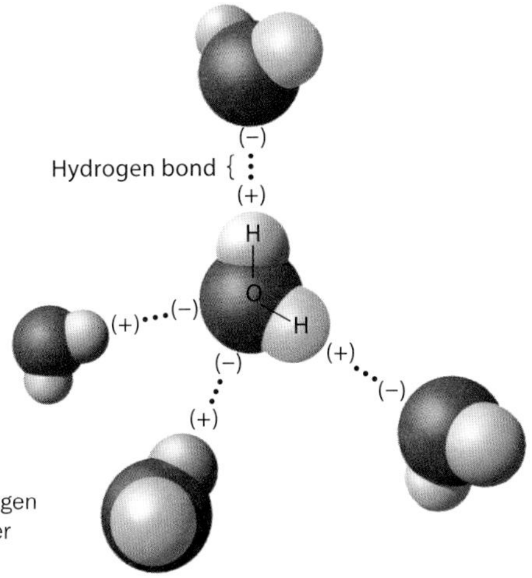

▶ **Figure 8** Hydrogen bonds between water molecules

can form a hydrogen bond with a nearby partial negative oxygen atom of another water molecule. Thus, each H_2O molecule can hydrogen-bond to as many as four partners.

Hydrogen bonds help to create a protein's shape (and thus its function) and hold the two strands of a DNA molecule together. Later in this chapter, we explore how water's polarity and hydrogen bonds give it unique, life-supporting properties. But first we discuss how the making and breaking of bonds change the composition of matter.

? What enables neighboring water molecules to hydrogen-bond to one another?

The molecules are polar, with the negative end (oxygen end) of one molecule attracted to a positive end (hydrogen end) of its neighbor.

9 Chemical reactions make and break chemical bonds

The basic chemistry of life has an overarching theme: The structure of atoms and molecules determines the way they behave. As we have seen, the chemical behavior of an atom is determined by the number and arrangement of its subatomic particles, particularly its electrons. Other properties emerge when atoms combine to form molecules and when molecules interact. Water is a good example, because its emergent properties sustain all life on Earth.

Hydrogen and oxygen gases can react to form water:

$$2\,H_2 + O_2 \longrightarrow 2\,H_2O$$

This is a **chemical reaction**, the breaking and making of chemical bonds, leading to changes in the composition of matter. In this case, two molecules of hydrogen ($2\,H_2$) react with one molecule of oxygen (O_2) to produce two molecules of water ($2\,H_2O$). The arrow indicates the conversion of the starting materials, called the **reactants**, to the **product**, the material resulting from the chemical reaction. Notice that the same *numbers* of hydrogen and oxygen atoms appear on the left and right sides of the arrow, although they are grouped differently. Chemical reactions do not create or destroy matter; they only rearrange it in various ways. As shown in **Figure 9**, the covalent bonds (represented here as white "sticks" between atoms) holding hydrogen atoms together in H_2 and holding oxygen atoms together in O_2 are broken, and new bonds are formed to yield the H_2O product molecules.

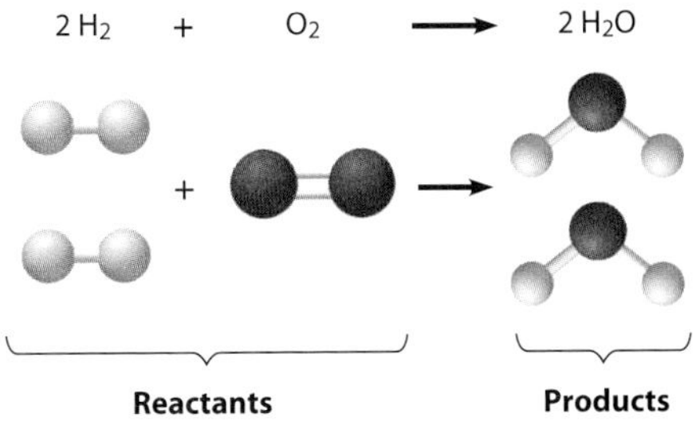

▲ **Figure 9** Breaking and making of bonds in a chemical reaction

Organisms cannot make water from H_2 and O_2, but they do carry out a great number of chemical reactions that rearrange matter in significant ways. Let's examine a chemical reaction that is essential to life on Earth: photosynthesis. The raw materials of photosynthesis are carbon dioxide (CO_2), which is taken from the air, and water (H_2O), which plants absorb from the soil. Within green plant cells, sunlight powers the conversion of these reactants to the sugar product glucose ($C_6H_{12}O_6$) and oxygen (O_2), a by-product that the plant releases into the air. The following chemical shorthand summarizes the process:

$$6\,CO_2 + 6\,H_2O \longrightarrow C_6H_{12}O_6 + 6\,O_2$$

Although photosynthesis is actually a sequence of many chemical reactions, we see that we end up with the same number and kinds of atoms we started with. Matter has simply been rearranged, with an input of energy provided by sunlight.

The chemistry of life is dynamic. Living cells routinely carry out thousands of chemical reactions. These reactions take place in the watery environment of a cell, and we look at the life-supporting properties of water next.

? Fill in the blanks with the correct numbers in the following chemical process:

$$C_6H_{12}O_6 + \_\_O_2 \longrightarrow \_\_CO_2 + \_\_H_2O$$

What process do you think this reaction represents? (*Hint:* Think about how your cells use these reactants to produce energy.)

$C_6H_{12}O_6 + 6\,O_2 \rightarrow 6\,CO_2 + 6\,H_2O$; the breakdown of sugar in the presence of oxygen to carbon dioxide and water, with the release of energy that the cell can use

Water's Life-Supporting Properties

10 Hydrogen bonds make liquid water cohesive

We can trace water's life-supporting properties to the structure and interactions of its molecules—their polarity and resulting hydrogen bonding between molecules (review Figure 8).

Hydrogen bonds between molecules of liquid water last for only a few trillionths of a second, yet at any instant, many molecules are hydrogen-bonded to others. This tendency of molecules of the same kind to stick together, called **cohesion**, is much stronger for water than for most other liquids. The cohesion of water is important in the living world. Trees, for example, depend on cohesion to help transport water and nutrients from their roots to their leaves. The evaporation of water from a leaf exerts a pulling force on water within the veins of the leaf. Because of cohesion, the force is relayed all the way down to the roots. **Adhesion**, the clinging of one substance to another, also plays a role. The adhesion of water to the cell walls of a plant's thin veins helps counter the downward pull of gravity.

Related to cohesion is **surface tension**, a measure of how difficult it is to stretch or break the surface of a liquid.

Herman Eisenbeiss/
Photo Researchers

◀ **Figure 10** Surface tension allows a water strider to walk on water.

Hydrogen bonds give water unusually high surface tension, making it behave as though it were coated with an invisible film. You can observe the surface tension of water by slightly overfilling a glass; the water will stand above the rim. The water strider in **Figure 10** takes advantage of the high surface tension of water to "stride" across ponds without breaking the surface.

? After a hard workout, you may see "beads" of sweat on your face. Can you explain what holds the sweat in droplet form?

● The cohesion of water molecules and its high surface tension hold water in droplets. The adhesion of water to your skin helps hold the beads in place.

11 Water's hydrogen bonds moderate temperature

If you have ever burned your finger on a metal pot while waiting for the water in it to boil, you know that water heats up much more slowly than metal. In fact, because of hydrogen bonding, water has a stronger resistance to temperature change than most other substances.

Temperature and heat are related but different. A swimmer crossing San Francisco Bay has a higher temperature than the water, but the bay contains far more heat because of its immense volume. **Heat** is the amount of energy associated with the movement of atoms and molecules in a body of matter. **Temperature** measures the intensity of heat—that is, the *average* speed of molecules rather than the *total* amount of heat energy in a body of matter.

Heat must be absorbed in order to break hydrogen bonds, and heat is released when hydrogen bonds form. To raise the temperature of water, heat energy must first disrupt hydrogen bonds before water molecules can move faster. Thus, water absorbs a large amount of heat while warming up only a few degrees. Conversely, when water cools, more hydrogen bonds form, and a considerable amount of heat is released.

Earth's giant water supply moderates temperatures, helping to keep them within limits that permit life. Oceans, lakes, and rivers store a huge amount of heat from the sun during warm periods. Heat given off from gradually cooling water warms the air. That's why coastal areas generally have milder climates than inland regions. Water's resistance to temperature change also stabilizes ocean temperatures, creating a favorable environment for marine life. Since water accounts for approximately 66% of your body weight, it also helps moderate your temperature.

Novastock/Photolibrary

◀ **Figure 11** Evaporative cooling occurs as sweat dries.

When a substance evaporates (changes physical state from a liquid to a gas), the surface of the liquid that remains behind cools down. This **evaporative cooling** occurs because the molecules with the greatest energy (the "hottest" ones) leave. It's as if the 10 fastest runners on the track team left school, lowering the average speed of the remaining team. Evaporative cooling helps prevent some land-dwelling organisms from overheating. Evaporation from a plant's leaves keeps them from becoming too warm in the sun, just as sweating helps dissipate our excess body heat (**Figure 11**). On a much larger scale, the evaporation of surface waters cools tropical seas.

? Explain the popular adage "It's not the heat, it's the humidity."

● High humidity hampers cooling by slowing the evaporation of sweat.

12 Ice is less dense than liquid water

Water exists on Earth in the form of a gas (water vapor), liquid, and solid. Unlike most substances, water is less dense as a solid than as a liquid. And as you might guess, this unusual property is due to hydrogen bonds.

As water freezes, each molecule forms stable hydrogen bonds with its neighbors, holding them at "arm's length" and creating a three-dimensional crystal. In **Figure 12**, compare the spaciously arranged molecules in the ice crystal with the more tightly packed molecules in the liquid water. The ice crystal has fewer molecules than an equal volume of liquid water. Therefore, ice is less dense and floats on top of liquid water.

If ice sank, then eventually ponds, lakes, and even oceans would freeze solid. Instead, when a deep body of water cools, the floating ice insulates the water below from colder air above. This "blanket" prevents the water below from freezing and allows fish and many other aquatic forms of life to survive under the frozen surface.

In the Arctic, this frozen surface serves as the winter hunting ground for polar bears (Figure 12). The shrinking of this ice cover as a result of global warming may doom these bears.

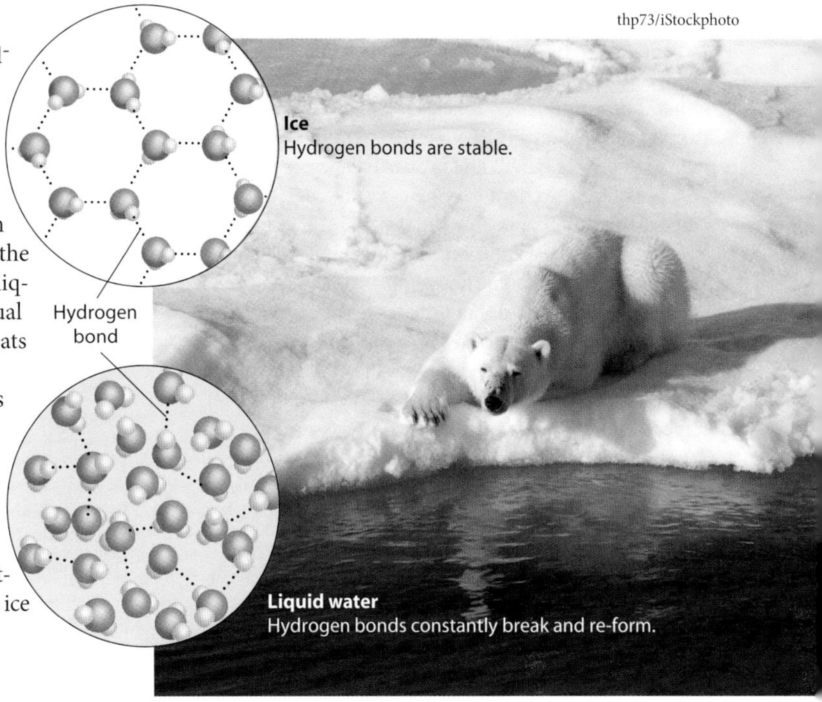

▲ **Figure 12** Hydrogen bonds between water molecules in ice and water

? Explain how the freezing of water can crack boulders.

Water in the crevices of a boulder expands as it freezes because the water molecules become spaced farther apart in forming ice crystals, cracking the rock.

13 Water is the solvent of life

If you add a teaspoon of table salt to a glass of water, the salt will dissolve and eventually become evenly mixed with the water, forming a solution. A **solution** is a liquid consisting of a uniform mixture of two or more substances. The dissolving agent (in this case, water) is the **solvent**, and a substance that is dissolved (salt) is a **solute**. An **aqueous solution** (from the Latin *aqua*, water) is one in which water is the solvent.

Water's versatility as a solvent results from the polarity of its molecules. **Figure 13** shows how a teaspoon of salt dissolves in water. At the surface of each grain, or crystal, the sodium and chloride ions are exposed to water. These ions and the water molecules are attracted to each other due to their opposite charges. The oxygen ends (red) of the water molecules have a partial negative charge and cling to the positive sodium ions (Na^+). The hydrogens of the water molecules, with their partial positive charge, are attracted to the negative chloride ions (Cl^-). Working inward from the surface of each salt crystal, water molecules eventually surround and separate all the ions. Water dissolves other ionic compounds as well. Seawater, for instance, contains a great variety of dissolved ions, as do your cells.

A compound doesn't need to be ionic to dissolve in water. A spoonful of sugar will also dissolve in a glass of water. Polar molecules such as sugar dissolve as water molecules surround them and form hydrogen bonds with their polar regions. Even large molecules, such as proteins, can dissolve if they have ionic or polar regions on their surface. As the solvent inside all cells, in blood, and in plant sap, water dissolves an enormous variety of solutes necessary for life.

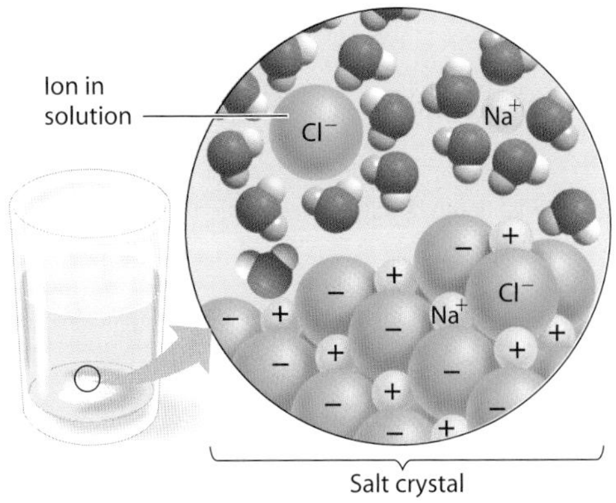

▲ **Figure 13** A crystal of salt (NaCl) dissolving in water

? Why are blood and most other biological fluids classified as aqueous solutions?

The solvent is water.

14 The chemistry of life is sensitive to acidic and basic conditions

In aqueous solutions, a very small percentage of the water molecules actually break apart (dissociate) into ions. The ions formed are called hydrogen ions (H^+) and hydroxide ions (OH^-). Hydrogen and hydroxide ions are very reactive. The proper balance of these ions is critical for the chemical processes that occur within an organism.

Some chemical compounds contribute additional H^+ to an aqueous solution, whereas others remove H^+ from it. A compound that donates hydrogen ions to solutions is called an **acid**. One example of a strong acid is hydrochloric acid (HCl), the acid in the gastric juice in your stomach. In solution, HCl dissociates completely into H^+ and Cl^-. An acidic solution has a higher concentration of H^+ than OH^-.

A **base** is a compound that accepts hydrogen ions and removes them from solution. Some bases, such as sodium hydroxide (NaOH), do this by donating OH^-; the OH^- combines with H^+ to form H_2O, thus reducing the H^+ concentration. Sodium hydroxide, also called lye, is a common ingredient in oven cleaners. The more basic a solution, the higher its OH^- concentration and the lower its H^+ concentration. Basic solutions are also called alkaline solutions.

We use the **pH scale** to describe how acidic or basic a solution is (pH stands for potential of hydrogen). As shown in **Figure 14**, the scale ranges from 0 (most acidic) to 14 (most basic). Each pH unit represents a 10-fold change in the concentration of H^+ in a solution. For example, lemon juice at pH 2 has 10 times more H^+ than an equal amount of a cola at pH 3 and 100 times more H^+ than tomato juice at pH 4.

Pure water and aqueous solutions that are neither acidic nor basic are said to be neutral; they have a pH of 7. They do contain some hydrogen and hydroxide ions, but the concentrations of the two kinds of ions are equal. The pH of the solution inside most living cells is close to 7. Even a slight change in pH can be harmful because the proteins and other complex molecules in cells are extremely sensitive to the concentrations of H^+ and OH^-.

The pH of human blood plasma (the fluid portion of the blood) is very close to 7.4. A person cannot survive for more than a few minutes if the blood pH drops to 7.0 or rises to 7.8. If you add a small amount of a strong acid to a liter of pure water, the pH drops from 7.0 to 0. If the same amount of acid is added to a liter of blood, however, the pH decrease is only from 7.4 to 7.3. How can the acid have so much less effect on the pH of blood? Biological fluids contain **buffers**, substances that minimize changes in pH. They do so by accepting H^+ when it is in excess and donating H^+ when it is depleted. There are several types of buffers that contribute to the pH stability in blood and many other internal solutions.

In the next module, we explore how changes in acidity can have environmental consequences.

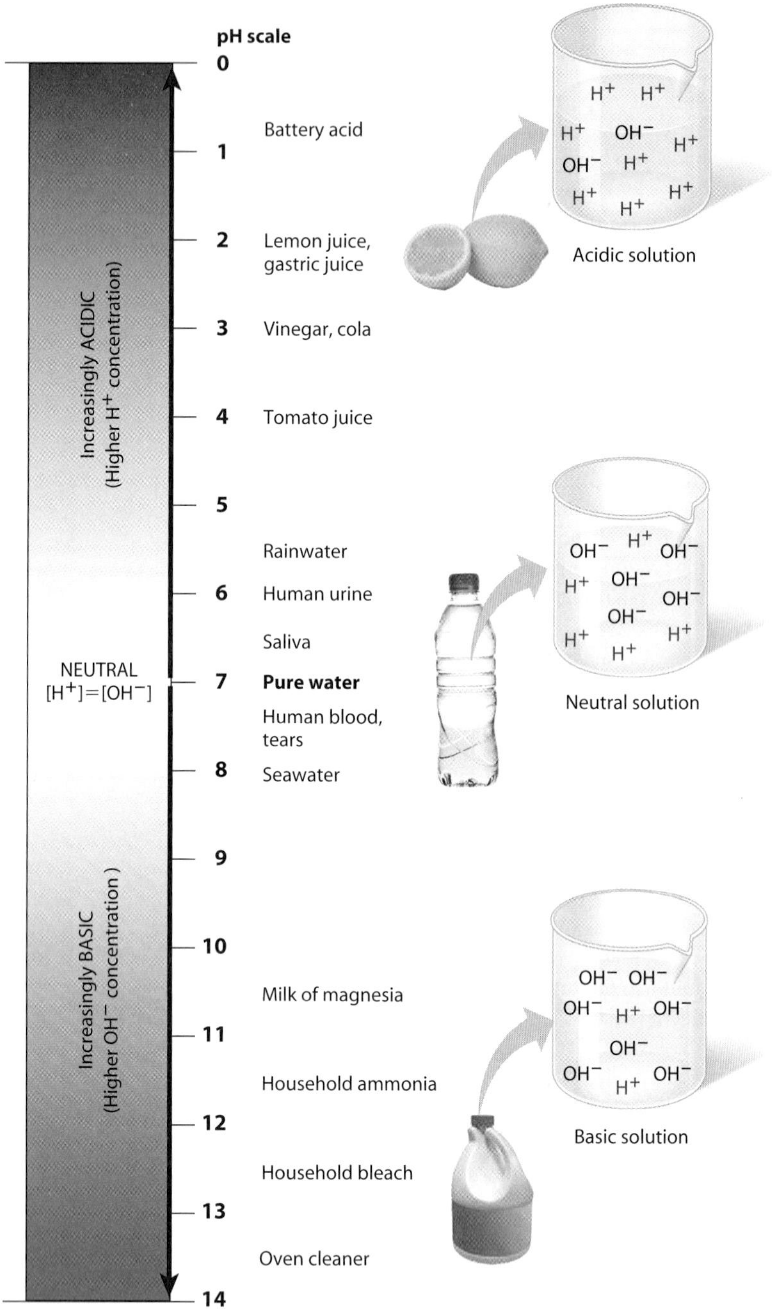

Jakub Semeniuk/iStockphoto; VR Photos/Shutterstock; VBeth Van Trees/Shutterstock

▲ **Figure 14** The pH scale reflects the relative concentrations of H^+ and OH^-.

? Compared to a basic solution at pH 9, the same volume of an acidic solution at pH 4 has ______ times more hydrogen ions (H^+).

● 100,000

CONNECTION

15 Acid precipitation and ocean acidification threaten the environment

Considering the dependence of all life on water, contamination of rivers, lakes, seas, and precipitation poses serious environmental problems. The burning of fossil fuels (coal, oil, and gas), which releases air-polluting compounds and large amounts of CO_2 into the atmosphere, is among the many threats to water quality posed by human activities. Chemical reactions of these compounds with water increase acidity and alter the delicate balance of conditions for life on Earth.

Sulfur oxides and nitrogen oxides released by burning fossil fuels react with water in the air to form strong acids, which fall to Earth with rain or snow. **Acid precipitation** refers to rain, snow, or fog with a pH lower than 5.2 (the pH of uncontaminated rain is 5.6). Acid precipitation has damaged life in lakes and streams and adversely affected plants through changes in soil chemistry. In the United States, amendments made in 1990 to the Clean Air Act have reduced acid precipitation and improved the health of most North American lakes and forests.

Carbon dioxide is the main product of fossil fuel combustion, and its steadily increasing release into the atmosphere is linked to global climate change (see Module 38.6). About 25% of human-generated CO_2 is absorbed by the oceans. An increase in CO_2 absorption is expected to change ocean chemistry and harm marine life and ecosystems.

In **ocean acidification,** CO_2 dissolving in seawater lowers ocean pH. That's because CO_2 reacts with water to produce carbonic acid (H_2CO_3). The change in acidity decreases the concentration of carbonate ions, which are required by corals and other organisms to produce their skeletons or shells, a process called calcification. Coral reef ecosystems act as havens for a great diversity of organisms (**Figure 15**). Other calcifying organisms are important food sources for salmon, herring, and other ocean fishes. Decreased calcification is likely to affect marine food webs and may substantially alter the productivity and biodiversity of Earth's oceans.

Shutterstock

▲ **Figure 15** Coral reefs are threatened by ocean acidification.

? What is the relationship between fossil fuel consumption and coral reefs?

Some of the increased CO_2 released by burning fossil fuels dissolves in and lowers the pH of the oceans. A lower pH reduces levels of carbonate ions, which then lowers the rate of calcification by coral animals.

EVOLUTION CONNECTION

16 The search for extraterrestrial life centers on the search for water

When astrobiologists search for signs of extraterrestrial life on distant planets, they look for evidence of water. Why? As we've seen in this chapter, the emergent properties of water (its cohesion, ability to moderate temperature and insulate, and versatility as a solvent) support life on Earth in many ways. Is it possible that some form of life has evolved on other planets that have water in their environment?

Researchers with the National Aeronautics and Space Administration (NASA) have found evidence that water was once abundant on Mars. In January 2004, NASA succeeded in landing two golf-cart-sized rovers, named *Spirit* and *Opportunity*, on Mars. These robotic geologists, using sophisticated instruments to determine the composition of rocks, detected a mineral that is formed only in the presence of water. And pictures sent back from the rovers and various orbiting Mars spacecraft have revealed physical evidence of past water.

In 2008, an analysis of Martian soil by the *Phoenix Mars Lander* provided evidence that components of the soil had at one time been dissolved in water. **Figure 16** shows a robotic arm of *Phoenix* that has gathered samples and scraped trenches that exposed ice just below the surface on Mars.

▶ **Figure 16** Robotic arm of the *Phoenix Mars Lander* probing for evidence of water

NASA

No evidence for life on Mars has yet been found. But is it possible that life is not unique to planet Earth? Finding evidence of life elsewhere would support the hypothesis that the chemical evolution of life is possible.

? Why is the presence of water important in the search for extraterrestrial life?

Water plays important roles in life as we know it, from moderating temperatures on the planet to functioning as the solvent of life.

REVIEW

For Practice Quizzes, BioFlix, MP3 Tutors, and Activities, go to www.masteringbiology.com.

Reviewing the Concepts

Elements, Atoms, and Compounds (1–4)

1 Organisms are composed of elements, in combinations called compounds. Oxygen, carbon, hydrogen, and nitrogen make up about 96% of living matter.

2 Trace elements are common additives to food and water.

3 Atoms consist of protons, neutrons, and electrons.

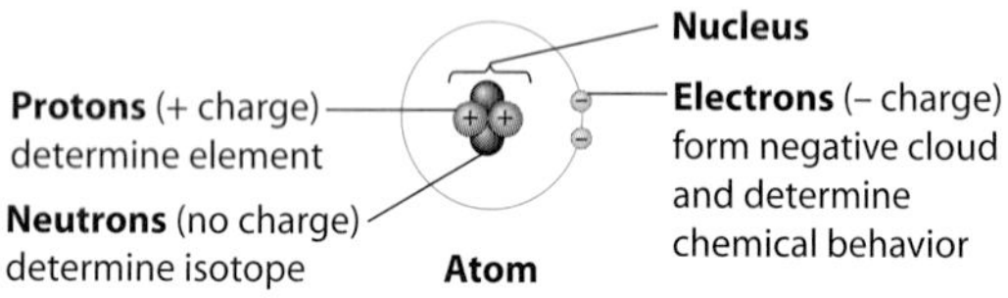

4 Radioactive isotopes can help or harm us. Radioactive isotopes are valuable in basic research and medicine.

Chemical Bonds (5–9)

5 The distribution of electrons determines an atom's chemical properties. An atom whose outer electron shell is not full tends to interact with other atoms and share, gain, or lose electrons, resulting in attractions called chemical bonds.

6 Covalent bonds join atoms into molecules through electron sharing. In a nonpolar covalent bond, electrons are shared equally. In polar covalent bonds, such as those found in water, electrons are pulled closer to the more electronegative atom.

7 Ionic bonds are attractions between ions of opposite charge. Electron gain and loss create charged atoms, called ions.

8 Hydrogen bonds are weak bonds important in the chemistry of life. The slightly positively charged H atoms in one polar molecule may be attracted to the partial negative charge of an O or N atom in a neighboring molecule.

9 Chemical reactions make and break chemical bonds. The composition of matter is changed as bonds are broken and formed to convert reactants to products.

Water's Life-Supporting Properties (10–16)

10 Hydrogen bonds make water molecules cohesive. Cohesion creates surface tension and allows water to move from plant roots to leaves.

11 Water's hydrogen bonds moderate temperature. Heat is absorbed when hydrogen bonds break and released when hydrogen bonds form. This helps keep temperatures relatively steady. As the most energetic water molecules evaporate, the surface of a substance cools.

12 Ice is less dense than liquid water. Floating ice protects lakes and oceans from freezing solid.

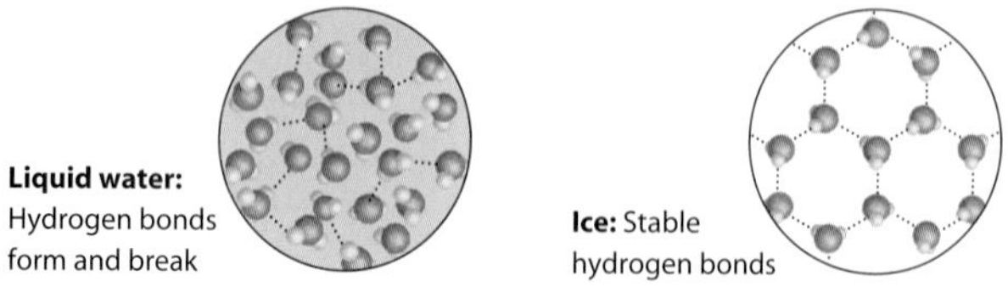

13 Water is the solvent of life. Polar or charged solutes dissolve when water molecules surround them, forming aqueous solutions.

14 The chemistry of life is sensitive to acidic and basic conditions. A compound that releases H^+ in solution is an acid, and one that accepts H^+ is a base. The pH scale ranges from 0 (most acidic) to 14 (most basic). The pH of most cells is close to 7 (neutral) and kept that way by buffers.

15 Acid precipitation and ocean acidification threaten the environment. The burning of fossil fuels increases the amount of CO_2 in the atmosphere and dissolved in the oceans. The acidification of the ocean threatens coral reefs and other marine organisms.

16 The search for extraterrestrial life centers on the search for water. The emergent properties of water support life on Earth and may contribute to the potential for life to have evolved on other planets.

Connecting the Concepts

1. Fill in the blanks in this concept map to help you tie together the key concepts concerning elements, atoms, and molecules.

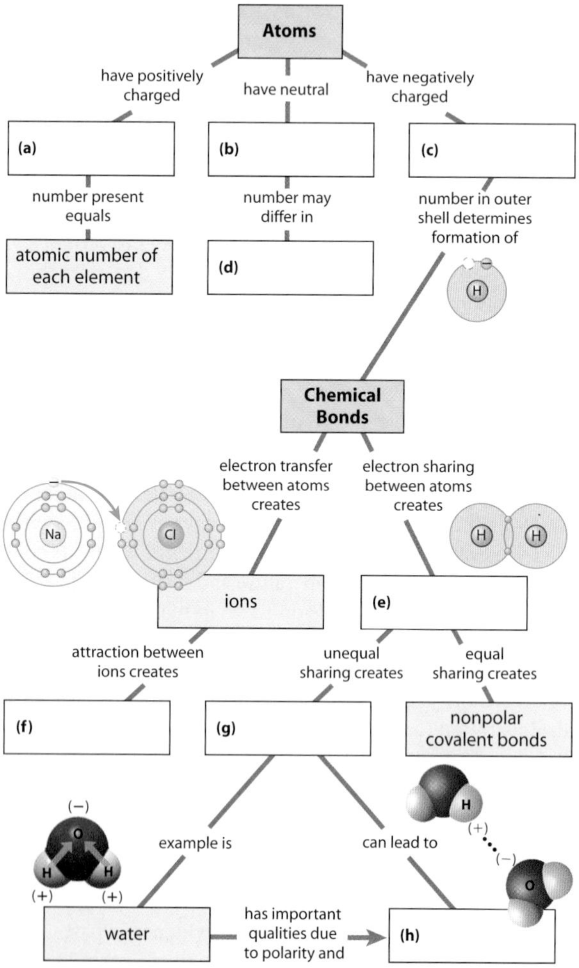

2. Create a concept map to organize your understanding of the life-supporting properties of water. A sample map is in the answer section, but the value of this exercise is in the thinking and integrating you must do to create your own map.

Testing Your Knowledge

Multiple Choice

3. Changing the ______ would change it into an atom of a different element.
   a. number of electrons surrounding the nucleus of an atom
   b. number of bonds formed by an atom
   c. number of protons in the nucleus of an atom
   d. electrical charge of an atom
   e. number of neutrons in the nucleus of an atom
4. Your body contains the smallest amount of which of the following elements?
   a. nitrogen c. carbon e. hydrogen
   b. phosphorus d. oxygen
5. A solution at pH 6 contains ______ than the same amount of a solution at pH 8.
   a. 2 times more H^+
   b. 4 times more H^+
   c. 100 times more H^+
   d. 4 times less H^+
   e. 100 times less H^+
6. Most of the unique properties of water result from the fact that water molecules
   a. are very small.
   b. are held together by covalent bonds.
   c. easily separate from one another.
   d. are constantly in motion.
   e. are polar and form hydrogen bonds.
7. A sulfur atom has 6 electrons in its outer shell. As a result, it forms ______ covalent bonds with other atoms. (*Explain your answer.*)
   a. two c. four e. eight
   b. three d. six
8. What does the word *trace* mean when you're talking about a trace element?
   a. The element is required in very small amounts.
   b. The element can be used as a label to trace atoms through an organism's body.
   c. The element is very rare on Earth.
   d. The element enhances health but is not essential for the organism's long-term survival.
   e. The element passes rapidly through the organism.
9. A can of cola consists mostly of sugar dissolved in water, with some carbon dioxide gas that makes it fizzy and makes the pH less than 7. In chemical terms, you could say that cola is an aqueous solution where water is the ______, sugar is a ______, and carbon dioxide makes the solution ______.
   a. solvent ... solute ... basic
   b. solute ... solvent ... basic
   c. solvent ... solute ... acidic
   d. solute ... solvent ... acidic
   e. not enough information to say
10. Radioactive isotopes can be used in medical studies because
   a. they allow researchers to time how long processes take.
   b. they are more reactive than nonradioactive isotopes.
   c. the cell does not recognize the extra protons in the nucleus, so isotopes are readily used in cellular processes.
   d. their location or quantity can be determined because of their radioactivity.
   e. their extra neutrons produce different colors that can be traced through the body.

True/False (Change false statements to make them true.)

11. Table salt, water, and carbon are compounds.
12. The smallest unit of an element is a molecule.
13. A bathtub full of lukewarm water may hold more heat than a teakettle full of boiling water.
14. If the atoms in a molecule share electrons equally, the molecule is said to be nonpolar.
15. Ice floats because water molecules in ice are more tightly packed than in liquid water.
16. Atoms in a water molecule are held together by the sharing of electrons.
17. Most acid precipitation results from the presence of pollutants from aerosol cans and air conditioners.
18. An atom that has gained or lost electrons is called an ion.

Describing, Comparing, and Explaining

19. Make a sketch that shows how water molecules hydrogen-bond with one another. Why do water molecules form hydrogen bonds? What unique properties of water result from water's tendency to form hydrogen bonds?
20. Describe two ways in which the water in your body helps stabilize your body temperature.
21. Compare covalent and ionic bonds.
22. What is an acid? A base? How is the acidity of a solution described?

Applying the Concepts

23. The diagram below shows the arrangement of electrons around the nucleus of a fluorine atom (left) and a potassium atom (right). What kind of bond do you think would form between these two atoms?

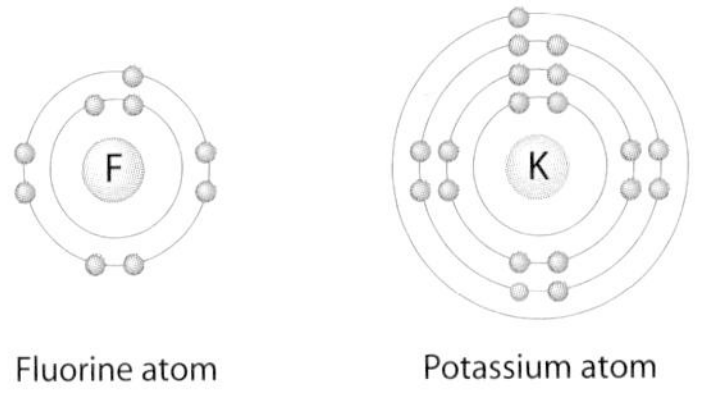

24. Look back at the abbreviated periodic table of the elements in Figure 5. If two or more elements are in the same row, what do they have in common? If two elements are in the same column, what do they have in common?
25. This chapter explains how the emergent properties of water contribute to the suitability of the environment for life. Until fairly recently, scientists assumed that other physical requirements for life included a moderate range of temperature, pH, and atmospheric pressure, as well as low levels of toxic chemicals. That view has changed with the discovery of organisms known as extremophiles, which have been found flourishing in hot, acidic sulfur springs, around hydrothermal vents deep in the ocean, and in soils with high levels of toxic metals. Why would astrobiologists be interested in studying extremophiles? What does the existence of life in such extreme environments say about the possibility of life on other planets?

Review Answers

1. a. protons; b. neutrons; c. electrons; d. different isotopes; e. covalent bonds; f. ionic bonds; g. polar covalent bonds; h. hydrogen bonding

2.

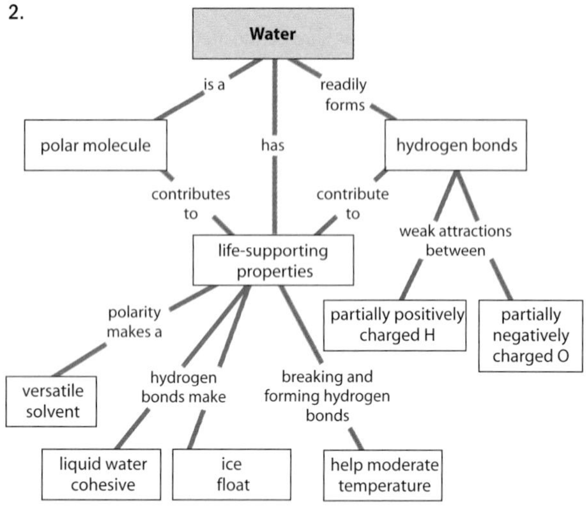

3. c 4. b 5. c 6. e 7. a (It needs to share 2 more electrons for a full outer shell of 8.) 8. a 9. c 10. d 11. F (Only salt and water are compounds; carbon is an element.) 12. F (The smallest unit of an element is an atom.) 13. T 14. T 15. F (Water molecules in ice are farther apart.) 16. T 17. F (Most acid precipitation results from burning fossil fuels.) 18. T

19. For a diagram, see Figure 8. Water molecules form hydrogen bonds because their slightly negative O atoms are attracted to the slightly positive H atoms of neighboring molecules. The unique properties of water that result from hydrogen bonding are cohesion, adhesion, surface tension, the ability to absorb and store large amounts of heat and to release heat as water cools, a solid form (ice) that is less dense than liquid water, and solvent properties.

20. First: Because increasing the temperature of water (the average speed of its molecules) requires breaking hydrogen bonds, a process that absorbs heat, a large amount of heat can be added to water before the water's temperature starts to rise. Conversely, when the surrounding temperature falls, new hydrogen bonds form in water, with the release of heat that slows the cooling process. Second: When the body becomes overheated, water evaporating from its surface decreases the body's temperature (evaporative cooling) because the hotter water molecules leave.

21. A covalent bond forms when atoms complete their outer shells by sharing electrons. Atoms can also complete their outer shells by gaining or losing electrons. This leaves the atoms as ions, with + or – charges. The oppositely charged ions are attracted to each other, forming an ionic bond.

22. An acid is a compound that donates hydrogen ions (H^+) to a solution. A base is a compound that accepts hydrogen ions and removes them from solution. Acidity is described by the pH scale, which measures H^+ concentration on a scale of 0 (most acidic) to 14 (most basic).

23. Fluorine needs 1 electron for a full outer shell of 8, and if potassium loses 1 electron, its outer shell will have 8. Potassium will lose an electron (becoming a + ion), and fluorine will pick it up (becoming a – ion). The ions can form an ionic bond.

24. The elements in a row all have the same number of electron shells. In a column, all the elements have the same number of electrons in their outer shell.

25. These extreme environments may be similar to those found on other planets. The fact that life may have evolved and continues to flourish in such extreme environments here on Earth suggests that some form of life may have evolved on other planets. Astrobiologists would want to study these extreme habitats and the adaptations to such challenging conditions that have evolved in extremophiles. In addition to seeking evidence for the past or current presence of water on Mars or other planets, scientists now know to search in environments that previously would have been thought incapable of supporting life.

The Molecules of Cells

From Chapter 3 of *Campbell Biology: Concepts & Connections*, Seventh Edition. Jane B. Reece, Martha R. Taylor, Eric J. Simon, Jean L. Dickey.

The Molecules of Cells

BIG IDEAS

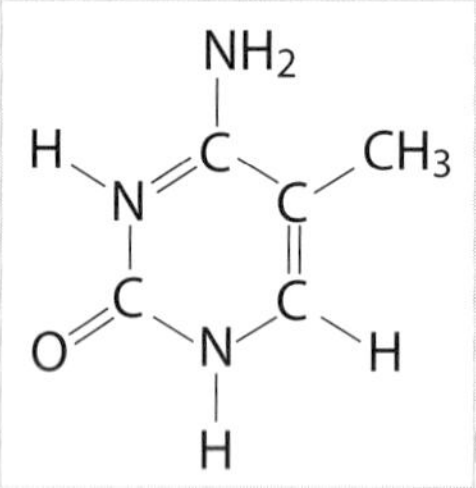

Introduction to Organic Compounds
(1–3)

Carbon-containing compounds are the chemical building blocks of life.

Alexandru Magurean/iStockphoto

Carbohydrates
(4–7)

Carbohydrates serve as a cell's fuel and building material.

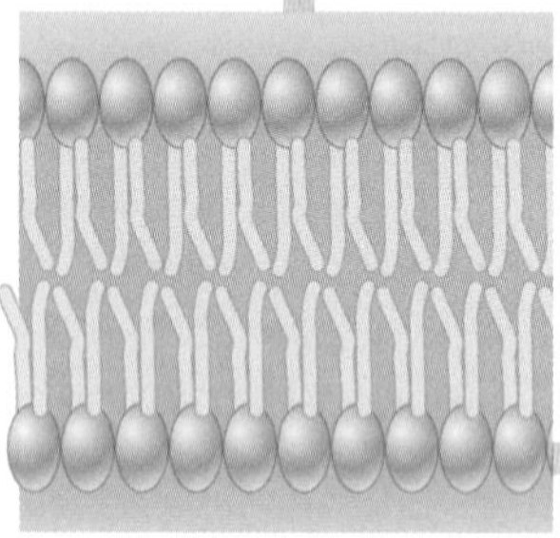

Lipids
(8–10)

Lipids are hydrophobic molecules with diverse functions.

Proteins
(11–13)

Proteins are essential to the structures and functions of life.

Fotosearch/age fotostock

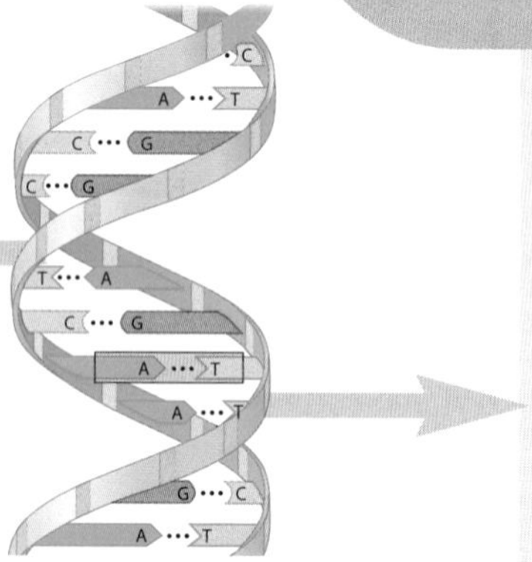

Nucleic Acids
(14–16)

Nucleic acids store, transmit, and help express hereditary information.

Is a big glass of milk a way to a healthy diet—or an upset stomach? Quite often, the answer is the latter. Most of the world's adult populations cannot easily digest milk-based foods. Such people suffer from lactose intolerance, the inability to properly break down lactose, the main sugar found in milk. Almost all infants are able to drink breast milk or other dairy products, benefiting from the proteins, fats, and sugars in this nutritious food. But as they grow older, many people find that drinking milk comes with a heavy dose of digestive discomfort.

The brightly colored ribbon model pictured above shows the three-dimensional structure of a large biological molecule called a protein. This protein is lactase, the enzyme that speeds the digestion of lactose into smaller sugars that can be absorbed by cells in the intestine. In most human populations, the production of this enzyme begins to decline after the age of 2. In the United States, as many as 80% of African Americans and Native Americans and 90% of Asian Americans are lactase-deficient once they reach their teenage years. Americans of northern European descent make up one of the few groups in which lactase production continues into adulthood. As a result, only about 10% of people in this group are lactose intolerant.

In people who easily digest milk, lactose (a sugar) is broken down by lactase (a protein), which is coded for by a gene made of DNA (a nucleic acid). Such molecular interactions, repeated in countless variations, drive all biological processes. In this chapter, we explore the structure and function of sugars, proteins, fats, and nucleic acids—the biological molecules that are essential to life. We begin with a look at carbon, the versatile atom at the center of life's molecules.

Introduction to Organic Compounds

1 Life's molecular diversity is based on the properties of carbon

When it comes to making molecules, carbon usually takes center stage. Almost all the molecules a cell makes are composed of carbon atoms bonded to one another and to atoms of other elements. Carbon is unparalleled in its ability to form large and complex molecules, which build the structures and carry out the functions required for life.

Carbon-based molecules are called **organic compounds**. Why are carbon atoms the lead players in the chemistry of life? The number of electrons in the outermost shell of its atoms determines an element's chemical properties. A carbon atom has 4 electrons in a valence shell that holds 8. Carbon completes its outer shell by sharing electrons with other atoms in four covalent bonds. Thus, each carbon atom is a connecting point from which a molecule can branch in up to four directions.

Figure 1A illustrates three representations of methane (CH_4), one of the simplest organic molecules. The structural formula shows that covalent bonds link four hydrogen atoms to the carbon atom. Each of the four lines in the formula represents a pair of shared electrons. The two models help you see that methane is three-dimensional, with the space-filling model on the right better portraying its overall shape. The ball-and-stick model shows that carbon's four bonds (the gray "sticks") angle out toward the corners of an imaginary tetrahedron (an object with four triangular sides). The red lines trace this shape, which occurs wherever a carbon atom participates in four single bonds. Different bond angles and shapes occur when carbon atoms form double bonds. Large organic molecules can have very elaborate shapes. And as we will see many times, a molecule's shape often determines its function.

Compounds composed of only carbon and hydrogen are called **hydrocarbons**. Methane and propane are examples of hydrocarbon fuels. As components of fats, longer hydrocarbons provide fuel to your body cells. **Figure 1B** illustrates some of the variety of hydrocarbon structures. The chain of carbon atoms in an organic molecule is called a **carbon skeleton** (shaded in gray in the figure). Carbon skeletons can vary in length and can be unbranched or branched. Carbon skeletons may also include double bonds, which can vary in number and location. Some carbon skeletons are arranged in rings.

The two compounds in the second row of Figure 1B, butane and isobutane, have the same molecular formula, C_4H_{10}, but

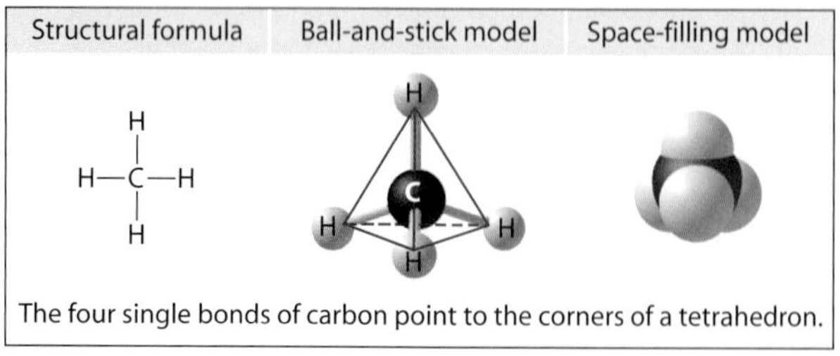

▲ **Figure 1A** Three representations of methane (CH_4)

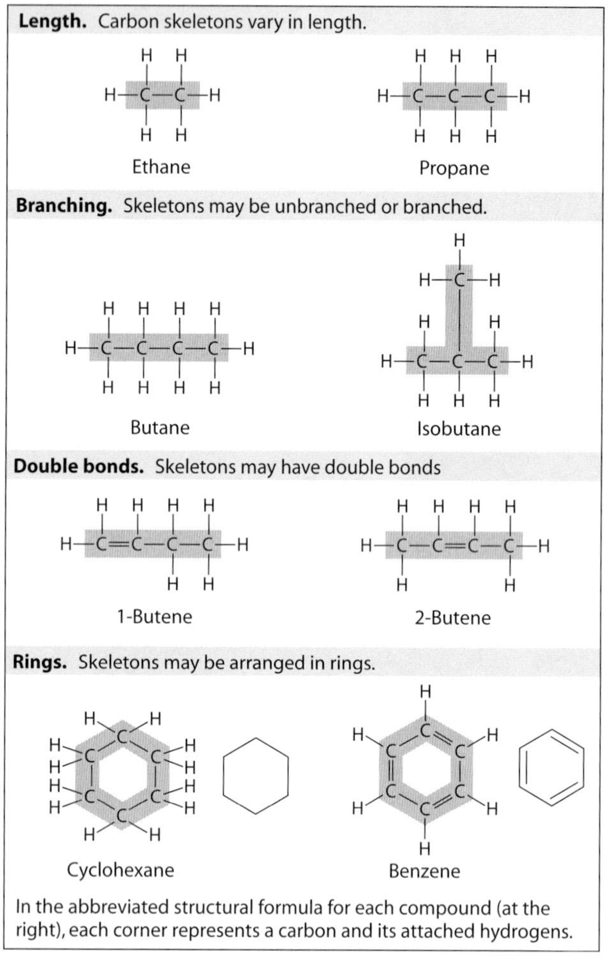

▲ **Figure 1B** Four ways that carbon skeletons can vary

differ in the bonding pattern of their carbon skeleton. The two molecules in the third row also have the same numbers of atoms, but they have different three-dimensional shapes because of the location of the double bond. Compounds with the same formula but different structural arrangements are called **isomers**. Isomers can also result from different spatial arrangements of the four partners bonded to a carbon atom. This type of isomer is important in the pharmaceutical industry, because the two isomers of a drug may not be equally effective or may have different (and sometimes harmful) effects. The different shapes of isomers result in unique properties and add greatly to the diversity of organic molecules.

? One isomer of methamphetamine is the addictive illegal drug known as "crank." The other is a medicine for sinus congestion. How can you explain the differing effects of the two isomers?

Isomers have different structures, or shapes, and the shape of a molecule usually helps determine the way it functions in the body.

2 A few chemical groups are key to the functioning of biological molecules

The unique properties of an organic compound depend not only on the size and shape of its carbon skeleton but also on the groups of atoms that are attached to that skeleton.

Table 2 illustrates six chemical groups important in the chemistry of life. The first five are called **functional groups**. They affect a molecule's function by participating in chemical reactions in characteristic ways. These groups are polar, because oxygen or nitrogen atoms exert a strong pull on shared electrons. This polarity tends to make compounds containing these groups **hydrophilic** (water-loving) and therefore soluble in water—a necessary condition for their roles in water-based life. The sixth group, a methyl group, is nonpolar and not reactive, but it affects molecular shape and thus function.

A **hydroxyl group** consists of a hydrogen atom bonded to an oxygen atom, which in turn is bonded to the carbon skeleton. Ethanol, shown in the table, and other organic compounds containing hydroxyl groups are called alcohols.

In a **carbonyl group**, a carbon atom is linked by a double bond to an oxygen atom. If the carbonyl group is at the end of a carbon skeleton, the compound is called an aldehyde; if it is within the chain, the compound is called a ketone. Sugars contain a carbonyl group and several hydroxyl groups.

A **carboxyl group** consists of a carbon double-bonded to an oxygen atom and also bonded to a hydroxyl group. The carboxyl group acts as an acid by contributing an H^+ to a solution and thus becoming ionized. Compounds with carboxyl groups are called carboxylic acids. Acetic acid, shown in the table, gives vinegar its sour taste.

An **amino group** has a nitrogen bonded to two hydrogens and the carbon skeleton. It acts as a base by picking up an H^+ from a solution. Organic compounds with an amino group are called amines. The building blocks of proteins are called amino acids because they contain an amino and a carboxyl group.

A **phosphate group** consists of a phophorus atom bonded to four oxygen atoms. It is usually ionized and attached to the carbon skeleton by one of its oxygen atoms. This structure is abbreviated as Ⓟ in this text. Compounds with phosphate groups are called organic phosphates and are often involved in energy transfers, as is the energy-rich compound ATP, shown in the table.

A **methyl group** consists of a carbon bonded to three hydrogens. Compounds with methyl groups are called methylated compounds. The addition of a methyl group to the component of DNA shown in the table affects the expression of genes.

Figure 2 shows how a small difference in chemical groups can lead to a big difference in body form and behavior. The male and female sex hormones shown here differ only in the groups highlighted with colored boxes. These subtle differences result in the different actions of these molecules, which help produce the contrasting features of males and females in lions and other vertebrates. Keeping in mind this basic scheme—carbon skeletons with chemical groups—we are now ready to see how our cells make large molecules out of smaller ones.

? **Identify the chemical groups that do *not* contain carbon.**

The hydroxyl, amino, and phosphate groups

TABLE 2 IMPORTANT CHEMICAL GROUPS OF ORGANIC COMPOUNDS

Chemical Group	Examples	
Hydroxyl group $—OH$	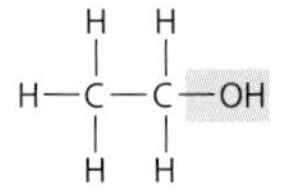 Alcohol	
Carbonyl group $>C=O$	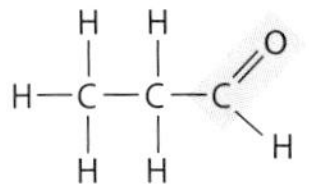 Aldehyde	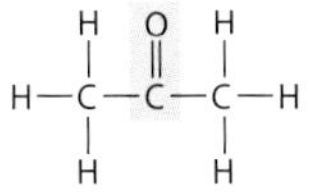 Ketone
Carboxyl group $—COOH$	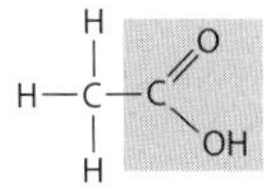 Carboxylic acid	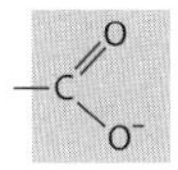 Ionized
Amino group $—NH_2$	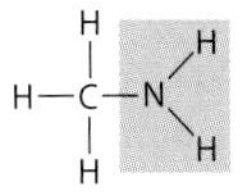 Amine	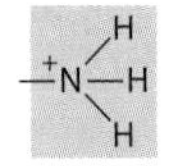 Ionized
Phosphate group $—OPO_3^{2-}$	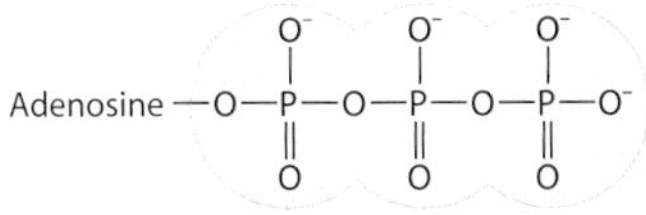 Organic phosphate	
Methyl group $—CH_3$	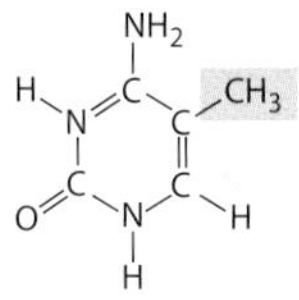 Methylated compound	

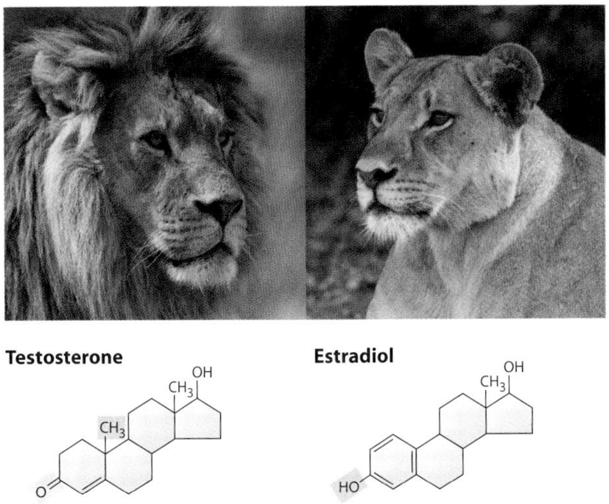

▲ **Figure 2** Differences in the chemical groups of sex hormones

3 Cells make a huge number of large molecules from a limited set of small molecules

Given the rich complexity of life on Earth, we might expect there to be an enormous diversity of types of molecules. Remarkably, however, the important molecules of all living things—from bacteria to elephants—fall into just four main classes: carbohydrates, lipids, proteins, and nucleic acids. On a molecular scale, molecules of three of these classes—carbohydrates, proteins, and nucleic acids—may be gigantic; in fact, biologists call them **macromolecules**. For example, a protein may consist of thousands of atoms. How does a cell make such a huge molecule?

Cells make most of their macromolecules by joining smaller molecules into chains called **polymers** (from the Greek *polys*, many, and *meros*, part). A polymer is a large molecule consisting of many identical or similar building blocks strung together, much as a train consists of a chain of cars. The building blocks of polymers are called **monomers**.

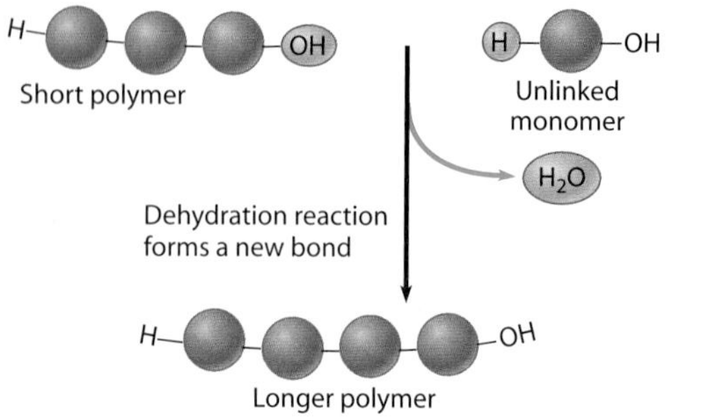

▲ **Figure 3A** Dehydration reaction building a polymer chain

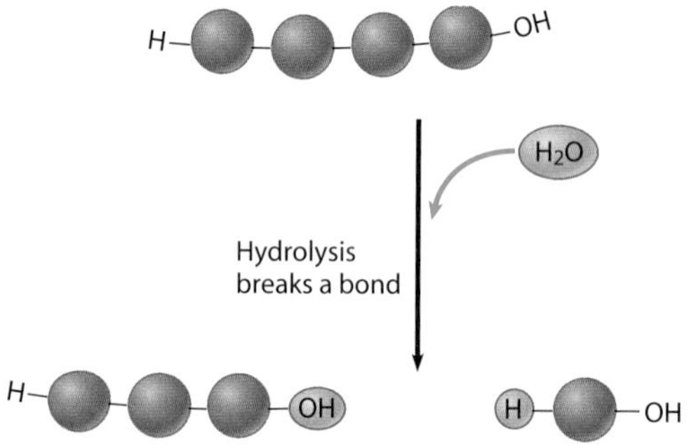

▲ **Figure 3B** Hydrolysis breaking down a polymer

Making Polymers Cells link monomers together to form polymers by a **dehydration reaction**, a reaction that removes a molecule of water. As you can see in **Figure 3A**, an unlinked monomer has a hydrogen atom (—H) at one end and a hydroxyl group (—OH) at the other. For each monomer added to a chain, a water molecule (H_2O) is released. Notice in Figure 3A that one monomer (the one at the right end of the short polymer in this example) loses a hydroxyl group and the other monomer loses a hydrogen atom to form H_2O. As this occurs, a new covalent bond forms, linking the two monomers. Dehydration reactions are the same regardless of the specific monomers and the type of polymer the cell is producing.

Breaking Polymers Cells not only make macromolecules but also have to break them down. For example, most of the organic molecules in your food are in the form of polymers that are much too large to enter your cells. You must digest these polymers to make their monomers available to your cells. This digestion process is called **hydrolysis**. Essentially the reverse of a dehydration reaction, hydrolysis means to break (*lyse*) with water (*hydro-*). As **Figure 3B** shows, the bond between monomers is broken by the addition of a water molecule, with the hydroxyl group from the water attaching to one monomer and a hydrogen attaching to the adjacent monomer.

The lactose-intolerant individuals you learned about in the chapter introduction are unable to hydrolyze such a bond in the sugar lactose because they lack the enzyme lactase. Both dehydration reactions and hydrolysis require the help of enzymes to make and break bonds. **Enzymes** are specialized macromolecules that speed up chemical reactions in cells.

The Diversity of Polymers The diversity of macromolecules in the living world is vast. Remarkably, a cell makes all its thousands of different macromolecules from a small list of ingredients—about 40 to 50 common components and a few others that are rare. Proteins, for example, are built from only 20 kinds of amino acids. Your DNA is built from just four kinds of monomers called nucleotides. The key to the great diversity of polymers is arrangement—variation in the sequence in which monomers are strung together.

The variety in polymers accounts for the uniqueness of each organism. The monomers themselves, however, are essentially universal. Your proteins and those of a tree or an ant are assembled from the same 20 amino acids. Life has a simple yet elegant molecular logic: Small molecules common to all organisms are ordered into large molecules, which vary from species to species and even from individual to individual in the same species.

In the remainder of the chapter, we explore each of the four classes of large biological molecules. Like water and simple organic molecules, large biological molecules have unique emergent properties arising from the orderly arrangement of their atoms. As you will see, for these molecules of life, structure and function are inseparable.

? **Suppose you eat some cheese. What reactions must occur for the protein of the cheese to be broken down into its amino acid monomers and then for these monomers to be converted to proteins in your body?**

● In digestion, the proteins are broken down into amino acids by hydrolysis. New proteins are formed in your body cells from these monomers in dehydration reactions.

Carbohydrates

4 Monosaccharides are the simplest carbohydrates

The name **carbohydrate** refers to a class of molecules ranging from the small sugar molecules dissolved in soft drinks to large polysaccharides, such as the starch molecules we consume in pasta and potatoes.

The carbohydrate monomers (single-unit sugars) are **monosaccharides** (from the Greek *monos*, single, and *sacchar*, sugar). The honey shown in **Figure 4A** consists mainly of monosaccharides called glucose and fructose. These and other single-unit sugars can be hooked together by dehydration reactions to form more complex sugars and polysaccharides.

Monosaccharides generally have molecular formulas that are some multiple of CH_2O. For example, the formula for glucose, a common monosaccharide of central importance in the chemistry of life, is $C_6H_{12}O_6$. **Figure 4B** illustrates the molecular structure of glucose, with its carbons numbered 1 to 6. This structure also shows the two trademarks of a sugar: a number of hydroxyl groups (—OH) and a carbonyl group (>C=O, highlighted in blue). The hydroxyl groups make a sugar an alcohol, and the carbonyl group, depending on its location, makes it either an aldose (an aldehyde sugar) or a ketose (a ketone sugar). As you see in Figure 4B, glucose is an aldose and fructose is a ketose. (Note that most names for sugars end in *-ose*. Also, as you saw with the enzyme lactase that digests lactose, the names for most enzymes end in *-ase*.)

If you count the numbers of different atoms in the fructose molecule in Figure 4B, you will find that its molecular formula is $C_6H_{12}O_6$, identical to that of glucose. Thus, glucose and fructose are isomers; they differ only in the arrangement of their atoms (in this case, the positions of the carbonyl groups). Seemingly minor differences like this give isomers different properties, such as how they react with other molecules. These differences also make fructose taste considerably sweeter than glucose.

The carbon skeletons of both glucose and fructose are six carbon atoms long. Other monosaccharides may have three to seven carbons. Five-carbon sugars, called pentoses, and six-carbon sugars, called hexoses, are among the most common.

It is convenient to draw sugars as if their carbon skeletons were linear, but in aqueous solutions, many monosaccharides form rings, as shown for glucose in **Figure 4C**. To form the glucose ring, carbon 1 bonds to the oxygen attached to carbon 5. As shown in the middle representation, the ring diagram of glucose and other sugars may be abbreviated by not showing the carbon atoms at the corners of the ring. Also, the bonds in the ring are often drawn with varied thickness, indicating that the ring is a relatively flat structure with attached atoms extending above and below it. The simplified ring symbol on the right is often used in this book to represent glucose.

Monosaccharides, particularly glucose, are the main fuel molecules for cellular work. Because cells release energy from glucose when they break it down, an aqueous solution of glucose (often called dextrose) may be injected into the bloodstream of sick or injured patients; the glucose provides an immediate energy source to tissues in need of repair. Cells also use the carbon skeletons of monosaccharides as raw material for making other kinds of organic molecules, such as amino acids and fatty acids. Sugars not used in these ways may be incorporated into disaccharides and polysaccharides, as we see next.

? Write the formula for a monosaccharide that has three carbons.

$C_3H_6O_3$

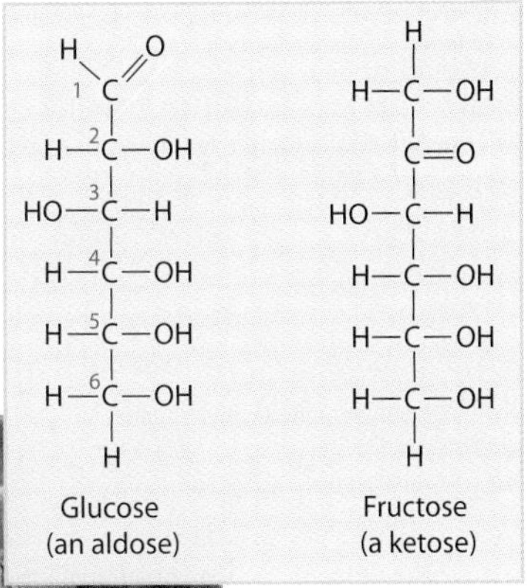

◀ **Figure 4B** Structures of glucose and fructose

▲ **Figure 4A** Bees with honey, a mixture of two monosaccharides

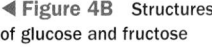

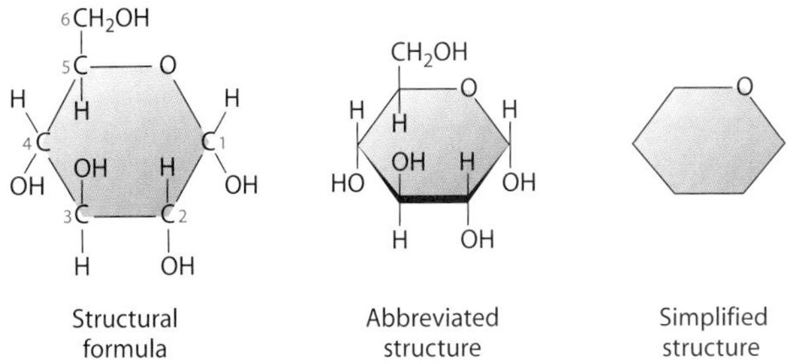

▲ **Figure 4C** Three representations of the ring form of glucose

5 Two monosaccharides are linked to form a disaccharide

Cells construct a **disaccharide** from two monosaccharide monomers by a dehydration reaction. **Figure 5** shows how maltose, also called malt sugar, is formed from two glucose monomers. One monomer gives up a hydroxyl group and the other gives up a hydrogen atom from a hydroxyl group. As H_2O is released, an oxygen atom is left, linking the two monomers. Maltose, which is common in germinating seeds, is used in making beer, malted milk shakes, and malted milk candy.

The most common disaccharide is sucrose, which is made of a glucose monomer linked to a fructose monomer. Transported in plant sap, sucrose provides a source of energy and raw materials to all the parts of the plant. We extract it from the stems of sugarcane or the roots of sugar beets to use as table sugar.

? Lactose, as you read in the chapter introduction, is the disaccharide sugar in milk. It is formed from glucose and galactose. The formula for both these monosaccharides is $C_6H_{12}O_6$. What is the formula for lactose?

$C_{12}H_{22}O_{11}$

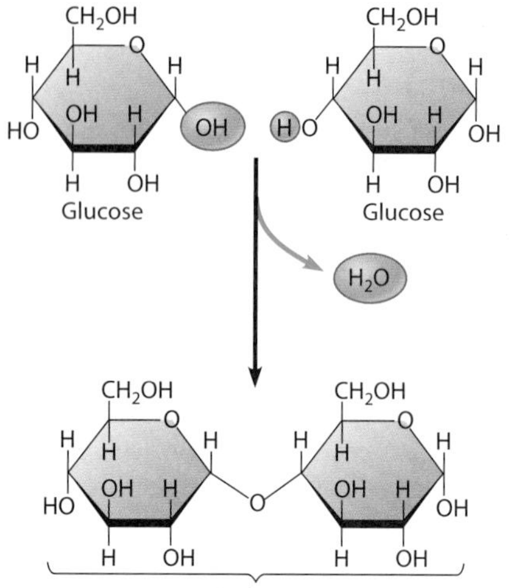

▲ **Figure 5** Disaccharide formation by a dehydration reaction

CONNECTION

6 What is high-fructose corn syrup, and is it to blame for obesity?

Kristin Piljay

▲ **Figure 6** High-fructose corn syrup (HFCS), a main ingredient of soft drinks and processed foods

If you want to sweeten your coffee or tea, you probably reach for sugar—the disaccharide sucrose. But if you drink sodas or fruit drinks, you're probably consuming the monosaccharides of sucrose in the form of high-fructose corn syrup. In fact, if you look at the label of almost any processed food, you will see high-fructose corn syrup listed as one of the ingredients (**Figure 6**). And you have probably heard reports linking high-fructose corn syrup to the "obesity epidemic."

What is high-fructose corn syrup (HFCS)? Let's start with the corn syrup part. The main carbohydrate in corn is starch, a polysaccharide. Industrial processing hydrolyzes starch into its component monomers, glucose, producing corn syrup. Glucose, however, does not taste as sweet to us as sucrose. Fructose, on the other hand, tastes much sweeter than both glucose and sucrose. When a new process was developed in the 1970s that used an enzyme to rearrange the atoms of glucose into the sweeter isomer, fructose (see Figure 4B), the high-fructose corn syrup industry was born. (High-fructose corn syrup is a bit of a misnomer, however, because the fructose is combined with regular corn syrup to produce a mixture of about 55% fructose and 45% glucose, not much different from the proportions in sucrose.)

This clear, goopy liquid is cheaper than sucrose and easier to mix into drinks and processed food. And it contains the same monosaccharides as sucrose, the disaccharide it is replacing. So is there a problem with HFCS? Some point to circumstantial evidence. From 1980 to 2000, the incidence of obesity doubled in the United States. In that same time period, the consumption of HFCS more than tripled, whereas the consumption of refined cane and beet sugar decreased 21%. Overall, the combined per capita consumption of HFCS and refined sugars increased 25% in that period. A 2001–2004 national health survey showed the average intake of added sugars and sweeteners was 22.2 teaspoons a day, with soft drinks and other beverages sweetened with HFCS the number one source.

So, is high-fructose corn syrup to blame for increases in obesity, type 2 diabetes, high blood pressure, and other chronic diseases associated with increased weight? Scientific studies are ongoing, and the jury is still out. There is consensus, however, that overconsumption of sugar or HFCS along with dietary fat and decreased physical activity contribute to weight gain. In addition, high sugar consumption also tends to replace eating more varied and nutritious foods. Sugars have been described as "empty calories" because they contain only negligible amounts of other nutrients. For good health, you require proteins, fats, vitamins, and minerals, as well as complex carbohydrates, the topic of the next module.

? How is high-fructose corn syrup made from corn?

Corn starch is hydrolyzed to glucose; then enzymes convert glucose to fructose. This fructose is combined with corn syrup to produce HFCS.

7 Polysaccharides are long chains of sugar units

Polysaccharides are macromolecules, polymers of hundreds to thousands of monosaccharides linked together by dehydration reactions. Polysaccharides may function as storage molecules or as structural compounds. **Figure 7** illustrates three common types of polysaccharides: starch, glycogen, and cellulose.

Starch, a storage polysaccharide in plants, consists entirely of glucose monomers. Starch molecules coil into a helical shape and may be unbranched (as shown in the figure) or branched. Starch granules serve as carbohydrate "banks" from which plant cells can withdraw glucose for energy or building materials. Humans and most other animals have enzymes that can hydrolyze plant starch to glucose. Potatoes and grains, such as wheat, corn, and rice, are the major sources of starch in the human diet.

Animals store glucose in a different form of polysaccharide, called **glycogen**. Glycogen is more highly branched than starch, as shown in the figure. Most of your glycogen is stored as granules in your liver and muscle cells, which hydrolyze the glycogen to release glucose when it is needed.

Cellulose, the most abundant organic compound on Earth, is a major component of the tough walls that enclose plant cells. Cellulose is also a polymer of glucose, but its monomers are linked together in a different orientation. (Carefully compare the oxygen "bridges" highlighted in yellow between glucose monomers in starch, glycogen, and cellulose in the figure.) Arranged parallel to each other, cellulose molecules are joined by hydrogen bonds, forming cable-like microfibrils. Layers of microfibrils combine with other polymers, producing strong support for trees and structures we build with lumber.

Animals do not have enzymes that can hydrolyze the glucose linkages in cellulose. Therefore, cellulose is not a nutrient for humans, although it does contribute to digestive system health. The cellulose that passes unchanged through your digestive tract is referred to as "insoluble fiber." Fresh fruits, vegetables, and grains are rich in fiber.

Some microorganisms do have enzymes that can hydrolyze cellulose. Cows and termites house such microorganisms in their digestive tracts and are thus able to derive energy from cellulose. Decomposing fungi also digest cellulose, helping to recycle its chemical elements within ecosystems.

Another structural polysaccharide, **chitin**, is used by insects and crustaceans to build their exoskeleton, the hard case enclosing the animal. Chitin is also found in the cell walls of fungi. Humans use chitin to make a strong and flexible surgical thread that decomposes after a wound or incision heals.

Almost all carbohydrates are hydrophilic owing to the many hydroxyl groups attached to their sugar monomers (see Figure 4B). Thus, cotton bath towels, which are mostly cellulose, are quite water absorbent due to the water-loving nature of cellulose. Next we look at a class of macromolecules that are not hydrophilic.

? Compare and contrast starch and cellulose, two plant polysaccharides.

Both are polymers of glucose, but the bonds between glucose monomers have different shapes. Starch functions mainly for sugar storage. Cellulose is a structural polysaccharide that is the main material of plant cell walls.

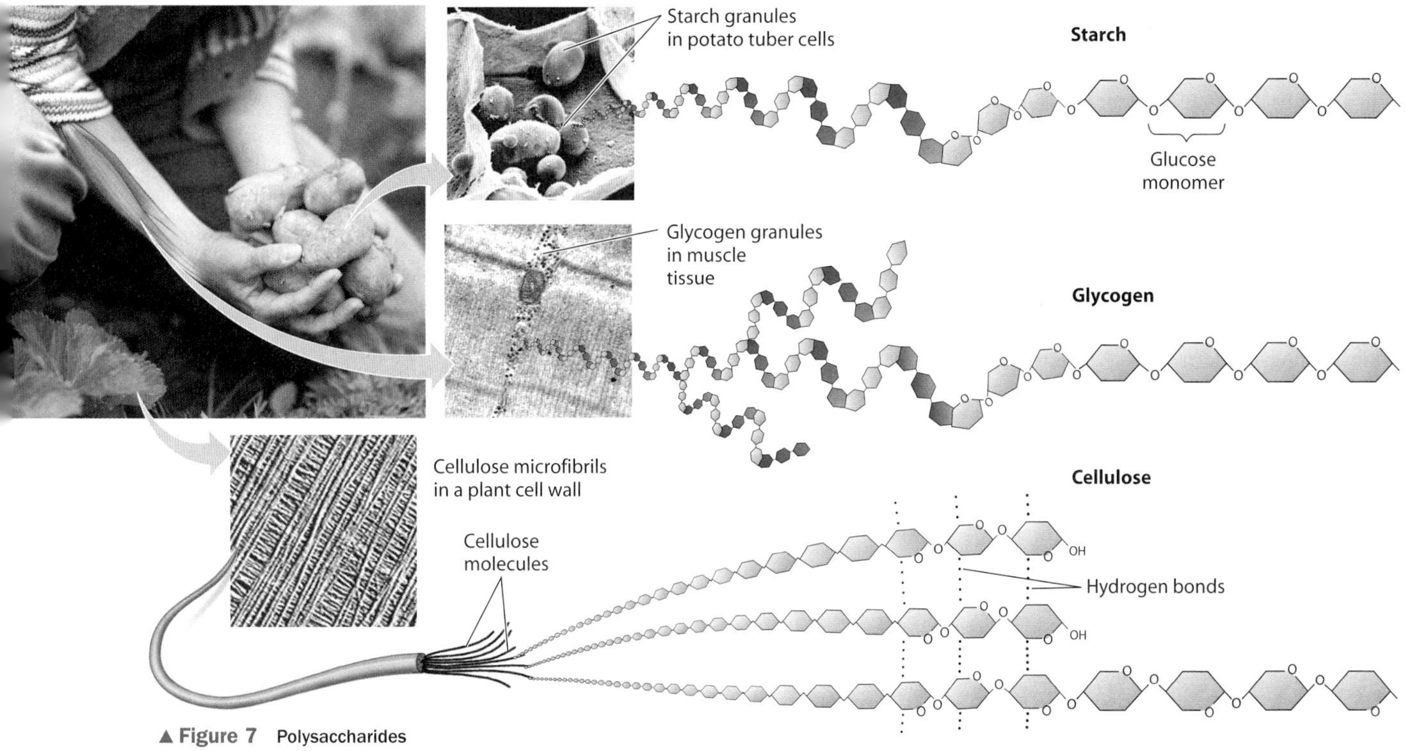

▲ **Figure 7** Polysaccharides

Dougal Waters/Getty Images; Biophoto Associates/Photo Researchers; Courtesy of Dr. L. M. Beidler; Biophoto Associates/Photo Researchers

Lipids

8 Fats are lipids that are mostly energy-storage molecules

Lipids are diverse compounds that are grouped together because they share one trait: They do not mix well with water. Lipids consist mainly of carbon and hydrogen atoms linked by nonpolar covalent bonds. In contrast to carbohydrates and most other biological molecules, lipids are **hydrophobic** (water-fearing). You can see this chemical behavior in an unshaken bottle of salad dressing: The oil (a type of lipid) separates from the vinegar (which is mostly water). The oils that ducks spread on their feathers make the feathers repel water **(Figure 8A)**, which helps such waterfowl stay afloat.

Lipids also differ from carbohydrates, proteins, and nucleic acids in that they are neither huge macromolecules nor polymers built from similar monomers. You will see that lipids vary a great deal in structure and function. In this and the next two modules, we will consider three types of lipids: fats, phospholipids, and steroids.

A **fat** is a large lipid made from two kinds of smaller molecules: glycerol and fatty acids. Shown at the top in **Figure 8B**, glycerol is an alcohol with three carbons, each bearing a hydroxyl group (—OH). A fatty acid consists of a carboxyl group (the functional group that gives these molecules the name fatty *acid*, —COOH) and a hydrocarbon chain, usually 16 or 18 carbon atoms in length. The nonpolar hydrocarbon chains are the reason fats are hydrophobic.

Figure 8B shows how one fatty acid molecule can link to a glycerol molecule by a dehydration reaction. Linking three fatty acids to glycerol produces a fat, as illustrated in **Figure 8C**. A synonym for fat is *triglyceride*, a term you may see on food labels or on medical tests for fat in the blood.

Some fatty acids contain one or more double bonds, which cause kinks (or bends) in the carbon chain. See the third fatty acid in Figure 8C. Such an **unsaturated fatty acid** has one fewer hydrogen atom on each carbon of the double bond. Fatty acids with no double bonds in their hydrocarbon chain have the maximum number of hydrogen atoms (are "saturated" with hydrogens) and are called **saturated fatty acids**. The kinks in unsaturated fatty acids prevent fats containing them from packing tightly together and solidifying at room temperature. Corn oil, olive oil, and other vegetable oils are called unsaturated fats.

Most animal fats are saturated. Their unkinked fatty acid chains pack closely together, making butter and beef fat solid at room temperature. When you see "hydrogenated vegetable oils" on a margarine label, it means that unsaturated fats have been converted to saturated fats by adding hydrogen. Unfortunately, hydrogenation also creates **trans fats**, a form of fat that recent research associates with health risks. Diets rich in saturated fats and trans fats may contribute to cardiovascular disease by promoting atherosclerosis. In this condition, lipid-containing deposits called plaques build up within the walls of blood vessels, reducing blood flow. Unsaturated fatty acids called omega-3 fatty acids are found in certain nuts, plant oils, and fatty fish and appear to protect against cardiovascular disease.

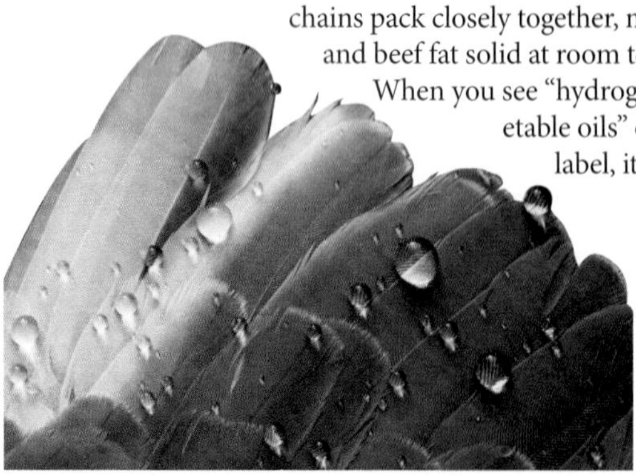

◄ **Figure 8A** Water beading on the oily coating of feathers

Tony Hamblin/Frank Lane Picture Agency/Corbis

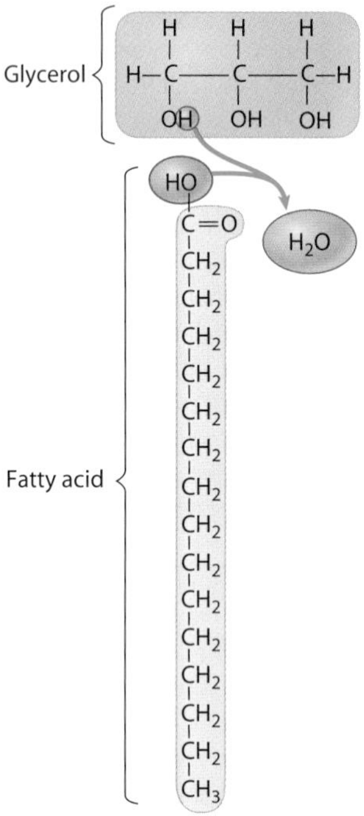

▲ **Figure 8B** A dehydration reaction linking a fatty acid molecule to a glycerol molecule

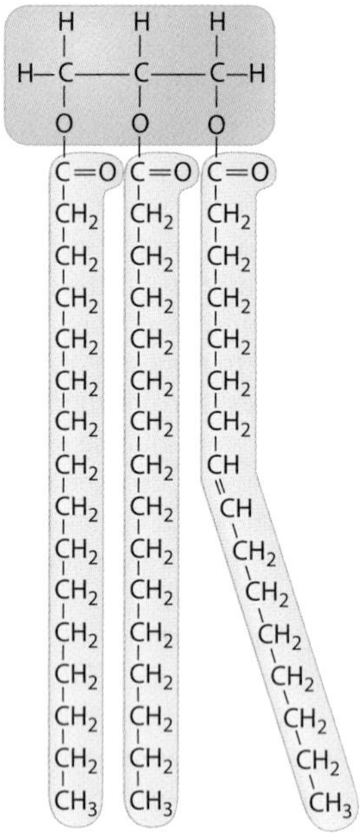

▲ **Figure 8C** A fat molecule (triglyceride) consisting of three fatty acids linked to glycerol

The main function of fats is long-term energy storage. A gram of fat stores more than twice as much energy as a gram of polysaccharide. For immobile plants, the bulky energy storage form of starch is not a problem. (Vegetable oils are generally obtained from seeds, where more compact energy storage is a benefit.) A mobile animal, such as a duck or a human, can get around much more easily carrying its energy stores in the form of fat. Of course, the downside of this energy-packed storage form is that it takes more effort for a person to "burn off" excess fat. In addition to storing energy, fatty tissue cushions vital organs and insulates the body.

? How do you think the structure of a monounsaturated fat differs from a polyunsaturated fat?

A monounsaturated fat has a fatty acid with a single double bond in its carbon chain. A polyunsaturated fat has a fatty acid with several double bonds.

9 Phospholipids and steroids are important lipids with a variety of functions

Cells could not exist without **phospholipids**, the major component of cell membranes. Phospholipids are structurally similar to fats, but they contain only two fatty acids attached to glycerol instead of three. As shown in **Figure 9A**, a negatively charged phosphate group (shown as a yellow circle in the figure and linked to another small molecule) is attached to glycerol's third carbon. (Note that glycerol is shown in orange.) The structure of phospholipids provides a classic example of how form fits function. The hydrophilic and hydrophobic ends of multiple molecules assemble in a bilayer of phospholipids to form a membrane **(Figure 9B)**. The hydrophobic tails of the fatty acids cluster in the center, and the hydrophilic phosphate heads face the watery environment on either side of the membrane. Each gray-headed, yellow-tailed structure in the membrane shown here represents a phospholipid; this symbol is used throughout in this book.

Steroids are lipids in which the carbon skeleton contains four fused rings, as shown in the structural formula of cholesterol in **Figure 9C**. (The diagram omits the carbons making up the rings and most of the chain and also their attached hydrogens.) **Cholesterol** is a common component in animal cell membranes, and animal cells also use it as a starting material for making other steroids, including sex hormones. Different steroids vary in the chemical groups attached to the rings, as you saw in Figure 2. Too much cholesterol in the blood may contribute to atherosclerosis.

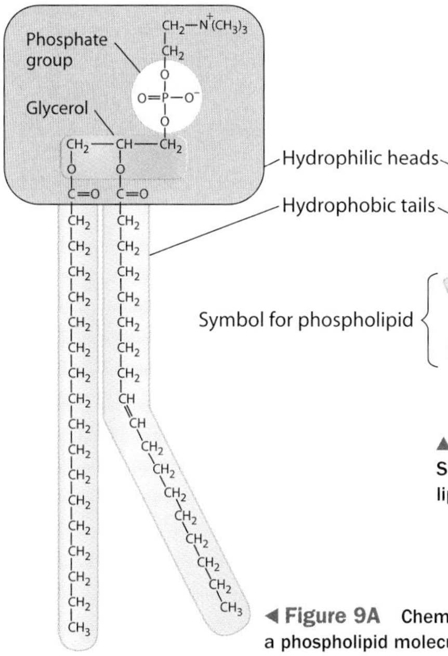

◄ **Figure 9A** Chemical structure of a phospholipid molecule

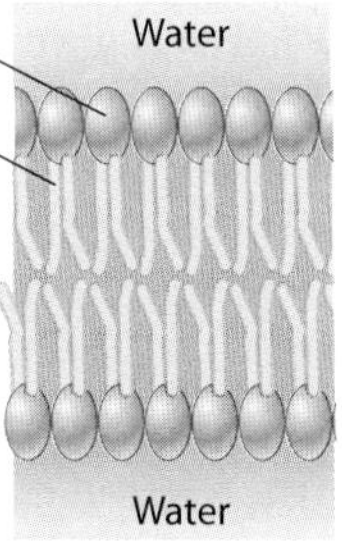

▲ **Figure 9B** Section of a phospholipid membrane

? **Compare the structure of a phospholipid with that of a fat (triglyceride).**

A phospholipid has two fatty acids and a phosphate group attached to glycerol. Three fatty acids are attached to the glycerol of a fat molecule.

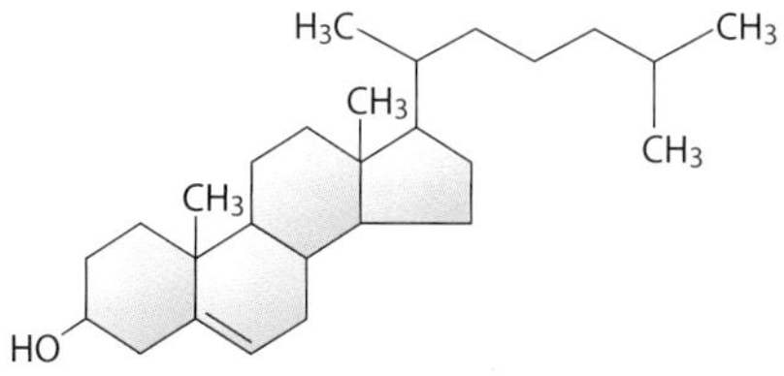

▲ **Figure 9C** Cholesterol, a steroid

CONNECTION

10 Anabolic steroids pose health risks

Anabolic steroids are synthetic variants of the male hormone testosterone. Testosterone causes a general buildup of muscle and bone mass in males during puberty and maintains masculine traits throughout life. Because anabolic steroids structurally resemble testosterone, they also mimic some of its effects. (The word *anabolic* comes from *anabolism*, the building of substances by the body.)

As prescription drugs, anabolic steroids are used to treat general anemia and diseases that destroy body muscle. However, some athletes use these drugs to build up their muscles quickly and enhance their performance. But at what cost? Steroid abuse may cause violent mood swings ("roid rage"), depression, liver damage or cancer, and high cholesterol levels and blood pressure. Use of these drugs often makes the body reduce its output of natural male sex hormones, which can cause shrunken testicles, reduced sex drive, infertility, and breast enlargement in men. Use in women has been linked to menstrual cycle disruption and development of masculine characteristics. A serious effect in teens is that bones may stop growing, stunting growth.

Shutterstock

Despite the risks, some athletes continue to use steroids, and unscrupulous chemists, trainers, and coaches try to find ways to avoid their detection. Meanwhile, the U.S. Congress, professional sports authorities, and high school and college athletic programs ban the use of anabolic steroids, implement drug testing, and penalize violators in an effort to keep the competition fair and protect the health of athletes.

? **How are dietary fats and anabolic steroids similar?**

Both fats and steroids are lipids, grouped together because they are hydrophobic molecules.

Proteins

11 Proteins are made from amino acids linked by peptide bonds

Nearly every dynamic function in your body depends on proteins. You have tens of thousands of different proteins, each with a specific structure and function. Of all of life's molecules, proteins are structurally the most elaborate and diverse. A **protein** is a polymer of amino acids. Protein diversity is based on differing arrangements of a common set of just 20 amino acid monomers.

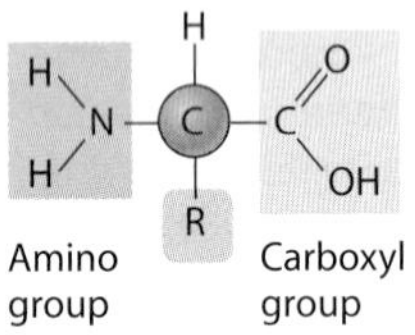

▲ **Figure 11A** General structure of an amino acid

Amino acids all have an amino group and a carboxyl group (which makes it an acid, hence the name amino *acid*). As you can see in the general structure shown in **Figure 11A**, both of these functional groups are covalently bonded to a central carbon atom, called the alpha carbon. Also bonded to the alpha carbon is a hydrogen atom and a chemical group symbolized by the letter R. The R group, also called the side chain, differs with each amino acid. In the simplest amino acid (glycine), the R group is just a hydrogen atom. In all others, such as those shown in **Figure 11B**, the R group consists of one or more carbon atoms with various chemical groups attached. The composition and structure of the R group determines the specific properties of each of the 20 amino acids that are found in proteins.

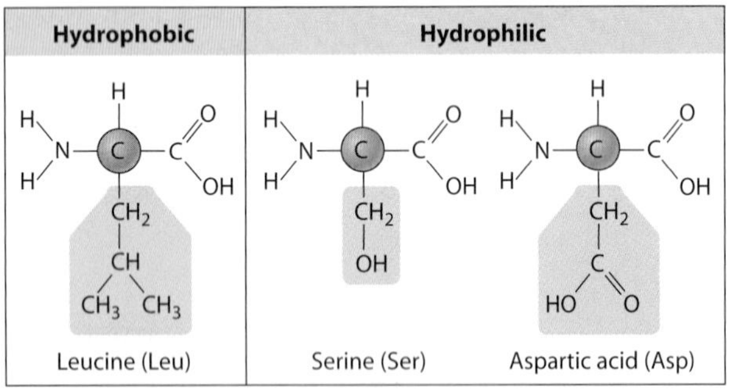

▲ **Figure 11B** Examples of amino acids with hydrophobic and hydrophilic R groups

The amino acids in Figure 11B represent two main types, hydrophobic and hydrophilic. Leucine (abbreviated Leu) is an example of an amino acid in which the R group is nonpolar and hydrophobic. Serine (Ser), with a hydroxyl group in its R group, is an example of an amino acid with a polar, hydrophilic R group. Aspartic acid (Asp) is acidic and negatively charged at the pH of a cell. (Indeed, all the amino and carboxyl groups of amino acids are usually ionized at cellular pH, as shown in Table 2.) Other amino acids have basic R groups and are positively charged. Amino acids with polar and charged R groups help proteins dissolve in the aqueous solutions inside cells.

Now that we have examined amino acids, let's see how they are linked to form polymers. Can you guess? Cells join amino acids together in a dehydration reaction that links the carboxyl group of one amino acid to the amino group of the next amino acid as a water molecule is removed (**Figure 11C**). The resulting covalent linkage is called a **peptide bond**. The product of the reaction shown in the figure is called a *di*peptide, because it was made from *two* amino acids. Additional amino acids can be added by the same process to form a chain of amino acids, a **polypeptide**. To release amino acids from the polypeptide by hydrolysis, a molecule of H_2O must be added back to break each peptide bond.

How is it possible to make thousands of different kinds of proteins from just 20 amino acids? The answer has to do with sequence. You know that thousands of English words can be made by varying the sequence of letters and word length. Although the protein "alphabet" is slightly smaller (just 20 "letters," rather than 26), the "words" are much longer. Most polypeptides are at least 100 amino acids in length; some are 1,000 or more. Each polypeptide has a unique sequence of amino acids. But a long polypeptide chain of specific sequence is not the same as a protein, any more than a long strand of yarn is the same as a sweater that can be knit from that yarn. A functioning protein is one or more polypeptide chains precisely coiled, twisted, and folded into a unique three-dimensional shape.

? In what way is the production of a dipeptide similar to the production of a disaccharide?

In both cases, the monomers are joined by a dehydration reaction.

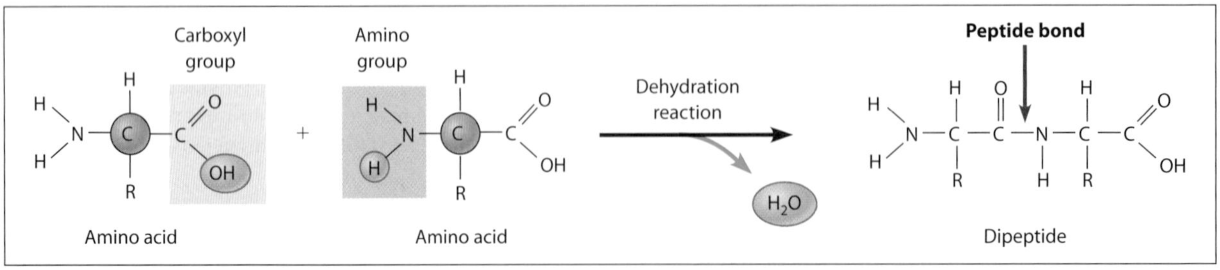

▲ **Figure 11C** Peptide bond formation

12 A protein's specific shape determines its function

What do the tens of thousands of different proteins in your body do? Probably their most important role is as *enzymes*, the chemical catalysts that speed and regulate virtually all chemical reactions in cells. Lactase, which you read about in the chapter introduction, is just one of thousands of different enzymes that may be produced by cells.

Fotosearch/age fotostock

▲ **Figure 12A** Structural proteins make up hair, tendons, and ligaments; contractile proteins are found in muscles.

In **Figure 12A**, you can see examples of two other types of proteins. *Structural proteins* are found in hair and the fibers that make up connective tissues such as tendons and ligaments. Muscle cells are packed with *contractile proteins.* Other types of proteins include *defensive proteins*, such as the antibodies of the immune system, and *signal proteins*, such as many of the hormones and other chemical messengers that help coordinate body activities by facilitating communication between cells. *Receptor proteins* may be built into cell membranes and transmit signals into cells. Hemoglobin in red blood cells is a *transport protein* that delivers O_2 to working muscles and tissues throughout the body. Other transport proteins move sugar molecules into cells for energy. Some proteins are *storage proteins*, such as ovalbumin, the protein of egg white, which serves as a source of amino acids for developing embryos. Milk proteins provide amino acids for baby mammals, and plant seeds contain storage proteins that nourish developing plant embryos.

The functions of all these different types of proteins depend on their specific shape. **Figure 12B** shows a ribbon model of lysozyme, an enzyme found in your sweat, tears, and saliva. Lysozyme consists of one long polypeptide, represented by the purple ribbon. Lysozyme's general shape is called globular. This overall shape is more apparent in **Figure 12C**, a space-filling model of lysozyme. In that model, the colors represent the different atoms of carbon, oxygen, nitrogen, and hydrogen. The barely visible yellow balls are sulfur atoms that form the stabilizing bonds shown as yellow lines in the ribbon model. Most enzymes and other proteins are globular. Structural proteins, such as those making up hair, tendons, and ligaments, are typically long and thin and are called fibrous proteins.

Descriptions such as globular and fibrous refer to a protein's general shape. Each protein also has a much more specific shape. The coils and twists of lysozyme's polypeptide ribbon appear haphazard, but they represent the molecule's specific, three-dimensional shape, and this shape is what determines its specific function. Nearly all proteins must recognize and bind to some other molecule to function. Lysozyme, for example, can destroy bacterial cells, but first it must bind to specific molecules on the bacterial cell surface. Lysozyme's specific shape enables it to recognize and attach to its molecular target, which fits into the groove you see on the right in the figures.

The dependence of protein function on a protein's specific shape becomes clear when proteins are altered. In a process called **denaturation**, polypeptide chains unravel, losing their specific shape and, as a result, their function. Changes in salt concentration and pH can denature many proteins, as can excessive heat. For example, visualize what happens when you fry an egg. Heat quickly denatures the clear proteins surrounding the yolk, making them solid, white, and opaque. One of the reasons why extremely high fevers are so dangerous is that some proteins in the body become denatured and cannot function.

Given the proper cellular environment, a newly synthesized polypeptide chain spontaneously folds into its functional shape. We examine the four levels of a protein's structure next.

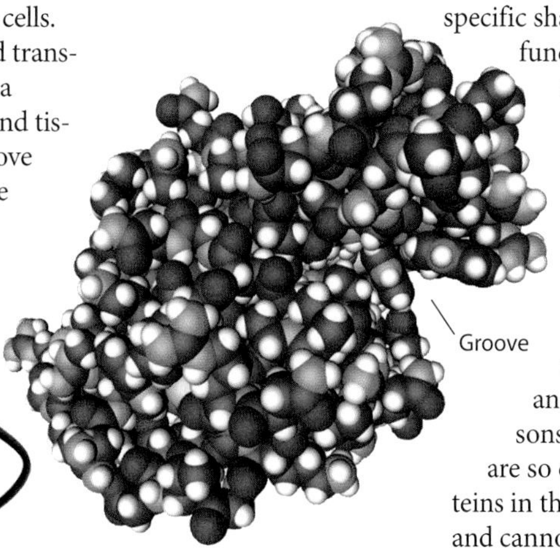

▲ **Figure 12C** Space-filling model of the protein lysozyme

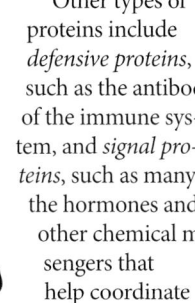

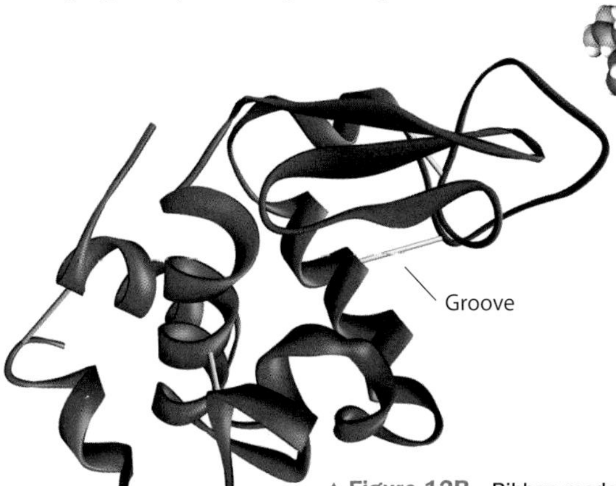

▲ **Figure 12B** Ribbon model of the protein lysozyme

Adapted from D. W. Heinz et al. 1993. How amino-acid insertions are allowed in an alpha-helix of T4 lysozyme. *Nature* 361: 561.

? Why does a denatured protein no longer function normally?

The function of each protein is a consequence of its specific shape, which is lost when a protein denatures.

13 A protein's shape depends on four levels of structure

Primary Structure The **primary structure** of a protein is its unique sequence of amino acids. As an example, let's consider transthyretin, an important transport protein found in your blood. Its specific shape enables it to transport vitamin A and one of the thyroid hormones throughout your body. A complete molecule of transthyretin has four identical polypeptide chains, each made up of 127 amino acids. **Figure 13A**, on the next page, shows part of one of these chains unraveled for a closer look at its primary structure. The three-letter abbreviations represent the specific amino acids that make up the chain.

In order for transthyretin or any other protein to perform its specific function, it must have the correct amino acids arranged in a precise order. The primary structure of a protein is determined by inherited genetic information. Even a slight change in primary structure may affect a protein's overall shape and thus its ability to function. For instance, a single amino acid change in hemoglobin, the oxygen-carrying blood protein, causes sickle-cell disease, a serious blood disorder.

Secondary Structure In the second level of protein structure, parts of the polypeptide coil or fold into local patterns called **secondary structure**. Coiling of a polypeptide chain results in a secondary structure called an alpha helix; a certain kind of folding leads to a secondary structure called a beta pleated sheet. Both of these patterns are maintained by regularly spaced hydrogen bonds between hydrogen atoms and oxygen atoms along the backbone of the polypeptide chain. Each hydrogen bond is represented in **Figure 13B** by a row of dots. Because the R groups of the amino acids are not involved in forming these secondary structures, they are omitted from the diagrams.

Peter Zaharov/iStockphoto

Transthyretin has only one alpha helix region (see **Figure 13C**). In contrast, some fibrous proteins, such as the structural protein of hair, have the alpha helix structure over most of their length.

Beta pleated sheets make up the core of many globular proteins, as is the case for transthyretin. Pleated sheets also dominate some fibrous proteins, including the silk protein of a spider's web, shown to the left. The combined strength of so many hydrogen bonds makes each silk fiber stronger than a steel strand of the same weight. Potential uses of spider silk proteins include surgical thread, fishing line, and bulletproof vests.

Tertiary Structure The term **tertiary structure** refers to the overall three-dimensional shape of a polypeptide, which, as we've said, determines the function of a protein. As shown in Figure 13C, a transthyretin polypeptide has a globular shape, which results from the compact arrangement of its alpha helix region and beta pleated sheet regions.

Here the R groups of the amino acids making up the polypeptide get involved in creating a protein's shape. Tertiary structure results from interactions between these R groups. For example, transthyretin and other proteins found in aqueous solutions are folded so that the hydrophobic R groups are on the inside of the molecule and the hydrophilic R groups on the outside, exposed to water. In addition to the clustering of hydrophobic groups, hydrogen bonding between polar side chains and ionic bonding of some of the charged (ionized) R groups help maintain the tertiary structure. A protein's shape may be reinforced further by covalent bonds called disulfide bridges. You saw disulfide bridges as the yellow lines in the ribbon model of lysozyme in Figure 12B.

Quaternary Structure Many proteins consist of two or more polypeptide chains aggregated into one functional macromolecule. Such proteins have a **quaternary structure**, resulting from the association of these polypeptides, which are known as "subunits." **Figure 13D** shows a complete transthyretin molecule with its four identical globular subunits.

Another example of a protein with quaternary structure is collagen, shown to the right. Collagen is a fibrous protein with three helical polypeptides intertwined into a larger triple helix. This arrangement gives the long fibers great strength, suited to their function as the girders of connective tissue in skin, bone, tendons, and ligaments. Collagen accounts for 40% of the protein in your body.

Collagen

Illustration of collagen structure by Irving Geis. Image from Irving Geis Collection, HHMI. Rights owned by Howard Hughes Medical Institute (HHMI). Not to be reproduced without permission.

Many other proteins have subunits that are different from one another. For example, the oxygen-transporting molecule hemoglobin has four polypeptides of two distinct types. Each polypeptide has a nonprotein attachment, called a heme, with an iron atom that binds oxygen.

What happens if a protein folds incorrectly? Many diseases, such as Alzheimer's and Parkinson's, involve an accumulation of misfolded proteins. Prions are infectious misshapen proteins that are associated with serious degenerative brain diseases such as mad cow disease. Such diseases reinforce the theme that structure fits function: A protein's unique three-dimensional shape determines its proper functioning.

? If a genetic mutation changes the primary structure of a protein, how might this destroy the protein's function?

Primary structure, the amino acid sequence, affects the secondary structure, which affects the tertiary structure, which affects the quaternary structure (if any). Thus, primary structure determines the shape of a protein, and the function of a protein depends on its shape. A shape change could eliminate function.

Four Levels of Protein Structure

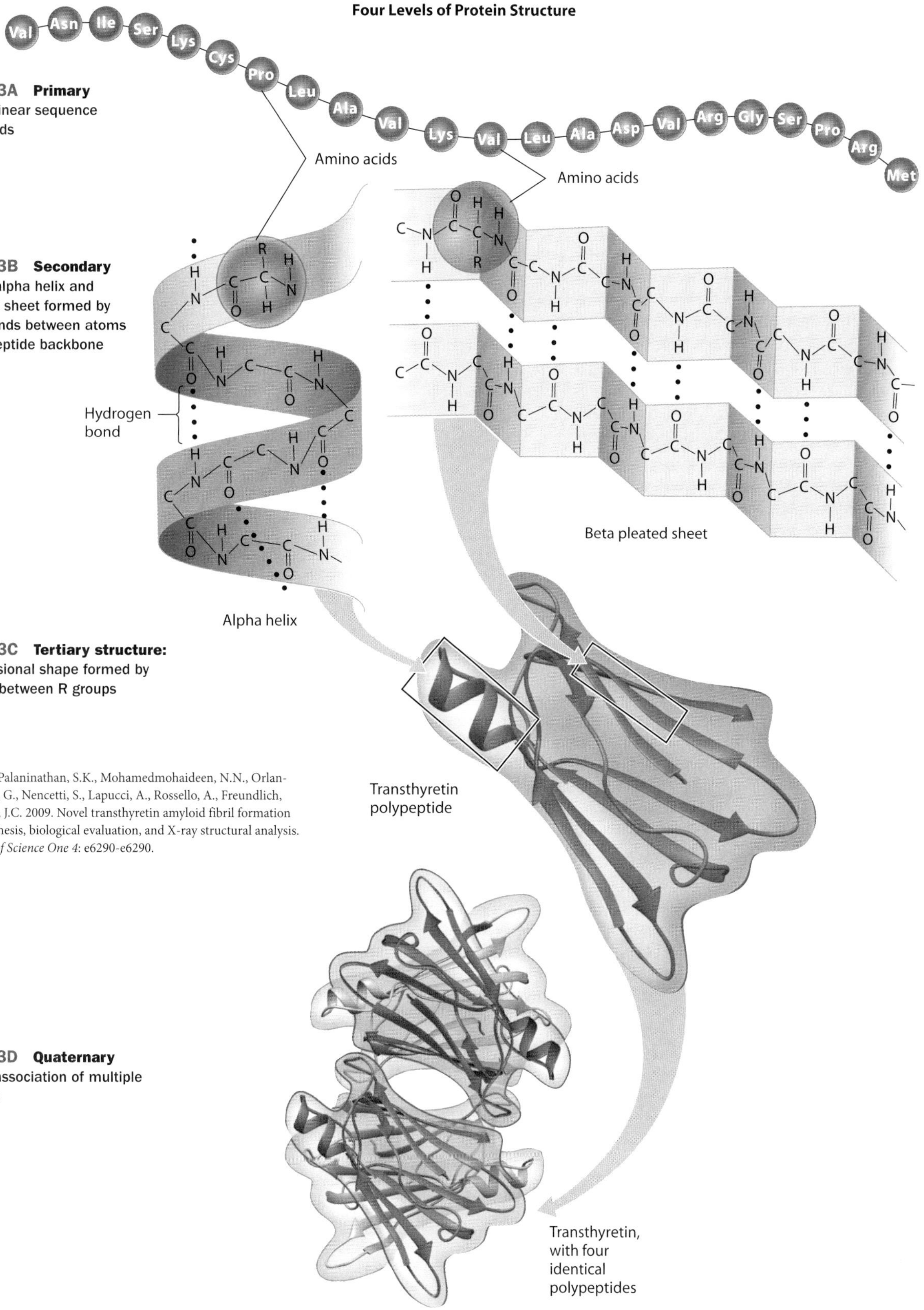

▶ **Figure 13A Primary structure:** linear sequence of amino acids

▶ **Figure 13B Secondary structure:** alpha helix and beta pleated sheet formed by hydrogen bonds between atoms of the polypeptide backbone

▶ **Figure 13C Tertiary structure:** three-dimensional shape formed by interactions between R groups

PDB ID 3GS0: Palaninathan, S.K., Mohamedmohaideen, N.N., Orlandini, E., Ortore, G., Nencetti, S., Lapucci, A., Rossello, A., Freundlich, J.S., Sacchettini, J.C. 2009. Novel transthyretin amyloid fibril formation inhibitors: synthesis, biological evaluation, and X-ray structural analysis. *Public Library of Science One 4*: e6290-e6290.

▶ **Figure 13D Quaternary structure:** association of multiple polypeptides

Nucleic Acids

14 DNA and RNA are the two types of nucleic acids

As we just saw, the primary structure of a polypeptide determines the shape of a protein. But what determines the primary structure? The amino acid sequence of a polypeptide is programmed by a discrete unit of inheritance known as a **gene**. Genes consist of **DNA (deoxyribonucleic acid)**, one of the two types of polymers called **nucleic acids**. The name *nucleic* comes from their location in the nuclei of eukaryotic cells. The genetic material that humans and other organisms inherit from their parents consists of DNA. Unique among molecules, DNA provides directions for its own replication. Thus, as a cell divides, its genetic instructions are passed to each daughter cell. These instructions program all of a cell's activities by directing the synthesis of proteins.

The genes present in DNA do not build proteins directly. They work through an intermediary—the second type of nucleic acid, known as **ribonucleic acid (RNA)**. **Figure 14** illustrates the main roles of these two types of nucleic acids in the production of proteins. In the nucleus of a eukaryotic cell, a gene directs the synthesis of an RNA molecule. We say that DNA is transcribed into RNA. The RNA molecule moves out of the nucleus and interacts with the protein-building machinery of the cell. There, the gene's instructions, written in "nucleic acid language," are translated into "protein language," the amino acid sequence of a polypeptide. (In prokaryotic cells, which lack nuclei, both transcription and translation take place within the cytoplasm of the cell.)

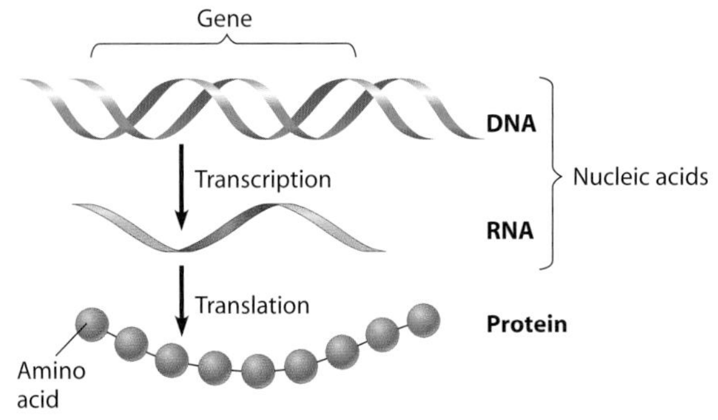

▲ **Figure 14** The flow of genetic information in the building of a protein

Recent research has found previously unknown types of RNA molecules that play many other roles in the cell. We return to the functions of DNA and RNA later in the book.

? How are the two types of nucleic acids functionally related?

The hereditary material of DNA contains the instructions for the primary structure of polypeptides. RNA is the intermediary that conveys those instructions to the protein-making machinery that assembles amino acids in the designated order.

15 Nucleic acids are polymers of nucleotides

The monomers that make up nucleic acids are **nucleotides**. As indicated in **Figure 15A**, each nucleotide contains three parts. At the center of a nucleotide is a five-carbon sugar (blue); the sugar in DNA is deoxyribose (shown in Figure 15A), whereas RNA has a slightly different sugar called ribose. Linked to one side of the sugar in both types of nucleotides is a negatively charged phosphate group (yellow). Linked to the sugar's other side is a nitrogenous base (green), a molecular structure containing nitrogen and carbon. (The nitrogen atoms tend to take up H^+ in aqueous solutions, which explains why it is called a nitrogenous *base*.) Each DNA nucleotide has one of four different nitrogenous bases: adenine (A), thymine (T), cytosine (C), and guanine (G). Thus, all genetic information is written in a four-letter alphabet. RNA nucleotides also contain the bases A, C, and G; but the base uracil (U) is found instead of thymine.

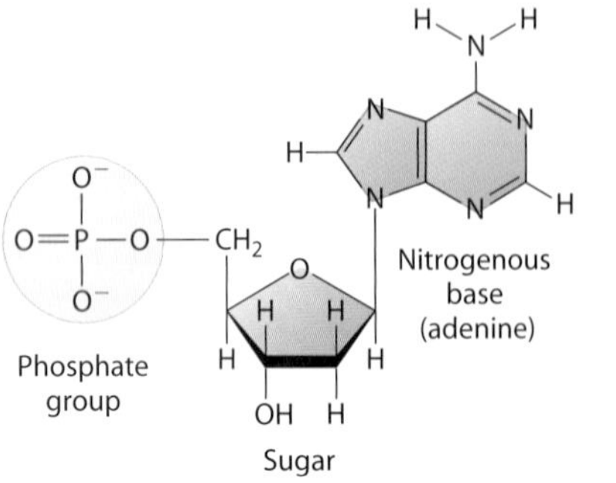

▲ **Figure 15A** A nucleotide, consisting of a phosphate group, a sugar, and a nitrogenous base

Like polysaccharides and polypeptides, a nucleic acid polymer—a polynucleotide—is built from its monomers by dehydration reactions. In this process, the sugar of one nucleotide bonds to the phosphate group of the next monomer. The result is a repeating sugar-phosphate backbone in the polymer, as represented by the blue and yellow ribbon in **Figure 15B**. (Note that the nitrogenous bases are not part of the backbone.)

RNA usually consists of a single polynucleotide strand, but DNA is a

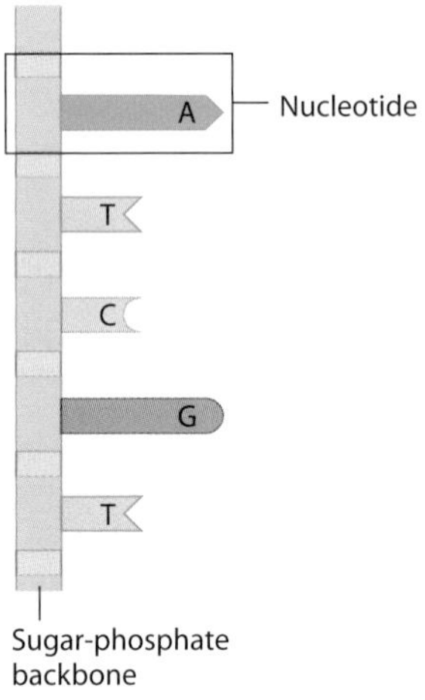

▲ **Figure 15B** Part of a polynucleotide

double helix, in which two polynucleotides wrap around each other **(Figure 15C)**. The nitrogenous bases protrude from the two sugar-phosphate backbones and pair in the center of the helix. As shown by their diagrammatic shapes in the figure, A always pairs with T, and C always pairs with G. The two DNA chains are held together by hydrogen bonds (indicated by the dotted lines) between their paired bases. These bonds are individually weak, but collectively they zip the two strands together into a very stable double helix. Most DNA molecules have thousands or even millions of base pairs.

Because of the base-pairing rules, the two strands of the double helix are said to be *complementary*, each a predictable counterpart of the other. Thus, if a stretch of nucleotides on one strand has the base sequence –AGCACT–, then the same stretch on the other strand must be –TCGTGA–. Complementary base pairing is the key to how a cell makes two identical copies of each of its DNA molecules every time it divides. Thus, the structure of DNA accounts for its function of transmitting genetic information whenever a cell reproduces. The same base-pairing rules (with the exception that U nucleotides of RNA pair with A nucleotides of DNA) also account for the precise transcription of information from DNA to RNA.

An organism's genes determine the proteins and thus the structures and functions of its body. Let's return to the subject of the chapter introduction—lactose intolerance—to conclude our study of biological molecules. In the next chapter, we move up in the biological hierarchy to the level of the cell.

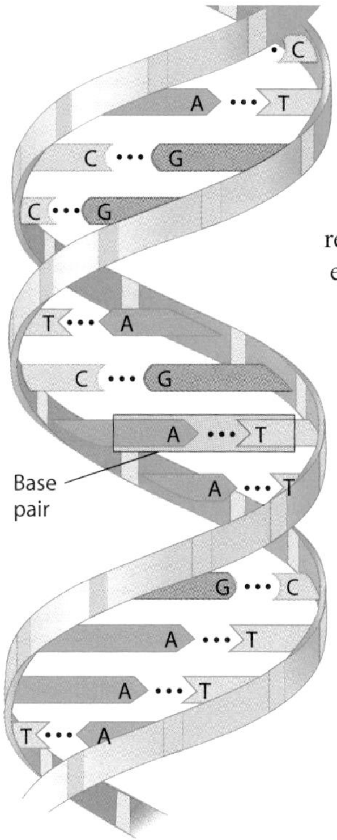

▲ **Figure 15C** DNA double helix

? What roles do complementary base pairing play in the functioning of nucleic acids?

Complementary base pairing makes possible the precise replication of DNA, ensuring that genetic information is faithfully transmitted every time a cell divides. It also ensures that RNA molecules carry accurate instructions for the synthesis of proteins.

EVOLUTION CONNECTION

16 Lactose tolerance is a recent event in human evolution

As you'll recall from the chapter introduction, the majority of people stop producing the enzyme lactase in early childhood and thus do not easily digest the milk sugar lactose. Researchers were curious about the genetic and evolutionary basis for the regional distribution of lactose tolerance and intolerance. In 2002, a group of scientists completed a study of the genes of 196 lactose-intolerant adults of African, Asian, and European descent. They determined that lactose intolerance is actually the human norm. It is "lactose tolerance" that represents a relatively recent mutation in the human genome.

The ability to make lactase into adulthood is concentrated in people of northern European descent, and the researchers speculated that lactose tolerance became widespread among this group because it offered a survival advantage. In northern Europe's relatively cold climate, only one harvest a year is possible. Therefore, animals were a main source of food for early humans in that region. Cattle were first domesticated in northern Europe about 9,000 years ago **(Figure 16)**. With milk and other dairy products at hand year-round, natural selection would have favored anyone with a mutation that kept the lactase gene switched on.

Researchers wondered whether the lactose tolerance mutation found in Europeans might be present in other cultures who kept dairy herds. Indeed, a 2006 study compared the genetic makeup and lactose tolerance of 43 ethnic groups in East Africa. The researchers identified three mutations, all different from each other and from the European mutation, that keep the lactase gene permanently turned on. The mutations appear to have occurred beginning around 7,000 years ago, around the time that archaeological evidence shows the domestication of cattle in these African regions.

Serge de Sazo/Photo Researchers

▲ **Figure 16** A prehistoric European cave painting of cattle

Mutations that conferred a selective advantage, such as surviving cold winters or withstanding drought by drinking milk, spread rapidly in these early pastoral peoples. Their evolutionary and cultural history is thus recorded in their genes and in their continuing ability to digest milk.

? Explain how lactose tolerance involves three of the four major classes of biological macromolecules.

Lactose, milk sugar, is a carbohydrate that is hydrolyzed by the enzyme lactase, a protein. The ability to make this enzyme and the regulation of when it is made is coded for in DNA, a nucleic acid.

REVIEW

For Practice Quizzes, BioFlix, MP3 Tutors, and Activities, go to www.masteringbiology.com.

Reviewing the Concepts

Introduction to Organic Compounds (1–3)

1 Life's molecular diversity is based on the properties of carbon. Carbon's ability to bond with four other atoms is the basis for building large and diverse organic compounds. Hydrocarbons are composed of only carbon and hydrogen. Isomers have the same molecular formula but different structures.

2 A few chemical groups are key to the functioning of biological molecules. Hydrophilic functional groups give organic molecules specific chemical properties.

3 Cells make a huge number of large molecules from a limited set of small molecules.

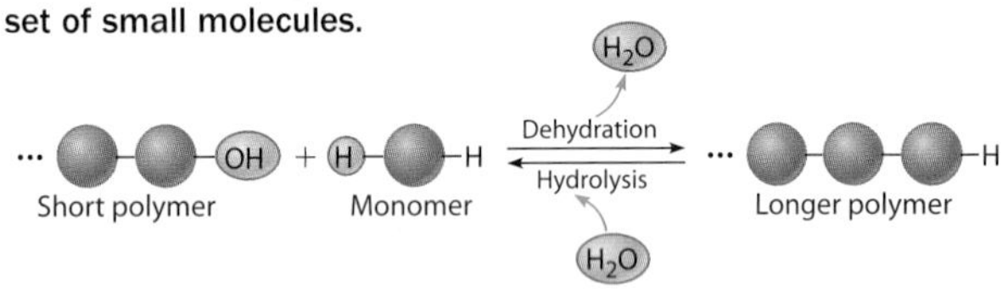

Carbohydrates (4–7)

4 Monosaccharides are the simplest carbohydrates. A monosaccharide has a formula that is a multiple of CH_2O and contains hydroxyl groups and a carbonyl group.

5 Two monosaccharides are linked to form a disaccharide.

6 What is high-fructose corn syrup, and is it to blame for obesity? HFCS, a mixture of glucose and fructose derived from corn, is commonly added to drinks and processed foods.

7 Polysaccharides are long chains of sugar units. Starch and glycogen are storage polysaccharides; cellulose is structural, found in plant cell walls. Chitin is a component of insect exoskeletons and fungal cell walls.

Lipids (8–10)

8 Fats are lipids that are mostly energy-storage molecules. Lipids are diverse, hydrophobic compounds composed largely of carbon and hydrogen. Fats (triglycerides) consist of glycerol linked to three fatty acids. Saturated fatty acids are found in animal fats; unsaturated fatty acids are typical of plant oils.

9 Phospholipids and steroids are important lipids with a variety of functions. Phospholipids are components of cell membranes. Steroids include cholesterol and some hormones.

10 Anabolic steroids pose health risks.

Proteins (11–13)

11 Proteins are made from amino acids linked by peptide bonds. Protein diversity is based on different sequences of amino acids, monomers that contain an amino group, a carboxyl group, an H, and an R group, all attached to a central carbon. The R groups distinguish 20 amino acids, each with specific properties.

12 A protein's specific shape determines its function. Proteins are involved in almost all of a cell's activities; as enzymes, they regulate chemical reactions.

13 A protein's shape depends on four levels of structure. A protein's primary structure is the sequence of amino acids in its polypeptide chain. Its secondary structure is the coiling or folding of the chain, stabilized by hydrogen bonds. Tertiary structure is the overall three-dimensional shape of a polypeptide, resulting from interactions among R groups. Proteins made of more than one polypeptide have quaternary structure.

Nucleic Acids (14–16)

14 DNA and RNA are the two types of nucleic acids. DNA and RNA serve as the blueprints for proteins and thus control the life of a cell.

15 Nucleic acids are polymers of nucleotides. Nucleotides are composed of a sugar, a phosphate group, and a nitrogenous base. DNA is a double helix; RNA is a single polynucleotide chain.

16 Lactose tolerance is a recent event in human evolution. Mutations in DNA have led to lactose tolerance in several human groups whose ancestors raised dairy cattle.

Connecting the Concepts

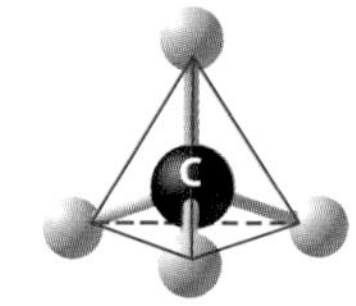

1. The diversity of life is staggering. Yet the molecular logic of life is simple and elegant: Small molecules common to all organisms are ordered into unique macromolecules. Explain why carbon is central to this diversity of organic molecules. How do carbon skeletons, chemical groups, monomers, and polymers relate to this molecular logic of life?
2. Complete the table to help review the structures and functions of the four classes of organic molecules.

Classes of Molecules and Their Components	Functions	Examples
Carbohydrates 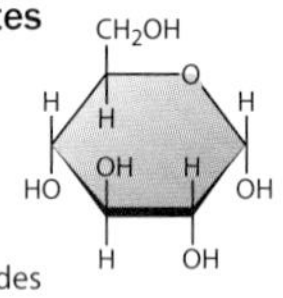 Monosaccharides	Energy for cell, raw material	a. ________
	b. ________	Starch, glycogen
	Plant cell support	c. ________
Lipids (don't form polymers) 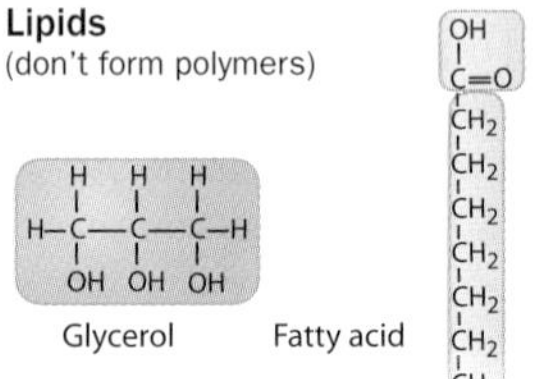 Glycerol / Fatty acid; Components of a fat molecule	Energy storage	d. ________
	e. ________	Phospholipids
	Hormones	f. ________
Proteins g. ________ h. ________ 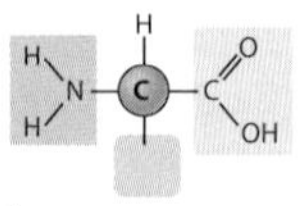 i. ________ Amino acid	j. ________	Lactase
	k. ________	Hair, tendons
	l. ________	Muscles
	Transport	m. ________
	Communication	Signal proteins
	n. ________	Antibodies
	Storage	Egg albumin
	Receive signals	Receptor protein
Nucleic Acids p. ________ o. ________ 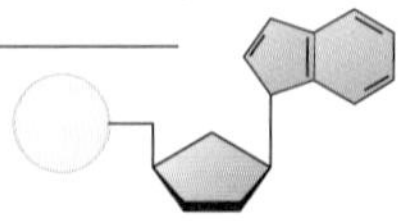 Nucleotide q. ________	Heredity	r. ________
	s. ________	DNA and RNA

Testing Your Knowledge

Multiple Choice

3. A glucose molecule is to starch as (*Explain your answer.*)
   a. a steroid is to a lipid.
   b. a protein is to an amino acid.
   c. a nucleic acid is to a polypeptide.
   d. a nucleotide is to a nucleic acid.
   e. an amino acid is to a nucleic acid.
4. What makes a fatty acid an acid?
   a. It does not dissolve in water.
   b. It is capable of bonding with other molecules to form a fat.
   c. It has a carboxyl group that donates an H^+ to a solution.
   d. It contains only two oxygen atoms.
   e. It is a polymer made of many smaller subunits.
5. Where in the tertiary structure of a water-soluble protein would you most likely find an amino acid with a hydrophobic R group?
   a. at both ends of the polypeptide chain
   b. on the outside, next to the water
   c. covalently bonded to another R group
   d. on the inside, away from water
   e. hydrogen-bonded to nearby amino acids
6. Cows can derive nutrients from cellulose because
   a. they produce enzymes that recognize the shape of the glucose-glucose bonds and hydrolyze them.
   b. they rechew their cud to break down cellulose fibers.
   c. one of their stomachs contains prokaryotes that can hydrolyze the bonds of cellulose.
   d. their intestinal tract contains termites, which produce enzymes to hydrolyze cellulose.
   e. they convert cellulose to starch and can digest starch.
7. A shortage of phosphorus in the soil would make it especially difficult for a plant to manufacture
   a. DNA.
   b. proteins.
   c. cellulose.
   d. fatty acids.
   e. sucrose.
8. Lipids differ from other large biological molecules in that they
   a. are much larger.
   b. are not polymers.
   c. do not have specific shapes.
   d. are nonpolar and therefore hydrophilic.
   e. contain nitrogen atoms.
9. Of the following functional groups, which is/are polar, tending to make organic compounds hydrophilic?
   a. carbonyl
   b. amino
   c. hydroxyl
   d. carboxyl
   e. all of the above
10. Unsaturated fats
   a. are more common in animals than in plants.
   b. have fewer fatty acid molecules per fat molecule.
   c. are associated with greater health risks than are saturated fats.
   d. have double bonds in their fatty acid chains.
   e. are usually solid at room temperature.

Describing, Comparing, and Explaining

11. List three different kinds of lipids and describe their functions.
12. Explain why heat, pH changes, and other environmental changes can interfere with a protein's function.
13. How can a cell make many different kinds of protein out of only 20 amino acids? Of the myriad possibilities, how does the cell "know" which proteins to make?
14. Briefly describe the various functions performed by proteins in a cell.
15. Explain how DNA controls the functions of a cell.
16. Sucrose is broken down in your intestine to the monosaccharides glucose and fructose, which are then absorbed into your blood. What is the name of this type of reaction? Using this diagram of sucrose, show how this would occur.

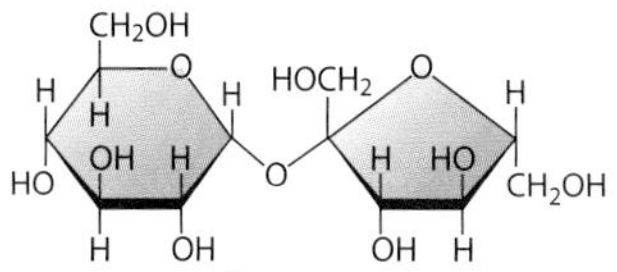

Sucrose

17. Circle and name the functional groups in this organic molecule. What type of compound is this? For which class of macromolecules is it a monomer?

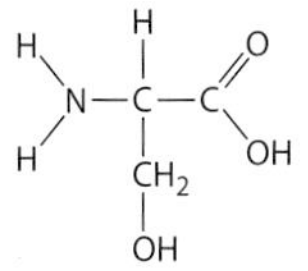

Applying the Concepts

18. Enzymes usually function best at an optimal pH and temperature. The following graph shows the effectiveness of two enzymes at various temperatures.

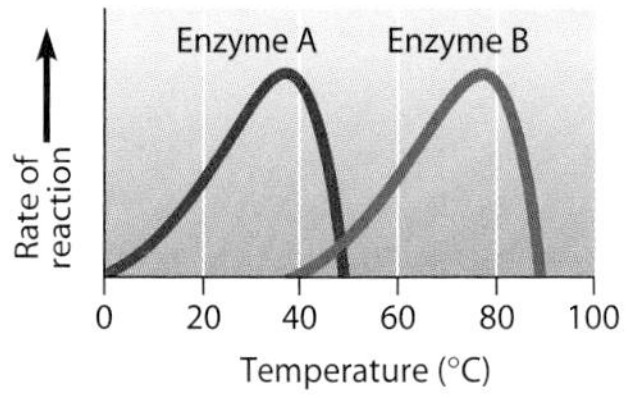

   a. At which temperature does enzyme A perform best? Enzyme B?
   b. One of these enzymes is found in humans and the other in thermophilic (heat-loving) bacteria. Which enzyme would you predict comes from which organism?
   c. From what you know about enzyme structure, explain why the rate of the reaction catalyzed by enzyme A slows down at temperatures above 40°C (140°F).
19. Some scientists hypothesize that life elsewhere in the universe might be based on the element silicon rather than on carbon. Look at the electron shell diagrams in the figure below. What properties does silicon share with carbon that would make silicon-based life more likely than, for example, neon-based or sulfur-based life?

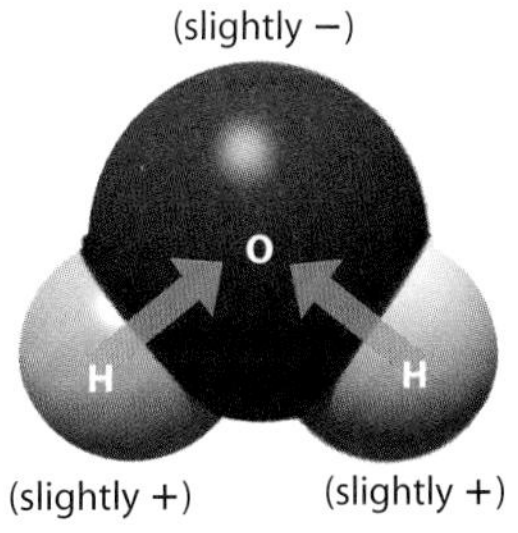

Review Answers

1. Carbon forms four covalent bonds, either with other carbon atoms, producing chains or rings of various lengths and shapes, or with other atoms, such as characteristic chemical groups that confer specific properties on a molecule. This is the basis for the incredible diversity of organic compounds. Organisms can link a small number of monomers into different arrangements to produce a huge variety of polymers.
2. a. glucose; b. energy storage; c. cellulose; d. fats; e. cell membrane component; f. steroids; g. amino group; h. carboxyl group; i. R group; j. enzyme; k. structural protein; l. movement; m. hemoglobin; n. defense; o. phosphate group; p. nitrogenous base; q. ribose or deoxyribose; r. DNA; s. code for proteins
3. d (The second kind of molecule is a polymer of the first.) 4. c 5. d 6. c 7. a 8. b 9. e 10. d
11. Fats (triglycerides)—energy storage. Phospholipids—major components of membranes. Steroids—cholesterol is a component of animal cell membranes; other steroids function as hormones.
12. Weak bonds that stabilize the three-dimensional structure of a protein are disrupted, and the protein unfolds. Function depends on shape, so if the protein is the wrong shape, it won't function properly.
13. Proteins are made of 20 amino acids arranged in many different sequences into chains of many different lengths. Genes, defined stretches of DNA, dictate the amino acid sequences of proteins in the cell.
14. Proteins function as enzymes, which catalyze chemical reactions. They also function in structure, movement, transport, defense, signaling, signal reception, and storage of amino acids (see Module 12).
15. The sequence of nucleotides in DNA is transcribed into a sequence of nucleotides in RNA, which determines the sequence of amino acids that will be used to build a polypeptide. Proteins mediate all the activities of a cell; thus, by coding for proteins, DNA controls the functions of a cell.
16. This is a hydrolysis reaction, which consumes water. It is essentially the reverse of the diagram in Figure 5, except that fructose has a different shape than glucose.
17. Circle NH_2, an amino group; COOH, a carboxyl group; and OH, a hydroxyl group on the R group. This is an amino acid, a monomer of proteins. The OH group makes it a polar amino acid.
18. a. A: at about 37°C; B: at about 78°C.
    b. A: from humans (human body temperature is about 37°C); B: from thermophilic bacteria.
    c. Above 40°C, the human enzyme denatures and loses its shape and thus its function. The increased thermal energy disrupts the weak bonds that maintain secondary and tertiary structure in an enzyme.
19. Silicon has four electrons in its outer electron shell, as does carbon. One would predict that silicon could thus form complex molecules by binding with four partners. Neon has a filled outer shell and is nonreactive. Sulfur can only form two covalent bonds and thus would not have the versatility of carbon or silicon.

A Tour of the Cell

A Tour of the Cell

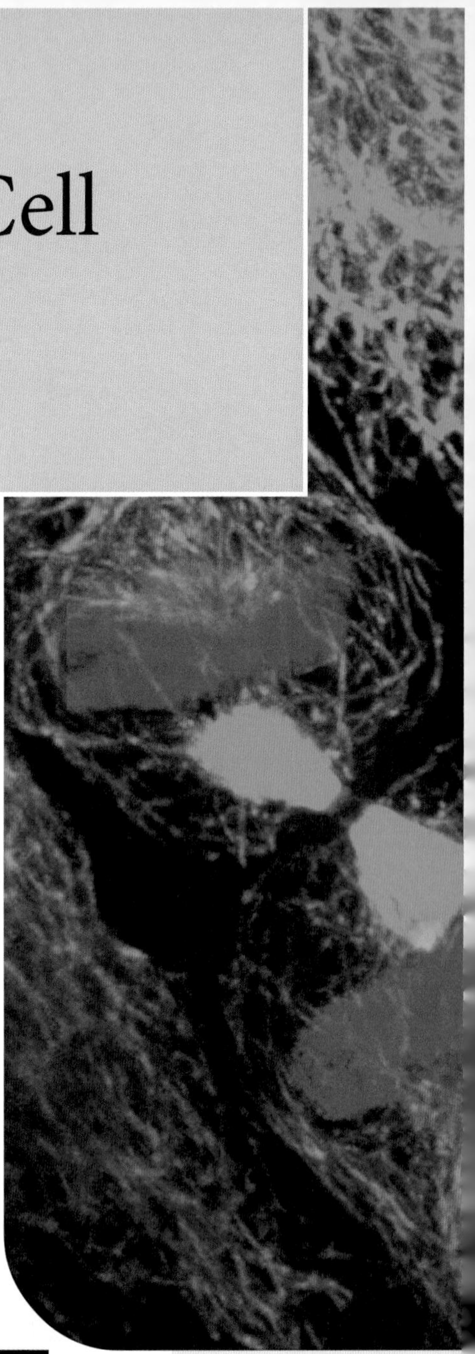

BIG IDEAS

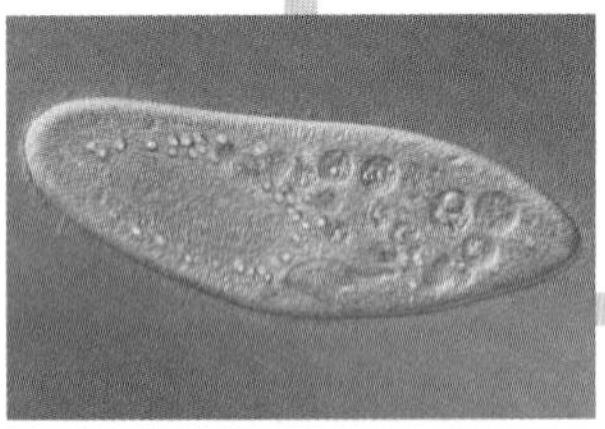

M.I. Walker/Photo Researchers

Introduction to the Cell
(1–4)

Microscopes reveal the structures of cells—the fundamental units of life.

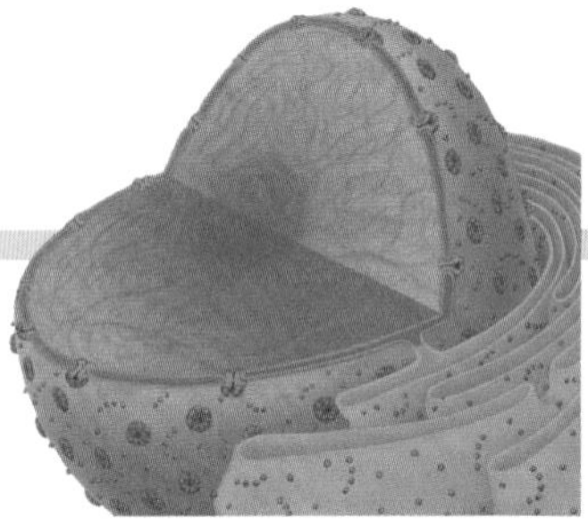

The Nucleus and Ribosomes
(5–6)

A cell's genetic instructions are housed in the nucleus and carried out by ribosomes.

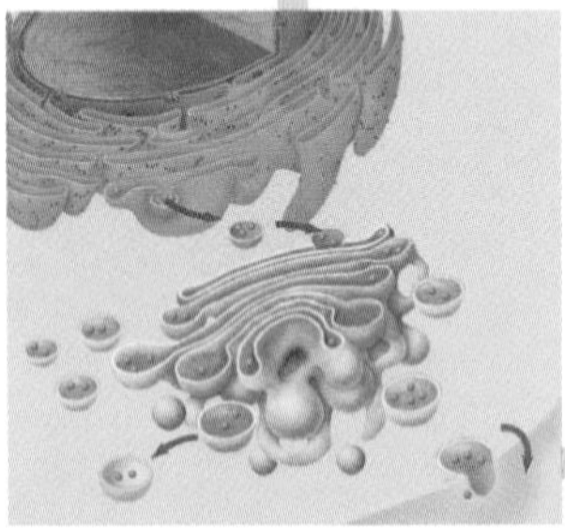

The Endomembrane System
(7–12)

The endomembrane system participates in the manufacture, distribution, and breakdown of materials.

Energy-Converting Organelles
(13–15)

Mitochondria in all cells and chloroplasts in plant cells function in energy processing.

Dr. Frank Solomon

The Cytoskeleton and Cell Surfaces
(16–22)

The cytoskeleton and extracellular components provide support, motility, and functional connections.

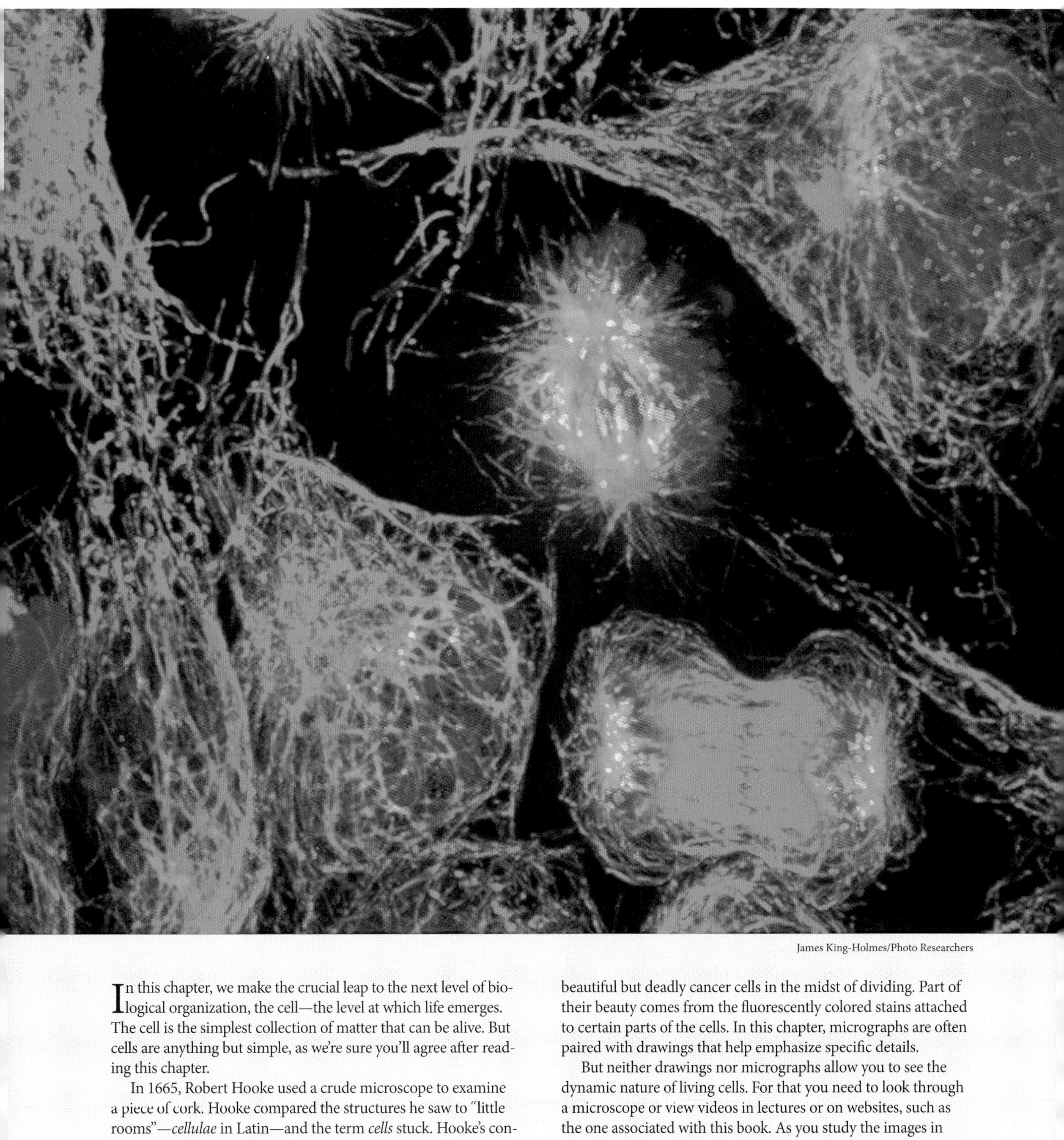

James King-Holmes/Photo Researchers

In this chapter, we make the crucial leap to the next level of biological organization, the cell—the level at which life emerges. The cell is the simplest collection of matter that can be alive. But cells are anything but simple, as we're sure you'll agree after reading this chapter.

In 1665, Robert Hooke used a crude microscope to examine a piece of cork. Hooke compared the structures he saw to "little rooms"—*cellulae* in Latin—and the term *cells* stuck. Hooke's contemporary, Antoni van Leeuwenhoek, working with more refined lenses, examined numerous subjects, from blood and sperm to pond water. His reports to the Royal Society of London included drawings and enthusiastic descriptions of his discoveries.

Since the days of Hooke and Leeuwenhoek, improved microscopes have vastly expanded our view of the cell. This micrograph (photo taken through a microscope) shows beautiful but deadly cancer cells in the midst of dividing. Part of their beauty comes from the fluorescently colored stains attached to certain parts of the cells. In this chapter, micrographs are often paired with drawings that help emphasize specific details.

But neither drawings nor micrographs allow you to see the dynamic nature of living cells. For that you need to look through a microscope or view videos in lectures or on websites, such as the one associated with this book. As you study the images in this chapter, keep in mind that cellular parts are not static; they are constantly moving and interacting.

This chapter focuses on cellular structures and functions. As you learn about the parts, however, remember that the phenomenon we call life emerges from the arrangement and interactions of the many components of a cell.

Introduction to the Cell

1 Microscopes reveal the world of the cell

Our understanding of nature often goes hand in hand with the invention and refinement of instruments that extend human senses. Before microscopes were first used in the 17th century, no one knew that living organisms were composed of cells. The first microscopes were light microscopes, like the ones you may use in a biology laboratory. In a **light microscope (LM)**, visible light is passed through a specimen, such as a microorganism or a thin slice of animal or plant tissue, and then through glass lenses. The lenses bend the light in such a way that the image of the specimen is magnified as it is projected into your eye or a camera.

Magnification is the increase in the apparent size of an object. **Figure 1A** shows a single-celled protist called *Paramecium*. The notation "LM230×" printed along the right edge of this **micrograph** tells you that the photograph was taken through a light microscope and that this image is 230 times the actual size of the organism.

The actual size of this *Paramecium* is about 0.33 millimeter (mm) in length. **Figure 1B** shows the size range of cells compared with objects both larger and smaller. The most common units of length that biologists use are listed at the bottom of the figure. Notice that the scale along the left side of the figure is logarithmic to accommodate the range of sizes shown. Starting at the top of the scale with 10 meters (m) and going down, each reference measurement marks a 10-fold decrease in length. Most cells are between 1 and 100 micrometers (μm) in diameter (yellow region of the figure) and are therefore visible only with a microscope. Certain bacteria are as small as 0.2 μm in diameter and can barely be seen with a light microscope, whereas bird eggs are large enough to be seen with the unaided eye. A single nerve cell running from the base of your spinal cord to your big toe may be 1 m in length, although it is so thin you would still need a microscope to see it.

Light microscopes can effectively magnify objects about 1,000 times. Greater magnification does not show more details clearly; indeed, the image becomes blurry. Thus, another important factor in microscopy is *resolution*, a measure of the clarity of an image. Resolution is the ability of an optical instrument to show two nearby objects as separate. For example, what looks to your unaided eye like a single star in the sky may be resolved as twin stars with a telescope. Just as the resolution of the human eye is limited, the light microscope cannot resolve detail finer than about 0.2 μm, about the size of the smallest bacterium. No matter how many times its image of such a bacterium is magnified, the light microscope cannot show the details of this small cell's structure.

From the time that Hooke discovered cells in 1665 until the middle of the 20th century, biologists had only light microscopes for viewing cells. With these microscopes and various staining techniques to increase contrast and highlight

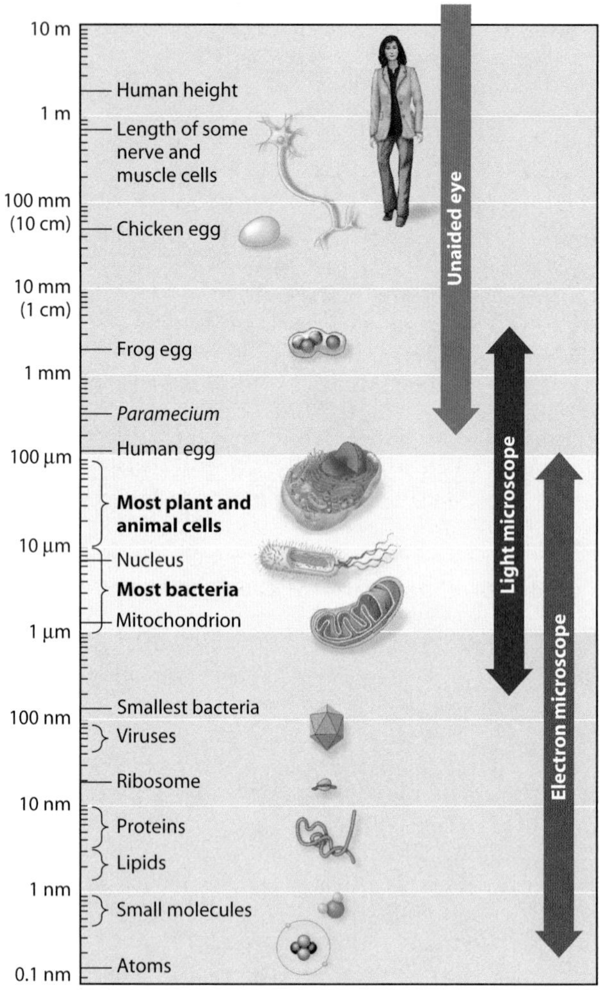

1 meter (m) = 10^0 meter (m) = 39.4 inches
1 centimeter (cm) = 10^{-2} m = 0.4 inch
1 millimeter (mm) = 10^{-3} m
1 micrometer (μm) = 10^{-3} mm = 10^{-6} m
1 nanometer (nm) = 10^{-3} μm = 10^{-9} m

▲ **Figure 1B** The size range of cells and related objects

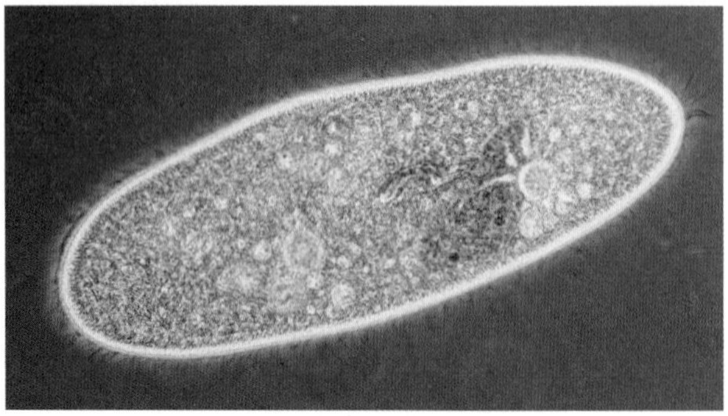
Michael Abbey/Photo Researchers

LM 230×

▲ **Figure 1A** Light micrograph of a protist, *Paramecium*

parts of the sample, these early biologists discovered a great deal—microorganisms, animal and plant cells, and even some of the structures within cells. By the mid-1800s, these discoveries led to the **cell theory**, which states that all living things are composed of cells and that all cells come from other cells.

Our knowledge of cell structure took a giant leap forward as biologists began using the electron microscope in the 1950s. Instead of using light, an **electron microscope (EM)** focuses a beam of electrons through a specimen or onto its surface. Electron microscopes can distinguish biological structures as small as about 2 nanometers (nm), a 100-fold improvement over the light microscope. This high resolution has enabled biologists to explore cell ultrastructure, the complex internal anatomy of a cell.

Figures 1C and **1D** show images produced by two kinds of electron microscopes. Biologists use the **scanning electron microscope (SEM)** to study the detailed architecture of cell surfaces. The SEM uses an electron beam to scan the surface of a cell or other sample, which is usually coated with a thin film of gold. The beam excites electrons on the surface, and these electrons are then detected by a device that translates their pattern into an image projected onto a video screen. The scanning electron micrograph in Figure 1C highlights the numerous cilia on *Paramecium*, projections it uses for movement. Notice the indentation, called the oral groove, through which food enters the cell. As you can see, the SEM produces images that look three-dimensional.

The **transmission electron microscope (TEM)** is used to study the details of internal cell structure. The TEM aims an electron beam through a very thin section of a specimen, just as a light microscope aims a beam of light through a specimen. The section is stained with atoms of heavy metals, which attach to certain cellular structures more than others. Electrons are scattered by these more dense parts, and the image is created by the pattern of transmitted electrons. Instead of using glass lenses, the TEM uses electromagnets as lenses to bend the paths of the electrons, magnifying and focusing an image onto a viewing screen or photographic film. The transmission electron micrograph in Figure 1D shows internal details of a protist called *Toxoplasma*. SEMs and TEMs are initially black and white but are often artificially colorized, as they are in these figures, to highlight or clarify structural features.

Electron microscopes have truly revolutionized the study of cells and their structures. Nonetheless, they have not replaced the light microscope. One problem is that electron microscopes cannot be used to study living specimens because the methods used to prepare the specimen kill the cells. For a biologist studying a living process, such as the movement of *Paramecium*, a light microscope equipped with a video camera is more suitable than either an SEM or a TEM.

There are different types of light microscopy. **Figure 1E** shows *Paramecium* as seen using differential interference contrast microscopy. This optical technique amplifies differences in density so that the structures in living cells appear almost three-dimensional. Other techniques use fluorescent stains that selectively bind to various cellular molecules (see the chapter introduction). In the last decade or two, light microscopy has seen significant and exciting technical advances that have increased magnification, resolution, and contrast. You will see many beautiful and illuminating examples of microscopy in this textbook.

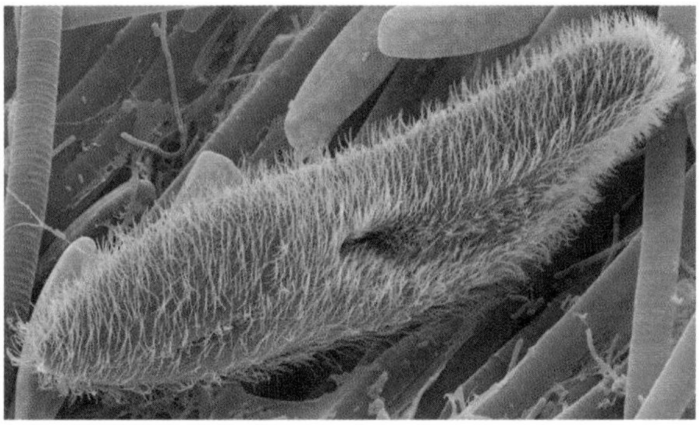

Andrew Syred/Photo Researchers

Colorized SEM 580×

▲ **Figure 1C** Scanning electron micrograph of *Paramecium*

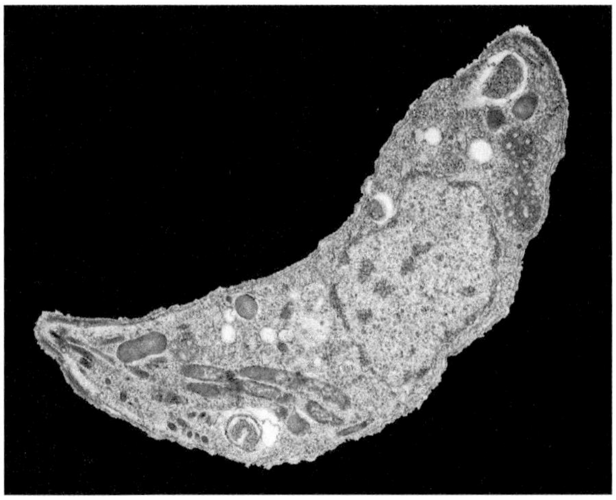

Dr. Klaus Boller/Photo Researchers

Colorized TEM 9,375×

▲ **Figure 1D** Transmission electron micrograph of *Toxoplasma*

? Which type of microscope would you use to study (a) the changes in shape of a living human white blood cell; (b) the finest details of surface texture of a human hair; (c) the detailed structure of an organelle in a liver cell?

(a) Light microscope; (b) scanning electron microscope; (c) transmission electron microscope

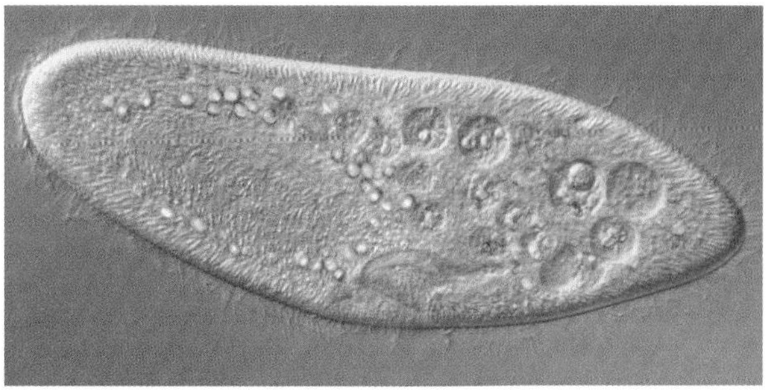

LM 380×

M.I. Walker/Photo Researchers

▲ **Figure 1E** Differential interference contrast micrograph of *Paramecium*

2 The small size of cells relates to the need to exchange materials across the plasma membrane

As you saw in Figure 1B, most cells are microscopic—unable to be seen without a microscope. Are there advantages to being so small? The logistics of carrying out a cell's functions appear to set both lower and upper limits on cell size. At minimum, a cell must be large enough to house enough DNA, protein molecules, and structures to survive and reproduce. But why aren't most cells as large as chicken eggs? The maximum size of a cell is influenced by geometry—the need to have a surface area large enough to service the volume of a cell. Active cells have a huge amount of traffic across their outer surface. A chicken's egg cell isn't very active, but once a chick embryo starts to develop, the egg is divided into many microscopic cells, each bounded by a membrane that allows the essential flow of oxygen, nutrients, and wastes across its surface.

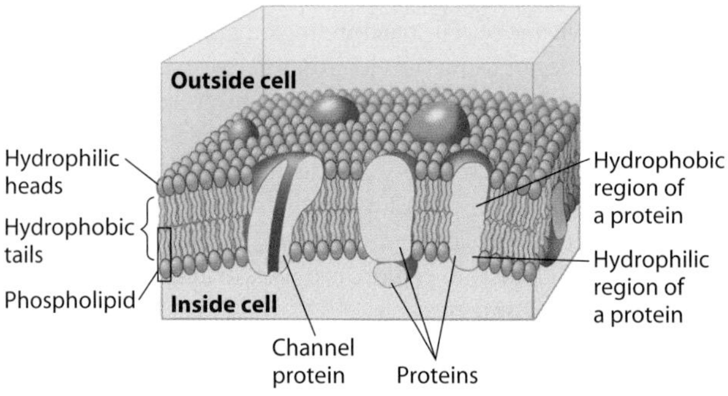

▲ **Figure 2B** A plasma membrane: a phospholipid bilayer with associated proteins

Surface-to-Volume Ratio Large cells have more surface area than small cells, but they have much less surface area relative to their volume than small cells. **Figure 2A** illustrates this relationship by comparing one large cube to 27 small ones. Using arbitrary units of measurement, the total volume is the same in both cases: 27 units^3 (height × width × length). The total surface areas, however, are quite different. A cube has six sides; thus, its surface area is six times the area of each side (height × width). The surface area of the large cube is 54 units^2, while the total surface area of all 27 cubes is 162 units^2 ($27 \times 6 \times 1 \times 1$), three times greater than the surface area of the large cube. Thus, we see that the smaller cubes have a much greater surface-to-volume ratio than the large cube. How about those neurons that extend from the base of your spine to your toes? Very thin, elongated shapes also provide a large surface area relative to a cell's volume.

The Plasma Membrane So what is a cell's surface like? And how does it control the traffic of molecules across it? The **plasma membrane** forms a flexible boundary between the living cell and its surroundings. For a structure that separates life from nonlife, this membrane is amazingly thin. It would take a stack of more than 8,000 of them to equal the thickness of this page. And, as you have come to expect with all things biological, the structure of the plasma membrane correlates with its function.

The structure of phospholipid molecules is well suited to their role as the main components of biological membranes. Composed of two distinct regions—a head with a negatively charged phosphate group and two nonpolar fatty acid tails, phospholipids form a two-layer sheet called a phospholipid bilayer. As you can see in **Figure 2B**, the phospholipids' hydrophilic (water-loving) heads face outward, exposed to the aqueous solutions on both sides of a membrane. Their hydrophobic (water-fearing) tails point inward, mingling together and shielded from water. Embedded in this lipid bilayer are diverse proteins, floating like icebergs in a phospholipid sea. The regions of the proteins within the center of the membrane are hydrophobic; the exterior sections exposed to water are hydrophilic.

Now let's see how the properties of the phospholipid bilayer and the proteins embedded in it relate to the plasma membrane's job as a traffic cop, regulating the flow of material into and out of the cell. Nonpolar molecules, such as O_2 and CO_2, can easily move across the membrane's hydrophobic interior. Some of the membrane's proteins form channels (tunnels) that shield ions and polar molecules as they pass through the hydrophobic center of the membrane. Still other proteins serve as pumps, using energy to actively transport molecules into or out of the cell.

In the next module, we consider other features common to all cells and take a closer look at the prokaryotic cells of domains Bacteria and Archaea.

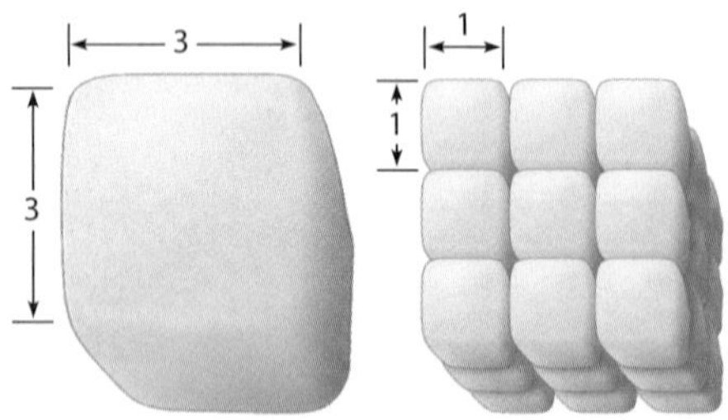

Total volume	27 units^3	27 units^3
Total surface area	54 units^2	162 units^2
Surface-to-volume ratio	2	6

▲ **Figure 2A** Effect of cell size on surface area

? To convince yourself that a small cell has more surface area relative to volume than a large cell, compare the surface-to-volume ratios of the large cube and one of the small cubes in Figure 2A.

Large cube: 54/27 = 2; small cube: 6/1 = 6 (surface area is 1 × 1 × 6 sides = 6 units^2; volume is 1 × 1 × 1 unit^3)

3 Prokaryotic cells are structurally simpler than eukaryotic cells

Two kinds of cells, which differ in size and structure, have evolved over time. Bacteria and archaea consist of **prokaryotic cells**, whereas all other forms of life (protists, fungi, plants, and animals) are composed of **eukaryotic cells**. Eukaryotic cells are distinguished by having a membrane-enclosed nucleus, which houses most of their DNA. The word *eukaryote* means "true nucleus" (from the Greek *eu*, true, and *karyon*, kernel, referring to the nucleus). The word *prokaryote* means "before nucleus" (from the Greek *pro*, before), reflecting the fact that prokaryotic cells evolved before eukaryotic cells. They are also, as you shall see, structurally much simpler than eukaryotic cells while sharing some common characteristics.

All cells have several basic features in common. In addition to being bounded by a plasma membrane, all cells have one or more **chromosomes** carrying genes made of DNA. And all cells contain **ribosomes**, tiny structures that make proteins according to instructions from the genes. The interior of both types of cell is called the **cytoplasm**. However, in eukaryotic cells, this term refers only to the region between the nucleus and the plasma membrane. The cytoplasm of a eukaryotic cell contains many membrane-enclosed organelles that perform specific functions.

The cutaway diagram in **Figure 3** reveals the structure of a generalized prokaryotic cell. Notice that the DNA is coiled into a region called the **nucleoid** (nucleus-like), but in contrast to the nucleus of eukaryotic cells, no membrane surrounds the DNA. The ribosomes of prokaryotes (shown here in brown) are smaller and differ somewhat from those of eukaryotes. These molecular differences are the basis for the action of some antibiotics, such as tetracycline and streptomycin, which target prokaryotic ribosomes. Thus, protein synthesis can be blocked for the bacterium that's invaded you, but not for you, the eukaryote who is taking the drug.

Outside the plasma membrane (shown here in gray) of most prokaryotes is a fairly rigid, chemically complex cell wall (orange). The wall protects the cell and helps maintain its shape. Some antibiotics, such as penicillin, prevent the formation of these protective walls. Again, since your cells don't have such walls, these antibiotics can kill invading bacteria without harming your cells. Certain prokaryotes have a sticky outer coat called a capsule (yellow) around the cell wall, helping to glue the cells to surfaces, such as sticks and rocks in fast-flowing streams or tissues within the human body. In addition to capsules, some prokaryotes have surface projections. Short projections help attach prokaryotes to each other or their substrate. Longer projections called **flagella** (singular, *flagellum*) propel a prokaryotic cell through its liquid environment.

It takes an electron microscope to see the details of any cell, and this is especially true of prokaryotic cells (Figure 3, right side). Most prokaryotic cells are about one-tenth the size of a typical eukaryotic cell. Eukaryotic cells are the main focus of this chapter, so we turn to these next.

? List three features that are common to prokaryotic and eukaryotic cells. List three features that differ.

Both types of cells have plasma membranes, chromosomes containing DNA, and ribosomes. Prokaryotic cells are smaller, do not have a nucleus that houses their DNA or other membrane-enclosed organelles, and have smaller, somewhat different ribosomes.

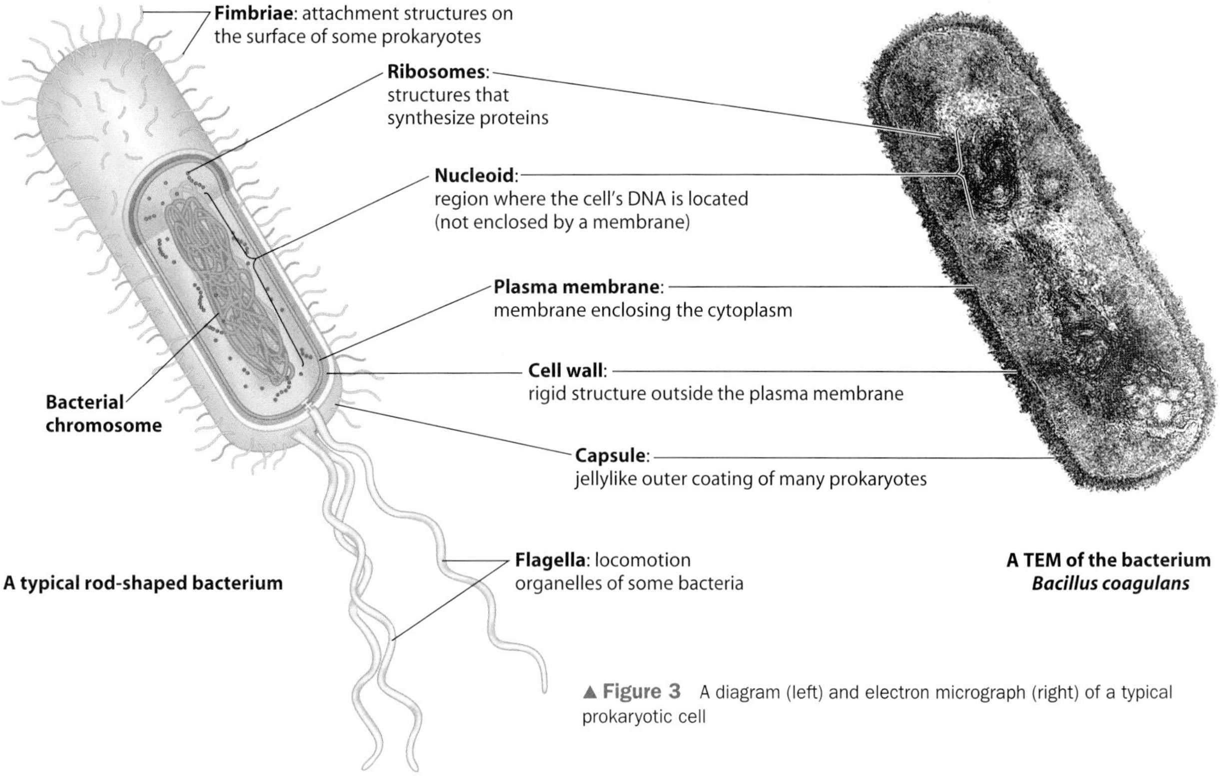

S. C. Holt, University of Texas Health Center/Biological Photo Service

▲ **Figure 3** A diagram (left) and electron micrograph (right) of a typical prokaryotic cell

4 Eukaryotic cells are partitioned into functional compartments

All eukaryotic cells—whether from animals, plants, protists, or fungi—are fundamentally similar to one another and profoundly different from prokaryotic cells. Let's look at an animal cell and a plant cell as representatives of the eukaryotes.

Figure 4A is a diagram of an idealized animal cell. No cell would look exactly like this. We color-code the various organelles and other structures in the diagrams for easier identification. And recall from the chapter introduction that in living cells many of these structures are moving and interacting.

The nucleus is the most obvious difference between a prokaryotic and eukaryotic cell. A eukaryotic cell also contains various other **organelles** ("little organs"), which perform specific functions in the cell. Just as the cell itself is wrapped in a membrane made of phospholipids and proteins that perform various functions, each organelle is bounded by a membrane with a lipid and protein composition that suits its function.

The organelles and other structures of eukaryotic cells can be organized into four basic functional groups as follows: (1) The nucleus and ribosomes carry out the genetic control of the cell. (2) Organelles involved in the manufacture, distribution, and breakdown of molecules include the endoplasmic reticulum, Golgi apparatus, lysosomes, vacuoles, and peroxisomes. (3) Mitochondria in all cells and chloroplasts in plant cells function in energy processing. (4) Structural support, movement, and communication between cells are the functions of the cytoskeleton, plasma membrane, and plant cell wall. These cellular components are identified in the figures on these two pages and will be examined in greater detail in the remaining modules of this chapter.

In essence, the internal membranes of a eukaryotic cell partition it into compartments. Many of the chemical activities of cells—activities known collectively as **cellular metabolism**—occur within organelles. In fact, many enzymatic proteins essential for metabolic processes are built into the membranes of organelles. The fluid-filled spaces within organelles are important as sites where specific chemical conditions are maintained. These conditions vary from one organelle to another and favor the metabolic processes occurring in each kind of organelle.

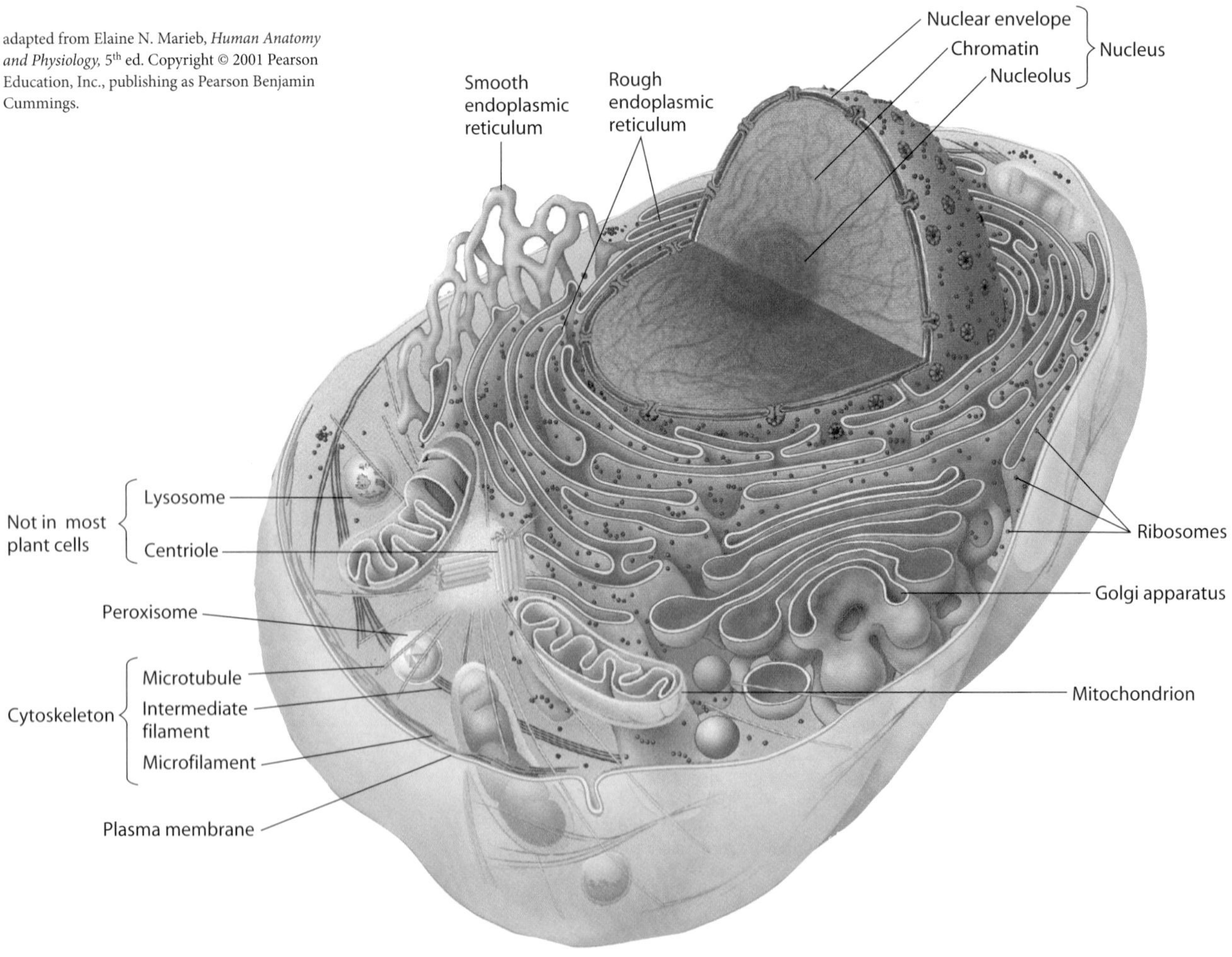

▲ **Figure 4A** An animal cell

For example, while a part of the endoplasmic reticulum is engaged in making steroid hormones, neighboring peroxisomes may be detoxifying harmful compounds and making hydrogen peroxide (H_2O_2) as a poisonous by-product of their activities. But because the H_2O_2 is confined within peroxisomes, where it is quickly converted to H_2O by resident enzymes, the rest of the cell is protected from destruction.

Almost all of the organelles and other structures of animal cells are also present in plant cells. As you can see in Figure 4A, however, there are a few exceptions: Lysosomes and centrioles are not found in plant cells. Also, although some animal cells have flagella or cilia (not shown in Figure 4A), among plants, only the sperm cells of a few species have flagella. (The flagella of prokaryotic cells differ in both structure and function from eukaryotic flagella.)

A plant cell **(Figure 4B)** also has some structures that an animal cell lacks. For example, a plant cell has a rigid, rather thick cell wall (as do the cells of fungi and many protists). Cell walls protect cells and help maintain their shape. Chemically different from prokaryotic cell walls, plant cell walls contain the polysaccharide cellulose. Plasmodesmata (singular, plasmodesma) are cytoplasmic channels through cell walls that connect adjacent cells. An important organelle found in plant cells is the chloroplast, where photosynthesis occurs. (Chloroplasts are also found in algae and some other protists.) Unique to plant cells is a large central vacuole, a compartment that stores water and a variety of chemicals.

Although we have emphasized organelles, eukaryotic cells contain nonmembranous structures as well. The cytoskeleton is composed of different types of protein fibers that extend throughout the cell. These networks provide for support and movement. As you can see by the many brown dots in both figures, ribosomes occur throughout the cytoplasm, as they do in prokaryotic cells. In addition, eukaryotic cells have many ribosomes attached to parts of the endoplasmic reticulum (making it appear "rough") and to the outer membrane of the nucleus.

Let's begin our in-depth tour of the eukaryotic cell, starting with the nucleus.

? Which of the following cellular structures differs from the others in the list: mitochondrion, chloroplast, ribosome, lysosome, vacuole? How does it differ?

Ribosome, because it is the only structure in the list that is not bounded by a membrane

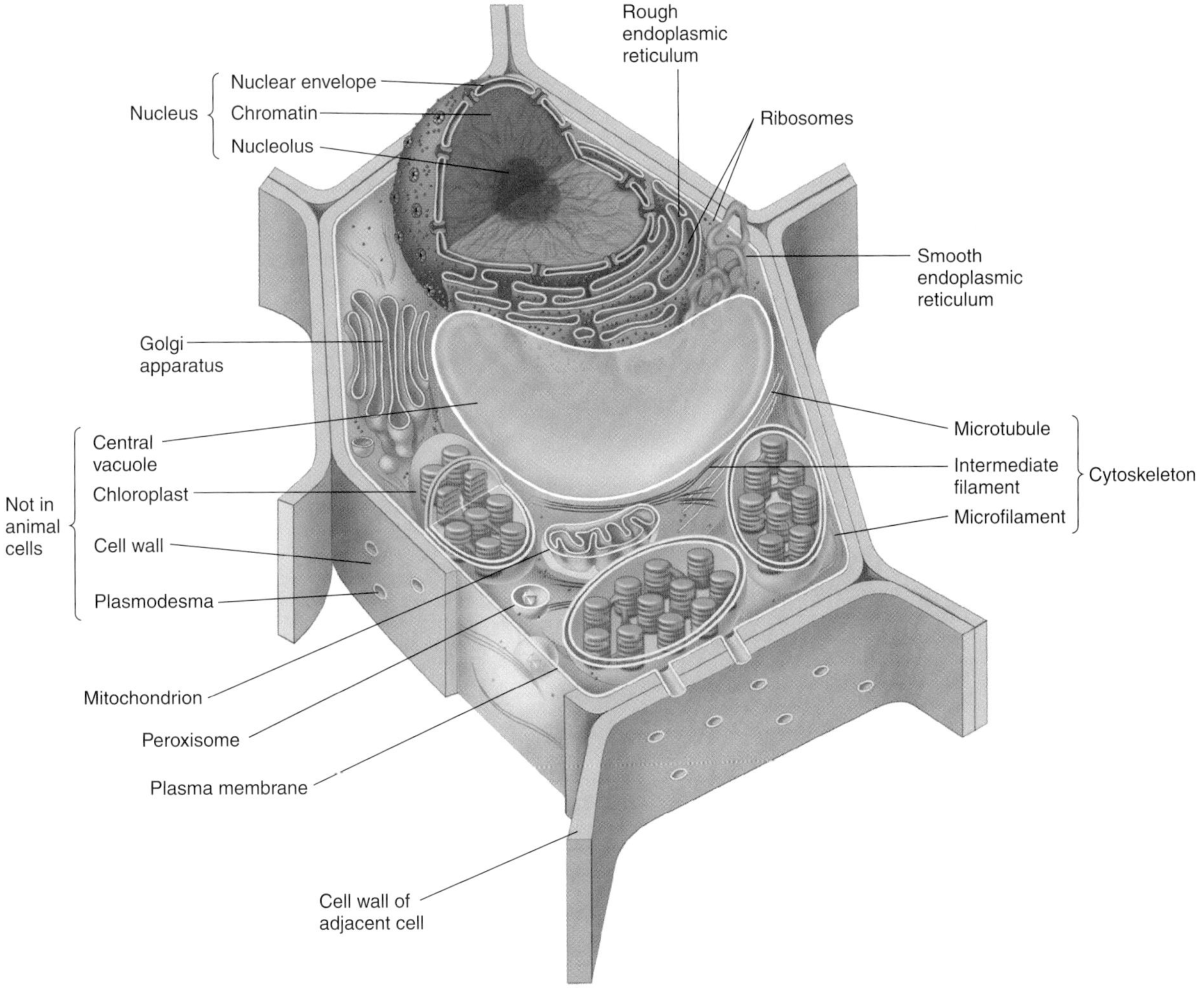

▲ **Figure 4B** A plant cell

The Nucleus and Ribosomes

5 The nucleus is the cell's genetic control center

You just saw a preview of the many intricate structures that can be found in a eukaryotic cell. A cell must build and maintain these structures and process energy for the work of transport, import, export, movement, and communication. But who is in charge of this bustling factory? Who stores the master plans, gives the orders, changes course in response to environmental input, and, when called upon, makes another factory just like itself? The cell's nucleus is this command center.

The **nucleus** contains most of the cell's DNA—its master plans—and controls the cell's activities by directing protein synthesis. The DNA is associated with many proteins in the structures called chromosomes. The proteins help organize and coil the long DNA molecule. Indeed, the DNA of the 46 chromosomes in one of your cells laid end to end would stretch to a length of over 2 m, but it must coil up to fit into a nucleus only 5 μm in diameter. When a cell is not dividing, this complex of proteins and DNA, called **chromatin**, appears as a diffuse mass, as shown in the TEM (left) and diagram (right) of a nucleus in **Figure 5**.

As a cell prepares to divide, the DNA is copied so that each daughter cell can later receive an identical set of genetic instructions. Just prior to cell division, the thin chromatin fibers coil up further, becoming thick enough to be visible with a light microscope as the familiar separate structures you would probably recognize as chromosomes.

Enclosing the nucleus is a double membrane called the **nuclear envelope**. Each of the membranes is a separate phospholipid bilayer with associated proteins. Similar in function to the plasma membrane, the nuclear envelope controls the flow of materials into and out of the nucleus. As you can see in the diagram in Figure 5, the nuclear envelope is perforated with protein-lined pores that regulate the movement of large molecules and also connects with the cell's network of membranes called the endoplasmic reticulum.

The **nucleolus**, a prominent structure in the nucleus, is the site where a special type of RNA called *ribosomal RNA* (rRNA) is synthesized according to instructions in the DNA. Proteins brought in through the nuclear pores from the cytoplasm are assembled with this rRNA to form the subunits of ribosomes. These subunits then exit through the pores to the cytoplasm, where they will join to form functional ribosomes.

The nucleus directs protein synthesis by making another type of RNA, *messenger RNA* (mRNA). Essentially, mRNA is a transcription of protein-synthesizing instructions written in a gene's DNA. The mRNA moves through the pores in the nuclear envelope to the cytoplasm. There it is translated by ribosomes into the amino acid sequences of proteins. Let's look at ribosomes next.

? What are the main functions of the nucleus?

To house and copy DNA and pass it on to daughter cells in cell division; to build ribosomal subunits; to transcribe DNA instructions into RNA and thereby control the cell's functions

adapted from Elaine N. Marieb, *Human Anatomy and Physiology*, 5th ed. Copyright © 2001 Pearson Education, Inc., publishing as Pearson Benjamin Cummings.

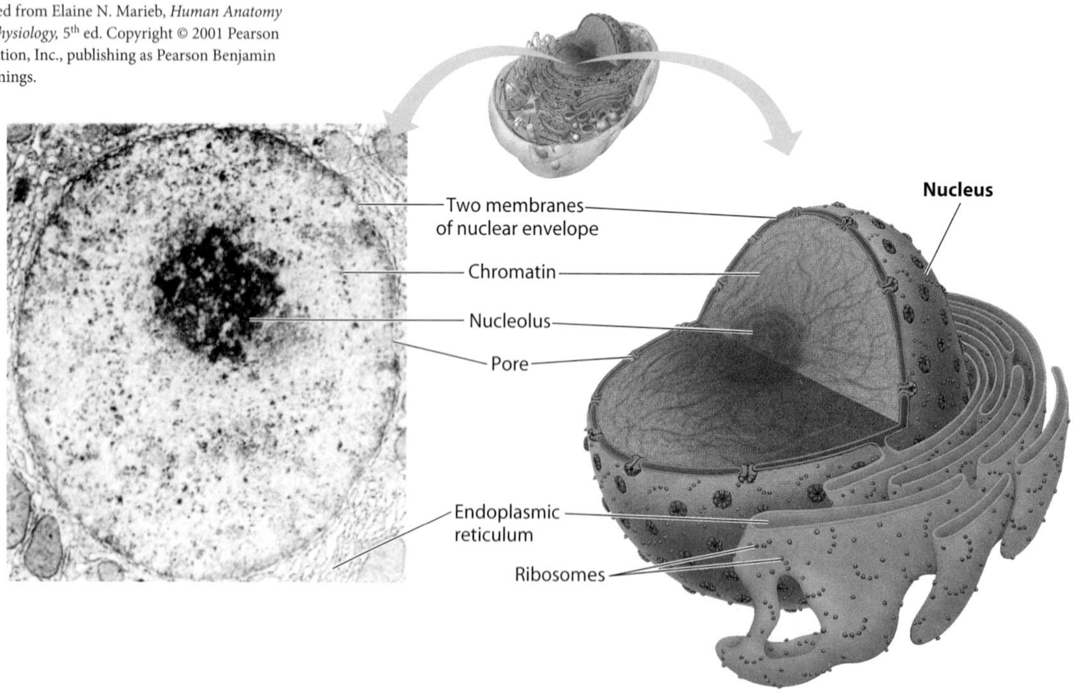

Courtesy of Richard Rodewald/Biological Photo Service

▲ **Figure 5** Transmission electron micrograph (left) and diagram (right) of the nucleus

6 Ribosomes make proteins for use in the cell and for export

If the nucleus is the command center, then ribosomes are the machines on which those commands are carried out. Ribosomes are the cellular components that use instructions sent from the nucleus to carry out protein synthesis. Cells that make a lot of proteins have a large number of ribosomes. For example, a human pancreas cell producing digestive enzymes may contain a few million ribosomes. What other structure would you expect to be prominent in cells that are active in protein synthesis? As you just learned, nucleoli assemble the subunits of ribosomes out of ribosomal RNA and protein.

As shown in the colorized TEM in **Figure 6**, ribosomes are found in two locations in the cell. *Free ribosomes* are suspended in the fluid of the cytoplasm, while *bound ribosomes* are attached to the outside of the endoplasmic reticulum or nuclear envelope. Free and bound ribosomes are structurally identical, and ribosomes can alternate between the two locations.

Most of the proteins made on free ribosomes function within the cytoplasm; examples are enzymes that catalyze the first steps of sugar breakdown. In Module 8, you will see how bound ribosomes make proteins that will be inserted into membranes, packaged in certain organelles, or exported from the cell.

At the bottom right in Figure 6, you see how ribosomes interact with messenger RNA (carrying the instructions from a gene) to build a protein. The nucleotide sequence of an mRNA molecule is translated into the amino acid sequence of a polypeptide. Next let's look at more of the manufacturing equipment of the cell.

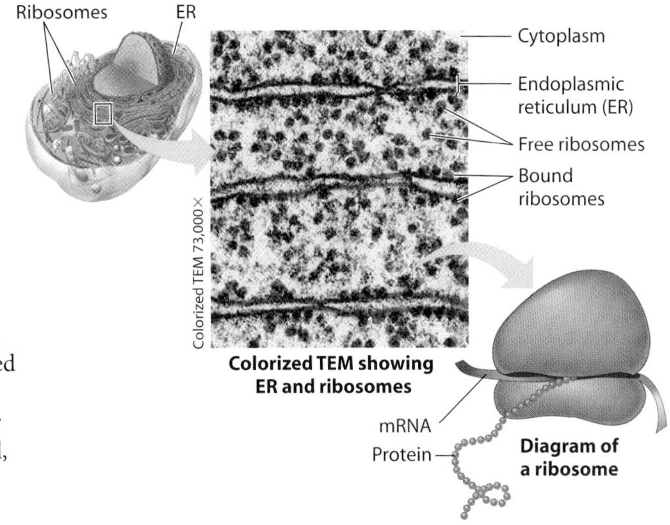

D. W. Fawcett/Photo Researchers

▲ **Figure 6** The locations and structure of ribosomes

? What role do ribosomes play in carrying out the genetic instructions of a cell?

Ribosomes synthesize proteins according to the instructions carried by messenger RNA from the DNA in the nucleus.

The Endomembrane System

7 Overview: Many cell organelles are connected through the endomembrane system

Ribosomes may be a cell's protein-making machines, but running a factory as complex as a cell requires infrastructure and many different departments that perform separate but related functions. Internal membranes, a distinguishing feature of eukaryotic cells, are involved in most of a cell's functions. Many of the membranes of the eukaryotic cell are part of an **endomembrane system**. Some of these membranes are physically connected and some are related by the transfer of membrane segments by tiny **vesicles**, sacs made of membrane.

The endomembrane system includes the nuclear envelope, endoplasmic reticulum, Golgi apparatus, lysosomes, vacuoles, and the plasma membrane. (The plasma membrane is not exactly an *endo*membrane in physical location, but it is related to the other membranes by the transfer of vesicles). Many of these organelles work together in the synthesis, distribution, storage, and export of molecules. We focus on these interrelated membranes in Modules 8–11.

The extensive network of flattened sacs and tubules called the **endoplasmic reticulum (ER)** is a prime example of the direct interrelatedness of parts of the endomembrane system. (The term *endoplasmic* means "within the cytoplasm," and *reticulum* is Latin for "little net.") As shown in Figure 5 on the facing page, membranes of the ER are continuous with the nuclear envelope. As we discuss next, there are two regions of ER—smooth ER and rough ER—that differ both in structure and in function. The membranes that form them, however, are connected.

The tubules and sacs of the ER enclose an interior space that is separate from the cytoplasmic fluid. Dividing the cell into separate functional compartments is an important aspect of the endomembrane system.

? Which structure includes all others in the list: rough ER, smooth ER, endomembrane system, nuclear envelope?

Endomembrane system

8 The endoplasmic reticulum is a biosynthetic factory

One of the major manufacturing sites in a cell is the endoplasmic reticulum. The diagram in **Figure 8A** shows a cutaway view of the interconnecting membranes of the smooth and rough ER. These two types of ER can be distinguished in the electron micrograph. **Smooth endoplasmic reticulum** is called *smooth* because it lacks attached ribosomes. **Rough endoplasmic reticulum** has ribosomes that stud the outer surface of the membrane; thus, it appears *rough* in the electron micrograph.

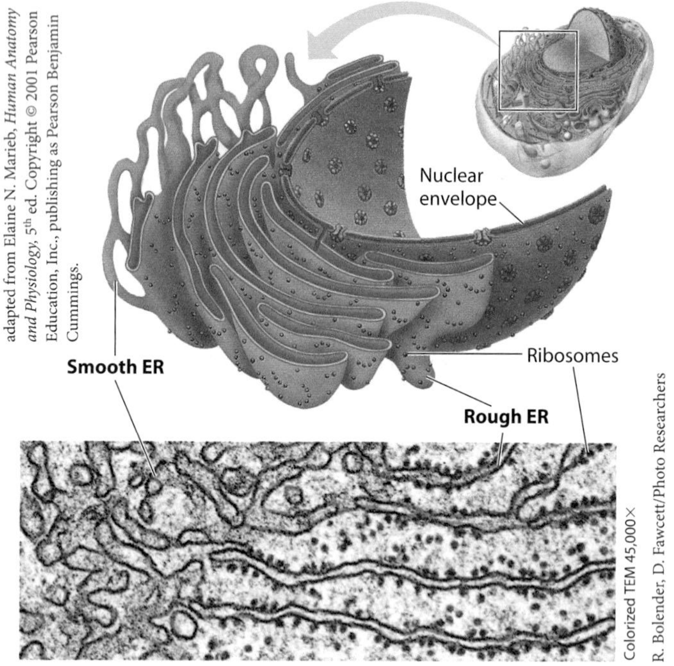

▲ **Figure 8A** Smooth and rough endoplasmic reticulum

Smooth ER The smooth ER of various cell types functions in a variety of metabolic processes. Enzymes of the smooth ER are important in the synthesis of lipids, including oils, phospholipids, and steroids. In vertebrates, for example, cells of the ovaries and testes synthesize the steroid sex hormones. These cells are rich in smooth ER, a structural feature that fits their function by providing ample machinery for steroid synthesis.

Our liver cells also have large amounts of smooth ER, with other important functions. Certain enzymes in the smooth ER of liver cells help process drugs, alcohol, and other potentially harmful substances. The sedative phenobarbital and other barbiturates are examples of drugs detoxified by these enzymes. As liver cells are exposed to such chemicals, the amount of smooth ER and its detoxifying enzymes increases, thereby increasing the rate of detoxification and thus the body's tolerance to the drugs. The result is a need for higher and higher doses of a drug to achieve a particular effect, such as sedation. Also, because detoxifying enzymes often cannot distinguish among related chemicals, the growth of smooth ER in response to one drug can increase tolerance to other drugs. Barbiturate abuse, for example, can decrease the effectiveness of certain antibiotics and other useful drugs.

Smooth ER has yet another function, the storage of calcium ions. In muscle cells, for example, a specialized smooth ER membrane pumps calcium ions into the interior of the ER. When a nerve signal stimulates a muscle cell, calcium ions rush from the smooth ER into the cytoplasmic fluid and trigger contraction of the cell.

Rough ER One of the functions of rough ER is to make more membrane. Phospholipids made by enzymes of the rough ER are inserted into the ER membrane. Thus, the ER membrane grows, and portions of it are transferred to other components of the endomembrane system as vesicles.

The bound ribosomes attached to rough ER produce proteins that will be inserted into the growing ER membrane, transported to other organelles, or secreted by the cell. An example of a secretory protein is insulin, a hormone secreted by specialized cells in the pancreas. Type 1 diabetes results when these cells are destroyed and a lack of insulin disrupts glucose metabolism in the body.

Figure 8B follows the synthesis, modification, and packaging of a secretory protein. ❶ As the polypeptide is synthesized by a bound ribosome following the instructions of an mRNA, it is threaded into the cavity of the rough ER. As it enters, the new protein folds into its three-dimensional shape. ❷ Short chains of sugars are often linked to the polypeptide, making the molecule a **glycoprotein** (*glyco* means "sugar"). ❸ When the molecule is ready for export from the ER, it is packaged in a **transport vesicle**, a vesicle that moves from one part of the cell to another. ❹ This vesicle buds off from the ER membrane.

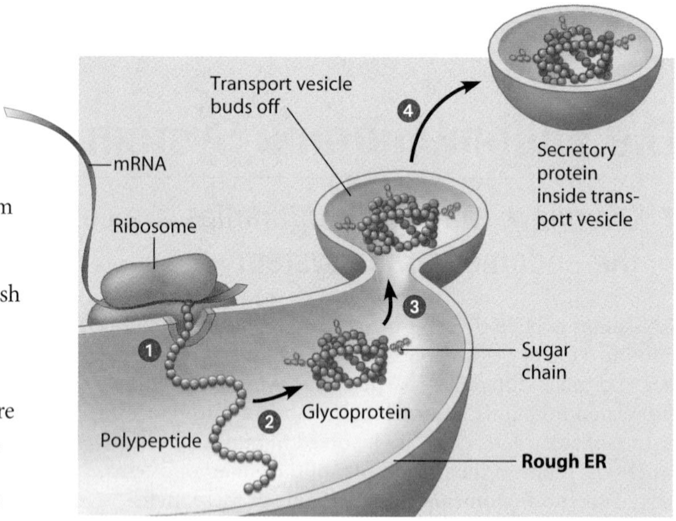

▲ **Figure 8B** Synthesis and packaging of a secretory protein by the rough ER

The vesicle now carries the protein to the Golgi apparatus (described in the next module) for further processing. From there, a transport vesicle containing the finished molecule makes its way to the plasma membrane and releases its contents from the cell.

? Explain why we say that the endoplasmic reticulum is a biosynthetic factory.

The ER produces a huge variety of molecules, including phospholipids for cell membranes, steroid hormones, and proteins (synthesized by bound ribosomes) for membranes, other organelles, and secretion by the cell.

9 The Golgi apparatus finishes, sorts, and ships cell products

After leaving the ER, many transport vesicles travel to the **Golgi apparatus**. Using a light microscope and a staining technique he developed, Italian scientist Camillo Golgi discovered this membranous organelle in 1898. The electron microscope confirmed his discovery more than 50 years later, revealing a stack of flattened sacs, looking much like a pile of pita bread. A cell may contain many, even hundreds, of these stacks. The number of Golgi stacks correlates with how active the cell is in secreting proteins—a multistep process that, as we have just seen, is initiated in the rough ER.

The Golgi apparatus serves as a molecular warehouse and finishing factory for products manufactured by the ER. You can follow this process in **Figure 9**. Note that the flattened Golgi sacs are not connected, as are ER sacs. ❶ One side of a Golgi stack serves as a receiving dock for transport vesicles produced by the ER. ❷ A vesicle fuses with a Golgi sac, adding its membrane and contents to the receiving side. ❸ Products of the ER are modified during their transit through the Golgi. ❹ The other side of the Golgi, the shipping side, gives rise to vesicles, which bud off and travel to other sites.

How might ER products be processed during their transit through the Golgi? Various Golgi enzymes modify the carbohydrate portions of the glycoproteins made in the ER, removing some sugars and substituting others. Molecular identification tags, such as phosphate groups, may be added that help the Golgi sort molecules into different batches for different destinations.

Until recently, the Golgi was viewed as a static structure, with products in various stages of processing moved from sac to sac by transport vesicles. Recent research has given rise to a new *maturation model* in which entire sacs "mature" as they move from the receiving to the shipping side, carrying and modifying their cargo as they go. The shipping side of the Golgi stack serves as a depot from which finished secretory products, packaged in transport vesicles, move to the plasma membrane for export from the cell. Alternatively, finished products may become part of the plasma membrane itself or part of another organelle, such as a lysosome, which we discuss next.

? What is the relationship of the Golgi apparatus to the ER in a protein-secreting cell?

The Golgi receives transport vesicles that bud from the ER and that contain proteins synthesized by ribosomes attached to the ER. The Golgi finishes processing the proteins and then dispatches transport vesicles that secrete the proteins to the outside of the cell.

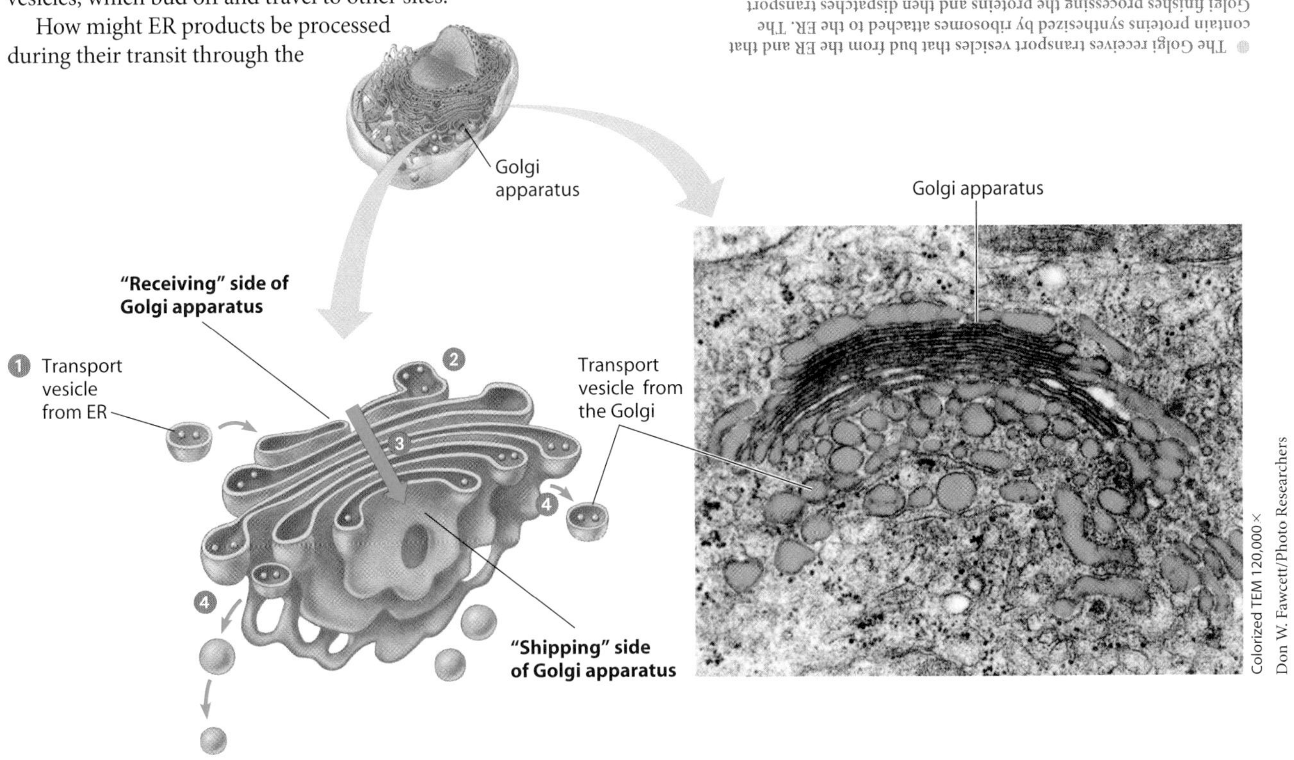

▲ **Figure 9** The Golgi apparatus

10 Lysosomes are digestive compartments within a cell

A **lysosome** is a membranous sac of digestive enzymes. The name *lysosome* is derived from two Greek words meaning "breakdown body." The enzymes and membranes of lysosomes are made by rough ER and processed in the Golgi apparatus. Illustrating a main theme of eukaryotic cell structure—compartmentalization—a lysosome provides an acidic environment for its enzymes, while safely isolating them from the rest of the cell.

Lysosomes have several types of digestive functions. Many protists engulf food particles into membranous sacs called food vacuoles. As **Figure 10A** shows, lysosomes fuse with food vacuoles and digest the food. The nutrients are then released into the cell fluid. Our white blood cells engulf and destroy bacteria using lysosomal enzymes. Lysosomes also serve as recycling centers for animal cells. Damaged organelles or small amounts of cell fluid become surrounded by a membrane. A lysosome fuses with such a vesicle (**Figure 10B**) and dismantles its contents, making organic molecules available for reuse. With the help of lysosomes, a cell continually renews itself.

The cells of people with inherited lysosomal storage diseases lack one or more lysosomal enzymes. The lysosomes become engorged with undigested material, eventually interfering with cellular function. In Tay-Sachs disease, for example, a lipid-digesting enzyme is missing, and brain cells become impaired by an accumulation of lipids. Lysosomal storage diseases are often fatal in early childhood.

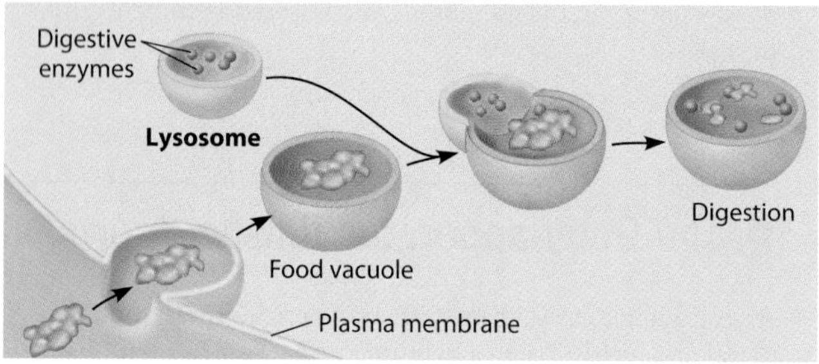

▲ **Figure 10A** Lysosome fusing with a food vacuole and digesting food

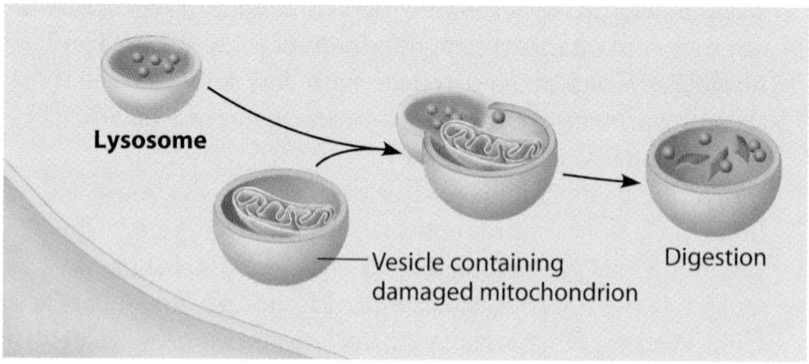

▲ **Figure 10B** Lysosome fusing with a vesicle containing a damaged organelle and digesting and recycling its contents

? **How is a lysosome like a recycling center?**

It breaks down damaged organelles and recycles their molecules.

11 Vacuoles function in the general maintenance of the cell

Vacuoles are large vesicles that have a variety of functions. In Figure 10A, you saw how a food vacuole forms as a cell ingests food. **Figure 11A** shows two contractile vacuoles in the protist *Paramecium*, looking somewhat like wheel hubs with radiating spokes. The "spokes" collect water from the cell, and the hub expels it to the outside. Freshwater protists constantly take in water from their environment. Without a way to get rid of the excess water, the cell would swell and burst.

In plants, some vacuoles have a digestive function similar to that of lysosomes in animal cells. Vacuoles in flower petals contain pigments that attract pollinating insects. Vacuoles may also contain poisons or unpalatable compounds that protect the plant against herbivores; examples include nicotine, caffeine, and various chemicals we use as pharmaceutical drugs. **Figure 11B** shows a plant cell's large **central vacuole**, which helps the cell grow in size by absorbing water and enlarging. It also stockpiles vital chemicals and acts as a trash can, safely storing toxic waste products.

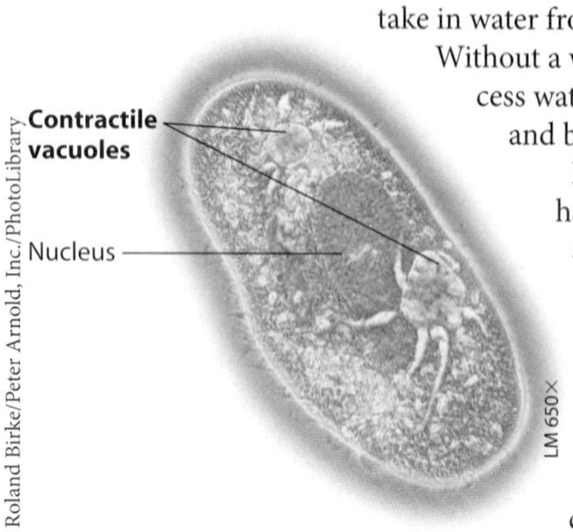

Roland Birke/Peter Arnold, Inc./PhotoLibrary

▲ **Figure 11A** Contractile vacuoles in *Paramecium*, a single-celled organism

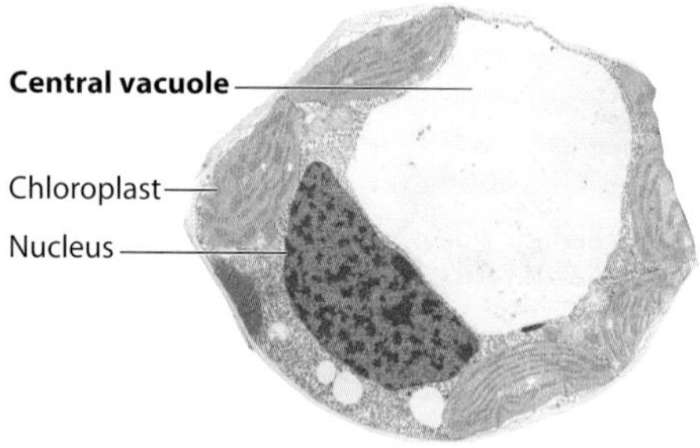

Colorized TEM 9,840×
Biophoto Associates/Photo Researchers

▲ **Figure 11B** Central vacuole in a plant cell

? **Is a food vacuole part of the endomembrane system?**

Yes; it forms by pinching in from the plasma membrane, which is part of the endomembrane system.

12 A review of the structures involved in manufacturing and breakdown

Figure 12 summarizes the relationships within the endomembrane system. You can see the direct *structural* connections between the nuclear envelope, rough ER, and smooth ER. The red arrows show the *functional* connections, as membranes and proteins produced by the ER travel in transport vesicles to the Golgi and on to other destinations. Some vesicles develop into lysosomes or vacuoles. Others transport products to the outside of the cell. When these vesicles fuse with the plasma membrane, their contents are secreted from the cell and their membrane is added to the plasma membrane.

Peroxisomes (see Figures 4A and B) are metabolic compartments that do not originate from the endomembrane system. In fact, how they are related to other organelles is still unknown. Some peroxisomes break down fatty acids to be used as cellular fuel. In your liver, peroxisomes detoxify harmful compounds, including alcohol. In these processes, enzymes transfer hydrogen from various compounds to oxygen, producing hydrogen peroxide (H_2O_2). Other enzymes in the peroxisome quickly convert this toxic product to water—another example of the importance of a cell's compartmental structure.

A cell requires a continuous supply of energy to perform the work of life. Next we consider two organelles that act as cellular power stations—mitochondria and chloroplasts.

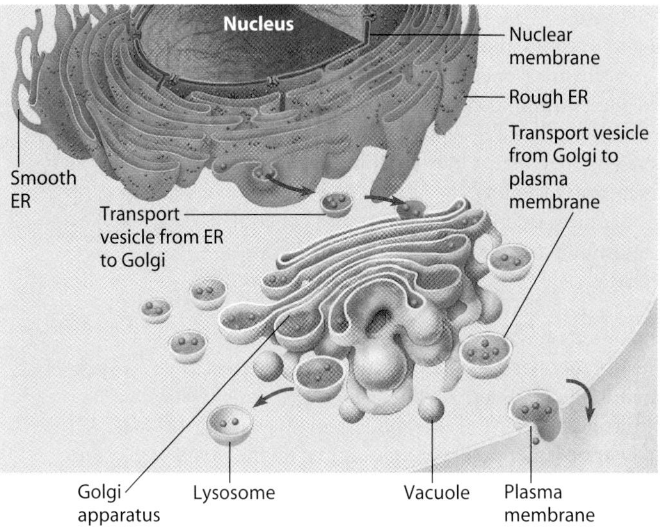

▲ Figure 12 Connections among the organelles of the endomembrane system

? How do transport vesicles help tie together the endomembrane system?

Transport vesicles move membranes and substances they enclose between components of the endomembrane system.

Energy-Converting Organelles

13 Mitochondria harvest chemical energy from food

Mitochondria (singular, *mitochondrion*) are organelles that carry out cellular respiration in nearly all eukaryotic cells, converting the chemical energy of foods such as sugars to the chemical energy of the molecule called ATP (adenosine triphosphate). ATP is the main energy source for cellular work.

As you have come to expect, a mitochondrion's structure suits its function. It is enclosed by two membranes, each a phospholipid bilayer with a unique collection of embedded proteins (Figure 13). The mitochondrion has two internal compartments. The first is the intermembrane space, the narrow region between the inner and outer membranes. The inner membrane encloses the second compartment, the **mitochondrial matrix**, which contains mitochondrial DNA and ribosomes, as well as many enzymes that catalyze some of the reactions of cellular respiration. The inner membrane is highly folded and contains many embedded protein molecules that function in ATP synthesis. The folds, called **cristae**, increase the membrane's surface area, enhancing the mitochondrion's ability to produce ATP.

? What is cellular respiration?

A process that converts the chemical energy of sugars and other food molecules to the chemical energy of ATP

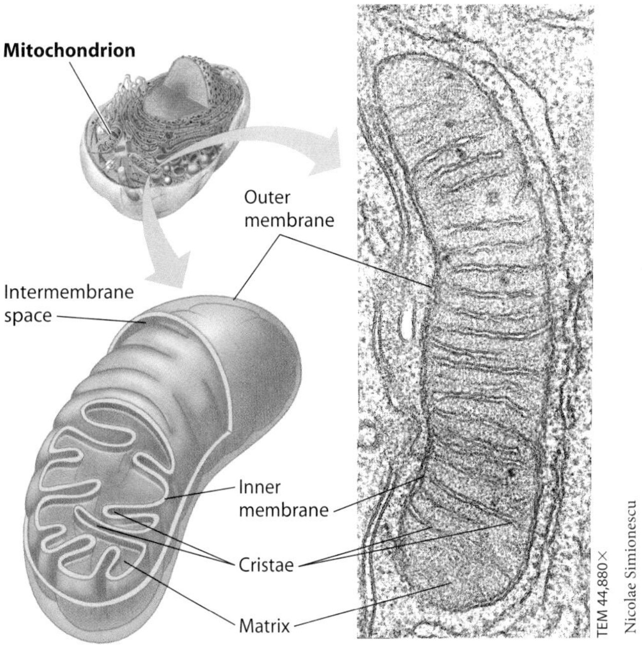

▲ Figure 13 The mitochondrion

14 Chloroplasts convert solar energy to chemical energy

Most of the living world runs on the energy provided by photosynthesis, the conversion of light energy from the sun to the chemical energy of sugar molecules. **Chloroplasts** are the photosynthesizing organelles of all photosynthetic eukaryotes. The chloroplast's solar power system is much more efficient than anything yet produced by human ingenuity.

Befitting an organelle that carries out complex, multistep processes, internal membranes partition the chloroplast into compartments **(Figure 14)**. The chloroplast is enclosed by an inner and outer membrane separated by a thin intermembrane space. The compartment inside the inner membrane holds a thick fluid called **stroma**, which contains chloroplast DNA and ribosomes as well as many enzymes. A network of interconnected sacs called **thylakoids** is inside the chloroplast. The compartment inside these sacs is called the thylakoid space. In some regions, thylakoids are stacked like poker chips; each stack is called a **granum** (plural, *grana*). The grana are the chloroplast's solar power packs—the sites where the green chlorophyll molecules embedded in thylakoid membranes trap solar energy. In the next module, we explore the surprising origin of mitochondria and chloroplasts.

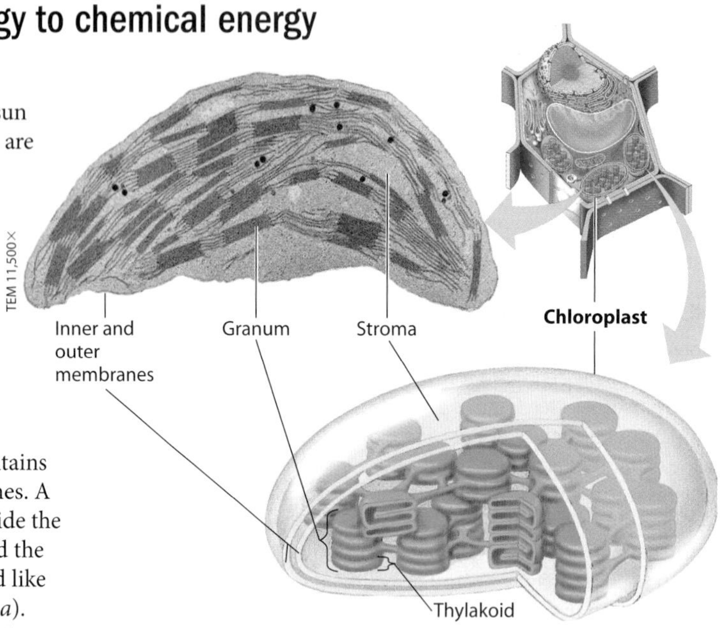

E.H. Newcomb & W.P. Wergin/Biological Photo Service

▲ **Figure 14** The chloroplast

? Which membrane in a chloroplast appears to be the most extensive? Why might this be so?

The thylakoid membranes are most extensive, providing a large area of membrane that contains chlorophyll for photosynthesis.

EVOLUTION CONNECTION

15 Mitochondria and chloroplasts evolved by endosymbiosis

Does it seem odd to you that mitochondria and chloroplasts contain DNA and ribosomes? Indeed, they have a single circular DNA molecule, similar in structure to the chromosome of prokaryotes. And their ribosomes are more similar to prokaryotic ribosomes than to eukaryotic ribosomes. Another interesting observation is that mitochondria and chloroplasts reproduce by a splitting process that is similar to that of certain prokaryotes.

The widely accepted **endosymbiont theory** states that mitochondria and chloroplasts were formerly small prokaryotes that began living within larger cells. The term *endosymbiont* refers to a cell that lives within another cell, called the host cell. These small prokaryotes may have gained entry to the larger cell as undigested prey or internal parasites **(Figure 15)**.

By whatever means the relationship began, we can hypothesize how the symbiosis could have become beneficial. In a world that was becoming increasingly aerobic from the oxygen-generating photosynthesis of prokaryotes, a host would have benefited from an endosymbiont that was able to use oxygen to release large amounts of energy from organic molecules by cellular respiration. And a host cell could derive nourishment from a photosynthetic endosymbiont. Over time, the host and endosymbionts would have become increasingly interdependent, eventually becoming a single organism.

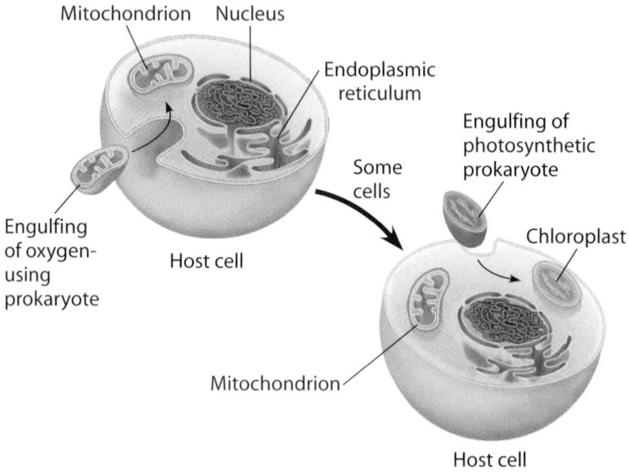

▲ **Figure 15** Endosymbiotic origin of mitochondria and chloroplasts

? All eukaryotes have mitochondria, but not all eukaryotes have chloroplasts. Can you propose an evolutionary explanation for this observation?

The first endosymbiotic event would have given rise to eukaryotic cells containing mitochondria. At least one of these cells may have then taken up a photosynthetic prokaryote, giving rise to eukaryotic cells that contain chloroplasts in addition to mitochondria.

The Cytoskeleton and Cell Surfaces

16 The cell's internal skeleton helps organize its structure and activities

Biologists once thought that organelles floated freely in the cell. But improvements in both light microscopy and electron microscopy have allowed scientists to uncover a network of protein fibers, collectively called the **cytoskeleton**, extending throughout the cytoplasm of a cell. These fibers function like a skeleton in providing for both structural support and cell motility. Cell motility includes both the internal movement of cell parts and the locomotion of a cell. These movements generally require the interaction of the cytoskeleton with proteins called motor proteins.

Three main kinds of fibers make up the cytoskeleton: microfilaments, the thinnest fiber; microtubules, the thickest; and intermediate filaments, in between in thickness. **Figure 16** shows three micrographs of cells of the same type, each stained with a different fluorescent dye that selectively highlights one of these types of fibers.

Microfilaments, also called actin filaments, are solid rods composed mainly of globular proteins called actin, arranged in a twisted double chain (bottom left of Figure 16). Microfilaments form a three-dimensional network just inside the plasma membrane that helps support the cell's shape. This is especially important for animal cells, which lack cell walls.

Microfilaments are also involved in cell movements. Actin filaments and thicker filaments made of the motor protein myosin interact to cause contraction of muscle cells. Localized contractions brought about by actin and myosin are involved in the amoeboid (crawling) movement of the protist *Amoeba* and some of our white blood cells.

Intermediate filaments may be made of various fibrous proteins that supercoil into thicker cables. Intermediate filaments serve mainly to reinforce cell shape and to anchor certain organelles. The nucleus is held in place by a cage of intermediate filaments. While microfilaments may be disassembled and reassembled elsewhere, intermediate filaments are often more permanent fixtures in the cell. The outer layer of your skin consists of dead skin cells full of intermediate filaments made of keratin proteins.

Microtubules are straight, hollow tubes composed of globular proteins called tubulins. As indicated in the bottom right of Figure 16, microtubules elongate by the addition of tubulin proteins, which consist of two subunits. Microtubules are readily disassembled in a reverse manner, and the tubulin proteins can be reused elsewhere in the cell. In many animal cells, microtubules grow out from a "microtubule-organizing center" near the nucleus. Within this region is a pair of **centrioles** (see Figure 4A).

Microtubules shape and support the cell and also act as tracks along which organelles equipped with motor proteins move. For example, a lysosome might "walk" along a microtubule to reach a food vacuole. Microtubules also guide the movement of chromosomes when cells divide, and as we see next, they are the main components of cilia and flagella—the locomotive appendages of cells.

? Which component of the cytoskeleton is most important in (a) holding the nucleus in place within the cell; (b) guiding transport vesicles from the Golgi to the plasma membrane; (c) contracting muscle cells?

(a) Intermediate filaments; (b) microtubules; (c) microfilaments

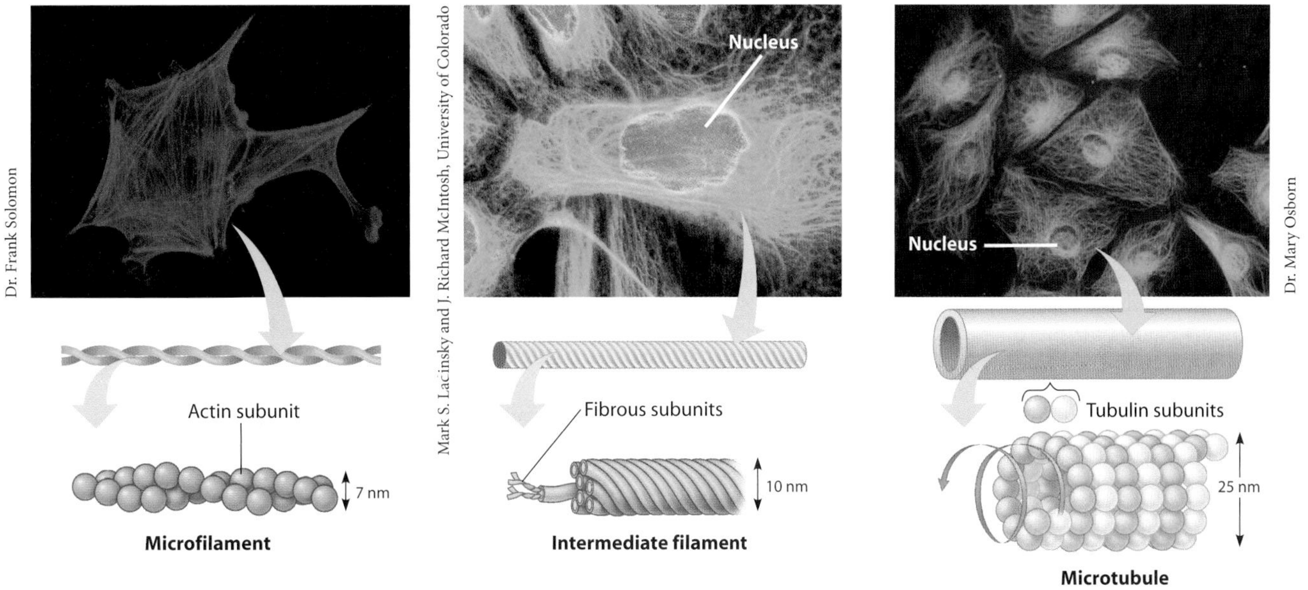

▲ **Figure 16** Three types of fibers of the cytoskeleton: microfilaments are stained red (left), intermediate filaments yellow-green (center), and microtubules green (right)

17 Cilia and flagella move when microtubules bend

The role of the cytoskeleton in movement is clearly seen in the motile appendages that protrude from certain cells. The short, numerous appendages that propel protists such as *Paramecium* (see Figure 1C) are called **cilia** (singular, *cilium*). Other protists may move using flagella, which are longer than cilia and usually limited to one or a few per cell.

Some cells of multicellular organisms also have cilia or flagella. For example, **Figure 17A** shows cilia on cells lining the human windpipe. In this case, the cilia sweep mucus containing trapped debris out of your lungs. (This cleaning function is impaired by cigarette smoke, which paralyzes the cilia.) Most animals and some plants have flagellated sperm. A flagellum, shown in **Figure 17B**, propels the cell by an undulating whip-like motion. In contrast, cilia work more like the coordinated oars of a rowing team.

Science Photo Library/Photo Researchers

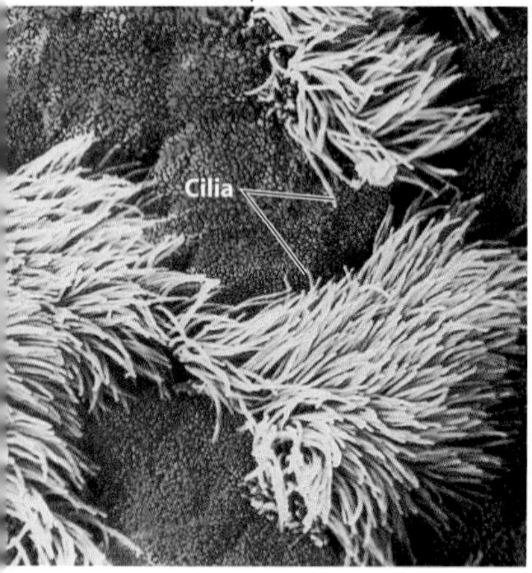

▲ **Figure 17A** Cilia on cells lining the respiratory tract

Though different in length and beating pattern, cilia and flagella have a common structure and mechanism of movement (**Figure 17C**). Both are composed of microtubules wrapped in an extension of the plasma membrane. In nearly all eukaryotic cilia and flagella, a ring of nine microtubule doublets surrounds a central pair of microtubules. This arrangement is called the 9 + 2 pattern. The microtubule assembly extends into an anchoring structure called a basal body (not shown in the figure), which consists of a ring of nine microtubule triplets. Basal bodies are very similar in structure to centrioles, which are found in the microtubule-organizing center of animal cells.

How does this microtubule assembly produce the bending movement of cilia and flagella? Bending involves large motor proteins called dyneins (red in the figure) that are attached along each outer microtubule doublet. A dynein protein has two "feet" that "walk" along an adjacent doublet, one foot maintaining contact while the other releases and reattaches one step farther along its neighboring microtubule. The outer doublets and two central microtubules are held together by flexible cross-linking proteins and radial spokes (purple in the diagram). If the doublets were not held in place, the walking action would make them slide past each other. Instead, the movements of the dynein feet cause the microtubules—and consequently the cilium or flagellum—to bend.

A cilium may also serve as a signal-receiving "antenna" for the cell. Cilia with this function are generally nonmotile (they lack the central pair of microtubules) and there is only one per cell. In fact, in vertebrate animals, it appears that almost all cells have what is called a *primary cilium*. Although the primary cilium was discovered over a century ago, its importance to embryonic development, sensory reception, and cell function is only now being recognized. Defective primary cilia have been linked to polycystic kidney disease and other human disorders.

? **Compare and contrast cilia and flagella.**

● Both cilia and flagella have the same 9 + 2 pattern of microtubules and mechanism for bending. Cilia are shorter, are more numerous, and beat in a coordinated oar-like pattern. The longer flagella, which are limited to one or a few per cell, undulate like a whip.

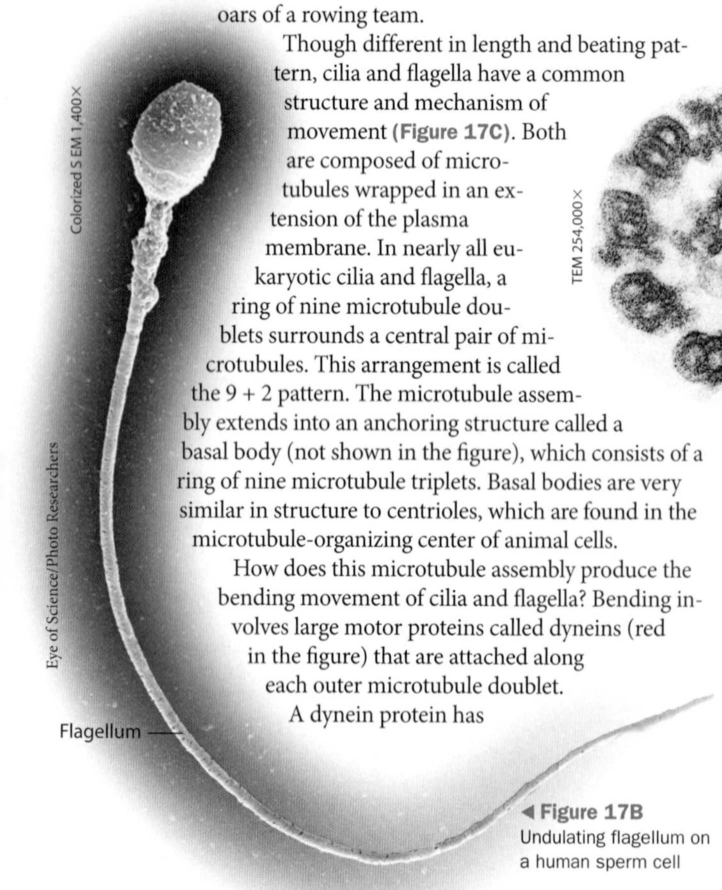

◄ **Figure 17B** Undulating flagellum on a human sperm cell

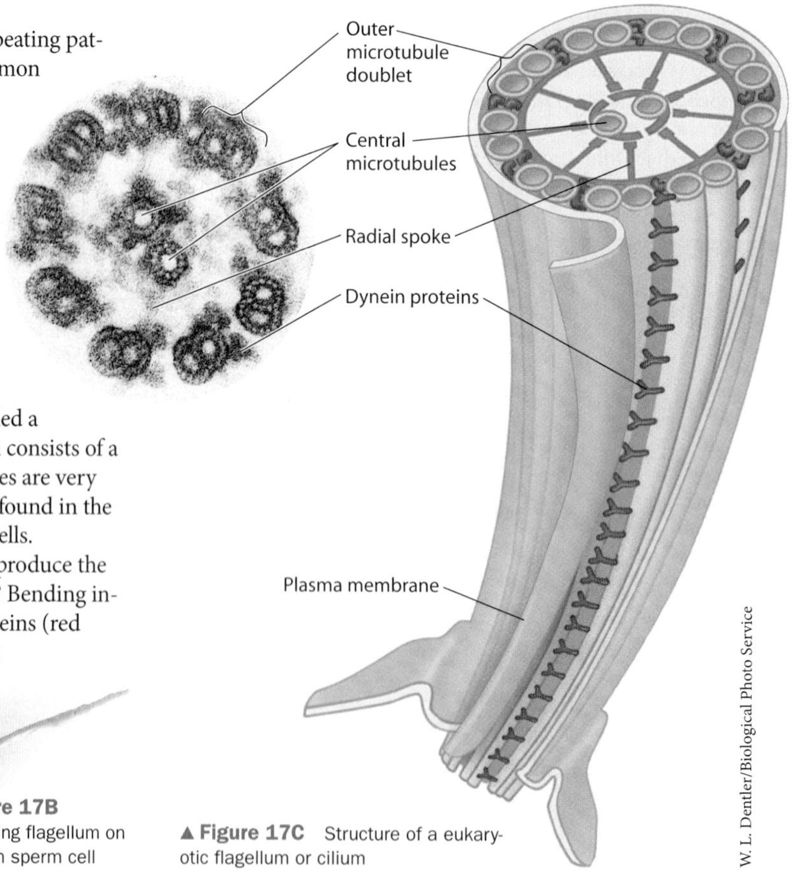

▲ **Figure 17C** Structure of a eukaryotic flagellum or cilium

CONNECTION 18 Problems with sperm motility may be environmental or genetic

Human sperm quality varies among men and in different geographic areas. In developed countries over the last 50 years, there has been an apparent decline in sperm quality—lower sperm counts, higher proportions of malformed sperm, and reduced motility. Various environmental factors are being studied as possible causes. One hypothesis links this trend to an increase in hormonally active chemicals in the environment.

One common group of chemicals that may be implicated is phthalates. These chemicals are used in cosmetics and are found in many types of plastics, including those used in medical tubing and, until banned by a U.S. congressional act that took effect in 2009, children's toys. Research has indicated that phthalates interfere with sex hormones and adversely affect sperm quality in rodents. Critics of these studies have claimed that the environmental exposures of people are beneath the levels causing effects in rodents. Several new studies, however, have indicated that normal levels of human exposure to these chemicals may result in impaired sperm quality.

Results of a five-year study of 463 men who had come to a hospital for infertility treatment found that the men with higher concentrations of breakdown products of phthalates in their urine had lower sperm counts and motility. But does a statistical correlation indicate cause and effect? Research continues on the potential reproductive health risks of hormone-disrupting chemicals in the environment.

Other problems with sperm motility are clearly genetic. Primary ciliary dyskinesia (PCD), also known as immotile cilia syndrome, is a fairly rare disease characterized by recurrent infections of the respiratory tract and immotile sperm. Compare the cross section of the flagellum of a sperm of a male with PCD in **Figure 18** with the TEM in Figure 17C. Do you notice the absence of the dynein proteins? How does that explain the seemingly unrelated symptoms of PCD? If microtubules cannot bend (see Module 17), then cilia cannot help cleanse the respiratory tract and sperm cannot swim.

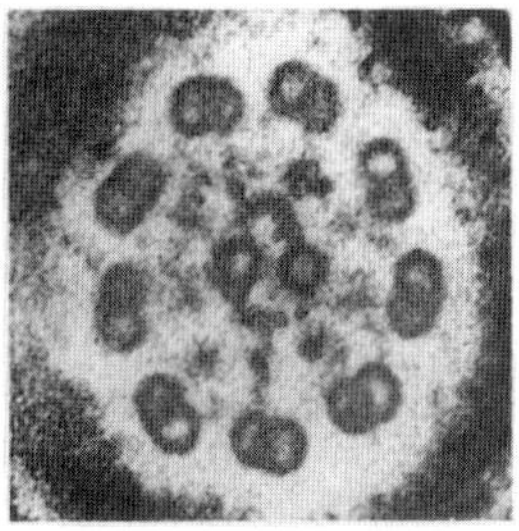
TEM 147,500×
Bjorn Afzelius

▲ **Figure 18** Cross section of immotile sperm flagellum

? Why does a lack of dynein proteins affect the action of both cilia and flagella?

Both cilia and flagella have the same arrangement of microtubule doublets with attached dynein motor proteins that cause them to bend.

19 The extracellular matrix of animal cells functions in support and regulation

The plasma membrane is usually regarded as the boundary of the living cell, but most cells synthesize and secrete materials that remain outside the plasma membrane. These extracellular structures are essential to many cell functions.

Animal cells produce an elaborate **extracellular matrix (ECM)** (**Figure 19**). This layer helps hold cells together in tissues and protects and supports the plasma membrane. The main components of the ECM are glycoproteins, proteins bonded with carbohydrates. The most abundant glycoprotein is collagen, which forms strong fibers outside the cell. In fact, collagen accounts for about 40% of the protein in your body. The collagen fibers are embedded in a network woven from other types of glycoproteins. Large complexes form when hundreds of small glycoproteins connect to a central long polysaccharide molecule (green in the figure). The ECM may attach to the cell through other glycoproteins that then bind to membrane proteins called integrins. **Integrins** span the membrane, attaching on the other side to proteins connected to microfilaments of the cytoskeleton.

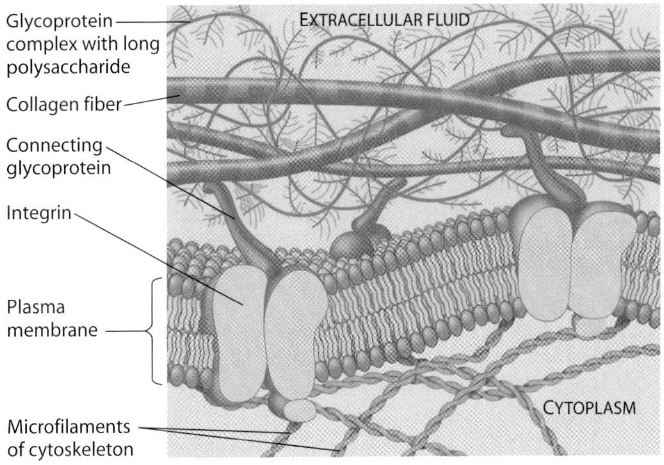

▲ **Figure 19** The extracellular matrix (ECM) of an animal cell

As their name implies, integrins have the function of integration: They transmit signals between the ECM and the cytoskeleton. Thus, the cytoskeleton can influence the organization of the ECM and vice versa. For example, research shows that the ECM can regulate a cell's behavior, directing the path along which embryonic cells move and even influencing the activity of genes through the signals it relays. Genetic changes in cancer cells may result in a change in the composition of the ECM they produce, causing such cells to lose their connections and spread to other tissues.

? Referring to Figure 19, describe the structures that provide support to the plasma membrane.

The membrane is attached through membrane proteins (integrins) to the microfilaments in the cytoskeleton and the glycoproteins and collagen fibers of the ECM.

20 Three types of cell junctions are found in animal tissues

Neighboring cells in animal tissues often adhere, interact, and communicate through specialized junctions between them. **Figure 20** uses cells lining the digestive tract to illustrate three types of cell junctions. (The projections at the top of the cells increase the surface area for absorption of nutrients.)

At *tight junctions*, the plasma membranes of neighboring cells are tightly pressed against each other and knit together by proteins. Forming continuous seals around cells, tight junctions prevent leakage of fluid across a layer of cells. The green arrows in the figure show how tight junctions prevent the contents of the digestive tract from leaking into surrounding tissues.

Anchoring junctions function like rivets, fastening cells together into strong sheets. Intermediate filaments made of sturdy keratin proteins anchor these junctions in the cytoplasm. Anchoring junctions are common in tissues subject to stretching or mechanical stress, such as skin and heart muscle.

Gap junctions, also called communicating junctions, are channels that allow small molecules to flow through protein-lined pores between cells. The flow of ions through gap junctions in the cells of heart muscle coordinates their contraction. Gap junctions are common in embryos, where communication between cells is essential for development.

? A muscle tear injury would probably involve the rupture of which type of cell junction?

Anchoring junction

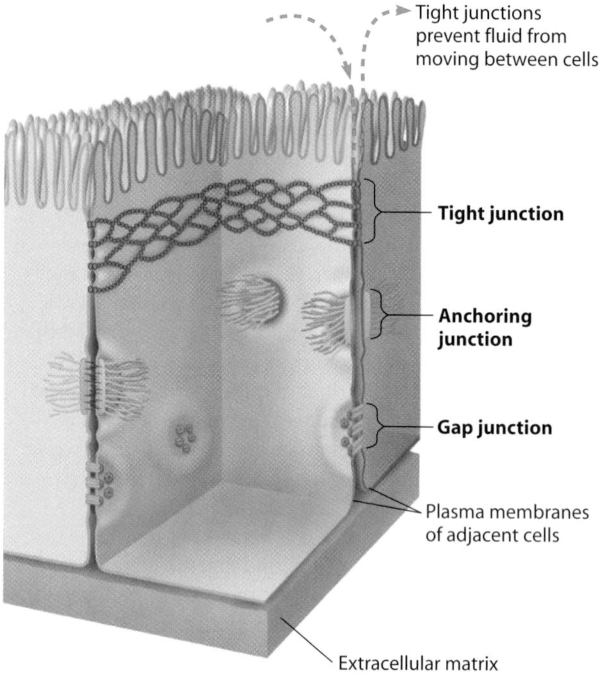

▲ **Figure 20** Three types of cell junctions in animal tissues

21 Cell walls enclose and support plant cells

The **cell wall** is one feature that distinguishes plant cells from animal cells. This rigid extracellular structure not only protects the cells but provides the skeletal support that keeps plants upright on land. Plant cell walls consist of fibers of cellulose embedded in a matrix of other polysaccharides and proteins. This fibers-in-a-matrix construction resembles that of fiberglass, a manufactured product also noted for its strength.

Figure 21 shows the layered structure of plant cell walls. Cells initially lay down a relatively thin and flexible primary wall, which allows the growing cell to continue to enlarge. Some cells then add a secondary wall deposited in laminated layers. Wood consists mainly of secondary walls, which are strengthened with rigid molecules called lignin. Between adjacent cells is a layer of sticky polysaccharides called pectins (shown here in dark brown), which glue the cells together. (Pectin is used to thicken jams and jellies.)

Despite their thickness, plant cell walls do not totally isolate the cells from each other. To function in a coordinated way as part of a tissue, the cells must have cell junctions, structures that connect them to one another. Figure 21 shows the numerous channels between adjacent plant cells, called **plasmodesmata** (singular, *plasmodesma*). Notice that the plasma membrane and the cytoplasm of the cells extend through the plasmodesmata, so that water and other small molecules can readily pass from cell to cell. Through plasmodesmata, the cells of a plant tissue share water, nourishment, and chemical messages.

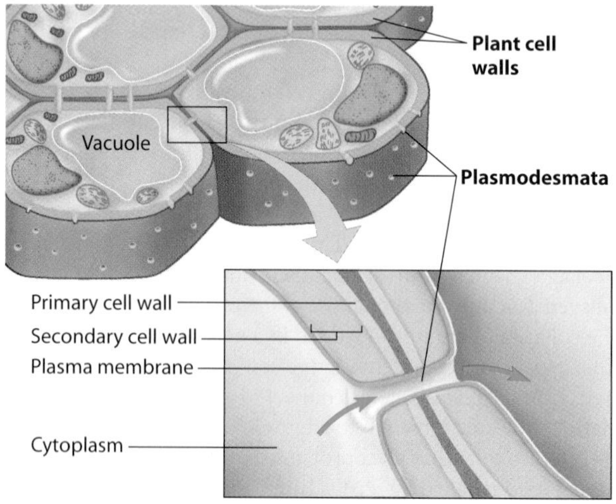

▲ **Figure 21** Plant cell walls and plasmodesmata

? Which animal cell junction is analogous to a plasmodesma?

A gap junction

22 Review: Eukaryotic cell structures can be grouped on the basis of four main functions

Congratulations, you have completed the grand tour of the cell. In the process, you have been introduced to many important cell structures. To provide a framework for this information and reinforce the theme that structure is correlated with function, we have grouped the eukaryotic cell organelles into four categories by general function, as reviewed in Table 22.

The first category is genetic control. Here we include the nucleus that houses a cell's genetic instructions and the ribosomes that produce the proteins coded for in those instructions. The second category includes organelles of the endomembrane system that are involved in the manufacture, distribution, and breakdown of materials. (Vacuoles have other functions as well.) The third category includes the two energy-processing organelles. The fourth category is structural support, movement, and intercellular communication. These functions are related because there must be rigid support against which force can be applied for movement to occur. In addition, a supporting structure that forms the cell's outer boundary is necessarily involved in communication with neighboring cells.

TABLE 22 | EUKARYOTIC CELL STRUCTURES AND FUNCTIONS

1. Genetic Control	
Nucleus	DNA replication, RNA synthesis; assembly of ribosomal subunits (in nucleoli)
Ribosomes	Polypeptide (protein) synthesis
2. Manufacturing, Distribution, and Breakdown	
Rough ER	Synthesis of membrane lipids and proteins, secretory proteins, and hydrolytic enzymes; formation of transport vesicles
Smooth ER	Lipid synthesis; detoxification in liver cells; calcium ion storage
Golgi apparatus	Modification and sorting of macromolecules; formation of lysosomes and transport vesicles
Lysosomes (in animal cells and some protists)	Digestion of ingested food, bacteria, and a cell's damaged organelles and macromolecules for recycling
Vacuoles	Digestion (food vacuole); storage of chemicals and cell enlargement (central vacuole); water balance (contractile vacuole)
Peroxisomes (not part of endomembrane system)	Diverse metabolic processes, with breakdown of toxic hydrogen peroxide by-product
3. Energy Processing	
Mitochondria	Conversion of chemical energy in food to chemical energy of ATP
Chloroplasts (in plants and some protists)	Conversion of light energy to chemical energy of sugars
4. Structural Support, Movement, and Communication Between Cells	
Cytoskeleton (microfilaments, intermediate filaments, and microtubules)	Maintenance of cell shape; anchorage for organelles; movement of organelles within cells; cell movement (crawling, muscle contraction, bending of cilia and flagella)
Extracellular matrix (in animals)	Support; regulation of cellular activities
Cell junctions	Communication between cells; binding of cells in tissues
Cell walls (in plants, fungi, and some protists)	Support and protection; binding of cells in tissues

Within most of these categories, a structural similarity underlies the general function of each component. Manufacturing depends heavily on a network of structurally and functionally connected membranes. All the organelles involved in the breakdown or recycling of materials are membranous sacs, inside of which enzymatic digestion can safely occur. In the energy-processing category, expanses of metabolically active membranes and intermembrane compartments within the organelles enable chloroplasts and mitochondria to perform the complex energy conversions that power the cell. Even in the diverse fourth category, there is a common structural theme in the various protein fibers of these cellular systems.

We can summarize further by emphasizing that these cellular structures form an integrated team—with the property of life emerging at the level of the cell from the coordinated functions of the team members. And finally we note that the overall structure of a cell is closely related to its specific function. Thus, cells that produce proteins for export contain a large quantity of ribosomes and rough ER, while muscle cells are packed with microfilaments, myosin motor proteins, and mitochondria.

All organisms share the fundamental feature of consist-ing of cells, each enclosed by a membrane that maintains internal conditions different from the surroundings and each carrying out metabolism, which involves the interconversion of different forms of energy and of chemical compounds.

? How do mitochondria, smooth ER, and the cytoskeleton all contribute to the contraction of a muscle cell?

Mitochondria supply energy in the form of ATP. The smooth ER helps regulate contraction by the uptake and release of calcium ions. Microfilaments function in the actual contractile apparatus.

REVIEW

For Practice Quizzes, BioFlix, MP3 Tutors, and Activities, go to www.masteringbiology.com.

Reviewing the Concepts

Introduction to the Cell (1–4)

1 Microscopes reveal the world of the cell. The light microscope can magnify up to 1,000 times. The greater magnification and resolution of the scanning and transmission electron microscopes reveal the ultrastructure of cells.

2 The small size of cells relates to the need to exchange materials across the plasma membrane. The microscopic size of most cells provides a large surface-to-volume ratio. The plasma membrane is a phospholipid bilayer with embedded proteins.

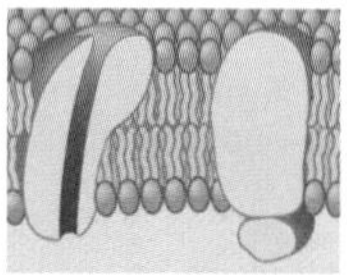

3 Prokaryotic cells are structurally simpler than eukaryotic cells. All cells have a plasma membrane, DNA, ribosomes, and cytoplasm. Prokaryotic cells are smaller than eukaryotic cells and lack a membrane-enclosed nucleus.

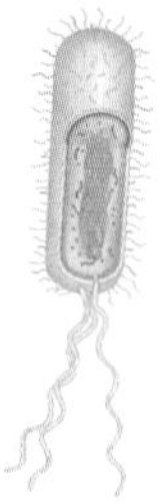

4 Eukaryotic cells are partitioned into functional compartments. Membranes form the boundaries of organelles, compartmentalizing a cell's activities.

The Nucleus and Ribosomes (5–6)

5 The nucleus is the cell's genetic control center. The nucleus houses the cell's DNA and directs protein synthesis by making messenger RNA. The nucleolus makes the subunits of ribosomes.

6 Ribosomes make proteins for use in the cell and for export. Composed of ribosomal RNA and proteins, ribosomes synthesize proteins according to directions from DNA.

The Endomembrane System (7–12)

7 Overview: Many cell organelles are connected through the endomembrane system.

8 The endoplasmic reticulum is a biosynthetic factory. The ER is a membranous network of tubes and sacs. Smooth ER synthesizes lipids and processes toxins. Rough ER manufactures membranes, and ribosomes on its surface produce membrane and secretory proteins.

9 The Golgi apparatus finishes, sorts, and ships cell products. The Golgi apparatus consists of stacks of sacs that modify ER products and then ship them to other organelles or to the cell surface.

10 Lysosomes are digestive compartments within a cell. Lysosomes house enzymes that function in digestion and recycling within the cell.

11 Vacuoles function in the general maintenance of the cell. Some protists have contractile vacuoles. Plant cells contain a large central vacuole that stores molecules and wastes and facilitates growth.

12 A review of the structures involved in manufacturing and breakdown. The organelles of the endomembrane system are interconnected structurally and functionally.

Energy-Converting Organelles (13–15)

13 Mitochondria harvest chemical energy from food.

14 Chloroplasts convert solar energy to chemical energy.

15 Mitochondria and chloroplasts evolved by endosymbiosis. These organelles originated from prokaryotic cells that became residents in a host cell.

The Cytoskeleton and Cell Surfaces (16–22)

16 The cell's internal skeleton helps organize its structure and activities. The cytoskeleton is a network of protein fibers. Microfilaments of actin enable cells to change shape and move. Intermediate filaments reinforce the cell and anchor certain organelles. Microtubules give the cell rigidity and act as tracks for organelle movement.

17 Cilia and flagella move when microtubules bend. Eukaryotic cilia and flagella are locomotor appendages made of microtubules in a 9 + 2 arrangement.

18 Problems with sperm motility may be environmental or genetic. Environmental chemicals or genetic disorders may interfere with movement of sperm and cilia.

19 The extracellular matrix of animal cells functions in support and regulation. The ECM consists mainly of glycoproteins, which bind tissue cells together, support the membrane, and communicate with the cytoskeleton.

20 Three types of cell junctions are found in animal tissues. Tight junctions bind cells to form leakproof sheets. Anchoring junctions rivet cells into strong tissues. Gap junctions allow substances to flow from cell to cell.

21 Cell walls enclose and support plant cells. Plant cell walls are made largely of cellulose. Plasmodesmata are connecting channels between cells.

22 Review: Eukaryotic cell structures can be grouped on the basis of four main functions. These functions are (1) genetic control; (2) manufacturing, distribution, and breakdown; (3) energy processing; and (4) structural support, movement, and communication between cells.

Connecting the Concepts

1. Label the structures in this diagram of an animal cell. Review the functions of each of these organelles.

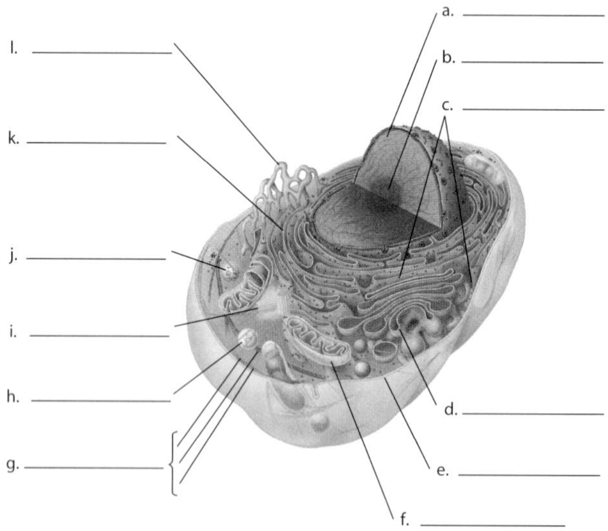

2. List some structures found in animal cells but not in plant cells.
3. List some structures found in plant cells but not in animal cells.

Testing Your Knowledge

Multiple Choice

4. The ultrastructure of a chloroplast is best studied using a
   a. light microscope.
   b. telescope.
   c. scanning electron microscope.
   d. transmission electron microscope.
   e. light microscope and fluorescent dyes.
5. The cells of an ant and a horse are, on average, the same small size; a horse just has more of them. What is the main advantage of small cell size?
   a. Small cells are less likely to burst than large cells.
   b. A small cell has a larger plasma membrane surface area than does a large cell.
   c. Small cells can better take up sufficient nutrients and oxygen to service their cell volume.
   d. It takes less energy to make an organism out of small cells.
   e. Small cells require less oxygen than do large cells.
6. Which of the following clues would tell you whether a cell is prokaryotic or eukaryotic?
   a. the presence or absence of a rigid cell wall
   b. whether or not the cell is partitioned by internal membranes
   c. the presence or absence of ribosomes
   d. whether or not the cell carries out cellular metabolism
   e. whether or not the cell contains DNA
7. Which statement correctly describes bound ribosomes?
   a. Bound ribosomes are enclosed in a membrane.
   b. Bound and free ribosomes are structurally different.
   c. Bound ribosomes are most commonly found on the surface of the plasma membrane.
   d. Bound ribosomes generally synthesize membrane proteins and secretory proteins.
   e. Bound ribosomes produce the subunits of microtubules, microfilaments, and intermediate filaments.

Choose from the following cells for questions 8–12:
   a. muscle cell in thigh of long-distance runner
   b. pancreatic cell that secretes digestive enzymes
   c. ovarian cell that produces the steroid hormone estrogen
   d. cell in tissue layer lining digestive tract
   e. white blood cell that engulfs bacteria

8. In which cell would you find the most lysosomes?
9. In which cell would you find the most mitochondria?
10. In which cell would you find the most smooth ER?
11. In which cell would you find the most rough ER?
12. In which cell would you find the most tight junctions?
13. A type of cell called a lymphocyte makes proteins that are exported from the cell. Which of the following traces the path of a protein from the site where its polypeptides are made to its export?
   a. chloroplast . . . Golgi . . . lysosomes . . . plasma membrane
   b. Golgi . . . rough ER . . . smooth ER . . . transport vesicle
   c. rough ER . . . Golgi . . . transport vesicle . . . plasma membrane
   d. smooth ER . . . Golgi . . . lysosome . . . plasma membrane
   e. nucleus . . . rough ER . . . Golgi . . . plasma membrane
14. Which of the following structures is *not* directly involved in cell support or movement?
   a. microfilament
   b. flagellum
   c. microtubule
   d. gap junction
   e. cell wall

Describing, Comparing, and Explaining

15. What four cellular components are shared by prokaryotic and eukaryotic cells?
16. Briefly describe the three kinds of junctions that can connect animal cells, and compare their functions.
17. What general function do the chloroplast and mitochondrion have in common? How are their functions different?
18. In what ways do the internal membranes of a eukaryotic cell contribute to the functioning of the cell?
19. Describe two different ways in which the motion of cilia can function in organisms.
20. Explain how a protein inside the ER can be exported from the cell without ever crossing a membrane.
21. Is this statement true or false? "Animal cells have mitochondria; plant cells have chloroplasts." Explain your answer.
22. Describe the structure of the plasma membrane of an animal cell. What would be found directly inside and outside the membrane?

Applying the Concepts

23. Imagine a spherical cell with a radius of 10 μm. What is the cell's surface area in μm^2? Its volume, in μm^3? What is the ratio of surface area to volume for this cell? Now do the same calculations for a second cell, this one with a radius of 20 μm. Compare the surface-to-volume ratios of the two cells. How is this comparison significant to the functioning of cells? (*Note*: For a sphere of radius r, surface area = $4\pi r^2$ and volume = πr^3. Remember that the value of π is 3.14.)
24. Cilia are found on cells in almost every organ of the human body, and the malfunction of cilia is involved in several human disorders. During embryological development, for example, cilia generate a leftward flow of fluid that initiates the left-right organization of the body organs. Some individuals with primary ciliary dyskinesia (see Module 18) exhibit *situs inversus*, in which internal organs such as the heart are on the wrong side of the body. Explain why this reversed arrangement may be a symptom of PCD.
25. The cells of plant seeds store oils in the form of droplets enclosed by membranes. Unlike typical biological membranes, this oil droplet membrane consists of a single layer of phospholipids rather than a bilayer. Draw a model for a membrane around such an oil droplet. Explain why this arrangement is more stable than a bilayer of phospholipids.
26. Doctors at a California university removed a man's spleen, standard treatment for a type of leukemia, and the disease did not recur. Researchers kept the spleen cells alive in a nutrient medium. They found that some of the cells produced a blood protein that showed promise as a treatment for cancer and AIDS. The researchers patented the cells. The patient sued, claiming a share in profits from any products derived from his cells. The California Supreme Court ruled against the patient, stating that his suit "threatens to destroy the economic incentive to conduct important medical research." The U.S. Supreme Court agreed. Do you think the patient was treated fairly? Is there anything else you would like to know about this case that might help you decide?

Review Answers

1. a. nucleus; b. nucleolus; c. ribosomes; d. Golgi apparatus; e. plasma membrane; f. mitochondrion; g. cytoskeleton; h. peroxisome; i. centriole; j. lysosome; k. rough endoplasmic reticulum; l. smooth endoplasmic reticulum. For functions, see Table 22.
2. flagellum or cilia (some plant sperm cells have flagella), lysosome, centriole (involved with microtubule formation)
3. chloroplast, central vacuole, cell wall
4. d 5. c (Small cells have a greater ratio of surface area to volume.) 6. b 7. d 8. e 9. a 10. c 11. b 12. d 13. c 14. d
15. DNA as genetic material, ribosomes, plasma membrane, and cytoplasm
16. Tight junctions form leakproof sheets of cells. Anchoring junctions link cells to each other; they form strong sheets of cells. Gap junctions are channels through which small molecules can move from cell to cell.
17. Both process energy. A chloroplast converts light energy to chemical energy (sugar molecules). A mitochondrion converts chemical energy (food molecules) to another form of chemical energy (ATP).
18. Different conditions and conflicting processes can occur simultaneously within separate, membrane-enclosed compartments. Also, there is increased area for membrane-attached enzymes that carry out metabolic processes.
19. Cilia may propel a cell through its environment or sweep a fluid environment past the cell.
20. A protein inside the ER is packaged inside transport vesicles that bud off the ER and then join to the Golgi apparatus. A transport vesicle containing the finished protein product then buds off the Golgi and travels to and joins with the plasma membrane, expelling the protein from the cell.
21. Part true, part false. All animal and plant cells have mitochondria; plant cells but not animal cells have chloroplasts.
22. The plasma membrane is a phospholipid bilayer with the hydrophilic heads facing the aqueous environment on both sides and the hydrophobic fatty acid tails mingling in the center of the membrane. Proteins are embedded in and attached to this membrane. Microfilaments form a three-dimensional network just inside the plasma membrane. The extracellular matrix outside the membrane is composed largely of glycoproteins, which may be attached to membrane proteins called integrins. Integrins can transmit information from the ECM to microfilaments on the other side of the membrane.
23. Cell 1: $S = 1{,}256\ \mu m^2$; $V = 4{,}187\ \mu m^3$; $S/V = 0.3$. Cell 2: $S = 5{,}024\ \mu m^2$; $V = 33{,}493\ \mu m^3$; $S/V = 0.15$. The smaller cell has a larger surface area relative to volume for absorbing food and oxygen and excreting waste. Small cells thus perform these activities more efficiently.
24. Individuals with PCD have nonfunctional cilia and flagella due to a lack of dynein motor proteins. This defect would also mean that the cilia involved in left-right pattern formation in the embryo would not be able to set up the fluid flow that initiates the normal arrangement of organs.
25. A single layer of phospholipids surrounding the oil droplet would have their hydrophobic fatty acid tails associated with the hydrophobic oil and their hydrophilic heads facing the aqueous environment of the cell outside the droplet.

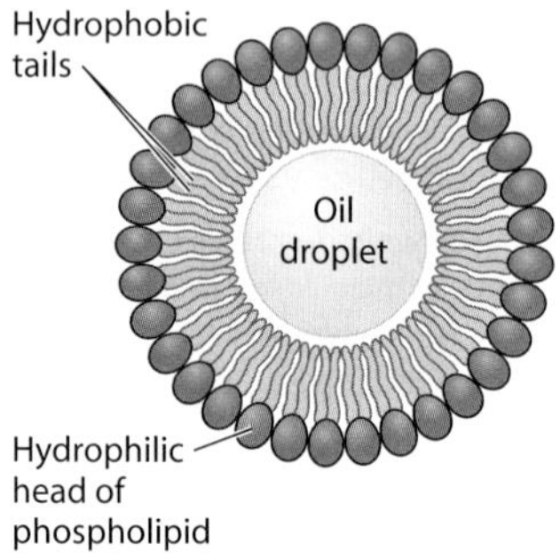

26. Some issues and questions to consider: Were the cells the patient's property, a gift, or just surplus? Was he asked to donate the cells? Was he informed about how the cells might be used? Is it important to ask permission or inform the patient in such a case? How much did the researchers modify the cells? What did they have to do to them to sell the product? Do the researchers and the university have a right to make money from patients' cells? Is the fact that they saved the patient's life a factor? Does the patient have the right to sell his cells? Would he have been able to sell the cells without the researchers' help?

The Working Cell

The Working Cell

BIG IDEAS

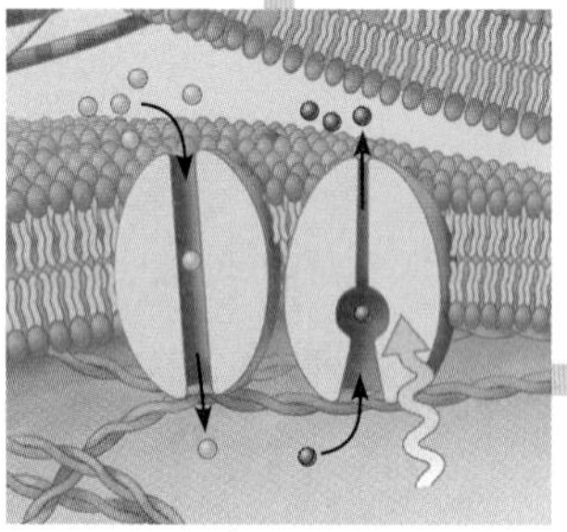

Membrane Structure and Function
(1–9)

The phospholipid and protein structure of cell membranes enables their many important functions.

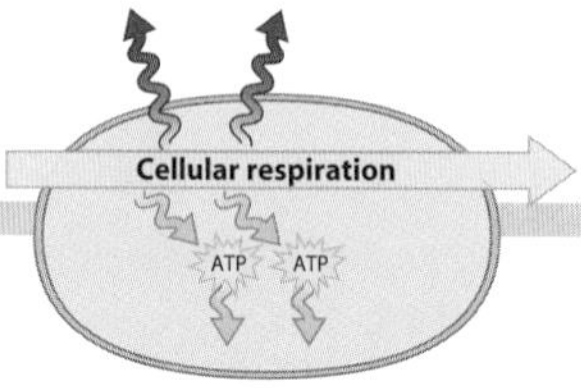

Energy and the Cell
(10–12)

A cell's metabolic reactions transform energy, producing ATP, which drives cellular work.

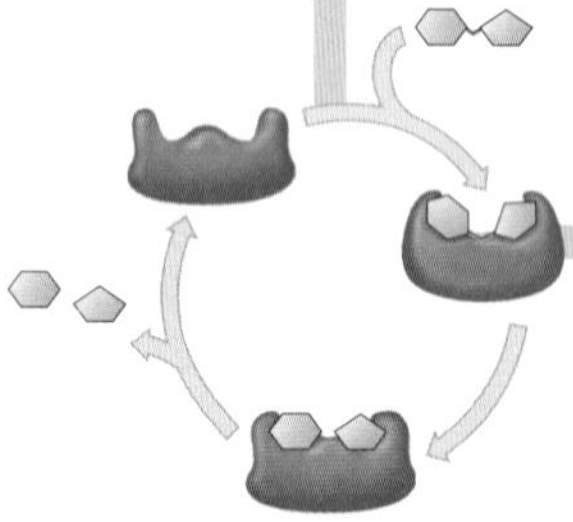

How Enzymes Function
(13–16)

Enzymes speed up a cell's chemical reactions and provide precise control of metabolism.

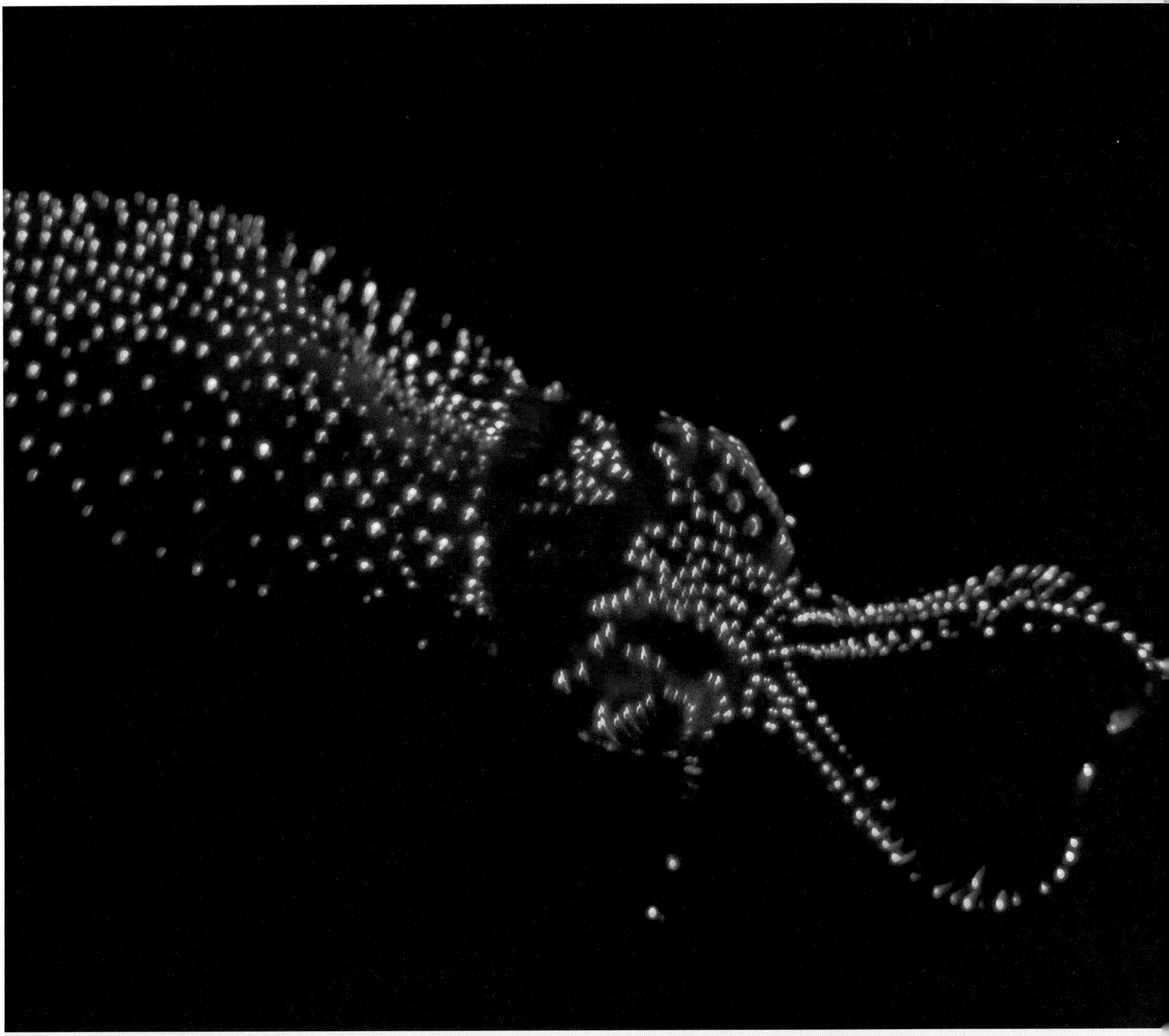

Dante Fenolio/Photo Researchers

Would you believe that this squid's glowing blue lights are a form of camouflage? Ocean predators often hunt by looking up, searching for a silhouette of their prey above them. But can an animal hide its silhouette in the open ocean? The answer is yes, if it turns on the lights. The firefly squid (*Watasenia scintillans*) shown here has light-producing organs called photophores, which emit a soft glow that matches the light filtering down from above. This counter-illumination masks the squid's silhouette. It turns out that many marine invertebrates and fishes hide from predators by producing such light, a process called bioluminescence.

You may be familiar with bioluminescence if you've seen fireflies. While such light production is fairly rare for land animals, it is quite common in the ocean. An estimated 90% of deep-sea marine life bioluminesce. For example, some microorganisms light up when attacked, drawing the attention of larger predators that may feed on their attackers. Some squids expel a cloud of glowing material instead of ink to confuse predators. And a deep-sea anglerfish uses a glowing glob of bacteria on a lure hanging above its huge mouth to attract both mates and prey.

The light these organisms produce comes from a chemical reaction that converts chemical energy to visible light. Bioluminescence is just one example of the multitude of energy conversions that a cell can perform. Many of a cell's reactions take place in organelles, such as those in the light-producing cells of a squid. And the enzymes that control these reactions are often embedded in the membranes of the organelle. Indeed, everything that happens when a squid turns on the lights to hide has some relation to the topics of this chapter: how working cells use membranes, energy, and enzymes.

Membrane Structure and Function

1 Membranes are fluid mosaics of lipids and proteins with many functions

The plasma membrane is the edge of life, the boundary that encloses a living cell. In eukaryotic cells, internal membranes partition the cell into specialized compartments. Membranes are composed of a bilayer of phospholipids with embedded and attached proteins. Biologists describe such a structure as a **fluid mosaic**.

In the cell, a membrane remains about as "fluid" as salad oil, with most of its components able to drift about like partygoers moving through a crowded room. Double bonds in the unsaturated fatty acid tails of some phospholipids produce kinks that prevent phospholipids from packing too tightly. In animal cell membranes, the steroid cholesterol helps stabilize the membrane at warm temperatures but also helps keep the membrane fluid at lower temperatures.

A membrane is a "mosaic" in having diverse protein molecules embedded in its fluid framework. The word *mosaic* can also refer to the varied functions of these proteins. Different types of cells have different membrane proteins, and the various membranes within a cell each contain a unique collection of proteins.

Figure 1, which diagrams the plasma membranes of two adjacent cells, illustrates six major functions performed by membrane proteins, represented by the purple oval structures. Some proteins help maintain cell shape and coordinate changes inside and outside the cell through their attachment to the cytoskeleton and extracellular matrix (ECM). Other proteins function as receptors for chemical messengers (signaling molecules) from other cells. The binding of a signaling molecule triggers a change in the protein, which relays the message into the cell, activating molecules that perform specific functions.

Some membrane proteins are enzymes, which may be grouped in a membrane to carry out sequential steps of a metabolic pathway. Membrane glycoproteins may be involved in cell-cell recognition. Their attached carbohydrates function as identification tags that are recognized by membrane proteins of other cells. This recognition allows cells in an embryo to sort into tissues and enables cells of the immune system to recognize and reject foreign cells, such as infectious bacteria. Membrane proteins also participate in the intercellular junctions that attach adjacent cells.

A final critical function is in transport of substances across the membrane. Membranes exhibit **selective permeability**; that is, they allow some substances to cross more easily than others. Many essential ions and molecules, such as glucose, require transport proteins to enter or leave the cell.

? Review the six different types of functions that proteins in a plasma membrane can perform.

Attachment to the cytoskeleton and ECM, signal transduction, enzymatic activity, cell-cell recognition, intercellular joining, and transport.

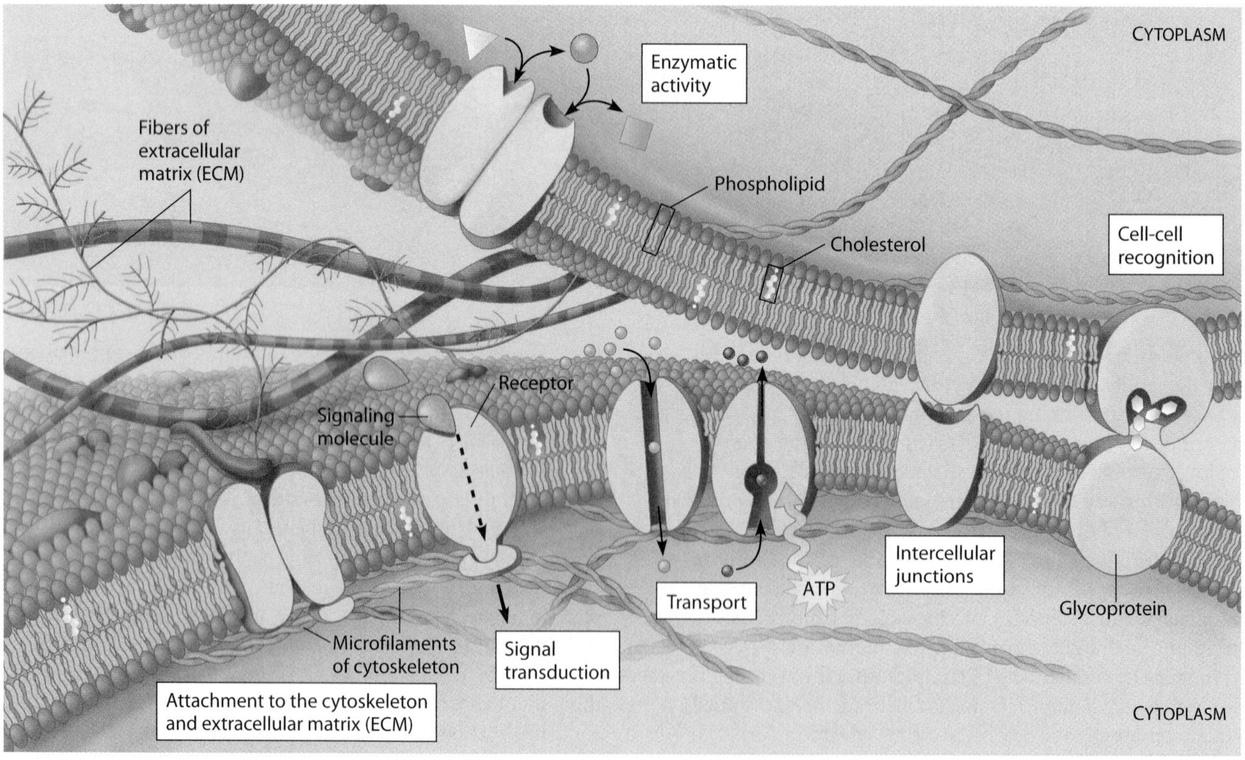

▲ **Figure 1** Some functions of membrane proteins

EVOLUTION CONNECTION

2 Membranes form spontaneously, a critical step in the origin of life

Phospholipids, the key ingredients of biological membranes, were probably among the first organic molecules that formed from chemical reactions on early Earth. These lipids could spontaneously self-assemble into simple membranes, as we can demonstrate in a test tube. When a mixture of phospholipids and water is shaken, the phospholipids organize into bilayers surrounding water-filled bubbles (**Figure 2**). This assembly requires neither genes nor other information beyond the properties of the phospholipids themselves.

The formation of membrane-enclosed collections of molecules was probably a critical step in the evolution of the first cells. A membrane can enclose a solution that is different in composition from its surroundings. A plasma membrane that allows cells to regulate their chemical exchanges with the environment is a basic requirement for life. Indeed, all cells are enclosed by a plasma membrane that is similar in structure and function—illustrating the evolutionary unity of life.

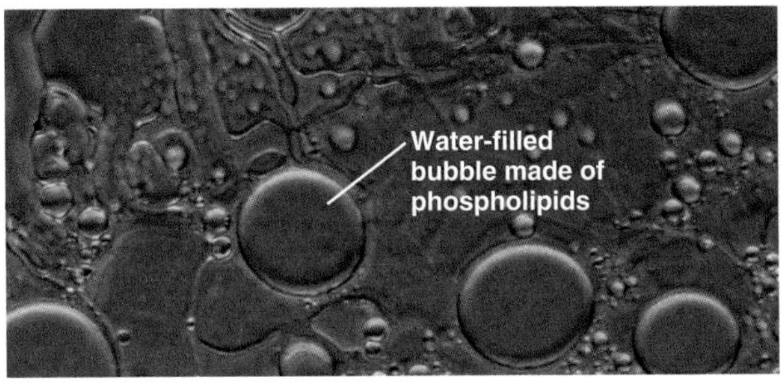

Peter B. Armstrong

▲ **Figure 2** Artificial membrane-bounded sacs

? **This is a diagram of a section of one of the membrane sacs shown in Figure 2. Describe its structure.**

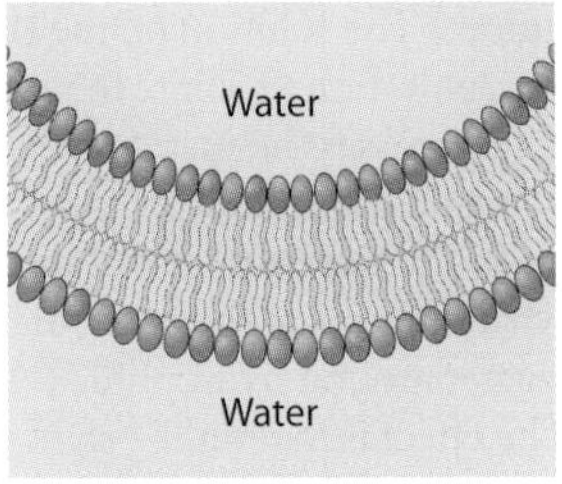

The phospholipids form a bilayer. The hydrophobic fatty acid tails cluster in the center, and the hydrophilic phosphate heads face the water on both sides.

3 Passive transport is diffusion across a membrane with no energy investment

Molecules vibrate and move randomly as a result of a type of energy called thermal motion (heat). One result of this motion is **diffusion**, the tendency for particles of any kind to spread out evenly in an available space. How might diffusion affect the movement of substances into or out of a cell?

The figures to the right will help you to visualize diffusion across a membrane. **Figure 3A** shows a solution of green dye separated from pure water by a membrane. Assume that this membrane has microscopic pores through which dye molecules can move. Thus, we say it is permeable to the dye. Although each molecule moves randomly, there will be a *net* movement from the side of the membrane where dye molecules are more concentrated to the side where they are less concentrated. Put another way, the dye diffuses down its **concentration gradient**. Eventually, the solutions on both sides will have equal concentrations of dye. At this dynamic equilibrium, molecules still move back and forth, but there is no *net* change in concentration on either side of the membrane.

Figure 3B illustrates the important point that two or more substances diffuse independently of each other; that is, each diffuses down its own concentration gradient.

Because a cell does not have to do work when molecules diffuse across its membrane, such movement across a membrane is called **passive transport**. Much of the traffic across cell membranes occurs by diffusion. For example, diffusion down concentration gradients is the sole means by which oxygen (O_2), essential for metabolism, enters your cells and carbon dioxide (CO_2), a metabolic waste, passes out of them.

Both O_2 and CO_2 are small, nonpolar molecules that diffuse easily across the phospholipid bilayer of a membrane. But can ions and polar molecules also diffuse across the hydrophobic interior of a membrane? They can if they are moving down their concentration gradients and if they have transport proteins to help them cross.

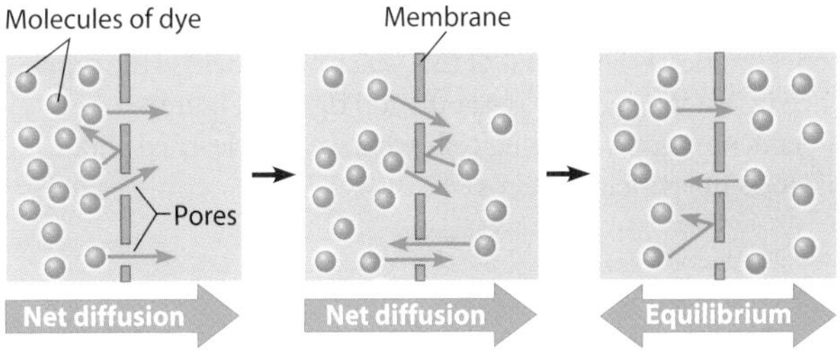

▲ **Figure 3A** Passive transport of one type of molecule

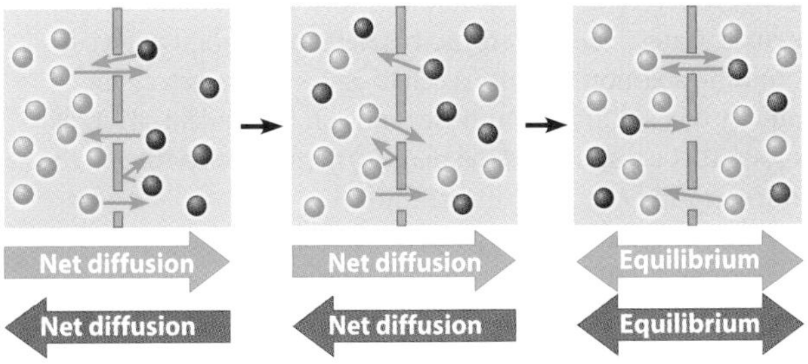

▲ **Figure 3B** Passive transport of two types of molecules

? **Why is diffusion across a membrane called passive transport?**

The cell does not expend energy to transport substances that are diffusing down their concentration gradients.

4 Osmosis is the diffusion of water across a membrane

One of the most important substances that crosses membranes by passive transport is water. In the next module, we consider the critical balance of water between a cell and its environment. But first let's explore a physical model of the diffusion of water across a selectively permeable membrane, a process called **osmosis**. Remember that a selectively permeable membrane allows some substances to cross more easily than others.

The top of **Figure 4** shows what happens if a membrane permeable to water but not to a solute (such as glucose) separates two solutions with different concentrations of solute. (A solute is a substance that dissolves in a liquid solvent, producing a solution.) The solution on the right side initially has a higher concentration of solute than that on the left. As you can see, water crosses the membrane until the solute concentrations are equal on both sides.

In the close-up view at the bottom of Figure 4, you can see what happens at the molecular level. Polar water molecules cluster around hydrophilic (water-loving) solute molecules. The effect is that on the right side, there are fewer water molecules available to cross the membrane. The less concentrated solution on the left, with fewer solute molecules, has more water molecules *free* to move. There is a net movement of water down its own concentration gradient, from the solution with more free water molecules (and lower solute concentration) to that with fewer free water molecules (and higher solute concentration). The result of this water movement is the difference in water levels you see at the top right of Figure 4.

Let's now apply to living cells what we have learned about osmosis in artificial systems.

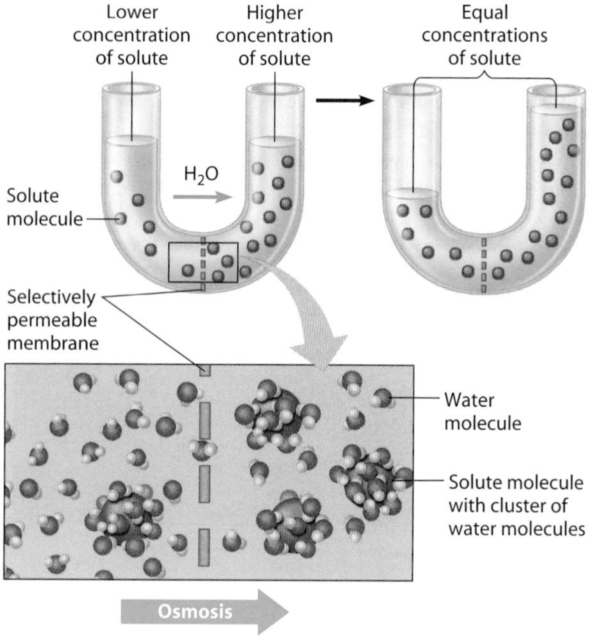

▲ **Figure 4** Osmosis, the diffusion of water across a membrane

? Indicate the direction of net water movement between two solutions—a 0.5% sucrose solution and a 2% sucrose solution—separated by a membrane not permeable to sucrose.

From the 0.5% sucrose solution (lower solute concentration) to the 2% sucrose solution (higher solute concentration)

5 Water balance between cells and their surroundings is crucial to organisms

Biologists use a special vocabulary to describe the relationship between a cell and its surroundings with regard to the movement of water. The term **tonicity** refers to the ability of a surrounding solution to cause a cell to gain or lose water. The tonicity of a solution mainly depends on its concentration of solutes that cannot cross the plasma membrane relative to the concentration of solutes inside the cell.

Figure 5, on the facing page, illustrates how the principles of osmosis and tonicity apply to cells. The effects of placing an animal cell in solutions of various tonicities are shown in the top row of the illustration; the effects of the same solutions on a plant cell are shown in the bottom row.

As shown in the top center of the figure, when an animal cell is immersed in a solution that is **isotonic** to the cell (*iso*, same, and *tonos*, tension), the cell's volume remains constant. The solute concentration of a cell and its isotonic environment are essentially equal, and the cell gains water at the same rate that it loses it. In your body, red blood cells are transported in the isotonic plasma of the blood. Intravenous (IV) fluids administered in hospitals must also be isotonic to blood cells. The body cells of most animals are bathed in an extracellular fluid that is isotonic to the cells. And seawater is isotonic to the cells of many marine animals, such as sea stars and crabs.

The upper left of the figure shows what happens when an animal cell is placed in a **hypotonic** solution (*hypo*, below), a solution with a solute concentration lower than that of the cell. (Can you figure out in which direction osmosis will occur? Where are there more free water molecules available to move?) The cell gains water, swells, and may burst (lyse) like an overfilled balloon. The upper right shows the opposite case—an animal cell placed in a **hypertonic** solution (*hyper*, above), a solution with a higher solute concentration. The cell shrivels and can die from water loss.

For an animal to survive in a hypotonic or hypertonic environment, it must have a way to prevent excessive uptake or excessive loss of water. The control of water balance is called **osmoregulation**. For example, a freshwater fish, which lives in a hypotonic environment, has kidneys and gills that work constantly to prevent an excessive buildup of water in the body.

Water balance issues are somewhat different for the cells of plants, prokaryotes, and fungi because of their cell walls.

As shown in the bottom center of Figure 5, a plant cell immersed in an isotonic solution is flaccid (limp). In contrast, a plant cell is turgid (very firm), which is the healthy state for most plant cells, in a hypotonic environment (bottom left). To become turgid, a plant cell needs a net inflow of water. Although the somewhat elastic cell wall expands a bit, the pressure it exerts prevents the cell from taking in too much water and bursting, as an animal cell would in a hypotonic environment. Plants that are not woody, such as most houseplants, depend on their turgid cells for mechanical support.

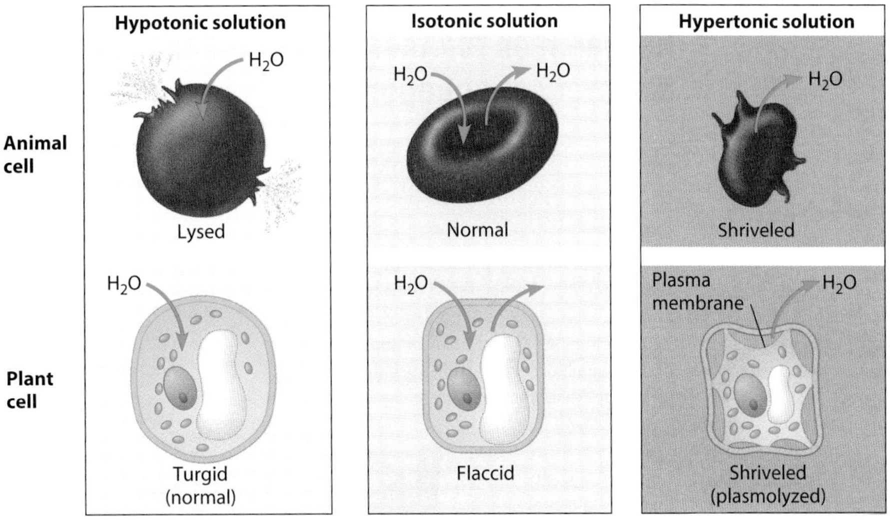

▲ **Figure 5** How animal and plant cells react to changes in tonicity. (Deepening shades of blue reflect increasing concentrations of solutes in the surrounding solutions.)

In a hypertonic environment (bottom right), a plant cell is no better off than an animal cell. As a plant cell loses water, it shrivels, and its plasma membrane pulls away from the cell wall. This process, called plasmolysis, causes the plant to wilt and can be lethal to the cell and the plant. The walled cells of bacteria and fungi also plasmolyze in hypertonic environments. Thus, meats and other foods can be preserved with concentrated salt solutions because the cells of food-spoiling bacteria or fungi become plasmolyzed and eventually die.

In the next module, we explore how water and other polar solutes move across cell membranes.

? Explain the function of the contractile vacuoles in the freshwater *Paramecium* shown in Figure 4.11A in terms of what you have just learned about water balance in cells.

The pond water in which *Paramecium* lives is hypotonic to the cell. The contractile vacuoles expel the water that constantly enters the cell by osmosis.

6 Transport proteins can facilitate diffusion across membranes

Recall that nonpolar, hydrophobic molecules can dissolve in the lipid bilayer of a membrane and cross it with ease. Polar or charged substances, meanwhile, can move across a membrane with the help of specific transport proteins in a process called **facilitated diffusion**. Without the transport protein, the substance cannot cross the membrane or it diffuses across it too slowly to be useful to the cell. Facilitated diffusion is a type of passive transport because it does not require energy. As in all passive transport, the driving force is the concentration gradient.

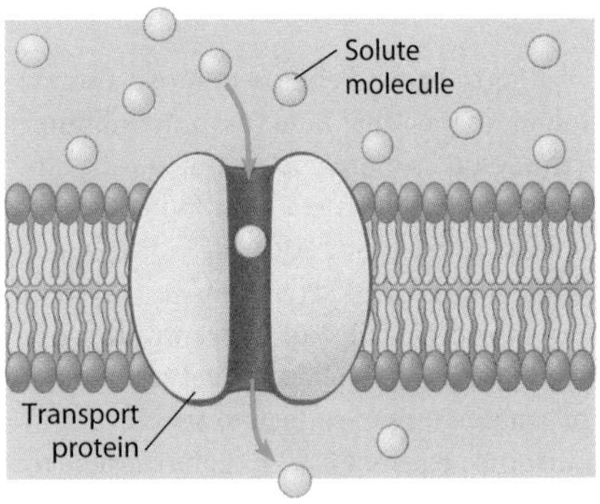

◀ **Figure 6** Transport protein providing a channel for the diffusion of a specific solute across a membrane

Figure 6 shows a common type of transport protein, which provides a hydrophilic channel that some molecules or ions use as a tunnel through the membrane. Another type of transport protein binds its passenger, changes shape, and releases its passenger on the other side. In both cases, the transport protein is specific for the substance it helps move across the membrane. The greater the number of transport proteins for a particular solute in a membrane, the faster the solute's rate of diffusion across the membrane.

Substances that use facilitated diffusion for crossing cell membranes include a number of sugars, amino acids, ions—and even water. The water molecule is very small, but because it is polar, its diffusion through a membrane's hydrophobic interior is relatively slow. The very rapid diffusion of water into and out of certain cells, such as plant cells, kidney cells, and red blood cells, is made possible by a protein channel called an **aquaporin**. A single aquaporin allows the entry or exit of up to 3 billion water molecules per second—a tremendous increase in water transport over simple diffusion.

? How do transport proteins contribute to a membrane's selective permeability?

Because they are specific for the solutes they transport, the numbers and kinds of transport proteins affect a membrane's permeability to various solutes.

SCIENTIFIC DISCOVERY

7 Research on another membrane protein led to the discovery of aquaporins

Peter Agre received the 2003 Nobel Prize in Chemistry for his discovery of aquaporins. In a recent interview, Dr. Agre described his research that led to this discovery:

> I'm a blood specialist (hematologist), and my particular interest has been proteins found in the plasma membrane of red blood cells. When I joined the faculty at the John Hopkins School of Medicine, I began to study the Rh blood antigens. Rh is of medical importance because of Rh incompatibility, which occurs when Rh-negative mothers have Rh-positive babies. Membrane-spanning proteins are really messy to work with. But we worked out a method to isolate the Rh protein. Our sample seemed to consist of two proteins, but we were sure that the smaller one was just a breakdown product of the larger one. We were completely wrong.
>
> Using antibodies we made to the smaller protein, we showed it to be one of the most abundant proteins in red cell membranes—200,000 copies per cell!—and even more abundant in certain kidney cells.

We asked Dr. Agre why cells have aquaporins.

> Not all cells do. Before our discovery, however, many physiologists thought that diffusion was enough for getting water into and out of *all* cells. Others said this couldn't be enough, especially for cells whose water permeability needs to be very high or regulated. For example, our kidneys must filter and reabsorb many liters of water every day. . . . People whose kidney cells have defective aquaporin molecules need to drink 20 liters of water a day to prevent dehydration. In addition, some patients make too much aquaporin, causing them to retain too much fluid. Fluid retention in pregnant women is caused by the synthesis of too much aquaporin. Knowledge of aquaporins may in the future contribute to the solution of medical problems.

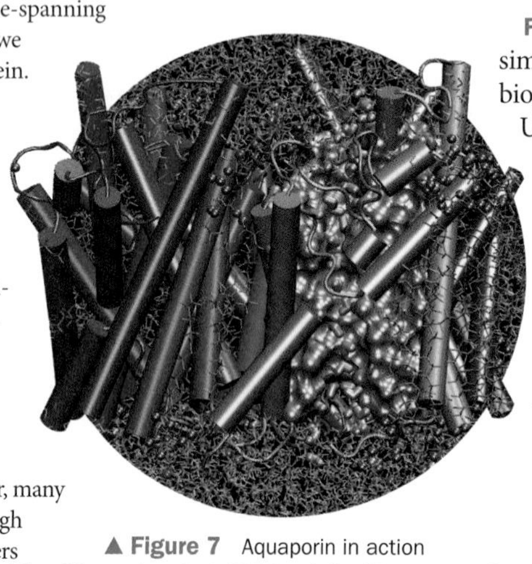

▲ **Figure 7** Aquaporin in action

"This water permeation through aquaporin was made by Drs. Emad Tajkhorshid and Klaus Schulten using VMD and is owned by the Theoretical and Computational Biophysics Group, NIH Resource for Macromolecular Modeling and Bioinformatics, at the Beckman Institute, University of Illinois at Urbana-Champaign."

Adapted from an interview in Neil Campbell and Jane Reece, *Biology* 7th ed. Copyright © 2005 Pearson Education, Inc., publishing as Pearson Benjamin Cummings.

Figure 7 is an image taken from a simulation produced by computational biophysicists at the University of Illinois, Urbana. Their model included four aquaporin channels spanning a membrane. You can see a line of blue water molecules flipping their way single file through the gold aquaporin. The simulation of this flipping movement allowed researchers to discover how aquaporins selectively allow only water molecules to pass through them.

? Why are aquaporins important in kidney cells?

Kidney cells must reabsorb a large amount of water when producing urine.

8 Cells expend energy in the active transport of a solute

In **active transport**, a cell must expend energy to move a solute *against* its concentration gradient—that is, across a membrane toward the side where the solute is more concentrated. The energy molecule ATP (described in more detail in Module 12) supplies the energy for most active transport.

Figure 8 shows a simple model of an active transport system that pumps a solute out of the cell against its concentration gradient. ❶ The process begins when solute molecules on the cytoplasmic side of the plasma membrane attach to specific binding sites on the transport protein. ❷ ATP then transfers a phosphate group to the transport protein, ❸ causing the protein to change shape in such a way that the solute is released on the other side of the membrane. ❹ The phosphate group detaches, and the transport protein returns to its original shape.

Active transport allows a cell to maintain internal concentrations of small molecules and ions that are different from concentrations in its surroundings. For example, the inside of an animal cell has a higher concentration of potassium ions (K^+) and a lower concentration of sodium ions (Na^+) than the solution outside the cell. The generation of nerve signals depends on these concentration differences, which a transport protein called the sodium-potassium pump helps maintain by shuttling Na^+ and K^+ against their concentration gradients.

? Cells actively transport Ca^{2+} out of the cell. Is calcium more concentrated inside or outside of the cell? Explain.

Outside: Active transport moves calcium against its concentration gradient.

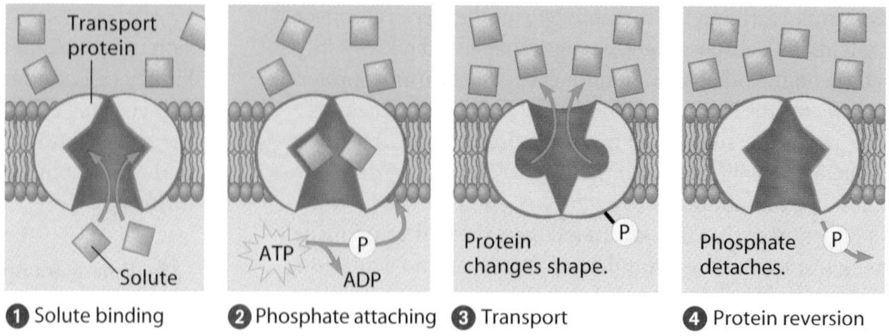

▲ **Figure 8** Active transport of a solute across a membrane

9 Exocytosis and endocytosis transport large molecules across membranes

So far, we've focused on how water and small solutes enter and leave cells. The story is different for large molecules.

A cell uses the process of **exocytosis** (from the Greek *exo*, outside, and *kytos*, cell) to export bulky materials such as proteins or polysaccharides. A transport vesicle filled with macromolecules buds from the Golgi apparatus and moves to the plasma membrane. Once there, the vesicle fuses with the plasma membrane, and the vesicle's contents spill out of the cell when the vesicle membrane becomes part of the plasma membrane. When we weep, for instance, cells in our tear glands use exocytosis to export a salty solution containing proteins. In another example, certain cells in the pancreas manufacture the hormone insulin and secrete it into the bloodstream by exocytosis.

Endocytosis (*endo*, inside) is a transport process that is the opposite of exocytosis. In endocytosis, a cell takes in large molecules. A depression in the plasma membrane pinches in and forms a vesicle enclosing material that had been outside the cell.

Figure 9 shows three kinds of endocytosis. The top diagram illustrates **phagocytosis**, or "cellular eating." A cell engulfs a particle by wrapping extensions called pseudopodia around it and packaging it within a membrane-enclosed sac large enough to be called a vacuole. The vacuole then fuses with a lysosome, whose hydrolytic enzymes digest the contents of the vacuole. The micrograph on the top right shows an amoeba taking in a food particle via phagocytosis.

The center diagram shows **pinocytosis**, or "cellular drinking." The cell "gulps" droplets of fluid into tiny vesicles. Pinocytosis is not specific; it takes in any and all solutes dissolved in the droplets. The micrograph in the middle shows pinocytosis vesicles forming (arrows) in a cell lining a small blood vessel.

In contrast to pinocytosis, **receptor-mediated endocytosis** is highly selective. Receptor proteins for specific molecules are embedded in regions of the membrane that are lined by a layer of coat proteins. The bottom diagram shows that the plasma membrane has indented to form a coated pit, whose receptor proteins have picked up particular molecules from the surroundings. The coated pit then pinches closed to form a vesicle that carries the molecules into the cytoplasm. The micrograph shows material bound to receptor proteins inside a coated pit.

Your cells use receptor-mediated endocytosis to take in cholesterol from the blood for synthesis of membranes and as a precursor for other steroids. Cholesterol circulates in the blood in particles called low-density lipoproteins (LDLs). LDLs bind to receptor proteins and then enter cells by endocytosis. In humans with the inherited disease familial hypercholesterolemia, LDL receptor proteins are defective and cholesterol accumulates to high levels in the blood, leading to atherosclerosis.

? As a cell grows, its plasma membrane expands. Does this involve endocytosis or exocytosis? Explain.

Exocytosis: When a transport vesicle fuses with the plasma membrane, its contents are released and the vesicle membrane adds to the plasma membrane.

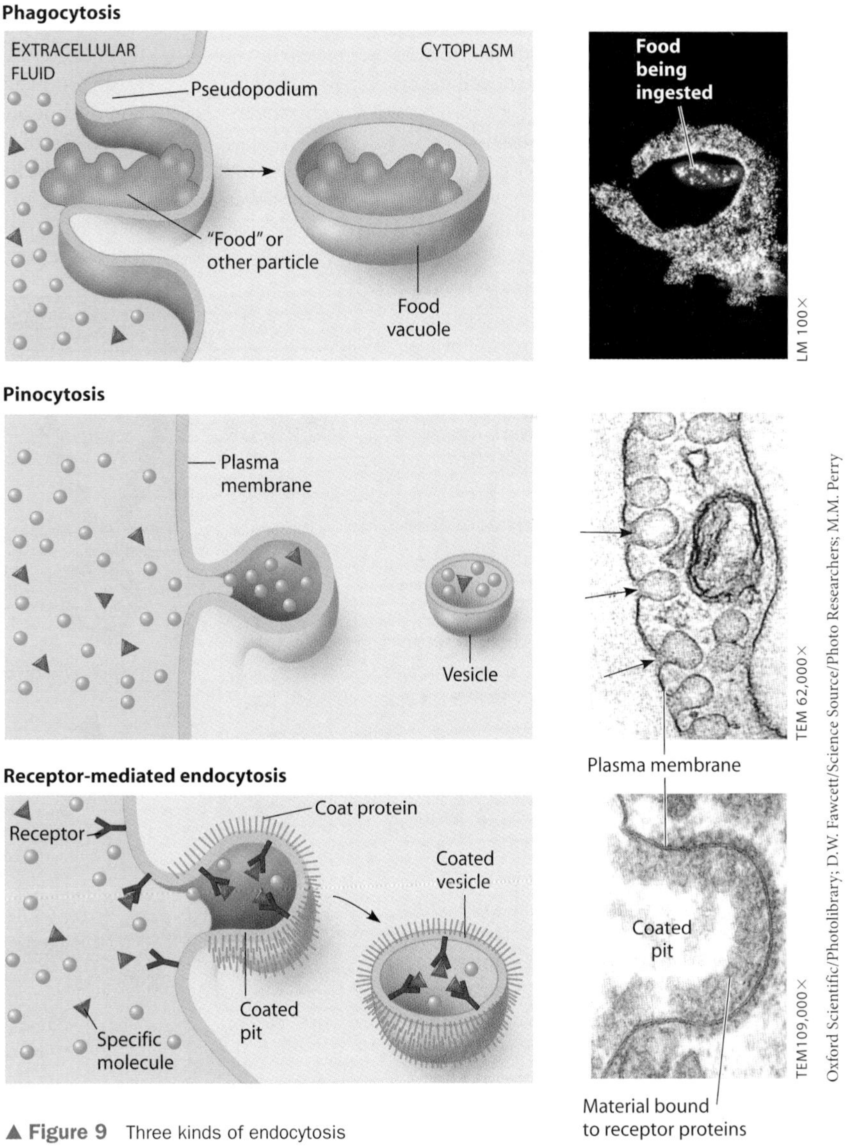

▲ **Figure 9** Three kinds of endocytosis

Energy and the Cell

10 Cells transform energy as they perform work

The title of this chapter is "The Working Cell." But just what type of work does a cell do? You just learned that a cell actively transports substances across membranes. The cell also builds those membranes and the proteins embedded in them. A cell is a miniature chemical factory in which thousands of reactions occur within a microscopic space. Some of these reactions release energy; others require energy. To understand how the cell works, you must have a basic knowledge of energy.

Forms of Energy We can define **energy** as the capacity to cause change or to perform work. There are two basic forms of energy: kinetic energy and potential energy. **Kinetic energy** is the energy of motion. Moving objects can perform work by transferring motion to other matter. For example, the movement of your legs can push bicycle pedals, turning the wheels and moving you and your bike up a hill. **Heat**, or thermal energy, is a type of kinetic energy associated with the random movement of atoms or molecules. Light, also a type of kinetic energy, can be harnessed to power photosynthesis.

Potential energy, the second main form of energy, is energy that matter possesses as a result of its location or structure. Water behind a dam and you on your bicycle at the top of a hill possess potential energy. Molecules possess potential energy because of the arrangement of electrons in the bonds between their atoms. **Chemical energy** is the potential energy available for release in a chemical reaction. Chemical energy is the most important type of energy for living organisms; it is the energy that can be transformed to power the work of the cell.

Energy Transformations **Thermodynamics** is the study of energy transformations that occur in a collection of matter. Scientists use the word *system* for the matter under study and refer to the rest of the universe—everything outside the system—as the *surroundings*. A system can be an electric power plant, a single cell, or the entire planet. An organism is an open system; that is, it exchanges both energy and matter with its surroundings.

The **first law of thermodynamics**, also known as the law of energy conservation, states that the energy in the universe is constant. Energy can be transferred and transformed, but it cannot be created or destroyed. A power plant does not create energy; it merely converts it from one form (such as the energy stored in coal) to the more convenient form of electricity. A plant cell converts light energy to chemical energy; it, too, is an energy transformer, not an energy producer.

If energy cannot be destroyed, then why can't organisms simply recycle their energy? It turns out that during every transfer or transformation, some energy becomes unusable—unavailable to do work. In most energy transformations, some energy is converted to heat, a disordered form of energy. Scientists use a quantity called **entropy** as a measure of disorder, or randomness. The more randomly arranged a collection of matter is, the greater its entropy. According to the **second law of thermodynamics**, energy conversions increase the entropy (disorder) of the universe.

Figure 10 compares a car and a cell to show how energy can be transformed and how entropy increases as a result. Automobile engines and living cells use the same basic process to make the chemical energy of their fuel available for work. The engine mixes oxygen with gasoline in an explosive chemical reaction that pushes the pistons, which eventually move the wheels. The waste products emitted from the exhaust pipe are mostly carbon dioxide and water, energy-poor, simple molecules. Only about 25% of the chemical energy stored in gasoline is converted to the kinetic energy of the car's movement; the rest is lost as heat.

Cells also use oxygen in reactions that release energy from fuel molecules. In the process called **cellular respiration**, the

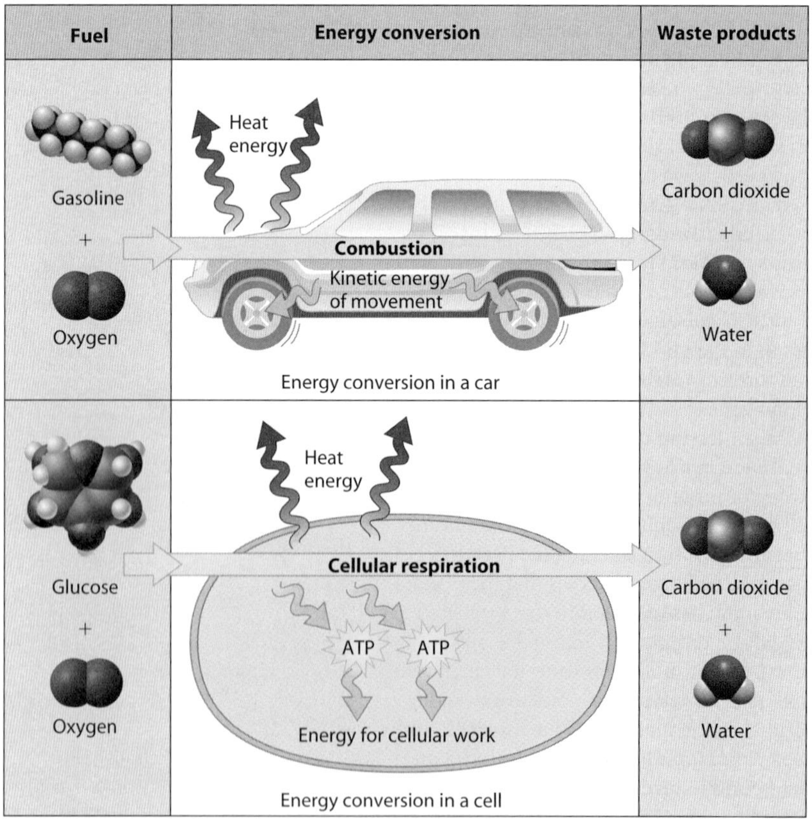

▲ **Figure 10** Energy transformations (with an increase in entropy) in a car and a cell

chemical energy stored in organic molecules is converted to a form that the cell can use to perform work. Just like for the car, the waste products are mostly carbon dioxide and water. Cells are more efficient than car engines, however, converting about 34% of the chemical energy in their fuel to energy for cellular work. The other 66% generates heat, which explains why vigorous exercise makes you so warm.

According to the second law of thermodynamics, energy transformations result in the universe becoming more disordered. How, then, can we account for biological order? A cell creates intricate structures from less organized materials. Although this increase in order corresponds to a decrease in entropy, it is accomplished at the expense of ordered forms of matter and energy taken in from the surroundings. As shown in Figure 10, cells extract the chemical energy of glucose and return disordered heat and lower-energy carbon dioxide and water to the surroundings. In a thermodynamic sense, a cell is an island of low entropy in an increasingly random universe.

? How does the second law of thermodynamics explain the diffusion of a solute across a membrane?

Diffusion across a membrane results in equal concentrations of solute, which is a more disordered arrangement (higher entropy) than a high concentration on one side and a low concentration on the other.

11 Chemical reactions either release or store energy

Chemical reactions are of two types: Either they release energy or they require an input of energy and store energy.

An **exergonic reaction** is a chemical reaction that releases energy (*exergonic* means "energy outward"). As shown in **Figure 11A**, an exergonic reaction begins with reactants whose covalent bonds contain more energy than those in the products. The reaction releases to the surroundings an amount of energy equal to the difference in potential energy between the reactants and the products.

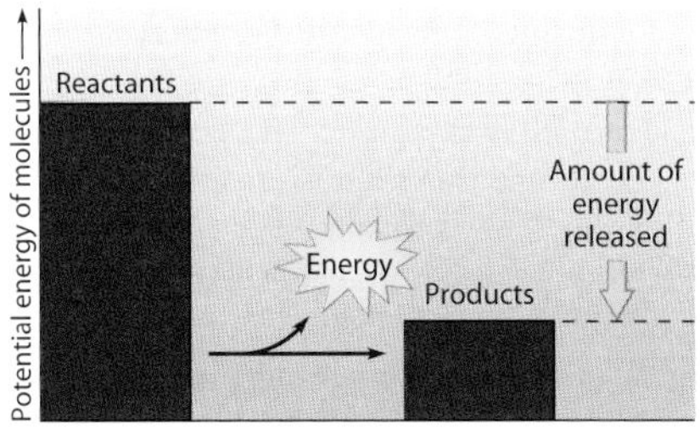

◀ Figure 11A Exergonic reaction, energy released

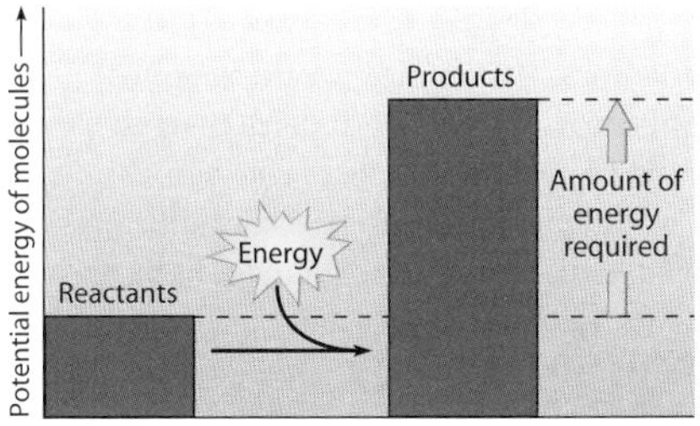

◀ Figure 11B Endergonic reaction, energy required

As an example of an exergonic reaction, consider what happens when wood burns. One of the major components of wood is cellulose, a large energy-rich carbohydrate composed of many glucose monomers. The burning of wood releases the energy of glucose as heat and light. Carbon dioxide and water are the products of the reaction.

As you learned in Module 10, cells release energy from fuel molecules in the process called cellular respiration. Burning and cellular respiration are alike in being exergonic. They differ in that burning is essentially a one-step process that releases all of a substance's energy at once. Cellular respiration, on the other hand, involves many steps, each a separate chemical reaction; you can think of it as a "slow burn." Some of the energy released from glucose by cellular respiration escapes as heat, but a substantial amount is converted to the chemical energy of ATP. Cells use ATP as an immediate source of energy.

The other type of chemical reaction requires a net input of energy. **Endergonic reactions** yield products that are rich in potential energy (*endergonic* means "energy inward"). As shown in **Figure 11B**, an endergonic reaction starts out with reactant molecules that contain relatively little potential energy. Energy is absorbed from the surroundings as the reaction occurs, so the products of an endergonic reaction contain more chemical energy than the reactants did. And as the graph shows, the amount of additional energy stored in the products equals the difference in potential energy between the reactants and the products.

Photosynthesis, the process by which plant cells make sugar, is an example of an endergonic process. Photosynthesis starts with energy-poor reactants (carbon dioxide and water molecules) and, using energy absorbed from sunlight, produces energy-rich sugar molecules.

Every working cell in every organism carries out thousands of exergonic and endergonic reactions. The total of an organism's chemical reactions is called **metabolism** (from the Greek *metabole*, change). We can picture a cell's metabolism as a road map of thousands of chemical reactions arranged as intersecting metabolic pathways. A **metabolic pathway** is a series of chemical reactions that either builds a complex molecule or breaks down a complex molecule into simpler compounds. The "slow burn" of cellular respiration is an example of a metabolic pathway in which a sequence of reactions slowly releases the potential energy stored in sugar.

All of an organism's activities require energy, which is obtained from sugar and other molecules by the exergonic reactions of cellular respiration. Cells then use that energy in endergonic reactions to make molecules and do the work of the cell. **Energy coupling**—the use of energy released from exergonic reactions to drive essential endergonic reactions—is a crucial ability of all cells. ATP molecules are the key to energy coupling. In the next module, we explore the structure and function of ATP.

? Cellular respiration is an exergonic process. Remembering that energy must be conserved, what becomes of the energy extracted from food during cellular respiration?

Some of it is stored in ATP molecules; the rest is released as heat.

12 ATP drives cellular work by coupling exergonic and endergonic reactions

ATP powers nearly all forms of cellular work. The structure of ATP, or adenosine triphosphate, is shown below in **Figure 12A**. The adenosine part of ATP consists of adenine, a nitrogenous base, and ribose, a five-carbon sugar. The triphosphate part is a chain of three phosphate groups (each symbolized by (P)). All three phosphate groups are negatively charged. These like charges are crowded together, and their mutual repulsion makes the triphosphate chain of ATP the chemical equivalent of a compressed spring.

As a result, the bonds connecting the phosphate groups are unstable and can readily be broken by hydrolysis, the addition of water. Notice in Figure 12A that when the bond to the third group breaks, a phosphate group leaves ATP—which becomes ADP (adenosine diphosphate)—and energy is released.

Thus, the hydrolysis of ATP is exergonic—it releases energy. How does the cell couple this reaction to an endergonic one? It usually does so by transferring a phosphate group from ATP to some other molecule. This phosphate transfer is called **phosphorylation**, and most cellular work depends on ATP energizing molecules by phosphorylating them.

There are three main types of cellular work: chemical, mechanical, and transport. As **Figure 12B** shows, ATP drives all three types of work. In chemical work, the phosphorylation of reactants provides energy to drive the endergonic synthesis of products. In an example of mechanical work, the transfer of phosphate groups to special motor proteins in muscle cells causes the proteins to change shape and pull on protein filaments, in turn causing the cells to contract. In transport work, as discussed in Module 8, ATP drives the active transport of solutes across a membrane against their concentration gradient by phosphorylating transport proteins.

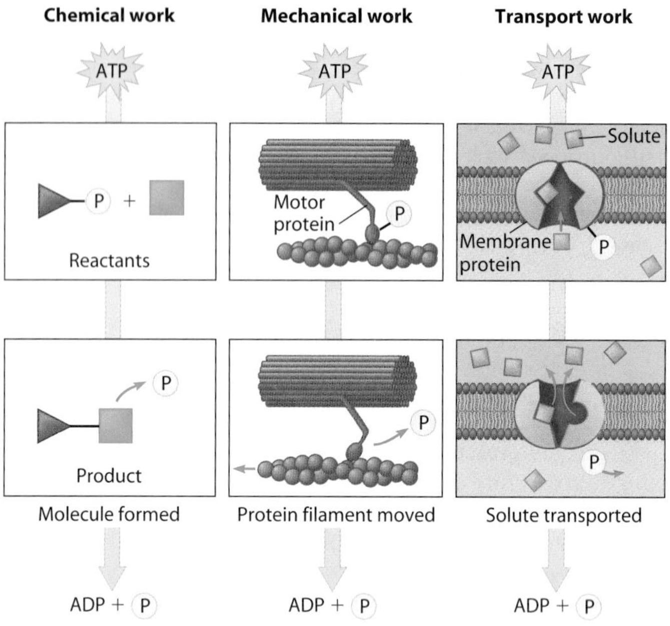

▲ **Figure 12B** How ATP powers cellular work

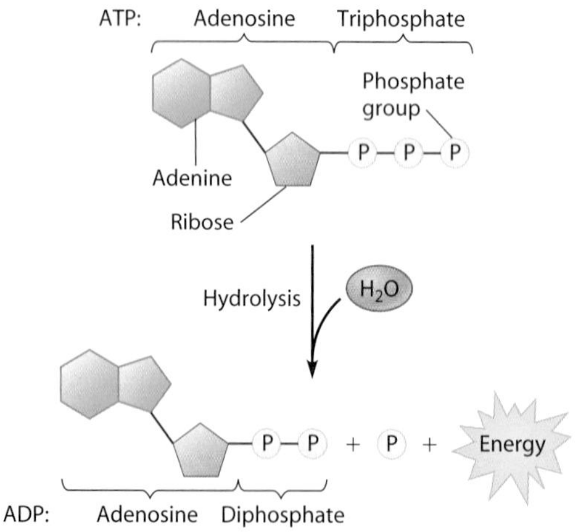

▲ **Figure 12A** The structure and hydrolysis of ATP. The reaction of ATP and water yields ADP, a phosphate group, and energy.

Work can be sustained because ATP is a renewable resource that cells regenerate. **Figure 12C**, below, shows the ATP cycle. Each side of this cycle illustrates energy coupling. Energy released in exergonic reactions, such as the breakdown of glucose during cellular respiration, is used to regenerate ATP from ADP. In this endergonic (energy-storing) process, a phosphate group is bonded to ADP. The hydrolysis of ATP releases energy that drives endergonic reactions. A cell at work uses ATP continuously, and the ATP cycle runs at an astonishing pace. In fact, a working muscle cell may consume and regenerate 10 million ATP molecules each second.

But even with a constant supply of energy, few metabolic reactions would occur without the assistance of enzymes. We explore these biological catalysts next.

? Explain how ATP transfers energy from exergonic to endergonic processes in the cell.

Exergonic processes phosphorylate ADP to form ATP. ATP transfers energy to endergonic processes by phosphorylating other molecules.

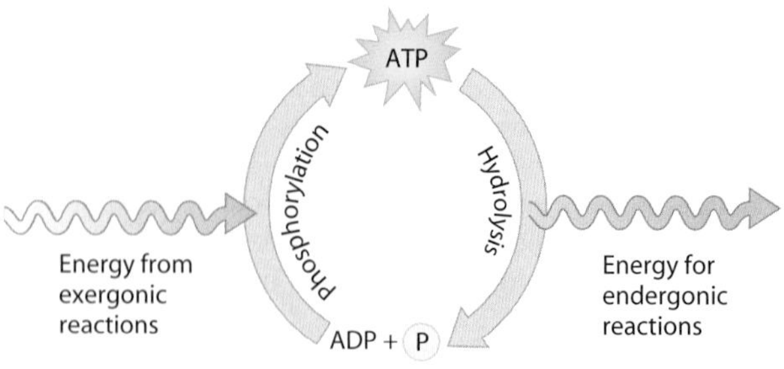

▲ **Figure 12C** The ATP cycle

How Enzymes Function

13 Enzymes speed up the cell's chemical reactions by lowering energy barriers

Your room gets messier; water flows downhill; sugar crystals dissolve in your coffee. Ordered structures tend toward disorder, and high-energy systems tend to change toward a more stable state of low energy. Proteins, DNA, carbohydrates, lipids—most of the complex molecules of your cells are rich in potential energy. Why don't these high-energy, ordered molecules spontaneously break down into less ordered, lower-energy molecules? They remain intact for the same reason that wood doesn't normally burst into flames or the gas in an automobile's gas tank doesn't spontaneously explode.

There is an energy barrier that must be overcome before a chemical reaction can begin. Energy must be absorbed to contort or weaken bonds in reactant molecules so that they can break and new bonds can form. We call this the **activation energy** (abbreviated E_A for energy of activation). We can think of E_A as the amount of energy needed for reactant molecules to move "uphill" to a higher-energy, unstable state so that the "downhill" part of a reaction can begin.

The energy barrier of E_A protects the highly ordered molecules of your cells from spontaneously breaking down. But now we have a dilemma. Life depends on countless chemical reactions that constantly change a cell's molecular makeup. Most of the essential reactions of metabolism must occur quickly and precisely for a cell to survive. How can the specific reactions that a cell requires get over that energy barrier?

One way to speed reactions is to add heat. Heat speeds up molecules and agitates atoms so that bonds break more easily and reactions can proceed. Certainly, adding a match to kindling will start a fire, and the firing of a spark plug ignites gasoline in an engine. But heating a cell would speed up all chemical reactions, not just the necessary ones, and too much heat would kill the cell.

The answer to our dilemma lies in **enzymes**—molecules that function as biological catalysts, increasing the rate of a reaction without being consumed by the reaction. Almost all enzymes are proteins, although some RNA molecules can also function as enzymes. An enzyme speeds up a reaction by lowering the E_A needed for a reaction to begin. **Figure 13** compares a reaction without (left) and with (right) an enzyme. Notice how much easier it is for the reactant to get over the activation energy barrier when an enzyme is involved. In the next module, we explore how the structure of an enzyme enables it to lower the activation energy, allowing a reaction to proceed.

? The graph below illustrates the course of a reaction with and without an enzyme. Which curve represents the enzyme-catalyzed reaction? What energy changes are represented by the lines labeled a, b, and c?

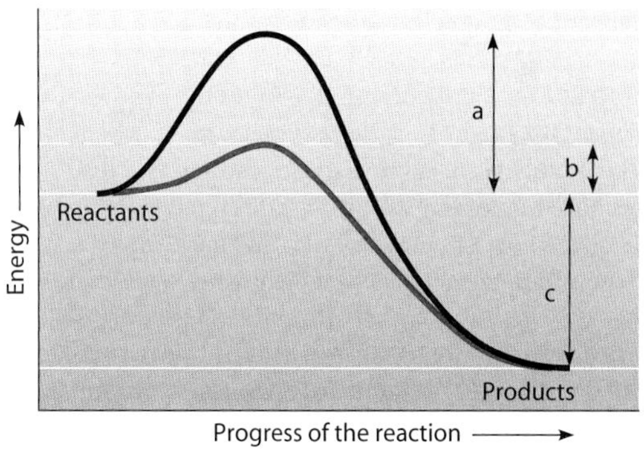

The red (lower) curve is the enzyme-catalyzed reaction. Line a is E_A without enzyme; b is E_A with enzyme; c is the change in energy between reactants and products, which is the same for both the catalyzed and uncatalyzed reactions.

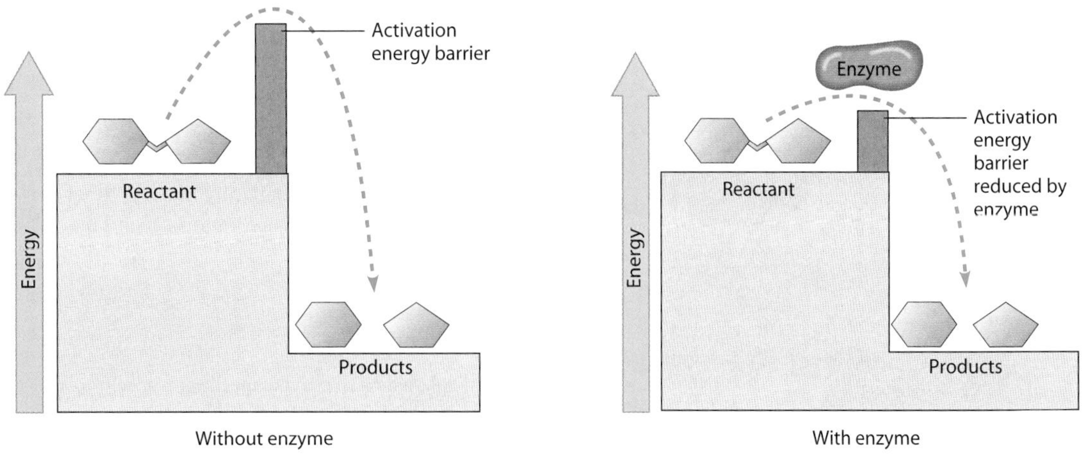

▲ **Figure 13** The effect of an enzyme in lowering E_A

14 A specific enzyme catalyzes each cellular reaction

You just learned that an enzyme catalyzes a reaction by lowering the E_A barrier. How does it do that? With the aid of an enzyme, the bonds in a reactant are contorted into the higher-energy, unstable state from which the reaction can proceed. Without an enzyme, the energy of activation might never be reached. For example, a solution of sucrose (table sugar) can sit for years at room temperature with no appreciable hydrolysis into its components glucose and fructose. But if we add a small amount of an enzyme to the solution, all the sucrose will be hydrolyzed within seconds.

An enzyme is very selective in the reaction it catalyzes. As a protein, an enzyme has a unique three-dimensional shape, and that shape determines the enzyme's specificity. The specific reactant that an enzyme acts on is called the enzyme's **substrate**. A substrate fits into a region of the enzyme called an **active site**. An active site is typically a pocket or groove on the surface of the enzyme formed by only a few of the enzyme's amino acids. The rest of the protein maintains the shape of the active site. Enzymes are specific because their active sites fit only specific substrate molecules.

The Catalytic Cycle **Figure 14** illustrates the catalytic cycle of an enzyme. Our example is the enzyme sucrase, which catalyzes the hydrolysis of sucrose to glucose and fructose. (Most enzymes have names that end in *-ase*, and many are named for their substrate.) ❶ The enzyme starts with an empty active site. ❷ Sucrose enters the active site, attaching by weak bonds. The active site changes shape slightly, embracing the substrate more snugly, like a firm handshake. This **induced fit** may contort substrate bonds or place chemical groups of the amino acids making up the active site in position to catalyze the reaction. (In reactions involving two or more reactants, the active site holds the substrates in the proper orientation for a reaction to occur.) ❸ The strained bond of sucrose reacts with water, and the substrate is converted (hydrolyzed) to the products glucose and fructose. ❹ The enzyme releases the products and emerges unchanged from the reaction. Its active site is now available for another substrate molecule, and another round of the cycle can begin. A single enzyme molecule may act on thousands or even millions of substrate molecules per second.

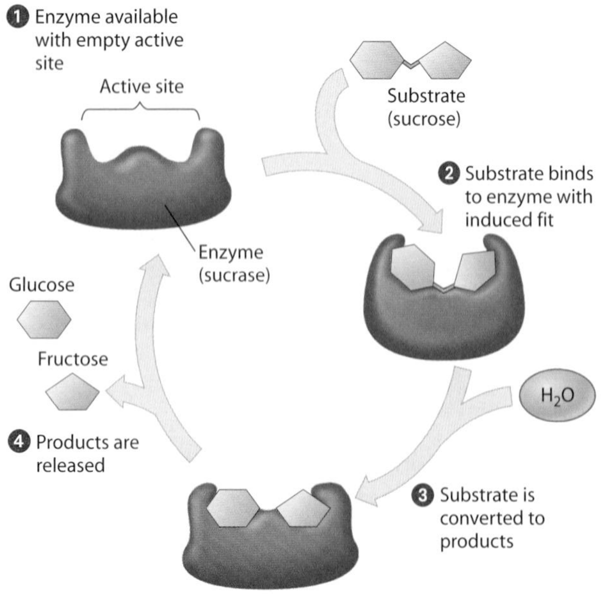

▲ **Figure 14** The catalytic cycle of an enzyme

Optimal Conditions for Enzymes As with all proteins, an enzyme's shape is central to its function, and this three-dimensional shape is affected by the environment. For every enzyme, there are optimal conditions under which it is most effective. Temperature, for instance, affects molecular motion, and an enzyme's optimal temperature produces the highest rate of contact between reactant molecules and the enzyme's active site. Higher temperatures denature the enzyme, altering its specific shape and destroying its function. Most human enzymes work best at 35–40°C (95–104°F), close to our normal body temperature of 37°C. Prokaryotes that live in hot springs, however, contain enzymes with optimal temperatures of 70°C (158°F) or higher. The enzymes of these bacteria are used in a technique that rapidly replicates DNA sequences from small samples.

The optimal pH for most enzymes is near neutrality, in the range of 6–8. There are exceptions, however. Pepsin, a digestive enzyme in the stomach, works best at pH 2. Such an environment would denature most enzymes, but the structure of pepsin is most stable and active in the acidic environment of the stomach.

Cofactors Many enzymes require nonprotein helpers called **cofactors**, which bind to the active site and function in catalysis. The cofactors of some enzymes are inorganic, such as the ions of zinc, iron, and copper. If the cofactor is an organic molecule, it is called a **coenzyme**. Most vitamins are important in nutrition because they function as coenzymes or raw materials from which coenzymes are made. For example, folic acid is a coenzyme for a number of enzymes involved in the synthesis of nucleic acids.

Chemical chaos would result if all of a cell's metabolic pathways were operating simultaneously. A cell must tightly control when and where its various enzymes are active. It does this either by switching on or off the genes that encode specific enzymes or by regulating the activity of enzymes once they are made. We explore this second mechanism in the next module.

? **Explain how an enzyme speeds up a specific reaction.**

An enzyme lowers the activation energy needed for a reaction when its specific substrate enters its active site. With an induced fit, the enzyme strains bonds that need to break or positions substrates in an orientation that aids the conversion of reactants to products.

15 Enzyme inhibitors can regulate enzyme activity in a cell

A chemical that interferes with an enzyme's activity is called an inhibitor. Scientists have learned a great deal about enzyme function by studying the effects of these chemicals. Some inhibitors resemble the enzyme's normal substrate and compete for entry into the active site. As shown in the lower left of **Figure 15A**, such a **competitive inhibitor** reduces an enzyme's productivity by blocking substrate molecules from entering the active site. Competitive inhibition can be overcome by increasing the concentration of the substrate, making it more likely that a substrate molecule rather than an inhibitor will be nearby when an active site becomes vacant.

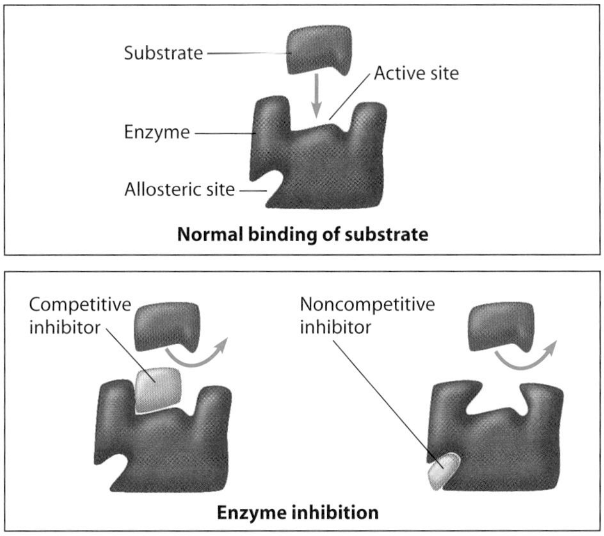

▲ **Figure 15A** How inhibitors interfere with substrate binding

In contrast, a **noncompetitive inhibitor** does not enter the active site. Instead, it binds to the enzyme somewhere else, a place called an allosteric site, and its binding changes the shape of the enzyme so that the active site no longer fits the substrate (lower right of Figure 15A).

Although enzyme inhibition sounds harmful, cells use inhibitors as important regulators of cellular metabolism. Many of a cell's chemical reactions are organized into metabolic pathways in which a molecule is altered in a series of steps, each catalyzed by a specific enzyme, to form a final product. If a cell is producing more of that product than it needs, the product may act as an inhibitor of one of the enzymes early in the pathway. **Figure 15B** illustrates this sort of inhibition, called **feedback inhibition**. Because only weak interactions bind inhibitor and enzyme, this inhibition is reversible. When the product is used up by the cell, the enzyme is no longer inhibited and the pathway functions again.

In the next module, we explore some uses that people make of enzyme inhibitors.

? Explain an advantage of feedback inhibition to a cell.

It prevents the cell from wasting valuable resources by synthesizing more of a particular product than is needed.

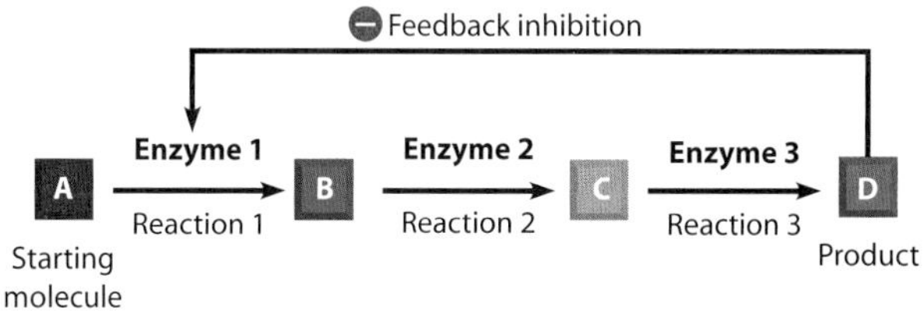

▲ **Figure 15B** Feedback inhibition of a biosynthetic pathway in which product D acts as an inhibitor of enzyme 1

CONNECTION

16 Many drugs, pesticides, and poisons are enzyme inhibitors

Many beneficial drugs act as enzyme inhibitors. Ibuprofen **(Figure 16)** is a common drug that inhibits an enzyme involved in the production of prostaglandins—messenger molecules that increase the sensation of pain and inflammation. Other drugs that function as enzyme inhibitors include some blood pressure medicines and antidepressants. Many antibiotics work by inhibiting enzymes of disease-causing bacteria. Penicillin, for example, blocks the active site of an enzyme that many bacteria use in making cell walls. Protease inhibitors are HIV drugs that target a key viral enzyme. And many cancer drugs are inhibitors of enzymes that promote cell division.

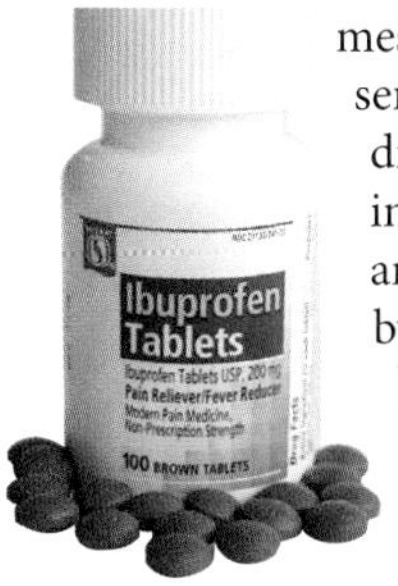

Krista Kennell/Newscom

▲ **Figure 16** Ibuprofen, an enzyme inhibitor

Humans have developed enzyme inhibitors as pesticides, and occasionally as deadly poisons for use in warfare. Poisons often attach to an enzyme by covalent bonds, making the inhibition irreversible. Poisons called nerve gases bind in the active site of an enzyme vital to the transmission of nerve impulses. The inhibition of this enzyme leads to rapid paralysis of vital functions and death. Pesticides such as malathion and parathion are toxic to insects (and dangerous to the people who apply them) because they also irreversibly inhibit this enzyme. Interestingly, some drugs reversibly inhibit this same enzyme and are used in anesthesia and treatment of certain diseases.

? What determines whether enzyme inhibition is reversible or irreversible?

If the inhibitor binds to the enzyme with covalent bonds, the inhibition is usually irreversible. When weak chemical interactions bind inhibitor and enzyme, the inhibition is reversible.

REVIEW

For Practice Quizzes, BioFlix, MP3 Tutors, and Activities, go to www.masteringbiology.com.

Reviewing the Concepts

Membrane Structure and Function (1–9)

1 Membranes are fluid mosaics of lipids and proteins with many functions. The proteins embedded in a membrane's phospholipid bilayer perform various functions.

2 Membranes form spontaneously, a critical step in the origin of life.

3 Passive transport is diffusion across a membrane with no energy investment. Solutes diffuse across membranes down their concentration gradients.

4 Osmosis is the diffusion of water across a membrane.

5 Water balance between cells and their surroundings is crucial to organisms. Cells shrink in a hypertonic solution and swell in a hypotonic solution. In isotonic solutions, animal cells are normal, but plant cells are flaccid.

6 Transport proteins can facilitate diffusion across membranes.

7 Research on another membrane protein led to the discovery of aquaporins. Aquaporins are water channels in cells with high water transport needs.

8 Cells expend energy in the active transport of a solute.

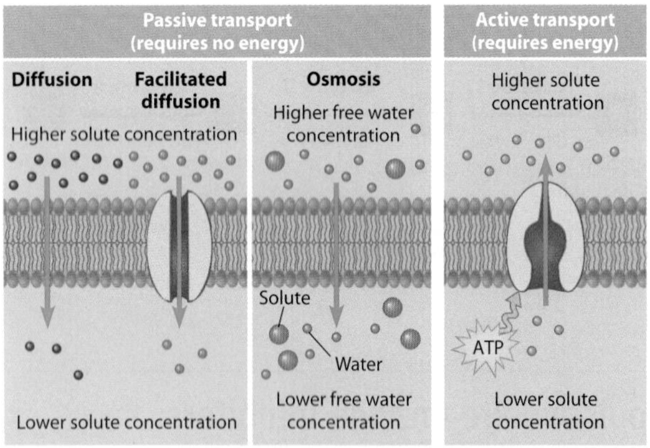

9 Exocytosis and endocytosis transport large molecules across membranes. A vesicle may fuse with the membrane and expel its contents (exocytosis), or the membrane may fold inward, enclosing material from the outside (endocytosis).

Energy and the Cell (10–12)

10 Cells transform energy as they perform work. Kinetic energy is the energy of motion. Potential energy is energy stored in the location or structure of matter. Chemical energy is potential energy available for release in a chemical reaction. According to the laws of thermodynamics, energy can change form but cannot be created or destroyed, and energy transformations increase disorder, or entropy, with some energy being lost as heat.

11 Chemical reactions either release or store energy. Exergonic reactions release energy. Endergonic reactions require energy and yield products rich in potential energy. Metabolism encompasses all of a cell's chemical reactions.

12 ATP drives cellular work by coupling exergonic and endergonic reactions. The transfer of a phosphate group from ATP is involved in chemical, mechanical, and transport work.

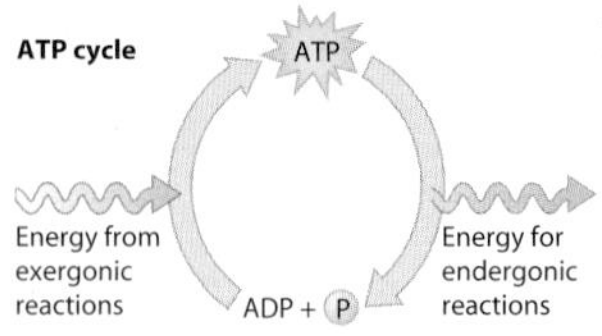

How Enzymes Function (13–16)

13 Enzymes speed up the cell's chemical reactions by lowering energy barriers. Enzymes are protein catalysts that decrease the activation energy (E_A) needed to begin a reaction.

14 A specific enzyme catalyzes each cellular reaction. An enzyme's substrate binds specifically to its active site.

15 Enzyme inhibitors can regulate enzyme activity in a cell. A competitive inhibitor competes with the substrate for the active site. A noncompetitive inhibitor alters an enzyme's function by changing its shape. Feedback inhibition helps regulate metabolism.

16 Many drugs, pesticides, and poisons are enzyme inhibitors.

Connecting the Concepts

1. Fill in the following concept map to review the processes by which molecules move across membranes.

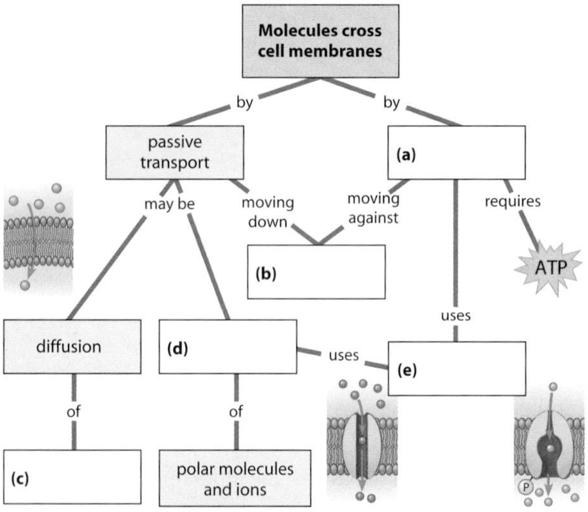

2. Label the parts of the following diagram illustrating the catalytic cycle of an enzyme.

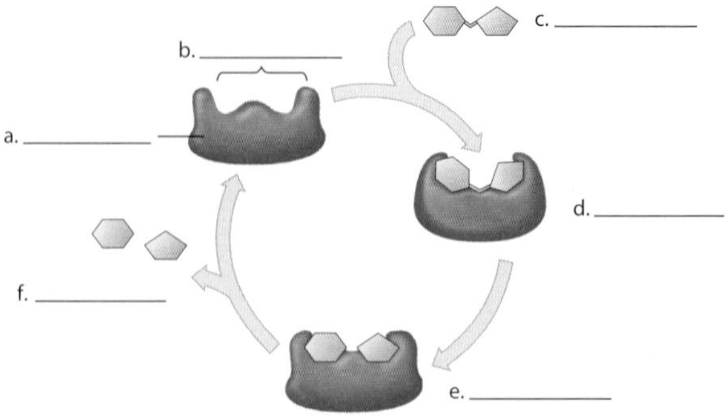

Testing Your Knowledge

Multiple Choice

3. Which best describes the structure of a cell membrane?
   a. proteins between two bilayers of phospholipids
   b. proteins embedded in a bilayer of phospholipids
   c. a bilayer of protein coating a layer of phospholipids
   d. phospholipids between two layers of protein
   e. cholesterol embedded in a bilayer of phospholipids

4. Consider the following: chemical bonds in the gasoline in a car's gas tank and the movement of the car along the road; a biker at the top of a hill and the ride he took to get there. The first parts of these situations illustrate ______, and the second parts illustrate ______.
   a. the first law of thermodynamics ... the second law
   b. kinetic energy ... potential energy
   c. an exergonic reaction ... an endergonic reaction
   d. potential energy ... kinetic energy
   e. the second law of thermodynamics ... the first law

5. A plant cell placed in distilled water will ______; an animal cell placed in distilled water will ______.
   a. burst ... burst
   b. become flaccid ... shrivel
   c. become flaccid ... be normal in shape
   d. become turgid ... be normal in shape
   e. become turgid ... burst

6. The sodium concentration in a cell is 10 times less than the concentration in the surrounding fluid. How can the cell move sodium out of the cell? (*Explain.*)
   a. passive transport
   b. diffusion
   c. active transport
   d. osmosis
   e. any of these processes

7. The synthesis of ATP from ADP and Ⓟ
   a. is an exergonic process.
   b. involves the hydrolysis of a phosphate bond.
   c. transfers a phosphate, priming a protein to do work.
   d. stores energy in a form that can drive cellular work.
   e. releases energy.

8. Facilitated diffusion across a membrane requires ______ and moves a solute ______ its concentration gradient.
   a. transport proteins ... up (against)
   b. transport proteins ... down
   c. energy ... up
   d. energy and transport proteins ... up
   e. energy and transport proteins ... down

Describing, Comparing, and Explaining

9. What are aquaporins? Where would you expect to find them?
10. How do the two laws of thermodynamics apply to living organisms?
11. What are the main types of cellular work? How does ATP provide the energy for this work?
12. Why is the barrier of the activation energy beneficial for organic molecules? Explain how enzymes lower E_A.
13. How do the components and structure of cell membranes relate to the functions of membranes?
14. Sometimes inhibitors can be harmful to a cell; often they are beneficial. Explain.

Applying the Concepts

15. Explain how each of the following food preservation methods would interfere with a microbe's enzyme activity and ability to break down food: canning (heating), freezing, pickling (soaking in acetic acid), salting.

16. A biologist performed two series of experiments on lactase, the enzyme that hydrolyzes lactose to glucose and galactose. First, she made up 10% lactose solutions containing different concentrations of enzyme and measured the rate at which galactose was produced (grams of galactose per minute). Results of these experiments are shown in Table A below. In the second series of experiments (Table B), she prepared 2% enzyme solutions containing different concentrations of lactose and again measured the rate of galactose production.

Table A: Rate and Enzyme Concentration

Lactose concentration	10%	10%	10%	10%	10%
Enzyme concentration	0%	1%	2%	4%	8%
Reaction rate	0	25	50	100	200

Table B: Rate and Substrate Concentration

Lactose concentration	0%	5%	10%	20%	30%
Enzyme concentration	2%	2%	2%	2%	2%
Reaction rate	0	25	50	65	65

   a. Graph and explain the relationship between the reaction rate and the enzyme concentration.
   b. Graph and explain the relationship between the reaction rate and the substrate concentration. How and why did the results of the two experiments differ?

17. The following graph shows the rate of reaction for two different enzymes: One is pepsin, a digestive enzyme found in the stomach; the other is trypsin, a digestive enzyme found in the intestine. As you may know, gastric juice in the stomach contains hydrochloric acid. Which curve belongs to which enzyme?

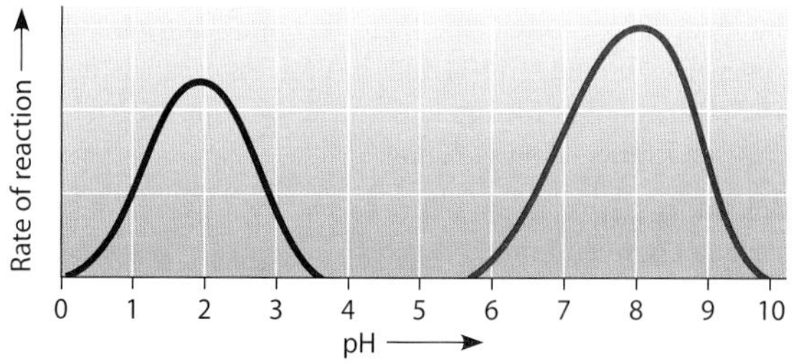

A lysosome, a digestive organelle in a cell, has an internal pH of around 4.5. Draw a curve on the graph that you would predict for a lysosomal enzyme, labeling its optimal pH.

18. Organophosphates (organic compounds containing phosphate groups) are commonly used as insecticides to improve crop yield. Organophosphates typically interfere with nerve signal transmission by inhibiting the enzymes that degrade transmitter molecules. They affect humans and other vertebrates as well as insects. Thus, the use of organophosphate pesticides poses some health risks. On the other hand, these molecules break down rapidly upon exposure to air and sunlight. As a consumer, what level of risk are you willing to accept in exchange for an abundant and affordable food supply?

Review Answers

1. a. active transport; b. concentration gradient; c. small nonpolar molecules; d. facilitated diffusion; e. transport proteins
2. a. enzyme; b. active site of enzyme; c. substrate; d. substrate in active site; induced fit strains substrate bonds; e. substrate converted to products; f. product molecules released; enzyme is ready for next catalytic cycle
3. b **4.** d **5.** e **6.** c (Only active transport can move solute against a concentration gradient.) **7.** d **8.** b
9. Aquaporins are water transport channels that allow for very rapid diffusion of water through a cell membrane. They are found in cells that have high water transport needs, such as blood cells, kidney cells, and plant cells.
10. Energy is neither created nor destroyed but can be transferred and transformed. Plants transform the energy of sunlight into chemical energy stored in organic molecules. Almost all organisms rely on the products of photosynthesis for the source of their energy. In every energy transfer or transformation, disorder increases as some energy is lost to the random motion of heat.
11. The work of cells falls into three main categories: mechanical, chemical, and transport. ATP provides the energy for cellular work by transferring a phosphate group to a protein (movement and transport) or to a substrate (chemical).
12. Energy is stored in the chemical bonds of organic molecules. The barrier of E_A prevents these molecules from spontaneously breaking down and releasing that energy. When a substrate fits into an enzyme's active site with an induced fit, its bonds may be strained and thus easier to break, or the active site may orient two substrates in such a way as to facilitate the reaction.
13. Cell membranes are composed of a phospholipid bilayer with embedded proteins. The bilayer creates the hydrophobic boundary between cells and their surroundings (or between organelles and the cytoplasm). The proteins perform the many functions of membranes, such as enzyme action, transport, attachment, and signaling.
14. Inhibitors that are toxins or poisons irreversibly inhibit key cellular enzymes. Inhibitors that are designed as drugs are beneficial, such as when they interfere with the enzymes of bacterial or viral invaders or cancer cells. Cells use feedback inhibition of enzymes in metabolic pathways as important mechanisms that conserve resources.
15. Heating, pickling, and salting denature enzymes, changing their shapes so they do not fit substrates. Freezing decreases the kinetic energy of molecules, so enzymes are less likely to interact with their substrates.
16. a. The more enzyme present, the faster the rate of reaction, because it is more likely that enzyme and substrate molecules will meet.

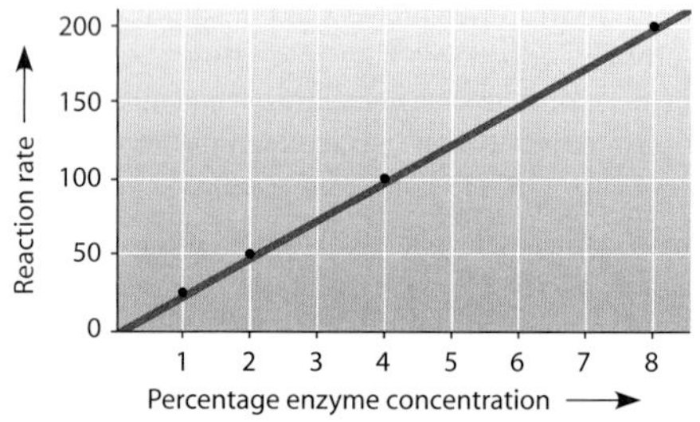

b. The more substrate present, the faster the reaction, for the same reason, but only up to a point. An enzyme molecule can work only so fast; once it is saturated (working at top speed), more substrate does not increase the rate.

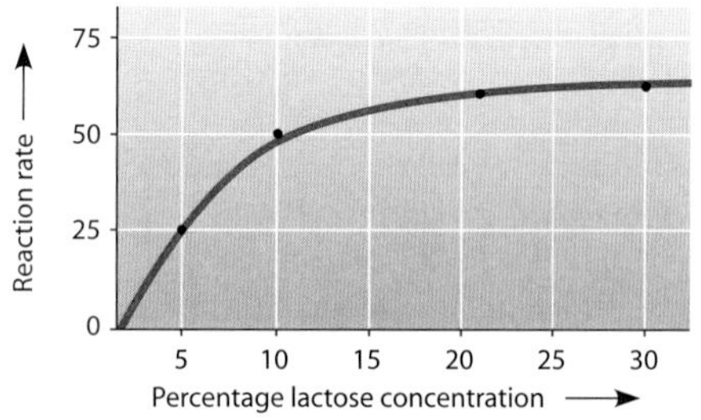

17. The black curve on the left would correspond to the stomach enzyme pepsin, which has a lower optimal pH, as is found in the stomach; the red curve on the right would correspond to trypsin, which has a higher optimal pH. The curve for a lysosomal enzyme should have an optimal pH at 4.5.

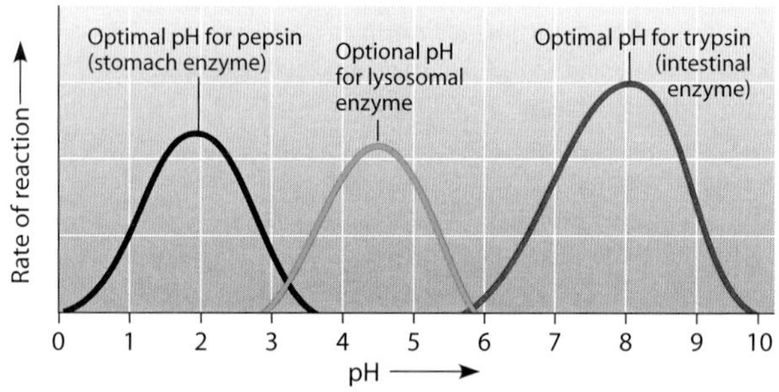

18. Some issues and questions to consider: Is improving crop yields of paramount importance in a world where many people can't get enough food? Does the fact that these compounds rapidly break down indicate that the risk to humans is low? How about the risks to people who work in agriculture or to other organisms, such as bees and other pollinating insects, birds, and small mammals? Might there be negative effects on ecosystems that are impossible to predict?

How Cells Harvest Chemical Energy

How Cells Harvest Chemical Energy

BIG IDEAS

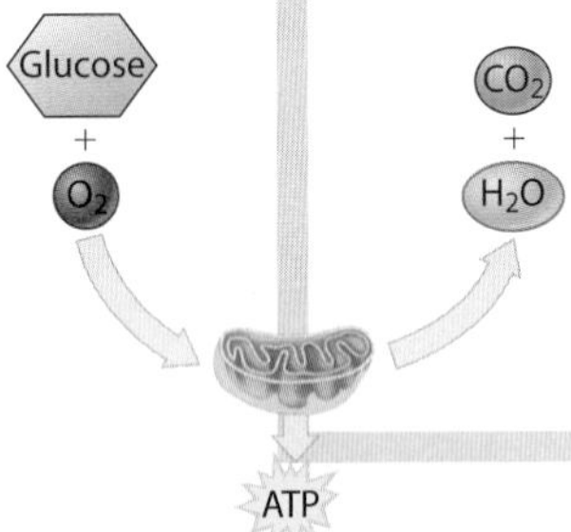

Cellular Respiration: Aerobic Harvesting of Energy
(1–5)

Cellular respiration oxidizes fuel molecules and generates ATP for cellular work.

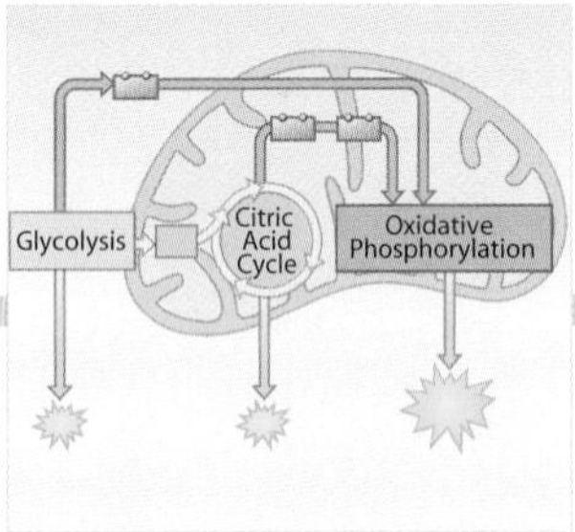

Stages of Cellular Respiration
(6–12)

The main stages of cellular respiration are glycolysis, the citric acid cycle, and oxidative phosphorylation.

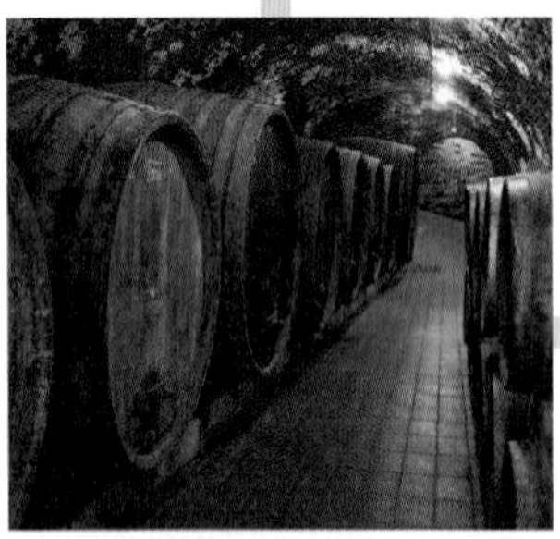

StockLite/Shutterstock

Fermentation: Anaerobic Harvesting of Energy
(13–14)

Fermentation regenerates NAD^+, allowing glycolysis and ATP production to continue without oxygen.

Fiona Rogers/naturepl.com

Connections Between Metabolic Pathways
(15–16)

The breakdown pathways of cellular respiration intersect with biosynthetic pathways.

Arto Hakola/Shutterstock

Potential energy stored in the chemical bonds of fuel molecules can be converted to kinetic energy, such as the movement of a car or, as you can see here, the leaping of a lemur. Where did the lemur get the energy for this dramatic leap? The obvious answer is, of course, from its food. But the more complete answer, as you will learn in this chapter, is from the harvesting of energy from food molecules that takes place in every cell in an animal's body. In fact, this process, called cellular respiration, also occurs in the cells of plants, fungi, and protists. And a similar process takes place in most prokaryotic organisms.

Cellular respiration is the breakdown of sugars and other food molecules in the presence of oxygen to carbon dioxide and water, generating a large amount of ATP, the energy currency that "pays for" cellular work. In the muscle cells of this leaping lemur, ATP powers the contraction of its muscles. The lemur's cells also use the energy of ATP to build and maintain cell structure, transport materials across membranes, manufacture products, grow, and divide.

In this chapter, we present some basic concepts about cellular respiration and then focus on the key stages of the process: glycolysis, the citric acid cycle, and oxidative phosphorylation. We'll also consider fermentation, an extended version of glycolysis that has deep evolutionary roots. We complete the chapter with a look at how the metabolic pathways that break down organic molecules connect to those that build such molecules.

But first let's take a step back and consider the original source of the energy for most cellular work on Earth today.

Cellular Respiration: Aerobic Harvesting of Energy

1 Photosynthesis and cellular respiration provide energy for life

Life requires energy. In almost all ecosystems, that energy ultimately comes from the sun. Photosynthesis is the process by which the sun's energy is captured. A brief overview here of the relationship between photosynthesis and cellular respiration will illustrate how these two processes provide energy for life (**Figure 1**). In photosynthesis, which takes place in a plant cell's chloroplast, the energy of sunlight is used to rearrange the atoms of carbon dioxide (CO_2) and water (H_2O) to produce glucose and oxygen (O_2). The lemur in Figure 1 obtains energy for leaping from tree to tree by eating plants. In **cellular respiration**, O_2 is consumed as glucose is broken down to CO_2 and H_2O; the cell captures the energy released in ATP. Cellular respiration takes place in the mitochondria of all eukaryotic cells.

This figure also shows that in these energy conversions, some energy is lost as heat. Life on Earth is solar powered, and energy makes a one-way trip through an ecosystem. Chemicals, however, are recycled. The CO_2 and H_2O released by cellular respiration are converted through photosynthesis to glucose and O_2, which are then used in respiration.

? What is misleading about the following statement? "Plant cells perform photosynthesis, and animal cells perform cellular respiration."

The statement implies that cellular respiration does not occur in plant cells. In fact, almost all eukaryotic cells use cellular respiration to obtain energy for their cellular work.

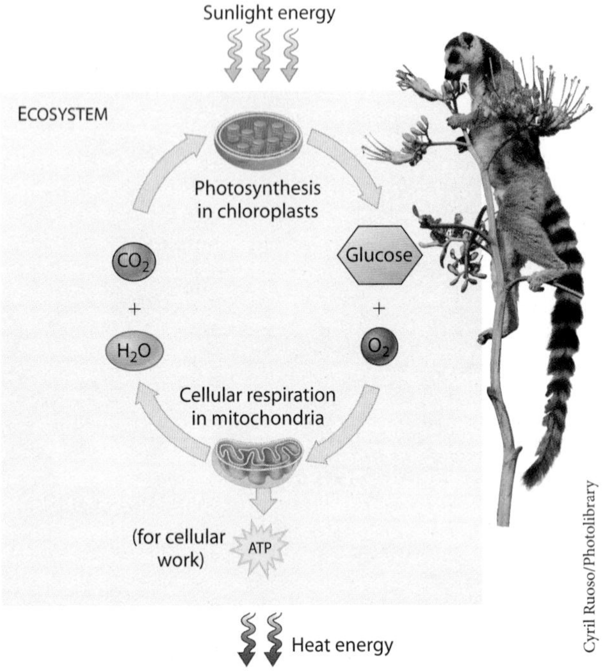

▲ **Figure 1** The connection between photosynthesis and cellular respiration

2 Breathing supplies O_2 for use in cellular respiration and removes CO_2

We often use the word *respiration* as a synonym for "breathing," the meaning of its Latin root. In that sense of the word, respiration refers to an exchange of gases: An organism obtains O_2 from its environment and releases CO_2 as a waste product. Biologists also define respiration as the aerobic (oxygen-requiring) harvesting of energy from food molecules by cells. This process is called cellular respiration to distinguish it from breathing.

Breathing and cellular respiration are closely related. As the runner in **Figure 2** breathes in air, her lungs take up O_2 and pass it to her bloodstream. The bloodstream carries the O_2 to her muscle cells. Mitochondria in the muscle cells use the O_2 in cellular respiration to harvest energy from glucose and other organic molecules and generate ATP. Muscle cells use ATP to contract. The runner's bloodstream and lungs also perform the vital function of disposing of the CO_2 waste , which, as you can see by the equation at the bottom of the figure, is produced in cellular respiration.

? How is your breathing related to your cellular respiration?

In breathing, CO_2 and O_2 are exchanged between your lungs and the air. In cellular respiration, cells use the O_2 obtained through breathing to break down fuel, releasing CO_2 as a waste product.

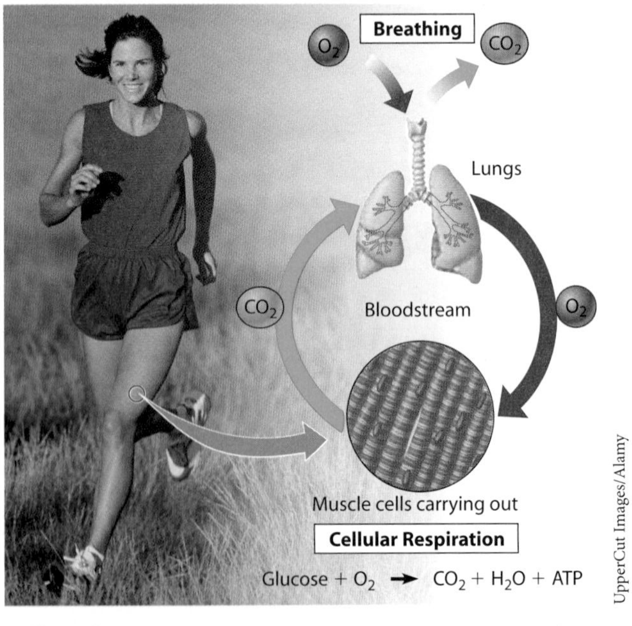

▲ **Figure 2** The connection between breathing and cellular respiration

3 Cellular respiration banks energy in ATP molecules

As the runner example in Figure 2 implies, oxygen usage is only a means to an end. Generating ATP for cellular work is the fundamental function of cellular respiration.

The balanced chemical equation in **Figure 3** summarizes cellular respiration as carried out by cells that use O_2 in harvesting energy from glucose. The simple sugar glucose ($C_6H_{12}O_6$) is the fuel that cells use most often, although other organic molecules can also be "burned" in cellular respiration. The equation tells us that the atoms of the reactant molecules $C_6H_{12}O_6$ and O_2 are rearranged to form the products CO_2 and H_2O. In this exergonic process, the chemical energy of the bonds in glucose is released and stored (or "banked") in the chemical bonds of ATP. The series of arrows in Figure 3 indicates that cellular respiration consists of many steps, not just a single reaction.

Cellular respiration can produce up to 32 ATP molecules for each glucose molecule, a capture of about 34% of the energy originally stored in glucose. The rest of the energy is released as heat. This may seem inefficient, but it compares very well with

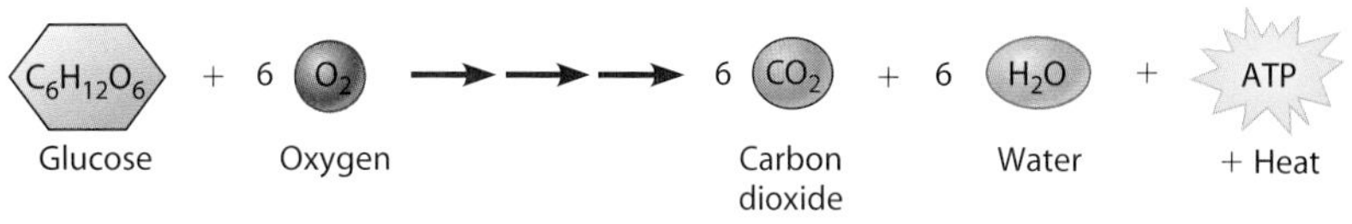

▲ **Figure 3** Summary equation for cellular respiration:
$C_6H_{12}O_6 + 6\ O_2 \longrightarrow 6\ CO_2 + 6\ H_2O + \text{Energy (ATP + Heat)}$

the efficiency of most energy-conversion systems. For instance, the average automobile engine is able to convert only about 25% of the energy in gasoline to the kinetic energy of movement.

How great are the energy needs of a cell? If ATP could not be regenerated through cellular respiration, you would use up nearly your body weight in ATP each day. Let's consider the energy requirements for various human activities next.

? Why are sweating and other body-cooling mechanisms necessary during vigorous exercise?

● The demand for ATP is supported by an increased rate of cellular respiration, but about 66% of the energy from food produces heat instead of ATP.

CONNECTION | # 4 The human body uses energy from ATP for all its activities

Your body requires a continuous supply of energy just to stay alive—to keep the heart pumping, to breathe, and to maintain body temperature. Your brain requires a huge amount of energy; its cells burn about 120 grams (g)—a quarter of a pound!—of glucose a day, accounting for about 15% of total oxygen consumption. Maintaining brain cells and other life-sustaining activities uses as much as 75% of the energy a person takes in as food during a typical day.

Above and beyond the energy you need for body maintenance, cellular respiration provides energy for voluntary activities. **Figure 4** shows the amount of energy it takes to perform some of these activities. The energy units are **kilocalories (kcal)**, the quantity of heat required to raise the temperature of 1 kilogram (kg) of water by 1°C. (The "Calories" listed on food packages are actually kilocalories, usually signified by a capital C.) The values shown do not include the energy the body consumes for its basic life-sustaining activities. Even sleeping or lying quietly requires energy for metabolism.

The U.S. National Academy of Sciences estimates that the average adult human needs to take in food that provides about 2,200 kcal of energy per day. This includes the energy expended in both maintenance and voluntary activity. Now we begin the study of how cells liberate the energy stored in fuel molecules to produce the ATP used to power the work of cells and thus the activities of the body.

Activity	kcal consumed per hour by a 67.5-kg (150-lb) person*
Running (8–9 mph)	979
Dancing (fast)	510
Bicycling (10 mph)	490
Swimming (2 mph)	408
Walking (4 mph)	341
Walking (3 mph)	245
Dancing (slow)	204
Driving a car	61
Sitting (writing)	28

*Not including kcal needed for body maintenance

▲ **Figure 4** Energy consumed by various activities

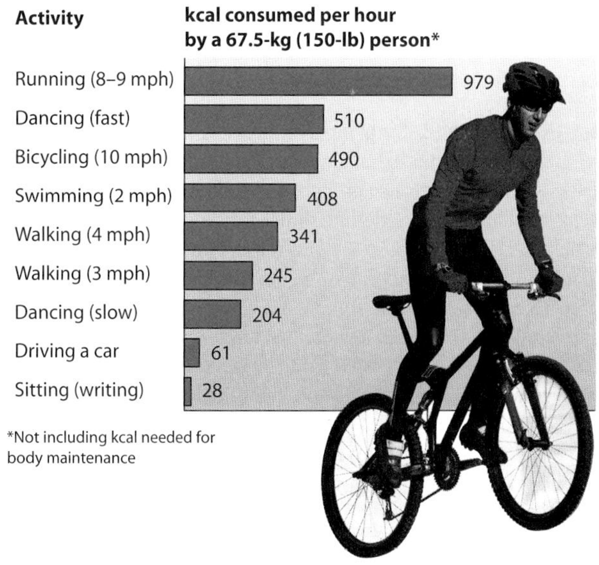

Corbis RF/Getty Images

? Walking at 3 mph, how far would you have to travel to "burn off" the equivalent of an extra slice of pizza, which has about 475 kcal? How long would that take?

● You would have to walk about 6 miles, which would take you about 2 hours. (Now you understand why the most effective exercise for losing weight is pushing away from the table!)

Data from C. M. Taylor and G. M. McLeod, *Rose's Laboratory Handbook for Dietetics*, 5th ed. (New York: Macmillan, 1949), p. 18; J. V. G. A. Durnin and R. Passmore, *Energy and Protein Requirements in FAO/WHO Technical Report* No. 522, 1973; W. D. McArdle, F. I. Katch, and V. L. Katch, *Exercise Physiology* (Philadelphia, PA: Lea & Feibiger, 1981); R. Passmore and J. V. G. A. Durnin, *Physiological Reviews* vol. 35, pp. 801–840 (1955).

5 Cells tap energy from electrons "falling" from organic fuels to oxygen

How do your cells extract energy from glucose? The answer involves the transfer of electrons during chemical reactions.

Redox Reactions During cellular respiration, electrons are transferred from glucose to oxygen, releasing energy. Oxygen attracts electrons very strongly, and an electron loses potential energy when it "falls" to oxygen. If you burn a cube of sugar, this electron fall happens very rapidly, releasing energy in the form of heat and light. Cellular respiration is a more controlled descent of electrons—more like stepping down an energy staircase, with energy released in small amounts that can be stored in the chemical bonds of ATP.

The movement of electrons from one molecule to another is an oxidation-reduction reaction, or **redox reaction** for short. In a redox reaction, the loss of electrons from one substance is called **oxidation**, and the addition of electrons to another substance is called **reduction**. A molecule is said to become oxidized when it loses one or more electrons and reduced when it gains one or more electrons. Because an electron transfer requires both a donor and an acceptor, oxidation and reduction always go together.

In the cellular respiration equation in **Figure 5A** below, you cannot see any electron transfers. What you do see are changes in the location of hydrogen atoms. These hydrogen movements represent electron transfers because each hydrogen atom consists of an electron (e^-) and a proton (hydrogen ion, or H^+). Glucose ($C_6H_{12}O_6$) loses hydrogen atoms (electrons) as it becomes oxidized to CO_2; simultaneously, O_2 gains hydrogen atoms (electrons) as it becomes reduced to H_2O. As they pass from glucose to oxygen, the electrons lose potential energy.

NADH and Electron Transport Chains An important player in the process of oxidizing glucose is a coenzyme called NAD^+, which accepts electrons and becomes reduced to NADH. NAD^+ (nicotinamide adenine dinucleotide) is an organic molecule that cells make from the vitamin niacin and use to shuttle electrons in redox reactions. The top equation in **Figure 5B** depicts the oxidation of an organic molecule. We show only its three carbons (●) and a few of its other atoms. An enzyme called dehydrogenase strips two hydrogen atoms from this molecule. Simultaneously, as shown in the lower equation, NAD^+ picks up the two electrons (⊖) and becomes reduced to NADH. One hydrogen ion (H^+) is released. (NADH is represented throughout this chapter as a light brown box carrying two blue electrons.)

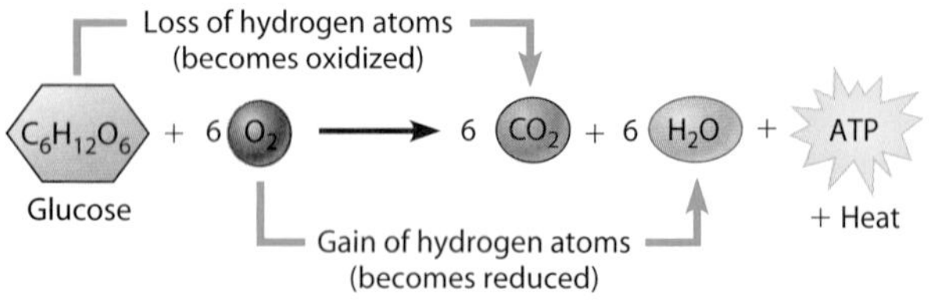

▲ **Figure 5A** Rearrangement of hydrogen atoms (with their electrons) in the redox reactions of cellular respiration

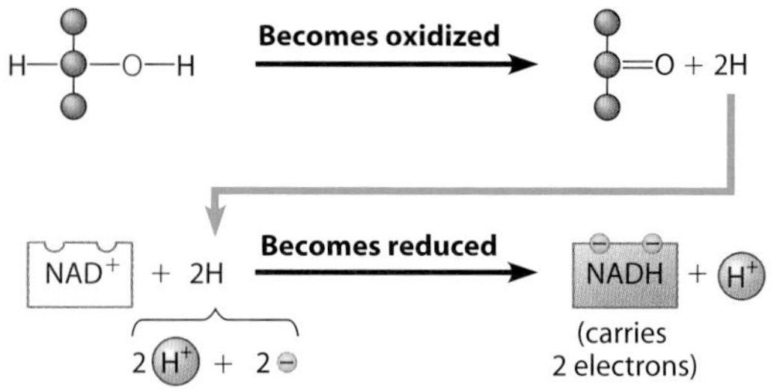

▲ **Figure 5B** A pair of redox reactions occuring simultaneously

Using the energy staircase analogy for electrons falling from glucose to oxygen, the transfer of electrons from an organic molecule to NAD^+ represents the first step. **Figure 5C** shows NADH delivering these electrons to the rest of the staircase—an **electron transport chain**. The steps in the chain are electron carrier molecules, shown here as purple ovals, built into the inner membrane of a mitochondrion. At the bottom of the staircase is O_2, the final electron acceptor.

The electron transport chain undergoes a series of redox reactions in which electrons pass from carrier to carrier down to oxygen. The redox steps in the staircase release energy in amounts small enough to be used by the cell to make ATP.

With an understanding of this basic mechanism of electron transfer and energy release, we can now explore cellular respiration in more detail.

? What chemical characteristic of the element oxygen accounts for its function in cellular respiration?

● Oxygen is very electronegative (see Module 2.6), making it very powerful in pulling electrons down the electron transport chain.

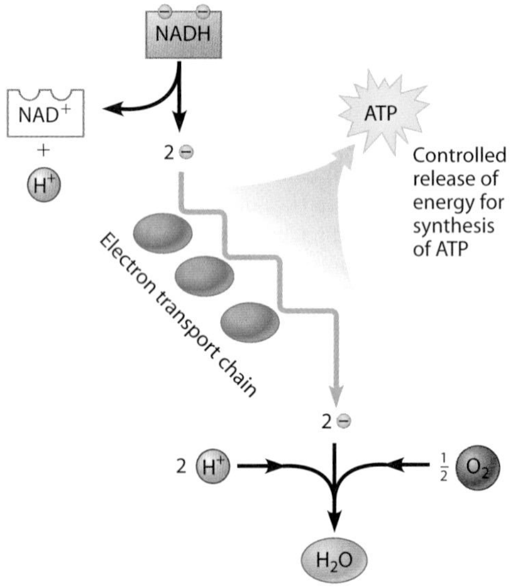

▲ **Figure 5C** In cellular respiration, electrons fall down an energy staircase and finally reduce O_2.

Stages of Cellular Respiration

6 Overview: Cellular respiration occurs in three main stages

Cellular respiration consists of a sequence of steps that can be divided into three main stages. **Figure 6** gives an overview of the three stages and shows where they occur in a eukaryotic cell. (In prokaryotic cells that use aerobic respiration, these steps occur in the cytoplasm, and the electron transport chain is built into the plasma membrane.)

Stage 1: Glycolysis (shown with an aqua background throughout this chapter) occurs in the cytoplasmic fluid of the cell. Glycolysis begins cellular respiration by breaking glucose into two molecules of a three-carbon compound called pyruvate.

Stage 2: Pyruvate oxidation and the citric acid cycle (shown in a salmon color) take place within the mitochondria. Pyruvate is oxidized to a two-carbon compound. The citric acid cycle then completes the breakdown of glucose to carbon dioxide. As suggested by the smaller ATP symbols in the diagram, the cell makes a small amount of ATP during glycolysis and the citric acid cycle. The main function of these first two stages, however, is to supply the third stage of respiration with electrons (shown with gold arrows).

Stage 3: Oxidative phosphorylation (purple background) requires an electron transport chain and a process known as chemiosmosis. NADH and a related electron carrier, $FADH_2$ (flavin adenine dinucleotide), shuttle electrons to an electron transport chain embedded in the inner mitochondrion membrane. Most of the ATP produced by cellular respiration is generated by oxidative phosphorylation, which uses the energy released by the downhill fall of electrons from NADH and $FADH_2$ to O_2 to phosphorylate ADP.

What couples the electron transport chain to ATP synthesis? As the electron transport chain passes electrons down the energy staircase, it also pumps hydrogen ions (H^+) across the inner mitochondrial membrane into the narrow intermembrane space. The result is a concentration gradient of H^+ across the membrane. In **chemiosmosis**, the potential energy of this concentration gradient is used to make ATP. The details of this process are explored in Module 10. In 1978, British biochemist Peter Mitchell was awarded the Nobel Prize for developing the theory of chemiosmosis.

The small amount of ATP produced in glycolysis and the citric acid cycle is made by substrate-level phosphorylation, a process we discuss in the next module. In the next several modules, we look more closely at the three stages of cellular respiration and the two mechanisms of ATP synthesis.

? Of the three main stages of cellular respiration represented in Figure 6, which one uses oxygen to extract chemical energy from organic compounds?

Oxidative phosphorylation, using the electron transport chain, which eventually transfers electrons to oxygen

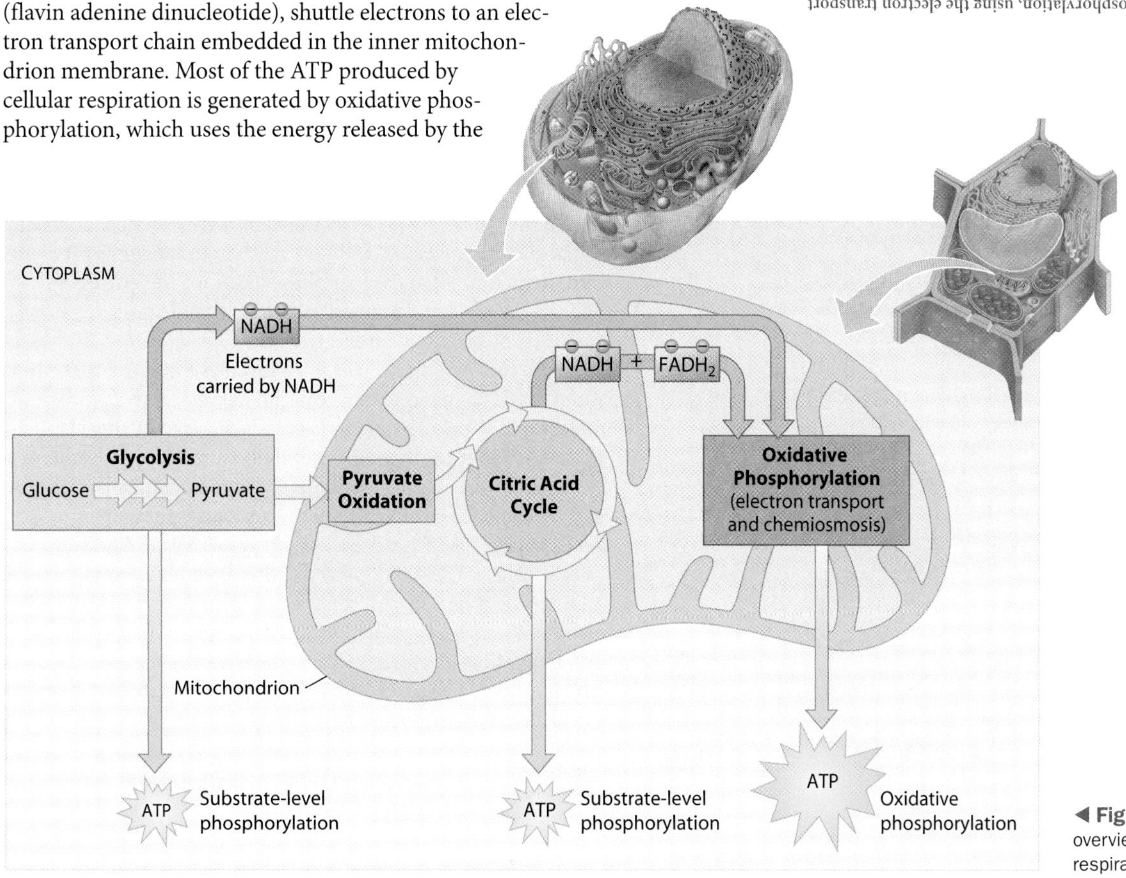

◀ **Figure 6** An overview of cellular respiration

7 Glycolysis harvests chemical energy by oxidizing glucose to pyruvate

Now that you have been introduced to the major players and processes, it's time to focus on the individual stages of cellular respiration. The term for the first stage, *glycolysis*, means "splitting of sugar" (*glyco*, sweet, and *lysis*, split), and that's exactly what happens during this phase.

Figure 7A below gives an overview of glycolysis in terms of input and output. Glycolysis begins with a single molecule of glucose and concludes with two molecules of pyruvate. (Pyruvate is the ionized form of pyruvic acid.) The gray balls represent the carbon atoms in each molecule; glucose has six carbons, and these same six carbons end up in the two molecules of pyruvate (three carbons in each). The straight arrow from glucose to pyruvate actually represents nine chemical steps, each catalyzed by its own enzyme. As these reactions occur, two molecules of NAD^+ are reduced to two molecules of NADH, and a net gain of two molecules of ATP is produced.

Figure 7B illustrates how ATP is formed in glycolysis by the process called **substrate-level phosphorylation**. In this process, an enzyme transfers a phosphate group from a substrate molecule directly to ADP, forming ATP. You will meet substrate-level phosphorylation again in the citric acid cycle, where a small amount of ATP is generated by this process.

The energy extracted from glucose during glycolysis is banked in a combination of ATP and NADH. The cell can use the energy in ATP immediately, but for it to use the energy in NADH, electrons from NADH must pass down an electron transport chain located in the inner mitochondrial membrane. And the pyruvate molecules still hold most of the energy of glucose; these molecules will be oxidized in the citric acid cycle.

Let's take a closer look at glycolysis. **Figure 7C**, on the next page, shows all the organic compounds that form in the nine chemical reactions of glycolysis. Commentary on the left

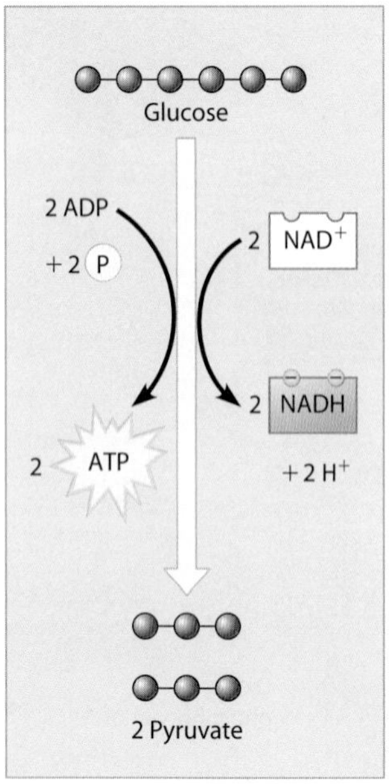

▶ **Figure 7A** An overview of glycolysis

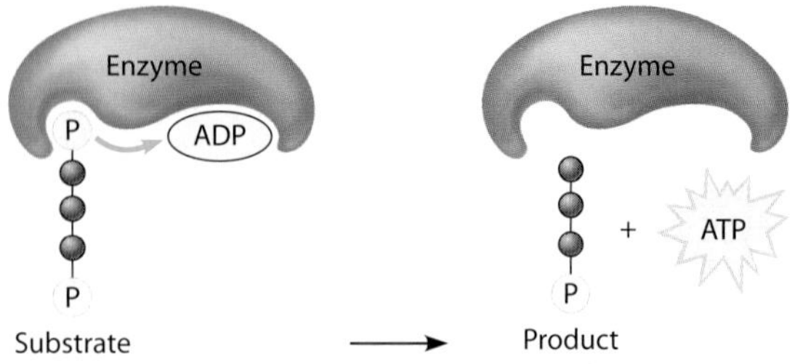

▲ **Figure 7B** Substrate-level phosphorylation: transfer of a phosphate group (P) from a substrate to ADP, producing ATP

highlights the main features of these reactions. The gray balls represent the carbon atoms in each of the compounds named on the right side of the figure.

The compounds that form between the initial reactant, glucose, and the final product, pyruvate, are known as **intermediates**. Glycolysis is an example of a *metabolic pathway*, in which each chemical step leads to the next one. For instance, the intermediate glucose 6-phosphate is the product of step 1 and the reactant for step 2. Similarly, fructose 6-phosphate is the product of step 2 and the reactant for step 3. Also essential are the specific enzymes that catalyze each chemical step; however, to help keep the figure simple, we have not included the enzymes.

As indicated in Figure 7C, the steps of glycolysis can be grouped into two main phases. Steps ❶–❹, the energy investment phase, actually *consume* energy. In this phase, ATP is used to energize a glucose molecule, which is then split into two small sugars that are now primed to release energy. The figure follows both of these three-carbon sugars through the second phase.

Steps ❺–❾, the energy payoff phase, *yield* energy for the cell. In this phase, two NADH molecules are produced for each initial glucose molecule, and four ATP molecules are generated. Remember that the first phase used two molecules of ATP, so the net gain to the cell is two ATP molecules for each glucose molecule that enters glycolysis.

These two ATP molecules from glycolysis account for only about 6% of the energy that a cell can harvest from a glucose molecule. The two NADH molecules generated during step 5 represent about another 16%, but their stored energy is not available for use in the absence of O_2. Some organisms—yeasts and certain bacteria, for instance—can satisfy their energy needs with the ATP produced by glycolysis alone. And some cells, such as muscle cells, may use this anaerobic production of ATP for short periods. Most cells and organisms, however, have far greater energy demands. The stages of cellular respiration that follow glycolysis release much more energy. In the next modules, we see what happens in most organisms after glycolysis oxidizes glucose to pyruvate.

? For each glucose molecule processed, what are the net molecular products of glycolysis?

Two molecules of pyruvate, two molecules of ATP, and two molecules of NADH

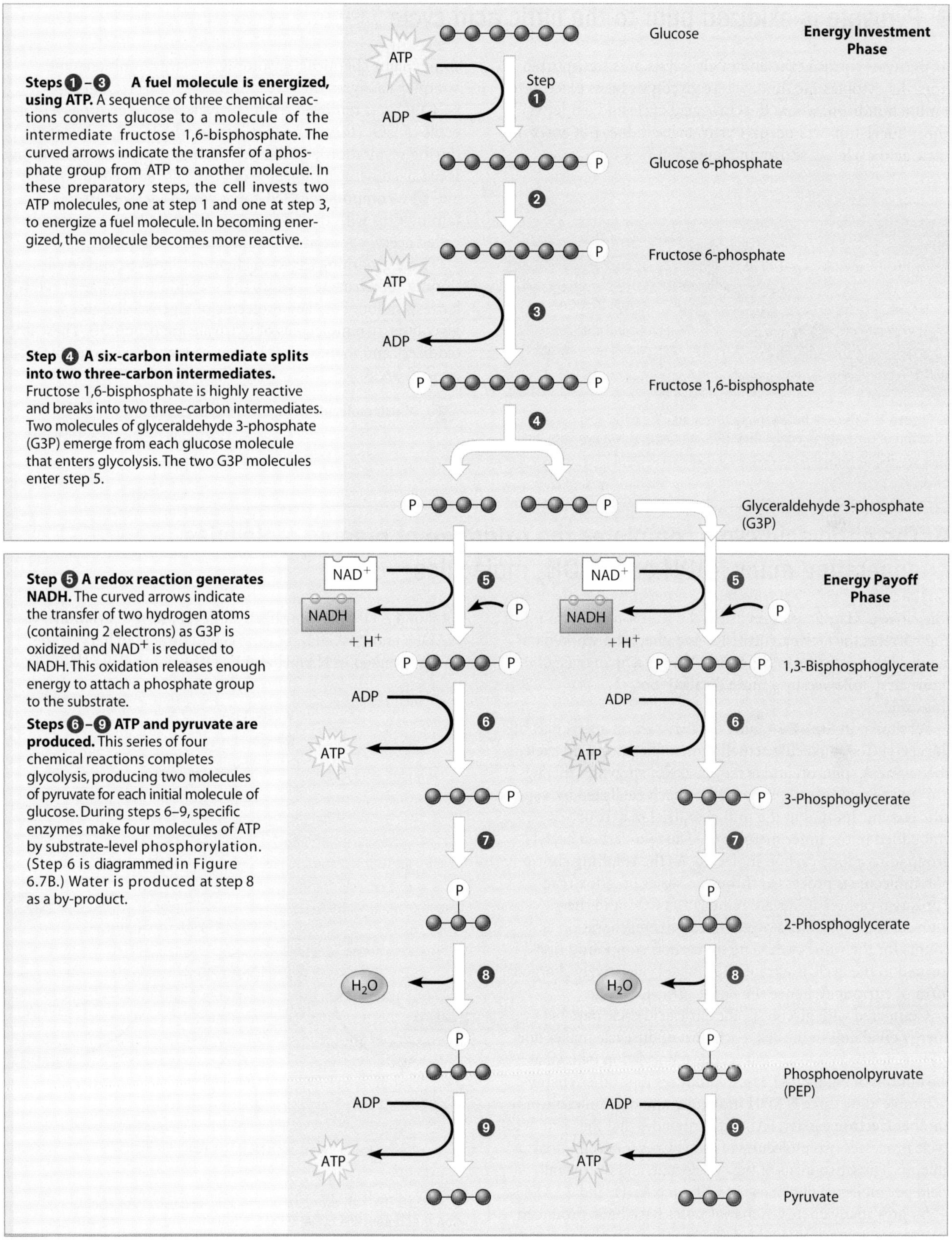

Energy Investment Phase
Glucose
ATP
ADP
Step 1
Glucose 6-phosphate
2
Fructose 6-phosphate
ATP
ADP
3
Fructose 1,6-bisphosphate
4
Glyceraldehyde 3-phosphate (G3P)
Steps 1–3 A fuel molecule is energized, using ATP. A sequence of three chemical reactions converts glucose to a molecule of the intermediate fructose 1,6-bisphosphate. The curved arrows indicate the transfer of a phosphate group from ATP to another molecule. In these preparatory steps, the cell invests two ATP molecules, one at step 1 and one at step 3, to energize a fuel molecule. In becoming energized, the molecule becomes more reactive.
Step 4 A six-carbon intermediate splits into two three-carbon intermediates. Fructose 1,6-bisphosphate is highly reactive and breaks into two three-carbon intermediates. Two molecules of glyceraldehyde 3-phosphate (G3P) emerge from each glucose molecule that enters glycolysis. The two G3P molecules enter step 5.
Energy Payoff Phase
NAD^+
NADH
$+ H^+$
5
P
1,3-Bisphosphoglycerate
ADP
ATP
6
3-Phosphoglycerate
7
2-Phosphoglycerate
H_2O
8
Phosphoenolpyruvate (PEP)
ADP
ATP
9
Pyruvate
Step 5 A redox reaction generates NADH. The curved arrows indicate the transfer of two hydrogen atoms (containing 2 electrons) as G3P is oxidized and NAD^+ is reduced to NADH. This oxidation releases enough energy to attach a phosphate group to the substrate.
Steps 6–9 ATP and pyruvate are produced. This series of four chemical reactions completes glycolysis, producing two molecules of pyruvate for each initial molecule of glucose. During steps 6–9, specific enzymes make four molecules of ATP by substrate-level phosphorylation. (Step 6 is diagrammed in Figure 6.7B.) Water is produced at step 8 as a by-product.

▲ **Figure 7C** Details of glycolysis

8 Pyruvate is oxidized prior to the citric acid cycle

As pyruvate forms at the end of glycolysis, it is transported from the cytoplasmic fluid, where glycolysis takes place, into a mitochondrion, where the citric acid cycle and oxidative phosphorylation will occur. Pyruvate itself does not enter the citric acid cycle. As shown in **Figure 8**, it first undergoes some major chemical "grooming." A large, multienzyme complex catalyzes three reactions: ❶ A carboxyl group ($—COO^-$) is removed from pyruvate and given off as a molecule of CO_2 (this is the first step in which CO_2 is released during respiration); ❷ the two-carbon compound remaining is oxidized while a molecule of NAD^+ is reduced to NADH; and ❸ a compound called coenzyme A, derived from a B vitamin, joins with the two-carbon group to form a molecule called acetyl coenzyme A, abbreviated **acetyl CoA**.

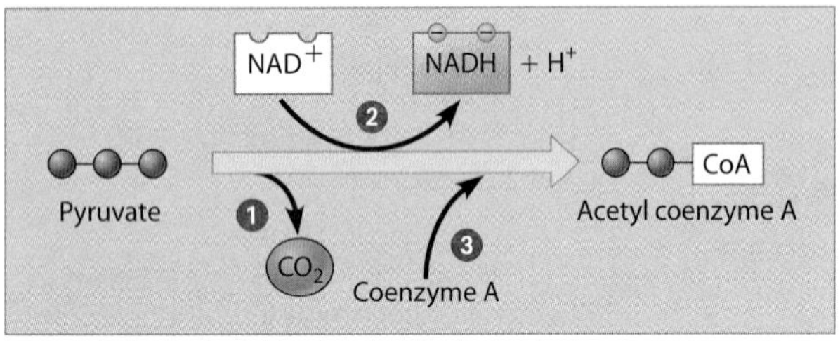

▲ **Figure 8** The link between glycolysis and the citric acid cycle: the oxidation of pyruvate to acetyl CoA. (Remember that two pyruvate, and thus two acetyl CoA, are produced from each glucose.)

These grooming steps—a chemical "haircut and conditioning" of pyruvate—set up the second major stage of cellular respiration. For each molecule of glucose that enters glycolysis, two molecules of pyruvate are produced. These are oxidized, and then two molecules of acetyl CoA enter the citric acid cycle.

? Which molecule in Figure 8 has been reduced?

NAD^+ has been reduced to NADH.

9 The citric acid cycle completes the oxidation of organic molecules, generating many NADH and $FADH_2$ molecules

The citric acid cycle is often called the Krebs cycle in honor of Hans Krebs, the German-British researcher who worked out much of this pathway in the 1930s. We present an overview figure first, followed by a more detailed look at this cycle.

As shown in **Figure 9A**, only the two-carbon acetyl part of the acetyl CoA molecule actually enters the citric acid cycle; coenzyme A splits off and is recycled. Not shown in this figure are the multiple steps that follow, each catalyzed by a specific enzyme located in the mitochondrial matrix or embedded in the inner membrane. The two-carbon acetyl group joins a four-carbon molecule. As the resulting six-carbon molecule is processed through a series of redox reactions, two carbon atoms are removed as CO_2, and the four-carbon molecule is regenerated; this regeneration accounts for the word cycle. The six-carbon compound first formed in the cycle is citrate, the ionized (negatively charged) form of citric acid; hence the name *citric acid cycle.*

Compared with glycolysis, the citric acid cycle pays big energy dividends to the cell. Each turn of the cycle makes one ATP molecule by substrate-level phosphorylation (shown at the bottom of Figure 9A). It also produces four other energy-rich molecules: three NADH molecules and one molecule of another electron carrier, $FADH_2$. Remember that the citric acid cycle processes two molecules of acetyl CoA for each initial glucose. Thus, two turns of the cycle occur, and the overall yield per molecule of glucose is 2 ATP, 6 NADH, and 2 $FADH_2$.

So how many energy-rich molecules have been produced by processing one molecule of glucose through glycolysis and the citric acid cycle? Up to this point, the cell has gained a total of 4 ATP (all from substrate-level phosphorylation), 10 NADH, and 2 $FADH_2$. For the cell to be able to harvest the energy banked in NADH and $FADH_2$, these molecules must

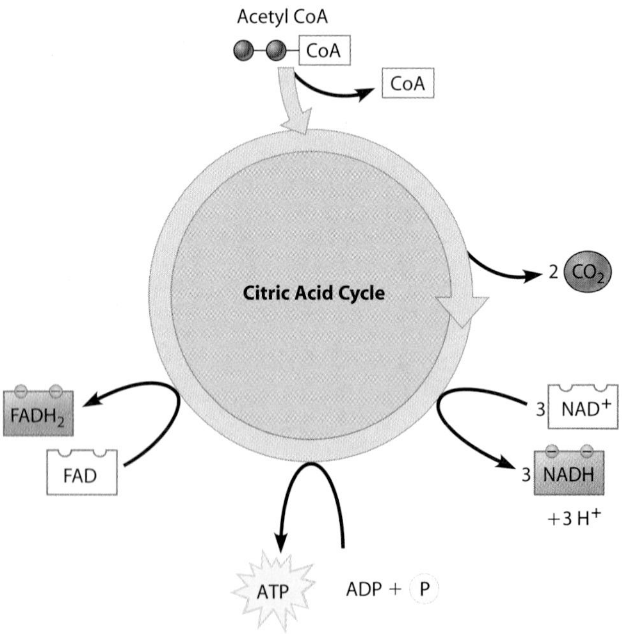

▲ **Figure 9A** An overview of the citric acid cycle: For each acetyl CoA that enters the cycle, 2 CO_2, 3 NADH, 1 $FADH_2$, and 1 ATP are produced. (Remember that 2 acetyl CoA are produced from glucose, so multiply by 2 to calculate a per-glucose return.)

shuttle their high-energy electrons to an electron transport chain. There the energy from the *oxidation* of organic molecules is used to *phosphorylate* ADP to ATP—hence the name *oxidative phosphorylation*. Before we look at how oxidative phosphorylation works, you may want to examine the inner workings of the citric acid cycle in **Figure 9B**, below.

? What is the total number of NADH molecules generated during the complete breakdown of one glucose molecule to six molecules of CO_2? (*Hint:* Combine the outputs of Modules 7–9.)

10 NADH: 2 from glycolysis; 2 from the oxidation of pyruvate; 6 from the citric acid cycle. (Did you remember to double the output due to the sugar-splitting step of glycolysis?)

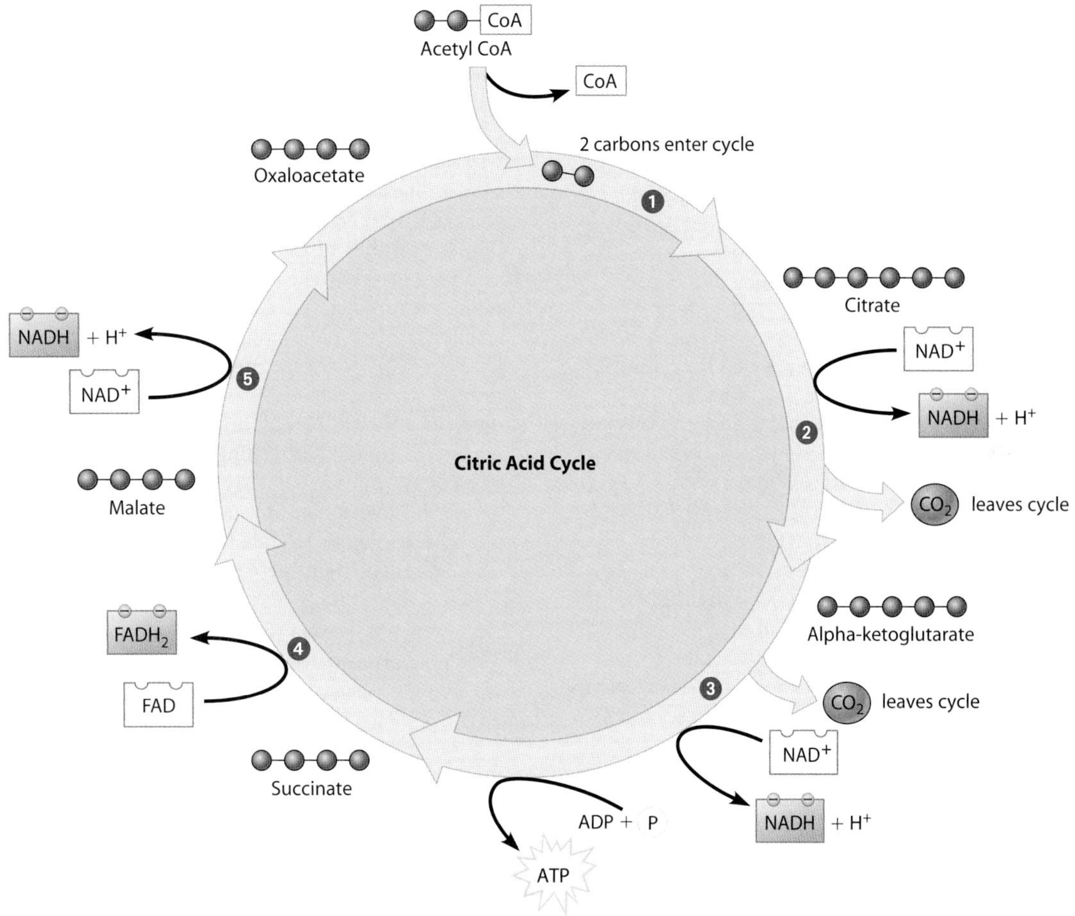

Step 1
Acetyl CoA stokes the furnace.

A turn of the citric acid cycle begins (top center) as enzymes strip the CoA portion from acetyl CoA and combine the remaining two-carbon acetyl group with the four-carbon molecule oxaloacetate (top left) already present in the mitochondrion. The product of this reaction is the six-carbon molecule citrate. Citrate is the ionized form of citric acid. All the acid compounds in this cycle exist in the cell in their ionized form, hence the suffix *-ate*.

Steps 2–3
NADH, ATP, and CO_2 are generated during redox reactions.

Successive redox reactions harvest some of the energy of the acetyl group by stripping hydrogen atoms from organic acid intermediates (such as alpha-ketoglutarate) and producing energy-laden NADH molecules. In two places, an intermediate compound loses a CO_2 molecule. Energy is harvested by substrate-level phosphorylation of ADP to produce ATP. A four-carbon compound called succinate emerges at the end of step 3.

Steps 4–5
Further redox reactions generate $FADH_2$ and more NADH.

Enzymes rearrange chemical bonds, eventually completing the citric acid cycle by regenerating oxaloacetate. Redox reactions reduce the electron carriers FAD and NAD^+ to $FADH_2$ and NADH, respectively. One turn of the citric acid cycle is completed with the conversion of a molecule of malate to oxaloacetate. This compound is then ready to start the next turn of the cycle by accepting another acetyl group from acetyl CoA.

▲ **Figure 9B** A closer look at the citric acid cycle. (Each glucose molecule yields two molecules of acetyl CoA, so the cycle runs two times for each glucose molecule oxidized.)

10 Most ATP production occurs by oxidative phosphorylation

Your main objective in this chapter is to learn how cells harvest the energy of glucose to make ATP. But so far, you've seen the production of only 4 ATP per glucose molecule. Now it's time for the big energy payoff. The final stage of cellular respiration is oxidative phosphorylation, which uses the electron transport chain and chemiosmosis—a process introduced in Module Oxidative phosphorylation clearly illustrates the concept of structure fitting function: The arrangement of electron carriers built into a membrane makes it possible to create an H^+ concentration gradient across the membrane and then use the energy of that gradient to drive ATP synthesis.

Figure 10 shows how an electron transport chain is built into the inner membrane of the mitochondrion. The folds (cristae) of this membrane enlarge its surface area, providing space for thousands of copies of the chain. Also embedded in the membrane are multiple copies of an enzyme complex called **ATP synthase**, which synthesizes ATP.

Electron Transport Chain Starting on the left in Figure 10, the gold arrow traces the path of electron flow from the shuttle molecules NADH and $FADH_2$ through the electron transport chain to oxygen, the final electron acceptor. It is in this end stage of cellular respiration that oxygen finally steps in to play its critical role. Each oxygen atom ($\frac{1}{2}O_2$) accepts 2 electrons from the chain and picks up 2 H^+ from the surrounding solution, forming H_2O.

Most of the carrier molecules of the chain reside in four main protein complexes (labeled I to IV in the diagram), while two mobile carriers transport electrons between the complexes. All of the carriers bind and release electrons in redox reactions, passing electrons down the "energy staircase." Three of the protein complexes use the energy released from these electron transfers to actively transport H^+ across the membrane, from where H^+ is less concentrated to where it is more concentrated. The green vertical arrows show H^+ being transported from the matrix of the mitochondrion (its innermost compartment) into the narrow intermembrane space.

Chemiosmosis Recall that chemiosmosis is a process that uses the energy stored in a hydrogen ion gradient across a membrane to drive ATP synthesis. How does that work? The H^+ concentration gradient stores potential energy, much the way a dam stores energy by holding back the elevated water behind it. The energy stored by a dam can be harnessed to do work (such as generating electricity) when the water is allowed to rush downhill, turning giant wheels called turbines. The ATP synthases built into the inner mitochondrial membrane act like miniature turbines. Indeed, ATP synthase is considered the smallest rotary motor known. The energy of the concentration gradient of H^+ across the membrane drives hydrogen ions through a channel in ATP synthase, as shown on the far right of the figure. The rush of H^+ through the channel spins a component of the complex, activating catalytic sites that attach phosphate groups to ADP to generate ATP.

We will account for how much ATP is made by oxidative phosphorylation in Module 12. But first, let's see what happens if something disrupts these processes.

? What effect would an absence of oxygen (O_2) have on the process illustrated in Figure 10?

Without oxygen to "pull" electrons down the electron transport chain, the energy stored in NADH could not be harvested for ATP synthesis.

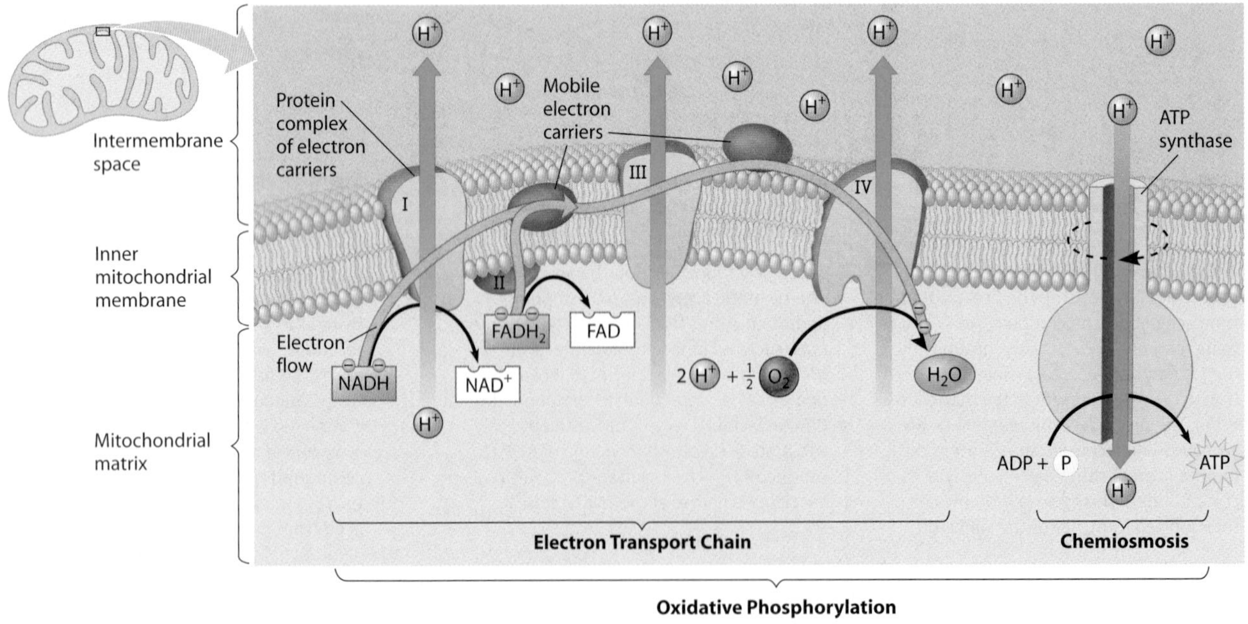

▲ **Figure 10** Oxidative phosphorylation: electron transport and chemiosmosis in a mitochondrion

CONNECTION 11 Interrupting cellular respiration can have both harmful and beneficial effects

A number of poisons produce their deadly effects by interfering with some of the events of cellular respiration that we have just discussed. **Figure 11** shows the places where three different categories of poisons obstruct the process of oxidative phosphorylation.

Poisons in one category block the electron transport chain. A substance called rotenone, for instance, binds tightly with one of the electron carrier molecules in the first protein complex, preventing electrons from passing to the next carrier molecule. By blocking the electron transport chain near its start and thus preventing ATP synthesis, rotenone literally starves an organism's cells of energy. Rotenone is often used to kill pest insects and fish.

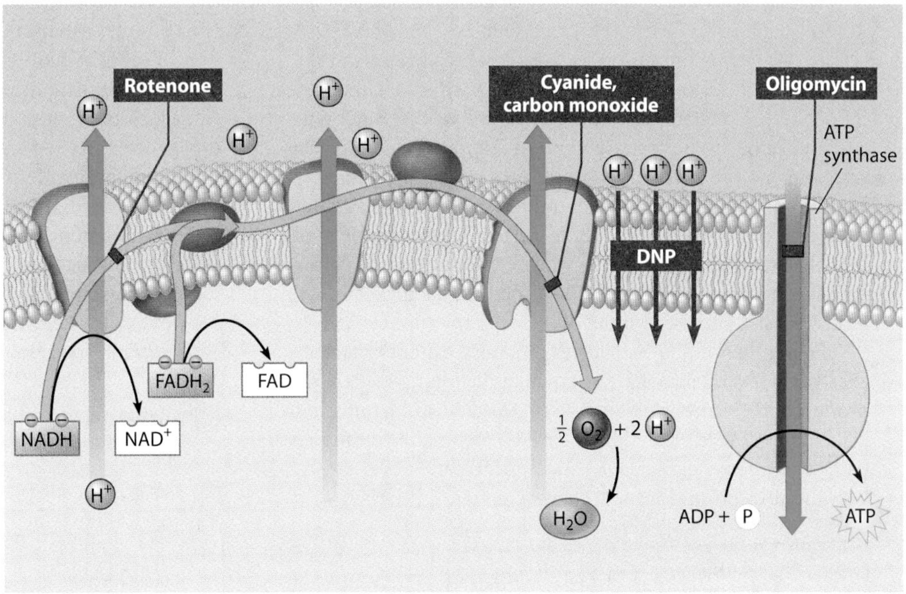

▲ **Figure 11** How some poisons affect the electron transport chain and chemiosmosis

Two other poisons in this category, cyanide and carbon monoxide, bind to an electron carrier in the fourth protein complex, where they block the passage of electrons to oxygen. This blockage is like turning off a faucet; electrons can no longer flow through the "pipe." The result is the same as with rotenone: No H^+ gradient is generated, and no ATP can be made. Cyanide was the lethal agent in an infamous case of product tampering: the Tylenol murders of 1982. Seven people in the Chicago area died after ingesting Tylenol capsules that had been laced with cyanide.

A second category of respiratory poison inhibits ATP synthase. An example of this type of poison is the antibiotic oligomycin, a compound used on the skin to combat fungal infections there. The right side of the figure shows that oligomycin blocks the passage of H^+ through the channel in ATP synthase, preventing fungal cells from making ATP and thereby killing them. The drug does not harm human cells because it cannot get through our outer layer of dead skin cells.

Poisons in a third category, called uncouplers, make the membrane of the mitochondrion leaky to hydrogen ions. Electron transport continues normally, but ATP cannot be made because leakage of H^+ through the membrane destroys the H^+ gradient. Cells continue to burn fuel and consume oxygen, often at a higher than normal rate, but to no avail, for they cannot make any ATP through chemiosmosis because no H^+ gradient exists.

A highly toxic uncoupler called dinitrophenol (DNP) is shown in the figure. When DNP is present, all steps of cellular respiration except chemiosmosis continue to run, even though almost all the energy is lost as heat. DNP poisoning produces an enormous increase in metabolic rate, profuse sweating as the body attempts to dissipate excess heat, and finally death.

For a short time in the 1930s, some physicians prescribed DNP in low doses as weight loss pills, but fatalities soon made it clear that there were far safer ways to lose weight.

Under certain conditions, however, an uncoupler that generates heat by abolishing the H^+ gradient in a mitochondrion may be beneficial. A remarkable adaptation is found in hibernating mammals and newborn infants in a tissue called brown fat. The cells of brown fat are packed full of mitochondria. The inner mitochondrial membrane contains an uncoupling protein, which allows H^+ to flow back down its concentration gradient without generating ATP. Activation of these uncouplers results in ongoing oxidation of stored fuel stores (fats) and the generation of heat, which protects hibernating mammals and newborns from dangerous drops in body temperature. In 2009, scientists discovered that adults retain deposits of brown fat. These tissues were found to be more active in colder weather, not surprisingly, and also in individuals who were thinner. This second finding suggests that brown fat may cause lean individuals to burn calories faster than other people do, and prompted some researchers to propose that these tissues could be a target for obesity-fighting drugs.

In the next module, we take a final look at the stages of cellular respiration and calculate the total amount of ATP harvested as a cell oxidizes glucose in the presence of oxygen to carbon dioxide and water.

? Looking at Figure 11, explain where uncoupling proteins would be found in the mitochondria of brown fat cells.

They would span the inner mitochondrial membrane, providing a channel through which H^+ would diffuse down its concentration gradient back into the matrix.

12 Review: Each molecule of glucose yields many molecules of ATP

Let's review what the cell accomplishes by oxidizing a molecule of glucose. **Figure 12** shows where each stage of cellular respiration occurs in a eukaryotic cell and how much ATP it produces. Starting on the left, glycolysis, which occurs in the cytoplasmic fluid, and the citric acid cycle, which occurs in the mitochondrial matrix, contribute a net total of 4 ATP per glucose molecule by substrate-level phosphorylation. The cell harvests much more energy than this via the carrier molecules NADH and $FADH_2$, which are produced in glycolysis and the citric acid cycle. The energy of the electrons they carry is used to make (according to current estimates) about 28 molecules of ATP by oxidative phosphorylation. Thus, the total yield of ATP molecules per glucose is about 32.

The number of ATP molecules produced cannot be stated exactly for several reasons. As shown in the figure, the NADH produced in glycolysis passes its electrons across the mitochondrial membrane to either NAD^+ or FAD, depending on the type of shuttle system used. Because $FADH_2$ adds its electrons later in the electron transport chain (see Figure 10), it contributes less to the H^+ gradient and thus generates less ATP. In addition, some of the energy of the H^+ gradient may be used for work other than ATP production, such as the active transport of pyruvate into the mitochondrion.

More important than the actual number of ATP molecules is the point that a cell can harvest a great deal of energy from glucose—up to about 34% of the molecule's potential energy. Because most of the ATP generated by cellular respiration results from oxidative phosphorylation, the ATP yield depends on an adequate supply of oxygen to the cell. Without oxygen to function as the final electron acceptor, electron transport and ATP production stop. But as we see next, some cells can oxidize organic fuel and generate ATP *without* oxygen.

? What would a cell's net ATP yield per glucose be in the presence of the poison DNP? (See Module 11.)

4 ATP, all from substrate-level phosphorylation. The uncoupler would destroy the H^+ concentration gradient necessary for chemiosmosis.

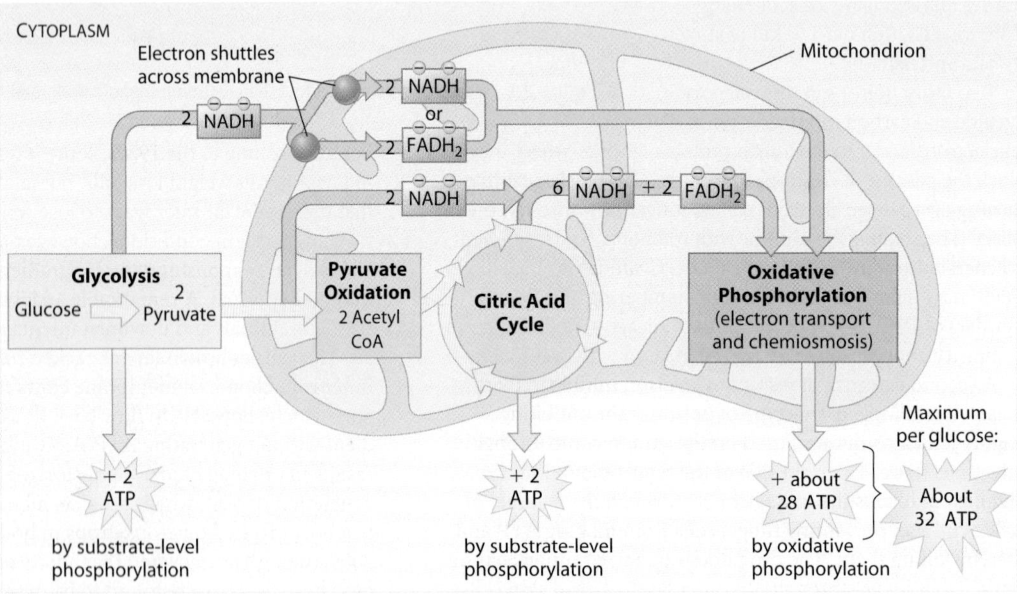

▶ Figure 12 An estimated tally of the ATP produced by substrate-level and oxidative phosphorylation in cellular respiration

Fermentation: Anaerobic Harvesting of Energy

13 Fermentation enables cells to produce ATP without oxygen

Fermentation is a way of harvesting chemical energy that does not require oxygen. The metabolic pathway that generates ATP during fermentation is glycolysis, the same pathway that functions in the first stage of cellular respiration. Remember that glycolysis uses no oxygen; it simply generates a net gain of 2 ATP while oxidizing glucose to two molecules of pyruvate and reducing NAD^+ to NADH. The yield of 2 ATP is certainly a lot less than the possible 32 ATP per glucose generated during aerobic respiration, but it is enough to keep your muscles contracting for a short period of time when oxygen is scarce. And many microorganisms supply all their energy needs with the 2 ATP per glucose yield of glycolysis.

There is more to fermentation, however, than just glycolysis. To oxidize glucose in glycolysis, NAD^+ must be present as an electron acceptor. This is no problem under aerobic conditions, because the cell regenerates its pool of NAD^+ when NADH passes its electrons into the mitochondrion, to be transported to the electron transport chain. Fermentation provides an anaerobic path for recycling NADH back to NAD^+.

Lactic Acid Fermentation One common type of fermentation is called **lactic acid fermentation**. Your muscle cells and certain bacteria can regenerate NAD^+ by this process, as illustrated in **Figure 13A**. You can see that NADH is oxidized to NAD^+ as pyruvate is reduced to lactate (the ionized form of lactic acid). Muscle cells can switch to lactic acid fermentation when the need for ATP outpaces the delivery of O_2 via the bloodstream. The lactate that builds up in muscle cells during strenuous exercise was previously thought to cause muscle fatigue and pain, but research now indicates that other factors are to blame. In any case, the lactate is carried in the blood to the liver, where it is converted back to pyruvate and oxidized in the mitochondria of liver cells.

The dairy industry uses lactic acid fermentation by bacteria to make cheese and yogurt. Other types of microbial fermentation turn soybeans into soy sauce and cabbage into sauerkraut.

Alcohol Fermentation For thousands of years, people have used **alcohol fermentation** in brewing, winemaking, and baking. Yeasts are single-celled fungi that normally use aerobic respiration to process their food. But they are also able to survive in anaerobic environments. Yeasts and certain bacteria recycle their NADH back to NAD^+ while converting pyruvate to CO_2 and ethanol **(Figure 13B)**. The CO_2 provides the bubbles in beer and champagne. Bubbles of CO_2 generated by baker's yeast cause bread dough to rise. Ethanol (ethyl alcohol), the two-carbon end product, is toxic to the organisms that produce it. Yeasts release their alcohol wastes to their surroundings, where it usually diffuses away. When yeasts are confined in a wine vat, they die when the alcohol concentration reaches 14%.

Types of Anaerobes Unlike muscle cells and yeasts, many prokaryotes that live in stagnant ponds and deep in the soil are *obligate anaerobes*, meaning they require anaerobic conditions and are poisoned by oxygen. Yeasts and many other bacteria are facultative anaerobes. A *facultative anaerobe* can make ATP either by fermentation or by oxidative phosphorylation, depending on whether O_2 is available. On the cellular level, our muscle cells behave as facultative anaerobes.

For a facultative anaerobe, pyruvate is a fork in the metabolic road. If oxygen is available, the organism will always use the more productive aerobic respiration. Thus, to make wine and beer, yeasts must be grown anaerobically so that they will ferment sugars and produce ethanol. For this reason, the wine barrels and beer fermentation vats in **Figure 13C** are designed to keep air out.

? A glucose-fed yeast cell is moved from an aerobic environment to an anaerobic one. For the cell to continue generating ATP at the same rate, how would its rate of glucose consumption need to change?

● The cell would have to consume glucose at a rate about 16 times the consumption rate in the aerobic environment (2 ATP per glucose molecule is made by fermentation versus 32 ATP by cellular respiration).

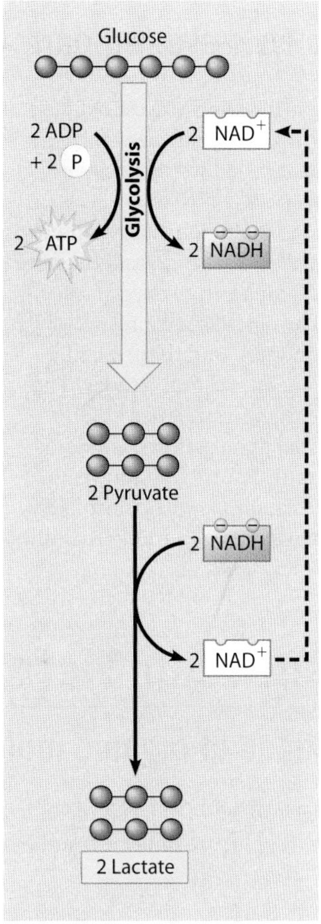

▲ **Figure 13A** Lactic acid fermentation: NAD^+ is regenerated as pyruvate is reduced to lactate.

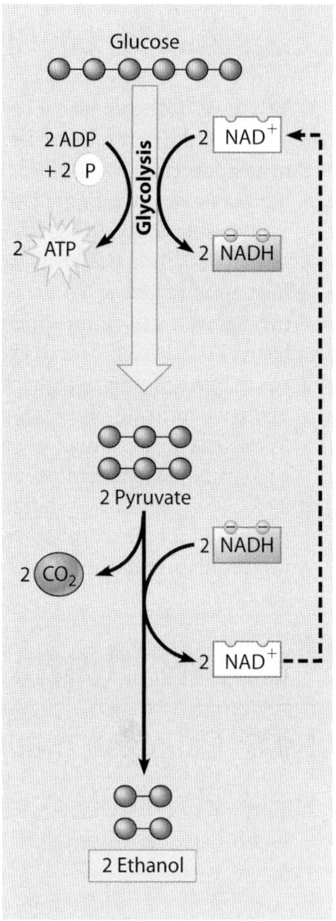

▲ **Figure 13B** Alcohol fermentation: NAD^+ is regenerated as pyruvate is broken down to CO_2 and ethanol.

StockLite/Shutterstock

Sean Lower

▲ **Figure 13C** Wine barrels and beer fermentation vats

EVOLUTION CONNECTION

14 Glycolysis evolved early in the history of life on Earth

Glycolysis is the universal energy-harvesting process of life. If you looked inside a bacterial cell, inside one of your body cells, or inside virtually any other living cell, you would find the metabolic machinery of glycolysis.

The role of glycolysis in both fermentation and respiration has an evolutionary basis. Ancient prokaryotes are thought to have used glycolysis to make ATP long before oxygen was present in Earth's atmosphere. The oldest known fossils of bacteria date back over 3.5 billion years, and they resemble some types of photosynthetic bacteria still found today. The evidence indicates, however, that significant levels of O_2, formed as a by-product of bacterial photosynthesis, did not accumulate in the atmosphere until about 2.7 billion years ago. Thus, for almost a billion years, prokaryotes most likely generated ATP exclusively from glycolysis, a process that does not require oxygen.

The fact that glycolysis is the most widespread metabolic pathway found in Earth's organisms today suggests that it evolved very early in the history of life. The location of glycolysis within the cell also implies great antiquity; the pathway does not require any of the membrane-bounded organelles of the eukaryotic cell, which evolved more than a billion years after the prokaryotic cell. Glycolysis is a metabolic heirloom from early cells that continues to function in fermentation and as the first stage in the breakdown of organic molecules by cellular respiration.

? List some of the characteristics of glycolysis that indicate that it is an ancient metabolic pathway.

Glycolysis occurs universally (functioning in both fermentation and respiration), does not require oxygen, and does not occur in a membrane-bounded organelle.

Connections Between Metabolic Pathways

15 Cells use many kinds of organic molecules as fuel for cellular respiration

Throughout this chapter, we have spoken of glucose as the fuel for cellular respiration. But free glucose molecules are not common in your diet. You obtain most of your calories as carbohydrates (such as sucrose and other disaccharide sugars and starch, a polysaccharide), fats, and proteins. You consume all three of these classes of organic molecules when you eat a handful of peanuts, for instance.

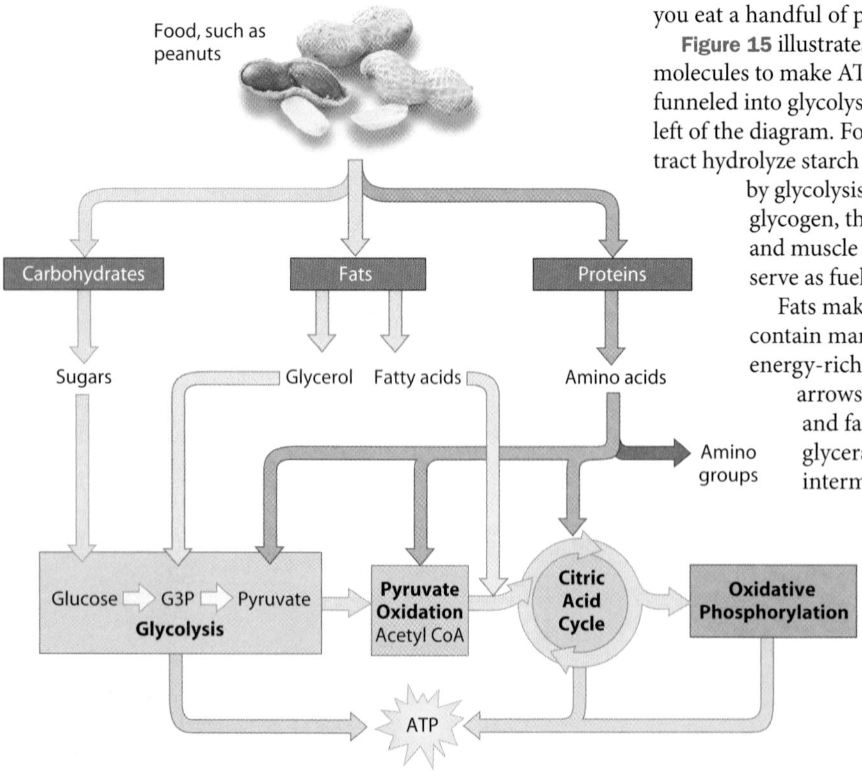

▲ **Figure 15** Pathways that break down various food molecules

Simon Smith/Dorling Kindersley

Figure 15 illustrates how a cell can use these three types of molecules to make ATP. A wide range of carbohydrates can be funneled into glycolysis, as indicated by the arrows on the far left of the diagram. For example, enzymes in your digestive tract hydrolyze starch to glucose, which is then broken down by glycolysis and the citric acid cycle. Similarly, glycogen, the polysaccharide stored in your liver and muscle cells, can be hydrolyzed to glucose to serve as fuel between meals.

Fats make excellent cellular fuel because they contain many hydrogen atoms and thus many energy-rich electrons. As the diagram shows (tan arrows), a cell first hydrolyzes fats to glycerol and fatty acids. It then converts the glycerol to glyceraldehyde 3-phosphate (G3P), one of the intermediates in glycolysis. The fatty acids are broken into two-carbon fragments that enter the citric acid cycle as acetyl CoA. A gram of fat yields more than twice as much ATP as a gram of carbohydrate. Because so many calories are stockpiled in each gram of fat, you must expend a large amount of energy to burn fat stored in your body. This helps explain why it is so difficult for a dieter to lose excess fat.

Proteins (purple arrows in Figure 15) can also be used for fuel, although your body preferentially burns sugars and fats first. To be oxidized as fuel, proteins must first be digested to their constituent amino acids. Typically, a cell will use most of these amino acids to make its own proteins. Enzymes can convert excess amino acids to intermediates of glycolysis or the citric acid cycle, and their energy is then harvested by cellular respiration. During the conversion, the amino groups are stripped off and later disposed of in urine.

? Animals store most of their energy reserves as fats, not as polysaccharides. What is the advantage of this mode of storage for an animal?

Most animals are mobile and benefit from a compact and concentrated form of energy storage. Also, because fats are hydrophobic, they can be stored without extra water associated with them (see Module 3.8).

16 Food molecules provide raw materials for biosynthesis

Not all food molecules are destined to be oxidized as fuel for making ATP. Food also provides the raw materials your cells use for biosynthesis—the production of organic molecules using energy-requiring metabolic pathways. A cell must be able to make its own molecules to build its structures and perform its functions. Some raw materials, such as amino acids, can be incorporated directly into your macromolecules. However, your cells also need to make molecules that are not present in your food. Indeed, glycolysis and the citric acid cycle function as metabolic interchanges that enable your cells to convert some kinds of molecules to others as you need them.

Figure 16 outlines the pathways by which your cells can make three classes of organic molecules using some of the intermediate molecules of glycolysis and the citric acid cycle. By comparing Figures 15 and 16, you can see clear connections between the energy-harvesting processes of cellular respiration and the biosynthetic pathways used to construct the organic molecules of the cell.

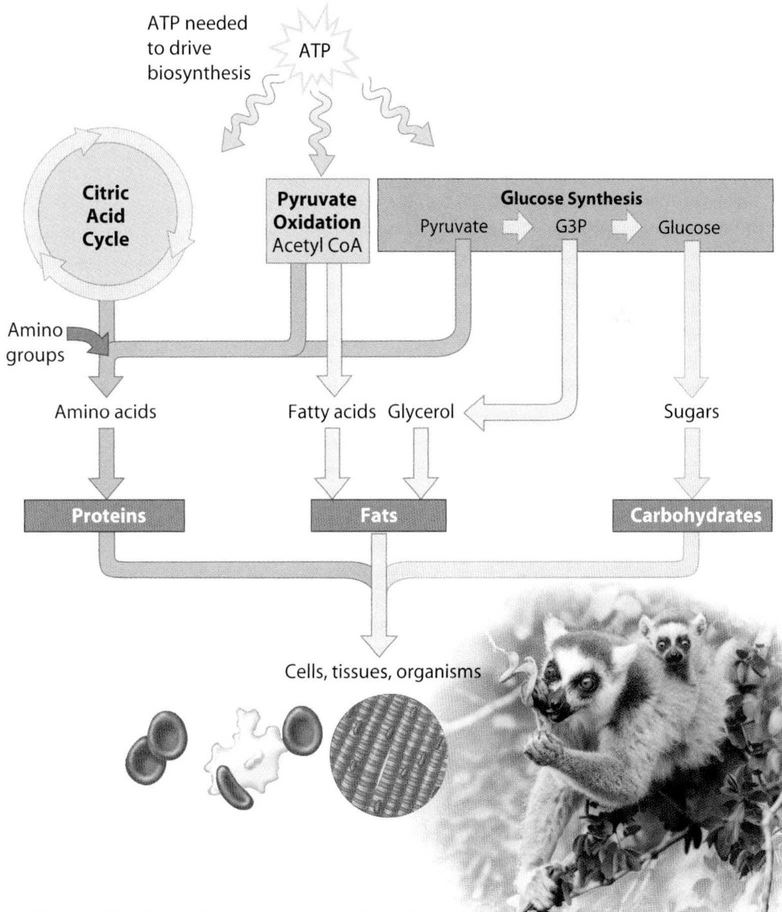

▲ **Figure 16** Biosynthesis of large organic molecules from intermediates of cellular respiration

Fiona Rogers/naturepl.com

Basic principles of supply and demand regulate these pathways. If there is an excess of a certain amino acid, for example, the pathway that synthesizes it is switched off. The most common mechanism for this control is feedback inhibition: The end product inhibits an enzyme that catalyzes an early step in the pathway. Feedback inhibition also controls cellular respiration. If ATP accumulates in a cell, it inhibits an early enzyme in glycolysis, slowing down respiration and conserving resources. On the other hand, the same enzyme is activated by a buildup of ADP in the cell, signaling the need for more energy.

The cells of all living organisms—including those of the lemurs shown in Figure 16 and the plants they eat—have the ability to harvest energy from the breakdown of organic molecules. When the process is cellular respiration, the atoms of the starting materials end up in carbon dioxide and water. In contrast, the ability to make organic molecules from carbon dioxide and water is not universal. Animal cells lack this ability, but plant cells can actually produce organic molecules from inorganic ones using the energy of sunlight.

? Explain how someone can gain weight and store fat even when on a low-fat diet. (*Hint:* Look for G3P and acetyl CoA in Figures 15 and 16.)

If caloric intake is excessive, body cells use metabolic pathways to convert the excess to fat. The glycerol and fatty acids of fats are made from G3P and acetyl CoA, respectively, both produced from the oxidation of carbohydrates.

REVIEW

For Practice Quizzes, BioFlix, MP3 Tutors, and Activities, go to www.masteringbiology.com.

Reviewing the Concepts

Cellular Respiration: Aerobic Harvesting of Energy (1–5)

1 Photosynthesis and cellular respiration provide energy for life. Photosynthesis uses solar energy to produce glucose and O_2 from CO_2 and H_2O. In cellular respiration, O_2 is consumed during the breakdown of glucose to CO_2 and H_2O, and energy is released.

2 Breathing supplies O_2 for use in cellular respiration and removes CO_2.

3 Cellular respiration banks energy in ATP molecules. The summary equation for cellular respiration is $C_6H_{12}O_6 + 6\ O_2 \longrightarrow 6\ CO_2 + 6\ H_2O$ + Energy (ATP + Heat).

4 The human body uses energy from ATP for all its activities.

5 Cells tap energy from electrons "falling" from organic fuels to oxygen. Electrons removed from fuel molecules (oxidation) are transferred to NAD^+ (reduction). NADH passes electrons to an electron transport chain. As electrons "fall" from carrier to carrier and finally to O_2, energy is released.

Stages of Cellular Respiration (6–12)

6 Overview: Cellular respiration occurs in three main stages.

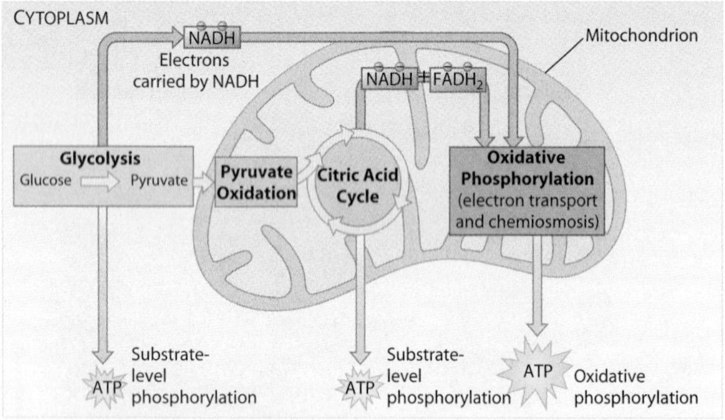

7 Glycolysis harvests chemical energy by oxidizing glucose to pyruvate. ATP is used to prime a glucose molecule, which is split in two. These three-carbon intermediates are oxidized to two molecules of pyruvate, yielding a net of 2 ATP and 2 NADH. ATP is formed by substrate-level phosphorylation, in which a phosphate group is transferred from an organic molecule to ADP.

8 Pyruvate is oxidized prior to the citric acid cycle. In the oxidation of pyruvate to acetyl CoA, CO_2 and NADH are produced.

9 The citric acid cycle completes the oxidation of organic molecules, generating many NADH and $FADH_2$ molecules. For each turn of the cycle, two carbons from acetyl CoA are added and 2 CO_2 are released; the energy yield is 1 ATP, 3 NADH, and 1 $FADH_2$.

10 Most ATP production occurs by oxidative phosphorylation. In mitochondria, electrons from NADH and $FADH_2$ travel down the electron transport chain to O_2, which picks up H^+ to form water. Energy released by these redox reactions is used to pump H^+ from the mitochondrial matrix into the intermembrane space. In chemiosmosis, the H^+ diffuses back across the inner membrane through ATP synthase complexes, driving the synthesis of ATP.

11 Interrupting cellular respiration can have both harmful and beneficial effects. Poisons can block electron flow, block the movement of H^+ through ATP synthase, or allow H^+ to leak through the membrane. Uncouplers in brown fat produce body heat.

12 Review: Each molecule of glucose yields many molecules of ATP. Substrate-level phosphorylation and oxidative phosphorylation produce up to 32 ATP molecules for every glucose molecule oxidized in cellular respiration.

Fermentation: Anaerobic Harvesting of Energy (13–14)

13 Fermentation enables cells to produce ATP without oxygen. Under anaerobic conditions, muscle cells, yeasts, and certain bacteria produce ATP by glycolysis. NAD^+ is recycled from NADH as pyruvate is reduced to lactate (lactic acid fermentation) or, in microbes, alcohol and CO_2 (alcohol fermentation).

14 Glycolysis evolved early in the history of life on Earth. Glycolysis occurs in the cytoplasm of nearly all organisms and is thought to have evolved in ancient prokaryotes.

Connections Between Metabolic Pathways (15–16)

15 Cells use many kinds of organic molecules as fuel for cellular respiration. Carbohydrates, fats, and proteins can all fuel cellular respiration.

16 Food molecules provide raw materials for biosynthesis. Cells use intermediates from cellular respiration and ATP for biosynthesis of other organic molecules. Metabolic pathways are often regulated by feedback inhibition.

Connecting the Concepts

1. Fill in the blanks in this summary map to help you review the key concepts of cellular respiration.

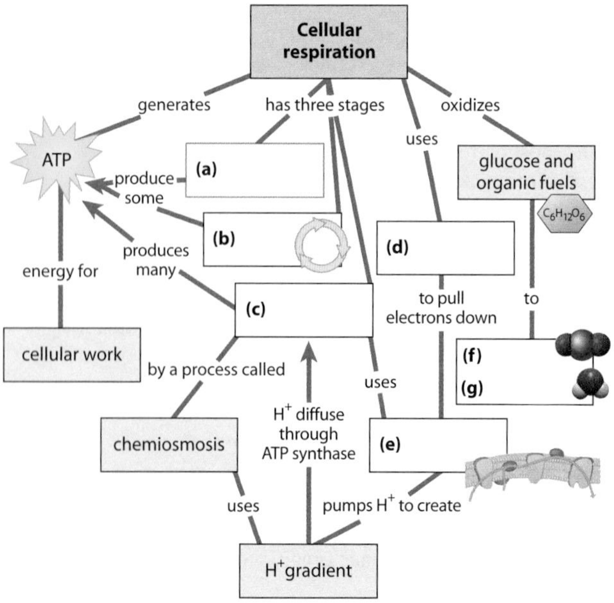

Testing Your Knowledge

Multiple Choice

2. What is the role of oxygen in cellular respiration?
   a. It is reduced in glycolysis as glucose is oxidized.
   b. It provides electrons to the electron transport chain.
   c. It combines with the carbon removed during the citric acid cycle to form CO_2.
   d. It is required for the production of heat and light.
   e. It accepts electrons from the electron transport chain.
3. When the poison cyanide blocks the electron transport chain, glycolysis and the citric acid cycle soon grind to a halt as well. Why do you think they stop?
   a. They both run out of ATP.
   b. Unused O_2 interferes with cellular respiration.
   c. They run out of NAD^+ and FAD.
   d. Electrons are no longer available.
   e. They run out of ADP.
4. A biochemist wanted to study how various substances were used in cellular respiration. In one experiment, he allowed a mouse to breathe air containing O_2 "labeled" by a particular isotope. In the mouse, the labeled oxygen first showed up in
   a. ATP.
   b. glucose ($C_6H_{12}O_6$).
   c. NADH.
   d. CO_2.
   e. H_2O.
5. In glycolysis, ______ is oxidized and ______ is reduced.
   a. NAD^+ . . . glucose
   b. glucose . . . oxygen
   c. ATP . . . ADP
   d. glucose . . . NAD^+
   e. ADP . . . ATP
6. Which of the following is the most immediate source of energy for making most of the ATP in your cells?
   a. the reduction of oxygen
   b. the transfer of (P) from intermediate substrates to ADP
   c. the movement of H^+ across a membrane down its concentration gradient
   d. the splitting of glucose into two molecules of pyruvate
   e. electrons moving through the electron transport chain
7. In which of the following is the first molecule becoming reduced to the second molecule?
   a. pyruvate $\longrightarrow$ acetyl CoA
   b. pyruvate $\longrightarrow$ lactate
   c. glucose $\longrightarrow$ pyruvate
   d. $NADH + H^+ \longrightarrow NAD^+ + 2\,H$
   e. $C_6H_{12}O_6 \longrightarrow 6\,CO_2$
8. Which of the following is a true distinction between cellular respiration and fermentation?
   a. NADH is oxidized by the electron transport chain in respiration only.
   b. Only respiration oxidizes glucose.
   c. Fermentation is an example of an endergonic reaction; cellular respiration is an exergonic reaction.
   d. Substrate-level phosphorylation is unique to fermentation; cellular respiration uses oxidative phosphorylation.
   e. Fermentation is the metabolic pathway found in prokaryotes; cellular respiration is unique to eukaryotes.

Describing, Comparing, and Explaining

9. Which of the three stages of cellular respiration is considered the most ancient? Explain your answer.
10. Explain in terms of cellular respiration why you need oxygen and why you exhale carbon dioxide.
11. Compare and contrast fermentation as it occurs in your muscle cells and as it occurs in yeast cells.
12. Explain how your body can convert excess carbohydrates in the diet to fats. Can excess carbohydrates be converted to protein? What else must be supplied?

Applying the Concepts

13. An average adult human requires 2,200 kcal of energy per day. Suppose your diet provides an average of 2,300 kcal per day. How many hours per week would you have to walk to burn off the extra calories? Swim? Run? (See Figure 4.)
14. Your body makes NAD^+ and FAD from two B vitamins, niacin and riboflavin. The Recommended Dietary Allowance for niacin is 20 mg and for riboflavin, 1.7 mg. These amounts are thousands of times less than the amount of glucose your body needs each day to fuel its energy needs. Why is the daily requirement for these vitamins so small?
15. In a detail of the citric acid cycle not shown in Figure 9B, an enzyme converts succinate to a compound called fumarate, with the release of H^+. You are studying this reaction using a suspension of bean cell mitochondria and a blue dye that loses its color as it takes up H^+. You know that the higher the concentration of succinate, the more rapid the decolorization of the dye. You set up reaction mixtures with mitochondria, dye, and three different concentrations of succinate (0.1 mg/L, 0.2 mg/L, and 0.3 mg/L). Which of the following graphs represents the results you would expect, and why?

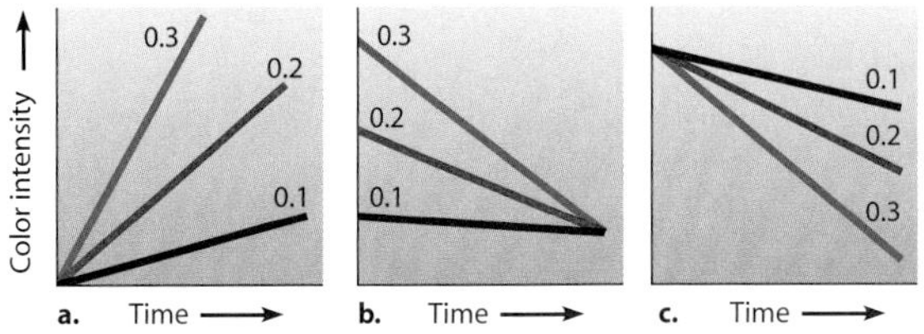

16. ATP synthase enzymes are found in the prokaryotic plasma membrane and in the inner membrane of a mitochondrion. What does this suggest about the evolutionary relationship of this eukaryotic organelle to prokaryotes?
17. Excess consumption of alcohol by a pregnant woman can cause a complex of birth defects called fetal alcohol syndrome (FAS). Symptoms of FAS include head and facial irregularities, heart defects, mental retardation, and behavioral problems. The U.S. Surgeon General's Office recommends that pregnant women abstain from drinking alcohol, and the government has mandated a warning label on liquor bottles: "Women should not drink alcoholic beverages during pregnancy because of the risk of birth defects." Imagine you are a server in a restaurant. An obviously pregnant woman orders several alcoholic drinks. How would you respond? Is it the woman's right to make those decisions about her unborn child's health? Do you bear any responsibility in the matter? Is a restaurant responsible for monitoring the health habits of its customers?

Review Answers

1. a. glycolysis; b. citric acid cycle; c. oxidative phosphorylation; d. oxygen; e. electron transport chain; f. CO_2; g. H_2O

2. e 3. c (NAD^+ and FAD, which are recycled by electron transport, are in limited supply in a cell.) 4. e 5. d 6. c 7. b (at the same time NADH is oxidized to NAD^+) 8. a

9. Glycolysis is considered the most ancient because it occurs in all living cells and doesn't require oxygen or membrane-enclosed organelles.

10. Oxygen picks up electrons from the oxidation of glucose at the end of the electron transport chain. Carbon dioxide results from the oxidation of glucose. It is released in the oxidation of pyruvate and in the citric acid cycle.

11. In lactic acid fermentation (in muscle cells), pyruvate is reduced by NADH to form lactate, and NAD^+ is recycled. In alcohol fermentation, pyruvate is broken down to CO_2 and ethanol as NADH is oxidized to NAD^+. Both types of fermentation allow glycolysis to continue to produce 2 ATP per glucose by recycling NAD^+.

12. As carbohydrates are broken down in glycolysis and the oxidation of pyruvate, glycerol can be made from G3P and fatty acids can be made from acetyl CoA. Amino groups, containing N atoms, must be supplied to various intermediates of glycolysis and the citric acid cycle to produce amino acids.

13. 100 kcal per day is 700 kcal per week. On the basis of Figure 4, walking 3 mph would require $\frac{700}{245}$ = about 2.8 hr; swimming, 1.7 hr; running, 0.7 hr.

14. NAD^+ and FAD are coenzymes that are not used up during the oxidation of glucose. NAD^+ and FAD are recycled when NADH and $FADH_2$ pass the electrons they are carrying to the electron transport chain. We need a small additional supply to replace those that are damaged.

15. a. No, this shows the blue color getting more intense. The reaction decolorizes the blue dye.
    b. No, this shows the dye being decolorized, but it also shows the three mixtures with different initial color intensities. The intensities should have started out the same, since all mixtures used the same concentration of dye.
    c. Correct. The mixtures all start out the same, and then the ones with more succinate (reactant) decolorize faster.

16. The presence of ATP synthase enzymes in prokaryotic plasma membranes and the inner membrane of mitochondria provides support for the theory of endosymbiosis—that mitochondria evolved from an engulfed prokaryote that used aerobic respiration (see Module 4.15).

17. Some issues and questions to consider: Is your customer aware of the danger? Do you have an obligation to protect the customer, even against her wishes? Does your employer have the right to dismiss you for informing the customer? For refusing to serve the customer? Could you or the restaurant later be held liable for injury to the fetus? Or is the mother responsible for willfully disregarding warnings about drinking?

Photosynthesis: *Using Light to Make Food*

Photosynthesis: *Using Light to Make Food*

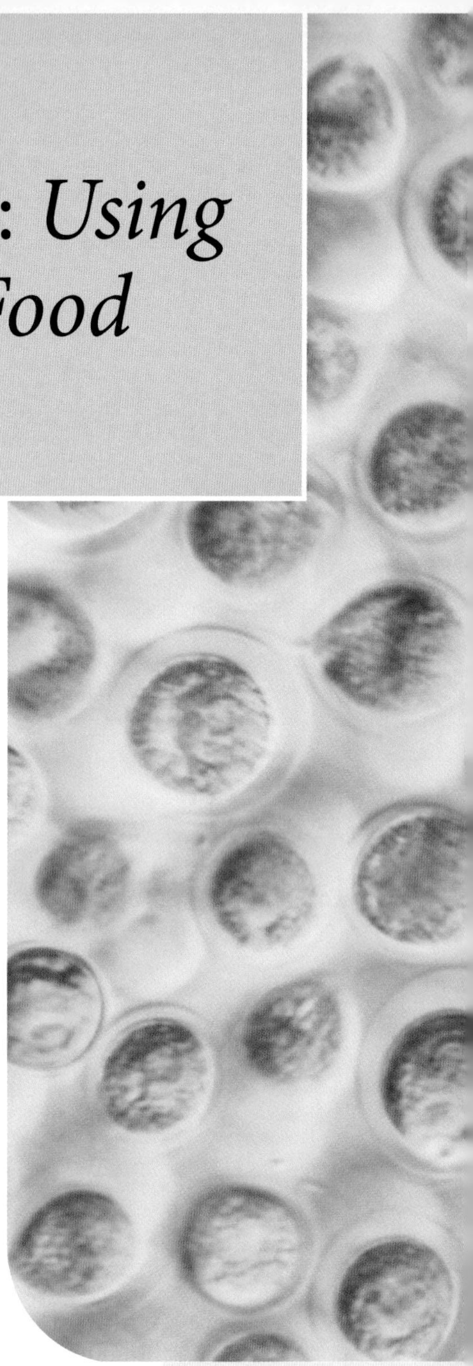

BIG IDEAS

An Overview of Photosynthesis
(1–5)

Plants and other photoautotrophs use the energy of sunlight to convert CO_2 and H_2O to sugar and O_2.

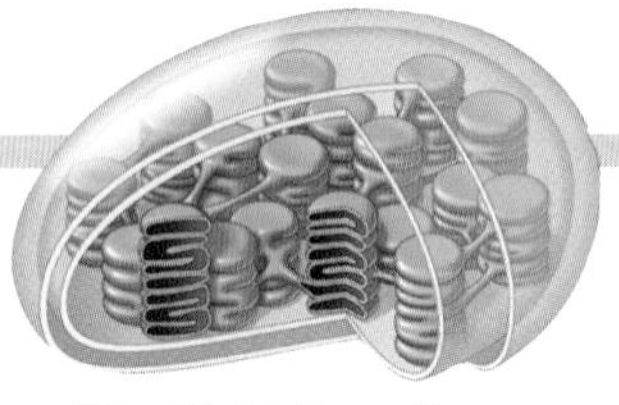

rodho/Shutterstock

The Light Reactions: Converting Solar Energy to Chemical Energy
(6–9)

In the thylakoids of a chloroplast, the light reactions generate ATP and NADPH.

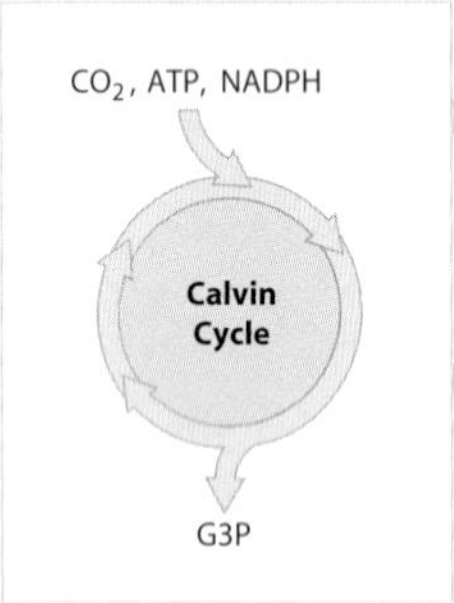

The Calvin Cycle: Reducing CO_2 to Sugar
(10–11)

The Calvin cycle, which takes place in the stroma of the chloroplast, uses ATP and NADPH to reduce CO_2 to sugar.

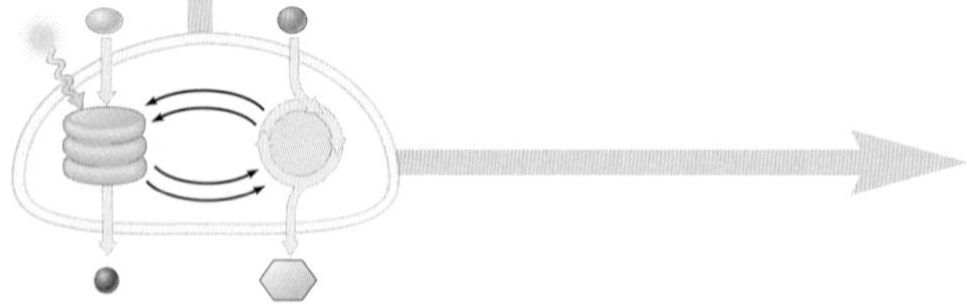

Photosynthesis Reviewed and Extended
(12–14)

Photosynthesis provides the energy and building material for ecosystems. It also affects global climate and the ozone layer.

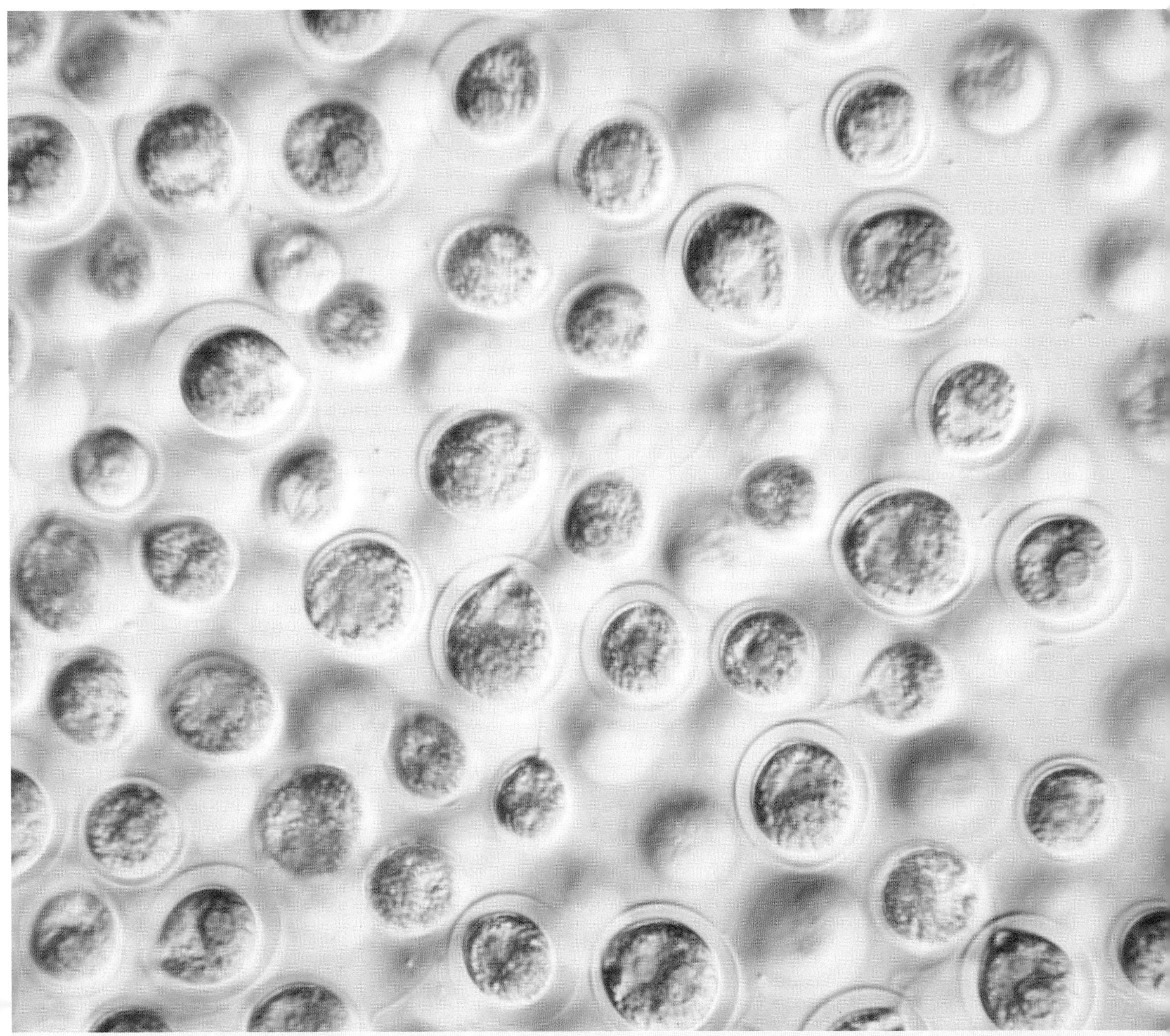

M. I. Walker/Photo Researchers

What do you think of when you see brilliant green cells like these? You might think about photosynthesis (especially if you are studying for a biology test). Others might simply admire nature's palette. But a new group of entrepreneurs has a different kind of green in mind—money.

The cells of plants, algae, and certain prokaryotes produce energy by the process of photosynthesis, which converts light energy to chemical energy, and then stores the chemical energy in sugar made from carbon dioxide and water. This stored energy is a valuable commodity. You purchase it—in the form of fruits and vegetables—to supply your body with energy. But now, enterprises ranging from small start-ups to giant oil companies are researching methods of harnessing photosynthesis for other energy needs, using single-celled algae as their "factories." For example, some algal species produce lipids that could be extracted and converted to biodiesel—green oil. Alternatively, carbohydrates produced by algae could be used to make ethanol. Unlike biofuel crops such as corn or oil palms, algae don't require large areas of fertile soil, and they grow very fast. Some designs even include using the CO_2 emissions from coal-fired power plants—a major contributor to climate change—as the carbon source for their photosynthesis.

In this chapter, you will learn how photosynthesis works. Because photosynthesis is a complex process, we begin with some basic concepts. Then we look more closely at the two stages of photosynthesis: the light reactions, in which solar energy is transformed into chemical energy, and the Calvin cycle, in which that chemical energy is used to make organic molecules. Finally, we explore ways in which photosynthesis affects our global environment.

An Overview of Photosynthesis

1 Autotrophs are the producers of the biosphere

Plants are **autotrophs** (meaning "self-feeders" in Greek) in that they make their own food and thus sustain themselves without consuming organic molecules derived from any other organisms. Plant cells capture light energy that has traveled 150 million kilometers from the sun and convert it to chemical energy. Because they use the energy of light, plants are specifically called *photoautotrophs*. Through the process of **photosynthesis**, plants convert CO_2 and H_2O to their own organic molecules and release O_2 as a by-product. Photoautotrophs are the ultimate source of organic molecules for almost all other organisms. They are often referred to as the producers of the biosphere because they produce its food supply. Producers feed the consumers of the biosphere—the **heterotrophs** that consume other plants or animals or decompose organic material (*hetero* means "other").

The photographs on this page illustrate some of the diversity among photoautotrophs. On land, plants, such as those in the forest scene in **Figure 1A**, are the predominant producers. In aquatic environments, there are several types of photoautotrophs. **Figure 1B** is a micrograph of unicellular photosynthetic protists. **Figure 1C** shows kelp, a large alga that forms extensive underwater "forests" off the coast of California. **Figure 1D** is a micrograph of cyanobacteria, abundant and important producers in freshwater and marine ecosystems.

In this chapter, we focus on photosynthesis in plants, which takes place in chloroplasts. The remarkable ability to harness light energy and use it to drive the synthesis of organic compounds emerges from the structural organization of these organelles: Photosynthetic pigments, enzymes, and other molecules are grouped together in membranes, allowing the sequences of reactions to be carried out efficiently. The process of photosynthesis most likely originated in a group of bacteria that had infolded regions of the plasma membrane containing such clusters of enzymes and other molecules. In fact, according to the widely accepted theory of endosymbiosis, chloroplasts originated from a photosynthetic prokaryote that took up residence inside a eukaryotic cell.

Let's begin our study of photosynthesis with an overview of the location and structure of plant chloroplasts.

? What do "self-feeding" photoautotrophs require from the environment in order to make their own food?

Light, carbon dioxide, and water. (Minerals are also required; you'll learn about the needs of plants in Chapter 32.)

Theresa de Salis/Peter Arnold, Inc./Photo Library

▲ **Figure 1A** Forest plants

M. I. Walker/Photo Researchers

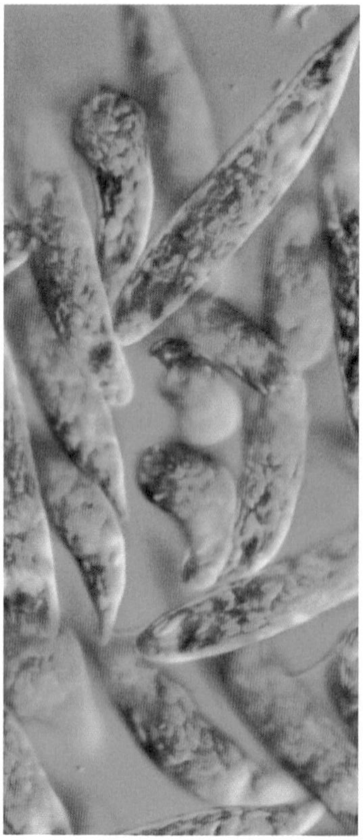

LM 1,400×

▲ **Figure 1B** Photosynthetic protists

Jeff Rotman/Nature Picture Library

▲ **Figure 1C** Kelp, a multicellular alga

Susan M. Barns, Ph.D.

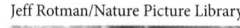

LM 460×

▲ **Figure 1D** Cyanobacteria (photosynthetic prokaryotes)

2 Photosynthesis occurs in chloroplasts in plant cells

All green parts of a plant have chloroplasts in their cells and can carry out photosynthesis. In most plants, however, the leaves have the most chloroplasts (about half a million in a square millimeter surface area of a leaf) and are the major sites of photosynthesis. Their green color comes from **chlorophyll**, a light-absorbing pigment in the chloroplasts that plays a central role in converting solar energy to chemical energy.

Figure 2 zooms in on a leaf to show the actual sites of photosynthesis. The leaf cross section shows a slice through a leaf. Chloroplasts are concentrated in the cells of the **mesophyll**, the green tissue in the interior of the leaf. Carbon dioxide enters the leaf, and oxygen exits, by way of tiny pores called **stomata** (singular, *stoma*, meaning "mouth"). Water absorbed by the roots is delivered to the leaves in veins. Leaves also use veins to export sugar to roots and other parts of the plant.

As you can see in the light micrograph of a single mesophyll cell, each cell has numerous chloroplasts. A typical mesophyll cell has about 30 to 40 chloroplasts. The bottom drawing and electron micrograph show the structures in a single chloroplast. Membranes in the chloroplast form the framework where many of the reactions of photosynthesis occur, just as mitochondrial membranes are the site for much of the energy-harvesting machinery we discussed in Chapter 6. In the chloroplast, an envelope of two membranes encloses an inner compartment, which is filled with a thick fluid called **stroma**. Suspended in the stroma is a system of interconnected membranous sacs, called **thylakoids**, which enclose another internal compartment, called the thylakoid space. (As you will see later, this thylakoid space plays a role analogous to the intermembrane space of a mitochondrion in the generation of ATP.) In many places, thylakoids are concentrated in stacks called **grana** (singular, *granum*). Built into the thylakoid membranes are the chlorophyll molecules that capture light energy. The thylakoid membranes also house much of the machinery that converts light energy to chemical energy, which is used in the stroma to make sugar.

Later in the chapter, we examine the function of these structures in more detail. But first, let's look more closely at the general process of photosynthesis.

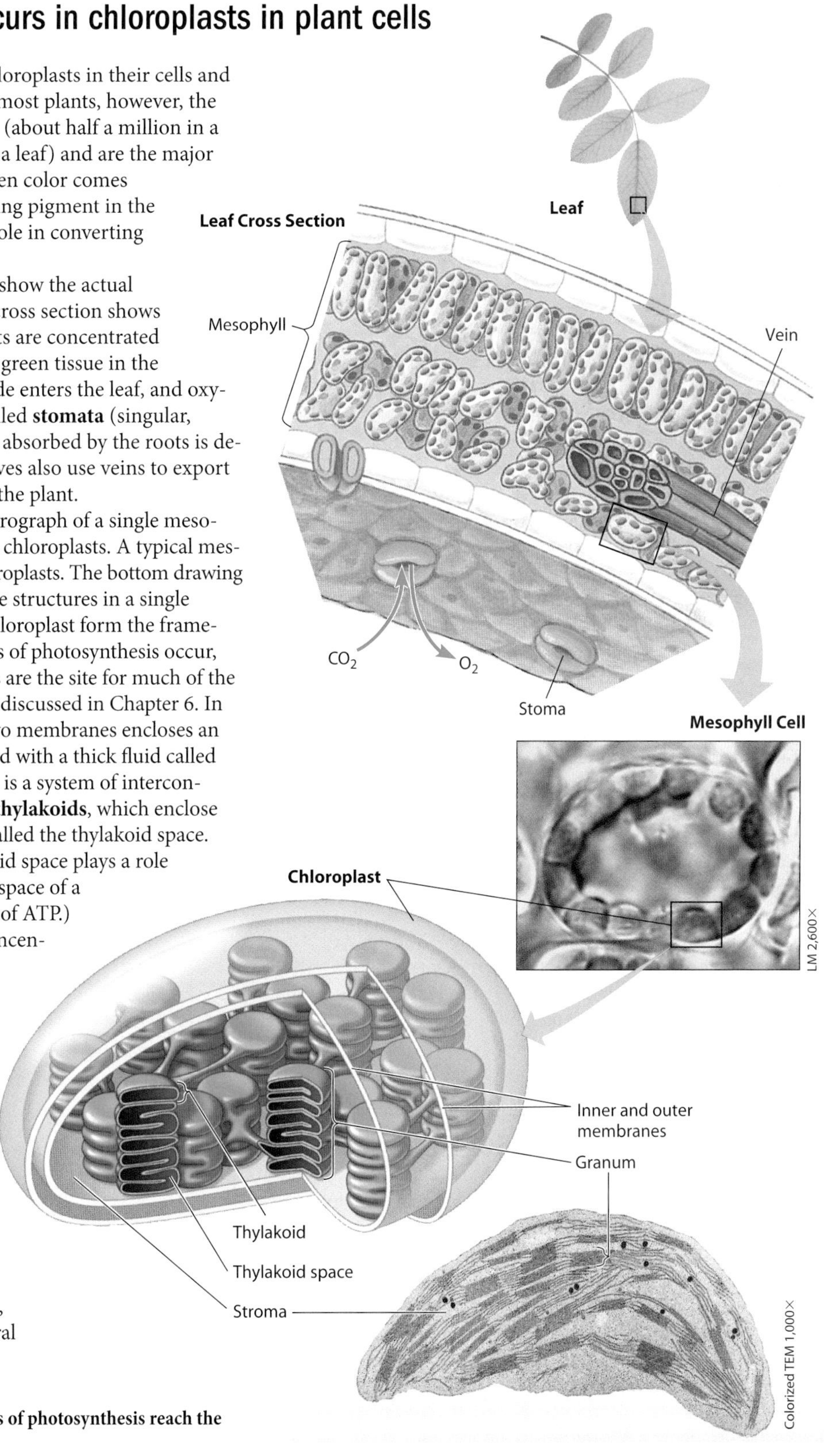

▲ **Figure 2** Zooming in on the location and structure of chloroplasts

rodho/Shutterstock; Graham Kent; E.H. Newcomb & W.P. Wergin/Biological Photo Service

? How do the reactant molecules of photosynthesis reach the chloroplasts in leaves?

● CO_2 enters leaves through stomata, and H_2O enters the roots and is carried to leaves through veins.

SCIENTIFIC DISCOVERY

3 Scientists traced the process of photosynthesis using isotopes

Martin Shields/Alamy

▲ **Figure 3A** Oxygen bubbles on the leaves of an aquatic plant

The leaves of plants that live in lakes and ponds are often covered with bubbles like the ones shown in **Figure 3A**. The bubbles are oxygen gas (O_2) produced during photosynthesis. But where does this O_2 come from?

The overall process of photosynthesis has been known since the 1800s: In the presence of light, green plants produce sugar and oxygen from carbon dioxide and water. Consider the basic equation for photosynthesis:

$$6\,CO_2 + 6\,H_2O \longrightarrow C_6H_{12}O_6 + 6\,O_2$$

Looking at this equation, you can understand why scientists hypothesized that photosynthesis first splits carbon dioxide ($CO_2 \longrightarrow C + O_2$), releasing oxygen gas, and then adds water (H_2O) to the carbon to produce sugar. In the 1930s, this idea was challenged by C. B. van Niel, who was working with photosynthesizing bacteria that produce sugar from CO_2 but do not release O_2. He hypothesized that in plants, H_2O is split, with the hydrogen becoming incorporated into sugar and the O_2 released as gas.

In the 1950s, scientists confirmed van Niel's hypothesis by using a heavy isotope of oxygen, ^{18}O, to follow the fate of oxygen atoms during photosynthesis. (This was one of the first uses of isotopes as tracers in biological research. Remember that isotopes are atoms with differing numbers of neutrons.) The photosynthesis equation you will see in the description of these experiments is slightly more detailed than the summary equation written above. It shows that water is actually both a reactant and a product in the reaction. (As is often done, glucose is shown as a product, although the direct product of photosynthesis is a three-carbon sugar that can be used to make glucose.) In these equations, the red type denotes the labeled oxygen—^{18}O.

Experiment 1: $6\,CO_2 + 12\,H_2O \longrightarrow C_6H_{12}O_6 + 6\,H_2O + 6\,O_2$

Experiment 2: $6\,CO_2 + 12\,H_2O \longrightarrow C_6H_{12}O_6 + 6\,H_2O + 6\,O_2$

In experiment 1, a plant given CO_2 containing ^{18}O gave off no labeled (^{18}O-containing) oxygen gas. But in experiment 2, a plant given H_2O containing ^{18}O did produce labeled O_2. These experiments show that the O_2 produced during photosynthesis comes from water and not from CO_2.

Additional experiments have revealed that the oxygen atoms in CO_2 and the hydrogen atoms in the reactant H_2O molecules end up in the sugar molecule and in water that is formed as a product. **Figure 3B** summarizes the fates of all the atoms that start out in the reactant molecules of photosynthesis.

The synthesis of sugar in photosynthesis involves numerous chemical reactions. Working out the details of these reactions also involved the use of isotopes, in this case, radioactive isotopes. In the mid-1940s, American biochemist Melvin Calvin and his colleagues began using radioactive ^{14}C to trace the sequence of intermediates formed in the cyclic pathway that produces sugar from CO_2. They worked for 10 years to elucidate this cycle, which is now called the Calvin cycle. Calvin received the Nobel Prize in 1961 for this work.

? Photosynthesis produces billions of tons of carbohydrate a year. Where does most of the mass of this huge amount of organic matter come from?

● Mostly from CO_2 in the air, which provides both the carbon and oxygen in carbohydrate. Water supplies only the hydrogen.

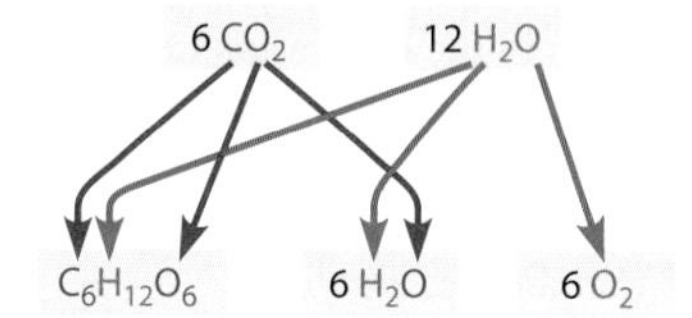

▲ **Figure 3B** Fates of all the atoms in photosynthesis

4 Photosynthesis is a redox process, as is cellular respiration

What actually happens when CO_2 and water are converted to sugar and O_2? Photosynthesis is a redox (oxidation-reduction) process, just as cellular respiration is. As indicated in the summary equation for photosynthesis (**Figure 4A**), CO_2 becomes reduced to sugar as electrons, along with hydrogen ions (H^+) from water, are added to it. Meanwhile, water molecules are oxidized; that is, they lose electrons, along with hydrogen ions. Recall that oxidation and reduction always go hand in hand.

Now compare the food-producing equation for photosynthesis with the energy-releasing equation for cellular respiration (**Figure 4B**). Overall, cellular respiration harvests energy

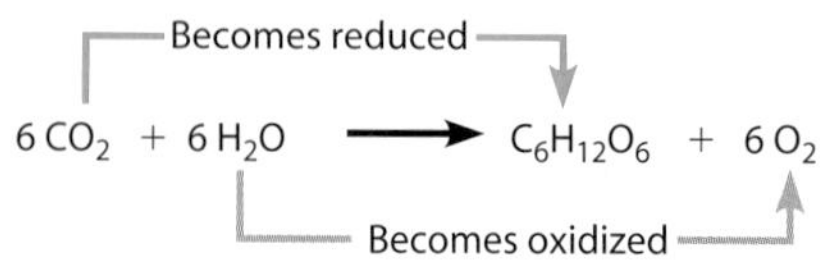

▲ **Figure 4A** Photosynthesis (uses light energy)

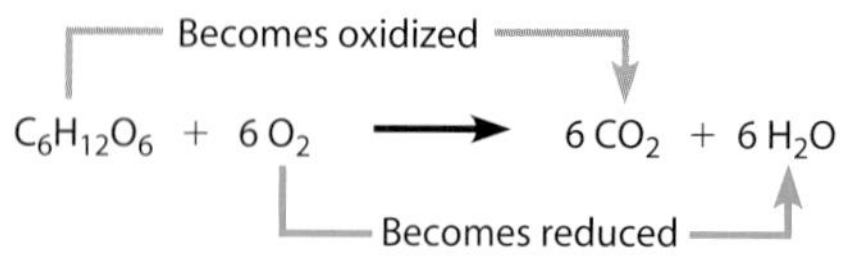

▲ **Figure 4B** Cellular respiration (releases chemical energy)

stored in a glucose molecule by oxidizing the sugar and reducing O_2 to H_2O. This process involves a number of energy-releasing redox reactions, with electrons losing potential energy as they "fall" down an electron transport chain to O_2. Along the way, the mitochondrion uses some of the energy to synthesize ATP.

In contrast, the food-producing redox reactions of photosynthesis require energy. The potential energy of electrons increases as they move from H_2O to CO_2 during photosynthesis. The light energy captured by chlorophyll molecules in the chloroplast provides this energy boost. Photosynthesis converts light energy to chemical energy and stores it in the chemical bonds of sugar molecules, which can provide energy for later use or raw materials for biosynthesis.

? Which redox process, photosynthesis or cellular respiration, is endergonic?

Photosynthesis

5 Overview: The two stages of photosynthesis are linked by ATP and NADPH

The equation for photosynthesis is a simple summary of a very complex process. Actually, photosynthesis occurs in two stages, each with multiple steps. **Figure 5** shows the inputs and outputs of the two stages and how the stages are related.

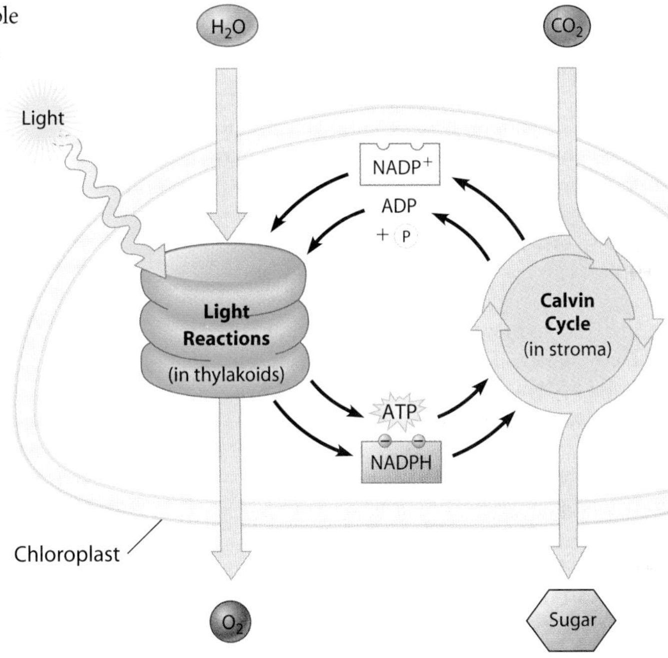

▲ **Figure 5** An overview of the two stages of photosynthesis in a chloroplast

The **light reactions** include the steps that convert light energy to chemical energy and release O_2. As shown in the figure, the light reactions occur in the thylakoid membranes. Water is split, providing a source of electrons and giving off O_2 as a by-product. Light energy absorbed by chlorophyll molecules built into the membranes is used to drive the transfer of electrons and H^+ from water to the electron acceptor **$NADP^+$**, reducing it to NADPH. This electron carrier is first cousin to NADH, which transports electrons in cellular respiration; the two differ only in the extra phosphate group in NADPH. NADPH temporarily stores electrons and hydrogen ions and provides "reducing power" to the Calvin cycle. The light reactions also generate ATP from ADP and a phosphate group.

In summary, the light reactions absorb solar energy and convert it to chemical energy stored in both ATP and NADPH. Notice that these reactions produce no sugar; sugar is not made until the Calvin cycle, which is the second stage of photosynthesis.

The **Calvin cycle** occurs in the stroma of the chloroplast (see Figure 5). It is a cyclic series of reactions that assembles sugar molecules using CO_2 and the energy-rich products of the light reactions. The incorporation of carbon from CO_2 into organic compounds, shown in the figure as CO_2 entering the Calvin cycle, is called **carbon fixation**. After carbon fixation, enzymes of the cycle make sugars by further reducing the carbon compounds.

As the figure suggests, it is NADPH produced by the light reactions that provides the electrons for reducing carbon in the Calvin cycle. And ATP from the light reactions provides chemical energy that powers several of the steps of the Calvin cycle. The Calvin cycle is sometimes referred to as the dark reactions, or light-independent reactions, because none of the steps requires light directly. However, in most plants, the Calvin cycle occurs during daylight, when the light reactions power the cycle's sugar assembly line by supplying it with NADPH and ATP.

The word *photosynthesis* encapsulates the two stages. *Photo*, from the Greek word for "light," refers to the light reactions; *synthesis*, meaning "putting together," refers to sugar construction by the Calvin cycle. In the next several modules, we look at these two stages in more detail. But first, let's consider some of the properties of light, the energy source that powers photosynthesis.

? For chloroplasts to produce sugar from carbon dioxide in the dark, they would need to be supplied with _____ and _____.

ATP . . . NADPH

The Light Reactions: Converting Solar Energy to Chemical Energy

6 Visible radiation absorbed by pigments drives the light reactions

What do we mean when we say that photosynthesis is powered by light energy from the sun? Sunlight is a type of energy called electromagnetic energy or electromagnetic radiation.

The Nature of Sunlight Electromagnetic energy travels in space as rhythmic waves analogous to those made by a pebble dropped in a puddle of water. **Figure 6A** shows the **electromagnetic spectrum**, the full range of electromagnetic wavelengths from the very short gamma rays to the very long-wavelength radio waves. As you can see in the center of the figure, visible light—the radiation your eyes see as different colors—is only a small fraction of the spectrum. It consists of wavelengths from about 380 nm to about 750 nm. The distance between the crests of two adjacent waves is called a **wavelength** (illustrated at the bottom right of the figure).

The model of light as waves explains many of light's properties. However, light also behaves as discrete packets of energy called photons. A **photon** has a fixed quantity of energy, and the shorter the wavelength of light, the greater the energy of its photons. In fact, the photons of wavelengths that are shorter than those of visible light have enough energy to damage molecules such as proteins and nucleic acids. This is why ultraviolet (UV) radiation can cause sunburns and skin cancer.

Photosynthetic Pigments **Figure 6B** shows what happens to visible light in the chloroplast. Light-absorbing molecules called *pigments*, built into the thylakoid membranes, absorb some wavelengths of light and reflect or transmit other wavelengths. We do not see the absorbed wavelengths; their energy has been absorbed by pigment molecules. What we see when we look at a leaf are the green wavelengths that the pigments transmit and reflect.

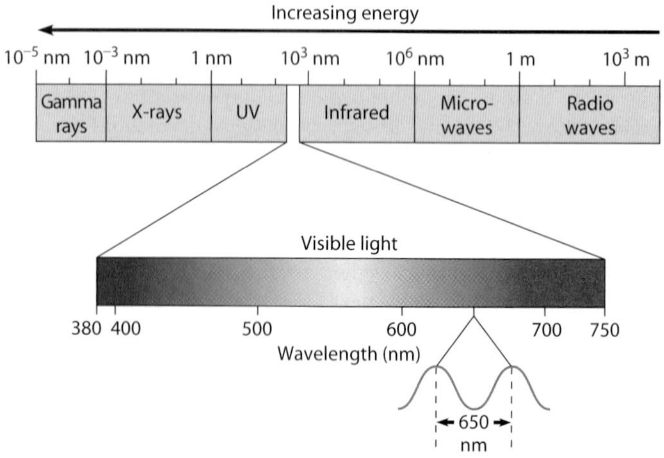

▲ **Figure 6A** The electromagnetic spectrum and the wavelengths of visible light. (A wavelength of 650 nm is illustrated.)

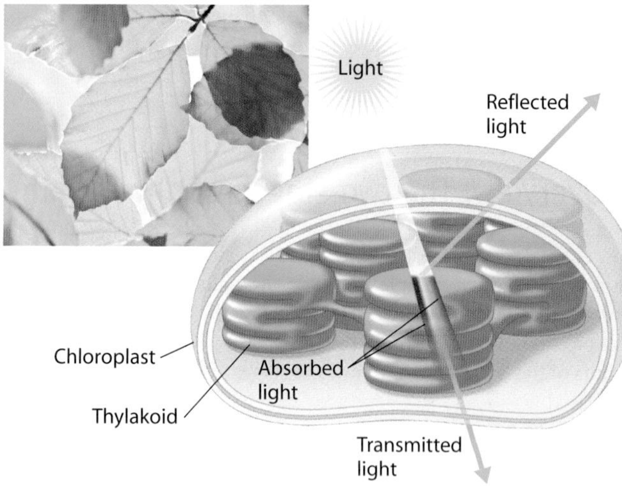

▲ **Figure 6B** The interaction of light with a chloroplast

Different pigments absorb light of different wavelengths, and chloroplasts contain more than one type of pigment. **Chlorophyll *a***, which participates directly in the light reactions, absorbs mainly blue-violet and red light. It looks blue-green because it reflects mainly green light. A very similar molecule, chlorophyll *b*, absorbs mainly blue and orange light and reflects (appears) yellow-green. Chlorophyll *b* broadens the range of light that a plant can use by conveying absorbed energy to chlorophyll *a*, which then puts the energy to work in the light reactions.

Chloroplasts also contain pigments called carotenoids, which are various shades of yellow and orange. The spectacular colors of fall foliage in certain parts of the world are due partly to the yellow-orange hues of longer-lasting carotenoids that show through once the green chlorophyll breaks down. Carotenoids may broaden the spectrum of colors that can drive photosynthesis. However, a more important function seems to be *photoprotection*: Some carotenoids absorb and dissipate excessive light energy that would otherwise damage chlorophyll or interact with oxygen to form reactive oxidative molecules that can damage cell molecules. Similar carotenoids, which we obtain from carrots and some other plants, have a photoprotective role in our eyes.

Each type of pigment absorbs certain wavelengths of light because it is able to absorb the specific amounts of energy in those photons. Next we see what happens when a pigment molecule such as chlorophyll absorbs a photon of light.

? **You may hear about the proposed health benefits of "phytochemicals" found in deep orange or red fruits and vegetables. How might such chemicals benefit a cell?**

As antioxidants that protect from reactive forms of oxidative molecules

7 Photosystems capture solar energy

Energy cannot be destroyed, but it can be transformed. Let's see how light energy can be transformed to other types of energy. When a pigment molecule absorbs a photon of light, one of the pigment's electrons jumps to an energy level farther from the nucleus. In this location, the electron has more potential energy, and we say that the electron has been raised from a ground state to an excited state. The excited state, like all high-energy states, is unstable. Generally, when isolated pigment molecules absorb light, their excited electrons drop back down to the ground state in a billionth of a second, releasing their excess energy as heat. This conversion of light energy to heat is what makes a black car so hot on a sunny day (black pigments absorb all wavelengths of light).

Some isolated pigments, including chlorophyll, emit light as well as heat after absorbing photons. We can demonstrate this phenomenon in the laboratory with a chlorophyll solution, as shown on the left in **Figure 7A**. When illuminated, the chlorophyll emits photons of light that produce a reddish afterglow called fluorescence. The right side of Figure 7A illustrates what happens in fluorescence: An absorbed photon boosts an electron of chlorophyll to an excited state, from which it falls back to the ground state, emitting its energy as heat and light.

But chlorophyll behaves very differently in isolation than it does in an intact chloroplast. In their native habitat of the thylakoid membrane, chlorophyll and other pigment molecules that absorb photons transfer the energy to other pigment molecules and eventually to a special pair of chlorophyll molecules. This pair passes off an excited electron to a neighboring molecule before it has a chance to drop back to the ground state.

In the thylakoid membrane, chlorophyll molecules are organized along with other pigments and proteins into clusters called photosystems **(Figure 7B)**. A **photosystem** consists of a number of light-harvesting complexes surrounding a reaction-center complex. A *light-harvesting complex* contains various pigment molecules bound to proteins. Collectively, the light-harvesting complexes function as a light-gathering antenna. The pigments absorb photons and pass the energy from molecule to molecule (thin yellow arrows) until it reaches the reaction center. The *reaction-center complex* contains the pair of special chlorophyll *a* molecules and a molecule called the *primary electron acceptor*, which is capable of accepting electrons and becoming reduced. The solar-powered transfer of an electron from the reaction-center chlorophyll *a* to the primary electron acceptor is the first step in the transformation of light energy to chemical energy in the light reactions.

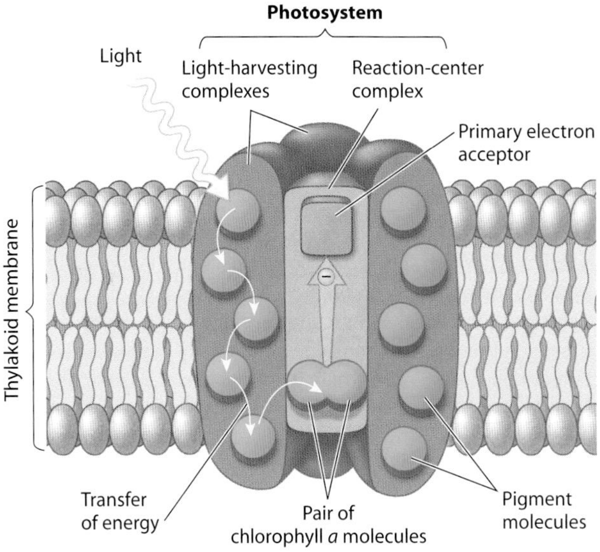

▲ **Figure 7B** A light-excited pair of chlorophyll molecules in the reaction center of a photosystem passing an excited electron to a primary electron acceptor

Two types of photosystems have been identified, and they cooperate in the light reactions. They are referred to as photosystem I and photosystem II, in order of their discovery, although photosystem II actually functions first in the sequence of steps that make up the light reactions. Each photosystem has a characteristic reaction-center complex. In photosystem II, the chlorophyll *a* of the reaction-center complex is called P680 because the light it absorbs best is red light with a wavelength of 680 nm. The reaction-center chlorophyll of photosystem I is called P700 because the wavelength of light it absorbs best is 700 nm (in the far-red part of the spectrum). Now let's see how the two photosystems work together in the light reactions to generate ATP and NADPH.

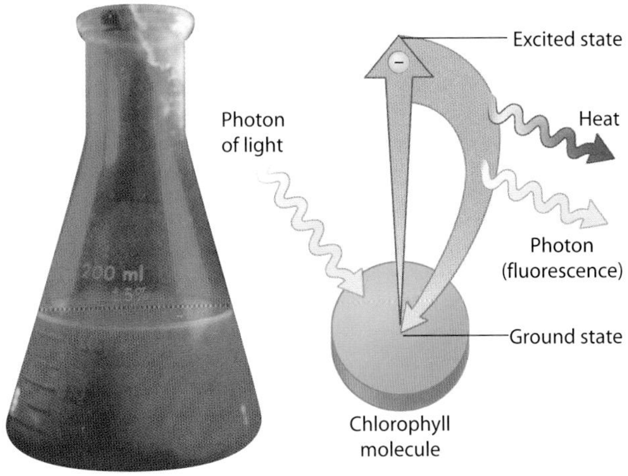

Christine Case

▲ **Figure 7A** A solution of chlorophyll glowing red when illuminated (left); a diagram of an isolated, light-excited chlorophyll molecule that releases a photon of red light (right)

? **Compared with a solution of isolated chlorophyll, why do intact chloroplasts release less heat and fluorescence when illuminated?**

In the chloroplasts, energy is passed from pigment molecule to pigment molecule, and eventually the light-excited electrons of reaction-center chlorophyll molecules are trapped by a primary electron acceptor rather than giving up their energy as heat and light.

8 Two photosystems connected by an electron transport chain generate ATP and NADPH

In the light reactions, light energy is transformed into the chemical energy of ATP and NADPH. In this process, electrons removed from H_2O pass from photosystem II to photosystem I to $NADP^+$. Between the two photosystems, the electrons move down an electron transport chain (similar to the one in cellular respiration) and provide energy for the synthesis of ATP.

Let's follow the flow of electrons (represented by gold arrows) in **Figure 8A**, which shows the two photosystems embedded in a thylakoid membrane. ❶ A pigment molecule in a light-harvesting complex absorbs a photon of light. The energy is passed to other pigment molecules and finally to the reaction center of photosystem II, where it excites an electron of chlorophyll P680 to a higher energy state. ❷ This electron is captured by the primary electron acceptor. ❸ Water is split, and its electrons are supplied one by one to P680, each replacing an electron lost to the primary electron acceptor. The oxygen atom combines with a second oxygen from another split water molecule, forming O_2.

❹ Each photoexcited electron passes from photosystem II to photosystem I via an electron transport chain. The exergonic "fall" of electrons provides energy for the synthesis of ATP by pumping H^+ across the membrane (not shown here). ❺ Meanwhile, light energy excites an electron of chlorophyll P700 in the reaction center of photosystem I. An adjacent primary electron acceptor captures the electron, and an electron that reaches the bottom of the electron transport chain from photosystem II replaces the lost electron in P700. ❻ Photoexcited electrons of photosystem I are passed through a short electron transport chain to $NADP^+$, reducing it to NADPH.

Admitedly, the scheme shown in Figure 8A is complicated. **Figure 8B** provides a mechanical analogy to help you focus on the key point: how the two photosystems cooperate in generating ATP and NADPH. The input of light energy, represented by the large yellow mallets, boosts electrons in the reaction-center complexes of both photosystems up to the excited state. The electrons are caught by the primary electron acceptor on top of the platform in each photosystem. Photosystem II passes the electrons through an ATP mill. Photosystem I hands its electrons off to reduce $NADP^+$ to NADPH.

NADPH, ATP, and O_2 are the products of the light reactions. Next we look in more detail at how ATP is formed.

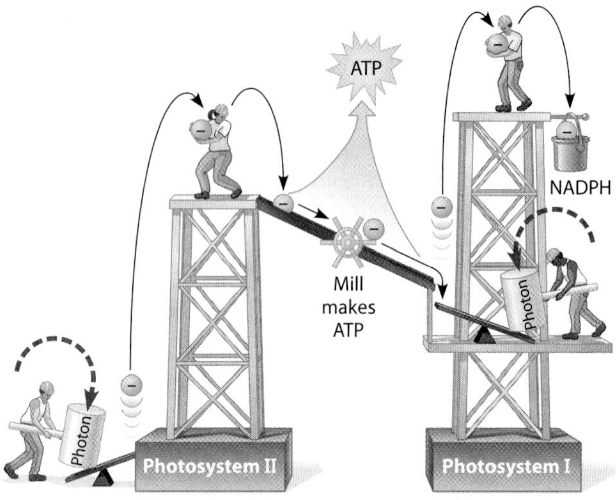

▲ **Figure 8B** A mechanical analogy of the light reactions

Adapted from Richard and David Walker. *Energy, Plants and Man*, fig. 4.1, p. 69. Sheffield: University of Sheffield. © Richard Walker. Used courtesy of Oxygraphics.

? Tracing the light reactions in Figure 8A, there is a flow of electrons from _____ to _____, which is reduced to _____, the source of electrons for sugar synthesis in the _____ cycle.

● water . . . $NADP^+$ . . . NADPH . . . Calvin

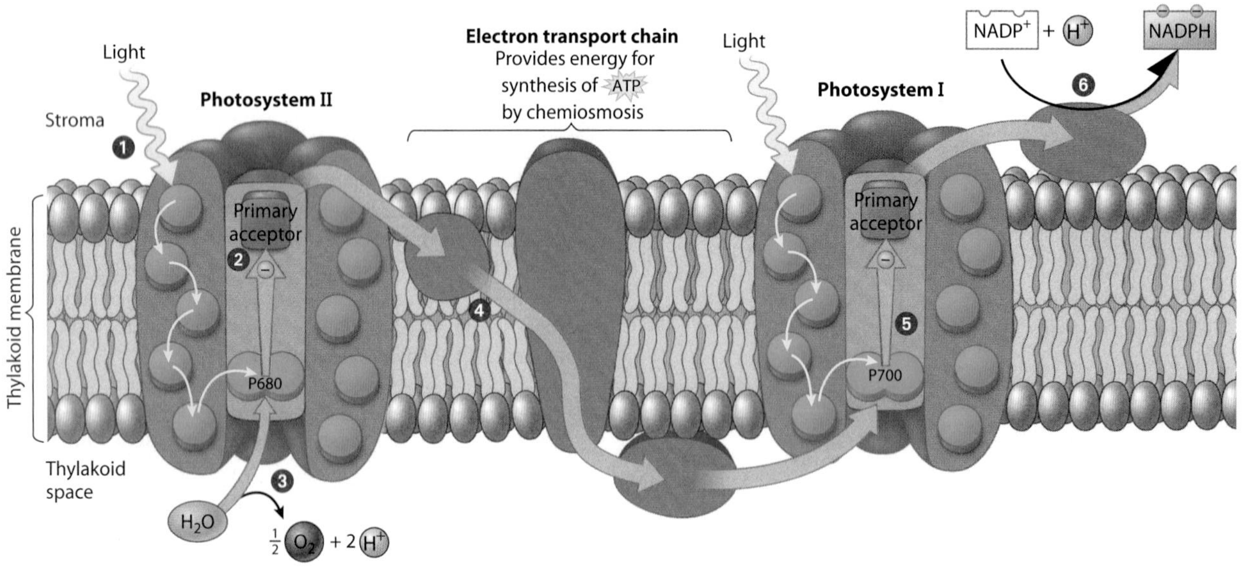

▲ **Figure 8A** Electron flow in the light reactions: light energy driving electrons from water to NADPH

9 Chemiosmosis powers ATP synthesis in the light reactions

Chemiosmosis is the mechanism of oxidative phosphorylation (ATP formation) in cellular respiration in a mitochondrion. Chemiosmosis is also the mechanism that generates ATP in a chloroplast. Recall that in chemiosmosis, the potential energy of a concentration gradient of hydrogen ions (H^+) across a membrane powers ATP synthesis. This gradient is created when an electron transport chain uses the energy released as it passes electrons down the chain to pump H^+ across a membrane.

Figure 9 illustrates the relationship between chloroplast structure and function in the light reactions. As in Figure 8A, we show the two photosystems and electron transport chains, all located within the thylakoid membrane of a chloroplast. Here you can see that as photoexcited electrons are passed down the electron transport chain connecting the two photosystems, hydrogen ions are pumped across the membrane from the stroma into the thylakoid space (inside the thylakoid sacs). This generates a concentration gradient across the membrane.

The flask-shaped structure in the figure represents an ATP synthase complex, which is very similar to the ones found in a mitochondrion. The energy of the concentration gradient drives H^+ back across the membrane through ATP synthase, which couples the flow of H^+ to the phosphorylation of ADP. Because the initial energy input is light, this chemiosmotic production of ATP in photosynthesis is called **photophosphorylation.**

How does photophosphorylation compare with oxidative phosphorylation? In cellular respiration, the high-energy electrons passed down the electron transport chain come from the oxidation of organic molecules. In photosynthesis, light energy is used to drive electrons that originally came from water to the top of the transport chain. Mitochondria transfer chemical energy from food to ATP; chloroplasts transform light energy into the chemical energy of ATP.

Notice that in the light-driven flow of electrons through the two photosystems, the final electron acceptor is $NADP^+$, not O_2 as in cellular respiration. Electrons do not end up at a low energy level in H_2O, as they do in respiration. Instead, they are stored at a high state of potential energy in NADPH.

In summary, the light reactions provide the chemical energy (ATP) and reducing power (NADPH) for the next stage of photosynthesis, the Calvin cycle. In the next module we see how that cycle makes sugar.

? What is the advantage of the light reactions producing NADPH and ATP on the stroma side of the thylakoid membrane?

The Calvin cycle, which uses the NADPH and ATP, occurs in the stroma.

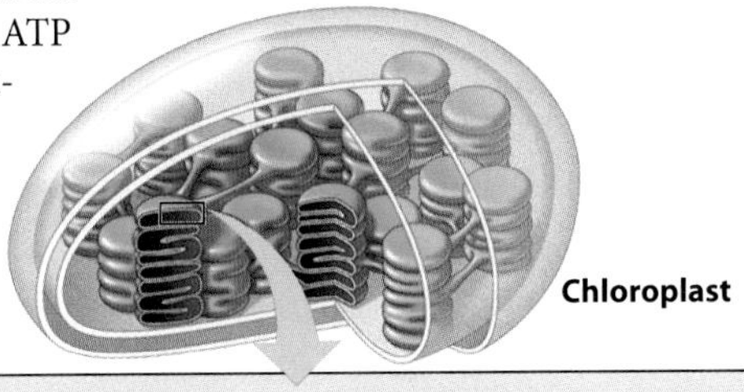

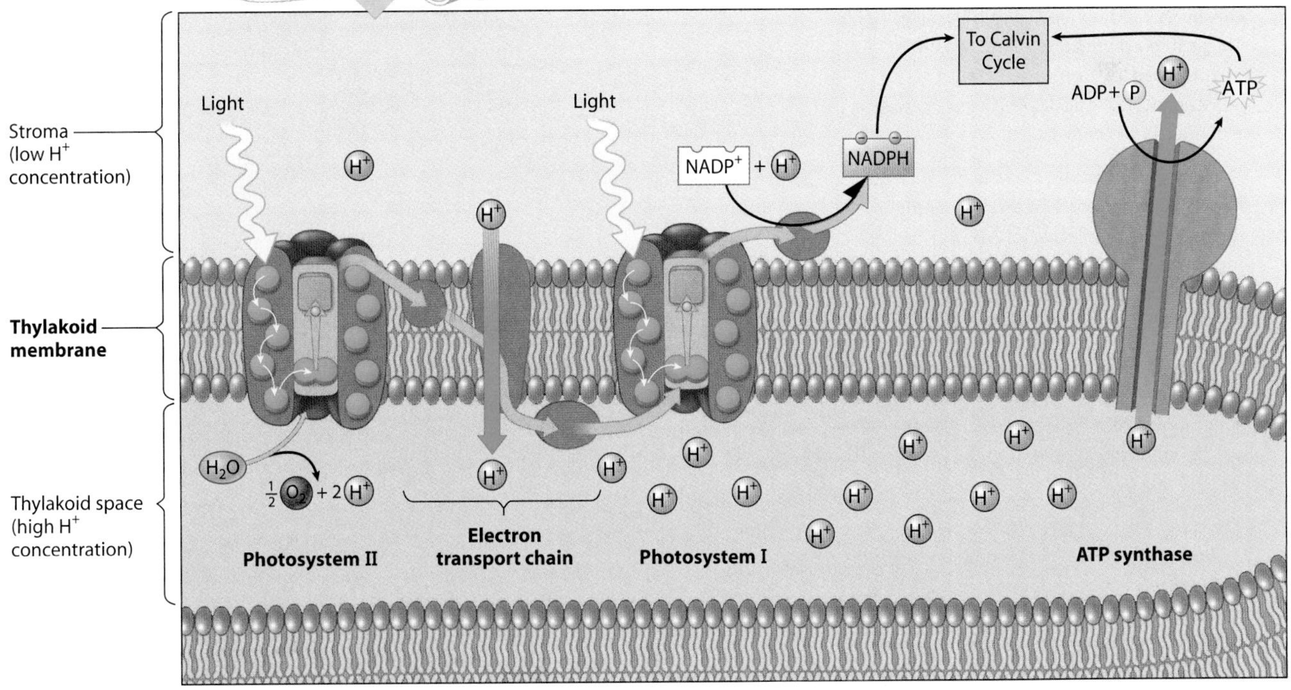

▲ **Figure 9** The production of ATP by chemiosmosis (numerous copies of these components present in each thylakoid)

The Calvin Cycle: Reducing CO_2 to Sugar

10 ATP and NADPH power sugar synthesis in the Calvin cycle

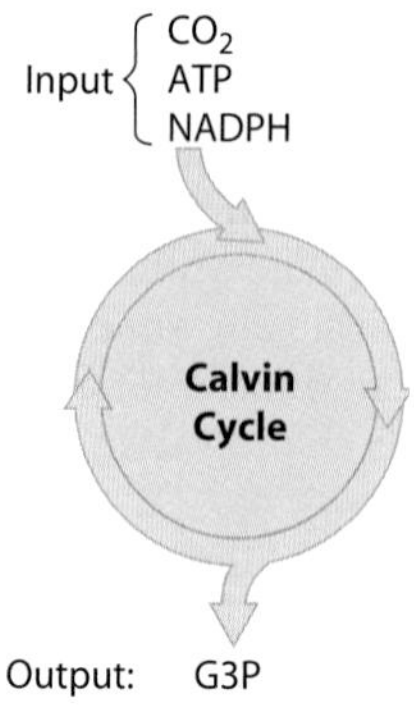

▲ **Figure 10A** An overview of the Calvin cycle

The Calvin cycle functions like a sugar factory within a chloroplast. As **Figure 10A** shows, inputs to this all-important food-making process are CO_2 (from the air) and ATP and NADPH (both generated by the light reactions). Using CO_2, energy from ATP, and high-energy electrons from NADPH, the Calvin cycle constructs an energy-rich, three-carbon sugar, glyceraldehyde 3-phosphate (G3P). A plant cell can use G3P to make glucose and other organic molecules as needed. (You already met G3P in glycolysis: It is the three-carbon sugar formed by the splitting of glucose.)

Figure 10B presents the details of the Calvin cycle. It is called a cycle because, like the citric acid cycle in cellular respiration, the starting material is regenerated with each turn of the cycle. In this case, the starting material is a five-carbon sugar named ribulose bisphosphate (RuBP). ❶ In the carbon fixation step, the enzyme rubisco attaches CO_2 to RuBP. (Rubisco is thought to be the most abundant protein on Earth.) ❷ In the next step, a reduction reaction, NADPH reduces the organic acid 3-PGA to G3P using the energy of ATP. To make a molecule of G3P, the cycle must incorporate the carbon atoms from three molecules of CO_2. The cycle actually incorporates one carbon at a time, but we show it starting with three CO_2 molecules so that we end up with a complete G3P molecule.

For this to be a cycle, RuBP must be regenerated. ❸ For every three CO_2 molecules fixed, one G3P molecule leaves the cycle as product, and the remaining five G3P molecules are rearranged, ❹ using energy from ATP to regenerate three molecules of RuBP.

Note that for the net synthesis of one G3P molecule, the Calvin cycle consumes nine ATP and six NADPH molecules, which were provided by the light reactions.

? **To synthesize one glucose molecule, the Calvin cycle uses ______ CO_2, ______ ATP, and ______ NADPH. Explain why this high number of ATP and NADPH molecules is consistent with the value of glucose as an energy source.**

6 . . . 18 . . . 12. Glucose is a valuable energy source because it is highly reduced, storing lots of potential energy in its electrons. The more energy a molecule stores, the more energy and reducing power required to produce that molecule.

Step ❶ Carbon fixation. An enzyme called rubisco combines CO_2 with a five-carbon sugar called ribulose bisphosphate (abbreviated RuBP). The unstable product splits into two molecules of the three-carbon organic acid, 3-phosphoglyceric acid (3-PGA). For three CO_2 entering, six 3-PGA result.

Step ❷ Reduction. Two chemical reactions (indicated by the two blue arrows) consume energy from six molecules of ATP and oxidize six molecules of NADPH. Six molecules of 3-PGA are reduced, producing six molecules of the energy-rich three-carbon sugar, G3P.

Step ❸ Release of one molecule of G3P. Five of the G3Ps from step 2 remain in the cycle. The single molecule of G3P you see leaving the cycle is the net product of photosynthesis. A plant cell uses G3P to make glucose and other organic compounds.

Step ❹ Regeneration of RuBP. A series of chemical reactions uses energy from ATP to rearrange the atoms in the five G3P molecules (15 carbons total), forming three RuBP molecules (15 carbons). These can start another turn of the cycle.

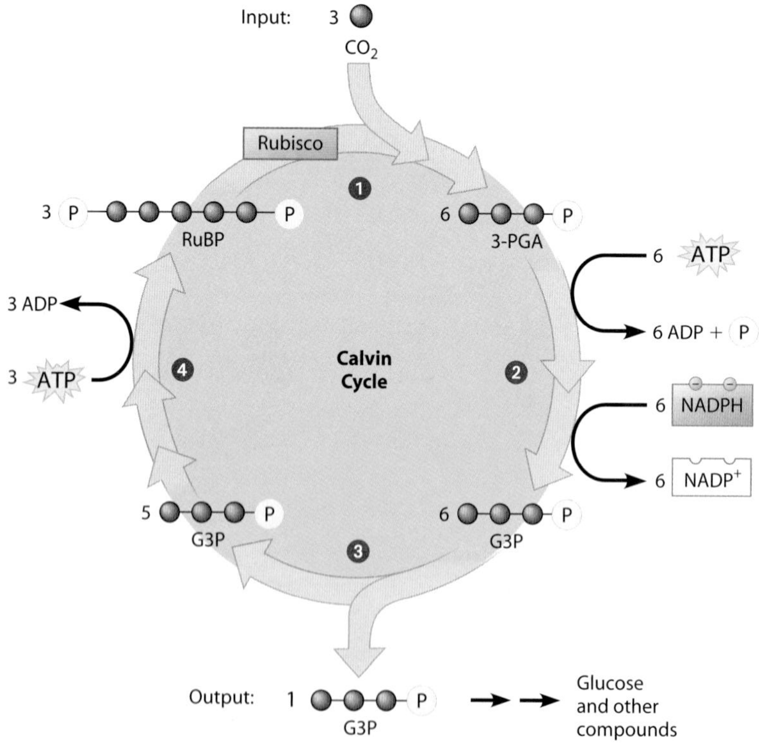

▲ **Figure 10B** Details of the Calvin cycle, which takes place in the stroma of a chloroplast

EVOLUTION CONNECTION

11 Other methods of carbon fixation have evolved in hot, dry climates

As you just learned in the previous module, the first step of the Calvin cycle is carbon fixation. Most plants use CO_2 directly from the air, and carbon fixation occurs when the enzyme rubisco adds CO_2 to RuBP (see step 1 of Figure 10B). Such plants are called **C_3 plants** because the first product of carbon fixation is the three-carbon compound 3-PGA. C_3 plants are widely distributed; they include such important agricultural crops as soybeans, oats, wheat, and rice. One problem that farmers face in growing C_3 plants is that hot, dry weather can decrease crop yield. In response to such conditions, plants close their stomata, the pores in their leaves. This adaptation reduces water loss and helps prevent dehydration, but it also prevents CO_2 from entering the leaf and O_2 from leaving. As a result, CO_2 levels get very low in the leaf and photosynthesis slows. And the O_2 released from the light reactions begins to accumulate, creating another problem.

As O_2 builds up in a leaf, rubisco adds O_2 instead of CO_2 to RuBP. A two-carbon product of this reaction is then broken down in the cell. This process is called **photorespiration** because it occurs in the light and, like respiration, it consumes O_2 and releases CO_2. But unlike cellular respiration, it uses ATP instead of producing it; and unlike photosynthesis, it yields no sugar. Photorespiration can, however, drain away as much as 50% of the carbon fixed by the Calvin cycle.

According to one hypothesis, photorespiration is an evolutionary relic from when the atmosphere had less O_2 than it does today. In the ancient atmosphere that prevailed when rubisco first evolved, the inability of the enzyme's active site to exclude O_2 would have made little difference. It is only after O_2 became so concentrated in the atmosphere that the "sloppiness" of rubisco presented a problem. New evidence also indicates that photorespiration may play a protective role when the products of the light reactions build up in a cell (as occurs when the Calvin cycle slows due to a lack of CO_2).

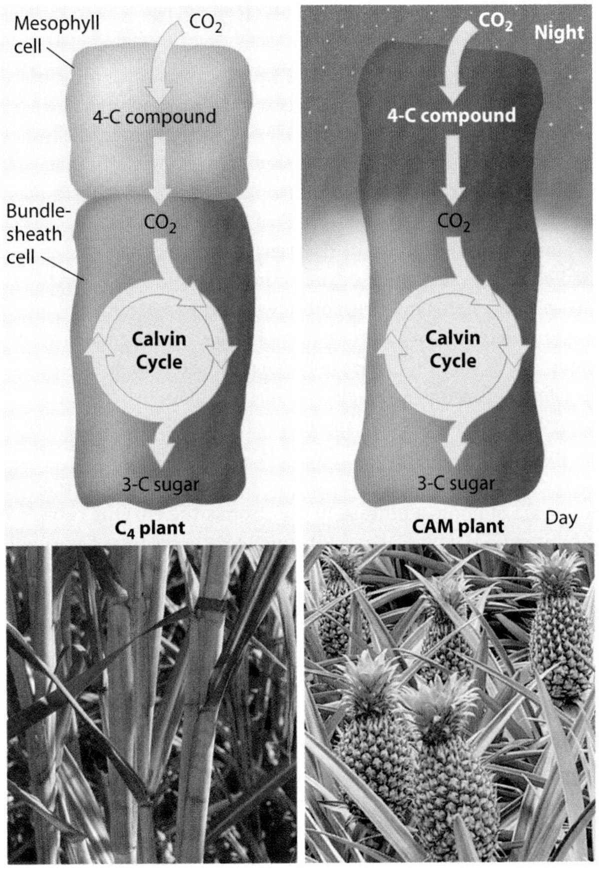

▲ Figure 11 Comparison of C_4 and CAM photosynthesis: The fixing of CO_2 into a four-carbon compound and the Calvin cycle occur in different cells in C_4 plants and at different times of day in CAM plants.

left Dinodia/Pixtal/AGE Fotostock
right ImageDJ/Jupiter Images/Getty Images

C_4 Plants In some plant species found in hot, dry climates, alternate modes of carbon fixation have evolved that minimize photorespiration and optimize the Calvin cycle. **C_4 plants** are so named because they first fix CO_2 into a four-carbon compound. When the weather is hot and dry, a C_4 plant keeps its stomata mostly closed, thus conserving water. It continues making sugars by photosynthesis using the pathway and the two types of cells shown on the left side of **Figure 11**. An enzyme in the mesophyll cells has a high affinity for CO_2 and can fix carbon even when the CO_2 concentration in the leaf is low. The resulting four-carbon compound then acts as a carbon shuttle; it moves into bundle-sheath cells, which are packed around the veins of the leaf, and releases CO_2. Thus, the CO_2 concentration in these cells remains high enough for the Calvin cycle to make sugars and avoid photorespiration. Corn and sugarcane are examples of agriculturally important C_4 plants.

CAM Plants A second photosynthetic adaptation has evolved in pineapples, many cacti, and other succulent (water-storing) plants, such as aloe and jade plants. Called **CAM plants**, these species are adapted to very dry climates. A CAM plant (right side of Figure 11) conserves water by opening its stomata and admitting CO_2 only at night. CO_2 is fixed into a four-carbon compound, which banks CO_2 at night and releases it during the day. Thus, the Calvin cycle can operate, even with the leaf's stomata closed during the day.

In C_4 plants, carbon fixation and the Calvin cycle occur in different types of cells. In CAM plants, these processes occur in the same cells, but at different times of the day. Keep in mind that CAM, C_4, and C_3 plants all eventually use the Calvin cycle to make sugar from CO_2. The C_4 and CAM pathways are two evolutionary adaptations that minimize photorespiration and maximize photosynthesis in hot, dry climates.

? Why would you expect photorespiration on a hot, dry day to occur less in C_4 and CAM plants than in C_3 plants?

Because of their initial fixing of carbon, both C_4 and CAM plants can supply rubisco with CO_2. When a C_3 plant closes its stomata, CO_2 levels drop and O_2 rises, making it more likely that rubisco will add O_2 to RuBP.

Photosynthesis Reviewed and Extended

12 Review: Photosynthesis uses light energy, carbon dioxide, and water to make organic molecules

Life on Earth is solar powered. As we have discussed, most of the living world depends on the food-making machinery of photosynthesis. **Figure 12** summarizes this vital process. The production of sugar from CO_2 is an emergent property that arises from the structure of a chloroplast—a structural arrangement that integrates the two stages of photosynthesis.

Starting on the left in the diagram, you see a summary of the light reactions, which occur in the thylakoid membranes. Two photosystems in the membranes capture solar energy, energizing electrons in chlorophyll molecules. Simultaneously, water is split, and O_2 is released. The photoexcited electrons are transferred through an electron transport chain, where energy is harvested to make ATP, and finally to $NADP^+$, reducing it to the high-energy compound NADPH.

The chloroplast's sugar factory is the Calvin cycle, the second stage of photosynthesis. In the stroma, the enzyme rubisco combines CO_2 with RuBP. ATP and NADPH are used to reduce 3-PGA to G3P. Sugar molecules made from G3P serve as a plant's own food supply.

About 50% of the carbohydrate made by photosynthesis is consumed as fuel for cellular respiration in the mitochondria of plant cells. Sugars also serve as starting material for making other organic molecules, such as a plant's proteins and lipids. Many glucose molecules are linked together to make cellulose, the main component of cell walls. Cellulose is the most abundant organic molecule in a plant—and probably on the surface of the planet. Most plants make much more food each day than they need. They store the excess in roots, tubers, seeds, and fruits.

Plants (and other photosynthesizers) not only feed themselves but also are the ultimate source of food for virtually all other organisms. Humans and other animals make none of their own food and are totally dependent on the organic matter made by photosynthesizers. Even the energy we acquire when we eat meat was originally captured by photosynthesis. The energy in a hamburger, for instance, came from sunlight that was originally converted to a chemical form in the grasses eaten by cattle.

The collective productivity of the tiny chloroplasts is truly amazing: Photosynthesis makes an estimated 160 billion metric tons of carbohydrate per year (about 176 billion tons). That's equivalent in mass to a stack of about 100 trillion copies of this textbook. No other chemical process on Earth can match the output of photosynthesis.

This review of photosynthesis is an appropriate place to reflect on the metabolic ground we have covered in this chapter. Virtually all organisms, plants included, use cellular respiration to obtain the energy they need from fuel molecules such as glucose. We followed the chemical pathways of glycolysis and the citric acid cycle, which oxidize glucose and release energy from it. We have now come full circle, seeing how plants trap sunlight energy and use it to reduce carbon dioxide to make glucose.

In tracing glucose synthesis and its breakdown, we have also seen that cells use several of the same mechanisms—electron transport, redox reactions, and chemiosmosis—in energy storage (photosynthesis) and energy harvest (cellular respiration).

? Explain this statement: No process is more important than photosynthesis to the welfare of life on Earth.

Photosynthesis is the ultimate source of the food for almost all organisms and the oxygen they need for cellular respiration.

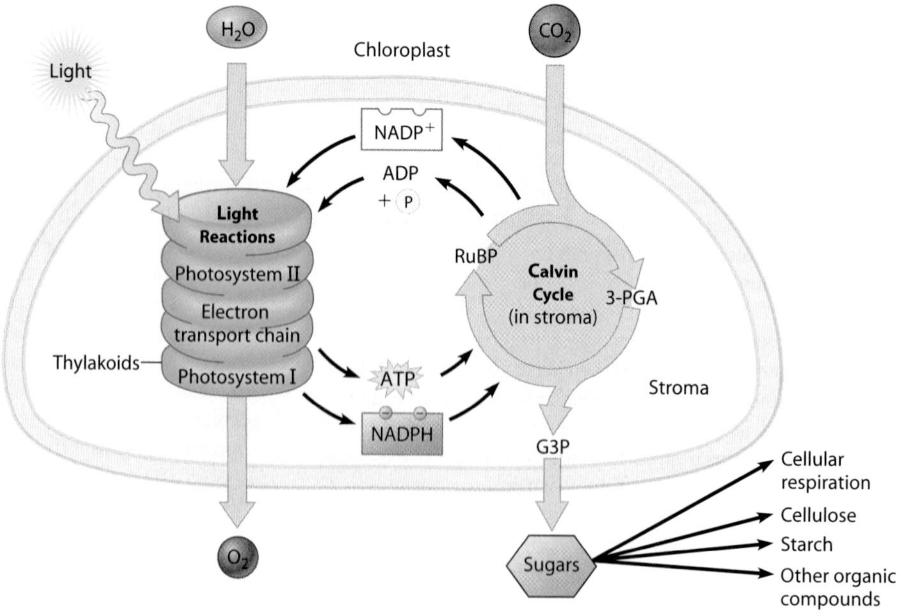

▶ **Figure 12** A summary of photosynthesis

CONNECTION 13 Photosynthesis may moderate global climate change

The greenhouse in **Figure 13A** is used to grow plants when the weather outside is too cold. The glass or plastic walls of a greenhouse allow solar radiation to pass through. The sunlight heats the soil, which in turn warms the air. The walls trap the warm air, raising the temperature inside.

An analogous process, called the **greenhouse effect**, operates on a global scale (**Figure 13B**). Solar radiation reaching Earth's atmosphere includes ultraviolet radiation and visible light. As we discuss in the next module, the ozone layer filters out most of the damaging UV radiation. Visible light passes through and is absorbed by the planet's surface, warming it. Heat radiating from the warmed planet is absorbed by gases in the atmosphere, which then reflect some of the heat back to Earth. This natural heating effect is highly beneficial. Without it, Earth would be much colder, and most life as we know it could not exist.

The gases in the atmosphere that absorb heat radiation are called greenhouse gases. Some occur naturally, such as water vapor, carbon dioxide, and methane, while others are synthetic, such as the chlorofluorocarbons we discuss in the next module. Human activities add to the levels of these greenhouse gases.

Carbon dioxide is one of the most important greenhouse gases. You have just learned that CO_2 is a raw material for photosynthesis and a waste product of cellular respiration. These two processes, taking place in microscopic chloroplasts and mitochondria, keep carbon cycling between CO_2 and more complex organic compounds on a global scale. Photosynthetic organisms absorb billions of tons of CO_2 each year. Most of that fixed carbon returns to the atmosphere via cellular respiration, the action of decomposers, and fires. But much of it remains locked in large tracts of forests and undecomposed organisms. And large amounts of carbon are in long-term storage in fossil fuels buried deep under Earth's surface.

Since 1850, the start of the Industrial Revolution, the atmospheric concentration of CO_2 has increased about 40%, mostly due to the combustion of fossil fuels, such as coal, oil, and gasoline. Increasing concentrations of greenhouse gases have been linked to **global climate change**, of which the major aspect is *global warming*. The predicted consequences of this slow but steady increase in average global temperature include melting of polar ice, rising sea levels, extreme weather patterns, droughts, increased extinction rates, and the spread of tropical diseases. Indeed, many of these effects are already being documented.

Unfortunately, the rise in atmospheric CO_2 levels during the last century coincided with widespread deforestation, which aggravated the global warming problem by reducing an effective CO_2 sink. As forests are cleared for lumber or agriculture, and as population growth increases the demand for fossil fuels, CO_2 levels will continue to rise.

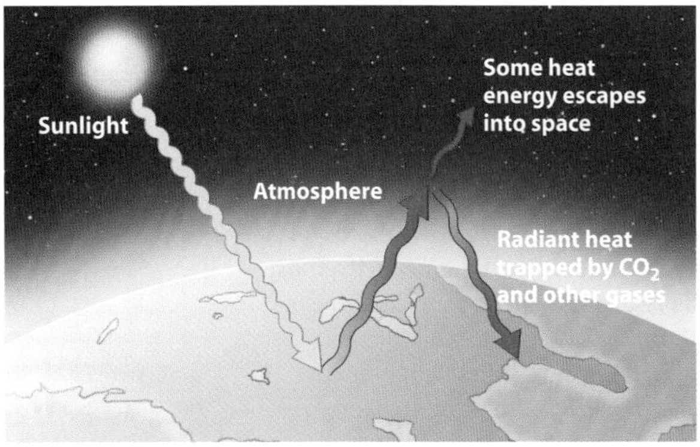

▲ **Figure 13B** CO_2 in the atmosphere and the greenhouse effect

Can photosynthesis offset this increase in atmospheric CO_2? Certainly, slowing the destruction of our forests will sustain their photosynthetic and carbon-storing contributions. Taking a lesson from plants, we can explore technologies that utilize solar energy for some of our energy needs. And as you read in the chapter introduction, biofuels hold out the promise of a renewable fuel source. As the plants or algae used for biofuels grow, their photosynthesis removes CO_2 from the atmosphere. The burning of these fuels releases CO_2 to the atmosphere, just as fossils fuels do. A key difference, however, is that fossil fuels come from the remains of ancient organisms, and their burning releases CO_2 that had been removed from the atmosphere by photosynthesis over the course of hundreds of millions of years. Growing and using alternative fuels could keep the cycle of CO_2 removal in photosynthesis balanced with CO_2 release in fuel burning.

? **Explain the greenhouse effect.**

Sunlight warms Earth's surface, which radiates heat to the atmosphere. CO_2 and other greenhouse gases absorb and radiate some heat back to Earth.

Doug Martin/Photo Researchers

▲ **Figure 13A** Plants growing in a greenhouse

SCIENTIFIC DISCOVERY

14 Scientific study of Earth's ozone layer has global significance

The process and importance of scientific discovery are illustrated by the story of how synthetic chemicals were destroying Earth's protective ozone layer and how the work of many scientists led to changes in worldwide environmental policies. As you now know, photosynthesis produces the O_2 on which almost all organisms depend for cellular respiration. This O_2 has another benefit: High in the atmosphere, high-energy solar radiation converts it to ozone (O_3). Acting as sunscreen for the planet, the ozone layer shields Earth from ultraviolet radiation. The balance between ozone formation and its natural destruction in the atmosphere, however, has been upset by human actions.

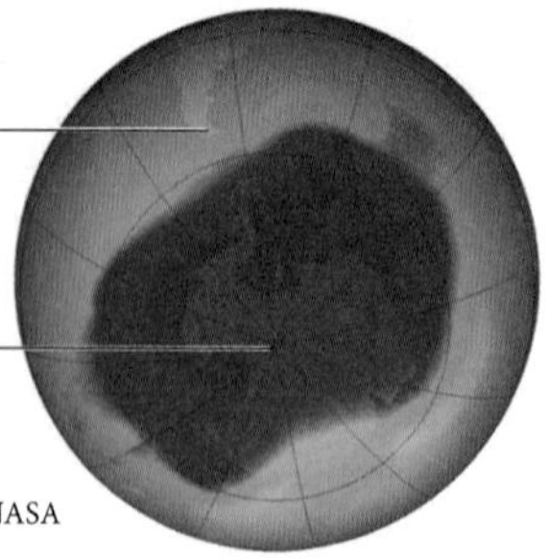

▲ **Figure 14A** The ozone hole in the Southern Hemisphere, spring 2006

Chlorofluorocarbons (CFCs) are chemicals developed in the 1930s that became widely used in aerosol sprays, refrigerators, and Styrofoam production. In 1970, a scientist wondered whether CFCs were accumulating in the environment and sent a homemade detector on a boat trip to Antarctica. He found CFCs in the air all along the journey. When he reported his findings at a scientific meeting in 1972, two chemists, Sherwood Rowland and Mario Molina, wondered what happened to CFCs once they entered the atmosphere. A search of the literature found that these chemicals were not broken down in the lower atmosphere.

But the intense solar radiation in the upper atmosphere could break down CFCs, releasing chlorine atoms. Molina learned that chlorine reacts with ozone, reducing it to O_2. Other reactions liberate the chlorine, allowing it to destroy more ozone. In 1974, Molina and Rowland published their work predicting that the release of CFCs would damage the ozone layer.

As other researchers tested the CFC-ozone depletion hypothesis, the evidence accumulated, and more people became concerned about ozone depletion. Others worried about the economic impact of banning CFCs, and CFC manufacturers mounted a campaign to cast doubt on the Molina-Rowland hypothesis in any way they could. A timetable for phasing out CFCs in aerosols was announced by the U.S. government in 1977, but worldwide production continued to increase.

Then, in 1982, a researcher noted a dramatic dip in the ozone layer over Antarctica. At first he suspected an instrument malfunction. After recording this decline for two more years, he published his observations. A reanalysis of data collected by NASA (National Aeronautics and Space Administration) over that period confirmed a gigantic hole in the ozone layer. This hole has appeared every spring over Antarctica since the late 1970s and continues today. **Figure 14A** shows an image produced from atmospheric data from 2006. Blue and purple colors show where there is the least ozone. The ozone depletion was much greater than had been predicted by Molina and Rowland. How could it be explained?

Susan Solomon (**Figure 14B**), with the National Oceanic and Atmospheric Administration (NOAA), developed a hypothesis that the unusual ice clouds that appear during early spring in Antarctica could speed up the reactions that destroy ozone. The then 30-year-old scientist led two research expeditions to the Antarctic. The data that she and her team collected indicated that chemical reactions occurring on the icy particles made CFCs hundreds of times more damaging than they normally are.

Further laboratory tests and field measurements taken from the ground, balloons, and airplanes supported Solomon's hypothesis. In response to these scientific findings, the first treaty to address Earth's environment was signed in 198 In the Montreal Protocol, many nations agreed to phase out CFCs. As research continued to establish the extent and danger of ozone depletion, these agreements were strengthened in 1990. In 1995, Molina and Rowland shared a Nobel Prize for their work on how CFCs were damaging the atmosphere.

Global emissions of CFCs are near zero now, but because these compounds are so stable, recovery of the ozone layer is not expected until around 2060. Meanwhile, unblocked UV radiation is predicted to increase skin cancer and cataracts, as well as damage crops and phytoplankton in the oceans.

In addition to being ozone destroyers, CFCs are also potent greenhouse gases. The phaseout of CFCs has avoided what would have been the equivalent of adding 10 gigatons of CO_2 to the atmosphere. (For comparison, the Kyoto Protocol of 1997 set a 2012 target of a reduction of 2 gigatons of CO_2 emissions.)

Whether an environmental problem involves CFCs or CO_2, the scientific research is often complicated and the solutions complex. The connections between science, technology, and society, so clearly exemplified by the work of the scientists studying the ozone layer, are a major theme of this chapter.

? Where does the ozone layer come from, and why is it so important to life on Earth?

● High in the atmosphere, radiation from the sun converts O_2 to ozone. The ozone layer absorbs potentially damaging UV radiation.

▲ **Figure 14B** Susan Solomon with a globe showing Antarctica

REVIEW

For Practice Quizzes, BioFlix, MP3 Tutors, and Activities, go to www.masteringbiology.com.

Reviewing the Concepts

An Overview of Photosynthesis (1–5)

1 Autotrophs are the producers of the biosphere. Plants, algae, and some protists and bacteria are photoautotrophs, the producers of food consumed by virtually all heterotrophic organisms.

2 Photosynthesis occurs in chloroplasts in plant cells. Chloroplasts are surrounded by a double membrane and contain stacks of thylakoids and a thick fluid called stroma.

3 Scientists traced the process of photosynthesis using isotopes. Experiments using both heavy and radioactive isotopes helped determine the details of the process of photosynthesis.

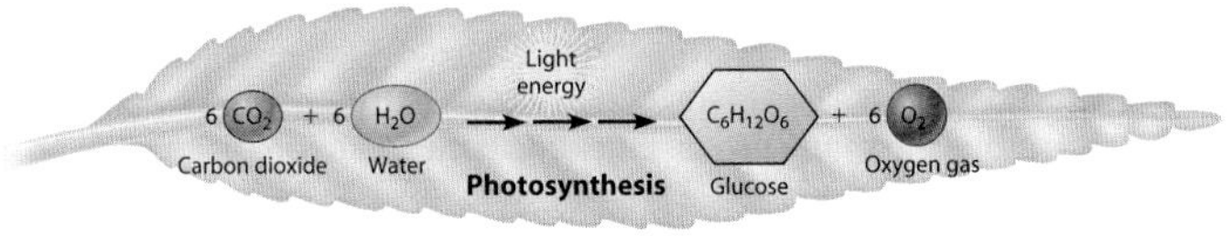

4 Photosynthesis is a redox process, as is cellular respiration. In photosynthesis, H_2O is oxidized and CO_2 is reduced.

5 Overview: The two stages of photosynthesis are linked by ATP and NADPH. The light reactions occur in the thylakoids, producing ATP and NADPH for the Calvin cycle, which takes place in the stroma.

The Light Reactions: Converting Solar Energy to Chemical Energy (6–9)

6 Visible radiation absorbed by pigments drives the light reactions. Certain wavelengths of visible light are absorbed by chlorophyll and other pigments. Carotenoids also function in photoprotection from excessive light.

7 Photosystems capture solar energy. Thylakoid membranes contain photosystems, each consisting of light-harvesting complexes and a reaction-center complex. A primary electron acceptor receives photoexcited electrons from chlorophyll.

8 Two photosystems connected by an electron transport chain generate ATP and NADPH. Electrons shuttle from photosystem II to photosystem I, providing energy to make ATP, and then reduce $NADP^+$ to NADPH. Photosystem II regains electrons as water is split and O_2 released.

9 Chemiosmosis powers ATP synthesis in the light reactions. In photophosphorylation, the electron transport chain pumps H^+ into the thylakoid space. The concentration gradient drives H^+ back through ATP synthase, driving the synthesis of ATP.

The Calvin Cycle: Reducing CO_2 to Sugar (10–11)

10 ATP and NADPH power sugar synthesis in the Calvin cycle. The steps of the Calvin cycle include carbon fixation, reduction, release of G3P, and regeneration of RuBP. Using carbon from CO_2, electrons from NADPH, and energy from ATP, the cycle constructs G3P, which is used to build glucose and other organic molecules.

11 Other methods of carbon fixation have evolved in hot, dry climates. In C_3 plants, a drop in CO_2 and rise in O_2 when stomata close divert the Calvin cycle to photorespiration. C_4 plants and CAM plants first fix CO_2 into a four-carbon compound that provides CO_2 to the Calvin cycle even when stomata close on hot, dry days.

Photosynthesis Reviewed and Extended (12–14)

12 Review: Photosynthesis uses light energy, carbon dioxide, and water to make organic molecules.

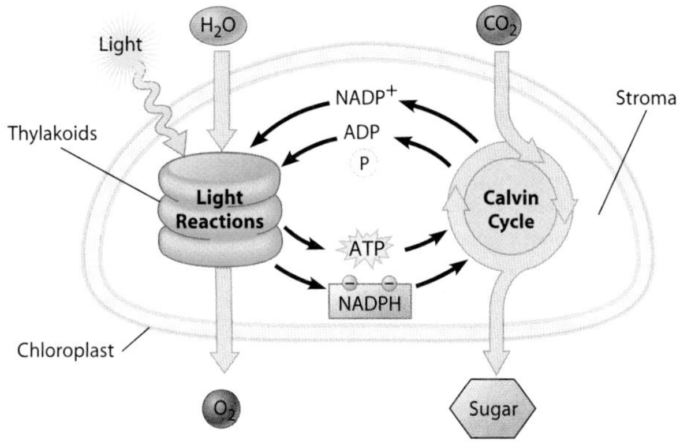

13 Photosynthesis may moderate global climate change. Reducing deforestation and fossil fuel use and growing biofuel crops, which remove CO_2 from the atmosphere, may help moderate the global warming caused by increasing CO_2 levels.

14 Scientific study of Earth's ozone layer has global significance. Solar radiation converts O_2 high in the atmosphere to ozone (O_3), which shields organisms from damaging UV radiation. Industrial chemicals called CFCs have caused dangerous thinning of the ozone layer, but international restrictions on CFC use are allowing a slow recovery.

Connecting the Concepts

1. The following diagram compares the chemiosmotic synthesis of ATP in mitochondria and chloroplasts. Fill in the blanks, which label the molecules shared by both processes and the regions in the chloroplast. Then, for both organelles, indicate which side of the membrane has the higher H^+ concentration.

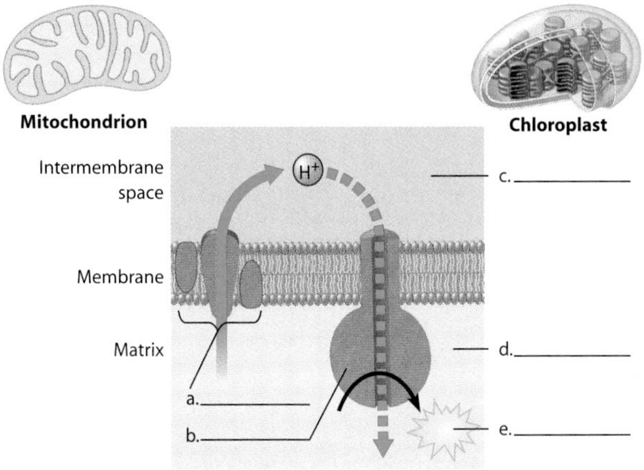

2. Continue your comparison of chemiosmosis and electron transport in mitochondria and chloroplasts. In each case,
   a. where do the electrons come from?
   b. how do the electrons get their high potential energy?
   c. what picks up the electrons at the end of the chain?
   d. how is the energy given up by the electrons used?

3. Complete this summary map of photosynthesis.

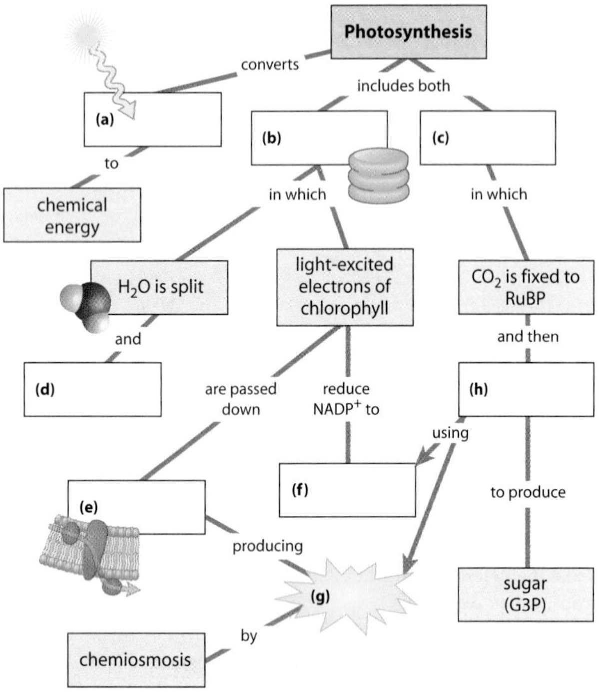

Testing Your Knowledge

Multiple Choice

4. In photosynthesis, ______ is oxidized and ______ is reduced.
   a. glucose . . . oxygen
   b. carbon dioxide . . . water
   c. water . . . carbon dioxide
   d. glucose . . . carbon dioxide
   e. water . . . oxygen

5. Which of the following are produced by reactions that take place in the thylakoids and consumed by reactions in the stroma?
   a. CO_2 and H_2O
   b. $NADP^+$ and ADP
   c. ATP and NADPH
   d. ATP, NADPH, and CO_2
   e. O_2 and ATP

6. When light strikes chlorophyll molecules in the reaction-center complex, they lose electrons, which are ultimately replaced by
   a. splitting water.
   b. breaking down ATP.
   c. oxidizing NADPH.
   d. fixing carbon.
   e. oxidizing glucose.

7. The reactions of the Calvin cycle are not directly dependent on light, but they usually do not occur at night. Why? (*Explain.*)
   a. It is often too cold at night for these reactions to take place.
   b. Carbon dioxide concentrations decrease at night.
   c. The Calvin cycle depends on products of the light reactions.
   d. Plants usually close their stomata at night.
   e. Most plants do not make four-carbon compounds, which they would need for the Calvin cycle at night.

8. How many "turns" of the Calvin cycle are required to produce one molecule of glucose? (Assume one CO_2 is fixed in each turn of the cycle.)
   a. 1
   b. 2
   c. 3
   d. 6
   e. 12

9. Which of the following does *not* occur during the Calvin cycle?
   a. carbon fixation
   b. oxidation of NADPH
   c. consumption of ATP
   d. regeneration of RuBP, the CO_2 acceptor
   e. release of oxygen

10. Why is it difficult for most plants to carry out photosynthesis in very hot, dry environments such as deserts?
    a. The light is too intense and destroys the pigment molecules.
    b. The closing of stomata keeps CO_2 from entering and O_2 from leaving the plant.
    c. They must rely on photorespiration to make ATP.
    d. Global warming is intensified in a desert environment.
    e. CO_2 builds up in the leaves, blocking carbon fixation.

11. How is photosynthesis similar in C_4 plants and CAM plants?
    a. In both cases, the light reactions and the Calvin cycle are separated in both time and location.
    b. Both types of plants make sugar without the Calvin cycle.
    c. In both cases, rubisco is not used to fix carbon initially.
    d. Both types of plants make most of their sugar in the dark.
    e. In both cases, thylakoids are not involved in photosynthesis.

Describing, Comparing, and Explaining

12. What are the major inputs and outputs of the two stages of photosynthesis?
13. Explain why a poison that inhibits an enzyme of the Calvin cycle will also inhibit the light reactions.
14. What do plants do with the sugar they produce in photosynthesis?

Applying the Concepts

15. Most experts agree that global warming is already occurring and will increase rapidly in this century. Many countries have made a commitment to reduce CO_2 emissions. Recent international negotiations, however, including a 2009 meeting in Copenhagen, Denmark, have yet to reach a consensus on a global strategy to reduce greenhouse gas emissions. Some countries have resisted taking action because a few scientists and policymakers think that the warming trend may be just a random fluctuation in temperature and/or not related to human activities or that cutting CO_2 emissions would sacrifice economic growth. Do you think we need more evidence before taking action? Or is it better to act now to reduce CO_2 emissions? What are the possible costs and benefits of each of these two strategies?
16. The use of biofuels (see chapter introduction) avoids many of the problems associated with gathering, refining, transporting, and burning fossil fuels. Yet biofuels are not without their own set of problems. What challenges do you think would arise from a large-scale conversion to biofuels? How do these challenges compare with those encountered with fossil fuels? Do you think any other types of energy sources have more benefits and fewer costs than the others? Which ones, and why?

Review Answers

1. a. electron transport chain; b. ATP synthase; c. thylakoid space; d. stroma; e. ATP. The higher H^+ concentration is found in the intermembrane space of the mitochondrion and in the thylakoid space of the chloroplast.

2. In mitochondria: a. Electrons come from food molecules. b. Electrons have high potential energy in the bonds in organic molecules. c. Electrons are passed to oxygen, which picks up H^+ and forms water.

   In chloroplasts: a. Electrons come from splitting of water. b. Light energy excites the electrons to a higher energy level. c. Electrons flow from water to the reaction-center chlorophyll in photosystem II to the reaction-center chlorophyll in photosystem I to $NADP^+$, reducing it to NADPH.

   In both processes: d. Energy released by redox reactions in the electron transport chain is used to transport H^+ across a membrane. The flow of H^+ down its concentration gradient back through ATP synthase drives the phosphorylation of ADP to make ATP.

3. a. light energy; b. light reactions; c. Calvin cycle; d. O_2 released; e. electron transport chain; f. NADPH; g. ATP; h. 3-PGA is reduced.

4. c **5.** c **6.** a **7.** c (NADPH and ATP from the light reactions are required by the Calvin cycle.) **8.** d **9.** e **10.** b **11.** c

12. Light reactions: Light and water are inputs; ATP, NADPH, and O_2 are outputs. Calvin cycle: CO_2, ATP, and NADPH are inputs; G3P is the output. Also, ADP and $NADP^+$ are inputs to the light reactions and outputs of the Calvin cycle.

13. The light reactions require ADP and $NADP^+$, which are not recycled from ATP and NADPH when the Calvin cycle stops.

14. Plants can break down the sugar for energy in cellular respiration or use the sugar as a raw material for making other organic molecules. Excess sugar is stored as starch.

15. Some issues and questions to consider: What are the risks that we take and costs we must pay if global warming continues? How certain do we have to be that warming is caused by human activities before we act? What can we do to reduce CO_2 emissions? Is it possible that the costs and sacrifices of reducing CO_2 emissions might actually improve our lifestyle?

16. Some issues and questions to consider: How much land would be required for large-scale conversion to biofuel production, and would that detract from land needed to produce food? Are there fertilizer needs and waste disposal issues with biofuel production? Is there a difference in net input and output of CO_2 between production and use of fossil fuels and production and use of biofuels? Which one offers a more long-term solution to energy needs? How do other alternative energy sources compare in cost, potential problems and pollution, and benefit?

The Cellular Basis of Reproduction and Inheritance

The Cellular Basis of Reproduction and Inheritance

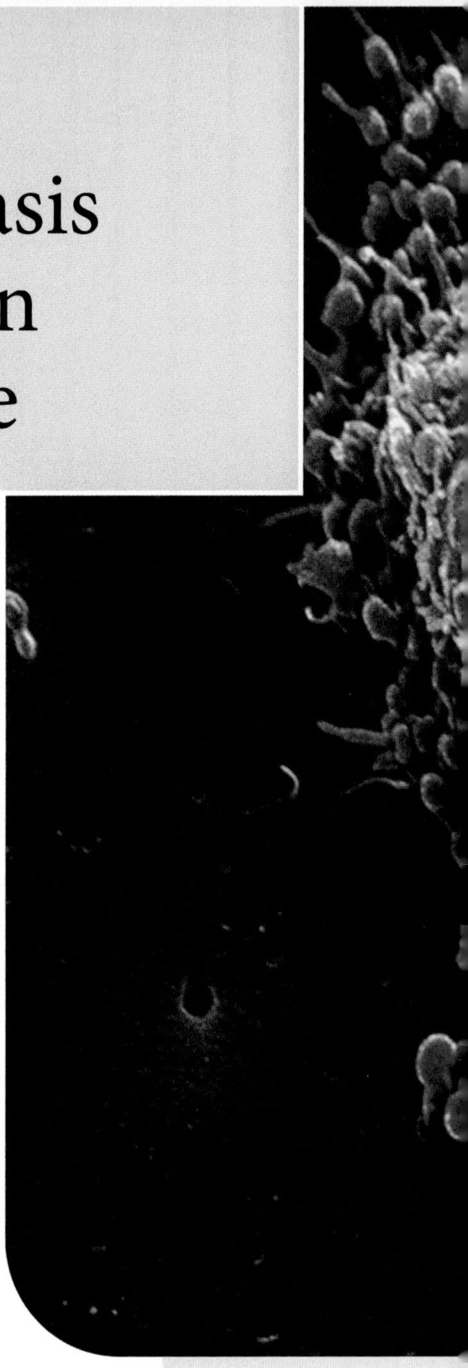

BIG IDEAS

Cell Division and Reproduction (1–2)

Cell division underlies many of life's important processes.

Dr. Yorgos Nikas/Photo Researchers

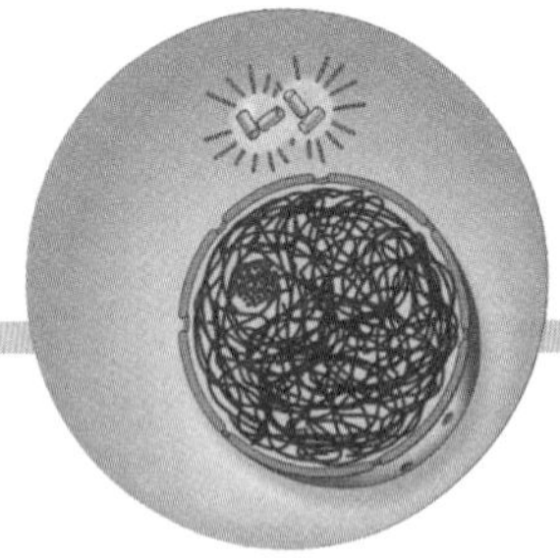

The Eukaryotic Cell Cycle and Mitosis (3–10)

Cells produce genetic duplicates through an ordered, tightly controlled series of steps.

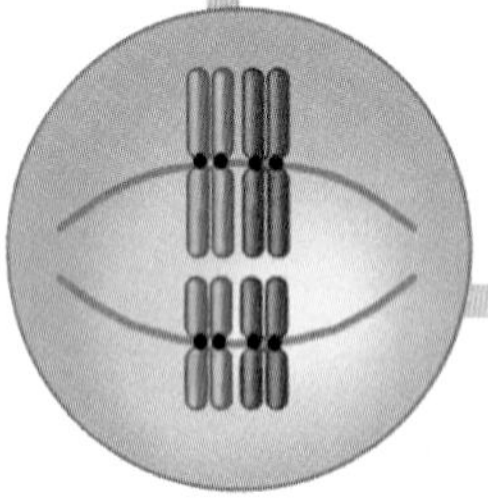

Meiosis and Crossing Over (11–17)

The process of meiosis produces genetically varied haploid gametes from diploid cells.

Michelle Gilders/Alamy

Alterations of Chromosome Number and Structure (18–23)

Errors in cell division can produce organisms with abnormal numbers of chromosomes.

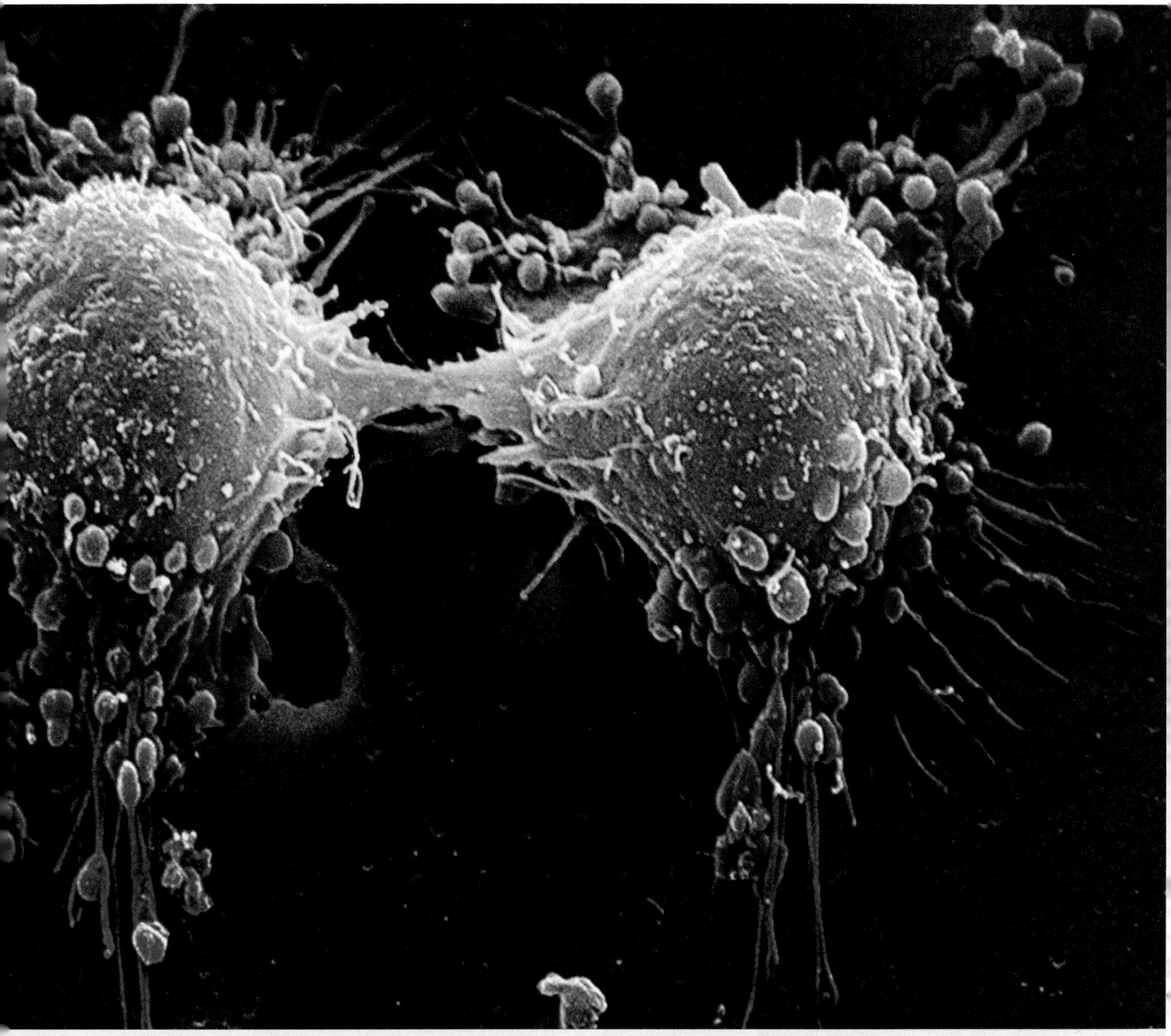

SPL/Photo Researchers

The photo above shows a cancer cell undergoing cell division, the creation of two cells from one. Cancer cells start as normal body cells that, because of genetic mutations, lose the ability to control the tempo of their own division. The result is rapid cell division that is no longer under the control of the host body—cell growth run amok! If left untreated, cancer cells may continue to divide and spread, invading other tissues and eventually killing the host. Most cancer treatments seek to prevent this outcome by disrupting one or more steps in cell division. Some anticancer drugs target dividing DNA; others disrupt cellular structures that assist in cell division. The goal of cancer treatment is to slow the spread of cancerous cells to the point that the body's immune system can overtake the growth, destroying the abnormal cells and restoring proper control of cell division.

Although cell division is harmful when it happens in a cancer cell, it is a necessary process for all forms of life. Why must cells divide? Some organisms, such as single-celled prokaryotes, reproduce themselves by splitting a single parent cell via cell division, creating two genetically identical offspring. In your body and the bodies of all other multicellular organisms, cell division allows for growth, replacement of damaged cells, and development of an embryo into an adult. Furthermore, in sexually reproducing organisms, eggs and sperm result from a particular type of cell division.

These examples illustrate the main point of this chapter: The perpetuation of life, including all aspects of reproduction and inheritance, is based on the reproduction of cells, or cell division. In this chapter, we discuss the two main types of cell division—mitosis and meiosis—and how they function within organisms.

Cell Division and Reproduction

1 Cell division plays many important roles in the lives of organisms

The ability of organisms to reproduce their own kind is the one characteristic that best distinguishes living things from nonliving matter. Only amoebas produce more amoebas, only people make more people, and only maple trees produce more maple trees. These simple facts of life have been recognized for thousands of years and are summarized by the age-old saying, "Like begets like."

However, the biological concept of reproduction includes more than just the birth of new organisms: Reproduction actually occurs much more often at the cellular level. When a cell undergoes reproduction, or **cell division**, the two "daughter" cells that result are genetically identical to each other and to the original "parent" cell. (Biologists traditionally use the word *daughter* in this context; it does not imply gender.) Before the parent cell splits into two, it duplicates its **chromosomes**, the structures that contain most of the cell's DNA. Then, during the division process, one set of chromosomes is distributed to each daughter cell. As a rule, the daughter cells receive identical sets of chromosomes from the lone, original parent cell. Each offspring cell will thus be genetically identical to the other and to the original parent cell.

Sometimes, cell division results in the reproduction of a whole organism. Many single-celled organisms, such as prokaryotes or the eukaryotic yeast cell in Figure 1A, reproduce by dividing in half, and the offspring are genetic replicas. This is an example of **asexual reproduction**, the creation of genetically identical offspring by a single parent, without the participation of sperm and egg. Many multicellular organisms can reproduce asexually as well. For example, some sea star species have the ability to grow new individuals from fragmented pieces (Figure 1B). And if you've ever grown a houseplant from a clipping, you've observed asexual reproduction in plants (Figure 1C). In asexual reproduction, there is one simple principle of inheritance: The lone parent and each of its offspring have identical genes.

London School of Hygiene & Tropical Medicine/Photo Researchers

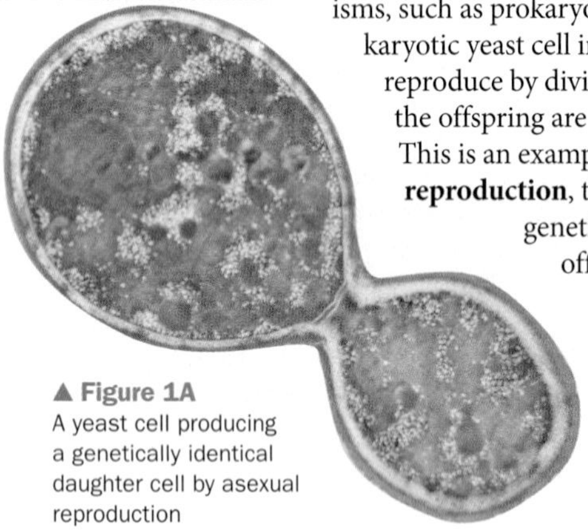

▲ **Figure 1A** A yeast cell producing a genetically identical daughter cell by asexual reproduction

Roger Steene/Image Quest Marine

▲ **Figure 1B** A sea star reproducing asexually

Sexual reproduction is different; it requires fertilization of an egg by a sperm. The production of gametes—egg and sperm—involves a special type of cell division that occurs only in reproductive organs (such as testes and ovaries in humans). As you'll learn later in the chapter, a gamete has only half as many chromosomes as the parent cell that gave rise to it, and these chromosomes contain unique combinations of genes. Therefore, in sexually reproducing species, like does not precisely beget like (Figure 1D). Offspring produced by sexual reproduction generally resemble their parents more closely than they resemble unrelated individuals of the same species, but they are not identical to their parents or to each other. Each offspring inherits a unique combination of genes from its two

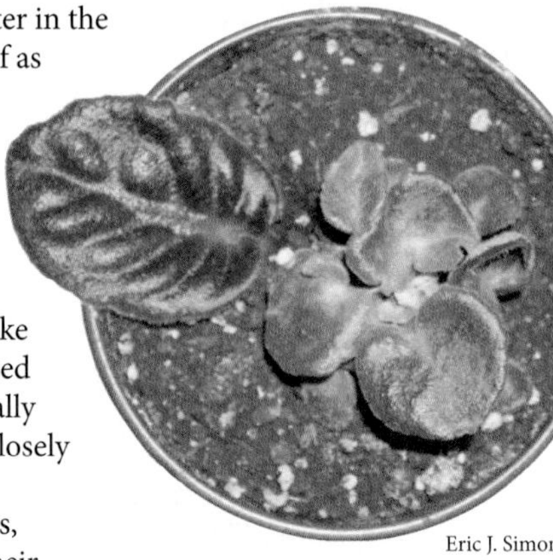

Eric J. Simon

▲ **Figure 1C** An African violet reproducing asexually from a cutting (the large leaf on the left)

▼ **Figure 1D** Sexual reproduction produces offspring with unique combinations of genes

Bob Thomas/Photographers Choice/Getty Images

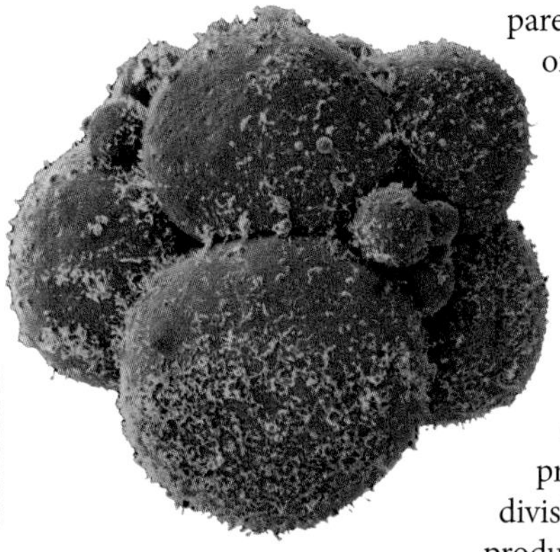

Dr. Yorgos Nikas/Photo Researchers

▲ **Figure 1E** Dividing cells in an early human embryo

parents, and this one-and-only set of genes programs a unique combination of traits. As a result, sexual reproduction can produce great variation among offspring.

In multicellular organisms, cell division plays other important roles, in addition to the production of gametes. Cell division enables sexually reproducing organisms to develop from a single cell—the fertilized egg, or zygote—into an adult organism (**Figure 1E**). All of the trillions of cells in your body arose via repeated cell divisions that began in your mother's body with a single fertilized egg cell. After an organism is fully grown, cell division continues to function in renewal and repair, replacing cells that die from normal wear and tear or accidents. Within your body, millions of cells must divide every second to replace damaged or lost cells (**Figure 1F**).

The type of cell division responsible for asexual reproduction and for the growth and maintenance of multicellular organisms involves a process called called mitosis. The production of egg and sperm cells involves a special type of cell division called meiosis. In the remainder of this chapter, you will learn the details of both types of cell division. To start, we'll look briefly at prokaryotic cell division in the next module.

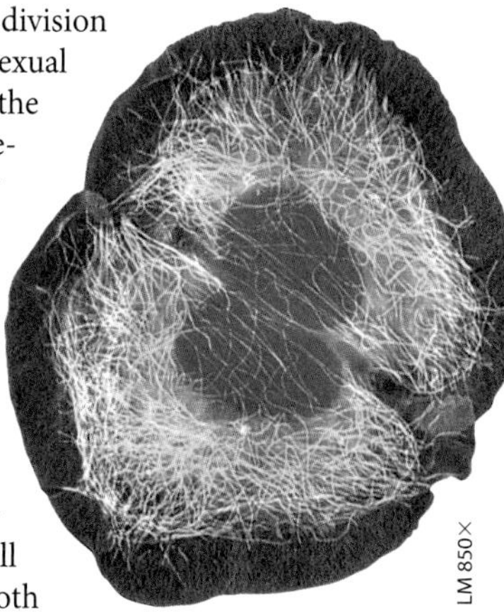

Dr. Torsten Wittmann/Photo Researchers

▲ **Figure 1F** A human kidney cell dividing

? What function does cell division play in an amoeba? What functions does it play in your body?

Reproduction; development, growth, and repair

2 Prokaryotes reproduce by binary fission

Prokaryotes (bacteria and archaea) reproduce by a type of cell division called **binary fission** ("dividing in half"). In typical prokaryotes, the majority of genes are carried on a single circular DNA molecule that, with associated proteins, constitutes the organism's chromosome. Although prokaryotic chromosomes are much smaller than those of eukaryotes, duplicating them in an orderly fashion and distributing the copies equally to two daughter bacteria is still a formidable task. Consider, for example, that when stretched out, the chromosome of the bacterium *Escherichia coli* (*E. coli*) is about 500 times longer than the cell itself. Accurately replicating this molecule when it is coiled and packed inside the cell is no small achievement.

Figure 2A illustrates binary fission in a prokaryote. ❶ As the chromosome is duplicating, the copies move toward the opposite ends of the cell. ❷ Meanwhile, the cell elongates. ❸ When chromosome duplication is complete and the cell has reached about twice its initial size, the plasma membrane grows inward and more cell wall is made, dividing the parent cell into two daughter cells. **Figure 2B** is an electron micrograph of a dividing bacterium (this cell is at a stage similar to the third illustration in Figure 2A).

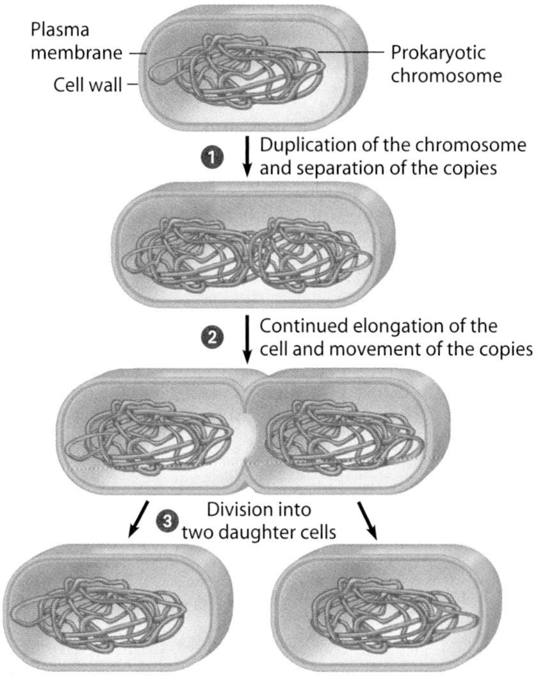

▲ **Figure 2A** Binary fission of a prokaryotic cell

? Why is binary fission classified as asexual reproduction?

Because the genetically identical offspring inherit their DNA from a single parent

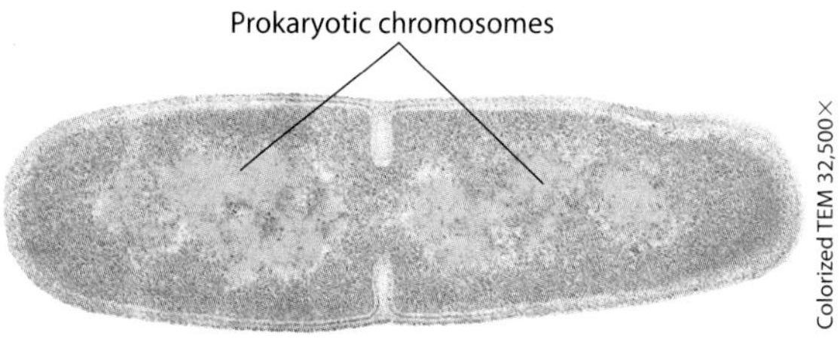

Lee D. Simon/Photo Researchers

▲ **Figure 2B** An electron micrograph of a dividing bacterium

The Eukaryotic Cell Cycle and Mitosis

3 The large, complex chromosomes of eukaryotes duplicate with each cell division

Eukaryotic cells, in general, are more complex and much larger than prokaryotic cells, and they have many more genes. Human cells, for example, carry around 21,000 genes, versus about 3,000 for a typical bacterium. Almost all the genes in the cells of humans, and in all other eukaryotes, are found in the cell nucleus, grouped into multiple chromosomes. (The exceptions include genes on the small DNA molecules of mitochondria and, in plants, chloroplasts.)

Most of the time, chromosomes exist as a diffuse mass of long, thin fibers that are far too long to fit in a cell's nucleus. In fact, if stretched out, the DNA in just one of your cells would be taller than you! DNA in this loose state is called **chromatin**, fibers composed of roughly equal amounts of DNA and protein molecules. Chromatin is too thin to be seen using a light microscope.

As a cell prepares to divide, its chromatin coils up, forming tight, distinct chromosomes that are visible under a light microscope. Why is it necessary for a cell's chromosomes to be compacted in this way? Imagine a situation from your own life: Your belongings are spread out over a considerable area of your home, but as you prepare to move to a new home, you need to gather them all up and pack them into small containers. Similarly, before a cell can undergo division, it must compact all its DNA into manageable packages. **Figure 3A** is a micrograph of a plant cell that is about to divide; each thick purple thread is a tightly packed individual chromosome.

Like a prokaryotic chromosome, each eukaryotic chromosome contains one long DNA molecule bearing hundreds or thousands of genes and, attached to the DNA, a number of protein molecules. However, the eukaryotic chromosome has a much more complex structure than the prokaryotic chromosome. The eukaryotic chromosome includes many more protein molecules, which help maintain the chromosome's structure and control the activity of its genes. The number of chromosomes in a eukaryotic cell depends on the species. For example, human body cells generally have 46 chromosomes, while the body cells of a dog have 78.

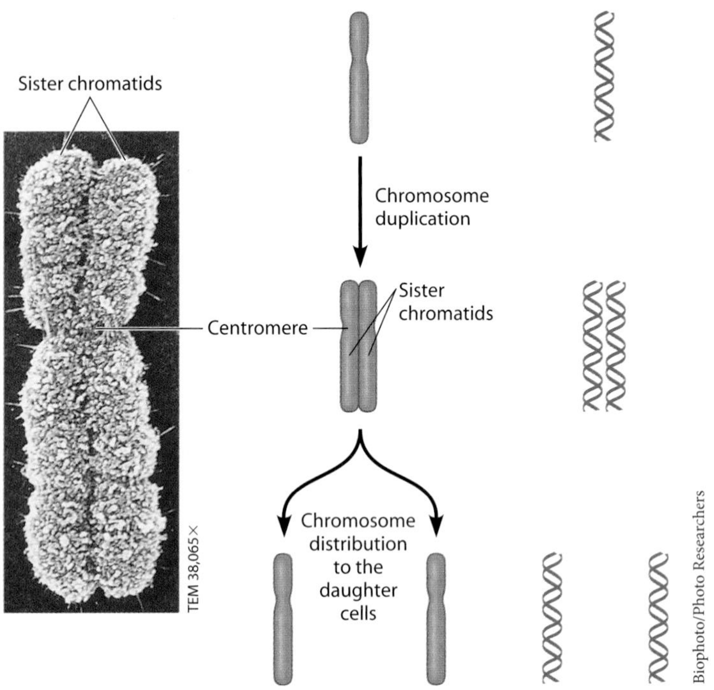

▲ **Figure 3B** Chromosome duplication and distribution

The chromosomes of a eukaryotic cell are duplicated before they condense and the cell divides. The DNA molecule of each chromosome is replicated, and new protein molecules attach as needed. Each chromosome now consists of two copies called **sister chromatids**, which contain identical copies of the DNA molecule (**Figure 3B**). The two sister chromatids are joined together especially tightly at a narrow "waist" called the **centromere** (visible near the center of each chromosome shown in the figure).

When the cell divides, the sister chromatids of a duplicated chromosome separate from each other. Once separated from its sister, each chromatid is called a chromosome, and it is identical to the chromosome the cell started with. One of the new chromosomes goes to one daughter cell, and the other goes to the other daughter cell. In this way, each daughter cell receives a complete and identical set of chromosomes. In humans, for example, a typical dividing cell has 46 duplicated chromosomes (or 92 chromatids), and each of the two daughter cells that results from it has 46 single chromosomes.

▶ **Figure 3A** A plant cell (from an African blood lily) just before cell division

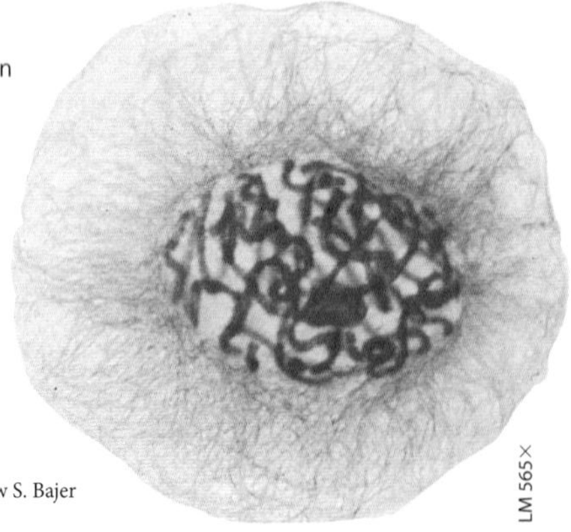

Andrew S. Bajer

? **When does a chromosome consist of two identical chromatids?**

When the cell is preparing to divide and has duplicated its chromosomes, but before the duplicates actually separate

4 The cell cycle multiplies cells

How do chromosome duplication and cell division fit into the life of a cell—and the life of an organism? As discussed in Module 1, cell division is essential to all life. Cell division is the basis of reproduction for every organism. It enables a multicellular organism to grow to adult size. It also replaces worn-out or damaged cells, keeping the total cell number in a mature individual relatively constant. In your own body, for example, millions of cells must divide every second to maintain the total number of about 10 trillion cells. Some cells divide once a day, others less often, and highly specialized cells, such as our mature muscle cells, not at all. The fact that some mature cells never divide explains why some kinds of damage—such as the death of cardiac muscle cells during a heart attack or the death of brain cells during a stroke—can never be reversed.

The process of cell division is a key component of the **cell cycle**, an ordered sequence of events that extends from the time a cell is first formed from a dividing parent cell until its own division into two cells. The cell cycle consists of two main stages: a growing stage (called interphase), during which the cell roughly doubles everything in its cytoplasm and precisely replicates its chromosomal DNA, and the actual cell division (called the mitotic phase).

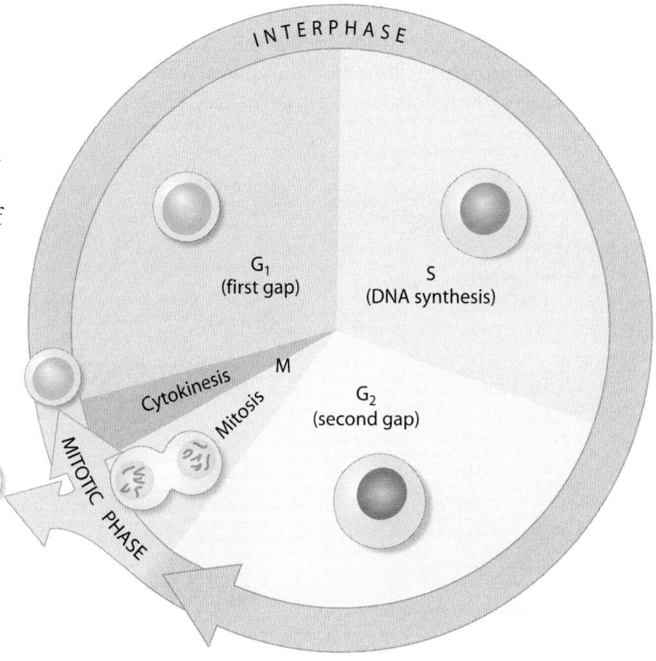

▲ **Figure 4** The eukaryotic cell cycle

As **Figure 4** indicates, most of the cell cycle is spent in **interphase**. This is a time when a cell's metabolic activity is very high and the cell performs its various functions within the organism. For example, a cell in your intestine might release digestive enzymes and absorb nutrients, while a white blood cell might circulate in your bloodstream, interacting with invading microbes. During interphase, a cell makes more cytoplasm. It increases its supply of proteins, creates more cytoplasmic organelles (such as mitochondria and ribosomes), and grows in size. Additionally, the chromosomes duplicate during this period. Typically, interphase lasts for at least 90% of the total time required for the cell cycle.

Interphase can be divided into three subphases: the G_1 phase ("first gap"), the S phase, and the G_2 phase ("second gap"). During all three subphases, the cell grows. However, chromosomes are duplicated only during the S phase. S stands for *synthesis* of DNA—also known as DNA replication. At the beginning of the S phase, each chromosome is single. At the end of this phase, after DNA replication, the chromosomes are double, each consisting of two sister chromatids. To summarize interphase: A cell grows (G_1), continues to grow as it copies its chromosomes (S), and then grows more as it completes preparations for cell division (G_2).

The **mitotic phase (M phase**; the blue area of the figure), the part of the cell cycle when the cell actually divides, accounts for only about 10% of the total time required for the cell cycle. The mitotic phase is itself divided into two overlapping stages, called mitosis and cytokinesis. In **mitosis**, the nucleus and its contents—most importantly the duplicated chromosomes—divide and are evenly distributed, forming two daughter nuclei. During **cytokinesis**, which usually begins before mitosis ends, the cytoplasm is divided in two. The combination of mitosis and cytokinesis produces two genetically identical daughter cells, each with a single nucleus, surrounding cytoplasm stocked with organelles, and a plasma membrane. Each newly produced daughter cell may then proceed through G_1 and repeat the cycle.

Mitosis is unique to eukaryotes and is the evolutionary solution to the problem of allocating identical copies of all the separate chromosomes to two daughter cells. Mitosis is a remarkably accurate mechanism. Experiments with yeast, for example, indicate that an error in chromosome distribution occurs only once in about 100,000 cell divisions.

The extreme accuracy of mitosis is essential to the development of your own body. All of us began as a single cell. Mitotic cell division ensures that all your body cells receive copies of the 46 chromosomes that were found in this original cell. Thus, every one of the trillions of cells in your body today can trace its ancestry back through mitotic divisions to that first cell produced when your father's sperm and mother's egg fused about nine months before your birth.

During the mitotic phase, a living cell viewed through a light microscope undergoes dramatic changes in the appearance of the chromosomes and other structures. In the next module, we'll use these visible changes as a guide to the stages of mitosis.

? A researcher treats cells with a chemical that prevents DNA synthesis from starting. This treatment would trap the cells in which part of the cell cycle?

G_1.

5 Cell division is a continuum of dynamic changes

Figure 5 illustrates the cell cycle for an animal cell using micrographs, drawings (simplified to include just four chromosomes), and text. The micrographs show cells from a newt, with chromosomes in blue and the mitotic spindle in green. Interphase is included, but the emphasis is on the dramatic changes that occur during cell division, the mitotic phase. Mitosis is a continuum, but biologists distinguish five main stages: **prophase**, **prometaphase**, **metaphase**, **anaphase**, and **telophase**.

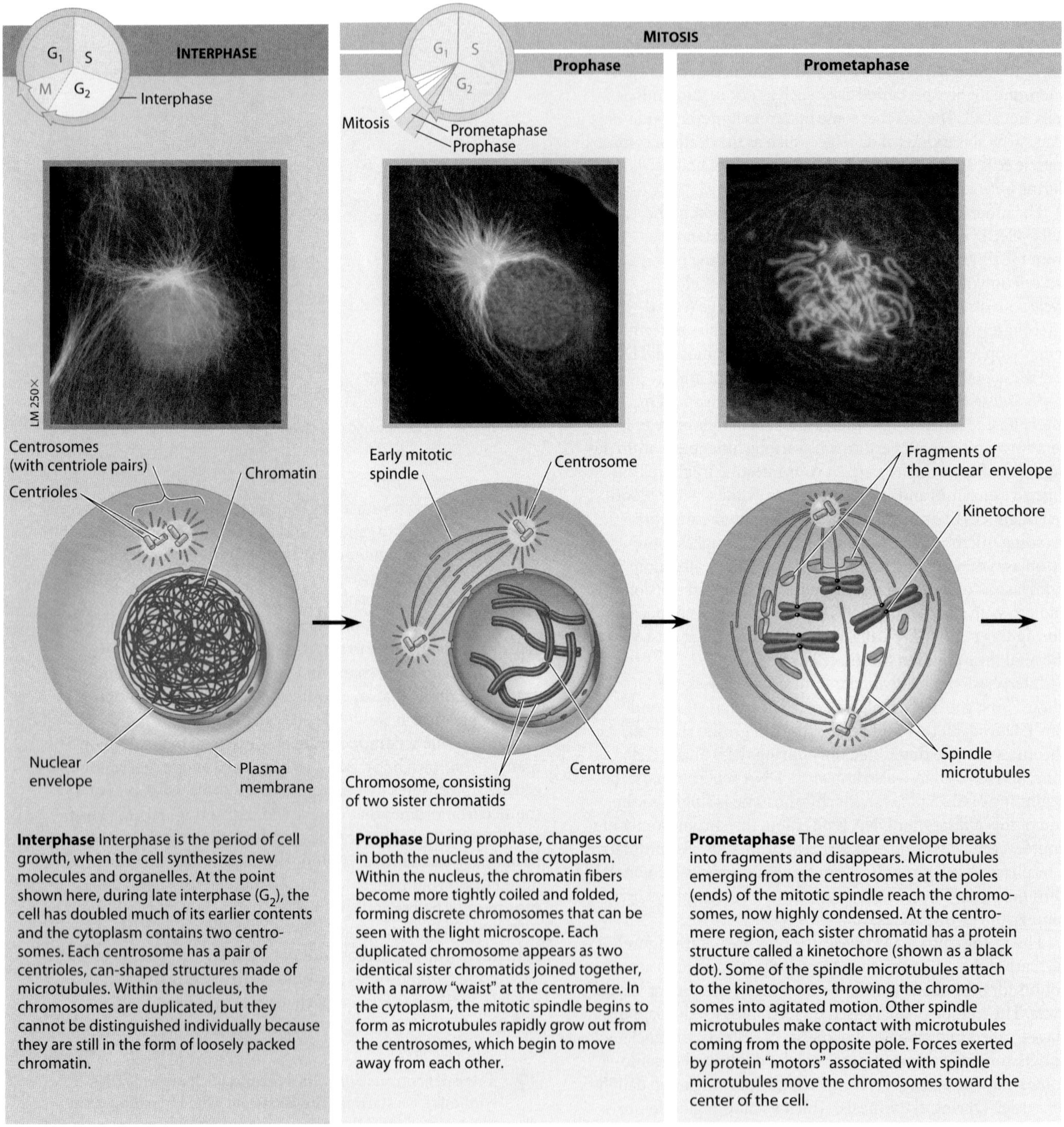

Interphase Interphase is the period of cell growth, when the cell synthesizes new molecules and organelles. At the point shown here, during late interphase (G_2), the cell has doubled much of its earlier contents and the cytoplasm contains two centrosomes. Each centrosome has a pair of centrioles, can-shaped structures made of microtubules. Within the nucleus, the chromosomes are duplicated, but they cannot be distinguished individually because they are still in the form of loosely packed chromatin.

Prophase During prophase, changes occur in both the nucleus and the cytoplasm. Within the nucleus, the chromatin fibers become more tightly coiled and folded, forming discrete chromosomes that can be seen with the light microscope. Each duplicated chromosome appears as two identical sister chromatids joined together, with a narrow "waist" at the centromere. In the cytoplasm, the mitotic spindle begins to form as microtubules rapidly grow out from the centrosomes, which begin to move away from each other.

Prometaphase The nuclear envelope breaks into fragments and disappears. Microtubules emerging from the centrosomes at the poles (ends) of the mitotic spindle reach the chromosomes, now highly condensed. At the centromere region, each sister chromatid has a protein structure called a kinetochore (shown as a black dot). Some of the spindle microtubules attach to the kinetochores, throwing the chromosomes into agitated motion. Other spindle microtubules make contact with microtubules coming from the opposite pole. Forces exerted by protein "motors" associated with spindle microtubules move the chromosomes toward the center of the cell.

Conly L. Rieder, Ph.D.

▲ **Figure 5** The stages of cell division by mitosis

The chromosomes are the stars of the mitotic drama. Their movements depend on the **mitotic spindle**, a football-shaped structure of microtubules that guides the separation of the two sets of daughter chromosomes. The spindle microtubules emerge from two **centrosomes**, clouds of cytoplasmic material that in animal cells contain pairs of centrioles. Centrosomes are also known as *microtubule-organizing centers*, a term describing their function.

? You view an animal cell through a microscope and observe dense, duplicated chromosomes scattered throughout the cell. Which state of mitosis are you looking at?

Prophase (since the chromosomes are condensed but not yet aligned)

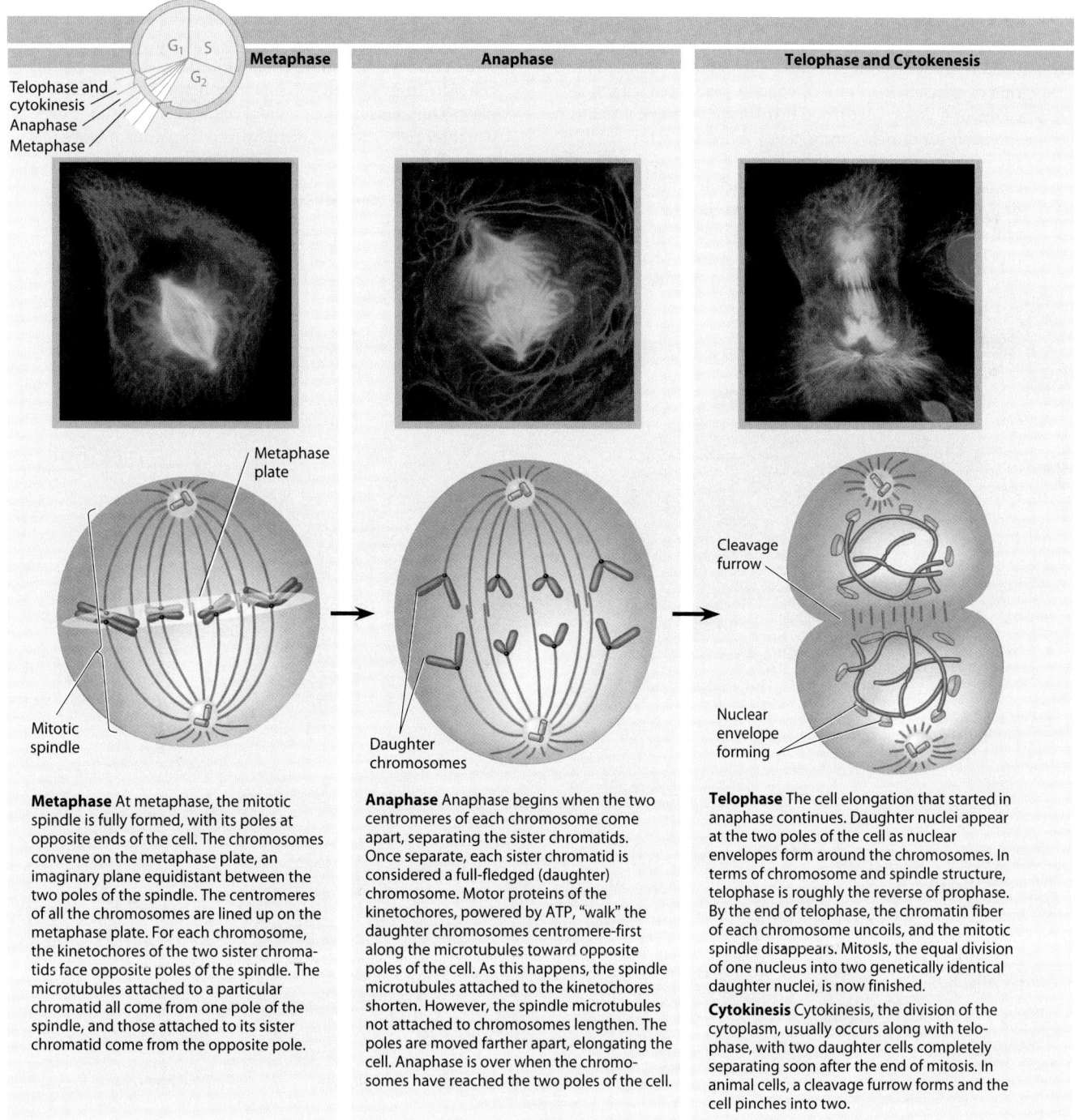

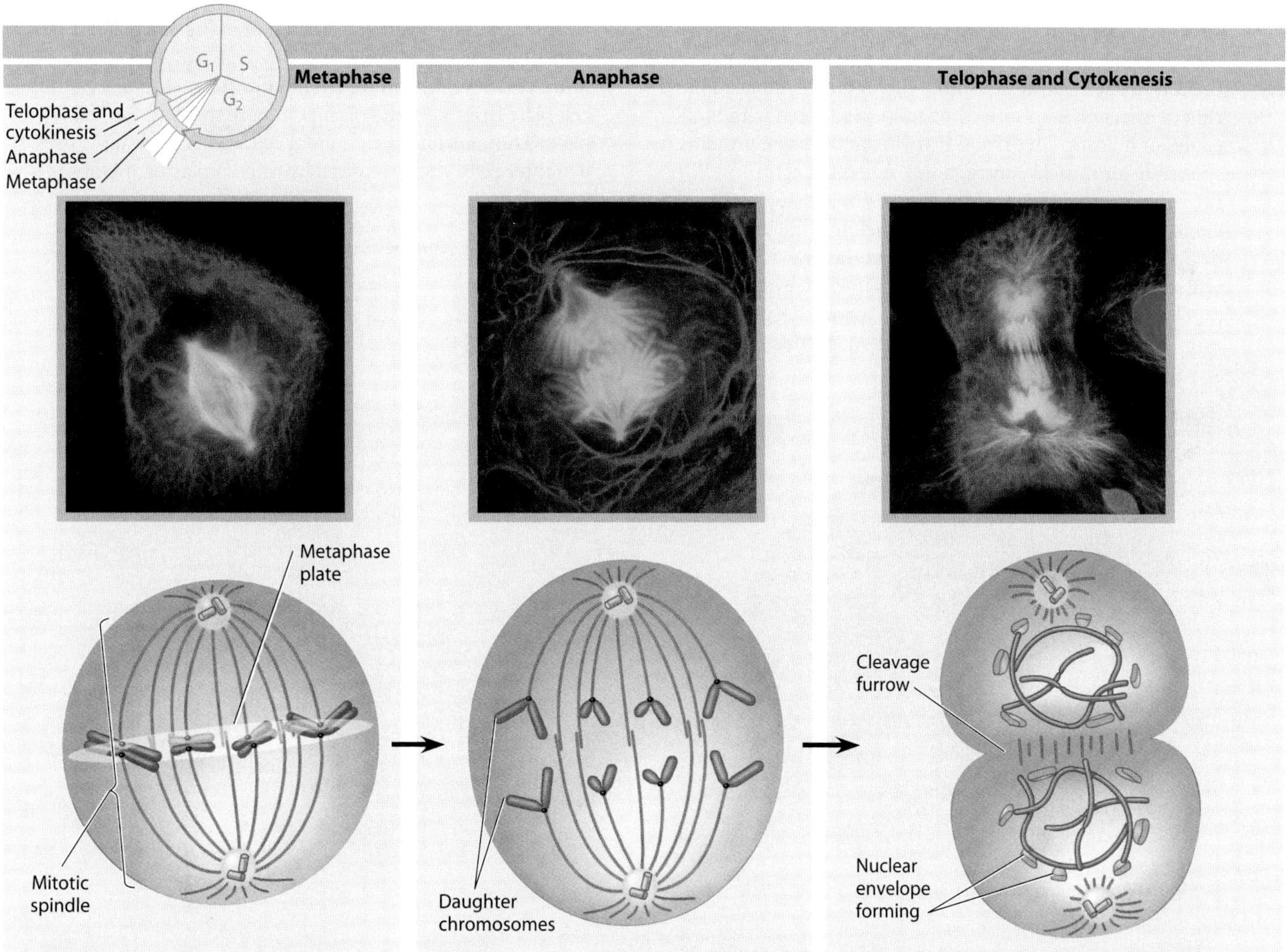

Metaphase At metaphase, the mitotic spindle is fully formed, with its poles at opposite ends of the cell. The chromosomes convene on the metaphase plate, an imaginary plane equidistant between the two poles of the spindle. The centromeres of all the chromosomes are lined up on the metaphase plate. For each chromosome, the kinetochores of the two sister chromatids face opposite poles of the spindle. The microtubules attached to a particular chromatid all come from one pole of the spindle, and those attached to its sister chromatid come from the opposite pole.

Anaphase Anaphase begins when the two centromeres of each chromosome come apart, separating the sister chromatids. Once separate, each sister chromatid is considered a full-fledged (daughter) chromosome. Motor proteins of the kinetochores, powered by ATP, "walk" the daughter chromosomes centromere-first along the microtubules toward opposite poles of the cell. As this happens, the spindle microtubules attached to the kinetochores shorten. However, the spindle microtubules not attached to chromosomes lengthen. The poles are moved farther apart, elongating the cell. Anaphase is over when the chromosomes have reached the two poles of the cell.

Telophase The cell elongation that started in anaphase continues. Daughter nuclei appear at the two poles of the cell as nuclear envelopes form around the chromosomes. In terms of chromosome and spindle structure, telophase is roughly the reverse of prophase. By the end of telophase, the chromatin fiber of each chromosome uncoils, and the mitotic spindle disappears. Mitosis, the equal division of one nucleus into two genetically identical daughter nuclei, is now finished.

Cytokinesis Cytokinesis, the division of the cytoplasm, usually occurs along with telophase, with two daughter cells completely separating soon after the end of mitosis. In animal cells, a cleavage furrow forms and the cell pinches into two.

6 Cytokinesis differs for plant and animal cells

Cytokinesis, the division of the cytoplasm into two cells, typically begins during telophase. Given the differences between animal and plant cells (particularly the presence of a stiff cell wall in plant cells), it isn't surprising that cytokinesis proceeds differently for these two types of eukaryotic cells.

In animal cells, cytokinesis occurs by a process known as cleavage. As shown in **Figure 6A**, the first sign of cleavage is the appearance of a **cleavage furrow**, a shallow indentation in the cell surface. At the site of the furrow, the cytoplasm has a ring of microfilaments made of actin, associated with molecules of myosin. (Actin and myosin are the same proteins responsible for muscle contraction.) When the actin microfilaments interact with the myosin, the ring contracts. Contraction of the myosin ring is much like the pulling of a drawstring on a hooded sweatshirt: As the drawstring is pulled, the ring of the hood contracts inward, eventually pinching shut. Similarly, the cleavage furrow deepens and eventually pinches the parent cell in two, producing two completely separate daughter cells, each with its own nucleus and share of cytoplasm.

Cytokinesis is markedly different in plant cells, which possess cell walls (**Figure 6B**). During telophase, membranous vesicles containing cell wall material collect at the middle of the parent cell. The vesicles fuse, forming a membranous disk called the **cell plate**. The cell plate grows outward, accumulating more cell wall materials as more vesicles fuse with it. Eventually, the membrane of the cell plate fuses with the plasma membrane, and the cell plate's contents join the parental cell wall. The result is two daughter cells, each bounded by its own plasma membrane and cell wall.

? Contrast cytokinesis in animals with cytokinesis in plants.

In animals, cytokinesis involves a cleavage furrow in which contracting microfilaments pinch the cell in two. In plants, it involves formation of a cell plate, a fusion of vesicles that forms new plasma membranes and new cell walls between the cells.

Cytokinesis

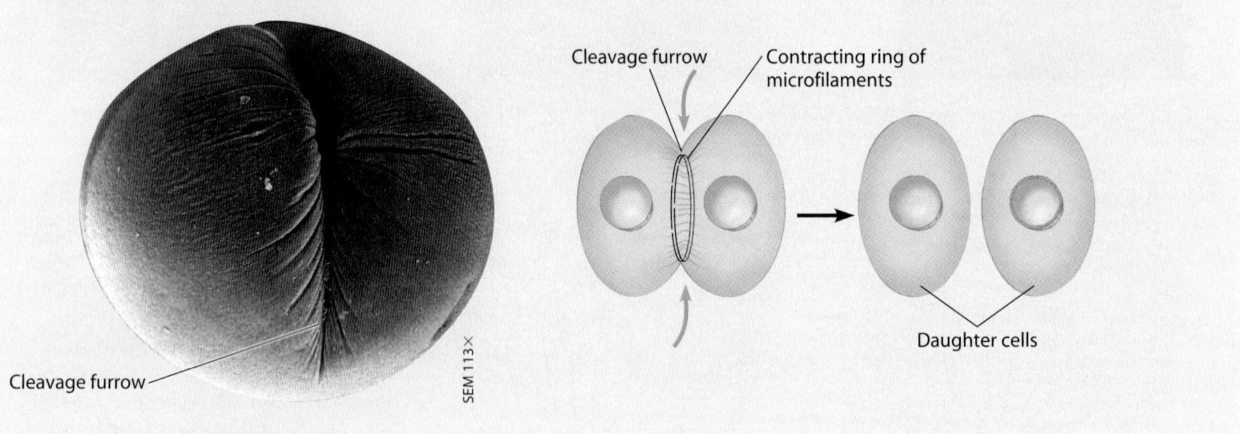

Don W. Fawcett/Photo Researchers

▲ **Figure 6A** Cleavage of an animal cell

Cytokinesis

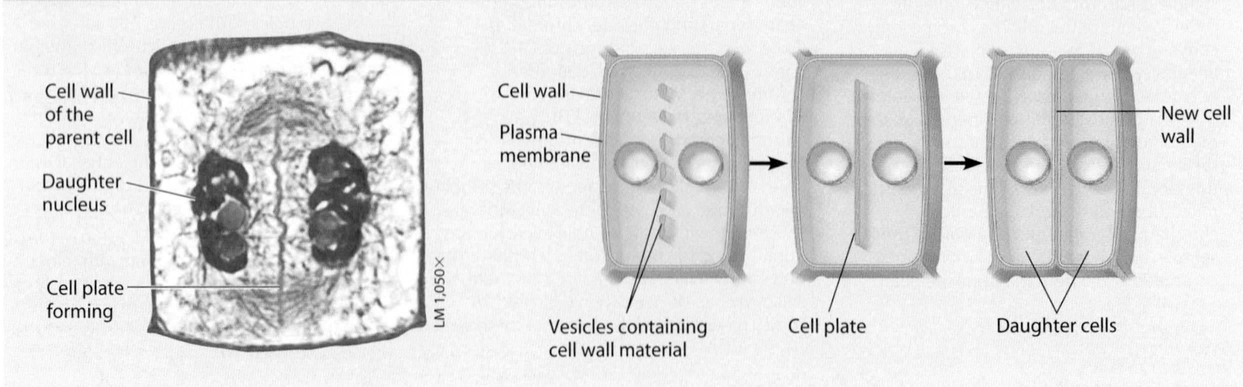

Eldon H. Newcomb

▲ **Figure 6B** Cell plate formation in a plant cell

7 Anchorage, cell density, and chemical growth factors affect cell division

For a plant or an animal to grow and develop normally and maintain its tissues once fully grown, it must be able to control the timing of cell division in different parts of its body. For example, in the adult human, skin cells and the cells lining the digestive tract divide frequently throughout life, replacing cells that are constantly being abraded and sloughed off. In contrast, cells in the human liver usually do not divide unless the liver is damaged.

By growing animal cells in culture, researchers have been able to identify many factors, both physical and chemical, that can influence cell division. For example, cells fail to divide if an essential nutrient is left out of the culture medium. And most types of mammalian cells divide in culture only if certain specific growth factors are included. A **growth factor** is a protein secreted by certain body cells that stimulates other cells to divide (**Figure 7A**). Researchers have discovered at least 50 different growth factors that can trigger cell division. Different cell types respond specifically to certain growth factors or a combination of growth factors. For example, injury to the skin causes blood platelets to release a protein called platelet-derived growth factor. This protein promotes the rapid growth of connective tissue cells that help seal the wound. Another well-studied example is a protein called vascular endothelial growth factor (VEGF), which stimulates the growth of new blood vessels during fetal development and after injury. Interestingly, VEGF overproduction is a hallmark of many dangerous cancers; several anticancer drug therapies work by inhibiting the action of VEGF.

The effect of a physical factor on cell division is clearly seen in **density-dependent inhibition**, a phenomenon in which crowded cells stop dividing. For example, animal cells growing on the surface of a dish multiply to form a single layer and usually stop dividing when they fill the space and touch one another (**Figure 7B**). If some cells are removed, those bordering the open space begin dividing again and continue until the vacancy is filled. What actually causes the cessation of growth? Studies of cultured cells suggest that physical contact of cell-surface proteins between adjacent cells is responsible for inhibiting cell division. One of the characteristics that distinguishes cancerous cells from normal body cells is their failure to exhibit density-dependent inhibition; cancer cells continue to divide even at high densities, piling up on one another.

Most animal cells also exhibit **anchorage dependence**; they must be in contact with a solid surface—such as the inside of a culture dish or the extracellular matrix of a tissue—to divide. Anchorage dependence, density-dependent inhibition, and the availability of growth factors are all important regulatory mechanisms controlling the division of the body's cells. How exactly do growth factors work? We pursue answers to this question in the next module.

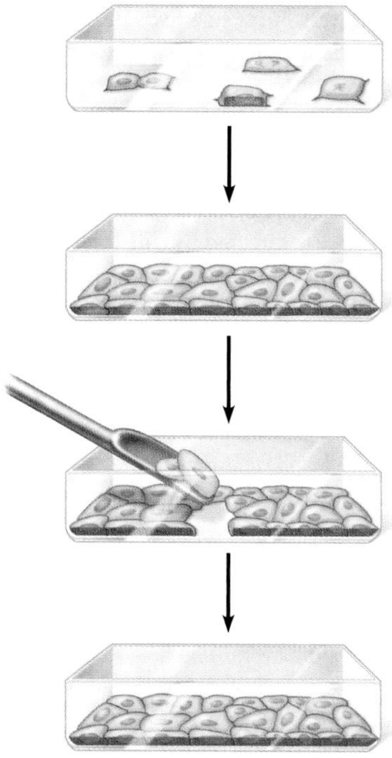
Cells anchor to the dish surface and divide.
When the cells have formed a complete single layer, they stop dividing (density-dependent inhibition).
If some cells are scraped away, the remaining cells divide to fill the dish with a single layer and then stop (density-dependent inhibition).

▲ **Figure 7B** An experiment demonstrating density-dependent inhibition, using animal cells grown in culture

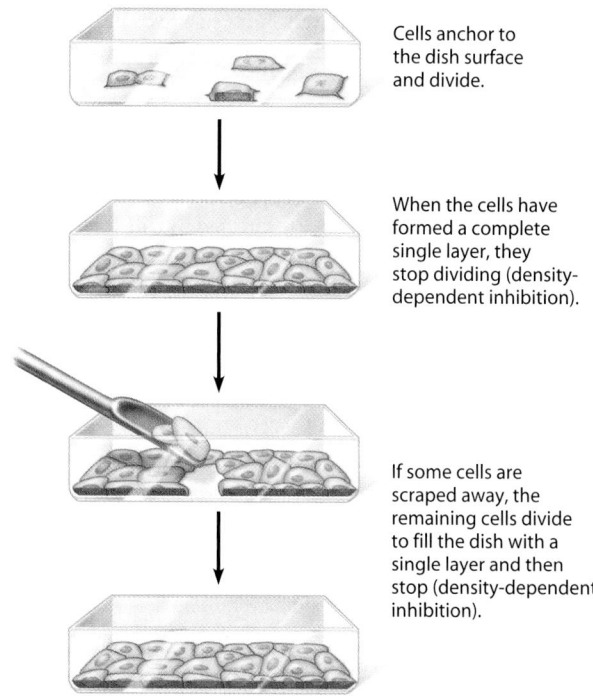

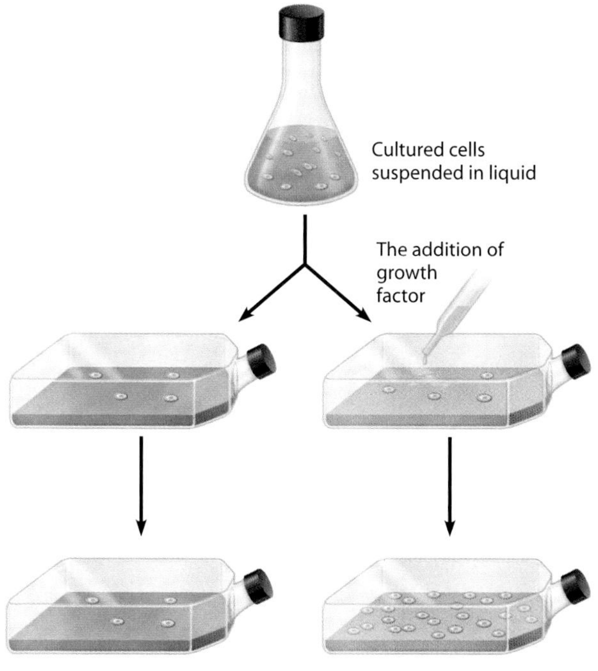
Cultured cells suspended in liquid
The addition of growth factor

▲ **Figure 7A** An experiment demonstrating the effect of growth factors on the division of cultured animal cells

? **Compared to a control culture, the cells in an experimental culture are fewer but much larger in size when they cover the dish surface and stop growing. What is a reasonable hypothesis for this difference?**

The experimental culture is deficient in one or more growth factors.

8 Growth factors signal the cell cycle control system

In a living animal, most cells are anchored in a fixed position and bathed in a solution of nutrients supplied by the blood, yet they usually do not divide unless they are signaled by other cells to do so. Growth factors are the main signals, and their role in promoting cell division leads us back to our earlier discussion of the cell cycle.

The sequential events of the cell cycle, represented by the circle of flat blocks in **Figure 8A**, are directed by a distinct cell cycle control system, represented by the knob in the center. The gray bar extending from the center represents the current position in the cell cycle. The **cell cycle control system** is a cyclically operating set of molecules in the cell that both triggers and coordinates key events in the cell cycle. The cell cycle is *not* like a row of falling dominoes, with each event causing the next one in line. Within the M phase, for example, metaphase does not automatically lead to anaphase. Instead, proteins of the cell cycle control system must trigger the separation of sister chromatids that marks the start of anaphase.

A checkpoint in the cell cycle is a critical control point where stop and go-ahead signals (represented by stop/go lights in the figure) can regulate the cycle. The default state in most animal cells is to halt the cell cycle at these checkpoints unless overridden by specific go-ahead signals.

The red and white gates in Figure 8A represent three major checkpoints in the cell cycle: during the G_1 and G_2 subphases of interphase and in the M phase. Intracellular signals detected by the control system tell it whether key cellular processes up to each point have been completed and thus whether or not the cell cycle should proceed past that point. The control system also receives messages from outside the cell, indicating both general environmental conditions and the presence of specific signal molecules from other cells. For many cells, the G_1 checkpoint seems to be the most important. If a cell receives a go-ahead signal—for example, from a growth factor—at the G_1 checkpoint, it will usually enter the S phase, eventually going on to complete its cycle and divide. If such a signal never arrives, the cell will switch to a permanently nondividing state called the G_0 phase. Many cells in the human body, such as mature nerve cells and muscle cells, are in the G_0 phase.

Figure 8B shows a simplified model for how a growth factor might affect the cell cycle control system at the G_1 checkpoint. A cell that responds to a growth factor (▼) has molecules of a specific receptor protein in its plasma membrane. Binding of the growth factor to the receptor triggers a signal transduction pathway in the cell. A signal transduction pathway is a series of protein molecules that conveys a message. In this case, that message leads to cell division. The "signals" are changes that each protein molecule induces in the next molecule in the pathway. Via a series of relay proteins, a signal finally reaches the cell cycle control system and overrides the brakes that otherwise prevent progress of the cell cycle. In Figure 8B, the cell cycle is set off from the cell in a separate diagram because the proteins making up the control system in the cell are not actually located together in one place.

Research on the control of the cell cycle is one of the hottest areas in biology today. This research is leading to a better understanding of cancer, which we discuss next.

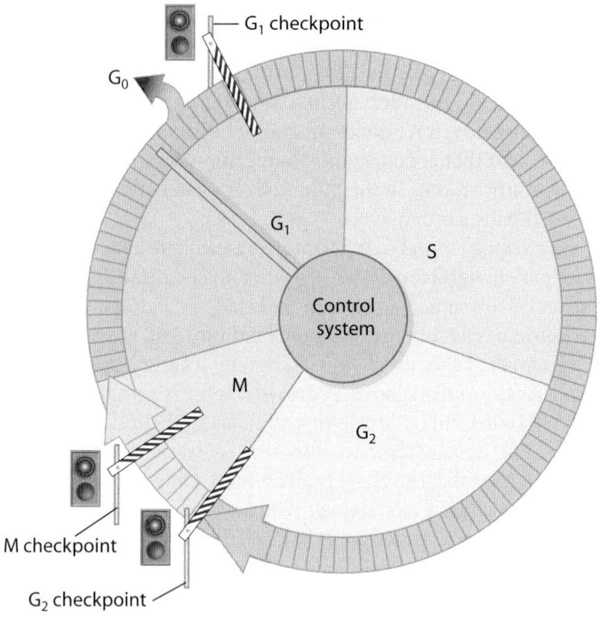

▲ **Figure 8A** A schematic model for the cell cycle control system

? **At which of the three checkpoints described in this module do the chromosomes exist as duplicated sister chromatids?**

● G_2 and the first half of M

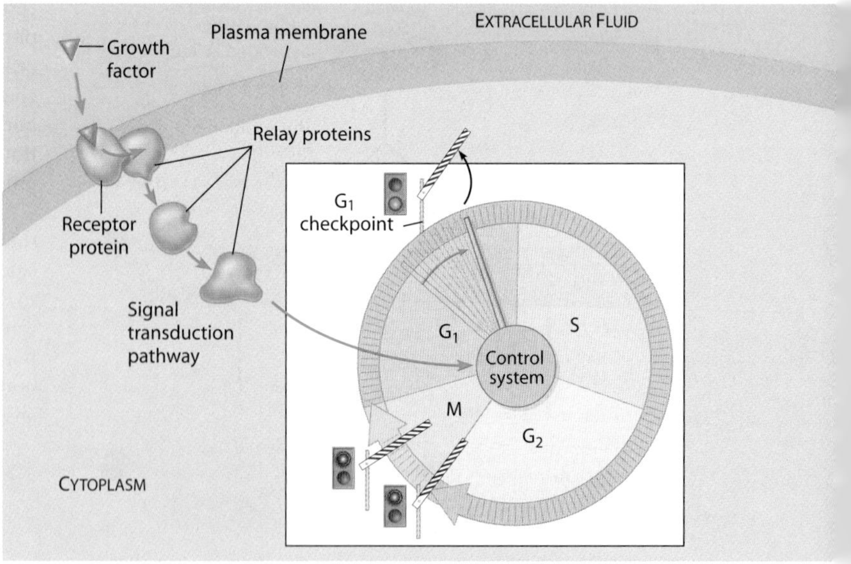

▲ **Figure 8B** How a growth factor signals the cell cycle control system

CONNECTION 9 Growing out of control, cancer cells produce malignant tumors

Cancer, which currently claims the lives of one out of every five people in the United States and other industrialized nations, is a disease of the cell cycle. Cancer cells do not heed the normal signals that regulate the cell cycle; they divide excessively and may invade other tissues of the body. If unchecked, cancer cells may continue to grow until they kill the organism.

The abnormal behavior of cancer cells begins when a single cell undergoes transformation, a process that converts a normal cell to a cancer cell. Transformation occurs following a mutation in one or more genes that encode for proteins in the cell cycle control system. Because a transformed cell grows abnormally, the immune system usually recognizes it and destroys it. However, if the cell evades destruction, it may proliferate to form a **tumor**, an abnormally growing mass of body cells. If the abnormal cells remain at the original site, the lump is called a **benign tumor**. Benign tumors can cause problems if they grow in and disrupt certain organs, such as the brain, but often they can be completely removed by surgery.

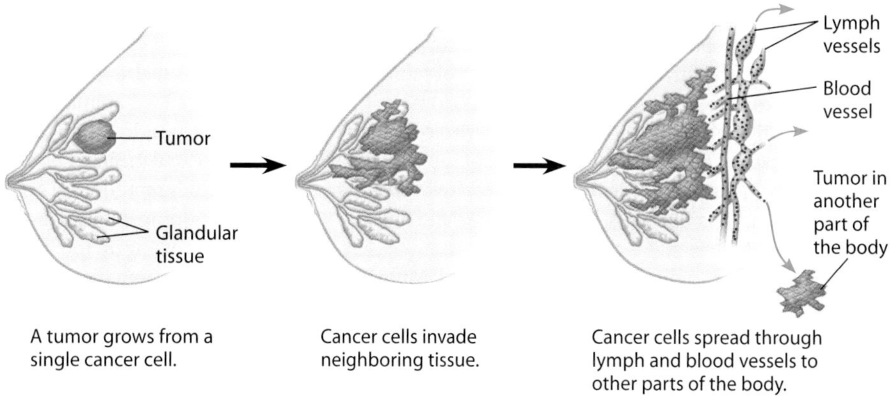

▲ **Figure 9** Growth and metastasis of a malignant (cancerous) tumor of the breast

In contrast, a **malignant tumor** can spread into neighboring tissues and other parts of the body, displacing normal tissue and interrupting organ function as it goes (**Figure 9**). An individual with a malignant tumor is said to have **cancer**. Cancer cells may separate from the original tumor or secrete signal molecules that cause blood vessels to grow toward the tumor. A few tumor cells may then enter the blood and lymph vessels and move to other parts of the body, where they may proliferate and form new tumors. The spread of cancer cells beyond their original site is called **metastasis**.

Cancers are named according to the organ or tissue in which they originate. Liver cancer, for example, starts in liver tissue and may or may not spread from there. Based on their site of origin, cancers are grouped into four categories. **Carcinomas** are cancers that originate in the external or internal coverings of the body, such as the skin or the lining of the intestine. **Sarcomas** arise in tissues that support the body, such as bone and muscle. Cancers of blood-forming tissues, such as bone marrow, spleen, and lymph nodes, are called **leukemias** and **lymphomas**.

From studying cancer cells in culture, researchers have learned that these cells do not heed the normal signals that regulate the cell cycle. For example, many cancer cells have defective cell cycle control systems that proceed past checkpoints even in the absence of growth factors. Other cancer cells synthesize growth factors themselves, making them divide continuously. If cancer cells do stop dividing, they seem to do so at random points in the cell cycle, rather than at the normal cell cycle checkpoints. Moreover, in culture, cancer cells are "immortal"; they can go on dividing indefinitely, as long as they have a supply of nutrients (whereas normal mammalian cells divide only about 20 to 50 times before they stop).

Luckily, many tumors can be successfully treated. A tumor that appears to be localized may be removed surgically. Alternatively, it can be treated with concentrated beams of high-energy radiation, which usually damages DNA in cancer cells more than it does in normal cells, perhaps because cancer cells have lost the ability to repair such damage. However, there is sometimes enough damage to normal body cells to produce harmful side effects. For example, radiation damage to cells of the ovaries or testes can lead to sterility.

To treat widespread or metastatic tumors, chemotherapy is used. During periodic chemotherapy treatments, drugs are administered that disrupt specific steps in the cell cycle. For instance, the drug paclitaxel (trade name Taxol) freezes the mitotic spindle after it forms, which stops actively dividing cells from proceeding past metaphase. (Interestingly, Taxol was originally discovered in the bark of the Pacific yew tree, found mainly in the northwestern United States.) Vinblastin, a chemotherapeutic drug first obtained from the periwinkle plant (found in the rain forests of Madagascar), prevents the mitotic spindle from forming.

The side effects of chemotherapy are due to the drugs' effects on normal cells that rapidly divide. Nausea results from chemotherapy's effects on intestinal cells, hair loss from effects on hair follicle cells, and susceptibility to infection from effects on immune cell production.

? What is metastasis?

Metastasis is the spread of cancer cells from their original site of formation to other sites in the body.

10 Review: Mitosis provides for growth, cell replacement, and asexual reproduction

The three micrographs in Figures 10A–10C summarize the roles that mitotic cell division plays in the lives of multicellular organisms. **Figure 10A** shows some of the cells from the tip of a rapidly growing onion plant root. Notice the two cells in the middle row, whose nuclei are in the process of mitosis, as evidenced by the visibly compact chromosomes. Cell division in the root tip produces new cells, which elongate and bring about growth of the root.

Figure 10B shows a dividing bone marrow cell. Mitotic cell division within the red marrow of your body's bones—particularly within your ribs, vertebrae, breastbone, and hip—continuously creates new blood cells that replace older ones. Similar processes replace cells throughout your body. For example, dividing cells within your epidermis continuously replace dead cells that slough off the surface of your skin.

Figure 10C is a micrograph of a hydra, a cnidarian (relative of a sea jelly) that is a common inhabitant of freshwater lakes. A hydra is a tiny multicellular animal that reproduces by either sexual or asexual means. This individual is reproducing asexually by budding. A bud starts out as a mass of mitotically dividing cells growing on the side of the parent. Eventually, the offspring detaches from the parent and takes up life on its own. The offspring is literally a "chip off the old block," being genetically identical to (a clone of) its parent.

In all of the examples described here, the new cells have exactly the same number and types of chromosomes as the parent cells because of the way duplicated chromosomes divide during mitosis. Mitosis makes it possible for organisms to grow, regenerate and repair tissues, and reproduce asexually by producing cells that carry the same genes as the parent cells.

If we examine the cells of any individual organism or those from individuals of any one species, we see that almost all of them contain the same number and types of chromosomes. In the next module, we take a look at how genetic material is organized in chromosomes.

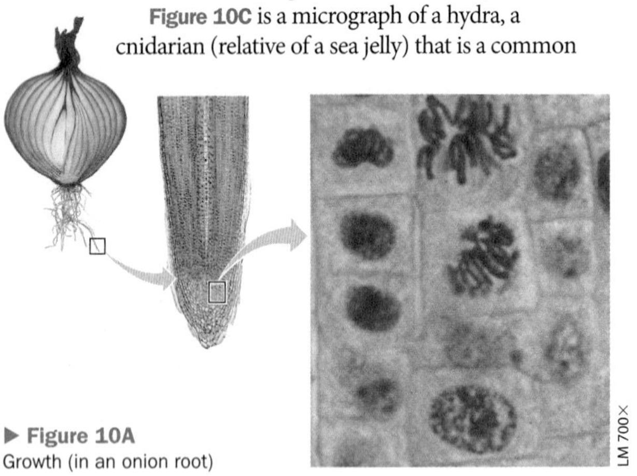

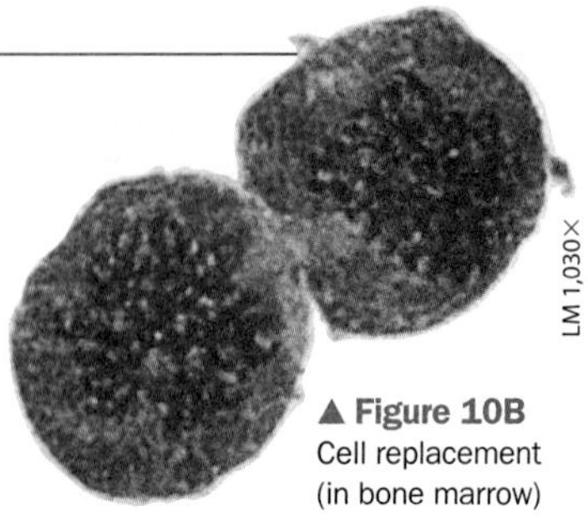

▲ **Figure 10B**
Cell replacement
(in bone marrow)

LM 1,030× Biophoto/Science Source/ Photo Researchers

LM 22× Biophoto/Science Source/Photo Researchers

▲ **Figure 10C**
Asexual reproduction
(of a hydra)

? The body cells of elephants have 56 chromosomes. If an elephant skin cell with 56 chromosomes divides by mitosis, each daughter cell will have _____ chromosomes.

56 ●

Garcia/photocuisine/Corbis; Joseph F. Gennaro Jr./ Photo Researchers; Brian Capon

▶ **Figure 10A**
Growth (in an onion root)

LM 700×

Meiosis and Crossing Over

11 Chromosomes are matched in homologous pairs

In humans, a typical body cell, called a **somatic cell**, has 46 chromosomes. If we use a microscope to examine human chromosomes in metaphase of mitosis, we see that the chromosomes, each consisting of two sister chromatids, can be arranged into matching pairs; **Figure 11** illustrates one pair of metaphase chromosomes. A human cell at metaphase contains 23 sets of duplicated chromosomes. Other species have different numbers of chromosomes, but these, too, usually occur in matched pairs. Moreover, when treated with special dyes, the chromosomes of a pair display matching staining patterns (represented by colored stripes in Figure 11).

Notice that each chromosome is duplicated, consisting of two sister chromatids joined at the centromere. Every (or almost every) chromosome has a twin that resembles it in length and centromere position. The two chromosomes of such a matching pair are called **homologous chromosomes**

(or homologs) because they both carry genes controlling the same inherited characteristics. For example, if a gene that determines whether a person has freckles is located at a particular place, or **locus** (plural, *loci*), on one chromosome—within the narrow orange band in our drawing, for instance—then the homologous chromosome has that same gene at that same locus. However, the two chromosomes of a homologous pair may have different versions of the same gene.

Pair of homologous chromosomes

Locus

Centromere

Sister chromatids

One duplicated chromosome

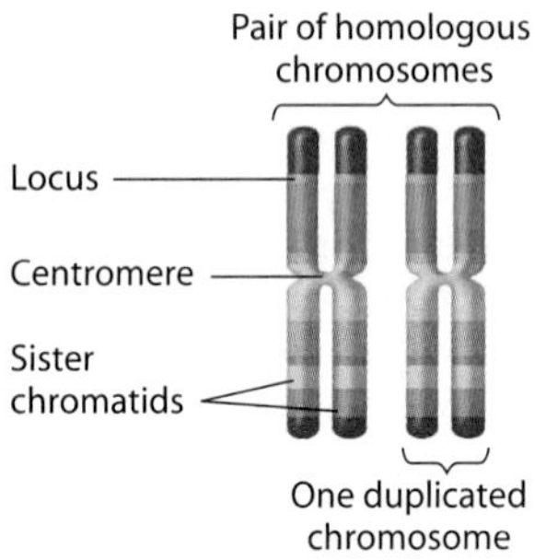

▲ **Figure 11** A pair of homologous chromosomes

In human females, the 46 chromosomes fall neatly into 23 homologous pairs. For a male, however, the chromosomes in one pair do not look alike. The nonidentical pair, only partly homologous, is the male's sex chromosomes. These **sex chromosomes** determine an individual's sex (male versus female), although these chromosomes carry genes that perform other functions as well. In mammals, males have one X chromosome and one Y chromosome. Females have two X chromosomes. The 22 remaining pairs of chromosomes, found in both males and females, are called **autosomes**. In the next module, we discuss how chromosomes are inherited.

? Are all of *your* chromosomes fully homologous?

If you are female, then yes. If you are male, then no (your X and Y are only partly homologous).

12 Gametes have a single set of chromosomes

The development of a fertilized egg into a new adult organism is one phase of a multicellular organism's **life cycle**, the sequence of stages leading from the adults of one generation to the adults of the next (**Figure 12A**). Having two sets of chromosomes, one inherited from each parent, is a key factor in the life cycle of humans and all other species that reproduce sexually.

Humans, as well as most other animals and many plants, are said to be **diploid** organisms because all body cells contain pairs of homologous chromosomes. The total number of chromosomes is called the diploid number (abbreviated $2n$). For humans, the diploid number is 46; that is, $2n = 46$. The exceptions are the egg and sperm cells, collectively known as **gametes**. Each gamete has a single set of chromosomes: 22 autosomes plus a sex chromosome, either X or Y. A cell with a single chromosome set is called a **haploid** cell; it has only one member of each homologous pair. For humans, the haploid number (abbreviated n) is 23; that is, $n = 23$.

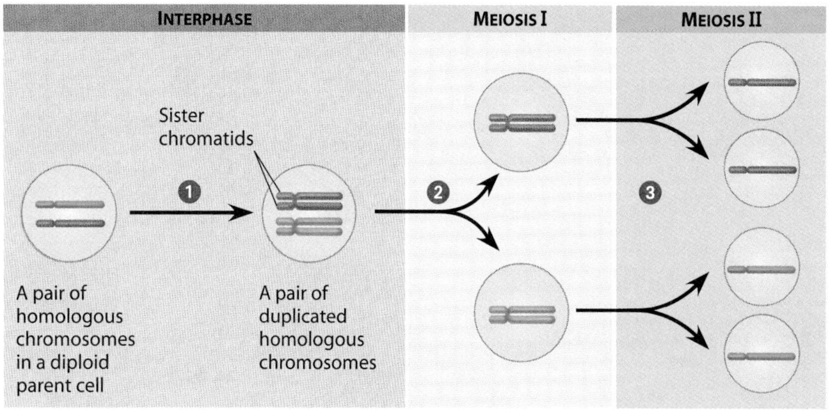

▲ **Figure 12B** How meiosis halves chromosome number

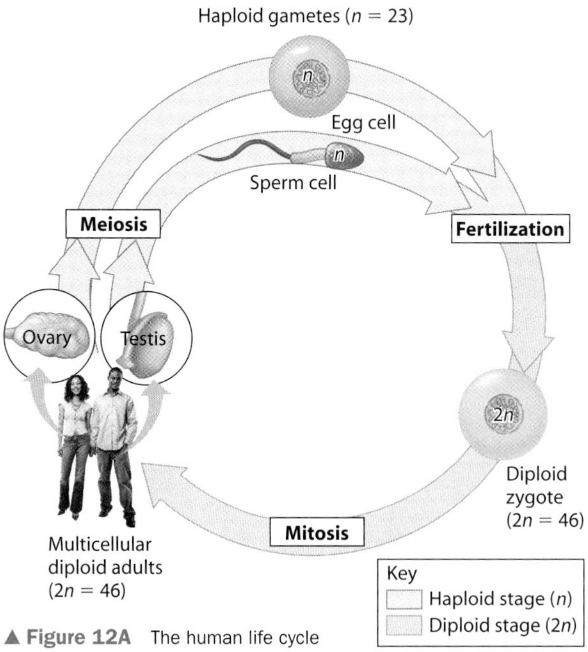

Ron Chapple Stock/Alamy

▲ **Figure 12A** The human life cycle

In the human life cycle, a haploid sperm cell from the father fuses with a haploid egg cell from the mother in the process of **fertilization**. The resulting fertilized egg, called a **zygote**, has two sets of homologous chromosomes and so is diploid. One set of homologous chromosomes comes from each parent. The life cycle is completed as a sexually mature adult develops from the zygote.

All sexual life cycles, including our own, involve an alternation of diploid and haploid stages. Producing haploid gametes prevents the chromosome number from doubling in every generation. Gametes are made by a special sort of cell division called meiosis, which occurs only in reproductive organs (ovaries and testes in animals). Whereas mitosis produces daughter cells with the same number of chromosomes as the parent cell, meiosis reduces the chromosome number by half. **Figure 12B** tracks one pair of homologous chromosomes. ❶ Each of the chromosomes is duplicated during interphase (before meiosis). ❷ The first division, meiosis I, segregates the two chromosomes of the homologous pair, packaging them in separate (haploid) daughter cells. But each chromosome is still doubled. ❸ Meiosis II separates the sister chromatids. Each of the four daughter cells is haploid and contains only a single chromosome from the homologous pair. We turn to meiosis next.

? Imagine you stain a human cell and view it under a microscope. You observe 23 chromosomes, including a Y chromosome. You could conclude that this must be a ______ cell taken from the organ called the ______ .

sperm (since it is haploid) . . . testis (since it is from a male)

13 Meiosis reduces the chromosome number from diploid to haploid

Meiosis is a type of cell division that produces haploid gametes in diploid organisms. Two haploid gametes may then combine via fertilization to restore the diploid state in the zygote. Were it not for meiosis, each generation would have twice as much genetic material as the generation before!

Many of the stages of meiosis closely resemble corresponding stages in mitosis. Meiosis, like mitosis, is preceded by the duplication of chromosomes. However, this single duplication is followed by two consecutive cell divisions, called meiosis I and meiosis II. Because one duplication of the chromosomes is followed by two divisions, each of the four daughter cells resulting from meiosis has a haploid set of chromosomes—half as many chromosomes as the parent cell. The drawings in **Figure 13** show the two meiotic divisions for an animal cell with a diploid number of 6. The members of a pair of homologous chromosomes in Figure 13 (and later figures) are colored red and blue to help distinguish them. (Imagine that the red chromosomes were inherited from the mother and the blue chromosomes from the father.) One of the most important events in meiosis occurs during

INTERPHASE: Chromosomes duplicate

MEIOSIS I: Homologous chromosomes separate

Prophase I | **Metaphase I** | **Anaphase I**

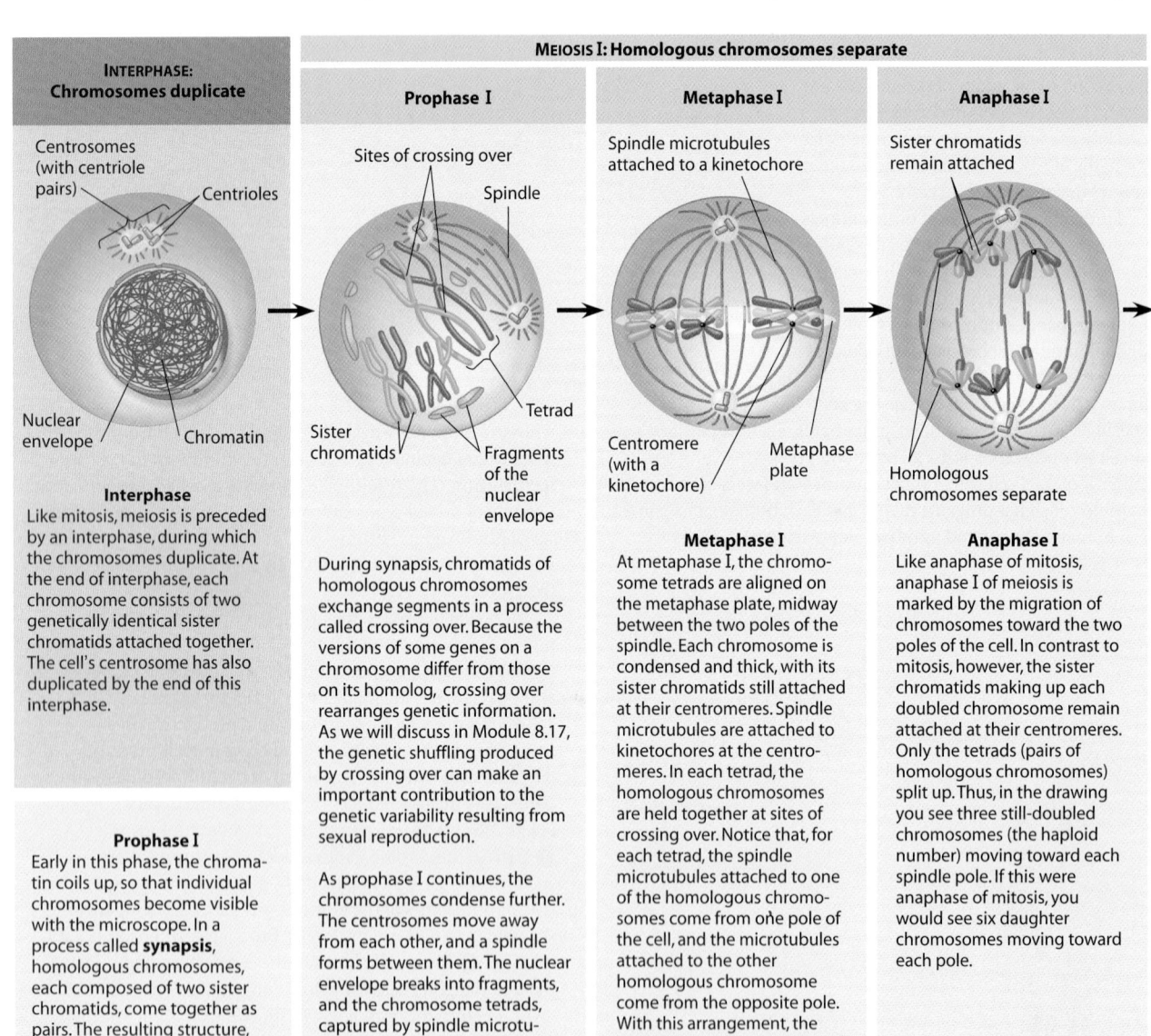

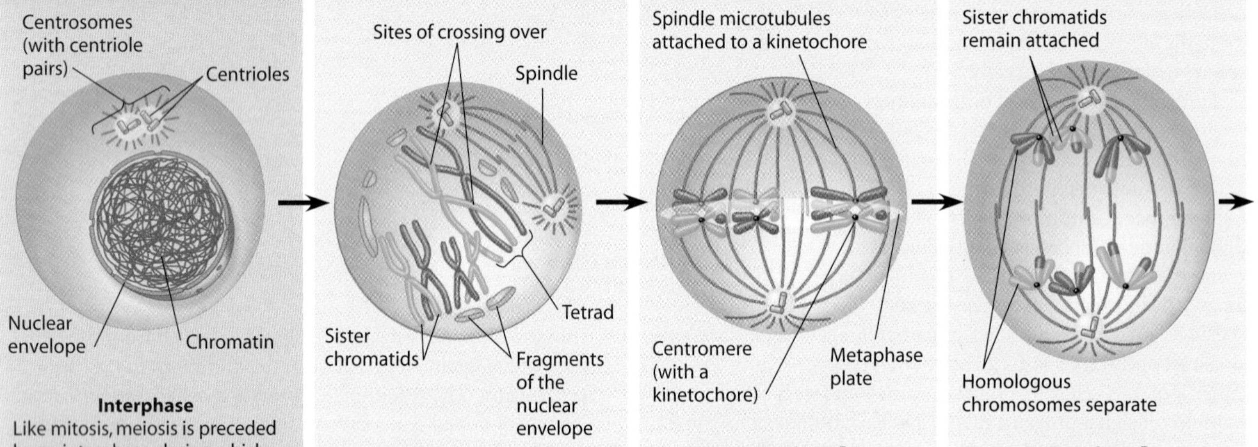

Interphase
Like mitosis, meiosis is preceded by an interphase, during which the chromosomes duplicate. At the end of interphase, each chromosome consists of two genetically identical sister chromatids attached together. The cell's centrosome has also duplicated by the end of this interphase.

Prophase I
Early in this phase, the chromatin coils up, so that individual chromosomes become visible with the microscope. In a process called **synapsis**, homologous chromosomes, each composed of two sister chromatids, come together as pairs. The resulting structure, consisting of four chromatids, is called a tetrad.

During synapsis, chromatids of homologous chromosomes exchange segments in a process called crossing over. Because the versions of some genes on a chromosome differ from those on its homolog, crossing over rearranges genetic information. As we will discuss in Module 8.17, the genetic shuffling produced by crossing over can make an important contribution to the genetic variability resulting from sexual reproduction.

As prophase I continues, the chromosomes condense further. The centrosomes move away from each other, and a spindle forms between them. The nuclear envelope breaks into fragments, and the chromosome tetrads, captured by spindle microtubules, are moved toward the center of the cell.

Metaphase I
At metaphase I, the chromosome tetrads are aligned on the metaphase plate, midway between the two poles of the spindle. Each chromosome is condensed and thick, with its sister chromatids still attached at their centromeres. Spindle microtubules are attached to kinetochores at the centromeres. In each tetrad, the homologous chromosomes are held together at sites of crossing over. Notice that, for each tetrad, the spindle microtubules attached to one of the homologous chromosomes come from one pole of the cell, and the microtubules attached to the other homologous chromosome come from the opposite pole. With this arrangement, the homologous chromosomes of each tetrad are poised to move toward opposite poles of the cell.

Anaphase I
Like anaphase of mitosis, anaphase I of meiosis is marked by the migration of chromosomes toward the two poles of the cell. In contrast to mitosis, however, the sister chromatids making up each doubled chromosome remain attached at their centromeres. Only the tetrads (pairs of homologous chromosomes) split up. Thus, in the drawing you see three still-doubled chromosomes (the haploid number) moving toward each spindle pole. If this were anaphase of mitosis, you would see six daughter chromosomes moving toward each pole.

Ed Reschke/Peter Arnold, Inc./PhotoLibrary

▲ **Figure 13** The stages of meiosis

prophase I. At this stage, four chromosomes (two sets of sister chromatids) are aligned and physically touching each other. When in this configuration, nonsister chromatids may trade segments. As you will learn in Module 17, this exchange of chromosome segments—called crossing over—is a key step in the generation of genetic diversity that occurs during sexual reproduction.

? A cell has the haploid number of chromosomes, but each chromosome has two chromatids. The chromosomes are arranged singly at the center of the spindle. What is the meiotic stage?

Metaphase II (since the chromosomes line up two by two in metaphase I)

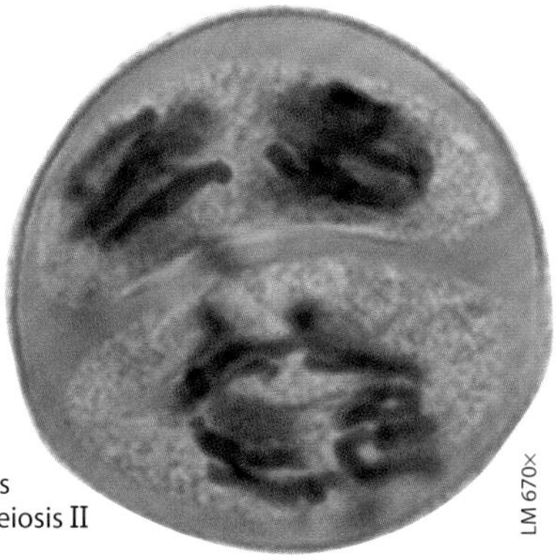

Two lily cells undergo meiosis II

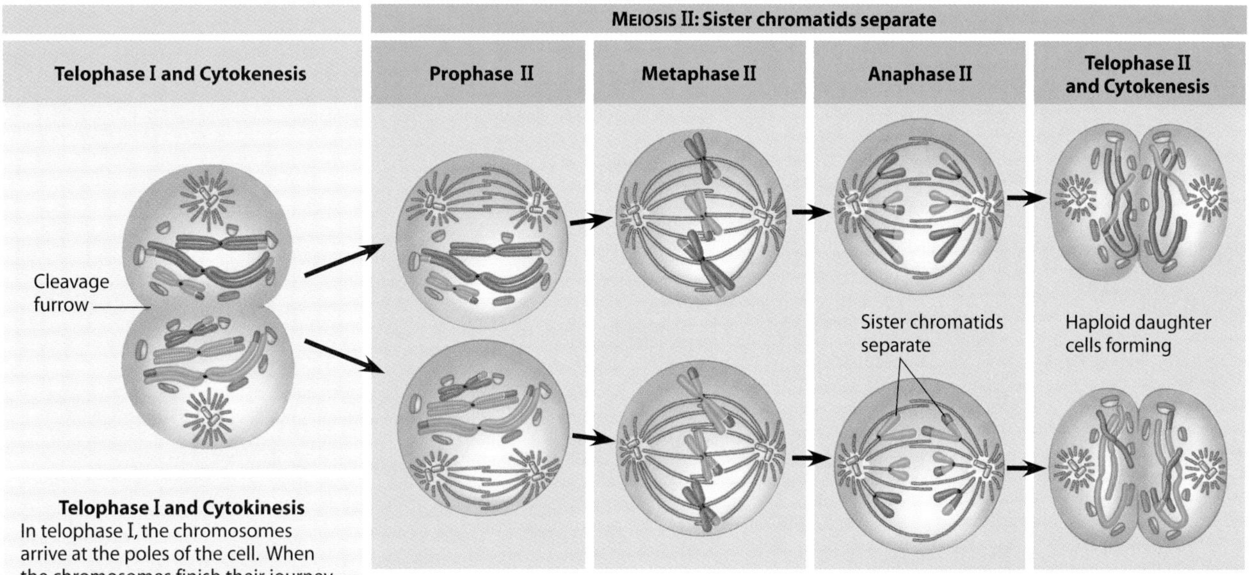

Telophase I and Cytokinesis

In telophase I, the chromosomes arrive at the poles of the cell. When the chromosomes finish their journey, each pole of the cell has a haploid chromosome set, although each chromosome is still in duplicate form at this point. In other words, each chromosome still consists of two sister chromatids. Usually, cytokinesis occurs along with telophase I, and two haploid daughter cells are formed.

Following telophase I in some organisms, the chromosomes uncoil and the nuclear envelope re-forms, and there is an interphase before meiosis II begins. In other species, daughter cells produced in the first meiotic division immediately begin preparation for the second meiotic division. In either case, no chromosome duplication occurs between telophase I and the onset of meiosis II.

Meiosis II

Meiosis II is essentially the same as mitosis. The important difference is that meiosis II starts with a haploid cell.

During prophase II, a spindle forms and moves the chromosomes toward the middle of the cell. During metaphase II, the chromosomes are aligned on the metaphase plate as they are in mitosis, with the kinetochores of the sister chromatids of each chromosome pointing toward opposite poles. In anaphase II, the centromeres of sister chromatids finally separate, and the sister chromatids of each pair, now individual daughter chromosomes, move toward opposite poles of the cell. In telophase II, nuclei form at the cell poles, and cytokinesis occurs at the same time. There are now four daughter cells, each with the haploid number of (single) chromosomes.

14 Mitosis and meiosis have important similarities and differences

You have now learned the two ways that cells of eukaryotic organisms divide. Mitosis, which provides for growth, tissue repair, and asexual reproduction, produces daughter cells that are genetically identical to the parent cell. Meiosis, needed for sexual reproduction, yields genetically unique haploid daughter cells—cells with only one member of each homologous chromosome pair.

For both mitosis and meiosis, the chromosomes duplicate only once, during the S phase of the preceding interphase. Mitosis involves one division of the nucleus, and it is usually accompanied by cytokinesis, producing two identical diploid cells. Meiosis entails two nuclear and cytoplasmic divisions, yielding four haploid cells.

Figure 14 compares mitosis and meiosis, tracing these two processes for a diploid parent cell with four chromosomes. Homologous chromosomes are those matching in size.

Notice that all the events unique to meiosis occur during meiosis I. In prophase I, duplicated homologous chromosomes pair to form **tetrads**, sets of four chromatids, with each pair of sister chromatids joined at their centromeres. In metaphase I, tetrads (not individual chromosomes) are aligned at the metaphase plate. During anaphase I, pairs of homologous chromosomes separate, but the sister chromatids of each chromosome stay together. At the end of meiosis I, there are two haploid cells, but each chromosome still has two sister chromatids.

Meiosis II is virtually identical to mitosis in that it separates sister chromatids. But unlike mitosis, each daughter cell produced by meiosis II has only a *haploid* set of chromosomes.

? Explain how mitosis conserves chromosome number while meiosis reduces the number from diploid to haploid.

In mitosis, the duplication of chromosomes is followed by one division of the cell. In meiosis, homologous chromosomes separate in the first of two cell divisions; after the second division, each new cell ends up with just a single haploid set.

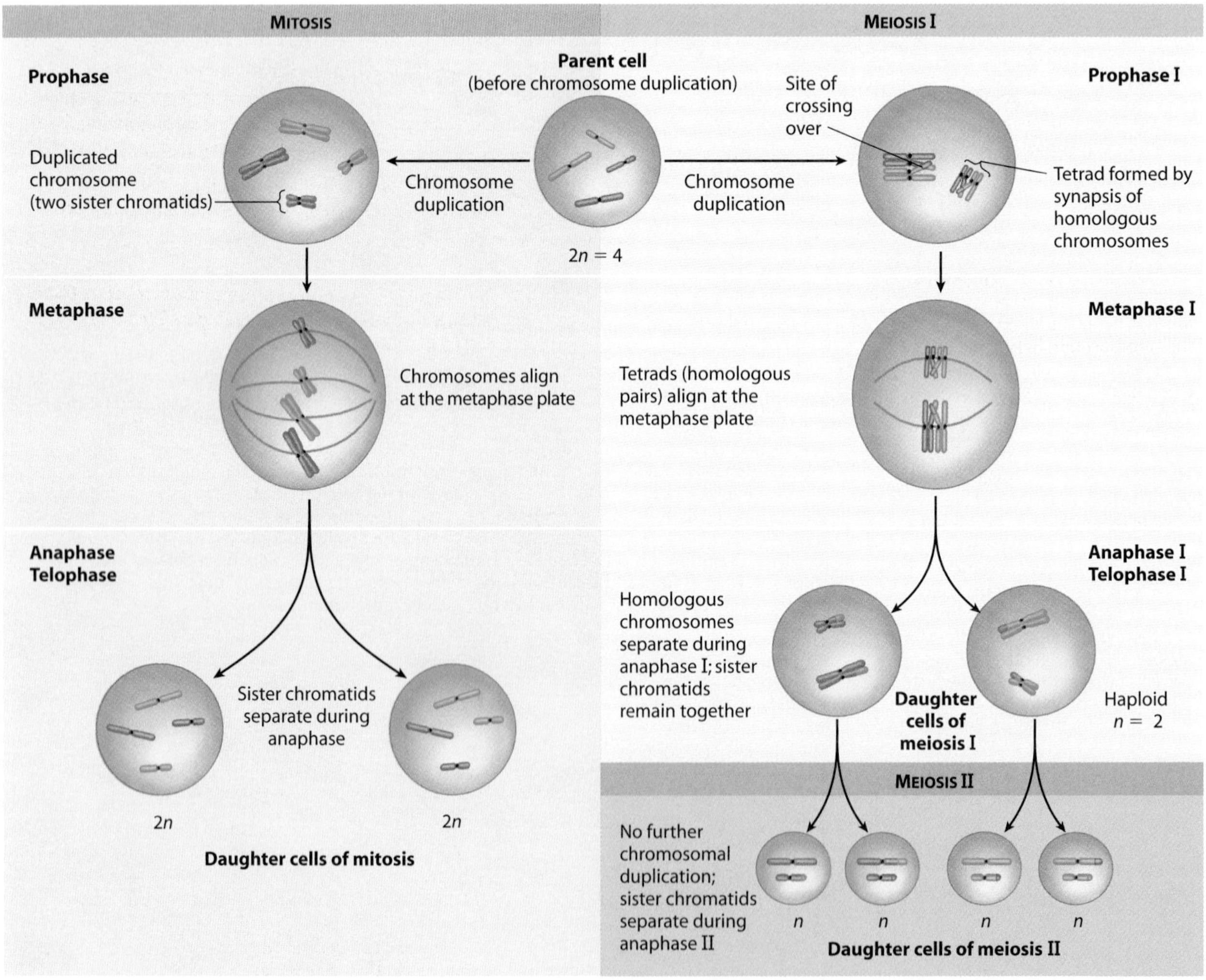

▲ **Figure 14** Comparison of mitosis and meiosis

15 Independent orientation of chromosomes in meiosis and random fertilization lead to varied offspring

As we discussed in Module 1, offspring that result from sexual reproduction are highly varied; they are genetically different from their parents and from one another. How does this genetic variation account result from meiosis?

Figure 15 illustrates one way in which the process of meiosis contributes to genetic differences in gametes. The figure shows how the arrangement of homologous chromosome pairs at metaphase of meiosis I affects the resulting gametes. Once again, our example is from a diploid organism with four chromosomes (two homologous pairs), with colors used to differentiate homologous chromosomes (red for chromosomes inherited from the mother and blue for those from the father).

The orientation of the pairs of homologous chromosomes (tetrads) at metaphase I—whether the maternal (red) or paternal (blue) chromosome is closer to a given pole—is as random as the flip of a coin. Thus, there is a 50% chance that a particular daughter cell will get the maternal chromosome of a certain homologous pair and a 50% chance that it will receive the paternal chromosome. In this example, there are two possible ways that the two tetrads can align during metaphase I. In possibility A, the tetrads are oriented with both red chromosomes on the same side of the metaphase plate. Therefore, the gametes produced in possibility A can each have either two red *or* two blue chromosomes (bottom row, combinations 1 and 2).

In possibility B, the tetrads are oriented differently (blue/red and red/blue). This arrangement produces gametes that each have one red and one blue chromosome. Furthermore, half the gametes have a big blue chromosome and a small red one (combination 3), and half have a big red chromosome and a small blue one (combination 4).

So we see that for this example, four chromosome combinations are possible in the gametes, and in fact the organism will produce gametes of all four types in equal quantities. For a species with more than two pairs of chromosomes, such as humans, *all* the chromosome pairs orient independently at metaphase I. (Chromosomes X and Y behave as a homologous pair in meiosis.)

For any species, the total number of combinations of chromosomes that meiosis can package into gametes is 2^n, where n is the haploid number. For the organism in this figure, $n = 2$, so the number of chromosome combinations is 2^2, or 4. For a human ($n = 23$), there are 2^{23}, or about 8 million possible chromosome combinations! This means that each gamete you produce contains one of roughly 8 million possible combinations of chromosomes inherited from your mother and father.

How many possibilities are there when a gamete from one individual unites with a gamete from another individual in fertilization? In humans, the random fusion of a single sperm with a single ovum during fertilization will produce a zygote with any of about 64 trillion (8 million $\times$ 8 million) combinations of chromosomes! While the random nature of fertilization adds a huge amount of potential variability to the offspring of sexual reproduction, there is in fact even more variety created during meiosis, as we see in the next two modules.

? **A particular species of worm has a diploid number of 10. How many chromosomal combinations are possible for gametes formed by meiosis?**

32; $2n = 10$, so $n = 5$ and $2^n = 32$

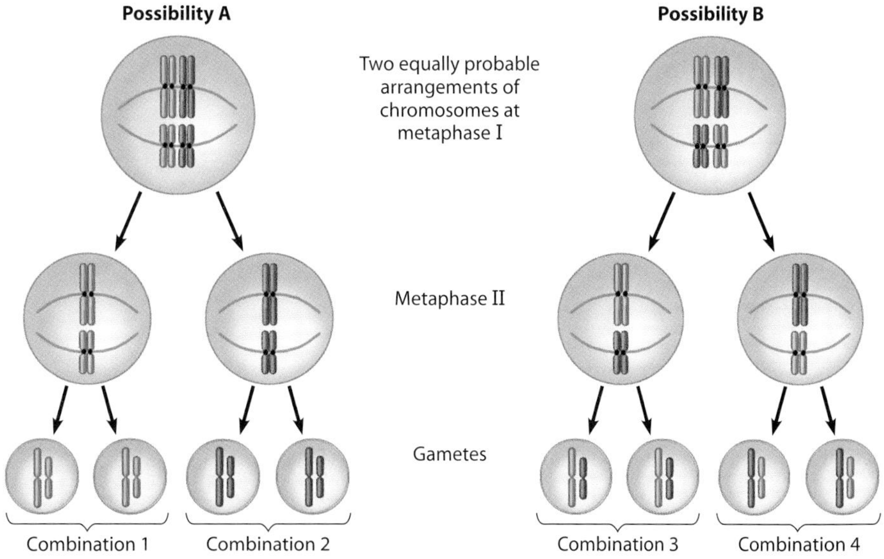

▲ **Figure 15** Results of the independent orientation of chromosomes at metaphase I

16 Homologous chromosomes may carry different versions of genes

So far, we have focused on genetic variability in gametes and zygotes at the whole-chromosome level. We have yet to discuss the actual genetic information—the genes—contained in the chromosomes. The question we need to answer now is this: What is the significance of the independent orientation of metaphase chromosomes at the level of genes?

Let's take a simple example, the single tetrad in **Figure 16**. The letters on the homologous chromosomes represent genes. Recall that homologous chromosomes have genes for the same characteristic at corresponding loci. Our example involves hypothetical genes controlling the appearance of mice. *C* and *c* are different versions of a gene for one characteristic, coat color; *E* and *e* are different versions of a gene for another characteristic, eye color. (Different versions of a gene contain slightly different nucleotide sequences in the chromosomal DNA.)

Let's say that *C* represents the gene for a brown coat and that *c* represents the gene for white coat. In the chromosome diagram, notice that *C* is at the same locus on the red homolog as *c* is on the blue one. Likewise, gene *E* (for black eyes) is at the same locus as *e* (pink eyes).

The fact that homologous chromosomes can bear two different kinds of genetic information for the same characteristic (for instance, coat color) is what really makes gametes—and therefore offspring—different from one another. In our example, a gamete carrying a red chromosome would have genes specifying brown coat color and black eye color, while a gamete with the homologous blue chromosome would have genes for white coat and pink eyes. Thus, we see how a tetrad with genes shown for only two characteristics can yield two genetically different kinds of gametes. In the next module, we go a step further and see how this same tetrad can actually yield *four* different kinds of gametes.

? In the tetrad of Figure 16, use labels to distinguish the pair of homologous chromosomes from sister chromatids.

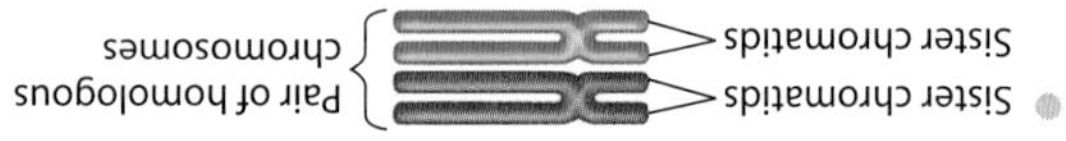

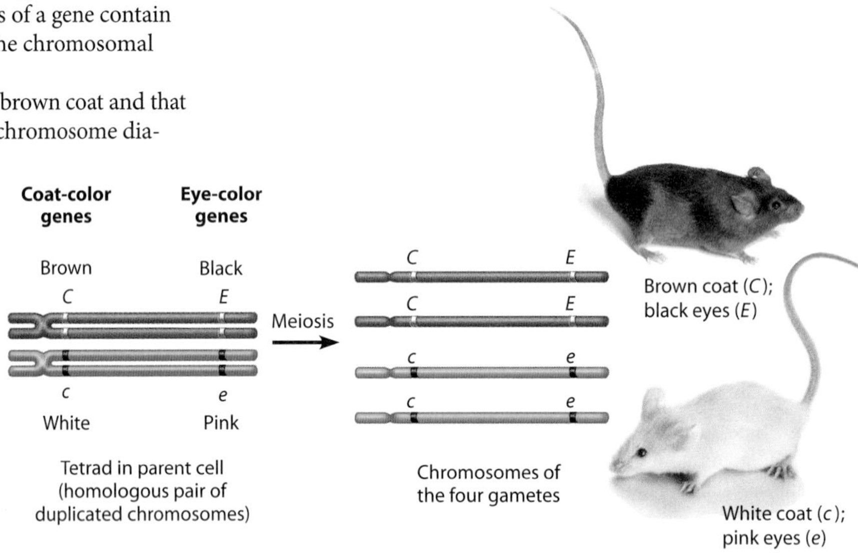

▲ **Figure 16** Differing genetic information (coat color and eye color) on homologous chromosomes

F. Schussler/Getty Images

Chris Collins/Bettmann/Corbis

17 Crossing over further increases genetic variability

Crossing over is an exchange of corresponding segments between nonsister chromatids of homologous chromosomes. The micrograph and drawing in **Figure 17A** show the results of crossing over between two homologous chromosomes during prophase I of meiosis. The chromosomes are a tetrad—a set of four chromatids, with each pair of sister chromatids joined together. The sites of crossing over appear as X-shaped regions; each is called a **chiasma** (Greek for "cross"). A chiasma (plural, *chiasmata*) is a place where two homologous (nonsister) chromatids are attached to each other. **Figure 17B** illustrates how crossing over can produce new combinations of genes, using as examples the hypothetical mouse genes mentioned in the previous module.

Crossing over begins very early in prophase I of meiosis. At that time, homologous chromosomes are paired all along their

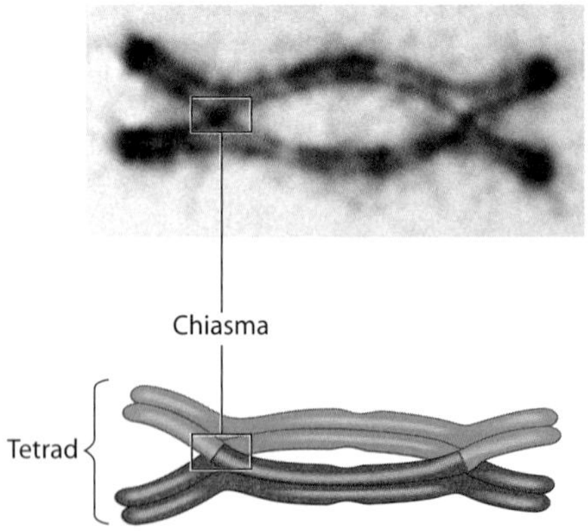

TEM 5,060×

Mark Petronczki and Maria Siomos

▲ **Figure 17A** Chiasmata, the sites of crossing over

lengths, with a precise gene-by-gene alignment. At the top of the figure is a tetrad with coat-color (*C, c*) and eye-color (*E, e*) genes labeled. ❶ The DNA molecules of two nonsister chromatids—one maternal (red) and one paternal (blue)—break at the same place. ❷ Immediately, the two broken chromatids join together in a new way (red to blue and blue to red). In effect, the two homologous segments trade places, or cross over, producing hybrid chromosomes (red/blue and blue/red) with new combinations of maternal and paternal genes. ❸ When the homologous chromosomes separate in anaphase I, each contains a new segment originating from its homolog. ❹ Finally, in meiosis II, the sister chromatids separate, each going to a different gamete.

In this example, if there were no crossing over, meiosis could produce only two genetic types of gametes. These would be the ones ending up with the "parental" types of chromosomes (either all blue or all red), carrying either genes *C* and *E* or genes *c* and *e*. These are the same two kinds of gametes we saw in Figure 16. With crossing over, two other types of gametes can result, ones that are part blue and part red. One of these carries genes *C* and *e*, and the other carries genes *c* and *E*. Chromosomes with these combinations of genes would not exist if not for crossing over. They are called "recombinant" because they result from **genetic recombination**, the production of gene combinations different from those carried by the original parental chromosomes.

In meiosis in humans, an average of one to three crossover events occur per chromosome pair. Thus, if you were to examine a chromosome from one of your gametes, you would most likely find that it is not exactly like any one of your own chromosomes. Rather, it is probably a patchwork of segments derived from a pair of homologous chromosomes, cut and pasted together to form a chromosome with a unique combination of genes.

We have now examined three sources of genetic variability in sexually reproducing organisms: independent orientation of chromosomes at metaphase I, random fertilization, and crossing over during prophase I of meiosis. The different versions of genes that homologous chromosomes may have at each locus originally arise from mutations, so mutations are ultimately responsible for genetic diversity in living organisms. Once these differences arise, reshuffling of the different versions during sexual reproduction increases genetic variation. When we discuss natural selection and evolution in Unit III, we will see that this genetic variety in offspring is the raw material for natural selection.

Our discussion of meiosis to this point has focused on the process as it normally occurs. In the next, and last, major section of the chapter, we consider some of the consequences of errors in the process.

? Describe how crossing over and the random alignment of homologous chromosomes on the metaphase I plate account for the genetic variation among gametes formed by meiosis.

● Crossing over creates recombinant chromosomes having a combination of genes that were originally on different, though homologous, chromosomes. Homologous chromosome pairs are oriented randomly at metaphase of meiosis I.

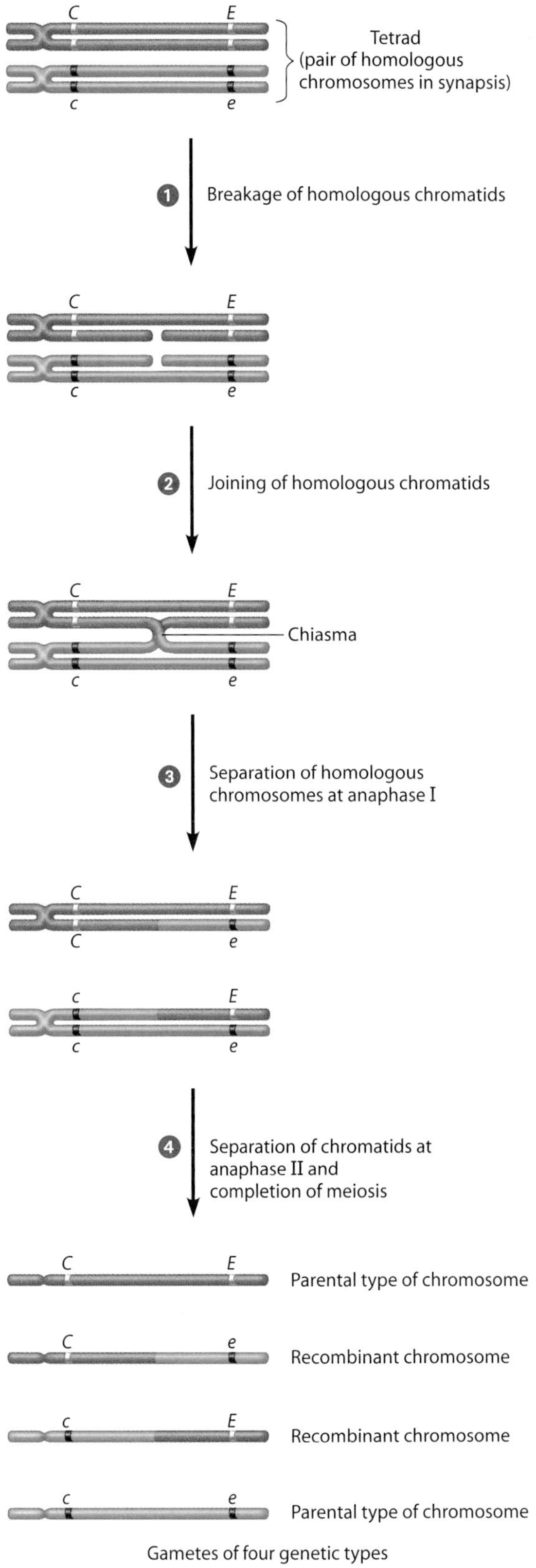

▲ **Figure 17B** How crossing over leads to genetic recombination

Alterations of Chromosome Number and Structure

18 A karyotype is a photographic inventory of an individual's chromosomes

Errors in meiosis can lead to gametes containing chromosomes in abnormal numbers or with major alterations in their structure. Fertilization involving such abnormal gametes results in offspring with chromosomal abnormalities. Such conditions can be readily detected in a **karyotype**, an ordered display of magnified images of an individual's chromosomes arranged in pairs. A karyotype shows the chromosomes condensed and doubled, as they appear in metaphase of mitosis.

To prepare a karyotype, medical scientists often use lymphocytes, a type of white blood cell. A blood sample is treated with a chemical that stimulates mitosis. After growing in culture for several days, the cells are treated with another chemical to arrest mitosis at metaphase, when the chromosomes, each consisting of two joined sister chromatids, are most highly condensed. Figure 18 outlines the steps in one method for the preparation of a karyotype from a blood sample.

The photograph in step 5 shows the karyotype of a normal human male. Images of the 46 chromosomes from a single diploid cell are arranged in 23 homologous pairs: autosomes numbered from 1 to 22 (starting with the largest) and one pair of sex chromosomes (X and Y in this case). The chromosomes had been stained to reveal band patterns, which are helpful in

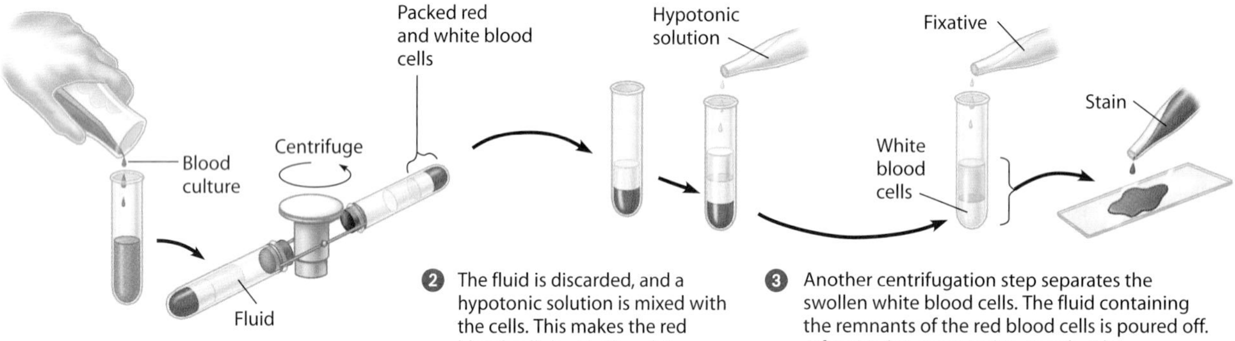

❶ A blood culture is centrifuged to separate the blood cells from fluid.

❷ The fluid is discarded, and a hypotonic solution is mixed with the cells. This makes the red blood cells burst. The white blood cells swell but do not burst, and their chromosomes spread out.

❸ Another centrifugation step separates the swollen white blood cells. The fluid containing the remnants of the red blood cells is poured off. A fixative (preservative) is mixed with the white blood cells. A drop of the cell suspension is spread on a microscope slide, dried, and stained.

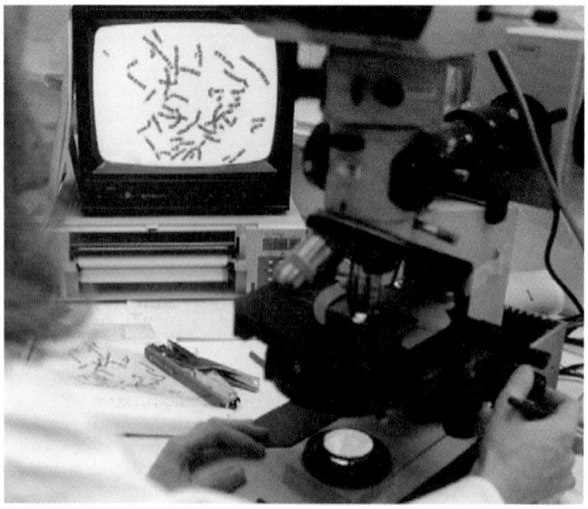

Veronique Burger/Phanie Agency/ Photo Researchers

❹ The slide is viewed with a microscope equipped with a digital camera. A photograph of the chromosomes is entered into a computer, which electronically arranges them by size and shape.

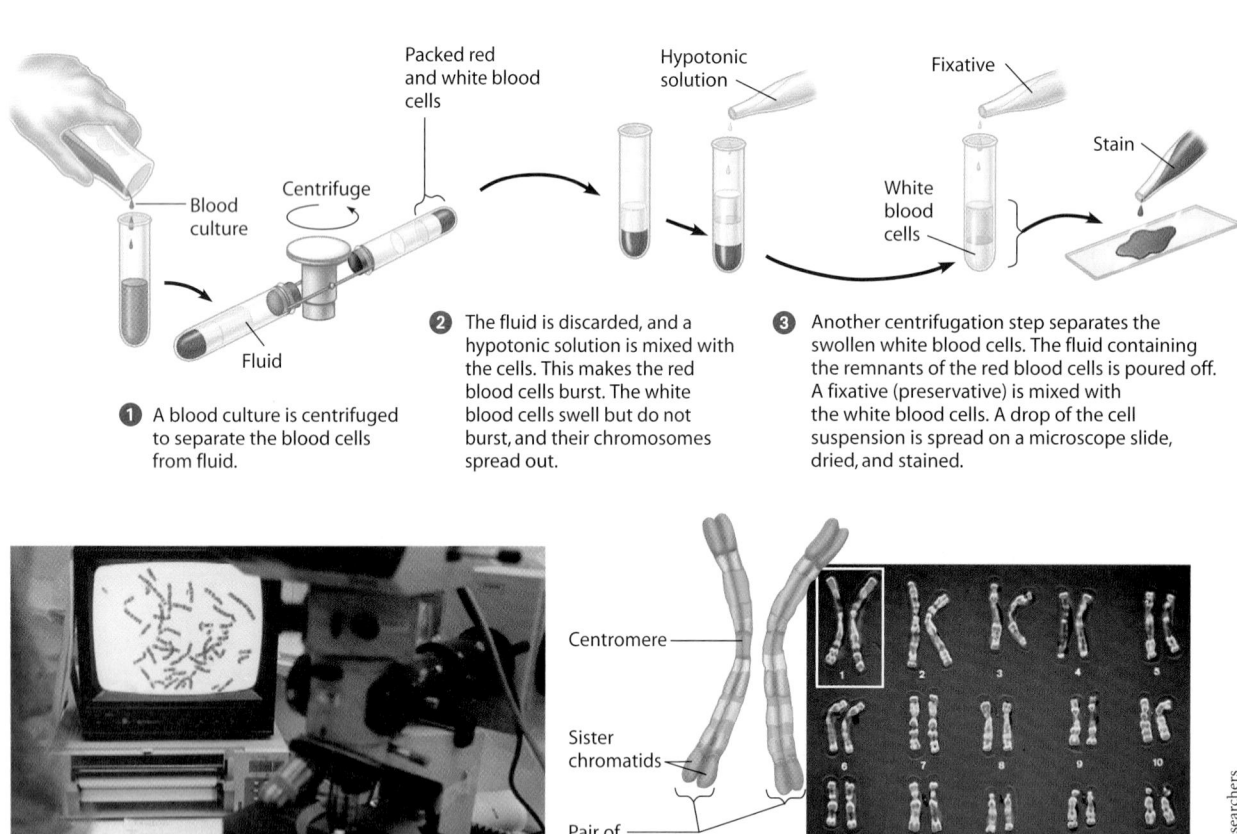

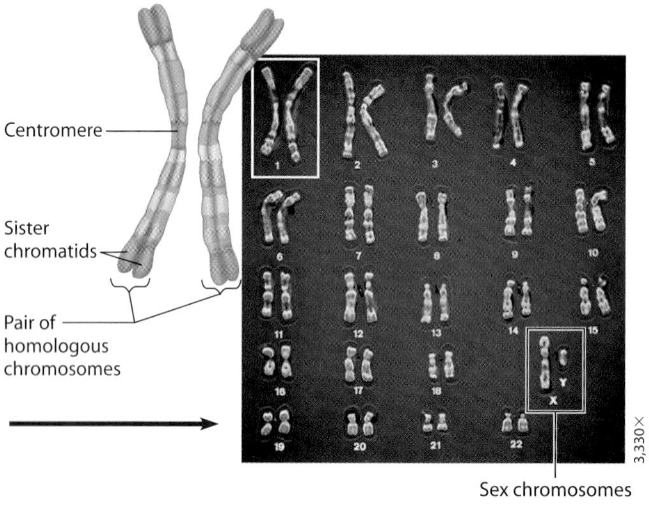

CNRI/SPL/Photo Researchers

❺ The resulting display is the karyotype. The 46 chromosomes here include 22 pairs of autosomes and two sex chromosomes, X and Y. Although difficult to discern in the karyotype, each of the chromosomes consists of two sister chromatids lying very close together (as shown in the diagram).

▲ **Figure 18** Preparation of a karyotype from a blood sample

differentiating the chromosomes and in detecting structural abnormalities. Among the alterations that can be detected by karyotyping is trisomy 21, the basis of Down syndrome, which we discuss next.

? How would the karyotype of a human female differ from the male karyotype in Figure 18?

Instead of an XY combination for the sex chromosomes, there would be a homologous pair of X chromosomes (XX).

CONNECTION 19 An extra copy of chromosome 21 causes Down syndrome

The karyotype in Figure 18 shows the normal human complement of 23 pairs of chromosomes. Compare it with **Figure 19A**; besides having two X chromosomes (because it's from a female), notice that there are three number 21 chromosomes, making 47 chromosomes in total. This condition is called **trisomy 21**.

In most cases, a human embryo with an abnormal number of chromosomes is spontaneously aborted (miscarried) long before birth. But some aberrations in chromosome number, including trisomy 21, appear to upset the genetic balance less drastically, and individuals carrying them can survive. These people have a characteristic set of symptoms, called a syndrome. A person with trisomy 21, for instance, has a condition called **Down syndrome**, named after John Langdon Down, who described the syndrome in 1866.

Trisomy 21 is the most common chromosome number abnormality. Affecting about one out of every 700 children born, it is also the most common serious birth defect in the United States. Chromosome 21 is one of our smallest chromosomes, but an extra copy produces a number of effects. Down syndrome includes characteristic facial features—frequently a round face, a skin fold at the inner corner of the eye, a flattened nose bridge, and small, irregular teeth—as well as short stature, heart defects, and susceptibility to respiratory infections, leukemia, and Alzheimer's disease.

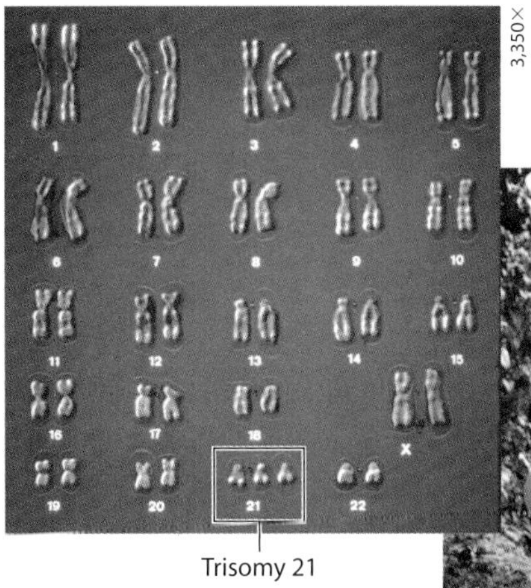

▲ **Figure 19A** A karyotype showing trisomy 21, and an individual with Down syndrome

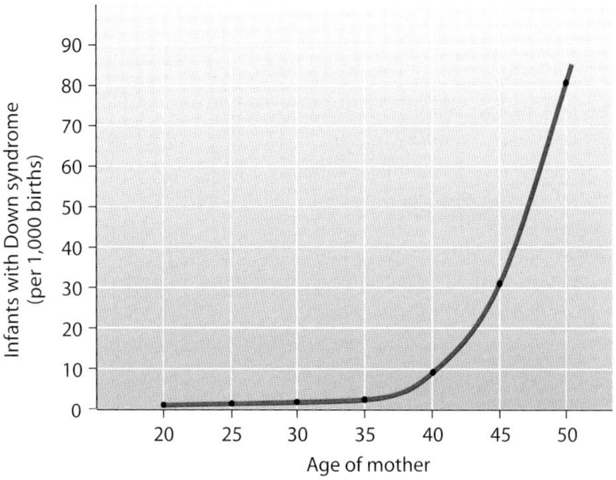

▲ **Figure 19B** Maternal age and incidence of Down syndrome

People with Down syndrome usually have a life span shorter than normal. They also exhibit varying degrees of mental retardation. However, some individuals with the syndrome live to middle age or beyond, and many are socially adept and able to hold jobs. A few women with Down syndrome have had children, though most people with the syndrome are sexually underdeveloped and sterile. Half the eggs produced by a woman with Down syndrome will have the extra chromosome 21, so there is a 50% chance that she will transmit the syndrome to her child.

As indicated in **Figure 19B**, the incidence of Down syndrome in the offspring of normal parents increases markedly with the age of the mother. Down syndrome affects less than 0.05% of children (fewer than one in 2,000) born to women under age 30. The risk climbs to 1% (ten in 1,000) for mothers at age 40 and is even higher for older mothers. Because of this relatively high risk, pregnant women over 35 are candidates for fetal testing for trisomy 21 and other chromosomal abnormalities.

What causes trisomy 21? We address that question in the next module.

? For mothers of age 47, the risk of having a baby with Down syndrome is about ______ per thousand births, or ______ %.

40 . . . 4

20 Accidents during meiosis can alter chromosome number

Within the human body, meiosis occurs repeatedly as the testes or ovaries produce gametes. Almost always, the meiotic spindle distributes chromosomes to daughter cells without error. But there is an occasional mishap in which the members of a chromosome pair fail to separate. Such an error is called a **nondisjunction**. Figures 20A and 20B illustrate two ways that a nondisjunction can occur. For simplicity, we use a hypothetical organism whose diploid chromosome number is 4. In both figures, the cell at the top is diploid ($2n = 4$), with two pairs of homologous chromosomes undergoing anaphase of meiosis I.

Sometimes, as in Figure 20A, a pair of homologous chromosomes do not separate during meiosis I. In this case, even though the rest of meiosis occurs normally, all the resulting gametes end up with abnormal numbers of chromosomes. Two of the gametes have three chromosomes; the other two gametes have only one chromosome each. In Figure 20B, meiosis I is normal, but one pair of sister chromatids fail to move apart during meiosis II. In this case, two of the resulting gametes are abnormal; the other two gametes are normal.

If an abnormal gamete produced by nondisjunction unites with a normal gamete during fertilization, the result is a zygote with an abnormal number of chromosomes. Mitosis will then transmit the mistake to all embryonic cells. If this were a real organism and it survived, it would have an abnormal karyotype and probably a syndrome of disorders caused by the abnormal number of genes. For example, if there is nondisjunction affecting human chromosome 21, some resulting gametes will carry an extra chromosome 21. If one of these gametes unites with a normal gamete, trisomy 21 (Down syndrome) will result. Nondisjunction can also affect chromosomes other than 21, as we see next.

? Explain how nondisjunction could result in a diploid gamete.

A diploid gamete would result if the nondisjunction affected all the chromosomes during one of the meiotic divisions.

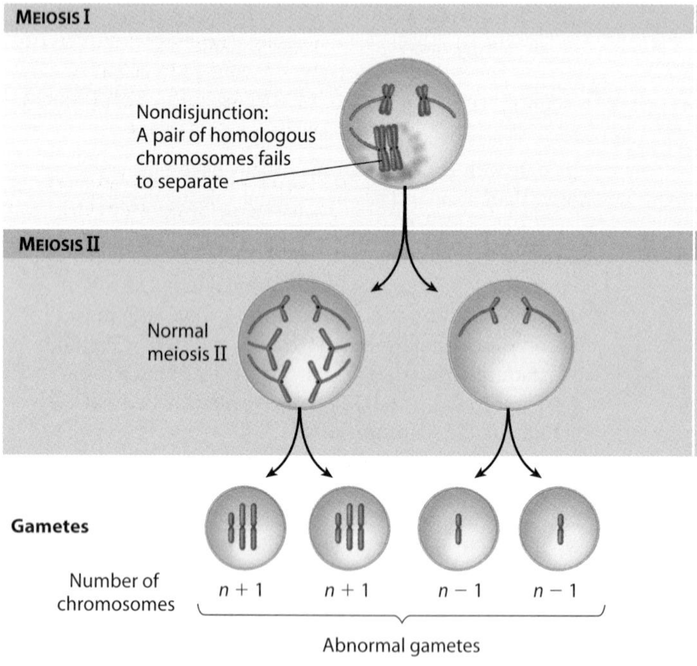

▲ **Figure 20A** Nondisjunction in meiosis I

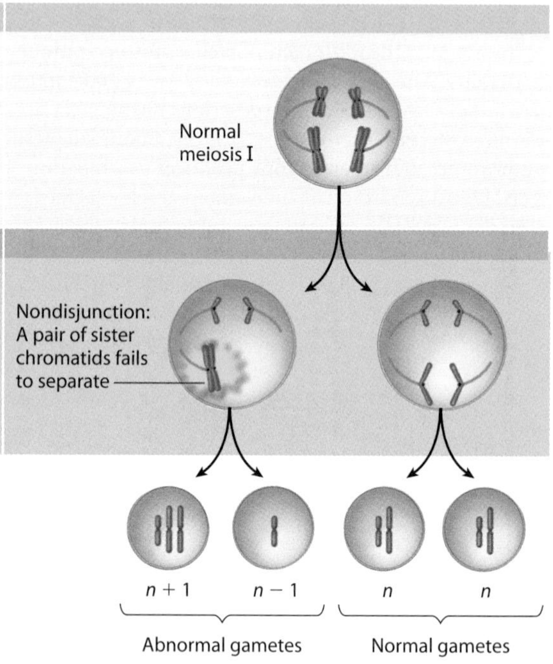

▲ **Figure 20B** Nondisjunction in meiosis II

CONNECTION

21 Abnormal numbers of sex chromosomes do not usually affect survival

Nondisjunction can result in abnormal numbers of sex chromosomes, X and Y. Unusual numbers of sex chromosomes seem to upset the genetic balance less than unusual numbers of autosomes. This may be because the Y chromosome is very small and carries relatively few genes. Furthermore, mammalian cells usually operate with only one functioning X chromosome because other copies of the chromosome become inactivated in each cell.

TABLE 21 ABNORMALITIES OF SEX CHROMOSOME NUMBER IN HUMANS

Sex Chromosomes	Syndrome	Origin of Nondisjunction	Frequency in Population
XXY	Klinefelter syndrome (male)	Meiosis in egg or sperm formation	$\frac{1}{2{,}000}$
XYY	None (normal male)	Meiosis in sperm formation	$\frac{1}{2{,}000}$
XXX	None (normal female)	Meiosis in egg or sperm formation	$\frac{1}{1{,}000}$
XO	Turner syndrome (female)	Meiosis in egg or sperm formation	$\frac{1}{5{,}000}$

Table 21 lists the most common human sex chromosome abnormalities. An extra X chromosome in a male, making him XXY, occurs approximately once in every 2,000 live births (once in every 1,000 male births). Men with this disorder, called Klinefelter syndrome, have male sex organs and normal intelligence, but the testes are abnormally small, the individual is sterile, and he often has breast enlargement and other female body characteristics. Klinefelter syndrome is also found in individuals with more than three sex chromosomes, such as XXYY, XXXY, or XXXXY. These abnormal numbers of sex chromosomes result from multiple nondisjunctions; such men are more likely to have developmental disabilities than XY or XXY individuals.

Human males with an extra Y chromosome (XYY) do not have any well-defined syndrome, although they tend to be taller than average. Females with an extra X chromosome (XXX) cannot be distinguished from XX females except by karyotype.

Females who lack an X chromosome are designated XO; the O indicates the absence of a second sex chromosome. These women have Turner syndrome. They have a characteristic appearance, including short stature and often a web of skin extending between the neck and the shoulders. Women with Turner syndrome are sterile because their sex organs do not fully mature at adolescence. If left untreated, girls with Turner syndrome have poor development of breasts and other secondary sexual characteristics. Artificial administration of estrogen can alleviate these symptoms. Women with Turner syndrome have normal intelligence. The XO condition is the sole known case where having only 45 chromosomes is not fatal in humans.

The sex chromosome abnormalities described here illustrate the crucial role of the Y chromosome in determining sex. In general, a single Y chromosome is enough to produce "maleness," even in combination with several X chromosomes. The absence of a Y chromosome yields "femaleness."

? What is the total number of autosomes you would expect to find in the karyotype of a female with Turner syndrome?

44 (plus one sex chromosome)

EVOLUTION CONNECTION

22 New species can arise from errors in cell division

Errors in meiosis or mitosis do not always lead to problems. In fact, biologists hypothesize that such errors have been instrumental in the evolution of many species. Numerous plant species, in particular, seem to have originated from accidents during cell division that resulted in extra sets of chromosomes. The new species is polyploid, meaning that it has more than two sets of homologous chromosomes in each somatic cell. At least half of all species of flowering plants are polyploid, including such useful ones as wheat, potatoes, apples, and cotton.

Let's consider one scenario by which a diploid ($2n$) plant species might generate a tetraploid ($4n$) plant. Imagine that, like many plants, our diploid plant produces both sperm and egg cells and can self-fertilize. If meiosis fails to occur in the plant's reproductive organs and gametes are instead produced by mitosis, the gametes will be diploid. The union of a diploid ($2n$) sperm with a diploid ($2n$) egg during self-fertilization will produce a tetraploid ($4n$) zygote, which may develop into a mature tetraploid plant that can itself reproduce by self-fertilization. The tetraploid plants will constitute a new species, one that has evolved in just one generation.

Although polyploid animal species are less common than polyploid plants, they are known to occur among the fishes and amphibians (**Figure 22**). Moreover, researchers in Chile have identified the first candidate for polyploidy among the mammals—a rat whose cells seem to be tetraploid. Tetraploid organisms are sometimes strikingly different from their recent diploid ancestors; they may be larger, for example. Scientists don't yet understand exactly how polyploidy brings about such differences.

? What is a polyploid organism?

One with more than two sets of homologous chromosomes in its body cells.

▶ **Figure 22** The gray tree frog (*Hyla versicolor*), a tetraploid organism

Michelle Gilders/Alamy

CONNECTION

23 Alterations of chromosome structure can cause birth defects and cancer

Even if all chromosomes are present in normal numbers, abnormalities in chromosome structure may cause disorders. Breakage of a chromosome can lead to a variety of rearrangements affecting the genes of that chromosome (Figure 23A). If a fragment of a chromosome is lost, the remaining chromosome will then have a **deletion**. If a fragment from one chromosome joins to a sister chromatid or homologous chromosome, it will produce a **duplication**. If a fragment reattaches to the original chromosome but in the reverse orientation, an **inversion** results.

Inversions are less likely than deletions or duplications to produce harmful effects, because in inversions all genes are still present in their normal number. Many deletions in human chromosomes, however, cause serious physical and mental problems. One example is a specific deletion in chromosome 5 that causes *cri du chat* ("cry of the cat") syndrome. A child born with this syndrome is mentally retarded, has a small head with unusual facial features, and has a cry that sounds like the mewing of a distressed cat. Such individuals usually die in infancy or early childhood.

Another type of chromosomal change is chromosomal **translocation**, the attachment of a chromosomal fragment to a nonhomologous chromosome. As shown in the figure, a translocation may be reciprocal; that is, two nonhomologous chromosomes may exchange segments. Like inversions, translocations may or may not be harmful. Some people with Down syndrome have only part of a third chromosome 21; as the result of a translocation, this partial chromosome is attached to another (nonhomologous) chromosome.

Whereas chromosomal changes present in sperm or egg can cause congenital disorders, such changes in a somatic cell may contribute to the development of cancer. For example, a chromosomal translocation in somatic cells in the bone marrow is associated with chronic myelogenous leukemia (CML). CML is one of the most common types of leukemia. (Leukemias are cancers affecting cells that give rise to white blood cells, or leukocytes.) In the cancerous cells of most CML patients, a part of chromosome 22 has switched places with a small fragment from a tip of chromosome 9 (Figure 23B). This reciprocal translocation creates a hybrid gene that codes for an abnormal protein. This protein stimulates cell division, leading to leukemia. The chromosome ending up with the activated cancer-causing gene is called the "Philadelphia chromosome," after the city where it was discovered.

Because the chromosomal changes in cancer are usually confined to somatic cells, cancer is not usually inherited.

? How is reciprocal translocation different from normal crossing over?

● Reciprocal translocation swaps chromosome segments between nonhomologous chromosomes. Crossing over normally exchanges corresponding segments between homologous chromosomes.

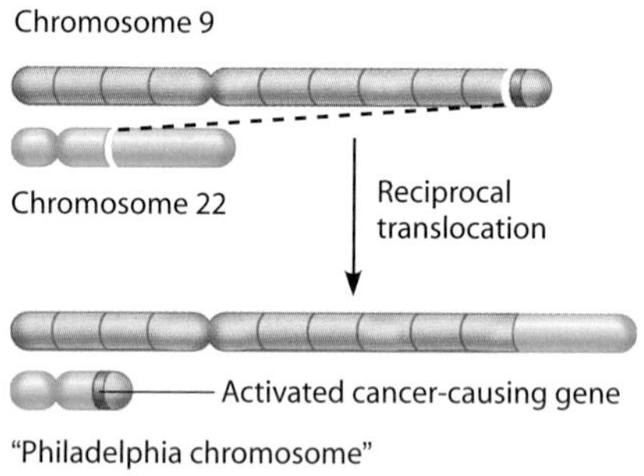
Chromosome 9
Chromosome 22
Reciprocal translocation
Activated cancer-causing gene
"Philadelphia chromosome"

▲ **Figure 23B** The translocation associated with chronic myelogenous leukemia

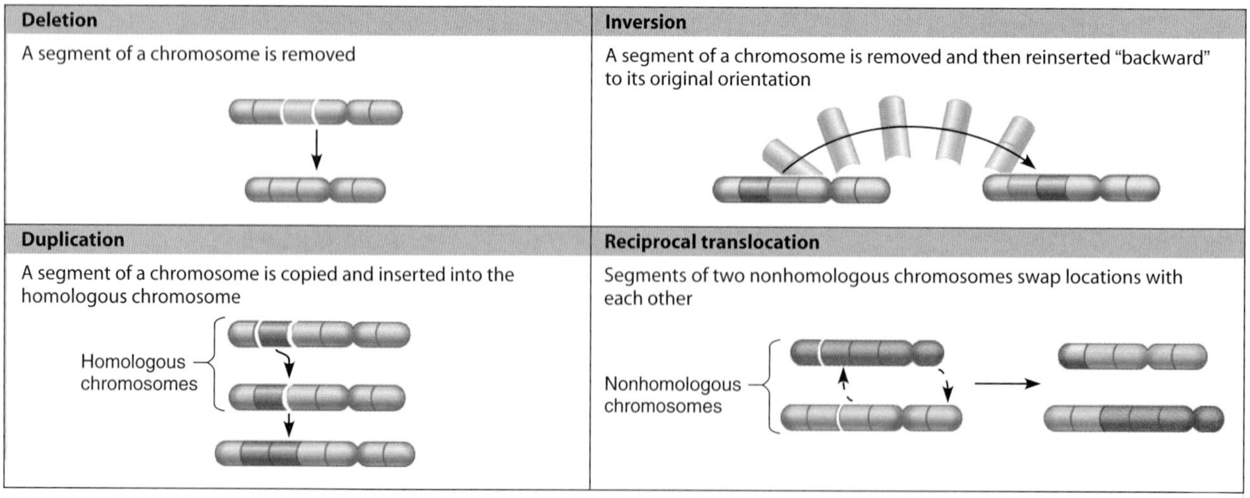
Deletion
A segment of a chromosome is removed
Inversion
A segment of a chromosome is removed and then reinserted "backward" to its original orientation
Duplication
A segment of a chromosome is copied and inserted into the homologous chromosome
Homologous chromosomes
Reciprocal translocation
Segments of two nonhomologous chromosomes swap locations with each other
Nonhomologous chromosomes

▲ **Figure 23A** Alterations of chromosome structure

REVIEW

For Practice Quizzes, BioFlix, MP3 Tutors, and Activities, go to www.masteringbiology.com.

Reviewing the Concepts

Cell Division and Reproduction (1–2)

1 Cell division plays many important roles in the lives of organisms. Cell division is at the heart of the reproduction of cells and organisms because cells come only from preexisting cells. Some organisms reproduce through asexual reproduction, and their offspring are all genetic copies of the parent and of each other. Others reproduce through sexual reproduction, creating a variety of offspring, each with a unique combination of traits.

2 Prokaryotes reproduce by binary fission. Prokaryotic cells reproduce asexually by cell division. As the cell replicates its single chromosome, the copies move apart; the growing membrane then divides the cell.

The Eukaryotic Cell Cycle and Mitosis (3–10)

3 The large, complex chromosomes of eukaryotes duplicate with each cell division. A eukaryotic cell has many more genes than a prokaryotic cell, and they are grouped into multiple chromosomes in the nucleus. Each chromosome contains one very long DNA molecule associated with proteins. Individual chromosomes are visible only when the cell is in the process of dividing; otherwise, they are in the form of thin, loosely packed chromatin fibers. Before a cell starts dividing, the chromosomes duplicate, producing sister chromatids (containing identical DNA) that are joined together. Cell division involves the separation of sister chromatids and results in two daughter cells, each containing a complete and identical set of chromosomes.

4 The cell cycle multiplies cells.

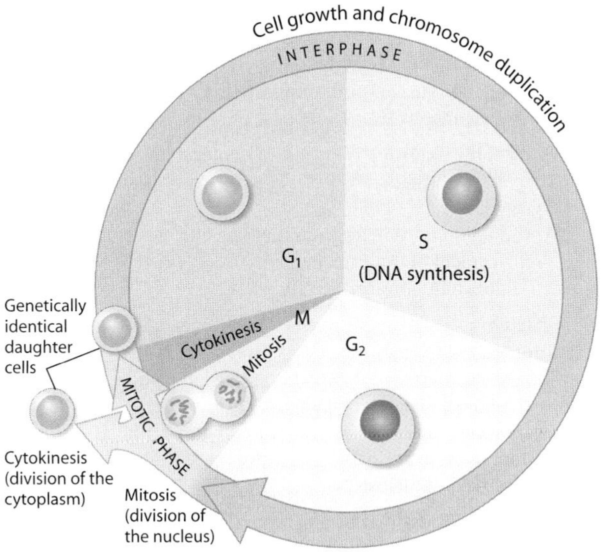

5 Cell division is a continuum of dynamic changes. Mitosis distributes duplicated chromosomes into two daughter nuclei. After the chromosomes are coiled up, a mitotic spindle made of microtubules moves them to the middle of the cell. The sister chromatids then separate and move to opposite poles of the cell, where two new nuclei form. Cytokinesis, in which the cell divides in two, overlaps the end of mitosis.

6 Cytokinesis differs for plant and animal cells. In animals, cytokinesis occurs by a constriction of the cell (cleavage). In plants, a membranous cell plate splits the cell in two.

7 Anchorage, cell density, and chemical growth factors affect cell division. Most animal cells divide only when stimulated by growth factors, and some not at all. Growth factors are proteins secreted by cells that stimulate other cells to divide. In laboratory cultures, most normal cells divide only when attached to a surface. They continue dividing until they touch one another.

8 Growth factors signal the cell cycle control system. A set of proteins within the cell controls the cell cycle. Signals affecting critical checkpoints in the cell cycle determine whether a cell will go through the complete cycle and divide. The binding of growth factors to specific receptors on the plasma membrane is usually necessary for cell division.

9 Growing out of control, cancer cells produce malignant tumors. Cancer cells divide excessively to form masses called tumors. Malignant tumors can invade other tissues. Radiation and chemotherapy are effective as cancer treatments because they interfere with cell division.

10 Review: Mitosis provides for growth, cell replacement, and asexual reproduction. When the cell cycle operates normally, mitotic cell division functions in growth, replacement of damaged and lost cells, and asexual reproduction.

Meiosis and Crossing Over (11–17)

11 Chromosomes are matched in homologous pairs. The somatic (body) cells of each species contain a specific number of chromosomes; for example, human cells have 46, consisting of 23 pairs (two sets) of homologous chromosomes. The chromosomes of a homologous pair of autosomes carry genes for the same characteristics at the same place, or locus. X and Y chromosomes are only partly homologous.

12 Gametes have a single set of chromosomes. Cells with two sets of homologous chromosomes are diploid. Gametes—eggs and sperm—are haploid cells with a single set of chromosomes. Sexual life cycles involve the alternation of haploid and diploid stages.

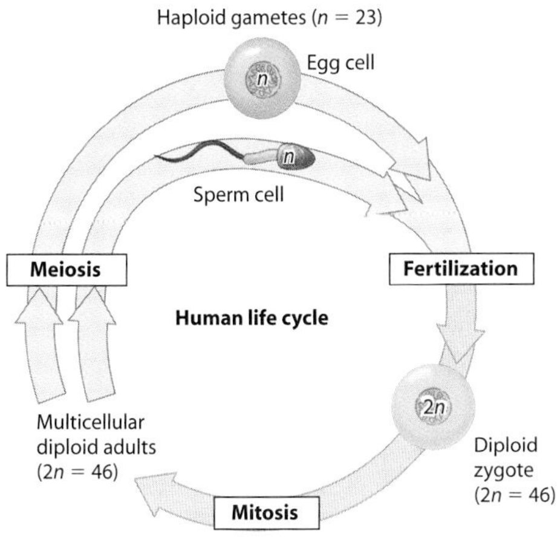

13 Meiosis reduces the chromosome number from diploid to haploid. Meiosis, like mitosis, is preceded by chromosome duplication, but in meiosis, the cell divides twice to form four daughter cells. The first division, meiosis I, starts with synapsis, the pairing of homologous chromosomes. In crossing over, homologous chromosomes exchange corresponding segments. Meiosis I separates the members of each homologous pair and produces two daughter cells, each with one set of chromosomes. Meiosis II is essentially the same as mitosis: In each of the cells, the sister chromatids of each chromosome separate. The result is a total of four haploid cells.

14 Mitosis and meiosis have important similarities and differences. Both mitosis and meiosis begin with diploid parent cells that have chromosomes duplicated during the previous interphase. But mitosis produces two genetically identical diploid somatic daughter cells, while meiosis produces four genetically unique haploid gametes.

15 Independent orientation of chromosomes in meiosis and random fertilization lead to varied offspring. Each chromosome of a homologous pair differs at many points from the other member of the pair. Random arrangements of chromosome pairs at metaphase I of meiosis leads to many different combinations of chromosomes in eggs and sperm. Random fertilization of eggs by sperm greatly increases this variation.

16 Homologous chromosomes may carry different versions of genes. The differences between homologous chromosomes are based on the fact that they can bear different versions of genes at corresponding loci.

17 Crossing over further increases genetic variability. Genetic recombination, which results from crossing over during prophase I of meiosis, increases variation still further.

Alterations of Chromosome Number and Structure (18–23)

18 A karyotype is a photographic inventory of an individual's chromosomes. To prepare a karyotype, white blood cells are isolated, stimulated to grow, arrested at metaphase, and photographed under a microscope. The chromosomes are arranged into ordered pairs so that any chromosomal abnormalities can be detected.

19 An extra copy of chromosome 21 causes Down syndrome. Trisomy 21, the most common chromosome number abnormality in the United States, results in a condition called Down syndrome.

20 Accidents during meiosis can alter chromosome number. An abnormal chromosome count is the result of nondisjunction, which can result from the failure of a pair of homologous chromosomes to separate during meiosis I or from the failure of sister chromatids to separate during meiosis II.

21 Abnormal numbers of sex chromosomes do not usually affect survival. Nondisjunction of the sex chromosomes during meiosis can result in individuals with a missing or extra X or Y chromosome. In some cases (such as XXY), this leads to syndromes; in other cases (such as XXX), the body is normal.

22 New species can arise from errors in cell division. Nondisjunction can produce polyploid organisms, ones with extra sets of chromosomes. Such errors in mitosis can be important in the evolution of new species.

23 Alterations of chromosome structure can cause birth defects and cancer. Chromosome breakage can lead to rearrangements—deletions, duplications, inversions, and translocations—that can produce genetic disorders or, if the changes occur in somatic cells, cancer.

Connecting the Concepts

1. Complete the following table to compare mitosis and meiosis.

	Mitosis	Meiosis
Number of chromosomal duplications		
Number of cell divisions		
Number of daughter cells produced		
Number of chromosomes in the daughter cells		
How the chromosomes line up during metaphase		
Genetic relationship of the daughter cells to the parent cell		
Functions performed in the human body		

Testing Your Knowledge

Multiple Choice

2. If an intestinal cell in a grasshopper contains 24 chromosomes, then a grasshopper sperm cell contains _______ chromosomes.
   a. 3
   b. 6
   c. 12
   d. 24
   e. 48

3. Which of the following phases of mitosis is essentially the opposite of prophase in terms of nuclear changes?
   a. telophase
   b. metaphase
   c. S phase
   d. interphase
   e. anaphase

4. A biochemist measured the amount of DNA in cells growing in the laboratory and found that the quantity of DNA in a cell doubled
   a. between prophase and anaphase of mitosis.
   b. between the G_1 and G_2 phases of the cell cycle.
   c. during the M phase of the cell cycle.
   d. between prophase I and prophase II of meiosis.
   e. between anaphase and telophase of mitosis.

5. Which of the following is *not* a function of mitosis in humans?
   a. repair of wounds
   b. growth
   c. production of gametes from diploid cells
   d. replacement of lost or damaged cells
   e. multiplication of somatic cells

6. A micrograph of a dividing cell from a mouse showed 19 chromosomes, each consisting of two sister chromatids. During which of the following stages of cell division could such a picture have been taken? (*Explain your answer.*)
   a. prophase of mitosis
   b. telophase II of meiosis
   c. prophase I of meiosis
   d. anaphase of mitosis
   e. prophase II of meiosis

7. Cytochalasin B is a chemical that disrupts microfilament formation. This chemical would interfere with
   a. DNA replication.
   b. formation of the mitotic spindle.
   c. cleavage.
   d. formation of the cell plate.
   e. crossing over.

8. It is difficult to observe individual chromosomes during interphase because
   a. the DNA has not been replicated yet.
   b. they are in the form of long, thin strands.
   c. they leave the nucleus and are dispersed to other parts of the cell.
   d. homologous chromosomes do not pair up until division starts.
   e. the spindle must move them to the metaphase plate before they become visible.
9. A fruit fly somatic cell contains 8 chromosomes. This means that ______ different combinations of chromosomes are possible in its gametes.
   a. 4
   b. 8
   c. 16
   d. 32
   e. 64
10. If a fragment of a chromosome breaks off and then reattaches to the original chromosome but in the reverse direction, the resulting chromosomal abnormality is called
    a. a deletion.
    b. an inversion.
    c. a translocation.
    d. a nondisjunction.
    e. a reciprocal translocation.
11. Why are individuals with an extra chromosome 21, which causes Down syndrome, more numerous than individuals with an extra chromosome 3 or chromosome 16?
    a. There are probably more genes on chromosome 21 than on the others.
    b. Chromosome 21 is a sex chromosome and chromosomes 3 and 16 are not.
    c. Down syndrome is not more common, just more serious.
    d. Extra copies of the other chromosomes are probably fatal.
    e. Chromosome 21 is more likely to produce a nondisjunction error than other chromosomes.

Describing, Comparing, and Explaining

12. An organism called a plasmodial slime mold is one large cytoplasmic mass with many nuclei. Explain how such a "megacell" could form.
13. Briefly describe how three different processes that occur during a sexual life cycle increase the genetic diversity of offspring.
14. Discuss the factors that control the division of eukaryotic cells grown in the laboratory. Cancer cells are easier to grow in the lab than other cells. Why do you suppose this is?
15. In the light micrograph below of dividing cells near the tip of an onion root, identify a cell in interphase, prophase, metaphase, anaphase, and telophase. Describe the major events occurring at each stage.

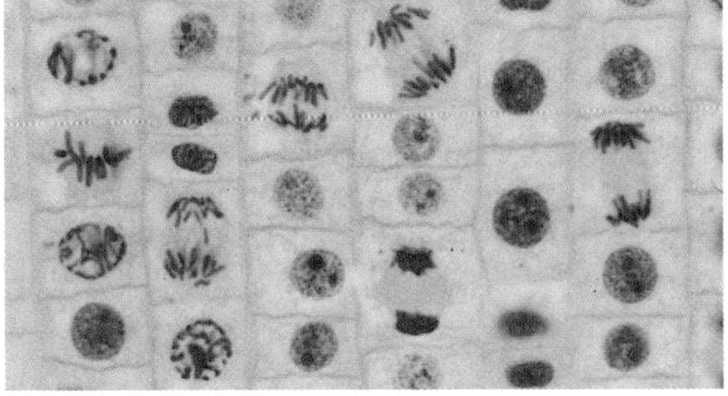

USDA/ARS/Agricultural Research Service

LM 500×

16. Compare cytokinesis in plant and animal cells. In what ways are the two processes similar? In what ways are they different?
17. Sketch a cell with three pairs of chromosomes undergoing meiosis, and show how nondisjunction can result in the production of gametes with extra or missing chromosomes.

Applying the Concepts

18. Suppose you read in the newspaper that a genetic engineering laboratory has developed a procedure for fusing two gametes from the same person (two eggs or two sperm) to form a zygote. The article mentions that an early step in the procedure prevents crossing over from occurring during the formation of the gametes in the donor's body. The researchers are in the process of determining the genetic makeup of one of their new zygotes. Which of the following predictions do you think they would make? Justify your choice, and explain why you rejected each of the other choices.
    a. The zygote would have 46 chromosomes, all of which came from the gamete donor (its one parent), so the zygote would be genetically identical to the gamete donor.
    b. The zygote *could* be genetically identical to the gamete donor, but it is much more likely that it would have an unpredictable mixture of chromosomes from the gamete donor's parents.
    c. The zygote would not be genetically identical to the gamete donor, but it would be genetically identical to one of the donor's parents.
    d. The zygote would not be genetically identical to the gamete donor, but it would be genetically identical to one of the donor's grandparents.
19. Bacteria are able to divide on a much faster schedule than eukaryotic cells. Some bacteria can divide every 20 minutes, while the minimum time required by eukaryotic cells in a rapidly developing embryo is about once per hour, and most cells divide much less often than that. State several testable hypotheses explaining why bacteria can divide at a faster rate than eukaryotic cells.
20. Red blood cells, which carry oxygen to body tissues, live for only about 120 days. Replacement cells are produced by cell division in bone marrow. How many cell divisions must occur each second in your bone marrow just to replace red blood cells? Here is some information to use in calculating your answer: There are about 5 million red blood cells per cubic millimeter (mm^3) of blood. An average adult has about 5 L (5,000 cm^3) of blood. (*Hint*: What is the total number of red blood cells in the body? What fraction of them must be replaced each day if all are replaced in 120 days?)
21. A mule is the offspring of a horse and a donkey. A donkey sperm contains 31 chromosomes and a horse egg cell 32 chromosomes, so the zygote contains a total of 63 chromosomes. The zygote develops normally. The combined set of chromosomes is not a problem in mitosis, and the mule combines some of the best characteristics of horses and donkeys. However, a mule is sterile; meiosis cannot occur normally in its testes (or ovaries). Explain why mitosis is normal in cells containing both horse and donkey chromosomes but the mixed set of chromosomes interferes with meiosis.

Review Answers

1.

	Mitosis	Meiosis
Number of chromosomal duplications	*1*	*1*
Number of cell divisions	*1*	*2*
Number of daughter cells produced	*2*	*4*
Number of chromosomes in the daughter cells	*Diploid (2n)*	*Haploid (n)*
How the chromosomes line up during metaphase	*Singly*	*In tetrads (metaphase I), then singly (metaphase II)*
Genetic relationship of the daughter cells to the parent cell	*Genetically identical*	*Genetically unique*
Functions performed in the human body	*Growth, development, and repair*	*Production of gametes*

2. c 3. a 4. b 5. c 6. e (A diploid cell would have an even number of chromosomes; the odd number suggests that meiosis I has been completed. Sister chromatids are together only in prophase and metaphase of meiosis II.) 7. c 8. b 9. c 10. b 11. d

12. Mitosis without cytokinesis would result in a single cell with two nuclei. Multiple rounds of cell division like this could produce such a "megacell."

13. Various orientations of chromosomes at metaphase I of meiosis lead to different combinations of chromosomes in gametes. Crossing over during prophase I results in an exchange of chromosome segments and new combinations of genes. Random fertilization of eggs by sperm further increases possibilities for variation in offspring.

14. In culture, normal cells usually divide only when they are in contact with a surface but not touching other cells on all sides (the cells usually grow to form only a single layer). The density-dependent inhibition of cell division apparently results from local depletion of substances called growth factors. Growth factors are proteins secreted by certain cells that stimulate other cells to divide; they act via signal transduction pathways to signal the cell cycle control system of the affected cell to proceed past its checkpoints. The cell cycle control systems of cancer cells do not function properly. Cancer cells generally do not require externally supplied growth factors to complete the cell cycle, and they divide indefinitely (in contrast to normal mammalian cells, which stop dividing after 20 to 50 generations)—two reasons why they are relatively easy to grow in the lab. Furthermore, cancer cells can often grow without contacting a solid surface, making it possible to culture them in suspension in a liquid medium.

15. Interphase (for example, third column from left in micrograph, third cell from top): Growth; metabolic activity; DNA synthesis. Prophase (for example, second column, cell at bottom): Chromosomes shorten and thicken; mitotic spindle forms. Metaphase (for example, first column, middle cell): Chromosomes line up on a plane going through the cell's equator. Anaphase (for example, third column, second cell from top): Sister chromatids separate and move to the poles of the cell. Telophase (for example, fourth column, fourth complete cell from top): Daughter nuclei form around chromosomes; cytokinesis usually occurs.

16. A ring of microfilaments pinches an animal cell in two, a process called cleavage. In a plant cell, membranous vesicles form a disk called the cell plate at the midline of the parent cell, cell plate membranes fuse with the plasma membrane, and a cell wall grows in the space, separating the daughter cells.

17. See Figures 20A and 20B.

18. a. No. For this to happen, the chromosomes of the two gametes that fused would have to represent, together, a complete set of the donor's maternal chromosomes (the ones that originally came from the donor's mother) and a complete set of the donor's paternal chromosomes (from the donor's father). It is much more likely that the zygote would be missing one or more maternal chromosomes and would have an excess of paternal chromosomes, or vice versa.
   b. Correct. Consider what would have to happen to produce a zygote genetically identical to the gamete donor: The zygote would have to have a complete set of the donor's maternal chromosomes and a complete set of the donor's paternal chromosomes. The first gamete in this union could contain any mixture of maternal and paternal chromosomes, but once that first gamete was "chosen," the second one would have to have one particular combination of chromosomes—the combination that supplies whatever the first gamete did not supply. So, for example, if the first three chromosomes of the first gamete were maternal, maternal, and paternal, the first three of the second gamete would have to be paternal, paternal, and maternal. The chance that all 23 chromosome pairs would be complementary in this way is only one in 22^3 (that is, one in 8,388,608). Because of independent assortment, it is much more likely that the zygote would have an unpredictable combination of chromosomes from the donor's father and mother.
   c. No. First, the zygote could not be genetically identical to the gamete donor (see b). Second, the zygote could not be identical to either of the gamete donor's parents because the donor only has half the genetic material of each of his or her parents. For example, even if the zygote were formed by two gametes containing only paternal chromosomes, the combined set of chromosomes could not be identical to that of the donor's father because it would still be missing half of the father's chromosomes.
   d. No. See answer c.

19. Some possible hypotheses: The replication of the DNA of the bacterial chromosome takes less time than the replication of the DNA in a eukaryotic cell. The time required for a growing bacterium to roughly double its cytoplasm is much less than for a eukaryotic cell. Bacteria have a cell cycle control system much simpler than that of eukaryotes.

20. 1 cm^3 = 1,000 mm^3, so 5,000 mm^3 of blood contains $5{,}000 \times 1{,}000 \times 5{,}000{,}000 = 25{,}000{,}000{,}000{,}000$, or 2.5×10^{13}, red blood cells. The number of cells replaced each day = $2.5 \times 10^{13}/120 = 2.1 \times 10^{11}$ cells. There are $24 \times 60 \times 60 = 86{,}400$ seconds in a day. Therefore, the number of cells replaced each second = $2.1 \times 10^{11}/86{,}400$ = about 2×10^6, or 2 million. Thus, about 2 million cell divisions must occur each second to replace red blood cells that are lost.

21. Each chromosome is on its own in mitosis; chromosome replication and the separation of sister chromatids occur independently for each horse or donkey chromosome. Therefore, mitotic divisions, starting with the zygote, are not impaired. In meiosis, however, homologous chromosomes must pair in prophase I. This process of synapsis cannot occur properly because horse and donkey chromosomes do not match in number or content.

Patterns of Inheritance

From Chapter 9 of *Campbell Biology: Concepts & Connections*, Seventh Edition. Jane B. Reece, Martha R. Taylor, Eric J. Simon, Jean L. Dickey.

Patterns of Inheritance

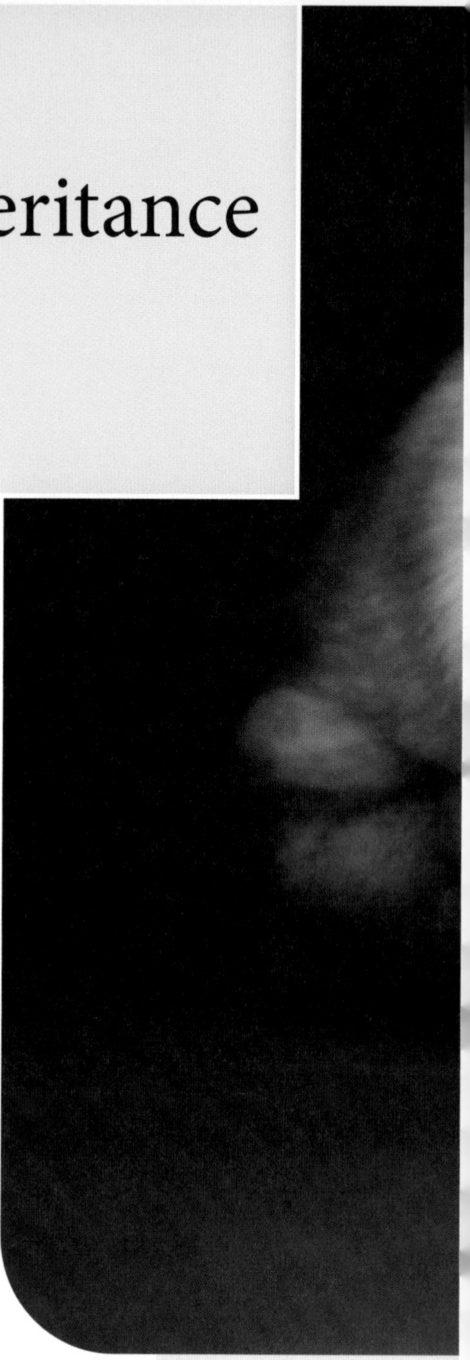

BIG IDEAS

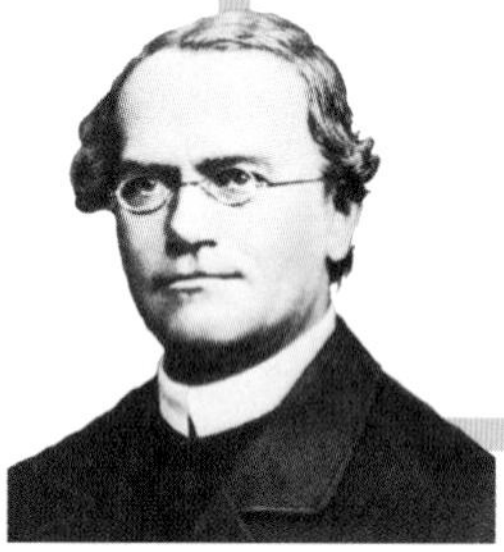
Hulton Archive/Getty Images

Mendel's Laws (1–10)

A few simple and long-established rules explain many aspects of heredity.

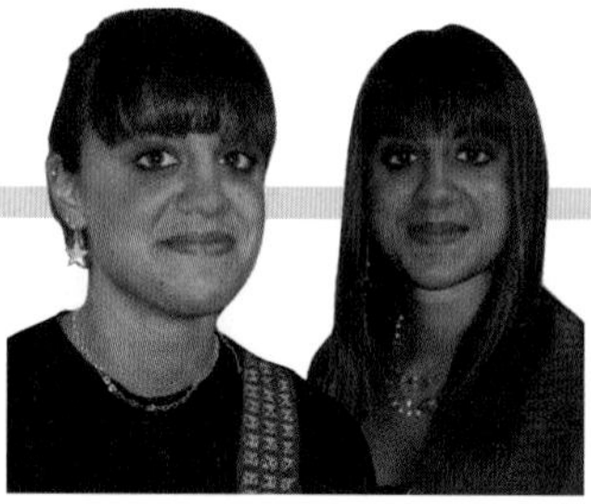
Eric J. Simon

Variations on Mendel's Laws (11–15)

Some inheritance patterns are more complex than the ones described by Mendel.

The Chromosomal Basis of Inheritance (16–19)

Hereditary rules can be understood by following the behavior of chromosomes.

Andrew Syred/Photo Researchers

Sex Chromosomes and Sex-Linked Genes (20–23)

Genes found on sex chromosomes display unique patterns of inheritance.

Charles Orrico/SuperStock

The cute canine shown above is a purebred Labrador retriever. Displaying the traits that distinguish this breed—such as a narrow muzzle and gentle eyes—this dog, when bred to other Labradors, is expected to produce puppies that look very similar. This is a reasonable expectation because a purebred dog has a well-documented pedigree that includes several generations of ancestors with similar genetic makeup and appearance. But similarities among purebred Labs extend beyond mere appearance; Labradors are generally kind, outgoing, intelligent, and docile. Such behavioral similarities suggests that breeders can select for temperament as well as physical traits.

Purebred pooches are living proof that dogs are more than man's best friend—they are also one of our longest-running genetic experiments. For thousands of years, different groups of people have chosen and mated dogs with specific traits. Continued over millennia, the result of such genetic tinkering is an incredibly diverse array of dog body types and behaviors.

Although humans have been applying genetics for thousands of years—by breeding food crops as well as domesticated animals—the biological basis of selective breeding has only recently been understood. In this chapter, we examine the rules that govern how inherited traits are passed from parents to offspring. We look at several kinds of inheritance patterns and how they allow us to predict the ratios of offspring with particular traits. Most importantly, we uncover a basic biological concept: how the behavior of chromosomes during gamete formation and fertilization accounts for the patterns of inheritance we observe. Along the way, we'll return several times to the Labrador retriever as an example of how basic genetic principles can help us understand the world around us.

Mendel's Laws

1 The science of genetics has ancient roots

Attempts to explain inheritance date back at least to ancient Greece. The physician Hippocrates (approximately 460–370 BCE) suggested an explanation called pangenesis. According to this idea, particles called pangenes travel from each part of an organism's body to the eggs or sperm and then are passed to the next generation; moreover, changes that occur in the body during an organism's life are passed on in this way. The Greek philosopher Aristotle (384–322 BCE; **Figure 1**) rejected this idea as simplistic, saying that what is inherited is the potential to produce body features rather than particles of the features themselves.

Actually, pangenesis proves incorrect in several respects. The reproductive cells are not composed of particles from somatic (body) cells, and changes in somatic cells do not influence eggs and sperm. For instance, no matter how much you enlarge your biceps by lifting weights, muscle cells in your arms do not transmit genetic information to your gametes, and your offspring will not be changed by your weight-lifting efforts. This may seem like common sense today, but the pangenesis hypothesis and the idea that traits acquired during an individual's lifetime are passed on to offspring prevailed well into the 19th century.

frog-traveller/Shutterstock

▲ **Figure 1** Aristotle

By observing inheritance patterns in ornamental plants, biologists of the early 19th century established that offspring inherit traits from both parents. The favored explanation of inheritance then became the "blending" hypothesis, the idea that the hereditary materials contributed by the male and female parents mix in forming the offspring similar to the way that blue and yellow paints blend to make green. For example, according to this hypothesis, after the genetic information for the colors of black and chocolate brown Labrador retrievers is blended, the colors should be as inseparable as paint pigments. But this is not what happens: Instead, the offspring of a purebred black Lab and a purebred brown Lab will all be black, but some of the dogs in the next generation will be brown (you'll learn why in Module 6). The blending hypothesis was finally rejected because it does not explain how traits that disappear in one generation can reappear in later ones.

? Horse breeders sometimes speak of "mixing the bloodlines" of two pedigrees. In what way is this phrase inaccurate?

It implies the blending hypothesis—that offspring are a blend of two parents, as in a liquid mixture.

2 Experimental genetics began in an abbey garden

Heredity is the transmission of traits from one generation to the next. The field of **genetics**, the scientific study of heredity, began in the 1860s, when an Augustinian monk named Gregor Mendel **(Figure 2A)** deduced the fundamental principles of genetics by breeding garden peas. Mendel lived and worked in an abbey in Brunn, Austria (now Brno, in the Czech Republic). Strongly influenced by his study of physics, mathematics, and chemistry at the University of Vienna, his research was both experimentally and mathematically rigorous, and these qualities were largely responsible for his success.

Hulton Archive/Getty Images

▲ **Figure 2A** Gregor Mendel

In a paper published in 1866, Mendel correctly argued that parents pass on to their offspring discrete "heritable factors." (It is interesting to note that Mendel's landmark publication came just seven years after Darwin's 1859 publication of *The Origin of Species*, making the 1860s a banner decade in the development of modern biology.) In his paper, Mendel stressed that the heritable factors, today called genes, retain their individuality generation after generation. That is, genes are like playing cards; a deck may be shuffled, but the cards always retain their original identities, and no card is ever blended with another. Similarly, genes may be rearranged but each gene permanently retains its identity.

Mendel probably chose to study garden peas because they had short generation times, produced large numbers of offspring from each mating, and came in many readily distinguishable varieties. For example, one variety has purple flowers, and another variety has white flowers. A heritable feature that varies among individuals, such as flower color, is called a **character**. Each variant for a character, such as purple or white flowers, is called a **trait**.

Perhaps the most important advantage of pea plants as an experimental model was that Mendel could strictly control matings. As **Figure 2B** shows, the petals

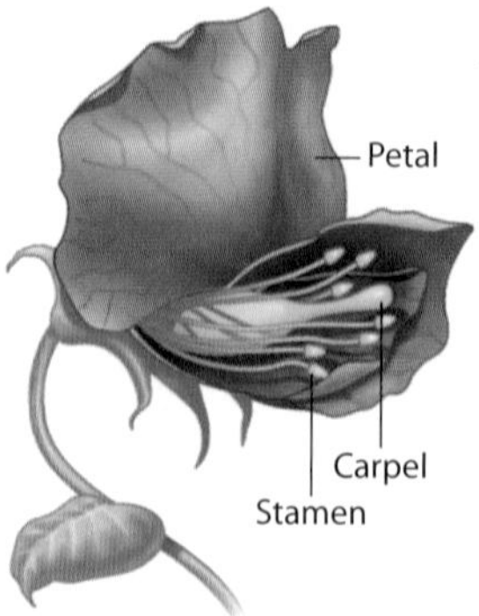

◀ **Figure 2B** The anatomy of a garden pea flower (with one petal removed to improve visibility)

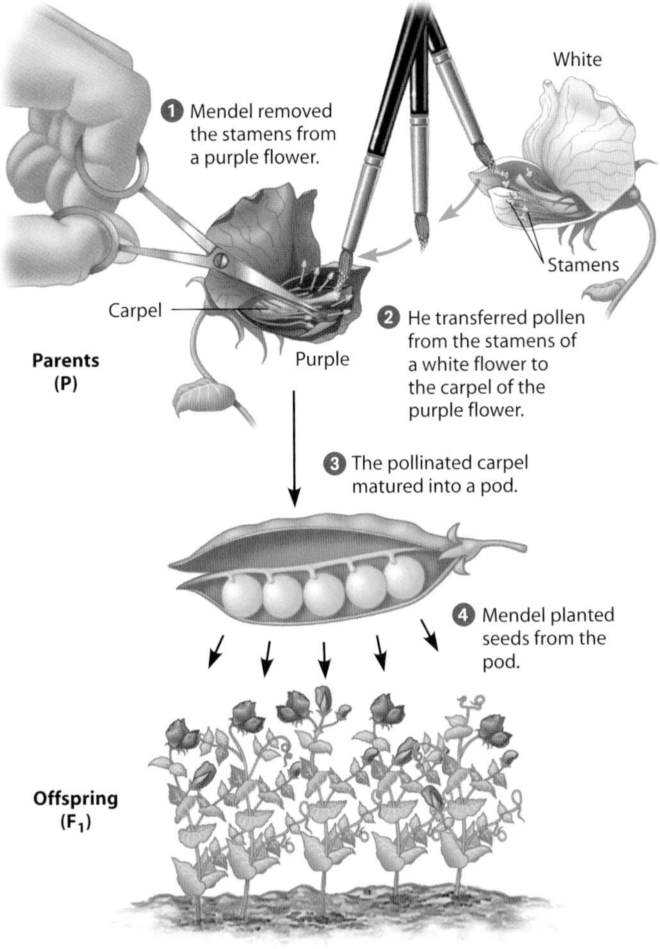

▲ **Figure 2C** Mendel's technique for cross-fertilization of pea plants

of the pea flower almost completely enclose the reproductive organs: the stamens and carpel. Consequently, pea plants usually are able to **self-fertilize** in nature. That is, sperm-carrying pollen grains released from the stamens land on the egg-containing carpel of the same flower. Mendel could ensure self-fertilization by covering a flower with a small bag so that no pollen from another plant could reach the carpel. When he wanted **cross-fertilization** (fertilization of one plant by pollen from a different plant), he used the method shown in **Figure 2C**. ❶ He prevented self-fertilization by cutting off the immature stamens of a plant before they produced pollen. ❷ To cross-fertilize the stamenless flower, he dusted its carpel with pollen from another plant. After pollination, ❸ the carpel developed into a pod, containing seeds (peas) that ❹ he planted. The seeds grew into offspring plants. Through these methods, Mendel could always be sure of the parentage of new plants.

Mendel's success was due not only to his experimental approach and choice of organism but also to his selection of characteristics to study. He chose to observe seven characters, each of which occurred as two distinct traits **(Figure 2D)**. Mendel worked with his plants until he was sure he had **true-breeding** varieties—that is, varieties for which self-fertilization produced offspring all identical to the parent. For instance, he identified a purple-flowered variety that, when self-fertilized, produced offspring plants that all had purple flowers.

Now Mendel was ready to ask what would happen when he crossed his different true-breeding varieties with each other. For example, what offspring would result if plants with purple flowers and plants with white flowers were cross-fertilized? The offspring of two different varieties are called **hybrids**, and the cross-fertilization itself is referred to as a hybridization, or simply a genetic **cross**. The true-breeding parental plants are called the **P generation** (P for parental), and their hybrid offspring are called the **F_1 generation** (F for filial, from the Latin word for "son"). When F_1 plants self-fertilize or fertilize each other, their offspring are the **F_2 generation**. We turn to Mendel's results next.

? Why was the development of true-breeding varieties critical to the success of Mendel's experiments?

True-breeding varieties allowed Mendel to predict the outcome of specific crosses and thus to run controlled experiments.

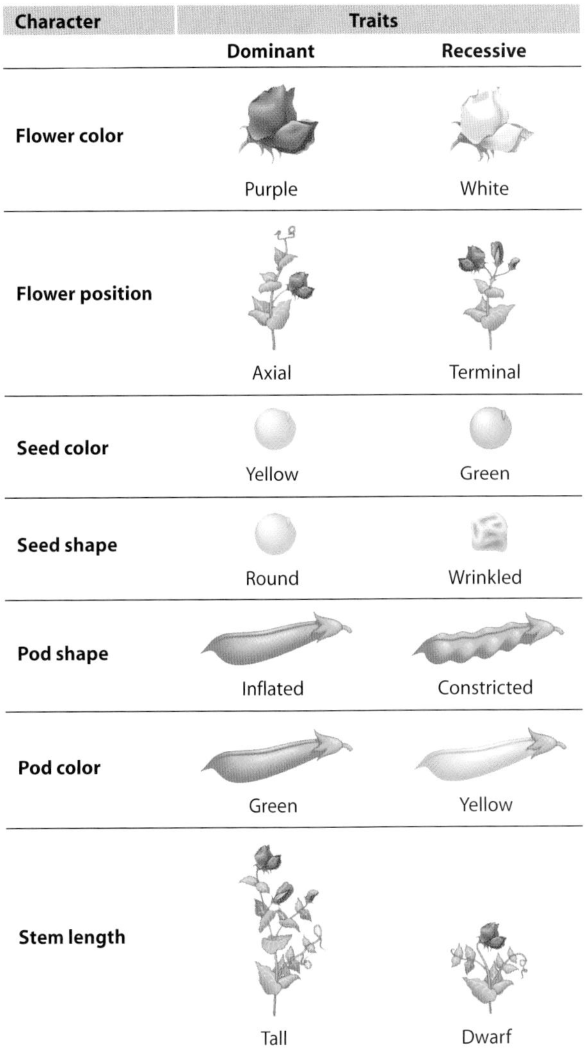

Character
Traits
Dominant
Recessive
Flower color
Purple
3 The pollinated carpel matured into a pod.
4 Mendel planted seeds from the pod.
White
Flower position
Axial
Terminal
Seed color
Yellow
Green
Seed shape
Round
Wrinkled
Pod shape
Inflated
Constricted
Pod color
Green
Yellow
Stem length
Tall
Dwarf

▲ **Figure 2D** The seven pea characters studied by Mendel

3 Mendel's law of segregation describes the inheritance of a single character

Mendel performed many experiments in which he tracked the inheritance of characters that occur in two forms, such as flower color. The results led him to formulate several hypotheses about inheritance. Let's look at some of his experiments and follow the reasoning that led to his hypotheses.

Figure 3A starts with a cross between a true-breeding pea plant with purple flowers and a true-breeding pea plant with white flowers. This is called a **monohybrid cross** because the parent plants differ in only one character—flower color. Mendel observed that F_1 plants all had purple flowers; they were not light purple, as predicted by the blending hypothesis. Was the white-flowered plant's genetic contribution to the hybrids lost? By mating the F_1 plants with each other, Mendel found the answer to be no. Out of 929 F_2 plants, 705 (about $\frac{3}{4}$) had purple flowers and 224 (about $\frac{1}{4}$) had white flowers, a ratio of about three plants with purple flowers to every one with white flowers (abbreviated as 3:1). Mendel reasoned that the heritable factor for white flowers did not disappear in the F_1 plants, but was masked when the purple-flower factor was present. He also deduced that the F_1 plants must have carried two factors for the flower-color character, one for purple and one for white.

Mendel observed these same patterns of inheritance for six other pea plant characters (see Figure 2D). From his results, he developed four hypotheses, described here using modern terminology, such as "gene" instead of "heritable factor."

1. *There are alternative versions of genes that account for variations in inherited characters.* For example, the gene for flower color in pea plants exists in two versions: one for purple and the other for white. The alternative versions of a gene are called **alleles**.
2. *For each character, an organism inherits two alleles, one from each parent.* These alleles may be the same or different. An organism that has two identical alleles for a gene is said to be **homozygous** for that gene (and is a "homozygote" for that trait). An organism that has two different alleles for a gene is said to be **heterozygous** for that gene (and is a "heterozygote").
3. *If the two alleles of an inherited pair differ, then one determines the organism's appearance and is called the* **dominant** *allele; the other has no noticeable effect on the organism's appearance and is called the* **recessive** *allele.* We use uppercase letters to represent dominant alleles and lowercase letters to represent recessive alleles.
4. *A sperm or egg carries only one allele for each inherited character because allele pairs separate (segregate) from each other during the production of gametes.* This statement is called the **law of segregation**. When sperm and egg unite at fertilization, each contributes its allele, restoring the paired condition in the offspring.

Figure 3B explains the results in Figure 3A. In this example, the letter *P* represents the dominant allele (for purple flowers),

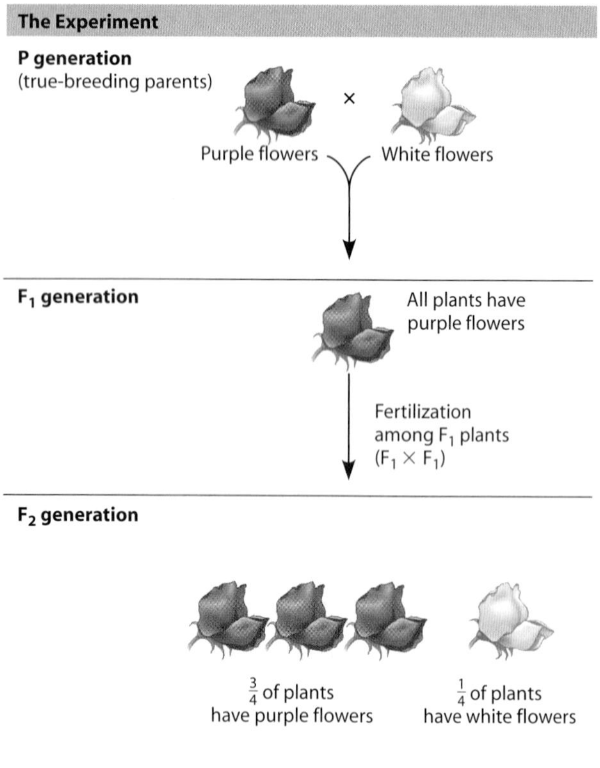

▲ **Figure 3A** Crosses tracking one character (flower color)

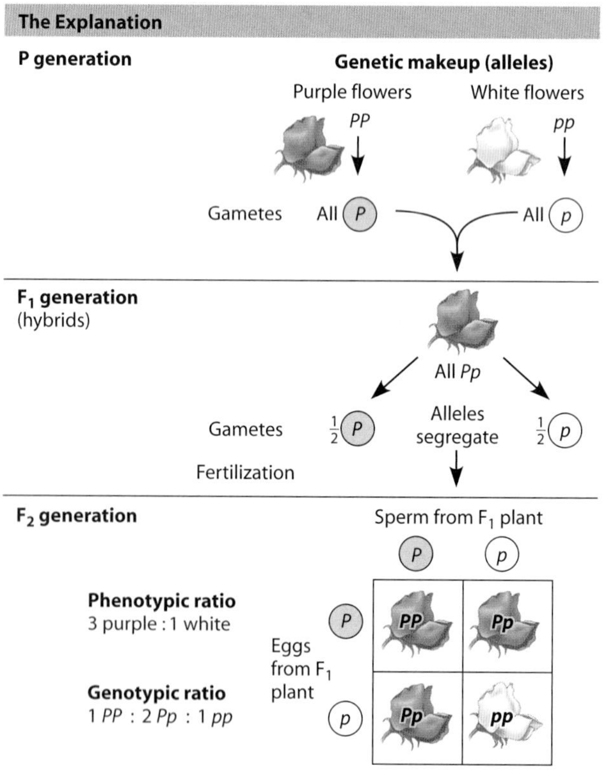

▲ **Figure 3B** An explanation of the crosses in Figure 3A

and p stands for the recessive allele (for white flowers). Both parental plants (at the top of the figure) were true-breeding, and Mendel's first two hypotheses propose that one parental variety had two alleles for purple flowers (*PP*) and the other had two alleles for white flowers (*pp*).

Consistent with hypothesis 4, the gametes of Mendel's parental plants each carried one allele; thus, the parental gametes in Figure 3B are either *P* or *p*. As a result of fertilization, the F_1 hybrids each inherited one allele for purple flowers and one for white. Hypothesis 3 explains why all of the F_1 hybrids (*Pp*) had purple flowers: The dominant *P* allele has its full effect in the heterozygote, while the recessive *p* allele has no effect.

Mendel's hypotheses also explain the 3:1 ratio in the F_2 generation. Because the F_1 hybrids are *Pp*, they make gametes *P* and *p* in equal numbers. The bottom diagram in Figure 3B, called a **Punnett square**, shows the four possible combinations of alleles that could occur when these gametes combine.

The Punnett square shows the proportions of F_2 plants predicted by Mendel's hypotheses. If a sperm carrying allele *P* fertilizes an egg carrying allele *P*, the offspring (*PP*) will produce purple flowers. Mendel's hypotheses predict that this combination will occur in $\frac{1}{4}$ of the offspring. As shown in the Punnett square, the hypotheses also predict that $\frac{1}{2}$ (or two of four) of the offspring will inherit one *P* allele and one *p* allele. These offspring (*Pp*) will all have purple flowers because *P* is dominant. The remaining $\frac{1}{4}$ of F_2 plants will inherit two *p* alleles and will have white flowers.

Because an organism's appearance does not always reveal its genetic composition, geneticists distinguish between an organism's physical traits, called its **phenotype** (such as purple or white flowers), and its genetic makeup, its **genotype** (in this example, *PP*, *Pp*, or *pp*). So now we see that Figure 3A shows just phenotypes while Figure 3B shows both phenotypes and genotypes in our sample crosses. For the F_2 plants, the ratio of plants with purple flowers to those with white flowers (3:1) is called the phenotypic ratio. The genotypic ratio, as shown by the Punnett square, is 1 *PP*:2 *Pp*:1 *pp*.

Mendel found that each of the seven characteristics he studied exhibited the same inheritance pattern: One parental trait disappeared in the F_1 generation, only to reappear in $\frac{1}{4}$ of the F_2 offspring. The mechanism underlying this inheritance pattern is stated by Mendel's law of segregation: Pairs of alleles segregate (separate) during gamete formation. The fusion of gametes at fertilization creates allele pairs once again. Research since Mendel's time has established that due to the separation of homologous chromosomes during meiosis II (see Modules 8.12–14), the law of segregation applies to all sexually reproducing organisms, including humans. We'll return to Mendel and his experiments with pea plants in Module 5.

? How can two plants with different genotypes for a particular inherited character be identical in phenotype?

One could be homozygous for the dominant allele and the other heterozygous.

4 Homologous chromosomes bear the alleles for each character

Figure 4 shows the locations of three genes on a pair of homologous chromosomes (homologs)—chromosomes that carry alleles of the same gene. Every diploid cell, whether from pea plant or human, has chromosomes in homologous pairs. One member of each pair comes from the organism's female parent, while the other member of the pair comes from the male parent.

Each labeled band on the chromosomes in the figure represents a gene **locus** (plural, *loci*), a specific location of a gene along the chromosome. The matching colors of the three corresponding loci on the two homologs highlight the fact that homologous chromosomes have genes for the same characters located at the same positions along their lengths. However, as the uppercase and lowercase letters next to the loci indicate, two homologous chromosomes may bear either the same alleles or different ones. This is the connection between Mendel's laws and homologous chromosomes: Alleles (alternative versions) of a gene reside at the same locus on homologous chromosomes.

The diagram here also serves as a review of some of the genetic terms we have encountered to this point. We will return to the chromosomal basis of inheritance in more detail beginning with Module 16.

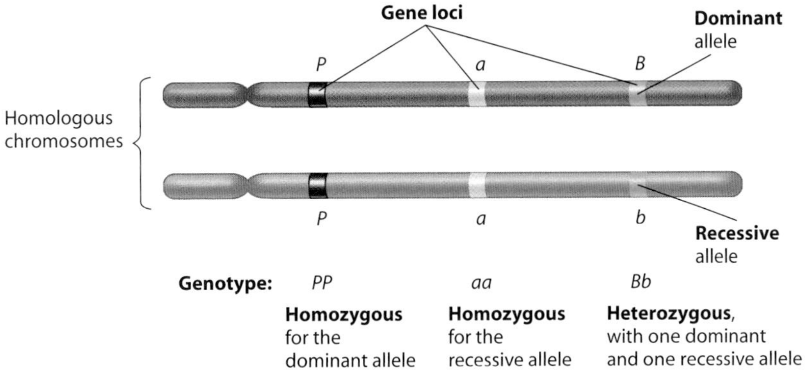

▲ **Figure 4** Three gene loci on homologous chromosomes

? An individual is heterozygous, *Bb*, for a gene. According to the law of segregation, each gamete formed by this individual will have *either* the *B* allele *or* the *b* allele. Recalling meiosis, explain the physical basis for this segregation of alleles.

The *B* and *b* alleles are located at the same gene locus on homologous chromosomes, which separate during meiosis I and are packaged in separate gametes.

5 The law of independent assortment is revealed by tracking two characters at once

Recall from Module 3 that Mendel established his law of segregation by following one character from the P generation through the F_1 and F_2 generations. From such monohybrid crosses, Mendel knew that the allele for round seed shape (designated *R*) was dominant to the allele for wrinkled seed shape (*r*) and that the allele for yellow seed color (*Y*) was dominant to the allele for green seed color (*y*). Mendel wondered: What would happen if he crossed plants that differ in both seed shape and seed color?

To find out, Mendel set up a **dihybrid cross**, a mating of parental varieties differing in two characters. Mendel crossed homozygous plants having round yellow seeds (genotype *RRYY*) with plants having wrinkled green seeds (*rryy*). Mendel knew that an *RRYY* plant would produce only gametes with *RY* alleles; an *rryy* plant would produce only gametes with *ry* alleles. Therefore, Mendel knew there was only one possible outcome for the F_1 generation: the union of *RY* and *ry* gametes would yield hybrids heterozygous for both characters (*RrYy*)—that is, dihybrids. All of these *RrYy* offspring would have round yellow seeds, the double dominant phenotype.

The F_2 generation is a bit trickier. Would the genes for seed color and seed shape be transmitted from parent to offspring as a package, or would they be inherited separately? To find out, Mendel crossed the *RrYy* F_1 plants with each other. He hypothesized two possible outcomes from this experiment: Either the dihybrid cross would exhibit *dependent* assortment, with the genes for seed color and seed shape inherited together as a set, or it would exhibit *independent* assortment, with the genes inherited independently from each other.

As shown on the left side of **Figure 5A**, the hypothesis of dependent assortment leads to the prediction that each F_2 plant would inherit one of two possible sperm (*RY* or *ry*) and one of two possible eggs (*RY* or *ry*), for a total of four combinations. The Punnett square shows that there could be only two F_2 phenotypes—round yellow or wrinkled green—in a 3:1 ratio. However, when Mendel actually performed this cross, he did not obtain these results, thus refuting the hypothesis of dependent assortment.

The alternative hypothesis—that the genes would exhibit independent assortment—is shown on the right side of Figure 5A. This leads to the prediction that the F_1 plants would produce four different gametes: *RY*, *Ry*, *rY*, and *ry*. Each F_2 plant would inherit one of four possible sperm and one of four possible eggs, for a total of 16 possible combinations. The Punnett square shows that fertilization among these gametes would lead to four different seed phenotypes—round yellow, round green, wrinkled yellow, or wrinkled green—in a 9:3:3:1 ratio. In fact, Mendel observed such a ratio in the F_2 plants, indicating that each pair of alleles segregates independently of the other.

The Punnett square on the right-hand side of Figure 5A also reveals that a dihybrid cross is equivalent to two monohybrid crosses—one for seed color and one for seed shape—occurring simultaneously. From the 9:3:3:1 ratio, we can see that there are 12 plants with round seeds to 4 with wrinkled seeds and 12 yellow-seeded plants to 4 green-seeded ones. These 12:4 ratios each reduce to 3:1, which is the F_2 ratio for a monohybrid cross; a monohybrid cross is occurring for each of the two traits. Mendel tried his seven pea characters in various dihybrid combinations and always obtained data close to the predicted 9:3:3:1 ratio. These results supported the hypothesis that each pair of alleles segregates (assorts) independently of other pairs of alleles during gamete formation. In other words, the inheritance of one character has no effect on the inheritance of another. This is called Mendel's **law of independent assortment**.

Figure 5B shows how this law applies to the inheritance of two characters in Labrador retrievers: black versus chocolate coat color and normal vision versus progressive retinal atrophy (PRA), an eye disorder that leads to blindness. As you would expect, these characters are controlled by separate genes. Black Labs have at

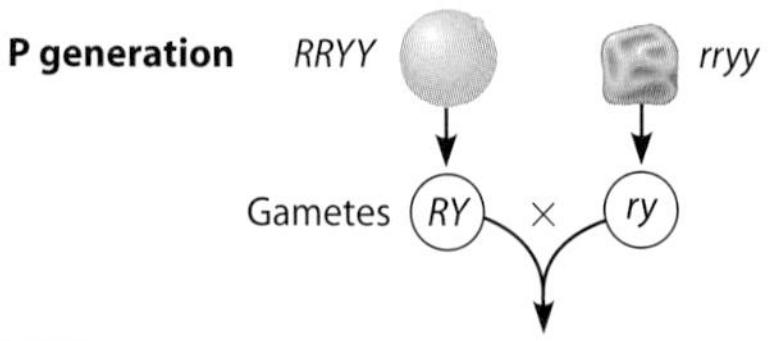

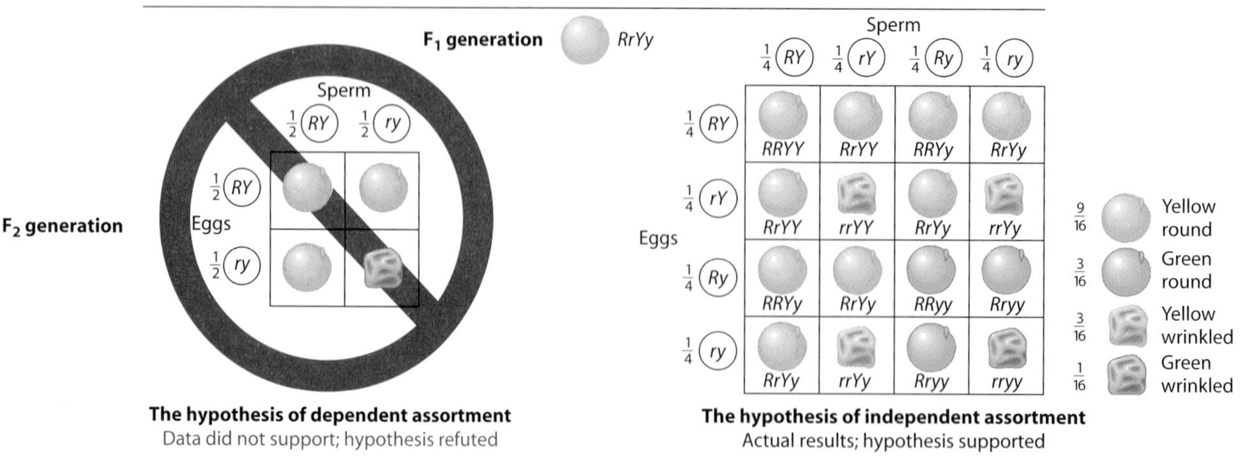

▶ **Figure 5A** Two hypotheses for segregation in a dihybrid cross

least one copy of an allele called *B*, which gives their hairs densely packed granules of a dark pigment. The *B* allele is dominant to *b*, which leads to a less tightly packed distribution of pigment granules. As a result, the coats of dogs with genotype *bb* are chocolate in color. (If you're wondering about yellow Labs, their color is controlled by a different gene altogether.) The allele that causes PRA, called *n*, is recessive to allele *N*, which is necessary for normal vision. Thus, only dogs of genotype *nn* become blind from PRA. (In the figure, blanks in the genotypes are used where a particular phenotype may result from multiple genotypes. For example, a black Lab may have either genotype *BB* or *Bb*, which we abbreviate as *B*_.)

The lower part of Figure 5B shows what happens when we mate two heterozygous Labs, both of genotype *BbNn*. The F_2 phenotypic ratio will be nine black dogs with normal eyes to three black with PRA to three chocolate with normal eyes to one chocolate with PRA. These results are analogous to the results in Figure 5A and demonstrate that the *B* and *N* genes are inherited independently.

? Predict the phenotypes of offspring obtained by mating a black Lab homozygous for both coat color and normal eyes with a chocolate Lab that is blind from PRA.

All offspring would be black with normal eyes (*BBNN* × *bbnn* → *BbNn*).

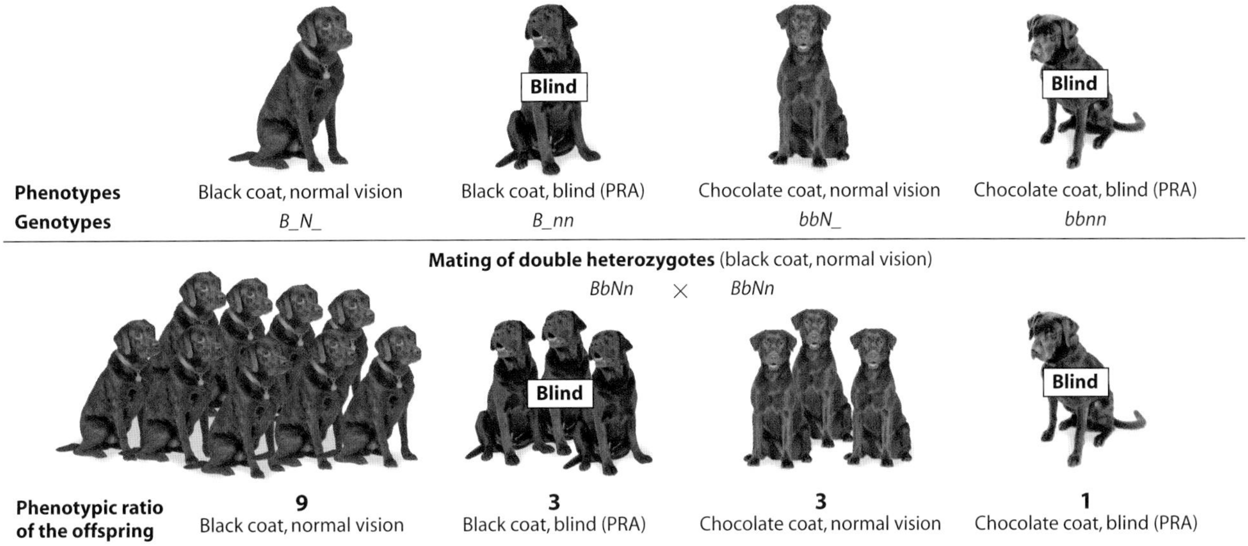

▲ **Figure 5B** Independent assortment of two genes in the Labrador retriever

6 Geneticists can use the testcross to determine unknown genotypes

Suppose you have a Labrador retriever with a chocolate coat. Referring to Figure 5B, you can tell that its genotype must be *bb*, the only combination of alleles that produces the chocolate-coat phenotype. But what if you had a black Lab? It could have one of two possible genotypes—*BB* or *Bb*—and there is no way to tell which is correct simply by looking at the dog. To determine your dog's genotype, you could perform a **testcross**, a mating between an individual of unknown genotype (your black Lab) and a homozygous recessive (*bb*) individual—in this case, a chocolate Lab.

Figure 6 shows the offspring that could result from such a mating. If, as shown on the bottom left, the black parent's genotype is *BB*, we would expect all the offspring to be black because a cross between genotypes *BB* and *bb* can produce only *Bb* offspring. On the other hand, if the black parent is *Bb*, as shown on the bottom right, we would expect both black (*Bb*) and chocolate (*bb*) offspring. Thus, the appearance of the offspring reveals the original black dog's genotype.

Mendel used testcrosses to verify that he had true-breeding varieties of plants. The testcross continues to be an important tool of geneticists for determining genotypes.

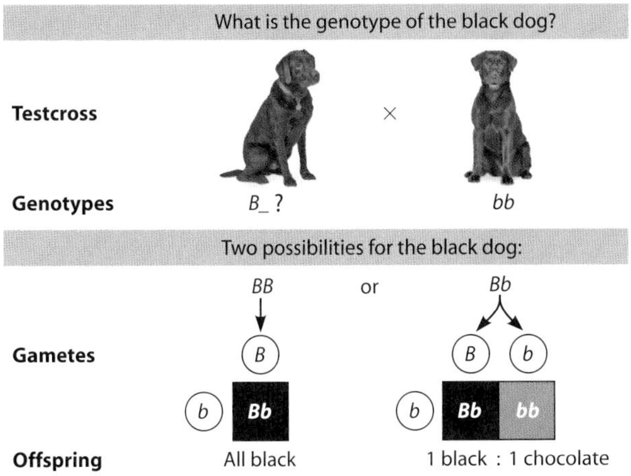

▲ **Figure 6** Using a testcross to determine genotype

? You use a testcross to determine the genotype of a Lab with normal eyes. Half of the offspring of the testcross are normal and half develop PRA. What is the genotype of the normal parent?

Heterozygous (*Nn*)

7 Mendel's laws reflect the rules of probability

Mendel's strong background in mathematics served him well in his studies of inheritance. He understood, for instance, that the segregation of allele pairs during gamete formation and the reforming of pairs at fertilization obey the rules of probability—the same rules that apply to the tossing of coins, the rolling of dice, and the drawing of cards. Mendel also appreciated the statistical nature of inheritance. He knew that he needed to obtain large samples—to count many offspring from his crosses—before he could begin to interpret inheritance patterns.

Let's see how the rules of probability apply to inheritance. The probability scale ranges from 0 to 1. An event that is certain to occur has a probability of 1, whereas an event that is certain *not* to occur has a probability of 0. For example, a tossed coin has a $\frac{1}{2}$ chance of landing heads and a $\frac{1}{2}$ chance of landing tails. These two possibilities add up to 1; the probabilities of all possible outcomes for an event to occur must always add up to 1. In another example, in a standard deck of 52 playing cards, the chance of drawing a jack of diamonds is $\frac{1}{52}$ and the chance of drawing any card other than the jack of diamonds is $\frac{51}{52}$, which together add up to 1.

An important lesson we can learn from coin tossing is that for each and every toss of the coin, the probability of heads is $\frac{1}{2}$. Even if heads has landed five times in a row, the probability of the next toss coming up heads is still $\frac{1}{2}$. In other words, the outcome of any particular toss is unaffected by what has happened on previous attempts. Each toss is an independent event.

If two coins are tossed simultaneously, the outcome for each coin is an independent event, unaffected by the other coin. What is the chance that both coins will land heads up? The probability of such a compound event is the product of the probabilities of each independent event; for the two tosses of coins, $\frac{1}{2} \times \frac{1}{2} = \frac{1}{4}$. This statistical principle is called the **rule of multiplication**, and it holds true for genetics as well as coin tosses.

Figure 7 represents a cross between F_1 Labrador retrievers that have the *Bb* genotype for coat color. The genetic cross is portrayed by the tossing of two coins that stand in for the two gametes (a dime for the egg and a penny for the sperm); the heads side of each coin stands for the dominant *B* allele and the tails side of each coin the recessive *b* allele. What is the probability that a particular F_2 dog will have the *bb* genotype? To produce a *bb* offspring, both egg and sperm must carry the *b* allele. The probability that an egg will have the *b* allele is $\frac{1}{2}$, and the probability that a sperm will have the *b* allele is also $\frac{1}{2}$. By the rule of multiplication, the probability that the two *b* alleles will come together at fertilization is $\frac{1}{2} \times \frac{1}{2} = \frac{1}{4}$. This is exactly the answer given by the Punnett square in Figure 7. If we know the genotypes of the parents, we can predict the probability for any genotype among the offspring.

Now consider the probability that an F_2 Lab will be heterozygous for the coat-color gene. As Figure 7 shows, there are two ways in which F_1 gametes can combine to produce a heterozygous offspring. The dominant (*B*) allele can come from the egg and the recessive (*b*) allele from the sperm, or vice versa. The probability that an event can occur in two or more alternative ways is the sum of the separate probabilities of the different ways; this is known as the **rule of addition**. Using this rule, we can calculate the probability of an F_2 heterozygote as $\frac{1}{4} + \frac{1}{4} = \frac{1}{2}$.

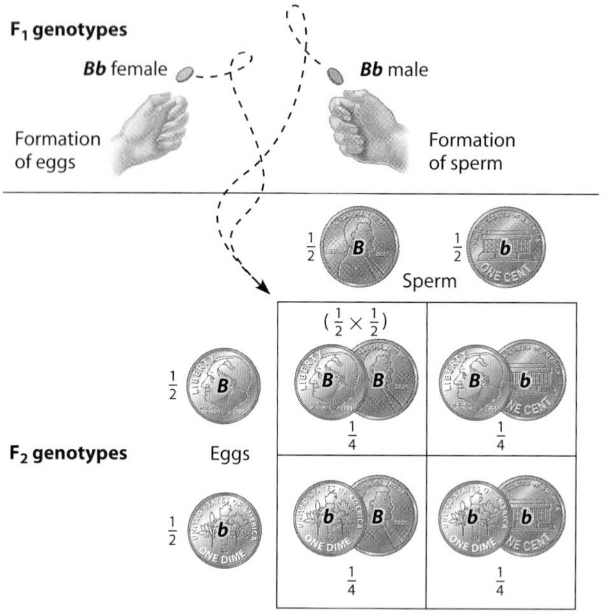

▲ **Figure 7** Segregation and fertilization as chance events

By applying the rules of probability to segregation and independent assortment, we can solve some rather complex genetics problems. For instance, we can predict the results of trihybrid crosses, in which three different characters are involved. Consider a cross between two organisms that both have the genotype *AaBbCc*. What is the probability that an offspring from this cross will be a recessive homozygote for all three genes (*aabbcc*)? Since each allele pair assorts independently, we can treat this trihybrid cross as three separate monohybrid crosses:

Aa × *Aa*: Probability of *aa* offspring = $\frac{1}{4}$

Bb × *Bb*: Probability of *bb* offspring = $\frac{1}{4}$

Cc × *Cc*: Probability of *cc* offspring = $\frac{1}{4}$

Because the segregation of each allele pair is an independent event, we use the rule of multiplication to calculate the probability that the offspring will be *aabbcc*:

$$\frac{1}{4}aa \times \frac{1}{4}bb \times \frac{1}{4}cc = \frac{1}{64}$$

We could reach the same conclusion by constructing a 64-section Punnett square, but that would take a lot of space!

? A plant of genotype *AABbCC* is crossed with an *AaBbCc* plant. What is the probability of an offspring having the genotype *AABBCC*?

● $\frac{1}{16}$ (that is, $\frac{1}{2} \times \frac{1}{4} \times \frac{1}{2}$)

CONNECTION 8 Genetic traits in humans can be tracked through family pedigrees

While much of the classic work that led to our understanding of genetics was performed on model organisms (such as peas and fruit flies), Mendel's laws apply to the inheritance of many human traits as well. **Figure 8A** illustrates alternative forms of three human characters that are each thought to be determined by simple dominant-recessive inheritance at one gene locus. (The genetic basis of many other human characters, such as eye and hair color, involves several genes and is not well understood.)

If we call the dominant allele of a gene under consideration *A*, the dominant phenotype results from either the homozygous genotype *AA* or the heterozygous genotype *Aa*. Recessive phenotypes always result from the homozygous genotype *aa*.

In genetics, the word *dominant* does not imply that a phenotype is either normal or more common than a recessive phenotype; **wild-type traits**—those prevailing in nature—are not necessarily specified by dominant alleles. Rather, dominance means that a heterozygote (*Aa*), carrying only a single copy of a dominant allele, displays the dominant phenotype. By contrast, the phenotype of the corresponding recessive allele is seen only in a homozygote (*aa*). Recessive traits may in fact be more common in the population than dominant ones. For example, the absence of freckles is more common than their presence.

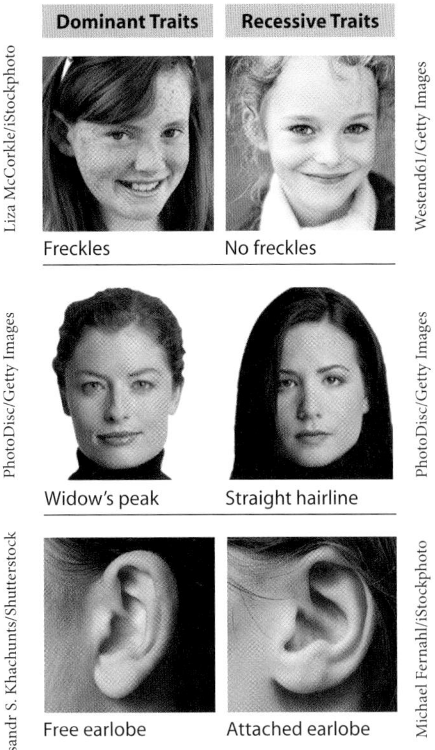

▲ **Figure 8A** Examples of single-gene inherited traits in humans

How can we determine the inheritance pattern of a particular human trait? Since we obviously cannot control human matings and run a testcross, geneticists must analyze the results of matings that have already occurred. First, a geneticist collects as much information as possible about a family's history for the trait of interest. The next step is to assemble this information into a family tree—the family **pedigree**. (You may associate pedigrees with purebred animals—such as racehorses or championship dogs—but they can represent human matings just as well.) To analyze the pedigree, the geneticist applies logic and Mendel's laws.

Let's apply this approach to the example in **Figure 8B**, a pedigree that traces the incidence of free versus attached earlobes in a hypothetical family. The letter *F* stands for the allele for free earlobes, and *f* symbolizes the allele for attached earlobes. In the pedigree, □ represents a male, ○ represents a female, and colored symbols (■ and ●) indicate that the person is affected by the trait under investigation (in this case, attached earlobes). The earliest generation studied is at the top of the pedigree, and the most recent generation is at the bottom.

Even if we didn't already know that *f* was recessive to *F*, we could apply Mendel's laws to deduce it from the pedigree. That is the only way that one of the third-generation sisters (at the bottom left) could have attached earlobes when both her parents did not; in the pedigree, this is represented by an affected individual (●) with two parents who are unaffected (□ and ○). We can therefore label all the individuals with attached earlobes in the pedigree—that is, all those with colored circles or squares—as homozygous recessive (*ff*).

Mendel's laws enable us to deduce the genotypes for most of the people in the pedigree. For example, both of the second-generation parents must have carried the *f* allele (which they passed on to the affected daughter) along with the *F* allele that gave them free earlobes. The same must be true of the first set of grandparents (the top left couple in the pedigree) because they both had free earlobes but their two sons had attached earlobes.

Notice that we cannot deduce the genotype of every member of the pedigree. For example, the sister with free earlobes in the bottom right of the pedigree must have at least one *F* allele, but she could be either *FF* or *Ff*. We cannot distinguish between these two possibilities using the available data.

? Look at Figure 8B and imagine that the sister with attached earlobes married a man with free earlobes and they had a daughter. What phenotype of this daughter would allow us to deduce all three genotypes in that family?

If the daughter had free earlobes, then she and her father must be *ff* and her mother must be *Ff*.

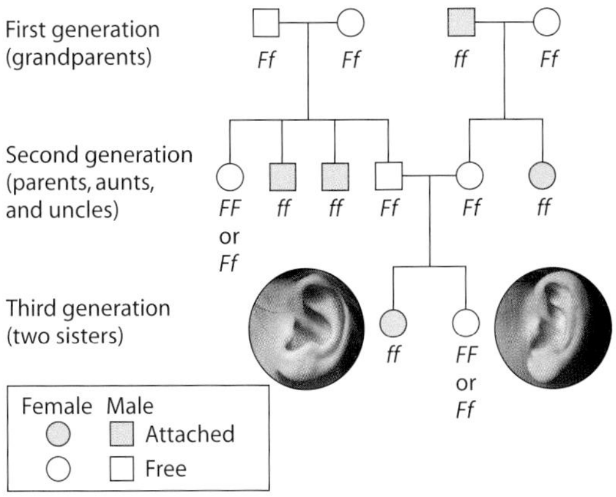

Michael Fernahl/iStockphoto
Aleksandr S. Khachunts/Shutterstock

▲ **Figure 8B** A pedigree showing the inheritance of attached versus free earlobes in a hypothetical family

CONNECTION

9 Many inherited disorders in humans are controlled by a single gene

The genetic disorders listed in **Table 9** are known to be inherited as dominant or recessive traits controlled by a single gene locus. These human disorders therefore show simple inheritance patterns like the traits Mendel studied in pea plants. The genes discussed in this module are all located on autosomes, chromosomes other than the sex chromosomes X and Y.

Recessive Disorders Most human genetic disorders are recessive. They range in severity from relatively mild, such as albinism (lack of pigmentation), to invariably fatal, such as Tay-Sachs disease. Most people who have recessive disorders are born to normal parents who are both heterozygotes—that is, who are **carriers** of the recessive allele for the disorder but are phenotypically normal.

Using Mendel's laws, we can predict the fraction of affected offspring likely to result from a mating between two carriers. Consider a form of inherited deafness caused by a recessive allele (**Figure 9A**). Suppose two heterozygous carriers (*Dd*) had a child. What is the probability that this child would be deaf? As the Punnett square in Figure 9A shows, each child of two carriers has a $\frac{1}{4}$ chance of inheriting two recessive alleles. To put it another way, we can say that about one-fourth of the children from such a mating are likely to be deaf. We can also say that a hearing ("normal") child from such a marriage has a $\frac{2}{3}$ chance of being a *Dd* carrier; that is, two out of three of the offspring with the hearing phenotype will be carriers. We can apply this same method of

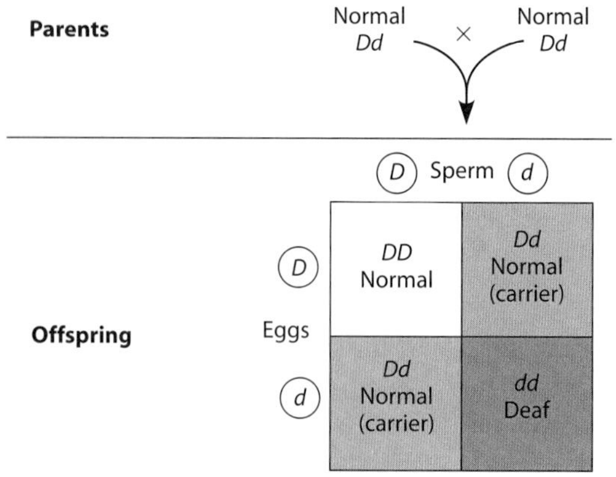

▲ **Figure 9A** Offspring produced by parents who are both carriers for a recessive disorder, a type of deafness

TABLE 9 SOME AUTOSOMAL DISORDERS IN HUMANS

Disorder	Major Symptoms	Incidence	Comments
Recessive disorders			
Albinism	Lack of pigment in the skin, hair, and eyes	$\frac{1}{22,000}$	Prone to skin cancer
Cystic fibrosis	Excess mucus in the lungs, digestive tract, liver; increased susceptibility to infections; death in early childhood unless treated	$\frac{1}{2,500}$ Caucasians	See Module 9
Galactosemia	Accumulation of galactose in tissues; mental retardation; eye and liver damage	$\frac{1}{100,000}$	Treated by eliminating galactose from diet
Phenylketonuria (PKU)	Accumulation of phenylalanine in blood; lack of normal skin pigment; mental retardation	$\frac{1}{10,000}$ in U.S. and Europe	See Module 10
Sickle-cell disease	Sickled red blood cells; damage to many tissues	$\frac{1}{400}$ African-Americans	See Module 13
Tay-Sachs disease	Lipid accumulation in brain cells; mental deficiency; blindness; death in childhood	$\frac{1}{3,500}$ Jews from central Europe	See Module 4.10
Dominant disorders			
Achondroplasia	Dwarfism	$\frac{1}{25,000}$	See Module 9
Alzheimer's disease (one type)	Mental deterioration; usually strikes late in life	Not known	Familial (inherited) Alzheimer's is a rare form of the disease
Huntington's disease	Mental deterioration and uncontrollable movements; strikes in middle age	$\frac{1}{25,000}$	See Module 9
Hypercholesterolemia	Excess cholesterol in the blood; heart disease	$\frac{1}{500}$ are heterozygous	See Module 11

pedigree analysis and prediction to any genetic trait controlled by a single gene locus.

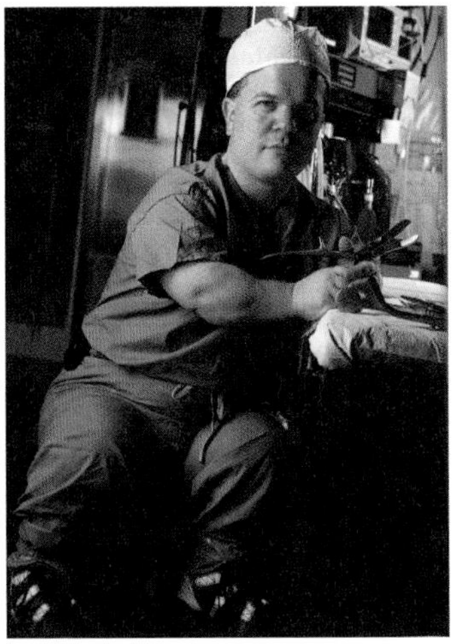

Michael Ciesielski Photography

▲ **Figure 9B** Dr. Michael C. Ain, a specialist in the repair of bone defects caused by achondroplasia and related disorders

The most common life-threatening genetic disease in the United States is **cystic fibrosis (CF)**. Affecting about 30,000 Americans and 70,000 people worldwide, the recessive CF allele is carried by about one in 31 Americans. A person with two copies of this allele has cystic fibrosis, which is characterized by an excessive secretion of very thick mucus from the lungs, pancreas, and other organs. This mucus can interfere with breathing, digestion, and liver function and makes the person vulnerable to recurrent bacterial infections. Although there is no cure for this disease, strict adherence to a daily health regimen—including gentle pounding on the chest and back to clear the airway, inhaled antibiotics, and a special diet—can have a profound impact on the health of the affected person. CF was once invariably fatal in childhood, but tremendous advances in treatment have raised the median survival age of Americans with CF to 37.

Most genetic disorders are not evenly distributed across all ethnic groups. CF, for example, is most common in Caucasians. Such uneven distribution is the result of prolonged geographic isolation of certain populations. For example, the isolated lives of the early inhabitants of Martha's Vineyard (an island off the coast of Massachusetts) led to frequent **inbreeding**, matings between close blood relatives. This caused the frequency of an allele that causes deafness to be high within the community, which led to a high incidence of deafness. Because the community was geographically isolated, the deafness allele was rarely transmitted to the public beyond Martha's Vineyard.

With the increased mobility in most societies today, it is relatively unlikely that two carriers of a rare, harmful allele will meet and mate. However, the probability increases greatly if close blood relatives have children. People with recent common ancestors are more likely to carry the same recessive alleles than are unrelated people. Inbreeding is therefore likely to produce offspring homozygous for recessive traits. Geneticists have observed increased incidence of harmful recessive traits among many types of inbred animals. For example, purebred Labrador retrievers are known to have high incidences of certain genetic defects, such as weak hip, knee, and elbow joints, in addition to eye problems. The detrimental effects of inbreeding are also seen in some endangered species.

Dominant Disorders Although many harmful alleles are recessive, a number of human disorders are caused by dominant alleles. Some are harmless conditions, such as extra fingers and toes (called polydactyly) or webbed fingers and toes.

A more serious dominant disorder is **achondroplasia**, a form of dwarfism in which the head and torso of the body develop normally, but the arms and legs are short (**Figure 9B**). About one out of every 25,000 people have achondroplasia. The homozygous dominant genotype (*AA*) causes death of the embryo, and therefore only heterozygotes (*Aa*), individuals with a single copy of the defective allele, have this disorder. (This also means that a person with achondroplasia has a 50% chance of passing the condition on to any children.) Therefore, all those who do not have achondroplasia, more than 99.99% of the population, are homozygous for the recessive allele (*aa*). This example makes it clear that a dominant allele is not necessarily more common in a population than a corresponding recessive allele.

Dominant alleles that cause lethal diseases are much less common than lethal recessives. One reason for this difference is that the dominant lethal allele cannot be carried by heterozygotes without affecting them. Many lethal dominant alleles result from mutations in a sperm or egg that subsequently kill the embryo. And if the afflicted individual is born but does not survive long enough to reproduce, he or she will not pass on the lethal allele to future generations. This is in contrast to lethal recessive mutations, which are perpetuated from generation to generation by healthy heterozygous carriers.

A lethal dominant allele can escape elimination, however, if it does not cause death until a relatively advanced age. One such example is the allele that causes **Huntington's disease**, a degenerative disorder of the nervous system that usually does not appear until middle age. Once the deterioration of the nervous system begins, it is irreversible and inevitably fatal. Because the allele for Huntington's disease is dominant, any child born to a parent with the allele has a 50% chance of inheriting the allele and the disorder. This example makes it clear that a dominant allele is not necessarily "better" than a corresponding recessive allele.

Until relatively recently, the onset of symptoms was the only way to know if a person had inherited the Huntington's allele. This is no longer the case. By analyzing DNA samples from a large family with a high incidence of the disorder, geneticists tracked the Huntington's allele to a locus near the tip of chromosome 4, and the gene has been sequenced. This information led to development of a test that can detect the presence of the Huntington's allele in an individual's genome. This is one of several genetic tests currently available. We'll explore the topic of personal genetic screening in the next module.

? **Peter is a 30-year-old man whose father died of Huntington's disease. Neither Peter's mother nor a much older sister show any signs of Huntington's. What is the probability that Peter has inherited Huntington's disease?**

Since his father had the disease, there is a $\frac{1}{2}$ chance that Peter received the gene. (The genotype of his sister is irrelevant.)

CONNECTION 10 New technologies can provide insight into one's genetic legacy

Some prospective parents are aware that they have an increased risk of having a baby with a genetic disease. For example, many pregnant women over age 35 know that they have a heightened risk of bearing children with Down syndrome, and some couples are aware that certain genetic diseases run in their families. These potential parents may want to learn more about their own and their baby's genetic makeup. Modern technologies offer ways to obtain such information before conception, during pregnancy, and after birth.

Genetic Testing Because most children with recessive disorders are born to healthy parents, the genetic risk for many diseases is determined by whether the prospective parents are carriers of the recessive allele. For an increasing number of genetic disorders, including Tay-Sachs disease, sickle-cell disease, and one form of cystic fibrosis, tests are available that can distinguish between individuals who have no disease-causing alleles and those who are heterozygous carriers. Other parents may know that a dominant but late-appearing disease, such as Huntington's disease, runs in their family. Such people may benefit from genetic tests for dominant alleles. Information from genetic testing (also called genetic screening) can inform decisions about whether to have a child.

Fetal Testing Several technologies are available for detecting genetic conditions in a fetus. Genetic testing before birth requires the collection of fetal cells. In **amniocentesis**, performed between weeks 14 and 20 of pregnancy, a physician carefully inserts a needle through the abdomen and into the mother's uterus while watching an ultrasound imager to help avoid the fetus (**Figure 10A**, left). The physician extracts about 10 milliliters (2 teaspoonsful) of the amniotic fluid that bathes the developing fetus. Tests for genetic disorders can be performed on fetal cells (mostly from shed skin) that have been isolated from the fluid. Before testing, these cells are usually cultured in the laboratory for several weeks. By then, enough dividing cells can be harvested to allow karyotyping and the detection of chromosomal abnormalities such as Down syndrome.

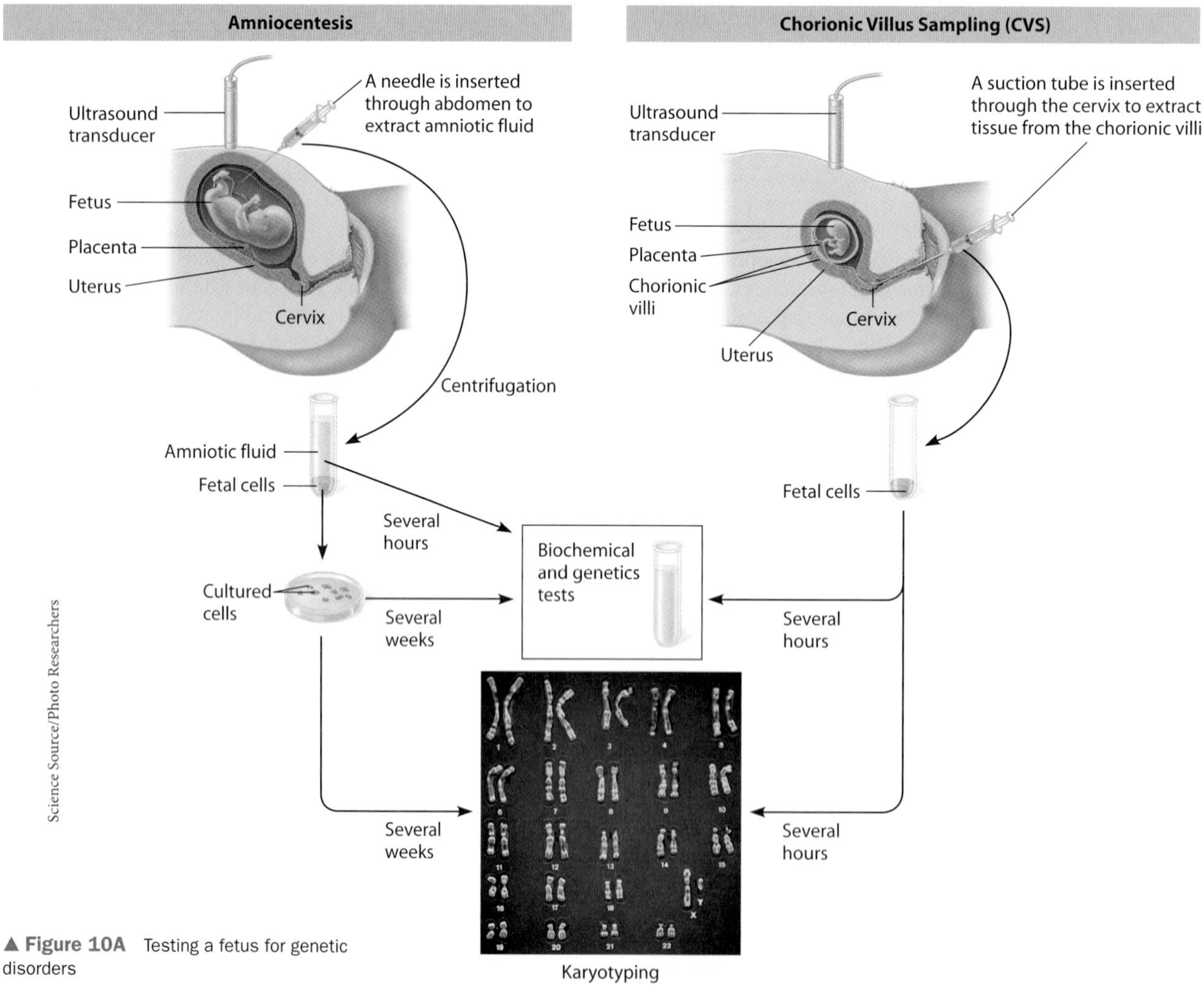

▲ **Figure 10A** Testing a fetus for genetic disorders

Biochemical tests can also be performed on the cultured cells, revealing conditions such as Tay-Sachs disease.

In another procedure, **chorionic villus sampling (CVS)**, a physician extracts a tiny sample of chorionic villus tissue from the placenta, the organ that carries nourishment and wastes between the fetus and the mother. The tissue can be obtained using a narrow, flexible tube inserted through the mother's vagina and cervix into the uterus (Figure 10A, right). Results of karyotyping and some biochemical tests can be available within 24 hours. The speed of CVS is an advantage over amniocentesis. Another advantage is that CVS can be performed as early as the 8th week of pregnancy.

Unfortunately, both amniocentesis and CVS pose some risk of complications, such as maternal bleeding, miscarriage, or premature birth. Complication rates for amniocentesis and CVS are about 1% and 2%, respectively. Because of the risks, these procedures are usually reserved for situations in which the possibility of a genetic disease is significantly higher than average. Newer genetic screening procedures involve isolating tiny amounts of fetal cells or DNA released into the mother's bloodstream. Although few reliable tests are yet available using this method, this promising and complication-free technology may soon replace more invasive procedures.

Blood tests on the mother at 15 to 20 weeks of pregnancy can help identify fetuses at risk for certain birth defects—and thus candidates for further testing that may require more invasive procedures (such as amniocentesis). The most widely used blood test measures the mother's blood level of alpha-fetoprotein (AFP), a protein produced by the fetus. High levels of AFP may indicate a neural tube defect in the fetus. (The neural tube is an embryonic structure that develops into the brain and spinal cord.) Low levels of AFP may indicate Down syndrome. For a more complete risk profile, a woman's doctor may order a "triple screen test," which measures AFP as well as two other hormones produced by the placenta. Abnormal levels of these substances in the maternal blood may also point to a risk of Down syndrome.

Fetal Imaging Other techniques enable a physician to examine a fetus directly for anatomical deformities. The most common procedure is **ultrasound imaging**, which uses sound waves to produce a picture of the fetus. **Figure 10B** shows an ultrasound scanner, which emits high-frequency sounds, beyond the range of hearing. When the sound waves bounce off the fetus, the echoes produce an image on the monitor. The inset image in Figure 10B shows a fetus at about 20 weeks. Ultrasound imaging is noninvasive—no foreign objects are inserted into the mother's body—and has no known risk. In another imaging method, fetoscopy, a needle-thin tube containing a fiber optic viewing scope is inserted into the uterus. Fetoscopy can provide highly detailed images of the fetus but, unlike ultrasound, carries risk of complications.

Newborn Screening Some genetic disorders can be detected at birth by simple tests that are now routinely performed in most hospitals in the United States. One common screening program is for phenylketonuria (PKU), a recessively inherited disorder that occurs in about one out of every 10,000 births in the United States. Children with this disease cannot properly break down the naturally occurring amino acid phenylalanine; and an accumulation of phenylalanine may lead to mental retardation. However, if the deficiency is detected in the newborn, a special diet low in phenylalanine can usually prevent retardation. Unfortunately, few other genetic disorders are currently treatable.

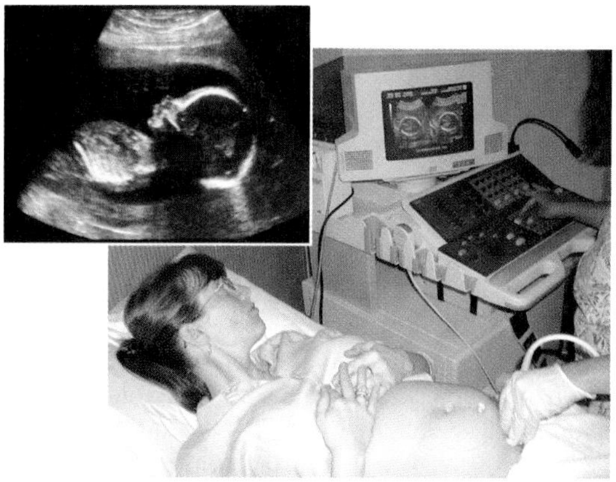
Gusto/Photo Researchers
Eric J. Simon

▲ **Figure 10B** Ultrasound scanning of a fetus

Ethical Considerations As new technologies such as fetal imaging and testing become more widespread, geneticists are working to make sure that they do not cause more problems than they solve. Consider the tests for identifying carriers of recessive diseases. Such information may enable people with family histories of genetic disorders to make informed decisions about having children. But these new methods for genetic screening pose problems, too. If confidentiality is breached, will carriers be stigmatized? For example, will they be denied health or life insurance, even though they themselves are healthy? Will misinformed employers equate "carrier" with disease? Geneticists stress that patients seeking genetic testing should receive counseling both before and after to clarify their family history, to explain the test, and to help them cope with the results. But with a wealth of genetic information becoming available, a full discussion of the meaning of the results might be time-consuming and costly, raising the question of who should pay for it.

Couples at risk for conceiving children with genetic disorders may now learn a great deal about their unborn children. In particular, CVS gives parents a chance to become informed very early in pregnancy. What is to be done with such information? If fetal tests reveal a serious disorder, the parents must choose between terminating the pregnancy and preparing themselves for a baby with severe problems. Identifying a genetic disease early can give families time to prepare—emotionally, medically, and financially.

Advances in biotechnology offer possibilities for reducing human suffering, but not before key ethical issues are resolved. The dilemmas posed by human genetics reinforce one of this book's central themes: the immense social implications of biology.

? What is the primary benefit of genetic screening by CVS? What is the primary risk?

CVS allows genetic screening to be performed very early in pregnancy and provides quick results, but it carries a risk of miscarriage.

Variations on Mendel's Laws

11 Incomplete dominance results in intermediate phenotypes

Mendel's laws explain inheritance in terms of discrete factors—genes—that are passed along from generation to generation according to simple rules of probability. Mendel's laws are valid for all sexually reproducing organisms, including garden peas, Labradors, and human beings. But just as the basic rules of musical harmony cannot account for all the rich sounds of a symphony, Mendel's laws stop short of explaining some patterns of genetic inheritance. In fact, for most sexually reproducing organisms, cases where Mendel's laws can strictly account for the patterns of inheritance are relatively rare. More often, the inheritance patterns are more complex, as we will see in this and the next four modules.

The F_1 offspring of Mendel's pea crosses always looked like one of the two parental varieties. In this situation—called **complete dominance**—the dominant allele has the same phenotypic effect whether present in one or two copies. But for some characters, the appearance of F_1 hybrids falls between the phenotypes of the two parental varieties, an effect called **incomplete dominance**. For instance, as **Figure 11A** illustrates, when red snapdragons are crossed with white snapdragons, all the F_1 hybrids have pink flowers. This third phenotype results from flowers of the heterozygote having less red pigment than the red homozygotes.

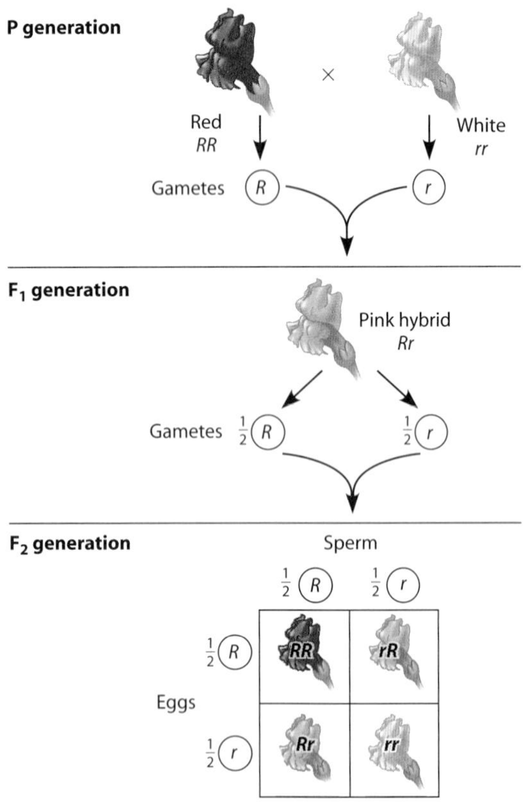

▲ **Figure 11A** Incomplete dominance in snapdragon flower color

Incomplete dominance does *not* support the blending hypothesis described in Module 1, which would predict that the red and white traits could never be retrieved from the pink hybrids. As the Punnett square at the bottom of Figure 11A shows, the F_2 offspring appear in a phenotypic ratio of one red to two pink to one white, because the red and white alleles segregate during gamete formation in the pink F_1 hybrids. In incomplete dominance, the phenotypes of heterozygotes differ from the two homozygous varieties, and the genotypic ratio and the phenotypic ratio are both 1:2:1 in the F_2 generation.

We also see examples of incomplete dominance in humans. One case involves a recessive allele (*h*) that can cause hypercholesterolemia, dangerously high levels of cholesterol in the blood. Normal individuals are *HH*. Heterozygotes (*Hh*; about one in 500 people) have blood cholesterol levels about twice normal. They are unusually prone to atherosclerosis, the blockage of arteries by cholesterol buildup in artery walls, and they may have heart attacks from blocked heart arteries by their mid-30s. This form of the disease can often be controlled through changes in diet and by taking statins, a class of medications that can significantly lower blood cholesterol. Hypercholesterolemia is even more serious in homozygous individuals (*hh*; about one in a million people). Homozygotes have about five times the normal amount of blood cholesterol and may have heart attacks as early as age 2. Homozygous hypercholesterolemia is harder to treat; options include high doses of statin drugs, organ surgeries or transplants, or physically filtering lipids from the blood.

Figure 11B illustrates the molecular basis for hypercholesterolemia. The dominant allele (*H*), which normal individuals carry in duplicate (*HH*), specifies a cell-surface receptor protein called an LDL receptor. Low-density lipoprotein (LDL, sometimes called "bad cholesterol") is transported in the blood. In certain cells, the LDL receptors mop up excess LDL particles from the blood and promote their breakdown. This process helps prevent the accumulation of cholesterol in

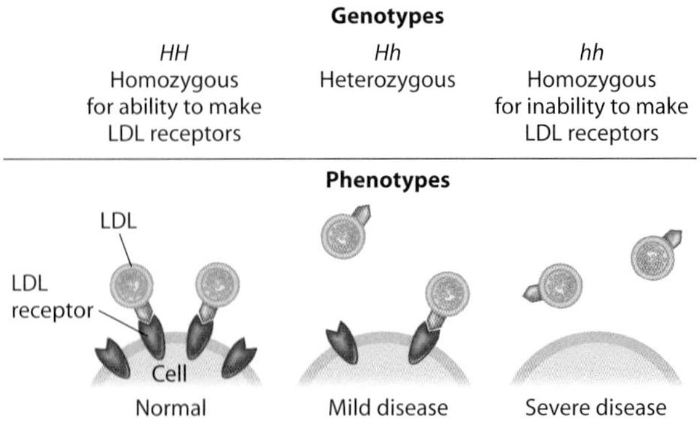

▲ **Figure 11B** Incomplete dominance in human hypercholesterolemia

arteries. Without the receptors, lethal levels of LDL build up in the blood. Heterozygotes (*Hh*) have only half the normal number of LDL receptors, and homozygous recessives (*hh*) have none.

? Why is a testcross unnecessary to determine whether a snapdragon with red flowers is homozygous or heterozygous?

Because the homozygotes and heterozygotes differ in phenotype: red flowers for the dominant homozygote and pink flowers for the heterozygote

12 Many genes have more than two alleles in the population

So far, we have discussed inheritance patterns involving only two alleles per gene (*H* versus *h*, for example). But most genes can be found in populations in more than two versions, known as multiple alleles. Although any particular individual carries, at most, two different alleles for a particular gene, in cases of multiple alleles, more than two possible alleles exist in the population.

For instance, the **ABO blood group** phenotype in humans involves three alleles of a single gene. Various combinations of three alleles for the ABO blood type produce four phenotypes: A person's blood type may be A, B, AB, or O (**Figure 12**). These letters refer to two carbohydrates, designated A and B, that may be found on the surface of red blood cells. A person's red blood cells may be coated with carbohydrate A (in which case they are said to have type A blood), carbohydrate B (type B), both carbohydrates (type AB), or neither carbohydrate (type O). (In case you are wondering, the "positive" and "negative" notations on blood types—referred to as the Rh blood group system—are due to inheritance of a separate, unrelated gene.)

Matching compatible blood types is critical for safe blood transfusions. If a donor's blood cells have a carbohydrate (A or B) that is foreign to the recipient, then the recipient's immune system produces blood proteins called antibodies (see Module 24.9) that bind specifically to the foreign carbohydrates and cause the donor blood cells to clump together, potentially killing the recipient. The clumping reaction is also the basis of a blood-typing test performed in the laboratory. In Figure 12, notice that AB individuals can receive blood from anyone without fear of clumping, making them "universal recipients," while donated type O blood never causes clumping, making those with type O blood "universal donors".

The four blood groups result from various combinations of the three different alleles: I^A (for an enzyme referred to as *I*, which adds carbohydrate A to red blood cells), I^B (for B), and *i* (for neither A nor B). Each person inherits one of these alleles from each parent. Because there are three alleles, there are six possible genotypes, as listed in the figure. Both the I^A and I^B alleles are dominant to the *i* allele. Thus, I^AI^A and I^Ai people have type A blood, and I^BI^B and I^Bi people have type B. Recessive homozygotes, *ii*, have type O blood, with neither carbohydrate. The I^A and I^B alleles are **codominant**: Both alleles are expressed in heterozygous individuals (I^AI^B), who have type AB blood. Note that codominance (the expression of both alleles) is different from incomplete dominance (the expression of one intermediate trait).

? Maria has type O blood, and her sister has type AB blood. The girls know that both of their maternal grandparents are type A. What are the genotypes of the girls' parents?

Their mother is I^Ai; their father is I^Bi.

Blood Group (Phenotype)	Genotypes	Carbohydrates Present on Red Blood Cells	Antibodies Present in Blood
A	I^AI^A or I^Ai	Carbohydrate A	Anti-B
B	I^BI^B or I^Bi	Carbohydrate B	Anti-A
AB	I^AI^B	Carbohydrate A and Carbohydrate B	*None*
O	*ii*	Neither	Anti-A, Anti-B

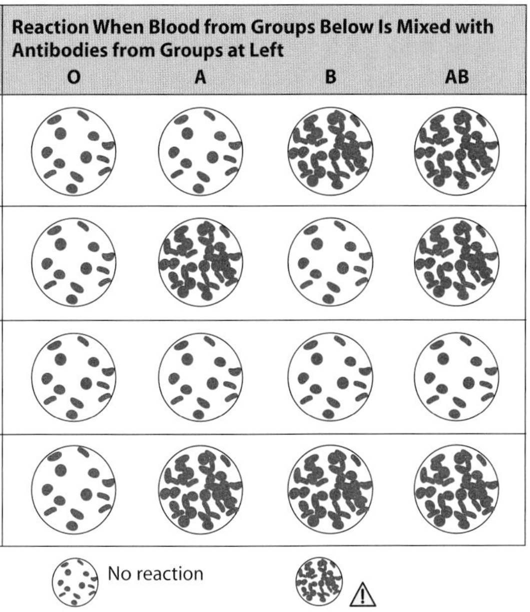
Reaction When Blood from Groups Below Is Mixed with Antibodies from Groups at Left
O
A
B
AB
No reaction

▲ **Figure 12** Multiple alleles for the ABO blood groups

13 A single gene may affect many phenotypic characters

All of our genetic examples to this point have been cases in which each gene specifies only one hereditary character. In most cases, however, one gene influences multiple characters, a property called **pleiotropy**.

An example of pleiotropy in humans is **sickle-cell disease** (sometimes called sickle-cell anemia). The direct effect of the sickle-cell allele is to make red blood cells produce abnormal hemoglobin proteins. These molecules tend to link together and crystallize, especially when the oxygen content of the blood is lower than usual because of high altitude, overexertion, or respiratory ailments. As the hemoglobin crystallizes, the normally disk-shaped red blood cells deform to a sickle shape with jagged edges (**Figure 13A**). Sickled cells are destroyed rapidly by the body, and their destruction may seriously lower the individual's red cell count, causing anemia and general weakening of the body. Also, because of their angular shape, sickled cells do not flow smoothly in the blood and tend to accumulate and clog tiny blood vessels. Blood flow to body parts is reduced, resulting in periodic fever, severe pain, and damage to various organs, including the heart, brain, and kidneys. Sickled cells also accumulate in the spleen, damaging it. The overall result is a disorder characterized by the cascade of symptoms shown in **Figure 13B**. Blood transfusions and certain drugs may relieve some of the symptoms, but there is no cure, and sickle-cell disease kills about 100,000 people in the world each year.

In most cases, only people who are homozygous for the sickle-cell allele suffer from the disease. Heterozygotes, who have one sickle-cell allele and one nonsickle allele, are usually healthy, although in rare cases they may experience some effects when oxygen in the blood is severely reduced, such as at very high altitudes. These effects may occur because the nonsickle and sickle-cell alleles are codominant at the molecular level: Both alleles are expressed in heterozygous individuals, and their red blood cells contain both normal and abnormal hemoglobin.

Sickle-cell disease is the most common inherited disorder among people of African descent, striking one in 400 African-Americans. About one in ten African-Americans is a carrier (an unaffected heterozygote). Among Americans of other ancestry, the sickle-cell allele is extremely rare.

One in ten is an unusually high frequency of carriers for an allele with such harmful effects in homozygotes. We might expect that the frequency of the sickle-cell allele in the population would be much lower because many homozygotes die before passing their genes to the next generation. The high frequency appears to be a vestige of the roots of African-Americans. Sickle-cell disease is most common in tropical Africa, where the deadly disease malaria is also prevalent. The parasite that causes malaria spends part of its life cycle inside red blood cells. When it enters those of a person with the sickle-cell allele, it triggers sickling. The body destroys most of the sickled cells, and the parasite does not grow well in those that remain. Consequently, sickle-cell carriers are resistant to malaria, and in many parts of Africa, they live longer and have more offspring than noncarriers who are exposed to malaria. In this way, malaria has kept the frequency of the sickle-cell allele relatively high in much of the African continent. To put it in evolutionary terms, as long as malaria is a danger, individuals with the sickle-cell allele have a selective advantage.

? How does sickle-cell disease exemplify the concept of pleiotropy?

● Homozygosity for the sickle-cell allele causes abnormal hemoglobin, and the impact of the abnormal hemoglobin on the shape of red blood cells leads to a cascade of symptoms in multiple organs of the body.

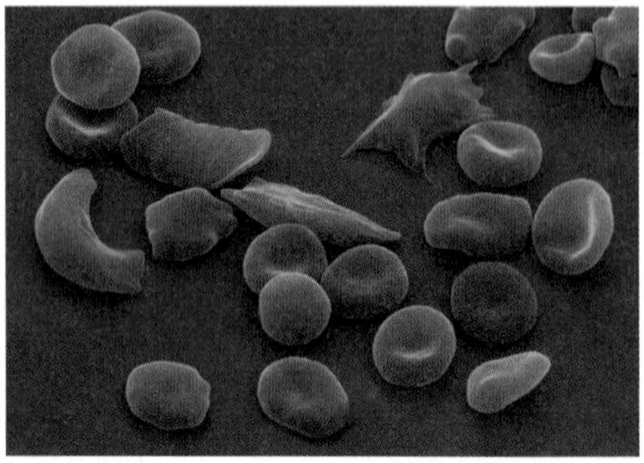

An individual homozygous for the sickle-cell allele

↓

Produces sickle-cell (abnormal) hemoglobin

↓

The abnormal hemoglobin crystallizes, causing red blood cells to become sickle-shaped

Sickled cell

SEM 1,045×

The mutiple effects of sickled cells

Damage to organs	Other effects
Kidney failure	Pain and fever
Heart failure	Joint problems
Spleen damage	Physical weakness
Brain damage (impaired mental function, paralysis)	Anemia
	Pneumonia and other infections

Eye of Science/Photo Researchers

▲ **Figure 13B** Sickle-cell disease, an example of pleitropy

Eye of Science/Photo Researchers

SEM 1,045×

▲ **Figure 13A** In this micrograph, you can see several jagged sickled cells in the midst of normal red blood cells.

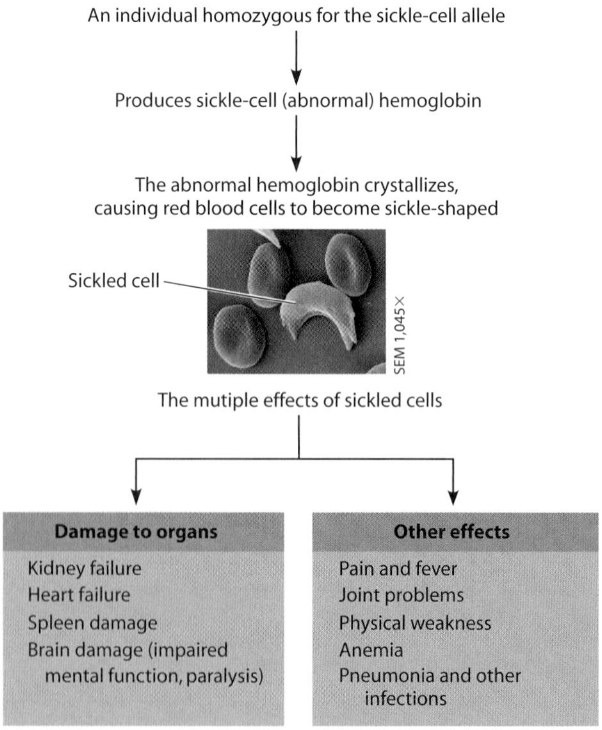

14 A single character may be influenced by many genes

Mendel studied genetic characters that could be classified on an either-or basis, such as purple or white flower color. However, many characteristics, such as human skin color and height, vary in a population along a continuum. Many such features result from **polygenic inheritance**, the additive effects of two or more genes on a single phenotypic character. (This is the converse of pleiotropy, in which a single gene affects several characters.)

Let's consider a hypothetical example. Assume that the continuous variation in human skin color is controlled by three genes that are inherited separately, like Mendel's pea genes. (Actually, genetic evidence indicates that *at least* three genes control this character.) The "dark-skin" allele for each gene (*A*, *B*, or *C*) contributes one "unit" of darkness to the phenotype and is incompletely dominant to the other allele (*a*, *b*, or *c*). A person who is *AABBCC* would be very dark, whereas an *aabbcc* individual would be very light. An *AaBbCc* person (resulting, for example, from a mating between an *AABBCC* person and an *aabbcc* person) would have skin of an intermediate shade. Because the alleles have an additive effect, the genotype *AaBbCc* would produce the same skin color as any other genotype with just three dark-skin alleles, such as *AABbcc,* since both of these individuals have three "units" of darkness.

The Punnett square in **Figure 14** shows all possible genotypes of offspring from a mating of two triple heterozygotes, the F_1 generation here. The row of squares below the Punnett square shows the seven skin pigmentation phenotypes that would theoretically result from this mating. The seven bars in the graph at the bottom of the figure depict the relative numbers of each of the phenotypes in the F_2 generation. This hypothetical example shows how inheritance of three genes could lead to a wide variety of pigmentation phenotypes. In real human populations, skin color has even more variations than shown in the figure, in part for reasons we discuss in the next module.

Up to this point in the chapter, we have presented four types of inheritance patterns that are extensions of Mendel's laws of inheritance: incomplete dominance, codominance, pleiotropy, and polygenic inheritance. It is important to realize that these patterns are extensions of Mendel's model, rather than exceptions to it. From Mendel's garden pea experiments came data supporting a particulate theory of inheritance, with the particles (genes) being transmitted according to the same rules of chance that govern the tossing of coins. The particulate theory holds true for all inheritance patterns, even the patterns that are more complex than the ones originally considered by Mendel. In the next module, we consider another important source of deviation from Mendel's standard model: the effect of the environment.

? Based on the model for skin color in Figure 14, an *AaBbcc* individual would be indistinguishable in phenotype from which of the following individuals: *AAbbcc, aaBBcc, AabbCc, Aabbcc,* or *aaBbCc*?

All except *Aabbcc*

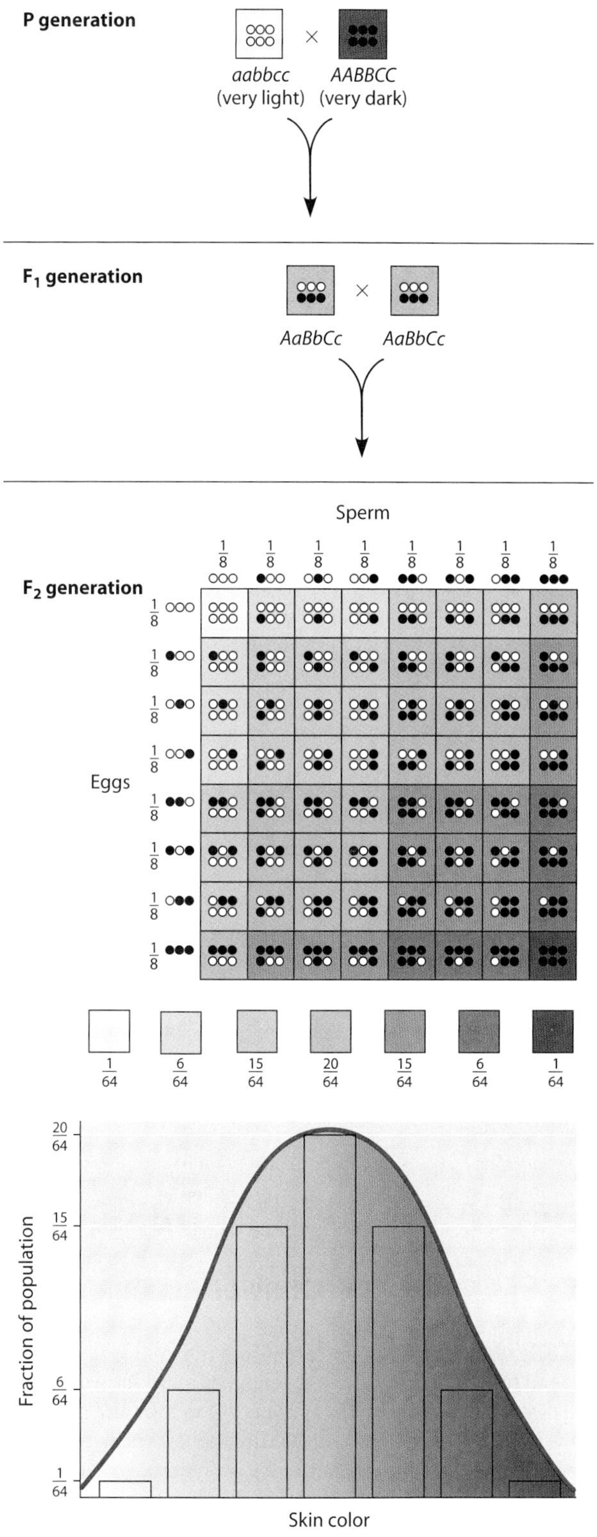

▲ **Figure 14** A model for polygenic inheritance of skin color

15 The environment affects many characters

In the previous module, we saw how a set of three hypothetical human skin-color genes could produce seven different phenotypes for skin color. If we examine a real human population for skin color, we might see more shades than just seven. The true range might be similar to the entire spectrum of color under the bell-shaped curve in Figure 14. In fact, no matter how carefully we characterize the genes for skin color, a purely genetic description will always be incomplete. This is because some intermediate shades of skin color result from the effects of environmental factors, such as exposure to the sun **(Figure 15A)**.

Many characters result from a combination of heredity and environment. For example, the leaves of a tree all have the same genotype, but they may vary greatly in phenotype such as size, shape, and color, depending on exposure to wind and sun and the tree's nutritional state. For humans, nutrition influences height; exercise alters build; sun-tanning darkens the skin; and experience improves performance on intelligence tests. As geneticists learn more and more about our genes, it is becoming clear that many human phenotypes—such as risk of heart disease and cancer and susceptibility to alcoholism and schizophrenia—are influenced by both genes and environment.

Whether human characters are more influenced by genes or by the environment—nature or nurture—is a very old and hotly contested debate. For some characters, such as the ABO blood group, a given genotype absolutely mandates a very specific phenotype, and the environment plays no role whatsoever. In contrast, a person's counts of red and white blood cells (the numbers of blood cells per milliliter of blood) are influenced greatly by environmental factors such as the altitude, the customary level of physical activity, and the presence of infectious agents.

It is important to realize that the individual features of any organism arise from a combination of genetic and environmental factors. Simply spending time with identical twins will convince anyone that environment, and not just genes, affects a person's traits **(Figure 15B)**. However, there is an important difference between these two sources of variation: Only genetic influences are inherited. Any effects of the environment are generally not passed on to the next generation.

? If most characters result from a combination of environment and heredity, why was Mendel able to ignore environmental influences on his pea plants?

● The characters he chose for study were all entirely genetically determined.

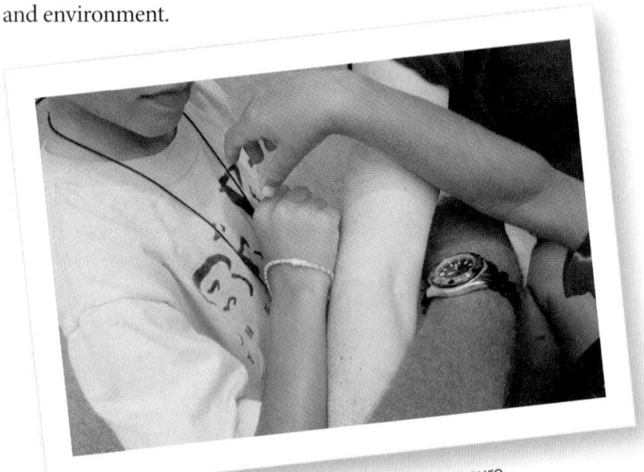

Eric J. Simon

▲ **Figure 15A** The effect of genes and sun exposure on the skin of one of this book's authors and his family

Eric J. Simon

▲ **Figure 15B** Varying phenotypes due to environmental factors in genetically identical twins

The Chromosomal Basis of Inheritance

16 Chromosome behavior accounts for Mendel's laws

Mendel published his results in 1866, but not until long after he died did biologists understand the significance of his work. Cell biologists worked out the processes of mitosis and meiosis in the late 1800s. Then, around 1900, researchers began to notice parallels between the behavior of chromosomes and the behavior of Mendel's heritable factors. Eventually, one of biology's most important concepts emerged. By combining these observations, the **chromosome theory of inheritance** states that genes occupy specific loci (positions) on chromosomes, and it is the chromosomes that undergo segregation and independent assortment during meiosis. Thus, it is the behavior of chromosomes during meiosis and fertilization that accounts for inheritance patterns.

We can see the chromosomal basis of Mendel's laws by following the fates of two genes during meiosis and fertilization in pea plants. In **Figure 16**, we show the genes for seed shape (alleles *R* and *r*) and seed color (*Y* and *y*) as black bars on different chromosomes. Notice that the Punnett square is repeated from Figure 5A; we will now follow the chromosomes to see how they account for the results of the dihybrid cross shown in the Punnett square. We start with the F_1 generation, in which all plants have the *RrYy* genotype. To simplify the diagram, we show only two of the seven pairs of pea chromosomes and three of the stages of meiosis: metaphase I, anaphase I, and metaphase II.

To see the chromosomal basis of the law of segregation (which states that pairs of alleles separate from each other during gamete formation via meiosis), let's follow just the homologous pair of long chromosomes, the ones carrying *R* and *r*, taking either the left or the right branch from the F_1 cell. Whichever arrangement the chromosomes assume at metaphase I, the two alleles segregate as the homologous chromosomes separate in anaphase I. And at the end of meiosis II, a single long chromosome ends up in each of the gametes. Fertilization then recombines the two alleles at random, resulting in F_2 offspring that are $\frac{1}{4}$ *RR*, $\frac{1}{2}$ *Rr*, and $\frac{1}{4}$ *rr*. The ratio of round to wrinkled phenotypes is thus 3:1 (12 round to 4 wrinkled), the ratio Mendel observed, as shown in the Punnett square in the figure.

To see the chromosomal basis of the law of independent assortment (which states that each pair of alleles sorts independently of other pairs of alleles during gamete formation; see Module 5), follow both the long and short (nonhomologous) chromosomes through the figure below. Two alternative, equally likely arrangements of tetrads can occur at metaphase I. The nonhomologous chromosomes (and their genes) assort independently, leading to four gamete genotypes. Random fertilization leads to the 9:3:3:1 phenotypic ratio in the F_2 generation.

? Which of Mendel's laws have their physical basis in the following phases of meiosis: (a) the orientation of homologous chromosome pairs in metaphase I; (b) the separation of homologs in anaphase I?

(a) The law of independent assortment; (b) the law of segregation

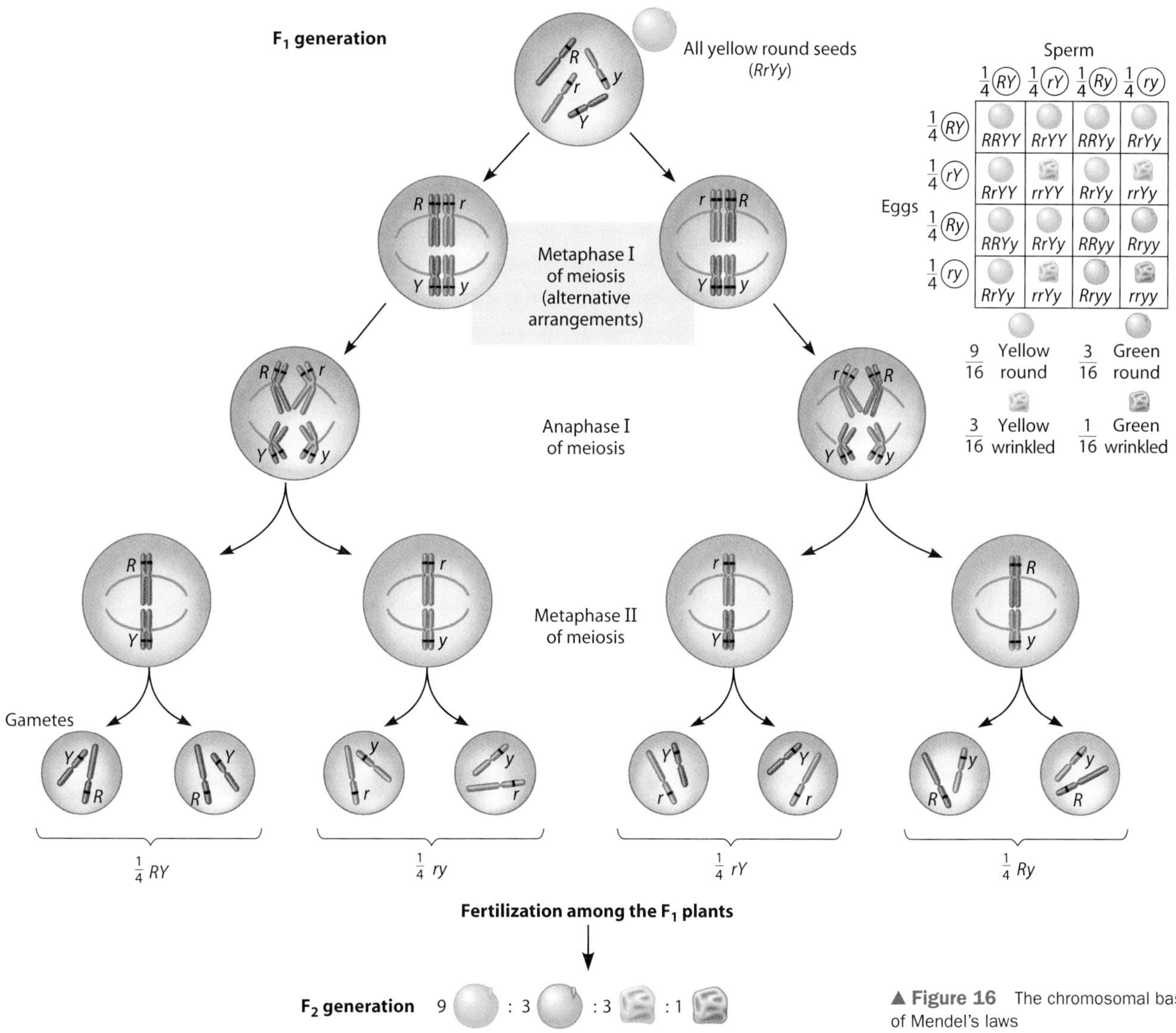

▲ **Figure 16** The chromosomal basis of Mendel's laws

SCIENTIFIC DISCOVERY 17 Genes on the same chromosome tend to be inherited together

In 1908, British biologists William Bateson and Reginald Punnett (originator of the Punnett square) observed an inheritance pattern that seemed inconsistent with Mendelian laws. Bateson and Punnett were working with two characters in sweet peas: flower color and pollen shape. They crossed doubly heterozygous plants (*PpLl*) that exhibited the dominant traits: purple flowers (expression of the *P* allele) and long pollen grains (expression of the *L* allele). The corresponding recessive traits are red flowers (in *pp* plants) and round pollen (in *ll* plants).

The top part of **Figure 17** illustrates Bateson and Punnett's experiment. When they looked at just one of the two characters (that is, either cross $Pp \times Pp$ or cross $Ll \times Ll$), they found that the dominant and recessive alleles segregated, producing a phenotypic ratio of approximately 3:1 for the offspring, in agreement with Mendel's law of segregation. However, when the biologists combined their data for the two characters, they did not see the predicted 9:3:3:1 ratio. Instead, as shown in the table, they found a disproportionately large number of plants with just two of the predicted phenotypes: purple long (almost 75% of the total) and red round (about 14%). The other two phenotypes (purple round and red long) were found in far fewer numbers than expected. What can account for these results?

The number of genes in a cell is far greater than the number of chromosomes; in fact, each chromosome has hundreds or thousands of genes. Genes located close together on the same chromosome tend to be inherited together and are called **linked genes**. Linked genes generally do not follow Mendel's law of independent assortment.

As shown in the explanation section of the figure, sweet-pea genes for flower color and pollen shape are located on the same chromosome. Thus, meiosis in the heterozygous (*PpLl*) sweet-pea plant yields mostly two genotypes of gametes (*PL* and *pl*) rather than equal numbers of the four types of gametes that would result if the flower-color and pollen-shape genes were not linked. The large numbers of plants with purple long and red round traits in the Bateson-Punnett experiment resulted from fertilization among the *PL* and *pl* gametes. But what about the smaller numbers of plants with purple round and red long traits? As you will see in the next module, the phenomenon of crossing over accounts for these offspring.

The Experiment

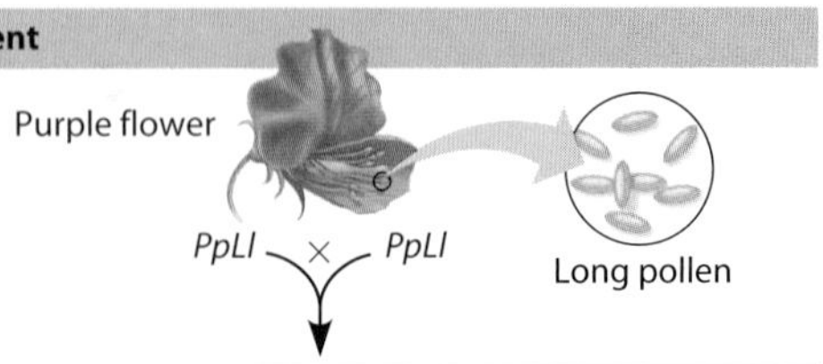

Phenotypes	Observed offspring	Prediction (9:3:3:1)
Purple long	284	215
Purple round	21	71
Red long	21	71
Red round	55	24

The Explanation: Linked Genes

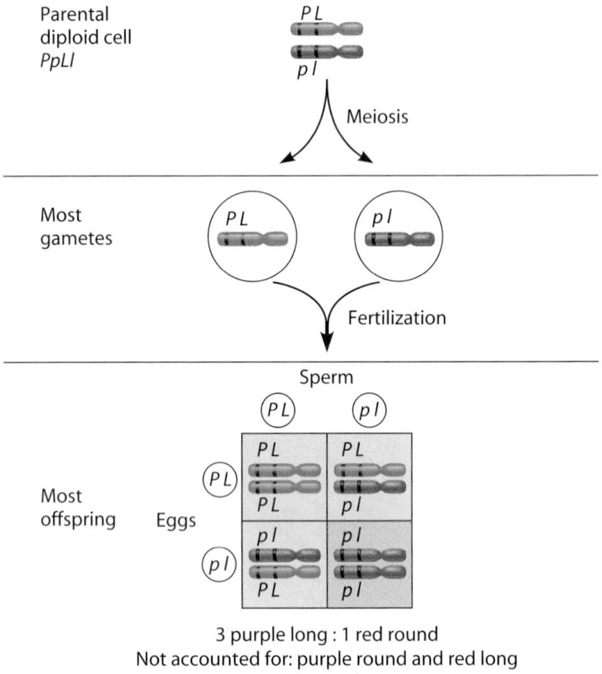

▲ **Figure 17** The experiment revealing linked genes in the sweet pea

? Why do linked genes tend to be inherited together?

Because they are located close together on the same chromosome

SCIENTIFIC DISCOVERY 18 Crossing over produces new combinations of alleles

During meiosis, crossing over between homologous chromosomes produces new combinations of alleles in gametes. Using the experiment shown in Figure 17 as an example, **Figure 18A** reviews this process, showing that two linked genes can give rise to four different gamete genotypes. Gametes with genotypes *PL* and *pl* carry parental-type chromosomes that have not been altered by crossing over. In contrast, gametes with genotypes *Pl* and *pL* are recombinant gametes. The exchange of chromosome segments during crossing over has produced new combinations of alleles. We can now understand the results of the Bateson-Punnett experiment presented in the previous module: The small fraction of offspring with recombinant phenotypes (purple round and red long) must have resulted from fertilization involving recombinant gametes.

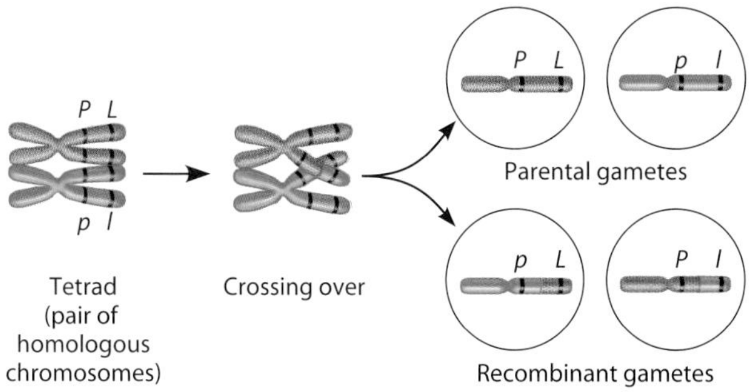

▲ **Figure 18A** Review: the production of recombinant gametes

The discovery of how crossing over creates gamete diversity confirmed the relationship between chromosome behavior and heredity. Some of the most important early studies of crossing over were performed in the laboratory of American embryologist Thomas Hunt Morgan in the early 1900s. Morgan and his colleagues used the fruit fly *Drosophila melanogaster* in many of their experiments (**Figure 18B**). Often seen flying around ripe fruit, *Drosophila* is a good research animal for genetic studies because it can be easily and inexpensively bred, producing each new generation in a couple of weeks.

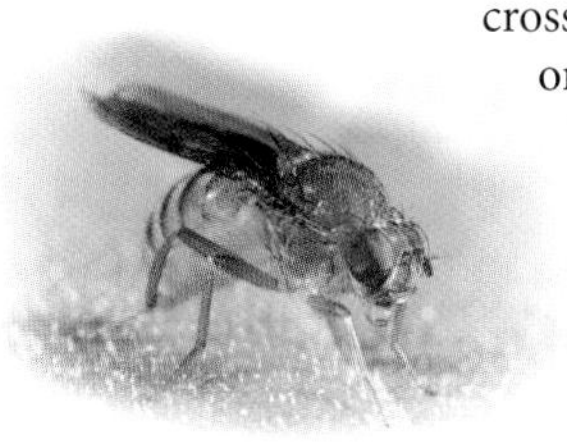

Graphic Science/Alamy

▲ **Figure 18B** *Drosophila melanogaster*

Figure 18C shows one of Morgan's experiments, a cross between a wild-type fruit fly (recall from Module 8 that "wild-type" refers to the traits most common in nature, in this case, gray body and long wings) and a fly with a black body and undeveloped, or vestigial, wings. Morgan knew the genotypes of these flies from previous studies. Here we use the following gene symbols:

G = gray body (dominant)
g = black body (recessive)
L = long wings (dominant)
l = vestigial wings (recessive)

In mating a heterozygous gray fly with long wings (genotype *GgLl*) with a black fly with vestigial wings (genotype *ggll*), Morgan performed a testcross (see Module 6). If the genes were not linked, then independent assortment would produce offspring in a phenotypic ratio of 1:1:1:1 ($\frac{1}{4}$ gray body, long wings; $\frac{1}{4}$ black body, vestigial wings; $\frac{1}{4}$ gray body, vestigial wings; and $\frac{1}{4}$ black body, long wings). But because these genes are linked, Morgan obtained the results shown in the top part of Figure 18C: Most of the offspring had parental phenotypes, but 17% of the offspring flies were recombinants. The percentage of recombinants is called the **recombination frequency**.

When Morgan first obtained these results, he did not know about crossing over. To explain the ratio of offspring, he hypothesized that the genes were linked and that some mechanism occasionally broke the linkage. Tests of the hypothesis proved him correct, establishing that crossing over was the mechanism that "breaks linkages" between genes.

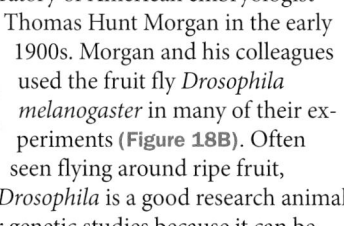

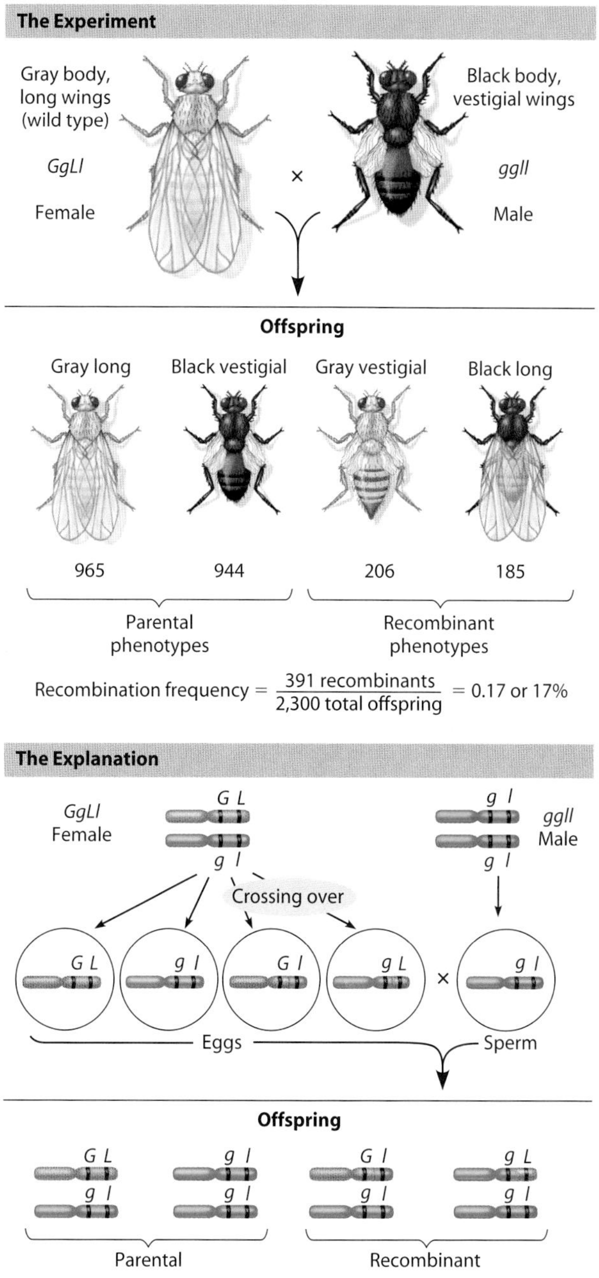

▲ **Figure 18C** A fruit fly experiment demonstrating the role of crossing over in inheritance

The lower part of Figure 18C explains Morgan's results in terms of crossing over. A crossover between chromatids of homologous chromosomes in parent *GgLl* broke linkages between the *G* and *L* alleles and between the *g* and *l* alleles, forming the recombinant chromosomes *Gl* and *gL*. Later steps in meiosis distributed the recombinant chromosomes to gametes, and random fertilization produced the four kinds of offspring Morgan observed.

? Return to the data in Figure 17. What is the recombination frequency between the flower-color and pollen-length genes?

11% ($\frac{42}{381}$)

19 Geneticists use crossover data to map genes

Working with *Drosophila*, T. H. Morgan and his students—Alfred H. Sturtevant in particular—greatly advanced our understanding of genetics during the early 20th century. One of Sturtevant's major contributions to genetics was an approach for using crossover data to map gene loci. His reasoning was elegantly simple: The greater the distance between two genes, the more points there are between them where crossing over can occur. With this principle in mind, Sturtevant began using recombination data from fruit fly crosses to assign relative positions of the genes on the chromosomes—that is, to map genes.

Figure 19A represents a part of the chromosome that carries the linked genes for black body (*g*) and vestigial wings (*l*) that we described in Module 18. This same chromosome also carries a gene that has a recessive allele (we'll call it *c*) determining cinnabar eye color, a brighter red than the wild-type color. Figure 19A shows the actual crossover (recombination) frequencies between these alleles, taken two at a time: 17% between the *g* and *l* alleles, 9% between *g* and *c*, and 5% between *c* and *l*. Sturtevant reasoned that these values represent the relative distances between the genes. Because the crossover frequencies between *g* and *c* and between *l* and *c* are approximately half that between *g* and *l*, gene *c* must lie roughly midway between *g* and *l*. Thus, the sequence of these genes on one of the fruit fly chromosomes must be *g*-*c*-*l*. Such a diagram of relative gene locations is called a **linkage map**.

Sturtevant started by assuming that the chance of crossing over is approximately equal at all points along a chromosome. We now know that this assumption is only approximately true; some locations along the chromosome are more prone to crossing over than others. Still, his method of mapping genes worked well, and it proved extremely valuable in establishing the relative positions of many other fruit fly genes. Eventually, enough data were accumulated to reveal that *Drosophila* has four groups of genes, corresponding to its four pairs of homologous chromosomes. **Figure 19B** is a genetic map showing just five of the gene loci on part of one chromosome: the loci labeled *g*, *c*, and *l* and two others. Notice that eye color is a character affected by more than one gene. Here we see the cinnabar-eye and brown-eye genes; still other eye-color genes are found elsewhere (see Module 21). For each of these genes, however, the wild-type allele specifies red eyes.

The linkage-mapping method has proved extremely valuable in establishing the relative positions of many genes in many organisms. The real beauty of the technique is that a wealth of information about genes can be learned simply by breeding and observing the organisms; no fancy equipment is required.

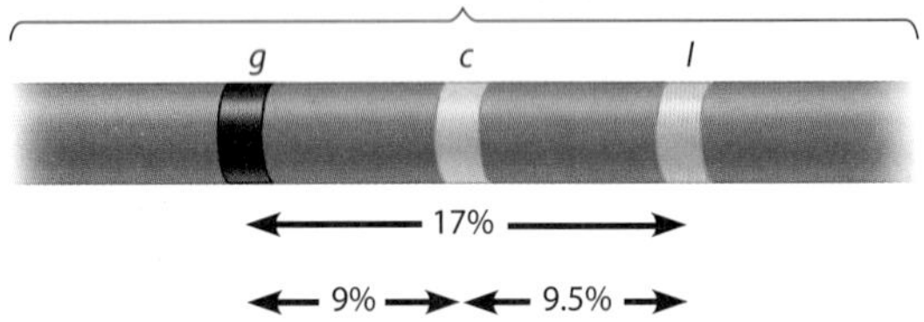

▲ **Figure 19A** Mapping genes from crossover data

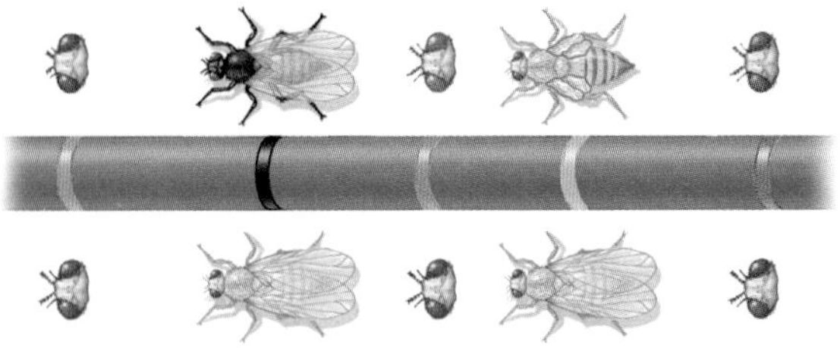

▲ **Figure 19B** A partial linkage map of a fruit fly chromosome

? You design *Drosophila* crosses to provide recombination data for a gene not included in Figure 19A. The gene has recombination frequencies of 3% with the vestigial-wing (*l*) locus and 7% with the cinnabar-eye (*c*) locus. Where is it located on the chromosome?

● The gene is located between the vestigial and cinnabar loci, a bit closer to the vestigial-wing locus (since it has a lower recombination frequency).

Sex Chromosomes and Sex-Linked Genes

20 Chromosomes determine sex in many species

Many animals, including fruit flies and all mammals, have a pair of **sex chromosomes**, designated X and Y, that determine an individual's sex (**Figure 20A**). **Figure 20B** reviews sex determination in humans. Individuals with one X chromosome and one Y chromosome are males; XX individuals are females.

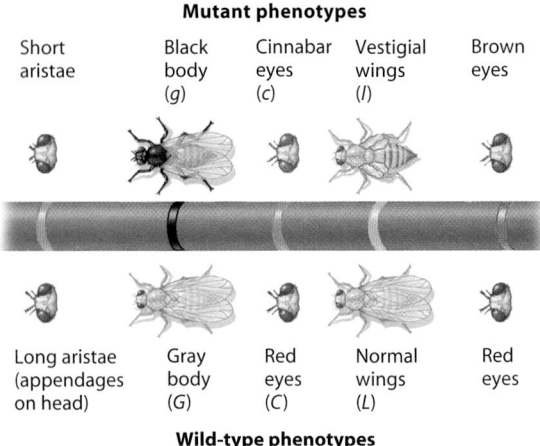

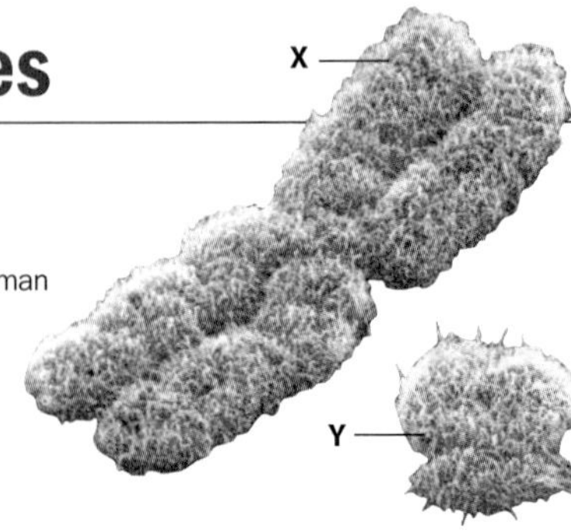

► **Figure 20A** The human sex chromosomes

Colorized SEM 31,955×
Andrew Syred/Photo Researchers

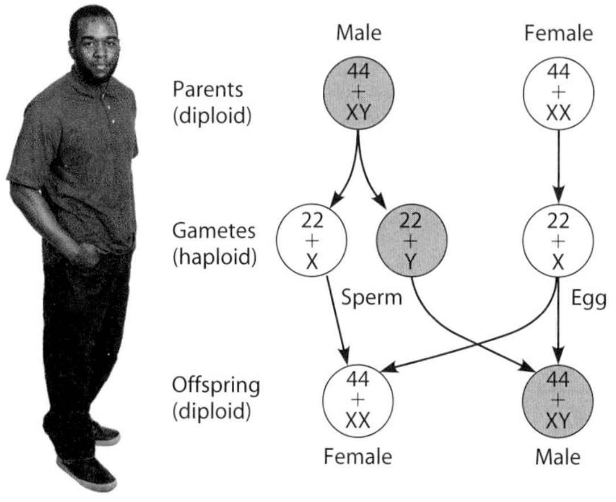

© Rob Byron-Fotolia

▲ **Figure 20B** The X-Y system

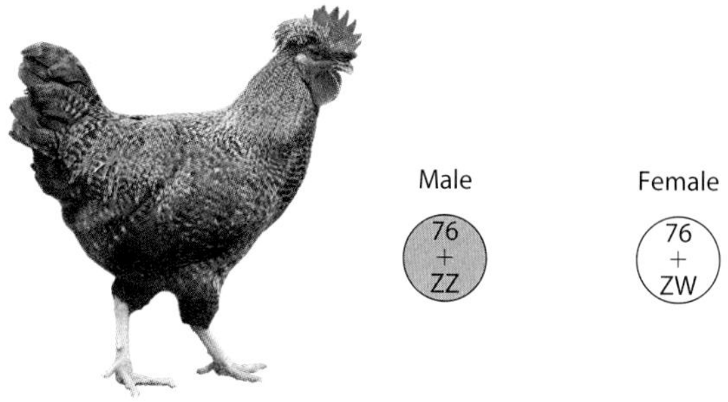

kosam/Shutterstock

▲ **Figure 20D** The Z-W system

Human males and females both have 44 autosomes (nonsex chromosomes). As a result of chromosome segregation during meiosis, each gamete contains one sex chromosome and a haploid set of autosomes (22 in humans). All eggs contain a single X chromosome. Of the sperm cells, half contain an X chromosome and half contain a Y chromosome. An offspring's sex depends on whether the sperm cell that fertilizes the egg bears an X chromosome or a Y chromosome. In the fruit fly's X-Y system, sex is determined by the ratio between the number of X chromosomes and the number of autosome sets, although the Y chromosome is essential for sperm formation.

The genetic basis of sex determination in humans is not yet completely understood, but one gene on the Y chromosome plays a crucial role. This gene, discovered by a British research team in 1990, is called *SRY* (for sex-determining region of Y) and triggers testis development. In the absence of *SRY*, an individual develops ovaries rather than testes. *SRY* codes for proteins that regulate other genes on the Y chromosome. These genes in turn produce proteins necessary for normal testis development.

The X-Y system is only one of several sex-determining systems. For example, grasshoppers, roaches, and some other insects have an X-O system, in which O stands for the absence of a sex chromosome **(Figure 20C)**. Females have two X chromosomes (XX); males have only one sex chromosome (XO). Males produce two classes of sperm (X-bearing and lacking any sex chromosome), and sperm cells determine the sex of the offspring at fertilization.

In contrast to the X-Y and X-O systems, eggs determine sex in certain fishes, butterflies, and birds **(Figure 20D)**. The sex chromosomes in these animals are designated Z and W. Males have the genotype ZZ; females are ZW. In this system, sex is determined by whether the egg carries a Z or a W.

Some organisms lack sex chromosomes altogether. In most ants and bees, sex is determined by chromosome number rather than by sex chromosomes **(Figure 20E)**. Females develop from fertilized eggs and thus are diploid. Males develop from unfertilized eggs—they are fatherless—and are haploid.

Most animals have two separate sexes; that is, individuals are either male or female. Many plant species have sperm-bearing and egg-bearing flowers borne on different individuals. Some plant species, such as date palms, have the X-Y system of sex determination; others, such as the wild strawberry, have the Z-W system. However, most plant species and some animal species have individuals that produce both sperm and eggs. In such species, all individuals have the same complement of chromosomes.

In Module 15, we discussed the role that environment plays in determining many characters. Among some animals, environment can even determine sex. For some species of reptiles, the temperature at which eggs are incubated during a specific period of embryonic development determines whether that embryo will develop into a male or female. For example, if green sea turtle hatchlings incubate above 30°C (86°F), nearly all the resulting turtles will be males. (Some worry that global climate change might have the unexpected consequence of affecting the makeup of turtle populations.) Such temperature-dependent sex determination is an extreme example of the environment affecting the phenotype of an individual.

? **During fertilization in humans, what determines the sex of the offspring?**

● Whether the egg is fertilized by a sperm bearing an X chromosome (producing a female offspring) or by a sperm with a Y chromosome (producing a male)

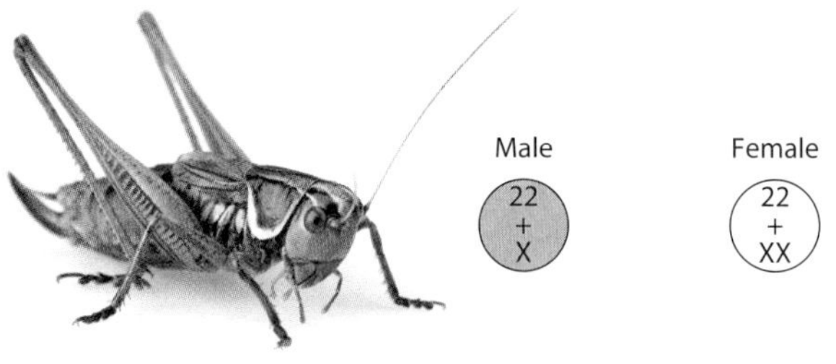

Tomasz Zachariasz/iStockphoto

▲ **Figure 20C** The X-O system

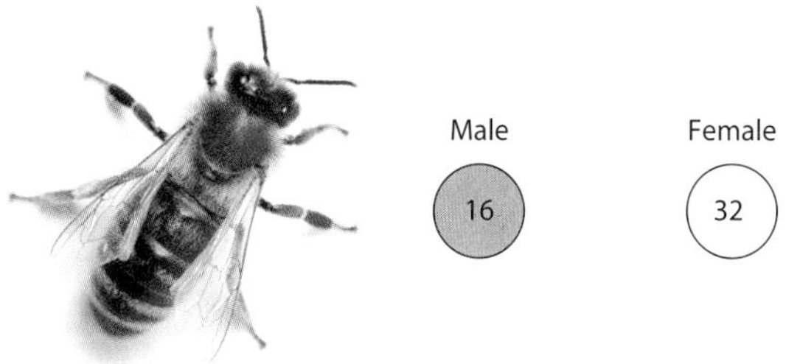

Tomasz Zachariasz/iStockphoto

▲ **Figure 20E** Sex determination by chromosome number

21 Sex-linked genes exhibit a unique pattern of inheritance

Besides bearing genes that determine sex, the sex chromosomes also contain genes for characters unrelated to femaleness or maleness. A gene located on either sex chromosome is called a **sex-linked gene**. Because the human X chromosome contains many more genes than the Y, the term has historically referred specifically to genes on the X chromosome. (Be careful not to confuse the term *sex-linked gene*, which refers to a gene on a sex chromosome, with the term *linked genes*, which refers to genes on the same chromosome that tend to be inherited together.)

The figures in this module illustrate inheritance patterns for white eye color in the fruit fly, an X-linked recessive trait. Wild-type fruit flies have red eyes; white eyes are very rare (**Figure 21A**). We use the uppercase letter *R* for the dominant, wild-type, red-eye allele and *r* for the recessive white-eye allele. Because these alleles are carried on the X chromosome, we show them as superscripts to the letter X. Thus, red-eyed male fruit flies have the genotype X^RY; white-eyed males are X^rY. The Y chromosome does not have a gene locus for eye color; therefore, the male's phenotype results entirely from his single X-linked gene. In the female, X^RX^R and X^RX^r flies have red eyes, and X^rX^r flies have white eyes.

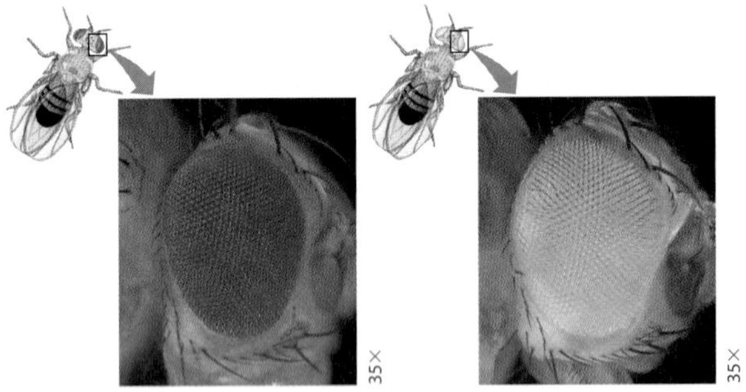

From: "Learning to Fly: Phenotypic Markers in Drosophila." A poster of common phenotypic markers used in Drosophila genetics. Jennifer Childress, Richard Behringer, and Georg Halder. 2005. Genesis 43(1). Cover illustration

▲ **Figure 21A** Fruit fly eye color determined by sex-linked gene

A white-eyed male (X^rY) will transmit his X^r to all of his female offspring but to none of his male offspring. This is because his female offspring, in order to be female, must inherit his X chromosome, but his male offspring must inherit his Y chromosome.

As shown in **Figure 21B**, when the female parent is a dominant homozygote (X^RX^R) and the male parent is X^rY, all the offspring have red eyes, but the female offspring are all carriers of the allele for white eyes (X^RX^r). When those offspring are bred to each other, the classic 3:1 phenotypic ratio of red eyes to white eyes appears among the offspring (**Figure 21C**). However, there is a twist: The white-eyed trait shows up only in males. All the females have red eyes, whereas half the males have red eyes and half have white eyes. All females inherit at least one dominant allele (from their male parent); half of them are homozygous dominant, whereas the other half are heterozygous carriers, like their female parent. Among the males, half of them inherit the recessive allele their mother was carrying, producing the white-eye phenotype.

Because the white-eye allele is recessive, a female will have white eyes only if she receives that allele on both X chromosomes. For example, if a heterozygous female mates with a white-eyed male, there is a 50% chance that each offspring will have white eyes (resulting from genotype X^rX^r or X^rY), regardless of sex (**Figure 21D**). Female offspring with red eyes are heterozygotes, whereas red-eyed male offspring completely lack the recessive allele.

? A white-eyed female *Drosophila* is mated with a red-eyed (wild-type) male. What result do you predict for the numerous offspring?

All female offspring will be red-eyed but heterozygous (X^RX^r); all male offspring will be white-eyed (X^rY).

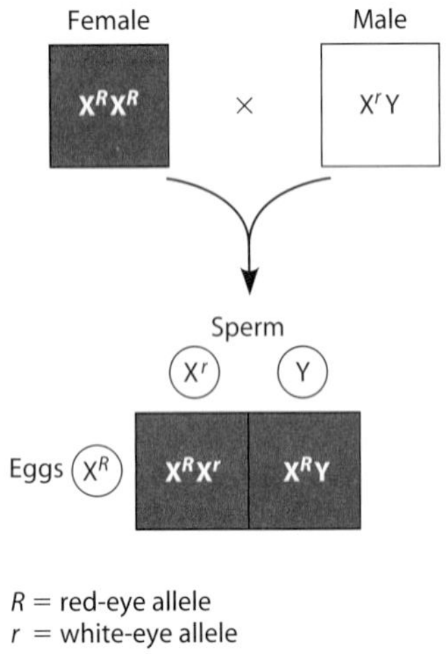

▲ **Figure 21B** A homozygous, red-eyed female crossed with a white-eyed male

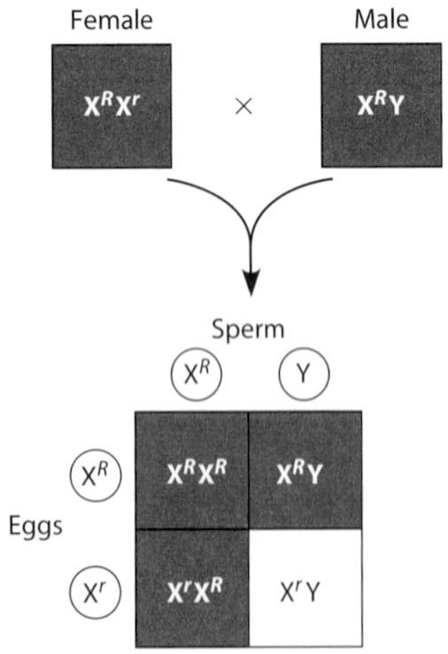

▲ **Figure 21C** A heterozygous female crossed with a red-eyed male

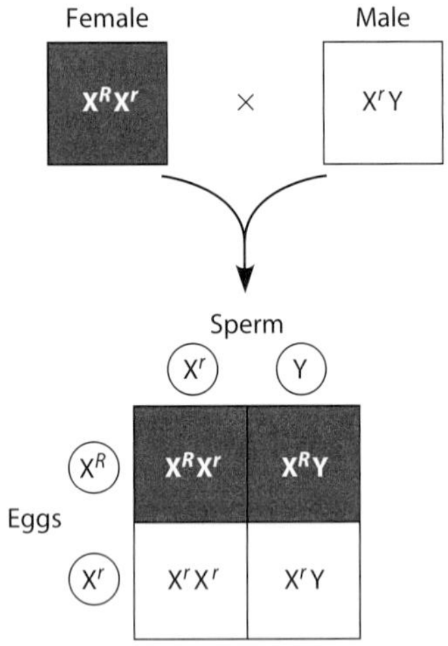

▲ **Figure 21D** A heterozygous female crossed with a white-eyed male

CONNECTION 22 Human sex-linked disorders affect mostly males

As in fruit flies, a number of human conditions result from sex-linked (X-linked) recessive alleles. Like a male fruit fly, if a man inherits only one X-linked recessive allele—from his mother—the allele will be expressed. In contrast, a woman has to inherit two such alleles—one from each parent—in order to exhibit the trait. Thus, recessive X-linked traits are expressed much more frequently in men than in women.

Hemophilia is an X-linked recessive trait with a long, well-documented history. Hemophiliacs bleed excessively when injured because they lack one or more of the proteins required for blood clotting. A high incidence of hemophilia plagued the royal families of Europe. Queen Victoria (1819–1901) of England was a carrier of the hemophilia allele. She passed it on to one of her sons and two of her daughters. Through marriage, her daughters then introduced the disease into the royal families of Prussia, Russia, and Spain. Thus, the age-old practice of strengthening international alliances by marriage effectively spread hemophilia through the royal families of several nations. The pedigree in Figure 22 traces the disease through one branch of the royal family. As you can see in the pedigree, Alexandra, like her mother and grandmother, was a carrier, and Alexis had the disease.

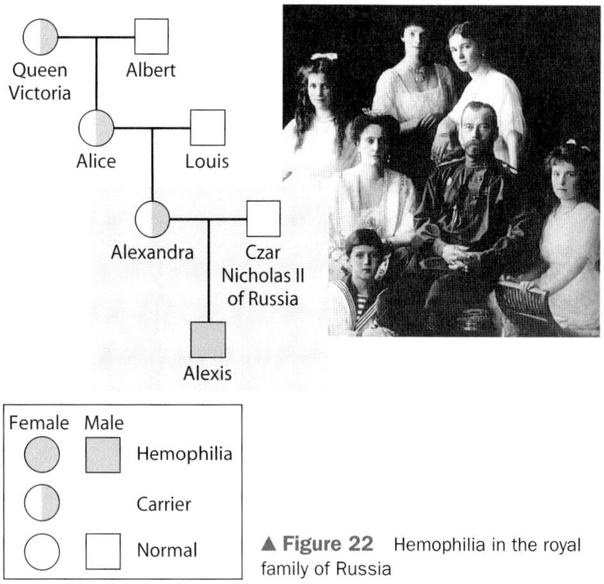

Taxi/Getty Images

▲ **Figure 22** Hemophilia in the royal family of Russia

Another recessive disorder in humans that is sex-linked is **red-green colorblindness**, a malfunction of light-sensitive cells in the eyes. Colorblindness is actually a class of disorders that involves several X-linked genes. A person with normal color vision can see more than 150 colors. In contrast, someone with red-green colorblindness can see fewer than 25. Mostly males are affected, but heterozygous females have some defects.

Duchenne muscular dystrophy, a condition characterized by a progressive weakening of the muscles and loss of coordination, is another human X-linked recessive disorder. The first symptoms appear in early childhood, when the child begins to have difficulty standing up. He is inevitably wheelchair-bound by age 12. Eventually, muscle tissue becomes severely wasted, and normal breathing becomes difficult. Affected individuals rarely live past their early 20s.

? Neither Tom nor Sue has hemophilia, but their first son does. If the couple has a second child, what is the probability that he or she will also have the disease?

$\frac{1}{4}$ ($\frac{1}{2}$ chance of a male child × $\frac{1}{2}$ chance that he will inherit the mutant X)

EVOLUTION CONNECTION 23 The Y chromosome provides clues about human male evolution

Barring mutations, the human Y chromosome passes essentially intact from father to son. By analyzing Y DNA, researchers can learn about the ancestry of human males.

In 2003, geneticists discovered that about 8% of males currently living in central Asia have Y chromosomes of striking genetic similarity. Further analysis traced their common genetic heritage to a single man living about 1,000 years ago. In combination with historical records, the data led to the speculation that the Mongolian ruler Genghis Kahn (Figure 23) may be responsible for the spread of the unusual chromosome to nearly 16 million men living today. A similar study of Irish men in 2006 suggested that nearly 10% of them were descendants of Niall of the Nine Hostages, a warlord who lived during the 5th century.

Another study of Y DNA seemed to confirm the claim by the Lemba people of southern Africa that they are descended from ancient Jews. Sequences of Y DNA distinctive of the Jewish priestly caste called Cohanim (descendants of Moses' brother Aaron, according to the Bible) are found at high frequencies among the Lemba.

▲ **Figure 23** Genghis Kahn

Dschingis Khan und seine Erben (exhibition catalogue), München 2005, p. 304

The discovery of the sex chromosomes and their pattern of inheritance was one of many breakthroughs in understanding how genes are passed from one generation to the next. During the first half of the 20th century, geneticists rediscovered Mendel's work, reinterpreted his laws in light of chromosomal behavior during meiosis, and firmly established the chromosome theory of inheritance. The chromosome theory set the stage for another explosion of experimental work in the second half of the 20th century.

? Why is the Y chromosome particularly useful in tracing recent human heritage?

Because it is passed directly from father to son, forming an unbroken chain of male lineage

REVIEW

For Practice Quizzes, BioFlix, MP3 Tutors, and Activities, go to www.masteringbiology.com.

Reviewing the Concepts

Mendel's Laws (1–10)

1 The science of genetics has ancient roots.

2 Experimental genetics began in an abbey garden. The science of genetics began with Gregor Mendel's quantitative experiments. Mendel crossed pea plants and traced traits from generation to generation. He hypothesized that there are alternative versions of genes (alleles), the units that determine heritable traits.

3 Mendel's law of segregation describes the inheritance of a single character. Mendel's law of segregation predicts that each set of alleles will separate as gametes are formed:

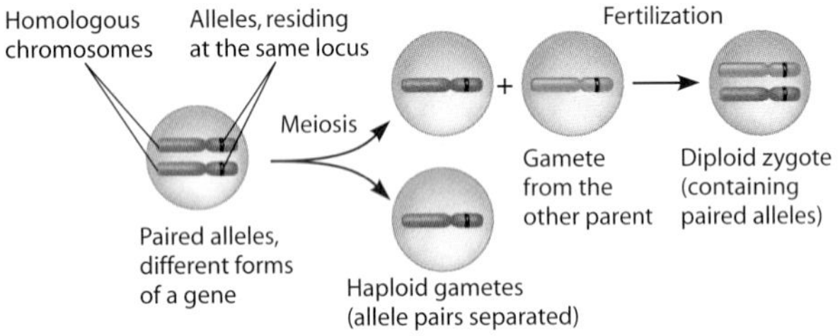

4 Homologous chromosomes bear the alleles for each character. When the two alleles of a gene in a diploid individual are different, the dominant allele determines the inherited trait, whereas the recessive allele has no effect.

5 The law of independent assortment is revealed by tracking two characters at once. Mendel's law of independent assortment states that the alleles of a pair segregate independently of other allele pairs during gamete formation.

6 Geneticists can use the testcross to determine unknown genotypes. The offspring of a testcross, a mating between an individual of unknown genotype and a homozygous recessive individual, can reveal the unknown's genotype.

7 Mendel's laws reflect the rules of probability. The rule of multiplication calculates the probability of two independent events both occurring. The rule of addition calculates the probability of an event that can occur in alternative ways.

8 Genetic traits in humans can be tracked through family pedigrees. The inheritance of many human traits follows Mendel's laws. Family pedigrees can help determine individual genotypes.

9 Many inherited disorders in humans are controlled by a single gene.

10 New technologies can provide insight into one's genetic legacy. Carrier screening, fetal testing, fetal imaging, and newborn screening can provide information for reproductive decisions but may create ethical dilemmas.

Variations on Mendel's Laws (11–15)

11 Incomplete dominance results in intermediate phenotypes. Mendel's laws are valid for all sexually reproducing species, but genotype often does not dictate phenotype in the simple way his laws describe, as shown in the figure at the top of the next column.

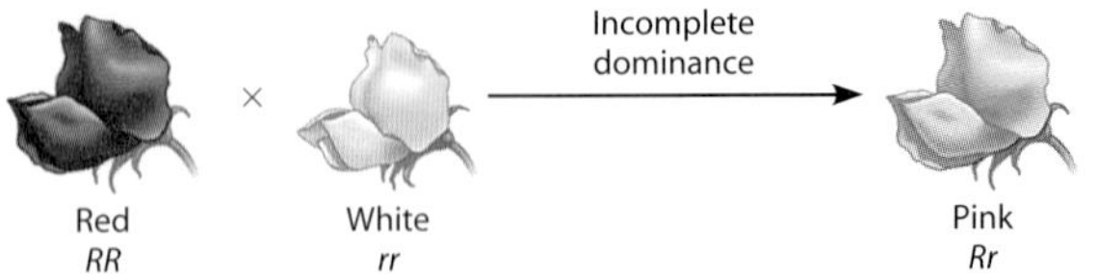

12 Many genes have more than two alleles in the population. For example, the ABO blood group phenotype in humans is controlled by three alleles that produce a total of four phenotypes.

13 A single gene may affect many phenotypic characters:

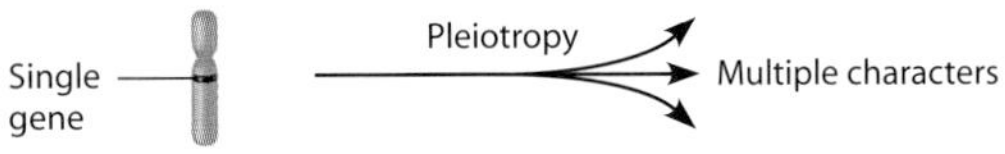

14 A single character may be influenced by many genes:

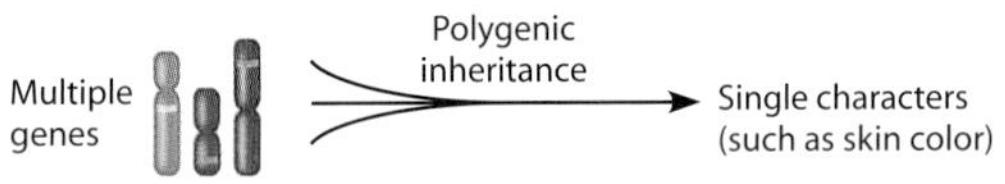

15 The environment affects many characters. Many traits are affected, in varying degrees, by both genetic and environmental factors.

The Chromosomal Basis of Inheritance (16–19)

16 Chromosome behavior accounts for Mendel's laws. Genes are located on chromosomes, whose behavior during meiosis and fertilization accounts for inheritance patterns.

17 Genes on the same chromosome tend to be inherited together. Such genes are said to be linked; they display non-Mendelian inheritance patterns.

18 Crossing over produces new combinations of alleles. Crossing over can separate linked alleles, producing gametes with recombinant chromosomes.

19 Geneticists use crossover data to map genes. Recombination frequencies can be used to map the relative positions of genes on chromosomes.

Sex Chromosomes and Sex-Linked Genes (20–23)

20 Chromosomes determine sex in many species. In mammals, a male has XY sex chromosomes, and a female has XX. The Y chromosome has genes for the development of testes, whereas an absence of the Y allows ovaries to develop. Other systems of sex determination exist in other animals and plants.

21 Sex-linked genes exhibit a unique pattern of inheritance. All genes on the sex chromosomes are said to be sex-linked. However, the X chromosome carries many genes unrelated to sex.

22 Human sex-linked disorders affect mostly males. Most sex-linked (X-linked) human disorders are due to recessive alleles and are seen mostly in males. A male receiving a single X-linked recessive allele from his mother will have the disorder; a female must receive the allele from both parents to be affected.

23 The Y chromosome provides clues about human male evolution. Because they are passed on intact from father to son, Y chromosomes can provide data about recent human evolutionary history.

Connecting the Concepts

1. Complete this concept map to help you review some key concepts of genetics.

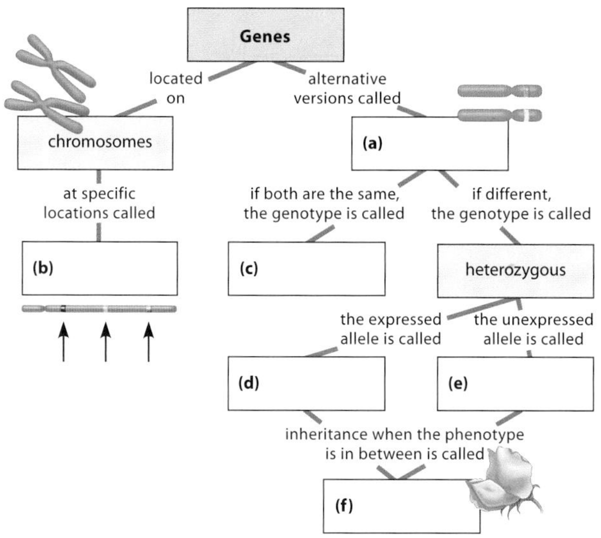

Testing Your Knowledge

Multiple Choice

2. Edward was found to be heterozygous (*Ss*) for sickle-cell trait. The alleles represented by the letters *S* and *s* are
   a. on the X and Y chromosomes.
   b. linked.
   c. on homologous chromosomes.
   d. both present in each of Edward's sperm cells.
   e. on the same chromosome but far apart.

3. Whether an allele is dominant or recessive depends on
   a. how common the allele is, relative to other alleles.
   b. whether it is inherited from the mother or the father.
   c. which chromosome it is on.
   d. whether it or another allele determines the phenotype when both are present.
   e. whether or not it is linked to other genes.

4. Two fruit flies with eyes of the usual red color are crossed, and their offspring are as follows: 77 red-eyed males, 71 ruby-eyed males, 152 red-eyed females. The allele for ruby eyes is
   a. autosomal (carried on an autosome) and dominant.
   b. autosomal and recessive.
   c. sex-linked and dominant.
   d. sex-linked and recessive.
   e. impossible to determine without more information.

5. A man who has type B blood and a woman who has type A blood could have children of which of the following phenotypes?
   a. A or B only
   b. AB only
   c. AB or O
   d. A, B, or O
   e. A, B, AB, or O

Additional Genetics Problems

6. Why do more men than women have colorblindness?

7. In fruit flies, the genes for wing shape and body stripes are linked. In a fly whose genotype is *WwSs*, *W* is linked to *S*, and *w* is linked to *s*. Show how this fly can produce gametes containing four different combinations of alleles. Which are parental-type gametes? Which are recombinant gametes? How are the recombinants produced?

8. Adult height in humans is at least partially hereditary; tall parents tend to have tall children. But humans come in a range of sizes, not just tall and short. Which extension of Mendel's model accounts for the hereditary variation in human height?

9. Tim and Jan both have freckles (see Module 8), but their son Mike does not. Show with a Punnett square how this is possible. If Tim and Jan have two more children, what is the probability that both will have freckles?

10. Both Tim and Jan (problem 9) have a widow's peak (see Module 8), but Mike has a straight hairline. What are their genotypes? What is the probability that Tim and Jan's next child will have freckles and a straight hairline?

11. In rabbits, black hair depends on a dominant allele, *B*, and brown hair on a recessive allele, *b*. Short hair is due to a dominant allele, *S*, and long hair to a recessive allele, *s*. If a true-breeding black, short-haired male is mated with a brown, long-haired female, describe their offspring. What will be the genotypes of the offspring? If two of these F_1 rabbits are mated, what phenotypes would you expect among their offspring? In what proportions?

12. A fruit fly with a gray body and red eyes (genotype *BbPp*) is mated with a fly having a black body and purple eyes (genotype *bbpp*). What ratio of offspring would you expect if the body-color and eye-color genes are on different chromosomes (unlinked)? When this mating is actually carried out, most of the offspring look like the parents, but 3% have a gray body and purple eyes, and 3% have a black body and red eyes. Are these genes linked or unlinked? What is the recombination frequency?

13. A series of matings shows that the recombination frequency between the black-body gene (problem 12) and the gene for dumpy (shortened) wings is 36%. The recombination frequency between purple eyes and dumpy wings is 41%. What is the sequence of these three genes on the chromosome?

14. A couple are both phenotypically normal, but their son suffers from hemophilia, a sex-linked recessive disorder. What fraction of their children are likely to suffer from hemophilia? What fraction are likely to be carriers?

15. Heather was surprised to discover she suffered from red-green colorblindness. She told her biology professor, who said, "Your father is colorblind too, right?" How did her professor know this? Why did her professor not say the same thing to the colorblind males in the class?

Applying the Concepts

16. In 1981, a stray black cat with unusual rounded, curled-back ears was adopted by a family in Lakewood, California. Hundreds of descendants of this cat have since been born, and cat fanciers hope to develop the "curl" cat into a show breed. The curl allele is apparently dominant and autosomal (carried on an autosome). Suppose you owned the first curl cat and wanted to breed it to develop a true-breeding variety. Describe tests that would determine whether the curl gene is dominant or recessive and whether it is autosomal or sex-linked. Explain why you think your tests would be conclusive. Describe a test to determine that a cat is true-breeding.

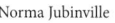

Norma Jubinville

Review Answers

1. a. alleles; b. loci; c. homozygous; d. dominant; e. recessive; f. incomplete dominance
2. c 3. d 4. d (Neither parent is ruby-eyed, but some offspring are, so it is recessive. Different ratios among male and female offspring show that it is sex-linked.) 5. e
6. Genes on the single X chromosome in males are always expressed because there are no corresponding genes on the Y chromosome to mask them. A male needs only one recessive colorblindness allele (from his mother) to show the trait; a female must inherit the allele from both parents, which is less likely.
7. The parental gametes are *WS* and *ws*. Recombinant gametes are *Ws* and *wS*, produced by crossing over.
8. Height appears to be a quantitative trait resulting from polygenic inheritance, like human skin color. See Module 14.
9. The trait of freckles is dominant, so Tim and Jan must both be heterozygous. There is a $\frac{3}{4}$ chance that they will produce a child with freckles and a $\frac{1}{4}$ chance that they will produce a child without freckles. The probability that the next two children will have freckles is $\frac{3}{4} \times \frac{3}{4} = \frac{9}{16}$.
10. As in problem 9, both Tim and Jan are heterozygous, and Mike is homozygous recessive. The probability of the next child having freckles is $\frac{3}{4}$. The probability of the next child having a straight hairline is $\frac{1}{4}$. The probability that the next child will have freckles and a straight hairline is $\frac{3}{4} \times \frac{1}{4} = \frac{3}{16}$.
11. The genotype of the black short-haired parent rabbit is *BBSS*. The genotype of the brown long-haired parent rabbit is *bbss*. The F_1 rabbits will all be black and short-haired, *BbSs*. The F_2 rabbits will be $\frac{9}{16}$ black short-haired, $\frac{3}{16}$ black long-haired, $\frac{3}{16}$ brown short-haired, and $\frac{1}{16}$ brown long-haired.
12. If the genes are not linked, the proportions among the offspring will be 25% gray red, 25% gray purple, 25% black red, 25% black purple. The actual percentages show that the genes are linked. The recombination frequency is 6%.
13. The recombination frequencies are black dumpy 36%, purple dumpy 41%, and black purple 6% (see problem 12). Since these recombination frequencies reflect distances between the genes, the sequence must be purple-black-dumpy (or dumpy-black-purple).
14. $\frac{1}{4}$ will be boys suffering from hemophilia, and $\frac{1}{4}$ will be female carriers. (The mother is a heterozygous carrier, and the father is normal.)
15. For a woman to be colorblind, she must inherit X chromosomes bearing the colorblindness allele from both parents. Her father has only one X chromosome, which he passes on to all his daughters, so he must be colorblind. A male need only inherit the colorblindness allele from a carrier mother; both his parents are usually phenotypically normal.
16. Start out by breeding the cat to get a population to work with. If the curl allele is recessive, two curl cats can have only curl kittens. If the allele is dominant, curl cats can have "normal" kittens. If the curl allele is sex-linked, ratios will differ in male and female offspring of some crosses. If the curl allele is autosomal, the same ratios will be seen among males and females. Once you have established that the curl allele is dominant and autosomal, you can determine if a particular curl cat is true-breeding (homozygous) by doing a testcross with a normal cat. If the curl cat is homozygous, all offspring of the testcross will be curl; if heterozygous, half of the offspring will be curl and half normal.

Molecular Biology of the Gene

Molecular Biology of the Gene

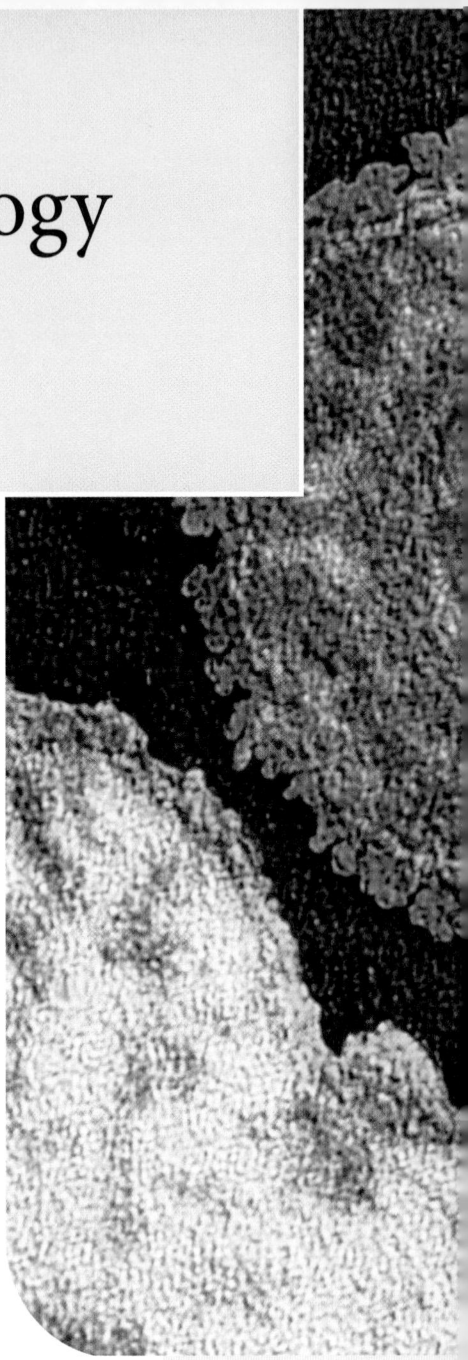

BIG IDEAS

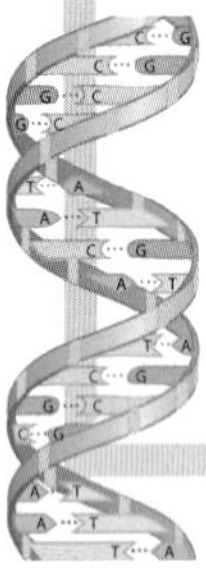

The Structure of the Genetic Material
(1–3)

A series of experiments established DNA as the molecule of heredity.

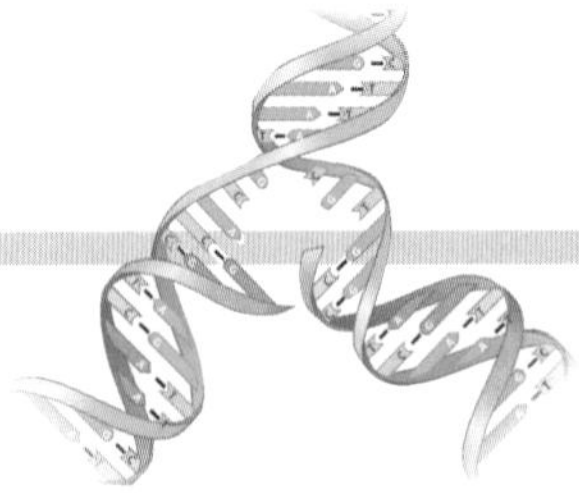

DNA Replication
(4–5)

Each DNA strand can serve as a template for another.

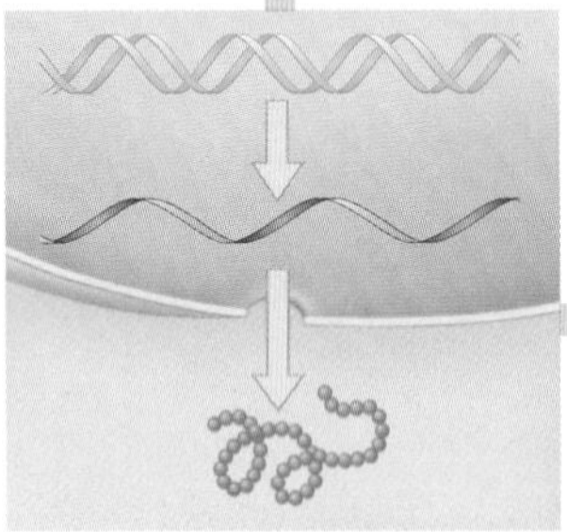

The Flow of Genetic Information from DNA to RNA to Protein
(6–16)

Genotype controls phenotype through the production of proteins.

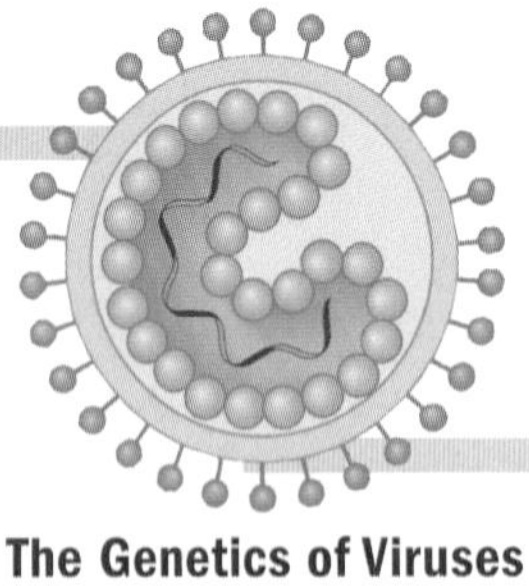

The Genetics of Viruses and Bacteria
(17–23)

Viruses and bacteria are useful model systems for the study of molecular biology.

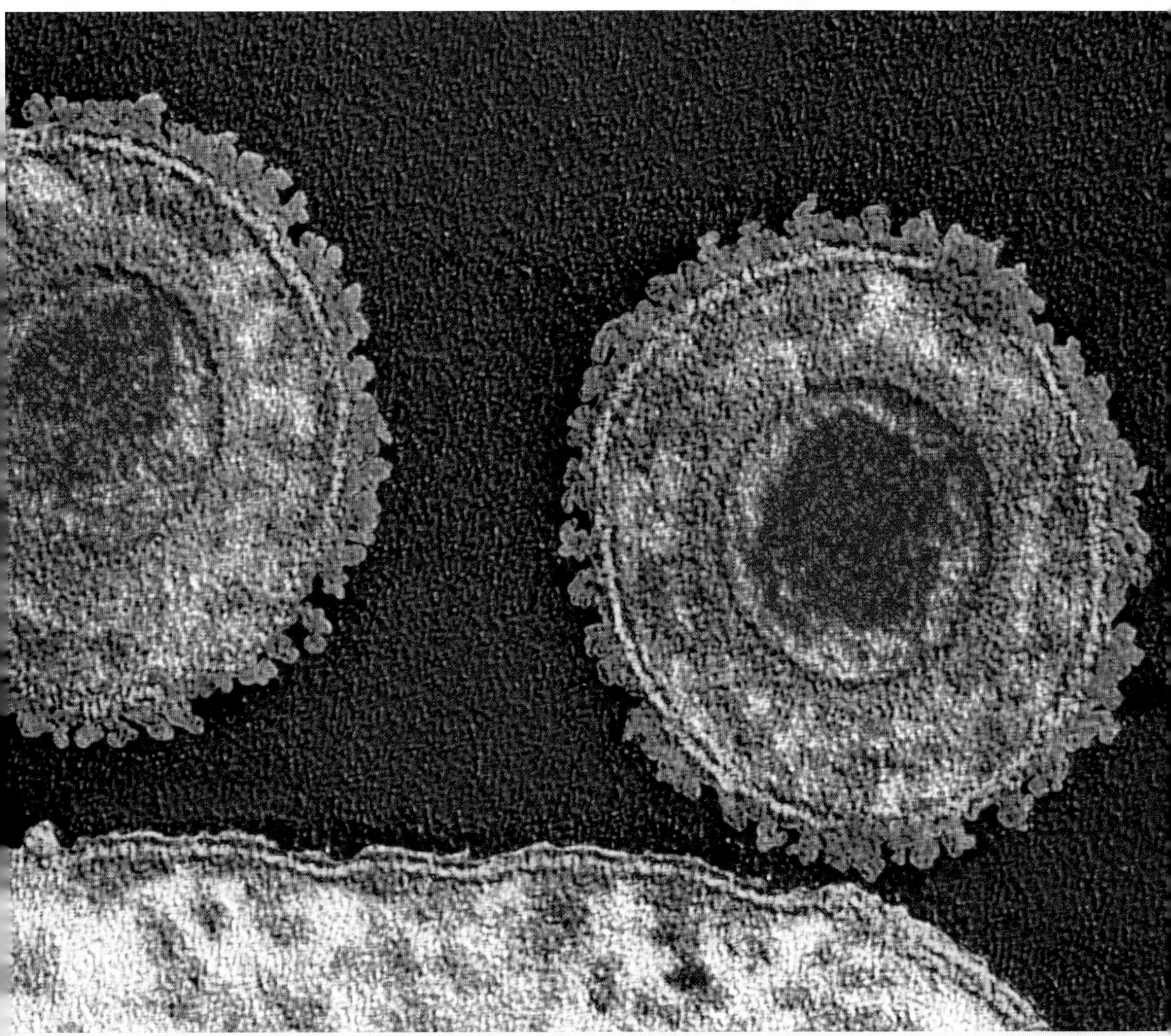

Eye of Science/Photo Researchers

The electron micrograph above shows herpesvirus, an infectious microbe that causes cold sores, genital herpes, chicken pox, and other human diseases. In the micrograph, protein spikes protrude from the exterior of the virus, while the genetic material, colored orange, is visible inside the cell.

Once it enters the human body, a herpesvirus tumbles along until it finds a suitable target cell, recognized when the virus's spikes bind to protein receptor molecules on the cell's surface. The outer membrane of the virus then fuses with the plasma membrane of the cell, and the inner part of the virus enters the cell. The virus DNA, its genetic material, soon enters the nucleus. In the nuclei of certain nerve cells, the viral DNA can remain dormant for long periods of time. Once activated, often under conditions of physical or emotional stress, the viral DNA hijacks the cell's own molecules and organelles and uses them to produce new copies of the virus. Virus production eventually causes host cells to burst. Such destruction causes the sores that are characteristic of herpes diseases. The released viruses can then infect other cells.

Viruses share some of the characteristics of living organisms, but are generally not considered alive because they are not cellular and cannot reproduce on their own. Because viruses have much less complex structures than cells, they are relatively easy to study on the molecular level. For this reason, we owe our first glimpses of the functions of DNA, the molecule that controls hereditary traits, to the study of viruses.

This chapter is about molecular biology—the study of DNA and how it serves as the basis of heredity. We'll explore the structure of DNA, how it replicates, and how it controls the cell by directing RNA and protein synthesis. We end with an examination of the genetics of viruses and bacteria.

The Structure of the Genetic Material

SCIENTIFIC DISCOVERY 1 Experiments showed that DNA is the genetic material

Today, even schoolchildren have heard of DNA, and scientists routinely manipulate DNA in the laboratory and use it to change the heritable characteristics of cells. Early in the 20th century, however, the precise identity of the molecular basis for inheritance was unknown. Biologists knew that genes were located on chromosomes and that the two chemical components of chromosomes were DNA and protein. Therefore, DNA and protein were the likely candidates to be the genetic material. Until the 1940s, the case for proteins seemed stronger because proteins appeared to be more structurally complex: Proteins were known to be made from 20 different amino acid building blocks, whereas DNA was known to be made from a mere four kinds of nucleotides. It seemed to make sense that the more complex molecule would serve as the hereditary material. Biologists finally established the role of DNA in heredity through experiments with bacteria and the viruses that infect them. This breakthrough ushered in the field of **molecular biology**, the study of heredity at the molecular level.

We can trace the discovery of the genetic role of DNA back to 1928. British medical officer Frederick Griffith was studying two strains (varieties) of a bacterium: a harmless strain and a pathogenic (disease-causing) strain that causes pneumonia. Griffith was surprised to find that when he killed the pathogenic bacteria and then mixed the bacterial remains with living harmless bacteria, some living bacterial cells were converted to the disease-causing form. Furthermore, all of the descendants of the transformed bacteria inherited the newly acquired ability to cause disease. Clearly, some chemical component of the dead bacteria could act as a "transforming factor" that brought about a heritable change in live bacteria.

Griffith's work set the stage for a race to discover the identity of the transforming factor. In 1952, American biologists Alfred Hershey and Martha Chase performed a very convincing set of experiments that showed DNA to be the genetic material of T2, a virus that infects the bacterium *Escherichia coli* (*E. coli*). Viruses that exclusively infect bacteria are called **bacteriophages** ("bacteria-eaters"), or **phages** for short. **Figure 1A** shows the

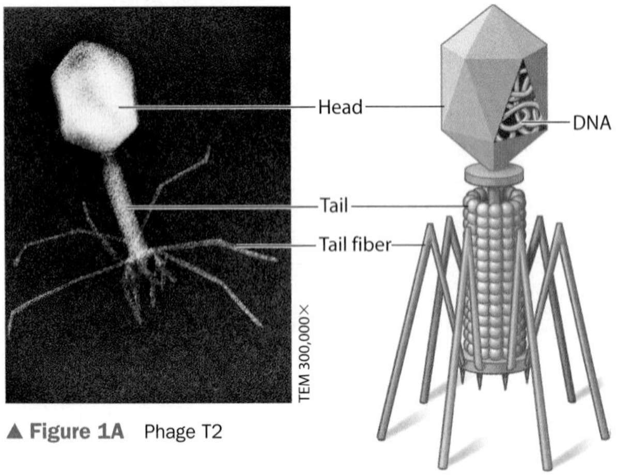

Robley C. Williams, University of California Berkeley/ Biological Photo Services

▲ **Figure 1A** Phage T2

structure of phage T2, which consists solely of DNA (blue) and protein (gold). Resembling a lunar landing craft, T2 has a DNA-containing head and a hollow tail with six jointed protein fibers extending from it. The fibers attach to the surface of a susceptible bacterium. Hershey and Chase knew that T2 could reprogram its host cell to produce new phages, but they did not know which component—DNA or protein—was responsible for this ability.

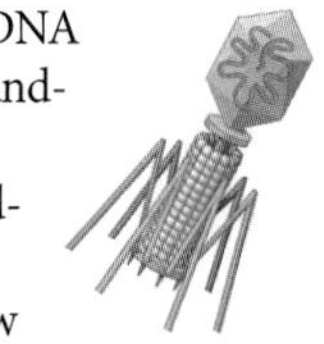

Hershey and Chase found the answer by devising an experiment to determine what kinds of molecules the phage transferred to *E. coli* during infection. Their experiment used only a few relatively simple tools: chemicals containing radioactive isotopes; a radioactivity detector; a kitchen blender; and a centrifuge, a device that spins test tubes to separate particles of different weights. (These are still basic tools of molecular biology.)

Hershey and Chase used different radioactive isotopes to label the DNA and protein in T2. First, they grew T2 with *E. coli* in a solution containing radioactive sulfur (bright yellow in **Figure 1B**). Protein contains sulfur but DNA does not, so as new phages were made, the radioactive sulfur atoms were incorporated only into the proteins of the bacteriophage. The researchers grew a separate batch of phages in a solution containing radioactive phosphorus (green). Because nearly all the phage's phosphorus is in DNA, this labeled only the phage DNA.

Armed with the two batches of labeled T2, Hershey and Chase were ready to perform the experiment outlined in Figure 1B. ❶ They allowed the two batches of T2 to infect separate samples of nonradioactive bacteria. ❷ Shortly after the onset of infection, they agitated the cultures in a blender to shake loose any parts of the phages that remained outside the bacterial cells. ❸ Then, they spun the mixtures in a centrifuge. The cells were deposited as a pellet at the bottom of the centrifuge tubes, but phages and parts of phages, being lighter, remained suspended in the liquid. ❹ The researchers then measured the radioactivity in the pellet and in the liquid.

Hershey and Chase found that when the bacteria had been infected with T2 phages containing labeled protein, the radioactivity ended up mainly in the liquid within the centrifuge tube, which contained phages but not bacteria. This result suggested that the phage protein did not enter the cells. But when the bacteria had been infected with phages whose DNA was tagged, then most of the radioactivity was in the pellet of bacterial cells at the bottom of the centrifuge tube. Furthermore, when these bacteria were returned to liquid growth medium, they soon lysed, or broke open, releasing new phages with radioactive phosphorus in their DNA but no radioactive sulfur in their proteins.

Figure 1C outlines our current understanding—as originally outlined by Hershey and Chase—of the replication cycle of phage T2. After the virus ❶ attaches to the host bacterial cell, it ❷ injects its DNA into the host. Notice that virtually all of

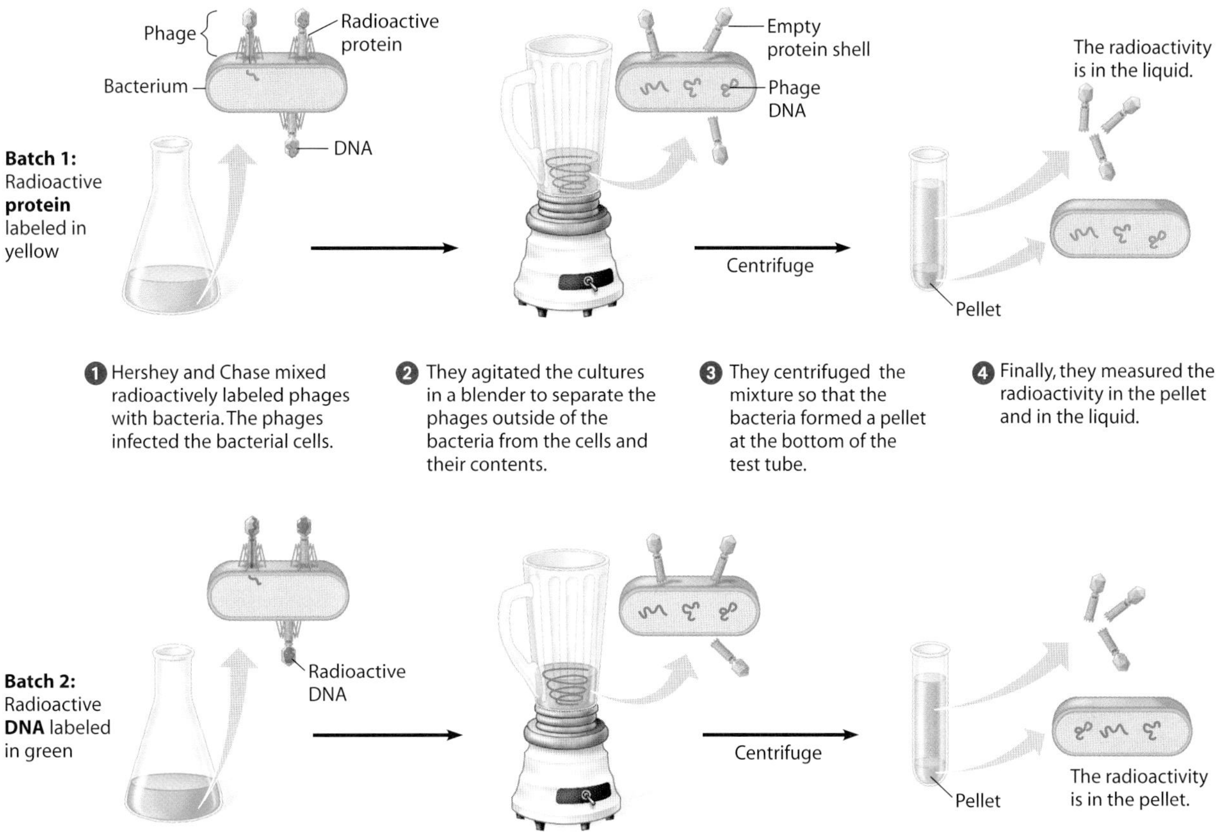

▲ **Figure 1B** The Hershey-Chase experiment

the viral protein (yellow) is left outside (which is why the radioactive protein did not show up in the host cells during the experiment shown at the top of Figure 1B). Once injected, the viral DNA causes the bacterial cells to 3 produce new phage proteins and DNA molecules—indeed, complete new phages—which soon 4 cause the cell to lyse, releasing the newly produced phages. In agreement with the experimental results of Hershey and Chase, it is the viral DNA that contains the instructions for making phages.

Once DNA was shown to be the molecule of heredity, understanding its structure became the most important quest in biology. In the next two modules, we'll review the structure of DNA and discuss how it was discovered.

? What convinced Hershey and Chase that DNA, rather than protein, is the genetic material of phage T2?

Radioactively labeled phage DNA, but not labeled protein, entered the host cell during infection and directed the synthesis of new viruses.

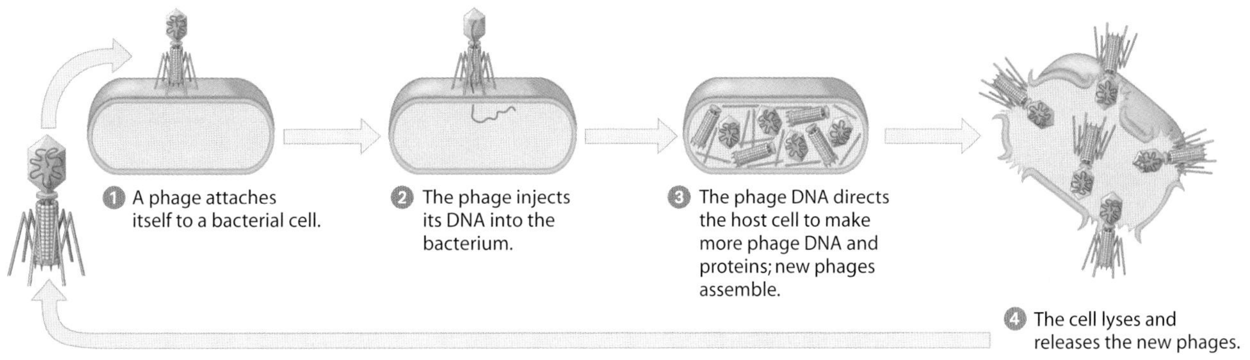

1 A phage attaches itself to a bacterial cell.
2 The phage injects its DNA into the bacterium.
3 The phage DNA directs the host cell to make more phage DNA and proteins; new phages assemble.
4 The cell lyses and releases the new phages.

▲ **Figure 1C** A phage replication cycle

2 DNA and RNA are polymers of nucleotides

By the time Hershey and Chase performed their experiments, much was already known about DNA. Scientists had identified all its atoms and knew how they were covalently bonded to one another. What was not understood was the specific arrangement of atoms that gave DNA its unique properties—the capacity to store genetic information, copy it, and pass it from generation to generation. However, only one year after Hershey and Chase published their results, scientists figured out the three-dimensional structure of DNA and the basic strategy of how it works. We will examine that momentous discovery in Module 3, but first, let's look at the underlying chemical structure of DNA and its chemical cousin RNA.

Recall from Module 3.15 that DNA and RNA are nucleic acids, consisting of long chains (polymers) of chemical units (monomers) called **nucleotides**. **Figure 2A** shows four representations of various parts of the same molecule. At left is a view of a DNA double helix. One of the strands is opened up (center) to show two different views of an individual DNA **polynucleotide**, a nucleotide polymer (chain). The view on the far right zooms into a single nucleotide from the chain. Each type of DNA nucleotide has a different nitrogen-containing base: adenine (A), cytosine (C), thymine (T), or guanine (G). Because nucleotides can occur in a polynucleotide in any sequence and polynucleotides vary in length from long to very long, the number of possible polynucleotides is enormous. The chain shown in this figure has the sequence ACTGG, only one of many possible arrangements of the four types of nucleotides that make up DNA.

Looking more closely at our polynucleotide, we see in the center of Figure 2A that each nucleotide consists of three components: a nitrogenous base (in DNA: A, C, T, or G), a sugar (blue), and a phosphate group (yellow). The nucleotides are joined to one another by covalent bonds between the sugar of one nucleotide and the phosphate of the next. This results in a **sugar-phosphate backbone**, a repeating pattern of sugar-phosphate-sugar-phosphate. The nitrogenous bases are arranged like ribs that project from the backbone.

Examining a single nucleotide in even more detail (on the right in Figure 2A), you can see the chemical structure of its three components. The phosphate group has a phosphorus atom (P) at its center and is the source of the word *acid* in *nucleic acid*. The sugar has five carbon atoms, shown in red here for emphasis—four in its ring and one extending above the ring. The ring also includes an oxygen atom. The sugar is called deoxyribose because, compared with the sugar ribose, it is missing an oxygen atom. (Notice that the C atom in the lower right corner of the ring is bonded to an H atom instead of to an —OH group, as it is in ribose; see Figure 2C. Hence, DNA is "deoxy"—which means "without an oxygen"—compared to RNA.)

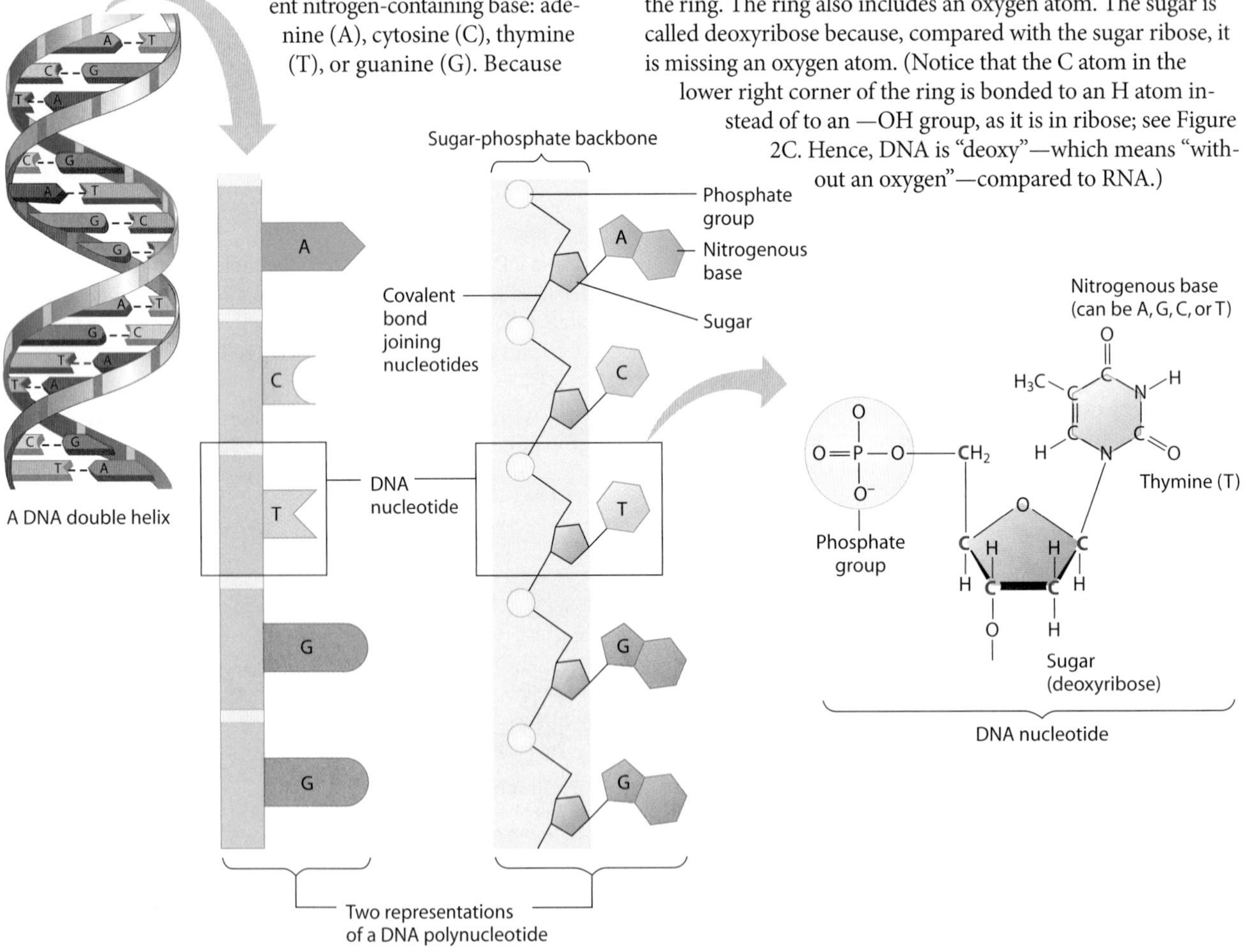

▲ **Figure 2A** The structure of a DNA polynucleotide

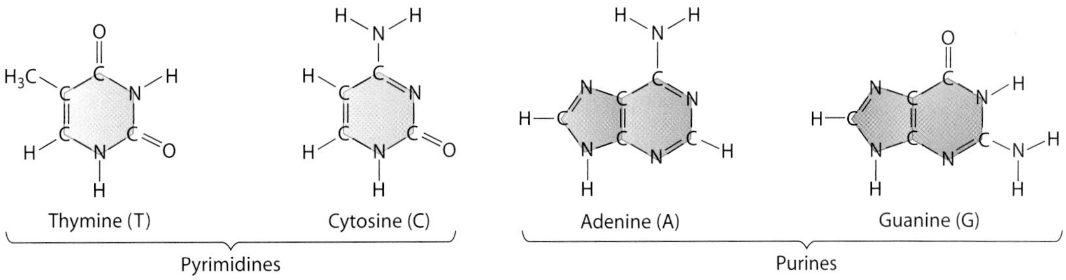

▲ **Figure 2B** The nitrogenous bases of DNA

The full name for **DNA** is **deoxyribonucleic acid**, with the *nucleic* portion of the word referring to DNA's location in the nuclei of eukaryotic cells. Each nitrogenous base (thymine, in our example at the right in Figure 2A) has a single or double ring consisting of nitrogen and carbon atoms with various functional groups attached. Recall from Module 3.2 that a functional group is a chemical group that affects a molecule's function by participating in specific chemical reactions. In the case of DNA, the main role of the functional groups is to determine which other kind of bases each base can hydrogen-bond with. For example, the NH_2 group hanging off cytosine is capable of forming a hydrogen bond to the C=O group hanging off guanine, but not with the NH_2 group protruding from adenine. The chemical groups of the bases are therefore responsible for DNA's most important property, which you will learn more about in the next module. In contrast to the acidic phosphate group, nitrogenous bases are basic, hence their name.

The four nucleotides found in DNA differ only in the structure of their nitrogenous bases (**Figure 2B**). At this point, the structural details are not as important as the fact that the bases are of two types. **Thymine (T)** and **cytosine (C)** are single-ring structures called pyrimidines. **Adenine (A)** and **guanine (G)** are larger, double-ring structures called purines. The one-letter abbreviations can be used either for the bases alone or for the nucleotides containing them.

What about RNA (**Figure 2C**)? As its name—ribonucleic acid—implies, its sugar is ribose rather than deoxyribose. Notice the ribose in the RNA nucleotide in Figure 2C; unlike deoxyribose, the sugar ring has an —OH group attached to the C atom at its lower-right corner. Another difference between RNA and DNA is that instead of thymine, RNA has a nitrogenous base called **uracil (U)**. (You can see the structure of uracil in Figure 2C; it is very similar to thymine.) Except for the presence of ribose and uracil, an RNA polynucleotide chain is identical to a DNA polynucleotide chain. **Figure 2D** is a computer graphic of a piece of RNA polynucleotide about 20 nucleotides long. In this 3-D view, each sphere represents an atom, and notice that the color scheme is the same as in the other figures in this module. The yellow phosphate groups and blue ribose sugars make it easy to spot the sugar-phosphate backbone.

In this module, we reviewed the structure of the nucleic acids DNA and RNA. In the next module, we'll see how two DNA polynucleotides join together in a molecule of DNA.

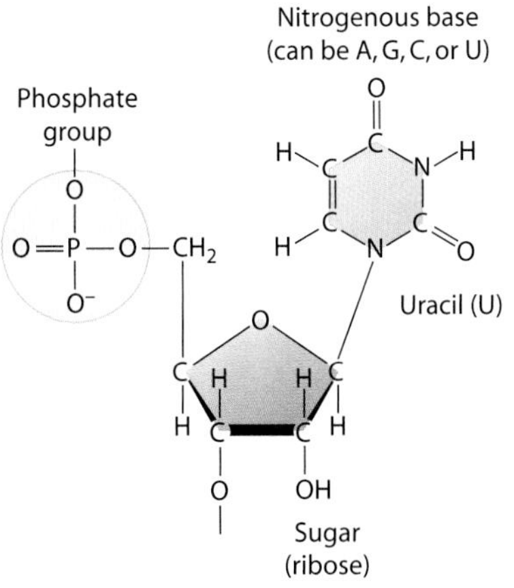

▲ **Figure 2C** An RNA nucleotide

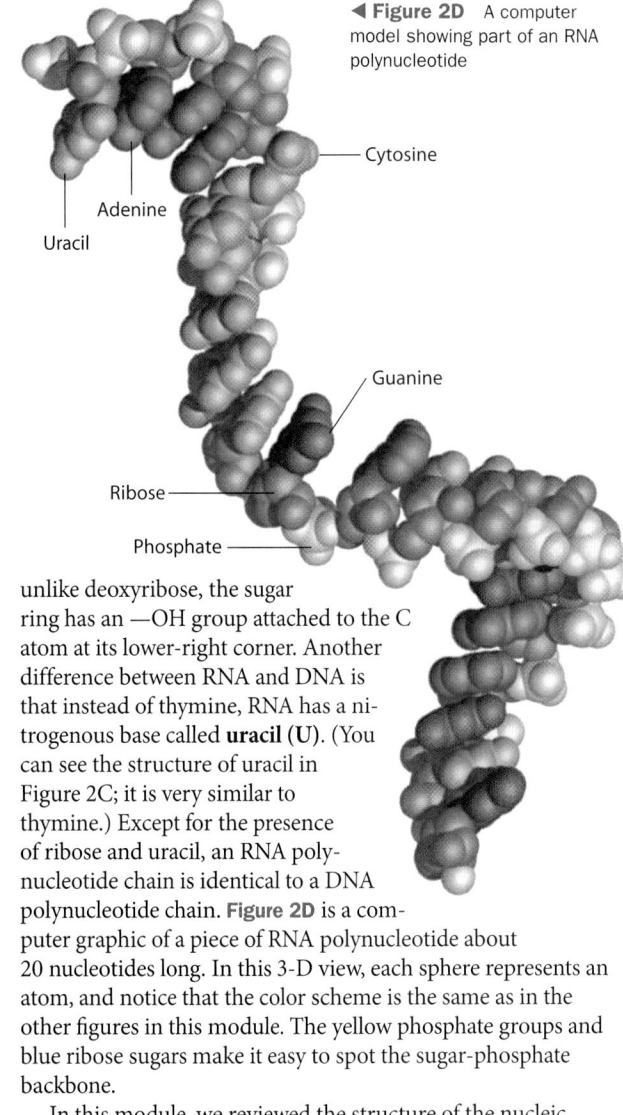

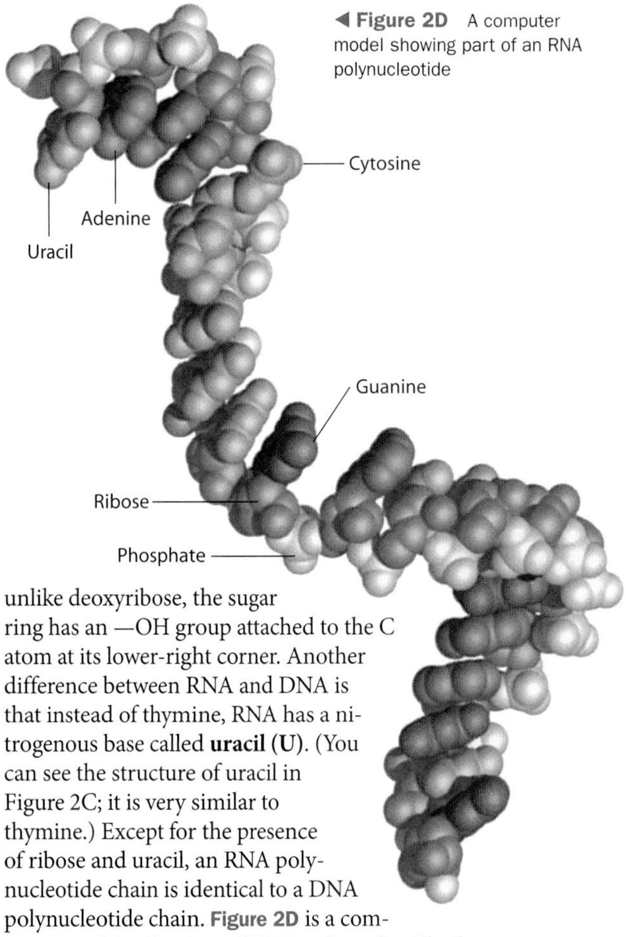

◀ **Figure 2D** A computer model showing part of an RNA polynucleotide

? **Compare and contrast DNA and RNA polynucleotides.**

Both are polymers of nucleotides consisting of a sugar, a nitrogenous base, and a phosphate. In RNA, the sugar is ribose; in DNA, it is deoxyribose. Both RNA and DNA have the bases A, G, and C, but DNA has a T and RNA has a U.

SCIENTIFIC DISCOVERY

3 DNA is a double-stranded helix

After the 1952 Hershey-Chase experiment convinced most biologists that DNA was the material that stored genetic information, a race was on to determine how the structure of this molecule could account for its role in heredity. By that time, the arrangement of covalent bonds in a nucleic acid polymer was well established, and researchers focused on discovering the three-dimensional shape of DNA. First to the finish line were two scientists who were relatively unknown at the time—American James D. Watson and Englishman Francis Crick.

The brief but celebrated partnership that solved the puzzle of DNA structure began soon after Watson, a 23-year-old newly minted Ph.D., journeyed to Cambridge University in England, where the more senior Crick was studying protein structure with a technique called X-ray crystallography. While visiting the laboratory of Maurice Wilkins at King's College in London, Watson saw an X-ray crystallographic image of DNA produced by Wilkins's colleague Rosalind Franklin (**Figure 3A**). A careful study of the image enabled Watson to deduce the basic shape of DNA to be a helix with a uniform diameter of 2 nanometers (nm), with its nitrogenous bases stacked about one-third of a nanometer apart. (For comparison, the plasma membrane of a cell is about 8 nm thick.) The diameter of the helix suggested that it was made up of two polynucleotide strands, a **double helix**.

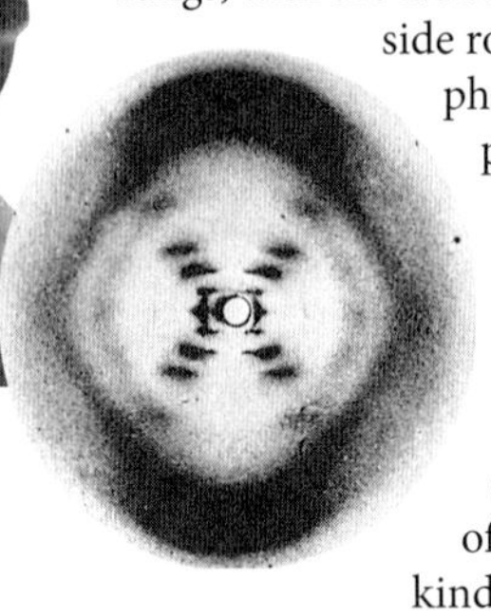

▲ **Figure 3A** Rosalind Franklin and her X-ray image of DNA

Watson and Crick began trying to construct a wire model of a double helix that would conform both to Franklin's data and to what was then known about the chemistry of DNA (**Figure 3B**). They had concluded that the sugar-phosphate backbones must be on the outside of the double helix, forcing the nitrogenous bases to swivel to the interior of the molecule. But how were the bases arranged in the interior of the double helix?

At first, Watson and Crick imagined that the bases paired like with like—for example, A with A and C with C. But that kind of pairing did not fit the X-ray data, which suggested that the DNA molecule has a uniform diameter. An A-A pair, with two double-ring bases, would be almost twice as wide as a C-C pair. It soon became apparent that a double-ringed base (purine) must always be paired with a single-ringed base (pyrimidine) on the opposite strand. Moreover, Watson and Crick realized that the individual structures of the bases dictated the pairings even more specifically. As discussed in the previous module, each base has functional groups protruding from its six-sided ring that can best form hydrogen bonds with one appropriate partner. Adenine can best form hydrogen bonds with thymine, and guanine with cytosine. In the biologist's shorthand, A pairs with T, and G pairs with C. A is also said to be "complementary" to T, and G to C.

Watson and Crick's pairing scheme not only fit what was known about the physical attributes and chemical bonding of DNA, but also explained some data obtained several years earlier by American biochemist Erwin Chargaff. Chargaff had discovered that the amount of adenine in the DNA of any one species was equal to the amount of thymine and that the amount of guanine was equal to that of cytosine. Chargaff's rules, as they are called, are explained by the fact that A on one of DNA's polynucleotide chains always pairs with T on the other polynucleotide chain, and G on one chain pairs only with C on the other chain.

You can picture the model of the DNA double helix proposed by Watson and Crick as a rope ladder with wooden rungs, with the ladder twisting into a spiral (**Figure 3C**). The side ropes are the equivalent of the sugar-phosphate backbones, and the rungs represent pairs of nitrogenous bases joined by hydrogen bonds.

Figure 3D shows three representations of the double helix. The shapes of the base symbols in the ribbonlike diagram on the left indicate the bases' complementarity; notice that the shape of any kind of base matches only one other kind of base. In the center of the diagram is an atomic-level version showing four base pairs, with the helix untwisted and the hydrogen bonds specified by dotted lines. Notice that a C-G base pair has functional groups that form three hydrogen bonds,

▲ **Figure 3B** Watson and Crick in 1953 with their model of the DNA double helix

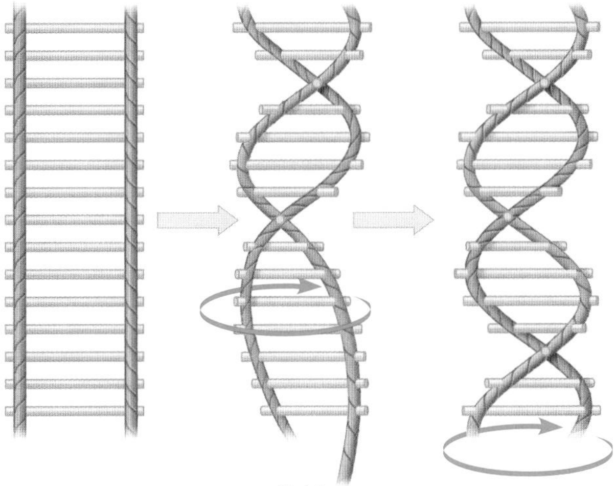

▲ **Figure 3C** A rope ladder model for the double helix

while an A-T base pair has functional groups that form two hydrogen bonds. This difference means that C-G base pairs are somewhat stronger than A-T base pairs. You can see that the two sugar-phosphate backbones of the double helix are oriented in opposite directions. (Notice that the sugars on the two strands are upside down with respect to each other.) On the right is a computer graphic showing most of the atoms of part of a double helix. The atoms that compose the deoxyribose sugars are shown as blue, phosphate groups as yellow, and nitrogenous bases as shades of green and orange.

Although the Watson-Crick base-pairing rules dictate the side-by-side combinations of nitrogenous bases that form the rungs of the double helix, they place no restrictions on the sequence of nucleotides along the length of a DNA strand. In fact, the sequence of bases can vary in countless ways, and each gene has a unique order of nucleotides, or base sequence.

In April 1953, Watson and Crick rocked the scientific world with a succinct paper explaining their molecular model for DNA in the journal *Nature*. In 1962, Watson, Crick, and Wilkins received the Nobel Prize for their work. (Rosalind Franklin probably would have received the prize as well but for her death from cancer in 1958; Nobel Prizes are never awarded posthumously.) Few milestones in the history of biology have had as broad an impact as the discovery of the double helix, with its A-T and C-G base pairing.

The Watson-Crick model gave new meaning to the words *genes* and *chromosomes*—and to the chromosome theory of inheritance. With a complete picture of DNA, we can see that the genetic information in a chromosome must be encoded in the nucleotide sequence of the molecule. One powerful aspect of the Watson-Crick model is that the structure of DNA suggests a molecular explanation for genetic inheritance, as we will see in the next module.

? Along one strand of a double helix is the nucleotide sequence GGCATAGGT. What is the complementary sequence for the other DNA strand?

CCGTATCCA

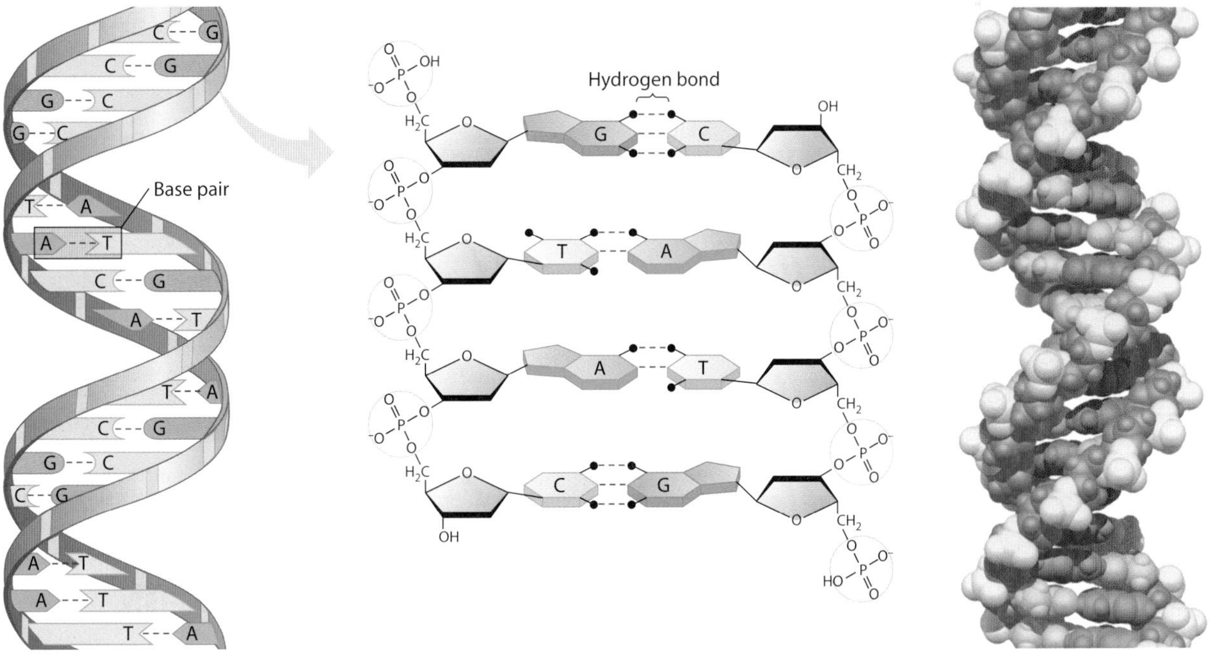

▲ **Figure 3D** Three representations of DNA

DNA Replication

4 DNA replication depends on specific base pairing

One of biology's overarching themes—the relationship between structure and function—is evident in the double helix. The idea that there is specific pairing of bases in DNA was the flash of inspiration that led Watson and Crick to the correct structure of the double helix. At the same time, they saw the functional significance of the base-pairing rules. They ended their classic 1953 paper with this statement: "It has not escaped our notice that the specific pairing we have postulated immediately suggests a possible copying mechanism for the genetic material."

The logic behind the Watson-Crick proposal for how DNA is copied—by specific pairing of complementary bases—is quite simple. You can see this by covering one of the strands in the parental DNA molecule in **Figure 4A**. You can determine the sequence of bases in the covered strand by applying the base-pairing rules to the unmasked strand: A pairs with T (and T with A), and G pairs with C (and C with G).

Watson and Crick predicted that a cell applies the same rules when copying its genes. As shown in Figure 4A, the two strands of parental DNA (blue) separate. Each then becomes a template for the assembly of a complementary strand from a supply of free nucleotides (gray) that is always available within the nucleus. The nucleotides line up one at a time along the template strand in accordance with the base-pairing rules. Enzymes link the nucleotides to form the new DNA strands. The completed new molecules, identical to the parental molecule, are known as daughter DNA (although no gender should be inferred).

Watson and Crick's model predicts that when a double helix replicates, each of the two daughter molecules will have one old strand, which was part of the parental molecule, and one newly created strand. This model for DNA replication is known as the **semiconservative model** because half of the parental molecule is maintained (conserved) in each daughter molecule. The semiconservative model of replication was confirmed by experiments performed in the 1950s.

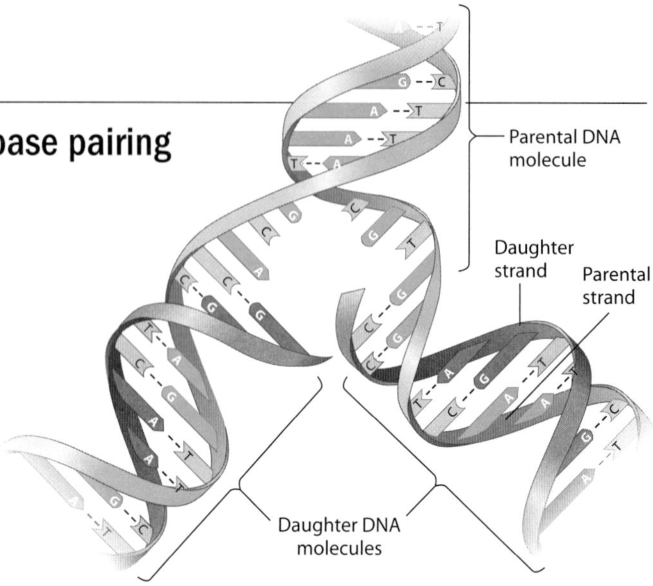

▲ **Figure 4B** The untwisting and replication of DNA

Although the general mechanism of DNA replication is conceptually simple, the actual process is complex, requiring the coordination of more than a dozen enzymes and other proteins. Some of the complexity arises from the need for the helical DNA molecule to untwist as it replicates and for the two new strands to be made roughly simultaneously **(Figure 4B)**. Another challenge is the speed of the process. *E. coli*, with about 4.6 million DNA base pairs, can copy its entire genome in less than an hour. Humans, with over 6 billion base pairs in 46 diploid chromosomes, require only a few hours. And yet, the process is amazingly accurate; typically, only about one DNA nucleotide per several billion is incorrectly paired. In the next module, we take a closer look at the mechanisms of DNA replication that allow it to proceed with such speed and accuracy.

? How does complementary base pairing make possible the replication of DNA?

● When the two strands of the double helix separate, free nucleotides can base-pair along each strand, leading to the synthesis of new complementary strands.

▶ **Figure 4A** A template model for DNA replication

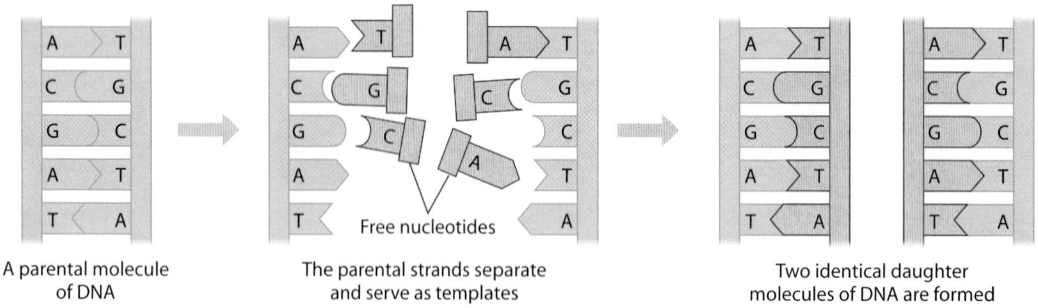

5 DNA replication proceeds in two directions at many sites simultaneously

Replication of a DNA molecule begins at special sites called *origins of replication*, short stretches of DNA having a specific sequence of nucleotides where proteins attach to the DNA and separate the strands. As shown in **Figure 5A**, replication then proceeds in both directions, creating replication "bubbles." The parental DNA strands (blue) open up as daughter strands

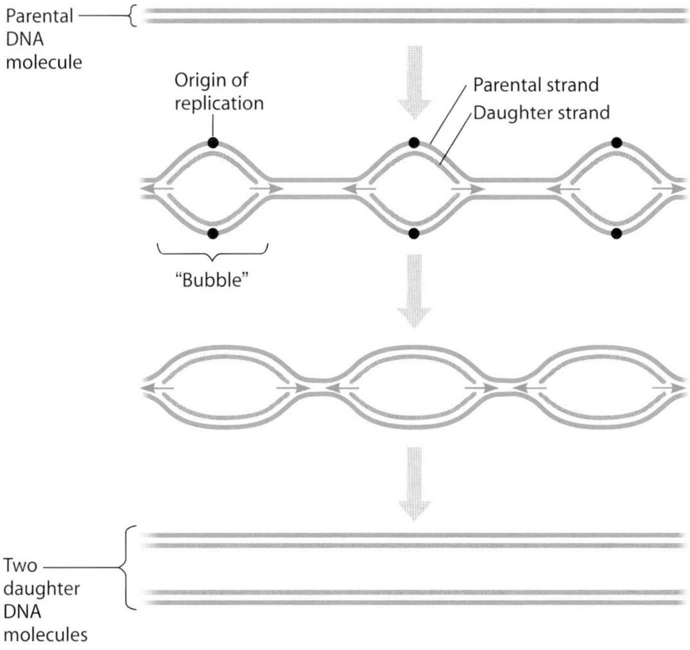

▲ **Figure 5A** Multiple bubbles in replicating DNA

(gray) elongate on both sides of each bubble. The DNA molecule of a eukaryotic chromosome has many origins where replication can start simultaneously. Thus, hundreds or thousands of bubbles can be present at once, shortening the total time needed for replication. Eventually, all the bubbles fuse, yielding two completed daughter DNA molecules (see the bottom of Figure 5A).

Figure 5B shows the molecular building blocks of a tiny segment of DNA, reminding us that the DNA's sugar-phosphate backbones run in opposite directions. Notice that each strand has a 3' ("three-prime") end and a 5' ("five-prime") end. The primed numbers refer to the carbon atoms of the nucleotide sugars. At one end of each DNA strand, the sugar's 3' carbon atom is attached to an —OH group; at the other end, the sugar's 5' carbon is attached to a phosphate group.

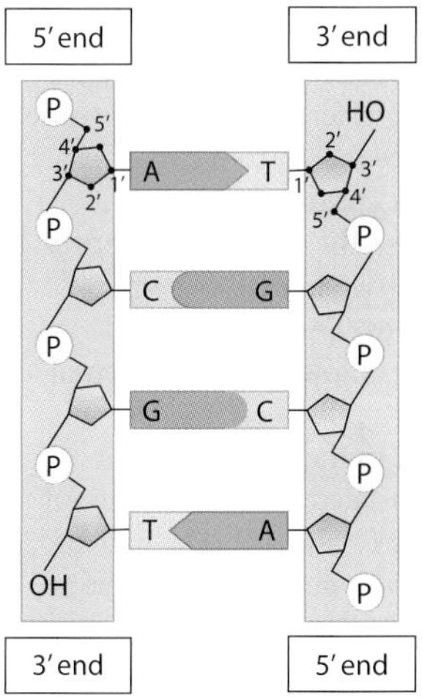

▲ **Figure 5B** The opposite orientations of DNA strands

The opposite orientation of the strands is important in DNA replication. The enzymes that link DNA nucleotides to a growing daughter strand, called **DNA polymerases**, add nucleotides only to the 3' end of the strand, never to the 5' end. Thus, a daughter DNA strand can only grow in the 5' → 3' direction. You see the consequences of this enzyme specificity in **Figure 5C**, where the forked structure represents one side of a replication bubble. One of the daughter strands (shown in gray) can be synthesized in one continuous piece by a DNA polymerase working toward the forking point of the parental DNA. However, to make the other daughter strand, polymerase molecules must work outward from the forking point. The only way this can be accomplished is if the new strand is synthesized in short pieces as the fork opens up. These pieces are called Okazaki fragments, after the Japanese husband-and-wife team of molecular biologists who discovered them. Another enzyme, called **DNA ligase**, then links, or ligates, the pieces together into a single DNA strand.

In addition to their roles in adding nucleotides to a DNA chain, DNA polymerases carry out a proofreading step that quickly removes nucleotides that have base-paired incorrectly during replication. DNA polymerases and DNA ligase are also involved in repairing DNA damaged by harmful radiation, such as ultraviolet light and X-rays, or toxic chemicals in the environment, such as those found in tobacco smoke.

DNA replication ensures that all the somatic cells in a multicellular organism carry the same genetic information. It is also the means by which genetic instructions are copied for the next generation of the organism. In the next module, we begin to pursue the connection between DNA instructions and an organism's phenotypic traits.

? What is the function of DNA polymerase in DNA replication?

As free nucleotides base-pair to a parental DNA strand, the enzyme covalently bonds them to the 3' end of a growing daughter strand.

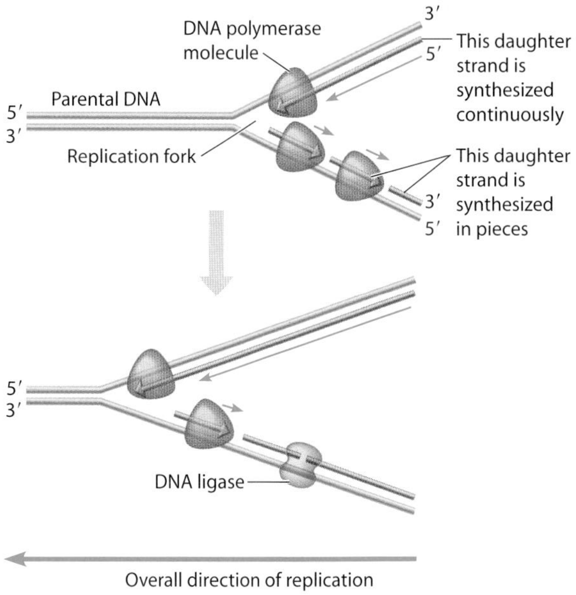

▲ **Figure 5C** How daughter DNA strands are synthesized

The Flow of Genetic Information from DNA to RNA to Protein

6 The DNA genotype is expressed as proteins, which provide the molecular basis for phenotypic traits

With our knowledge of DNA, we can now define genotype and phenotype more precisely. An organism's genotype, its genetic makeup, is the heritable information contained in its DNA. The phenotype is the organism's physical traits. So what is the molecular connection between genotype and phenotype?

The answer is that the DNA inherited by an organism specifies traits by dictating the synthesis of proteins. In other words, proteins are the links between the genotype and the phenotype. However, a gene does not build a protein directly. Rather, a gene dispatches instructions in the form of RNA, which in turn programs protein synthesis. This fundamental concept in biology, termed the "central dogma" by Francis Crick, is summarized in **Figure 6A**. The molecular "chain of command" is from DNA in the nucleus of the cell to RNA to protein synthesis in the cytoplasm. The two main stages are **transcription**, the synthesis of RNA under the direction of DNA, and **translation**, the synthesis of protein under the direction of RNA.

The relationship between genes and proteins was first proposed in 1909, when English physician Archibald Garrod suggested that genes dictate phenotypes through enzymes, the proteins that catalyze chemical reactions. Garrod hypothesized that an inherited disease reflects a person's inability to make a particular enzyme, and he referred to such diseases as "inborn errors of metabolism." He gave as one example the hereditary condition called alkaptonuria, in which the urine is dark because it contains a chemical called alkapton. Garrod reasoned that individuals without the disorder have an enzyme that breaks down alkapton, whereas alkaptonuric individuals cannot make the enzyme. Garrod's hypothesis was ahead of its time, but research conducted decades later proved him right.

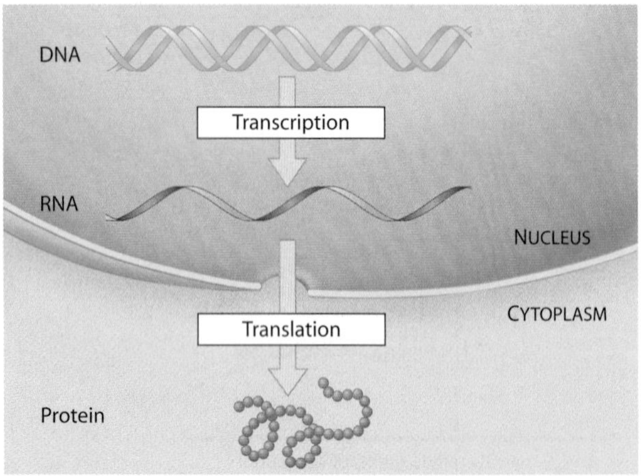

▲ **Figure 6A** The flow of genetic information in a eukaryotic cell

In the intervening years, biochemists accumulated evidence that cells make and break down biologically important molecules via metabolic pathways, as in the synthesis of an amino acid or the breakdown of a sugar. Each step in a metabolic pathway is catalyzed by a specific enzyme. Therefore, individuals lacking one of the enzymes for a pathway are unable to complete it.

The major breakthrough in demonstrating the relationship between genes and enzymes came in the 1940s from the work of American geneticists George Beadle and Edward Tatum with the bread mold *Neurospora crassa* (**Figure 6B**). Beadle and Tatum studied strains of the mold that were unable to grow on a simple growth medium. Each of these so-called nutritional mutants turned out to lack an enzyme in a metabolic pathway that produced some molecule the mold needed, such as an amino acid. Beadle and Tatum also showed that each mutant was defective in a single gene. This result suggested the one gene–one enzyme hypothesis—the idea that the function of a gene is to dictate the production of a specific enzyme.

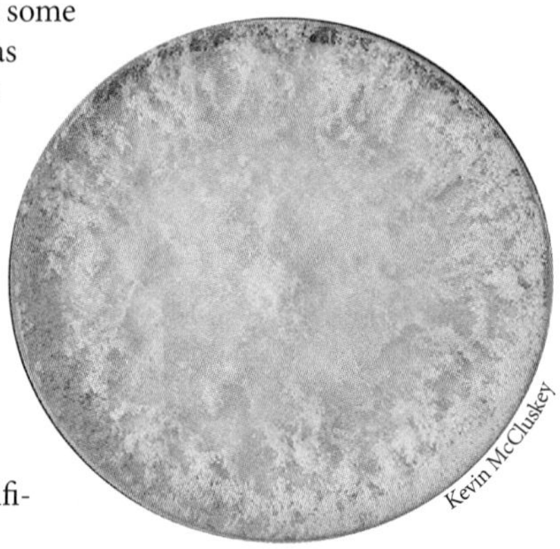

▲ **Figure 6B** The bread mold *Neurospora crassa* growing in a culture dish

The one gene–one enzyme hypothesis has been amply confirmed, but with important modifications. First, it was extended beyond enzymes to include *all* types of proteins. For example, keratin (the structural protein of hair) and the hormone insulin are two examples of proteins that are not enzymes. So biologists began to think in terms of one gene–one protein. However, many proteins are made from two or more polypeptide chains, with each polypeptide specified by its own gene. For example, hemoglobin, the oxygen-transporting protein in your blood, is built from two kinds of polypeptides, encoded by two different genes. Thus, Beadle and Tatum's hypothesis is now stated as follows: The function of a gene is to dictate the production of a polypeptide. Even this description is not entirely accurate, in that the RNA transcribed from some genes is not translated (you'll learn about two such kinds of RNA in Modules 11 and 12). The flow of information from genotype to phenotype continues to be an active research area.

? What are the functions of transcription and translation?

Transcription is the transfer of information from DNA to RNA. Translation is the use of the information in RNA to make a polypeptide.

7 Genetic information written in codons is translated into amino acid sequences

Genes provide the instructions for making specific proteins. But a gene does not build a protein directly. As you have learned, the bridge between DNA and protein synthesis is the nucleic acid RNA: DNA is transcribed into RNA, which is then translated into protein. Put another way, information within the cell flows as DNA → RNA → protein. This is sometimes stated as: "DNA makes RNA makes protein."

Transcription and translation are linguistic terms, and it is useful to think of nucleic acids and proteins as having languages. To understand how genetic information passes from genotype to phenotype, we need to see how the chemical language of DNA is translated into the different chemical language of proteins.

What, exactly, is the language of nucleic acids? Both DNA and RNA are polymers (long chains) made of nucleotide monomers (the individual units that make up the polymer). In DNA, there are four types of nucleotides, which differ in their nitrogenous bases (A, T, C, and G). The same is true for RNA, although it has the base U instead of T.

Figure 7 focuses on a small region of one gene (gene 3, shown in light blue) carried by a DNA molecule. DNA's language is written as a linear sequence of nucleotide bases on a polynucleotide, a sequence such as the one you see on the enlarged DNA segment in the figure. Specific sequences of bases, each with a beginning and an end, make up the genes on a DNA strand. A typical gene consists of hundreds or thousands of nucleotides in a specific sequence.

The pink strand underneath the enlarged DNA segment represents the results of transcription: an RNA molecule. The process is called transcription because the nucleic acid language of DNA has been rewritten (transcribed) as a sequence of bases on RNA. Notice that the language is still that of nucleic acids, although the nucleotide bases on the RNA molecule are complementary to those on the DNA strand. As we will see in Module 9, this is because the RNA was synthesized using the DNA as a template.

The purple chain represents the results of translation, the conversion of the nucleic acid language to the polypeptide language (recall that proteins consist of one or more polypeptides). Like nucleic acids, polypeptides are polymers, but the monomers that compose them are the 20 amino acids common to all organisms. Again, the language is written in a linear sequence, and the sequence of nucleotides of the RNA molecule dictates the sequence of amino acids of the polypeptide. The RNA acts as a messenger carrying genetic information from DNA.

During translation, there is a change in language from the nucleotide sequence of the RNA to the amino acid sequence of the polypeptide. How is this translation achieved? Recall that there are only four different kinds of nucleotides in DNA (A, G, C, T) and RNA (A, G, C, U). In translation, these four nucleotides must somehow specify all 20 amino acids. Consider whether each single nucleotide base were to specify one amino acid. In this case, only four of the 20 amino acids could be accounted for, one for each type of base. What if the language consisted of two-letter code words? If we read the bases of a gene two at a time—AG, for example, could specify one amino acid, whereas AT could designate a different amino acid—then only 16 arrangements would be possible (4^2), which is still not enough to specify all 20 amino acids. However, if the code word in DNA consists of a triplet, with each arrangement of three consecutive bases specifying an amino acid—AGT specifies one amino acid, for example, while AGA specifies a different one—then there can be 64 (that is, 4^3) possible code words, more than enough to specify the 20 amino acids. Indeed, there are enough triplets to allow more than one coding for each amino acid. For example, the base triplets AAT and AAC could both code for the same amino acid. Thus, triplets of bases are the smallest "words" of uniform length that can specify all the amino acids (see the brackets below the strand of RNA in Figure 7).

Experiments have verified that the flow of information from gene to protein is based on a **triplet code**: The genetic instructions for the amino acid sequence of a polypeptide chain are written in DNA and RNA as a series of nonoverlapping three-base "words" called **codons**. Notice in the figure that three-base codons in the DNA are transcribed into complementary three-base codons in the RNA, and then the RNA codons are translated into amino acids that form a polypeptide. We turn to the codons themselves in the next module.

? What is the minimum number of nucleotides necessary to code for 100 amino acids?

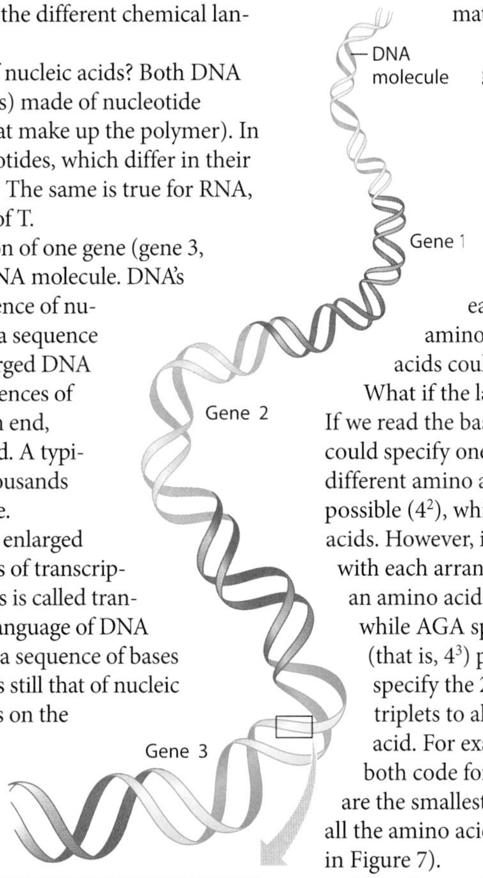

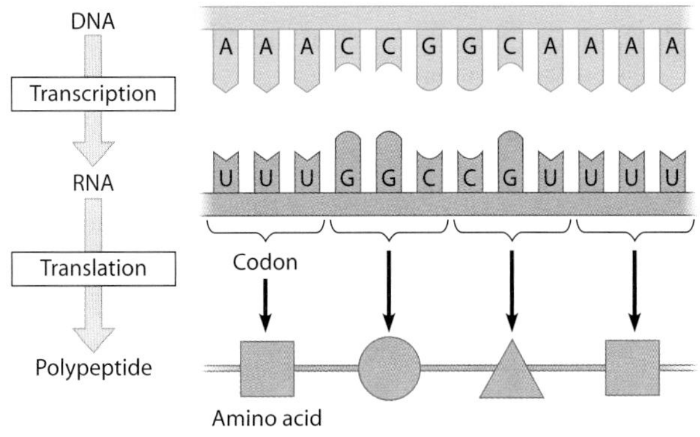

▲ **Figure 7** Transcription and translation of codons

8 The genetic code dictates how codons are translated into amino acids

During the 1960s, scientists cracked the **genetic code**, the set of rules that relate codons in RNA to amino acids in proteins. The rules were established by a series of elegant experiments that disclosed the amino acid translations of each of the nucleotide-triplet code words. The first codon was deciphered in 1961 by American biochemist Marshall Nirenberg. He synthesized an artificial RNA molecule by linking together identical RNA nucleotides having uracil as their only base. No matter where this message started or stopped, it could contain only one type of triplet codon: UUU. Nirenberg added this "poly-U" to a test-tube mixture containing ribosomes and the other ingredients required for polypeptide synthesis. This mixture translated the poly-U into a polypeptide containing a single kind of amino acid, phenylalanine (Phe). Thus, Nirenberg learned that the RNA codon UUU specifies the amino acid phenylalanine. By variations on this method, the amino acids specified by all the codons were soon determined.

As shown in **Figure 8A**, 61 of the 64 triplets code for amino acids. The triplet AUG (green box in the figure) has a dual function: It codes for the amino acid methionine (Met) and also can provide a signal for the start of a polypeptide chain. Three codons (red) do not designate amino acids. They are the stop codons that mark the end of translation.

Notice in Figure 8A that there is redundancy in the code but no ambiguity. For example, although codons UUU and UUC both specify phenylalanine (redundancy), neither of them ever represents any other amino acid (no ambiguity). The codons in the figure are the triplets found in RNA. They have a straightforward, complementary relationship to the codons in DNA, with UUU in the RNA matching AAA in the DNA, for example. The nucleotides making up the codons occur in a linear order along the DNA and RNA, with no gaps or "punctuation" separating the codons.

First base	Second base: U		Second base: C		Second base: A		Second base: G		Third base
U	UUU	Phe	UCU	Ser	UAU	Tyr	UGU	Cys	U
	UUC		UCC		UAC		UGC		C
	UUA	Leu	UCA		UAA	Stop	UGA	Stop	A
	UUG		UCG		UAG	Stop	UGG	Trp	G
C	CUU	Leu	CCU	Pro	CAU	His	CGU	Arg	U
	CUC		CCC		CAC		CGC		C
	CUA		CCA		CAA	Gln	CGA		A
	CUG		CCG		CAG		CGG		G
A	AUU	Ile	ACU	Thr	AAU	Asn	AGU	Ser	U
	AUC		ACC		AAC		AGC		C
	AUA		ACA		AAA	Lys	AGA	Arg	A
	AUG	Met or start	ACG		AAG		AGG		G
G	GUU	Val	GCU	Ala	GAU	Asp	GGU	Gly	U
	GUC		GCC		GAC		GGC		C
	GUA		GCA		GAA	Glu	GGA		A
	GUG		GCG		GAG		GGG		G

▲ **Figure 8A** Dictionary of the genetic code (RNA codons)

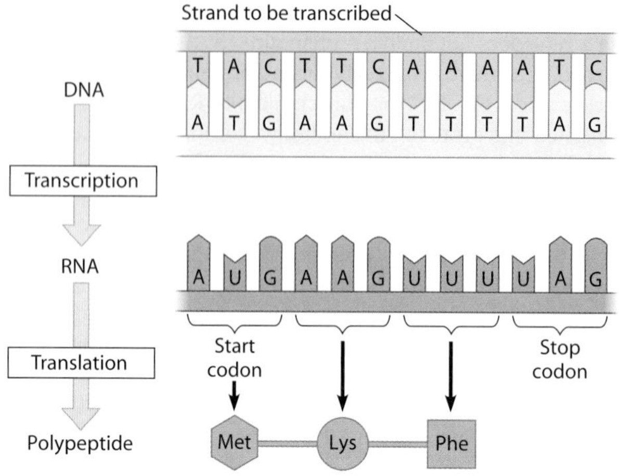

▲ **Figure 8B** Deciphering the genetic information in DNA

As an exercise in translating the genetic code, consider the 12-nucleotide segment of DNA in **Figure 8B**. Let's read this as a series of triplets. Using the base-pairing rules (with U in RNA instead of T), we see that the RNA codon corresponding to the first transcribed DNA triplet, TAC, is AUG. As you can see in Figure 8A, AUG specifies, "Place Met as the first amino acid in the polypeptide." The second DNA triplet, TTC, dictates RNA codon AAG, which designates lysine (Lys) as the second amino acid. We continue until we reach a stop codon (UAG in this example).

The genetic code is nearly universal, shared by organisms from the simplest bacteria to the most complex plants and animals. Such universality is key to modern DNA technologies because it allows scientists to mix and match genes from various species **(Figure 8C)**. A language shared by all living things must have evolved early enough in the history of life to be present in the common ancestors of all modern organisms. A shared genetic vocabulary is a reminder of the kinship that connects all life on Earth.

? Translate the RNA sequence CCAUUUACG into the corresponding amino acid sequence.

Pro-Phe-Thr

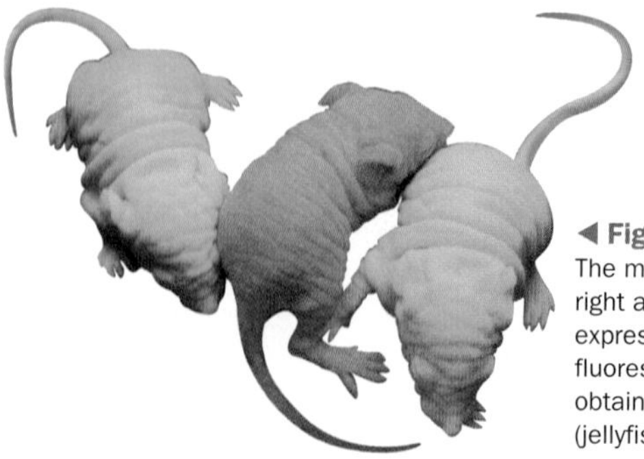

◀ **Figure 8C** The mice to the left and right are engineered to express a green fluorescence protein obtained from a jelly (jellyfish)

Dr. Masaru Okabe, Research Institute for Microbial Diseases

9 Transcription produces genetic messages in the form of RNA

In eukaryotic cells, transcription, the transfer of genetic information from DNA to RNA, occurs in the nucleus. (The nucleus, after all, contains the DNA; see Figure 6A for a review.) An RNA molecule is transcribed from a DNA template by a process that resembles the synthesis of a DNA strand during DNA replication (see Module 4).

Figure 9A is a close-up view of the process of transcription. As with replication, the two DNA strands must first separate at the place where the process will start. In transcription, however, only one of the DNA strands serves as a template for the newly forming RNA molecule; the other strand is unused. The nucleotides that make up the new RNA molecule take their place one at a time along the DNA template strand by forming hydrogen bonds with the nucleotide bases there.
Notice that the RNA nucleotides follow the same base-pairing rules that govern DNA replication, except that U, rather than T, pairs with A. The RNA nucleotides are linked by the transcription enzyme **RNA polymerase**, symbolized in the figure by the large gray shape.

Figure 9B is an overview of the transcription of an entire prokaryotic gene. (We focus on prokaryotes here; eukaryotic transcription occurs via a similar but more complex process.) Specific sequences of nucleotides along the DNA mark where transcription of a gene begins and ends. The "start transcribing" signal is a nucleotide sequence called a **promoter**. A promoter is a specific binding site for RNA polymerase and determines which of the two strands of the DNA double helix is used as the template in transcription.

❶ The first phase of transcription, called initiation, is the attachment of RNA polymerase to the promoter and the start of RNA synthesis. ❷ During a second phase of transcription, elongation, the RNA grows longer. As RNA synthesis continues, the RNA strand peels away from its DNA template, allowing the two separated DNA strands to come back together in the region already transcribed. ❸ Finally, in the third phase, termination, the RNA polymerase reaches a sequence of bases in the DNA template called a **terminator**. This sequence signals the end of the gene; at that point, the polymerase molecule detaches from the RNA molecule and the gene.

In addition to producing RNA that encodes amino acid sequences, transcription makes two other kinds of RNA that are involved in building polypeptides. We discuss these three kinds of RNA—messenger RNA, transfer RNA, and ribosomal RNA—in the next three modules.

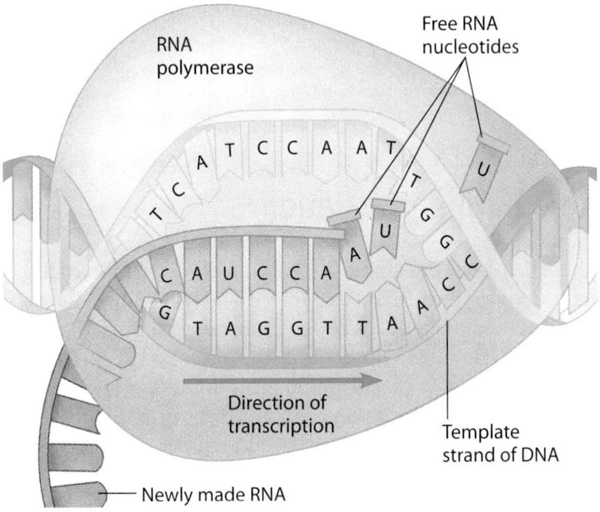

▲ **Figure 9A** A close-up view of transcription

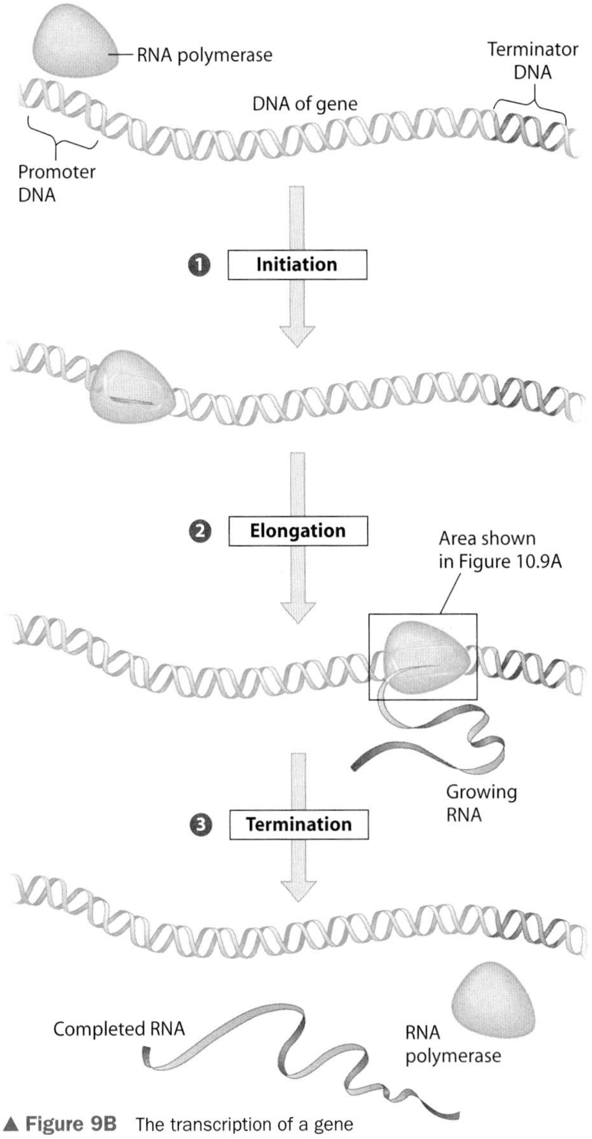

▲ **Figure 9B** The transcription of a gene

? What is a promoter? What molecule binds to it?

A promoter is a specific nucleotide sequence at the start of a gene where RNA polymerase attaches and begins transcription.

10 Eukaryotic RNA is processed before leaving the nucleus as mRNA

The kind of RNA that encodes amino acid sequences is called **messenger RNA (mRNA)** because it conveys genetic messages from DNA to the translation machinery of the cell. Messenger RNA is transcribed from DNA, and the information in the mRNA is then translated into polypeptides. In prokaryotic cells, which lack nuclei, transcription and translation occur in the same place: the cytoplasm. In eukaryotic cells, however, mRNA molecules must exit the nucleus via the nuclear pores and enter the cytoplasm, where the machinery for polypeptide synthesis is located.

Before leaving the nucleus as mRNA, eukaryotic transcripts are modified, or processed, in several ways (**Figure 10**). One kind of RNA processing is the addition of extra nucleotides to the ends of the RNA transcript. These additions include a small cap (a single G nucleotide) at one end and a long tail (a chain of 50 to 250 A nucleotides) at the other end. The cap and tail (yellow in the figure) facilitate the export of the mRNA from the nucleus, protect the mRNA from attack by cellular enzymes, and help ribosomes bind to the mRNA. The cap and tail themselves are not translated into protein.

Another type of RNA processing is made necessary in eukaryotes by noncoding stretches of nucleotides that interrupt the nucleotides that actually code for amino acids. It is as if unintelligible sequences of letters were randomly interspersed in an otherwise intelligible document. Most genes of plants and animals, it turns out, include such internal noncoding regions, which are called **introns** ("intervening sequences"). The coding regions—the parts of a gene that are expressed—are called **exons**. As Figure 10 shows, both exons (darker color) and introns (lighter color) are transcribed from DNA into RNA. However, before the RNA leaves the nucleus, the introns are removed, and the exons are joined to produce an mRNA molecule with a continuous coding sequence. (The short noncoding regions just inside the cap and tail are considered parts of the first and last exons.) This cutting-and-pasting process is called **RNA splicing**. In most cases, RNA splicing is catalyzed by a complex of proteins and small RNA molecules, but sometimes the RNA transcript itself catalyzes the process. In other words, RNA can sometimes act as an enzyme that removes its own introns! RNA splicing also provides a means to produce multiple polypeptides from a single gene.

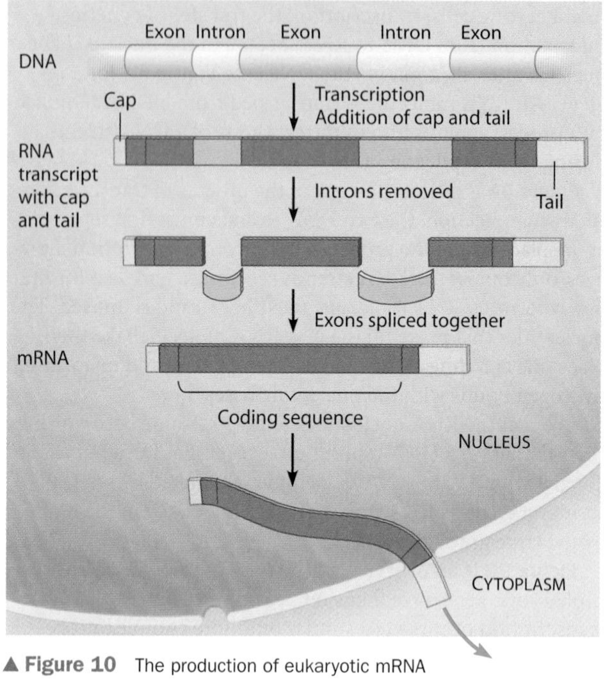

▲ **Figure 10** The production of eukaryotic mRNA

As we have discussed, translation is a conversion between different languages—from the nucleic acid language to the protein language—and it involves more elaborate machinery than transcription. The first important ingredient required for translation is the processed mRNA. Once it is present, the machinery used to translate mRNA requires enzymes and sources of chemical energy, such as ATP. In addition, translation requires two heavy-duty components: ribosomes and a kind of RNA called transfer RNA, the subject of the next module.

? Explain why most eukaryotic genes are longer than the mRNA that leaves the nucleus.

These genes have introns, noncoding sequences of nucleotides that are spliced out of the initial RNA transcript, to produce mRNA.

11 Transfer RNA molecules serve as interpreters during translation

Translation of any language requires an interpreter, someone or something that can recognize the words of one language and convert them to another. Translation of a genetic message carried in mRNA into the amino acid language of proteins also requires an interpreter. To convert the words of nucleic acids (codons) to the amino acid words of proteins, a cell employs a molecular interpreter, a special type of RNA called **transfer RNA (tRNA)**.

A cell that is producing proteins has in its cytoplasm a supply of amino acids, either obtained from food or made from other chemicals. But amino acids themselves cannot recognize the codons in the mRNA. The amino acid tryptophan, for example, is no more attracted by codons for tryptophan than by any other codons. It is up to the cell's molecular interpreters, tRNA molecules, to match amino acids to the appropriate codons to form the new polypeptide. To perform this task, tRNA molecules must carry out two functions: (1) picking up the appropriate amino acids and (2) recognizing the appropriate codons in the mRNA. The unique structure of tRNA molecules enables them to perform both tasks.

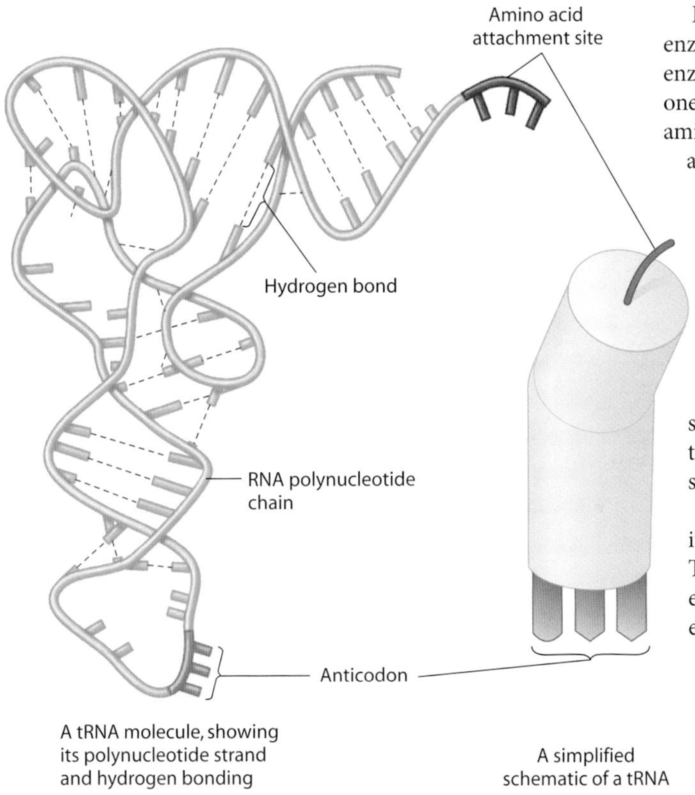

▲ **Figure 11A** The structure of tRNA

Figure 11A shows two representations of a tRNA molecule. The structure on the left shows the backbone and bases, with hydrogen bonds between bases shown as dashed magenta lines. The structure on the right is a simplified schematic that emphasizes the most important parts of the structure. Notice from the structure on the left that a tRNA molecule is made of a single strand of RNA—one polynucleotide chain—consisting of about 80 nucleotides. By twisting and folding upon itself, tRNA forms several double-stranded regions in which short stretches of RNA base-pair with other stretches via hydrogen bonds. A single-stranded loop at one end of the folded molecule contains a special triplet of bases called an **anticodon**. The anticodon triplet is complementary to a codon triplet on mRNA. During translation, the anticodon on tRNA recognizes a particular codon on mRNA by using base-pairing rules. At the other end of the tRNA molecule is a site where one specific kind of amino acid can attach.

In the modules that follow, we represent tRNA with the simplified shape shown on the right in Figure 11A. This shape emphasizes the two parts of the molecule—the anticodon and the amino acid attachment site—that give tRNA its ability to match a particular nucleic acid word (a codon in mRNA) with its corresponding protein word (an amino acid). Although all tRNA molecules are similar, there is a slightly different variety of tRNA for each amino acid.

Each amino acid is joined to the correct tRNA by a specific enzyme. There is a family of 20 versions of these enzymes, one enzyme for each amino acid. Each enzyme specifically binds one type of amino acid to all tRNA molecules that code for that amino acid, using a molecule of ATP as energy to drive the reaction. The resulting amino acid–tRNA complex can then furnish its amino acid to a growing polypeptide chain, a process that we describe in Module 12.

The computer graphic in **Figure 11B** shows a tRNA molecule (green) and an ATP molecule (purple) bound to the enzyme molecule (blue). (To help you see the two distinct molecules, the tRNA molecule is shown with a stick representation, while the enzyme is shown as space-filling spheres.) In this figure, you can see the proportional sizes of these three molecules. The amino acid that would attach to the tRNA is not shown; it would be less than half the size of the ATP.

Once an amino acid is attached to its appropriate tRNA, it can be incorporated into a growing polypeptide chain. This is accomplished within ribosomes, the cellular organelles directly responsible for the synthesis of protein. We examine ribosomes in the next module.

? What is an anticodon, and what is its function?

It is the base triplet of a tRNA molecule that couples the tRNA to a complementary codon in the mRNA. This is a key step in translating mRNA to polypeptide.

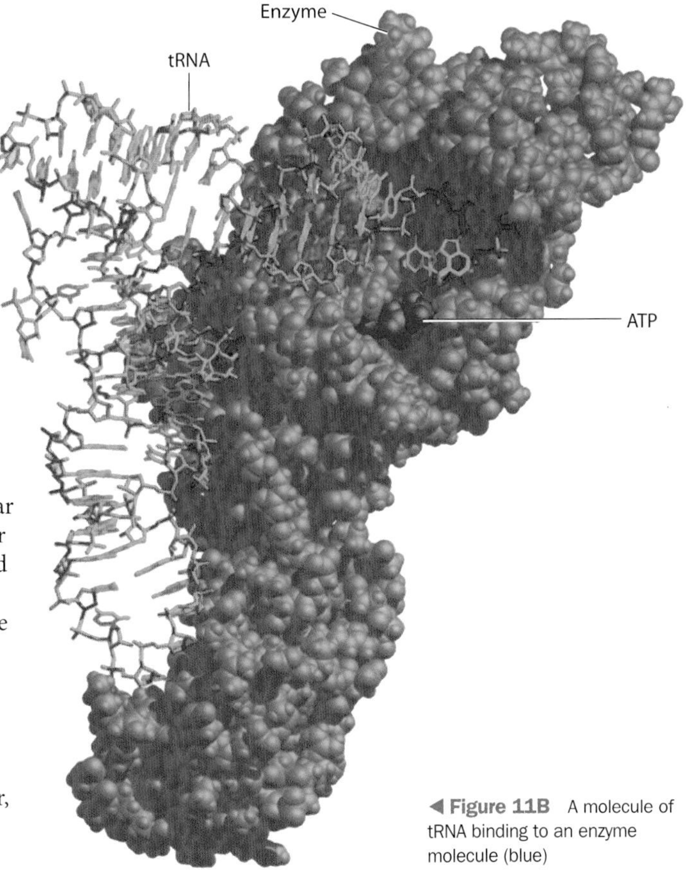

◀ **Figure 11B** A molecule of tRNA binding to an enzyme molecule (blue)

12 Ribosomes build polypeptides

We have now looked at many of the components a cell needs to carry out translation: instructions in the form of mRNA molecules, tRNA to interpret the instructions, a supply of amino acids and enzymes (for attaching amino acids to tRNA), and ATP for energy. The final components are the **ribosomes**, structures in the cytoplasm that position mRNA and tRNA close together and catalyze the synthesis of polypeptides.

A ribosome consists of two subunits, each made up of proteins and a kind of RNA called **ribosomal RNA (rRNA)**. In **Figure 12A**, you can see the actual shapes and relative sizes of the ribosomal subunits. You can also see where mRNA, tRNA, and the growing polypeptide are located during translation.

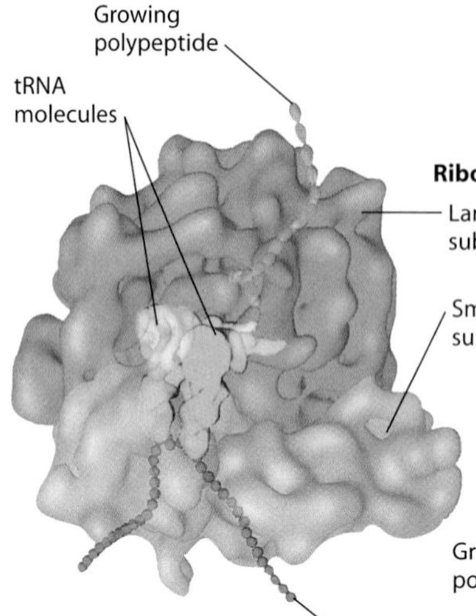

▲ **Figure 12A** The true shape of a functioning ribosome

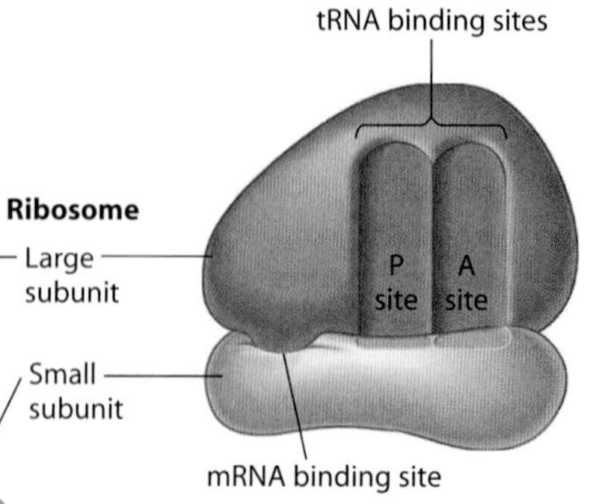

▲ **Figure 12B** A ribosome with empty binding sites

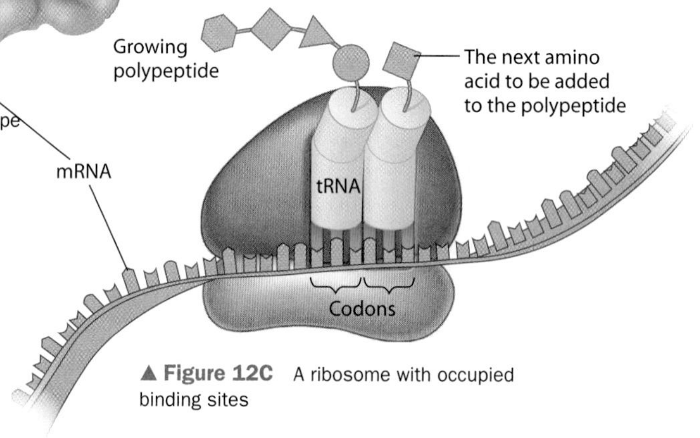

▲ **Figure 12C** A ribosome with occupied binding sites

The ribosomes of prokaryotes and eukaryotes are very similar in function, but those of eukaryotes are slightly larger and different in composition. The differences are medically significant. Certain antibiotic drugs can inactivate prokaryotic ribosomes while leaving eukaryotic ribosomes unaffected. These drugs, such as tetracycline and streptomycin, are used to combat bacterial infections.

The simplified drawings in Figures 12B and 12C indicate how tRNA anticodons and mRNA codons fit together on ribosomes. As **Figure 12B** shows, each ribosome has a binding site for mRNA and the two main binding sites (P and A) for tRNA. **Figure 12C** shows tRNA molecules occupying these two sites. The subunits of the ribosome act like a vise, holding the tRNA and mRNA molecules close together, allowing the amino acids carried by the tRNA molecules to be connected into a polypeptide chain. In the next two modules, we examine the steps of translation in detail.

? How does a ribosome facilitate protein synthesis?

A ribosome holds mRNA and tRNAs together and connects amino acids from the tRNAs to the growing polypeptide chain.

13 An initiation codon marks the start of an mRNA message

Translation can be divided into the same three phases as transcription: initiation, elongation, and termination. The process of polypeptide initiation brings together the mRNA, a tRNA bearing the first amino acid, and the two subunits of a ribosome.

As shown in **Figure 13A**, an mRNA molecule is longer than the genetic message it carries. The light pink nucleotides at either end of the molecule are not part of the message, but help the mRNA to bind to the ribosome. The initiation process establishes exactly where translation will begin, ensuring that the mRNA codons are translated into the correct sequence of amino acids.

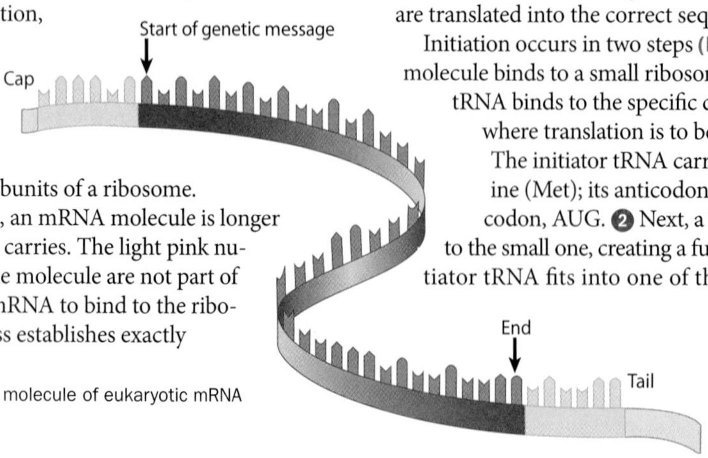

▶ **Figure 13A** A molecule of eukaryotic mRNA

Initiation occurs in two steps **(Figure 13B)**. ❶ An mRNA molecule binds to a small ribosomal subunit. A special initiator tRNA binds to the specific codon, called the **start codon**, where translation is to begin on the mRNA molecule. The initiator tRNA carries the amino acid methionine (Met); its anticodon, UAC, binds to the start codon, AUG. ❷ Next, a large ribosomal subunit binds to the small one, creating a functional ribosome. The initiator tRNA fits into one of the two tRNA binding sites on the ribosome. This site, called the **P site**, will hold the growing polypeptide. The other tRNA binding site, called the **A site**, is vacant and ready for the next amino-acid-bearing tRNA.

? What would happen if a genetic mutation changed a start codon to some other codon?

The messenger RNA transcribed from the mutated gene would be nonfunctional because ribosomes could not initiate translation correctly.

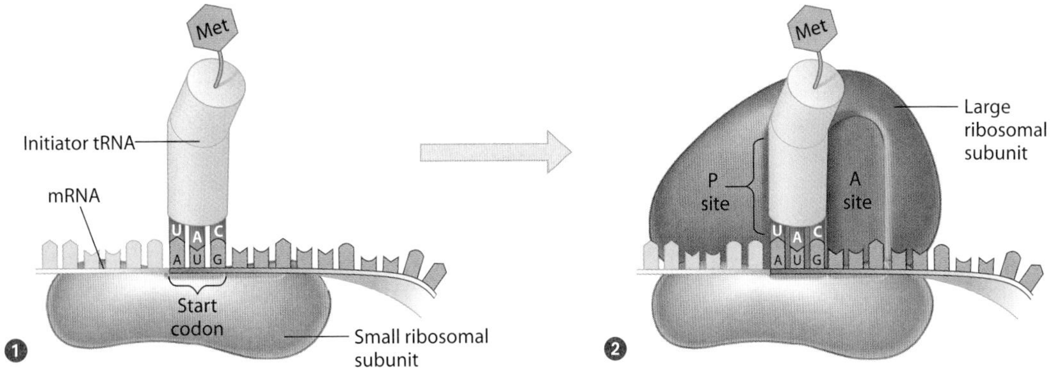

▲ **Figure 13B** The initiation of translation

14 Elongation adds amino acids to the polypeptide chain until a stop codon terminates translation

Once initiation is complete, amino acids are added one by one to the first amino acid. Each addition occurs in a three-step elongation process **(Figure 14)**:

❶ **Codon recognition.** The anticodon of an incoming tRNA molecule, carrying its amino acid, pairs with the mRNA codon in the A site of the ribosome.

❷ **Peptide bond formation.** The polypeptide separates from the tRNA in the P site and attaches by a new peptide bond to the amino acid carried by the tRNA in the A site. The ribosome catalyzes formation of the peptide bond, adding one more amino acid to the growing polypeptide chain.

❸ **Translocation.** The P site tRNA now leaves the ribosome, and the ribosome translocates (moves) the remaining tRNA in the A site, with the growing polypeptide, to the P site. The codon and anticodon remain hydrogen-bonded, and the mRNA and tRNA move as a unit. This movement brings into the A site the next mRNA codon to be translated, and the process can start again with step 1.

Elongation continues until a **stop codon** reaches the ribosome's A site. Stop codons—UAA, UAG, and UGA—do not code for amino acids but instead act as signals to stop translation. This is the termination stage of translation. The completed polypeptide is freed from the last tRNA, and the ribosome splits back into its separate subunits.

? What happens as a tRNA passes through the A and P binding sites on the ribosome?

In the A site, its amino acid receives the growing polypeptide from the tRNA that precedes it. In the P site, it gives up the polypeptide to the tRNA that follows it.

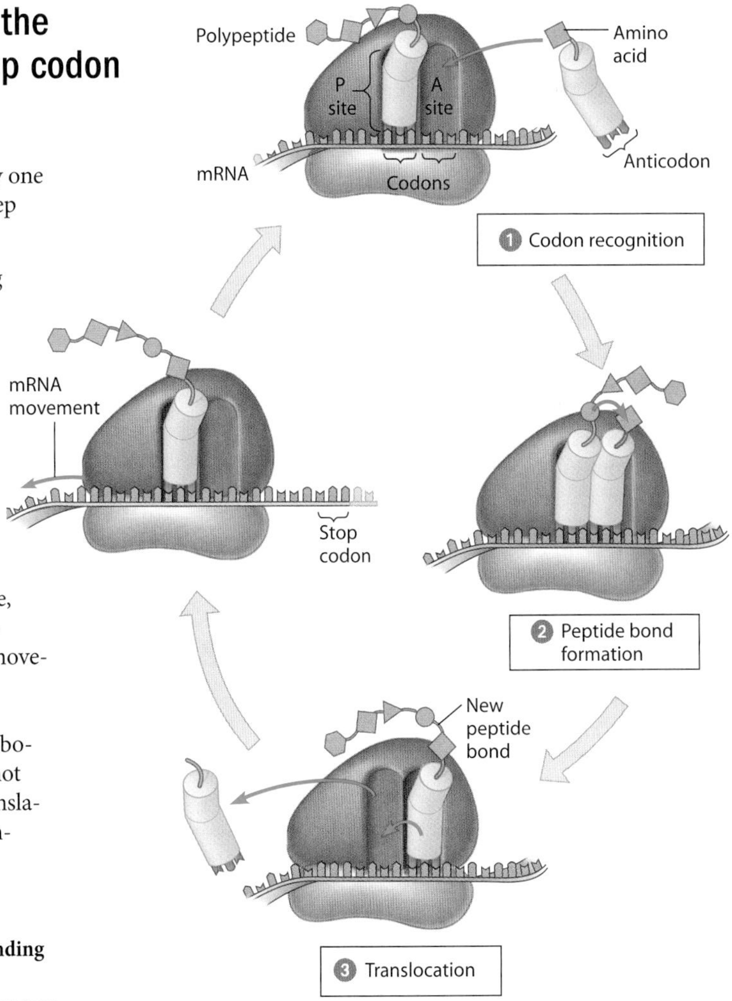

▲ **Figure 14** Polypeptide elongation; the small green arrows indicate movement

15 Review: The flow of genetic information in the cell is DNA → RNA → protein

Figure 15 summarizes the main stages in the flow of genetic information from DNA to RNA to protein. ❶ In transcription (DNA → RNA), the mRNA is synthesized on a DNA template. In eukaryotic cells, transcription occurs in the nucleus, and the messenger RNA must travel from the nucleus to the cytoplasm. In prokaryotes, transcription occurs in the cytoplasm.

❷–❺ Translation (RNA → protein) can be divided into four steps, all of which occur in the cytoplasm in eukaryotic cells. When the polypeptide is complete at the end of step 5, the two ribosomal subunits come apart, and the tRNA and mRNA are released (not shown in this figure). Translation is rapid; a single ribosome can make an average-sized polypeptide in less than a minute. Typically, an mRNA molecule is translated simultaneously by a number of ribosomes. Once the start codon emerges from the first ribosome, a second ribosome can attach to it; thus, several ribosomes may trail along on the same mRNA molecule.

As it is made, a polypeptide coils and folds, assuming a three-dimensional shape, its tertiary structure. Several polypeptides may come together, forming a protein with quaternary structure.

What is the overall significance of transcription and translation? These are the main processes whereby genes control the structures and activities of cells—or, more broadly, the way the genotype produces the phenotype. The chain of command originates with the information in a gene, a specific linear sequence of nucleotides in DNA. The gene serves as a template, dictating transcription of a complementary sequence of nucleotides in mRNA. In turn, mRNA dictates the linear sequence in which amino acids assemble to form a specific polypeptide. Finally, the proteins that form from the polypeptides determine the appearance and the capabilities of the cell and organism.

? Which of the following molecules or structures does not participate directly in translation: ribosomes, transfer RNA, messenger RNA, DNA?

DNA ●

Transcription

DNA

mRNA

RNA polymerase

❶ **Transcription**
mRNA is transcribed from a DNA template.

Translation CYTOPLASM

❷ **Amino acid attacment**
Each amino acid attaches to its proper tRNA with the help of a specific enzyme and ATP.

Amino acid

Enzyme

tRNA

ATP

Anticodon

Initiator tRNA

Large ribosomal subunit

Start Codon

mRNA

Small ribosomal subunit

❸ **Initiation of polypeptide synthesis**
The mRNA, the first tRNA, and the ribosomal subunits come together.

Growing polypeptide

New peptide bond forming

Codons

mRNA

❹ **Elongation**
A succession of tRNAs add their amino acids to the polypeptide chain as the mRNA is moved through the ribosome, one codon at a time.

Polypeptide

❺ **Termination**
The ribosome recognizes a stop codon. The polypeptide is terminated and released.

Stop codon

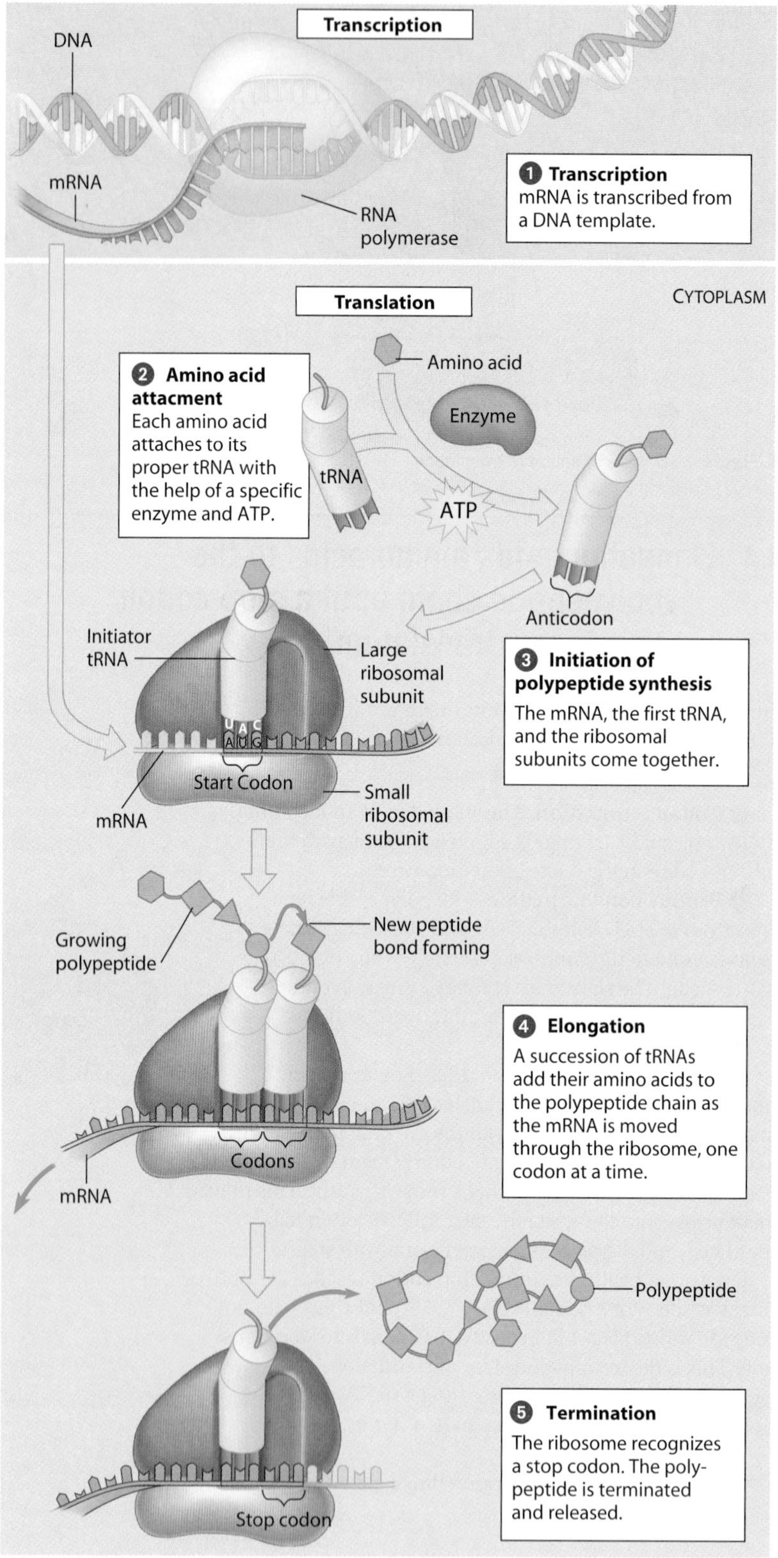

▲ **Figure 15** A summary of transcription and translation

16 Mutations can change the meaning of genes

Many inherited traits can be understood in molecular terms. For instance, sickle-cell disease can be traced through a difference in a protein to one tiny change in a gene. In one of the two kinds of polypeptides in the hemoglobin protein, an individual with sickle-cell disease has a single different amino acid—valine (Val) instead of glutamate (Glu). This difference is caused by the change of a single nucleotide in the coding strand of DNA **(Figure 16A)**. In the double helix, one nucleotide *pair* is changed.

Any change in the nucleotide sequence of DNA is called a **mutation**. Mutations can involve large regions of a chromosome or just a single nucleotide pair, as in sickle-cell disease. Here we consider how mutations involving only one or a few nucleotide pairs can affect gene translation.

Mutations within a gene can be divided into two general categories: nucleotide substitutions, and nucleotide insertions or deletions **(Figure 16B)**. A nucleotide substitution is the replacement of one nucleotide and its base-pairing partner with another pair of nucleotides. For example, in the second row in Figure 16B, A replaces G in the fourth codon of the mRNA. What effect can a substitution have? Because the genetic code is redundant, some substitution mutations have no effect at all. For example, if a mutation causes an mRNA codon to change from GAA to GAG, no change in the protein product would result because GAA and GAG both code for the same amino acid (Glu; see Figure 8A). Such a change is called a **silent mutation**.

Other substitutions, called **missense mutations**, do change the amino acid coding. For example, if a mutation causes an mRNA codon to change from GGC to AGC, as in the second row of Figure 16B. The resulting protein will have a serine (Ser) instead of a glycine (Gly) at this position. Some missense mutations have little or no effect on the shape or function of the resulting protein, but others, as in the case of sickle-cell disease, prevent the protein from performing its normal function.

Occasionally, a nucleotide substitution leads to an improved protein that enhances the success of the mutant organism and its descendants. Much more often, though, mutations are harmful. Some substitutions, called **nonsense mutations**, change an amino acid codon into a stop codon. For example, if an AGA (Arg) codon is mutated to a UGA (stop) codon, the result will be a prematurely terminated protein, which probably will not function properly.

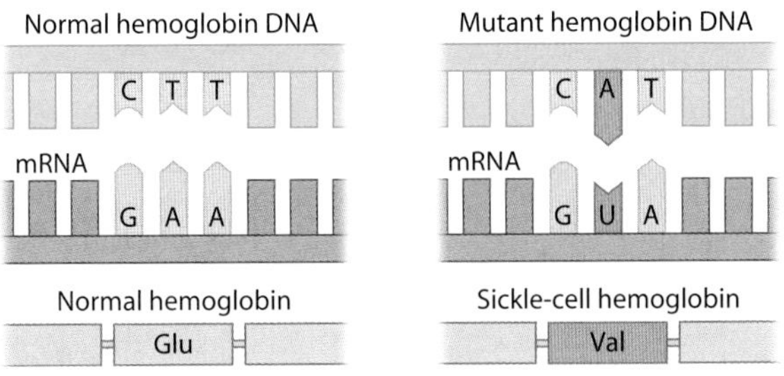

▲ **Figure 16A** The molecular basis of sickle-cell disease

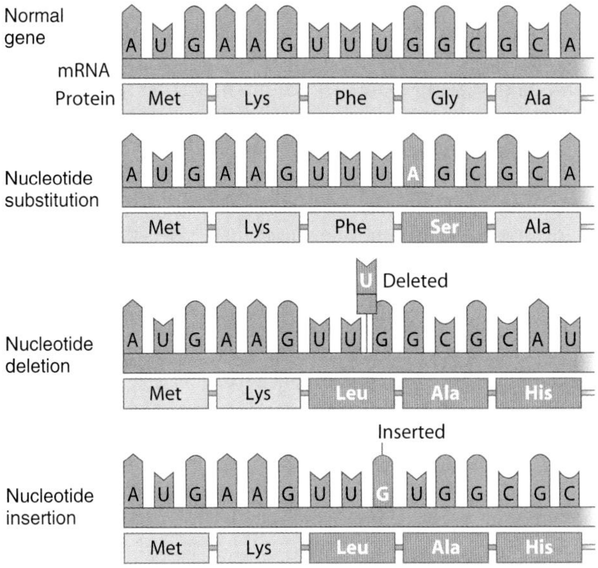

▲ **Figure 16B** Types of mutations and their effects

Mutations involving the insertion or deletion of one or more nucleotides in a gene often have disastrous effects. Because mRNA is read as a series of nucleotide triplets (codons) during translation, adding or subtracting nucleotides may alter the **reading frame** (triplet grouping) of the message. All the nucleotides that are "downstream" of the insertion or deletion will be regrouped into different codons (Figure 16B, bottom two rows). The result will most likely be a nonfunctional polypeptide.

The production of mutations, called **mutagenesis**, can occur in a number of ways. Spontaneous mutations are due to errors that occur during DNA replication or recombination. Other mutations are caused by physical or chemical agents, called **mutagens**. High-energy radiation, such as X-rays or ultraviolet light, is a physical mutagen. One class of chemical mutagens consists of chemicals that are similar to normal DNA bases but pair incorrectly or are otherwise disruptive when incorporated into DNA. For example, the anti-AIDS drug AZT works because its structure is similar enough to thymine that viral polymerases incorporate it into newly synthesized DNA, but different enough that the drug blocks further replication.

Although mutations are often harmful, they are also extremely useful, both in nature and in the laboratory. It is because of mutations that there is such a rich diversity of genes in the living world, a diversity that makes evolution by natural selection possible. Mutations are also essential tools for geneticists. Whether naturally occurring (as in Mendel's peas) or created in the laboratory (Morgan used X-rays to make most of his fruit fly mutants), mutations create the different alleles needed for genetic research.

? How could a single nucleotide substitution result in a shortened protein product?

A substitution that changed an amino acid codon into a stop codon would produce a prematurely terminated polypeptide.

The Genetics of Viruses and Bacteria

17 Viral DNA may become part of the host chromosome

As we discussed in Module 1, bacteria and viruses served as models in experiments that uncovered the molecular details of heredity. Now let's take a closer look at viruses, focusing on the relationship between viral structure and the processes of nucleic acid replication, transcription, and translation.

In a sense, a **virus** is nothing more than "genes in a box": an infectious particle consisting of a bit of nucleic acid wrapped in a protein coat called a **capsid** and, in some cases, a membrane envelope. Viruses are parasites that can replicate (reproduce) only inside cells. In fact, the host cell provides most of the components necessary for replicating, transcribing, and translating the viral nucleic acid.

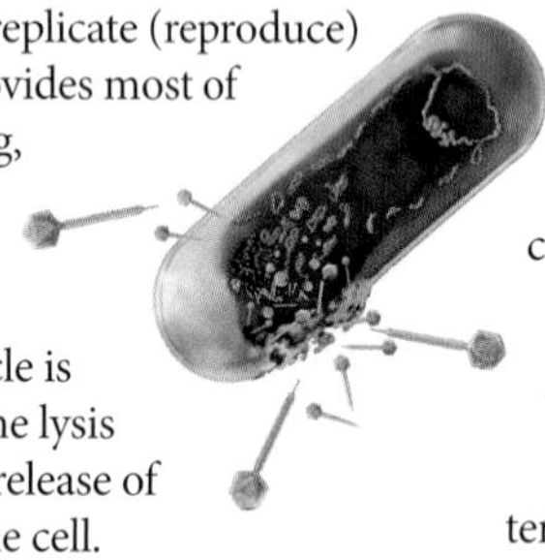

In Figure 1C, we described the replication cycle of phage T2. This sort of cycle is called a **lytic cycle** because it results in the lysis (breaking open) of the host cell and the release of the viruses that were produced within the cell. Some phages can also replicate by an alternative route called the lysogenic cycle. During a **lysogenic cycle**, viral DNA replication occurs without destroying the host cell.

In **Figure 17**, you see the two kinds of cycles for a phage called lambda that infects *E. coli*. Both cycles begin when the phage DNA ❶ enters the bacterium and ❷ forms a circle. The DNA then embarks on one of the two pathways. In the lytic cycle (left), ❸ lambda's DNA immediately turns the cell into a virus-producing factory, and ❹ the cell soon lyses and releases its viral products.

In the lysogenic cycle, however, ❺ viral DNA is inserted by genetic recombination into the bacterial chromosome. Once inserted, the phage DNA is referred to as a **prophage**, and most of its genes are inactive. ❻ Every time the *E. coli* cell prepares to divide, it replicates the phage DNA along with its own chromosome and passes the copies on to daughter cells. A single infected cell can thereby quickly give rise to a large population of bacterial cells that all carry prophage. The lysogenic cycle enables viruses to spread without killing the host cells on which they depend. The prophages may remain in the bacterial cells indefinitely. ❼ Occasionally, however, an environmental signal—typically, one that indicates an unfavorable turn in the environment, such as an increase in radiation, drought, or certain toxic chemicals—triggers a switchover from the lysogenic cycle to the lytic cycle. This so-called genetic switch causes the viral DNA to be excised from the bacterial chromosome, eventually leading to death of the host cell.

Sometimes, the few prophage genes active in a lysogenic bacterium can cause medical problems. For example, the bacteria that cause diphtheria, botulism, and scarlet fever would be harmless to humans if it were not for the prophage genes they carry. Certain of these genes direct the bacteria to produce the toxins responsible for making people ill. In the next module, we will explore viruses that infect animals and plants.

? Describe one way a virus can perpetuate its genes without destroying its host cell. What is this type of replication cycle called?

Some viruses can insert their DNA into a chromosome of the host cell, which replicates the viral genes when it replicates its own DNA prior to cell division. This is called the lysogenic cycle.

▶ **Figure 17** Two types of phage replication cycles

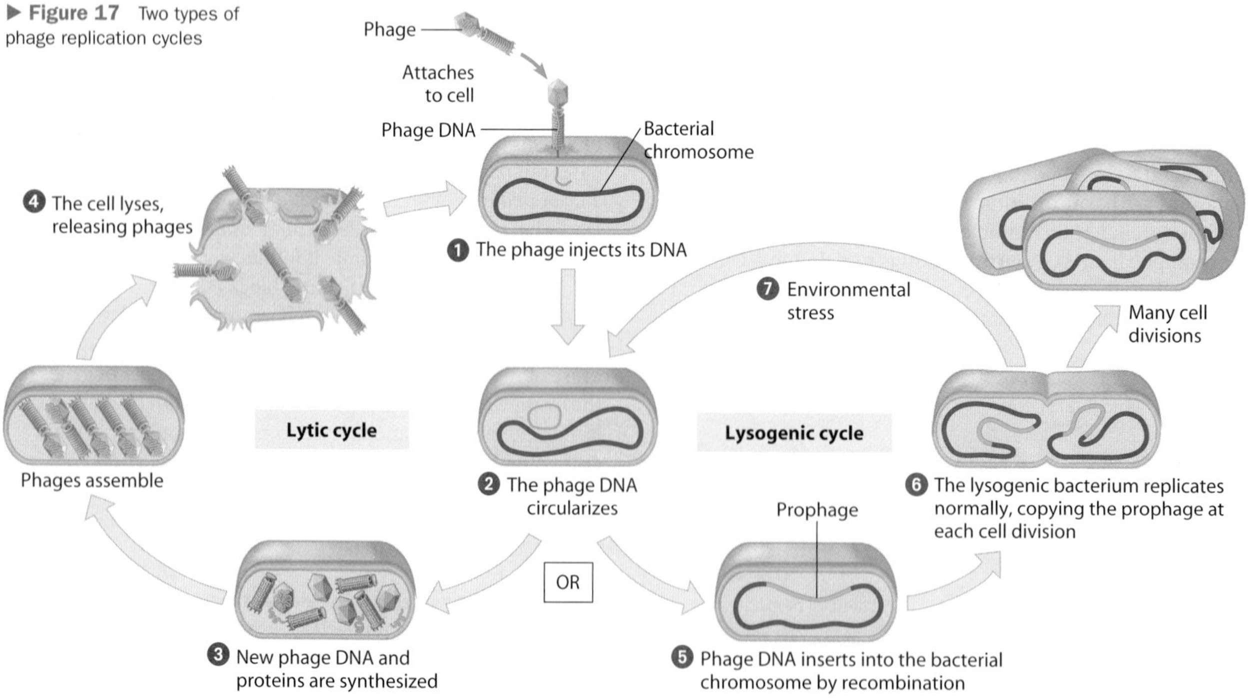
Phage
Attaches to cell
Phage DNA
Bacterial chromosome
❶ The phage injects its DNA
❹ The cell lyses, releasing phages
Phages assemble
Lytic cycle
❷ The phage DNA circularizes
❸ New phage DNA and proteins are synthesized
OR
❼ Environmental stress
Many cell divisions
Lysogenic cycle
Prophage
❻ The lysogenic bacterium replicates normally, copying the prophage at each cell division
❺ Phage DNA inserts into the bacterial chromosome by recombination

Russell Kightley/SPL/Photo Researchers

CONNECTION 18 Many viruses cause disease in animals and plants

Viruses can cause disease in both animals and plants. A typical animal virus has a membranous outer envelope and projecting spikes of glycoprotein (protein molecules with attached sugars). The envelope helps the virus enter and leave the host cell. Many animal viruses have RNA rather than DNA as their genetic material. Examples of RNA viruses include those that cause the common cold, measles, mumps, polio, and AIDS. Examples of diseases caused by DNA viruses include hepatitis, chicken pox, and herpes infections.

Figure 18 shows the replication cycle of a typical enveloped RNA virus: the mumps virus. (Once a common childhood disease characterized by fever and painful swelling of the salivary glands, mumps has become quite rare in industrialized nations.) When the virus contacts a host cell, the glycoprotein spikes attach to receptor proteins on the cell's plasma membrane. The envelope fuses with the cell's membrane, allowing the protein-coated RNA to ❶ enter the cytoplasm. ❷ Enzymes then remove the protein coat. ❸ An enzyme that entered the cell as part of the virus uses the virus's RNA genome as a template for making complementary strands of RNA (pink). The new strands have two functions: ❹ They serve as mRNA for the synthesis of new viral proteins, and ❺ they serve as templates for synthesizing new viral genome RNA. ❻ The new coat proteins assemble around the new viral RNA. ❼ Finally, the viruses leave the cell by cloaking themselves in the host cell's plasma membrane. Thus, the virus obtains its envelope from the host cell, leaving the cell without necessarily lysing it.

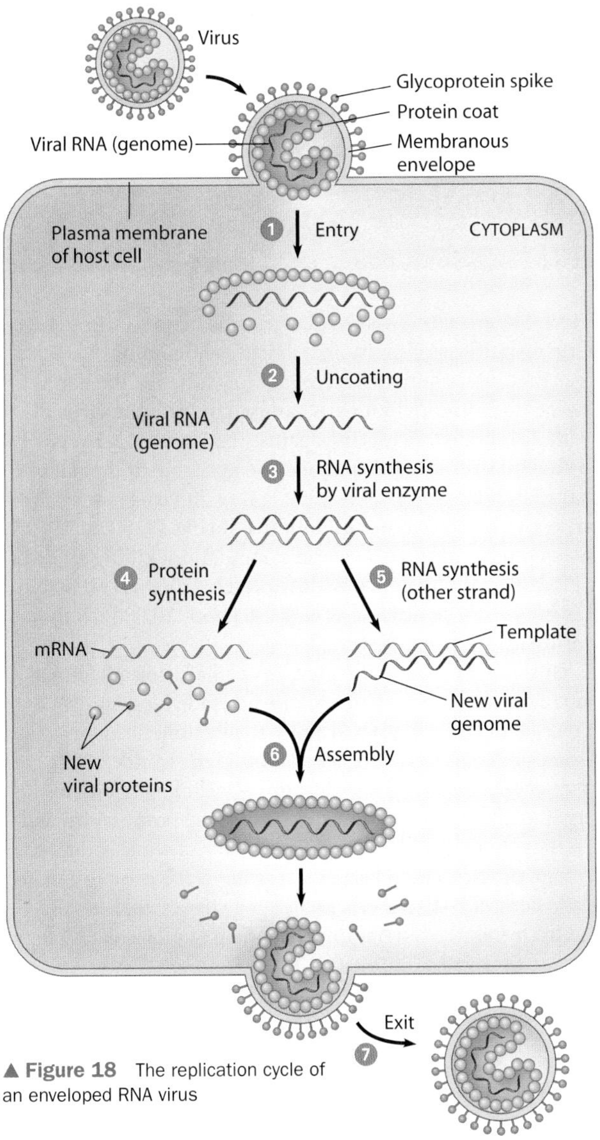

▲ **Figure 18** The replication cycle of an enveloped RNA virus

Not all animal viruses replicate in the cytoplasm. For example, herpesviruses, which you read about in the chapter introduction, are enveloped DNA viruses that replicate in the host cell's nucleus; they acquire their envelopes from the cell's nuclear membranes. While inside the nuclei of certain nerve cells, herpesvirus DNA may remain permanently dormant, without destroying these cells. From time to time, physical stress, such as a cold or sunburn, or emotional stress may stimulate the herpesvirus DNA to begin production of the virus, which then infects cells at the body's surface and brings about cold sores or genital sores.

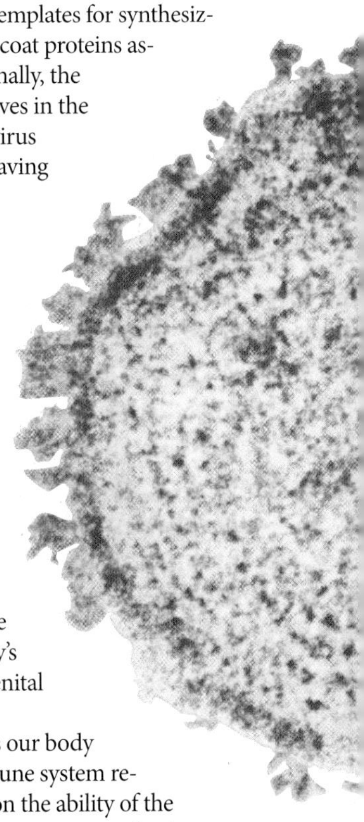

Hazel Appleton, Centre for Infections/Health Protection Agency/Photo Researchers

The amount of damage a virus causes our body depends partly on how quickly our immune system responds to fight the infection and partly on the ability of the infected tissue to repair itself. We usually recover completely from colds because our respiratory tract tissue can efficiently replace damaged cells by mitosis. In contrast, the poliovirus attacks nerve cells, which are not usually replaceable. The damage to such cells, unfortunately, is permanent. In such cases, we try to prevent the disease with vaccines.

Viruses that infect plants can stunt plant growth and diminish crop yields. Most known plant viruses are RNA viruses. To infect a plant, a virus must first get past the plant's outer protective layer of cells (the epidermis). Once a virus enters a plant cell and begins replicating, it can spread throughout the entire plant through plasmodesmata, the cytoplasmic connections that penetrate the walls between adjacent plant cells. Plant viruses may spread to other plants by insects, herbivores, humans, or farming tools. As with animal viruses, there are no cures for most viral diseases of plants. Agricultural scientists focus instead on preventing infections and on breeding resistant varieties of crop plants.

? Explain how some viruses replicate without having DNA.

The genetic material of these viruses is RNA, which is replicated inside the host cell by special enzymes encoded by the virus. The viral genome (or its complement) serves as mRNA for the synthesis of viral proteins.

EVOLUTION CONNECTION

19 Emerging viruses threaten human health

Viruses that appear suddenly or are new to medical scientists are called **emerging viruses**. There are many familiar examples. HIV, the AIDS virus, is a classic example: This virus appeared in New York and California in the early 1980s, seemingly out of nowhere. The deadly Ebola virus, recognized initially in 1976 in central Africa, is one of several emerging viruses that cause hemorrhagic fever, an often fatal illness characterized by fever, vomiting, massive bleeding, and circulatory system collapse. A number of other dangerous new viruses cause encephalitis, an inflammation of the brain. One example is the West Nile virus, which appeared for the first time in North America in 1999 and has since spread to all 48 contiguous U.S. states. West Nile virus is spread primarily by mosquitoes, which carry the virus in blood sucked from one victim and can transfer it to another victim. Severe acute respiratory syndrome (SARS) first appeared in China in 2002. Within eight months, about 8,000 people were infected, of whom some 10% died. Researchers quickly identified the infectious agent as a previously unknown, single-stranded RNA coronavirus, so named for its crown-like "corona" of spikes.

▼ **Figure 19** People in Mexico City wearing masks in an attempt to prevent spread of the 2009 H1N1 virus (shown in the inset)

left Health Protection Agency Centre for Infections/Photo Researchers, Inc. **right** Mario Guzman/epa/Corbis

From where and how do such viruses burst on the human scene, giving rise to rare or previously unknown diseases? Three processes contribute to the emergence of viral diseases: mutation, contact between species, and spread from isolated populations.

The mutation of existing viruses is a major source of new viral diseases. RNA viruses tend to have unusually high rates of mutation because errors in replicating their RNA genomes are not subject to the kind of proofreading and repair mechanisms that help reduce errors in DNA replication. Some mutations enable existing viruses to evolve into new strains (genetic varieties) that can cause disease in individuals who have developed immunity to ancestral strains. That is why we need yearly flu vaccines: Mutations create new influenza virus strains to which previously vaccinated people have no immunity.

New viral diseases often arise from the spread of existing viruses from one host species to another. Scientists estimate that about three-quarters of new human diseases have originated in other animals. For example, in 1997, at least 18 people in Hong Kong were infected with a strain of flu virus called H5N1, which was previously seen only in birds. A mass culling of all of Hong Kong's 1.5 million domestic birds appeared to stop that outbreak. Beginning in 2002, however, new cases of human infection by this bird strain began to crop up around southeast Asia. As of 2009, the disease caused by this virus, now called "avian flu," has killed more than 250 people, and more than 100 million birds have either died from the disease or been killed to prevent the spread of infection.

In 2009, scientists in Mexico and then the United States became aware of a rapidly spreading new strain of flu called H1N1 (**Figure 19**). This particular virus evolved through genetic reshuffling of multiple flu viruses, including ones that infect humans, birds, and pigs (hence the name "swine flu," although humans cannot contract the virus directly from pigs). A vaccine against the 2009 H1N1 was rushed into production and became available in the fall of that year. Interestingly, the 2009 H1N1 flu virus is very similar to a virus that, in just 18 months during 1918 and 1919, infected one-third of the world's population, killing an estimated 50 million people worldwide. The 2009 H1N1 strain was not nearly as deadly; as of 2010, about 20,000 deaths worldwide had been reported.

The spread of a viral disease from a small, isolated human population can also lead to widespread epidemics. For instance, AIDS went unnamed and virtually unnoticed for decades before it began to spread around the world. In this case, technological and social factors—including affordable international travel, blood transfusions, sexual promiscuity, and the abuse of intravenous drugs—allowed a previously rare human disease to become a global scourge. If we ever do manage to control HIV and other emerging viruses, that success will likely follow from our understanding of molecular biology.

? Why doesn't a flu shot one year give us immunity to flu in subsequent years?

Influenza viruses evolve rapidly by frequent mutation; thus, the strains that infect us later will most likely be different from the ones to which we've been vaccinated.

20 The AIDS virus makes DNA on an RNA template

The devastating disease **AIDS** (acquired immunodeficiency syndrome) is caused by **HIV** (human immunodeficiency virus), an RNA virus with some special properties. In outward appearance, HIV resembles the flu or mumps virus (**Figure 20A**). Its membranous envelope and glycoprotein spikes enable HIV to enter and leave a host cell much the way the mumps virus does (see Figure 18). Notice, however, that HIV contains two identical copies of its RNA instead of one. HIV also has a different mode of replication. HIV carries molecules of an enzyme called **reverse transcriptase**, which catalyzes reverse transcription, the synthesis of DNA on an RNA template. This unusual process, which is opposite the usual DNA → RNA flow of genetic information, characterizes **retroviruses** (retro means "backward").

Figure 20B illustrates what happens after HIV RNA is uncoated in the cytoplasm of a host cell. ❶ Reverse transcriptase (●) uses the RNA as a template to make a DNA strand and then ❷ adds a second, complementary DNA strand. ❸ The resulting viral DNA enters the cell's nucleus and inserts itself into the chromosomal DNA, becoming a provirus (analogous to a prophage). The host's RNA polymerase ❹ transcribes the proviral DNA into RNA, which can then be ❺ translated by ribosomes into viral proteins. ❻ New viruses assembled from these components leave the cell and can infect other cells.

HIV infects and kills white blood cells that play important roles in the body's immune system. The loss of such cells causes the body to become susceptible to other infections that it would normally be able to fight off. Such secondary infections cause the syndrome (a collection of symptoms) that can kill an AIDS patient.

? Why is HIV classified as a retrovirus?

It synthesizes DNA from its RNA genome. This is the reverse ("retro") of the usual DNA → RNA information flow.

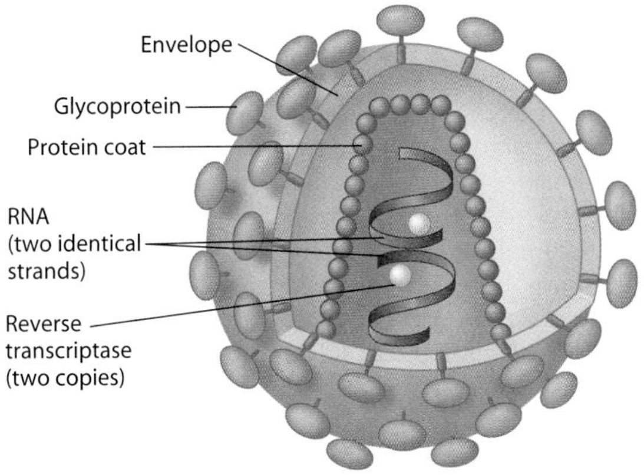

▲ **Figure 20A** A model of HIV structure

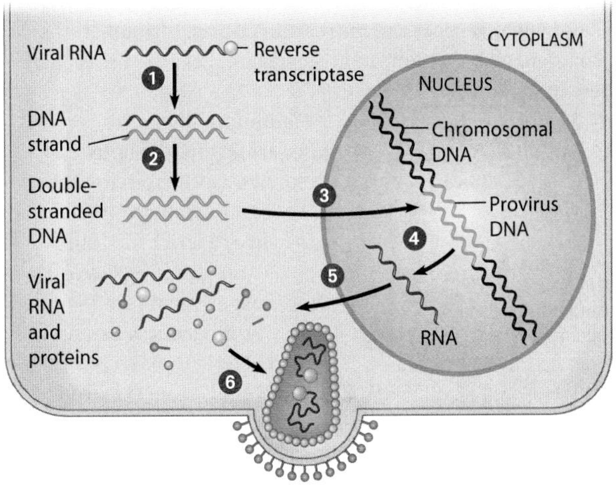

▲ **Figure 20B** The behavior of HIV nucleic acid in a host cell

21 Viroids and prions are formidable pathogens in plants and animals

Viruses may be small and simple, but they dwarf another class of pathogens: viroids. **Viroids** are small circular RNA molecules that infect plants. Unlike the nucleic acid of a virus, viroids do not encode proteins but can replicate in host plant cells, apparently using cellular enzymes. These small RNA molecules seem to cause errors in the regulatory systems that control plant growth. The typical signs of viroid diseases are abnormal development and stunted growth.

An important lesson to learn from viroids is that a single molecule can be an infectious agent. Viroids consist solely of nucleic acid, whose ability to be replicated is well known. Even more surprising are infectious proteins called **prions**, which cause a number of degenerative brain diseases in various animal species, including scrapie in sheep and goats, chronic wasting disease in deer and elk, mad cow disease (formally called bovine spongiform encephalopathy, or BSE), and Creutzfeldt-Jakob disease in humans. (The human disease is exceedingly rare, with at most a few hundred cases per year in the United States.)

A prion is thought to be a misfolded form of a protein normally present in brain cells. When the prion enters a cell containing the normal form of protein, the prion somehow converts the normal protein molecules to the misfolded prion versions. The abnormal proteins clump together, which may lead to loss of brain tissue (although *how* this occurs is the subject of much debate and ongoing research). To date, there is no cure for prion diseases, and the only hope for developing effective treatments lies in understanding and preventing the process of infection.

? What makes prions different from all other known infectious agents?

Prions are proteins and have no nucleic acid.

22 Bacteria can transfer DNA in three ways

By studying viral replication, researchers also learn about the mechanisms that regulate DNA replication and gene expression in living cells. Bacteria are equally valuable as microbial models in genetics research. As prokaryotic cells, bacteria allow researchers to investigate molecular genetics in the simplest living organisms.

Most of a bacterium's DNA is found in a single chromosome, a closed loop of DNA with associated proteins. In the diagrams here, we show the chromosome much smaller than it actually is relative to the cell. A bacterial chromosome is hundreds of times longer than its cell; it fits inside the cell because it is tightly folded.

Bacterial cells reproduce by replication of the bacterial chromosome followed by binary fission. Because binary fission is an asexual process involving only a single parent, the bacteria in a colony are genetically identical to the parental cell. But this does not mean that bacteria lack ways to produce new combinations of genes. In fact, in the bacterial world, there are three mechanisms by which genes can move from one cell to another: transformation, transduction, and conjugation. Let's discuss each of these in turn.

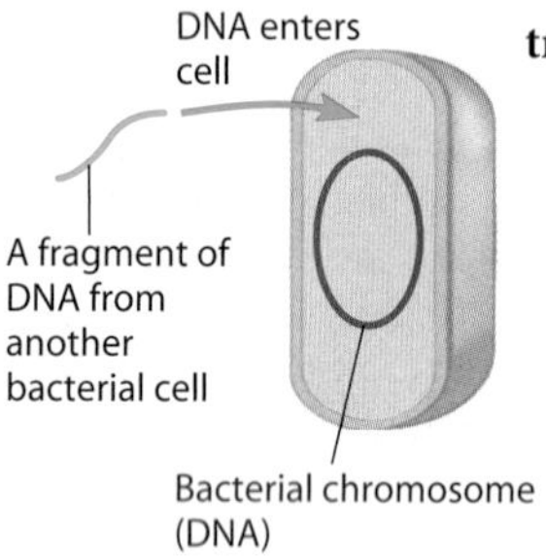

▲ **Figure 22A** Transformation

Figure 22A illustrates **transformation**, the uptake of foreign DNA from the surrounding environment. In Frederick Griffith's "transforming factor" experiment (see Module 1), a harmless strain of bacteria took up pieces of DNA left from the dead cells of a disease-causing strain. The DNA from the pathogenic bacteria carried a gene that made the cells resistant to an animal's defenses, and when the previously harmless bacteria acquired this gene, it could cause pneumonia in infected animals.

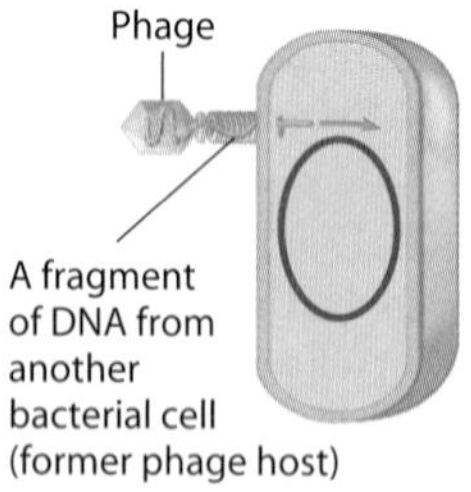

▲ **Figure 22B** Transduction

Bacteriophages, the viruses that infect bacteria, provide the second means of bringing together genes of different bacteria. The transfer of bacterial genes by a phage is called **transduction**. During a lytic infection, when new viruses are being assembled in an infected bacterial cell, a fragment of DNA belonging to the host cell may be mistakenly packaged within the phage's coat instead of the phage's DNA. When the phage infects a new bacterial cell, the DNA stowaway from the former host cell is injected into the new host **(Figure 22B)**.

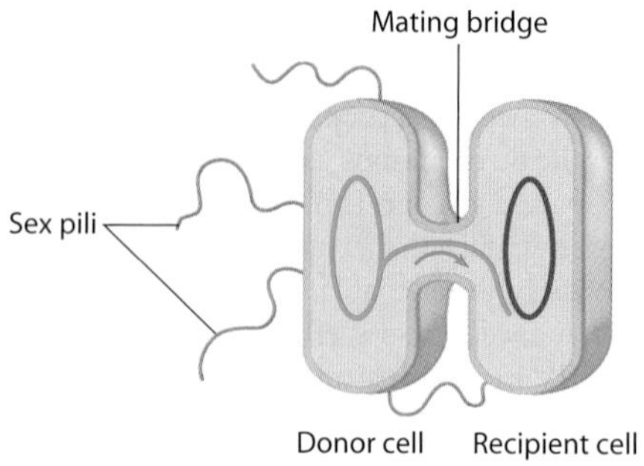

▲ **Figure 22C** Conjugation

Figure 22C is an illustration of what happens at the DNA level when two bacterial cells "mate." This physical union of two bacterial cells—of the same or different species—and the DNA transfer between them is called **conjugation**. The donor cell has hollow appendages called sex pili, one of which is attached to the recipient cell in the figure. The outside layers of the cells have fused, and a cytoplasmic bridge has formed between them. Through this mating bridge, donor cell DNA (light blue in the figure) passes to the recipient cell. The donor cell replicates its DNA as it transfers it, so the cell doesn't end up lacking any genes. The DNA replication is a special type that allows one copy to peel off and transfer into the recipient cell.

Once new DNA gets into a bacterial cell, by whatever mechanism, part of it may then integrate into the recipient's chromosome. As **Figure 22D** indicates, integration occurs by crossing over between the donor and recipient DNA molecules, a process similar to crossing over between eukaryotic chromosomes. Here we see that two crossovers result in a piece of the donated DNA replacing part of the recipient cell's original DNA. The leftover pieces of DNA are broken down and degraded, leaving the recipient bacterium with a recombinant chromosome.

As we'll see in the next module, the transfer of genetic material between bacteria has important medical consequences.

? **The three modes of gene transfer between bacteria are _______, which is transfer via a virus; _______, which is the uptake of DNA from the surrounding environment; and _______, which is bacterial "mating."**

transduction . . . transformation . . . conjugation

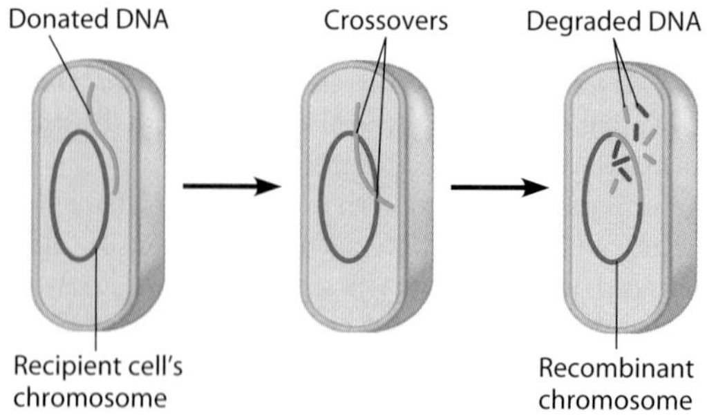

▲ **Figure 22D** The integration of donated DNA into the recipient cell's chromosome

23 Bacterial plasmids can serve as carriers for gene transfer

The ability of a donor *E. coli* cell to carry out conjugation is usually due to a specific piece of DNA called the **F factor** (F for *fertility*). The F factor carries genes for making sex pili and other requirements for conjugation; it also contains an origin of replication, where DNA replication starts.

Let's see how the F factor behaves during conjugation. In **Figure 23A**, the F factor (light blue) is integrated into the donor bacterium's chromosome. When this cell conjugates with a recipient cell, the donor chromosome starts replicating at the F factor's origin of replication, indicated by the blue dot on the DNA. The growing copy of the DNA peels off the chromosome and heads into the recipient cell. Thus, part of the F factor serves as the leading end of the transferred DNA, but right behind it are genes from the donor's original chromosome. The rest of the F factor stays in the donor cell. Once inside the recipient cell, the transferred donor genes can recombine with the corresponding part of the recipient chromosome by crossing over. If crossing over occurs, the recipient cell may be genetically changed, but it usually remains a recipient because the two cells break apart before the rest of the F factor transfers.

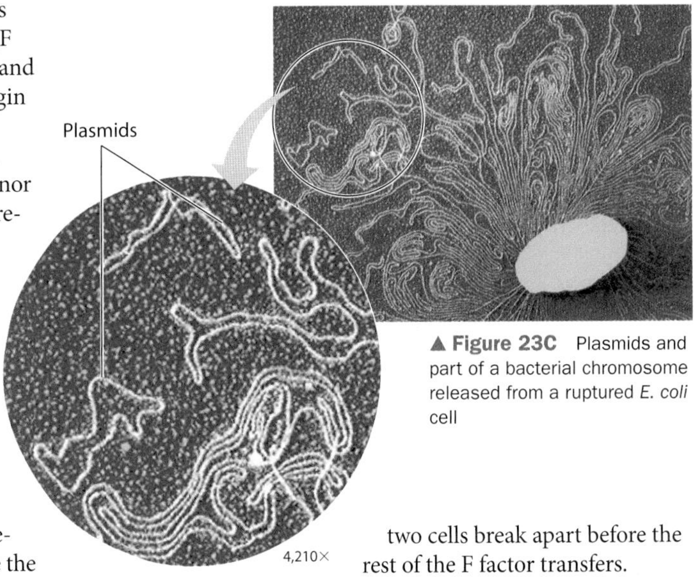

▲ **Figure 23C** Plasmids and part of a bacterial chromosome released from a ruptured *E. coli* cell

Colorized TEM 1,730×

Huntington Potter, University of South Florida College of Medicine

Alternatively, as **Figure 23B** shows, an F factor can exist as a **plasmid**, a small, circular DNA molecule separate from the bacterial chromosome. Every plasmid has an origin of replication, required for its replication within the cell. Some plasmids, including the F factor plasmid, can bring about conjugation and move to another cell. When the donor cell in Figure 23B mates with a recipient cell, the F factor replicates and at the same time transfers one whole copy of itself, in linear rather than circular form, to the recipient cell. The transferred plasmid re-forms a circle in the recipient cell, and the cell becomes a donor.

E. coli and other bacteria have many different kinds of plasmids. You can see several from one cell in **Figure 23C**, along with part of the bacterial chromosome, which extends in loops from the ruptured cell. Some plasmids carry genes that can affect the survival of the cell. Plasmids of one class, called **R plasmids**, pose serious problems for human medicine. Transferable R plasmids carry genes for enzymes that destroy antibiotics such as penicillin and tetracycline. Bacteria containing R plasmids are resistant (hence the designation R) to antibiotics that would otherwise kill them. The widespread use of antibiotics in medicine and agriculture has tended to kill off bacteria that lack R plasmids, whereas those with R plasmids have multiplied. As a result, an increasing number of bacteria that cause human diseases, such as food poisoning and gonorrhea, are becoming resistant to antibiotics.

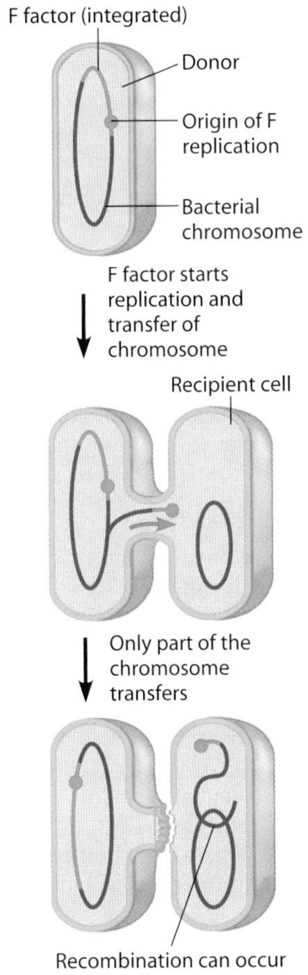

▲ **Figure 23A** Transfer of chromosomal DNA by an integrated F factor

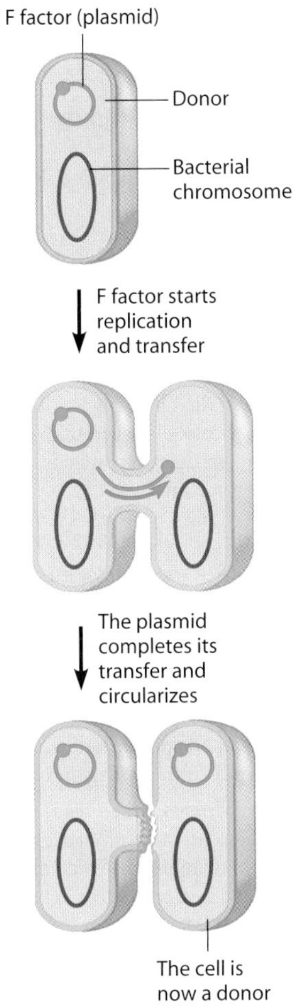

▲ **Figure 23B** Transfer of an F factor plasmid

? Plasmids are useful tools for genetic engineering. Can you guess why?

Scientists can take advantage of the ability of plasmids to carry foreign genes, to replicate, and to be inherited by progeny cells.

REVIEW

For Practice Quizzes, BioFlix, MP3 Tutors, and Activities, go to www.masteringbiology.com.

Reviewing the Concepts

The Structure of the Genetic Material (1–3)

1 Experiments showed that DNA is the genetic material. One key experiment demonstrated that certain phages (bacterial viruses) reprogram host cells to produce more phages by injecting their DNA.

2 DNA and RNA are polymers of nucleotide.

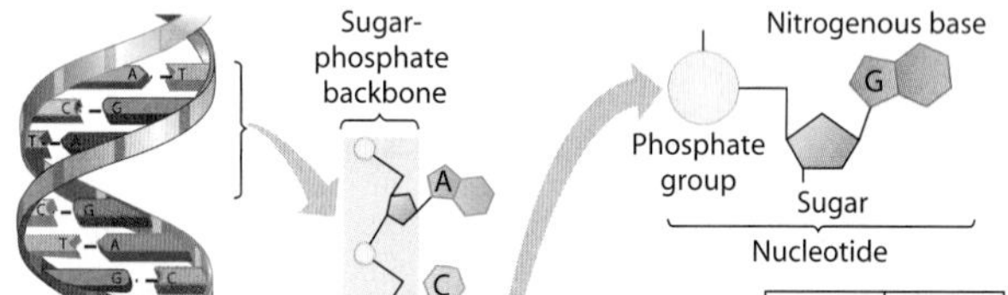

	DNA	RNA
Nitrogenous bases	C G A T	C G A U
Sugar	Deoxyribose	Ribose

3 DNA is a double-stranded helix. Watson and Crick worked out the three-dimensional structure of DNA: two polynucleotide strands wrapped around each other in a double helix. Hydrogen bonds between bases hold the strands together. Each base pairs with a complementary partner: A with T, G with C.

DNA Replication (4–5)

4 DNA replication depends on specific base pairing. DNA replication starts with the separation of DNA strands. Enzymes then use each strand as a template to assemble new nucleotides into a complementary strand.

5 DNA replication proceeds in two directions at many sites simultaneously. Using the enzyme DNA polymerase, the cell synthesizes one daughter strand as a continuous piece. The other strand is synthesized as a series of short pieces, which are then connected by the enzyme DNA ligase.

The Flow of Genetic Information from DNA to RNA to Protein (6–16)

6 The DNA genotype is expressed as proteins, which provide the molecular basis for phenotypic traits. The DNA of a gene—a linear sequence of many nucleotides—is transcribed into RNA, which is translated into a polypeptide.

7 Genetic information written in codons is translated into amino acid sequences. Codons are base triplets.

8 The genetic code dictates how codons are translated into amino acids. Nearly all organisms use an identical genetic code to convert the codons of a gene to the amino acid sequence of a polypeptide.

9 Transcription produces genetic messages in the form of RNA. In the nucleus, the DNA helix unzips, and RNA nucleotides line up and hydrogen-bond along one strand of the DNA, following the base-pairing rules.

10 Eukaryotic RNA is processed before leaving the nucleus as mRNA. Noncoding segments of RNA called introns are spliced out, and a cap and tail are added to the ends of the mRNA.

11 Transfer RNA molecules serve as interpreters during translation. Translation takes place in the cytoplasm. A ribosome attaches to the mRNA and translates its message into a specific polypeptide, aided by transfer RNAs (tRNAs). Each tRNA is a folded molecule bearing a base triplet called an anticodon on one end; a specific amino acid is added to the other end.

12 Ribosomes build polypeptides. Made of rRNA and proteins, ribsomes have binding sites for tRNAs and mRNA.

13 An initiation codon marks the start of an mRNA message.

14 Elongation adds amino acids to the polypeptide chain until a stop codon terminates translation. As the mRNA moves one codon at a time relative to the ribosome, a tRNA with a complementary anticodon pairs with each codon, adding its amino acid to the growing polypeptide chain.

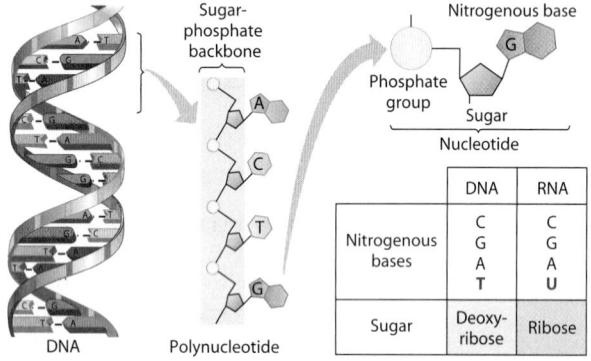

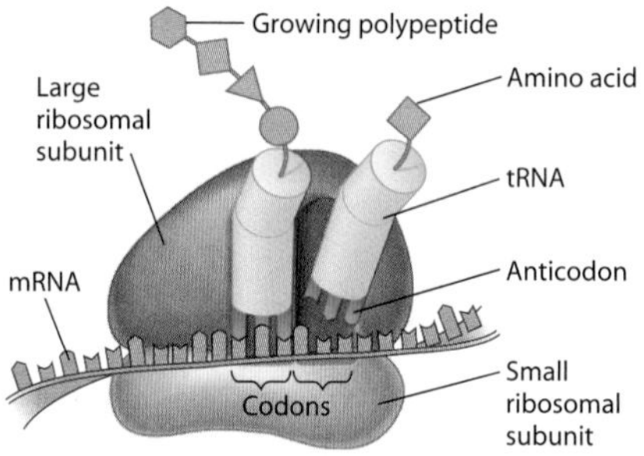

15 Review: The flow of genetic information in the cell is DNA → RNA → protein. The sequence of codons in DNA, via the sequence of codons in mRNA, spells out the primary structure of a polypeptide.

16 Mutations can change the meaning of genes. Mutations are changes in the DNA nucleotide sequence, caused by errors in DNA replication or recombination, or by mutagens. Substituting, inserting, or deleting nucleotides alters a gene, with varying effects on the organism.

The Genetics of Viruses and Bacteria (17–23)

17 Viral DNA may become part of the host chromosome. Viruses can be regarded as genes packaged in protein. When phage DNA enters a lytic cycle inside a bacterium, it is replicated, transcribed, and translated; the new viral DNA and protein molecules then assemble into new phages, which burst from the host cell. In the lysogenic cycle, phage DNA inserts into the host chromosome and is passed on to generations of daughter cells. Much later, it may initiate phage production.

18 Many viruses cause disease in animals and plants. Flu viruses and most plant viruses have RNA, rather than DNA, as their genetic material. Some animal viruses steal a bit of host cell membrane as a protective envelope. Some viruses can remain latent in the host's body for long periods.

19 Emerging viruses threaten human health.

20 The AIDS virus makes DNA on an RNA template. HIV is a retrovirus: It uses RNA as a template for making DNA, which then inserts into a host chromosome.

21 Viroids and prions are formidable pathogens in plants and animals. Viroids are RNA molecules that can infect plants. Prions are infectious proteins that can cause brain diseases in animals.

22 Bacteria can transfer DNA in three ways. Bacteria can transfer genes from cell to cell by transformation, transduction, or conjugation.

23 Bacterial plasmids can serve as carriers for gene transfer. Plasmids are small circular DNA molecules separate from the bacterial chromosome.

Connecting the Concepts

1. Check your understanding of the flow of genetic information through a cell by filling in the blanks.

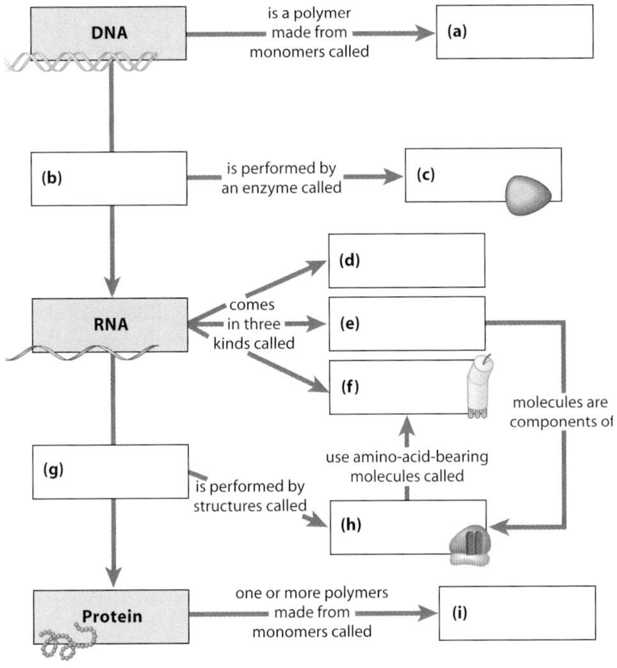

Testing Your Knowledge

Multiple Choice

2. Scientists have discovered how to put together a bacteriophage with the protein coat of phage T2 and the DNA of phage lambda. If this composite phage were allowed to infect a bacterium, the phages produced in the host cell would have ____________. (*Explain your answer.*)
   a. the protein of T2 and the DNA of lambda
   b. the protein of lambda and the DNA of T2
   c. a mixture of the DNA and proteins of both phages
   d. the protein and DNA of T2
   e. the protein and DNA of lambda

3. A geneticist found that a particular mutation had no effect on the polypeptide encoded by a gene. This mutation probably involved
   a. deletion of one nucleotide.
   b. alteration of the start codon.
   c. insertion of one nucleotide.
   d. deletion of the entire gene.
   e. substitution of one nucleotide.

4. Which of the following correctly ranks the structures in order of size, from largest to smallest?
   a. gene-chromosome-nucleotide-codon
   b. chromosome-gene-codon-nucleotide
   c. nucleotide-chromosome-gene-codon
   d. chromosome-nucleotide-gene-codon
   e. gene-chromosome-codon-nucleotide

5. The nucleotide sequence of a DNA codon is GTA. A messenger RNA molecule with a complementary codon is transcribed from the DNA. In the process of protein synthesis, a transfer RNA pairs with the mRNA codon. What is the nucleotide sequence of the tRNA anticodon?
   a. CAT
   b. CUT
   c. GUA
   d. CAU
   e. GT

Describing, Comparing, and Explaining

6. Describe the process of DNA replication: the ingredients needed, the steps in the process, and the final product.

7. Describe the process by which the information in a eukaryotic gene is transcribed and translated into a protein. Correctly use these words in your description: tRNA, amino acid, start codon, transcription, RNA splicing, exons, introns, mRNA, gene, codon, RNA polymerase, ribosome, translation, anticodon, peptide bond, stop codon.

Applying the Concepts

8. A cell containing a single chromosome is placed in a medium containing radioactive phosphate so that any new DNA strands formed by DNA replication will be radioactive. The cell replicates its DNA and divides. Then the daughter cells (still in the radioactive medium) replicate their DNA and divide, and a total of four cells are present. Sketch the DNA molecules in all four cells, showing a normal (nonradioactive) DNA strand as a solid line and a radioactive DNA strand as a dashed line.

9. The base sequence of the gene coding for a short polypeptide is CTACGCTAGGCGATTGACT. What would be the base sequence of the mRNA transcribed from this gene? Using the genetic code in Figure 8A, give the amino acid sequence of the polypeptide translated from this mRNA. (*Hint*: What is the start codon?)

   Researchers on the Human Genome Project have determined the nucleotide sequences of human genes and in many cases identified the proteins encoded by the genes. Knowledge of the nucleotide sequences of genes might be used to develop lifesaving medicines or treatments for genetic defects. In the United States, both government agencies and biotechnology companies have applied for patents on their discoveries of genes. In Britain, the courts have ruled that a naturally occurring gene cannot be patented. Do you think individuals and companies should be able to patent genes and gene products? Before answering, consider the following: What are the purposes of a patent? How might the discoverer of a gene benefit from a patent? How might the public benefit? What might be some positive and negative results of patenting genes?

Review Answers

1. a. nucleotides; b. transcription; c. RNA polymerase; d. mRNA; e. rRNA; f. tRNA; g. translation; h. ribosomes; i. amino acids
2. e (Only the phage DNA enters a host cell; lambda DNA determines both DNA and protein.) 3. e 4. b 5. c
6. Ingredients: Original DNA, nucleotides, several enzymes and other proteins, including DNA polymerase and DNA ligase. Steps: Original DNA strands separate at a specific site (origin of replication), free nucleotides hydrogen-bond to each strand according to base-pairing rules, and DNA covalently bonds the nucleotides to form new strands. New nucleotides are added only to the 3′ end of a growing strand. One new strand is made in one continuous piece; the other new strand is made in a series of short pieces that are then joined by DNA ligase. Product: Two identical DNA molecules, each with one old strand and one new strand.
7. A gene is the polynucleotide sequence with information for making one polypeptide. Each codon—a triplet of bases in DNA or RNA—codes for one amino acid. Transcription occurs when RNA polymerase produces RNA using one strand of DNA as a template. In prokaryotic cells, the RNA transcript may immediately serve as mRNA. In eukaryotic cells, the RNA is processed: A cap and tail are added, and RNA splicing removes introns and links exons together to form a continuous coding sequence. A ribosome is the site of translation, or polypeptide synthesis, and tRNA molecules serve as interpreters of the genetic code. Each folded tRNA molecule has an amino acid attached at one end and a three-base anticodon at the other end. Beginning at the start codon, mRNA is moved relative to the ribosome a codon at a time. A tRNA with a complementary anticodon pairs with each codon, adding its amino acid to the polypeptide chain. The amino acids are linked by peptide bonds. Translation stops at a stop codon, and the finished polypeptide is released. The polypeptide folds to form a functional protein, sometimes in combination with other polypeptides.
8.

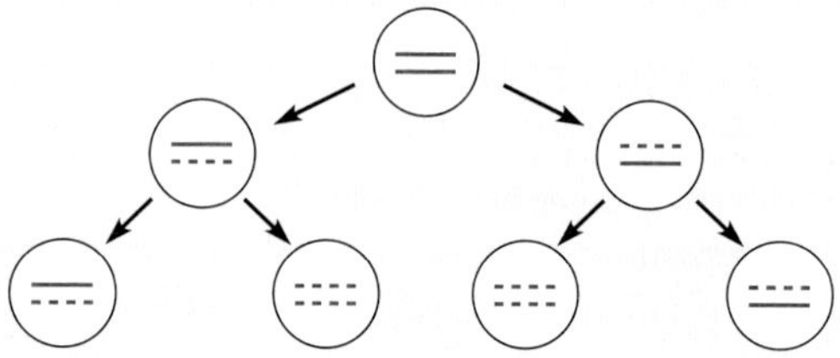

9. mRNA: GAUGCGAUCCGCUAACUGA. Amino acids: Met-Arg-Ser-Ala-Asn.
10. Some issues and questions to consider: Is it fair to issue a patent for a gene or gene product that occurs naturally in every human being? Or should a patent be issued only for something new that is invented rather than found? Suppose another scientist slightly modifies the gene or protein. How different does the gene or protein have to be to avoid patent infringement? Might patents encourage secrecy and interfere with the free flow of scientific information? What are the benefits to the holder of a patent? When research discoveries cannot be patented, what are the scientists' incentives for doing the research? What are the incentives for the institution or company that is providing financial support?

How Genes Are Controlled

How Genes Are Controlled

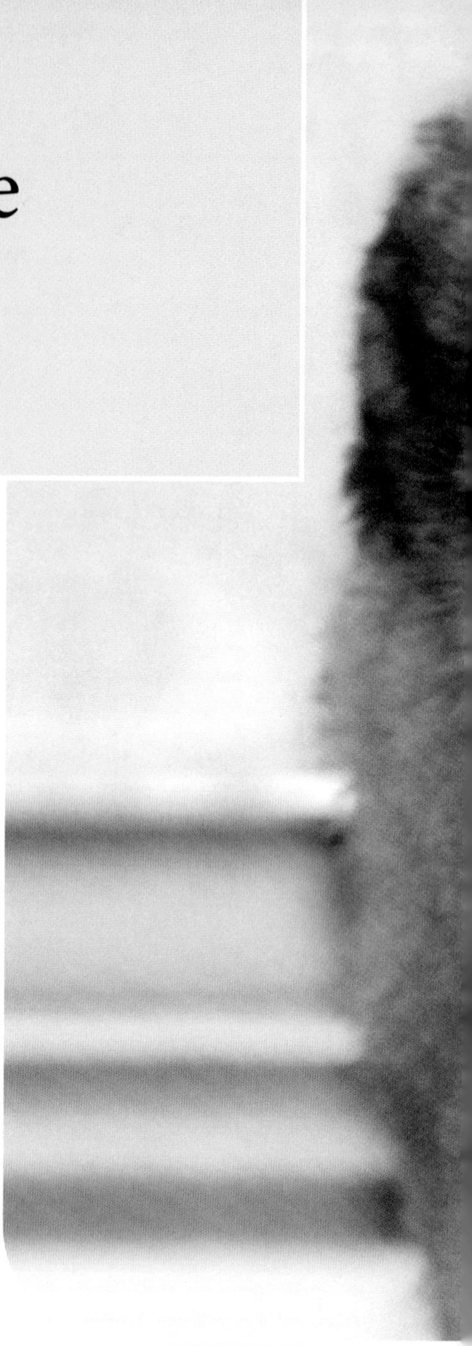

BIG IDEAS

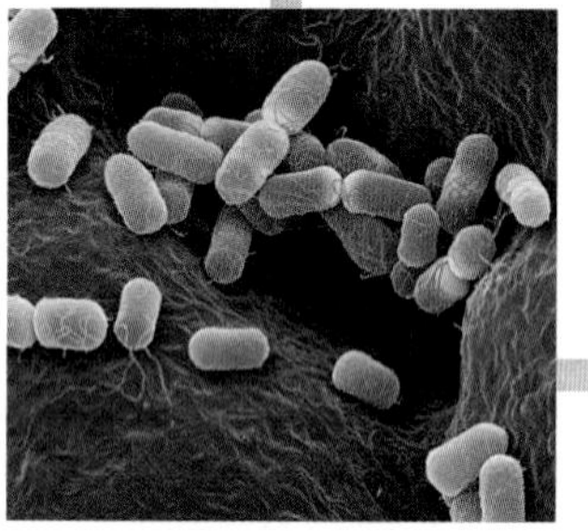

Martin Oeggerli/Photo Researchers

Control of Gene Expression (1–11)

Cells can turn genes on and off through a variety of mechanisms.

Pat Sullivan/Associated Press

Cloning of Plants and Animals (12–15)

Cloning demonstrates that many body cells retain their full genetic potential.

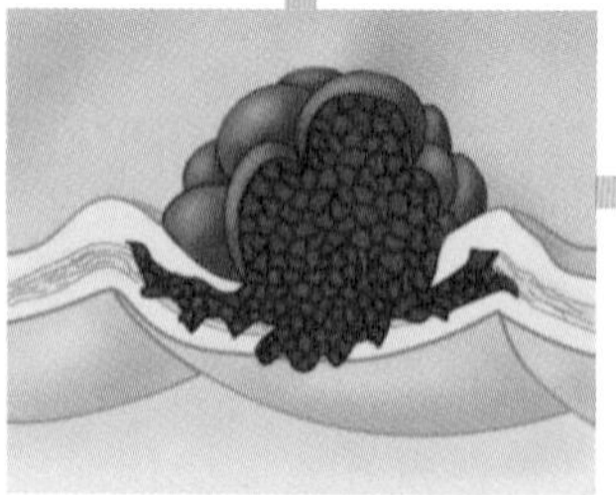

The Genetic Basis of Cancer (16–19)

Changes in genes that control gene expression can lead to out-of-control cell growth.

Lee Jae Won/Reuters

The smiling canine shown above is a cloned grey wolf (*Canis lupus*), born in a South Korean lab in 2007. In this context, the term *clone* refers to an individual created by asexual reproduction (that is, reproduction of a single individual that does not involve fusion of sperm and egg). First demonstrated in the 1950s with frogs, animal cloning became much more commonplace after 1997, when Scottish researchers announced the first successful cloning of a mammal: the world-famous Dolly, a sheep cloned from a mammary cell extracted from an adult ewe.

Cloning efforts to date have focused on farm animals (such as sheep), important research organisms (such as mice), and endangered species (such as wolves). Indeed, cloning may be the only way to repopulate some highly endangered species. However, conservationists argue that cloning may trivialize the tragedy of extinction and detract from efforts to preserve natural habitats. They correctly point out that cloning does not increase genetic diversity and is therefore not as beneficial to endangered species as natural reproduction.

The cloning of an animal from a single body cell demonstrates that the starting cell contained a complete genome capable of directing the production of all the cell types in an organism. The development of a multicellular organism with many different kinds of cells thus depends on the turning on and off of different genes in different cells—the control of gene expression.

Whether an organism is unicellular or multicellular, its cells may alter their patterns of gene expression in response to the organism's needs. We begin this chapter with examples of how and where this may occur. Next we look at the methods and applications of plant and animal cloning. Finally, we discuss cancer, a disease that can be caused by changes in gene expression.

Control of Gene Expression

1 Proteins interacting with DNA turn prokaryotic genes on or off in response to environmental changes

Picture an *Escherichia coli* (*E. coli*) bacterium living in your intestine (**Figure 1A**). Its environment changes continuously, depending on your dietary whims. For example, if you eat a sweet roll for breakfast, the bacterium will be bathed in sugars and broken-down fats. Later, if you have a salad for lunch, the *E. coli*'s environment will change drastically. How can a bacterium cope with such a constantly shifting flow of resources?

The answer is that **gene regulation**—the turning on and off of genes—can help organisms respond to environmental changes. What does it mean to turn a gene on or off? Genes determine the nucleotide sequences of specific mRNA molecules, and mRNA in turn determines the sequences of amino acids in protein molecules (DNA → RNA → protein). Thus, a gene that is turned on is being transcribed into RNA, and that message is being translated into specific protein molecules. The overall process by which genetic information flows from genes to proteins—that is, from genotype to phenotype—is called **gene expression**. The control of gene expression makes it possible for cells to produce specific kinds of proteins when and where they are needed.

It's no coincidence that we used *E. coli* as our example. Our earliest understanding of gene control came from studies of this bacterium by French biologists François Jacob and Jacques Monod. *E. coli* has a remarkable ability to change its metabolic activities in response to changes in its environment. For example, *E. coli* produces enzymes needed to metabolize a specific nutrient only when that nutrient is available. Bacterial cells that can conserve resources and energy have an advantage over cells that are unable to do so. Thus, natural selection has favored bacteria that express only the genes whose products are needed by the cell. Let's look at how the regulation of gene transcription helps *E. coli* efficiently use available resources.

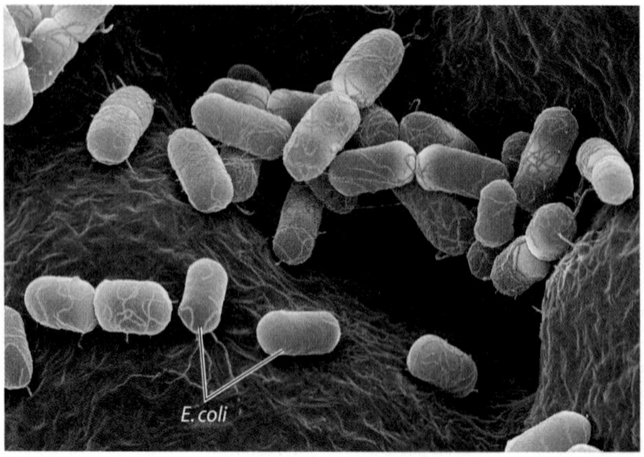

Martin Oeggerli/Photo Researchers

Colorized SEM 5,000×

▲ **Figure 1A** Cells of *E. coli* bacteria

The *lac* Operon Imagine the bacterium in your intestine soon after you drink a glass of milk. One of the main nutrients in milk is the disaccharide sugar lactose. When lactose is plentiful in the intestine, *E. coli* makes the enzymes necessary to absorb the sugar and use it as an energy source. Conversely, when lactose is not plentiful, *E. coli* does not waste its energy producing these enzymes.

Recall that enzymes are proteins; their production is an outcome of gene expression. *E. coli* can make lactose-utilization enzymes because it has genes that code for these enzymes. **Figure 1B** presents a model (first proposed in 1961 by Jacob and Monod) to explain how an *E. coli* cell can turn genes coding for lactose-utilization enzymes off or on, depending on whether lactose is available.

E. coli uses three enzymes to take up and start metabolizing lactose, and the genes coding for these three enzymes are regulated as a single unit. The DNA at the top of Figure 1B represents a small segment of the bacterium's chromosome. Notice that the three genes that code for the lactose-utilization enzymes (light blue) are next to each other in the DNA.

Adjacent to the group of lactose enzyme genes are two control sequences, short sections of DNA that help control the enzyme genes. One control sequence is a **promoter**, a site where the transcription enzyme, RNA polymerase, attaches and initiates transcription—in this case, transcription of all three lactose enzyme genes (as depicted in the bottom panel of Figure 1B). Between the promoter and the enzyme genes, a DNA control sequence called an **operator** acts as a switch. The operator determines whether RNA polymerase can attach to the promoter and start transcribing the genes.

Such a cluster of genes with related functions, along with the control sequences, is called an **operon**; with rare exceptions, operons exist only in prokaryotes. The key advantage to the grouping of related genes into operons is that a single "on-off switch" can control the whole cluster. The operon discussed here is called the *lac* operon, short for lactose operon. When an *E. coli* bacterium encounters lactose, all the enzymes needed for its metabolism are made at once because the operon's genes are all controlled by a single switch, the operator. But what determines whether the operator switch is on or off?

The top panel of Figure 1B shows the *lac* operon in "off" mode, its status when there is no lactose in the cell's environment. Transcription is turned off by a protein called a **repressor** ( ), a protein that functions by binding to the operator ( ) and physically blocking the attachment of RNA polymerase ( ) to the promoter ( ). On the left side of the figure, you can see where the repressor comes from. A gene called a **regulatory gene** (dark blue), located outside the operon, codes for the repressor. The regulatory gene is expressed continually, so the cell always has a small supply of repressor molecules.

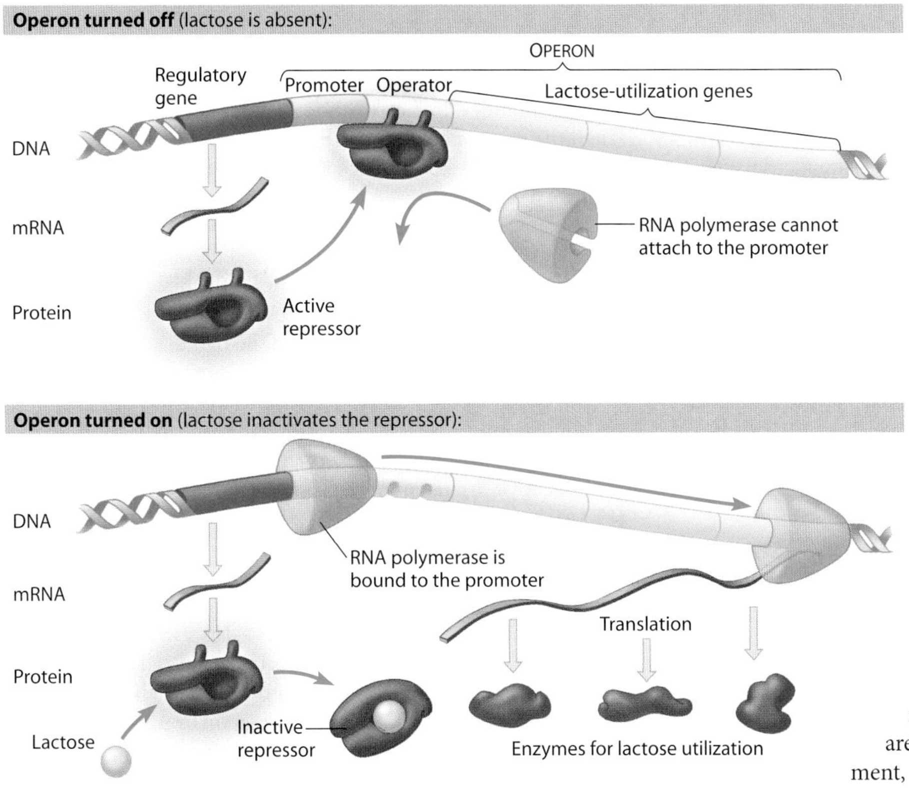

▲ **Figure 1B** The *lac* operon

How can an operon be turned on if its repressor is always present? As the bottom panel of Figure 1B indicates, lactose interferes with the attachment of the *lac* repressor to the operator by binding to the repressor and changing its shape. With its new shape, the repressor cannot bind to the operator, and the operator switch remains on. RNA polymerase can now bind to the promoter (since it is no longer being blocked) and from there transcribes the genes of the operon. The resulting mRNA carries coding sequences for all three enzymes needed for lactose metabolism. The cell can translate the message in this single mRNA into three separate polypeptides because the mRNA has multiple codons signaling the start and stop of translation.

The *lac* operon is so efficient that the addition of lactose to a bacterium's environment results in a thousandfold increase in lactose utilization enzymes in just 15 minutes. The newly produced mRNA and protein molecules will remain intact for only a short time before cellular enzymes break them down. When the synthesis of mRNA and protein stops because lactose is no longer present, the molecules quickly disappear.

Other Kinds of Operons The *lac* operon is only one type of operon in bacteria. Other types also have a promoter, an operator, and several adjacent genes, but they differ in the way the operator switch is controlled. **Figure 1C** shows two types of repressor-controlled operons. The *lac* operon's repressor is active when alone and inactive when bound to lactose. A second type of operon, represented here by the *trp* operon, is controlled by a repressor that is *inactive* alone. To be active, this type of repressor must combine with a specific small molecule. In our example, the small molecule is tryptophan (Trp), an amino acid essential for protein synthesis. *E. coli* can make tryptophan from scratch, using enzymes encoded in the *trp* operon. But it will stop making tryptophan and simply absorb it in prefabricated form from the surroundings whenever possible. When *E. coli* is swimming in tryptophan in the intestines (as occurs when you eat foods such as milk and poultry), the tryptophan binds to the repressor of the *trp* operon. This activates the *trp* repressor, enabling it to switch off the operon. Thus, this type of operon allows bacteria to stop making certain essential molecules when the molecules are already present in the environment, saving materials and energy for the cells.

Another type of operon control involves **activators**, proteins that turn operons *on* by binding to DNA. These proteins act by making it easier for RNA polymerase to bind to the promoter, rather than by blocking RNA polymerase, as repressors do. Activators help control a wide variety of operons.

Armed with a variety of operons regulated by repressors and activators, *E. coli* and other prokaryotes can thrive in frequently changing environments. Next we examine how more complex eukaryotes regulate their genes.

? A certain mutation in *E. coli* impairs the ability of the *lac* repressor to bind to the *lac* operator. How would this affect the cell?

The cell would wastefully produce the enzymes for lactose metabolism continuously, even when lactose is not present.

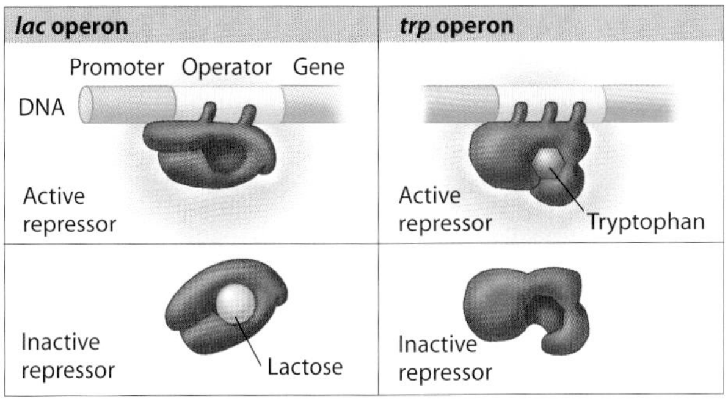

▲ **Figure 1C** Two types of repressor-controlled operons

2 Chromosome structure and chemical modifications can affect gene expression

The cells of all organisms, whether prokaryotes or eukaryotes, must be able to turn genes on and off in response to signals from their external and internal environments. All multicellular eukaryotes also require an additional level of gene control: During the repeated cell divisions that lead from a zygote to an adult in a multicellular organism, individual cells must undergo **differentiation**—that is, they must become specialized in structure and function, with each type of cell fulfilling a distinct role. Your body, for example, contains hundreds of different types of cells. What makes a kidney cell different from, say, a bone cell?

To perform its specialized role, each cell type must maintain a specific program of gene expression in which some genes are expressed and others are not. Almost all the cells in an organism contain an identical genome, yet the subset of genes expressed in each cell type is unique, reflecting its specific function. Each adult human cell expresses only a small fraction of its total genes at any given time. And even one particular cell type can change its pattern of gene expression over time in response to developmental signals or other changes in the environment.

The differences between cell types, therefore, are due not to different genes being present, but to selective gene expression. In this module, we begin our exploration of gene regulation in eukaryotes by looking at the chromosomes, where almost all of a cell's genes are located.

DNA Packing The DNA in just a single human chromosome would, if stretched out, average 4 cm in length, thousands of times greater than the diameter of the nucleus. All of this DNA can fit within the nucleus because of an elaborate, multilevel system of packing—coiling and folding—of the DNA in each chromosome. A crucial aspect of DNA packing is the association of the DNA with small proteins called **histones**. In fact, histone proteins account for about half the mass of eukaryotic chromosomes. (Prokaryotes have analogous proteins, but lack the degree of DNA packing seen in eukaryotes.)

Figure 2A shows a model for the main levels of DNA packing. At the left, notice that the unpacked double-helical molecule of DNA has a diameter of 2 nm. At the first level of packing, histones attach to the DNA double helix. In the electron micrograph near the top left of the figure, notice how the DNA-histone complex has the appearance of beads on a string. Each "bead," called a **nucleosome**, consists of DNA wound around a protein core of eight histone molecules. Short stretches of DNA, called linkers, are the "strings" that join consecutive "beads" of nucleosomes.

At the next level of packing, the beaded string is wrapped into a tight helical fiber. This fiber coils further into a thick supercoil with a diameter of about 300 nm. Looping and folding can further compact the DNA, as you can see in the metaphase chromosome at the right of the figure. Viewed as a whole, Figure 2A gives a sense of how successive levels of coiling and folding enable a huge amount of DNA to fit into a cell nucleus.

DNA packing can block gene expression by preventing RNA polymerase and other transcription proteins from contacting the DNA. Cells seem to use higher levels of packing for long-term inactivation of genes. Highly compacted chromatin, which is found not only in mitotic chromosomes—such as the duplicated chromosome shown below—but also in varying regions of interphase chromosomes, is generally not expressed at all.

Chemical Modifications and Epigenetic Inheritance In addition to being able to change their level of packing, the cells of many eukaryotic organisms have the capability to establish and maintain chemical modifications to their chromosomes in

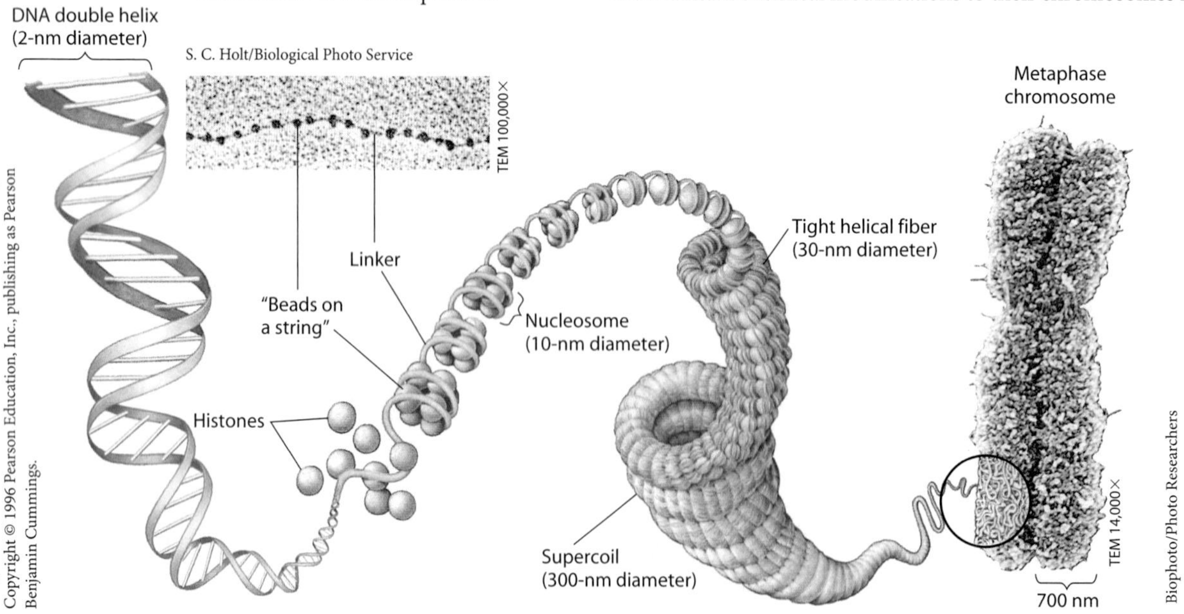

▲ **Figure 2A** DNA packing in a eukaryotic chromosome

ways that help regulate gene expression. For example, the addition of chemical groups to some of the amino acids in histone proteins, or their removal, can cause the proteins to bind DNA more tightly or loosely, altering the ability of transcription machinery to reach those genes.

DNA itself can also be the target of chemical modification. In one such type of modification, certain enzymes add a methyl group (CH_3) to DNA bases, usually cytosine, without changing the actual sequence of the bases. Individual genes are usually more heavily methylated in cells in which they are not expressed, and removal of the extra methyl groups can turn on some of these genes. Thus, DNA methylation appears to play a role in turning genes off. At least in some species, DNA methylation seems to be essential for the long-term inactivation of genes. Such modifications are a normal and necessary mechanism for the regulation of gene expression, and improper methylation can lead to problems for the organism. For example, insufficient DNA methylation can lead to abnormal embryonic development in many species.

Once methylated, genes usually stay that way through successive cell divisions in a given individual. At DNA sites where one strand is already methylated, enzymes methylate the corresponding daughter strand after each round of DNA replication. Methylation patterns are therefore passed on, and cells forming specialized tissues keep a chemical record of what occurred during embryonic development. In this way, modifications to the DNA and histones can be passed along to future generations of cells—that is, they can be inherited. Inheritance of traits transmitted by mechanisms not directly involving the nucleotide sequence is called **epigenetic inheritance**. Whereas mutations in the DNA are permanent changes, modifications to the chromatin, which do not affect the sequence of DNA itself, can be reversed by processes that are not yet fully understood.

Researchers are amassing more and more evidence for the importance of epigenetic information in the regulation of gene expression. Epigenetic variations might help explain differences in identical twins. It is often the case that one identical twin acquires a genetically influenced disease, such as schizophrenia, but the other does not, despite their identical genomes. Researchers suspect that epigenetics may be behind such differences. Alterations in normal patterns of DNA methylation are seen in some cancers, where they are associated with inappropriate gene expression. Evidently, enzymes that modify chromatin structure are integral parts of the eukaryotic cell's machinery for regulating transcription.

X Inactivation Female mammals, including humans, inherit two X chromosomes, whereas males inherit only one. So why don't females make twice as much of the proteins encoded by genes on the X chromosome compared to the amounts in males? It turns out that in female mammals, one X chromosome in each somatic (body) cell is chemically modified and highly compacted, rendering it almost entirely inactive. Inactivation of an X chromosome involves modification of the DNA (by, for example, methylation) and the histone proteins that help compact it. A specific gene on the X chromosomes ensures that one and only one of them will be inactivated. This **X chromosome inactivation** is initiated early in embryonic development, when one of the two X chromosomes in each cell is inactivated at random. As a result, the cells of females and males have the same effective dose (one copy) of these genes. The inactive X in each cell of a female condenses into a compact object called a **Barr body**.

Which X chromosome is inactivated is a matter of chance in each embryonic cell, but once an X chromosome is inactivated, all descendant cells have the same copy turned off—an example of epigenetic inheritance. Consequently, females consist of a mosaic of two types of cells: those with the active X derived from the father and those with the active X derived from the mother. If a female is heterozygous for a gene on the X chromosome (a sex-linked gene), about half her cells will express one allele, while the others will express the alternate allele.

A striking example of this mosaic phenomenon is the tortoiseshell cat, which has orange and black patches of fur (**Figure 2B**). The relevant fur-color gene is on the X chromosome, and the tortoiseshell phenotype requires the presence of two different alleles, one for orange fur and one for black fur. Normally, only females can have both alleles because only they have two X chromosomes. If a female is heterozygous for the tortoiseshell gene, she will have the tortoiseshell phenotype.

Orange patches are formed by populations of cells in which the X chromosome with the orange allele is active; black patches have cells in which the X chromosome with the black allele is active.

In this module, we have seen how the physical structure of chromosomes can affect which genes are expressed in a cell. In the next module, we discuss mechanisms for regulating genes in active, unpacked chromosomes.

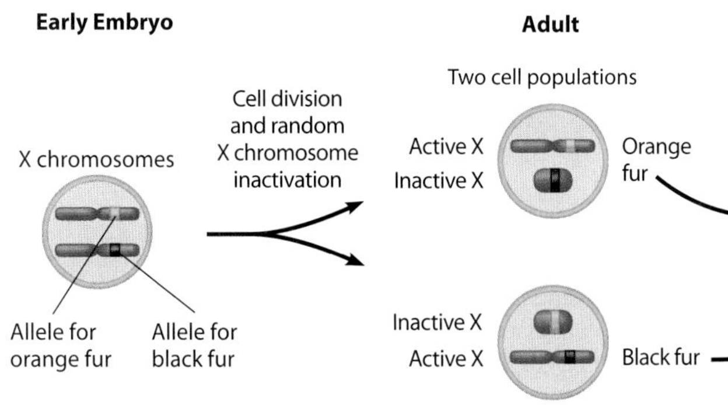

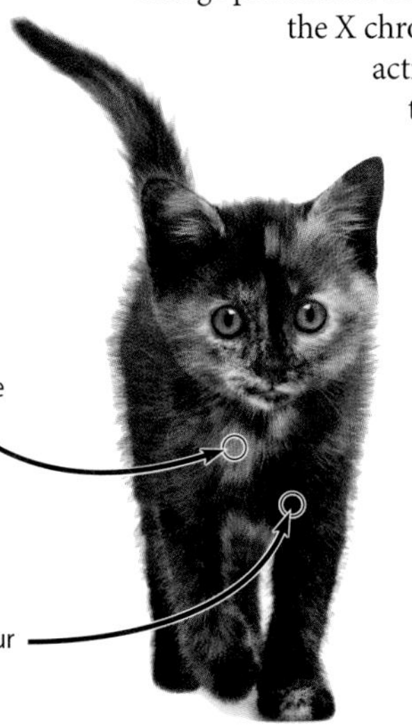

Eric Isselée/Shutterstock

▲ **Figure 2B** A tortoiseshell pattern on a female cat, a result of X chromosome inactivation

? If a nerve cell and a skin cell in your body have the same genes, how can the cells be so different?

Each cell type must be expressing certain genes that are present in, but not expressed in, the other cell type.

3 Complex assemblies of proteins control eukaryotic transcription

The packing and unpacking of chromosomal DNA provide a coarse adjustment for eukaryotic gene expression by making a region of DNA either more or less available for transcription. The fine-tuning begins with the initiation of RNA synthesis—transcription. In both prokaryotes and eukaryotes, the initiation of transcription (whether transcription starts or not) is the most important stage for regulating gene expression.

Like prokaryotes (see Module 1), eukaryotes employ regulatory proteins—activators and repressors—that bind to specific segments of DNA and either promote or block the binding of RNA polymerase, turning the transcription of genes on or off. However, most eukaryotic genes have individual promoters and other control sequences and are not clustered together as in operons.

The current model for the initiation of eukaryotic transcription features an intricate array of regulatory proteins that interact with DNA and with one another to turn genes on or off. In eukaryotes, activator proteins seem to be more important than repressors. That is, in multicellular eukaryotes, the "default" state for most genes seems to be "off." A typical animal or plant cell needs to turn on (transcribe) only a small percentage of its genes, those required for the cell's specialized structure and function. Housekeeping genes, those continually active in virtually all cells for routine activities such as glycolysis, may be in an "on" state by default.

In order to function, eukaryotic RNA polymerase requires the assistance of proteins called **transcription factors.** In the model depicted in **Figure 3**, the first step in initiating gene transcription is the binding of activator proteins ( )to DNA control sequences called **enhancers** ( ). In contrast to the operators of prokaryotic operons, enhancers are usually far away on the chromosome from the gene they help regulate. The binding of activators to enhancers leads to bending of the DNA. Once the DNA is bent, the bound activators interact with other transcription factor proteins ( ), which then bind as a complex at the gene's promoter ( ). This large assembly of proteins facilitates the correct attachment of RNA polymerase to the promoter and the initiation of transcription. Only when the complete complex of proteins has assembled can the polymerase begin to move along the gene, producing an RNA strand. As shown in the figure, several enhancers and activators may be involved. Not shown are *repressor* proteins called **silencers** that may bind to DNA sequences and *inhibit* the start of transcription.

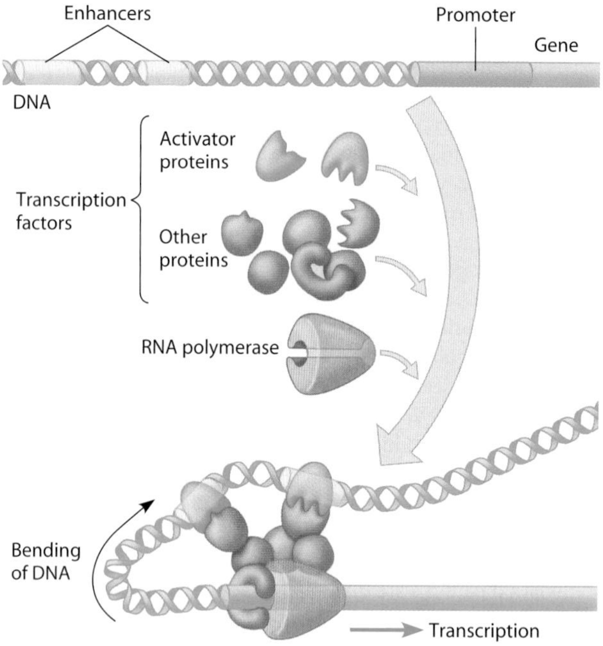

▲ **Figure 3** A model for the turning on of a eukaryotic gene

If eukaryotic genomes only rarely have operons, how does a eukaryotic cell deal with genes of related function that all need to be turned on or off at the same time? Making the situation even more complex, genes coding for the enzymes of a metabolic pathway are often scattered across different chromosomes. The key to coordinated gene expression in eukaryotes is often the association of a specific combination of control sequences with every gene of a particular metabolic pathway. Copies of the activators that recognize these control sequences bind to them all at once (since they are all identical), promoting simultaneous transcription of the genes, no matter where they are in the genome. In the next module, we consider another method of gene regulation that is unique to eukaryotes.

? What must occur before RNA polymerase can bind to a promoter and transcribe a specific eukaryotic gene?

Enhancers must bind to transcription factors to facilitate the attachment of RNA polymerase to the promoter.

4 Eukaryotic RNA may be spliced in more than one way

Although regulation of transcription is the most important step in gene regulation in most cells, transcription alone does not equal gene expression. Several other points along the path from DNA to protein can be regulated. Within a eukaryotic cell, for example, RNA transcripts are processed into mRNA before moving to the cytoplasm for translation by the ribosomes. RNA processing includes the addition of a cap and a tail, as well as the removal of any introns—noncoding DNA segments that interrupt the genetic message—and the splicing together of the remaining exons.

Some scientists think that the splicing process may help control the flow of mRNA from nucleus to cytoplasm because until splicing is completed, the RNA is attached to the molecules of the splicing machinery and cannot pass through the nuclear pores. Moreover, in some cases, the cell can carry out splicing in more than one way, generating different mRNA molecules from the same RNA transcript. Notice in **Figure 4**, for example, that one mRNA molecule ends up with the green exon and the other with the brown exon. With this sort of **alternative RNA splicing**, an organism can produce more than one type of polypeptide from a single gene.

One interesting example of two-way splicing is found in the fruit fly, where the differences between males and females are largely due to different patterns of RNA splicing. Results from the Human Genome Project suggest that alternative splicing is very common in humans. Included among the many instances already known is one gene whose transcript can be spliced to encode *seven* alternative versions of a protein, each of which is made in a different type of cell.

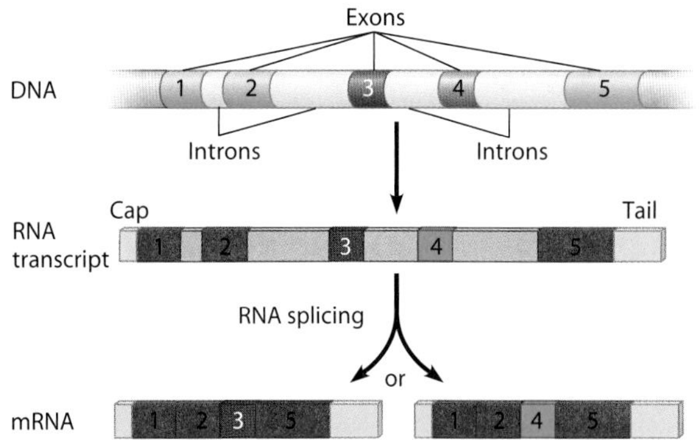

▲ **Figure 4** The production of two different mRNAs from the same gene

? How does alternative RNA splicing enable a single gene to encode more than one kind of polypeptide?

Each kind of polypeptide is encoded by an mRNA molecule containing a different combination of exons.

5 Small RNAs play multiple roles in controlling gene expression

Recall that only 1.5% of the human genome—and a similarly small percentage of the genomes of many other multicellular eukaryotes—codes for proteins. Another very small fraction of DNA consists of genes for ribosomal RNA and transfer RNA. Until recently, most of the remaining DNA was thought to be untranscribed and therefore considered to be lacking any genetic information. However, a flood of recent data has contradicted this view. It turns out that a significant amount of the genome is transcribed into functioning but non-protein-coding RNAs, including a variety of small RNAs. While many questions about the functions of these RNAs remain unanswered, researchers are uncovering more evidence of their biological roles every day.

In 1993, researchers discovered small RNA molecules, called **microRNAs (miRNAs)**, that can bind to complementary sequences on mRNA molecules (**Figure 5**). Each miRNA, typically about 20 nucleotides long, ❶ forms a complex with protein. The miRNA-protein complex can ❷ bind to any mRNA molecule with the complementary sequence. Then the complex either ❸ degrades the target mRNA or ❹ blocks its translation. It has been estimated that miRNAs may regulate the expression of up to one-third of all human genes, a striking figure given that miRNAs were unknown a mere 20 years ago.

Researchers can take advantage of miRNA to artificially control gene expression. For example, injecting miRNA into a cell can turn off expression of a gene with a sequence that matches the miRNA, a procedure called **RNA interference (RNAi).** The RNAi pathway may have evolved as a natural defense against infection by certain viruses with RNA genomes. In 2006, two American researchers were awarded a Nobel Prize for their discovery and categorization of RNA interference.

Biologists are excited about these recent discoveries, which hint at a large, diverse population of RNA molecules in the cell that play crucial roles in regulating gene expression—and have gone largely unnoticed until now. Our new understanding may lead to important clinical applications. For example, in 2009, researchers discovered a particular microRNA that is essential to the proper functioning of the pancreas. Without it, insulin-producing beta cells die off, which can lead to diabetes. Clearly, we must revise the long-standing view that because they code for proteins, messenger RNAs are the most important RNAs in terms of cellular function.

? If a gene has the sequence AATTCGCG, what would be the sequence of an miRNA that turns off the gene?

The gene will be transcribed as the mRNA sequence UUAAGCGC; an miRNA of sequence AAUUCGCG would bind to and disable this mRNA.

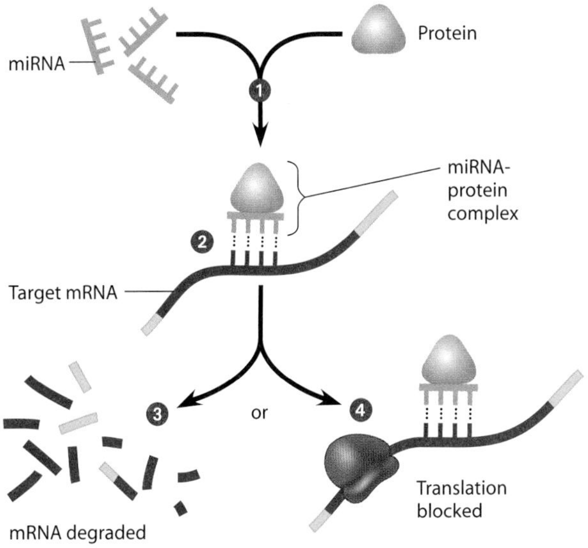

▲ **Figure 5** Mechanisms of RNA interference

6 Later stages of gene expression are also subject to regulation

Even after a eukaryotic mRNA is fully processed and transported to the cytoplasm, there are several additional opportunities for regulation. Such control points include mRNA breakdown, initiation of translation, protein activation, and protein breakdown.

Breakdown of mRNA Molecules of mRNA do not remain intact forever. Enzymes in the cytoplasm eventually break them down, and the timing of this event is an important factor regulating the amounts of various proteins that are produced in the cell. Long-lived mRNAs can be translated into many more protein molecules than short-lived ones. Prokaryotic mRNAs have very short lifetimes; they are typically degraded by enzymes within a few minutes after their synthesis. This is one reason bacteria can change their protein production so quickly in response to environmental changes. In contrast, the mRNA of eukaryotes can have lifetimes of hours or even weeks.

A striking example of long-lived mRNA is found in vertebrate red blood cells, which manufacture large quantities of the protein hemoglobin. In most species of vertebrates, the mRNAs for hemoglobin are unusually stable. They probably last as long as the red blood cells that contain them—about a month or a bit longer in reptiles, amphibians, and fishes—and are translated again and again. Mammals are an exception. When their red blood cells mature, they lose their ribosomes (along with their other organelles) and thus cease to make new hemoglobin. However, mammalian hemoglobin itself lasts about as long as the red blood cells last, around four months.

Initiation of Translation The process of translating an mRNA into a polypeptide also offers opportunities for regulation. Among the molecules involved in translation are a great many proteins that control the start of polypeptide synthesis. Red blood cells, for instance, have an inhibitory protein that prevents translation of hemoglobin mRNA unless the cell has a supply of heme, the iron-containing chemical group essential for hemoglobin function. (It is the iron atom of the heme group to which oxygen molecules actually attach.) By controlling the start of protein synthesis, cells can avoid wasting energy if the needed components are currently unavailable.

Protein Activation After translation is complete, some polypeptides require alterations before they become functional. Post-translational control mechanisms in eukaryotes often involve the cleavage (cutting) of a polypeptide to yield a smaller final product that is the active protein, able to carry out a specific function in the organism. In **Figure 6**, we see the example of the hormone insulin, which is a protein. Insulin is synthesized in the cells of the pancreas as one long polypeptide that has no hormonal activity. After translation is completed, the polypeptide folds up, and covalent bonds form between the sulfur (S) atoms of sulfur-containing amino acids (see Figure 3.12B, which shows S—S bonds in another protein). Two H atoms are lost as each S—S bond forms, linking together parts of the polypeptide in a specific way. Finally, a large center portion is cut away, leaving two shorter chains held together by the sulfur linkages. This combination of two shorter polypeptides is the form of insulin that functions as a hormone. By controlling the timing of such protein modifications, the rate of insulin synthesis can be fine-tuned.

Protein Breakdown The final control mechanism operating after translation is the selective breakdown of proteins. Though mammalian hemoglobin may last as long as the red blood cell housing it, the lifetimes of many other proteins are closely regulated. Some of the proteins that trigger metabolic changes in cells are broken down within a few minutes or hours. This regulation allows a cell to adjust the kinds and amounts of its proteins in response to changes in its environment. It also enables the cell to maintain its proteins in prime working order. Indeed, when proteins are damaged, they are usually broken down right away and replaced by new ones that function properly.

Over the last five modules, you have learned about several ways that eukaryotes can control gene expression. The next module summarizes all of these processes.

? Review Figure 6. If the enzyme responsible for cleaving inactive insulin is deactivated, what effect will this have on the form and function of insulin?

The final molecule will have a shape different from that of active insulin and therefore will not be able to function as a hormone.

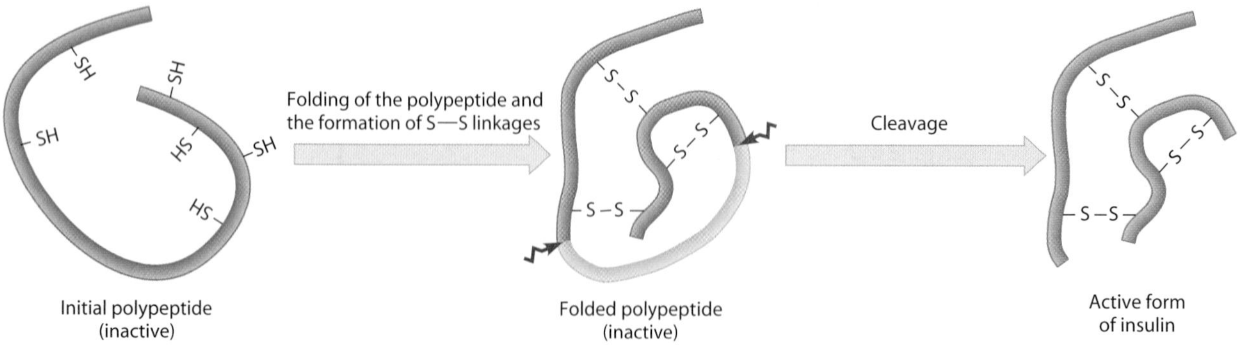

▲ **Figure 6** Protein activation: the role of polypeptide cleavage in producing the active insulin protein

7 Review: Multiple mechanisms regulate gene expression in eukaryotes

Figure 7 provides a review of eukaryotic gene expression and highlights the multiple control points where the process can be turned on or off, speeded up, or slowed down. Picture the series of pipes that carry water from your local water supply, perhaps a reservoir, to a faucet in your home. At various points, valves control the flow of water. We use this model in the figure to illustrate the flow of genetic information from a chromosome—a reservoir of genetic information—to an active protein that has been synthesized in the cell's cytoplasm. The multiple mechanisms that control gene expression are analogous to the control valves in water pipes. In the figure, each gene expression "valve" is indicated by a control knob. Note that these knobs represent *possible* control points; for most proteins, only a few control points may be important. The most important control point, in both eukaryotes and prokaryotes, is usually the start of transcription. In the diagram, the large yellow knob represents the mechanisms that regulate the start of transcription.

Although the initiation of transcription is the most important control point, there are several other opportunities for regulation. RNA processing in the nucleus adds nucleotides to the ends of the RNA (cap and tail) and splices out introns. As we discussed in Module 4, a growing body of evidence suggests the importance of control at this stage. Once mRNA reaches the cytoplasm, additional stages that can be regulated include mRNA translation and eventual breakdown, possible alteration of the polypeptide to activate it, and the eventual breakdown of the protein.

Despite its numerous steps, Figure 7 actually oversimplifies the control of gene expression. What it does not show is the web of control that connects different genes, often through their products. We have seen examples in both prokaryotes and eukaryotes of the actions of gene products (usually proteins) on other genes or on other gene products within the same cell. The genes of operons in *E. coli*, for instance, are controlled by repressor or activator proteins encoded by regulatory genes on the same DNA molecule. In eukaryotes, many genes are controlled by proteins encoded by regulatory genes on other chromosomes. The numerous interactions of these various proteins, taken in total, result in flexible yet precise control of gene expression.

In eukaryotes, cellular differentiation results from the selective turning on and off of genes. In the next module, we examine the stage in the life cycle of a multicellular eukaryote when cellular differentiation by selective gene expression is most vital: the development of a multicellular embryo from a unicellular zygote.

? Of the nine regulatory "valves" in Figure 7, which five can also operate in a prokaryotic cell?

(1) Control of transcription; (2) control of mRNA breakdown; (3) control of translation; (4) control of protein activation; and (5) control of protein breakdown

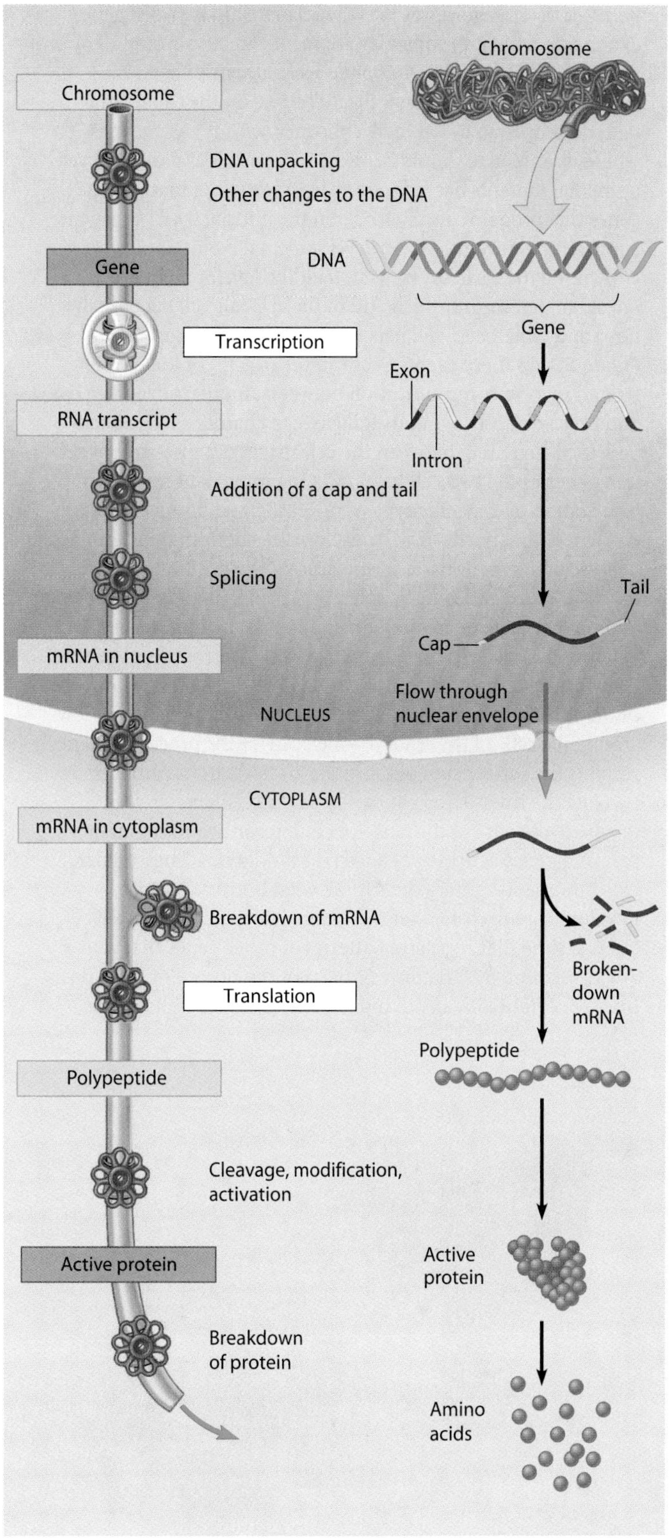

▲ **Figure 7** The gene expression "pipeline" in a eukaryotic cell

8 Cell signaling and cascades of gene expression direct animal development

Some of the first glimpses into the relationship between gene expression and embryonic development came from studies of mutants of the fruit fly *Drosophila melanogaster*. **Figure 8A** shows the heads of two fruit flies. The one on the right, a mutant, developed in a strikingly abnormal way: It has two legs where its antennae should be! Research on this and other developmental mutants has led to the identification of many of the genes that program development in the normal fly. This genetic approach has revolutionized developmental biology.

Among the earliest events in fruit fly development are ones that determine which end of the egg cell will become the head and which end will become the tail. As you can see in **Figure 8B**, ❶ these events occur in the ovaries of the mother fly and involve communication between an unfertilized egg cell and cells adjacent to it in its follicle (egg chamber). The back-and-forth signaling between the cells triggers expression of certain genes in the two cell types. ❷ One important result is the localization of a specific type of mRNA (pink) at the end of the egg where the fly's head will develop, thus defining the animal's head-to-tail axis. (Similar events lead to the positioning of the top-to-bottom and side-to-side axes.) Molecular interactions and further gene expression bring about growth of the egg.

After the egg is fertilized and laid, repeated rounds of mitosis transform the zygote into an embryo. The early embryo makes proteins that diffuse through its cell layers. Cell signaling—now among the cells of the embryo—helps drive the process. ❸ The result is the subdivision of the embryo's body into segments.

Now the finer details of the fly can take shape. Protein products of some of the axis-specifying genes and segment-forming genes activate yet another set of genes. These genes, called homeotic genes, determine what body parts will develop from each segment. A **homeotic gene** is a master control gene that regulates batteries of other genes that actually determine the anatomy of parts of the body. For example, one set of homeotic genes in fruit flies instructs cells in the segments of the head and thorax (midbody) to form antennae and legs, respectively. Elsewhere, these homeotic genes remain turned off, while others are turned on. ❹ The eventual outcome is an adult fly. Notice that the adult's body segments correspond to those of the embryo in step 3. It was mutation of a homeotic gene that was responsible for the abnormal fly in Figure 8A.

Cascades of gene expression, with the protein products of one set of genes activating other sets of genes, are a common theme in development. Next we look at how DNA technology can help elucidate gene expression in any cell.

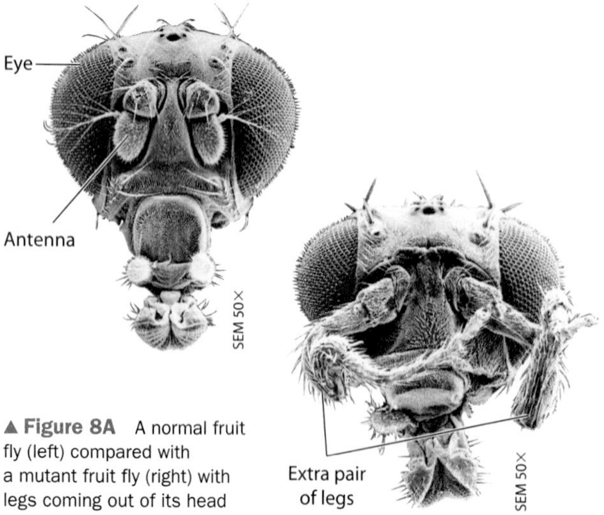

▲ **Figure 8A** A normal fruit fly (left) compared with a mutant fruit fly (right) with legs coming out of its head

? What determines which end of a developing fruit fly will become the head?

A specific kind of mRNA localizes at the end of the unfertilized egg that will become the head.

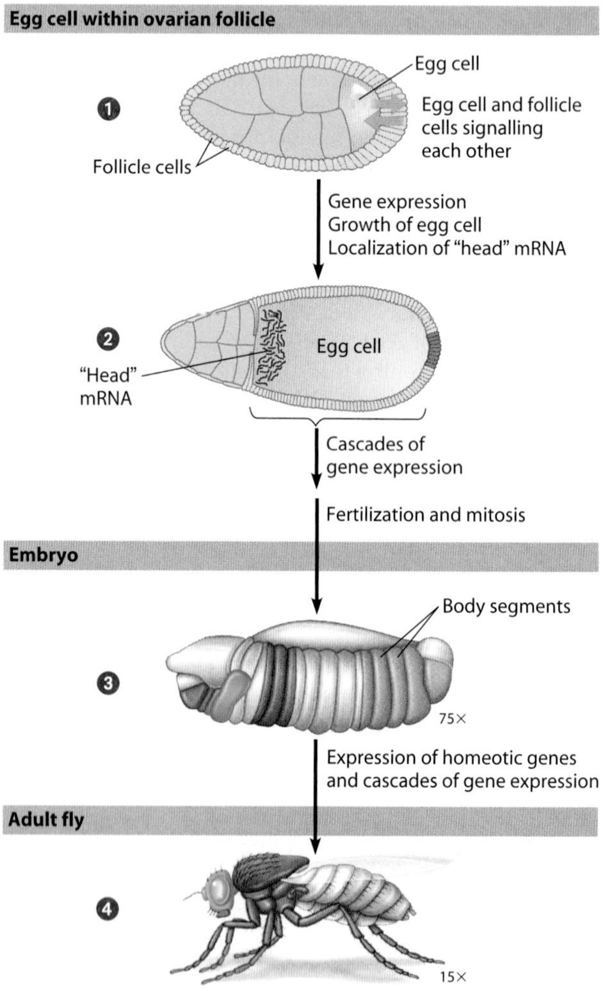

▲ **Figure 8B** Key steps in the early development of head-tail axis in a fruit fly

CONNECTION 9 DNA microarrays test for the transcription of many genes at once

A major goal of biologists is to learn how genes act together within a functioning organism. Now that a number of whole genomes have been sequenced, it is possible to study the expression of large groups of genes. Researchers can use gene sequences as probes to investigate which genes are transcribed in different situations, such as in different tissues or at different stages of development. They also look for groups of genes that are expressed in a coordinated manner, with the aim of identifying networks of gene expression across an entire genome.

Genome-wide expression studies are made possible by DNA microarrays. A **DNA microarray** is a glass slide with tiny amounts of thousands of different kinds of single-stranded DNA fragments fixed to it in tiny wells in a tightly spaced array, or grid. (A DNA microarray is also called a DNA chip or gene chip by analogy to a computer chip.) Each fixed DNA fragment is obtained from a particular gene; a single microarray thus carries DNA from thousands of genes, perhaps even all the genes of an organism.

Figure 9 outlines how microarrays are used. ❶ A researcher collects all of the mRNA transcribed from genes in a particular type of cell at the current moment. ❷ This collection of mRNAs is mixed with reverse transcriptase (an enzyme that produces DNA from an RNA template to produce a mixture of DNA fragments. These fragments are called cDNAs (complementary DNAs) because each one is complementary to one of the mRNAs. The cDNAs are produced in the presence of nucleotides that have been modified to fluoresce (glow). The fluorescent cDNA collection thus represents all of the genes that are being actively transcribed in the cell. ❸ A small amount of the fluorescently labeled cDNA mixture is added to each of the wells in the microarray. If a molecule in the cDNA mixture is complementary to a DNA fragment at a particular location on the grid, the cDNA molecule binds to it, becoming fixed there. ❹ After unbound cDNA is rinsed away, the remaining cDNA produces a detectable glow in the microarray. The pattern of glowing spots enables the researcher to determine which genes were being transcribed in the starting cells.

DNA microarrays are a potential boon to medical research. For example, a 2002 study showed that DNA microarray data can classify different types of leukemia into specific subtypes based on the activity of 17 genes. This information can be used to predict which of several available regimens of chemotherapy is likely to be most effective. Further research suggests that many cancers have a variety of subtypes with different patterns of gene expression that can be identified with DNA microarrays. Indeed, some oncologists predict that DNA microarrays will usher in a new era where medical treatment is customized to each patient.

DNA microarrays can also reveal general profiles of gene expression over the lifetime of an organism. In one example of a global expression study using this technique, researchers performed DNA microarray experiments on more than 90% of the genes of the nematode worm *Caenorhabditis elegans* during every stage of its life cycle. The results showed that expression of nearly 60% of the *C. elegans* genes changed dramatically during development. This study supported the model held by most developmental biologists that embryonic development of multicellular eukaryotes involves a complex and elaborate program of gene expression, rather than simply the expression of a small number of important genes.

? What is learned from a DNA microarray?

Which genes are active (transcribed) in a particular sample of cells

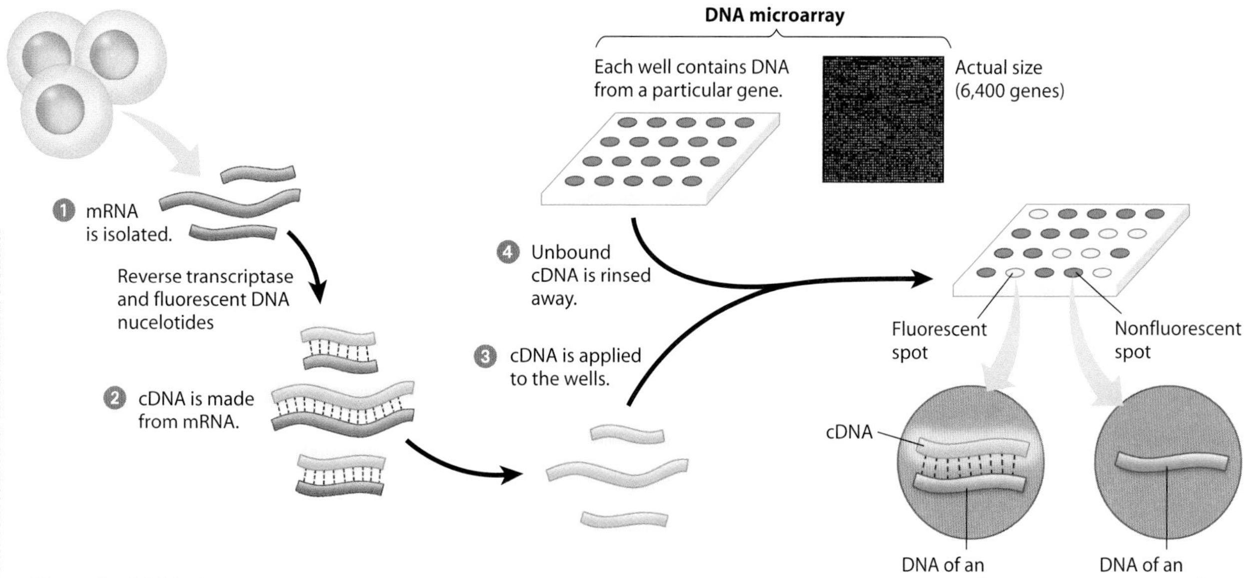

▲ **Figure 9** A DNA microarray

10 Signal transduction pathways convert messages received at the cell surface to responses within the cell

Cell-to-cell signaling, with proteins or other kinds of molecules carrying messages from signaling cells to receiving (target) cells, is a key mechanism in the coordination of cellular activities. In most cases, a signaling molecule acts by binding to a receptor protein in the plasma membrane of the target cell and initiating a signal transduction pathway in the cell. A **signal transduction pathway** is a series of molecular changes that converts a signal on a target cell's surface to a specific response inside the cell.

Figure 10 shows the main elements of a signal transduction pathway in which the target cell's response is the expression of a gene. ❶ The signaling cell secretes a signaling molecule. ❷ This molecule binds to a receptor protein embedded in the target cell's plasma membrane. ❸ The binding activates the first in a series of relay proteins within the target cell. Each relay molecule activates another. ❹ The last relay molecule in the series activates a transcription factor that ❺ triggers transcription of a specific gene. ❻ Translation of the mRNA produces a protein.

Signal transduction pathways are crucial to many cellular functions. Throughout your study of biology, you'll see their importance again and again. We'll revisit them when we discuss cancer later in this chapter (see, for example, Module 18); and we'll see how they relate to hormone function in animals and plants.

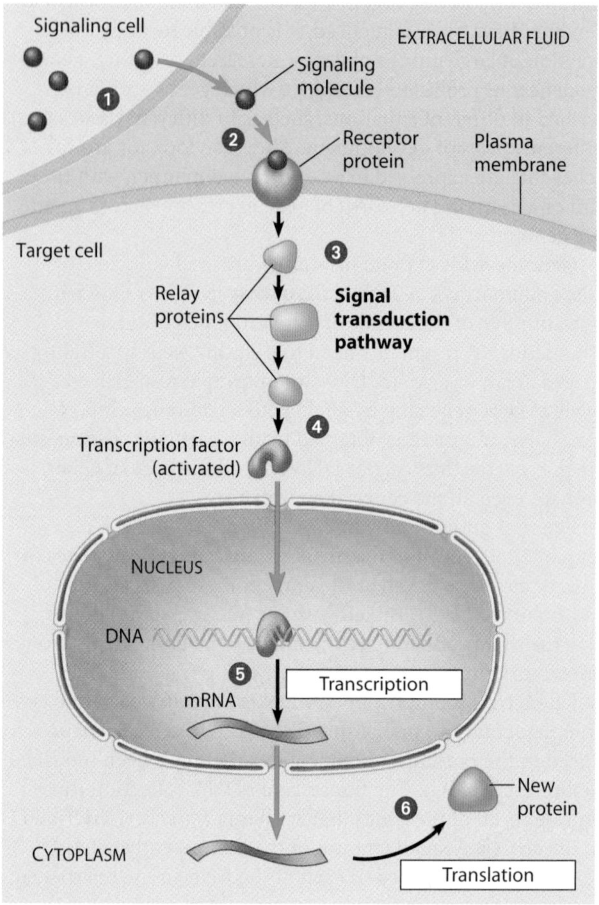

▲ **Figure 10** A signal transduction pathway that turns on a gene

? **How can a signaling molecule from one cell alter gene expression in a target cell without even entering the target cell?**

By binding to a receptor protein in the membrane of the target cell and triggering a signal transduction pathway that activates transcription factors

EVOLUTION CONNECTION

11 Cell-signaling systems appeared early in the evolution of life

As explained in Module 10, one cell can communicate with another by secreting molecules that bind to surface proteins on a target cell. How ancient and widespread are such signaling systems among Earth's organisms? To answer these questions, we can look at communication between microorganisms, for modern microbes are a window on the role of cell signaling in the evolution of life on Earth.

One topic of cell "conversation" is sex—at least for the yeast *Saccharomyces cerevisiae,* which people have used for millennia to make bread, wine, and beer. Researchers have learned that cells of this yeast identify their mates by chemical signaling. There are two sexes, or mating types, called **a** and **α** (**Figure 11**). Cells of mating type **a** secrete a chemical signal called **a** factor, which can bind to specific receptor proteins on nearby **α** cells. At the same time, **α** cells secrete **α** factor, which binds to receptors on **a** cells. Without actually

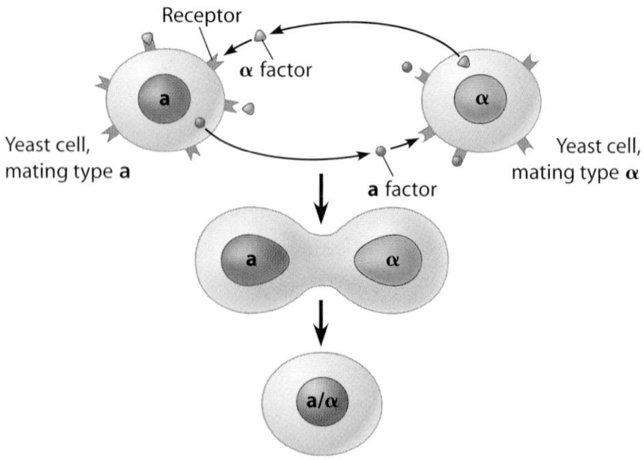

▲ **Figure 11** Communication between mating yeast cells

entering the target cells, the two mating factors cause the cells to grow toward each other and bring about other cellular changes. The result is the fusion, or mating, of two cells of opposite type. The new **a/α** cell contains all the genes of both original cells, a combination of genetic resources that provides advantages to the cell's descendants, which arise by subsequent cell divisions.

The signal transduction pathways involved in this yeast mating system have been extensively studied, as have many other pathways in yeasts and other organisms. Amazingly, the molecular details of signal transduction in yeast and mammals are strikingly similar, even though the last common ancestor of these two groups of organisms lived over a billion years ago. These similarities—and others more recently uncovered between signaling systems in bacteria and plants—suggest that early versions of the cell-signaling mechanisms used today evolved well before the first multicellular creatures appeared on Earth. Scientists think that signaling mechanisms evolved first in ancient prokaryotes and single-celled eukaryotes and then became adapted for new uses in their multicellular descendants.

? In what sense is the joining of yeast mating types "sex"?

The process results in the creation of a diploid cell that is a genetic blend of two parental haploid cells.

Cloning of Plants and Animals

12 Plant cloning shows that differentiated cells may retain all of their genetic potential

One of the most important "take home lessons" from this chapter is that differentiated cells express only a small percentage of their genes. So then how do we know that all the genes are still present? And if all the genes are still there, do differentiated cells retain the potential to express them?

One way to approach these questions is to see if a differentiated cell can dedifferentiate, or reverse its differentiation, and then be stimulated to generate a whole new organism. In plants, this ability is common. In fact, if you have ever grown a plant from a small cutting, you've seen evidence that a differentiated plant cell can undergo cell division and give rise to all the tissues of an adult plant. On a larger scale, the technique described in **Figure 12** can be used to produce hundreds or thousands of genetically identical plants from the cells of a single plant. For example, when cells from a carrot are transferred to a culture medium, a single cell can begin dividing and eventually grow into an adult plant, a genetic replica of the parent plant. Such an organism, produced through asexual reproduction from a single parent, is called a **clone**. The fact that a mature plant cell can dedifferentiate and then give rise to all the different kinds of specialized cells of a new plant shows that differentiation does not necessarily involve irreversible changes in the plant's DNA.

Plant cloning is now used extensively in agriculture. For some plants, such as orchids, cloning is the only commercially practical means of reproducing plants. In other cases, cloning has been used to reproduce a plant with desirable traits, such as high fruit yield or the ability to resist a plant pathogen.

But is this sort of cloning possible in animals? A good indication that differentiation need not impair an animal cell's genetic potential is the natural process of **regeneration**, the regrowth of lost body parts. When a salamander loses a leg, for example, certain cells in the leg stump dedifferentiate, divide, and then redifferentiate, giving rise to a new leg. Many animals, especially among the invertebrates, can regenerate lost parts, and in a few relatively simple animals, isolated differentiated cells can dedifferentiate and then develop into an entire organism. Further evidence for the complete genetic potential of animal cells comes from cloning experiments, our next topic.

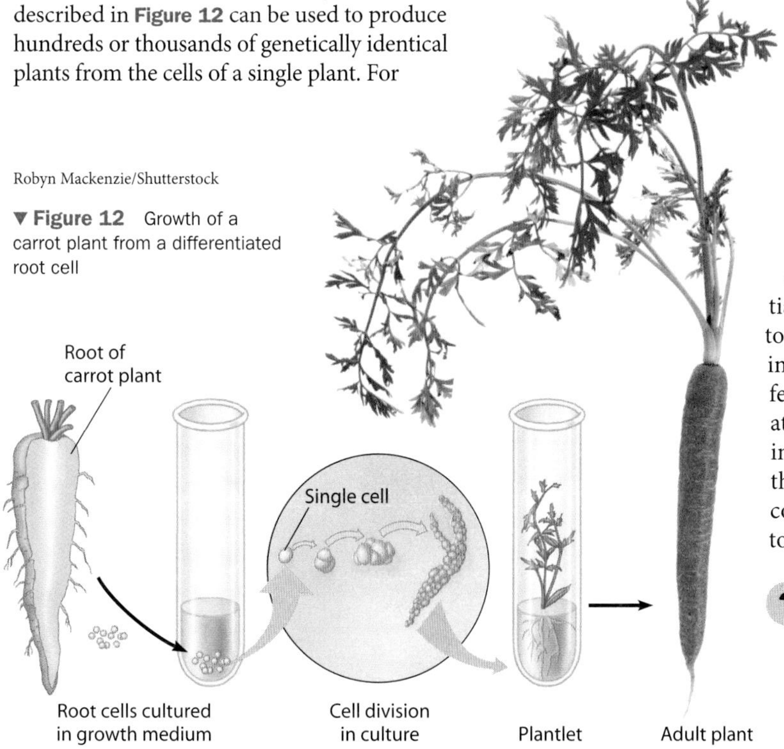

Robyn Mackenzie/Shutterstock

▼ Figure 12 Growth of a carrot plant from a differentiated root cell

? How does the cloning of plants from differentiated cells support the view that differentiation is based on the control of gene expression rather than on irreversible changes in the genome?

Cloning shows that all the genes of a fully differentiated plant are still present, but some may be turned off.

13 Nuclear transplantation can be used to clone animals

Animal cloning can be achieved through a procedure called **nuclear transplantation** (**Figure 13**). First performed in the 1950s using cells from frog embryos, nuclear transplantation involves replacing the nucleus of an egg cell or a zygote with the nucleus of an adult somatic cell. The recipient cell may then begin to divide. By about 5 days later, repeated cell divisions have formed a blastocyst, a hollow ball of about 100 cells. At this point, the blastocyst may be used for different purposes, as indicated by the two branches in Figure 13.

If the animal to be cloned is a mammal, further development requires implanting the blastocyst into the uterus of a surrogate mother (Figure 13, upper branch). The resulting animal will be genetically identical to the donor of the nucleus—a "clone" of the donor. This type of cloning, which results in the birth of a new living individual, is called **reproductive cloning**. Scottish researcher Ian Wilmut and his colleagues used reproductive cloning to produce the world-famous sheep Dolly in 1997. The researchers used an electric shock to fuse 277 specially treated adult sheep udder cells with eggs from which they had removed the nuclei. After several days of growth, 29 of the resulting embryos were implanted in the uteruses of surrogate mothers. One of the embryos developed into Dolly. As expected, Dolly resembled her genetic parent, the nucleus donor, not the egg donor or the surrogate mother.

In a different cloning procedure (Figure 13, lower branch), **embryonic stem cells (ES cells)** are harvested from the blastocyst. In nature, embryonic stem cells give rise to all the different kinds of specialized cells of the body. In the laboratory, embryonic stem cells are easily grown in culture, where, given the right conditions, they can perpetuate themselves indefinitely. When the major aim is to produce embryonic stem cells for therapeutic treatments, the process is called **therapeutic cloning**. In the next two modules, we discuss applications of reproductive and therapeutic cloning, respectively.

? What are the intended products of reproductive cloning and therapeutic cloning?

Reproductive cloning is used to produce new individuals. Therapeutic cloning is used to harvest embryonic stem cells.

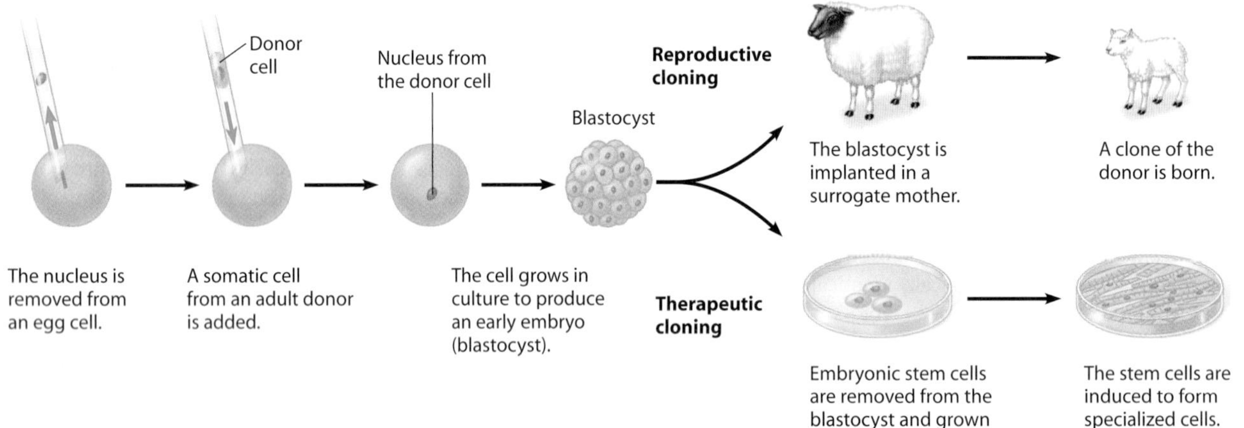
Donor cell
Nucleus from the donor cell
Blastocyst
Reproductive cloning
The blastocyst is implanted in a surrogate mother.
A clone of the donor is born.
The nucleus is removed from an egg cell.
A somatic cell from an adult donor is added.
The cell grows in culture to produce an early embryo (blastocyst).
Therapeutic cloning
Embryonic stem cells are removed from the blastocyst and grown in culture.
The stem cells are induced to form specialized cells.

▲ **Figure 13** Nuclear transplantation for cloning

CONNECTION

14 Reproductive cloning has valuable applications, but human reproductive cloning raises ethical issues

Since Dolly's landmark birth in 1997, researchers have cloned many other mammals, including mice, cats, horses, cows, mules, pigs, rabbits, ferrets, and dogs. We have already learned much from such experiments. For example, cloned animals of the same species do not always look or behave identically. In a herd of cows cloned from the same line of cultured cells, certain cows are dominant and others are more submissive. Another example is the first cloned cat, named CC for Carbon Copy (**Figure 14**). She has a calico coat, like her single female parent, but the color and pattern are different due to random X chromosome inactivation (see Module 2). Moreover, CC and her lone parent behave differently; CC is playful, while her mother is more

▶ **Figure 14** CC, the world's first cloned cat (right), and her lone parent (left)

Pat Sullivan/Associated Press

reserved. You've probably observed that identical human twins, which are naturally occurring "clones," are always slightly different. Clearly, environmental influences and random phenomena can play a significant role during development.

Reproductive cloning has potential for many practical applications. On an experimental basis, agricultural scientists are cloning farm animals with specific sets of desirable traits in the hope of creating high-yielding, genetically identical herds. The pharmaceutical industry is experimenting with cloning mammals for the production of potentially valuable drugs. For example, researchers have produced piglet clones that lack one of two copies of a gene for a protein that can cause immune system rejection in humans. Such pigs may one day provide organs for transplant into humans. Some wildlife biologists hope that reproductive cloning can be used to restock the populations of endangered animals. Among the rare animals that have been cloned are a wild mouflon (a small European sheep), a banteng (a Javanese cow), a gaur (an Asian ox), and gray wolves.

The successful cloning of various mammals has heightened speculation that humans could be cloned. Critics point out that there are many obstacles—both practical and ethical—to human cloning. Practically, animal cloning is extremely difficult and inefficient. Only a small percentage of cloned embryos develop normally. Ethically, the discussion about whether or not humans should be cloned—and if so, under what circumstances—is far from settled. Meanwhile, the research and the debate continue.

? **If you cloned your dog, would you expect the original and the clone to look and act exactly alike?**

No. While cloning produces genetically identical dogs, appearance and behavior are affected by environment.

CONNECTION

15 Therapeutic cloning can produce stem cells with great medical potential

Therapeutic cloning produces embryonic stem cells, cells that in the early animal embryo differentiate to give rise to all the cell types in the body. When grown in laboratory culture, embryonic stem cells can divide indefinitely (like cancer cells). Furthermore, the right conditions can induce changes in gene expression that cause differentiation into a variety of cell types (**Figure 15**).

The adult body also has stem cells, which serve to replace nonreproducing specialized cells as needed. In contrast to embryonic stem cells, **adult stem cells** are able to give rise to many but not all cell types in the organism. For example, bone marrow contains several types of stem cells, including one that can generate all the different kinds of blood cells. Although adult animals have only tiny numbers of stem cells, scientists are learning to identify and isolate these cells from various tissues and, in some cases, to grow them in culture.

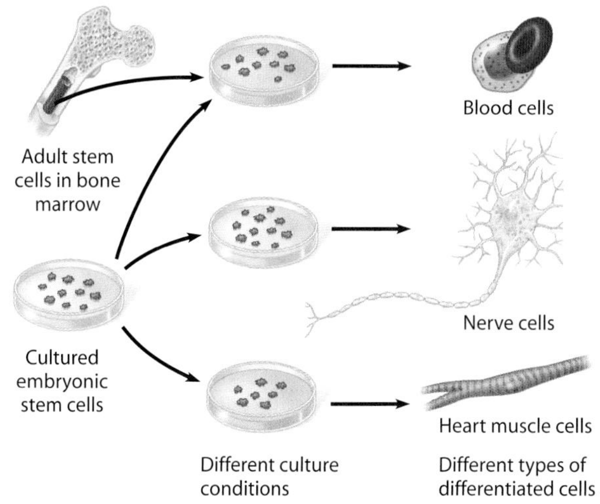

▲ **Figure 15** Differentiation of stem cells in culture

Recently, researchers have discovered stem cells in human skin, hair, eyes, and oral tissues. However, the developmental potential of adult stem cells is limited to certain cell types, so embryonic stem cells are considered more promising than adult stem cells for medical applications, at least for now. In 2007, three research groups reported transforming mouse skin cells into embryonic stem cells simply by causing the skin cells to express four "stem cell" master regulatory genes. The researchers used retroviruses as vectors to introduce extra copies of these genes into the skin cells.

The ultimate aim of therapeutic cloning is to supply cells for the repair of damaged or diseased organs: for example, insulin-producing pancreatic cells for people with diabetes or certain kinds of brain cells for people with Parkinson's disease or Alzheimer's disease. Adult stem cells from donor bone marrow have long been used as a source of immune system cells in patients whose own immune systems have been destroyed by genetic disorders or radiation treatments for cancer. More recently, clinical trials using bone marrow stem cells have shown slight success in promoting regeneration of heart tissue in patients whose hearts have been damaged by heart attacks. In the future, a donor nucleus from a patient with a particular disease could allow production of embryonic stem cells for treatment that match the patient and are thus not rejected by his or her immune system.

While many people believe that reproductive cloning of humans is unethical, opinions vary more widely about the morality of therapeutic cloning using embryonic stem cells. As with reproductive cloning, the research and the debate continue.

? **In nature, how do embryonic stem cells differ from adult stem cells?**

Embryonic cells give rise to all the different kinds of cells in the body. Adult stem cells generate only a few related types of cells.

The Genetic Basis of Cancer

16 Cancer results from mutations in genes that control cell division

Cancerous cells have escaped from the control mechanisms that normally limit their growth. Scientists have learned that such escape is often due to changes in gene expression.

The abnormal behavior of cancer cells was observed years before anything was known about the cell cycle, its control, or the role genes play in making cells cancerous. One of the earliest clues to the cancer puzzle was the discovery, in 1911, of a virus that causes cancer in chickens. Recall that viruses are simply molecules of DNA or RNA surrounded by pro-tein and in some cases a membranous envelope. Viruses that cause cancer can become permanent residents in host cells by inserting their nucleic acid into the DNA of host chromosomes.

The genes that a cancer-causing virus inserts into a host cell can make the cell cancerous. Such a gene, which can cause cancer when present in a single copy in the cell, is called an **oncogene** (from the Greek *onkos*, tumor). Over the last century, researchers have identified a number of viruses that harbor cancer-causing genes. One example is the human papillomavirus (HPV), which is associated with several types of cancer, most frequently cervical cancer.

Proto-oncogenes In 1976, American molecular biologists J. Michael Bishop, Harold Varmus, and their colleagues made a startling discovery. They found that the cancer-causing chicken virus discovered in 1911 contains an oncogene that is an altered version of a normal gene found in chicken cells. Subsequent research has shown that the chromosomes of many animals, including humans, contain genes that can be converted to oncogenes. A normal gene that has the potential to become an oncogene is called a **proto-oncogene**. (These terms can be confusing, so they bear repeating: a *proto-oncogene* is a normal, healthy gene that, if changed, can become a cancer-causing *oncogene*.) Thus, a cell can acquire an oncogene either from a virus or from the mutation of one of its own genes.

The cancer research conducted by Bishop and Varmus focused on proto-oncogenes. Searching for the normal role of these genes, researchers found that many proto-oncogenes code for growth factors—proteins that stimulate cell division—or for other proteins that somehow affect growth factor function or some other aspect of the cell cycle. When all these proteins are functioning normally, in the right amounts at the right times, they help properly control cell division and cellular differentiation.

How might a proto-oncogene—a gene that has an essential function in normal cells—become an oncogene, a cancer-causing gene? In general, an oncogene arises from a genetic change that leads to an increase either in the amount of the proto-oncogene's protein product or in the activity of each protein molecule. **Figure 16A** illustrates three kinds of changes in DNA that can produce oncogenes. Let's assume that the starting proto-oncogene codes for a protein that stimulates cell division. On the left in the figure, a mutation (green) in the proto-oncogene itself creates an oncogene that codes for a hyperactive protein, which is produced in the usual amount but whose stimulating effect is stronger than normal. In the center, an error in DNA replication or recombination generates multiple copies of the gene, which are all transcribed and translated; the result is an excess of the normal stimulatory protein. On the right, the proto-oncogene has been moved from its normal location in the cell's DNA to another location. At its new site, the gene is under the control of a different promoter, one that causes it to be transcribed more often than normal; the normal protein is again made in excess. So in all three cases, normal gene expression is changed, and the cell is stimulated to divide excessively.

▶ **Figure 16A** Alternative ways to make oncogenes from a proto-oncogene (all leading to excessive cell growth)

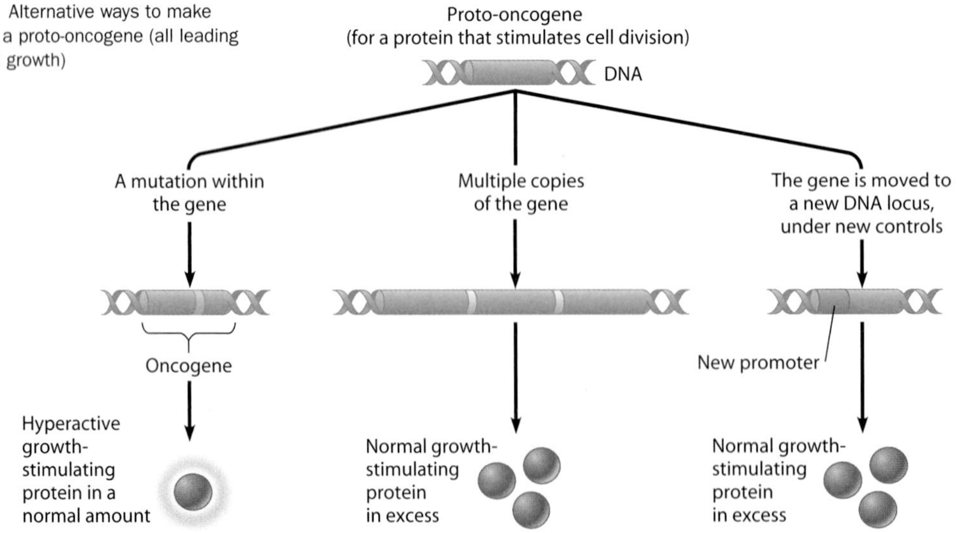

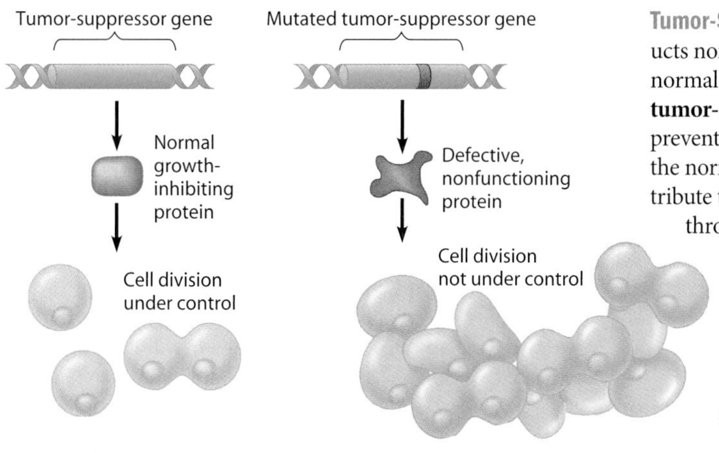

▲ **Figure 16B** The effect of a mutation in a tumor-suppressor gene

Tumor-Suppressor Genes In addition to genes whose products normally *promote* cell division, cells contain genes whose normal products *inhibit* cell division. Such genes are called **tumor-suppressor genes** because the proteins they encode help prevent uncontrolled cell growth. Any mutation that decreases the normal activity of a tumor-suppressor protein may contribute to the onset of cancer, in effect stimulating growth through the absence of suppression (**Figure 16B**). Scientists have also discovered a class of tumor-suppressor genes that function in the repair of damaged DNA. When these genes are mutated, other cancer-causing mutations are more likely to accumulate.

? How do proto-oncogenes relate to oncogenes?

A proto-oncogene is a normal gene that, if mutated, can become a cancer-causing oncogene.

17 Multiple genetic changes underlie the development of cancer

Nearly 150,000 Americans will be stricken by cancer of the colon (the main part of the large intestine) this year. One of the best-understood types of human cancer, colon cancer illustrates an important principle about how cancer develops: More than one somatic mutation is needed to produce a full-fledged cancer cell. As in many cancers, the development of malignant (spreading) colon cancer is gradual.

Figure 17A illustrates this idea using colon cancer as an example. ❶ Colon cancer begins when an oncogene arises or is activated through mutation, causing unusually frequent division of apparently normal cells in the colon lining. ❷ Later, one or more additional DNA mutations, such as the inactivation of a tumor-suppressor gene, cause the growth of a small benign tumor (a polyp) in the colon wall. ❸ Still more mutations eventually lead to formation of a malignant tumor, a tumor that has the potential to metastasize (spread). The requirement for several mutations—the actual number is usually four or more—explains why cancers can take a long time to develop.

Thus, the development of a malignant tumor is paralleled by a gradual accumulation of mutations that convert proto-oncogenes to oncogenes and knock out tumor-suppressor genes. Multiple changes must occur at the DNA level for a cell to become fully cancerous. Such changes usually include the appearance of at least one active oncogene and the mutation or loss of several tumor-suppressor genes. In **Figure 17B**, colors distinguish the normal cell (tan) from cells with one or more mutations leading to increased cell division and cancer (red). Once a cancer-promoting mutation occurs (the red band on the chromosome), it is passed to all the descendants of the cell carrying it.

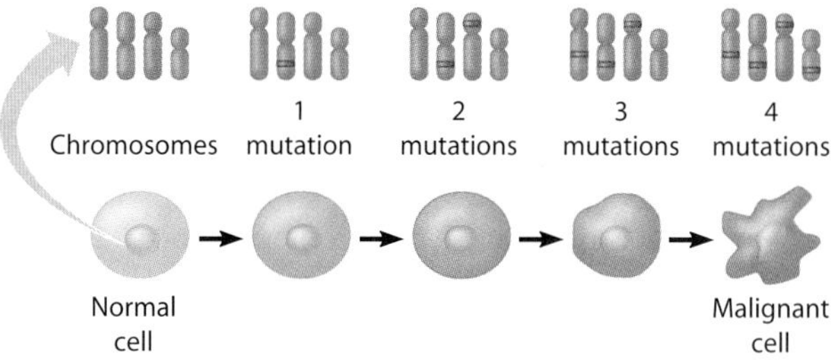

▲ **Figure 17B** Accumulation of mutations in the development of a cancer cell

The fact that more than one somatic mutation is generally needed to produce a full-fledged cancer cell may help explain why the incidence of cancer increases with age. If cancer results from an accumulation of mutations that occur throughout life, then the longer we live, the more likely we are to develop cancer. Cancer researchers hope to learn about mutations that cause cancer through the Cancer Genome Project, a 10-year effort to map all human cancer-causing genes.

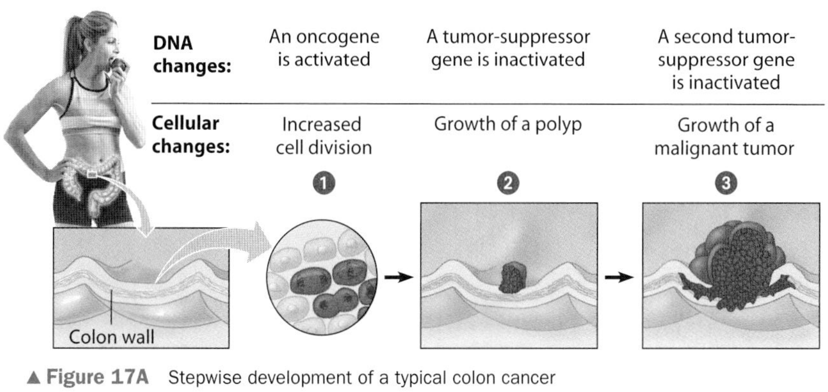

▲ **Figure 17A** Stepwise development of a typical colon cancer

? Epithelial cells, those that line body cavities, are frequently replaced and so divide more often than most other types of body cells. Will epithelial cells become cancerous more or less frequently than other types of body cells?

More frequent cell divisions will result in more frequent mutation and thus a greater chance of cancer.

18 Faulty proteins can interfere with normal signal transduction pathways

To understand how oncogenes and defective tumor-suppressor genes can contribute to uncontrolled cell growth, we need to look more closely at the normal functions of proto-oncogenes and tumor-suppressor genes. Genes in both categories often code for proteins involved in signal transduction pathways leading to gene expression (see Module 10).

The figures below (excluding, for the moment, the white boxes) illustrate two types of signal transduction pathways leading to the synthesis of proteins that influence the cell cycle. In **Figure 18A**, the pathway leads to the stimulation of cell division. The initial signal is a growth factor (●), and the target cell's ultimate response is the production of a protein that stimulates the cell to divide. By contrast, **Figure 18B** shows an inhibitory pathway, in which a growth-*inhibiting* factor (▼) causes the target cell to make a protein that inhibits cell division. In both cases, the newly made proteins function by interacting with components of the cell cycle control system, although the figures here do not show these interactions.

Now, let's see what can happen when the target cell undergoes a cancer-causing mutation. The white box in Figure 18A shows the protein product of an oncogene resulting from mutation of a proto-oncogene called *ras*. The normal product of *ras* is a relay protein. Ordinarily, a stimulatory pathway like this will not operate unless the growth factor is available. However, an oncogene protein that is a hyperactive version of a protein in the pathway may trigger the pathway even in the absence of a growth factor. In this example, the oncogene protein is a hyperactive version of the *ras* relay protein that issues signals on its own. In fact, abnormal versions or amounts of any of the pathway's components—from the growth factor itself to the transcription factor—could have the same final effect: overstimulation of cell division.

The white box in Figure 18B indicates how a mutant tumor-suppressor protein can affect cell division. In this case, the mutation affects a gene called *p53*, which codes for an essential transcription factor. This mutation leads to the production of a faulty transcription factor, one that the signal transduction pathway cannot activate. As a result, the gene for the inhibitory protein at the bottom of the figure remains turned off, and excessive cell division may occur.

Mutations of the *ras* and *p53* genes have been implicated in many kinds of cancer. In fact, mutations in *ras* occur in about 30% of human cancers, and mutations in *p53* occur in more than 50%. As we see next, carcinogens are responsible for many mutations that lead to cancer.

? Contrast the action of an oncogene with that of a cancer-causing mutation in a tumor-suppressor gene.

● An oncogene encodes an abnormal protein that stimulates cell division via a signal transduction pathway; a mutant tumor-suppressor gene encodes a defective protein unable to function in a pathway that normally inhibits cell division.

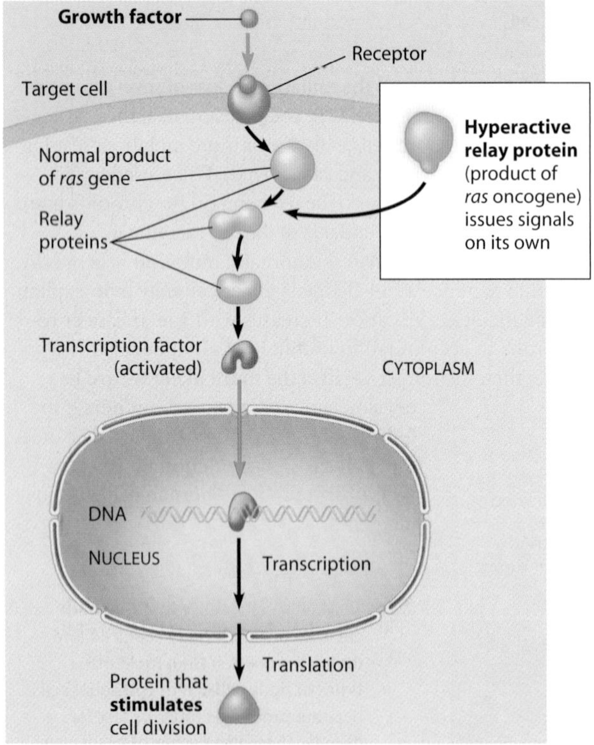

▲ **Figure 18A** A stimulatory signal transduction pathway and the effect of an oncogene protein

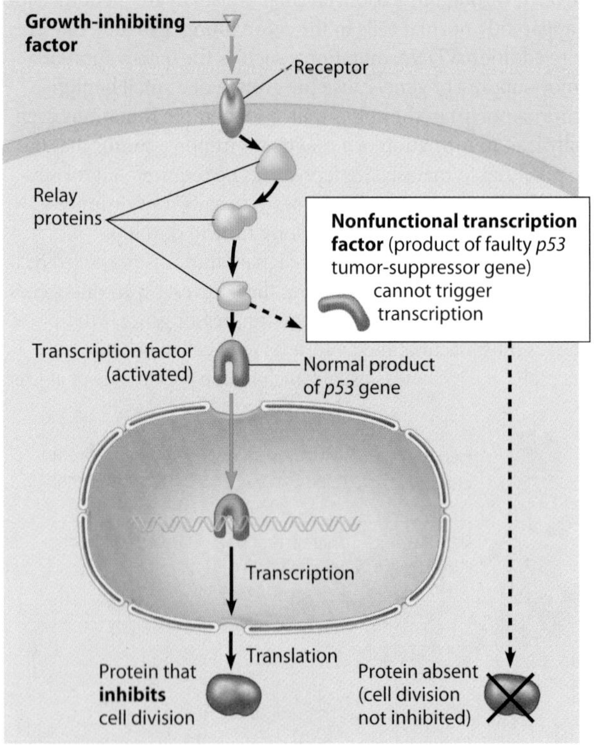

▲ **Figure 18B** An inhibitory signal transduction pathway and the effect of a faulty tumor-suppressor protein

CONNECTION 19 Lifestyle choices can reduce the risk of cancer

Cancer is the second-leading cause of death in most industrialized nations, after heart disease. Death rates due to certain forms of cancer—including stomach, cervical, and uterine cancers—have decreased in recent years, but the overall cancer death rate is on the rise, currently increasing at about 1% per decade. Table 19 lists the most common cancers in the United States and associated risk factors for each.

The fact that multiple genetic changes are required to produce a cancer cell helps explain the observation that cancers can run in families. An individual inheriting an oncogene or a mutant allele of a tumor-suppressor gene is one step closer to accumulating the necessary mutations for cancer to develop than an individual without any such mutations.

But the majority of cancers are not associated with a mutation that is passed from parent to offspring; they arise from new mutations caused by environmental factors. Cancer-causing agents, factors that alter DNA and make cells cancerous, are called **carcinogens**. Most mutagens, substances that cause mutations, are carcinogens. Two of the most potent mutagens are X-rays and ultraviolet radiation in sunlight. X-rays are a significant cause of leukemia and brain cancer. Exposure to UV radiation from the sun is known to cause skin cancer, including a deadly type called melanoma.

The one substance known to cause more cases and types of cancer than any other single agent is tobacco. More people die of lung cancer (nearly 160,000 Americans in 2010) than any other form of cancer. Most tobacco-related cancers come from smoking, but the passive inhalation of secondhand smoke is also a risk. As Table 19 indicates, tobacco use, sometimes in combination with alcohol consumption, causes a number of other types of cancer in addition to lung cancer. In nearly all cases, cigarettes are the main culprit, but smokeless tobacco products, such as snuff and chewing tobacco, are linked to cancer of the mouth and throat.

How do carcinogens cause cancer? In many cases, the genetic changes that cause cancer result from decades of exposure to the mutagenic effects of carcinogens. Carcinogens can also produce their effect by promoting cell division. Generally, the higher the rate of cell division, the greater the chance for mutations resulting from errors in DNA replication or recombination. Some carcinogens seem to have both effects. For instance, the hormones linked to breast and uterine cancers promote cell division and may also cause genetic changes that lead to cancer. In other cases, several different agents, such as viruses and one or more carcinogens, may together produce cancer.

Avoiding carcinogens is not the whole story, for there is growing evidence that some food choices significantly reduce cancer risk. For instance, eating 20–30 grams (g) of plant fiber daily—roughly equal to the amount of fiber in four slices of whole-grain bread, 1 cup of bran flakes, one apple, and 1/2 cup of carrots combined—and at the same time reducing animal fat intake may help prevent colon cancer. There is also evidence that other substances in fruits and vegetables, including vitamins C and E and certain compounds related to vitamin A, may offer protection against a variety of cancers. Cabbage and its relatives, such as broccoli and cauliflower, are thought to be especially rich in substances that help prevent cancer, although the identities of these substances are not yet established. Determining how diet influences cancer has become a major research goal.

The battle against cancer is being waged on many fronts, and there is reason for optimism in the progress being made. It is especially encouraging that we can help reduce our risk of acquiring and increase our chance of surviving some of the most common forms of cancer by the choices we make in our daily life. Not smoking, exercising adequately, avoiding overexposure to the sun, and eating a high-fiber, low-fat diet can all help prevent cancer. Furthermore, seven types of cancer can be easily detected: cancers of the skin and oral cavity (via physical exam), breast (via self-exams and mammograms for higher-risk women), prostate (via rectal exam), cervix (via Pap smear), testes (via self-exam), and colon (via colonoscopy). Regular visits to the doctor can help identify tumors early, thereby significantly increasing the possibility of successful treatment.

TABLE 19 CANCER IN THE UNITED STATES

Cancer	Risk Factors	Estimated Number of Cases in 2010
Lung	Tobacco smoke	222,520
Prostate	African heritage; possibly dietary fat	217,730
Breast	Estrogen	209,060
Colon, rectum	High dietary fat; tobacco smoke; alcohol	142,570
Lymphomas	Viruses (for some types)	74,030
Urinary bladder	Tobacco smoke	70,530
Melanoma of the skin	Ultraviolet light	68,130
Kidney	Tobacco smoke	58,240
Uterus	Estrogen	43,470
Pancreas	Tobacco smoke; obesity	43,140
Leukemias	X-rays; benzene; virus (for one type)	43,050
Oral cavity	Tobacco in various forms; alcohol	36,540
Liver	Alcohol; hepatitis viruses	24,120
Brain and nerve	Trauma; X-rays	22,020
Ovary	Obesity; many ovulation cycles	21,880
Stomach	Table salt; tobacco smoke	21,000
Cervix	Sexually transmitted viruses; tobacco smoke	12,200
All others		199,330

Data from *American Cancer Society. Cancer Facts & Figures 2010.* Atlanta: American Cancer Society; 2010.

? Which of the most common cancers affect primarily males? Which affect primarily females?

Males: prostate; females: breast, uterus, cervix

REVIEW

For Practice Quizzes, BioFlix, MP3 Tutors, and Activities, go to www.masteringbiology.com.

Reviewing the Concepts

Control of Gene Expression (1–11)

1 Proteins interacting with DNA turn prokaryotic genes on or off in response to environmental changes. In prokaryotes, genes for related enzymes are often controlled together in units called operons. Regulatory proteins bind to control sequences in the DNA and turn operons on or off in response to environmental changes.

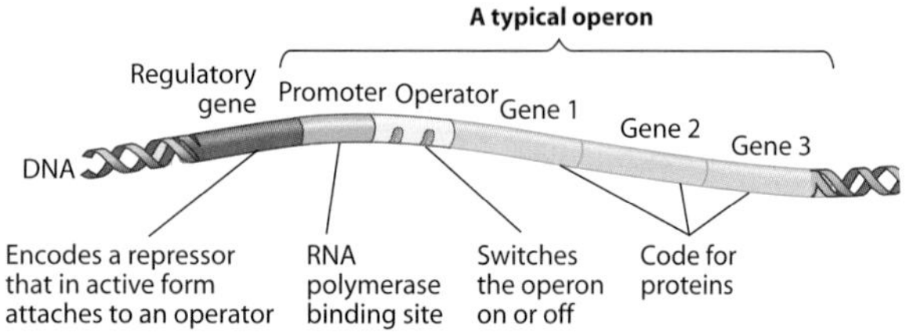

2 Chromosome structure and chemical modifications can affect gene expression. In multicellular eukaryotes, different types of cells make different proteins because different combinations of genes are active in each type. A chromosome contains DNA wound around clusters of histone proteins, forming a string of beadlike nucleosomes. DNA packing tends to block gene expression by preventing access of transcription proteins to the DNA. One example of DNA packing is X chromosome inactivation in the cells of female mammals. Chemical modification of DNA bases or histone proteins can result in epigenetic inheritance.

3 Complex assemblies of proteins control eukaryotic transcription. A variety of regulatory proteins interact with DNA and with each other to turn the transcription of eukaryotic genes on or off.

4 Eukaryotic RNA may be spliced in more than one way. After transcription, alternative RNA splicing may generate two or more types of mRNA from the same transcript.

5 Small RNAs play multiple roles in controlling gene expression. MicroRNAs, bound to proteins, can prevent gene expression by forming complexes with mRNA molecules.

6 Later stages of gene expression are also subject to regulation. The lifetime of an mRNA molecule helps determine how much protein is made, as do factors involved in translation. A protein may need to be activated in some way, and eventually the cell will break it down.

7 Review: Multiple mechanisms regulate gene expression in eukaryotes. Figure 7 reviews the multiple stages of eukaryotic gene expression, each stage offering opportunities for regulation.

8 Cell signaling and cascades of gene expression direct animal development. A series of RNAs and proteins produced in the embryo control the development of an animal from a fertilized egg.

9 DNA microarrays test for the transcription of many genes at once. Scientists can use a DNA microarray to gather data about which genes are turned on or off in a particular cell.

10 Signal transduction pathways convert messages received at the cell surface to responses within the cell. A glass slide containing DNA fragments from thousands of genes can be used to test which of those genes are being produced in a particular cell type.

11 Cell-signaling systems appeared early in the evolution of life. Similarities among organisms suggest that signal transduction pathways evolved early in the history of life on Earth.

Cloning of Plants and Animals (12–15)

12 Plant cloning shows that differentiated cells may retain all of their genetic potential. A clone is an individual created by asexual reproduction and thus genetically identical to a single parent.

13 Nuclear transplantation can be used to clone animals. Inserting DNA from a host cell into a nucleus-free egg can result in an early embryo that is a clone of the DNA donor. Implanting a blastocyst into a surrogate mother allows for the birth of a cloned mammal.

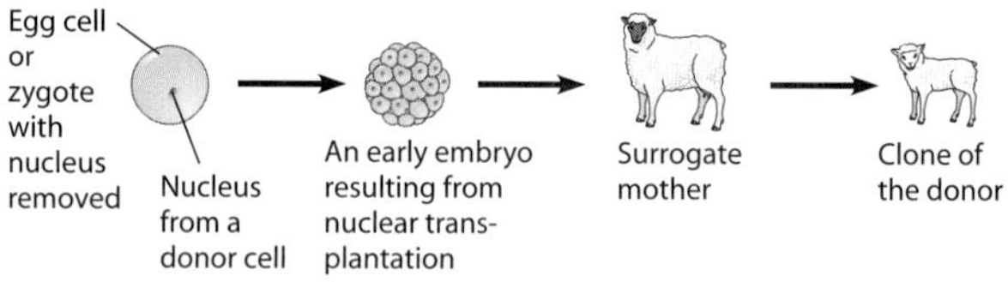

14 Reproductive cloning has valuable applications, but human reproductive cloning raises ethical issues.

15 Therapeutic cloning can produce stem cells with great medical potential. The goal of therapeutic cloning is to produce embryonic stem cells. Such cells may eventually be used for a variety of therapeutic purposes. Like embryonic stem cells, adult stem cells can both perpetuate themselves in culture and give rise to differentiated cells. Unlike embryonic stem cells, adult stem cells normally give rise to only a limited range of cell types.

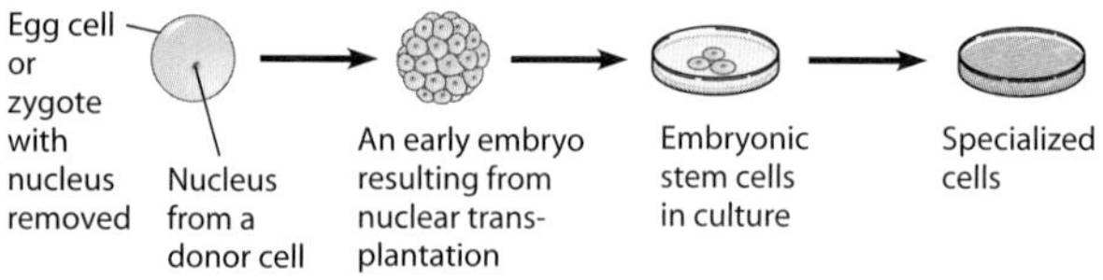

The Genetic Basis of Cancer (16–19)

16 Cancer results from mutations in genes that control cell division. Cancer cells, which divide uncontrollably, result from mutations in genes whose protein products affect the cell cycle. A mutation can change a proto-oncogene, a normal gene that helps control cell division, into an oncogene, which causes cells to divide excessively. Mutations that inactivate tumor-suppressor genes have similar effects.

17 Multiple genetic changes underlie the development of cancer. Cancers result from a series of genetic changes.

18 Faulty proteins can interfere with normal signal transduction pathways. Many proto-oncogenes and tumor-suppressor genes code for proteins active in signal transduction pathways regulating cell division.

19 Lifestyle choices can reduce the risk of cancer. Reducing exposure to carcinogens, which induce cancer-causing mutations, and making other lifestyle choices can help reduce cancer risk.

Connecting the Concepts

1. Complete the following concept map to test your knowledge of gene regulation.

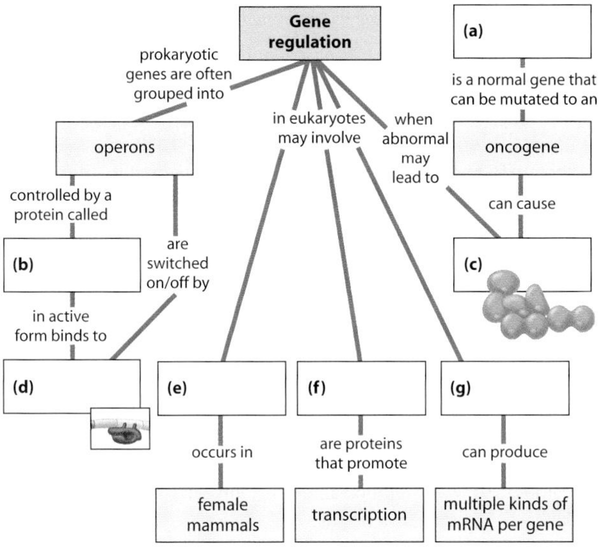

Testing Your Knowledge

Multiple Choice

2. The control of gene expression is more complex in multicellular eukaryotes than in prokaryotes because ______. (*Explain your answer.*)
   a. eukaryotic cells are much smaller
   b. in a multicellular eukaryote, different cells are specialized for different functions
   c. prokaryotes are restricted to stable environments
   d. eukaryotes have fewer genes, so each gene must do several jobs
   e. eukaryotic genes code for proteins
3. Your bone cells, muscle cells, and skin cells look different because
   a. each cell contains different kinds of genes.
   b. they are present in different organs.
   c. different genes are active in each kind of cell.
   d. they contain different numbers of genes.
   e. each cell has different mutations.
4. Which of the following methods of gene regulation do eukaryotes and prokaryotes have in common?
   a. elaborate packing of DNA in chromosomes
   b. activator and repressor proteins, which attach to DNA
   c. the addition of a cap and tail to mRNA after transcription
   d. *lac* and *trp* operons
   e. the removal of noncoding portions of RNA
5. A homeotic gene does which of the following?
   a. It serves as the ultimate control for prokaryotic operons.
   b. It regulates the expression of groups of other genes during development.
   c. It represses the histone proteins in eukaryotic chromosomes.
   d. It helps splice mRNA after transcription.
   e. It inactivates one of the X chromosomes in a female mammal.
6. All your cells contain proto-oncogenes, which can change into cancer-causing genes. Why do cells possess such potential time bombs?
   a. Viruses infect cells with proto-oncogenes.
   b. Proto-oncogenes are genetic "junk" with no known function.
   c. Proto-oncogenes are unavoidable environmental carcinogens.
   d. Cells produce proto-oncogenes as a by-product of mitosis.
   e. Proto-oncogenes normally control cell division.
7. Which of the following is a valid difference between embryonic stem cells and the stem cells found in adult tissues?
   a. In laboratory culture, only adult stem cells are immortal.
   b. In nature, only embryonic stem cells give rise to all the different types of cells in the organism.
   c. Only adult stem cells can differentiate in culture.
   d. Embryonic stem cells are generally more difficult to grow in culture than adult stem cells.
   e. Only embryonic stem cells are found in every tissue of the adult body.

Describing, Comparing, and Explaining

8. A mutation in a single gene may cause a major change in the body of a fruit fly, such as an extra pair of legs or wings. Yet it probably takes the combined action of hundreds or thousands of genes to produce a wing or leg. How can a change in just one gene cause such a big change in the body?

Applying the Concepts

9. You obtain an egg cell from the ovary of a white mouse and remove the nucleus from it. You then obtain a nucleus from a liver cell from an adult black mouse. You use the methods of nuclear transplantation to insert the nucleus into the empty egg. After some prompting, the new zygote divides into an early embryo, which you then implant into the uterus of a brown mouse. A few weeks later, a litter of mice is born. What color will they be? Why?
10. Mutations can alter the function of the *lac* operon (see Module 1). Predict how the following mutations would affect the function of the operon in the presence and absence of lactose:
    a. Mutation of regulatory gene; repressor cannot bind to lactose.
    b. Mutation of operator; repressor will not bind to operator.
    c. Mutation of regulatory gene; repressor will not bind to operator.
    d. Mutation of promoter; RNA polymerase will not attach to promoter.
11. A chemical called dioxin is produced as a by-product of some chemical manufacturing processes. This substance was present in Agent Orange, a defoliant sprayed on vegetation during the Vietnam War. There has been a continuing controversy over its effects on soldiers exposed to it during the war. Animal tests have suggested that dioxin can be lethal and can cause birth defects, cancer, organ damage, and immune system suppression. But its effects on humans are unclear, and even animal tests are inconclusive. Researchers have discovered that dioxin enters a cell and binds to a protein that in turn attaches to the cell's DNA. How might this mechanism help explain the variety of dioxin's effects? How might you determine whether a particular individual became ill as a result of exposure to dioxin?

Review Answers

1. a. proto-oncogene; b. repressor (or activator); c. cancer; d. operator; e. X inactivation; f. transcription factors; g. alternative RNA splicing
2. b (Different genes are active in different kinds of cells.) 3. c 4. b 5. b 6. e 7. b
8. A mutation in a single gene can influence the actions of many other genes if the mutated gene is a control gene, such as a homeotic gene. A single control gene may encode a protein that affects (activates or represses) the expression of a number of other genes. In addition, some of the affected genes may themselves be control genes that in turn affect other batteries of genes. Cascades of gene expression are common in embryonic development.
9. Black, because the DNA of the cell was obtained from a black mouse.
10. a. If the mutated repressor could still bind to the operator on the DNA, it would continuously repress the operon; enzymes for lactose utilization would not be made, whether or not lactose was present.
    b. The *lac* genes would continue to be transcribed and the enzymes made, whether or not lactose was present.
    c. Same predicted result as for b.
    d. RNA polymerase would not be able to transcribe the genes; no proteins would be made, whether or not lactose was present.
11. The protein to which dioxin binds in the cell is probably a transcription factor that regulates multiple genes (see Module 3). If the binding of dioxin influences the activity of this transcription factor—either activating or inactivating it—dioxin could thereby affect multiple genes and thus have a variety of effects on the body. The differing effects in different animals might be explained by differing genetic details in the different species. It would be extremely difficult to demonstrate conclusively that dioxin exposure was the cause of illness in a particular individual, even if dioxin had been shown to be present in the person's tissues. However, if you had detailed information about how dioxin affects patterns of gene expression in humans and were able to show dioxin-specific abnormal patterns in the patient (perhaps using DNA microarrays; see Module 9), you might be able to establish a strong link between dioxin and the illness.

DNA Technology and Genomics

From Chapter 12 of *Campbell Biology: Concepts & Connections*, Seventh Edition. Jane B. Reece, Martha R. Taylor, Eric J. Simon, Jean L. Dickey.

DNA Technology and Genomics

BIG IDEAS

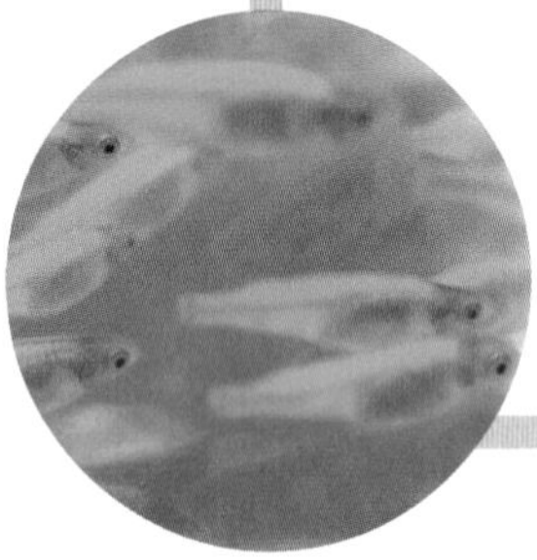

AP Wide World Photos

Gene Cloning
(1–5)

A variety of laboratory techniques can be used to copy and combine DNA molecules.

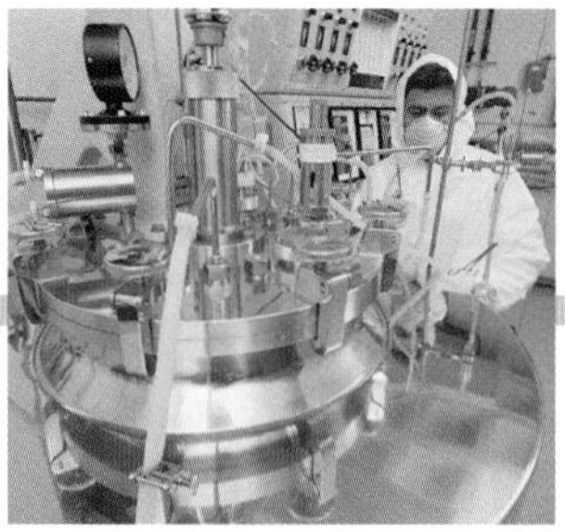

Hank Morgan/Photo Researchers

Genetically Modified Organisms
(6–10)

Transgenic cells, plants, and animals are used in agriculture and medicine.

Steve Helber/AP Wide World Photos

DNA Profiling
(11–16)

Genetic markers can be used to definitively match a DNA sample to an individual.

Philippe Plailly & Atelier Daynes/Photo Researchers

Genomics
(17–21)

The study of complete DNA sets helps us learn about evolutionary history.

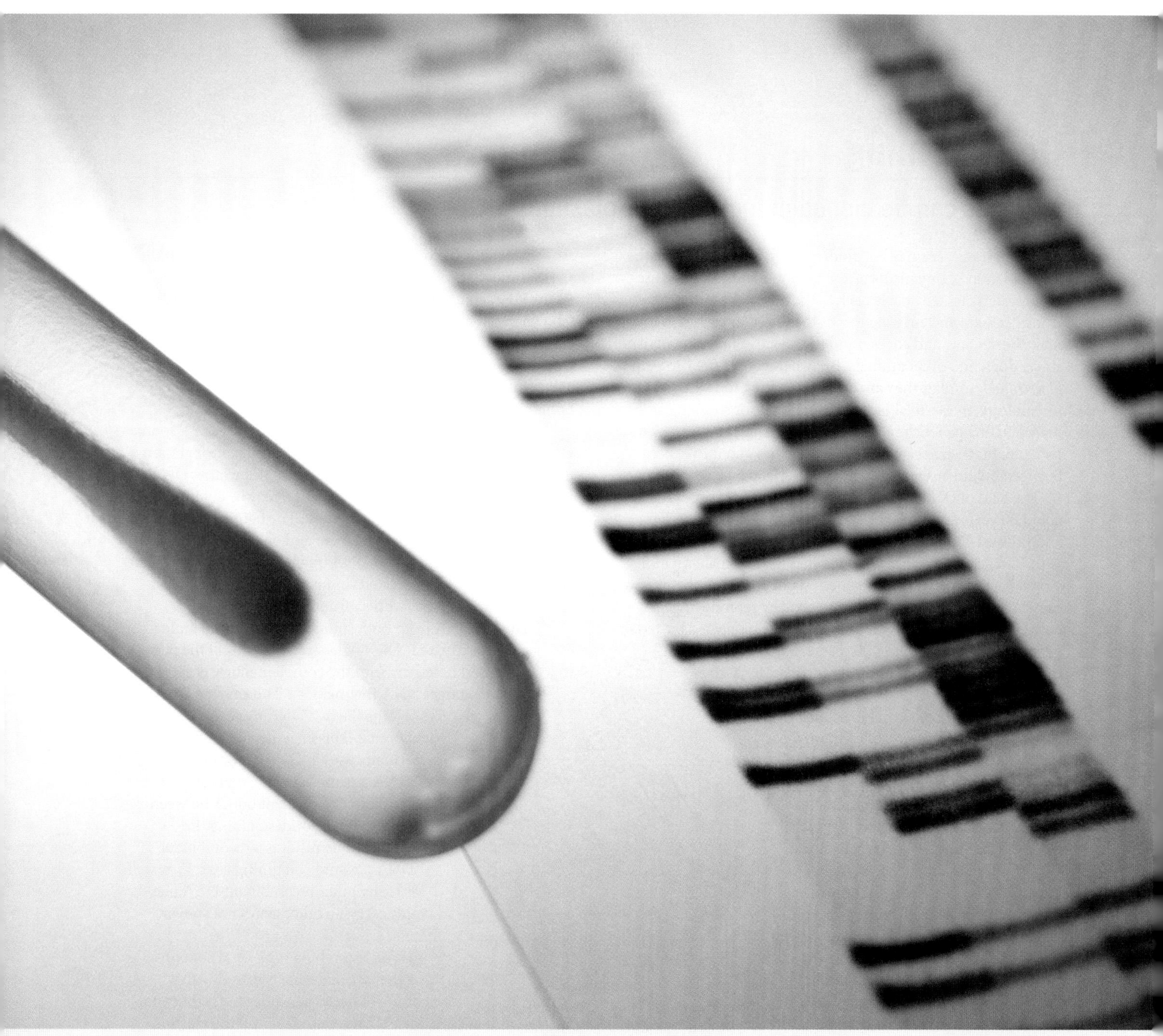

Tek Image/Photo Researchers

DNA technology—a set of methods for studying and manipulating genetic material—has rapidly revolutionized the field of forensics, the scientific analysis of evidence for legal investigations. Since its introduction, DNA analysis has become a standard law enforcement tool. The photograph above shows a swab containing DNA for analysis. The data on the printout beneath it is part of a DNA profile. Because the DNA sequence of every person is unique (except for identical twins), DNA profiling can be used to determine with near certainty whether two DNA samples are from the same individual. And due to its unbiased nature—a DNA profile is able to prove innocence as well as guilt—DNA profiling has provided crucial evidence in many famous cases.

You will learn in this chapter that DNA technology has many practical applications beyond the courtroom. In fact, DNA technology has led to some of the most remarkable scientific advances in recent years. Applications of DNA technology include the use of gene cloning to produce medical and industrial products, such as human insulin made by bacteria; the development of genetically modified organisms for agriculture, such as crop plants that produce their own insecticide; and even the investigation of historical questions, such as famous mysteries of paternity, from Thomas Jefferson to the modern day. Equally important, DNA technology is invaluable in many areas of biological research, including cancer and evolution.

As we discuss these various applications throughout this chapter, we'll consider the specific techniques involved, how they are applied, and some of the social, legal, and ethical issues that are raised by the new technologies.

Gene Cloning

1 Genes can be cloned in recombinant plasmids

Although it may seem like a modern field, **biotechnology**, the manipulation of organisms or their components to make useful products, actually dates back to the dawn of civilization. Consider such ancient practices as the use of microbes to make beer, wine, and cheese, and the selective breeding of livestock, dogs, and other animals. But when people use the term *biotechnology* today, they are usually referring to **DNA technology**, modern laboratory techniques for studying and manipulating genetic material. Using these techniques, scientists can, for instance, modify specific genes and move them between organisms as different as bacteria, plants, and animals.

The field of DNA technology grew out of discoveries made about 60 years ago by American geneticists Joshua Lederberg and Edward Tatum. They performed a series of experiments with *Escherichia coli* (*E. coli*) that demonstrated that two individual bacteria can combine genes—a phenomenon that was previously thought to be limited to sexually reproducing eukaryotic organisms. With this work, they pioneered bacterial genetics, a field that within 20 years made *E. coli* the most thoroughly studied and understood organism at the molecular level.

In the 1970s, the field of biotechnology exploded with the invention of methods for making recombinant DNA in a test tube. **Recombinant DNA** is formed when scientists combine nucleotide sequences (pieces of DNA) from two different sources—often different species—to form a single DNA molecule. Today, recombinant DNA technology is widely used in the field of **genetic engineering**, the direct manipulation of genes for practical purposes. Scientists have genetically engineered bacteria to mass-produce a variety of useful chemicals, from cancer drugs to pesticides. Scientists have also transferred genes from bacteria into plants and from one animal species into another (**Figure 1A**).

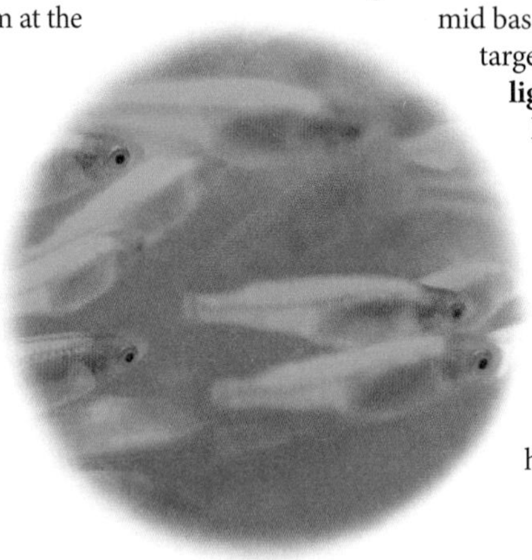

AP Wide World Photos

▲ **Figure 1A** Glowing fish produced by transferring a gene originally obtained from a jelly (cnidarian)

To manipulate genes in the laboratory, biologists often use bacterial **plasmids**, which are small, circular DNA molecules that replicate (duplicate) separately from the much larger bacterial chromosome. Because plasmids can carry virtually any gene and are passed from one generation of bacteria to the next, they are key tools for **gene cloning**, the production of multiple identical copies of a gene-carrying piece of DNA. Gene-cloning methods are central to the production of useful products via genetic engineering.

Consider a typical genetic engineering challenge: A molecular biologist at a pharmaceutical company has identified a gene that codes for a valuable product, a hypothetical substance called protein V. The biologist wants to manufacture large amounts of protein V. **Figure 1B** illustrates how the techniques of gene cloning can be used to accomplish this goal.

To begin, the biologist isolates two kinds of DNA: ❶ a bacterial plasmid that will serve as the **vector**, or gene carrier, and ❷ the DNA containing the gene of interest—in this case, gene *V* (shown in red in the figure)—along with other, unwanted genes. Often, the plasmid comes from the bacterium *E. coli*. The DNA containing gene *V* could come from a variety of sources, such as a different bacterium, a plant, a nonhuman animal, or even human tissue cells growing in laboratory culture.

The researcher treats both the plasmid and the gene *V* source DNA with an enzyme that cuts DNA. ❸ An enzyme is chosen that cleaves the plasmid in only one place. ❹ The other DNA, which is usually much longer in sequence, may be cut into many fragments, one of which carries gene *V*. The figure shows the processing of just one DNA fragment and one plasmid, but actually, millions of plasmids and DNA fragments, most of which do not contain gene *V*, are treated simultaneously. The cuts leave single-stranded ends, as we'll explain in Module 2.

❺ The cut DNA from both sources—the plasmid and target gene—are mixed. The single-stranded ends of the plasmid base-pair with the complementary ends of the target DNA fragment. ❻ The enzyme **DNA ligase** joins the two DNA molecules by covalent bonds. This enzyme, which the cell normally uses in DNA replication, is a "DNA pasting" enzyme that catalyzes the formation of covalent bonds between adjacent nucleotides, joining the strands. The result is a recombinant DNA plasmid containing gene *V*, as well as many other recombinant DNA plasmids carrying other genes not shown here.

❼ The recombinant plasmid containing the targeted gene is mixed with a culture of bacteria. Under the right conditions, a bacterium takes up the plasmid DNA by transformation. ❽ This recombinant bacterium then reproduces to form a **clone** of cells, a group of identical cells descended from a single ancestral cell, each carrying a copy of gene *V*. This step is the actual gene cloning. In our example, the biologist will eventually grow a cell clone large enough to produce protein V in marketable quantities.

❾ Gene cloning can be used to produce a variety of desirable products. Copies of the gene itself can be the immediate product, to be used in further genetic engineering projects. For example, a pest-resistance gene present in one plant species might be cloned and transferred into plants of another species. Other times, the protein product of the cloned gene is harvested and used. For example, an enzyme that creates a faded look in blue jeans can be harvested in large quantities from

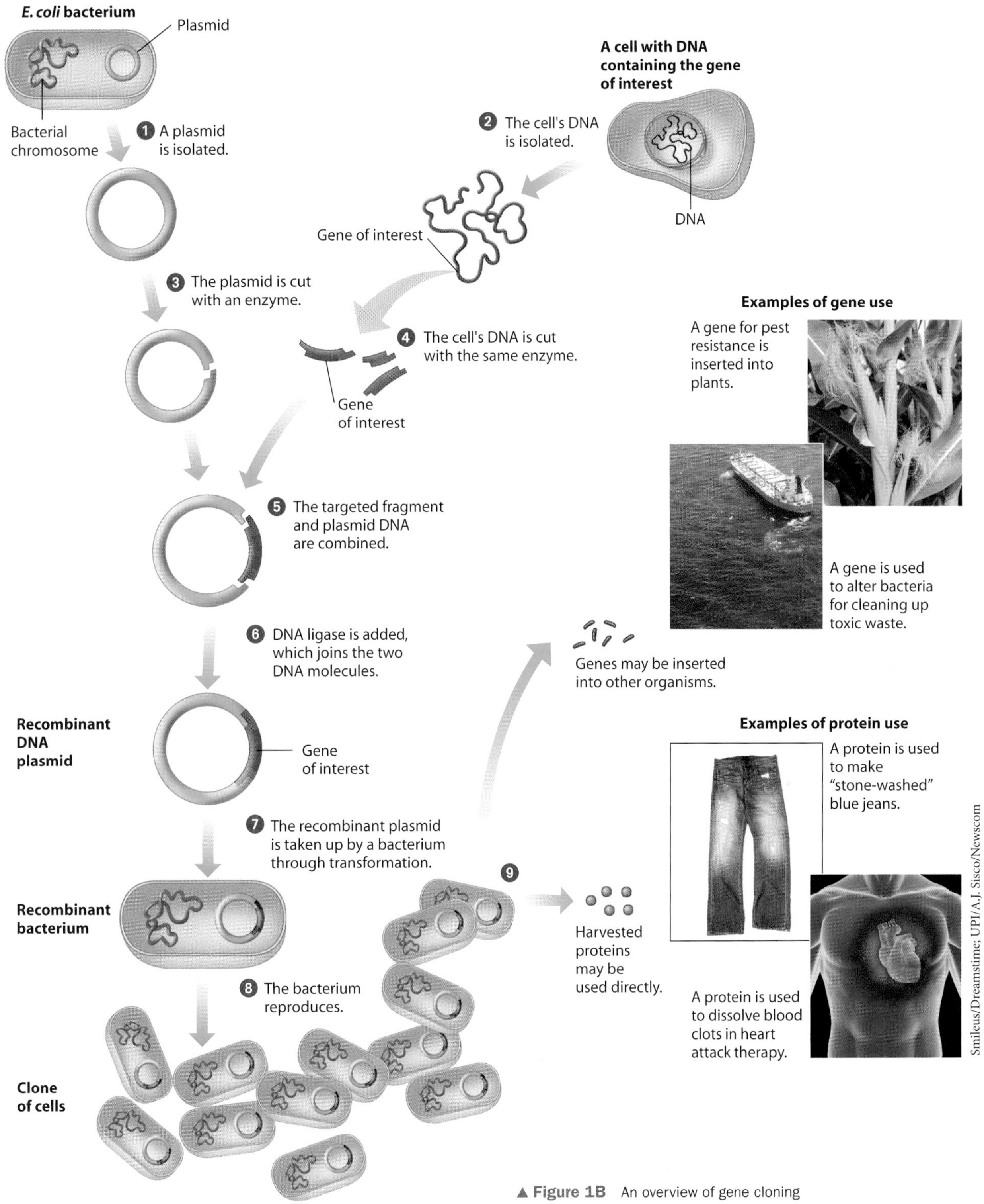

E. coli bacterium
Plasmid
Bacterial chromosome
1 A plasmid is isolated.
A cell with DNA containing the gene of interest
2 The cell's DNA is isolated.
DNA
Gene of interest
3 The plasmid is cut with an enzyme.
4 The cell's DNA is cut with the same enzyme.
Gene of interest
5 The targeted fragment and plasmid DNA are combined.
6 DNA ligase is added, which joins the two DNA molecules.
Recombinant DNA plasmid
Gene of interest
7 The recombinant plasmid is taken up by a bacterium through transformation.
Recombinant bacterium
8 The bacterium reproduces.
Clone of cells
9
Genes may be inserted into other organisms.
Examples of gene use
A gene for pest resistance is inserted into plants.
A gene is used to alter bacteria for cleaning up toxic waste.
Harvested proteins may be used directly.
Examples of protein use
A protein is used to make "stone-washed" blue jeans.
A protein is used to dissolve blood clots in heart attack therapy.
Smileus/Dreamstime; UPI/A.J. Sisco/Newscom

▲ **Figure 1B** An overview of gene cloning

bacteria carrying the cloned gene (that's right: No stones are used to make stone-washed jeans!).

In the next four modules, we discuss the methods outlined in Figure 1B. You may find it useful to turn back to this summary figure as each technique is discussed.

? Why does the rapid reproduction of bacteria make them a good choice for cloning a foreign gene?

A foreign gene located within plasmid DNA inside a bacterium is replicated each time the cell divides, resulting in rapid accumulation of many copies of the gene.

2 Enzymes are used to "cut and paste" DNA

In the gene-cloning procedure outlined in Figure 1B, a recombinant DNA molecule is created by combining two ingredients: a bacterial plasmid and the gene of interest. To understand how these DNA molecules are spliced together, you need to learn how enzymes cut and paste DNA. The cutting tools are bacterial enzymes called **restriction enzymes**. In nature, these enzymes protect bacterial cells against intruding DNA from other organisms or viruses. They work by chopping up the foreign DNA, a process that restricts the ability of the invader to do harm to the bacterium. (The bacterial cell's own DNA is protected from restriction enzymes through chemical modification by other enzymes.)

Biologists have identified hundreds of different restriction enzymes. Each restriction enzyme is specific, recognizing a particular short DNA sequence, usually four to eight nucleotides long. For example, a restriction enzyme called *Eco*RI (found naturally in *E. coli*) only recognizes the DNA sequence GAATTC, whereas the enzyme called *Bam*HI only recognizes GGATCC. The DNA sequence recognized by a particular restriction enzyme is called a **restriction site**. Once a restriction site is recognized, the restriction enzyme cuts both strands of the DNA at specific points within the sequence. All copies of a particular DNA molecule always yield the same set of DNA fragments when exposed to the same restriction enzyme. In other words, a restriction enzyme cuts a DNA molecule in a precise, reproducible way.

Figure 2 ❶ shows a piece of DNA containing one recognition sequence for the restriction enzyme *Eco*RI. In this case, the restriction enzyme cuts each DNA strand between the bases A and G within the sequence, producing pieces of DNA called **restriction fragments.** ❷ Notice that the DNA is cut unevenly; the staggered cuts yield two double-stranded DNA fragments with single-stranded ends, called "sticky ends." Sticky ends are the key to joining DNA restriction fragments originating from different sources because these short extensions can form hydrogen-bonded base pairs with complementary single-stranded stretches of DNA.

❸ A piece of DNA (gray, with the red area showing the gene of interest) from another source is now added. Notice that the gray DNA has single-stranded ends identical in base sequence to the sticky ends on the blue DNA. The gray, "foreign" DNA has ends with this particular base sequence because it was cut from a larger molecule by the same restriction enzyme used to cut the blue DNA. ❹ The complementary ends on the blue and gray fragments allow them to stick together by base pairing. (The hydrogen bonds are not shown in the figure.) This union between the blue and gray DNA fragments is temporary; it can be made permanent by the "pasting" enzyme DNA ligase. ❺ The final outcome is a stable molecule of recombinant DNA.

The ability to cut DNA with restriction enzymes and then paste it back together with DNA ligase is the key to the gene-cloning procedure outlined in Figure 1B and most other modern genetic engineering methods. This particular cloning procedure, which uses a mixture of fragments from the entire genome of an organism, is called a "shotgun" approach. Thousands of different recombinant plasmids are produced, and a clone of each is made. The complete set of plasmid clones, each carrying copies of a particular segment from the initial genome, is a type of library. The next three modules discuss such libraries in more detail.

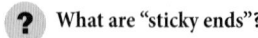

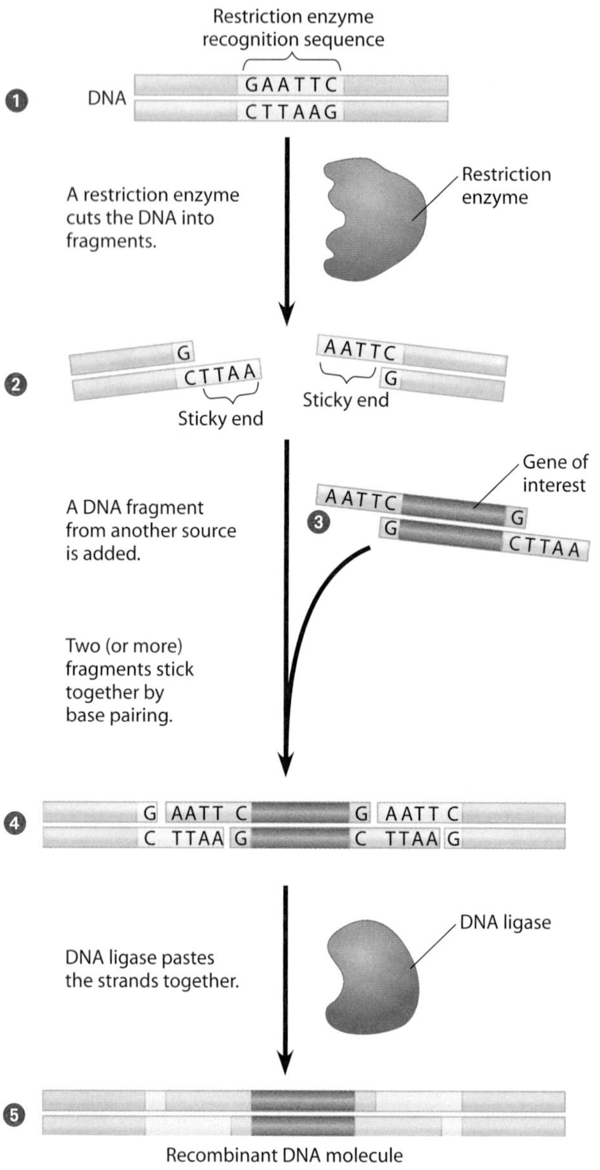

▲ **Figure 2** Creating recombinant DNA using a restriction enzyme and DNA ligase

? **What are "sticky ends"?**

Single-stranded regions whose unpaired bases can hydrogen-bond to the complementary sticky ends of other fragments created by the same restriction enzyme

3 Cloned genes can be stored in genomic libraries

Each bacterial clone from the procedure in Figure 1B consists of identical cells with plasmids carrying one particular fragment of target DNA. The entire collection of all the cloned DNA fragments from a genome is called a **genomic library**. On the left side of **Figure 3**, the red, yellow, and green DNA segments represent three of the thousands of different library "books" that are "shelved" in plasmids inside bacterial cells. A typical cloned DNA fragment is big enough to carry one or a few genes, and together, the fragments include the entire genome of the organism from which the DNA was derived.

Bacteriophages (also called phages)—viruses that infect bacteria—can also serve as vectors when cloning genes (Figure 3, right). When a phage is used, the DNA fragments are inserted into phage DNA molecules. The recombinant phage DNA can then be introduced into a bacterial cell through the normal infection process. Inside the cell, phage DNA is replicated, producing new phage particles, each carrying the foreign DNA. Another type of vector commonly used in library construction is a bacterial artificial chromosome (BAC). BACs are essentially large plasmids containing only the genes necessary to ensure replication. The primary advantage of BACs is that they can carry more foreign DNA than other vectors. In the next module, we look at another source of DNA for cloning: eukaryotic mRNA.

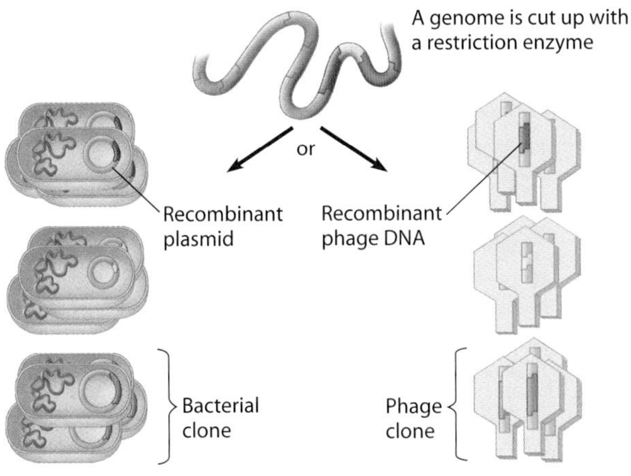

▲ **Figure 3** Genomic libraries

? In what sense does a genomic library have multiple copies of each "book"?

Each "book"—a piece of DNA from the genome that was the source of the library—is present in every recombinant bacterium or phage in a clone.

4 Reverse transcriptase can help make genes for cloning

Rather than starting with an entire eukaryotic genome, a researcher can focus on the genes expressed in a particular kind of cell by using its mRNA as the starting material for cloning. As shown in **Figure 4**, ① the chosen cells transcribe their genes and ② process the transcripts, removing introns and splicing exons together, producing mRNA. ③ The researcher isolates the mRNA and makes single-stranded DNA transcripts from it using the enzyme **reverse transcriptase** (gold in the figure). ④ Another enzyme is added to break down the mRNA, and ⑤ DNA polymerase (the enzyme that replicates DNA) is used to synthesize a second DNA strand.

The double-stranded DNA that results from such a procedure, called **complementary DNA (cDNA)**, represents only the subset of genes that had been transcribed into mRNA in the starting cells. Among other purposes, such a cDNA library is useful for studying the genes responsible for the specialized functions of a particular cell type, such as brain or liver cells. Because cDNAs lack introns, they are shorter than the full versions of the genes and therefore easier to work with.

In the next module, you will learn how to find one particular piece of DNA from among the thousands that are stored in a genomic or cDNA library.

? Why is a cDNA gene made using reverse transcriptase often shorter than the natural form of the gene?

Because cDNAs are made from spliced mRNAs, which lack introns.

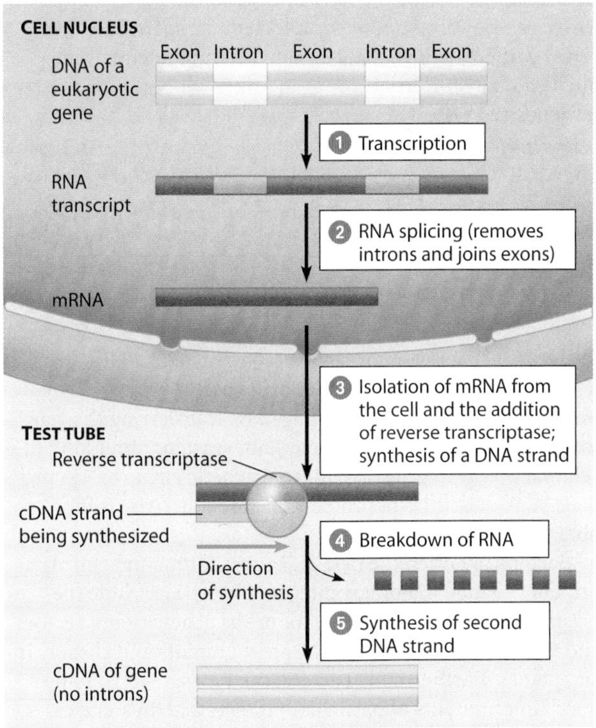

▲ **Figure 4** Making an intron-lacking gene from eukaryotic mRNA

5 Nucleic acid probes identify clones carrying specific genes

Often, the most difficult task in gene cloning is finding the right "books" in a genomic library—that is, identifying clones containing a desired gene from among all those created. For example, a researcher might want to pull out just the clone of bacteria carrying the red gene in Figure 3. If bacterial clones containing a specific gene actually translate the gene into protein, they can be identified by testing for the protein product. However, not every desired gene produces detectable proteins. In such cases, researchers can also test directly for the gene itself.

Methods for detecting a gene directly depend on base pairing between the gene and a complementary sequence on another nucleic acid molecule, either DNA or RNA. When at least part of the nucleotide sequence of a gene is known, this information can be used to a researcher's advantage. Taking a simplified example, if we know that a hypothetical gene contains the sequence TAGGCT, a biochemist can synthesize a short single strand of DNA with the complementary sequence (ATCCGA) and label it with a radioactive isotope or fluorescent tag. This labeled, complementary molecule is called a **nucleic acid probe** because it is used to find a specific gene or other nucleotide sequence within a mass of DNA. (In practice, probe molecules are usually considerably longer than six nucleotides.)

Figure 5 shows how a probe works. The DNA sample to be tested is treated with heat or chemicals to separate the DNA strands. When the radioactive DNA probe is added to these strands, it tags the correct molecules—that is, it finds the correct books in the genomic library—by hydrogen-bonding to the complementary sequence in the gene of interest. Such a probe can be simultaneously applied to many bacterial clones to screen all of them at once for a desired gene.

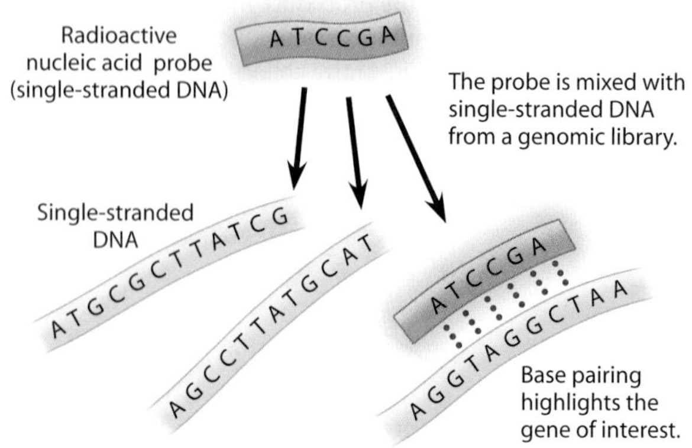

▲ **Figure 5** How a DNA probe tags a gene by base pairing

In one technique, a piece of filter paper is pressed against bacterial colonies (clones) growing on a petri dish. The filter paper picks up cells from each colony. A chemical treatment is used to break open the cells and separate the DNA strands. The DNA strands are then soaked in probe solution. Any bacterial colonies carrying the gene of interest will be tagged on the filter paper, marking them for easy identification. Once the researcher identifies a colony carrying the desired gene, the cells can be grown further, and the gene of interest, or its protein product, can be collected in large amounts.

? How does a probe consisting of radioactive DNA or RNA enable a researcher to find the bacterial clones carrying a particular gene?

The probe molecules bind to and label DNA only from the cells containing the gene of interest, which has a complementary DNA sequence.

Genetically Modified Organisms

6 Recombinant cells and organisms can mass-produce gene products

Recombinant cells and organisms constructed by DNA technology are used to manufacture many useful products, chiefly proteins (**Table 6**, on the next page). By transferring the gene for a desired protein into a bacterium, yeast, or other kind of cell that is easy to grow in culture, a genetic engineer can produce large quantities of proteins that are otherwise difficult to obtain.

Bacteria are often the best organisms for manufacturing a protein product. Major advantages of bacteria include the plasmids and phages available for use as gene-cloning vectors and the fact that bacteria can be grown rapidly and cheaply in large tanks. Furthermore, bacteria can be engineered to produce large amounts of particular proteins and, in some cases, to secrete the proteins directly into their growth medium, simplifying the task of collecting and purifying the products. As Table 6 shows, a number of proteins of importance in human medicine and agriculture are being produced in the bacterium *Escherichia coli*.

Despite the advantages of using bacteria, it is sometimes desirable or necessary to use eukaryotic cells to produce a protein product. Often, the first-choice eukaryotic organism for protein production is the yeast used in making bread and beer, *Saccharomyces cerevisiae*. As bakers and brewers have recognized for centuries, yeast cells are easy to grow. And like *E. coli*, yeast cells can take up foreign DNA and integrate it into their genomes. Yeast cells also have plasmids that can be used as gene vectors, and yeast is often better than bacteria at synthesizing and secreting eukaryotic proteins.

TABLE 6 SOME PROTEIN PRODUCTS OF RECOMBINANT DNA TECHNOLOGY

Product	Made In	Use
Human insulin	*E. coli*	Treatment for diabetes
Human growth hormone (HGH)	*E. coli*	Treatment for growth defects
Epidermal growth factor (EGF)	*E. coli*	Treatment for burns, ulcers
Interleukin-2 (IL-2)	*E. coli*	Possible treatment for cancer
Bovine growth hormone (BGH)	*E. coli*	Improving weight gain in cattle
Cellulase	*E. coli*	Breaking down cellulose for animal feeds
Taxol	*E. coli*	Treatment for ovarian cancer
Interferons (alpha and gamma)	*S. cerevisiae*, *E. coli*	Possible treatment for cancer and viral infections
Hepatitis B vaccine	*S. cerevisiae*	Prevention of viral hepatitis
Erythropoietin (EPO)	Mammalian cells	Treatment for anemia
Factor VIII	Mammalian cells	Treatment for hemophilia
Tissue plasminogen activator (TPA)	Mammalian cells	Treatment for heart attacks and some strokes

S. cerevisiae is currently used to produce a number of proteins. In certain cases, the same product—for example, interferons used in cancer research—can be made in either yeast or bacteria. In other cases, such as the hepatitis B vaccine, yeast alone is used.

The cells of choice for making some gene products come from mammals. Many proteins that mammalian cells normally secrete are glycoproteins, proteins with chains of sugars attached. Because only mammalian cells can attach the sugars correctly, mammalian cells must be used for making these products. For example, recombinant mammalian cells growing in laboratory cultures are currently used to produce human erythropoietin (EPO), a hormone that stimulates the production of red blood cells. EPO is used as a treatment for anemia, but the drug is also abused by some athletes who seek the advantage of artificially high levels of oxygen-carrying red blood cells (called "blood doping"; see Module 23.13).

Recently, pharmaceutical researchers have been exploring the mass production of gene products by whole animals or plants rather than cultured cells. Genetic engineers have used recombinant DNA technology to insert genes for desired human proteins into other mammals, where the protein encoded by the recombinant gene may secreted in the animal's milk. For example, a gene for antithrombin—a human protein that helps prevent inappropriate blood clotting—has been inserted into the genome of a goat (**Figure 6A**); isolated from the milk, the protein can be administered to patients with a rare hereditary disorder in which this protein is lacking. The pig in **Figure 6B** has been genetically modified to produce human hemoglobin; this vital blood protein can be supplied to patients via blood transfusion.

Brad DeCecco Photography

▲ **Figure 6A** A goat carrying a gene for a human blood protein that is secreted in the milk

However, genetically engineered animals are difficult and costly to produce. Typically, a biotechnology company starts by injecting the desired DNA into a large number of embryos, which are then implanted into surrogate mothers. With luck, one or a few recombinant animals may result; success rates for such procedures are very low. Once a recombinant organism is successfully produced, it may be cloned. The result can be a genetically identical herd—a grazing pharmaceutical "factory" of "pharm" animals that produce otherwise rare biological substances for medical use.

We continue an exploration of the medical applications of DNA technology in the next module.

? Why can't glycoproteins be mass-produced by engineered bacteria or yeast cells?

Because bacteria and yeast cells cannot correctly attach the sugar groups to the protein of glycoproteins

DNX/Photolibrary

▲ **Figure 6B** A pig that has been genetically modified to produce a useful human protein

CONNECTION

7 DNA technology has changed the pharmaceutical industry and medicine

DNA technology, and gene cloning in particular, is widely used to produce medicines and to diagnose diseases.

Therapeutic Hormones Consider the first two products in Table 6 on the previous page—human insulin and human growth hormone. About 2 million people with diabetes in the United States depend on insulin treatment. Before 1982, the main sources of this hormone were slaughtered pigs and cattle. Insulin extracted from these animals is chemically similar, but not identical, to human insulin, and it causes harmful side effects in some people. Genetic engineering has largely solved this problem by developing bacteria that synthesize and secrete the human form of insulin. In 1982, Humulin (**Figure 7A**)—human insulin produced by bacteria—became the first recombinant DNA drug approved by the U.S. Food and Drug Administration.

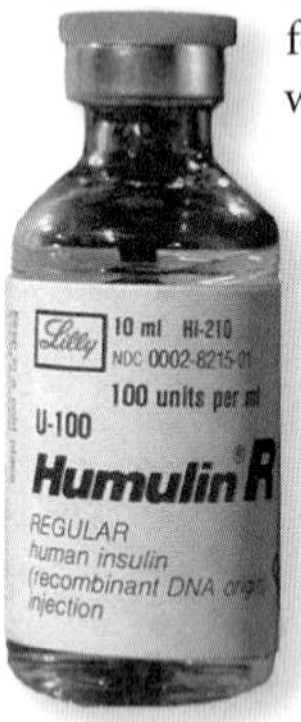

Eli Lilly and Company. Used with permission

▲ **Figure 7A** Human insulin produced by bacteria

Treatment with human growth hormone (HGH) is a boon to children born with a form of dwarfism caused by inadequate amounts of HGH. Because growth hormones from other animals are not effective in humans, children with HGH deficiency historically have had to rely on scarce supplies from human cadavers or else face dwarfism. In 1985, however, molecular biologists made an artificial gene for HGH by joining a human DNA fragment to a chemically synthesized piece of DNA; using this gene, they were able to produce HGH in *E. coli*. HGH from recombinant bacteria is now widely used.

Another important pharmaceutical product produced by genetic engineering is tissue plasminogen activator (TPA). If administered soon after a heart attack, this protein helps dissolve blood clots and reduces the risk of subsequent heart attacks.

Diagnosis of Disease DNA technology is being used increasingly to diagnose disease. Among the hundreds of genes for human diseases that have been identified are those for sickle-cell disease, hemophilia, cystic fibrosis, and Huntington's disease. Affected individuals with such diseases often can be identified before the onset of symptoms, even before birth. It is also possible to identify symptomless carriers of potentially harmful recessive alleles. Additionally, DNA technology can pinpoint infections. For example, DNA analysis can help track down and identify elusive viruses such as HIV, the virus that causes AIDS.

Vaccines DNA technology is also helping medical researchers develop vaccines. A **vaccine** is a harmless variant (mutant) or derivative of a pathogen—usually a bacterium or virus—that is used to stimulate the immune system to mount a lasting defense against that pathogen. For many viral diseases, prevention by vaccination is the only medical way to prevent illness.

Genetic engineering can be used in several ways to make vaccines. One approach is to use genetically engineered cells or organisms to produce large amounts of a protein molecule that is found on the pathogen's outside surface. This method has been used to make the vaccine against the hepatitis B virus. Hepatitis is a disabling and sometimes fatal liver disease, and the hepatitis B virus may also cause liver cancer. **Figure 7B** shows a tank for growing yeast cells that have been engineered to carry the gene for the virus's surface protein. Made by the yeast, this protein will be the main ingredient of the vaccine.

Another way to use DNA technology in vaccine development is to make a harmless artificial mutant of the pathogen by altering one or more of its genes. When a harmless mutant is used as a so-called "live vaccine," it multiplies in the body and may trigger a strong immune response. Artificial-mutant vaccines may cause fewer side effects than vaccines that have traditionally been made from natural mutants.

Yet another method for making vaccines employs a virus related to the one that causes smallpox. Smallpox was once a dreaded human disease, but it was eradicated worldwide in the 1970s by widespread vaccination with a harmless variant of the smallpox virus. Using this harmless virus, genetic engineers could replace some of the genes encoding proteins that induce immunity to smallpox with genes that induce immunity to other diseases. In fact, the virus could be engineered to carry genes needed to vaccinate against several diseases simultaneously. In the future, one inoculation may prevent a dozen diseases.

Genetic engineering rapidly transformed the field of medicine and continues to do so today. But genetically modified organisms affect our lives in other ways, as we'll see next.

? Human growth hormone and insulin produced by DNA technology are used in the treatment of _______ and _______, respectively.

dwarfism . . . diabetes

Hank Morgan/Photo Researchers

▲ **Figure 7B** Equipment used in the production of a vaccine against hepatitis B

CONNECTION

8 Genetically modified organisms are transforming agriculture

Since ancient times, people have selectively bred agricultural crops to make them more useful. Today, DNA technology is quickly replacing traditional breeding programs as scientists work to improve the productivity of agriculturally important plants and animals. Genetic engineers have produced many varieties of **genetically modified (GM) organisms**, organisms that have acquired one or more genes by artificial means. If the newly acquired gene is from another organism, typically of another species, the recombinant organism is called a **transgenic organism**.

The most common vector used to introduce new genes into plant cells is a plasmid from the soil bacterium *Agrobacterium tumefaciens* called the **Ti plasmid** (**Figure 8A**). ❶ With the help of a restriction enzyme and DNA ligase, the gene for the desired trait (indicated in red in the figure) is inserted into a modified version of the plasmid. ❷ Then the recombinant plasmid is put into a plant cell, where the DNA carrying the new gene integrates into the plant chromosome. ❸ Finally, the recombinant cell is cultured and grown into a plant.

With an estimated 1 billion people facing malnutrition, GM crops may be able to help a great many hungry people by improving food production, pest resistance, and the nutritional value of crops. For example, in India, the insertion of a salinity-resistance gene has enabled new varieties of rice to grow in water three times as salty as seawater. Similar research is under way in Australia to help improve wheat yields in salty soil. In Hawaii, the ring spot virus seemed poised to devastate the papaya industry until a GM variety resistant to the virus was introduced in 1992. Golden Rice, a transgenic variety created in 2000 with a few daffodil genes, produces yellow grains containing beta-carotene, which our body uses to make vitamin A. A new strain (Golden Rice 2) uses corn genes to boost beta-carotene levels even higher (**Figure 8B**). This rice could help prevent vitamin A deficiency—and the resulting blindness—among the half of the world's people who depend on rice as their staple food.

In addition to agricultural applications, genetic engineers are now creating plants that make human proteins for medical use. A recently developed transgenic rice strain harbors genes for milk proteins that can be used in rehydration formulas to treat infant diarrhea, a serious problem in developing countries. Other pharmaceutical trials currently under way involve using modified corn to treat cystic fibrosis, safflower to treat diabetes, and duckweed to treat hepatitis. Although these trials seem promising, no plant-made drugs intended for use by humans have yet to be approved or sold.

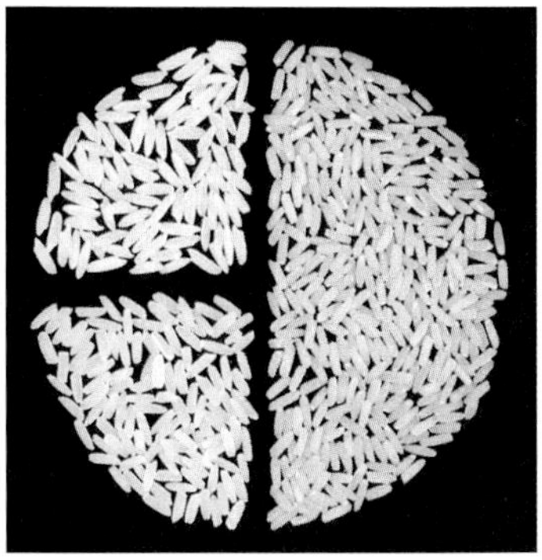
Syngenta Corporate Communications

▲ **Figure 8B** A mix of conventional rice (white), the original Golden Rice (light gold), and Golden Rice 2 (dark gold)

Agricultural researchers are also producing transgenic animals, as mentioned in Module 6. To do this, scientists remove egg cells from a female and fertilize them. They then inject a previously cloned gene directly into the nuclei of the fertilized eggs. Some of the cells integrate the foreign DNA into their genomes. The engineered embryos are then surgically implanted in a surrogate mother. If an embryo develops successfully, the result is an animal containing a gene from a third "parent," which may even be of another species.

The goals in creating a transgenic animal are often the same as the goals of traditional breeding—for instance, to make a sheep with better quality wool or a cow that will mature in less time. In 2006, researchers succeeded in transferring a fat metabolism gene from a roundworm into a pig. Meat from the resulting swine had levels of healthy omega-3 fatty acids—which are believed to reduce the risk of heart disease—four to five times higher than meat from normal pigs. Atlantic salmon have been genetically modified by the addition of a more active growth hormone gene from Chinook salmon. Such fish can grow to market size in about half the time of conventional salmon. As of 2010, the FDA was considering whether to grant approval for the modified salmon to be sold as food. To date, the vast majority of the GM organisms that contribute to our food supply are not animals, but crop plants. As we'll discuss next, some people question whether such genetically modified organisms are beneficial to our society.

? What is the function of the Ti plasmid in the creation of transgenic plants?

It is used as the vector for introducing foreign genes into a plant cell.

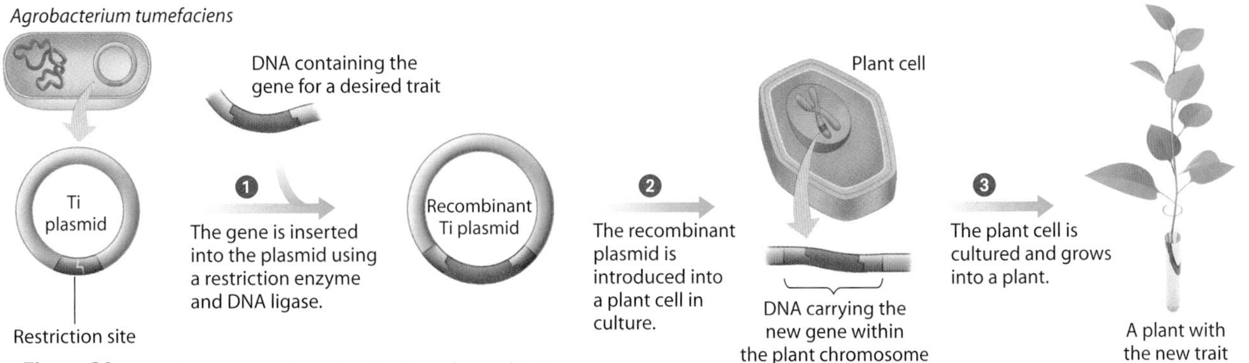

Agrobacterium tumefaciens
DNA containing the gene for a desired trait
Ti plasmid
Restriction site
❶ The gene is inserted into the plasmid using a restriction enzyme and DNA ligase.
Recombinant Ti plasmid
❷ The recombinant plasmid is introduced into a plant cell in culture.
Plant cell
DNA carrying the new gene within the plant chromosome
❸ The plant cell is cultured and grows into a plant.
A plant with the new trait
Vladimir Nikitin/Shutterstock

▲ **Figure 8A** Using the Ti plasmid to genetically engineer plants

9 Genetically modified organisms raise concerns about human and environmental health

As soon as scientists realized the power of DNA technology, they began to worry about potential dangers. Early concerns focused on the possibility that recombinant DNA technology might create new pathogens. What might happen, for instance, if cancer cell genes were transferred into infectious bacteria or viruses? To guard against such rogue microbes, scientists developed a set of guidelines that were adopted as formal government regulations in the United States and some other countries. One safety measure is a set of strict laboratory procedures designed to protect researchers from infection by engineered microbes and to prevent the microbes from accidentally leaving the laboratory (**Figure 9A**). In addition, strains of microorganisms to be used in recombinant DNA experiments are genetically crippled to ensure that they cannot survive outside the laboratory. Finally, certain obviously dangerous experiments have been banned.

Today, most public concern about possible hazards centers not on recombinant microbes but on genetically modified organisms (GMOs) used for food. Some fear that crops carrying genes from other species might be hazardous to human health or the environment. Others fear that the protein products of transplanted genes might lead to allergic reactions.

About a decade ago, negotiators from 130 countries, including the United States, agreed on a Biosafety Protocol that requires all exporters to identify GM organisms present in bulk food shipments and allows importing countries to decide whether they pose environmental or health risks. Although the majority of several staple crops grown in the United States—including corn and soybeans—are genetically modified, products made from GMOs are not required to be labeled in any way. Chances are, you eat a food containing GMOs nearly every day, but the lack of labeling means there is little chance that you would be able to say for certain. However, labeling of foods containing more than trace amounts of GMOs is required in Europe, Japan, Australia, and some other countries. Critics of GM crops point out that labeling would allow consumers to decide for themselves whether or not they wish to be exposed to GM foods. Some biotechnology advocates, however, respond that similar demands were not made when "transgenic" crop plants produced by traditional breeding techniques were put on the market. For example, triticale was created decades ago by combining the genomes of wheat and rye—two plants that do not interbreed in nature. Triticale is now grown worldwide.

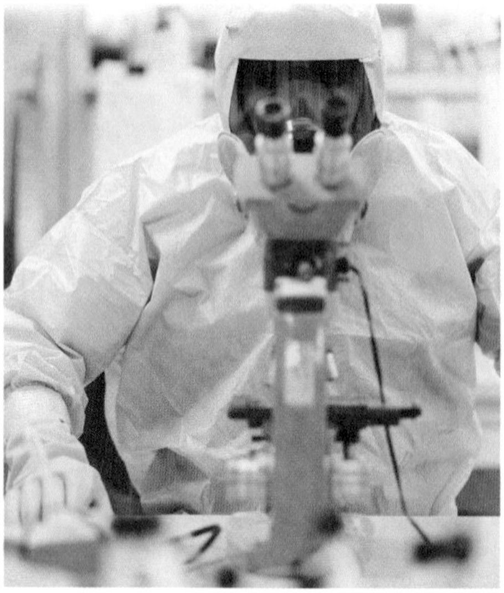

▶ Figure 9A A maximum-security laboratory at the Pasteur Institute in Paris

Ocean/Corbis

Advocates of a cautious approach toward GM crops also fear that transgenic plants might pass their new genes to close relatives in nearby wild areas (**Figure 9B**). We know that lawn and crop grasses, for example, commonly exchange genes with wild relatives via pollen transfer. If crop plants carrying genes for resistance to herbicides, diseases, or insect pests pollinated wild ones, the offspring might become "superweeds" that would be very difficult to control. In 2003, the U.S. Department of Agriculture imposed multimillion-dollar fines and tightened rules for tests involving GM plants after leftover corn plants engineered to make a pig vaccine popped up in a soybean field in Nebraska. Concern has also been raised that the widespread use of GM seeds may reduce natural genetic diversity, leaving crops susceptible to catastrophic die-offs in the event of a sudden change to the environment or introduction of a new pest.

Dave Porter/Alamy

▲ Figure 9B Genetically engineered crop plants growing near their wild relatives

Today, governments and regulatory agencies throughout the world are grappling with how to facilitate the use of biotechnology in agriculture, industry, and medicine while ensuring that new products and procedures are safe. In the United States, such applications of biotechnology are evaluated for potential risks by multiple government agencies. Meanwhile, these same agencies and the public must consider the ethical implications of biotechnology.

In the case of GM plants and certain other applications of DNA technology, zero risk is probably unattainable. Scientists and the public need to weigh the possible benefits versus risks on a case-by-case basis. The best scenario would be for us to proceed with caution, basing our decisions on sound scientific information rather than on either irrational fear or blind optimism.

? What is one of the concerns about engineering crop plants by adding genes for herbicide resistance?

● The possibility that the genes could escape via cross-pollination to weeds that are closely related to the crop species

CONNECTION 10 Gene therapy may someday help treat a variety of diseases

In this chapter, we have discussed transgenic viruses, bacteria, yeast, plants, and animals. What about transgenic humans? Why would anyone want to insert genes into a living person?

One reason to tamper with the human genome is the potential for treating a variety of diseases by **gene therapy**—alteration of an afflicted individual's genes for therapeutic purposes. In people with disorders traceable to a single defective gene, it might be possible to replace or supplement the defective gene by inserting a normal allele into cells of the tissue affected by the disorder. Once there, the normal allele may be expressed, potentially curing the disease after just a single treatment.

For gene therapy to be permanent, the normal allele would have to be transferred to cells that multiply throughout a person's life. Bone marrow cells, which include the stem cells that give rise to all the cells of the blood and immune system, are prime candidates. **Figure 10** outlines one possible procedure for gene therapy. ❶ The normal gene is cloned, converted to an RNA version, and then inserted into the RNA genome of a harmless retrovirus vector. ❷ Bone marrow cells are taken from the patient and infected with the virus. ❸ The virus inserts a DNA version of its genome, including the normal human gene, into the cells' DNA ❹ The engineered cells are then injected back into the patient. If the procedure succeeds, the cells will multiply throughout the patient's life and produce a steady supply of the missing protein, curing the patient.

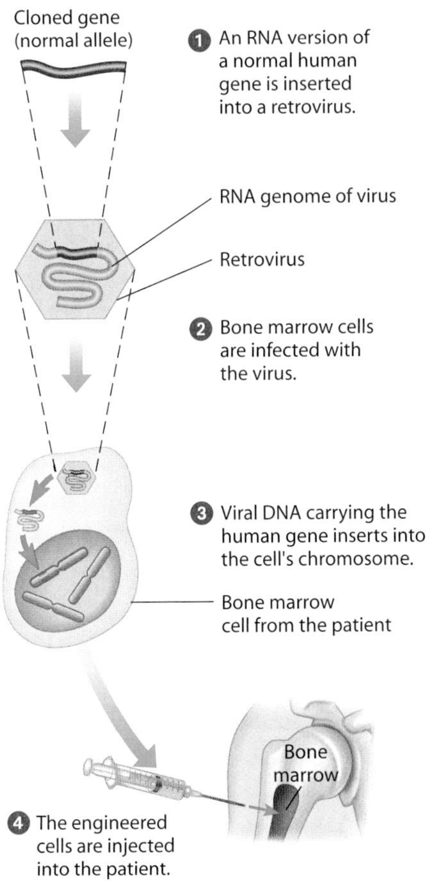
Cloned gene
(normal allele)
❶ An RNA version of a normal human gene is inserted into a retrovirus.
RNA genome of virus
Retrovirus
❷ Bone marrow cells are infected with the virus.
❸ Viral DNA carrying the human gene inserts into the cell's chromosome.
Bone marrow cell from the patient
Bone marrow
❹ The engineered cells are injected into the patient.

▲ **Figure 10** One type of gene therapy procedure

The first successful human gene therapy trial, begun in 2000, used this method to treat 10 young children with severe combined immunodeficiency disease (SCID), a disorder in which the patient lacks a functional immune system. Nine of these patients showed significant improvement, providing the first definitive success of gene therapy. However, three of the patients subsequently developed leukemia, a cancer of the blood cells, and one died. Researchers discovered that in two of the cases, the inserted DNA appeared to disrupt a gene involved in proliferation and development of blood cells. This insertion somehow caused the leukemia. Active research into treating SCID continues with new, tougher safety guidelines.

A 2009 gene therapy trial involved a disease called Leber's congenital amaurosis (LCA). People with one form of LCA have a defective version of a gene needed to produce rhodopsin, a pigment that enables the eye to detect light. In such people, photoreceptor cells gradually die, causing progressive blindness. An international research team found that a single injection—containing a virus carrying the normal gene—into one eye of affected children improved vision in that eye, sometimes enough to allow normal functioning.

The use of gene therapy raises several technical questions. For example, how can researchers build in gene control mechanisms to ensure that cells with the transferred gene make appropriate amounts of the gene product at the right time and in the right parts of the body? And how can they be sure that the gene's insertion does not harm the cell's normal function?

In addition to technical challenges, gene therapy raises difficult ethical questions. Some critics suggest that tampering with human genes in any way will inevitably lead to the practice of eugenics, the deliberate effort to control the genetic makeup of human populations. Other observers see no fundamental difference between the transplantation of genes into somatic cells and the transplantation of organs.

The implications of genetically manipulating gamete-forming cells or zygotes (already accomplished in lab animals) are more problematic. This possibility raises the most difficult ethical questions of all: Should we try to eliminate genetic defects in our children and their descendants? Should we interfere with evolution in this way? From a biological perspective, the elimination of unwanted alleles from the gene pool could backfire. Genetic variety is a necessary ingredient for the survival of a species as environmental conditions change with time. Genes that are damaging under some conditions may be advantageous under others (one example is the sickle-cell allele). Are we willing to risk making genetic changes that could be detrimental to our species in the future? We may have to face this question soon.

? What characteristic of bone marrow makes those cells good targets for gene therapy?

They multiply throughout a person's life.

DNA Profiling

11 The analysis of genetic markers can produce a DNA profile

Modern DNA technology methods have rapidly transformed the field of **forensics**, the scientific analysis of evidence for crime scene investigations and other legal proceedings. The most important application to forensics is **DNA profiling**, the analysis of DNA samples to determine whether they came from the same individual.

How do you prove that two samples of DNA come from the same person? You could compare the entire genomes found in the two samples, but such an approach would be extremely impractical, requiring a lot of time and money. Instead, scientists compare genetic markers, sequences in the genome that vary from person to person. Like a gene, which is one type of genetic marker, a genetic marker within a noncoding stretch of DNA is more likely to be a match between relatives than between unrelated individuals.

Figure 11 summarizes the basic steps in creating a DNA profile. ❶ First, DNA samples are isolated from the crime scene, suspects, victims, or stored evidence. ❷ Next, selected markers from each DNA sample are amplified (copied many times), producing an adequate supply for testing. ❸ Finally, the amplified DNA markers are compared, proving which samples were derived from the same individual. In the next four modules, we'll explore the methods behind these steps in detail.

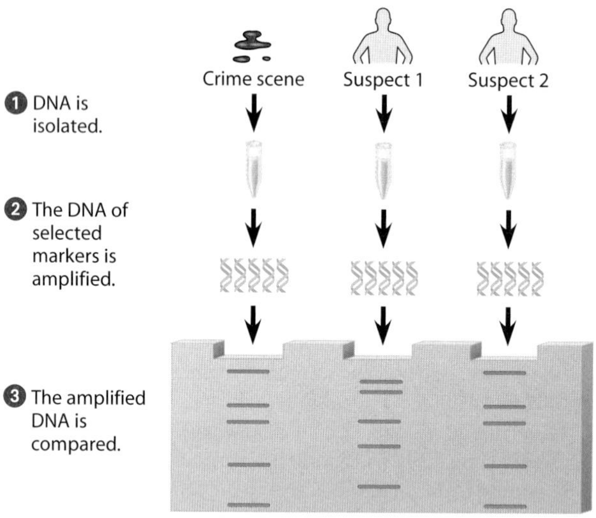
Crime scene
Suspect 1
Suspect 2
❶ DNA is isolated.
❷ The DNA of selected markers is amplified.
❸ The amplified DNA is compared.

▲ **Figure 11** An overview of DNA profiling

? According to the data presented in Figure 11, which suspect left DNA at the crime scene?

Suspect 2; notice that the number and location of the DNA markers match between suspect 2's DNA and the crime scene DNA.

12 The PCR method is used to amplify DNA sequences

Cloning DNA in host cells is often the best method for preparing large quantities of DNA from a particular gene (see Module 1). However, when the source of DNA is scant or impure, the polymerase chain reaction is a much better method. The **polymerase chain reaction (PCR)** is a technique by which a specific segment of a DNA molecule can be targeted and quickly amplified in the laboratory. Starting with a minute sample of blood or other tissue, automated PCR can generate billions of copies of a DNA segment in just a few hours, producing enough DNA to allow a DNA profile to be constructed.

In principle, PCR is fairly simple (**Figure 12**). A repeated, three-step cycle brings about a chain reaction that doubles the population of identical DNA molecules during each round. The key to amplifying one particular segment of DNA and no others

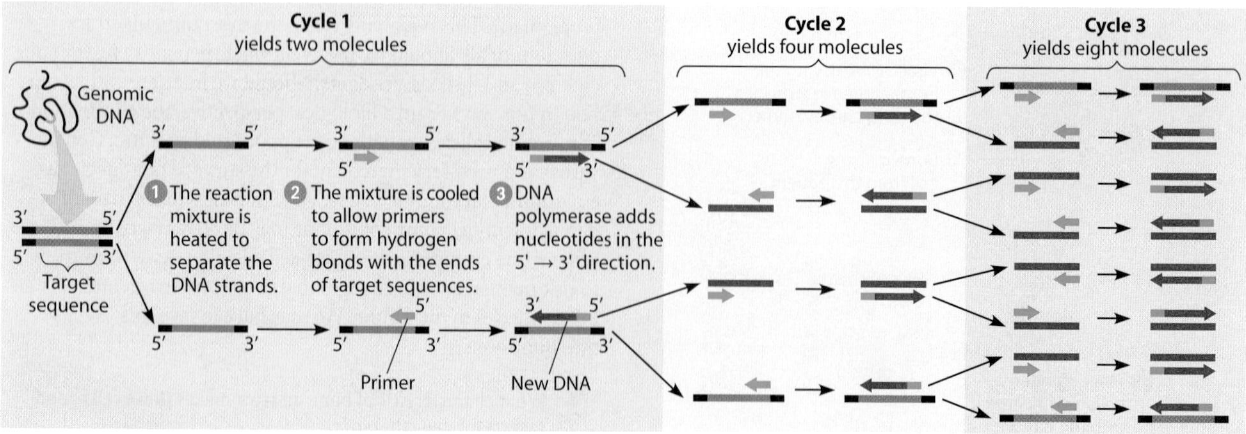
Cycle 1
yields two molecules
Cycle 2
yields four molecules
Cycle 3
yields eight molecules
Genomic DNA
Target sequence
❶ The reaction mixture is heated to separate the DNA strands.
❷ The mixture is cooled to allow primers to form hydrogen bonds with the ends of target sequences.
❸ DNA polymerase adds nucleotides in the 5' → 3' direction.
Primer
New DNA

▲ **Figure 12** DNA amplification by PCR

is the use of **primers**, short (usually 15 to 20 nucleotides long), chemically synthesized single-stranded DNA molecules with sequences that are complementary to sequences at each end of the target sequence. One primer is complementary to one strand at one end of the target sequence; the second primer is complementary to the other strand at the other end of the sequence. The primers thus bind to sequences that flank the target sequence, marking the start and end points for the segment of DNA being amplified.

1 In the first step of each PCR cycle, the reaction mixture is heated to separate the strands of the DNA double helices. 2 Next, the strands are cooled. As they cool, primer molecules hydrogen-bond to their target sequences on the DNA. 3 In the third step, a heat-stable DNA polymerase builds new DNA strands by extending the primers in the $5' \rightarrow 3'$ direction. These three steps are repeated over and over, doubling the amount of DNA after each three-step cycle. A key prerequisite for automating PCR was the discovery of an unusual DNA polymerase, first isolated from a bacterium living in hot springs, that could withstand the heat at the start of each cycle. Without such a heat-stable polymerase, PCR would not be possible because standard DNA polymerases would denature (unfold) during the heating step of each cycle.

Just as impressive as the speed of PCR is its sensitivity. Only minute amounts of DNA need to be present in the starting material, and this DNA can even be in a partially degraded state. The key to the high sensitivity is the primers. Because the primers only bind the sequences associated with the target, the DNA polymerase duplicates only the desired segments of DNA. Other DNA will not be bound by primers and thus not copied by the DNA polymerase.

Devised in 1985, PCR has had a major impact on biological research and biotechnology. PCR has been used to amplify DNA from a wide variety of sources: fragments of ancient DNA from a mummified human, a 40,000-year-old frozen woolly mammoth, and a 30-million-year-old plant fossil; DNA from fingerprints or from tiny amounts of blood, tissue, or semen found at crime scenes; DNA from single embryonic cells for rapid prenatal diagnosis of genetic disorders; and DNA of viral genes from cells infected with viruses that are difficult to detect, such as HIV.

? Why is amplification of DNA from a crime scene often necessary?

The DNA from a crime scene is often scant and impure.

13 Gel electrophoresis sorts DNA molecules by size

Many approaches for studying DNA molecules in the lab make use of **gel electrophoresis**. A gel is a thin slab of jellylike material often made from agarose, a carbohydrate polymer extracted from seaweed. Because agarose contains a dense tangle of cable-like threads (similar to the structure of fiberglass, and resembling a jungle dense with vines), it can act as a molecular sieve that separates macromolecules—usually proteins or nucleic acids—on the basis of size, electrical charge, or other physical properties.

Figure 13 outlines how gel electrophoresis can be used to separate mixtures of DNA fragments obtained from three different sources. A DNA sample from each source is placed in a separate well (or hole) at one end of a flat, rectangular gel. A negatively charged electrode from a power supply is attached near the end of the gel containing the DNA, and a positive electrode is attached near the other end. Because all nucleic acid molecules carry negative charges on their phosphate groups (PO_4^-), the DNA molecules all travel through the gel toward the positive pole. However, longer DNA fragments are held back by the thicket of polymer fibers within the gel, so they move more slowly than the shorter fragments. Over time, shorter molecules move farther through the gel than longer fragments. Gel electrophoresis thus separates DNA fragments by length, with shorter molecules migrating toward the bottom faster than longer molecules.

When the current is turned off, a series of bands is left in each "lane" of the gel. Each band is a collection of DNA fragments of the same length. The bands can be made visible by staining, by exposure onto photographic film (if the DNA is radioactively labeled), or by measuring fluorescence (if the DNA is labeled with a fluorescent dye).

? What causes DNA molecules to move toward the positive pole during electrophoresis? Why do large molecules move more slowly than smaller ones?

The negatively charged phosphate groups of the DNA are attracted to the positive pole; the gel restricts the movement of longer fragments more.

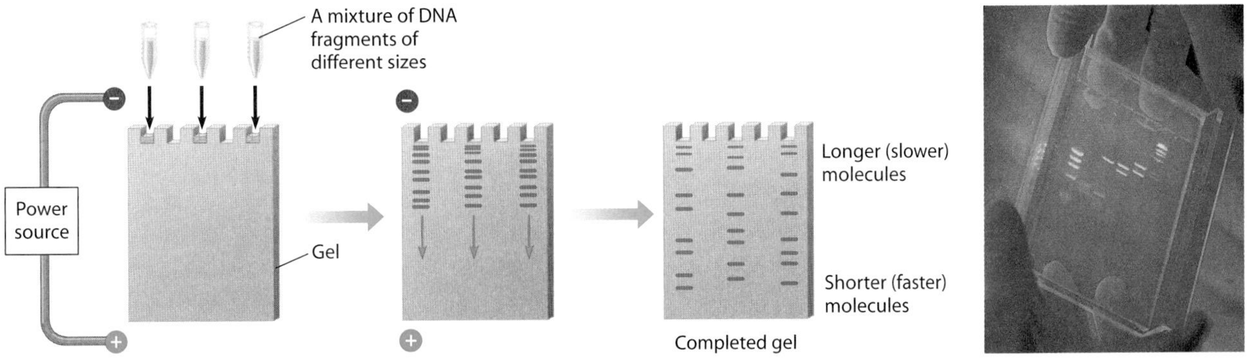

Repligen Corporation

▲ **Figure 13** Gel electrophoresis of DNA

14 STR analysis is commonly used for DNA profiling

If you take another look at the overview of DNA profiling in Figure 11, you will see that we have learned about DNA amplification by PCR (step 2) and gel electrophoresis (step 3). Now, let's put them together to see how a DNA profile is made.

To create a DNA profile, a forensic scientist must compare genetic markers from two or more DNA samples. The genetic markers most often used in DNA profiling are inherited variations in the lengths of repetitive DNA segments. **Repetitive DNA** consists of nucleotide sequences that are present in multiple copies in the genome; much of the DNA that lies between genes in humans is of this type. Some regions of repetitive DNA vary considerably from one individual to the next.

For DNA profiling, the relevant type of repetitive DNA consists of short sequences repeated many times in a row; such a series of repeats is called a **short tandem repeat (STR)**. For example, one person might have the sequence AGAT repeated 12 times in a row at one place in the genome, the sequence GATA repeated 45 times in a row at a second place, and so on. Another person is likely to have the same sequences at the same places but with different numbers of repeats. These stretches of repetitive DNA, like any genetic marker, are more likely to be an exact match between relatives than between unrelated individuals.

STR analysis is a method of DNA profiling that compares the lengths of STR sequences at specific sites in the genome. Most commonly, STR analysis compares the number of repeats of specific four-nucleotide DNA sequences at 13 sites scattered throughout the genome. Each of these repeat sites, which typically contain from 3 to 50 four-nucleotide repeats in a row, vary widely from person to person. In fact, some of the short tandem repeats used in the standard procedure can be found in up to 80 different variations in the human population.

Consider the two samples of DNA shown in **Figure 14A**, where the top DNA was obtained at a crime scene and the bottom DNA from a suspect's blood. The two segments have the same number of repeats at the first site: 7 repeats of the four-nucleotide DNA sequence AGAT (shown in orange). Notice, however, that they differ in the number of repeats at the second site: 8 repeats of GATA (shown in purple) in the crime scene DNA, compared with 13 repeats in the suspect's DNA. To create a DNA profile, a scientist uses PCR to specifically amplify the regions of DNA that include these STR sites. This can be done by using primers matching nucleotide sequences known to flank the STR sites. The resulting DNA molecules are then compared by gel electrophoresis.

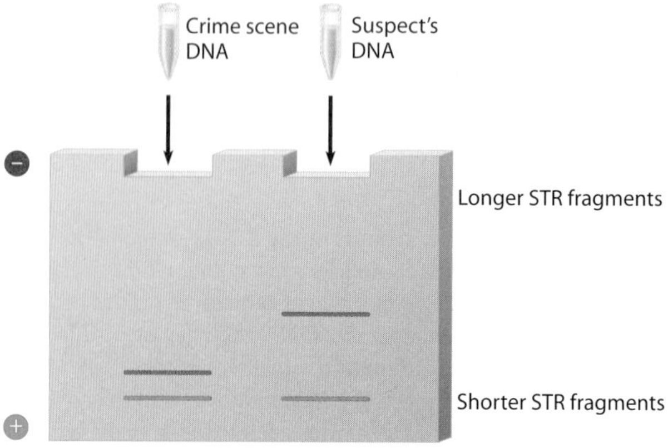

▲ **Figure 14B** DNA profiles generated from the STRs in Figure 14A

Figure 14B shows a gel that could have resulted from the STR fragments in Figure 14A. The differences in the locations of the bands reflect the different lengths of the DNA fragments. (A gel from an actual DNA profile would typically contain more than just two bands in each lane.) This gel would provide evidence that the crime scene DNA did not come from the suspect. Notice that electrophoresis allows us to see similarities as well as differences between mixtures of DNA molecules. Thus, data from DNA profiling can provide evidence of either innocence or guilt.

Although other methods have been used in the past, STR analysis of 13 predetermined STR sites is the current standard for DNA profiling in forensic and legal systems. Once determined, the number of repeats at each of these sites can be entered into the Combined DNA Index System (CODIS) database, administered by the Federal Bureau of Investigation. Within the human population, so much variation exists within the 13 standard sites that a DNA profile made from them can definitely identify a single person from within the entire human population. In the next module, we'll examine several real-world examples of how this technology has been used.

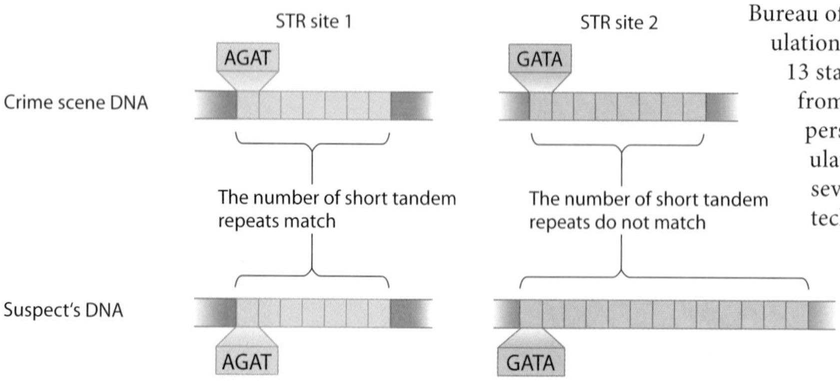

▲ **Figure 14A** Two representative STR sites from crime scene DNA samples

? **What are STRs? What is STR analysis?**

STRs are regions of the genome that contain varying numbers of in-a-row repeats of a short nucleotide sequence; STR analysis is a technique for determining whether two DNA samples have identical STRs.

CONNECTION

15 DNA profiling has provided evidence in many forensic investigations

When a violent crime is committed, body fluids or small pieces of tissue may be left at the crime scene or on the clothes of the victim or assailant. If rape has occurred, semen may be recovered from the victim. DNA profiling can match such samples to the person they came from with a high degree of certainty because the DNA sequence of every person is unique, except for identical twins. And with PCR amplification of DNA, a tissue sample as small as 20 cells can be sufficient for testing.

Since its introduction in 1986, DNA profiling has become a standard tool of forensics and has provided crucial evidence in many famous cases. In the O. J. Simpson murder trial, DNA analysis proved that blood in Simpson's car belonged to the victims and that blood at the crime scene belonged to Simpson. (The jury in this case did not find the DNA evidence alone to be sufficient and Simpson was found not guilty.) During the investigation that led to his impeachment, President Bill Clinton repeatedly denied having sexual relations with Monica Lewinsky—until DNA profiling proved that his semen was on her dress.

Of course, DNA evidence can prove innocence as well as guilt. The Innocence Project, a nonprofit organization dedicated to overturning wrongful convictions, has used DNA technology and legal work to exonerate over 260 convicted criminals since 1989, including 15 on death row (**Figure 15A**). In more than a third of these cases, DNA profiling also identified the true perpetrators.

Steve Helber/AP Wide World Photos

▲ **Figure 15A** STR analysis proved that convicted murderer Earl Washington was innocent, freeing him after 17 years in prison.

The use of DNA profiling extends beyond crimes. For instance, a comparison of the DNA of a child and the purported father can conclusively settle a question of paternity. Sometimes, paternity is of historical interest: DNA profiling proved that Thomas Jefferson or a close male relative fathered a child with his slave Sally Hemings. Going back much further, one of the strangest cases of DNA profiling is that of Cheddar Man, a 9,000-year-old skeleton found in a cave near Cheddar, England (**Figure 15B**). DNA was extracted from his tooth and analyzed. The DNA profile showed that Cheddar Man was a direct ancestor—through approximately 300 generations—of a present-day schoolteacher who lived only a half mile from the cave!

South West News Service, Bristol, England

▲ **Figure 15B** Cheddar Man and one of his modern-day descendants

DNA profiling can also be used to identify victims. The largest such effort in history occurred after the World Trade Center attack of September 11, 2001. Forensic scientists, under the coordination of the Office of the Chief Medical Examiner of New York City, worked for years to identify over 20,000 samples of victims' remains. DNA profiles of tissue samples from the disaster site were matched to DNA profiles from tissue known to be from the victims. If no sample of a victim's DNA was available, blood samples from close relatives were used to confirm identity through near matches. Over half of the identified victims at the World Trade Center site were identified solely by DNA evidence, providing closure to many grieving families.

Just how reliable is DNA profiling? When the standard CODIS set of 13 STR sites (see Module 14) is used, the probability of finding the same DNA profile in randomly selected, unrelated individuals is less than one in 10 billion. Put another way, a standard DNA profile can provide a statistical match of a particular DNA sample to just one living human. For this reason, DNA analyses are now accepted as compelling evidence by legal experts and scientists alike. In fact, DNA analysis on stored forensic samples has provided the evidence needed to solve many "cold cases" in recent years.

DNA analysis has also been used to probe the origin of nonhuman materials. In 1998, the U.S. Fish and Wildlife Service began testing the DNA in caviar to determine if the fish eggs originated from the species claimed on the label. By conclusively proving the origin of contraband animal products, DNA profiling could help protect endangered species. In another example, a 2005 study determined that DNA extracted from a 27,000-year-old Siberian mammoth was 98.6% identical to DNA from modern African elephants.

Although DNA profiling has provided definitive evidence in many investigations, the method is far from foolproof. Problems can arise from insufficient data, human error, or flawed evidence. While the science behind DNA profiling is irrefutable, the human element remains a possible confounding factor.

? In what way is DNA profiling valuable for determining innocence as well as guilt?

A DNA profile can prove with near certainty that a sample of DNA does not come from a particular individual. DNA profiling therefore can provide evidence in support of guilt or innocence.

16 RFLPs can be used to detect differences in DNA sequences

Recall that a genetic marker is a DNA sequence that varies in a population. Like different alleles of a gene, the DNA sequence at a specific place on a chromosome may exhibit small nucleotide differences, or polymorphisms (from the Greek for "many forms"). Geneticists have cataloged many single-base-pair variations in the genome. Such a variation found in at least 1% of the population is called a **single nucleotide polymorphism (SNP**, pronounced "snip"). SNPs occur on average about once in 100 to 300 base pairs in the human genome either in the coding sequence of a gene or in a noncoding sequence.

SNPs may alter a restriction site—the sequence recognized by a restriction enzyme. Such alterations change the lengths of the restriction fragments formed by that enzyme when it cuts the DNA. A sequence variation of this type is called a **restriction fragment length polymorphism (RFLP**, pronounced "rif-lip"). Thus, RFLPs can serve as genetic markers for particular loci in the genome. RFLPs have many uses. For example, disease-causing alleles can be diagnosed with reasonable accuracy if a closely linked RFLP marker has been found. Alleles for a number of genetic diseases were first detected by means of RFLPs in this indirect way.

Restriction fragment analysis involves two of the methods you have learned about: DNA fragments produced by restriction enzymes (see Module 2) are sorted by gel electrophoresis (see Module 13). The number of restriction fragments and their sizes reflect the specific sequence of nucleotides in the starting DNA.

At the top of **Figure 16**, you can see corresponding segments of DNA from two DNA samples prepared from human tissue. Notice that the DNA sequences differ by a single base pair (highlighted in gold). In this case, the restriction enzyme cuts DNA between two cytosine (C) bases in the sequence CCGG and in its complement, GGCC. Because DNA from the first sample has two recognition sequences for the restriction enzyme, it is cleaved in two places, yielding three restriction fragments (labeled *w*, *x*, and *y*). DNA from the second sample, however, has only one recognition sequence and yields only two restriction fragments (*z* and *y*). Notice that the lengths of restriction fragments, as well as the number of fragments, differ, depending on the exact sequence of bases in the DNA.

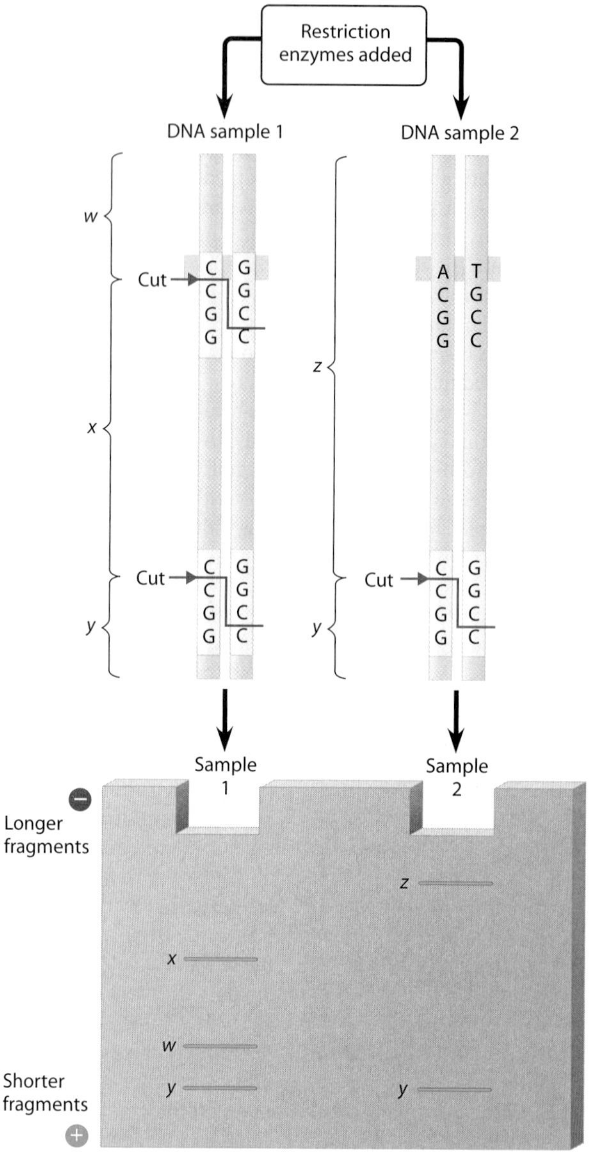

▲ **Figure 16** RFLP analysis

To detect the differences between the collections of restriction fragments, we need to separate the restriction fragments in the two mixtures and compare their lengths. This process, called RFLP analysis, is accomplished through gel electrophoresis. As shown in the bottom of the figure, the three kinds of restriction fragments from sample 1 separate into three bands in the gel, while those from sample 2 separate into only two bands. Notice that the shortest fragment from sample 1 (*y*) produces a band at the same location as the identical short fragment from the sample 2. So you can see that electrophoresis allows us to see similarities as well as differences between mixtures of restriction fragments—and similarities as well as differences between the base sequences in DNA from two individuals. The restriction fragment analysis in Figure 16 clearly shows that the two DNA samples differ in sequence. While RFLP analysis is rarely used today for identification, this method was vital in some of the earliest discoveries of disease-causing genes. For example, the gene for Huntington's disease was found after researchers used RFLPs to track a genetic marker that was closely associated with the disorder.

? **You use a restriction enzyme to cut a DNA molecule that has three copies of the enzyme's recognition sequence clustered near one end. When you separate the restriction fragments by gel electrophoresis, how do you expect the bands to look?**

Three bands near the positive pole at the bottom of the gel (small fragments) and one band near the negative pole at the top of the gel (large fragment)

Genomics

17 Genomics is the scientific study of whole genomes

By the 1980s, biologists were using RFLPs to help map important genes in humans and some other organisms. But it didn't take long for biologists to think on a larger scale. In 1995, a team of scientists announced that they had determined the nucleotide sequence of the entire genome of *Haemophilus influenzae*, a bacterium that can cause several human diseases, including pneumonia and meningitis. **Genomics**, the science of studying a complete set of genes (a genome) and their interactions, was born.

Since 1995, researchers have used the tools and techniques of DNA technology to develop more and more detailed maps of the genomes of a number of species. The first targets of genomics research were bacteria, which have relatively little DNA. The genome of *H. influenzae*, for example, contains only 1.8 million nucleotides and 1,709 genes. But soon, the attention of genomics researchers turned toward more complex organisms with much larger genomes. As of 2009, the genomes of over 1,000 species have been published, and thousands more are in progress. Table 17 lists some of the completed genomes; for diploids, the size refers to the haploid genome. The vast majority of genomes under study are from prokaryotes, including *Escherichia coli* and several hundred other bacteria (some of medical importance), and a few dozen archaea. Over 100 eukaryotic species have been sequenced, including vertebrates, invertebrates, fungi, and plants.

Baker's yeast (*Saccharomyces cerevisiae*) was the first eukaryote to have its full sequence determined, and the roundworm *Caenorhabditis elegans* was the first multicellular organism. Other sequenced animals include the fruit fly (*Drosophila melanogaster*) and the lab mouse (*Mus musculus*), both model organisms for genetics. Plants, such as one type of mustard (*Arabidopsis thaliana*, an important research organism) and rice (*Oryza sativa*, one of the world's most economically important crops), have also been completed. Other recently completed eukaryotic genomes include sorghum (another important commercial crop) and the honeybee, dog, chicken, and sea urchin.

In 2005, researchers completed the genome sequence for our closest living relative on the evolutionary tree of life, the chimpanzee (*Pan troglodytes*). Comparisons with human DNA revealed that we share 96% of our genome with our closest animal relative. As you will see in Module 21, genomic scientists are currently finding and studying the important differences, shedding scientific light on the age-old question of what makes us human.

Why map so many genomes? Not only are all genomes of interest in their own right, but comparative analysis provides invaluable insights into the evolutionary relationships among organisms. Also, having maps of a variety of genomes helps scientists interpret the human genome. For example, when scientists find a nucleotide sequence in the human genome similar to a yeast gene whose function is known, they have a valuable clue to the function of the human sequence. Indeed, several yeast protein-coding genes are so similar to certain human disease-causing genes that researchers have figured out the functions of the disease genes by studying their normal yeast counterparts. Many genes of disparate organisms are turning out to be astonishingly similar, to the point that one researcher has joked that he now views fruit flies as "little people with wings."

? **Why is it useful to sequence nonhuman genomes?**

● Besides their value in understanding evolution, comparative analysis of nonhuman genes helps scientists interpret human data.

TABLE 17 | SOME IMPORTANT COMPLETED GENOMES

Organism	Year Completed	Size of Haploid Genome (in Base Pairs)	Approximate Number of Genes
Haemophilus influenzae (bacterium)	1995	1.8 million	1,700
Saccharomyces cerevisiae (yeast)	1996	12 million	6,300
Escherichia coli (bacterium)	1997	4.6 million	4,400
Caenorhabditis elegans (nematode)	1998	100 million	20,100
Drosophila melanogaster (fruit fly)	2000	165 million	13,700
Arabidopsis thaliana (mustard plant)	2000	120 million	27,000
Mus musculus (mouse)	2001	2.6 billion	22,000
Oryza sativa (rice)	2002	430 million	42,000
Homo sapiens (humans)	2003	3.0 billion	21,000
Rattus norvegius (lab rat)	2004	2.8 billion	25,000
Pan troglodytes (chimpanzee)	2005	3.1 billion	22,000
Macaca mulatta (macaque)	2007	2.9 billion	22,000
Xenopus tropicalis (frog)	2010	1.7 billion	20,000

CONNECTION

18 The Human Genome Project revealed that most of the human genome does not consist of genes

The **Human Genome Project (HGP)** had the goals of determing the nucleotide sequence of all DNA in the human genome and identifying the location and sequence of every gene. The HGP began in 1990 at 20 government-funded research centers in six countries. Several years into the project, private companies, chiefly Celera Genomics, in the United States, joined the effort. At the completion of the final draft of the sequence, over 99% of the genome had been determined to 99.999% accuracy. (There remain a few hundred gaps of unknown sequences within the human genome that will require special methods to figure out.) The DNA sequences determined by the HGP have been deposited in a publicly available database called Genbank.

The chromosomes in the human genome—22 autosomes plus the X and Y sex chromosomes—contain approximately 3.0 billion nucleotide pairs of DNA. To try to get a sense of this much DNA, imagine that its nucleotide sequence is printed in letters (A, T, C, and G) like the letters in this book. At this size, the sequence would fill a stack of books 18 stories high! The biggest surprise from the HGP is the small number of human genes. The current estimate is about 21,000 genes—very close to the number found in a nematode worm. How, then, do we account for human complexity? Part of the answer may lie in alternative RNA splicing; scientists think that a typical human gene specifies several polypeptides.

In humans, as in most complex eukaryotes, only a small amount of our total DNA (about 1.5%) is contained in genes that code for proteins, tRNAs, or rRNAs (**Figure 18**). Most multicellular eukaryotes have a huge amount of noncoding DNA; about 98.5% of human DNA is of this type. About one-quarter of our DNA consists of introns and gene control sequences such as promoters and enhancers. The remaining noncoding DNA has been dubbed "junk DNA," a tongue-in-cheek way of saying that scientists don't fully understand its functions.

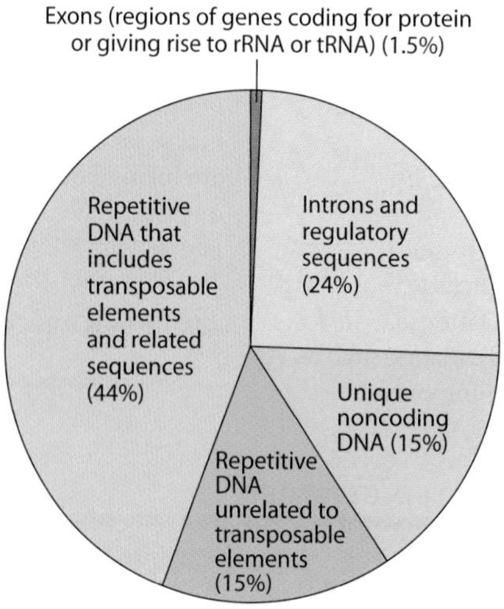

▲ **Figure 18** Composition of the human genome

Much of the DNA between genes consists of repetitive DNA, nucleotide sequences present in many copies in the genome. The repeated units of some of this DNA, such as the STRs used in DNA profiling, are short (see Module 14). Stretches of DNA with thousands of short repetitions are also prominent at the centromeres and ends of chromosomes—called **telomeres**—suggesting that this DNA plays a role in chromosome structure.

In the second main type of repetitive DNA, each repeated unit is hundreds of nucleotides long, and the copies are scattered around the genome. Most of these sequences seem to be associated with **transposable elements** ("jumping genes"), DNA segments that can move or be copied from one location to another in a chromosome and even between chromosomes. Researchers believe that transposable elements, through their copy-and-paste mechanism, are responsible for the proliferation of dispersed repetitive DNA in the human genome.

The potential benefits of having a complete map of the human genome are enormous, especially to medicine. For instance, hundreds of disease-associated genes have already been identified. One example is the gene that is mutated in an inherited type of Parkinson's disease, a debilitating brain disorder that causes motor problems of increasing severity. Until recently, Parkinson's disease was not known to have a hereditary component. But data from the Human Genome Project mapped a small number of cases of Parkinson's disease to a specific gene. Interestingly, an altered version of the protein encoded by this gene has also been tied to Alzheimer's disease, suggesting a previously unknown link between these two brain disorders. Moreover, the same gene is also found in rats, where it plays a role in the sense of smell, and in zebra finches, where it is thought to be involved in song learning. Cross-species comparisons such as these may uncover clues about the role played by the normal version of the protein in the human brain. And such knowledge could eventually lead to treatment for the half a million Americans with Parkinson's disease.

One interesting question about the Human Genome Project is: Whose genome was sequenced? The human genome sequenced by the public consortium was actually a reference genome compiled from a group of individuals. The genome sequenced by Celera consisted primarily of DNA sampled from the company's president. These representative sequences will serve as standards so that comparisons of individual differences and similarities can be made. Starting in 2007, the genomes of a number of other individuals—the first was James Watson, codiscoverer of the structure of DNA—have also been sequenced. These sequences are part of a larger effort to collect information on all of the genetic variations that affect human characteristics. As the amount of sequence data multiplies, the small differences that account for individual variation within our species will come to light.

? **The haploid human genome consists of about _______ base pairs and _______ genes spread over _______ different chromosomes (provide three numbers).**

3 billion . . . 21,000 . . . 24 (22 autosomes plus 2 sex chromosomes)

19 The whole-genome shotgun method of sequencing a genome can provide a wealth of data quickly

Sequencing an entire genome is a complex task that requires careful work. The Human Genome Project proceeded through three stages that provided progressively more detailed views of the human genome. First, geneticists combined pedigree analyses of large families to map over 5,000 genetic markers (mostly RFLPs) spaced throughout all of the chromosomes. The resulting low-resolution *linkage map* provided a framework for mapping other markers and for arranging later, more detailed maps of particular regions. Next, researchers determined the number of base pairs between the markers in the linkage map. These data helped them construct a *physical map* of the human genome. Finally came the most arduous part of the project: determining the nucleotide sequences of the set of DNA fragments that had been mapped. Advances in automated DNA sequencing were crucial to this endeavor.

This three-stage approach is logical and thorough. However, in 1992, molecular biologist J. Craig Venter proposed an alternative strategy called the **whole-genome shotgun method** and set up the company Celera Genomics to implement it. His idea was essentially to skip the genetic and physical mapping stages and start directly with the sequencing step. In the whole-genome shotgun method, an entire genome is chopped by restriction enzymes into fragments that are cloned and sequenced in just one stage (**Figure 19**). High-performance computers running specialized mapping software can assemble the millions of overlapping short sequences into a single continuous sequence for every chromosome—an entire genome.

Today, the whole-genome shotgun approach is the method of choice for genomic researchers because it is fast and relatively inexpensive. However, recent research has revealed some limitations of this method, such as difficulties with repetitive sequences, suggesting that a hybrid approach that combines whole-genome shotgunning with physical or genetic maps may prove to be the most useful method in the long run.

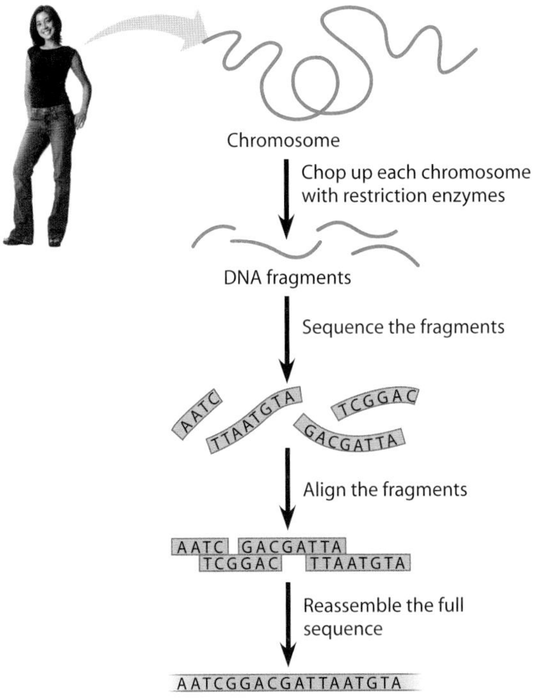

Rubberball/Getty Images

▲ **Figure 19** The whole-genome shotgun method

? What are the primary advantages of the whole-genome shotgun method?

It is faster and cheaper than the three-stage method of genome sequencing.

20 Proteomics is the scientific study of the full set of proteins encoded by a genome

The successes in the field of genomics have encouraged scientists to attempt a similar systematic study of the full protein sets (proteomes) encoded by genomes, an approach called **proteomics**. The number of different proteins in humans far exceeds the number of genes—about 100,000 proteins versus about 21,000 genes. And since proteins, not genes, actually carry out most of the activities of the cell, scientists must study when and where proteins are produced in an organism and how they interact in order to understand the functioning of cells and organisms. Given the huge number of proteins and the myriad ways that their production can be controlled, assembling and analyzing proteomes pose many experimental challenges. Ongoing advances are beginning to provide the tools to meet those challenges.

Genomics and proteomics are enabling biologists to approach the study of life from an increasingly holistic perspective. Biologists are now in a position to compile complete catalogs of genes and proteins—that is, a full listing of all the "parts" that contribute to the operation of cells, tissues, and organisms. With such catalogs in hand, researchers are shifting their attention from the individual parts to how they function together in biological systems.

? If every protein is encoded by a gene, how can humans have many more proteins than genes? *Hint*: See Module 11.4.

The RNA transcribed from one gene may be spliced several different ways to produce different mRNAs that are translated into different proteins.

EVOLUTION CONNECTION

21 Genomes hold clues to human evolution

Comparisons of genome sequences from different species allow geneticists to evaluate the evolutionary relationships between those species. The more similar in sequence the same gene is in two species, the more closely related those species are in their evolutionary history. Comparing genes of closely related species sheds light on recent evolutionary events, whereas comparing those of distantly related species helps us understand more ancient evolutionary history.

The small number of genetic differences between closely related species makes it easier to correlate phenotypic differences between the species with particular genetic differences. The completion of the chimpanzee genome in 2005 has allowed us to compare our genome with that of our primate cousins. Such an analysis revealed that these two genomes differ by 1.2% in single-base substitutions. Researchers were surprised when they found a further 2.7% difference due to insertions or deletions of larger regions in the genome of one or the other species, with many of the insertions being duplications or other repetitive DNA. In fact, a third of the human duplications are not present in the chimpanzee genome, and some of these duplications contain regions associated with human diseases. All of these observations provide clues to the forces that might have swept the two genomes along different paths, but we don't have a complete picture yet.

What about specific genes and types of genes that differ between humans and chimpanzees? Using evolutionary analyses, biologists have identified a number of genes that have evolved faster in humans. Among them are genes involved in defense against malaria and tuberculosis and a gene regulating brain size. One gene that changed rapidly in the human lineage is *FOXP2*, a gene implicated in speech and vocalization. Differences between the *FOXP2* gene in humans and chimpanzees may play a role in the ability of humans, but not chimpanzees, to communicate by speech.

Neanderthals (*Homo neanderthalensis*) were humans' closest relatives (**Figure 21**). First appearing at least 300,000 year ago, Neanderthals lived in Europe and Asia until suddenly going extinct a mere 30,000 years ago. Modern humans (*Homo sapiens*) first appeared in Africa around 200,000 years ago and spread into Europe and Asia around 50,000 years ago—meaning that modern humans and Neanderthals most likely comingled for some time.

Philippe Plailly & Atelier Daynes/Photo Researchers

▲ **Figure 21**
Reconstruction of a Neanderthal female, based on a 36,000-year-old skull

A 2009 rough draft of a 60%-complete Neanderthal genome has, for the first time, allowed detailed genomic comparisons between two species in the genus *Homo*. Using 38,000-year-old thigh bone fossils of two *Homo neanderthalensis* females discovered in a Croatian cave, genomic analysis confirmed Neanderthals as a separate species and as our closest relatives (much closer than chimpanzees). Further comparisons, completed in 2010, suggested that Neanderthals and some *H. sapiens* probably did interbreed. Analysis of the sequence of the *FOXP2* gene showed that Neanderthals had the same allele as modern humans, hinting that Neanderthals may have had the same ability to speak as we do. Other genetic analyses of less complete Neanderthal genomes revealed one male to have an unusual allele for a pigment gene that would have given him pale skin and red hair. And, interestingly, analysis of the lactase gene suggests that Neanderthals, like the majority of modern humans, were lactose intolerant as adults.

Comparisons with Neanderthals and chimpanzees are part of a larger effort to learn more about the human genome. Other research efforts are extending genomic studies to many more species. These studies will advance our understanding of all aspects of biology, including health, ecology, and evolution. In fact, comparisons of the completed genome sequences of bacteria, archaea, and eukaryotes supported the theory that these are the three fundamental domains of life.

? How can cross-species comparisons of the nucleotide sequences of a gene provide insight into evolution?

● Similarities in gene sequences correlate with evolutionary relatedness; greater genetic similarities reflect a more recent shared ancestry.

REVIEW

MB For Practice Quizzes, BioFlix, MP3 Tutors, and Activities, go to www.masteringbiology.com.

Reviewing the Concepts

Gene Cloning (1–5)

1 Genes can be cloned in recombinant plasmids. Gene cloning is one application of biotechnology, the manipulation of organisms or their components to make useful products. Researchers can create plasmids containing recombinant DNA and insert those plasmids into bacteria. If the recombinant bacteria multiply into a clone, the foreign genes are also duplicated and copies of the gene or its protein product can be harvested.

2 Enzymes are used to "cut and paste" DNA. Restriction enzymes cut DNA at specific sequences, forming restriction fragments. DNA ligase "pastes" DNA fragments together.

3 Cloned genes can be stored in genomic libraries. Genomic libraries, sets of DNA fragments containing all of an organism's genes, can be constructed and stored using cloned bacterial plasmids, phages, or bacterial artificial chromosomes (BACs).

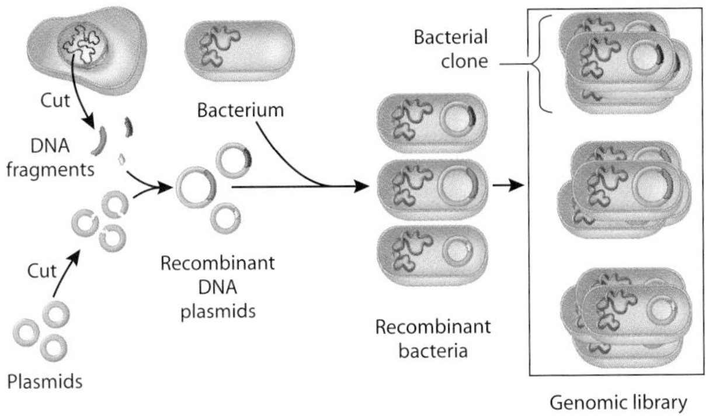

4 Reverse transcriptase can help make genes for cloning. cDNA libraries contain only the genes that are transcribed by a particular type of cell.

5 Nucleic acid probes identify clones carrying specific genes. A short, single-stranded molecule of labeled DNA or RNA can tag a desired gene in a library.

Genetically Modified Organisms (6–10)

6 Recombinant cells and organisms can mass-produce gene products. Bacteria, yeast, cell cultures, and whole animals can be used to make products for medical and other uses.

7 DNA technology has changed the pharmaceutical industry and medicine. Researchers use gene cloning to produce hormones, diagnose diseases, and produce vaccines.

8 Genetically modified organisms are transforming agriculture. A number of important crop plants are genetically modified.

9 Genetically modified organisms raise concerns about human and environmental health. Genetic engineering involves risks, such as ecological damage from GM crops.

10 Gene therapy may someday help treat a variety of diseases.

DNA Profiling (11–16)

11 The analysis of genetic markers can produce a DNA profile. DNA technology—methods for studying and manipulating genetic material—has revolutionized the field of forensics. DNA profiling—the analysis of DNA fragments—can determine whether two samples of DNA come from the same individual.

12 The PCR method is used to amplify DNA sequences. The polymerase chain reaction (PCR) can be used to amplify a DNA sample. The use of specific primers that flank the desired sequence ensures that only a particular subset of the DNA sample will be copied.

13 Gel electrophoresis sorts DNA molecules by size.

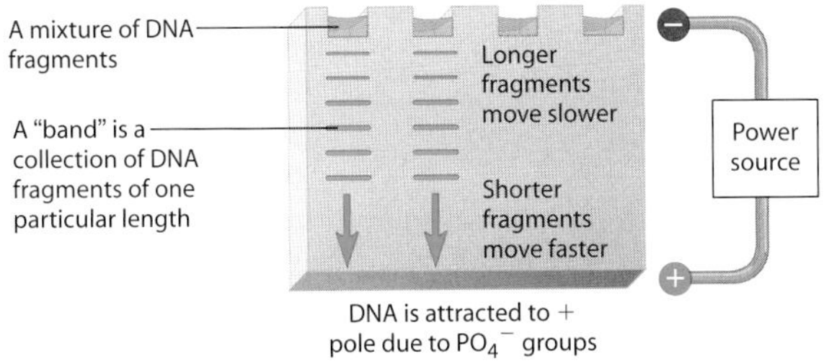

14 STR analysis is commonly used for DNA profiling. Short tandem repeats (STRs) are stretches of DNA that contain short nucleotide sequences repeated many times in a row. DNA profiling by STR analysis involves amplifying a set of 13 STRs.

15 DNA profiling has provided evidence in many forensic investigations. The applications of DNA profiling include helping to solve crimes and establishing paternity.

16 RFLPs can be used to detect differences in DNA sequences. Restriction fragment length polymorophisms (RFLPs) reflect differences in the sequences of DNA samples.

Genomics (17–21)

17 Genomics is the scientific study of whole genomes. Genomics researchers have sequenced many prokaryotic and eukaryotic genomes. Besides being of interest in their own right, nonhuman genomes can be compared with the human genome.

18 The Human Genome Project revealed that most of the human genome does not consist of genes. Data from the Human Genome Project (HGP) revealed that the human genome contains about 21,000 genes and a huge amount of noncoding DNA, much of which consists of repetitive nucleotide sequences and transposable elements that can move about within the genome.

19 The whole-genome shotgun method of sequencing a genome can provide a wealth of data quickly. The HGP uses genetic and physical mapping of chromosomes followed by DNA sequencing. Modern genomic analysis often uses the faster whole-genome shotgun method.

20 Proteomics is the scientific study of the full set of proteins encoded by a genome.

21 Genomes hold clues to human evolution.

Connecting the Concepts

1. Imagine you have found a small quantity of DNA. Fill in the following diagram, which outlines a series of DNA technology experiments you could perform to study this DNA.

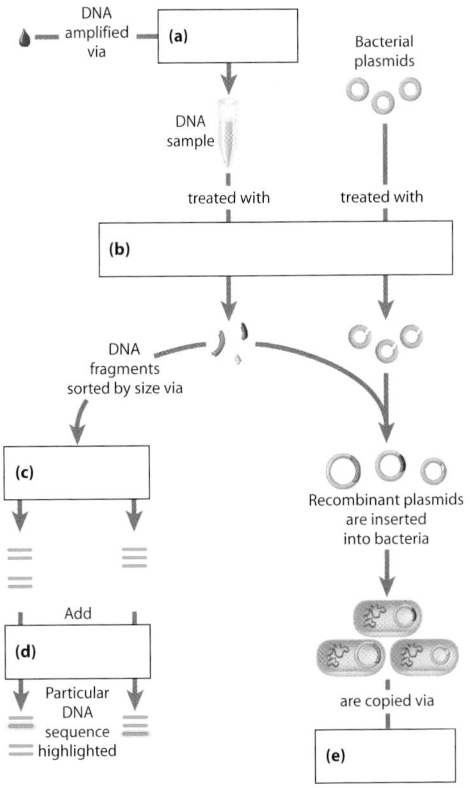

Testing Your Knowledge

Multiple Choice

2. Which of the following would be considered a transgenic organism?
   a. a bacterium that has received genes via conjugation
   b. a human given a corrected human blood-clotting gene
   c. a fern grown in cell culture from a single fern root cell
   d. a rat with rabbit hemoglobin genes
   e. a human treated with insulin produced by bacteria
3. When a typical restriction enzyme cuts a DNA molecule, the cuts are uneven, giving the DNA fragments single-stranded ends. These ends are useful in recombinant DNA work because
   a. they enable a cell to recognize fragments produced by the enzyme.
   b. they serve as starting points for DNA replication.
   c. the fragments will bond to other fragments with complementary ends.
   d. they enable researchers to use the fragments as molecular probes.
   e. only single-stranded DNA segments can code for proteins.
4. The DNA profiles used as evidence in a murder trial look something like supermarket bar codes. The pattern of bars in a DNA profile shows
   a. the order of bases in a particular gene.
   b. the presence of various-sized fragments of DNA.
   c. the presence of dominant or recessive alleles for particular traits.
   d. the order of genes along particular chromosomes.
   e. the exact location of a specific gene in a genomic library.
5. A biologist isolated a gene from a human cell, attached it to a plasmid, and inserted the plasmid into a bacterium. The bacterium made a new protein, but it was nothing like the protein normally produced in a human cell. Why? (*Explain your answer.*)
   a. The bacterium had undergone transformation.
   b. The gene did not have sticky ends.
   c. The gene contained introns.
   d. The gene did not come from a genomic library.
   e. The biologist should have cloned the gene first.
6. A paleontologist has recovered a tiny bit of organic material from the 400-year-old preserved skin of an extinct dodo. She would like to compare DNA from the sample with DNA from living birds. Which of the following would be most useful for increasing the amount of DNA available for testing?
   a. restriction fragment analysis
   b. polymerase chain reaction
   c. molecular probe analysis
   d. electrophoresis
   e. Ti plasmid technology
7. How many genes are there in a human sperm cell?
   a. 23
   b. 46
   c. 5,000–10,000
   d. about 21,000
   e. about 3 billion

Describing, Comparing, and Explaining

8. Why does DNA profiling rely on comparing specific genetic markers rather than the entire genome?
9. Explain how you might engineer *E. coli* to produce human growth hormone (HGH) using the following: *E. coli* containing a plasmid, DNA carrying the gene for HGH, DNA ligase, a restriction enzyme, equipment for manipulating and growing bacteria, a method for extracting and purifying the hormone, an appropriate DNA probe. (Assume that the human HGH gene lacks introns.)
10. Recombinant DNA techniques are used to custom-build bacteria for two main purposes: to obtain multiple copies of certain genes and to obtain useful proteins produced by certain genes. Give an example of each of these applications in medicine and agriculture.

Applying the Concepts

11. A biochemist hopes to find a gene in human liver cells that codes for an important blood-clotting protein. She knows that the nucleotide sequence of a small part of the blood-clotting gene is CTGGACTGACA. Briefly outline a possible method she might use to isolate the desired gene.
12. What is left for genetic researchers to do now that the Human Genome Project has determined nearly complete nucleotide sequences for all of the human chromosomes? Explain.
13. Today, it is fairly easy to make transgenic plants and animals. What are some important safety and ethical issues raised by this use of recombinant DNA technology? What are some of the possible dangers of introducing genetically engineered organisms into the environment? What are some reasons for and against leaving decisions in these areas to scientists? To business owners and executives? What are some reasons for and against more public involvement? How might these decisions affect you? How do you think these decisions should be made?
14. In the not-too-distant future, gene therapy may be an option for the treatment and cure of some inherited disorders. What do you think are the most serious ethical issues that must be dealt with before human gene therapy is used on a large scale? Why do you think these issues are important?
15. The possibility of extensive genetic testing raises questions about how personal genetic information should be used. For example, should employers or potential employers have access to such information? Why or why not? Should the information be available to insurance companies? Why or why not? Is there any reason for the government to keep genetic files? Is there any obligation to warn relatives who might share a defective gene? Might some people avoid being tested for fear of being labeled genetic outcasts? Or might they be compelled to be tested against their wishes? Can you think of other reasons to proceed with caution?

Review Answers

1. a. PCR; b. a restriction enzyme; c. gel electrophoresis; d. nucleic acid probe; e. cloning
2. d 3. c 4. b 5. c (Bacteria lack the RNA-splicing machinery needed to delete eukaryotic introns.) 6. b 7. d
8. Because it would be too expensive and time consuming to compare whole genomes. By choosing STR sites that vary considerably from person to person, investigators can get the necessary degree of specificity without sequencing DNA.
9. Isolate plasmids from a culture of *E. coli*. Cut the plasmids and the human DNA containing the HGH gene with the restriction enzyme to produce molecules with sticky ends. Join the plasmids and the fragments of human DNA with ligase. Allow *E. coli* to take up recombinant plasmids. Bacteria will then replicate plasmids and multiply, producing clones of bacterial cells. Identify a clone carrying and expressing the HGH gene using a nucleic acid probe. Grow large amounts of the bacteria and extract and purify HGH from the culture.
10. Medicine: Genes can be used to produce transgenic lab animals for AIDS research or for research related to human gene therapy. Proteins can be hormones, enzymes, blood-clotting factor, or the active ingredient of vaccines. Agriculture: Foreign genes can be inserted into plant cells or animal eggs to produce transgenic crop plants or farm animals. Animal growth hormones are examples of agriculturally useful proteins that can be made using recombinant DNA technology.
11. She could start with DNA isolated from liver cells (the entire genome) and carry out the procedure outlined in Module 1 to produce a collection of recombinant bacterial clones, each carrying a small piece of liver cell DNA. To find the clone with the desired gene, she could then make a probe of radioactive RNA with a nucleotide sequence complementary to part of the gene: GACCUGACUGU. This probe would bind to the gene, labeling it and identifying the clone that carries it. Alternatively, the biochemist could start with mRNA isolated from liver cells and use it as a template to make cDNA (using reverse transcriptase). Cloning this DNA rather than the entire genome would yield a smaller library of genes to be screened—only those active in liver cells. Furthermore, the genes would lack introns, making the desired gene easier to manipulate after isolation.
12. Determining the nucleotide sequences is just the first step. Once researchers have written out the DNA "book," they will have to try to figure out what it means—what the nucleotide sequences code for and how they work.
13. Some issues and questions to consider: What are some of the unknowns in recombinant DNA experiments? Do we know enough to anticipate and deal with possible unforeseen and negative consequences? Do we want this kind of power over evolution? Who should make these decisions? If scientists doing the research were to make the decisions about guidelines, what factors might shape their judgment? What might shape the judgment of business executives in the decision-making process? Does the public have a right to a voice in the direction of scientific research? Does the public know enough about biology to get involved in this decision-making process? Who represents "the public," anyway?
14. Some issues and questions to consider: What kinds of impact will gene therapy have on the individuals who are treated? On society? Who will decide what patients and diseases will be treated? What costs will be involved, and who will pay them? How do we draw the line between treating disorders and "improving" the human species?
15. Some issues and questions to consider: Should genetic testing be mandatory or voluntary? Under what circumstances? Why might employers and insurance companies be interested in genetic data? Since genetic characteristics differ among ethnic groups and between the sexes, might such information be used to discriminate? Which of these questions do you think is most important? Which issues are likely to be the most serious in the future?

The Origin of Species

The Origin of Species

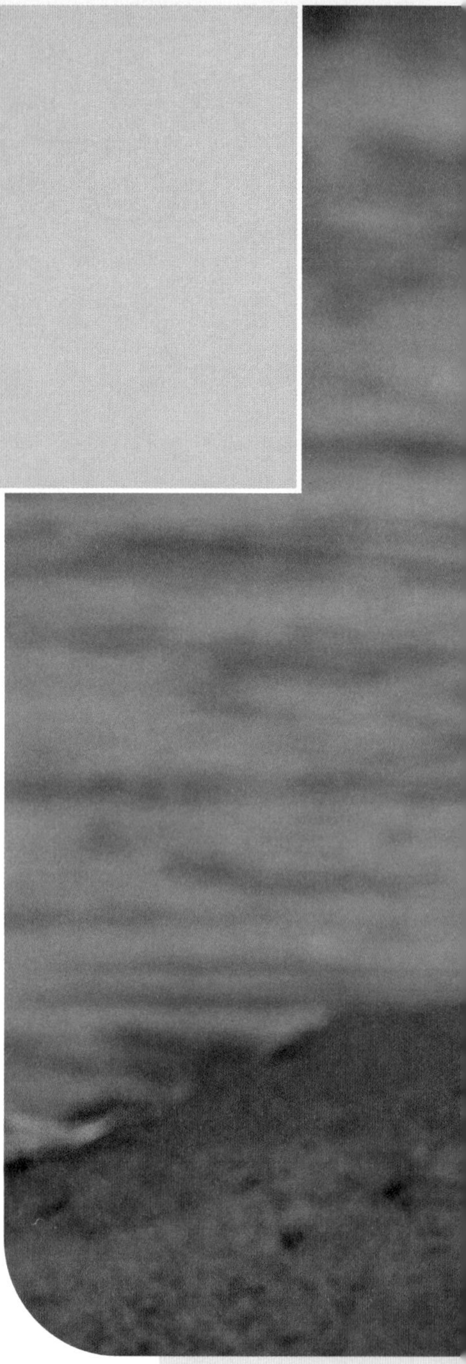

BIG IDEAS

Defining Species

(1–3)

A species can be defined as a group of populations whose members can produce fertile offspring.

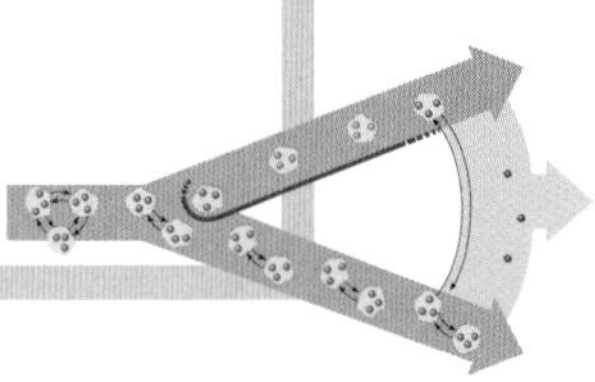

Mechanisms of Speciation

(4–11)

Speciation can take place with or without geographic isolation, as long as reproductive barriers evolve that keep species separate.

David Hosking/FLPA

What happened to this bird's wings? The flightless cormorant (*Phalacrocorax harrisi*) pictured above is found only on two islands in the Galápagos. There are many other cormorant species found elsewhere in the world that *can* fly. Like its flying relatives, this bird feeds in the water, diving for octopuses, eels, and bottom-dwelling fishes. Here we see it drying its wings after such a dive, a behavior common to all cormorants because their feathers are not waterproof.

How did this cormorant get to these isolated islands if it cannot fly? And why are these birds found nowhere else in the world? An ancestral cormorant species is thought to have flown from the Americas to the Galápagos Islands more than 3 million years ago. Terrestrial mammals could not make the trip over the wide expanse of ocean, and thus predatory mammals were absent as these immigrants took up residence on the islands. The sea was readily accessible from the rocky shoreline, and nesting sites among the rocks were easily reached. Thus, without the need to fly to escape predators or reach breeding grounds, natural selection may no longer have favored the large wings, muscles, and keel bone needed for flight. Indeed, birds with smaller wings would have used less energy to maintain these structures, perhaps instead channeling resources to the production of offspring.

In this chapter, we explore how such changes may lead to the origin of new species, the process that is responsible for the amazing diversity of present and past life on Earth. We begin with the biological definition of a species and describe the mechanisms through which new species may evolve. We also explore some of the evidence for speciation and how scientists study this evolutionary process.

Defining Species

1 The origin of species is the source of biological diversity

Darwin was eager to explore landforms newly emerged from the sea when he came to the Galápagos Islands. He noted that these volcanic islands, despite their geologic youth, were teeming with plants and animals found nowhere else in the world, including the flightless cormorant pictured in the chapter introduction. He realized that these species, like the islands, were relatively new. He wrote in his diary: "Both in space and time, we seem to be brought somewhat near to that great fact—that mystery of mysteries—the first appearance of new beings on this Earth."

Even though Darwin titled his seminal work *On the Origin of Species by Means of Natural Selection*, most of his theory of evolution focused on the role of natural selection in the gradual adaptation of a population to its environment. We call this process microevolution—changes in the gene pool of a population from one generation to the next. But if microevolution were *all* that happened, then Earth would be inhabited only by a highly adapted version of the first form of life.

The "mystery of mysteries" that fascinated Darwin is **speciation**, the process by which one species splits into two or more species. He saw evidence of speciation in the flightless cormorants of the Galápagos, which he could compare with the flying species he saw elsewhere in his travels. Indeed, more than 40 living species of cormorants have now been identified, with others known from the fossil record. **Figure 1** shows one common species found along the Florida coastline.

Each time speciation occurs, the diversity of life increases. Over the course of 3.5 billion years, an ancestral species first gave rise to two or more different species, which then branched to new lineages, which branched again, until we arrive at the millions of species that live, or once lived, on Earth. This origin of species explains both the diversity and the unity of life. When one species splits into two, the new species share many characteristics because they are descended from a common ancestor. Thus, we see not only the uniqueness of the flightless cormorant of the Galápagos but also the similarities between this scrawny-winged bird and its broad-winged relatives.

? How does microevolution differ from speciation?

● Microevolution involves evolutionary changes within a population; speciation occurs when a population changes enough that it diverges from its parent species and becomes a new species.

AKrieger/Photolibrary

▲ **Figure 1** Great cormorant (*Phalacrocorax carbo*) drying its broad wings

2 There are several ways to define a species

The word *species* is from the Latin for "kind" or "appearance," and indeed, even young children learn to distinguish between kinds of plants and animals—between dogs and cats, or roses and dandelions—from differences in their appearance. Although the basic idea of species as distinct life-forms seems intuitive, devising a more formal definition is not so easy.

In many cases, the differences between two species are obvious. A quick comparison of the great cormorant shown in Figure 1 and the flightless cormorant in the chapter introduction would lead you to conclude that these are two different species. In other cases, the physical differences between two species are not so obvious. The two birds in **Figure 2A** look much the same—the one on the left is an eastern meadowlark (*Sturnella magna*); the bird on the right is a western meadowlark (*Sturnella neglecta*). They are distinct species, however, because their songs and other behaviors are different enough that each type of meadowlark breeds only with individuals of its own species.

Malcolm Schuyl/Alamy

David Kjaer/Nature Picture Library

▲ **Figure 2A** Similarity between two species: the eastern meadowlark (left) and western meadowlark (right)

How similar are members of the same species? Whereas the individuals of many species exhibit fairly limited variation in physical appearance, certain other species—our own, for example—seem extremely varied. The physical diversity within

▲ **Figure 2B** Diversity within one species

our species (partly illustrated in **Figure 2B**) might lead you to guess that there are several human species. Despite these outward appearances, however, humans all belong to the same species, *Homo sapiens*.

The Biological Species Concept How then, do biologists define a species? And what keeps one species distinct from others? The primary definition of species used in this book is called the **biological species concept**. It defines a species as a group of populations whose members have the potential to interbreed in nature and produce fertile offspring (offspring that themselves can reproduce). A businesswoman in Manhattan may be unlikely to meet a dairy farmer in Mongolia, but if the two should happen to meet and mate, they could have viable babies that develop into fertile adults. Thus, members of a biological species are united by being reproductively compatible, at least potentially.

Members of different species do not usually mate with each other. In effect, **reproductive isolation** prevents genetic exchange (gene flow) and maintains the gap between species. But there are some pairs of clearly distinct species that do occasionally interbreed. The resulting offspring are called **hybrids**. An example is the grizzly bear (*Ursus arctos*) and the polar bear (*Ursus maritimus*), whose hybrid offspring have been called "grolar bears" (**Figure 2C**). The two species have been known to interbreed in zoos, and DNA testing confirmed that a bear shot in 2006 in the Canadian Arctic was a wild polar bear–grizzly offspring. Another grolar bear was shot in 2010, and scientists predict an increase in such hybrids as melting polar sea ice brings the two species into contact more often. Clearly, identifying species solely on the basis of reproductive isolation can be more complex than it may seem.

There are other instances in which applying the biological species concept is problematic. For example, there is no way to determine whether organisms that are now known only through fossils were once able to interbreed. Also, this criterion is useless for organisms such as prokaryotes that reproduce asexually. Because of such limitations, alternative species concepts are useful in certain situations.

▲ **Figure 2C** Hybridization between two species of bears

Other Definitions of Species For most organisms—sexual, asexual, and fossils alike—classification is based mainly on physical traits such as shape, size, and other features of morphology (form). This **morphological species concept** has been used to identify most of the 1.8 million species that have been named to date. The advantages of this concept are that it can be applied to asexual organisms and fossils and does not require information on possible interbreeding. The disadvantage, however, is that this approach relies on subjective criteria, and researchers may disagree on which features distinguish a species.

Another species definition, the **ecological species concept**, identifies species in terms of their ecological niches, focusing on unique adaptations to particular roles in a biological community. For example, two species of fish may be similar in appearance but distinguishable based on what they eat or the depth of water in which they are usually found.

Finally, the **phylogenetic species concept** defines a species as the smallest group of individuals that share a common ancestor and thus form one branch on the tree of life. Biologists trace the phylogenetic history of such a species by comparing its characteristics, such as morphology or DNA sequences, with those of other organisms. These sorts of analyses can distinguish groups that are generally similar yet different enough to be considered separate species. Of course, agreeing on the amount of difference required to establish separate species remains a challenge.

Each species definition is useful, depending on the situation and the questions being asked. The biological species concept, however, helps focus on how these discrete groups of organisms arise and are maintained by reproductive isolation. Because reproductive isolation is an essential factor in the evolution of many species, we look at it more closely next.

? Which species concepts could you apply to both asexual and sexual species? Explain.

The morphological, ecological, and phylogenetic species concepts could all be used because they do not rely on the criterion of reproductive isolation.

3 Reproductive barriers keep species separate

Clearly, a fly will not mate with a frog or a fern. But what prevents species that are closely related from interbreeding? Reproductive isolation depends on one or more types of reproductive barriers—biological features of the organism that prevent individuals of different species from interbreeding. As shown in **Figure 3A**, the various types of reproductive barriers that isolate the gene pools of species can be categorized as either prezygotic or postzygotic, depending on whether they function before or after zygotes (fertilized eggs) form.

Prezygotic Barriers **Prezygotic barriers** prevent mating or fertilization between species. There are five main types of prezygotic barriers. In the first, called habitat isolation, two species that live in the same area but not in the same kinds of habitats encounter each other rarely, if at all. Two species of garter snake in the genus *Thamnophis* are found in western North America, but one lives mainly in water and the other on land **(Figure 3B)**. Habitat isolation also affects parasites that are confined to certain host species and thus do not have the opportunity to interbreed.

In a second type of prezygotic barrier, temporal isolation, species breed at different times—different times of the day, different seasons, or even different years. The skunks shown in **Figure 3C** are an example. The breeding season of the eastern spotted skunk (*Spilogale putoris*) is in the late winter, while the western spotted skunk (*Spilogale gracilis*) breeds in the fall. Thus, even though their geographic ranges overlap in the Great Plains, each type of skunk breeds only with individuals of its own species. Many plants also exhibit seasonal differences in reproduction. Some flowering plants are temporally isolated because their flowers open at different times of the day, so pollen cannot be transferred between them.

In behavioral isolation, a third type of prezygotic barrier, there is little or no "mate recognition" between females and males of different species. Special signals that attract mates and elaborate mating behaviors that are unique to a species are probably the most important reproductive barriers between closely related animals. For example, male fireflies of various species signal to females of their kind by blinking their lights in particular rhythms. Females respond only to signals of their own species, flashing back and attracting the males.

Figure 3D, on the next page, shows part of a courtship ritual. Many species will not mate until the male and female have performed an elaborate ritual that is unlike that of any other species. These blue-footed boobies are involved in a courtship

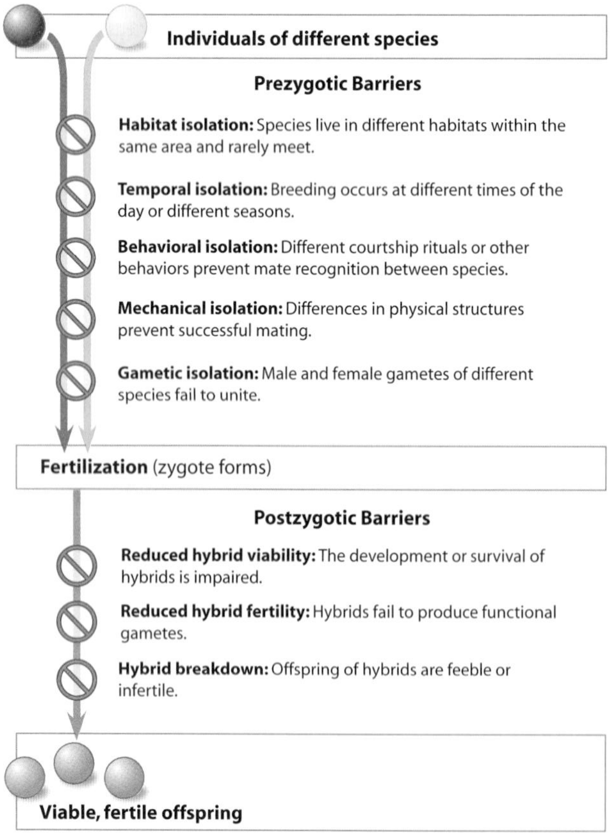

Individuals of different species
Prezygotic Barriers
Habitat isolation: Species live in different habitats within the same area and rarely meet.
Temporal isolation: Breeding occurs at different times of the day or different seasons.
Behavioral isolation: Different courtship rituals or other behaviors prevent mate recognition between species.
Mechanical isolation: Differences in physical structures prevent successful mating.
Gametic isolation: Male and female gametes of different species fail to unite.
Fertilization (zygote forms)
Postzygotic Barriers
Reduced hybrid viability: The development or survival of hybrids is impaired.
Reduced hybrid fertility: Hybrids fail to produce functional gametes.
Hybrid breakdown: Offspring of hybrids are feeble or infertile.
Viable, fertile offspring

▲ **Figure 3A** Reproductive barriers between species

Joe McDonald/Photoshot Holdings Ltd./Bruce Coleman

Joe McDonald/Corbis

▲ **Figure 3B** **Habitat isolation:** Two species of garter snake do not mate because one lives mainly in water and the other lives on land.

USDA/APHIS Animal and Plant Health Inspection Service

Stephen Krasemann/Photo Researchers

▲ **Figure 3C** **Temporal isolation:** The eastern spotted skunk (top) and western spotted skunk (bottom) breed at different times of the year.

© Sohns/imagebroker/Alamy

▲ **Figure 3D** **Behavioral Isolation:** The courtship ritual in blue-footed boobies is a behavioral barrier to mating between species.

Joseph Dovala/Image Quest Marine

▲ **Figure 3F** **Gametic isolation:** Gametes of these red and purple urchins are unable to fuse because surface proteins do not match.

dance in which the male points his beak, tail, and wing tips to the sky. Part of the "script" also calls for the male to high-step, a dance that advertises his bright blue feet.

Mechanical isolation, a fourth type of prezygotic barrier, occurs when female and male sex organs are not compatible; for instance, the male copulatory organs of many insect species have a unique and complex structure that fits the female parts of only one species. **Figure 3E** shows how the spiraling of the shells of two species of snails in different directions prevents their genital openings from aligning, resulting in a mechanical barrier to mating.

Mechanical barriers can also contribute to reproductive isolation of plants. Many species have flower structures that are adapted to specific insect or animal pollinators that transfer pollen only between plants of the same species.

Gametic isolation is a fifth type of prezygotic barrier. Sperm of one species may not be able to fertilize the eggs of another species. Gametic isolation is very important when fertilization is external. Male and female sea urchins **(Figure 3F)** of many different species release eggs and sperm into the sea, but fertilization occurs only if species-specific molecules on the surfaces of egg and sperm attach to each other. A similar mechanism of molecular recognition enables a flower to discriminate between pollen of its own species and pollen of a different species.

Postzygotic Barriers In contrast to prezygotic barriers, **postzygotic barriers** operate after hybrid zygotes have formed. In some cases, there is reduced hybrid viability—most hybrid offspring do not survive. For example, certain salamanders that live in the same habitats may occasionally hybridize, but most hybrids do not complete development.

Another type of postzygotic barrier is reduced hybrid fertility, in which hybrid offspring reach maturity and are vigorous but sterile. A mule, for example, is the robust but sterile offspring of a female horse and a male donkey **(Figure 3G)**. Because infertile hybrids cannot produce offspring with either parent species, genes do not flow between the parent species.

In a third type of postzygotic barrier, called hybrid breakdown, the first-generation hybrid offspring are viable and fertile, but when these hybrids mate with one another or with either parent species, the offspring are feeble or sterile. For example, different species of cotton plants can produce fertile hybrids, but the offspring of the hybrids do not survive.

Next we examine situations that make reproductive isolation and speciation possible.

Ueshima R, Asami T. Evolution: single-gene speciation by left-right reversal. Nature. 2003 Oct 16;425(6959):679 Fig. 1.

▲ **Figure 3E** **Mechanical isolation:** The genital openings (indicated by arrows) of these snails are not aligned, and mating cannot be completed.

? **Two closely related fish live in the same lake, but one feeds along the shoreline and the other is a bottom feeder in deep water. This is an example of ______ isolation, which is a ______ reproductive barrier.**

habitat . . . prezygotic

photosbyjohn/Shutterstock

Photodisc/Getty Images

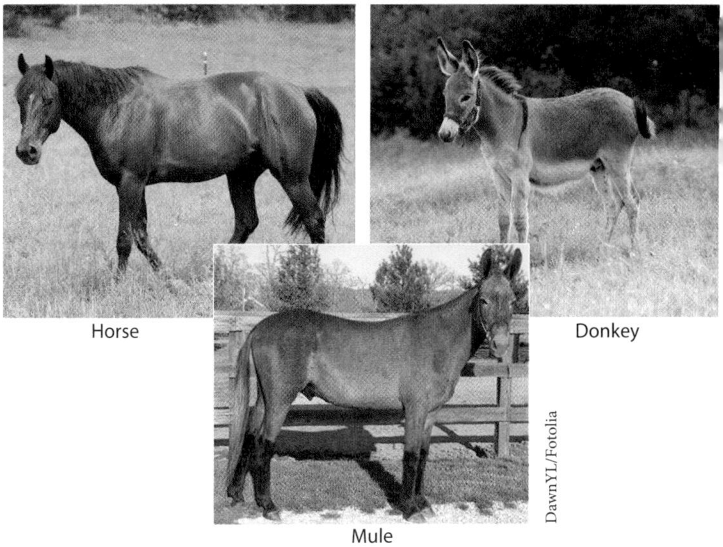

DawnYL/Fotolia

▲ **Figure 3G** **Reduced hybrid fertility:** A horse and a donkey may produce a hybrid mule.

Mechanisms of Speciation

4 In allopatric speciation, geographic isolation leads to speciation

A key event in the origin of a new species is the separation of a population from other populations of the same species. With its gene pool isolated, the splinter population can follow its own evolutionary course. Changes in allele frequencies caused by natural selection, genetic drift, and mutation will not be diluted by alleles entering from other populations (gene flow). The initial block to gene flow may come from a geographic barrier that isolates a population. This mode of speciation is called **allopatric speciation** (from the Greek *allos*, other, and *patra*, fatherland). Populations separated by a geographic barrier are known as allopatric populations.

Geographic Barriers Several geologic processes can isolate populations. A mountain range may emerge and gradually split a population of organisms that can inhabit only lowlands. A large lake may subside until there are several smaller lakes, isolating certain fish populations. Or, continents themselves can split and move apart.

Allopatric speciation can also occur when individuals colonize a remote area and become geographically isolated from the parent population. The flightless cormorant (see chapter introduction) most likely originated in this way from an ancestral flying species that reached the Galápagos Islands.

How large must a geographic barrier be to keep allopatric populations apart? The answer depends on the ability of the organisms to move. Birds, mountain lions, and coyotes can easily cross mountain ranges. The windblown pollen of trees is not hindered by such barriers, and the seeds of many plants may be carried back and forth by animals. In contrast, small rodents may find a canyon or a wide river a formidable barrier. The Grand Canyon and Colorado River **(Figure 4A)** separate two species of antelope squirrels. Harris's antelope squirrel (*Ammospermophilus harrisii*) inhabits the south rim. Just a few kilometers away on the north rim, but separated by the deep and wide canyon, lives the closely related white-tailed antelope squirrel (*Ammospermophilus leucurus*).

Evidence of Allopatric Speciation Many studies provide evidence that speciation has occurred in allopatric populations.

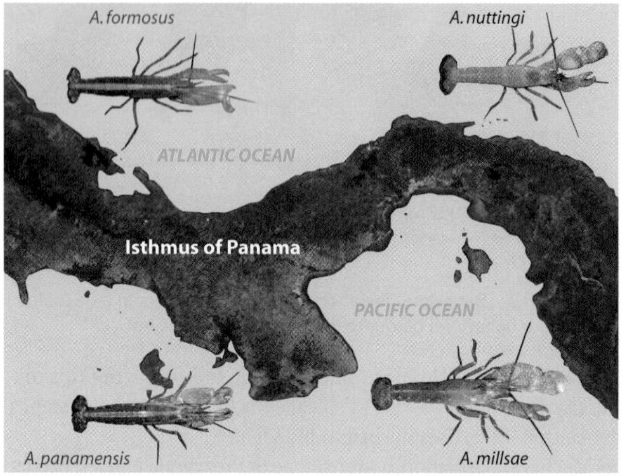

Arthur Anker

▲ **Figure 4B** Allopatric speciation in snapping shrimp: two of the 15 pairs of shrimp that are separated by the Isthmus of Panama

An interesting example is the 30 species of snapping shrimp in the genus *Alpheus* that live off the Isthmus of Panama, the land bridge that connects South and North America **(Figure 4B)**. Snapping shrimp are named for the snapping together of their single oversized claw, which creates a high-pressure blast that stuns their prey. Morphological and genetic data group these shrimp into 15 pairs, with the members of each pair being each other's closest relative. In each case, one member of the pair lives on the Atlantic side of the isthmus, while the other lives on the Pacific side, strongly suggesting that geographic separation of the ancestral species of these snapping shrimp led to allopatric speciation.

? Geologic evidence indicates that the Isthmus of Panama gradually closed about 3 million years ago. Genetic analyses indicate that the various species of snapping shrimp originated from 9 to 3 million years ago, with the species pairs that live in deepest water diverging first. How would you interpret these data?

The deeper species would have been separated into two isolated populations first, which enabled them to diverge into new species first.

John Shaw/Photoshot Holdings Ltd./Bruce Coleman

Michael Fogden/Photoshot Holdings Ltd./Bruce Coleman

▲ **Figure 4A** Allopatric speciation of geographically isolated antelope squirrels

Corbis

5 Reproductive barriers can evolve as populations diverge

Geographic isolation creates opportunities for speciation, but it does not necessarily lead to new species. Speciation occurs only when the gene pool undergoes changes that establish reproductive barriers such as those described in Module 3. What might cause such barriers to arise? The environment of an isolated population may include different food sources, different types of pollinators, and different predators. As a result of natural selection acting on preexisting variations (or as a result of genetic drift or mutation), a population's traits may change in ways that also establish reproductive barriers.

Researchers have successfully documented the evolution of reproductive isolation with laboratory experiments. While at Yale University, Diane Dodd tested the hypothesis that reproductive barriers can evolve as a by-product of changes in populations as they adapt to different environments. Dodd raised fruit flies on different food sources. Some populations were fed starch; others were fed maltose. After about 40 generations, populations raised on starch digested starch more efficiently, and those raised on maltose digested maltose more efficiently.

Dodd then combined flies from various populations in mating experiments. **Figure 5A** shows some of her results. When flies from "starch populations" were mixed with flies from "maltose populations," the flies mated more frequently with like partners (left grid), even when the like partners came from different populations. In one of the control tests (right grid), flies taken from different populations adapted to starch were about as likely to mate with each other as with flies from their own populations. The mating preference shown in the experimental group is an example of a prezygotic barrier. The reproductive barrier was not absolute—some mating between maltose flies and starch flies did occur—but reproductive isolation was under way as these allopatric populations became adapted to different environments.

A reproductive barrier in plants is often pollinator choice. Perhaps populations of an ancestral species became separated in environments that had either more hummingbirds than bees or vice versa. Flower color and shape would evolve through natural selection in ways that attracted the most common pollinator, and these changes would help separate the species should they later share the same region. Two species of closely related monkey flower are found in the same area of the Sierra Nevada, but they rarely interbreed. Bumblebees prefer the pink-flowered *Mimulus lewisii*, and hummingbirds prefer the red-flowered *Mimulus cardinalis*. When scientists experimentally exchanged the alleles for flower color between these two species, the color of their flowers changed **(Figure 5B)**. In a field experiment, the now light orange *M. lewisii* received many more visits from hummingbirds than did the normal pink-flowered *M. lewisii*, and the now pinker *M. cardinalis* flowers received many more visits from bumblebees than the normal red-flowered plants. Thus, a change in flower color influenced pollinator preference, which normally provides a reproductive barrier between these two species.

Sometimes reproductive barriers can arise even when populations are not geographically separated, as we see next.

? Females of the Galápagos finch *Geospiza difficilis* respond to the songs of males from their island but ignore songs of males from other islands. How would you interpret these findings?

Behavioral barriers to reproduction have begun to develop in these allopatric (geographically separated) finch populations.

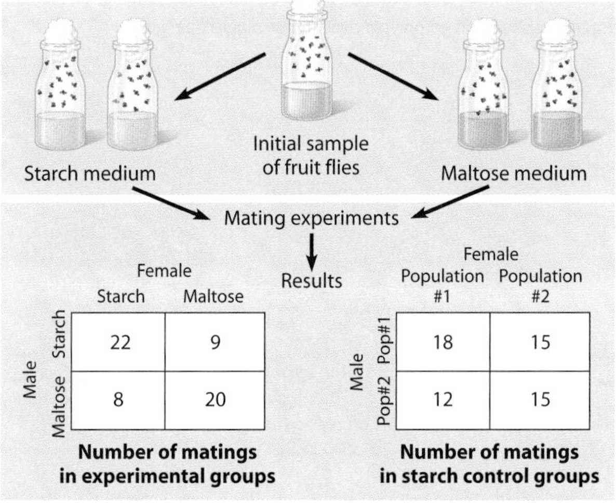

▲ **Figure 5A** Evolution of reproductive barriers in laboratory populations of fruit flies adapted to different food sources

Adapted from D. M. B. Dodd, *Evolution*, vol. 11, pp. 1308-1311. Reprinted by permission of the Society for the Study of Evolution.

▲ **Figure 5B** Transferring an allele between monkey flowers changes flower color and influences pollinator choice.

6 Sympatric speciation takes place without geographic isolation

In **sympatric speciation** (from the Greek *syn*, together, and *patra*, fatherland), a new species arises within the same geographic area as its parent species. How can reproductive isolation develop when members of sympatric populations remain in contact with each other? Sympatric speciation may occur when mating and the resulting gene flow between populations are reduced by factors such as polyploidy, habitat differentiation, and sexual selection.

Many plant species have originated from accidents during cell division that resulted in extra sets of chromosomes. New species formed in this way are **polyploid**, meaning that their cells have more than two complete sets of chromosomes. Figure 6A shows one way in which a tetraploid plant ($4n$, with four sets of chromosomes) can arise from a parent species that is diploid. ❶ A failure of cell division after chromosome duplication could double a cell's chromosomes. ❷ If this $4n$ cell gives rise to a tetraploid branch, flowers produced on this branch would produce diploid gametes. ❸ If self-fertilization occurs, as it commonly does in plants, the resulting tetraploid zygotes would develop into plants that can produce fertile tetraploid offspring by self-pollination or by mating with other tetraploids.

A tetraploid cannot, however, produce fertile offspring by mating with a parent plant. The fusion of a diploid ($2n$) gamete from the tetraploid plant and a haploid (n) gamete from the diploid parent would produce triploid ($3n$) offspring. Triploid individuals are sterile; they cannot produce normal gametes because the odd number of chromosomes cannot form homologous pairs and separate normally during meiosis. Thus, the formation of a tetraploid ($4n$) plant is an instantaneous speciation event: A new species, reproductively isolated from its parent species, is produced in just one generation.

Most polyploid species, however, arise from hybridization of two different species. Figure 6B illustrates one way in which this can happen. ❶ When haploid gametes from two different species combine, the resulting hybrid is normally sterile because its chromosomes cannot pair during meiosis. ❷ However, the hybrid may reproduce asexually, as many plants can do. ❸ Subsequent errors in cell division may produce chromosome duplications that result in a diploid set of chromosomes ($2n = 10$). Now chromosomes *can* pair in meiosis, and haploid gametes will be produced; thus, a fertile polyploid species has formed. The new species has a chromosome number equal to the sum of the diploid chromosome numbers of its parent species. Again, this new species is reproductively isolated, this time from both parent species.

Does polyploid speciation occur in animals? It appears to happen occasionally. For example, the gray tree frog is thought to have originated in this way.

Sympatric speciation in animals is more likely to happen through habitat differentiation or sexual selection. Both these factors may have been involved in the origin of as many as 600 species of small fish called cichlids in Lake Victoria in East Africa. Adaptations for exploiting different food sources may have evolved in different subgroups of the original cichlid population. If those sources were in different habitats, mating between the populations would become rare, isolating their gene pools as each population becomes adapted to a different resource. As you will learn in Module 10, speciation in these brightly colored fish may also have been driven by the type of sexual selection in which females choose mates based on coloration. Such mate choice can contribute to reproductively isolating populations, keeping the gene pools of newly forming species separate. Of course, both habitat differentiation and sexual selection can also contribute to the formation of reproductive barriers between allopatric populations.

? Revisit the reproductive barriers in Figure 3A and choose the barrier that isolates a viable, fertile polyploid plant from its parental species.

Reduced hybrid fertility

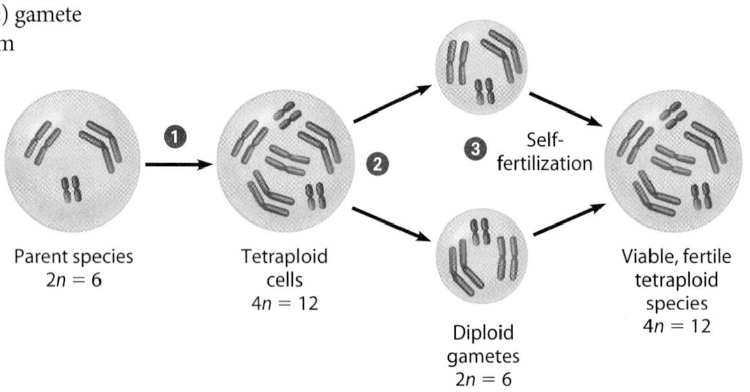

▲ **Figure 6A** Sympatric speciation by polyploidy within a single species

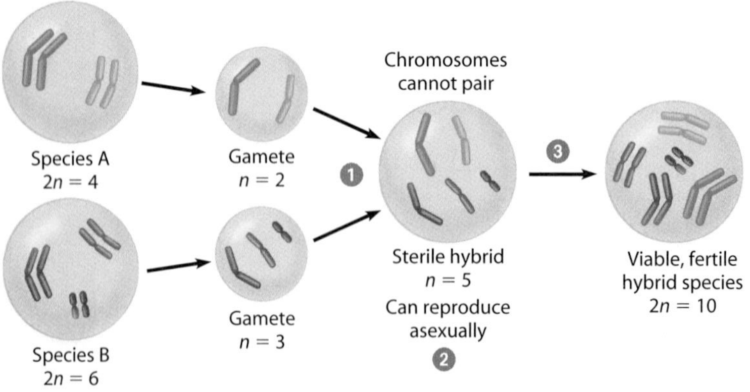

▲ **Figure 6B** Sympatric speciation producing a hybrid polyploid from two different species

EVOLUTION CONNECTION

7 Most plant species trace their origin to polyploid speciation

Plant biologists estimate that 80% of living plant species are descendants of ancestors that formed by polyploid speciation. Hybridization between two species accounts for most of these species, perhaps because of the adaptive advantage of the diverse genes a hybrid inherits from different parental species.

Many of the plants we grow for food are polyploids, including oats, potatoes, bananas, peanuts, barley, plums, apples, sugarcane, coffee, and wheat. Cotton, also a polyploid, provides one of the world's most popular clothing fibers.

Wheat, the most widely cultivated plant in the world, occurs as 20 different species of *Triticum*. We know that humans were cultivating wheat at least 10,000 years ago because wheat grains of *Triticum monococcum* ($2n = 14$) have been found in the remains of Middle Eastern farming villages from that time. This species has small seed heads and is not highly productive, but some varieties are still grown in the Middle East.

Our most important wheat species is bread wheat (*Triticum aestivum*), a polyploid with 42 chromosomes. **Figure 7** illustrates how this species may have evolved; the uppercase letters represent not genes but *sets of chromosomes* that have been traced through the lineage.

❶ The process may have begun with hybridization between two wheats, one the cultivated species *T. monococcum* (AA), the other one of several wild species that probably grew as weeds at the edges of fields (BB). Chromosome sets A and B of the two species would not have been able to pair at meiosis, making the AB hybrid sterile. ❷ However, an error in cell division and self-fertilization would have produced a new species (AABB) with 28 chromosomes. Today, we know this species as emmer wheat (*T. turgidum*), varieties of which are grown widely in Eurasia and western North America. It is used mainly for making macaroni and other noodle products because its proteins hold their shape better than bread-wheat proteins.

The final steps in the evolution of bread wheat are thought to have occurred in early farming villages on the shores of European lakes more than 8,000 years ago. ❸ The cultivated emmer wheat, with its 28 chromosomes, hybridized spontaneously with the closely related wild species *T. tauschii* (DD), which has 14 chromosomes. The hybrid (ABD, with 21 chromosomes) was sterile, ❹ but a cell division error in this hybrid and self-fertilization doubled the chromosome number to 42. The result was bread wheat, with two each of the three ancestral sets of chromosomes (AABBDD).

Today, plant geneticists generate new polyploids in the laboratory by using chemicals that induce meiotic and mitotic cell division errors. Researchers can produce new hybrids with special qualities, such as a hybrid combining the high yield of wheat with the hardiness of rye.

? Why are errors in mitosis or meiosis a necessary part of speciation by hybridization between two species?

If a hybrid has a single copy of the chromosomes from two species, homologous pairs cannot join and separate during meiosis to produce gametes. Errors in mitosis or meiosis must somehow duplicate chromosomes so that there is a diploid number of each set and normal gametes can form.

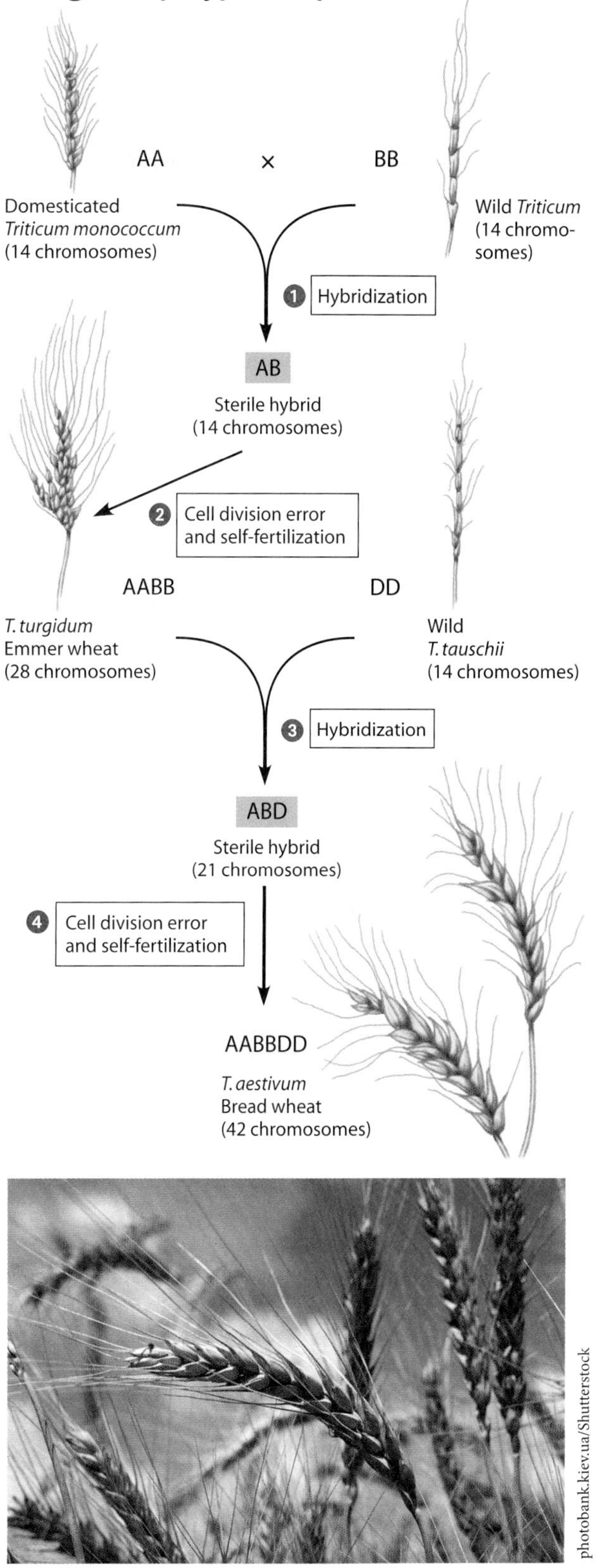

▲ **Figure 7** The evolution of bread wheat, *Triticum aestivum*

8 Isolated islands are often showcases of speciation

Although sympatric speciation is known to occur in plants and occasionally in animals, most of the diverse life-forms on Earth are thought to have originated by allopatric speciation. Isolated island chains offer some of the best evidence of this type of speciation. These groups of islands are often inhabited by unique collections of species. Islands that have physically diverse habitats and that are far enough apart to permit populations to evolve in isolation but close enough to allow occasional dispersions to occur are often the sites of multiple speciation events. The evolution of many diverse species from a common ancestor is known as **adaptive radiation**.

The Galápagos Archipelago, located about 900 km (560 miles) west of Ecuador, is one of the world's great showcases of adaptive radiation. Each island was born naked from underwater volcanoes and was gradually clothed by plants, animals, and microorganisms derived from strays that rode the ocean currents and winds from other islands and the South American mainland.

The Galápagos Islands today have numerous plants, snails, reptiles, and birds that are found nowhere else on Earth. For example, they have 14 species of closely related finches, which are often called Darwin's finches because Darwin collected them during his around-the-world voyage on the *Beagle*. These birds share many finchlike traits, but they differ in their feeding habits and their beaks, which are specialized for what they eat. Their various foods include insects, large or small seeds, cactus fruits, even eggs of other species. The woodpecker finch uses cactus spines or twigs as tools to pry insects from trees. The "vampire" finch is noted for pecking wounds on the backs of seabirds and drinking their blood. **Figure 8** illustrates some of these birds, with their distinctive beaks adapted for their specific diet.

How might Darwin's finch species have evolved from a small population of ancestral birds that colonized one of the islands? Completely isolated on the island, the founder population may have changed significantly as natural selection adapted it to the new environment, and thus it became a new species. Later, a few individuals of this species may have migrated to a neighboring island, where, under different conditions, this new founder population was changed enough through natural selection to become another new species. Some of these birds then may have recolonized the first island and coexisted there with the original ancestral species if reproductive barriers kept the species distinct. Multiple colonizations and speciations on the many separate islands of the Galápagos probably followed.

Today, each of the Galápagos Islands has several species of finches, with as many as 10 on some islands. The effects of the adaptive radiation of Darwin's finches are evident not just in their many types of beaks but also in their different habitats—some live in trees and others spend most of their time on the ground. Reproductive isolation due to species-specific songs helps keep the species separate.

The closely related finches of the Galápagos provide evidence not only of past speciation events. In the next module we see how scientists studying these birds are documenting current evolutionary changes.

? **Explain why isolated island chains provide opportunities for adaptive radiations.**

The chance colonization of an island often presents a species with new resources and an absence of predators. Through natural selection acting on existing variation, the island population becomes adapted to this new habitat and may evolve into a new species. Subsequent colonizations of nearby islands would provide additional opportunities for adaptation and genetic drift, which could lead to further speciations.

INTERFOTO/Alamy

Cactus-seed-eater (cactus finch)

Mary Plage/Photolibrary

Tool-using insect-eater (woodpecker finch)

Frans Lanting/DanitaDelimont.com

Seed-eater (medium ground finch)

▲ **Figure 8** Examples of differences in beak shape and size in Galápagos finches, each adapted for a specific diet

SCIENTIFIC DISCOVERY 9 A long-term field study documents evolution in Darwin's finches

Peter and Rosemary Grant have been conducting field research on Darwin's finches in the Galápagos for more than three decades. In 2009, the Grants received the prestigious Kyoto Prize for their long-term studies of natural selection and evolution in response to environmental changes.

How did the Grants come to work with these finches? They were looking for a pristine, undisturbed place to study variation within populations. In 1973, Peter banded about 60 medium ground finches (*Geospiza fortis*) on Daphne Major, an isolated, uninhabited island in the Galápagos. Returning eight months later with Rosemary and their young daughters, he found all but two of the banded birds. With such an opportunity to work with a small, isolated population, the Grants decided to study these birds for 3 years. One evolutionary question led to another, and for more than 34 years the Grants have done research on Daphne Major. Each year, the Grants and their student assistants have captured, tagged, measured, and studied every finch on this desolate, 100-acre island.

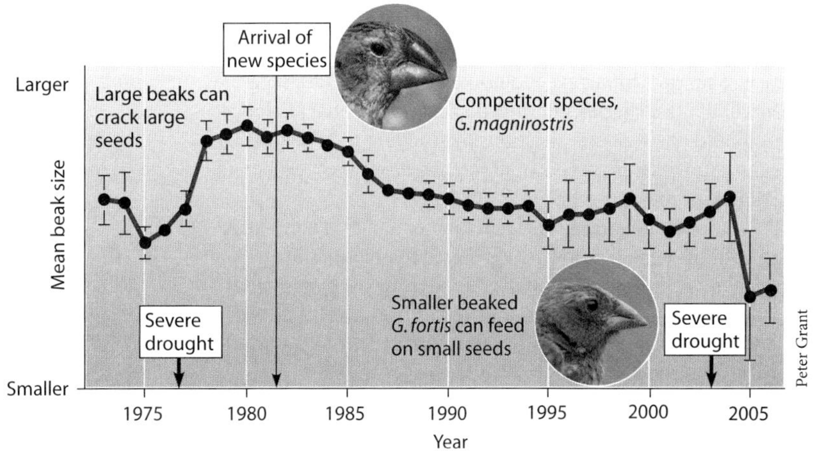

▲ **Figure 9** Changes in mean beak size in the medium ground finch (*G. fortis*)

What were some of the questions that kept the Grants returning for so many years? In addition to their study of beak size and shape and natural selection (see Module 13.3), the Grants focused on other evolutionary questions: How might competition for food from another species affect beak size? What keeps two finch species distinct despite their ability to interbreed? How might a new finch species originate?

During their long years of study, the Grants found opportunities to answer some of their questions. As the graph in **Figure 9** shows, they documented an increase in mean (average) beak size in the medium ground finch population following a severe drought in 1977, when the only seeds available were much tougher to crack than the small seeds that are abundant in wetter years. During the drought, 84% of the population disappeared. The survivors had much larger beaks, and this beak size was inherited by their offspring.

Then, in 1982, a new competitor species, the large ground finch (*G. magnirostris*), arrived on the island. It fed on large, hard seeds, taking more than were harvested by the resident medium ground finches. When another drought occurred in 2003–2004, the populations of both ground finches declined sharply as all seed supplies dwindled. As you can see in Figure 9, mean beak size decreased abruptly in the following year. Due to the depletion of large seeds by the competitive large ground finch, the medium ground finches that survived the drought had small beaks that most likely helped them feed more efficiently on very small seeds. Thus, over the 30 years of study, the average beak size of the medium ground finch has shifted in two different directions in response to two different selection pressures—drought and competition.

The Grants also study reproductive barriers between species. They found that occasional interbreeding between the medium ground finch and the cactus finch happens when a male learns to sing the song of a different species. For example, a cactus finch nestling whose father dies or does not sing much may learn a neighbor's song, even if the neighbor is a medium ground finch. Thus, a medium ground finch female might breed with a cactus finch because he sings her song.

To find out whether these interspecies couples would create a new hybrid species, the Grants followed their offspring. They found that the hybrids have intermediate bill sizes and thus can only survive during wet years when there are plenty of small seeds. During dry years, the hybrids can't crack the larger, harder seeds that the medium ground finches can eat and can't compete with cactus finches for cactus seeds. Thus, the severe selection during drought years helps keep the medium ground finch and the cactus finch on separate evolutionary paths.

Hybrids, however, can introduce more genetic variation on which natural selection can act. The Grants have documented increases in the variation within the medium ground finch and cactus finch populations when hybrids have bred with members of the parent species.

Peter Grant conjectures about hybrid finches:

> Perhaps hybrids occasionally disperse . . . to another island that has neither parent species. The hybrids could start a new population with a range of genetic variation different from the parent species. . . . I see no reason why hybridization hasn't been important right from the beginning, from the first divergence of the ancestral finch stock that reached the islands.

The research goes on. The Grants have teamed up with molecular scientists to identify the genes involved in beak development. The Galápagos continue to provide a laboratory in which varying natural conditions allow scientists to make and test predictions about evolution, natural selection, and speciation.

? What were the advantages of Daphne Major as a research site?

● The resident finch populations were small and isolated, and individual birds and their offspring could be followed over many years.

10 Hybrid zones provide opportunities to study reproductive isolation

What happens when separated populations of closely related species come back into contact with one another? Will reproductive barriers be strong enough to keep the species separate? Or will the two species interbreed and become one? Biologists attempt to answer such questions by studying **hybrid zones**, regions in which members of different species meet and mate, producing at least some hybrid offspring.

Figure 10A illustrates the formation of a hybrid zone, starting with the ancestral species. ❶ Three populations are connected by gene flow. ❷ A barrier to gene flow separates one population. ❸ This population diverges from the other two. ❹ Gene flow is reestablished in the hybrid zone. Let's consider possible outcomes for this hybrid zone over time.

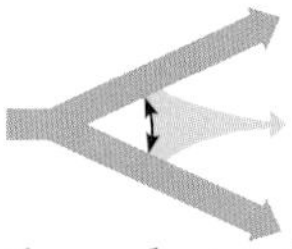

Reinforcement When hybrid offspring are less fit than members of both parent species, we might expect natural selection to strengthen, or *reinforce*, reproductive barriers, thus reducing the formation of unfit hybrids. And we would predict that barriers between species should be stronger where the species overlap (that is, where the species are sympatric).

As an example, consider the closely related pied flycatcher and collared flycatcher illustrated in **Figure 10B**. When populations of these two species do not overlap (that is, when they are allopatric), males closely resemble each other, with similar black and white coloration (see left side of Figure 10B). However, when populations of the two species are sympatric, male collared flycatchers are still black but with enlarged patches of white, whereas male pied flycatchers are a dull brown (see right side of Figure 10B). The photographs at the bottom of the figure show two pied flycatchers, the one on the left from a population that has no overlap with collared flycatchers and the one on the right from a population in an area where both species coexist. When scientists performed mate choice experiments, they found that female flycatchers frequently made mistakes when presented with males from allopatric populations, which look similar. But females never selected mates from the other species when presented with males from sympatric populations, which look different. Thus, reproductive barriers are reinforced when populations of these two species overlap.

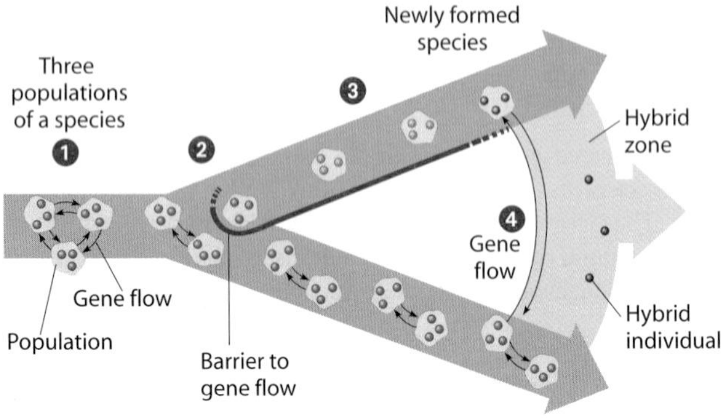

▲ **Figure 10A** Formation of a hybrid zone

▲ **Figure 10B** Reinforcement of reproductive barriers

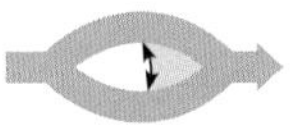

Fusion What happens when the reproductive barriers between species are not strong and the species come into contact in a hybrid zone? So much gene flow may occur that the speciation process reverses, causing the two hybridizing species to fuse into one.

Such a situation may be occurring among the cichlid species in Lake Victoria that we discussed in Module 6. Many species are reproductively isolated by female mate choice based on male coloration. As shown in **Figure 10C**, males of *Pundamilia nyererei* have a bright red back and dorsal fin, whereas males of the closely related species, *P. pundamilia*, have a bright blue dorsal fin.

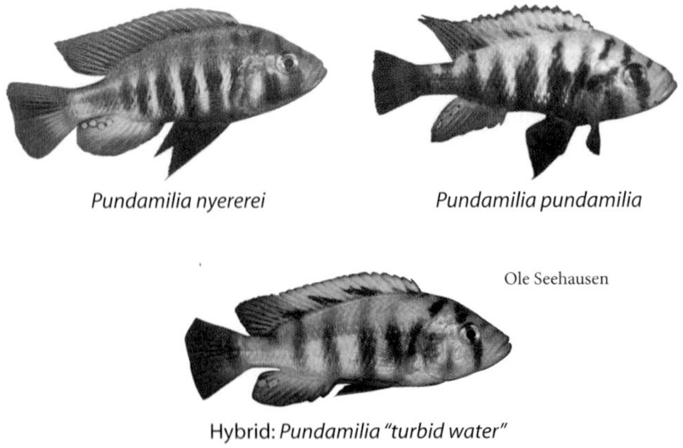

▲ **Figure 10C** Fusion: males of *Pundamilia nyererei* and *Pundamilia pundamilia* contrasted with a hybrid from an area with turbid water

Various mate choice experiments have shown that females prefer brightly colored—and rightly colored—males.

In the past 30 years, as many as 200 species of cichlids have disappeared from Lake Victoria. Some species were driven to extinction by an introduced predator, the Nile perch. But many species not eaten by Nile perch are also disappearing. The waters of Lake Victoria have become increasingly murky due to pollution over the past 30 years. Researchers have found that female cichlids from turbid water are less choosy with regard to male coloration, and males from these areas are less brightly colored. Many viable hybrid offspring are produced by interbreeding, and the once isolated gene pools of the parent species are combining—two species fusing into a single hybrid species (see bottom of Figure 10C).

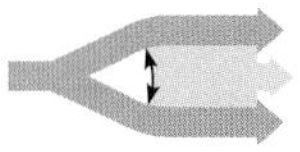

Stability One might predict that either reinforcement of reproductive barriers or fusion of gene pools would occur in a hybrid zone. However, many hybrid zones are fairly stable, and hybrids continue to be produced. Although these hybrids allow for some gene flow between populations, each species maintains its own integrity. The island inhabited by medium ground finches and cactus finches, with their occasional hybrid offspring, is an example of a stable hybrid zone.

? Why might hybrid zones be called "natural laboratories" in which to study speciation?

By studying the fate of hybrids over time, scientists can directly observe factors that cause (or fail to cause) reproductive isolation.

11 Speciation can occur rapidly or slowly

Biologists continue to make field observations and devise experiments to study evolution in progress. However, much of the evidence for evolution comes from the fossil record. What does this record say about the process of speciation?

Many fossil species appear suddenly in a layer of rock and persist essentially unchanged through several layers (strata) until disappearing just as suddenly. Paleontologists Niles Eldredge and the late Stephen Jay Gould coined the term **punctuated equilibria** to describe these long periods of little change, or equilibrium, punctuated by abrupt episodes of speciation. The top of **Figure 11** illustrates the evolution of two lineages of butterflies in a punctuated pattern. Notice that the butterfly species change little, if at all, once they appear.

Other fossil species appear to have diverged gradually over long periods of time. As shown at the bottom of Figure 11, differences gradually accumulate, and new species (represented by the two butterflies at the far right) evolve gradually from the ancestral population.

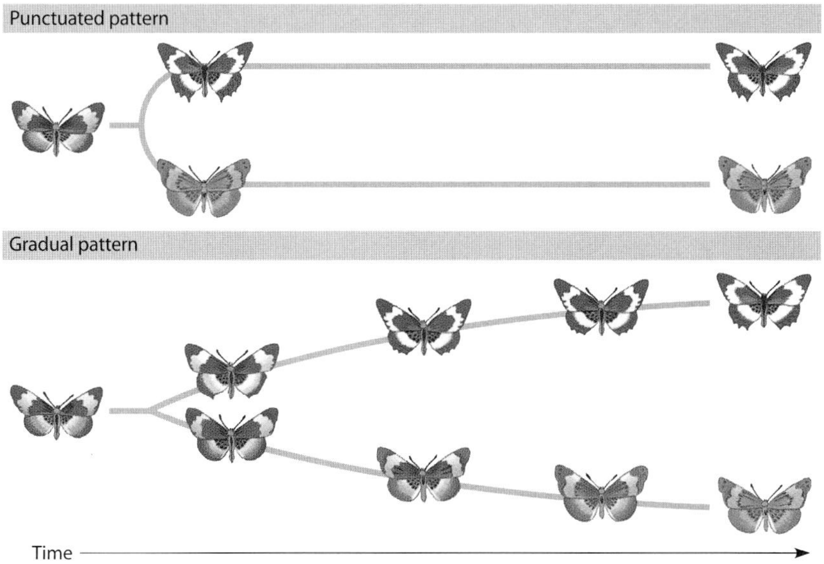

▲ **Figure 11** Two models for the tempo of speciation

Even when fossil evidence points to a punctuated pattern, species may not have originated as rapidly as it appears. Suppose that a species survived for 5 million years but that most of the changes in its features occurred during the first 50,000 years of its existence. Time periods this short often cannot be distinguished in fossil strata. And should a new species originate from a small, isolated population—as no doubt many species have—the chances of fossils being found are low.

But what about the total length of time between speciation events—between when a new species forms and when its populations diverge enough to produce another new species? In one survey of 84 groups of plants and animals, this time ranged from 4,000 to 40 million years. Overall, the time between speciation events averaged 6.5 million years. What are the implications of such long time frames? They tell us that it has taken vast spans of time for life on Earth to evolve and that it takes a long time for life to recover from mass extinctions, as have occurred in the past and may be occurring now.

As you've seen, speciation may begin with small differences. However, as species diverge and speciate again and again, these changes may eventually lead to new groups that differ greatly from their ancestors. The cumulative effects of multiple speciations, as well as extinctions, have shaped the dramatic changes documented in the fossil record.

? How does the punctuated equilibrium model account for the relative rarity of transitional fossils linking newer species to older ones?

If speciation takes place in a relatively short time or in a small isolated population, the transition of one species to another may be difficult to find in the fossil record.

REVIEW

For Practice Quizzes, BioFlix, MP3 Tutors, and Activities, go to www.masteringbiology.com.

Reviewing the Concepts

Defining Species (1–3)

1 The origin of species is the source of biological diversity. Speciation, the process by which one species splits into two or more species, accounts for both the unity and diversity of life.

2 There are several ways to define a species. The biological species concept holds that a species is a group of populations whose members can interbreed and produce fertile offspring with each other but not with members of other species. This concept emphasizes reproductive isolation. Most organisms are classified based on observable traits—the morphological species concept.

3 Reproductive barriers keep species separate. Such barriers isolate a species' gene pool and prevent interbreeding.

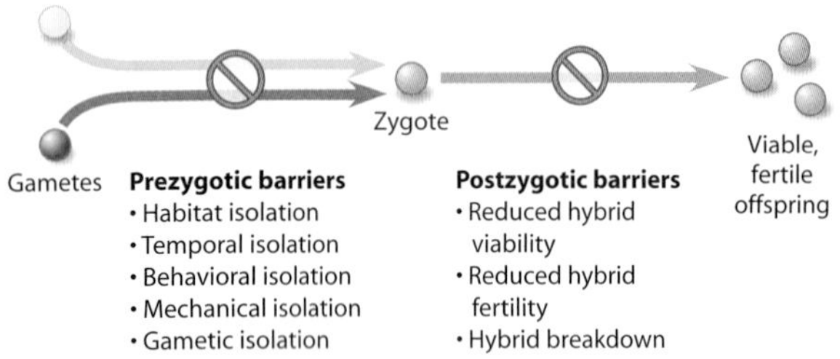

Mechanisms of Speciation (4–11)

4 In allopatric speciation, geographic isolation leads to speciation. Geographically separated from other populations, a small population may become genetically unique as its gene pool is changed by natural selection, mutation, or genetic drift.

5 Reproductive barriers can evolve as populations diverge. Researchers have documented the beginning of reproductive isolation in fruit fly populations adapting to different food sources and have identified a gene for flower color involved in the pollinator choice that helps separate monkey flower species.

6 Sympatric speciation takes place without geographic isolation. Many plant species have evolved by polyploidy, duplication of the chromosome number due to errors in cell division. Habitat differentiation and sexual selection, usually involving mate choice, can lead to sympatric (and allopatric) speciation.

7 Most plant species trace their origin to polyploid speciation. Many plants, including food plants such as bread wheat, are the result of hybridization and polyploidy.

8 Isolated islands are often showcases of speciation. Repeated isolation, speciation, and recolonization events on isolated island chains have led to adaptive radiations of species, many of which are found nowhere else in the world.

9 A long-term field study documents evolution in Darwin's finches. Peter and Rosemary Grant have documented changes in beak size resulting from natural selection due to drought and competition, as well as interbreeding between some Galápagos finch species.

10 Hybrid zones provide opportunities to study reproductive isolation. Hybrid zones are regions in which populations of different species overlap and produce at least some hybrid offspring. Over time, reinforcement may strengthen barriers to reproduction, or fusion may reverse the speciation process as gene flow between species increases. In stable hybrid zones, a limited number of hybrid offspring continue to be produced.

11 Speciation can occur rapidly or slowly. The punctuated equilibria model draws on the fossil record, where species change most as they arise from an ancestral species and then change relatively little for the rest of their existence. Other species appear to have evolved more gradually. The time interval between speciation events varies from a few thousand years to tens of millions of years.

Connecting the Concepts

1. Name the two types of speciation represented by this diagram. For each type, describe how reproductive barriers may develop between the new species.

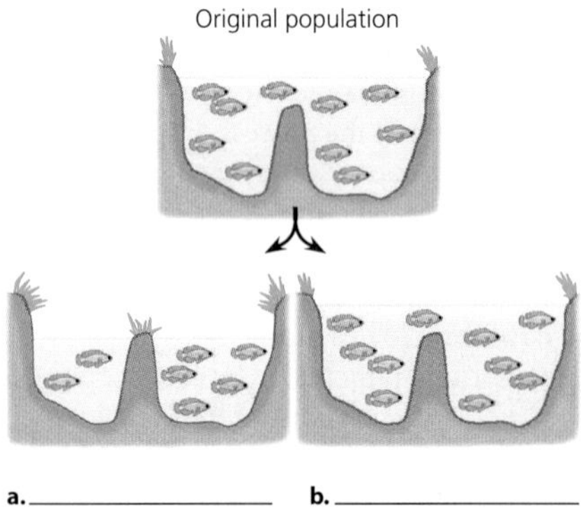

2. Fill in the blanks in the following concept map.

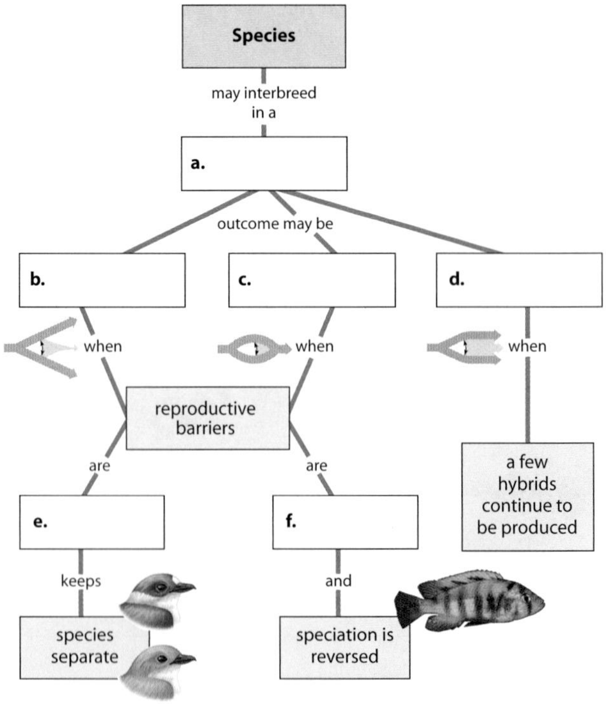

Testing Your Knowledge

Multiple Choice

3. Which concept of species would be most useful to a field biologist identifying new plant species in a tropical forest?
   a. biological
   b. ecological
   c. morphological
   d. phylogenetic
   e. a and b combined
4. The *largest* unit within which gene flow can readily occur is a
   a. population.
   b. species.
   c. genus.
   d. phylum.
   e. hybrid offspring.
5. Bird guides once listed the myrtle warbler and Audubon's warbler as distinct species that lived side by side in parts of their ranges. However, recent books show them as eastern and western forms of a single species, the yellow-rumped warbler. Most likely, it has been found that these two kinds of warblers
   a. live in similar habitats and eat similar foods.
   b. interbreed often in nature, and the offspring are viable and fertile.
   c. are almost identical in appearance.
   d. have many genes in common.
   e. sing similar songs.
6. Which of the following is an example of a postzygotic reproductive barrier?
   a. One *Ceanothus* shrub lives on acid soil, another on alkaline soil.
   b. Mallard and pintail ducks mate at different times of year.
   c. Two species of leopard frogs have different mating calls.
   d. Hybrid offspring of two species of jimsonweeds always die before reproducing.
   e. Pollen of one species of tobacco cannot fertilize a different species of tobacco.
7. Biologists have found more than 500 species of fruit flies on the various Hawaiian Islands, all apparently descended from a single ancestor species. This example illustrates
   a. polyploidy.
   b. temporal isolation.
   c. adaptive radiation.
   d. sympatric speciation.
   e. postzygotic barriers.
8. A new plant species C, which formed from hybridization of species A ($2n = 16$) with species B ($2n = 12$), would probably produce gametes with a chromosome number of
   a. 12.
   b. 14.
   c. 16.
   d. 28.
   e. 56.
9. A horse ($2n = 64$) and a donkey ($2n = 62$) can mate and produce a mule. How many chromosomes would there be in a mule's body cells?
   a. 31
   b. 62
   c. 63
   d. 126
   e. 252
10. What prevents horses and donkeys from hybridizing to form a new species?
   a. limited hybrid fertility
   b. limited hybrid viability
   c. hybrid breakdown
   d. gametic isolation
   e. prezygotic barrier
11. When hybrids produced in a hybrid zone can breed with each other and with both parent species, and they survive and reproduce as well as members of the parent species, one would predict that
   a. the hybrid zone would be stable.
   b. sympatric speciation would occur.
   c. reinforcement of reproductive barriers would keep the parent species separate.
   d. reproductive barriers would lessen and the two parent species would fuse.
   e. a new hybrid species would form through allopatric speciation.
12. Which of the following factors would *not* contribute to allopatric speciation?
   a. A population becomes geographically isolated from the parent population.
   b. The separated population is small, and genetic drift occurs.
   c. The isolated population is exposed to different selection pressures than the parent population.
   d. Different mutations begin to distinguish the gene pools of the separated populations.
   e. Gene flow between the two populations continues to occur.

Describing, Comparing, and Explaining

13. Explain how each of the following makes it difficult to clearly define a species: variation within a species, geographically isolated populations, asexual species, fossil organisms.
14. Explain why allopatric speciation would be less likely on an island close to a mainland than on a more isolated island.
15. What does the term *punctuated equilibria* describe?
16. Can factors that cause sympatric speciation also cause allopatric speciation? Explain.

Applying the Concepts

17. Cultivated American cotton plants have a total of 52 chromosomes ($2n = 52$). In each cell, there are 13 pairs of large chromosomes and 13 pairs of smaller chromosomes. Old World cotton plants have 26 chromosomes ($2n = 26$), all large. Wild American cotton plants have 26 chromosomes, all small. Propose a testable hypothesis to explain how cultivated American cotton probably originated.
18. Explain how the murky waters of Lake Victoria may be contributing to the decline in cichlid species. How might these polluted waters affect the formation of new species? How might the loss of cichlid species be slowed?
19. The red wolf, *Canis rufus,* which was once widespread in the southeastern and southcentral United States, was declared extinct in the wild by 1980. Saved by a captive breeding program, the red wolf has been reintroduced in areas of eastern North Carolina. The current wild population estimate is about 100 individuals. It is presently being threatened with extinction due to hybridization with coyotes, *Canis latrans*, which have become more numerous in the area. Red wolves and coyotes differ in terms of morphology, DNA, and behavior, although these differences may disappear if interbreeding continues. Although designated as an endangered species under the Endangered Species Act, some people think that the endangered status of red wolves should be withdrawn and resources should not be spent to protect what is not a "pure" species. Do you agree? Why or why not?

Review Answers

1. a. Allopatric speciation: Reproductive barriers may evolve between these two geographically separated populations as a by-product of the genetic changes associated with each population's adaptation to its own environment or as a result of genetic drift or mutation.
   b. Sympatric speciation: Some change, perhaps in resource use or female mate choice, may lead to a reproductive barrier that isolates the gene pools of these two populations, which are not separated geographically. Once the gene pools are separated, each species may go down its own evolutionary path. If speciation occurs by polyploidy—which is common in plants but unusual in animals—then the new species is instantly isolated from the parent species.
2. a. hybrid zone; b. reinforcement; c. fusion; d. stability; e. strengthened; f. weakened or eliminated
3. c **4.** b **5.** b **6.** d **7.** c **8.** b **9.** c **10.** a **11.** d **12.** e
13. Different physical appearances may indicate that organisms belong in different species; but they may just be physical differences within a species. Isolated populations may or may not be able to interbreed; breeding experiments would need to be performed to determine this. Organisms that reproduce only asexually and fossil organisms do not have the potential to interbreed and produce fertile offspring; therefore, the biological species concept cannot apply to them.
14. There is more chance for gene flow between populations on a mainland and nearby island. This interbreeding would make it more difficult for reproductive isolation to develop and separate the two populations.
15. The term *punctuated equilibria* refers to a common pattern seen in the fossil record, in which most species diverge relatively quickly as they arise from an ancestral species and then remain fairly unchanged for the rest of their existence as a species.
16. Yes. Factors such as polyploidy, sexual selection, and habitat specialization can lead to reproductive barriers that would separate the gene pools of allopatric as well as sympatric populations.
17. A broad hypothesis would be that cultivated American cotton arose from a sequence of hybridization, mistakes in cell division, and self-fertilization. We can divide this broad statement into at least three hypotheses. *Hypothesis 1*: The first step in the origin of cultivated American cotton was hybridization between a wild American cotton plant (with 13 pairs of small chromosomes) and an Old World cotton plant (with 13 pairs of large chromosomes). If this hypothesis is correct, we would predict that the hybrid offspring would have had 13 small chromosomes and 13 large chromosomes. *Hypothesis 2*: The second step in the origin of cultivated American cotton was a failure of cell division in the hybrid offspring, such that all chromosomes were duplicated (now 26 small and 26 large). If this hypothesis is true, we would expect the resulting gametes to each have had 13 large chromosomes and 13 small chromosomes. *Hypothesis 3*: The third step in the origin of cultivated American cotton was self-fertilization of these gametes. If this hypothesis is true, we would expect the outcome of self-fertilization to be a hybrid plant with 52 chromosomes: 13 pairs of large ones and 13 pairs of small ones. Indeed, this is the genetic makeup of cultivated American cotton.
18. By decreasing the ability of females to distinguish males of their own species, the polluted turbid waters have increased the frequency of mating between members of species that had been reproductively isolated from one another. As the number of hybrid fish increase, the parent species' gene pools may fuse, resulting in a loss of the two separate parent species and the formation of a new hybrid species. Future speciation events in Lake Victoria cichlids are less likely to occur in turbid water because females are less able to base mate choice on male breeding color. Reducing the pollution in the lake may help reverse this trend.
19. Some issues and questions to consider: One could look at this question in two ways: If the biological species concept is followed strictly, one could argue that red wolves and coyotes are the same species, since they can interbreed. Because coyotes are not rare, this line of argument would suggest that red wolves should not be protected. On the other hand, because red wolves and coyotes differ in many ways, they can be viewed as distinct species by other species concepts. Protecting the remaining red wolves from hybridizing with coyotes can preserve their distinct species status. The rationale behind protecting all endangered groups is the desire to preserve genetic diversity. Questions for society in general are: What is the value of any particular species and its genetically distinct subgroups? And how far are we willing to go to preserve a rare and distinct group of organisms? How should the costs of preserving genetic diversity compare with the costs of other public projects?

Tracing Evolutionary History

Tracing Evolutionary History

BIG IDEAS

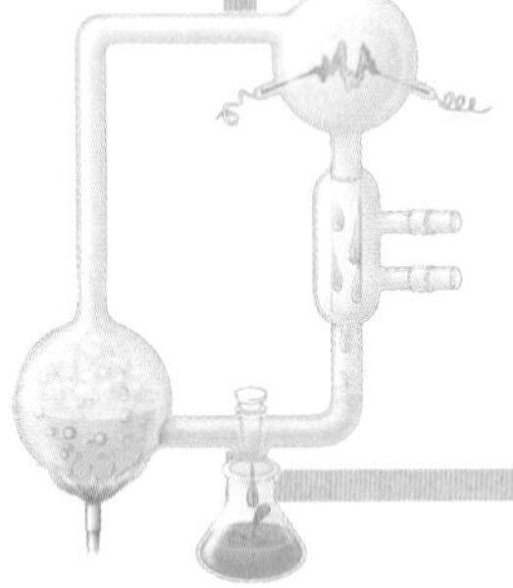

Early Earth and the Origin of Life (1–3)

Scientific experiments can test the four-stage hypothesis of how life originated on early Earth.

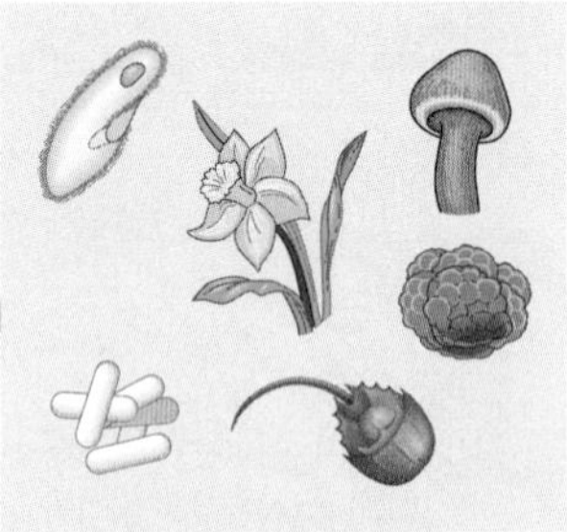

Major Events in the History of Life (4–6)

The fossil record and radiometric dating establish a geologic record of key events in life's history.

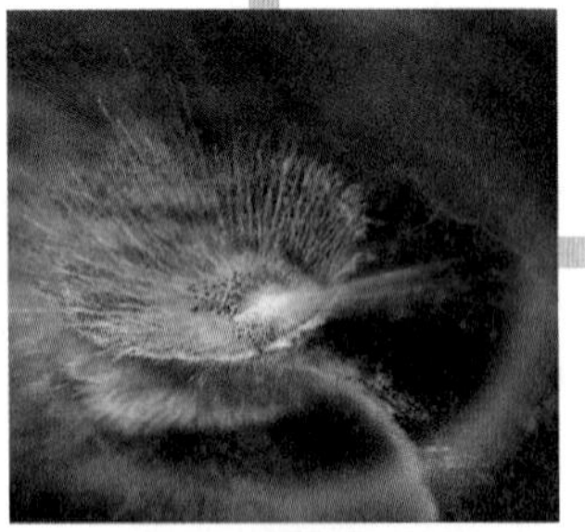

Joe Tucciarone/Photo Researchers

Mechanisms of Macroevolution (7–13)

Continental drift, mass extinctions, adaptive radiations, and changes in developmental genes have all contributed to macroevolution.

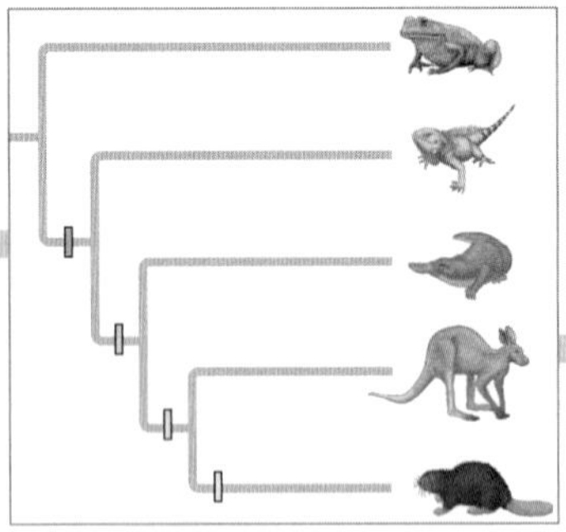

Phylogeny and the Tree of Life (14–19)

The evolutionary history of a species is reconstructed using fossils, homologies, and molecular systematics.

Theo Allofs/CorbisS

BIOS Huguet Pierre/Peter Arnold, Inc./PhotoLibrary

What do the three flying vertebrates pictured above have in common? Over the course of evolution, the walking forelimbs of their ancestors have become remodeled as flying wings.

Pterosaurs were the first to take to the air. These flying reptiles existed for about 140 million years and then went extinct about 65 million years ago, along with most other dinosaurs. Birds evolved from feathered dinosaurs about 150 million years ago. Bats diverged from insectivorous, tree-dwelling mammals some 50 million years ago. Different times, different ancestors, same ability—powered flight. So are the wings of pterosaurs, birds, and bats the same or different?

As you can see in the artist's depiction of a pterosaur, its membranous wing was supported along much of its length by one greatly elongated finger. A bird wing is composed of feathers and is supported by an elongated forearm and modified wrist and hand bones. Finger bones are reduced and fused (imagine the bones of a chicken wing). Bat wings consist of a membrane supported by the arm bones and four very elongated fingers.

As different as the architecture of these wings are, however, all three structures represent the remodeling of the same ancestral forelimb. These examples of Darwin's "descent with modification" illustrate again that evolution is a remodeling process, which can gradually adapt existing structures to new functions.

In this chapter, we turn to macroevolution, the major changes recorded in the history of life over vast tracts of time. We also explore some of the mechanisms responsible for such changes. Finally, we consider how scientists organize the amazing diversity of life according to evolutionary relationships. To approach these wide-ranging topics, we begin with the most basic of questions: How did life first arise on planet Earth?

Early Earth and the Origin of Life

1 Conditions on early Earth made the origin of life possible

Earth is one of eight planets orbiting the sun, which is one of billions of stars in the Milky Way. The Milky Way, in turn, is one of billions of galaxies in the universe. The star closest to our sun is 40 trillion kilometers away.

The universe has not always been so spread out. Physicists have evidence that before the universe existed in its present form, all matter was concentrated in one mass. The mass seems to have blown apart with a "big bang" sometime between 12 and 14 billion years ago and has been expanding ever since.

Scientific evidence indicates that Earth formed about 4.6 billion years ago from a vast swirling cloud of dust that surrounded the young sun. As gases, dust, and rocks collided and stuck together, larger bodies formed, and the gravity of the larger bodies in turn attracted more matter, eventually forming Earth and other planets.

Conditions on Early Earth Immense heat would have been generated by the impact of meteorites and compaction by gravity, and young planet Earth probably began as a molten mass. The mass then sorted into layers of varying densities, with the least dense material on the surface, solidifying into a thin crust.

As the bombardment of early Earth slowed about 3.9 billion years ago, conditions on the planet were extremely different from those of today. The first atmosphere was probably thick with water vapor, along with various compounds released by volcanic eruptions, including nitrogen and its oxides, carbon dioxide, methane, ammonia, hydrogen, and hydrogen sulfide. As Earth slowly cooled, the water vapor condensed into oceans. Not only was the atmosphere of young Earth very different from the atmosphere we know today, but lightning, volcanic activity, and ultraviolet radiation were much more intense.

When Did Life Begin? The earliest evidence of life on Earth comes from fossils that are about 3.5 billion years old. One of these fossils is pictured in the inset in **Figure 1**; the larger illustration is an artist's rendition of what Earth may have looked like at that time. Life is already present in this painting, as shown by the "stepping stones" that dominate the shoreline. These rocks, called **stromatolites**, were built up by ancient photosynthetic prokaryotes. As evident in the fossil stromatolite shown in the inset, the rocks are layered. The prokaryotes that built them bound thin films of sediment together, then migrated to the surface and started the next layer. Similar layered mats are still being formed today by photosynthetic prokaryotes in a few shallow, salty bays, such as Shark Bay, in western Australia.

Photosynthesis is not a simple process, so it is likely that significant time had elapsed before life as complex as the organisms that formed the ancient stromatolites had evolved. The evidence that these prokaryotes lived 3.5 billion years ago is strong support for the hypothesis that life in a simpler form arose much earlier, perhaps as early as 3.9 billion years ago.

How Did Life Arise? From the time of the ancient Greeks until well into the 1800s, it was commonly believed that life arises from nonliving matter. Many people believed, for instance, that flies come from rotting meat and fish from ocean mud. Experiments by the French scientist Louis Pasteur in 1862, however, confirmed that all life arises only by the reproduction of preexisting life.

Pasteur ended the argument over spontaneous generation of present-day organisms, but he did not address the question of how life arose in the first place. To attempt to answer that question, for which there is no fossil evidence available, scientists develop hypotheses and test their predictions.

Scientists hypothesize that chemical and physical processes on early Earth could have produced very simple cells through a sequence of four main stages:

1. The abiotic (nonliving) synthesis of small organic molecules, such as amino acids and nitrogenous bases
2. The joining of these small molecules into polymers, such as proteins and nucleic acids
3. The packaging of these molecules into "protocells," membrane-enclosed droplets that maintained an internal chemistry different from that of their surroundings
4. The origin of self-replicating molecules that eventually made inheritance possible

In the next two modules, we examine some of the experimental evidence for each of these four stages.

? Why do 3.5-billion-year-old stromatolites suggest that life originated *before* 3.5 billion years ago?

If photosynthetic prokaryotes existed by 3.5 billion years ago, a simpler, nonphotosynthetic cell probably originated well before that time.

Francois Gohier/Photo Researchers

Peter Sawyer/Chip Clark

▲ **Figure 1** A depiction of Earth about 3 billion years ago (inset: photo of a cross section of a fossilized stromatolite)

SCIENTIFIC DISCOVERY

2 Experiments show that the abiotic synthesis of organic molecules is possible

In 1953, when Stanley Miller was a 23-year-old graduate student in the laboratory of Harold Urey at the University of Chicago, he performed experiments that attracted global attention. Miller was the first to show that amino acids and other organic molecules could be formed under conditions thought to simulate those of early Earth.

Miller's experiments were a test of a hypothesis about the origin of life developed in the 1920s by Russian chemist A. I. Oparin and British scientist J. B. S. Haldane. Oparin and Haldane independently proposed that conditions on early Earth could have generated organic molecules. They reasoned that present-day conditions on Earth do not allow the spontaneous synthesis of organic compounds simply because the atmosphere is now rich in oxygen. As a strong oxidizing agent, O_2 tends to disrupt chemical bonds. However, before the early photosynthetic prokaryotes added O_2 to the air, Earth may have had a reducing (electron-adding) atmosphere. The energy for this abiotic synthesis of organic compounds could have come from lightning and intense UV radiation.

Figure 2A is a diagram of the apparatus used in Miller's experiment. A flask of warmed water represented the primeval sea. ❶ The water was heated so that some vaporized and moved into a second, higher flask. ❷ The "atmosphere" in this higher flask consisted of water vapor, hydrogen gas (H_2), methane (CH_4), and ammonia (NH_3)—the gases that scientists at the time thought prevailed in the ancient world. Electrodes discharged sparks into the flask to mimic lightning. ❸ A condenser with circulating cold water cooled the atmosphere, raining water and any dissolved compounds back down into the miniature sea. ❹ As material cycled through the apparatus, Miller periodically collected samples for chemical analysis.

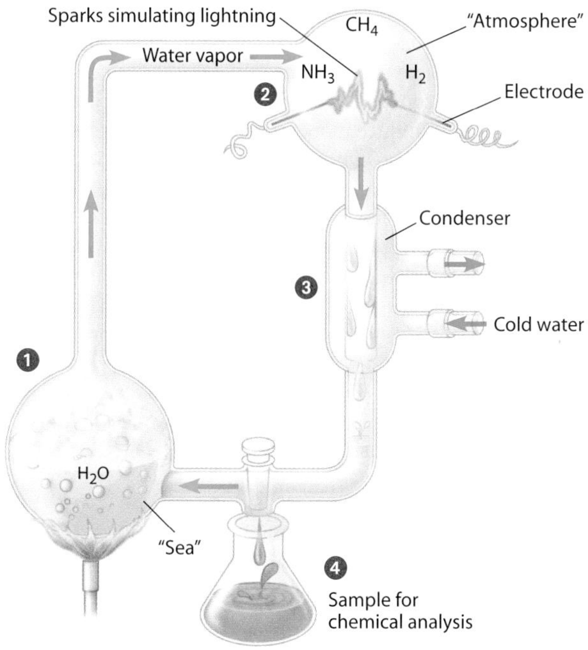

▲ **Figure 2A** The synthesis of organic compounds in Miller's 1953 experiment

Miller identified a variety of organic molecules that are common in organisms, including hydrocarbons (long chains of carbon and hydrogen) and some of the amino acids that make up proteins. Many laboratories have since repeated Miller's classic experiment using various atmospheric mixtures and produced organic compounds.

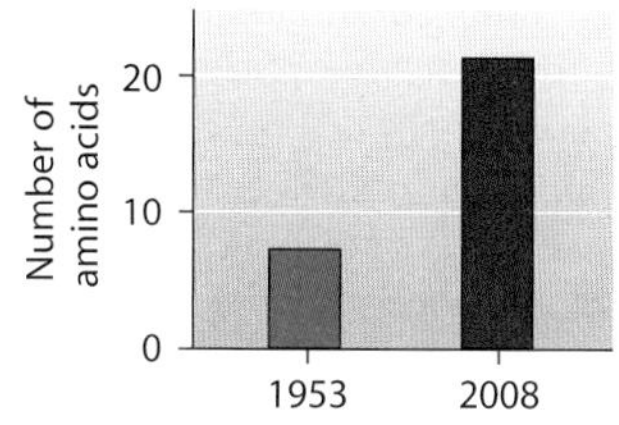

▲ **Figure 2B** Numbers of amino acids synthesized in Miller's original 1953 experiment and found in a 2008 reanalysis of samples from his simulated volcanic eruption

There is some evidence that the early atmosphere was made up primarily of nitrogen and carbon dioxide and was not as reducing as once assumed. Nevertheless, recent experiments using such atmospheres have also produced organic molecules. And it is possible that small "pockets" of the early atmosphere—perhaps near volcanic openings—were reducing. In 2008, a former graduate student of Miller's discovered some samples from an experiment that Miller had designed to mimic volcanic conditions. Reanalyzing these samples using modern equipment, he identified additional organic compounds that had been synthesized. Indeed, as **Figure 2B** shows, many more amino acids had been produced under Miller's simulated volcanic conditions than were produced in his original 1953 experiment.

Alternatively, submerged volcanoes or deep-sea hydrothermal vents—gaps in Earth's crust where hot water and minerals gush into deep oceans—may have provided the initial chemical resources for life. Such environments are among the most extreme in which life exists today, and some researchers are exploring the hypothesis that life may have begun in similar regions on early Earth.

Another source of organic molecules may have been meteorites. Fragments of a 4.5-billion-year-old meteorite that fell to Earth in Australia in 1969 contain more than 80 types of amino acids, some in large amounts. Recent studies have shown that this meteorite also contains other key organic molecules, including lipids, simple sugars, and nitrogenous bases such as uracil. Chemical analyses show that these organic compounds are not contaminants from Earth. Research will continue on the possible origins of organic molecules on early Earth.

? Which of the four stages in the hypothetical scenario of the origin of simple cells was Stanley Miller testing with his experiments?

Stage 1: Conditions on early Earth favored synthesis of organic molecules important to life, such as amino acids, from simpler ingredients.

Scientific Discovery: Adapted from an interview in Neil Campbell, *Biology,* 2nd ed.

3 Stages in the origin of the first cells probably included the formation of polymers, protocells, and self-replicating RNA

The abiotic synthesis of small organic molecules would have been a first step in the origin of life. But what is the evidence that the next three stages could have occurred on early Earth?

Abiotic Synthesis of Polymers In a cell, enzymes catalyze the joining of monomers to build polymers. Could this happen without enzymes? Scientists have produced polymers in the laboratory by dripping dilute solutions of amino acids or RNA monomers onto hot sand, clay, or rock. The heat vaporizes the water and concentrates the monomers, some of which then spontaneously bond together in chains. On early Earth, waves may have splashed organic monomers onto fresh lava or other hot rocks and then rinsed polypeptides and other polymers back into the sea.

Formation of Protocells A key step in the origin of life would have been the isolation of a collection of organic molecules within a membrane-enclosed compartment. Laboratory experiments demonstrate that small membrane-enclosed sacs or vesicles form when lipids are mixed with water. When researchers add to the mixture a type of clay thought to have been common on early Earth, such vesicles form at a faster rate. Organic molecules become concentrated on the surface of this clay and thus more easily interact. As shown by the smaller droplets forming in **Figure 3A**, these abiotically created vesicles are able to grow and divide (reproduce). Researchers have shown that these vesicles can absorb clay particles to which RNA and other molecules are attached. In a similar fashion, protocells on early Earth may have been able to form, reproduce, and create and maintain an internal environment different from their surroundings.

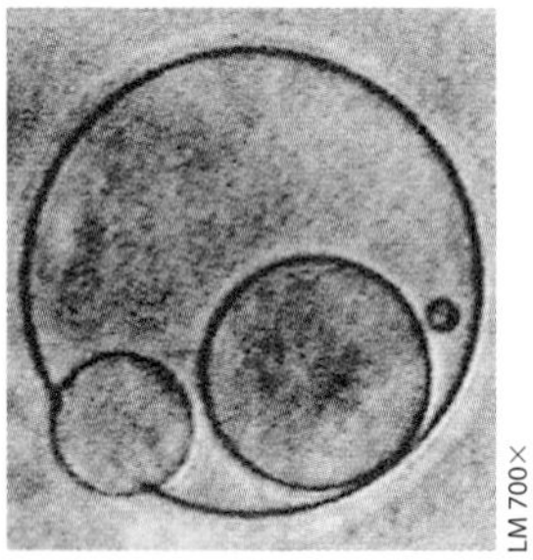
LM 700×

Fred M. Menger and Kurt Gabrielson, Emory University

▲ **Figure 3A** Microscopic vesicle, with membranes made of lipids, "giving birth" to smaller vesicles

Self-Replicating RNA Today's cells store their genetic information as DNA, transcribe the information into RNA, and then translate RNA messages into proteins. As we have seen in Chapter 10, this DNA → RNA → protein assembly system is extremely intricate. Most likely, it emerged gradually through a series of refinements of much simpler processes.

What were the first genes like? One hypothesis is that they were short strands of self-replicating RNA. Laboratory experiments have shown that short RNA molecules can assemble spontaneously from nucleotide monomers. Furthermore, when RNA is added to a solution containing a supply of RNA monomers, new RNA molecules complementary to parts of the starting RNA sometimes assemble. So we can imagine a scenario on early Earth like the one in **Figure 3B**: ❶ RNA monomers adhere to clay particles and become concentrated. ❷ Some monomers spontaneously join, forming the first small "genes." ❸ Then an RNA chain complementary to one of these genes assembles. If the new chain, in turn, serves as a template for another round of RNA assembly, the result is a replica of the original gene.

This replication process could have been aided by the RNA molecules themselves, acting as catalysts for their own replication. The discovery that some RNAs, which scientists call **ribozymes**, can carry out enzyme-like functions supports this hypothesis. Thus, the "chicken and egg" paradox of which came first, genes or enzymes, may be solved if the chicken and egg came together in the same RNA molecules. Scientists use the term "RNA world" for the hypothetical period in the evolution of life when RNA served as both rudimentary genes and catalytic molecules.

Once some protocells contained self-replicating RNA molecules, natural selection would have begun to shape their properties. Those that contained genetic information that helped them grow and reproduce more efficiently than others would have increased in number, passing their abilities on to subsequent generations. Mutations, errors in copying RNA "genes," would result in additional variation on which natural selection could work. At some point during millions of years of selection, DNA, a more stable molecule, replaced RNA as the repository of genetic information, and protocells passed a fuzzy border to become true cells. The stage was then set for the evolution of diverse life-forms, changes that we see documented in the fossil record.

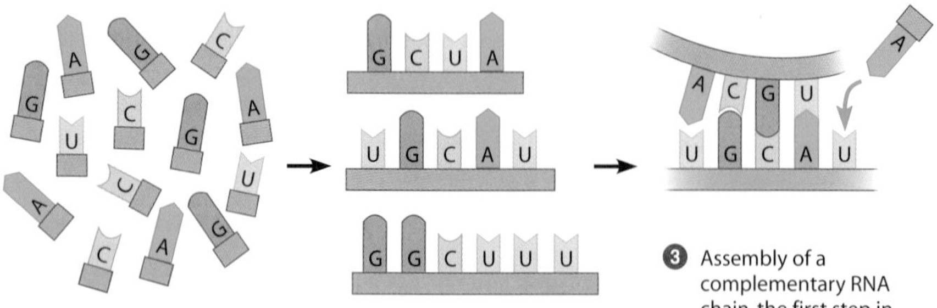
❶ Collection of monomers
❷ Formation of short RNA polymers: simple "genes"
❸ Assembly of a complementary RNA chain, the first step in the replication of the original "gene"

▲ **Figure 3B** A hypothesis for the origin of the first genes

? **Why would the formation of protocells represent a key step in the evolution of life?**

Segregating mixtures of molecules within compartments could concentrate organic molecules and facilitate chemical reactions. Natural selection could act on protocells once self-replicating "genes" evolved.

Major Events in the History of Life

4 The origins of single-celled and multicellular organisms and the colonization of land were key events in life's history

We now begin our study of **macroevolution**, the broad pattern of changes in life on Earth. **Figure 4** shows a timeline from the origin of Earth 4.6 billion years ago to the present. The colored bands represent some key origin points in the history of life. Earth's history can be divided into three eons of geologic time. The Archaean and the Proterozoic eons together lasted about 4 billion years. The Phanerozoic eon includes the last half billion years.

Origin of Prokaryotes As we discussed, the earliest evidence of life comes from fossil stromatolites formed by ancient photosynthetic prokaryotes (see Figure 1). Prokaryotes (the gold band in Figure 4) were Earth's sole inhabitants from at least 3.5 billion years ago to about 2 billion years ago. During this time, prokaryotes transformed the biosphere. As a result of prokaryotic photosynthesis, oxygen saturated the seas and began to appear in the atmosphere 2.7 billion years ago (the teal band). By 2.2 billion years ago, atmospheric O_2 began to increase rapidly, causing an "oxygen revolution." Many prokaryotes were unable to live in this aerobic environment and became extinct, while some species survived in anaerobic habitats, where we find their descendants living today. The evolution of cellular respiration, which uses O_2 in harvesting energy from organic molecules, allowed other prokaryotes to flourish.

Origin of Single-celled Eukaryotes The oldest widely accepted fossils of eukaryotes are about 2.1 billion years old (the salmon band). The more complex eukaryotic cell originated when small prokaryotic cells capable of aerobic respiration or photosynthesis took up life inside larger cells. After the first eukaryotes appeared, a great range of unicellular forms evolved, giving rise to the diversity of single-celled eukaryotes that continue to flourish today.

Origin of Multicellular Eukaryotes Another wave of diversification followed: the origin of multicellular forms whose descendants include a variety of algae, plants, fungi, and animals. Molecular comparisons suggest that the common ancestor of multicellular eukaryotes arose about 1.5 billion years ago (the light blue band). The oldest known fossils of multicellular eukaryotes are of relatively small algae that lived 1.2 billion years ago.

Larger and more diverse multicellular organisms do not appear in the fossil record until about 575 million years ago. A great increase in the diversity of animal forms occurred 535–525 million years ago, in a period known as the Cambrian explosion. Animals are shown on the timeline by the bright blue band.

Colonization of Land There is fossil evidence that photosynthetic prokaryotes coated damp terrestrial surfaces well over a billion years ago. However, larger forms of life did not begin to colonize land until about 500 million years ago (the purple band).

Plants colonized land in the company of fungi. Even today, the roots of most plants are associated with fungi that aid in absorption of water and minerals; the fungi receive nutrients in return. Such mutually beneficial associations are evident in some of the oldest plant fossils.

The most widespread and diverse land animals are arthropods (particularly insects and spiders) and tetrapods (vertebrates with four appendages). Tetrapods include humans, but we are late arrivals on the scene—the human lineage diverged from other primates around 6 to 7 million years ago, and our own species originated about 195,000 years ago. If the clock of Earth's history were rescaled to represent an hour, humans appeared less than 0.2 second ago!

In the next two modules, we see how scientists have determined when these key episodes in Earth's history have occurred in geologic time.

? For how long did life on Earth consist solely of single-celled organisms?

More than 2 billion years: Molecular estimates date the common ancestor of multicellular eukaryotes at 1.5 billion years ago; the oldest known fossils are about 1.2 billion years old.

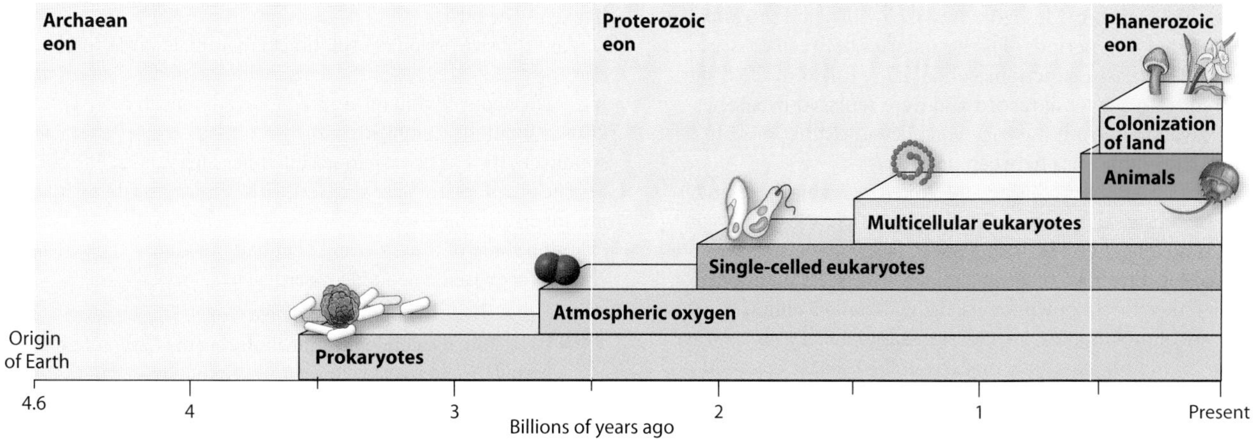

▲ **Figure 4** Some key events in the history of life on Earth

5 The actual ages of rocks and fossils mark geologic time

Geologists use several techniques to determine the ages of rocks and the fossils they contain. The method most often used, called **radiometric dating**, is based on the decay of radioactive isotopes (unstable forms of an element). Fossils contain isotopes of elements that accumulated when the organisms were alive. For example, a living organism contains both the common isotope carbon-12 and the radioactive isotope carbon-14 in the same ratio as is present in the atmosphere. Once an organism dies, it stops accumulating carbon, and the stable carbon-12 in its tissues does not change. Its carbon-14, however, starts to decay to another element. The rate of decay is expressed as a half-life, the time required for 50% of the isotope in a sample to decay. With a half-life of 5,730 years, half the carbon-14 in a specimen decays in about 5,730 years, half the remaining carbon-14 decays in the next 5,730 years, and so on (**Figure 5**). Knowing both the half-life of a radioactive isotope and the ratio of radioactive to stable isotope in a fossil enables us to determine the age of the fossil.

Carbon-14 is useful for dating relatively young fossils—up to about 75,000 years old. Radioactive isotopes with longer half-lives are used to date older fossils.

There are indirect ways to estimate the age of much older fossils. For example, potassium-40, with a half-life of 1.3 billion years, can be used to date volcanic rocks hundreds of millions of years old. A fossil's age can be inferred from the ages of the rock layers above and below the stratum in which it is found.

By dating rocks and fossils, scientists have established a geologic record of Earth's history.

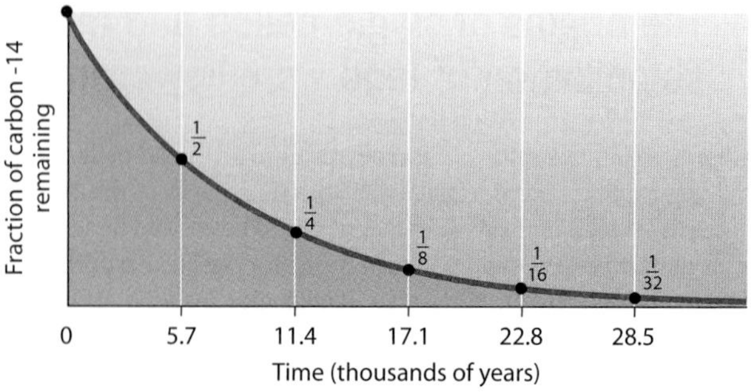

▲ **Figure 5** Radiometric dating using carbon-14

? Estimate the age of a fossil found in a sedimentary rock layer between two layers of volcanic rock that are determined to be 530 and 520 million years old.

● We can infer that the organism lived approximately 525 million years ago.

6 The fossil record documents the history of life

The fossil record, the sequence in which fossils appear in rock strata, is an archive of evolutionary history. Based on this sequence and the ages of rocks and fossils, geologists have established a **geologic record**, as shown in **Table 6**, on the facing page. As you saw in Figure 4, Earth's history is divided into three eons, the Archaean, Proterozoic, and Phanerozoic. The timeline in Table 6 shows the lengths and ages (in millions of years ago) of these eons. Note that the Phanerozoic eon, which is only the last 542 million years, is expanded in the table to show the key events in the evolution of multicellular eukaryotic life. This eon is divided into three eras: the Paleozoic, Mesozoic, and Cenozoic, and the eras are subdivided into periods. The boundaries between eras are marked by mass extinctions, when many forms of life disappeared from the fossil record and were replaced by species that diversified from the survivors. Lesser extinctions often mark the boundaries between periods.

Rocks from the Archaean and Proterozoic eons have undergone extensive change over time, and much of their fossil content is no longer visible. Nonetheless, paleontologists have pieced together ancient events in life's history. As mentioned earlier, the oldest known fossils, dating from 3.5 billion years ago, are of prokaryotes; the oldest fossils of eukaryotic cells are from 2.1 billion years ago. Strata from the Ediacaran period (635–542 million years ago) bear diverse fossils of multicellular algae and soft-bodied animals.

Dating from about 542 million years ago, rocks of the Paleozoic ("ancient animal") era contain fossils of lineages that gave rise to present-day organisms, as well as many lineages that have become extinct. During the early Paleozoic, virtually all life was aquatic, but by about 400 million years ago, plants and animals were well established on land.

The Mesozoic ("middle animal") era is also known as the age of reptiles because of its abundance of reptilian fossils, including those of the dinosaurs. The Mesozoic era also saw the first mammals and flowering plants (angiosperms). By the end of the Mesozoic, dinosaurs had become extinct except for one lineage—the birds.

An explosive period of evolution of mammals, birds, insects, and angiosperms began at the dawn of the Cenozoic ("recent animal") era, about 65 million years ago. Because much more is known about the Cenozoic era than about earlier eras, our table subdivides the Cenozoic periods into finer intervals called epochs.

The chapters in Unit IV describe the enormous diversity of life-forms that have evolved on Earth. In the next section, we examine some of the processes that have produced the distinct changes seen in the geologic record.

? What were the dominant animals during the Carboniferous period? When were gymnosperms the dominant plants? (*Hint*: Look at Table 6.)

● Amphibians. Gymnosperms were dominant during the Triassic and Jurassic periods (251–145.5 million years ago).

TABLE 6 | THE GEOLOGIC RECORD

Relative Duration of Eons	Era	Period	Epoch	Age (millions of years ago)	Important Events in the History of Life
Phanerozoic	Cenozoic	Quaternary	Holocene	0.01	Historical time 
			Pleistocene	2.6	Ice ages; origin of genus *Homo* 
		Tertiary	Pliocene	5.3	Appearance of bipedal human ancestors 
			Miocene	23	Continued radiation of mammals and angiosperms; earliest direct human ancestors 
			Oligocene	33.9	Origins of many primate groups 
			Eocene	55.8	Angiosperm dominance increases; continued radiation of most present-day mammalian orders 
			Paleocene	65.5	Major radiation of mammals, birds, and pollinating insects 
	Mesozoic	Cretaceous		145.5	Flowering plants (angiosperms) appear and diversify; many groups of organisms, including most dinosaurs, become extinct at end of period 
		Jurassic		199.6	Gymnosperms continue as dominant plants; dinosaurs abundant and diverse 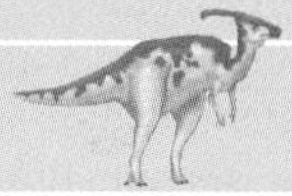
		Triassic		251	Cone-bearing plants (gymnosperms) dominate landscape; dinosaurs evolve and radiate; origin of mammals 
	Paleozoic	Permian		299	Radiation of reptiles; origin of most present-day groups of insects; extinction of many marine and terrestrial organisms at end of period 
		Carboniferous		359	Extensive forests of vascular plants form; first seed plants appear; origin of reptiles; amphibians dominant 
		Devonian		416	Diversification of bony fishes; first tetrapods and insects appear 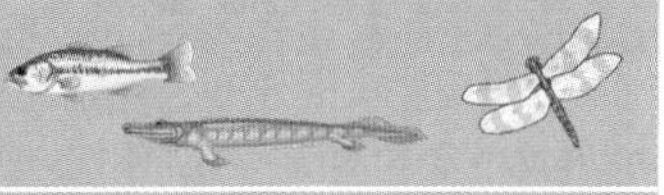
		Silurian		444	Diversification of early vascular plants 
		Ordovician		488	Marine algae abundant; colonization of land by diverse fungi, plants, and animals 
		Cambrian		542	Sudden increase in diversity of many animal phyla (Cambrian explosion) 
Proterozoic		Ediacaran		635	Diverse algae and soft-bodied invertebrate animals appear 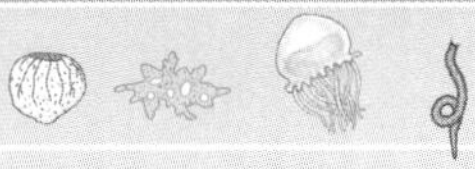
				2,100	Oldest fossils of eukaryotic cells appear 
				2,500	
Archaean				2,700	Concentration of atmospheric oxygen begins to increase
				3,500	Oldest fossils of cells (prokaryotes) appear
				3,800	Oldest known rocks on Earth's surface
				Approx. 4,600	Origin of Earth

Mechanisms of Macroevolution

7 Continental drift has played a major role in macroevolution

The fossil record documents macroevolution, the major events in the history of life on Earth. In this section, we explore some of the factors that helped shape these evolutionary changes, such as plate tectonics, mass extinctions, and adaptive radiations.

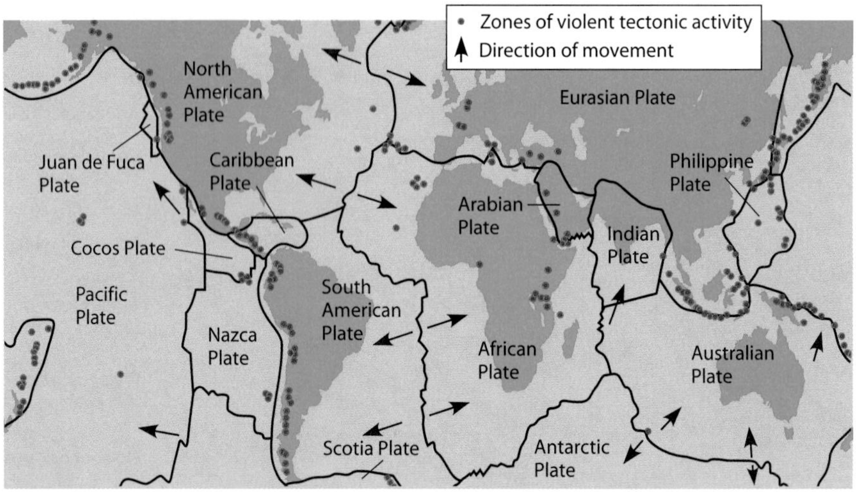

▲ **Figure 7B** Earth's tectonic plates

Plate Tectonics If photographs of Earth were taken from space every 10,000 years and then spliced together, it would make a remarkable movie. The seemingly "rock solid" continents we live on move over time. Since the origin of multicellular eukaryotes roughly 1.5 billion years ago, there have been three occasions in which the landmasses of Earth came together to form a supercontinent, then later broke apart. Each time the landmasses split, they yielded a different configuration of continents. Geologists estimate that the continents will come together again and form a new supercontinent roughly 250 million years from now.

The continents and seafloors form a thin outer layer of planet Earth, called the crust, which covers a mass of hot, viscous material called the mantle. The outer core is liquid and the inner core is solid (**Figure 7A**). According to the theory of **plate tectonics**, Earth's crust is divided into giant, irregularly shaped plates (outlined in black in **Figure 7B**) that essentially float on the underlying mantle. In a process called continental drift, movements in the mantle cause the plates to move (black arrows in the figure). In some cases, the plates are moving away from each other. North America and Europe, for example, are drifting apart at a rate of about 2 cm per year. In other cases, two plates are sliding past each other, forming regions where earthquakes are common. In still other cases, two plates are colliding. Massive upheavals may occur, forming mountains along the plate boundaries. The red dots in Figure 7B indicate zones of violent geologic activity, most of which are associated with plate boundaries.

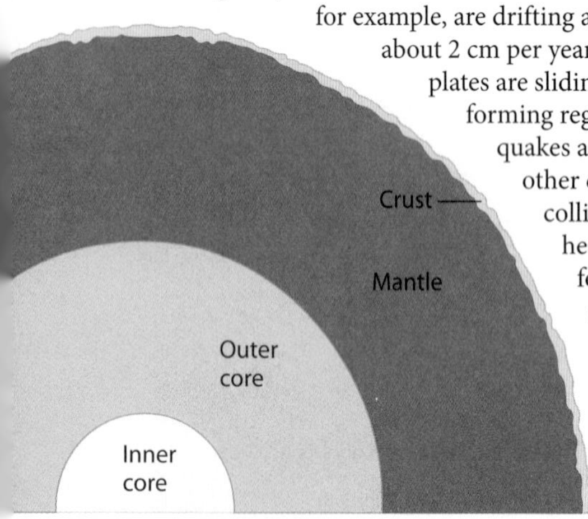

▲ **Figure 7A** Cross-sectional view of Earth (with the thickness of the crust exaggerated)

Consequences of Continental Drift Throughout Earth's history, continental drift has reshaped the physical features of the planet and altered the habitats in which organisms live. **Figure 7C**, on the facing page, shows continental movements that greatly influenced life during the Mesozoic and Cenozoic eras. ❶ About 250 million years ago, near the end of the Paleozoic era, plate movements brought all the previously separated landmasses together into a supercontinent we call **Pangaea**, meaning "all land." When the landmasses fused, ocean basins became deeper, lowering the sea level and draining the shallow coastal seas. Then, as now, most marine species inhabited shallow waters, and much of that habitat was destroyed. The interior of the vast continent was cold and dry. Overall, the formation of Pangaea had a tremendous impact on the physical environment and climate. As the fossil record documents, biological diversity was reshaped. Many species were driven to extinction, and new opportunities arose for organisms that survived the crisis.

During the Mesozoic era, Pangaea started to break apart, causing a geographic isolation of colossal proportions. As the continents drifted apart, each became a separate evolutionary arena—a huge island on which organisms evolved in isolation from their previous neighbors. ❷ At first, Pangaea split into northern and southern landmasses, which we call Laurasia and Gondwana, respectively. ❸ By the end of the Mesozoic era, some 65 million years ago, the modern continents were beginning to take shape. Note that at that time Madagascar (home of the lemurs) became isolated and India was still a large island. Then, around 45 million years ago, the India plate collided with the Eurasian plate, and the slow, steady buckling at the plate boundary formed the Himalayas, the tallest and youngest of Earth's mountain

Map adapted from http://geology.er. usgs.gov/eastern/plates.html.

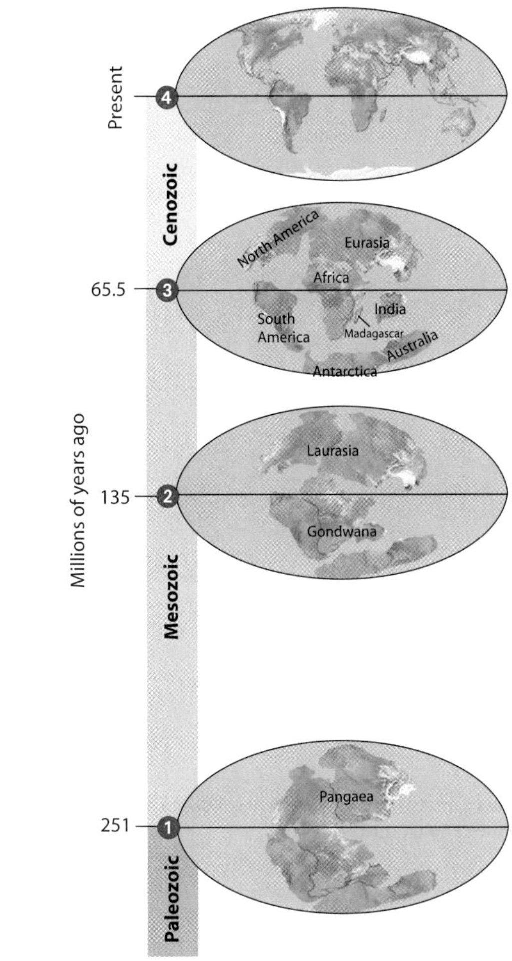

▲ **Figure 7C** Continental drift during the Phanerozoic eon

ranges. ❹ The continents continue to drift today, and the Himalayas are still growing (about 1 cm per year).

The pattern of continental mergings and separations solves many biogeographic puzzles, including Australia's great diversity of marsupials (pouched mammals). Fossil evidence suggests that marsupials originated in what is now Asia and reached Australia via South America and Antarctica while the continents were still joined. The subsequent breakup of continents set Australia "afloat" like a great ark of marsupials. The few early eutherians (placental mammals) that lived there became extinct, while on other continents, most marsupials became extinct. Isolated on Australia, marsupials evolved and diversified, filling ecological roles analogous to those filled by eutherians on other continents.

The history of continental drift also explains the distribution of a group of ancient vertebrates called lungfishes (**Figure 7D**). Today, there are six species of lungfishes in the world, four in Africa and one each in Australia and South America (yellow areas in Figure 7D). What is the evolutionary history of these animals? Although their present-day distribution may indicate that lungfishes evolved after separation of the supercontinent Gondwana (see Figure 7C), their fossil distribution tells a different story. As the orange triangles in Figure 7D indicate, fossil lungfishes have been found on all continents except Antarctica. This widespread fossil record indicates that lungfishes evolved when Pangaea was intact.

In the next module, we consider some of the perils associated with the movements of Earth's crustal plates.

? Paleontologists have discovered fossils of the same species of Permian freshwater reptiles in West Africa and Brazil, regions that are separated by 3,000 km of ocean. How could you explain such finds?

West Africa and Brazil were connected during the early Mesozoic era, and these reptiles must have ranged across both areas.

▶ **Figure 7D**
Lungfish distribution, a result of continental drift

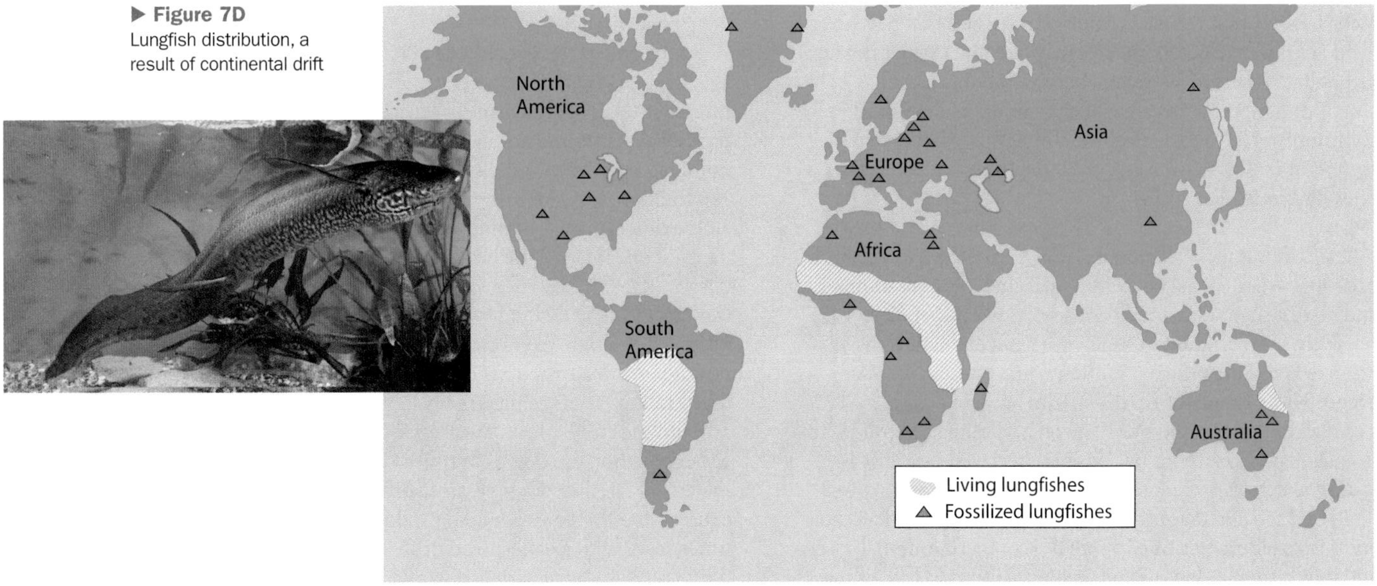

Wildlife/D.Harms/Still Pictures/ Specialiststock.com

Adapted from W. K. Purves and G. H. Orians, *Life: The Science of Biology*, 2nd ed., fig. 16.5c, p. 1180 (New York: Sinauer Associates, 1987). Used with permission.

CONNECTION

8 Plate tectonics may imperil human life

Not only do moving crustal plates cause continents to collide, pile up, and build mountain ranges; they also produce volcanoes and earthquakes. The boundaries of plates are hot spots of such geologic activity. California's frequent earthquakes are a result of movement along the infamous San Andreas Fault, part of the border where the Pacific and North American plates grind together and gradually slide past each other (**Figure 8**)—in what we can call a strike-slip fault. Two major earthquakes have occurred in the region in the past century: the San Francisco earthquake of 1906 and the 1989 Loma Prieta earthquake, also near San Francisco.

In such a strike-slip fault, the two plates do not slide smoothly past each other. They often stick in one spot until enough pressure builds along the fault that the landmasses suddenly jerk forward, releasing massive amounts of energy and causing the surrounding area to move or shake. A strike-slip fault runs under Haiti and is responsible for the devastating magnitude 7.0 earthquake of January 2010. In Haiti, the North American plate is moving west past the Caribbean plate (see Figure 7B).

Undersea earthquakes can cause giant waves, such as the massive 2004 tsunamis that resulted when a large area of a fault in the Indian Ocean ruptured near the meeting point of the Indian, Eurasian, and Australian plates.

A volcano is a rupture that allows hot, molten rock, ash, and gases to escape from beneath Earth's crust. Volcanoes are often found where tectonic plates are diverging or converging, as opposed to sliding past each other. Volcanoes can cause

▲ **Figure 8** The San Andreas Fault (an aerial view near San Luis Obispo County), a boundary between two crustal plates

tremendous devastation, as when Mt. Vesuvius in southern Italy erupted in 79 AD, burying Pompeii in a layer of ash. But sometimes volcanoes imperil more than just local life, as we see in the next module.

? **Volcanoes usually destroy life. How might undersea volcanoes create new opportunities for life?**

● By creating new landmasses on which life can evolve, such as the Galápagos and Hawaiian Islands

9 During mass extinctions, large numbers of species are lost

Extinction is inevitable in a changing world. Indeed, the fossil record shows that the vast majority of species that have ever lived are now extinct. A species may become extinct because its habitat has been destroyed, because of unfavorable climatic changes, or because of changes in its biological community, such as the evolution of new predators or competitors. Extinctions occur all the time, but extinction rates have not been steady.

Mass Extinctions The fossil record chronicles a number of occasions when global environmental changes were so rapid and disruptive that a majority of species were swept away in a relatively short amount of time. Five mass extinctions have occurred over the past 500 million years. In each of these events, 50% or more of Earth's species became extinct.

Of all the mass extinctions, the ones marking the ends of the Permian and Cretaceous periods have received the most attention. The Permian extinction, which occurred about 251 million years ago and defines the boundary between the Paleozoic and Mesozoic eras, claimed about 96% of marine animal species and took a tremendous toll on terrestrial life as well. This mass extinction occurred in less than 500,000 years, and possibly in just a few thousand years—an instant in the context of geologic time.

At the end of the Cretaceous period, about 65 million years ago, the world again lost an enormous number of species—more than half of all marine species and many lineages of terrestrial plants and animals. At that point, dinosaurs had dominated the land and pterosaurs had ruled the air for some 150 million years. After the Cretaceous mass extinction, almost all the dinosaurs were gone, leaving behind only the descendants of one lineage, the birds.

Causes of Mass Extinctions The Permian mass extinction occurred at a time of enormous volcanic eruptions in what is now Siberia. Vast stretches of land were covered with lava hundreds to thousands of meters thick. Besides spewing lava and ash into the atmosphere, the eruptions may have produced enough carbon dioxide to warm the global climate by an estimated 6°C. Reduced temperature differences between the equator and the poles would have slowed the mixing of ocean water, leading to a widespread drop in oxygen concentration in the water. This oxygen deficit would have killed many marine organisms and promoted the growth of anaerobic bacteria that

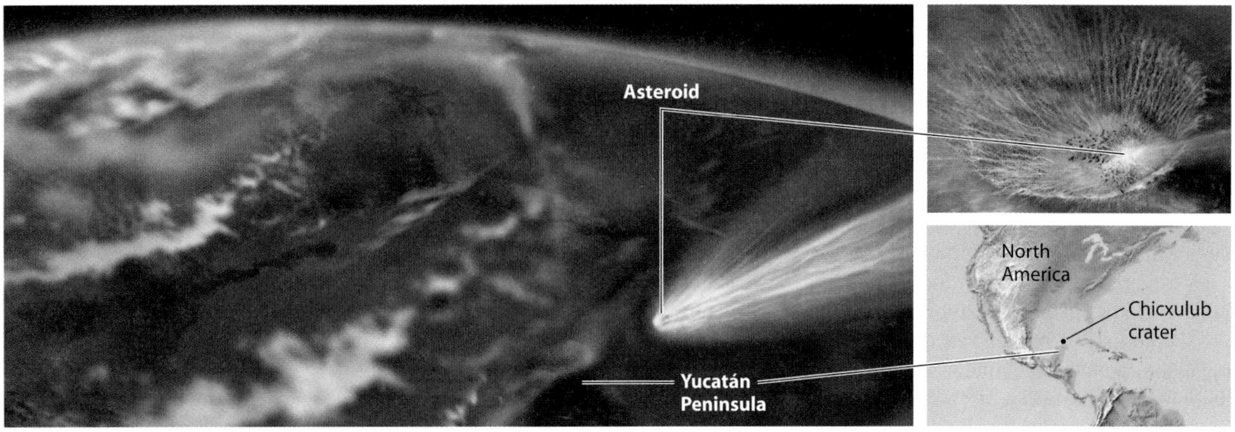

▲ **Figure 9** The impact hypothesis for the Cretaceous mass extinction

emit a poisonous by-product, hydrogen sulfide. As this gas bubbled out of the water, it would have killed land plants and animals and initiated chemical reactions that would destroy the protective ozone layer. Thus, a cascade of factors may have contributed to the Permian extinction.

One clue to a possible cause of the Cretaceous mass extinction is a thin layer of clay enriched in iridium that separates sediments from the Mesozoic and Cenozoic eras. Iridium is an element very rare on Earth but common in meteorites and other extraterrestrial objects that occasionally fall to Earth. The rocks of the Cretaceous boundary layer have many times more iridium than normal Earth levels. Most paleontologists conclude that the iridium layer is the result of fallout from a huge cloud of dust that billowed into the atmosphere when an asteroid or large comet hit Earth. The cloud would have blocked light and severely disturbed the global climate for months.

Is there evidence of such an asteroid? A large crater, the 65-million-year-old Chicxulub impact crater, has been found in the Caribbean Sea near the Yucatán Peninsula of Mexico (**Figure 9**). About 180 km wide (about 112 miles), the crater is the right size to have been caused by an object with a diameter of 10 km (about 6 miles). The horseshoe shape of the crater and the pattern of debris in sedimentary rocks indicate that an asteroid or comet struck at a low angle from the southeast. The artist's interpretation in Figure 9 represents the impact and its immediate effect—a cloud of hot vapor and debris that could have killed most of the plants and animals in North America within hours. The collision is estimated to have released more than a billion times the energy of the nuclear bombs dropped in Japan during World War II.

In March 2010, an international team of scientists reviewed two decades' worth of research on the Cretaceous extinction and endorsed the asteroid hypothesis as the triggering event. Nevertheless, research will continue on other contributing causes and the multiple and interrelated effects of this major ecological disaster.

Consequences of Mass Extinctions Whatever their causes, mass extinctions affect biological diversity profoundly. By removing large numbers of species, a mass extinction can decimate a thriving and complex ecological community. Mass extinctions are random events that act on species indiscriminately. They can permanently remove species with highly advantageous features and change the course of evolution forever. Consider what would have happened if our early primate ancestors living 65 million years ago had died out in the Cretaceous mass extinction—or if a few large, predatory dinosaurs had *not* become extinct!

How long does it take for life to recover after a mass extinction? The fossil record shows that it typically takes 5–10 million years for the diversity of life to return to previous levels. In some cases, it has taken much longer: It took about 100 million years for the number of marine families to recover after the Permian mass extinction.

Is a Sixth Mass Extinction Under Way? Human actions that result in habitat destruction and climate change are modifying the global environment to such an extent that many species are currently threatened with extinction. In the past 400 years, more than a thousand species are known to have become extinct. Scientists estimate that this rate is 100 to 1,000 times the normal rate seen in the fossil record. Does this represent the beginning of a sixth mass extinction?

This question is difficult to answer, partly because it is hard to document both the total number of species on Earth and the number of extinctions that are occurring. It is clear that losses have not reached the level of the other "big five" extinctions. Monitoring, however, does show that many species are declining at an alarming rate, suggesting that a sixth (human-caused) mass extinction could occur within the next few centuries or millennia. And as seen with prior mass extinctions, it may take millions of years for life on Earth to recover.

But the fossil record also shows a creative side to the destruction. Mass extinctions can pave the way for adaptive radiations in which new groups rise to prominence, as we see next.

? The Permian and Cretaceous mass extinctions mark the ends of the ______ and ______ eras, respectively. (*Hint*: Refer back to Table 6.)

● Paleozoic . . . Mesozoic

10 Adaptive radiations have increased the diversity of life

Adaptive radiations are periods of evolutionary change in which many new species evolve from a common ancestor, often following the colonization of new, unexploited areas. Adaptive radiations have also followed each mass extinction, when survivors became adapted to the many vacant ecological roles, or niches, in their communities.

For example, fossil evidence indicates that mammals underwent a dramatic adaptive radiation after the extinction of terrestrial dinosaurs 65 million years ago **(Figure 10)**. Although mammals originated 180 million years ago, fossils older than 65 million years indicate that they were mostly small and not very diverse. Early mammals may have been eaten or outcompeted by the larger and more diverse dinosaurs. With the disappearance of the dinosaurs (except for the bird lineage), mammals expanded greatly in both diversity and size, filling the ecological roles once occupied by dinosaurs.

The history of life has also been altered by radiations that followed the evolution of new adaptations, such as the wings of pterosaurs, birds, and bats (see chapter introduction). Major new adaptations facilitated the colonization of land by plants, insects, and tetrapods. The radiation of land plants, for example, was associated with features such as stems that supported the plant against gravity and a waxy coat that protected leaves from water loss. Finally, note that organisms that arise in an adaptive radiation can serve as a new source of food for still other organisms. In this way, the diversification of land plants stimulated a series of adaptive radiations in insects that ate or pollinated plants—helping to make insects the most diverse group of animals on Earth today.

Now that we've looked at geologic and environmental influences, let's consider how genes can affect macroevolution.

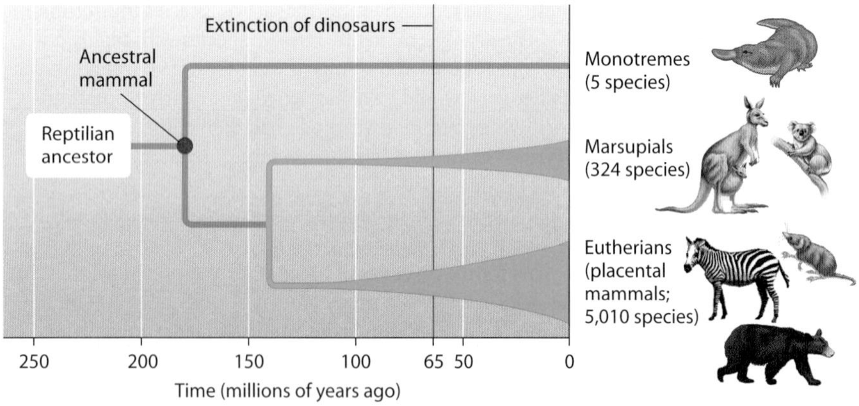

▲ **Figure 10** Adaptive radiation of mammals (width of line reflects numbers of species).

? In addition to the new resources of plants, what other factors likely promoted the adaptive radiation of insects on land?

● Many unfilled niches on land and the evolution of wings and a supportive, protective, and waterproof exoskeleton

11 Genes that control development play a major role in evolution

The fossil record can tell us *what* the great events in the history of life have been and *when* they occurred. Continental drift, mass extinctions, and adaptive radiation provide a big-picture view of *how* those changes came about. But now we are increasingly able to understand the basic biological mechanisms that underlie the changes seen in the fossil record.

Scientists working at the interface of evolutionary biology and developmental biology—the research field abbreviated "**evo-devo**"—are studying how slight genetic changes can become magnified into major morphological differences between species. Genes that program development control the rate, timing, and spatial pattern of change in an organism's form as it develops from a zygote into an adult. A great many of these genes appear to have been conserved throughout evolutionary history: The same or very similar genes are involved in the development of form across multiple lineages.

Changes in Rate and Timing Many striking evolutionary transformations are the result of a change in the rate or timing of developmental events. **Figure 11A** shows a photograph of an axolotl, a salamander that illustrates a phenomenon called **paedomorphosis** (from the Greek *paedos*, of a child, and *morphosis*, formation), the retention in the adult of body structures that were juvenile features in an ancestral species. Most salamander species have aquatic larvae (with gills) that undergo metamorphosis in becoming terrestrial adults (with lungs). The axolotl is a salamander that grows to a sexually mature adult while retaining gills and other larval features.

▲ **Figure 11A** An axolotl, a paedomorphic salamander

Stephen Dalton/Photo Researchers

Slight changes in the relative growth of different body parts can change an adult form substantially. As the skulls and photo in **Figure 11B**, on the next page, show, humans and chimpanzees are much more alike as fetuses than they are as adults. As development proceeds,

Jean Kern

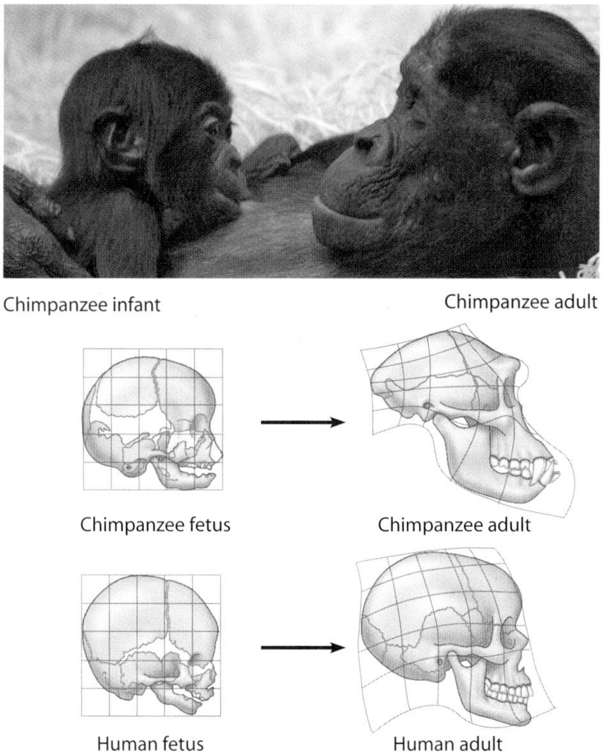

▲ **Figure 11B** Chimpanzee and human skull shapes compared

accelerated growth in the jaw produces the elongated skull, sloping forehead, and massive jaws of an adult chimpanzee. In the human lineage, genetic changes that slowed the growth of the jaw relative to other parts of the skull produced an adult whose head proportions still resembled that of a child (and that of a baby chimpanzee). Our large skull and complex brain are among our most distinctive features. Compared to the slow growth of a chimpanzee brain after birth, our brain continues to grow at the rapid rate of a fetal brain for the first year of life.

Changes in Spatial Pattern Homeotic genes, the master control genes described in Module 11.8, determine such basic features as where a pair of wings or legs will develop on a fruit fly. Changes in homeotic genes or in how or where such genes are expressed can have a profound impact on body form. Consider, for example, the evolution of snakes from a four-limbed lizard-like ancestor. Researchers have found that one pattern of expression of two homeotic genes in tetrapods results in the formation of forelimbs and of vertebrae with ribs, whereas a different pattern of expression of these two genes results in the development of vertebrae with ribs but no limbs, as in snakes.

New Genes and Changes in Genes New developmental genes that arose as a result of gene duplications may have facilitated the origin of new body forms. For example, a fruit fly (an invertebrate) has a single cluster of several homeotic genes that direct the development of major body parts. A mouse (a vertebrate) has four clusters of very similar genes that occur in the same linear order on chromosomes and direct the development of the same body regions as the fly genes. Two duplications of these gene clusters appear to have occurred in the evolution of vertebrates from invertebrate animals. Mutations in these duplicated genes may then have led to the origin of novel vertebrate characteristics, such as a backbone, jaws, and limbs.

Changes in Gene Regulation Researchers are finding that changes in the form of organisms often are caused by mutations that affect the regulation of developmental genes. As we just discussed, such a change in gene expression was shown to correlate with the lack of forelimbs in snakes.

Additional evidence for this type of change in gene regulation is seen in studies of the threespine stickleback fish. In western Canada, these fish live in the ocean and also in lakes that formed when the coastline receded during the past 12,000 years. Ocean populations have bony plates that make up a kind of body armor and a large set of pelvic spines that help deter predatory fish. The body armor and pelvic spines are reduced or absent in threespine sticklebacks that live in lakes. The loss of the pelvic spine appears to have been driven by natural selection, because freshwater predators such as dragonfly larvae capture juvenile sticklebacks by grasping onto the spines. **Figure 11C** shows specimens of an ocean and a lake stickleback, which have been stained to highlight their bony plates and spines.

Researchers have identified a key gene that influences the development of these spines. Was the reduction of spines in lake populations due to changes in the gene itself or to changes in how the gene is expressed? It turns out that the gene is identical in the two populations, and it is expressed in the mouth region and other tissues of embryos from both populations. Studies have shown, however, that while the gene is also expressed in the developing pelvic region of ocean sticklebacks, it is not turned on in the pelvic region in lake sticklebacks. This example shows how morphological change can be caused by altering the expression of a developmental gene in some parts of the body but not others.

? **Research shows that many differences in body form are caused by changes in gene regulation and not changes in the nucleotide sequence of the developmental gene itself. Why might this be the case?**

A change in sequence may affect a gene's function wherever that gene is expressed—with potentially harmful effects. Changes in the regulation of gene expression can be limited to specific areas in a developing embryo.

Dr. William A. Cresko

▲ **Figure 11C** Stickleback fish from ocean (top) and lake (bottom), stained to show bony plates and spines. (Arrow indicates the absence of the pelvic spine in the lake fish.)

12 Evolutionary novelties may arise in several ways

Let's see how the Darwinian theory of gradual change can account for the evolution of intricate structures such as eyes or of new body structures such as wings. Most complex structures have evolved in increments from simpler versions having the same basic function—a process of refinement. But sometimes we can trace the origin of evolutionary novelties to the gradual adaptation of existing structures to new functions.

As an example of the process of gradual refinement, consider the amazing camera-like eyes of vertebrates and squids. Although the eyes of vertebrates evolved independently of those of squids, both evolved from a simple ancestral patch of photoreceptor cells through a series of incremental modifications that benefited their owners at each stage. Indeed, there appears to have been a single evolutionary origin of light-sensitive cells, and all animals with eyes—vertebrates and invertebrates alike—share the same master genes that regulate eye development.

Figure 12 illustrates the range of complexity in the structure of eyes among molluscs living today. Simple patches of pigmented cells enable limpets to distinguish light from dark, and they cling more tightly to their rock when a shadow falls on them—a behavioral adaptation that reduces the risk of being eaten. Other molluscs have eyecups that have no lenses or other means of focusing images but can indicate light direction. In those molluscs that do have complex eyes, the organs probably evolved in small steps of adaptation. You can see examples of such small steps in Figure 12.

Although eyes have retained their basic function of vision throughout their evolutionary history, evolutionary novelty can also arise when structures that originally played one role gradually acquire a different one. Structures that evolve in one context but become co-opted for another function are sometimes called *exaptations*. This term suggests that a structure can become adapted to alternative functions; it does not mean that a structure evolves in anticipation of future use. Natural selection cannot predict the future; it can only improve an existing structure in the context of its current use. Novel features can arise gradually via a series of intermediate stages, each of which has some function in the organism's current context.

Consider the evolution of birds from a feathered dinosaur ancestor. Feathers could not have evolved as an adaptation for upcoming flights. Their first utility may have been for insulation. It is possible that longer, winglike forelimbs and feathers, which increased the surface area of these forelimbs, were co-opted for flight after functioning in some other capacity, such as mating displays, thermoregulation, or camouflage (all functions that feathers still serve today). The first flights may have been only short glides to the ground or from branch to branch in tree-dwelling species. Once flight itself became an advantage, natural selection would have gradually remodeled feathers and wings to fit their additional function.

The flippers of penguins are another example of the modification of existing structures for different functions. Penguins cannot fly, but their modified wings are powerful oars that make them strong, fast underwater swimmers.

? **Explain why the concept of exaptation does not imply that a structure evolves in anticipation of some future environmental change.**

Although a structure is co-opted for new or additional functions in a new environment, the structure existed because it worked as an adaptation in the old environment.

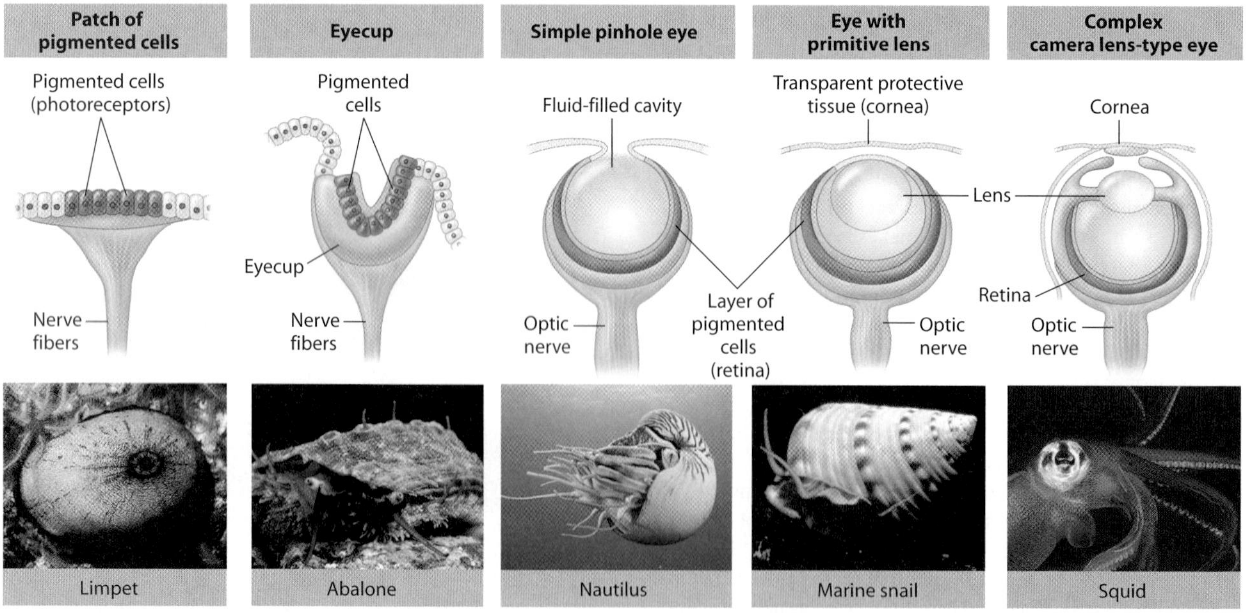

▲ **Figure 12** A range of eye complexity among molluscs

Hal Beral/V & W/Image Quest Marine; Christophe Courteau/Biosphoto/Photo Researchers; Reinhard Dirscherl/Photolibrary; Jim Greenfield/Image Quest Marine; Image Quest Marine

EVOLUTION CONNECTION

13 Evolutionary trends do not mean that evolution is goal directed

The fossil record seems to reveal trends in the evolution of many species. An example is the modern horse (genus *Equus*), a descendant of an ancestor about the size of a large dog that lived some 55 million years ago. Named *Hyracotherium*, this ancestor had four toes on its front feet, three toes on its hind feet, and teeth adapted to browsing on shrubs and trees. In contrast, the present-day horse is larger, has only one toe on each foot, and has teeth modified for grazing on grasses.

Figure 13 shows the fossil record of horses, with the vertical bars representing the period of time each group persisted in the record. If we select only certain species (those highlighted in yellow), it is possible to arrange a sequence of animals that were intermediate in form between *Hyracotherium* and living horses. If these were the only fossils known, they could create the illusion of a single trend in an unbranched lineage, progressing toward larger size, reduced number of toes, and teeth modified for grazing. However, if we consider *all* fossil horses known today, this apparent trend vanishes. Clearly, there was no "trend" toward grazing. Only a few lineages were grazers (green area of figure); the other lineages (tan area), all of which are now extinct, remained multi-toed browsers. The genus *Equus* is the only surviving twig of an evolutionary tree that is so branched that it is more like a bush.

Branching evolution *can* lead to a real evolutionary trend, however. One model of long-term trends compares species to individuals: Speciation is their birth, extinction their death, and new species that diverge from them are their offspring. According to this model of species selection, unequal survival of species and unequal generation of new species play a role in macroevolution similar to the role of unequal reproduction in microevolution. In other words, the species that generate the greatest number of new species determine the direction of major evolutionary trends.

Evolutionary trends can also result directly from natural selection. For example, when horse ancestors invaded the grasslands that spread during the mid-Cenozoic, there was strong selection for grazers that could escape predators by running faster. This trend would not have occurred without open grasslands.

Whatever its cause, it is important to recognize that an evolutionary trend does not imply that evolution is goal directed. Evolution is the result of interactions between organisms and the current environment. If conditions change, an apparent trend may cease or even reverse itself.

In the final section, we explore how biologists arrange life's astounding diversity into an evolutionary tree of life.

? A trend in the evolution of mammals was toward a larger brain size. Use the species selection model to explain how such a trend could occur.

Those species with larger brains persisted longer before extinction and gave rise to more "offspring" species than did species with smaller brains.

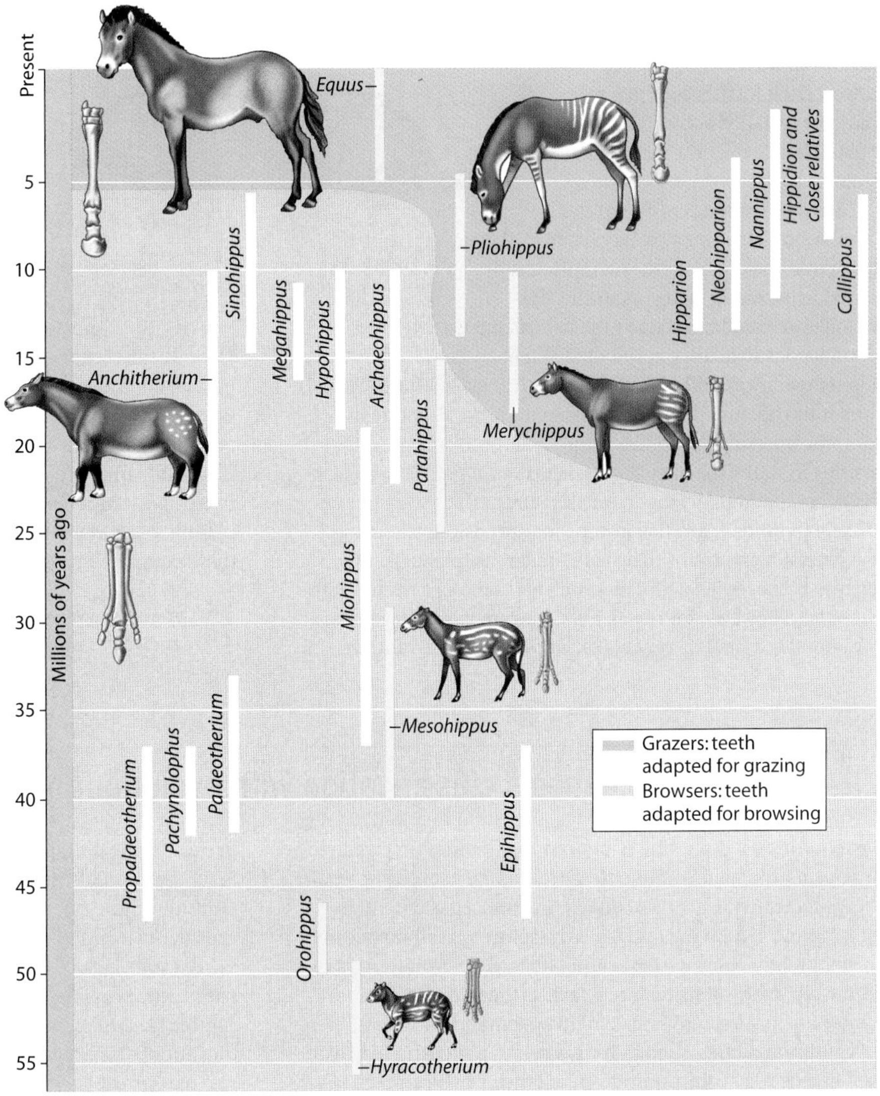

▶ **Figure 13** The branched evolution of horses

Phylogeny and the Tree of Life

14 Phylogenies based on homologies reflect evolutionary history

So far in this chapter, we have looked at the major evolutionary changes that have occurred during the history of life on Earth and explored some of the mechanisms that underlie the process of macroevolution. Now we shift our focus to how biologists use the pattern of evolution to distinguish and categorize the millions of species that live, and have lived, on Earth.

The evolutionary history of a species or group of species is called **phylogeny** (from the Greek *phylon*, tribe, and *genesis*, origin). The fossil record provides a substantial chronicle of evolutionary change that can help trace the phylogeny of many groups. It is, however, an incomplete record, as many of Earth's species probably never left any fossils; many fossils that formed were probably destroyed by later geologic processes; and only a fraction of existing fossils have been discovered. Even with its limitations, however, the fossil record is a remarkably detailed account of biological change over the vast scale of geologic time.

In addition to evidence from the fossil record, phylogeny can also be inferred from morphological and molecular homologies among living organisms. Homologies are similarities due to shared ancestry. Homologous structures may look different and function differently in different species, but they exhibit fundamental similarities because they evolved from the same structure in a common ancestor. For instance, the whale limb is adapted for steering in water; the bat wing is adapted for flight. Nonetheless, the bones that support these two structures, which were present in their common mammalian ancestor, are basically the same.

Generally, organisms that share similar morphologies are likely to be closely related. The search for homologies is not without pitfalls, however, for not all likenesses are inherited from a common ancestor. In a process called **convergent evolution**, species from different evolutionary branches may come to resemble one another if they live in similar environments and natural selection has favored similar adaptations. In such cases, body structures and even whole organisms may resemble each other.

▲ **Figure 14** Australian "mole" (top) and North American mole (bottom)

Similarity due to convergent evolution is called **analogy**. For example, the two mole-like animals shown in **Figure 14** are very similar in external appearance. They both have enlarged front paws, small eyes, and a pad of protective thickened skin on the nose. However, the Australian "mole" (top) is a marsupial, meaning that its young complete their embryonic development in a pouch outside the mother's body. The North American mole (bottom) is a eutherian, which means that its young complete development in the mother's uterus. Genetic and fossil evidence indicate that the last common ancestor of these two animals lived 140 million years ago. And in fact, that ancestor and most of its descendants were not mole-like. Analogous traits evolved independently in these two mole lineages as they each became adapted to burrowing lifestyles.

In addition to molecular comparisons and fossil evidence, another clue to distinguishing homology from analogy is to consider the complexity of the structure being compared. For instance, the skulls of a human and a chimpanzee (see Figure 11B) consist of many bones fused together, and the composition of these skulls matches almost perfectly, bone for bone. It is highly improbable that such complex structures have separate origins. More likely, the genes involved in the development of both skulls were inherited from a common ancestor, and these complex structures are homologous.

? Human forearms and a bat's wings are ______. A bat's wings and a bee's wings are ______.

homologous . . . analogous

15 Systematics connects classification with evolutionary history

Systematics is a discipline of biology that focuses on classifying organisms and determining their evolutionary relationships. In the 18th century, Carolus Linnaeus introduced a system of naming and classifying species, a discipline we call **taxonomy**. Although Linnaeus's system was not based on evolutionary relationship, many of its features, such as the two-part Latin names for species, remain useful in systematics.

Common names, such as bird, bat, and pterosaur, may work well in everyday communication, but they can be ambiguous because there are many species of each of these kinds of organisms. And some common names are downright misleading. Consider these three "fishes": jellyfish (a cnidarian), crayfish (a crustacean), and silverfish (an insect).

To avoid such confusion, biologists assign each species a two-part scientific name, or **binomial**. The first part is the **genus** (plural, *genera*) to which a species belongs. The second part of the binomial is unique for each species within the genus. The two parts must be used together to name a species. For example,

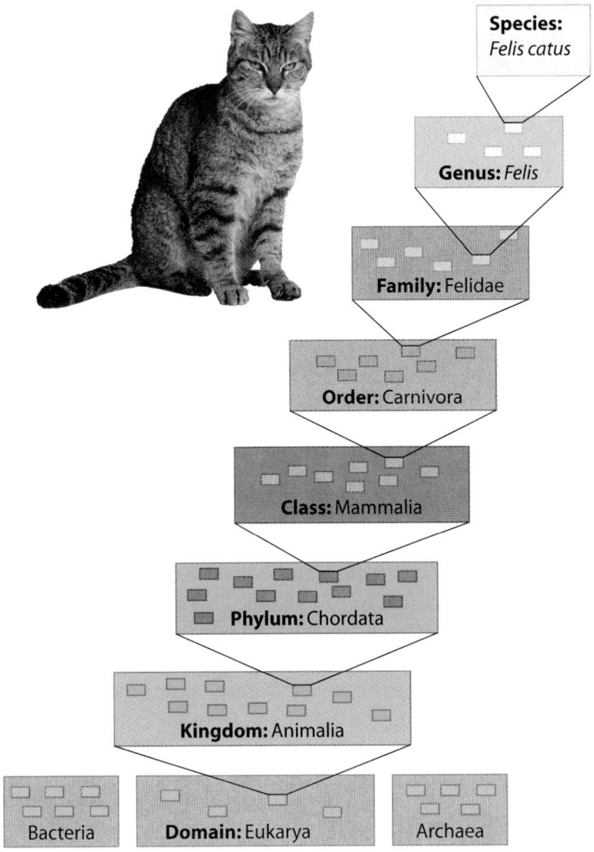

▲ **Figure 15A** Hierarchical classification of the domestic cat

the scientific name for the gray squirrel is *Sciurus carolinensis* and for the Carolina chickadee is *Poecile carolinensis*. Notice that the first letter of the genus name is capitalized and that the binomial is italicized and latinized.

In addition to naming species, Linnaeus also grouped them into a hierarchy of categories. Beyond the grouping of species within genera (as indicated by the binomial), the Linnaean system extends to progressively broader categories of classification. It places similar genera in the same **family**, puts families into **orders**, orders into **classes**, classes into **phyla** (singular, *phylum*), phyla into **kingdoms**, and, more recently, kingdoms into **domains**.

Figure 15A uses the domestic cat (*Felis catus*) to illustrate this progressively broader classification system. The genus *Felis* includes the domestic cat and several closely related species of small wild cats (represented by small yellow boxes in the figure). The genus *Felis* is placed in the cat family, Felidae, along with other genera of cats, such as the genus *Panthera*, which includes the tiger, leopard, jaguar, and African lion. Family Felidae belongs to the order Carnivora, which also includes the family Canidae (for example, the wolf and coyote) and several other families. Order Carnivora is grouped with many other orders in the class Mammalia, the mammals. Class Mammalia is one of the classes belonging to the phylum Chordata in the kingdom Animalia, which is one of several kingdoms in the domain Eukarya. Each taxonomic unit at any level—family Felidae or class Mammalia, for instance—is called a **taxon** (plural, *taxa*).

Grouping organisms into more inclusive categories seems to come naturally to humans—it is a way to structure our world. Classifying species into higher (broader) taxa, however, is ultimately arbitrary. Higher classification levels are generally defined by morphological characteristics chosen by taxonomists rather than by quantitative measurements that could apply to the same taxon level across all lineages. Because of such difficulties with determining higher taxa, some biologists propose that classification be based entirely on evolutionary relationships, using a PhyloCode, as it is called, which only names groups that include a common ancestor and all its descendants. While PhyloCode would change the way taxa are defined, the names of most groups would remain the same, just without "ranks," such as family, order, or class, attached to them.

Ever since Darwin, systematics has had a goal beyond simple organization: to have classification reflect evolutionary relationships. Biologists traditionally use **phylogenetic trees** to depict hypotheses about the evolutionary history of species. These branching diagrams reflect the hierarchical classification of groups nested within more inclusive groups. **Figure 15B** illustrates the connection between classification and phylogeny. This tree shows the classification of some of the taxa in the order Carnivora and the probable evolutionary relationships among these groups. Note that such a phylogenetic tree does not indicate when a particular species evolved but only the pattern of descent from the last common ancestors of the species shown.

Next we explore how phylogenetic trees are constructed.

? **How much of the classification in Figure 15A do we share with the domestic cat?**

● We are classified the same from the domain to the class level: Both cats and humans are mammals. We do not belong to the same order.

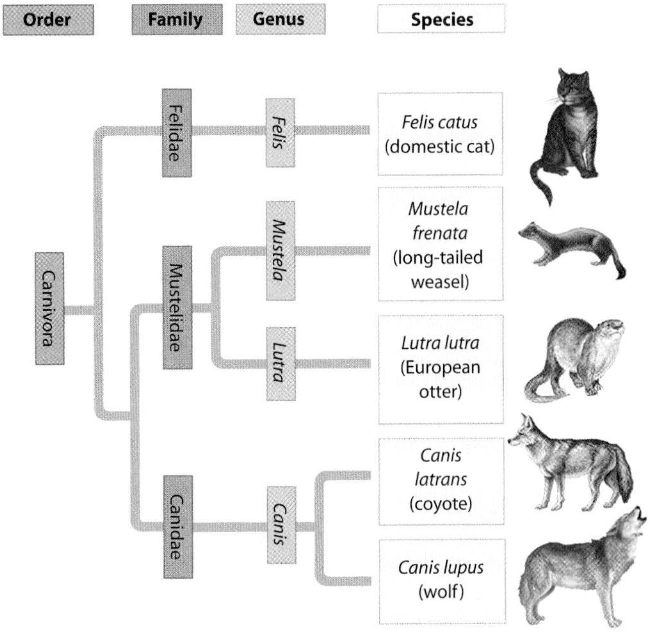

▲ **Figure 15B** Relating classification to phylogeny

16 Shared characters are used to construct phylogenetic trees

In reconstructing a group's evolutionary history, biologists first sort homologous features, which reflect evolutionary relationship, from analogous features, which do not. They then infer phylogeny using these homologous characters.

Cladistics The most widely used method in systematics is called **cladistics**. Common ancestry is the primary criterion used to group organisms into **clades** (from the Greek *clados*, branch). A clade is a group of species that includes an ancestral species and all its descendants. Such an inclusive group of ancestor and descendants, be it a genus, family, or some broader taxon, is said to be **monophyletic** (meaning "single tribe"). Clades reflect the branching pattern of evolution and can be used to construct phylogenetic trees.

Cladistics is based on the Darwinian concept that organisms both share characteristics with their ancestors and differ from them. For example, all mammals have backbones, but the presence of a backbone does not distinguish mammals from other vertebrates. The backbone predates the branching of the mammalian clade from other vertebrates. Thus, we say that for mammals, the backbone is a **shared ancestral character** that originated in an ancestor of all vertebrates. In contrast, hair, a character shared by all mammals but not found in their ancestors, is considered a **shared derived character**, an evolutionary novelty unique to mammals. Shared derived characters distinguish clades and thus the branch points in the tree of life.

Inferring Phylogenies Using Shared Characters The simplified example in **Figure 16A** illustrates that the sequence in which shared derived characters appear can be used to construct a phylogenetic tree. The figure compares five animals according to the presence or absence of a set of characters. An important part of cladistics is a comparison between a so-called ingroup and an outgroup. The **outgroup** (in this example, the frog) is a species from a lineage that is known to have diverged before the lineage that includes the species we are studying, the **ingroup**.

In our example, the frog (representing amphibians, the outgroup) and the other four animals (collectively the ingroup) are all related in that they are tetrapods (vertebrates with limbs). By comparing members of the ingroup with each other and with the outgroup, we can determine which characters are the derived characters—evolutionary innovations—that define the sequence of branch points in the phylogeny of the ingroup.

In the character table in Figure 16A, 0 indicates that a particular character is not present in a group; 1 indicates that the character is present. Let's work through this example step by step. All the animals in the ingroup have an amnion, a membrane that encloses the embryo in a fluid-filled sac. The outgroup does not have this character. Now consider the next character—hair and mammary glands. This character is present in all three mammals (the duck-billed platypus, kangaroo, and beaver) but not the iguana or frog. The third character in the table is gestation, the carrying of developing offspring within the uterus of the female parent. Both the outgroup and iguanas do not exhibit gestation. Instead, frogs release their eggs into the water, and iguanas and most other reptiles lay eggs with a shell. One of the mammals, the duck-billed platypus, also lays eggs with a shell; and from this we might infer that the duck-billed platypus represents an early branch point in the mammalian clade. In fact, this hypothesis is strongly supported by structural, fossil, and molecular evidence. The final character is long gestation, in which an offspring completes its embryonic development within the uterus. This is the case for a beaver, but a kangaroo has a very short gestation

TAXA

CHARACTERS	Frog	Iguana	Duck-billed platypus	Kangaroo	Beaver
Amnion	0	1	1	1	1
Hair, mammary glands	0	0	1	1	1
Gestation	0	0	0	1	1
Long gestation	0	0	0	0	1

Character Table

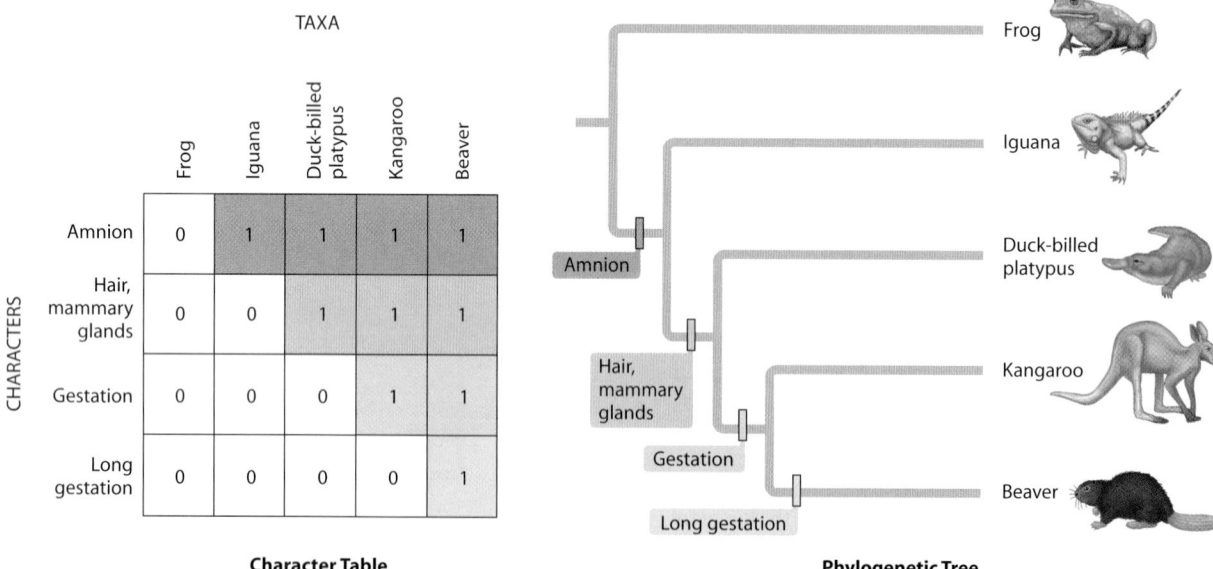

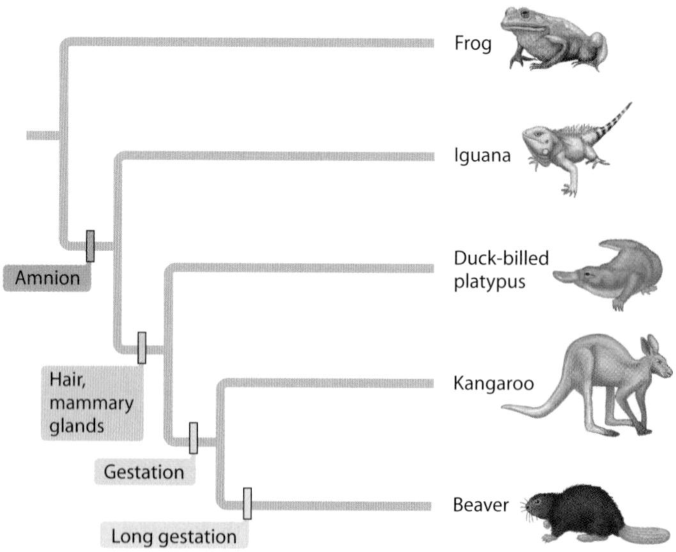

Phylogenetic Tree

▲ **Figure 16A** Constructing a phylogenetic tree using cladistics

period and completes its embryonic development while nursing in its mother's pouch.

We can now translate the data in our table of characters into a phylogenetic tree. Such a tree is constructed from a series of two-way branch points. Each branch point (also called a node) represents the divergence of two groups from a common ancestor and the emergence of a lineage possessing a new set of derived characters. By tracing the distribution of shared derived characters, you can see how we inferred the sequence of branching and the evolutionary relationships of this group of animals.

Parsimony Useful in many areas of science, **parsimony** is the adoption of the simplest explanation for observed phenomena. Systematists use the principle of parsimony to construct phylogenetic trees that require the smallest number of evolutionary changes. For instance, parsimony leads to the hypothesis that a beaver is more closely related to a kangaroo than to a platypus, because in both the beaver and the kangaroo, embryos begin development within the female uterus. It is possible that gestation evolved twice, once in the kangaroo lineage and independently in the beaver lineage, but this explanation is more complicated and therefore less likely. Typical cladistic analyses involve much more complex data sets than we presented in Figure 16A (often including comparisons of DNA sequences) and are usually handled by computer programs designed to construct parsimonious trees.

Phylogenetic Trees as Hypotheses Systematists use many kinds of evidence, such as structural and developmental features, molecular data, and behavioral traits, to reconstruct evolutionary histories. However, even the best tree represents only the most likely hypothesis based on available evidence. As new data accumulate, hypotheses are revised and new trees drawn.

An example of a redrawn tree is shown in **Figure 16B**. In traditional vertebrate taxonomy, crocodiles, snakes, lizards, and other reptiles were classified in the class Reptilia, while birds were placed in the separate class Aves. However, such a

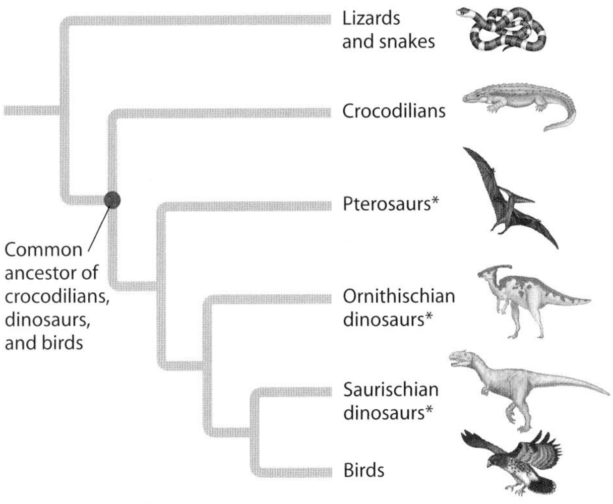

▲ **Figure 16B** A phylogenetic tree of reptiles (* indicates extinct lineages)

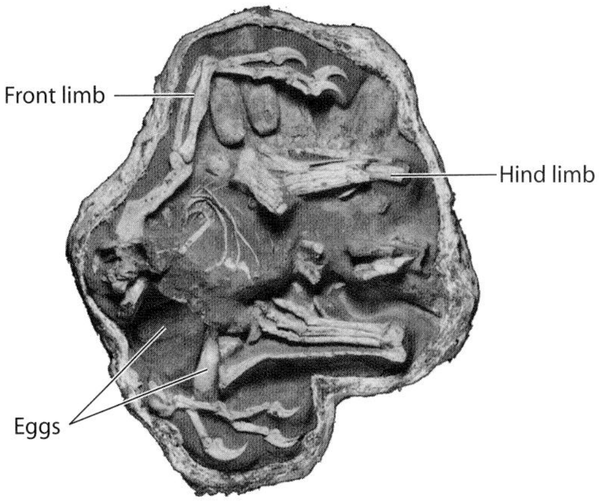

Neg./Transparency no.5789. Courtesy Dept. of Library Services, American Museum of Natural History

▲ **Figure 16C** Fossil remains of *Oviraptor* and eggs. The orientation of the bones, which surround the eggs, suggests that the dinosaur died while incubating or protecting its eggs.

reptilian clade is not monophyletic—in other words, it does not include an ancestral species and all of its descendants, one group of which includes the birds. Many lines of evidence support the tree shown in Figure 16B, showing that birds belong to the clade of reptiles.

Thinking of phylogenetic trees as hypotheses allows us to use them to make and test predictions. For example, if our phylogeny is correct, then features shared by two groups of closely related organisms should be present in their common ancestor. Using this reasoning, consider the novel predictions that can be made about dinosaurs. As seen in the tree in Figure 16B, the closest *living* relatives of birds are crocodiles. Birds and crocodiles share numerous features: They have four-chambered hearts, they "sing" to defend territories and attract mates (although a crocodile "song" is more like a bellow), and they build nests. Both birds and crocodiles care for and warm their eggs by brooding. Birds brood by sitting on their eggs, whereas crocodiles cover their eggs with their neck. Reasoning that any feature shared by birds and crocodiles is likely to have been present in their common ancestor (denoted by the red circle in Figure 16B) and all of its descendants, biologists hypothesize that dinosaurs had four-chambered hearts, sang, built nests, and exhibited brooding.

Internal organs such as hearts rarely fossilize, and it is, of course, difficult to determine whether dinosaurs sang. However, fossilized dinosaur nests have been found. **Figure 16C** shows a fossil of an *Oviraptor* dinosaur thought to have died in a sandstorm while incubating or protecting its eggs. The hypothesis that dinosaurs built nests and exhibited brooding has been further supported by additional fossil discoveries that show other species of dinosaurs caring for their eggs.

The more we know about an organism and its relatives, the more accurately we can portray its phylogeny. In the next module, we consider how molecular biology is providing valuable data for tracing evolutionary history.

? To distinguish a particular clade of mammals within the larger clade that corresponds to class Mammalia, why is hair not a useful characteristic?

Hair is a shared ancestral character common to all mammals and thus is not helpful in distinguishing different mammalian subgroups.

17 An organism's evolutionary history is documented in its genome

The more recently two species have branched from a common ancestor, the more similar their DNA sequences should be. The longer two species have been on separate evolutionary paths, the more their DNA is expected to have diverged.

Molecular Systematics Using DNA or other molecules to infer relatedness, a method called **molecular systematics** is a valuable approach for tracing phylogeny. Scientists have sequenced more than 110 billion bases of DNA from thousands of species. This enormous database has fueled a boom in the study of phylogeny and clarified many evolutionary relationships.

For example, consider the red panda, an endangered, Southeast Asian tree-dwelling mammal that feeds mostly on bamboo. It was initially classified as a close relative of the giant panda, then as a member of the raccoon family. Recent molecular studies, however, suggest that the red panda represents a separate group, which diverged from the lineage that led to the raccoon and the weasel families. Molecular evidence has also begun to sort out the relationships among the species of bears. Figure 17 presents a phylogenetic hypothesis for the families that include bears, raccoons, weasels, and the red panda. Notice that this phylogenetic tree includes a timeline, which is based on fossil evidence and molecular data that can estimate when many of these divergences occurred. Most of the phylogenetic trees we have seen so far indicate only the relative order in which lineages diverged; they do not show the timing of those events.

Bears, raccoons, and the red panda are closely related mammals that diverged fairly recently. But biologists can also use DNA analyses to assess relationships between groups of organisms that are so phylogenetically distant that structural similarities are absent. It is also possible to reconstruct phylogenies among groups of present-day prokaryotes and other microorganisms for which we have no fossil record at all. Molecular biology has helped to extend systematics to the extremes of evolutionary relationships far above and below the species level, ranging from major branches of the tree of life to its finest twigs.

The ability of molecular trees to encompass both short and long periods of time is based on the observation that different genes evolve at different rates. The DNA specifying ribosomal RNA (rRNA) changes relatively slowly, so comparisons of DNA sequences in these genes are useful for investigating relationships between taxa that diverged hundreds of millions of years ago. Studies of the genes for rRNA have shown, for example, that fungi are more closely related to animals than to green plants—something that certainly could not have been deduced from morphological comparisons alone.

In contrast, the DNA in mitochondria (mtDNA) evolves relatively rapidly and can be used to investigate more recent evolutionary events. For example, researchers have used mtDNA sequences to study the relationships between Native American groups. Their studies support earlier evidence that the Pima of Arizona, the Maya of Mexico, and the Yanomami of Venezuela are closely related, probably descending from the first wave of immigrants to cross the Bering Land Bridge from Asia to the Americas about 15,000 years ago.

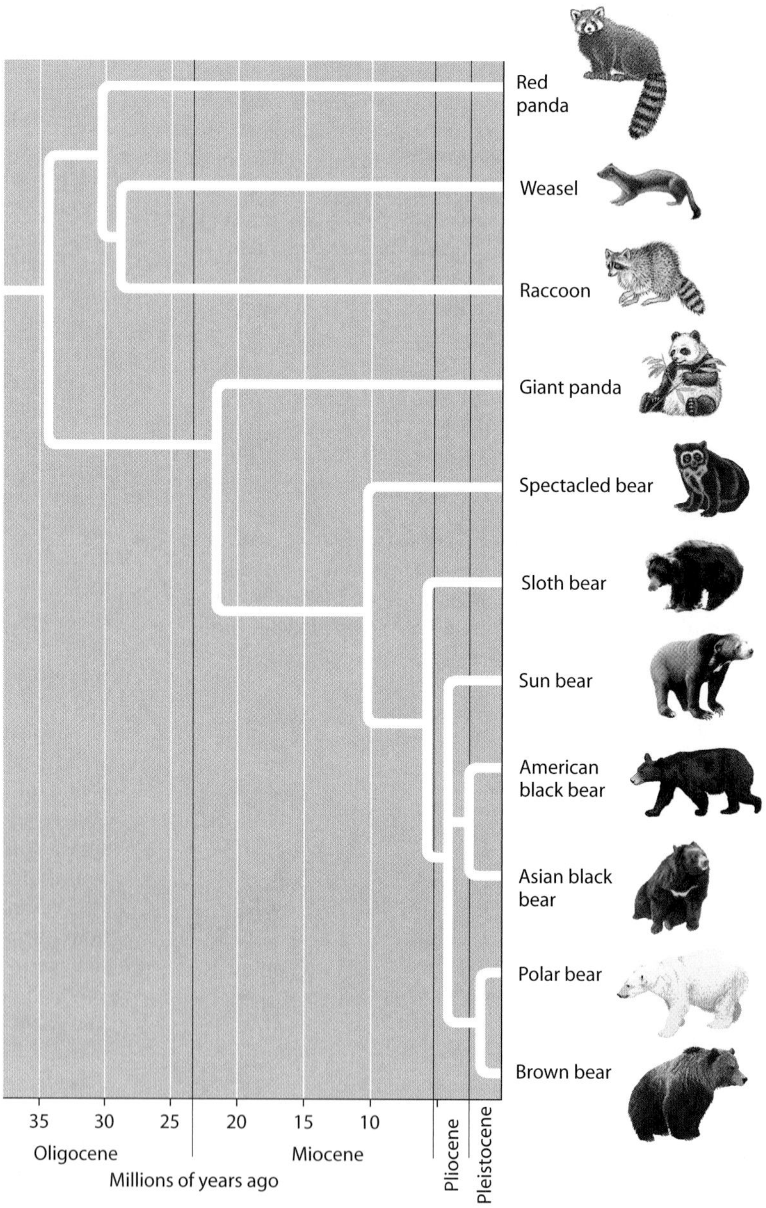

▲ **Figure 17** A phylogenetic tree based on molecular data

Genome Evolution Now that we can compare entire genomes, including our own, some interesting facts have emerged. As you may have heard, the genomes of humans and chimpanzees are strikingly similar. An even more remarkable fact is that homologous genes (similar genes that species share because of descent from a common ancestor) are widespread and can extend over huge evolutionary distances. While the genes of humans and mice are certainly not identical, 99% of them are detectably homologous. And 50% of human genes are homologous with those of yeast. This remarkable commonality demonstrates that all living organisms share many biochemical and developmental pathways and provides overwhelming support for Darwin's theory of "descent with modification."

Gene duplication has played a particularly important role in evolution because it increases the number of genes in the genome, providing additional opportunities for further evolutionary changes (see Module 11). Molecular techniques now allow scientists to trace the evolutionary history of such duplications—in which lineage they occurred and how the multiple copies of genes have diverged from each other over time.

Another interesting fact evident from genome comparisons is that the number of genes has not increased at the same rate as the complexity of organisms. Humans have only about four times as many genes as yeasts. Yeasts are simple, single-celled eukaryotes; humans have a complex brain and a body that contains more than 200 different types of tissues. Evidence is emerging that many human genes are more versatile than those of yeast, but explaining the mechanisms of such versatility remains an exciting scientific challenge.

? What types of molecules should be compared to help determine whether fungi are more closely related to plants or to animals?

Because these organisms diverged so long ago, scientists should compare molecules that change or evolve very slowly, such as the DNA that specifies rRNA.

18 Molecular clocks help track evolutionary time

Some regions of genomes appear to accumulate changes at constant rates. Comparisons of certain homologous DNA sequences for taxa known to have diverged during a certain time period have shown that the number of nucleotide substitutions is proportional to the time that has elapsed since the lineages branched. For example, homologous genes of bats and dolphins are much more alike than are homologous genes of sharks and tuna. This is consistent with the fossil evidence that sharks and tuna have been on separate evolutionary paths much longer (more than 420 million years) than have bats and dolphins (perhaps 60 million years). In this case, molecular divergence has kept better track of time than have changes in morphology.

For a gene shown to have a reliable average rate of change, a **molecular clock** can be calibrated in actual time by graphing the number of nucleotide differences against the dates of evolutionary branch points known from the fossil record. The graph line can then be used to estimate the dates of other evolutionary episodes not documented in the fossil record.

Figure 18 shows how a molecular clock has been used to date the origin of HIV infection in humans. HIV, the virus that causes AIDS, is descended from viruses that infect chimpanzees and other primates. When did HIV jump to humans? The virus has spread to humans more than once, but the most widespread strain in humans is HIV-1 M. To pinpoint the earliest infection by this strain, researchers compared samples of the virus from various times during the epidemic, including a sample from 1959. The samples showed that the virus has evolved in a clock-like fashion. Extrapolating backward from these data indicates that HIV-1 M first spread to humans during the 1930s.

Some biologists are skeptical about the accuracy of molecular clocks because the rate of molecular change may vary at different times, in different genes, and in different groups of organisms. In some cases, problems may be avoided by calibrating molecular clocks with many genes rather than just one or a few genes. One group of researchers used sequence data

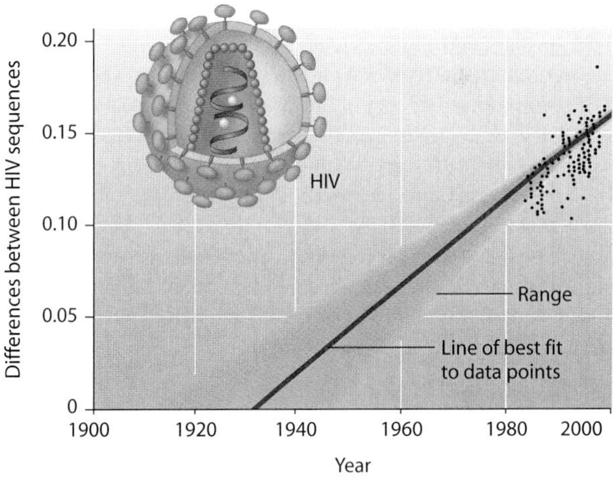

▲ **Figure 18** Dating the origin of HIV-1 M with a molecular clock. The data points in the upper-right corner represent different HIV samples taken at known times.

from 658 genes to construct a molecular clock that covered almost 600 million years of vertebrate evolution. Their estimates of divergence times agreed closely with fossil-based estimates. An abundant fossil record extends back only about 550 million years, and molecular clocks have been used to date evolutionary divergences that occurred a billion or more years ago. But the estimates assume that the clocks have been constant for all that time. Thus, such estimates are highly uncertain.

Evolutionary theory holds that all of life has a common ancestor. Molecular systematics is helping to link all living organisms into a comprehensive tree of life, as we see next.

? What is a molecular clock? What assumption underlies the use of such a clock?

A molecular clock estimates the actual time of evolutionary events based on the number of DNA changes. It is based on the assumption that some regions of genomes evolve at constant rates.

19 Constructing the tree of life is a work in progress

Phylogenetic trees are hypotheses about evolutionary history. Like all hypotheses, they are revised, or in some cases rejected, in accordance with new evidence. As you have learned, molecular systematics and cladistics are remodeling some trees.

Over the years, many schemes have been proposed for classifying all of life. Historically, a two-kingdom system divided all organisms into plants and animals. But it was beset with problems. Where do bacteria fit? Or photosynthetic unicellular organisms that move? And what about the fungi?

By the late 1960s, many biologists recognized five kingdoms: Monera (prokaryotes), Protista (a diverse kingdom consisting mostly of unicellular eukaryotes), Plantae, Fungi, and Animalia. However, molecular studies highlighted fundamental flaws in the five-kingdom system. Biologists have since adopted a **three-domain system**, which recognizes three basic groups: two domains of prokaryotes, Bacteria and Archaea, and one domain of eukaryotes, called Eukarya. Kingdoms Fungi, Plantae, and Animalia are still recognized, but kingdoms Monera and Protista are obsolete because they are not monophyletic.

Molecular and cellular evidence indicates that the two lineages of prokaryotes (bacteria and archaea) diverged very early in the evolutionary history of life. Molecular evidence also suggests that archaea are more closely related to eukaryotes than to bacteria. **Figure 19A** is an evolutionary tree based largely on rRNA genes. As you just learned, rRNA genes have evolved so slowly that homologies between distantly related organisms can still be detected. This tree shows that ❶ the first major split in the history of life was the divergence of the bacteria from the other two domains, followed by the divergence of domains Archaea and Eukarya.

Comparisons of complete genomes from the three domains, however, show that, especially during the early history of life, there have been substantial interchanges of genes between organisms in the different domains. These took place through **horizontal gene transfer**, a process in which genes are transferred from one genome to another through mechanisms such as plasmid exchange and viral infection and even through the fusion of different organisms. Figure 19A shows two major episodes of horizontal gene transfer: ❷ gene transfer between a mitochondrial ancestor and the ancestor of eukaryotes and ❸ gene transfer between a chloroplast ancestor and the ancestor of green plants. Module 4.15 describes the endosymbiont theory for the origin of mitochondria and chloroplasts.

Some scientists have argued that horizontal gene transfers were so common that the early history of life should be represented as a tangled network of connected branches. Others have suggested that the early history of life is best represented by a ring, not a tree (**Figure 19B**). Based on an analysis of hundreds of genes, some researchers have hypothesized that the eukaryote lineage (gold in the figure) arose when an early archaean (teal) fused with an early bacterium (purple). In this model, eukaryotes are as closely related to bacteria as they are to archaea—an evolutionary relationship that can best be shown in a ring of life. As new data and new methods for analyzing that data emerge, constructing a comprehensive tree of life will continue to challenge and intrigue scientists.

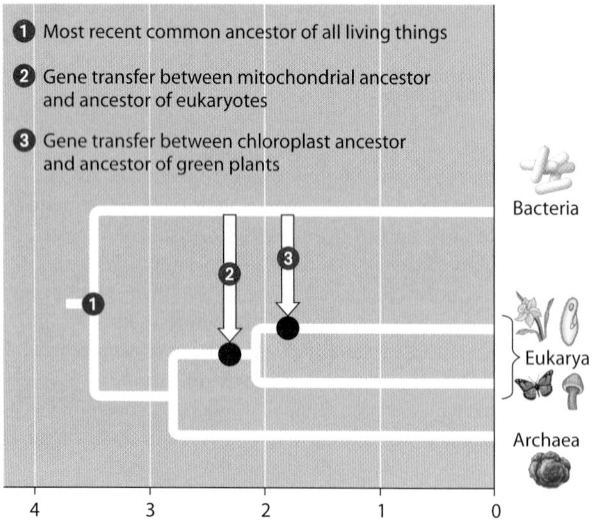

▲ **Figure 19A** Two major episodes of horizontal gene transfer in the history of life (dates are uncertain)

Adapted from S. Blair Hedges. The origin and evolution of model organisms. *Nature Reviews Genetics* 3: 838-848, fig. 1, p. 840.

? **Why might the evolutionary history of the earliest organisms be best represented by a ring of life?**

There appear to have been multiple horizontal gene transfers among these earliest organisms before the three domains of life eventually emerged from the ring to give rise to Earth's tremendous diversity of life.

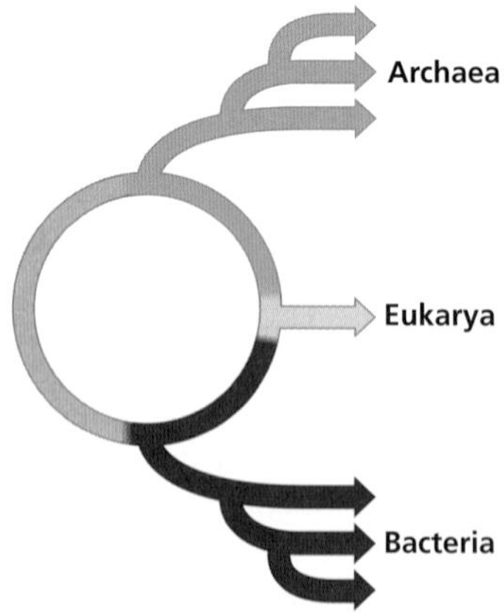

▲ **Figure 19B** Is the tree of life really a ring of life? In this model, eukaryotes arose when an early archaean fused with an early bacterium.

REVIEW

For Practice Quizzes, BioFlix, MP3 Tutors, and Activities, go to www.masteringbiology.com.

Reviewing the Concepts

Early Earth and the Origin of Life (1–3)

1 Conditions on early Earth made the origin of life possible. Earth formed some 4.6 billion years ago. Fossil stromatolites formed by prokaryotes date back 3.5 billion years.

2 Experiments show that the abiotic synthesis of organic molecules is possible.

3 Stages in the origin of the first cells probably included the formation of polymers, protocells, and self-replicating RNA. Macromolecules may have polymerized on hot rocks. The first genes may have been catalytic RNA molecules. Protocells containing self-replicating molecules could have been acted on by natural selection.

Major Events in the History of Life (4–6)

4 The origins of single-celled and multicellular organisms and the colonization of land were key events in life's history. The following timeline is based on the fossil record.

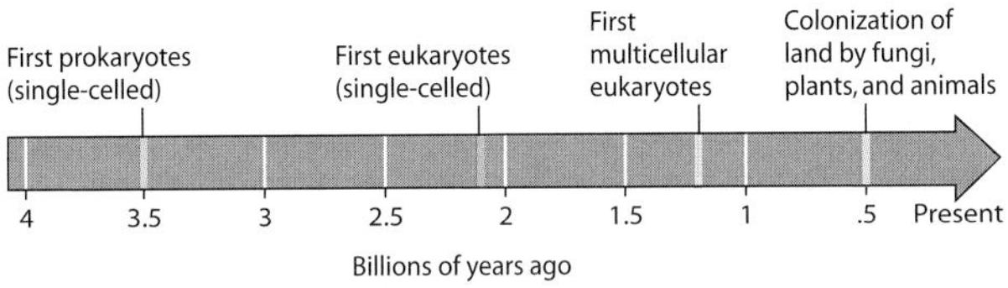

5 The actual ages of rocks and fossils mark geologic time. Radiometric dating can date rocks and fossils.

6 The fossil record documents the history of life. In the geologic record, eras and periods are separated by major transitions in lifeforms, often caused by extinctions.

Mechanisms of Macroevolution (7–13)

7 Continental drift has played a major role in macroevolution. The formation and split-up of Pangaea affected the distribution and diversification of organisms.

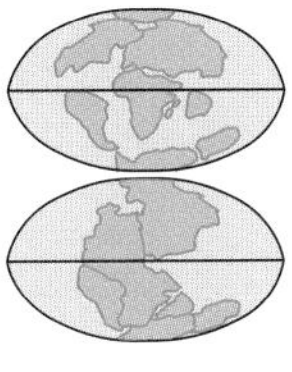

8 Plate tectonics may imperil human life. Volcanoes and earthquakes often occur at the boundaries of Earth's plates.

9 During mass extinctions, large numbers of species are lost. The Permian extinction is linked to the effects of extreme volcanic activity, and the Cretaceaous extinction, which included most dinosaurs, may have been caused by the impact of an asteroid.

10 Adaptive radiations have increased the diversity of life. The origin of many new species often follows mass extinctions, colonization of new habitats, and the evolution of new adaptations.

11 Genes that control development play a major role in evolution. "Evo-devo" combines evolutionary and developmental biology. New forms can evolve by changes in the number, sequences, or regulation of developmental genes.

12 Evolutionary novelties may arise in several ways. Complex structures may evolve in stages from simpler versions with the same basic function or from the gradual adaptation of existing structures to new functions.

13 Evolutionary trends do not mean that evolution is goal directed. An evolutionary trend may be a result of species selection or natural selection in changing environments.

Phylogeny and the Tree of Life (14–19)

14 Phylogenies based on homologies reflect evolutionary history. Homologous structures and molecular sequences provide evidence of common ancestry.

15 Systematics connects classification with evolutionary history. Taxonomists assign each species a binomial—a genus and species name. Genera are grouped into progressively broader categories. A phylogenetic tree is a hypothesis of evolutionary relationships.

16 Shared characters are used to construct phylogenetic trees. Cladistics uses shared derived characters to define clades. A parsimonious tree requires the fewest evolutionary changes.

17 An organism's evolutionary history is documented in its genome. Molecular systematics uses molecular comparisons to build phylogenetic trees. Homologous genes are found across distantly related species.

18 Molecular clocks help track evolutionary time. Regions of DNA that change at a constant rate can provide estimated dates of past events.

19 Constructing the tree of life is a work in progress. Evidence of multiple horizontal gene transfers suggests that the early history of life may be best represented by a ring, from which domains Bacteria, Archaea, and Eukarya emerge.

Connecting the Concepts

1. Using the figure below, describe the stages that may have led to the origin of life.

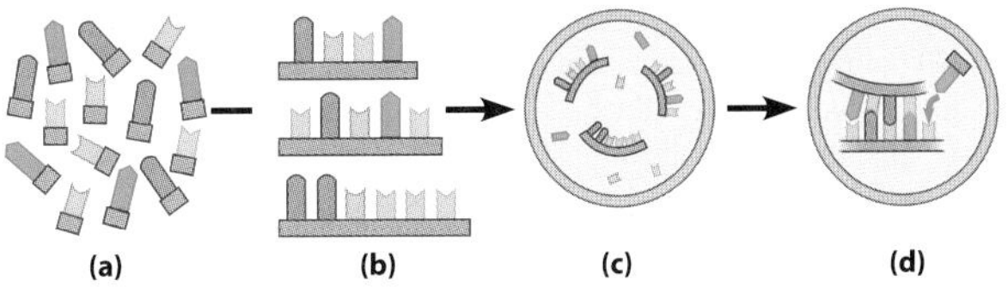

2. Fill in this concept map about systematics.

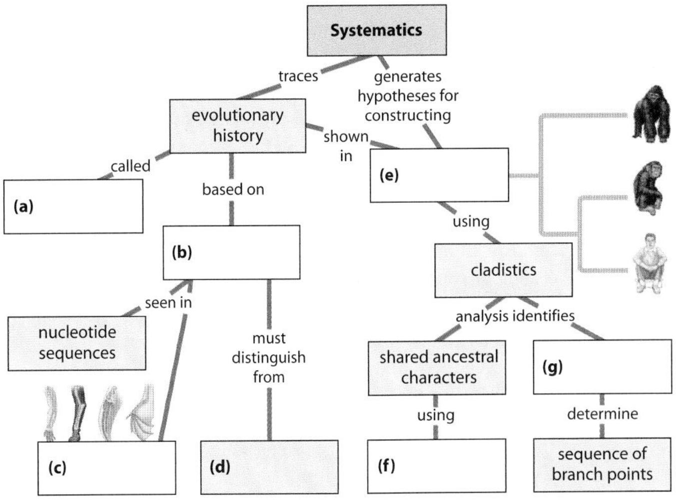

Testing Your Knowledge

Multiple Choice

3. You set your time machine for 3 billion years ago and push the start button. When the dust clears, you look out the window. Which of the following describes what you would probably see?
   a. plants and animals very different from those alive today
   b. a cloud of gas and dust in space
   c. green scum in the water
   d. land and water sterile and devoid of life
   e. an endless expanse of red-hot molten rock
4. Ancient photosynthetic prokaryotes were very important in the history of life because they
   a. were probably the first living things to exist on Earth.
   b. produced the oxygen in the atmosphere.
   c. are the oldest known archaea.
   d. were the first multicellular organisms.
   e. showed that life could evolve around deep-sea vents.
5. The animals and plants of India are very different from the species in nearby Southeast Asia. Why might this be true?
   a. India was once covered by oceans and Asia was not.
   b. The climates of the two regions are different.
   c. India is in the process of separating from the rest of Asia.
   d. Life in India was wiped out by ancient volcanic eruptions.
   e. India was a separate continent until about 45 mya.
6. Adaptive radiations may be promoted by all of the following *except* one. Which one?
   a. mass extinctions that result in vacant ecological niches
   b. colonization of an isolated region with few competitors
   c. a gradual change in climate
   d. a novel adaptation
   e. another adaptive radiation providing new food sources
7. A swim bladder is a gas-filled sac that helps fish maintain buoyancy. Evidence indicates that early fish gulped air into primitive lungs, helping them survive in stagnant waters. The evolution of the swim bladder from lungs of an ancestral fish is an example of
   a. an evolutionary trend.
   b. paedomorphosis.
   c. changes in homeotic gene expression.
   d. the gradual refinement of a structure with the same function.
   e. exaptation.
8. If you were using cladistics to build a phylogenetic tree of cats, which would be the best choice for an outgroup?
   a. kangaroo c. domestic cat e. lion
   b. leopard d. iguana
9. Which of the following could provide the best data for determining the phylogeny of very closely related species?
   a. the fossil record
   b. a comparison of embryological development
   c. their morphological differences and similarities
   d. a comparison of nucleotide sequences in homologous genes and mitochondrial DNA
   e. a comparison of their ribosomal DNA sequences
10. Major divisions in the geologic record are marked by
   a. radioactive dating.
   b. distinct changes in the types of fossilized life.
   c. continental drift.
   d. regular time intervals measured in millions of years.
   e. the appearance, in order, of prokaryotes, eukaryotes, protists, animals, plants, and fungi.

Describing, Comparing, and Explaining

11. Distinguish between microevolution and macroevolution.
12. Which are more likely to be closely related: two species with similar appearance but divergent gene sequences or two species with different appearances but nearly identical genes? Explain.
13. How can the Darwinian concept of descent with modification explain the evolution of such complex structures as an eye?
14. Explain why changes in the regulation of developmental genes may have played such a large role in the evolution of new forms.
15. What types of molecular comparisons are used to determine the very early branching of the tree of life? Explain.

Applying the Concepts

16. Measurements indicate that a fossilized skull you unearthed has a carbon-14/carbon-12 ratio about one-sixteenth that of the skulls of present-day animals. What is the approximate age of the fossil? (The half-life of carbon-14 is 5,730 years.)
17. A paleontologist compares fossils from three dinosaurs and *Archaeopteryx*, the earliest known bird. The following table shows the distribution of characters for each species, where 1 means that the character is present and 0 means it is not. The outgroup (not shown in the table) had none of the characters. Arrange these species on the phylogenetic tree below and indicate the derived character that defines each branch point.

Trait	*Velociraptor*	*Coelophysis*	*Archaeopteryx*	*Allosaurus*
Hollow bones	1	1	1	1
Three-fingered hand	1	0	1	1
Half-moon-shaped wrist bone	1	0	1	0
Reversed first toe	0	0	1	0

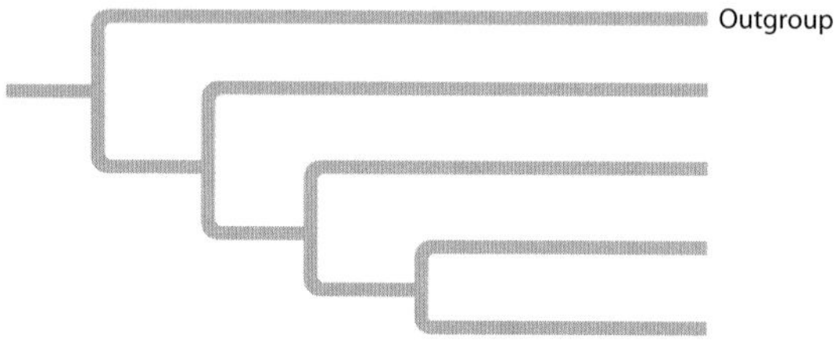

18. Experts estimate that human activities cause the extinction of hundreds of species every year. The natural rate of extinction is thought to average only a few species per year. As we continue to alter the environment, especially by destroying tropical rain forests and altering Earth's climate, the resulting wave of extinctions may rival previous mass extinctions. Considering that life has endured five mass extinctions before, should we be concerned that we may cause a sixth? How would such an extinction differ from previous extinctions? What might be the consequences for the surviving species, including ourselves?

Review Answers

1. a. Abiotic synthesis of important molecules from simpler chemicals in atmosphere, with lightning or UV radiation as energy source
   b. Polymerization of monomers, perhaps on hot rocks
   c. Enclosure within a lipid membrane, which maintained a distinct internal environment
   d. Beginnings of heredity as RNA molecules replicated themselves. Natural selection could have acted on protocells that enclosed self-replicating RNA.
2. a. phylogeny; b. homologies; c. morphology; d. analogies; e. phylogenetic tree; f. outgroup; g. shared derived characters
3. c 4. b 5. e 6. c 7. e 8. a 9. d 10. b
11. Microevolution is the change in the gene pool of a population from one generation to the next. Macroevolution involves the pattern of evolutionary changes over large time spans and includes the origin of new groups and evolutionary novelties as well as mass extinctions.
12. The latter are more likely to be closely related, because even small genetic changes can produce divergent physical appearances. But if genes have diverged greatly, it implies that lineages have been separate for some time, and the similar appearances may be analogous, not homologous.
13. Complex structures can evolve by the gradual refinement of earlier versions of those structures, all of which served a useful function in each ancestor.
14. Where and when key developmental genes are expressed in a developing embryo can greatly affect the final form and arrangement of body parts. The regulation of gene expression allows these genes to continue to be expressed in some areas, turned off in other areas, and/or expressed at different times during development.
15. The ribosomal RNA genes, which specify the RNA parts of ribosomes, have evolved so slowly that homologies between even distantly related organisms can still be detected. Analysis of other homologous genes is also used.
16. 22,920 years old, a result of four half-life reductions
17.

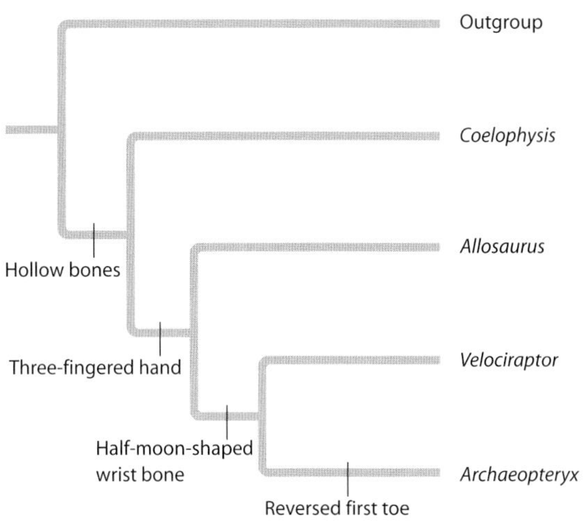

18. Some issues and questions to consider: Whereas previous mass extinctions have resulted from catastrophic events, such as asteroid collisions or volcanism, a sixth mass extinction would be caused by a single species—the result of human-caused environmental alteration. Mass extinctions can reduce complex ecological communities to much simpler ones. It can take millions of years for diversity to recover from a mass extinction. Do we have an ethical responsibility to preserve other species? By disrupting ecological communities throughout the world, a sixth mass extinction would have great consequences for all species alive today—including humans.

How Populations Evolve

From Chapter 13 of *Campbell Biology: Concepts & Connections*, Seventh Edition. Jane B. Reece, Martha R. Taylor, Eric J. Simon, Jean L. Dickey.

How Populations Evolve

BIG IDEAS

ARCHIV/Photo Researchers

Darwin's Theory of Evolution

(1–6)

Darwin's theory of evolution explains the adaptations of organisms and the unity and diversity of life.

The Evolution of Populations

(7–10)

Genetic variation makes evolution possible within a population.

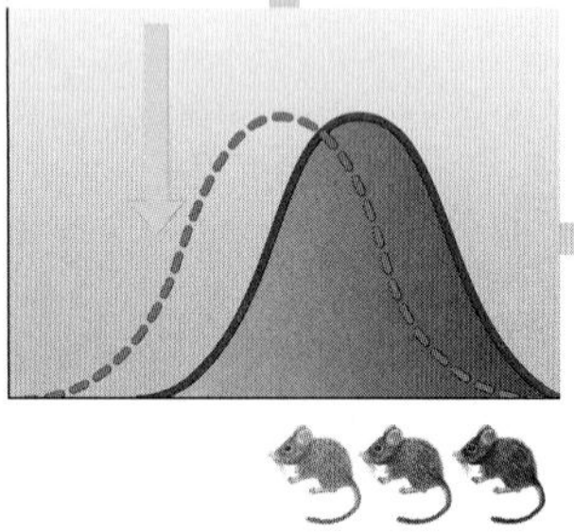

Mechanisms of Microevolution

(11–17)

Natural selection, genetic drift, and gene flow can alter gene pools; natural selection leads to adaptive evolution.

tbkmedia.de/Alamy

Charles Darwin no doubt encountered the birds with the bright blue feet pictured above in his travels to the Galápagos Islands. As we'll see in this chapter, Darwin's observations on these isolated, volcanic islands contributed greatly to his theory of evolution. Like the other species Darwin observed, the blue-footed booby (*Sula nebouxii*) has physical features that make it suited to its environment. Its large, webbed feet make great flippers, propelling the bird through the water at high speeds—a huge advantage when hunting fish. Other traits also serve it well in its sea-faring life. The booby's body and bill are streamlined, like a torpedo, minimizing friction as it dives from heights up to 24 m (over 75 feet) into the shallow water below. To pull out of this high-speed dive once it hits the water, the booby uses its large tail as a brake.

The booby's webbed feet, streamlined shape, and large tail are examples of adaptations—inherited traits that enhance an organism's ability to survive and reproduce in a particular environment. In this chapter, we examine how such adaptations evolve through the process of natural selection.

We thus begin our study of evolution—what Darwin called descent with modification. Evolution explains both the unity and diversity of life—how Earth's many species are related through descent from common ancestors and how these species have diverged over time.

It is the changes in species over time that have transformed life on Earth from its earliest forms to what Darwin called life's "endless forms most beautiful." Let's begin this grand journey with Darwin's sea voyage.

Darwin's Theory of Evolution

1 A sea voyage helped Darwin frame his theory of evolution

If you visit the Galápagos Islands, a chain of volcanic islands located about 900 km (560 miles) off the Pacific coast of South America, you will see many of the same sights that fascinated Darwin more than a century ago: blue-footed boobies waddling around; lumbering giant tortoises (whose Spanish name is *galápagos*) (**Figure 1A**); and marine iguanas basking on dark lava rocks or feeding on algae in the ocean (**Figure 1B**). You will also observe some of the finches Darwin collected—closely related small birds that represent more than a dozen species (distinct types of organisms). A major difference between these birds is the shape of their beaks, which are adapted for crushing seeds, feeding on cactus flowers, or snagging insects. Some finches even find their meals on tortoises, who raise themselves up and stretch out their necks to allow the birds to hop aboard and gobble up parasites.

One of Darwin's lasting contributions is the scientific explanation for the striking ways in which organisms are suited for life in their environment. Let's trace the path Darwin took to his theory of **evolution**, the idea that Earth's many species are descendants of ancestral species that were different from those living today.

Darwin's Cultural and Scientific Context Some early Greek philosophers suggested that life might change gradually over time. But the Greek philosopher Aristotle, whose views had an enormous impact on Western culture, generally viewed species as perfect and permanent. Judeo-Christian culture reinforced this idea with a literal interpretation of the biblical book of Genesis, which holds that species were individually designed by a divine creator. The idea that all living species are unchanging in form and inhabit an Earth that is only about 6,000 years old dominated the intellectual and cultural climate of the Western world for centuries.

In the mid-1700s, the study of **fossils**—the imprints or remains of organisms that lived in the past—revealed a succession of fossil forms in layers of sedimentary rock that differed from current life-forms. In the early 1800s, French naturalist Jean Baptiste Lamarck suggested that the best explanation for the relationship of fossils to current organisms is that life evolves. Today, we remember Lamarck mainly for his erroneous view of *how* species evolve. He proposed that by using or not using its body parts, an individual may change its traits and then pass those changes on to its offspring. He suggested, for instance, that the ancestors of the giraffe had lengthened their necks by stretching higher and higher into the trees to reach leaves. Our understanding of genetics refutes Lamarck's idea of the inheritance of acquired characteristics. But the fact remains that by strongly advocating evolution and by proposing that species evolve as a result of interactions with their environment, Lamarck helped set the stage for Darwin.

Darwin's Sea Voyage Charles Darwin was born in 1809, on the same day as Abraham Lincoln. Even as a boy, Darwin had a consuming interest in nature. When not reading books about nature, he was fishing, hunting, and collecting insects. His father, an eminent physician, could see no future for his son as a naturalist and sent him to medical school. But Darwin found medicine boring and surgery before the days of anesthesia horrifying. He quit medical school and enrolled at Cambridge University, intending to become a clergyman. At that time, many scholars of science belonged to the clergy.

Soon after graduation, Darwin's botany professor and mentor recommended him to the captain of the HMS *Beagle*, who was preparing a voyage to chart poorly known stretches of the South American coast. In December 1831, at the age of 22, Darwin began the round-the-world voyage that profoundly

Gerry Ellis/Getty Images

▲ **Figure 1A** A giant tortoise, one of the unique inhabitants of the Galápagos Islands

Photovolcanica.com/Shutterstock

▲ **Figure 1B** A marine iguana feeding on algae in the waters around the Galápagos Islands

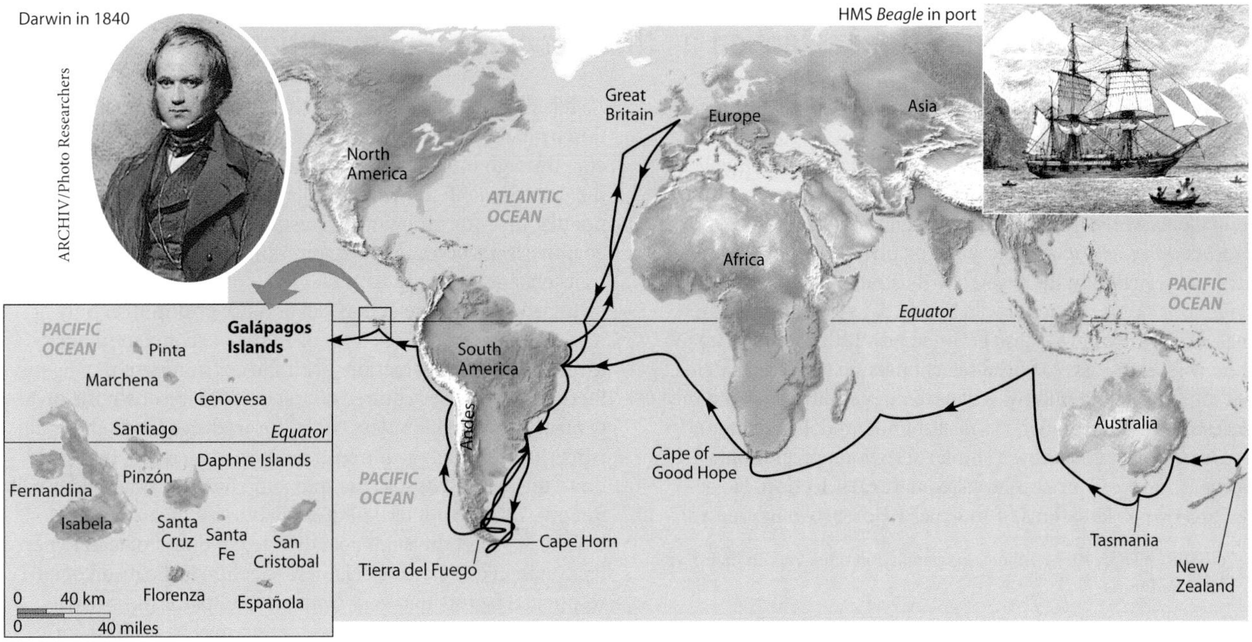

ARCHIV/Photo Researchers

National Maritime Museum Picture Library, London, England

▲ **Figure 1C** The voyage of the *Beagle* (1831–1836), with insets showing a young Charles Darwin and the ship on which he sailed

influenced his thinking and eventually the thinking of the world **(Figure 1C)**.

While the ship's crew surveyed the coast, Darwin spent most of his time on shore, collecting thousands of plants and animals as well as fossils. He noted the characteristics of plants and animals that made them well suited to such diverse environments as the jungles of Brazil, the grasslands of Argentina, and the desolate and frigid lands at the southern tip of South America. Darwin asked himself why fossils found on the South American continent were more similar to present-day South American species than to fossils found on other continents. During his visit to the Galápagos Islands, Darwin observed many unusual organisms, most of which were not known from anywhere else in the world. Many of these were similar to, but different from, the plants and animals of the nearest mainland. Even the individual islands had species different from those on other islands.

While on the voyage, Darwin was strongly influenced by Charles Lyell's newly published *Principles of Geology*. Having read Lyell's book and witnessed an earthquake that raised part of the coastline of Chile almost a meter, Darwin realized that natural forces gradually changed Earth's surface and that these forces are still operating in modern times. Thus, the growth of mountains as a result of earthquakes could account for the presence of the marine snail fossils that he had collected on mountaintops in the Andes.

By the time Darwin returned to Great Britain five years after the *Beagle* first set sail, his experiences and reading had led him to seriously doubt that Earth and all its living organisms had been specially created only a few thousand years earlier. Darwin had come to realize that Earth was very old and constantly changing. Once in Great Britain, Darwin began to analyze his collections and to discuss them with colleagues. He continued to read, correspond with other scientists, and maintain extensive journals of his observations, studies, and thoughts.

Darwin's Writings By the early 1840s, Darwin had composed a long essay describing the major features of his theory of evolution. He realized that his ideas would cause an uproar, however, and he delayed publishing his essay. Even as he procrastinated, Darwin continued to compile evidence in support of his hypothesis. In 1858, Alfred Wallace, a British naturalist doing fieldwork in Indonesia, conceived a hypothesis almost identical to Darwin's. Wallace asked Darwin to evaluate the manuscript he had written to see if it merited publication. Darwin wrote to Lyell, "So all my originality, whatever it may amount to, will be smashed." Lyell and another colleague, however, presented Wallace's paper and excerpts of Darwin's earlier essay together to the scientific community.

In 1859, Darwin published his book titled *On the Origin of Species by Means of Natural Selection*. Commonly referred to as *The Origin of Species*, Darwin's book presented the world with a logical and well-supported explanation for evolution. Darwin provided evidence that present-day species arose from a succession of ancestors. He called this evolutionary history of life "descent with modification." As the descendants of a remote ancestor spread into various habitats over millions and millions of years, they accumulated diverse modifications, or **adaptations**, that fit them to specific ways of life in their environment. Darwin's proposed mechanism for this descent with modification is natural selection, the topic of our next module.

? **What was Darwin's phrase for evolution? What does it mean?**

Descent with modification. An ancestral species could diversify into many descendant species by the accumulation of adaptations to various environments.

2 Darwin proposed natural selection as the mechanism of evolution

Darwin devoted much of *The Origin of Species* to exploring the adaptation of organisms to their environment. First, he discussed familiar examples of domesticated plants and animals. Humans have modified other species by selecting and breeding individuals with desired traits over many generations—a process called **artificial selection**. You can see evidence of artificial selection in the vegetables illustrated in **Figure 2**, all varieties of a single species of wild mustard. Crop plants and animals bred as livestock or as pets often bear little resemblance to their wild ancestors. Artificial selection has led to greater yields of crops, meat, and milk, as well as to dogs as different from the ancestral wolf as Chihuahuas, dachshunds, and Afghan hounds.

Darwin explained how a similar selection process could occur in nature, a process he called **natural selection**. He described two observations from which he drew two inferences:

OBSERVATION #1: Members of a population often vary in their inherited traits.

OBSERVATION #2: All species are capable of producing more offspring than the environment can support.

INFERENCE #1: Individuals whose inherited traits give them a higher probability of surviving and reproducing in a given environment tend to leave more offspring than other individuals.

INFERENCE #2: This unequal production of offspring will lead to the accumulation of favorable traits in a population over generations.

Darwin recognized the connection between natural selection and the capacity of organisms to "overreproduce." He had read an essay written in 1798 by economist Thomas Malthus, who contended that much of human suffering—disease, famine, and war—was the consequence of human populations increasing faster than food supplies and other resources. Darwin deduced that the production of more individuals than the limited resources can support leads to a struggle for existence, with only some offspring surviving in each generation. Of the many eggs laid, young born, and seeds spread, only a tiny fraction complete development and leave offspring. The rest are eaten, starved, diseased, unmated, or unable to reproduce for other reasons. The essence of natural selection is this unequal reproduction. Individuals whose traits better enable them to obtain food or escape predators or tolerate physical conditions will survive and reproduce more successfully, passing these adaptive traits to their offspring.

Darwin reasoned that if artificial selection can bring about so much change in a relatively short period of time, then natural selection could modify species considerably over hundreds or thousands of generations. Over vast spans of time, many traits that adapt a population to its environment will accumulate. If the environment changes, however, or if individuals move to a new environment, natural selection will select for adaptations to these new conditions, sometimes producing changes that result in the origin of a completely new species in the process.

It is important to emphasize three key points about evolution by natural selection. First, although natural selection occurs through interactions between individual organisms and the environment, individuals do not evolve. Rather, it is the population—the group of organisms—that evolves over time as adaptive traits become more common in the group and other traits change or disappear.

Second, natural selection can amplify or diminish only heritable traits. Certainly, an organism may become modified through its own interactions with the environment during its lifetime, and those acquired characteristics may help the organism survive. But unless coded for in the genes of an organism's gametes, such acquired characteristics cannot be passed on to offspring. Thus, a championship female bodybuilder will not give birth to a muscle-bound baby.

Third, evolution is not goal directed; it does not lead to perfectly adapted organisms. Natural selection is the result of environmental factors that vary from place to place and over time. A trait that is favorable in one situation may be useless—or even detrimental—in different circumstances. And as you will see, adaptations are often compromises: A blue-footed booby's webbed feet may be efficient in water, but they make for awkward walking on land.

Now let's look at some examples of natural selection.

? Compare artificial selection and natural selection.

In artificial selection, humans choose the desirable traits and breed only organisms with those traits. In natural selection, the environment does the choosing: Individuals with traits best suited to the environment survive and reproduce most successfully, passing those adaptive traits to offspring.

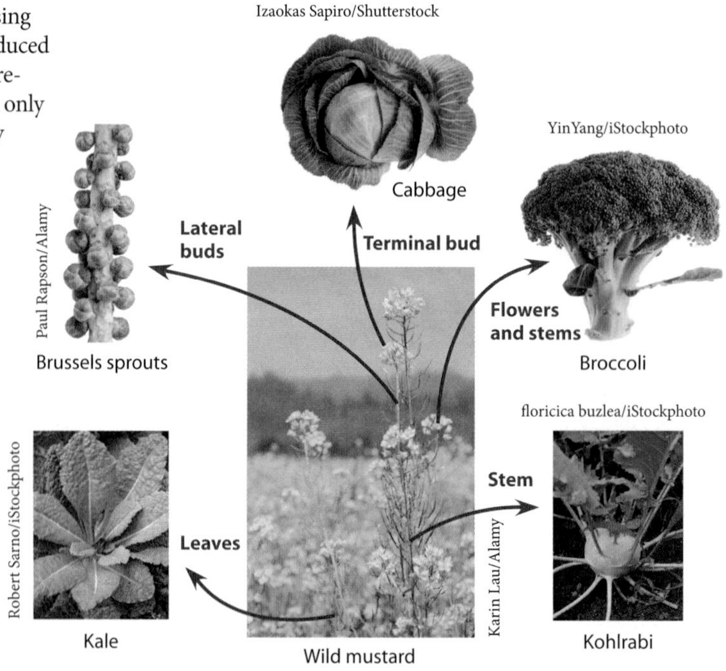

▲ **Figure 2** Artificial selection: different vegetables produced as humans have selected for variations in different parts of the wild mustard plant

3 Scientists can observe natural selection in action

The exquisite camouflage adaptations shown in **Figure 3A** by insects that evolved in different environments illustrate the results of natural selection. But do we have examples of natural selection in action?

Indeed, biologists have documented evolutionary change in thousands of scientific studies. A classic example comes from Peter and Rosemary Grant's work with finches in the Galápagos Islands over more than 30 years. As part of their research, they measured changes in beak size in a population of a ground finch species. These birds eat mostly small seeds. In dry years, when all seeds are in short supply, birds must eat more large seeds. Birds with larger, stronger beaks have a feeding advantage and greater reproductive success, and the Grants measured an increase in the average beak depth for the population. During wet years, smaller beaks are more efficient for eating the now abundant small seeds, and they found a decrease in average beak depth.

An unsettling example of natural selection in action is the evolution of pesticide resistance in hundreds of insect species. Whenever a new type of pesticide is used to control agricultural pests, the story is similar **(Figure 3B)**: A relatively small amount of poison dusted onto a crop may kill 99% of the insects, but subsequent sprayings are less and less effective. The few survivors of the first pesticide wave are insects that are genetically resistant, carrying an allele (alternative form of a gene, colored red in the figure) that somehow enables them to resist the chemical attack. So the poison kills most members of the population, leaving the resistant individuals to reproduce and pass the alleles for pesticide resistance to their offspring. The proportion of pesticide-resistant individuals thus increases in each generation. Like the finches, the insect population has adapted to environmental change through natural selection.

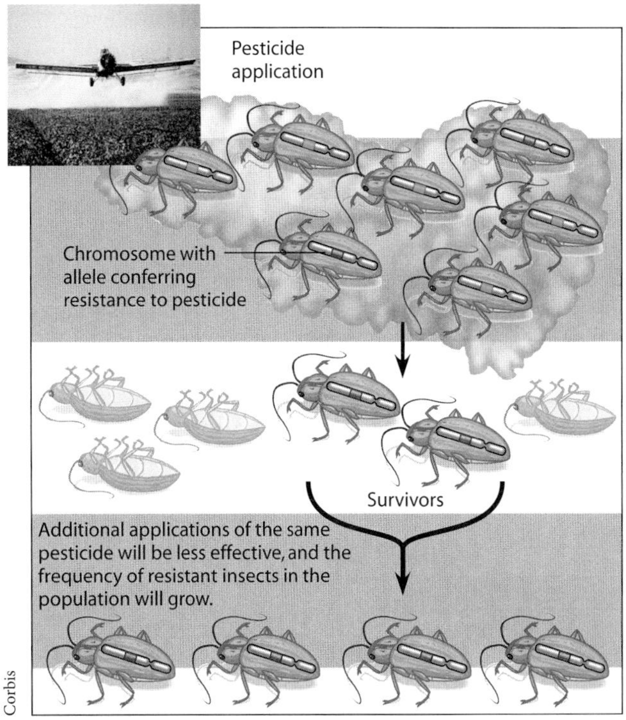

Corbis

▲ **Figure 3B** Evolution of pesticide resistance in an insect population

These examples of evolutionary adaptation highlight two important points about natural selection. First, natural selection is more an editing process than a creative mechanism. A pesticide does not create alleles that allow insects to resist it. Rather, the presence of the pesticide leads to natural selection for insects already in the population that have those alleles. Second, natural selection is contingent on time and place: It favors those heritable traits in a varying population that fit the current, local environment. For instance, mutations that endow houseflies with resistance to the pesticide DDT also reduce their growth rate. Before DDT was introduced, such mutations were a handicap to the flies that had them. But once DDT was part of the environment, the mutant alleles were advantageous, and natural selection increased their frequency in fly populations.

Now that we have seen some direct observations of natural selection, let's turn to the many other types of scientific evidence that support evolution.

Edward S. Ross

Michael & Patricia Fogden/Corbis

▲ **Figure 3A** Camouflage as an example of evolutionary adaptation

? In what sense is natural selection more an editing process than a creative process?

● Natural selection cannot create beneficial traits on demand but instead "edits" variation in a population by selecting for individuals with those traits that are best suited to the current environment.

4 The study of fossils provides strong evidence for evolution

Darwin developed his theory of descent with modification using evidence from artificial selection and from comparisons of the geographic distribution of species and the body structures of different species. (These comparisons are described in the next module.) He also relied on the fossil record. Fossils document differences between past and present organisms and the fact that many species have become extinct.

The photographs on this page illustrate a number of fossils, each of which formed in a somewhat different way. The organic substances of a dead organism usually decay rapidly, but the hard parts of an animal that are rich in minerals, such as the bones and teeth of dinosaurs and the shells of clams and snails, may remain as fossils. The fossilized skull in **Figure 4A** is from one of our early relatives, *Homo erectus*, who lived some 1.5 million years ago in Africa.

Many of the fossils that **paleontologists** (scientists who study fossils) find in their digs are not the actual remnants of organisms at all but are replicas of past organisms. Such fossils result when a dead organism captured in sediment decays and leaves an empty mold that is later filled by minerals dissolved in water. The casts that form when the minerals harden are replicas of the organism, as seen in the 375-million-year-old casts of ammonites shown in **Figure 4B**. Ammonites were shelled marine organisms.

Trace fossils are footprints, burrows, and other remnants of an ancient organism's behavior. The boy in **Figure 4C** is standing in a 150-million-year-old dinosaur track in Colorado.

Some fossils actually retain organic material. The leaf in **Figure 4D** is about 40 million years old. It is a thin film pressed in rock, still greenish with remnants of its chlorophyll and well enough preserved that biologists can analyze its molecular and cellular structure. In rare instances, an entire organism, including its soft parts, is fossilized. That can happen only when the individual is buried in a medium that prevents bacteria and fungi from decomposing the corpse. The insect in **Figure 4E** got stuck in the resin of a tree about 35 million years ago. The resin hardened into amber (fossilized resin), preserving the insect. Other media can also preserve organisms. Explorers

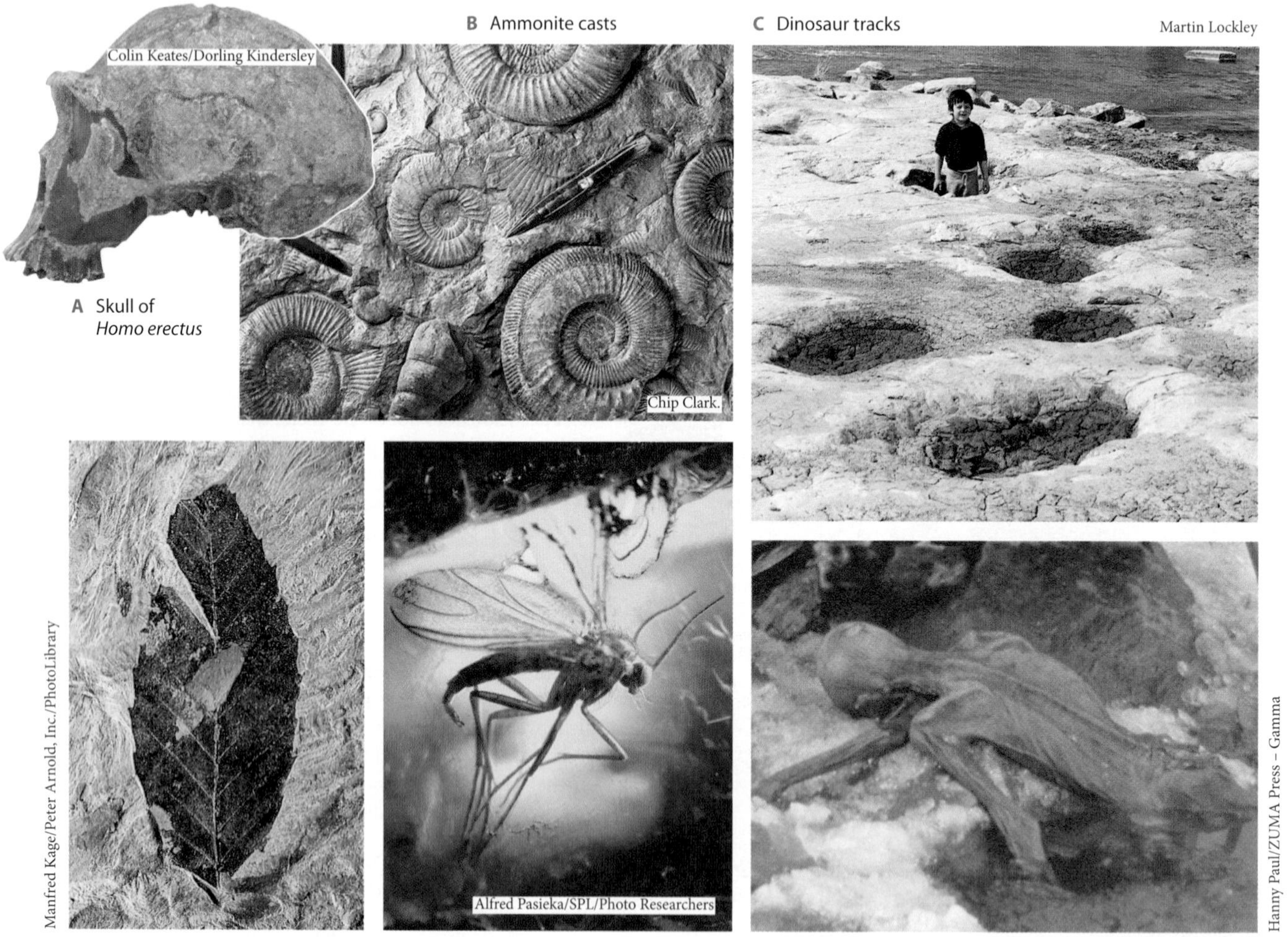
Colin Keates/Dorling Kindersley
Chip Clark.
Martin Lockley
Manfred Kage/Peter Arnold, Inc./PhotoLibrary
Alfred Pasieka/SPL/Photo Researchers
Hanny Paul/ZUMA Press – Gamma

A Skull of *Homo erectus*
B Ammonite casts
C Dinosaur tracks
D Fossilized organic matter of a leaf
E Insect in amber
F "Ice Man"

▲ **Figure 4A–F** A gallery of fossils

Albo/Shutterstock

▲ **Figure 4G** Strata of sedimentary rock at the Grand Canyon

have discovered mammoths, bison, and even prehistoric humans frozen in ice or preserved in acid bogs. Such rare discoveries make the news, as did the 1991 discovery of the "Ice Man" in **Figure 4F**, who died 5,000 years ago. However, biologists rely mainly on the fossils formed in sedimentary rock to reconstruct the history of life.

The **fossil record**—the sequence in which fossils appear within layers of sedimentary rocks—provides some of the strongest evidence of evolution. Sedimentary rocks form from layers of sand and mud that settle to the bottom of seas, lakes, and swamps. Over millions of years, deposits pile up and compress the older sediments below into rock. When organisms die, they settle along with the sediments and may be preserved as fossils. The rate of sedimentation and the types of particles that settle vary over time. As a result, the rock forms in **strata**, or layers. Younger strata are on top of older ones; thus, the relative ages of fossils can be determined by the layer in which they are found. **Figure 4G** (above) shows strata of sedimentary rock at the Grand Canyon. The Colorado River has cut through more than 2,000 m of rock, exposing sedimentary layers that can be read like huge pages from the book of life. Scan the canyon wall from rim to floor and you look back through hundreds of millions of years. Each layer entombs fossils that represent some of the organisms from that period of Earth's history. Of course, the fossil record is incomplete: Not all organisms live in areas that favor fossilization, many rocks are distorted by geologic processes, and not all fossils that have been preserved will be found.

The fossil record reveals the historical sequence in which organisms evolved, and radiometric dating helps determine the ages of rocks and fossils. The oldest known fossils, dating from about 3.5 billion years ago, are prokaryotes. Molecular and cellular evidence also indicates that prokaryotes were the ancestors of all life. More than a billion years passed before fossils of single-celled eukaryotes formed, and approximately a billion more before multicellular eukaryotes appeared. Fossils in younger strata reveal the sequential evolution of various groups of eukaryotic organisms.

Darwin lamented the lack of fossils showing how existing groups of organisms gave rise to new groups. In the 150 years since publication of *The Origin of Species*, however, many fossil discoveries illustrate the transition of existing groups to new groups. For example, a series of fossils traces the gradual modification of jaws and teeth in the evolution of mammals from a reptilian ancestor.

Another series of fossils traces the evolution of whales from four-legged land mammals. Whales living today have forelegs in the form of flippers but lack hind legs. In the past few decades, remarkable fossils of extinct mammals have been discovered in Pakistan, Egypt, and North America that document the transition from life on land to life in the sea **(Figure 4H)**. Additional fossils show that *Pakicetus* and *Rodhocetus* (shown in the figure) had a type of ankle bone that is otherwise unique to the group of mammals that includes pigs, hippos, cows, camels, and deer. The ankle bone similarity strongly suggests that whales (and dolphins and porpoises) are most closely related to this group of land mammals.

Next we look at other sources of evidence for evolution.

? **What types of animals do you think would be most represented in the fossil record? Explain your answer.**

Animals with hard parts, such as shells or bones that readily fossilize, and those that lived in areas where sedimentary rock may form

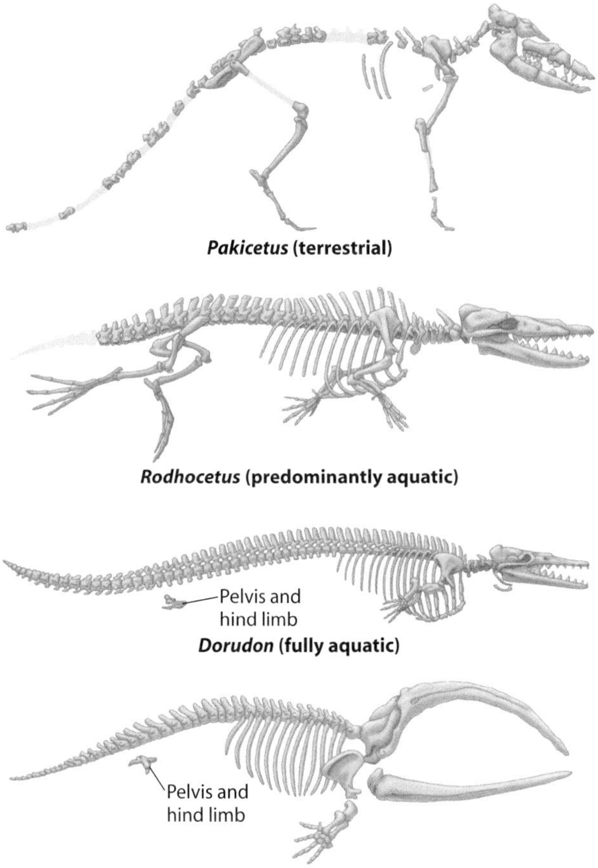

▲ **Figure 4H** A transition to life in the sea: fossils of four species that link whales with their land-dwelling ancestors

5 Many types of scientific evidence support the evolutionary view of life

Darwin's wide range of evidence convinced many of the scientists of his day that organisms do indeed evolve. Let's continue to explore Darwin's evidence as well as some new types of data that provide overwhelming support for evolution.

Biogeography It was the geographic distribution of species, known as **biogeography**, that first suggested to Darwin that organisms evolve from ancestral species. Darwin noted that Galápagos animals resembled species of the South American mainland more than they resembled animals on islands that were similar but much more distant. The logical explanation was that the Galápagos species evolved from animals that migrated from South America. These immigrants eventually gave rise to new species as they became adapted to their new environments. Among the many examples from biogeography that make sense only in the historical context of evolution are the marsupials in Australia. This unique collection of pouched mammals, such as kangaroos and koalas, evolved in isolation after geologic activities separated that island continent from the landmasses on which placental mammals diversified.

Comparative Anatomy Also providing support for evolution and cited extensively by Darwin is comparative anatomy. Anatomical similarities between species give signs of common descent. Similarity in characteristics that results from common ancestry is known as **homology**. As **Figure 5A** shows, the same skeletal elements make up the forelimbs of humans, cats, whales, and bats. The functions of these forelimbs differ. A whale's flipper does not do the same job as a bat's wing, so if these structures had been uniquely engineered, we would expect that their basic designs would be very different. The logical explanation is that the arms, forelegs, flippers, and wings of these different mammals are variations on an anatomical structure of an ancestral organism that over millions of years has become adapted to different functions. Biologists call such anatomical similarities in different organisms **homologous structures**—features that often have different functions but are structurally similar because of common ancestry.

Comparative anatomy illustrates that evolution is a remodeling process in which ancestral structures that originally functioned in one capacity become modified as they take on new functions—the kind of process that Darwin called descent with modification.

Comparing early stages of development in different animal species reveals additional homologies not visible in adult organisms. For example, at some point in their development, all vertebrate embryos have a tail posterior to the anus, as well as structures called pharyngeal (throat) pouches. These homologous pouches ultimately develop into structures with very different functions, such as gills in fishes and parts of the ears and throat in humans. Note the pharyngeal pouches and tails of the bird embryo (left) and the human embryo (right) in **Figure 5B**, on the facing page.

Some of the most interesting homologies are "leftover" structures that are of marginal or perhaps no importance to the organism. These **vestigial structures** are remnants of features that served important functions in the organism's ancestors. For example, the small pelvis and hind-leg bones of ancient whales (see Figure 4H) are vestiges of their walking ancestors. Another example are the eye remnants that are buried under scales in blind species of cave fishes—a vestige of their sighted ancestors.

Molecular Biology Anatomical homology is not helpful in linking very distantly related organisms such as plants and animals and microorganisms. But in recent decades, advances in **molecular biology** have enabled biologists to read a molecular history of evolution in the DNA sequences of organisms. The hereditary background of an organism is documented in its DNA and in the proteins encoded there. Siblings have greater similarity in their DNA and proteins than do unrelated individuals of the same species. And if two species have homologous genes with sequences that match closely, biologists conclude that these sequences must have been inherited from a relatively recent common ancestor. Conversely, the greater the number of sequence differences between species, the more distant is their last common ancestor. Molecular comparisons between diverse organisms have allowed biologists to develop hypotheses about the evolutionary divergence of major branches on the tree of life.

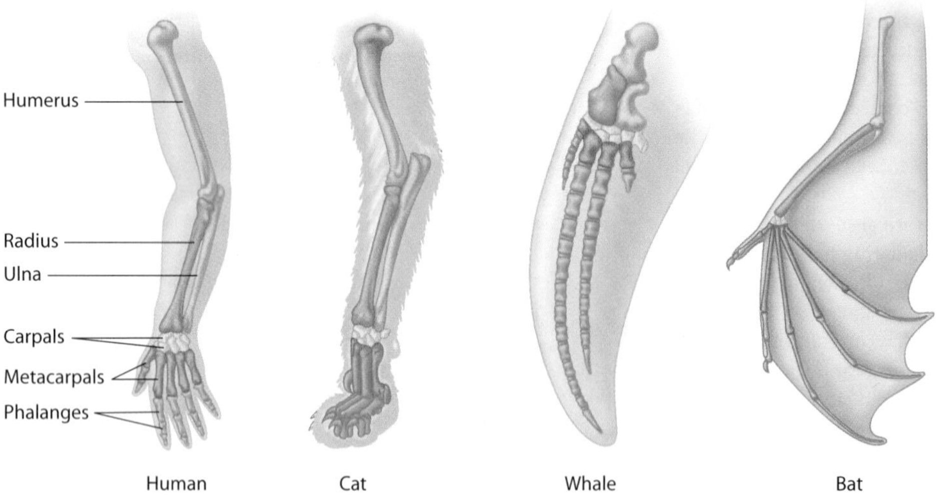

▲ **Figure 5A** Homologous structures: vertebrate forelimbs

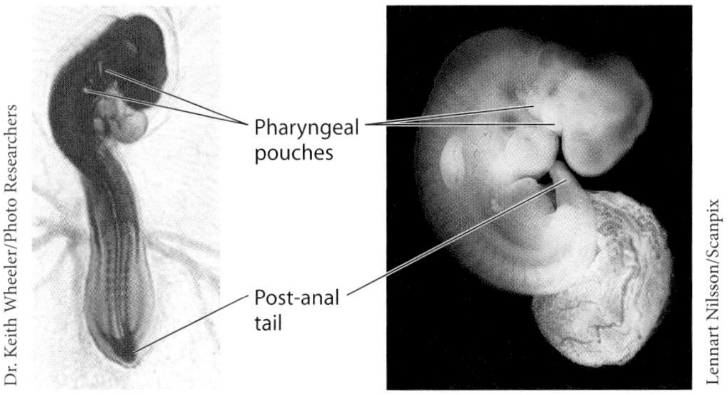

Chick embryo Human embryo

▲ **Figure 5B** Homologous structures in vertebrate embryos

Darwin's boldest hypothesis was that all life-forms are related. Molecular biology provides strong evidence for this claim: All forms of life use the same genetic language of DNA and RNA, and the genetic code—how RNA triplets are translated into amino acids—is essentially universal. Thus, it is likely that all species descended from common ancestors that used this code. But molecular homologies go beyond a shared code. For example, organisms as dissimilar as humans and bacteria share homologous genes that have been inherited from a very distant common ancestor.

? What is homology? How does the concept of homology relate to molecular biology?

Homology is similarity in different species due to evolution from a common ancestor. Similarities in DNA sequences or proteins reflect the evolutionary relationship that is the basis of homology.

6 Homologies indicate patterns of descent that can be shown on an evolutionary tree

Darwin was the first to view the history of life as a tree, with multiple branchings from a common ancestral trunk to the descendant species at the tips of the twigs. Biologists represent these patterns of descent with an **evolutionary tree**, although today they often turn the trees sideways, as in **Figure 6**.

Homologous structures, both anatomical and molecular, can be used to determine the branching sequence of such a tree. Some homologous characteristics, such as the genetic code, are shared by all species because they date to the deep ancestral past. In contrast, characteristics that evolved more recently are shared only within smaller groups of organisms. For example, all tetrapods (from the Greek *tetra*, four, and *pod*, foot) possess the same basic limb bone structure illustrated in Figure 5A, but their ancestors do not.

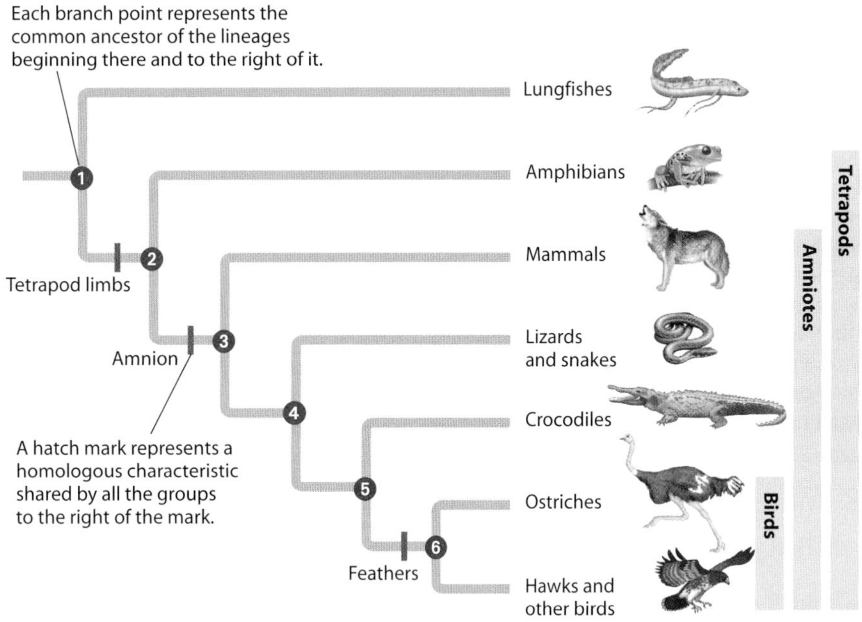

▲ **Figure 6** An evolutionary tree for tetrapods and their closest living relatives, the lungfishes

Figure 6 is an evolutionary tree of tetrapods (amphibians, mammals, and reptiles, including birds) and their closest living relatives, the lungfishes. In this diagram, each branch point represents the common ancestor of all species that descended from it. For example, lungfishes and all tetrapods descended from ancestor ❶. Three homologies are shown by the purple hatch marks on the tree—tetrapod limbs, the amnion (a protective embryonic membrane), and feathers. Tetrapod limbs were present in ancestor ❷ and hence are found in all of its descendants (the tetrapods). The amnion was present only in ancestor ❸ and thus is shared only by mammals and reptiles. Feathers were present only in ancestor ❻ and hence are found only in birds.

Evolutionary trees are hypotheses reflecting our current understanding of patterns of evolutionary descent. Some trees are more speculative because less data may be available. Others, such as the one in Figure 6, are supported by strong combinations of fossil, anatomical, and DNA sequence data.

In the next section, we consider the smallest scale of evolution—genetic changes within a population over time.

? Refer to the evolutionary tree in Figure 6. Are crocodiles more closely related to lizards or birds?

Look for the most recent common ancestor of these groups. Crocodiles are more closely related to birds because they share a more recent common ancestor with birds (ancestor ❺) than with lizards (ancestor ❹).

The Evolution of Populations

7 Evolution occurs within populations

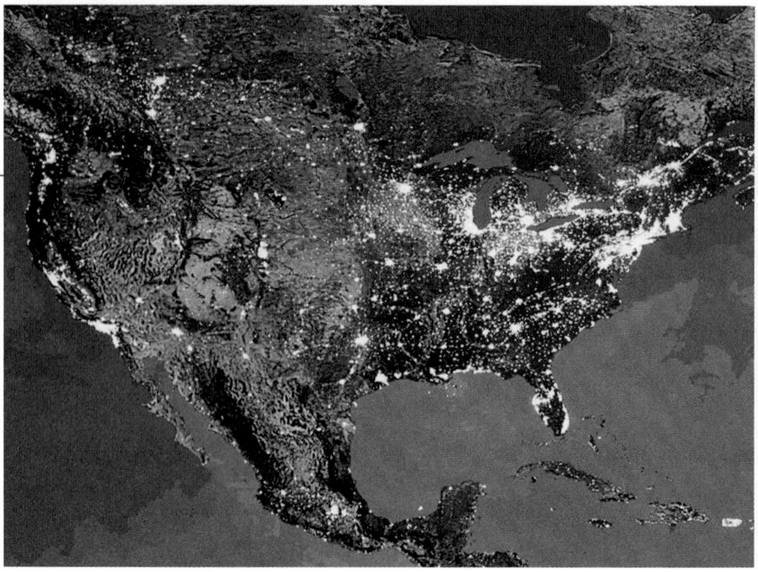

Earth Imaging/Getty Images

▲ **Figure 7** Human population centers in North America

One common misconception about evolution is that individual organisms evolve during their lifetimes. It is true that natural selection acts on individuals: Each individual's combination of traits affects its survival and reproductive success. But the evolutionary impact of natural selection is only apparent in the changes in a population of organisms over time.

A **population** is a group of individuals of the same species that live in the same area and interbreed. We can measure evolution as a change in the prevalence of certain heritable traits in a population over a span of generations. The increasing proportion of resistant insects in areas sprayed with pesticide is one example (see Module 3). Natural selection favored insects with alleles for pesticide resistance; these insects left more offspring, and the population changed, or evolved.

Let's examine some key features of populations. Different populations of the same species may be isolated from one another, with little interbreeding and thus little exchange of genes between them. Such isolation is common for populations confined to widely separated islands or in different lakes. However, populations are not usually so isolated, and they rarely have sharp boundaries. **Figure 7** is a nighttime satellite photograph showing the lights of population centers in North America. We know that these populations are not really isolated; people move around, and there are suburban and rural communities between cities. Nevertheless, people are most likely to choose mates locally. Thus, for humans and other species, members of one population are generally more closely related to each other than to members of other populations.

In studying evolution at the population level, biologists focus on what is called the **gene pool**, the total collection of genes in a population at any one time. The gene pool consists of all the alleles in all the individuals making up a population. For many genes, there are two or more alleles in the gene pool. For example, in an insect population, there may be two alleles relating to pesticide breakdown, one that codes for an enzyme that breaks down a certain pesticide and one for a version of the enzyme that does not. In populations living in fields sprayed with pesticide, the allele for the enzyme conferring resistance will increase in frequency and the other allele will decrease in frequency. When the relative frequencies of alleles in a population change like this over a number of generations, evolution is occurring on its smallest scale. Such a change in a gene pool is often called **microevolution**.

Next let's look at the genetic variation that makes microevolution possible.

? Why can't an individual evolve?

● Evolution involves changes in the genetic makeup of a population over time. An individual's genetic makeup rarely changes during its lifetime.

8 Mutation and sexual reproduction produce the genetic variation that makes evolution possible

In *The Origin of Species*, Darwin provided evidence that life on Earth has evolved over time, and he proposed that natural selection, in favoring some heritable traits over others, was the primary mechanism for that change. But he could not explain the cause of variation among individuals, nor could he account for how those variations passed from parents to offspring.

Just a few years after the publication of *The Origin of Species*, Gregor Mendel wrote a groundbreaking paper on inheritance in pea plants. By breeding peas in his abbey garden, Mendel discovered the very hereditary processes required for natural selection. Although the significance of Mendel's work was not recognized during his or Darwin's lifetime, it set the stage for understanding the genetic differences on which evolution is based.

Genetic Variation You have no trouble recognizing your friends in a crowd. Each person has a unique genome, reflected in individual variations in appearance and other traits. Indeed, individual variation occurs in all species, as illustrated on the next page in **Figure 8** by the lady beetles and the garter snakes. All four of these snakes were captured in one Oregon field. In addition to obvious physical differences, most populations have a great deal of variation that can be observed only at the molecular level.

Of course, not all variation in a population is heritable. The phenotype results from a combination of the genotype, which is inherited, and many environmental influences. For instance, if you undertake a strength-training program to build up your muscle mass, you will not pass your environmentally produced physique to your offspring. Only the genetic component of variation is relevant to natural selection.

Many of the characters that vary in a population result from the combined effect of several genes. Polygenic inheritance produces characters that vary more or less continuously—in

Laura Jesse, Extension Entomologist, Iowa State University

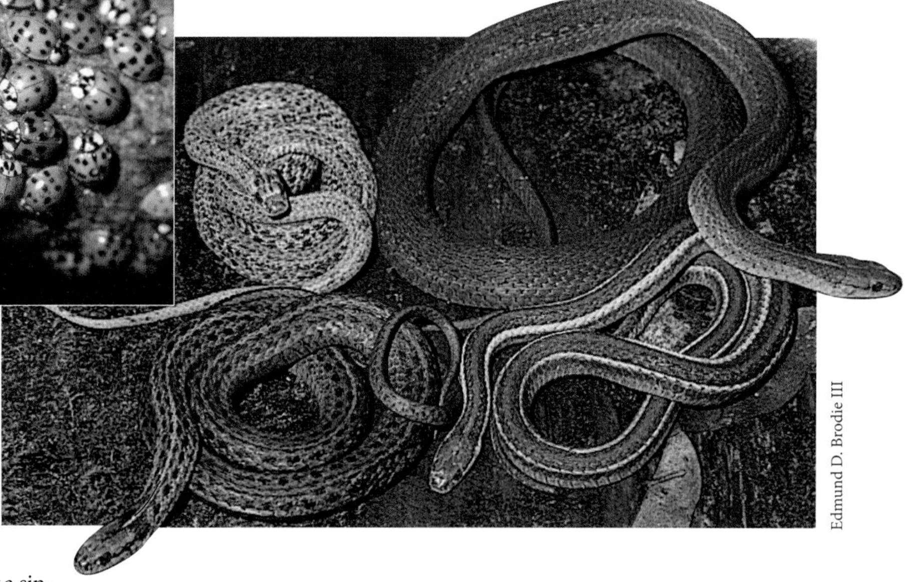

Edmund D. Brodie III

▲ **Figure 8** Variation within a species of Asian lady beetles (above) and within a species of garter snakes (right)

human height, for instance, from very short individuals to very tall ones. By contrast, other features, such as Mendel's purple and white pea flowers or human blood types, are determined by a single gene locus, with different alleles producing distinct phenotypes. But where do these alleles come from?

Mutation New alleles originate by mutation, a change in the nucleotide sequence of DNA. Thus, mutation is the ultimate source of the genetic variation that serves as raw material for evolution. In multicellular organisms, however, only mutations in cells that produce gametes can be passed to offspring and affect a population's genetic variability.

A mutation that affects a protein's function will probably be harmful. An organism is a refined product of thousands of generations of past selection, and a random change in its DNA is not likely to improve its genome any more than randomly changing some words on a page is likely to improve a story.

On rare occasions, however, a mutant allele may actually improve the adaptation of an individual to its environment and enhance its reproductive success. This kind of effect is more likely when the environment is changing in such a way that mutations that were once disadvantageous are favorable under the new conditions. The evolution of DDT-resistant houseflies (see Module 3) illustrates this point.

Chromosomal mutations that delete, disrupt, or rearrange many gene loci at once are almost certain to be harmful. But duplication of a gene or small pieces of DNA through errors in meiosis can provide an important source of genetic variation. If a repeated segment of DNA can persist over the generations, mutations in duplicated genes may accumulate, eventually leading to new genes with novel functions. For example, the remote ancestors of mammals carried a single gene for detecting odors, which has been duplicated repeatedly. As a result, mice have about 1,300 different olfactory receptor genes. It is likely that such dramatic increases helped early mammals by enabling them to distinguish among many different smells. And duplications of genes that control development are linked to the origin of vertebrate animals from an invertebrate ancestor.

In prokaryotes, mutations can quickly generate genetic variation. Because bacteria multiply so rapidly, a beneficial mutation can increase in frequency in a matter of hours or days. And because bacteria are haploid, with a single allele for each gene, a new allele can have an effect immediately.

Mutation rates in animals and plants average about one in every 100,000 genes per generation. For these organisms, low mutation rates, long time spans between generations, and diploid genomes prevent most mutations from significantly affecting genetic variation from one generation to the next.

Sexual Reproduction In organisms that reproduce sexually, most of the genetic variation in a population results from the unique combination of alleles that each individual inherits. (Of course, the origin of those allele variations is past mutations.)

Fresh assortments of existing alleles arise every generation from three random components of sexual reproduction: crossing over, independent orientation of homologous chromosomes at metaphase I of meiosis, and random fertilization. During prophase I, pairs of homologous chromosomes, one set inherited from each parent, trade some of their genes by crossing over. Then each pair of chromosomes separates into gametes independently of other chromosome pairs. Thus, gametes from any individual vary extensively in their genetic makeup. Finally, each zygote made by a mating pair has a unique assortment of alleles resulting from the random union of sperm and egg.

Genetic variation is necessary for a population to evolve, but variation alone does not guarantee that microevolution will occur. In the next module, we'll explore how to test whether evolution is occurring in a population.

? **What is the ultimate source of genetic variation? What is the source of most genetic variation in a population that reproduces sexually?**

Mutation; unique combinations of alleles resulting from sexual reproduction

9 The Hardy-Weinberg equation can test whether a population is evolving

To understand how microevolution works, let's first examine a simple population in which evolution is not occurring and thus the gene pool is not changing. Consider an imaginary population of blue-footed boobies with individuals that differ in foot webbing (**Figure 9A**). Let's assume that foot webbing is controlled by a single gene and that the allele for nonwebbed feet (W) is completely dominant to the allele for webbed feet (w). The term *dominant* may seem to suggest that over many generations, the W allele will somehow come to "dominate," becoming more and more common at the expense of the recessive allele. In fact, this is not what happens. The shuffling of alleles that accompanies sexual reproduction does not alter the genetic makeup of the population. In other words, no matter how many times alleles are segregated into different gametes and united in different combinations by fertilization, the frequency of each allele in the gene pool will remain constant unless other factors are operating. This equilibrium is known as the **Hardy-Weinberg principle**, named for the two scientists who derived it independently in 1908.

To test the Hardy-Weinberg principle, let's look at two generations of our imaginary booby population. **Figure 9B** shows the frequencies of alleles in the gene pool of the original population. We have a total of 500 birds; of these, 320 have the genotype WW (nonwebbed feet), 160 have the heterozygous genotype, Ww (also nonwebbed feet, because the nonwebbed allele W is dominant), and 20 have the genotype ww (webbed feet). The proportions or frequencies of the three genotypes are shown in the middle of Figure 9B: 0.64 for WW ($\frac{320}{500}$), 0.32 for Ww ($\frac{160}{500}$), and 0.04 for ww ($\frac{20}{500}$).

From these genotype frequencies, we can calculate the frequency of each allele in the population. Because these are diploid organisms, this population of 500 has a total of 1,000 alleles for foot type. To determine the number of W alleles, we add the number in the WW boobies, $2 \times 320 = 640$, to the number in the Ww boobies, 160. The total number of W alleles is thus 800. The frequency of the W allele, which we will call p, is $\frac{800}{1,000}$, or 0.8. We can calculate the frequency of the w allele in a similar way; this frequency, called q, is 0.2. The letters p and q are often used to represent allele frequencies. (You will notice

▶ **Figure 9A** Imaginary blue-footed boobies, with and without foot webbing

Phenotypes			
Genotypes	WW	Ww	ww
Number of animals (total = 500)	320	160	20
Genotype frequencies	$\frac{320}{500} = 0.64$	$\frac{160}{500} = 0.32$	$\frac{20}{500} = 0.04$
Number of alleles in gene pool (total = 1,000)	640 W	160 W + 160 w	40 w
Allele frequencies	$\frac{800}{1,000} = 0.8$ W		$\frac{200}{1,000} = 0.2$ w

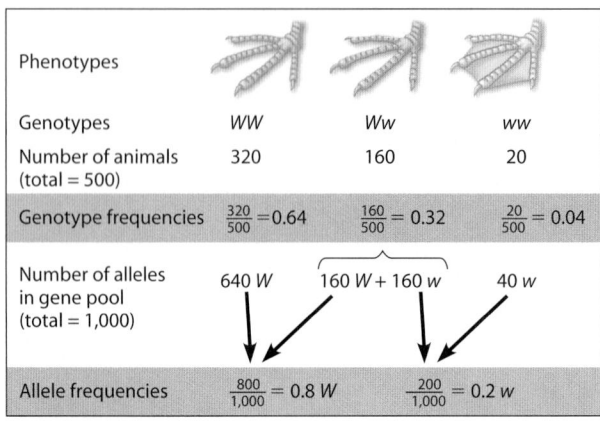

▲ **Figure 9B** Gene pool of the original population of imaginary blue-footed boobies

that $p + q = 1$. The combined frequencies of all alleles for a gene in a population must equal 1. If there are only two alleles and you know the frequency of one allele, you can calculate the frequency of the other.)

What happens when the boobies of this parent population form gametes? At the end of meiosis, each gamete has one allele for foot type, either W or w. The frequency of the two alleles in the gametes will be the same as their frequencies in the gene pool of the parental population, 0.8 for W and 0.2 for w.

Figure 9C, on the next page, shows a Punnett square that uses these gamete allele frequencies and the rule of multiplication to calculate the frequencies of the three genotypes in the next generation. The probability of producing a WW individual (by combining two W alleles from the pool of gametes) is $p \times p = p^2$, or $0.8 \times 0.8 = 0.64$. Thus, the frequency of WW boobies in the next generation would be 0.64. Likewise, the frequency of ww individuals would be $q^2 = 0.04$. For heterozygous individuals, Ww, the genotype can form in two ways, depending on whether the sperm or egg supplies the dominant allele. In other words, the frequency of Ww would be $2pq = 2 \times 0.8 \times 0.2 = 0.32$. Do these frequencies look familiar? Notice that the three genotypes have the same frequencies in the next generation as they did in the parent generation.

Finally, what about the frequencies of the alleles in this new generation? Because the genotype frequencies are the same as in the parent population, the allele frequencies p and q are also the same. In fact, we could follow the frequencies of alleles and genotypes through many generations, and the results would continue to be the same. Thus, the gene pool of this population is in a state of equilibrium—Hardy-Weinberg equilibrium.

Now let's write a general formula for calculating the frequencies of genotypes in a population from the frequencies of alleles in the gene pool. In our imaginary blue-footed booby population, the frequency of the W allele (p) is 0.8, and the frequency of the w allele (q) is 0.2. Again note that $p + q = 1$. Also notice

Gametes reflect allele frequencies of parental gene pool	Sperm	
	W $p = 0.8$	w $q = 0.2$
Eggs W $p = 0.8$	WW $p^2 = 0.64$	Ww $pq = 0.16$
w $q = 0.2$	wW $qp = 0.16$	ww $q^2 = 0.04$

Next generation:

Genotype frequencies	0.64 WW	0.32 Ww	0.04 ww
Allele frequencies		0.8 W	0.2 w

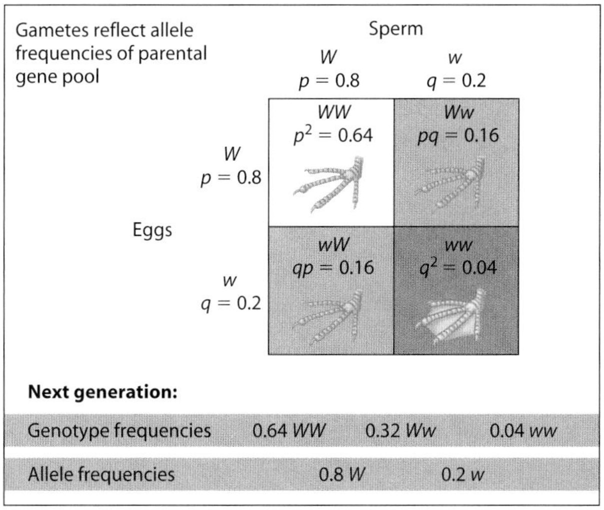

▲ **Figure 9C** Gene pool of next generation of imaginary boobies

in Figures 9B and 9C that the frequencies of the three possible genotypes in the populations also add up to 1 (that is, 0.64 + 0.32 + 0.04 = 1). We can represent these relationships symbolically with the Hardy-Weinberg equation:

$$ \underset{\substack{\text{Frequency} \\ \text{of homozygous} \\ \text{dominants}}}{p^2} + \underset{\substack{\text{Frequency} \\ \text{of heterozygotes}}}{2pq} + \underset{\substack{\text{Frequency} \\ \text{of homozygous} \\ \text{recessives}}}{q^2} = 1 $$

If a population is in Hardy-Weinberg equilibrium, allele and genotype frequencies will remain constant generation after generation. The Hardy-Weinberg principle tells us that something other than the reshuffling processes of sexual

reproduction is required to change allele frequencies in a population. One way to find out what factors *can* change a gene pool is to identify the conditions that must be met if genetic equilibrium is to be maintained. For a population to be in Hardy-Weinberg equilibrium, it must satisfy five main conditions:

1. Very large population
2. No gene flow between populations
3. No mutations
4. Random mating
5. No natural selection

Let's expand on these conditions: (1) The smaller the population, the more likely that allele frequencies will fluctuate by chance from one generation to the next. (2) When individuals move into or out of populations, they add or remove alleles, altering the gene pool. (3) By changing alleles or deleting or duplicating genes, mutations modify the gene pool. (4) If individuals mate preferentially, such as with close relatives (inbreeding), random mixing of gametes does not occur, and genotype frequencies change. (5) The unequal survival and reproductive success of individuals (natural selection) can alter allele frequencies.

Rarely are all five conditions met in real populations, and thus, allele and genotype frequencies often do change. The Hardy-Weinberg equation can be used to test whether evolution is occurring in a population. The equation also has medical applications, as we see next.

? Which is *least* likely to alter allele and genotype frequencies in a few generations of a large, sexually reproducing population: gene flow, mutation, or natural selection? Explain.

● Mutation. Because mutations are rare, their effect on allele and genotype frequencies from one generation to the next is likely to be small.

CONNECTION **10 The Hardy-Weinberg equation is useful in public health science**

Public health scientists use the Hardy-Weinberg equation to estimate how many people carry alleles for certain inherited diseases. Consider the case of phenylketonuria (PKU), which is an inherited inability to break down the amino acid phenylalanine. PKU occurs in about one out of 10,000 babies born in the United States, and if untreated results in severe mental retardation. Newborns are now routinely tested for PKU, and symptoms can be lessened if individuals living with the disease follow a diet that strictly regulates the intake of phenylalanine. Packaged foods with ingredients such as aspartame, a common artificial sweetener that contains phenylalanine, must be labeled clearly (**Figure 10**).

PKU is due to a recessive allele, so the frequency of individuals born with PKU corresponds to the q^2 term in the Hardy-Weinberg equation. Given one PKU occurrence per 10,000 births, $q^2 = 0.0001$. Therefore, the frequency of the recessive allele for PKU in the population, q, equals the square root of 0.0001, or 0.01. And the frequency of the dominant allele, p, equals 1 − q, or 0.99. The frequency of carriers,

heterozygous people who do not have PKU but may pass the PKU allele on to offspring, is $2pq$, which equals $2 \times 0.99 \times 0.01$, or 0.0198. Thus, the equation tells us that about 2% (actually 1.98%) of the U.S. population are carriers of the PKU allele. Estimating the frequency of a harmful allele is part of any public health program dealing with genetic diseases.

▶ **Figure 10** A warning to individuals with PKU

INGREDIENTS: SORBITOL, MAGNESIUM STEARATE, ARTIFICIAL FLAVOR, **ASPARTAME† (SWEETENER),** ARTIFICIAL COLOR (YELLOW 5 LAKE, BLUE 1 LAKE), ZINC GLUCONATE. **†PHENYLKETONURICS: CONTAINS PHENYLALANINE**

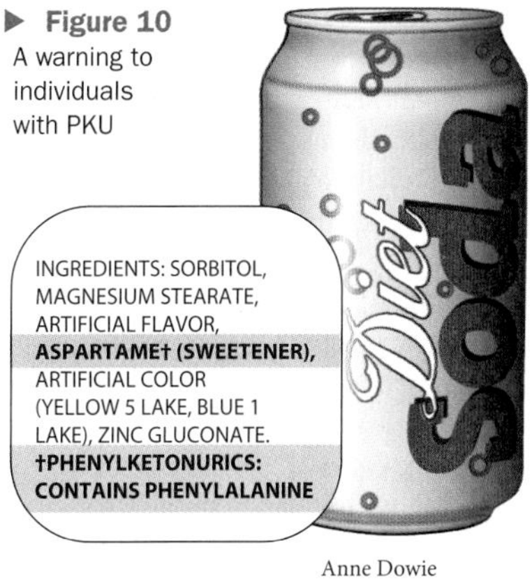

Anne Dowie

? Which term in the Hardy-Weinberg equation—p^2, $2pq$, or q^2—corresponds to the frequency of individuals who have no alleles for the disease PKU?

● The frequency of individuals with no PKU alleles is p^2.

Mechanisms of Microevolution

11 Natural selection, genetic drift, and gene flow can cause microevolution

Deviations from the five conditions named in Module 9 for Hardy-Weinberg equilibrium can alter allele frequencies in a population (microevolution). Although new genes and new alleles originate by mutation, these random and rare events probably change allele frequencies little within a population of sexually reproducing organisms. Nonrandom mating can affect the frequencies of homozygous and heterozygous genotypes, but by itself usually does not affect allele frequencies. The three main causes of evolutionary change are natural selection, genetic drift, and gene flow.

Natural Selection The condition for Hardy-Weinberg equilibrium that there be no natural selection—that all individuals in a population be equal in ability to reproduce—is probably never met in nature. Populations consist of varied individuals, and some variants leave more offspring than others. In our imaginary blue-footed booby population, birds with webbed feet (genotype *ww*) might survive better and produce more offspring because they are more efficient at swimming and catching food than birds without webbed feet. Genetic equilibrium would be disturbed as the frequency of the *w* allele increased in the gene pool from one generation to the next.

Genetic Drift Flip a coin a thousand times, and a result of 700 heads and 300 tails would make you suspicious about that coin. But flip a coin 10 times, and an outcome of 7 heads and 3 tails would seem within reason. The smaller the sample, the more likely that chance alone will cause a deviation from an idealized result—in this case, an equal number of heads and tails. Let's apply that logic to a population's gene pool. The frequencies of alleles will be more stable from one generation to the next when a population is large. In a process called **genetic drift**, chance events can cause allele frequencies to fluctuate unpredictably from one generation to the next. The smaller the population, the more impact genetic drift is likely to have. In fact, an allele can be lost from a small population by such chance fluctuations. Two situations in which genetic drift can have a significant impact on a population are those that produce the bottleneck effect and the founder effect.

Earthquakes, floods, or fires may kill large numbers of individuals, leaving a small surviving population that is unlikely to have the same genetic makeup as the original population. Such a drastic reduction in population size is called a **bottleneck effect**. Analogous to shaking just a few marbles through a bottleneck (**Figure 11A**), certain alleles (purple marbles) may be present at higher frequency in the surviving population than in the original population, others (green marbles) may be present at lower frequency, and some (orange marbles) may not be present at all. After a bottlenecking event, genetic drift may continue for many generations until the population is again large enough for fluctuations due to chance to have less of an impact. Even if a population that has passed through a bottleneck ultimately recovers its size, it may have low levels of genetic variation—a legacy of the genetic drift that occurred when the population was small.

▲ **Figure 11A** The bottleneck effect

One reason it is important to understand the bottleneck effect is that human actions may create severe bottlenecks for other species, such as the endangered Florida panther and the African cheetah (see Figure 12). The greater prairie chicken (**Figure 11B**) is another example. Millions of these birds once lived on the prairies of Illinois. But as the prairies were converted to farmland and other uses during the 19th and 20th centuries, the number of greater prairie chickens plummeted. By 1993, only two Illinois populations remained, with a total of fewer than 50 birds. Less than 50% of the eggs of these birds hatched. Researchers compared the DNA of the 1993 population with DNA extracted from museum specimens dating back to the 1930s. They surveyed six gene loci and found that the modern birds had lost 30% of the alleles that were present in the museum specimens. Thus, genetic drift as a result of the bottleneck reduced the genetic variation of the population and may have increased the frequency of harmful alleles, leading to the low egg-hatching rate.

▼ **Figure 11B** Greater prairie chicken (*Tympanuchus cupido*)

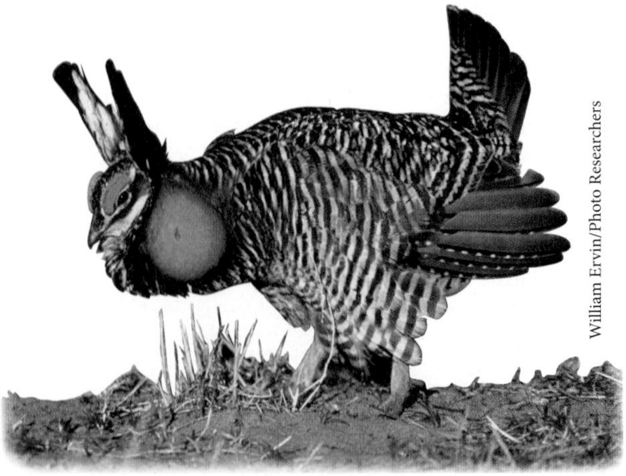

Genetic drift is also likely when a few individuals colonize an island or other new habitat, producing what is called the **founder effect**. The smaller the group, the less likely the genetic makeup of the colonists will represent the gene pool of the larger population they left.

The founder effect explains the relatively high frequency of certain inherited disorders among some human populations established by small numbers of colonists. In 1814, 15 people founded a colony on Tristan da Cunha, a group of small islands in the middle of the Atlantic Ocean. Apparently, one of the colonists carried a recessive allele for retinitis pigmentosa, a progressive form of blindness. Of the 240 descendants who still lived on the islands in the 1960s, four had retinitis pigmentosa, and at least nine others were known to be heterozygous carriers of the allele. The frequency of this allele is 10 times higher on Tristan da Cunha than in the British population from which the founders came.

Gene Flow Allele frequencies in a population can also change as a result of **gene flow**, where a population may gain or lose alleles when fertile individuals move into or out of a population or when gametes (such as plant pollen) are transferred between populations. Gene flow tends to reduce differences between populations. For example, humans today move more freely about the world than in the past, and gene flow has become an important agent of evolutionary change in previously isolated human populations.

Let's return to the Illinois greater prairie chickens and see how gene flow improved their fate. To counteract the lack of genetic diversity, researchers added a total of 271 birds from neighboring states to the Illinois populations. This strategy worked. New alleles entered the population, and the egg-hatching rate improved to more than 90%.

? How might gene flow between populations living in different habitats actually interfere with each population's adaptation to its local environment?

The introduction of alleles that may not be beneficial in a particular habitat prevents the population living there from becoming fully adapted to its local conditions.

12 Natural selection is the only mechanism that consistently leads to adaptive evolution

Genetic drift, gene flow, and even mutation can cause microevolution. But only by chance could these events result in improving a population's fit to its environment. Evolution by natural selection, on the other hand, is a blend of chance and "sorting": chance in the random collection of genetic variation packaged in gametes and combined in offspring, and sorting in that some alleles are favored over others in a given environment. Because of this sorting effect, only natural selection consistently leads to adaptive evolution—evolution that results in a better fit between organisms and their environment.

The adaptations of organisms include many striking examples. Remember the adaptations of blue-footed boobies to their marine environment and the amazing camouflage of mantids shown in Figure 3A. Or consider the remarkable speed of the cheetah (**Figure 12**). The cheetah can accelerate from 0 to 40 miles per hour in three strides and to a full speed of 70 miles per hour in seconds. Its entire body—skeleton, muscles, joints, heart, and lungs—is built for speed.

Such adaptations for speed are the result of natural selection. By consistently favoring some alleles over others, natural selection improves the match between organisms and their environment. However, the environment may change over time. As a result, what constitutes a "good match" between an organism and its environment is a moving target, making adaptive evolution a continuous, dynamic process.

Let's take a closer look at natural selection. The commonly used phrases "struggle for existence" and "survival of the fittest" are misleading if we take them to mean direct competition between individuals. There *are* animal species in which individuals lock horns or otherwise do combat to determine mating privilege. But reproductive success is generally more subtle and passive. In a varying population of moths, certain individuals may produce more offspring than others because their wing colors hide them from predators better. Plants in a wildflower population may differ in reproductive success because some attract more pollinators, owing to slight variations in flower color, shape, or fragrance. In a given environment, such traits can lead to greater **relative fitness**: the contribution an individual makes to the gene pool of the next generation relative to the contributions of other individuals. The fittest individuals in the context of evolution are those that produce the largest number of viable, fertile offspring and thus pass on the most genes to the next generation.

? Explain how the phrase "survival of the fittest" differs from the biological definition of relative fitness.

Survival alone does not guarantee reproductive success. An organism's relative fitness is determined by its number of fertile offspring and thus its relative contribution to the gene pool of the next generation.

▼ **Figure 12** The flexible spine of a cheetah stretching out in the middle of its 7-m (23-foot) stride

Tom Brakefield/Corbis

13 Natural selection can alter variation in a population in three ways

Evolutionary fitness is related to genes, but it is an organism's phenotype—its physical traits, metabolism, and behavior—that is directly exposed to the environment. Let's see how natural selection can affect the distribution of phenotypes using an imaginary mouse population that has a heritable variation in fur coloration. The bell-shaped curve in the top graph of **Figure 13** depicts the frequencies of individuals in an initial population in which fur color varies along a continuum from very light (only a few individuals) through various intermediate shades (many individuals) to very dark (a few individuals). The bottom graphs show three ways in which natural selection can alter the phenotypic variation in the mouse population. The blue downward arrows symbolize the pressure of natural selection working against certain phenotypes.

Stabilizing selection favors intermediate phenotypes. In the mouse population depicted in the graph on the bottom left, stabilizing selection has eliminated the extremely light and dark individuals, and the population has a greater number of intermediate phenotypes, which may be best suited to an environment with medium gray rocks. Stabilizing selection typically reduces variation and maintains the status quo for a particular character. For example, this type of selection keeps the majority of human birth weights in the range of 3–4 kg (6.5–9 pounds). For babies a lot smaller or larger than this, infant mortality may be greater.

Directional selection shifts the overall makeup of the population by acting against individuals at *one* of the phenotypic extremes. For the mouse population in the bottom center graph, the trend is toward darker fur color, as might occur if a fire darkened the landscape and now darker fur would more readily camouflage the animal. Directional selection is most common during periods of environmental change or when members of a species migrate to some new habitat with different environmental conditions. The changes we described in populations of insects exposed to pesticides are an example of directional selection. Another example is the increase in beak depth in a population of Galápagos finches following a drought, when the birds had to be able to eat more large seeds to survive (see Module 3).

Disruptive selection typically occurs when environmental conditions vary in a way that favors individuals at *both* ends of a phenotypic range over individuals with intermediate phenotypes. For the mice in the graph on the bottom right, individuals with light and dark fur have increased numbers. Perhaps the mice colonized a patchy habitat where a background of light soil was studded with areas of dark rocks. Disruptive selection can lead to two or more contrasting phenotypes in the same population. For example, in a population of African black-bellied seedcracker finches, large-billed birds, which specialize in cracking hard seeds, and small-billed birds, which feed mainly on soft seeds, survive better than birds with intermediate-sized bills, which are fairly inefficient at cracking both types of seeds.

Next we consider a special case of selection, one that leads to phenotypic differences between males and females.

? **What type of selection probably resulted in the color variations evident in the garter snakes in Figure 8?**

Disruptive selection

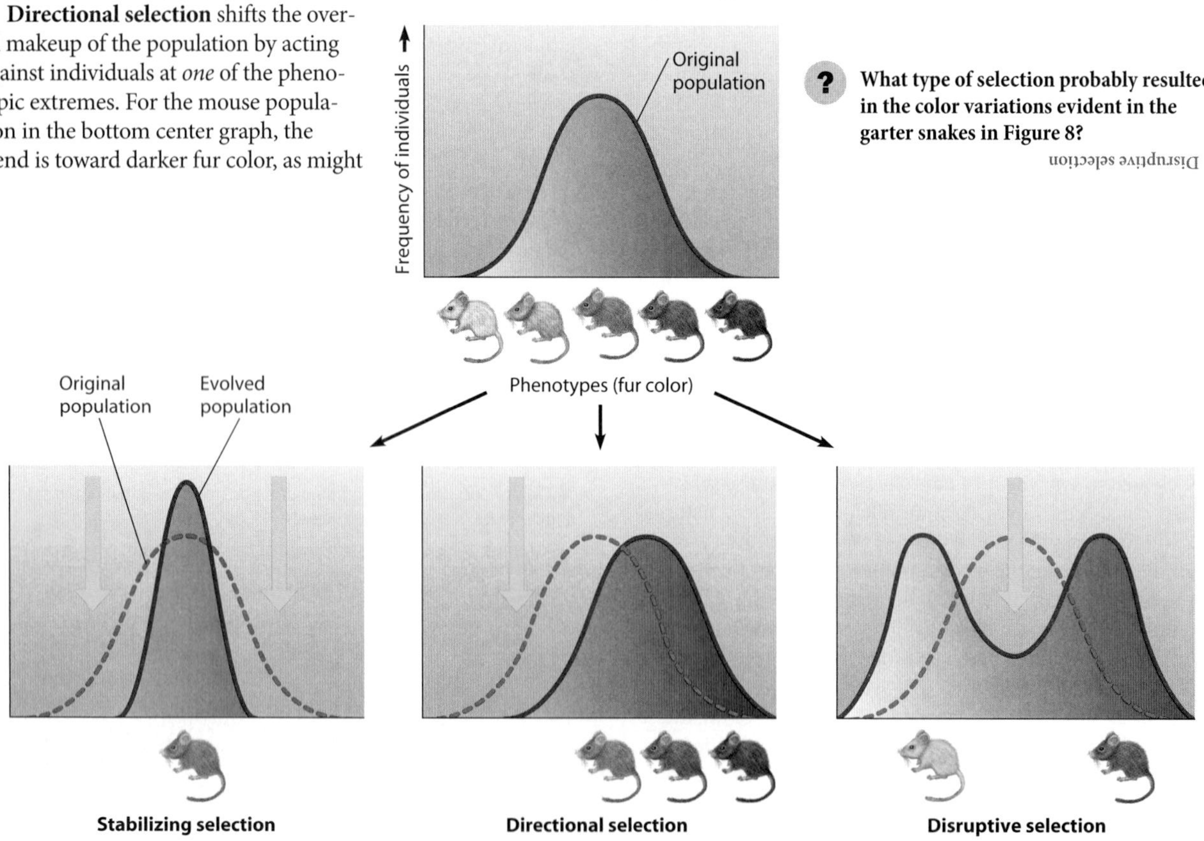

▲ **Figure 13** Three possible effects of natural selection on a phenotypic character

14 Sexual selection may lead to phenotypic differences between males and females

Darwin was the first to examine **sexual selection**, a form of natural selection in which individuals with certain traits are more likely than other individuals to obtain mates. The males and females of an animal species obviously have different reproductive organs. But they may also have secondary sexual characteristics, noticeable differences not directly associated with reproduction or survival. This distinction in appearance, called **sexual dimorphism**, is often manifested in a size difference, but it can also include forms of adornment, such as manes on lions or colorful plumage on birds (**Figure 14A**). Males are usually the showier sex, at least among vertebrates.

In some species, individuals compete directly with members of the same sex for mates (**Figure 14B**). This type of sexual selection is called *intrasexual selection* (within the same sex, most often the males). Contests may involve physical combat, but are more often ritualized displays. Intrasexual selection is frequently found in species where the winning individual acquires a harem of mates.

In a more common type of sexual selection, called *intersexual selection* (between sexes) or *mate choice*, individuals of one sex (usually females) are choosy in selecting their mates. Males with the largest or most colorful adornments are often the most attractive to females. The extraordinary feathers of a peacock's tail are an example of this sort of "choose me" statement. What intrigued Darwin is that some of these mate-attracting features do not seem to be otherwise adaptive and may in fact pose some risks. For example, showy plumage may make male birds more visible to predators. But if such secondary sexual characteristics help a male gain a mate, then they will be reinforced over the generations for the most Darwinian of reasons—because they enhance reproductive success. Every time a female chooses a mate based on a certain appearance or behavior, she perpetuates the alleles that influenced her to make that choice and allows a male with that particular phenotype to perpetuate his alleles.

What is the advantage to females of being choosy? One hypothesis is that females prefer male traits that are correlated with "good genes." In several bird species, research has shown that traits preferred by females, such as bright beaks or long tails, are related to overall male health. The "good genes" hypothesis was also tested in gray tree frogs. Female frogs prefer to mate with males that give long mating calls (**Figure 14C**). Researchers collected eggs from wild gray tree frogs. Half of each female's eggs were fertilized with sperm from long-calling males, the others with sperm from short-calling males. The offspring of long-calling male frogs grew bigger, faster, and survived better than their half-siblings fathered by short-calling males. The duration of a male's mating call was shown to be indicative of the male's overall genetic quality, supporting the hypothesis that female mate choice can be based on a trait that indicates whether the male has "good genes."

Next we return to the concept of directional selection, focusing on the evolution of antibiotic resistance in bacteria.

George D. Lepp/Corbis

▲ **Figure 14B** A contest for access to mates between two male elks

Dave Blackey/PhotoLibrary

▲ **Figure 14A** Extreme sexual dimorphism (peacock and peahen)

? Males with the most elaborate ornamentation may garner the most mates. How might choosing such a mate be advantageous to a female?

An elaborate display may signal good health and therefore good genes, which in turn could be passed along to the female's offspring.

Barry Mansell/ Nature Picture Library

▲ **Figure 14C** A male gray tree frog calling for mates

EVOLUTION CONNECTION

15 The evolution of antibiotic resistance in bacteria is a serious public health concern

Antibiotics are drugs that kill infectious microorganisms. Many antibiotics are naturally occurring chemicals derived from soil-dwelling fungi or bacteria. For these microbes, antibiotic production is an adaptation that helps them compete with bacteria for space and nutrients. Penicillin, originally isolated from a fungus, has been widely prescribed since the 1940s. A revolution in human health followed its introduction, rendering many previously fatal diseases easily curable. During the 1950s, some doctors even predicted the end of human infectious diseases.

Why hasn't that optimistic prediction come true? It did not take into account the force of evolution. In the same way that pesticides select for resistant insects (see Module 3), antibiotics select for resistant bacteria. A gene that codes for a protein that breaks down an antibiotic or a mutation that alters the site where an antibiotic binds can make a bacterium and its offspring resistant to that antibiotic. Again we see the chance and sorting aspects of natural selection: the chance variations in bacteria and the sorting effect of antibiotics as nonresistant bacteria are killed and resistant strains are left to flourish.

In what ways do we contribute to the problem of antibiotic resistance? Livestock producers add antibiotics to animal feed as a growth promoter and to prevent illness. These practices may select for bacteria that are resistant to standard antibiotics. Doctors may overprescribe antibiotics—for example, to patients with viral infections, which do not respond to antibiotic treatment. And patients may misuse prescribed antibiotics by prematurely stopping the medication because they feel better. This allows mutant bacteria that may be killed more slowly by the drug to survive and multiply. Subsequent mutations in such bacteria may lead to full-blown antibiotic resistance.

Difficulty in treating infections is a serious public health concern. Penicillin is virtually useless today in its original form. New drugs have been developed, but they are rendered ineffective as resistant bacteria evolve. Natural selection for antibiotic resistance is particularly strong in hospitals, where antibiotic use is extensive. Nearly 100,000 people die each year in the United States from infections they contract in the hospital, often because the bacteria are resistant to multiple antibiotics. A formidable "superbug" known as MRSA (methicillin-resistant *Staphylococcus aureus*) can cause "flesh-eating disease" **(Figure 15)** and potentially fatal systemic (whole-body) infections. Incidents of MRSA infections in both hospital and community settings have increased dramatically in the past decade. The medical community and pharmaceutical companies are engaged in an ongoing race against the powerful force of bacterial evolution.

? **Explain why the following statement is incorrect: "Antibiotics have created resistant bacteria."**

● The use of antibiotics has increased the frequency of genes for resistance that were already naturally present in bacterial populations.

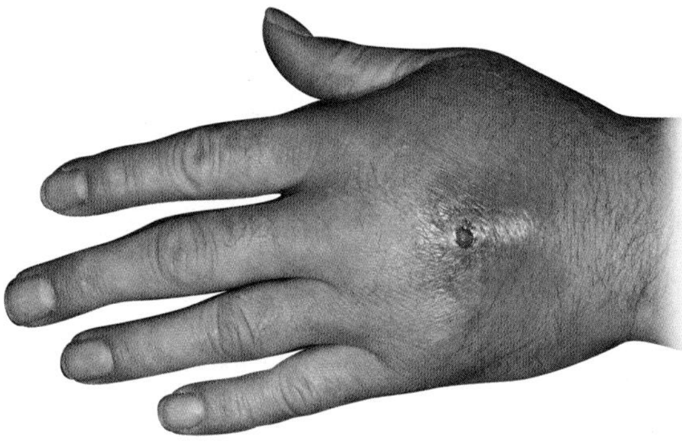

Scott Camazine/Photo Researchers

▲ **Figure 15** A MRSA skin infection

16 Diploidy and balancing selection preserve genetic variation

Natural selection acting on some variants within a population adapts that population to its environment. But what prevents natural selection from eliminating all variation as it selects against unfavorable genotypes? Why aren't less adaptive alleles eliminated as the "best" alleles are passed to the next generation? It turns out that the tendency for natural selection to reduce variation in a population is countered by mechanisms that maintain variation.

Most eukaryotes are diploid. Having two sets of chromosomes helps to prevent populations from becoming genetically uniform. The effects of recessive alleles may not often be displayed in a diploid population. A recessive allele is subject to natural selection only when it influences the phenotype, as only occurs in homozygous recessive individuals. In a heterozygote, a recessive allele is, in effect, protected from natural selection. The "hiding" of recessive alleles in the presence of dominant ones can maintain a huge pool of alleles that may not be favored under present conditions but that could be advantageous if the environment changes.

Genetic variation may also be preserved by natural selection, the same force that generally reduces it. **Balancing selection** occurs when natural selection maintains stable frequencies of two or more phenotypic forms in a population.

Heterozygote advantage is a type of balancing selection in which heterozygous individuals have greater reproductive success than either type of homozygote, with the result that two or more alleles for a gene are maintained in the population. An example of heterozygote advantage is the protection from malaria conferred by the sickle-cell allele. The frequency of the sickle-cell allele in Africa is generally highest in areas where malaria is a major cause of death. The environment in those areas favors heterozygotes, who are protected from the most severe effects of malaria. Less favored are individuals homozygous for the normal hemoglobin allele, who are susceptible to malaria, and individuals homozygous for the sickle-cell allele, who develop sickle-cell disease.

Frequency-dependent selection is a type of balancing selection that maintains two different phenotypic forms in a population. In this case, selection acts against either phenotypic form if it becomes too common in the population. An example of frequency-dependent selection is a scale-eating fish in Lake Tanganyika, Africa, which attacks other fish from behind, darting in to remove a few scales from the side of its prey. As you can see in **Figure 16**, some of these fish are "left-mouthed" and others are "right-mouthed." Because their mouth twists to the left, left-mouthed fish always attack their prey's right side—try twisting your lower jaw and lips to the left and imagine which side of a fish you could take a bite from. Similarly, right-mouthed fish attack from the left. Prey fish guard more effectively against attack from whichever phenotype is most common. Thus, scale-eating fish with the less common phenotype have a feeding advantage that enhances survival and reproductive success. As you can see on the graph, the frequency of left-mouthed fish oscillates over time, as frequency-dependent selection keeps each phenotype close to 50%.

Some of the DNA variation in populations probably has little or no impact on reproductive success, such as some nucleotide differences found in noncoding sequences of DNA (see Module 10.10) or changes in protein-coding genes that have little effect on protein function. But even if only a fraction of the variation in a gene pool affects reproductive success, that is still an enormous resource of raw material for natural selection and the adaptive evolution it brings about.

From M. Hori. 1993. Frequency-dependent natural selection in the handedness of scale-eating Cichlid fish. *Science* 260: 216–219, fig. 2a. Copyright ©1993 American Association for the Advancement of Science. Reprinted with permission from AAAS.

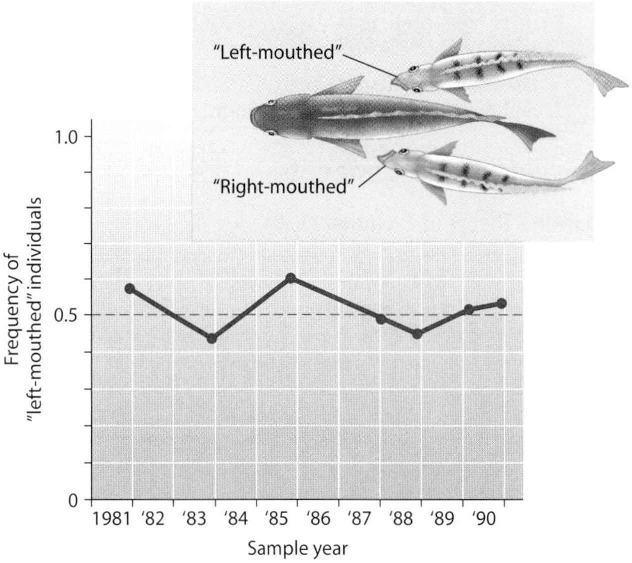

▲ **Figure 16** Frequency-dependent selection for left-mouthed and right-mouthed scale-eating fish (*Perissodus microlepis*)

? Why would natural selection tend to reduce genetic variation more in populations of haploid organisms than in populations of diploid organisms?

All alleles in a haploid organism are phenotypically expressed and are hence screened by natural selection.

17 Natural selection cannot fashion perfect organisms

Though natural selection leads to adaptation, there are several reasons why nature abounds with organisms that seem to be less than ideally "engineered" for their lifestyles.

1. *Selection can act only on existing variations.* Natural selection favors only the fittest variants from the phenotypes that are available, which may not be the ideal traits. New, advantageous alleles do not arise on demand.
2. *Evolution is limited by historical constraints.* Each species has a legacy of descent with modification from ancestral forms. Evolution does not scrap ancestral anatomy and build each new complex structure from scratch; it co-opts existing structures and adapts them to new situations. If the ability to fly were to evolve in a terrestrial animal, it might be best to grow an extra pair of limbs that would serve as wings. However, evolution operates on the traits an organism already has. Thus, as birds and bats evolved from four-legged ancestors, their existing forelimbs took on new functions for flight and they were left with only two limbs for walking.
3. *Adaptations are often compromises.* Each organism must do many different things. A blue-footed booby uses its webbed feet to swim after prey in the ocean, but these same feet make for clumsy travel on land.
4. *Chance, natural selection, and the environment interact.* Chance events probably affect the genetic structure of populations to a greater extent than was once recognized. When a storm blows insects over an ocean to an island, the wind does not necessarily transport the individuals that are best suited to the new environment. In small populations, genetic drift can result in the loss of beneficial alleles. In addition, the environment may change unpredictably from year to year, again limiting the extent to which adaptive evolution results in a close match between organisms and the environment.

With all these constraints, we cannot expect evolution to craft perfect organisms. Natural selection operates on a "better than" basis. We can see evidence for evolution in the imperfections of the organisms it produces.

? Humans owe much of their physical versatility and athleticism to their flexible limbs and joints. But we are prone to sprains, torn ligaments, and dislocations. Why?

Adaptations are compromises: Structural reinforcement has been compromised as agility was selected for.

REVIEW

For Practice Quizzes, BioFlix, MP3 Tutors, and Activities, go to www.masteringbiology.com.

Reviewing the Concepts

Darwin's Theory of Evolution (1–6)

1 A sea voyage helped Darwin frame his theory of evolution. Darwin's theory differed greatly from the long-held notion of a young Earth inhabited by unchanging species. Darwin called his theory descent with modification, which explains that all of life is connected by common ancestry and that descendants have accumulated adaptations to changing environments over vast spans of time.

2 Darwin proposed natural selection as the mechanism of evolution.

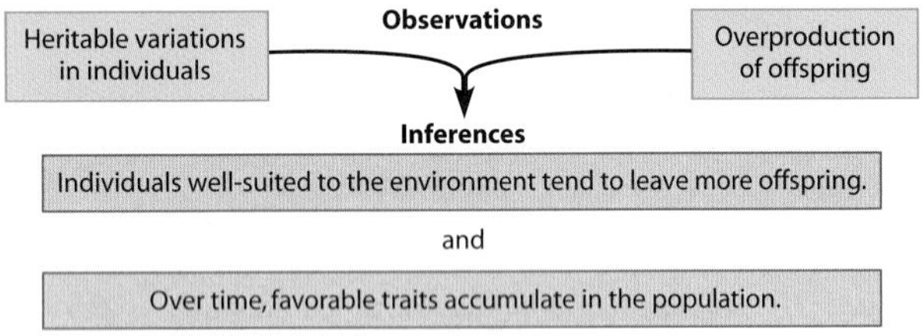

3 Scientists can observe natural selection in action.

4 The study of fossils provides strong evidence for evolution. The fossil record reveals the historical sequence in which organisms have evolved.

5 Many types of scientific evidence support the evolutionary view of life. Biogeography, comparative anatomy, and molecular biology all document evolution. Homologous structures and DNA sequences reveal evolutionary relationships.

6 Homologies indicate patterns of descent that can be shown on an evolutionary tree.

The Evolution of Populations (7–10)

7 Evolution occurs within populations. Microevolution is a change in the frequencies of alleles in a population's gene pool.

8 Mutation and sexual reproduction produce the genetic variation that makes evolution possible.

9 The Hardy-Weinberg equation can test whether a population is evolving. The Hardy-Weinberg principle states that allele and genotype frequencies will remain constant if a population is large, mating is random, and there is no mutation, gene flow, or natural selection.

Allele frequencies	$p + q = 1$
Genotype frequencies	$p^2 + 2pq + q^2 = 1$

Dominant homozygotes (p^2), Heterozygotes ($2pq$), Recessive homozygotes (q^2)

10 The Hardy-Weinberg equation is useful in public health science.

Mechanisms of Microevolution (11–17)

11 Natural selection, genetic drift, and gene flow can cause microevolution. The bottleneck effect and founder effect lead to genetic drift.

12 Natural selection is the only mechanism that consistently leads to adaptive evolution. Relative fitness is the relative contribution an individual makes to the gene pool of the next generation. As a result of natural selection, favorable traits increase in a population.

13 Natural selection can alter variation in a population in three ways.

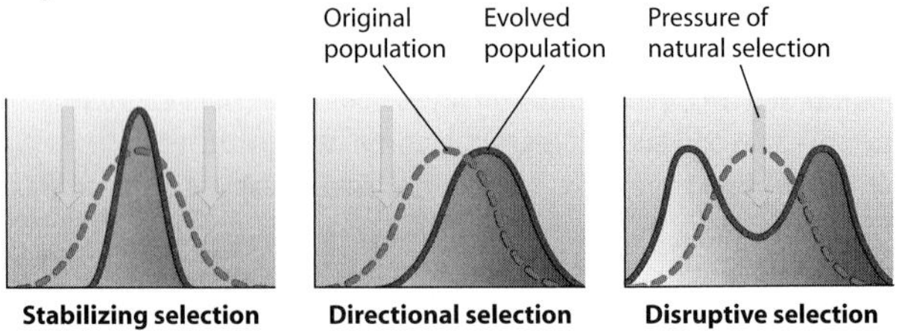

14 Sexual selection may lead to phenotypic differences between males and females. Secondary sex characteristics can give individuals an advantage in mating.

15 The evolution of antibiotic resistance in bacteria is a serious public health concern.

16 Diploidy and balancing selection preserve genetic variation. Diploidy preserves variation by "hiding" recessive alleles. Balancing selection may result from heterozygote advantage or frequency-dependent selection.

17 Natural selection cannot fashion perfect organisms. Natural selection can act only on available variation; anatomical structures result from modified ancestral forms; adaptations are often compromises; and chance, natural selection, and the environment interact.

Connecting the Concepts

1. Summarize the key points of Darwin's theory of descent with modification, including his proposed mechanism of evolution.
2. Complete this concept map describing potential causes of evolutionary change within populations.

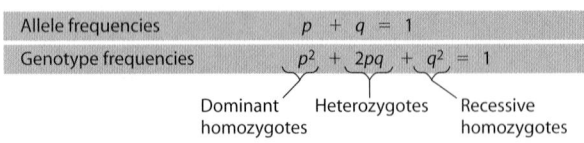

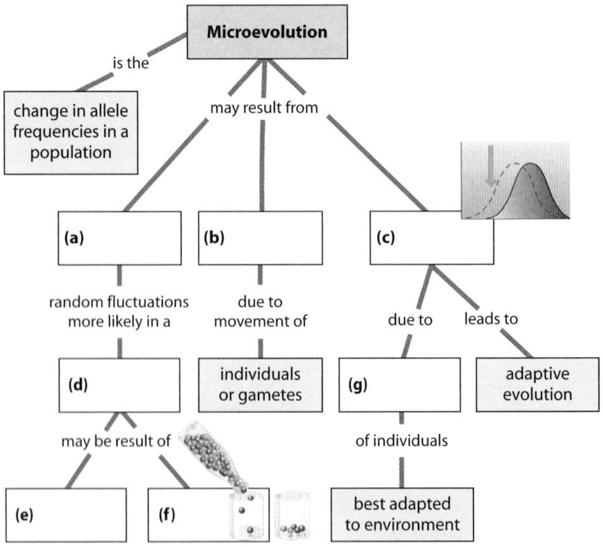

Testing Your Knowledge

Multiple Choice

3. Which of the following did not influence Darwin as he synthesized the theory of evolution by natural selection?
   a. examples of artificial selection that produce large and relatively rapid changes in domesticated species
   b. Lyell's *Principles of Geology*, on gradual geologic changes
   c. comparisons of fossils with living organisms
   d. the biogeographic distribution of organisms, such as the unique species on the Galápagos Islands
   e. Mendel's paper describing the laws of inheritance
4. Natural selection is sometimes described as "survival of the fittest." Which of the following best measures an organism's fitness?
   a. how many fertile offspring it produces
   b. its mutation rate
   c. how strong it is when pitted against others of its species
   d. its ability to withstand environmental extremes
   e. how much food it is able to make or obtain
5. Mutations are rarely the cause of evolution in populations of plants and animals because
   a. they are often harmful and do not get passed on.
   b. they do not directly produce most of the genetic variation present in a diploid population.
   c. they occur very rarely.
   d. they are passed on only when in cells that lead to gametes.
   e. all of the above
6. In an area of erratic rainfall, a biologist found that grass plants with alleles for curled leaves reproduced better in dry years, and plants with alleles for flat leaves reproduced better in wet years. This situation would tend to_____. *(Explain your answer.)*
   a. cause genetic drift in the grass population.
   b. preserve genetic variation in the grass population.
   c. lead to stabilizing selection in the grass population.
   d. lead to uniformity in the grass population.
   e. cause gene flow in the grass population.
7. Which of the following pairs of structures is *least* likely to represent homology?
   a. the hemoglobin of a human and that of a baboon
   b. the mitochondria of a plant and those of an animal
   c. the wings of a bird and those of an insect
   d. the tail of a cat and that of an alligator
   e. the foreleg of a pig and the flipper of a whale
8. If an allele is recessive and lethal in homozygotes before they reproduce,
   a. the allele is present in the population at a frequency of 0.001.
   b. the allele will be removed from the population by natural selection in approximately 1,000 years.
   c. the allele will likely remain in the population at a low frequency because it cannot be selected against in heterozygotes.
   d. the fitness of the homozygous recessive genotype is 0.
   e. both c and d are correct.
9. Darwin's claim that all life is descended from a common ancestor is best supported with evidence from _____. *(Explain your answer.)*
   a. the fossil record.
   b. molecular biology.
   c. evolutionary trees.
   d. comparative anatomy.
   e. comparative embryology.
10. In a population with two alleles, B and b, the allele frequency of b is 0.4. B is dominant to b. What is the frequency of individuals with the dominant phenotype if the population is in Hardy-Weinberg equilibrium?
    a. 0.16
    b. 0.36
    c. 0.48
    d. 0.84
    e. You cannot tell from this information.
11. Within a few weeks of treatment with the drug 3TC, a patient's HIV population consists entirely of 3TC-resistant viruses. How can this result best be explained?
    a. HIV can change its surface proteins and resist vaccines.
    b. The patient must have become reinfected with a resistant virus.
    c. A few drug-resistant viruses were present at the start of treatment, and natural selection increased their frequency.
    d. The drug caused the HIV genes to change.
    e. HIV began making drug-resistant versions of its enzymes in response to the drug.

Describing, Comparing, and Explaining

12. Write a paragraph briefly describing the kinds of scientific evidence for evolution.
13. Sickle-cell disease is caused by a recessive allele. Roughly one out of every 400 African Americans (0.25%) is afflicted with sickle-cell disease. Use the Hardy-Weinberg equation to calculate the percentage of African Americans who are carriers of the sickle-cell allele. (*Hint*: $q^2 = 0.0025$.)
14. It seems logical that natural selection would work toward genetic uniformity; the genotypes that are most fit produce the most offspring, increasing the frequency of adaptive alleles and eliminating less adaptive alleles. Yet there remains a great deal of genetic variation within populations. Describe some of the factors that contribute to this variation.

Applying the Concepts

15. A population of snails is preyed on by birds that break the snails open on rocks, eat the soft bodies, and leave the shells. The snails occur in both striped and unstriped forms. In one area, researchers counted both live snails and broken shells. Their data are summarized below:

	Striped	Unstriped	Total	Percent Striped
Living	264	296	560	47.1
Broken	486	377	863	56.3

    Which snail form seems better adapted to this environment? Why? Predict how the frequencies of striped and unstriped snails might change in the future.

16. Advocates of "scientific creationism" and "intelligent design" lobby school districts for such things as a ban on teaching evolution, equal time in science classes to teach alternative versions of the origin and history of life, or disclaimers in textbooks stating that evolution is "just a theory." They argue that it is only fair to let students evaluate both evolution and the idea that all species were created by God as the Bible relates or that, because organisms are so complex and well adapted, they must have been created by an intelligent designer. Do you think that alternative views of evolution should be taught in science courses? Why or why not?

Review Answers

1. According to Darwin's theory of descent with modification, all life has descended from a common ancestral form as a result of natural selection. Individuals in a population have hereditary variations. The overproduction of offspring in the face of limited resources leads to a struggle for existence. Individuals that are well suited to their environment tend to leave more offspring than other individuals, leading to the gradual accumulation of adaptations to the local environment in the population.
2. a. genetic drift; b. gene flow; c. natural selection; d. small population; e. founder effect; f. bottleneck effect; g. unequal reproductive success
3. e **4.** a **5.** e **6.** b (Erratic rainfall and unequal reproductive success would ensure that a mixture of both forms remained in the population.) **7.** c **8.** e **9.** b (All of these provide evidence of evolution, but DNA and the nearly universal genetic code are best able to connect all of life's diverse forms through common ancestry.) **10.** d **11.** c
12. Your paragraph should include such evidence as fossils and the fossil record, biogeography, comparative anatomy, comparative embryology, DNA and protein comparisons, artificial selection, and examples of natural selection.
13. If $q^2 = 0.0025$, $q = 0.05$. Since $p + q = 1$, $p = 1 - q = 0.95$. The proportion of heterozygotes is $2pq = 2 \times 0.95 \times 0.05 = 0.095$. About 9.5% of African Americans are carriers.
14. Genetic variation is retained in a population by diploidy and balanced selection. Recessive alleles are hidden from selection when in the heterozygote; thus, less adaptive or even harmful alleles are maintained in the gene pool and are available should environmental conditions change. Both heterozygote advantage and frequency-dependent selection tend to maintain alternate alleles in a population.
15. The unstriped snails appear to be better adapted. Striped snails make up 47% of the living population but 56% of the broken shells. Assuming that all the broken shells result from the meals of birds, we would predict that bird predation would reduce the frequency of striped snails and the frequency of unstriped individuals would increase.
16. Some issues and questions to consider: Who should decide curriculum, scientific experts in a field or members of the community? Are these alternative versions scientific ideas? Who judges what is scientific? If it is fairer to consider alternatives, should the door be open to all alternatives? Are constitutional issues (separation of church and state) involved here? Can a teacher be compelled to teach an idea he or she disagrees with? Should a student be required to learn an idea he or she thinks is wrong?

Microbial Life: *Prokaryotes and Protists*

From Chapter 16 of *Campbell Biology: Concepts & Connections*, Seventh Edition. Jane B. Reece, Martha R. Taylor, Eric J. Simon, Jean L. Dickey.

Microbial Life: *Prokaryotes and Protists*

BIG IDEAS

Dr. Tony Brain/David Parker/Science Photo Library/Photo Researchers, Inc.

Prokaryotes
(1–12)

Prokaryotes, the smallest organisms known, are extraordinarily diverse.

Manfred Kage/Peter Arnold/PhotoLibrary

Protists
(13–21)

Protists are eukaryotes. Though most are unicellular, microscopic organisms, some protists are multicellular.

Andre Seale/AGE fotostock

Coral reefs are vivid displays of biological diversity. Exploring the reef shown above, you would encounter many species of brightly colored fish and a multitude of sea anemones, sponges, sea stars, urchins, crustaceans, and, of course, corals. Now, you may be wondering what this glorious variety of animals has to do with a chapter about microbes, organisms too small to be seen without a microscope. Without microbes, the reef would not exist. Photosynthesis by prokaryotes and protists feeds all of the animals. The photosynthesizers, in turn, depend on prokaryotes to convert dead organic material to fertilizer that nourishes their growth. And without the supply of sugars made by the protists that reside in their cells, corals would lack the energy to build and maintain the enormous structures that support and shelter the reef community.

The vital roles of prokaryotes and protists are not limited to coral reefs. They are essential to the health of every ecosystem—including the ecosystems they inhabit in and on your body! In this chapter, you will learn about some of these ecological functions and discover how they benefit natural ecosystems as well as human endeavors. You will also sample a bit of the remarkable diversity of prokaryotes and protists, including some of the undesirable species, the microbes that cause disease.

With this chapter, we begin our exploration of the diversity of life. We focus mostly on current diversity, but we also retrace evolutionary journeys that changed the face of our planet, such as the colonization of land by plants and vertebrates. And so it is fitting that we begin with the prokaryotes, Earth's first life-form, and the protists, the bridge between unicellular eukaryotes and multicellular plants, fungi, and animals.

Prokaryotes

1 Prokaryotes are diverse and widespread

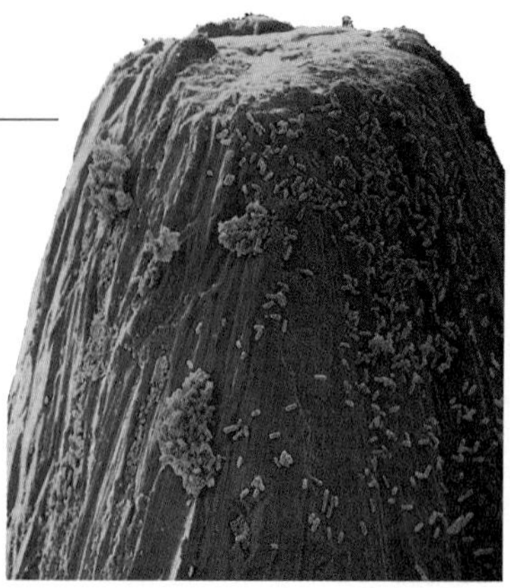

▶ **Figure 1** Bacteria on the point of a pin

Dr. Tony Brain/David Parker/Science Photo Library/Photo Researchers

In the first half of this chapter, you will learn about prokaryotes, organisms that have a cellular organization fundamentally different from that of eukaryotes. Whereas eukaryotic cells have a membrane-enclosed nucleus and numerous other membrane-enclosed organelles, prokaryotic cells lack these structural features. Prokaryotes are also typically much smaller than eukaryotes. You can get an idea of the size of most prokaryotes from **Figure 1**, a colorized scanning electron micrograph of the point of a pin (purple) covered with numerous bacteria (orange). Most prokaryotic cells have diameters in the range of 1–5 μm, much smaller than most eukaryotic cells (typically 10–100 μm).

Despite their small size, prokaryotes have an immense impact on our world. They are found wherever there is life, including in and on the bodies of multicellular organisms. The collective biological mass (biomass) of prokaryotes is at least 10 times that of all eukaryotes! Prokaryotes also thrive in habitats too cold, too hot, too salty, too acidic, or too alkaline for any eukaryote. And scientists are just beginning to investigate the extensive prokaryotic diversity in the oceans.

Although prokaryotes are a constant presence in our environment, we hear most about the relatively few species that cause illnesses. We focus on bacterial **pathogens**, disease-causing agents, in Module 10. But benign or beneficial prokaryotes are far more common than harmful prokaryotes. For example, each of us harbors several hundred species of prokaryotes in and on our body. Many of the bacteria that live on our skin perform helpful housekeeping functions such as decomposing dead skin cells. Some of our intestinal residents supply essential vitamins and enable us to obtain nutrition from food molecules that we can't otherwise digest. Bacteria also guard the body against pathogenic intruders.

Prokaryotes are also essential to the health of the environment. They help to decompose dead organisms and other organic waste material, returning vital chemical elements to the environment. They are indispensable components of the chemical cycle that makes nitrogen available to plants and other organisms. If prokaryotes were to disappear, the chemical cycles that sustain life would halt, and all forms of eukaryotic life would also be doomed. In contrast, prokaryotic life would undoubtedly persist in the absence of eukaryotes, as it once did for billions of years.

There are two very different kinds of prokaryotes, which are classified in the domains **Archaea** and **Bacteria**. In the next several modules, we describe the features that have made prokaryotes so successful, followed by a look at the diversity of each domain.

? The number of bacterial cells that live in and on our body is greater than the number of eukaryotic cells that make up the body. Why aren't we aware of these trillions of cells?

● We can't sense our own eukaryotic cells individually, and bacterial cells are much smaller than that. Also, our "companion" microbes are adapted for coexisting with us.

2 External features contribute to the success of prokaryotes

Some of the diversity of prokaryotes is evident in their external features, including shape, cell walls, and projections such as flagella. These features are useful for identifying prokaryotes as well as helping the organisms survive in their environments.

Cell Shape Determining cell shape by microscopic examination is an important step in identifying prokaryotes. The micrographs in **Figure 2A** show three of the most common prokaryotic cell shapes. Spherical prokaryotic cells are called **cocci** (singular, *coccus*). Cocci that occur in chains, like the ones in Figure 2A, are called streptococci (from the Greek *streptos*, twisted). The bacterium that causes strep throat in humans is a streptococcus. Other cocci occur in clusters; they are called staphylococci (from the Greek *staphyle*, cluster of grapes).

Rod-shaped prokaryotes are called **bacilli** (singular, *bacillus*). Most bacilli occur singly, like the *Escherichium*

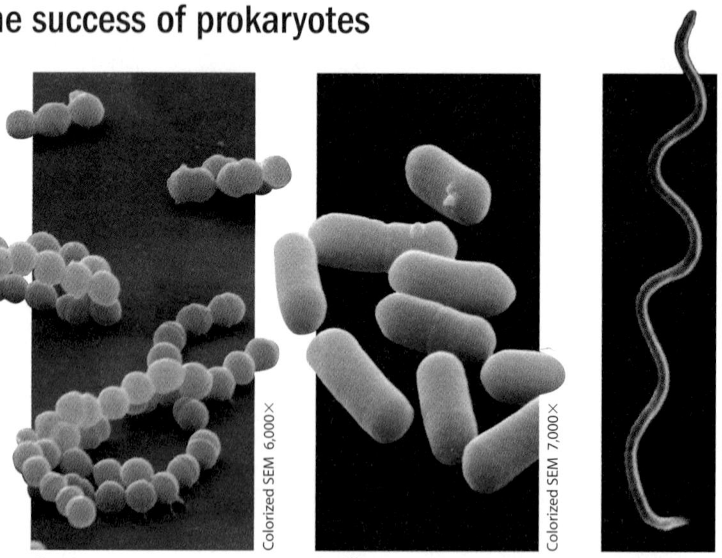

▲ **Figure 2A** Prokaryote shapes

Eye of Science/Photo Researchers; David McCarthy/Photo Researchers; Stem Jems/Photo Researchers

coli cells in the middle photo in Figure 2A. (*E. coli* resides in our digestive tract.) However, the cells of some species occur in pairs or chains. Bacilli may also be threadlike, or filamentous.

A third prokaryotic cell shape is spiral, like a corkscrew. Spiral prokaryotes that are relatively short and rigid are called *spirilla*; those with longer, more flexible cells, like the one shown on the right in Figure 2A, which causes Lyme disease, are called *spirochetes*. The bacterium that causes syphilis is also a spirochete. Spirochetes include some giants by prokaryotic standards—cells 0.5 mm long (though very thin).

Cell Wall Nearly all prokaryotes have a cell wall, a feature that enables them to live in a wide range of environments. The cell wall provides physical protection and prevents the cell from bursting in a hypotonic environment (see Module 5.5). The cell walls of bacteria fall into two general types, which scientists can identify with a technique called the **Gram stain** (**Figure 2B**). Gram-positive bacteria have simpler walls with a relatively thick layer of a unique material called **peptidoglycan**, a polymer of sugars cross-linked by short polypeptides. The walls of gram-negative bacteria stain differently. They have less peptidoglycan and are more complex, with an outer membrane that contains lipids bonded to carbohydrates. The cell walls of archaea do not contain peptidoglycan, but can also be gram-positive or gram-negative.

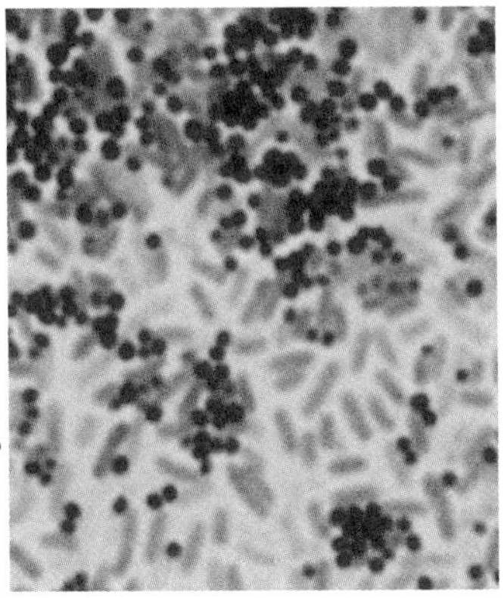

▲ **Figure 2B** Gram-positive (purple) and gram-negative (pink) bacteria

In medicine, Gram stains are often used to detect the presence of bacteria and indicate the type of antibiotic to prescribe. Among disease-causing bacteria, gram-negative species are generally more threatening than gram-positive species because lipid molecules of the outer membrane of gram-negative bacteria are often toxic. The membrane also protects the gram-negative bacteria against the body's defenses and hinders the entry of antibiotic drugs into the bacterium.

The cell wall of many prokaryotes is covered by a capsule, a sticky layer of polysaccharide or protein. The capsule enables prokaryotes to adhere to a surface or to other individuals in a colony. Capsules can also shield pathogenic prokaryotes from attacks by their host's immune system. The capsule surrounding the *Streptococcus* bacterium shown in **Figure 2C** enables it to attach to cells that line the human respiratory tract—in this image, a tonsil cell.

Projections Some prokaryotes have external structures that extend beyond the cell wall. Many bacteria and archaea are equipped with flagella, adaptations that enable them to move about in response to chemical or physical signals in their

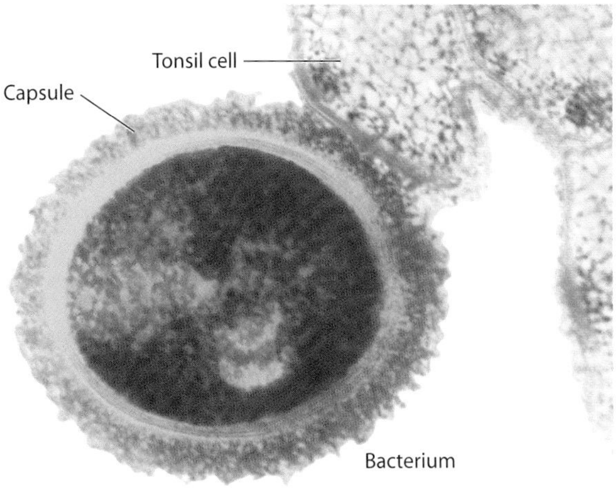

▲ **Figure 2C** Capsule

environment. For example, prokaryotes can move toward nutrients or other members of their species or away from a toxic substance. Flagella may be scattered over the entire cell surface or concentrated at one or both ends of the cell. Unlike the flagellum of eukaryotic cells, the prokaryotic flagellum is a naked protein structure that lacks microtubules. The flagellated bacterium in **Figure 2D** is *E. coli,* as seen in a TEM.

Figure 2D also illustrates the hairlike projections called **fimbriae** that enable some prokaryotes to stick to a surface or to one another. Fimbriae allow many pathogenic bacteria to latch onto the host cells they colonize. For example, *Neisseria gonorrheae*, which causes the sexually transmitted infection gonorrhea, uses fimbriae to attach to cells in the reproductive tract. During sexual intercourse, *N. gonorrheae* bacteria may also attach to sperm cells and travel to a woman's oviducts; an infection in these narrow tubes can impair fertility.

? How could a microscope help you distinguish the cocci that cause "staph" infections from those that cause "strep" throat?

It would show clusters of cells for staphylococcus and chains of cells for streptococcus.

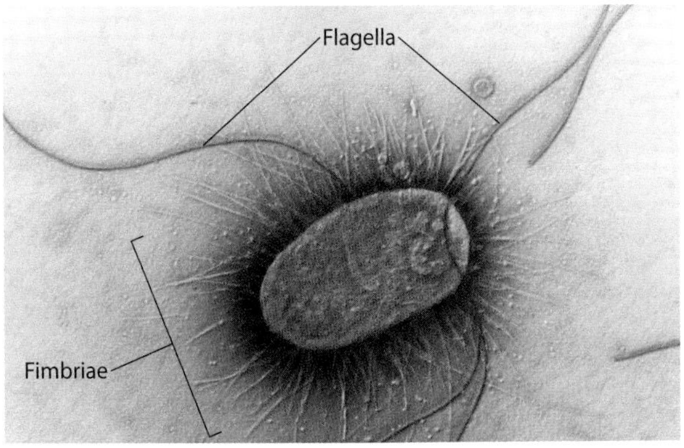

▲ **Figure 2D** Flagella and fimbriae

3 Populations of prokaryotes can adapt rapidly to changes in the environment

Certainly a large part of the success of prokaryotes is their potential to reproduce quickly in a favorable environment. Dividing by binary fission, a single prokaryotic cell becomes 2 cells, which then become 4, 8, 16, and so on. While many prokaryotes produce a new generation within 1–3 hours, some species can produce a new generation in only 20 minutes under optimal conditions. If reproduction continued unchecked at this rate, a single prokaryote could give rise to a colony outweighing Earth in only three days!

Salmonella bacteria, which cause food poisoning, are commonly found on raw poultry and eggs, but the bacterial population is often too small to cause symptoms. Refrigeration slows (but does not stop) bacterial reproduction. However, when raw poultry is left in the warm environment of the kitchen, bacteria multiply rapidly and can quickly reach a risky population size. Similarly, bacteria that remain on the counter, cutting board, or kitchen implements may continue to reproduce. So be sure to cook poultry thoroughly (an internal temperature of 165°F is considered safe), and clean anything that has come into contact with the raw poultry with soap and hot water or an antimicrobial cleaner.

Each time DNA is replicated prior to binary fission, spontaneous mutations occur. As a result, rapid reproduction generates a great deal of genetic variation in a prokaryote population. If the environment changes, an individual that possesses a beneficial gene can quickly take advantage of the new conditions. For example, exposure to antibiotics may select for antibiotic resistance in a bacterial population.

The amount of DNA in a prokaryotic cell is on average only about one-thousandth as much as that in a eukaryotic cell. The genome of a typical prokaryote is one long, circular chromosome (**Figure 3A**). (In an intact cell, it is packed into a distinct region.) Many prokaryotes also have additional small, circular DNA molecules called plasmids, which replicate independently of the chromosome. Some plasmids carry genes that enhance survival under certain conditions. For example, plasmids may provide resistance to antibiotics, direct the metabolism of rarely encountered nutrients, or have other "contingency" functions. The ability of many prokaryotes to transfer plasmids within and even between species provides another rapid means of adaptation to changes in the environment.

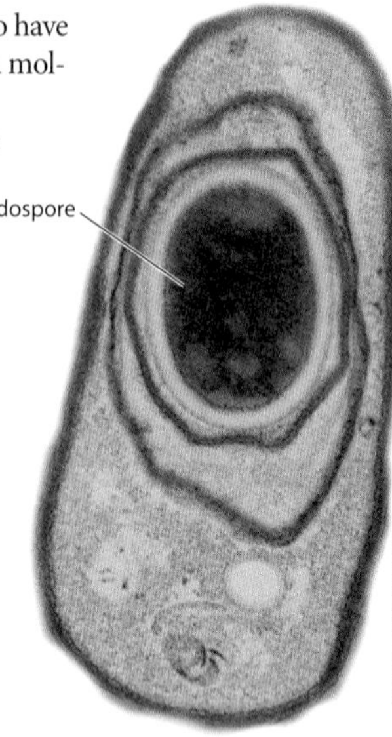

▲ **Figure 3B** An endospore within an anthrax bacterium cell

If environmental conditions become too harsh to sustain active metabolism—for example, when food or moisture is depleted—some prokaryotes form specialized resistant cells. **Figure 3B** shows an example of such an organism, *Bacillus anthracis*, the bacterium that causes a disease called anthrax in cattle, sheep, and humans. There are actually two cells here, one inside the other. The outer cell, which will later disintegrate, produced the specialized inner cell, called an **endospore**. The endospore, which has a thick, protective coat, dehydrates and becomes dormant. It can survive all sorts of trauma, including extreme heat or cold. When the endospore receives cues that the environmental conditions have improved, it absorbs water and resumes growth.

Some endospores can remain dormant for centuries. Not even boiling water kills most of these resistant cells, making it difficult to get rid of spores in a contaminated area. An island off the coast of Scotland that was used for anthrax testing in 1942 was finally declared safe 48 years later, after tons of formaldehyde were applied and huge amounts of top soil were removed. The food-canning industry kills endospores of dangerous bacteria such as *Clostridium botulinum*, the source of the potentially fatal disease botulism, by heating the food to a temperature of 110–150°C (230–300°F) with high-pressure steam.

Another feature that contributes to the success of prokaryotes is the diversity of ways in which they obtain their nourishment, which we consider in the next module.

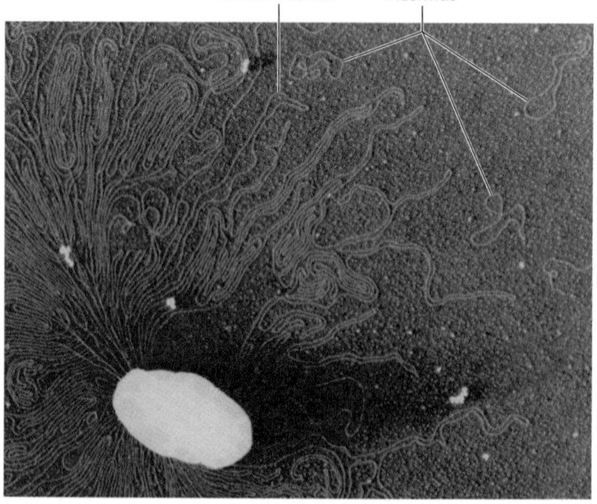

▲ **Figure 3A** DNA released from a ruptured bacterial cell

? Why does rapid reproduction produce high genetic variation in populations of prokaryotes?

Each time DNA replicates, spontaneous mutations occur.

4 Prokaryotes have unparalleled nutritional diversity

One way to organize the vast diversity of prokaryotes is by their mode of nutrition—how they obtain energy for cellular work and carbon to build organic molecules. Prokaryotes exhibit much more nutritional diversity than eukaryotes. This allows them to inhabit almost every nook and cranny on Earth.

Source of Energy As shown in **Figure 4**, two sources of energy are used by prokaryotes. Like plants, prokaryotic *phototrophs* capture energy from sunlight. Prokaryotic cells do not have chloroplasts, but some prokaryotes have thylakoid membranes where photosynthesis takes place.

Prokaryotes called *chemotrophs* harness the energy stored in chemicals, either organic molecules or inorganic chemicals such as hydrogen sulfide (H_2S), elemental sulfur (S), iron (Fe)-containing compounds, or ammonia (NH_3).

Source of Carbon Organisms that make their own organic compounds from inorganic sources are autotrophic. **Autotrophs**, including plants and some prokaryotes and protists, obtain their carbon atoms from carbon dioxide (CO_2). Most prokaryotes, as well as animals, fungi, and some protists, are **heterotrophs**, meaning they obtain their carbon atoms from the organic compounds of other organisms.

Mode of Nutrition The terms used to describe how an organism obtains energy and carbon are combined to describe its mode of nutrition (see Figure 4).

Photoautotrophs harness sunlight for energy and use CO_2 for carbon. Cyanobacteria, such as the example shown in Figure 4, are photoautotrophs. As in plants, photosynthesis in cyanobacteria uses chlorophyll *a* and produces O_2 as a by-product.

Photoheterotrophs obtain energy from sunlight but get their carbon atoms from organic sources. This unusual mode of nutrition is found in only a few types of bacteria called purple nonsulfur bacteria. Many of them, including the example shown in Figure 4, are found in aquatic sediments.

Chemoautotrophs harvest energy from inorganic chemicals and use carbon from CO_2 to make organic molecules. Because they don't depend on sunlight, chemoautotrophs can thrive in conditions that seem totally inhospitable to life. Near hydrothermal vents, where scalding water and hot gases surge into the sea more than a mile below the surface, chemoautotrophic bacteria use sulfur compounds as a source of energy. The organic molecules they produce using CO_2 from the seawater support diverse animal communities. The chemoautotrophs shown in Figure 4 live between layers of rocks buried 100 m below Earth's surface. Chemoautotrophs are also found in more predictable habitats, such as the soil.

Chemoheterotrophs, which acquire both energy and carbon from organic molecules, are by far the largest and most diverse group of prokaryotes. Almost any organic molecule is food for some species of chemoheterotrophic prokaryote.

? **Which term would describe your mode of nutrition?**

Chemoheterotrophy

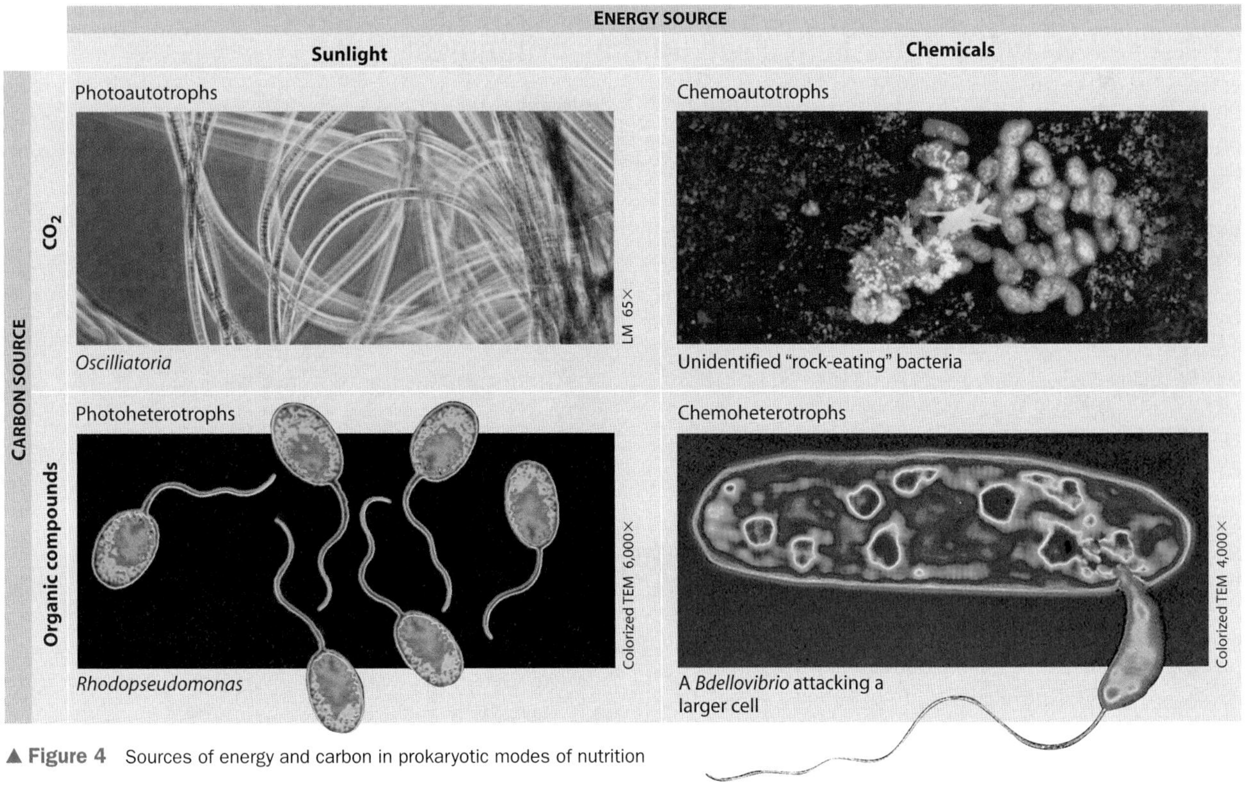

▲ **Figure 4** Sources of energy and carbon in prokaryotic modes of nutrition

Sinclair Stammers/Photo Researchers

Pasieka/SPL/Photo Researchers

T. Stevens & P. McKinley, PNNL/Photo Researchers

Alfred Pasieka/Peter Arnold/Photolibrary

CONNECTION 5 Biofilms are complex associations of microbes

In many natural environments, prokaryotes attach to surfaces in highly organized colonies called **biofilms**. A biofilm may consist of one or several species of prokaryotes, and it may include protists and fungi as well. Biofilms can form on almost any support, including rocks, soil, organic material (including living tissue), metal, and plastic. You have a biofilm on your teeth—dental plaque is a biofilm that can cause tooth decay. Biofilms can even form without a solid foundation—for example, on the surface of stagnant water.

Biofilm formation begins when prokaryotes secrete signaling molecules that attract nearby cells into a cluster. Once the cluster becomes sufficiently large, the cells produce a gooey coating that glues them to the support and to each other, making the biofilm extremely difficult to dislodge. For example, if you don't scrub your shower, you could find a biofilm growing around the drain—running water alone is not strong enough to wash it away. As the biofilm gets larger and more complex, it becomes a "city" of microbes. Communicating by chemical signals, members of the community coordinate the division of labor, defense against invaders, and other activities. Channels in the biofilm allow nutrients to reach cells in the interior and allow wastes to leave, and a variety of environments develop within it.

Biofilms are common among bacteria that cause disease in humans. For instance, ear infections and urinary tract infections are often the result of biofilm-forming bacteria. Cystic fibrosis patients are vulnerable to pneumonia caused by bacteria that form biofilms in their lungs. Biofilms of harmful bacteria can also form on implanted medical devices such as catheters, replacement joints, or pacemakers. The complexity of biofilms makes these infections especially difficult to defeat. Antibiotics may not be able to penetrate beyond the outer layer of cells, leaving much of the community intact. For example, some biofilm bacteria produce an enzyme that breaks down penicillin faster than it can diffuse inward.

Biofilms that form in the environment can be difficult to eradicate, too. A variety of industries spend billions of dollars every year trying to get rid of biofilms that clog and corrode pipes, gum up filters and drains, and coat the hulls of ships (**Figure 5**). Biofilms in water distribution pipes may survive chlorination, the most common method of ensuring that drinking water does not contain any harmful microorganisms. For example, biofilms of *Vibrio cholera*, the bacterium that causes cholera, found in water pipes were capable of withstanding levels of chlorine 10 to 20 times higher than the concentrations routinely used to chlorinate drinking water.

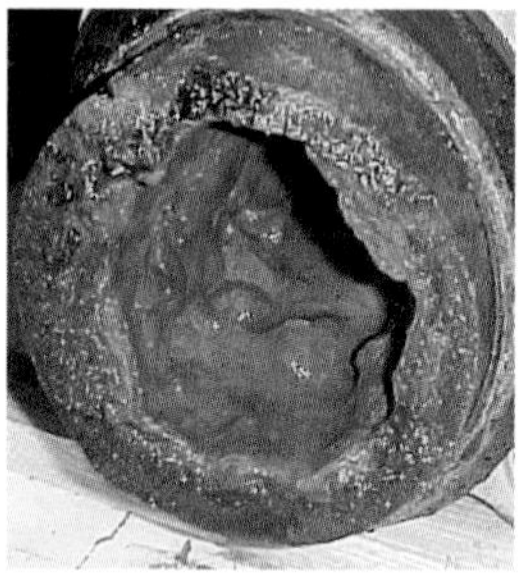
Garry Palmateer

▲ **Figure 5** A biofilm fouling the insides of a pipe

? **Why are biofilms difficult to eradicate?**

The biofilm sticks to the surface it resides on, and the cells that make up the biofilm stick to each other; the outer layer of cells may prevent antimicrobial substances from penetrating into the interior of the biofilm.

CONNECTION 6 Prokaryotes help clean up the environment

The characteristics that have made prokaryotes so widespread and successful—their nutritional diversity, adaptability, and capacity for forming biofilms—also make them useful for cleaning up contaminants in the environment. **Bioremediation** is the use of organisms to remove pollutants from soil, air, or water.

Prokaryotic decomposers are the mainstays of sewage treatment facilities. Raw sewage is first passed through a series of screens and shredders, and solid matter settles out from the liquid waste. This solid matter, called sludge, is then gradually added to a culture of anaerobic prokaryotes, including both bacteria and archaea. The microbes decompose the organic matter in the sludge into material that can be placed in a landfill or used as fertilizer.

Liquid wastes are treated separately from the sludge. In **Figure 6A**, you can see a trickling filter system, one type of mechanism for treating liquid wastes. The long horizontal pipes rotate slowly, spraying liquid wastes through the air onto a thick bed of rocks, the filter. Biofilms of aerobic bacteria and fungi growing on the rocks remove much of the organic material dissolved in the waste. Outflow from

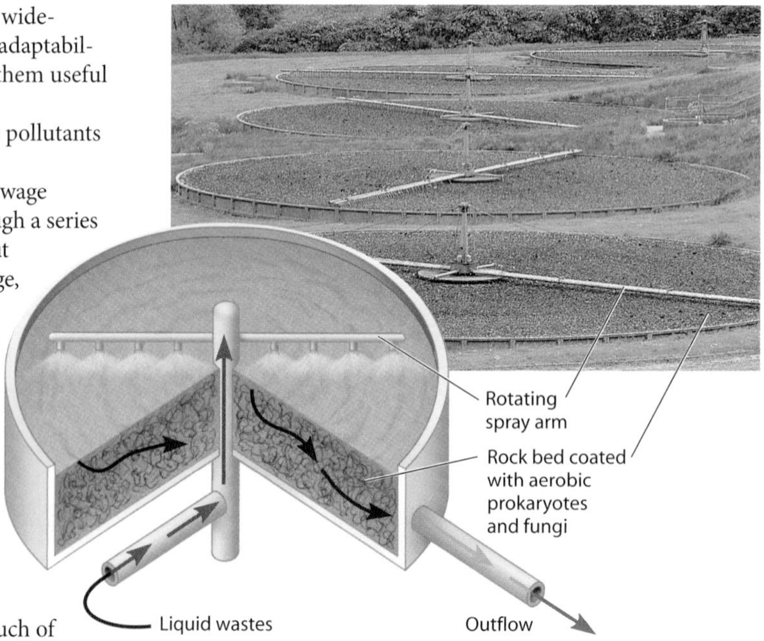

Photo Researchers

▲ **Figure 6A** The trickling filter system at a sewage treatment plant the rock bed is sterilized and then released, usually into a river or ocean.

Bioremediation has also become a useful tool for cleaning up toxic chemicals released into the soil and water. Naturally occurring prokaryotes capable of degrading pollutants such as oil, solvents, and pesticides are often present in contaminated soil and water, but their activity is limited by environmental factors such as nutrient availability. In **Figure 6B**, workers are spraying an oil-polluted beach in Alaska with fertilizers to stimulate soil bacteria to speed up the breakdown of oil. Chemical dispersants were used in the massive Gulf of Mexico oil spill in 2010. Like detergents that help clean greasy dishes, these chemicals break oil into smaller droplets that offer more surface area for microbial attack.

▲ **Figure 6B** Treatment of an oil spill in Alaska

ExxonMobil Corporation

? What is bioremediation?

The use of organisms to clean up pollution

7 Bacteria and archaea are the two main branches of prokaryotic evolution

As you learned in Module 15.19, researchers recently discovered that many prokaryotes once classified as bacteria are actually more closely related to eukaryotes and belong in a domain of their own. As a result, prokaryotes are now classified in two domains: Bacteria and Archaea (from the Greek *archaios,* ancient). Many bacterial and archaeal genomes have now been sequenced. When compared with each other and with the genomes of eukaryotes, these genome sequences strongly support the three-domain view of life. Some genes of archaea are similar to bacterial genes, others to eukaryotic genes, and still others seem to be unique to archaea.

Table 7 summarizes some of the main differences between the three domains. Differences between the ribosomal RNA (rRNA) sequences provided the first clues of a deep division among prokaryotes. Other differences in the cellular machinery for gene expression include differences in RNA polymerases (the enzymes that catalyze the synthesis of RNA) and in the presence of introns within genes. The cell walls and membranes of bacteria and archaea are also distinctive. Bacterial cell walls contain peptidoglycan (see Module 2), while archaea do not. Furthermore, the lipids forming the backbone of plasma membranes differ between the two domains. Intriguingly, for most of the features listed in the table, archaea have at least as much in common with eukaryotes as they do with bacteria.

Now that you are familiar with the general characteristics of prokaryotes and the features underlying their spectacular success, let's take a look at prokaryotic diversity. We begin with domain Archaea.

? As different as bacteria and archaea are, both groups are characterized by ______ cells, which lack nuclei and other membrane-enclosed organelles.

prokaryotic

TABLE 7 | DIFFERENCES BETWEEN THE DOMAINS BACTERIA, ARCHAEA, AND EUKARYA

Characteristic	Bacteria	Archaea	Eukarya
rRNA sequences	Some unique to bacteria	Some unique to archaea; some match eukaryotic sequences	Some unique to eukaryotes; some match archaeal sequences
RNA polymerase	One kind; relatively small and simple	Several kinds; complex	Several kinds; complex
Introns	Rare	In some genes	Present
Peptidoglycan in cell wall	Present	Absent	Absent
Histones associated with DNA	Absent	Present in some species	Present

8 Archaea thrive in extreme environments—and in other habitats

Archaea are abundant in many habitats, including places where few other organisms can survive. The archaeal inhabitants of extreme environments have unusual proteins and other molecular adaptations that enable them to metabolize and reproduce effectively. Scientists are only beginning to learn about these adaptations.

A group of archaea called the **extreme halophiles** ("salt lovers") thrive in very salty places, such as the Great Salt Lake in Utah, the Dead Sea, and seawater-evaporating ponds used to produce salt. Many species flourish when the salinity of the water is 15–30% and can tolerate even higher salt concentrations. Because seawater, with a salt concentration of about 3%, is hypertonic enough to shrivel most cells, these archaea have very little competition from other organisms. Extremely salty environments may turn red, purple, or yellow as a result of the dense growth and colorful pigments of halophilic archaea.

Another group of archaea, the **extreme thermophiles** ("heat lovers"), thrive in very hot water; some even live near deep-ocean vents, where temperatures are above 100°C, the boiling point of water at sea level! One such habitat is the Nevada geyser shown in **Figure 8A**. Other thermophiles thrive in acid. Many hot, acidic pools in Yellowstone National Park harbor such archaea, which give the pools a vivid greenish color. One of these organisms, *Sulfolobus*, can obtain energy by oxidizing sulfur or a compound of sulfur and iron; the mechanisms involved may be similar to those used billions of years ago by the first cells.

A third group of archaea, the **methanogens**, live in anaerobic (oxygen-lacking) environments and give off methane as a waste product. Many thrive in anaerobic mud at the bottom of lakes and swamps. You may have seen methane, also called marsh gas, bubbling up from a swamp. A large amount of methane is generated in solid waste landfills, where methanogens flourish in the anaerobic conditions. Many municipalities collect this methane and use it as a source of energy **(Figure 8B)**. Great numbers of methanogens also inhabit the digestive tracts of cattle, deer, and other animals that depend heavily on cellulose for their nutrition. Because methane is a greenhouse gas, landfills and livestock contribute significantly to global warming.

Accustomed to thinking of archaea as inhabitants of extreme environments, scientists have been surprised to discover their abundance in more moderate conditions, especially in the oceans. Archaea live at all depths, making up a substantial fraction of the prokaryotes in waters more than 150 m beneath the surface and half of the prokaryotes that live below 1,000 m. Archaea are thus one of the most abundant cell types in Earth's largest habitat.

Because bacteria have been the subject of most prokaryotic research for over a century, much more is known about them than about archaea. Now that the evolutionary and ecological importance of archaea has come into focus, we can expect research on this domain to turn up many more surprises about the history of life and the roles of microbes in ecosystems.

Jack Dykinga/Stone/Getty Images

▲ **Figure 8A** Orange and yellow colonies of heat-loving archaea growing in a Nevada geyser

Jim West/Alamy

▲ **Figure 8B** Pipes for collecting gas from a landfill

? Some archaea are referred to as "extremophiles." Why?

Because they can thrive in extreme environments too hot, too salty, or too acidic for other organisms

9 Bacteria include a diverse assemblage of prokaryotes

Domain Bacteria is currently divided into five groups based on comparisons of genetic sequences. In this module, we sample some of the diversity in each group.

Proteobacteria are all gram-negative and share a particular rRNA sequence. With regard to other characteristics, however, this large group encompasses enormous diversity. For example, all four modes of nutrition are represented.

Chemoheterotrophic proteobacteria include pathogens such as *Salmonella*, one cause of food poisoning; *Vibrio cholerae*, which causes cholera; and species of *Bordetella* that cause pertussis (whooping cough). *Escherichia coli*, which is a common resident of the intestines of humans and other mammals and a favorite research organism, is also a member of this group. Proteobacteria even include bacteria that attack other bacteria.

Bdellovibrio (see Figure 4) charge their prey at up to 100 μm/sec (comparable to a human running 600 km/hr) and bore into the prey by spinning at 100 revolutions per second.

Thiomargarita namibiensis (**Figure 9A**), an example of a photoautotrophic species of proteobacteria, uses H_2S to generate organic molecules from CO_2. The small greenish globules you see in the photo are sulfur wastes. *T. namibiensis* is a giant among prokaryotes. Single cells are typically 100–300 μm in diameter, roughly 100 times the length of an *E. coli* cell and large enough to be seen with the unaided eye. Other proteobacteria, including *Rhodopseudomonas* (see Figure 4), are photoheterotrophs; they cannot convert CO_2 to sugars.

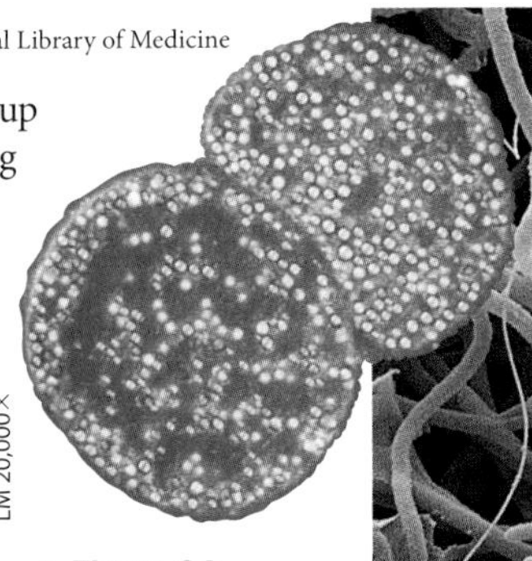

National Library of Medicine

LM 20,000×

▲ **Figure 9A** *Thiomargarita namibiensis*

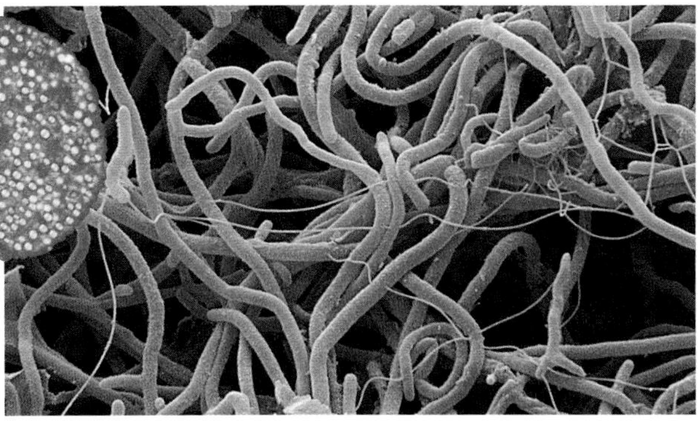

Colorized SEM 2,000×

© David Scharf/RGB Ventures LLC dba SuperStock/Alamy

▲ **Figure 9B** *Streptomyces*, the source of many antibiotics

Chemoautotrophic soil bacteria such as *Nitrosomonas* obtain energy by oxidizing inorganic nitrogen compounds. These and related species of proteobacteria are essential to the chemical cycle that makes nitrogen available to plants.

Proteobacteria also include *Rhizobium* species that live symbiotically in root nodules of legumes such as soybeans and peas. **Symbiosis** is a close association between organisms of two or more species, and *endo*symbiosis refers to one species, called the endosymbiont, living *within* another. *Rhizobium* endosymbionts fix nitrogen; that is, they convert atmospheric nitrogen gas to a form usable by their legume host.

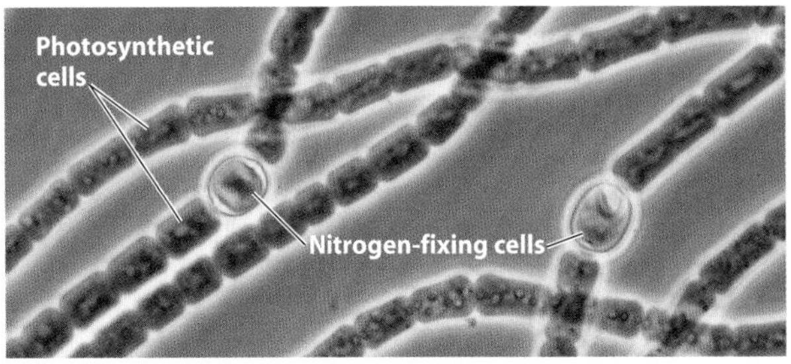

LM 650×

Susan M. Barns, Ph.D.

▲ **Figure 9C** *Anabaena*, a filamentous cyanobacterium

A second major group of bacteria, **gram-positive bacteria**, rivals the proteobacteria in diversity. One subgroup, the actinomycetes (from the Greek *mykes*, fungus, for which these bacteria were once mistaken), forms colonies of branched chains of cells. Actinomycetes are very common in the soil, where they decompose organic matter. Soil-dwelling species in the genus *Streptomyces*, shown in **Figure 9B**, are cultured by pharmaceutical companies as a source of many antibiotics, including streptomycin. Gram-positive bacteria also include many solitary species, such as *Bacillus anthracis* (see Figure 3B). The pathogens *Staphylococcus* and *Streptococcus* are also gram-positive bacteria. Mycoplasmas are gram-positive bacteria that lack cell walls and are the tiniest of all known cells, with diameters as small as 0.1 μm, only about five times as large as a ribosome.

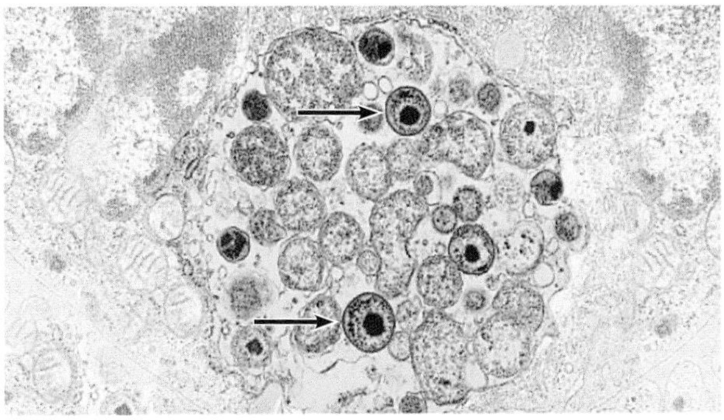

Colorized TEM 7,000×

Moredon Animal Health/SPL/Photo Researchers

▲ **Figure 9D** *Chlamydia* cells (arrows) inside an animal cell

The **cyanobacteria** are the only group of prokaryotes with plantlike, oxygen-generating photosynthesis. Ancient cyanobacteria generated the oxygen that changed Earth's atmosphere hundreds of millions of years ago. Today, cyanobacteria provide an enormous amount of food for freshwater and marine ecosystems. Some species, such as the cyanobacterium *Anabaena* in **Figure 9C**, have specialized cells that fix nitrogen. Many species of cyanobacteria have symbiotic relationships with organisms such as fungi, mosses, and a variety of marine invertebrates.

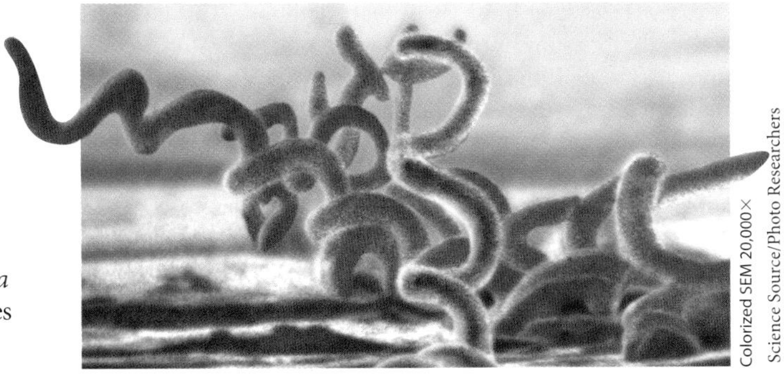

Colorized SEM 20,000×

Science Source/Photo Researchers

▲ **Figure 9E** *Treponema pallidum*, the spirochete that causes syphilis

The **chlamydias**, which live inside eukaryotic host cells, form a fourth bacterial group (**Figure 9D**). *Chlamydia trachomatis* is a common cause of blindness in developing countries and also causes nongonococcal urethritis, the most common sexually transmitted disease in the United States.

Spirochetes, the fifth group, are helical bacteria that spiral through their environment by means of rotating, internal filaments. Some spirochetes are notorious pathogens: *Treponema pallidum*, shown in **Figure 9E**, causes syphilis, and *Borrelia burgdorferi* (see Figure 2A) causes Lyme disease.

? **How are *Thiomargarita namibiensis* similar to the cyanobacteria?**

● They are both photoautotrophic.

CONNECTION

10 Some bacteria cause disease

All organisms, people included, are almost constantly exposed to pathogenic bacteria. Most often, our body's defenses prevent pathogens from affecting us. Occasionally, however, a pathogen establishes itself in the body and causes illness. Even some of the bacteria that are normal residents of the human body can make us ill when our defenses have been weakened by poor nutrition or by a viral infection.

Most bacteria that cause illness do so by producing a poison—either an exotoxin or an endotoxin. **Exotoxins** are proteins that bacterial cells secrete into their environment. They include some of the most powerful poisons known. For example, *Staphylococcus aureus*, shown in **Figure 10**, produces several exotoxins. Although *S. aureus* is commonly found on the skin and in the nasal passages, if it enters the body through a wound, it can cause serious disease. One of its exotoxins destroys the white blood cells that attack invading bacteria, resulting in the pus-filled skin bumps characteristic of methicillin-resistant *S. aureus* infections (MRSA). Food may also be contaminated with *S. aureus* exotoxins, which are so potent that less than a millionth of a gram causes vomiting and diarrhea.

Endotoxins are lipid components of the outer membrane of gram-negative bacteria that are released when the cell dies or is digested by a defensive cell. All endotoxins induce the same general symptoms: fever, aches, and sometimes a dangerous drop in blood pressure (septic shock). Septic shock triggered by an endotoxin of *Neisseria meningitides*, which causes bacterial meningitis, can kill a healthy person in a matter of days or even hours. Because the bacteria are easily transmitted among people living in close contact, many colleges require students to be vaccinated against this disease. Other examples of endotoxin-producing bacteria include the species of *Salmonella* that cause food poisoning and typhoid fever.

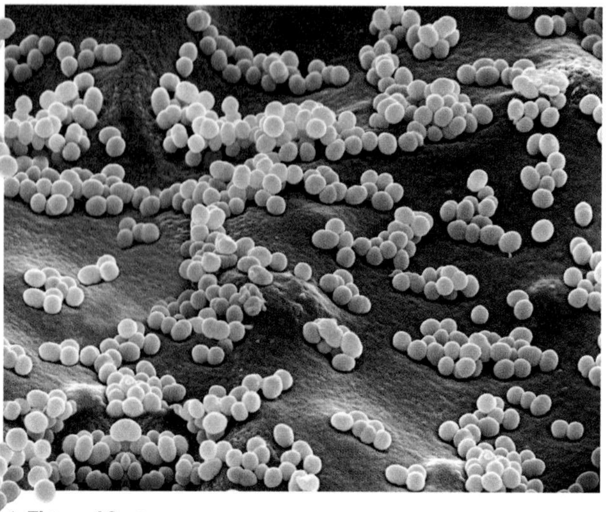
▲ **Figure 10** *Staphylococcus aureus*, an exotoxin producer

Colorized SEM 12,000×
David M. Phillips/Photo Researchers

? Contrast exotoxins with endotoxins.

Exotoxins are proteins secreted by pathogenic bacteria; endotoxins are components of the outer membrane of pathogenic bacteria.

SCIENTIFIC DISCOVERY

11 Koch's postulates are used to prove that a bacterium causes a disease

To test the hypothesis that a certain bacterium is the cause of a disease, a researcher must satisfy four conditions. This method of hypothesis testing, formulated by microbiologist Robert Koch in the late 19th century, is known as *Koch's postulates*. For a human disease, the researcher must be able to (1) find the candidate bacterium in every case of the disease; (2) isolate the bacterium from a person who has the disease and grow it in pure culture; (3) show that the cultured bacterium causes the disease when transferred to a healthy subject (usually an animal); and (4) isolate the bacterium from the experimentally infected subject. So when Australian microbiologist Barry Marshall hypothesized that chronic gastritis (an inflammation of the stomach lining that can lead to ulcers) was caused by a bacterium called *Helicobacter pylori*, he knew he would need to fulfill these criteria.

Over the course of several years, Marshall satisfied the first two requirements, but his efforts to infect animals failed to produce results. Taking another approach to the problem, he designed an experiment to test whether antibiotics would cure

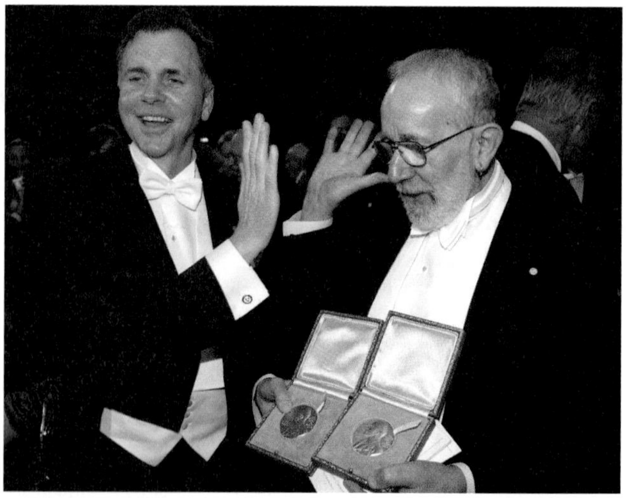
▲ **Figure 11** Marshall (left) and colleague Robin Warren celebrating the success of Marshall's unorthodox experiment

Anders Wiklund/Scanpix/Reuters

ulcers. However, the scientific community was highly skeptical of Marshall's hypothesis. Doctors and scientists had long believed that chronic gastritis was caused by lifestyle factors such as stress, diet, and smoking and saw no reason to pursue alternative explanations. As a result, Marshall had difficulty obtaining funding, and his research stalled.

Marshall didn't give up, continuing to accumulate evidence supporting his hypothesis. But he was frustrated watching so many patients suffer life-threatening complications from peptic (stomach) ulcers when his research might yield a simple cure. At last, Marshall decided to take a radical course of action—he would experiment on himself. He concocted a nasty brew of *H. pylori* and swallowed it. Several days later, he became ill from gastritis (step 3 of Koch's postulates). His stomach lining proved to be teeming with *H. pylori* (step 4). Marshall then cleared up his infection with antibiotics. He subsequently received funding to continue his research, and other scientists followed up with further studies. Several years after Marshall's big gulp, an international conference of disease specialists finally accepted his conclusions about *H. pylori*.

In 1993, before Marshall established the pathogenicity of *H. pylori*, peptic ulcers resulted in 1 million hospitalizations, 6,500 deaths, and $6 billion in health-care costs in the United States. Today, most ulcer patients are cured with antibiotics shortly after diagnosis. Barry Marshall and collaborator Robin Warren were awarded the 2005 Nobel Prize in Medicine for their discovery of *H. pylori* and its role in peptic ulcers (**Figure 11**, on previous page).

? How would Marshall's proposed experiment with antibiotics test his hypothesis that ulcers are caused by bacteria?

If ulcers are caused by bacteria, then treatment with antibiotics should kill the bacteria and cure the ulcer.

CONNECTION 12 Bacteria can be used as biological weapons

In October 2001, endospores of *Bacillus anthracis*, the bacterium that causes anthrax, were mailed to members of the news media and the U.S. Senate. These attacks, which resulted in five deaths, raised public awareness of the potential for biological agents such as viruses and bacteria to be used as weapons. *B. anthracis*, *Yersinia pestis* (the bacterium that causes plague), and the exotoxin of *Clostridium botulinum* are among the biological agents that are considered the highest-priority threats today.

B. anthracis forms hardy endospores (see Figure 3B) that are commonly found in the soil of agricultural regions, where large grazing animals can become infected. People who work in agriculture, leather tanning, or wool processing occasionally catch the easily treated cutaneous (skin) form of anthrax from infected animals. Inhalation anthrax, which affects the lungs, is the deadly form of the disease. "Weaponizing" anthrax involves manufacturing a preparation of endospores that disperses easily in the air, where they will be inhaled by the target population. Endospores that enter the lungs germinate, and the bacteria multiply, producing an exotoxin that eventually accumulates to lethal levels in the blood. *B. anthracis* can be controlled by certain antibiotics, but the antibiotics only kill the bacteria—they can't eliminate the toxin already in the body. As a result, inhalation anthrax has a very high death rate.

Yersinia pestis bacteria are carried by rodents in many parts of the world and are transmitted by fleas. People contract the disease when they are bitten by infected fleas. When diagnosed in time, plague can be treated by antibiotics.

In the bubonic form of plague, egg-size swellings called buboes grow in the groin, armpits, and neck, and black blotches appear where the bacteria proliferate in clots under the skin (**Figure 12**). If bacteria spill into the bloodstream, death from the bacterial toxins is certain. Inhalation of *Y. pestis* causes the pneumonic form of plague. Bacteria infect cells in the lungs, becoming airborne as the disintegrating lung tissue is coughed out in droplets. As a result, pneumonic plague is easily transmitted from person to person and can become epidemic, the worst scenario for an attack by bioweapons.

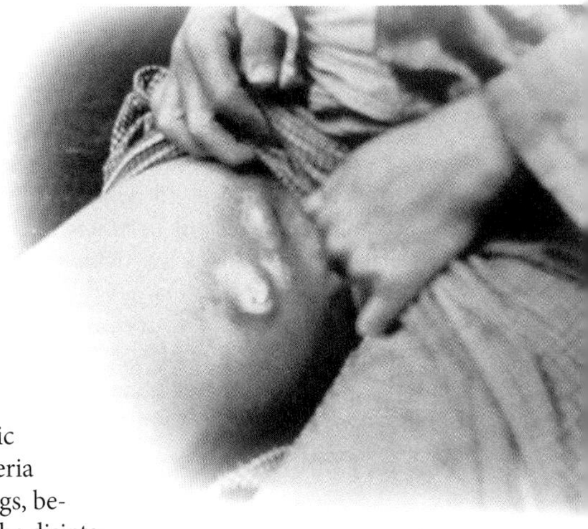

▶ **Figure 12** Swellings (buboes) characteristic of the bubonic form of plague

Centers for Disease Control and Prevention

Unlike other biological agents, the weapon form of *C. botulinum* is the exotoxin it produces, botulinum, rather than the living microbes. Botulinum is the deadliest poison known. Thirty grams of pure toxin, a bit more than an ounce, could kill every person in the United States. Botulinum blocks transmission of the nerve signals that cause muscle contraction, resulting in paralysis of the muscles required for breathing. As a weapon, it would most likely be broadcast as an aerosol.

? Why is *Bacillus anthracis* an effective bioweapon?

It is easy to obtain from the soil and forms potentially deadly endospores that resist destruction and can be easily dispersed.

Protists

13 Protists are an extremely diverse assortment of eukaryotes

Protists are a diverse collection of mostly unicellular eukaryotes. Biologists used to classify all protists in a kingdom called Protista, but now it is thought that these organisms constitute multiple kingdoms within domain Eukarya. While our knowledge of the evolutionary relationships among these diverse groups remains incomplete, *protist* is still useful as a convenient term to refer to eukaryotes that are not plants, animals, or fungi.

Protists obtain their nutrition in a variety of ways (**Figure 13A**). Some protists are autotrophs, producing their food by photosynthesis; these are called **algae** (another useful term that is not taxonomically meaningful). Many algae, including the one shown on the left in Figure 13A, are multicellular. Other protists, informally called **protozoans**, are heterotrophs, eating bacteria and other protists. Some heterotrophic protists are fungus-like and obtain organic molecules by absorption, and some are parasitic. **Parasites** derive their nutrition from a living host, which is harmed by the interaction. *Giardia*, shown in the middle of Figure 13A, is a human parasite. Still other protists are **mixotrophs**, capable of both photosynthesis and heterotrophy, depending on availability of light and nutrients. An example is *Euglena*, shown on the right in Figure 13A.

Protist habitats are also diverse. Most protists are aquatic, and they are found almost anywhere there is moisture, including terrestrial habitats such as damp soil and leaf litter. Others inhabit the bodies of various host organisms. For example, **Figure 13B** shows one of the protists that are endosymbionts in the intestinal tract of termites. Termite endosymbionts digest the tough cellulose in the wood eaten by their host. Some of these protists even have endosymbionts of their own—prokaryotes that metabolize the cellulose.

As eukaryotes, protists are more complicated than any prokaryotes. Their cells have a membrane-enclosed nucleus (containing multiple chromosomes) and other organelles characteristic of eukaryotic cells. The flagella and cilia of protistan cells have a 9 + 2 pattern of microtubules, another typical eukaryotic trait.

Because most protists are unicellular, they are justifiably considered the simplest eukaryotes. However, the cells of many protists are among the most elaborate in the world. This level of cellular complexity is not really surprising, for each unicellular protist is a complete eukaryotic organism analogous to an entire animal or plant.

With their extreme diversity, protists are difficult to categorize. Recent molecular and cellular studies have shaken the foundations of protistan taxonomy as much as they have that of the prokaryotes. Intuitive groupings such as protozoans and algae are phylogenetically meaningless because

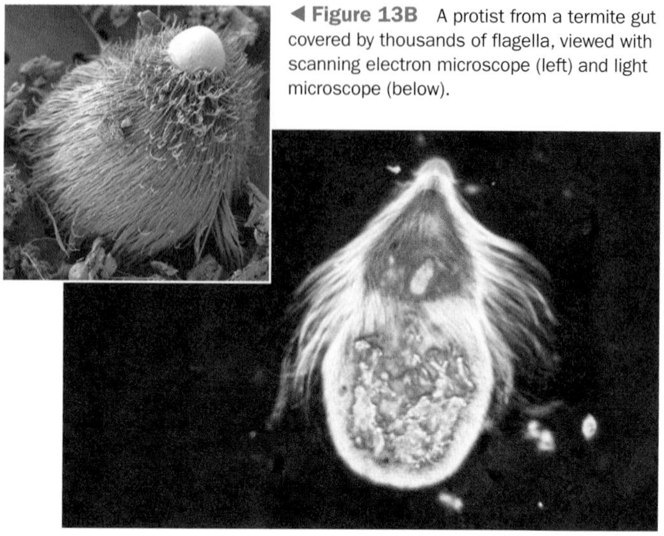

◀ **Figure 13B** A protist from a termite gut covered by thousands of flagella, viewed with scanning electron microscope (left) and light microscope (below).

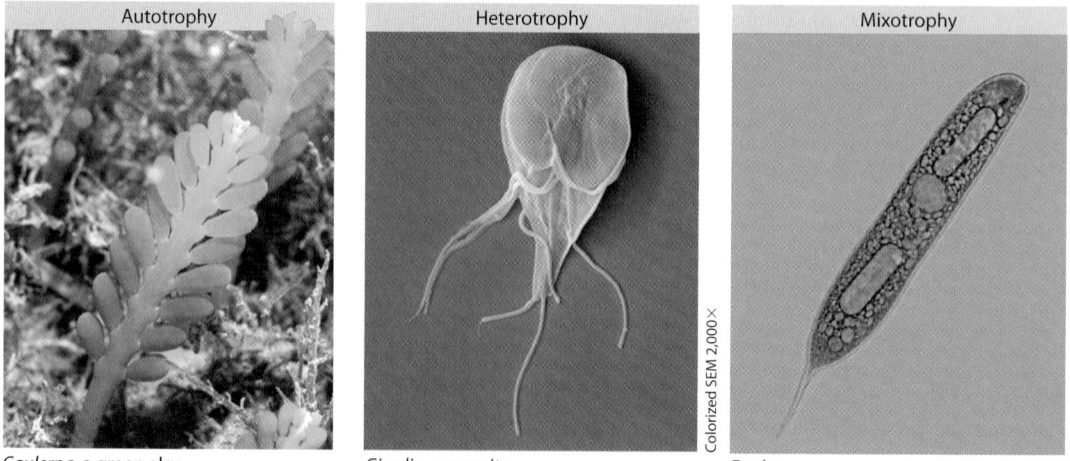

Caulerpa, a green alga — *Giardia*, a parasite — *Euglena*

▶ **Figure 13A** Protist modes of nutrition

the nutritional modes used to categorize them are spread across many different lineages. It is now clear that there are multiple clades of protists, with some lineages more closely related to plants, fungi, or animals than they are to other protists. We have chosen to organize our brief survey of protist diversity using one current hypothesis of protist phylogeny, which proposes five monophyletic "supergroups": Chromalveolata, Rhizaria, Excavata, Unikonta, and Archaeplastida. While there is general agreement on some of these groupings, others are hotly debated—the classification of protists is very much a work in progress. Before embarking on our tour of protists, however, let's consider how their extraordinary diversity originated.

? What is a general definition for "protist"?

A eukaryote that is not an animal, fungus, or plant

EVOLUTION CONNECTION

14 Secondary endosymbiosis is the key to much of protist diversity

As Module 13 indicates, protists are bewilderingly diverse. What is the origin of this enormous diversity? To explain, let's first review the endosymbiont theory for the origin of mitochondria and chloroplasts. According to that theory, eukaryotic cells evolved when prokaryotes with special capabilities established residence within other, larger prokaryotes. The evidence for this theory includes the structural and molecular similarities between prokaryotic cells and present-day mitochondria and chloroplasts. Mitochondria and chloroplasts even replicate their own DNA and reproduce by a process similar to that of prokaryotes.

Scientists think that heterotrophic eukaryotes evolved first; these had mitochondria but not chloroplasts. As shown in **Figure 14**, autotrophic eukaryotes are thought to have arisen later from a lineage of heterotrophic eukaryotes descended from ❶ an individual that engulfed an autotrophic cyanobacterium through primary endosymbiosis. If the cyanobacterium continued to function within its host cell, its photosynthesis would have provided a steady source of food for the heterotrophic host and thus given it a significant selective advantage. And because the cyanobacterium had its own DNA, it could reproduce to make multiple copies of itself within the host cell. In addition, cyanobacteria could be passed on when the host reproduced. Over time, ❷ the descendants of the original cyanobacterium evolved into chloroplasts. The chloroplast-bearing lineage of eukaryotes later diversified into ❸ the autotrophs green algae and red algae (see Module 20).

On subsequent occasions during eukaryotic evolution, ❹ green algae and red algae themselves became endosymbionts following ingestion by heterotrophic eukaryotes. Heterotrophic host cells enclosed the algal cells in food vacuoles ❺ but the algae—or parts of them—survived and became cellular organelles. The presence of the endosymbionts, which also had the ability to replicate themselves, gave their hosts a selective advantage. This process, in which an autotrophic eukaryotic protist became endosymbiotic in a heterotrophic eukaryotic protist, is called **secondary endosymbiosis**. Secondary endosymbiosis appears to be a major key to protist diversity. Figure 14 shows how secondary endosymbiosis of green algae could give rise to mixotrophs, such as the *Euglena* in Figure 13A. Secondary endosymbiosis of red algae led to nutritional diversity in other groups of protists.

? Distinguish between primary and secondary endosymbiosis.

In primary endosymbiosis, the endosymbiont is a prokaryote. In secondary endosymbiosis, the endosymbiont is a eukaryote.

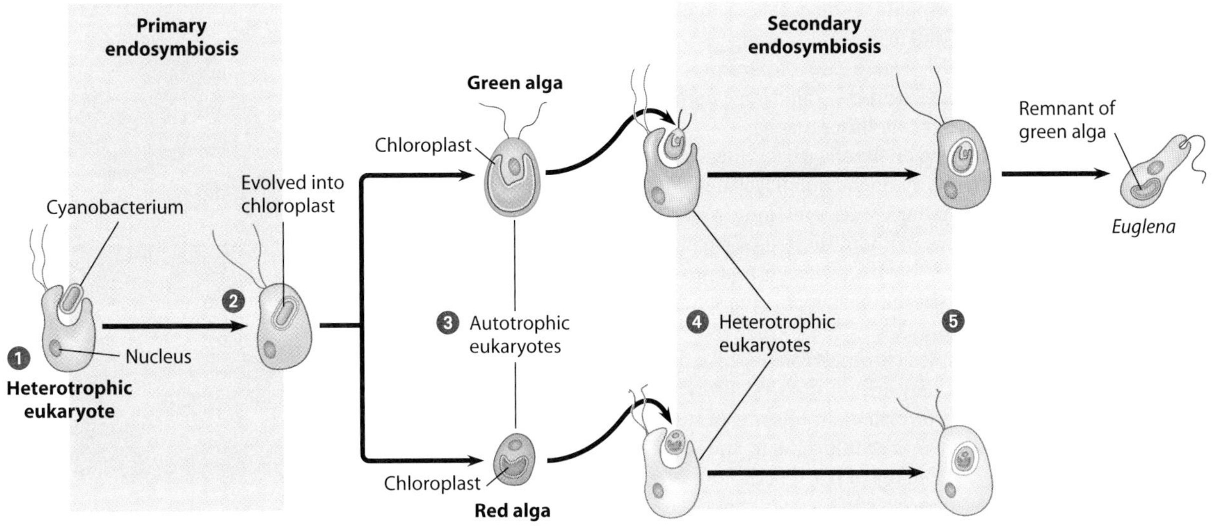

▲ **Figure 14** The theory of the origin of protistan diversity through endosymbiosis (mitochondria not shown)

15 Chromalveolates represent the range of protist diversity

Our sample of protist diversity begins with the supergroup **Chromalveolata** (the chromalveolates). This large, extremely diverse group includes autotrophic, heterotrophic, and mixotrophic species and multicellular as well as unicellular species. Although some evidence supports Chromalveolata as monophyletic, it is a controversial grouping.

Autotrophic chromalveolates include **diatoms**, unicellular algae that have a unique glassy cell wall containing silica. The cell wall of a diatom consists of two halves that fit together like the bottom and lid of a shoe box (**Figure 15A**). Both freshwater and marine environments are rich in diatoms, and the organic molecules these microscopic algae produce are a key source of food in all aquatic environments. Some diatoms store food reserves in the form of lipid droplets as well as carbohydrates. In addition to being a rich source of energy, the lipids make the diatoms buoyant, which keeps them floating near the surface in the sunlight. Massive accumulations of fossilized diatoms make up thick sediments known as diatomaceous earth, which is mined for use as a filtering medium and as a grinding and polishing agent.

Colorized SEM 200×

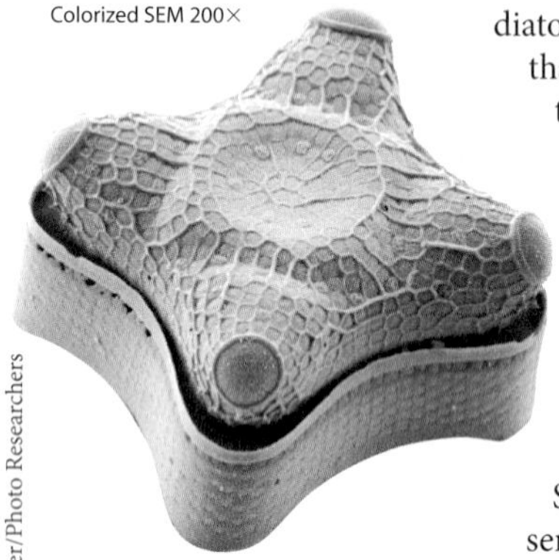

Steve Gschmeissner/Photo Researchers

▲ **Figure 15A** Diatom, a unicellular alga

Dinoflagellates, a group that includes unicellular autotrophs, heterotrophs, and mixotrophs, are also very common components of marine and freshwater plankton (communities of microorganisms that live near the water's surface). Blooms—population explosions—of autotrophic dinoflagellates sometimes cause warm coastal waters to turn pinkish orange, a phenomenon known as "red tide" (**Figure 15B**). Toxins produced by some red-tide dinoflagellates have killed large numbers of fish. Humans who consume molluscs that have accumulated the toxins by feeding on dinoflagellates may be affected as well. One genus of photosynthetic dinoflagellates resides within the cells of reef-building corals, providing at least half the energy used by the corals. Without these algal partners, corals could not build and sustain the massive reefs that provide the food, living space, and shelter that support the splendid diversity of the reef community.

Miriam Godfrey, NIWA

▲ **Figure 15B** A red tide caused by *Gymnodinium*, a dinoflagellate

Brown algae are large, complex chromalveolates. Like diatoms and some dinoflagellates, they are autotrophic. Brown algae owe their characteristic brownish color to some of the pigments in their chloroplasts. All are multicellular, and most are marine. Brown algae include many of the species commonly called seaweeds. We use the word *seaweeds* here to refer to marine algae that have large multicellular bodies but lack the roots, stems, and leaves found in most plants. (Some red and green algae are also referred to as seaweeds.) **Figure 15C** shows an underwater bed of brown algae called **kelp** off the coast of California. Anchored to the seafloor by rootlike structures, kelp may grow to heights of 60 m, taller than a 15-story building. Fish, sea lions, sea otters, and gray whales regularly use these kelp "forests" as their feeding grounds.

Georgie Holland/AGE fotostock

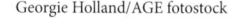

▲ **Figure 15C** Brown algae: a kelp "forest"

Water molds are heterotrophic unicellular chromalveolates that typically decompose dead plants and animals in freshwater habitats. Because many species resemble fungi (**Figure 15D**, top of facing page), water molds were classified as fungi until molecular comparisons revealed their kinship to protists. Parasitic water molds sometimes grow on the skin or gills of fish. Water molds also include plant parasites called downy mildews. "Late blight" of potatoes, a disease caused by a downy mildew, led to a devastating famine in Ireland in the mid-1800s. A closely related pathogen has swept through tomato crops in the eastern United States since 2009, depriving fast-food burgers of a standard topping and home gardeners of a favorite summertime treat.

Fred Rhoades

▲ **Figure 15D** Water mold (white threads) decomposing a goldfish

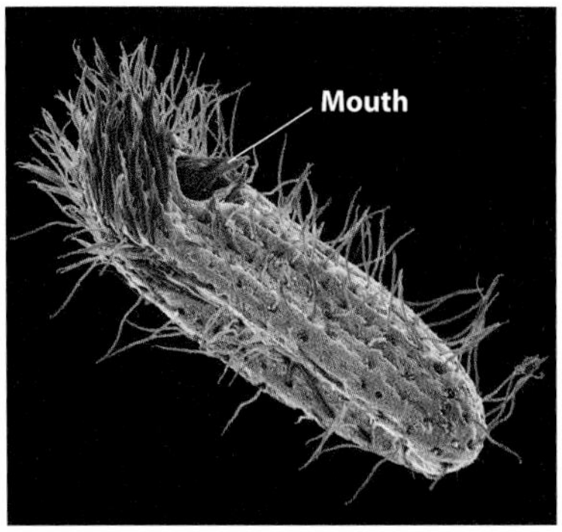

Colorized SEM

Steve Gscheissner/Photo Researchers

▲ **Figure 15E** A freshwater ciliate showing cilia distributed over the cell surface and around the mouth

Ciliates, named for their use of cilia to move and to sweep food into their mouth, are a group of unicellular protists that includes heterotrophs and mixotrophs. You may have seen the common freshwater protist *Paramecium* in a biology lab. Like many ciliates, the example shown in **Figure 15E** swims by beating its cilia in a wavelike motion. Other ciliates "crawl" over a surface using cilia that are arranged in bundles along the length of the cell.

Another subgroup of Chromalveolata is made up of parasites, including some that cause serious diseases in humans. For example, *Plasmodium*, which causes malaria, kills nearly a million people a year. Some stages of *Plasmodium*'s complex life cycle take place in certain species of mosquitoes, which transmit the parasite to humans.

? Which subgroups of chromalveolates include autotrophs?

Diatoms, dinoflagellates, and brown algae

CONNECTION

16 Can algae provide a renewable source of energy?

Have you ever wondered what the "fossils" are in fossil fuels? They are organic remains of organisms that lived hundreds of millions of years ago. Diatoms are thought to be the main source of oil, while coal was formed from primitive plants. However, rapid consumption is depleting the world's supply of readily accessible fossil fuels.

Entrepreneurs are now eying the lipid droplets in diatoms and other algae as a renewable source of energy. After all, the energy we extract from fossil fuels was originally stored in organisms through the process of photosynthesis. Why wait millions of years? If unicellular algae could be grown on a large scale, the oil could be harvested and processed into biodiesel. When supplied with light, carbon dioxide, and nutrients, unicellular algae reproduce rapidly. In one scenario, algae could be grown indoors in closed "bioreactor" vessels under tightly controlled environmental conditions (**Figure 16**). Outdoor systems using closed bioreactors or open-air ponds are also being developed.

There are numerous technical hurdles to overcome before the industrial-scale production of biofuel from algae becomes a reality. Investigators must identify the most productive of the hundreds of algal species and test whether they are suitable for mass culturing methods. With further research, scientists may be able to improve desirable characteristics such as growth rate or oil yield through genetic engineering. In addition, manufacturers need to develop cost-effective methods of harvesting the algae and extracting and processing the oil. Nevertheless, there might be an alga-powered vehicle in your future.

? What characteristics of unicellular algae make them attractive candidates for the production of biofuels?

Rapid reproduction; would not occupy farmland needed to grow food crops

Xinhua/Photoshot/Newscom

▲ **Figure 16** Green algae in a bioreactor

17 Rhizarians include a variety of amoebas

The clade **Rhizaria** was recently proposed based on similarities in DNA, although some scientists disagree and think that rhizarians should be placed in Chromalveolata. The two largest groups in Rhizaria, foraminiferans and radiolarians, are among the organisms referred to as amoebas. **Amoebas** move and feed by means of **pseudopodia** (singular, *pseudopodium*), which are temporary extensions of the cell. Molecular systematics now indicates that amoebas are dispersed across many taxonomic groups. Most of the amoebas in Rhizaria are distinguished from other amoebas by their threadlike (rather than lobe-shaped) pseudopodia.

Foraminiferans (forams) (**Figure 17A**) are found both in the ocean and in fresh water. They have porous shells, called *tests*, composed of organic material hardened by calcium carbonate. The pseudopodia, which function in feeding and locomotion, extend through small pores in the test (see Figure 17A inset). Ninety percent of forams that have been identified are fossils. The fossilized tests, which are a component of sedimentary rock, are excellent markers for correlating the ages of rocks in different parts of the world.

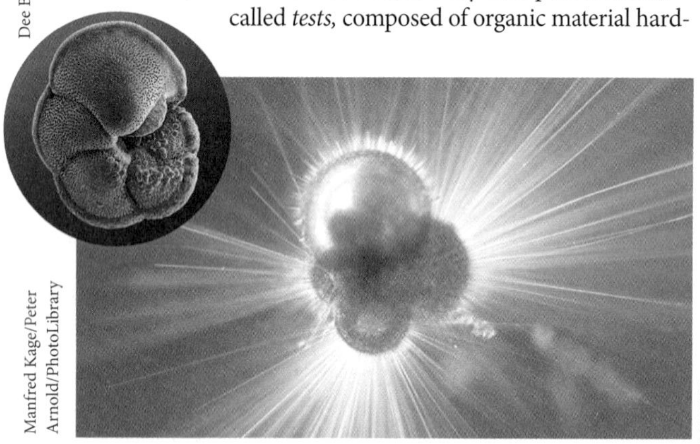

▲ **Figure 17A** A foraminiferan (inset SEM shows a foram test)

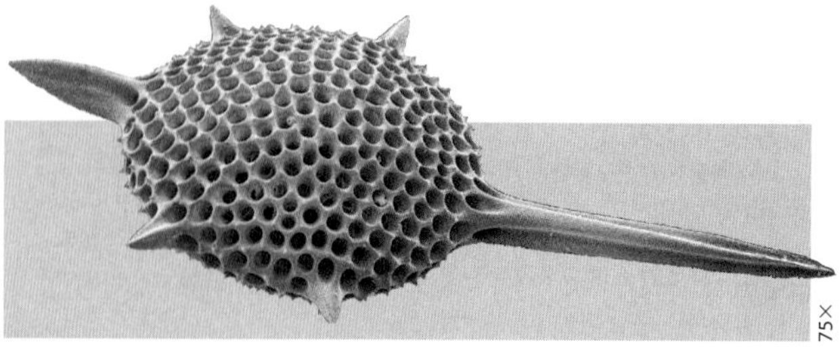

▲ **Figure 17B** A radiolarian skeleton

Like forams, **radiolarians** produce a mineralized support structure, in this case an internal skeleton made of silica (**Figure 17B**). The cell is also surrounded by a test composed of organic material. Most species of radiolarians are marine. When they die, their hard parts, like those of forams, settle to the bottom of the ocean and become part of the sediments. In some areas, radiolarians are so abundant that sediments, known as radiolarian ooze, are hundreds of meters thick.

? What minerals compose foram tests and radiolarian skeletons?

Forams, calcium carbonate; radiolarians, silica

18 Some excavates have modified mitochondria

Excavata (the excavates) has recently been proposed as a clade on the basis of molecular and morphological similarities. The name refers to an "excavated" feeding groove possessed by some members of the group. Many excavates have modified mitochondria that lack functional electron transport chains and use anaerobic pathways such as glycolysis to extract energy. Heterotrophic excavates include the termite endosymbiont shown in Figure 13B. There are also autotrophic species and mixotrophs, such as *Euglena* (see Figure 13A).

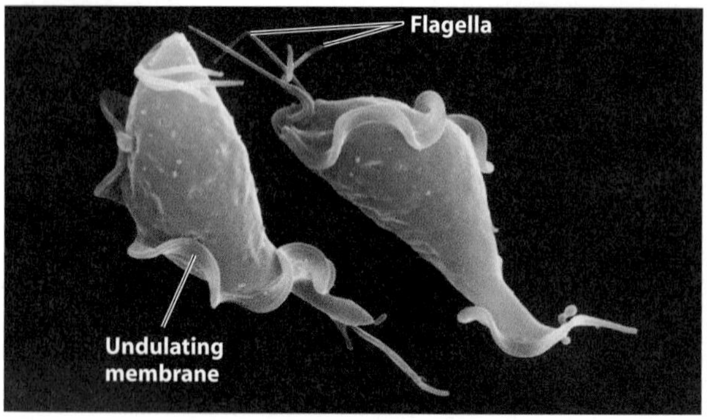

▲ **Figure 18A** A parasitic excavate: *Trichomonas vaginalis*

Some excavates are parasites. *Giardia intestinalis* (see Figure 13A) is a common waterborne parasite that causes severe diarrhea. People most often pick up *Giardia* by drinking water contaminated with feces containing the parasite. For example, a swimmer in a lake or river might accidentally ingest water contaminated with feces from infected animals, or a hiker might drink contaminated water from a seemingly pristine stream. (Boiling the water first will kill *Giardia.*)

Another excavate, *Trichomonas vaginalis* (**Figure 18A**), is a common sexually transmitted parasite that causes an estimated 5 million new infections each year. The parasite travels through the reproductive tract by moving its flagella and undulating part of its membrane. In women, the protists feed on white blood cells and bacteria living on the cells lining the vagina. *T. vaginalis* also infects the cells lining the male reproductive tract, but limited availability of food results in very small population sizes. Consequently, males typically have no symptoms of infection, although they can repeatedly infect their female partners.

The only treatment available is a drug called metronidazole. Disturbingly, drug resistance seems to be evolving in *T. vaginalis*, especially on college campuses.

Members of the excavate genus *Trypanosoma* are parasites that can be transmitted to humans by insects. For instance, the trypanosome shown in **Figure 18B** causes sleeping sickness, a potentially fatal disease spread by the African tsetse fly. The squiggly "worms" in the photo are cells of *Trypanosoma;* the circular cells are human red blood cells.

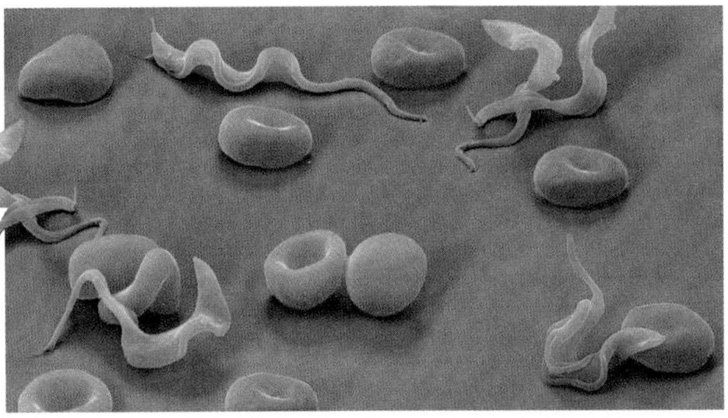

▲ **Figure 18B** A parasitic excavate: *Trypanosoma* (with blood cells)

Colorized SEM 1,500×

Eye of Science/Science Source/Photo Researchers

? How do the nutritional modes of *Euglena* and *Trichomonas* differ?

Euglena is mixotrophic; *Trichomonas* is strictly heterotrophic.

---

19 Unikonts include protists that are closely related to fungi and animals

Unikonta is a controversial grouping that joins two well-established clades: **amoebozoans**, which are protists, and a second clade that includes animals and fungi. You'll learn about the amoebozoans in this module, then return to the second clade in the last module of this chapter.

Amoebozoans, including many species of free-living amoebas, some parasitic amoebas, and the slime molds, have lobe-shaped pseudopodia. The amoeba in **Figure 19A** is poised to ingest an alga. Its pseudopodia arch around the prey and will enclose it in a food vacuole. Free-living amoebas creep over rocks, sticks, or mud at the bottom of a pond or ocean. A parasitic species of amoeba causes amoebic dysentery, a potentially fatal diarrheal disease.

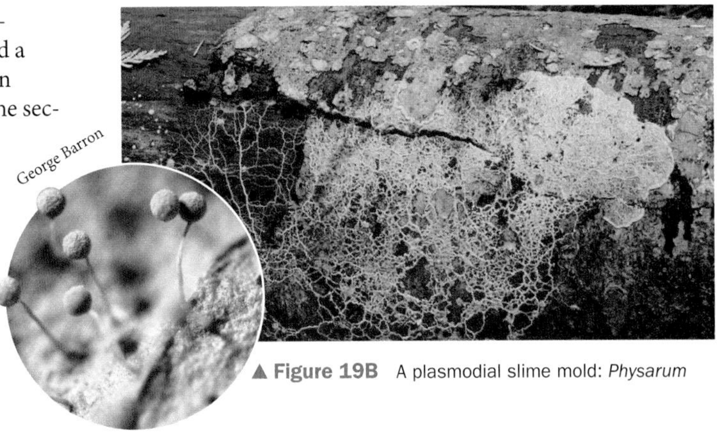

▲ **Figure 19B** A plasmodial slime mold: *Physarum*

George Barron

Ray Simons/Photo Researchers

The yellow, branching growth on the dead log in **Figure 19B** is an amoebozoan called a **plasmodial slime mold**. These protists are common where there is moist, decaying organic matter and are often brightly pigmented. Although it is large and has many extensions, the organism in Figure 19B is not multicellular. Rather, it is a **plasmodium**, a single, multinucleate mass of cytoplasm undivided by plasma membranes. (Don't confuse this word with the chromalveolate *Plasmodium*, which causes malaria.) Because most of the nuclei go through mitosis at the same time, plasmodial slime molds are used to study molecular details of the cell cycle.

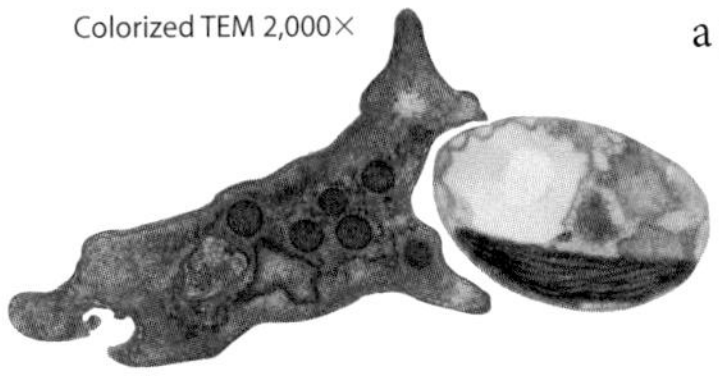

Colorized TEM 2,000×

▲ **Figure 19A** An amoeba beginning to ingest an algal cell

Biophoto Associates/Photo Researchers

The plasmodium extends pseudopodia through soil and rotting logs, engulfing food by phagocytosis as it grows. Within the fine channels of the plasmodium, cytoplasm streams first one way and then the other in pulsing flows that probably help distribute nutrients and oxygen. When food and water are in short supply, the plasmodium stops growing and differentiates into reproductive structures (shown in the inset in Figure 19B) that produce spores. When conditions become favorable, the spores release haploid cells that fuse to form a zygote, and the life cycle continues.

Cellular slime molds are also common on rotting logs and decaying organic matter. Most of the time, these organisms exist as solitary amoeboid cells. When food is scarce, the amoeboid cells swarm together, forming a slug-like aggregate that wanders around for a short time. Some of the cells then dry up and form a stalk supporting an asexual reproductive structure in which yet other cells develop into spores. The cellular slime mold *Dictyostelium*, shown in **Figure 19C**, is a useful model for researchers studying the genetic mechanisms and chemical changes underlying cellular differentiation.

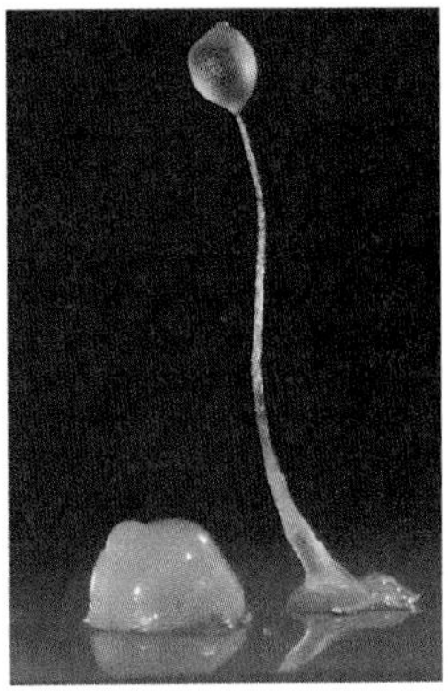

15×

Robert Kay

▲ **Figure 19C** An aggregate of amoeboid cells (left) and the reproductive structure of a cellular slime mold, *Dictyostelium*

? Contrast the plasmodium of a plasmodial slime mold with the slug-like stage of a cellular slime mold.

A plasmodium is not multicellular, but is one cytoplasmic mass with many nuclei; the slug-like stage of a cellular slime mold consists of many cells.

20 Archaeplastids include red algae, green algae, and land plants

Almost all the members of the supergroup **Archaeplastida** are autotrophic. As you learned in Module 14, autotrophic eukaryotes are thought to have arisen by primary endosymbiosis of a cyanobacterium that evolved into chloroplasts. The descendants of this ancient protist evolved into red algae and green algae, which are key photosynthesizers in aquatic food webs. Archaeplastida also includes land plants, which evolved from a group of green algae.

The warm coastal waters of the tropics are home to the majority of species of **red algae**. Their red color comes from an accessory pigment that masks the green of chlorophyll. Although a few species are unicellular, most red algae are multicellular. Multicellular red algae are typically soft-bodied, but some have cell walls encrusted with hard, chalky deposits. Encrusted species, such as the one in **Figure 20A**, are common on coral reefs, and their hard parts are important in building and maintaining the reef. Other red algae are commercially important. Carrageenan, a gel that is used to stabilize many products, including ice cream, chocolate milk, and pudding, is derived from species of red algae. Sheets of a red alga, known as nori, are used to wrap sushi. Agar, a polysaccharide used as a substrate for growing bacteria, also comes from red algae.

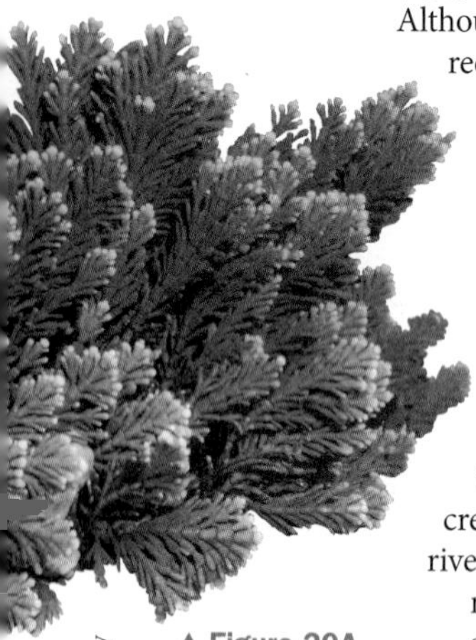

Paul Kay/Photolibrary

▲ **Figure 20A** An encrusted red alga on a coral reef

Green algae, which are named for their grass-green chloroplasts, include unicellular and colonial species as well as multicellular seaweeds. The micrograph on the right in **Figure 20B** shows *Chlamydomonas,* a unicellular alga common in freshwater lakes and ponds. It is propelled through the water by two flagella. (Such cells are said to be biflagellated.) *Volvox*, shown on the left, is a colonial green alga. Each *Volvox* colony is a hollow ball composed of hundreds or thousands of biflagellated cells. As the flagella move, the colony tumbles slowly through the water. Some of the large colonies shown here contain small daughter colonies that will eventually be released.

Aaron J. Bell/Photo Researchers

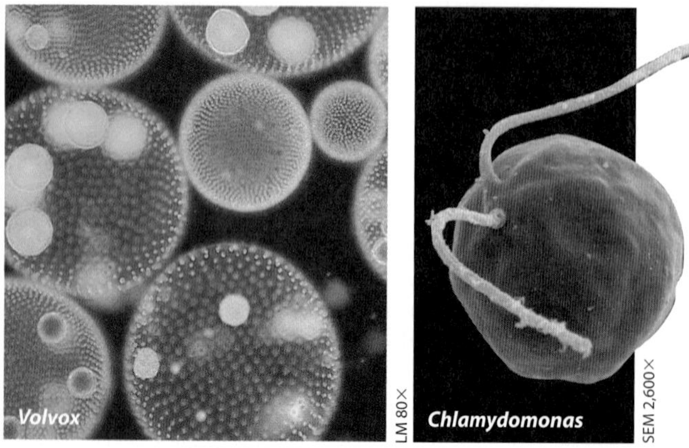

Manfred Kage/Peter Arnold, Inc./PhotoLibrary

▲ **Figure 20B** Green algae, colonial (left) and unicellular (right)

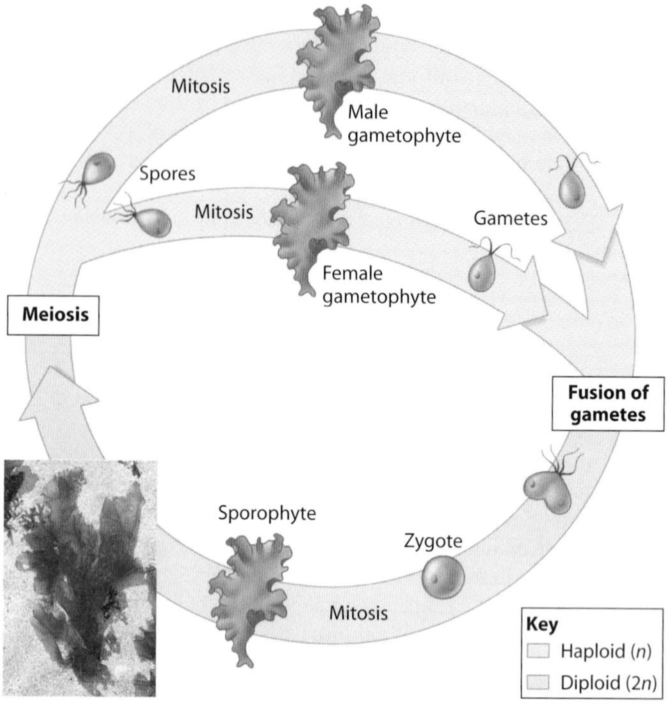

D.P. Wilson/Eric and David Hosking/Photo Researchers

▲ **Figure 20C** The life cycle of *Ulva*, a multicellular green alga

Ulva, or sea lettuce, is a multicellular green alga. Like many multicellular algae and all land plants, *Ulva* has a complex life cycle that includes an **alternation of generations** (**Figure 20C**). In this type of life cycle, a multicellular diploid ($2n$) form alternates with a multicellular haploid (n) form. Notice in the figure that multicellular diploid forms are called **sporophytes**, because they produce spores. The sporophyte generation alternates with a haploid generation that features a multicellular haploid form called a **gametophyte**, which produces gametes. In *Ulva*, the gametophyte and sporophyte organisms are identical in appearance; both look like the one in the photograph, although they differ in chromosome number. The haploid gametophyte produces gametes by mitosis, and fusion of the gametes begins the sporophyte generation. In turn, cells in the sporophyte undergo meiosis and produce haploid, flagellated spores. The life cycle is completed when a spore settles to the bottom of the ocean and develops into a gametophyte.

? How does chromosome number differ in the gametophyte and sporophyte in the alternation of generations life cycle?

The gametophyte is haploid (n); the sporophyte is diploid ($2n$).

EVOLUTION CONNECTION **21** Multicellularity evolved several times in eukaryotes

Increased complexity often makes more variations possible. Thus, the origin of the eukaryotic cell led to an evolutionary radiation of new forms of life. As you have seen in this chapter, unicellular protists, which are structurally complex eukaryotic cells, are much more diverse in form than the simpler prokaryotes. The evolution of multicellular bodies broke through another threshold in structural organization.

Multicellular organisms—seaweeds, plants, animals, and most fungi—are fundamentally different from unicellular ones. In a unicellular organism, all of life's activities occur within a single cell. In contrast, a multicellular organism has various specialized cells that perform different functions and are dependent on one another. For example, some cells give the organism its shape, while others make or procure food, transport materials, enable movement, or reproduce.

As you have seen in this chapter, multicellular organisms have evolved from three different ancestral lineages: chromalveolates (brown algae), unikonts (fungi and animals), and archaeplastids (red algae and green algae). **Figure 21A** summarizes some current hypotheses for the early phylogeny of land plants and animals, which are all multicellular, and fungi, which are mostly multicellular.

According to one hypothesis, two separate unikont lineages led to fungi and animals. Based on molecular clock calculations, scientists estimate that the ancestors of animals and fungi diverged more than 1 billion years ago. A combination of morphological and molecular evidence suggests that a group of unikonts called *choanoflagellates* are the closest living protist relatives of animals. The bottom half of **Figure 21B** shows that the cells of choanoflagellates strongly resemble the "collar cells" with which sponges, the group that is closest to the root of the animal tree, obtain food. Similar cells have been found in other animals, but not in fungi or plants. Some species of choanoflagellates live as colonies, federations of independent cells sticking loosely together. Scientists hypothesize that the common ancestor of living animals may have been a stationary colonial choanoflagellate similar to the one shown in Figure 21B.

A different group of unikont protists is thought to have given rise to the fungi. Molecular evidence suggests that a group of single-celled protists called *nucleariids*, amoebas that feed on algae and bacteria, are the closest living relatives of fungi (top of Figure 21B).

A group of green algae called *charophytes* are the closest living relatives of land plants. Around 500 million years ago, the move onto land began, probably as green algae living along the edges of lakes gave rise to primitive plants.

? **In what way do multicellular organisms differ fundamentally from unicellular ones?**

In unicellular organisms, all the functions of life are carried out within a single cell. Multicellular organisms have specialized cells that perform different functions.

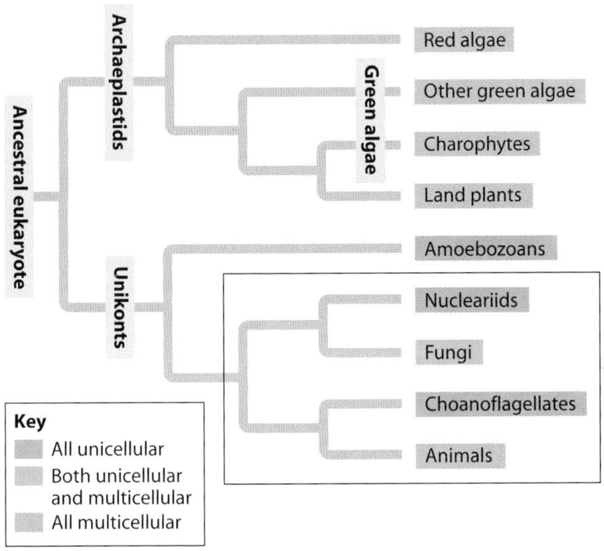

▲ **Figure 21A** A hypothesis for the phylogeny of plants, fungi, and animals

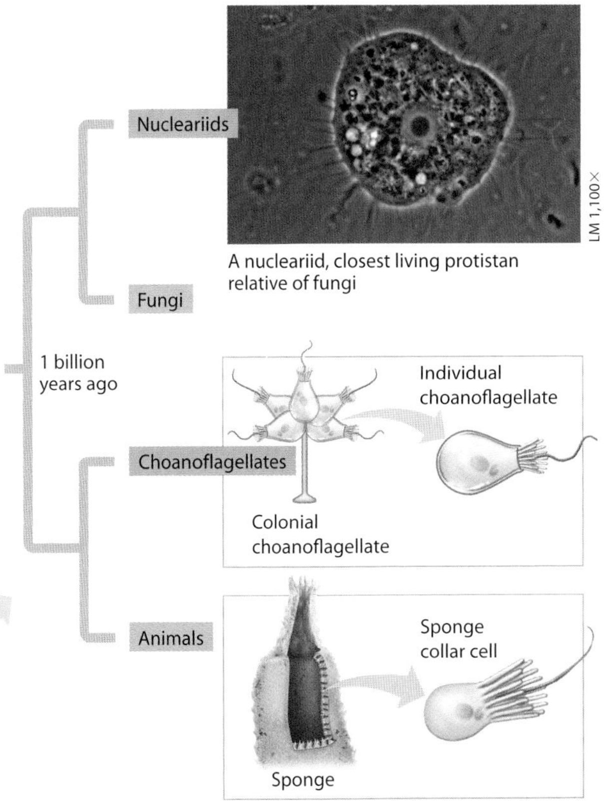

▲ **Figure 21B** The closest living protist relatives of fungi (top) and animals (bottom)

David Patterson

REVIEW

For Practice Quizzes, BioFlix, MP3 Tutors, and Activities, go to www.masteringbiology.com.

Reviewing the Concepts

Prokaryotes (1–12)

1 Prokaryotes are diverse and widespread. Prokaryotes are the most numerous organisms. Although small, they have an immense impact on the environment and on our own health.

2 External features contribute to the success of prokaryotes. Prokaryotes can be classified by shape and by reaction to a Gram stain. Almost all prokaryotes have a cell wall. Other external features may include a s ticky capsule, flagella that provide motility, or fimbriae that attach to surfaces. These external features enable prokaryotes to be successful in a wide range of environments.

3 Populations of prokaryotes can adapt rapidly to changes in the environment. Rapid prokaryote population growth by binary fission generates a great deal of genetic variation, increasing the likelihood that some members of the population will survive changes in the environment. Some prokaryotes form specialized cells called endospores that remain dormant through harsh conditions.

4 Prokaryotes have unparalleled nutritional diversity.

Nutritional mode	Energy source	Carbon source
Photoautotroph	Sunlight	CO_2
Chemoautotroph	Inorganic chemicals	
Photoheterotroph	Sunlight	Organic compounds
Chemoheterotroph	Organic compounds	

5 Biofilms are complex associations of microbes. Prokaryotes attach to surfaces and form biofilm communities that are difficult to eradicate, causing both medical and environmental problems.

6 Prokaryotes help clean up the environment. Prokaryotes are often used for bioremediation, including in sewage treatment facilities.

7 Bacteria and archaea are the two main branches of prokaryotic evolution.

8 Archaea thrive in extreme environments—and in other habitats. Domain Archaea includes extreme halophiles ("salt lovers"), extreme thermophiles ("heat lovers"), and methanogens that thrive in anaerobic conditions. They are also a major life-form in the ocean.

9 Bacteria include a diverse assemblage of prokaryotes. Domain Bacteria is currently organized into five major groups: proteobacteria, gram-positive bacteria, cyanobacteria, chlamydias, and spirochetes.

10 Some bacteria cause disease. Pathogenic bacteria often cause disease by producing exotoxins or endotoxins.

Exotoxin
Secreted by cell

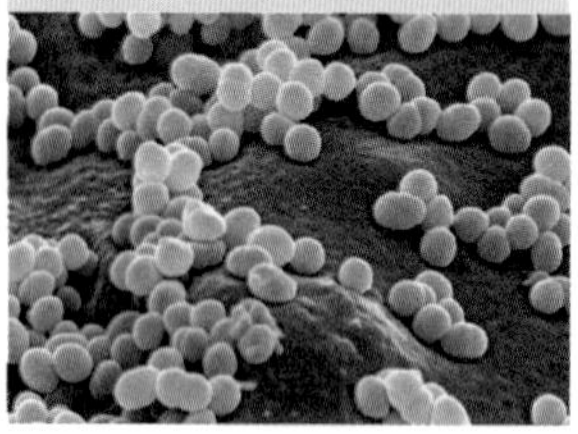

David M. Phillips/Photo Researchers
SEM 12,400×

Staphylococcus aureus

Endotoxin
Component of gram-negative plasma membrane

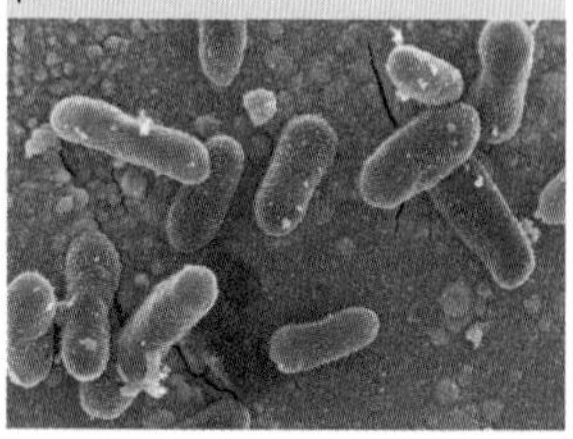

SEM 4,000×
Dr. Gary Gaugler/Photo Researchers

Salmonella enteritidis

11 Koch's postulates are used to prove that a bacterium causes a disease. Barry Marshall used them to show that peptic ulcers are usually caused by a bacterium, *Helicobacter pylori*.

12 Bacteria can be used as biological weapons. Bacteria, such as the species that cause anthrax and plague, and bacterial toxins, such as botulinum, can be used as biological weapons.

Protists (13–21)

13 Protists are an extremely diverse assortment of eukaryotes. Protists are mostly unicellular eukaryotes that are found in a variety of aquatic or moist habitats. They may be autotrophic, heterotrophic, or mixotrophic. Molecular systematists are exploring protistan phylogeny, but at present it is highly tentative.

14 Secondary endosymbiosis is the key to much of protist diversity. Endosymbiosis of prokaryotic cells resulted in the evolution of eukaryotic cells containing mitochondria. By a similar process, heterotrophic eukaryotic cells engulfed cyanobacteria, which became chloroplasts. Secondary endosymbiosis of eukaryotic cells by red and green algae gave rise to diverse lineages of protists.

15 Chromalveolates represent the range of protist diversity. Chromalveolates include diatoms, dinoflagellates, brown algae water molds, ciliates, and a group whose members are all parasites.

16 Can algae provide a renewable source of energy? Researchers are working on methods of growing diatoms and other algae as a source of biofuels.

17 Rhizarians include a variety of amoebas. Rhizarian amoebas have threadlike pseudopodia and include forams, which have calcium carbonate shells, and radiolarians, which have internal skeletons made of silica.

18 Some excavates have modified mitochondria. Some excavates are anaerobic protists that have modified mitochondria; they include parasitic *Giardia*, *Trichomonas vaginalis*, and *Trypanosomas*. Other excavates include *Euglena*, a mixotroph, and termite endosymbionts.

19 Unikonts include protists that are closely related to fungi and animals. Amoebozoans, the protistan unikonts, include amoebas with lobe-shaped pseudopodia, plasmodial slime molds, and cellular slime molds. Fungi and animals are also unikonts.

20 Archaeplastids include red algae, green algae, and land plants. Red algae, which are mostly multicellular, include species that contribute to the structure of coral reefs and species that are commercially valuable. Green algae may be unicellular, colonial, or multicellular. The life cycles of many algae involve the alternation of haploid gametophyte and diploid sporophyte generations. Archaeplastida also includes land plants, which arose from a group of green algae called charophytes.

21 Multicellularity evolved several times in eukaryotes. Multicellular organisms have cells specialized for different functions. Multicellularity evolved in ancestral lineages of chromalveolates (brown algae), unikonts (fungi and animals), and archaeplastids (red and green algae).

Connecting the Concepts

1. Explain how each of the following characteristics contributes to the success of prokaryotes: cell wall, capsule, flagella, fimbriae, endospores.

2. Fill in the blanks on the phylogenetic tree to show current hypotheses for the origin of multicellular organisms.

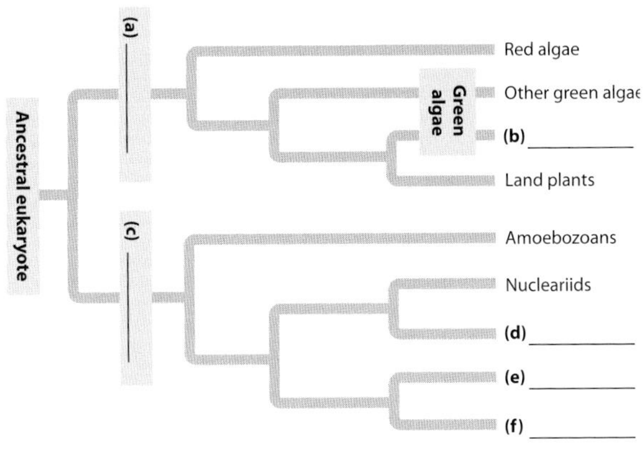

Testing Your Knowledge

Multiple Choice

3. In terms of nutrition, autotrophs are to heterotrophs as
   a. kelp are to diatoms.
   b. archaea are to bacteria.
   c. slime molds are to algae.
   d. algae are to slime molds.
   e. pathogenic bacteria are to harmless bacteria.

4. The bacteria that cause tetanus can be killed only by prolonged heating at temperatures considerably above boiling. This suggests that tetanus bacteria
   a. have cell walls containing peptidoglycan.
   b. protect themselves by secreting antibiotics.
   c. secrete endotoxins.
   d. are autotrophic.
   e. produce endospores.

5. Glycolysis is the only metabolic pathway common to nearly all organisms. To scientists, this suggests that it
   a. evolved many times during the history of life.
   b. was first seen in early eukaryotes.
   c. appeared early in the history of life.
   d. must be very complex.
   e. appeared rather recently in the evolution of life.

6. A new organism has been discovered. Tests have revealed that it is unicellular, is autotrophic, and has a cell wall that contains peptidoglycan. Based on this evidence, it should be classified as a(n)
   a. alga.
   b. archaean.
   c. diatom.
   d. bacterium.
   e. excavate.

7. Which pair of protists has support structures composed of silica?
   a. dinoflagellates and diatoms
   b. diatoms and radiolarians
   c. radiolarians and forams
   d. forams and amoebozoans
   e. amoebozoans and dinoflagellates

8. Which of the following groups is incorrectly paired with an example of that group?
   a. excavates—trypanosome causing sleeping sickness
   b. chromalveolates—parasites such as *Plasmodium*
   c. chromalveolates—brown algae
   d. unikonts—water mold
   e. archaeplastids—red algae

9. Which of the following prokaryotes is not pathogenic?
   a. *Chlamydia*
   b. *Rhizobium*
   c. *Streptococcus*
   d. *Salmonella*
   e. *Bacillus anthracis*

Describing, Comparing, and Explaining

10. Explain why prokaryote populations can adapt rapidly to changes in their environment.
11. Explain how protist diversity may have arisen through secondary endosymbiosis.
12. *Chlamydomonas* is a unicellular green alga. How does it differ from a photosynthetic bacterium, which is also single-celled? How does it differ from a protozoan, such as an amoeba? How does it differ from larger green algae, such as sea lettuce (*Ulva*)?
13. What characteristic distinguishes true multicellularity from colonies of cells?

Applying the Concepts

14. Imagine you are on a team designing a moon base that will be self-contained and self-sustaining. Once supplied with building materials, equipment, and organisms from Earth, the base will be expected to function indefinitely. One of the team members has suggested that everything sent to the base be sterilized so that no bacteria of any kind are present. Do you think this is a good idea? Predict some of the consequences of eliminating all bacteria from an environment.
15. The buildup of CO_2 in the atmosphere resulting from the burning of fossil fuels is regarded as a major contributor to global warming. Diatoms and other microscopic algae in the oceans counter this buildup by using large quantities of atmospheric CO_2 in photosynthesis, which requires small quantities of iron. Experts suspect that a shortage of iron may limit algal growth in the oceans. Some scientists have suggested that one way to reduce CO_2 buildup might be to fertilize the oceans with iron. The iron would stimulate algal growth and thus the removal of more CO_2 from the air. A single supertanker of iron dust, spread over a wide enough area, might reduce the atmospheric CO_2 level significantly. Do you think this approach would be worth a try? Why or why not?

Review Answers

1. Cell wall: maintains cell shape; provides physical protection; prevents cell from bursting in a hypotonic environment
   Capsule: enables cell to stick to substrate or to other individuals in a colony; shields pathogens from host's defensive cells
   Flagella: provide motility, enabling cell to respond to chemical or physical signals in the environment that lead to nutrients or other members of their species and away from toxic substances
   Fimbriae: allow cells to attach to surfaces, including host cells, or to each other
   Endospores: withstand harsh conditions
2. a. Archaeplastids; b. Charophytes; c. Unikonts; d. Fungi; e. Choanoflagellates; f. Animals
3. d (Algae are autotrophs; slime molds are heterotrophs.) 4. e 5. c 6. d 7. b 8. d 9. b
10. Rapid rate of reproduction enables prokaryotes to colonize favorable habitats quickly. The production of large numbers of cells by binary fission results in a great deal of genetic variation, making it more likely that some individuals will survive—and be able to recolonize the habitat—if the environment changes again.
11. Small, free-living prokaryotes were probably engulfed by a larger cell and took up residence inside. A symbiotic relationship developed between the host cell and engulfed cells, which became mitochondria. By a similar process, heterotrophic eukaryotic cells engulfed cyanobacteria, which became chloroplasts. Lineages of these autotrophic cells diverged into red and green algae. Secondary endosymbiosis of eukaryotic cells by red and green algae gave rise to diverse lineages of protists.
12. *Chlamydomonas* is a eukaryotic cell, much more complex than a prokaryotic bacterium. It is autotrophic, while amoebas are heterotrophic. It is unicellular, unlike multicellular sea lettuce.
13. Multicellular organisms have a greater extent of cellular specialization and more interdependence of cells. New organisms are produced from a single cell, either an egg or an asexual spore.
14. Not a good idea; all life depends on bacteria. You could predict that eliminating all bacteria from an environment would result in a buildup of toxic wastes and dead organisms (both of which bacteria decompose), a shutdown of all chemical cycling, and the consequent death of all organisms.
15. Some issues and questions to consider: Could we determine beforehand whether the iron would really have the desired effect? How? Would the "fertilization" need to be repeated? Could it be a cure for the problem, or would it merely treat the symptoms? Might the iron treatment have side effects? What might they be?

The Evolution of Plant and Fungal Diversity

From Chapter 17 of *Campbell Biology: Concepts & Connections*, Seventh Edition. Jane B. Reece, Martha R. Taylor, Eric J. Simon, Jean L. Dickey.

The Evolution of Plant and Fungal Diversity

BIG IDEAS

BlueMoon Stock/Alamy

Plant Evolution and Diversity
(1–2)

A variety of adaptations enable plants to live on land.

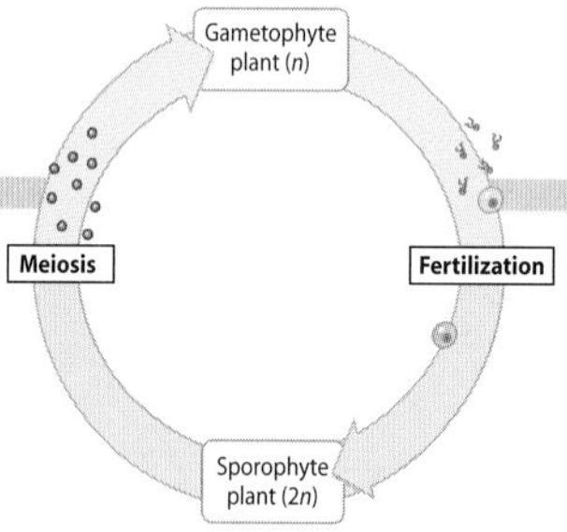

Alternation of Generations and Plant Life Cycles
(3–13)

Plant life cycles alternate haploid (gametophyte) and diploid (sporophyte) generations.

Diversity of Fungi
(14–21)

Fungi are a diverse group of organisms that acquire nutrients through absorption. Many fungi have complex life cycles.

J S Sira/age fotostock

To an insect, the lovely plants in the photo might appear to offer an inviting platform for resting in the warm sun. Woe to the insect that tries it! The plant's name, Venus flytrap, is fitting. Shortly after the insect touches down, the leaves will snap shut and seal themselves around the victim. Digestive juices will begin to eat away at the insect's body, until all that remains is the prey's tough exoskeleton. Finally, the leaves will reabsorb the fluids and relax their grasp, allowing the prey's ghostly remains to blow away.

More than 600 species of plants are carnivores—they feed on animals. Although most are small and consume only insects, some carnivorous plants are large enough to capture frogs and even small mammals. If you are intrigued by these "meat-eaters," you have something in common with Charles Darwin, who published a book about them. Through his investigations, Darwin answered a question that you may also have: Why would plants have adaptations to entrap and digest animals? The surprising answer is that carnivorous plants absorb and use nutrients from this food, just as we do. Animals are a good source of protein, a nitrogen-rich macromolecule. Thus, carnivory enables plants to thrive in poor soil, especially where nitrogen is scarce.

As you'll learn in this chapter, evolution in the plant kingdom began with structural and reproductive adaptations for life on land. From a modest beginning, plants have diversified into 290,000 present-day species with adaptations that enable them to live in all kinds of environmental conditions. However, it is unlikely that plants would be so successful without the fungal partners that help them acquire vital mineral nutrients from the soil. Thus, our exploration of plant evolution in this chapter is accompanied by a look at the diversity of their hidden allies, the fungi.

Plant Evolution and Diversity

1 Plants have adaptations for life on land

Plants and green algae called charophytes are thought to have evolved from a common ancestor. Like plants, charophytes are photosynthetic eukaryotes, and many species have complex, multicellular bodies (**Figure 1A**). In this module, you will learn about the characteristics that distinguish plants from their algal relatives. As you will see, many of the adaptations that evolved after plants diverged from algae facilitated survival and reproduction on dry land. Some plant groups, including water weeds such as *Anacharis* that are used in aquariums, returned to aquatic habitats during their evolution. However, most present-day plants live in terrestrial environments.

▲ **Figure 1A** *Chara*, an elaborate charophyte

The algal ancestors of plants may have carpeted moist fringes of lakes or coastal salt marshes over 500 million years ago. These shallow-water habitats were subject to occasional drying, and natural selection would have favored algae that could survive periodic droughts. Some species accumulated adaptations that enabled them to live permanently above the water line. The modern-day green alga *Coleochaete* (**Figure 1B**), which grows at the edges of lakes as disklike, multicellular colonies, may resemble one of these early plant ancestors.

Adaptations making life on dry land possible had accumulated by about 475 million years ago, the age of the oldest known plant fossils. The evolutionary novelties of these first land plants opened the new frontier of a terrestrial habitat. Early plant life would have thrived in the new environment. Bright sunlight was virtually limitless on land; the atmosphere had an abundance of carbon dioxide (CO_2); and at first there were relatively few pathogens and plant-eating animals.

Linda Graham, University of Wisconsin-Madison

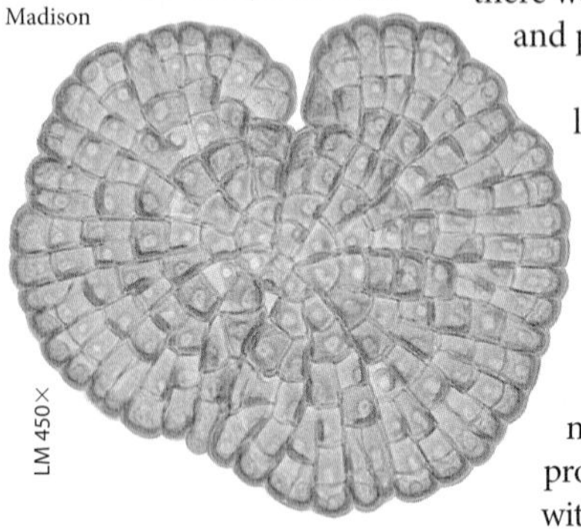

▲ **Figure 1B** *Coleochaete*, a simple charophyte

On the other hand, life on land poses a number of challenges. Because terrestrial organisms are surrounded by air rather than water, they must be able to maintain moisture inside their cells, support the body in a nonbuoyant medium, and reproduce and disperse offspring without water. As nonmotile organisms, plants must also anchor their bodies in the soil and obtain resources from both soil and air. Thus, the water-to-land transition required fundamental changes in algal structure and life cycle.

Figure 1C compares how plants such as moss, ferns, and pines differ from multicellular algae like *Chara*. Many algae are anchored by a holdfast, but generally they have no rigid tissues and are supported by the surrounding water. The whole algal body obtains CO_2 and minerals directly from the water. Almost all of the organism receives light and can perform photosynthesis. For reproduction, flagellated sperm swim to fertilize an egg. The offspring are dispersed by water as well.

Maintaining Moisture The aboveground parts of most land plants are covered by a waxy cuticle that prevents water loss. Gas exchange cannot occur directly through the cuticle, but CO_2 and O_2 diffuse across the leaf surfaces through the tiny pores called stomata. Two surrounding cells regulate each stoma's opening and closing. Stomata are usually open during sunlight hours, allowing gas exchange, and closed at other times, preventing water loss by evaporation.

Obtaining Resources from Two Locations A typical plant must obtain chemicals from both soil and air, two very different media. Water and mineral nutrients are mainly found in the soil; light and CO_2 are available above ground. Most plants have discrete organs—roots, stems, and leaves—that help meet this resource challenge.

Plant roots provide anchorage and absorb water and mineral nutrients from the soil. Above ground, a plant's stems bear leaves, which obtain CO_2 from the air and light from the sun, enabling them to perform photosynthesis. Growth-producing regions of cell division, called **apical meristems**, are found near the tips of stems and roots. The elongation and branching of a plant's stems and roots maximize exposure to the resources in the soil and air.

A plant must be able to connect its subterranean and aerial parts, conducting water and minerals upward from its roots to its leaves and distributing sugars produced in the leaves throughout its body. Most plants, including ferns, pines, and flowering plants, have **vascular tissue**, a network of thick-walled cells joined into narrow tubes that extend throughout the plant body (traced in red in Figure 1C). The photograph of part of an aspen leaf in **Figure 1D**, at the bottom of the next page, shows the leaf's network of veins, which are fine branches of the vascular tissue. There are two types of vascular tissue. **Xylem** includes dead cells that form microscopic pipes conveying water and minerals up from the roots. **Phloem**, which consists entirely of living cells, distributes sugars throughout the plant. In contrast to plants with elaborate vascular tissues, mosses lack a complex transport system (although some mosses do have simple vascular tissue). With limited means for distributing water and minerals from soil

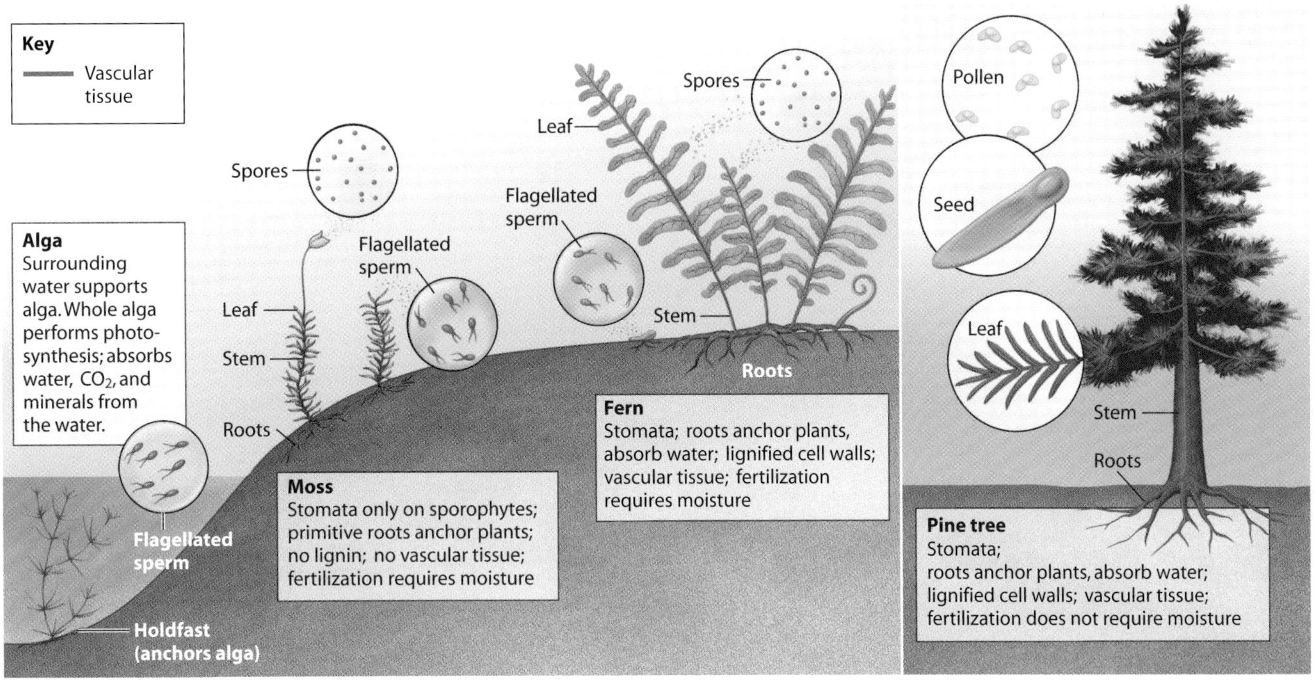

▲ **Figure 1C** Comparing the terrestrial adaptations of moss, fern, and pine with *Chara*, a multicellular green alga

to the leaves, the height of nonvascular plants is severely restricted.

Supporting the Plant Body Because air provides much less support than water, plants must be able to hold themselves up against the pull of gravity. The cell walls of some plant tissues, including xylem, are thickened and reinforced by a chemical called **lignin.** The absence of lignified cell walls in mosses and other plants that lack vascular tissue is another limitation on their height.

Reproduction and Dispersal Reproduction on land presents complex challenges. For *Chara* and other algae, the surrounding water ensures that gametes and offspring stay moist. Plants, however, must keep their gametes and developing embryos from drying out in the air. Like the earliest land plants, mosses and ferns still produce gametes in male and female structures called **gametangia** (singular, *gametangium*), which consist of protective jackets of cells surrounding the gamete-producing cells. The egg remains in the female gametangium and is fertilized there by a sperm that swims through a film of water. As a result, mosses and ferns can only reproduce in a moist environment. Pines and flowering plants have **pollen grains**, structures that contain the sperm-producing cells. Pollen grains are carried close to the egg by wind or animals; moisture is not required for bringing sperm and egg together.

In all plants, the fertilized egg (zygote) develops into an embryo while attached to and nourished by the parent plant. This multicellular, dependent embryo is the basis for designating plants as **embryophytes** (*phyte* means "plant"), distinguishing them from algae.

The life cycles of all plants involve the alternation of a haploid generation, which produces eggs and sperm, and a diploid generation, which produces spores within protective structures called **sporangia** (singular, *sporangium*). A **spore** is a cell that can develop into a new organism without fusing with another cell. The earliest land plants relied on tough-walled spores for dispersal, a trait retained by mosses and ferns today. Pines and flowering plants have seeds for launching their offspring. Seeds are elaborate embryo-containing structures that are well protected from the elements and are dispersed by wind or animals (see Module 10). To emphasize this important reproductive difference, plants that disperse their offspring as spores are often referred to as seedless plants. Now let's take a closer look at the evolution of the major plant groups.

? What adaptations enable plants to grow tall?

Vascular tissues to transport materials from belowground parts to aboveground parts and vice versa; lignified cell walls to provide structural support

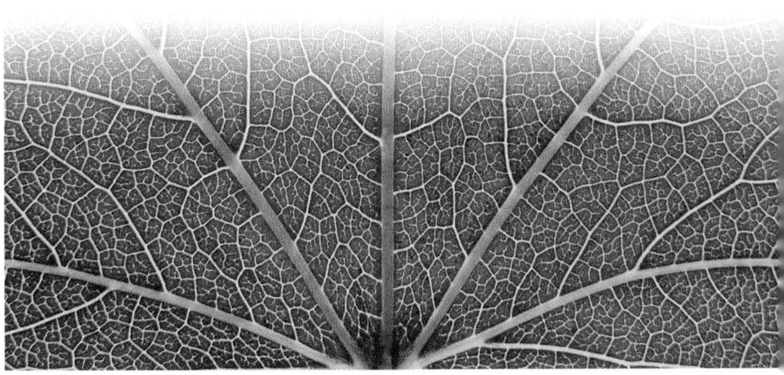

Corbis

▲ **Figure 1D** The network of veins in a leaf

2 Plant diversity reflects the evolutionary history of the plant kingdom

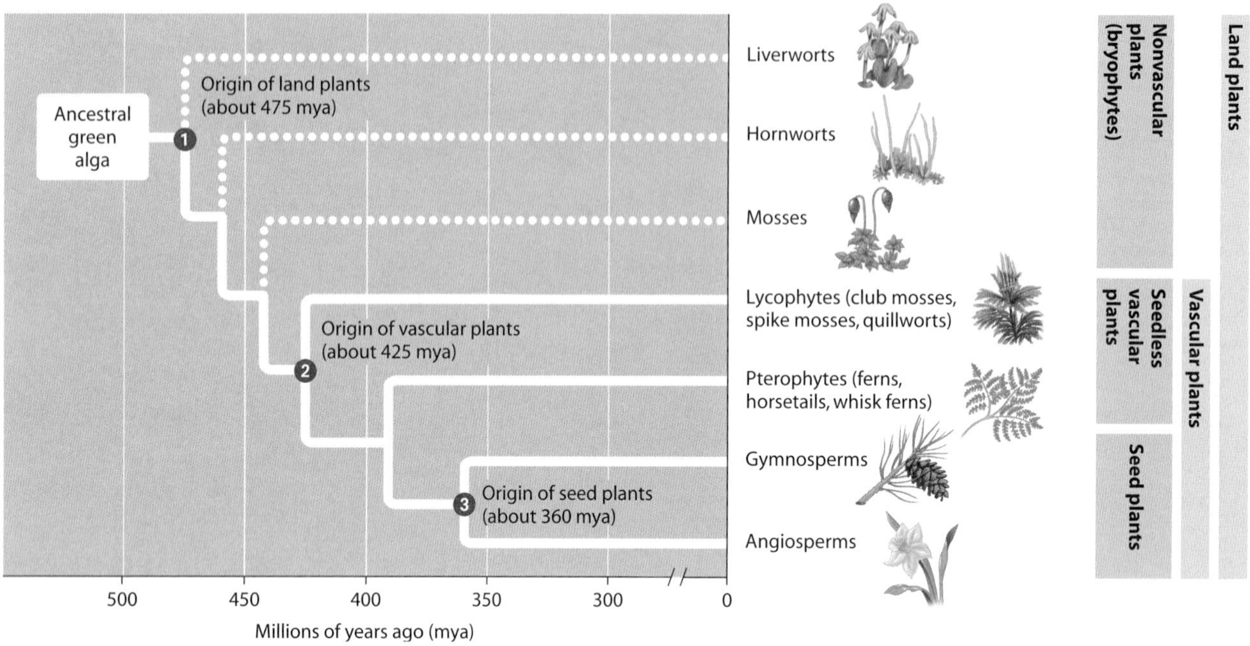

▲ **Figure 2A** Some highlights of plant evolution (Dotted lines indicate uncertain evolutionary relationships.)

Figure 2A highlights some of the major events in the history of the plant kingdom and presents a widely held view of the relationships between surviving lineages of plants.

❶ After plants originated from an algal ancestor approximately 475 million years ago, early diversification gave rise to seedless, nonvascular plants, including mosses, liverworts, and hornworts **(Figure 2B)**. These plants, which are informally called **bryophytes**, resemble other plants in having apical meristems and embryos that are retained on the parent plant, but they lack true roots and leaves. Without lignified cell walls, bryophytes with an upright growth habit lack support. A mat of moss actually consists of many plants growing in a tight pack. Like people crowded together in a small space, their bodies hold one another upright. The mat is spongy and retains water. Other bryophytes grow flat against the ground. Growth in dense mats facilitates fertilization by flagellated sperm swimming through a film of water left by rain or dew.

❷ The origin of **vascular plants** occurred about 425 million years ago. Their lignin-hardened vascular tissues provide strong support, enabling stems to stand upright and grow tall on land. Two clades of vascular plants are informally called **seedless vascular plants** (**Figure 2C**, next page): the lycophytes (such as club mosses) and the widespread pterophytes (ferns and their relatives). A fern has well-developed roots and rigid stems. Ferns are common in temperate forests, but they are most diverse in the tropics. In some tropical species, called tree ferns, upright stems can grow several meters tall. Like bryophytes, however, ferns and club mosses require moist conditions for fertilization, and they disperse their offspring as spores that are carried by air currents.

❸ The first vascular plants with seeds evolved about 360 million years ago. Today, the seed plant lineage accounts for over 90% of the approximately 290,000 species of living plants. Seeds and pollen are key adaptations that improved the ability of plants to diversify in terrestrial habitats. A **seed**

Tony Wharton/Frank Lane Picture Agency/Corbis

Moss

Dr. Jeremy Burgess/Photo Researchers

Liverwort

Hornwort

Hidden Forest

▲ **Figure 2B** Bryophytes

Colin Walton/Dorling Kindersley

Fern (a pterophyte)

◀ **Figure 2C** Seedless vascular plants

Ed Reschke/Photolibrary

Club moss (a lycophyte). Spores are produced in the upright tan-colored structures.

Koos Van Der Lende/AGE fotostock

A jacaranda tree

Terry W. Eggers/ Corbis

Barley, a grass

▲ **Figure 2E** Angiosperms

consists of an embryo packaged with a food supply within a protective covering. This survival packet facilitates wide dispersal of plant embryos (see Module 10). As you learned in Module 1, pollen brings sperm-producing cells into contact with egg-producing parts without water. And unlike flagellated sperm, which can swim a few centimeters at most, pollen can travel great distances.

Gymnosperms (from the Greek *gymnos*, naked, and *sperma*, seed) were among the earliest seed plants. Seeds of gymnosperms are said to be "naked" because they are not produced in specialized chambers. The largest clade of gymnosperms is the conifers, consisting mainly of cone-bearing trees, such as pine, spruce, and fir. (The term *conifer* means "cone-bearing.") Some examples of gymnosperms that are less common are the ornamental ginkgo tree, desert shrubs in the genus *Ephedra*, and the palmlike cycads **(Figure 2D)**. Gymnosperms flourished alongside the dinosaurs in the Mesozoic era.

The most recent major episode in plant evolution was the appearance of flowering plants, or **angiosperms** (from the Greek *angion*, container, and *sperma*, seed), at least 140 million years ago. Flowers are complex reproductive structures that develop seeds within protective chambers. The great majority of living plants—some 250,000 species—are angiosperms, which include a wide variety of plants, such as grasses, flowering shrubs, and flowering trees **(Figure 2E)**.

In summary, four key adaptations for life on land distinguish the main lineages of the plant kingdom. (1) Dependent embryos are present in all plants. (2) Lignified vascular tissues mark a lineage that gave rise to most living plants. (3) Seeds are found in a lineage that includes all living gymnosperms and angiosperms and that dominates the plant kingdom today. (4) Flowers mark the angiosperm lineage. As you will see in the next several modules, the life cycles of living plants reveal additional details about plant evolution.

? **Identify which of the following traits is shared by all plants: flowers, seeds, retained embryo, vascular tissue.**

An embryo retained on parent plant

▼ **Figure 2D** Gymnosperms

blickwinkel/Alamy

Ginkgo

John Cancalosi/Photolibrary

Cycad

Bob Gibbons/Alamy

Ephedra (Mormon tea)

Pichugin Dmitry/shutterstock

A conifer

Alternation of Generations and Plant Life Cycles

3 Haploid and diploid generations alternate in plant life cycles

Plants have life cycles that are very different from ours. Humans are diploid individuals. That is, each of us has two sets of chromosomes, one from each parent. The only haploid stages in the human life cycle are sperm and eggs. Plants have an **alternation of generations**: The diploid and haploid stages are distinct, multicellular bodies. The haploid generation of a plant produces gametes and is called the **gametophyte**. The diploid generation produces spores and is called the **sporophyte**. In a plant's life cycle, these two generations alternate in producing each other.

The life cycles of all plants follow the pattern shown in **Figure 3**. Be sure that you understand this diagram; then review it after studying each life cycle to see how the pattern applies. Starting at the top of the figure, a haploid gametophyte plant produces gametes (sperm and eggs) by mitosis—not by meiosis, as in animals. Fertilization results in a diploid zygote. The zygote grows by mitotic cell division and develops into the multicellular diploid sporophyte plant. The sporophyte produces haploid spores by meiosis. A spore then develops by mitosis into a multicellular haploid gametophyte.

Although some algae exhibit alternation of generations, the closest algal relatives of plants, the charophytes, do not. Thus, this life cycle appears to have evolved independently in plants; it is a derived character of land plants.

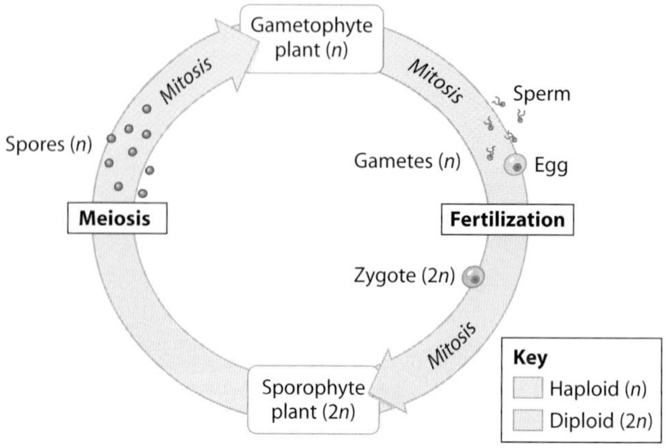

▲ **Figure 3** Alternation of generations

The next module highlights the life cycle of mosses. In mosses, as in all nonvascular plants, the gametophyte is the larger, more obvious stage of the life cycle.

? How does gamete production differ in plants and animals?

Animals produce gametes by meiosis; in plants, gametes are produced by mitosis.

4 The life cycle of a moss is dominated by the gametophyte (See text at top of next page.)

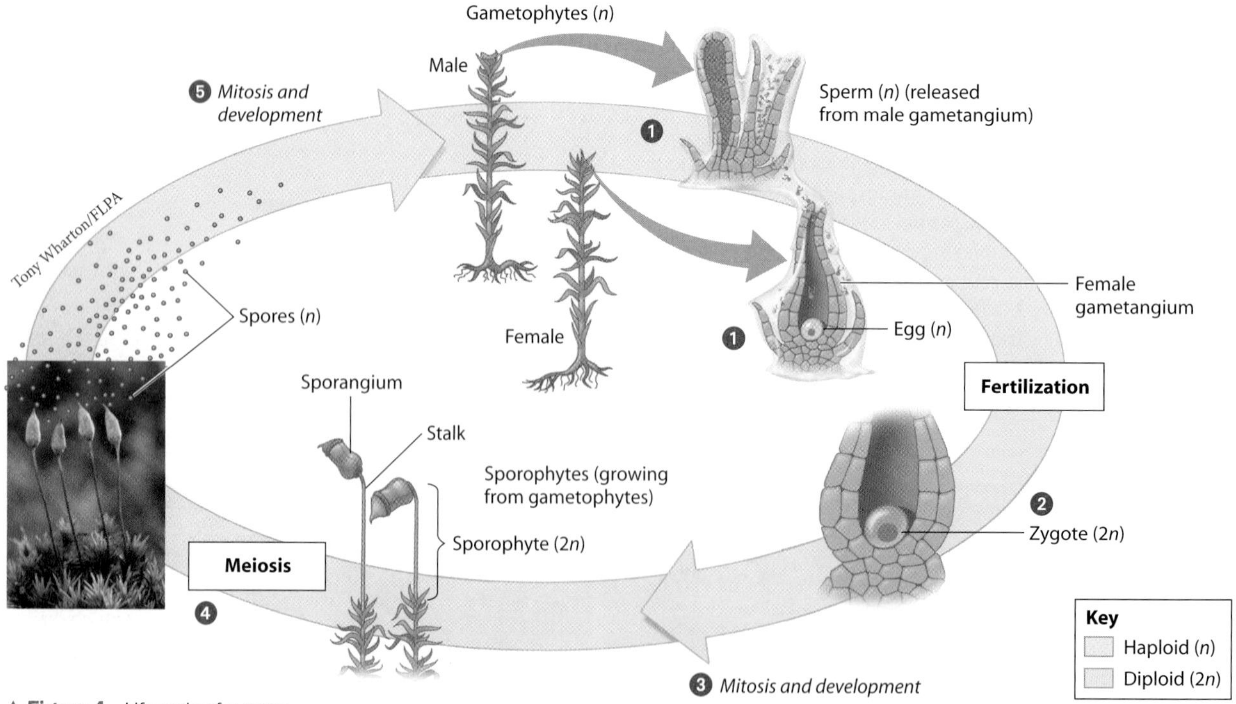

▲ **Figure 4** Life cycle of a moss

In a moss, the green, cushiony growth we see consists of gametophytes. Follow the moss life cycle in **Figure 4** at the bottom of the previous page. ❶ Gametes develop in male and female gametangia, usually on separate plants. The flagellated sperm are released from the male gametangium and swim through a film of water to the egg in the female gametangium. After fertilization, ❷ the zygote remains in the gametangium. ❸ There it divides by mitosis, developing into a sporophyte embryo and then a mature sporophyte, which remains attached to the gametophyte. ❹ Meiosis occurs in the sporangia at the tips of the sporophyte stalks. After meiosis, haploid spores are released from the sporangium. ❺ The spores undergo mitosis and develop into gametophyte plants.

? How do moss sperm travel from male gametangia to female gametangia, where fertilization of eggs occurs?

The flagellated sperm swim through a film of water.

5 Ferns, like most plants, have a life cycle dominated by the sporophyte

All we usually see of a fern is the sporophyte. But let's start with the gametophyte in the fern life cycle in **Figure 5**. ❶ Fern gametophytes often have a distinctive heart-like shape, but they are quite small (about 0.5 cm across) and inconspicuous. ❷ Like mosses, ferns have flagellated sperm that require moisture to reach an egg. Although eggs and sperm are usually produced in separate locations on the same gametophyte, a variety of mechanisms promote cross-fertilization between gametophytes. ❸ The zygote remains on the gametophyte as it develops into a new sporophyte. ❹ Unlike moss, the gametophyte dies and the sporophyte becomes an independent plant. ❺ The black dots in the photograph are clusters of sporangia, in which cells undergo meiosis, producing haploid spores. The spores are released and develop into gametophytes by mitosis.

Today, about 95% of all plants, including all seed plants, have a dominant sporophyte in their life cycle. But before we resume this story, let's look back at a time in plant history, before seed plants rose to dominance, when ferns and other seedless vascular plants covered much of the land surface.

? What is the major difference between the moss and fern life cycles?

In mosses, the dominant plant body is the gametophyte. In ferns, the sporophyte is dominant and is independent of the gametophyte.

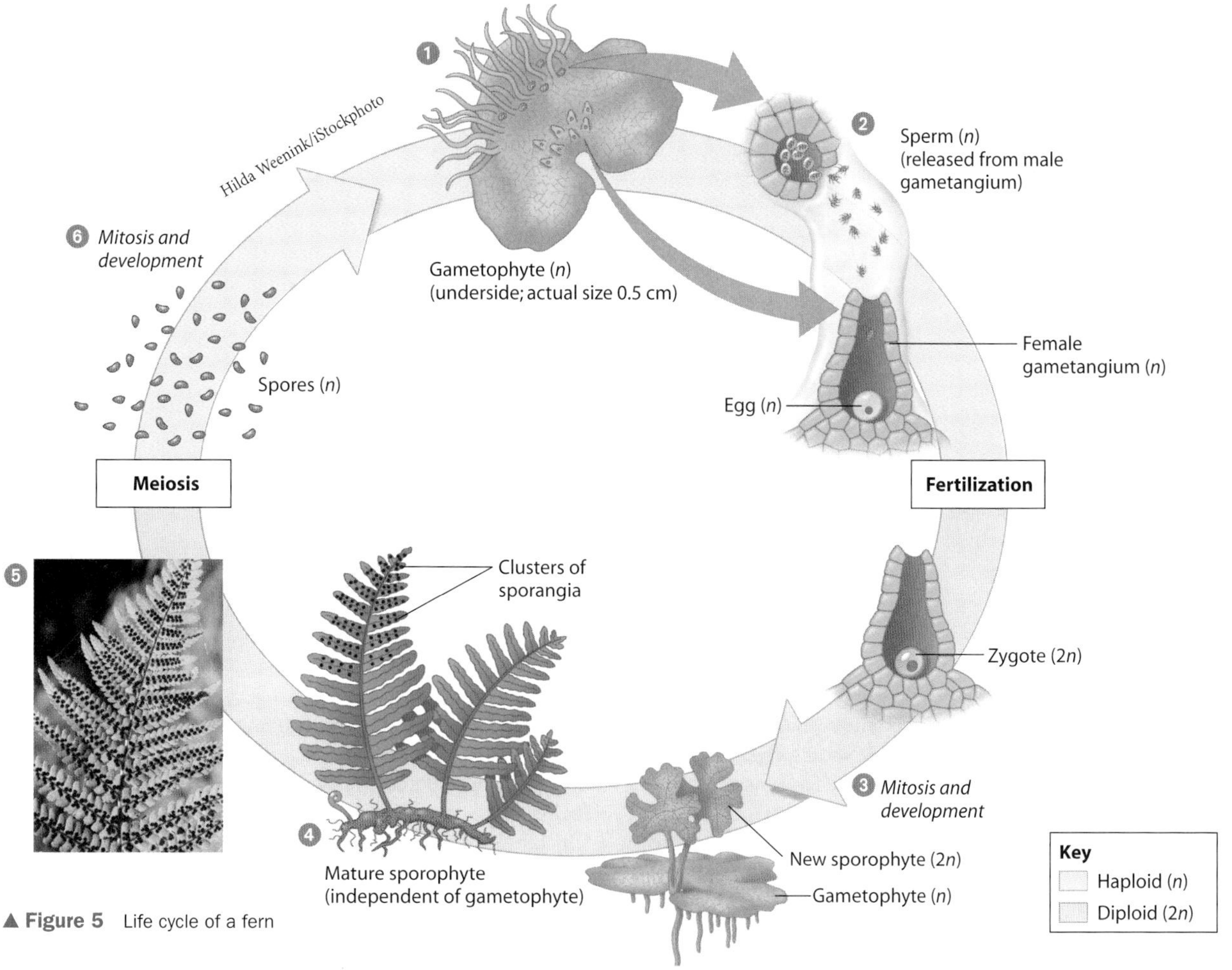

▲ **Figure 5** Life cycle of a fern

6 Seedless vascular plants dominated vast "coal forests"

During the Carboniferous period (about 360–299 million years ago), the two clades of seedless vascular plants, lycophytes (for example, club mosses) and pterophytes (such as ferns and their relatives), grew in vast forests in the low-lying wetlands of what is now Eurasia and North America. At that time, these continents were close to the equator and had warm, humid climates that supported broad expanses of lush vegetation.

Figure 6 shows an artist's reconstruction of one of these forests based on fossil evidence. Tree ferns are visible in the foreground. Most of the large trees are lycophytes, giants that grew as tall as a 12-story building, with diameters of more than 2 m (6 feet). (For a sense of scale, dragonflies such as the one shown in the foreground had wing spans of up to 1 m.) Vertebrates were adapting to terrestrial habitats in parallel with plants; amphibians and early reptiles lived among these trees.

Photosynthesis in these immense swamp forests fixed large amounts of carbon from CO_2 into organic molecules, dramatically reducing CO_2 levels in the atmosphere. Because atmospheric CO_2 traps heat, this change caused global cooling. Photosynthesis generated great quantities of organic matter. As the plants died, they fell into stagnant swamps and did not decay completely. Their remains formed thick organic deposits called peat. Later, seawater covered the swamps, marine sediments covered the peat, and pressure and heat gradually converted the peat to coal, black sedimentary rock made up of fossilized plant material. Coal deposits from the Carboniferous period are the most extensive ever formed. (The name Carboniferous comes from the Latin *carbo*, coal, and *fer*, bearing.) Coal, oil, and natural gas are called **fossil fuels** because they were formed from the remains of ancient organisms. (Oil and natural gas were formed from marine organisms.) Since the Industrial Revolution, coal has been a crucial source of energy for human society. However, burning these fossil fuels releases CO_2 and other greenhouse gases into the atmosphere, which are now causing a warming climate.

As temperatures dropped during the late Carboniferous period, glaciers formed. The global climate turned drier, and the vast swamps and forests began to disappear. The climate change provided an opportunity for the early seed plants, which grew along with the seedless plants in the Carboniferous swamps. With their wind-dispersed pollen and protective seeds, seed plants could complete their life cycles on dry land.

▲ **Figure 6** A reconstruction of an extinct forest dominated by seedless plants

? How did the tropical swamp forests contribute to global cooling in the Carboniferous period?

Photosynthesis by the abundant plant life reduced atmospheric CO_2, a gas that traps heat.

7 A pine tree is a sporophyte with gametophytes in its cones

In seed plants, a specialized structure within the sporophyte houses all reproductive stages, including spores, eggs, sperm, zygotes, and embryos. In gymnosperms such as pines and other conifers, this structure is called a cone. If you look at the longitudinal section of the cones in **Figure 7**, you'll see that the cone resembles a short stem bearing thick leaves. The resemblance is not surprising—cones are shoots that were modified by natural selection to serve a different function. Each "leaf," or scale, of the cone contains sporangia that produce spores by meiosis. Unlike seedless plants, however, the spores are not released. Rather, spores give rise to gametophytes within the shelter of the cone. The gametophytes later produce gametes, which unite to form a new sporophyte.

A pine tree bears two types of cones, which produce spores that develop into male and female gametophytes. The smaller ones, called pollen cones (❶ in Figure 7), produce the male gametophytes. Pollen cones contain many sporangia, each of which makes numerous haploid spores by meiosis. Male gametophytes, or pollen grains, develop from the spores. Mature pollen cones release millions of microscopic pollen grains in great clouds. You may have seen yellowish conifer pollen covering cars or floating on puddles after a spring rain. Pollen grains, which are carried by the wind, house cells that will develop into sperm if by chance they land on a cone that contains a female gametophyte.

❷ An ovulate cone, which produces the female gametophyte, is larger than a pollen cone. Each of its stiff scales bears a pair of **ovules**. (Only one is shown in the diagram.) Each ovule contains a sporangium surrounded by a protective covering called the integument. ❸ **Pollination** occurs when a pollen grain lands on an ovulate scale and enters an ovule. After pollination, the scales of the cone grow together, sealing it until the seeds are mature. Meiosis then occurs in a spore "mother cell" within the ovule. Over the course of many months, ❹ one surviving haploid spore cell develops into the female gametophyte, which makes eggs. ❺ Meanwhile, a tiny tube grows out of each pollen grain, digests its way through the ovule, and eventually releases sperm, which are simply nuclei rather than independent cells, near an egg. Fertilization typically occurs more than a year after pollination.

❻ Usually, all eggs in an ovule are fertilized, but ordinarily only one zygote develops fully into a sporophyte embryo. ❼ The ovule matures into a seed, which contains the embryo's food supply (the remains of the female gametophyte) and has a tough **seed coat** (the ovule's integument). In a typical pine, seeds are shed about two years after pollination. ❽ The seed is dispersed by wind, and when conditions are favorable, it germinates, and its embryo grows into a pine seedling.

In summary, all the reproductive stages of conifers are housed in cones borne on sporophytes. The ovule is a key adaptation—a protective device for all the female stages in the life cycle, as well as the site of pollination, fertilization, and embryonic development. The ovule becomes the seed, an important terrestrial adaptation and a major factor in the success of the conifers and flowering plants.

Next we consider flowering plants, the most diverse and geographically widespread of all plants. Angiosperms dominate most landscapes today, and it is their flowers that account for their unparalleled success.

? What key adaptations have contributed to the overwhelming success of seed plants?

Pollen transfers sperm to eggs without the need for water. Seeds protect, nourish, and help disperse plant embryos. The female gametophytes of seed plants are protected on the sporophyte plant.

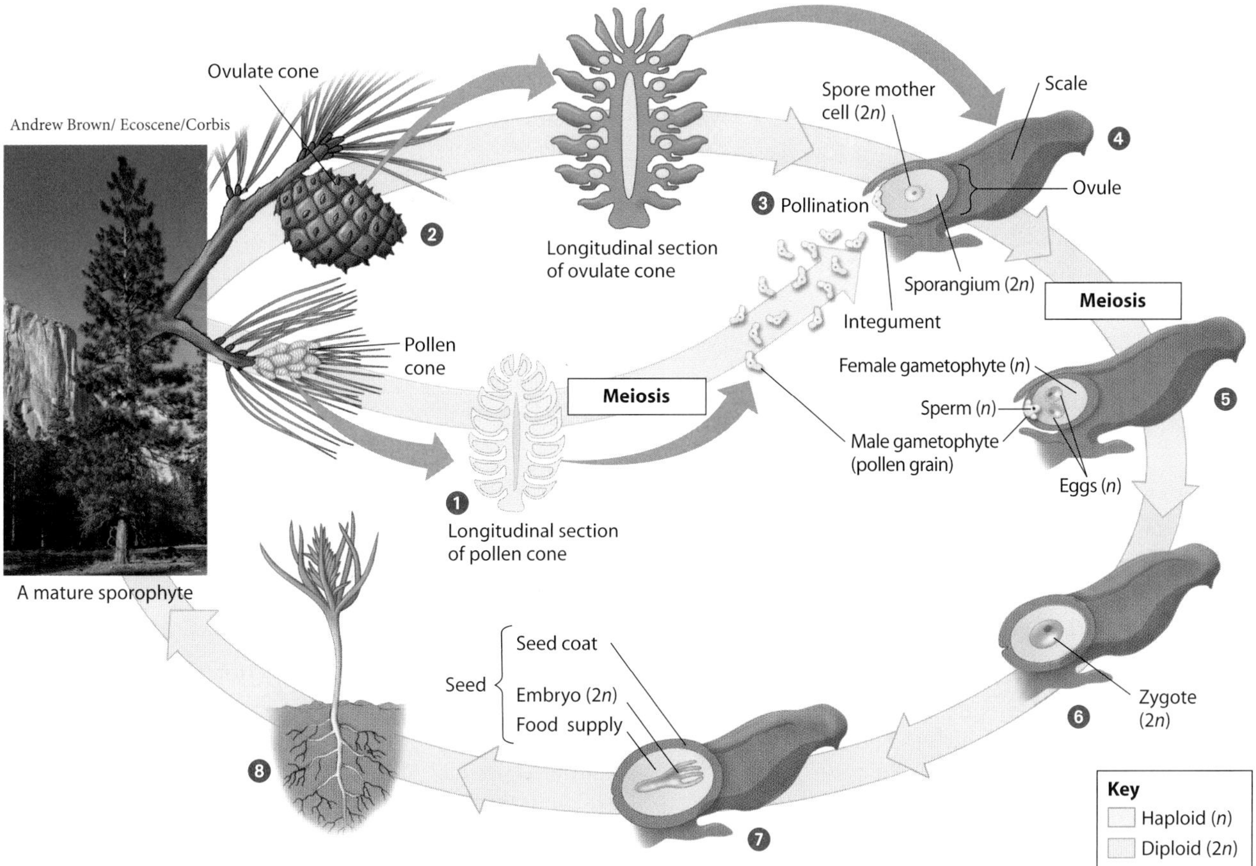

▲ **Figure 7** Life cycle of a pine tree

8 The flower is the centerpiece of angiosperm reproduction

No organisms make a showier display of their sex life than angiosperms (**Figure 8A**). From roses to cherry blossoms, flowers are the sites of pollination and fertilization. Like pine cones, flowers house separate male and female sporangia and gametophytes, and the mechanisms of sexual reproduction, including pollination and fertilization, are similar. And like cones, flowers are also short stems bearing modified leaves. However, as you can see in **Figure 8B**, the modifications are quite different from the scales of a pine cone. Each floral structure is highly specialized for a different function, and the structures are attached in a circle to a receptacle at the base of the flower. The outer layer of the circle consists of the **sepals**, which are usually green. They enclose the flower before it opens. When the sepals are peeled away, the next layer is the **petals**, which are conspicuous and attract animal pollinators. As we'll explore further in Module 12, showy petals are a major reason for the overwhelming success of angiosperms.

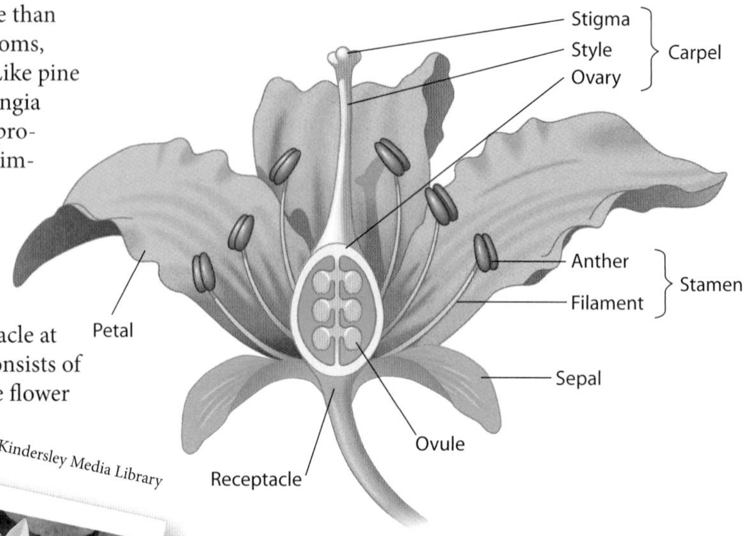

▲ **Figure 8B** The parts of a flower

▲ **Figure 8A** Some examples of floral diversity

Plucking off a flower's petals reveals the filaments of the **stamens**. The **anther**, a sac at the top of each filament, contains male sporangia and will eventually release pollen. At the center of the flower is the **carpel**, the female reproductive structure. It includes the **ovary**, a unique angiosperm adaptation that encloses the ovules. If you cut open the ovary of a flower, you can see its white, egg-shaped ovules. As in pines, each ovule contains a sporangium that will produce a female gametophyte and eventually become a seed. The ovary matures into a fruit, which aids in seed dispersal, as we'll discuss shortly. In the next module, you will learn how the alternation of generations life cycle proceeds in angiosperms.

? Where are the male and female sporangia located in flowering plants?

Anthers contain the male sporangia; ovules contain the female sporangia.

9 The angiosperm plant is a sporophyte with gametophytes in its flowers

In angiosperms, as in gymnosperms, the sporophyte generation is dominant and produces the gametophyte generation within its body. **Figure 9** illustrates the life cycle of a flowering plant and highlights features that have been especially important in angiosperm evolution. Starting at the "Meiosis" box at the top of Figure 9, ❶ meiosis in the anthers of the flower produces haploid spores that undergo mitosis and form the male gametophytes, or pollen grains. ❷ Meiosis in the ovule produces a haploid spore that undergoes mitosis and forms the few cells of the female gametophyte, one of which becomes an egg. ③ Pollination occurs when a pollen grain, carried by the wind or an animal, lands on the stigma. As in gymnosperms, a tube grows from the pollen grain to the ovule, and a sperm fertilizes the egg, ④ forming a zygote. Also as in gymnosperms, ⑤ a seed develops from each ovule. Each seed consists of an embryo (a new sporophyte) surrounded by a food supply and a seed coat derived from the integuments. While the seeds develop, ⑥ the ovary's wall thickens, forming the fruit that encloses the seeds. When conditions are favorable, ⑦ a seed germinates. As the embryo begins to grow, it uses the food supply from the seed until it can begin to photosynthesize. Eventually, it develops into a mature sporophyte plant, completing the life cycle.

The evolution of flowers that attract animals, which carry pollen more reliably than the wind, was a key adaptation of angiosperms. The success of angiosperms was also enhanced by their ability to reproduce rapidly. Fertilization in angiosperms usually occurs about 12 hours after pollination, making it possible for the plant to produce seeds in only a few days or weeks. As we mentioned in Module 7, a typical pine takes years to produce seeds. Rapid seed production is especially advantageous in harsh environments such as deserts, where growing seasons are extremely short.

Another feature contributing to the success of angiosperms is the development of fruits, which protect and help disperse the seeds, as we see in the next module.

? What is the difference between pollination and fertilization?

Pollination is the transfer of pollen by wind or animals from stamens to the tips of carpels. Fertilization is the union of egg and sperm; the sperm are released from the pollen tube after the tube grows and makes contact with an ovule.

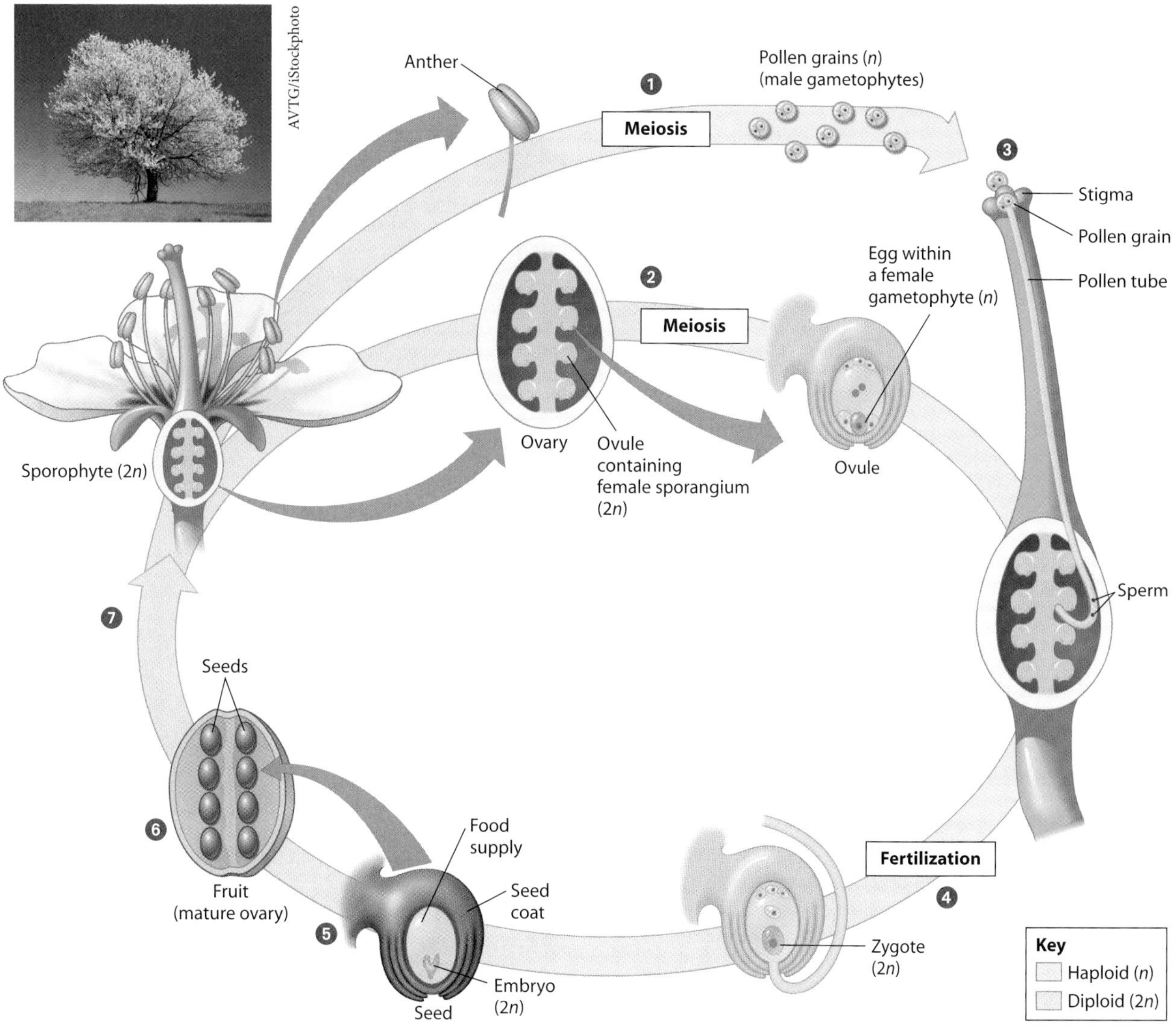

▲ **Figure 9** Life cycle of an angiosperm

10 The structure of a fruit reflects its function in seed dispersal

A **fruit**, the ripened ovary of a flower, is an adaptation that helps disperse seeds. Some angiosperms depend on wind for seed dispersal. For example, the fruit of a dandelion (**Figure 10A**) has a parachute-like extension that carries the tiny seed away from the parent plant on wind currents. Hook-like modifications of the outer layer of the fruit or seed coat allow some angiosperms to hitch a ride on animals. The fruits of the cocklebur plant (**Figure 10B**), for example, may be carried for miles before they open and release their seeds.

Many angiosperms produce fleshy, edible fruits that are attractive to animals as food. While the seeds are developing, these fruits are green and effectively camouflaged against green foliage. When ripe, the fruit turns a bright color, such as red or yellow, advertising its presence to birds and mammals. When the catbird in **Figure 10C** eats a berry, it digests the fleshy part of the fruit, but most of the tough seeds pass unharmed through its digestive tract. The bird may then deposit the seeds, along with a supply of natural fertilizer, some distance from where it ate the fruit.

The dispersal of seeds in fruits is one of the main reasons that angiosperms are so successful. Humans have also made extensive use of fruits and seeds, as we see next.

? What is a fruit?

A ripened ovary of a flower, which contains, protects, and aids in the dispersal of seeds

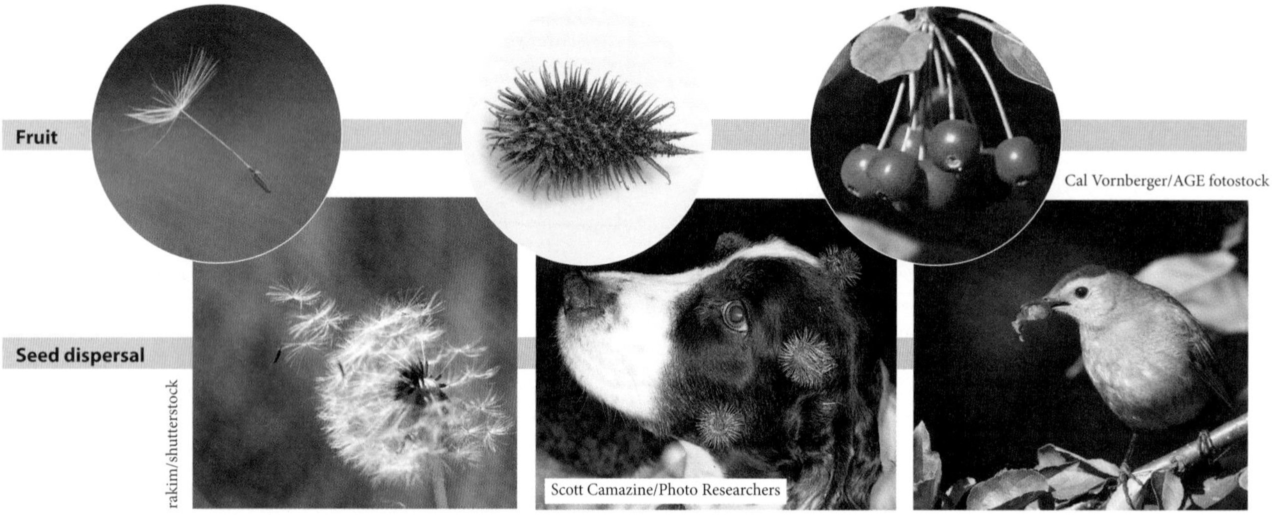

▲ **Figure 10A** Dandelion seeds can be launched into the air by a light breeze.

▲ **Figure 10B** Cocklebur fruits may be carried by animal fur.

▲ **Figure 10C** Seeds within edible fruits are often dispersed in animal feces.

CONNECTION 11 Angiosperms sustain us—and add spice to our diets

We depend on the fruits and seeds of angiosperms for much of our food. Corn, rice, wheat, and other grains, the main food sources for most of the world's people and their domesticated animals, are dry fruits. Many food crops are fleshy fruits, including apples, cherries, oranges, tomatoes, squash, and cucumbers. (In scientific terms, a fruit is an angiosperm structure containing seeds, so some vegetables are also fruits.) While most people can easily recognize grains and fleshy fruits as plant products, fewer realize that spices such as nutmeg, cinnamon, cumin, cloves, ginger, and licorice come from angiosperms. **Figure 11** shows the source of a condiment found on most American dinner tables: black pepper. The pepper fruits are harvested before ripening, then dried and ground into powder or sold whole as "peppercorns." In medieval Europe, peppercorns were so valuable that they were used as currency. Rent and taxes could be paid in peppercorns; as a form of wealth, peppercorns were included in dowries and left in wills. The search for a sea route to obtain pepper and other precious spices from India and southeast Asia led to the Age of Exploration and had a lasting impact on European history.

? Suppose you found a cluster of pepper berries like the ones in Figure 11. How would you know that they are fruits?

Each berry has seeds inside it.

▲ **Figure 11** Berries (fruits) on *Piper nigrum*

EVOLUTION CONNECTION

12 Pollination by animals has influenced angiosperm evolution

Most of us associate flowers with colorful petals and sweet fragrances, but not all flowers have these accessories. **Figure 12A**, for example, shows flowers of a red maple, which have many stamens but no petals (carpels are borne on separate flowers). Compare those flowers to the large, vibrantly colored columbine in **Figure 12B**. Such an elaborate flower costs the columbine an enormous amount of energy to produce, but the investment pays off when a pollinator, attracted by the flower's color or scent, carries the plant's pollen to another flower of the same species. Red maple, on the other hand, devotes substantial energy to making massive amounts of pollen for release into the wind, a far less certain method of pollination. Both species have adaptations to achieve pollination, which is necessary for reproductive success, but they allocate their resources differently.

MShieldsPhotos/Alamy

▲ **Figure 12A** Flowers of red maple, whose pollen is carried by the wind

Plant scientists estimate that about 90% of angiosperms employ animals to transfer their pollen. Birds, bats, and many different species of insects (notably bees, butterflies, moths, and beetles) serve as pollinators. These animals visit flowers in search of a meal, which the flowers provide in the form of nectar, a high-energy fluid. For pollinators, the colorful petals and alluring odors are signposts that mark food resources.

The cues that flowers offer are keyed to the sense of sight and smell of certain types of pollinators. For example, birds are attracted by bright red and orange flowers but not to particular scents, while most beetles are drawn to fruity odors but are indifferent to color. The petals of bee-pollinated flowers may be marked with guides in contrasting colors that lead to the nectar. In some flowers, the nectar guides are pigments that reflect ultraviolet light, a part of the electromagnetic spectrum that is invisible to us and most other animals, but readily apparent to bees. Flowers that are pollinated by night-flying animals such as bats and moths typically have large, light-colored, highly scented flowers that can easily be found at night. Some flowers even produce an enticing imitation of the smell of rotting flesh, thereby attracting pollinators such as carrion flies and beetles.

Many flowers have additional adaptations that improve pollen transfer, and thus reproductive success, once a pollinator arrives. The location of the nectar, for example, may manipulate the visitor's position in a way that maximizes pollen pickup and deposition. In **Figure 12C**, the pollen-bearing stamens of a scotch broom flower arch over the bee as it harvests nectar. Some of the pollen the bee picks up here will rub off onto the stigmas of other flowers it visits. In the columbine, as well as in many other flowers, the nectar can only be reached by pollinators with long tongues, a group that includes butterflies, moths, birds, and some bees.

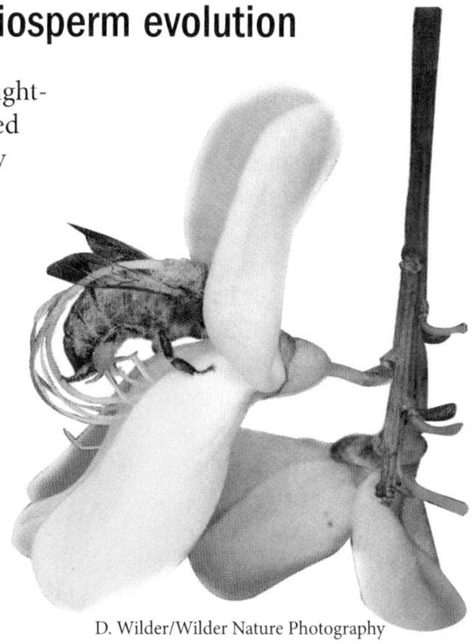

D. Wilder/Wilder Nature Photography

▲ **Figure 12C** A bee picking up pollen from a scotch broom flower as it feeds on nectar

Pollination is only effective if the pollen transfer occurs between members of the same species, but relatively few pollinators visit one species of flower exclusively. Biologists hypothesize that it takes time, through trial and error, for a pollinator to learn to extract nectar from a flower. Insects, for instance, can only remember one extraction technique at a time. Thus, pollinators are most successful at obtaining food if they visit another flower with the same cues immediately after mastering a technique for nectar extraction.

Although floral characteristics are adaptations that attract pollinators, they are a source of enjoyment to people, as well. Humans use flowering plants, including many species of trees, for a variety of purposes. As we consider in the next module, however, the most essential role of plants in our lives is as food.

Greg Vaughn/Photolibrary

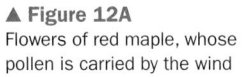

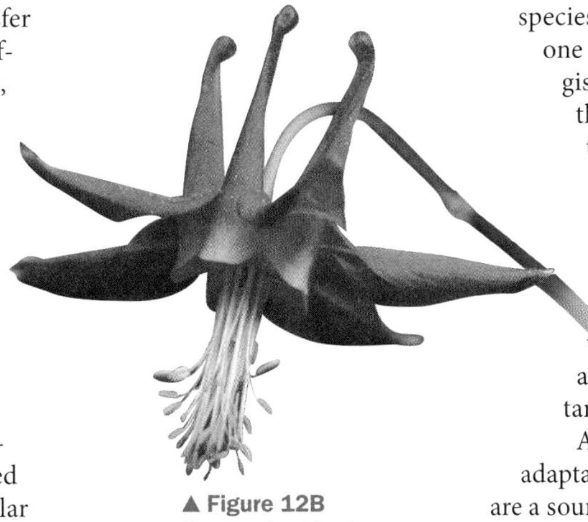

▲ **Figure 12B** Showy columbine flower

? **What type of pollinator do you think would be attracted to the columbine in Figure 12B?**

Long-tongued birds, because the flower is red and has long floral tubes

CONNECTION 13 Plant diversity is vital to the future of the world's food supply

Plants have always been a mainstay of the human diet, but the way we obtain our plant food has changed enormously. Early humans made use of any edible plant species that was available, probably eating different plants as the seasons changed, as do present-day hunter-gatherers. During the development of agriculture, people in different parts of the world domesticated the tastiest, most easily cultivated species, gradually increasing their productivity through generation after generation of artificial selection. In modern agriculture, plant-breeding techniques have further narrowed the pool of food plant diversity by creating a select few genotypes possessing the most desirable characteristics. Most of the world's population is now fed by varieties of rice, wheat, corn, and soybeans that require specific cultivation techniques.

Agriculture has also changed the landscape. Over thousands of years, the expanding human population created farms by clear-cutting or burning forests. More recently, deforestation has accelerated to replace the vast expanses of cropland that have been severely degraded by unsustainable agricultural practices. But converting more land to farms is not the only way to ensure an adequate food supply for the future. Plant diversity offers possibilities for developing new crops and improving existing ones.

Some new crops may come from the hundreds of species of nutritious fruits, nuts, and grains that people gather and use locally. In a recent study of potential food sources in Africa, scientists identified dozens of wild plants that might be suitable for domestication and regional production (**Figure 13A**). Promising candidates include intriguingly named fruits such as chocolate berries, gingerbread plums, and monkey oranges. In addition, some regions already have unique domesticated or semi-domesticated crops with the potential for greater production, especially in marginal farmland. For example, some African grains tolerate heat and drought, and many grow better on infertile soil than grains cultivated elsewhere. One species is so tough that it grows on sand dunes where the annual rainfall is less than 70 mm (about 2.5 inches) per year!

Bart Wursten

Breaking Ground

▲ **Figure 13A** Sugar plums (left) and safou (right), two wild fruits that may be ripe for domestication

UNEP/GRID - Sioux Falls (DEWA)

▲ **Figure 13B** Satellite photo of Amazonian rain forest. The "fishbone" pattern marks a network of roads carved through the forest when farmers and loggers came to the area.

Modifying crops to enable them to thrive in less than ideal conditions—either through traditional breeding methods or biotechnology—is another approach. Genes that enable plants to grow in salty soil, to resist pests, or to tolerate heat and drought would also be useful. But where might such genes be found? All crop plants were originally derived from wild ancestors. Those ancestors and their close relatives are a rich source of genetic diversity that could be used to bolster existing crops.

Both of these approaches to crop diversification are undermined by the ongoing loss of natural plant diversity caused by habitat destruction. Besides clear-cutting to create farmland, forests are being lost to logging, mining, and air pollution. Roads slice vast tracts of lands into fragments that are too small to support a full array of species (**Figure 13B**). As a result, species that could potentially be domesticated are instead being lost. So are many wild relatives of crop species, and with them a priceless pool of genetic diversity. These losses are just one example of the impact of declining biodiversity on the future of our species.

? Name two ways that the loss of plant diversity might affect the world's future supply of food.

Potential crop plants could be lost; potentially useful genes carried by wild ancestors or close relatives of crop plants could be lost.

Diversity of Fungi

14 Fungi absorb food after digesting it outside their bodies

You have probably seen members of the kingdom **Fungi** at some time, whether they were mushrooms sprouting in a lawn, bracket fungi attached to a tree like small shelves, or fuzzy patches of mold on leftover food. Despite the differences in their visible body forms, each of these fungi is obtaining food from its substrate in the same way. All fungi are heterotrophs that acquire their nutrients by **absorption**. They secrete powerful enzymes that digest macromolecules into monomers and then absorb the small nutrient molecules into their cells. For example, enzymes secreted by fungi growing on a loaf of bread digest the bread's starch into glucose molecules, which the fungal cells absorb. Some fungi produce enzymes that digest cellulose and lignin, the major structural components of plants. Consequently, they are essential decomposers in most ecosystems.

The feeding structures of a fungus are a network of threadlike filaments called **hyphae** (singular *hypha*). Hyphae branch repeatedly as they grow, forming a mass known as a **mycelium** (plural, *mycelia*) (**Figure 14A**). The "umbrellas" that you recognize as fungi, such as the ones in **Figure 14B**, are reproductive structures. Like the rest of the fungal body, the reproductive structures are made up of hyphae. In the type of fungus shown, mushrooms arise as small buds on a mycelium that extends throughout the food source, hidden from view. When a bud has developed sufficiently, the rapid absorption of water (for example, after a rainfall) creates enough hydraulic pressure to pop the mushroom to the surface. Above ground, the mushroom produces tiny reproductive cells called spores at the tips of specialized hyphae, and the spores are then dispersed on air currents.

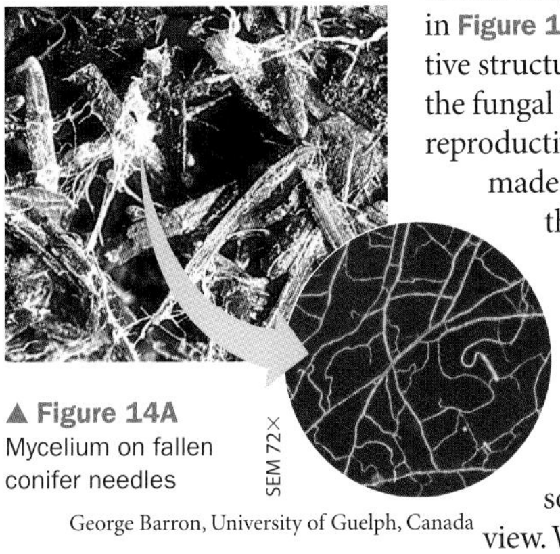

▲ **Figure 14A** Mycelium on fallen conifer needles

Fred Rhoades, Mycena Consulting

George Barron, University of Guelph, Canada

Fungal hyphae are surrounded by a cell wall. Unlike plants, which have cellulose cell walls, most fungi have cell walls made of **chitin**, a strong, flexible nitrogen-containing polysaccharide, identical to the chitin found in the external skeletons of insects. In most fungi, the hyphae consist of chains of cells separated by cross-walls that have pores large enough to allow ribosomes, mitochondria, and even nuclei to flow from cell to cell. Some fungi lack cross-walls entirely and have many nuclei within a single mass of cytoplasm.

Fungi cannot run or fly in search of food. But their mycelium makes up for the lack of mobility by being able to grow at a phenomenal rate, branching throughout a food source and extending its hyphae into new territory. If you were to break open the log beneath the mushrooms in Figure 14B, for example, you would see strands of hyphae throughout the wood. Because its hyphae grow longer without getting thicker, the fungus develops a huge surface area from which it can secrete digestive enzymes and through which it can absorb food.

Not all fungi are decomposers. In Module 18, you will learn about fungi that are **parasites**, obtaining their nutrients at the expense of living plants or animals. Also in this chapter, you will encounter fungi that live symbiotically with other organisms. Of special significance is the symbiosis between fungi and plant roots called a **mycorrhiza** (meaning "fungus root"; plural, *mycorrhizae*). The hyphae of some mycorrhizal fungi branch into the root cells; other species surround the root but don't penetrate its living cells. Both types of mycorrhizae absorb phosphorus and other essential minerals from the soil and make them available to the plant. Sugars produced by the plant through photosynthesis nourish the fungus, making the relationship mutually beneficial. The mycorrhizal partnership is thought to have played a crucial role in the success of early land plants. It remains important today, as mycorrhizal associations are present in nearly all vascular plants.

? Contrast how fungi digest and absorb their food with your own digestion.

A fungus digests its food externally by secreting enzymes onto the food and then absorbing the small nutrients that result from digestion. In contrast, humans and most other animals eat relatively large pieces of food and digest the food within their bodies.

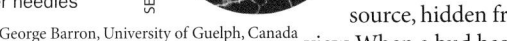

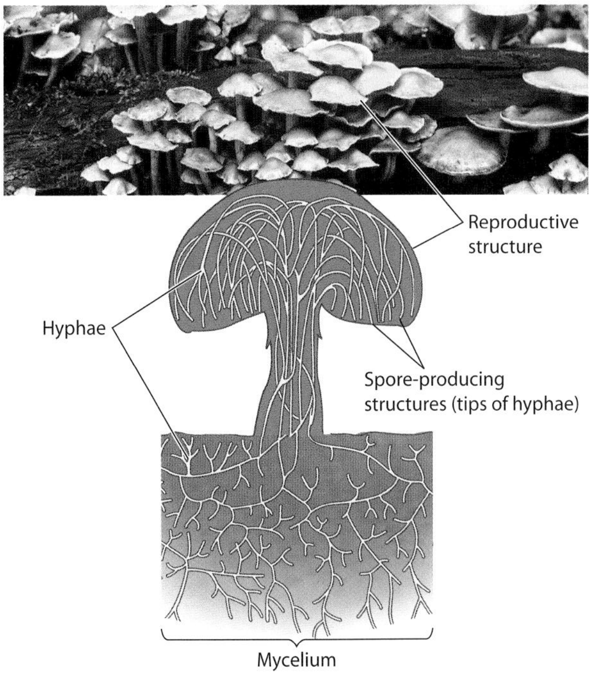

BlueMoon Stock/Alamy

▲ **Figure 14B** Fungal reproductive and feeding structures

15 Fungi produce spores in both asexual and sexual life cycles

Fungal reproduction typically involves the release of vast numbers of haploid spores, which are transported easily over great distances by wind or water. A spore that lands in a moist place where food is available germinates and produces a new haploid fungus. As you can see in **Figure 15**, however, spores can be produced either sexually or asexually.

In many fungi, sexual reproduction involves mycelia of different mating types. Hyphae from each mycelium release signaling molecules and grow toward each other. ❶ When the hyphae meet, their cytoplasms fuse. But this fusion of cytoplasm is often not followed immediately by the fusion of "parental" nuclei. Thus, many fungi have what is called a **heterokaryotic stage** (from the Greek, meaning "different nuclei"), in which cells contain two genetically distinct haploid nuclei. Hours, days, or even centuries may pass before the parental nuclei fuse, ❷ forming the usually short-lived diploid phase. ❸ Zygotes undergo meiosis, producing haploid spores. As you'll learn in the next module, the specialized structures in which these spores are formed are used to classify fungi.

In asexual reproduction, ❹ spore-producing structures arise from haploid mycelia that have undergone neither a heterokaryotic stage nor meiosis. Many fungi that reproduce sexually can also produce spores asexually. In addition, asexual reproduction is the only known means of spore production in some fungi, informally known as **imperfect fungi**. Many species commonly called molds and yeasts are imperfect fungi. The term **mold** refers to any rapidly growing fungus that reproduces asexually by producing spores, often at the tips of specialized hyphae. These familiar furry carpets often appear on aging fruit and bread, and in seldom-cleaned shower stalls. The term **yeast** refers to any single-celled fungus. Yeasts reproduce asexually by cell division, often by budding—pinching off small "buds" from a parent cell. Yeasts inhabit liquid or moist habitats, such as plant sap and animal tissues.

We will examine fungal reproduction more closely in Module 17, but first let's look at the classification of fungi.

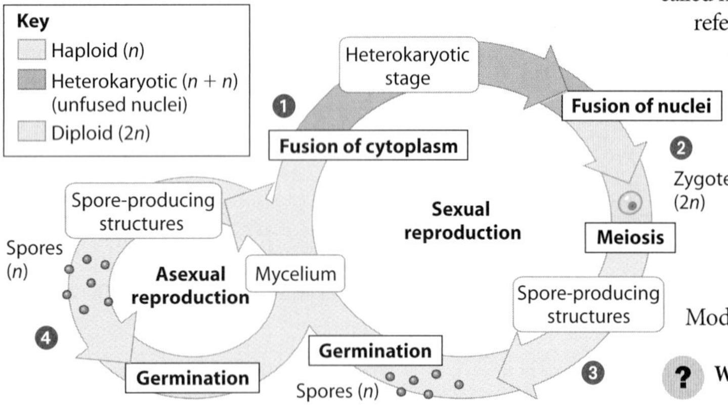

▲ **Figure 15** Generalized life cycle of a fungus

? What is the heterokaryotic stage of a fungus?

The stage in which each cell has two different nuclei (from two different parents), with the nuclei not yet fused

16 Fungi are classified into five groups

Biologists who study fungi have described over 100,000 species, but there may be as many as 1.5 million. Sexual reproductive structures are often used to classify species. Molecular analysis has helped scientists understand the phylogeny of fungi. The lineages that gave rise to fungi and animals are thought to have diverged from a flagellated unikont ancestor more than 1 billion years ago. The oldest undisputed fossils of fungi, however, are only about 460 million years old, perhaps because the ancestors of terrestrial fungi were microscopic and fossilized poorly.

Figure 16A shows a current hypothesis of fungal phylogeny. The multiple lines leading to the chytrids and the zygomycetes indicate that these groups are probably not monophyletic. For now, though, most biologists still talk in terms of the five groups of fungi shown here.

The **chytrids**, the only fungi with flagellated spores, are thought to represent the earliest lineage of fungi. They are common in lakes, ponds, and soil. Some species are decomposers; others parasitize protists, plants, or animals. Some researchers have linked the widespread decline of amphibian species to a highly infectious fungal disease caused by a species of chytrid. Populations of frogs in mountainous regions of Central America and Australia have suffered massive mortality from this emerging disease.

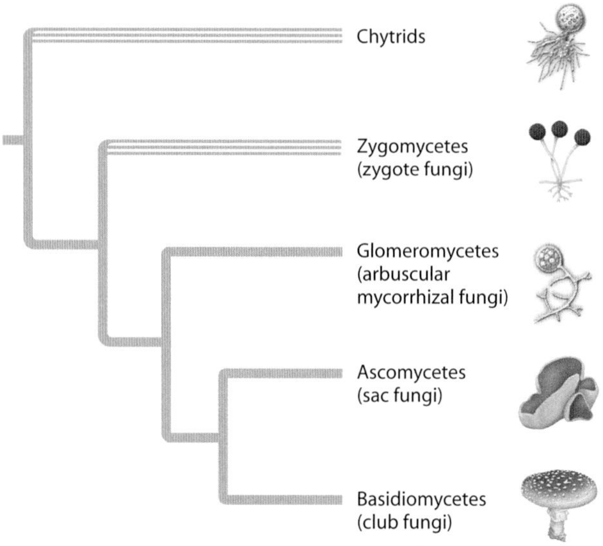

▲ **Figure 16A** A proposed phylogenetic tree of fungi

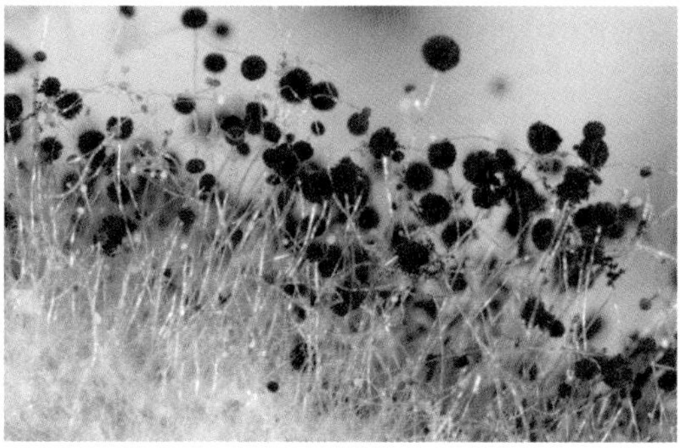
Gregory G. Dimijian/Photo Researchers

▲ **Figure 16B** Zygomycete: *Rhizopus stolonifer*, black bread mold

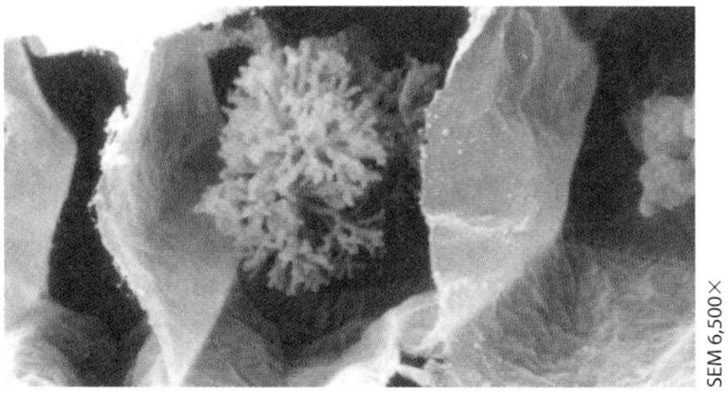

M. F. Brown/Biological Photo Service

▲ **Figure 16C** Glomeromycete: an arbuscule in a root cell

The **zygomycetes**, or **zygote fungi**, are characterized by their protective zygosporangium, where zygotes produce haploid spores by meiosis. This diverse group includes fast-growing molds, such as black bread mold **(Figure 16B)** and molds that rot produce such as peaches, strawberries, and sweet potatoes. Some zygote fungi are parasites on animals.

The **glomeromycetes** (from the Latin *glomer*, ball) form a distinct type of mycorrhiza in which hyphae that invade plant roots branch into tiny treelike structures known as arbuscules **(Figure 16C)**. About 90% of all plants have such symbiotic partnerships with glomeromycetes, which deliver phosphate and other minerals to plants while receiving organic nutrients in exchange.

The **ascomycetes**, or **sac fungi**, are named for saclike structures called asci (from the Greek *asco*, pouch) that produce spores in sexual reproduction. They live in a variety of marine, freshwater, and terrestrial habitats and range in size from unicellular yeasts to elaborate morels and cup fungi **(Figure 16D)**. Ascomycetes include some of the most devastating plant pathogens. Other species of ascomycetes live with green algae or cyanobacteria in symbiotic associations called lichens, which we discuss in Module 21.

Phil Dotson/Photo Researchers

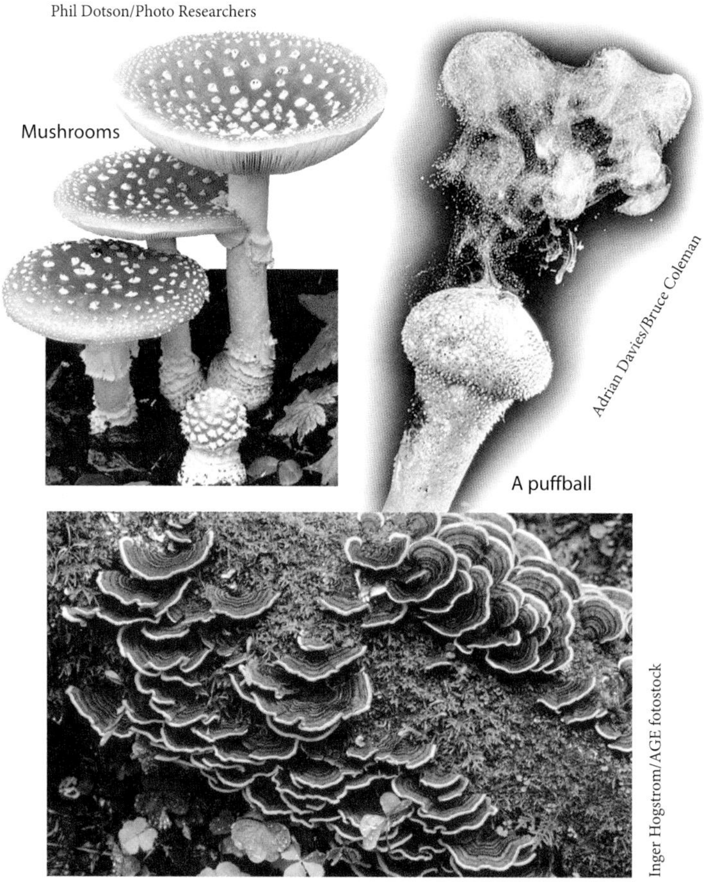

Adrian Davies/Bruce Coleman

Inger Hogstrom/AGE fotostock

▲ **Figure 16E** Basidiomycetes (club fungi)

When you think of fungi, you probably picture mushrooms, puffballs, or shelf fungi **(Figure 16E)**. These are examples of **basidiomycetes**, or **club fungi**. They are named for their club-shaped, spore-producing structure, called a basidium (meaning "little pedestal" in Latin; plural, *basidia*; see Figure 17B). Many species excel at breaking down the lignin found in wood and thus play key roles as decomposers. For example, shelf fungi often break down the wood of weak or damaged trees and continue to decompose the wood after the tree dies. The basidiomycetes also include two groups of particularly destructive plant parasites, the rusts and smuts, which we discuss in Module 18.

In the next module, we explore the life cycles of some representative fungi.

Frank Young/Papilio/Corbis

ARCO/H Reinhard/age fotostock

▲ **Figure 16D** Ascomycetes

? What is one reason that chytrids are thought to have diverged earliest in fungal evolution?

Chytrids are the only fungi that have flagellated spores, a characteristic of the ancestor of fungi.

17 Fungal groups differ in their life cycles and reproductive structures

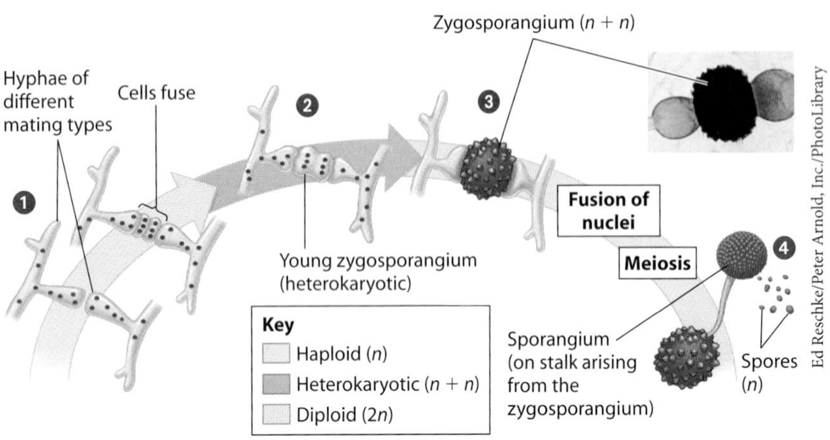

▲ **Figure 17A** Sexual reproduction of a zygote fungus

The life cycle of black bread mold (see Figure 16B) is typical of zygomycetes. As hyphae grow through its food, the fungus reproduces asexually, forming spores in sporangia at the tips of upright hyphae. When the food is depleted, the fungus reproduces sexually. As shown in **Figure 17A**, ❶ hyphae from mycelia of different mating types fuse and ❷ produce a cell containing multiple nuclei from two parents. ❸ This young zygosporangium develops into a thick-walled structure that can tolerate dry or harsh environments. When conditions are favorable, the parental nuclei fuse to form diploid zygotes, which undergo meiosis, ❹ producing haploid spores.

Like the zygomycetes, ascomycetes, such as those shown in Figure 16D, reproduce asexually when conditions are suitable. In many species, sexual reproduction takes place in the fall, with the haploid spores maturing in the early spring. The genetic diversity of these sexually produced spores increases the likelihood that at least one genotype will successfully establish itself in the environment encountered in the new season. The surviving individuals then reproduce asexually for several generations through the spring and summer before undergoing sexual reproduction.

Now let's follow the life cycle of a mushroom, a basidiomycete, starting at the center bottom of **Figure 17B**. ❶ The heterokaryotic stage begins when hyphae of two different mating types fuse, ❷ forming a heterokaryotic mycelium, which grows and produces the mushroom. In the club-shaped cells called basidia, which line the gills of a mushroom, ❸ the haploid nuclei fuse, forming diploid nuclei. Each diploid nucleus then undergoes meiosis, producing haploid spores. ❹ A mushroom can release as many as a billion spores. If spores land on moist matter that can serve as food, ❺ they germinate and grow into haploid mycelia.

Much of the success of fungi is due to their reproductive capacity, both in the asexual production of spores and in sexual reproduction, which increases genetic variation and facilitates adaptation to harsh or changing conditions.

? Under what conditions is asexual reproduction advantageous? Under what conditions is sexual reproduction advantageous?

In asexual reproduction, a successful genotype can be propagated to take advantage of consistent, favorable conditions. Sexual reproduction reshuffles alleles, providing numerous genotypes that will be "tested" by the environment. This is an advantage when environmental conditions are changing.

▶ **Figure 17B** Life cycle of a mushroom

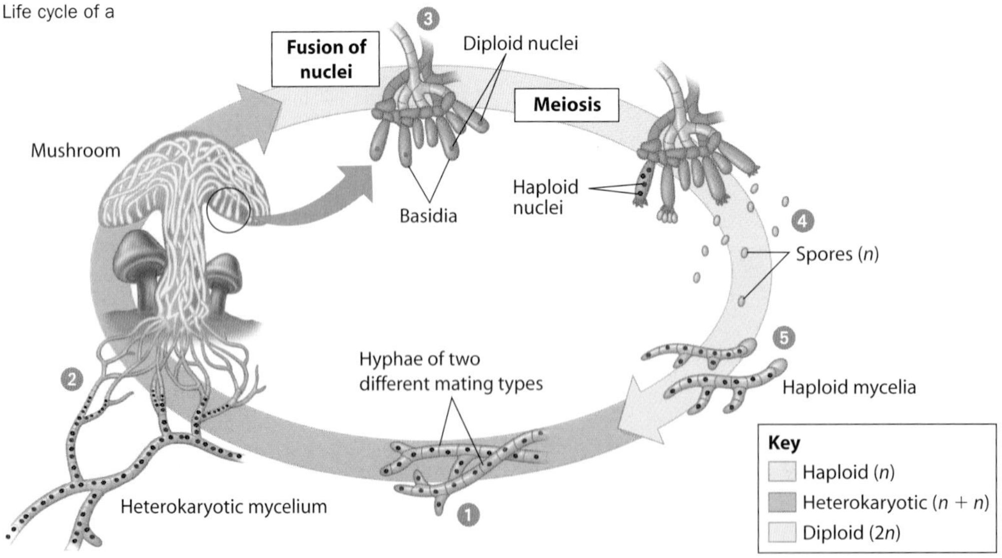

CONNECTION

18 Parasitic fungi harm plants and animals

Of the 100,000 known species of fungi, about 30% make their living as parasites or pathogens, mostly in or on plants. In some cases, fungi have literally changed the landscape. In 1926, a fungus that causes Dutch elm disease was accidentally introduced into the United States on logs sent from Europe to make furniture. (The name refers to the Netherlands, where the disease was first identified; the fungus originated in Asia.) Over the course of several decades, the fungus destroyed around 70% of the elm trees across the eastern United States. English elms (a different species), such as those in **Figure 18A**, fared even worse. They were completely annihilated. Recently, scientists studying the DNA of English elms have found evidence that they were all genetically identical, derived by asexual reproduction from a single ancestor brought to England by the Romans 2,000 years ago. As a result, they were all equally susceptible to the ravages of Dutch elm disease.

Fungi are a serious problem as agricultural pests. Crop fields typically contain genetically identical individuals of a single species planted close together—ideal conditions for the spread of disease. About 80% of plant diseases are caused by fungi, causing tremendous economic losses each year. Between 10% and 50% of the world's fruit harvest is lost each year to fungal attack. A variety of fungi, including smuts and rusts, are common on grain crops. The ear of corn shown in **Figure 18B** is infected with a club fungus called corn smut. The grayish growths, known as galls, are made up of heterokaryotic hyphae that invade a developing corn kernel and eventually displace it. When a gall matures, it breaks open and releases thousands of blackish spores. In parts of Central America, the smutted ears are cooked and eaten as a delicacy, but generally corn smut is regarded as a scourge.

Some of the fungi that attack food crops are toxic to humans. The seed heads of many kinds of grain, including rye, wheat, and oats, are sometimes infected with fungal growths called ergots, the dark structures on the seed head of rye shown in **Figure 18C**. Consumption of flour made from ergot-infested grain can cause gangrene, nervous spasms, burning sensations, hallucinations, temporary insanity, and death. Several toxins have been isolated from ergots. One, called lysergic acid, is the raw material from which the hallucinogenic drug LSD is made. Certain others are medically useful in small doses. For instance, an ergot compound is useful in treating high blood pressure and in stopping maternal bleeding after childbirth.

MT Media/LPI

▲ **Figure 18A** Stately English elms in Australia, unaffected by Dutch elm disease

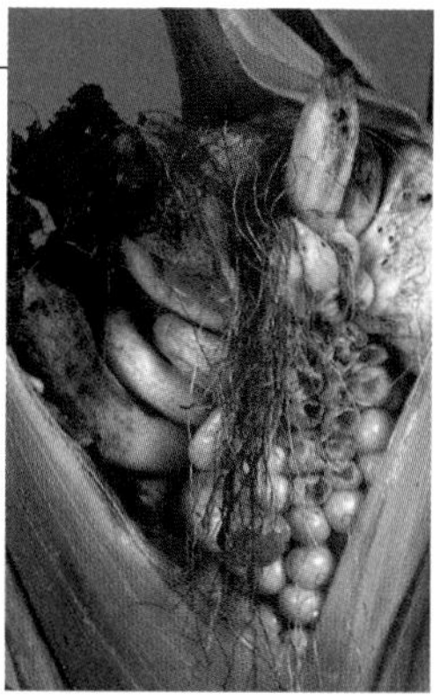

Astrid & Hanns-Frieder Michler/Photo Researchers

▲ **Figure 18B** Corn smut

Animals are much less susceptible to parasitic fungi than are plants. Only about 50 species of fungi are known to be parasitic in humans and other animals. In humans, fungi cause infections ranging from annoyances such as athlete's foot to deadly lung diseases. The general term for a fungal infection is **mycosis**. Skin mycoses include the disease called ringworm, so named because it can appear as circular red areas on the skin. The ringworm fungus can infect virtually any skin surface. Most commonly, it attacks the feet, causing the intense itching and sometimes blistering known as athlete's foot. Systemic mycoses are fungal infections that spread throughout the body, usually from spores that are inhaled. These can be very serious diseases. Coccidioidomycosis is a systemic mycosis that produces tuberculosis-like symptoms in the lungs. It is so deadly that it is considered a potential biological weapon.

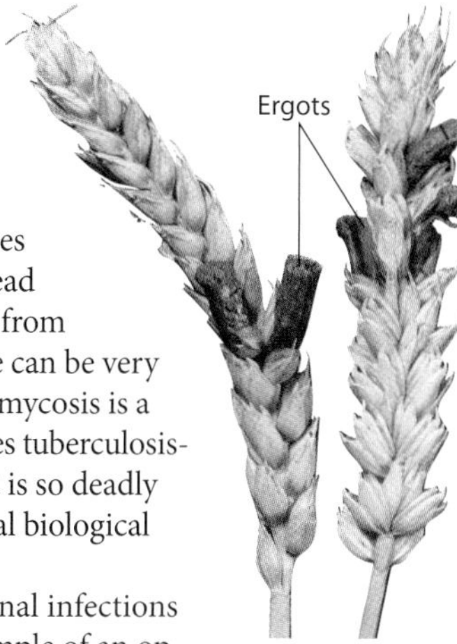

Nigel Cattlin/Photo Researchers

▲ **Figure 18C** Ergots on rye

The yeast that causes vaginal infections (*Candida albicans*) is an example of an opportunistic pathogen—a normal inhabitant of the body that causes problems only when some change in the body's microbiology, chemistry, or immunology allows the yeast to grow unchecked. Many other opportunistic mycoses have increased in recent decades, in part because of AIDS, which compromises the immune system.

Despite the prominent role of fungi as plant pathogens and occasionally human pathogens, most species of fungi are beneficial. In the next two modules, we consider the ecological and practical benefits of fungi.

? What is a mycosis? What is an opportunistic pathogen?

A mycosis is a fungal infection. An opportunistic pathogen is a normal inhabitant of the body that grows out of control when there is a change in the body's microbiology, chemistry, or immunology.

CONNECTION 19 Fungi have enormous ecological benefits

As you have read, fungi have been major players in terrestrial communities ever since they moved onto land in the company of plants. As symbiotic partners in mycorrhizae, fungi supply essential nutrients to plants and are enormously important in natural ecosystems and agriculture.

Fungi, along with prokaryotes, are essential decomposers in ecosystems, breaking down organic matter and restocking the environment with vital nutrients essential for plant growth. So many fungal spores are in the air that as soon as a leaf falls or an insect dies, it is covered with spores and is soon infiltrated by fungal hyphae (**Figure 19**). If fungi and prokaryotes in a forest suddenly stopped decomposing, leaves, logs, feces, and dead animals would pile up on the forest floor. Plants—and the animals that eat plants—would starve because elements taken from the soil would not be replenished through decomposition.

Almost any organic (carbon-containing) substance can be consumed by fungi. During World War II, the moist tropical heat of Southeast Asia and islands in the Pacific Ocean provided ideal conditions for fungal decomposition of wood and natural fibers such as canvas and cotton. Packing crates, military uniforms, and tents quickly disintegrated, causing supply problems for the military forces. Synthetic substances are more resistant to fungal attack, but some fungi have the useful ability to break down toxic pollutants, including the pesticide DDT and certain chemicals that cause cancer. Scientists are also investigating the possibility of using fungi that can digest petroleum products to clean up oil spills and other chemical messes.

Emmanuel LATTES/Alamy

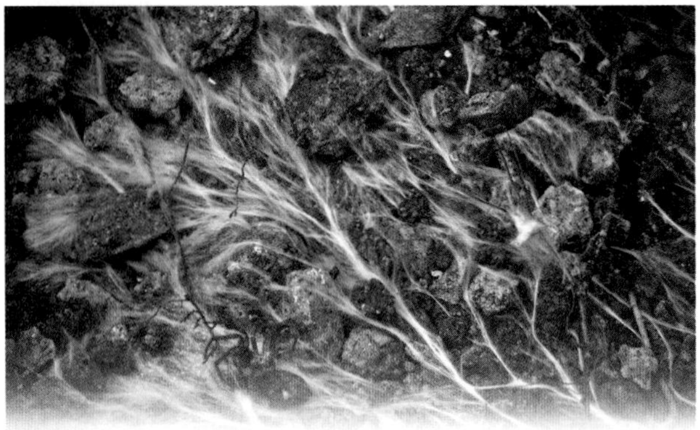

▲ **Figure 19** A fungal mycelium

? Name two essential roles that fungi play in terrestrial ecosystems.

In mycorrhizae, fungi help plants acquire nutrients from the soil. When soil fungi decompose dead animals, fallen leaves, and other organic materials, they release nutrients that fertilize plant growth.

CONNECTION 20 Fungi have many practical uses

Fungi have a number of culinary uses. Most of us have eaten mushrooms, although we may not have realized that we were ingesting reproductive structures of subterranean fungi. The distinctive flavors of certain cheeses, including Roquefort and blue cheese (**Figure 20A**), come from fungi used to ripen them. Truffles, which are produced by certain mycorrhizal fungi associated with tree roots, are highly prized by gourmets. And humans have used yeasts for thousands of years to produce alcoholic beverages and cause bread to rise.

Fungi are medically valuable as well. Like the bacteria called actinomycetes, some fungi produce antibiotics that we use to treat bacterial diseases. In fact, the first antibiotic discovered was penicillin, which is made by the common mold called *Penicillium*. In **Figure 20B**, the clear area between the mold and the bacterial growth is where the antibiotic produced by *Penicillium* has inhibited the growth of the bacteria (*Staphylococcus aureus*).

Fungi also figure prominently in research in molecular biology and in biotechnology. Researchers often use yeasts to study the molecular genetics of eukaryotes because they are easy to culture and manipulate. Yeasts have also been genetically modified to produce human proteins for research and for medical use.

Fungi may play a major role in the future production of biofuels from plants. Ideally, the biofuels would be derived from plants or plant parts that could not be used to feed people or livestock, such as straw, certain grasses, and wood. These plant materials are primarily made up of cellulose and lignin, large

FILATOV ALEXEY/Shutterstock

▲ **Figure 20A** Blue cheese

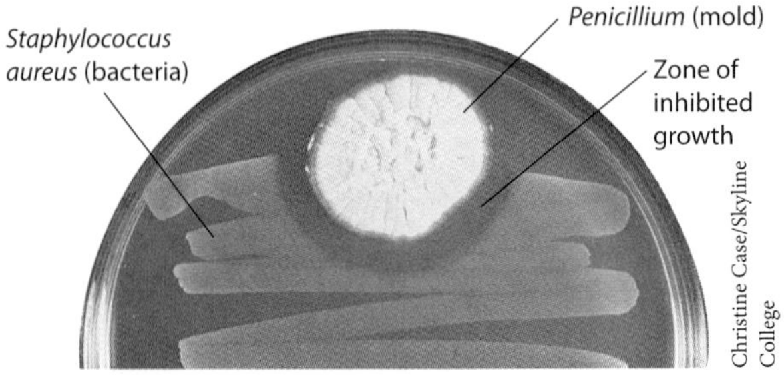

Christine Case/Skyline College

▲ **Figure 20B** A culture of *Penicillium* and bacteria

molecules that are difficult to decompose. Researchers are currently investigating a variety of fungi that produce enzymes capable of digesting the toughest plant parts. Basidiomycetes called white rot fungi (because the enzymatic breakdown of lignin bleaches the brown color out of wood) are promising candidates for this application (**Figure 20C**).

? What do you think is the function of the antibiotics that fungi produce in their natural environments?

The antibiotics probably block the growth of microorganisms, especially prokaryotes that compete with the fungi for nutrients and other resources.

▲ **Figure 20C** White rot fungus

Huetter, C/AGE fotostock

21 Lichens are symbiotic associations of fungi and photosynthetic organisms

The rock in **Figure 21A** is covered with a living crust of **lichens**, symbiotic associations of millions of unicellular green algae or cyanobacteria held in a mass of fungal hyphae. The partners are so closely entwined that they appear to be a single organism. How does this merger come about? When the growing hyphal tips of a lichen-forming fungus come into contact with a suitable partner, the hyphae quickly fork into a network of tendrils that encircle and overgrow the algal cells (**Figure 21B**). The fungus invariably benefits from the symbiosis, receiving food from its photosynthetic partner. In fact, fungi with the ability to form lichens rarely thrive on their own in nature. In many lichens that have been studied, the alga or cyanobacterium also benefits, as the fungal mycelium provides a suitable habitat that helps it absorb and retain water and minerals. In other lichens, it is not clear whether the relationship benefits the photosynthetic partner or is only advantageous to the fungus.

© William Mullins/Alamy

▲ **Figure 21A** Several of the 200 to 300 species of lichen that live in Antarctica

Lichens are rugged and able to live where there is little or no soil. As a result, they are important pioneers on new land. Lichens grow into tiny rock crevices, where the acids they secrete help to break down the rock to soil, paving the way for future plant growth. Some lichens can tolerate severe cold, and carpets of them cover the arctic tundra. Caribou feed on lichens known as reindeer "moss" (**Figure 21C**) in their winter feeding grounds in Alaska.

Lichens can also withstand severe drought. They are opportunists, growing in spurts when conditions are favorable. When it rains, a lichen quickly absorbs water and photosynthesizes at a rapid rate. In dry air, it dehydrates and photosynthesis may stop, but the lichen remains alive more or less indefinitely. Some lichens are thousands of years old, rivaling the oldest plants and fungi as the oldest organisms on Earth.

Eye of Science/Photo Researchers

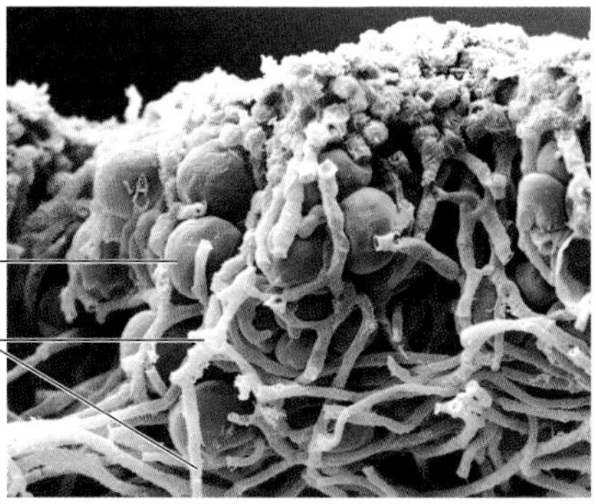

▲ **Figure 21B** The close relationship between fungal and algal partners in a lichen

Brian & Sophia Fuller/Alamy

▲ **Figure 21C** Reindeer moss, a lichen

As tough as lichens are, many do not withstand air pollution. Because they get most of their minerals from the air, in the form of dust or compounds dissolved in raindrops, lichens are very sensitive to airborne pollutants. The death of lichens is often a sign that air quality in an area is deteriorating.

Fungi are the third group of eukaryotes we have surveyed so far. (Protists and plants were the first two groups.) Strong evidence suggests that they evolved from unikont protists, a group that also gave rise to the fourth and most diverse group of eukaryotes, the animals.

? What benefit do fungi in lichens receive from their partners?

Access to the sugar molecules that algae or cyanobacteria produce by photosynthesis

REVIEW

For Practice Quizzes, MP3 Tutors, and Activities, go to www.masteringbiology.com.

Reviewing the Concepts

Plant Evolution and Diversity (1–2)

1 Plants have adaptations for life on land.

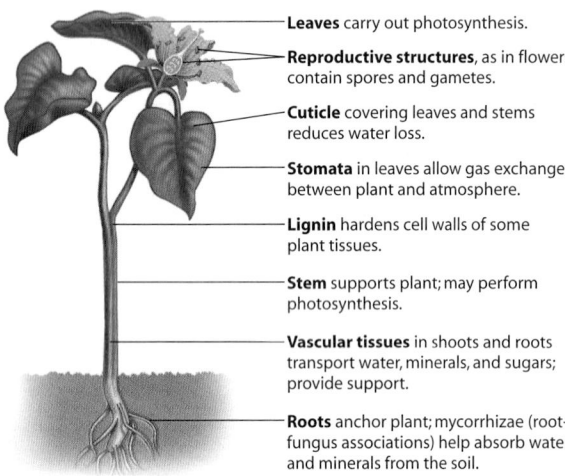

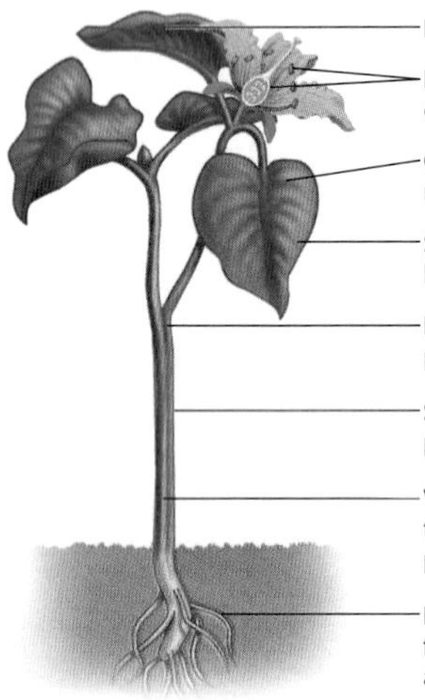

Leaves carry out photosynthesis.
Reproductive structures, as in flowers, contain spores and gametes.
Cuticle covering leaves and stems reduces water loss.
Stomata in leaves allow gas exchange between plant and atmosphere.
Lignin hardens cell walls of some plant tissues.
Stem supports plant; may perform photosynthesis.
Vascular tissues in shoots and roots transport water, minerals, and sugars; provide support.
Roots anchor plant; mycorrhizae (root-fungus associations) help absorb water and minerals from the soil.

2 Plant diversity reflects the evolutionary history of the plant kingdom. Nonvascular plants (bryophytes) include the mosses, hornworts, and liverworts. Vascular plants have supportive conducting tissues. Ferns are seedless vascular plants with flagellated sperm. Seed plants have sperm-transporting pollen grains and protect embryos in seeds. Gymnosperms, such as pines, produce seeds in cones. The seeds of angiosperms develop within protective ovaries.

Alternation of Generations and Plant Life Cycles (3–13)

3 Haploid and diploid generations alternate in plant life cycles. The haploid gametophyte produces eggs and sperm by mitosis. The zygote develops into the diploid sporophyte, in which meiosis produces haploid spores. Spores grow into gametophytes.

4 The life cycle of a moss is dominated by the gametophyte. A mat of moss is mostly gametophytes, which produce eggs and swimming sperm. The zygote develops on the gametophyte into the smaller sporophyte.

5 Ferns, like most plants, have a life cycle dominated by the sporophyte. The sporophyte is the dominant generation. Sperm, produced by the gametophyte, swim to the egg.

6 Seedless vascular plants dominated vast "coal forests."

7 A pine tree is a sporophyte with gametophytes in its cones. A pine tree is a sporophyte; tiny gametophytes grow in its cones. A sperm from a pollen grain fertilizes an egg in the female gametophyte. The zygote develops into a sporophyte embryo, and the ovule becomes a seed, with stored food and a protective coat.

8 The flower is the centerpiece of angiosperm reproduction. A flower usually consists of sepals, petals, stamens (produce pollen), and carpels (produce ovules).

9 The angiosperm plant is a sporophyte with gametophytes in its flowers. The sporophyte is independent, with tiny, dependent gametophytes protected in flowers. Ovules become seeds, and ovaries become fruits.

10 The structure of a fruit reflects its function in seed dispersal.

11 Angiosperms sustain us—and add spice to our diets.

12 Pollination by animals has influenced angiosperm evolution. Flowers attract pollinators by color and scent. Visiting pollinators are rewarded with nectar and pollen.

13 Plant diversity is vital to the future of the world's food supply. As plant biodiversity is lost through extinction and habitat destruction, potential crop species and valuable genes are lost.

Diversity of Fungi (14–21)

14 Fungi absorb food after digesting it outside their bodies. Fungi are heterotrophic eukaryotes that digest their food externally and absorb the resulting nutrients. A fungus usually consists of a mass of threadlike hyphae, called a mycelium.

15 Fungi produce spores in both asexual and sexual life cycles.

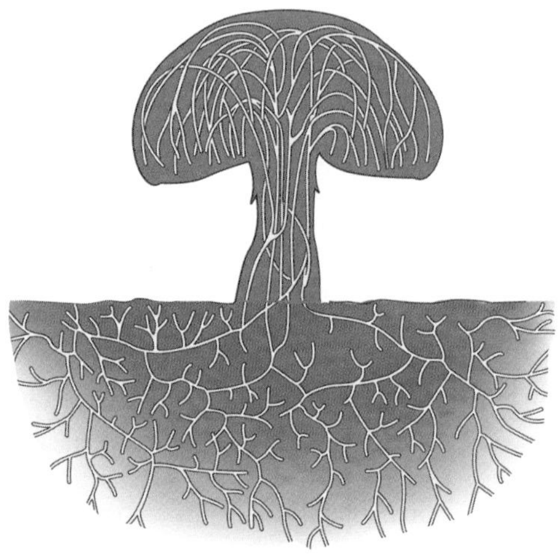

In some fungi, fusion of haploid hyphae produces a heterokaryotic stage containing nuclei from two parents. After the nuclei fuse, meiosis produces haploid spores.

16 Fungi are classified into five groups. Fungi evolved from a protist ancestor. Fungal groups include chytrids, zygomycetes, glomeromycetes, ascomycetes, and basidiomycetes.

17 Fungal groups differ in their life cycles and reproductive structures.

18 Parasitic fungi harm plants and animals.

19 Fungi have enormous ecological benefits. Fungi are essential decomposers and also participate in mycorrhizae.

20 Fungi have many practical uses. Some fungi provide food or antibiotics.

21 Lichens are symbiotic associations of fungi and photosynthetic organisms. The photosynthesizers are algae or cyanobacteria.

Connecting the Concepts

1. In this abbreviated diagram, identify the four major plant groups and the key terrestrial adaptation associated with each of the three major branch points.

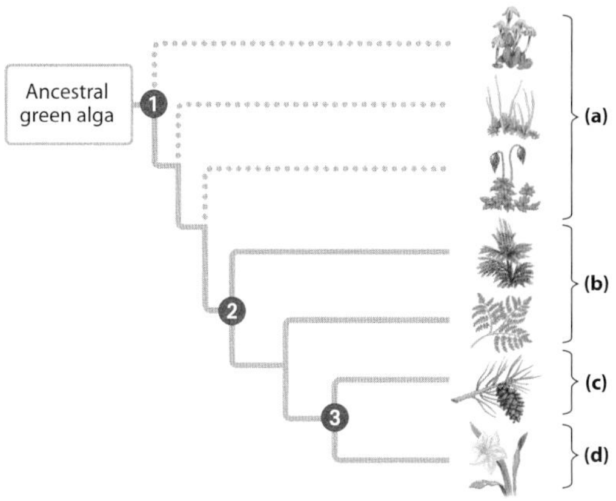

2. Identify the cloud seen in each photograph. Describe the life cycle events associated with each cloud.

R. J. Erwin/Photo Researchers

(a) Pine tree, a gymnosperm

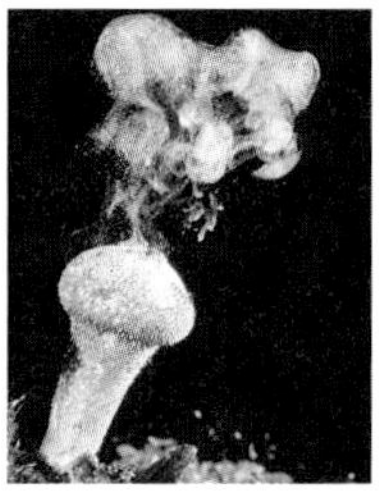

Adrian Davies/Bruce Coleman

(b) Puffball, a club fungus

Testing Your Knowledge

Multiple Choice

3. Angiosperms are different from all other plants because only they have
   a. a vascular system.
   b. flowers.
   c. a life cycle that involves alternation of generations.
   d. seeds.
   e. a dominant sporophyte phase.
4. Which of the following produce eggs and sperm? (*Explain your answer.*)
   a. the sexual reproductive structures of a fungus
   b. fern sporophytes
   c. moss gametophytes
   d. the anthers of a flower
   e. moss sporangia
5. The eggs of seed plants are fertilized within ovules, and the ovules then develop into
   a. seeds.
   b. spores.
   c. gametophytes.
   d. fruit.
   e. sporophytes.
6. The diploid sporophyte stage is dominant in the life cycles of all of the following *except*
   a. a pine tree.
   b. a dandelion.
   c. a rose bush.
   d. a fern.
   e. a moss.
7. Under a microscope, a piece of a mushroom would look most like
   a. jelly.
   b. a tangle of string.
   c. grains of sugar or salt.
   d. a piece of glass.
   e. foam.
8. Which of the following is an opportunistic pathogen that can cause a mycosis?
   a. HIV, the AIDS virus
   b. the fungus that produces ergots on rye, which can cause serious symptoms if milled into flour
   c. smuts, serious pathogens of grain crops
   d. *Bacillus anthracis*, which causes anthrax
   e. *Candida albicans*, which causes vaginal yeast infections
9. Which of the following terms includes all the others?
   a. angiosperm
   b. gymnosperm
   c. vascular plant
   d. fern
   e. seed plant
10. Which of the following is a plant with flagellated sperm and a sporophyte-dominated life cycle?
    a. chytrid
    b. moss
    c. charophyte
    d. fern
    e. liverwort

Describing, Comparing, and Explaining

11. Compare a seed plant with an alga in terms of adaptations for life on land versus life in the water.
12. How do animals help flowering plants reproduce? How do the animals benefit?
13. Why are fungi and plants classified in different kingdoms?

Applying the Concepts

14. Many fungi produce antibiotics, such as penicillin, which are valuable in medicine. But of what value might the antibiotics be to the fungi? Similarly, fungi often produce compounds with unpleasant tastes and odors as they digest their food. What might be the value of these chemicals to the fungi? How might production of antibiotics and odors have evolved?
15. In April 1986, an accident at a nuclear power plant in Chernobyl, Ukraine, scattered radioactive fallout for hundreds of miles. In assessing the biological effects of the radiation, researchers found mosses to be especially valuable as organisms for monitoring the damage. As mentioned in Module 10.16, radiation damages organisms by causing mutations. Explain why it is faster to observe the genetic effects of radiation on mosses than on plants from other groups. Imagine that you are conducting tests shortly after a nuclear accident. Using potted moss plants as your experimental organisms, design an experiment to test the hypothesis that the frequency of mutations decreases with the organism's distance from the source of radiation.

Review Answers

1. a. bryophytes (nonvascular plants); b. seedless vascular plants (ferns and relatives); c. gymnosperms; d. angiosperms; 1. apical meristems and embryos retained in the parent plant; 2. lignin-hardened vascular tissue; 3. seeds that protect and disperse embryos
2. a. This is a cloud of pollen being released from a pollen cone of a pine tree. In pollen cones, spores produce millions of the male gametophytes—the pollen grains.
   b. This is a cloud of haploid spores produced by a puffball fungus. Each spore may germinate to produce a haploid mycelium.
3. b 4. c (It is the only gametophyte among the possible answers.) 5. a 6. e 7. b 8. e 9. c 10. d
11. The alga is surrounded and supported by water, and it has no supporting tissues, vascular system, or special adaptations for obtaining or conserving water. Its whole body is photosynthetic, and its gametes and embryos are dispersed into the water. The seed plant has lignified vascular tissues that support it against gravity and carry food and water. The seed plant also has specialized organs that absorb water and minerals (roots), provide support (stems and roots), and photosynthesize (leaves and stems). It is covered by a waterproof cuticle and has stomata for gas exchange. Its sperm are carried by pollen grains, and embryos develop on the parent plant and are then protected and provided for by seeds.
12. Animals carry pollen from flower to flower and thus help fertilize the plants' eggs. They also disperse seeds by consuming fruit or carrying fruit that clings to their fur. In return, they get food (nectar, pollen, fruit).
13. Plants are autotrophs; they have chlorophyll and make their own food by photosynthesis. Fungi are heterotrophs that digest food externally and absorb nutrient molecules. There are also many structural differences; for example, the threadlike fungal mycelium is different from the plant body, and their cell walls are made of different substances. Plants evolved from green algae, which belong to the protist supergroup Archaeplastida; the ancestor of fungi was in the protist supergroup Unikonta. Molecular evidence indicates that fungi are more closely related to animals than to plants.
14. Antibiotics probably kill off bacteria that compete with fungi for food. Similarly, bad tastes and odors deter animals that eat or compete with fungi. Those fungi that produce antibiotics or bad-smelling and bad-tasting chemicals would survive and reproduce more successfully than fungi unable to inhibit competitors. Animals that could recognize the smells and tastes also would survive and reproduce better than their competitors. Thus, natural selection would favor fungi that produce the chemicals and, to some extent, the competitors deterred by them.
15. Moss gametophytes, the dominant stage in the moss life cycle, are haploid plants. The diploid (sporophyte) generation is dominant in most other plants. Recessive mutations are not expressed in a diploid organism unless both homologous chromosomes carry the mutation. In haploid organisms, recessive mutations are apparent in the phenotype of the organism because haploid organisms have only one set of chromosomes. Some factors to consider in designing your experiment: What are the advantages/disadvantages of performing the experiment in the laboratory? In the field? What variables would be important to control? How many potted plants should you use? At what distances from the radiation source should you place them? What would serve as a control group for the experiment? What age of plants should you use?

The Evolution of Invertebrate Diversity

From Chapter 18 of *Campbell Biology: Concepts & Connections*, Seventh Edition. Jane B. Reece, Martha R. Taylor, Eric J. Simon, Jean L. Dickey.

The Evolution of Invertebrate Diversity

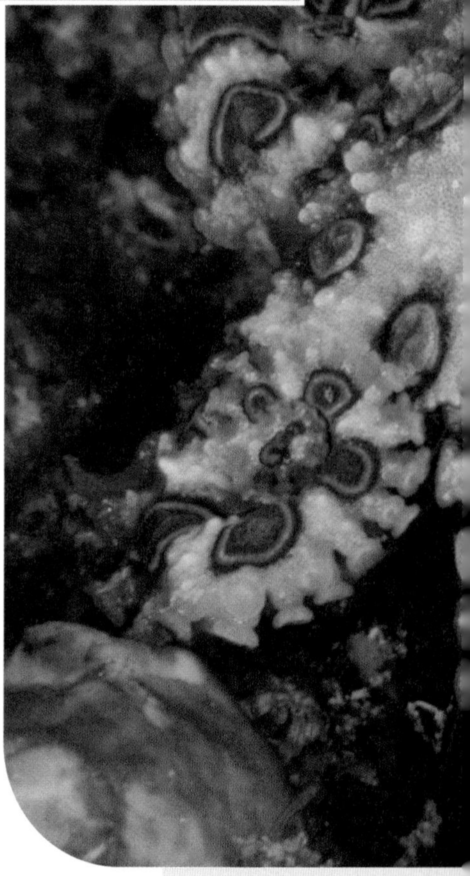

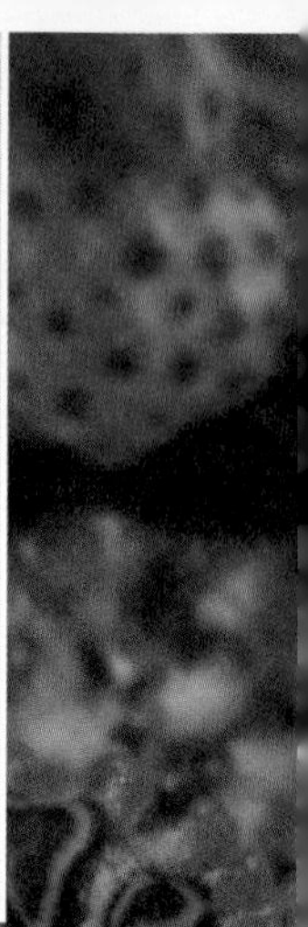

BIG IDEAS

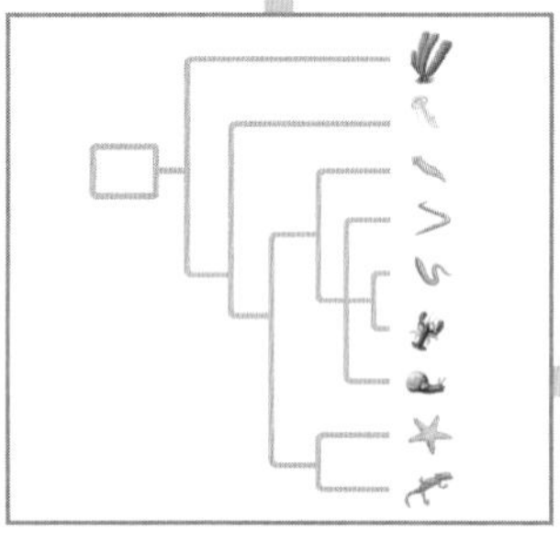

Animal Evolution and Diversity
(1–4)

Animal body plans can be used to build a phylogenetic tree.

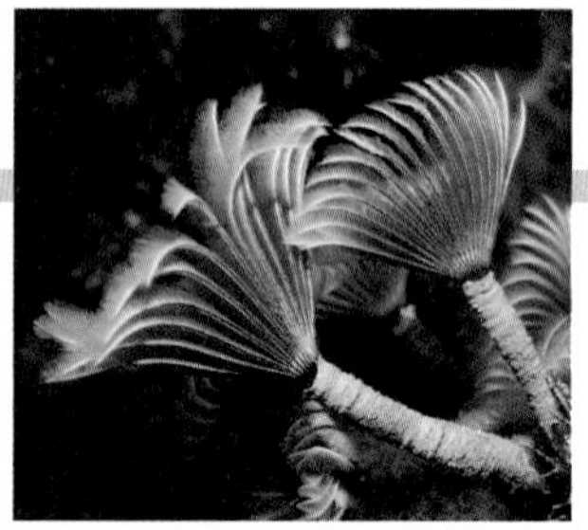

Jurgen Freund/Nature Picture Library

Invertebrate Diversity
(5–14)

In this chapter, you will learn about animals without backbones.

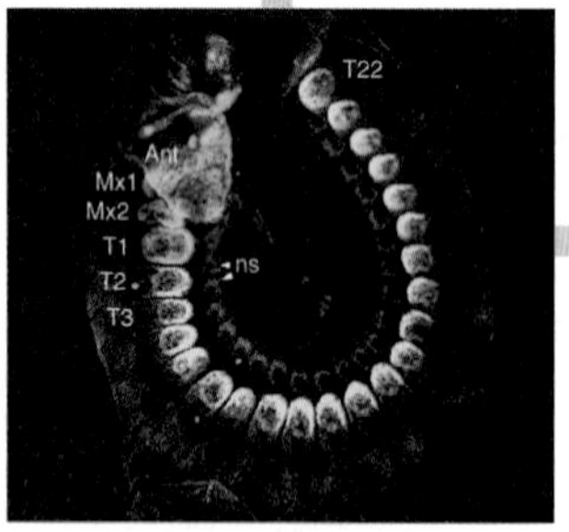

Steve Paddock, Ph.D

Animal Phylogeny and Diversity Revisited
(15–16)

Scientists are using molecular data to revise the phylogenetic tree for animals.

David & Debi Henshaw/Alamy

The shallow rock pools of Sydney Harbor, Australia, offer a fascinating glimpse of invertebrate diversity. The tentacles of colorful sea anemones sway lazily in the water. A shrimp darts out from behind a rock but quickly retreats to the safety of its hiding place. The large blue sea star draped over a rock would be spotted easily by a casual observer, but it takes a sharp eye to see its tiny cousin, camouflaged against the background and surrounded by its even tinier offspring. It would also take a keen eye to see the pale yellow-brown octopus the size of a golf ball that is loitering inconspicuously on the bottom of the pool. But this octopus changes color instantly when threatened. A sudden fierce glow of sapphire-blue circles proclaims its identity—a blue-ringed octopus, one of the deadliest creatures in the ocean (see photo above). The flash of color is not an empty threat—it packs a toxin 10,000 times more lethal than cyanide.

The blue-ringed octopus is unusual, however. Most octopuses tend to rely on nonaggressive defense mechanisms such as camouflage. In fact, many cephalopods, the group that includes octopuses, employ an impressive color palette for communication and camouflage, and they display a remarkable assortment of behaviors. With their shifting colors and fluid bodies, these are among the most intriguing members of the kingdom Animalia. But you can decide for yourself. This chapter is a brief tour of the vast diversity found in the animal kingdom—diversity that has been evolving for perhaps a billion years. We will encounter a spectacular variety of forms ranging from corals to cockroaches to chordates. We will look at 9 major phyla of the roughly 35 phyla in the kingdom. And we will see that identifying, classifying, and arranging this diversity remain a work in progress. But first let's define what an animal is!

Animal Evolution and Diversity

1 What is an animal?

Animals are multicellular, heterotrophic eukaryotes that (with a few exceptions) obtain nutrients by ingestion. Now that's a mouthful—and speaking of mouthfuls, **Figure 1A** shows a rock python just beginning to ingest a gazelle. **Ingestion** means eating food. This mode of nutrition contrasts animals with fungi, which absorb nutrients after digesting food outside their body. Animals digest food within their body after ingesting other organisms, dead or alive, whole or by the piece.

Animals also have cells with distinctive structures and specializations. Animal cells lack the cell walls that provide strong support in the bodies of plants and fungi. Animal cells are held together by extracellular structural proteins, the most abundant of which is collagen, and by unique types of intercellular junctions. In addition, all but the simplest animals have muscle cells for movement and nerve cells for conducting impulses.

Other unique features are seen in animal reproduction and development. Most animals are diploid and reproduce sexually; eggs and sperm are the only haploid cells, as shown in the life cycle of a sea star in **Figure 1B**. ❶ Male and female adult animals make haploid gametes by meiosis, and ❷ an egg and a sperm fuse, producing a zygote. ❸ The zygote divides by mitosis, ❹ forming an early embryonic stage called a **blastula**, which is usually a hollow ball of cells. ❺ In the sea star and most other animals, one side of the blastula folds inward, forming a stage called a **gastrula**. ❻ The internal sac formed by gastrulation becomes the digestive tract, lined by a cell layer called the **endoderm**. The embryo also has an **ectoderm**, an outer cell layer that gives rise to the outer covering of the animal and, in some phyla, to the central nervous system. Most animals have a third embryonic layer, known as the **mesoderm**, which forms the muscles and most internal organs.

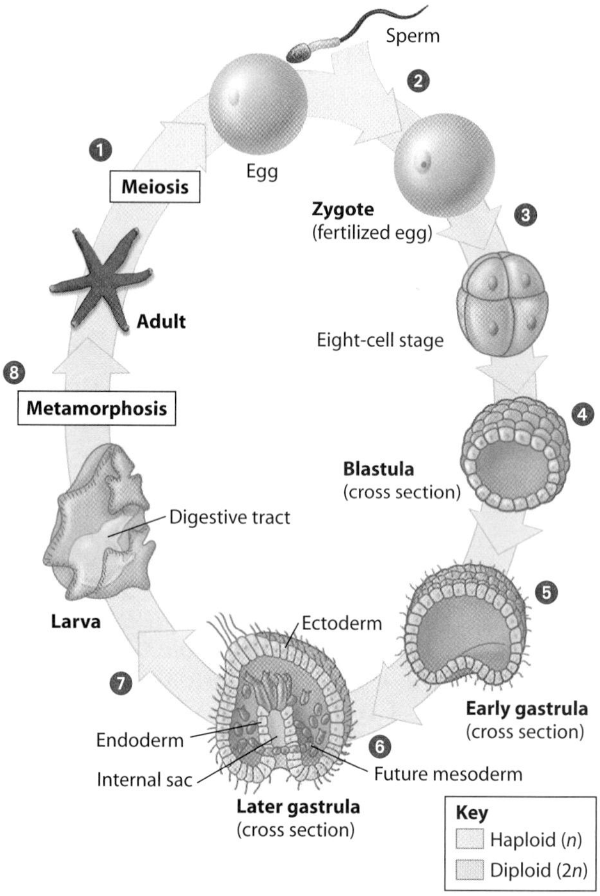

Georgette Douwma/Nature Picture Library

▲ **Figure 1B** The life cycle of a sea star

▲ **Figure 1A** Ingestion, the animal way of life

Gunter Ziesler/Peter Arnold, Inc./PhotoLibrary

After the gastrula stage, many animals develop directly into adults. Others, such as the sea star, ❼ develop into one or more larval stages first. A **larva** is an immature individual that looks different from the adult animal. ❽ The larva undergoes a major change of body form, called **metamorphosis**, in becoming an adult capable of reproducing sexually.

This transformation of a zygote into an adult animal is controlled by clusters of homeotic genes. The study of these master control genes has helped scientists investigate the phylogenetic relationships among the highly diverse animal forms we are about to survey.

? List the distinguishing characteristics of animals.

● Bodies composed of multiple eukaryotic cells; ingestion of food (heterotrophic nutrition); absence of cell walls; unique cell junctions; nerve and muscle cells (generally); sexual reproduction and life cycles with unique embryonic stages; unique developmental genes; gametes alone representing the haploid stage of the life cycle

2 Animal diversification began more than half a billion years ago

As you learned in Module 16.21, the lineage that gave rise to animals is thought to have diverged from a flagellated unikont ancestor more than 1 billion years ago. This ancestor may have resembled modern choanoflagellates, colonial protists that are the closest living relatives of animals. But despite the molecular data indicating this early origin of animals, the oldest generally accepted animal fossils that have yet been found are 575–550 million years old, from the late Ediacaran period. These fossils were first discovered in the 1940s, in the Ediacara Hills of Australia (hence the name). Similar fossils have since been found in Asia, Africa, and North America. All are impressions of soft-bodied animals that varied in shape and ranged in length from 1 cm to 1 m (**Figure 2A**). Although some of the fossils may belong to groups of invertebrates that still exist today, such as sponges and cnidarians, most do not appear to be related to any living organism.

Animal diversification appears to have accelerated rapidly from 535 to 525 million years ago, during the Cambrian period. Because many animal body plans and new phyla appear in the fossils from such an evolutionarily short time span, biologists call this episode the Cambrian explosion. The most celebrated source of Cambrian fossils is a fine-grained deposit of sedimentary rock in British Columbia. The Burgess Shale, as it is known, provided a cornucopia of perfectly preserved animal fossils. In contrast to the uniformly soft-bodied Ediacaran animals, many Cambrian animals had hard body parts such as shells and spikes. Many of these fossils are clearly related to existing animal groups. For example, scientists have classified more than a third of the species found in the Burgess Shale as arthropods, including the one labeled in **Figure 2B**. (Present-day arthropods include crabs, shrimp, and insects.) Another striking fossil represented in this reconstruction is an early member of our own phylum, Chordata. Other fossils are more difficult to place, and some are downright weird, such as the spiky creature called *Hallucigenia* and *Anomalocaris* (dominating the left half of the illustration), a formidable predator that grew to an estimated 2 feet in length. The circular structure on the underside of the animal's head is its mouth.

Lisa-Ann Gershwin, University of California-Berkeley, Museum of Paleontology

The Museum Board of South Australia 2004
Photographer: Dr. J. Gehling

Spriggina floundersi (about 3 cm long)

Dickinsonia costata (about 8 cm across)

▲ **Figure 2A** Ediacaran fossils

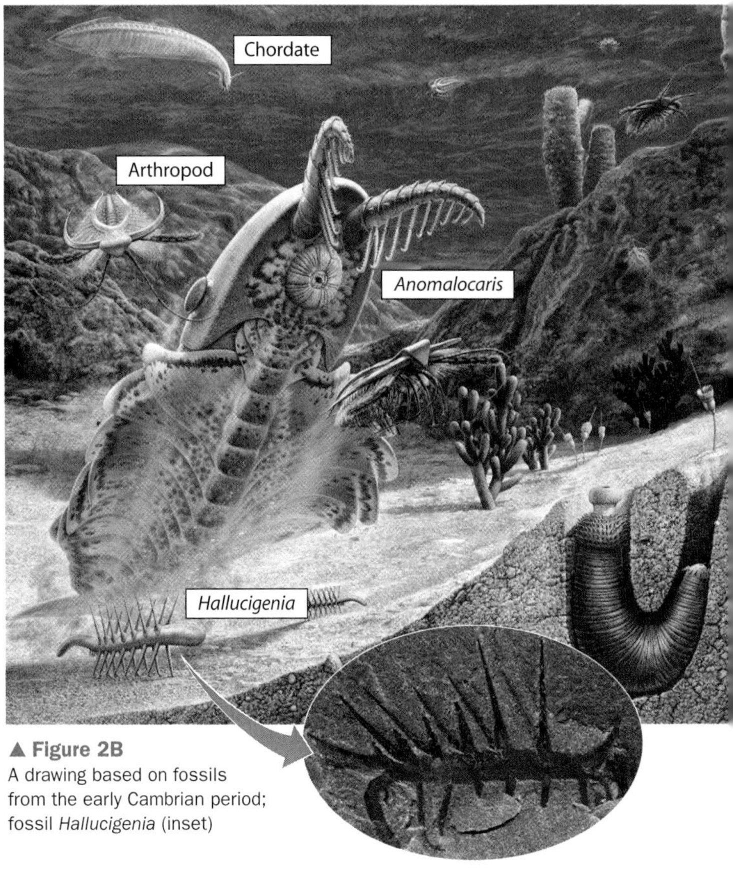

The Natural History Museum, London

▲ **Figure 2B** A drawing based on fossils from the early Cambrian period; fossil *Hallucigenia* (inset)

Chip Clark

What ignited the Cambrian explosion? Scientists have proposed several hypotheses, including increasingly complex predator-prey relationships and an increase in atmospheric oxygen. But whatever the cause of the rapid diversification, it is highly probable that the set of homeotic genes—the genetic framework for complex bodies—was already in place. Much of the diversity in body form among the animal phyla is associated with variations in where and when homeotic genes are expressed within developing embryos. The role of these master control genes in the evolution of animal diversity will be discussed further in Module 16.

Most animals are **invertebrates**, so called because they lack a vertebral column (backbone): Of the 35 or so animal phyla (systematists disagree on the precise number), all of the animals in all but one phylum are invertebrates. Let's look at some of the anatomical features biologists use to classify this vast animal diversity.

? What are two major differences between the fossil animals from the Ediacaran and the Cambrian periods?

Ediacaran animals were all soft-bodied; many Cambrian animals had hard parts such as shells. Few Ediacaran animals can be classified as members of present-day groups; many Cambrian animals are clearly related to present-day groups.

3 Animals can be characterized by basic features of their "body plan"

One way that biologists categorize the diversity of animals is by certain general features of body structure, which together describe what is called an animal's "body plan." Distinctions between body plans help biologists infer the phylogenetic relationships between animal groups.

Symmetry is a prominent feature of the body plan. Some animals have **radial symmetry**: As illustrated by the sea anemone on the left in Figure 3A, the body parts radiate from the center like the spokes of a bicycle wheel. Any imaginary slice through the central axis divides a radially symmetric animal into mirror images. Thus, the animal has a top and a bottom, but not right and left sides. As shown by the lobster in the figure, an animal with **bilateral symmetry** has mirror-image right and left sides; a distinct head, or **anterior**, end; a tail, or **posterior**, end; a back, or **dorsal**, surface; and a bottom, or **ventral**, surface.

The symmetry of an animal reflects its lifestyle. A radial animal is typically sedentary or passively drifting, meeting its environment equally on all sides. In bilaterally symmetric animals, the brain, sense organs, and mouth are usually located in the head. This arrangement facilitates mobility. As the animal travels headfirst through the environment, its eyes and other sense organs contact the environment first.

Body plans also vary in the organization of tissues. True tissues are collections of specialized cells, usually isolated from other tissues by membrane layers, that perform specific functions. An example is the nervous tissue of your brain and spinal cord. Sponges lack true tissues, but in other animals, the cell layers formed during gastrulation (see Figure 1B) give rise to true tissues and to organs. Some animal embryos have only ectoderm and endoderm; most also have mesoderm, making a body with three tissue layers.

Animals arising from embryos with three tissue layers may be characterized by the presence or absence of a **body cavity**. This fluid-filled space between the digestive tract and outer body wall cushions the internal organs and enables them to grow and move independently of the body wall. In soft-bodied animals, a noncompressible fluid in the body cavity forms a **hydrostatic skeleton** that provides a rigid structure against which muscles contract, moving the animal.

In the figures to the right, colors indicate the tissue layers from which the adult arose: ectoderm (blue), mesoderm (red), and endoderm (yellow). The cross section of a segmented worm in Figure 3B shows a body cavity called a **true coelom**, which is completely lined by tissue derived from mesoderm. A roundworm (Figure 3C) has a body cavity called a **pseudocoelom.** A pseudocoelom is not completely lined by tissue derived from mesoderm. Despite the name, pseudocoeloms function just like coeloms. A flatworm (Figure 3D) reveals a body that is solid except for the cavity of the digestive sac—it lacks a coelom.

Animals that arise from three tissue layers can be separated into two groups based on details of their embryonic development, such as the fate of the opening formed during gastrulation that leads to the developing digestive tract. In **protostomes** (from the Greek *protos*, first, and *stoma*, mouth), this opening becomes the mouth; in **deuterostomes** (from the Greek *deutero*, second), this opening becomes the anus, and the mouth forms from a second opening. Other differences between protostomes and deuterostomes include the pattern of early cell divisions and the way the coelom forms.

Next we see how these general features of body plan are used to infer relationships between animal groups.

? List four features that can describe an animal's body plan.

Symmetry, number of embryonic tissue layers, body cavity type, patterns of embryonic development

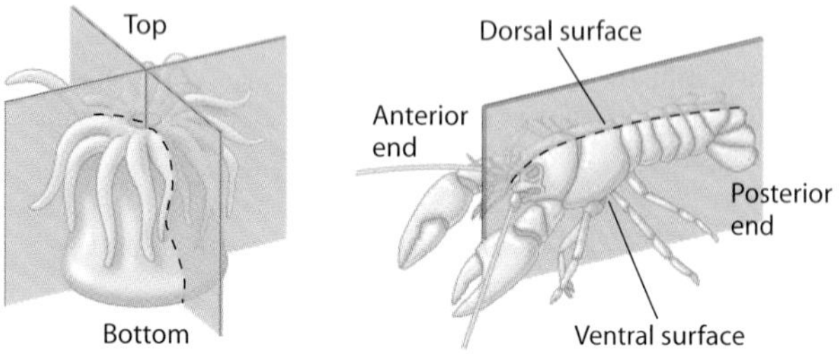

▲ **Figure 3A** Radial (left) and bilateral (right) symmetry

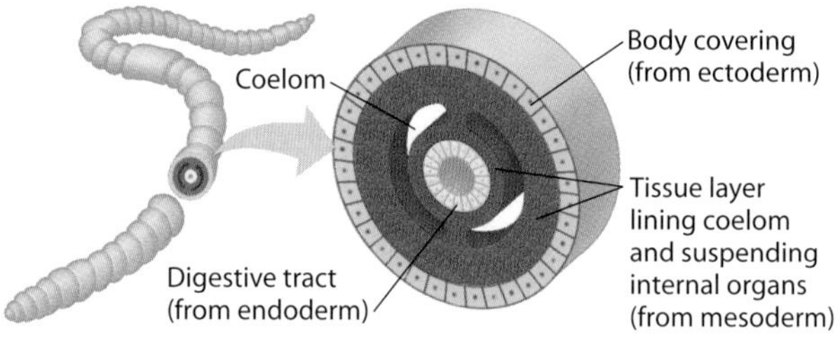

▲ **Figure 3B** True coelom (a segmented worm)

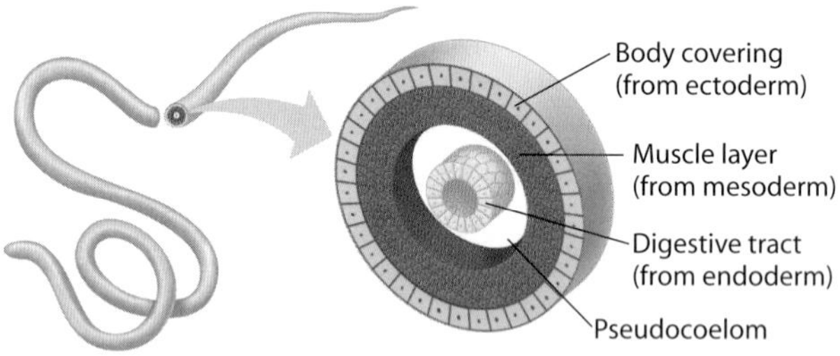

▲ **Figure 3C** Pseudocoelom (a roundworm)

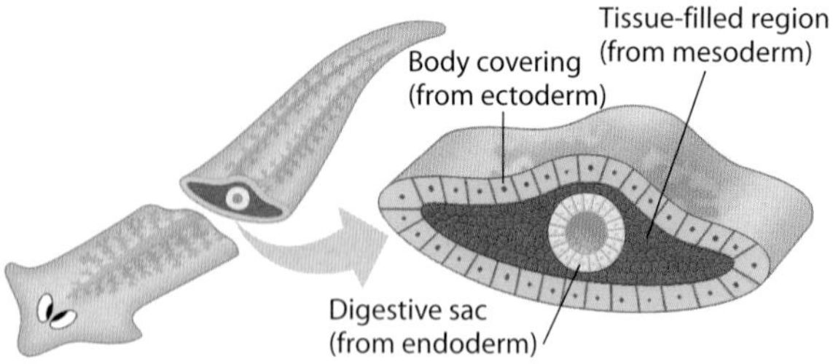

▲ **Figure 3D** No body cavity (a flatworm)

4 The body plans of animals can be used to build phylogenetic trees

Because animals diversified so rapidly on the scale of geologic time, it is difficult to sort out the evolutionary relationships among the various phyla using only the fossil record. Traditionally, biologists have proposed hypotheses about animal phylogeny based on morphological studies of living animals as well as fossils, often using the characteristics of body plan and embryonic development described in the preceding module.

Figure 4 presents a morphology-based phylogenetic tree of the major phyla of the animal kingdom. At the far left is the hypothetical ancestral colonial protist. The tree has a series of branch points that represent shared derived characters. The first branch point splits the sponges from the clade of **eumetazoans** ("true animals"), the animals with true tissues. The next branch point separates the animals with radial symmetry from those with bilateral symmetry. Notice that most animal phyla have bilateral symmetry and thus belong to the clade called **bilaterians**. This morphology-based tree then divides the bilaterians into two clades based on embryology: protostomes and deuterostomes.

All phylogenetic trees are hypotheses for the key events in the evolutionary history that led to the animal phyla now living on Earth. Increasingly, researchers are adding molecular comparisons to their data sets for identifying clades. In Module 15, we will see that such data are leading to new hypotheses for grouping animal phyla. But first let's look at the unique characteristics of each phylum and meet some examples.

? What shared derived character separates bilaterians from cnidarians?

Bilateral symmetry

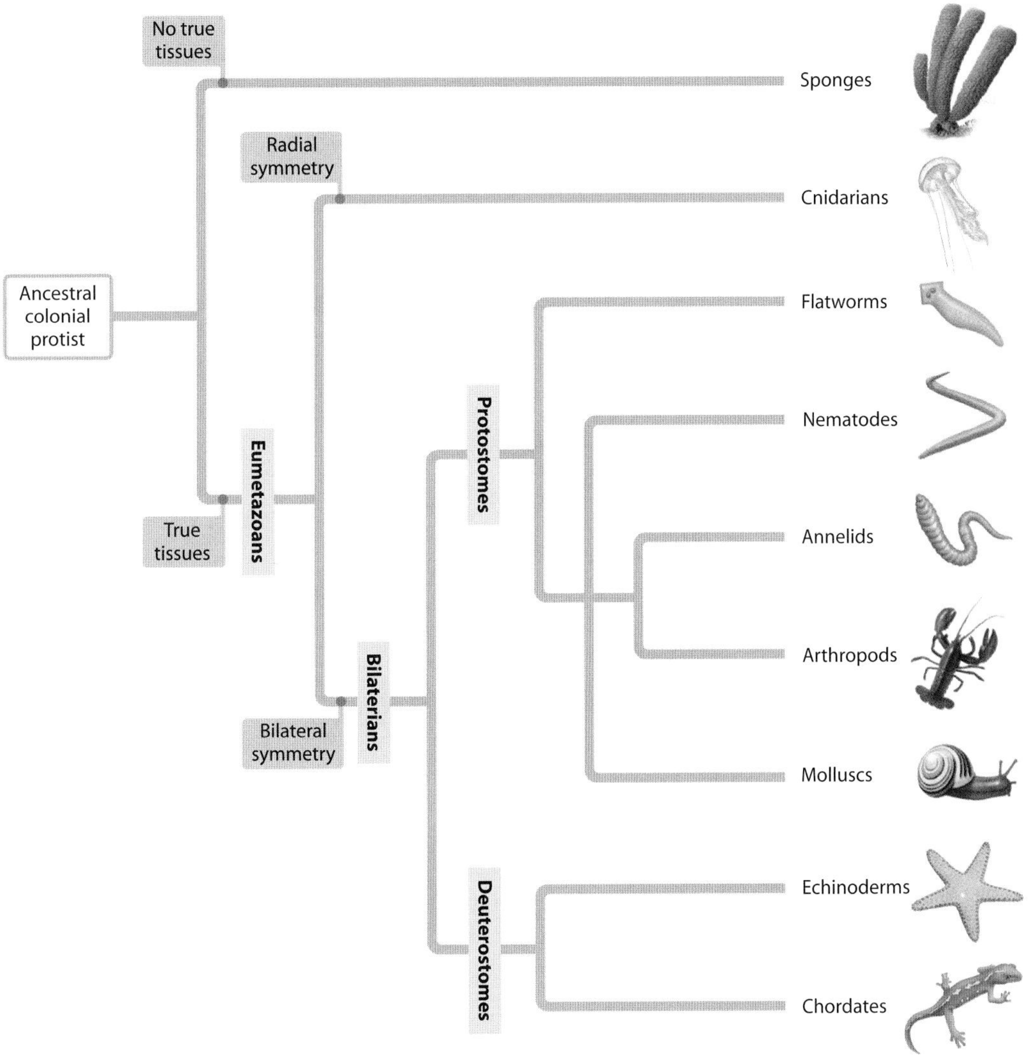

▲ **Figure 4** A hypothesis of animal phylogency based on morphological comparisons

Invertebrate Diversity

5 Sponges have a relatively simple, porous body

Sponges (phylum Porifera) are the simplest of all animals. They have no nerves or muscles, though their individual cells can sense and react to changes in the environment. The majority of species are marine, although some are found in fresh water. Some sponges are radially symmetric, but most lack body symmetry. There is variation in the size and internal structure of sponges. For example, the purple tube sponge shown in **Figure 5A** can reach heights of 1.5 m (about 5 feet). The body of *Scypha*, a small sponge only about 1–3 cm tall resembles a simple sac. Other sponges, such as the azure vase sponge, have folded body walls and irregular shapes.

A simple sponge resembles a thick-walled sac perforated with holes. (*Porifera* means "pore-bearer" in Latin.) Water enters through the pores into a central cavity, then flows out through a larger opening **(Figure 5B)**. More complex sponges have branching water canals.

The body of a sponge consists of two layers of cells separated by a gelatinous region. Since the cell layers are loose associations of cells, they are not considered true tissues. The inner cell layer consists of flagellated "collar" cells called **choanocytes** (tan in Figure 5B), which help to sweep water through the sponge's body. **Amoebocytes** (blue), which wander through the middle body region, produce supportive skeletal fibers (yellow) composed of a flexible protein called spongin and mineralized particles called spicules. Most sponges have both types of skeletal components, but some, including those used as bath sponges, only contain spongin.

Sponges are examples of **suspension feeders** (also known as filter feeders), animals that collect food particles from water passed through some type of food-trapping equipment. Sponges feed by filtering food particles suspended in the water that streams through their porous bodies. To obtain enough food to grow by 100 g (about 3 ounces), a sponge must filter roughly 1,000 kg (about 275 gallons) of seawater. Choanocytes trap food particles in mucus on the membranous collars that surround the base of their flagella and then engulf the food by phagocytosis. Amoebocytes pick up food packaged in food vacuoles from choanocytes, digest it, and carry the nutrients to other cells.

Adult sponges are **sessile**, meaning they are anchored in place—they cannot escape from predators. Researchers have found that sponges produce defensive compounds such as toxins and antibiotics that deter pathogens, parasites, and predators. Some of these compounds may prove useful to humans as new drugs.

Biologists hypothesize that sponge lineages arose very early from the multicellular organisms that gave rise to the animal kingdom. The choanocytes of sponges and the cells of living choanoflagellates are similar, supporting the molecular evidence that animals evolved from a flagellated protist ancestor. Sponges are the only animal phylum covered in this book that lack true tissues and thus are not members of the clade Eumetazoa ("true animals"). You will learn about the simplest Eumetazoans in the next module.

? Why is it thought that sponges represent the earliest branch of the animal kingdom?

Sponges lack true tissues, and their choanocytes resemble certain flagellated protists.

Image Quest Marine

Charles R. Wyttenbach/ Biological Photo Service

▲ **Figure 5A** Sponges

Andrew J. Martinez/Photo Researchers

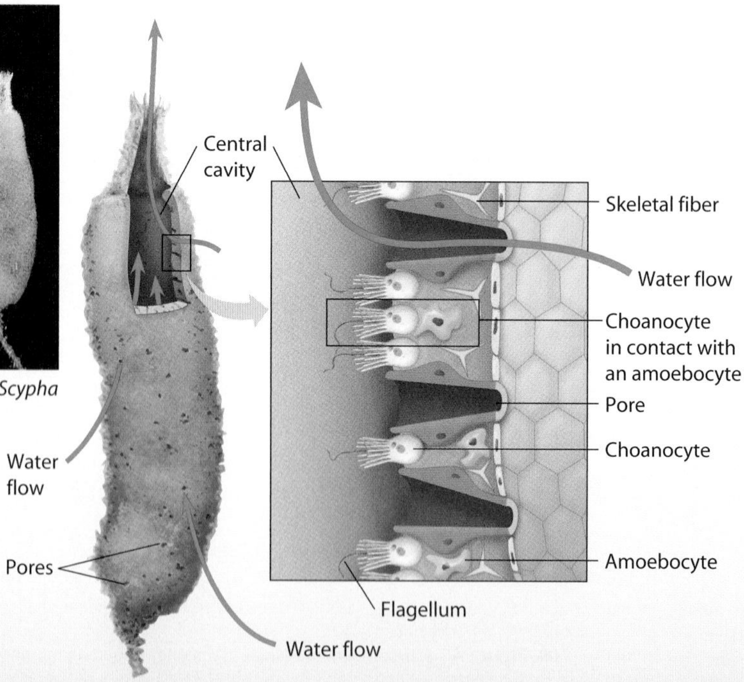

▲ **Figure 5B** The structure of a simple sponge

6 Cnidarians are radial animals with tentacles and stinging cells

Among eumetazoans, one of the oldest groups is phylum Cnidaria, which includes the hydras, sea anemones, corals, and jellies (also called "jellyfish"). **Cnidarians** are characterized by radial symmetry and bodies arising from only two tissue layers. The simple body of most cnidarians has an outer epidermis and an inner cell layer that lines the digestive cavity. A jelly-filled middle region may contain scattered amoeboid cells. Contractile tissues and nerves occur in their simplest forms in cnidarians.

Cnidarians exhibit two kinds of radially symmetric body forms. Hydras, common in freshwater ponds and lakes, and sea anemones have a cylindrical body with tentacles projecting from one end. This body form is a **polyp** (Figure 6A). The other type of cnidarian body is the **medusa**, exemplified by the marine jelly in Figure 6B. While polyps are mostly stationary, medusae move freely about in the water. They are shaped like an umbrella with a fringe of tentacles around the lower edge. A few jellies have tentacles 60–70 m long dangling from an umbrella up to 2 m in diameter, but the diameter of most jellies ranges from 2 to 40 cm.

Some cnidarians pass sequentially through both a polyp stage and a medusa stage in their life cycle. Others exist only as medusae; still others, including hydras and sea anemones, exist only as polyps.

Despite their flowerlike appearance, cnidarians are carnivores that use their tentacles to capture small animals and protists and to push the prey into their mouths. In a polyp, the mouth is on the top of the body, at the hub of the radiating tentacles. In a medusa, the mouth is in the center of the undersurface. The mouth leads into a multifunctional compartment called a **gastrovascular cavity** (from the Greek *gaster*, belly, and Latin *vas*, vessel), where food is digested. The mouth is the only opening in the body, so it is also the exit for undigested food and other wastes. Fluid in the gastrovascular cavity circulates nutrients and oxygen to internal cells and removes metabolic wastes (hence the "vascular" in gastrovascular;). The fluid also acts as a hydrostatic skeleton, supporting the body and helping to give a cnidarian its shape, much like water can give shape to a balloon. When the animal closes its mouth, the volume of the cavity is fixed. Then contraction of selected cells changes the shape of the animal and can produce movement.

Phylum Cnidaria (from the Greek *cnide*, nettle) is named for its unique stinging cells, called **cnidocytes**, that function in defense and in capturing prey. Each cnidocyte contains a fine thread coiled within a capsule (Figure 6C). When it is discharged, the thread can sting or entangle prey. Some large marine cnidarians use their stinging threads to catch fish. A group of cnidarians called cubozoans have highly toxic cnidocytes. The sea wasp, a cubozoan found off the coast of northern Australia, is the deadliest organism on Earth: One animal may produce enough poison to kill as many as 60 people.

Coral animals are polyp-form cnidarians that secrete a hard external skeleton. Each generation builds on top of the skeletons of previous generations, constructing the characteristic shapes of "rocks" we call coral. Reef-building corals depend on sugars produced by symbiotic algae to supply them with enough energy to maintain the reef structure in the face of erosion and reef-boring animals.

? What are three functions of a cnidarian's gastrovascular cavity?

Digestion, circulation (transport of oxygen, nutrients, and wastes), and physical support and movement

Gwen Fidler/Comstock Images

A hydra (about 2–25 mm tall)

Jonathan Bird/Photolibrary

A sea anemone (about 6 cm in diameter)

▲ **Figure 6A** Polyp body form

Reinhard Dirscherl/Photolibrary

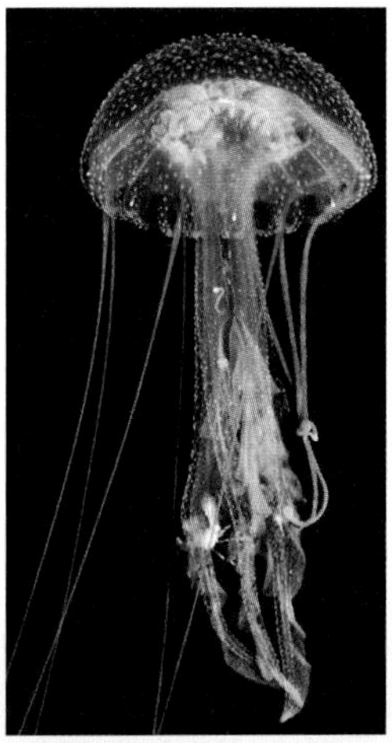

A marine jelly (about 6 cm in diameter)

▲ **Figure 6B** Medusa body form

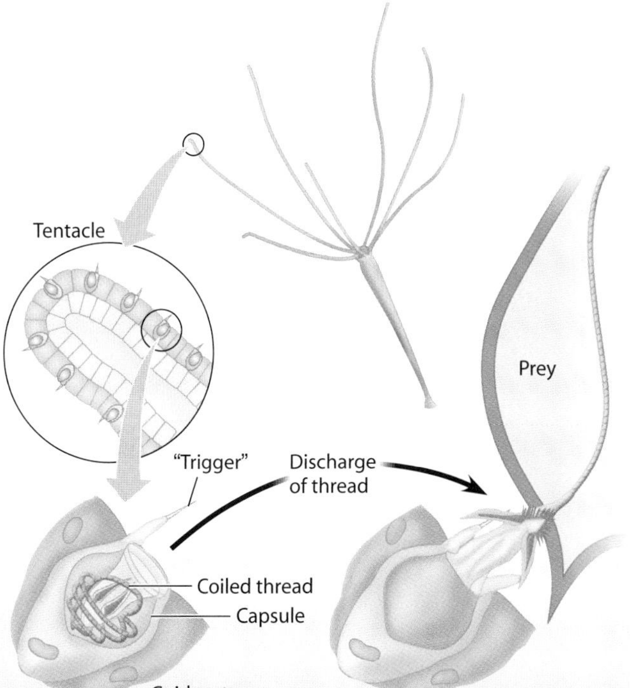

▲ **Figure 6C** Cnidocyte action

7 Flatworms are the simplest bilateral animals

The vast majority of animal species belong to the clade Bilateria, which consists of animals with bilateral symmetry and three embryonic tissue layers. In Modules 7 through 12, you'll learn about phyla belonging to the protostome group of bilaterians. **Flatworms**, phylum Platyhelminthes (from the Greek *platys*, flat, and *helmis*, worm), are the simplest of the bilaterians. These thin, often ribbonlike animals range in length from about 1 mm to 20 m and live in marine, freshwater, and damp terrestrial habitats. In addition to free-living forms, there are many parasitic species. Like cnidarians, most flatworms have a gastrovascular cavity with only one opening. Fine branches of the gastrovascular cavity distribute food throughout the animal.

There are three major groups of flatworms. Worms called planarians represent the **free-living flatworms** (**Figure 7A**). A planarian has a head with a pair of light-sensitive eyecups and a flap at each side that detects chemicals. Dense clusters of nerve cells form a simple brain, and a pair of nerve cords connect with small nerves that branch throughout the body.

When a planarian feeds, it sucks food in through a mouth at the tip of a muscular tube that projects from a surprising location—the mid-ventral surface of the body (as shown in the figure). Planarians live on the undersurfaces of rocks in freshwater ponds and streams. Using cilia on their ventral surface, they crawl about in search of food. They also have muscles that enable them to twist and turn.

A second group of flatworms, the **flukes**, live as parasites in other animals. Many flukes have suckers that attach to their host and a tough protective covering. Reproductive organs occupy nearly the entire interior of these worms.

Many flukes have complex life cycles that facilitate dispersal of offspring to new hosts. Larvae develop in an intermediate host. The larvae then infect the final host in which they live as adults. For example, blood flukes called schistosomes spend part of their life cycle in snails. The final hosts are humans, who suffer symptoms such as abdominal pain, bloody diarrhea, and liver problems as a result of the parasite's eggs lodging in their organs and blood capillaries. Although schistosomes are not found in the United States, more than 200 million people around the world are infected by these parasites each year.

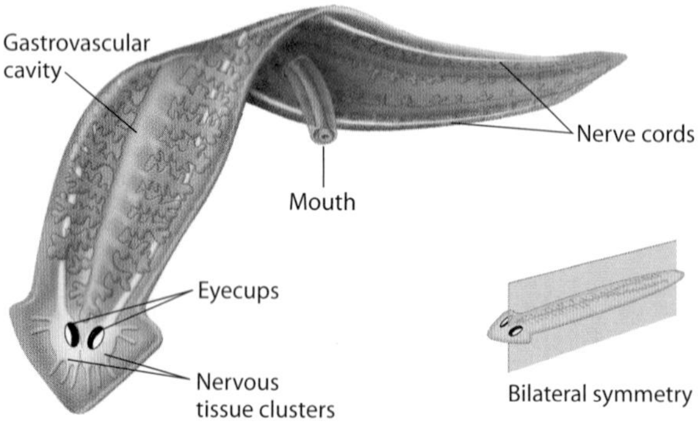

▲ **Figure 7A** A free-living flatworm, the planarian (most are about 5–10 mm long)

Tapeworms are another parasitic group of flatworms. Adult tapeworms inhabit the digestive tracts of vertebrates, including humans. In contrast with planarians and flukes, most tapeworms have a very long, ribbonlike body with repeated units. As **Figure 7B** shows, the anterior end, called the scolex, is armed with hooks and suckers that grasp the host. Notice that there is no mouth. Bathed in the partially digested food in the intestines of their hosts, they simply absorb nutrients across their body surface and have no digestive tract. Because of this adaptation to their parasitic lifestyle, tapeworms are an exception to our definition of animals in Module 1; other animals ingest nutrients. Behind the scolex is a long ribbon of repeated units filled with both male and female reproductive structures. The units at the posterior end, which are full of ripe eggs, break off and pass out of the host's body in feces.

Like parasitic flukes, tapeworms have a complex life cycle, usually involving more than one host. Most species take advantage of the predator-prey relationships of their hosts. A prey species—a sheep or a rabbit, for example—may become infected by eating grass contaminated with tapeworm eggs. Larval tapeworms develop in these hosts, and a predator—a coyote or a dog, for instance—becomes infected when it eats an infected prey animal. The adult tapeworms then develop in the predator's intestines. Humans can be infected with tapeworms by eating undercooked beef, pork, or fish infected with tapeworm larvae. The larvae are microscopic, but the adults of some species can reach lengths of 2 m in the human intestine.

The term *worm* is commonly applied to any slender, elongated invertebrate. Thus, flatworms are just one of three major animal phyla known as worms. You'll learn about roundworms in the next module and segmented worms in Module 10.

? **Flatworms and cnidarians differ in symmetry, with flatworms being ______ and cnidarians being ______, but the animals of both phyla have a ______.**

bilateral . . . radial . . . gastrovascular cavity

Eye of Science/Photo Researchers

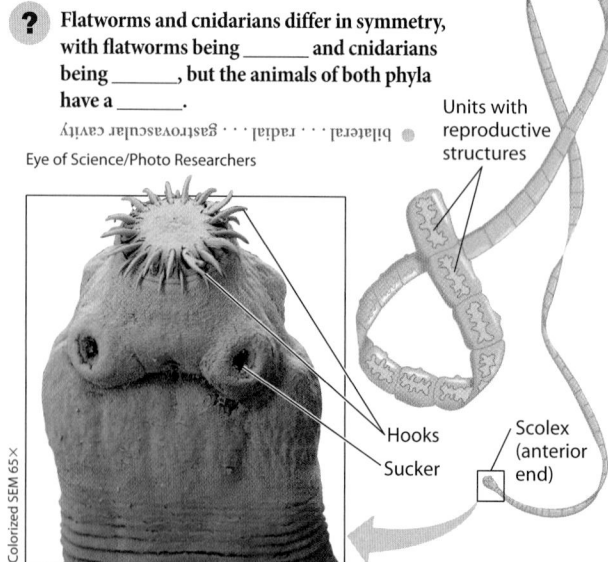

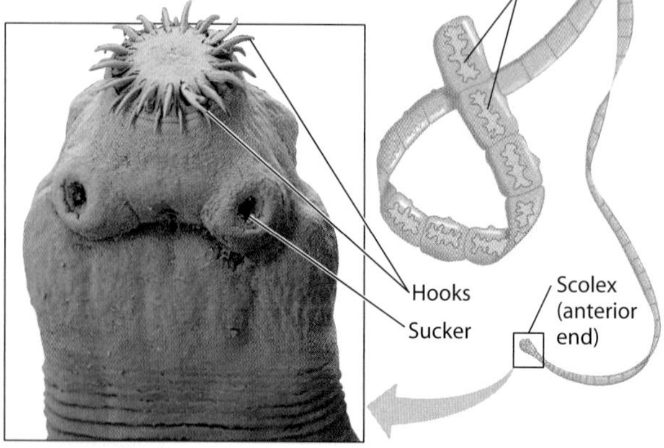

▲ **Figure 7B** A tapeworm, a parasitic flatworm

8 Nematodes have a pseudocoelom and a complete digestive tract

Nematodes, also called roundworms, make up the phylum Nematoda. As bilaterians, these animals have bilateral symmetry and an embryo with three tissue layers. In contrast with flatworms, roundworms have a fluid-filled body cavity (a pseudocoelom, not completely lined with mesoderm) and a digestive tract with two openings.

Nematodes are cylindrical with a blunt head and tapered tail. They range in size from less than 1 mm to more than a meter. Several layers of tough, nonliving material called a **cuticle** cover the body and prevent the nematode from drying out. In parasitic species, the cuticle protects the nematode from the host's digestive system. When the worm grows, it periodically sheds its cuticle (molts) and secretes a new, larger one. What looks like a corduroy coat on the nematode in **Figure 8A** is its cuticle.

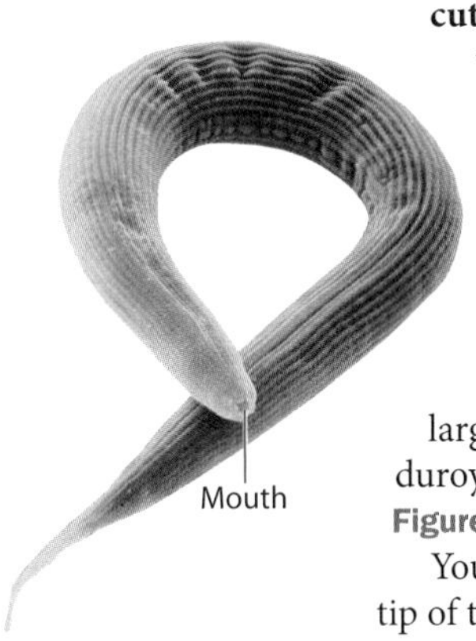

▲ **Figure 8A**
A free-living nematode

You can also see the mouth at the tip of the blunt anterior end of the nematode in Figure 8A. Nematodes have a **complete digestive tract** extending as a tube from the mouth to the anus near the tip of the tail. Food travels only one way through the system and is processed as it moves along. In animals with a complete digestive tract, the anterior regions of the tract churn and mix food with enzymes, while the posterior regions absorb nutrients and then dispose of wastes. This division of labor makes the process more efficient and allows each part of the digestive tract to be specialized for its particular function.

Fluid in the pseudocoelom of a nematode distributes nutrients absorbed from the digestive tract throughout the body. The pseudocoelom also functions as a hydrostatic skeleton, and contraction of longitudinal muscles produces a whiplike motion that is characteristic of nematodes.

Although about 25,000 species of nematodes have been named, estimates of the total number of species range as high as 500,000. Free-living nematodes live virtually everywhere there is rotting organic matter, and their numbers are huge. Ninety thousand individuals were found in a single rotting apple lying on the ground; an acre of topsoil contains billions of nematodes. Nematodes are important decomposers in soil and on the bottom of lakes and oceans. Some are predators, eating other microscopic animals.

Little is known about most free-living nematodes. A notable exception is the soil-dwelling species *Caenorhabditis elegans*, an important research organism. A *C. elegans* adult consists of only about 1,000 cells—in contrast with the human body, which consists of some 10 trillion cells. By following every cell division in the developing embryo, biologists have been able to trace the lineage of every cell in the adult worm. The genome of *C. elegans* has been sequenced, and ongoing research contributes to our understanding of how genes control animal development, the functioning of nervous systems, and even some of the mechanisms of aging.

Other nematodes thrive as parasites in the moist tissues of plants and in the body fluids and tissues of animals. The largest known nematodes are parasites of whales and measure more than 7 m (23 feet) long! Many species are serious agricultural pests that attack the roots of plants or parasitize animals. The dog heartworm (**Figure 8B**), a common parasite, is deadly to dogs and can also infect other pets such as cats and ferrets. It is spread by mosquitoes, which pick up heartworm eggs in the blood of an infected host and transmit them when sucking the blood of another animal. Although dog heartworms were once found primarily in the southeastern United States, they are now common throughout the contiguous United States. Regular doses of a preventive medication can protect dogs from heartworm.

Humans are host to at least 50 species of nematodes, including a number of disease-causing organisms. *Trichinella spiralis* causes a disease called trichinosis in a wide variety of mammals, including humans. People usually acquire the worms by eating undercooked pork or wild game containing the juvenile worms. Cooking meat until it is no longer pink kills the worms. Hookworms, which grapple onto the intestinal wall and suck blood, infect millions of people worldwide. Although hookworms are small (about 10 mm long), a heavy infestation can cause severe anemia.

You might expect that an animal group as numerous and widespread as nematodes would include a great diversity of body form. In fact, the opposite is true. Most species look very much alike. In sharp contrast, animals in the phylum Mollusca, which we examine next, exhibit enormous diversity in body form.

? What is the advantage of a complete digestive tract?

Different parts of the digestive tract can be specialized for different functions.

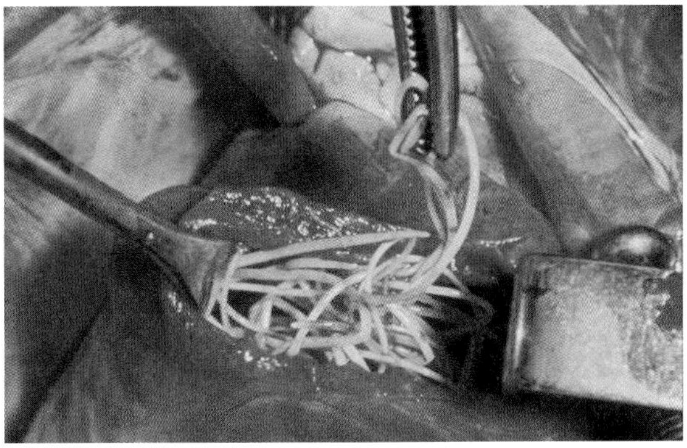

▲ **Figure 8B** A dog heart infested by heartworm, a parasitic nematode

9 Diverse molluscs are variations on a common body plan

Snails, slugs, oysters, clams, octopuses, and squids are just a few of the great variety of animals known as **molluscs** (phylum Mollusca). Molluscs are soft-bodied animals (from the Latin *molluscus*, soft), but most are protected by a hard shell.

You may wonder how animals as different as octopuses and clams could belong in the same phylum, but these and other molluscs have inherited several common features from their ancestors. Figure 9A illustrates the basic body plan of a mollusc, consisting of three main parts: a muscular **foot** (gray in the drawing), which functions in locomotion; a **visceral mass** (orange) containing most of the internal organs; and a **mantle** (purple), a fold of tissue that drapes over the visceral mass and secretes a shell in molluscs such as clams and snails. In many molluscs, the mantle extends beyond the visceral mass, producing a water-filled chamber called the mantle cavity, which houses the gills (left side in Figure 9A).

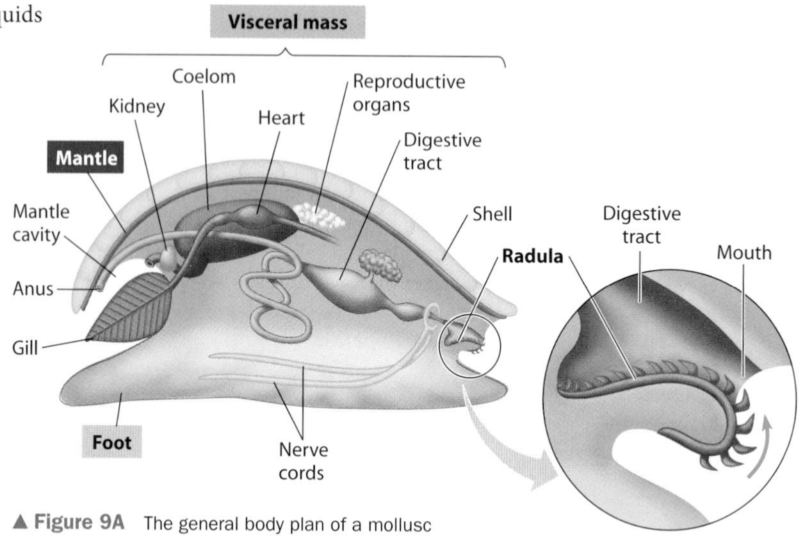

▲ **Figure 9A** The general body plan of a mollusc

Figure 9A shows yet another body feature found in many molluscs—a unique rasping organ called a **radula**, which is used to scrape up food. In a snail, for example, the radula extends from the mouth and slides back and forth like a backhoe, scraping and scooping algae off rocks. You can observe a radula in action by watching a snail graze on the glass wall of an aquarium.

Most molluscs have separate sexes, with reproductive organs located in the visceral mass. The life cycle of many marine molluscs includes a ciliated larva called a trochophore (Figure 9B). As you will learn in Module 15, trochophore larvae are a significant characteristic in determining phylogenetic relationships among the invertebrate phyla.

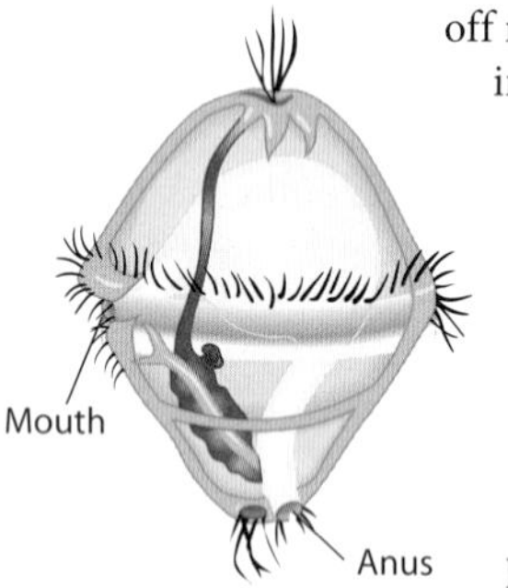

▲ **Figure 9B** Trochophore larva

In contrast with flatworms, which have no body cavity, and nematodes, which have a pseudocoelom, molluscs have a true coelom (brown in Figure 9A). Also, unlike flatworms and nematodes, molluscs have a **circulatory system**—an organ system that pumps blood and distributes nutrients and oxygen throughout the body.

These basic body features have evolved in markedly different ways in different groups of molluscs. The three most diverse groups are the gastropods (including snails and slugs), bivalves (including clams, scallops, and oysters), and cephalopods (including squids and octopuses).

Gastropods The largest group of molluscs is called the **gastropods** (from the Greek *gaster*, belly, and *pod*, foot), found in fresh water, salt water, and terrestrial environments. In fact, they include the only molluscs that live on land. Most gastropods are protected by a single, spiraled shell into which the animal can retreat when threatened. Many gastropods have a distinct head with eyes at the tips of tentacles, like the land snail in Figure 9C. Terrestrial snails lack the gills typical of aquatic molluscs; instead, the lining of the mantle cavity functions as a lung, exchanging gases with the air.

Most gastropods are marine, and shell collectors delight in their variety. Slugs, however, are unusual molluscs in that they have lost their mantle and shell during their evolution. The long, colorful projections on the sea slug in Figure 9C function as gills.

▲ **Figure 9C** Gastropods

Marevision/AGE fotostock

Harold W. Pratt/Biological Photo Service

A scallop (about 10 cm in diameter)

Mussels (each about 6 cm long)

▲ **Figure 9D** Bivalves

Bivalves The **bivalves** (from the Latin *bi*, double, and *valva*, leaf of a folding door) include numerous species of clams, oysters, mussels, and scallops. They have shells divided into two halves that are hinged together. Most bivalves are suspension feeders. The mantle cavity contains gills that are used for feeding as well as gas exchange. The mucus-coated gills trap fine food particles suspended in the water, and cilia sweep the particles to the mouth. Most bivalves are sedentary, living in sand or mud. They may use their muscular foot for digging and anchoring. Mussels are sessile, secreting strong threads that attach them to rocks, docks, and boats **(Figure 9D)**. The scallop in Figure 9D can skitter along the seafloor by flapping its shell, rather like the mechanical false teeth sold in novelty shops. Notice the many bluish eyes peering out between the two halves of its hinged shell. The eyes are set into the fringed edges of the animal's mantle.

Cephalopods The **cephalopods** (from the Greek *kephale*, head, and *pod*, foot) differ from gastropods and bivalves in being adapted to the lifestyle of fast, agile predators. The chambered nautilus in **Figure 9E** is a descendant of ancient groups with external shells, but in other cephalopods, the shell is small and internal (as in squids) or missing altogether (as in octopuses). If you have a pet bird, you may have hung the internal shell of another cephalopod, the cuttlefish, in its cage. Such "cuttlebones" are commonly given to caged birds as a source of calcium. Cephalopods use beak-like jaws and a radula to crush or rip prey apart. The mouth is at the base of the foot, which is drawn out into several long tentacles for catching and holding prey.

The squid in Figure 9E ranks with fishes as a fast, streamlined predator. It darts about by drawing water into its mantle cavity and then forcing a jet of water out through a muscular siphon. It steers by pointing the siphon in different directions. Octopuses, such as the ones described in the chapter introduction, live on the seafloor, where they prowl about in search of crabs and other food.

All cephalopods have large brains and sophisticated sense organs that contribute to their success as mobile predators. Cephalopod eyes are among the most complex sense organs in the animal kingdom. Each eye contains a lens that focuses light and a retina on which clear images form. Octopuses are considered among the most intelligent invertebrates and have shown remarkable learning abilities in laboratory experiments.

The so-called colossal squid, which lives in the ocean depths near Antarctica, is the largest of all invertebrates. Specimens are rare, but in 2007, a male colossal squid measuring 10 m and weighing 450 kg was hauled in by a fishing boat. Females are generally even larger, and scientists estimate that the colossal squid averages around 13 m in overall length. The giant squid, which rivals the colossal in length, is thought to average about 10 m. Most information about these impressive animals has been gleaned from dead specimens that washed ashore. In 2004, however, Japanese zoologists succeeded in capturing the first video footage of a giant squid in its natural habitat.

? Identify the mollusc group that includes each of these examples: garden snail, clam, squid.

Gastropod, bivalve, cephalopod

Inga Ivanova/iStockphoto

A squid (internal shell)

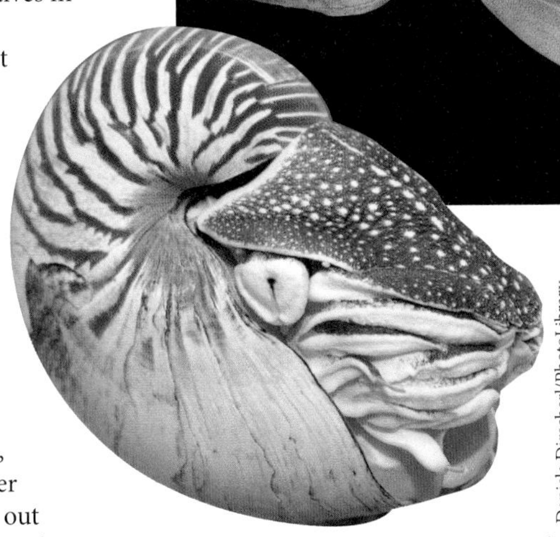

Daniela Dirscherl/PhotoLibrary

A chambered nautilus (about 21 cm in diameter)

◀ **Figure 9E** Cephalopods

10 Annelids are segmented worms

A segmented body resembling a series of fused rings is the hallmark of phylum Annelida (from the Latin *anellus*, ring). **Segmentation**, the subdivision of the body along its length into a series of repeated parts (segments), played a central role in the evolution of many complex animals. A segmented body allows for greater flexibility and mobility, and it probably evolved as an adaptation facilitating movement. An earthworm, a typical **annelid**, uses its flexible, segmented body to crawl and burrow rapidly into the soil.

Annelids range in length from less than 1 mm to 3 m, the length of some giant Australian earthworms. They are found in damp soil, in the sea, and in most freshwater habitats. Some aquatic annelids swim in pursuit of food, but most are bottom-dwelling scavengers that burrow in sand and mud. There are three main groups of annelids: earthworms and their relatives, polychaetes, and leeches.

Earthworms and Their Relatives **Figure 10A** illustrates the segmented anatomy of an earthworm. Internally, the coelom is partitioned by membrane walls (only a few are fully shown here). Many of the internal body structures are repeated within each segment. The nervous system (yellow) includes a simple brain and a ventral nerve cord with a cluster of nerve cells in each segment. Excretory organs (green), which dispose of fluid wastes, are also repeated in each segment (only a few are shown in this diagram). The digestive tract, however, is not segmented; it passes through the segment walls from the mouth to the anus.

Many invertebrates, including most molluscs and all arthropods (which you will meet next), have what is called an **open circulatory system**, in which blood is pumped through vessels that open into body cavities, where organs are bathed directly in blood. Annelids and vertebrates, in contrast, have a **closed circulatory system**, in which blood remains enclosed in vessels as it distributes nutrients and oxygen throughout the body. As you can see in the diagram at the lower left, the main vessels of the earthworm circulatory system—a dorsal blood vessel and a ventral blood vessel—are connected by segmental vessels. The pumping organ, or "heart," is simply an enlarged region of the dorsal blood vessel plus five pairs of segmental vessels near the anterior end.

Each segment is surrounded by longitudinal and circular muscles. Earthworms move by coordinating the contraction of these two sets of muscles. The muscles work against the coelomic fluid in each segment, which acts as a hydrostatic skeleton. Each segment also has four pairs of stiff bristles that provide traction for burrowing.

Earthworms are hermaphrodites; that is, they have both male and female reproductive structures. However, they do not fertilize their own eggs. Mating earthworms align their bodies facing in opposite directions and exchange sperm. Fertilization occurs some time later, when a specialized organ, visible as the

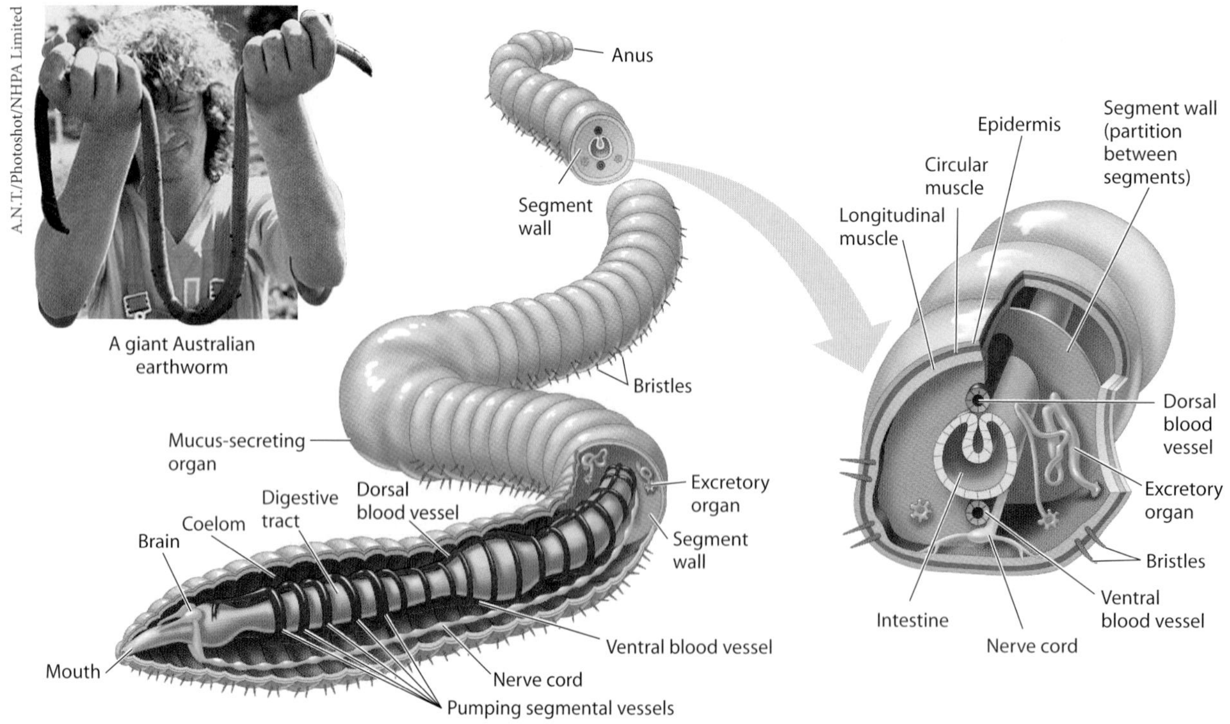

▲ **Figure 10A** Segmentation and internal anatomy of an earthworm

Jurgen Freund/Nature Picture Library

Alexander Semenov/Photo Researchers.

A sandworm

Peter Herring/Image Quest Marine

▲ **Figure 10B** Polychaetes

thickened region of the worm in Figure 10A, secretes a cocoon made of mucus. The cocoon slides along the worm, picking up the worm's own eggs and the sperm it received from its partner. The cocoon then slips off the worm into the soil, where the embryos develop.

Earthworms eat their way through the soil, extracting nutrients as soil passes through their digestive tube. Undigested material, mixed with mucus secreted into the digestive tract, is eliminated as castings (feces) through the anus. Farmers and gardeners value earthworms because the animals aerate the soil and their castings improve the soil's texture. Charles Darwin estimated that a single acre of British farmland had about 50,000 earthworms, producing 18 tons of castings per year.

Polychaetes The **polychaetes** (from the Greek *polys*, many, and *chaeta*, hair), which are mostly marine, form the largest group of annelids.

Many polychaetes live in tubes and extend feathery appendages coated with mucus that trap suspended food particles. Tube-dwellers usually build their tubes by mixing mucus with bits of sand and broken shells. Some species of tube-dwellers are colonial, such as the group shown in **Figure 10B**. The circlet of feathery appendages seen at the mouth of each tube extends from the head of the worm inside. The free-swimming polychaete in Figure 10B travels in the open ocean by moving the paddle-like appendages on each segment. In polychaetes such as the sandworm that live in the sediments, stiff bristles (called *chaetae*) on the appendages help the worm wriggle about in search of small invertebrates to eat. In many polychaetes, the appendages are richly supplied with blood vessels and are either associated with the gills or function as gills themselves.

Leeches The third main group of annelids is the **leeches**, which are notorious for their bloodsucking habits. However, most species are free-living carnivores that eat small invertebrates such as snails and insects. The majority of leeches inhabit fresh water, but there are also marine species and a few terrestrial species that inhabit moist vegetation in the tropics. Leeches range in length from 1 to 30 cm.

Some bloodsucking leeches use razor-like jaws to slit the skin of an animal. The host is usually oblivious to this attack because the leech secretes an anesthetic as well as an anticoagulant into the wound. The leech then sucks as much blood as it can hold, often more than 10 times its own weight. After this gorging, a leech can last for months without another meal.

Until the 1920s, physicians used leeches for bloodletting. For centuries, illness was thought to result from an imbalance in the body's fluids, and the practice of bloodletting was originally conceived to restore the natural balance. Later, physicians viewed bloodletting as a kind of spring cleaning for the body to remove any toxins or "bad blood" that had accumulated. Leeches are still occasionally applied to remove blood from bruised tissues (**Figure 10C**) and to help relieve swelling in fingers or toes that have been sewn back on after accidents. Blood tends to accumulate in a reattached finger or toe until small veins have a chance to grow back into it.

Astrid & Hanns-Frieder Michler/ SPL/Photo Researchers

▲ **Figure 10C**
A medicinal leech applied to drain blood from a patient

The anticoagulant produced by leeches has also proved to be medically useful. It is used to dissolve blood clots that form during surgery or as a result of heart disease. Because it is difficult to obtain this chemical from natural sources, it is now being produced through genetic engineering.

The segments of an annelid are all very similar. In the next group we explore, the arthropods, body segments and their appendages have become specialized, serving a variety of functions.

? Tapeworms and bloodsucking leeches are parasites. What are the key differences between these two?

Whereas both are composed of repeated segments, the segments of a tapeworm are filled mostly with reproductive organs and are shed from the posterior end of the animal. Tapeworms are flatworms with no body cavity and, in their parasitic lifestyle, not even a gastrovascular cavity. Leeches have a true coelom and a complete digestive tract.

11 Arthropods are segmented animals with jointed appendages and an exoskeleton

Over a million species of **arthropods**—including crayfish, lobsters, crabs, barnacles, spiders, ticks, and insects—have been identified. Biologists estimate that the arthropod population of the world numbers about a billion billion (10^{18}) individuals! In terms of species diversity, geographic distribution, and sheer numbers, phylum Arthropoda must be regarded as the most successful animal phylum.

The diversity and success of arthropods are largely related to their segmentation, their hard exoskeleton, and their jointed appendages, for which the phylum is named (from the Greek *arthron*, joint, and *pod*, foot). As indicated in the drawing of a lobster in **Figure 11A**, the appendages are variously adapted for sensory reception, defense, feeding, walking, and swimming. The arthropod body, including the appendages, is covered by an **exoskeleton**, an external skeleton that protects the animal and provides points of attachment for the muscles that move the appendages. This exoskeleton is a cuticle, a nonliving covering that in arthropods is hardened by layers of protein and chitin, a polysaccharide. The exoskeleton is thick around the head, where its main function is to house and protect the brain. It is paper-thin and flexible in other locations, such as the joints of the legs. As it grows, an arthropod must periodically shed its old exoskeleton and secrete a larger one, a complex process called **molting**.

In contrast with annelids, which have similar segments along their body, the body of most arthropods arises from several distinct groups of segments that fuse during development: the head, thorax, and abdomen. In some arthropods, including the lobster, the exoskeleton of the head and thorax is partly fused, forming a body region called the cephalothorax. Each of the segment groups is specialized for a different function. In a lobster, the head bears sensory antennae, eyes, and jointed mouthparts on the ventral side. The thorax bears a pair of defensive appendages (the pincers) and four pairs of legs for walking. The abdomen has swimming appendages.

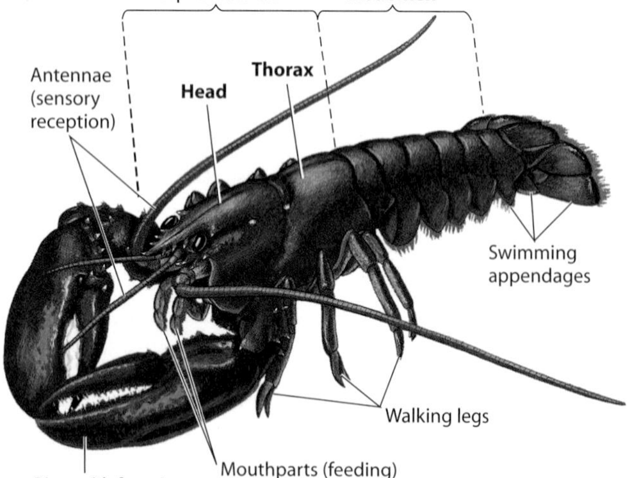

▲ **Figure 11A** The structure of a lobster, an arthropod

▲ **Figure 11B** A horseshoe crab (up to about 30 cm wide)

Angelo Giampiccolo/bluegreenpictures.com/naturepl.com.

Like molluscs, arthropods have an open circulatory system in which a tube-like heart pumps blood through short arteries into spaces surrounding the organs. A variety of gas exchange organs have evolved. Most aquatic species have gills. Terrestrial insects have internal air sacs that branch throughout the body.

Fossils and molecular evidence suggest that living arthropods represent four major lineages that diverged early in the evolution of arthropods. The figures in this module illustrate representatives of three of these lineages. The fourth, the insects, will be discussed in Module 12.

Chelicerates **Figure 11B** shows a **horseshoe crab**, a species that has survived with little change for hundreds of millions of years. This "living fossil" is the only surviving member of a group of marine **chelicerates** (from the Greek *chele*, claw, and *keras*, horn) that were abundant in the sea some 300 million years ago. One member of this group, the water scorpion, could grow up to 3 m (almost 10 feet) long. Horseshoe crabs are common on the Atlantic and Gulf coasts of the United States.

Living chelicerates also include the scorpions, spiders, ticks, and mites, collectively called **arachnids**. Most arachnids live on land. Scorpions (**Figure 11C**, left, on facing page) are nocturnal hunters. Their ancestors were among the first terrestrial carnivores, preying on other arthropods that fed on early land plants. Scorpions have a large pair of pincers for defense and capturing prey. The tip of the tail bears a poisonous stinger. Scorpions eat mainly insects and spiders and attack people only when prodded or stepped on. Only a few species are dangerous to humans, but the sting is painful nonetheless.

William Dow/Corbis

Scott Camazine/Alamy

Oliver Meckes/Ottawa/Photo Reseachers

Colorized SEM

A scorpion

A black widow spider
(about 1 cm wide)

A dust mite
(about 0.4 mm long)

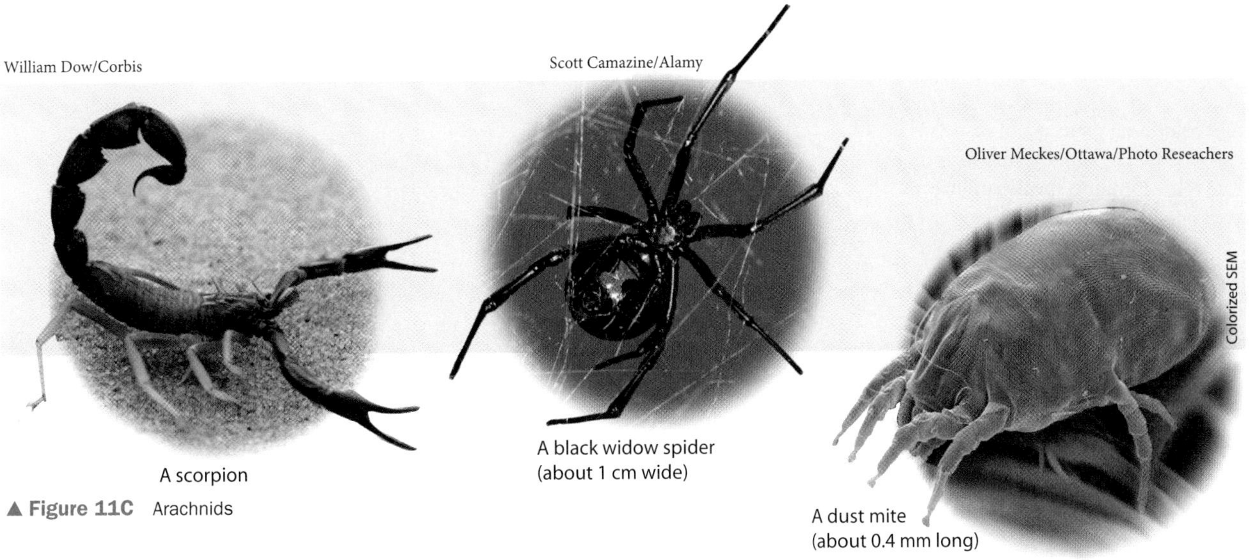

▲ Figure 11C Arachnids

Spiders, a diverse group of arachnids, hunt insects or trap them in webs of silk that they spin from specialized glands on their abdomen (see Figure 11C, center). Mites make up another large group of arachnids. On the right in Figure 11C is a micrograph of a dust mite, a ubiquitous scavenger in our homes. Thousands of these microscopic animals can thrive in a few square centimeters of carpet or in one of the dust balls that form under a bed. Dust mites do not carry infectious diseases, but many people are allergic to them.

Millipedes and Centipedes The animals in this lineage have similar segments over most of their body and superficially resemble annelids; however, their jointed legs identify them as arthropods. **Millipedes (Figure 11D)** are wormlike terrestrial creatures that eat decaying plant matter. They have two pairs of short legs per body segment. **Centipedes (Figure 11E)** are terrestrial carnivores with a pair of poison claws used in defense and to paralyze prey such as cockroaches and flies. Each of their body segments bears a single pair of long legs.

Crustaceans The **crustaceans** are nearly all aquatic. Lobsters and crayfish are in this group, along with numerous barnacles, crabs, and shrimps. **(Figure 11F)**. Barnacles are marine crustaceans with a cuticle that is hardened into a shell containing calcium carbonate, which may explain why they were once classified as molluscs. Their jointed appendages project from their shell to strain food from the water. Most barnacles anchor themselves to rocks, boat hulls, pilings, or even whales. The adhesive they produce is as strong as any glue ever invented. Other crustaceans include small copepods and krill, which serve as food sources for many fishes and whales.

We turn next to the fourth lineage of arthropods, the insects, whose numbers dwarf all other groups combined.

? **List the characteristics that arthropods have in common.**

Segmentation, exoskeleton, specialized jointed appendages ◖

Matthew Cole/Shutterstock

◀ Figure 11D
A millipede (about
7 cm long)

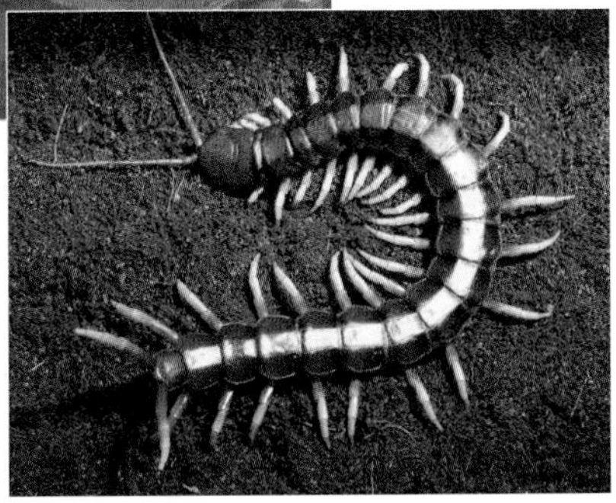

▶ Figure 11E
A Peruvian giant
centipede (can reach
30 cm in length)

Mark Smith/Photo Researchers

▶ Figure 11F
Crustaceans

Colin Milkins/Photolibrary

A ghost crab
(body about
2.5 cm across)

Goose barnacles
(about 2 cm high)

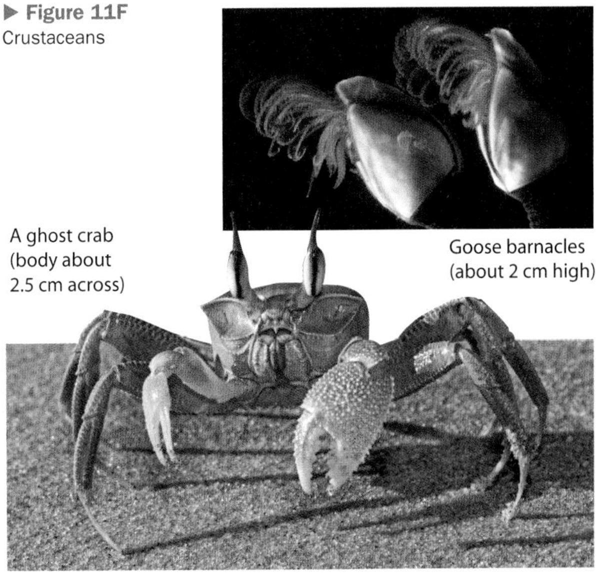

Maximilian Weinzierl/Alamy

EVOLUTION CONNECTION

12 Insects are the most successful group of animals

The evolutionary success of insects is unrivaled by any other group of animals. More than a million species of insects have been identified, comprising over 70% of all animal species. Entomologists (scientists who study insects) think that fewer than half the total number of insect species have been identified, and some believe there could be as many as 30 million. Insects are distributed worldwide and have a remarkable ability to survive challenging terrestrial environments. Although they have also flourished in freshwater habitats, insects are rare in the seas, where crustaceans are the dominant arthropods.

What characteristics account for the extraordinary success of insects? One answer lies in the features they share with other arthropods—body segmentation, an exoskeleton, and jointed appendages. Other key features include flight, a waterproof coating on the cuticle, and a complex life cycle. In addition, many insects have short generation times and large numbers of offspring. For example, *Culex pipiens*, the most widely distributed species of mosquito, has a generation time of roughly 10 days, and a single female can lay many hundreds of eggs over the course of her lifetime. Thus, natural selection acts rapidly, and alleles that offer a reproductive advantage can quickly be established in a population.

Life Cycles One factor in the success of insects is a life cycle that includes metamorphosis, during which the animal takes on different body forms as it develops from larva to adult. Only the adult insect is sexually mature and has wings. More than 80% of insect species, including beetles, flies, bees, and moths and butterflies, undergo **complete metamorphosis**. The larval stage (such as caterpillars, which are the larvae of moths and butterflies, and maggots, which are fly larvae) is specialized for eating and growing. A larva typically molts several times as it grows, then exists as an encased, nonfeeding pupa while its body rebuilds from clusters of embryonic cells that have been held in reserve. The insect then emerges as an adult that is specialized for reproduction and dispersal. Adults and larvae eat different foods, permitting the species to make use of a wider range of resources and avoiding intergenerational competition. Figure 12A shows the larva, pupa, and adult of the rhinoceros beetle (*Oryctes nasicornis*), named for the horn on the male's head.

Other insect species undergo **incomplete metamorphosis**, in which the transition from larva to adult is achieved through multiple molts, but without forming a pupa. In some species, including grasshoppers and cockroaches, the juvenile forms resemble the adults. In others, the body forms and lifestyles are very different. The larvae of dragonflies, for example, are aquatic, but the adults live on land.

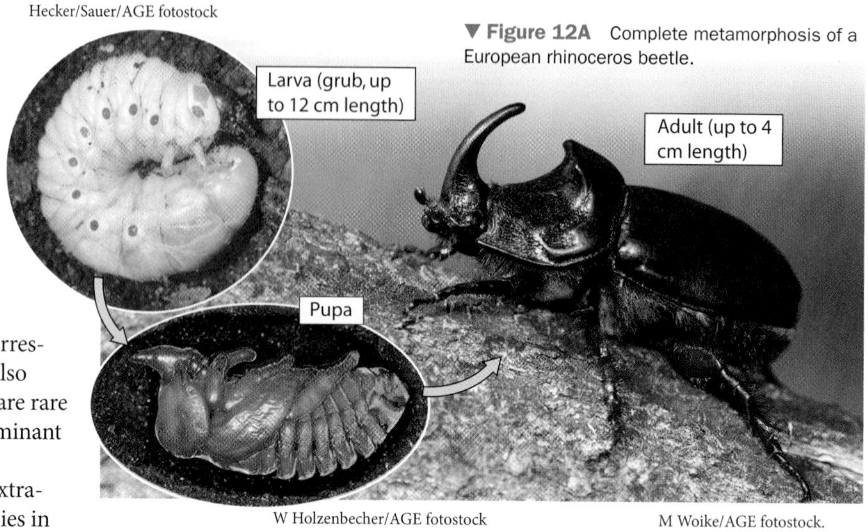

▼ **Figure 12A** Complete metamorphosis of a European rhinoceros beetle.

Modular Body Plan Like other arthropods, insects have specialized body regions—a head, a thorax, and an abdomen (**Figure 12B**). These regions arise from the fusion of embryonic segments during development. Early in development, the embryonic segments are identical to each other. However, they soon diverge as different genes are expressed in different segments, giving rise to the three distinct body parts and to a variety of appendages, including antennae, mouthparts, legs, and wings. The insect body plan is essentially modular: Each embryonic segment is a separate building block that develops independently of the other segments. As a result, a mutation that changes homeotic gene expression can change the structure of one segment or its appendages without affecting any of the others. In the evolution of the grasshopper, for example, genetic changes in one thoracic segment produced the specialized jumping legs but did not affect the other two leg-producing segments. Wings, antennae, and mouthparts have all evolved in a similar fashion, by the specialization of independent segments through changes in the timing and location of homeotic gene expression. Much of the extraordinary diversification of insects resulted from modifications of the appendages that adapted them for specialized functions.

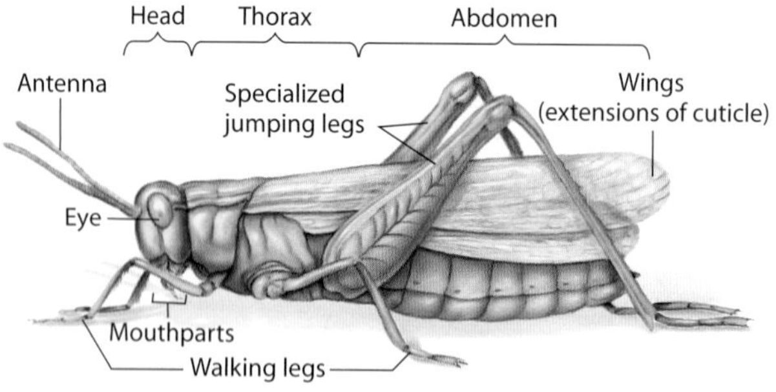

▲ **Figure 12B** Modular body plan of insects, as seen in a grasshopper

Fred Bruemmer/Photolibrary

Martin Shields/Alamy

Ingo Arndt/Nature Picture Library

▲ **Figure 12C** Remarkable resemblances

The head typically bears a pair of sensory antennae, a pair of eyes, and several pairs of mouthparts. The mouthparts are adapted for particular kinds of eating—for example, for chewing plant material (in grasshoppers); for biting and tearing prey (praying mantis); for lapping up fluids (houseflies); or for piercing into and sucking the fluids of plants (aphids) or animals (mosquitoes). When flowering plants appeared, adaptations for nectar feeding became advantageous. As a result of this variety in mouthparts, insects have adaptations that exploit almost every conceivable food source.

Most adult insects have three pairs of legs, which may be adapted for walking, jumping, grasping prey, digging into the soil, or even paddling on water. Insects are the only invertebrates that can fly; most adult insects have one or two pairs of wings. (Some insects, such as fleas, are wingless.) Flight, which is an effective means of dispersal and escape from predators, was a major factor in the success of insects. And because the wings are extensions of the cuticle, insects have acquired the ability to fly without sacrificing any legs. By contrast, the wings of birds and bats are modified limbs. With a single pair of walking legs, those animals are generally clumsy when on the ground.

Protective Color Patterns In many groups of insects, adaptations of body structures have been coupled with protective coloration. Many different animals, including insects, have camouflage, color patterns that blend into the background. But insects also have elaborate disguises that include modifications to their antennae, legs, wings, and bodies. For instance, there are insects that resemble twigs, leaves, and bird droppings (**Figure 12C**). Some even do a passable imitation of vertebrates. The "snake" in **Figure 12D** is actually a hawk moth caterpillar. The colors of its dorsal side are an effective camouflage. When disturbed, however, it flips over to reveal the snake-like eyes of its ventral side, even puffing out its thorax to enhance the deception. "Eyespots" that resemble vertebrate eyes are common in several groups of moths and butterflies. **Figure 12E** shows a member of a genus known informally as owl butterflies. A flash of these large "eyes" startles would-be predators. In other species, eyespots deflect the predator's attack away from vital body parts.

Stephen J. Krasemann/Photo Researchers

▲ **Figure 12D** A hawk moth caterpillar

How could evolution have produced these complex color patterns? It turns out that the genetic mechanism by which eyespots evolve is very similar to the mechanism by which specialized appendages evolve. Butterfly wings have a modular construction similar to that of embryonic body segments. Each section can change independently of the others and can therefore have a unique pattern. And like the specialization of appendages, eyespots result from different patterns of homeotic gene expression during development.

The study of insects has given evolutionary developmental biologists valuable insight into the genetic mechanisms that have generated the amazing diversity of life. We'll return to this topic in Module 16. But first let's complete our survey of the invertebrate phyla.

? Contrast incomplete and complete metamorphosis.

In complete metamorphosis, there is a pupal stage; in incomplete metamorphosis, there is not.

Bruce Coleman Inc./Alamy Images

Purestock/Getty Images

▲ **Figure 12E** An owl butterfly (left) and a long-eared owl (right)

13 Echinoderms have spiny skin, an endoskeleton, and a water vascular system for movement

Echinoderms, such as sea stars, sand dollars, and sea urchins, are slow-moving or sessile marine animals. Most are radially symmetric as adults. Both the external and the internal parts of a sea star, for instance, radiate from the center like spokes of a wheel. The bilateral larval stage of echinoderms, however, tells us that echinoderms are not closely related to cnidarians or other animals that never show bilateral symmetry.

The phylum name Echinodermata (from the Greek *echin*, spiny, and *derma*, skin) refers to the prickly bumps or spines of a sea star or sea urchin. These are extensions of the hard calcium-containing plates that form the **endoskeleton**, or internal skeleton, under the thin skin of the animal.

Unique to echinoderms is the **water vascular system**, a network of water-filled canals that branch into extensions called tube feet **(Figure 13A)**. Tube feet function in locomotion, feeding, and gas exchange. A sea star pulls itself slowly over the seafloor using its suction-cup-like tube feet. Its mouth is centrally located on its undersurface. When the sea star shown in **Figure 13B** encounters an oyster or clam, its favorite food, it grips the bivalve with its tube feet and pulls until the mollusc's muscle tires enough to create a narrow opening between the two valves of the shell. The sea star then turns its stomach inside out, pushing it through its mouth and into the opening. The sea star's stomach digests the soft parts of its prey inside the mollusc's shell. When the meal is completed, the sea star withdraws its stomach from the empty shell.

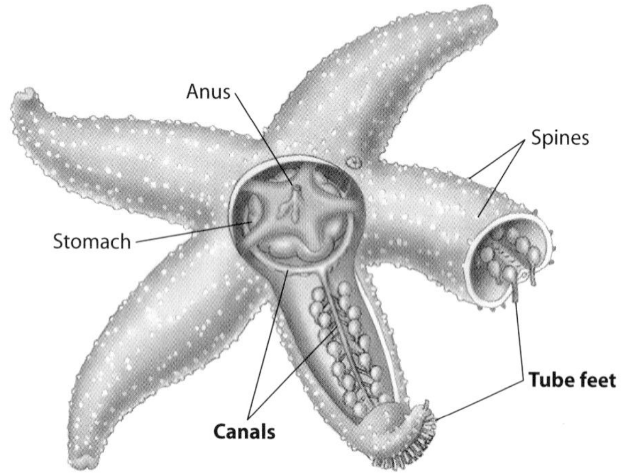

▲ **Figure 13A** The water vascular system (canals and tube feet) of a sea star (top view)

Andrew J Martinez/Photo Researchers/Getty Images

▲ **Figure 13B** A sea star feeding on a clam

Louise Murray/AGE fotostock

▲ **Figure 13C** A sea urchin

Sea stars and some other echinoderms are capable of regeneration. Arms that are lost are readily regrown.

In contrast with sea stars, sea urchins are spherical and have no arms. They do have five rows of tube feet that project through tiny holes in the animal's globe-like case. If you look carefully in **Figure 13C**, you can see the long, threadlike tube feet projecting among the spines of the sea urchin. Sea urchins move by pulling with their tube feet. They also have muscles that pivot their spines, which can aid in locomotion. Unlike the carnivorous sea stars, most sea urchins eat algae.

Other echinoderm groups include brittle stars, which move by thrashing their long, flexible arms; sea lilies, which live attached to the substrate by a stalk; and sea cucumbers, odd elongated animals that resemble their vegetable namesake more than they resemble other echinoderms.

Though echinoderms have many unique features, we see evidence of their relation to other animals in their embryonic development. As we discussed in Modules 3 and 4, differences in patterns of development have led biologists to identify echinoderms and chordates (which include vertebrates) as a clade of bilateral animals called deuterostomes. Thus, echinoderms are more closely related to our phylum, the chordates, than to the protostome animals, such as molluscs, annelids, and arthropods. We examine the chordates next.

? Contrast the skeleton of an echinoderm with that of an arthropod.

An echinoderm has an endoskeleton; an arthropod has an exoskeleton.

14 Our own phylum, Chordata, is distinguished by four features

You may be surprised to find the phylum that includes humans in a chapter on invertebrate diversity. However, vertebrates evolved from invertebrate ancestors and continue to share the distinctive features that identify members of the phylum Chordata. The embryos, and often the adults, of chordates possess: (1) a **dorsal, hollow nerve cord**; (2) a **notochord**, a flexible, supportive, longitudinal rod located between the digestive tract and the nerve cord; (3) **pharyngeal slits** located in the pharynx, the region just behind the mouth; and (4) a muscular **post-anal tail** (a tail posterior to the anus). You can see these four features in the diagrams in Figures 14A and 14B. The two chordates shown, a tunicate and a lancelet, are called invertebrate chordates because they do not have a backbone.

Adult **tunicates** are stationary and look more like small sacs than anything we usually think of as a chordate (**Figure 14A**). Tunicates often adhere to rocks and boats, and they are common on coral reefs. The adult has no trace of a notochord, nerve cord, or tail, but it does have prominent pharyngeal slits that function in feeding. The tunicate larva, however, is a swimming, tadpole-like organism that exhibits all four distinctive chordate features.

Tunicates are suspension feeders. Seawater enters the adult animal through an opening at the top, passes through the pharyngeal slits into a large cavity in the animal, and exits back into the ocean via an excurrent siphon on the side of the body (see the photo in Figure 14A). Food particles are trapped in a net made of mucus and then transported to the intestine, where they are digested. Because they shoot a jet of water through their excurrent siphon when threatened, tunicates are often called sea squirts.

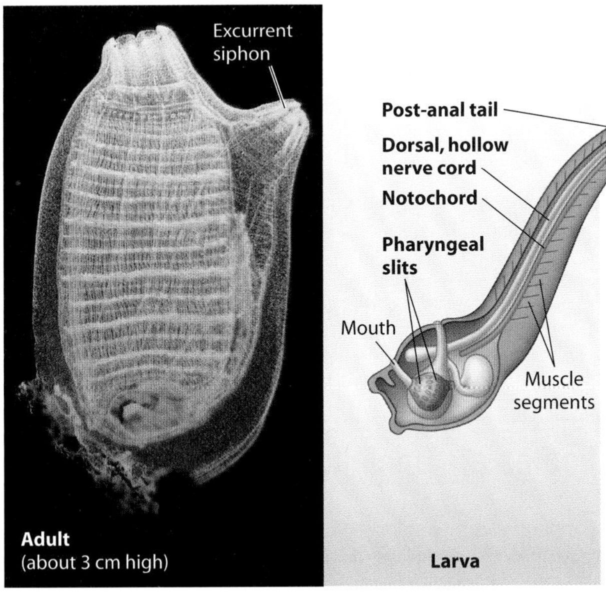

Robert Brons/Biological Photo Service

▲ **Figure 14A** A tunicate

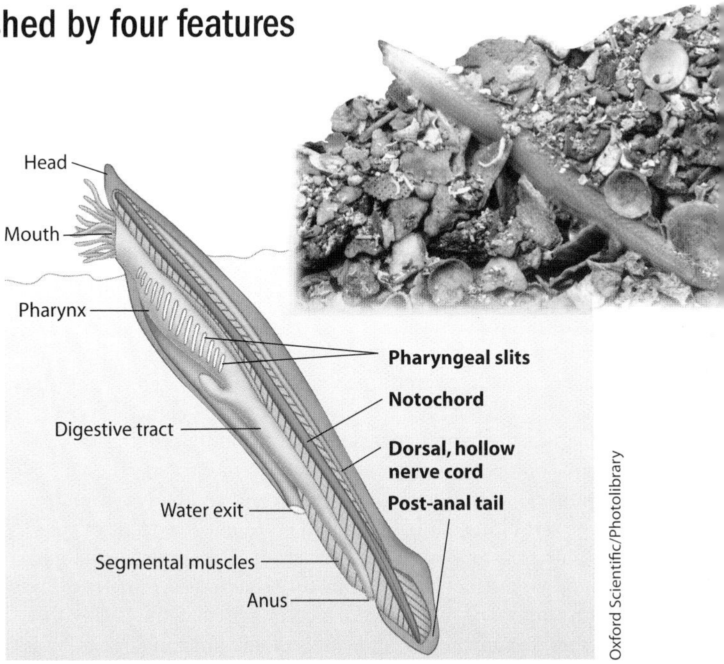

Oxford Scientific/Photolibrary

▲ **Figure 14B** A lancelet (5–15 cm long)

Lancelets, another group of marine invertebrate chordates, also feed on suspended particles. Lancelets are small, bladelike chordates that live in marine sands (**Figure 14B**). When feeding, a lancelet wriggles backward into the sand with its head sticking out. As in tunicates, a net of mucus secreted across the pharyngeal slits traps food particles. Water flowing through the slits exits via an opening in front of the anus.

Lancelets clearly illustrate the four chordate features. They also have segmental muscles that flex their body from side to side, producing slow swimming movements. These serial muscles are evidence of the lancelet's segmentation. Although not unique to chordates, body segmentation is also a chordate characteristic.

What is the relationship between the invertebrate chordates and the vertebrates? The tunicates likely represent the earliest branch of the chordate lineage. Molecular evidence indicates that the lancelets are the closest living nonvertebrate relatives of vertebrates. Research has shown that the same genes that organize the major regions of the vertebrate brain are expressed in the same pattern at the anterior end of the lancelet nerve cord.

The invertebrate chordates have helped us identify the four chordate hallmarks. We'll look at the vertebrate members of this phylum in the next chapter. Next, we review the invertebrates by revisiting the phylogenetic tree presented in Module 4.

? What four features do we share with invertebrate chordates, such as lancelets?

Human embryos and invertebrate chordates all have (1) a dorsal, hollow nerve cord; (2) a notochord; (3) pharyngeal slits; and (4) a post-anal tail.

Animal Phylogeny and Diversity Revisited

15 An animal phylogenetic tree is a work in progress

As you learned in Module 4, biologists have used evidence from the fossil record, morphology, and embryology to make hypotheses about the evolutionary history of animal groups. Recently, scientists have accumulated molecular data, chiefly DNA sequences, that shed new light on these phylogenetic relationships. **Figure 15** presents a slightly revised tree based on this new evidence.

Let's see how the new hypotheses represented by the tree compare with the traditional, morphology-based tree. At the base (far left) of both trees is the ancestral colonial protist that was the hypothetical ancestor of animals. Both trees agree on the early divergence of phylum Porifera (the sponges), with their lack of true tissues and body symmetry. The rest of the animal kingdom represents the clade of eumetazoans, animals with true tissues. The eumetazoans split into two distinct lineages that differ in body symmetry and the number of cell layers formed in gastrulation. The hydras, jellies, sea anemones, and corals of phylum Cnidaria are radially symmetric and have two cell layers. The other lineage consists of bilateral animals, the bilaterians. The two trees agree on all of these early branchings. Both trees also recognize the deuterostomes, which include the echinoderms and chordates, as a clade.

The purple branches in Figure 15 show how molecular data have changed the hypotheses represented by Figure 4. As you can see, the revised tree distinguishes two clades within the protostomes: lophotrochozoans and ecdysozoans. The lophotrochozoans, while grouped on the basis of molecular similarities, are named for the feeding apparatus (called a lophophore) of some phyla in the group (which we did not discuss) and for the trochophore larva found in molluscs and annelids (see Figure 9B). Lophotrochozoans include the flatworms, molluscs, annelids, and many other phyla that we did not survey. The ecdysozoans include the nematodes and arthropods. Both of these phyla have exoskeletons that must be shed for the animal to grow. This molting process is called ecdysis, the basis for the name ecdysozoan.

Both the tree here and the one in Figure 4 represent hypotheses for the key events in the evolutionary history that led to the animal phyla now living on Earth. Like other phylogenetic trees, this latest one serves to stimulate research and discussion, and it is subject to revision as new information is acquired. As biologists continue to revise some branch points, however, the tree's overall message remains the same: The animal kingdom's great diversity arose through the process of evolution, and all animals exhibit features reminiscent of their evolutionary history. In the next module, we look at one of the common threads of evolutionary history, the generation of diversity from shared ancestral genes.

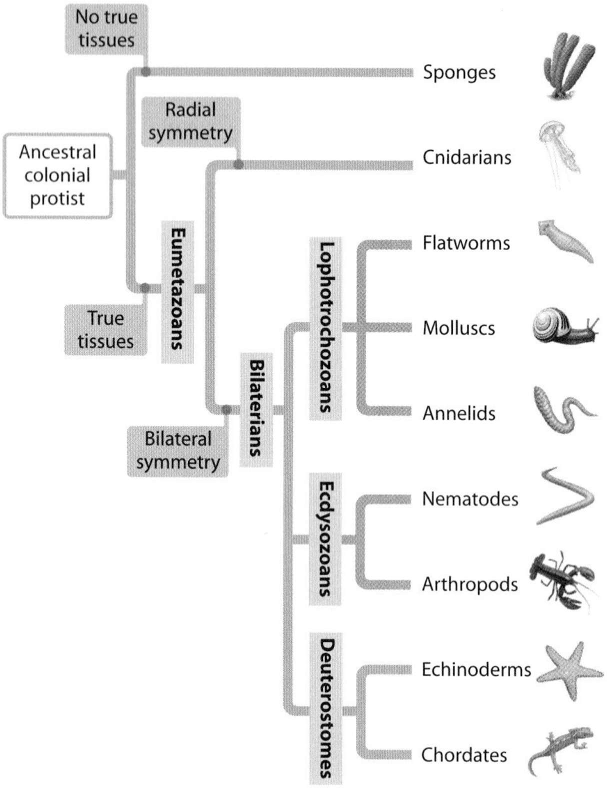

▲ **Figure 15** A molecular-based phylogenetic tree

? Compare the placement of annelids, arthropods, and molluscs in Figures 4 and 15. How do they differ?

In the morphology-based tree, annelids and arthropods are hypothesized to be more closely related to each other than to molluscs, largely based on their segmented bodies. In the molecular-based tree, arthropods are separated from both annelids and molluscs and are placed in the ecdysozoan clade.

EVOLUTION CONNECTION

16 The genes that build animal bodies are ancient

This chapter, a brief tour of invertebrate life, has introduced you to some of the immense variety of animal bodies—flower-like sea anemones, echinoderms in the shapes of stars and cucumbers, and arthropods with countless configurations of appendages and color patterns, to name just a few. How can we explain the evolution of such strikingly different forms from a common protistan ancestor? As exciting new discoveries in the genetics and evolution of development (evo-devo) are integrated with existing lines of evidence from the fossil record, comparative anatomy, and molecular biology, a new understanding of the evolution of animal diversity is beginning to emerge.

Until recently, scientists assumed that the construction of complex organs such as eyes and hearts required more complex instructions than for simple organs. Consequently, they expected the results of the Human Genome Project to reflect the complexity of the human body, estimating that we have about 100,000 genes. As you may recall, geneticists were taken aback to learn how few genes humans have (about 21,000), especially in comparison with other animals (about 14,000 in fruit flies).

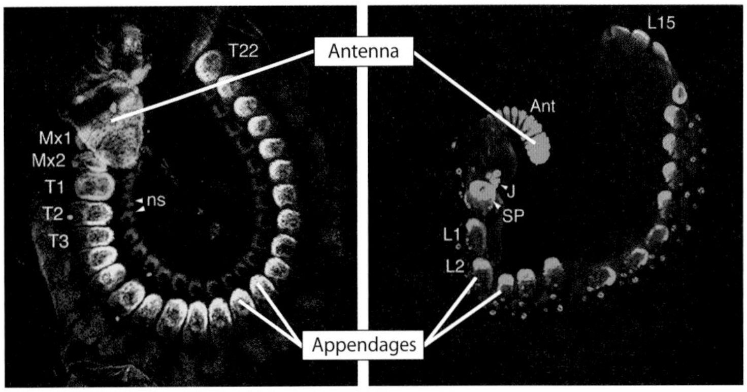

Steve Paddock, Ph.D

▲ **Figure 16B** Expression of homeotic genes in the embryos of a centipede (left) and a velvet worm (right). (The labels identify distinct body segments.)

It is now clear that important genes responsible for building animal bodies are shared by virtually every member of the animal kingdom—they have existed for at least half a billion years. These ancient genes are the master control genes called homeotic genes, the body-building genes you learned about earlier.

Let's look at one investigation into the involvement of homeotic genes in the evolution of animal diversity, in this case, the origin of the arthropod body plan. How did the enormous variety of distinct body segments bearing specialized appendages evolve? According to one hypothesis, the ancestors of arthropods had a small number of body segment types whose development was controlled by a small number of homeotic genes. New genes that originated on the arthropod branch of the tree, perhaps by duplication of existing homeotic genes, expanded the diversity of segment and appendage types.

To test the hypothesis, a team of scientists compared homeotic genes in arthropods with those of their closest living relatives, known as velvet worms. Velvet worms are one of the small animal phyla (containing 110 species) that we omitted from our coverage of animal diversity. The wormlike bodies of velvet worms bear fleshy antennae and numerous short, identical appendages that are used for walking **(Figure 16A)**.

Michael Fogden/Photolibrary

▲ **Figure 16A** *Peripatus*, a velvet worm

If velvet worms have only a subset of the homeotic genes found in arthropods, then the hypothesis is supported—arthropods acquired additional genes after the lineages diverged. However, results showed that velvet worms *do* have all of the arthropod homeotic genes. The researchers concluded that body segment diversity did *not* result from new homeotic genes that arose in arthropods.

What other explanation might account for arthropod diversity? Changes in the regulation of developmental gene expression—when and where the master control genes are transcribed and translated into proteins—can result in morphological changes. **Figure 16B** shows differences in gene expression patterns in a centipede embryo (left; also see Figure 11E) and a velvet worm embryo (right). The green stain indicates expression of a homeotic gene that is involved in the formation of appendages in a wide range of taxa. Two homeotic genes unique to arthropods and their close relatives are expressed in the body regions stained red or yellow. As you can see, the velvet worm deploys these unique genes only in the posterior tip of its body. In the centipede, and in other groups of arthropods that have been studied, the locations where these genes are expressed correspond with the boundaries between one segment type and the next.

Experiments such as this have demonstrated that the evolution of new structures and new types of animals does not require new genes. Rather, the genetic differences that result in new forms arise in the segments of DNA that control when and where ancient homeotic genes are expressed. To put it simply, building animal bodies is not just about which genes are present—it's about how they are used.

The work of evo-devo researchers is deepening our understanding of how Earth's enormous wealth of biological diversity evolved.

? Researchers found that velvet worms and arthropods share the same set of homeotic genes. What conclusion did they draw from this result?

The evolution of diverse arthropod body segment types was not the result of new genes in arthropods.

REVIEW

For Practice Quizzes, BioFlix, MP3 Tutors, and Activities, go to www.masteringbiology.com.

Reviewing the Concepts

Animal Evolution and Diversity (1–4)

1 What is an animal? Animals are multicellular eukaryotes that have distinctive cell structures and specializations and obtain their nutrients by ingestion. Animal life cycles and embryonic development also distinguish animals from other groups of organisms.

2 Animal diversification began more than half a billion years ago. The oldest animal fossils are from the late Ediacaran period and are 575–550 million years old. Animal diversification accelerated rapidly during the "Cambrian explosion" from 535–525 million years ago.

3 Animals can be characterized by basic features of their "body plan." Body plans may vary in symmetry (radial or bilateral), body cavity (none, pseudocoelom, or true coelom), and embryonic development (protostomes or deuterostomes).

4 The body plans of animals can be used to build phylogenetic trees.

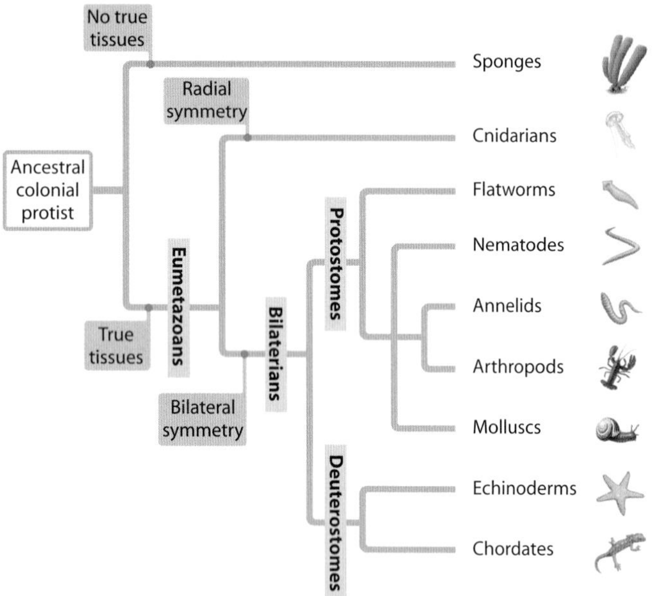

Invertebrate Diversity (5–14)

5 Sponges have a relatively simple, porous body. Members of the phylum Porifera have no true tissues. Their flagellated choanocytes filter food from water passing through pores in the body.

6 Cnidarians are radial animals with tentacles and stinging cells. Members of the phylum Cnidaria have true tissues and a gastrovascular cavity. Their two body forms are polyps (such as hydras) and medusae (jellies).

7 Flatworms are the simplest bilateral animals. Members of the phylum Platyhelminthes are bilateral animals with no body cavity. A planarian has a gastrovascular cavity and a simple nervous system. Flukes and tapeworms are parasitic flatworms with complex life cycles.

8 Nematodes have a pseudocoelom and a complete digestive tract. Members of the phylum Nematoda are covered by a protective cuticle. Many nematodes (roundworms) are free-living decomposers; others are plant or animal parasites.

9 Diverse molluscs are variations on a common body plan. Members of the phylum Mollusca include gastropods (such as snails and slugs), bivalves (such as clams), and cephalopods (such as octopuses and squids). All have a muscular foot and a mantle, which encloses the visceral mass and may secrete a shell. Many molluscs feed with a rasping radula.

10 Annelids are segmented worms. Members of the phylum Annelida include earthworms, polychaetes, and leeches.

11 Arthropods are segmented animals with jointed appendages and an exoskeleton. The four lineages of the phylum Arthropoda are chelicerates (arachnids such as spiders), the lineage of millipedes and centipedes, the aquatic crustaceans (lobsters and crabs), and the terrestrial insects.

12 Insects are the most successful group of animals. Their development often includes metamorphosis. Insects have a three-part body (head, thorax, and abdomen) and three pairs of legs; most have wings. Specialized appendages and protective color patterns, which frequently result from evolutionary changes in the timing and location of homeotic gene expression, have played a major role in this group's success.

13 Echinoderms have spiny skin, an endoskeleton, and a water vascular system for movement. Members of the phylum Echinodermata, such as sea stars, are radially symmetric as adults.

14 Our own phylum, Chordata, is distinguished by four features. Chordates have a dorsal, hollow nerve cord, a stiff notochord, pharyngeal slits, and a muscular post-anal tail. The simplest chordates are tunicates and lancelets, marine invertebrates that use their pharyngeal slits for suspension feeding.

Animal Phylogeny and Diversity Revisited (15–16)

15 An animal phylogenetic tree is a work in progress. A revised phylogenetic tree based on molecular data distinguishes two protostome clades: the lophotrochozoans and the ecdysozoans.

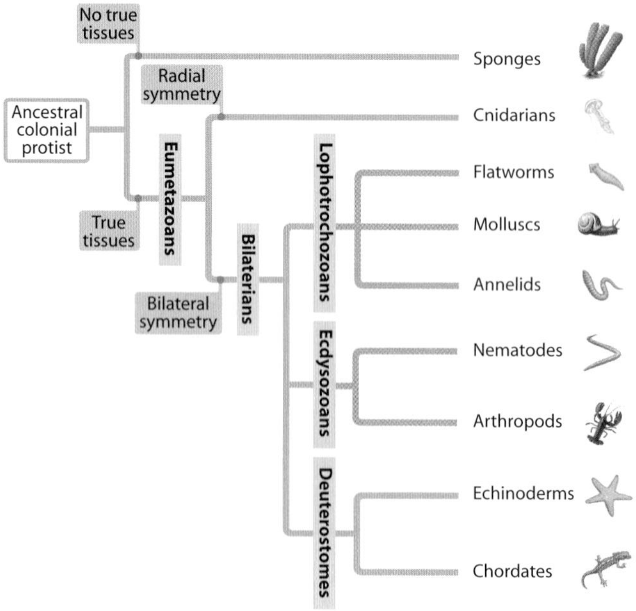

16 The genes that build animal bodies are ancient. Changes in the regulation of homeotic gene expression have been significant factors in the evolution of animal diversity.

Connecting the Concepts

1. The table below lists the common names of the nine animal phyla surveyed in this chapter. For each phylum, list the key characteristics and some representatives.

Phylum	Characteristics	Representatives
Sponges		
Cnidarians		
Flatworms		
Nematodes		
Molluscs		
Annelids		
Arthropods		
Echinoderms		
Chordates		

Testing Your Knowledge

Multiple Choice

2. Bilateral symmetry in animals is best correlated with
   a. an ability to see equally in all directions.
   b. the presence of a skeleton.
   c. motility and active predation and escape.
   d. development of a true coelom.
   e. adaptation to terrestrial environments.
3. Jon found an organism in a pond, and he thinks it's a freshwater sponge. His friend Liz thinks it looks more like an aquatic fungus. How can they decide whether it is an animal or a fungus?
   a. See if it can swim.
   b. Figure out whether it is autotrophic or heterotrophic.
   c. See if it is a eukaryote or a prokaryote.
   d. Look for cell walls under a microscope.
   e. Determine whether it is unicellular or multicellular.
4. Which of the following groupings includes the largest number of species? (*Explain your answer.*)
   a. invertebrates
   b. chordates
   c. arthropods
   d. insects
   e. vertebrates
5. Which of the following animal groups does not have tissues derived from mesoderm?
   a. annelids
   b. amphibians
   c. echinoderms
   d. cnidarians
   e. flatworms
6. Molecular comparisons place nematodes and arthropods in clade Ecdysozoa. What characteristic do they share that is the basis for the name Ecdysozoa?
   a. a complete digestive tract
   b. body segmentation
   c. molting of an exoskeleton
   d. bilateral symmetry
   e. a true coelom

Matching

7. Include the vertebrates	a. annelids
8. Medusa and polyp body forms	b. nematodes
9. The simplest bilateral animals	c. sponges
10. The most primitive animal group	d. arthropods
11. Earthworms, polychaetes, and leeches	e. flatworms
12. Largest phylum of all	f. cnidarians
13. Closest relatives of chordates	g. molluscs
14. Body cavity is a pseudocoelom	h. echinoderms
15. Have a muscular foot and a mantle	i. chordates

Describing, Comparing, and Explaining

16. Compare the structure of a planarian (a flatworm) and an earthworm with regard to the following: digestive tract, body cavity, and segmentation.
17. Name two phyla of animals that are radially symmetric and two that are bilaterally symmetric. How do the general lifestyles of radial and bilateral animals differ?
18. One of the key characteristics of arthropods is their jointed appendages. Describe four functions of these appendages in four different arthropods.
19. Compare the phylogenetic tree from Module 4 with the one from Module 15 (see previous page for both phylogenetic trees). What are the similarities and differences? Why have scientists revised the tree?

Applying the Concepts

20. A marine biologist has dredged up an unknown animal from the seafloor. Describe some of the characteristics that could be used to determine the animal phylum to which the creature should be assigned.

Review Answers

1. Sponges: sessile, saclike body with pores, suspension feeder; sponges
Cnidarians: radial symmetry, gastrovascular cavity, cnidocytes, polyp or medusa body form; hydras, sea anemones, jellies, corals
Flatworms: bilateral symmetry, gastrovascular cavity, no body cavity; free-living planarians, flukes, tapeworms
Nematodes: pseudocoelom, covered with cuticle, complete digestive tract, ubiquitous, free-living and parasitic; roundworms, heartworms, hookworms, trichinosis worms
Molluscs: muscular foot, mantle, visceral mass, circulatory system, many with shells, radula in some; snails and slugs, bivalves, cephalopods (squids and octopuses)
Annelids: segmented worms, closed circulatory system, many organs repeated in each segment; earthworms, polychaetes, leeches
Arthropods: exoskeleton, jointed appendages, segmentation, open circulatory system; chelicerates (spiders), crustaceans (lobsters, crabs), millipedes and centipedes, insects
Echinoderms: radial symmetry as adult, water vascular system with tube feet, endoskeleton, spiny skin; sea stars, sea urchins
Chordates: (1) notochord, (2) dorsal, hollow nerve cord, (3) pharyngeal slits, (4) post-anal tail; tunicates, lancelets, hagfish, and all the vertebrates (lampreys, sharks, ray-finned fishes, lobe-fins, amphibians, reptiles (including birds), mammals

2. c **3.** d **4.** a (The invertebrates include all animals except the vertebrates.) **5.** d **6.** c **7.** i **8.** f **9.** e **10.** c **11.** a **12.** d **13.** h **14.** b **15.** g

16. The gastrovascular cavity of a flatworm is an incomplete digestive tract; the worm takes in food and expels waste through the same opening. An earthworm has a complete digestive tract; food travels one way, and different areas are specialized for different functions. The flatworm's body is solid and unsegmented. The earthworm has a coelom, allowing its internal organs to grow and move independently of its outer body wall. Fluid in the coelom cushions internal organs, acts as a skeleton, and aids circulation. Segmentation of the earthworm, including its coelom, allows for greater flexibility and mobility.

17. Cnidarians and most adult echinoderms are radially symmetric, while most other animals, such as arthropods and chordates, are bilaterally symmetric. Most radially symmetric animals stay in one spot or float passively. Most bilateral animals are more active and move headfirst through their environment.

18. For example, the legs of a horseshoe crab are used for walking, while the antennae of a grasshopper have a sensory function. Some appendages on the abdomen of a lobster are used for swimming, while the scorpion catches prey with its pincers. (Note that the scorpion stinger and insect wings are not considered jointed appendages.)

19. Both trees agree on the early branching of eumetazoans into two groups based on body symmetry and the number of cell layers formed in gastrulation. Both trees recognize deuterostomes as a clade of bilaterians. In the morphological tree, protostomes are a second clade of bilaterians. The molecular tree distinguishes two clades within the protostomes, lophotrochozoans and ecdysozoans.

20. Important characteristics include symmetry, the presence and type of body cavity, segmentation, type of digestive tract, type of skeleton, and appendages.

The Evolution of Vertebrate Diversity

From Chapter 19 of *Campbell Biology: Concepts & Connections*, Seventh Edition. Jane B. Reece, Martha R. Taylor, Eric J. Simon, Jean L. Dickey.

The Evolution of Vertebrate Diversity

BIG IDEAS

Publiphoto/Photo Researchers

Vertebrate Evolution and Diversity
(1–8)

The major clades of chordates are distinguished by traits such as hinged jaws, two pairs of limbs, fluid-filled eggs with shells, and milk.

David TIipling/naturepl.com.

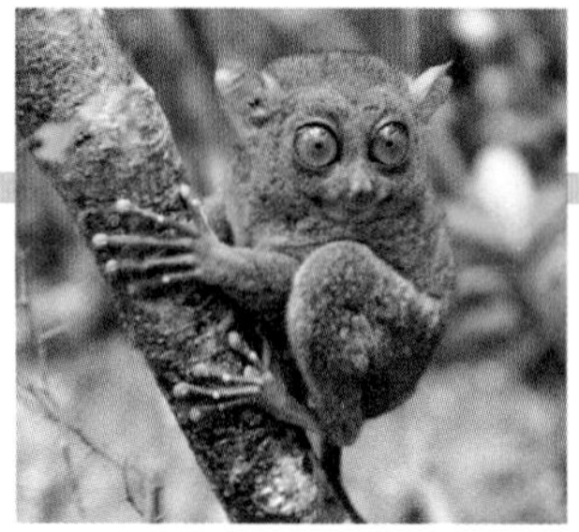

Primate Diversity
(9–10)

Humans have many characteristics in common with other primates, including forward-facing eyes, limber shoulder and hip joints, and opposable thumbs.

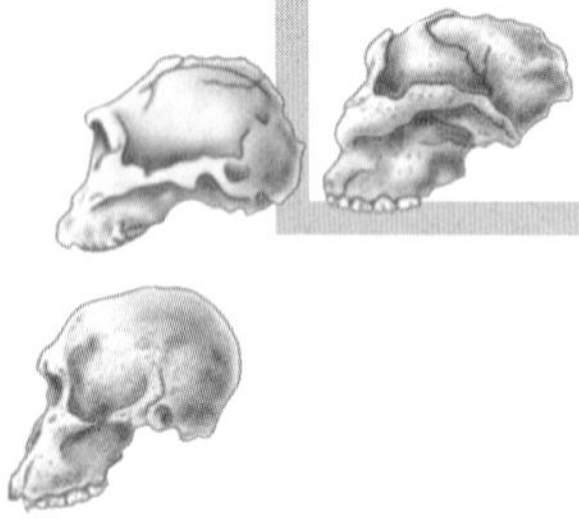

Hominin Evolution
(11–17)

Hominins, species that are on the human branch of the evolutionary tree, include approximately 20 extinct species.

Dave Watts/Alamy

Scientists have identified more than 1.3 million species of animals. But classifying a new animal isn't always easy. Imagine being the first zoologist to encounter the animal above in its native Australia. What would you make of it? It has a bill and webbed feet similar to a duck's, but the rest of its furry body looks very much like that of a muskrat or other aquatic rodent. To make the case even more confusing, this animal lays eggs. Is it a bird or a mammal? The decision is easier once you study it a little more closely. The animal, called a duck-billed platypus, has mammary glands that produce milk for its young. That trait, along with its hair, identifies it as a mammal.

Part of our interest in studying animals is exploring their many fascinating adaptations. Scientists investigating the platypus bill found that it is not composed of hard, inert materials like a duck's bill but rather is covered by soft skin filled with sensitive nerve endings. While the duck and the platypus both use their bills to dig for food in muddy waters, the platypus's bill also serves as a sensory organ to help it locate food and avoid obstacles underwater. When a platypus dives, it closes its eyes, so it relies on its bill to "see" its surroundings. Indeed, biologists have found that a large portion of the platypus brain is devoted to processing sensory information from its bill.

The incredible diversity of animal life arose through hundreds of millions of years of evolution, as natural selection shaped adaptations to Earth's diverse and changing environments. In this chapter, we continue our tour of the animal kingdom by exploring our own group, the vertebrates. We end the chapter, on the diversity of life, with a look at our predecessors—the primates who first walked on two legs, evolved a large, sophisticated brain, and eventually dominated Earth.

Vertebrate Evolution and Diversity

1 Derived characters define the major clades of chordates

Using a combination of anatomical, molecular, and fossil evidence, biologists have developed hypotheses for the evolution of chordate groups. **Figure 1** illustrates a current view of the major clades of chordates and lists some of the derived characters that define the clades. You can see that the tunicates are thought to be the first group to branch from the chordate lineage. Unlike tunicates, all other chordates have a brain, albeit a small one in the lancelets (only a swollen tip of the nerve cord).

The next transition was the development of a head that consists of a brain at the anterior end of the dorsal nerve cord, eyes and other sensory organs, and a skull. These innovations opened up a completely new way of feeding for chordates: active predation. All chordates with a head are called **craniates** (from the word *cranium*, meaning "skull").

The origin of a backbone came next. The **vertebrates** are distinguished by a more extensive skull and a backbone, or **vertebral column**, composed of a series of bones called **vertebrae** (singular, *vertebra*). These skeletal elements enclose the main parts of the nervous system. The skull forms a case for the brain, and the vertebrae enclose the nerve cord. The vertebrate skeleton is an endoskeleton, made of either flexible cartilage or a combination of hard bone and cartilage. Bone and cartilage are mostly nonliving material. But because there are living cells that secrete the nonliving material, the endoskeleton can grow with the animal.

The next major transition was the origin of jaws, which opened up new feeding opportunities. The evolution of lungs or lung derivatives, followed by muscular lobed fins with skeletal support, opened the possibility of life on land. **Tetrapods**, jawed vertebrates with two pairs of limbs, were the first vertebrates on land. The evolution of **amniotes**, tetrapods with a terrestrially adapted egg, completed the transition to land.

In the next several modules, we'll survey the vertebrates, from the jawless lampreys to the fishes to the tetrapods to the amniotes.

? **List the hierarchy of clades to which mammals belong.**

Chordates, craniates, vertebrates, jawed vertebrates, tetrapods, amniotes

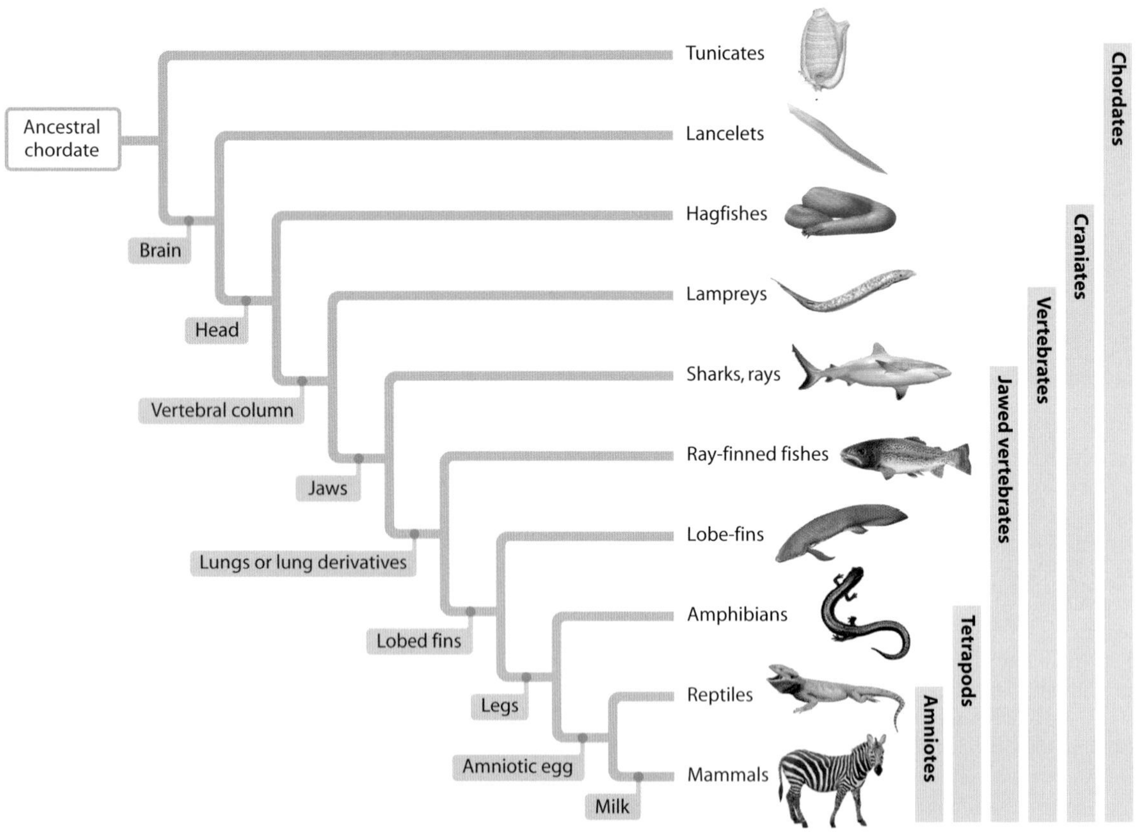

▲ **Figure 1** A phylogenetic tree of chordates, showing key derived characters

2 Hagfishes and lampreys lack hinged jaws

The most primitive surviving craniates (chordates with heads) are hagfishes and lampreys. In hagfishes **(Figure 2A)**, the notochord—a strong, flexible rod that runs most of the length of the body—is the body's main support in the adult. The notochord also persists in the adult lamprey, but rudimentary vertebral structures are also present. Consequently, lampreys are considered vertebrates, but hagfishes are not. Neither hagfishes nor lampreys have jaws.

Present-day hagfishes (roughly 40 species) scavenge dead or dying vertebrates on the cold, dark seafloor. Although nearly blind, they have excellent senses of smell and touch. They feed by entering the animal through an existing opening or by creating a hole using sharp, tooth-like structures on the tongue that grasp and tear flesh. For leverage, the hagfish may tie its tail in a knot, then slide the knot forward to tighten it against the prey's body. The knot trick is also part of its antipredator behavior. When threatened, a hagfish exudes an enormous amount of slime from special glands on the sides of its body (see Figure 2A, inset). The slime may make the hagfish difficult to grasp, or it may repel the predator. After the danger has passed, the hagfish ties itself into a knot and slides the knot forward, peeling off the layer of slime.

Fisherman who use nets have long been familiar with hagfishes. With their keen chemical senses, hagfishes are quick to detect bait and entrapped fish. Many fishermen have hauled in a net filled with feasting hagfishes, unsalable fish, and bucketfuls of slime. But hagfishes have gained economic importance recently. Both the meat and the skin, which is used to make faux-leather "eel-skin" belts, purses, and boots, are valuable commodities. Asian fisheries have been harvesting hagfish for decades. As Asian fishing grounds have been depleted, the industry has moved to North America. Some populations of hagfish have been eradicated along the West Coast, and fisheries are now looking to the East Coast and South America for fresh stocks.

TOM MCHUGH/Photo Researchers/Getty Images

Brandon D. Cole/Corbis

▲ **Figure 2A** Hagfish and slime (inset)

Gary Meszaros/Photo Researchers

Marevision/AGE fotostock

▲ **Figure 2B** A sea lamprey, with its rasping mouth (inset)

Lampreys represent the oldest living lineage of vertebrates **(Figure 2B)**. Lamprey larvae resemble lancelets. They are suspension feeders that live in freshwater streams, where they spend much of their time buried in sediment. Most lampreys migrate to the sea or lakes as they mature into adults.

Most species of lamprey are parasites, and just seeing the mouth of a sea lamprey (see Figure 2B, inset) suggests what it can do. The lamprey attaches itself to the side of a fish, uses its rasping tongue to penetrate the skin, and feeds on its victim's blood and tissues. After invading the Great Lakes via canals, these voracious vertebrates multiplied rapidly, decimating fish populations as they spread. Since the 1960s, streams that flow into the lakes have been treated with a chemical that reduces lamprey numbers, and fish populations have been recovering.

? Why are hagfishes described as craniates rather than vertebrates?

They have a head but no vertebrae; their body is supported by a notochord.

3 Jawed vertebrates with gills and paired fins include sharks, ray-finned fishes, and lobe-finned fishes

Jawed vertebrates appeared in the fossil record in the mid-Ordovician period, about 470 million years ago, and steadily became more diverse. Their success probably relates to their paired fins and tail, which allowed them to swim after prey, as well as to their jaws, which enabled them to catch and eat a wide variety of prey instead of feeding as mud-suckers or suspension feeders. Sharks, fishes, amphibians, reptiles (including birds), and mammals—the vast majority of living vertebrates—have jaws supported by two skeletal parts held together by a hinge. Where did these hinged jaws come from? According to one hypothesis, they evolved by modification of skeletal supports of the anterior pharyngeal (gill) slits. The first part of **Figure 3A** shows the skeletal rods supporting the gill slits in a hypothetical ancestor. The main function of these gill slits was trapping suspended food particles. The other two parts of the figure show changes that may have occurred as jaws evolved. By following the red and green structures, you can see how two pairs of skeletal rods near the mouth have become the jaws and their supports. The remaining gill slits, no longer required for suspension feeding, remained as sites of gas exchange.

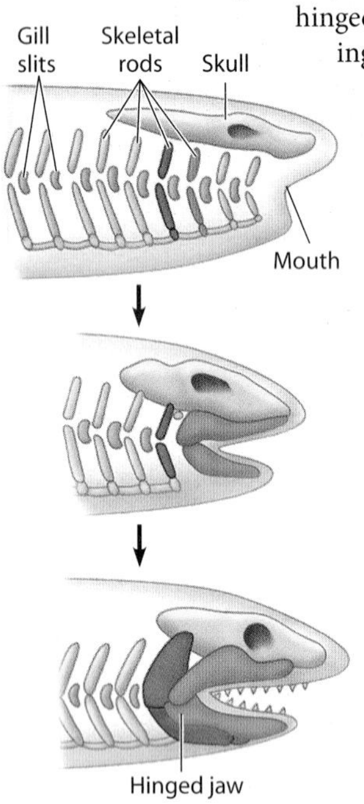

▲ **Figure 3A** A hypothesis for the origin of vertebrate jaws

Three lineages of jawed vertebrates with gills and paired fins are commonly called fishes. The sharks and rays of the class Chondrichthyes, which means "cartilage fish," have changed little in over 300 million years. As shown in Figure 1, lungs or lung derivatives are the key derived character of the clade that includes the ray-finned fishes and the lobe-fins. Muscular fins supported by stout bones further characterize the lobe-fins.

Chondrichthyans Sharks and rays, the **chondrichthyans**, have a flexible skeleton made of cartilage. The largest sharks are suspension feeders that eat small, floating plankton. Most sharks, however, are adept predators—fast swimmers with a streamlined body, powerful jaws, and knifelike teeth (**Figure 3B**). A shark has sharp vision and a keen sense of smell. On its head it has electrosensors, organs that can detect the minute electric fields produced by muscle contractions in nearby animals. Sharks and most other aquatic vertebrates have a **lateral line system**, a row of sensory organs running along each side that are sensitive to changes in water pressure and can detect minor vibrations caused by animals swimming nearby.

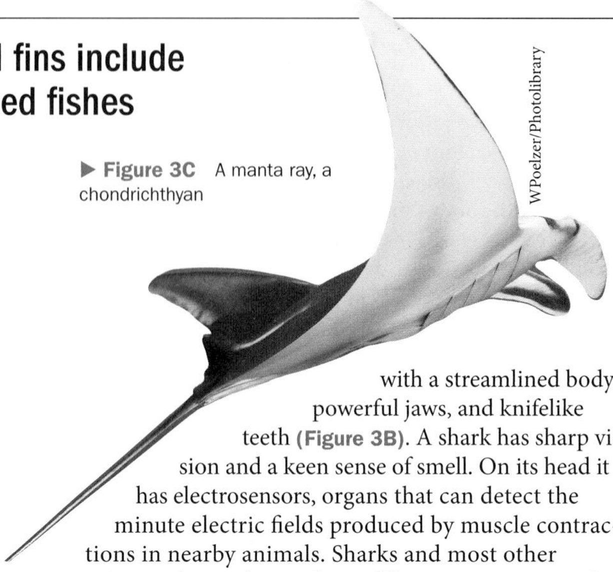

▶ **Figure 3C** A manta ray, a chondrichthyan

While the bodies of sharks are streamlined for swimming in the open ocean, rays are adapted for life on the bottom. Their bodies are dorsoventrally flattened, with the eyes on the top of the head. Once settled, they flip sand over their bodies with their broad pectoral fins and lie half-buried for much of the day. The tails of stingrays bear sharp spines with venom glands at the base. Where stingrays are common, swimmers and divers must take care not to step on or swim too closely over a concealed ray. The sting is painful and in rare cases fatal. Steve Irwin, a wildlife expert and television personality (*The Crocodile Hunter*), died when the 10-inch barb of a stingray pierced his heart while he was filming on the Great Barrier Reef in Australia.

The largest rays swim through the open ocean filtering plankton (**Figure 3C**). Some of these fishes are truly gigantic, measuring up to 6 m (19 feet) across the fins. The fin extensions in front of the mouth, which led to the common name devilfish, help funnel in water for suspension feeding.

Ray-finned Fishes In **ray-finned fishes**, which include the familiar tuna, trout, and goldfish, the skeleton is made of bone—cartilage reinforced with a hard matrix of calcium phosphate. Their fins are supported by thin, flexible skeletal rays. Most have flattened scales covering their skin and secrete a coating of mucus that reduces drag during

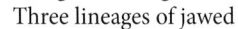

▶ **Figure 3B** A sand bar shark, a chondrichthyan

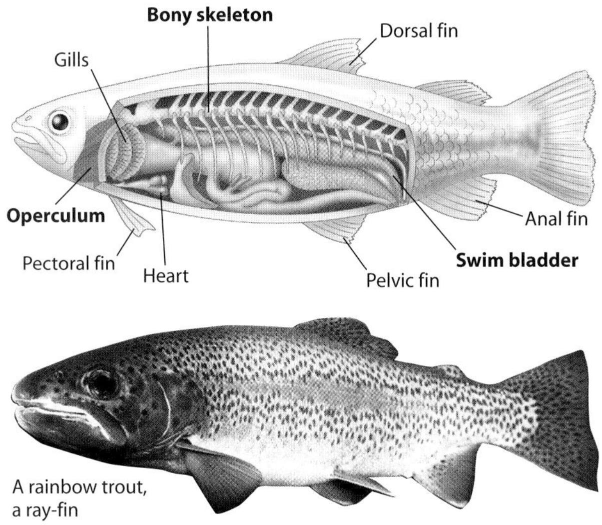

Tom and Pat Leeson/Photo Researchers

▲ **Figure 3D** The anatomical features of a ray-finned fish

swimming. **Figure 3D** highlights key features of a ray-finned fish such as the rainbow trout shown in the photograph. On each side of the head, a protective flap called an **operculum** covers a chamber housing the gills. Movement of the operculum allows the fish to breathe without swimming. (By contrast, sharks must generally swim to pass water over their gills.) Ray-finned fishes also have a lung derivative that helps keep them buoyant—the **swim bladder**, a gas-filled sac. Swim bladders evolved from balloon-like lungs, which the ancestral fishes may have used in shallow water to supplement gas exchange by their gills.

Ray-finned fishes, which emerged during the Devonian period along with the lobe-fins, include the greatest number of species of any vertebrate group, more than 27,000, and more species are discovered all the time. They have adapted to virtually every aquatic habitat on Earth. From the basic structural adaptations that gave them great maneuverability, speed, and feeding efficiency, various groups have evolved specialized body forms, fins, and feeding adaptations. **Figure 3E** shows a sample of the variety. The flounder's flattened body is nearly invisible on the seabed. Pigment cells in its skin match the background for excellent camouflage. Notice that both eyes are on the top of its head. The larvae of flounders and other flatfishes have eyes on both sides of the head. During development, one eye migrates to join the other on the side that will become the top. The balloon fish doesn't always look like a spiky beach ball. It raises its spines and inflates its body to deter predators. The small fins of the seahorse help it maneuver in dense vegetation, and the long tail is used for grasping onto a support. Seahorses have an unusual method of reproduction. The female deposits eggs in the male's abdominal brood pouch. His sperm fertilize the eggs, which develop inside the pouch.

Lobe-finned Fishes The key derived character of the **lobe-fins** is a series of rod-shaped bones in their muscular pectoral and pelvic fins. During the Devonian, they lived in coastal wetlands and may have used their lobed fins to "walk" underwater. Today, three lineages of lobe-fins survive: The coelacanth is a deep-sea dweller once thought to be extinct. The lungfishes are represented by a few Southern Hemisphere genera that generally inhabit stagnant waters and gulp air into lungs connected to the pharynx (**Figure 3F**). And the third lineage, the tetrapods, adapted to life on land during the mid-Devonian and gave rise to terrestrial vertebrates, as we see next.

? **From what structure might the swim bladder of ray-finned fishes have evolved?**

Simple lungs of an ancestral species

Marevision/AGE Fotostock

Jung Hsuan/Shutterstock

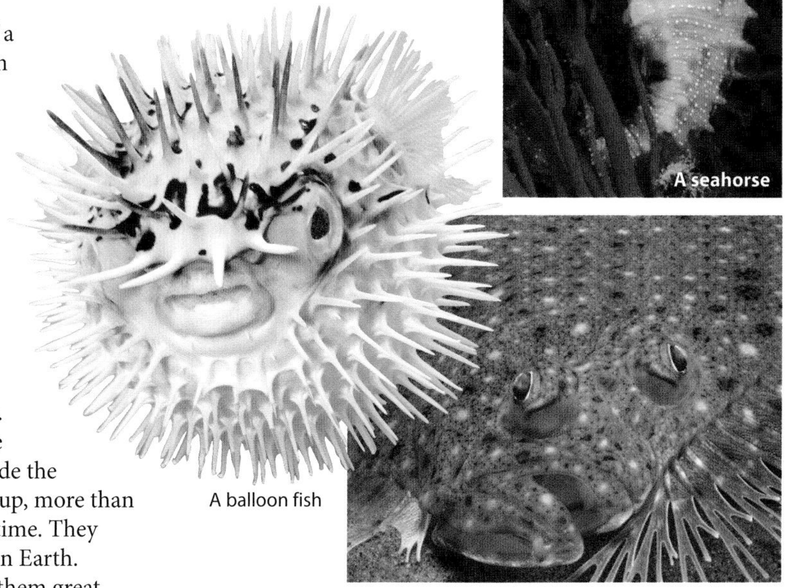

A balloon fish

A flounder

Jeff Rotman/Getty Images

▲ **Figure 3E** A variety of ray-finned fishes

Rudie Kuiter/OceanwideImages.com.

▲ **Figure 3F** A lobe-finned lungfish (about 1 m long)

EVOLUTION CONNECTION

4 New fossil discoveries are filling in the gaps of tetrapod evolution

During the late Devonian period, a line of lobe-finned fishes gave rise to tetrapods (meaning "four feet" in Greek), which today are defined as jawed vertebrates that have limbs and feet that can support their weight on land. Adaptation to life on land was a key event in vertebrate history; all subsequent groups of vertebrates—amphibians, mammals, and reptiles (including birds)—are descendants of these early land-dwellers.

The dramatic differences between aquatic and terrestrial environments shaped plant bodies and life cycles. Like plants, vertebrates faced obstacles on land in regard to gas exchange, water conservation, structural support, and reproduction. But vertebrates had other challenges as well. Sensory organs that worked in water had to be adapted or replaced by structures that received stimuli transmitted through air. And, crucially, a new means of locomotion was required.

Lobe-finned fishes were long considered the most likely immediate ancestors of tetrapods. Their fleshy paired fins contain bones that appear to be homologous to tetrapod limb bones, and some of the modern lobe-fins have lungs that extract oxygen from the air. Alfred Romer, a renowned paleontologist, hypothesized that these features enabled lobe-fins to survive by moving from one pool of water to another as aquatic habitats shrank during periods of drought. With Romer's gift for vivid imagery, it was easy to imagine the fish dragging themselves short distances across the Devonian landscape, those with the best locomotor skills surviving such journeys to reproduce. In this way, according to the hypothesis, vertebrates gradually became fully adapted to a terrestrial existence. But fossil evidence of the transition was scarce.

For decades, the most informative fossils available were *Eusthenopteron*, a 385-million-year-old specimen that was clearly a fish, and *Ichthyostega*, which lived 365 million years ago and had advanced tetrapod features (**Figure 4A**). The ray-finned tail and flipper-like hind limbs of *Ichthyostega* indicated that it spent considerable time in the water, but its well-developed front limbs with small, fingerlike bones and powerful shoulders showed that it was capable of locomotion on land. Unlike the shoulder bones of lobe-finned fishes, which are connected directly to the skull, *Ichthyostega* had a neck, a feature advantageous for terrestrial life. *Eusthenopteron* and *Ichthyostega* represented two widely separated points in the transition from fins to limbs. But what happened in between?

Recent fossil finds have begun to fill in the gap. Scientists have discovered lobe-fin fossils that are more similar to tetrapods than to *Eusthenopteron*, including a 380-million-year-old fish called *Panderichthys* (see Figure 4A). With its long snout, flattened body shape, and eyes on top of its head, *Panderichthys* looked a bit like a crocodile. It had lungs as well as gills and an opening that allowed water to enter through the top of the skull, a possible indication of a shallow-water habitat. Its paired fins had fishlike rays, but the dorsal and anal fins had been lost, and the tail fin was much smaller than in *Eusthenopteron*. Certain features of its skull were more like those of a tetrapod. Although it had no neck, the bones connecting forelimb to skull were intermediate in shape between that of a fish and a tetrapod. *Panderichthys* could have been capable of leveraging its fins against the bottom as it propelled itself through shallow water. It was a fish, but a tetrapod-like fish.

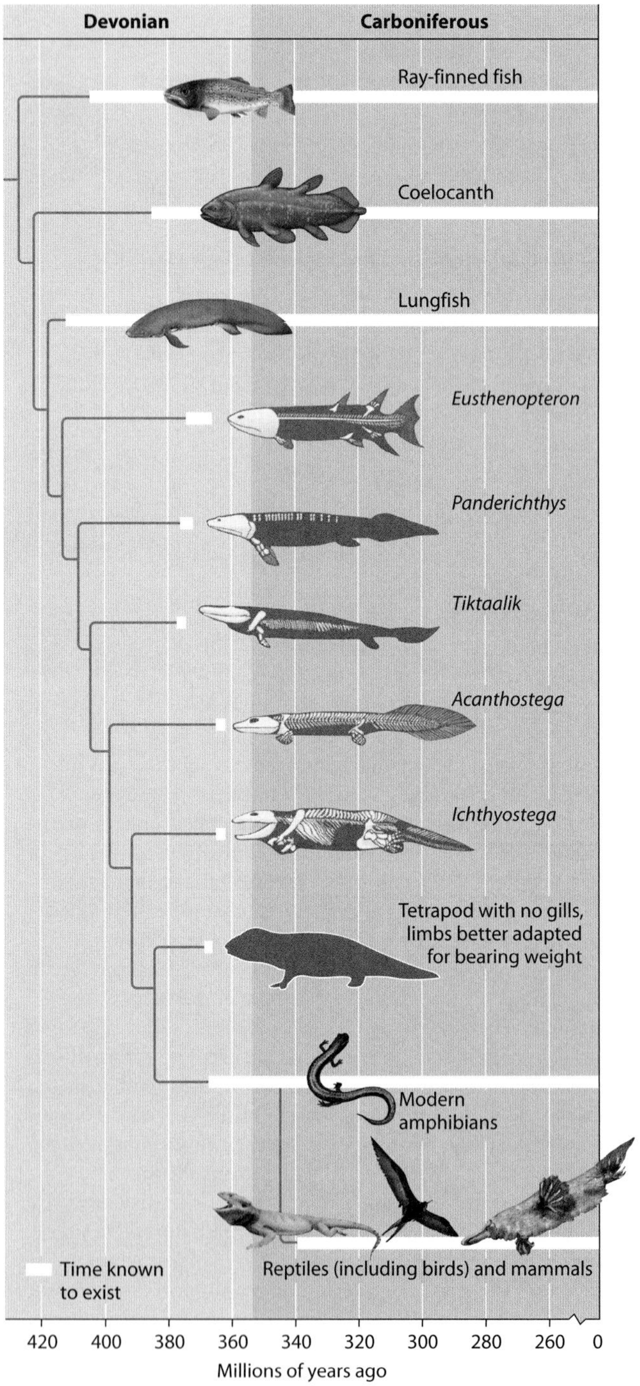

▲ **Figure 4A** Some of the transitional forms in tetrapod evolution

Adapted from B. Holmes. Sept. 9, 2006. Meet your ancestor – the fish that crawled. *New Scientist*, issue 2568: 35-39. By permission of *New Scientist* magazine.

On the other hand, *Acanthostega* (**Figure 4B**) was more of a fishlike tetrapod, and it turned scientists' ideas about tetrapod evolution upside down. Like *Ichthyostega*, *Acanthostega* had a neck, structural modifications that strengthened its backbone and skull, and four limbs with toes. But its limbs could not have supported the animal on land, nor could its ribs have prevented its lungs from collapsing out of water. The startling conclusion was that the first tetrapods were not fish with lungs that had gradually evolved legs as they dragged themselves from pool to pool in search of water. Instead, they were fish with necks and four limbs that raised their heads above water and could breathe oxygen from the air. **Figure 4C** shows an artist's rendering, with fanciful colors.

In 2006, a team of scientists added another important link to the chain of evidence. Using information from the dates and habitats of previous specimens, they predicted that transitional forms might be found in rock formed from the sediments of shallow river environments during a particular time period in the late Devonian. They found a suitable area to search in Arctic Canada which was located near the equator during that time. There they discovered several remarkable fossils of an animal they named *Tiktaalik* ("large freshwater fish" in the language of the Nunavut Inuit tribe from that region). The specimens were exquisitely preserved; even the fishlike scales are clearly visible. *Tiktaalik* straddled the border between *Panderichthys* and *Acanthostega* (see Figure 4A). Its paddle-like forelimbs were part fin, part foot. The fin rays had not been replaced by toes. The joints would have served to prop the animal up, but not enable it to walk. It had well-developed gills like a fish, but a tetrapod-like neck. It was a perfectly intermediate form.

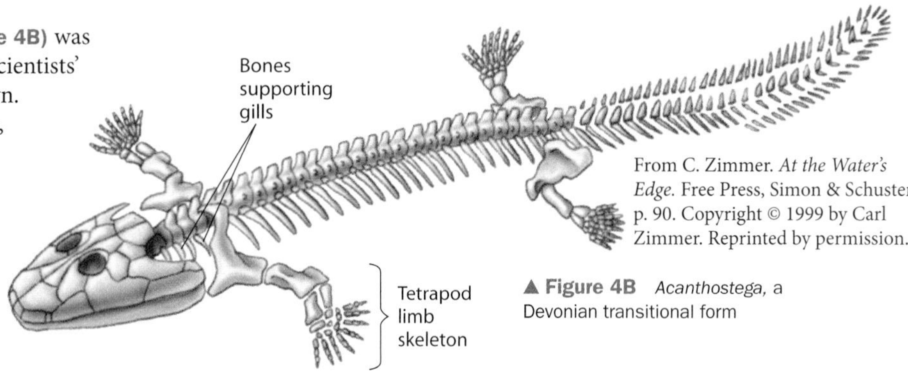

From C. Zimmer. *At the Water's Edge.* Free Press, Simon & Schuster, p. 90. Copyright © 1999 by Carl Zimmer. Reprinted by permission.

▲ **Figure 4B** *Acanthostega*, a Devonian transitional form

With these images of early tetrapods gleaned from the fossil record in mind, let's look at the environmental conditions that drove their evolution. Plants had colonized the land 100 million years earlier, followed by arthropods. By the time *Tiktaalik* appeared, shallow water would have been a complex environment, with fallen trees and other debris from land plants, along with rooted aquatic plants, providing food and shelter for a variety of organisms. Even a meter-long predator like *Tiktaalik* would have found plenty to eat. But warm, stagnant water is low in oxygen. The ability to supplement its oxygen intake by air breathing—by lifting the head out of the water—would have been an advantage.

Once tetrapods had adaptations that enabled them to leave the water for extended periods of time, they diversified rapidly. Food and oxygen were plentiful in the Carboniferous swamp forests. From one of the many lines of tetrapods that settled ashore, modern amphibians evolved.

? **How did *Acanthostega* change scientists' concept of tetrapod evolution?**

It showed that the first tetrapods were more fishlike than previously thought. They did not spend time on land.

▼ **Figure 4C** The first tetrapod, a four-limbed fish that lived in shallow water and could breathe air

Publiphoto/Photo Researchers

5 Amphibians are tetrapods—vertebrates with two pairs of limbs

© Piotr Naskrecki (MYN)/Nature Picture Library/Corbis

▲ **Figure 5A** A redback salamander

A.Noellert/WILDLIFE/Peter Arnold, Inc./Photo Library

Amphibians include salamanders, frogs, and caecilians. Some present-day salamanders are entirely aquatic, but those that live on land walk with a side-to-side bending of the body that probably resembles the swagger of early terrestrial tetrapods (**Figure 5A**). Frogs are more specialized for moving on land, using their powerful hind legs to hop along the terrain. Caecilians (**Figure 5B**) are nearly blind and are legless, adaptations that suit their burrowing lifestyle. However, they evolved from a legged ancestor.

Most amphibians are found in damp habitats, where their moist skin supplements their lungs for gas exchange. Amphibian skin usually has poison glands that may play a role in defense. Poison dart frogs have particularly deadly poisons, and their vivid coloration warns away potential predators (**Figure 5C**).

In Greek, the word *amphibios* means "living a double life," a reference to the metamorphosis of many frogs. A frog spends much of its time on land, but it lays its eggs in water. During the breeding season, many species fill the air with their mating calls. As you can see in **Figure 5D**, frog eggs are encapsulated in a jellylike material. Consequently, they must be surrounded by moisture to prevent them from drying out. The larval stage, called a tadpole, is a legless, aquatic algae-eater with gills, a lateral line system resembling that of fishes, and a long, finned tail (**Figure 5E**). In changing into a frog, the tadpole undergoes a radical metamorphosis. When a young frog crawls onto shore and continues life as a terrestrial insect-eater, it has four legs and air-breathing lungs instead of gills (see Figure 5C). Not all amphibians live such a double life, however. Some species are strictly terrestrial, and others are exclusively aquatic. *Toad* is a term generally used to refer to frogs that have rough skin and live entirely in terrestrial habitats.

For the past 25 years, zoologists have been documenting a rapid and alarming decline in amphibian populations throughout the world. In 2009, roughly 30% of all known species were at risk for extinction. Multiple causes are contributing to the decline, including habitat loss, climate change, and the spread of a pathogenic fungus.

Amphibians were the first vertebrates to colonize the land. The early amphibians probably feasted on insects and other invertebrates in the lush forests of the Carboniferous period. As a result, amphibians became so widespread and diverse that the Carboniferous period is sometimes called the age of amphibians. However, the distribution of amphibians was limited by their vulnerability to dehydration. Adaptations that evolved in the next clade of vertebrates we discuss, the amniotes, enabled them to complete their life cycles entirely on land.

? In what ways are amphibians not completely adapted for terrestrial life?

● Their eggs are not well protected against dehydration; many species have an aquatic larval form; their skin is not waterproof and must remain moist to permit gas exchange.

Michael Fogden/Bruce Coleman

▲ **Figure 5C** An adult poison dart frog

Hans Pfletschinger/Peter Arnold, Inc./PhotoLibrary

▶ **Figure 5E** A tadpole undergoing metamorphosis

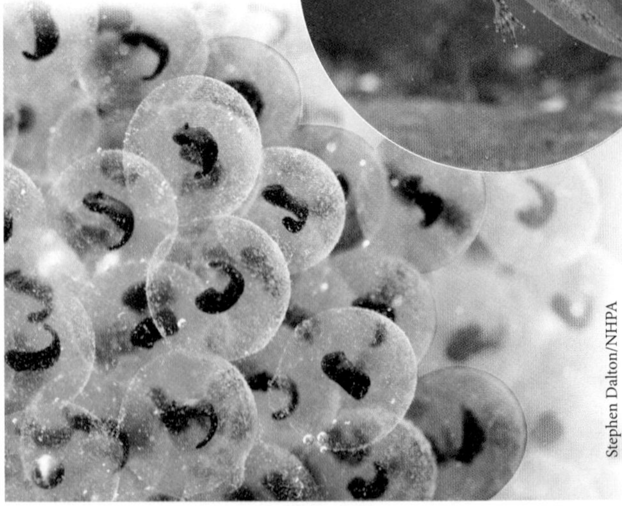

Stephen Dalton/NHPA

▲ **Figure 5D** Frog eggs; a tadpole is developing in the center of each ball of jelly

6 Reptiles are amniotes—tetrapods with a terrestrially adapted egg

Reptiles (including birds) and mammals are **amniotes**. The major derived character of this clade is the amniotic egg **(Figure 6A)**. The **amniotic egg** contains specialized extraembryonic membranes, so called because they are not part of the embryo's body. The *amnion*, for which the amniotic egg is named, is a fluid-filled sac surrounding the embryo **(Figure 6B)**. The *yolk sac* contains a rich store of nutrients for the developing embryo, like the yellow yolk of a chicken egg. Additional nutrients are available from the albumen ("egg white"). The *chorion* and the membrane of the *allantois* enable the embryo to obtain oxygen from the air for aerobic respiration and dispose of the carbon dioxide produced. The allantois is also a disposal sac for other metabolic waste products. With a waterproof shell to enclose the embryo and its life-support system, reptiles were the first vertebrates to be able to complete their life cycles on land. The seed played a similar role in the evolution of plants.

The clade of amniotes called **reptiles** includes lizards, snakes, turtles, crocodilians, and birds, along with a number of extinct groups such as most of the dinosaurs. Lizards are the most numerous and diverse reptiles other than birds. Snakes, which are closely related to lizards, may have become limbless as their ancestors adapted to a burrowing lifestyle. Turtles have changed little since they evolved, although their ancestral lineage is still uncertain. Crocodiles and alligators (crocodilians) are the largest living reptiles—the saltwater crocodiles measure up to 6.3 m (as long as a stretch limousine) in length and weigh up to a ton. Crocodilians spend most of the time in water, breathing air through upturned nostrils.

In addition to an amniotic egg protected in a waterproof shell, reptiles have several other adaptations for terrestrial living not found in amphibians. Reptilian skin, covered with scales waterproofed with the tough protein keratin, keeps the body from drying out. Reptiles cannot breathe through their dry skin and obtain most of their oxygen with their lungs, using their rib cage to help ventilate their lungs.

Lizards, snakes, crocodilians, and turtles are sometimes said to be "cold-blooded" because they do not use their metabolism to produce body heat. Nonetheless, these animals may regulate their temperature through their behavior. The bearded dragon of the Australian outback **(Figure 6C)** commonly warms up in the morning by sitting on warm rocks and basking in the sun. If the lizard gets too hot, it seeks shade. Animals that absorb external heat rather than generating much of their own are said to be **ectothermic** (from the Greek *ektos*, outside, and *therme*, heat), a term that is more appropriate than the term *cold-blooded*. Because the energy demands of ectothermic animals are low, reptiles are well suited to deserts, where food is scarce.

Like the amphibians from which they evolved, reptiles were once much more prominent than they are today. Following the decline of amphibians, reptilian lineages expanded rapidly, creating a dynasty that lasted 200 million years. Most dinosaurs died out during the period of mass extinctions about 65 million years ago. Descendants of one dinosaur lineage, however, survive today as the reptilian group we know as birds.

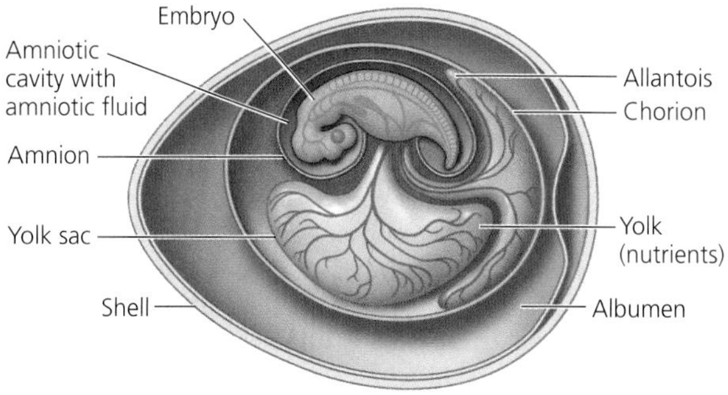

▲ **Figure 6B** An amniotic egg

? **What is an amniotic egg?**

A shelled egg in which an embryo develops in a fluid-filled amniotic sac and is nourished by a yolk.

Sylvain Cordier/Photolibrary

▲ **Figure 6A** A bull snake laying eggs

Steven David Miller/naturepl.com

▲ **Figure 6C** A bearded dragon basking in the sun

7 Birds are feathered reptiles with adaptations for flight

Almost all **birds** can fly, and nearly every part of a bird's body reflects adaptations that enhance flight. The forelimbs have been remodeled as feather-covered wings that act as airfoils, providing lift and maneuverability in the air. Large flight muscles anchored to a central ridge along the breastbone provide power. Some species, such as the seagoing frigate bird in **Figure 7A**, have wings adapted to soaring on air currents, and they flap their wings only occasionally. Others, such as hummingbirds, excel at maneuvering but must flap almost continuously to stay aloft. The few flightless groups of birds include the ostrich, which is the largest bird in the world, and the emu, the largest native bird in Australia.

Many features help reduce weight for flight: Present-day birds lack teeth; their tail is supported by only a few small vertebrae; their feathers have hollow shafts; and their bones have a honeycombed structure, making them strong but light. For example, the frigate bird has a wingspan of more than 2 m, but its whole skeleton weighs only about 113 g (4 ounces).

sodapix/photolibrary

▲ **Figure 7A** A soaring frigate bird

Flying requires a great amount of energy, and present-day birds have a high rate of metabolism. Unlike other living reptiles, they are **endothermic**, using heat generated by metabolism to maintain a warm, steady body temperature. Insulating feathers help to maintain their warm body temperature. In support of their high metabolic rate, birds have a highly efficient circulatory system, and their lungs are even more efficient at extracting oxygen from the air than are the lungs of mammals.

Frans Lanting/Corbis

▲ **Figure 7B** A male wandering albatross performing a courtship display for a potential mate

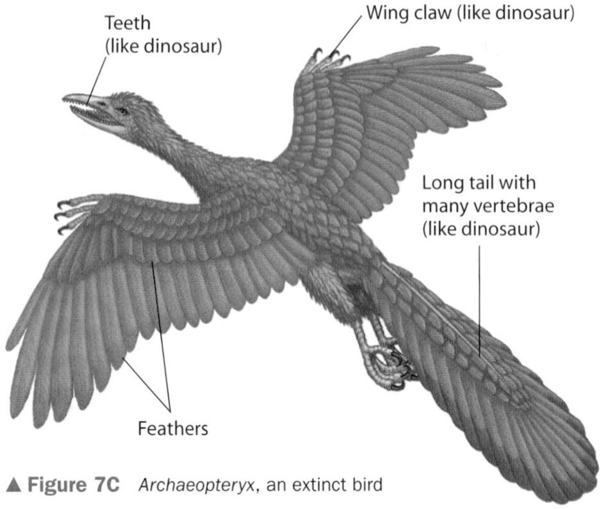

▲ **Figure 7C** *Archaeopteryx*, an extinct bird

Flying also requires acute senses and fine muscle control. Birds have excellent eyesight, perhaps the best of all vertebrates, and the visual and motor areas of the brain are well developed.

Birds typically display very complex behaviors, particularly during breeding season. Courtship often involves elaborate rituals. The male frigate bird in Figure 7A, for example, inflates its red throat pouch like an enormous balloon to attract females. The wandering albatross in **Figure 7B** employs a different kind of courtship display. In many species of birds, males and females take turns incubating the eggs and then feeding the young. Some birds migrate great distances each year to different feeding or breeding grounds.

Strong evidence indicates that birds evolved from a lineage of small, two-legged dinosaurs called theropods. **Figure 7C** is an artist's reconstruction based on a 150-million-year-old fossil of the oldest known, most primitive bird, called *Archaeopteryx* (from the Greek *archaios*, ancient, and *pteryx*, wing). Like living birds, it had feathered wings, but otherwise it was more like a small two-legged dinosaur of its era—with teeth, wing claws, and tail with many vertebrae. Over the past decade, Chinese paleontologists have excavated fossils of many feathered theropods, including specimens that predate *Archaeopteryx* by 5–10 million years. Such findings imply that feathers, which are homologous to reptilian scales, evolved long before powered flight. Early feathers may have functioned in insulation or courtship displays. As more of these fossils are discovered, scientists are gaining new insight into the evolution of flight.

? List some adaptations of birds that enhance flight.

Reduced weight, endothermy with high metabolism, efficient respiratory and circulatory systems, feathered wings shaped like airfoils, good eyesight

8 Mammals are amniotes that have hair and produce milk

There are two major lineages of amniotes: one that led to the reptile clade and one that produced the mammals. Two features—hair and mammary glands that produce milk—are the distinguishing traits of **mammals**. Like birds, mammals are endothermic. Hair provides insulation that helps maintain a warm body temperature. Efficient respiratory and circulatory systems (including a four-chambered heart) support the high rate of metabolism characteristic of endotherms. Differentiation of teeth adapted for eating many kinds of foods is also characteristic of mammals; different kinds of teeth specialize in cutting, piercing, crushing, or grinding. The three major lineages of mammals—monotremes (egg-laying mammals), marsupials (mammals with a pouch), and eutherians (placental mammals)—differ in their reproductive patterns.

The only existing egg-laying mammals, known as **monotremes**, are echidnas (spiny anteaters) and the duck-billed platypus, described in the chapter introduction. The female platypus usually lays two eggs and incubates them in a nest. After hatching, the young lick up milk secreted onto the mother's fur **(Figure 8A)**.

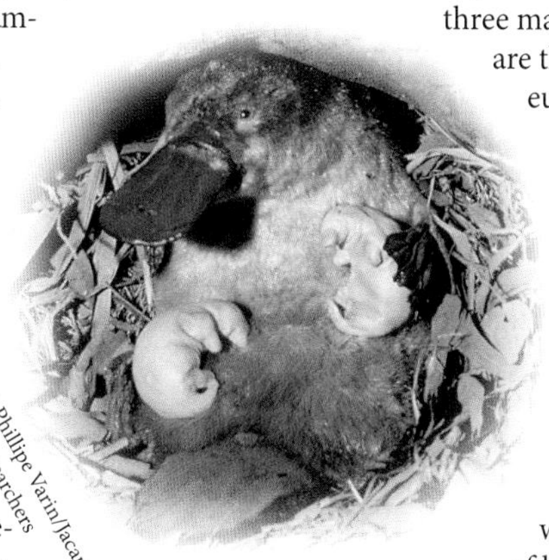

Jean Phillipe Varin/Jacana/ Photo Researchers

▲ **Figure 8A** Monotremes: a duck-billed platypus with newly hatched young

All other mammals are born rather than hatched. During development, the embryos remain inside the mother and receive their nourishment directly from her blood. Mammalian embryos produce extraembryonic membranes that are homologous to those found in the amniotic egg, including the amnion, which retains its function as a protective fluid-filled sac. The chorion, yolk sac, and allantois have different functions in mammals than in reptiles. Membranes from the embryo join with the lining of the uterus to form a **placenta**, a structure in which nutrients from the mother's blood diffuse into the embryo's blood.

Marsupials have a brief gestation and give birth to tiny, embryonic offspring that complete development while attached to the mother's nipples. The nursing young are usually housed in an external pouch **(Figure 8B)**. Nearly all marsupials live in Australia, New Zealand, and Central and South America. The opossum is the only North American marsupial. **Eutherians** are mammals that bear fully developed live young. They are commonly called **placental mammals** because their placentas are more complex than those of marsupials, and the young complete their embryonic development in the mother's uterus attached to the placenta. The large silvery membrane still clinging to the newborn zebra in **Figure 8C** is the amniotic sac. Elephants, rodents, rabbits, dogs, cows, whales, bats, and humans are all examples of eutherians.

The first true mammals arose about 200 million years ago and were probably small, nocturnal insect-eaters. Of the three main groups of living mammals, monotremes are the oldest lineage. Marsupials diverged from eutherians about 180 million years ago. During the Mesozoic era, mammals remained about the size of today's shrews, which are very small insectivores. After the extinction of large dinosaurs at the end of the Cretaceous period, however, mammals underwent an adaptive radiation, giving rise to large terrestrial herbivores and predators, as well as bats and aquatic mammals such as porpoises and whales. Humans belong to the mammalian order Primates, along with monkeys and apes. We begin our study of human evolution with the next module.

? What are the two distinguishing features of mammals?

Hair and mammary glands, which produce milk

Photolibrary/Corbis

▲ **Figure 8B** Marsupials: a gray kangaroo with her young in her pouch

Mitch Reardon/Photo Researchers

▲ **Figure 8C** Eutherians: a zebra with newborn

Primate Diversity

9 The human story begins with our primate heritage

The mammalian order Primates includes the lemurs, tarsiers, monkeys, and apes. The earliest primates were probably small arboreal (tree-dwelling) mammals that arose before 65 million years ago, when dinosaurs still dominated the planet. Most living primates are still arboreal, and the primate body has a number of features that were shaped, through natural selection, by the demands of living in trees. Although humans never lived in trees, the human body retains many of the traits that evolved in our arboreal ancestors.

▼ **Figure 9A** A slender loris

E. H. Rao/Photo Researchers

The squirrel-sized slender loris in **Figure 9A** illustrates a number of distinguishing primate features. It has limber shoulder and hip joints, enabling it to climb and to swing from one branch to another. The five digits of its grasping feet and hands are highly mobile; the separation of its big toe from the other toes and its flexible thumb give the loris the ability to hang onto branches and manipulate food. The great sensitivity to touch of the hands and feet also aids in manipulation. Lorises have a short snout and eyes set close together on the front of the face. The position of the eyes makes their two fields of vision overlap, enhancing depth perception, an important trait for maneuvering in trees. We humans share all these basic primate traits with the slender loris except for the widely spaced big toe.

As shown in the phylogenetic tree in **Figure 9B**, the slender loris belongs to one of three main groups of living primates. The lorises and pottos of tropical Africa and southern Asia are placed in one group along with the lemurs. Ranging from the pygmy mouse lemur, which weighs 25 g (1 ounce), to the sifakas **(Figure 9C)**, which may weigh as much as 8 kg (17.6 pounds), lemurs are a diverse group. But of about 50 species of lemurs originally present on the island of Madagascar (see back cover), 18 have become extinct since humans first colonized the island about 2,000 years ago. Because of continuing deforestation and hunting, all lemurs, including the ring-tailed lemur on the book's cover, are currently endangered species.

The tarsiers form a second group of primates. Limited to Southeast Asia, these small, nocturnal tree-dwellers have flat faces with large eyes **(Figure 9D)**. Fossil evidence indicates that tarsiers are more closely related to anthropoids, the third group of primates, than to the lemur-loris-potto group.

The **anthropoids** (from the Greek *anthropos*, man, and *eidos*, form) include monkeys and apes. As shown in Figure 9B,

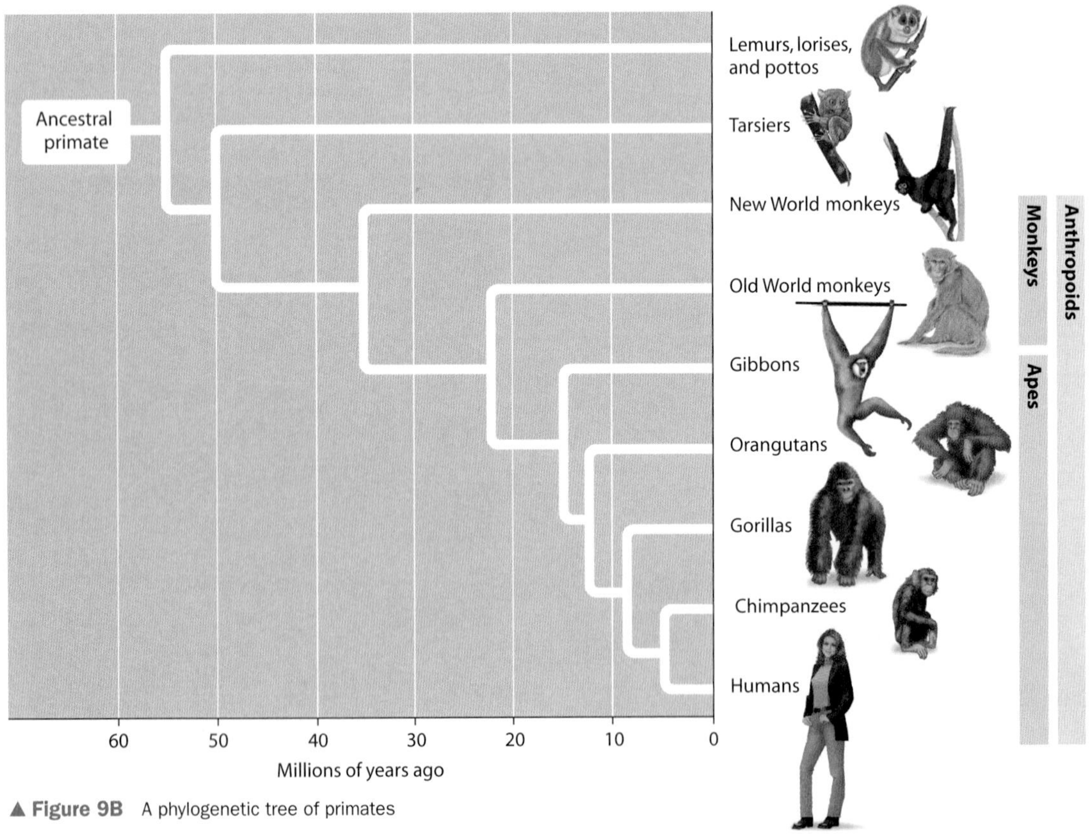

▲ **Figure 9B** A phylogenetic tree of primates

the fossil record indicates that anthropoids began diverging from other primates about 50 million years ago. Anthropoids have a fully **opposable thumb**; that is, they can touch the tip of all four fingers with their thumb. In monkeys and most apes, the opposable thumb functions in a grasping "power grip," but in humans, a distinctive bone structure at the base of the thumb allows it to be used for more precise manipulation.

Notice that monkeys do not constitute a monophyletic group. The first monkeys probably evolved in the Old World (Africa and Asia) and may have reached South America by rafting on logs from Africa. The monkeys of the Old World and New World (the Americas) have been evolving separately for over 30 million years. Both New and Old World monkeys are active during the day and usually live in bands held together by social behavior.

New World monkeys, found in Central and South America, are all arboreal. Their nostrils are wide open and far apart, and many, such as the black spider monkey, an inhabitant of rain forests in eastern South America (**Figure 9E**, right), have a long tail that is prehensile—specialized for grasping tree limbs. The squirrel-sized golden lion tamarin (Figure 9E, left) is a New World monkey that inhabits lowland rain forests of eastern Brazil. In the 1970s, habitat destruction reduced this species to the brink of extinction. Thanks to an intense international conservation effort, numbers have rebounded to about 1,200 in the wild.

Old World monkeys lack a prehensile tail, and their nostrils open downward. They include macaques (**Figure 9F**), mandrills, baboons, and rhesus monkeys. Many species are arboreal, but some, such as the baboons of the African savanna, are ground dwelling.

Monkeys differ from most apes in having forelimbs that are about equal in length to their hind limbs. Old World monkeys and apes, which you will learn about in the next module, diverged about 20–25 million years ago. The human lineage probably diverged from an ancestor shared with chimpanzees sometime between 5 and 7 million years ago.

? **To which mammalian order do we belong? What are the three main groups of this order?**

Primates: (1) lorises, pottos, and lemurs; (2) tarsiers; and (3) anthropoids

Kevin Schafer/Corbis

A golden lion tamarin (note nostrils that open to the side)

Kevin Schafer/Photo Researchers

A black spider monkey (note prehensile tail)

▲ **Figure 9E** New World monkeys

Bildagentur/Photolibrary

▲ **Figure 9C** A Coquerel's sifaka (pronounced "she-fa'-ka")

David Tipling/naturepl.com.

▶ **Figure 9D** A tarsier, member of a distinct primate group

Digital Vision/Getty Images

▲ **Figure 9F** Old World monkeys: a macaque with its young

10 Humans and four other groups of apes are classified as anthropoids

In addition to monkeys, the anthropoid group includes apes: gibbons, orangutans, gorillas, chimpanzees (and bonobos), and humans (**Figure 10**). Apes lack a tail and have relatively long arms and short legs. Compared to other primates, they have larger brains relative to body size, and consequently their behavior is more flexible. Gorillas, chimpanzees, and humans have a high degree of social organization. Except for the human lineage, the apes have a smaller geographic range than the monkeys; they evolved and diversified only in Africa and Southeast Asia and are confined to tropical regions, mainly rain forests.

The nine species of gibbons, all found in Southeast Asia, are entirely arboreal. Gibbons are the smallest, lightest, and most acrobatic of the apes. They are also the only nonhuman apes that are monogamous, with mated pairs remaining together for life.

The orangutan is a shy, solitary ape that lives in the rain forests of Sumatra and Borneo. The largest living arboreal mammal, it moves rather slowly through the trees, supporting its stocky body with all four limbs. Orangutans may occasionally venture onto the forest floor.

The gorilla is the largest ape: Some males are almost 2 m (6.5 feet) tall and weigh about 200 kg (440 pounds). Found only in African rain forests, gorillas usually live in groups of up to about 20 individuals. They spend nearly all their time on the ground. Gorillas can stand upright on their hind legs, but when they walk on all fours, their knuckles contact the ground.

Like the gorilla, the chimpanzee and a closely related species called the bonobo are knuckle walkers. These apes spend as much as a quarter of their time on the ground. Both species inhabit tropical Africa. Chimpanzees have been studied extensively, and many aspects of their behavior resemble human behavior. For example, chimpanzees make and use simple tools. The individual in Figure 10 is using a blade of grass to "fish" for termites. Chimpanzees also raid other social groups of their own species, exhibiting behavior formerly thought to be uniquely human.

Molecular evidence indicates that chimpanzees and gorillas are more closely related to humans than they are to other apes. Humans and chimpanzees are especially closely related; their genomes are 99% identical. Nevertheless, human and chimpanzee genomes have been evolving separately since the two lineages diverged from their last common ancestor between 5 and 7 million years ago. Because fossil apes are extremely rare, we know little about that ancestor. However, researchers studying the skeletal features of a 4.4-million-year-old species called *Ardipithecus ramidus* recently concluded that present-day apes such as chimpanzees are the result of substantial evolution since the lineages diverged.

? **What are the five groups of apes within the anthropoid category?**

Gibbons, orangutans, gorillas, chimpanzees, and humans

Steve Bloom/Alamy Images

A gibbon

Anup Shah/NaturePL

An orangutan

Figure 10 Nonhuman anthropoids

Gerry Ellis/Digital Vision/Getty Images

A chimpanzee

Image Source/Alamy Images

A gorilla and offspring

Hominin Evolution

11 The hominin branch of the primate tree includes species that coexisted

Paleoanthropology, the study of human origins and evolution, focuses on the tiny slice of biological history that has occurred since the divergence of human and chimpanzee lineages from their common ancestor. Paleoanthropologists have unearthed fossils of approximately 20 species of extinct **hominins**, species that are more closely related to humans than to chimpanzees and are therefore on the human branch of the evolutionary tree. (The older term *hominid* is still used by some anthropologists.) These fossils have shown that many of the characters that distinguish humans from other apes first emerged long before *Homo sapiens* appeared.

Thousands of hominin fossils have been discovered, and each new find sheds light on the story of human evolution. However, paleoanthropologists are still vigorously debating hominin classification and phylogenetic relationships. Therefore, **Figure 11** presents some of the known hominins in a timeline rather than in a tree diagram like the one in Figure 9B. The vertical bars indicate the approximate time period when each species existed, as currently known from the fossil record.

One inference about human phylogeny can immediately be made from Figure 11: Hominins did not evolve in a straight line leading directly to *Homo sapiens*. At times several hominin species coexisted, and some must have been dead ends that did not give rise to new lineages.

The oldest hominin yet discovered, *Sahelanthropus tchadensis*, lived from about 7 to 6 million years ago, around the time when the human and chimpanzee lineages diverged. However, most of the hominin fossils that have been found are less than 4 million years old. Thus, the 4.4-million-year-old fossils of *Ardipithecus ramidus*, painstakingly uncovered and reconstructed by international teams of scientists over the past 15 years, represent an unprecedented perspective on early hominin evolution. *Ardipithecus* was a woodland creature that moved in the trees by walking along branches on the flat parts of its hands and feet. It was equally capable of moving on the ground, and its skeletal features suggest that it walked upright.

Hominin diversity increased dramatically in the period between 4 and 2 million years ago. The first fossil member of our own genus, *Homo*, dates from that time. By 1 million years ago, only species of *Homo* existed. Eventually, all *Homo* species except one—our own—ended in extinction.

? Based on the fossil evidence represented in Figure 11, how many hominin species coexisted 1.7 million years ago?

Five: *P. boisei, P. robustus, H. habilis, H. ergaster, H. erectus*

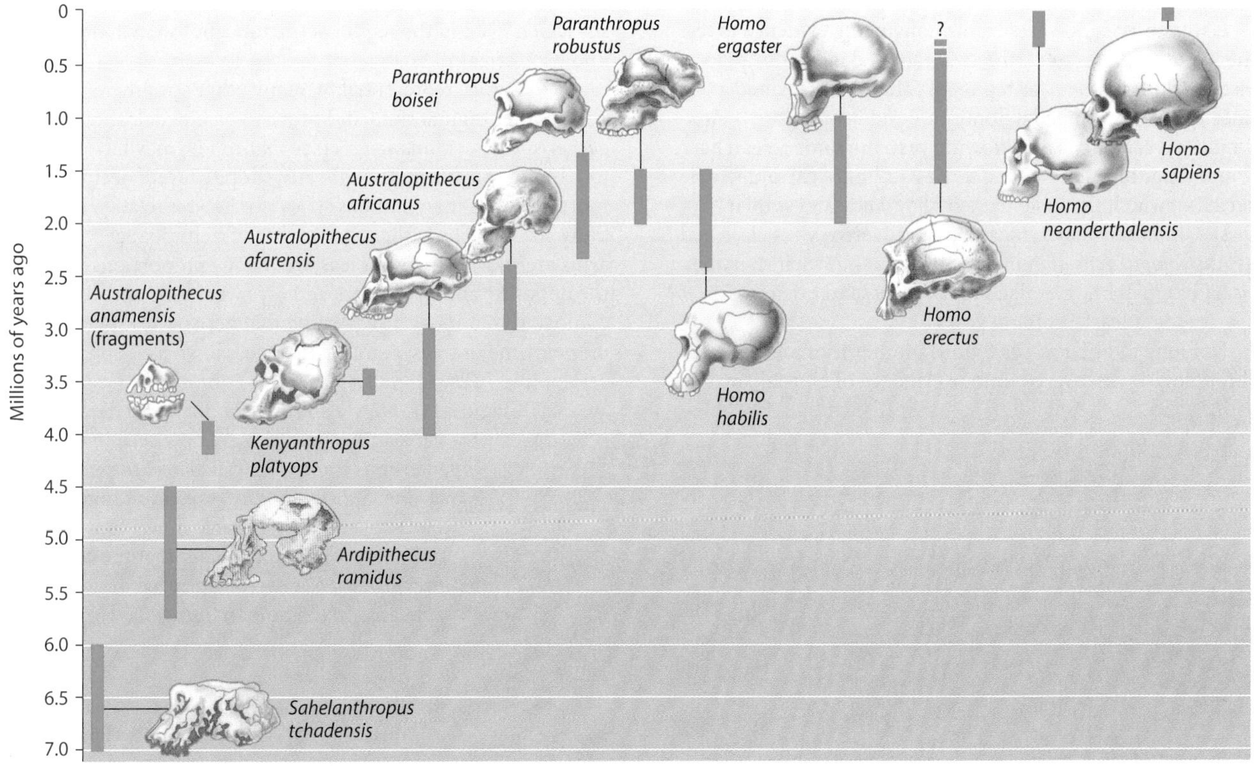

▲ **Figure 11** A timeline for some hominin species

Drawn from photos of fossils: *A. ramidus* adapted from www.age-of-the-sage.org/evolution/ardi_fossilized_skeleton.htm. *A. anamensis* and *H. neanderthalensis* adapted from *The Human Evolution Coloring Book*. *K. platyops* drawn from photo in Meave Leakey et al., New hominid genus from eastern Africa shows diverse middle Pliocene lineages, *Nature*, March 22, 2001, 410:433. *P. boisei* drawn from a photo by David Bill. *H. ergaster* drawn from a photo at www.inhandmuseum.com. *S. tchadensis* drawn from a photo in Michel Brunet et al., A new hominid from the Upper Miocene of Chad, Central Africa, *Nature*, July 11, 2002, 418:147, fig. 1b.

12 Australopiths were bipedal and had small brains

Present-day humans and chimpanzees clearly differ in two major features: Humans are bipedal (walk upright) and have much larger brains. When did these features emerge? In the early 20th century, paleoanthropologists hypothesized that increased brain size was the initial change that separated hominins from apes. Bipedalism and other adaptations came later as hominin intelligence led to changes in food-gathering methods, parental care, and social interactions.

The evidence needed to test this hypothesis would come from hominin fossils. Hominin skulls would reveal brain size. Evidence of upright stance might be found by examining the limb and pelvic structures. Another clue to bipedalism is the location of the opening in the base of the skull through which the spinal cord exits. In chimpanzees and other species that are primarily quadrupeds, the spinal cord exits toward the rear of the skull, at an angle that allows the eyes to face forward (**Figure 12A**, left). In bipeds, including humans, the spinal cord emerges from the floor of the braincase, so the head can be held directly over the body (Figure 12A, right).

In 1924, a scientist working in South Africa discovered a skull that he interpreted as having both apelike and human characteristics. He claimed that the skull represented a new genus and species of bipedal human relatives, which he named *Australopithecus africanus* ("southern ape of Africa"). However, the skull had been blasted out of its geologic context before the scientist received it, making it difficult to establish the fossil's age.

It was another 50 years before convincing evidence to test the hypothesis was unearthed. A team of paleoanthropologists working in the Afar region of Ethiopia discovered a knee joint from a bipedal hominin—and it was more than 3 million years old. The following year, the same researchers found a significant portion of a 3.24-million-year-old female skeleton, which they nicknamed Lucy. Lucy and similar fossils (hundreds have since been discovered) were classified as *Australopithecus afarensis.* The fossils show that *A. afarensis* had a small brain, walked on two legs, and existed as a species for at least 1 million years.

Not long after Lucy was found, another team of paleoanthropologists discovered unique evidence of bipedalism in ancient hominins. While working in what is now Tanzania in East Africa, they found a 3.6-million-year-old layer of hardened volcanic ash crisscrossed with tracks of hyenas, giraffes, and several extinct species of mammals—including upright-walking hominins (**Figure 12B**). After the ash had settled, rain had dampened it. The feet of two hominins, one large and one small, walking close together, made impressions in the ash as if it were wet sand on a beach. The ash, composed of a cement-like material, solidified soon after and was buried by more ash from a later volcanic eruption. The hominins strolling across that ancient landscape may have been *A. afarensis*, which lived in the region at the time, but we will never know for certain.

John Reader/SPL/Photo Researchers

▲ **Figure 12B** Evidence of bipedalism in early hominins: footprints in ancient ash

A. africanus, represented by many other specimens besides the original skull, did prove to be bipedal, and an earlier species, *A. anamensis* (see Figure 11), may have been, too. In other features, though, australopiths were decidedly more like apes than humans. Lucy's brain size relative to her body size was about the same as that of a chimpanzee. Our arms are shorter than our legs, but Lucy's proportions were the opposite, and her fingers and toes were long and curved compared to ours—all suggesting that her species spent some of their time in trees. Another lineage known as "robust" australopiths—*Paranthropus boisei* and *Paranthropus robustus* in Figure 11—were also small-brained bipeds.

Paleoanthropologists are now certain that bipedalism is a very old trait. There is evidence that even *Sahelanthropus*, the oldest hominin yet discovered, was capable of walking upright. It was only much later that the other major human trait—an enlarged brain—appeared in the human lineage.

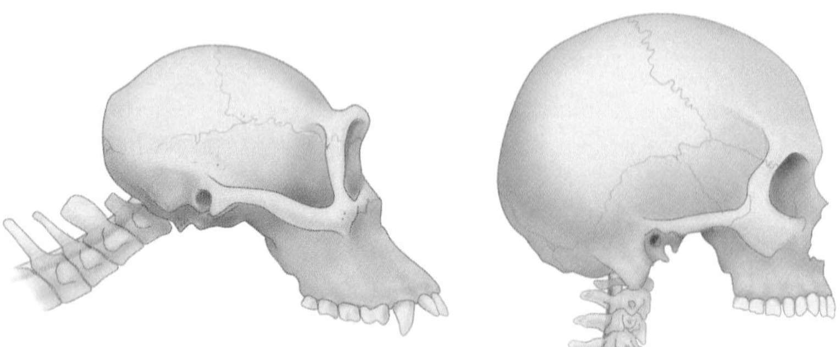

▲ **Figure 12A** The angle of spinal cord exit from skull in chimpanzee (left) and human (right)

? How can paleoanthropologists conclude that a species was bipedal based on only a fossil skull?

By the location of the opening where the spinal cord exits the skull

13 Larger brains mark the evolution of *Homo*

By measuring the capacity of fossil skulls, paleoanthropologists can estimate the size of the brain, which, relative to body size, roughly indicates the animal's intelligence. The brain volume of *Homo sapiens*, at an average 1,300 cm^3, is approximately triple that of australopiths **(Figure 13A)**. As evolutionary biologist Stephan J. Gould put it, "Mankind stood up first and got smart later."

At 400–450 cm^3, the brains of australopiths were too small to qualify them as members of the genus *Homo*, but how big is big enough? What distinguishes humanlike from apelike brain capacity? When a team of paleoanthropologists found crude stone tools along with hominin fossils, they decided that the toolmaker must be one of us and dubbed their find *Homo habilis* ("handy man"). Its brain volume of 510–690 cm^3 was a significant jump from australopiths, but some scientists did not consider this large enough to be included in the genus *Homo*. Many *H. habilis* fossils ranging in age from about 1.6 to 2.4 million years have since been found, some appearing more humanlike than others.

Homo ergaster, dating from 1.9 to 1.6 million years ago, marks a new stage in hominin evolution. With a further increase in brain size, ranging from 750 to 850 cm^3, *H. ergaster* was associated with more sophisticated stone tools. Its limb proportions and rib cage were similar to those of modern humans, and its short, straight fingers indicate that it did not climb trees. Its long, slender legs with hip joints were well adapted for long-distance walking.

Fossils of *H. ergaster* were originally thought to come from early members of another species, *Homo erectus*. In *H. erectus* ("upright man"), average brain volume had increased to 940 cm^3; the range of sizes overlaps that of *H. ergaster*. Members of *H. erectus* were the first hominins to extend their range beyond Africa. The oldest known fossils of hominins outside Africa, discovered in 2000 in the former Soviet Republic of Georgia, are *H. erectus* dating back 1.8 million years. Others have been found in China and Indonesia. Most fossil evidence indicates that *H. erectus* became extinct at some point about 200,000 years ago, but recent discoveries raise the possibility that a population survived until 50,000 years ago.

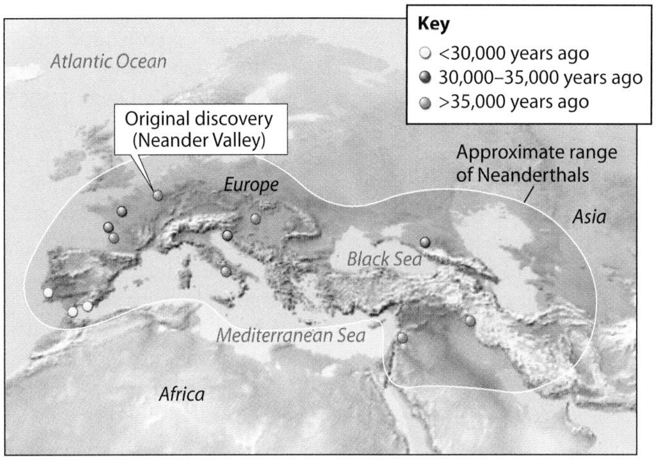

Adapted from D. Jones. Nov. 11, 2006. Blueprint for a Neanderthal. *New Scientist*, issue 2577. By permission of *New Scientist* magazine.

▲ **Figure 13B** Sites where Neanderthal fossils have been found

Homo neanderthalensis, commonly called Neanderthals, are perhaps the best known hominins. They had a brain even larger than ours and hunted big game with tools made from stone and wood. Neanderthals were living in Europe as long as 350,000 years ago and later spread to the Near East **(Figure 13B)**, but by 28,000 years ago, the species was extinct.

Since the discovery of fossilized remains of *H. neanderthalensis* in the Neander Valley in Germany 150 years ago, people have wondered how Neanderthals are related to us. Were they the ancestors of Europeans? Close cousins? Or part of a different branch of evolution altogether? By comparing mitochondrial DNA sequences from Neanderthals and living humans, researchers showed that Neanderthals are not the ancestors of Europeans. Rather, the last common ancestor of humans and Neanderthals lived around 500,000 years ago. However, a comparison of the nuclear genome sequence of *Homo sapiens* with that from Neanderthal fossils, completed in 2010, suggests that Neanderthals and some *H. sapiens* that had left Africa probably did interbreed. Further analysis of Neanderthal DNA may reveal details about the physical appearance, physiology, and brain structure of these intriguing hominins.

In the next module, we look at the origin and worldwide spread of our own species, *Homo sapiens*.

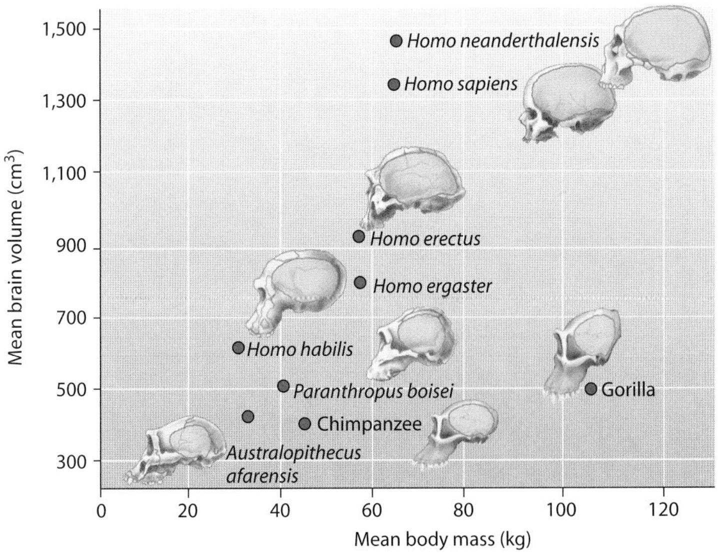

▲ **Figure 13A** Brain volume versus body mass in anthropoids

? Place the following hominins in order of increasing brain volume: *Australopithecus*, *H. erectus*, *H. ergaster*, *H. habilis*, *H. sapiens*.

Australopithecus, *H. habilis*, *H. ergaster*, *H. erectus*, *H. sapiens*

14 From origins in Africa, *Homo sapiens* spread around the world

Evidence from fossils and DNA studies is coming together to support a compelling hypothesis about how our own species, *Homo sapiens*, emerged and spread around the world.

The ancestors of humans originated in Africa. Older species (perhaps *H. ergaster* or *H. erectus*) gave rise to newer species and ultimately to *H. sapiens*. The oldest known fossils of our own species have been discovered in Ethiopia and include fossils that are 160,000 and 195,000 years old. These early humans lacked the heavy browridges of *H. erectus* and Neanderthals (*H. neanderthalensis*), and they were more slender.

Molecular evidence about the origin of humans supports the conclusions drawn from fossils. In addition to showing that living humans are more closely related to one another than to Neanderthals, DNA studies indicate that Europeans and Asians share a more recent common ancestor and that many African lineages represent earlier branches on the human tree. These findings strongly suggest that all living humans have ancestors that originated as *H. sapiens* in Africa.

This conclusion is further supported by analyses of mitochondrial DNA, which is maternally inherited, and Y chromosomes, which are transmitted from fathers to sons—as well as recently sequenced autosomal DNA. Such studies suggest that all living humans inherited their mitochondrial DNA from a common ancestral woman who lived approximately 160,000–200,000 years ago. Mutations on the Y chromosomes can serve as markers for tracing the ancestry and relationships among males alive today. By comparing the Y chromosomes of males from various geographic regions, researchers were able to infer divergence from a common African ancestor.

These lines of evidence suggest that our species emerged from Africa in one or more waves, spreading first into Asia 50,000–60,000 years ago, then to Europe, Southeast Asia, and Australia (**Figure 14**). The date of the first arrival of humans in the New World is hotly debated.

The rapid expansion of *H. sapiens* may have been spurred by the evolution of human cognition as our species evolved in Africa. Although Neanderthals and other hominins were able to produce sophisticated tools, they showed little creativity and not much capability for symbolic thought, as far as we can tell. In contrast, researchers are beginning to find evidence of increasingly sophisticated thought as *H. sapiens* evolved. In 2002, researchers reported the discovery in South Africa of 77,000-year-old art, and by 36,000 years ago, *H. sapiens* were producing spectacular cave paintings.

As *H. sapiens* spread around the globe, populations adapted to the new environments they encountered. Consequently, some differences among people are attributable to their deep ancestry. We'll look at one such adaptation in Module 16.

? What types of evidence indicate that *Homo sapiens* originated in Africa?

Fossils and analyses of mitochondrial DNA and chromosomal DNA

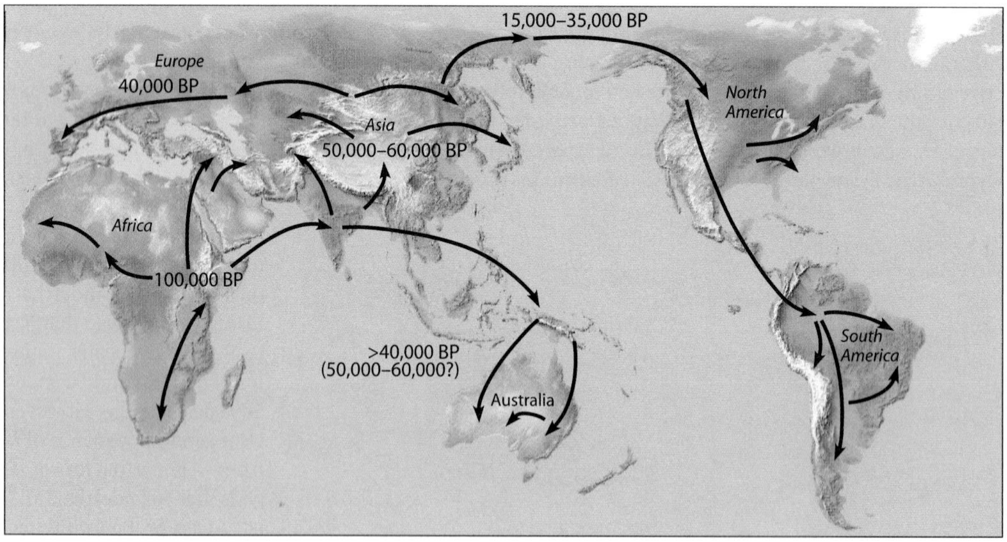

▶ **Figure 14** The spread of modern *Homo sapiens* (dates given as years before present, BP)

Reprinted by permission from Macmillan Publishers Ltd.: *Nature Genetics*. From L. Luca Cavalli-Sforza and M. W. Feldman. 2003. The application of molecular genetic approaches to the study of human evolution. Vol. 33: 266-275, fig. 3, copyright 2003.

15 Who were the "hobbits"?

New findings continually update our understanding of human evolution. Perhaps the most surprising was the 2004 discovery of the nearly complete skeleton of a tiny hominin on Flores, one of the Indonesian islands. In size and brain capacity, the adult female was comparable to Lucy—about 1 m tall, with a chimp-sized brain—and quickly acquired the nickname "hobbit." Despite its diminutive size, its skull (**Figure 15**) displays some humanlike traits, and it apparently made and used stone tools. Most astonishingly, it lived as recently as 18,000 years ago.

The researchers who discovered the "hobbit" proposed that it was a previously unknown human species that evolved from a population of *Homo erectus* into a dwarf form on the island.

Beawiharta/Reuters/Corbis

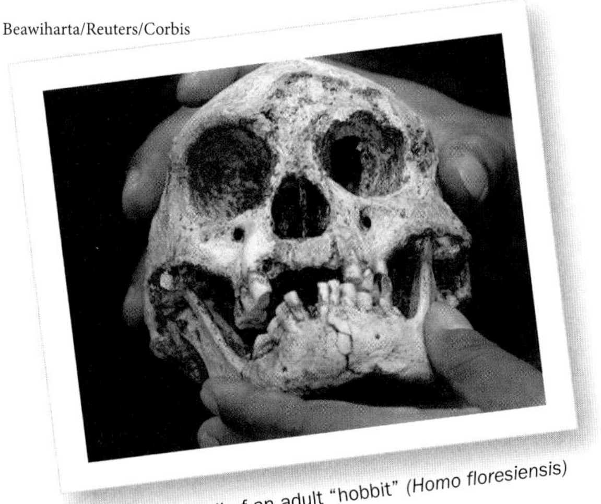

▲ **Figure 15** The skull of an adult "hobbit" (*Homo floresiensis*)

They named it *Homo floresiensis*. Some scientists agreed with them, but others vehemently disputed their conclusions. One counterargument suggested that the hobbit was simply a *Homo sapiens* adult with a genetic disorder.

Since the initial discovery, researchers have unearthed the bones of an estimated 14 *H. floresiensis* individuals, but no additional skulls have been located. Numerous scientists have pored over the fossils, and the more this species is studied, the more mysterious it becomes. For example, the feet are very long in proportion to the legs, the toes are long and curved, and the foot lacks an arch—all apelike characteristics. On the other hand, the big toe aligns with the others, unlike the ape toe, which diverges at an angle. But if *H. floresiensis* is determined to be a very primitive *Homo*, more like *H. habilis* than *H. erectus*, then new questions are raised. It must have left Africa even earlier than *H. erectus* and managed to extend its range thousands of miles without the long, striding legs and sophisticated tools of that species. If that was the case, then undiscovered hominin fossils much older than *H. erectus* must exist somewhere between Africa and Indonesia.

As researchers target new sites to explore and excavation continues in the cave where the hobbits were found, the mystery continues. The final answers await the discovery of more fossils.

? What characteristics prompted the discoverers of the "hobbit" to classify it in the genus *Homo*?

Humanlike skull characteristics and the apparent use of stone tools

EVOLUTION CONNECTION

16 Human skin color reflects adaptations to varying amounts of sunlight

In today's diverse society, skin color is one of the most striking differences among individuals. For centuries, people assumed that these differences reflected more fundamental genetic distinctions, but modern genetic analysis has soundly disproved those assumptions. Is there an evolutionary explanation for skin color differences? To develop hypotheses, scientists began with the observation that human skin color varies geographically. People indigenous to tropical regions have darker skin pigmentation than people from more northerly latitudes.

Skin color results from a pigment called melanin that is produced by specialized skin cells. We all have melanin-producing cells, but the cells are less active in people who have light-colored skin. In addition to absorbing visible light, and therefore appearing dark-colored, melanin absorbs ultraviolet wavelengths. We know that ultraviolet (UV) radiation causes mutations, but it has other effects in skin as well.

UV radiation helps catalyze the synthesis of vitamin D in the skin. This vitamin is essential for proper bone development, so it is especially important for pregnant women and small children to receive adequate amounts. By blocking UV radiation, melanin prevents vitamin D synthesis. Dark-skinned humans evolving in equatorial Africa received sufficient UV radiation to make vitamin D, but northern latitudes receive less sunlight. The loss of skin pigmentation in humans that migrated north from Africa probably helped their skin receive adequate UV radiation to produce enough vitamin D.

Why did dark skin evolve in humans in the first place? UV radiation degrades folate (folic acid), a vitamin that is vital for fetal development and spermatogenesis. Researchers hypothesize that dark skin was selected for because melanin protects folate from the intense tropical sunlight. The evolution of differing skin pigmentations likely provided a balance between folate protection and vitamin D production (Table 16).

Because skin color was the product of natural selection, similar environments produced similar degrees of pigmentation. Widely separated populations may have the same adaptation, regardless of how they are related. As a result, skin color is not a useful characteristic for judging phylogenetic relationships.

TABLE 16 | CORRELATION OF UV RADIATION WITH RISK OF VITAMIN D AND FOLATE DEFICIENCIES

Latitude	UV Radiation	Risk of Vitamin D Deficiency	Risk of Folate Deficiency
Tropical latitudes 0–23.5°	High	Low	High
Higher latitudes 23.5–90°	Low	High	Low

? Why didn't folate degradation select against lightly pigmented people in northern latitudes?

UV radiation is less intense in northern latitudes than in the tropics, so it did not have an adverse effect on folate levels.

CONNECTION 17 Our knowledge of animal diversity is far from complete

When an Englishman sent home the skin of a duck-billed platypus more than 200 years ago, it was one of thousands of previously unknown species of organisms pouring into Europe from naturalists exploring Africa, Asia, and North and South America, as well as Australia. You might think that after centuries of scientific exploration, only tiny organisms such as microbes and insects remain to be found. But the days of exploring new ecosystems and discovering new species are not over. In fact, better access to remote areas, coupled with new mapping technologies, has renewed the pace of discovery. According to a report issued in 2009, 18,516 species were described for the first time in 2007. As you might expect, over half of them were insects, but the list also includes more than 1,200 vertebrates.

The Mekong region of Southeast Asia, an area of diverse landscapes surrounding the Mekong River as it flows from southern China to the China Sea, is one of many treasure troves of previously unknown species that are currently being explored. Over the past decade, more than 1,000 new species have been identified in the region, including the leopard gecko in **Figure 17A**, one of more than 400 new species of vertebrates that scientists have turned up there. To the southeast, remote mountains on the island of New Guinea are also yielding hundreds of discoveries, including a frog with a droopy nose (**Figure 17B**) and the tiny wallaby shown in **Figure 17C**. The first new species of monkey found in more than a century was discovered in the eastern Himalayas. Discoveries of new primate species are extremely rare, but four others have also been made recently: two new lemur species in Madagascar and two new monkeys, an Old World species in Tanzania and a New World species in Bolivia.

Previously undescribed species are being reported almost daily from every continent and a wide variety of habitats. And researchers are just beginning to explore the spectacular diversity of the oceans. The Census of Marine Life, a decade-long collaboration among scientists from 80 nations, has reported the discovery of more than 5,000 new species. Thousands more are expected to be found as new technology enables scientists to investigate deep-sea habitats. Recent expeditions have also gleaned hundreds of new species from the seas surrounding Antarctica, and the collapse of Antarctic ice shelves has allowed researchers their first glimpse of life on a seafloor that had previously been hidden from view. Even places that are regularly visited by people offer surprises. For example, over 100 new marine species were identified recently on a coral reef near Australia.

When a new species is described, taxonomists learn as much as possible about its physical and genetic characteristics and assign it to the appropriate groups in the Linnaean system. As a result, most new species automatically acquire a series of names from domain through genus. But every species also has a unique identifier, and the honor of choosing it belongs to the discoverer. Species are often named for their habitat or a notable feature.

In a new twist, naming rights for recently discovered species have been auctioned off to raise money for conservation organizations, which undertake many of the projects that survey biological diversity. The right to name a new species of monkey cost the winning bidder $650,000, and donors spent more than $2 million for the honor of naming 10 new species of fish. Naming rights are available for smaller budgets, too—the top bid to name a new species of shrimp was $2,900. The proceeds from these auctions go toward funding new expeditions and preserving the habitats of the newly discovered species. In many cases, such discoveries are made as roads and settlements reach farther into new territory. Consequently, many species are endangered soon after they are discovered.

? What factors are responsible for the recent increase in the number of new species found?

Technology; encroachment of human activities into wilderness areas

AFP Photo/WWF GMPO HO/Newscom.

▶ Figure 17A Leopard gecko, a newly discovered lizard from northern Vietnam

▶ Figure 17B Pinocchio frog, recently discovered in New Guinea

Tim Laman/National Geographic Stock

Tim Laman/National Geographic Stock

▲ Figure 17C Dwarf wallaby, a rabbit-sized member of the kangaroo family

REVIEW

Reviewing the Concepts

For Practice Quizzes, BioFlix, MP3 Tutors, and Activities, go to www.masteringbiology.com.

Vertebrate Evolution and Diversity (1–8)

1 Derived characteristics define the major clades of chordates.

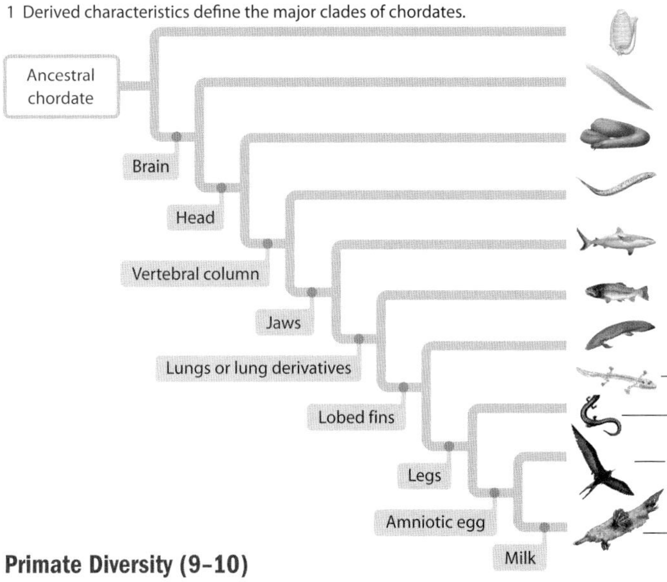

2 Hagfishes and lampreys lack hinged jaws.

3 Jawed vertebrates with gills and paired fins include sharks, ray-finned fishes, and lobe-finned fishes.

4 New fossil discoveries are filling in the gaps of tetrapod evolution.

5 Amphibians are tetrapods—vertebrates with two pairs of limbs.

6 Reptiles are amniotes—tetrapods with a terrestrially adapted egg.

7 Birds are feathered reptiles with adaptations for flight.

8 Mammals are amniotes that have hair and produce milk.

Primate Diversity (9–10)

9 The human story begins with our primate heritage. Primates had evolved as small arboreal mammals by 65 million years ago. Primate characters include limber joints, grasping hands and feet with flexible digits, a short snout, and forward-pointing eyes that enhance depth perception. The three groups of living primates are the lorises, pottos, and lemurs; the tarsiers; and the anthropoids (monkeys and apes).

10 Humans and four other groups of apes are classified as anthropoids. Apes, which have larger brains than other primates and lack tails, include gibbons, orangutans, gorillas, chimpanzees, and humans. We humans share 99% of our genome with chimpanzees, our closest living relatives.

Hominin Evolution (11–17)

11 The hominin branch of the primate tree includes species that coexisted. Paleoanthropologists have found about 20 species of extinct hominins, species that are more closely related to humans than to chimpanzees. Some of these species lived at the same time.

12 Australopiths were bipedal and had small brains.

13 Larger brains mark the evolution of *Homo*. The genus *Homo* includes hominins with larger brains and evidence of tool use. *Homo ergaster* had a larger brain than *H. habilis*. *H. erectus*, with a larger brain than *H. ergaster*, was the first hominin to spread out of Africa. *H. neanderthalensis* lived in Europe as recently as 30,000 years ago.

14 From origins in Africa, *Homo sapiens* spread around the world. Evidence from fossils and DNA studies has enabled scientists to trace early human history.

15 Who were the "hobbits"? Fossils of small hominins named *Homo floresiensis* that were found in Indonesia are controversial. Scientists are trying to determine their relationship to other hominins.

16 Human skin color reflects adaptations to varying amounts of sunlight. Human skin color variations probably resulted from natural selection balancing the body's need for folate with the need to synthesize vitamin D.

17 Our knowledge of animal diversity is far from complete. Thousands of new species are discovered each year.

Connecting the Concepts

1. In the primate phylogenetic tree below, fill in groups (a)–(e). Which of the groups are anthropoids? Which of the groups are apes?

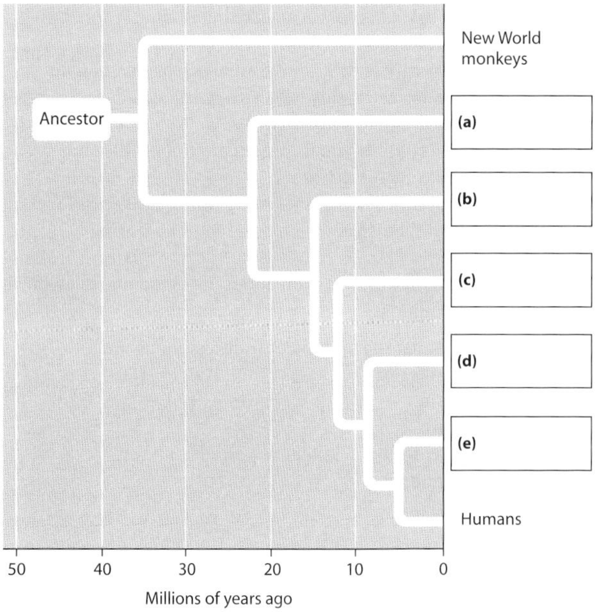

2. In the chordate phylogenetic tree below, fill in the key derived character that defines each clade.

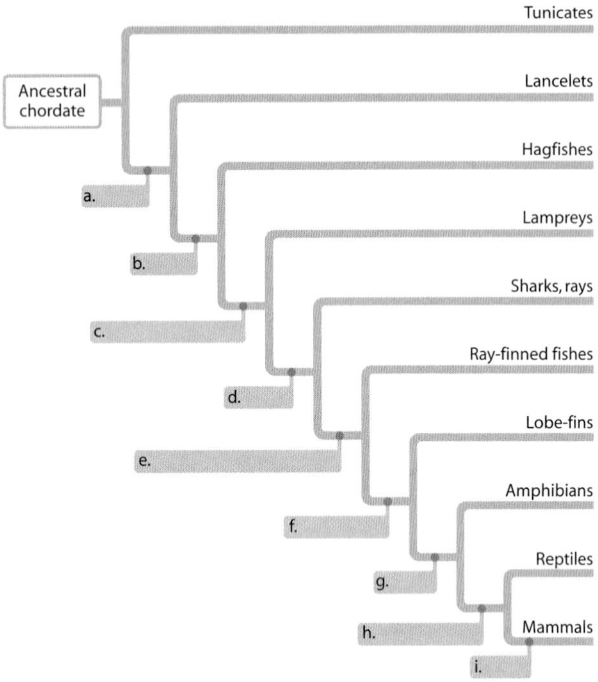

Testing Your Knowledge

Multiple Choice

3. A lamprey, a shark, a lizard, and a rabbit share all the following characteristics except
   a. pharyngeal slits in the embryo or adult.
   b. vertebrae.
   c. hinged jaws.
   d. a dorsal, hollow nerve cord.
   e. a post-anal tail.
4. Why were the *Tiktaalik* fossils an exciting discovery for scientists studying tetrapod evolution?
   a. They are the earliest frog-like animal discovered to date.
   b. They show that tetrapods successfully colonized land much earlier than previously thought.
   c. They have a roughly equal combination of fishlike and tetrapod-like characteristics.
   d. They demonstrate conclusively that limbs evolved as lobe-fins dragged themselves from pond to pond during droughts.
   e. They reveal new information about the breeding behavior of early tetrapods.
5. Fossils suggest that the first major trait distinguishing hominins from other primates was
   a. a larger brain.
   b. erect posture.
   c. forward-facing eyes with depth perception.
   d. grasping hands.
   e. toolmaking.
6. Which of the following correctly lists possible ancestors of humans from the oldest to the most recent?
   a. *Homo erectus, Australopithecus, Homo habilis*
   b. *Australopithecus, Homo habilis, Homo erectus*
   c. *Australopithecus, Homo erectus, Homo habilis*
   d. *Homo ergaster, Homo erectus, Homo neanderthalensis*
   e. *Homo habilis, Homo erectus, Australopithecus*
7. Which of these is not a member of the anthropoids?
   a. chimpanzee   b. ape
   c. tarsier   d. human
   e. New World monkey
8. Studies of DNA support which of the following?
   a. Members of the group called australopiths were the first to migrate from Africa.
   b. Neanderthals are more closely related to humans in Europe than to humans in Africa.
   c. *Homo sapiens* originated in Africa.
   d. *Sahelanthropus* was the earliest hominin.
   e. Chimpanzees are more similar to gorillas and orangutans than to humans.
9. The earliest members of the genus *Homo*
   a. had a larger brain compared to other hominins.
   b. probably hunted dinosaurs.
   c. lived in Europe.
   d. lived about 4 million years ago.
   e. were the first hominins to be bipedal.

Describing, Comparing, and Explaining

10. Compare the adaptations of amphibians and reptiles for terrestrial life.
11. Birds and mammals are both endothermic, and both have four-chambered hearts. Most reptiles are ectothermic and have three-chambered hearts. Why don't biologists group birds with mammals? Why do most biologists now consider birds to be reptiles?
12. What adaptations inherited from our primate ancestors enable humans to make and use tools?
13. Summarize the hypotheses that explain variation in human skin color as adaptations to variation in UV radiation.

Applying the Concepts

14. A good scientific hypothesis is based on existing evidence and leads to testable predictions. What hypothesis did the paleontologists who discovered *Tiktaalik* test? What evidence did they use to predict where they would find fossils of transitional forms?
15. Explain some of the reasons why humans have been able to expand in number and distribution to a greater extent than most other animals.
16. Anthropologists are interested in locating areas in Africa where fossils 4–8 million years old might be found. Why?

Review Answers

1. a. Old World monkeys; b. gibbons; c. orangutans; d. gorillas; e. chimpanzees. All are anthropoids; gibbons, orangutans, gorillas, chimpanzees, and humans are apes.
2. a. brain; b. head; c. vertebral column; d. jaws; e. lungs or lung derivatives; f. lobed fins; g. legs; h. amniotic egg; i. milk
3. c 4. c 5. b 6. b 7. c 8. c 9. a
10. Amphibians have four limbs adapted for locomotion on land, a skeletal structure that supports the body in a nonbuoyant medium, and lungs. However, most amphibians are tied to water because they obtain some of their oxygen through thin, moist skin and they require water for fertilization and development. Reptiles are completely adapted to life on land. They have amniotic eggs that contain food and water for the developing embryo and a shell to protect it from dehydration. Reptiles are covered by waterproof scales that enable them to resist dehydration (more efficient lungs eliminate the need for gas exchange through the skin).
11. Fossil evidence supports the evolution of birds from a small, bipedal, feathered dinosaur, which was probably endothermic. The last common ancestor that birds and mammals shared was the ancestral amniote. The four-chambered hearts of birds and mammals must have evolved independently.
12. Several primate characteristics make it easy for us to make and use tools—mobile digits, opposable fingers and thumb, and great sensitivity of touch. Primates also have forward-facing eyes, which enhances depth perception and eye-hand coordination, and a relatively large brain.
13. UV radiation is most intense in tropical regions and decreases farther north. Skin pigmentation is darkest in people indigenous to tropical regions and much lighter in northern latitudes. Scientists hypothesize that depigmentation was an adaptation to permit sufficient exposure to UV radiation, which catalyzes the production of vitamin D, a vitamin that permits the calcium absorption needed for both maternal and fetal bones. Dark pigmentation is hypothesized to protect against degradation of folate, a vitamin essential to normal embryonic development.
14. The paleontologists who discovered *Tiktaalik* hypothesized the existence of transitional forms between fishlike tetrapods such as *Panderichthys* and tetrapod-like fish such as *Acanthostega.* From the available evidence, they knew the time periods when fishlike tetrapods and tetrapod-like fish lived. From the rocks in which the fossils had been found, they knew the geographic region and the type of habitat these creatures occupied. With this knowledge, they predicted the type of rock formation where transitional fossils might be found.
15. Our intelligence and culture—accumulated and transmitted knowledge, beliefs, arts, and products—have enabled us to overcome our physical limitations and alter the environment to fit our needs and desires.
16. Most anthropologists think that humans and chimpanzees diverged from a common ancestor 5–7 million years ago. Primate fossils 4–8 million years old might help us understand how the human lineage first evolved.

Unifying Concepts of Animal Structure and Function

Unifying Concepts of Animal Structure and Function

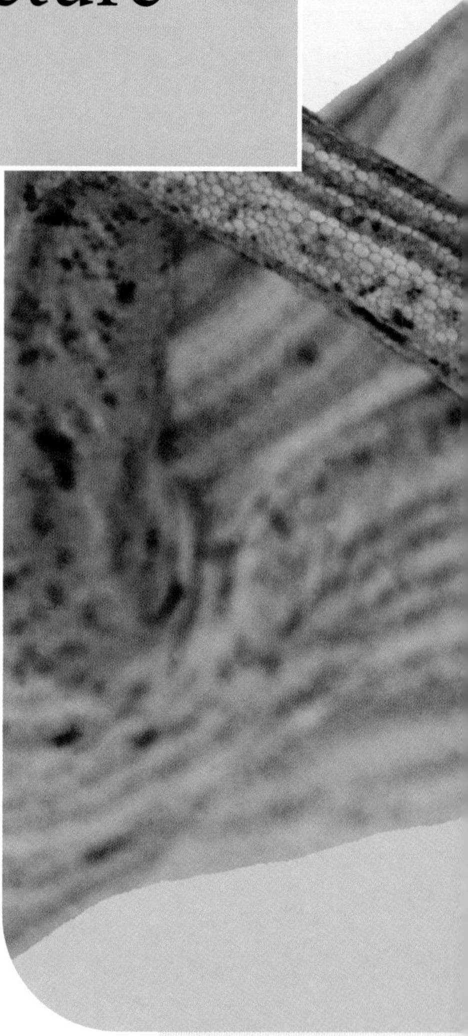

BIG IDEAS

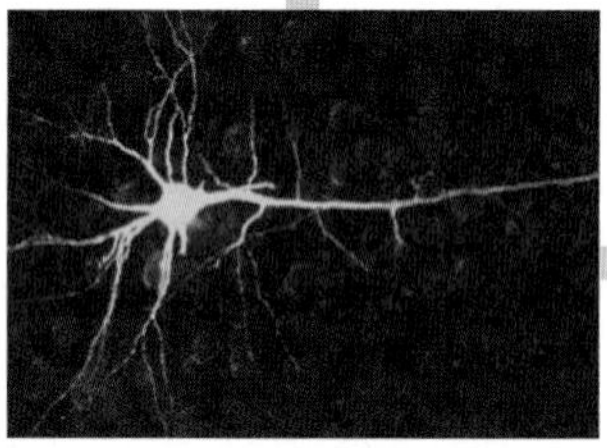

Ulrich Gartner

Structure and Function in Animal Tissues
(1–7)

The structural hierarchy in an animal begins with cells and tissues, whose forms correlate with their functions.

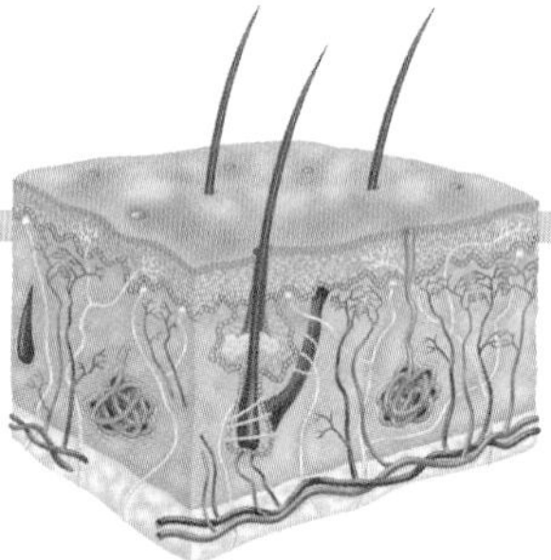

Organs and Organ Systems
(8–12)

Tissues are arranged into organs, which may be functionally coordinated in organ systems.

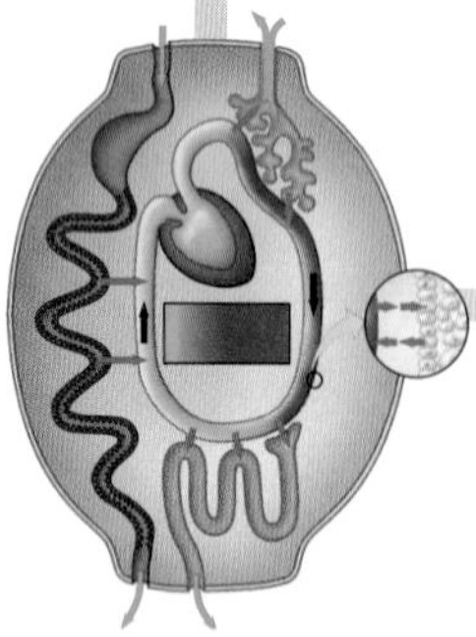

External Exchange and Internal Regulation
(13–15)

Complex animals have internal surfaces that facilitate exchange with the environment. Feedback control maintains homeostasis in many animals.

Martin Harvey/Photolibrary

Spiderman is known for his ability to climb walls, but few vertebrates have similar talents. One exception is the gecko, a small lizard commonly found in the tropics. Geckos can walk up walls and even across ceilings. How do they do it?

The explanation relates to hairs on the gecko's toes (see the photo above)—specifically, to short-lived positive and negative charges that arise on these hairs and on the surfaces they touch. It took a multidisciplinary team of biologists and engineers and a specially designed apparatus to determine the force of attraction between these fleeting charges. Each instance of attraction is weak, but there are so many microscopic hairs—about half a million on each toe, each with hundreds of split ends—that the combined strength of these forces would allow a gecko to support eight times its weight hanging from a single toe!

The gecko's ability to walk on walls is a function that emerges from special structural adaptations of the hairs on its toes. The correlation of structure and function is an overarching theme of biology. The chapters in this unit explore form and function in the context of the various challenges animals face: how to obtain nutrients and oxygen and distribute them throughout the body, fight infection, excrete wastes, reproduce, and sense and respond to the environment. The adaptations that represent the various solutions to these problems have been fashioned by natural selection, fitting structure to function over the course of evolution.

This chapter opens with an overview of animal structure and function. We'll explore the hierarchy of cells, tissues, organs, and organ systems. And we'll see how an animal's structure enhances exchange with the external environment as well as regulation of the internal environment.

Structure and Function in Animal Tissues

1 Structure fits function at all levels of organization in the animal body

When discussing structure and function, biologists distinguish anatomy from physiology. **Anatomy** is the study of the form of an organism's structures; **physiology** is the study of the functions of those structures. A biologist interested in anatomy, for instance, might focus on the arrangement of muscles and bones in a gecko's legs. A physiologist might study how a gecko's muscles function. Despite their different approaches, both scientists are working toward a better understanding of the connection between structure and function, such as how the structural adaptations of the hairs on its toes give the gecko its remarkable ability to walk on walls.

Structure in the living world is organized in hierarchical levels. We followed the progression from molecules to cells in Unit I. Now, let's trace the hierarchy in animals from cells to organisms. (In Unit VI, we follow the same hierarchy in plants. And in Unit VII, we pick up the trail again, moving from organisms to ecosystems.) Emergent properties—novel properties that were not present at the preceding level of the hierarchy of life—arise as a result of the structural and functional organization of each level's component parts.

Figure 1 illustrates structural hierarchy in a ring-tailed lemur. **Part A** shows a single muscle cell in the lemur's heart. This cell's main function is to contract, and the stripes in the cell indicate the precise alignment of strands of proteins that perform that function. Each muscle cell is also branched, providing for multiple connections to other cells that ensure coordinated contractions of all the muscle cells in the heart.

Together, these heart cells make up a tissue **(Part B),** the second structural level. A **tissue** is an integrated group of similar cells that perform a common function. The cells of a tissue are specialized, and their structure enables them to perform a specific task—in this instance, coordinated contraction.

Part C, the heart itself, illustrates the organ level of the hierarchy. An **organ** is made up of two or more types of tissues that together perform a specific task. In addition to muscle tissue, the heart includes nervous, epithelial, and connective tissue.

Part D shows the circulatory system, the organ system of which the heart is a part. An **organ system** consists of multiple organs that together perform a vital body function. Other parts of the circulatory system include the blood and the blood vessels: arteries, veins, and capillaries.

In **Part E,** the lemur itself forms the final level of this hierarchy. An organism contains a number of organ systems, each specialized for certain tasks and all functioning together as an integrated, coordinated unit. For example, the lemur's circulatory system cannot function without oxygen supplied by the respiratory system and nutrients supplied by the digestive system. And it takes the coordination of several other organ systems to enable this animal to walk or climb trees.

The ability to climb trees or walls emerges from the specific arrangement of specialized structures. As we see throughout our study of the anatomy and physiology of animals, form fits function at each level of the structural hierarchy. In several modules to come, we focus on the tissue level of this biological hierarchy. But first, let's explore how evolution shapes an animal's body form.

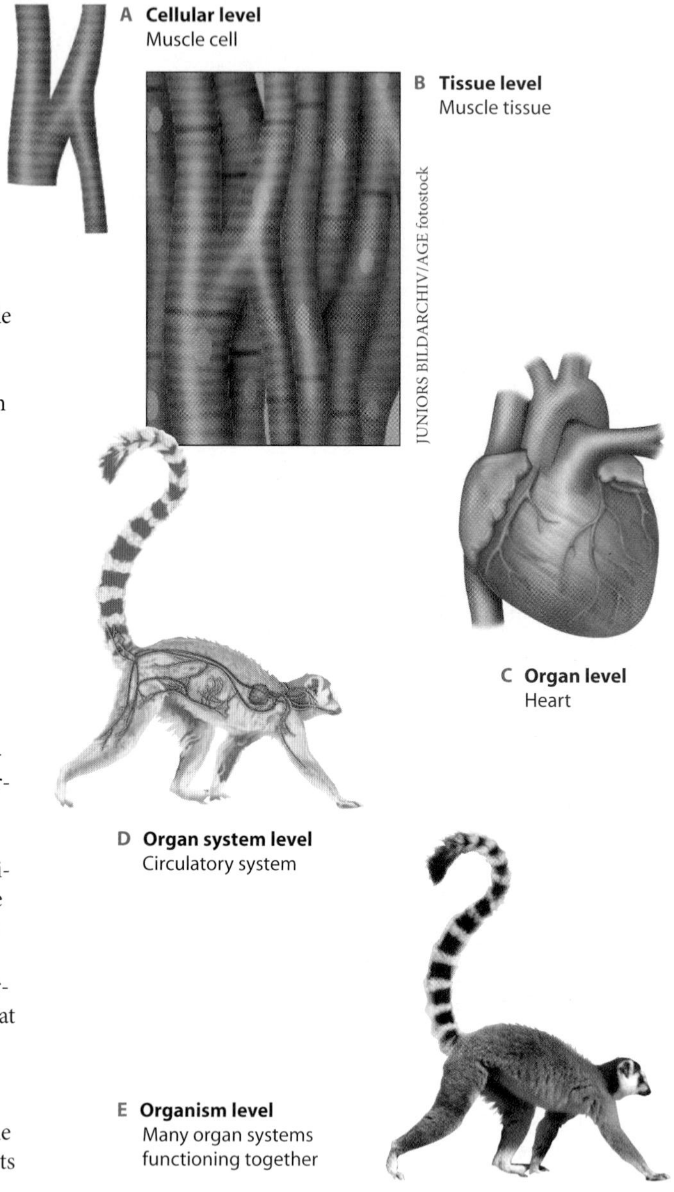

▲ **Figure 1** A structural hierarchy in a ring-tailed lemur

? Explain how the ability to pump blood is an emergent property of a heart, which is at the organ level of the biological hierarchy.

The specific structural organization and integration of the individual muscle, connective, epithelial, and nervous tissues of a heart enable the function of pumping blood.

EVOLUTION CONNECTION

2 An animal's form reflects natural selection

An animal's size and shape are fundamental aspects of form that affect the way an animal interacts with its environment. Although biologists often refer to "body plan" or "design" in discussing animal form, they are not implying a process of conscious invention. The body plan of an animal is the result of millions of years of evolution. Natural selection fits form to function by selecting the variations that best meet the challenges of an animal's environment.

How might physical laws affect the evolution of an animal's form? Consider some of the properties of water that limit the possible shapes for fast swimmers. Water is about a thousand times denser than air. Thus, any bump on the body surface that causes drag slows a swimmer much more than it would a runner or a flyer. What shape, then, is best suited for rapid movement through the dense medium of water?

Tuna and other fast fishes can swim at speeds up to 80 km/hr (about 50 miles per hour). Sharks, penguins, dolphins, and seals are also fast swimmers. As illustrated by the examples in **Figure 2**, such animals share a streamlined, tapered body form. The similar shape found in speedy fishes, sharks, and aquatic birds and mammals is an example of convergent evolution. Natural selection often shapes similar adaptations when diverse organisms face the same environmental challenge, such as the resistance of water to fast swimming.

Physical laws also influence body form with regard to maximum size. As body dimensions increase, thicker skeletons are required to support the body. In addition, as bodies increase in size, the muscles required for movement become an ever-larger proportion of the total body mass. At some point, mobility becomes limited. By considering the fraction of body mass in leg muscles and the force such muscles can generate, scientists can estimate maximum running speed for a wide range of body plans. Such calculations suggest that *Tyrannosaurus rex*, a dinosaur that was more than 6 m tall (about 20 feet), could probably only lumber from place to place, a far cry from the thunderous sprint depicted in the movie *Jurassic Park*.

? Explain how a seal, a penguin, and a shark illustrate convergent evolution.

All three have a streamlined, tapered shape, the result of natural selection adapting each to fast swimming in its dense, aquatic environment.

▲ **Figure 2** Convergent evolution of body shape in fast swimmers

kp Wolf/Photolibrary

3 Tissues are groups of cells with a common structure and function

In almost all animals, the cells of the body are organized into tissues. The term *tissue* is from a Latin word meaning "weave," and some tissues resemble woven cloth in that they consist of a meshwork of nonliving fibers and other extracellular substances surrounding living cells. Other tissues are held together by a sticky glue that coats the cells or by special junctions between adjacent plasma membranes. The structure of tissues relates to their specific functions.

The specialization of complex body parts such as organs and organ systems is largely based on varied combinations of a limited set of cells and tissue types. For example, your lungs and blood vessels have very distinct functions, but they are lined by tissues that are of the same basic type.

Your body is built from four main types of tissues: epithelial, connective, muscle, and nervous. We examine the structure and function of these tissue types in the next four modules.

? How is a tissue different from a cell and an organ?

Tissues are collections of similar cells that perform a common function. Several different tissue types usually produce the structure of an organ.

4 Epithelial tissue covers the body and lines its organs and cavities

Epithelial tissues, or epithelia (singular, *epithelium*), are sheets of closely packed cells that cover your body surface and line your internal organs and cavities. The tightly knit cells form a protective barrier and, in some cases, a surface for exchange with the fluid or air on the other side. One side of an epithelium is attached to a basal lamina, a dense mat of extracellular matrix consisting of fibrous proteins and sticky polysaccharides that separates the epithelium from the underlying tissues. The other side, called the apical surface, faces the outside of an organ or the inside of a tube or passageway.

Epithelial tissues are named according to the number of cell layers they have and according to the shape of the cells on their apical surface. A simple epithelium has a single layer of cells, whereas a stratified epithelium has multiple layers. A pseudostratified epithelium has a single layer but appears stratified because the cells vary in length. The shape of the cells may be squamous (flat like floor tiles), cuboidal (like dice), or columnar (like bricks on end). **Figure 4** shows examples of different types of epithelia. In each case, the pink color identifies the cells of the epithelium itself.

The structure of each type of epithelium fits its function. Simple squamous epithelium **(Part A)** is thin and leaky and thus suitable for exchanging materials by diffusion. You would find it lining your capillaries and the air sacs of your lungs.

Both cuboidal and columnar epithelia have cells with a relatively large amount of cytoplasm, facilitating their role of secretion or absorption of materials. **Part B** shows a cuboidal epithelium forming a tube in the kidney. Such epithelia are also found in glands, such as the thyroid and salivary glands. A simple columnar epithelium **(Part C)** lines your intestines, where it secretes digestive juices and absorbs nutrients.

The many layers of the stratified squamous epithelium in **Part D** make it well suited for lining surfaces subject to abrasion, such as your outer skin and the linings of your mouth and esophagus. Stratified squamous epithelium regenerates rapidly by division of the cells near the basal lamina. New cells move toward the apical surface as older cells slough off.

The pseudostratified ciliated columnar epithelium in **Part E** forms a mucous membrane that lines portions of your respiratory tract and helps keep your lungs clean. Dust, pollen, and other particles are trapped in the mucus it secretes and then swept up and out of your respiratory tract by the beating of the cilia on its cells.

? **Epithelial tissues are named according to the _______ of cells on their apical surface and the number of cell _______.**

shape . . . layers

Image Source/Getty Images

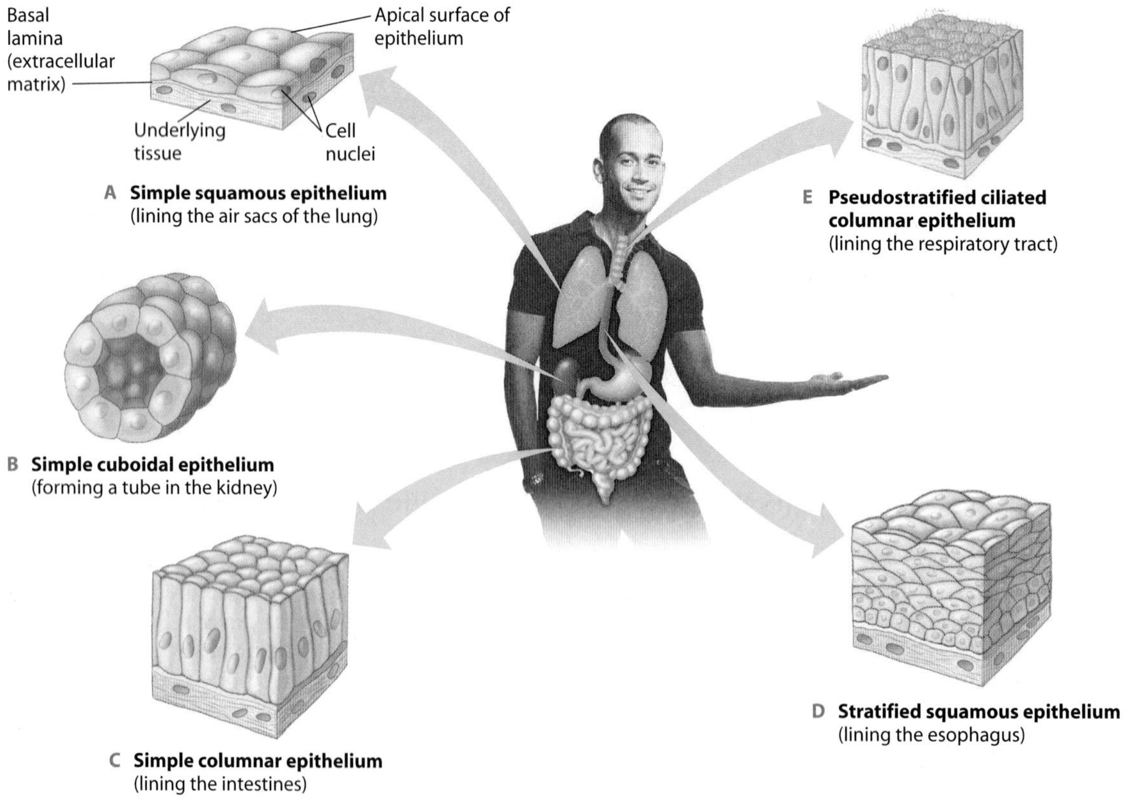

▲ **Figure 4** Types of epithelial tissue

5 Connective tissue binds and supports other tissues

In contrast to epithelium, **connective tissue** consists of a sparse population of cells scattered throughout an extracellular material called a matrix. The cells produce and secrete the matrix, which usually consists of a web of fibers embedded in a liquid, jelly, or solid. Connective tissues may be grouped into six major types. **Figure 5** shows micrographs of each type and illustrates where each would be found in your arm, for example.

The most widespread connective tissue in your body is called **loose connective tissue** (**Part A**) because its matrix is a loose weave of fibers. Many of the fibers consist of the strong, ropelike protein collagen. Other fibers are more elastic, making the tissue resilient as well as strong. Loose connective tissue serves mainly to bind epithelia to underlying tissues and hold organs in place. In the figure, we show the loose connective tissue that lies directly under the skin.

Fibrous connective tissue (**Part B**) has densely packed parallel bundles of collagen fibers, an arrangement that maximizes its strength. This tissue forms tendons, which attach your muscles to bone, and ligaments, which connect your bones at joints.

Adipose tissue (**Part C**) stores fat in large, closely packed adipose cells held in a matrix of fibers. This tissue pads and insulates your body and stores energy. Each adipose cell contains a large fat droplet that swells when fat is stored and shrinks when fat is used as fuel.

The matrix of **cartilage** (**Part D**), a connective tissue that forms a strong but flexible skeletal material, consists of collagen fibers embedded in a rubbery material. Cartilage commonly surrounds the ends of bones, providing a shock-absorbing surface. It also supports your ears and nose and forms the cushioning disks between your vertebrae.

Bone (**Part E**) has a matrix of collagen fibers embedded in a hard mineral substance made of calcium, magnesium, and phosphate. The combination of fibers and minerals makes bone strong without being brittle. The microscopic structure of compact regions of bones contains repeating circular layers of matrix, each with a central canal containing blood vessels and nerves. Like other tissues, bone contains living cells and can therefore grow as you grow and mend when broken.

Blood (**Part F**) transports substances throughout your body and thus functions differently from other connective tissues. Its extensive extracellular matrix is a liquid called plasma, which consists of water, salts, and dissolved proteins. Suspended in the plasma are red blood cells, which carry oxygen; white blood cells, which function in defense against disease; and platelets, which aid in blood clotting.

? Why does blood qualify as a type of connective tissue?

Because it consists of a relatively sparse population of cells surrounded by a noncellular matrix, which in this case is a fluid called plasma

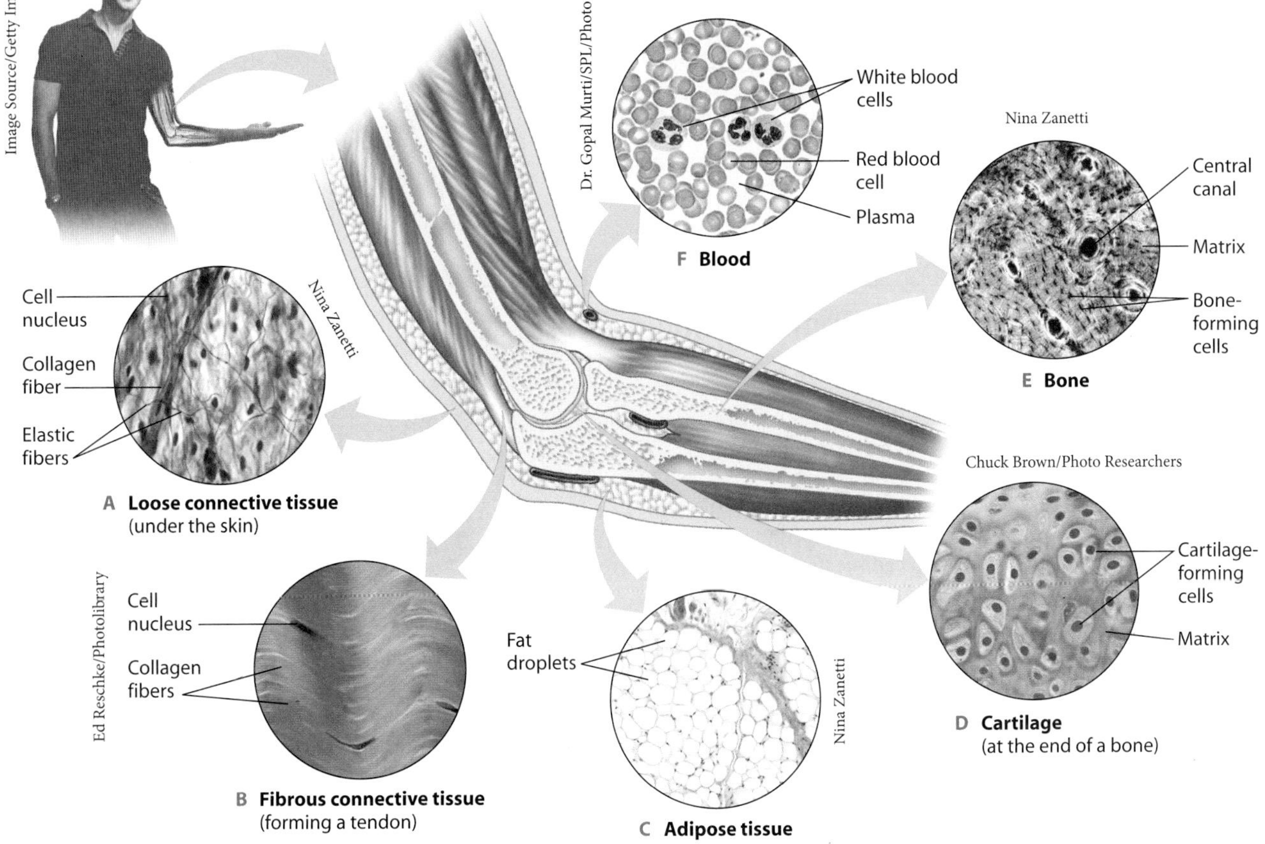

▲ **Figure 5** Types of connective tissue

6 Muscle tissue functions in movement

Muscle tissue is the most abundant tissue in most animals. It consists of long cells called muscle fibers, each containing many molecules of contractile proteins. **Figure 6** shows micrographs of the three types of vertebrate muscle tissue.

Skeletal muscle (**Part A**) is attached to your bones by tendons and is responsible for voluntary movements of your body, such as walking or bouncing a ball. The arrangement of the contractile units along the length of muscle fibers gives the cells a striped, or striated, appearance, as you can see in the micrograph below.

Cardiac muscle (**Part B**) forms the contractile tissue of your heart. It is striated like skeletal muscle, but it is under involuntary control, meaning that you cannot consciously control its contraction. Cardiac muscle fibers are branched, interconnecting at specialized junctions that rapidly relay the signal to contract from cell to cell during your heartbeat.

Smooth muscle (**Part C**) gets its name from its lack of striations. Smooth muscle is found in the walls of your digestive tract, arteries, and other internal organs. It is responsible for involuntary body activities, such as the movement of food through your intestines. Smooth muscle cells are spindle-shaped and contract more slowly than skeletal muscles, but can sustain contractions for a longer period of time.

? **The muscles responsible for a gecko climbing a wall are _______________ muscles.**

skeletal

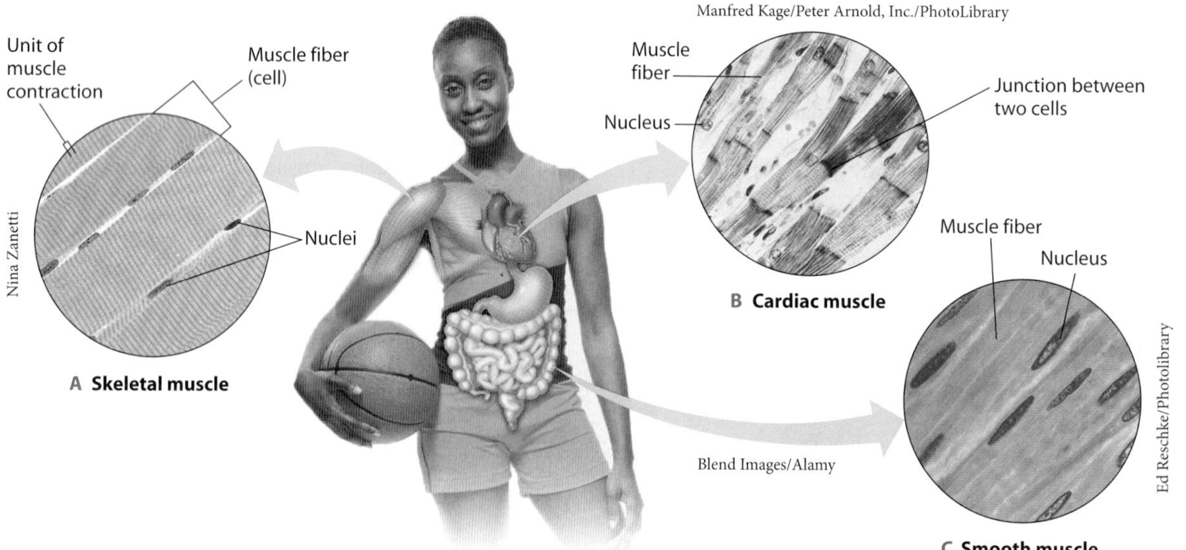

▲ **Figure 6** The three types of muscle tissue

7 Nervous tissue forms a communication network

Nervous tissue senses stimuli and rapidly transmits information. Nervous tissue is found in your brain and spinal cord, as well as in the nerves that transmit signals throughout your body.

The structural and functional unit of nervous tissue is the nerve cell, or **neuron**, which is uniquely specialized to conduct electrical nerve impulses. As you can see in the micrograph in **Figure 7**, a neuron consists of a cell body (containing the cell's nucleus and other organelles) and a number of slender extensions. Dendrites and the cell body receive nerve impulses from other neurons. Axons, which are often bundled together into nerves, transmit signals toward other neurons or to an effector, such as a muscle cell.

Nervous tissue actually contains many more supporting cells than neurons. Some of these cells surround and insulate axons, promoting faster transmission of signals. Others help nourish neurons and regulate the fluid around them.

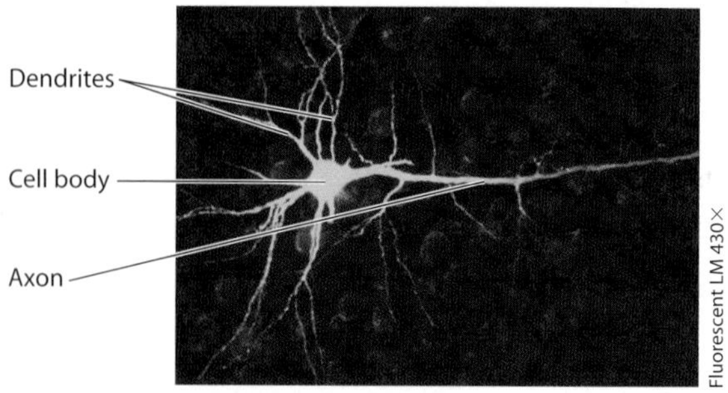

▲ **Figure 7** A neuron   Ulrich Gartner

? **How does the long length of some axons (such as those that extend from your lower spine to your toes) relate to the function of a neuron?**

It allows for the transmission of a nerve signal over a long distance directly to specific muscle cells.

Organs and Organ Systems

8 Organs are made up of tissues

In all but the simplest animals, tissues are arranged into organs that perform specific functions. The heart, for example, while mostly muscle, also has epithelial, connective, and nervous tissues. Epithelial tissue lining the heart chambers prevents leakage and provides a smooth surface over which blood can flow. Connective tissue makes the heart elastic and strengthens its walls. Neurons regulate the contractions of cardiac muscle.

In some organs, tissues are organized in layers, as you can see in the diagram of the small intestine in Figure 8. The lumen, or interior space, of the small intestine is lined by a columnar epithelium that secretes digestive juices and absorbs nutrients. Notice the finger-like projections that increase the surface area of this lining. Underneath the epithelium (and extending into the projections) is connective tissue, which contains blood vessels. The two layers of smooth muscle, oriented in different directions, propel food through the intestine. The smooth muscle, in turn, is surrounded by another layer of connective tissue and epithelial tissue.

An organ represents a higher level of structure than the tissues composing it, and it performs functions that none of its component tissues can carry out alone. These functions emerge from the coordinated interactions of tissues.

? Explain why a disease that damages connective tissue can impair most of the body's organs.

Connective tissue is a component of most organs.

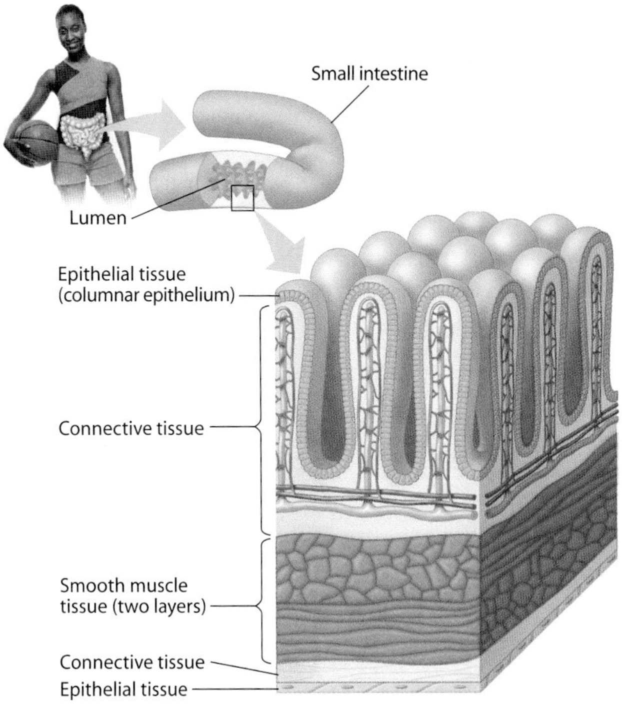

▲ **Figure 8** Tissue layers of the wall of the small intestine

CONNECTION

9 Bioengineers are learning to produce tissues and organs for transplants

Injuries, diseases, and birth defects often leave individuals with disfiguring or life-threatening conditions. Scientists are increasingly turning to bioengineering in their search for ways to repair or replace damaged tissues and organs.

One of the most successful tissue-engineering advances has come in the form of artificial skin, a type of human-engineered tissue designed for everyone from burn victims to diabetic patients with skin ulcers. The tissue is grown from human fibroblasts, tissue-generating cells often harvested from newborn foreskin tissue. These cells are applied along a tiny protein scaffolding, where they multiply and produce a three-dimensional skin substitute containing active living cells.

In 2006, researchers reported the first successful transplantation and long-term functioning of laboratory-grown bladders (Figure 9). Such bladders have now been implanted in almost 30 children and adults. These organs were grown from the patients' own cells, thus avoiding the risk of rejection. Organs grown from a patient's cells may someday reduce the shortage of organs available for transplants.

In another remarkable advance in tissue engineering, some imaginative researchers filled recycled inkjet print cartridges with a cell solution instead of ink and, using desktop printers, actually "printed" cells in layers on a biodegradable gel. The gel can be removed, leaving a three-dimensional cellular structure. Sheets of artificial tissues or even pieces of organs may someday be produced by such printing techniques. Researchers may also use such cellular systems for testing drugs or for basic research into cell growth and interactions.

Brian Walker/AP Photo

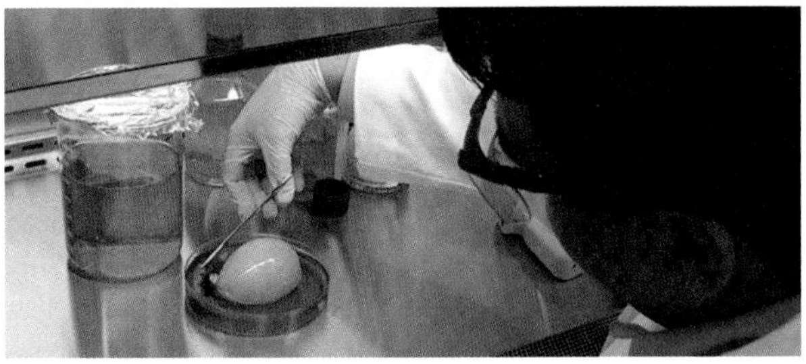

▲ **Figure 9** A laboratory-grown bladder

? Why is it beneficial to grow replacement tissues or organs from a patient's own cells?

The patient's body would not reject the tissue as foreign.

10 Organ systems work together to perform life's functions

Just as it takes several different tissues to build an organ, it takes the integration of several organs into organ systems to perform the body's functions. **Figure 10** illustrates the organ systems found in humans and other mammals. As you read through the brief descriptions of these systems and study their components in the figure, remember that all of the organ systems are interdependent and work together to create a functional organism.

A The **circulatory system** delivers O_2 and nutrients to your body cells and transports CO_2 to the lungs and metabolic wastes to the kidneys.

B The **respiratory system** exchanges gases with the environment, supplying your blood with O_2 and disposing of CO_2.

C The **integumentary system** protects your body against physical injury, infection, excessive heat or cold, and drying out.

D The **skeletal system** supports your body, protects organs such as your brain and lungs, and provides the framework for muscles to produce movement.

E The **muscular system** moves your body, maintains posture, and produces heat.

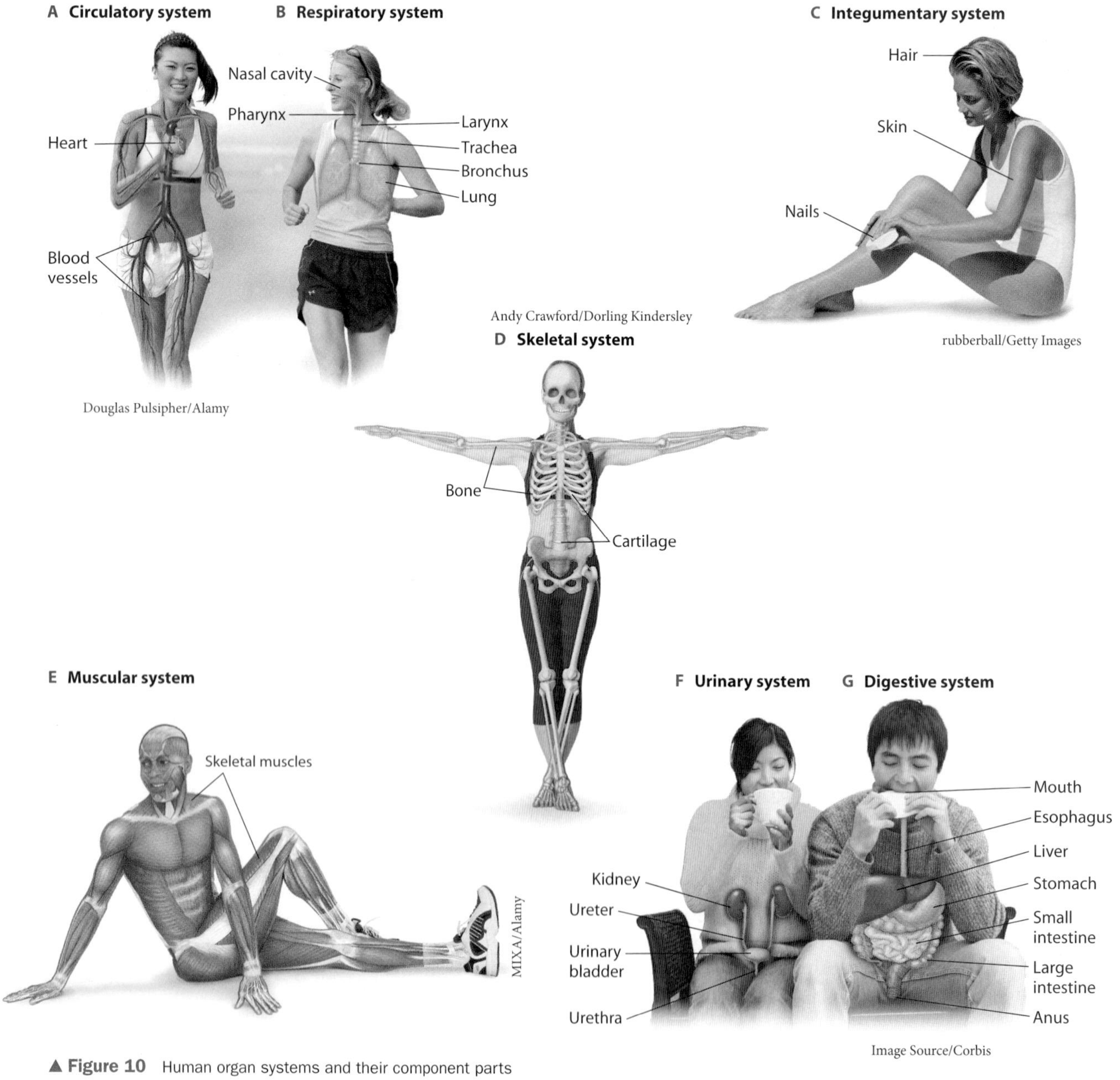

▲ **Figure 10** Human organ systems and their component parts

F The **urinary system** removes waste products from your blood and excretes urine. It also regulates the chemical makeup, pH, and water balance of your blood.

G The **digestive system** ingests and digests your food, absorbs nutrients, and eliminates undigested material.

H The **endocrine system** secretes hormones that regulate the activities of your body, thus maintaining an internal steady state called homeostasis.

I The **lymphatic system** returns excess body fluid to the circulatory system and functions as part of the immune system.

J The **immune system** defends your body against infections and cancer.

K The **nervous system** coordinates your body's activities by detecting stimuli, integrating information, and directing the body's responses.

L The **reproductive system** produces gametes and sex hormones. The female system supports a developing embryo and produces milk.

The ability to perform life's functions emerges from the organization and coordination of all the body's organ systems. Indeed, the whole is greater than the sum of its parts.

? Which two organ systems are most directly involved in regulating all other systems?

The nervous system and the endocrine system

H Endocrine system

Rolf Bruderer/Blend Images/Corbis

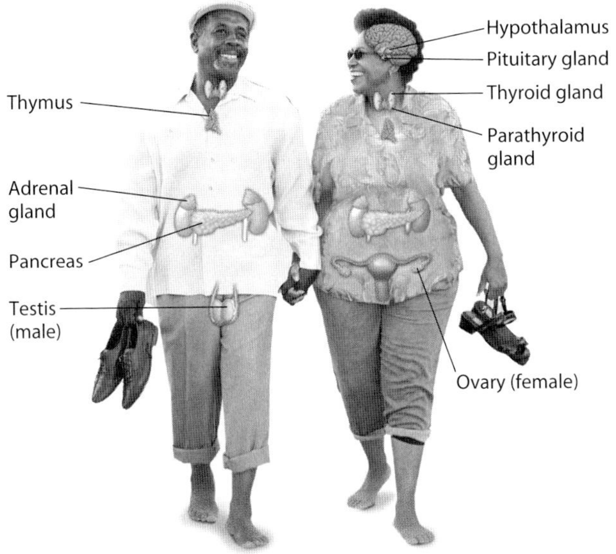

I, J Lymphatic and immune systems

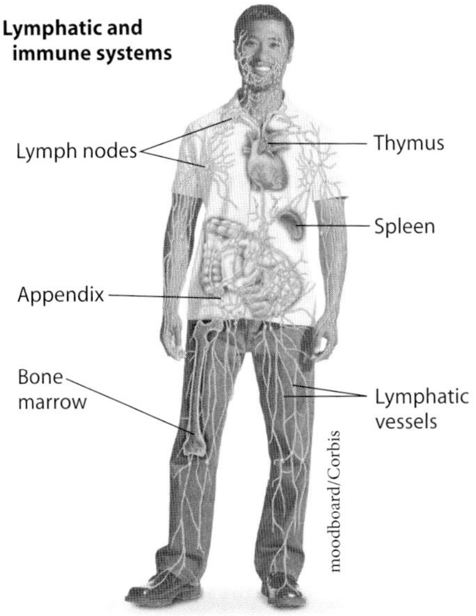

Jeremy Woodhouse/Photolibrary

K Nervous system

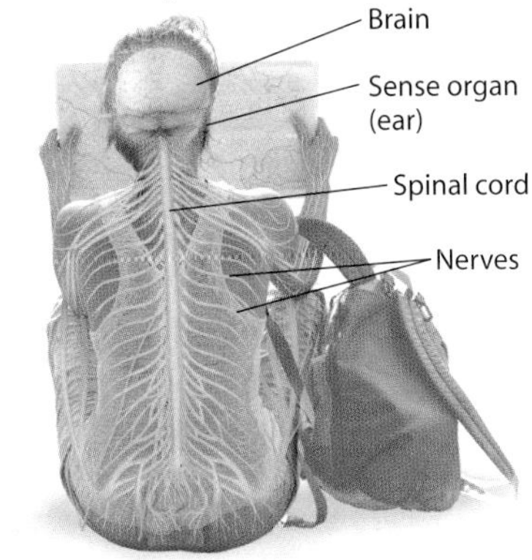

L Reproductive system

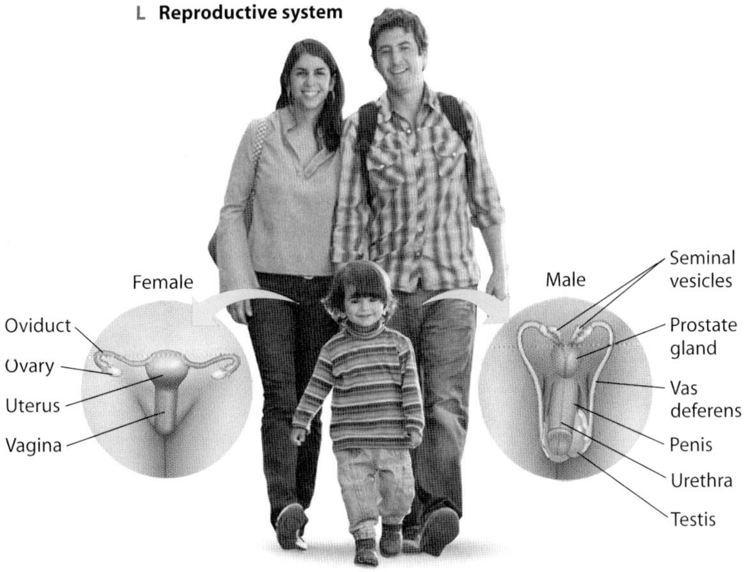

Ryan McVay/Getty Images

CONNECTION 11 New imaging technology reveals the inner body

Let's explore some imaging techniques that allow physicians to "see" the body's organs without resorting to surgery.

X-rays X-rays, discovered in 1895, were the first means of producing a photographic image of internal organs. X-rays are a type of high-energy radiation that passes readily through soft tissues. The features that show up most distinctly on X-rays are shadows of hard structures, such as bones and dense tumors that block the rays. Conventional X-rays are used routinely to check for broken bones and tooth cavities.

CT A newer X-ray technology is called computed tomography (CT). This computer-assisted technique produces images of a series of thin cross sections through the body. The patient is often given a special liquid to improve the contrast of the images and is then slowly moved through a doughnut-shaped CT machine as the X-ray source circles around the body, illuminating successive sections from many angles. The CT scanner's computer then produces high-resolution video images of the cross sections, which can be studied individually or combined into various three-dimensional views.

CT scans are excellent diagnostic tools. They can detect small differences between normal and abnormal tissues in many organs, but are especially useful for evaluating problems that affect the abdomen and brain—areas where conventional X-ray procedures are of little help.

Another useful technique uses ultra-fast CT scanners to show the actual movements of body organs, such as the heart beating and blood flowing through vessels. Physicians use this technique to identify heart defects and constricted blood vessels, as well as to monitor the status of coronary bypass grafts.

MRI A completely different technique, magnetic resonance imaging (MRI), takes advantage of the behavior of the hydrogen atoms in water molecules. MRI uses powerful magnets to align the hydrogen nuclei and then knocks them out of alignment with a brief pulse of radio waves. In response, the hydrogen atoms give out faint radio signals of their own, which are picked up by the MRI scanner and translated into an image. Since water is a major component of all of our soft tissues, MRI visualizes them well. For example, MRI scans are commonly done to assess joint injuries. **Figure 11A** is an MRI scan of a knee that reveals a tear in the outer meniscus, one of the cushioning disks of cartilage between the bones.

PET Positron-emission tomography (PET) is an imaging technology that differs from both CT and MRI in its ability to yield information about metabolic processes at specific locations in the body. In preparation for a PET scan, the patient is injected with a small amount of a molecule—glucose, for example—labeled with a radioactive isotope. Metabolically active cells take up more of the labeled glucose than less active cells. A PET scanner pinpoints metabolic "hot spots" by highlighting the sites of most intense radiation. Cancerous tissue that is actively growing is readily identified in a PET scan.

PET is proving most valuable for measuring the metabolic activity of various parts of the brain. This technique is providing insights into brain activity in people affected by stroke or by disorders such as schizophrenia, epilepsy, depression, and Alzheimer's disease. PET is also used to learn about the healthy brain. It has been used to map the different regions of the brain involved in language—those that are active during speaking, reading, and hearing words.

A recent technological advance is a combination CT-PET scanner, which performs both types of scans at the same time. Combining the images benefits from the better anatomical representation of a CT scan and the metabolic information of a PET scan. The CT-PET scan in **Figure 11B** reveals the location of two tumors (bright white) in a patient with lung cancer.

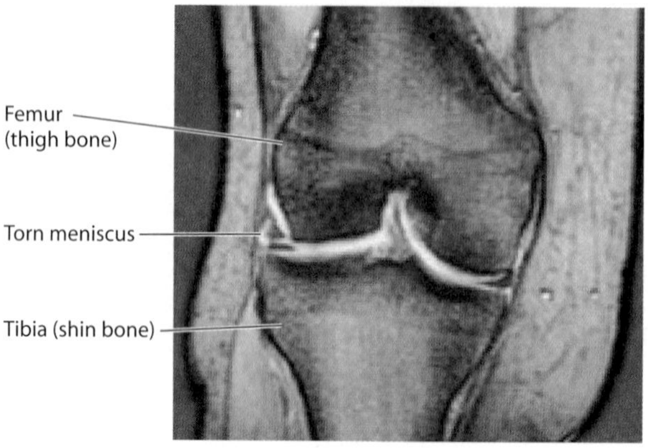

SPL/Photo Researchers

▲ **Figure 11A** A colorized MRI scan of a knee joint, showing a tear (small red areas) in the outer meniscus

? Why are the imaging techniques described in this module regarded as relatively noninvasive in contrast to such diagnostic methods as exploratory surgery or biopsy?

Although they may involve injections, these techniques require no surgery.

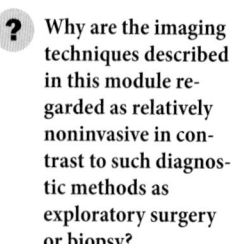

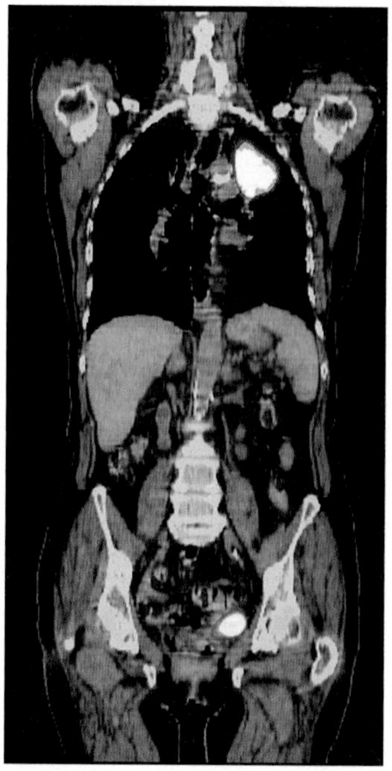

Simon Fraser/Photo Researchers

▲ **Figure 11B** A combined CT-PET scan showing tumors (bright white) in the upper left lung and lower abdomen

12 The integumentary system protects the body

Most of the organ systems introduced in Module 10 are examined in more detail in the chapters of this unit. Here we take a brief look at the integumentary system, which consists of the skin, hair, and nails.

Structure of Skin As shown in **Figure 12**, your skin consists of two layers: the epidermis and the dermis. The epidermis is a stratified squamous epithelium (many layers of flat cells; see Module 4). Rapid cell division near the base of the epidermis serves to replenish the skin cells that are constantly abraded from your body surface. As these new cells are pushed upward in the epidermis by the addition of new cells below them, they fill with the fibrous protein keratin, release a waterproofing glycolipid, and eventually die. These dead, tightly joined cells remain at the surface of your skin for up to two weeks. This continuous process means that you get a brand new epidermis every few weeks.

The dermis is the inner layer of the skin. It consists of a fairly dense connective tissue with many resilient elastic fibers and strong collagen fibers. The dermis contains hair follicles, oil and sweat glands, muscles, nerves, sensory receptors, and blood vessels. Beneath the skin lies the hypodermis, a layer of adipose tissue. (The hypodermis is the site of many of the injections you receive with a hypodermic needle.)

Functions of Skin The structure of the skin corresponds to its functions. The keratin-filled, tightly joined cells at the surface of the epidermis provide a waterproof covering that protects your body from dehydration and prevents penetration by microbes. And as close encounters with the environment abrade the epidermis, replacement cells move up.

The sensory receptors in the skin provide important environmental information to your brain: Is this surface too hot or cold or sharp? Your touch receptors help you to manipulate food and tools and to feel your way around in the dark. The profusion of small blood vessels and the 2.5 million sweat glands in the dermis also facilitate the important function of temperature regulation, as Figure 15 illustrates.

One of the metabolic functions of the skin is the synthesis of vitamin D, which is required for absorbing calcium. Ultraviolet (UV) light catalyzes the conversion of a derivative of cholesterol to vitamin D in the cells in the lower layers of your epidermis. Adequate sunlight is needed for this synthesis. But sunlight also has damaging effects on the skin. In response to ultraviolet radiation, skin cells make more melanin, which moves to the outer layers of the skin and is visible as a tan. A tan is not indicative of good health; it signals that skin cells have been damaged. DNA changes caused by the UV rays can lead to premature aging of the skin, cataracts in the eyes, and skin cancers.

Hair and Nails In mammals, hair is an important component of the integumentary system. Hair is a flexible shaft of flattened, keratin-filled dead cells, which were produced by a hair follicle. Associated with hair follicles are oil glands, whose secretions lubricate the hair, condition the surrounding skin, and inhibit the growth of bacteria. Look at the hair follicle in Figure 12 and you will see that it is wrapped in nerve endings. Hair plays important sensory functions, as its slightest movement is relayed to the nervous system. (You can get a sense of this sensitivity by lightly touching the hair on your head.)

Hair insulates the bodies of most mammals—although in humans this insulation is limited to the head. Land mammals react to cold by raising their fur, which traps a layer of air and increases the insulating power of the fur. Look again at the hair follicle in the figure. The muscle attached to it is responsible for raising the hair when you get cold. The resulting "goose bumps" are a vestige of hair raising inherited from our furry ancestors.

Fingernails and toenails are the final component of your integumentary system. These protective coverings are also composed of keratin. Fingernails facilitate fine manipulation (and are useful for chewing when nervous). In other mammals, the digits may end in claws or hooves.

The integumentary system encloses and protects an animal from its environment. But animals cannot be isolated systems; they must exchange gases, take in food, and eliminate wastes. In the next section, we explore some of the structural adaptations that provide for these functions.

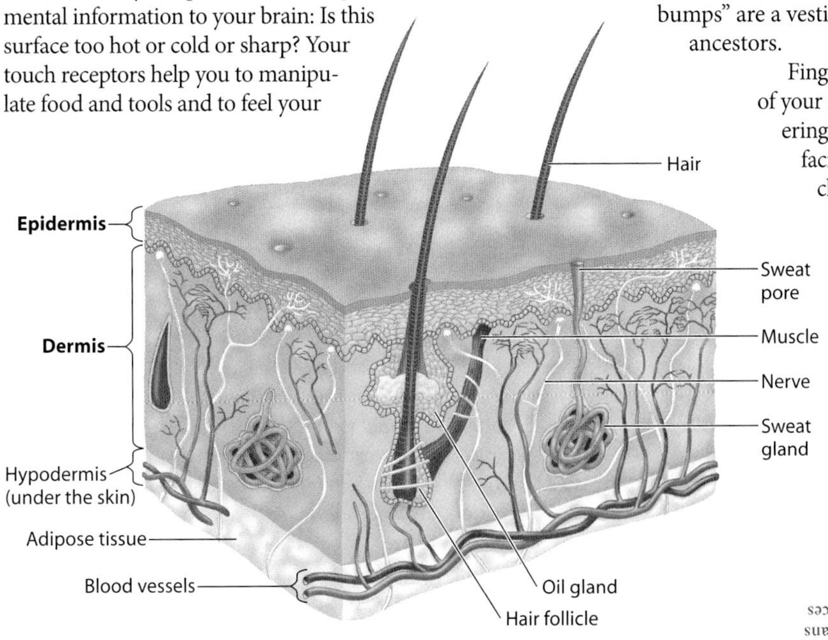

▲ **Figure 12** A section of skin, the major organ of the integumentary system

? Describe three structures associated with a hair follicle that contribute to the functions of hair.

The nerve endings sense when the hair is moved; the muscle raises the hair (producing goose bumps in humans but warming other mammals); and the oil gland produces lubricating and antibacterial secretions.

External Exchange and Internal Regulation

13 Structural adaptations enhance exchange with the environment

Every living organism is an open system, meaning that it exchanges matter and energy with its surroundings. You, for example, take in O_2 and water and food, and in exchange, you breathe out CO_2, urinate, defecate, sweat, and radiate heat. The exchange of materials with the environment must extend to the level of each individual cell. Exchange of gases, nutrients, and wastes occurs as substances dissolved in an aqueous solution move across the plasma membrane of every cell.

A freshwater hydra has a body wall only two cell layers thick. The outside layer is in contact with its water environment; the inner layer is bathed by fluid in its saclike body cavity. This internal fluid circulates in and out of the hydra's mouth. Thus, every body cell exchanges materials directly with an aqueous environment.

Another common body plan that maximizes exchange with the environment is a flat, thin shape. For instance, a parasitic tapeworm may be several meters long, but because it is very thin, most of its cells are bathed in the intestinal fluid of its host—the source of its nutrients.

The saclike body of a hydra or the paper-thin one of the tapeworm works well for animals with a simple body structure. However, most animals are composed of compact masses of cells and have an outer surface that is relatively small compared with the animal's overall volume. As an extreme example, the ratio of a whale's outer surface area to its volume is hundreds of thousands of times smaller than that of a small animal like a hydra. Still, every cell in the whale's body must be bathed in fluid, have access to oxygen and nutrients, and be able to dispose of its wastes. How is all this accomplished?

In whales and other complex animals, the evolutionary adaptation that provides for sufficient exchange with the environment is in the form of extensively branched or folded surfaces. In almost all cases, these surfaces lie within the body, protected by the integumentary system from dehydration or damage and allowing for a streamlined body shape. In humans, the digestive, respiratory, and circulatory systems rely on exchange surfaces within the body that each have a surface area more than 25 times that of the skin. Indeed, if you lined up all the tiny capillaries across which exchange between the blood and body cells occurs within your body, they would circle the globe!

Figure 13A is a schematic model illustrating four of the organ systems of a compact, complex animal. Each system has a large, specialized internal exchange surface. We have placed the circulatory system in the middle because of its central role in transporting substances between the other three systems. The blue arrows indicate exchange of materials between the circulatory system and the other systems.

Actually, direct exchange does not occur between the blood and the cells of tissues and organs. Body cells are bathed in a solution called **interstitial fluid** (see the circular enlargement in Figure 13A). Exchange takes place through this fluid. In other words, to get from the blood to body cells, or vice versa, materials pass through the interstitial fluid.

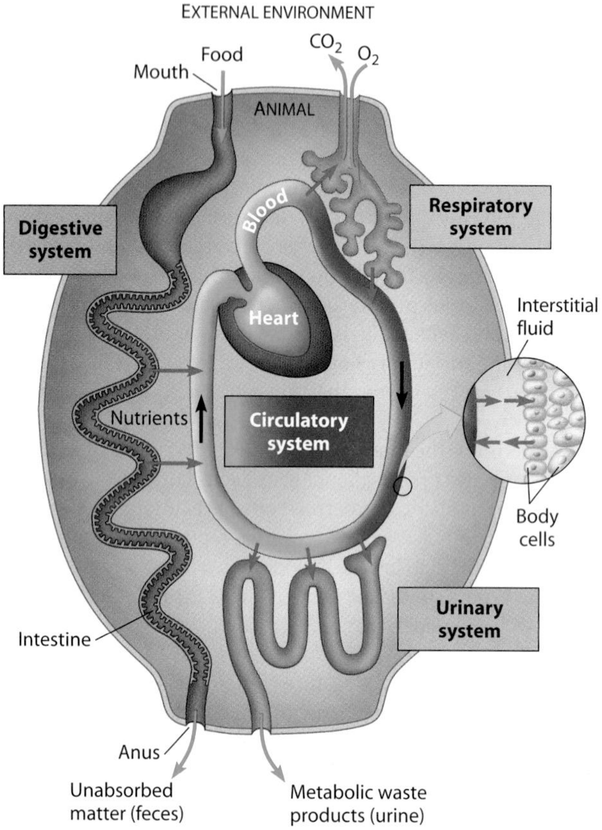

▲ **Figure 13A** A schematic representation showing indirect exchange between the environment and the cells of a complex animal

The digestive system, especially the intestine, has an expanded surface area resulting from folds and projections of its inner lining (see Figure 8). Nutrients are absorbed from the lumen of the intestine into the cells lining this large surface area. They then pass through the interstitial fluid and into capillaries, tiny branched blood vessels that form an exchange network with the digestive surfaces. This system is so effective that enough nutrients move into the circulatory system to support the rest of the cells in the body.

The extensive, epithelium-lined tubes of the urinary system are equally effective at increasing the surface area for exchange. Enmeshed in capillaries, excretory tubes extract metabolic wastes that the blood brings from throughout the body. The wastes move out of the blood into the excretory tubes and pass out of the body in urine.

The respiratory system also has an enormous internal surface area across which gases are exchanged. Your lungs are not shaped like big balloons, as they are sometimes depicted, but rather like millions of tiny balloons at the tips of finely branched

Science Photo Library/Photo Researchers

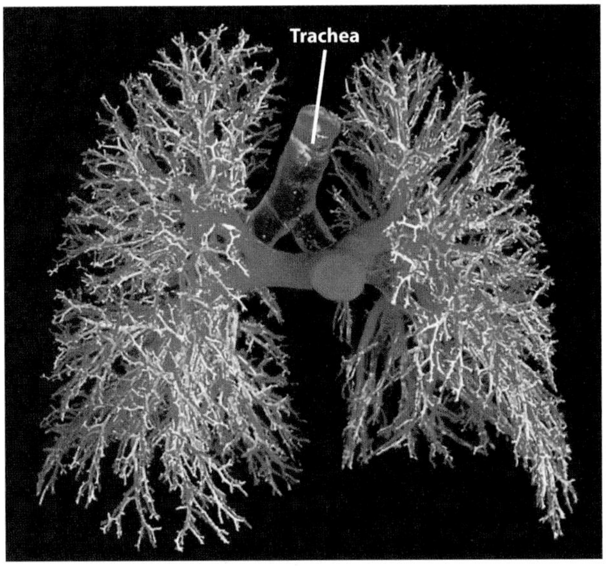

▲ **Figure 13B** A resin model of the finely branched air tubes (white) and blood vessels (red) of the human lungs

air tubes. **Figure 13B** is a resin model of the human lungs. Blood vessels that convey blood from the heart to the lungs divide into tiny blood vessels that radiate throughout the lungs, shown in the model as the red branches. The white branches represent tiny air tubes that end in multilobed sacs lined with thin epithelium and surrounded by capillaries. Oxygen readily moves from the air in the lungs across this epithelium and into the blood in the capillaries. The blood returns to the heart and is then pumped throughout the body to supply all cells with oxygen.

Both Figures 13A and 13B highlight two basic concepts in animal biology: First, any animal with a complex body—one with most of its cells not in direct contact with its external environment—must have internal structures that provide sufficient surface area to service those cells. Second, the organ systems of the body are interdependent; it takes their coordinated actions to produce a functional organism.

? How do the structures of the lungs, small intestine, and kidneys relate to the function of exchange with the environment?

These organs all have a huge number of sacs, projections, or tubes that greatly increase the surface area across which exchange of materials can occur.

14 Animals regulate their internal environment

More than a century ago, French physiologist Claude Bernard recognized that *two* environments are important to an animal: the external environment surrounding the animal and the internal environment, where its cells actually live. The internal environment of a vertebrate is the interstitial fluid that fills the spaces around the cells. Many animals maintain relatively constant conditions in their internal environment. Your own body maintains the salt and water balance of your internal fluids and also keeps your body temperature at about 37°C (98.6°F). A bird, such as the ptarmigan shown in **Figure 14**, also maintains salt and water balance and a body temperature of about 40°C, or 104°F, even in winter. The bird uses energy from its food to generate body heat, and it has a thick, insulating coat of down feathers. A gecko does not generate its own body heat, but it can maintain a fairly constant body temperature by basking in the sun or resting in the shade. It does, however, regulate the salt and water balance of its body fluids.

Cre8tive Images/Shutterstock

Today, Bernard's concept of the internal environment is included in the principle of **homeostasis**, which means "a steady state." As Figure 14 illustrates, conditions may fluctuate widely in the external environment, but homeostatic mechanisms regulate internal conditions, resulting in much smaller changes in the animal's internal environment. Both birds and mammals have control systems that keep body temperature within a narrow range, despite large changes in the temperature of the external environment.

MVaresvuo/Photolibrary

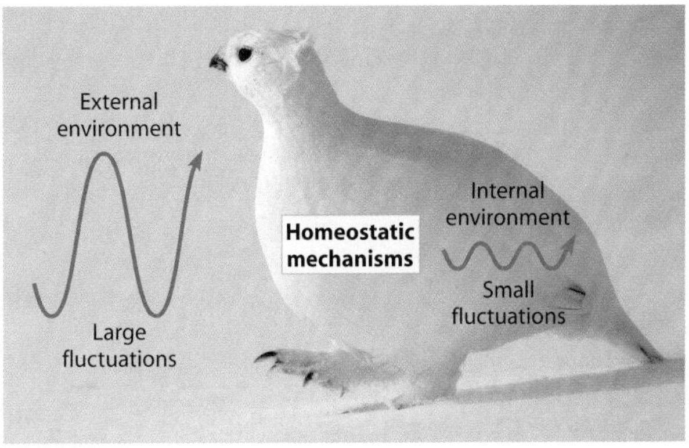

▲ **Figure 14** A model of homeostasis in a white-tailed ptarmigan in its snowy habitat

The internal environment of an animal always fluctuates slightly. Homeostasis is a dynamic state, an interplay between outside forces that tend to change the internal environment and internal control mechanisms that oppose such changes. An animal's homeostatic control systems maintain internal conditions within a range where life's metabolic processes can occur. In the next module, we explore a general mechanism for how many of these control systems function.

? Look back at Figure 13A. What are some ways in which the circulatory system contributes to homeostasis?

By its exchanges with the digestive, respiratory, and urinary systems, the blood helps maintain the proper balance of materials in the interstitial fluid surrounding body cells.

15 Homeostasis depends on negative feedback

Most of the control mechanisms of homeostasis are based on **negative feedback**, in which a change in a variable triggers mechanisms that reverse that change. To identify the components of a negative-feedback system, consider the simple example of the regulation of room temperature. You set the thermostat at a comfortable temperature—call this its set point. When a sensor in the thermostat detects that the temperature has dropped below this set point, the thermostat turns on the furnace. The response (heat) reverses the drop in temperature. Then, when the temperature rises to the set point, the thermostat turns the furnace off. Physiologists would call the thermostat a control center, which senses a stimulus (room temperature below the set point) and activates a response.

Many of the control centers that maintain homeostasis in animals are located in the brain. For example, your "thermostat" operates by negative feedback to switch on and off mechanisms that maintain your body temperature around 37°C (98.6°F). As shown in the upper part of **Figure 15**, when the thermostat senses a rise in temperature above the set point, it activates cooling mechanisms, such as the dilation of blood vessels in your skin and sweating. Once body temperature returns to normal, the thermostat shuts off these cooling mechanisms. When your body temperature falls below the set point (lower part of the figure), the thermostat activates warming mechanisms, such as constriction of blood vessels to reduce heat loss and shivering to generate heat. Again, a return to normal temperature shuts off these mechanisms.

As you examine the body's organ systems in detail, you will encounter many examples of homeostatic control and negative feedback, as well as constant reminders of the relationship between structure and function.

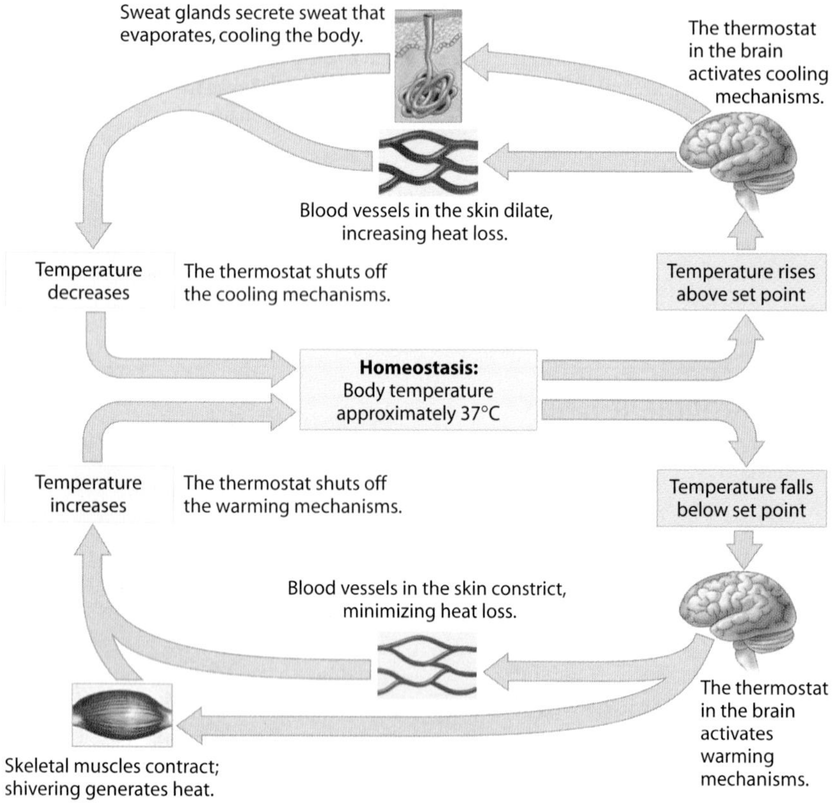

▲ **Figure 15** Feedback control of body temperature

? How would adding an air conditioner contribute to the homeostatic control of room temperature?

The air conditioner would cool the room when temperature exceeded the set point. Such opposing, or antagonistic, pairs of control circuits are common homeostatic mechanisms in the body.

REVIEW

For Practice Quizzes, BioFlix, MP3 Tutors, and Activities, go to www.masteringbiology.com.

Reviewing the Concepts

Structure and Function in Animal Tissues (1–7)

1 Structure fits function at all levels of organization in the animal body.

2 An animal's form reflects natural selection. Physical laws affect the evolution of an animal's size and shape.

3 Tissues are groups of cells with a common structure and function.

Function	**4 Epithelial tissue covers the body and lines its organs and cavities.**	**5 Connective tissue binds and supports other tissues.**	**6 Muscle tissue functions in movement.**	**7 Nervous tissue forms a communication network.**
Structure	Sheets of closely packed cells	Sparse cells in extracellular matrix	Long cells (fibers) with contractile proteins	Neurons with branching extensions; supporting cells
Example	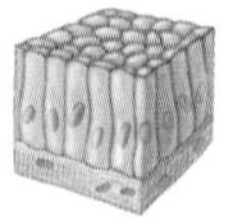 Columnar epithelium	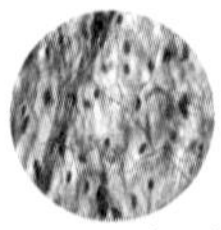 Loose connective tissue	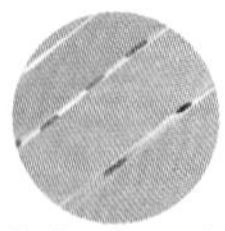 Skeletal muscle	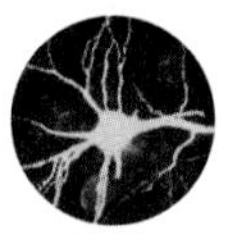 Neuron

Organs and Organ Systems (8–12)

8 Organs are made up of tissues.

9 Bioengineers are learning to produce tissues and organs for transplants.

10 Organ systems work together to perform life's functions. The integumentary system covers the body. Skeletal and muscular systems support and move it. The digestive and respiratory systems obtain food and oxygen, respectively, and the circulatory system transports these materials. The urinary system disposes of wastes, and the immune and lymphatic systems protect the body from infection. The nervous and endocrine systems control and coordinate body functions. The reproductive system produces offspring.

11 New imaging technology reveals the inner body. CT, MRI, and PET are used in medical diagnosis and research.

12 The integumentary system protects the body.

External Exchange and Internal Regulation (13–15)

13 Structural adaptations enhance exchange with the environment. Complex animals have specialized internal structures that increase surface area. Exchange of materials between blood and body cells takes place through the interstitial fluid.

14 Animals regulate their internal environment.

15 Homeostasis depends on negative feedback. Control systems detect change and direct responses. Negative-feedback mechanisms keep internal variables fairly constant, with small fluctuations around set points.

Connecting the Concepts

1. There are several key concepts introduced in this chapter: Structure correlates with function; an animal's body has a hierarchy of organization with emergent properties at each level; and complex bodies have structural adaptations that increase surface area for exchange. Label the tissue layers shown in this section of the small intestine, and describe how this diagram illustrates these three concepts.

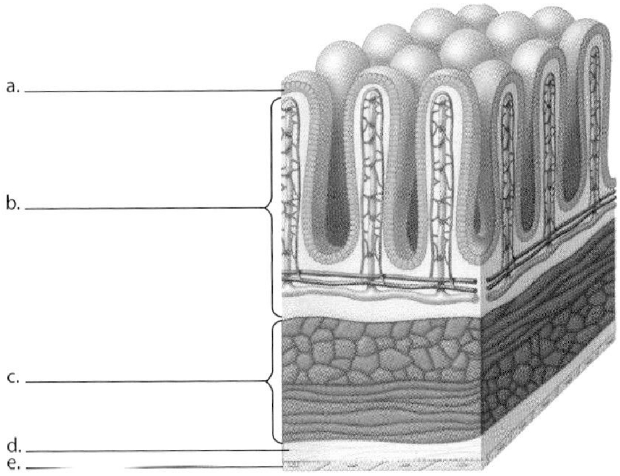

Testing Your Knowledge

Multiple Choice

2. The cells in the human body are in contact with an internal environment consisting of
   a. blood.
   b. connective tissue.
   c. interstitial fluid.
   d. matrix.
   e. mucous membranes.

3. Which of the following body systems facilitates (but doesn't regulate) the functions of the other systems?
   a. respiratory system
   b. endocrine system
   c. digestive system
   d. urinary system
   e. circulatory system

4. Negative-feedback mechanisms are
   a. most often involved in maintaining homeostasis.
   b. activated only when a variable rises above a set point.
   c. analogous to a furnace that produces heat.
   d. found only in birds and mammals.
   e. all of the above.

5. Which of the following best illustrates homeostasis? (*Explain your answer.*)
   a. Most adult humans are between 5 and 6 feet tall.
   b. The lungs and intestines have large surface areas.
   c. All the cells of the body are about the same size.
   d. When the salt concentration of the blood goes up, the kidneys expel more salt.
   e. When oxygen in the blood decreases, you feel dizzy.

Matching *(Terms in the right-hand column may be used more than once.)*

6. Closely packed cells covering a surface
7. Neurons
8. Adipose tissue, blood, and cartilage
9. May be simple or stratified
10. Scattered cells embedded in matrix
11. Senses stimuli and transmits signals
12. Cells are called fibers
13. Cells may be squamous, cuboidal, or columnar
14. Skeletal, cardiac, or smooth

a. connective tissue
b. muscle tissue
c. nervous tissue
d. epithelial tissue

Describing, Comparing, and Explaining

15. Briefly explain how the structure of each of these tissues is well suited to its function: stratified squamous epithelium in the skin, neurons in the brain, simple squamous epithelium lining the lung, bone in the skull.
16. Describe ways in which the bodies of complex animals are structured for exchanging materials with the environment. Do all animals share such features?

Applying the Concepts

17. Suppose at the end of a hard run on a hot day you find that there are no drinks left in the cooler. If, out of desperation, you dunk your head into the cooler, how might the ice-cold water affect the rate at which your body temperature returns to normal?
18. Some companies promote whole-body CT scans as health screens that consumers may purchase. Many parents request CT scans when they take an injured child to the ER. There is growing concern that widespread and repeated use of this technology is exposing adults and children to potentially high levels of radiation from procedures that may not be medically necessary. How can consumers evaluate the health risks versus benefits of these procedures? What is the government's role in regulating these commercial ventures?

Review Answers

1. a. epithelial tissue; b. connective tissue; c. smooth muscle tissue; d. connective tissue; e. epithelial tissue

   The structure of the specialized cells in each type of tissue fits their function. For example, columnar epithelial cells are specialized for absorption and secretion; the fibers and cells of connective tissue provide support and connect the tissues. The hierarchy from cell to tissue to organ is evident in this diagram. The functional properties of a tissue or organ emerge from the structural organization and coordination of its component parts. The many projections of the lining of the small intestine greatly increase the surface area for absorption of nutrients.

2. c 3. e 4. a 5. d (Expelling salt opposes the increase in blood salt concentration, thereby maintaining a constant internal environment.) 6. d 7. c 8. a 9. d 10. a 11. c 12. b 13. d 14. b

15. Stratified squamous epithelium consists of many cell layers. The outer cells are flattened, filled with the protein keratin, and dead, providing a protective, waterproof covering for the body. Neurons are cells with long extensions that conduct signals to other cells, making multiple connections in the brain. Simple squamous epithelium is a single, thin layer of cells that allows for diffusion of gases across the lining of the lung. Bone cells are surrounded by a matrix that consists of fibers and mineral salts, forming a hard protective covering around the brain.

16. Extensive exchange surfaces are often located within the body. The surfaces of the intestine, urinary system, and lungs are highly folded and divided, increasing their surface area for exchange. These surfaces interface with many blood capillaries. Not all animals have such extensive exchange surfaces. Animals with small, simple bodies or thin, flat bodies have a greater surface-to-volume ratio, and their cells are closer to the surface, enabling direct exchange between cells and the outside environment.

17. The ice water would cool the blood in your head, which would then circulate throughout your body. This effect would accelerate the return to a normal body temperature. If, however, the ice water cooled the blood vessel that supplies the thermostat in your brain so that it sensed a decrease in temperature, this control center would respond by inhibiting sweating and constricting blood vessels in the skin, thereby slowing the cooling of your body.

18. Some issues and questions to consider: Should a doctor's prescription be required for a whole-body CT scan? Should such scans be available only to those who can pay for them? Are CT scan machines calibrated so that they expose children or small adults to less radiation? Have there been research studies to test the effect of repeated exposures to radiation from CT scans? Whose responsibility is it to perform such studies and then publicize the results?

Nutrition and Digestion

Nutrition and Digestion

BIG IDEAS

Shutterstock

Obtaining and Processing Food
(1–3)

Animals ingest food, digest it in specialized compartments, absorb nutrients, and eliminate wastes.

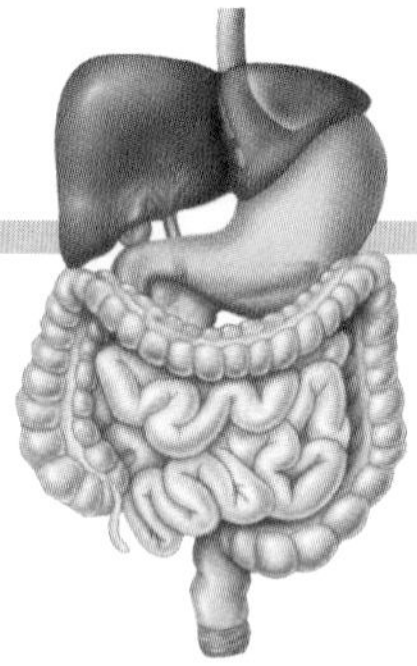

The Human Digestive System
(4–13)

Food is processed sequentially in the mouth, stomach, and small intestine, where nutrients are absorbed.

Steve Allen/Brand X Pictures/Jupiter Images

Nutrition
(14–22)

A healthy diet fuels activities, provides organic building blocks, and provides nutrients a body cannot manufacture.

SuperStock

Does your diet include junk food? If you consume the kinds of items pictured here, are they a steady part of your diet or occasional splurges? Do you try to choose nutritious foods and eat balanced meals, or does your diet fall victim to a hectic schedule or cravings for sweets? What role does food play in your life?

All animals must eat. Unlike plants, animals must consume food to provide both energy and the organic building blocks used to assemble new molecules, cells, and tissues. Animals also need essential nutrients such as vitamins and minerals, which are required for proper metabolic function. Throughout evolutionary history, animal feeding mechanisms and digestive anatomy and physiology have been shaped by natural selection to fit the available food sources and provide adequate nutrition. However, in the United States and in some other industrialized nations, some of us have begun to take in vastly more food than we need, thanks to relatively cheap, readily available, and calorie-dense foods. This, combined with sedentary jobs and inactive lifestyles, has landed us in the middle of an obesity crisis. An estimated 68% of people in the United States are considered overweight, and of these, almost 100 million are categorized as obese. Obesity has been associated with increased risk for heart disease, certain cancers, and diabetes. Thus, our modern diet appears to be contributing to shorter, less healthy lives.

In this chapter, we consider nutrition and diet in the context of the human digestive system. This discussion will help you determine if you are eating the right foods in the proper proportions needed to provide your body with the nutrients it needs for good health. But first, let's look at the variety of ways in which animals obtain and process food.

Obtaining and Processing Food

1 Animals obtain and ingest their food in a variety of ways

All animals eat other organisms—dead or alive, whole or by the piece. In general, animals fall into one of three dietary categories. **Herbivores**, such as cattle, gorillas, sea urchins, and snails, eat mainly autotrophs—plants and algae. **Carnivores**, such as lions, hawks, spiders, and whales, mostly eat other animals. Animals that consume *both* plants and animals are called **omnivores** (from the Latin *omni*, all, and *vorus*, devouring). Omnivores include humans, as well as crows, cockroaches, and raccoons.

▲ **Figure 1A** A suspension feeder: a tube worm filtering food through its tentacles

Several types of feeding mechanisms have evolved among animals. Many aquatic animals are **suspension feeders**, which sift small organisms or food particles from water. For example, humpback whales use comb-like plates called baleen to strain krill and small fish from the enormous volumes of water they take into their mouths. Clams and oysters are also suspension feeders. A film of mucus on their gills traps tiny morsels suspended in the water, and cilia on the gills sweep the food to the mouth. **Figure 1A** shows the feathery tentacles of a suspension-feeding tube worm.

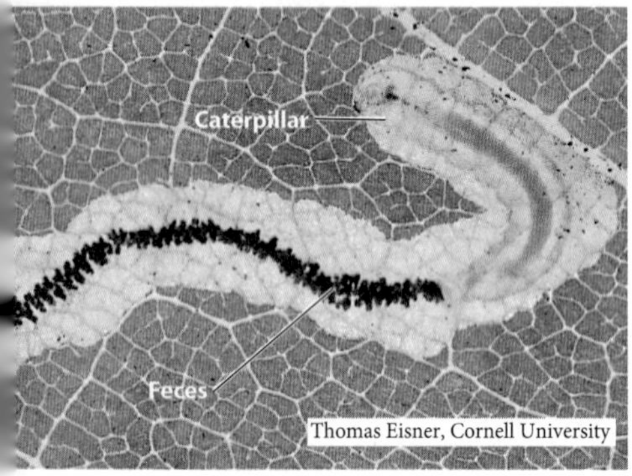

▲ **Figure 1B** A substrate feeder: a caterpillar eating its way through the soft tissues inside an oak leaf

Substrate feeders live in or on their food source and eat their way through it. **Figure 1B** shows a leaf miner caterpillar, the larva of a moth. The dark spots on the leaf are a trail of feces that the caterpillar leaves in its wake. Other substrate feeders include maggots (fly larvae), which burrow into animal carcasses, and earthworms, which eat their way through soil, digesting partially decayed organic material and helping to aerate and fertilize the soil as they go.

Fluid feeders suck nutrient-rich fluids from a living host. Aphids, for example, tap into the sugary sap in plants. Bloodsuckers, such as mosquitoes and ticks, pierce animals with needlelike mouthparts. The female mosquito in **Figure 1C** has just filled her abdomen with a meal of human blood. (Only female mosquitoes suck blood; males live on plant nectar.) In contrast to such parasitic fluid feeders, some fluid feeders actually benefit their hosts. For example, hummingbirds and bees move pollen between flowers as they fluid-feed on nectar.

Roger Eritja/Photolibrary

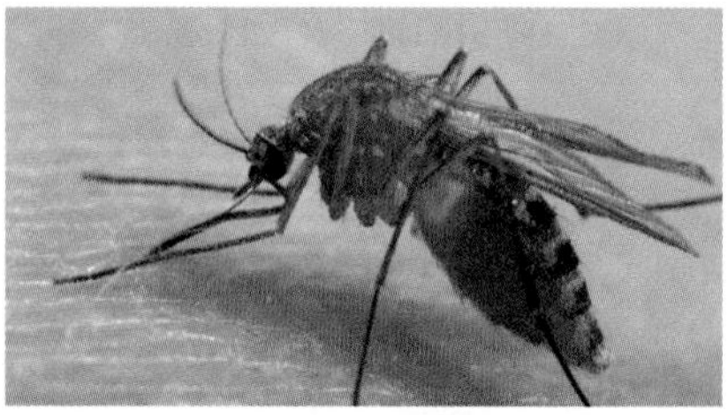

▲ **Figure 1C** A fluid feeder: a mosquito sucking blood

Rather than filtering food from water, eating their way through a substrate, or sucking fluids from other animals or plants, most animals are **bulk feeders**, meaning they ingest large pieces of food. **Figure 1D** shows a gray heron preparing to swallow its prey whole. A bulk feeder may use such utensils as tentacles, pincers, claws, poisonous fangs, or jaws and teeth to kill its prey or to tear off pieces of meat or vegetation. Whatever the type of food or feeding mechanism, the processing of food involves four stages, as we see next.

? Ring-tailed lemurs eat fruit, leaves, insects, and even small birds. Name their diet category and type of feeding mechanism.

Omnivore and bulk feeder

Jennette van Dyk/Getty Images

▲ **Figure 1D** A bulk feeder: a gray heron preparing to swallow a fish headfirst

2 Overview: Food processing occurs in four stages

So far we have discussed what animals eat and how they feed. As shown in **Figure 2A**, ❶ **ingestion**, the act of eating, is only the first of four main stages of food processing. The second stage, ❷ **digestion**, is the breaking down of food into molecules small enough for the body to absorb. Digestion typically occurs in two phases. First, food may be mechanically broken into smaller pieces. In animals with teeth, the process of chewing or tearing breaks large chunks of food into smaller ones. The second phase of digestion is the chemical breakdown process called hydrolysis. Catalyzed by specific enzymes, hydrolysis breaks the chemical bonds in food molecules by adding water to them.

Most of the organic matter in food consists of proteins, fats, and carbohydrates—all large molecules. Animals cannot use these materials directly for two reasons. First, these molecules are too large to pass through plasma membranes into the cells. Second, an animal needs small components to make the molecules of its own body. Most food molecules, for instance, the proteins in the cat's food shown in the figure below, are different from those that make up an animal's body.

All organisms use the same building blocks to make their macromolecules. For instance, cats, caterpillars, and humans all make their proteins from the same 20 amino acids. Digestion breaks the polymers in food into monomers. As shown in **Figure 2B**, proteins are split into amino acids, polysaccharides and disaccharides are broken down into monosaccharides, and nucleic acids are split into nucleotides (and their components). Fats are not polymers, but they are split into their components, glycerol and fatty acids. The animal can then use these small molecules to make the specific large molecules it needs.

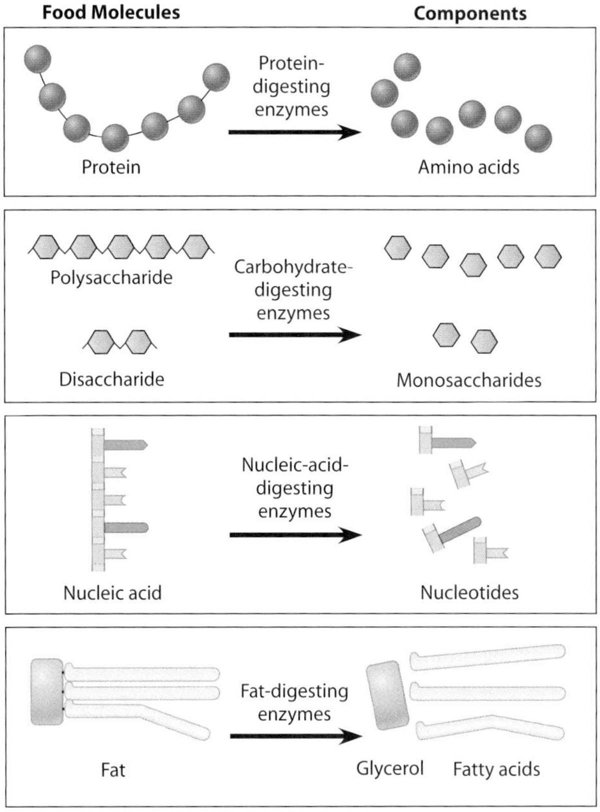

▲ **Figure 2B** Chemical digestion: the breakdown of large organic molecules to their components

In the third stage of food processing, ❸ **absorption**, the cells lining the digestive tract take up, or absorb, the products of digestion—small molecules such as amino acids and simple sugars. From the digestive tract, these nutrients travel in the blood to body cells, where they are used to build a cell's large molecules or broken down further to provide energy. In an animal that eats much more than its body immediately uses, many of the nutrient molecules are converted to fat for storage. In the fourth and last stage of food processing, ❹ **elimination**, undigested material passes out of the digestive tract.

How can an animal digest food without digesting its own cells and tissues? After all, digestive enzymes hydrolyze the same biological molecules that animals are made of—and it is obviously important to avoid digesting ourselves! The evolutionary adaptation found in most animal species is the chemical digestion of food within specialized compartments. We discuss digestive compartments in the next module.

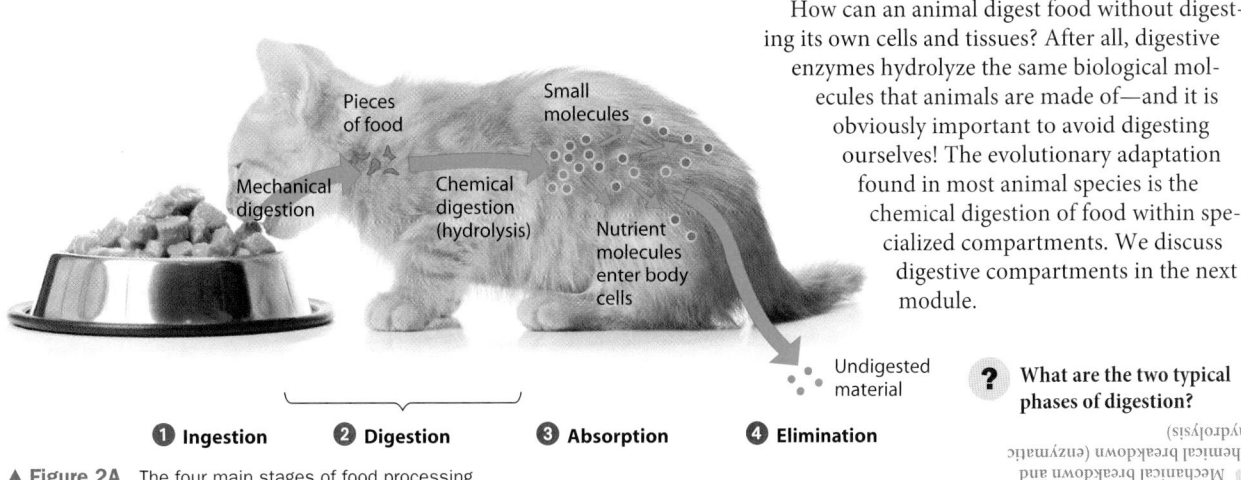

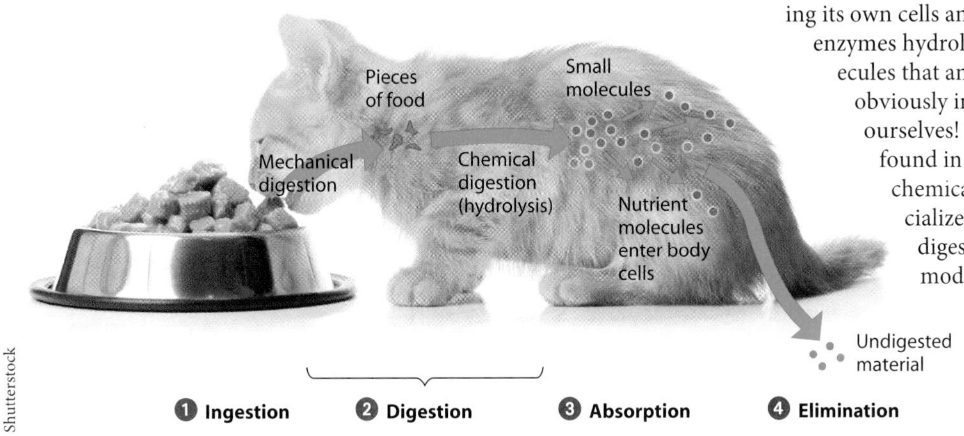

▲ **Figure 2A** The four main stages of food processing

? What are the two typical phases of digestion?

Mechanical breakdown and chemical breakdown (enzymatic hydrolysis)

3 Digestion occurs in specialized compartments

The simplest type of digestive compartment is a food vacuole, in which digestion occurs within a cell. After a cell engulfs a food particle, the newly formed food vacuole fuses with a lysosome containing hydrolytic enzymes. This type of digestion is common in single-celled protists, but sponges are the only animals that digest their food entirely in food vacuoles. In contrast, most animals have a digestive compartment that is surrounded by, rather than within, body cells. Such compartments enable an animal to devour much larger pieces of food than could fit in a food vacuole.

Simpler animals such as cnidarians and flatworms have a **gastrovascular cavity**, a digestive compartment with a single opening, the **mouth**. Figure 3A shows a hydra digesting a water flea, which it has captured with its tentacles and stuffed into its mouth. ❶ Gland cells lining the gastrovascular cavity secrete digestive enzymes that ❷ break down the food into smaller particles. ❸ Other cells engulf these small food particles, and ❹ digestion is completed in food vacuoles. Undigested material is expelled through the mouth.

Most animals have an **alimentary canal**, a digestive tract with two openings, a mouth and an anus. Because food moves in one direction, specialized regions of the tube can carry out digestion and absorption of nutrients in sequence.

Food entering the mouth usually passes into the **pharynx**, or throat. Depending on the species, the **esophagus** may channel food to a crop, gizzard, or stomach. A **crop** is a pouch-like organ in which food is softened and stored. **Stomachs** and **gizzards** may also store food temporarily, but they are more muscular and they churn and grind the food. Chemical digestion and nutrient absorption occur mainly in the **intestine**. Undigested materials are expelled through the **anus**.

Figure 3B illustrates three examples of alimentary canals. The digestive tract of an earthworm includes a muscular pharynx that sucks food in through the mouth. Food passes through the esophagus and is stored in the crop. Mechanical digestion takes place in the muscular gizzard, which pulverizes food with the aid of small bits of sand and gravel. Chemical digestion and absorption occur in the intestine, and undigested material is expelled through the anus.

A grasshopper also has a crop where food is stored. Most digestion in a grasshopper occurs in the midgut region, where projections called gastric pouches function in digestion and absorption. The hindgut mainly reabsorbs water and compacts wastes. Many birds have three separate chambers: a crop, a stomach, and a gravel-filled gizzard, in which food is pulverized. Chemical digestion and absorption occur in the intestine.

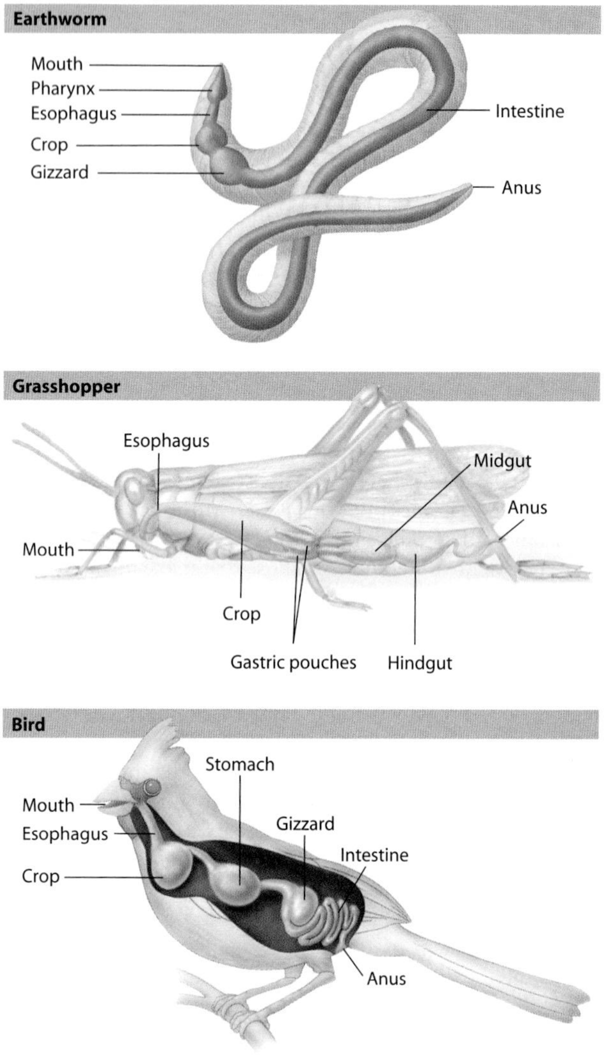

▲ **Figure 3B** Three examples of alimentary canals

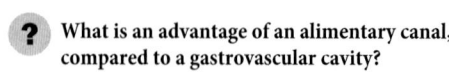

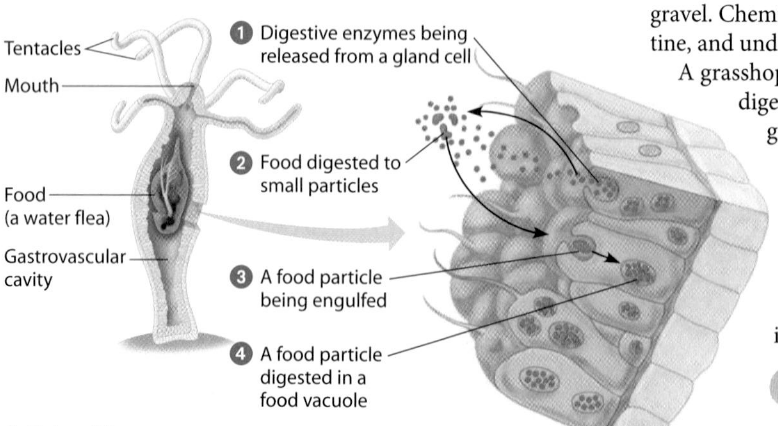

▲ **Figure 3A** Digestion in the gastrovascular cavity of a hydra

? What is an advantage of an alimentary canal, compared to a gastrovascular cavity?

● An alimentary canal has specialized regions, which can carry out digestion and absorption sequentially.

The Human Digestive System

4 The human digestive system consists of an alimentary canal and accessory glands

As an introduction to our own digestive system, **Figure 4** illustrates the human alimentary canal and its accessory glands. The schematic diagram on the left gives you an overview of the sequence of the organs and the locations of the accessory glands. These glands—the salivary glands, gallbladder, liver, and pancreas—are labeled in blue on the figure. They secrete digestive juices that enter the alimentary canal through ducts.

Food is ingested and chewed in the mouth, or **oral cavity**, and then pushed by the tongue into the pharynx. Once food is swallowed, muscles propel it through the alimentary canal by **peristalsis**, alternating waves of contraction and relaxation of the smooth muscles lining the canal (see Module 6). It is peristalsis that enables you to process and digest food even while lying down. After chewing a bite of food, it only takes 5–10 seconds for it to pass from the pharynx down the esophagus and into your stomach.

As shown in the enlargement, below right, muscular ringlike valves, called **sphincters**, regulate the passage of food into and out of the stomach. The sphincter controlling the passage out of the stomach works like a drawstring to close the stomach off, keeping food there for about 2–6 hours, long enough for stomach acids and enzymes to begin digestion. The final steps of digestion and nutrient absorption occur in the small intestine over a period of 5–6 hours. Undigested material moves slowly through the large intestine (taking 12–24 hours), and feces are stored in the rectum and then expelled through the anus.

In the next several modules, we follow a snack—an apple and some crackers and cheese—through your alimentary canal to see in more detail what happens to the food in each of the processing stations along the way.

? By what process does food move from the pharynx to the stomach of an astronaut in the weightless environment of a space station?

Peristalsis

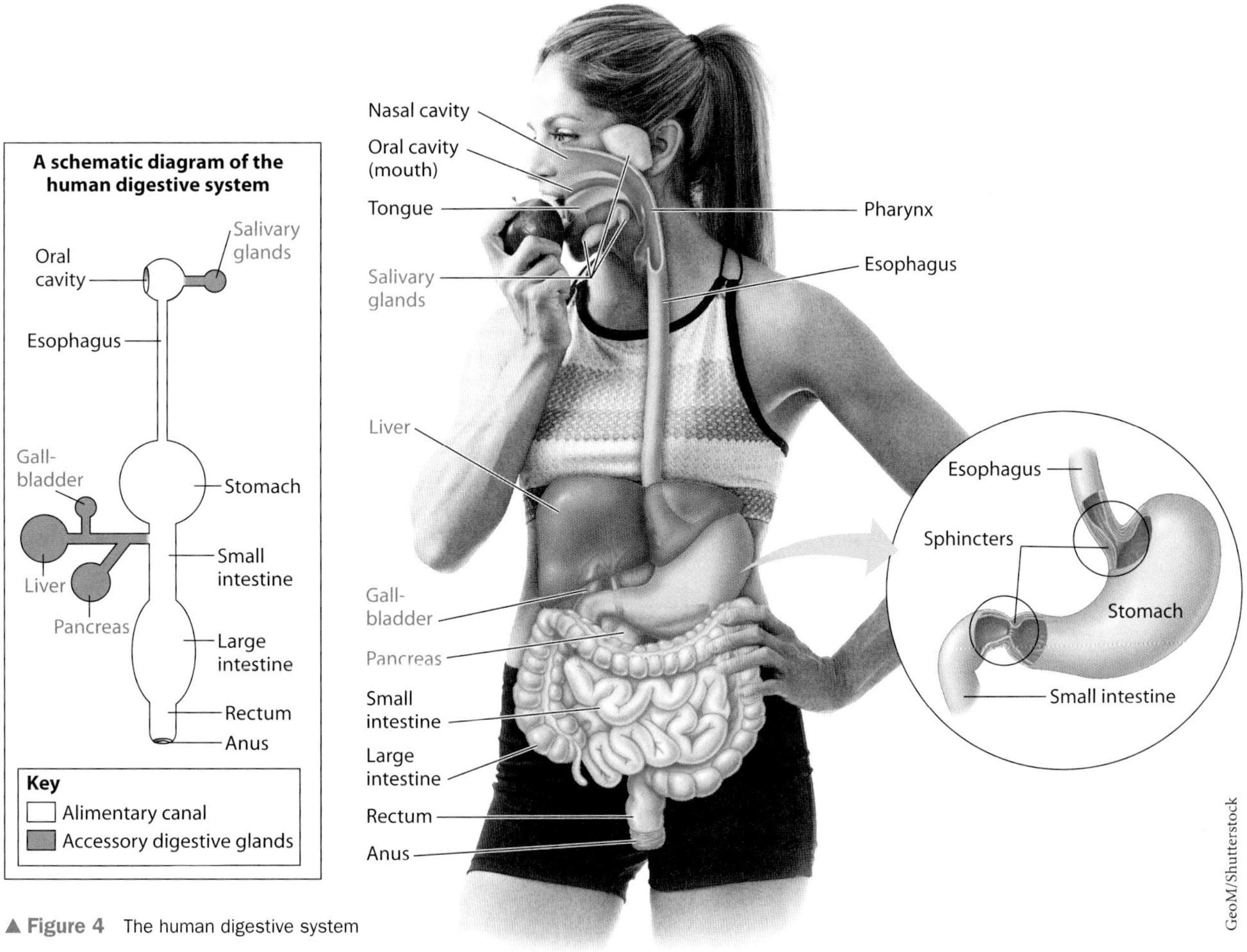

▲ **Figure 4** The human digestive system

5 Digestion begins in the oral cavity

Both mechanical digestion and chemical digestion begin in your oral cavity. Chewing cuts, smashes, and grinds food, making it easier to swallow and exposing more food surface to digestive enzymes. As **Figure 5** shows, you have several kinds of teeth. Starting at the front on one side of the upper or lower jaw, there are two bladelike incisors. You use these for biting into your apple. Behind the incisors is a single pointed canine tooth. (Canine teeth are much bigger in carnivores, which use them to kill and rip apart prey.) Next come two premolars and three molars, which grind and crush your food. You use these to pulverize your apple, cheese, and crackers. The third molar, a "wisdom" tooth, does not appear in all people, and in some people it pushes against the other teeth and must be removed.

As you anticipate your snack, your **salivary glands** may start delivering saliva through ducts to the oral cavity even before you take a bite. This is a response to the sight or smell or even thought of food. The presence of food in the oral cavity continues to stimulate salivation. In a typical day, your salivary glands secrete more than a liter (1 L) of saliva. You can see the duct opening in Figure 5. All three pairs of salivary glands are visible in Figure 4.

Saliva contains several substances important in food processing. A slippery glycoprotein (carbohydrate-protein complex) protects the soft lining of your mouth and lubricates food for easier swallowing. Buffers neutralize food acids, helping prevent tooth decay. Antibacterial agents kill many of the bacteria that enter your mouth with food. Saliva also contains the digestive enzyme amylase, which begins hydrolyzing the starch in your cracker into the disaccharide maltose.

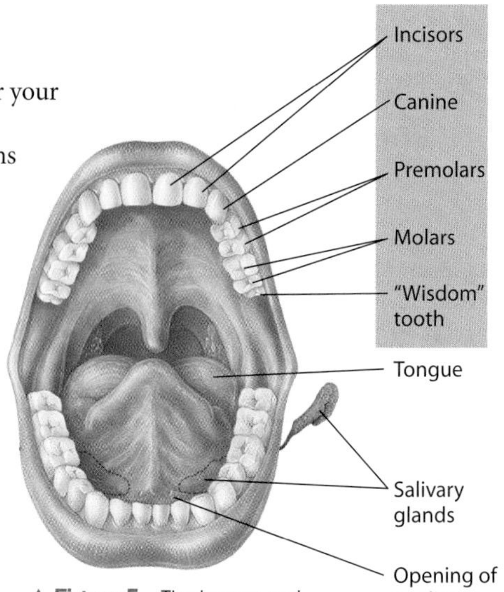

▲ **Figure 5** The human oral cavity

Also prominent in the oral cavity is the tongue, a muscular organ covered with taste buds. Besides enabling you to taste your meal, the tongue manipulates food and helps shape it into a ball called a **bolus**. When swallowing, the tongue pushes the bolus to the back of your oral cavity and into the pharynx.

? **Chewing functions in the ____ phase of digestion. What does the amylase in saliva do?**

mechanical; begins the chemical digestion of starch

6 After swallowing, peristalsis moves food through the esophagus to the stomach

Openings into both the esophagus and the trachea (windpipe) are in the pharynx, the region you call your throat. Most of the time, as shown on the left in **Figure 6A**, the esophageal opening is closed off by a sphincter (blue arrows). Air enters the larynx and flows past the vocal cords in your voice box, through the trachea, to your lungs (black arrows).

This situation changes when you start to swallow. The tongue pushes the bolus of food into the pharynx, triggering the swallowing reflex (center of Figure 6A). The larynx, the upper part of the respiratory tract, moves upward. This movement tips a flap of cartilage called the epiglottis over the opening to the larynx, preventing food from entering the trachea. You can see this motion in the bobbing of your larynx (also called your Adam's apple) during swallowing. The esophageal sphincter relaxes, and the bolus enters the esophagus (green arrow). As shown on the right side of Figure 6A,

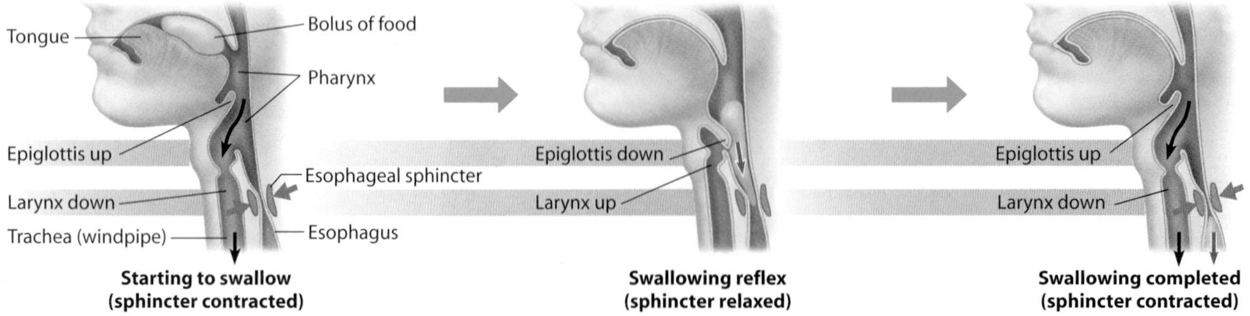

▲ **Figure 6A** The human swallowing reflex

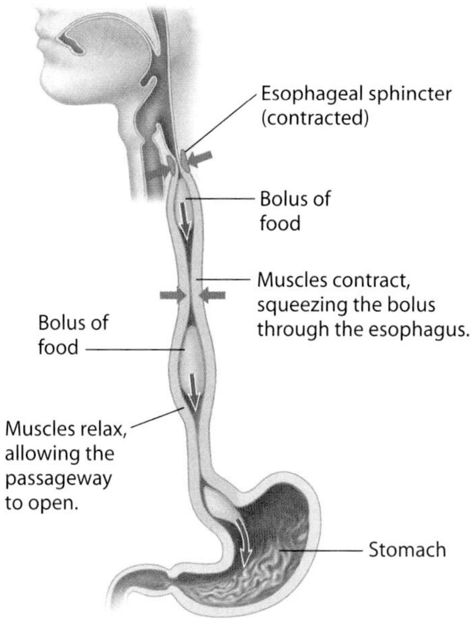

▲ **Figure 6B** Peristalsis moving a food bolus through the esophagus

the esophageal sphincter then contracts above the bolus, and the epiglottis tips up again, reopening the breathing tube.

The esophagus is a muscular tube that conveys food from the pharynx to the stomach. The muscles at the top of the esophagus are under voluntary control; thus, you begin the act of swallowing voluntarily. But then involuntary contractions of smooth muscles in the rest of the esophagus take over. **Figure 6B** shows how muscle contractions—peristalsis—squeeze a bolus toward the stomach. Muscle contractions continue in waves until the bolus enters the stomach. Peristalsis also moves digesting food through the intestines.

The structure of the esophagus fits its function. It has tough yet elastic connective tissues that allow it to stretch to accommodate a bolus, layers of circular and longitudinal smooth muscles for peristalsis, and a stratified epithelial lining that replenishes cells abraded off during swallowing. The length of the esophagus varies by species. For example, fishes have no lungs to bypass and have a very short esophagus. And it will come as no surprise that giraffes have a very long esophagus.

Kleer/Shutterstock.

Usually our breathing and swallowing are carefully coordinated. Next we consider what happens when they are not.

? What prevents food from going down the wrong tube?

The epiglottis tips over the opening to the trachea when swallowing.

CONNECTION 7 The Heimlich maneuver can save lives

Sometimes our swallowing mechanism goes awry. A person may eat too quickly or fail to chew food thoroughly. Or a young child may swallow an object too big to pass through the esophagus. Such mishaps can lead to a blocked pharynx or trachea. The blockage may prevent air from flowing into the trachea, causing the person to choke. If breathing is not restored within minutes, brain damage or death may result.

To save someone who is choking, you need to dislodge any foreign objects in the throat and get air flowing. This quick assistance can come through the use of the Heimlich maneuver. The procedure, invented by Dr. Henry Heimlich in the 1970s, allows people with little medical training to step in and aid a choking victim.

The maneuver is usually performed on someone who is seated or standing up. Stand behind the victim and place your arms around the victim's waist. Make a fist with one hand, and place it against the victim's upper abdomen, well below the rib cage. Then place the other hand over the fist and press into the victim's abdomen with a quick upward thrust. When done correctly, the diaphragm is forcibly elevated, pushing air into the trachea. Repeat this procedure until the object is forced out of the victim's airway (**Figure 7**). The maneuver can also be performed on drowning victims to clear the lungs of water before beginning CPR. You can even use your own fist or the back of a chair to force air upward and dislodge an object from your own trachea.

? If food is stuck in the pharynx, what effect could it have on nearby structures? (*Hint:* See Figure 6A.)

The epiglottis may be tipped down, blocking the opening to the trachea.

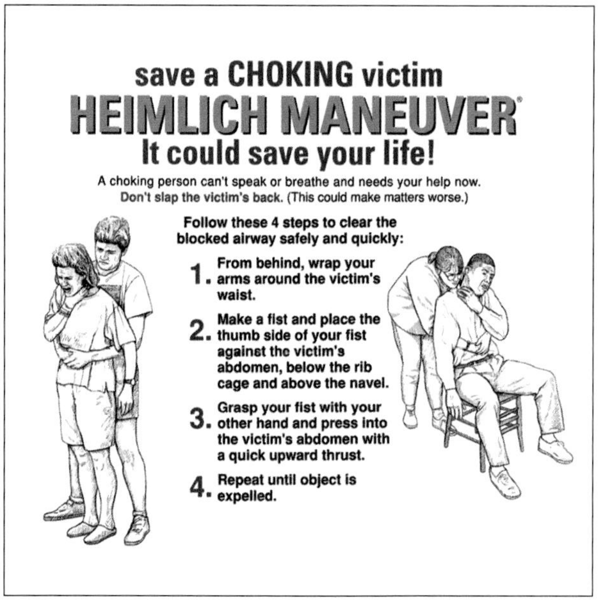

▲ **Figure 7** The Heimlich maneuver for helping choking victims

8 The stomach stores food and breaks it down with acid and enzymes

Having a stomach is the main reason you do not need to eat constantly. With its accordion-like folds and highly elastic wall, your stomach can stretch to accommodate about 2 L of food and drink, usually enough to satisfy your needs for hours.

Some chemical digestion occurs in the stomach. The stomach secretes **gastric juice**, which is made up of a protein-digesting enzyme, mucus, and strong acid. The pH of gastric juice is about 2, acidic enough to dissolve iron nails. One function of the acid is to break apart the cells in food and denature (unravel) proteins. The acid also kills most bacteria and other microbes that are swallowed with food.

The interior surface of the stomach wall is highly folded, and as the diagram in **Figure 8** shows, it is dotted with pits leading into tubular gastric glands. The gastric glands have three types of cells that secrete different components of gastric juice. Mucous cells (shown here in dark pink) secrete mucus, which lubricates and protects the cells lining the stomach. Parietal cells (yellow) secrete hydrogen and chloride ions, which combine in the lumen (cavity) of the stomach to form hydrochloric acid (HCl). Chief cells (pink) secrete pepsinogen, an inactive form of the enzyme pepsin.

The diagram on the far right of the figure indicates how active pepsin is formed. ❶ Pepsinogen, H^+, and Cl^- are secreted into the lumen of the stomach. ❷ Next, the HCl converts some pepsinogen to pepsin. ❸ Pepsin itself then helps activate more pepsinogen, starting a chain reaction. This series of events is an example of positive feedback, in which the end product of a process promotes the formation of more end product.

Now what does all this active pepsin do? Pepsin begins the chemical digestion of proteins—those in your cheese snack, for instance. It splits the polypeptide chains of proteins into smaller polypeptides, which will be broken down further in the small intestine.

What prevents gastric juice from digesting away the stomach lining? Secreting pepsin in the inactive form of pepsinogen helps protect the cells of the gastric glands, and mucus helps protect the stomach lining from both pepsin and acid. Regardless, the epithelium of the stomach is constantly eroded. But don't worry, enough new cells are generated by mitosis to replace your stomach lining completely about every three days.

Another protection for the stomach is that gastric glands do not secrete acidic gastric juice constantly. Their activity is regulated by a combination of nerve signals and hormones. When you see, smell, or taste food, a signal from your brain stimulates your gastric glands. And as food arrives in your stomach, it stretches the stomach walls and triggers the release of the hormone **gastrin**. Gastrin circulates in the bloodstream, returning to the stomach (green dashed line in the top section of Figure 8), where it stimulates additional secretion of gastric juice. As much as 3 L of gastric juice may be secreted in a day.

What prevents too much gastric juice from being secreted? A negative-feedback mechanism inhibits the secretion of

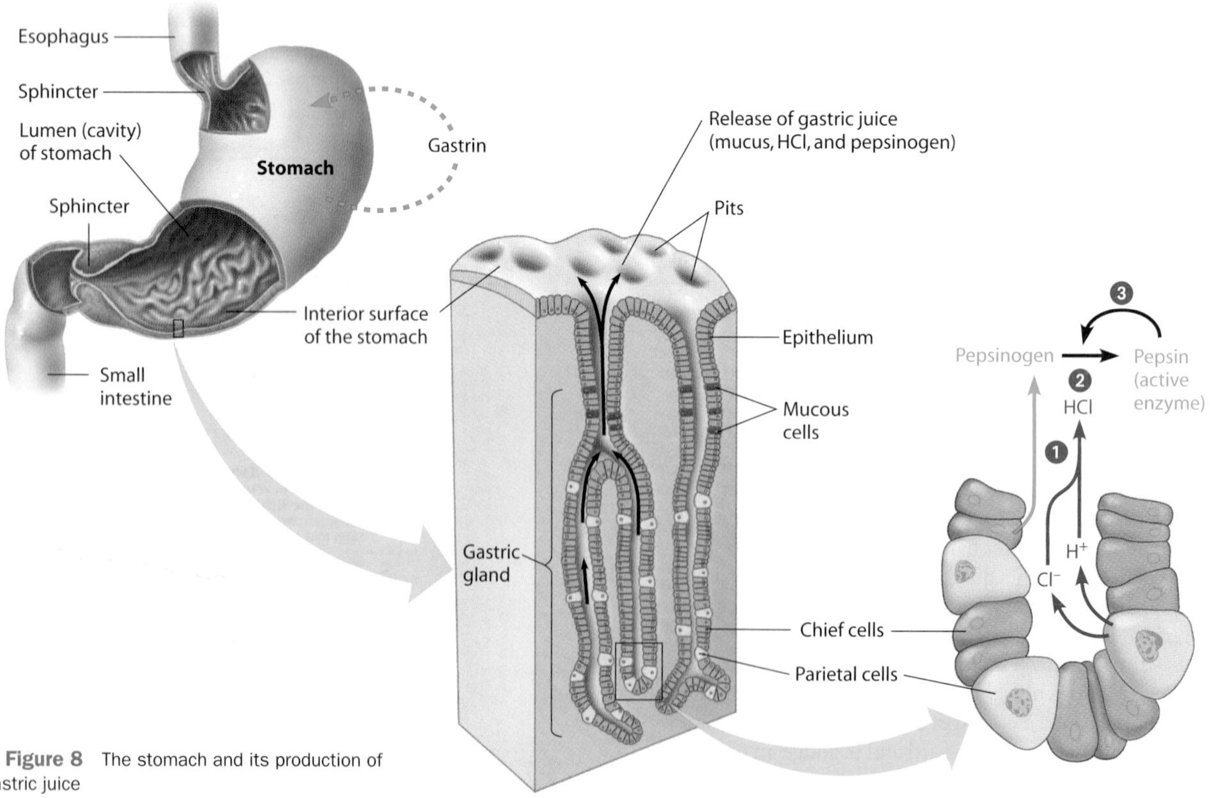

▲ **Figure 8** The stomach and its production of gastric juice

gastric juice when the stomach contents become too acidic. The acid inhibits the release of gastrin, and with less gastrin in the blood, the gastric glands secrete less gastric juice.

About every 20 seconds, the stomach contents are mixed by the churning action of muscles in the stomach wall. You may feel hunger pangs when your stomach has been empty for hours and strongly contracts. As a result of mixing and enzyme action, what begins in the stomach as a recently swallowed apple, cracker, and cheese snack becomes an acidic, nutrient-rich broth known as **chyme**. The sphincter between the stomach and the small intestine regulates the passage of chyme, which leaves the stomach and enters the small intestine a squirt at a time.

We'll continue with the digestion of your snack in Module 10. But first, let's consider some digestive problems.

? If you add pepsinogen to a test tube containing protein dissolved in distilled water, not much protein will be digested. What inorganic chemical could you add to the tube to accelerate protein digestion? What effect will it have?

HCl or some other acid will convert inactive pepsinogen to active pepsin, which will begin digestion of the protein and also activate more pepsinogen.

CONNECTION 9 Digestive ailments include acid reflux and gastric ulcers

A stomachful of digestive juice laced with strong acid breaks apart the cells in your food, kills bacteria, and begins the digestion of proteins. At the same time, these chemicals can be dangerous to unprotected cells. The opening between the esophagus and the stomach is usually closed until a bolus arrives. Occasionally, however, acid reflux occurs. This backflow of chyme into the lower end of the esophagus causes the feeling we call heartburn—actually "esophagus-burn."

Some people suffer acid reflux frequently and severely enough to harm the lining of the esophagus, a condition called GERD (gastroesophageal reflux disease). GERD can often be treated with lifestyle changes. Doctors usually recommend that patients stop smoking, avoid alcohol, lose weight, eat small meals, refrain from lying down for 2–3 hours after eating, and sleep with the head of the bed raised. Medications to treat GERD include antacids, which reduce stomach acidity, and drugs called H2 blockers, such as Pepcid AC or Zantac, which impede acid production. Proton pump inhibitors, such as Prilosec, have a different mechanism of action and are very effective at stopping acid production. When lifestyle changes and medications fail to alleviate the symptoms, surgery to strengthen the lower esophageal sphincter may be an option.

Can all that acid also cause problems in the stomach? A gel-like coat of mucus normally protects the stomach wall from the corrosive effect of digestive juice. When it fails, open sores called gastric ulcers can develop in the stomach wall. The symptoms are usually a gnawing pain in the upper abdomen, which may occur a few hours after eating.

Gastric ulcers were formerly thought to be caused by too much acid or too little mucus and linked to factors such as aspirin, smoking, alcohol, coffee, and, in particular, psychological stress. However, in 1982, researchers Barry Marshall and Robin Warren reported that an acid-tolerant bacterium called *Helicobacter pylori* causes ulcers (**Figure 9**). Their hypothesis was poorly received. Marshall then experimented on himself by drinking a mixture laced with *H. pylori* bacteria—and soon developed gastritis, a mild inflammation of the stomach. Although Marshall and Warren were awarded the 2005 Nobel Prize for their discovery, we do not recommend this kind of experiment!

The low pH of the stomach kills most microbes, but not *H. pylori*. This bacterium burrows beneath the mucus and releases harmful chemicals. Growth of *H. pylori* seems to result in a localized loss of protective mucus and damage to the cells lining the stomach. Numerous white blood cells move into the stomach wall to fight the infection, and their presence is associated with gastritis. Gastric ulcers develop when pepsin and hydrochloric acid destroy cells faster than the cells can regenerate. Eventually, the stomach wall may erode to the point that it actually has a hole in it. This hole can lead to a life-threatening infection within the abdomen or internal bleeding. *H. pylori* is found in up to 90% of ulcer and gastritis sufferers. Some studies also link *H. pylori* to certain kinds of stomach cancer.

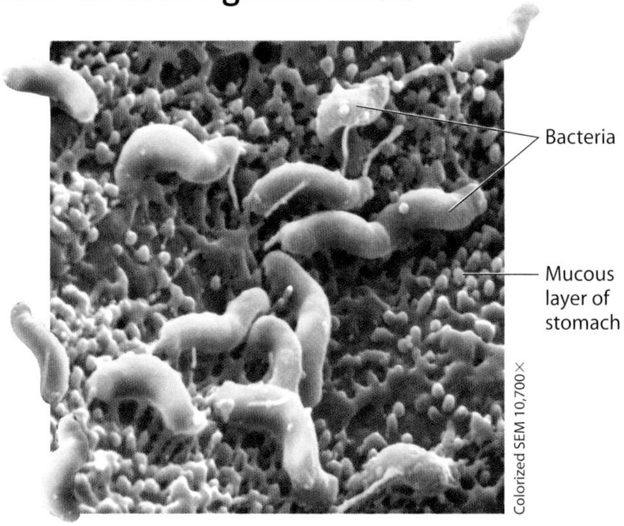

Oliver Meckes/Ottawa/Photo Researchers

▲ **Figure 9** An ulcer-causing bacterium, *Helicobacter pylori*

Gastric ulcers usually respond to a combination of antibiotics and bismuth (the active ingredient in Pepto-Bismol), which eliminates the bacteria and promotes healing.

When partially digested food leaves the stomach, it is accompanied by gastric juices, and so the first section of the small intestine—the duodenum—is also susceptible to ulcers, as is the esophagus in cases of severe GERD.

? In contrast to most microbes, the species that causes ulcers thrives in an environment with a very low ____.

pH

10 The small intestine is the major organ of chemical digestion and nutrient absorption

So what is the status of your snack as it passes out of the stomach and into the small intestine? At this point in its journey through the digestive tract, the food has been mechanically reduced to smaller pieces and mixed with digestive juices; it now resembles a thick soup. Chemically, the digestion of starch in the cracker began in the mouth, and the breakdown of protein in the cheese began in the stomach. The rest of the digestion of the large molecules in your snack occurs in the **small intestine**. The nutrients that result from this digestion are absorbed into the blood from the small intestine. With a length of more than 6 m, the small intestine is the longest organ of the alimentary canal. Its name does not come from its length but from its diameter, which is only about 2.5 cm; the large intestine is much shorter but has twice the diameter.

Sources of Digestive Enzymes and Bile The first 25 cm (10 inches) or so of the small intestine is called the **duodenum**. This is where chyme squirted from the stomach mixes with digestive juices from the pancreas, liver, and gallbladder, as well as from gland cells in the intestinal wall itself (**Figure 10A**). The **pancreas** produces pancreatic juice, a mixture of digestive enzymes and an alkaline solution rich in bicarbonate. The bicarbonate acts as a buffer to neutralize the acidity of chyme as it enters the small intestine. The pancreas also produces hormones that regulate blood glucose levels.)

In addition to its many other functions, the **liver** produces bile. **Bile** contains bile salts that break up fats into small droplets that are more susceptible to attack by digestive enzymes. The **gallbladder** stores bile until it is needed in the small intestine. In response to chyme, hormones produced by the duodenum stimulate the release of bile from the liver, as well as digestive juices from the pancreas.

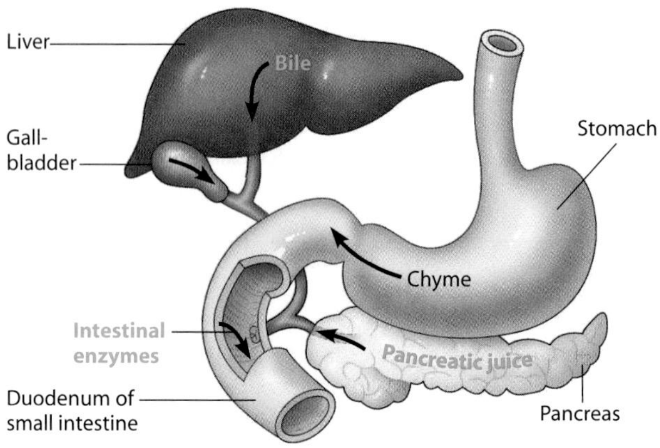

▲ **Figure 10A** The duodenum and associated digestive organs

As already mentioned, the intestinal wall itself also produces digestive enzymes. Some of these enzymes are secreted into the lumen of the small intestine; others are bound to the surface of epithelial cells.

Digestion in the Small Intestine **Table 10** summarizes the processes of enzymatic digestion that occur in the small intestine. All four types of large molecules—carbohydrates, proteins, nucleic acids, and fats—are digested in the small intestine. As we discuss the digestion of each, the table will help you keep track of the enzymes involved (in blue type).

Remember that the digestion of carbohydrates, such as those in your cracker, began in the oral cavity. Their digestion is completed in the small intestine. An enzyme called pancreatic amylase hydrolyzes polysaccharides into the disaccharide maltose. The enzyme maltase then splits maltose into

TABLE 10 | ENZYMATIC DIGESTION IN THE SMALL INTESTINE

Carbohydrates				
Polysaccharides	Pancreatic amylase →	Maltose (and other disaccharides)	Maltase, sucrase, lactase, etc. →	Monosaccharides
Proteins				
Polypeptides	Trypsin, chymotrypsin →	Smaller polypeptides	Various peptidases →	Amino acids
Nucleic Acids				
DNA and RNA	Nucleases →	Nucleotides	Other enzymes →	Nitrogenous bases, sugars, and phosphates
Fats				
Fat globules	Bile salts →	Fat droplets (emulsified)	Lipase →	Fatty acids and glycerol

the monosaccharide glucose. Maltase is one of a family of enzymes, each specific for the hydrolysis of a different disaccharide. For example, sucrase hydrolyzes table sugar (sucrose), and lactase digests lactose. Lactose is common in milk and cheese. Children generally produce much more lactase than do adults; as a result, many adults have difficulty digesting milk products. If you are lactose intolerant, you will likely experience digestive discomfort shortly after eating the cheese in your snack.

The small intestine also completes the digestion of proteins that was begun in the stomach. The pancreas and the duodenum produce enzymes that completely dismantle polypeptides into amino acids. The enzymes trypsin and chymotrypsin break polypeptides into smaller polypeptides. Several types of enzymes called peptidases then split off one amino acid at a time from these smaller polypeptides. Working together, this enzyme team digests proteins much faster than any single enzyme could.

Yet another team of enzymes, the nucleases, hydrolyzes nucleic acids. Nucleases from the pancreas split DNA and RNA (which are present in the cells of food sources) into their component nucleotides. The nucleotides are then broken down into nitrogenous bases, sugars, and phosphates by other enzymes.

Digestion of fats is a special problem because fats are insoluble in water and tend to clump together in large globules. How is this problem solved? First, bile salts in bile separate and coat smaller fat droplets, a process called emulsification. When there are many small droplets, a larger surface area of fat is exposed to lipase, a pancreatic enzyme that breaks fat molecules down into fatty acids and glycerol.

By the time the mixture of chyme and digestive juices has moved through your duodenum, chemical digestion of your snack is just about complete. The main function of the rest of the small intestine is to absorb nutrients and water.

Absorption in the Small Intestine Structurally, the small intestine is well suited for its task of absorbing nutrients. As **Figure 10B** shows, the inner wall of the small intestine has large circular folds with numerous small, finger-like projections called **villi** (singular, *villus*). Each of the epithelial cells on the surface of a villus has many tiny projections, called **microvilli**. This combination of folds and projections greatly increases the surface area across which nutrients are absorbed. Indeed, the lining of your small intestine has a huge surface area—roughly 300 m^2, about the size of a tennis court.

Some nutrients are absorbed by simple diffusion; others are pumped against concentration gradients into the epithelial cells. Notice that a small lymph vessel (shown in yellow in the figure) penetrates the core of each villus. After fatty acids and glycerol are absorbed by an epithelial cell, these building blocks are recombined into fats, which are then coated with proteins and transported into a lymph vessel. These vessels are part of the lymphatic system—a network of vessels filled with a clear fluid called lymph that eventually empties into large veins near the heart.

Notice that each villus also has a network of capillaries. Many absorbed nutrients, such as amino acids and sugars, pass directly out of the intestinal epithelium and then across the thin walls of the capillaries into the blood.

Where does this nutrient-laden blood go? To the liver, where we also head in the next module.

? **Amylase is to ____ as ____ is to DNA.**

polysaccharides ... nuclease

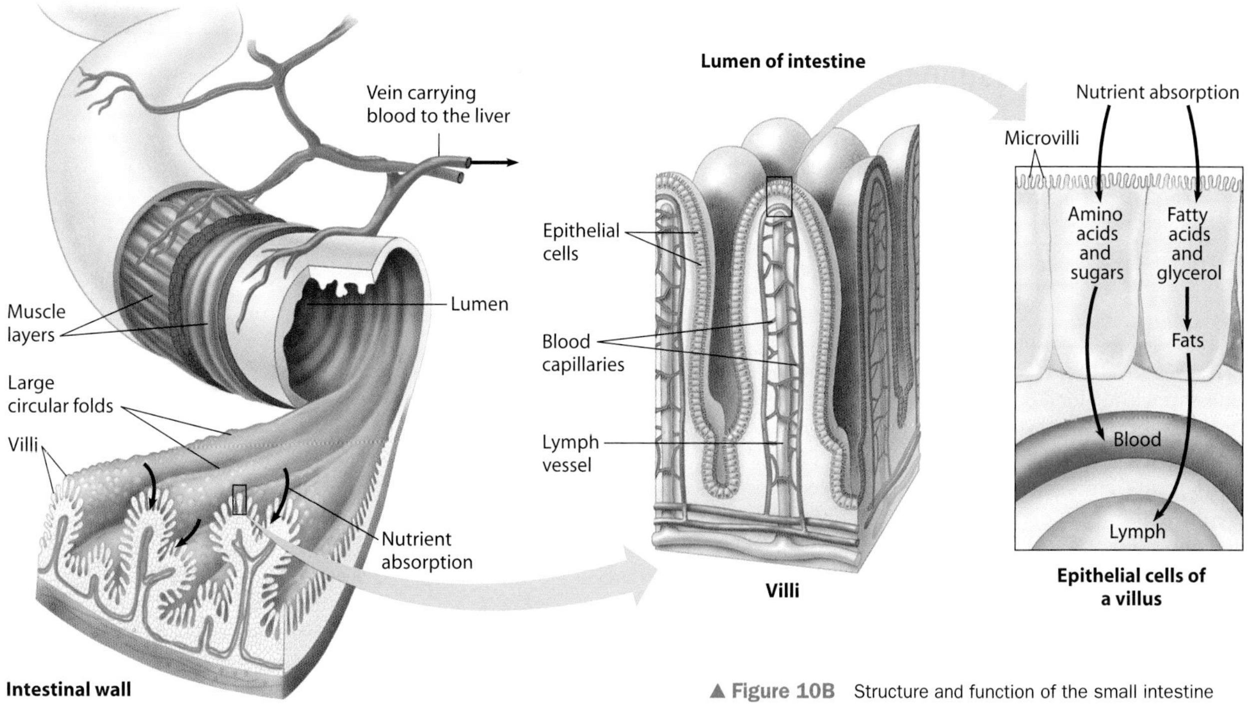

▲ **Figure 10B** Structure and function of the small intestine

11 One of the liver's many functions is processing nutrient-laden blood from the intestines

The liver has a strategic location in your body—between the intestines and the heart. As shown in **Figure 11**, capillaries from the small and large intestines converge into veins that lead into the **hepatic portal vein**. This large vessel transports blood to the liver, thus giving the liver first access to nutrients absorbed in the intestines. The liver removes excess glucose from the blood and converts it to glycogen (a polysaccharide), which is stored in liver cells. In balancing the storage of glycogen with the release of glucose to the blood, your liver plays a key role in regulating metabolism.

The liver also converts many of the nutrients it receives to new substances. Liver cells synthesize many essential proteins. Among these are plasma proteins important in blood clotting and in maintaining the osmotic balance of the blood, as well as lipoproteins that transport fats and cholesterol to body cells. If your diet includes too much junk food and too many calories, the liver converts the excess to fat, which is stored in your body.

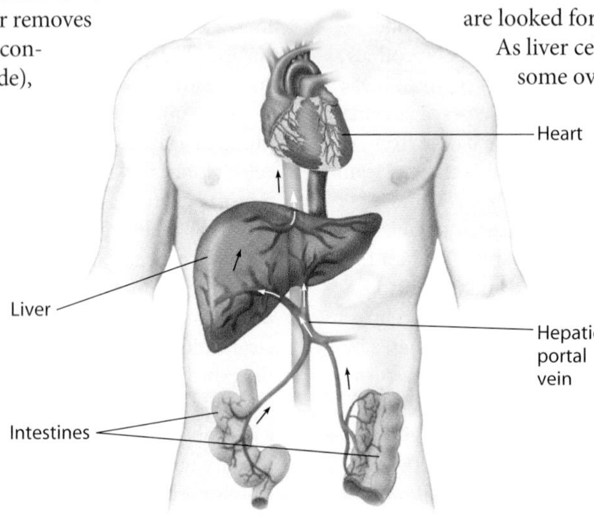

▲ **Figure 11** The hepatic portal vein carrying blood from the intestines to the liver

The liver has a chance to modify and detoxify substances absorbed by the digestive tract before the blood carries these materials to the rest of the body. It converts toxins such as alcohol or other drugs to inactive products that are excreted in the urine. These breakdown products are what are looked for in urine tests for various drugs. As liver cells detoxify alcohol or process some over-the-counter and prescription drugs, however, they can be damaged. The combination of alcohol and certain drugs, such as acetaminophen, is particularly harmful.

As you already learned, the liver produces bile, which aids in the digestion of fats. It processes nitrogen-containing wastes from the breakdown of proteins for disposal in urine.

? What two functions of the liver relate to the hepatic portal vein?

As blood is delivered directly from the intestines, the liver can process and regulate the absorbed nutrients and remove toxic substances.

12 The large intestine reclaims water and compacts the feces

Most of the nutrients from your snack have been absorbed, and the large intestine now processes whatever remains. The **large intestine**, or **colon**, is about 1.5 m long and 5 cm in diameter. As the enlargement in **Figure 12** shows, at the junction where the small intestine leads into the large intestine there is a small pouch called the **cecum**. Compared with many other mammals, we humans have a small cecum. The **appendix**, a small, finger-like extension of the cecum, contains a mass of white blood cells that make a minor contribution to immunity. Despite this role, the appendix itself is prone to infection (appendicitis). If this occurs, emergency surgery is usually required to remove the appendix and prevent the spread of infection.

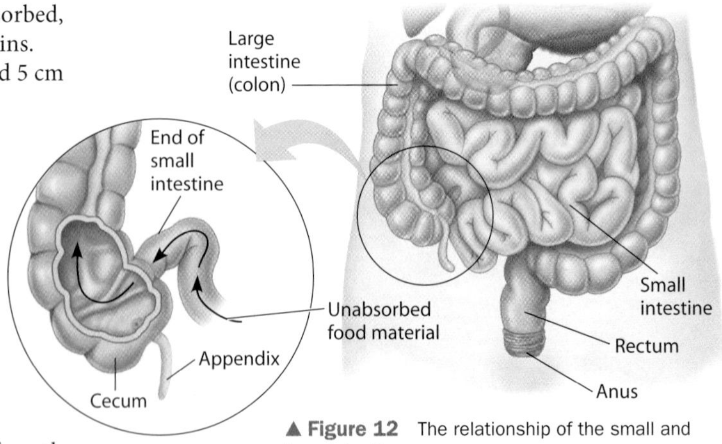

▲ **Figure 12** The relationship of the small and large intestines

One major function of the colon is to absorb water from the alimentary canal. Altogether, about 7 L of fluid enters the lumen of your digestive tract each day as the solvent of the various digestive juices. About 90% of this water is absorbed back into the blood and tissue fluids by the small intestine and colon. The wastes of the digestive system, called **feces**, become more solid as water is reabsorbed and they move along the colon by peristalsis. The feces consist mainly of indigestible plant fibers—cellulose from your apple, for instance—and prokaryotes that normally live in the colon. Some of your colon bacteria, such as *E. coli*, produce important vitamins, including several B vitamins and vitamin K. These substances are absorbed into the bloodstream and supplement your dietary intake of vitamins.

Feces are stored in the final portion of the colon, the **rectum**, until they can be eliminated. Strong contractions of the colon create the urge to defecate. Two rectal sphincters, one voluntary and the other involuntary, regulate the opening of the anus.

If the lining of the colon is irritated—by a viral or bacterial infection, for instance—the colon is less effective in reclaiming water, and diarrhea may result. The opposite problem, constipation, occurs when peristalsis moves the feces along too slowly; the colon reabsorbs too much water, and the feces become too compacted. Constipation often results from a diet that does not include enough plant fiber.

? Explain why treatment with antibiotics for an extended period may cause a vitamin K deficiency.

The antibiotics may kill the bacteria that synthesize vitamin K in the colon.

EVOLUTION CONNECTION

13 Evolutionary adaptations of vertebrate digestive systems relate to diet

Natural selection has favored adaptations that fit the structure of an animal's digestive system to the function of digesting the kind of food the animal eats. Large, expandable stomachs are common adaptations in carnivores, which may go a long time between meals and must eat as much as they can when they do catch prey. A 200-kg lion can consume 40 kg (almost 90 pounds) of meat in one meal! After such a feast, the lion may not hunt again for a few days.

The length of an animal's digestive tract also correlates with diet. In general, herbivores and omnivores have longer alimentary canals, relative to their body size, than carnivores. Vegetation is more difficult to digest than meat because it contains plant cell walls. A longer canal provides more time for digestion and more surface area for the absorption of nutrients.

Most herbivorous animals have special chambers that house great numbers of bacteria and protists. The animals lack the enzymes needed to digest the cellulose in plants. The microbes break down cellulose to simple sugars, which the animals then absorb or obtain by digesting the microbes.

Many herbivorous mammals—horses, elephants, and koalas, for example—house cellulose-digesting microbes in a large cecum. **Figure 13** compares the digestive tract of a carnivore, the coyote, with that of an herbivore, the koala. These two mammals are about the same size, but the koala's intestine is much longer and includes the longest cecum (about 2 m) of any animal of its size. With the aid of bacteria in its cecum, the koala gets almost all its food and water from the leaves of eucalyptus trees.

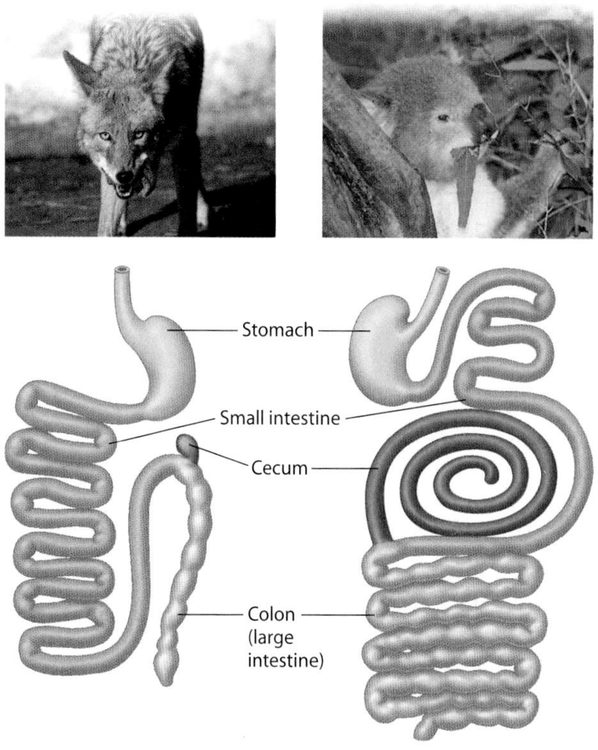

Kristin Piljay/Alamy

EyeWire Collection/ Photodisc/Getty Images

▲ **Figure 13** The alimentary canal in a carnivore (coyote) and an herbivore (koala)

In rabbits and some rodents, cellulose-digesting bacteria live in the large intestine as well as in the cecum. Many of the nutrients produced by these microbes are initially lost in the feces because they do not go through the small intestine, the main site of nutrient absorption. Rabbits and rodents recover these nutrients by eating some of their feces, thus passing the food through the alimentary canal a second time. The feces from the second round of digestion, the familiar rabbit "pellets," are more compact and are not reingested.

The most elaborate adaptations for an herbivorous diet have evolved in the mammals called **ruminants**, which include cattle, sheep, and deer. The stomach of a ruminant has four chambers containing symbiotic microbes. A cow periodically regurgitates food from the first two chambers and "chews its cud," exposing more plant fibers to its microbes for digestion. The cud is then swallowed and moves to the last stomach chambers, where digestion is completed. A cow actually obtains many of its nutrients by digesting the microbes along with the nutrients they produce. The microbes reproduce so rapidly that their numbers remain stable despite this constant loss.

? Name two advantages of a long alimentary canal in herbivores.

It provides increased time for processing of difficult-to-digest plant material and increased surface area for absorption of nutrients.

Nutrition

14 Overview: An animal's diet must satisfy three needs

All animals—whether herbivores like koalas, carnivores like coyotes, or omnivores like humans—have the same basic nutritional needs. All animals must obtain (1) fuel to power all body activities; (2) organic molecules to build the animal's own molecules; and (3) essential nutrients, substances the animal cannot make for itself.

We have seen that digestion dismantles the large molecules in food. Cells can then use the resulting small molecules for energy or assemble them into their own complex molecules—the proteins, carbohydrates, lipids, and nucleic acids needed to build and maintain cell structure and function.

Eating too little food, too much food, or the wrong mixture of foods can endanger an animal's health. Starting with the need for fuel and paying particular attention to humans, we discuss basic nutritional needs in the rest of this chapter.

? What are the three needs that an adequate diet fills?

Fuel, organic building materials, and essential nutrients

15 Chemical energy powers the body

It takes energy to read this book. It also takes energy to digest your snack, walk to class, and perform all the other activities of your body. Cellular respiration produces the body's energy currency, ATP, by oxidizing organic molecules obtained from food. Usually, cells use carbohydrates and fats as fuel sources. Fats are especially rich in energy: The oxidation of a gram of fat liberates more than twice the energy liberated from a gram of carbohydrate or protein. The energy content of food is measured in **kilocalories** (**1 kcal** = 1,000 calories). The calories listed on food labels or referred to in regard to nutrition are actually kilocalories and are often written as Calories (capital C).

The rate of energy consumption by an animal is called its **metabolic rate**. It is the sum of all the energy used by biochemical reactions over a given time interval. Cellular metabolism must continuously drive several processes for an animal to remain alive. These include cell maintenance, breathing, the beating of the heart, and, in birds and mammals, the maintenance of body temperature. The number of kilocalories a resting animal requires to fuel these essential processes for a given time is called the **basal metabolic rate (BMR)**. The BMR for humans averages 1,300–1,500 kcal per day for adult females and about 1,600–1,800 kcal per day for adult males. This is about equivalent to the rate of energy use by a 75-watt light bulb. But this is only a basal (base) rate—the amount of energy you "burn" lying motionless. Any activity, even working at your desk, consumes kilocalories in addition to the BMR. The more active you are, the greater your actual metabolic rate and the greater the number of kilocalories your body uses per day.

Table 15 gives you an idea of the amount of activity it takes for a 68-kg (150-pound) person to use up the kilocalories contained in several common foods. What happens when you take in more Calories than you use? Rather than discarding the extra energy, your cells store it in various forms. Your liver and muscles store energy in the form of glycogen, a polymer of glucose molecules. Most of us store enough glycogen to supply about a day's worth of basal metabolism. Your body also stores excess energy as fat. This happens even if your diet contains little fat because the liver converts excess carbohydrates and proteins to fat. The average human's energy needs can be fueled by the oxidation of only 0.3 kg of fat per day. Most healthy people have enough stored fat to sustain them through several weeks of starvation. We discuss fat storage and its consequences in Module 20. But first let's consider the essential nutrients that must be supplied in the diet.

? What is the difference between metabolic rate and basal metabolic rate?

Metabolic rate is the total energy used for all activities in a unit of time; BMR is the minimum number of kilocalories that a resting animal needs to maintain life's basic processes for a unit of time.

TABLE 15 EXERCISE REQUIRED TO "BURN" THE CALORIES (KCAL) IN COMMON FOODS

	Jogging	Swimming	Walking
Speed of exercise	**9 min/mi**	**30 min/mi**	**20 min/mi**
kcal "burned" per hour	**775**	**408**	**245**
Cheeseburger (quarter-pound), 417 kcal	32 min	1 hr, 1 min	1 hr, 42 min
Pepperoni pizza (1 large slice), 280 kcal	22 min	42 min	1 hr, 8 min
Non-diet soft drink (12 oz), 152 kcal	12 min	22 min	37 min
Whole wheat bread (1 slice), 65 kcal	5 min	10 min	16 min

These data are for a person weighing 68 kg (150 pounds).

16 An animal's diet must supply essential nutrients

Besides providing fuel and organic raw materials, an animal's diet must also supply **essential nutrients**. These are materials that must be obtained in preassembled form because the animal's cells cannot make them from *any* raw material. Some nutrients are essential for all animals, whereas others are needed only by certain species. For example, vitamin C is an essential nutrient for humans and other primates, guinea pigs, and some birds and snakes, but most animals can make vitamin C as needed. There are four classes of essential nutrients: essential fatty acids, essential amino acids, vitamins, and minerals.

Essential Fatty Acids Our cells make fats and other lipids by combining fatty acids with other molecules, such as glycerol. We can make most of the fatty acids we need. Those we cannot make, called **essential fatty acids**, must be obtained from our diet. One essential fatty acid, linoleic acid, is especially important because it is needed to make some of the phospholipids of cell membranes. Because seeds, grains, and vegetables generally provide ample amounts of essential fatty acids, deficiencies are rare.

Essential Amino Acids Proteins are built from 20 different kinds of amino acids. Adult humans can make 12 of these amino acids from other compounds. The remaining eight, known as the **essential amino acids**, must be obtained from the diet. Infants also require a ninth, histidine. A deficiency of a single essential amino acid impairs protein synthesis and can lead to protein deficiency.

The simplest way to get all the essential amino acids is to eat meat and animal by-products such as eggs, milk, and cheese. The proteins in these products are said to be "complete," meaning they provide all the essential amino acids in the proportions needed by the human body. In contrast, most plant proteins are "incomplete," or deficient in one or more essential amino acids. If you are vegetarian (by choice, or, as for much of the world's population, by economic necessity), the key to good nutrition is to eat a varied diet of plant proteins that together supply sufficient quantities of all the essential amino acids.

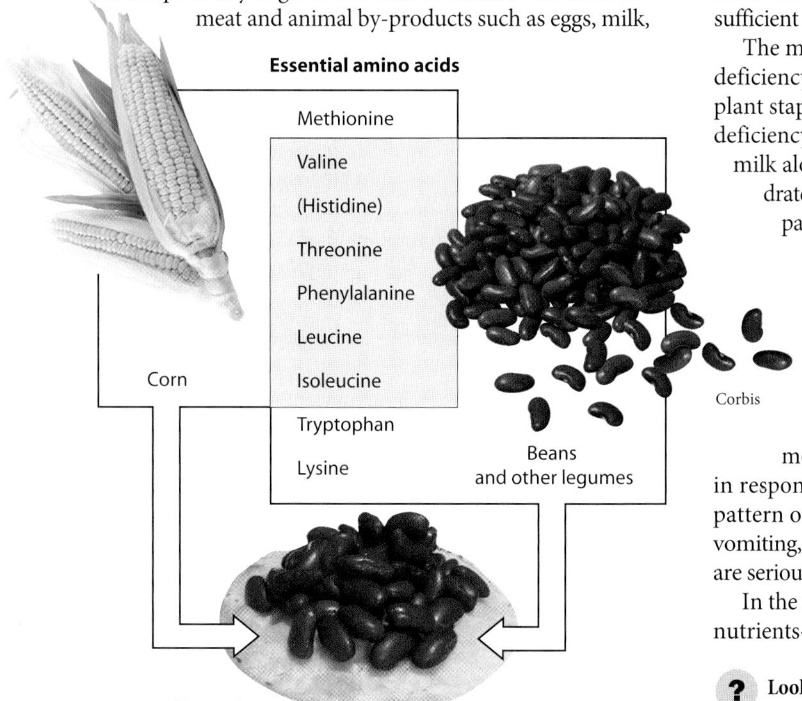

Yasonya/iStockphoto

Corbis

Debbie Hardin

▲ **Figure 16** Essential amino acids from a vegetarian diet

Simply by eating a combination of beans and corn, for example, vegetarians can get all the essential amino acids (**Figure 16**). The combination of a legume (such as beans, peanuts, or soybeans) and a grain often provides the right balance. Most societies have, by trial and error, developed balanced diets that prevent protein deficiency. The Latin American staple of rice and beans is an example.

Dietary Deficiencies A diet that is chronically deficient in calories or lacks one or more essential nutrients results in **malnutrition**. Failing to obtain adequate nutrition can have serious health consequences.

Living in an industrialized country where food is plentiful and most people can afford it, you may find it hard to relate to starvation. But more than 1 billion people in the world do not have enough to eat, and an estimated 4.5 million die of hunger each year, most of them children. Undernutrition, which occurs when diets do not supply sufficient chemical energy, may occur when food supplies are disrupted by drought, wars, or other crises and when severe poverty prevents people from obtaining sufficient food.

The most common type of human malnutrition is protein deficiency resulting from a diet that must depend on a single plant staple—just corn, rice, or potatoes, for instance. Protein deficiency often begins when a child's diet shifts from breast milk alone to food that is mostly starch or other carbohydrates. Such children, if they survive infancy, often have impaired physical and mental development.

Depending on food availability and choices, it is even possible for an overnourished (obese) individual to be malnourished. For example, malnutrition can result from a steady diet of junk food, which offers little nutritional value.

Another cause of undernutrition is anorexia nervosa, an eating disorder in which individuals, most often females, starve themselves compulsively, in response to an intense fear of gaining weight. Bulemia is a pattern of binge eating followed by purging through induced vomiting, abuse of laxatives, or excessive exercise. Both disorders are serious health problems and can even lead to death.

In the next module, we continue our look at essential nutrients—this time vitamins and minerals.

? Look carefully at Figure 16. A diet consisting strictly of corn would probably result in a deficiency of which essential amino acids?

Tryptophan and lysine

17 A healthy human diet includes 13 vitamins and many essential minerals

A **vitamin** is an organic nutrient required in your diet, but only in very small amounts. For example, 1 tablespoon of vitamin B_{12} could provide the daily requirement for nearly a million people. Depending on the vitamin, the required daily amount ranges from about 0.01 to 100 mg. To help you imagine how small these amounts are, consider that a small peanut weighs about 1 g, so 100 mg would be one-tenth of a small peanut. And some vitamin requirements are one-ten-thousandth of that!

Table 17A lists 13 essential vitamins and their major dietary sources. As you can see from the functions in the body and symptoms of deficiency listed in the table, vitamins are absolutely necessary to your health—helping to keep your skin healthy, your nervous system in good working order, and your vision clear, among dozens of other actions.

Vitamins are classified as water-soluble or fat-soluble. Water-soluble vitamins include the B vitamins and vitamin C. Many B vitamins function in the body as coenzymes, enabling the catalytic functions of enzymes that are used over and over in metabolic reactions. Vitamin C is required in the production of connective tissue.

Fat-soluble vitamins include vitamins A, D, E, and K. Vitamin A is a component of the visual pigments in your eyes. Among populations subsisting on simple rice diets, people are often afflicted with vitamin A deficiency, which can cause blindness or death. Vitamin D aids in calcium absorption and bone formation. Your dietary requirement for vitamin D is variable because you synthesize this vitamin from other molecules when your skin is exposed to sunlight. Vitamin E functions as an antioxidant that helps prevent damage to your cells. Vitamin K is necessary for proper blood clotting.

Minerals are simple inorganic nutrients, also required in small amounts—from less than 1 mg to about 2,500 mg per day. Table 17B, on the facing page, lists your daily mineral requirements. As you can see from the table, you need the first seven

TABLE 17A | VITAMIN REQUIREMENTS OF HUMANS

Vitamin	Major Dietary Sources	Functions in the Body	Symptoms of Deficiency
Water-Soluble Vitamins			
Vitamin B_1 (thiamine)	Pork, legumes, peanuts, whole grains  Steve Gorton/Dorling Kindersley	Coenzyme used in removing CO_2 from organic compounds	Beriberi (tingling, poor coordination, reduced heart function)
Vitamin B_2 (riboflavin)	Dairy products, meats, enriched grains, vegetables	Component of coenzyme FAD	Skin lesions, such as cracks at corners of mouth
Vitamin B_3 (niacin)	Nuts, meats, grains	Component of coenzymes NAD^+ and $NADP^+$	Skin and gastrointestinal lesions, delusions, confusion
Vitamin B_5 (pantothenic acid)	Meats, dairy products, whole grains, fruits, vegetables  Chic Type/iStockphoto	Component of coenzyme A	Fatigue, numbness, tingling of hands and feet
Vitamin B_6 (pyridoxine)	Meats, vegetables, whole grains	Coenzyme used in amino acid metabolism	Irritability, convulsions, muscular twitching, anemia
Vitamin B_7 (biotin)	Legumes, other vegetables, meats	Coenzyme in synthesis of fats, glycogen, and amino acids	Scaly skin inflammation, neuromuscular disorders
Vitamin B_9 (folic acid)	Green vegetables, oranges, nuts, legumes, whole grains	Coenzyme in nucleic acid and amino acid metabolism	Anemia, birth defects
Vitamin B_{12} (cobalamin)	Meats, eggs, dairy products	Production of nucleic acids and red blood cells	Anemia, numbness, loss of balance
Vitamin C (ascorbic acid)	Citrus fruits, broccoli, tomatoes  Magnet Creative/iStockphoto	Used in collagen synthesis; antioxidant	Scurvy (degeneration of skin, teeth), delayed wound healing
Fat-Soluble Vitamins			
Vitamin A (retinol)	Dark green and orange vegetables and fruits, dairy products 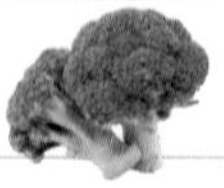 Motorolka/Shutterstock	Component of visual pigments; maintenance of epithelial tissues	Blindness, skin disorders, impaired immunity
Vitamin D	Dairy products, egg yolk	Aids in absorption and use of calcium and phosphorus	Rickets (bone deformities) in children; bone softening in adults
Vitamin E (tocopherol)	Vegetable oils, nuts, seeds  NatUlrick/Shutterstock	Antioxidant; helps prevent damage to cell membranes	Nervous system degeneration
Vitamin K	Green vegetables, tea; also made by colon bacteria	Important in blood clotting	Defective blood clotting

Data from RDA Subcommittee, *Recommended Dietary Allowances* (Washington, D. C.: National Academy Press, 1989); M. E. Shils and V. R. Young, *Modern Nutrition in Health and Disease* (Philadelphia, PA: Lea & Feibiger, 1988)

Steve Gorton/Dorling Kindersley; Chic Type/iStockphoto; Magnet Creative/iStockphoto; Motorolka/Shutterstock; NatUlrick/Shutterstock

minerals in amounts greater than 200 mg per day (about two-tenths of that small peanut). You need the rest in much smaller quantities. The table includes the dietary sources for each mineral, and lists the functions and the symptoms of deficiency for most of these minerals.

Along with other vertebrates, we humans require relatively large amounts of calcium and phosphorus to construct and maintain our skeleton. Too little calcium, especially before the age of 30, can result in the degenerative bone disease osteoporosis. Calcium is also necessary for the normal functioning of nerves and muscles, and phosphorus is an ingredient of ATP and nucleic acids.

Iron is a component of hemoglobin, the oxygen-carrying protein found in your red blood cells. Vertebrates need iodine to make thyroid hormones, which regulate metabolic rate. Worldwide, iodine deficiency is a serious human health problem and is ranked as the leading cause of preventable mental retardation.

Sodium, potassium, and chlorine are important in nerve function and help maintain the osmotic balance of your cells. Most of us ingest far more salt (sodium chloride) than we need. The average U.S. citizen eats enough salt to provide about 20 times the required amount of sodium. Packaged (prepared) foods and most junk foods contain large amounts of sodium chloride, even if they don't taste very salty. Ingesting too much sodium has been associated with high blood pressure and other health risks.

A varied diet usually includes enough vitamins and minerals and is considered the best source of these nutrients. Such diets meet the **Recommended Dietary Allowances (RDAs)**, minimum amounts of nutrients that are needed each day, as determined by a national scientific panel. The U.S. Department of Agriculture makes specific recommendations for certain population groups, such as additional B_{12} for people over age 50, folic acid for pregnant women, and extra vitamin D for people with dark skin (which blocks the synthesis of this vitamin) and for those exposed to insufficient sunlight.

The subject of vitamin dosage has led to heated scientific and popular debate. Some people argue that RDAs are set too low, and some believe, probably mistakenly, that massive doses of vitamins confer health benefits. In general, any excess water-soluble vitamins consumed will be eliminated in urine. But high doses of niacin have been shown to cause liver damage, and large doses of vitamin C can result in gastrointestinal upset. Excessive amounts of fat-soluble vitamins accumulate in body fat. Thus, overdoses may have toxic effects. Excess vitamin A and K are both linked to liver damage. In addition to the dangers of high salt intake, excessive consumption of all the listed minerals may be harmful. For example, in some regions of Africa where the water supply is especially iron-rich, as much as 10% of the population have liver damage as a result of iron overload.

Nevertheless, when we discuss health concerns relating to vitamins and minerals, most of the time we are concerned with deficiencies. Remember that a diet that doesn't include adequate quantities of fresh fruits and vegetables, either as a result of poor food choices or limited supplies or resources, is unlikely to provide the nutrients needed for good health.

? Which of the vitamins and minerals listed in these tables are involved with the formation or maintenance of bones and teeth?

Vitamin C, vitamin D, calcium, phosphorus, and fluorine

TABLE 17B | MINERAL REQUIREMENTS OF HUMANS

	Mineral*	Dietary Sources	Functions in the Body	Symptoms of Deficiency
Greater than 200 mg per day required	Calcium (Ca)	Dairy products, dark green vegetables, legumes	Bone and tooth formation, blood clotting, nerve and muscle function	Impaired growth, loss of bone mass
	Phosphorus (P)	Dairy products, meats, grains	Bone and tooth formation, acid-base balance, nucleotide synthesis	Weakness, loss of minerals from bone, calcium loss
	Sulfur (S)	Proteins from many sources	Component of certain amino acids	Impaired growth, fatigue, swelling
	Potassium (K)	Meats, dairy products, many fruits and vegetables, grains	Acid-base balance, water balance, nerve function	Muscular weakness, paralysis, nausea, heart failure
	Chlorine (Cl)	Table salt	Acid-base balance, water balance, nerve function, formation of gastric juice	Muscle cramps, reduced appetite
	Sodium (Na)	Table salt	Acid-base balance, water balance, nerve function	Muscle cramps, reduced appetite
	Magnesium (Mg)	Whole grains, green leafy vegetables	Enzyme cofactor; ATP bioenergetics	Nervous system disturbances
	Iron (Fe)	Meats, eggs, legumes, whole grains, green leafy vegetables	Component of hemoglobin and of electron carriers; enzyme cofactor	Iron-deficiency anemia, weakness, impaired immunity
	Fluorine (F)	Drinking water, tea, seafood	Maintenance of tooth structure	Higher frequency of tooth decay
	Iodine (I)	Seafood, iodized salt	Component of thyroid hormones	Goiter (enlarged thyroid gland)

*Additional minerals required in trace amounts are chromium (Cr), cobalt (Co), copper (Cu), manganese (Mn), molybdenum (Mo), selenium (Se), and zinc (Zn). All of these minerals, as well as those in the table, are harmful when consumed in excess.

Data from (1) M. E. Shils, "Magnesium," in M. E. Shiels and V. R. Young, eds. *Modern Nutrition in Health and Disease* (Philadelphia, PA: Lea & Feibiger, 1988); (2) V. F. Fairbanks and E. Beutler, "Iron" [same as 1]; (3) N. W. Solomons, "Zinc and Copper" [same as 1]; (4) RDA Subcommittee, *Recommended Dietary Allowances* (Washington, D. C.: National Academy Press, 1989); (5) E. J. Underwood, Trace Elements in Human and Animal Nutrition (New York: Academic Press, 1977).

SCIENTIFIC DISCOVERY

18 Scientists use observations and experiments to determine nutritional needs

How did scientists figure out that to stay healthy we need the vitamins and minerals you just learned about? Many insights into human nutrition have come from *epidemiology*, the study of human health and disease within populations. This type of research often looks for links between health and diet. A classic example involved the high incidence of scurvy in sailors on long sea voyages. This disease could decimate a crew with gradual weakening, loss of teeth, open wounds, and eventual death. In 1747, a British naval surgeon treated scurvy-ridden sailors with citrus fruits and obtained dramatic cures. British sailors were soon provided rations of lime juice on long voyages, earning them the nickname "limeys." The active agent for the cure and prevention of scurvy was later identified as vitamin C.

Another example of nutritional research relates to birth defects. Researchers in the 1970s observed that children born to women of low socioeconomic status were more likely to have neural tube defects. These birth defects occur when tissue fails to enclose the developing brain or spinal cord. The English scientist Richard Smithells thought that malnutrition might be responsible. He designed an experiment in which some women (the experimental group) took a multivitamin starting at least four weeks before conception and other women (the control group) did not take a multivitamin.

As **Table 18** illustrates, the women taking supplements gave birth to fewer children with neural tube defects, providing evidence that vitamin supplementation protects against such defects. In later studies, Smithells obtained evidence that folic acid was the vitamin responsible for this protection, a finding confirmed by other researchers. Based on these studies, the United States in 1998 began to require that folic acid be added to foods such as bread and cereals. Using data from birth defect registries, the Centers for Disease Control and Prevention (CDC) has estimated that neural tube defects have since dropped by 25% in the United States.

Scientists also use animals to study human nutrition, as you'll learn in Module 20. But first let's see what types of nutritional information you can get from food labels.

TABLE 18 RESULTS OF STUDY OF VITAMINS AND NEURAL TUBE DEFECTS

Group	Number of Infants/Fetuses Studied	Infants/Fetuses with a Neural Tube Defect
Vitamin supplements (experimental group)	141	1 (0.7%)
No vitamin supplements (control group)	204	12 (5.9%)

? **What is the difference between an epidemiological/observational study and an experimental study of nutritional needs?**

An epidemiological study looks for correlations between health and diet, which may suggest experiments to test if a certain nutrient affects health.

CONNECTION

19 Food labels provide nutritional information

The U.S. Food and Drug Administration (FDA) requires packaged-food labels to list ingredients from the greatest amount by weight to the least, and to provide "nutrition facts," as shown in **Figure 19**. For selected ingredients, we see the amount per serving and as the percentage of a daily value (requirement or limit) based on a 2,000-kcal diet.

Food labels emphasize nutrients believed to be associated with disease risks (fats, cholesterol, and sodium) and with a healthy diet (dietary fiber, protein, and certain vitamins and minerals). Note that this bread supplies 6% of the daily value of folic acid. FDA regulations change from time to time; for example, trans fat levels must now be listed (see Module 3.8).

Food labels also provide information on total daily needs. For example, less than 20 g of saturated fat and at least 25 g of dietary fiber are recommended for those with a 2,000-kcal daily diet. Reading food labels can help you make informed choices about what you eat.

? **What percentage of the daily requirements for the fat-soluble vitamins is provided by a slice of the bread in Figure 19?**

0%

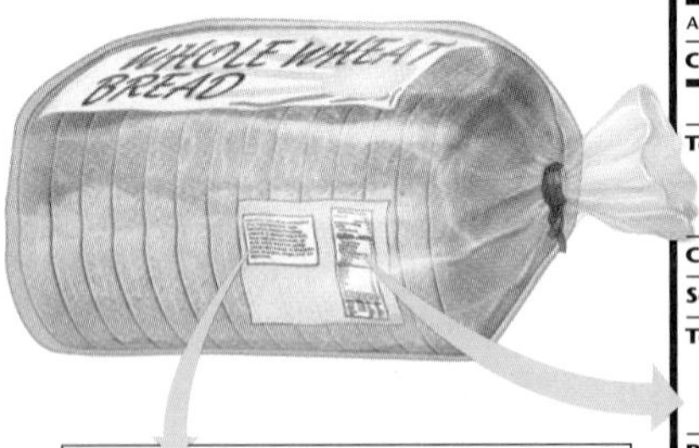

Ingredients: whole wheat flour, water, high fructose corn syrup, wheat gluten, soybean or canola oil, molasses, yeast, salt, cultured whey, vinegar, soy flour, calcium sulfate (source of calcium).

Nutrition Facts

Serving Size 1 slice (43g)
Servings Per Container 16

Amount Per Serving

Calories 100	Calories from Fat 10
	% Daily Value*
Total Fat 1.5g	2%
Saturated Fat 0g	0%
Trans Fat 0g	0%
Cholesterol 0mg	0%
Sodium 190mg	8%
Total Carbohydrate 19g	6%
Dietary Fiber 3g	12%
Sugars 3g	
Protein 4g	

Vitamin A 0%	Vitamin C 0%
Calcium 2%	Iron 4%
Thiamine 6%	Riboflavin 2%
Niacin 6%	Folic Acid 6%

* Percent Daily Values are based on a 2,000 calorie diet. Your daily values may be higher or lower depending on your calorie needs:

	Calories:	2,000	2,500
Total Fat	Less than	65g	80g
Sat. Fat	Less than	20g	25g
Cholesterol	Less than	300mg	300mg
Sodium	Less than	2,400mg	2,400mg
Total Carbohydrate		300g	375g
Dietary Fiber		25g	30g

Calories per gram:
Fat 9 • Carbohydrate 4 • Protein 4

▲ **Figure 19** A whole wheat bread label

EVOLUTION CONNECTION

20 The human health problem of obesity may reflect our evolutionary past

Overnourishment, consuming more food energy than the body needs for normal metabolism, causes **obesity**, the excessive accumulation of fat. The World Health Organization now recognizes obesity as a major global health problem. As noted in the introduction to this chapter, in many countries the increased availability of fattening foods and large portions, combined with more sedentary lifestyles, puts excess weight on bodies. In the United States, the percentage of obese (very overweight) people has doubled to more than 30% in the past two decades, and another 35% are overweight. Weight problems often begin at a young age: About 15% of children and adolescents in the United States are obese; another 17% are overweight.

Obesity contributes to a number of health problems, including type 2 diabetes, cancer of the colon and breast, and cardiovascular disease. Obesity is estimated to be a factor in 300,000 deaths per year in the United States. The results of a 15-year study, published in 2010, indicates that obesity now surpasses smoking in its contribution to disease and the shortening of healthy life spans.

The obesity epidemic has stimulated an increase in scientific research on the causes and possible treatments for weight-control problems. Inheritance is one factor in obesity, which helps explain why certain people have to struggle harder than others to control their weight. Dozens of genes have been identified that code for weight-regulating hormones. Scientists continue to study the signaling pathways that regulate both long-term and short-term appetite and the body's storage of fat. This research gives us reason to be somewhat optimistic that obese people who have inherited defects in these weight-controlling mechanisms may someday be treated with a new generation of drugs. But so far, the diversity of defects in these complex systems has made it difficult to develop drugs that are effective and free from serious side effects.

The complexity of weight control in humans is evident from studies of the hormone leptin, one of the key long-term appetite regulators in mammals. Leptin is produced by adipose (fat) cells (**Figure 20A**). As the amount of adipose tissue increases, leptin levels in the blood rise, which normally cues the brain to suppress appetite. This is one of the feedback mechanisms that usually keep people from becoming obese in spite of access to an abundance of food. Conversely, loss of body fat decreases leptin levels, signaling the brain to increase appetite.

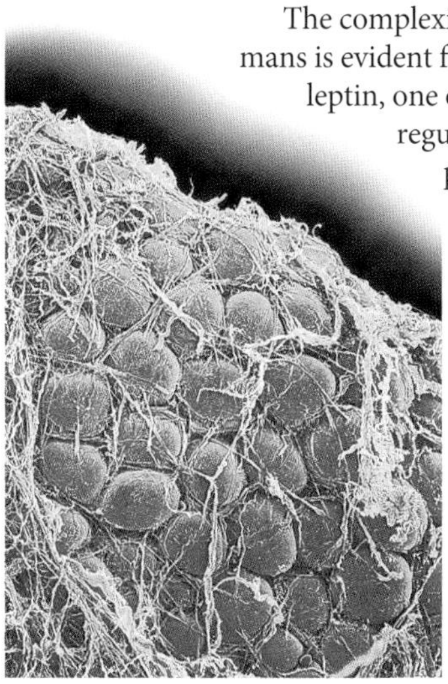

Susumu Nishinaga/SPL/Photo Researchers

▲ **Figure 20A** Fat cells from the abdomen of a human

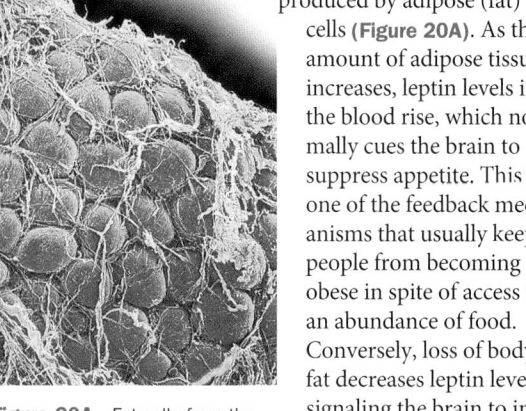

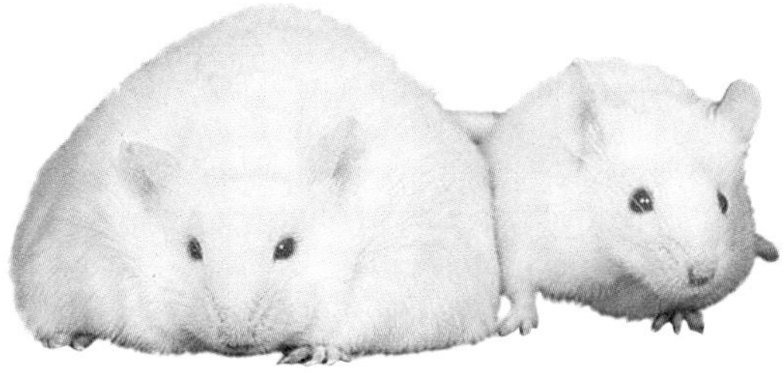

Photo courtesy of The Jackson Laboratory Bar Harbor, Maine

▲ **Figure 20B** A mouse with a defect in a gene for leptin, an appetite-suppressing hormone (left); a normal mouse (right)

Researchers found that mice that inherit a defect in the gene for leptin become very obese (**Figure 20B**). They then discovered that they could treat these obese mice by injecting them with leptin.

The discovery of the leptin-deficiency mutation in mice made headlines and initially generated excitement because humans also have a leptin gene. And indeed, obese children who have inherited a mutant form of the leptin gene do lose weight after leptin treatments. However, relatively few obese people have such deficiencies. In fact, most obese humans have an abnormally high level of leptin, which, after all, is produced by adipose tissue. For some reason, the brain's satiety center does not respond to their high leptin levels—they are leptin resistant. Thus, the search is on for drugs that may reverse this leptin resistance or target some of the other hormonal pathways that suppress appetite.

Some of our current struggles with obesity may be a consequence of our evolutionary history. Most of us crave foods that are fatty: fries, chips, burgers, cheese, and ice cream. Though fat hoarding can be a health liability today, it may actually have been advantageous in our evolutionary past. Only in the past few centuries have large numbers of people had access to a reliable supply of high-calorie food. Our ancestors on the African savanna were hunter-gatherers who probably survived mainly on seeds and other plant products, a diet only occasionally supplemented by hunting game or scavenging meat from animals killed by other predators. In such a feast-and-famine existence, natural selection may have favored those individuals with a physiology that induced them to gorge on rich, fatty foods on those rare occasions when such treats were available. Individuals with genes promoting the storage of fat during feasts may have been more likely than their thinner peers to survive famines.

So perhaps our modern taste for fats and sugars reflects the selective advantage it conveyed in our evolutionary history. Although we know it is unhealthful, many of us find it difficult to overcome the ancient survival behavior of stockpiling for the next famine.

? In what two ways does the hormone leptin regulate appetite? In which of these ways does leptin apparently not function in obese humans?

A drop in leptin due to a loss of adipose tissue stimulates appetite; a high level of leptin, produced by increased body fat, depresses appetite. The second mechanism does not seem to function in some people.

CONNECTION 21 What are the health risks and benefits of weight loss plans?

Is it our evolutionary past or our sedentary lifestyle? Is it super-sized fast food or the addition of high-fructose corn syrup (HFCS) to processed foods? Why are so many people overweight, and how do you know if you are one of them? A standard method of determining healthy weight is body mass index (BMI), a ratio of weight to height (**Figure 21**). A BMI of 25–29 is considered overweight, and above 30 is obese.

Whatever is fueling the dramatic increases in overweight citizens is also igniting our interest in ways to shed body fat. According to some estimates, the U.S. market for weight loss products and services, worth about $30 billion in 1992, has expanded to more than $60 billion a year. But has this huge increase in expenditures bought us thinner, healthier bodies? Not yet.

Thousands of diet books have been written. Magazines, the Internet, even several weight loss reality TV shows bombard us with advice and promises. How can we know which diet plan is fastest, safest, most long-lasting?

In recent years, many popular weight loss schemes have focused on reduced intake of carbohydrates. People following “low-carb” diets often drop sugar, bread, fruits, and potatoes from their diet, swapping in cheese, nuts, and meat instead. Because of the success stories of people who have lost weight and the fatty foods these diets allow, this approach has surged in popularity. Americans spend as much as $15 billion a year on “low-carb” diet aids and foods. Although some studies have found these diets to be effective, others have found that they offer only short-lived benefits. The fatty foods encouraged in such diets may contribute to health problems, and reductions in fruits and vegetables cut a person’s intake of vitamins, minerals, and fiber. As a result, few doctors recommend low-carbohydrate diets as a healthy way to long-term weight loss.

Low-carb diets unseated low-fat diets, an earlier dieting trend with its own flood of low-fat (but often high-sugar) processed foods and attendant health concerns about inadequate fatty acids or protein. Among the many other types of diets are prepackaged meal plans and group programs where dieters attend meetings or join online chats for diet and exercise plans and support.

Losing weight is certainly big business, with some plans costing thousands of dollars. However, an online program from the USDA, called MyPyramid Tracker, is free. It asks you to enter your food intake and physical activity daily. The program then analyzes your nutrient and energy intake as well as kilocalories expended by your activities. An energy balance summary indicates the weight gain or loss you can expect from your data.

Clynt Garnham Lifestyle/Alamy

Some severely obese individuals may be candidates for weight loss surgery. Gastric bypass surgery, which reduces the size of the stomach and the length of the small intestine, is an increasingly popular weight loss solution, with about 150,000 operations performed a year. This surgery limits food intake capacity and nutrient absorption. Although it has documented risks, studies show that gastric bypass surgery reduces some obesity-related health risks. A new lap band surgery involves an adjustable ring placed around the upper part of the stomach, which restricts the amount of food that can be eaten at one time. This laparoscopic procedure is less invasive and reversible, and its use is increasing.

Scientific studies of weight loss diets indicate that sustainability is the major shortcoming of all diets. There appears to be no silver bullet for losing weight and keeping it off without lifestyle changes. These changes involve a combination of increased exercise and a restricted but balanced diet that provides at least 1,200 kcal per day and adequate amounts of all essential nutrients. Such a combination can trim the body gradually and keep the extra pounds off.

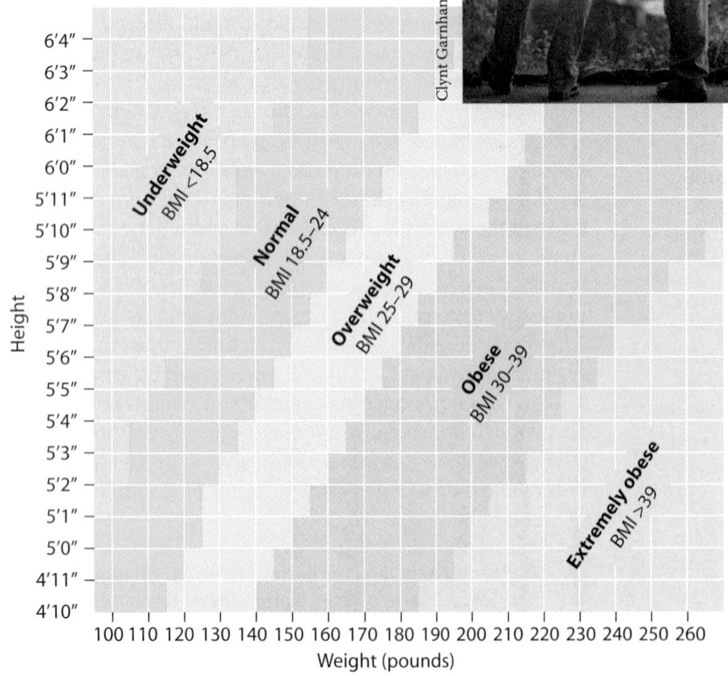

▲ **Figure 21** Body mass index (BMI): one measure of healthy weight

? In what sense is maintaining a stable body weight a matter of caloric bookkeeping?

● When your metabolism burns as many kilocalories a day as you take in with your food, a stable body weight will result.

Data adapted from *Clinical Guidelines on the Identification, Evaluation, and Treatment of Overweight and Obesity in Adults: The Evidence Report.* Data found at the National Heart, Lung, and Blood Institute: http://www.nhlbi.nih.gov/guidelines/obesity/bmi_tbl.htm.

CONNECTION 22 Diet can influence risk of cardiovascular disease and cancer

Food influences far more than your size and appearance. Diet also plays an important role in a person's risk of developing serious illnesses, including cardiovascular disease and cancer. Some risk factors associated with cardiovascular disease, such as family history, are unavoidable, but others, such as smoking and lack of exercise, you can influence through your behavior. Diet is another behavioral factor that affects cardiovascular health. For instance, a diet rich in saturated fats is linked to high blood cholesterol, which in turn is linked to cardiovascular disease.

Cholesterol travels through the body in particles made up of thousands of molecules of cholesterol and other lipids bound to a protein. High blood levels of one type of particle called **low-density lipoproteins (LDLs)** generally correlate with a tendency to develop blocked blood vessels, high blood pressure, and consequent heart attacks. In contrast to LDLs, cholesterol particles called **high-density lipoproteins (HDLs)** may decrease the risk of vessel blockage, perhaps because HDLs convey excess cholesterol to the liver, where it is broken down. Some research indicates that reducing LDLs while maintaining or increasing HDLs lowers the risk of cardiovascular disease. How do you increase your levels of "good" cholesterol? You can exercise more, which tends to increase HDL levels. And you can abstain from smoking, because smoking has been shown to lower HDL levels.

How do you decrease your levels of "bad" cholesterol? You can avoid a diet high in saturated fats, which tend to increase LDL levels. Saturated fats (see Module 3.8) are found in eggs, butter, and most meats. Saturated fats are also found in artificially saturated ("hydrogenated") vegetable oils. The hydrogenation process, which solidifies vegetable oils, also produces a type of fat called trans fat. Trans fats tend not only to increase LDL levels but also to lower HDL levels, a two-pronged attack on cardiovascular health. By contrast, eating mainly unsaturated fats, such as found in fatty fish like salmon, certain nuts, and most liquid vegetable oils, including corn, soybean, and olive oils, tends to lower LDL levels and raise HDL levels. These oils are also important sources of vitamin E, whose antioxidant effect may help prevent blood vessel blockage, and omega-3 fatty acids, which appear to protect against cardiovascular disease.

Diet also seems to influence our risk for certain cancers. Some research suggests a link between diets heavy in fats or carbohydrates and the incidence of breast cancer. The incidence of colon cancer and prostate cancer may be linked to a diet rich in saturated fat or red meat. Other foods may help fight cancer. For example, some fruits and vegetables (**Figure 22**) are rich in antioxidants, chemicals that help protect cells from damaging molecules known as free radicals. Antioxidants may help prevent cancer, although this link is still debated by scientists. Foods that are particularly high in antioxidants include berries, beans, nuts, and dried fruit.

Despite the progress researchers have made in studying nutrition and health, it is often difficult to design controlled experiments that establish the link between the two. Experiments that may damage participants' health are clearly unethical. Some studies rely on self-reported food intake, and the accuracy of participants' memories may influence the outcome. As you learned in Module 18, scientists often do studies that correlate certain health characteristics with groups that have particular diets or lifestyles. For example, many French people eat high-fat diets and drink wine, yet have lower rates of obesity and heart disease than do Americans. When researchers notice apparent contradictions like these, they attempt to control for other variables and isolate the factors responsible for such observations, such as the fact that the French eat smaller portions; eat more unprocessed, fresh foods; and snack infrequently.

Steve Allen/Brand X Pictures/Jupiter Images

▲ **Figure 22** Foods that contribute to good health

Even with large, controlled intervention trials, results may be contradictory or inconclusive. For instance, an 8-year study of almost 49,000 postmenopausal women found that low-fat diets failed to reduce the risk of breast and colon cancer and did not affect the incidence of cardiovascular disease. LDL and cholesterol levels did decrease slightly in the low-fat group, however.

The relationship between foods and health is complex, and we have much to learn. The American Cancer Society (ACS) suggests that following the dietary guidelines in **Table 22**, in combination with physical activity, can help lower cancer risk. The ACS's main recommendation is to "eat a variety of healthful foods, with an emphasis on plant sources."

? If you are trying to minimize the damaging effects of blood cholesterol on your cardiovascular system, your goal is to ____ your LDLs and ____ your HDLs.

decrease . . . increase

TABLE 22 DIETARY GUIDELINES FOR REDUCING CANCER RISK

Maintain a healthy weight throughout life.
Eat five or more servings of a variety of fruits and vegetables daily.
Choose whole grains over processed (refined) grains.
Limit consumption of processed and red meats.
If you drink alcoholic beverages, limit yourself to a maximum of one or two drinks a day (a drink = 12 ounces of beer, 5 ounces of wine, or 1.5 ounces of 80% distilled spirits).

CHAPTER 21 REVIEW

For Practice Quizzes, BioFlix, MP3 Tutors, and Activities, go to www.masteringbiology.com.

Reviewing the Concepts

Obtaining and Processing Food (1–3)

1 Animals obtain and ingest their food in a variety of ways. Animals may be herbivores, carnivores, or omnivores and may obtain food by suspension, substrate, fluid, or bulk feeding.

2 Overview: Food processing occurs in four stages. The stages are ingestion, digestion, absorption, and elimination.

3 Digestion occurs in specialized compartments. Food may be digested in food vacuoles, gastrovascular cavities, or alimentary canals, which run from mouth to anus with specialized regions.

The Human Digestive System (4–13)

4 The human digestive system consists of an alimentary canal and accessory glands. The rhythmic muscle contractions of peristalsis squeeze food along the alimentary canal.

5 Digestion begins in the oral cavity. The teeth break up food, saliva moistens it, and an enzyme in saliva begins the hydrolysis of starch. The tongue pushes the bolus of food into the pharynx.

6 After swallowing, peristalsis moves food through the esophagus to the stomach. The swallowing reflex moves food into the esophagus and keeps it out of the trachea.

7 The Heimlich maneuver can save lives. This procedure can dislodge food from the pharynx or trachea during choking.

8 The stomach stores food and breaks it down with acid and enzymes. Pepsin in gastric juice begins to digest protein.

9 Digestive ailments include acid reflux and gastric ulcers.

10 The small intestine is the major organ of chemical digestion and nutrient absorption. Enzymes from the pancreas and cells of the intestinal wall digest food molecules. Bile, made in the liver and stored in the gallbladder, emulsifies fat for attack by enzymes. Folds of the intestinal lining and finger-like villi (with microscopic microvilli) increase the area across which absorbed nutrients move into capillaries and lymph vessels.

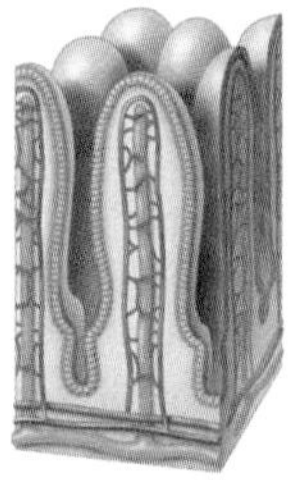

11 One of the liver's many functions is processing nutrient-laden blood from the intestines. The liver regulates nutrient levels in the blood, produces bile, detoxifies alcohol and drugs, and synthesizes blood proteins.

12 The large intestine reclaims water and compacts the feces. Some bacteria in the colon produce vitamins. Feces are stored in the rectum before elimination.

13 Evolutionary adaptations of vertebrate digestive systems relate to diet. Herbivores may have longer alimentary canals than carnivores and compartments that house cellulose-digesting microbes.

Nutrition (14–22)

14 Overview: An animal's diet must satisfy three needs. The diet must provide chemical energy, raw materials for biosynthesis, and essential nutrients.

15 Chemical energy powers the body. Metabolic rate, the rate of energy consumption, includes the basal metabolic rate (BMR) plus the energy used for other activities.

16 An animal's diet must supply essential nutrients. Essential fatty acids are easily obtained from the diet. The eight essential amino acids can be obtained from animal protein or the proper combination of plant foods. Malnutrition results from a diet lacking in sufficient calories or essential nutrients.

17 A healthy human diet includes 13 vitamins and many essential minerals. Most vitamins function as coenzymes. Minerals are inorganic nutrients that play a variety of roles. A varied diet usually meets the RDAs for these nutrients.

18 Scientists use observations and experiments to determine nutritional needs. Epidemiology relates diets to health characteristics in populations. Controlled experiments can identify essential nutrients.

19 Food labels provide nutritional information.

20 The human health problem of obesity may reflect our evolutionary past. The dramatic rise in obesity is linked to a lack of exercise and abundance of fattening foods and may partly stem from an evolutionary advantage of fat hoarding.

21 What are the health risks and benefits of weight loss plans? Diets that restrict important nutrients may help individuals lose weight but have health risks.

22 Diet can influence risk of cardiovascular disease and cancer. The ratio of HDLs to LDLs is influenced by diet.

Connecting the Concepts

1. Label the parts of the human digestive system below and indicate the functions of these organs and glands.

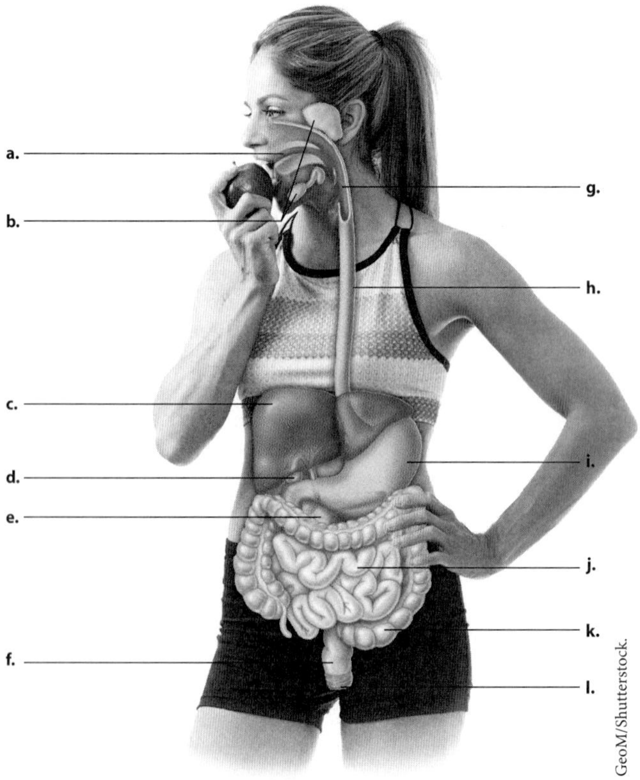

GeoM/Shutterstock.

2. Complete the following map summarizing the nutritional needs of animals that are met by a healthy diet.

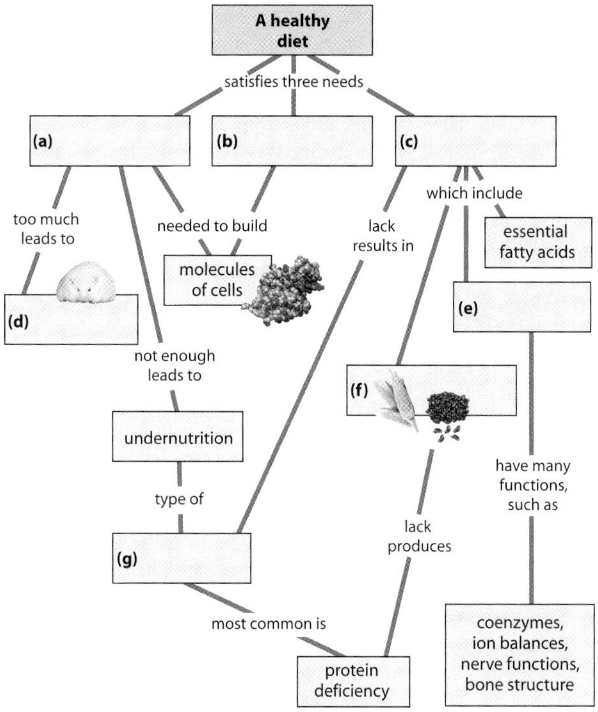

Testing Your Knowledge

Multiple Choice

3. Earthworms, which are substrate feeders,
   a. feed mostly on mineral substrates.
   b. filter small organisms from the soil.
   c. are bulk feeders.
   d. are herbivores that eat autotrophs.
   e. eat their way through the soil, feeding on partially decayed organic matter.
4. The energy content of fats
   a. is released by bile salts.
   b. may be lost unless an herbivore eats some of its feces.
   c. is, per gram, twice that of carbohydrates or proteins.
   d. can reverse the effects of undernutrition.
   e. Both c and d are correct.
5. Which of the following statements is false?
   a. A healthy human has enough stored fat to supply calories for several weeks.
   b. An increase in leptin levels leads to an increase in appetite and weight gain.
   c. The interconversion of glucose and glycogen takes place in the liver.
   d. After glycogen stores are filled, excessive calories are stored as fat, regardless of their original food source.
   e. Carbohydrates and fats are normally used as fuel before proteins are used.
6. Which of the following is mismatched with its function?
   a. most B vitamins—coenzymes
   b. vitamin E—antioxidant
   c. vitamin K—blood clotting
   d. iron—component of thyroid hormones
   e. phosphorus—bone formation, nucleotide synthesis
7. Why is it necessary for healthy vegetarians to combine different plant foods or eat some eggs or milk products?
   a. to make sure they obtain sufficient calories
   b. to provide sufficient vitamins
   c. to make sure they ingest all essential fatty acids
   d. to make their diet more interesting
   e. to provide all essential amino acids for protein synthesis

Describing, Comparing, and Explaining

8. A peanut butter and jelly sandwich contains carbohydrates, proteins, and fats. Describe what happens to the sandwich when you eat it. Discuss ingestion, digestion, absorption, and elimination.

Applying the Concepts

9. How might our craving for fatty foods, which is helping to fuel the obesity crisis, have evolved through natural selection?
10. Use the Nutrition Facts label to the right to answer these questions:
    a. What percentage of the total Calories in this product is from fat?
    b. Is this product a good source of vitamin A and calcium? Explain.
    c. Each gram of fat supplies 9 Calories. Based on the grams of saturated fat and its % Daily Value, calculate the upper limit of saturated fat (in grams and Calories) that an individual on a 2,000-Calorie/day diet should consume.

Nutrition Facts	
Serving Size 1/2 Cup (83g)	
Servings Per Container 8	
Amount Per Serving	
Calories 190	Calories from Fat 110
	% Daily Value*
Total Fat 12g	18%
Saturated Fat 8g	40%
Trans Fat 0g	0%
Cholesterol 45mg	15%
Sodium 75mg	3%
Total Carbohydrate 18g	6%
Dietary Fiber 0g	0%
Sugars 17g	
Protein 3g	
Vitamin A 10%	Vitamin C 8%
Calcium 10%	Iron 0%
*Percent Daily Values (DV) are based on a 2,000 calorie diet.	

11. One common piece of dieting advice is to replace energy-dense food with nutrient-dense food. What does this mean?
12. The media report numerous claims and counterclaims about the benefits and dangers of certain foods, dietary supplements, and diets. Have you modified your eating habits on the basis of nutritional information disseminated by the media? Why or why not? How should we evaluate whether such nutritional claims are valid?
13. It is estimated that 15% of Americans do not always have access to enough food. Worldwide, more than 1 billion people go to bed hungry most nights, and millions of people have starved to death in recent decades. In some cases, war, poor crop yields, and disease epidemics strip people of food. Many say instead that it is not inadequate food production but unequal food distribution that causes food shortages. What responsibility do nations have for feeding their citizens? For feeding the people of other countries? What do you think you can do to lessen world hunger?

Review Answers

1. a. oral cavity—ingests and chews food; b. salivary glands—produce saliva; c. liver—produces bile and processes nutrient-laden blood from intestines; d. gallbladder—stores bile; e. pancreas—produces digestive enzymes and bicarbonate; f. rectum—stores feces before elimination; g. pharynx—site of openings into esophagus and trachea; h. esophagus—transports bolus to stomach by peristalsis; i. stomach—stores food, mixes food with acid, begins digestion of proteins; j. small intestine—digestion and absorption; k. large intestine—absorbs water, compacts feces; l. anus—eliminates feces
2. a. fuel, chemical energy; b. raw materials, monomers; c. essential nutrients; d. overnutrition or obesity; e. vitamins and minerals; f. essential amino acids; g. malnutrition
3. e 4. e 5. b 6. d 7. e
8. You ingest the sandwich one bite at a time. In the oral cavity, chewing begins mechanical digestion, and salivary amylase action on starch begins chemical digestion. When you swallow, food passes through the pharynx and esophagus to the stomach. Mechanical and chemical digestion continues in the stomach, where HCl in gastric juice breaks apart food cells and pepsin begins protein digestion. In the small intestine, enzymes from the pancreas and intestinal wall break down starch, protein, and nucleic acids to monomers. Bile from the liver and gallbladder emulsifies fat droplets for attack by enzymes. Most nutrients are absorbed into the bloodstream through the villi of the small intestine. Fats travel through lymph vessels. In the large intestine, absorption of water is completed, and undigested material and intestinal bacteria are compacted into feces, which are eliminated through the anus.
9. Our craving for fatty foods may have evolved from the feast-and-famine existence of our ancestors. Natural selection may have favored individuals who gorged on and stored high-energy molecules, as they were more likely to survive famines.
10. a. 58% (110/190)
    b. Based on a 2,000-Calorie diet, this product supplies about 9.5% of daily Calories, and it supplies 10% of vitamin A and calcium. If all food consumed supplied a similar quantity, the daily requirement for these two nutrients would be met.
    c. The 8 g of saturated fat in this product represents 40% of the daily value. Thus, the daily value must be 20 g (8/0.4 = 20). This represents 180 Calories from saturated fat per day.
11. Sodas, chips, cookies, and candy provide many calories (high energy) but few vitamins, minerals, proteins, or other nutrients. Unprocessed, fresh foods such as fruits and vegetables are considered nutrient dense; they provide substantial amounts of vitamins, minerals, and other nutrients and relatively few calories.
12. Some issues and questions to consider: What are the roles of family, school, advertising, media, and government in providing nutritional information? How might the available information be improved? What types of scientific studies form the foundation of various nutritional claims?
13. Some issues and questions to consider: In wealthy countries, what are the factors that make it difficult for some people to get enough food? In your community, what types of help exist to feed hungry people? Think of two recent food crises in other countries and what caused them. Did other countries or international organizations provide aid? Which ones, and how did they help? Did that aid address the underlying causes of malnutrition and starvation in the stricken area or only provide temporary relief? How might that aid be changed to offer more permanent solutions to food shortages?

Gas Exchange

Gas Exchange

BIG IDEAS

Mechanisms of Gas Exchange
(1–5)

Gas exchange occurs across thin, moist surfaces in respiratory organs such as gills, tracheal systems, and lungs.

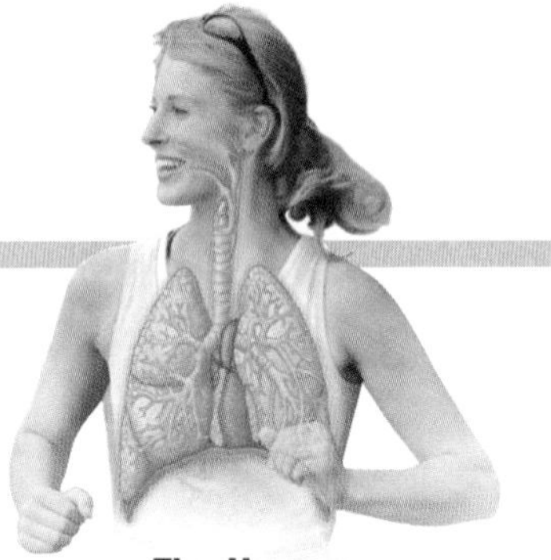

Douglas Pulsipher/Alamy

The Human Respiratory System
(6–9)

Air travels through branching tubes to the lungs, where gases are exchanged with the blood.

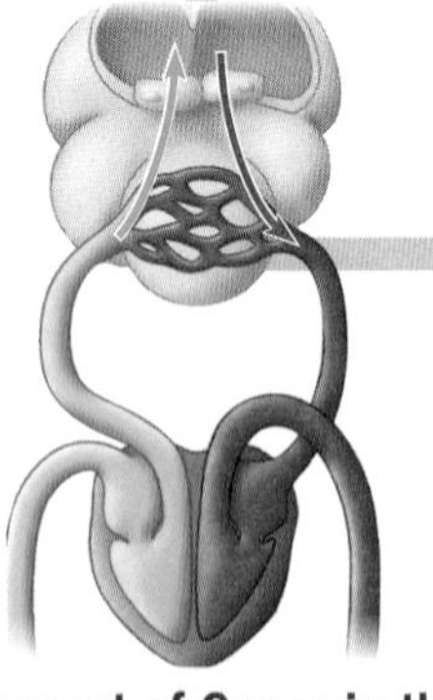

Transport of Gases in the Human Body
(10–12)

The circulatory system transports O_2 to body tissues and returns CO_2 to the lungs.

John Holmes/FLPA

These bar-headed geese are flying over the highest mountains in the world, the Himalayas, on their migration between winter quarters in India and summer breeding grounds in Russia. These same peaks, however, have claimed the lives of some of the world's most experienced mountain climbers—the journey into thin air can weaken muscles, cloud minds, and sometimes fill lungs with fluid. The air at the top of Mount Everest is so low in oxygen that most people would quickly pass out if exposed to it.

So how do these birds manage to fly at such heights? One factor is the efficiency of their lungs. In addition, their blood contains a form of hemoglobin that has a very high affinity for oxygen. Their circulatory system has a large number of capillaries that provide oxygen-rich blood to their flight muscles, and the muscles pack a lot of myoglobin, a protein that stores a ready supply of oxygen. These adaptations allow these high-flying birds to travel even where oxygen is in very short supply.

Without O_2, the metabolic machinery that releases energy from food molecules shuts down. It is the continuous supply of O_2 to body cells that makes the difference between life and death in all the environments that animals inhabit—water, land, and the thin air of the Himalayas.

Respiratory systems provide for the exchange of O_2 and the waste product CO_2 between an animal and its environment. In this chapter, we explore the various types of gas exchange systems that have evolved in animals. We then take a closer look at the structures and functions of the human respiratory system. We conclude with a preview of the circulatory system, which delivers the oxygen essential for life to all body cells.

Mechanisms of Gas Exchange

1 Overview: Gas exchange in humans involves breathing, transport of gases, and exchange with body cells

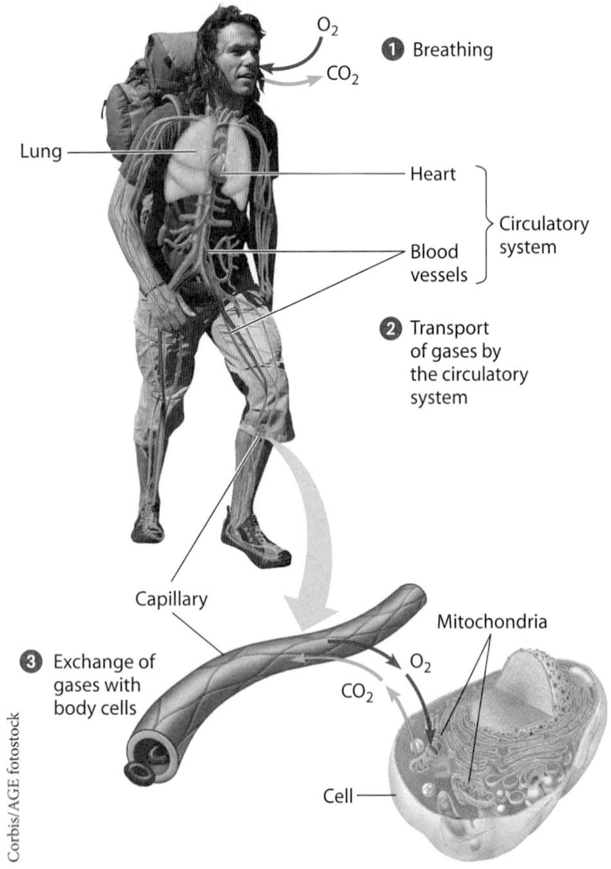

▲ **Figure 1** The three phases of gas exchange in a human

Gas exchange makes it possible for you to put to work the food molecules the digestive system provides. **Figure 1** presents an overview of the three phases of gas exchange in humans and other animals with lungs. 1 Breathing: As you inhale, a large, moist internal surface is exposed to the air entering the lungs. Oxygen (O_2) diffuses across the cells lining the lungs and into surrounding blood vessels. At the same time, carbon dioxide (CO_2) diffuses from the blood into the lungs. As you exhale, CO_2 leaves your body.

2 Transport of gases by the circulatory system: The O_2 that diffused into the blood attaches to hemoglobin in red blood cells. The red vessels in the figure are transporting O_2-rich blood from the lungs to capillaries in the body's tissues. CO_2 is also transported in blood, from the tissues back to the lungs, carried in the blue vessels shown here.

3 Exchange of gases with body cells: Your cells take up O_2 from the blood and release CO_2 to the blood. O_2 functions in cellular respiration in the mitochondria as the final electron acceptor in the stepwise breakdown of fuel molecules. H_2O and CO_2 are waste products, and ATP is produced that will power cellular work. The gas exchange occurring as we breathe is often called respiration; do not confuse this exchange with cellular respiration.

Cellular respiration requires a continuous supply of O_2 and the disposal of CO_2. Gas exchange involves both the respiratory and circulatory systems in servicing your body's cells.

? Humans cannot survive for more than a few minutes without O_2. Why?

Cells require a steady supply of O_2 for cellular respiration to produce enough ATP to function. Without enough ATP, cells and the organism die.

2 Animals exchange O_2 and CO_2 across moist body surfaces

The part of an animal's body where gas exchange with the environment occurs is called the respiratory surface. Respiratory surfaces are made up of living cells, and like all cells, their plasma membranes must be wet to function properly. Thus, respiratory surfaces are always moist.

Gas exchange takes place by diffusion. The surface area of the respiratory surface must be large enough to take up sufficient O_2 for every cell in the body. Usually, a single layer of cells forms the respiratory surface. This thin, moist layer allows O_2 to diffuse rapidly into the circulatory system or directly into body tissues and also allows CO_2 to diffuse out.

The four figures on the facing page illustrate, in simplified form, four types of respiratory organs, structures in which gas exchange with the external environment occurs. In each of these figures, the circle represents a cross section of the animal's body through the respiratory surface. The yellow areas represent the respiratory surfaces; the green outer circles represent body surfaces with little or no role in gas exchange. The boxed enlargements show gas exchange occurring across the respiratory surface.

Some animals use their entire outer skin as a gas exchange organ. The earthworm in **Figure 2A** is an example. The cross-sectional diagram shows its whole body surface as yellow; there are no specialized gas exchange surfaces. Oxygen diffuses into a dense network of thin-walled capillaries lying just beneath the skin. Earthworms and other skin-breathers must live in damp places or in water because their whole body surface has to stay moist. Animals that breathe only through their skin are

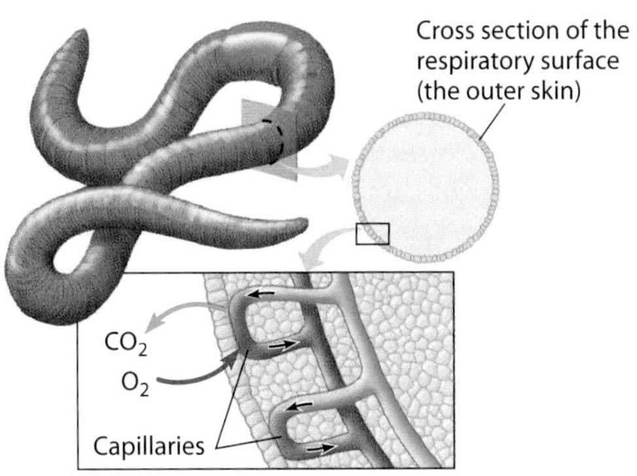

▲ **Figure 2A** The skin: the outer body surface

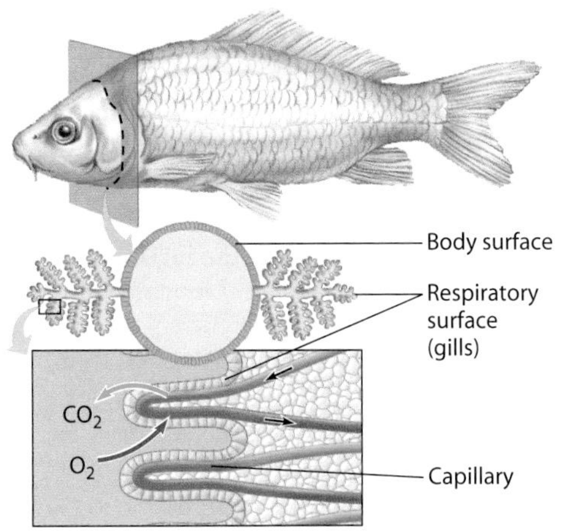

▲ **Figure 2B** Gills: extensions of the body surface

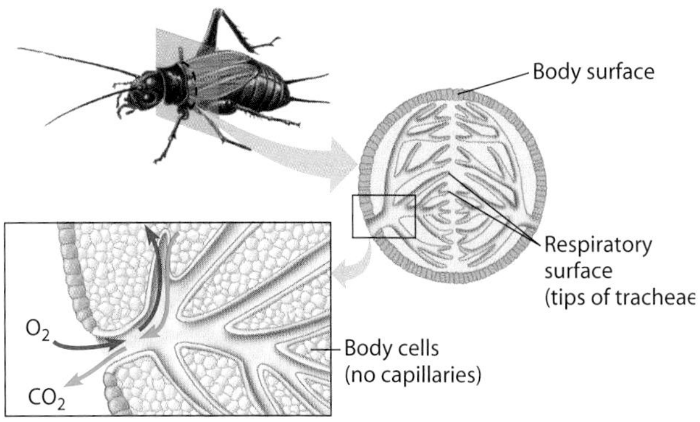

▲ **Figure 2C** A tracheal system: air tubes that extend throughout the body

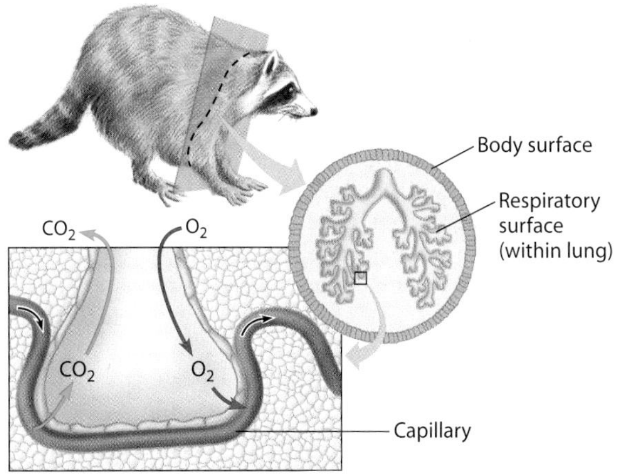

▲ **Figure 2D** Lungs: internal thin-walled sacs

generally small, and many are long and thin or flattened. These shapes provide a high ratio of respiratory surface to body volume, allowing for sufficient gas exchange for all the cells in the body.

In most animals, the skin surface is not extensive enough to exchange gases for the whole body. Consequently, certain parts of the body have become adapted as highly branched respiratory surfaces with large surface areas. Such gas exchange organs include gills, tracheal systems, and lungs.

Gills have evolved in most aquatic animals. **Gills** are extensions, or outfoldings, of the body surface specialized for gas exchange. Many marine worms have flap-like gills that extend from each body segment. The gills of clams and crayfish are clustered in one body location. A fish **(Figure 2B)** has a set of feather-like gills on each side of its head. As indicated in the enlargement, gases diffuse across the gill surface between the water and the blood. Because the respiratory surfaces of aquatic animals extend into the surrounding water, keeping the surface moist is not a problem.

In most terrestrial animals, the respiratory surface is folded into the body rather than projecting from it. The infolded surface opens to the air only through narrow tubes, an arrangement that helps retain the moisture that is essential for the cells of the respiratory surfaces to function.

The **tracheal system** of insects **(Figure 2C)** is an extensive system of branching internal tubes called tracheae, with a moist, thin epithelium forming the respiratory surface at their tips. As you will see in Module 4, the smallest branches exchange gases directly with body cells. Thus, gas exchange in insects requires no assistance from the circulatory system.

Most terrestrial vertebrates have **lungs (Figure 2D)**, which are internal sacs lined with moist epithelium. As the diagram indicates, the inner surfaces of the lungs are extensively subdivided, forming a large respiratory surface. Gases are carried between the lungs and the body cells by the circulatory system.

We examine gills, tracheae, and lungs more closely in the next several modules.

? How does the structure of the respiratory surface of a gill or lung fit its function?

These respiratory surfaces are moist and thin so that gases can easily diffuse across them and into or out of the closely associated capillaries. They are highly branched or subdivided, providing a large surface area for exchange.

3 Gills are adapted for gas exchange in aquatic environments

Oceans, lakes, and other bodies of water contain O_2 as a dissolved gas. The gills of fishes and many invertebrate animals, including lobsters and clams, tap this source of O_2.

An advantage of exchanging gases in water is that there is no problem keeping the respiratory surface moist. A disadvantage, however, is that the concentration of oxygen dissolved in water is only about 3% of that in an equivalent volume of air. And the warmer and saltier the water, the less O_2 it holds. Thus, gills—especially those of large, active fishes and squids in warm oceans—must be very efficient to obtain enough O_2 from the surrounding water.

Structure of Fish Gills The drawings in **Figure 3** show the architecture of fish gills, which are among the most efficient gas exchange organs in the aquatic world. There are four supporting gill arches on each side of the body. Two rows of gill filaments project from each gill arch. Each filament bears many flattened plates called lamellae (singular, lamella), which are the actual respiratory surfaces. A lamella is full of tiny capillaries that are so narrow that blood cells pass through them in single file. Thus, every red blood cell comes in close contact with O_2 dissolved in the surrounding water.

What you can't see in the drawings are the movements that ventilate the gills. We use the term **ventilation** to refer to any mechanism that increases the flow of water or air over the respiratory surface (in gills, tracheae, or lungs). Increasing this flow ensures a fresh supply of O_2 and the removal of CO_2. Blue arrows in the drawings represent the one-way flow of water into the mouth, across the gills, and out the side of the fish's body. Swimming fish simply open their mouths and let water flow over their gills. Fish also pump water across the gills by the coordinated opening and closing of the mouth and operculum, the stiff flap that covers and protects the gills. Because water is dense and contains so little oxygen, most fish expend considerable energy in ventilating their gills.

Countercurrent Exchange The flow of blood inside a gill in a direction opposite the movement of water past the gill makes it possible for oxygen to diffuse into the blood by an efficient process called **countercurrent exchange**. The name reflects the fact that substances (or heat) are exchanged between two fluids that are moving *counter* to each other. Let's see how this arrangement enhances the exchange of gases in a fish gill.

In the enlargement on the top right of Figure 3, notice that the direction that water flows over the surface of the lamellae (blue arrows) is opposite that of the blood flow within each lamella (black arrows). The countercurrent exchange diagram in the lower right illustrates the transfer of O_2 from water to

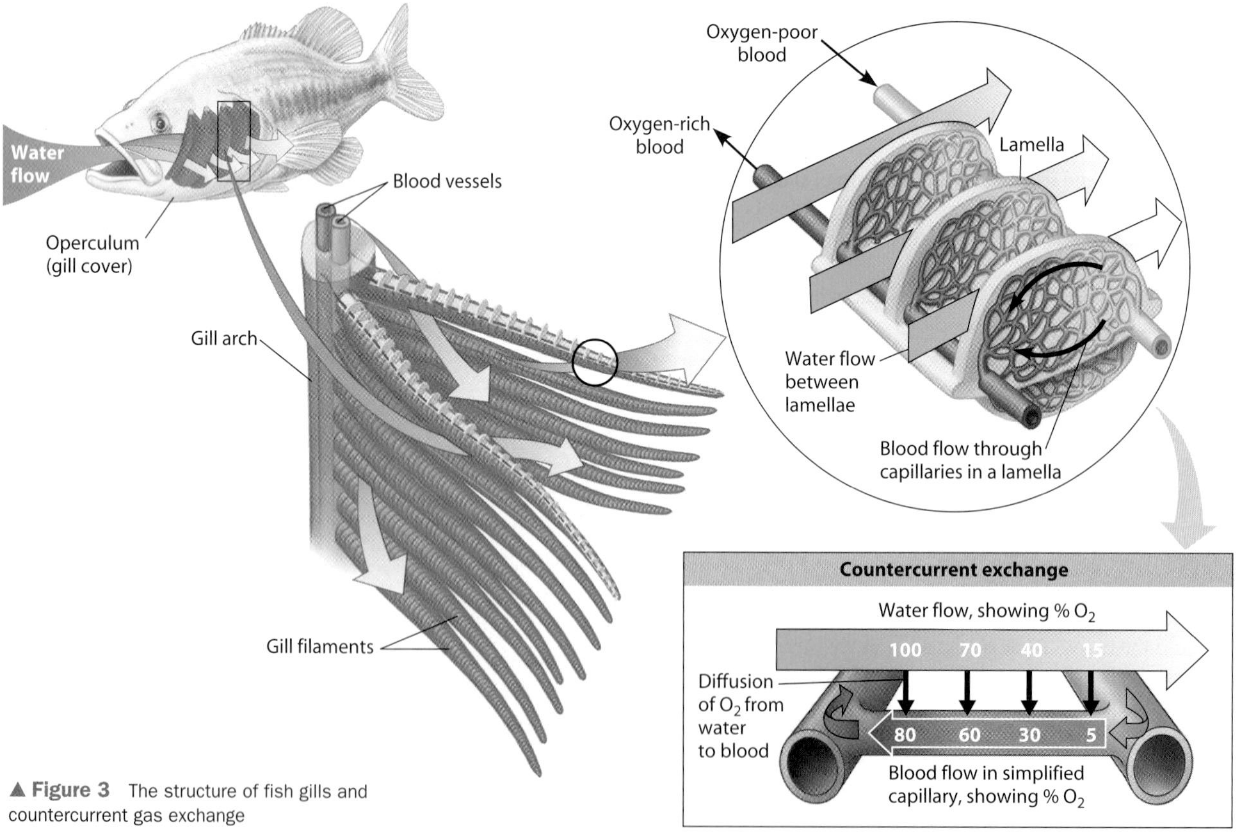

▲ **Figure 3** The structure of fish gills and countercurrent gas exchange

blood. The percentages shown indicate the changing amount of O_2 dissolved in each fluid. Notice that as blood flows through a lamella and picks up more and more O_2, it comes in contact with water that is closer to beginning its passage over the gills and thus has more O_2 available. As a result, a concentration gradient is maintained that favors the diffusion of O_2 from the water to the blood along the entire length of the capillary.

This countercurrent exchange mechanism is so efficient that fish gills can remove more than 80% of the O_2 dissolved in the water flowing through them. The mechanism of countercurrent exchange is also important in temperature regulation,

No matter how efficient they may be, gills are unsuitable for an animal living on land. An expansive surface of wet membrane extending out from the body and exposed to air would lose too much water to evaporation. Therefore, most terrestrial animals house their respiratory surfaces within the body, opening to the atmosphere through narrow tubes, as we see next.

? What would be the maximum percentage of the water's O_2 a gill could extract if its blood flowed in the same direction as the water instead of counter to it? (This is a challenging one! It may help to sketch it out.)

50%. As O_2 diffuses from the water into the blood as they flow in the same direction, the concentration gradient becomes less and less steep, until there is an equal amount of O_2 in both, and O_2 can no longer diffuse from water to blood.

4 The tracheal system of insects provides direct exchange between the air and body cells

There are two big advantages to breathing air: Air contains a much higher concentration of O_2 than does water, and air is much lighter and easier to move than water. Thus, a terrestrial animal expends much less energy than an aquatic animal ventilating its respiratory surface. The main problem facing an air-breathing animal, however, is the loss of water to the air by evaporation.

The tracheal system of insects, with respiratory surfaces at the tips of tiny branching tubes inside the body, greatly reduces evaporative water loss. **Figure 4A** illustrates the tracheal system in a grasshopper. The largest tubes, called tracheae, open to the outside and are reinforced by rings of chitin, as shown in the blowup on the bottom right of the figure. (An insect's tough exoskeleton is also made of chitin.) Enlarged portions of tracheae form air sacs (shown in pink) near organs that require a large supply of O_2.

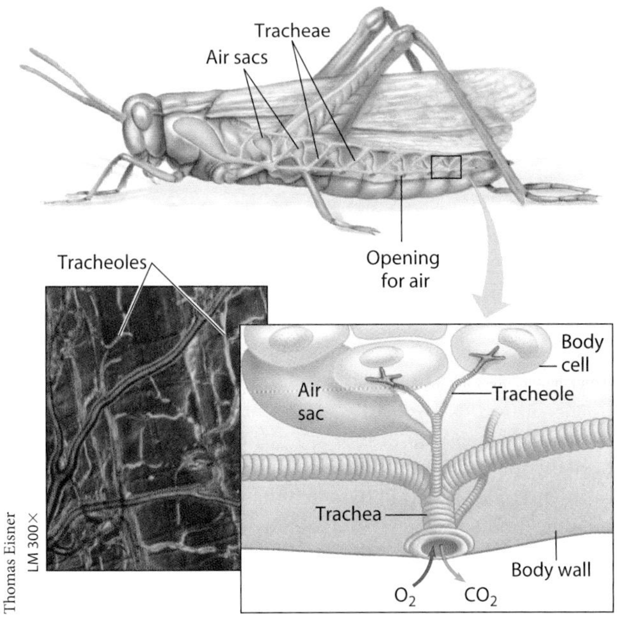

▲ **Figure 4A** The tracheal system of an insect

▲ **Figure 4B** A grasshopper in flight

The micrograph on the left in Figure 4A shows how these tubes branch repeatedly. The smallest branches, called tracheoles, extend to nearly every cell in the insect's body. Their tiny tips have closed ends and contain fluid (blue in the drawing). Gas is exchanged with body cells by diffusion across the moist epithelium that lines these tips. The structure of a tracheal system matches its function of exchanging gases directly with body cells. Thus, the circulatory system of insects is not involved in transporting gases.

For a small insect, diffusion through the tracheae brings in enough O_2 to support cellular respiration. Larger insects may ventilate their tracheal systems with rhythmic body movements that compress and expand the air tubes like bellows. An insect in flight **(Figure 4B)** has a very high metabolic rate and consumes 10 to 200 times more O_2 than it does at rest. In many insects, alternating contraction and relaxation of the flight muscles rapidly pumps air through the tracheal system.

? In what basic way does the process of gas exchange in insects differ from that in both fishes and humans?

The circulatory system of insects is not involved in transporting gases to and from the body cells.

EVOLUTION CONNECTION

5 The evolution of lungs facilitated the movement of tetrapods onto land

The colonization of land by vertebrates was one of the pivotal milestones in the history of life. The evolution of legs from fins may be the most obvious change in body design, but the refinement of lung breathing was just as important. And although skeletal changes were undoubtedly required in the transition from fins to legs, the evolution of lungs for breathing on land also required skeletal changes. Interestingly, current fossil evidence supports the hypothesis that the earliest changes in the front fins and shoulder girdle of tetrapod ancestors may actually have been breathing adaptations that enabled a fish in shallow water to push itself up to gulp in air.

Paleontologists have uncovered numerous transitional forms in tetrapod evolution. It now seems clear that tetrapods first evolved in shallow water from what some researchers jokingly call "fishapods." These ancient forms had both gills and lungs. The adaptations for air breathing evident in their fossils include a flat skull with a strong, elongated snout, as well as a muscular neck and shoulders that enabled the animal to lift the head clear of water and into the unsupportive air. Strengthening of the lower jaw may have facilitated the pumping motion presumed to be used by early air-breathing tetrapods and still employed by frogs to inflate their lungs. The recently discovered 375-million-year-old fossil of *Tiktaalik* (**Figure 5**) illustrates some of these air-breathing adaptations.

The first tetrapods on land diverged into three major lineages: amphibians, reptiles (including birds), and mammals. Most amphibians have small lungs and rely heavily on the diffusion of gases across body surfaces. Reptiles and mammals rely on lungs for gas exchange. In general, the size and complexity of lungs are correlated with an animal's metabolic rate and thus oxygen need. For example, the lungs of birds and mammals, whose high body temperatures are maintained by a high metabolic rate, have a greater area of exchange surface than the lungs of similar-sized amphibians and nonbird reptiles, which have a much lower metabolic rate.

We explore the mammalian respiratory system next.

Ted Daeschler/AFP/Getty Images

▲ **Figure 5** A cast of a fossil of *Tiktaalik*

? How might adaptations for breathing air be linked to the evolution of tetrapod limbs?

Fossil evidence indicates that changes in the neck, shoulder girdle, and limb bones may have helped early tetrapod ancestors lift their heads above water to gulp air.

The Human Respiratory System

6 In mammals, branching tubes convey air to lungs located in the chest cavity

As in all mammals, your lungs are located in your chest, or thoracic cavity, and are protected by the supportive rib cage. The thoracic cavity is separated from the abdominal cavity by a sheet of muscle called the **diaphragm**. You will see how the diaphragm helps ventilate your lungs in Module 8.

Figure 6A, on the facing page, shows the human respiratory system (along with the esophagus and heart, for orientation). Air enters your respiratory system through the nostrils. It is filtered by hairs and warmed, humidified, and sampled for odors as it flows through a maze of spaces in the nasal cavity. You can also draw in air through your mouth, but mouth breathing does not allow the air to be processed by your nasal cavity.

From the nasal cavity or mouth, air passes to the **pharynx**, where the paths for air and food cross. As you will remember from the previous chapter, when you swallow food, the **larynx** (the upper part of the respiratory tract) moves upward and tips the epiglottis over the opening of your **trachea**, or windpipe. The rest of the time, the air passage in the pharynx is open for breathing.

The larynx is often called the voice box. When you exhale, the outgoing air rushes by a pair of **vocal cords** in the larynx, and you can produce sounds by voluntarily tensing muscles that stretch the cords so they vibrate. You produce high-pitched sounds when your vocal cords are tightly stretched and vibrating very fast. When the cords are less tense, they vibrate slowly and produce low-pitched sounds.

From the larynx, air passes into your trachea. Rings of cartilage (shown in the figure in blue) reinforce the walls of the larynx and trachea, keeping this part of the airway open. The trachea forks into two **bronchi** (singular, *bronchus*), one leading to each lung. Within the lung, the bronchus branches repeatedly into finer and finer tubes called **bronchioles**. Bronchitis is a condition in which these small tubes become inflamed and constricted, making breathing difficult.

As the enlargement on the right of Figure 6A shows, the bronchioles dead-end in grapelike clusters of air sacs called **alveoli** (singular, *alveolus*). Each of your lungs contains millions of these tiny sacs. Together they have a surface area of about

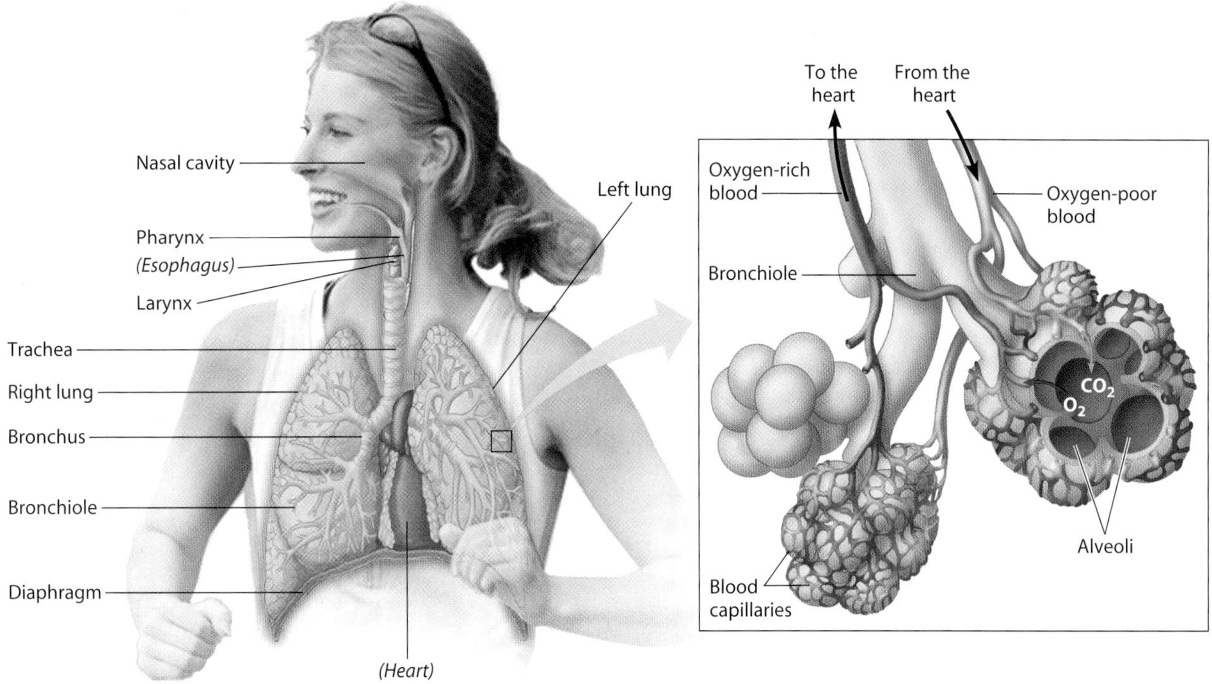

Douglas Pulsipher/Alamy

▲ **Figure 6A** The anatomy of the human respiratory system (left) and details of the alveoli (right)

100 m^2, 50 times that of your skin. The inner surface of each alveolus is lined with a thin layer of epithelial cells. The O_2 in inhaled air dissolves in a film of moisture on the epithelial cells. It then diffuses across the epithelium and into the dense web of blood capillaries that surrounds each alveolus. **Figure 6B** is a scanning electron micrograph showing the network of capillaries enclosing the alveoli. (The alveoli in this micrograph appear as empty spaces because the blood vessels were injected with a solution that hardened to form casts of the capillaries, and the tissues of the alveoli were then dissolved.) This close association between capillaries and alveoli also enables CO_2 to diffuse the opposite way—from the capillaries, across the epithelium of the alveolus, into the air space, and finally out in the exhaled air.

The major branches of your respiratory system are lined by a moist epithelium covered by cilia and a thin film of mucus. The cilia and mucus are the respiratory system's cleaning system. The beating cilia move mucus with trapped dust, pollen, and other contaminants upward to the pharynx, where it is usually swallowed.

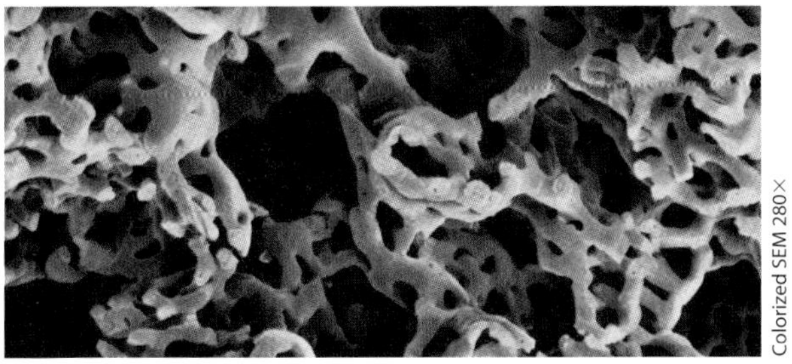

Motta & Macchiarelli/Anatomy Dept./Univ. "LA Sapienza", Rome/ Photo Researchers

▲ **Figure 6B** A colorized electron micrograph showing the network of capillaries that surround the alveoli in the lung

Respiratory Problems Alveoli are so small that specialized secretions called **surfactants** are required to keep them from sticking shut from the surface tension of their moist surface. Respiratory distress syndrome due to a lack of lung surfactant is a common disease seen in babies born 6 weeks or more before their due dates. Surfactants typically appear in the lungs after 33 weeks of embryonic development; birth normally occurs at 38 weeks. Artificial surfactants are now administered through a breathing tube to treat such preterm infants.

Alveoli are highly susceptible to airborne contaminants. Defensive white blood cells patrol them and engulf foreign particles. However, if too much particulate matter reaches the alveoli, the delicate lining of these small sacs becomes damaged and the efficiency of gas exchange drops. Studies have shown a significant association between exposure to fine particles and premature death. Air pollution and tobacco smoke are two sources of these lung-damaging particles.

Exposure to such pollutants can cause continual irritation and inflammation of the lungs and lead to chronic obstructive pulmonary disease (COPD). COPD includes two main conditions: emphysema and chronic bronchitis. In emphysema, the delicate walls of alveoli become permanently damaged and the lungs lose the elasticity that helps expel air during exhalation. With COPD, both lung ventilation and gas exchange are severely impaired. Patients experience labored breathing, coughing, and frequent lung infections. COPD is a major cause of disability and death in the United States.

? How does the structure of alveoli match their function?

Alveoli have a thin, moist epithelium across which dissolved O_2 and CO_2 can easily diffuse into or out of the surrounding capillaries. The huge collective surface area of all the alveoli enables the passage of many gas molecules.

CONNECTION

7 Smoking is a serious assault on the respiratory system

One of the worst sources of lung-damaging air pollutants is tobacco smoke, which is mainly microscopic particles of carbon coated with toxic chemicals. A single drag on a cigarette exposes a person to more than 4,000 chemicals.

Remember that cilia on cells lining the respiratory tract sweep contaminant-laden mucus up and out of the airways. Tobacco smoke irritates these cells, inhibiting or destroying their cilia. Frequent coughing—common in heavy smokers—is the respiratory system's attempt to clear the mucus no longer moved by the cilia. Smoke's toxins also kill the white blood cells that reside in the respiratory tract and engulf foreign particles. Thus, smoking disables the normal cleansing and protective mechanisms of the respiratory system.

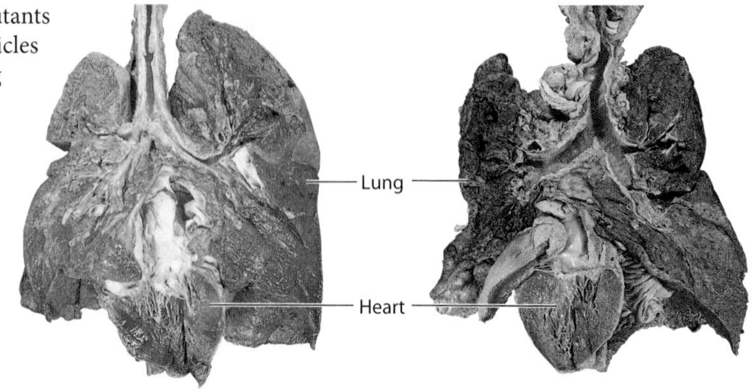

Martin M. Rotker

▲ **Figure 7** Healthy lungs (left) and cancerous lungs (right)

Smoking is a leading cause of emphysema (see Module 6), which causes breathlessness and constant fatigue as the body is forced to spend more and more energy just breathing.

Some of the toxins in tobacco smoke cause lung cancer. The photographs in **Figure 7** show a cutaway view of a pair of healthy human lungs (left) and the lungs of a cancer victim (right), whose lungs are black from the long-term buildup of smoke particles. Smokers account for 90% of all lung cancer cases. Most victims die within one year of diagnosis. Smoking is also linked to an increased risk of numerous other cancers.

Cardiovascular disease is the second highest cause of smoking-related deaths. Smokers have a higher rate of heart attacks and stroke. Smoking raises blood pressure and increases harmful cholesterol levels in the blood.

Every year in the United States, smoking kills about 440,000 people, more than all the deaths caused by accidents, alcohol and drug abuse, HIV, and murders combined. On average, adults who smoke die 13 to 14 years earlier than nonsmokers. Studies show that nonsmokers exposed to secondary cigarette smoke are also at risk. Young children are particularly susceptible, with increased risk of asthma, bronchitis, and pneumonia.

About 15 years after quitting, a former smoker's risk of most diseases linked to smoking is similar to that of people who have never smoked. No lifestyle choice has a more positive impact on the long-term health of you and those with whom you live than not smoking.

? **What causes "smoker's cough"?**

Smoke damages cilia, inhibiting their ability to sweep mucus and trapped particles from the respiratory tract. The body tries to compensate by coughing.

8 Negative pressure breathing ventilates your lungs

Breathing is the alternate inhalation and exhalation of air. This ventilation of your lungs maintains high O_2 and low CO_2 concentrations at the respiratory surface.

Figure 8 shows the changes that occur during breathing. Put your hands on your rib cage and inhale: You can feel your ribs move upward and out as muscles between the ribs contract. At the same time, your diaphragm contracts and moves downward, expanding the chest cavity. The volume of your lungs increases with the expanding chest cavity, which lowers the air pressure in the alveoli to less than atmospheric pressure. Flowing from a region of higher pressure to one of lower pressure, air is pulled through the nostrils and down the breathing tubes to the alveoli. This type of ventilation is called **negative pressure breathing**.

The diagram on the right in Figure 8 shows what happens when you exhale. Your rib muscles and diaphragm both relax, decreasing the volume of the rib cage and chest cavity, which increases the air pressure inside the lungs, forcing air out.

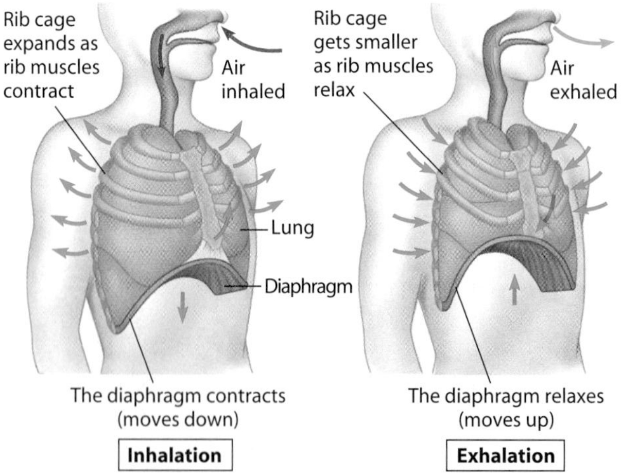

▲ **Figure 8** Negative pressure breathing

Notice that the diaphragm curves farther upward into the chest cavity when relaxed.

Each year, you take between 4 million and 10 million breaths. The volume of air in each breath is about 500 milliliters (mL) when you breathe quietly. The volume of air breathed during maximal inhalation and exhalation is called **vital capacity**. It averages about 3.4 L and 4.8 L for college-age females and males, respectively. (Women tend to have smaller rib cages and lungs.) The lungs actually hold more air than the vital capacity. Because the alveoli do not completely collapse, a residual volume of "dead" air remains in the lungs even after you blow out as much air as you can. As lungs lose resilience (springiness) with age or as the result of disease, such as emphysema, less air exits on exhalation and residual volume increases at the expense of vital capacity.

Because the lungs do not completely empty, each inhalation mixes fresh air with oxygen-depleted air. Thus, you can extract only about 25% of the O_2 in the air you inhale. As mentioned in the chapter introduction, the gas exchange system of birds is much more efficient than that of humans. Unlike the in-and-out flow of air in the human alveoli, the air in the lungs of birds moves in one direction through tiny passageways where gas exchange occurs. Because of this one-way flow of air, oxygen-depleted air does not remain in a bird's lungs after exhalation. Thus, birds can extract more oxygen from a volume of inhaled air, enough oxygen to fly over the Himalayas.

? Explain how negative pressure breathing ventilates your lungs.

● The lungs expand along with the expanding chest cavity during inhalation. This increase in volume lowers the pressure in the lungs, which draws air into the lungs. During exhalation, the process is reversed.

9 Breathing is automatically controlled

Although you can voluntarily hold your breath or breathe faster and deeper, most of the time your breathing is under involuntary control. Figure 9 illustrates how a **breathing control center** in a part of the brain called the medulla oblongata ensures that your breathing rate is coordinated with your body's need for oxygen. ❶ Nerves from the breathing control center signal the diaphragm and rib muscles to contract, causing you to inhale. When you are at rest, these nerve signals result in about 10 to 14 inhalations per minute. Between inhalations, the muscles relax, and you exhale.

❷ The control center regulates breathing rate in response to changes in the CO_2 level of the blood. When you exercise vigorously, for instance, your metabolism speeds up and your body cells generate more CO_2 as a waste product. The CO_2 goes into the blood, where it reacts with water to form carbonic acid. The acid slightly lowers the pH of the blood and the fluid bathing the brain, the cerebrospinal fluid. When the medulla senses this pH drop, its breathing control center increases both the rate and depth of your breathing. As a result, more CO_2 is eliminated in the exhaled air, and the pH of the blood returns to normal.

The O_2 concentration in the blood usually has little effect on the breathing control center. Because the same process that consumes O_2—cellular respiration—also produces CO_2, a rise in CO_2 is generally a good indication of a decrease in blood oxygen. Thus, by responding to lowered pH, the breathing control center increases blood oxygen level.

❸ Secondary control over breathing is exerted by sensors in the aorta and carotid arteries that monitor concentrations of O_2 as well as CO_2. When the O_2 level in the blood is severely depressed, these sensors signal the control center via nerves to increase the rate and depth of breathing. This response may occur, for example, at high altitudes, where the air is so thin that you cannot get enough O_2 by breathing normally.

The breathing control center responds to a variety of nervous and chemical signals that serve to keep the rate and depth of your breathing in tune with the changing metabolic needs of your body. Breathing rate must also be coordinated with the activity of the circulatory system, which transports blood to and from the alveolar capillaries. We examine the role of the circulatory system in gas exchange more closely in the next module.

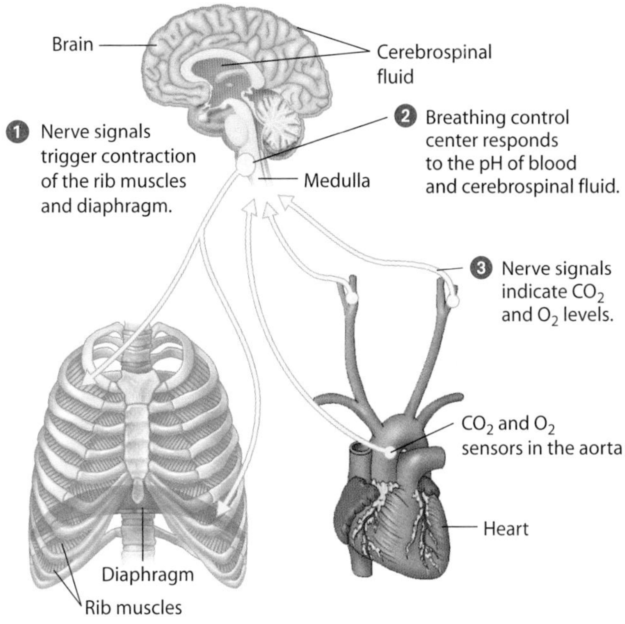
Brain
Cerebrospinal fluid
❶ Nerve signals trigger contraction of the rib muscles and diaphragm.
❷ Breathing control center responds to the pH of blood and cerebrospinal fluid.
Medulla
❸ Nerve signals indicate CO_2 and O_2 levels.
CO_2 and O_2 sensors in the aorta
Heart
Diaphragm
Rib muscles

▲ **Figure 9** How the breathing control center regulates breathing

? How is the increased need for O_2 during exercise accommodated by the breathing control center?

● During exercise, cells release more CO_2 to the blood, which forms carbonic acid, lowering the pH of the blood. The breathing center senses the decrease in pH and sends impulses to increase breathing rate, thus supplying more O_2.

Transport of Gases in the Human Body

10 Blood transports respiratory gases

How does oxygen get from your lungs to all the other tissues in your body, and how does carbon dioxide travel from the tissues to your lungs? To answer these questions, we must look at the basic organization of the human circulatory system.

Figure 10 is a diagram showing the main components of your circulatory system and their roles in gas exchange. Let's start with the heart, in the middle of the diagram. One side of the heart handles oxygen-poor blood (colored blue). The other side handles oxygen-rich blood (red). As indicated in the lower left of the diagram, oxygen-poor blood returns to the heart from capillaries in body tissues. The heart pumps this blood to the alveolar capillaries in the lungs. Gases are exchanged between air in the alveoli and blood in the capillaries (top of diagram). Blood that has lost CO_2 and gained O_2 returns to the heart and is then pumped out to body tissues.

The exchange of gases between capillaries and the cells around them occurs by the diffusion of gases down gradients of pressure. A mixture of gases, such as air, exerts pressure. You see evidence of gas pressure whenever you open a can of soda, releasing the pressure of the CO_2 it contains. Each kind of gas in a mixture accounts for a portion of the total pressure of the mixture. Thus, each gas has what is called a **partial pressure**. Molecules of each kind of gas will diffuse down a gradient of their own partial pressure independently of the other gases. At the bottom of the figure, for instance, O_2 moves from oxygen-rich blood, through the interstitial fluid, and into tissue cells because it diffuses from a region of higher partial pressure to a region of lower partial pressure. The tissue cells maintain this gradient as they consume O_2 in cellular respiration. The CO_2 produced as a waste product of cellular respiration diffuses down its own partial pressure gradient out of tissue cells and into the capillaries. Diffusion down partial pressure gradients also accounts for gas exchange in the alveoli.

? What is the physical process underlying gas exchange?

Diffusion of each gas down its partial pressure gradient

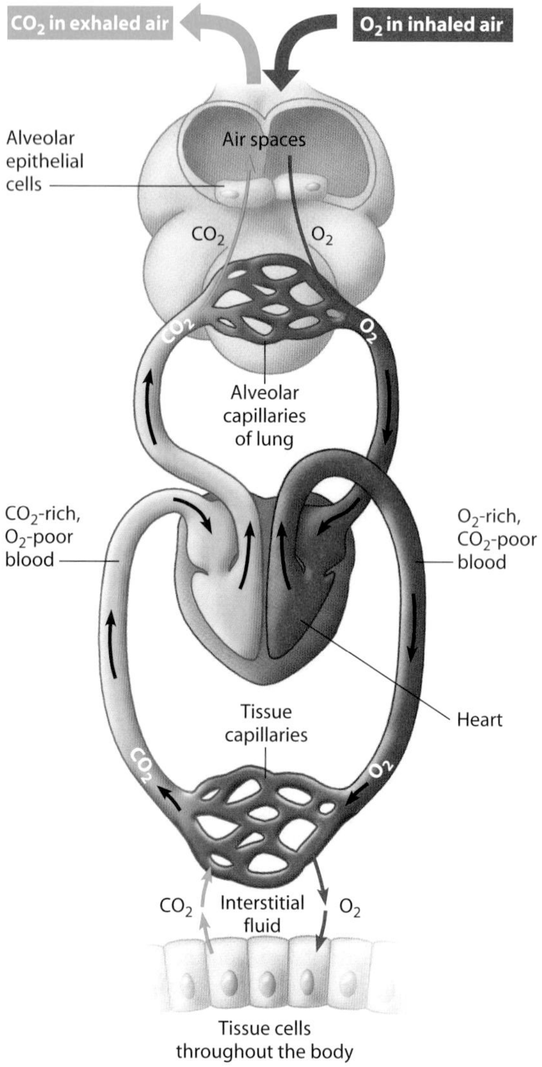

▲ **Figure 10** Gas transport and exchange in the body

Adapted from Elaine N. Marieb, *Human Anatomy and Physiology*, 4th ed. Copyright © 1998 Pearson Education, Inc., publishing as Pearson Benjamin Cummings

11 Hemoglobin carries O_2, helps transport CO_2, and buffers the blood

Oxygen is not highly soluble in water, and most animals transport O_2 bound to proteins called respiratory pigments. These molecules have distinctive colors, hence the name pigment. Many molluscs and arthropods use a blue, copper-containing pigment. Almost all vertebrates and many invertebrates use **hemoglobin**, an iron-containing pigment that turns red when it binds O_2.

Each of your red blood cells is packed with about 250 million molecules of hemoglobin. A hemoglobin molecule consists of four polypeptide chains of two different types, depicted with two shades of purple in **Figure 11**, on the next page. Attached to each polypeptide is a chemical group called a heme (colored blue in the figure), at the center of which is an iron atom (black). Each iron atom binds one O_2 molecule. Thus, every hemoglobin molecule can carry up to four O_2 molecules. Hemoglobin loads up with O_2 in the lungs and transports it to the body's tissues. There, hemoglobin unloads some or all of its cargo, depending on the O_2 needs of the cells. The partial pressure of O_2 in the tissue reflects how much O_2 the cells are using and determines how much O_2 is unloaded.

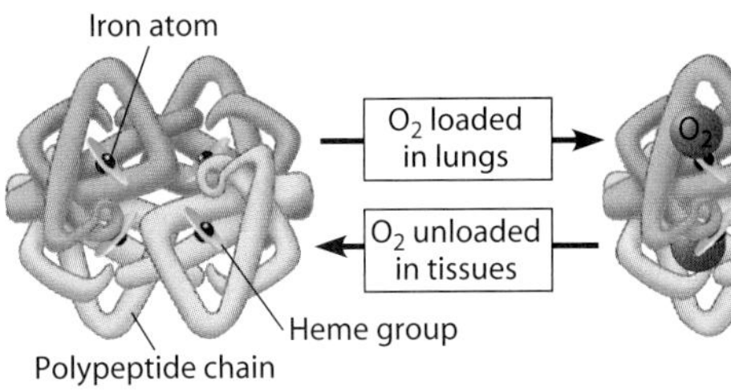

▲ **Figure 11** Hemoglobin loading and unloading O_2

Hemoglobin is a multipurpose molecule. It also helps transport CO_2 and assists in buffering the blood. Most of the CO_2 that diffuses from tissue cells into a capillary enters red blood cells, where some of it combines with hemoglobin. The rest reacts with water, forming carbonic acid (H_2CO_3), which then breaks apart into a hydrogen ion (H^+) and a bicarbonate ion (HCO_3^-). This reversible reaction is shown below:

$$\underset{\text{Carbon dioxide}}{CO_2} + \underset{\text{Water}}{H_2O} \rightleftharpoons \underset{\text{Carbonic acid}}{H_2CO_3} \rightleftharpoons \underset{\text{Hydrogen ion}}{H^+} + \underset{\text{Bicarbonate ion}}{HCO_3^-}$$

Hemoglobin binds most of the H^+ produced by this reaction, minimizing the change in blood pH. (As discussed in Module 9, the slight drop in pH due to the increased production of CO_2 during exercise is the stimulus to increase breathing rate.) The bicarbonate ions diffuse into the plasma, where they are carried to the lungs.

As blood flows through capillaries in the lungs, the reaction is reversed. Bicarbonate ions combine with H^+ to form carbonic acid; carbonic acid is converted to CO_2 and water; and CO_2 diffuses from the blood to the alveoli and leaves the body in exhaled air.

We have seen how O_2 and CO_2 are transported between your lungs and body tissue cells via the bloodstream. In the next module, we consider a special case of gas exchange between two circulatory systems.

? O_2 in the blood is transported bound to ______ within ______ ______ cells, and CO_2 is mainly transported as ______ ions within the plasma.

hemoglobin . . . red blood . . . bicarbonate

CONNECTION 12 The human fetus exchanges gases with the mother's blood

Figure 12 is a drawing of a human fetus inside the mother's uterus. The fetus literally swims in a protective watery bath, the amniotic fluid. Its nonfunctional lungs are full of fluid. How does the fetus exchange gases with the outside world? It does this by way of the placenta, a composite organ that includes tissues from both fetus and mother. A large network of capillaries fans out into the placenta from blood vessels in the umbilical cord of the fetus. These capillaries exchange gases with the maternal blood that circulates in the placenta, and the mother's circulatory system transports the gases to and from her lungs. Aiding O_2 uptake by the fetus is fetal hemoglobin, which attracts O_2 more strongly than does adult hemoglobin.

One of the reasons that smoking is considered a health risk during pregnancy is because it reduces, perhaps by as much as 25%, the supply of oxygen reaching the placenta. Lower oxygen levels delay fetal development and growth, resulting in a higher incidence of premature birth, low birth weight, and brain and lung defects. Avoiding cigarette smoke protects the health of both mothers and babies.

Let's move on to what happens when a baby is born. Very soon after delivery, placental gas exchange with the mother ceases, and the baby's lungs must begin to work. Carbon dioxide acts as the signal. As soon as CO_2 stops diffusing from the fetus into the placenta, CO_2 levels rise in the fetal blood. The resulting drop in blood pH stimulates the breathing control center in the infant's brain, and the newborn gasps and takes its first breath.

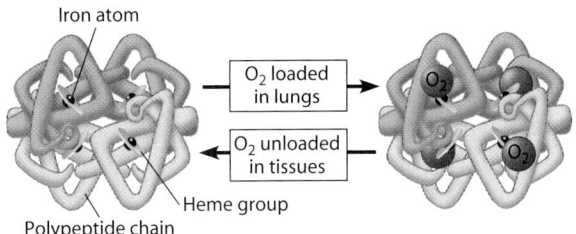

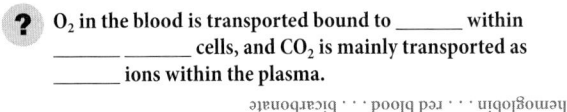

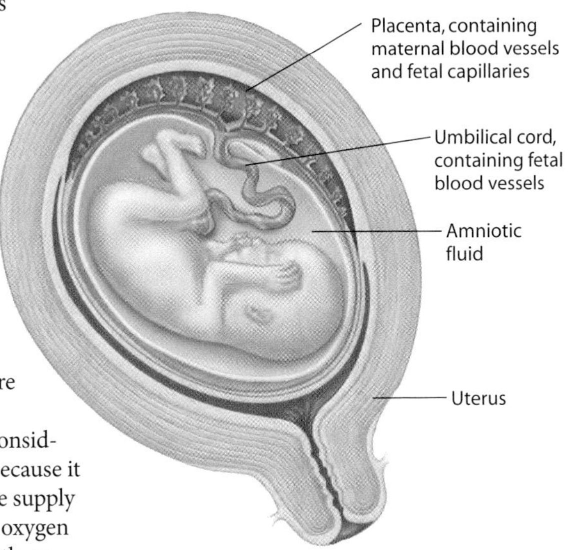

▲ **Figure 12** A human fetus and placenta in the uterus

A human birth and the radical changes in gas exchange mechanisms that accompany it are extraordinary events. Resulting from millions of years of evolutionary adaptation, these events are on a par with the remarkable flying ability of the geese we discussed in the chapter introduction. For a goose to breathe the thin air and fly great distances high above Earth or for a human baby to switch almost instantly from living in water and exchanging gases with maternal blood to breathing air directly requires truly remarkable adaptations in the organism's respiratory system. Also required are adaptations of the circulatory system, which, as we have seen, supports the respiratory system in its gas exchange function.

? How does fetal hemoglobin enhance oxygen transfer from mother to fetus across the placenta?

Because fetal hemoglobin has a greater affinity for O_2 than does adult hemoglobin, it helps "pull" the O_2 from maternal blood to fetal blood.

REVIEW

For Practice Quizzes, BioFlix, MP3 Tutors, and Activities, go to www.masteringbiology.com.

Reviewing the Concepts

Mechanisms of Gas Exchange (1–5)

1 Overview: Gas exchange in humans involves breathing, transport of gases, and exchange with body cells. Gas exchange, the interchange of O_2 and CO_2 between an organism and its environment, provides O_2 for cellular respiration and removes its waste product, CO_2.

2 Animals exchange O_2 and CO_2 across moist body surfaces. Respiratory surfaces must be thin and moist for diffusion of O_2 and CO_2 to occur. Some animals use their entire skin as a gas exchange organ. In most animals, gills, a tracheal system, or lungs provide large respiratory surfaces for gas exchange.

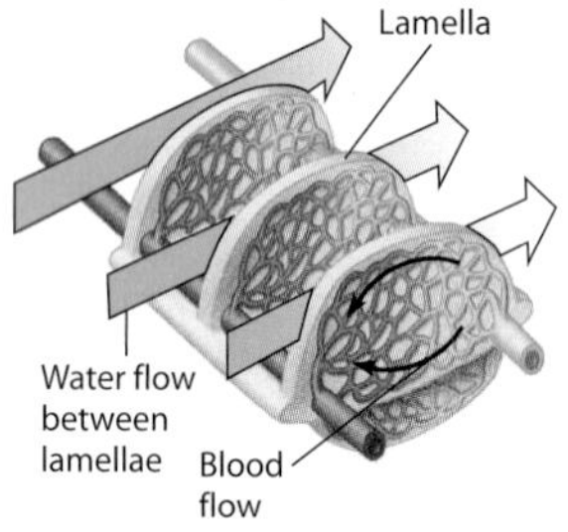

3 Gills are adapted for gas exchange in aquatic environments. Gills absorb O_2 dissolved in water. In a fish, gas exchange is enhanced by ventilation and the countercurrent flow of water and blood.

4 The tracheal system of insects provides direct exchange between the air and body cells. A network of finely branched tubes transports O_2 directly to body cells and moves CO_2 from them.

5 The evolution of lungs facilitated the movement of tetrapods onto land. Skeletal adaptations of air-breathing fish may have helped early tetrapods move onto land.

The Human Respiratory System (6–9)

6 In mammals, branching tubes convey air to lungs located in the chest cavity. Inhaled air passes through the pharynx and larynx into the trachea, bronchi, and bronchioles to the alveoli. Mucus and cilia in the respiratory passages protect the lungs.

7 Smoking is a serious assault on the respiratory system. Smoking causes lung cancer, heart disease, and COPD.

8 Negative pressure breathing ventilates your lungs. The contraction of rib muscles and diaphragm expands the chest cavity, reducing air pressure in the alveoli and drawing air into the lungs.

9 Breathing is automatically controlled. A breathing control center in the brain keeps breathing in tune with body needs, sensing and responding to the CO_2 level in the blood. A drop in blood pH triggers an increase in the rate and depth of breathing.

Transport of Gases in the Human Body (10–12)

10 Blood transports respiratory gases. The heart pumps oxygen-poor blood to the lungs, where it picks up O_2 and drops off CO_2. Oxygen-rich blood returns to the heart and is pumped to body cells, where it drops off O_2 and picks up CO_2.

11 Hemoglobin carries O_2, helps transport CO_2, and buffers the blood.

12 The human fetus exchanges gases with the mother's blood. Fetal hemoglobin enhances oxygen transfer from maternal blood in the placenta. At birth, rising CO_2 in fetal blood stimulates the breathing control center to initiate breathing.

Connecting the Concepts

1. Complete the following concept map to review some of the concepts of gas exchange.

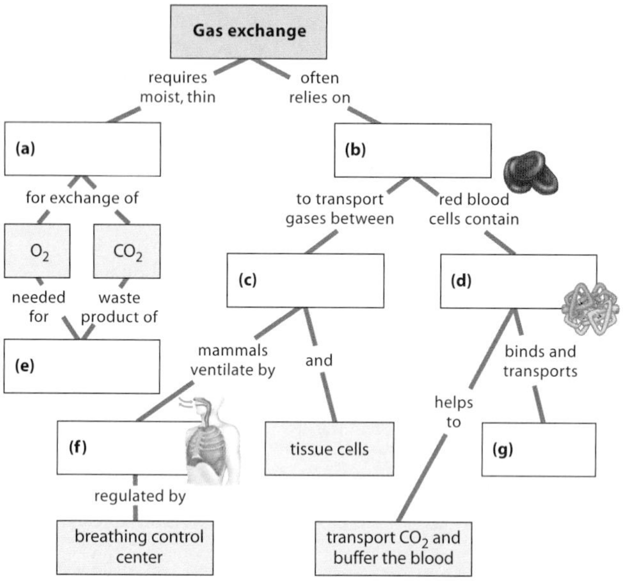

2. Label the parts of the human respiratory system.

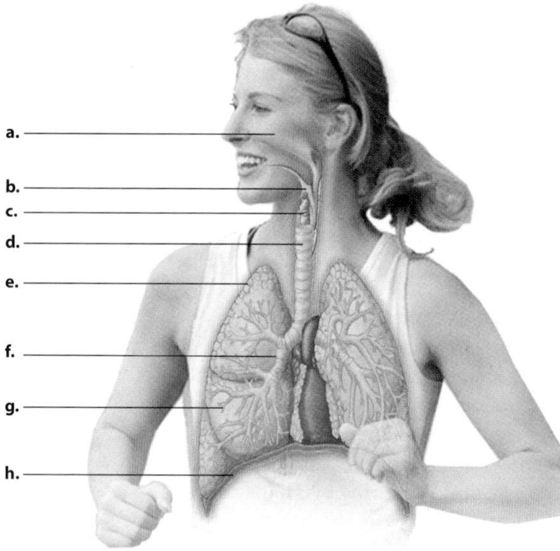

Douglas Pulsipher/Alamy

Testing Your Knowledge

Multiple Choice

3. When you hold your breath, which of the following first leads to the urge to breathe?
   a. falling CO_2
   b. falling O_2
   c. rising CO_2
   d. rising pH of the blood
   e. both c and d

4. Countercurrent gas exchange in the gills of a fish
   a. speeds up the flow of water through the gills.
   b. maintains a gradient that enhances diffusion.
   c. enables the fish to obtain oxygen without swimming.
   d. means that blood and water flow at different rates.
   e. allows O_2 to diffuse against its partial pressure gradient.
5. When you inhale, the diaphragm
   a. relaxes and moves upward.
   b. relaxes and moves downward.
   c. contracts and moves upward.
   d. contracts and moves downward.
   e. is not involved in the breathing movements.
6. In which of the following organisms does oxygen diffuse directly across a respiratory surface to cells, without being carried by the blood?
   a. a grasshopper
   b. a whale
   c. an earthworm
   d. a sparrow
   e. a mouse
7. What is the function of the cilia in the trachea and bronchi?
   a. to sweep air into and out of the lungs
   b. to increase the surface area for gas exchange
   c. to vibrate when air is exhaled to produce sounds
   d. to dislodge food that may have slipped past the epiglottis
   e. to sweep mucus with trapped particles up and out of the respiratory tract
8. What do the alveoli of mammalian lungs, the gill filaments of fish, and the tracheal tubes of insects have in common?
   a. use of a circulatory system to transport gases
   b. respiratory surfaces that are infoldings of the body wall
   c. countercurrent exchange
   d. a large, moist surface area for gas exchange
   e. all of the above
9. Which of the following is the best explanation for why birds can fly over the Himalayas while most humans require oxygen masks to climb these mountains?
   a. Birds are much smaller and require less oxygen.
   b. Birds use positive pressure breathing, whereas humans use negative pressure breathing.
   c. With their one-way flow of air and efficient ventilation, the lungs of birds extract more O_2 from the air.
   d. The circulatory system of birds is much more efficient at delivering oxygen to tissues than is that of humans.
   e. Humans are endotherms and thus require more oxygen than do birds, which are ectotherms.

Describing, Comparing, and Explaining

10. What are two advantages of breathing air, compared with obtaining dissolved oxygen from water? What is a comparative disadvantage of breathing air?
11. Trace the path of an oxygen molecule in its journey from the air to a muscle cell in your arm, naming all the structures involved along the way.
12. Carbon monoxide (CO) is a colorless, odorless gas found in furnace and automobile engine exhaust and cigarette smoke. CO binds to hemoglobin 210 times more tightly than does O_2. (CO binds with an electron transport protein and disrupts cellular respiration.) Explain why CO is such a deadly gas.

Applying the Concepts

13. Partial pressure reflects the relative amount of gas in a mixture and is measured in millimeters of mercury (mm Hg). Llamas are native to the Andes Mountains in South America. The partial pressure of O_2 (abbreviated P_{O_2}) in the atmosphere where llamas live is about half of the P_{O_2} at sea level. As a result, the P_{O_2} in the lungs of llamas is about 50 mm Hg, whereas the P_{O_2} in human lungs at sea level is about 100 mm Hg.

    A dissociation curve for hemoglobin shows the percentage of saturation (the amount of O_2 bound to hemoglobin) at increasing values of P_{O_2}. As you see in the graph below, the dissociation curves for llama and human hemoglobin differ. Compare these two curves and explain how the hemoglobin of llamas is an adaptation to living where the air is "thin."

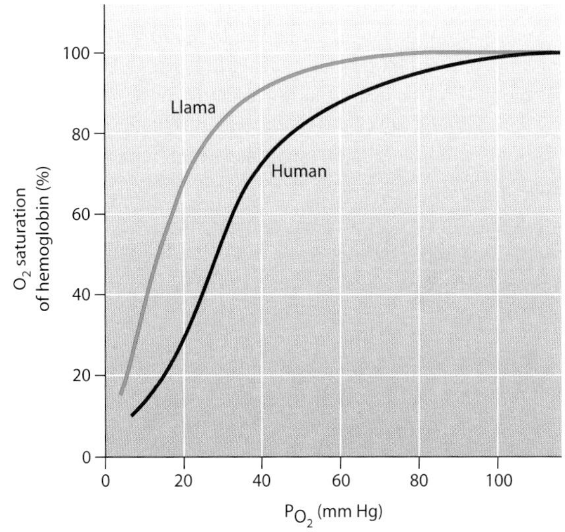

14. Mountain climbers often spend weeks adjusting to the low oxygen concentration at high altitudes before and during their ascent of high peaks. During that time, their bodies begin to produce more red blood cells. Some runners and cyclists prepare for competition by training at high altitudes or by sleeping in a tent in which P_{O_2} is kept artificially low. Explain why this training strategy may improve an athlete's performance.
15. One of the many mutant opponents that the movie monster Godzilla contends with is Mothra, a giant mothlike creature with a wingspan of 7–8 m. Science fiction creatures like these can be critiqued on the grounds of biomechanical and physiological principles. Focusing on the principles of gas exchange that you learned about in this chapter, what problems would Mothra face? Why do you think truly giant insects are improbable?
16. Hundreds of studies have linked smoking with cardiovascular and lung disease. According to health authorities, smoking is the leading cause of preventable, premature death in the United States. Antismoking and health groups have proposed that cigarette advertising in all media be banned. What are some arguments in favor of such a ban on cigarette advertising? In opposition? Do you favor or oppose such a ban? Defend your position.

Review Answers

1. a. respiratory surface; b. circulatory system; c. lungs; d. hemoglobin; e. cellular respiration; f. negative pressure breathing; g. O_2
2. a. nasal cavity; b. pharynx; c. larynx; d. trachea; e. right lung; f. bronchus; g. bronchiole; h. diaphragm
3. c 4. b 5. d 6. a 7. e 8. d 9. c
10. Advantages of breathing air: It has a higher concentration of O_2 than water and is easier to move over the respiratory surface. Disadvantage of breathing air: Living cells on the respiratory surface must remain moist, but breathing air dries out this surface.
11. Nasal cavity, pharynx, larynx, trachea, bronchus, bronchiole, alveolus, through wall of alveolus into blood vessel, blood plasma, into red blood cell, attaches to hemoglobin, carried by blood through heart, blood vessel in muscle, dropped off by hemoglobin, out of red blood cell, into blood plasma, through capillary wall, through interstitial fluid, and into muscle cell.
12. Both these effects of carbon monoxide interfere with cellular respiration and the production of ATP. By binding more tightly to hemoglobin, CO would decrease the amount of O_2 picked up in the lungs and delivered to body cells. Without sufficient O_2 to act as the final electron acceptor, cellular respiration would slow. And by blocking electron flow in the electron transport chain, cellular respiration and ATP production would cease. Without ATP, cellular work stops and cells and organisms die.
13. Llama hemoglobin has a higher affinity for O_2 than does human hemoglobin. The dissociation curve shows that its hemoglobin becomes saturated with O_2 at the lower P_{O_2} of the high altitudes to which llamas are adapted. At that P_{O_2}, human hemoglobin is only 80% saturated.
14. The athlete's body would respond to training at high altitudes or sleeping in an artificial atmosphere with lower P_{O_2} by producing more red blood cells. Thus, the athlete's blood would carry more O_2, and this increase in aerobic capacity may improve endurance and performance.
15. Insects have a tracheal system for gas exchange. To provide O_2 to all the body cells in such a huge moth, the tracheal tubes would have to be wider (to provide enough ventilation across longer distances) and very extensive (to service large flight muscles and other tissues), thus presenting problems of water loss and increased weight. Both the tracheal system and the weight of the exoskeleton limit the size of insects.
16. Some issues and questions to consider: Would a total ban on advertising decrease the number of cigarette smokers? If cigarettes are legal, can the right of cigarette manufacturers to advertise their product be restricted in this manner? Have similar bans on advertising of other legal but potentially deadly products (such as alcohol) been tried? How have they worked? Do health concerns outweigh commercial concerns? If cigarettes are so bad, should they be declared illegal?

Circulation

From Chapter 23 of *Campbell Biology: Concepts & Connections*, Seventh Edition. Jane B. Reece, Martha R. Taylor, Eric J. Simon, Jean L. Dickey.

Circulation

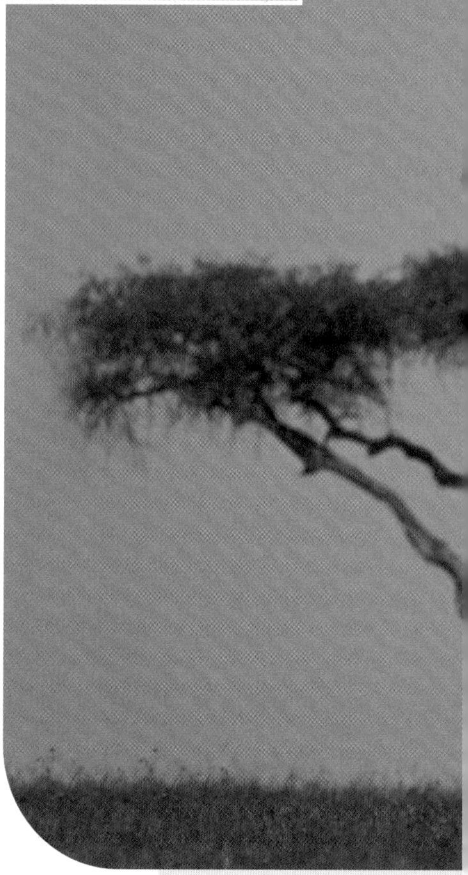

BIG IDEAS

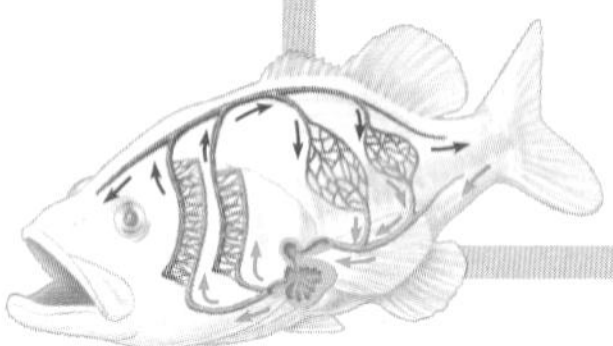

Circulatory Systems
(1–2)

Internal transport systems carry materials between exchange surfaces and body cells.

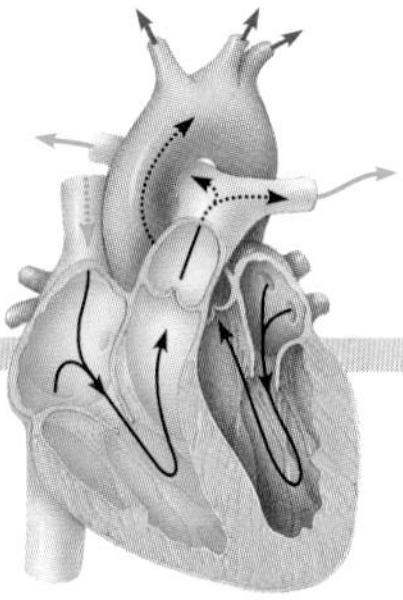

The Human Cardiovascular System and Heart
(3–6)

The heart pumps blood through the pulmonary circuit and the systemic circuit.

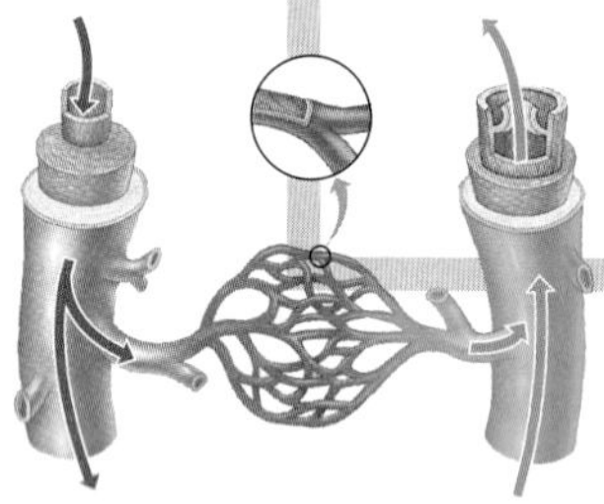

Structure and Function of Blood Vessels
(7–11)

Blood flows through arteries to capillaries, where exchange occurs with body cells, and returns to the heart in veins.

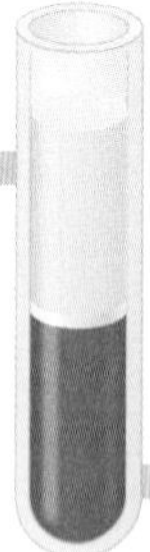

Structure and Function of Blood
(12–15)

Red blood cells carry oxygen, white blood cells fight infections, and platelets function in blood clotting.

Frank Stober/Photolibrary

You may not think that you have much in common with the giraffes pictured above. But as mammals, you share many characteristics, such as hair, female mammary glands that produce milk, an efficient respiratory system, and a four-chambered heart. As land animals, you and a giraffe also share the challenges presented by the persistent, unwavering force of gravity.

Your circulatory system is strongly affected by gravity, which tends to pull blood downward into the lower parts of the body. When you are upright, your heart must pump blood against gravity from your heart to your brain. The challenge is even greater for a giraffe eating leaves from a tall tree. It takes a high blood pressure (about twice that of a human) to pump blood up that long neck. But when a giraffe bends down to drink, the pull of gravity almost doubles the blood pressure in the arteries leading to its head. Special valves, saclike sinuses, and other mechanisms protect the giraffe's brain from this potentially dangerous spike in blood pressure.

How does blood travel uphill in the veins of a giraffe's long legs, or in your own legs, to return to the heart? As you or a giraffe walks, leg muscles squeeze the veins and force the blood upward. Veins also have one-way valves that prevent blood from flowing back down the legs.

These adaptations of the circulatory system facilitate the distribution of blood in spite of the pull of gravity. Most animals have a circulatory system that connects organs involved in gas exchange, digestion, and waste processing. We begin this chapter with a survey of some of the solutions to transport that natural selection has favored in different animals. We then turn to the human cardiovascular system and explore the structures and functions of the heart, blood vessels, and blood.

Circulatory Systems

1 Circulatory systems facilitate exchange with all body tissues

To sustain life, an animal must acquire nutrients, exchange gases, and dispose of waste products, and these needs ultimately extend to every cell in the body. In most animals, these functions are facilitated by a **circulatory system**. A circulatory system is necessary in any animal whose body is too large or too complex for such exchange to occur by diffusion alone. Diffusion is inadequate for transporting materials over distances greater than a few cell widths—far less than the distance oxygen must travel between your lungs and brain or the distance nutrients must go between your small intestine and the muscles in your arms and legs. An internal transport system must bring resources close enough to cells for diffusion to be effective.

Several types of internal transport have evolved in animals. For example, in cnidarians and most flatworms, a central gastrovascular cavity serves both in digestion and in distribution of substances throughout the body. The body wall of a hydra is only two or three cells thick, so all the cells can exchange materials directly with the water surrounding the animal or with the fluid in its gastrovascular cavity. Nutrients and other materials have only a short distance to diffuse between cell layers.

A gastrovascular cavity is not adequate for animals with thick, multiple layers of cells. Such animals require a true circulatory system, which consists of a muscular pump **(heart),** a circulatory fluid, and a set of tubes (vessels) to carry the circulatory fluid.

Two basic types of circulatory systems have evolved in animals. Many invertebrates, including most molluscs and all arthropods, have an **open circulatory system**. The system is called "open" because fluid is pumped through open-ended vessels and flows out among the tissues; there is no distinction between the circulatory fluid and interstitial fluid. In an insect, such as the grasshopper **(Figure 1A)**, pumping of the tubular heart drives body fluid into the head and the rest of the body (black arrows). Body movements help circulate the fluid as materials are exchanged with body cells. When the heart relaxes, fluid enters through several pores. Each pore has a valve that closes when the heart contracts, preventing backflow of the circulating fluid. In insects, respiratory gases are conveyed to and from body cells by the tracheal system (not shown here), not by the circulatory system.

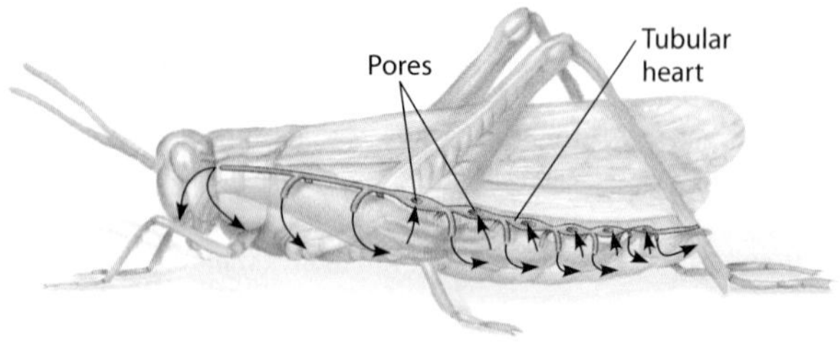

▲ **Figure 1A** The open circulatory system of a grasshopper

Earthworms, squids, octopuses, and vertebrates (such as ourselves and giraffes) all have a **closed circulatory system**. It is called "closed" because the circulatory fluid, **blood**, is confined to vessels, keeping it distinct from the interstitial fluid. There are three kinds of vessels: **Arteries** carry blood away from the heart to body organs and tissues; **veins** return blood to the heart; and **capillaries** convey blood between arteries and veins within each tissue. The vertebrate circulatory system is often called a **cardiovascular system** (from the Greek *kardia*, heart, and Latin *vas*, vessel). How extensive are the vessels in your cardiovascular system? If all your blood vessels were lined up end to end, they would circle Earth's equator twice.

The cardiovascular system of a fish **(Figure 1B)** illustrates some key features of a closed circulatory system. The heart of a fish has two main chambers. The **atrium** (plural, *atria*) receives blood from the veins, and the **ventricle** pumps blood to the gills via large arteries. As in all figures depicting closed circulatory systems in this book, red represents oxygen-rich blood and blue represents oxygen-poor blood. After passing through the gill capillaries, the blood, now oxygen-rich, flows into large arteries that carry it to all other parts of the body. The large arteries branch into **arterioles**, small vessels that give rise to capillaries. Networks of capillaries called **capillary beds** infiltrate every organ and tissue in the body. The thin walls of the capillaries allow chemical exchange between the blood and the interstitial fluid. The capillaries converge into **venules**, which in turn converge into larger veins that return blood to the heart.

In the next module, we compare the cardiovascular systems of different vertebrate groups.

? What are the key differences between an open circulatory system and a closed circulatory system?

The vessels in an open circulatory system do not form an enclosed circuit from the heart, through the body, and back to the heart, and the circulatory fluid is not distinct from interstitial fluid, as is the blood in a closed circulatory system.

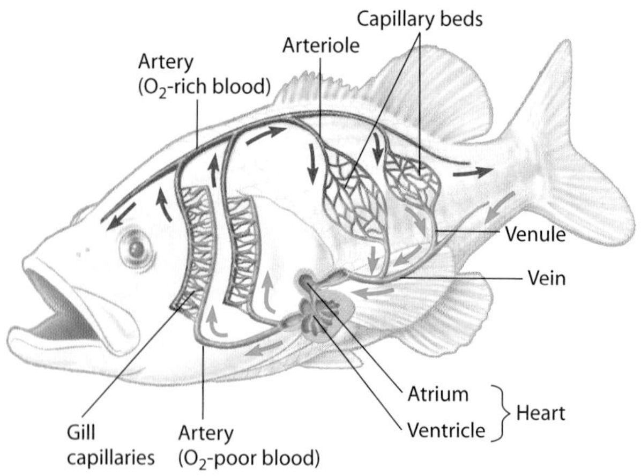

▲ **Figure 1B** The closed circulatory system of a fish

EVOLUTION CONNECTION

2 Vertebrate cardiovascular systems reflect evolution

The colonization of land by vertebrates was a major episode in the history of life. As aquatic vertebrates became adapted for terrestrial life, nearly all of their organ systems underwent major changes. One of these was the change from gill breathing to lung breathing, and this switch was accompanied by important changes in the cardiovascular system.

As illustrated in Figure 1B and diagrammed in **Figure 2A**, blood passes through the heart of a fish once in each circuit through the body, an arrangement called **single circulation**. Blood pumped from the ventricle travels first to the gill capillaries. Blood pressure drops considerably as blood flows through the numerous, narrow gill capillaries. An artery carries the oxygen-rich blood from the gills to capillaries in the tissues and organs, from which the blood returns to the atrium of the heart. The animal's swimming movements help to propel the blood through the body.

A single circuit would not supply enough pressure to move blood through the capillaries of the lungs and then to the body capillaries of a terrestrial vertebrate. The evolutionary adaptation that resulted in a more vigorous flow of blood to body organs is called **double circulation**, in which blood is pumped a second time after it loses pressure in the lungs. The **pulmonary circuit** carries blood between the heart and gas exchange tissues in the lungs, and the **systemic circuit** carries blood between the heart and the rest of the body.

You can see an example of these two circuits in **Figure 2B**. (Notice that the right side of the animal's heart is on the left in the diagram. It is customary to draw the system as though in a body facing you from the page.) Frogs and other amphibians have a three-chambered heart. The right atrium receives blood returning from the systemic capillaries in the body's organs. The ventricle pumps blood to capillary beds in the lungs and skin. Because gas exchange occurs both in the lungs and across the thin, moist skin, this is called a *pulmocutaneous circuit*. Oxygen-rich blood returns to the left atrium. Although blood from the two atria mixes in the single ventricle, a ridge diverts most of the oxygen-poor blood to the pulmocutaneous circuit and most of the oxygen-rich blood to the systemic circuit.

In the three-chambered heart of turtles, snakes, and lizards, the ventricle is partially divided, and less mixing of blood occurs. The ventricle is completely divided in crocodilians.

In all birds and mammals (for example, the lemur in **Figure 2C**), the heart has four chambers: two atria and two ventricles. The right side of the heart handles only oxygen-poor blood; the left side receives and pumps only oxygen-rich blood. The evolution of a powerful four-chambered heart was an essential adaptation to support the high metabolic rates of birds and mammals, which are endothermic. Endotherms use about 10 times as much energy as equal-sized ectotherms. Therefore, their circulatory system needs to deliver much more fuel and oxygen to body tissues. This requirement is met by a large heart that is able to pump a large volume of blood through separate systemic and pulmonary circulations. Birds and mammals descended from different reptilian ancestors, and their four-chambered hearts evolved independently—an example of convergent evolution, in which natural selection favors the same adaptation in response to similar environmental challenges.

? What is the difference between the single circulation of a fish and the double circulation of a land vertebrate?

In a fish, blood travels from gill capillaries to body capillaries before returning to the heart. In a land vertebrate, blood returns to the heart and is pumped a second time between the pulmonary and systemic circuits.

George Peters/iStockphoto

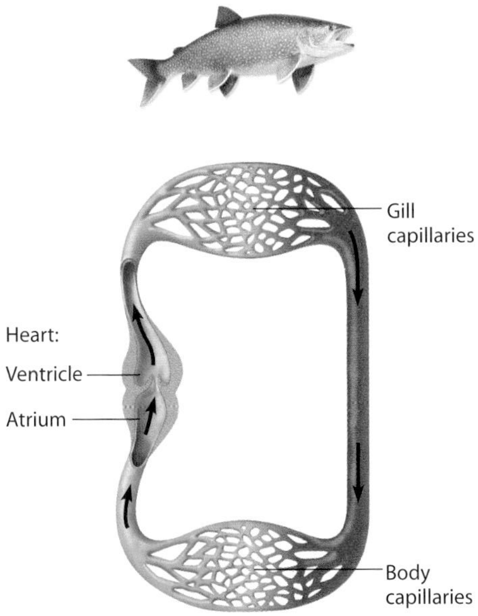

▲ **Figure 2A** The single circulation and two-chambered heart of a fish

Matt Antonino/Shutterstock

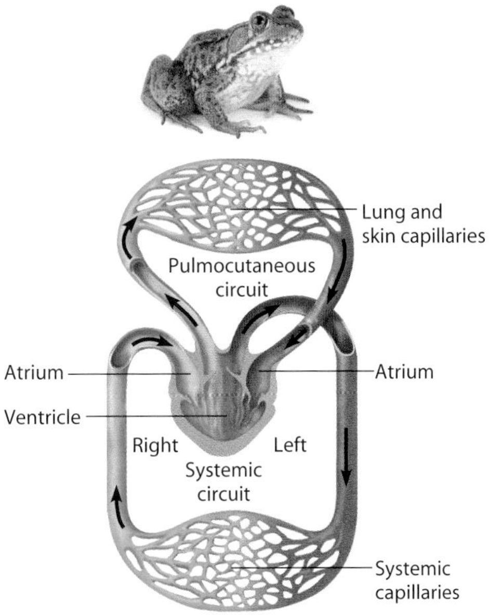

▲ **Figure 2B** The double circulation and three-chambered heart of an amphibian

Chris Mattison/Alamy

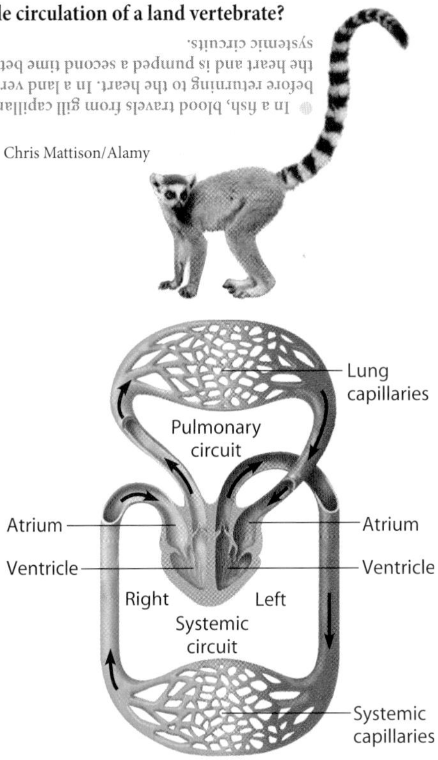

▲ **Figure 2C** The double circulation and four-chambered heart of a bird or mammal

The Human Cardiovascular System and Heart

3 The human cardiovascular system illustrates the double circulation of mammals

Let's follow the flow of blood through the human circulatory system. Starting in the right ventricle in **Figure 3A**, we trace the pulmonary circuit first. ❶ The right ventricle pumps oxygen-poor blood to the lungs via ❷ the **pulmonary arteries**. As blood flows through ❸ capillaries in the lungs, it takes up O_2 and unloads CO_2. Oxygen-rich blood flows back through ❹ the **pulmonary veins** to ❺ the left atrium. Next, the oxygen-rich blood flows from the left atrium into ❻ the left ventricle.

Now let's trace the systemic circuit. As Figure 3A shows, the left ventricle pumps oxygen-rich blood into ❼ the **aorta**. The aorta is our largest blood vessel, with a diameter of about 2.5 cm, roughly equal to the diameter of a quarter. The first branches from the aorta are the coronary arteries (not shown), which supply blood to the heart muscle itself. Next there are large branches leading to ❽ the head, chest, and arms, and the abdominal regions and legs. For simplicity, Figure 3A does not show the individual organs, but within each organ, arteries lead to arterioles that branch into capillaries. The capillaries rejoin as venules, which lead to veins. ❾ Oxygen-poor blood from the upper

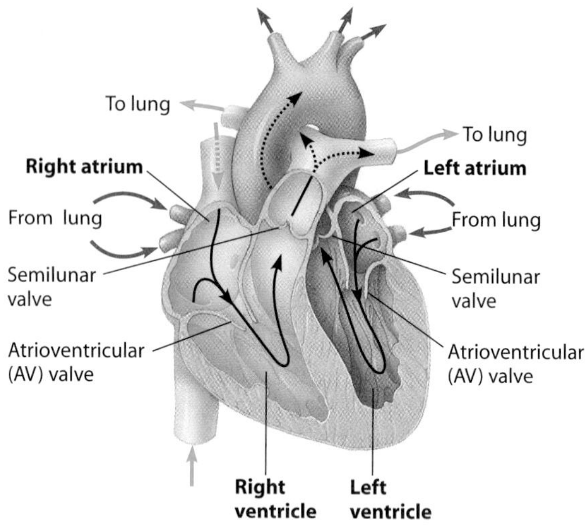

▲ **Figure 3B** Blood flow through the human heart

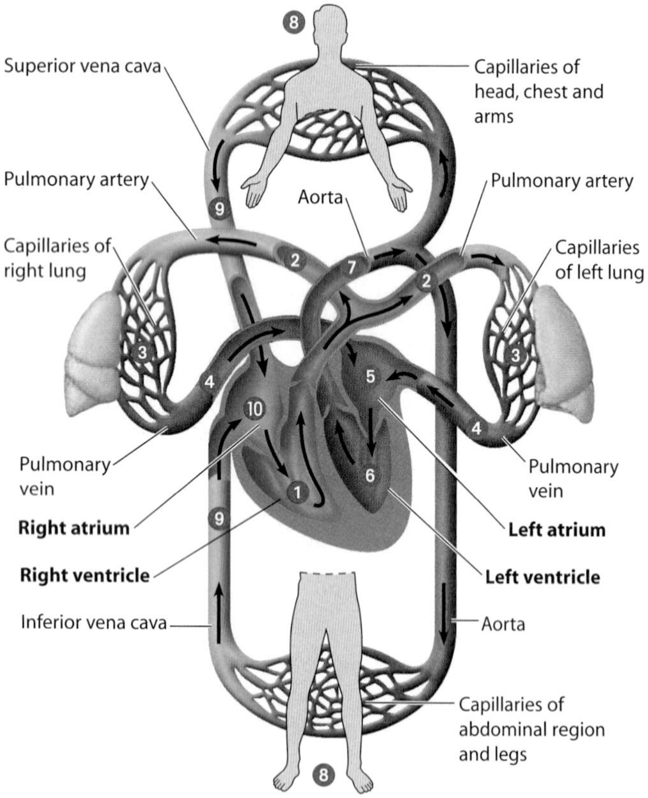

▲ **Figure 3A** Blood flow through the double circulation of the human cardiovascular system

part of the body is channeled into a large vein called the **superior vena cava,** and from the lower part of the body it flows through the **inferior vena cava**. The two venae cavae empty into ❿ the right atrium. As the blood flows from the right atrium into the right ventricle, we complete our journey, only to start the pulmonary circuit again at the right ventricle.

Remember that the path of any single red blood cell is always heart to lung capillaries to heart to body tissue capillaries and back to heart. In one systemic circuit, a blood cell may travel to the brain; in the next (after a pulmonary circuit), it may travel to the legs. A red blood cell never travels from the brain to the legs without first returning to the heart and being pumped to the lungs to be recharged with oxygen.

Figure 3B shows the path of blood through the human heart. About the size of a clenched fist, your heart is enclosed in a sac just under the sternum (breastbone). The heart is formed mostly of cardiac muscle tissue. Its thin-walled atria collect blood returning to the heart. The thicker-walled ventricles pump blood to the lungs and to all other body tissues. Notice that the left ventricle walls are thicker than the right, a reflection of how much farther it pumps blood in the body. Flap-like valves between the atria and ventricles and at the openings to the pulmonary artery and the aorta regulate the direction of blood flow. We'll look at these valves and the functioning of the heart in the next module.

? Why does blood in the pulmonary veins have more O_2 than blood in the venae cavae, which are also veins?

● Pulmonary veins carry blood from the lungs, where it picks up O_2, to the heart. The venae cavae carry blood returning to the heart after delivering O_2 to body tissues.

4 The heart contracts and relaxes rhythmically

The four-chambered heart is the hub of the circulatory system. It separately but simultaneously pumps oxygen-poor blood to the lungs and oxygen-rich blood to the body. Its pumping action occurs as a rhythmic sequence of contraction and relaxation, called the **cardiac cycle.** When the heart contracts, it pumps blood; when it relaxes, blood fills its chambers.

The Cardiac Cycle How long does a cardiac cycle take? If you have a heart rate of 72 beats per minute, your cardiac cycle takes about 0.8 second. **Figure 4** shows that when the heart is relaxed, in the phase called ❶ **diastole**, blood flows into all four of its chambers. Blood enters the right atrium from the venae cavae and the left atrium from the pulmonary veins (see Figure 3A). The valves between the atria and the ventricles (atrioventricular, or AV, valves) are open. The valves leading from the ventricles to the aorta or pulmonary artery (semilunar valves) are closed. Diastole lasts about 0.4 second, during which the ventricles nearly fill with blood.

The contraction phase of the cardiac cycle is called **systole.** ❷ Systole begins with a very brief (0.1-second) contraction of the atria that completely fills the ventricles with blood (atrial systole). ❸ Then the ventricles contract for about 0.3 second (ventricular systole). The force of their contraction closes the AV valves, opens the semilunar valves located at the exit from each ventricle, and pumps blood into the large arteries. Blood flows into the relaxed atria during the second part of systole, as the green arrows in step 3 indicate.

Because it pumps blood to your whole body, the left ventricle contracts with greater force than the right. Both ventricles, however, pump the same volume of blood. The volume of blood that each ventricle pumps per minute is called **cardiac output**. This volume is equal to the amount of blood pumped each time a ventricle contracts (about 70 mL, or a little more than $\frac{1}{4}$ cup, for the average person) times the **heart rate** (number of beats per minute). At an average resting heart rate of 72 beats per minute, cardiac output would be calculated as 70 mL/beat $\times$ 72 beats/min = about 5 L/min, roughly equivalent to the total volume of blood in your body. Thus, a drop of blood can travel the entire systemic circuit in just 1 minute.

Heart rate and cardiac output vary, depending on age, fitness, and other factors. Both increase, for instance, during heavy exercise, in which cardiac output can increase fivefold, enabling the circulatory system to provide the additional oxygen needed by hardworking muscles. A well-trained athlete's heart may strengthen and enlarge with a resulting increase in the volume of blood a ventricle pumps. Thus, a resting heart rate of an athlete may be as low as 40 beats/min and still produce a normal cardiac output of about 5 L/min. During competition, a trained athlete's cardiac output may increase sevenfold.

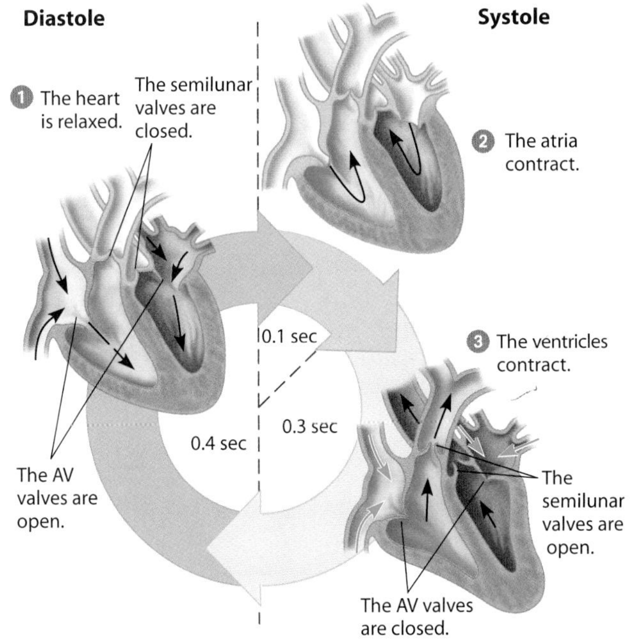
Diastole
Systole
❶ The heart is relaxed.
The semilunar valves are closed.
❷ The atria contract.
0.1 sec
0.3 sec
0.4 sec
❸ The ventricles contract.
The AV valves are open.
The semilunar valves are open.
The AV valves are closed.

▲ **Figure 4** A cardiac cycle in a human with a heart rate of about 72 beats a minute

Heart Valves Notice again in Figure 4 how the heart valves act as one-way doors at the exits of the atria and ventricles during a cardiac cycle. Made of flaps of connective tissue, these valves open when pushed from one side and close when pushed from the other. The powerful contraction of the ventricles forces blood against the AV valves, which closes them and keeps blood from flowing back into the atria. The semilunar valves are pushed open when the ventricles contract. When the ventricles relax, blood in the arteries starts to flow back toward the heart, causing the flaps of the semilunar valves to close and preventing blood from flowing back into the ventricles.

You can follow the closing of the two sets of heart valves either with a stethoscope or by pressing your ear tightly against the chest of a friend (or a friendly dog). The sound pattern is "lub-dup, lub-dup, lub-dup." The "lub" sound comes from the recoil of blood against the closed AV valves. The "dup" is produced by the recoil of blood against the closed semilunar valves.

Someone who is trained can detect the hissing sound of a **heart murmur**, which may indicate a defect in one or more of the heart valves. A murmur occurs when a stream of blood squirts backward through a valve. Some people are born with murmurs, while others have their valves damaged by infection (from rheumatic fever, for instance). Most valve defects do not reduce the efficiency of blood flow enough to warrant surgery. Those that do can be corrected by replacing the damaged valves with synthetic ones or with valves taken from an organ donor (human or other animal, usually a pig).

The next module explores the control of the cardiac cycle.

? **During a cardiac cycle of 0.8 second, the atria are generally relaxed for ______ second.**

0.7

5 The SA node sets the tempo of the heartbeat

In vertebrates, the cardiac cycle originates in the heart itself. Your cardiac muscle cells have their own beat. Each contracts and relaxes without any signal from the nervous system. But if each of these cells has its own intrinsic rhythm, how are their contractions coordinated so that your heart beats as an integrated unit? The answer lies with a group of cells that make up the pacemaker, or **SA (sinoatrial) node**, which sets the rate at which all the muscle cells of your heart contract.

The SA node, situated in the upper wall of the right atrium, generates electrical impulses much like those produced by nerve cells. These signals spread rapidly through the specialized junctions between cardiac muscle cells. **Figure 5A** shows the sequence of electrical events in the heart. ❶ Signals (shown in the figure in yellow) from the SA node spread quickly through both atria, making them contract in unison. ❷ The impulses pass to a relay point called the **AV (atrioventricular) node**, located between the right atrium and right ventricle. Here the signals are delayed about 0.1 second, which ensures that the atria empty completely before the ventricles contract. ❸ Specialized muscle fibers (orange) then relay the signals to the apex of the heart and ❹ up through the walls of the ventricles, triggering the strong contractions that drive the blood out of the heart.

The electrical impulses in the heart generate electrical changes in the skin, which can be detected by electrodes and recorded as an electrocardiogram (ECG or EKG). The yellow portion of the graphs under each heart in Figure 5A indicates the part of an ECG that matches the electrical event shown in the heart. In step 4, the portion of the ECG to the right of the yellow "spike" is the electrical activity that readies the ventricles for the next round of contraction.

An ECG can provide data about heart health, such as the existence of arrhythmias. These are abnormal heart rhythms, including heart rates that are too slow or too fast and fibrillations (flutterings) of the atria or ventricles. In certain kinds of heart disease, the heart's self-pacing system fails to maintain a normal heart rhythm. In such cases, doctors can implant in the chest an artificial pacemaker **(Figure 5B)**, a device that emits electrical signals to trigger normal heartbeats.

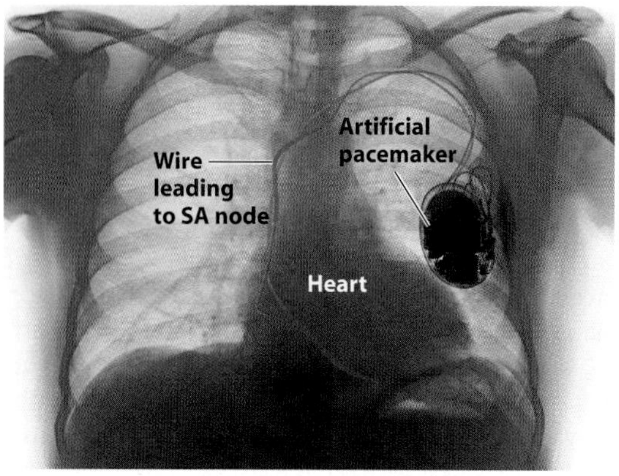
Artificial pacemaker
Wire leading to SA node
Heart

Gustoimages/Photo Researchers

▲ **Figure 5B** An artificial pacemaker implanted in the chest

During a heart attack, the SA node is often unable to maintain a normal rhythm. Electrical shocks applied to the chest by a defibrillator may reset the SA node and restore proper cardiac function. The availability of automatic external defibrillators (AEDs) has saved thousands of lives. Unlike hospital defibrillators, AEDs are designed to be used by laypeople and are placed in public places (such as airports, movie theaters, and shopping malls) where they are quickly accessible.

A variety of cues help regulate the SA node. Two sets of nerves with opposite effects can direct this pacemaker to speed up or slow down, depending on physiological needs and emotional cues. Heart rate is also influenced by hormones, such as epinephrine, the "fight-or-flight" hormone released at times of stress. An increased heart rate provides more blood to muscles that may be needed to "fight or flee."

? A slight decrease in blood pH causes the SA node to increase the heart rate. How would this control mechanism benefit a person during strenuous exercise?

More CO_2 in the blood, as would occur with increased exercise, causes pH to drop. An accelerated heart rate enhances delivery of O_2-rich blood to body tissues and CO_2-rich blood to the lungs for removal of CO_2. (Breathing rate is also speeded up by this mechanism.)

▶ **Figure 5A** The sequence of electrical events in a heartbeat (top) and the corresponding electrocardiogram (bottom). (Yellow represents electrical signals.)

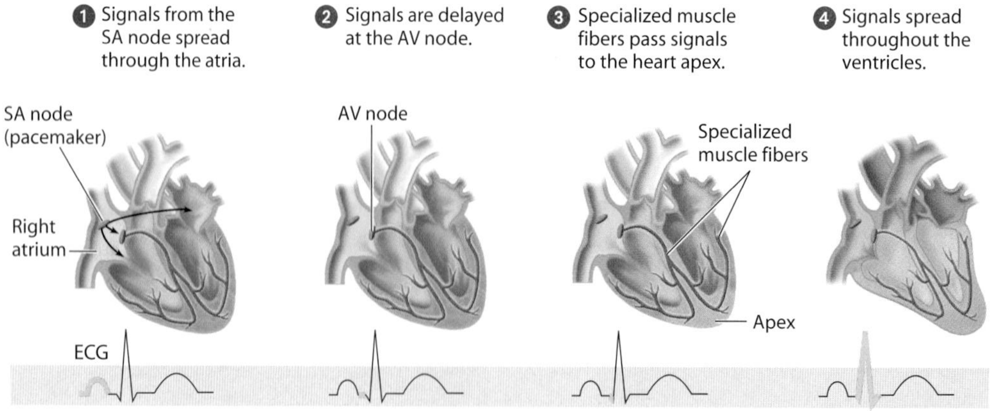
❶ Signals from the SA node spread through the atria.
❷ Signals are delayed at the AV node.
❸ Specialized muscle fibers pass signals to the heart apex.
❹ Signals spread throughout the ventricles.
SA node (pacemaker)
Right atrium
AV node
Specialized muscle fibers
Apex
ECG

CONNECTION 6 What is a heart attack?

Like all of your cells, your heart muscle cells require nutrients and oxygen-rich blood to survive. Indeed, their needs are high, as your heart contracts more than 100,000 times a day. Where blood exits the heart via the aorta, several coronary arteries (shown in red in **Figure 6A**) immediately branch off to feed the heart muscle. If one or more of these blood vessels become blocked, heart muscle cells will quickly die (gray area in Figure 6A). A **heart attack**, also called a myocardial infarction, is the damage or death of cardiac muscle tissue, usually as a result of such blockage. (Rarely, a severe spasm of a coronary artery, possibly related to the use of a drug such as cocaine, may trigger a heart attack.) Approximately one-third of heart attack victims die almost immediately, as their damaged heart can no longer provide the brain and other vital tissues with sufficient oxygen. For those who survive, the ability of the damaged heart to pump blood may be seriously impaired.

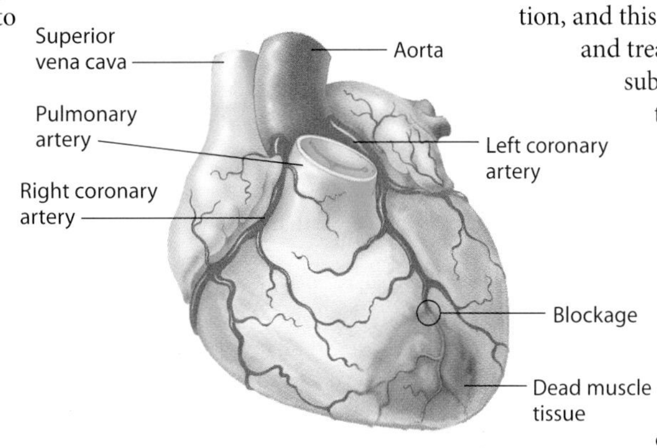

▲ **Figure 6A** Blockage of a coronary artery, resulting in a heart attack

More than half of all deaths in the United States are caused by **cardiovascular disease**—disorders of the heart and blood vessels. A **stroke** is the death of brain tissue due to the lack of O_2 resulting from the rupture or blockage of arteries in the head. The suddenness of a heart attack or stroke belies the fact that the arteries of most victims became impaired gradually by a chronic cardiovascular disease known as **atherosclerosis** (from the Greek *athero*, paste, and *sclerosis*, hardness). During the course of this disease, fatty deposits called plaques develop in the inner walls of arteries, narrowing the passages through which blood can flow (**Figure 6B**).

As a plaque grows, it incorporates fibrous connective tissue and cholesterol, and the walls of the artery become thick and stiff. A blood clot is more likely to become trapped in a vessel that has been narrowed in this way. Furthermore, plaques may rupture and cause blood clots to form, or fragments of the ruptured plaque may become lodged in other narrowed arteries.

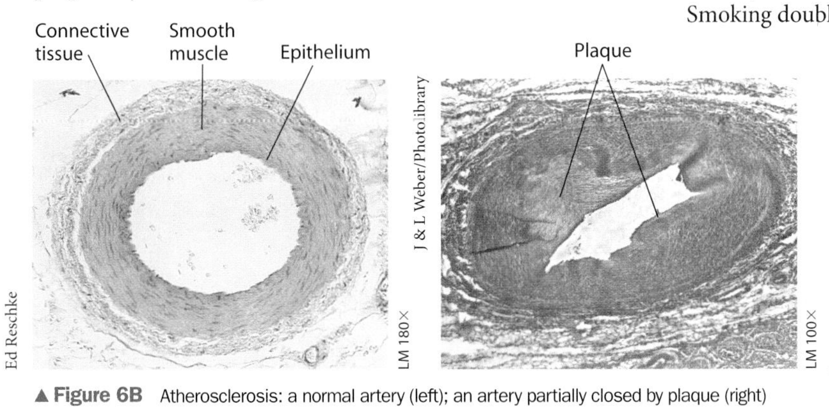

▲ **Figure 6B** Atherosclerosis: a normal artery (left); an artery partially closed by plaque (right)

There are treatments available for cardiovascular disease, and more continue to be developed. Heart attack victims are treated with clot-dissolving drugs, which stop many heart attacks and help prevent damage. Diagnostic tests ranging from cholesterol and blood pressure measurements to sophisticated imaging techniques such as CT and MRI help identify those at risk. Drugs are available that can lower cholesterol and blood pressure, risk factors we discuss in Module 9.

Recently, researchers have found that inflammation plays a central role in atherosclerosis and blood clot formation, and this has led to new ways of diagnosing and treating cardiovascular disease. A substance known as C-reactive protein (CRP) is produced by the liver during episodes of acute inflammation, and significant levels of CRP in the blood appear to be a predictor of cardiovascular disease. Aspirin, which blocks the body's inflammatory response, is typically given during a cardiac episode and is often prescribed in low doses to help prevent the recurrence of heart attacks and stroke.

Angioplasty (insertion of a tiny catheter with a small balloon that is inflated to compress plaques and widen clogged arteries) and stents (small wire mesh tubes that prop arteries open) are common treatments for atherosclerosis. In bypass surgery, a more drastic remedy, blood vessels removed from a patient's legs are sewn into the heart to detour blood around clogged arteries. In extreme cases, a heart transplant may be necessary. With the severe shortage of donor hearts, various artificial pumping devices are being developed.

Fortunately, the U.S. death rate from cardiovascular disease has been cut in half over the past 50 years. Health education, early diagnosis, and reduction of risk factors, particularly smoking, have contributed to this decline.

To some extent, the tendency to develop cardiovascular disease appears to be inherited. There are three behaviors under your control, however, that significantly affect the risk. Smoking doubles the risk of heart attack and harms the circulatory system in other ways. Exercise can cut the risk of heart disease in half. Eating a heart-healthy diet, low in cholesterol and trans and saturated fats, can reduce the risk of atherosclerosis.

? Why is it important to identify a person's risk of cardiovascular disease?

Although lifestyle choices (exercise, not smoking, healthy diet) are beneficial for everyone, they are particularly important for people with an increased risk of cardiovascular disease. Also, such a person may benefit from drugs to reduce the risk of heart attack or the resulting damage.

Structure and Function of Blood Vessels

7 The structure of blood vessels fits their functions

Now that we have explored the structure and function of the heart, let's look at the amazingly extensive series of vessels that transport blood throughout your body.

Functions of Blood Vessels The blood vessels of the circulatory system must have an intimate connection with all the body's tissues. The micrograph in **Figure 7A** shows a capillary that carries oxygenated, nutrient-rich blood to smooth muscle cells. Notice that red blood cells pass single file through the capillary, coming close enough to the surrounding tissue that O_2 can diffuse out of them into the muscle cells.

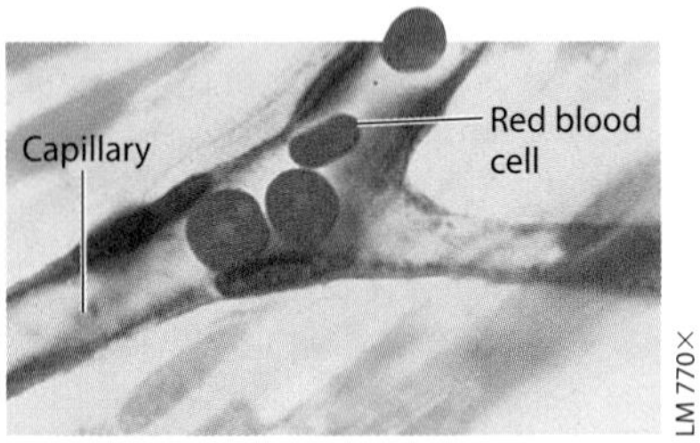

Lennart Nilsson/Scanpix

▲ **Figure 7A** A capillary in smooth muscle tissue

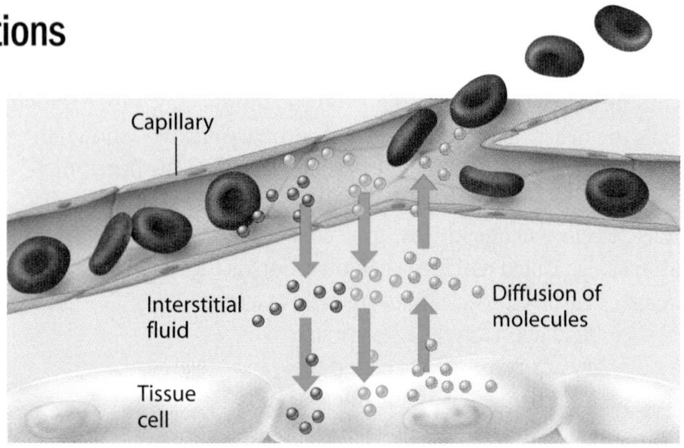

▲ **Figure 7B** Diffusion between blood and tissue cells

In **Figure 7B** (above right), the downward arrows show the route that molecules take in diffusing from blood in a capillary to tissue cells. As we discussed in Module 20.13, cells are immersed in interstitial fluid. Molecules such as O_2 (●) and nutrients (●) diffuse out of a capillary into the interstitial fluid and then from the fluid into a tissue cell.

In addition to transporting O_2 and nutrients, blood vessels convey metabolic wastes to waste disposal organs: CO_2 to the lungs and a variety of other metabolic wastes to the kidneys. The upward arrows in Figure 7B represent the diffusion of waste molecules (●) out of a tissue cell, through the interstitial fluid, and into the capillary.

The circulatory system plays several key roles in maintaining a constant internal environment (homeostasis). By exchanging molecules with the interstitial fluid, it helps control the makeup of the environment in which the tissue cells live. As we'll see in later chapters, the circulatory system is also involved in body defense, temperature regulation, and hormone distribution.

Structure of Blood Vessels **Figure 7C** illustrates the structures of the different kinds of blood vessels and how the vessels are connected. Look first at the capillaries (center). Appropriate to its function of exchanging materials, a capillary has a very thin wall of a single layer of epithelial cells, which is wrapped in a thin basal lamina. The inner surface of the capillary is smooth, which keeps the blood cells from being abraded as they tumble along.

Arteries, arterioles, veins, and venules have thicker walls than capillaries. Their walls have the same smooth epithelium but are reinforced by two other tissue layers. An outer layer of connective tissue with elastic fibers enables the vessels to stretch and recoil. The middle layer consists mainly of smooth muscle. Both these layers are thicker and sturdier in arteries, providing the strength and elasticity to accommodate the rapid flow and high pressure of blood pumped by the heart. Arteries are also able to regulate blood flow by constricting or relaxing their smooth muscle layer. The thinner-walled veins convey blood back to the heart at low velocity and pressure. Within large veins, flaps of tissue act as one-way valves, which permit blood to flow only toward the heart.

? **How does the structure of a capillary relate to its function?**

● The small diameter and thin walls of capillaries facilitate the exchange of substances between blood and interstitial fluid.

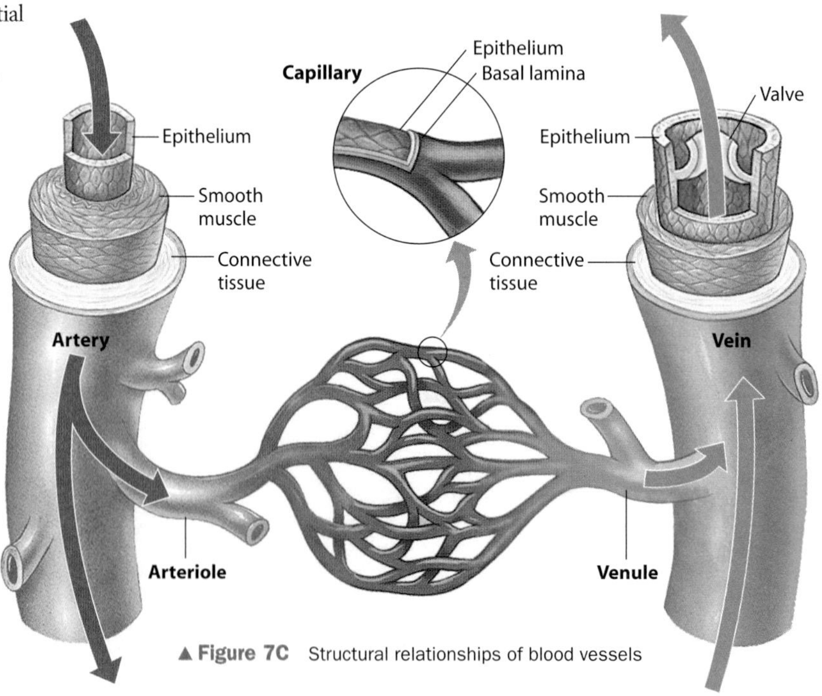

▲ **Figure 7C** Structural relationships of blood vessels

8 Blood pressure and velocity reflect the structure and arrangement of blood vessels

Now that you've looked at the structure of blood vessels, let's explore the forces that move your blood through these vessels.

Blood Pressure **Blood pressure** is the force that blood exerts against the walls of your blood vessels. Created by the pumping of the heart, blood pressure drives the flow of blood from the heart through arteries and arterioles to capillary beds.

When the ventricles contract, blood is forced into the arteries faster than it can flow into the arterioles. This stretches the elastic walls of the arteries. You can feel this rhythmic stretching of the arteries when you measure your heart rate by taking your **pulse**. You can see this surge in pressure (expressed in millimeters of mercury, mm Hg) as the pressure peaks in the top graph of **Figure 8A**. The pressure caused by ventricular contraction is called systolic pressure. The elastic arteries snap back during diastole, maintaining pressure on the blood and a continuous flow of blood into arterioles and capillaries. The dips in pressure in the top graph represent diastolic pressure.

The diagram at the center of Figure 8A shows the relative sizes and numbers of blood vessels as blood flows from the aorta through arteries to capillaries and back through veins to the venae cavae. Blood pressure is highest in the aorta and arteries and declines abruptly as the blood enters the arterioles. The pressure drop results mainly from the resistance to blood flow caused by friction between the blood and the walls of the millions of narrow arterioles and capillaries.

Blood pressure in the arteries depends on the volume of blood pumped into the aorta and also on the restriction of blood flow into the narrow openings of the arterioles. These openings are controlled by smooth muscles. When the muscles relax, the arterioles dilate, and blood flows through them more readily, causing a fall in blood pressure. Physical and emotional stress can raise blood pressure by triggering nervous and hormonal signals that constrict these blood vessels. Regulatory mechanisms coordinate cardiac output and changes in the arteriole openings to maintain adequate blood pressure as demands on the circulatory system change.

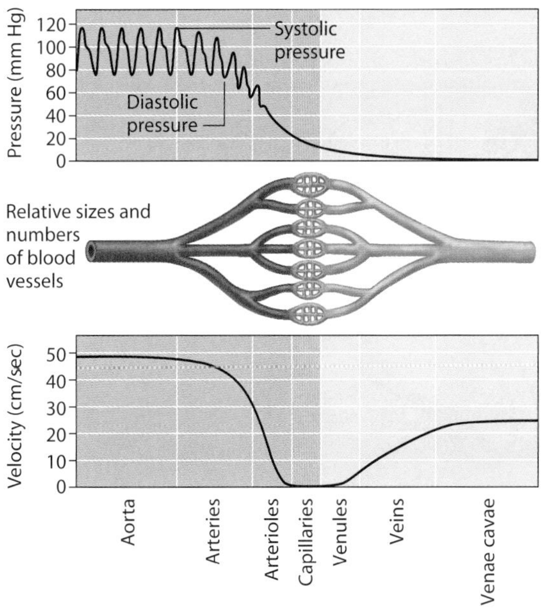

▲ **Figure 8A** Blood pressure and velocity in the blood vessels

Blood Velocity The blood's velocity (rate of flow, expressed in centimeters per second, cm/sec) is illustrated in the bottom graph of Figure 8A. As the figure shows, velocity declines rapidly in the arterioles, drops to almost zero in the capillaries, and then speeds up in the veins. What accounts for these changes? As larger arteries divide into smaller and more numerous arterioles, the total combined cross-sectional area of the many vessels is much greater than the diameter of the one artery that feeds into them. If there were only one arteriole per artery, the blood would actually flow faster through the arteriole, the way water does when you add a narrow nozzle to a garden hose. However, there are many arterioles per artery, so the effect is like taking the nozzle off the hose: As you increase the diameter of a pipe, the flow rate goes down.

The cross-sectional area is greatest in the capillaries, and the velocity of blood is slowest through them. The steady, leisurely flow of blood in the capillaries enhances the exchange of substances with body cells.

By the time blood reaches the veins, its pressure has dropped to near zero. The blood has encountered so much resistance as it passes through the millions of tiny arterioles and capillaries that the force from the pumping heart no longer propels it. How, then, does blood return to the heart, especially when it must travel up the legs of a tall giraffe against gravity? As we discussed in the chapter introduction, whenever the body moves, muscles squeeze blood through the veins. **Figure 8B** shows how veins are often sandwiched between skeletal muscles and how one-way valves allow the blood to flow only toward the heart. Breathing also helps return blood to the heart. When you inhale, the change in pressure within your chest cavity causes the large veins near your heart to expand and fill.

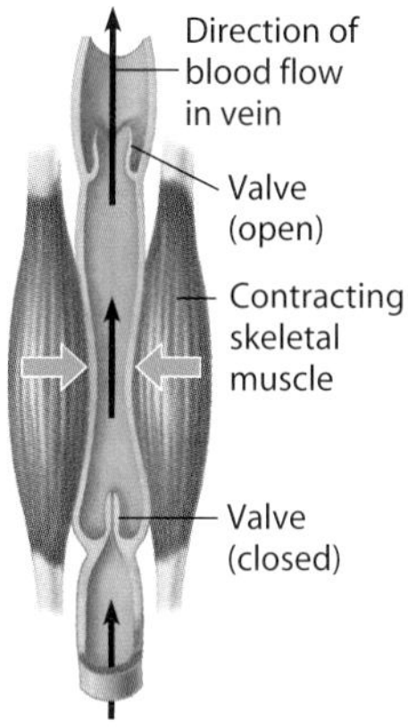

▲ **Figure 8B**
Blood flow in a vein

Next we look at how blood pressure is measured.

? If blood pressure in the veins drops to zero, why does blood velocity increase as blood flows from venules to veins?

The total diameter of the veins is less than the venules. The velocity increases, just as water flows faster when a nozzle narrows the opening of a hose.

CONNECTION 9 Measuring blood pressure can reveal cardiovascular problems

Blood pressure is generally measured in an artery in the arm at the same height as the heart. As indicated in Figure 9, ❶ a typical blood pressure for a healthy young adult is about 120/70. The first (sometimes referred to as the top) number is the systolic pressure; the second (or bottom) number is the diastolic pressure (see Module 8).

Figure 9 shows how blood pressure is measured using a sphygmomanometer, or blood pressure cuff. ❷ Once wrapped around the upper arm, where large arteries are accessible, the cuff is inflated until the pressure is strong enough to close the artery and cut off blood flow to the lower arm. ❸ A stethoscope is used to listen for sounds of blood flow below the cuff as the cuff is gradually deflated. The first sound of blood spurting through the constricted artery indicates that the pressure exerted by the cuff has fallen just below that in the artery. The pressure at this point is the systolic pressure. The sound of blood flowing unevenly through the artery continues until the pressure of the cuff falls below the pressure of the artery during diastole. ❹ Blood now flows continuously through the artery, and the sound of blood flow ceases. The reading at this point is the diastolic pressure.

Optimal blood pressure for adults is below 120 mm Hg for systolic pressure and below 80 mm Hg for diastolic pressure. Lower values are generally considered better, except in rare cases where low blood pressure may indicate a serious underlying condition (such as an endocrine disorder, malnutrition, or internal bleeding). Blood pressure higher than the normal range, however, may indicate a serious cardiovascular disorder.

High blood pressure, or **hypertension**, is persistent systolic blood pressure higher than 140 mm Hg and/or diastolic blood pressure higher than 90 mm Hg. Hypertension affects almost 30% of the adult population in the United States. It is sometimes called a "silent killer" because high blood pressure often displays no outward symptoms for years but may be leading to severe health problems.

High blood pressure harms the cardiovascular system in several ways. Elevated pressure requires the heart to work harder to pump blood throughout the body, and over time the left ventricle may enlarge as a result. When the coronary blood supply does not keep up with the demands of this increase in muscle mass, the heart muscle weakens. In addition, the increased force on arterial walls throughout the body causes tiny ruptures. The resulting inflammation promotes plaque formation, aggravating atherosclerosis (see Module 6) and increasing the risk of blood clot formation. Prolonged hypertension is the major cause of heart attack, heart disease, stroke, and also kidney failure, as renal arteries and arterioles may be damaged by high pressure.

In many patients, the exact cause of hypertension cannot be firmly established. Some predispositions to hypertension cannot be avoided. For example, males have a greater risk of high blood pressure up to age 55; females have a greater risk after menopause. African Americans are more prone to hypertension than Caucasians. Blood pressure generally increases with age, as does the prevalence of hypertension. Heredity also plays a role; children of parents with hypertension are twice as likely to develop the condition.

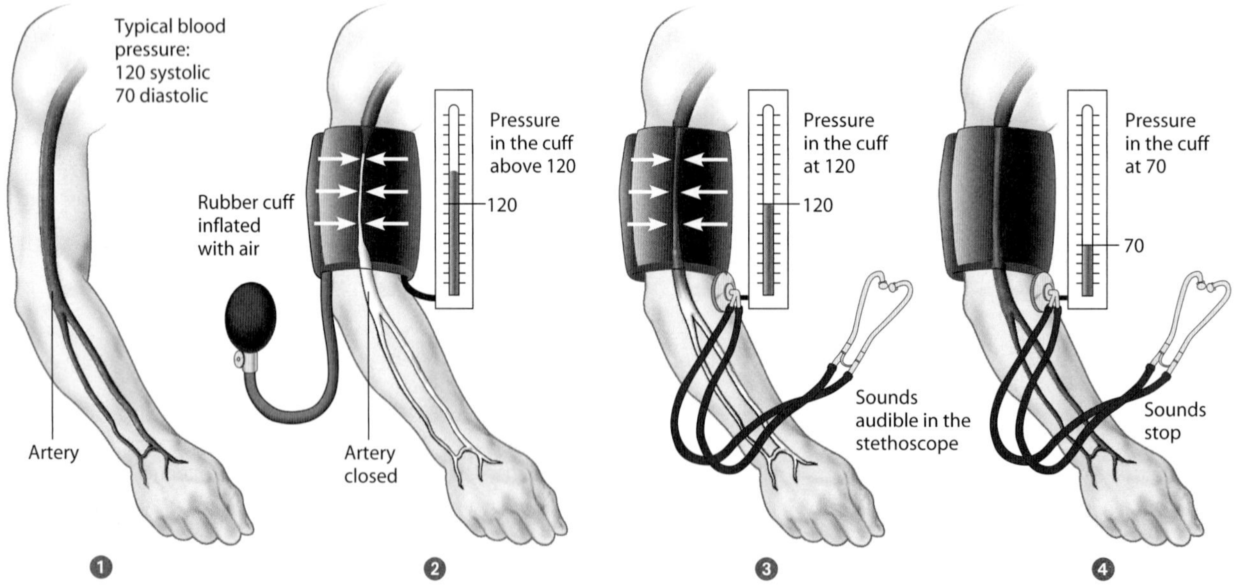

▲ Figure 9 Measuring blood pressure

However, no matter how many unavoidable predispositions a person may have, there are lifestyle changes that can prevent or control hypertension in just about everybody: eating a heart-healthy diet, not smoking, avoiding excess alcohol (more than two drinks per day), exercising regularly (30 minutes of moderate activity on most days), and maintaining a healthy weight. Many people associate salt with high blood pressure, but it is a contributing factor only in a small percentage of people. If lifestyle changes don't lower blood pressure, there are several effective antihypertensive medications.

? Listening with a stethoscope below a sphygmomanometer cuff, you hear sounds that begin at 140 mm Hg and cease at 95 mm Hg. What are the systolic and diastolic blood pressure readings for this person, and do they indicate a health risk?

Systolic = 140; diastolic = 95; blood pressure = 140/95; yes, this person has hypertension and may be at risk for cardiovascular disease and kidney failure.

10 Smooth muscle controls the distribution of blood

You learned in Module 8 that smooth muscles in arteriole walls can influence blood pressure by changing the resistance to blood flow out of the arteries and into arterioles. The smooth muscles in the arteriole walls also regulate the distribution of blood to the capillaries of the various organs. At any given time, only about 5–10% of your body's capillaries have blood flowing through them. However, each tissue has many capillaries, so every part of your body is supplied with blood at all times. Capillaries in a few organs, such as the brain, heart, kidneys, and liver, usually carry a full load of blood. In many other sites, however, blood supply varies as blood is diverted from one place to another, depending on need.

Figure 10 illustrates a second mechanism that regulates the flow of blood into capillaries. Notice that in both parts of this figure there is a capillary called a thoroughfare channel, through which blood streams directly from arteriole to venule. This channel is always open. Rings of smooth muscle located at the entrance to capillary beds, called precapillary sphincters, regulate the passage of blood into the branching capillaries. As you can see in the figure, ❶ blood flows through a capillary bed when its precapillary sphincters are relaxed. ❷ It bypasses the capillary bed when the sphincters are contracted.

After a meal, for instance, precapillary sphincters in the wall of your digestive tract relax, letting a large quantity of blood pass through the capillary beds. The products of digestion are absorbed into the blood, which delivers them to the rest of the body. During strenuous exercise, many of the capillaries in the digestive tract are closed off, and blood is supplied more generously to your skeletal muscles. This is one reason why heavy exercise right after eating may cause indigestion or abdominal cramps (and why you shouldn't swim too soon after eating—just like mom always said). Blood flow to your skin is regulated to help control body temperature. An increase in blood supply to the skin helps to release the excess heat generated by exercise.

Nerve impulses, hormones, and chemicals produced locally influence the contraction of the smooth muscles that regulate the flow of blood to capillary beds. For example, the chemical histamine released by cells at a wound site causes smooth muscle relaxation, increasing blood flow and the supply of infection-fighting white blood cells.

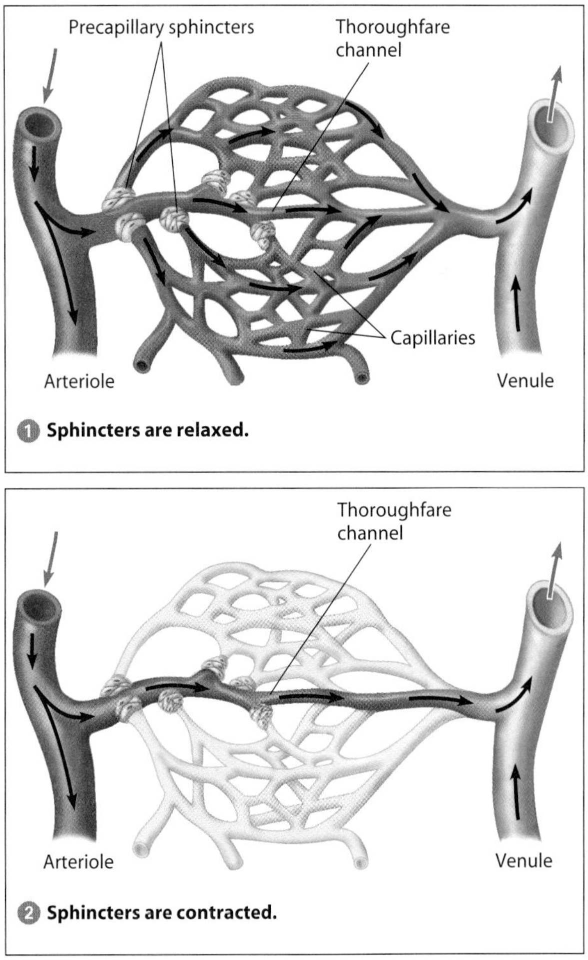

▲ **Figure 10** The control of capillary blood flow by precapillary sphincters

Next we consider how substances are exchanged when blood flows through a capillary.

? What two mechanisms restrict the distribution of blood to a capillary bed?

Constriction of an arteriole, so that less blood reaches a capillary bed, and contraction of precapillary sphincters, so that blood flows through thoroughfare channels only, not capillary beds

11 Capillaries allow the transfer of substances through their walls

Capillaries are the only blood vessels with walls thin enough for substances to cross between the blood and the interstitial fluid that bathes the body cells. This transfer of materials is the most important function of the circulatory system, so let's examine the process more carefully.

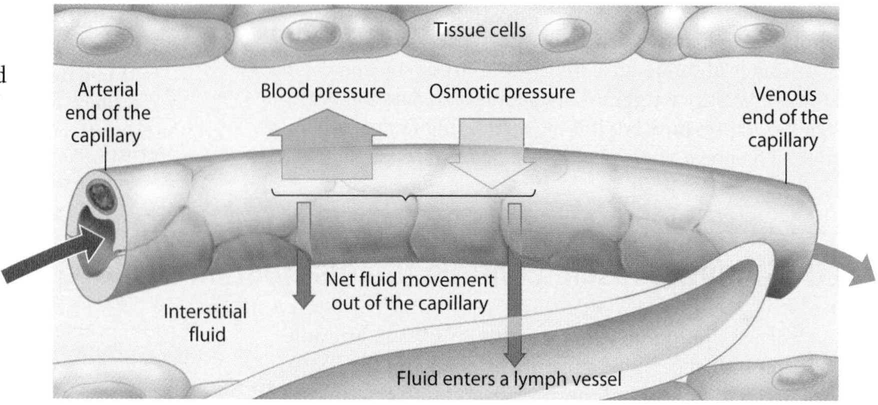

▲ **Figure 11B** The movement of fluid out of a capillary and into a lymph vessel

Capillary Structure and Function

Take a look at **Figure 11A** below. It shows a cross section of a capillary that serves skeletal muscle cells, illustrated with a micrograph and a labeled drawing. The drawing will help you interpret the micrograph. The capillary wall consists of adjoining epithelial cells that enclose a lumen, or interior space. The lumen is just large enough for red blood cells to tumble through in single file. The nucleus you see in the figure belongs to one of the two cells making up this portion of the capillary wall. (The other cell's nucleus does not appear in this particular cross section.) Interstitial fluid, shown in blue in the drawing, fills the space between the capillary and the muscle cells.

The exchange of substances between the blood and the interstitial fluid occurs in several ways. Some nonpolar molecules, such as O_2 and CO_2, simply diffuse through the epithelial cells of the capillary wall. Larger molecules may be carried across an epithelial cell in vesicles that form by endocytosis on one side of the cell and then release their contents by exocytosis on the other side.

In addition, the capillary wall is leaky; there are small pores in the wall and narrow clefts between the epithelial cells making up the wall (see Figure 11A). Water and small solutes, such as sugars and salts, move freely through these pores and clefts. Blood cells and dissolved proteins generally remain inside the capillary because they are too large to fit through these passageways. Much of the exchange between blood and interstitial fluid is the result of the pressure-driven flow of fluid (consisting of water and dissolved solutes) through these openings.

The diagram in **Figure 11B** above shows part of a capillary with blood flowing from its arterial end (near an arteriole) to its venous end (near a venule). What are the active forces that drive fluid into or out of the capillary? One of these forces is blood pressure, which tends to push fluid outward. The other is osmotic pressure, a force that tends to pull fluid back because the blood has a higher concentration of solutes than the interstitial fluid. Proteins dissolved in the blood account for much of this high solute concentration. On average, blood pressure is greater than the opposing forces, leading to a net loss of fluid from capillaries.

Fluid Return via the Lymphatic System Each day, you lose approximately 4–8 L of fluid from your capillaries to the surrounding tissues. The lost fluid is picked up by your lymphatic system, which includes a network of tiny vessels intermingled among the capillaries (see Figure 11B). After diffusing into these vessels, the fluid, now called lymph, is returned to your circulatory system through ducts that join with large veins in your neck. You learned in Module 21.10 that lymph vessels also transport fats from the small intestine to the blood.

Now that we have examined the structure and function of the heart and blood vessels, we turn our focus to the composition of the blood.

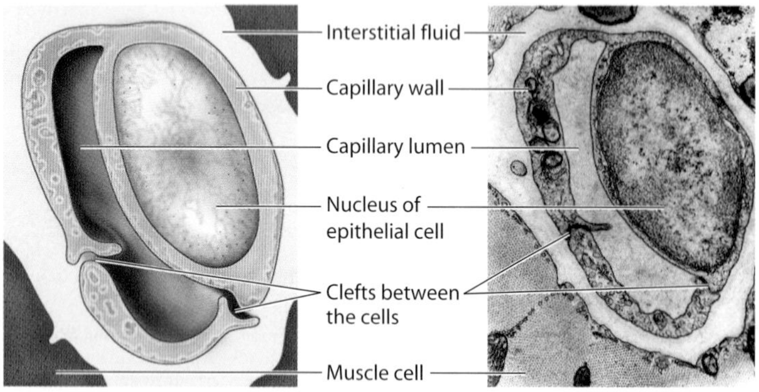

▲ **Figure 11A** A capillary in cross section

? **Explain how a severe protein deficiency in the diet that decreases the concentration of blood plasma proteins can cause edema, the accumulation of fluid in body tissues.**

Decreased blood protein concentration reduces the osmotic gradient across the capillary walls, thus reducing the pull of fluid back into the capillaries.

Structure and Function of Blood

12 Blood consists of red and white blood cells suspended in plasma

Your body has about 5 L of blood (about 5.3 quarts). Blood consists of several types of cells suspended in a liquid called **plasma**. When a blood sample is taken, the cells can be separated from the plasma by spinning the sample in a centrifuge, after adding a chemical to prevent the blood from clotting. The cellular elements (cells and cell fragments), which make up about 45% of the volume of blood, settle to the bottom of the centrifuge tube, underneath the transparent, straw-colored plasma (**Figure 12**).

Plasma is about 90% water. Among its many solutes are inorganic salts in the form of dissolved ions. The functions of these ions (also called electrolytes) include keeping the pH of blood at about 7.4 and maintaining the osmotic balance between blood and interstitial fluid.

Plasma proteins also help maintain osmotic balance. Some proteins act as buffers. Fibrinogen is a plasma protein that functions in blood clotting, and immunoglobulins are proteins important in body defense (immunity).

Plasma also contains a wide variety of substances in transit from one part of the body to another, such as nutrients, waste products, O_2, CO_2, and hormones.

There are two classes of cells suspended in blood plasma: **red blood cells** and **white blood cells**. Also suspended in plasma are **platelets**, cell fragments that are involved in the process of blood clotting.

Red blood cells are also called **erythrocytes**. The structure of a red blood cell suits its main function, which is to carry oxygen. Human red blood cells are small biconcave disks, thinner in the center than at the sides. Their small size and shape create a large surface area across which oxygen can diffuse. Red blood cells lack a nucleus, allowing more room to pack in hemoglobin. Each tiny red blood cell contains about 250 million molecules of hemoglobin and thus can transport about a billion oxygen molecules. A single drop of your blood contains about 25 million cells (a drop is about 50 mL), which means you have about 25 trillion red blood cells in your 5 L of blood. Think about how many molecules of oxygen are traveling through your bloodstream!

There are five major types of white blood cells, or **leukocytes**, as pictured in Figure 12: monocytes, neutrophils, basophils, eosinophils, and lymphocytes. Their collective function is to fight infections. For example, monocytes and neutrophils are **phagocytes**, which engulf and digest bacteria and debris from your own dead cells. White blood cells actually spend much of their time moving through interstitial fluid, where most of the battles against infection are waged. There are also great numbers of white cells in the lymphatic system.

? For every one white blood cell in normal human blood, there are about ______ to ______ red blood cells.

500 . . . 1,000

Plasma (55%)

Constituent	Major functions
Water	Solvent for carrying other substances
Ions (blood electrolytes) Sodium Potassium Calcium Magnesium Chloride Bicarbonate	Osmotic balance, pH buffering, and maintaining ion concentration of interstitial fluid
Plasma proteins	Osmotic balance and pH buffering
Fibrinogen	Clotting
Immunoglobulins (antibodies)	Defense
Substances transported by blood Nutrients (e.g., glucose, fatty acids, vitamins) Waste products of metabolism Respiratory gases (O_2 and CO_2) Hormones	

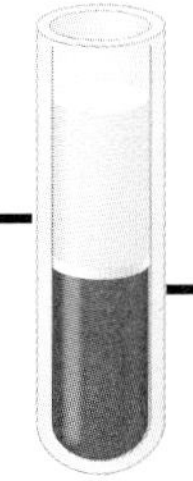

Centrifuged blood sample

Cellular elements (45%)

Cell type	Number per μL (mm^3) of blood	Functions
Red blood cells (erythrocytes) 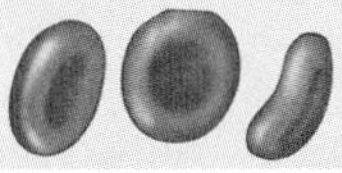	5–6 million	Transport of O_2 and some CO_2
White blood cells (leukocytes) 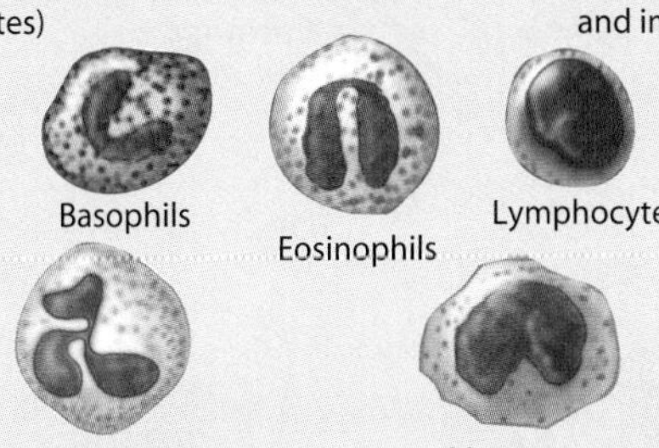	5,000–10,000	Defense and immunity
Platelets 	250,000–400,000	Blood clotting

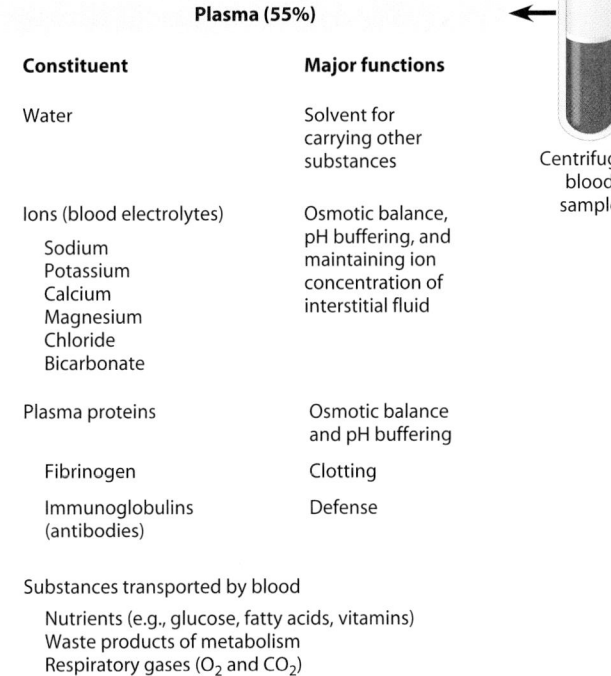

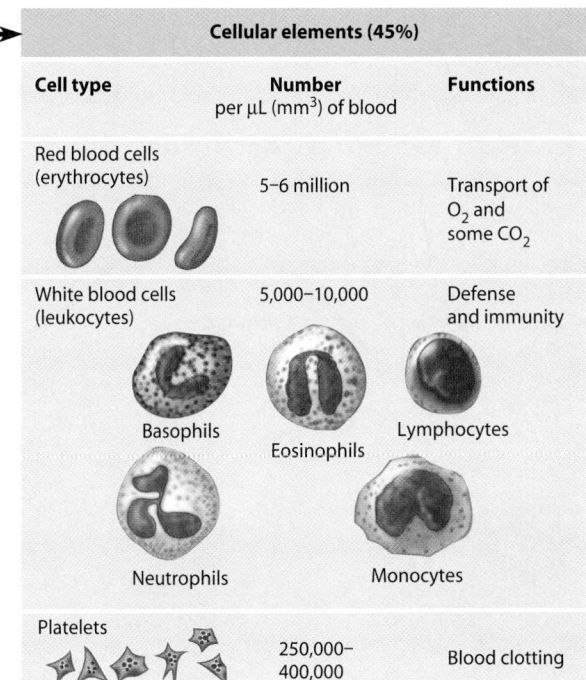

▲ **Figure 12** The composition of blood

CONNECTION 13 Too few or too many red blood cells can be unhealthy

Adequate numbers of red blood cells **(Figure 13)** are essential for healthy body function. After circulating for three or four months, red blood cells are broken down and their mol-ecules recycled. Much of the iron removed from the hemoglobin is returned to the bone marrow, where new red blood cells are formed at the amazing rate of 2 million per second.

An abnormally low amount of hemoglobin or a low number of red blood cells is a condition called **anemia**. An anemic person feels constantly tired because body cells do not get enough oxygen. Anemia can result from a variety of factors, including excessive blood loss, vitamin or mineral deficiencies, and certain forms of cancer. Iron deficiency is the most common cause. Women are more likely to develop iron deficiency than men because of blood loss during menstruation. Pregnant women are generally prescribed iron supplements to support the developing fetus and placenta.

The production of red blood cells in the bone marrow is controlled by a negative-feedback mechanism that is sensitive to the amount of oxygen reaching the tissues via the blood. If the tissues are not receiving enough oxygen, the kidneys produce a hormone called **erythropoietin (EPO)** that stimulates the bone marrow to produce more red blood cells. Patients on kidney dialysis often have very low red blood cell counts because their kidneys do not produce enough erythropoietin. Genetically engineered EPO has significantly helped these patients, as well as cancer and AIDS patients, who also often suffer from anemia.

Ingram Publishing/Photolibrary

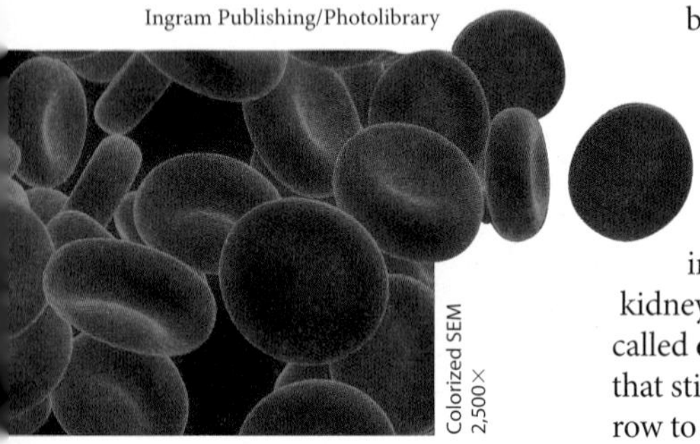

▲ **Figure 13**
Human red blood cells

One of the physiological adaptations of individuals who live at high altitudes, where oxygen levels are low, is the production of more red blood cells. Many athletes train at high altitudes to benefit from this effect. But other athletes take more drastic and illegal measures to increase the oxygen-carrying capacity of their blood and improve their performance. Injecting synthetic EPO can increase normal red blood cell volume from 45% to as much as 65%. Other athletes seek an unfair advantage by blood doping—withdrawing and storing their red blood cells and then reinjecting them before a competition. Athletic commissions test for these practices by measuring the percentage of red blood cells in the blood volume or by testing for EPO-related drugs. In recent years, a number of well-known runners and cyclists have tested positive for these drugs and have forfeited both their records and their right to compete in the future.

But there can be even more serious consequences. In some athletes, a combination of dehydration from a long race and blood already thickened by an increased number of red blood cells has led to severe medical problems, such as clotting, stroke, heart failure, and even death. Indeed, EPO-related drugs have been blamed for the deaths of dozens of athletes.

? Why might increasing the number of red blood cells result in greater endurance and speed?

The additional red blood cells increase the oxygen-carrying capacity of blood and thus the oxygen supply to working muscles.

14 Blood clots plug leaks when blood vessels are injured

You may get cuts and scrapes from time to time, yet you don't bleed to death from such minor injuries because your blood contains self-sealing materials that are activated when blood vessels are injured. These sealants are platelets and the previously mentioned plasma protein **fibrinogen**.

What happens when you sustain an injury? Your body's immediate response is to constrict the damaged blood vessel, thereby reducing blood loss and allowing time for repairs to begin. **Figure 14A** shows the stages of the clotting process. ❶ When the epithelium (shown as tan) lining a blood vessel is damaged, connective tissue in the vessel wall is exposed to blood. Platelets (purple) rapidly adhere to the exposed tissue and release chemicals that make nearby platelets sticky. ❷ Soon a cluster of sticky platelets forms a plug that quickly provides protection against additional blood loss. Clotting factors in the plasma and released from the clumped platelets set off a chain of reactions that culminate in the formation of a reinforced patch, called a scab when it's on the skin. In this complex process, an activated enzyme converts fibrinogen to a threadlike protein called **fibrin**. ❸ Threads of fibrin (white) reinforce the plug, forming a fibrin clot.

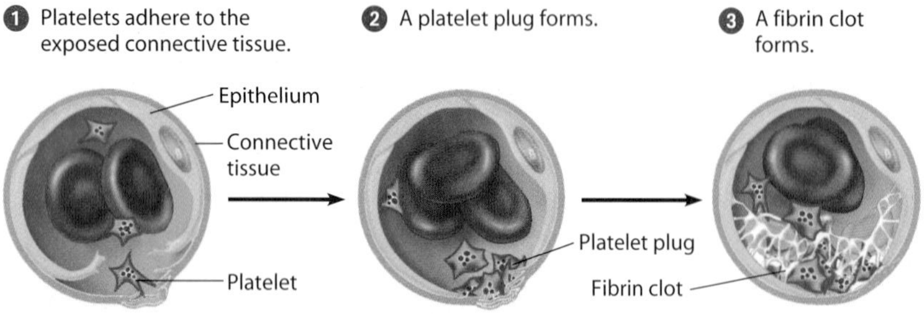

▲ **Figure 14A** The blood-clotting process

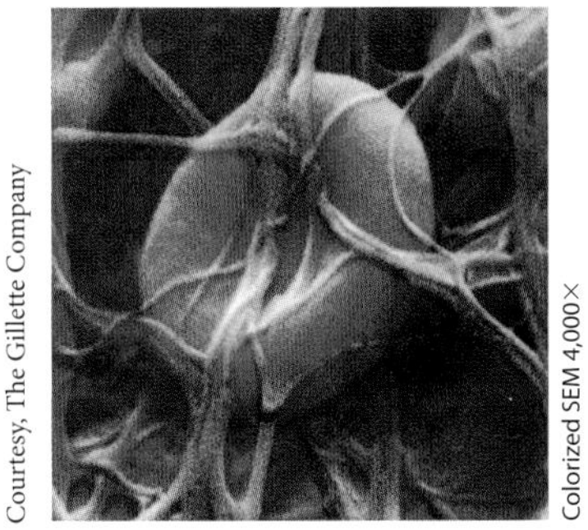

▲ **Figure 14B** A fibrin clot

Figure 14B is a micrograph of a fibrin clot. Within an hour after a fibrin clot forms, the platelets contract, pulling the torn edges closer together and reducing the size of the area in need of repair. Chemicals released by platelets also stimulate cell division in smooth muscle and connective tissue, initiating the healing process.

The clotting mechanism is so important that any defect in it can be life-threatening. In the inherited disease hemophilia, excessive, sometimes fatal bleeding occurs from even minor cuts or bruises. In other people, the formation of blood clots in the *absence* of injury can be a problem. Anticlotting factors in the blood normally prevent spontaneous clotting. If blood clots form within a vessel, they can block the flow of blood. Such a clot, called a thrombus, can be dangerous if it blocks a blood vessel of the heart or brain (see Module 6). Aspirin, heparin, and warfarin are anticoagulant drugs that work by different mechanisms to prevent undesirable clotting in patients at risk for heart attack or stroke.

? What is the role of platelets in blood clot formation?

Platelets adhere to exposed connective tissue and release various chemicals that help a platelet plug form and activate the pathway leading to a fibrin clot. (Also, some of the chemicals promote healing in other ways.)

CONNECTION

15 Stem cells offer a potential cure for blood cell diseases

The red marrow inside bones such as the ribs, vertebrae, sternum, and pelvis is a spongy tissue in which unspecialized cells called **multipotent stem cells** differentiate into blood cells. Multipotent stem cells are so named because they have the ability to form multiple types of cells, in this case all the blood cells and platelets. When a stem cell divides, one daughter cell remains a stem cell while the other can take on a specialized function.

As shown in **Figure 15**, multipotent stem cells in bone marrow give rise to two different types of stem cells: lymphoid stem cells and myeloid stem cells. Lymphoid stem cells produce two different types of lymphocytes, which function in the immune system. Myeloid stem cells can differentiate into other white blood cells, platelets, and erythrocytes. The stem cells continually produce all the blood cells needed throughout life.

Leukemia is cancer of the white blood cells, or leukocytes. Because cancerous cells grow uncontrollably, a person with leukemia has an unusually high number of leukocytes, most of which are defective. These overabundant cells crowd out the bone marrow cells that are developing into red blood cells and platelets, causing severe anemia and impaired clotting.

Leukemia is usually fatal unless treated, and not all cases respond to the standard cancer treatments—radiation and chemotherapy. An alternative treatment involves destroying the cancerous bone marrow completely and replacing it with healthy bone marrow. Injection of as few as 30 stem cells can repopulate the blood and immune system. Patients may be treated with their own bone marrow: Marrow from the patient is harvested, processed to remove as many cancerous cells as possible, and then reinjected. Alternatively, a suitable donor, often a sibling, may provide the marrow. Such a patient requires lifelong treatment with drugs that suppress the tendency of some of the transplanted marrow cells to "reject" the cells of the recipient.

Stem cell research holds great promise, and leukemia is just one of several blood diseases that may be treated by bone marrow stem cells. Recently, researchers have been able to isolate bone marrow stem cells and grow them in the laboratory. In a few cases, they have induced these stem cells to differentiate into more than just blood cells. Thus, these adult stem cells may eventually provide cells for human tissue and organ transplants.

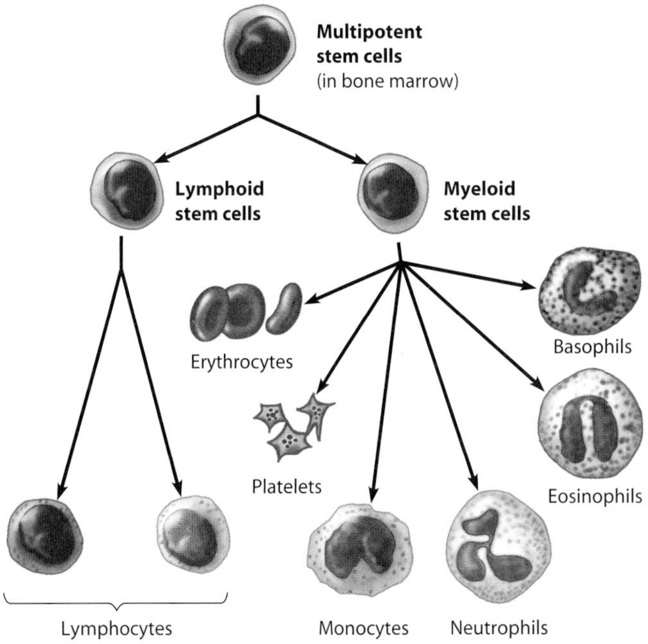

▲ **Figure 15** Differentiation of blood cells from stem cells

? Why would a leukemia patient's bone marrow need to be destroyed and then replaced with a transplant?

Destruction of the bone marrow would be necessary to kill the cancerous leukocytes. The patient would need replacement multipotent stem cells to continue making both red and white blood cells.

R E V I E W

For Practice Quizzes, BioFlix, MP3 Tutors, and Activities, go to www.masteringbiology.com.

Reviewing the Concepts

Circulatory Systems (1–2)

1 Circulatory systems facilitate exchange with all body tissues. Gastrovascular cavities function in both digestion and transport. In open circulatory systems, a heart pumps fluid through open-ended vessels to bathe tissue cells directly. In closed circulatory systems, a heart pumps blood, which travels through arteries to capillaries to veins and back to the heart.

2 Vertebrate cardiovascular systems reflect evolution. A fish's two-chambered heart pumps blood in a single circuit. Land vertebrates have double circulation with a pulmonary and a systemic circuit. Amphibians and many reptiles have three-chambered hearts; birds and mammals have four-chambered hearts.

The Human Cardiovascular System and Heart (3–6)

3 The human cardiovascular system illustrates the double circulation of mammals. The mammalian heart has two thin-walled atria and two thick-walled ventricles. The right side of the heart receives and pumps oxygen-poor blood; the left side receives oxygen-rich blood from the lungs and pumps it to all other organs.

4 The heart contracts and relaxes rhythmically. During diastole of the cardiac cycle, blood flows from the veins into the heart chambers; during systole, contractions of the atria push blood into the ventricles, and then stronger contractions of the ventricles propel blood into the large arteries. Cardiac output is the amount of blood per minute pumped by a ventricle. Heart valves prevent the backflow of blood.

5 The SA node sets the tempo of the heartbeat. The SA node, or pacemaker, generates electrical signals that trigger contraction of the atria. The AV node relays these signals to the ventricles. An electrocardiogram records the electrical changes.

6 What is a heart attack? A heart attack is damage or death of cardiac muscle, usually resulting from a blocked coronary artery.

Structure and Function of Blood Vessels (7–11)

7 The structure of blood vessels fits their functions.

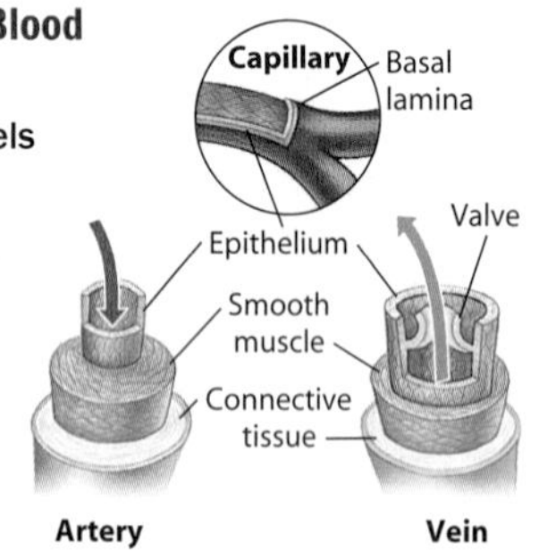

8 Blood pressure and velocity reflect the structure and arrangement of blood vessels. Blood pressure depends on cardiac output and the resistance of vessels. Pressure is highest in the arteries. Blood velocity is slowest in the capillaries. Muscle contractions and one-way valves keep blood moving through veins to the heart.

9 Measuring blood pressure can reveal cardiovascular problems. Blood pressure is measured as systolic and diastolic pressures. Hypertension is a serious cardiovascular problem that in most cases can be controlled.

10 Smooth muscle controls the distribution of blood. Constriction of arterioles and precapillary sphincters controls blood flow through capillary beds.

11 Capillaries allow the transfer of substances through their walls. Blood pressure forces fluid and small solutes out of the capillary at the arterial end. Excess fluid is returned to the circulatory system through lymph vessels.

Structure and Function of Blood (12–15)

12 Blood consists of red and white blood cells suspended in plasma. Plasma contains various inorganic ions, proteins, nutrients, wastes, gases, and hormones. Red blood cells (erythrocytes) transport O_2 bound to hemoglobin. White blood cells (leukocytes) fight infections.

13 Too few or too many red blood cells can be unhealthy. The hormone erythropoietin regulates red blood cell production.

14 Blood clots plug leaks when blood vessels are injured. Platelets adhere to connective tissue of damaged vessels and help convert fibrinogen to fibrin, forming a clot that plugs the leak.

15 Stem cells offer a potential cure for blood cell diseases. Stem cells in bone marrow produce all types of blood cells.

Connecting the Concepts

1. Use the following diagram to review the flow of blood through a human cardiovascular system. Label the indicated parts, highlight the vessels that carry oxygen-rich blood, and then trace the flow of blood by numbering the circles from 1 to 10, starting with 1 in the right ventricle. (When two locations are equivalent in the pathway, such as right and left lung capillaries or capillaries of top and lower portion of the body, assign them both the same number.)

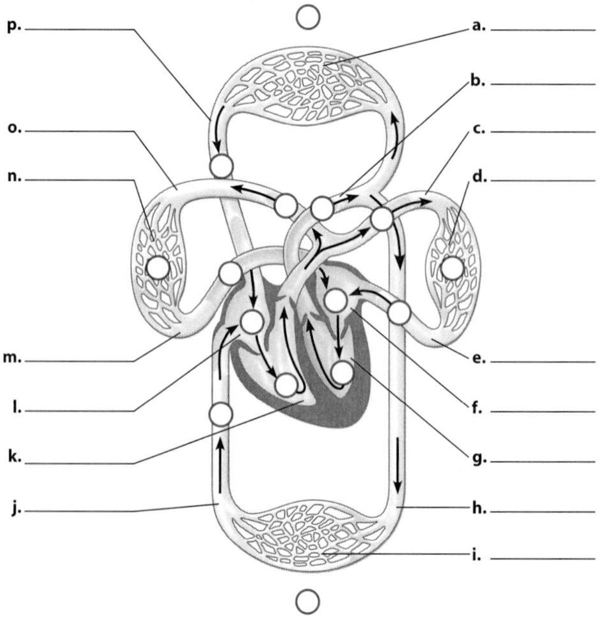

Testing Your Knowledge

Multiple Choice

2. Blood pressure is highest in ______, and blood moves most slowly in ______.
   a. veins; capillaries
   b. arteries; capillaries
   c. veins; arteries
   d. capillaries; arteries
   e. arteries; veins

3. When the doctor listened to Janet's heart, he heard "lub-hiss, lub-hiss" instead of the normal "lub-dup" sounds. The hiss is most likely due to ______. (*Explain your answer.*)
   a. a clogged coronary artery
   b. a defective atrioventricular (AV) valve
   c. a damaged pacemaker (SA node)
   d. a defective semilunar valve
   e. high blood pressure

4. Which of the following is the main difference between your cardiovascular system and that of a fish?
   a. In a fish, blood is oxygenated by passing through a capillary bed.
   b. Your heart has two chambers; a fish heart has four.
   c. Your circulation has two circuits; fish circulation has one.
   d. Your heart chambers are called atria and ventricles.
   e. Yours is a closed system; the fish's is an open system.

5. Paul's blood pressure is 150/90. The 150 indicates ______, and the 90 indicates ______.
   a. pressure in the left ventricle; pressure in the right ventricle
   b. arterial pressure; heart rate
   c. pressure during ventricular contraction; pressure during heart relaxation
   d. systemic circuit pressure; pulmonary circuit pressure
   e. pressure in the arteries; pressure in the veins

6. Which of the following *initiates* the process of blood clotting?
   a. damage to the lining of a blood vessel
   b. exposure of blood to the air
   c. conversion of fibrinogen to fibrin
   d. attraction of leukocytes to a site of infection
   e. conversion of fibrin to fibrinogen

7. Blood flows more slowly in the arterioles than in the artery that supplies them because the arterioles
   a. must provide opportunity for exchange with the interstitial fluid.
   b. have thoroughfare channels to venules that are often closed off, slowing the flow of blood.
   c. have sphincters that restrict flow to capillary beds.
   d. are narrower than the artery.
   e. collectively have a larger cross-sectional area than does the artery.

8. Which of the following is *not* a true statement about open and closed circulatory systems?
   a. Both systems have some sort of a heart that pumps a circulatory fluid through the body.
   b. A frog has an open circulatory system; other vertebrates have closed circulatory systems.
   c. The blood and interstitial fluid are separate in a closed system but are indistinguishable in an open system.
   d. The open circulatory system of an insect does not transport O_2 to body cells; closed circulatory systems do transport O_2.
   e. Some of the circulation of blood in both systems results from body movements.

9. If blood was supplied to all of the body's capillaries at one time,
   a. blood pressure would fall dramatically.
   b. resistance to blood flow would increase.
   c. blood would move too rapidly through the capillaries.
   d. the amount of blood returning to the heart would increase.
   e. the increased gas exchange in the lungs and in the supply of O_2 to muscles would allow for strenuous exercise.

Describing, Comparing, and Explaining

10. Trace the path of blood starting in a pulmonary vein, through the heart, and around the body, returning to the pulmonary vein. Name, in order, the heart chambers and types of vessels through which the blood passes.

11. Explain how the structure of capillaries relates to their function of exchanging substances with the surrounding interstitial fluid. Describe how that exchange occurs.

12. Here is a blood sample that has been spun in a centrifuge. List, as completely as you can, the components you would find in the straw-colored fluid at the top of this tube and in the dense red portion at the bottom.

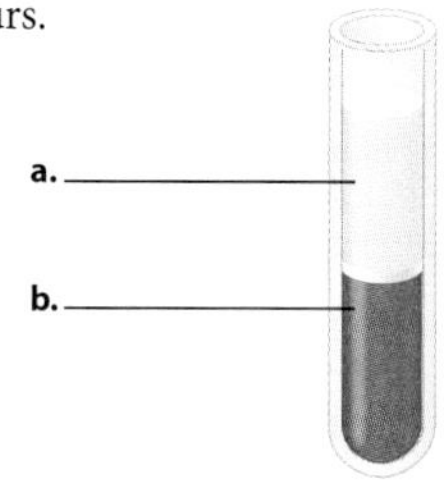

Applying the Concepts

13. Some babies are born with a small hole in the wall between the left and right ventricles. How might this affect the oxygen content of the blood pumped out of the heart into the systemic circuit?

14. Juan has a disease in which damaged kidneys allow some of his normal plasma proteins to be removed from the blood. How might this condition affect the osmotic pressure of blood in capillaries, compared with that of the surrounding interstitial fluid? One of the symptoms of this kidney malfunction is an accumulation of excess interstitial fluid, which causes Juan's arms and legs to swell. Can you explain why this occurs?

15. Some years ago, a 19-year-old woman received a bone marrow transplant from her 1-year-old sister. The woman was suffering from a deadly form of leukemia and was almost certain to die without a transplant. The parents had decided to have another child in a final attempt to provide their daughter with a matching donor. Although the ethics of the parents' decision were criticized, doctors report that this situation is not uncommon. In your opinion, is it acceptable to have a child to provide an organ or tissue donation? Why or why not?

16. Physiologists speculate about cardiovascular adaptations in dinosaurs—some of which had necks almost 10 m (33 feet) long. Such animals would have required a systolic pressure of nearly 760 mm Hg to pump blood to the brain when the head was fully raised. Some analyses suggest that dinosaurs' hearts were not powerful enough to generate such pressures, leading to the speculation that long-necked dinosaurs fed close to the ground rather than raising their heads to feed on high foliage. Scientists also debate whether dinosaurs had a "reptile-like" or "bird-like" heart. Most modern reptiles have a three-chambered heart with just one ventricle. Birds, which evolved from a lineage of dinosaurs, have a four-chambered heart. Some scientists believe that the circulatory needs of these long-necked dinosaurs provide evidence that dinosaurs must have had a four-chambered heart. Why might they conclude this?

Review Answers

1. a. capillaries of head, chest, and arms; b. aorta; c. pulmonary artery; d. capillaries of left lung; e. pulmonary vein; f. left atrium; g. left ventricle; h. aorta; i. capillaries of abdominal region and legs; j. inferior vena cava; k. right ventricle; l. right atrium; m. pulmonary vein; n. capillaries of right lung; o. pulmonary artery; p. superior vena cava

   See text Figure 3A for numbers and red vessels that carry oxygen-rich blood.
2. b 3. d (The second sound is the closing of the semilunar valves as the ventricles relax.) 4. c 5. c 6. a 7. e 8. b 9. a
10. Pulmonary vein, left atrium, left ventricle, aorta, artery, arteriole, body tissue capillary bed, venule, vein, vena cava, right atrium, right ventricle, pulmonary artery, capillary bed in lung, pulmonary vein
11. Capillaries are very numerous, producing a large surface area for exchange close to body cells. The capillary wall is only one epithelial cell thick. Pores in the wall and clefts between epithelial cells allow fluid with small solutes to move out of the capillary.
12. a. Plasma (the straw-colored fluid) would contain water, inorganic salts (ions such as sodium, potassium, calcium, magnesium, chloride, and bicarbonate), plasma proteins such as fibrinogen and immunoglobulins (antibodies), and substances transported by blood, such as nutrients (for example, glucose, amino acids, vitamins), waste products of metabolism, respiratory gases (O_2 and CO_2), and hormones.

    b. The red portion would contain erythrocytes (red blood cells), leukocytes (white blood cells—basophils, eosinophils, neutrophils, lymphocytes, and monocytes), and platelets.
13. Oxygen content is reduced as oxygen-poor blood returning to the right ventricle from the systemic circuit mixes with oxygen-rich blood of the left ventricle.
14. Proteins are important solutes in blood, accounting for much of the osmotic pressure that counters the flow of fluid out of a capillary. If protein concentration is reduced, the inward pull of osmotic pressure will fail to balance the outward push of blood pressure, and more fluid will leave the capillary and accumulate in the tissues.
15. Some issues and questions to consider: Is it ethical to have a child to save the life of another? Is it right to conceive a child as a means to an end—to produce a tissue or organ? Is this a less acceptable reason than most reasons parents have for bearing children? Do parents even need a reason for conceiving a child? Do parents have the right to make decisions like this for their young children? How will the donor (and recipient) feel about this when the donor is old enough to know what happened?
16. With a three-chambered heart, there is some mixing of oxygen-rich blood returning from the lungs with oxygen-poor blood returning from the systemic circulation. Thus, the blood of a dinosaur might not have supplied enough O_2 to support the higher metabolism and strong cardiac muscle contractions needed to generate such a high systolic blood pressure. Also, with a single ventricle pumping simultaneously to both pulmonary and systemic circuits, the blood pumped to the lungs would be at such a high pressure that it would damage the lungs.

The Immune System

The Immune System

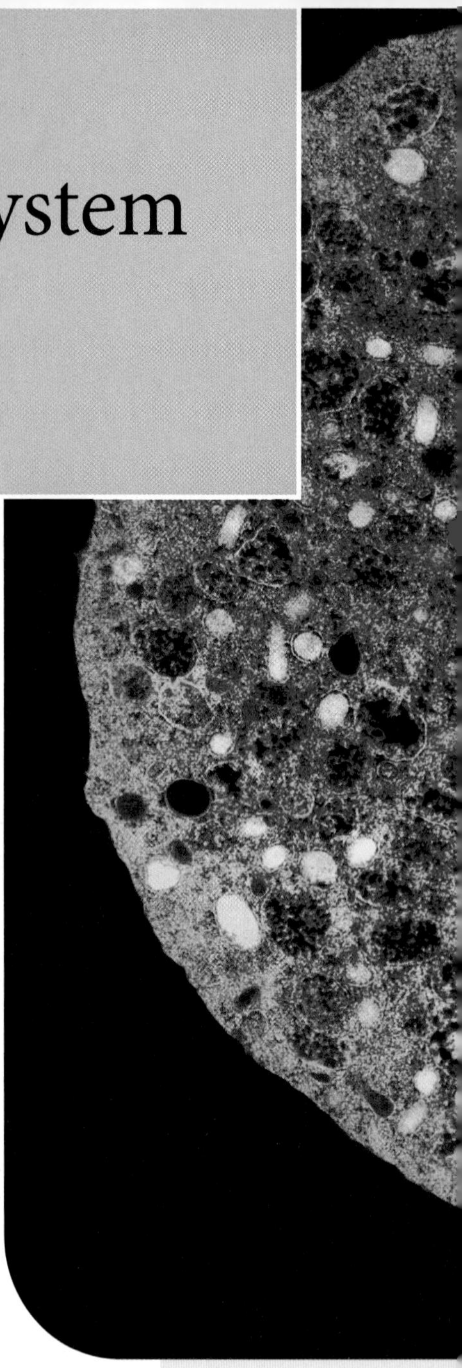

BIG IDEAS

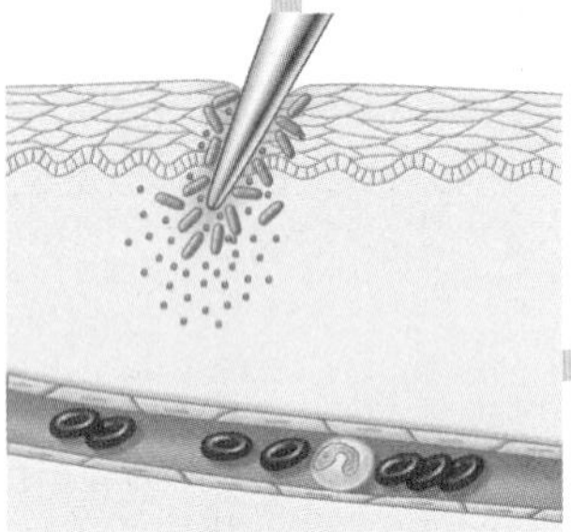

David McNew/Getty Images

Innate Immunity
(1–3)

All animals have immune defenses that are always at the ready.

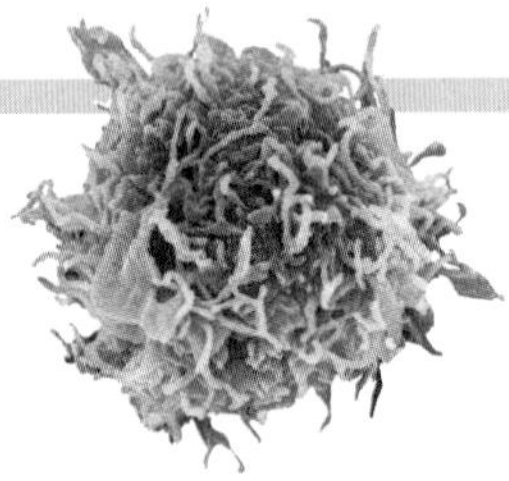

Adaptive Immunity
(4–15)

Vertebrates custom-tailor the immune response to specific pathogens.

Salisbury District Hospital/Photo Researchers

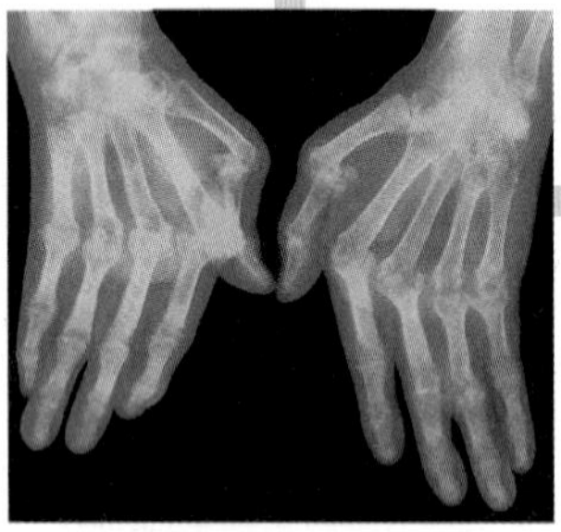

Disorders of the Immune System
(16–17)

Malfunctions of the immune response can cause problems that range from mild to severe.

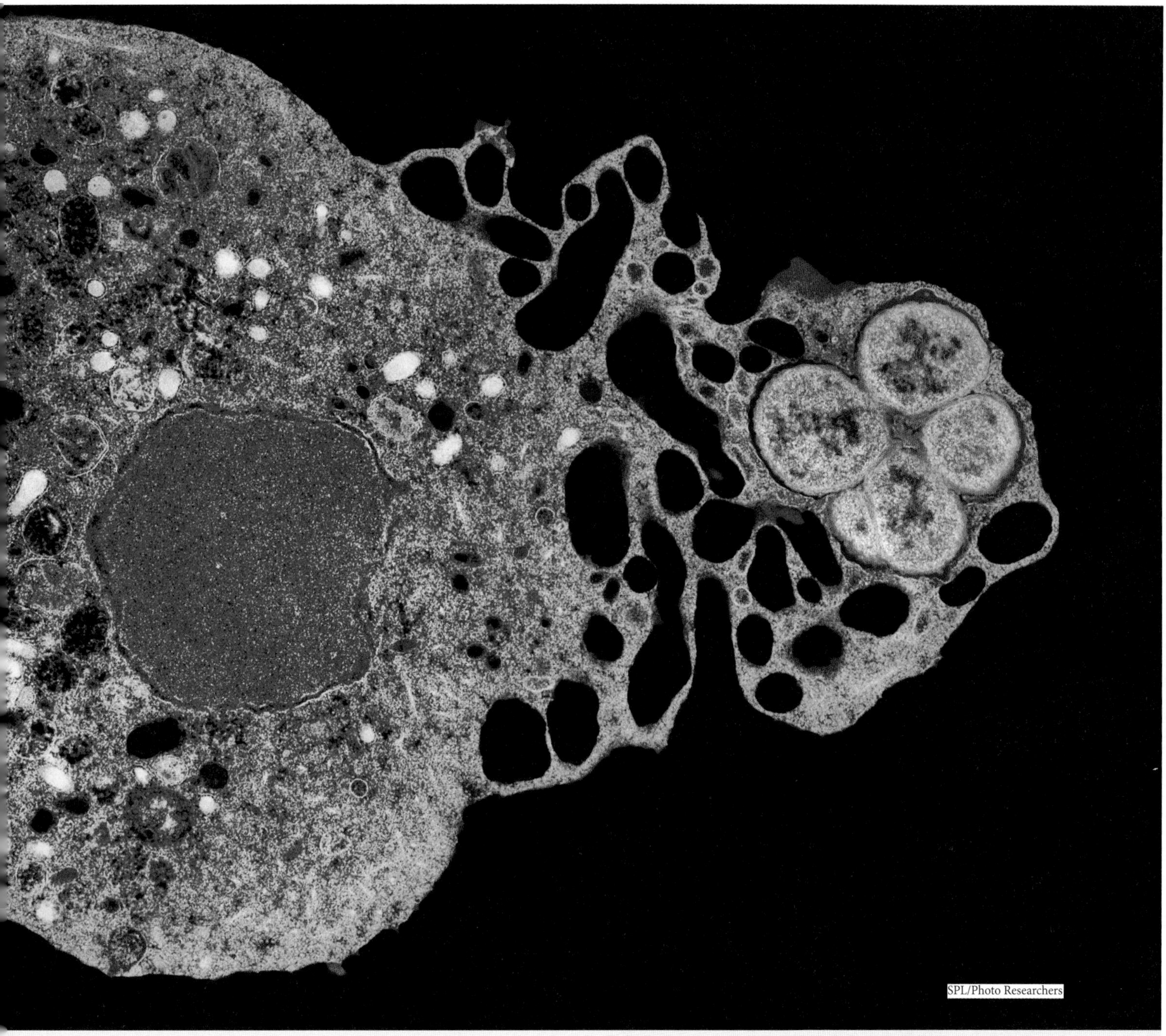
SPL/Photo Researchers

The large purple structure in the electron micrograph above is a neutrophil, a kind of white blood cell found in the human immune system. Neutrophils are capable of recognizing foreign invaders within the human body and destroying them by phagocytosis, or engulfment. In this micrograph, the neutrophil is in the process of engulfing four bacteria (the yellow spheres) of the species *Neisseria gonorrhoeae*. These bacteria cause the sexually transmitted disease gonorrhea. Once engulfed, the bacteria will be destroyed by the neutrophil.

As you will see in this chapter, your body's immune system recognizes and attacks agents that can cause disease. Some of the defenses of the immune system are innate; that is, they are always deployed and waiting to counter an invader. Other immune defenses are adaptive; they require recognition of a specific infectious agent. Once a pathogen is recognized, the immune system rallies large numbers of specific molecules and cells that can search out and destroy the invader. By combining multiple defenses—both innate and adaptive—into a coordinated immune system, animals avoid or limit many infections.

In this chapter, you will learn about the body's varied defenses and how they work to protect the body. Following a discussion of innate immunity in the first three modules, we concentrate on the mechanisms of adaptive immunity. We'll also look at how that knowledge has been applied to improve human health and the problems—from seasonal nuisances to fatal illnesses—that may arise when the intricate interplay of our body's defenses goes awry. Finally, we'll examine one of the most important applications of the study of the immune system: the continuing struggle against the deadly disease AIDS.

Innate Immunity

1 All animals have innate immunity

Nearly everything in the environment teems with **pathogens**, agents that cause disease. Yet we do not constantly become ill, thanks to the **immune system**, the body's system of defenses against agents that cause disease. The immune systems of all animals include **innate immunity**, a set of defenses that are active immediately upon infection and are the same whether or not the pathogen has been encountered previously (**Figure 1A**).

Invertebrate Innate Immunity Invertebrates rely solely on innate immunity. For example, insects have an exoskeleton, which is a tough, dry barrier that keeps out bacteria and viruses. Pathogens that breach these external defenses confront a set of internal defenses. Additional physical barriers, a low pH, and the secretion of lysozyme, an enzyme that breaks down bacterial cell walls, protect the insect digestive system. Circulating insect immune cells are capable of **phagocytosis**, cellular ingestion and digestion of foreign substances. The insect innate immune system also includes recognition proteins that bind to molecules found only on pathogens. Recognition of an invading microbe triggers the production of antimicrobial peptides that bring about the destruction of the invaders.

Vertebrate Innate Immunity In vertebrates, innate immunity coexists with the more recently evolved system of adaptive immunity (discussed later in this chapter). In mammals, innate defenses include barriers such as skin and mucous membranes that protect organ systems open to the external environment, such as the digestive, respiratory, reproductive, and urinary systems. Nostril hairs filter incoming air, and mucus in the respiratory tract traps most microbes and dirt that get past the nasal filter. Cilia on cells lining the respiratory tract then sweep the mucus and any trapped microbes upward and out, helping to prevent lung infections.

Innate immunity (1–3) The response is the same whether or not the pathogen has been previously encountered		Adaptive immunity (4–15) Found only in vertebrates; previous exposure to the pathogen enhances the immune response
External barriers (1)	**Internal defenses (1–2)**	
• Skin/exoskeleton • Acidic environment • Secretions • Mucous membranes • Hairs • Cilia	• Phagocytic cells • NK cells • Defensive proteins • Inflammatory response (2)	• Antibodies (8–10) • Lymphocytes (11–13)
	The lymphatic system (3)	

▲ **Figure 1A** An overview of animal immune systems (along with the module number where each topic is covered)

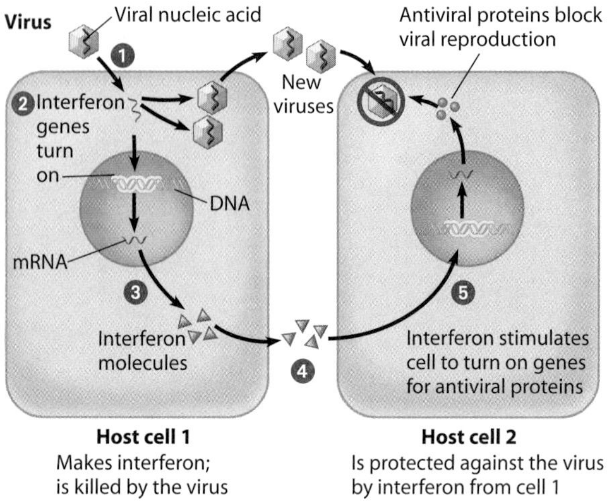

▲ **Figure 1B** The interferon mechanism against viruses

Microbes that breach a mammal's external barriers, such as those that enter through a cut in the skin, are confronted by innate immune cells. These are all classified as white blood cells, although they are found in interstitial fluid as well as in the blood. Most, such as abundant **neutrophils**, are phagocytes (phagocytic cells). **Macrophages** ("big eaters") are large phagocytic cells that wander through the interstitial fluid, "eating" any bacteria and virus-infected cells they encounter. **Natural killer (NK) cells** attack cancer cells and virus-infected cells by releasing chemicals that lead to cell death.

Other components of vertebrate innate immunity include proteins that either attack microbes directly or impede their reproduction. **Interferons** are proteins, produced by virus-infected cells, that help to limit the cell-to-cell spread of viruses (**Figure 1B**). ❶ The virus infects a cell, which causes ❷ interferon genes in the cell's nucleus to be turned on. ❸ The cell synthesizes interferon. The infected cell then dies, but ❹ its interferon molecules may diffuse to neighboring healthy cells, ❺ stimulating them to produce other proteins that inhibit viral reproduction.

Additional innate immunity in vertebrates is provided by the **complement system**, a group of about 30 different kinds of proteins that circulate in an inactive form in the blood. As you'll see later in the chapter, these proteins can act together (in complement) with other defense mechanisms. Substances on the surfaces of many microbes activate the complement system, resulting in a cascade of steps that may lead to the lysis, or bursting, of the invaders. Certain complement proteins also help trigger the inflammatory response, the subject of the next module.

? Which components of innate immunity described here actually help prevent infection? Which come into play only after infection has occurred?

Skin/exoskeleton, secretions, mucous membranes; phagocytic cells (macrophages, neutrophils), NK cells, defensive proteins (interferons, complement system)

2 Inflammation mobilizes the innate immune response

The **inflammatory response** is a major component of our innate immunity. Any damage to tissue, whether caused by microorganisms or by physical injury—even just a scratch or an insect bite—triggers this response. You may see signs of the inflammatory response when your skin is cut. The area becomes red, warm, and swollen. This reaction is inflammation, which literally means "setting on fire."

Figure 2 shows the chain of events when a pin has broken the skin, allowing infection by bacteria. ❶ The damaged cells soon release chemical alarm signals, such as **histamine**. ❷ The chemicals spark the mobilization of various defenses. Histamine, for instance, induces neighboring blood vessels to dilate and become leaky. Blood flow to the damaged area increases, and blood plasma passes out of the leaky vessels into the interstitial fluid of the affected tissues. Other chemicals, some that are part of the complement system, attract phagocytes to the area. Squeezing between the cells of the blood vessel wall, these phagocytic white blood cells (yellow in the figure) migrate out of the blood into the tissue spaces. The local increase in blood flow, fluid, and cells produces the redness, heat, and swelling characteristic of inflammation.

The major function of the inflammatory response is to disinfect and clean injured tissues. ❸ The white blood cells that migrate into the area engulf bacteria and the remains of any body cells killed by them or by the physical injury. Many of the white cells die in the process, and their remains are also engulfed and digested. The pus that often accumulates at the site of an infection consists mainly of dead white cells, fluid that has leaked from capillaries, and other tissue debris.

The inflammatory response also helps prevent the spread of infection to surrounding tissues. Clotting proteins present in blood plasma pass into the interstitial fluid during inflammation. Along with platelets, these substances form local clots that help seal off the infected region and allow healing to begin.

Inflammation may be localized, as we have just described, or widespread (systemic). Sometimes microorganisms such as bacteria or protozoans get into the blood or release toxins that are carried throughout the body in the bloodstream. The body may react with several inflammatory weapons. For instance, the number of white blood cells circulating in the blood may increase severalfold within just a few hours; an elevated "white cell count" is one way to diagnose certain infections. Another response to systemic infection is fever, an abnormally high body temperature. Toxins themselves may trigger the fever, or macrophages may release compounds that cause the body's thermostat to be set at a higher temperature. A very high fever is dangerous, but a moderate one may stimulate phagocytosis and hasten tissue repair. (*How* a moderate fever aids immune defenses is a subject of active debate within the medical community.) Anti-inflammatory drugs, such as aspirin and ibuprofen, dampen the normal inflammatory response and thus help reduce swelling and fever.

Sometimes bacterial infections bring about an overwhelming systemic inflammatory response leading to a condition called septic shock. Characterized by very high fever and low blood pressure, septic shock is a common cause of death in hospital intensive care units. Clearly, while local inflammation is an essential step toward healing, widespread inflammation can be devastating.

? Why is the inflammatory response considered a form of *innate* immunity?

Because the response is the same regardless of whether the invader has been previously encountered

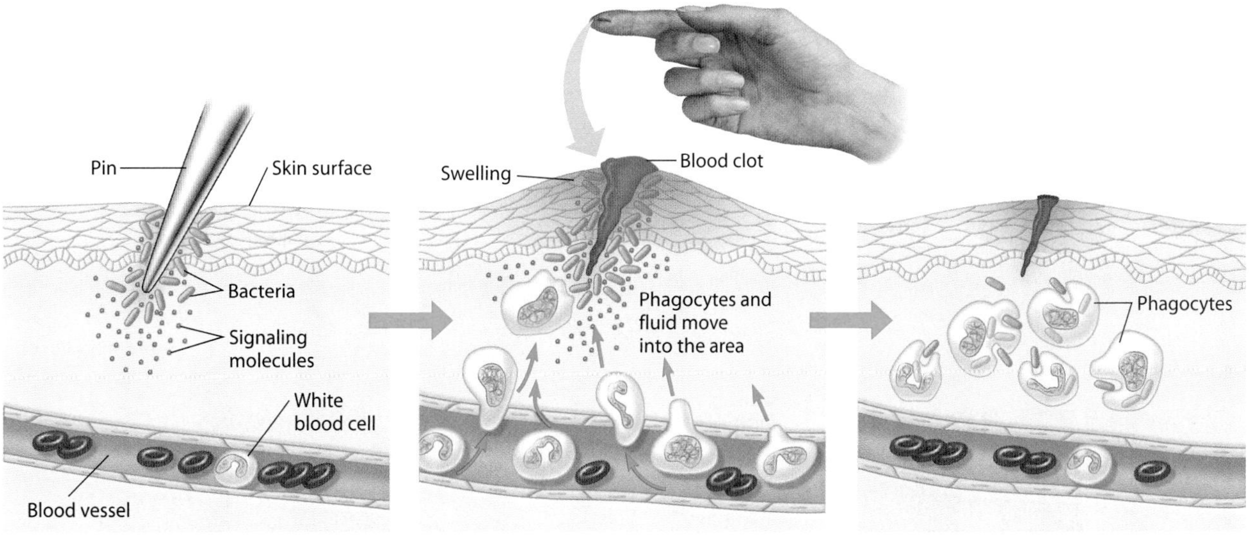

❶ Tissue injury; signaling molecules, such as histamine, are released.

❷ Dilation and increased leakiness of local blood vessels; phagocytes migrate to the area.

❸ Phagocytes (macrophages and neutrophils) consume bacteria and cellular debris; the tissue heals.

▲ **Figure 2** The inflammatory response

3 The lymphatic system becomes a crucial battleground during infection

The **lymphatic system**, which is involved in both innate and adaptive immunity, consists of a branching network of vessels, numerous **lymph nodes**—little round organs packed with macrophages and white blood cells called lymphocytes—the bone marrow, and several organs (**Figure 3**). The lymphatic vessels carry a fluid called **lymph**, which is similar to the interstitial fluid that surrounds body cells but contains less oxygen and fewer nutrients.

The lymphatic system is closely associated with the circulatory system. If an infectious agent gets inside the body, it usually winds up in the circulatory system. From there, it is carried into the lymphatic system, which can usually filter it out. The filtered fluid can then be recycled into the circulatory system. The lymphatic system thus has two main functions: to return tissue fluid back to the circulatory system and to fight infection.

A small amount of the fluid that enters the tissue spaces from the blood in a capillary bed does not reenter the blood capillaries. Instead, this fluid is returned to the blood via lymphatic vessels. The enlargement in Figure 3 (bottom right) shows a branched lymphatic vessel in the process of taking up fluid from tissue spaces in the skin. As shown here, fluid enters the lymphatic system by diffusing into tiny, dead-end lymphatic capillaries that are intermingled among the blood capillaries.

Lymph drains from the lymphatic capillaries into larger and larger lymphatic vessels. Eventually, the fluid reenters the circulatory system via two large lymphatic vessels that fuse with veins in the chest. As the close-up indicates, the lymphatic vessels resemble veins in having valves that prevent the backflow of fluid toward the capillaries (see Figure 23.7C). Also, like veins, lymphatic vessels depend mainly on the movement of skeletal muscles to squeeze their fluid along. The green arrows in the close-ups indicate the flow of lymph.

When your body is fighting an infection, the organs of the lymphatic system become a major battleground. As lymph circulates through the lymphatic organs—such as the lymph node shown in Figure 3, top right—it carries microbes, parts of microbes, and their toxins picked up from infection sites anywhere in the body. Once inside lymphatic organs, macrophages that reside there permanently may engulf the invaders as part of the innate immune response. Lymph nodes fill with huge numbers of defensive cells, causing the tender "swollen glands" in your neck and armpits that your doctor looks for as a sign of infection.

In addition to the functioning of macrophages and other components of the innate immune response, cells of the adaptive immune response, called lymphocytes, may be activated against specific invaders. We examine adaptive immunity next.

? What are the two main functions of the lymphatic system?

To return fluid from interstitial spaces to the circulatory system and to combat infection

▼ **Figure 3** The human lymphatic system

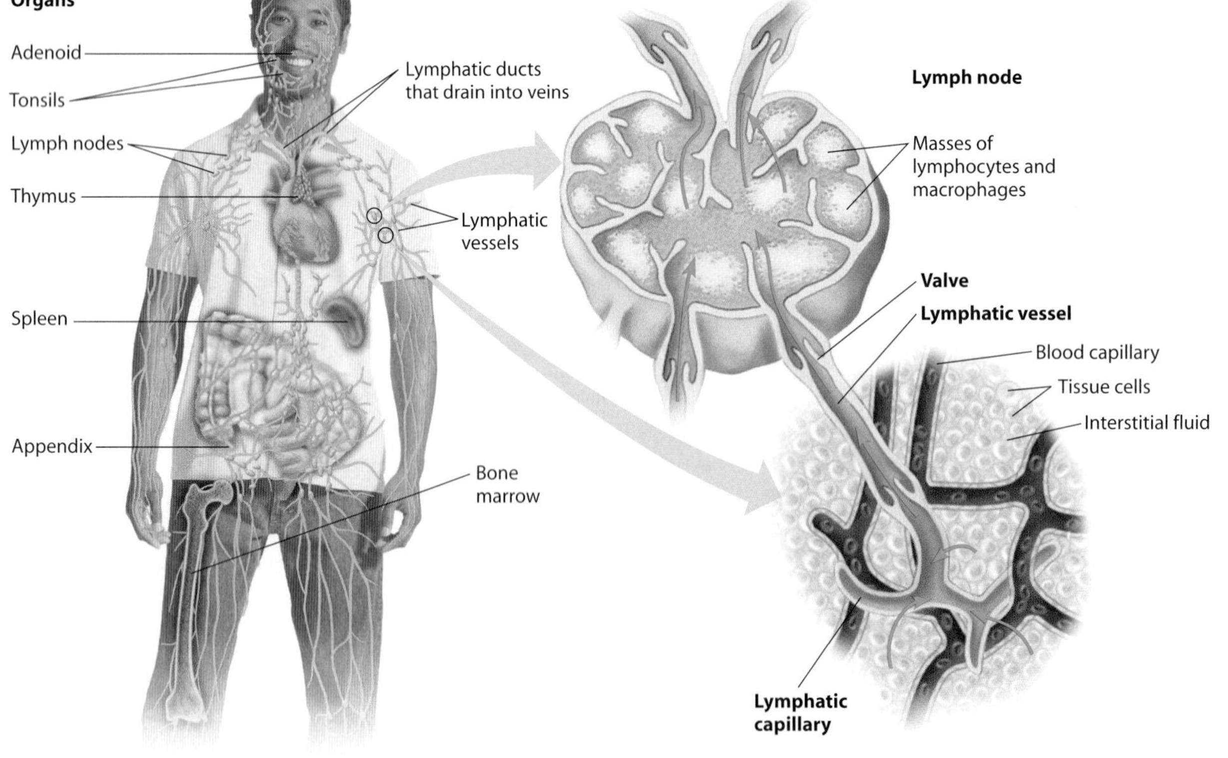

Ryan McVay/Getty Images

Adaptive Immunity

4 The adaptive immune response counters specific invaders

All the defenses you've learned about so far are called *innate* because they're ready "off the rack"; that is, innate defenses are always standing by, ready to be used in their current form. When the innate immune response fails to ward off a pathogen, a set of adaptive defenses, ones that are "custom-tailored" to each specific invader, provides a second line of defense. **Adaptive immunity**—also called acquired immunity—is a set of defenses, found only in vertebrates, that is activated only after exposure to specific pathogens. Thus, unlike innate immunity, adaptive immunity differs from individual to individual, depending on what pathogens they have been previously exposed to. Once activated, the adaptive immune response provides a strong defense against pathogens that is highly specific; that is, it acts against one infectious agent but not another. Moreover, adaptive immunity can amplify certain innate responses, such as inflammation and the complement system.

David McNew/Getty Images

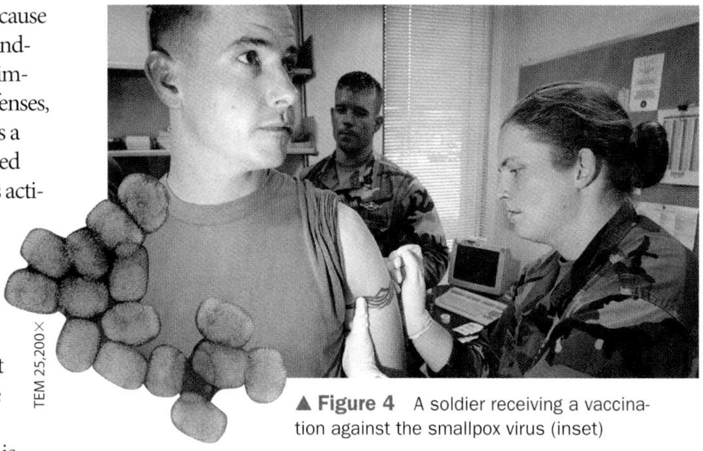

TEM 25,200×

Science Source/Photo Researchers

▲ **Figure 4** A soldier receiving a vaccination against the smallpox virus (inset)

Any molecule that elicits an adaptive immune response is called an **antigen**. Antigens may be molecules that protrude from pathogens or other particles, such as viruses, bacteria, mold spores, pollen, house dust, or the cell surfaces of transplanted organs. Antigens may also be substances released into the extracellular fluid, such as toxins secreted by bacteria. When the immune system detects an antigen, it responds with an increase in the number of cells that either attack the invader directly or produce immune proteins called antibodies. An **antibody** is a protein found in blood plasma that attaches to one particular kind of antigen and helps counter its effects. (The word *antigen* is a contraction of "*anti*body-*gen*erating," a reference to the fact that the foreign agent provokes an adaptive immune response.) The defensive cells and antibodies produced against a particular antigen are usually specific to that antigen; they are ineffective against any other foreign substance.

Adaptive immunity has a remarkable "memory"; it can "remember" antigens it has encountered before and react against them more quickly and vigorously on subsequent exposures. For example, if a person is infected by the varicella zoster virus and then contracts chicken pox, the immune system "remembers" certain molecules on the virus. Should the virus enter the body again, the adaptive immune response mounts a quick and decisive attack that usually destroys the virus before symptoms appear. Thus, in the adaptive immune response, unlike innate immunity, exposure to a foreign agent enhances future responses to that same agent.

Adaptive immunity is usually obtained by natural exposure to antigens (that is, by being infected), but it can also be achieved by **vaccination**, also known as immunization. In this procedure, the immune system is confronted with a **vaccine** composed of a harmless variant or part of a disease-causing microbe, such as an inactivated bacterial toxin, a dead or weakened microbe, or a piece of a microbe. The vaccine stimulates the immune system to mount defenses against this harmless antigen, defenses that will also be effective against the actual pathogen because it has similar antigens. Once you have been successfully vaccinated, your immune system will respond quickly if it is exposed to the actual microbe. Such protection may last for life.

In industrialized nations, routine childhood vaccination has virtually eliminated many viral diseases. In the United States, most children receive a series of shots starting soon after birth, including vaccinations against diphtheria/pertussis/tetanus (DPT, or "diptet"), polio, hepatitis, chicken pox, and measles/mumps/rubella (MMR). One of the major success stories of modern vaccination involves smallpox, a potentially fatal viral infection that affected over 50 million people per year worldwide in the 1950s. A massive vaccination effort has been so effective that there have been no cases of smallpox since 1977. Since 2001, however, the U.S. government has stockpiled hundreds of millions of doses of smallpox vaccine and has begun to vaccinate high-risk health-care and military workers in case the smallpox virus is used in a bioterrorist attack (**Figure 4**).

Whether antigens enter the body naturally (if you catch the flu) or artificially (if you get a flu shot), the resulting immunity is called **active immunity** because the person's own immune system actively produces antibodies. It is also possible to acquire **passive immunity** by receiving premade antibodies. For example, a fetus obtains antibodies from its mother's bloodstream; a baby receives antibodies in breast milk that protect the digestive tract (although these antibodies are broken down there and do not enter the bloodstream); and travelers sometimes get a shot containing antibodies to pathogens they are likely to encounter. In yet another example, the effects of a poisonous snakebite may be counteracted by injecting the victim with antibodies extracted from animals previously immunized against the venom. Passive immunity is temporary because the recipient's immune system is not stimulated by antigens. Immunity lasts only as long as the antibodies do; after a few weeks or months, these proteins break down and are recycled by the body.

? Why is protection resulting from a vaccination considered *active* immunity rather than *passive* immunity?

Because the body itself produces the immunity by mounting an immune response and generating antibodies, even though the stimulus consists of artificially introduced antigens.

5 Lymphocytes mount a dual defense

Lymphocytes, white blood cells that spend most of their time in the tissues and organs of the lymphatic system, are responsible for adaptive immunity. Like all blood cells, lymphocytes originate from stem cells in the bone marrow. As shown in Figure 5A, some immature lymphocytes continue developing in the bone marrow; these become specialized as B lymphocytes, or **B cells**. Other immature lymphocytes migrate to the thymus, a gland above the heart. There, the lymphocytes become specialized as T lymphocytes, or **T cells**. Both B cells and T cells eventually make their way via the blood to the lymph nodes, spleen, and other lymphatic organs.

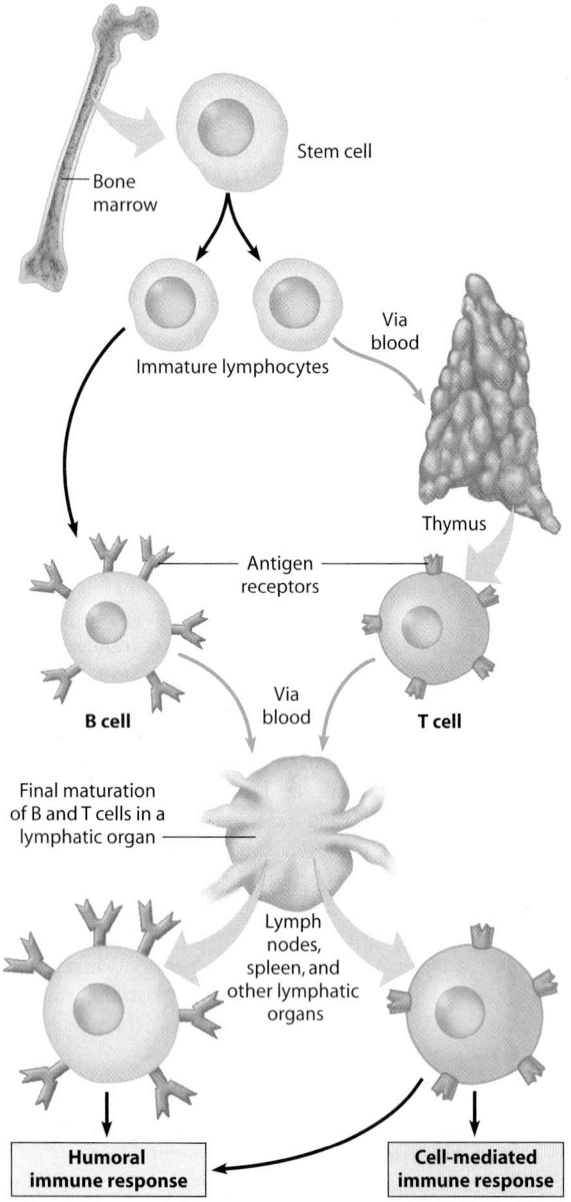

▲ **Figure 5A** The development of B cells and T cells

The B cells and T cells of the adaptive immune response together provide a dual defense. The first type of defense is called the **humoral immune response**. The humoral immune response involves the secretion of free-floating antibodies by B cells (Figure 5B, left) into the blood and lymph. (The humoral response is so named because blood and lymph were long ago called body "humors.") The humoral system defends primarily against bacteria and viruses present in body fluids. As discussed in the last paragraph of Module 4, the humoral immune response can be passively transferred by injecting antibody-containing blood plasma from an immune individual into a nonimmune individual. As you will see in Module 9, antibodies mark invaders by binding to them. The resulting antigen-antibody complexes are easily recognized for destruction and disposal by phagocytic cells.

The second type of adaptive immunity, produced by T cells (Figure 5B, right), is called the **cell-mediated immune response**. As its name implies, this defensive system results from the action of defensive cells, in contrast to the action of free-floating defensive antibody proteins of the humoral response. Certain T cells attack body cells infected with bacteria or viruses. Other T cells function indirectly by promoting phagocytosis by other white blood cells and by stimulating B cells to produce antibodies. Thus, T cells play a part in both the cell-mediated and humoral immune responses.

When a B cell develops in bone marrow or a T cell develops in the thymus, certain genes in the cell are turned on. This causes the cell to synthesize molecules of a specific protein, which are then incorporated into the plasma membrane. As indicated in Figure 5A, these protein molecules stick out from the cell's surface. The molecules are **antigen receptors**, capable of binding one specific type of antigen. Each B or T cell has about 100,000 antigen receptors, and all the receptors on a single cell are identical—they all recognize the same antigen. In the case of a B cell, the receptors are almost identical to the particular antibody that the B cell will secrete. Once a B cell or T cell has its surface proteins in place, it can recognize a specific antigen and mount an immune response against it. One cell may recognize an antigen on the mumps virus, for instance, while another detects a particular antigen on a tetanus-causing bacterium.

In Figure 5A, you can see that after the B cells and T cells have developed their antigen receptors, these lymphocytes leave the bone marrow and thymus and move via the bloodstream to the lymph nodes, spleen, and other parts of the lymphatic system. In these organs, many B and T cells take up residence and encounter infectious agents that have penetrated the body's outer defenses. Because lymphatic capillaries extend into virtually all the body's tissues, bacteria or viruses infecting nearly any part of the body eventually enter the lymph and are carried to the lymphatic organs. As we will describe in Module 7, when a B or T cell within a lymphatic organ first confronts the specific antigen that it is programmed to recognize, it differentiates further and becomes a fully mature component of the immune system.

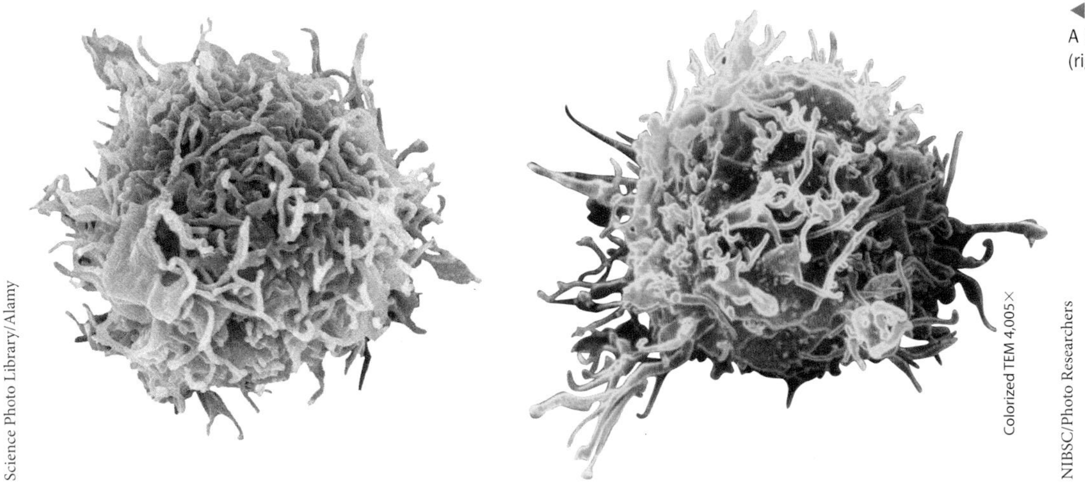

◀ **Figure 5B** A B cell (left) and T cell (right)

An enormous diversity of B cells and T cells develops in each individual. Researchers estimate that each of us has millions of different kinds—enough to recognize and bind to virtually every possible antigen. A small population of each kind of lymphocyte lies in wait in our body, genetically programmed to recognize and respond to a specific antigen. Only a tiny fraction of the immune system's lymphocytes will ever be used, but they are all available if needed. It is as if the immune system maintains a huge standing army of soldiers, each made to recognize one particular kind of invader. The majority of soldiers never encounter their target and remain idle. But when an invader does appear, chances are good that some lymphocytes will be able to recognize it, bind to it, and call in reinforcements. We'll take a closer look at the different types of T cells in Modules 11 and 12.

? Contrast the targets of the humoral immune response with those of the cell-mediated immune response.

The humoral immune response works against pathogens in the body fluids; the cell-mediated immune response attacks infected cells.

6 Antigens have specific regions where antibodies bind to them

As molecules that elicit the adaptive immune response, antigens usually do not belong to the host animal. Most antigens are proteins or large polysaccharides that protrude from the surfaces of viruses or foreign cells. Common examples are protein-coat molecules of viruses, parts of the capsules and cell walls of bacteria, and macromolecules on the surface cells of other kinds of organisms, such as protozoans and parasitic worms. (Sometimes a particular microbe is called an antigen, but this usage is misleading because the microbe will almost always have several kinds of antigenic molecules.) Other sources of antigenic molecules include blood cells or tissue cells from other individuals, of the same species or of a different species. Antigenic molecules are also found dissolved in body fluids; foreign molecules of this type include bacterial toxins and bee venom.

As shown in **Figure 6**, an antibody usually recognizes and binds to a small surface-exposed region of an antigen called an **antigenic determinant**, also known as an epitope. An antigen-binding site, a specific region on the antibody molecule, recognizes an antigenic determinant by the fact that the binding site and antigenic determinant have complementary shapes, like an enzyme and substrate or a lock and key. An antigen usually has several different determinants (there are three in the diagram here), so different antibodies (two, in this case) can bind to the same antigen. A single kind of antigen molecule may thus stimulate the immune system to make several distinct antibodies against it. Notice that each antibody molecule has two identical antigen-binding sites. We'll return to antibody structure in Module 8. But first let's see how the body produces large quantities of antibodies and defensive cells in response to specific infections.

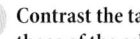

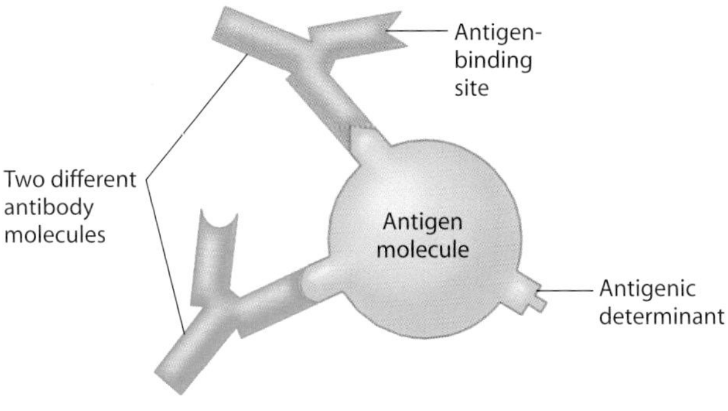

▲ **Figure 6** The binding of antibodies to antigenic determinants

? Why is it inaccurate to refer to a pathogen, such as a virus, as an antigen?

It is inaccurate because antigens are not whole pathogens; they are molecules, which may be chemical components of a pathogen's surface. One pathogen may have many antigens.

7 Clonal selection musters defensive forces against specific antigens

The immune system's ability to defend against a wide variety of antigens depends on a process known as **clonal selection**. Once inside the body, an antigen encounters a diverse pool of B and T lymphocytes. However, one particular antigen interacts only with the tiny fraction of lymphocytes bearing receptors specific to that antigen. Once activated by the antigen, these few "selected" cells proliferate, forming a clone—a genetically identical population—of thousands of cells all specific for the stimulating antigen. This antigen-driven cloning of lymphocytes—clonal selection—is a vital step in the adaptive immune response against infection.

The Steps of Clonal Selection Figure 7A indicates how clonal selection of B cells works in the humoral immune response. (A similar mechanism activates clonal selection for T cells in the cell-mediated immune response.) ❶ The row of three cells at the top of the figure represents a vast repertoire of B cells in a lymph node. Notice that each lymphocyte has its own specific type of antigen receptor embedded in its surface (represented by different colors in the figure). The cells' receptors are in place before they ever encounter an antigen.

The first time an antigen enters the body and is swept into a lymph node, ❷ antigenic determinants on its surface bind to the few B cells that happen to have complementary receptors. Other lymphocytes, without the appropriate binding sites (in the figure, the ones with the green and blue receptors), are not affected. Primed by the interaction with the antigen, ❸ the selected cell is activated: It grows, divides, and differentiates into two genetically identical yet physically distinct types of cells. Both newly produced types of cells are specialized for defending against the very antigen that triggered the response. ❹ One group of newly cloned cells is short-lived but fast-acting **effector cells**, which combat the antigen. Because the example in the figure involves B cells, the effector cells produced are **plasma cells**. Each plasma cell secretes antibody molecules into the blood and lymph, all of the same type. Each plasma cell makes as many as 2,000 copies of its antibody per second and thus requires large amounts of endoplasmic reticulum, a characteristic of cells actively synthesizing and secreting proteins. The secreted antibodies circulate in the blood and lymphatic fluid, contributing to the humoral immune response. Although highly effective at combating infection, each effector cell lasts only 4 or 5 days before dying off.

❺ A second group of cells produced by the activated B cells is a smaller number of **memory cells**, which differ from effector cells in both appearance and function. In contrast to short-lived effector cells, memory cells may last for decades. They remain in the lymph nodes, poised to be activated by a second exposure to the antigen. In fact, in some cases, memory cells confer lifetime immunity, as they may after vaccination against such childhood diseases as mumps and measles. Steps 1–5 show the initial phase of adaptive immunity, called the

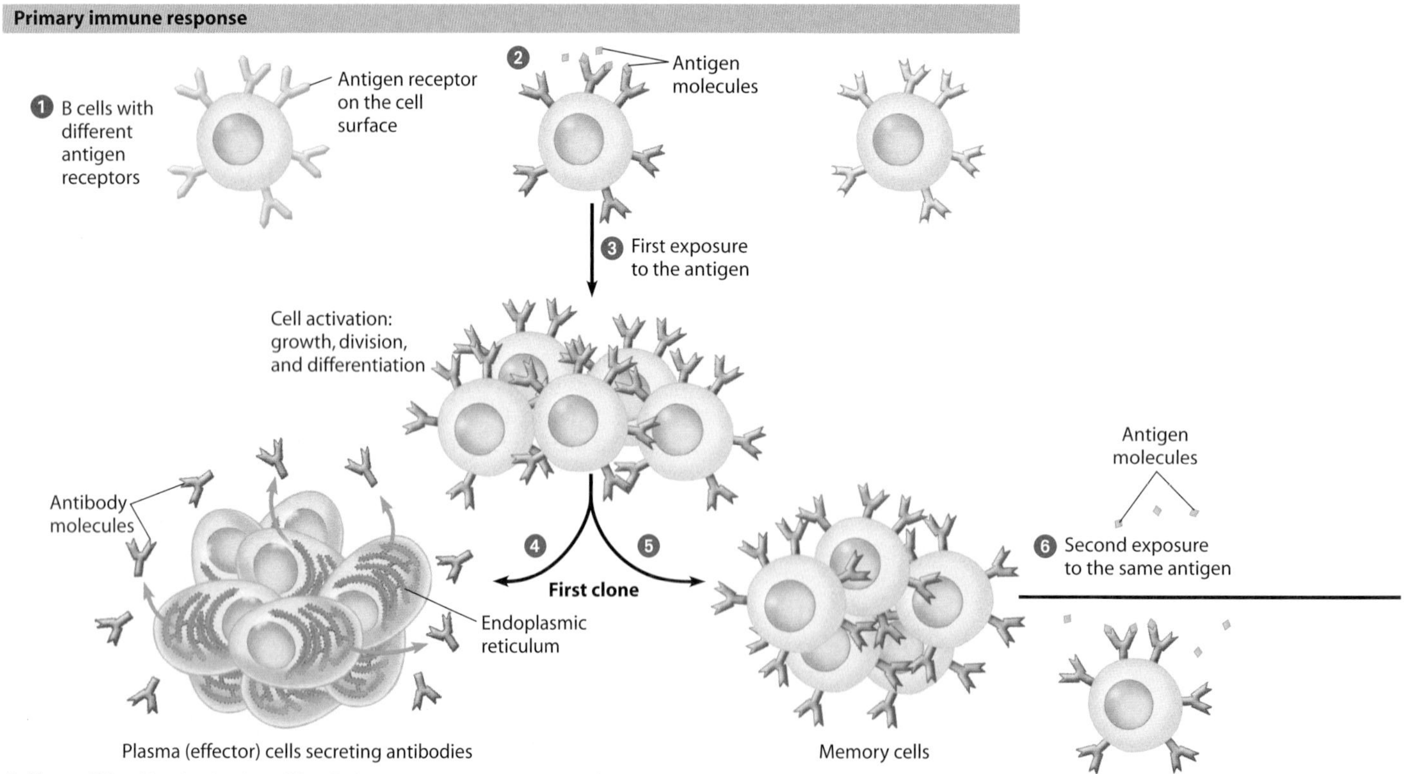

▲ **Figure 7A** Clonal selection of B cells in the primary and secondary immune responses

primary immune response. This phase occurs when lymphocytes are exposed to an antigen for the first time.

❻ When memory cells produced during the primary response are activated by a second exposure to the same antigen—which may occur soon or long after the primary immune response—they initiate the **secondary immune response**. This response is faster and stronger than the first. Another round of clonal selection ensues. The selected memory cells multiply quickly, producing a large second clone of lymphocytes that mount the secondary response. Like the first clone, the second clone includes effector cells that produce antibodies and memory cells capable of responding to future exposures to the antigen. In our example here with B cells, the secondary response produces very high levels of antibodies that, though they are short-lived, are often more effective against the antigen than those produced during the primary response.

The concept of clonal selection is so fundamental to understanding adaptive immunity that it is worth restating: Each antigen, by binding to specific receptors, selectively activates a tiny fraction of lymphocytes; these few selected cells then give rise to a clone of many cells, all specific for and dedicated to eliminating the antigen that started the response. Thus, we see that the versatility of the adaptive immune response depends on a great diversity of preexisting lymphocytes with different antigen receptors.

Primary versus Secondary Immune Responses Now that we have seen how clonal selection works, let's take a look at the two phases of the adaptive immune response in an individual. The

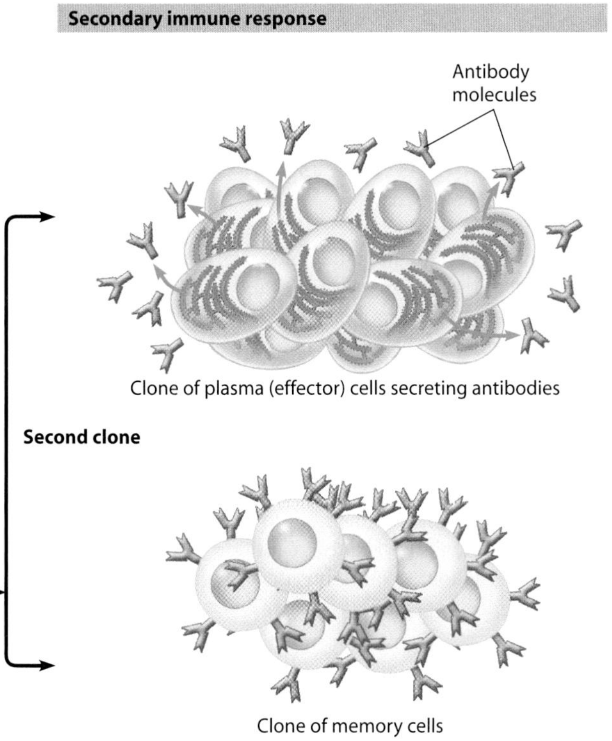

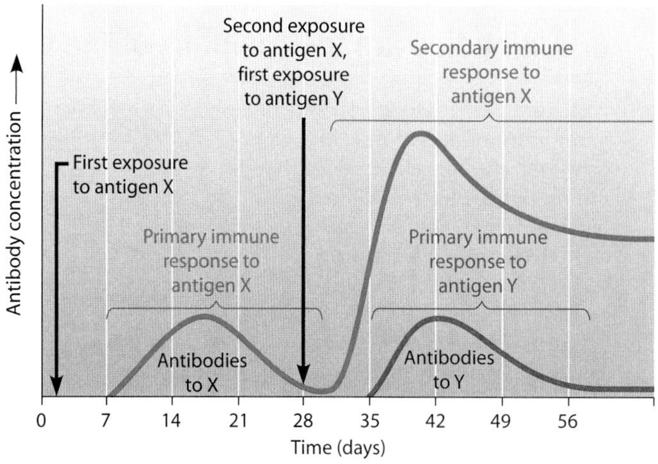

▲ **Figure 7B** The two phases of the adaptive immune response

blue curve in **Figure 7B** illustrates the difference between the two phases, triggered by two exposures to the same antigen. On the far left of the graph, you can see that the primary response does not start right away; it usually takes several days for the lymphocytes to become activated by an antigen (called X here) and form clones of effector cells. When the effector cell clone forms, antibodies start showing up in the blood, as the graph shows. During this delay, a stricken individual may become ill. The antibody level reaches its peak 2–3 weeks after initial exposure. As the antibody levels in the blood and lymph rise, the symptoms of the illness typically diminish and disappear. The primary response subsides as the effector cells die out.

The second exposure to antigen X (at day 28 in the graph) triggers the secondary immune response. Notice that this secondary response starts faster than the primary response, typically in 2–7 days, versus 10–17 days. As mentioned, the secondary response is also of greater magnitude, producing higher levels of antibodies, and is more prolonged. This is why vaccination is so effective: The vaccine induces a primary immune response that produces memory cells; an encounter with the actual pathogen then elicits a rapid and strong secondary immune response.

The red curve in Figure 7B illustrates the specificity of the immune response. If the body is exposed to a different antigen (Y), even after it has already responded to antigen X, it responds with another primary response, this one directed against antigen Y. The response to Y is not enhanced by the response to X; that is, adaptive immunity is specific.

Although we have focused on the humoral immune response (produced by B cells) in this module, clonal selection, effector cells, and memory cells are features of the cell-mediated immune response (produced by T cells) as well. In the next several modules, we discuss the humoral immune response further. After that, we focus on how the cell-mediated arm of the immune system helps defend the body against pathogens.

? What is the immunological basis for referring to certain diseases, such as mumps, as *childhood* diseases?

One bout with the pathogen, which most often occurs during childhood, is usually enough to confer immunity for the rest of that individual's life.

8 Antibodies are the weapons of the humoral immune response

B cells are the "frontline warriors" of the humoral immune response. Plasma cells—the effector cells produced during clonal selection of B cells (as shown in Figure 7A)—make and secrete antibodies, proteins that serve as molecular weapons of defense.

We have been using Y-shaped symbols to represent antibodies, and their shape actually does resemble a Y, as the computer-generated rendering of an antibody molecule in **Figure 8A** illustrates. **Figure 8B** is a simplified diagram explaining antibody structure. Each antibody molecule is made up of four polypeptide chains, two identical "heavy" chains and two identical "light" chains. In both figures, the parts colored in shades of pink represent the fairly long heavy chains of amino acids that give the molecule its Y shape. Bonds (the black lines in Figure 8B) at the fork of the Y hold these chains together. The two green regions in each figure are shorter chains of amino acids, the light chains. Each of the light chains is bonded to one of the heavy chains. As Figure 8A indicates, the bonded chains actually intertwine.

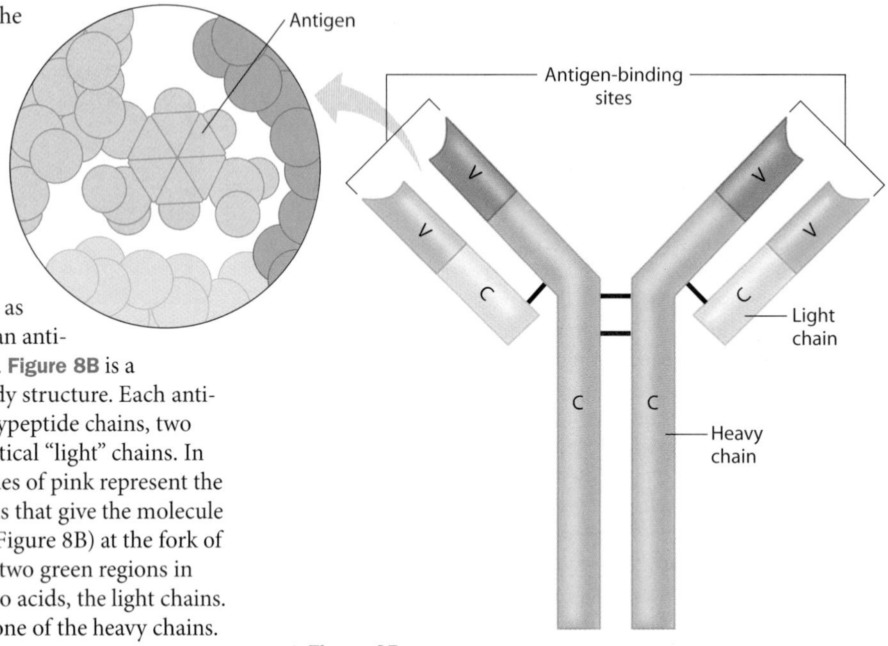

▲ **Figure 8B** Antibody structure and the binding of an antigen-binding site to its complementary antigen (enlargement)

An antibody molecule has two related functions in the humoral immune response: to recognize and bind to a certain antigen and to assist in neutralizing the antigen it recognizes. The structure of an antibody allows it to perform both of these functions. Notice in Figure 8B that each of the four chains of the molecule has a C (constant) region, where amino acid sequences vary little among different antibodies, and a V (variable) region, where the amino acid sequence varies extensively among antibodies. At the tip of each arm of the Y, a pair of V regions forms an **antigen-binding site**, a region of the molecule responsible for the antibody's recognition-and-binding function. A huge variety in the three-dimensional shapes of the binding sites of different antibody molecules arises from a similarly large variety in the amino acid sequences in the V regions; hence the term *variable*. The top left of Figure 8B illustrates such a fit, with the recognized antigen colored gold. The great structural variety of antigen-binding sites accounts for the diversity of lymphocytes and gives the humoral immune system the ability to react to virtually any kind of antigen.

The tail of the antibody molecule, formed by the constant regions of the heavy chains, helps mediate the disposal of the bound antigen. Antibodies with different kinds of heavy-chain C regions are grouped into different classes. Humans and other mammals have five major classes of antibodies, called IgA, IgD, IgE, IgG, and IgM (Ig stands for immunoglobulin, another name for antibody). Each of the five classes differs in where it's found in the body and how it works. However, all five classes of antibodies perform the same basic function: to mark invaders for elimination. We take a closer look at this process next.

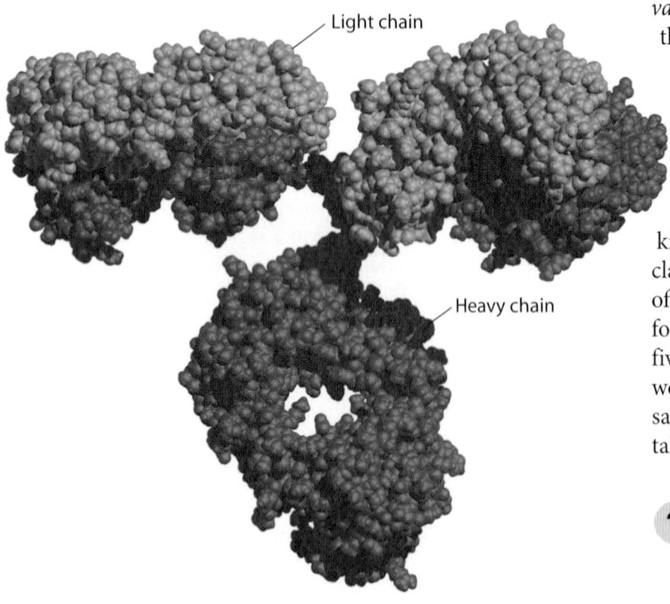

▲ **Figure 8A** A computer graphic of an antibody molecule

? How is the specificity of an antibody molecule for an antigen analogous to an enzyme's specificity for its substrate?

● Both antibodies and enzymes are proteins with binding sites of specific shape that recognize and bind to other molecules (antigens for antibodies, substrates for enzymes) with complementary shapes.

9 Antibodies mark antigens for elimination

Antibodies do not kill pathogens. Instead, antibodies mark a pathogen by combining with it to form an antigen-antibody complex. Weak chemical bonds between antigen molecules and the antigen-binding sites on antibody molecules hold the complex together. Once marked in this manner, other immune system components bring about the destruction of the antigen.

As **Figure 9** illustrates, the binding of antibodies to antigens can trigger several mechanisms that disable or destroy an invader. In neutralization, antibodies bind to surface proteins on a virus or bacterium, thereby blocking its ability to infect a host cell and presenting an easily recognized structure to macrophages. This increases the likelihood that the foreign cell will be engulfed by phagocytosis. Another antibody mechanism is the agglutination (clumping together) of viruses, bacteria, or foreign eukaryotic cells. Because each antibody molecule has at least two binding sites, antibodies can hold a clump of invading cells together. Agglutination makes the cells easy for phagocytes to capture. A third mechanism, precipitation, is similar to agglutination, except that the antibody molecules link dissolved antigen molecules together. This makes the antigen molecules precipitate; that is, they separate, in solid form, from the surrounding liquid. Precipitation, like the other effector mechanisms discussed so far, enhances engulfment by phagocytes.

One of the most important steps in the humoral immune response is the activation of the complement system (see Module 1) by antigen-antibody complexes. Activated complement system proteins (right side of the figure) can attach to a foreign cell. Once there, several activated proteins may form a complex that pokes a hole in the plasma membrance of the foreign cell, causing cell lysis, or rupture.

Taken as a whole, this figure illustrates a fundamental concept of adaptive immunity: All antibody mechanisms involve a *specific* recognition-and-attack phase followed by a *nonspecific* destruction phase. Thus, the antibodies of the humoral immune response, which identify and bind to foreign invaders, work with the components of innate immunity, such as phagocytes and complement, to form a complete defense system.

? How does adaptive humoral immunity interact with the body's innate immune system?

Antibodies mark specific antigens for destruction, but it is usually the complement system, phagocytes, or other components of innate immunity that destroy the antigens.

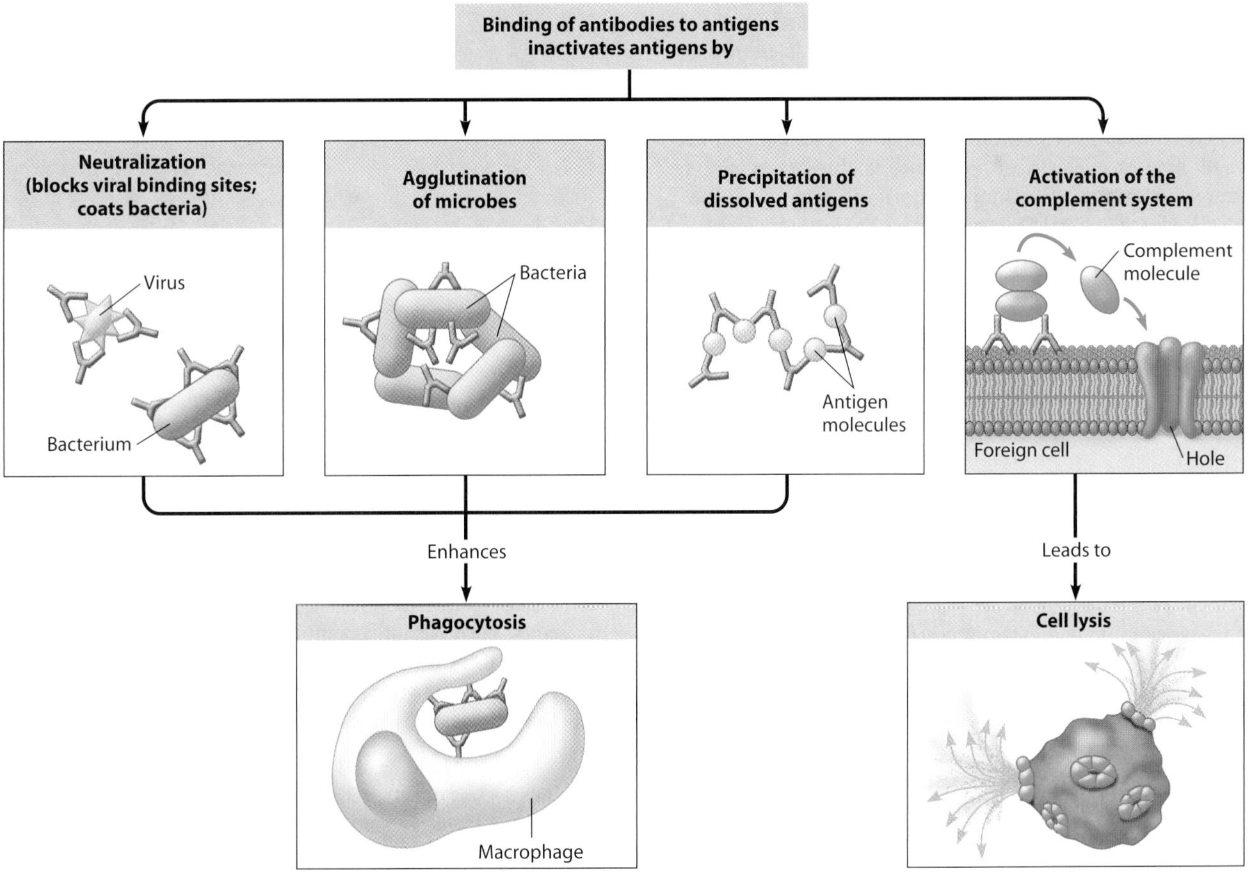

▲ **Figure 9** Effector mechanisms of the humoral immune response

CONNECTION 10 Monoclonal antibodies are powerful tools in the lab and clinic

Because of their ability to tag specific molecules or cells, antibodies are widely used in laboratory research, clinical diagnosis, and the treatment of disease. One way to prepare antibodies is to inject a small sample of antigen into a rabbit or mouse. In response to the antigen, the animal produces antibodies, which can be collected directly from its blood. However, because the antigen usually has many different antigenic determinants, the result is polyclonal: a mixture of different antibodies produced by different clones of cells.

In the late 1970s, a technique was developed for making **monoclonal antibodies (mAb)**. The term *monoclonal* means that all the cells producing the antibodies are descendants of a single cell; thus, they all produce identical antibody molecules that are specific for the same antigenic determinant. Monoclonal antibodies are harvested from cell cultures rather than from live animals.

The trick to making monoclonal antibodies is the fusion of two cells to form a hybrid cell with a combination of desirable properties. First, an animal is injected with an antigen that will stimulate B cells to make the desired antibody. At the same time, cancerous tumor cells, which can multiply indefinitely, are grown in a culture. The scientist then fuses a tumor cell with a normal antibody-producing B cell from the animal. The hybrid cell makes antibody molecules specific for a single antigenic determinant and is able to multiply indefinitely in a laboratory dish. The desired antibody can then be isolated from these cultured cells. Thus, large amounts of identical antibody molecules can be prepared.

One common use of monoclonal antibodies is in home pregnancy tests. The most popular type of test uses mAb to detect a hormone called human chorionic gonadotrophin (hCG), which is present in the urine of pregnant women. (hCG helps maintain the uterus during early pregnancy.) When urine is applied to the testing stick, it moves through a series of bands that contain different types of mAb (**Figure 10**). In the first band, any hCG ( ) that is present in the urine binds to an mAb ( ) specific for hCG. This first band also contains a second mAb ( ) that serves as a control. As the fluid moves into the second band, another mAb ( ) binds to the hCG-mAb complex, forming a "sandwich" that activates a colored dye visible as a stripe on the testing stick (in the figure, this is the blue stripe in the box at the center of the kit). If the woman is not pregnant, no hCG will be present in the urine, and thus no dye-activating complex will be formed. When the fluid moves into the third band, yet another type of monoclonal antibody ( ) binds to the control mAb, producing a different colored stripe (the red stripe in the bottom box). This positive control shows that the test strip is working.

Monoclonal antibodies are particularly useful for the diagnosis of diseases. For example, monoclonal antibodies can be prepared that will bind to a bacterium or virus that causes a sexually transmitted disease. If the antibodies have been labeled for easy detection (by a dye, for instance), they will reveal the presence of the bacterium or virus.

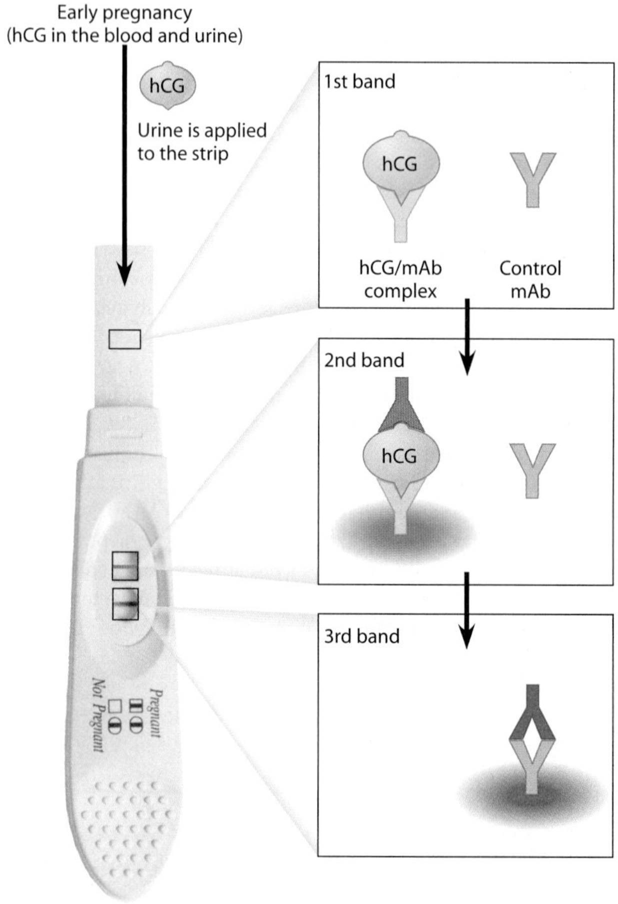

Raimund Koch/Photonica Amana America, Inc./Getty Images

▲ **Figure 10** Monoclonal antibodies used in a home pregnancy test

Monoclonal antibodies also have great promise for use in the treatment of certain diseases, including cancer. For example, Herceptin, a genetically engineered monoclonal antibody, is used to treat a common form of aggressive breast cancer. The Herceptin antibody molecules act by binding to growth factor receptors that are present in excess on the cancer cells. The drug helps slow the progress of the cancer by preventing the receptors from stimulating the cells to grow. Furthermore, certain types of cancer can be treated with tumor-specific monoclonal antibodies bound to toxin molecules. The toxin-linked antibodies carry out a precise search-and-destroy mission, selectively attaching to and killing tumor cells. The design of antibodies that target specific disease-associated molecules is an exciting and active area of research.

? In what way does the term *monoclonal* accurately reflect how these antibodies are made?

The prefix *mono* means "one," and *clonal* refers to the creation of an identical clone (group) of cells. Monoclonal antibodies are identical molecules produced by cells descended from a single cell.

11 Helper T cells stimulate the humoral and cell-mediated immune responses

The antibody-producing B cells of the humoral immune response make up one army of the adaptive immune response network. The humoral defense system identifies and helps destroy invaders that are in our blood, lymph, or interstitial fluid—in other words, outside our body cells. But many invaders, including all viruses, enter cells and reproduce there. It is the cell-mediated immune response produced by T cells that battles pathogens that have already entered body cells.

Whereas B cells respond to free antigens present in body fluids, T cells respond only to antigens present on the surfaces of the body's own cells. Recall from Module 7 that effector cells act quickly against an antigen. There are two main kinds of effector T cells. **Cytotoxic T cells** attack body cells that are infected with pathogens; we'll discuss these T cells in Module 12. **Helper T cells** play a role in many aspects of immunity. They help activate cytotoxic T cells and macrophages and even help stimulate B cells to produce antibodies. Other types of T cells include memory T cells, analogous to memory B cells.

Helper T cells interact with other white blood cells—including macrophages and B cells—that function as **antigen-presenting cells (APCs)**. All of the cell-mediated immune response and much of the humoral immune response depend on the precise interaction of antigen-presenting cells and helper T cells. This interaction activates the helper T cells, which can then go on to activate other cells of the immune system.

As its name implies, an antigen-presenting cell presents a foreign antigen to a helper T cell. Consider a typical antigen-presenting cell, a macrophage. As shown in **Figure 11**, ❶ the macrophage ingests a microbe or other foreign particle and breaks it into fragments—foreign antigens (▲). Then molecules of a special protein (Y) belonging to the macrophage, which we will call a **self protein** (because it belongs to the body), ❷ bind the foreign antigens—**nonself molecules**—and ❸ display them on the cell's surface. (Each of us has a unique set of self proteins, which serve as identity markers for our body cells.) ❹ Helper T cells recognize and bind to the combination of a self protein and a foreign antigen—called a self-nonself complex (Y)—displayed on an antigen-presenting cell. This double-recognition system is like the system banks use for safe-deposit boxes: Opening your box requires the banker's key along with your specific key.

The ability of a helper T cell to specifically recognize a unique self-nonself complex on an antigen-presenting cell depends on the receptors (purple) embedded in the T cell's plasma membrane. A T cell receptor actually has two binding sites: one for antigen and one for self protein. The two binding sites enable a T cell receptor to recognize the overall shape of a self-nonself complex on an antigen-presenting cell. The immune response is specific because the receptors on each helper T cell bind only one kind of self-nonself complex on an antigen-presenting cell.

The binding of a T cell receptor to a self-nonself complex activates the helper T cell. Several other kinds of signals can enhance this activation. For example, certain proteins secreted by the antigen-presenting cell, such as interleukin-1 (green arrow), diffuse to the helper T cell and stimulate it.

Activated helper T cells promote the immune response in several ways, with a major mechanism being the secretion of additional stimulatory proteins. One such protein, interleukin-2 (blue arrows), has three major effects. ❺ First, it makes the helper T cell itself grow and divide, producing both memory cells and additional active helper T cells. This positive-feedback loop amplifies the cell-mediated defenses against the antigen at hand. Second, interleukin-2 ❻ helps activate B cells, thus stimulating the humoral immune response. And third, ❼ it stimulates the activity of cytotoxic T cells, our next topic.

? How can one helper T cell stimulate both humoral and cell-mediated immunity?

By releasing stimulatory proteins that activate both cytotoxic T cells and B cells

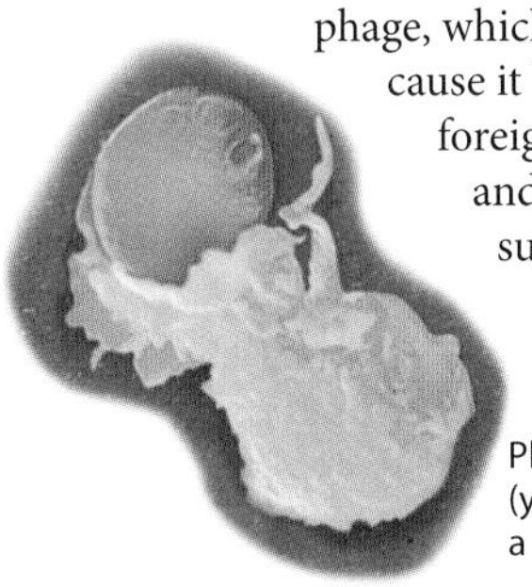

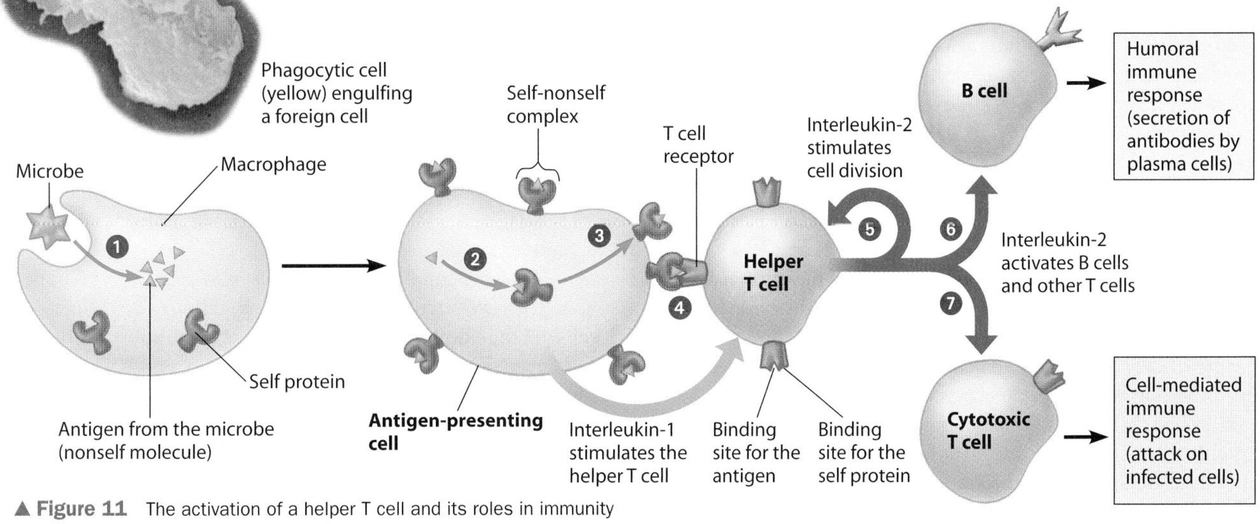

Biology Media/Science Source/Photo Researchers

▲ **Figure 11** The activation of a helper T cell and its roles in immunity

12 Cytotoxic T cells destroy infected body cells

As you have just learned, two types of T cells participate in the cell-mediated immune response: helper T cells and cytotoxic T cells. Helper T cells activate many kinds of cells, including cytotoxic T cells, the only T cells that actually kill infected cells.

Once activated, cytotoxic T cells identify infected cells in the same way that helper T cells identify antigen-presenting cells. An infected cell has foreign antigens—molecules belonging to the viruses or bacteria infecting it—attached to self proteins on its surface (Figure 12). Like a helper T cell, a cytotoxic T cell carries receptors that can bind with a self-nonself complex on the infected cell.

Cytotoxic T cells also play a role in protecting the body against the spread of some cancers. About 20% of human cancers are caused by viruses. Examples include the hepatitis B virus, which can trigger liver cancer, and the human papillomavirus (HPV), which can trigger cervical cancer. When a human cancer cell harbors such a virus, viral proteins may end up on the surface of the infected cell. If they do, they may be recognized by a cytotoxic T cell, which can then destroy the infected cell, halting the proliferation of that cancerous cell.

The self-nonself complex on an infected body cell is like a red flag to cytotoxic T cells that have matching receptors. As shown in the figure, ❶ a cytotoxic T cell binds to the infected cell. The binding activates the T cell, which then synthesizes several toxic proteins that act on the bound cell, including one called perforin. ❷ Perforin is discharged from the cytotoxic T cell and attaches to the infected cell's plasma membrane, making holes in it. T cell enzymes then enter the infected cell and promote its death by apoptosis, programmed cell death. ❸ The infected cell is destroyed, and the cytotoxic T cell may move on to destroy other cells infected with the same pathogen.

? Compare and contrast the T cell receptor with the antigen receptor on the surface of a B cell.

Both receptors bind to a specific antigen, but the T cell receptor only recognizes that antigen when it is presented along with a "self" marker on the surface of one of the body's own cells.

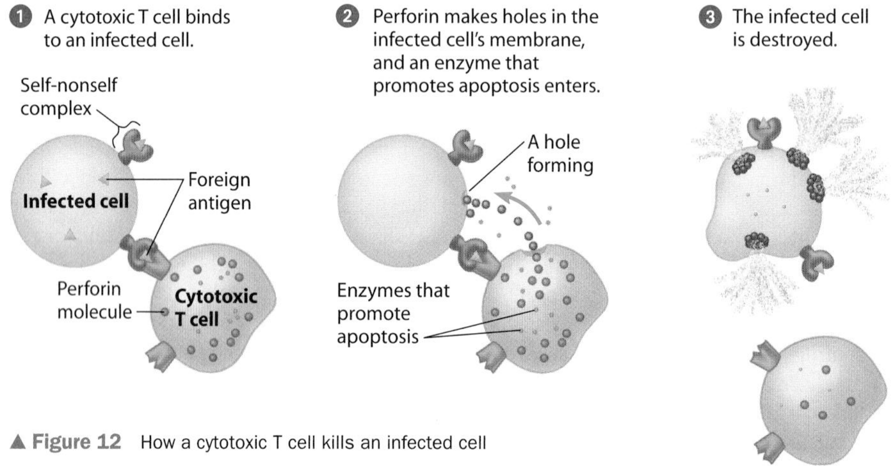
❶ A cytotoxic T cell binds to an infected cell.
Self-nonself complex
Infected cell
Foreign antigen
Perforin molecule
Cytotoxic T cell
❷ Perforin makes holes in the infected cell's membrane, and an enzyme that promotes apoptosis enters.
A hole forming
Enzymes that promote apoptosis
❸ The infected cell is destroyed.

▲ Figure 12 How a cytotoxic T cell kills an infected cell

CONNECTION 13 HIV destroys helper T cells, compromising the body's defenses

AIDS (acquired immunodeficiency syndrome) results from infection by **HIV**, the **human immunodeficiency virus**. Since the epidemic was first recognized in 1981, AIDS has killed more than 27 million people, and more than 33 million people are currently living with HIV. In 2009, 6 million people were newly infected with HIV, and over 1.8 million died, including over 300,000 children under the age of 15. The vast majority of HIV infections and AIDS deaths occur in the nonindustrialized nations of southern Asia and sub-Saharan Africa.

Although HIV can infect a variety of cells, it most often attacks helper T cells (Figure 13). As HIV depletes the body of helper T cells, both the cell-mediated and humoral immune responses are severely impaired. This drastically compromises the body's ability to fight infections.

How does HIV destroy helper T cells? Transmission of HIV requires the transfer of the virus from person to person via body fluids such as semen, blood, or breast milk. Once HIV is in the bloodstream, proteins on the surface of the virus can bind to proteins on the surface of a helper T cell. Attached to the T cell, HIV may enter and begin to reproduce. Inside the host helper T cell, the RNA genome of HIV is reverse-transcribed, and the newly produced DNA is integrated into the T cell's genome. This viral genome can now direct the production of new viruses from inside the T cell, generating up to 1,000 or more per day. Eventually, the host helper T cell dies from the damaging effects of virus production or from virus-triggered apoptosis.

After the new copies of the virus are released into the bloodstream, the HIV circulates, infecting and killing other helper T cells. As the number of T cells decreases, the body's ability to fight even the mildest infection is hampered, and AIDS eventually develops. It may take 10 years or more for full-blown AIDS symptoms to appear after the initial HIV infection.

Immune system impairment makes AIDS patients susceptible to cancers and **opportunistic infections**, ones that can be successfully fought off by a person with a healthy immune

Lennart Nilsson/Scanpix

Colorized EM 9,000×

▲ **Figure 13** A human helper T cell (red) under attack by HIV (blue dots)

system. For example, infection by a common fungus called *Pneumocystis carinii* rarely occurs among healthy individuals. In a person with AIDS, however, infection by *P. carinii* can cause severe pneumonia and death. Likewise, Kaposi's sarcoma, a very rare skin cancer, used to be seen exclusively among the elderly or patients receiving chemotherapy treatments. It is now most frequently seen among AIDS patients because their immune systems are unable to fight progression of the disease. In fact, it was an unusual cluster of Kaposi's sarcoma and pneumocystis pneumonia cases in 1981 that first brought AIDS to the attention of the medical community.

AIDS is incurable, although certain drugs can slow HIV reproduction and the progress of AIDS. One area of success has been in the development of drugs that can drastically lower transmission rates of HIV from mother to child. Unfortunately, these drugs are expensive and not available to all who need them. Combinations of drugs have proved effective at slowing the progress of AIDS, but the multidrug regimens are complicated and expensive, and may have debilitating side effects. Furthermore, any interruption of the treatment often results in the rapid decline of the health of the patient.

Despite decades of effort and billions of dollars spent, an AIDS vaccine remains elusive. A hint of success emerged from a six-year study involving 16,000 volunteers in Thailand. In 2009, researchers announced that the combination of two vaccines—neither of which is effective alone—appeared to offer some protection from HIV infection. Although the effect was small, just barely qualifying as statistically significant, it was the first success after 25 years of effort.

Until there is a vaccine or a cure, the best way to stop AIDS is to educate people about how the virus is transmitted: through unprotected sex, direct contact with blood (usually through shared intravenous drug needles), or from mother to baby. Safe sex behaviors, such as reducing promiscuity, using latex condoms, and avoiding intravenous drug use, could save millions of lives.

? **What is the function of the HIV enzyme reverse transcriptase?**

Reverse transcriptase catalyzes the synthesis of a DNA version of HIV's RNA genome.

EVOLUTION CONNECTION

14 The rapid evolution of HIV complicates AIDS treatment

As HIV reproduces, mutational changes occur that can generate new strains of the virus. In fact, the virus mutates at a very high rate during replication. Some of the newly mutated viruses will be less susceptible to destruction by the immune system. Such viruses will survive, proliferate, and mutate further. The virus thus evolves within the host body.

At one time, there was great hope that a "cocktail" of three anti-AIDS drugs (**Figure 14**), each of which attacks a different part of the HIV life cycle, could completely eliminate the virus in an infected person. It was hoped that virus strains resistant to one drug would be defeated by another. Such hope greatly underestimated the ability of HIV to evolve. Although people with access to such drugs do survive much longer and have a greatly improved quality of life, some HIV strains have evolved that are resistant even to multidrug regimens; thus, the virus is usually not totally eliminated from a patient's immune system. Current HIV treatment prolongs the life of HIV-positive people, but it is only a temporary fix and not a cure for AIDS.

Disturbingly, drug-resistant HIV strains are now being found in newly infected patients. This demonstrates that HIV readily adapts through natural selection to a changing environment—one in which drug treatments are widely available. In other words, the presence of anti-AIDS drugs in the environment has created a selection pressure that favors the spread of drug-resistant strains. Thus, the battle continues, with medical science on one side and the constantly evolving HIV on the other.

Impact Visuals/Newscom

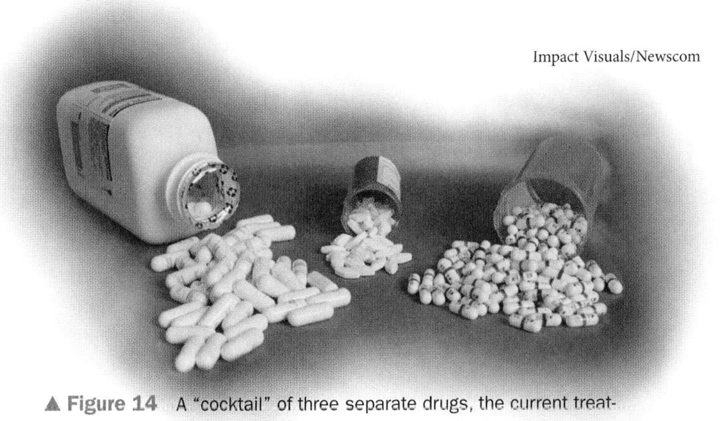

▲ **Figure 14** A "cocktail" of three separate drugs, the current treatment for people living with HIV

? **Why is it difficult to develop an AIDS vaccine?**

Because HIV evolves rapidly.

15 The immune system depends on our molecular fingerprints

As we have seen, the ability of lymphocytes to recognize the body's own molecules—that is, to distinguish self from non-self—enables our adaptive immune response to battle foreign invaders without harming healthy body cells. The key to this ability is that each person's cells have a unique collection of self proteins on the surface. These specific self proteins provide molecular "fingerprints" that can be recognized by the immune system. As lymphocytes mature, those with receptors that bind the body's own molecules are destroyed or deactivated, leaving only lymphocytes that react to foreign molecules. As a result, our lymphocytes do not attack our own molecules or cells.

The immune system not only distinguishes body cells from microbial cells, but also can tell your cells from those of other people. Genes at multiple chromosomal loci code for **major histocompatibility complex (MHC) molecules**, the main self proteins. (The green self proteins shown in Figures 11 and 12 are encoded by MHC genes.) Because there are hundreds of alleles in the human population for each MHC gene, it is extremely rare for any two people (except identical twins) to have completely matching sets of MHC self proteins.

The immune system's ability to recognize foreign antigens does not always work in our favor. For example, when a person receives an organ transplant or tissue graft, the person's immune system recognizes the MHC markers on the donor's cells as foreign and attacks them. To minimize rejection, doctors look for a donor with self proteins matching the recipient's as closely as possible. The best match is to transplant the patient's own tissue, as when a burn victim receives skin grafts removed from other parts of his or her body. Otherwise, identical twins provide the closest match, followed by nonidentical siblings. Sometimes doctors use drugs to suppress the immune response against the transplant. Unfortunately, these drugs may also reduce the ability to fight infections and cancer.

? In what sense is a cell's set of MHC surface markers analogous to a fingerprint?

The set of MHC ("self") markers is unique to each individual.

Disorders of the Immune System

CONNECTION 16 Malfunction or failure of the immune system causes disease

Our immune system is highly effective, protecting us against most potentially harmful invaders. But when the system fails to function properly, serious disease can result.

Autoimmune diseases result when the immune system goes awry and turns against some of the body's own molecules. In systemic lupus erythematosus (lupus), for example, B cells produce antibodies against a wide range of self molecules, such as histones and DNA released by the normal breakdown of body cells. Lupus is characterized by skin rashes, fever, arthritis, and kidney malfunction. Rheumatoid arthritis is another antibody-mediated autoimmune disease; it leads to damage and painful inflammation of the cartilage and bone of joints (**Figure 16**). In type 1 (insulin-dependent) diabetes mellitus, the insulin-producing cells of the pancreas are attacked by cytotoxic T cells. In multiple sclerosis (MS), T cells react against the myelin sheath that surrounds parts of many neurons, causing progressive muscle paralysis. Recent research suggests that Crohn's disease, a chronic inflammation of the digestive tract, may be caused by an autoimmune reaction against normal flora (bacteria) that inhabit the intestinal tract.

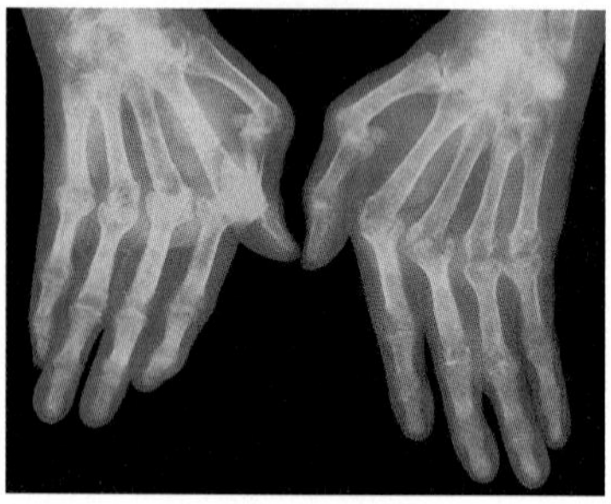

Salisbury District Hospital/Photo Researchers

▲ **Figure 16** An X-ray image of hands affected by rheumatoid arthritis

Gender, genetics, and environment all influence susceptibility to autoimmune disorders. For example, many autoimmune diseases afflict females more than males; women are two to three times more likely to suffer from MS and rheumatoid arthritis and nine times more likely to develop lupus. The cause of this sex bias is an area of active research and debate.

Most medicines for treating autoimmune diseases either suppress immunity in general or are limited to the alleviation of specific symptoms. However, as research scientists learn more about these diseases and about the normal operation of the immune system, they hope to develop more effective therapies.

In contrast to autoimmune diseases are a variety of defects called **immunodeficiency diseases** in which an immune response is defective or absent. People born immunodeficient are thus susceptible to frequent and recurrent infections. In the rare congenital disease severe combined immunodeficiency (SCID), both T cells and B cells are absent or inactive. People with SCID are extremely vulnerable to even minor infections. Until recently, their only hope for survival was to live behind protective barriers (providing inspiration for "bubble boy" stories in the popular media) or to receive a successful bone marrow transplant that would continue to supply functional lymphocytes. Since the early 1990s, medical researchers have been testing a gene therapy for this disease, with some success.

Immunodeficiency is not always an inborn condition; it may be acquired later in life. In addition to AIDS, another example is Hodgkin's disease, a type of cancer that damages the lymphatic system and can depress the immune system. Radiation therapy and the drug treatments used against many cancers can also disrupt the immune system.

There is growing evidence that physical and emotional stress can harm immunity. Hormones secreted by the adrenal glands during stress affect the numbers of white blood cells and may suppress the immune system in other ways. The association between emotional stress and immune function also involves the nervous system. Some neurotransmitters secreted when we are relaxed and happy may enhance immunity. In one study, college students were examined just after a vacation and again during final exams. Their immune systems were impaired in various ways during exam week; for example, interferon levels were lower. These and other observations indicate that general health and state of mind affect immunity.

? What is a probable side effect of autoimmune disease treatments that suppress the immune system?

Lowered resistance to infections

CONNECTION 17 Allergies are overreactions to certain environmental antigens

Allergies are hypersensitive (exaggerated) responses to otherwise harmless antigens in our surroundings. Antigens that cause allergies are called **allergens**. Common allergens include protein molecules on pollen grains and on the feces of tiny mites that live in house dust. Many people who are allergic to cats and dogs are actually allergic to proteins in the animal's saliva that get deposited on the fur when the animal licks itself. Allergic reactions typically occur very rapidly and in response to tiny amounts of an allergen. A person allergic to cat or dog saliva, for instance, may react to a few molecules of the allergen in a matter of minutes. Allergic reactions can occur in many parts of the body, including the nasal passages, bronchi, and skin. Symptoms may include sneezing, runny nose, coughing, wheezing, and itching.

The symptoms of an allergy result from a two-stage reaction sequence outlined in Figure 17. The first stage, called sensitization, occurs when a person is first exposed to an allergen—pollen, for example. ❶ After an allergen enters the bloodstream, it binds to effector B cells (plasma cells) with complementary receptors. ❷ The B cells then proliferate through clonal selection and secrete large amounts of antibodies to this allergen. ❸ Some of these antibodies attach by their base to the surfaces of mast cells, body cells that produce histamine and other chemicals that trigger the inflammatory response (see Module 2).

The second stage of an allergic response begins when the person is exposed to the same allergen later. The allergen enters the body and ❹ binds to the antibodies attached to mast cells. ❺ This causes the mast cells to release histamine, which triggers the allergic symptoms. As in inflammation, histamine causes blood vessels to dilate and leak fluid and so causes nasal irritation, itchy skin, and tears. **Antihistamines** are drugs that interfere with histamine's action and give temporary relief from an allergy.

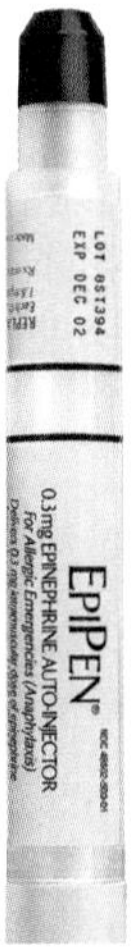
EpiPen

Allergies range from seasonal nuisances to severe, life-threatening responses. Anaphylactic shock is an especially dangerous type of allergic reaction. It may occur in people who are extremely sensitive to certain allergens, such as bee venom, penicillin, or allergens in peanuts or shellfish. Any contact with these allergens makes their mast cells release inflammatory chemicals very suddenly. As a result, their blood vessels dilate abruptly, causing a rapid, potentially fatal drop in blood pressure, a condition called shock. Fortunately, anaphylactic shock can be counteracted with injections of the hormone epinephrine. People with severe allergies often carry "epi pens" to protect themselves in case of exposure.

? How do antihistamines relieve allergy symptoms?

By blocking the action of histamine, which produces the symptoms of the inflammatory response

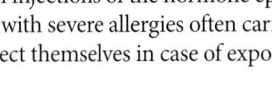

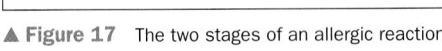

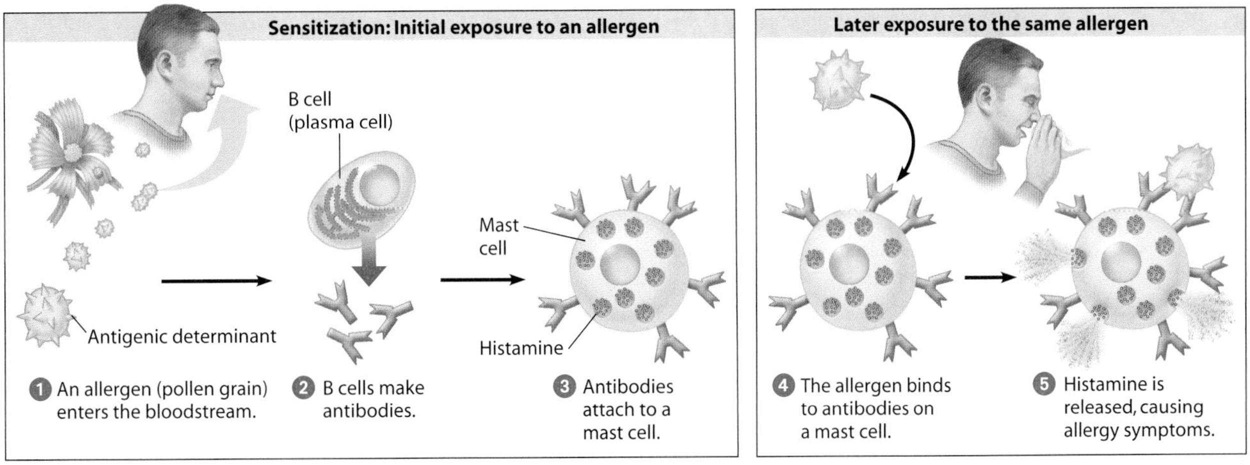
Sensitization: Initial exposure to an allergen
B cell (plasma cell)
Mast cell
Antigenic determinant
Histamine
❶ An allergen (pollen grain) enters the bloodstream.
❷ B cells make antibodies.
❸ Antibodies attach to a mast cell.
Later exposure to the same allergen
❹ The allergen binds to antibodies on a mast cell.
❺ Histamine is released, causing allergy symptoms.

▲ **Figure 17** The two stages of an allergic reaction

Author series Dey L. P. Pharmaceuticals

REVIEW

For Practice Quizzes, BioFlix, MP3 Tutors, and Activities, go to www.masteringbiology.com.

Reviewing the Concepts

Innate Immunity (1–3)

1 All animals have innate immunity. Innate defenses include the skin, mucous membranes, phagocytic cells, and antimicrobial proteins such as interferons and the complement system.

2 Inflammation mobilizes the innate immune response. Tissue damage triggers the inflammatory response, which can disinfect tissues and limit further infection.

3 The lymphatic system becomes a crucial battleground during infection. Lymphatic vessels collect fluid from body tissues and return it as lymph to the blood. Lymph organs, such as the spleen and lymph nodes, are packed with white blood cells that fight infections.

Adaptive Immunity (4–15)

4 The adaptive immune response counters specific invaders. Antigens from infections or vaccinations trigger adaptive immunity.

5 Lymphocytes mount a dual defense. Millions of kinds of B cells and T cells, each with different membrane receptors, wait in the lymphatic system, where they may respond to invaders.

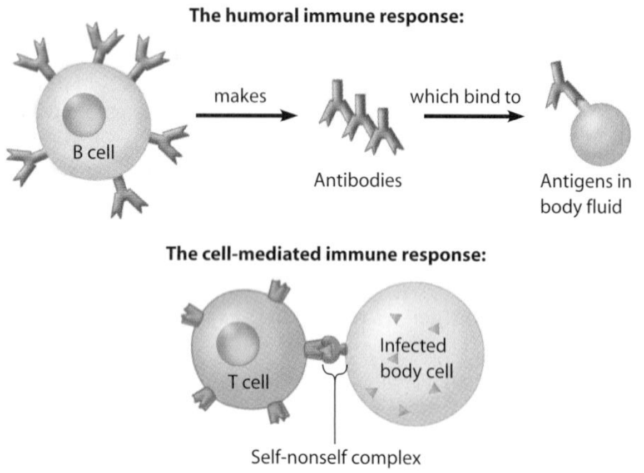

6 Antigens have specific regions where antibodies bind to them.

7 Clonal selection musters defensive forces against specific antigens. When an antigen enters the body, it activates only a small subset of lymphocytes that have complementary receptors. The selected cells multiply into clones of short-lived effector cells specialized for defending against the antigen that triggered the response and also into memory cells, which confer long-term immunity. Activated by subsequent exposure to the antigen, memory cells mount a rapid and strong secondary immune response.

8 Antibodies are the weapons of the humoral immune response. Antibodies are secreted by plasma (effector) B cells into the blood and lymph. An antibody has antigen-binding sites specific to the antigenic determinants that elicited its secretion.

9 Antibodies mark antigens for elimination.

10 Monoclonal antibodies are powerful tools in the lab and clinic.

11 Helper T cells stimulate the humoral and cell-mediated immune responses. In the cell-mediated immune response, an antigen-presenting cell displays a foreign antigen (a nonself molecule) and one of the body's own self proteins to a helper T cell. The helper T cell's receptors recognize the self-nonself complexes, and the interaction activates the helper T cell. In turn, the helper T cell can activate cytotoxic T cells and B cells.

12 Cytotoxic T cells destroy infected body cells.

13 HIV destroys helper T cells, compromising the body's defenses. The AIDS virus opens the way for opportunistic infections.

14 The rapid evolution of HIV complicates AIDS treatment.

15 The immune system depends on our molecular fingerprints. The immune system normally reacts only against nonself substances, not against self. Transplanted organs may be rejected because these cells lack the unique "fingerprint" of the recipient's self proteins.

Disorders of the Immune System (16–17)

16 Malfunction or failure of the immune system causes disease. In autoimmune diseases, the immune system turns against the body's own molecules. In immunodeficiency diseases, immune components are lacking and frequent infections occur.

17 Allergies are overreactions to certain environmental antigens.

Connecting the Concepts

1. Complete this concept map to summarize the key concepts concerning the body's defenses.

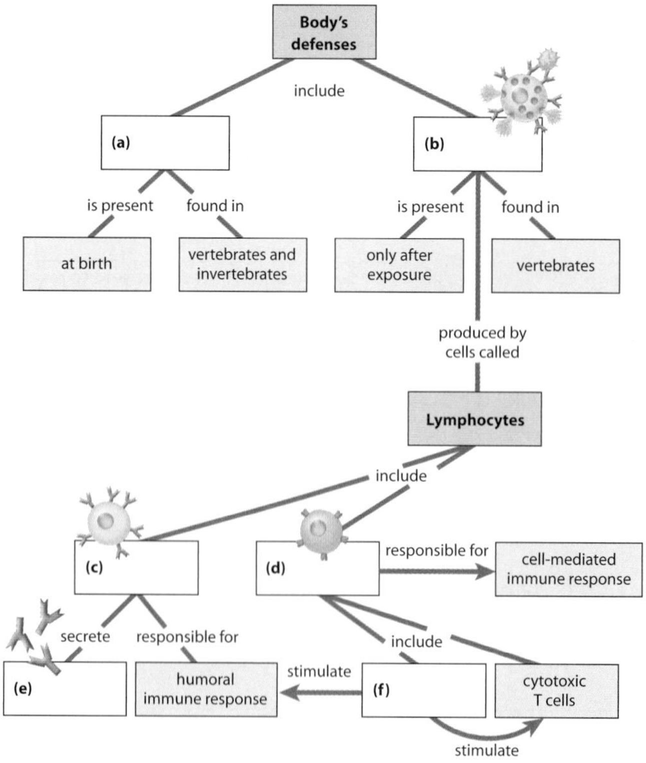

Testing Your Knowledge

Multiple Choice

2. Foreign molecules that elicit an immune response are called
   a. pathogens.
   b. antibodies.
   c. lymphocytes.
   d. histamines.
   e. antigens.

3. Which of the following is *not* part of the vertebrate innate defense system?
   a. natural killer cells
   b. antibodies
   c. interferons
   d. complement system
   e. inflammation

4. Which of the following best describes the difference in the way B cells and cytotoxic T cells deal with invaders?
   a. B cells confer active immunity; T cells confer passive immunity.
   b. B cells send out antibodies to attack; certain T cells can do the attacking themselves.
   c. T cells handle the primary immune response; B cells handle the secondary response.
   d. B cells are responsible for the cell-mediated immune response; T cells are responsible for the humoral immune response.
   e. B cells attack the first time the invader is present; T cells attack subsequent times.

5. The antigen-binding sites of an antibody molecule are formed from the molecule's variable regions. Why are these regions called variable?
   a. They can change their shapes on command to fit different antigens.
   b. They change their shapes when they bind to an antigen.
   c. Their specific shapes are unimportant.
   d. They have different shapes on antibodies to different antigens.
   e. Their sizes vary considerably from one antibody to another.

6. Cytotoxic T cells are able to recognize infected body cells because
   a. the infection changes the surfaces of infected cells.
   b. B cells help them.
   c. the infected cells produce antigens.
   d. infected cells release antibodies into the blood.
   e. helper T cells destroy them first.

Matching

7. Attacks infected body cells
8. Carries out the humoral immune response
9. Causes allergy symptoms
10. Phagocytic white blood cell
11. General name for a B or T cell
12. Required for initiating the secondary immune response
13. Cell most commonly attacked by HIV

a. lymphocyte
b. cytotoxic T cell
c. helper T cell
d. mast cell
e. macrophage
f. B cell
g. memory cell

Describing, Comparing, and Explaining

14. Describe how AIDS is transmitted and how immune system cells in an infected person are affected by HIV. Why is AIDS particularly deadly compared to other viral diseases? What are the most effective means of preventing HIV transmission?

15. What is inflammation? How does it protect the body? Why is inflammation considered part of the innate immune response?

16. Your roommate is rushed to the hospital after suffering a severe allergic reaction to a bee sting. After she is treated and released, she asks you (the local biology expert!) to explain what happened. She says, "I don't understand how this could have happened. I've been stung by bees before and didn't have a reaction." Suggest a hypothesis to explain what has happened to cause her severe allergic reaction and why she did not have the reaction after previous bee stings.

Applying the Concepts

17. Organ donation saves many lives each year. Even though some transplanted organs are derived from living donors, the majority come from patients who die but still have healthy organs that can be of value to a transplant recipient. Potential organ donors can fill out an organ donation card to specify their wishes. If the donor is in critical condition and dying, the donor's family is usually consulted to discuss the donation process. Generally, the next of kin must approve before donation can occur, regardless of whether the patient has completed an organ donation card. In some cases, the donor's wishes are overridden by a family member. Do you think that family members should be able to deny the stated intentions of the potential donor? Why or why not? Have you signed up to be an organ donor? Why or why not?

18. There is great concern about the rate of HIV infection among teenagers. Schools in some large cities have instituted programs to make condoms available to students, along with counseling about safer sex. These plans have divided school boards and communities. Some citizens and church groups are opposed to giving condoms to students on the grounds that it might appear to encourage sexual activity. By contrast, many school and public health officials view the situation as a health issue rather than a moral issue. The heart of the controversy seems to be whether the schools should take such a direct role in this part of student life. What are the reasons for and against distribution of condoms? What do you think the school's role should be?

19. One of the key difficulties in the development of anti-HIV drugs is the fact that HIV will only infect humans. This precludes testing of drugs in animals and instead requires that drugs be tested on human volunteer subjects. The developing world (particularly sub-Saharan Africa and Southeast Asia) has the highest rates of HIV infection. Consequently, drug companies frequently conduct studies in these regions. Some people decry such tests, fearing that drug companies may profit hugely from the use of economically disadvantaged people. Others counter that such tests are the only way to find new and cheaper drugs that will ultimately help everyone. What do you think are the ethical issues surrounding trials of anti-HIV drugs in the developing world? Which side do you think has the more morally compelling argument?

Review Answers

1. a. innate immunity; b. adaptive immunity; c. B cells; d. T cells; e. antibodies; f. helper T cells
2. e 3. b 4. b 5. d 6. a 7. b 8. f 9. d 10. e 11. a 12. g 13. c
14. AIDS is mainly transmitted in blood and semen. It enters the body through slight wounds during sexual contact or via needles contaminated with infected blood. AIDS is deadly because it infects helper T cells, crippling both the humoral and cell-mediated immune responses and leaving the body vulnerable to other infections. The most effective way to avoid HIV transmission is to prevent contact with body fluids by practicing safe sex and avoiding intravenous drugs.
15. Inflammation is triggered by tissue injury. Damaged cells release histamine and other chemicals, which cause nearby blood vessels to dilate and become leakier. Blood plasma leaves vessels, and phagocytes are attracted to the site of injury. An increase in blood flow, fluid accumulation, and increased cell population cause redness, heat, and swelling. Inflammation disinfects and cleans the area and curtails the spread of infection from the injured area. Inflammation is considered part of the innate immune response because similar defenses are presented in response to any infection.
16. One hypothesis is that your roommate's previous bee stings caused her to become sensitized to the allergens in bee venom. During sensitization, antibodies to allergens attach to receptor proteins on mast cells. During this sensitization stage, she would not have experienced allergy symptoms. When she was exposed to the bee venom again at a later time, the bee venom allergens bound to the mast cells, which triggered her allergic reaction.
17. There is no correct answer to this question. Some issues and questions to consider: Possible directions include the idea that if the donor felt strongly about the process, then his or her wishes should be respected. The opposite direction would be that the next of kin should be able to approve or deny the procedure. Other considerations may be appropriate, including religious beliefs.
18. Some issues and questions to consider: How important is it to protect students from HIV? Is this a function of schools? Do schools serve other such "noneducational" purposes? Should parents or citizens' and church groups—on either side of the issue—have a say in this, or is it a matter between the school and the student? Does the distribution of condoms condone or sanction sexual activity or promiscuity? Is a school legally liable if a school-issued condom fails to protect a student? Are there alternative measures, such as education, that might be as effective for slowing the spread of HIV?
19. Some issues and questions to consider: How much do people in various nations stand to gain by the development of new drugs, in terms of both lives saved and profits made? How can oversight be used to ensure that drug companies are acting in the best interests of all their patients, and not purely for profit? Can studies be modified so as to maximize the potential benefits to HIV-infected people while minimizing the risks to study participants? Or is such a trade-off impossible? Should studies on humans be banned altogether?

Control of Body Temperature and Water Balance

From Chapter 25 of *Campbell Biology: Concepts & Connections*, Seventh Edition. Jane B. Reece, Martha R. Taylor, Eric J. Simon, Jean L. Dickey.

Control of Body Temperature and Water Balance

BIG IDEAS

Four Oaks/Shutterstock

Thermoregulation
(1–3)

Animals use various homeostatic mechanisms to control body temperature.

George Peters/iStockphoto

Osmoregulation and Excretion
(4–10)

Animals regulate the movement of water, solutes, and wastes into and out of the body.

Lynn Rogers/Peter Arnold, Inc./PhotoLibrary

Bears, like the two grizzlies (*Ursus arctos horribilis*) shown in the photo, emerge in spring after a winter spent in their dens. For months hibernating bears hardly move. If you were to examine a dormant bear, you would find physiological processes at work that aid homeostasis, the maintenance of nearly constant internal conditions despite fluctuations in the external environment. These physiological processes are evolutionary adaptations that conserve energy as the animal taps into stored body fat.

To make it through winter, a bear must thermoregulate. Thermoregulation is the maintenance of internal temperature within narrow limits (in this case, about 5°C lower than normal). Thermoregulation is aided by several adaptations. For example, a bear's stored body fat and dense fur provide insulation, curling up keeps heat loss to a minimum, and reduced blood flow to the bear's extremities decreases heat loss.

Besides having to regulate body temperature, an overwintering bear faces other challenges. While dormant, bears do not eat or drink, nor do they expel waste. Their bodies compensate for this through osmoregulation, the control of the input and output of water and solutes, and by controlling excretion. A bear has adaptations that aid in these processes. For example, dormant bears can metabolize nitrogen-containing wastes that accumulate in the bloodstream, converting the waste molecules to harmless forms.

In this chapter, we explore the three homeostatis mechanisms we just discussed: thermoregulation, osmoregulation, and excretion. You'll see that, like the bear, most animals can survive fluctuations in the external environment because homeostatic control mechanisms allow the internal environment of animals to stay within a narrow, tolerable range. To begin, we'll survey some of the ways that animals regulate body temperature.

Thermoregulation

1 An animal's regulation of body temperature helps maintain homeostasis

Throughout their lives, animals continually exchange heat, water, and dissolved solutes with their environment. Several mechanisms have evolved that allow animals to make such exchanges while maintaining homeostasis. The first homeostatic mechanism that we examine is **thermoregulation**, the process by which animals maintain an internal temperature within a tolerable range. Thermoregulation is critical to survival because most of life's processes are sensitive to changes in body temperature. Each species of animal has an optimal temperature range. Thermoregulation helps keep body temperature within that range, enabling cells to function effectively even when the external temperature fluctuates.

Internal metabolism and the external environment provide heat for thermoregulation. Most mammals and birds, a few other reptiles, some fishes, and many insect species are **endotherms**, meaning they are warmed mostly by heat generated by their own metabolism. In contrast, most amphibians, lizards, many fishes, and most invertebrates are **ectotherms**, meaning they gain most of their heat by absorbing it from external sources.

Keep in mind, though, that endothermic and ectothermic modes of thermoregulation are not mutually exclusive. For example, a bird is mainly endothermic, but it may warm itself in the sun on a cold morning much as an ectothermic lizard does. In the following two modules, we examine some of the ways that animals regulate their body temperatures.

? A lizard warming itself on a hot rock is an example of an ______. Why?

Ectotherm. Lizards absorb most of their body heat from their surroundings.

2 Heat is gained or lost in four ways

An animal can exchange heat with the environment by four physical processes. *Conduction* is the direct transfer of thermal motion (heat) between molecules of objects in direct contact with each other, as when an animal is physically touching an object in its environment. Heat is always transferred from an object of higher temperature to one of lower temperature. In **Figure 2**, heat conducted from the warm rock (red arrows) elevates the lizard's body temperature.

Convection is the transfer of heat by the movement of air or liquid past a surface. In the figure, a breeze removes heat from the lizard's tail (orange arrows) by convection.

Radiation is the emission of electromagnetic waves. Radiation can transfer heat between objects that are not in direct contact, as when an animal absorbs heat radiating from the sun (yellow arrows). The lizard also radiates some of its own heat into the external environment.

Evaporation is the loss of heat from the surface of a liquid that is losing some of its molecules as a gas. A lizard loses heat as moisture evaporates from its nostrils (blue arrow).

Mike Wilkes/naturepl.com

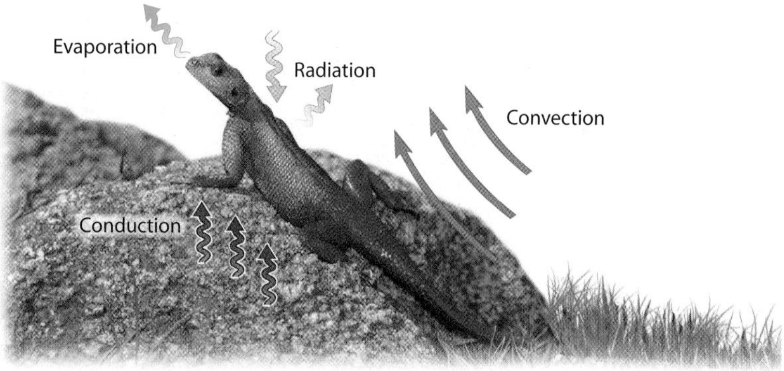

▲ **Figure 2** Mechanisms of heat exchange

? If you are sweating on a hot day and turn a fan on yourself, what two mechanisms contribute to your cooling?

Evaporation (of sweat) and convection (fan moving air)

3 Thermoregulation involves adaptations that balance heat gain and loss

Different animals are adapted to different environmental temperatures. Within their optimal temperature range, endotherms and many ectotherms maintain a fairly constant internal temperature despite external temperature fluctuations. Five general categories of adaptations help animals thermoregulate.

Cherkas/Shutterstock

Metabolic Heat Production In cold weather, hormonal changes tend to boost the metabolic rate of birds and mammals, increasing their heat production. Simply moving around more or shivering produces heat as a metabolic by-product of the contraction of skeletal muscles. Honeybees survive cold winters by clustering together and shivering in their hive. The metabolic activity of all the bees together generates enough heat to keep the cluster alive.

Insulation A major thermoregulatory adaptation in mammals and birds is insulation—hair (often called fur), feathers, and fat layers—which reduces the flow of heat between an animal and its environment. Most land mammals and birds react to cold by raising their fur or feathers, which traps a thicker layer of air next to the warm skin, improving the insulation. (In humans, our muscles raise our hair in the cold, causing goose bumps, a vestige from our furry ancestors.) Aquatic mammals (such as seals) and aquatic birds (such as penguins) are insulated by a thick layer of fat called blubber.

Circulatory Adaptations Heat loss can be altered by changing the amount of blood flowing to the skin. In a bird or mammal (and some ectotherms), nerves signals cause surface blood vessels to constrict or dilate, depending on the external temperature. When the vessels are constricted, less blood flows from the warm body core to the body surface, reducing the rate of heat loss through radiation. Conversely, dilation (opening) of surface blood vessels increases the rate of heat loss. Large, thin ears capable of radiating a lot of heat have evolved in elephants, helping to cool their large bodies in their native tropical climates (**Figure 3A**). Flapping their ears increases heat dissipation by convection. Dilation is what causes your face to become flushed after rigorous exercise or after sitting in a hot tub.

Figure 3B illustrates a circulatory adaptation found in many birds and mammals. In **countercurrent heat exchange**, warm and cold blood flow in opposite (countercurrent) directions in two adjacent blood vessels. Warm blood (red) from the body core cools as it flows down the goose's leg or the dolphin's flipper. But the arteries carrying the warm blood are in close contact with veins conveying cool blood (blue) back toward the body core. As shown in the figure by the black arrows, heat passes from the warmer blood to the cooler blood along the whole length of these side-by-side vessels (because heat always flows from where it is warmer to where it is cooler). By the time blood leaves the leg or flipper and returns to the body core it is almost as warm as the body core. Thus, even when the animal is standing on ice or swimming in frigid water, heat loss through the extremities is minimal.

Some endothermic bony fishes and sharks also have countercurrent exchange mechanisms. All fishes lose heat as blood passes through the gills. In large, powerful swimmers such as the great white shark, cold blood returning from the gills is

Jeff Lepore/Photo Researchers

O. Alamany & E. Vicens/Corbis

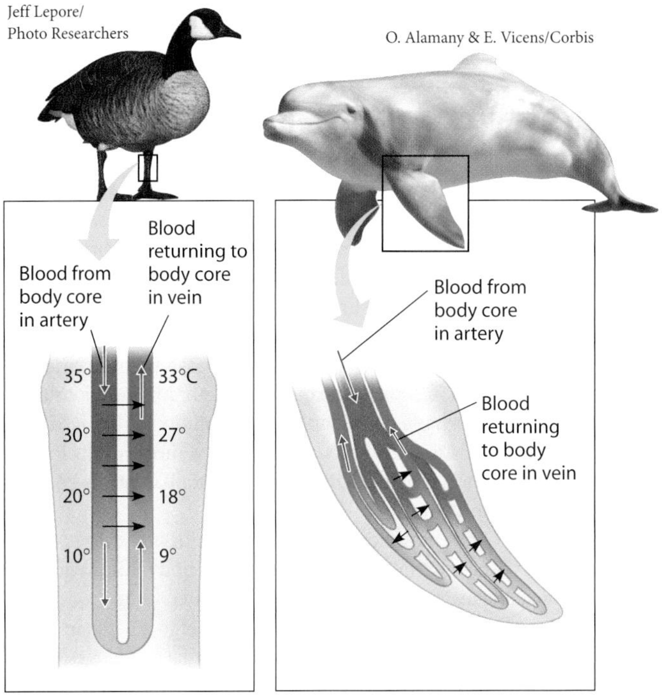

▲ **Figure 3B** Countercurrent heat exchange

transported in large vessels lying just under the skin. Small branches of these vessels deliver oxygenated blood to the deep muscles. Each branch runs side by side with a vessel carrying warm blood outward from the inner body. The resulting countercurrent heat exchange retains heat around the main swimming muscles and enables the vigorous, sustained activity of these endothermic animals.

Evaporative Cooling Many animals live in places where thermoregulation requires cooling as well as warming. Some animals have adaptations that greatly increase evaporative cooling, such as panting, sweating, or spreading saliva on body surfaces. As shown in Figure 3A, elephants spray water over their bodies to aid evaporation. Humans sweat and other animals (such as dogs) lose heat as moisture evaporates during panting.

Behavioral Responses Both endotherms and ectotherms control body temperature through behavioral responses. Some birds and butterflies migrate seasonally to more suitable climates. Other animals, such as desert lizards, warm themselves in the sun when it is cold and find cool, damp areas or burrows when it is hot. Many animals bathe, which brings immediate cooling by convection and continues to cool for some time by evaporation. We humans dress for warmth.

Although some animals can tolerate minor fluctuations in body temperature, few can withstand even small changes in the balance of water and solutes in body fluids. We consider this topic, osmoregulation, next.

? Compare countercurrent heat exchange with the countercurrent exchange of oxygen in fish gills.

In both cases, countercurrent exchange enhances transfer all along the length of a blood vessel—transfer of heat from one vessel to another in the case of a heat exchanger and transfer of oxygen between water and vessels in the case of gills.

▼ **Figure 3A** Heat dissipation via ear flapping (convection) and via water spray (evaporative cooling)

Four Oaks/Shutterstock

Osmoregulation and Excretion

4 Animals balance their level of water and solutes through osmoregulation

Osmoregulation is the homeostatic control of the uptake and loss of water and solutes, dissolved substances such as salt (NaCl) and other ions. Animal cells cannot survive substantial water gain or loss: They swell and burst if there is a significant net uptake of water; they shrivel and die if there is a substantial net loss of water. Osmosis is one process whereby animals regulate their uptake and loss of fluids. Osmosis is a movement of water across a selectively permeable membrane from a solution with lower solute concentration to a solution with higher solute concentration. So, because solute movement results in the movement of water by osmosis, osmoregulation controls both solute and water levels.

Some sea-dwelling animals—such as squids, sea stars, and most other marine invertebrates—have body fluids with a solute concentration equal to that of seawater. Called **osmoconformers**, such animals do not undergo a net gain or loss of water. They therefore face no substantial challenges in water balance. However, because they differ considerably from seawater in the concentrations of certain specific solutes, they must actively transport these solutes to maintain homeostasis. For example, the resting state of squid neurons requires a higher concentration of sodium ions (Na^+) outside the cell than inside.

Many animals—land animals (such as elephants), freshwater animals (such as trout), and marine vertebrates (such as sharks)—have body fluids whose solute concentration differs from that of their environment. Therefore, they must actively regulate water movement. Such animals are called **osmoregulators**.

The freshwater fish in **Figure 4A** has a much higher solute concentration in its internal fluids than does fresh water. Therefore, a freshwater fish constantly gains water by osmosis through its body surface, especially through its gills. It also loses salt by diffusion to its more dilute environment. Freshwater fish also actively take in salt through their gills. Although such fish do not drink water, food helps supply some additional ions. To dispose of excess water and to conserve solutes, the fish's kidneys produce large amounts of **urine**, the waste material produced by its urinary system.

The saltwater fish in **Figure 4B** has different osmoregulatory challenges. Because its internal fluids are lower in total solutes than seawater, a saltwater fish loses water by osmosis across its body surfaces. It also gains salt by diffusion and from the food it eats. The fish balances the water loss by taking in large amounts of seawater, and it balances solutes by pumping out excess salt through its gills. It also saves water by producing only small amounts of concentrated urine, in which it disposes of some excess ions.

Land animals face an additional homeostatic challenge. Because of the threat of dehydration, adaptations that prevent or reduce this danger provide important evolutionary advantages. Insects have tough exoskeletons impregnated with waterproof wax, which helps conserve water. Most terrestrial vertebrates, including humans, have an outer skin formed of multiple layers of dead, water-resistant cells, which also minimizes surface water loss. As we will see in Module 8, the kidney also plays a major role in conserving water. Essential to survival on land are adaptations that protect fertilized eggs and developing embryos from drying out. Many insects (such as pesky summertime mosquitoes) lay their eggs in moist areas, and the eggs of many species (such as birds, turtles, and other reptiles) are surrounded by a tough, watertight shell. Likewise, the embryos of reptiles and mammals develop in a water-filled amniotic sac surrounded by protective membranes. In addition to reproductive adaptations, land animals maintain water balance by drinking and eating moist foods and by producing water metabolically through cellular respiration. Despite such adaptations, however, most terrestrial animals lose water in many ways: in urine and feces, across their skin, and from the surfaces of gas exchange organs.

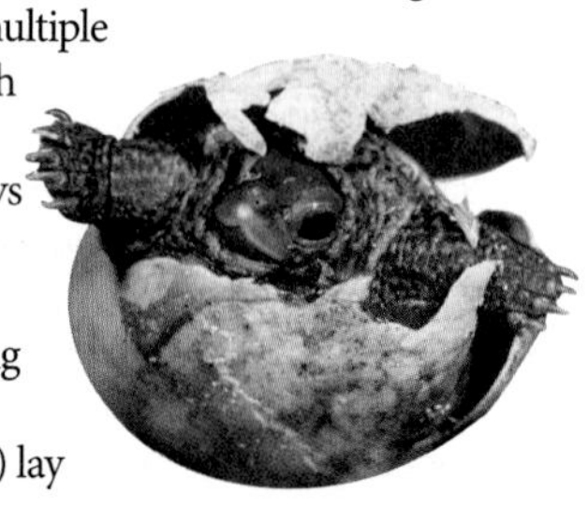

Zig Leszczynski/AA/PhotoLibrary

? **Why are no freshwater animals osmoconformers?**

● Osmoconformers have solute concentrations equal to that of the environment. The body fluid of a freshwater osmoconformer would be very low in ions and be too dilute to support life's processes.

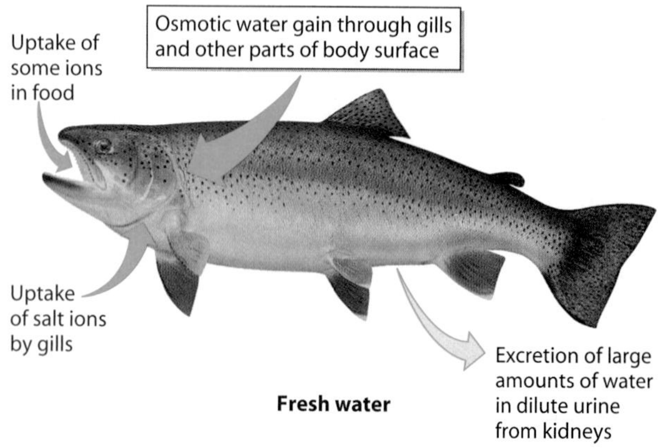

George Peters/iStockphoto

▲ **Figure 4A** Osmoregulation in a rainbow trout, a freshwater fish

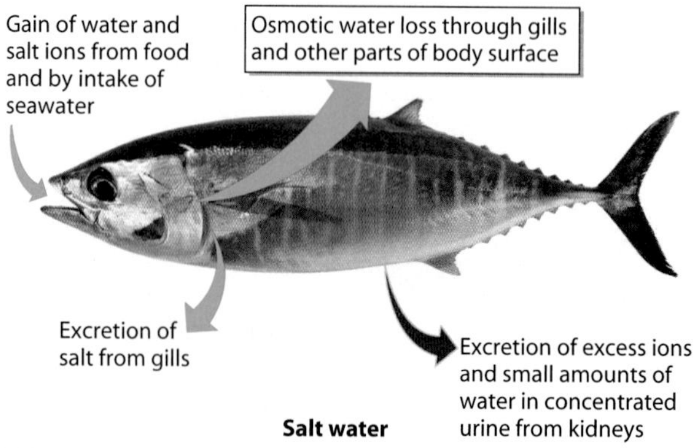

holbox/Shutterstock

▲ **Figure 4B** Osmoregulation in a bluefin tuna, a saltwater fish

EVOLUTION CONNECTION

5 A variety of ways to dispose of nitrogenous wastes has evolved in animals

Waste disposal is a crucial aspect of osmoregulation. Because most metabolic wastes must be dissolved in water to be removed from the body, the type and quantity of an animal's wastes will have a large impact on its water balance. Metabolism produces a number of toxic by-products, such as the nitrogenous (nitrogen-containing) wastes that result from the breakdown of proteins and nucleic acids. An animal disposes of these metabolic wastes through excretion, the third and final homeostatic control mechanism we discuss in this chapter.

The form of an animal's nitrogenous wastes depends on its evolutionary history and its habitat. As **Figure 5** indicates, most aquatic animals dispose of their nitrogenous wastes as **ammonia**. Among the most toxic of all metabolic by-products, ammonia (NH_3) is formed when amino groups ($—NH_2$) are removed from amino acids and nucleic acids. Ammonia is too toxic to be stored in the body, but it is highly soluble and diffuses rapidly across cell membranes. If an animal is surrounded by water, ammonia readily diffuses out of its cells and body. Small, soft-bodied invertebrates, such as planarians (flatworms), excrete ammonia across their whole body surface. Fishes excrete it mainly across their gills.

Ammonia excretion does not work well for land animals. Because it is so toxic, ammonia must be transported and excreted in large volumes of very dilute solutions. It is therefore more efficient for land animals to convert ammonia to less toxic compounds, either urea or uric acid, that can be safely transported and stored in the body and released periodically by the urinary system. The disadvantage of excreting urea or uric acid is that the animal must use energy to produce these compounds from ammonia.

As shown in Figure 5, mammals, most adult amphibians, sharks, some bony fishes, and turtles excrete **urea**. Urea is produced in the vertebrate liver by a metabolic cycle that combines ammonia with carbon dioxide. The circulatory system transports urea to the kidneys. Urea is highly soluble in water. It is also some 100,000 times less toxic than ammonia, so it can be held in a concentrated solution in the body and disposed of with relatively little water loss. Some animals can switch between excreting ammonia and urea, depending on environmental conditions. Certain toads, for example, excrete ammonia (thus saving energy) when in water as tadpoles, but they excrete mainly urea (reducing water loss) when they become land-dwelling adults.

Urea can be stored in a concentrated solution, but it must be diluted with water for disposal. By contrast, land animals that excrete **uric acid** (insects, land snails, and many reptiles, including birds) avoid the water loss problem almost completely. As you can see in the figure, uric acid is a considerably more complex molecule than either urea or ammonia. Like urea, uric acid is relatively nontoxic. But unlike either ammonia or urea, uric acid is largely insoluble in water, and thus water is not used to dilute it. In most cases, uric acid is excreted as a semisolid paste. (The white material in bird droppings is mostly uric acid—which explains why this substance can cause so much damage to the paint job on a car.) An animal must expend more energy to excrete uric acid than to excrete urea, but the higher energy cost is balanced by the great savings in body water.

An animal's type of reproduction also influences whether it excretes urea or uric acid. Urea can diffuse out of a shell-less amphibian egg or be carried away from a mammalian embryo in the mother's blood. However, the shelled eggs produced by birds and other reptiles are not permeable to liquids. The evolution of uric acid as a waste product therefore conveyed a selective advantage because it precipitates out of solution and can remain as a harmless solid that is left behind when the animal hatches.

In the next five modules, we focus on one specific example of a system that has evolved for osmoregulation and excretion: the human urinary system.

? Aquatic turtles excrete both urea and ammonia; land turtles excrete mainly uric acid. What could account for this difference?

Although uric acid as a waste product evolved in terrestrial reptiles with their shelled eggs, natural selection favored the energy savings of ammonia and urea for aquatic turtles.

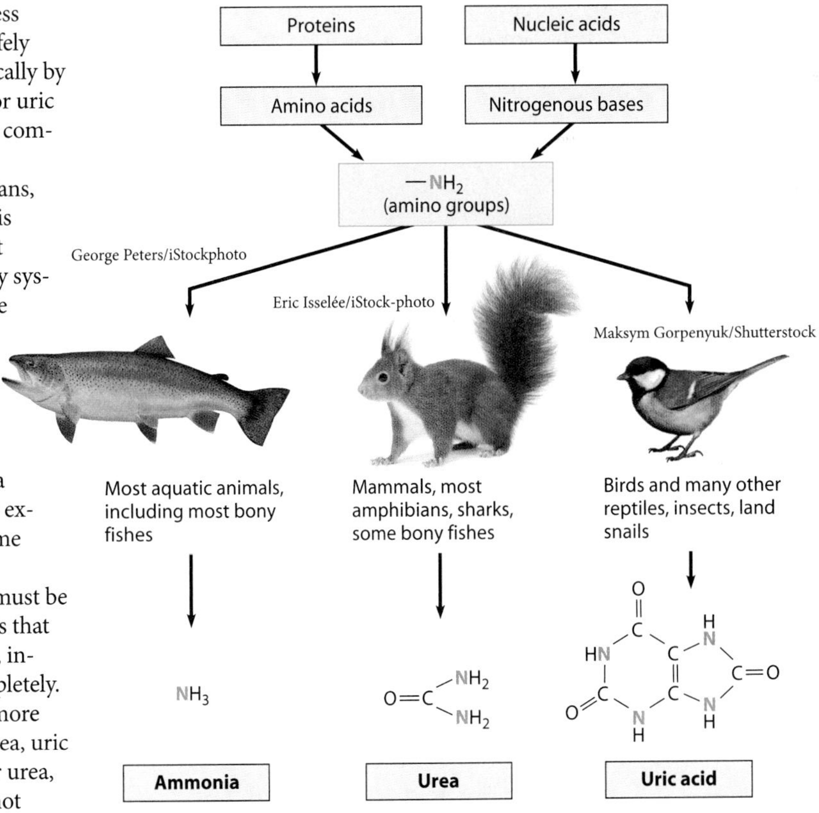

▲ **Figure 5** Nitrogen-containing metabolic waste products

6 The urinary system plays several major roles in homeostasis

Survival in any environment requires a precise balance between the competing demands of waste disposal and an animal's need for water. The **urinary system** plays a central role in homeostasis, forming and excreting urine while regulating the amount of water and solutes in body fluids.

In humans, the main processing centers of the urinary system are the two kidneys. Each is a compact organ a bit smaller than a fist, located on either side of the abdomen. The kidneys contain about 80 km of tubules (small tubes) and an intricate network of tiny blood capillaries. The human body contains only about 5 L of blood, but because this blood circulates repeatedly, about 1,100–2,000 L pass through the capillaries in our kidneys every day. From this enormous circulation of blood, every day our kidneys extract about 180 L of fluid, called **filtrate**, consisting of water, urea, and a number of valuable solutes, including glucose, amino acids, ions, and vitamins. If we excreted all the filtrate as urine, we would lose vital nutrients and dehydrate rapidly. Instead, our kidneys refine the filtrate, concentrating the urea and recycling most of the water and useful solutes to the blood. In a typical day, we excrete only about 1.5 L of urine.

In the human kidneys, numerous tubules are organized in close association with a dense network of capillaries (**Figure 6**). Starting with the whole system in **Part A**, blood to be filtered enters each kidney via a renal artery, shown in red; blood that has been filtered leaves the kidney in the renal vein, shown in blue. Urine leaves each kidney through a duct called a

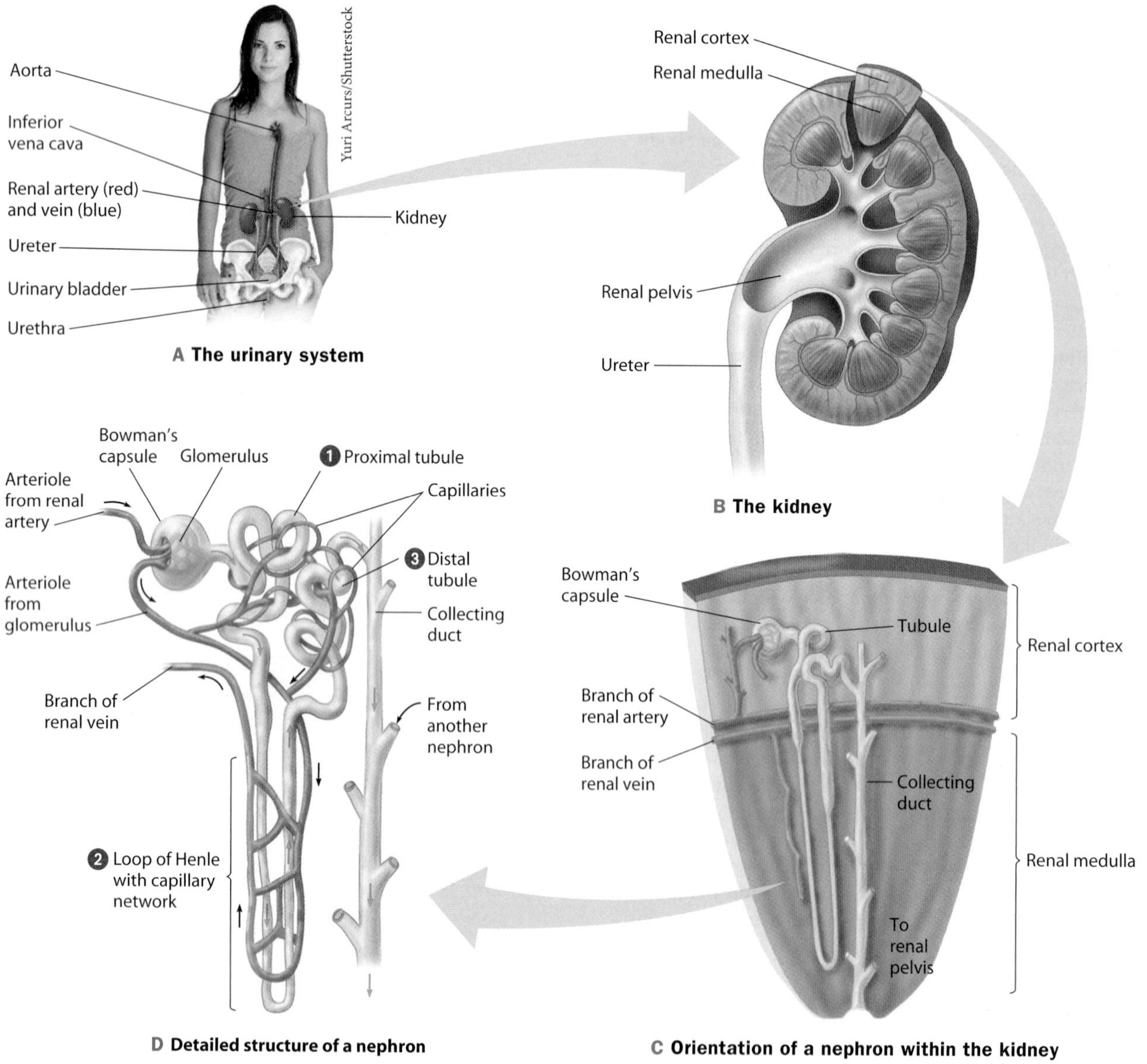

▲ **Figure 6** Anatomy of the human excretory system

ureter. Both ureters drain into the **urinary bladder**. During urination, urine is expelled from the bladder through a tube called the **urethra**, which empties to the outside near the vagina in females and through the penis in males. Sphincter muscles near the junction of the urethra and the bladder control urination.

As shown in **Part B**, the kidney has two main regions, an outer **renal cortex** and an inner **renal medulla**. Each kidney contains about a million tiny functional units called **nephrons**, one of which is shown in **Part C**. A nephron consists of a single folded tubule and associated blood vessels. Performing the kidney's functions in miniature, the nephron extracts a tiny amount of filtrate from the blood and then refines the filtrate into a much smaller quantity of urine. Each nephron starts and ends in the kidney's cortex; some extend into the medulla, as in Part C. The receiving end of the nephron is a cup-shaped swelling, called **Bowman's capsule**. At the other end of the nephron is the **collecting duct**, which carries urine to the renal pelvis.

Part D shows a nephron in more detail, along with its blood vessels. Bowman's capsule envelops a ball of capillaries called the **glomerulus** (plural, *glomeruli*). The glomerulus and Bowman's capsule make up the blood-filtering unit of the nephron. Here, blood pressure forces water and solutes from the blood in the glomerular capillaries across the wall of Bowman's capsule and into the nephron tubule. This process creates the filtrate, leaving blood cells and large molecules such as plasma proteins behind in the capillaries.

The rest of the nephron refines filtrate. The tubule has three sections: ❶ the **proximal tubule** (in the cortex); ❷ the **loop of Henle**, a hairpin loop with a capillary network that carries filtrate toward—and in some cases into—the medulla and then back toward the cortex; and ❸ the **distal tubule** (called distal because it is the most distant from Bowman's capsule). The distal tubule drains into a collecting duct, which receives filtrate from many nephrons. From the kidney's many collecting ducts, the processed filtrate, or urine, passes into a chamber called the renal pelvis and then into the ureter, from which it is then expelled.

The intricate association between blood vessels and tubules is the key to nephron function. As shown in Part D, the nephron has two distinct networks of capillaries. One network is the glomerulus, a finely divided portion of an arteriole that branches from the renal artery. Leaving the glomerulus, the arteriole reforms and carries blood to the second capillary network, which surrounds the proximal and distal tubules. This second network works with the tubule in refining the filtrate. Some of the vessels in this network parallel the loop of Henle, with blood flowing downward and then back up. Leaving the nephron, the capillaries converge to form a venule leading toward the renal vein.

With the structure of a nephron in mind, we focus next on what actually happens as our urinary system filters blood, refines the filtrate, and excretes urine.

? Place these parts of a nephron in the order in which filtrate moves through them: proximal tubule, Bowman's capsule, collecting duct, distal tubule, loop of Henle.

Bowman's capsule, proximal tubule, loop of Henle, distal tubule, collecting duct

7 Overview: The key processes of the urinary system are filtration, reabsorption, secretion, and excretion

Our urinary system produces and disposes of urine in four major processes, shown in **Figure 7**. First, during **filtration**, the pressure of the blood forces water and other small molecules through a capillary wall into the start of a kidney tubule, forming filtrate.

Two processes then refine the filtrate. In **reabsorption**, water and valuable solutes—including glucose, salt, other ions, and amino acids—are reclaimed from the filtrate and returned to the blood. In **secretion**, substances in the blood are transported into the filtrate. When there is an excess of H^+ in the blood, for example, these ions are secreted into the filtrate, thus keeping the blood from becoming acidic. Secretion also eliminates certain drugs and other toxic substances from the blood. In both reabsorption and secretion, water and solutes move between the tubule and capillaries by passing through the interstitial fluid.

Fourth, in **excretion**, urine—the waste-containing product of filtration, reabsorption, and secretion—passes from the kidneys to the outside via the ureters, urinary bladder, and urethra. In the next module, we examine these processes and their control in detail.

? Urine differs in composition from the fluid that enters a nephron tubule by filtration because of the processes of ______ and ______.

reabsorption . . . secretion

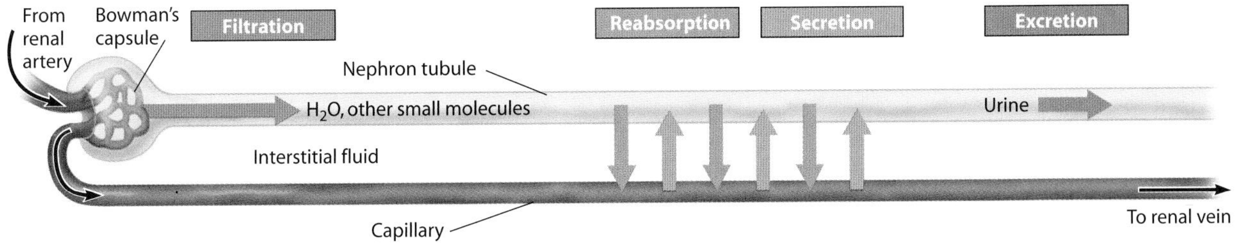

▲ **Figure 7** Major processes of the urinary system

8 Blood filtrate is refined to urine through reabsorption and secretion

Let's take a closer look at how a single nephron and collecting duct in the kidney produce urine from a blood filtrate that initially consists of both wanted substances and unwanted substances.

The broad arrows in **Figure 8** indicate where reabsorption and secretion occur along the nephron tubule. The purple arrows pointing out of the tubule represent reabsorption, which may occur by active transport, passive diffusion, or osmosis. The blue arrows pointing into the tubule represent secretion. For simplicity, this figure omits the capillary network that surrounds the tubule.

The pink area of the figure represents the interstitial fluid, through which solutes and water move between the tubule and capillaries. The intensity of the color corresponds to the concentration of solutes in the interstitial fluid: The concentration is lowest in the cortex of the kidney and becomes progressively greater toward the inner medulla. By maintaining this solute gradient, the kidney can extract and save most of the water from the filtrate. All along the tubule, wherever you see an indication that water is passing out of the filtrate into the interstitial fluid, the water is moving by osmosis. Water flows this way because the solute concentration of the interstitial fluid exceeds that of the filtrate.

To get an overview of the process, let's first discuss the activities of the proximal and distal tubules. ❶ The proximal tubule actively transports nutrients such as glucose and amino acids from the filtrate into the interstitial fluid to be reabsorbed into the capillaries. NaCl (salt) is reabsorbed from both the proximal and distal tubules. As NaCl is transported from the filtrate to the interstitial fluid, water follows by osmosis. Secretion of excess H^+ and reabsorption of HCO_3^- also occur at the proximal and distal tubules, helping to regulate the blood's pH. Potassium concentration in the blood is regulated by secretion of excess K^+ into the distal tubule. Drugs and poisons that were processed in the liver are secreted into the proximal tubule.

The loop of Henle and the collecting duct have one major function: water reabsorption. ❷ The long loop of Henle carries the filtrate deep into the medulla and then back to the cortex. The presence of NaCl and some urea in the interstitial fluid maintains the high concentration gradient in the medulla, which in turn increases water reabsorption by osmosis. Water leaves the tubule because the interstitial fluid in the medulla has a higher solute concentration than the filtrate. As soon as the water from the filtrate passes into the interstitial fluid, it moves into nearby blood capillaries and is carried away to be recycled in the body. This prompt removal is essential because the water would otherwise dilute the interstitial fluid surrounding the loop and destroy the concentration gradient necessary for water reabsorption.

Just after the filtrate rounds the hairpin turn in the loop of Henle, water reabsorption stops because the tubule there is impermeable to water. As the filtrate moves back toward the cortex, NaCl leaves the filtrate, first passively and then actively as the cells of the tubule pump NaCl into the interstitial fluid. It is primarily this movement of salt that creates the solute gradient in the interstitial fluid of the medulla.

❸ Final refining of the filtrate occurs in the collecting duct. Because it actively reabsorbs NaCl, the collecting duct is important in determining how much salt is excreted in the urine. In the medulla, the collecting duct becomes permeable to urea and some leaks out, adding to the high concentration gradient in the interstitial fluid. As the filtrate moves through the collecting duct, more water is reabsorbed before the final product, urine, passes into the renal pelvis. In sum, the nephron returns much of the water that filters into it from the blood. This water conservation is one of the major functions of the kidneys.

? What would happen if reabsorption in the proximal and distal tubules were to cease?

Needed solutes would pass into the urine, depriving the body of substances it requires.

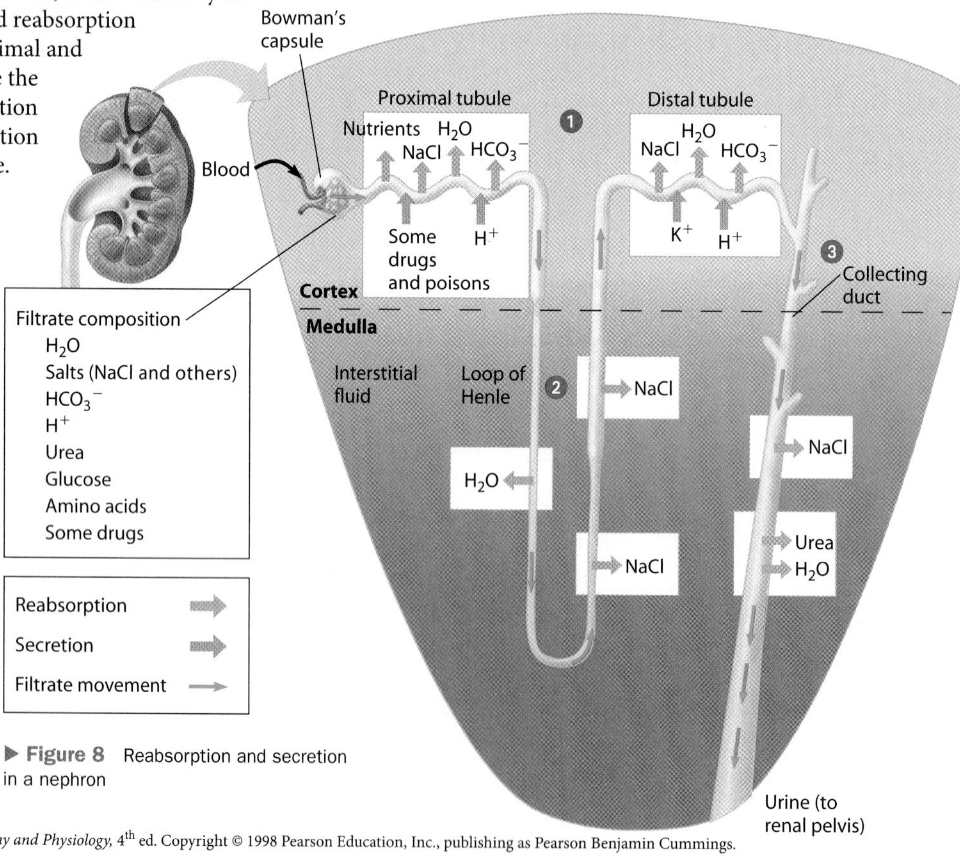

▶ **Figure 8** Reabsorption and secretion in a nephron

9 Hormones regulate the urinary system

The body controls its concentration of water and solutes by increasing or decreasing the levels of hormones that act on the kidney's nephrons. If you start to become dehydrated, the solute concentration of your body fluids rises. When this concentration gets too high, the brain increases levels of a hormone called **antidiuretic hormone (ADH)** in your blood. This hormone signals the nephrons in your kidneys to reabsorb more water from the filtrate. This reabsorption increases the amount of water returning to your blood (where it is needed) and decreases the amount of water excreted (resulting in concentrated urine). Dark-colored urine indicates that you have not been drinking enough water.

On the other hand, if you drink a lot of water, the solute concentration of your body fluids becomes too dilute. In response, blood levels of ADH drop, causing the nephrons to reabsorb less water from the filtrate, resulting in dilute, watery urine. This process explains why your urine is very clear after you drink a lot of water. (Increased urination is called diuresis, and it is because ADH acts against this state that it is called antidiuretic hormone.) Diuretics, such as alcohol, are substances that inhibit the release of ADH and therefore result in excessive urinary water loss. Drinking alcohol will make you urinate more frequently, and the resulting dehydration contributes to the symptoms of a hangover. Caffeine is also a diuretic; you may have noticed that drinking coffee, tea, or cola causes you to urinate soon afterward.

We see that our kidneys' regulatory functions are controlled by an elaborate system of checks and balances.

? Some of the drugs classified as diuretics make the epithelium of the collecting duct less permeable to water. How would this affect kidney function?

The collecting ducts would reabsorb less water, and thus the diuretic would increase water loss in the urine.

CONNECTION

10 Kidney dialysis can be lifesaving

A person can survive with one functioning kidney, but if both kidneys fail, the buildup of toxic wastes and the lack of regulation of blood ion concentrations, pH, and pressure will lead to certain and rapid death. More than 60% of all cases of kidney disease are caused by hypertension and diabetes, but the prolonged use of pain relievers (even common, over-the-counter ones), alcohol, and other drugs are also possible causes.

Knowing how the nephron works helps us understand how some of its functions can be performed artificially when the kidneys are damaged. In **dialysis**, blood is filtered by a machine, which mimics the action of a nephron (**Figure 10**). Like the nephrons of the kidney, the machine sorts small molecules of the blood, keeping some and discarding others. The patient's blood is pumped from an artery through a series of selectively permeable tubes. The tubes are immersed in a dialyzing solution that resembles the chemical makeup of the interstitial fluid that bathes the nephrons. As the blood circulates through the tubing, urea and excess ions diffuse out. Needed substances diffuse from the dialyzing solution into the blood. The machine continually discards the used dialyzing solution as wastes build up.

Dialysis treatment is life sustaining for people with kidney failure, but it is costly, takes a lot of time (about 4–6 hours three time a week), and must be continued for life—or until the patient is able to undergo kidney transplantation. In some cases, a kidney from a living compatible donor (usually a close relative) or a deceased organ donor can be transplanted into a person with kidney failure. Unfortunately, the number of people who need kidneys is much greater than the number of kidneys available, and the average wait for a kidney donation in the United States is three to five years.

? How does the composition of dialyzing solution compare with that of the patient's blood plasma?

Dialyzing solution has a solute concentration similar to that of interstitial fluid. The solution contains no urea, which allows urea from the patient's blood to diffuse into it.

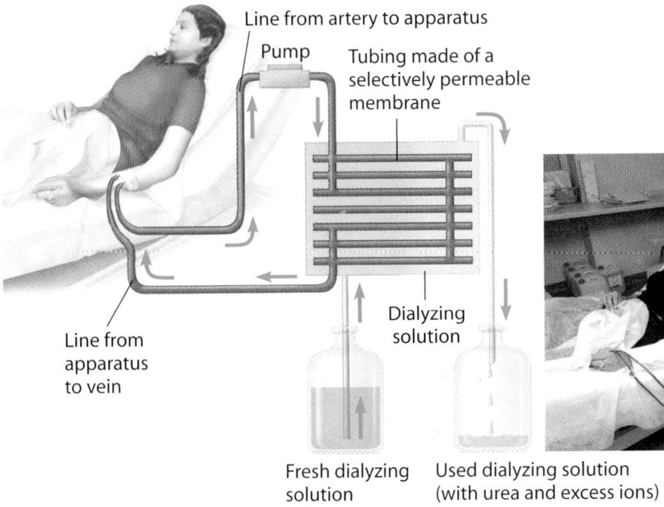

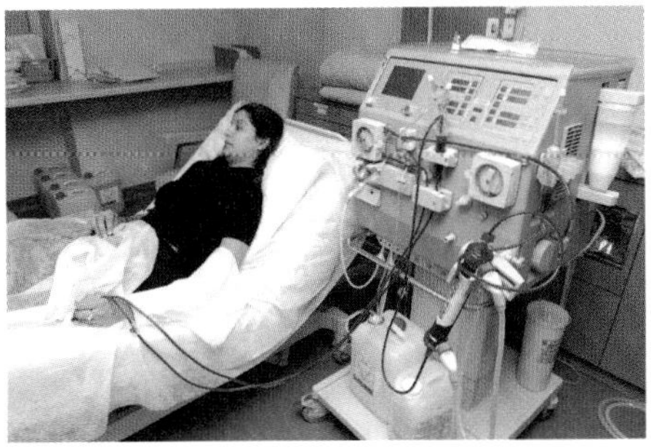

Phanie/Photo Researchers

▲ **Figure 10** Kidney dialysis

REVIEW

For Practice Quizzes, BioFlix, MP3 Tutors, and Activities, go to www.masteringbiology.com.

Reviewing the Concepts

Thermoregulation (1–3)

1 An animal's regulation of body temperature helps maintain homeostasis. Thermoregulation maintains body temperature within a tolerable range. Endotherms derive body heat mainly from their metabolism; ectotherms absorb heat from their surroundings.

2 Heat is gained or lost in four ways. Heat exchange with the environment occurs by conduction, convection, radiation, and evaporation.

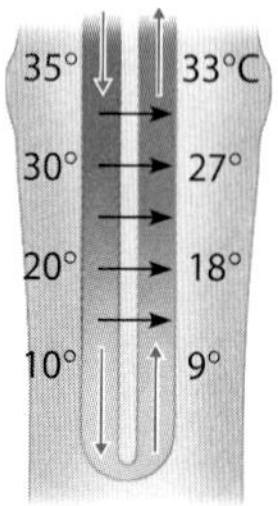

3 Thermoregulation involves adaptations that balance heat gain and loss. Adaptations for thermoregulation include increased metabolic heat production; insulation; circulatory adaptations; evaporative cooling; and behavioral responses.

Osmoregulation and Excretion (4–10)

4 Animals balance their level of water and solutes through osmoregulation. Osmoconformers have the same internal solute concentration as seawater. Osmoregulators control their solute concentrations.

	Gain Water	Lose Water	Salt
Freshwater Fish 	Osmosis	Excretion	Pump in
Saltwater Fish 	Drinking	Osmosis	Excrete, pump out
Land Animal 	Drinking, eating	Evaporation, urinary system	

Animals can conserve water by means of the kidneys, waterproof skin, and reproductive and behavioral adaptations.

5 A variety of ways to dispose of nitrogenous wastes has evolved in animals. Excretion is the disposal of toxic nitrogenous wastes. Ammonia (NH_3) is poisonous but soluble and is easily disposed of by aquatic animals. Urea and uric acid are less toxic and easier to store but require significant energy to produce.

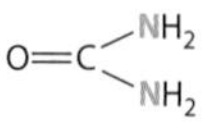

Urea

6 The urinary system plays several major roles in homeostasis. The urinary system expels wastes and regulates water and ion balance. Nephrons extract a filtrate from the blood and refine it to urine. Urine leaves the kidneys via the ureters, is stored in the urinary bladder, and is expelled through the urethra.

7 Overview: The key processes of the urinary system are filtration, reabsorption, secretion, and excretion. In filtration, blood pressure forces water and many small solutes into the nephron. In reabsorption, water and valuable solutes are reclaimed from the filtrate. In secretion, excess H^+ and toxins are added to the filtrate. In excretion, urine is expelled.

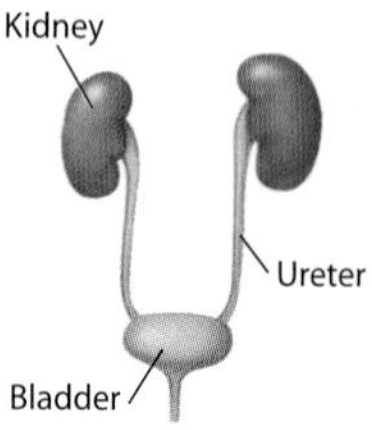

8 Blood filtrate is refined to urine through reabsorption and secretion. Nutrients, salt, and water are reabsorbed from the proximal and distal tubules within the nephron. Secretion of H^+ and reabsorption of HCO_3^- help regulate pH. High NaCl concentration in the medulla promotes reabsorption of water.

9 Hormones regulate the urinary system. Antidiuretic hormone (ADH) regulates the amount of water excreted by the kidneys.

10 Kidney dialysis can be lifesaving. A dialysis machine removes wastes from blood and maintains solute concentration.

Connecting the Concepts

1. Complete this map, which presents the three main topics of this chapter.

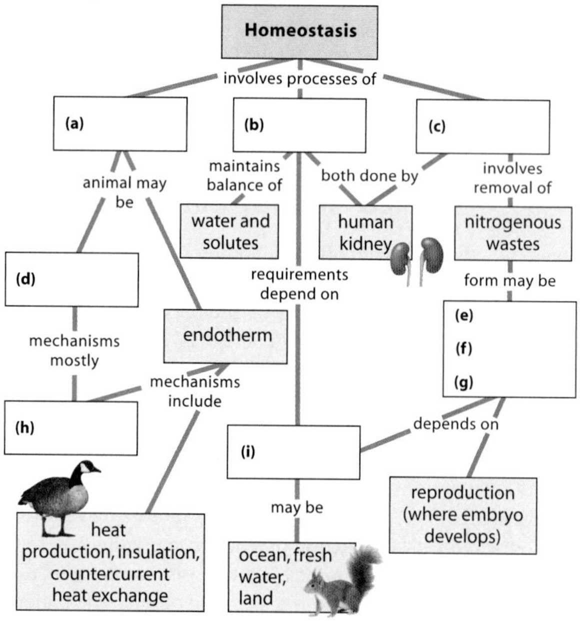

2. In this schematic of urine production in a nephron, label the four processes involved and list some of the substances that are moved in each process.

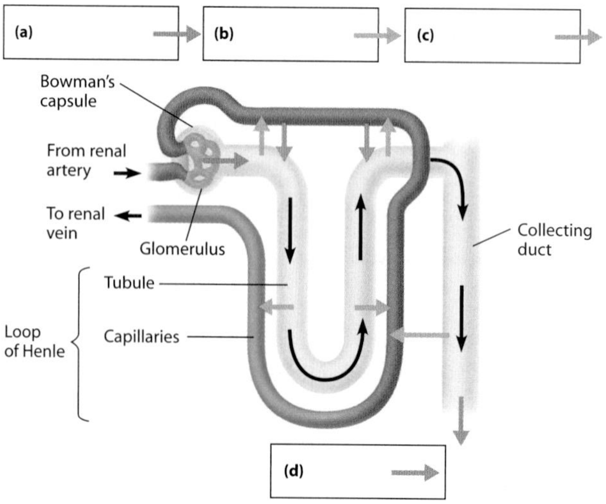

Testing Your Knowledge

Multiple Choice

3. The main difference between endotherms and ectotherms is
   a. how they conserve water.
   b. whether they are warm or cold.
   c. the source of most of their body heat.
   d. whether they live in a warm or cold environment.
   e. whether they maintain a fairly stable body temperature.
4. You place your hand on a black car hood in bright summer sunshine. The car hood was heated by the process of ______ and your hand was warmed by the process of ______.
   a. conduction . . . radiation
   b. radiation . . . convection
   c. radiation . . . conduction
   d. evaporation . . . radiation
5. Which of the following is not an adaptation for reducing the rate of heat loss to the environment?
   a. feathers or fur
   b. reducing blood flow to surface blood vessels
   c. contraction of flight muscles before a moth takes off
   d. countercurrent heat exchange
   e. thick layer of fat
6. In each nephron of the kidney, the glomerulus and Bowman's capsule
   a. filter the blood and capture the filtrate.
   b. reabsorb water into the blood.
   c. break down harmful toxins and poisons.
   d. reabsorb ions and nutrients.
   e. refine and concentrate the urine for excretion.
7. As filtrate passes through the loop of Henle, salt is removed and concentrated in the interstitial fluid of the medulla. This high concentration enables nephrons to
   a. excrete the maximum amount of salt.
   b. neutralize toxins that might be found in the kidney.
   c. control the pH of the interstitial fluid.
   d. excrete a large amount of water.
   e. reabsorb water very efficiently.
8. Birds and insects excrete uric acid, whereas mammals and most amphibians excrete mainly urea. What is the chief advantage of uric acid over urea as a waste product?
   a. Uric acid is more soluble in water.
   b. Uric acid is a much simpler molecule.
   c. It takes less energy to make uric acid.
   d. Less water is required to excrete uric acid.
   e. More solutes are removed excreting uric acid.
9. Which process in the nephron is least selective?
   a. secretion
   b. reabsorption
   c. filtration
   d. active transport of salt
   e. passive diffusion of salt
10. A freshwater fish would be expected to
    a. pump salt out through its gills.
    b. produce copious quantities of dilute urine.
    c. diffuse urea out through the gills.
    d. have scales and a covering of mucus that reduce water loss to the environment.
    e. do all of the above.

Matching

Match each of the following components of blood with what happens to it as the blood is processed by the kidney. Note that each lettered choice may be used more than once.

11. Water	a. passes into filtrate; almost all excreted in urine
12. Glucose	b. remains in blood
13. Plasma protein	c. passes into filtrate; mostly reabsorbed
14. Toxins or drugs	d. secreted and excreted
15. Red blood cell	
16. Urea	

Describing, Comparing, and Explaining

17. Compare the problems of water and salt regulation a salmon faces when it is swimming in the ocean and when it migrates into fresh water to spawn.
18. Can ectotherms have stable body temperatures? Explain.

Applying the Concepts

19. Assuming equal size, which of these organisms would produce the greatest amount of nitrogenous wastes? Explain.
    a. An endotherm or an ectotherm?
    b. A carnivore or an herbivore (assume both are endotherms)?
20. You are studying a large tropical reptile that has a high and relatively stable body temperature. How would you determine whether this animal is an endotherm or an ectotherm?
21. Passing by a lake in midwinter, you notice a small flock of geese standing on the frozen surface. Imagine what it would be like for you to stand there with no boots or warm pants. Propose a hypothesis to explain why the birds' legs do not freeze. You may assume you have equipment for measuring temperatures in the birds' legs. What results would you expect if your hypothesis is correct?
22. The kidneys remove many drugs from the blood, and these substances show up in the urine. Some employers require a urine drug test at the time of hiring and/or at intervals during the term of employment. Why do some employers feel that drug testing is necessary? Do you think that passing a drug test is a valid criterion for employment? If so, for what types of jobs? Would you take a drug test to get or keep a job? Why or why not?
23. Kidneys were the first organs to be transplanted successfully. A donor can live a normal life with a single kidney, making it possible for individuals to donate a kidney to an ailing relative or even an unrelated individual. In some countries, poor people sell kidneys to transplant recipients through organ brokers. What are some of the ethical issues associated with organ commerce?

Review Answers

1. a. thermoregulation; b. osmoregulation; c. excretion; d. ectotherm; e.–g. ammonia, urea, uric acid; h. behavioral responses; i. environment

2. a. filtration: water; NaCl, HCO_3^-, H^+, urea, glucose, amino acids, some drugs; b. reabsorption: nutrients, NaCl, water, HCO_3^-, urea; c. secretion: some drugs and toxins, H^+, K^+; d. excretion: urine containing water, urea, and excess ions

3. c 4. c 5. c 6. a 7. e 8. d 9. c 10. b 11. c 12. c 13. b 14. d 15. b 16. a

17. In salt water, the fish loses water by osmosis. It drinks salt water and disposes of salts through its gills. Its kidneys conserve water and excrete excess ions. In fresh water, it gains water by osmosis. Its kidneys excrete a lot of dilute urine. Its gills take up salt, and some ions are ingested with food.

18. Yes. Ectotherms that live in very stable environments, such as tropical seas or deep oceans, have stable body temperatures. And terrestrial ectotherms can maintain relatively stable temperatures by behavioral means.

19. a. An endotherm would produce more nitrogenous wastes because it must eat more food to maintain its higher metabolic rate.
    b. A carnivore, because it eats more protein and thus produces more breakdown products of protein digestion—nitrogenous wastes.

20. You could take it back to the laboratory and measure its body temperature under different ambient temperatures.

21. A countercurrent heat exchange in the birds' legs reduces the loss of heat from the body. You would expect the temperature of blood flowing back to the body from the legs to be only slightly cooler than the blood flowing from the body to the legs.

22. Some issues and questions to consider: Could drug use endanger the safety of the employee or others? Is drug testing relevant to jobs where safety is not a factor? Is drug testing an invasion of privacy, interfering in the private life of an employee? Is an employer justified in banning drug use off the job if it does not affect safety or ability to do the job? Do the same criteria apply to employers requiring the test? Could an employer use a drug test to regulate other employee behavior that is legal, such as smoking?

23. Some issues and questions to consider: Should human organs be sold? If the donor is poor, he or she may also not be in the best of health. What are the health risks to the donor? Will the recipient be assured of buying a healthy organ? How much does the organ broker make from this transaction? Both parties in such a transaction are extremely desperate. Should regulations be in place when people may not be able to make reasoned decisions? Or do people have a right to sell parts of their bodies, just as they can now sell other possessions or their labor?

Hormones and the Endocrine System

Hormones and the Endocrine System

BIG IDEAS

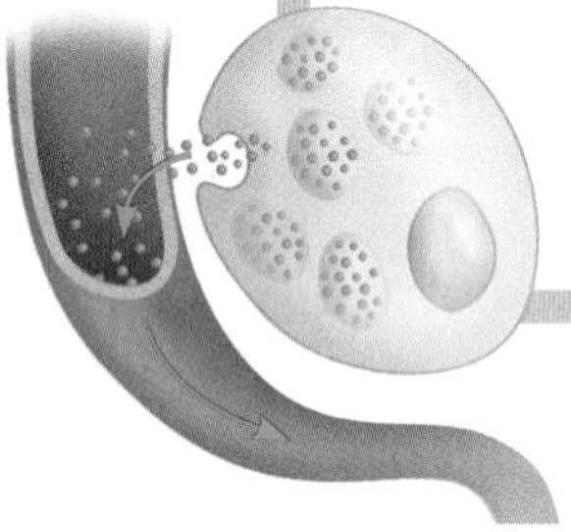

Jonathan Blair/Corbis

The Nature of Chemical Regulation
(1–2)

Hormones affect cells using two distinct mechanisms.

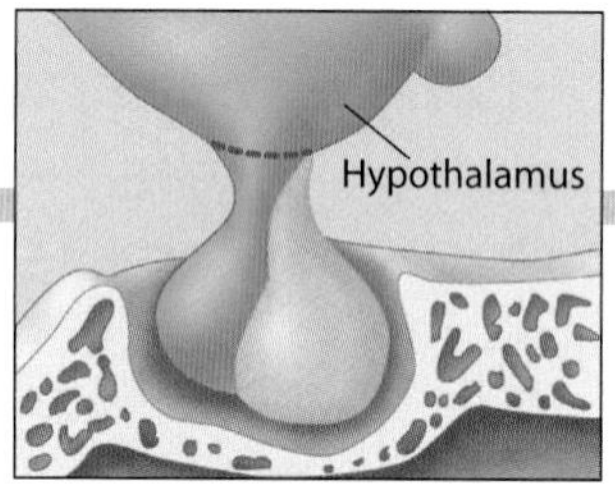

The Vertebrate Endocrine System
(3–4)

The hypothalamus exerts master control over many other endocrine glands.

Hormones and Homeostasis
(5–11)

Hormones, often in antagonistic pairs, regulate whole-body processes.

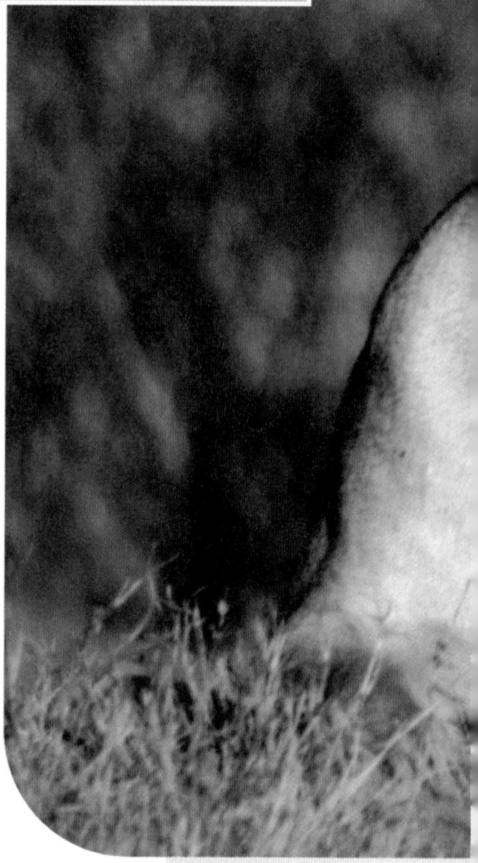

Mitch Reardon/Lonely Planet Images

Few sights in nature are as impressive as a male lion (*Panthera leo*) stalking prey across the savannah. Often weighing more than 500 pounds, males of the species are easily recognized by the thick mane about their necks. Females lack manes; the lion is thus the only feline species to display a distinct phenotypic difference between the sexes. Research has shown that the color of a lion's mane correlates with age and food intake. In general, the darker the mane, the healthier the lion. Females mate more frequently with darker-maned males (perhaps because this indicates a greater ability to obtain food), while males avoid other darker-maned males (perhaps because this indicates greater fighting prowess).

Manes develop at puberty when male lions—like the males of many mammal species—begin to produce high levels of the sex hormone testosterone. Hormones are chemicals that help regulate body functions such as energy use, metabolism, and growth. The primary role of testosterone and related male sex hormones is to promote the development and maintenance of uniquely male traits, such as male reproductive anatomy. Testosterone also affects many secondary sexual characteristics, such as growth and maintenance of a mane and the greater height and weight of males compared to females.

This chapter focuses on how hormones help coordinate activities in different parts of the body, enabling organ systems to function cooperatively. We explore the general theme of homeostasis. In this chapter, we explore how chemical signals maintain an animal's dynamic steady state. We begin with an overview of how hormones work in all vertebrates and then turn to the major components of the human endocrine system. Along the way, we consider many examples of the effects of hormonal imbalance.

The Nature of Chemical Regulation

1 Chemical signals coordinate body functions

Animals rely on many kinds of chemical signals to regulate their body activities. A **hormone** is a substance that acts as one such chemical signal. Hormones are carried by the circulatory system (usually in the blood) to other parts of the body, where they may communicate regulatory messages. Hormones are made and secreted mainly by organs called **endocrine glands**. Collectively, all of an animal's hormone-secreting cells constitute its **endocrine system**, one of two body systems for communication and chemical regulation.

The other system of internal communication and regulation is the nervous system. Unlike the endocrine system, which sends chemical signals through the bloodstream, the nervous system transmits electrical signals via nerve cells. These rapid messages control split-second responses. The flick of a frog's tongue as it catches a fly and the jerk of your hand away from a flame result from high-speed nerve signals. The endocrine system coordinates slower but longer-lasting responses. In some cases, the endocrine system takes hours or even days to act, partly because of the time it takes for hormones to be made and transported to all their target organs and partly because the cellular response may take time.

Hormones reach all cells of the body, and the endocrine system is especially important in controlling whole-body activities. For example, hormones coordinate the body's responses to stimuli such as dehydration, low blood glucose levels, or stress, as with the rapid increase in heartbeat and awareness that you feel when faced with a sudden threat. Hormones also regulate long-term developmental processes, such as the appearance of characteristics that distinguish an adult animal from a juvenile animal and bring about the physical and behavioral changes that underlie sexual maturity.

Figure 1A sketches the activity of a hormone-secreting cell. Membrane-enclosed secretory vesicles in the endocrine cell are full of molecules of the hormone (∴). The endocrine cell secretes the molecules directly into blood vessels. From there, hormones may travel via the circulatory system to all parts of the body, but only certain types of cells, called **target cells**, have receptors for that specific hormone. (A hormone is ignored by other, nontarget cells.) A single hormone molecule may dramatically alter a target cell's metabolism by turning on or off the production of a number of enzymes. A tiny amount of a hormone can govern the activities of enormous numbers of target cells in a variety of organs. (In the next module, we look at *how* hormones trigger responses in their target cells.)

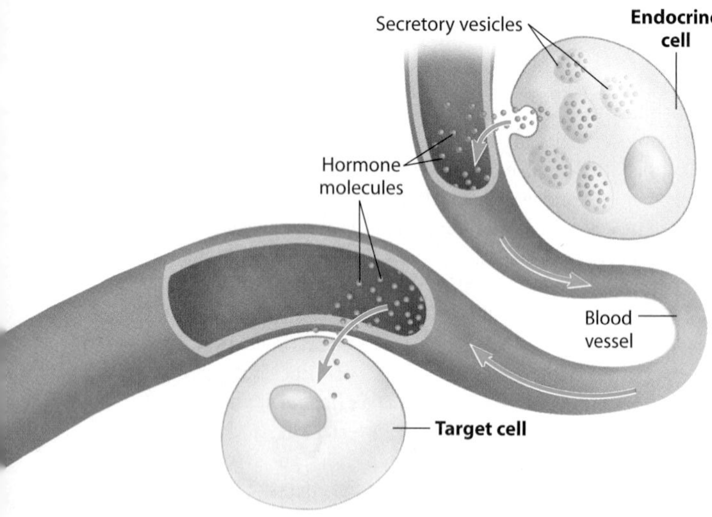

▲ **Figure 1A** Hormone from an endocrine cell

Chemical signals play a major role in coordinating the internal functioning of animals. Hormones are the body's long-distance chemical regulators and convey information via the bloodstream to target cells that may or may not be close to the cell that initiates the signal. Other chemical signals—local regulators—are secreted into the interstitial fluid and affect only nearby target cells. Still other chemical signals, called pheromones, carry messages between different individuals of a species, as in mate attraction.

Although it is convenient to distinguish between the endocrine and nervous systems, in reality the lines between these two regulatory systems are blurred. In particular, certain specialized nerve cells called **neurosecretory cells** perform functions of both systems. Like all nerve cells, neurosecretory cells conduct nerve signals, but they also make and secrete hormones into the blood. A neurosecretory cell thus functions similarly to the endocrine cell shown in Figure 1A.

A few chemicals serve both as hormones in the endocrine system and as chemical signals in the nervous system. Epinephrine (adrenaline), for example, functions in vertebrates as the "fight-or-flight" hormone (so called because it prepares the body for sudden action) and as a neurotransmitter. A neurotransmitter is a chemical that carries information from one nerve cell to another or from a nerve cell to another kind of cell that will react. When a nerve signal reaches the end of a nerve cell, it triggers the secretion of neurotransmitter molecules (∴ in **Figure 1B**). Unlike most hormones, however, neurotransmitters usually do not travel in the bloodstream but instead move between adjacent neurons. In the rest of this chapter, we explore the endocrine system and its hormones.

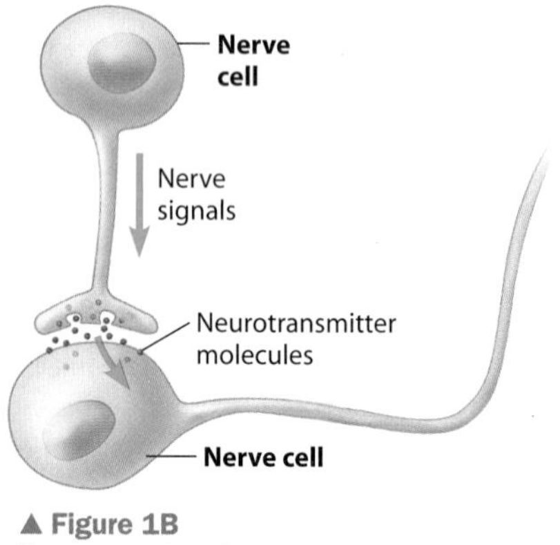

▲ **Figure 1B** Neurotransmitter from a nerve cell

? How do hormones usually travel between an endocrine gland and their target cells?

● Via the bloodstream

2 Hormones affect target cells using two main signaling mechanisms

Two major classes of molecules function as hormones in vertebrates. The first class, the **amino-acid-derived hormones**, includes proteins, peptides (polypeptides of only 3–30 amino acids), and amines (modified versions of single amino acids). They are almost all water-soluble. Hormones of the second main class, the **steroid hormones**, are small lipid-soluble molecules made from cholesterol.

Regardless of their chemical structure, signaling by any of these molecules involves three key events: reception, signal transduction, and response. *Reception* of the signal occurs when a hormone binds to a specific receptor protein on or in the target cell. Each signal molecule has a specific shape that can be recognized by its target cell receptors. The binding of a signal molecule to a receptor protein triggers events within the target cell—*signal transduction*—that convert the signal from one form to another. The result is a *response*, a change in the cell's behavior. Cells that lack receptors for a particular chemical signal do not respond to that signal.

Although both amino-acid-derived (water-soluble) and steroid (lipid-soluble) hormones carry out these three key steps, they do so by different mechanisms. We now take a closer look at how each type of hormone elicits cellular responses.

Water-soluble hormones cannot pass through the phospholipid bilayer of the plasma membrane, but they can bring about cellular changes indirectly, without entering their target cells. The receptor proteins for most water-soluble hormones are embedded in the plasma membrane of target cells and project outward from the cell surface (**Figure 2A**). ❶ A water-soluble hormone molecule (○) binds to the receptor protein, activating it. ❷ This initiates a signal transduction pathway, a series of changes in cellular proteins (relay molecules) that converts an extracellular chemical signal to a specific intracellular response. ❸ The final relay molecule (○) activates a protein (▲) that carries out the cell's response, either in the cytoplasm (such as activating an enzyme) or in the nucleus (turning on or off genes). One hormone may trigger a variety of responses in target cells because each cell may contain different receptors for that hormone, diverse signal transduction pathways, or several proteins that can initiate different responses.

While water-soluble hormones bind to receptors in the plasma membrane, steroid hormones pass through the phospholipid bilayer and bind to receptors inside the cell. As shown in **Figure 2B**, ❶ a steroid hormone (▽) enters a cell by diffusion. If the cell is a target cell, the hormone ❷ binds to an open receptor protein in the cytoplasm or nucleus. Rather than triggering a signal transduction pathway, as happens with a water-soluble hormone, the hormone-receptor complex itself usually carries out the transduction of the hormonal signal: The complex acts as a transcription factor—a gene activator or repressor. ❸ The hormone-receptor complex attaches to specific sites on the cell's DNA in the nucleus. (These sites are enhancers.) ❹ The binding to DNA stimulates gene regulation, turning genes either on (by promoting transcription of certain genes into RNA) or off.

Because a hormone can bind to a variety of receptors in various kinds of target cells, different kinds of cells can respond differently to the same hormone. The main effect of epinephrine on heart muscle cells, for example, is cellular contraction, which speeds up the heartbeat; its main effect on liver and muscle cells, however, is glycogen breakdown, providing glucose (an energy source) to body cells. Together, these effects speed the delivery of energy to cells throughout the body, allowing for a rapid response to the stress that triggered the release of the hormone.

? What are two major differences between the action of steroid hormones and the action of nonsteroid hormones?

(1) Steroid hormones bind to receptors inside the cell; other hormones bind to plasma membrane receptors. (2) Steroid hormones always affect gene expression; other hormones have this or other effects.

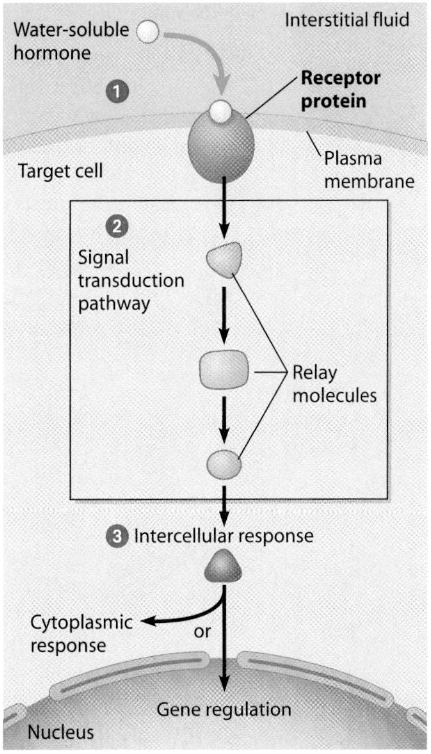

▲ **Figure 2A** A hormone that binds a plasma membrane receptor

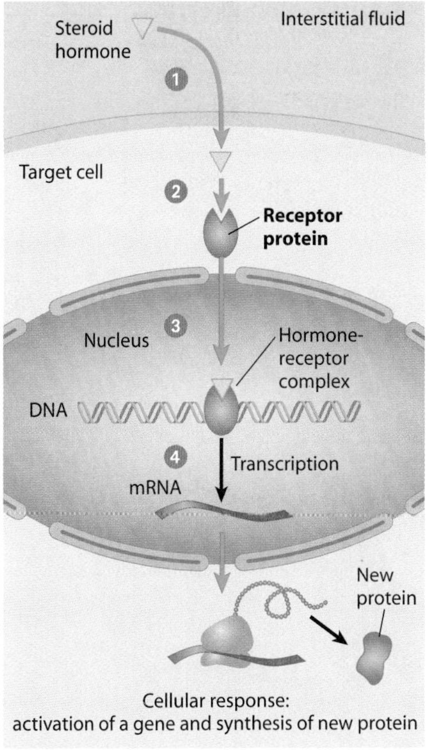

▲ **Figure 2B** A hormone that binds an intracellular receptor

The Vertebrate Endocrine System

3 Overview: The vertebrate endocrine system consists of more than a dozen major glands

Some endocrine glands (such as the thyroid) are specialists that primarily secrete hormones into the blood. Other glands (such as the pancreas) serve dual roles, having both endocrine and nonendocrine functions. Still other organs (such as the stomach) are primarily nonendocrine but have some cells that secrete hormones.

Figure 3 shows the locations of the major human endocrine glands. **Table 3** summarizes the actions of the main hormones they produce and how the glands are regulated. (When reviewing the information in this table, keep in mind that this chapter covers only the major endocrine glands and hormones; there are other hormone-secreting structures—the heart, liver, and stomach, for example—and other hormones that we will not discuss.)

Notice the distribution of the chemical classes of hormone (proteins, peptides, amines, and steroids) in Table 3. Only the sex organs and the cortex of the adrenal gland produce steroid hormones, the main type of hormone that actually enters target cells. Most of the endocrine glands produce water-soluble hormones, which generally bind to plasma membrane receptors and act via signal transduction.

Hormones have a wide range of targets. Some, like the sex hormones, which promote male and female characteristics, affect most of the tissues of the body. Other hormones, such as glucagon from the pancreas, have only a few kinds of target cells (liver and fat cells for glucagon). Some hormones have other endocrine glands as their targets. For example, the pituitary gland produces thyroid-stimulating hormone, which promotes activity of the thyroid gland.

The close association between the endocrine system and the nervous system is apparent in both Figure 3 and Table 3. For example, the hypothalamus, which is part of the brain, secretes many hormones that regulate other endocrine glands, especially the pituitary. We explore structural and functional connections between the endocrine system and the nervous system further in Module 4.

Endocrine glands that we do not discuss at length in this chapter include the pineal and the thymus. The **pineal gland** is a pea-sized mass of tissue near the center of the brain. The pineal gland synthesizes and secretes melatonin, a hormone that links environmental light conditions with biological rhythms, particularly the sleep/wake circadian rhythms. Melatonin is sometimes called "the dark hormone" because it is secreted at night. In diurnal (day-active) animals, melatonin production peaks in the middle of the night and then gradually falls. Some people ingest melatonin supplements as sleep aids, but the effectiveness of this treatment has not been established by scientists. Although there is good evidence that nightly increases in natural melatonin play a significant role in promoting sleep, we do not yet know exactly what effects melatonin has on body cells and precisely how sleep/wake cycles are controlled.

The **thymus gland** lies under the breastbone in humans and is quite large during childhood. But it wasn't until the 1960s that researchers discovered its importance in the immune system. Thymus cells secrete several important hormones, including a peptide that stimulates the development of T cells. Beginning at puberty, when the immune system reaches maturity, the thymus shrinks drastically. However, it continues to secrete its T-cell–stimulating hormones throughout life.

In the rest of this chapter, we explore several of the endocrine glands listed in Figure 3 and Table 3. We focus on the hormones produced by each organ and how they help to maintain homeostasis within the human body.

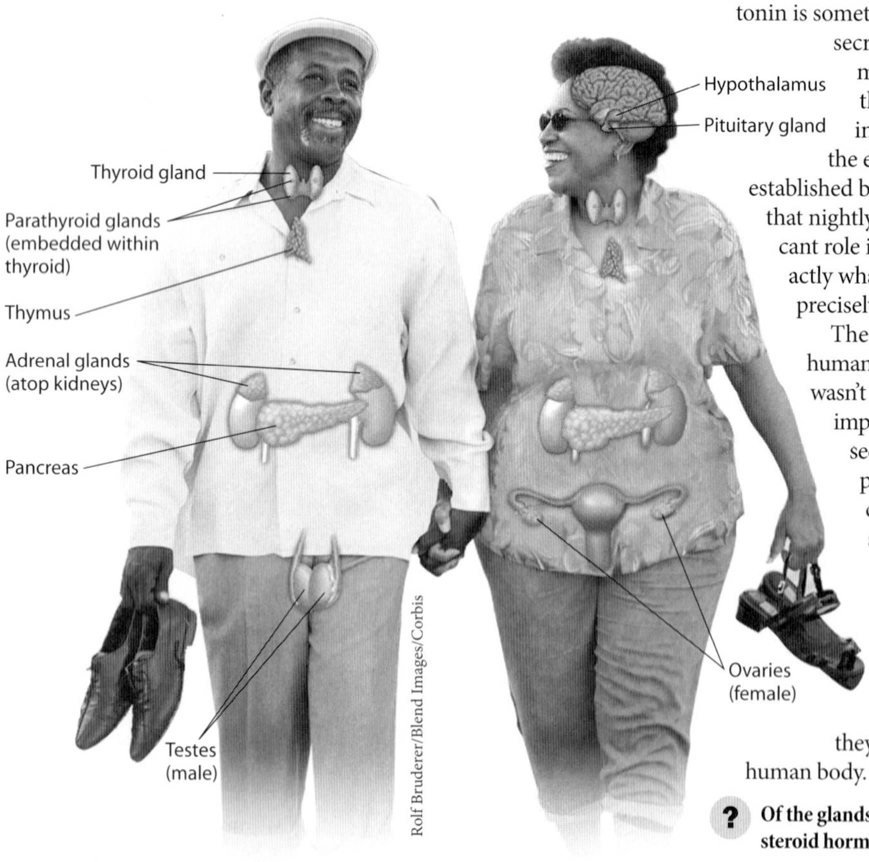

▲ **Figure 3** The major endocrine glands in humans

? Of the glands listed in Table 3, which are the only ones to secrete steroid hormones?

The testes, ovaries, and adrenal cortex

TABLE 3 | MAJOR HUMAN ENDOCRINE GLANDS AND SOME OF THEIR HORMONES

Gland (module)	Hormone	Chemical Class	Representative Actions	Regulated by
Hypothalamus (4) 	Hormones released by the posterior pituitary and hormones that regulate the anterior pituitary (see below)			
Pituitary gland (4)				
Posterior lobe (releases hormones made by hypothalamus) 	Oxytocin	Peptide	Stimulates contraction of uterus during labor and ejection of milk from mammary glands	Nervous system
	Antidiuretic hormone (ADH)	Peptide	Promotes retention of water by kidneys	Water/salt balance
Anterior lobe 	Growth hormone (GH)	Protein	Stimulates growth (especially bones) and metabolic functions	Hypothalamic hormones
	Prolactin (PRL)	Protein	Stimulates milk production and secretion in females	Hypothalamic hormones
	Follicle-stimulating hormone (FSH)	Protein	Stimulates production of ova and sperm	Hypothalamic hormones
	Luteinizing hormone (LH)	Protein	Stimulates ovaries and testes	Hypothalamic hormones
	Thyroid-stimulating hormone (TSH)	Protein	Stimulates thyroid gland	Thyroxine in blood; hypothalamic hormones
	Adrenocorticotropic hormone (ACTH)	Peptide	Stimulates adrenal cortex to secrete glucocorticoids	Glucocorticoids; hypothalamic hormones
Pineal gland (3) 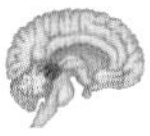	Melatonin	Amine	Involved in rhythmic activities (daily and seasonal)	Light/dark cycles
Thyroid gland (5–6) 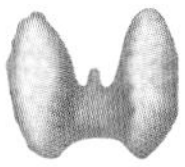	Thyroxine (T_4) and triiodothyronine (T_3)	Amine	Stimulate and maintain metabolic processes	TSH
	Calcitonin	Peptide	Lowers blood calcium level	Calcium in blood
Parathyroid glands (5–6) 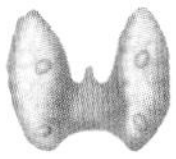	Parathyroid hormone (PTH)	Peptide	Raises blood calcium level	Calcium in blood
Thymus (3) 	Thymosin	Peptide	Stimulates T cell development	Not known
Adrenal gland (9)				
Adrenal medulla 	Epinephrine and norepinephrine	Amine	Increase blood glucose; increase metabolic activities; constrict certain blood vessels	Nervous system
Adrenal cortex 	Glucocorticoids	Steroid	Increase blood glucose	ACTH
	Mineralocorticoids	Steroid	Promote reabsorption of Na^+ and excretion of K^+ in kidneys	K^+ in blood
Pancreas (7–8) 	Insulin	Protein	Lowers blood glucose	Glucose in blood
	Glucagon	Protein	Raises blood glucose	Glucose in blood
Testes (10) 	Androgens	Steroid	Support sperm formation; promote development and maintenance of male secondary sex characteristics	FSH and LH
Ovaries (10) 	Estrogens	Steroid	Stimulate uterine lining growth; promote development and maintenance of female secondary sex characteristics	FSH and LH
	Progesterone	Steroid	Promotes uterine lining growth	FSH and LH

4 The hypothalamus, which is closely tied to the pituitary, connects the nervous and endocrine systems

As we saw earlier, the distinction between the endocrine system and the nervous system often blurs, especially when we consider the hypothalamus and its intricate association with the pituitary gland. The **hypothalamus (Figure 4A)** is the main control center of the endocrine system. As part of the brain, the hypothalamus receives information from nerves about the internal condition of the body and about the external environment. It then responds by sending out appropriate nervous or endocrine signals. Its hormonal signals directly control the pituitary gland, which in turn secretes hormones that influence numerous body functions. The hypothalamus thus exerts master control over the endocrine system by using the pituitary to relay directives to other glands.

As Figure 4A shows, the pea-sized **pituitary gland** consists of two distinct parts: a posterior lobe and an anterior lobe, both situated in a pocket of skull bone just under the hypothalamus. The **posterior pituitary** is composed of nervous tissue and is actually an extension of the hypothalamus. It stores and secretes two hormones that are made in the hypothalamus. In contrast, the **anterior pituitary** is composed of endocrine cells that synthesize and secrete numerous hormones directly into the blood. Several of these hormones control the activity of other endocrine glands. The hypothalamus exerts control over the anterior pituitary by secreting two kinds of hormones into short blood vessels that connect the two organs: releasing hormones and inhibiting hormones. A **releasing hormone** stimulates the anterior pituitary to secrete one or more specific hormones, and an **inhibiting hormone** induces the anterior pituitary to stop secreting one or more specific hormones.

Figures 4B and 4C emphasize the structural and functional connections between the hypothalamus and the pituitary. As **Figure 4B** indicates, a set of neurosecretory cells extends from the hypothalamus into the posterior pituitary. These cells synthesize the peptide hormones oxytocin and antidiuretic hormone (ADH). These hormones (▴▴) are channeled along the neurosecretory cells into the posterior pituitary. When released into the blood from the posterior pituitary, oxytocin causes uterine muscles to contract during childbirth and mammary glands to eject milk during nursing. ADH helps cells of the kidney tubules reabsorb water, thus decreasing urine volume when the body needs to retain water. When the body has too much water, the hypothalamus slows the release of ADH from the posterior pituitary.

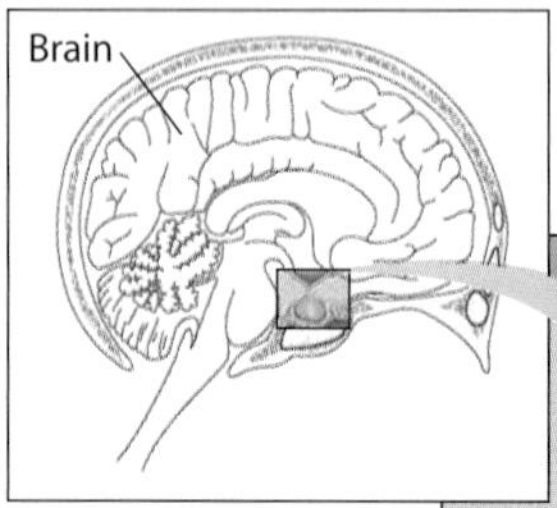
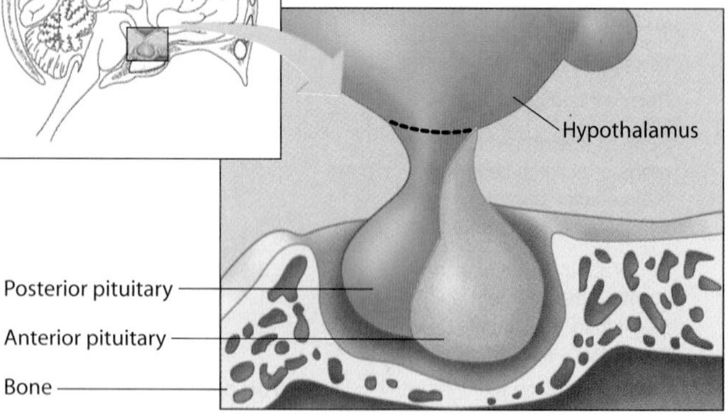

▲ **Figure 4A** Location of the hypothalamus and pituitary

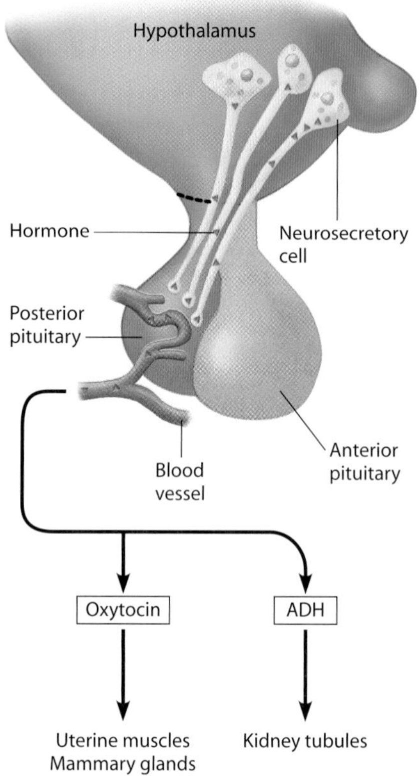

▲ **Figure 4B** Hormones of the posterior pituitary

Figure 4C shows a second set of neurosecretory cells in the hypothalamus. These cells secrete releasing and inhibiting hormones (•••) that control the anterior pituitary. A system of small blood vessels carries these hormones from the hypothalamus to the anterior pituitary. In response to hypothalamic releasing hormones, the anterior pituitary synthesizes and releases many different peptide and protein hormones (▴▴). These protein hormones, in turn, influence a broad range of body activities. **Thyroid-stimulating hormone (TSH)**, adrenocorticotropic hormone (ACTH), follicle-stimulating hormone (FSH), and luteinizing hormone (LH) all activate other endocrine glands. Feedback mechanisms control the secretion of these hormones by the anterior pituitary. Another anterior pituitary hormone, **prolactin (PRL)**, produces very different effects in different species (see Module 11); in mammals, prolactin stimulates the mammary glands.

Of all the pituitary secretions, none has a broader effect than the protein called **growth hormone (GH)**. GH promotes

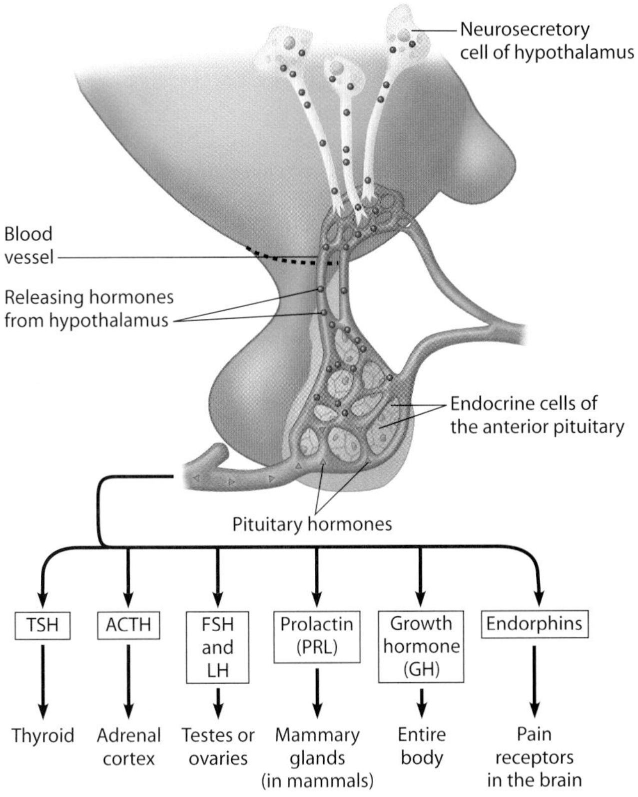

▲ **Figure 4C** Hormones of the anterior pituitary

protein synthesis and the use of body fat for energy metabolism in a wide variety of target cells. In young mammals, GH promotes the development and enlargement of all parts of the body. Abnormal production of GH can result in several human disorders. Excessive production of GH in adulthood, a condition known as acromegaly, stimulates bony growth in the face, hands, and feet. Too much GH during childhood, usually due to a pituitary tumor, can lead to gigantism (**Figure 4D**). In contrast, too little GH in childhood can lead to pituitary dwarfism. Administering growth hormone to children with GH deficiency can prevent this. Once extracted only in minute quantities from pituitaries of cadavers, human GH is now produced in large amounts by genetically engineered bacteria. Unfortunately, the increased availability of human GH has allowed some athletes to abuse it to bulk up their muscles. Such abuse is extremely dangerous and can lead to disfigurement, heart failure, and cancers.

The **endorphins**, hormones produced by the anterior pituitary as well as other parts of the brain, are the body's natural painkillers. These chemical signals bind to receptors in the brain and dull the perception of pain. The effect on the nervous system is similar to that of the drug morphine, earning endorphins the nickname "natural opiates" (although it would be more accurate to call opiates "artificial endorphins"). Some researchers speculate that the so-called runner's high results partly from the release of endorphins when stress and pain in the body reach critical levels. Some research also suggests that endorphins may be released during deep meditation, by acupuncture treatments, or even by eating spicy foods.

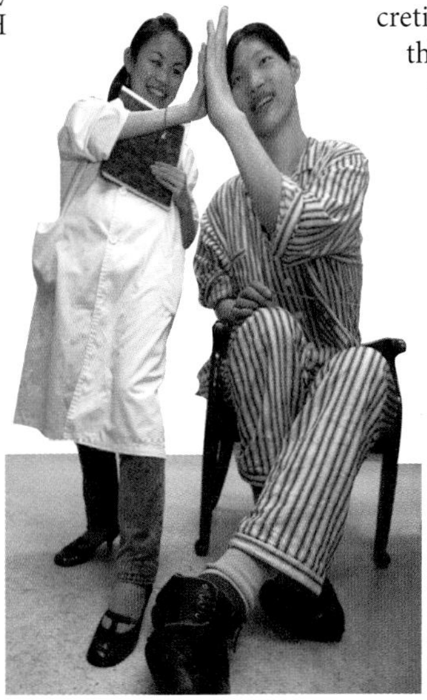

AP Wide World Photos

▲ **Figure 4D** Gigantism, caused by an excess of growth hormone during childhood

Figure 4E shows how the hypothalamus operates through the anterior pituitary to direct the activity of another endocrine organ, the thyroid gland. The hypothalamus secretes a releasing hormone known as **TRH (TSH-releasing hormone)**. In turn, TRH stimulates the anterior pituitary to produce thyroid-stimulating hormone (TSH). Under the influence of TSH, the thyroid secretes the hormone thyroxine into the blood. Thyroxine is converted to another hormone that increases the metabolic rate of most body cells, warming the body as a result.

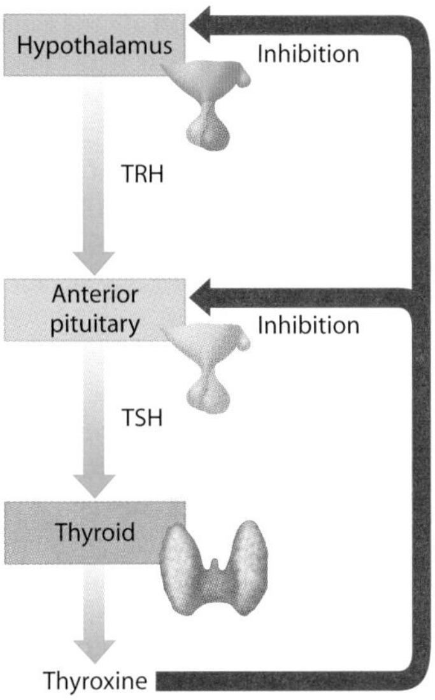

▲ **Figure 4E** Control of thyroxine secretion

Precise regulation of the TRH-TSH-thyroxine system keeps the hormones at levels that maintain homeostasis. The hypothalamus takes some cues from the environment; for instance, cold temperatures tend to increase its secretion of TRH. In addition, as the red arrows in Figure 4E indicate, negative-feedback mechanisms control the secretion of thyroxine. When thyroxine increases in the blood, it acts on the hypothalamus and anterior pituitary, inhibiting TRH and TSH secretion and consequently thyroxine synthesis.

To summarize what we discussed in this module: The anterior and posterior pituitary, directed by the hypothalamus, stimulate a number of other endocrine glands with specific functions. In the next seven modules, we explore several of these endocrine glands, the hormones they produce, and their effects on the vertebrate body.

? Alcohol inhibits secretion of ADH by the anterior pituitary. Predict how this action of alcohol would affect urination.

Since ADH stimulates reabsorption of water in the kidneys, alcohol increases the volume of urine produced.

Hormones and Homeostasis

5 The thyroid regulates development and metabolism

Your **thyroid gland** is located in your neck just under your larynx (voice box). Thyroid hormones perform several important homeostatic functions that affect virtually all the tissues of the body.

The thyroid produces two very similar amine hormones, both of which contain the element iodine. One of these, **thyroxine**, is called **T_4** because it contains four iodine atoms; the other, **triiodothyronine**, is called **T_3** because it contains three iodine atoms. In target cells, most T_4 is converted to T_3. One of the crucial roles of both hormones is in development and maturation. In a bullfrog, for example, these hormones trigger the profound reorganization of body tissues that occurs as a tadpole—a strictly aquatic organism—transforms into an adult frog, which may spend much of its time on land (**Figure 5A**). Thyroid hormones are equally important in mammals, especially in bone and nerve cell development. T_3 and T_4 also help maintain normal blood pressure, heart rate, muscle tone, digestion, and reproductive function. Throughout the body, these hormones tend to increase the rate of oxygen consumption and cellular metabolism.

Too much or too little of the thyroid hormones in the blood can result in serious metabolic disorders. An excess of T_3 and T_4 in the blood (*hyper*thyroidism) can make a person overheat, sweat profusely, become irritable, develop high blood pressure, and lose weight. The most common form of hyperthyroidism is Graves' disease; protuding eyes caused by fluid accumulation behind the eyeballs are a typical symptom. Conversely, insufficient amounts of T_3 and T_4 (*hypo*thyroidism) can cause weight gain, lethargy, and intolerance to cold.

Hypothyroidism can result from a defective thyroid gland or from dietary disorders. For example, severe iodine deficiency during childhood can cause cretinism, which results in retarded skeletal growth and poor mental development. And in adults, insufficient iodine in the diet can cause **goiter**, an enlargement of the thyroid. In such cases, the thyroid gland cannot synthesize adequate amounts of its T_3 and T_4 hormones. The lack of T_3 and T_4 interrupts the feedback loops that control thyroid activity (**Figure 5B**). The blood never carries enough of the T_3 and T_4 hormones to shut off the secretion of TRH (TSH-releasing hormone) or TSH. The thyroid enlarges because TSH continues to stimulate it.

Chris Howes/Wild Places Photography/Alamy

John M. Burnley/ Photo Researchers

▲ **Figure 5A** The maturation of a tadpole (below) into an adult frog (above), as regulated by thyroid hormones

Fortunately, both hypo- and hyperthyroidism can be successfully treated. For example, many cases of goiter can be treated simply by adding iodine to the diet. Seawater is a rich source of iodine, and goiter rarely occurs in people living near the seacoast, where the soil is iodine-rich and a lot of iodine-rich seafood is consumed. Goiter is less common today than in the past thanks to the incorporation of iodine into table salt, but it still affects thousands of people in developing nations. Treatment for hyperthyroidism (an overactive thyroid) takes advantage of the fact that the thyroid accumulates iodine: Patients drink a solution containing a small dose of radioactive iodine, which kills off some thyroid cells. This "radioactive cocktail" can kill off just enough cells to reduce thyroid output and relieve symptoms.

? By what mechanism does thyroxine switch off its own production?

By negative feedback: It inhibits the secretion of TRH from the hypothalamus and TSH from the pituitary.

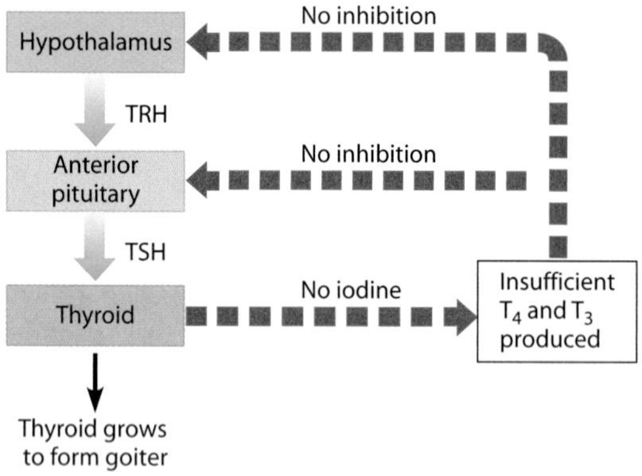

▲ **Figure 5B** How iodine deficiency causes goiter

6 Hormones from the thyroid and parathyroid glands maintain calcium homeostasis

To work properly, many body functions require an appropriate level of calcium in the blood and interstitial fluid. Without calcium, nerve signals cannot be transmitted from cell to cell, muscles cannot function properly, blood cannot clot, and cells cannot transport molecules across their membranes. The thyroid and parathyroid glands function in the homeostasis of calcium ions (Ca^{2+}), keeping the concentration of the ions within a narrow range.

You have four disk-shaped **parathyroid glands**, all embedded in the surface of the thyroid. Two peptide hormones, **calcitonin** from the thyroid and **parathyroid hormone (PTH)** from the parathyroids, regulate blood calcium levels.

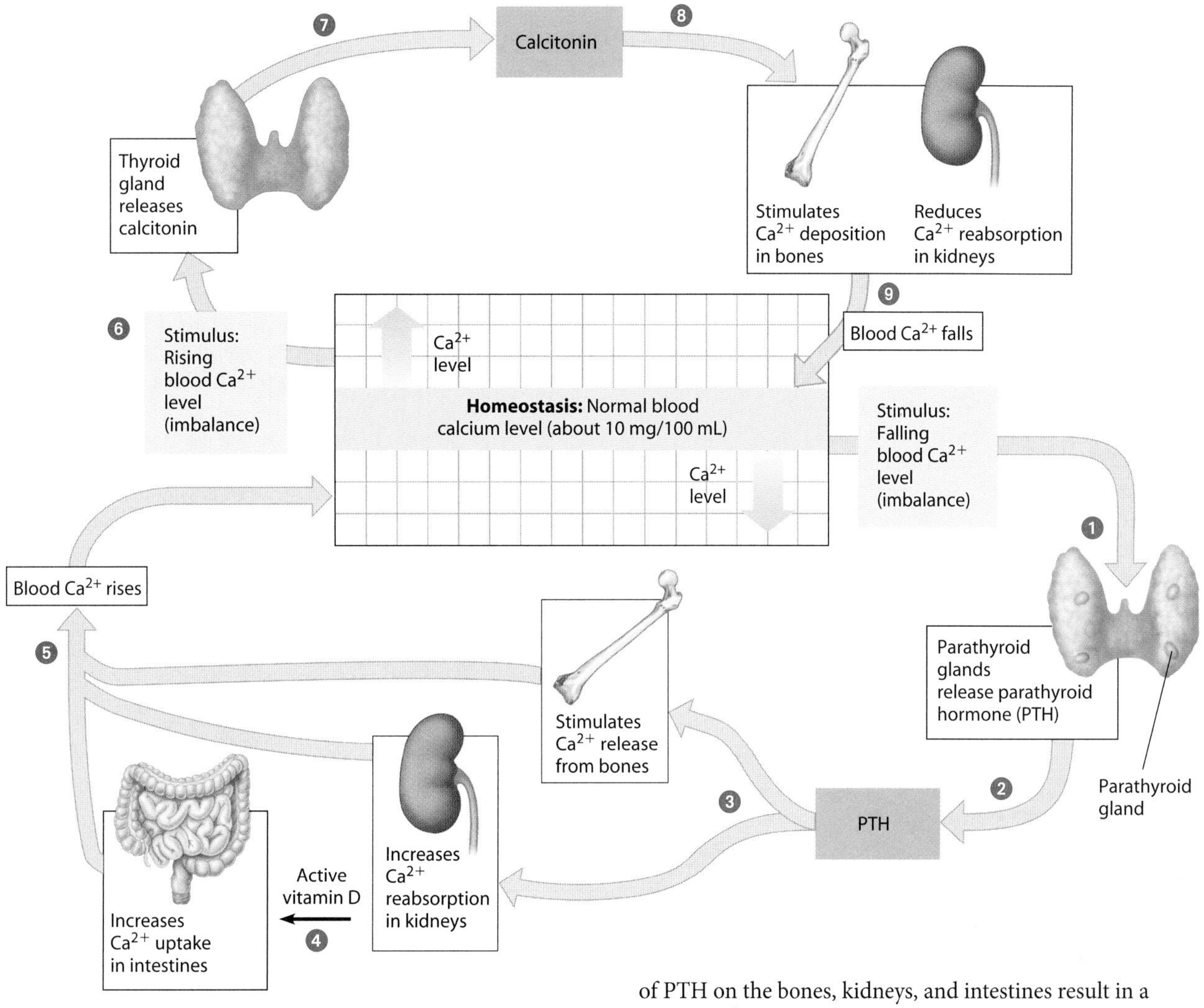

▲ **Figure 6** Calcium homeostasis

Calcitonin and PTH are said to be **antagonistic hormones** because the actions of one oppose the actions of the other: In this example, calcitonin lowers the calcium level in the blood, whereas PTH raises it. As **Figure 6** indicates, this pair of antagonistic hormones operate by means of feedback systems that keep the calcium level near the homeostatic set point of 10 mg of Ca^{2+} per 100 mL of blood.

❶ When the blood Ca^{2+} level drops below the set point, ❷ the parathyroids release PTH into the blood. ❸ PTH stimulates the release of calcium ions from bones and increases Ca^{2+} reabsorption by the kidneys. The kidneys also play an indirect role in calcium homeostasis, which involves vitamin D. We obtain this vitamin in inactive form from food and also from chemical reactions in our skin when it is exposed to sunlight. Transported in the blood, inactive vitamin D undergoes sequential steps of activation in the liver and kidneys. ❹ The active form of vitamin D, secreted by the kidneys, acts as a hormone. It stimulates the intestines to increase uptake of Ca^{2+} from food. ❺ The combined effects of PTH on the bones, kidneys, and intestines result in a higher Ca^{2+} level in the blood.

The top part of the diagram shows how calcitonin from the thyroid gland reverses the effects of PTH. ❻ A rise in blood Ca^{2+} above the set point ❼ induces the thyroid gland to secrete calcitonin. ❽ Calcitonin, in turn, has two main effects: It causes more Ca^{2+} to be deposited in the bones, and it makes the kidneys reabsorb less Ca^{2+} as they form urine. ❾ The result is a lower Ca^{2+} level in the blood.

In summary, a sensitive balancing system maintains calcium homeostasis. The system depends on feedback control by two antagonistic hormones. Failure of the system can have far-reaching effects. For example, a shortage of PTH causes the blood calcium level to drop dramatically, leading to convulsive contractions of the skeletal muscles. This condition, known as tetany, can be fatal. In such cases, a rapid intravenous infusion of calcium can be lifesaving.

? Why are calcitonin and PTH referred to as antagonistic hormones?

Antagonistic hormones have opposite effects on the body. In this case, calcitonin lowers blood calcium levels, whereas PTH raises blood calcium levels.

7 Pancreatic hormones regulate blood glucose levels

The **pancreas** produces two hormones that play a large role in managing the body's energy supplies. Clusters of endocrine cells, called islets of Langerhans, are scattered throughout the pancreas. Each islet has a population of beta cells, which produce the hormone **insulin**, and a population of alpha cells, which produce another hormone, **glucagon**. Insulin and glucagon—both protein hormones—are secreted directly into the blood.

As shown in Figure 7, insulin and glucagon are antagonistic hormones that regulate the concentration of glucose in the blood. The two hormones counter each other in a feedback circuit that precisely manages the amount of circulating glucose available to use as cellular fuel versus the amount of glucose stored as the polymer glycogen in body cells. Through such feedback, the concentration of glucose in the blood determines the relative amounts of insulin and glucagon secreted by the islet cells.

In the top half of the diagram, you see what happens when the glucose concentration of the blood rises above the set point of about 90 mg of glucose per 100 mL of blood, as it does shortly after we eat a carbohydrate-rich meal. ❶ The rising blood glucose level ❷ stimulates the beta cells in the pancreas to secrete more insulin into the blood. ❸ The insulin stimulates nearly all body cells to take up glucose from the blood, which decreases the blood glucose level. Liver cells (and skeletal muscle cells) take up much of the glucose and use it to form glycogen, which they store. Insulin also stimulates cells to metabolize the glucose for immediate energy use, for the storage of energy in fats, or for the synthesis of proteins. ❹ When the blood glucose level falls to the set point, the beta cells lose their stimulus to secrete insulin.

Following the bottom half of the diagram, you see what happens when ❺ the blood glucose level starts to dip below the set point, as it may between meals or during strenuous exercise. ❻ The pancreatic alpha cells respond by secreting more glucagon. ❼ Glucagon is a fuel mobilizer, signaling liver cells to break glycogen down into glucose, convert amino acids and

▼ **Figure 7** Glucose homeostasis

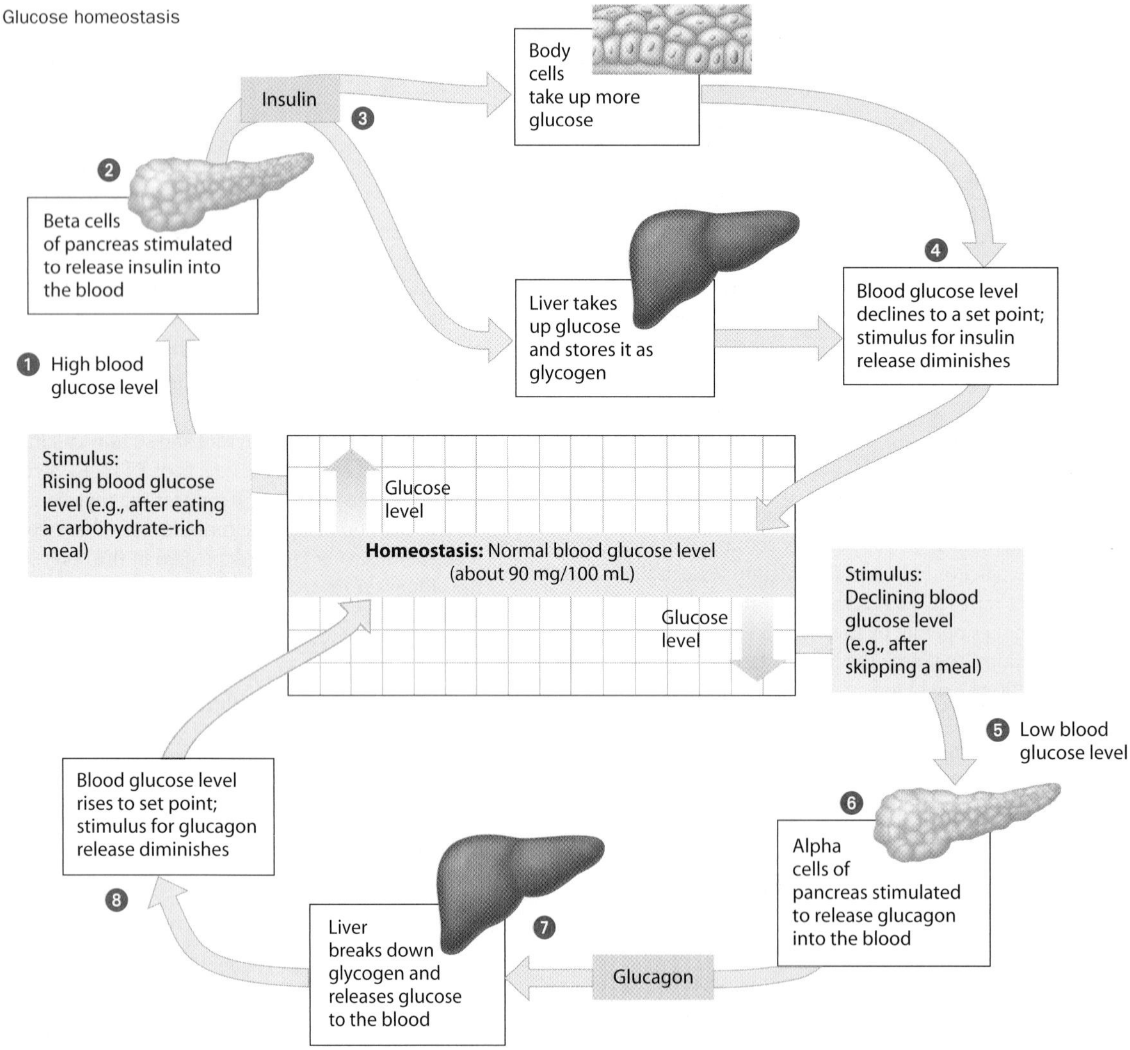

fat-derived glycerol to glucose, and release the glucose into the blood. 8 Then, when the blood glucose level returns to the set point, the alpha cells slow their secretion of glucagon.

In the next module, we see what can happen when this delicately balanced system breaks down.

? How is the insulin-glucagon relationship similar to the calcitonin-PTH relationship?

In both cases, the two hormones are antagonists that help maintain homeostasis by counteracting one another's effects. Their actions keep the blood concentration of a key chemical (glucose for insulin-glucagon; calcium ions for calcitonin-PTH) near the set point.

CONNECTION 8 Diabetes is a common endocrine disorder

Diabetes mellitus is a serious hormonal disease in which body cells are unable to absorb glucose from the blood. It affects about 24 million Americans—8% of the total population—and an estimated 6 million people with diabetes remain undiagnosed, unaware that they have a serious illness. Diabetes develops when there is not enough insulin in the blood or when target cells do not respond normally to blood insulin. In either case, the cells cannot obtain enough glucose from the blood even though there is plenty. Thus starved for fuel, cells are forced to burn the body's supply of fats and proteins. Meanwhile, because the digestive system continues to absorb glucose from ingested food, the glucose concentration in the blood can become extremely high—so high, in fact, that measurable amounts of glucose are excreted in the urine. (Normally, the kidneys leave no glucose in the urine.)

The exact causes of diabetes are elusive, but both genetic and environmental factors appear to be important. There are treatments for diabetes mellitus—insulin supplements and/or special diets—but no cure. Untreated diabetes can cause dehydration, blindness, cardiovascular and kidney disease, and nerve damage. Every year, more than 200,000 Americans die from the disease or its complications, making diabetes the seventh leading cause of death in the United States.

There are three types of diabetes mellitus. Type 1 (insulin-dependent) diabetes is an autoimmune disease in which white blood cells of the body's own immune system attack and destroy the pancreatic beta cells. As a result, the pancreas does not produce enough insulin, and glucose builds up in the blood. Type 1 diabetes generally develops during childhood. Type 1 patients are treated with injections, several times daily, of human insulin (**Figure 8A**), which is produced by genetically engineered bacteria.

In type 2 (non-insulin-dependent) diabetes, insulin is produced, but target cells fail to take up glucose from the blood, causing blood glucose levels to remain elevated. Type 2 diabetes is almost always associated with being overweight and underactive, although whether obesity causes diabetes (and if so, how) remains unknown. This form of diabetes generally appears after age 40, but even young people who are overweight and inactive can develop the disease. In the United States, more than 90% of diabetics are type 2. Many type 2 diabetics can manage their disease with exercise and diet; some require medication.

A third type of diabetes affects about 4% of pregnant women in the United States. Called gestational diabetes, it can affect any pregnant woman, even one who has never shown symptoms of diabetes before. The causes of gestational diabetes are not known, but if left untreated, gestational diabetes can lead to dangerously large babies, which can greatly complicate delivery. If diagnosed, diet and/or insulin injections can prevent most problems.

How is diabetes detected? Early signs include a lack of energy, a craving for sweets, frequent urination, and persistent thirst. The diagnostic test for diabetes is a glucose tolerance test: The person swallows a sugar solution and then has blood drawn at prescribed time intervals. Each blood sample is tested for glucose. If the blood glucose level exceeds 200 mg/100 mL 2 hours after eating, the person has diabetes. In **Figure 8B**, you can compare the glucose tolerance of a person with diabetes with that of a healthy individual. A healthy body can maintain a nearly constant concentration of blood glucose; the body of a diabetic experiences a broad range of blood glucose concentrations.

Diabetes is not the only disease that can result from problems with insulin. Some people have hyperactive beta cells that put too much insulin into the blood when sugar is eaten. As a result, their blood glucose level can drop well below normal. This condition, called **hypoglycemia**, usually occurs 2–4 hours after a meal and may be accompanied by hunger, weakness, sweating, and nervousness. In severe cases, when the brain receives inadequate amounts of glucose, a person may develop convulsions, become unconscious, and even die. Hypoglycemia is uncommon, and most forms of it can be controlled by reducing sugar intake and eating smaller, more frequent meals.

? Three hours after glucose ingestion, the person with diabetes whose test is shown in Figure 8B has a blood glucose concentration about ______ times that of the normal individual.

three

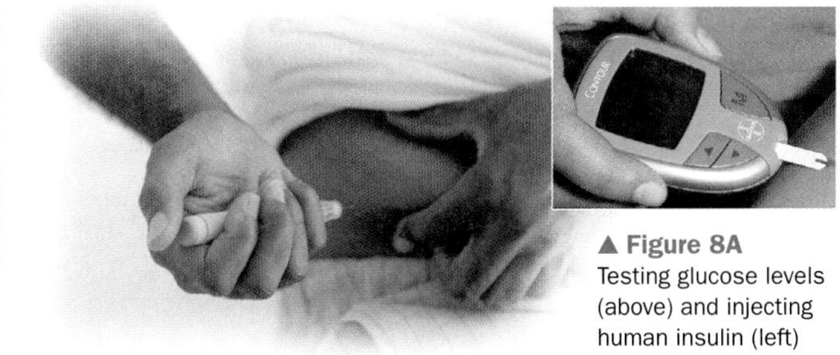

Ian Hooton/Photo Researchers

Edward Kinsman/Photo Researchers

▲ Figure 8A Testing glucose levels (above) and injecting human insulin (left)

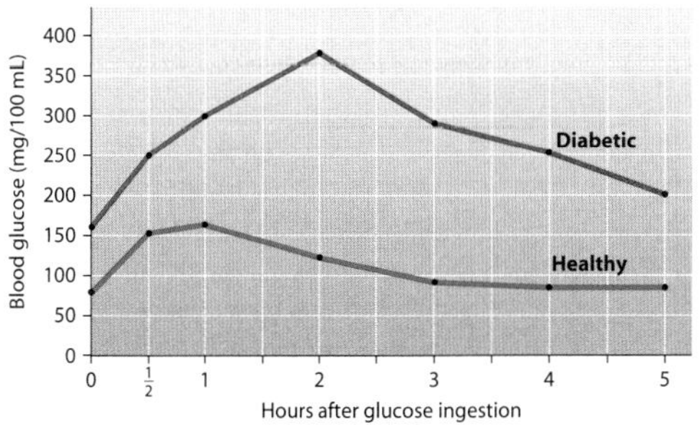

▲ Figure 8B Results of glucose tolerance tests

Data from C. J. Bryne, D. F. Saxton, et al., *Laboratory Tests: Implications for Nursing Care*, 2nd ed. Copyright © 1986 Pearson Education, Inc., publishing as Pearson Benjamin Cummings, 1301 Sansome St., San Francisco, CA 94111.

9 The adrenal glands mobilize responses to stress

The endocrine system includes two **adrenal glands**, one sitting on top of each kidney. As you can see in **Figure 9** (inset, top left), each adrenal gland is actually made up of two glands fused together: a central portion called the **adrenal medulla** and an outer portion called the **adrenal cortex**. Though the cells they contain and the hormones they produce are different, both the adrenal medulla and the adrenal cortex secrete hormones that enable the body to respond to stress.

The adrenal medulla produces the "fight-or-flight" hormones that we discussed earlier; they ensure a rapid, short-term response to stress. You've probably felt your heart beat faster and your skin develop goose bumps when sensing danger. Facing an unexpected threat, like a pop quiz, can cause these short-term stress symptoms. Positive emotions—extreme pleasure, for instance—can produce the same effects. These reactions are triggered by two amine hormones secreted by the adrenal medulla, **epinephrine** (adrenaline) and **norepinephrine** (noradrenaline).

Stressful stimuli, whether negative or positive, activate certain nerve cells in the hypothalamus. ❶ These cells send nerve signals via the spinal cord to the adrenal medulla, ❷ stimulating it to secrete epinephrine and norepinephrine into the blood. Epinephrine and norepinephrine have somewhat different effects on tissues, but both contribute to the short-term stress response. Both hormones stimulate liver cells to release glucose, thus making more fuel available for cellular work. They also prepare the body for action by raising the blood pressure, breathing rate, and metabolic rate. In addition, epinephrine and norepinephrine change blood flow patterns, making some organs more active and others less so. For example, epinephrine dilates blood vessels in the brain and skeletal muscles, thus increasing alertness and the muscles' ability to react to stress. At the same time, epinephrine and norepinephrine constrict blood vessels elsewhere, thereby reducing activities that are not immediately involved in the stress response, such as digestion. The short-term stress response prepares the body for a quick and decisive reaction—as would be needed to confront a sudden danger—but it occurs and subsides rapidly.

In contrast to epinephrine and norepinephrine secreted by the adrenal medulla, hormones secreted by the adrenal cortex can provide a slower, longer-lasting response to stress that can last for days. The adrenal cortex responds to endocrine signals—chemical signals in the blood—rather than to nerve cell signals. As shown in Figure 9, ❸ the hypothalamus secretes a releasing hormone that ❹ stimulates target cells in the anterior pituitary to secrete the hormone **adrenocorticotropic hormone (ACTH)**. ❺ In turn, ACTH stimulates cells of the adrenal cortex to synthesize and secrete a family of steroid hormones called the

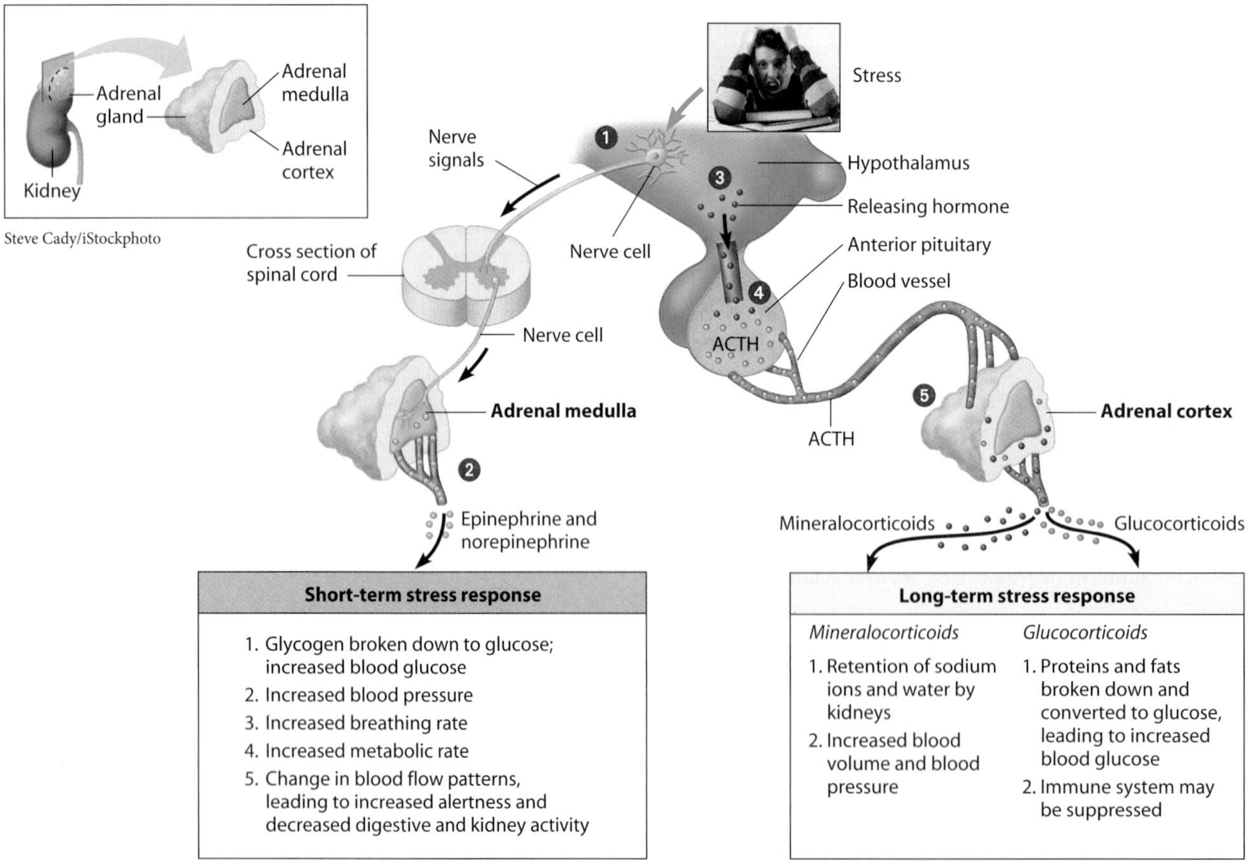

Steve Cady/iStockphoto

▲ **Figure 9** How the adrenal glands control our responses to stress

corticosteroids. The two main types in humans are the mineralocorticoids and the glucocorticoids. Both help maintain homeostasis when the body experiences long-term stress.

Mineralocorticoids act mainly on salt and water balance. One of these hormones (aldosterone) stimulates the kidneys to reabsorb sodium ions and water, with the overall effect of increasing the volume of the blood and raising blood pressure as a response to prolonged stress.

Glucocorticoids function mainly in mobilizing cellular fuel, thus reinforcing the effects of glucagon. Glucocorticoids promote the synthesis of glucose from noncarbohydrates such as proteins and fats. When the body cells consume more glucose than the liver can provide from glycogen stores, glucocorticoids stimulate the breakdown of muscle proteins, making amino acids available for conversion to glucose by the liver. This makes more glucose available in the blood as cellular fuel in response to stress.

Very high levels of glucocorticoids can suppress the body's defense system, including the inflammatory response that occurs at infection sites. For this reason, physicians may use glucocorticoids to treat excessive inflammation. The glucocorticoid cortisone, for example, was once regarded as a miracle drug for treating serious inflammatory conditions such as arthritis. Cortisone and other glucocorticoids can relieve swelling and pain from inflammation; but by suppressing immunity, they can also make a person highly susceptible to infection.

Physicians often prescribe oral glucocorticoids to relieve pain from athletic injuries. However, glucocorticoids are potentially very dangerous; prolonged use can depress the activity of the adrenal glands and cause side effects such as a weakened immune system, easy bruising, weak bones, weight gain, muscle breakdown, and increased risk of diabetes. It is safer, but still potentially dangerous, to inject a glucocorticoid at the site of injury. With this treatment, the pain usually subsides, but its underlying cause remains. Masking the pain covers up the pain's message—that tissue is damaged.

? Epinephrine and norepinephrine are responsible for the __________ stress response while the mineralocorticoids and glucocorticoids are responsible for the __________ stress response.

short-term . . . long-term

10 The gonads secrete sex hormones

The sex hormones are steroid hormones that affect growth and development and also regulate reproductive cycles and sexual behavior. The **gonads**, or sex glands (ovaries in the female and testes in the male), secrete sex hormones in addition to producing gametes (ova and sperm).

The gonads of mammals produce three major categories of sex hormones: estrogens, progestins, and androgens. Females and males have all three types, but in different proportions. Females have a high ratio of estrogens to androgens. In humans, **estrogens** maintain the female reproductive system and promote the development of female features like smaller body size, higher-pitched voice, breasts, and wider hips. In all mammals, **progestins**, such as progesterone, are primarily involved in preparing and maintaining the uterus to support an embryo.

In general, **androgens** stimulate the development and maintenance of the male reproductive system. Males have a high ratio of androgens to estrogens, with their main androgen being **testosterone**. In humans, androgens produced by male embryos during the seventh week of development stimulate the embryo to develop into a male rather than a female. During puberty, high concentrations of androgens trigger the development of male characteristics, such as a lower-pitched voice, facial hair, and large skeletal muscles. The muscle-building action of androgens and other anabolic steroids has enticed some athletes to abuse them at great risk to their health.

Research has established that the process of sex determination by androgens occurs in a highly similar manner in all vertebrates, suggesting that androgens had this role early in evolution. As discussed in the chapter opener, testosterone may play a role in the development of manes in male lions. In elephant seals, male androgens promote development of large bodies weighing 2 tons or more, an inflatable enlargement of the nasal cavity, a thick hide that can withstand bloody conflicts, and aggressive behavior toward other males. The two males in **Figure 10** are fighting by slamming their bodies against each other. In so doing one will establish dominance over the other and the right to mate with many females.

The synthesis of sex hormones by the gonads is regulated by the hypothalamus and anterior pituitary. In response to a releasing factor from the hypothalamus, the anterior pituitary secretes follicle-stimulating hormone (FSH) and luteinizing hormone (LH). These stimulate the ovaries or testes to synthesize and secrete the sex hormones, among other effects.

Jonathan Blair/Corbis

? Name the human male and female gonads and the primary hormone(s) that each produce(s).

Ovaries (estrogens and progestins) and testes (androgens)

▶ **Figure 10** Male elephant seals in combat

EVOLUTION CONNECTION

11 A single hormone can perform a variety of functions in different animals

Hormones play important roles in all vertebrates, and some of the same hormones can be found in vertebrates that are only distantly related. Interestingly, the same hormone can have different actions in different animals—a strong indication that hormonal regulation was an early evolutionary adaptation.

The peptide hormone prolactin (PRL), produced and secreted by the anterior pituitary under the direction of the hypothalamus, is a good example. Prolactin produces diverse effects in different vertebrate species. In humans, PRL performs several important functions related to childbirth. During late pregnancy, PRL stimulates mammary glands to grow and produce milk. (A brief surge in prolactin levels just before menstruation causes breast swelling and tenderness in some women.) Suckling by a newborn stimulates further release of PRL, which in turn increases the milk supply (Figure 11). High prolactin levels during nursing tend to prevent the ovaries from releasing eggs, decreasing the chances of a new pregnancy occurring during the time of breast-feeding. This may be an evolutionary adaptation that helps ensure that adequate care is given to newborns.

PRL plays a wide variety of other roles unrelated to childbirth. In some nonhuman mammals, PRL stimulates nest building. In birds, PRL regulates fat metabolism and reproduction. In amphibians, it stimulates movement toward water in preparation for breeding and affects metamorphosis. In fish that migrate between salt and fresh water (salmon, for example), PRL helps regulate salt and water balance in the gills and kidneys.

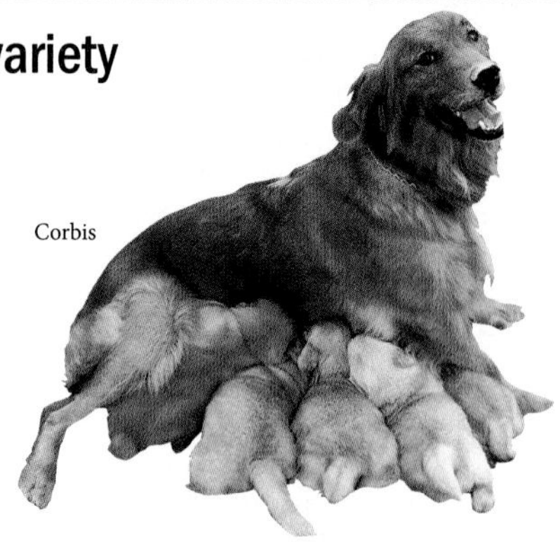

▲ **Figure 11** Suckling promotes prolactin production

Such diverse effects suggest that prolactin is an ancient hormone whose functions diversified through evolution. Over millions of years, the prolactin molecule stayed the same, but its role changed dramatically—a good example of how evolution can both preserve unity (in terms of the structure of the molecule itself) and promote diversity (in terms of the varying roles it plays).

? PRL promotes the production of milk. Newborn suckling promotes PRL production. This is an example of ______ feedback.

positive

R E V I E W

For Practice Quizzes, BioFlix, MP3 Tutors, and Activities, go to www.masteringbiology.com.

Reviewing the Concepts

The Nature of Chemical Regulation (1–2)

1 Chemical signals coordinate body functions. Hormones are signaling molecules, usually carried in the blood, that cause specific changes in target cells. All hormone-secreting cells make up the endocrine system, which works with the nervous system in regulating body activities.

2 Hormones affect target cells using two main signaling mechanisms.

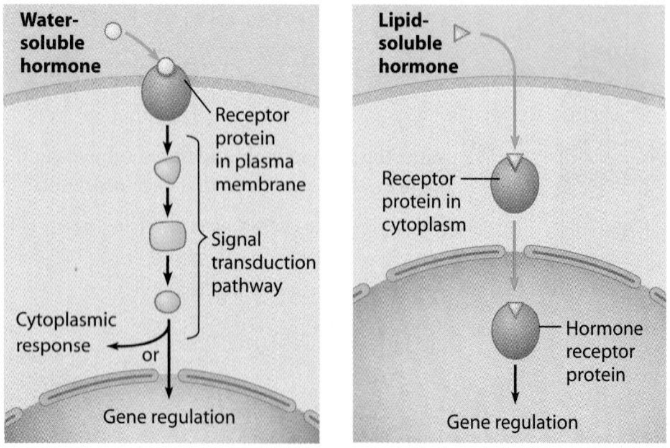

The Vertebrate Endocrine System (3–4)

3 Overview: The vertebrate endocrine system consists of more than a dozen major glands. Some glands are specialized for hormone secretion only; some also do other jobs. Some hormones have a very narrow range of targets and effects; others have numerous effects on many kinds of target cells.

4 The hypothalamus, which is closely tied to the pituitary, connects the nervous and endocrine systems. The releasing and inhibiting hormones from the hypothalamus control the secretion of several other hormones.

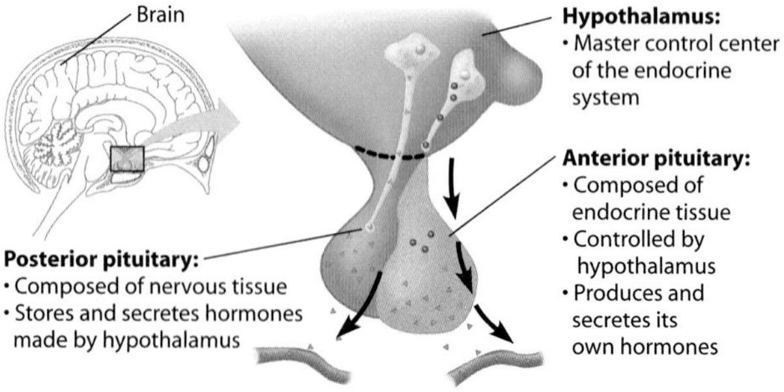

Hormones and Homeostasis (5–11)

5 The thyroid regulates development and metabolism. Two amine hormones from the thyroid gland, T_4 and T_3, regulate an animal's development and metabolism. Negative feedback maintains homeostatic levels of T_4 and T_3 in the blood. Thyroid imbalance can cause disease.

6 Hormones from the thyroid and parathyroid glands maintain calcium homeostasis. Blood calcium level is regulated by a well-balanced antagonism between calcitonin from the thyroid and parathyroid hormone from the parathyroid glands.

7 Pancreatic hormones regulate blood glucose levels.

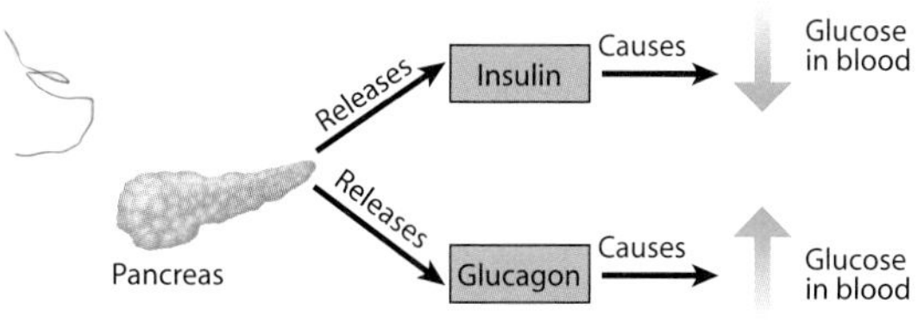

8 Diabetes is a common endocrine disorder. Diabetes mellitus results from a lack of insulin or a failure of cells to respond to it.

9 The adrenal glands mobilize responses to stress. Nerve signals from the hypothalamus stimulate the adrenal medulla to secrete epinephrine and norepinephrine, which quickly trigger the fight-or-flight response. ACTH from the pituitary causes the adrenal cortex to secrete glucocorticoids and mineralocorticoids, which boost blood pressure and energy in response to long-term stress.

10 The gonads secrete sex hormones. Estrogens, progestins, and androgens are steroid sex hormones produced by the gonads in response to signals from the hypothalamus and pituitary. Estrogens and progestins (mainly progesterone) stimulate the development of female characteristics and maintain the female reproductive system. Androgens, such as testosterone, trigger the development of male characteristics.

11 A single hormone can perform a variety of functions in different animals. A single hormone, such as prolactin, can assume diverse roles.

Connecting the Concepts

Match each hormone (top) with the gland where it is produced (center) and its effect on target cells (bottom).

1. thyroxine
2. epinephrine
3. androgens
4. insulin
5. melatonin
6. FSH
7. PTH
8. ADH

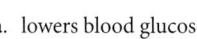

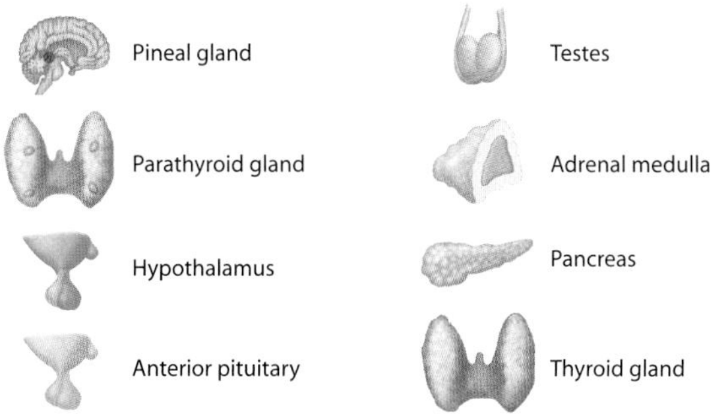

a. lowers blood glucose
b. stimulates ovaries
c. triggers fight-or-flight
d. promotes male characteristics
e. regulates metabolism
f. influences sleep/wake rhythms
g. raises blood calcium level
h. boosts water retention

Testing Your Knowledge

Multiple Choice

9. Which of the following controls the activity of all the others?
   a. thyroid gland
   b. pituitary gland
   c. adrenal cortex
   d. hypothalamus
   e. ovaries
10. The pancreas increases its output of insulin in response to
    a. an increase in body temperature.
    b. changing cycles of light and dark.
    c. a decrease in blood glucose.
    d. a hormone secreted by the anterior pituitary.
    e. an increase in blood glucose.
11. Which of the following hormones have antagonistic (opposing) effects? Choose all that apply.
    a. parathyroid hormone and calcitonin
    b. glucagon and thyroxine
    c. growth hormone and epinephrine
    d. ACTH and cortisone
    e. epinephrine and norepinephrine
12. The body is able to maintain a relatively constant level of thyroxine in the blood because
    a. thyroxine stimulates the pituitary to secrete thyroid-stimulating hormone (TSH).
    b. thyroxine inhibits the secretion of TSH-releasing hormone (TRH) from the hypothalamus.
    c. TRH inhibits the secretion of thyroxine by the thyroid gland.
    d. thyroxine stimulates the hypothalamus to secrete TRH.
    e. thyroxine stimulates the pituitary to secrete TRH.
13. Which of the following hormones has the broadest range of targets?
    a. ADA
    b. oxytocin
    c. TSH
    d. epinephrine
    e. ACTH

Describing, Comparing, and Explaining

14. Explain how the hypothalamus controls body functions through its action on the pituitary gland. How does control of the anterior and posterior pituitary differ?
15. Explain how the same hormone might have different effects on two different target cells and no effect on nontarget cells.

Applying the Concepts

16. A strain of transgenic mice remains healthy as long as you feed them regularly and do not let them exercise. After they eat, their blood glucose level rises slightly and then declines to a homeostatic level. However, if these mice fast or exercise at all, their blood glucose drops dangerously. Which hypothesis best explains their problem? (*Explain your choice.*)
    a. The mice have insulin-dependent diabetes.
    b. The mice lack insulin receptors on their cells.
    c. The mice lack glucagon receptors on their cells.
    d. The mice cannot synthesize glycogen from glucose.
17. How could a hormonal imbalance result in a person who is genetically male but physically female?

Review Answers

1. Testes: 3, d
   Pineal gland: 5, f
   Parathyroid gland: 7, g
   Adrenal medulla: 2, c
   Hypothalamus: 8, h
   Pancreas: 4, a
   Anterior pituitary: 6, b
   Thyroid gland: 1, e

9. d **10.** e **11.** a **12.** b (Negative feedback: When thyroxine increases, it inhibits TSH, which reduces thyroxine secretion.) **13.** d

14. The hypothalamus secretes releasing hormones and inhibiting hormones, which are carried by the blood to the anterior pituitary. In response to these signals from the hypothalamus, the anterior pituitary increases or decreases its secretion of a variety of hormones that directly affect body activities or influence other glands. Neurosecretory cells that extend from the hypothalamus into the posterior pituitary secrete hormones that are stored in the posterior pituitary until they are released into the blood.

15. Only cells with the proper receptors will respond to a hormone. For a steroid hormone, the presence (or absence) and types of receptor proteins inside the cell determine the hormone's effect. For a nonsteroid hormone, the types of receptors on the cell's plasma membrane are key, and the proteins of the signal transduction pathway may have different effects inside different cells.

16. a. No. Blood sugar level goes too low. Diabetes would tend to make the blood sugar level go too high after a meal.
    b. No. Insulin is working, as seen by the homeostatic blood sugar response to feeding.
    c. Correct. Without glucagon, exercise and fasting lower blood sugar, the cells cannot mobilize any sugar reserves, and blood sugar level drops. Insulin (which lowers blood sugar) has no effect.
    d. No. If this were true, blood sugar level would increase too much after a meal.

17. If cells within a male embryo do not secrete testosterone at the proper time during development, the embryo will develop into a female despite being genetically male.

Reproduction and Embryonic Development

Reproduction and Embryonic Development

BIG IDEAS

John R. Finnerty

Asexual and Sexual Reproduction
(1–2)

Some animals can reproduce asexually, but most reproduce by the fusion of egg and sperm.

C. Edelman/La Vilette/Photo Researchers

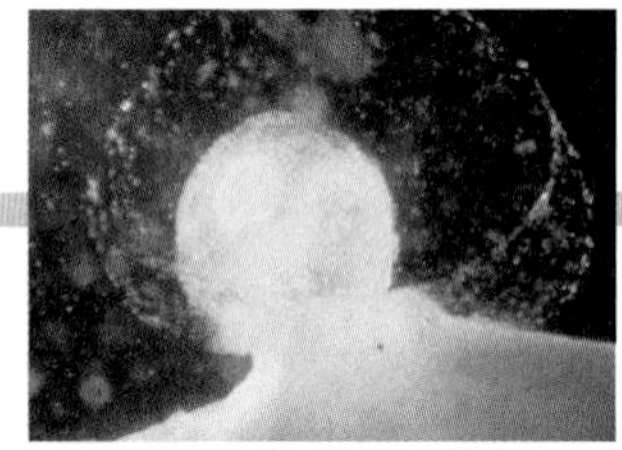

Human Reproduction
(3–8)

Human males and females have structures that produce, store, and deliver gametes.

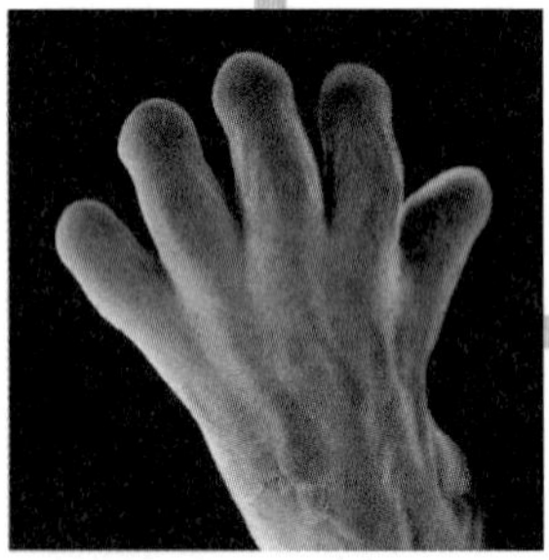

David Barlow

Principles of Embryonic Development
(9–14)

A zygote develops into an embryo through a series of carefully regulated processes.

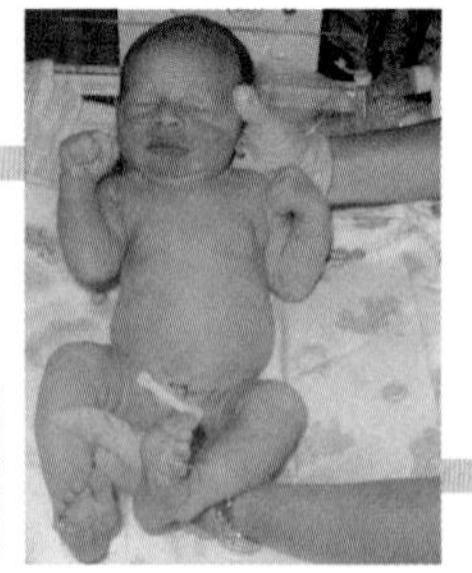

Eric J. Simon

Human Development
(15–18)

A human fetus develops within the uterus for 9 months.

Picture Partners/Photo Researchers

Chances are you know a family with twins. Indeed, the rate of twin births in the United States and other industrialized nations more than doubled between 1980 and 2006. What is the cause of this remarkable baby bonanza? One answer is the increased use of drugs to treat infertility.

One common cause of infertility among women is a failure to ovulate (release an immature egg). Fertility drugs that alter the levels of one or more hormones that control ovulation have allowed thousands of infertile couples to have babies. But an unintended consequence of fertility drugs is that they often promote the release of multiple eggs. Each egg may then fuse with a sperm, resulting in multiple embryos. In fact, over 10% of women taking fertility drugs become pregnant with more than one embryo.

Despite the successful births now enjoyed by previously infertile couples, multiple births are risky. Newborns from multiple births are premature more often than babies from single births. They also have lower birth weights, are less likely to survive (the mortality rate is five times higher among twins), and are more likely to suffer lifelong disabilities if they do survive. Thus, while modern medicine offers infertile couples new options, these options carry risks that require careful consideration.

Fertility drugs are just one example of how reproductive technologies can alter the normal reproductive cycle. We investigate more of these reproductive options at the end of this chapter.

We begin this chapter with a brief introduction to the diverse ways that animals reproduce, followed by a close look at the reproductive system of our own species. In the second half of the chapter, we discuss the processes of fertilization and embryonic development in vertebrates and then focus on human embryonic development and birth.

Asexual and Sexual Reproduction

1 Asexual reproduction results in the generation of genetically identical offspring

Individual animals have a finite life span. A species transcends this limit only by **reproduction**, the creation of new individuals from existing ones. Animals reproduce in a great variety of ways, but there are two principal modes: asexual and sexual.

Asexual reproduction (reproduction without sex) is the creation of genetically identical offspring by a lone parent. Because asexual reproduction proceeds without the fusion of egg and sperm, the resulting offspring are genetic copies of the one parent.

Several types of asexual reproduction are found among animals. Many invertebrates, such as the hydra in **Figure 1A**, reproduce asexually by **budding**, splitting off new individuals from outgrowths of existing ones. The sea anemone in the center of **Figure 1B** is undergoing **fission**, the separation of a parent into two or more individuals of about equal size. Asexual reproduction can also result from the process of **fragmentation**, the breaking of the parent body into several pieces, followed by **regeneration**, the regrowth of lost body parts. In sea stars (starfish) of the genus *Linckia*, for example, a whole new individual can develop from a broken-off arm plus a bit of the central body. Thus, a single animal with five arms, if broken apart, could potentially give rise to five offspring via asexual reproduction in a matter of weeks. In some species of sea sponges, if a single sponge is pushed through a wire mesh, each of the resulting clumps of cells can regrow into a new sponge. (Besides the natural means of asexual reproduction discussed here, many species have been the target of artificial asexual reproduction.

Oxford Scientific/PhotoLibrary

▲ **Figure 1A** Asexual reproduction via budding in a hydra

David Wrobel

▲ **Figure 1B** Asexual reproduction of a sea anemone (*Anthopleura elegantissima*) by fission

In nature, asexual reproduction has several potential advantages. For one, it allows animals that do not move from place to place or that live in isolation to produce offspring without finding mates. Another advantage is that it enables an animal to produce many offspring quickly; no time or energy is lost in production of eggs and sperm or in mating. Asexual reproduction perpetuates a particular genotype faithfully, precisely, and rapidly. Therefore, it can be an effective way for animals that are genetically well suited to an environment to quickly expand their populations and exploit available resources.

A potential disadvantage of asexual reproduction is that it produces genetically uniform populations. Genetically similar individuals may thrive in one particular environment, but if the environment changes and becomes less favorable, all individuals may be affected equally, and the entire population may die out.

? What kinds of environments would likely be advantageous to asexually reproducing organisms? Why?

Relatively unchanging environments favor asexual reproduction because well-suited individuals can rapidly multiply and use available resources.

2 Sexual reproduction results in the generation of genetically unique offspring

Sexual reproduction is the creation of offspring through the process of **fertilization**, the fusion of two haploid (n) sex cells, or **gametes**, to form a diploid ($2n$) **zygote** (fertilized egg). Vote that n refers to the haploid number of chromosomes and $2n$ refers to the diploid number; for humans, $n = 23$ and $2n = 46$.) The male gamete, the **sperm**, is a relatively small cell that moves by means of a flagellum. The female gamete, the **egg**, is a much larger cell that is not self-propelled. The zygote—and the new individual it develops into—contains a unique combination of genes inherited from the parents via the egg and sperm.

Most animals reproduce mainly or exclusively by sexual reproduction, which increases genetic variability among offspring. Meiosis and random fertilization can generate enormous genetic variation. And such variation is the raw material of evolution by natural selection. The variability produced by the reshuffling of genes in sexual reproduction may provide greater adaptability to changing environments. According to this hypothesis, when an environment changes suddenly or drastically, more offspring will survive and reproduce if they aren't all genetically very similar.

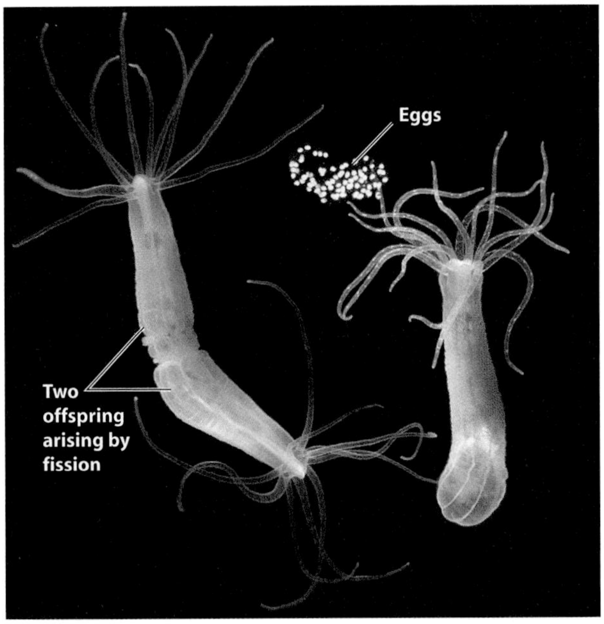

John R. Finnerty

▲ **Figure 2A** Asexual (left) and sexual (right) reproduction in the starlet sea anemone (*Nematostella vectensis*)

John Cancalosi/Nature Picture Library

▲ **Figure 2C** Frogs in an embrace that triggers the release of eggs and sperm (the sperm are too small to be seen)

Animals that can reproduce both asexually and sexually benefit from both modes. In **Figure 2A**, you can see two sea anemones of the same species; the one on the left is reproducing asexually (via fission) while the one on the right is releasing eggs. Many other marine invertebrates can also reproduce by both modes. Why would such dual reproductive capabilities be advantageous to an organism? From several well-studied cases, it is known that certain animals reproduce asexually when there is ample food and when water temperatures are favorable for rapid growth and development. Asexual reproduction usually continues until cold temperatures signal the approach of winter or until the food supply dwindles or the habitat starts to dry up. At that point, the animals switch to a sexual reproduction mode, resulting in a generation of genetically varied individuals with better potential to adapt to the changing conditions.

Although sexual reproduction has advantages, it presents a problem for nonmobile animals and for those that live solitary lives: how to find a mate. One solution that has evolved is **hermaphroditism**, in which each individual has both female and male reproductive systems. (The term comes from the Greek myth in which Hermaphroditus, the son of the gods Hermes and Aphrodite, fused with a woman to form one individual of both sexes.) Although some hermaphrodites, such as tapeworms, can fertilize their own eggs, most must mate with another member of the same species. When hermaphrodites mate (for example, the two earthworms in **Figure 2B**), each animal serves as both male and female, donating and receiving sperm. For hermaphrodites, there is only one sex, so every individual encountered is a potential mate. Mating can therefore result in twice as many offspring than if only one individual's eggs were fertilized.

Hans Pfletschinger/Peter Arnold, Inc. /Photo Library

▲ **Figure 2B** Hermaphroditic earthworms mating

The mechanics of fertilization play an important part in sexual reproduction. Many aquatic invertebrates and most fishes and amphibians exhibit **external fertilization**: The parents discharge their gametes into the water, where fertilization then occurs, often without the male and female even making physical contact. Timing is crucial because the eggs must be ready for fertilization when sperm contact them. For many species—certain clams that live in freshwater rivers and lakes, for instance—environmental cues such as temperature and day length cause a whole population to release gametes all at once. Males or females may also emit a chemical signal as they release their gametes. The signal triggers gamete release in members of the opposite sex. Most fishes and amphibians with external fertilization have specific courtship rituals that trigger simultaneous gamete release in the same vicinity by the female and male. An example of such a mating ritual is the clasping of a female frog by a male (**Figure 2C**).

In contrast to external fertilization, **internal fertilization** occurs when sperm are deposited in or close to the female reproductive tract and gametes unite within the tract. Nearly all terrestrial animals exhibit internal fertilization, which is an adaptation that enables sperm to reach an egg despite a dry external environment. Internal fertilization usually requires **copulation**, or sexual intercourse. It also requires complex reproductive systems, including organs for gamete storage and transport and organs that facilitate copulation. For examples of these complex structures, we turn next to the human female and male.

? In terms of genetic makeup, what is the most important difference between the outcome of sexual reproduction and that of asexual reproduction?

The offspring of sexual reproduction are genetically diverse, whereas the offspring of asexual reproduction are genetically identical.

Human Reproduction

3 Reproductive anatomy of the human female

Although we tend to focus on the anatomical differences between the human male and female reproductive systems, there are also some important similarities. Both sexes have a pair of **gonads**, the organs that produce gametes. Also, both sexes have ducts that store and deliver gametes as well as structures that facilitate copulation. In this and the next module, we examine the anatomical features of the human reproductive system, beginning with female anatomy.

A woman's gonads, her **ovaries**, are each about an inch long, with a bumpy surface (**Figure 3A**). The bumps are **follicles**, each consisting of one or more layers of cells that surround, nourish, and protect a single developing egg cell. In addition to producing egg cells, the ovaries produce hormones. Specifically, the follicle cells produce the female sex hormone estrogen. (In this chapter, we use the word *estrogen* to refer collectively to several closely related chemicals that affect the body similarly.)

A female is born with 1–2 million follicles, but only several hundred will release egg cells during her reproductive years. Starting at puberty, one follicle (or, rarely, more than one) matures and releases an immature egg cell about every 28 days. This monthly cycle continues until a female reaches menopause, which usually occurs around age 50. An immature egg cell is ejected from the follicle in a process called **ovulation**, shown in **Figure 3B**.

After ovulation, the follicular tissue that had been surrounding the egg that was just ejected grows within the ovary to form a solid mass called the **corpus luteum**; you can see one in the ovary on the left in Figure 3A. The corpus luteum secretes additional estrogen as well as progesterone, a hormone that helps maintain the uterine lining during pregnancy. If the released egg is not fertilized, the corpus luteum degenerates, and a new follicle matures during the next cycle. We discuss ovulation and female hormonal cycles further in later modules.

Notice in Figure 3A that each ovary lies next to the opening of an **oviduct**, also called a fallopian tube. The oviduct opening resembles a funnel fringed with finger-like projections. The projections touch the surface of the ovary, but the ovary is

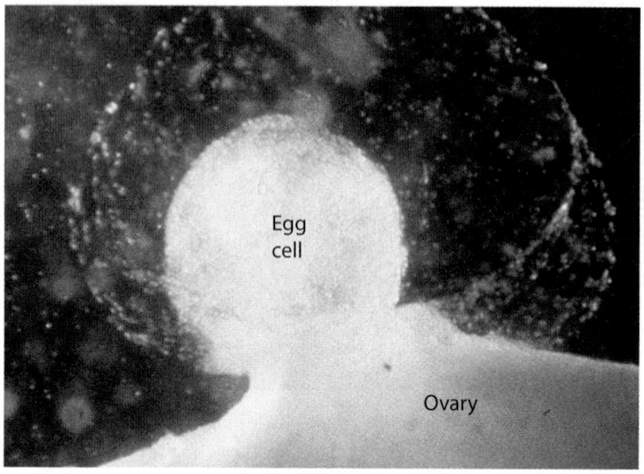

▲ **Figure 3B** Ovulation

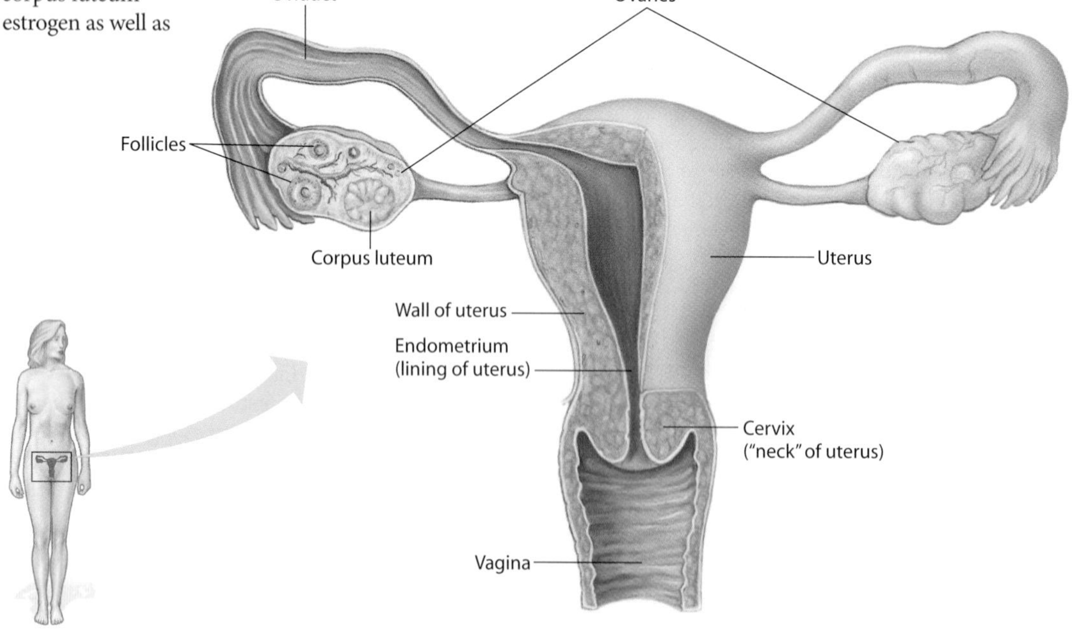

▲ **Figure 3A** Front view of female reproductive anatomy (upper portion)

actually separated from the opening of the oviduct by a tiny space. When ovulation occurs, the egg cell passes across the space and into the oviduct, where cilia sweep it toward the uterus. If sperm are present, fertilization may occur in the upper part of the oviduct. The resulting zygote starts to divide, thus becoming an embryo, as it moves along within the oviduct.

The **uterus**, also known as the womb, is the actual site of pregnancy. The uterus is only about 3 inches long in a woman who has never been pregnant, but during pregnancy it expands considerably as the baby develops. The uterus has a thick muscular wall, and its inner lining, the **endometrium**, is richly supplied with blood vessels. An embryo implants in the endometrium, and development is completed there. The term **embryo** is used for the stage in development from the first division of the zygote until body structures begin to appear, about the 9th week in humans. From the 9th week until birth, a developing human is called a **fetus**.

The uterus is the normal site of pregnancy. However, in about 1% of pregnancies, the embryo implants somewhere else, resulting in an **ectopic pregnancy**. Most ectopic pregnancies occur in the oviduct and are called tubal pregnancies. An ectopic pregnancy is a serious medical emergency that requires surgical intervention; otherwise, it can rupture surrounding tissues, causing severe bleeding and even death of the mother.

The narrow neck at the bottom of the uterus is the **cervix**, which opens into the vagina. It is recommended that women have a yearly Pap test in which cells are removed from around the cervix and examined under a microscope for signs of cervical cancer. Regular Pap smears greatly increase the chances of detecting cervical cancer early and therefore treating it successfully. The cervix opens to the **vagina**, a thin-walled, but strong, muscular chamber that serves as the birth canal through which the baby is born. The vagina is also the repository for sperm during sexual intercourse. Glands near the vaginal opening secrete mucus during sexual arousal, lubricating the vagina and facilitating intercourse.

You can see more features of female reproductive anatomy in **Figure 3C**, a side view. **Vulva** is the collective term for the external female genitalia. Notice that the vagina opens to the outside just behind the opening of the urethra, the tube through which urine is excreted. A pair of slender skin folds, the **labia minora**, border the openings, and a pair of thick, fatty ridges, the **labia majora**, protect the vaginal opening. Until sexual intercourse or vigorous physical activity ruptures it, a thin piece of tissue called the hymen partly covers the vaginal opening.

Several female reproductive structures are important in sexual arousal, and stimulation of them can produce highly pleasurable sensations. The vagina, labia minora, and a small erectile organ called the **clitoris** all engorge with blood and enlarge during sexual activity. The clitoris consists of a short shaft supporting a rounded **glans**, or head, covered by a small hood of skin called the **prepuce**. In Figure 3C, blue highlights the spongy erectile tissue within the clitoris that fills with blood during arousal. The clitoris, especially the glans, has an enormous number of nerve endings and is very sensitive to touch. Keep in mind the details of female reproductive anatomy as you read the next module, and you'll notice many similarities in the human male.

? Where does fertilization occur? In which organ does the fetus develop?

The oviduct; the uterus

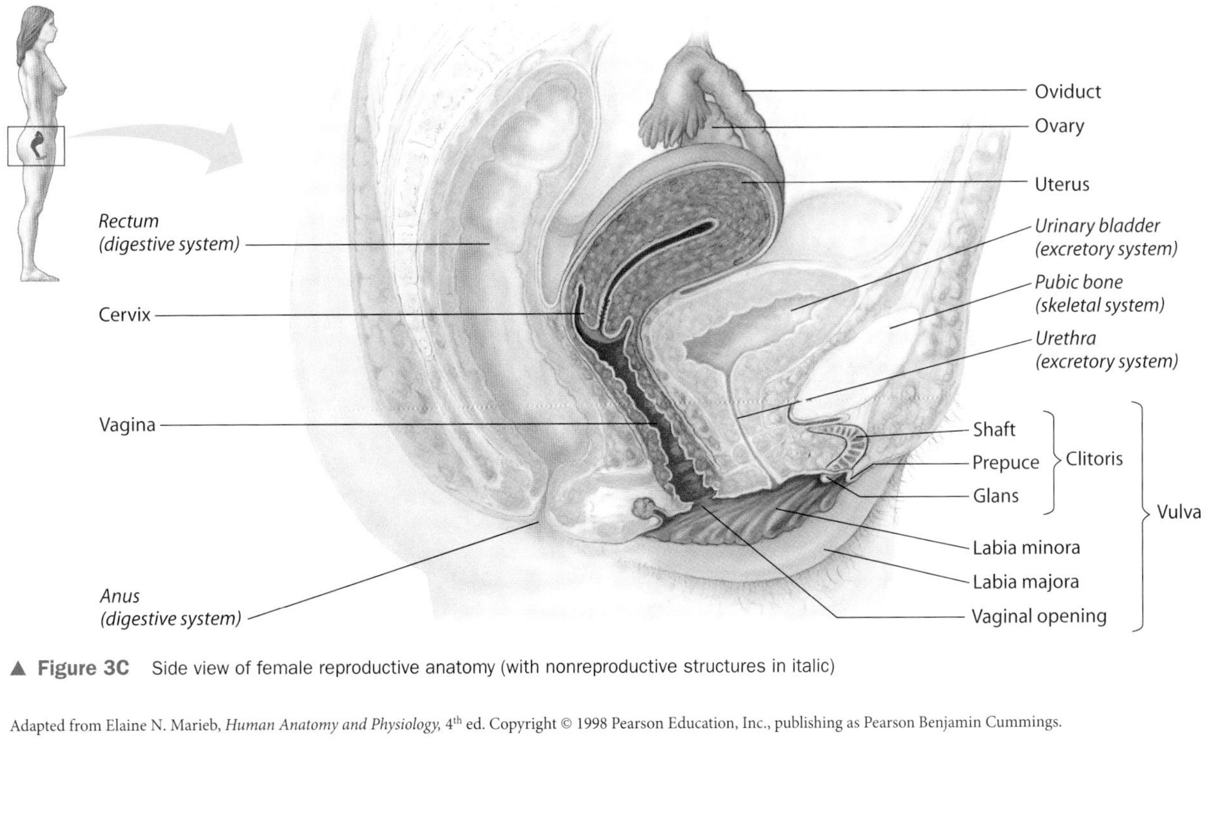

Edelman/La Vilette/Photo Researchers.

▲ **Figure 3C** Side view of female reproductive anatomy (with nonreproductive structures in italic)

4 Reproductive anatomy of the human male

Figures 4A and **4B** present front and side views of the male reproductive system. The male gonads, or **testes** (singular, *testis*), are each housed outside the abdominal cavity in a sac called the **scrotum**. A testis and scrotum together are called a **testicle**. Sperm cannot develop optimally at human core body temperature; the scrotum keeps the sperm-forming cells about 2°C cooler, which allows them to function normally. In cold conditions, muscles around the scrotum contract, pulling the testes toward the body, thereby maintaining the proper temperature.

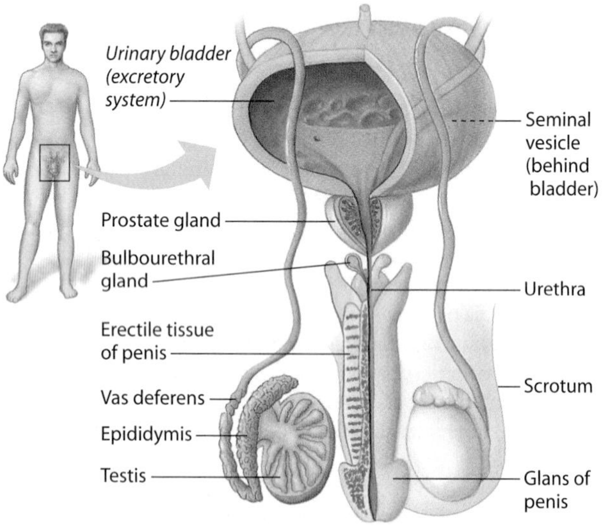

▲ **Figure 4A** Front view of male reproductive anatomy

Now let's track the path of sperm from one of the testes out of the male's body. From each testis, sperm pass into a coiled tube called the **epididymis**, which stores the sperm while they continue to develop. Sperm leave the epididymis during **ejaculation**, the expulsion of sperm-containing fluid from the penis. At that time, muscular contractions propel the sperm from the epididymis through another duct called the **vas deferens**. The vas deferens (which is the target of a vasectomy; see Module 8) passes upward into the abdomen and loops around the urinary bladder. Next to the bladder, the vas deferens joins a short duct from a gland, the seminal vesicle. The two ducts unite to form a short **ejaculatory duct**, which joins its counterpart conveying sperm from the other testis. Each ejaculatory duct empties into the urethra, which conveys both urine and sperm out through the penis, although not at the same time. Thus, unlike the female, the male has a direct connection between the reproductive and urinary systems.

In addition to the testes and ducts, the reproductive system of human males contains three sets of glands: the seminal vesicles, the prostate gland, and the bulbourethral glands. The two **seminal vesicles** secrete a thick fluid that contains fructose, which provides most of the energy used by the sperm as they propel themselves through the female reproductive tract. The **prostate gland** secretes a thin fluid that further nourishes the sperm. The two **bulbourethral glands** secrete a clear, alkaline mucus.

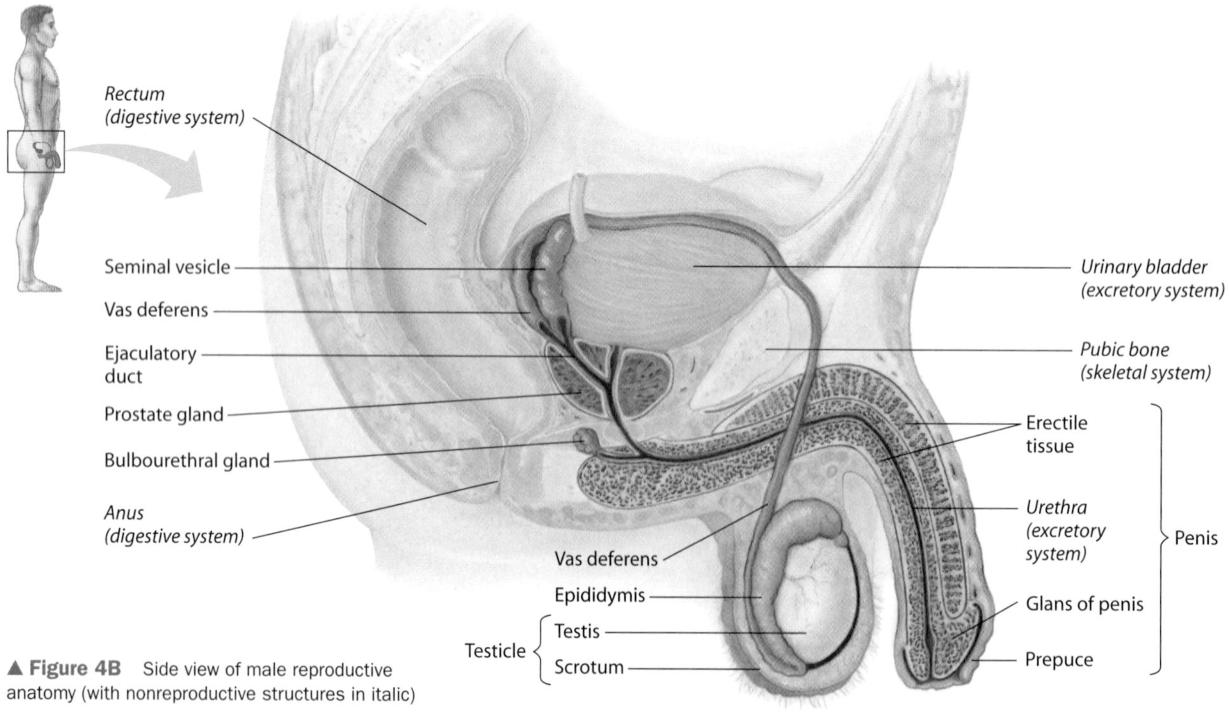

▲ **Figure 4B** Side view of male reproductive anatomy (with nonreproductive structures in italic)

Adapted from Elaine N. Marieb, *Human Anatomy and Physiology,* 4ᵗʰ ed. Copyright © 1998 Pearson Education, Inc., publishing as Pearson Benjamin Cummings.

Together, the sperm and the glandular secretions make up **semen**, the fluid ejaculated from the penis during **orgasm**, a series of rhythmic, involuntary contractions of the reproductive structures. About 2–5 mL (1 teaspoonful) of semen are discharged during a typical ejaculation. About 95% of the fluid consists of glandular secretions. The other 5% is made up of sperm (typically 200–500 million of them), only one of which may eventually fertilize an egg. The alkalinity of the semen balances the acidity of any traces of urine in the urethra and neutralizes the acidic environment of the vagina, protecting the sperm and increasing their motility.

The human **penis** consists mainly of erectile tissue (shown in blue in Figures 4A and 4B) that can fill with blood to cause an erection during sexual arousal. Erection is essential for insertion of the penis into the vagina. Like the clitoris, the penis consists of a shaft that supports the glans, or head. The glans is richly supplied with nerve endings and is highly sensitive to stimulation. A fold of skin called the prepuce, or foreskin, covers the glans. Circumcision, the surgical removal of the prepuce, arose from religious traditions. The majority of American males are circumcised, although the practice remains somewhat controversial because of a slight (less than 1%) chance of complication from the surgery and questions about its benefit. Recent studies have suggested that circumcision does have a medical advantage: The procedure significantly reduces a man's chance of contracting and passing on sexually transmitted diseases, including AIDS.

Figure 4C illustrates the process of ejaculation and summarizes the production of semen and its expulsion. Ejaculation occurs in two stages. ❶ At the peak of sexual arousal, muscle contractions in multiple glands force secretions into the urethra and propel sperm from the epididymis. At the same time, a sphincter muscle at the base of the bladder contracts, preventing urine from leaking into the urethra from the bladder. Another sphincter also contracts, closing off the entrance of the urethra into the penis. The section of the urethra between the two sphincters fills with semen and expands. ❷ In the second stage of ejaculation, the expulsion stage, the sphincter at the base of the penis relaxes, admitting semen into the penis. At the same time, a series of strong muscle contractions around the base of the penis and along the urethra expels the semen from the body.

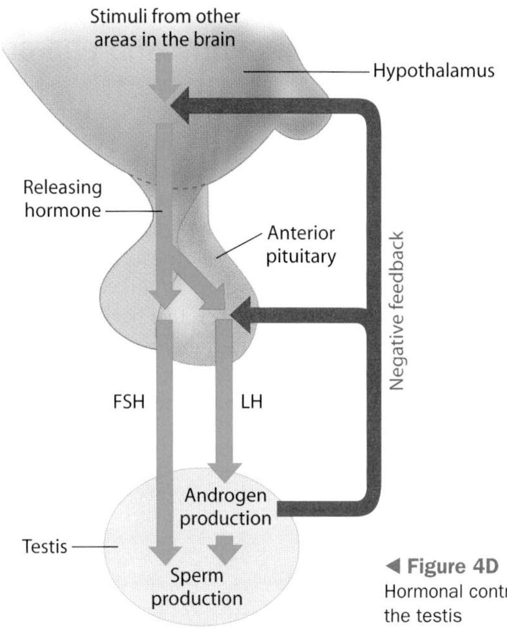

◀ **Figure 4D** Hormonal control of the testis

Figure 4D shows how hormones control sperm production by the testes. The hypothalamus secretes a releasing hormone that regulates release of follicle-stimulating hormone (FSH) and luteinizing hormone (LH) by the anterior pituitary. FSH increases sperm production by the testes, while LH promotes the secretion of androgens, mainly testosterone. Androgens stimulate sperm production. In addition, androgens carried in the blood help maintain homeostasis by a negative-feedback mechanism (red arrows), inhibiting secretion of both the releasing hormone and LH. Under the control of this chemical regulating system, the testes produce hundreds of millions of sperm every day, from puberty well into old age. Next we'll see how sperm and eggs are made.

? Arrange the following organs in the correct sequence for the travel of sperm: epididymis, testis, urethra, vas deferens.

Testis, epididymis, vas deferens, urethra

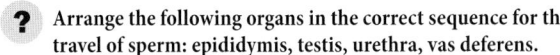

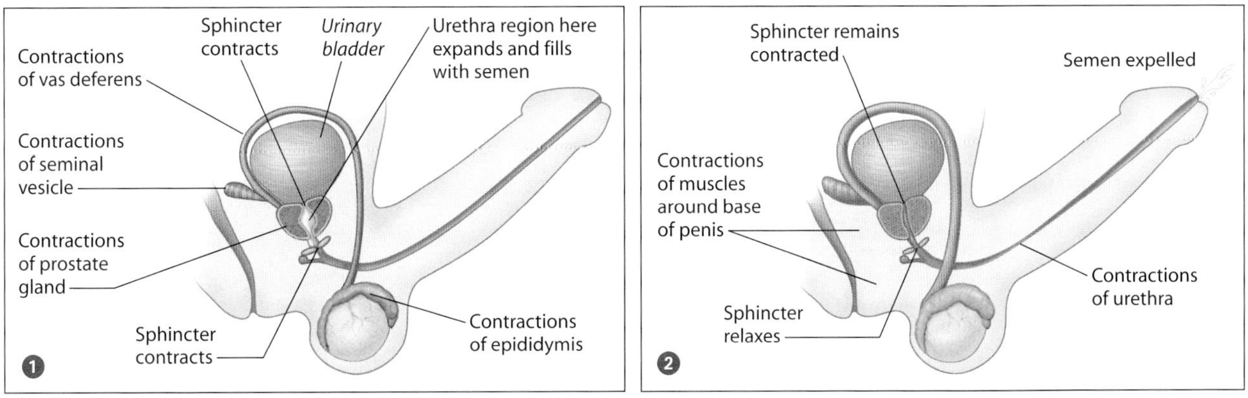

▲ **Figure 4C** The two stages of ejaculation

Adapted from Robert Crooks and Karla Baur, *Our Sexuality,* 5th ed. (Redwood City, CA: Benjamin Cummings, 1993). Copyright © 1993 The Benjamin Cummings Publishing Company.

5 The formation of sperm and egg cells requires meiosis

Both sperm and egg are haploid (n) cells that develop by meiosis from diploid ($2n$) cells in the gonads. Recall that the diploid chromosome number in humans is 46; that is, $2n = 46$. There are significant differences in **gametogenesis** between human males and females, so we'll examine the processes separately.

Figure 5A outlines **spermatogenesis**, the formation of sperm cells. Sperm develop in the testes in coiled tubes called the **seminiferous tubules**. Diploid cells that begin the process are located near the outer wall of the tubules (at the top of the enlarged wedge of tissue in Figure 5A). These cells multiply continuously by mitosis, and each day about 3 million of them differentiate into **primary spermatocytes**, the cells that undergo meiosis. Meiosis I of a primary spermatocyte produces two **secondary spermatocytes**, each with the haploid number of chromosomes ($n = 23$). Meiosis II then forms four cells, each with the haploid number of chromosomes. A sperm cell develops by differentiation of each of these haploid cells and is gradually pushed toward the center of the seminiferous tubule. From there it passes into the epididymis, where it matures, becomes motile, and is stored until ejaculation. In human males, spermatogenesis takes about 10 weeks.

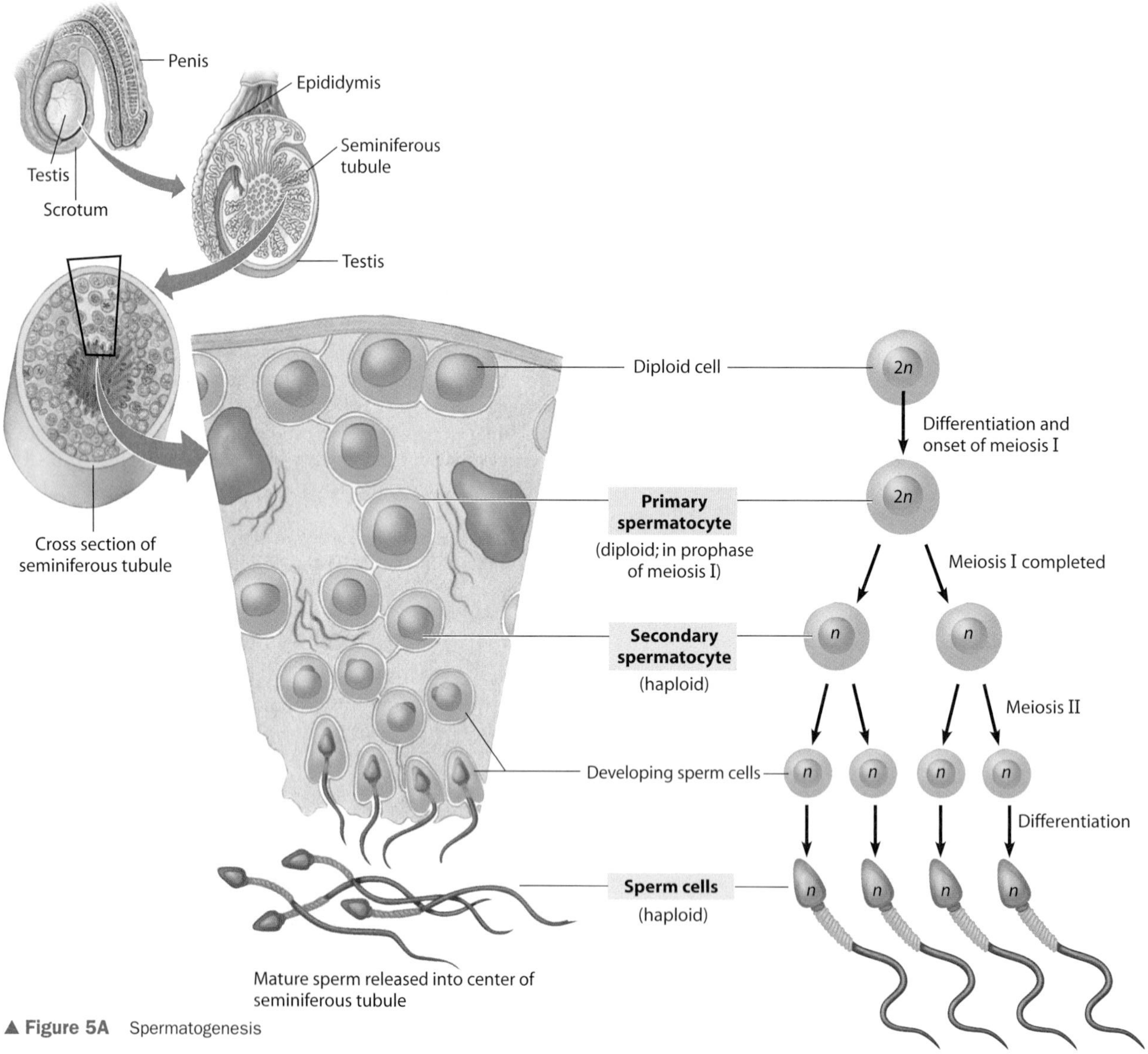

▲ **Figure 5A** Spermatogenesis

The right side of **Figure 5B** shows **oogenesis**, the development of a mature egg (also called an ovum; plural, *ova*). Most of the process occurs in the ovary. Oogenesis actually begins prior to birth, when a diploid cell in each developing follicle begins meiosis. At birth, each follicle contains a dormant **primary oocyte**, a diploid cell that is resting in prophase of meiosis I. A primary oocyte can be hormonally triggered to develop further. Between puberty and menopause, about every 28 days, FSH (follicle-stimulating hormone) from the pituitary stimulates one of the dormant follicles to develop. The follicle enlarges, and the primary oocyte completes meiosis I and begins meiosis II. Meiosis then halts again at metaphase II. In the female, the division of the cytoplasm in meiosis I is unequal, with a single **secondary oocyte** receiving almost all of it. The smaller of the two daughter cells, called the first polar body, receives almost no cytoplasm.

The secondary oocyte is released by the ovary during ovulation. It enters the oviduct, and if a sperm cell penetrates it, the secondary oocyte completes meiosis II. Meiosis II is also unequal, yielding a second polar body and the mature egg (ovum). The haploid nucleus of the mature egg can then fuse with the haploid nucleus of the sperm cell, producing a zygote.

Although not shown in Figure 5B, the first polar body may also undergo meiosis II, forming two cells. These and the second polar body receive virtually no cytoplasm and quickly degenerate, leaving the mature egg with nearly all the cytoplasm and thus the bulk of the nutrients contained in the original diploid cell.

The left side of Figure 5B is a cutaway view of an ovary. The series of follicles here represents the changes one follicle undergoes over time. An actual ovary would have thousands of dormant follicles, each containing a primary oocyte. Usually, only one follicle has a dividing oocyte at any one time. Meiosis I occurs as the follicle matures. About the time the secondary oocyte forms, the pituitary hormone LH (luteinizing hormone) triggers ovulation, the rupture of the follicle and expulsion of the secondary oocyte. The ruptured follicle then develops into a corpus luteum ("yellow body"). Unless fertilization occurs, the corpus luteum degenerates before another follicle starts to develop.

Oogenesis and spermatogenesis are alike in that they both produce haploid gametes. However, these two processes differ in some important ways. First, only one mature egg results from each diploid cell that undergoes meiosis. The other products of oogenesis, the polar bodies, degenerate. By contrast, in spermatogenesis, all four products of meiosis develop into mature gametes. Second, although the cells from which sperm develop continue to divide by mitosis throughout the male's life, this is probably not the case for the comparable cells in the human female. Third, oogenesis has long "resting" periods, whereas spermatogenesis produces mature sperm in an uninterrupted sequence.

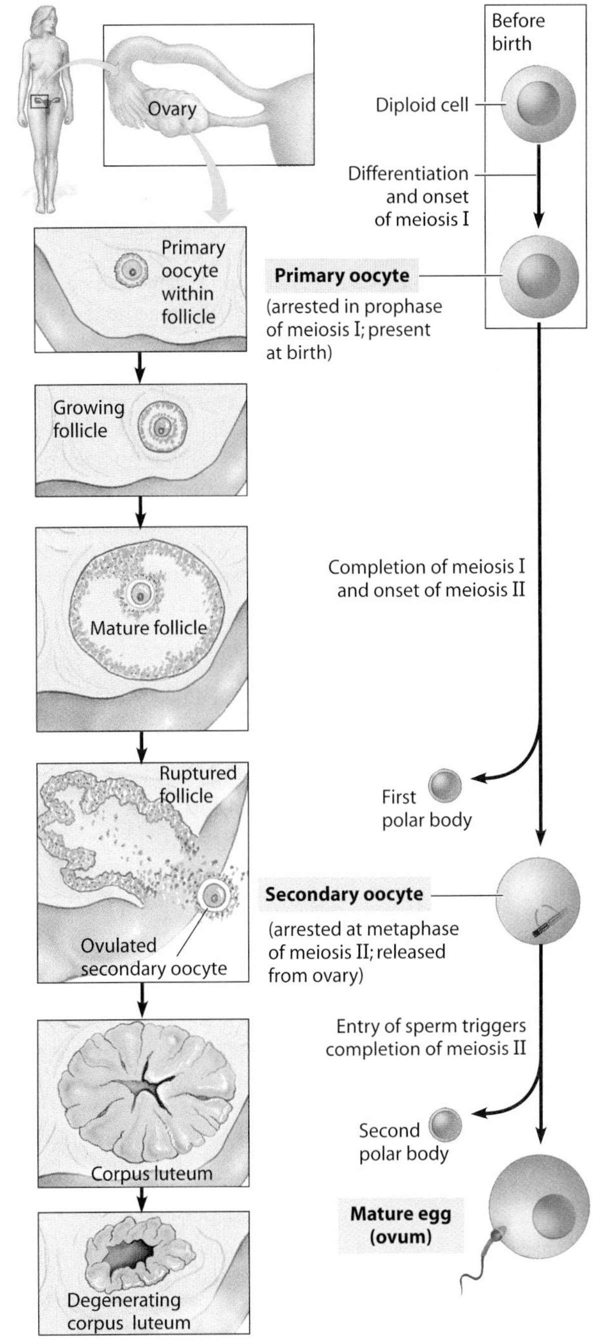

▲ **Figure 5B** Oogenesis and the development of an ovarian follicle

? Which process in the development of sperm and eggs is responsible for the genetic variation among gametes? (*Hint*: Review Module 8.15.)

The random alignment of homologous chromosomes during meiosis, specifically meiosis I

6 Hormones synchronize cyclic changes in the ovary and uterus

Oogenesis is one part of a female mammal's **reproductive cycle**, a recurring sequence of events that produces eggs, makes them available for fertilization, and prepares the body for pregnancy. The reproductive cycle is actually one integrated cycle involving cycles in two different reproductive organs: the ovaries and the uterus. In discussing oogenesis in the last module, we described the **ovarian cycle**, cyclic events that occur about every 28 days in the human ovary. Changes in the uterus define the **menstrual cycle**. Hormonal messages link the two cycles, synchronizing follicle growth in the ovaries with the establishment of a uterine lining that can support a growing embryo. The hormone story is complex and involves intricate feedback mechanisms. Table 6 lists the major hormones and their roles. Figure 6, on the facing page, shows how the events of the ovarian cycle (Part C) and menstrual cycle (Part E) are synchronized through the actions of multiple hormones (shown in Parts A, B, and D). Notice the time scale at the bottom of Part E; it also applies to Parts B–D.

An Overview of the Ovarian and Menstrual Cycles Let's begin with the structural events of the ovarian and menstrual cycles. For simplicity, we have divided the ovarian cycle (Part C of the figure) into two phases separated by ovulation: the pre-ovulatory phase, when a follicle is growing and a secondary oocyte is developing, and the post-ovulatory phase, after the follicle has become a corpus luteum.

Events in the menstrual cycle (Part E) are synchronized with the ovarian cycle. By convention, the first day of a woman's "period" is designated day 1 of the menstrual cycle. Uterine bleeding, called **menstruation**, usually lasts for 3–5 days. Notice that this corresponds to the beginning of the pre-ovulatory phase of the ovarian cycle. During menstruation, the endometrium (the blood-rich inner lining of the uterus) breaks down and leaves the body through the vagina. The menstrual discharge consists of blood, small clusters of endometrial cells, and mucus. After menstruation, the endometrium regrows. It continues to thicken through the time of ovulation, reaching a maximum thickness at about 20–25 days. If an embryo has not implanted in the uterine lining, menstruation begins again, marking the start of the next ovarian and menstrual cycles.

Now let's consider the hormones that regulate the ovarian and menstrual cycles. The ebb and flow of the hormones listed in Table 6 synchronize events in the ovarian cycle (the growth of the follicle and ovulation) with events in the menstrual cycle (preparation of the uterine lining for possible implantation of an embryo). A releasing hormone from the hypothalamus in the brain regulates secretion of the two pituitary hormones FSH and LH. Changes in the blood levels of FSH, LH, and two other hormones—estrogen and progesterone—coincide with specific events in the ovarian and menstrual cycles.

Hormonal Events Before Ovulation Focusing on Part A of Figure 6, we see that the releasing hormone from the hypothalamus stimulates the anterior pituitary ❶ to increase its output of FSH and LH. True to its name, ❷ FSH stimulates the growth of an ovarian follicle, in effect starting the ovarian cycle. In turn, the follicle secretes estrogen. Early in the pre-ovulatory phase, the follicle is small (Part C) and secretes relatively little estrogen (Part D). As the follicle grows, ❸ it secretes more and more estrogen, and the rising but still relatively low level of estrogen exerts negative feedback on the pituitary. This keeps the blood levels of FSH and LH low for most of the pre-ovulatory phase (Part B). As the time of ovulation approaches, hormone levels change drastically, with estrogen reaching a critical peak (Part D) just before ovulation. This high level of estrogen exerts positive feedback on the hypothalamus (green arrow in Part A), which then ❹ makes the pituitary secrete surges of FSH and LH. By comparing Parts B and D of the figure, you can see that the peaks in FSH and LH occur just after the estrogen peak. It may help to place a piece of paper over the figure and slide it slowly to the right. As you uncover the figure, you will see the follicle getting bigger and the estrogen level rising to its peak, followed almost immediately by the LH and FSH surges. ❺ Then, just to the right of the peaks, comes the dashed line representing ovulation.

Hormonal Events at Ovulation and After LH stimulates the completion of meiosis I, transforming the primary oocyte in the follicle into a secondary oocyte. It also signals

TABLE 6 | HORMONES OF THE OVARIAN AND MENSTRUAL CYCLES

Hormone	Secreted by	Major Roles
Releasing hormone	Hypothalamus	Regulates secretion of LH and FSH by pituitary
Follicle-stimulating hormone (FSH)	Pituitary	Stimulates growth of ovarian follicle
Leuteinizing hormone (LH)	Pituitary	Stimulates growth of ovarian follicle and production of secondary oocyte; promotes ovulation; promotes development of corpus luteum and secretion of other hormones
Estrogen	Ovarian follicle	Low levels inhibit pituitary; high levels stimulate hypothalamus; promotes growth of endometrium
Estrogen and progesterone	Corpus luteum	Maintain endometrium; high levels inhibit hypothalamus and pituitary; sharp drops promote menstruation

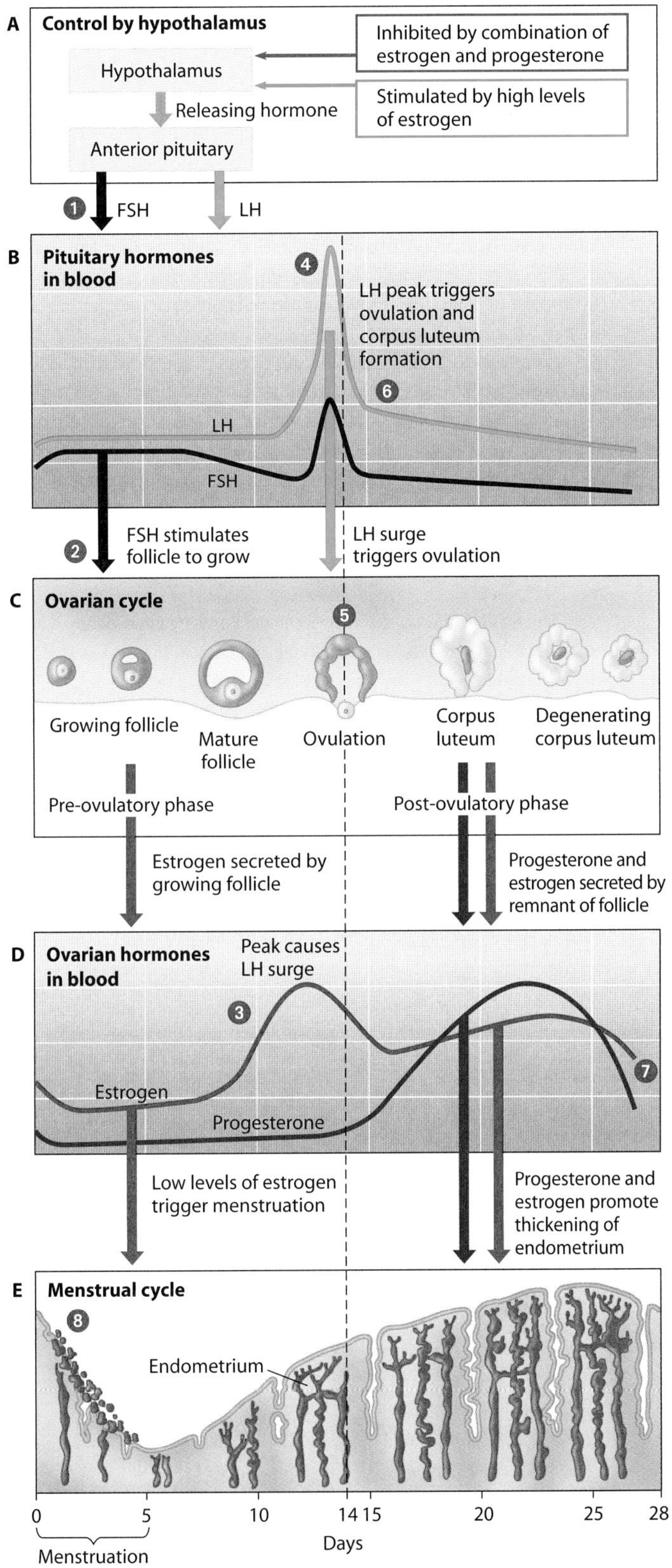

▲ **Figure 6** The reproductive cycle of the human female

enzymes to rupture the follicle, allowing ovulation to occur, and triggers the development of the corpus luteum from the ruptured follicle (hence its name, luteinizing hormone). LH also promotes the secretion of progesterone and estrogen by the corpus luteum. In Part D of the figure, you can see the progesterone peak and the second (lower and wider) estrogen peak after ovulation.

High levels of estrogen and progesterone in the blood following ovulation have a strong influence on both the ovary and uterus. The combination of the two hormones exerts negative feedback on the hypothalamus and pituitary, producing ❻ falling FSH and LH levels. The drop in FSH and LH prevents follicles from developing and ovulation from occurring during the post-ovulatory phase. Also, the LH drop is followed by the gradual degeneration of the corpus luteum. Near the end of the post-ovulatory phase, unless an embryo has implanted in the uterus, the corpus luteum stops secreting estrogen and progesterone. ❼ As blood levels of these hormones decline, the hypothalamus once again can stimulate the pituitary to secrete more FSH and LH, and a new cycle begins.

Control of the Menstrual Cycle Hormonal regulation of the menstrual cycle is simpler than that of the ovarian cycle. The menstrual cycle (Part E) is directly controlled by estrogen and progesterone alone. You can see the effects of these hormones by comparing Parts D and E of the figure. Starting around day 5 of the cycle, the endometrium thickens in response to the rising levels of estrogen and, later, progesterone. When the levels of these hormones drop, the endometrium begins to slough off. ❽ Menstrual bleeding begins soon after, on day 1 of a new cycle.

We have now described what happens in the human ovary and uterus in the absence of fertilization. As we'll see later, the ovarian and menstrual cycles are put on hold if fertilization and pregnancy occur. Early in pregnancy, the developing embryo, implanted in the endometrium, releases a hormone (human chorionic gonadotropin, or hCG). This hormone acts like LH in that it maintains the corpus luteum, which continues to secrete progesterone and estrogen, keeping the endometrium intact. (hCG also boosts testosterone production in males, so its use is banned by many sports organizations. In 2009, slugger Manny Ramirez was suspended from baseball after testing positive for hCG.) We'll return to the events of pregnancy in Modules 15 and 16.

? Which hormonal change triggers the onset of menstruation?

The drop in the levels of estrogen and progesterone. These changes are caused by negative feedback of these hormones on the hypothalamus and pituitary after ovulation.

CONNECTION

7 Sexual activity can transmit disease

Sexually transmitted diseases (STDs) are contagious diseases spread by sexual contact. Table 7 lists the most common STDs in the United States, organized by the type of infectious agent. Notice that bacteria, viruses, protists, and fungi can all cause STDs.

Bacterial STDs are generally curable, but treatment must be given early, before any permanent damage is done. The most common bacterial STD (with nearly a million new cases reported annually in the United States and over 90 million worldwide) is **chlamydia**. Chlamydia poses a public health challenge because it is frequently "silent," producing no visible symptoms. Health officials estimate that for every reported case of chlamydia, two other cases go unreported. The primary symptom of chlamydia is a burning sensation and genital discharge during urination, but half of infected men and three-quarters of infected women do not notice any symptoms. Long-term complications are rare among men, but up to 40% of infected women develop pelvic inflammatory disease (PID). The inflammation associated with PID may block the oviducts or scar the uterus, causing infertility. Luckily, treatment for chlamydia is easy: A single dose of an antibiotic usually cures the disease completely. But early screening is required to catch the disease before any scarring occurs. Sexually active women are encouraged to be screened for chlamydia and other STDs annually.

In contrast to bacterial STDs, viral STDs are not curable. They can be controlled by medications, but symptoms and the ability to infect others remain a possibility through a person's lifetime. One in five Americans is infected with **genital herpes**, caused by the herpes simplex virus type 2 (HSV-2), a variant of the virus that causes oral cold sores. Symptoms first appear about a week after exposure. Blisters form on the external genitalia. After a few days, the blisters change to scabs that fall off. Most outbreaks heal within a few weeks without leaving a scar. But the virus is not gone: It lies dormant within nearby nerve cells. Months or years later, the virus can reemerge, causing fresh sores that allow the virus to be spread to sexual partners. Abstinence during outbreaks, the use of condoms, and the use of antiviral medications that minimize symptoms can reduce the spread of infection. However, there is no cure for genital herpes, so infection lasts a lifetime.

AIDS, caused by HIV, poses one of the greatest health challenges in the world today, particularly among the developing nations of Africa and Asia. Yet even within the United States, there are 56,000 new infections each year, one-third of which result from heterosexual contact. Another sexually transmitted virus is the human papillomavirus (HPV). In 2006, a vaccine against HPV was approved that protects against genital warts and helps prevent infection by HPV strains that cause 70% of cervical cancers.

Many STDs can cause long-term problems or even death if left untreated. Anyone who is sexually active should have regular medical exams, be tested for STDs, and seek immediate help if any suspicious symptoms appear, even if they are mild. STDs are most prevalent among teenagers and young adults; nearly two-thirds of infections occur among people under 25. The best way to avoid the spread of STDs is, of course, abstinence. Alternatively, latex condoms provide the best protection for "safe sex."

? How are bacterial STDs different from viral STDs in terms of their long-term prognosis?

Bacterial STDs can be cured; viral STDs can be controlled but not cured.

TABLE 7 STDS COMMON IN THE UNITED STATES

Disease	Microbial Agent	Major Symptoms and Effects	Treatment
Bacterial			
Chlamydia	*Chlamydia trachomatis*	Genital discharge, itching and/or painful urination; often no symptoms in women; pelvic inflammatory disease (PID)	Antibiotics
Gonorrhea	*Neisseria gonorrhoeae*	Genital discharge; painful urination; sometimes no symptoms in women; PID	Antibiotics
Syphilis	*Treponema pallidum*	Ulcer (chancre) on genitalia in early stages; spreads throughout body and can be fatal if not treated	Antibiotics can cure in early stages
Viral			
Genital herpes	Herpes simplex virus type 2, occasionally type 1	Recurring symptoms: small blisters on genitalia, painful urination, skin inflammation; linked to cervical cancer, miscarriage, birth defects	Valacyclovir can prevent recurrences
Genital warts	Papillomaviruses	Painless growths on genitalia; some of the viruses linked to cancer	Removal by freezing
AIDS and HIV infection	HIV	Severe impairment of immune responses	Combination of drugs
Protozoan			
Trichomoniasis	*Trichomonas vaginalis*	Vaginal irritation, itching, and discharge; usually no symptoms in men	Antiprotozoal drugs
Fungal			
Candidiasis (yeast infections)	*Candida albicans*	Similar to symptoms of trichomoniasis; frequently acquired nonsexually	Antifungal drugs

C. Edelman/La Vilette/Photo Researchers.

CONNECTION | 8 Contraception can prevent unwanted pregnancy

Contraception is the deliberate prevention of pregnancy. Complete abstinence (avoiding intercourse) is the only totally effective method of birth control, but other methods are effective to varying degrees. Sterilization, surgery that prevents sperm from reaching an egg, is very reliable. A woman may have a **tubal ligation** ("having her tubes tied"), in which a doctor removes a short section from each oviduct, often tying (ligating) the remaining ends. A man may undergo a **vasectomy**, in which a doctor cuts a section out of each vas deferens to prevent sperm from reaching the urethra. Both forms of sterilization are relatively safe and free from side effects. Sterilization procedures are generally considered permanent, but can sometimes be surgically reversed. Surgical reversals of tubal ligations or vasectomies are becoming increasingly successful, but these major surgeries carry some risk.

The effectiveness of other methods of contraception depends on how they are used. Temporary abstinence, also called the **rhythm method** or **natural family planning**, depends on refraining from intercourse during the days around ovulation, when fertilization is most likely. In theory, the time of ovulation can be determined by monitoring changes in body temperature and the composition of cervical mucus, but careful monitoring and record keeping are required. Additionally, the length of the reproductive cycle can vary from month to month, and sperm can survive for 3–5 days within the female reproductive tract, making natural family planning quite unreliable in actual practice. Withdrawal of the penis from the vagina before ejaculation is also ineffective because sperm may be released before climax.

If used correctly, barrier methods can be quite effective at physically preventing the union of sperm and egg. Condoms are sheaths, usually made of latex, that fit over the penis. A diaphragm is a dome-shaped rubber cap that covers the cervix; it requires a doctor's visit for proper fitting. Barrier devices (including condoms) are more effective when used in combination with **spermicides**, sperm-killing cream, foam, or jelly; spermicides used alone are unreliable.

TABLE 8 | CONTRACEPTIVE METHODS

Method	Pregnancies per 100 Women per Year*: Used Correctly	Typically
Birth control pill (combination)	0.1	5
Vasectomy	0.1	0.15
Tubal ligation	0.2	0.5
Rhythm method	1–9	20
Withdrawal	4	19
Condom (male)	3	14
Diaphragm and spermicide	6	20
Spermicide alone	6	26

***Without contraception, about 85 pregnancies would occur.**

Data from R. Hatcher et al., *Contraceptive Technology: 1990-1992*, p. 134 (New York: Irvington, 1990).

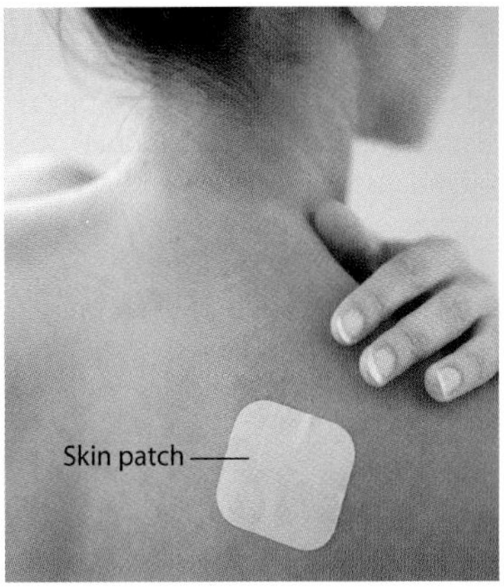

◀ **Figure 8** A contraceptive skin patch

Some of the most effective methods of contraception work by preventing the release of egg cells. **Oral contraceptives**, or **birth control pills**, come in several different forms that contain synthetic estrogen and/or progesterone (or a synthetic progesterone-like hormone called progestin). In addition to pills, various combinations of these hormones are also available as an injection (Depo-Provera), a ring inserted into the vagina, or a skin patch (**Figure 8**). Steady intake of these hormones simulates their constant levels during pregnancy. In response, the hypothalamus fails to send the signals that start development of an ovarian follicle. Ovulation ceases, preventing pregnancy.

Certain drugs can prevent fertilization or implantation even after intercourse has occurred. Combination birth control pills can be prescribed in high doses for emergency contraception, also called **morning after pills (MAPs)**. If taken within 3 days after unprotected intercourse, MAPs are about 75% effective at preventing pregnancy. Such treatments should only be used in emergencies because they have significant side effects.

If pregnancy has already occurred, the drug mifepristone, or RU486, can induce an abortion, the termination of a pregnancy in progress. If taken within the first 7 weeks, RU486 blocks progesterone receptors in the uterus, thus preventing progesterone from maintaining pregnancy. Mifepristone requires a doctor's prescription and several visits to a medical facility and may cause significant side effects.

Table 8 lists common methods of contraception, along with their failure rates when used correctly and when used typically. Note that these two rates are often quite different, emphasizing the importance of learning to use contraception correctly. It is also important to note that condoms are the only means of "safe sex" that can prevent (but not eliminate the risk of) both unwanted pregnancy and sexually transmitted diseases (STDs); other contraceptive methods do not prevent STDs.

? What is the fundamental difference between barrier methods (such as condoms) and oral contraceptives in terms of their means of preventing pregnancies?

Barrier methods prevent sperm from reaching an egg, while birth control pills prevent the release of eggs altogether.

Principles of Embryonic Development

9 Fertilization results in a zygote and triggers embryonic development

The last six modules focused on the anatomy and physiology of the human reproductive system. In the next six modules, we examine the results of reproduction: the formation and development of an embryo. Embryonic development begins with fertilization, the union of a sperm and an egg to form a diploid zygote. Fertilization combines haploid sets of chromosomes from two individuals and also activates the egg by triggering metabolic changes that start embryonic development.

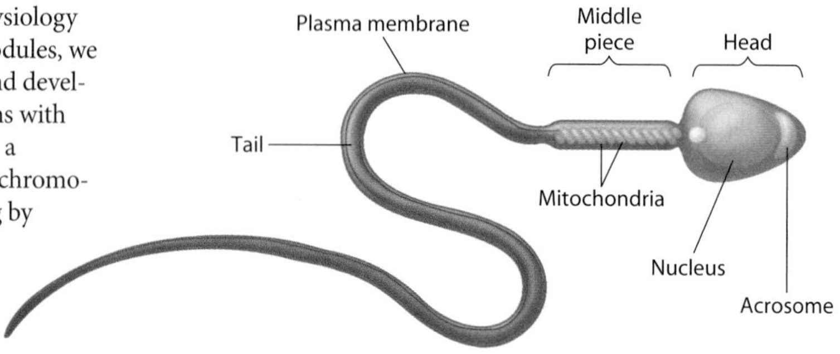

▲ **Figure 9B** The structure of a human sperm cell

The Properties of Sperm Cells **Figure 9A** is a micrograph of an unfertilized human egg that is surrounded by sperm. Among all of these sperm, only one will enter and fertilize the egg. All the other sperm—the ones shown here and millions more that were ejaculated with them—will die. The one sperm that penetrates the egg adds its unique set of genes to those of the egg and contributes to the next generation.

Figure 9B illustrates the structure of a mature human sperm cell, a clear case of form fitting function. The sperm's streamlined shape is an adaptation for swimming through fluids in the vagina, uterus, and oviduct of the female. Its thick head contains a haploid nucleus and is tipped with a vesicle, the **acrosome**, which lies just inside the plasma membrane. The acrosome contains enzymes that help the sperm penetrate the egg. The middle piece of the sperm contains mitochondria. The sperm absorbs high-energy nutrients, especially the sugar fructose, from the semen. Thus fueled, its mitochondria provide ATP for movement of the tail, which is actually a flagellum. By the time a sperm has reached the egg, it has consumed much of the energy available to it. But a successful sperm will have enough energy left to penetrate the egg and deposit its nucleus in the egg's cytoplasm.

The Process of Fertilization **Figure 9C**, on the facing page, illustrates the sequence of events in fertilization. This diagram is based on fertilization in sea urchins (phylum Echinodermata, on which a great deal of research has been done. Similar processes occur in other animals, including humans. The diagram traces one sperm through the successive activities of fertilization. Notice that to reach the egg nucleus, the sperm nucleus must pass through three barriers: the egg's jelly coat (yellow), a middle region of glycoproteins called the vitelline layer (pink), and the egg cell's plasma membrane.

Let's follow the steps shown in the figure. ❶ The contact of a sperm with the jelly coat of the egg triggers the release from

▶ **Figure 9A** A human egg cell surrounded by sperm

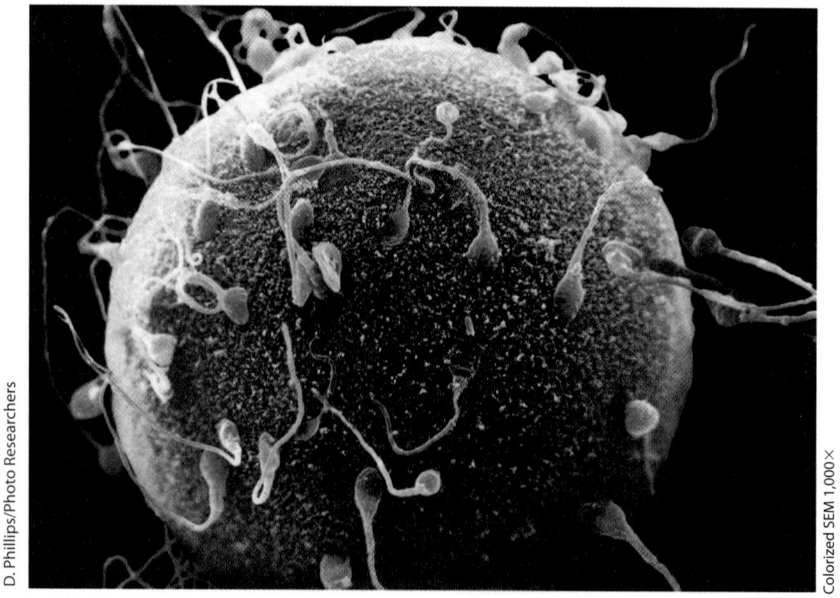

the sperm's acrosome of a cloud of enzyme molecules by exocytosis. (2) The enzyme molecules digest a cavity into the jelly. When the sperm head reaches the vitelline layer, (3) species-specific protein molecules on its surface bind with specific receptor proteins on the vitelline layer. The binding between these proteins ensures that sperm of other species cannot fertilize the egg. This specificity is especially important when fertilization is external because the sperm of other species may be present in the water. After the specific binding occurs, the sperm proceeds through the vitelline layer, and (4) the sperm's plasma membrane fuses with that of the egg. Fusion of the two membranes (5) makes it possible for the sperm nucleus to enter the egg.

Fusion of the sperm and egg plasma membranes triggers a number of important changes in the egg. Two such changes prevent other sperm from entering the egg. About 1 second after the membranes fuse, the entire egg plasma membrane becomes impenetrable to other sperm cells. Shortly thereafter, (6) the vitelline layer hardens and separates from the plasma membrane. The space quickly fills with water, and the vitelline layer becomes impenetrable to sperm. If these events did not occur and an egg were fertilized by more than one sperm, the resulting zygote nucleus would contain too many chromosomes, and the zygote could not develop normally.

(7) About 20 minutes after the sperm nucleus enters the egg, the sperm and egg nuclei fuse. Gearing up for the enormous growth and development that will soon follow, DNA synthesis and cellular respiration begin. The first cell division occurs after about 90 minutes, marking the end of the fertilization stage.

Note that the sperm provided chromosomes to the zygote, but little else. The zygote's cytoplasm and various organelles were all provided by the mother through the egg. In the next module, we begin to trace the development of the zygote into a new animal.

? Why is the vitelline layer particularly important among aquatic animals that use external fertilization?

Protein receptors on the vitelline layer match with species-specific proteins on the sperm; this ensures that sperm of a different species will not fertilize the egg.

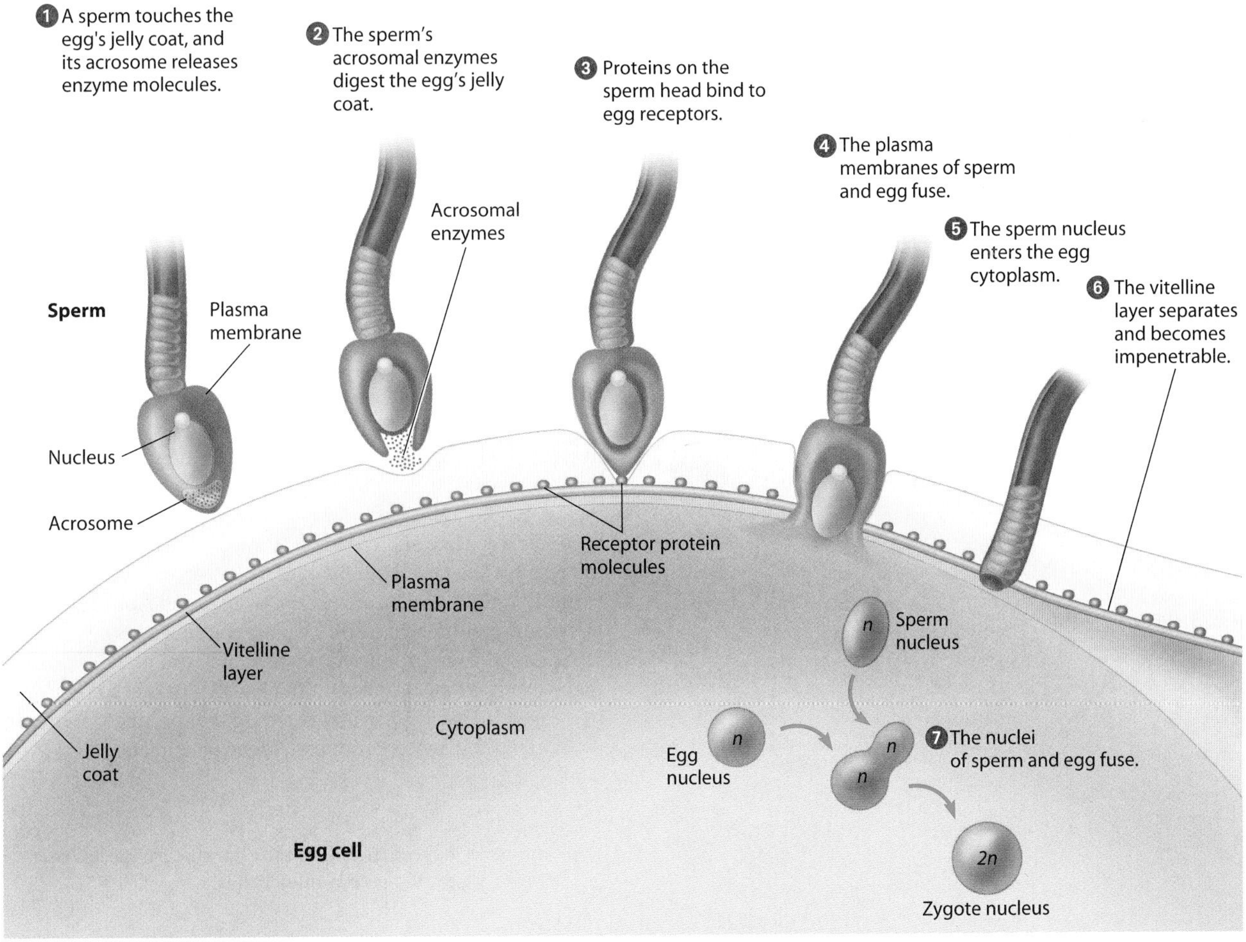

(1) A sperm touches the egg's jelly coat, and its acrosome releases enzyme molecules.
(2) The sperm's acrosomal enzymes digest the egg's jelly coat.
(3) Proteins on the sperm head bind to egg receptors.
(4) The plasma membranes of sperm and egg fuse.
(5) The sperm nucleus enters the egg cytoplasm.
(6) The vitelline layer separates and becomes impenetrable.
Sperm
Plasma membrane
Acrosomal enzymes
Nucleus
Acrosome
Receptor protein molecules
Plasma membrane
Vitelline layer
Jelly coat
Cytoplasm
Sperm nucleus
n
Egg nucleus
n
n
n
(7) The nuclei of sperm and egg fuse.
Egg cell
2n
Zygote nucleus

▲ **Figure 9C** The process of fertilization in a sea urchin

10 Cleavage produces a ball of cells from the zygote

An animal consists of many thousands, millions, even trillions of cells that are precisely organized into complex tissues and organs. The transformation from a zygote to this multicellular state is truly phenomenal. Order and precision are required at every step, and both are clearly displayed in the first two major phases of embryonic development: cleavage and gastrulation.

Cleavage is a rapid succession of cell divisions that produces a ball of cells—a multicellular embryo—from the zygote. Nutrients stored in the egg nourish the dividing cells. DNA replication, mitosis, and cytokinesis occur rapidly, but gene transcription virtually shuts down and few new proteins are synthesized. The embryo does not enlarge significantly; instead, cleavage partitions the cytoplasm of the one-celled zygote into many smaller cells, each with its own nucleus.

Figure 10 illustrates cleavage in a sea urchin. (A similar process occurs in humans.) As the first three steps show, the number of cells doubles with each cleavage division. In a sea urchin, a doubling occurs about every 20 minutes, and the whole cleavage process takes about 3 hours to produce a solid ball of cells. Notice that as cleavage proceeds, the total size of the ball remains constant even as the cells double. As a result, each cell in the ball is much smaller than the original cell that formed the zygote. As cleavage continues, a fluid-filled cavity called the **blastocoel** forms in the center of the embryo. At the completion of cleavage, there is a hollow ball of cells called the **blastula**.

Cells removed from a human blastocyst (the equivalent of the sea urchin blastula) are useful in research. Such cells are called embryonic stem cells. Because they have yet to become specialized, embryonic stem cells have great therapeutic potential to replace just about any kind of mature cells that have been lost to damage or illness. But harvesting the embryonic stem cells destroys the embryo, which raises ethical questions. Research using embryonic stem cells remains one of the hottest areas of biological study.

Cleavage makes two important contributions to early development. It creates a multicellular embryo, the blastula, from a single-celled zygote. Cleavage is also an organizing process, partitioning the multicellular embryo into developmental regions. The cytoplasm of the zygote contains a variety of chemicals that control gene expression during early development. During cleavage, regulatory chemicals become localized in particular groups of cells, where they later activate the genes that direct the formation of specific parts of the animal. Gastrulation, the next phase of development, further refines the embryo's cellular organization.

Rarely, and apparently at random, a cell in the early embryo may separate and "reset" as if it were the original zygote; the result is the development of identical (monozygotic) twins. (Nonidentical, or dizygotic, twins result from a completely different mechanism: Two separate eggs fuse with two separate sperm to produce two genetically unique zygotes that develop in the uterus simultaneously.) In exceedingly rare cases, the separation and resetting that produce identical twins can occur twice, producing identical triplets.

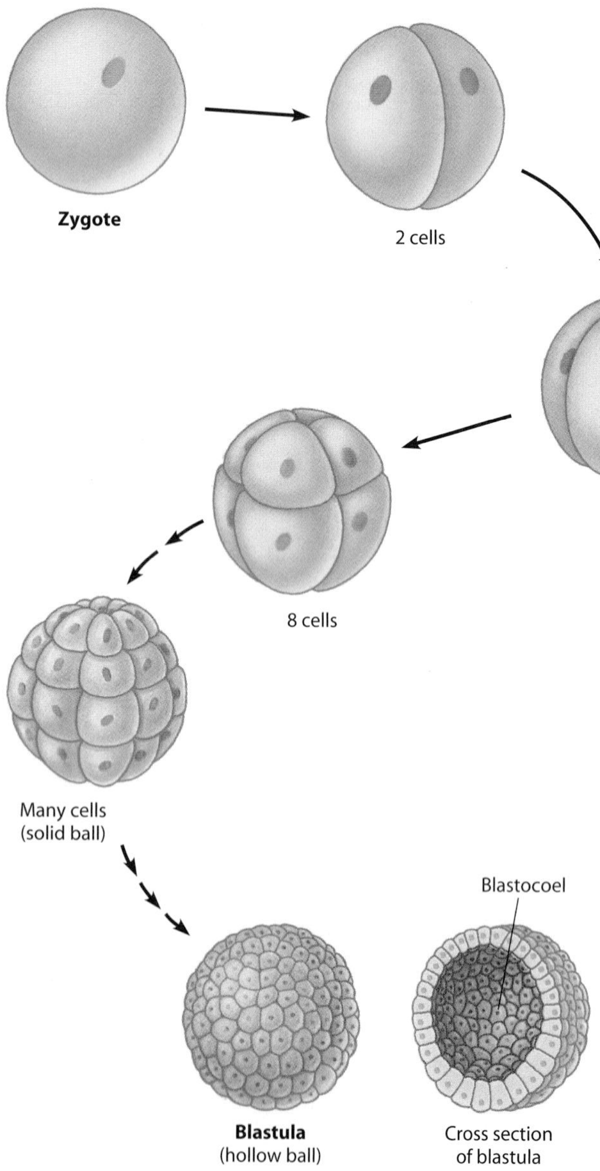

▲ **Figure 10** Cleavage in a sea urchin

? How does the reduction of cell size during cleavage increase oxygen supply to the cells' mitochondria?

Smaller cells have a greater plasma membrane surface area relative to cellular volume, and this facilitates diffusion of oxygen from the environment to the cells' cytoplasm.

11 Gastrulation produces a three-layered embryo

After cleavage, the rate of cell division slows dramatically. Groups of cells then undergo **gastrulation**, the second major phase of embryonic development. During gastrulation, cells take up new locations that will allow later formation of all the organs and tissues. As gastrulation proceeds, the embryo is organized into a three-layer stage called a **gastrula**.

The three layers produced by gastrulation are embryonic tissues called **ectoderm**, **endoderm**, and **mesoderm**. The ectoderm forms the outer layer (skin) of the gastrula. The endoderm forms an embryonic digestive tract. And the mesoderm lies between the ectoderm and endoderm. Eventually, these three cell layers develop into all the parts of the adult animal. For instance, our nervous system and the outer layer (epidermis) of our skin come from ectoderm; the innermost lining of our digestive tract arises from endoderm; and most other organs and tissues, such as the kidney, heart, muscles, and the inner layer of our skin (dermis), develop from mesoderm. **Table 11** lists the major organs and tissues that arise in most vertebrates from the three main embryonic tissue layers.

The mechanics of gastrulation vary somewhat, depending on the species. We have chosen the frog, a vertebrate that has long been a favorite of researchers, to demonstrate how gastrulation produces three cell layers. The top of **Figure 11** shows the frog blastula, formed by cleavage (as discussed in the previous module). The frog blastula is a partially hollow ball of unequally sized cells. The cells toward one end, called the animal pole, are smaller than those near the opposite end, the vegetal pole. The three colors on the blastula in the figure indicate regions of cells that will give rise to the primary cell layers in the gastrula at the bottom of the figure: ectoderm (blue), endoderm (yellow), and mesoderm (red). (Notice that each layer may be more than one cell thick.)

During gastrulation (shown in the center of Figure 11), cells migrate to new positions that will form the three layers. Gastrulation begins when a small groove, called the blastopore, appears on one side of the blastula. In the blastopore, cells of the future endoderm (yellow) move inward from the surface and fold over to produce a simple digestive cavity. Meanwhile, the cells that will form ectoderm (blue) spread downward over more of the surface of the embryo, and the cells that will form mesoderm (red) begin to spread into a thin layer inside the embryo, forming a middle layer between the other two.

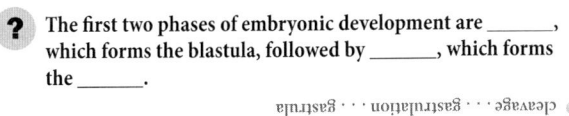

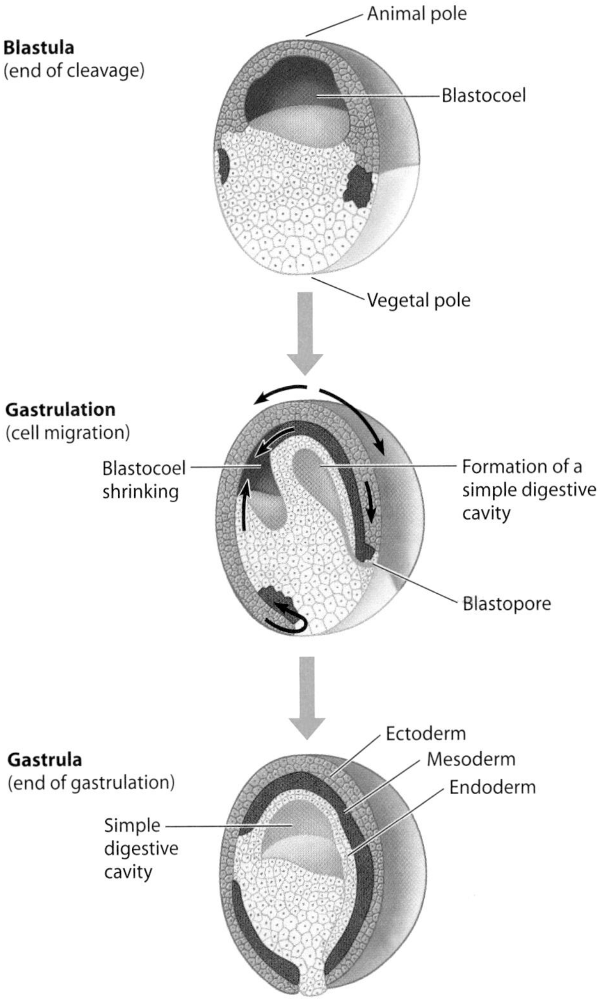

▲ **Figure 11** Development of the frog gastrula

As shown at the bottom of the figure, gastrulation is completed when cell migration has resulted in a three-layered embryo. Ectoderm covers most of the surface. Mesoderm forms a layer between the ectoderm and the endoderm.

Although gastrulation differs in detail from one animal group to another, the process is driven by the same general mechanisms in all species. The timing of these events also varies with the species. In many frogs, for example, cleavage and gastrulation together take about 15–20 hours.

? **The first two phases of embryonic development are ______, which forms the blastula, followed by ______, which forms the ______.**

cleavage · · · gastrulation · · · gastrula

TABLE 11 | DERIVATIVES OF THE THREE EMBRYONIC TISSUE LAYERS

Embryonic Layer	Organs and Tissues in the Adult
Ectoderm	Epidermis of skin; epithelial lining of mouth and rectum; sense receptors in epidermis; cornea and lens of eye; nervous system
Endoderm	Epithelial lining of digestive tract (except mouth and rectum); epithelial lining of respiratory system; liver; pancreas; thyroid; parathyroids; thymus; lining of urethra, urinary bladder, and reproductive system
Mesoderm	Skeletal system; muscular system; circulatory system; excretory system; reproductive system (except gamete-forming cells); dermis of skin; lining of body cavity

12 Organs start to form after gastrulation

In organizing the embryo into three layers, gastrulation sets the stage for the shaping of an animal. Once the ectoderm, endoderm, and mesoderm form, cells in each layer begin to differentiate into tissues and embryonic organs. The cutaway drawing in **Figure 12A** shows the developmental structures that appear in a frog embryo a few hours after the completion of gastrulation. The orientation drawing at the upper left of the figure indicates a corresponding cut through an adult frog.

We see two structures in the embryo in Figure 12A that were not present at the gastrula stage described in the last module. An organ called the notochord has developed in the mesoderm, and a structure that will become the hollow nerve cord is beginning to form in the ectoderm (in the region that is colored green). Recall that the notochord and the dorsal, hollow nerve cord are two of the hallmarks of the chordates.

Made of a substance similar to cartilage, the **notochord** extends for most of the embryo's length and provides support for other developing tissues. Later in development, the notochord will function as a core around which mesodermal cells gather and form the backbone.

The area shown in green in the cutaway drawing of Figure 12A is a thickened region of ectoderm called the neural plate. From it arises a pair of pronounced ectodermal ridges, called neural folds, visible in both the drawing and the micrograph below it. If you now look at the series of diagrams in **Figure 12B**, you will see what happens as the neural folds and neural plate develop further. The neural plate rolls up and forms a tube, which sinks beneath the surface of the embryo and is covered by an outer layer of ectoderm. If you look carefully at the figure, you'll see that cells of the ectoderm fold inward by changing shape, first elongating and then becoming wedge-shaped. The result is a tube of ectoderm—the **neural tube**—which is destined to become the brain and spinal cord.

Figure 12C, on the facing page, shows a later frog embryo (about 12 hours older than the one in Figure 12A), in which the neural tube has formed. Notice in the drawing that the neural tube lies directly above the notochord. The relative positions of the neural tube, notochord, and digestive cavity give us a preview of the basic body plan of a frog. The spinal cord will lie within extensions of the dorsal (upper) surface of the backbone (which will replace the notochord), and the

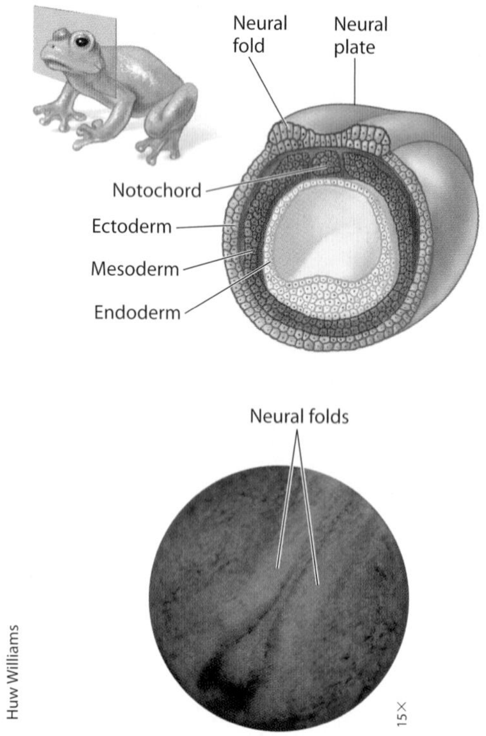

▲ **Figure 12A** The beginning of organ development in a frog: the notochord, neural folds, and neural plate

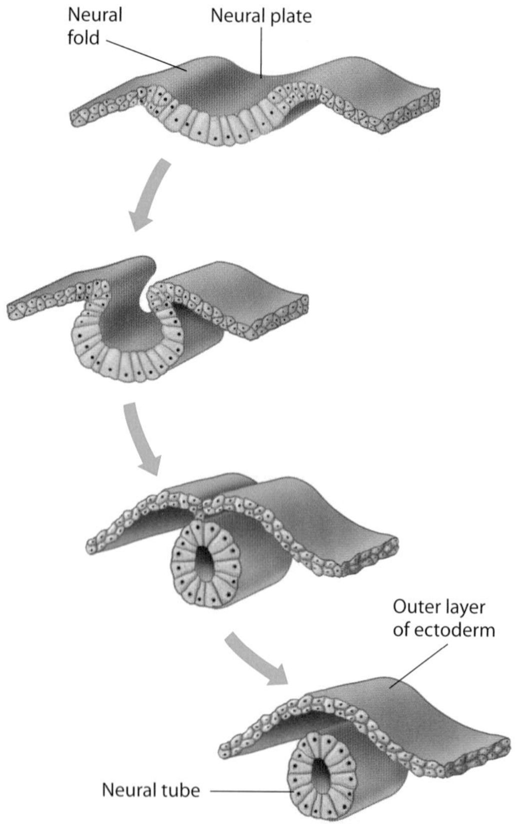

▲ **Figure 12B** Formation of the neural tube

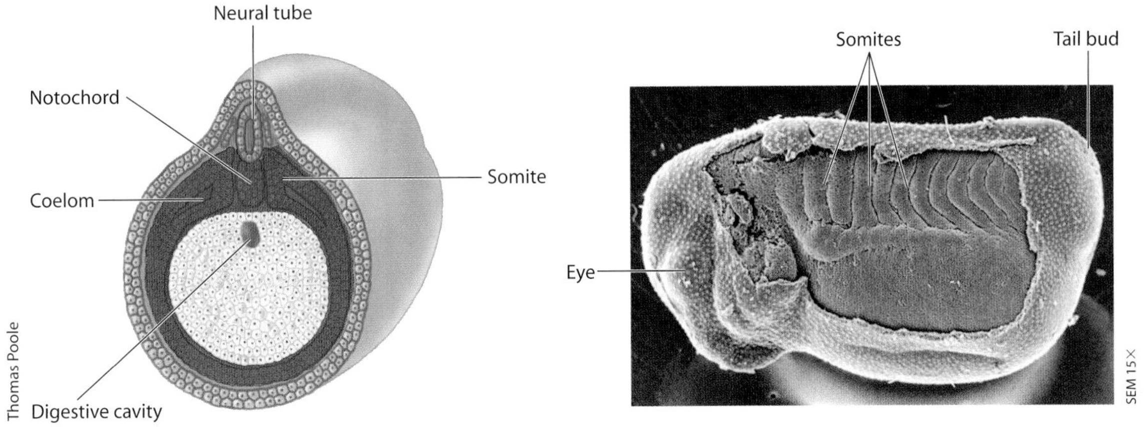

▲ **Figure 12C** An embryo with completed neural tube, somites, and coelom

digestive tract will be ventral to (beneath) the backbone. We see this same arrangement of organs in all vertebrates.

The importance of these processes is underscored by human birth defects that result from improper signaling between embryonic tissues. For example, spina bifida is a condition that results from the failure of the tube of ectoderm cells to close properly and form the spine during the first month of fetal development. Infants born with spina bifida often have permanent nerve damage that results in paralysis of the lower limbs.

Besides the appearance of the neural tube and digestive cavity, Figure 12C shows several other fundamental changes in the frog embryo. In the micrograph, which is a side view, you can see that the embryo is more elongated than the one in Figure 12A. You can also see the beginnings of an eye and a tail (called the tail bud). Part of the ectoderm has been removed to reveal a series of internal ridges called somites. The somites are blocks of mesoderm that will give rise to segmental structures (constructed of repeating units), such as the vertebrae and associated muscles of the backbone. In the cross-sectional drawing, notice that the mesoderm next to the somites is developing a hollow space—the body cavity, or **coelom**. Segmented body parts and a coelom are basic features of all chordates.

In this and the previous two modules, we have observed the sequence of changes that occur as an animal begins to take shape. To summarize, the key phases in embryonic development are cleavage (which creates a multicellular blastula from a zygote), gastrulation (which organizes the embryo into a gastrula with three discrete layers), and organ formation (which generates embryonic organs from the three embryonic tissue layers). These same three phases occur in nearly all animals.

If we followed a frog's development beyond the stage represented in Figure 12C, within a few hours we would be able to monitor muscular responses and a heartbeat and see a set of gills with blood circulating in them. A long tail fin would grow from the tail bud. The timing of the later stages in frog development varies enormously, but in many species, by 5–8 days after development begins, we would see all the body tissues and organs of a tadpole emerge from cells of the ectoderm, mesoderm, and endoderm. Eventually, the structures of the tadpole **(Figure 12D)** would transform into the tissues and organs of an adult frog.

Watching embryos develop helps us appreciate the enormous changes that occur as one tiny cell, the zygote, gives rise to a highly structured, many-celled animal. Your own body, for instance, is a complex organization of some 60 trillion cells, all of which arose from a zygote smaller than the period at the end of this sentence. Discovering how this incredibly intricate arrangement is achieved is one of biology's greatest challenges. Through research that combines the experimental manipulation of embryos with cell biology and molecular genetics, developmental biologists have begun to work out the mechanisms that underlie development. We examine several of these mechanisms in the next module.

? What is the embryonic basis for the dorsal, hollow nerve cord that is common to all members of our phylum?

The neural tube, which becomes the brain and spinal cord, develops from a dorsal ectodermal plate that folds to form an interior tube.

▲ **Figure 12D** A tadpole

13 Multiple processes give form to the developing animal

The development of an animal embryo depends on several cellular processes. For example, as you learned in the last module, changes in cell shape help form the neural tube (see Figure 12B).

Most developmental processes depend on signals passed between neighboring cells and cell layers, telling embryonic cells precisely what to do and when to do it. The mechanism by which one group of cells influences the development of an adjacent group of cells is called **induction**. Induction may be mediated by diffusible signals or, if the cells are in direct contact, by cell-surface interactions. Induction plays a major role in the early development of virtually all tissues and organs. Its effect is to switch on a set of genes whose expression makes the receiving cells differentiate into a specific tissue. Many inductions involve a sequence of inductive steps from different surrounding tissues that progressively determine the fate of cells. In the eye, for example, lens formation involves precisely timed inductive signals from ectodermal, mesodermal, and endodermal cells. In the developing animal, a sequence of inductive signals leads to increasingly greater specialization of cells as organs begin to take shape.

Cell migration is also essential in development. For example, during gastrulation, cells "crawl" within the embryo by extending and contracting cellular protrusions, similar to the pseudopodia of amoeboid cells. Migrating cells may follow inductive chemical trails secreted by cells near their specific destination. Once a migrating cell reaches its destination, surface proteins enable it to recognize similar cells. The cells join together and secrete glycoproteins that glue them in place.

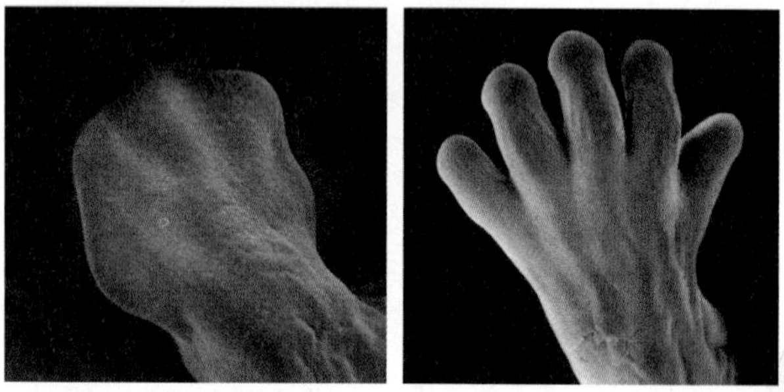

David Barlow

▲ **Figure 13A** Apoptosis in a developing human hand

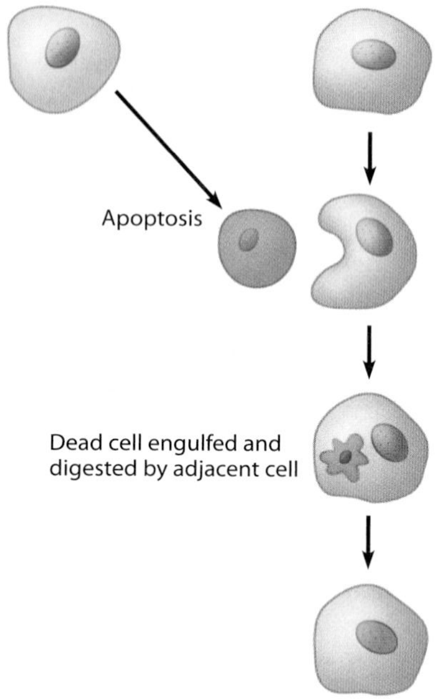

▲ **Figure 13B** Apoptosis at the cellular level

Finally, they differentiate, taking on the characteristics of a particular tissue.

Another important developmental process is **apoptosis**, the timely and tidy suicide of cells. Apoptosis is a type of **programmed cell death**. Animal cells make proteins that have the ability, when activated, to kill the cell that produces them. In humans, the timely death of specific cells in developing hands and feet creates the spaces between fingers and toes (**Figure 13A**). In **Figure 13B**, the cell on the left shrinks and dies because suicide proteins have been activated. Meanwhile, signals from the dying cell make an adjacent cell phagocytic. This cell engulfs and digests the dead cell, keeping the embryo free of harmful debris.

? Induction often involves signal transduction pathways. What do you suppose their role is?

● They mediate between the chemical signal received by the cell and the resulting changes in gene expression and other responses by the cell.

EVOLUTION CONNECTION

14 Pattern formation during embryonic development is controlled by ancient genes

So far, we have discussed the formation of individual organs. What directs the formation of large body features, such as the limbs? The shaping of an animal's major parts involves **pattern formation**, the emergence of a body form with specialized organs and tissues in the right places. Research indicates that master control genes respond to chemical signals that tell a cell where it is relative to other cells in the embryo. These positional signals determine which master control genes will be expressed and, consequently, which body parts will form. Research has shown that such control genes arose early in

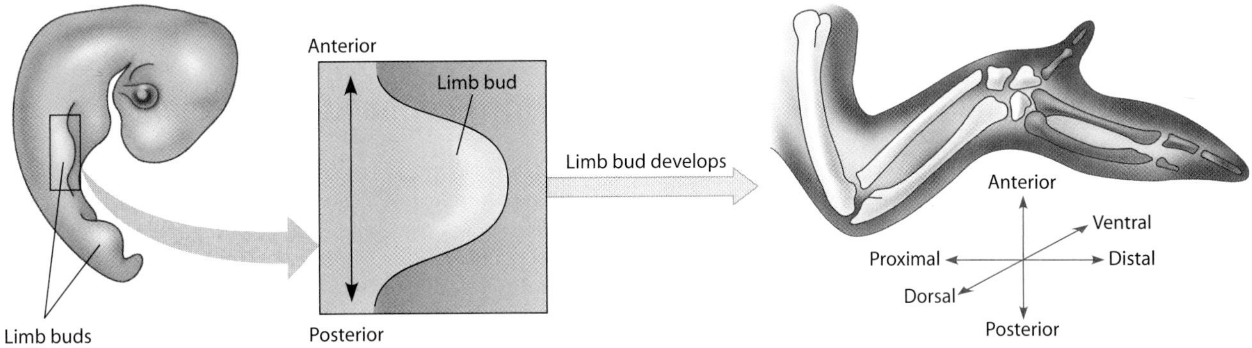

▲ **Figure 14A** The normal development of a wing

the evolution of animals and so play similar roles across diverse animal groups. The field of biology that studies the evolution of developmental processes is called
evo-devo.

Vertebrate limbs, such as bird wings, begin as embryonic structures called limb buds (**Figure 14A**). Each component of a chick wing, such as a specific bone or muscle, develops with a precise location and orientation relative to three axes: the proximal-distal axis (the "shoulder-to-fingertip" axis), the anterior-posterior axis (the "thumb-to-little finger" axis), and the dorsal-ventral axis (the "knuckle-to-palm" axis). The embryonic cells within a limb bud respond to positional information indicating location along these three axes. Only with this information will the cell's genes direct the synthesis of the proteins needed for normal differentiation in that cell's specific location.

Among the most exciting biological discoveries in recent years is that a class of similar genes—**homeotic genes**—help direct embryonic pattern formation in a wide variety of organisms. Researchers studying homeotic genes in fruit flies found a common structural feature: Every homeotic gene they looked at contained a common sequence of 180 nucleotides. Very similar sequences have since been found in virtually every eukaryotic organism examined so far, including yeasts, plants, and humans—and even some prokaryotes. These nucleotide sequences are called **homeoboxes**, and each is translated into a segment (60 amino acids long) of the protein product of the homeotic gene. The homeobox polypeptide segment binds to specific sequences in DNA, enabling homeotic proteins that contain it to turn groups of genes on or off during development.

Figure 14B highlights some striking similarities in the chromosomal locations and the developmental roles of some homeobox-containing homeotic genes in two quite different animals. The figure shows portions of chromosomes that carry homeotic genes in the fruit fly and the mouse. The colored boxes represent homeotic genes that are very similar in flies and mice. Notice that the order of genes on the fly chromosome is the same as on the four mouse chromosomes and that the gene order on the chromosomes corresponds to analogous body regions in both animals. These similarities suggest that the original version of these homeotic genes arose very early in the history of life and that the genes have remained remarkably unchanged for eons of animal evolution. By their presence in such diverse creatures, homeotic genes illustrate one of the central themes of biology: unity in diversity due to shared evolutionary history.

A major goal of developmental research is to learn how the one-dimensional information encoded in the nucleotide sequence of a zygote's DNA directs the development of the three-dimensional form of an animal. Pattern formation requires cells to receive and interpret environmental cues that vary from one location to another. These cues, acting together along three axes, tell cells where they are in the three-dimensional realm of a developing organ. In the next two modules, we'll see the results of this process as we watch an individual of our own species take shape.

? How is pattern formation already apparent at the gastrula stage? (*Hint*: Review Figure 11.)

● The major axes of the animal—anterior-posterior, dorsal-ventral, and proximal-distal—are already set at the gastrula stage.

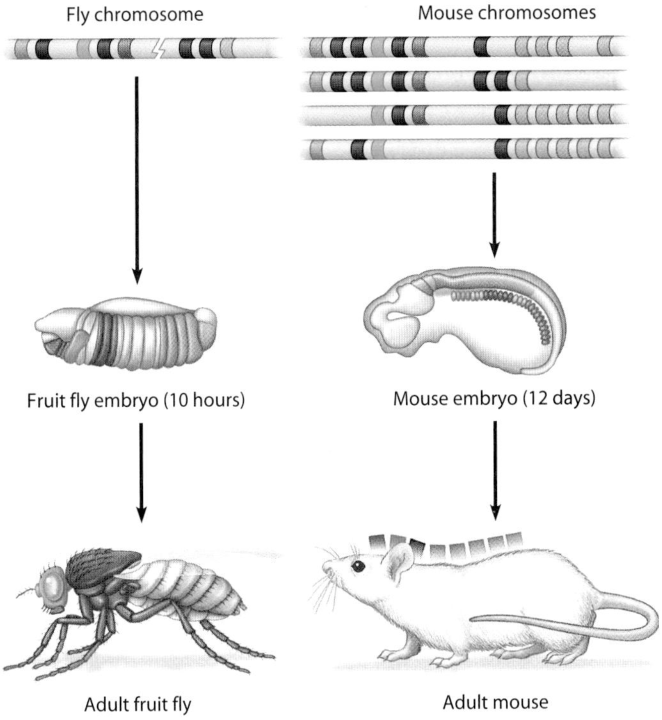

▲ **Figure 14B** Comparison of fruit fly and mouse homeotic genes

Adapted from an illustration by William McGinnis in Peter Radetsky, "The homeobox: Something very precious that we share with flies, from egg to adult." Bethesda, MD: Howard Hughes Medical Institute, 1992, p. 92. Reprinted by permission of William McGinnis.

Human Development

15 The embryo and placenta take shape during the first month of pregnancy

Pregnancy, or **gestation**, is the carrying of developing young within the female reproductive tract. It begins with the fertilization of the egg by a sperm and continues until birth. Duration of pregnancy varies considerably among animal species; gestation in mice lasts about 21 days, while elephants carry their young for 600 days. Human pregnancy averages 266 days (38 weeks) from fertilization (also called **conception** in humans), or 40 weeks (9 months) from the start of the last menstrual cycle.

An Overview of Developmental Events The figures in this module illustrate, in cross section, the changes that occur during the first month of human development. The insets at the lower right of Figures 15C–15F show the embryo's actual size at each stage.

Fertilization occurs in the oviduct (**Figure 15A**). Cleavage starts about 24 hours after fertilization and continues as the embryo moves down the oviduct toward the uterus. By the 6th or 7th day after fertilization, the embryo has reached the uterus, and cleavage has produced about 100 cells. The embryo is now a hollow sphere of cells called a **blastocyst** (the mammalian equivalent of the sea urchin blastula we saw in Figure 10).

The human blastocyst (**Figure 15B**) has a fluid-filled cavity, an inner cell mass that will actually form the baby, and an outer layer of cells called the **trophoblast**. The trophoblast secretes enzymes that enable the blastocyst to implant in the endometrium, the uterine lining (gray in all the figures).

The blastocyst starts to implant in the uterus about a week after conception. In **Figure 15C**, you can see extensions of the trophoblast spreading into the endometrium; these extensions consist of multiplying cells. The trophoblast cells eventually form part of the **placenta**, the organ that provides nourishment and oxygen to the embryo and helps dispose of its metabolic wastes. The placenta consists of both embryonic and maternal tissues.

In Figure 15C, the cells colored purple and yellow are derived from the inner cell mass. Most of the purple cells will give rise to the embryo. The yellow cells, some purple cells, and some trophoblast cells will give rise to four structures called the **extraembryonic membranes**, which develop as attachments to the embryo and help support it. You can see three of these membranes—the amnion (from purple cells), the yolk sac (from yellow cells), and the chorion (partly from the trophoblast)—starting to take shape in **Figure 15D**. A later stage (**Figure 15E**) shows the fourth extraembryonic membrane, the allantois, developing as an extension of the yolk sac.

Gastrulation, the stage shown in Figure 15D, is under way by 9 days after conception. There is already evidence of the three embryonic layers—ectoderm (blue), endoderm (yellow), and mesoderm (red). The embryo itself (not including the membranes) develops from the three inner cell layers shown in Figure 15E. The ectoderm layer will form the outer part of the embryo's skin. As indicated in the drawing, the ectoderm layer is continuous with the amnion. Similarly, the embryo's digestive tract will develop from the endoderm layer, which is continuous with the yolk sac. The bulk of most other organs will develop from the central layer of mesoderm.

Roles of the Extraembryonic Membranes **Figure 15F** shows the embryo about a month after fertilization, with its life-support system, made up largely of the four extraembryonic membranes. By this time, the **amnion** has grown to enclose the embryo. The amniotic cavity is filled with fluid, which protects the embryo. The amnion usually breaks just before childbirth,

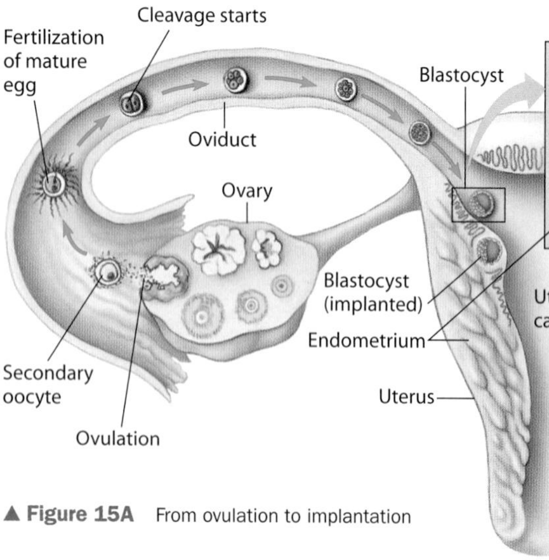

▲ **Figure 15A** From ovulation to implantation

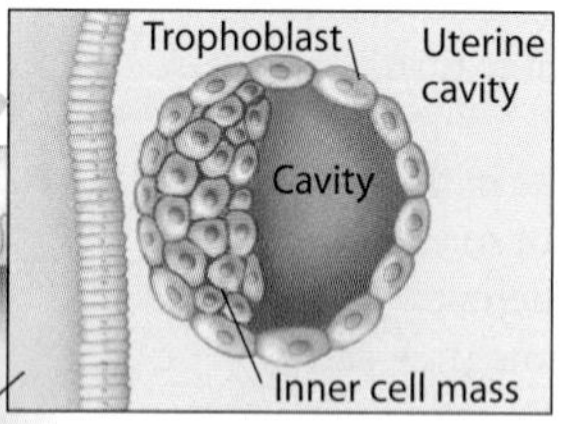

▲ **Figure 15B** Blastocyst (6 days after conception)

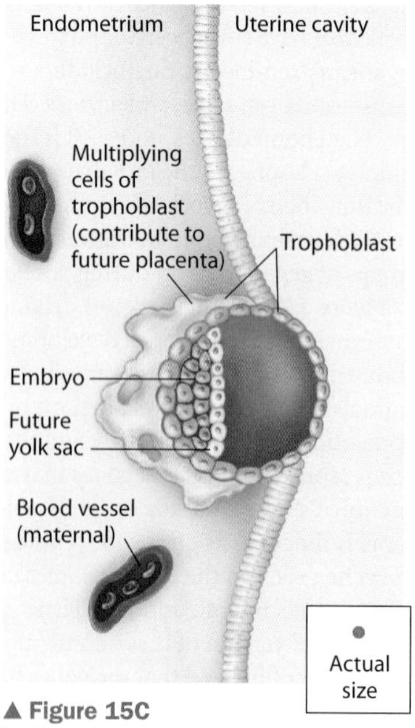

▲ **Figure 15C** Implantation under way (about 7 days)

and the amniotic fluid leaves the mother's body through her vagina ("her water broke").

In humans and most other mammals, the **yolk sac** contains no yolk, but is given the same name as the homologous structure in other vertebrates. In a bird egg, the yolk sac contains a large mass of yolk. Isolated within a shelled egg outside of the mother's body, a developing bird will obtain nourishment from the yolk rather than from a placenta. In mammals, the yolk sac, which remains small, has other important functions: It produces the embryo's first blood cells and its first germ cells, the cells that will give rise to the gamete-forming cells in the gonads.

The **allantois** also remains small in mammals. It forms part of the umbilical cord—the lifeline between the embryo and the placenta. It also forms part of the embryo's urinary bladder. In birds and other reptiles, the allantois expands around the embryo and functions in waste disposal.

The outermost extraembryonic membrane, the **chorion**, completely surrounds the embryo and other extraembryonic membranes. The chorion becomes part of the placenta, where it functions in gas exchange. Cells in the chorion secrete a hormone called **human chorionic gonadotropin (hCG)**, which maintains production of estrogen and progesterone by the corpus luteum of the ovary during the first few months of pregnancy. Without these hormones, menstruation would occur, and the embryo would abort spontaneously. Levels of hCG in maternal blood are so high that some is excreted in the urine, where it can be detected by pregnancy tests.

The Placenta Looking again at Figure 15D, notice the knobby outgrowths on the outside of the chorion. In Figure 15E, these outgrowths, now called **chorionic villi**, are larger and contain mesoderm. In Figure 15F, the mesoderm cells have formed into embryonic blood vessels in the chorionic villi. By this stage, the placenta is fully developed. Starting with the chorion and extending outward, the placenta is a composite organ consisting of chorionic villi closely associated with the blood vessels of the mother's endometrium. The villi are actually bathed in tiny pools of maternal blood. The mother's blood and the embryo's blood are not in direct contact. Instead, the chorionic villi absorb nutrients and oxygen from the mother's blood and pass these substances to the embryo via the chorionic blood vessels that are shown in red. The chorionic vessels shown in blue carry wastes away from the embryo. The wastes diffuse into the mother's bloodstream and are excreted by her kidneys.

The placenta is a vital organ with both embryonic and maternal parts that mediates exchange of nutrients, gases, and the products of excretion between the embryo and the mother. However, the placenta cannot always protect the embryo from substances circulating in the mother's blood. A number of viruses—the German measles virus and HIV, for example—can cross the placenta. German measles can cause serious birth defects; HIV-infected babies usually die of AIDS within a few years without treatment. Most drugs, both prescription and not, also cross the placenta, and many can harm the developing embryo. Alcohol and the chemicals in tobacco smoke, for instance, raise the risk of miscarriage and birth defects. Alcohol can cause a set of birth defects called fetal alcohol syndrome, which includes mental retardation.

? Why does testing for hCG in a woman's urine or blood work as an early test of pregnancy?

Because this hormone is secreted by the chorion of an embryo

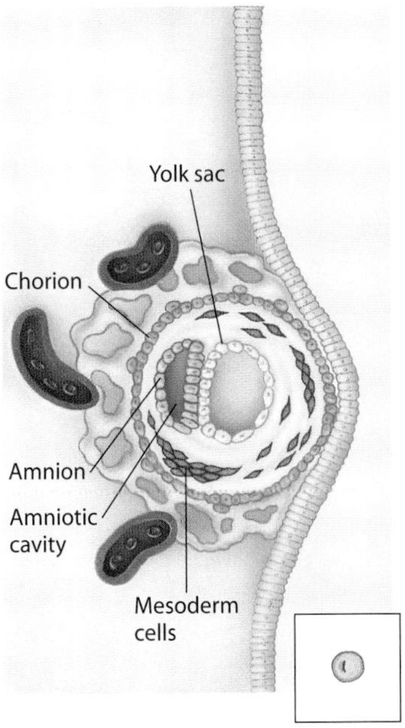

▲ **Figure 15D** Embryonic layers and extraembryonic membranes starting to form (9 days)

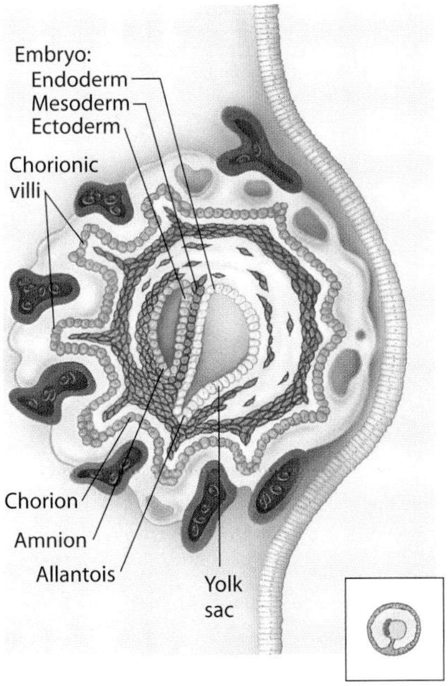

▲ **Figure 15E** Three-layered embryo and four extraembryonic membranes (16 days)

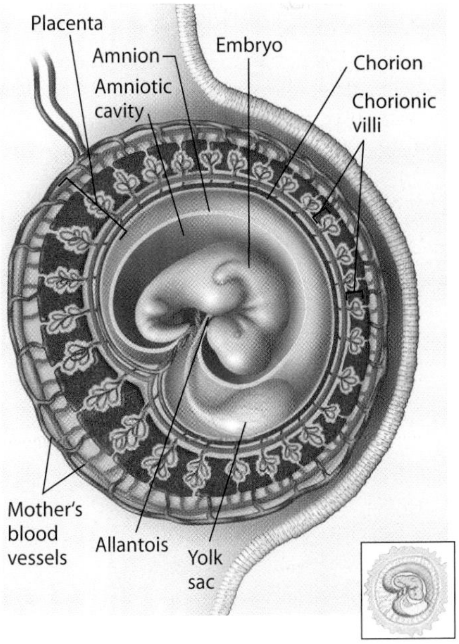

▲ **Figure 15F** Placenta formed (31 days)

16 Human development from conception to birth is divided into three trimesters

In the previous module, we followed human development through the first 4 weeks. In this module, we use photographs to illustrate the rest of human development in the uterus. For convenience, we divide the period of human development from conception to birth into three **trimesters** of about 3 months each.

First Trimester The first trimester is the time of the most radical change for both mother and embryo. **Figure 16A** shows a human embryo about 5 weeks after fertilization. In that brief time, this highly organized multicellular embryo has developed from a single cell. Not shown are the extraembryonic membranes that surround the embryo or most of the umbilical cord that attaches it to the placenta. This embryo is about 7 mm (0.28 inch) long and has a number of features in common with the somite stage of a frog embryo (see Figure 12C). The embryo has a notochord and a coelom, both formed from mesoderm. Its brain and spinal cord have begun to take shape from a tube of ectoderm. The human embryo also has four stumpy limb buds, a short tail, and elements of gill pouches. The gill pouches appear during embryonic development in all chordates; in land vertebrates, they eventually develop into parts of the throat and middle ear. Overall, a month-old human embryo is similar to other vertebrates at the somite stage of development.

Figure 16B shows a developing human, now called a fetus, about 9 weeks after fertilization. The large pinkish structure on the left is the placenta, attached to the fetus by the umbilical cord. The clear sac around the fetus is the amnion. By this time, the fetus looks decidedly human, rather than generally vertebrate. It is about 5.5 cm (2.2 inches) long, and all the major structures of the adult are present in rudimentary form. The somites have developed into the segmental muscles and bones of the back and ribs. The limb buds have become tiny arms and legs with fingers and toes. The first trimester is when most of the body's organs form. Because these important structures are taking shape, during this stage an embryo is particularly susceptible to damage by radiation, drugs, or alcohol, all of which can lead to birth defects.

Although only a few inches long, the fetus can move its arms and legs, turn its head, frown, and make sucking motions with its lips. By the end of the first trimester, the fetus

Timeline of Human Fetal Development

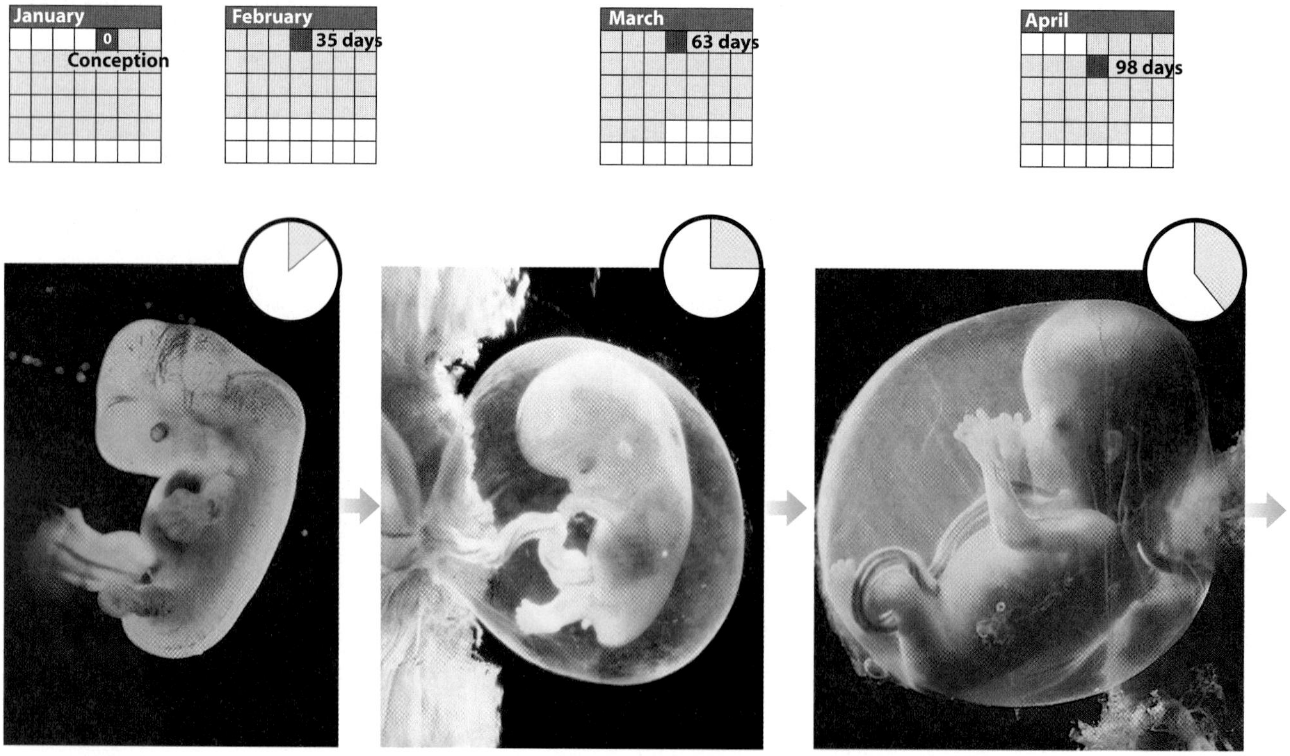

Lennart Nilsson/Scanpix

▲ **Figure 16A** 5 weeks (35 days)

▲ **Figure 16B** 9 weeks (63 days)

▲ **Figure 16C** 14 weeks (98 days)

looks like a miniature human being, although its head is still oversized for the rest of the body. The sex of the fetus is usually evident by this time, and its heartbeat can be detected.

Second Trimester The main developmental changes during the second and third trimesters include an increase in size and general refinement of the human features—nothing as dramatic as the changes of the first trimester. **Figure 16C** shows a fetus at 14 weeks, 2 weeks into the second trimester. The fetus is now about 6 cm (2.4 inches) long. During the second trimester, the placenta takes over the task of maintaining itself by secreting progesterone, rather than receiving it from the corpus luteum. At the same time, the placenta stops secreting hCG, and the corpus luteum, no longer needed to maintain pregnancy, degenerates.

At 20 weeks **(Figure 16D)**, well into the second trimester, the fetus is about 19 cm (7.6 inches) long, weighs about half a kilogram (1 pound), and has the face of an infant, complete with eyebrows and eyelashes. Its arms, legs, fingers, and toes have lengthened. It also has fingernails and toenails and is covered with fine hair. Fetuses of this age are usually quite active. The mother's abdomen has become markedly enlarged, and she may often feel her baby move. Because of the limited space in the uterus, the fetus flexes forward into the so-called fetal position. By the end of the second trimester, the fetus's eyes are open and its teeth are forming.

Third Trimester The third trimester (28 weeks to birth) is a time of rapid growth as the fetus gains the strength it will need to survive outside the protective environment of the uterus. Babies born prematurely—as early as 24 weeks—may survive, but they require special medical care after birth. During the third trimester, the fetus's circulatory system and respiratory system undergo changes that will allow the switch to air breathing. The fetus gains the ability to maintain its own temperature, and its bones begin to harden and its muscles thicken. It also loses much of its fine body hair, except on its head. The fetus becomes less active as it fills the space in the uterus. As the fetus grows and the uterus expands around it, the mother's abdominal organs become compressed and displaced, leading to frequent urination, digestive blockages, and strain in the back muscles. At birth **(Figure 16E)**, babies average about 50 cm (20 inches) in length and weigh 3–4 kg (6–8 pounds).

? Certain drugs cause their most serious damage to an embryo very early in pregnancy, often before the mother even realizes she is pregnant. Why?

Because organ systems, such as the circulatory and nervous systems, begin to develop early in the first trimester.

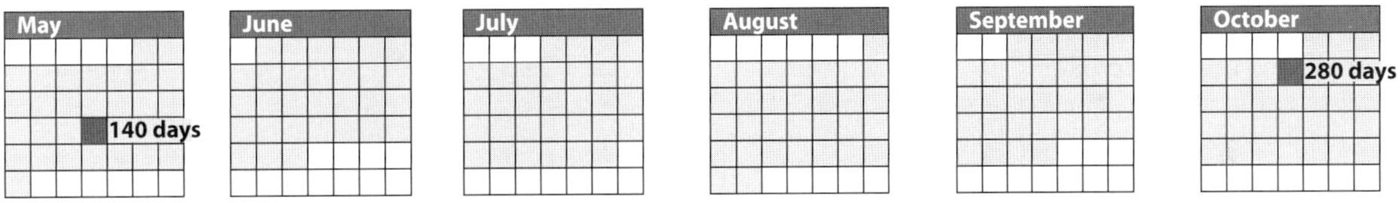

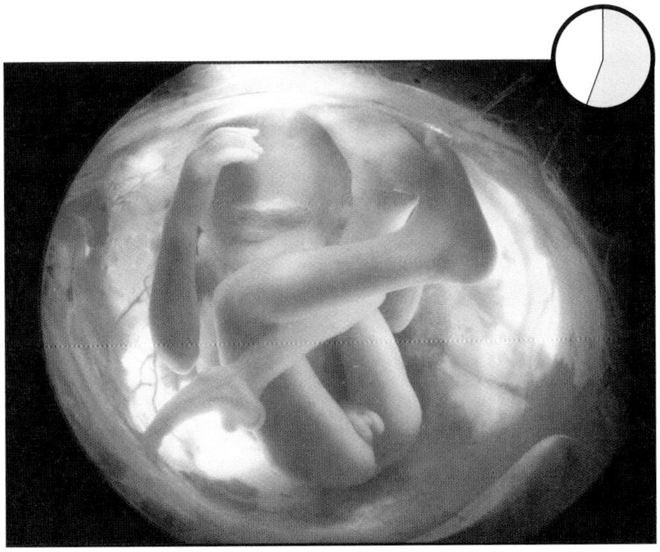

Lennart Nilsson/Scanpix

▲ **Figure 16D** 20 weeks (140 days)

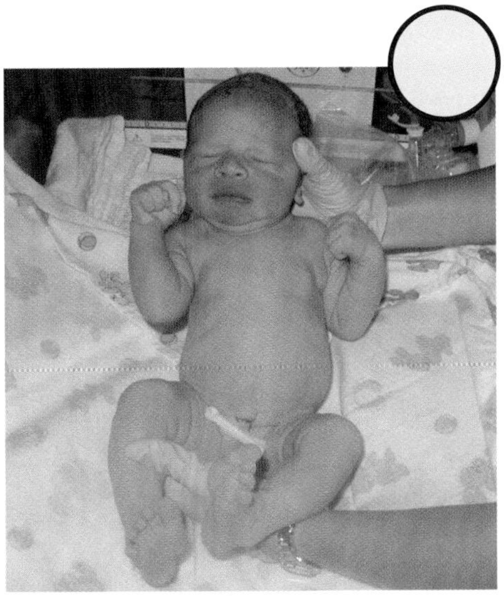

Eric J. Simon

▲ **Figure 16E** At birth (280 days)

17 Childbirth is induced by hormones and other chemical signals

The series of events that expel an infant from the uterus is called **labor**. Several hormones play key roles in this process **(Figure 17A)**. One hormone, estrogen, reaches its highest level in the mother's blood during the last weeks of pregnancy. An important effect of this estrogen is to trigger the formation of numerous oxytocin receptors on cells of the uterus. Cells of the fetus produce the hormone oxytocin, and late in pregnancy, the mother's pituitary gland secretes it in increasing amounts. Oxytocin stimulates the smooth muscles in the wall of the uterus, producing the series of increasingly strong, rhythmic contractions characteristic of labor. It also stimulates the placenta to make prostaglandins, local tissue regulators that stimulate the uterine muscle cells to contract even more.

The induction of labor involves **positive feedback**, a type of control in which a change triggers mechanisms that amplify that change. In this case, oxytocin and prostaglandins cause uterine contractions that in turn stimulate the release of more oxytocin and prostaglandins. The result is a steady increase in contraction intensity, climaxing in forceful muscle contractions that propel a baby from the uterus.

Figure 17B shows the three stages of labor. As the process begins, the cervix (neck of the uterus) gradually opens, or dilates. ❶ The first stage, dilation, is the time from the onset of labor until the cervix reaches its full dilation of about 10 cm. Dilation is the longest stage of labor, lasting 6–12 hours or even considerably longer.

❷ The period from full dilation of the cervix to delivery of the infant is called the expulsion stage. Strong uterine contractions, lasting about 1 minute each, occur every 2–3 minutes, and the mother feels an increasing urge to push or bear down with her abdominal muscles. Within a period of 20 minutes to an hour or so, the infant is forced down and out of the uterus and vagina. An attending physician or midwife (or nervous father!) clamps and cuts the umbilical cord after the baby is expelled. ❸ The final stage is the delivery of the placenta ("afterbirth"), usually within 15 minutes after the birth of the baby.

Hormones continue to be important after the baby and placenta are delivered. Decreasing levels of progesterone and estrogen allow the uterus to start returning to its state before

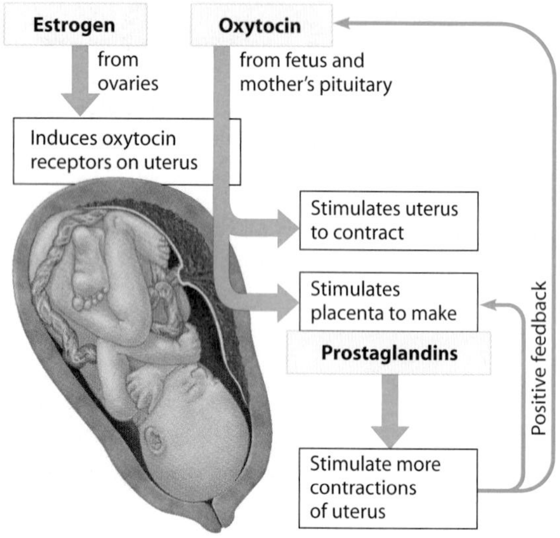

▲ **Figure 17A** The hormonal induction of labor

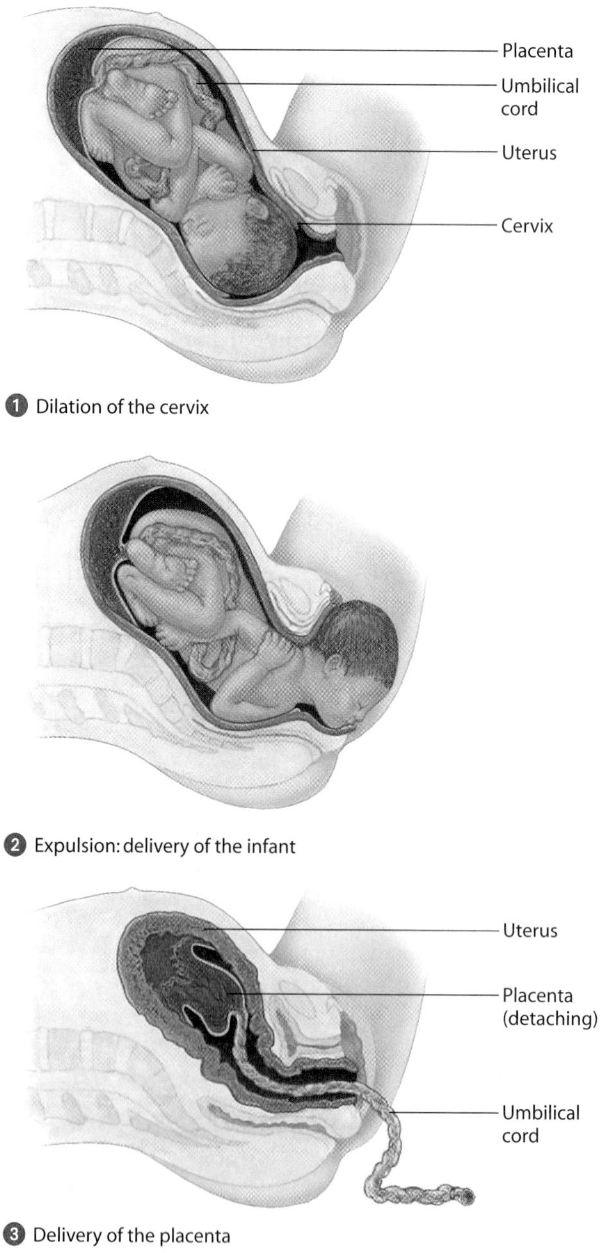

▲ **Figure 17B** The three stages of labor

pregnancy. In response to suckling by the newborn, as well as falling levels of progesterone after birth, the pituitary secretes prolactin and oxytocin. These two hormones promote milk production and release (called lactation) by the mammary glands. At first, a yellowish fluid called colostrum, rich in protein and antibodies, is secreted. After 2–3 days, the production of regular milk begins.

? The onset of labor is marked by dilation of the ______.

cervix

CONNECTION 18 Reproductive technologies increase our reproductive options

About 15% of couples who want children are unable to conceive, even after 12 months of unprotected intercourse. Such a condition, called **infertility**, can have many causes. A man's testes may not produce enough sperm (a "low sperm count"), or those that are produced may be defective. Underproduction of sperm is frequently caused by the man's scrotum being too warm, so a switch of underwear from briefs (which hold the scrotum close to the body) to boxers may help. In other cases, infertility is caused by **impotence**, also called erectile dysfunction, the inability to maintain an erection. Temporary impotence can result from alcohol or drug use or from psychological problems. Permanent impotence can result from nervous system or circulatory problems. Female infertility can result from a lack of eggs, a failure to ovulate, or blocked oviducts (often caused by scarring due to sexually transmitted diseases). Other women are able to conceive, but cannot support a growing embryo in the uterus. The resulting multiple miscarriages can take a heavy emotional toll.

Reproductive technologies can help many cases of infertility. Drug therapies (including Viagra) and penile implants can be used to treat impotence. If a man produces no functioning sperm, the couple may elect to use another man's sperm that has been donated to a sperm bank.

If a woman has normal eggs that are not being released properly, hormone injections can induce ovulation. Such treatments frequently result in multiple pregnancies. If a woman has no eggs of her own, they, too, can be obtained from a donor for fertilization and implantation into the uterus. While sperm can be collected without any danger to the donor, collection of eggs involves surgery and therefore some pain and risk for the donating woman.

If a woman produces eggs but is unable to support a growing fetus, she and her partner may hire a surrogate mother. In such cases, the couple enters into a legal contract with a woman who agrees to be implanted with the couple's embryo and carry it to birth. However, a number of states have laws restricting surrogate motherhood because of the serious ethical and legal problems that can arise.

Many infertile couples turn to fertilization procedures called **assisted reproductive technologies**. In these procedures, eggs (secondary oocytes) are surgically removed from a woman's ovaries after hormonal stimulation, fertilized, and returned to the woman's body. Eggs, sperm, and embryos from such procedures can be frozen for later pregnancy attempts.

With **in vitro fertilization (IVF)**, the most common assisted reproductive technology procedure, a woman's eggs are mixed with sperm in culture dishes (*in vitro* means "in glass") and incubated for several days to allow fertilized eggs to start developing (Figure 18). When they have developed into embryos of at least eight cells each, the embryos are carefully inserted into the woman's uterus.

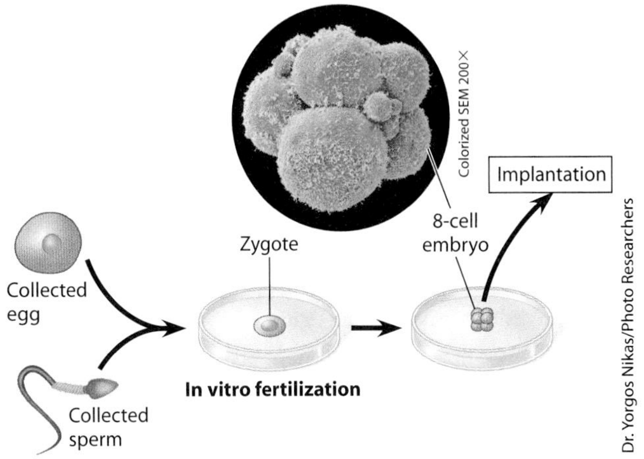

▲ **Figure 18** In vitro fertilization

If mature sperm are defective, of low number, or even absent, fertilization can often be achieved by a technique called intracytoplasmic sperm injection (ICSI). In this form of IVF, the head of a sperm is drawn up into a needle and injected directly into an egg to achieve fertilization.

Abnormalities arising as a consequence of an IVF procedure appear to be quite rare, although some research has shown small but significant risks of lower birth weights and higher rates of birth defects. Despite such risks and the high cost (typically $10,000 per attempt, whether it succeeds or not), IVF techniques are now performed at medical centers throughout the world and result in the birth of thousands of babies each year.

? Explain how IVF can involve up to three different people in the birth of a child.

One woman (1) may become pregnant with an embryo created using the sperm of a man (2) and the mature egg of a second woman (3).

In this chapter, we have watched a single-celled product of sexual reproduction, the zygote, become transformed into a new organism, complete with all organ systems. One of the first of those organ systems to develop is the nervous system.

REVIEW

For Practice Quizzes, BioFlix, MP3 Tutors, and Activities, go to www.masteringbiology.com.

Reviewing the Concepts

Asexual and Sexual Reproduction (1–2)

1 Asexual reproduction results in the generation of genetically identical offspring. Asexual reproduction can proceed by budding, fission, or fragmentation/regeneration and enables one individual to produce many offspring rapidly.

2 Sexual reproduction results in the generation of genetically unique offspring. Sexual reproduction involves the fusion of gametes from two parents, resulting in genetic variation among offspring. This may enhance survival of a population in a changing environment.

Human Reproduction (3–8)

3 Reproductive anatomy of the human female. The human reproductive system consists of a pair of ovaries (in females) or testes (in males), ducts that carry gametes, and structures for copulation. A woman's ovaries contain follicles that nurture eggs and produce sex hormones. Oviducts convey eggs to the uterus, where a fertilized egg develops. The uterus opens into the vagina, which receives the penis during intercourse and serves as the birth canal.

4 Reproductive anatomy of the human male. A man's testes produce sperm, which are expelled through ducts during ejaculation. Several glands contribute to the formation of fluid that nourishes and protects sperm.

5 The formation of sperm and egg cells requires meiosis. Spermatogenesis and oogenesis produce sperm and eggs, respectively. Primary spermatocytes are made continuously in the testes; these diploid cells undergo meiosis to form four haploid sperm. In females, each month, one primary oocyte forms a secondary oocyte, which, if penetrated by a sperm, completes meiosis and becomes a mature egg. The haploid nucleus of the mature egg then fuses with the haploid nucleus of the sperm, forming a diploid zygote.

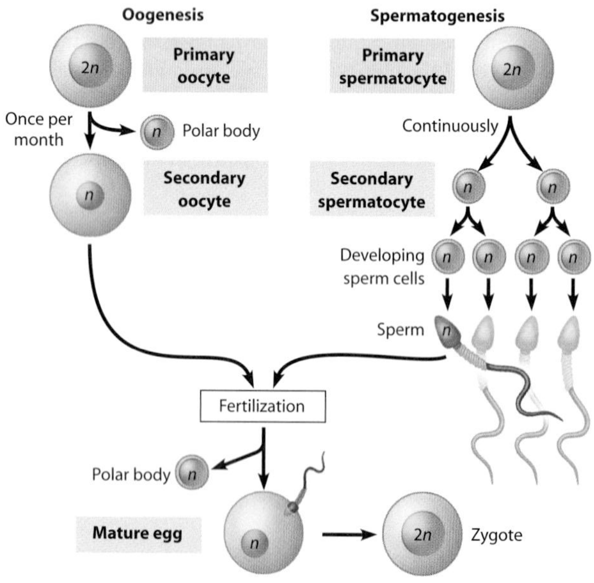

6 Hormones synchronize cyclic changes in the ovary and uterus. Approximately every 28 days, the hypothalamus signals the anterior pituitary to secrete FSH an d LH, which trigger the growth of a follicle and ovulation, the release of an egg. The follicle becomes the corpus luteum, which secretes both estrogen and progesterone. These two hormones stimulate the endometrium (the uterine lining) to thicken, preparing the uterus for implantation. They also inhibit the hypothalamus, reducing FSH and LH secretion. If the egg is not fertilized, the drop in LH shuts down the corpus luteum and its hormones. This triggers menstruation, the breakdown of the endometrium. The hypothalamus and pituitary then stimulate another follicle, starting a new cycle. If fertilization occurs, a hormone from the embryo maintains the uterine lining and prevents menstruation.

7 Sexual activity can transmit disease. STDs caused by bacteria can often be cured, but viral diseases can only be controlled.

8 Contraception can prevent unwanted pregnancy. Several forms of contraception can prevent pregnancy, with varying degrees of success.

Principles of Embryonic Development (9–14)

9 Fertilization results in a zygote and triggers embryonic development. During fertilization, a sperm releases enzymes that pierce the egg's coat. Sperm surface proteins bind to egg receptor proteins, sperm and egg plasma membranes fuse, and the two nuclei unite. Changes in the egg membrane prevent entry of additional sperm, and the fertilized egg (zygote) develops into an embryo.

10 Cleavage produces a ball of cells from the zygote. Cleavage is a rapid series of cell divisions that results in a blastula.

11 Gastrulation produces a three-layered embryo. In gastrulation, cells migrate and form a rudimentary digestive cavity and three layers of cells.

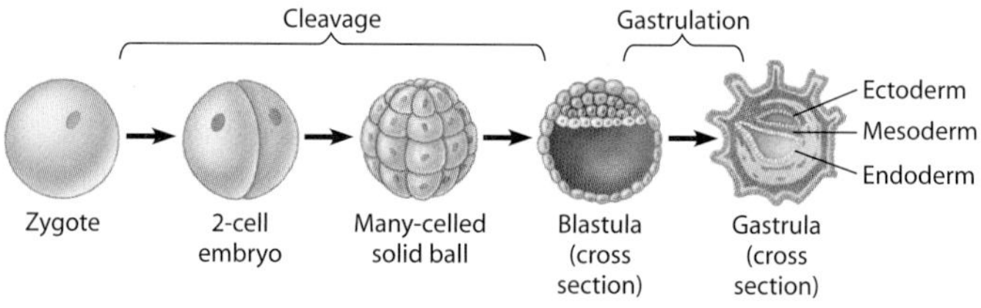

12 Organs start to form after gastrulation. After gastrulation, the three embryonic tissue layers give rise to specific organ systems.

13 Multiple processes give form to the developing animal. Tissues and organs take shape in a developing embryo as a result of cell shape changes, cell migration, and programmed cell death. Through induction, adjacent cells and cell layers influence each other's differentiation via chemical signals.

14 Pattern formation during embryonic development is controlled by ancient genes. Pattern formation, the emergence of the parts of a structure in their correct relative positions, involves the response of genes to spatial variations of chemicals in the embryo. Homeotic genes contain homeoboxes, nucleotide sequences that appeared early in the evolutionary history of animals.

Human Development (15–18)

15 The embryo and placenta take shape during the first month of pregnancy. Human development begins with fertilization in the oviduct. Cleavage produces a blastocyst, whose inner cell mass becomes the embryo. The blastocyst's outer layer, the trophoblast, implants in the uterine wall. Gastrulation occurs, and organs develop from the three embryonic layers. Meanwhile, the four extraembryonic membranes develop: the amnion, the chorion, the yolk sac, and the allantois. The embryo floats in the fluid-filled amniotic cavity, while the chorion and embryonic mesoderm form the embryo's part of the placenta. The placenta's chorionic villi absorb food and oxygen from the mother's blood.

16 Human development from conception to birth is divided into three trimesters. The most rapid changes occur during the first trimester. At 9 weeks, the embryo is called a fetus. The second and third trimesters are times of growth and preparation for birth.

17 Childbirth is induced by hormones and other chemical signals. Estrogen makes the uterus more sensitive to oxytocin, which acts with prostaglandins to initiate labor. The cervix dilates, the baby is expelled by strong muscular contractions, and the placenta follows.

18 Reproductive technologies increase our reproductive options. In in vitro fertilization, eggs are extracted and fertilized in the lab. The resulting embryo is implanted into a woman.

Connecting the Concepts

1. This graph plots the rise and fall of pituitary and ovarian hormones during the human ovarian cycle. Identify each hormone (A–D) and the reproductive events with which each one is associated (P–S). For A–D, choose from estrogen, LH, FSH, and progesterone. For P–S, choose from ovulation, growth of follicle, menstruation, and development of corpus luteum. How would the right-hand side of this graph be altered if pregnancy occurred? What other hormone is responsible for triggering this change?

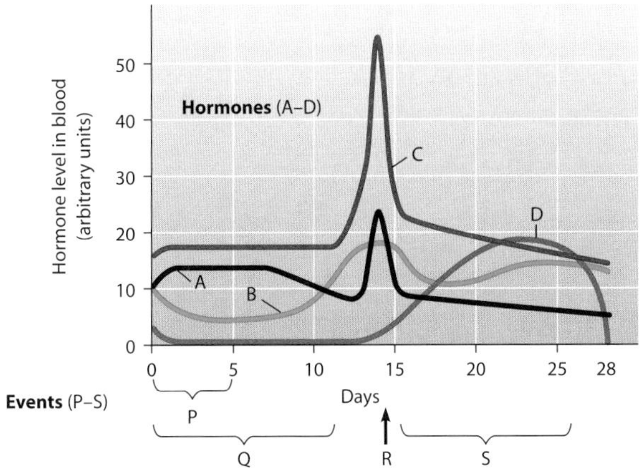

Testing Your Knowledge

Multiple Choice

2. After a sperm penetrates an egg, it is important that the vitelline layer separate from the egg so that it can
   a. secrete important hormones.
   b. enable the fertilized egg to implant in the uterus.
   c. prevent more than one sperm from entering the egg.
   d. attract additional sperm to the egg.
   e. activate the egg for embryonic development.
3. In an experiment, a researcher colored a bit of tissue on the outside of a frog gastrula with an orange fluorescent dye. The embryo developed normally. When the tadpole was placed under an ultraviolet light, which of the following glowed bright orange? (*Explain your answer.*)
   a. the heart
   b. the pancreas
   c. the brain
   d. the stomach
   e. the liver
4. How does a zygote differ from a mature egg?
   a. A zygote has more chromosomes.
   b. A zygote is smaller.
   c. A zygote consists of more than one cell.
   d. A zygote is much larger.
   e. A zygote divides by meiosis.
5. A woman had several miscarriages. Her doctor suspected that a hormonal insufficiency was causing the lining of the uterus to break down, as it does during menstruation, terminating her pregnancies. Treatment with which of the following might help her remain pregnant?
   a. oxytocin
   b. follicle-stimulating hormone
   c. testosterone
   d. luteinizing hormone
   e. prolactin

Matching

6. Turns into the corpus luteum	a. vas deferens
7. Female gonad	b. prostate gland
8. Site of spermatogenesis	c. endometrium
9. Site of fertilization in humans	d. testis
10. Site of human gestation	e. follicle
11. Sperm duct	f. uterus
12. Secretes seminal fluid	g. ovary
13. Lining of uterus	h. oviduct

Describing, Comparing, and Explaining

14. Compare sperm formation with egg formation. In what ways are the processes similar? In what ways are they different?
15. The embryos of reptiles (including birds) and mammals have systems of extraembryonic membranes. What are the functions of these membranes, and how do fish and frog embryos survive without them?
16. In an embryo, nerve cells grow out from the spinal cord and form connections with the muscles they will eventually control. What mechanisms described in this chapter might explain how these cells "know" where to go and which cells to connect with?

Applying the Concepts

17. As a frog embryo develops, the neural tube forms from ectoderm along what will be the frog's back, directly above the notochord. To study this process, a researcher extracted a bit of notochord tissue and inserted it under the ectoderm where the frog's belly would normally develop. What can the researcher hope to learn from this experiment? Predict the possible outcomes. What experimental control would you suggest?
18. Should parents undergoing in vitro fertilization have the right to choose which embryos to implant based on genetic criteria, such as the presence or absence of disease-causing genes? Should they be able to choose based on the sex of the embryo? How could you distinguish acceptable from unacceptable criteria? Do you think such options should be legislated?

Review Answers

1. A. FSH; B. estrogen; C. LH; D. progesterone; P. menstruation; Q. growth of follicle; R. ovulation; S. development of corpus luteum

   If pregnancy occurs, the embryo produces human chorionic gonadotrophin (hCG), which maintains the corpus luteum, keeping levels of estrogen and progesterone high.

2. c 3. c (The outer layer in a gastrula is the ectoderm; of the choices given, only the brain develops from ectoderm.) 4. a 5. d 6. e 7. g 8. d 9. h 10. f 11. a 12. b 13. c

14. Both produce haploid gametes. Spermatogenesis produces four small sperm; oogenesis produces one large egg. In humans, the ovary contains all the primary oocytes at birth, while testes can keep making primary spermatocytes throughout life. Oogenesis is not complete until fertilization, but sperm mature without eggs.

15. The extraembryonic membranes provide a moist environment for the embryos of terrestrial vertebrates and enable the embryos to absorb food and oxygen and dispose of wastes. Such membranes are not needed when an embryo is surrounded by water, as are those of fishes and amphibians.

16. The nerve cells may follow chemical trails to the muscle cells and identify and attach to them by means of specific surface proteins.

17. The researcher might find out whether chemicals from the notochord stimulate the nearby ectoderm to become the neural tube, a process called induction. Transplanted notochord tissue might cause ectoderm anywhere in the embryo to become neural tissue. Control: Transplant non-notochord tissue under the ectoderm of the belly area.

18. Some issues and questions to consider: What characteristics might parents like to select for? If parents had the right to choose embryos based on these characteristics, what are some of the possible benefits? What are potential pitfalls? Could an imbalance of the population result?

Nervous Systems

Nervous Systems

BIG IDEAS

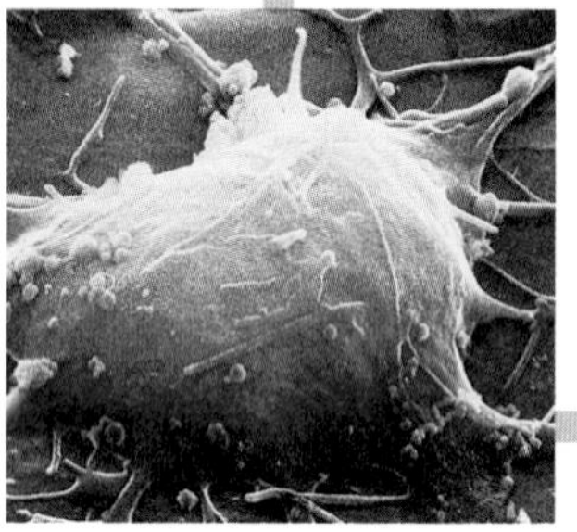

Manfred Kage/Peter Arnold, Inc./ PhotoLibrary

Nervous System Structure and Function (1–2)

Neurons receive and process inputs and communicate responses.

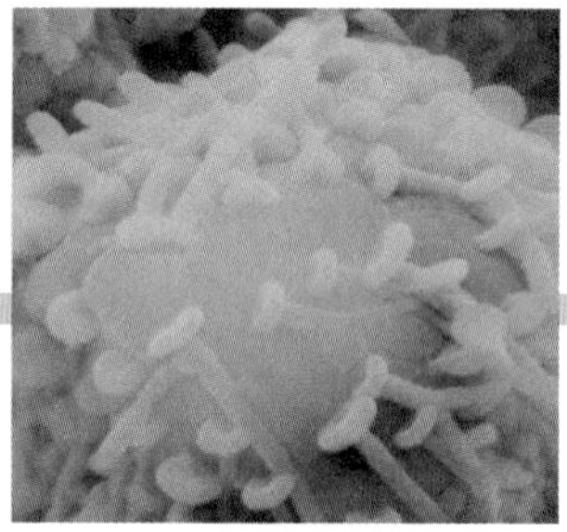

Edwin R. Lewis, Professor Emeritus

Nerve Signals and Their Transmission (3–9)

Nerve signals are electrical messages generated by the movement of ions across membranes.

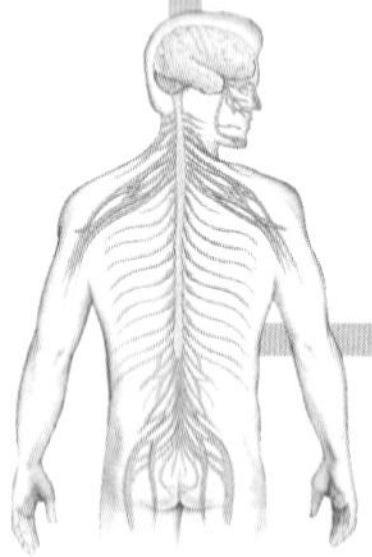

An Overview of Animal Nervous Systems (10–13)

The vertebrate nervous system can be understood as a structural and functional hierarchy.

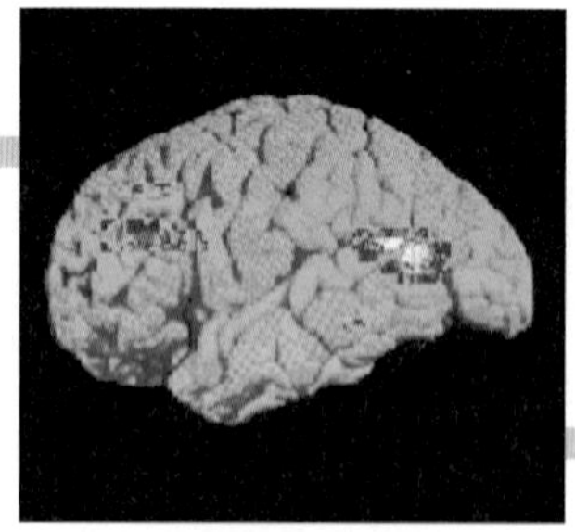

WDCN/Univ. College London/ Photo Researchers

The Human Brain (14–20)

Modern research techniques are illuminating how the interplay of brain regions controls our actions and behavior.

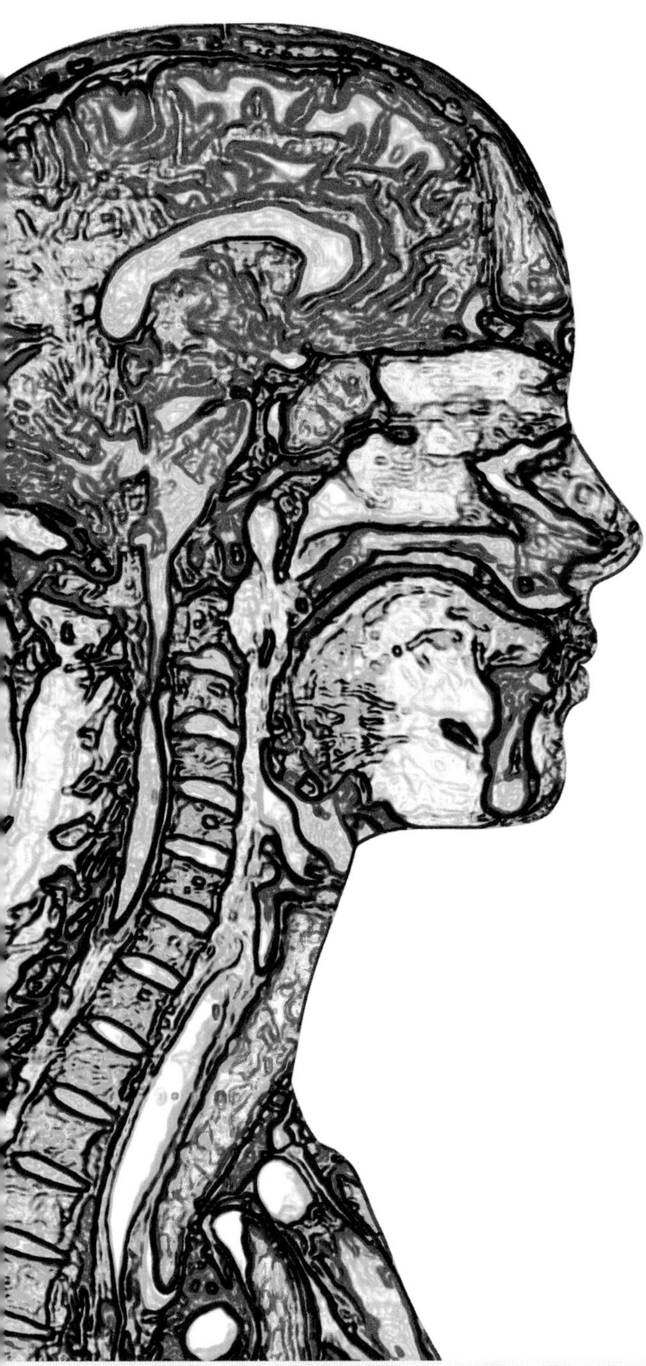

Mehau Kulyk/Science Source

Protected inside the bony vertebrae of the spine (your "backbone") is a gelatinous bundle of nervous tissue called the spinal cord. The inch-thick spinal cord acts as the central communication conduit between the brain and the rest of the body. Millions of nerve fibers carry motor information from the brain to the muscles, while other nerve fibers bring sensory information from the body to the brain. The spinal cord acts like a communication cable jam-packed with fibers, each of which carries high-speed messages between the central hub and an outlying area. These signals control the voluntary and involuntary actions of your body.

If a communications cable is cut, signals cannot get through. In humans, the spinal cord is rarely severed because the bony vertebrae provide rigid protection. But a traumatic blow to the spinal column and subsequent bleeding, swelling, and scarring can crush the delicate nerve bundles and prevent signals from passing.

Over a quarter million Americans are living with spinal cord injuries. About half have paraplegia—paralysis of the lower half of the body—while half have quadriplegia—paralysis from the neck down, which may necessitate permanent breathing assistance from an artificial respirator. Spinal cord injuries most often happen to men, and more than half of those injured are in their teens and 20s. The most common causes are vehicle accidents, gunshots, and falls—and nearly a thousand cases a year are sports related. Any injury to the spinal cord is usually permanent because the spinal cord cannot be repaired.

In this chapter, we explore the structure, function, and evolution of animal nervous systems. We'll start with an overview of nerve cells, which function similarly in nearly all animals. Then we'll focus on the vertebrate nervous system and the structure and function of the human brain.

Nervous System Structure and Function

1 Nervous systems receive sensory input, interpret it, and send out appropriate commands

Nervous systems are the most intricately organized data processing systems on Earth. Your brain, for instance, contains an estimated 100 billion **neurons**, nerve cells that transmit signals from one location in the body to another. A neuron consists of a cell body, containing the nucleus and other cell organelles, and long, thin extensions that convey signals. Each neuron may communicate with thousands of others, forming networks that enable us to learn, remember, perceive our surroundings, and move.

With few exceptions, nervous systems have two main anatomical divisions. The first anatomical division, called the **central nervous system (CNS)**, consists of the brain and, in vertebrates, the spinal cord. The other anatomical division of the nervous system, the **peripheral nervous system (PNS)**, is made up mostly of nerves that carry signals into and out of the CNS. A **nerve** is a communication line consisting of a bundle of neurons tightly wrapped in connective tissue. In addition to nerves, the PNS also has **ganglia** (singular, *ganglion*), clusters of neuron cell bodies.

A nervous system has three interconnected functions (**Figure 1A**). **Sensory input** is the conduction of signals from sensory receptors, such as light-detecting cells of the eye, to the CNS. **Integration** is the analysis and interpretation of the sensory signals and the formulation of appropriate responses. **Motor output** is the conduction of signals from the integration centers to **effector cells**, such as muscle cells or gland cells, which perform the body's responses. The integration of sensory input and motor output is not usually rigid and linear, but involves the continuous background activity symbolized by the circular arrow in Figure 1A.

The relationship between neurons and nervous system structure and function is easiest to see in the relatively simple circuits that produce **reflexes**, or automatic responses to stimuli (**Figure 1B**). The small blue, green, and purple balls in the figure represent neuron cell bodies; the thin colored lines represent neuron extensions. Three functional types of neurons correspond to a nervous system's three main functions: **Sensory neurons** convey signals from sensory receptors into the CNS. In many animals, **interneurons** are located entirely within the CNS. They integrate data and then relay appropriate signals to other interneurons or to motor neurons. Finally, **motor neurons** convey signals from the CNS to effector cells. (For simplicity, this figure shows only one neuron of each functional type, but nearly every body activity actually involves many neurons of each type.)

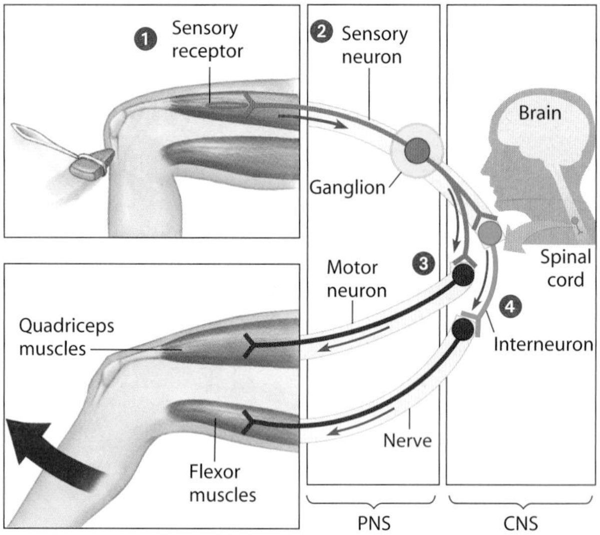

▲ **Figure 1B** The knee-jerk reflex

When the knee is tapped, ❶ a sensory receptor detects a stretch in the muscle, and ❷ a sensory neuron conveys this information into the spinal cord (part of the CNS). In the CNS, the information goes to ❸ a motor neuron and to ❹ one or more interneurons. One set of muscles (quadriceps) responds to motor signals conveyed by a motor neuron by contracting, jerking the lower leg forward. At the same time, another motor neuron, responding to signals from an interneuron, inhibits the flexor muscles, making them relax and not resist the action of the quadriceps.

As you can see from this example, the nervous system depends on the ability of neurons to receive and convey signals. Next, we'll examine how that is done.

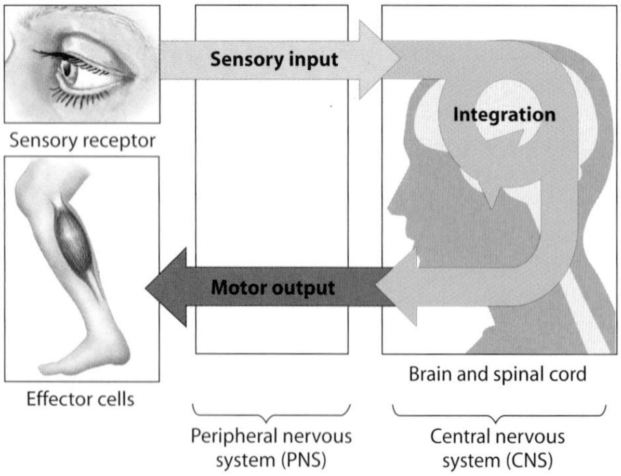

▲ **Figure 1A** Organization of a nervous system

? If someone tickles the bottom of your foot, your ankle automatically flexes. Arrange the following neurons in the correct sequence for information flow during this reflex: interneuron, sensory neuron, motor neuron.

Sensory neuron ⟶ interneuron ⟶ motor neuron

2 Neurons are the functional units of nervous systems

The ability of neurons to receive and transmit information depends on their structure. **Figure 2** depicts a motor neuron, like those that carry command signals from your spinal cord to your skeletal muscles.

Most of a neuron's organelles, including its nucleus, are located in the **cell body**. Arising from the cell body are two types of extensions: numerous dendrites and a single axon. **Dendrites** (from the Greek *dendron*, tree) are highly branched extensions that *receive* signals from other neurons and convey this information toward the cell body. Dendrites are often short. In contrast, the **axon** is typically a much longer extension that *transmits* signals to other cells, which may be other neurons or effector cells. Some axons, such as the ones that reach from your spinal cord to muscle cells in your feet, can be over a meter long.

Neurons make up only part of a nervous system. To function normally, neurons of all vertebrates and most invertebrates require supporting cells called **glia**. Depending on the type, glia may nourish neurons, insulate the axons of neurons, or help maintain homeostasis of the extracellular fluid surrounding neurons. In the mammalian brain, glia outnumber neurons by as many as 50 to 1.

Figure 2 shows one kind of glial cell, called a Schwann cell, which is found in the PNS. (Analogous cells are found in the CNS.) In many vertebrates, axons that convey signals rapidly are enclosed along most of their length by a thick insulating material, analogous to the plastic insulation that covers many electrical wires. This insulating material, called the **myelin sheath**, resembles a chain of oblong beads. Each bead is actually a Schwann cell, and the myelin sheath is essentially a chain of Schwann cells, each wrapped many times around the axon. The gaps between Schwann cells are called **nodes of Ranvier**, and they are the only points along the axon that require nerve signals to be regenerated, which is a time-consuming process. Everywhere else, the myelin sheath insulates the axon, preserving the signal and allowing it to propagate quickly. Thus, when a nerve signal travels along a myelinated axon, it needs to be rejuvenated only at the nodes. The resulting signal is much faster than one that must be regenerated constantly along the length of the axon. In the human nervous system, signals can travel along a myelinated axon about 150 m/sec (over 330 miles per hour), which means that a command from your brain can make your fingers move in just a few milliseconds. Without myelin sheaths, the signals would be over 10 times slower.

The debilitating autoimmune disease multiple sclerosis (MS) demonstrates the importance of myelin. MS leads to a gradual destruction of myelin sheaths by the individual's own immune system. The result is a progressive loss of signal conduction, muscle control, and brain function.

Notice in Figure 2 that the axon ends in a cluster of branches. A typical axon has hundreds or thousands of these branches, each with a **synaptic terminal** at the very end. The junction between a synaptic terminal and another cell is called a synapse. As we will see in Module 6, this is where infor-mation is passed between neurons. With the basic structure of a neuron in mind, let's take a closer look at the signals that neurons convey.

? What is the function of the myelin sheath? How does it accomplish this function?

It speeds up conduction of signals along axons by insulating the axon.

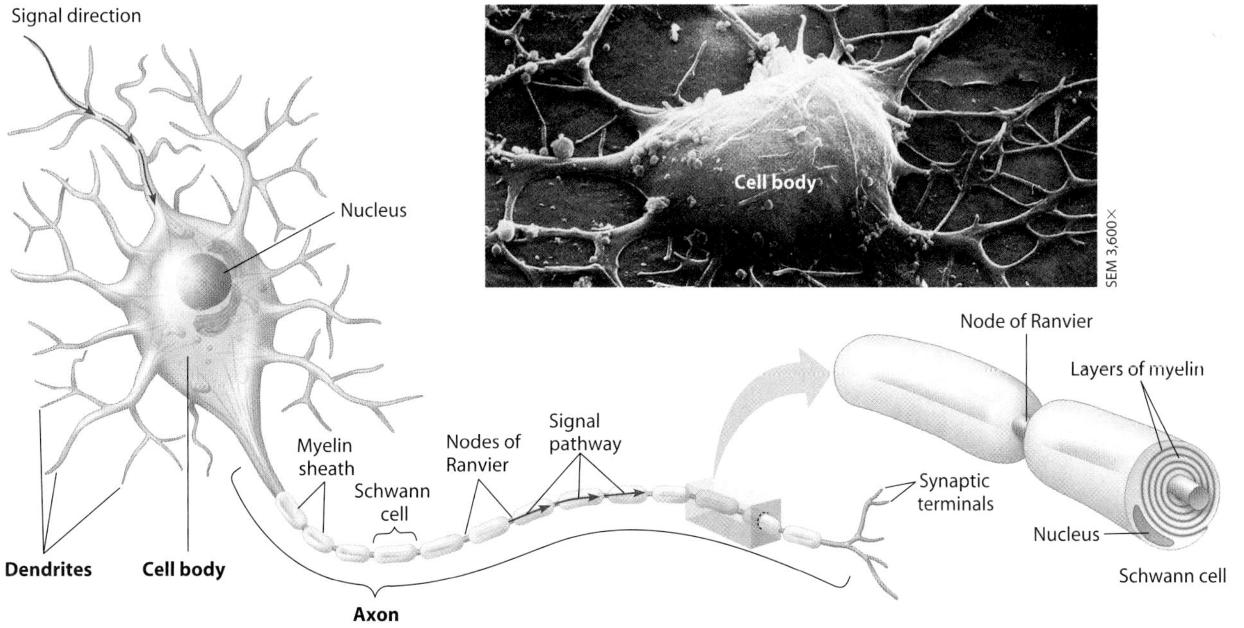

▲ **Figure 2** Structure of a myelinated motor neuron

Manfred Kage/Peter Arnold, Inc./PhotoLibrary

Nerve Signals and Their Transmission

3 Nerve function depends on charge differences across neuron membranes

To understand nerve signals, we must first study a resting neuron, one that is not transmitting a signal. Like all cells, a resting neuron has potential energy. In neurons, potential energy can be put to work sending signals from one part of the body to another. This potential energy, called the **membrane potential**, exists as an electrical charge difference across the neuron's plasma membrane: The inside of the cell is negatively charged relative to the outside as a result of unequal distribution of positively and negatively charged ions. Since opposite charges tend to move toward each other, a membrane stores energy by holding opposite charges apart, like a battery. The strength (voltage) of a neuron's stored energy can be measured with microelectrodes connected to a voltmeter. The voltage across the plasma membrane of a resting neuron is called the **resting potential**. A neuron's resting potential is about −70 millivolts (mV)—about 5% of the voltage in a flashlight battery. The minus sign indicates that the inside of the cell is negative relative to the outside.

The resting potential exists because of differences in ionic composition of the fluids inside and outside the neuron (**Figure 3**). The plasma membrane surrounding the neuron has protein channels and pumps that regulate the passage of inorganic ions. A resting membrane has many open potassium (K^+) channels but only a few open sodium (Na^+) channels, allowing much more potassium than sodium to diffuse across the membrane. Therefore, Na^+ is more concentrated outside the neuron than inside. (Mutations in genes for an Na^+ channel in the brain can cause groups of nerve cells to fire excessively, resulting in epileptic seizures.) But K^+, which is more concentrated inside, can flow out through the many open K^+ channels. As the positively charged potassium ions diffuse out, the inside of the neuron becomes less positive—that is, more negative—relative to the outside. Also helping maintain the resting potential are membrane proteins called **sodium-potassium (Na^+-K^+) pumps**. Using energy from ATP, these pumps actively transport Na^+ out of the neuron and K^+ in, thereby helping keep the concentration of Na^+ low in the neuron and K^+ high.

This ionic gradient (high K^+/low Na^+ concentrations inside coupled with low K^+/high Na^+ concentrations outside) produces an electrical potential difference, or voltage, across the membrane—the resting potential. Notice an important point: *The membrane potential can change from its resting value if the membrane's permeability to particular ions changes.* As we will see in the next module, this is the basis of nearly all electrical signals in the nervous system.

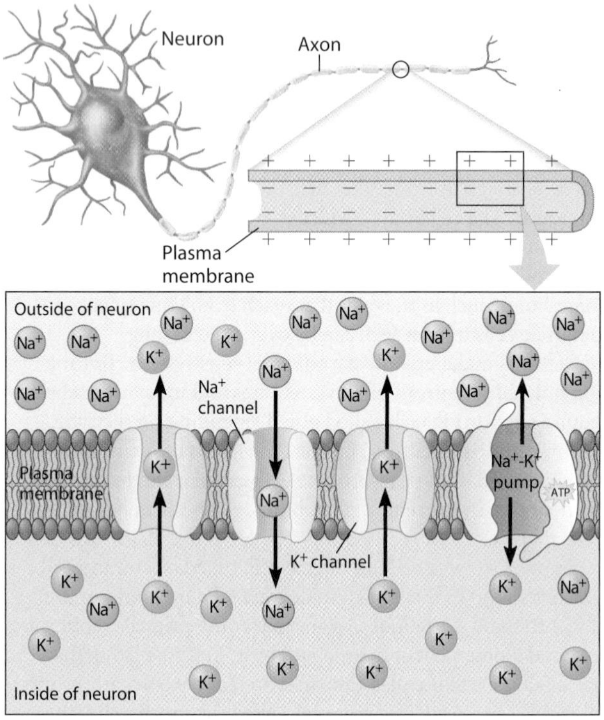

▲ **Figure 3** How the resting potential is generated

? If a neuron's membrane suddenly becomes more permeable to sodium ions, there is a rapid net movement of Na^+ into the cell. What are the two forces that drive the ions inward?

The greater concentration of Na^+ outside the cell than inside and the membrane potential (negatively charged inside versus outside) favor the inward diffusion of Na^+.

4 A nerve signal begins as a change in the membrane potential

Turning on a flashlight uses the potential energy stored in a battery to create light. In a similar way, stimulating a neuron's plasma membrane can trigger the use of the membrane's potential energy to generate a nerve signal. A **stimulus** is any factor that causes a nerve signal to be generated. Examples of stimuli include light, sound, a tap on the knee, or a chemical signal from another neuron.

The discovery of giant axons in squids (up to 1 mm in diameter) gave researchers their first chance to study how stimuli trigger signals in a living neuron. From microelectrode studies with squid neurons, British biologists A. L. Hodgkin and A. F. Huxley worked out the details of nerve signal transmission in the 1940s, earning a Nobel Prize for their findings. The nerve signal transmission process, summarized in

Figure 4, applies to neurons in nearly all animals, including humans.

The graph in the middle of the figure traces the electrical changes that make up an **action potential**, a change in membrane voltage that transmits a nerve signal along an axon. The graph records electrical events over time (in milliseconds) at a particular place on the membrane where a stimulus is applied. ❶ The graph starts out at the membrane's resting potential (–70 mV). ❷ The stimulus is applied. If it is strong enough, the voltage rises to what is called the *threshold* (–50 mV, in this case). The difference between the threshold and the resting potential is the minimum change in the membrane's voltage that must occur to generate the action potential (+20 mV, in this case). ❸ Once the threshold is reached, the action potential is triggered. The membrane polarity reverses abruptly, with the interior of the cell becoming positive with respect to the outside. ❹ The membrane then rapidly repolarizes as the voltage drops back down, ❺ undershoots the resting potential, and ❶ finally returns to it.

What actually causes the electrical changes of the action potential? The rapid flip-flop of the membrane potential is a result of the rapid movements of ions across the membrane at Na^+ and K^+ voltage-gated channels, so called because they have special gates that open and close, depending on changes in membrane potential. These are a kind of transport protein. The diagrams surrounding the graph show the ion movements.

Starting at lower left, ❶ the resting membrane separates a positively charged outside environment from a negatively charged inside environment. ❷ A stimulus triggers the opening of a few Na^+ voltage-gated channels in the membrane, and a tiny amount of Na^+ enters the axon. This makes the inside surface of the membrane slightly less negative. If the stimulus is strong enough, a sufficient number of these Na^+ channels open to raise the voltage to the threshold. ❸ Once the threshold is reached, more of these Na^+ channels open, allowing even more Na^+ to diffuse into the cell. As more and more Na^+ moves in, the voltage soars to its peak. ❹ The peak voltage triggers closing and inactivation of the Na^+ voltage-gated channels. Meanwhile, the K^+ voltage-gated channels open, allowing K^+ to diffuse out rapidly. These changes produce the downswing on the graph. ❺ A very brief undershoot of the resting potential results because these K^+ channels close slowly. ❶ The membrane then returns to its resting potential. In a typical mammalian neuron, this entire process takes just a few milliseconds, meaning that a neuron can produce hundreds of nerve signals per second.

? Is the generation of the peak of an action potential an example of positive feedback or negative feedback?

The opening of Na^+ gates caused by stimulation of the neuron changes the membrane potential, and this change causes more of the voltage-gated Na^+ channels to open. This is an example of a positive-feedback mechanism.

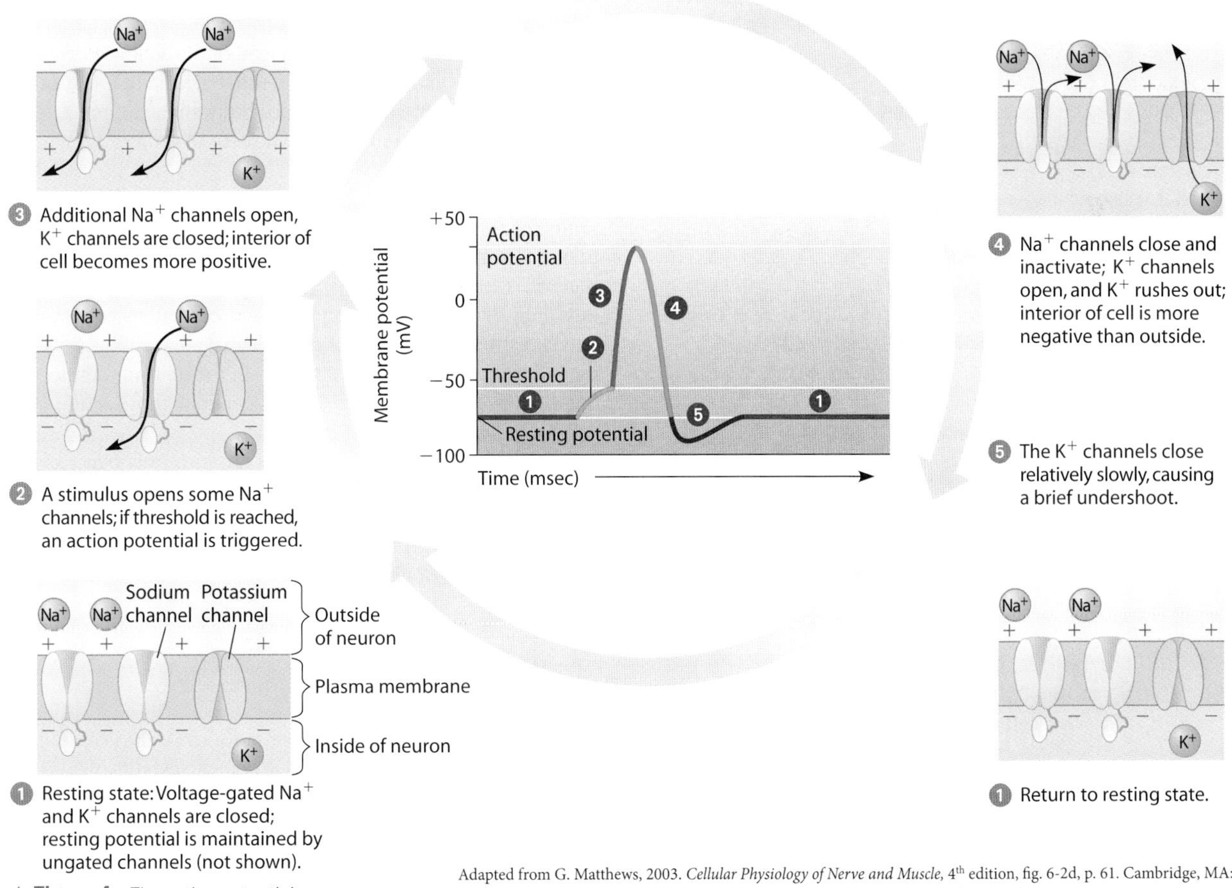

Na+
K+
+
–
❸ Additional Na+ channels open, K+ channels are closed; interior of cell becomes more positive.
❷ A stimulus opens some Na+ channels; if threshold is reached, an action potential is triggered.
Sodium channel
Potassium channel
Outside of neuron
Plasma membrane
Inside of neuron
❶ Resting state: Voltage-gated Na+ and K+ channels are closed; resting potential is maintained by ungated channels (not shown).
+50
0
–50
–100
Membrane potential (mV)
Action potential
Threshold
Resting potential
Time (msec)
❹ Na+ channels close and inactivate; K+ channels open, and K+ rushes out; interior of cell is more negative than outside.
❺ The K+ channels close relatively slowly, causing a brief undershoot.
❶ Return to resting state.

▲ **Figure 4** The action potential

Adapted from G. Matthews, 2003. *Cellular Physiology of Nerve and Muscle*, 4th edition, fig. 6-2d, p. 61. Cambridge, MA: Blackwell Scientific Publications. Reprinted by permission of Wiley Blackwell.

5 The action potential propagates itself along the axon

An action potential is a localized electrical event—a rapid change from the resting potential at a specific place along the neuron. A nerve signal starts out as an action potential generated in the axon, typically where the axon meets the cell body. To function as a long-distance signal, this local event must be passed along the axon from the cell body to the synaptic terminals. It does so by regenerating itself along the axon (**Figure 5**). The effect of this action potential is like tipping the first of a row of standing dominoes: The first domino does not travel along the row, but its fall is relayed to the end of the row, one domino at a time.

The three steps in Figure 5 show the changes that occur in an axon segment as a nerve signal passes from left to right. Let's first focus on the axon region on the far left. ❶ When this region of the axon (blue) has its Na^+ channels open, Na^+ rushes inward, and an action potential is generated. This corresponds to the upswing of the curve (step 2) in Figure 4. ❷ Soon, the K^+ channels in that same region open, allowing K^+ to diffuse out of the axon; at this time, its Na^+ channels are closed and inactivated at that point on the axon, and we would see the downswing of the action potential. ❸ A short time later, we would see no signs of an action potential at this (far-left) spot because the axon membrane here has restored itself and returned to its resting potential.

Now let's see how these events lead to the "domino effect" of a nerve signal. In step 1 of the figure, the blue arrows pointing sideways within the axon indicate local spreading of the electrical changes caused by the inflowing Na^+ associated with the first action potential. These changes are large enough to reach threshold in the neighboring regions, triggering the opening of Na^+ channels. As a result, a second action potential is generated, as indicated by the blue region in step 2. In the same way, a third action potential is generated in step 3, and each action potential generates another all the way down the axon. The net result is the movement of a nerve impulse from the cell body to the synaptic terminals.

So why are action potentials propagated in only one direction along the axon (left to right in the figure)? As the blue arrows indicate, local electrical changes do spread in both directions in the axon. However, these changes cannot open Na^+ channels and generate an action potential when the Na^+ channels are inactivated (step 4 in Figure 4). Thus, an action potential cannot be generated in the regions where K^+ is leaving the axon (green in the figure) and Na^+ channels are still inactivated. Consequently, the inward flow of Na^+ that depolarizes the axon membrane ahead of the action potential cannot produce another action potential behind it. Once an action potential starts where the cell body and axon meet, it moves along the axon in only one direction: toward the synaptic terminals.

Action potentials are all-or-none events; that is, they are the same no matter how strong or weak the stimulus that triggers them (as long as threshold is reached). How, then, do action potentials relay different intensities of information (such as a loud sound versus a soft sound) to your central nervous system? It is the *frequency* of action potentials that changes with the intensity of stimuli. For example, in the neurons connecting the ear to the brain, loud sounds generate more action potentials per second than quiet sounds.

In the past three modules, we've examined how nerve signals are conducted along a single neuron. In the next four modules, we focus on how signals pass from one neuron to another.

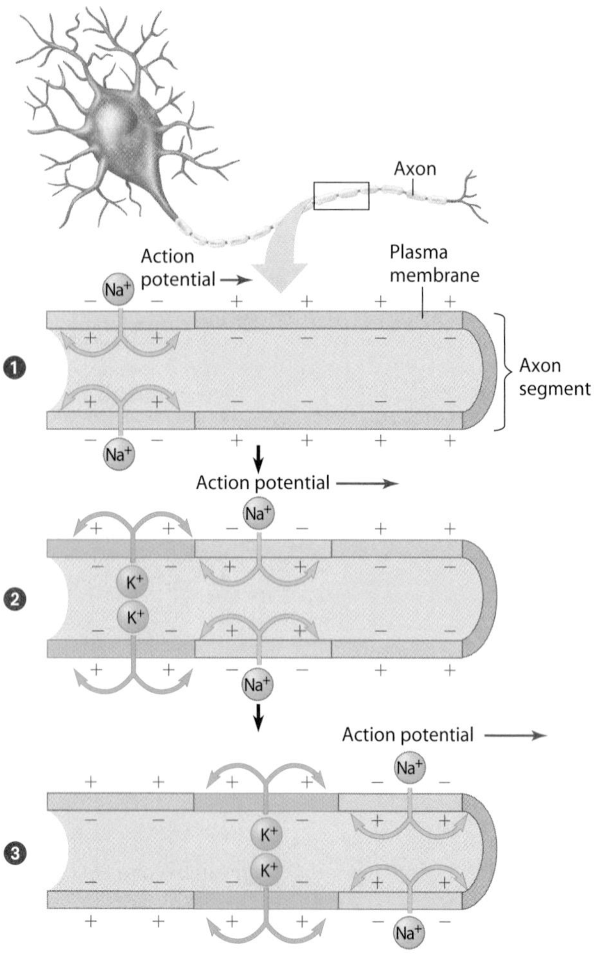

▲ **Figure 5** Propagation of the action potential along an axon

? During an action potential, ions move across the neuron plasma membrane in a direction perpendicular to the direction of the impulse along the neuron. What is it that actually travels along the neuron as the signal?

● The signal is the wavelike change in membrane potential, the self-perpetuated action potential that regenerates sequentially at points farther and farther away from the site of stimulation.

6 Neurons communicate at synapses

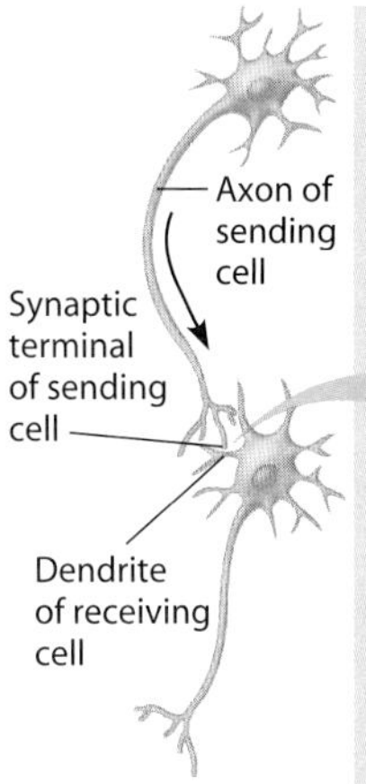

If an action potential travels in one direction along an axon, what happens when the signal arrives at the end of the neuron? To continue conveying information, the signal must be passed to another cell. This occurs at a **synapse**, or relay point, between a synaptic terminal of a sending neuron and a receiving cell. The receiving cell can be another neuron or an effector cell such as a muscle cell or endocrine cell.

Synapses come in two varieties: electrical and chemical. In an electrical synapse, electrical current flows directly from a neuron to a receiving cell. The receiving cell is stimulated quickly and at the same frequency of action potentials as the sending neuron. Lobsters and many fishes can flip their tails with lightning speed because the neurons that carry signals for these movements communicate by fast electrical synapses. In the human body, electrical synapses are found in the heart and digestive tract, where nerve signals maintain steady, rhythmic muscle contractions.

In contrast, when an action potential reaches a chemical synapse, it stops there. This is because, unlike electrical synapses, chemical synapses have a narrow gap, called the **synaptic cleft**, separating a synaptic terminal of the sending (presynaptic) cell from the receiving (postsynaptic) cell. The cleft is very narrow—only about 50 nm, about 1/1,000th the width of a human hair—but it prevents the action potential from spreading directly to the receiving cell. Instead, the action potential (an electrical signal) is first converted to a chemical signal consisting of molecules of **neurotransmitter**. The chemical signal may then generate an action potential in the receiving cell.

Let's follow the events that occur at a typical chemical synapse, shown in **Figure 6**. Molecules of the neurotransmitter (⁘) are in membrane-enclosed sacs called **synaptic vesicles** in the sending neuron's synaptic terminals. ❶ An action potential arrives at the synaptic terminal. ❷ The action potential causes some synaptic vesicles to fuse with the plasma membrane of the sending cell. ❸ The fused vesicles release their neurotransmitter molecules by exocytosis into the synaptic cleft, and the neurotransmitter rapidly diffuses across the cleft.

The subsequent steps in Figure 6 show one example of what can happen next at a chemical synapse. ❹ The released neurotransmitter binds to complementary receptors on ion channel proteins in the receiving cell's plasma membrane. ❺ In a typical chemical synapse, the binding of neurotransmitter to receptor opens these chemically gated ion channels in the receiving cell's membrane. With the channels open, ions can diffuse into the receiving cell and trigger new action potentials. ❻ The neurotransmitter is broken down by an enzyme or is transported back into the signaling cell, and the ion channels close. Step 6 ensures that the neurotransmitter's effect is brief and precise.

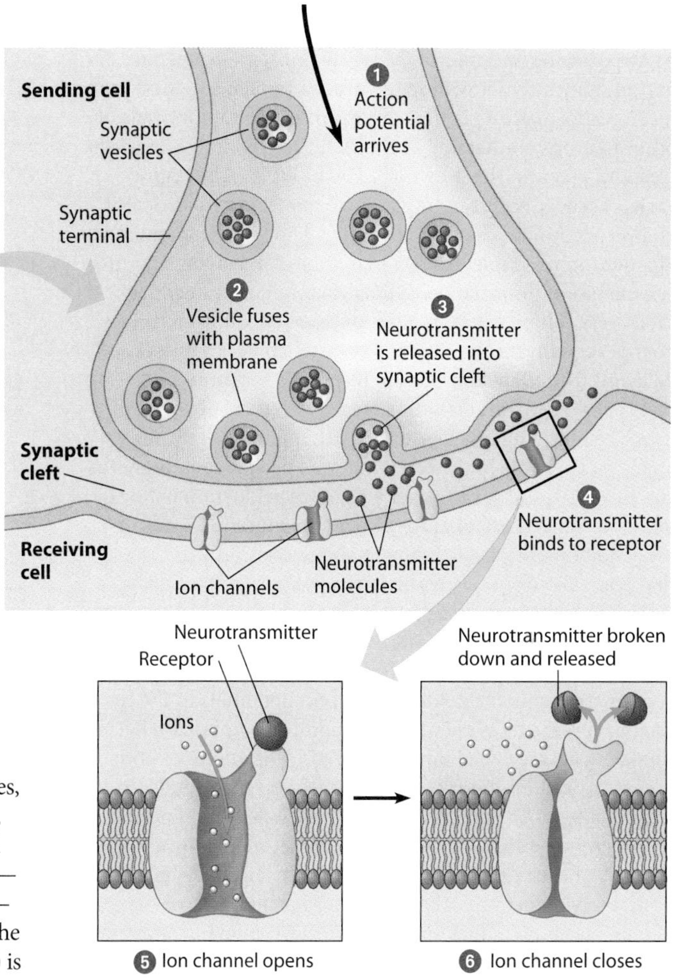

▲ **Figure 6** Neuron communication at a typical chemical synapse

Adapted from Becker, Kleinsmith, and Hardin, *The World of the Cell*, 6th ed. Copyright © 2006 Pearson Education, Inc., publishing as Pearson Benjamin Cummings.

You can review what we have covered so far in this chapter by thinking about what is happening right now in your own nervous system. Action potentials carrying information about the words on this page are streaming along sensory neurons from your eyes to your brain. Arriving at synapses with receiving cells (interneurons in the brain), the action potentials are triggering the release of neurotransmitters at the ends of the sensory neurons. The neurotransmitters are diffusing across synaptic clefts and triggering changes in some of your interneurons—changes that lead to integration of the signals and ultimately to what the signals actually mean (in this case, the meaning of words and sentences). Next, motor neurons in your brain will send out action potentials to muscle cells in your eyes, telling them to move to the next section of text.

? How does a synapse ensure that signals pass only in one direction, from a sending neuron to a receiving cell?

The signal can go only one way at any one synapse because only the sending neuron releases neurotransmitter, and only the receiving cell has receptors for the neurotransmitter.

7 Chemical synapses enable complex information to be processed

As the drawing and micrograph in **Figure 7** indicate, one neuron may interact with many others. In fact, a neuron may receive information via neurotransmitters from hundreds of other neurons connecting at thousands of synaptic terminals (red and green in the drawing). The inputs can be highly varied because each sending neuron may secrete a different quantity or kind of neurotransmitter. The plasma membrane of a neuron resembles a tiny circuit board, receiving and processing bits of information in the form of neurotransmitter molecules. These exceedingly sophisticated living circuit boards account for the nervous system's ability to process data and formulate appropriate responses to stimuli.

What do neurotransmitters actually do to receiving neurons? The binding of a neurotransmitter to a receptor may open ion channels in the receiving cell's plasma membrane, as you saw in Figure 6, or trigger a signal transduction pathway that does so. The effect of the neurotransmitter depends on the kind of membrane channel it opens. Neurotransmitters that open Na^+ channels, for instance, may trigger action potentials in the receiving cell. Such effects are referred to as excitatory (green in the drawing). In contrast, many neurotransmitters open membrane channels for ions that *decrease* the tendency to develop action potentials in the receiving cell—such as channels that admit Cl^- or release K^+. These effects are called inhibitory (red). The effects of both excitatory and inhibitory signals can vary in magnitude. In general, the more neurotransmitter molecules that bind to receptors on the receiving cell and the closer the synapse is to the base of the receiving cell's axon, the stronger the effect.

A receiving neuron's plasma membrane may receive signals—both excitatory and inhibitory—from many different sending neurons. If the excitatory signals are collectively strong enough to raise the membrane potential to threshold, an action potential will be generated in the receiving cell. That neuron then passes signals along its axon to other cells at a rate that represents a summation of all the information it has received. Signal frequency is key because action potentials are all-or-none events. Each new receiving cell, in turn, processes this information along with all its other inputs.

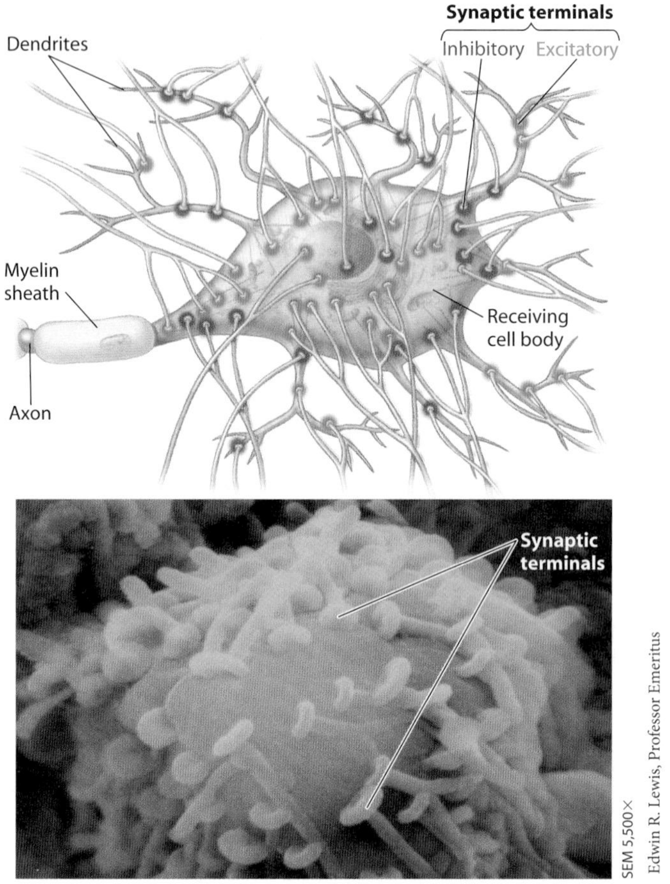

▲ **Figure 7** A neuron's multiple synaptic inputs

? Contrast how excitatory and inhibitory signals change a receiving cell's membrane potential relative to triggering an action potential.

Neurotransmitters in an excitatory signal open ion channels that move the receiving cell's membrane potential closer to threshold. Neurotransmitters in an inhibitory signal open ion channels that move the cell's membrane potential farther from threshold.

8 A variety of small molecules function as neurotransmitters

As discussed in Modules 6 and 7, the propagation of nerve signals across chemical synapses depends on neurotransmitters. A variety of small molecules serve this function.

Many neurotransmitters are small, nitrogen-containing organic molecules. One, called **acetylcholine**, is important in the brain and at synapses between motor neurons and muscle cells. Depending on the kind of receptors on receiving cells, acetylcholine may be excitatory or inhibitory. For instance, acetylcholine makes our skeletal muscles contract but slows the rate of contraction of cardiac muscles. Botulinum toxin (sold as Botox), made by the bacteria that cause botulism food poisoning, inhibits the release of acetylcholine. Botox injections disable the synapses that control certain facial muscles, thereby eliminating wrinkles around the eyes or mouth.

Four other neurotransmitters—aspartate, glutamate, glycine, and GABA (gamma aminobutyric acid)—are actually amino acids. All are known to be important in the central nervous system. Aspartate and glutamate act primarily at excitatory synapses, while glycine and GABA act at inhibitory synapses.

Biogenic amines are neurotransmitters derived from amino acids. Some examples of biogenic amines are epinephrine, norepinephrine, serotonin, and dopamine, the first two of which also function as hormones (see Chapter 26). Biogenic amines are important neurotransmitters in the central nervous system. Serotonin and dopamine affect sleep, mood, attention, and learning. Imbalances of biogenic amines are associated with various disorders. For example, the degenerative illness Parkinson's disease is associated with a lack of dopamine in the brain. Reduced levels of norepinephrine and serotonin seem to be linked with some types of depression. Some psychoactive drugs, including LSD and mescaline, apparently produce their hallucinatory effects by binding to serotonin and dopamine receptors in the brain.

Many neuropeptides, relatively short chains of amino acids, also serve as neurotransmitters. The endorphins are peptides that function as both neurotransmitters and hormones, decreasing our perception of pain during times of physical or emotional stress. Endorphins may be released in response to a wide variety of stimuli, including traumatic injury, muscle fatigue, and even eating very spicy foods.

Neurons also use some dissolved gases, notably nitric oxide (NO), as chemical signals. During sexual arousal in human males, certain neurons release NO into blood vessels in the erectile tissue of the penis, and the NO triggers an erection. (The erectile dysfunction drug Viagra promotes this effect of NO.) Neurons produce NO molecules on demand, rather than storing them in synaptic vesicles. The dissolved gas diffuses into neighboring cells, produces a change, and is broken down—all within a few seconds.

? What determines whether a neuron is affected by a specific neurotransmitter?

To be affected by a particular neurotransmitter, a neuron must have specific receptors for that neurotransmitter.

CONNECTION 9 Many drugs act at chemical synapses

Many psychoactive drugs, even common ones such as caffeine, nicotine, and alcohol, affect the action of neurotransmitters in the brain's billions of synapses (**Figure 9**). Caffeine, found in coffee, tea, chocolate, and many soft drinks, keeps us awake by countering the effects of inhibitory neurotransmitters, ones that normally suppress nerve signals. Nicotine acts as a stimulant by binding to and activating acetylcholine receptors. Alcohol is a strong depressant. Its precise effect is not yet known, but it seems to increase the inhibitory effects of the neurotransmitter GABA.

Many prescription drugs used to treat psychological disorders alter the effects of neurotransmitters (see Module 20). The most popular class of antidepressant medication works by affecting the action of serotonin. Called selective serotonin reuptake inhibitors (SSRIs), these medications block the removal of serotonin from synapses, increasing the amount of time this mood-altering neurotransmitter is available to receiving cells. Tranquilizers such as diazepam (Valium) and alprazolam (Xanax) activate the receptors for the neurotransmitter GABA, increasing its effect at inhibitory synapses.

PhotoDisc/Getty Images

Steve Gorton/Dorling Kindersley

▲ **Figure 9** Alcohol, nicotine, and caffeine—drugs that alter the effects of neurotransmitters

In other cases, a drug may bind to and block a receptor, reducing a neurotransmitter's effect. For instance, some drugs used to treat schizophrenia block receptors for the neurotransmitter dopamine. Some drugs used to treat attention deficit hyperactivity disorder (ADHD), such as methylphenidate (Ritalin) and Adderall (a combination drug), block the reuptake of dopamine and norepinephrine, which can increase alertness. Some students abuse Adderall as a "study drug," which can be quite dangerous due to significant risk of seizures and cardiac problems. Because the effects of these drugs are poorly understood, a physician should monitor anyone taking them.

What about illegal drugs? Stimulants such as amphetamines and cocaine increase the release and availability of norepinephrine and dopamine at synapses. Abuse of these drugs can cause symptoms resembling schizophrenia. The active ingredient in marijuana (tetrahydrocannabinol, or THC) binds to brain receptors normally used by other neurotransmitters that seem to play a role in pain, depression, appetite, memory, and fertility. Opiates—morphine, codeine, and heroin—bind to endorphin receptors, reducing pain and producing euphoria. Not surprisingly, opiates are used medicinally for pain relief. However, abuse of opiates may permanently change the brain's chemical synapses and reduce normal synthesis of neurotransmitters. As explained in Module 14, these drugs are also highly addictive.

The drugs discussed here are used for a variety of purposes, both medicinal and recreational. While they increase alertness and sense of well-being or reduce physical and emotional pain, they may disrupt the brain's finely tuned neural pathways, altering the chemical balances that are the product of millions of years of evolution.

? What is the biochemical basis for the pain-relieving properties of morphine?

Morphine binds to the same receptors as endorphins, the body's natural pain relievers.

An Overview of Animal Nervous Systems

EVOLUTION CONNECTION

10 The evolution of animal nervous systems reflects changes in body symmetry

To this point in the chapter, we have concentrated on the cellular mechanisms that are fundamental to nearly all animal nervous systems. There is remarkable uniformity throughout the animal kingdom in the way nerve cells function—a strong indication that the basic architecture of the neuron was an early evolutionary adaptation. But during subsequent animal evolution, great diversity emerged in the organization of nervous systems as a whole.

The ability to sense and react originated billions of years ago with prokaryotes that could detect changes in their environment and respond in ways that enhanced their survival and reproductive success. Later, modification of simple recognition and response processes provided multicellular organisms with a mechanism for communication between cells of the body. By 500 million years ago, systems of neurons allowing animals to sense and move rapidly were present in essentially their current forms.

Animals on the earliest branches of the animal evolutionary tree—sponges, for instance—lack a nervous system. The first modern phylum to evolve a nervous system was the cnidarians. Hydras, jellies, and other cnidarians have a **nerve net** (**Figure 10A**), a diffuse, weblike system of interconnected neurons extending throughout the body. Neurons of the nerve net control the contractions of the digestive cavity, movement of the tentacles, and other functions.

Cnidarians and adult echinoderms have radial, uncentralized nervous systems. Sea stars and many other echinoderms have a nerve net in each arm connected by radial nerves to a central nerve ring. This organization is better suited than a diffuse nerve net for controlling complex motion, such as the coordinated movements of hundreds of tube feet that allow a sea star to move in one direction.

The appearance of bilateral symmetry marks a key branch point in the evolution of animals and their nervous systems. Bilaterally symmetric animals tend to move through the environment with the head—usually equipped with sense organs and a brain—first encountering new stimuli. Flatworms are the simplest animals to have evolved two hallmarks of bilateral symmetry: **cephalization**, an evolutionary trend toward the concentration of the nervous system at the head end, and **centralization**, the presence of a central nervous system distinct from a peripheral nervous system. For example, the planarian worm in **Figure 10B** has a small brain composed of ganglia (clusters of nerve cell bodies) and two parallel **nerve cords**, elongated bundles of neurons that control the animal's movements. These elements constitute the simplest clearly defined CNS in the animal kingdom. Nerves that connect the CNS with the rest of the body make up the PNS.

In subsequent animal phyla, the CNS evolved in complexity. For instance, the brains of leeches (**Figure 10C**) have a greater concentration of neurons than those of flatworms, and leech ventral nerve cords contain segmentally arranged ganglia. The insect shown in **Figure 10D** has a brain composed of several fused ganglia, and its ventral nerve cord also has a ganglion in each body segment. Each of these ganglia directs the activity of muscles in its segment of the body.

Molluscs serve as a good illustration of how natural selection leads to correlation of the structure of a nervous system with an animal's interaction with the environment. Sessile or slow-moving molluscs, such as clams, have little or no cephalization and relatively simple sense organs. In contrast, the relatively large brain of a squid (**Figure 10E**), accompanied by complex eyes and rapid signaling along giant axons, correlates well with the active predatory life of these animals.

In the next several modules, we explore the complex nervous systems that evolved in our own subphylum, the vertebrates.

? Why is it advantageous for the brain of most bilateral animals to be located at the head end?

Cephalization places the brain close to major sense organs, which are concentrated on the end of the animal that leads the way as the animal moves through its environment.

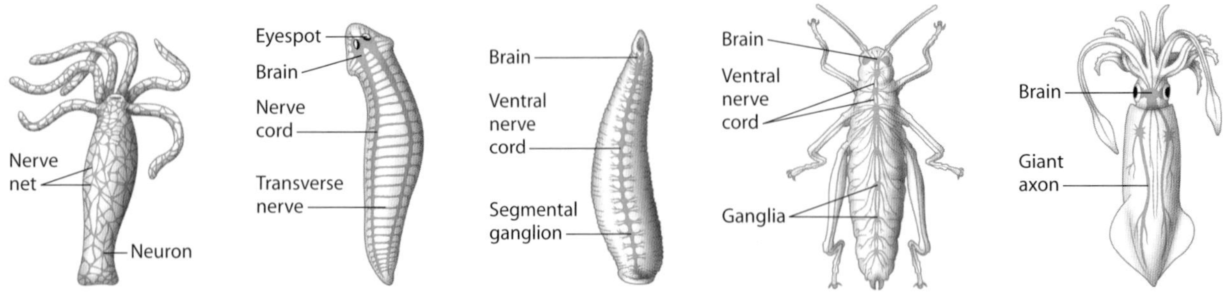

A Hydra (cnidarian) B Flatworm (planarian) C Leech (annelid) D Insect (arthropod) E Squid (mollusc)

▲ **Figure 10** Invertebrate nervous systems (not to scale)

11 Vertebrate nervous systems are highly centralized

Vertebrate nervous systems are diverse in structure and level of sophistication. For instance, those of dolphins and humans are much more complex structurally than those of frogs or fishes; they are also much more powerful integrators. However, all vertebrate nervous systems have fundamental similarities. All have distinct central and peripheral elements and are highly centralized. In all vertebrates, the brain and spinal cord make up the CNS, while the PNS comprises the rest of the nervous system (**Figure 11A**). The **spinal cord**, a jellylike bundle of nerve fibers that runs lengthwise inside the spine, conveys information to and from the brain and integrates simple responses to certain stimuli (such as the knee-jerk reflex). The master control center of the nervous system, the **brain**, includes homeostatic centers that keep the body functioning smoothly; sensory centers that integrate data from the sense organs; and (in humans, at least) centers of emotion and intellect. The brain also sends motor commands to muscles.

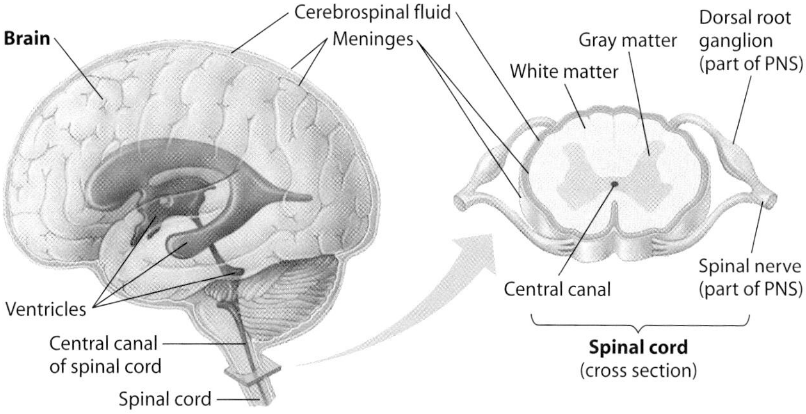

▲ **Figure 11B** Fluid-filled spaces of the vertebrate CNS

A vast network of blood vessels services the CNS. Brain capillaries are more selective than those elsewhere in the body. They allow essential nutrients and oxygen to pass freely into the brain, but keep out many chemicals, such as metabolic wastes from other parts of the body. This selective mechanism, called the **blood-brain barrier**, maintains a stable chemical environment for the brain.

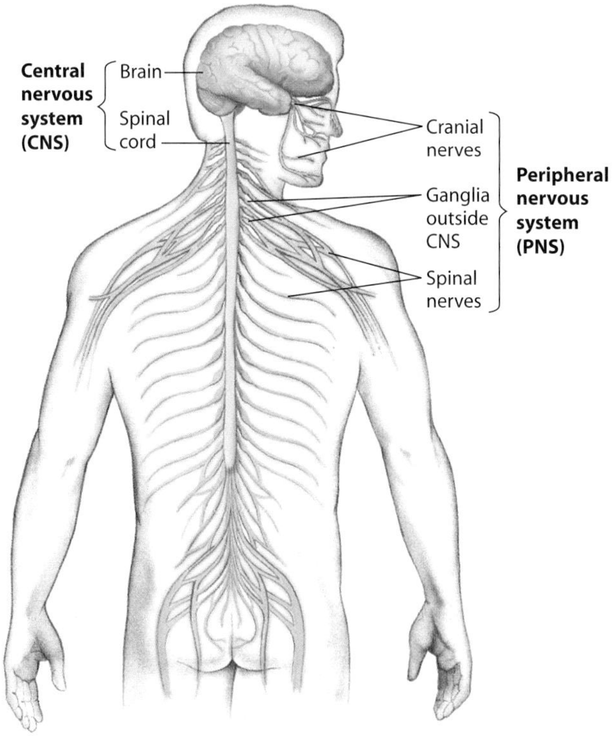

▲ **Figure 11A** A vertebrate nervous system (back view)

Fluid-filled spaces in the brain called **ventricles** are continuous with the narrow **central canal** of the spinal cord (**Figure 11B**). These cavities are filled with **cerebrospinal fluid**, which is formed within the brain by the filtration of blood. Circulating slowly through the central canal and ventricles (and then draining back into veins), the cerebrospinal fluid cushions the CNS and assists in supplying nutrients and hormones and removing wastes. Also protecting the brain and spinal cord are layers of connective tissue, called **meninges**. If the cerebrospinal fluid becomes infected by bacteria or viruses, the meninges may become inflamed, a condition called meningitis. Viral meningitis is generally not harmful, but bacterial meningitis can have serious consequences if not treated with antibiotics. A sample of cerebrospinal fluid can be collected for testing by a spinal tap, a procedure in which a narrow needle is inserted through the spinal column into the central canal. In mammals, cerebrospinal fluid circulates between layers of the meninges, providing an additional protective cushion for the CNS.

As shown on the right side of Figure 11B, the CNS has white matter and gray matter. **White matter** is composed mainly of axons (with their whitish myelin sheaths); **gray matter** consists mainly of nerve cell bodies and dendrites.

The ganglia and nerves of the vertebrate PNS are a vast communication network. Cranial nerves originate in the brain and usually end in structures of the head and upper body (your eyes, nose, and ears, for instance). Spinal nerves originate in the spinal cord and extend to parts of the body below the head. All spinal nerves and most cranial nerves contain sensory and motor neurons. Thousands of incoming and outgoing signals pass within the same nerves all the time.

In the next module, we look closely at the vertebrate PNS.

? A vertebrate's central nervous system consists of the ______ and the ______.

brain . . . spinal cord

12 The peripheral nervous system of vertebrates is a functional hierarchy

As shown in **Figure 12A**, the PNS can be divided into two functional components: the motor system and the autonomic nervous system. The **motor system** carries signals to and from skeletal muscles, mainly in response to external stimuli. When you read, for instance, these neurons carry commands that make your eye muscles move. The control of skeletal muscles can be voluntary, as when you raise your hand to ask a question, or involuntary, as in a knee-jerk reflex controlled by the spinal cord.

The **autonomic nervous system** regulates the internal environment by controlling smooth and cardiac muscles and the organs and glands of the digestive, cardiovascular, excretory, and endocrine systems. This control is generally involuntary.

As you can see in Figure 12A, the autonomic nervous system is composed of three divisions: parasympathetic, sympathetic, and enteric. The parasympathetic and sympathetic divisions have largely antagonistic (opposite) effects on most body organs. The **parasympathetic division** primes the body for activities that gain and conserve energy for the body ("rest and digest"). A sample of the effects of parasympathetic signals appears on the left in **Figure 12B**, on the facing page. These include stimulating the digestive organs, such as the salivary glands, stomach, and pancreas; decreasing the heart rate; and increasing glycogen production.

Neurons of the **sympathetic division** tend to have the opposite effect, preparing the body for intense, energy-consuming activities, such as fighting, fleeing, or competing in a strenuous game (the "fight-or-flight" response). You see some of the effects of sympathetic signals on the right side of the figure. The digestive organs are inhibited, the bronchi dilate so that more air can pass through them, the heart rate increases, the liver releases the energy compound glucose into the blood, and the adrenal glands secrete the hormones epinephrine and norepinephrine.

Fight-or-flight and relaxation are opposite extremes. Your body usually operates at intermediate levels, with most of your organs receiving both sympathetic and parasympathetic signals. The opposing signals adjust an organ's activity to a suitable level. In regulating some body functions, however, the two divisions complement rather than antagonize each other. For example, in regulating reproduction, erection is promoted by the parasympathetic division while ejaculation is promoted by the sympathetic division.

Sympathetic and parasympathetic neurons emerge from different regions of the CNS, represented in Figure 12B by black dots, and they use different neurotransmitters. As the dots in the figure indicate, neurons of the parasympathetic system emerge from the brain and the lower part of the spinal cord. Most parasympathetic neurons produce their effects by releasing the neurotransmitter acetylcholine at synapses within target organs. In contrast, neurons of the sympathetic system emerge from the middle regions of the

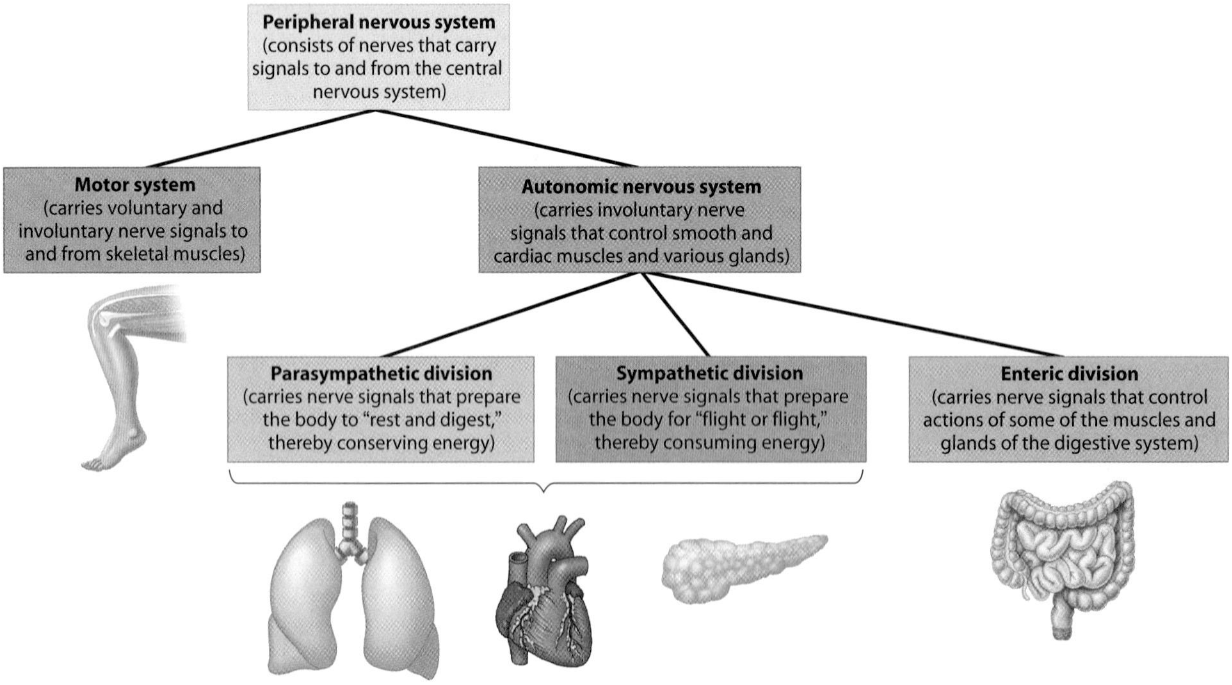

▲ **Figure 12A** Functional divisions of the vertebrate PNS

spinal cord. Most sympathetic neurons release the neurotransmitter norepinephrine at target organs.

While it is convenient to divide the PNS into motor and autonomic components, it is important to realize that these two divisions cooperate to maintain homeostasis. In response to a drop in body temperature, for example, the brain signals the autonomic nervous system to constrict surface blood vessels, which reduces heat loss. At the same time, the brain also signals the motor nervous system to cause shivering, which increases heat production.

The **enteric division** (not shown in Figure 12B) of the autonomic nervous system consists of networks of neurons in the digestive tract, pancreas, and gallbladder. Within these organs, neurons of the enteric division control secretion as well as activity of the smooth muscles that produce peristalsis. Although the enteric division can function independently, it is normally regulated by the sympathetic and parasympathetic divisions.

In the next several modules, we take a closer look at the highest level of the nervous system's structural hierarchy: the brain.

? How would a drug that inhibits the parasympathetic nervous system affect a person's pulse?

Since signals from the parasympathetic nervous system slow the heart rate, a drug that inhibits the parasympathetic division would result in an increased pulse.

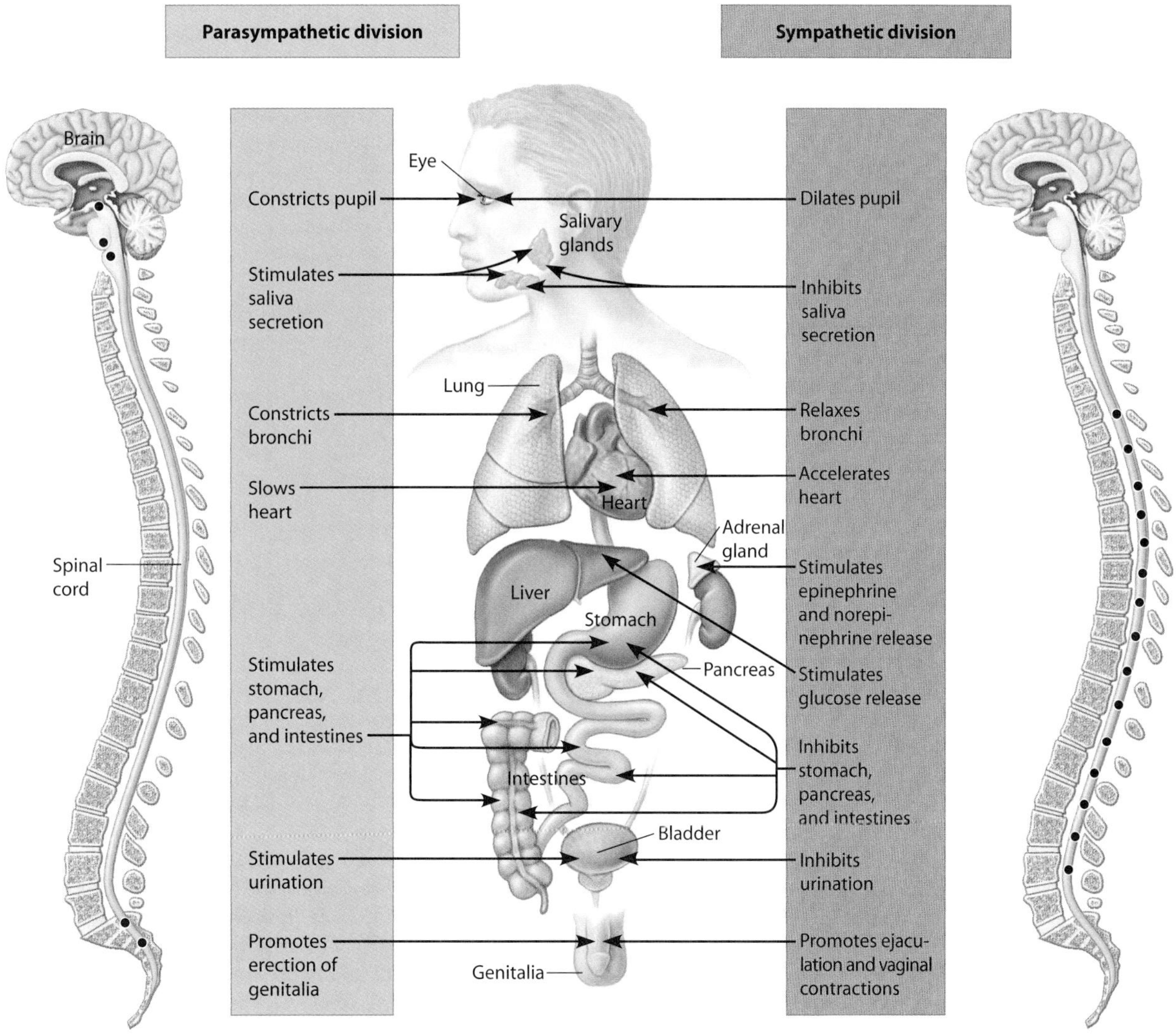

▲ **Figure 12B** The parasympathetic and sympathetic divisions of the autonomic nervous system

13 The vertebrate brain develops from three anterior bulges of the neural tube

Before we focus on the human brain in the next section, we close our overview of animal nervous systems by examining the embryonic development of the vertebrate nervous system from the dorsal hollow nerve cord, one of the four distinguishing features of chordates. During early embryonic development in all vertebrates, three bilaterally symmetric bulges—the **forebrain**, **midbrain**, and **hindbrain**—appear at the anterior end of the neural tube (Figure 13 left). In the course of vertebrate evolution, the forebrain and hindbrain gradually became subdivided—both structurally and functionally—into regions that assume specific responsibilities.

Another trend in brain evolution was the increasing integrative power of the forebrain. Evolution of the most complex vertebrate behavior paralleled the evolution of the **cerebrum**, an outgrowth of the forebrain that is the most sophisticated center of homeostatic control and integration. The cerebrum of birds and mammals is much larger relative to other parts of the brain than the cerebrum of fishes, amphibians, and other reptiles. Also, the sophisticated behavior of birds and mammals is directly correlated with their large cerebrum.

During the embryonic development of the human brain, the most profound changes occur in the region of the forebrain. Rapid, expansive growth of this region during the second and third months creates the cerebrum, which extends over and around much of the rest of the brain (Figure 13, right). By the sixth month of development, foldings increase the surface area of the cerebrum. This extensively convoluted outer region is called the **cerebral cortex**. (See Figure 14A, on the next page.) As you'll see, the cerebrum develops into two halves, called the left and right cerebral hemispheres.

Porpoises, whales, and primates also have large and complex cerebral cortices. Porpoises communicate using a large variety of sounds. They also have the ability to locate objects, such as prey, using sound echoes. Much of a porpoise's cerebral cortex may be devoted to processing information about its sound-oriented world. In contrast, the brains of humans and other primates are strongly oriented toward visual perceptions. Humans have the largest brain surface area, relative to body size, of all animals. Next we take a look at the main components of the human brain.

Embryonic Brain Regions	Brain Structures Present in Adult
Forebrain	Cerebrum (cerebral hemispheres; includes cerebral cortex, white matter, basal ganglia)
	Diencephalon (thalamus, hypothalamus, posterior pituitary, pineal gland)
Midbrain	Midbrain (part of brainstem)
Hindbrain	Pons (part of brainstem), cerebellum
	Medulla oblongata (part of brainstem)

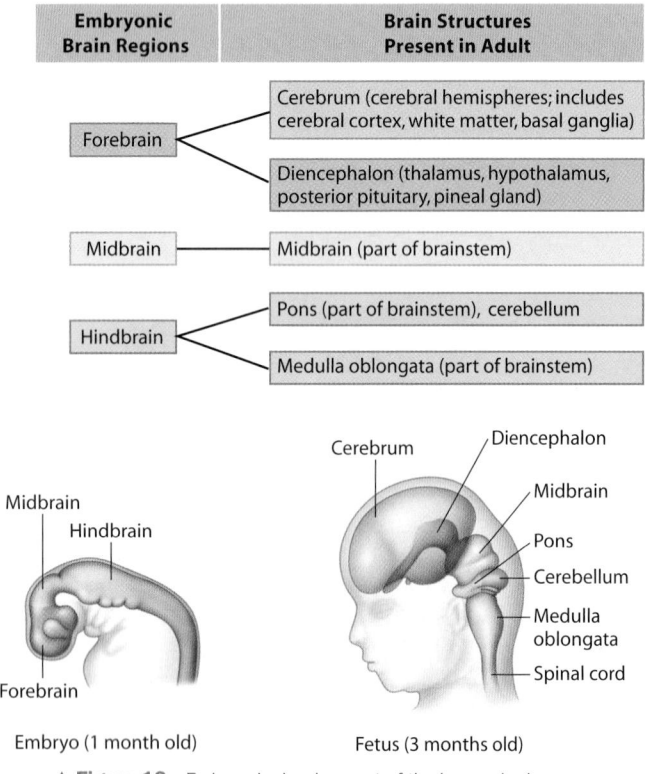

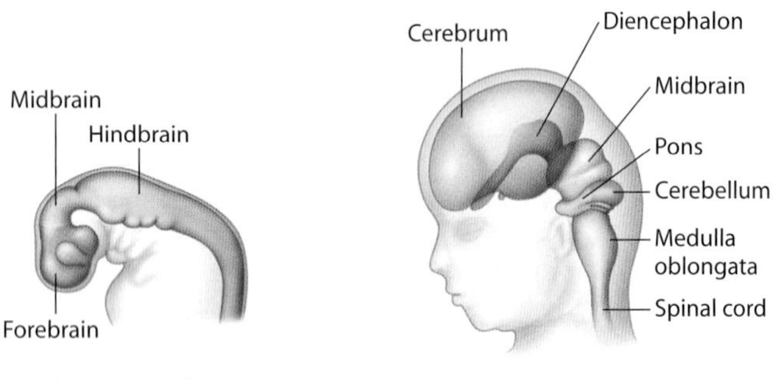

▲ **Figure 13** Embryonic development of the human brain

? Which region of the brain has changed the most during the course of vertebrate evolution?

The cerebrum, especially the cerebral cortex

The Human Brain

14 The structure of a living supercomputer: The human brain

Composed of up to 100 billion intricately organized neurons, with a much larger number of supporting cells, the human brain is more powerful than the most sophisticated computer. Figure 14A shows the three ancestral brain regions in their fully developed adult form. Table 14 summarizes the anatomy and physiology of the major structures of the brain.

Looking first at the lower brain centers, two sections of the hindbrain (blue), the **medulla oblongata** and **pons**, and the midbrain (purple) make up a functional unit called the **brainstem**. Consisting of a stalk with cap-like swellings at the anterior end of the spinal cord, the brainstem is, evolutionarily speaking, one of the older parts of the vertebrate brain. The brainstem coordinates and filters the conduction of information from sensory and motor neurons to the higher brain regions. It also regulates sleep and arousal and helps coordinate body movements, such as walking. Table 14 lists some of the individual functions of the medulla oblongata, pons, and midbrain.

Another part of the hindbrain, the **cerebellum** (light blue), is a planning center for body movements. It also plays a role in learning, decision making, and remembering motor responses. The cerebellum receives sensory information about the position of joints and the length of muscles, as well as information from

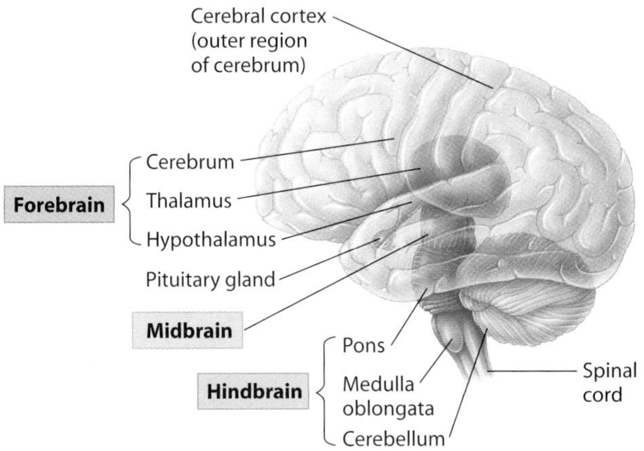

▲ **Figure 14A** The main parts of the human brain

the auditory and visual systems. It also receives input concerning motor commands issued by the cerebrum. The cerebellum uses this information to coordinate movement and balance. Hand-eye coordination is an example of such control by the cerebellum. If the cerebellum is damaged, the eyes can follow a moving object, but they will not stop at the same place as the object.

The most sophisticated integrating centers are those derived from the forebrain (orange and gold)—the thalamus, the hypothalamus, and the cerebrum. The **thalamus** contains most of the cell bodies of neurons that relay information to the cerebral cortex. The thalamus first sorts data into categories (all of the touch signals from a hand, for instance). It also suppresses some signals and enhances others. The thalamus then sends information on to the appropriate higher brain centers for further interpretation and integration.

That the hypothalamus controls the pituitary gland and the secretion of many hormones. The hypothalamus also regulates body temperature, blood pressure, hunger, thirst, sex drive, and fight-or-flight responses, and it helps us experience emotions such as rage and pleasure. A "pleasure center" in the hypothalamus could also be called an addiction center, for it is strongly affected by certain addictive drugs, such as amphetamines and cocaine. As described in Module 9, these drugs increase the effects of norepinephrine and dopamine at synapses in the pleasure center, producing a short-term high, often followed by depression. Cocaine addiction may involve chemical changes in the pleasure center and elsewhere in the hypothalamus.

A pair of hypothalmic structures called the suprachiasmatic nuclei function as an internal timekeeper, our **biological clock**. Receiving visual input from the eyes (light/dark cycles, in particular), the clock maintains our **circadian rhythms**—daily cycles of biological activity—such as the sleep/wake cycle. Research with many different species has shown that without environmental cues, biological clocks keep time in a free-running way. For example, when humans are placed in artificial settings that lack environmental cues, our biological clocks and circadian rhythms maintain a cycle of approximately 24 hours 11 minutes, with very little variation among individuals.

TABLE 14 | MAJOR STRUCTURES OF THE HUMAN BRAIN

Brain Structure	Major Functions
Brainstem	Conducts data to and from other brain centers; helps maintain homeostasis; coordinates body movement
Medulla oblongata	Controls breathing, circulation, swallowing, digestion
Pons	Controls breathing
Midbrain	Receives and integrates auditory data; coordinates visual reflexes; sends sensory data to higher brain centers
Cerebellum	Coordinates body movement; plays role in learning and in remembering motor responses
Thalamus	Serves as input center for sensory data going to the cerebrum, output center for motor responses leaving the cerebrum; sorts data
Hypothalamus	Functions as homeostatic control center; controls pituitary gland; serves as biological clock
Cerebrum	Performs sophisticated integration; plays major role in memory, learning, speech, emotions; formulates complex behavioral responses

The cerebrum, the largest and most complex part of our brain, consists of right and left **cerebral hemispheres** (**Figure 14B**), each responsible for the opposite side of the body. A thick band of nerve fibers called the **corpus callosum** facilitates communication between the hemispheres, enabling them to process information together. Under the corpus callosum, groups of neurons called the **basal nuclei** are important in motor coordination. If they are damaged, a person may be immobilized. Degeneration of the basal nuclei occurs in Parkinson's disease (see Module 20). The most extensive area of our cerebrum, the cerebral cortex, is the focus of the next module.

? Choosing from the structures in Table 14, identify the brain part most important in solving an algebra problem.

Cerebrum

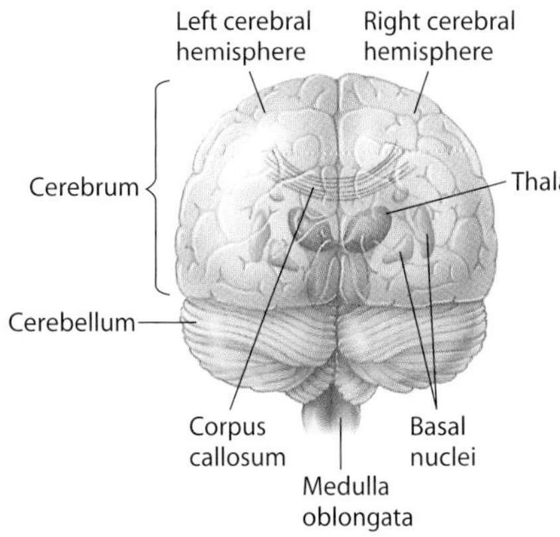

◀ **Figure 14B** A rear view of the brain

15 The cerebral cortex is a mosaic of specialized, interactive regions

Although less than 5 mm thick—less than the thickness of a pencil—the highly folded cerebral cortex accounts for about 80% of the total human brain mass. It contains some 10 billion neurons and hundreds of billions of synapses. Its intricate neural circuitry produces our most distinctive human traits: reasoning and mathematical abilities, language skills, imagination, artistic talent, and personality traits. Assembling information it receives from our eyes, ears, nose, taste buds, and touch sensors, the cerebral cortex also creates our sensory perceptions—what we are actually aware of when we see, hear, smell, taste, or touch something. In addition, the cerebral cortex regulates our voluntary movements.

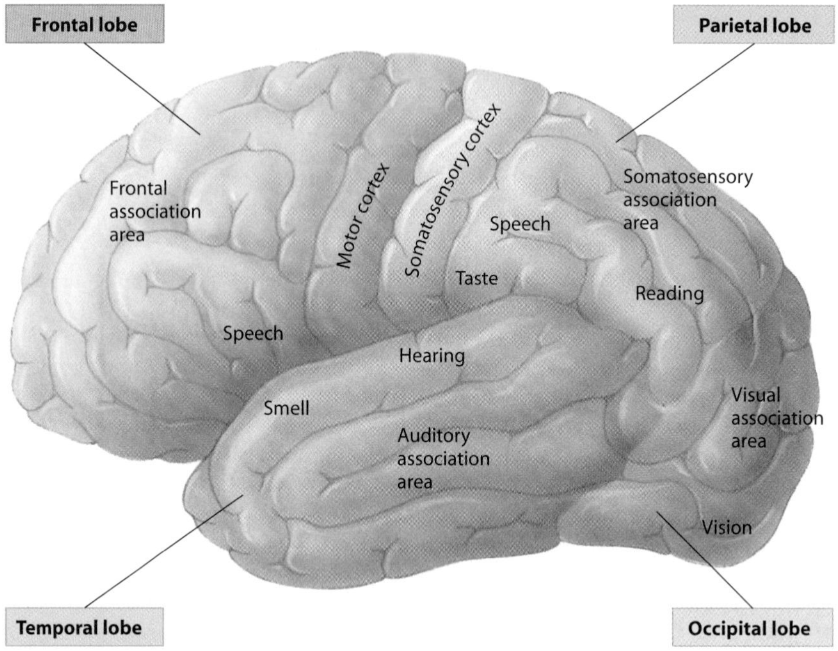

▲ **Figure 15** Functional areas of the left cerebral hemisphere

Like the rest of the cerebrum, the cerebral cortex is divided into right and left sides connected by the corpus callosum. Each side of the cerebral cortex has four lobes named for a nearby bone of the skull: the frontal, parietal, temporal, and occipital lobes, which are represented by different colors in **Figure 15**. Researchers have identified a number of functional areas within each lobe.

Two areas of known function form the boundary between the frontal and parietal lobes. One area, called the motor cortex, functions mainly in sending commands to skeletal muscles, signaling appropriate responses to sensory stimuli. Next to the motor cortex, the somatosensory cortex receives and partially integrates signals from touch, pain, pressure, and temperature receptors throughout the body. The cerebral cortex also has centers that receive and begin processing sensory information concerned with vision, hearing, taste, and smell.

Making up most of our cerebral cortex are numerous **association areas**, sites of higher mental activities—roughly, what we call thinking. Each sensory-receiving center of the cerebral cortex and the somatosensory cortex cooperates with an adjacent association area. In humans, a large association area in the frontal lobe uses varied inputs from many other areas of the brain to evaluate consequences, make considered judgments, and also plan for the future. Imaging techniques, such as PET scanning, are beginning to show how a complicated interchange of signals among the sensory-receiving centers and the association areas produces our sensory perceptions.

Language results from extremely complex interactions among several association areas. For instance, the parietal lobe of the cerebral cortex has association areas used for reading and speech. These areas obtain visual information (the appearance of words on a page) from the vision centers. Then, if the words are to be spoken aloud, they arrange the information into speech patterns and tell another speech center, in the frontal lobe, how to make the motor cortex move the tongue, lips, and other muscles to form words. When we hear words, the parietal areas perform similar functions using information from auditory centers of the cerebral cortex.

You may have heard people comment that they are "left-brained" or "right-brained." In a phenomenon known as **lateralization**, areas in the two cerebral hemispheres become specialized for different functions during brain development in infants and children. In most people, the left cerebral hemisphere becomes adept at language, logic, and mathematical operations, as well as detailed skeletal motor control and processing of fine visual and auditory details. The right cerebral hemisphere is stronger at spatial relations, pattern and face recognition, and nonverbal thinking. (In about 10% of us, these roles of the left and right cerebral hemispheres are reversed or the hemispheres are less specialized.)

How have researchers identified the functions of different parts of the brain? We explore this question in the next two modules.

? A stroke that causes loss of speech and numbness of the right side of the body has probably damaged brain tissue in the ______ lobe of the ______ cerebral hemisphere.

parietal . . . left

CONNECTION 16 Injuries and brain operations provide insight into brain function

The physiology of the human brain is exquisitely complex, making it one of the most difficult structures to study in all of biology. No animal model or computer simulation can accurately predict its complicated functions. New techniques, such as PET scans and fMRIs (see Modules 17), are allowing researchers to associate specific parts of the brain with various activities. Much of what has been learned about the brain, however, has come from rare individuals whose brains were altered through injury, illness, or surgery. By studying such "broken brains," researchers have gained insight into how healthy brains operate.

The first well-publicized case of this type involved a man named Phineas Gage. In 1848, while working as a railroad construction foreman in Vermont, Gage accidentally exploded a dynamite charge that propelled a 3-foot-long spike through his head. The 13-pound steel rod entered his left cheek and traveled upward behind his left eye and out the top of his skull, landing several yards away. Incredibly, Gage walked away from the accident and appeared to have an intact intellect. However, his associates soon noticed drastic changes in his personality, with new propensities toward meanness, vulgarity, irresponsibility, and an inability to control his behavior.

At the time, Gage's doctor was able to note these changes, but understanding of the brain was insufficient to explain them. Luckily, the doctor preserved Gage's skull and the spike, allowing a group of researchers in 1994 to produce a computer model of the injury **(Figure 16A)**. The modern analysis offered an explanation for Gage's bizarre behavior: The rod had pierced both frontal lobes of his brain. People with these sorts of injuries often exhibit irrational decision making and difficulty processing emotions. As you will learn in Module 19, the frontal lobes are part of the limbic system, a group of brain structures involved with emotions.

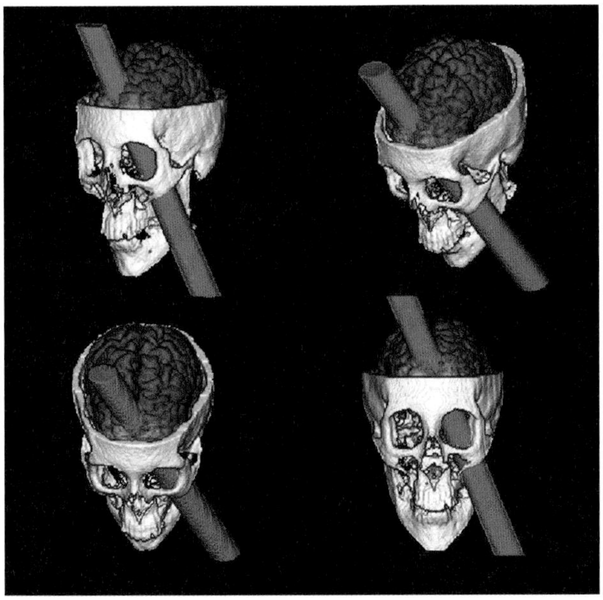

The Dornsife Neuroscience Imaging Center

▲ **Figure 16A** Computer model of Phineas Gage's injury

Beginning with the work of several neurosurgeons in the 1950s, many of the functional areas of the cerebral cortex have been identified during brain surgery. The cortex lacks cells that detect pain; thus, after anesthetizing the scalp, a neurosurgeon can operate on the cerebrum with the patient awake. Parts of the cortex can be stimulated with a harmless electrical current. Stimulation of specific areas can cause someone to experience different sensations or recall memories. Researchers can obtain information about the effects simply by questioning the conscious patient.

Neurophysiologists have also gained insight into the interrelatedness of the brain's two cerebral hemispheres. As discussed in Module 15, association areas in the left and right sides become specialized for different functions. Much of what we know about this lateralization stems from the work of Roger Sperry with patients whose corpus callosum (communicating fibers between the two cerebral hemispheres; see Module 14) had been surgically cut to treat severe epileptic seizures. In a series of ingenious experiments, Sperry demonstrated that his patients were unable to verbalize sensory information that was received by only the right cerebral hemisphere.

One of the most radical surgical alterations of the brain is a hemispherectomy **(Figure 16B)**—the removal of most of one-half of the brain, excluding deep structures such as the thalamus, brainstem, and basal nuclei. This procedure is performed to alleviate severe seizure disorders that originate from one of the cerebral hemispheres as a result of illness, abnormal development, or stroke. Incredibly, with just half a brain, hemispherectomy patients recover quickly, often leaving the hospital within a few weeks. Their intellectual capacities are undiminished, although the side of the body opposite the surgery remains partially paralyzed. Higher brain functions that previously originated from the missing half of the brain begin to be controlled by the opposite side. The younger the patient is, ideally less than 5 or 6 years old, the faster and more complete the recovery. Development after hemispherectomy is a striking example of the remarkable plasticity of the brain.

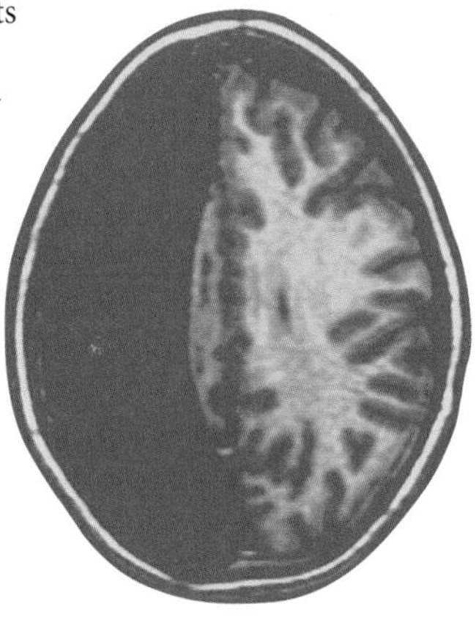

Johns Hopkins University

▲ **Figure 16B** X-ray of hemispherectomy patient after surgery

? How are researchers able to investigate brain function during brain surgery?

The cortex lacks pain receptors. Regions of the brain can be stimulated during surgery, and the conscious patient can report sensations or memories.

CONNECTION 17 fMRI scans provide insight into brain structure and function

Functional magnetic resonance imaging (fMRI) is a scanning and imaging technology that can "light up" metabolic processes as they occur within living tissue. Since the procedure can be performed on a conscious patient, fMRI scans can provide significant insights into brain structure and function.

MRI uses powerful magnets to align and then locate atoms within living tissue. During fMRI, a subject lies with his or her head in the hole of a large, doughnut-shaped magnet. When the brain is scanned with electromagnetic waves, changes in blood oxygen usage at sites of neuronal activity generate a signal that can be recorded. A computer uses the data to construct a three-dimensional map of the subject's brain activity.

Studies using fMRI confirm hypotheses based on older technologies about the roles of specific brain areas in movement and intention. Researchers have applied such techniques to correlate specific brain regions with nearly every aspect of human cognition, consciousness, and emotion.

Functional magnetic resonance imaging is proving to be a powerful diagnostic tool for a wide variety of illnesses. A 2010 study, for example, used fMRI to examine the brains of veterans suffering from Gulf War syndrome, a set of symptoms—including confusion, vertigo, mood swings, fatigue, and numbness—reported by over 175,000 U.S. troops since the first Gulf War in the early 1990s. A research team exposed test subjects to a variety of stimuli (including threatening pictures and word association games) and used fMRI to observe which parts of the brain "lit up." The researchers found that soldiers reporting different symptoms displayed abnormalities in different regions of the brain. For example, veterans who reported difficulties with attention often displayed abnormal fMRI measurements in their thalamus (which is associated with the ability to concentrate). The pair of fMRI scans in **Figure 17** shows that healthy veterans and veterans suffering from Gulf War syndrome had different areas of brain activity when performing a particular task. Overall, the researchers found a suite of brain differences that appeared to generally correlate with the symptoms reported by each individual. Such data may help accurately identify, diagnose, and treat people suffering from Gulf War syndrome and similar traumas. These types of diagnostic methods are at the forefront of neuroscience, one of biology's most fascinating and rapidly developing subdisciplines.

? **What does an fMRI scan actually measure?**

Changes in the use of oxygen by living tissues

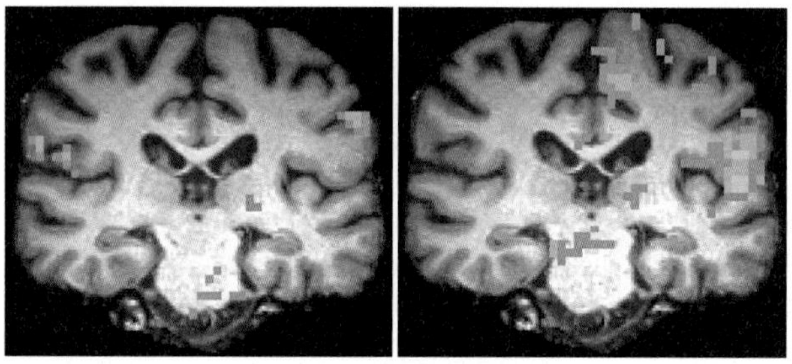

Robert W. Haley, M.D.

▲ **Figure 17** Comparative human fMRI images showing regions of brain activity in healthy veterans (left) and veterans with Gulf War syndrome (right) when performing the same task.

18 Several parts of the brain regulate sleep and arousal

As anyone who has drifted off to sleep during a lecture or while reading a book knows, attentiveness and mental alertness can change rapidly. Arousal is a state of awareness of the outside world. Its counterpart is sleep, a state when external stimuli are received but not consciously perceived.

The brainstem contains several centers for controlling arousal and sleep. One such center is the reticular formation, a diffuse network of neurons that extends through the core of the brainstem. Acting as a sensory filter, the reticular formation receives data from sensory receptors and selects which information reaches the cerebrum. The more information the cerebrum receives, the more alert and aware a person is, although the brain often ignores certain familiar and repetitive stimuli—the feel of your clothes against your skin, for example—while actively processing other inputs. Sleep and wakefulness are also regulated by specific parts of the brainstem: The pons and medulla contain centers that promote sleep when stimulated, and the midbrain has a center that promotes arousal.

Although we know very little about its function, sleep is essential for survival. Furthermore, sleep is an active state, at least for the brain. By placing electrodes at multiple sites on the scalp, researchers can record patterns of electrical activity called brain waves in an electroencephalogram, or EEG. These recordings reveal that brain wave frequency changes as the brain progresses through several distinct stages of sleep. During the stage called REM (rapid eye movement) sleep, the brain waves are rapid and irregular, more like those of the awake state. We have most of our dreams during REM sleep, which typically occurs about six times a night for periods of 5–50 minutes each.

Understanding the function of sleeping and dreaming remains a compelling research problem. One hypothesis is that sleep and dreams are involved in the consolidation of learning and memory, and experiments show that regions of the brain activated during a learning task can become active again during sleep.

? **What prevents the cerebral cortex from being overwhelmed by all the sensory stimuli arriving from sensory receptors?**

The reticular formation filters out unimportant stimuli.

19 The limbic system is involved in emotions, memory, and learning

Mapping the parts of the brain involved in human emotions, learning, and memory and studying the interactions between these parts are among the great challenges in biology today. Much of human emotion, learning, and memory depends on our **limbic system**. This functional unit includes parts of the thalamus and hypothalamus and two partial rings around them formed by portions of the cerebral cortex **(Figure 19)**. Two cerebral structures, the amygdala and the hippocampus, play key roles in memory, learning, and emotions.

The limbic system is central to such behaviors as nurturing infants and bonding emotionally to other people. Primary emotions that produce laughing and crying are mediated by the limbic system, and it also attaches emotional "feelings" to basic survival mechanisms of the brainstem, such as feeding, aggression, and sexuality. The intimate relationship between our feelings and our thoughts results from interactions between the limbic system and the prefrontal cortex, which is involved in complex learning, reasoning, and personality.

Memory, which is essential for learning, is the ability to store and retrieve information derived from experience. The **amygdala** is central in recognizing the emotional content of facial expressions and laying down emotional memories. Sensory data converge in the amygdala, which seems to act as a memory filter, somehow labeling information to be remembered by tying it to an event or emotion of the moment. The **hippocampus** is involved in both the formation of memories and their recall. Portions of the frontal lobes are involved in associating primary emotions with different situations.

We sense our limbic system's role in both emotion and memory when certain odors bring back "scent memories." Have you ever had a particular smell suddenly make you nostalgic for something that happened when you were a child? As indicated in Figure 19, signals from your nose enter your brain through the olfactory bulb, which connects with the limbic system. Thus, a specific scent can immediately trigger emotional reactions and memories.

Short-term memory, as the name implies, lasts only a short time—usually only a few minutes. It is short-term memory that allows you to dial a phone number just after looking it up. You may, however, be able to recall the number weeks after you originally looked it up, or even longer. This is because you have stored it in **long-term memory**. The transfer of information from short-term to long-term memory is enhanced by rehearsal, positive or negative emotional states mediated by the amygdala, and the association of new data with data previously learned and stored in long-term memory. For example, it's easier to learn a new card game if you already have "card sense" from playing other card games.

Factual memories, involving names, faces, words, and places, are different from skill or procedural memories. Skill memories usually involve motor activities that are learned by repetition without consciously remembering specific information. You perform skills, such as tying your shoes, riding a bicycle, or hitting a baseball, without consciously recalling the individual steps required to do these tasks correctly. Once a skill memory is learned, it is difficult to unlearn. For example, a person who has played tennis with a self-taught, awkward backhand has a tougher time learning the correct form than a beginner just learning the game. Bad habits, as we all know, are hard to break.

Information processing by the brain generally seems to involve a complex interplay of several integrating centers. By experimenting with animals, studying amnesia (memory loss) in humans, and using brain-imaging techniques, scientists have begun to map some of the major brain pathways involved in memory. Their proposed pathway involves the hippocampus and amygdala, which receive sensory information from the cortex and convey it to other parts of the limbic system and to the prefrontal cortex. The storage of a memory is completed when signals return to the area in the cortex where the sensory perception originated.

In the final module of this chapter, we'll discuss four major disorders of the nervous system, including their symptoms and treatments.

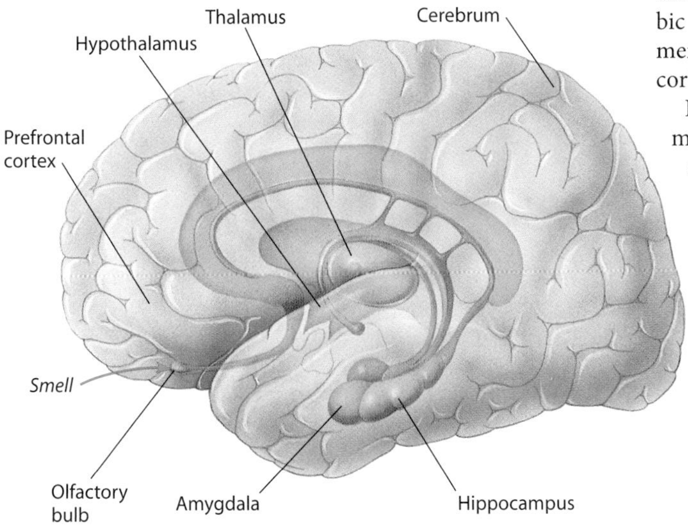

▲ **Figure 19** The limbic system (shown in shades of gold)

? Which three factors help transfer information from short-term to long-term memory?

Rehearsal, emotional associations, and connection with previously learned data

CONNECTION 20 Changes in brain physiology can produce neurological disorders

Neurological disorders (diseases of the nervous system) take an enormous toll on society. Examples of neurological disorders include schizophrenia, depression, Alzheimer's disease, and Parkinson's disease. While these conditions are not yet curable, there are a number of treatments available.

Schizophrenia About 1% of the world's population suffers from **schizophrenia**, a severe mental disturbance characterized by psychotic episodes in which patients have a distorted perception of reality. The symptoms of schizophrenia typically include hallucinations (such as "voices" that only the patient can hear); delusions (generally paranoid); blunted emotions; distractibility; lack of initiative; and difficulty with verbal expression. Contrary to commonly held belief, schizophrenics do not necessarily exhibit a "split personality." There seem to be several different forms of schizophrenia, and it is unclear whether they represent different disorders or variations of the same underlying disease.

The physiological causes of schizophrenia are unknown, although the disease has a strong genetic component. Studies of identical twins show that if one twin has schizophrenia, there is a 50% chance that the other twin will have it, too. Since identical twins share identical genes, this indicates that schizophrenia has an equally strong environmental component, the nature of which has not been identified.

There are several treatments for schizophrenia that can usually alleviate the major symptoms, but they often have major side effects. Such treatments focus on brain pathways that use dopamine as a neurotransmitter. Identification of the genetic mutations associated with schizophrenia may yield new insights about the causes of the disease, which may in turn lead to effective therapies with fewer drawbacks.

Depression Nearly 20 million American adults are affected by depression, about two-thirds of them women. Two broad forms of depressive illness are recognized: major depression and bipolar disorder.

People with **major depression** may experience persistent sadness, loss of interest in pleasurable activities, changes in body weight and sleep patterns, loss of energy, and suicidal thoughts. While all of us feel sad from time to time, major depression is extreme and more persistent, leaving a person unable to live a normal life for months at a time. One of the most common forms of mental illness, major depression affects about one in every seven individuals at some point in their adult lives. If left untreated, symptoms may become more frequent and severe over time.

Bipolar disorder, or manic-depressive disorder, is characterized by extreme mood swings and affects about 1% of the population. The manic phase is marked by high self-esteem, increased energy and flow of ideas, extreme talkativeness, inappropriate risk taking, promiscuity, and reckless spending. In its milder forms, this phase is sometimes associated with great creativity, and some well-known artists, musicians, and writers (including Van Gogh, Beethoven, and Hemingway) had periods of intense creative output during their manic phases. The depressive phase of bipolar disorder is characterized by lowered ability to feel pleasure, loss of motivation, sleep disturbances, and feelings of worthlessness. These symptoms can be so severe that some individuals attempt suicide.

Both bipolar disorder and major depression have a genetic component. As in schizophrenia, there is also a strong environmental influence; stress, especially severe stress in childhood, may be an important factor.

In recent years, researchers have begun to learn how brain physiology is involved in depression. The PET scans shown in **Figure 20A** compare the brain of a depressed patient (top) with that of a healthy patient (bottom). The red and yellow colors indicate areas of low brain activity. Note that the PET scan from the depressed patient shows decreased activity in certain areas of the brain.

Several kinds of medication for depression are available and can be quite effective. Many depressed people have an imbalance of neurotransmitters, particularly serotonin. Some medications are intended to correct such imbalances. The most commonly prescribed class of antidepressant drugs are selective serotonin reuptake inhibitors (SSRIs). As mentioned in Module 9, these medications increase the amount of time that serotonin is available to stimulate certain neurons in the brain, which appears to relieve some symptoms. In 1987, fluoxetine (best known as Prozac) became the first SSRI approved to treat depression. Other frequently prescribed SSRIs include paroxetine (Paxil) and sertraline (Zoloft). As shown in **Figure 20B**, the number of prescriptions for SSRIs in the United States tripled over a 10-year period, but began to level off in the last decade. SSRIs now represent one of the largest

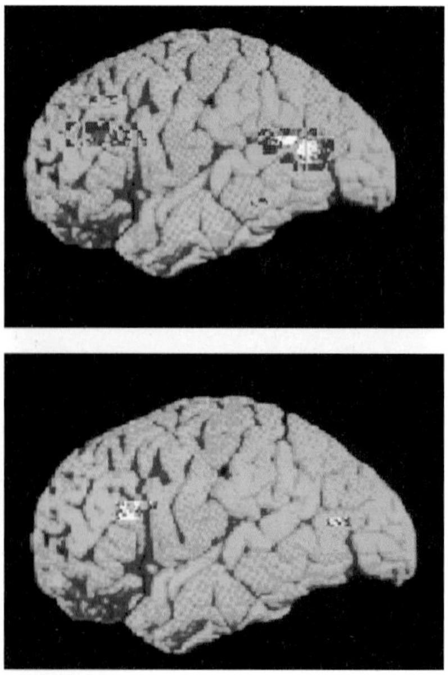

▶ **Figure 20A** PET scans showing brain activity in a depressed person (top) and healthy person (bottom)

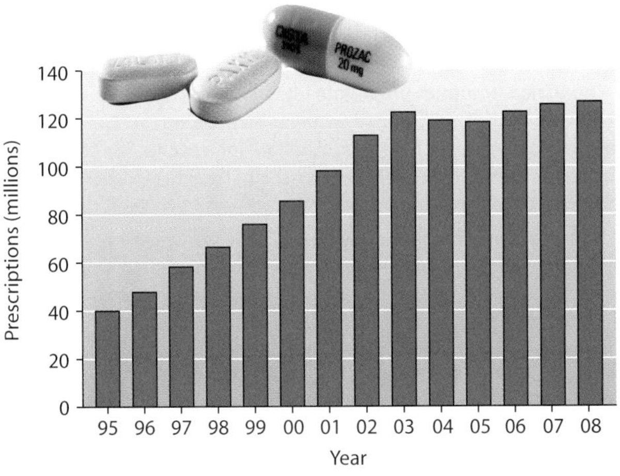

Jonathan Nourak/Stone/Getty Images

▲ **Figure 20B** SSRI prescriptions in the United States

classes of medication in the United States, in terms of both dollars spent and number of prescriptions.

Alzheimer's Disease A form of mental deterioration, or dementia, **Alzheimer's disease (AD)** is characterized by confusion and memory loss. Its incidence is usually age related, rising from about 10% at age 65 to about 35% at age 85. Thus, by helping humans live longer, modern medicine is increasing the proportion of AD patients in the population. The disease is progressive; patients gradually become less able to function and eventually need to be dressed, bathed, and fed by others. There are also personality changes, almost always for the worse. Patients often lose their ability to recognize people, even family members, and may treat them with suspicion and hostility.

At present, a firm diagnosis of AD is difficult to make while the patient is alive because there are other causes of dementia. However, AD results in a characteristic pathology of the brain: Neurons die in huge areas of the brain, and brain tissue often shrinks. The shrinkage is visible with brain imaging but is not enough to positively identify AD. What is diagnostic is the postmortem finding of two features—neurofibrillary tangles and senile plaques—in the remaining brain tissue. Neurofibrillary tangles are bundles of a protein called tau, a protein that normally helps regulate the movement of nutrients along microtubules within neurons. Senile plaques are aggregates of beta-amyloid, an insoluble peptide of 40 to 42 amino acids that is cleaved from a membrane protein normally present in neurons. The plaques appear to trigger the death of surrounding neurons, but whether amyloid plaques cause Alzheimer's or result from it remains unclear.

Parkinson's Disease Approximately 1 million people in the United States suffer from **Parkinson's disease** (**Figure 20C**), a motor disorder characterized by difficulty in initiating movements, slowness of movement, and rigidity. Patients often have a masked facial expression, muscle tremors, poor balance, a flexed posture, and a shuffling gait. Like Alzheimer's disease, Parkinson's is progressive, and the risk increases with age. The incidence of Parkinson's disease is about 1% at age 65 and about 5% at age 85.

The symptoms of Parkinson's disease result from the death of neurons in the basal nuclei. These neurons normally release dopamine from their synaptic terminals. The disease itself appears to be caused by a combination of environmental and genetic factors. Evidence for a genetic role includes the fact that some families with an increased incidence of Parkinson's disease carry a mutated form of the gene for a protein important in normal brain function.

At present, there is no cure for Parkinson's disease, although various treatments can help control the symptoms. Treatments include drugs such as L-dopa, a precursor of the neurotransmitter dopamine, and surgery. One potential cure is to develop stem cells into dopamine-secreting neurons in the laboratory and implant them in patients' brains. In laboratory experiments, transplantation of such cells into rats with a Parkinson's-like condition can lead to a recovery of motor control. Whether this kind of regenerative medicine will also work in humans is one of many important questions on the frontier of modern brain research.

Unraveling the biological basis of neurological disorders remains one of the most challenging tasks of modern biology. Recent advances in locating genes that correlate with CNS disorders and creating animal models for these diseases have provided new insights, but many aspects of our nervous system remain mysterious.

? What do the initials SSRI stand for? Relate the name to its mechanism of action.

SSRIs (selective serotonin reuptake inhibitors) are antidepressant drugs that specifically (*selectively*) prevent (*inhibit*) the reabsorption (*reuptake*) of the neurotransmitter *serotonin* in the brain.

Douglas Graham/Roll Call/Sygma/Corbis

▲ **Figure 20C** Actor Michael J. Fox (right) and boxer Muhammad Ali, both of whom suffer from Parkinson's disease, testifying before the United States Senate about funding for the disorder

REVIEW

MB For Practice Quizzes, BioFlix, MP3 Tutors, and Activities, go to www.masteringbiology.com.

Reviewing the Concepts

Nervous System Structure and Function (1–2)

1 Nervous systems receive sensory input, interpret it, and send out appropriate commands. The nervous system obtains and processes sensory information and sends commands to effector cells (such as muscle cells) that carry out appropriate responses.

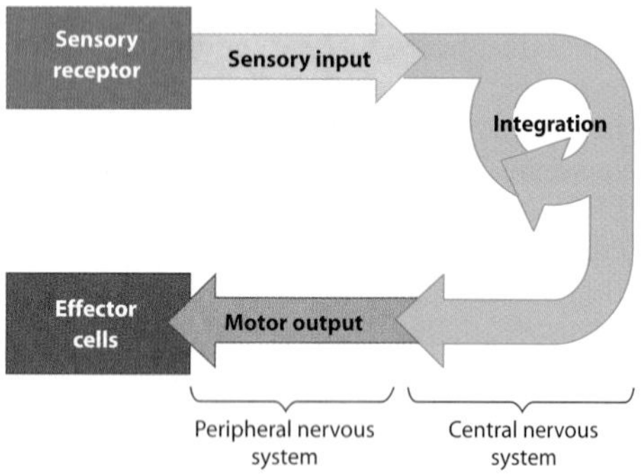

Sensory neurons conduct signals from sensory receptors to the central nervous system (CNS), which consists of the brain and, in vertebrates, the spinal cord. Interneurons in the CNS integrate information and send it to motor neurons. Motor neurons, in turn, convey signals to effector cells. Located outside the CNS, the peripheral nervous system (PNS) consists of nerves (bundles of sensory and motor neurons) and ganglia (clusters of cell bodies of the neurons).

2 Neurons are the functional units of nervous systems. Neurons are cells specialized for carrying signals.

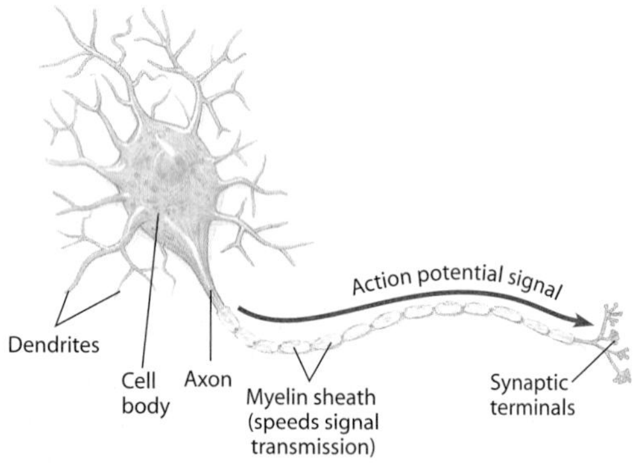

Nerve Signals and Their Transmission (3–9)

3 Nerve function depends on charge differences across neuron membranes. At rest, a neuron's plasma membrane has an electrical voltage called the resting potential. The resting potential is caused by the membrane's ability to maintain a positive charge on its outer surface opposing a negative charge on its inner (cytoplasmic) surface.

4 A nerve signal begins as a change in the membrane potential. A stimulus alters the permeability of a portion of the membrane, allowing ions to pass through and changing the membrane's voltage. A nerve signal, called an action potential, is a change in the membrane voltage from the resting potential to a maximum level and back to the resting potential.

5 The action potential propagates itself along the axon. Action potentials are self-propagated in a one-way chain reaction along an axon. An action potential is an all-or-none event. The frequency of action potentials (but not their strength) changes with the strength of the stimulus.

6 Neurons communicate at synapses. The transmission of signals between neurons or between neurons and effector cells occurs at junctions called synapses. At electrical synapses, electrical signals pass directly between cells. At chemical synapses, the sending (presynaptic) cell secretes a chemical signal, a neurotransmitter, which crosses the synaptic cleft and binds to a specific receptor on the surface of the receiving (postsynaptic) cell.

7 Chemical synapses enable complex information to be processed. Some neurotransmitters excite a receiving cell; others inhibit a receiving cell's activity by decreasing its ability to develop action potentials. A cell may receive differing signals from many neurons; the summation of excitation and inhibition determines whether or not it will transmit a nerve signal.

8 A variety of small molecules function as neurotransmitters. Many neurotransmitters are small, nitrogen-containing molecules.

9 Many drugs act at chemical synapses. Many common drugs can affect the brain's delicate chemistry.

An Overview of Animal Nervous Systems (10–13)

10 The evolution of animal nervous systems reflects changes in body symmetry. Radially symmetric animals have a nervous system arranged in a weblike system of neurons called a nerve net. Among bilaterally symmetric animals, nervous systems evolved to exhibit cephalization, the concentration of the nervous system in the head end, and centralization, the presence of a central nervous system.

11 Vertebrate nervous systems are highly centralized. The brain and spinal cord contain fluid-filled spaces. Cranial and spinal nerves make up the peripheral nervous system.

12 The peripheral nervous system of vertebrates is a functional hierarchy.

Central Nervous System (CNS)		**Peripheral Nervous System** (PNS)	
Brain	Spinal cord: nerve bundle that communicates with body	Motor system: voluntary control over muscles	Autonomic nervous system: involuntary control over organs • Parasympathetic division: rest and digest • Sympathetic division: fight or flight

13 The vertebrate brain develops from three anterior bulges of the neural tube. The vertebrate brain evolved by the enlargement and subdivision of the hindbrain, midbrain, and forebrain. The size and complexity of the cerebrum in birds and mammals correlate with their sophisticated behavior.

The Human Brain (14–20)

14 The structure of a living supercomputer: The human brain. The midbrain and subdivisions of the hindbrain, together with the thalamus and hypothalamus of the forebrain, function mainly in conducting information to and from higher brain centers. They regulate homeostatic functions, keep track of body position, and sort sensory information. The forebrain's cerebrum is the largest and most complex part of the brain. Most of the cerebrum's integrative power resides in the cerebral cortex of the two cerebral hemispheres.

15 The cerebral cortex is a mosaic of specialized, interactive regions. Specialized integrative regions of the cerebral cortex include the somatosensory cortex and centers for vision, hearing, taste, and smell. The motor cortex directs responses. Association areas, concerned with higher mental activities such as reasoning and language, make up most of the cerebrum. The right and left cerebral hemispheres tend to specialize in different mental tasks.

16 Injuries and brain operations provide insight into brain function. Brain injuries and surgeries have been used to study brain function.

17 fMRI scans provide insight into brain structure and function. The scans detect oxygen use at sites of neuronal activity.

18 Several parts of the brain regulate sleep and arousal. Sleep and arousal involve activity by the hypothalamus, medulla oblongata, pons, and neurons of the reticular formation.

19 The limbic system is involved in emotions, memory, and learning. The limbic system is a group of integrating centers in the cerebral cortex, thalamus, and hypothalamus. One key function is transfer of information from short-term to long-term memory.

20 Changes in brain physiology can produce neurological disorders. Treatment can often improve patients' lives, but cures are elusive.

Connecting the Concepts

1. Test your understanding of the nervous system by matching the following labels with their corresponding letters: CNS, effector cells, interneuron, motor neuron, PNS, sensory neuron, sensory receptor, spinal cord, synapse.

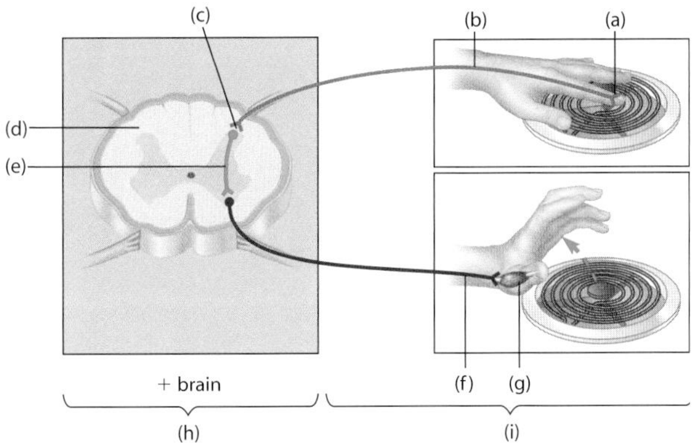

Testing Your Knowledge

Multiple Choice

2. Joe accidentally touched a hot pan. His arm jerked back, and an instant later, he felt a burning pain. How would you explain the fact that his arm moved before he felt the pain?
   a. His limbic system blocked the pain momentarily, but the important pain signals eventually got through.
   b. His response was a spinal cord reflex that occurred before the pain signals got to the brain.
   c. It took a while for his brain to search long-term memory and figure out what was going on.
   d. Motor neurons are myelinated; sensory neurons are not. The signals traveled faster to his muscles.
   e. This scenario is not actually possible. The brain must register pain before a person can react.

3. Which of the following mediates sleep and arousal?
   a. the reticular formation, along with the hypothalamus and thalamus
   b. the limbic system, which includes the amygdala and hippocampus
   c. the left hemisphere of the cerebral cortex
   d. the midbrain and cerebellum
   e. the parasympathetic and sympathetic divisions of the nervous system

4. Anesthetics block pain by blocking the transmission of nerve signals. Which of these three chemicals might work as anesthetics? (*Circle all that apply and explain your selections.*)
   a. a chemical that prevents the opening of sodium channels in membranes
   b. a chemical that inhibits the enzymes that degrade neurotransmitters
   c. a chemical that blocks neurotransmitter receptors

Describing, Comparing, and Explaining

5. As you hold this book, nerve signals are generated in nerve endings in your fingertips and sent to your brain. Once a touch has caused an action potential at one end of a neuron, what causes the nerve signal to move from that point along the length of the neuron to the other end? What is the nerve signal, exactly? Why can't it go backward? How is the nerve signal transmitted from one neuron to the next across a synapse? Write a short paragraph that answers these questions.

Applying the Concepts

6. Using microelectrodes, a researcher recorded nerve signals in four neurons in the brain of a snail. The neurons are called A, B, C, and D in the table below. A, B, and C all can transmit signals to D. In three experiments, the animal was stimulated in different ways. The number of nerve signals transmitted per second by each of the cells is recorded in the table. Write a short paragraph explaining the different results of the three experiments.

	Signals/sec			
	A	B	C	D
Experiment #1	50	0	40	30
Experiment #2	50	0	60	45
Experiment #3	50	30	60	0

7. Brain injuries tend to be severe because most neurons, once damaged, will not regenerate. The use of embryonic stem cells has been proposed as a potential treatment for many neurological diseases. Neurons developed from such stem cells in laboratory culture might be implanted in the brain of a person with Parkinson's disease, as mentioned in the text, or Alzheimer's. These neurons might be able to replace those damaged by the disease. Do you favor or oppose research along those lines? Explain your answer.

8. Alcohol's depressant effects on the nervous system cloud judgment and slow reflexes. Alcohol consumption is a factor in most fatal traffic accidents in the United States. What are some other impacts of alcohol abuse on society? What are some of the responses of people and society to alcohol abuse? Do you think this is primarily an individual or societal problem? Do you think our responses to alcohol abuse are appropriate and proportional to the seriousness of the problem?

Review Answers

1. (a) sensory receptor; (b) sensory neuron; (c) synapse; (d) spinal cord; (e) interneuron; (f) motor neuron; (g) effector cells; (h) CNS; (i) PNS

2. b 3. a 4. Both a and c would prevent action potentials from occurring; b could actually increase the generation of action potentials.

5. At the point where the action potential is triggered, sodium ions rush into the neuron. They diffuse laterally and cause sodium gates to open in the adjacent part of the membrane, triggering another action potential. The moving wave of action potentials, each triggering the next, is a moving nerve signal. Behind the action potential, sodium gates are temporarily inactivated, so the action potential can only go forward. At a synapse, the transmitting cell releases a chemical neurotransmitter, which binds to receptors on the receiving cell and may trigger a nerve signal in the receiving cell.

6. The results show the cumulative effect of all incoming signals on neuron D. Comparing experiments 1 and 2, we see that the more nerve signals D receives from C, the more it sends; C is excitatory. Because neuron A is not varied here, its action is unknown; it may be either excitatory or mildly inhibitory. Comparing experiments 2 and 3, we see that neuron B must release a strongly inhibitory neurotransmitter, because when B is transmitting, D stops.

7. Some issues and questions to consider: Some people might be against the use of embryonic stem cells for any disorder. This may be related to religious or moral beliefs. Potentially, some people may not be aware of the source of these stem cells and may be against their use because they believe that the cells come from elective abortions. Other people may agree with the use of stem cell research because of the potential to cure diseases that are currently fatal. Some people who may have a neutral opinion on the issue might be swayed by the thought of a loved one who might be helped by stem cell therapy.

8. Some issues and questions to consider: What is the role of alcohol in crime? What are its effects on families and in the workplace? In what ways is the individual responsible for alcohol abuse? The family? Society? Who is affected by alcohol abuse? How effective are treatment and punishment in curbing alcohol abuse? Who pays for alcohol abuse and consequent treatment or punishment? Is it possible to enjoy alcohol without abusing it?

The Senses

The Senses

BIG IDEAS

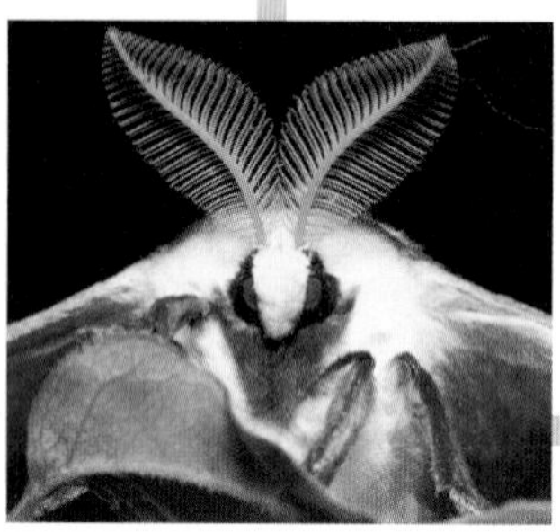

John Mitchell/Photo Researchers

Sensory Reception
(1–3)

Animal sensations begin as stimuli detected by sensory receptor cells.

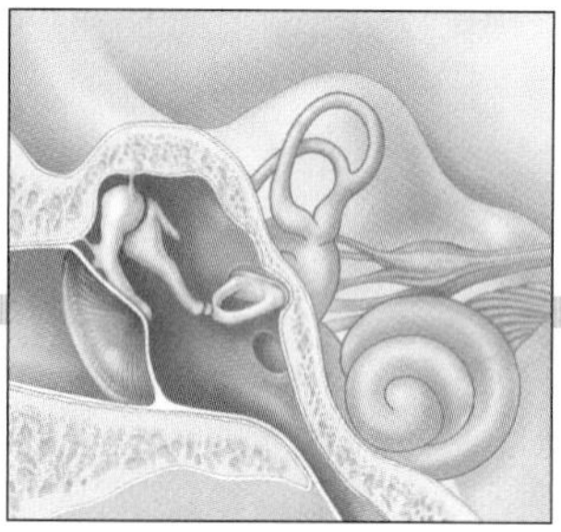

Hearing and Balance
(4–6)

The human ear contains structures that detect sound and body movement.

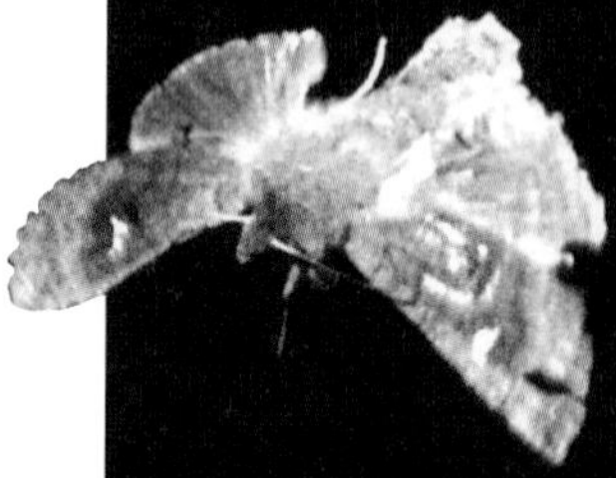

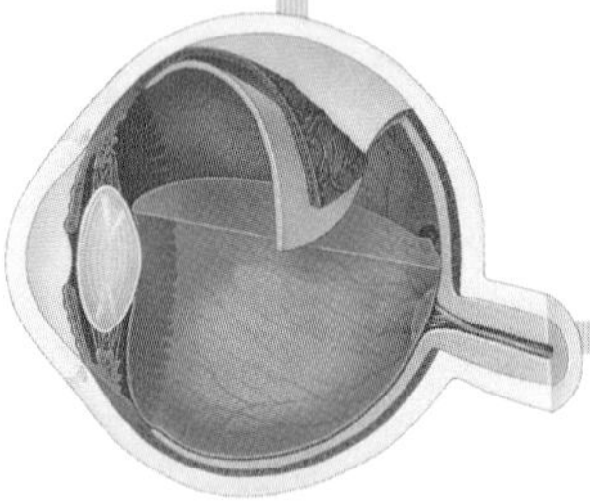

Vision
(7–10)

In the vertebrate eye, a single lens focuses an image onto photoreceptor cells.

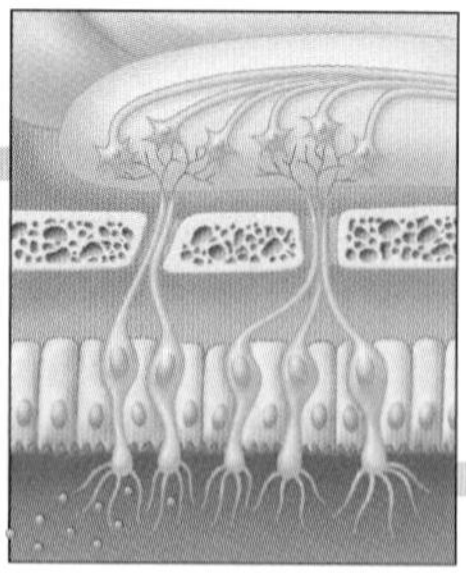

Taste and Smell
(11–13)

Chemoreceptors on the tongue and in the nose detect chemicals.

Dietmar Nill/Photolibrary

Many animals perceive the world in ways that we cannot. For example, the bat (*Plecotus auritus*) shown in the photo is navigating by echolocation, which allows it to identify and locate objects by detecting the echoes of emitted sound waves. Many bat species are able to produce high-pitched squeaks in their larynx and emit them through their mouths and noses, which are shaped in a way that focuses the sound waves. These signals are usually ultrasonic—that is, at frequencies above the range of human hearing. Bats receive the echoes of the emitted sounds via large ears. Their brains can process the time delay and spatial arrangement of the echoes to determine the size, shape, location, speed, and direction of objects in their environment. Bats emit about 15 pulses of sound per second while scanning for prey and up to 100 pulses per second while actively capturing insects.

In another example of echolocation, some marine mammals—including dolphins, porpoises, killer whales, and sperm whales—produce ultrasonic clicking sounds in their nasal passages. The sounds bounce off a skull bone and through an oil-filled structure in their forehead that focuses the sound. The animal then receives echoes via a narrow window of bone behind the jaw. Echolocation has also been observed in some species of cave-dwelling birds and forest-dwelling shrews (small mammals).

Although we humans cannot perceive our surroundings through echolocation, we can perceive the world in great detail using the senses we do have. In this chapter, we focus on four human senses: hearing, vision, taste, and smell. To begin, we look at the general principles of sensory reception that underlie all animal senses. Then we turn our attention to human sensory structures and discuss how they function.

Sensory Reception

1 Sensory organs share a common cellular basis

The operation of all animal senses—the ones used by humans as well as the nonhuman senses such as echolocation—originates in **sensory receptors**, specialized neurons or other cells that are tuned to the conditions of the external world or internal organs. Sensory cells detect stimuli: chemical flavorings in your food; light emitted by your TV screen; tension in a muscle as you grasp a computer mouse; sound waves produced by your MP3 player; touch sensations as you hold someone's hand; or other sensations such as electricity, cold, or heat. Once a stimulus is detected, a sensory receptor cell triggers action potentials (electrical signals) that send this information to the central nervous system (CNS), the brain and spinal cord.

An example of sensory receptors in action can be observed as a great hammerhead shark (*Sphyrna mokarran*) hunts for prey. It sweeps its wide head back and forth, like a beachcomber scanning the shore with a metal detector **(Figure 1)**. As the shark moves, sensory receptor cells in its head can detect prey—perhaps a flounder or stingray hiding on or just beneath the ocean floor. The shark is hunting with a sense that humans lack: electroreception, the ability to sense electric fields. Certain sensory cells in a hammerhead shark's head are electromagnetic receptors (see Module 3) that can detect the minute electrical activity of muscles or nerves in nearby prey—even if the prey animals are perfectly still—and these same cells can relay reports about this information to the shark's brain. Electroreception is common among fish and is found in some amphibians and the monotremes (egg-laying mammals such as the duck-billed platypus).

Andre Seale/Photolibrary

▲ **Figure 1** A hammerhead shark hunting by electroreception

Whether electroreception in hammerhead sharks or our own ability to taste or hear, the functioning of every sense starts with the detection of stimuli by sensory receptor cells. Detection leads to sensation, the identification of the particular type of stimulus—the sweet taste of ice cream or the shrill sound of an alarm clock. Sensation depends on the part of the brain that receives the action potential. Before we turn to the details of the human senses, we'll next discuss how sensory receptor cells convert stimuli to action potentials.

? How is the signal produced by an electroreceptive cell in a hammerhead shark similar to the signal produced by a cell in your eye?

● Both types of cells produce action potentials that are passed to the CNS.

2 Sensory receptors convert stimulus energy to action potentials

Sensory organs, such as your eyes or taste buds, contain sensory receptors that detect stimuli. The sensory receptors in your eyes detect light energy (photons); those in your taste buds detect chemicals, such as salt ions or sugar molecules. The sensory receptor's job is completed when it sends information to the central nervous system by triggering action potentials.

In stimulus detection, the receptor cell converts one type of signal (the stimulus) to another type, an electrical signal. This conversion, called **sensory transduction**, produces a change in the cell's membrane potential (the potential energy stored by the plasma membrane of the receptor cell. This change results from the opening or closing of ion channels in the sensory receptor's plasma membrane.

Figure 2A on the facing page shows sensory transduction occurring when sensory receptor cells in a taste bud detect sugar molecules, as when you lick an ice cream cone. ❶ The sugar molecules (⁘) first arrive at the taste bud, where ❷ they bind to sweet receptor proteins, specific protein molecules embedded in the plasma membrane of a taste receptor cell. The binding triggers ❸ a signal transduction pathway that causes ❹ some ion channels in the membrane to close and other ion channels to open. Changes in the flow of ions create a graded change in membrane potential called a **receptor potential**. Receptor potentials vary; the stronger the stimulus, the greater the receptor potential.

Once a stimulus is converted to a receptor potential, the receptor potential usually results in signals passing into the central nervous system. In our taste bud example, ❺ each receptor cell forms a synapse with a sensory neuron. In many cases, a receptor cell constantly secretes neurotransmitter (⁘) into this synapse at a set rate, triggering a steady stream of action potentials in the sensory neuron. ❻ The graph shows the rate at which the sensory neuron sends action potentials when the taste receptor is not detecting any sugar. The graph also shows what happens when there are enough sugar molecules to trigger a strong receptor potential, causing the receptor cell to release even more neurotransmitter than usual. This additional neurotransmitter increases the rate of action potential

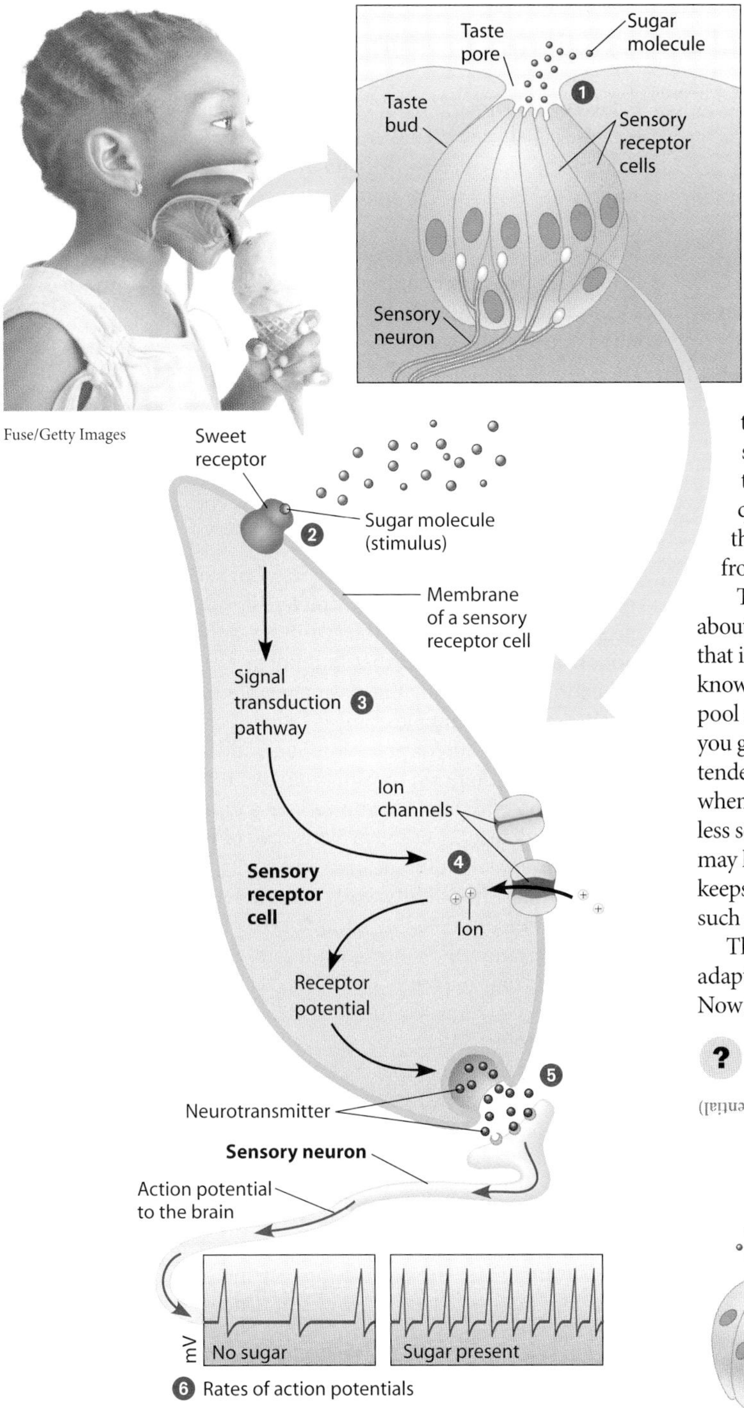

▲ **Figure 2A** Sensory transduction at a taste bud

generation in the sensory neuron, which in turn signals the brain that the sensory receptor detects a stimulus.

You can see that sensory receptor cells transduce (convert) stimuli to receptor potentials, which trigger action potentials that enter the central nervous system for processing. Because action potentials are the same no matter where or how they are produced, how do they communicate a sweet taste instead of a salty one? In **Figure 2B**, the taste bud on the left has sensory receptors that respond to sugar, and the taste bud on the right responds to salt. The sensory neurons from the sugar-detecting taste bud synapse with different interneurons in the brain than those contacted by neurons from the salt-detecting taste bud. The brain distinguishes stimulus types by the particular interneurons that are stimulated.

The graphs in Figure 2B also indicate how action potentials communicate information about the intensity of stimuli. In each case, the left part of the graph represents the rate at which the sensory neurons in the taste bud transmit action potentials when the taste receptors are not stimulated. The right side of each graph shows that the rate of transmission depends on the intensity of the stimulus. The stronger the stimulus, the more neurotransmitter released by the receptor cell and the more frequently the sensory neuron transmits action potentials to the brain. The brain interprets the intensity of the stimulus from the rate at which it receives action potentials.

There is an important qualification to what we have just said about stimulus intensity. Have you ever noticed how an odor that is strong at first seems to fade with time, even when you know the odorous substance is still there? Or how the water in a pool may seem shockingly cold when you dive in, but not when you get used to it? This effect is called **sensory adaptation**, the tendency of some sensory receptors to become less sensitive when they are stimulated repeatedly. When receptors become less sensitive, they trigger fewer action potentials, and the brain may lose its awareness of stimuli as a result. Sensory adaptation keeps the body from reacting to normal background stimuli, such as the feeling of clothes on your skin.

This overview of sensory transduction, transmission, and adaptation explains how sensory receptors work in general. Now let's look at some different kinds of receptors.

? **What is meant by sensory transduction?**

The conversion of a stimulus signal to an electrical signal (a receptor potential) by a sensory receptor cell

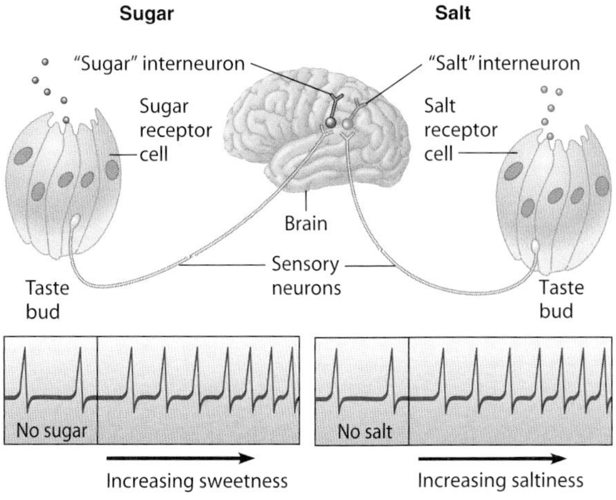

▲ **Figure 2B** Action potentials transmitting different taste sensations

3 Specialized sensory receptors detect five categories of stimuli

Based on the type of signals to which they respond, we can group sensory receptors into general categories: pain receptors, thermoreceptors (sensors for both heat and cold), mechanoreceptors (sensors for touch and pressure), chemoreceptors, and electromagnetic receptors. These receptor types work in various combinations to produce the human senses.

Figure 3A, showing a section of human skin, reveals why the surface of our body is sensitive to such a variety of stimuli. Our skin contains pain receptors, thermoreceptors, and mechanoreceptors. Each of these receptors is a modified dendrite of a sensory neuron. The neuron recognizes stimuli and sends action potentials to the central nervous system. In other words, each receptor serves as both a receptor cell and a sensory neuron. Most of the dendrites in the dermis (the underlying region of the skin) are wrapped in one or more layers of connective tissue (purple areas in the figure); however, the pain and touch receptors in the epidermis (outer skin layer) and the touch receptors around the base of hairs are naked dendrites.

Pain Receptors In humans and most other mammals, the skin has the highest density of **pain receptors** (see pink label in the figure). Pain receptors may respond to excess heat or pressure or to chemicals released from damaged or inflamed tissues. Pain often indicates injury or disease and usually makes an animal withdraw to safety. Chemicals produced in an animal's body sometimes increase pain.

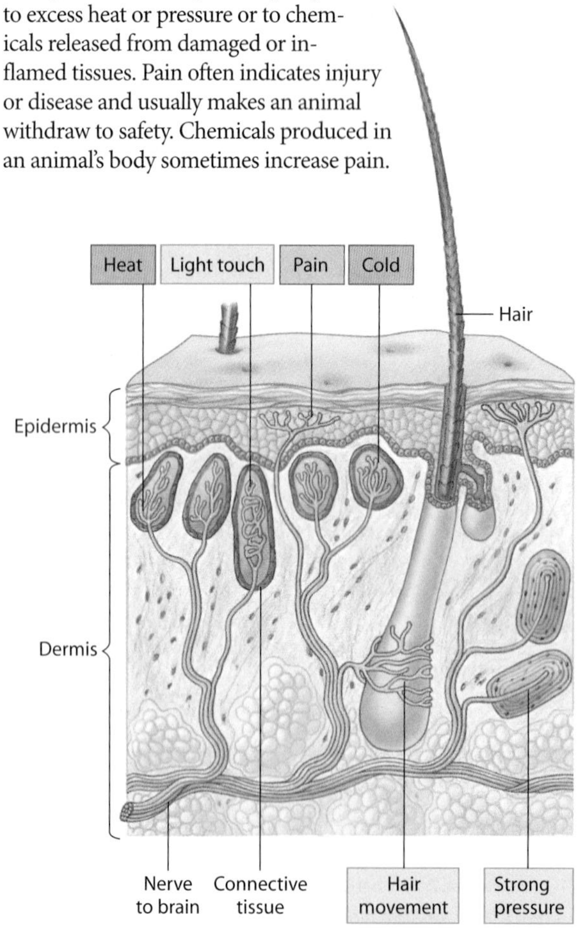

▲ **Figure 3A** Sensory receptors in the human skin

For example, damaged tissues produce prostaglandins, which increase pain by increasing the sensitivity of pain receptors. Aspirin and ibuprofen reduce pain by inhibiting prostaglandin synthesis.

Thermoreceptors **Thermoreceptors** in the skin (blue labels in the figure) detect either heat or cold. Other temperature sensors located deep in the body monitor the temperature of the blood. The hypothalamus acts as the body's thermostat: Receiving action potentials from both surface and deep sensors, the hypothalamus keeps a mammal's or bird's body temperature within a narrow range. Interestingly, scientists recently discovered that sensory receptors for high temperatures also respond to capsaicin, the chemical that makes chili peppers taste "spicy hot."

Mechanoreceptors Different types of **mechanoreceptors** (green labels in the figure) are stimulated by various forms of mechanical energy, such as touch and pressure, stretching, motion, and sound. All of these stimuli produce their effects by bending or stretching the plasma membrane of a receptor cell. When the membrane changes shape, it becomes more permeable to sodium or potassium ions, and the mechanical energy of the stimulus is transduced into a receptor potential.

At the top of Figure 3A, you can see a type of mechanoreceptor that detects light touch. This type of receptor can transduce very slight inputs of mechanical energy into action potentials. Another type of pressure sensor, lying deeper in the skin, is stimulated by strong pressure. A third type of mechanoreceptor, the touch receptor around the base of the hair, detects hair movements. Touch receptors at the base of the stiff whiskers on a cat are extremely sensitive and enable the animal to detect close objects by touch in the dark. Another type of mechanoreceptor (not shown) is found in our skeletal muscles. Sensitive to changes in muscle length, **stretch receptors** monitor the position of body parts.

A variety of mechanoreceptors collectively called **hair cells** detect sound waves and other forms of movement in water. The "hairs" on these sensors are either specialized types of cilia or cellular projections called microvilli. The sensory hairs project from the surface of a receptor cell into either the external environment, such as the water surrounding a fish, or an internal fluid-filled compartment, such as our inner ear.

Figure 3B shows how hair cells work, ❶ starting with a receptor cell at rest. ❷ When fluid movement bends the hairs in one direction, the hairs stretch the cell membrane, increasing its permeability to certain ions. This makes the hair cell secrete more neurotransmitter molecules and increases the rate of action potential production by a sensory neuron. ❸ When the hairs bend in the opposite direction, ion permeability decreases, the hair cell releases fewer neurotransmitter molecules, and the rate of action potentials decreases. We'll see later that hair cells are involved in both hearing and balance.

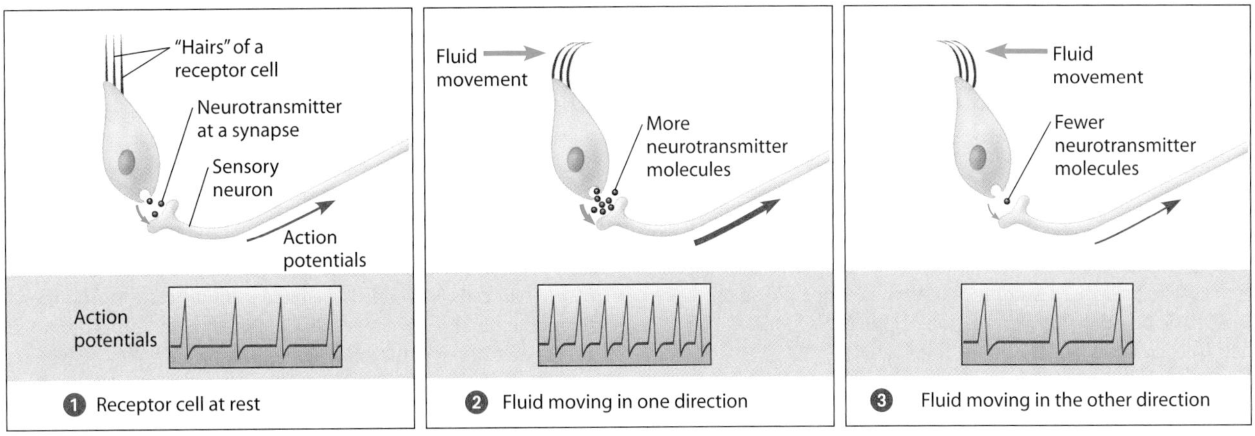

▲ **Figure 3B** Mechanoreception by a hair cell

Chemoreceptors **Chemoreceptors** include the sensory receptors in our nose and taste buds, which are attuned to chemicals in the external environment, as well as some receptors that detect chemicals in the body's internal environment. Internal chemoreceptors include sensors in some of our arteries that can detect changes in the amount of O_2 in the blood. Osmoreceptors in the brain detect changes in the total solute concentration of the blood and stimulate thirst when blood osmolarity increases.

One of the most sensitive chemoreceptors in the animal kingdom is on the antennae of the male silkworm moth *Bombyx mori* (**Figure 3C**). The antennae are covered with thousands of sensory hairs (visible in the micrograph). The hairs have chemoreceptors that detect a sex pheromone released by the female.

Electromagnetic Receptors Energy occurring as electricity, magnetism, or various wavelengths of light, may be detected by **electromagnetic receptors**. A platypus, for example, has electroreceptors on its bill that can detect electric fields generated by the muscles of prey, such as crustaceans, frogs, and small fishes.

John Mitchell/Photo Researchers

Dr. R.A. Steinbrecht

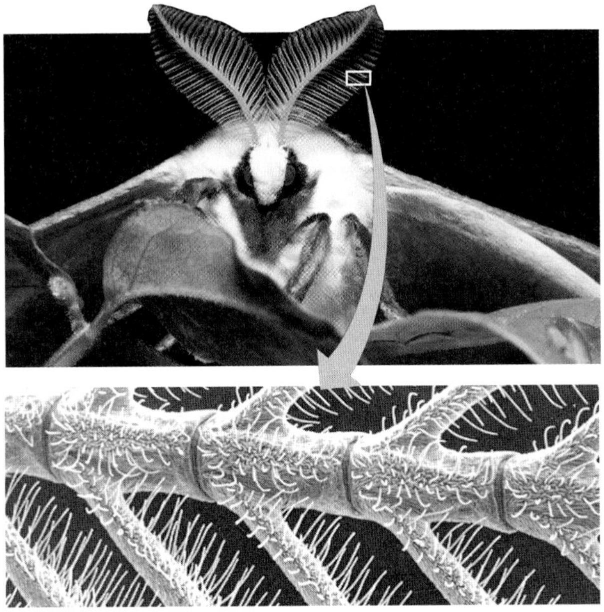

▲ **Figure 3C** Chemoreceptors on the antennae of a moth

Many animals appear to use Earth's magnetic field to orient themselves as they migrate. Migratory birds, fishes (such as salmon and trout), turtles, amphibians, and bees are believed to rely on magnetoreception to navigate relative to Earth's magnetic field—although the precise physiological basis for this sense remains to be discovered.

Photoreceptors, including eyes, are probably the most common type of electromagnetic receptor. Photoreceptors detect the electromagnetic energy of light, which may be in the visible or ultraviolet part of the electromagnetic spectrum. The rattlesnake in **Figure 3D** has prominent eyes that detect visible light. Near its eyes are two infrared receptor organs. These organs have specialized electromagnetic receptors that detect the body heat of the snake's preferred prey, small mammals and birds. The receptors can detect the infrared radiation emitted by a mouse a meter away.

James Gerholdt/Photolibrary

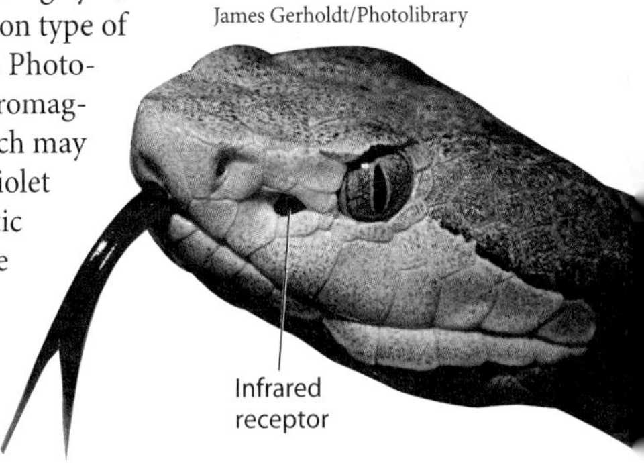

▲ **Figure 3D** Electromagnetic receptor organs in a snake

A variety of light detectors have evolved in the animal kingdom. Despite their differences, however, all photoreceptors contain similar pigment molecules that absorb light, and evidence indicates that all photoreceptors may be homologous, having a common ancestry. We consider some of the different types of eyes found in animals in Module 7. But first, let's consider the senses of hearing and balance.

? **For each of the following senses in humans, identify the type of receptor: seeing, tasting, hearing, smelling.**

Photoreceptor; chemoreceptor; mechanoreceptor; chemoreceptor

Hearing and Balance

4 The ear converts air pressure waves to action potentials that are perceived as sound

The human ear is really two separate organs, one for hearing and the other for maintaining balance. We look at the structure and function of our hearing organ in this module and then turn to our sense of balance in Module 5. Both organs operate on the same basic principle: the stimulation of long projections on hair cells (mechanoreceptors) in fluid-filled canals.

Structure The ear is complex, and it will help you to learn about its structure before studying how it functions (Figure 4A). The ear has three regions: the outer ear, the middle ear, and the inner ear. The **outer ear** consists of the flap-like **pinna**—the fleshy structure we commonly refer to as our "ear"—and the **auditory canal**. The pinna and the auditory canal collect sound waves and channel them to the **eardrum**, a sheet of tissue that separates the outer ear from the **middle ear** (Figure 4B). Both the outer ear and middle ear are common sites of childhood infections (called swimmer's ear and otitis media, respectively).

When sound pressure waves strike the eardrum, the eardrum vibrates and passes the vibrations to three small bones: the hammer (more formally, the malleus), anvil (incus), and stirrup (stapes). The stirrup is connected to the oval window, a membrane-covered hole in the skull bone through which vibrations pass into the inner ear. The middle ear also opens into a passage called the **Eustachian tube**, which connects with the pharynx (back of the throat), allowing air pressure to stay equal on either side of the eardrum. This tube is what enables you to "pop" your ears by swallowing hard or yawning to equalize pressure when changing altitude in an airplane.

The **inner ear** consists of fluid-filled channels in the bones of the skull. Sound vibrations or movements of the head set the fluid in motion. One of the channels, the **cochlea** (Latin for "snail"), is a long, coiled tube. The cross-sectional view of the cochlea in Figure 4C shows that inside it are three canals, each of which is filled with fluid. Our actual hearing organ, the **organ of Corti**, is located within the middle canal. The organ of Corti consists of an array of hair cells embedded in a **basilar membrane**, the floor of the middle canal. The hair cells are the sensory receptors of the ear. As you can see in the enlargement on the far right, a jellylike projection called the tectorial membrane extends from the wall of the middle canal (Figure 4D). Notice that the tips of the hair cells are in contact with the overlying tectorial membrane. Sensory neurons synapse with the base of the hair cells and carry action potentials to the brain via the auditory nerve.

Hearing Now let's see how the parts of the ear function in hearing. Sound waves, which move as pressure waves in the air, are collected by the pinna and auditory canal of the outer ear. These pressure waves make your eardrum vibrate with the same frequency as the sound (Figure 4E). The frequency, measured in hertz (Hz), is the number of vibrations per second.

From the eardrum, the vibrations are concentrated as they are transferred through the hammer, anvil, and stirrup in the middle ear to the oval window. Vibrations of the oval window then produce pressure waves in the fluid within the cochlea. The vibrations first pass from the oval window into the fluid in the upper canal of the cochlea. Pressure waves travel through the upper canal to the tip of the cochlea, at the coil's center. The pressure waves then enter the lower canal and gradually fade away.

As a pressure wave passes through the upper canal of the cochlea, it pushes downward on the middle canal, making the basilar membrane vibrate. Vibration of the basilar membrane makes the hairlike projections on the hair cells alternately brush against and draw away from the overlying tectorial membrane. When a hair cell's projections are bent, ion channels in its plasma

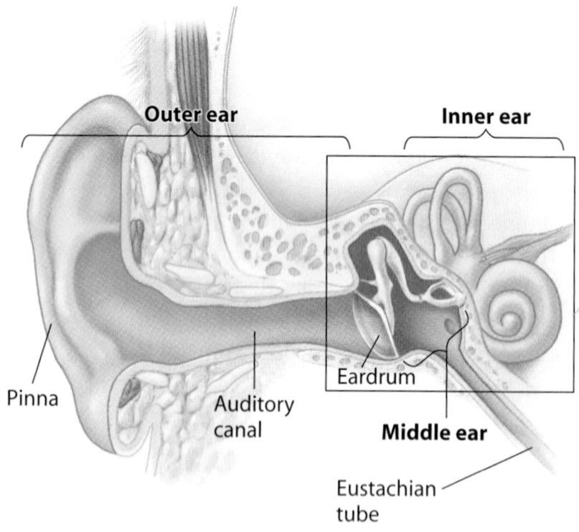

▲ **Figure 4A** An overview of the human ear

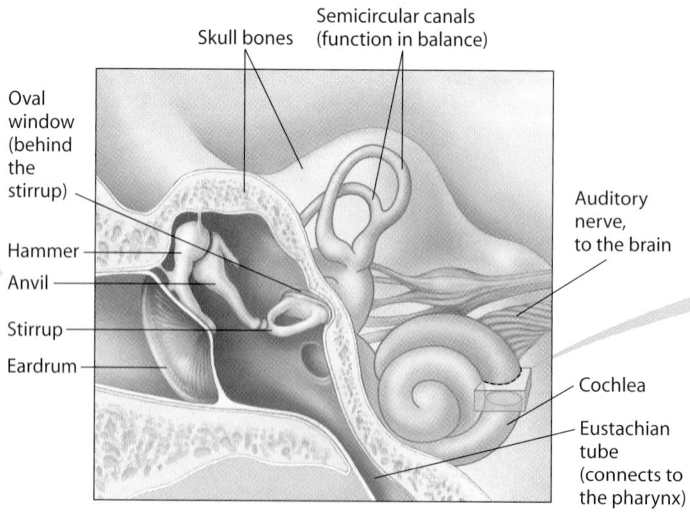

▲ **Figure 4B** The middle ear and the inner ear

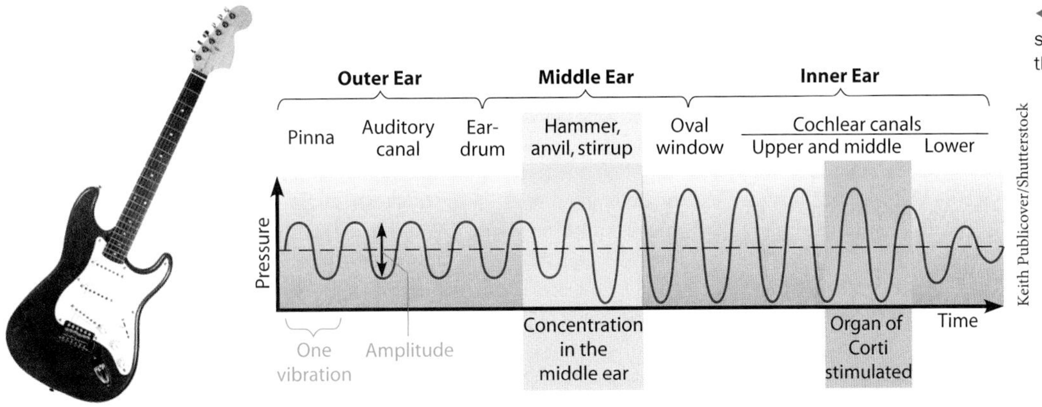

◀ **Figure 4E** The route of sound wave vibrations through the ear

membrane open, and positive ions enter the cell. As a result, the hair cell develops a receptor potential and releases more neurotransmitter molecules at its synapse with a sensory neuron. In turn, the sensory neuron sends more action potentials to the brain through the auditory nerve.

Volume and Pitch The brain senses a sound as an increase in the frequency of action potentials from the auditory nerve. But how is the quality of the sound determined? The higher the volume (loudness) of sound, the higher the amplitude (height) of the pressure wave it generates. In the ear, a higher amplitude of pressure waves produces more vigorous vibrations of fluid in the cochlea, more pronounced bending of the hair cells, and thus more action potentials generated in the sensory neurons.

The pitch of a sound depends on the frequency of the sound waves. High-pitched sounds, such as high notes sung by a soprano, generate high-frequency waves. Low-pitched sounds, like the low notes made by a bass, generate low-frequency waves. How does the cochlea distinguish sounds of different pitch? The key is that the basilar membrane is not uniform along its length. The end near the oval window is relatively narrow and stiff, while the other end, near the tip of the cochlea, is wider and more flexible. As a result, the basilar membrane varies in its sensitivity to particular frequencies of vibration, and the region vibrating most vigorously at any instant sends the most action potentials to auditory centers in the brain. The brain interprets the information and gives us a perception of pitch. Young people with healthy ears can hear pitches in the range of 20–20,000 Hz. We gradually lose the ability to hear the highest tones beginning around age 8, and women generally hear higher tones better than men do. Dogs can hear sounds as high as 40,000 Hz, and bats can emit and hear clicking sounds as high-pitched as 100,000 Hz.

Deafness can be caused by the inability of the ear to conduct sounds, resulting from middle-ear infections, a ruptured eardrum, or stiffening of the middle-ear bones (a common age-related problem). Deafness can also result from damage to sensory receptors or neurons. You may have noticed that a few hours in a very loud environment, such as a rock concert, leaves you with ringing or buzzing sounds in your ears, a condition called tinnitus. Frequent or prolonged exposure to sounds of more than 90 decibels (dB) can damage or destroy hair cells, which are never replaced. Ear plugs can provide protection to both rock musicians and their fans. Because few parts of our anatomy are more delicate than the organ of Corti, deafness is often progressive and permanent. In recent years, however, many deaf people have received cochlear implants, electronic devices that convert sounds to electrical impulses that stimulate the auditory nerve directly.

? How does the ear convert sound waves in the air to pressure waves of the fluid in the cochlea?

Sound waves in air cause the eardrum to vibrate. The small bones attached to the inside of the eardrum transmit the movement to the oval window on the wall of the inner ear. Vibrations of the oval window set in motion the fluid in the inner ear, which includes the cochlea.

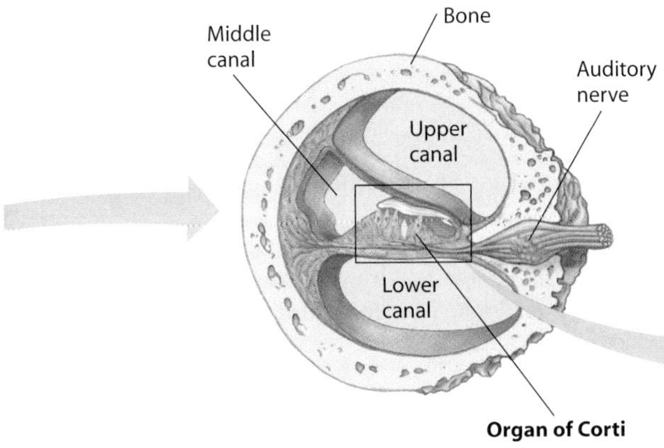

▲ **Figure 4C** A cross section through the cochlea

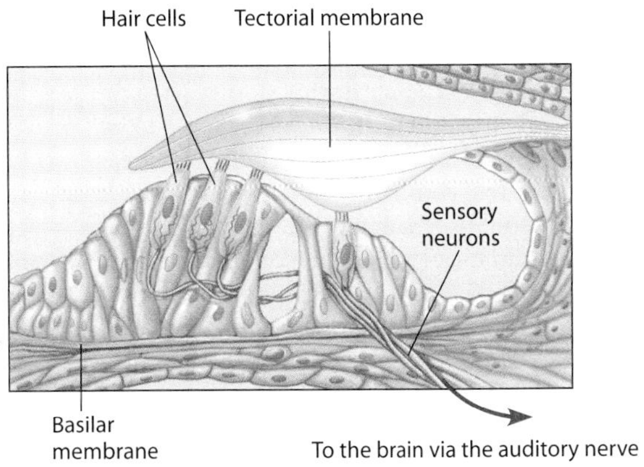

▲ **Figure 4D** The organ of Corti

5 The inner ear houses our organs of balance

Several organs in the inner ear detect body position and movement. These fluid-filled structures lie next to the cochlea **(Figure 5)** and include three semicircular canals and two chambers, the utricle and the saccule. All the equilibrium structures operate on the same principle: the bending of hairs on hair cells.

The ear contains three **semicircular canals** that detect changes in the head's rate of rotation or angular movement. The canals are arranged in three perpendicular planes and can therefore detect movement in all directions. A swelling at the base of each semicircular canal contains a cluster of hair cells with their hairs projecting into a gelatinous mass called a cupula (shown in the enlargements). When you rotate your head in any direction, the thick, sticky fluid in the canals moves more slowly than your head. Consequently, the fluid presses against the cupula, bending the hairs. The faster you rotate your head, the greater the pressure and the higher the frequency of action potentials sent to the brain. If you rotate your head at a constant speed, the fluid in the canals begins moving with the head, and the pressure on the cupula is reduced. But if you stop suddenly, the fluid continues to move just as coffee continues to rotate in a cup even after you stop stirring. The swirling fluid continues to stimulate the hair cells, which may make you feel dizzy. This explains why a situation that causes your inner-ear fluid to move—such as a whirling amusement park ride or even just twirling around quickly—makes you dizzy only after you try to stand still.

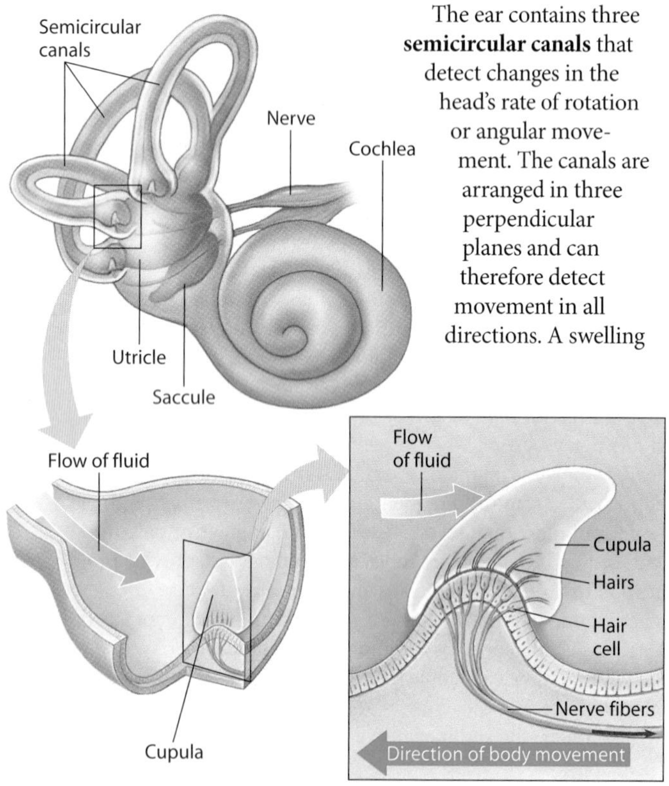

▲ **Figure 5** Equilibrium structures in the inner ear

Clusters of hair cells in the utricle and saccule detect the position of the head with respect to gravity. The hairs of these cells project into a gelatinous material containing many small particles of calcium carbonate. When the position of the head changes, this heavy material bends the hairs in a different direction, causing an increase or decrease in the rate at which action potentials are sent to the brain. The brain determines the new position of the head by interpreting the altered flow of action potentials.

The equilibrium receptors provide data the brain needs to determine the position and movement of the head. Using this information, the brain develops and sends out commands that enable the skeletal muscles to balance the body.

? What type of receptor cell is common to our senses of hearing and equilibrium?

Hair cells, which are mechanoreceptors

Adapted from Elaine N. Marieb, *Human Anatomy and Physiology*, 4ᵗʰ ed. Copyright © 1998 Pearson Education, Inc., publishing as Pearson Benjamin Cummings.

CONNECTION 6 What causes motion sickness?

Boating, flying, or even riding in a car can make us dizzy and nauseated, a condition called motion sickness. Some people can begin feeling ill simply at the thought of a boat or plane ride. Many others get sick only during storms at sea or during turbulence in flight. Motion sickness is believed to be caused when the brain receives signals from equilibrium receptors in the inner ear that conflict with visual signals from the eyes. When a susceptible person is inside a moving ship, for instance, signals from the equilibrium receptors in the inner ear indicate, correctly, that the body is moving (in relation to the environment outside the ship). In conflict with these signals, the eyes may tell the brain that the body is in a stationary environment, say, inside a cabin. Somehow the conflicting signals make the person feel ill. Symptoms may be relieved by closing the eyes, limiting head movements, or focusing on a stable horizon. Many sufferers of motion sickness take a sedative such as Dramamine or Bonine to relieve their symptoms. Long-lasting, drug-releasing skin patches prevent motion sickness by inhibiting input to the brain from the equilibrium sensors.

Because motion sickness can be a severe problem for astronauts, the National Aeronautics and Space Administration (NASA) conducts research on the problem. NASA has discovered that some people can learn to consciously control body functions, including the vomiting reflex. Astronauts receive intensive training in how to exert "mind over body" when zero gravity starts to induce motion sickness.

? Explain how someone could suffer motion sickness when watching a film shot from the front of a roller coaster.

There would be conflicting information between vision ("I'm moving") and the equilibrium sense ("I'm sitting still").

Vision

EVOLUTION CONNECTION

7 Several types of eyes have evolved independently among animals

The ability to detect light plays a central role in the lives of nearly all animals. Although there is great diversity in the organs animals use to perceive light, comparison across animal species demonstrates unity in the underlying mechanism of light detection. All animal light detectors—from simple clusters of cells that detect only the direction and intensity of light to complex organs that form images—are based on cells called photoreceptors, which contain pigment molecules that absorb light.

Most invertebrates have some kind of light-detecting organ. One of the simplest is the **eyecup**, the visual sensory organ of free-living (nonparasitic) flatworms called planarians (**Figure 7A**). Planarian eyecups contain photoreceptor cells that are partially shielded by darkly pigmented cells. Light can enter the eyecup only where there is no pigment and the openings of the two eyecups face opposite directions. The brain compares the rate of nerve impulses coming from the two eyecups. Although no distinct image is formed, the animal can turn until the sensations are equal and minimal, allowing it to move away from the light source to reach a dark hiding place.

Two major types of image-forming eyes have evolved in invertebrates. Compound eyes are found in insects and crustaceans. A **compound eye** consists of up to several thousand light-detectors called ommatidia (**Figure 7B**). Every ommatidium has its own light-focusing lens and several photoreceptor cells. Each ommatidium picks up light from a tiny portion of the field of view. The animal's brain then forms a mosaic visual image by assembling the data from all the ommatidia.

Compound eyes are extremely acute motion detectors, providing an important advantage for flying insects and other small animals often threatened by predators. The compound eyes of most insects also provide excellent color vision. Some species, such as honeybees, can see ultraviolet light (invisible to humans), which helps them locate certain nectar-bearing flowers.

The second type of image-forming eye to have evolved in invertebrates, such as squids, is the **single-lens eye**. The single-lens eye also evolved independently in vertebrates. **Figure 7C** shows a human eye. (The single-lens eye found in squids and other invertebrates differs in some of the details.) The human eye has a small opening at the center of the eye, the **pupil**, through which light enters. Analogous to a camera's shutter, an adjustable doughnut-shaped **iris** changes the diameter of the pupil to let in more or less light. After going through the pupil, light passes through a single disklike **lens**. The lens focuses light onto the **retina**, which consists of many photoreceptor cells.

Photoreceptor cells of the retina transduce light energy, and action potentials pass via sensory neurons in the optic nerve to the visual centers of the brain. Photoreceptor cells are highly concentrated at the retina's center of focus, called the **fovea**. There are no photoreceptor cells in the blind spot, the part of the retina where the optic nerve passes through the back of the eye. We cannot detect light that is focused on the blind spot, but having two eyes with overlapping fields of view enables us to perceive uninterrupted images. In the next module, we examine the human eye's image-forming structures in more detail.

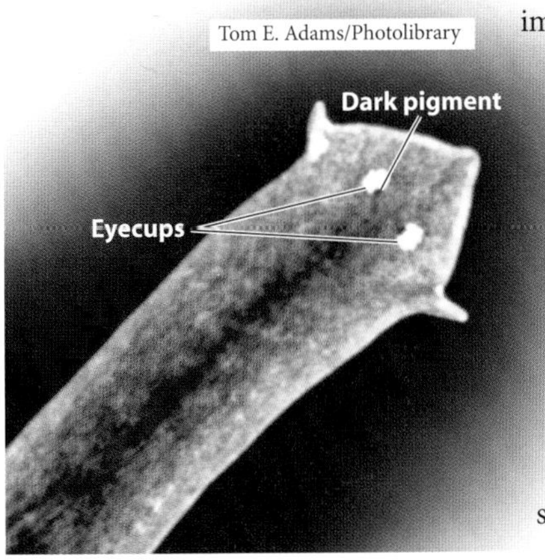

▲ **Figure 7A** The eyecups of a planarian

▲ **Figure 7B** The two compound eyes of a fly, each made up of thousands of ommatidia

? What key optical feature is found in the eyes of both insects and squids but is not present in planarians?

Lenses, which focus light onto photoreceptor cells

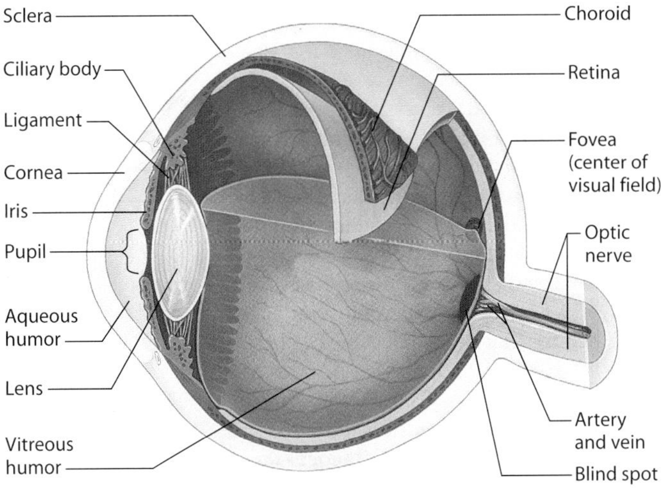

▲ **Figure 7C** The single-lens eye of a vertebrate

8 Humans have single-lens eyes that focus by changing position or shape

The outer surface of the human eyeball is a tough, whitish layer of connective tissue called the **sclera** (see Figure 7C). At the front of the eye, the sclera becomes the transparent **cornea**, which lets light into the eye and also helps focus light. The sclera surrounds a pigmented layer called the **choroid**. The anterior choroid forms the iris, which gives the eye its color. After going through the pupil, the opening at the center of the iris, light passes through the disklike lens, which is held in position by ligaments. At the back of the eyeball is the retina, a layer just inside the choroid that contains photoreceptor cells.

The lens and ciliary body divide the eye into two fluid-filled chambers. The large chamber behind the lens is filled with jelly-like **vitreous humor**. The much smaller chamber in front of the lens contains the thinner **aqueous humor**. The humors help maintain the shape of the eyeball. In addition, the aqueous humor circulates through its chamber, supplying nutrients and oxygen to the lens, iris, and cornea and carrying off wastes. Blockage of the ducts that drain this fluid can cause glaucoma, increased pressure inside the eye that may lead to blindness. If diagnosed early, glaucoma can be treated with medications that increase the circulation of aqueous humor.

A thin mucous membrane helps keep the outside of the eye moist. This membrane, called the **conjunctiva**, lines the inner surface of the eyelids and folds back over the white of the eye (but not the cornea). An infection or allergic reaction may cause inflammation of the conjunctiva, a condition called conjunctivitis, or "pink eye." Bacterial conjunctivitis usually clears up with antibiotic eyedrops. Viral conjunctivitis usually clears up on its own, although it is very contagious, especially among young children. If you are in contact with someone who has conjunctivitis, the best way to avoid spreading the disease is through careful hygiene: Don't touch the affected area, wash your hands frequently and thoroughly, and wash objects that may come into contact with the infection (such as linens and towels) with hot water and bleach.

A gland above the eye secretes tears, a dilute salt solution that is spread across the eyeball by blinking and that drains into ducts that lead into the nasal cavities. This fluid cleanses and moistens the eye surface. Excess secretion in response to eye irritation or strong emotions causes tears to spill over the eyelid and fill the nasal cavities, producing sniffles. Some scientists speculate that emotional tears play a role in reducing stress.

A lens focuses light onto a retina by bending light rays. Focusing can occur in two ways. The lens may be rigid, as in squids and many fishes. In this case, focusing occurs as muscles move the lens back or forth, as you might focus on an object using a magnifying glass. Or, as in the mammalian eye, the lens may be flexible, with focusing accomplished by changing the shape of the lens, depending on the distance to the object being viewed. The thicker the lens, the more sharply it bends light.

The shape of the mammalian lens is controlled by the muscles attached to the choroid and the ligaments that suspend the lens **(Figure 8)**. When the eye focuses on a nearby object, these muscles contract, pulling the choroid toward the lens, which reduces the tension on the ligaments. As these ligaments slacken, the elastic lens becomes thick and round, as shown in the top diagram of Figure 8. This change allows the diverging light rays from a close object to be bent and focused. The light rays from the object actually cross as they pass through the lens, resulting in an upside-down image striking the retina. (The brain interprets this inverted image so that we perceive it as right-side up.)

Light from distant objects approaches in parallel rays that require less bending for proper focusing on the retina. When the eye focuses on a distant object, the muscles controlling the lens relax, and the choroid moves away from the lens. This puts tension on the ligaments and flattens the elastic lens, as shown in the bottom diagram.

Flattening the lens makes it less thick, and so the light is not bent as much. This allows a distant object to be placed into focus. At least, that is how the human eye is *supposed* to work. In the next module, we consider ways that this mechanism can fail and ways to correct such failures.

? Arrange the following eye parts into the correct sequence encountered by photons of light traveling into the eye: pupil, retina, cornea, lens, vitreous humor, aqueous humor.

● Cornea → aqueous humor → pupil → lens → vitreous humor → retina

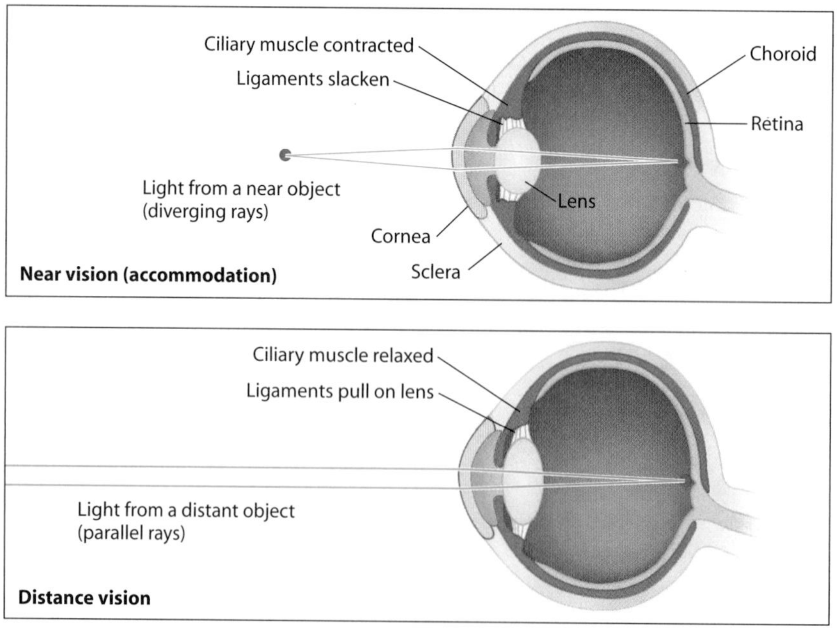

▲ **Figure 8** How human lenses focus light

CONNECTION | 9 Artificial lenses or surgery can correct focusing problems

Reading from an eye chart measures your **visual acuity**, the ability of your eyes to distinguish fine detail. When you have your eyes tested, the examiner asks you to read a line of letters sized for legibility at a distance of 20 feet, using one eye at a time. If you can do this accurately, you have normal (20/20) acuity in each eye. This means that from a distance of 20 feet, each of your eyes can read the chart's line of letters designated for 20 feet.

Visual acuity of 20/10 is better than normal and means that you can read letters from a distance of 20 feet that a person with 20/20 vision can only read at 10 feet. On the other hand, visual acuity of 20/50 is worse than normal. In this circumstance, you would have to stand at a distance of 20 feet to read what a person with normal acuity can read at 50 feet.

Three of the most common vision problems are nearsightedness, farsightedness, and astigmatism. All three are focusing problems, easily corrected with artificial lenses. People with **nearsightedness** cannot focus well on distant objects, although they can see well at short distances (the condition is named for the type of vision that is *unimpaired*). A nearsighted eyeball (**Figure 9A**) is longer than normal. The lens cannot flatten enough to compensate, and it focuses distant objects in front of the retina instead of on it. Nearsightedness (also known as myopia) is corrected by glasses or contact lenses that are thinner in the middle than at the outside edge. The lenses make the light rays from distant objects diverge as they enter the eye. The focal point formed by the lens in the eye then falls on the retina.

Farsightedness (also known as hyperopia) is the opposite of nearsightedness. It occurs when the eyeball is shorter than normal, causing the lens to focus images behind the retina (**Figure 9B**). Farsighted people see distant objects normally but cannot focus on close objects. Corrective lenses that are thicker in the middle than at the outside edge compensate for farsightedness by making light rays from nearby objects converge slightly before they enter the eye. Another type of farsightedness, called presbyopia, develops with age. Beginning around the mid-40s, the lens of the eye becomes less elastic. As a result, the lens gradually loses its ability to focus on nearby objects, and reading without glasses becomes difficult.

Astigmatism is blurred vision caused by a misshapen lens or cornea. This makes light rays converge unevenly and not focus at one point on the retina. Corrective lenses are asymmetric in a way that compensates for the asymmetry in the eye.

Surgical procedures are an option for treating vision disorders. In laser-assisted in situ keratomileusis (LASIK), a laser is used to reshape the cornea and change its focusing ability. More than 1 million LASIK procedures are performed each year to correct a variety of vision problems.

? A person with 20/100 vision in both eyes must stand at ______ feet to read what someone with normal vision can read at ______ feet. Is this better or worse than normal acuity?

20 . . . 100; worse

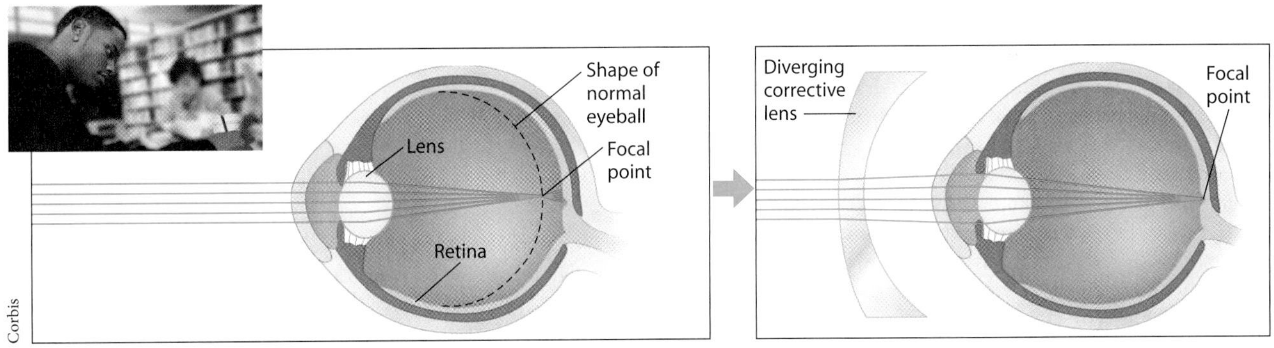

▲ **Figure 9A** A nearsighted eye (eyeball too long)

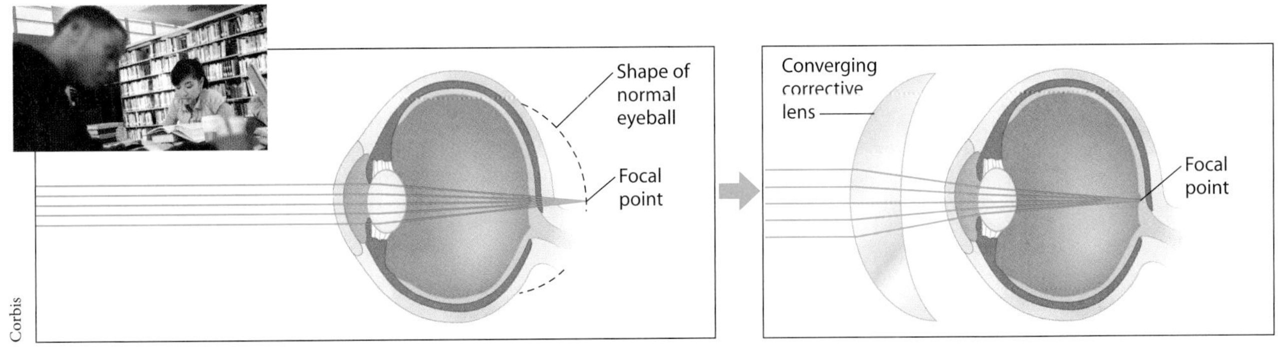

▲ **Figure 9B** A farsighted eye (eyeball too short)

10 The human retina contains two types of photoreceptors: rods and cones

The human retina contains two types of photoreceptors named for their shapes **(Figure 10A)**. **Cones** are stimulated by bright light and can distinguish color, but they contribute little to night vision. **Rods** are extremely sensitive to light and enable us to see in dim light, though only in shades of gray. The relative numbers of rods and cones an animal has correlates with whether an animal is most active during the day or night. Each human eye contains about 125 million rod cells and 6 million cone cells.

In humans, rods are found in greatest density at the outer edges of the retina and are completely absent from the fovea, the retina's center of focus. If you look directly toward a dim star in the night sky, the star is hard to see. Looking just to the side, however, makes your lens focus the starlight onto the parts of the retina with the most rods, and you can see the star better. By contrast, you achieve your sharpest day vision by looking straight at the object of interest. This is because cones are densest (about 150,000 per square millimeter) in the fovea.

As Figure 10A shows, each rod and cone consists of an array of membranous disks containing light-absorbing visual pigments. Rods contain a visual pigment called **rhodopsin**, which can absorb dim light. (Rhodopsin is derived from vitamin A, which is why vitamin A deficiency can cause "night blindness.") Interestingly, a change in a single amino acid of rhodopsin causes it to react to light in the ultraviolet range; this difference allows many species of birds to see beyond the range of human vision.

Cones contain visual pigments called **photopsins**, which absorb bright, colored light. We have three types of cones, each containing a different type of photopsin. These cells are called blue cones, green cones, and red cones, referring to the colors absorbed best by their photopsin. We can perceive a great number of colors because the light from each particular color triggers a unique pattern of stimulation among the three types of cones. Colorblindness results from a deficiency in one or more types of cones. The most common type is red-green colorblindness, in which red and green are seen as the same color—either red or green, depending on which type of cone is deficient.

Figure 10B shows the pathway of light into the eye and through the cell layers of the retina. Notice that the tips of the rods and cones are embedded in the back of the retina (pink cells). Light must pass through several relatively transparent layers of neurons before reaching the pigments in the rods and cones. Like all sensory receptors, rods and cones are stimulus transducers. When rhodopsin and photopsin absorb light, they change chemically, and the change alters the permeability of the cell's membrane. The resulting receptor potential triggers a change in the release of neurotransmitters from the synaptic terminals (see the left side of the rod and cone in Figure 10A). This release initiates a complex integration process in the retina. As shown in Figure 10B, visual information transduced by the rods and cones passes from the photoreceptor cells through the network of neurons (red arrows). Notice the numerous synapses between the photoreceptor cells and the neurons and among the neurons themselves. Integration in this maze of synapses helps sharpen images and increases the contrast between light and dark areas. Action potentials carry the partly integrated information to the brain via the optic nerve. Three-dimensional perceptions (what we actually see) result when visual input coming from the two eyes is integrated further in several processing centers of the cerebral cortex.

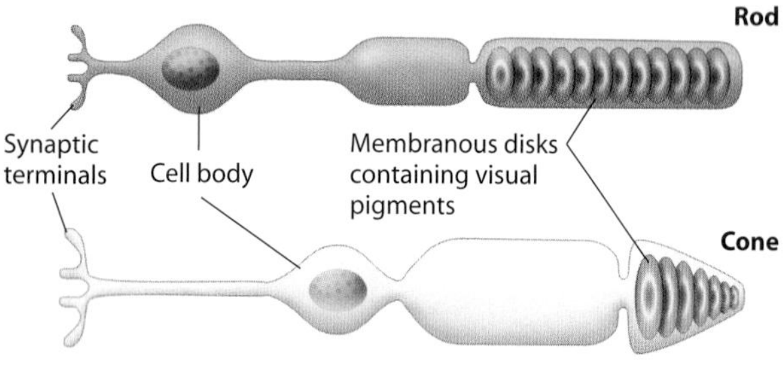

▲ **Figure 10A** Photoreceptor cells

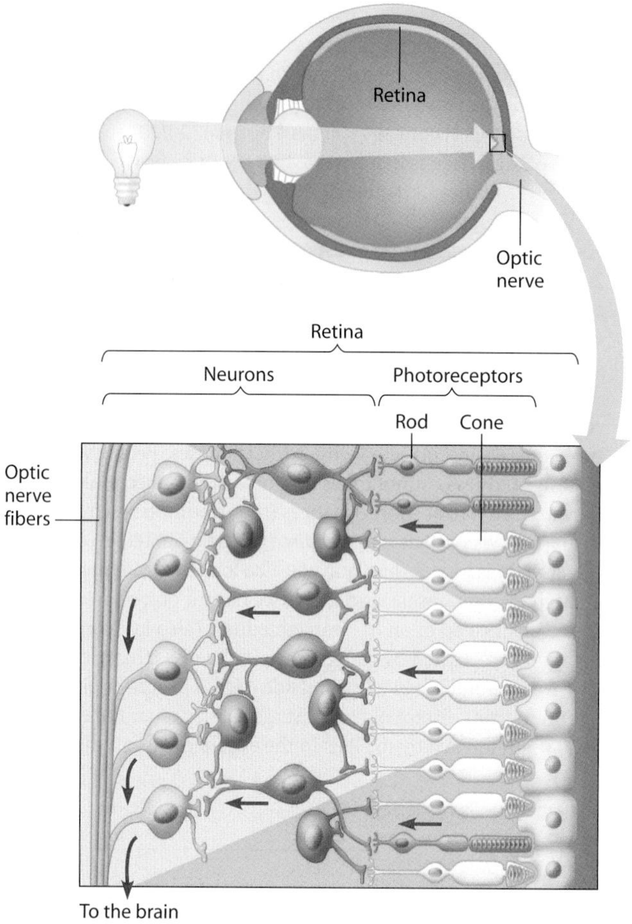

▲ **Figure 10B** The vision pathway from light source to optic nerve

? Explain why our night vision is mostly in shades of gray rather than in color.

Rods are more sensitive than cones to light, and thus low-intensity light stimulates far more rods than cones. Rods do not detect color.

Taste and Smell

11 Taste and odor receptors detect chemicals present in solution or air

Your senses of smell and taste depend on receptor cells that detect chemicals in the environment. Chemoreceptors in your taste buds detect molecules in solution; chemoreceptors in your nose detect airborne molecules.

Olfactory (smell) receptors are sensory neurons that line the upper portion of the nasal cavity and send impulses along their axons directly to the olfactory bulb of the brain (**Figure 11**). Notice the cilia extending from the tips of these chemoreceptors into the mucus that coats the nasal cavity. When an odorous substance (∴) diffuses into this region and dissolves in the mucus, it can bind to specific receptor proteins on the cilia. The binding triggers a membrane depolarization and generates action potentials. As the signals are integrated in the brain, we perceive odor. Humans can distinguish thousands of different odors.

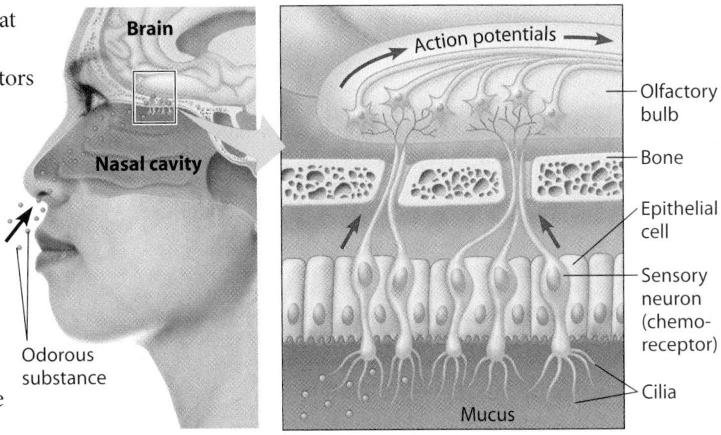

▲ **Figure 11** Smell in humans

szefei/Shutterstock

Many animals rely heavily on their sense of smell for survival. Most other mammals have a much more discriminating sense of smell than humans. Odors often provide more information than visual images about food, the presence of mates, or danger. In contrast, humans often pay more attention to sights and sounds than to smells. ("Seeing is believing" is thus a very human-centric idea!)

Receptor cells for taste in mammals are modified epithelial cells organized into taste buds on the tongue. In addition to the four familiar taste perceptions—sweet, sour, salty, and bitter—a fifth, called *umami* (Japanese for "delicious") is elicited by glutamate, an amino acid. Umami describes the savory flavor common in meats, cheeses, and other protein-rich foods, as well as the flavor-enhancing chemical monosodium glutamate (MSG). Any region of the tongue with taste buds can detect any of the five types of taste. ("Taste maps" of the tongue can thus be misleading.) An individual taste cell has a single type of receptor and thus detects only one of the five tastes.

Although the receptors and brain pathways for taste and smell are independent, the two senses do interact. Indeed, much of what we call taste is really smell, as you've probably noticed when a stuffy head cold dulls your perception of taste.

? What is the key structural difference between taste receptors and olfactory receptors in terms of cellular structure? (*Hint*: Review Module 2.)

Taste receptors release neurotransmitters, triggering action potentials carried to the brain by sensory neurons. Olfactory receptors are themselves modified sensory neurons.

CONNECTION

12 "Supertasters" have a heightened sense of taste

RubberBall/Alamy

If you don't like your vegetables (and we mean *really* don't like your vegetables), there may be a genetic basis behind your aversion. About 25% of us are "supertasters." For reasons that are not clearly understood, supertasters have three times the sensitivity to bitter tastes than other people. Supertasters tend to perceive bitter tastes in many more foods than do normal tasters. Hence, supertasters typically have more food dislikes, often avoiding coffee; alcoholic beverages; fatty foods; vegetables such as spinach, brussels sprouts, cabbage, kale, and broccoli; and other common foods.

Are you a supertaster? There are two signs: hypersensitivity to a bitter-tasting chemical called propylthiouracil and more than normal numbers of fungiform papillae, the structures on your tongue that house taste buds. Try sticking your tongue out, looking at it in a mirror, and comparing the number of taste buds you can see on your tongue with those you can see on a friend's tongue. (It helps if you coat your tongue with blue food dye first.)

There are consequences to being a supertaster beyond matters of food preference. An aversion to fatty foods may offer health benefits to supertasters (such as reduced risk of heart disease). But perhaps because they tend to avoid somewhat bitter but healthy vegetables and other foods, supertasters also have a higher risk of obesity, colon cancer, and other serious health problems. Thus, taste-related food choices can have health implications.

? Why might being a supertaster be harmful to your health?

Supertasters tend to avoid slightly bitter but healthful vegetables.

13 Review: The central nervous system couples stimulus with response

In this chapter and the previous one, we focused on information gathering and processing. Sensory receptors provide an animal's nervous system with vital data that enable the animal to avoid danger, find food and mates, and maintain homeostasis—in short, to survive.

A bat catching an insect helps us summarize the sequence of information flow in an animal. Ultrasonic squeaks emitted by the bat echo off an insect. Within milliseconds, receptor cells in the bat's ears transduce the sound, and action potentials representing the echo of the insect enter the brain. Before the insect can fly out of reach, a vast network of neurons in the bat's brain, with millions of synapses, integrates the information and sends out command signals, again in the form of action potentials. The commands go out via motor neurons to muscles, and the bat swoops and grabs its prey.

? What three general types of neurons are involved when a bat catches an insect?

Sensory neurons, interneurons, and motor neurons

REVIEW

For Activity Quizzes, BioFlix, MP3 Tutors, and Activities, go to www.masteringbiology.com.

Reviewing the Concepts

Sensory Reception (1–3)

1 Sensory organs share a common cellular basis. All animal senses (both human and nonhuman) arise from receptor cells communicating action potentials to the central nervous system.

2 Sensory receptors convert stimulus energy to action potentials. Sensory receptors are specialized cells that detect stimuli. Sensory transduction converts stimulus energy to receptor potentials, which trigger action potentials that are transmitted to the brain. Action potential frequency reflects stimulus strength:

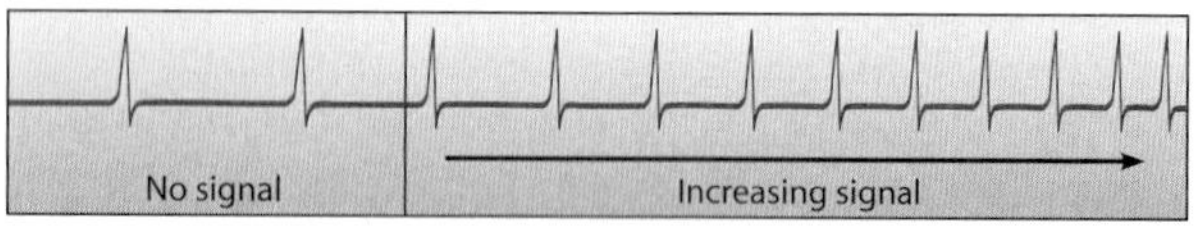

Repeated stimuli may lead to adaptation, a decrease in sensitivity.

3 Specialized sensory receptors detect five categories of stimuli. Pain receptors sense dangerous stimuli. Thermoreceptors detect heat or cold. Mechanoreceptors respond to mechanical energy (such as touch, pressure, and sound). Chemoreceptors respond to chemicals. Electromagnetic receptors respond to electricity, magnetism, and light (sensed by photoreceptors).

Hearing and Balance (4–6)

4 The ear converts air pressure waves to action potentials that are perceived as sound. The human ear channels sound waves through the outer ear to the eardrum to a chain of bones in the middle ear to the fluid in the coiled cochlea in the inner ear:

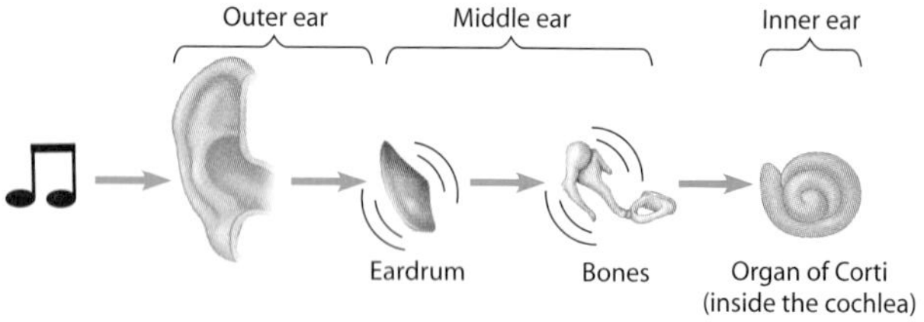

Pressure waves in the fluid bend hair cells of the organ of Corti against a membrane, triggering nerve signals to the brain. Louder sounds generate more action potentials; pitches stimulate different regions of the organ of Corti.

5 The inner ear houses our organs of balance. The semicircular canals and utricle and saccule located in the inner ear sense body position and movement.

6 What causes motion sickness? Conflicting signals from the inner ear and eyes may cause motion sickness.

Vision (7–10)

7 Several types of eyes have evolved independently among animals. Animal eyes range from simple eyecups that sense light intensity and direction to the many-lensed compound eyes of insects to the single-lens eyes of squids and vertebrates, including humans.

8 Humans have single-lens eyes that focus by changing position or shape. In the human eye, the cornea and flexible lens focus light on photoreceptor cells in the retina:

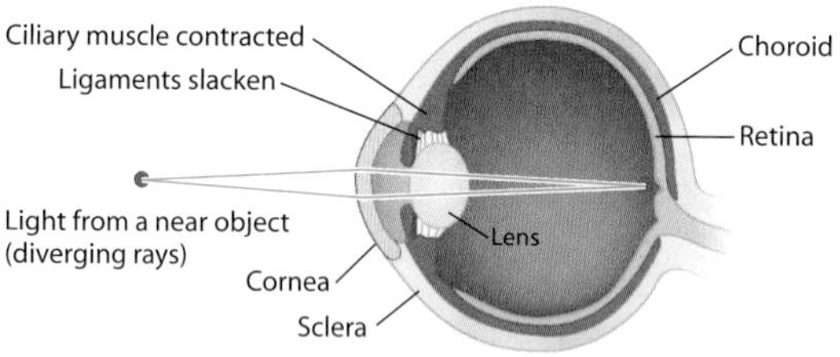

9 Artificial lenses or surgery can correct focusing problems. Focusing involves changing the shape of the lens. Nearsightedness and farsightedness result when the focal point is not on the retina. Corrective lenses bend the light rays to compensate.

10 The human retina contains two types of photoreceptors: rods and cones. Rods allow us to see shades of gray in dim light, and cones allow us to see color in bright light.

Taste and Smell (11–13)

11 Taste and odor receptors detect chemicals present in solution or air. Taste and smell depend on chemoreceptors that bind specific molecules. Taste receptors, located in taste buds on the tongue, produce five taste sensations. Olfactory (smell) sensory neurons line the nasal cavity. Various odors and tastes result from the integration of input from many receptors.

12 "Supertasters" have a heightened sense of taste.

13 Review: The central nervous system couples stimulus with response. The nervous system receives sensory information, integrates it, and commands appropriate muscle responses.

Connecting the Concepts

1. Complete this concept map summarizing sensory receptors.

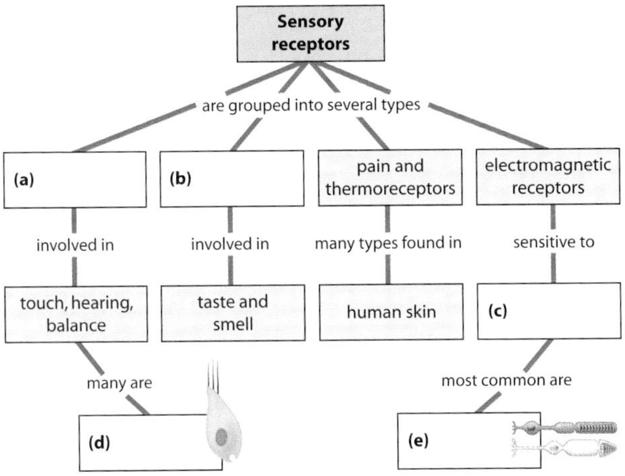

Testing Your Knowledge

Multiple Choice

2. Eighty-year-old Mr. Johnson was becoming slightly deaf. To test his hearing, his doctor held a vibrating tuning fork tightly against the back of Mr. Johnson's skull. This sent vibrations through the bones of the skull, setting the fluid in the cochlea in motion. Mr. Johnson could hear the tuning fork this way, but not when it was held away from the skull a few inches from his ear. The problem was probably in the ___________. (*Explain.*)
   a. auditory center in Mr. Johnson's brain.
   b. auditory nerve leading to the brain.
   c. hair cells in the cochlea.
   d. bones of the middle ear.
   e. fluid of the cochlea.

3. Which of the following correctly traces the path of light into your eye?
   a. lens, cornea, pupil, retina
   b. cornea, pupil, lens, retina
   c. cornea, lens, pupil, retina
   d. lens, pupil, cornea, retina
   e. pupil, cornea, lens, retina

4. If you look away from this book and focus your eyes on a distant object, the eye muscles ___________ and the lenses ___________ to focus images on the retinas.
   a. relax . . . flatten
   b. relax . . . become more rounded
   c. contract . . . flatten
   d. contract . . . become more rounded
   e. contract . . . relax

5. Which of the following are *not* present in human skin?
   a. thermoreceptors
   b. electromagnetic receptors
   c. touch receptors
   d. pressure receptors
   e. pain receptors

6. Jim had his eyes tested and found that he has 20/40 vision. This means that
   a. the muscles in his iris accommodate too slowly.
   b. he is farsighted.
   c. the vision in his left eye is normal, but his right eye is defective.
   d. he can see at 40 feet what a person with normal vision can see at 20 feet.
   e. he can see at 20 feet what a person with normal vision can see at 40 feet.

7. What do the receptor cells on the skin of a fish and the cochlea of your ear have in common?
   a. They use hair cells to sense sound or pressure waves.
   b. They are organs of equilibrium.
   c. They use electromagnetic receptors to sense pressure waves in fluid.
   d. They use granules that signal a change in position and stimulate their receptor cells.
   e. They are homologous structures that share a common evolutionary origin with all organs of hearing.

Describing, Comparing, and Explaining

8. How does your brain determine the volume and pitch of sounds?

9. As you read these words, the lenses of your eyes project patterns of light representing the letters onto your retinas. There the photoreceptors respond to the patterns of light and dark and transmit nerve signals to the brain. The brain then interprets the words. In this example, which processes relate to sensation and which to perception?

10. For what purposes do animals use their senses of taste and smell?

Applying the Concepts

11. Sensory organs tend to come in pairs. We have two eyes and two ears. Similarly, a planarian worm has two eyecups, a rattlesnake has two infrared receptors, and a butterfly has two antennae. Propose a testable hypothesis that could explain the advantage of having two ears or eyes instead of one.

12. Sea turtles bury their eggs on the beach above the high-tide line. When the baby turtles hatch, they dig their way to the surface of the sand and quickly head straight for the water. How do you think the turtles know which way to go? Outline an experiment to test your hypothesis.

13. Have you ever felt your ears ringing after listening to loud music from an iPod or at a concert? Can this music be loud enough to permanently impair your hearing? Do you think people are aware of the possible danger of prolonged exposure to loud music? Should anything be done to warn or protect them? If you think so, what action would you suggest? What effect might warnings have?

Review Answers

1. a. mechanoreceptors; b. chemoreceptors; c. electricity, magnetism, light; d. hair cells; e. photoreceptors
2. d (He could hear the tuning fork against his skull, so the cochlea, nerve, and brain are OK. Apparently, sounds are not being transmitted to the cochlea; therefore, the bones are the problem.) 3. b 4. a 5. b 6. e 7. a
8. Louder sounds create pressure waves with greater amplitude, moving hair cells more and generating a greater frequency of action potentials. Different pitches affect different parts of the basilar membrane; different hair cells stimulate different sensory neurons that transmit action potentials to different parts of the brain.
9. Sensation is the detection of stimuli (light) by the photoreceptors of the retina and transmission of action potentials to the brain. Perception is the interpretation of these nerve signals—sorting out the patterns of light and dark and determining their meaning.
10. Taste is used to sample food and determine its quality. Smell is used for many functions—communicating territories (scent marking), navigating (salmon), locating mates (moths), sensing danger (predators, fires), and finding food.
11. Some possible hypotheses: Paired sensory receptors enable an animal to determine the direction from which stimuli come. Paired receptors enable comparison of the intensity of stimuli on either side. Paired receptors enable comparison of slightly different images seen by the eyes or sounds heard by the ears (thus enabling the brain to perceive depth and distance).
12. Do the turtles hear the surf? Plug the ears of some turtles and not others. If turtles without earplugs head for the water and turtles with earplugs get lost, they probably hear the ocean. Or do they smell the water? Plug their nostrils and follow the same process.
13. Some issues and questions to consider: Assuming that the sound is loud enough to impair hearing, how long an exposure is necessary for this to occur? Does exposure have to occur all at once, or is damage cumulative? Who is responsible, concert promoters or listeners? Should there be regulations regarding sound exposure at concerts (as there are for job-related noise)? Are young people sufficiently mature and aware to heed such warnings?

How Animals Move

How Animals Move

BIG IDEAS

Photodisc/PhotoLibrary

Movement and Locomotion
(1–2)

Locomotion requires the collaboration of muscles and skeleton.

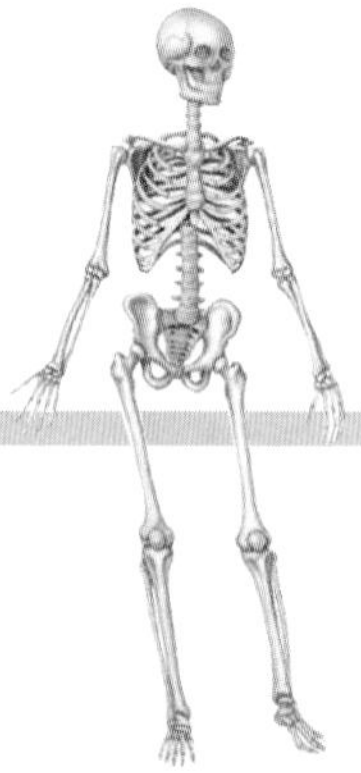

The Vertebrate Skeleton
(3–6)

Vertebrates have similar skeletons, but modifications of the ancestral body plan resulted in structural adaptations for different functions.

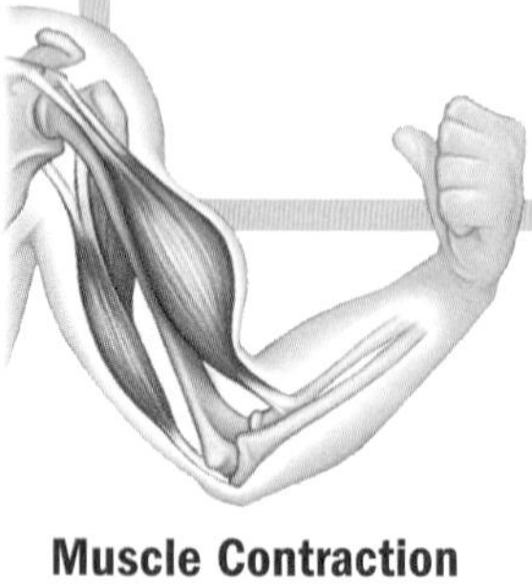

Muscle Contraction and Movement
(7–12)

Muscles work with the skeleton to move the body. Each muscle cell has its own contractile apparatus.

Manfred Grebler/Alamy

A horse at full gallop is a magnificent example of an animal that is born to run. The legs of horses are made up of the same skeletal elements as those of other vertebrates, including humans. But as the ancestors of modern horses evolved on the open plains and savanna, the original five toes were lost or reduced, until each leg rested on a single wear-resistant hoof. The hoof is separated from the shinbone by a long bone called the cannon, a highly modified version of a bone that is part of the foot of most other tetrapods. Thus, the joint that appears to be a horse's knee actually corresponds to our wrist or ankle. With these long, slender legs, a horse can cover as much as 7 m (23 feet) in a single stride. In most animals, including humans, the advantage of long limbs is offset by the added weight of the muscle required to move them—muscle is very heavy. But the cannon bone, like our hands and feet, carries only lightweight tendons. Other muscles in horses are reduced or redistributed as well. For example, our shoulders and hips have ball-and-socket joints, which enable the limbs to rotate freely but also require stabilizing muscles. Horses have hinge joints in their shoulders and hips. Hinge joints bend in a single plane, like our knees, and lack stabilizing muscles.

Most animals have the ability to move from one place to another. Some, like horses, move swiftly. Others—snails and tortoises, for example—are exceedingly slow. Modes of travel are diverse, too. In addition to animals that run or walk, there are animals that crawl, hop, fly, or swim. Almost all are variations on a common theme: muscles working in partnership with a skeletal system. In this chapter, we sample the types of animal movement and the skeletal systems that support it. We also learn about our own muscle and skeletal systems and how we can increase our strength and endurance through athletic training.

Movement and Locomotion

1 Locomotion requires energy to overcome friction and gravity

Movement is a distinguishing characteristic of animals. Even animals that are attached to a substrate move their body parts. All types of animal movement have underlying similarities. At the cellular level, every form of movement involves protein strands moving against one another, an energy-consuming process. In muscle cell contraction and amoeboid movement, the cellular system is based on microfilaments. Microtubules are the main components of cilia and flagella.

Most animals are fully mobile. **Locomotion**—active travel from place to place—requires energy to overcome two forces that tend to keep an animal stationary: friction and gravity. The relative importance of these two forces varies, depending on the environment. Water is dense and offers considerable resistance to a body moving through it, so an aquatic animal must expend energy overcoming friction. Gravity is not much of a problem, because water supports much or all of the animal's weight. On land, gravity is the main challenge—air provides no support for an animal's body. When a land animal walks, runs, or hops, its leg muscles expend energy both to propel it and to keep it from falling down. On the other hand, air offers very little resistance to an animal moving through it, at least at moderate speeds. Friction is limited to the points of contact between the animal's body and the ground. In the remainder of this module, we look at the major modes of animal locomotion.

Swimming Animals swim in diverse ways. Many insects, for example, swim the way we do, using their legs as oars to push against the water. Squids, scallops, and some jellies are jet-propelled, taking in water and squirting it out in bursts. Fishes swim by moving their body and tail from side to side (**Figure 1A**). Whales and other aquatic mammals move by undulating their body and tail from top to bottom. A sleek, streamlined shape, like that of seals, porpoises, penguins, and many fishes, is an adaptation that aids rapid swimming.

Danita Delimont/Alamy

▲ **Figure 1A** A fish swimming

Walking and Running A walking animal moves each leg in turn, overcoming friction between the foot and the ground with each step. To maintain balance, a four-legged animal usually keeps three feet on the ground at all times when walking slowly. Bipedal (two-footed) animals, such as birds and humans, are less stable on land and keep part of at least one foot on the ground when walking. A running four-legged animal may move two or three legs with each stride. At some gaits, all of its feet may be off the ground simultaneously (**Figure 1B**). At running speeds, momentum, more than foot contact, stabilizes the body's position, just as it keeps a moving bicycle upright.

Tamara Bauer/iStockphoto

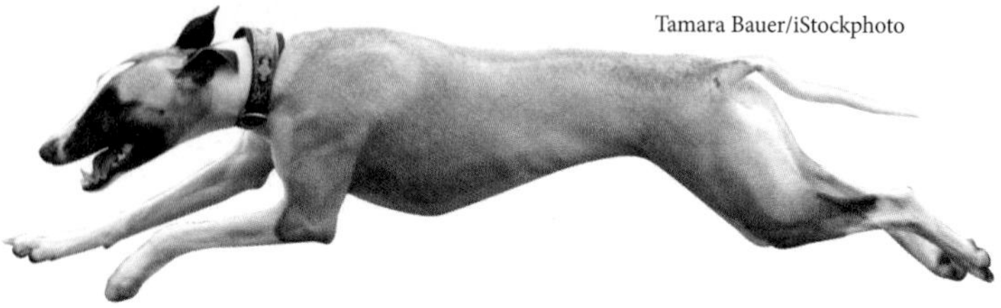

▲ **Figure 1B** A dog running at full speed

Hopping Some animals—for example, kangaroos—travel mainly by hopping (**Figure 1C**), a specialized mode of locomotion that has also evolved independently in several rodents. Large muscles in the hind legs of kangaroos generate a lot of power. Tendons (which connect muscle to bone) in the legs also momentarily store energy when the kangaroo lands—somewhat like the spring on a pogo stick. The higher the jump, the tighter the spring coils when a pogo stick lands and the greater the tension in the tendons when a kangaroo lands. In both cases, the stored energy is available for the next jump. For the kangaroo, the tension in its legs is a cost-free energy boost that reduces the total amount of energy the animal expends to travel. At rest, the kangaroo sits upright with its tail and both hind feet touching the ground. This position stabilizes the animal's body and costs little energy to maintain.

The pogo stick analogy applies to many other land animals as well. The legs of an insect, horse, or human, for instance, retain some spring during walking or running, although less than those of a hopping kangaroo.

Crawling Animals that have no limbs, or very short limbs, drag their bodies along the ground in a crawling movement. Because much of the animal's body is in contact with the ground, its energy is mainly expended to overcome friction

Dave Watts/NHPA/Photo Researchers

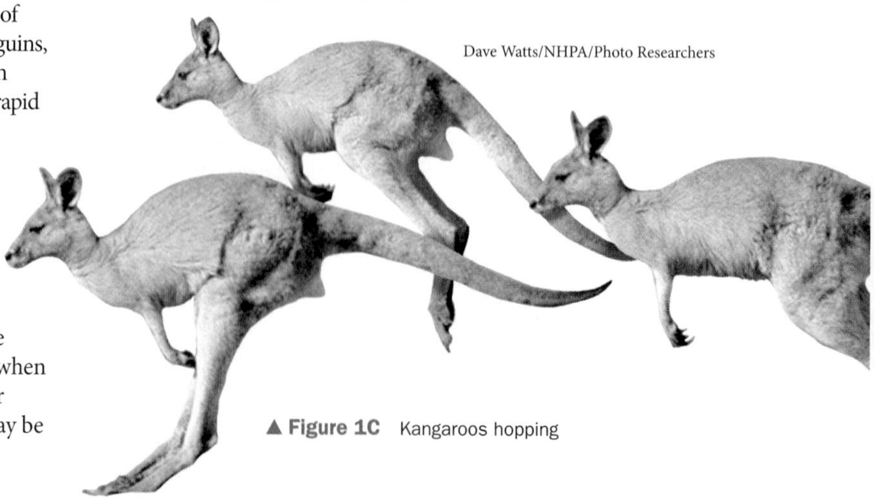

▲ **Figure 1C** Kangaroos hopping

rather than gravity. Many snakes crawl rapidly by undulating the entire body from side to side. Aided by large, movable scales on its underside, a snake's body pushes against the ground, driving the animal forward. Boa constrictors and pythons creep forward in a straight line, driven by muscles that lift belly scales off the ground, tilt them forward, and then push them backward against the ground.

Earthworms crawl by peristalsis, a type of movement produced by rhythmic waves of muscle contractions passing from head to tail. To move by peristalsis, an animal needs a set of muscles that elongates the body and another set that shortens it. Also required are a way to anchor its body to the ground and a hydrostatic skeleton, which we discuss further in Module 2. As illustrated in **Figure 1D**, the contraction of circular muscles, which encircle the circumference of the body, constricts and elongates some regions of the fluid-filled segments of a crawling earthworm. At the same time, longitudinal muscles that run the length of the body shorten and thicken other regions. Stiff bristles on the underside of the body grip the ground and provide traction, like the spikes on track shoes. (If you run your fingers along the belly of an earthworm, the bristles feel like whisker stubble.) In position ❶, segments at the head and tail ends of the worm are short and thick (longitudinal muscles contracted) and anchored to the ground by bristles. Just behind the head, a group of segments is thin and elongated (circular muscles contracted), with bristles held away from the ground. In position ❷, the head has moved forward because circular muscles in the head segments have contracted. Segments just behind the head and near the tail are now thick and anchored by bristles, thus preventing the head from slipping backward. In position ❸, the head segments are thick again and anchored to the ground in their new position, well ahead of their starting point. The rear segments of the worm now release their hold on the ground and are pulled forward.

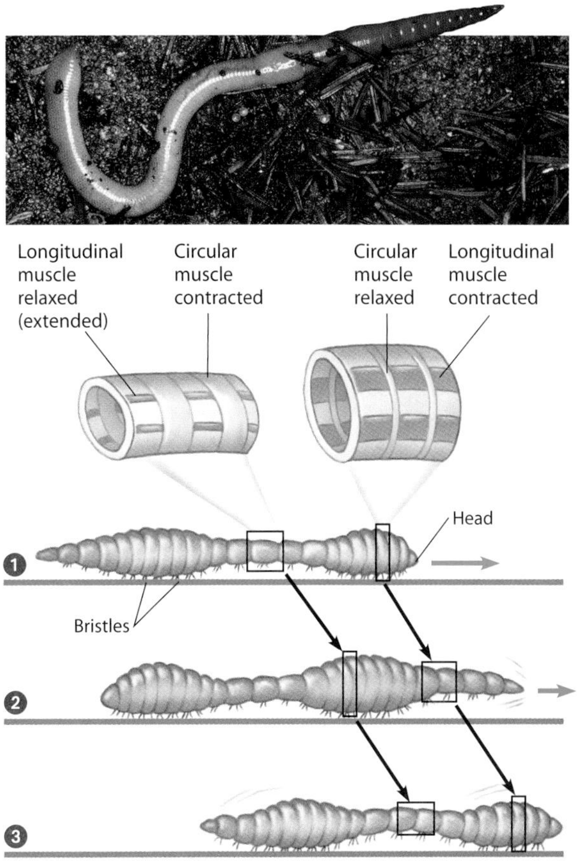

▲ **Figure 1D** An earthworm crawling by peristalsis

Flying Many phyla of animals include species that crawl, walk, or run, and almost all phyla include swimmers. But flying has evolved in only a few animal groups: insects, reptiles (including birds), and, among the mammals, bats. A group of large flying reptiles died out millions of years ago, leaving birds and bats as the only flying vertebrates.

For an animal to become airborne, its wings must develop enough "lift" to completely overcome the pull of gravity. The key to flight is the shape of wings. All types of wings, including those of airplanes, are airfoils—structures whose shape alters air currents in a way that creates lift. As **Figure 1E** shows, an airfoil has a leading edge that is thicker than the trailing edge. It also has an upper surface that is somewhat convex and a lower surface that is flattened or concave. This shape makes the air passing over the wing travel farther than the air passing under the wing. As a result, air molecules are spaced farther apart above the wing than below it, and the air pressure underneath the wing is greater. This pressure difference provides the lift for flight.

Birds can reach great speeds and cover enormous distances. Swifts, which can fly 170 km/hr (105 mph), are the fastest. The bird that migrates the farthest is the arctic tern, which flies round-trip between the North and South Poles each year.

An animal's muscle system provides the power to overcome friction and gravity. However, movement and locomotion result from a collaboration between muscles and a skeletal system. In the next module, you'll learn how skeletal systems are involved in movement. You'll also learn about some of the other functions of skeletal systems.

? Contrast swimming with walking in terms of the forces an animal must overcome to move.

Friction resists an animal moving through water, but gravity has little effect because of the animal's buoyancy; air poses little resistance to an animal walking on land, but the animal must support itself against the force of gravity.

▼ **Figure 1E** A bald eagle flying

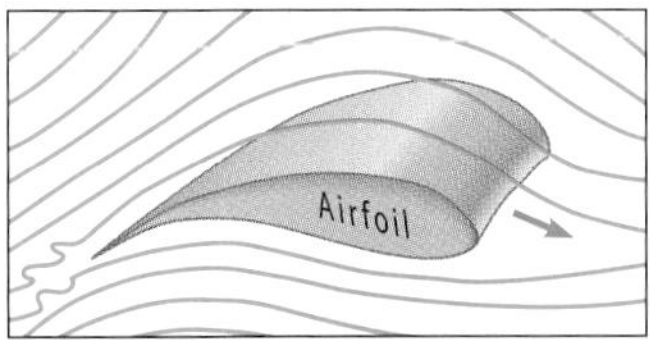

2 Skeletons function in support, movement, and protection

A skeleton has many functions. An animal could not move without its skeleton, and most land animals would sag from their own weight if they had no skeleton to support them. Even an animal in water would be a formless mass without a skeletal framework to maintain its shape. Skeletons also may protect an animal's soft parts. For example, the vertebrate skull protects the brain, and the ribs form a cage around the heart and lungs.

There are three main types of skeletons: hydrostatic skeletons, exoskeletons, and endoskeletons. All three types have multiple functions.

Hydrostatic Skeletons A **hydrostatic skeleton** consists of fluid held under pressure in a closed body compartment. This is very different from the more familiar skeletons made of hard materials. Nonetheless, a hydrostatic skeleton helps protect other body parts by cushioning them from shocks. It also gives the body shape and provides support for muscle action.

Earthworms have a fluid-filled internal body cavity, or coelom. As a segmented animal, the earthworm has its coelom divided into separate compartments. The fluid in these segments functions as a hydrostatic skeleton, and the action of circular and longitudinal muscles working against the hydrostatic skeleton produces the peristaltic movement described in Module 1.

Cnidarians, such as hydras and jellies, also have a hydrostatic skeleton. A hydra (**Figure 2A**), for example, has contractile cells in its body wall that enable it to alter its body shape by exerting pressure on the water-filled gastrovascular cavity. When a hydra closes its mouth and the contractile cells encircling its gastrovascular cavity contract, the body elongates, just as the earthworm elongates when its circular muscles contract (Figure 2A, left). The squeezing action also extends the tentacles. A hydra often sits in this position for hours, waiting for prey such as small worms or crustaceans that it can snare with its tentacles. If the hydra is disturbed, its mouth opens, allowing water to flow out. At the same time, contractile cells arranged longitudinally in the body wall contract, causing the body to shorten (Figure 2A, right).

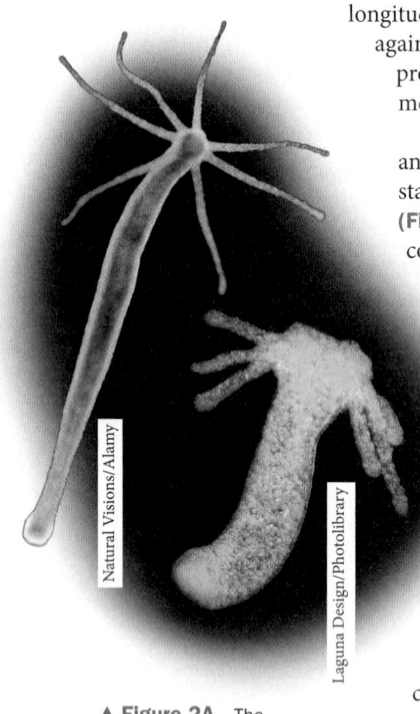

▲ **Figure 2A** The hydrostatic skeleton of a hydra in two states

Hydrostatic skeletons work well for many aquatic animals and for terrestrial animals that crawl or burrow by peristalsis. Most animals with hydrostatic skeletons are soft and flexible. In addition to extending its body and tentacles, for example, a hydra can expand its body around ingested prey that are larger than the gastrovascular cavity. An earthworm can burrow through soil because it is flexible and has a hydrostatic skeleton. Similarly, having an expandable body and a hydrostatic skeleton enables tube-dwelling animals such as feather duster worms to extend out of their tubes for feeding and gas exchange and then quickly squeeze back into the tube when threatened. However, a hydrostatic skeleton cannot support the forms of terrestrial locomotion in which an animal's body is held off the ground, such as walking.

Exoskeletons A variety of aquatic and terrestrial animals have a rigid external skeleton, or **exoskeleton**. Recall from Module 18.11 that the exoskeleton is a characteristic of the phylum Arthropoda, a group that includes insects, spiders, and crustaceans such as crabs. The arthropod exoskeleton is a tough covering composed of layers of protein and the polysaccharide chitin. The muscles are attached to knobs and plates on the inner surfaces of the exoskeleton. At the joints of legs, the exoskeleton is thin and flexible, allowing movement. If you have eaten crab legs, you cracked the exoskeleton to extract the tasty muscle within.

Because the exoskeleton is composed of nonliving material, it does not grow with the animal. It must be shed (molted) and replaced by a larger exoskeleton at intervals to allow for the animal's growth (**Figure 2B**). Depending on the species, most insects molt from four to eight times before

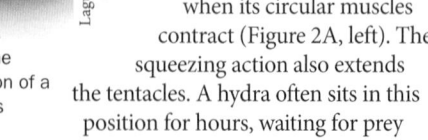

▲ **Figure 2B** The exoskeleton of an arthropod: a crab molting

reaching adult size. A few insect species and certain other arthropods, such as lobsters and crabs, molt at intervals throughout life.

An arthropod is never without an exoskeleton of some sort. For instance, a newly molted crab is covered by a soft, elastic exoskeleton that formed under the old one. Soon after molting, the crab expands its body by gulping air or water. Its new exoskeleton then hardens in the expanded position, and the animal has room for additional growth. If you have ever eaten soft-shell crab, you took advantage of this brief period when the new exoskeleton is tender enough to chew. As you can imagine, a newly molted arthropod is very susceptible to predation. Besides being weakly armored, it is usually less mobile, because the soft exoskeleton cannot support the full action of its muscles.

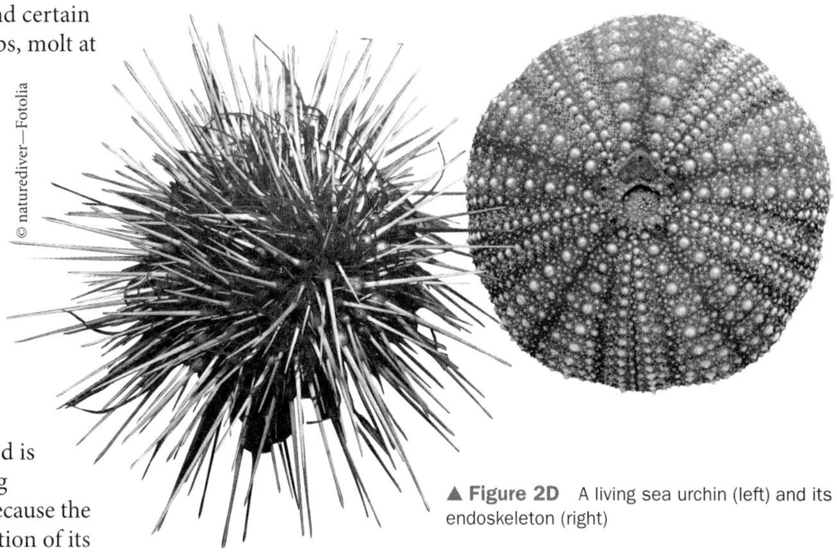

© naturediver—Fotolia

Kaj R. Svensson/Photo Researchers

▲ **Figure 2D** A living sea urchin (left) and its endoskeleton (right)

The shells of molluscs such as clams, snails, and cowries (**Figure 2C**)—shells you might find on a beach—are also exoskeletons. Unlike the chitinous arthropod exoskeleton, mollusc shells are made of a mineral, calcium carbonate. The mantle, a sheetlike extension of the animal's body wall, secretes the shell. As a mollusc grows, it does not molt; rather, it enlarges the diameter of its shell by adding to its outer edge.

Endoskeletons An **endoskeleton** consists of hard or leathery supporting elements situated among the soft tissues of an animal. Sponges, for example, are reinforced by a framework of tough protein fibers or by mineral-containing particles. Usually microscopic and sharp-pointed, the particles consist of inorganic material such as calcium salts or silica. Sea stars, sea urchins, and most other echinoderms have an endoskeleton of hard plates beneath their skin. In living sea urchins, about all you see are the movable spines, which are attached to the endoskeleton by muscles (**Figure 2D**). A dead urchin with its spines removed reveals the plates that form a rigid skeletal case (Figure 2D, right).

Carlos Villoch/Image Quest Marine

▲ **Figure 2C** The exoskeleton of a mollusc: a cowrie (a marine snail)

Vertebrates have endoskeletons consisting of cartilage or a combination of cartilage and bone. One major lineage of vertebrates, the sharks, have endoskeletons of cartilage reinforced with calcium. **Figure 2E** shows the more common condition for vertebrates. Bone makes up most of a frog's skeleton, as it does in bony fishes and land vertebrates. The frog skeleton and the skeletons of most other vertebrates also include some cartilage (blue in the figure), mainly in areas where flexibility is needed. Next, let's take a closer look at endoskeletons.

? What are the advantages and disadvantages of an exoskeleton as compared to an endoskeleton?

An exoskeleton may offer greater protection to body parts, but must usually be molted for the animal to grow.

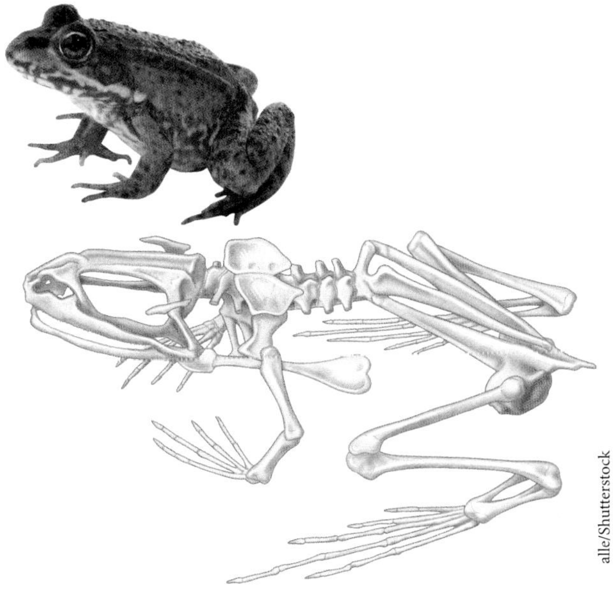

alle/Shutterstock

▲ **Figure 2E** Bone (tan) and cartilage (blue) in the endoskeleton of a vertebrate: a frog

Adapted from Lawrence G. Mitchell, John A. Mutchmor, and Warren D. Dolphin, *Zoology*.

The Vertebrate Skeleton

EVOLUTION CONNECTION

3 Vertebrate skeletons are variations on an ancient theme

The vertebrate skeletal system provided the structural support and means of locomotion that enabled tetrapods to colonize land. Subsequent evolution produced diverse groups of animals: amphibians, reptiles (including birds), and mammals. Each of those groups has diverse body forms whose skeletons are constructed from modified versions of the same parts. Even the skeletons of whales and dolphins, mammals that evolved from land-dwelling ancestors, are variations on the same theme.

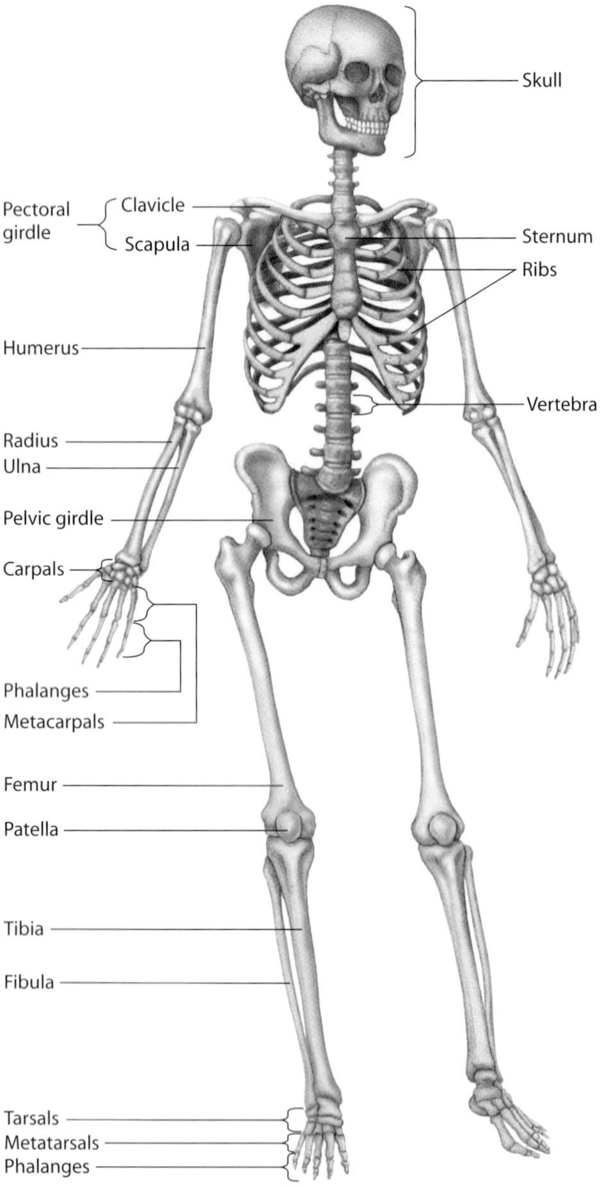

▲ **Figure 3A** The human skeleton

All vertebrates have an **axial skeleton** (orange in **Figure 3A**) supporting the axis, or trunk, of the body. The axial skeleton consists of the skull, enclosing and protecting the brain; the vertebral column (backbone), enclosing the spinal cord; and, in most vertebrates, a rib cage around the lungs and heart.

The backbone, the definitive characteristic of vertebrates, consists of a series of individual bones, the vertebrae, joined by pads of tough cartilage known as discs. The number of vertebrae varies among species. Pythons have 400, while an adult human has 24. All vertebrae have the same basic structure, with slight variations that reflect the position of each vertebra in the backbone. Anatomists divide the vertebral column into the regions shown in **Figure 3B**: cervical (neck), which support the head; thoracic (chest), which form joints with the ribs; lumbar (lower back); sacral (between the hips); and coccygeal (tail). In humans, the sacral vertebrae fuse into a single bone called the sacrum. Our small coccygeal vertebrae are partially fused into the coccyx or "tailbone."

Most vertebrates also have an **appendicular skeleton** (tan in Figure 3A), which is made up of the bones of the appendages and the bones that anchor the appendages to the axial skeleton. In a land vertebrate, the pectoral (shoulder) girdle and the pelvic girdle provide a base of support for the bones of the forelimbs and hind limbs. Modified versions of the same bones are found in all vertebrate limbs, whether they are arms, legs, fins, or wings. This variety of limbs equips vertebrates for every form of locomotion.

A few groups of vertebrates, including snakes, lost their limbs during their evolution. How did this happen? The identity of vertebrae is established during embryonic development by the pattern of master control (homeotic) genes expressed in the somites. (Somites are the blocks of embryonic tissue that give rise to the vertebral column.) Two of the homeotic genes that direct the differentiation of vertebrae are *Hoxc6* and *Hoxc8*. These

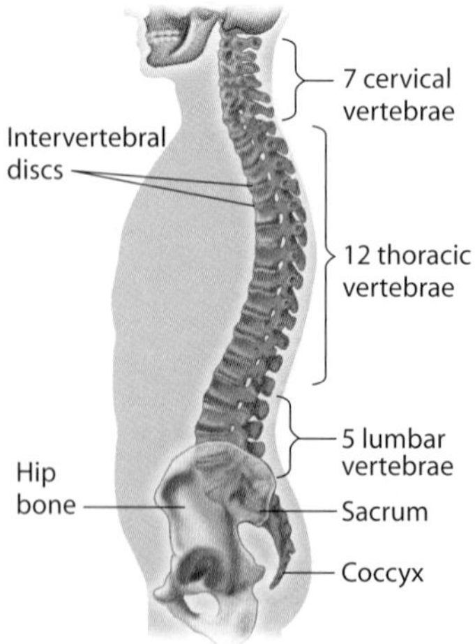

▲ **Figure 3B** The human backbone, showing the groups of vertebrae

genes are associated with the development of thoracic vertebrae, which support the ribs. **Figure 3C** shows the range of vertebrae formed by somites that expressed *Hoxc6* (red), *Hoxc8* (blue), or both (purple) in a chicken and a python.

In the python, both *Hoxc6* and *Hoxc8* are expressed in all somites for nearly the entire length of the vertebral column. As a result, the first rib-bearing thoracic vertebra is located immediately posterior to the head. Pythons have no cervical vertebrae. Chickens, on the other hand, have several cervical vertebrae, ending at the point where *Hoxc6* expression—and thoracic vertebrae—begin.

During the evolution of snakes, mutation in the DNA segments that control the expression of *Hoxc6* and *Hoxc8* changed cervical vertebrae to thoracic. In all vertebrates, the forelimbs originate at the boundary between cervical and thoracic vertebrae. Because this position does not exist in snakes, forelimbs do not form.

? **What type of vertebrae do chickens and humans both have that snakes do not have? How does this difference affect the appendicular skeleton of snakes?**

Chickens and humans have cervical vertebrae; snakes do not (we did not compare vertebrae posterior to thoracic vertebrae). Because forelimbs form at the boundary between cervical and thoracic vertebrae, snakes lack forelimbs.

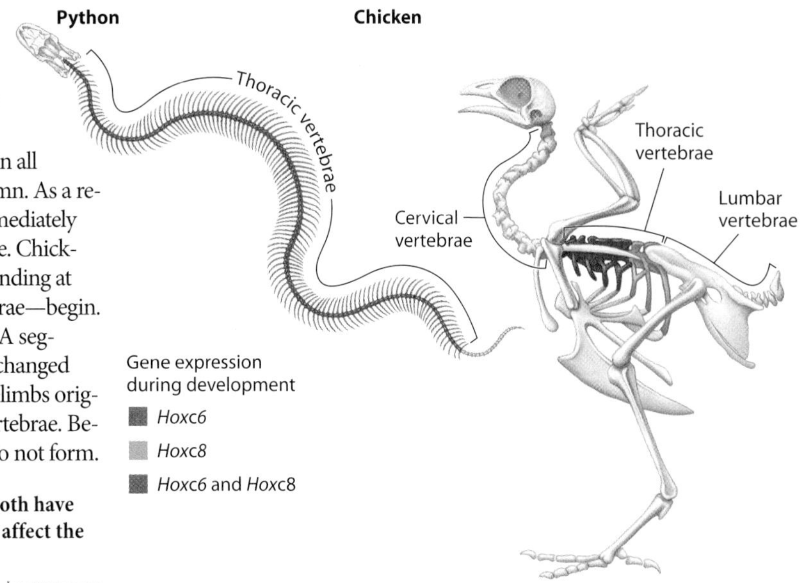

▲ **Figure 3C** Expression of two *Hox* genes in a python (left) and a chicken (right)

4 Bones are complex living organs

The expression "dry as a bone" should not be taken literally. Your bones are actually complex organs consisting of several kinds of moist, living tissues. **Figure 4** shows a human humerus (upper arm bone). A sheet of fibrous connective tissue, shown in pink (most visible in the enlargement on the lower right), covers most of the outside surface. This tissue helps form new bone in the event of a fracture. A thin sheet of cartilage (blue) forms a cushion-like surface for movable joints, protecting the ends of bones as they glide against one another. The bone itself contains living cells that secrete a surrounding material, or matrix. Bone matrix consists of flexible fibers of the protein collagen with crystals of a mineral made of calcium and phosphate bonded to them. The collagen keeps the bone flexible and nonbrittle, while the hard mineral matrix resists compression.

The shaft of this long bone is made of compact bone, a term referring to its dense structure. Notice that the compact bone surrounds a central cavity. The central cavity contains **yellow bone marrow**, which is mostly stored fat brought into the bone by the blood. The ends, or heads, of the bone have an outer layer of compact bone and an inner layer of spongy bone, so named because it is honeycombed with small cavities. The cavities contain **red bone marrow** (not shown in the figure), a specialized tissue that produces our blood cells.

Like all living tissues, bone cells carry out metabolism. Blood vessels that extend through channels in the bone transport nutrients and regulatory hormones to its cells and remove waste materials. Nerves running parallel to the blood vessels help regulate the traffic of materials between the bone and the blood.

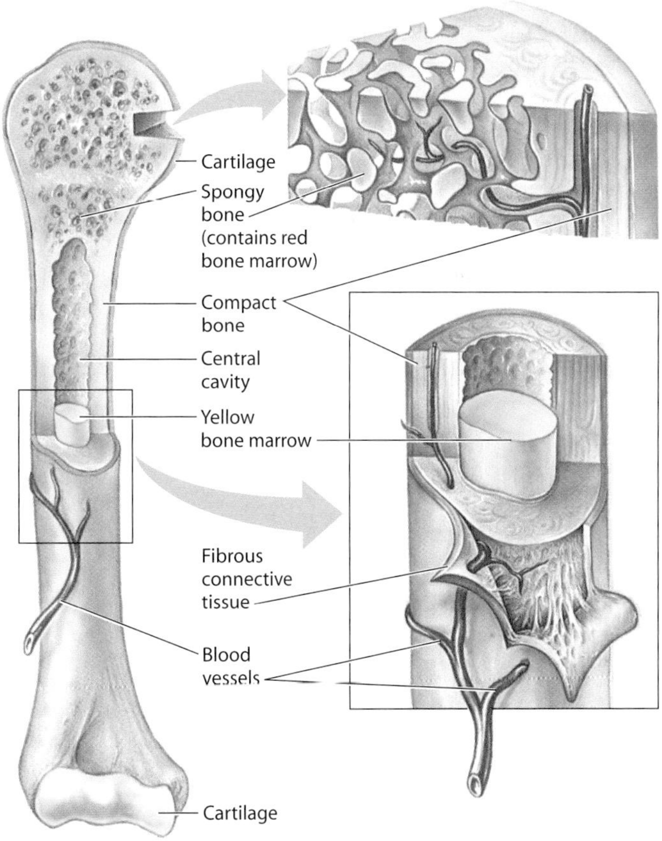

▲ **Figure 4** The structure of an arm bone

? **What causes the colors of yellow and red bone marrow?**

Stored fat and developing red blood cells, respectively.

CONNECTION 5 Healthy bones resist stress and heal from injuries

Bones are constantly subjected to stress as we go about our daily lives; exercise or physical labor causes additional stress. Excessive bone fatigue can lead to so-called stress fractures, hairline cracks in the bone, just as the accumulation of small amounts of stress on metals can cause a break. For example, when you bend a paper clip repeatedly, the metal fatigues and finally snaps. Unlike metal, however, bone is composed of living, dynamic tissue. Cells continually remove old bone matrix and replace it with new material. Stress fractures only occur if this repair process cannot keep up with the amount of stress placed on a bone. An athlete usually becomes aware of the problem and can allow time for healing, but in a racehorse, stress damage might go unnoticed until the fatigued bone breaks suddenly during a race.

A bone may also break when subjected to an external force that exceeds its resiliency. The average American will break two bones during his or her lifetime, most commonly the forearm or, for people over 75, the hip. Usually, this type of fracture occurs from a sudden impact, such as a fall or car accident. Wearing appropriate protective gear, such as seat belts, helmets, or padding, can protect your bones from high-force trauma.

Fortunately for active people, broken bones can heal themselves. A physician assists the process by putting the bone back into its natural alignment and then immobilizing it until the body's normal bone-building cells can repair the break. A splint or cast is used to protect the injured area and prevent movement. In severe cases, a fracture can be repaired surgically by inserting plates, rods, and/or screws that hold the broken pieces together (**Figure 5A**). In certain cases, however, severely injured or diseased bone is beyond repair and must be replaced. Broken hip joints, for example, can be replaced with artificial ones made of titanium or cobalt alloys. Researchers have recently developed new methods of bone replacement, including grafts (from the patient or from a cadaver) and the use of synthetic polymers.

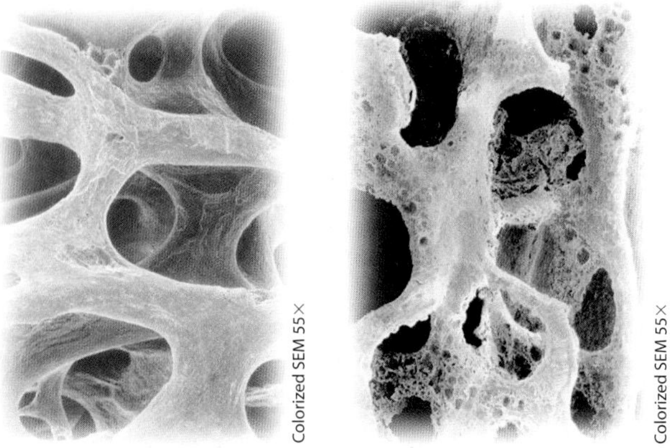

Professor P. Motta, Department of Anatomy, University 'la Sapienza" Rome/Photo Researchers, Inc.

▲ **Figure 5B** Healthy spongy bone tissue (left) and bone damaged by osteoporosis (right)

The risk of bone fracture increases if bones are porous and weak. **Figure 5B** contrasts healthy bone tissue (left) and bone eroded by osteoporosis (right). **Osteoporosis** is characterized by low bone mass and structural deterioration of bone tissue. This weakness emerges from an imbalance in the process of bone maintenance—the destruction of bone material exceeds the rate of replacement. Because the natural mechanism of bone maintenance responds to bone usage, weight-bearing exercise such as walking or running strengthens bones. On the other hand, disuse causes bones to become thinner. Strong bones also require an adequate intake of dietary calcium and enough vitamin D, which are both essential to bone replacement.

Until recently, osteoporosis was mostly considered a problem for women after menopause; estrogen contributes to normal bone maintenance. But while osteoporosis remains a serious health problem for older women, it is also becoming a concern for men and younger people. Doctors have noted a dramatic increase in bone fractures in children and teenagers in recent years. Many scientists believe that this is the result of exercising less and getting less calcium in the diet and less vitamin D from exposure to sunlight. Prevention of osteoporosis in later years begins with exercise and sufficient calcium and vitamin D while bones are still increasing in density (up until about age 30).

Other lifestyle habits, such as smoking, may also contribute to osteoporosis. There is a strong genetic component as well; young women whose mothers or grandmothers suffer from osteoporosis should be especially concerned with maintaining good bone health. Treatments for osteoporosis include calcium and vitamin D supplements and drugs that slow bone loss.

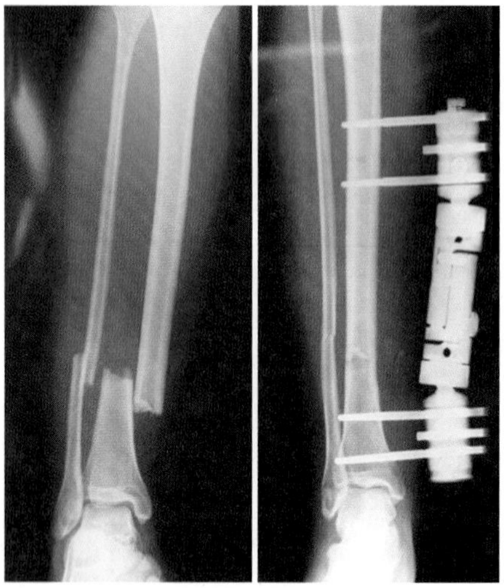

Jochen Tack Alamy

▲ **Figure 5A** X-rays of a broken leg (left) and the same leg after the bones were set with a plate and screws (right)

? How do exercise and adequate calcium intake help prevent osteoporosis?

Bone tissue responds to the stress of exercise by stepping up the repair process, which builds greater bone density. Calcium is the primary component of the rigid mineral matrix of bone.

6 Joints permit different types of movement

Much of the versatility of the vertebrate skeleton comes from its joints. Bands of strong fibrous connective tissue called **ligaments** hold together the bones of movable joints. **Ball-and-socket joints**, such as are found where the humerus joins the pectoral girdle (**Figure 6**, left), enable us to rotate our arms and legs and move them in several planes. A ball-and-socket joint also joins the femur to the pelvic girdle. **Hinge joints** permit movement in a single plane, just as the hinge on a door enables it to open and close. Our elbows (shown in Figure 6, center) and knees are hinge joints. Hinge joints are especially vulnerable to injury in sports like volleyball, basketball, and tennis that demand quick turns, which can twist the joint sideways. A **pivot joint** enables us to rotate the forearm at the elbow (Figure 6, right). A pivot joint between the first and second cervical vertebrae allows movement of the head from side to side, for example, the motion you make when you say "no." As you'll learn in the next module, muscles supply the force to move the bones of each joint.

? Where we have ball-and-socket joints, horses have hinge joints. How does this affect the movements they can perform?

Hinge joints restrict the movement of their legs to a single plane, making horses less flexible than humans.

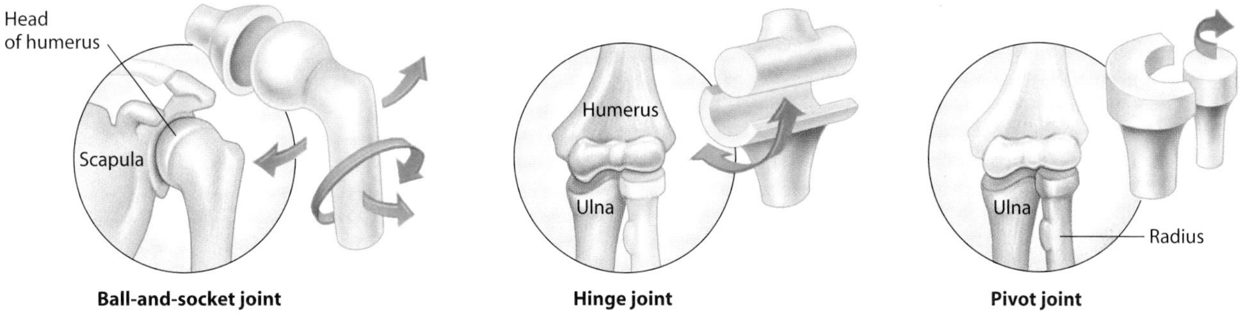

▲ **Figure 6** Three kinds of joints

Muscle Contraction and Movement

7 The skeleton and muscles interact in movement

Figure 7 shows how an animal's muscles interact with its bones to produce movement. Muscles are connected to bones by **tendons**. For example, the upper ends of the biceps and triceps muscles shown in the figure are anchored to bones in the shoulder. The lower ends of these muscles are attached to bones in the forearm. The action of a muscle is always to contract, or shorten. A muscle *pulls* the bone to which it is attached—it can only move the bone in one direction. A different muscle is needed to reverse the action. Thus, back-and-forth movement of body parts involves antagonists, a pair of muscles (or muscle groups) that can pull the same bone in opposite directions. While one antagonist contracts, the other relaxes.

The biceps and triceps muscles are an example of an antagonistic pair. Imagine that you are picking up a glass of water to drink. To raise the glass to your lips, your biceps muscle contracts, pulling the forearm bones toward you as your elbow bends. To put the glass back on the table, you must lower your forearm. Now the triceps muscle contracts, pulling the forearm bones down. The quadriceps, which extend the lower leg, and the hamstring, which flexes the lower leg, are additional examples of an antagonistic pair of muscles.

All animals—very small ones like ants and giant ones like elephants—have antagonistic pairs of muscles that apply opposite forces to move parts of their skeleton. Next we see how a muscle's structure explains its ability to contract.

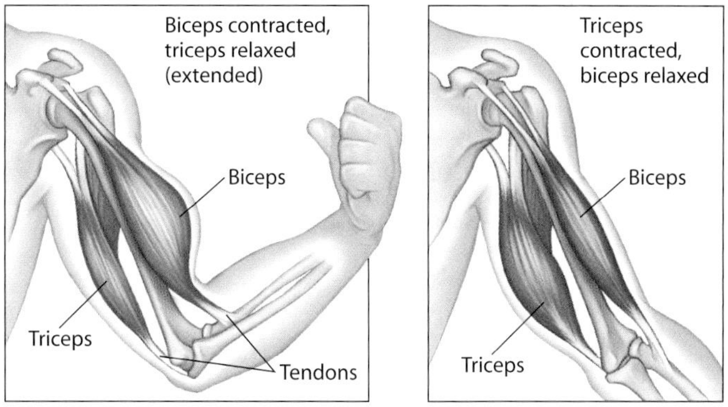

▲ **Figure 7** Antagonistic action of muscles to pull bones up or down in the human arm

? When exercising to strengthen muscles, why is it important to impose resistance while both flexing and extending the limbs?

This exercises both muscles of antagonistic pairs, which only do work when they are contracting.

8 Each muscle cell has its own contractile apparatus

The skeletal muscle system is a beautiful illustration of the relationship between structure and function. Each muscle in the body is made up of a hierarchy of smaller and smaller parallel strands, from the muscle itself down to the contractile protein molecules that produce body movements.

Figure 8 shows the levels of organization of skeletal muscle. As indicated at the top of the figure, a muscle consists of many bundles of **muscle fibers**—roughly 250,000 in a typical human biceps muscle—oriented parallel to each other. Each muscle fiber is a single long, cylindrical cell that has many nuclei. Most of its volume is occupied by hundreds or thousands of **myofibrils**, discrete bundles of proteins that include the contractile proteins **actin** and **myosin**. Skeletal muscle is also called striated (striped) muscle because the arrangement of the proteins creates a repeating pattern of stripes along the length of a myofibril that is visible under a light microscope. Beneath the drawing of a myofibril in Figure 8 is an electron micrograph that shows one unit of the pattern, which is called a **sarcomere**. Structurally, a sarcomere is the region between two dark, narrow lines, called Z lines, in the myofibril. Each myofibril consists of a long series of sarcomeres. Functionally, the sarcomere is the contractile apparatus in a myofibril—the muscle fiber's fundamental unit of action.

The diagram of the sarcomere at the bottom of Figure 8 explains the features visible in the micrograph. The pattern of horizontal stripes is the result of the alternating bands of **thin filaments**, composed primarily of actin molecules, and **thick filaments**, which are made up of myosin molecules. The Z lines consist of proteins that connect adjacent thin filaments. The light band surrounding each Z line contains only thin filaments. The dark band centered in the sarcomere is the location of the thick filaments. The actin molecules in the thin filaments are globular proteins arrayed in long strands. In addition to actin, thin filaments include proteins called troponin and tropomyosin that play a key role in regulating muscle contraction.

Next we examine the structure of a sarcomere in detail and see how it functions in muscle contraction.

? The two most abundant proteins of a myofibril are ______ and ______.

actin . . . myosin

▶ **Figure 8** The contractile apparatus of skeletal muscle

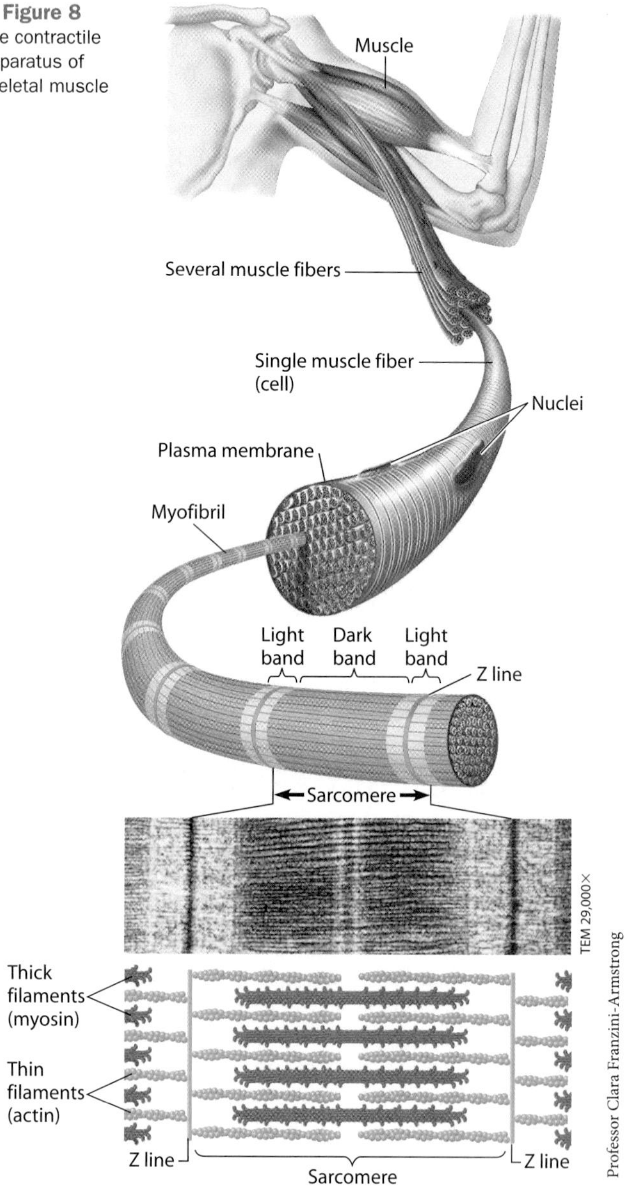

Adapted from Elaine N. Marieb, *Human Anatomy and Physiology*, 4th ed. Copyright © 1998 Pearson Education, Inc., publishing as Pearson Benjamin Cummings.

9 A muscle contracts when thin filaments slide along thick filaments

How does the structure of a sarcomere relate to its function? According to the sliding-filament model of muscle contraction, a sarcomere contracts (shortens) when its thin filaments slide along its thick filaments. Figure 9A, on the next page, is a simplified diagram that shows a sarcomere in a relaxed muscle, in a contracting muscle, and in a fully contracted muscle. Notice in the contracting sarcomere that the Z lines and the thin filaments (blue) have moved closer together. When the muscle is fully contracted, the thin filaments overlap in the middle of the sarcomere. Contraction shortens the sarcomere without changing the lengths of the thick and thin filaments. A whole muscle can shorten about 35% of its resting length when all sarcomeres contract.

Myosin acts as the engine of movement. Each myosin molecule has a long "tail" region and a globular "head" region. The tails of the myosin molecules in a thick filament lie parallel to each other, with their heads sticking out to the side. Each head has two binding sites. One of the binding

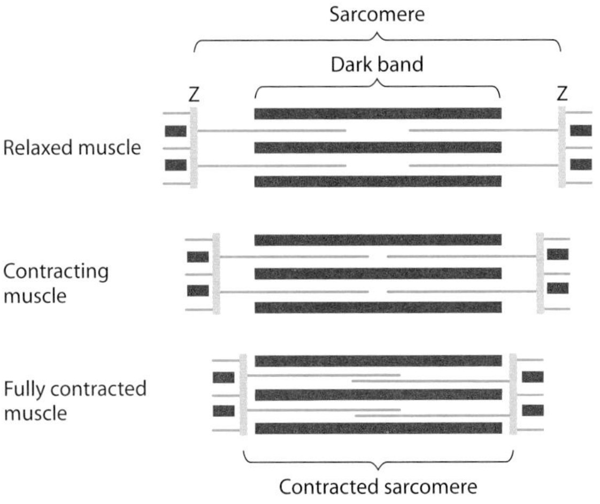

▲ **Figure 9A** The sliding-filament model of muscle contraction

sites matches a binding site on the actin molecules (subunits) of the thin filament. ATP binds at the other site, which is also capable of hydrolyzing the ATP to release its energy—the energy that powers muscle contraction.

Each myosin head pivots back and forth in a limited arc as it changes shape from a low-energy configuration to a high-energy configuration and back again. During these changes, the myosin head swings toward the thin filament, binds with an actin molecule, and drags the thin filament through the remainder of its arc. The myosin head then releases the actin molecule and returns to its starting position to repeat the same motion with a different actin molecule.

Let's follow the key events of this process in **Figure 9B**. The myosin head binds a molecule of ATP, as shown at ❶. At this point, the myosin head is in its low-energy position. ❷ Myosin hydrolyzes the ATP to ADP and phosphate P, releasing energy that extends the myosin head toward the thin filament. ❸ The myosin head extends further, and its other binding site latches on to the binding site of an actin. The result is a connection between the two filaments—a cross-bridge. ❹ ADP and P are released, and the myosin head pivots back to its low-energy configuration. This action, called the power stroke, pulls the thin filament toward the center of the sarcomere.

The cross-bridge remains intact until ❺ another ATP molecule binds to the myosin head, and the whole process repeats. On the next power stroke, the myosin head attaches to an actin molecule ahead of the previous one on the thin filament (closer to the Z line). This sequence—detach, extend, attach, pull, detach—occurs again and again in a contracting muscle. Though we show only one myosin head in the figure, a typical thick filament has about 350 heads, each of which can bind and unbind to a thin filament about five times per second. The combined action of hundreds of myosin heads on each thick filament ratchets the thin filament toward the center of the sarcomere, much like the people on one side of a tug-of-war. Each person (a myosin head) pulls hand over hand on the rope (the thin filament)—the rope moves, the people do not. As long as sufficient ATP is present, the process continues until the muscle is fully contracted or until the signal to contract stops.

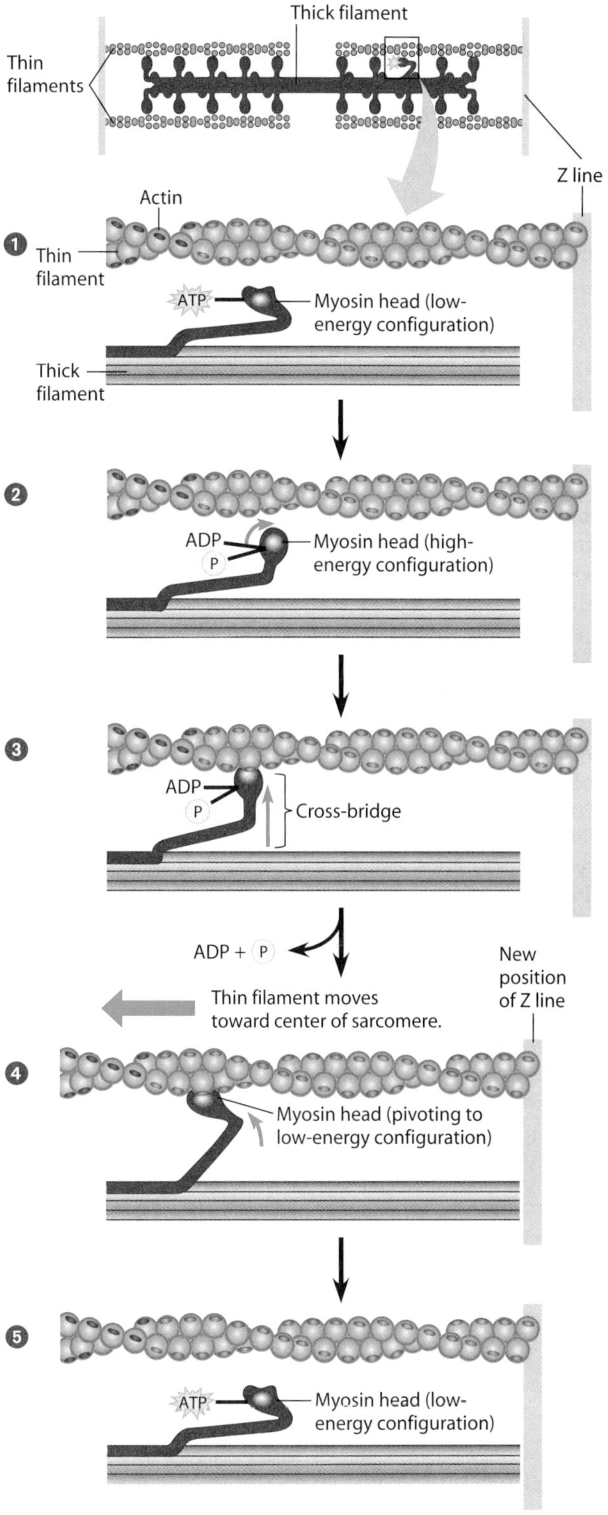

▲ **Figure 9B** The mechanism of filament sliding

? Which region of a sarcomere becomes shorter during contraction of a muscle?

The light bands shorten and even disappear as the thin filaments slide (are pulled) toward the center of the sarcomere.

10 Motor neurons stimulate muscle contraction

What prevents muscles from contracting whenever ATP is present? Signals from the central nervous system, conveyed by motor neurons, are required to initiate and sustain muscle contraction. When a motor neuron sends out an action potential, its synaptic terminals release the neurotransmitter acetylcholine, which diffuses across the synapse to the plasma membrane of the muscle fiber **(Figure 10A)**.

The plasma membrane of muscle fibers is unusual in two ways. Like the plasma membrane of neurons, the plasma membrane of a muscle fiber is electrically excitable—it can propagate action potentials. Also, the plasma membrane extends deep into the interior of the muscle fiber via infoldings called transverse (T) tubules. As a result, when a motor neuron triggers an action potential in a muscle fiber, it spreads throughout the entire volume of the cell, rather than only along the surface. The T tubules are in close contact with the endoplasmic reticulum (ER; blue in Figure 10A), a network of interconnected tubules within the muscle fiber. The action potential causes channels in the ER to open, releasing calcium ions (Ca^{2+}) into the cytoplasmic fluid. Now let's see why Ca^{2+} is necessary for muscle contraction.

When a muscle fiber is in a resting state, the regulatory proteins tropomyosin and troponin block the myosin binding sites on the actin molecules. As shown in **Figure 10B**, two strands of tropomyosin wrap around the thin filament, blocking access to the binding sites. The muscle fiber cannot contract while these sites are blocked. When Ca^{2+} binds to troponin, the tropomyosin moves away from the myosin-binding sites, allowing contraction to occur. As long as the cytoplasmic fluid is flooded with Ca^{2+}, contraction continues. When motor neurons stop sending action potentials to the muscle fibers, the ER pumps Ca^{2+} back out of the cytoplasmic fluid, binding sites on the actin molecules are again blocked, the sarcomeres stop contracting, and the muscle relaxes.

A large muscle such as the calf muscle is composed of roughly a million muscle fibers. However, only about 500 motor neurons run to the calf muscle. Each motor neuron has axons that branch out to synapse with many muscle fibers distributed throughout the muscle. Thus, an action potential from a single motor neuron in the calf causes the simultaneous contraction of roughly 2,000 muscle fibers. A motor neuron and all the muscle fibers it controls is called a **motor unit**. **Figure 10C** shows two motor units; one controls two muscle fibers (motor unit 1 in the figure), and the other (motor unit 2) controls three.

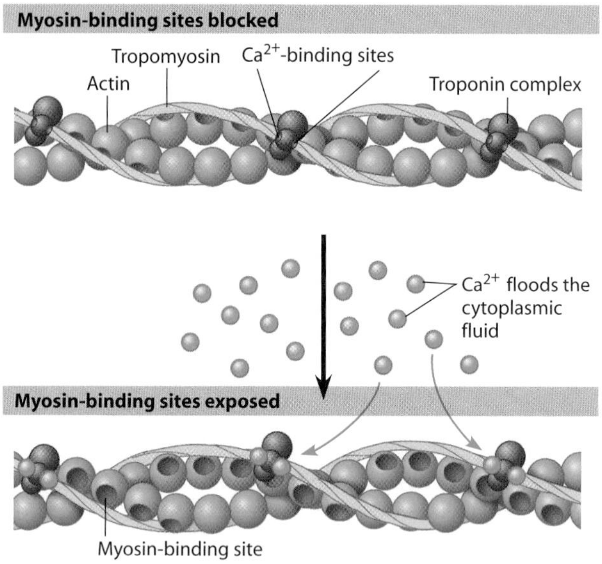

▲ **Figure 10B** Thin filament, showing the interactions among actin, regulatory proteins, and Ca^{2+}

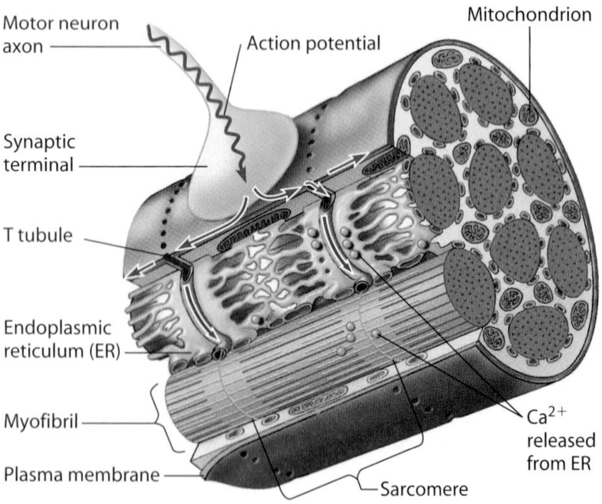

▲ **Figure 10A** How a motor neuron stimulates muscle contraction

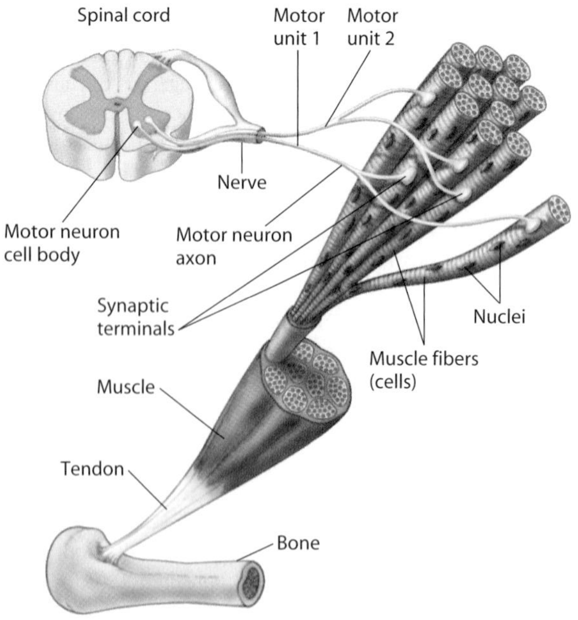

▲ **Figure 10C** Motor units

The organization of individual neurons and muscle cells into motor units is the key to the action of whole muscles. We can vary the amount of force our muscles develop: When you arm wrestle, for example, you might change the amount of force developed by the biceps and triceps several times in the course of a match. The ability to do this depends mainly on the nature of motor units. More forceful contractions result when additional motor units are activated. Thus, depending on how many motor units your brain commands to contract, you can apply a small amount of force to lift a fork or considerably more to lift, say, this textbook. In muscles requiring precise control, such as those controlling eye movements, a motor neuron may control only a single muscle fiber.

? How does the endoplasmic reticulum help regulate muscle contraction?

By reversibly taking up releasing Ca^{2+}, the ER regulates the cytoplasmic concentration of this ion, which is required in the cytoplasmic fluid for the binding of myosin to actin.

CONNECTION 11 Aerobic respiration supplies most of the energy for exercise

Many people exercise to stay in shape or to lose a few pounds before swimsuit season. Activities such as jogging, swimming, or other aerobic workouts are typical fitness routines (**Figure 11**). Aerobic exercise is an effective method of maintaining or losing weight—the goal is to burn at least as many calories as you consume. The number of calories burned varies with the intensity and duration of the exercise, as you saw in the examples listed in Figure 6.4. Most of the energy expended during exercise is used for muscle movement, specifically, to break the cross-bridges formed during sarcomere contraction. Here we refer to this energy in terms of ATP rather than calories.

Muscles have a very small amount of ATP on hand. ATP can also be obtained using a high-energy molecule called phosphocreatine (PCr, also known as creatine phosphate) that is stored in the muscles. The enzymatic transfer of a phosphate group from PCr to ADP makes ATP almost instantaneously. Together, ATP and PCr can provide enough energy for a 10- to 15-second burst of activity, enough for a 100-m sprint (see **Table 11**).

▲ **Figure 11** Running, a good form of aerobic exercise

Christopher Nuzzaco/iStockphoto

TABLE 11 SOURCES OF ATP FOR ATHLETIC ACTIVITIES

Athletic Activity	Energy Use	Main Source of ATP
100-m sprint; power lifting	10- to 15-second burst of activity	Stored ATP and PCr
200-m or 400-m race	Intense effort sustained over a short period of time	Stored ATP and PCr plus lactic acid fermentation
Jogging; long-distance running	Prolonged, low-level activity	Aerobic respiration
Tennis; squash; soccer	Prolonged, low-level activity with intermittent surges of intense effort	Aerobic respiration and lactic acid fermentation

The bulk of the ATP for aerobic exercise comes from the oxygen-requiring process of aerobic respiration, which derives ATP by the breakdown of the energy-rich sugar glucose. Muscles are richly supplied with blood vessels that bring O_2 and glucose and carry away CO_2, the by-product of aerobic respiration. Breathing and heart rate increase during exercise, facilitating the exchange of gases. Muscles can also get oxygen from a hemoglobin-like molecule called myoglobin.

If the demand for ATP outstrips the oxygen supply, muscle fibers can carry out the anaerobic process called lactic acid fermentation, which also uses glucose as a starting molecule. Fermentation supplies only a fraction of the ATP obtainable through aerobic respiration, but the process works $2\frac{1}{2}$ times faster.

Glucose for ATP production is available from the bloodstream, and muscle tissue stores glycogen, which can be broken down to provide more glucose. The liver stores glycogen, too, and can release glucose into the bloodstream. Although the body can also mobilize fats as a source of fuel, the process of harvesting ATP from fatty acid breakdown is too slow to keep pace with the demands of increasing exercise intensity. Casual athletes are in no danger of running out of metabolic fuel during exercise, though—glycogen stores are typically more than adequate.

When exercise begins, it takes a few minutes for the aerobic "machinery" to start producing enough ATP to meet the increased demand. After exercise, muscles must repay the "oxygen debt" that was incurred during the first few minutes of exercise. For example, after a run, you breathe rapidly, and your heart rate remains elevated for a time as your muscles replenish their supplies of ATP and PCr and restore myoglobin to its oxygenated state. Oxygen is also used to metabolize the lactic acid that was produced by fermentation.

In the next module, we consider some of the effects that exercise has on muscles.

? Compare the substances required, the ATP output, and the speed of aerobic respiration to lactic acid fermentation in muscles.

Aerobic respiration requires glucose and O_2; fermentation requires glucose, but not O_2. Aerobic respiration produces many more ATP molecules than fermentation, but fermentation produces ATP molecules faster.

CONNECTION 12 Characteristics of muscle fiber affect athletic performance

The fibers that make up a muscle are not all alike. The contractions of so-called "fast-twitch" fibers are rapid and powerful, but the fibers fatigue quickly. "Slow-twitch" fibers can sustain repeated contractions and are slow to fatigue, but their contractions are less forceful. The characteristics of slow and fast fibers are summarized in **Table 12** (below), along with the characteristics of intermediate fibers, which are also abundant in human muscle. Most of the features associated with fiber type reflect the pathway(s) the fiber preferentially uses to generate ATP from energy-rich molecules; the possible pathways are aerobic respiration and fermentation, which is anaerobic.

Each muscle typically has a mixture of fast, slow, and intermediate fiber types, broadly correlated to the type of work it does. Muscles that maintain the body's posture, for example, are constantly active; they have a high proportion of slow, fatigue-resistant fibers. In other muscles, such as those used for walking and running, the proportion of each fiber type varies widely among individuals. Although to some extent training can convert one type of fiber to another, most of this variation appears to be genetic. **Figure 12** (above right) compares the percentage of the three fiber types that are typical of the quadriceps (thigh) muscles of elite athletes and average adults.

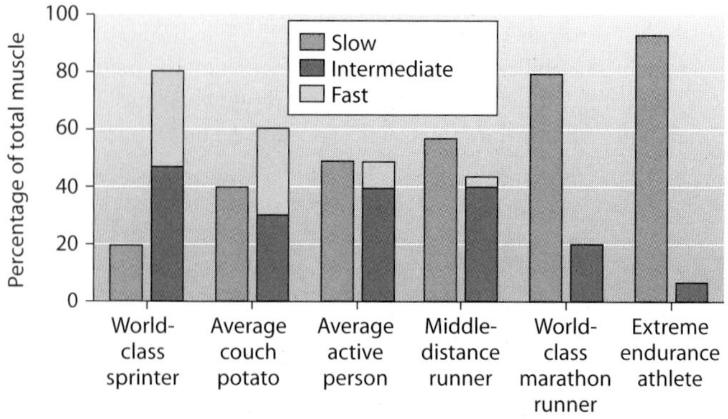

▲ **Figure 12** Percentage of slow, intermediate, and fast muscle fibers in quadriceps (thigh) muscles of different individuals

The differences among muscle fibers can be traced to the myosin that makes up the thick filaments. Fast, slow, and intermediate fibers contain different forms of myosin that hydrolyze ATP at different speeds. The faster ATP is consumed during muscle contraction, the faster the metabolic process needed to supply the ATP.

The fastest fibers cycle through cross-bridges rapidly, producing forceful contractions that power brief, explosive movements such as those that occur when hitting a baseball, hoisting a heavy weight, or bursting out of the blocks for a sprint. For such forceful contractions, the fibers use ATP at a breakneck pace. Therefore, the oxygen-independent process of fermentation is the primary mechanism for producing ATP. The combination of fast fibers with more fatigue-resistant intermediate fibers appears to be ideal for sprinters.

Hydrolysis of ATP occurs much more slowly in the myosin found in slow fibers. The slower pace at which cross-bridges are made and broken results in more sustained but less forceful contractions. Athletes who have a predominance of slow fibers excel at endurance events such as running, cycling, or swimming long distances. The slow fibers that give middle-distance runners aerobic endurance are complemented by intermediate fibers that sustain a higher power output over a longer period.

Exercise can change the composition of muscle fibers to some extent. Indeed, muscles have a remarkable ability to adapt to how they are used. Muscle fibers respond to a variety of signals generated within the cell during exercise, including mechanical stretching, change in calcium concentration, and low oxygen levels. Hormones may also trigger cellular changes. The specific responses of muscle fibers correspond to the stresses placed on them. For example, weight lifting stimulates muscle fibers to produce additional myofibrils. The thick filaments of the new myofibrils are composed of the fast form of myosin, increasing the force that the muscle can generate. Muscles adapt to aerobic exercise by synthesizing more myoglobin, increasing the number of mitochondria, and adding capillaries—changes that result in increased endurance and greater resistance to fatigue.

As athletes increase their strength and endurance through training, they continue to set new records. The limits of human performance have not yet been reached, but improved understanding of how animals move brings us closer to knowing what those limits are.

TABLE 12 CHARACTERISTICS OF MUSCLE FIBERS

Characteristic	Slow Fibers	Intermediate Fibers	Fast Fibers
Speed of contraction	Slow	Fast	Fast
Rate of fatigue	Fatigue slowly	Intermediate	Fatigue rapidly
Primary pathway for making ATP	Aerobic respiration	Aerobic (some fermentation)	Anaerobic (fermentation)
Myoglobin content	High	High	Low
Mitochondria and capillaries	Many	Many	Few

? **According to the graph in Figure 12, how does becoming active affect the percentage of muscle fibers that use aerobic respiration?**

The percentages of slow and intermediate fibers, which both use aerobic respiration, increase.

REVIEW

For Practice Quizzes, BioFlix, MP3 Tutors, and Activities, go to masteringbiology.com.

Reviewing the Concepts

Movement and Locomotion (1–2)

1 Locomotion requires energy to overcome friction and gravity. Animals that swim are supported by water but are slowed by friction. Animals that walk, hop, or run on land are less affected by friction but must support themselves against gravity. Burrowing or crawling animals must overcome friction. They may move by side-to-side undulation or by peristalsis. The wings of birds, bats, and flying insects are airfoils, which generate enough lift to overcome gravity.

Photodisc/Photolibrary; Tamara Bauer/iStockphoto; Danita Delimont/Alamy; Dave Watts/NHPA/Photo Researchers

2 Skeletons function in support, movement, and protection. Worms and cnidarians have hydrostatic skeletons—fluid held under pressure in closed body compartments. Exoskeletons are hard external cases, such as the chitinous, jointed skeletons of arthropods. The vertebrate endoskeleton is composed of cartilage and bone.

The Vertebrate Skeleton (3–6)

3 Vertebrate skeletons are variations on an ancient theme. Vertebrate skeletons consist of an axial skeleton (skull, vertebrae, and ribs) and an appendicular skeleton (shoulder girdle, upper limbs, pelvic girdle, and lower limbs). There are many variations on this basic body plan, which may have evolved through changes in gene regulation.

4 Bones are complex living organs. Cartilage at the ends of bones cushions the joints. Bone cells, serviced by blood vessels and nerves, reside in a matrix of flexible protein fibers and hard calcium salts. Long bones have a fat-storing central cavity and spongy bone at their ends. Spongy bone contains red marrow, where blood cells are made.

5 Healthy bones resist stress and heal from injuries. Bone cells continue to replace and repair bone throughout life. Osteoporosis, a bone disease characterized by weak, porous bones, occurs when bone destruction exceeds replacement.

6 Joints permit different types of movement.

Muscle Contraction and Movement (7–12)

7 The skeleton and muscles interact in movement. Antagonistic pairs of muscles produce opposite movements. Muscles perform work only when contracting.

8 Each muscle cell has its own contractile apparatus. Muscle fibers, or cells, consist of bundles of myofibrils, which contain bundles of overlapping thick (myosin) and thin (actin) protein filaments. Sarcomeres, repeating groups of thick and thin filaments, are the contractile units.

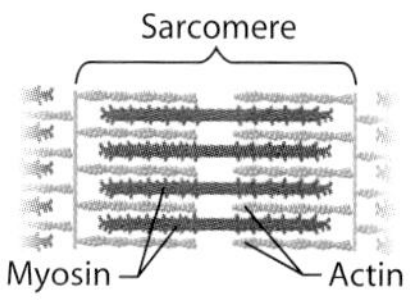

9 A muscle contracts when thin filaments slide along thick filaments. According to the sliding-filament model of muscle contraction, the myosin heads of the thick filaments bind ATP and extend to high-energy states. The heads then attach to binding sites on the actin molecules and pull the thin filaments toward the center of the sarcomere.

10 Motor neurons stimulate muscle contraction. Motor neurons carry action potentials that initiate muscle contraction. A neuron and the muscle fibers it controls constitute a motor unit. The neurotransmitter acetylcholine released at a synaptic terminal triggers an action potential that passes along T tubules into the center of the muscle cell. Calcium ions released from the endoplasmic reticulum initiate muscle contraction by moving the regulatory protein tropomyosin away from the myosin-binding sites on actin.

11 Aerobic respiration supplies most of the energy for exercise. Aerobic respiration requires a constant supply of glucose and oxygen. The anaerobic process of fermentation can start producing ATP faster than aerobic respiration can, but it produces less.

12 Characteristics of muscle fiber affect athletic performance. The classification of muscle fibers as slow, intermediate, and fast is based on the pathway(s) they use to generate ATP. Most muscles have a combination of fiber types, which can be affected by exercise. Exercise also affects other aspects of muscle composition.

Connecting the Concepts

1. Complete this concept map on animal movement.

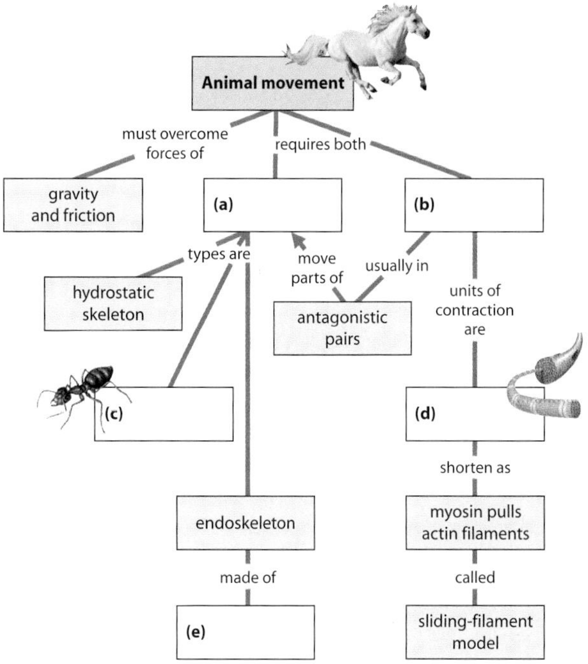

Testing Your Knowledge

Multiple Choice

2. A human's internal organs are protected mainly by the
   a. hydrostatic skeleton.
   b. motor unit.
   c. axial skeleton.
   d. exoskeleton.
   e. appendicular skeleton.
3. Arm muscles and leg muscles are arranged in antagonistic pairs. How does this affect their functioning?
   a. It provides a backup if one of the muscles is injured.
   b. One muscle of the pair pushes while the other pulls.
   c. A single motor neuron can control both of them.
   d. It allows the muscles to produce opposing movements.
   e. It doubles the strength of contraction.
4. Gravity would have the least effect on the movement of which of the following? *(Explain your answer.)*
   a. a salmon
   b. a human
   c. a snake
   d. a sparrow
   e. a grasshopper
5. Which of the following bones in the human arm corresponds to the femur in the leg?
   a. radius
   b. tibia
   c. humerus
   d. metacarpal
   e. ulna
6. Which of the following animals is correctly matched with its type of skeleton?
   a. fly—endoskeleton
   b. earthworm—exoskeleton
   c. dog—exoskeleton
   d. lobster—exoskeleton
   e. bee—hydrostatic skeleton
7. When a horse is running fast, its body position is stabilized by
   a. side-to-side undulation.
   b. energy stored in tendons.
   c. the lift generated by its movement through the air.
   d. foot contact with the ground.
   e. its momentum.
8. What is the role of calcium in muscle contraction?
   a. Its binding to a regulatory protein causes the protein to move, exposing actin binding sites to the myosin heads.
   b. It provides energy for contraction.
   c. It blocks contraction when the muscle relaxes.
   d. It is the neurotransmitter released by a motor neuron, and it initiates an action potential in a muscle fiber.
   e. It forms the heads of the myosin molecules in the thick filaments inside a muscle fiber.
9. Muscle A and muscle B have the same number of fibers, but muscle A is capable of more precise control than muscle B. Which of the following is likely to be true of muscle A? (*Explain your answer.*)
   a. It is controlled by more neurons than muscle B.
   b. It contains fewer motor units than muscle B.
   c. It is controlled by fewer neurons than muscle B.
   d. It has larger sarcomeres than muscle B.
   e. Each of its motor units consists of more cells than the motor units of muscle B.
10. Which of the following statements about skeletons is true?
    a. Hydrostatic skeletons are soft and do not protect body parts.
    b. Chitin is a major component of vertebrate skeletons.
    c. Evolution of bipedalism involved little change in the axial skeleton.
    d. Most cnidarians must shed their skeleton periodically in order to grow.
    e. Vertebrate bones contain living cells.

Describing, Comparing, and Explaining

11. In terms of both numbers of species and numbers of individuals, insects are the most successful land animals. Write a paragraph explaining how their exoskeletons help them live on land. Are there any disadvantages to having an exoskeleton?
12. A hawk swoops down, seizes a mouse in its talons, and flies back to its perch. Explain how its wings enable it to overcome the downward pull of gravity as it flies upward.
13. The greatest concentration of thoroughbred horse farms is in the bluegrass region of Kentucky. The grass in the limestone-based soil of this area is especially rich in calcium. How does this grass affect the development of championship horses?
14. Describe how you bend your arm, starting with action potentials and ending with the contraction of a muscle. How does a strong contraction differ from a weak one?
15. Using examples, explain this statement: "Vertebrate skeletons are variations on a theme."

Applying the Concepts

16. Drugs are often used to relax muscles during surgery. Which of the following chemicals do you think would make the best muscle relaxant, and why? Chemical A: Blocks acetylcholine receptors on muscle cells. Chemical B: Floods the cytoplasm of muscle cells with calcium ions.
17. An earthworm's body consists of a number of fluid-filled compartments, each with its own set of longitudinal and circular muscles. But in the roundworm, a single fluid-filled cavity occupies the body, and there are only longitudinal muscles that run its entire length. Predict how the movement of a roundworm would differ from the movement of an earthworm.
18. When a person dies, muscles become rigid and fixed in position —a condition known as rigor. Rigor mortis occurs because muscle cells are no longer supplied with ATP (when breathing stops, ATP synthesis ceases). Calcium also flows freely into dying cells. The rigor eventually disappears because the biological molecules break down. Explain, in terms of the mechanism of contraction described in Modules 9 and 10, why the presence of calcium and the lack of ATP would cause muscles to become rigid, rather than limp, soon after death.
19. A goal of the Americans with Disabilities Act is to allow people with physical limitations to fully participate in society. Perhaps you have a disability or know someone with a disability. Imagine that a neuromuscular disease or injury makes it impossible for you to walk. Think about your activities during the last 24 hours. How would your life be different if you had to get around in a wheelchair? What kinds of barriers or obstacles would you encounter? What kinds of changes would have to be made in your activities and surroundings to accommodate your change in mobility?

Review Answers

1. a. skeleton; b. muscles; c. exoskeleton; d. sarcomeres; e. bone and cartilage
2. c **3.** d **4.** a (Water supports aquatic animals, reducing the effects of gravity.) **5.** c **6.** d **7.** e **8.** a **9.** a (Each neuron controls a smaller number of muscle fibers.) **10.** e
11. Advantages of an insect exoskeleton include strength, good protection for the body, flexibility at joints, and protection from water loss. The major disadvantage is that the exoskeleton must be shed periodically as the insect grows, leaving the insect temporarily weak and vulnerable.
12. The bird's wings are airfoils, with convex upper surfaces and flat or concave lower surfaces. As the wings beat, air passing over them travels farther than air beneath. Air molecules above the wings are more spread out, lowering pressure. Higher pressure beneath the wings pushes them up.
13. Calcium is needed for healthy bone development. It strengthens bones and makes them less susceptible to stress fractures.
14. Action potentials from the brain travel down the spinal cord and along a motor neuron to the muscle. The neuron releases a neurotransmitter, which triggers action potentials in a muscle fiber membrane. These action potentials initiate the release of calcium ions from the ER of the cell. Calcium enables myosin heads of the thick filaments to bind with the actin of the thin filaments. ATP provides energy for the movement of myosin heads, which causes the thick and thin filaments to slide along one another, shortening the muscle fiber. The shortening of muscle fibers pulls on bones, bending the arm. If more motor units are activated, the contraction is stronger.
15. The fundamental vertebrate body plan includes an axial skeleton (skull, backbone, and rib cage) and an appendicular skeleton (bones of the appendages). Species vary in the numbers of vertebrae and the numbers of different types of vertebrae they possess. For example, pythons have no cervical vertebrae. Almost all mammals have seven cervical vertebrae but may have different numbers of other types. For example, human coccygeal vertebrae are small and fused together, but horses and other animals with long tails have many coccygeal vertebrae. Limb bones have been modified into a variety of appendages, such as wings, fins, and limbs. Snakes have no appendages.
16. Chemical A would work better, because acetylcholine triggers contraction. Blocking it would prevent contraction. Chemical B would actually increase contraction, because Ca^{2+} allows contraction to occur.
17. Circular muscles in the earthworm body wall decrease the diameter of each segment, squeezing internal fluid and lengthening the segment. Longitudinal muscles shorten and thicken each segment. Different parts of the earthworm can lengthen while others shorten, producing a crawling motion. The whole roundworm body moves at once because of a lack of segmentation. The body can only shorten or bend, not lengthen, because of a lack of circular muscles. Roundworms simply thrash from side to side.
18. The binding of calcium ions causes the regulatory protein tropomyosin to move out of the way, enabling myosin heads to bind to actin. This results in muscle contraction. ATP causes the myosin heads of the thick filaments to detach from the thin filaments (Figure 9B, step 1). If there is no ATP present, the myosin heads remain attached to the thin filaments, and the muscle fiber remains fixed in position.
19. Some issues and questions to consider: Are the places where you live, work, or attend class accessible to a person in a wheelchair? If you were in a wheelchair, would you have trouble with doors, stairs, drinking fountains, toilet facilities, and eating facilities? What kinds of transportation would be available to you, and how convenient would they be? What activities would you have to forgo? How might your disability alter your relationships with your friends and family? How well would you manage on your own?

Plant Structure, Growth, and Reproduction

From Chapter 31 of *Campbell Biology: Concepts & Connections*, Seventh Edition. Jane B. Reece, Martha R. Taylor, Eric J. Simon, Jean L. Dickey.

Plant Structure, Growth, and Reproduction

BIG IDEAS

zhuda/Shutterstock

Plant Structure and Function
(1–6)

Plant bodies contain specialized cells grouped into tissues, organs, and organ systems.

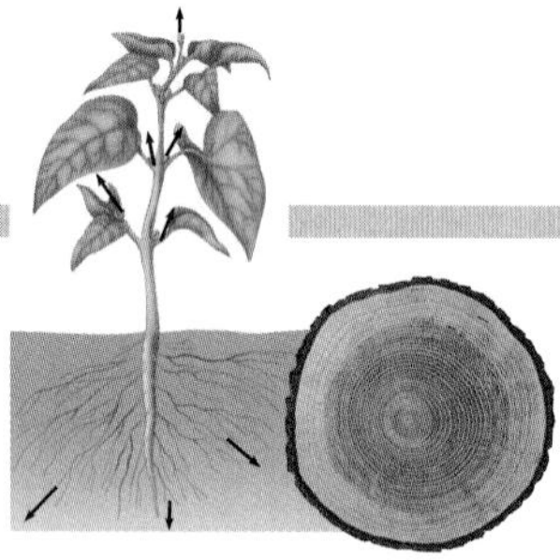

Plant Growth
(7–8)

All plants increase their length via primary growth, and woody plants thicken via secondary growth.

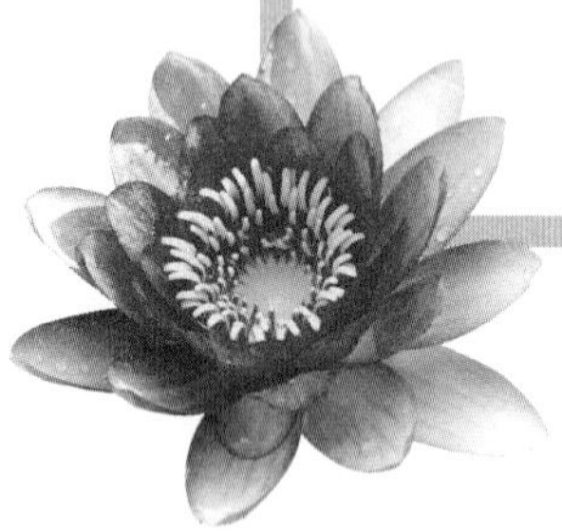

Reproduction of Flowering Plants
(9–15)

Sexual reproduction in angiosperms involves pollination, development of fruit and seeds, seed dispersal, germination, and growth.

Michael Nichols/National Geographic Stock

Found in northern California and southern Oregon, coast redwoods (*Sequoia sempervirens*) are the tallest trees in the world. The tallest one, named Hyperion, stands at 379.1 feet (115.5 m). Estimated to be over 2,000 years old, Hyperion is still growing. To protect this tree, only a few park workers and plant biologists know its location.

The only way to measure the height of a tree accurately is to climb it—talk about an extreme sport! Imagine yourself standing at the trunk of Hyperion. Its base stretches 15 feet to either side. You look up. The trunk rises 20 stories (60 m) before the first limbs appear. You strap on a helmet, a tree-climbing harness, and soft-soled boots to avoid damaging the bark. Using a crossbow, you shoot a rope over the first branch. You then pull yourself up the tree, foot by foot, using mechanical ascenders, which are rope-climbing devices used by rock climbers.

As you climb upward, you find that the redwood canopy supports its own unique ecosystem. Many species—such as golden-brown ants, pink earthworms, and salamanders—appear to spend their entire lives hundreds of feet in the air. The unique species of this redwood canopy are found nowhere else on Earth.

The coast redwood is one of about 600 living species of cone-bearing trees called conifers. Conifers are gymnosperms, one of two groups of seed plants. Gymnosperms bear seeds that are exposed, typically on cones, whereas plants of the other group, the angiosperms, produce seeds enclosed in fruits. Since angiosperms—the flowering plants—make up over 90% of the plant kingdom, we concentrate on them in this unit. By studying angiosperm structure in relation to growth and reproduction, we can gain insight into the structure-function relationships of all plants.

Plant Structure and Function

CONNECTION 1 People have manipulated plants since prehistoric times

To tell the story of human society, you have to talk about plants. From the dawn of civilization to the frontiers of genetic engineering, human progress has always depended on expanding our use of plants for food, fuel, clothing, and countless other trappings of modern life.

To illustrate, let's consider humankind's relationship with a single crop: wheat. Today, wheat accounts for about 20% of all calories consumed worldwide. In the United States, it is one of the most valuable cash crops, with American farmers exporting 30 million tons each year. And we do more than just eat wheat. For example, researchers at the USDA developed a new material made from wheat that can be used to make biodegradable packaging materials, and a protein that protects the wheat plant from cold can be added to ice cream to help prevent freezer burn.

The roots of today's prosperous relationship between people and wheat can be traced back to the earliest human settlements. Throughout much of prehistory, humans were nomadic hunter-gatherers, moving and foraging with the changing seasons. People gathered and ate seeds from wild grasses and cereals, but didn't plant them. About 10,000 years ago, a major shift occurred as people in several parts of the world began to domesticate wild plants. This domestication allowed for the production of surplus food and the formation of year-round farming villages. Farming, in turn, led to the establishment of cities and the emergence of modern civilizations.

When and where did this crucial shift take place? The origin of wheat cultivation has been traced to a region of the Middle East dubbed the "Fertile Crescent," near the upper reaches of the Tigris and Euphrates rivers **(Figure 1)**. Archaeological excavations of sites older than 10,000 years have found remnants of wild wheat. But in younger sites, cultivated wheat also appears, displacing wild varieties in the archaeological record over the course of more than a millennium.

ARCO/Diez, O/AGE fotostock

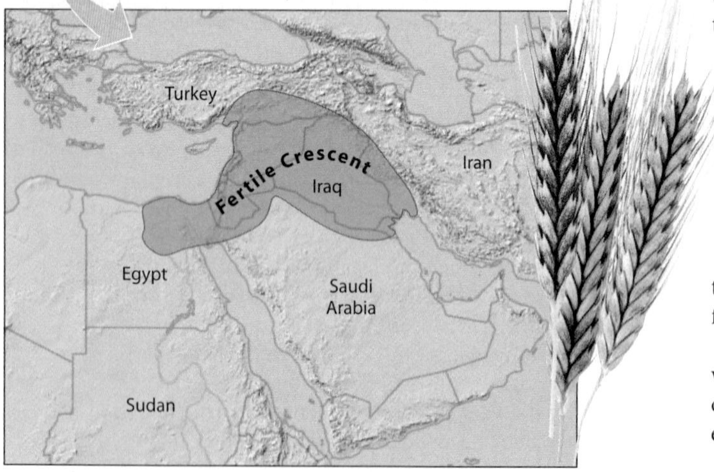

▲ **Figure 1** The Fertile Crescent region (shaded green), original site of the cultivation of wheat

Once begun, cultivation rapidly replaced gathering in the Middle East. These habits quickly spread. By 7,000 years ago, bread wheat was cultivated in Egypt, India, China, and northern Europe. Farming became the dominant way of life across Europe by 6,000 years ago. For most of the subsequent millennia, little changed in the production of wheat. Individual farmers tinkered to produce slightly superior varieties, but these were small, incremental improvements.

A major leap forward occurred in the latter half of the 20th century. Faced with the rapidly expanding human population and a dwindling availability of farmland, an international effort was undertaken to improve wheat and other staple crops. The methods were straightforward: controlled cross-pollinations between varieties with desirable traits, selection of the few plants possessing the best combinations of characteristics, cross-pollination of these plants, and so on. Through such repeated selective breeding experiments, today's wheat varieties were created. The improved traits of modern wheat include a short stature with double the grain per acre of full-sized varieties; the ability to grow in a wide variety of climates and soil conditions (such as dry or acidic soils); short generation times; and resistance to over 30 pests and pathogens. As a result of the so-called "green revolution," wheat yield (tons of grain per acre planted) more than doubled from 1940 to 1980 while the cost of wheat production was cut in half.

Wheat and other crops continue to be the target of considerable research today, primarily through the methods of genetic engineering and genomics. One goal is to enhance the nutritional value of food crops. In the case of wheat, some wild varieties are much more nutritious than domesticated varieties, with 10–15% higher concentrations of protein, iron, and zinc. In 2006, researchers identified and cloned a gene responsible for this difference. (Domestic wheat contains the gene, but at least one copy of it is inactive.) The hope is to transfer the active version of this gene into domesticated varieties, increasing their nutritional content.

Another group of researchers developed genetically modified (GM) wheat with triple the usual amount of the polysaccharide amylose, a component of dietary fiber. They achieved this by turning off a gene that normally breaks down amylose in wheat. High-fiber strains may be a health boon because eating more fiber has been linked to reduced colorectal cancer, heart disease, diabetes, and obesity. Other genetic engineers are attempting to identify genes that confer resistance to a new strain of wheat rust, a virulent fungal pathogen that threatens the world's wheat crops.

As we continue to refine modern DNA technology methods, we will continue to deepen our understanding of wheat and other crops. Keeping in mind this staple crop as one example, we will explore the structure of angiosperms in the next five modules.

? When and where did humans first cultivate wheat?

About 10,000 years ago in the Middle East

2 The two major groups of angiosperms are the monocots and the eudicots

Angiosperms have dominated the land for over 100 million years, and there are about 250,000 known species of flowering plants living today. Most of our foods come from a few hundred domesticated species of flowering plants. Among these foods are roots, such as beets and carrots; the fruits of trees and vines, such as apples, nuts, berries, and squashes; the fruits and seeds of legumes, such as peas and beans; and grains, the fruits of grasses such as wheat, rice, and corn (maize).

On the basis of several structural features, plant biologists (also called botanists) traditionally placed most angiosperms into two groups, called monocots and dicots. The names *monocot* and *dicot* refer to the first leaves on the plant embryo. These embryonic leaves are called seed leaves, or **cotyledons**. A **monocot** embryo has one seed leaf; a **dicot** embryo has two seed leaves. The great majority of dicots, called the **eudicots** ("true" dicots), are evolutionarily related, having diverged from a common ancestor about 125 million years ago; a few smaller groups of dicots have evolved independently. In this chapter, we will focus on monocots and eudicots (**Figure 2**).

Monocots are a large group of related plants that include the orchids, bamboos, palms, and lilies, as well as the grains and other grasses. You can see the single cotyledon inside the seed on the top left in Figure 2. The leaves, stems, flowers, and roots of monocots are also distinctive. Most monocots have leaves with parallel veins. Monocot stems have vascular tissues (internal tissues that transport water and nutrients) organized into bundles that are arranged in a scattered pattern. The flowers of most monocots have their petals and other parts in multiples of three. Monocot roots form a shallow fibrous system—a mat of threads—that spreads out below the soil surface. With most of their roots in the top few centimeters of soil, monocots, especially grasses, make excellent ground cover that reduces erosion. Fibrous root systems are well adapted to shallow soils where rainfall is light.

Most flowering shrubs and trees are eudicots, such as the majority of our ornamental plants and many of our food crops, including nearly all of our fruits and vegetables. You can see the two cotyledons of a typical eudicot in the seed on the lower left in Figure 2. Eudicot leaves have a multibranched network of veins, and eudicot stems have vascular bundles arranged in a ring. Eudicot flowers usually have petals and other parts in multiples of four or five. The large, vertical root of a eudicot, which is known as a taproot, goes deep into the soil, as you know if you've ever tried to pull up a dandelion. Taproots are well adapted to soils with deep groundwater.

A close look at a structure often reveals its function. Conversely, function provides insight into the "logic" of a structure. In the modules that follow, we'll take a detailed look at the correlation between plant structure and function.

? The terms *monocot* and *eudicot* refer to the number of_____ on the developing embryo in a seed.

cotyledons (seed leaves)

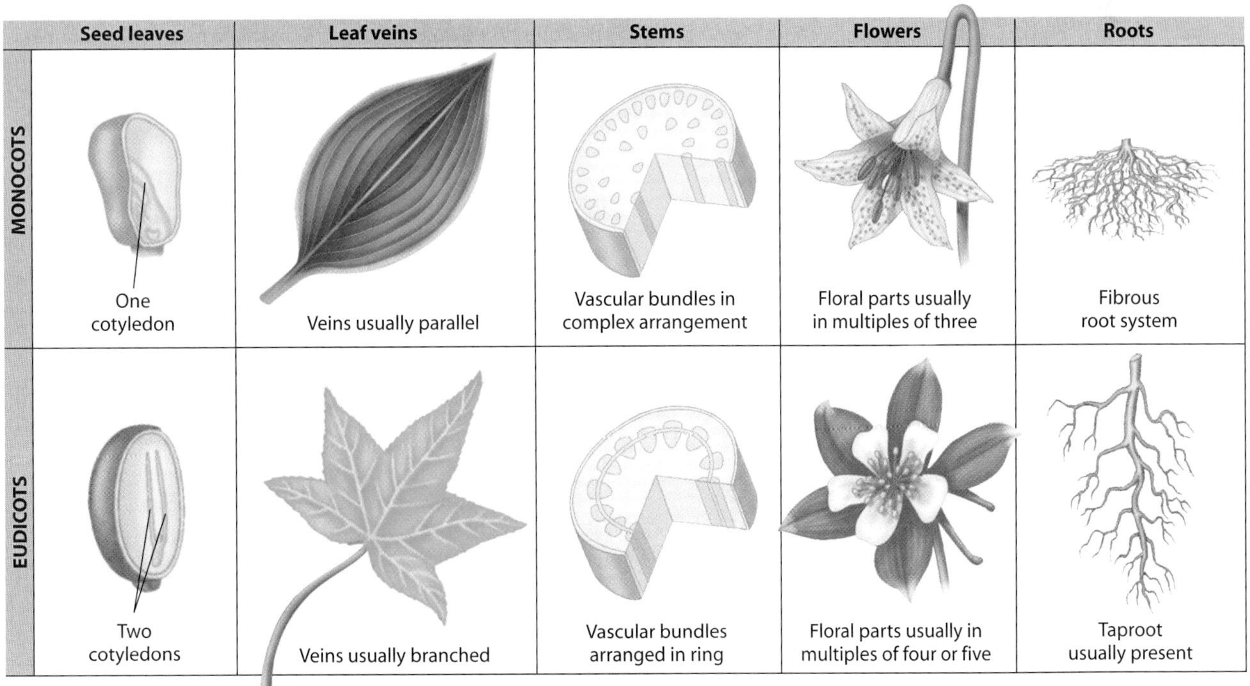

Seed leaves
Leaf veins
Stems
Flowers
Roots
MONOCOTS
One cotyledon
Veins usually parallel
Vascular bundles in complex arrangement
Floral parts usually in multiples of three
Fibrous root system
EUDICOTS
Two cotyledons
Veins usually branched
Vascular bundles arranged in ring
Floral parts usually in multiples of four or five
Taproot usually present

▲ **Figure 2** A comparison of monocots and eudicots

3 A typical plant body contains three basic organs: roots, stems, and leaves

Plants, like most animals, have organs composed of different tissues, which in turn are composed of one or more cell types. An **organ** consists of several types of tissues that together carry out particular functions. In this and the next module, we'll focus on plant organs. We will then work our way down the structural hierarchy and examine plant tissues (in Module 5) and cells (in Module 6).

The basic structure of plants reflects their evolutionary history as land-dwelling organisms. Most plants must draw resources from two very different environments: They must absorb water and minerals from the soil, while simultaneously obtaining CO_2 and light from above ground. The subterranean roots and aerial shoots (stems and leaves) of a typical land plant, such as the generalized flowering plant shown in Figure 3, perform these vital functions. Neither roots nor shoots can survive without the other. Lacking chloroplasts and living in the dark, most roots would starve without sugar and other organic nutrients transported from the photosynthetic leaves of the shoot system. Conversely, stems and leaves depend on the water and minerals absorbed by roots.

A plant's **root system** anchors it in the soil, absorbs and transports minerals and water, and stores food. Near the root tips, a vast number of tiny tubular projections called **root hairs** enormously increase the root surface area for absorption of water and minerals. As shown on the far right of the figure, each root hair is an extension of an epidermal cell (a cell in the outer layer of the root). It is difficult to move an established plant without injuring it because transplantation often damages the plant's delicate root hairs.

The **shoot system** of a plant is made up of stems, leaves, and adaptations for reproduction, which in angiosperms are the flowers. (We'll return to flowers in Module 9.) The **stems** are the parts of the plant that are generally above the ground and that support and separate the leaves (thereby promoting photosynthesis) and flowers (responsible for reproduction). In the case of a tree, the stems are the trunk and all the branches, including the smallest twigs. A stem has **nodes**, the points at which leaves are attached, and **internodes**, the portions of the stem between nodes. The **leaves** are the main photosynthetic organs in most plants, although green stems also perform photosynthesis. Most leaves consist of a flattened blade and a stalk, or petiole, which joins the leaf to a node of the stem.

The two types of buds in the figure are undeveloped shoots. When a plant stem is growing in length, the **terminal bud** (also called the apical bud), at the apex (tip) of the stem has developing leaves and a compact series of nodes and internodes. The **axillary buds**, one in each of the angles formed by a leaf and the stem, are usually dormant. In many plants, the terminal bud produces hormones that inhibit growth of the axillary buds, a phenomenon called **apical dominance**. By concentrating resources on growing taller, apical dominance is an evolutionary adaptation that increases the plant's exposure to light. This is especially important where vegetation is dense. However, branching is also important for increasing the exposure of the shoot system to the environment, and under certain conditions, the axillary buds begin growing. Some develop into shoots bearing flowers, and others become nonreproductive branches complete with their own terminal buds, leaves, and axillary buds. Removing the terminal bud usually stimulates the growth of axillary buds. This is why pruning fruit trees and "pinching back" houseplants makes them bushier.

The drawing in Figure 3 gives an overview of plant structure, but it by no means represents the enormous diversity of angiosperms. Next, let's look briefly at some variations on the basic themes of root and stem structure.

Adam Hart-Davis/SPL/Photo Researchers

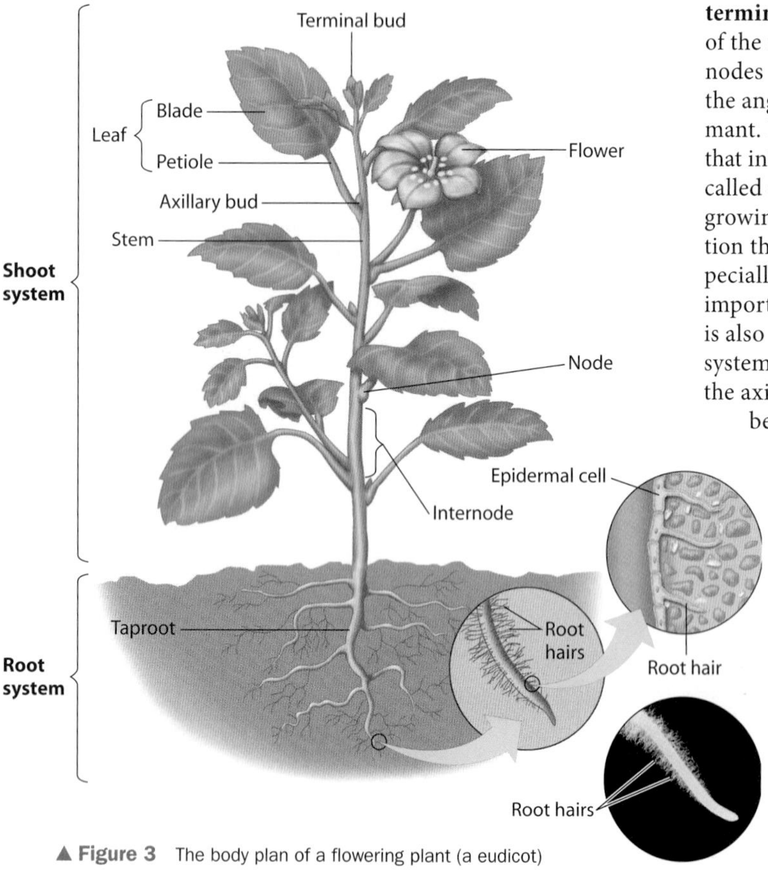

▲ **Figure 3** The body plan of a flowering plant (a eudicot)

? Name the two organ systems and three basic organs found in typical plants.

Root system and shoot system; roots, stems, leaves

4 Many plants have modified roots, stems, and leaves

▲ **Figure 4A** The modified root of a sugar beet plant

Over evolutionary history, the three basic plant organs—roots, stems, and leaves—have become adapted for a variety of functions. Carrots, turnips, sugar beets, and sweet potatoes, for instance, all have unusually large taproots that store food in the form of carbohydrates such as starch (**Figure 4A**). The plants consume the stored sugars during flowering and fruit production. For this reason, root crops are harvested before flowering.

Figure 4B shows three examples of modified stems. The strawberry plant has a horizontal stem called a stolon (or "runner") that grows along the ground. Stolons enable a plant to reproduce asexually, as plantlets form at nodes along their length. That is why strawberries, if left unchecked, can rapidly fill your garden. You've seen a different stem modification if you have ever dug up an iris plant or cooked with fresh ginger; the large, brownish, rootlike structures are actually **rhizomes**, horizontal stems that grow near the soil surface. Rhizomes store food and, having buds, can also form new plants. About every three years, gardeners can dig up iris rhizomes, split them, and replant to get multiple identical plants. A potato plant has rhizomes that end in enlarged structures specialized for storage called **tubers** (the potatoes we eat). Potato "eyes" are axillary buds on the tubers that can grow when planted, allowing potatoes to be easily propagated. Plant bulbs are underground shoots containing swollen leaves that store food. As you peel an onion, you are removing layers of leaves attached to a short stem.

▲ **Figure 4C** Modified leaves: the tendrils of a pea plant (left) and cactus spines (right)

Plant leaves, too, are highly varied. Grasses and many other monocots, for instance, have long leaves without petioles. Some eudicots, such as celery, have enormous petioles—the stalks we eat—which contain a lot of water and stored food. The left photograph in **Figure 4C** shows a modified leaf called a **tendril**, with its tips coiled around a support structure. Tendrils help plants climb. (Some tendrils, as in grapevines, are modified stems.) The spines of the barrel cactus (right photo) are modified leaves that protect the plant from being eaten by animals. The main part of the cactus is the stem, which is adapted for photosynthesis and water storage.

So far, we have examined plants as we see them with the unaided eye. Next, we move further down the structural hierarchy to explore plant tissues.

? In what sense do cactus spines deviate from the typical function of plant leaves?

Cactus spines are protective but are not the main sites of photosynthesis (as leaves are on most plants).

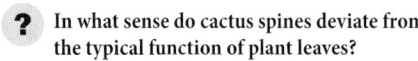

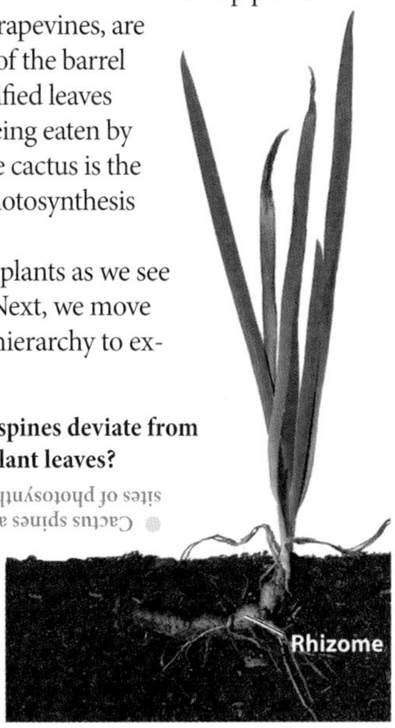

▲ **Figure 4B** Three kinds of modified stems: stolons, rhizomes, and tubers

5 Three tissue systems make up the plant body

Like the organs of most animals, the organs of plants contain tissues with characteristic functions. A **tissue** is a group of cells that together perform a specialized function. For example, **xylem** tissue contains water-conducting cells that convey water and dissolved minerals upward from the roots, while **phloem** tissue contains cells that transport sugars and other organic nutrients from leaves or storage tissues to other parts of the plant.

Each plant organ—root, stem, or leaf—has three types of tissues: dermal, vascular, and ground tissues. Each of these three categories forms a **tissue system**, a functional unit connecting all of the plant's organs. Each tissue system is continuous throughout the entire plant body, but the systems are arranged differently in leaves, stems, and roots (**Figure 5**). In this module, we examine the tissue systems of young roots and shoots. Later we will see that the tissue systems are somewhat different in older roots and stems.

The **dermal tissue system** (brown in the figure) is the plant's outer protective covering. Like our own skin, it forms the first line of defense against physical damage and infectious organisms. In many plants, the dermal tissue system consists of a single layer of tightly packed cells called the **epidermis**. The epidermis of leaves and most stems has a waxy coating called the **cuticle**, which helps prevent water loss. The second tissue system is the **vascular tissue system** (purple). It is made up of xylem and phloem tissues and provides support and long-distance transport between the root and shoot systems. Tissues that are neither dermal nor vascular make up the **ground tissue system** (yellow). It accounts for most of the bulk of a young plant, filling the spaces between the epidermis and vascular tissue system. Ground tissue internal to the vascular tissue is called **pith**, and ground tissue external to the vascular tissue is called **cortex**. The ground tissue system has diverse functions, including photosynthesis, storage, and support.

The close-up views on the right side of Figure 5 show how these three tissue systems are organized in typical plant roots, stems, and leaves. The view at the bottom left shows in cross section the three tissue systems in a young eudicot root. Water and minerals that are absorbed from the soil must enter through the epidermis. In the center of the root, the vascular tissue system forms a **vascular cylinder**, with the cross sections of xylem cells radiating from the center like spokes of a wheel and phloem cells filling in the wedges between the spokes. The ground tissue system of the root, the region between the vascular cylinder and epidermis, consists entirely of cortex. The cortex cells store food as starch and take up minerals that have entered the root through the epidermis. The innermost layer of the cortex is the **endodermis**, a cylinder one cell thick. The endodermis is a selective barrier, determining which substances pass between the rest of the cortex and the vascular tissue.

The bottom right of Figure 5 shows a cross section of a young monocot root. There are several similarities to the eudicot root: an outer layer of epidermis (dermal tissue) surrounding a large cortex (ground tissue), with a vascular cylinder (vascular tissue) at the center. But in a monocot root, the vascular tissue consists of a central core of cells surrounded by a ring of xylem and a ring of phloem.

As the center of Figure 5 indicates, the young stem of a eudicot looks quite different from that of a monocot. Both stems have their vascular tissue system arranged in numerous **vascular bundles**. However, in monocot stems the bundles are scattered, whereas in eudicots they are arranged in a ring. This ring separates the ground tissue into cortex and pith regions. The cortex fills the space between the vascular ring and the epidermis. The pith fills the center of the stem and is often important in food storage. In a monocot stem, the ground tissue is not separated into these regions because the vascular bundles don't form a ring.

The top of Figure 5 illustrates the arrangement of the three tissue systems in a typical eudicot leaf. The epidermis is interrupted by tiny pores called **stomata** (singular, *stoma*), which allow exchange of CO_2 and O_2 between the surrounding air and the photosynthetic cells inside the leaf. Also, most of the water vapor lost by a plant passes through stomata. Each stoma is flanked by two **guard cells**, which regulate the opening and closing of the stoma.

The ground tissue system of a leaf, called the **mesophyll**, is sandwiched between the upper and lower epidermis. Mesophyll consists mainly of cells specialized for photosynthesis. The green structures in the diagram are their chloroplasts. In this eudicot leaf, notice that cells in the lower area of mesophyll are loosely arranged, with a labyrinth of air spaces through which CO_2 and O_2 circulate. The air spaces are particularly large in the vicinity of stomata, where gas exchange with the outside air occurs. In many monocot leaves and in some eudicot leaves, the mesophyll is not arranged in distinct upper and lower areas.

In both monocots and eudicots, the leaf's vascular tissue system is made up of a network of veins. As you can see in Figure 5, each **vein** is a vascular bundle composed of xylem and phloem tissues surrounded by a protective sheath of cells. The veins' xylem and phloem, continuous with the vascular bundles of the stem, are in close contact with the leaf's photosynthetic tissues. This ensures that those tissues are supplied with water and mineral nutrients from the soil and that sugars made in the leaves are transported throughout the plant. The vascular structure also functions as a skeleton that reinforces the shape of the leaf.

In the last three modules, we have examined plant structure at the level of organs and tissues. In the next module, we will complete our descent into the structural hierarchy of plants by taking a look at plant cells.

? **For each of the following structures in your body, name the most analogous plant tissue system: circulatory system, skin, adipose tissue (body fat).**

Vascular tissue system, dermal tissue system, ground tissue system

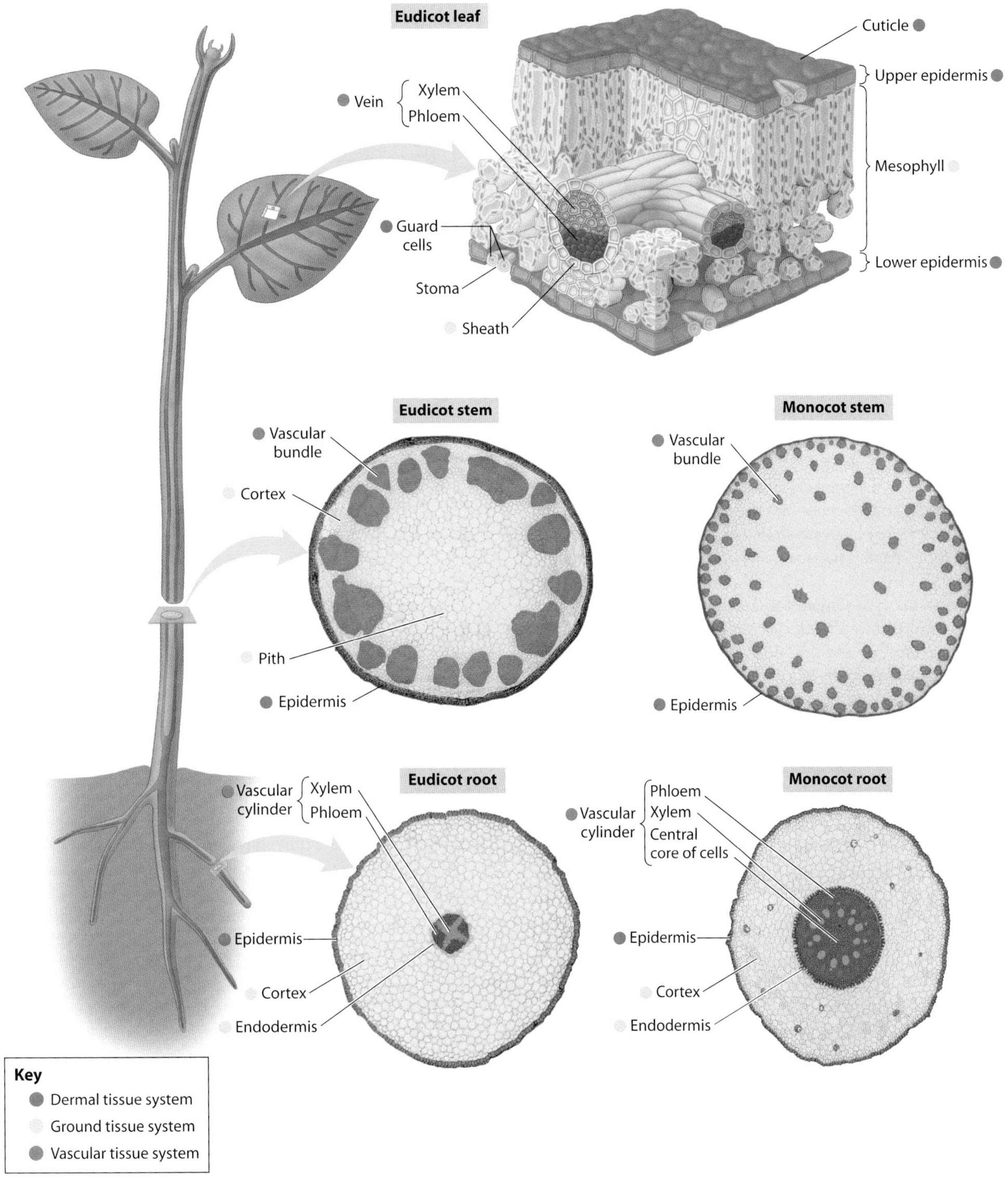
Eudicot leaf
Cuticle
Upper epidermis
Vein
Xylem
Phloem
Mesophyll
Guard cells
Stoma
Sheath
Lower epidermis
Eudicot stem
Vascular bundle
Cortex
Pith
Epidermis
Monocot stem
Vascular bundle
Epidermis
Eudicot root
Vascular cylinder
Xylem
Phloem
Epidermis
Cortex
Endodermis
Monocot root
Vascular cylinder
Phloem
Xylem
Central core of cells
Epidermis
Cortex
Endodermis
Key
Dermal tissue system
Ground tissue system
Vascular tissue system

▲ **Figure 5** The three tissue systems

6 Plant cells are diverse in structure and function

In addition to features shared with other eukaryotic cells, most plant cells have three unique structures (**Figure 6A**): chloroplasts, the sites of photosynthesis; a central vacuole containing fluid that helps maintain cell turgor (firmness); and a protective cell wall made from the structural carbohydrate cellulose surrounding the plasma membrane.

The enlargement on the right in Figure 6A highlights the adjoining cell walls of two cells. Many plant cells, especially those that provide structural support, have a two-part cell wall; a primary cell wall is laid down first, and then a more rigid secondary cell wall forms between the plasma membrane and the primary wall. The primary walls of adjacent cells in plant tissues are held together by a sticky layer called the middle lamella. Pits, where the cell wall is relatively thin, allow migration of water between adjacent cells. Plasmodesmata are open channels in adjacent cell walls through which cytoplasm and various molecules can flow from cell to cell.

The structure of a plant cell and the nature of its wall often correlate with the cell's main functions. As you consider the five major types of plant cells shown in Figures 6B–6F, notice the structural adaptations that make their specific functions possible.

Parenchyma cells (**Figure 6B**) are the most abundant type of cell in most plants. They usually have only primary cell walls, which are thin and flexible. Parenchyma cells perform most metabolic functions of a plant, such as photosynthesis, aerobic respiration, and food storage. Most parenchyma cells can divide and differentiate into other types of plant cells under certain conditions, such as during the repair of an injury. In the laboratory, it is even possible to regenerate an entire plant from a single parenchyma cell (as you'll see in Module 14).

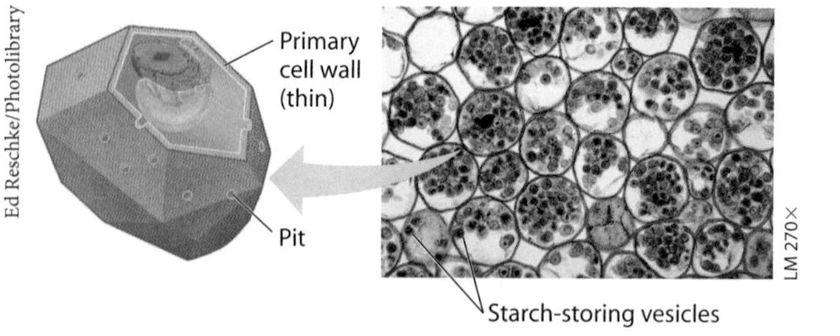

▲ **Figure 6B** Parenchyma cell

▲ **Figure 6C** Collenchyma cell

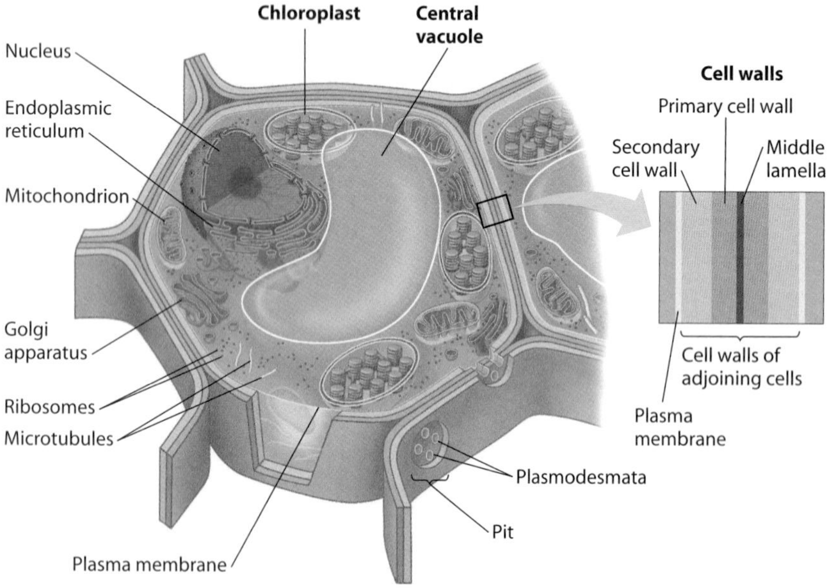

▲ **Figure 6A** The structure of a plant cell

Collenchyma cells (**Figure 6C**) resemble parenchyma cells in lacking secondary walls, but they have unevenly thickened primary walls. These cells provide flexible support in actively growing parts of the plant; young stems and petioles often have collenchyma cells just below their surface (the "string" of a celery stalk, for example). These living cells elongate as stems and leaves grow.

Sclerenchyma cells (**Figure 6D**) have thick secondary cell walls usually strengthened with lignin, the main chemical component of wood. Mature sclerenchyma cells cannot elongate and thus are found only in regions of the plant that have stopped growing in length. When mature, most sclerenchyma cells are dead, their cell walls forming a rigid "skeleton" that supports the plant much as steel beams do in the interior of a building.

Figure 6D shows two types of sclerenchyma cells. One, called a **fiber**, is long and slender and is usually arranged in bundles. Some plant tissues with abundant fiber cells are commercially important; hemp fibers, for example, are used to make rope and clothing. **Sclereids**, which are shorter than fiber cells, have thick, irregular, and very hard secondary walls. Sclereids impart the hardness to nutshells and seed coats and the gritty texture to the soft tissue of a pear.

The xylem tissue of angiosperms includes two types of water-conducting cells: tracheids and vessel elements. Both have rigid, lignin-containing secondary cell walls. As **Figure 6E** shows, the

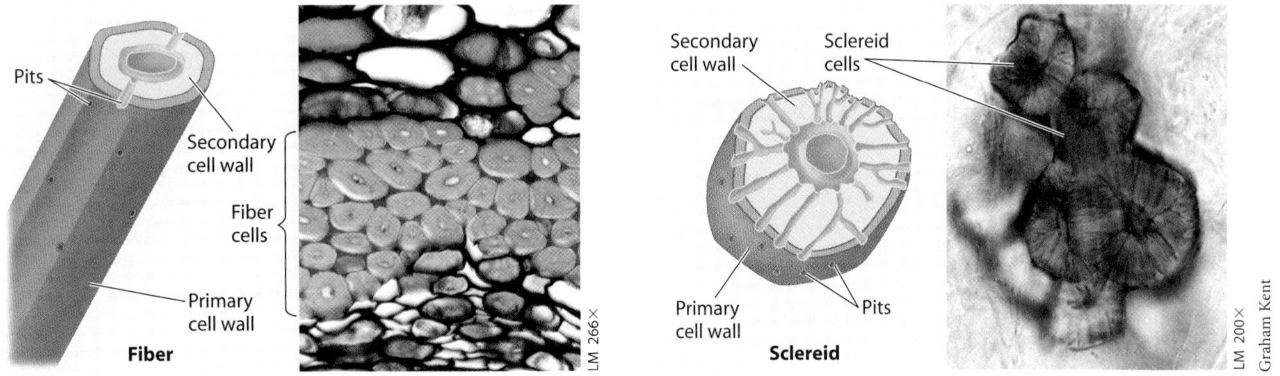

▲ **Figure 6D** Sclerenchyma cells: fiber (left) and sclereid (right)

tracheids are long, thin cells with tapered ends. **Vessel elements** are wider, shorter, and less tapered. Chains of tracheids or vessel elements with overlapping ends form a system of tubes that conveys water from the roots to the stems and leaves as part of xylem tissue. The tubes are hollow because both tracheids and vessel elements are dead when mature, with only their cell walls remaining. Water passes through pits in the walls of tracheids and vessel elements and through openings in the end walls of vessel elements. With their thick, rigid walls, these cells also function in support.

Food-conducting cells, known as **sieve-tube elements** (or sieve-tube members), are also arranged end to end, forming tubes as part of phloem tissue (**Figure 6F**). Unlike water-conducting cells, however, sieve-tube elements remain alive at maturity, though they lose most organelles, including the nucleus and ribosomes. This reduction in cell contents enables nutrients to pass more easily through the cell. The end walls between sieve-tube elements, called **sieve plates**, have pores that allow fluid to flow from cell to cell along the sieve tube. Alongside each sieve-tube element is at least one **companion cell**, which is connected to the sieve-tube element by numerous plasmodesmata. One companion cell may serve multiple sieve-tube elements by producing and transporting proteins to all of them.

Now that we have reached the lowest level in the structural hierarchy of plants—cells—let's review by moving back up. Cells of plants are grouped into tissues with characteristic functions. For example, xylem tissue contains water-conducting cells that convey water and dissolved minerals upward from the roots as well as sclerenchyma cells, which provide support, and parenchyma cells, which store various materials. Xylem and phloem tissues are organized into the vascular tissue system, which provides structural support and long-term transport throughout the plant body. The vascular, dermal, and ground tissue systems connect all the plant organs: roots, stems, and leaves. This review completes our survey of basic plant anatomy. Next, we examine how plants grow.

? Identify which of the following cell types can give rise to all others in the list: collenchyma, sclereid, parenchyma, vessel element, companion cell.

Parenchyma

N.C. Brown Center for Ultrastructure Studies, SUNY-Environmental Science & Forestry, Syracuse, NY

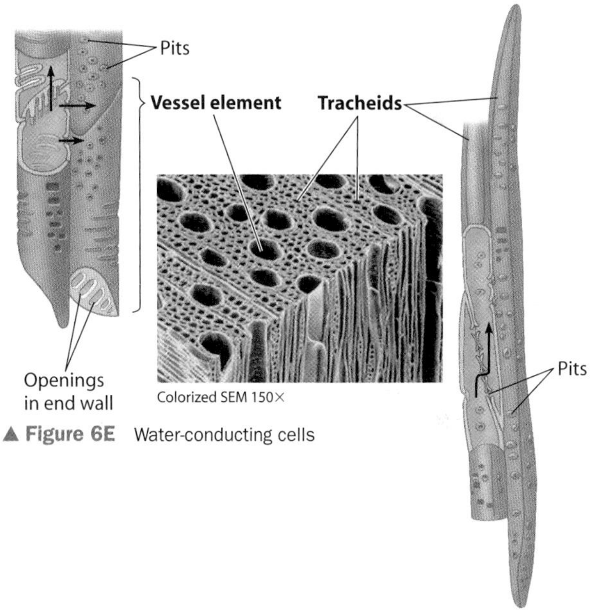

▲ **Figure 6E** Water-conducting cells

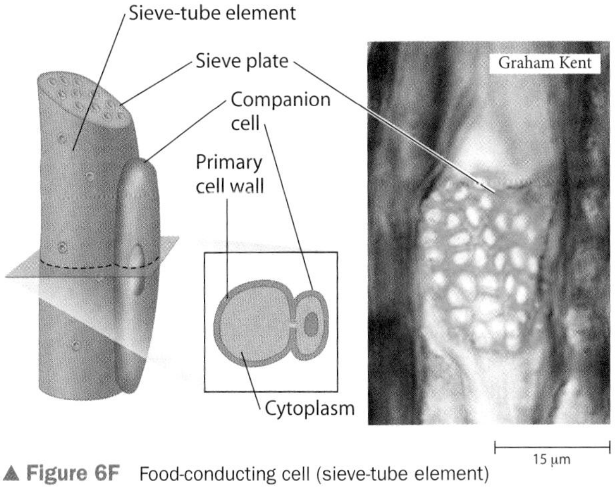

▲ **Figure 6F** Food-conducting cell (sieve-tube element)

Plant Growth

7 Primary growth lengthens roots and shoots

So far, we have looked at the structure of plant tissues and cells in mature organs. We will now consider how such organization arises through plant growth.

The growth of a plant differs from that of an animal in a fundamental way. Most animals are characterized by **determinate growth**; that is, they cease growing after reaching a certain size. Most species of plants, however, continue to grow as long as they live, a condition called **indeterminate growth**. Indeterminate growth allows a plant to continuously increase its exposure to sunlight, air, and soil.

Indeterminate growth does not mean that plants are immortal; they do, of course, die. Flowering plants are categorized as annuals, biennials, or perennials, based on the length of their life cycle, the time from germination through flowering and seed production to death. **Annuals** complete their life cycle in a single year or less. Our most important food crops (including grains and legumes) are annuals, as are a great number of wildflowers. **Biennials** complete their life cycle in two years, with flowering and seed production usually occurring during the second year. Beets, parsley, turnips, and carrots are biennials, but we usually harvest them in their first year and so miss seeing their flowers. **Perennials** are plants that live and reproduce for many years. They include trees, shrubs, and some grasses.

Some plants are among the oldest organisms alive. Some of the coast redwoods we discussed in the chapter introduction are estimated to be 2,000 to 3,000 years old, and another conifer, the bristlecone pine, can live thousands of years longer than that (as you'll see in Module 15). Some buffalo grass of the North American plains is believed to have been growing for 10,000 years from seeds that sprouted at the end of the last ice age. When a plant dies, it is not usually from old age, but from an infection or some environmental trauma, such as fire, severe drought, or consumption by animals.

Growth in all plants is made possible by tissues called meristems. A **meristem** consists of undifferentiated (unspecialized) cells that divide when conditions permit, generating additional cells. Some products of this division remain in the meristem and produce still more cells, while others differentiate and are incorporated into tissues and organs of the growing plant. Meristems at the tips of roots and in the buds of shoots are called **apical meristems** (Figure 7A). Cell division in the apical meristems produces the new cells that enable a plant to grow in length, a process called **primary growth**. Tissues produced by primary growth are called primary tissues. Primary growth enables roots to push through the soil and allows shoots to grow upward, increasing exposure to light and CO_2. Although apical meristems lengthen both roots and shoots, there are important differences in the mechanisms of primary growth in each system. We will examine them separately, starting at the bottom with roots.

Figure 7B illustrates primary growth in a longitudinal section through a growing onion root. The root tip is covered by a thimble-like **root cap** that protects the delicate, actively dividing cells of the apical meristem. Growth in length occurs just behind the root tip, where three zones of cells at successive stages of primary growth are located. Moving away from the root tip, they are called the zone of cell division, the zone of elongation, and the zone of differentiation. The three zones of cells overlap, with no sharp boundaries between them.

The zone of cell division includes the root apical meristem and cells that derive from it. New root cells are produced in this region, including the cells of the root cap. In the zone of elongation, root cells elongate, sometimes to more than 10 times their original length. It is cell elongation that pushes the root tip farther into the soil. The cells lengthen, rather than expand equally in all directions, because of the circular arrangement of cellulose fibers in parallel bands in their cell walls. The enlargement diagrams at the left of the root indicate how this works. The cells elongate by taking up water, and as they do, the cellulose fibers (shown in red) separate, somewhat like an expanding accordion. The cells cannot expand greatly in width because the cellulose fibers do not stretch much.

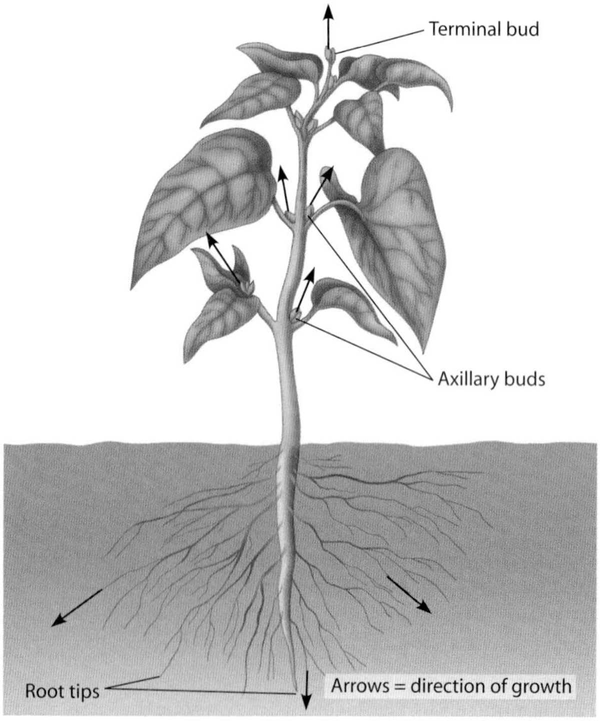

▲ **Figure 7A** Locations of apical meristems, which are responsible for primary growth

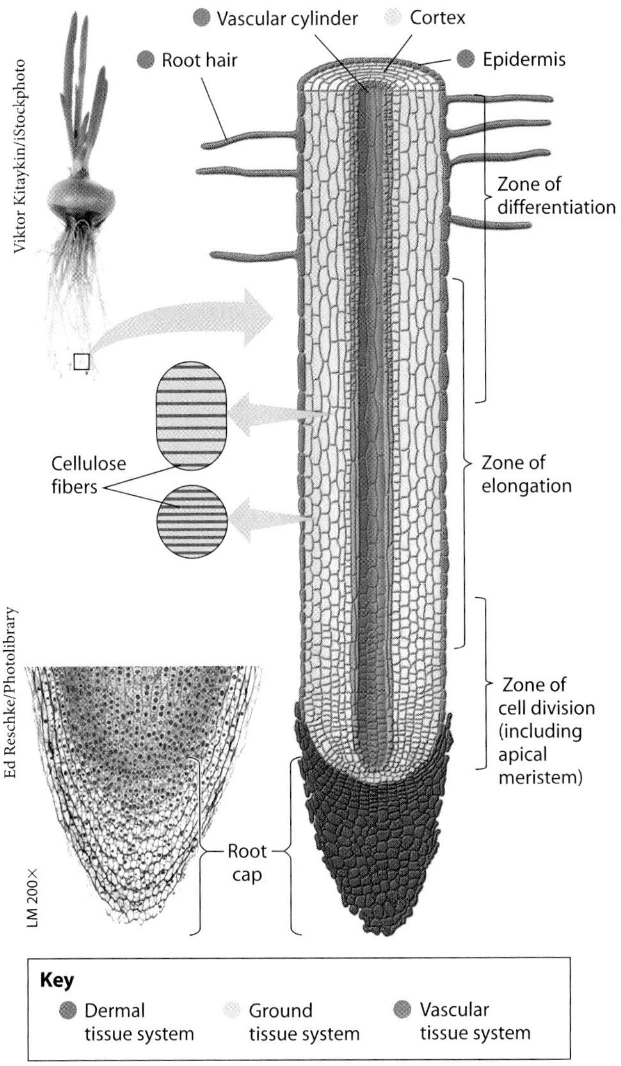

▲ **Figure 7B** Primary growth of a root

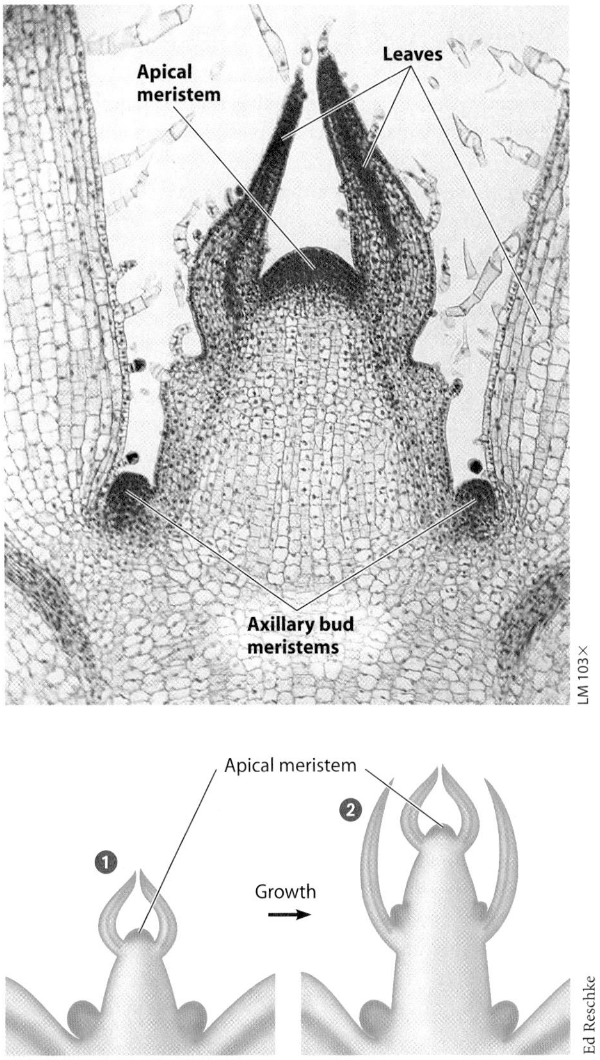

▲ **Figure 7C** Primary growth of a shoot

The three tissue systems of a mature plant (dermal, ground, and vascular) complete their development in the zone of differentiation. Cells of the vascular cylinder differentiate into **primary xylem** and **primary phloem**. Differentiation of cells—the specialization of their structure and function—results from differential gene expression. Cells in the vascular cylinder, for instance, develop into primary xylem or phloem cells because certain genes are turned on and are expressed as specific proteins, whereas other genes in these cells are turned off. Genetic control of differentiation is one of the most active areas of plant research.

The micrograph in **Figure 7C** shows a section through the end of a growing shoot that was cut lengthwise from its tip to just below its uppermost pair of axillary buds. You can see the apical meristem, which is a dome-shaped mass of dividing cells at the tip of the terminal bud. Elongation occurs just below this meristem, and the elongating cells push the apical meristem upward, instead of downward as in the root. As the apical meristem advances upward, some of its cells remain behind, and these become new axillary bud meristems at the base of the leaves.

The drawings in Figure 7C show two stages in the growth of a shoot. Stage ❶ is just like the micrograph. At the later stage shown in sketch ❷, the apical meristem has been pushed upward by elongating cells underneath.

Primary growth accounts for a plant's lengthwise growth. The stems and roots of many plants increase in thickness, too, and in the next module, we see how this usually happens.

? **A plant grows taller due to cell division within the _____ at the tips of the shoots. Such lengthening is called _____.**

apical meristem . . . primary growth

8 Secondary growth increases the diameter of woody plants

In nonwoody plants, primary growth produces nearly all of the plant body. Woody plants (such as trees, shrubs, and vines), however, grow in diameter in addition to length, thickening in older regions where primary growth has ceased. This increase in thickness of stems and roots, called **secondary growth**, is caused by the activity of dividing cells in tissues that are called **lateral meristems**. These dividing cells are arranged into two cylinders, known as the vascular cambium and the cork cambium, that extend along the length of roots and stems.

The **vascular cambium** is a cylinder of meristem cells one cell thick between the primary xylem and primary phloem. In **Figure 8A**, left, this region of the stem is just beginning secondary growth. Except for the vascular cambium, the stem at this stage of growth is virtually the same as a young stem undergoing primary growth (compare this figure with the eudicot stem in Figure 5). Secondary growth adds layers of vascular tissue on either side of the vascular cambium, as indicated by the black arrows.

The drawings at the center and the right show the results of secondary growth. Tissues produced by secondary growth are called secondary tissues. In the center drawing, the vascular cambium has given rise to two new tissues: **secondary xylem** to its interior and **secondary phloem** to its exterior. Each year, the vascular cambium gives rise to layers of secondary xylem and secondary phloem that are larger in circumference than the previous layer (see the drawing at the right). In this way, the vascular cambium thickens roots and stems.

Secondary xylem makes up the **wood** of a tree, shrub, or vine. Over the years, a woody stem gets thicker and thicker as its vascular cambium produces layer upon layer of secondary xylem. The cells of the secondary xylem have thick walls rich in lignin, giving wood its characteristic hardness and strength.

Annual growth rings, such as those in the cross section in **Figure 8B**, result from the layering of secondary xylem. The layers are visible as rings because of uneven activity of the vascular cambium during the year. In woody plants that live in temperate regions, such as most of the United States, the vascular cambium becomes dormant each year during winter, and secondary growth is interrupted. When secondary growth resumes in the spring, a cylinder of early wood forms. Early wood cells are made up of the first new xylem cells to develop and are usually larger in diameter and thinner-walled than those produced later in summer. The boundary between the large cells of early wood and the smaller cells of the late wood produced during the previous growing season is usually a distinct ring visible in cross sections of tree trunks and roots. Therefore, a tree's age can be estimated by counting its annual rings. Dendrochronology (*dendros* is Greek for "tree") is the science of analyzing tree ring growth patterns. The rings may have varying thicknesses, reflecting climate conditions and therefore the amount of seasonal growth in a given year. In fact, growth ring patterns in older trees are one source of evidence for recent global climate change.

Now let's return to Figure 8A and see what happens to the parts of the stem that are *external* to the vascular cambium. Unlike xylem, the external tissues do not accumulate over the years. Instead, they are sloughed off at about the same rate they are produced.

Notice at the left of the figure that the epidermis and cortex, both the result of primary growth, make up the young stem's external covering. When secondary growth begins (center drawing), the epidermis is sloughed off and replaced with a new outer layer called **cork** (brown). Mature cork cells are dead and have thick, waxy walls that protect the underlying tissues of the stem from water loss, physical damage, and pathogens.

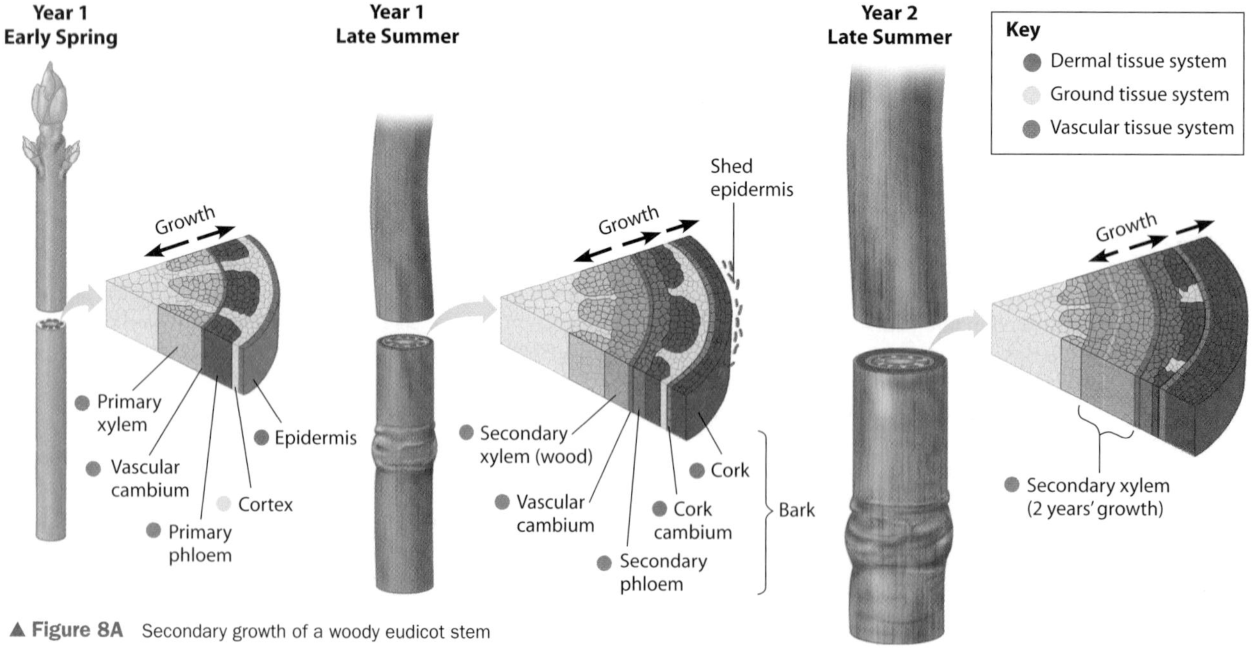

▲ **Figure 8A** Secondary growth of a woody eudicot stem

Cork is produced by meristem tissue called the **cork cambium** (light brown), which first forms from parenchyma cells in the cortex. As the stem thickens and the secondary xylem expands, the original cork and cork cambium are pushed outward and fall off, as is evident in the cracked, peeling bark of many tree trunks. A new cork cambium forms to the inside. When no cortex is left, it forms from parenchyma cells in the phloem.

Everything external to the vascular cambium is called **bark**. As indicated in the diagram at the right in Figure 8B, its main components are the secondary phloem, the cork cambium, and the cork. The youngest secondary phloem (next to the vascular cambium) functions in sugar transport. The older secondary phloem dies, as does the cork cambium you see here. Pushed outward, these tissues and cork produced by the cork cambium help protect the stem until they, too, are sloughed off as part of the bark. Keeping pace with secondary growth, cork cambium keeps regenerating from the younger secondary phloem and keeps producing a steady supply of cork.

The log on the left in Figure 8B shows the results of several decades of secondary growth. The bulk of a trunk like this is dead tissue. The living tissues in it are the vascular cambium, the youngest secondary phloem, the cork cambium, and cells in the wood rays, which you can see radiating from the center of the log in the drawing on the right. The **wood rays** consist of parenchyma cells that transport water and nutrients, store organic nutrients, and aid in wound repair. The **heartwood**, in the center of the trunk, consists of older layers of secondary xylem. These cells no longer transport water; they are clogged with resins and other compounds that make heartwood resistant to rotting. Because heartwood doesn't conduct water, a large tree can survive even if the center of its trunk is hollow (**Figure 8C**). The lighter-colored **sapwood** is younger secondary xylem that does conduct xylem fluid (sap).

Thousands of useful products are made from wood—from construction lumber to fine furniture, musical instruments, paper, insulation, and a long list of chemicals, including turpentine, alcohols, artificial vanilla flavoring, and preservatives. Among the qualities that make wood so useful are a unique combination of strength, hardness, lightness, high insulating properties, durability, and workability. In many cases, there is simply no good substitute for wood. A wooden oboe, for instance, produces far richer sounds than a plastic one. Fence posts made of locust tree wood actually last much longer in the ground than metal ones. Ball bearings are sometimes made of a very hard wood called lignum vitae. Unlike metal bearings, they require no lubrication because a natural oil completely penetrates the wood.

California Historical Society Collection (CHS-1177), University of Southern California on behalf of the USC Specialized Libraries and Archival Collections

▲ **Figure 8C** A giant sequoia that survived for 88 years with a tunnel cut through its heartwood.

In a sense, wood is analogous to the hard endoskeletons of many land animals. It is an evolutionary adaptation that enables a shrub or tree to remain upright and keep growing year after year on land—sometimes to attain enormous heights, as we saw in the chapter's introduction. In the next few modules, we discuss some additional adaptations that enable plants to live on land—those that facilitate reproduction.

? Why can some large trees survive after a tunnel has been cut through their center?

The center of an older tree consists of old layers of secondary xylem that no longer transport materials through the plant. It can thus be removed without harming the tree.

Don Mason/Corbis

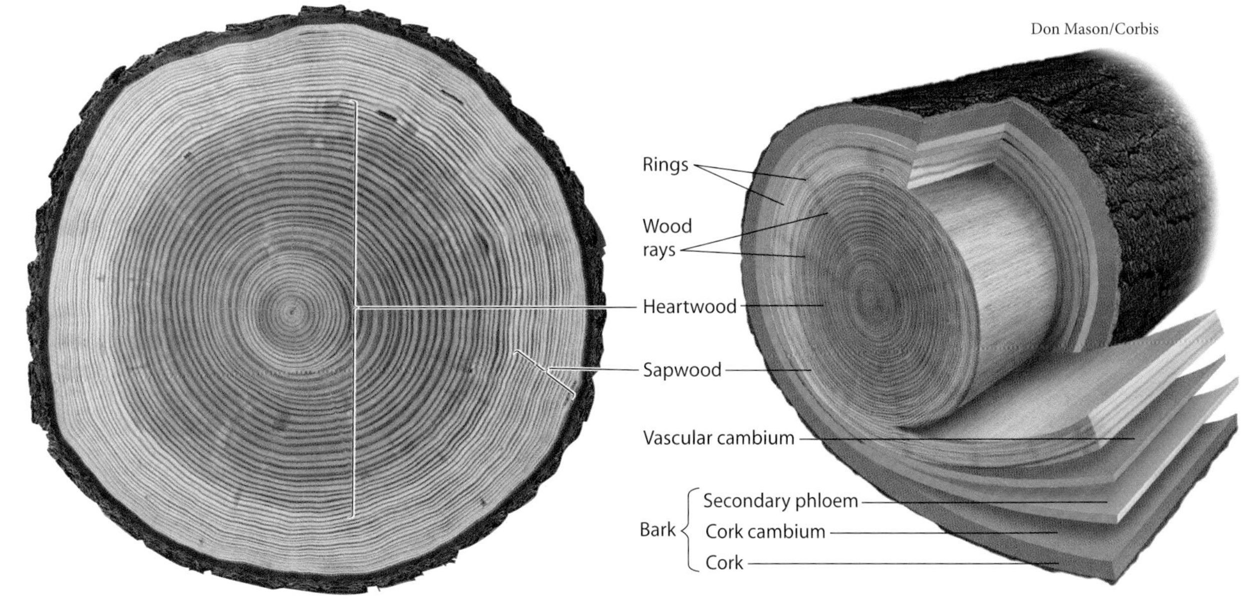

▲ **Figure 8B** Anatomy of a log

Reproduction of Flowering Plants

9 The flower is the organ of sexual reproduction in angiosperms

It has been said that an oak tree is merely an acorn's way of making more acorns. Indeed, evolutionary fitness for any organism is measured only by its ability to produce healthy, fertile offspring. Thus, from an evolutionary viewpoint, all the structures and functions of a plant can be interpreted as mechanisms contributing to reproduction. In the remaining modules, we explore the reproductive biology of angiosperms, beginning here with a brief overview.

▲ **Figure 9A** Some variations in flower shape

Flowers, the reproductive shoots of angiosperms, can vary greatly in shape (**Figure 9A**). Despite such variation, nearly all flowers contain four types of modified leaves called floral organs: sepals, petals, stamens, and carpels (**Figure 9B**). The **sepals**, which enclose and protect the flower bud, are usually green and more leaflike than the other floral organs (picture the green at the base of a rosebud). The **petals** are often colorful and advertise the flower to pollinators. The stamens and carpels are the reproductive organs, containing the sperm and eggs, respectively.

A **stamen** consists of a stalk (filament) tipped by an anther. Within the **anther** are sacs in which pollen is produced via meiosis. Pollen grains house the cells that develop into sperm.

A **carpel** has a long slender neck (style) with a sticky stigma at its tip. The **stigma** is the landing platform for pollen. The base of the carpel is the **ovary**, which contains one or more **ovules**, each containing a developing egg and supporting cells. The term **pistil** is sometimes used to refer to a single carpel or a group of fused carpels.

Figure 9C shows the life cycle of a generalized angiosperm. ❶ Fertilization occurs in an ovule within a flower. ❷ As the ovary develops into a fruit, ❸ the ovule develops into the seed containing the embryo. The fruit protects the seed and aids in dispersing it. Completing the life cycle, ❹ the seed then **germinates** (begins to grow) in a suitable habitat; ❺ the embryo develops into a seedling; and the seedling grows into a mature plant.

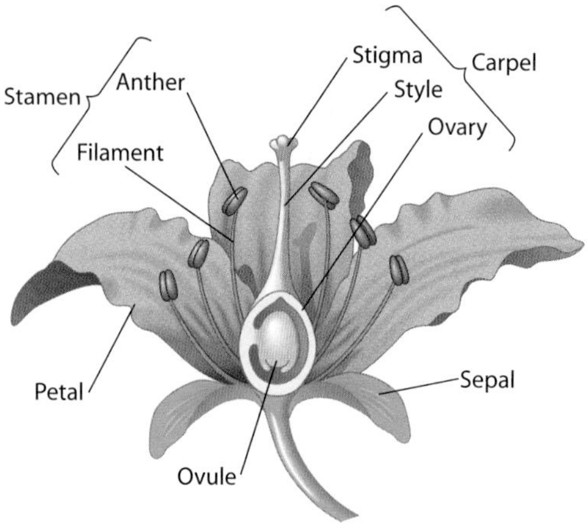

▲ **Figure 9B** The structure of a flower

In the next four modules, we examine key stages in the angiosperm sexual life cycle in more detail. We will see that there are a number of variations in the basic themes presented here.

? Pollen develops within the _____ of _____. Ovules develop within the _____ of _____.

anthers . . . stamens . . . ovaries . . . carpels

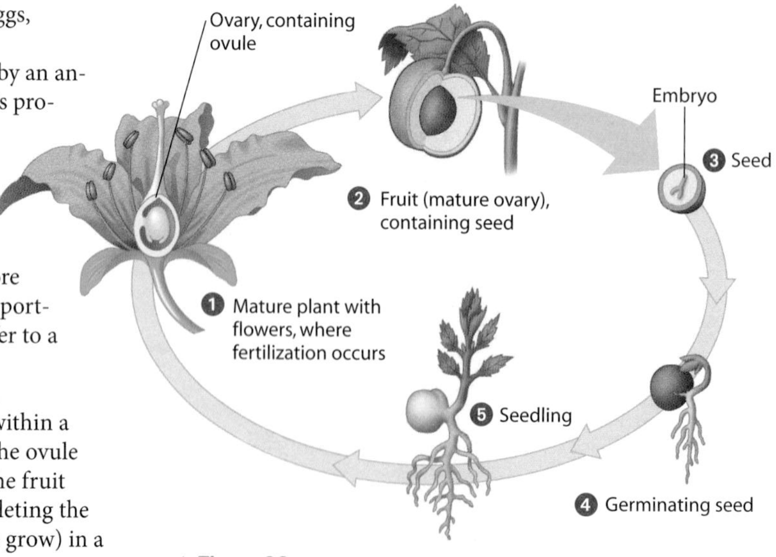

▲ **Figure 9C** Life cycle of a generalized angiosperm

10 The development of pollen and ovules culminates in fertilization

The life cycles of plants are characterized by an alternation of generations, in which haploid (n) and diploid ($2n$) generations take turns producing each other. The roots, stems, leaves, and most of the reproductive structures of angiosperms are diploid. The diploid plant body is called the **sporophyte**. A sporophyte produces special structures, the anthers and ovules, in which cells undergo meiosis to produce haploid cells called spores. Each spore then divides via mitosis and becomes a multicellular **gametophyte**, the plant's haploid generation. The gametophyte produces gametes by mitosis. At fertilization, gametes from the male and female gametophytes unite, producing a diploid zygote. The life cycle is completed when the zygote divides by mitosis and develops into a new sporophyte. In angiosperms, the sporophyte is the dominant generation: It is larger, more obvious, and longer-living than the gametophyte. In this module, we take a microscopic look at the gametophytes of a flowering plant.

Pollen grains are the male gametophytes. The cells that develop into pollen grains are found within a flower's anthers (top left in Figure 10). Each cell first undergoes meiosis, forming four haploid spores. Each spore then divides by mitosis, forming two haploid cells, called the tube cell and the generative cell. The generative cell passes into the tube cell, and a thick wall forms around them. The resulting pollen grain is ready for release from the anther.

Moving to the top right of the figure, we can follow the development of the flower parts that form the female gametophyte and eventually the egg. In most species, the ovary of a flower contains several ovules, but only one is shown here. An ovule contains a central cell (gold) surrounded by a protective covering of smaller cells (yellow). The central cell enlarges and undergoes meiosis, producing four haploid spores. Three of the spores usually degenerate, but the surviving one enlarges and divides by mitosis, producing a multicellular structure known as the **embryo sac**. Housed in several layers of protective cells (yellow) produced by the sporophyte plant, the embryo sac is the female gametophyte. The sac contains a large central cell with two haploid nuclei. One of its other cells is the haploid egg, ready to be fertilized.

The first step leading to fertilization is **pollination** (at the center of the figure), the delivery of pollen from anther to stigma. Most angiosperms depend on insects, birds, or other animals to transfer pollen. Most major crops rely on insects, mainly bees. But the pollen of some plants—such as grasses and many trees—is windborne (as anyone bothered by pollen allergies knows).

After pollination, the pollen grain germinates on the stigma. Its tube cell gives rise to the pollen tube, which grows downward into the ovary. Meanwhile, the generative cell divides by mitosis, forming two sperm. When the pollen tube reaches the base of the ovule, it enters the ovary and discharges its two sperm near the embryo sac. One sperm fertilizes the egg, forming the diploid zygote. The other contributes its haploid nucleus to the large diploid central cell of the embryo sac. This cell, now with a triploid ($3n$) nucleus, will give rise to a food-storing tissue called **endosperm**.

The union of two sperm cells with different nuclei of the embryo sac is called **double fertilization**, and the resulting production of endosperm is unique to angiosperms. Endosperm will develop only in ovules containing a fertilized egg, thereby preventing angiosperms from squandering nutrients.

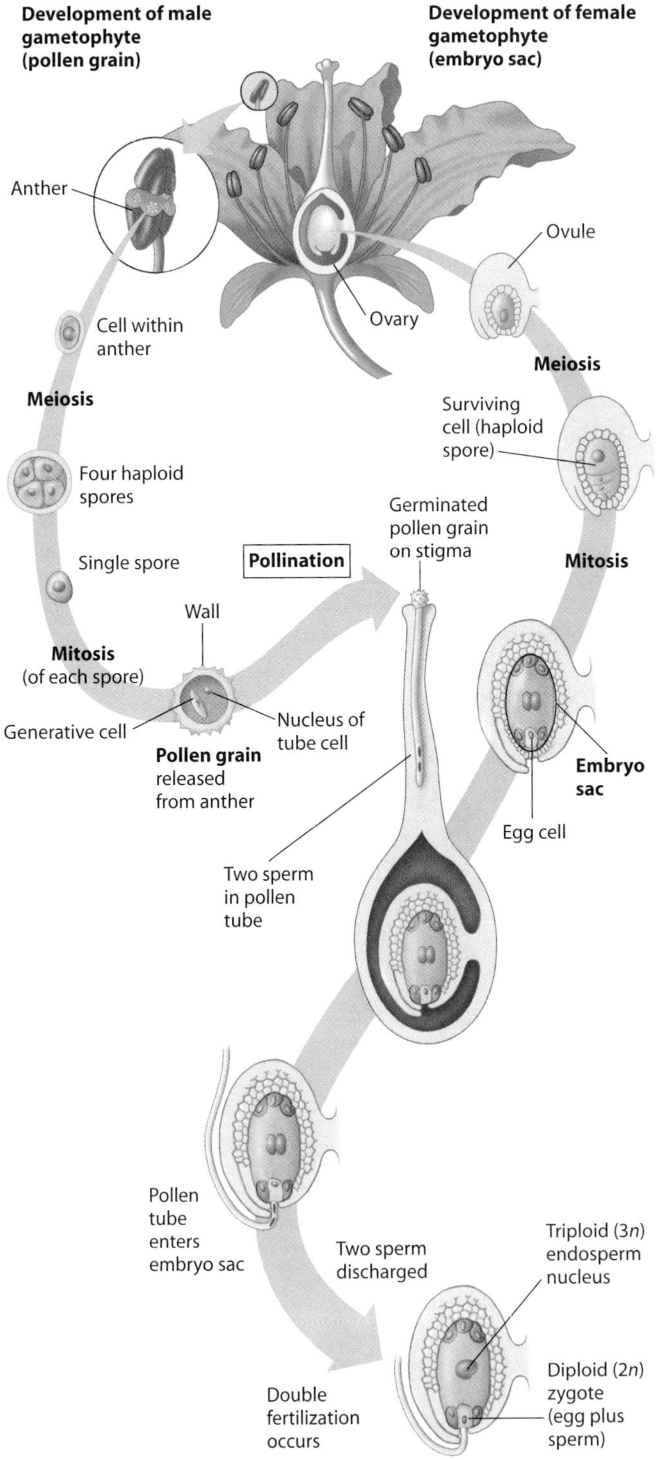

▲ Figure 10 Gametophyte development and fertilization in an angiosperm

? **Fertilization unites the sperm cell, which develops within the male gametophyte (or _____), with the egg cell, which develops within the female gametophyte (or _____).**

pollen grain . . . embryo sac

11 The ovule develops into a seed

After fertilization, the ovule, containing the triploid central cell and the diploid zygote, begins developing into a seed. As the embryo develops from the zygote, the seed stockpiles proteins, oils, and starch to varying degrees, depending on the species. This is what makes seeds such a major source of nutrition for many animals.

As shown in **Figure 11A**, embryonic development begins when the zygote divides by mitosis into two cells. Repeated division of one of the cells then produces a ball of cells that becomes the embryo. The other cell divides to form a thread of cells that pushes the embryo into the endosperm. The bulges you see on the embryo are the developing cotyledons. You can tell that the plant in this drawing is a eudicot, since it has two cotyledons.

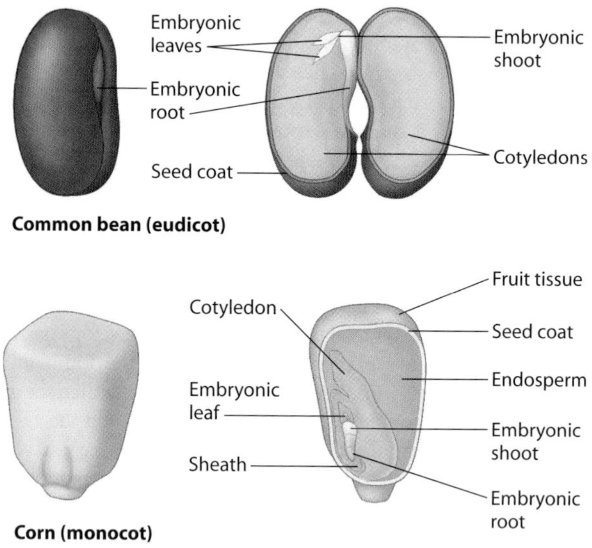

▲ **Figure 11B** Seed structure

The result of embryonic development in the ovule is a mature seed (Figure 11A, bottom right). Near the end of its maturation, the seed loses most of its water and forms a hard, resistant **seed coat** (brown). The embryo, surrounded by its endosperm food supply (gold), becomes dormant; it will not develop further until the seed germinates. Seed dormancy, a condition in which growth and development are suspended temporarily, is a key evolutionary adaptation. Dormancy allows time for a plant to disperse its seeds and increases the chance that a new generation of plants will begin growing only when environmental conditions, such as temperature and moisture, favor survival.

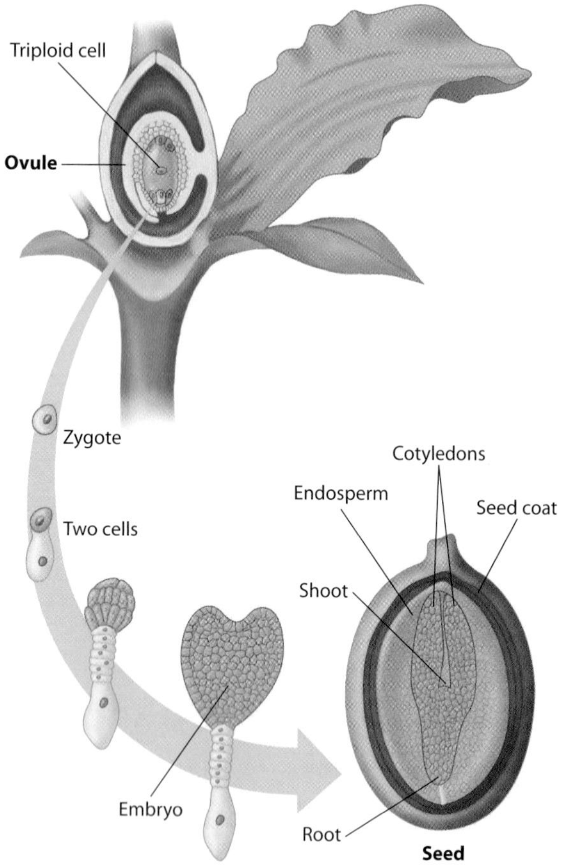

▲ **Figure 11A** Development of a eudicot plant embryo

The dormant embryo contains a miniature root and shoot, each equipped with an apical meristem. After the seed germinates, the apical meristems will sustain primary growth as long as the plant lives. Also present in the embryo are the three tissues that will form the epidermis, cortex, and primary vascular tissues.

Figure 11B contrasts the internal structures of eudicot and monocot seeds. In the eudicot (a bean), the embryo is an elongated structure with two thick cotyledons (tan). The embryonic root develops just below the point at which the cotyledons are attached to the rest of the embryo. The embryonic shoot, tipped by a pair of miniature embryonic leaves, develops just above the point of attachment. The bean seed contains no endosperm because its cotyledons absorb the endosperm nutrients as the seed forms. The nutrients start passing from the cotyledons to the embryo when it germinates.

A kernel of corn, an example of a monocot, is actually a fruit containing one seed. Everything you see in the drawing is the seed, except the kernel's outermost covering. The covering is dried fruit tissue, the former wall of the ovary, and is tightly bonded to the seed coat. Unlike the bean, the corn seed contains a large endosperm and a single, thin cotyledon. The cotyledon absorbs the endosperm's nutrients during germination. Also unlike the bean, the embryonic root and shoot in corn each have a protective sheath.

? What is the role of the endosperm in a seed?

The endosperm provides nutrients to the developing embryo.

12 The ovary develops into a fruit

In the previous two modules, we followed the angiosperm life cycle from the flower on the sporophyte plant through the transformation of an ovule into a seed. While the seeds are developing from ovules, hormonal changes triggered by fertilization cause the flower's ovary to thicken and mature into a fruit. A **fruit** is a specialized vessel that houses and protects seeds and helps disperse them from the parent plant. Although a fruit typically consists of a mature ovary, it can include other flower parts as well. A pea pod is a fruit, as is a peach, orange, tomato, cherry, or corn kernel.

The photographs in **Figure 12A** illustrate the changes in a pea plant leading to pod formation. ❶ Soon after pollination, ❷ the flower drops its petals, and the ovary starts to grow. The ovary expands tremendously, and its wall thickens, ❸ forming the pod, or fruit, which holds the peas, or seeds.

Figure 12B matches the parts of a pea flower with what they become in the pod. The wall of the ovary becomes the pod. The ovules, within the ovary, develop into the seeds. The small, threadlike structure at the end of the pod is what remains of the upper part of the flower's carpel. The sepals of the flower often stay attached to the base of the green pod. Peas are usually harvested at this stage of fruit development. If the pods are allowed to develop further, they become dry and brownish and will split open, releasing the seeds.

❶

❷

❸

Walter H. Hodge/Peter Arnold, Inc./Photo Library

▲ **Figure 12A** Development of a fruit, a pea pod

As shown in the examples in **Figure 12C**, mature fruits can be either fleshy or dry. Oranges, plums, and grapes are examples of fleshy fruits, in which the wall of the ovary becomes soft during ripening. Dry fruits include beans, nuts, and grains. The dry, wind-dispersed fruits of grasses, harvested while on the plant, are major staple foods for humans. The cereal grains of wheat, rice, corn, and other grasses, though easily mistaken for seeds, are each actually a fruit with a dry outer covering (the former wall of the ovary) that adheres to the seed coat of the seed within.

Various adaptations of fruits help disperse seeds. The seeds of some flowering plants, such as dandelions and maples, are contained within fruits that function like kites or propellers, adaptations that enhance dispersal by wind. Some fruits, such as coconuts, are adapted to dispersal by water. And many angiosperms rely on animals to carry seeds. Some of these plants have fruits modified as burrs that cling to animal fur (or the clothes of humans). Other angiosperms produce edible fruits, which are usually nutritious, sweet tasting, and vividly colored, advertising their ripeness. When an animal eats the fruit, it digests the fruit's fleshy part, but the tough seeds usually pass unharmed through the animal's digestive tract. Animals may deposit the seeds, along with a supply of fertilizer, kilometers from where the fruit was eaten.

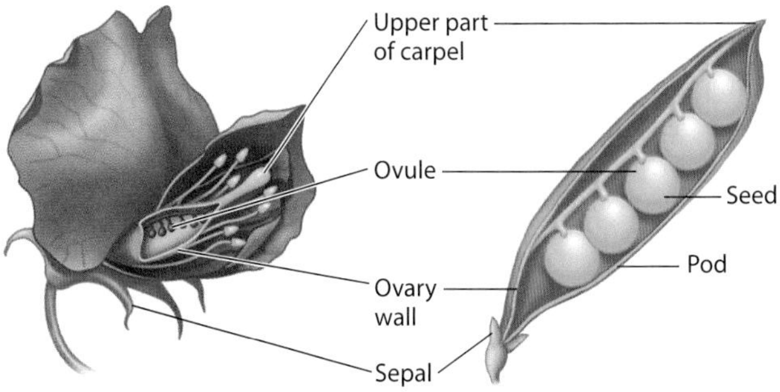

▲ **Figure 12B** The correspondence between flower and fruit in the pea plant

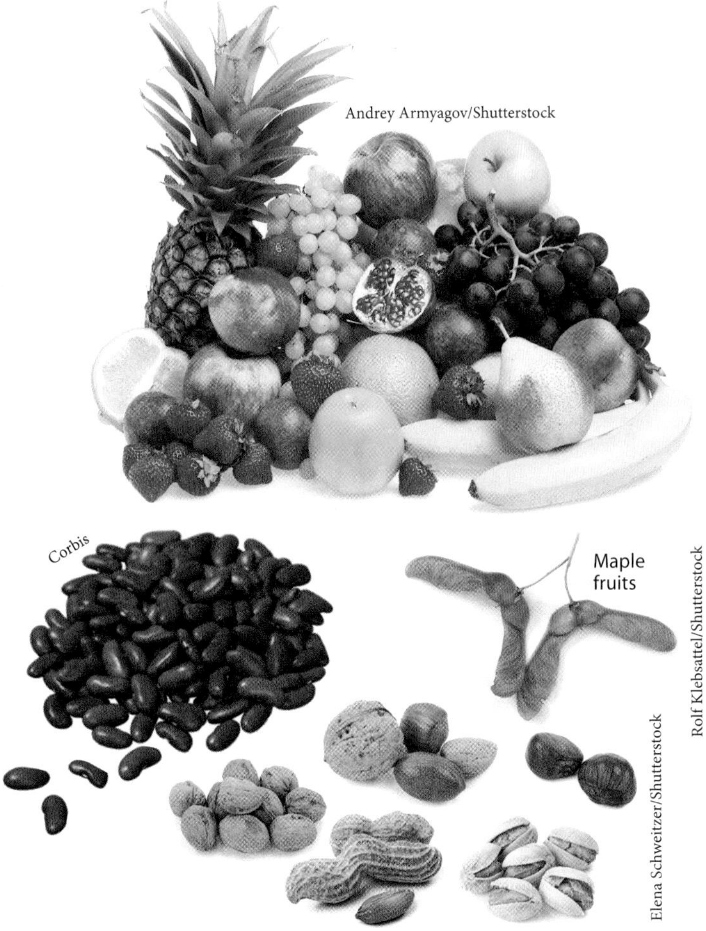

▲ **Figure 12C** A collection of fleshy (top) and dry (bottom) fruits

? Seed is to _____ as _____ is to ovary.

ovule . . . fruit

13 Seed germination continues the life cycle

The germination of a seed is often used to symbolize the beginning of life, but as we have seen, the seed already contains a miniature plant, complete with embryonic root and shoot. Thus, at germination, the plant does not begin life but rather resumes the growth and development that were temporarily suspended during seed dormancy.

Germination usually begins when the seed takes up water. The hydrated seed expands, rupturing its coat. The inflow of water triggers metabolic changes in the embryo that make it start growing again. Enzymes begin digesting stored nutrients in the endosperm or cotyledons, and these nutrients are transported to the growing regions of the embryo.

The figures below trace germination in a eudicot (a garden bean) and a monocot (corn). In **Figure 13A**, notice that the embryonic root of a bean emerges first and grows downward from the germinating seed. Next, the embryonic shoot emerges, and a hook forms near its tip. The hook protects the delicate shoot tip by holding it downward, rather than pushing it up through the abrasive soil. As the shoot breaks through the soil surface, its tip is lifted gently out of the soil as exposure to light stimulates the hook to straighten. The first foliage leaves then expand from the shoot tip and begin making food by photosynthesis. The cotyledons emerge from the soil and become leaflike photosynthetic structures. In many other plants, such as peas, the cotyledons remain behind in the soil and decompose.

Corn and other monocots use a different mechanism for breaking ground at germination (**Figure 13B**). A protective sheath surrounding the shoot pushes upward and breaks through the soil. The shoot tip then grows up through the tunnel provided by the sheath. The corn cotyledon remains in the soil and decomposes.

In the wild, only a small fraction of fragile seedlings endure long enough to reproduce. Production of enormous numbers of seeds compensates for the odds against individual survival. Asexual reproduction, generally simpler and less hazardous for offspring than sexual reproduction, is an alternative means of plant propagation, as we see next.

? Which meristems provide additional cells for early growth of a seedling after germination?

The apical meristems of the shoot and root

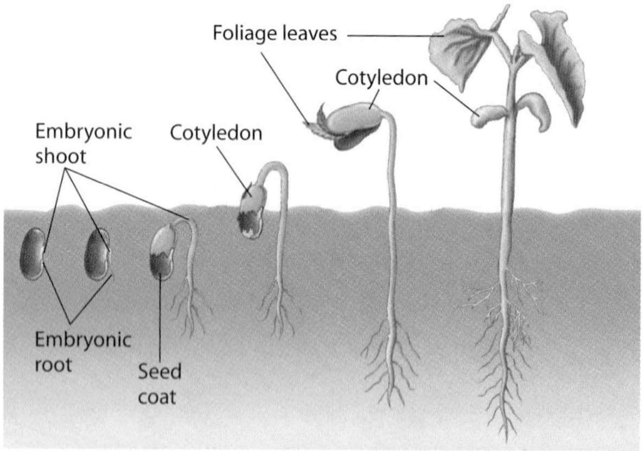

▲ **Figure 13A** Bean germination (a eudicot)

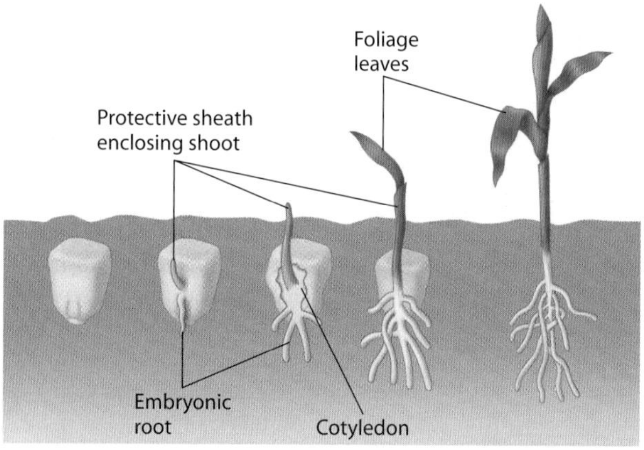

▲ **Figure 13B** Corn germination (a monocot)

14 Asexual reproduction produces plant clones

Imagine chopping off your finger and watching it grow and develop into an exact copy of you. This would be an example of asexual reproduction, which in plants is also called vegetative propagation. The resulting asexually produced offspring, often called a **clone**, is genetically identical to its single parent.

Asexual reproduction in angiosperms and other plants is an extension of their capacity to grow throughout life. A plant's meristematic tissues can sustain growth indefinitely. In addition, parenchyma cells can divide and differentiate into the various types of cells.

In nature, asexual reproduction in plants often involves **fragmentation**, the separation of a parent plant into parts that develop into whole plants. A garlic bulb (**Figure 14A**) is actually an underground stem that functions in storage. A single large bulb fragments into several parts, called cloves. Each clove can give rise to a separate plant, as indicated by the green shoots emerging from some of them. The white, paper-thin sheaths are leaves that are attached to the stem.

Tim Hill/Alamy

▲ **Figure 14A** Cloves of a garlic bulb

Each of the small trees you see in **Figure 14B** is a sprout from the roots of a coast redwood tree (discussed in the chapter introduction). Each small tree develops its own root system separate from the parent tree.

The ring of plants in **Figure 14C** is a clone of creosote bushes growing in the Mojave Desert in southern California. In this case, the word *clone* refers to a group of genetically identical organisms. All these bushes came from generations of asexual reproduction by roots. Making the oldest trees seem youthful, this clone apparently began with a single plant that germinated from a seed about 12,000 years ago. The original plant probably occupied the center of the ring.

Some aspen groves, such as those shown in **Figure 14D**, consist of thousands of trees descended by asexual reproduction from the root system of a single parent. (Genetic differences between groves that arose from different parents result in different timing for the development of fall color.)

Many plants can reproduce both sexually and asexually. What advantages can asexual reproduction offer? For one thing, a particularly fit parent plant can clone many copies of itself, all of which would be equally well suited to current conditions. Also, the offspring may not be as fragile as seedlings produced by sexual reproduction. And if a plant is isolated (and therefore unlikely to be pollinated), asexual reproduction may be the only means of reproduction.

The ability of plants to reproduce asexually provides many opportunities for growers to produce large numbers of plants with minimal effort and expense. For example, most of our fruit trees and houseplants are asexually propagated from cuttings. Several other plants are propagated from root sprouts (for example, raspberries) or pieces of underground stems (such as potatoes) or by grafting a bud from one plant onto a closely related plant (for example, most varieties of wine grapes).

Plants can also be asexually propagated in the laboratory. The test tube in **Figure 14E** contains a geranium plantlet that was grown from a few meristem cells cut from a mature plant and cultured in a growth medium containing nutrients and hormones. Using this method, a single plant can be cloned into thousands of copies. Orchids and certain pine trees used for mass plantings are commonly propagated this way.

Plant cell culture methods also enable researchers to grow plants from genetically engineered plant cells. Foreign genes are incorporated into a single parenchyma cell, and the cell is then cultured so that it multiplies and develops into a new plant. The resulting genetically modified (GM) plant may then be able to grow and reproduce normally. The commercial adoption of GM crops by farmers has been one of the most rapid cases of technology transfer in the history of agriculture, but it is not without controversy.

Rosenfeld Images Ltd/Photo Researchers

▲ **Figure 14E** Test-tube cloning

Aside from the issues raised by GM plants, modern agriculture faces some potentially serious problems. Nearly all of today's crop plants have very little genetic variability. In fact, we grow most crops in monocultures, cultivating a single plant variety on large areas of land. Given these conditions, plant scientists fear that a small number of diseases could devastate large crop areas. In response, plant breeders are working to maintain "gene banks," storage sites for seeds of many different plant varieties that can be used to breed new hybrids.

? **Which mode of reproduction (sexual or asexual) would generally be more advantageous in a location where the composition of the soil is constantly changing? Why?**

Sexual, because it generates genetic variation among the offspring, which enhances the potential for adaptation to a changing environment.

Frank Balthis

▲ **Figure 14B** Sprouts from the roots of coast redwood trees

Dennis Frates/Alamy

▲ **Figure 14D** Aspen trees

Dan Suzio/SPL/Photo Researchers

▶ **Figure 14C** A ring of creosote bushes

EVOLUTION CONNECTION

15 Evolutionary adaptations help some plants to live very long lives

Some plants can survive a very long time. As noted earlier, some coast redwoods, like the tree you read about in the chapter introduction, are estimated to be 2,000 to 3,000 years old. However, those redwoods are mere youngsters compared with some other trees. The tree shown in **Figure 15** is a 4,600-year-old bristlecone pine (*Pinus longaeva*). Bristlecone pines are found only in six western U.S. states (including the White Mountain region of California, pictured here). Another bristlecone pine, named Methuselah, is believed to be Earth's oldest living organism. Its location is kept secret to protect the tree.

A long life enhances evolutionary fitness by increasing the number of reproductive opportunities. What evolutionary adaptations in plants can help trees to live so long? Adult trees, like most adult plants, retain meristems, which allow for continued growth and repair throughout life. A tree can thus replace organs that have been lost or damaged by trauma or disease. Also, thick wood can protect against insects and disease.

? Why is the presence of meristems essential for the long life of many plants?

Plant meristem can give rise to multiple types of cells throughout life.

David Welling/Nature Picture Library

▲ **Figure 15** A bristlecone pine tree growing in California

REVIEW

For Practice Quizzes, BioFlix, MP3 Tutors, and Activities, go to www.masteringbiology.com.

Reviewing the Concepts

Plant Structure and Function (1–6)

1 People have manipulated plants since prehistoric times. Humans share a long and prosperous history with angiosperms (flowering plants).

2 The two major groups of angiosperms are the monocots and the eudicots. These two groups differ in the number of seed leaves and the structure of roots, stems, leaves, and flowers.

3 A typical plant body contains three basic organs: roots, stems, and leaves. The structure of a flowering plant allows it to draw resources from both soil and air.

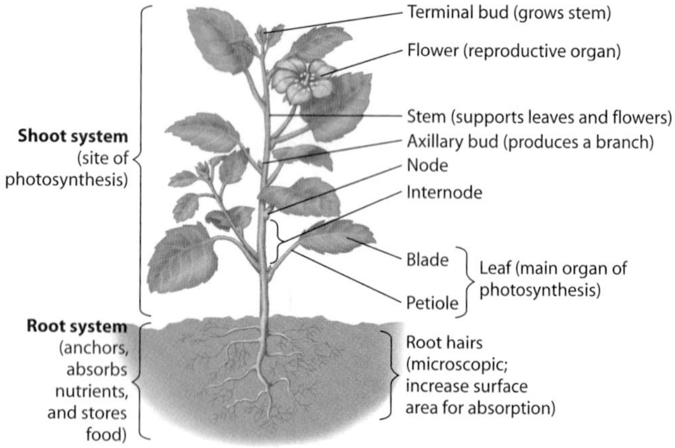

4 Many plants have modified roots, stems, and leaves. In addition to their primary functions, plant organs may store food, promote asexual reproduction, and provide protection.

5 Three tissue systems make up the plant body. Roots, stems, and leaves are each made up of dermal, vascular, and ground tissues. Dermal tissue covers and protects the plant. In leaves, dermal tissue has stomata, pores with guard cells that regulate exchange of gases and water vapor with the environment. The vascular tissue system contains xylem and phloem, which function in support and transport. Xylem conveys water and dissolved minerals, and phloem transports sugars. The ground tissue system functions in storage, photosynthesis, and support.

6 Plant cells are diverse in structure and function. The major types of plant cells are parenchyma, collenchyma, sclerenchyma (including fiber and sclereid cells), water-conducting cells (tracheids and vessel elements), and food-conducting cells (sieve-tube elements).

Plant Growth (7–8)

7 Primary growth lengthens roots and shoots. Meristems, areas of unspecialized, dividing cells, are where plant growth originates. Apical meristems at the tips of roots and in terminal buds and axillary buds of shoots initiate primary growth by producing new cells. A root or shoot lengthens as the cells elongate and differentiate.

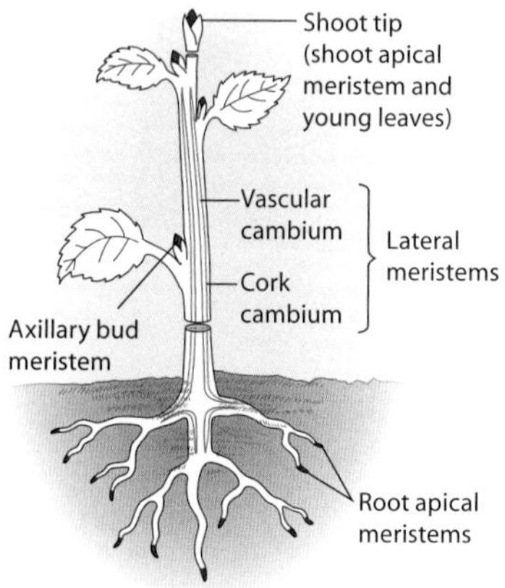

8 Secondary growth increases the diameter of woody plants. An increase in a plant's diameter, called secondary growth, arises from cell division in a cylinder of meristem cells called the vascular cambium. The vascular cambium thickens a stem by adding layers of secondary xylem, or wood, next to its inner surface. Outside the vascular cambium, the bark includes secondary phloem, cork cambium, and protective cork cells produced by the cork cambium.

Reproduction of Flowering Plants (9–15)

9 The flower is the organ of sexual reproduction in angiosperms. The angiosperm flower consists of sepals, petals, stamens, and carpels. Pollen grains develop in anthers, at tips of stamens. The tip of the carpel, the stigma, receives pollen grains. The ovary, at the carpel's base, houses the egg-producing ovule.

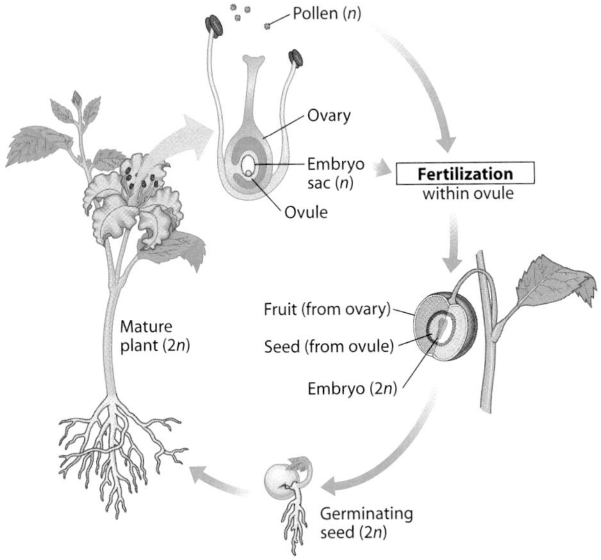

10 The development of pollen and ovules culminates in fertilization. Haploid spores are formed within ovules and anthers. The spores in anthers give rise to male gametophytes—pollen grains—which produce sperm. A spore in an ovule produces the embryo sac, the female gametophyte. Each embryo sac has an egg cell. Pollination is the arrival of pollen grains onto a stigma. A pollen tube grows into the ovule, and sperm pass through it and fertilize both the egg and a second cell. This process is called double fertilization.

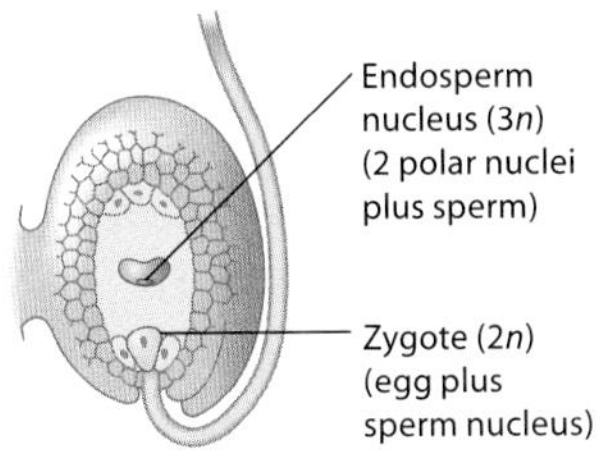

11 The ovule develops into a seed. After fertilization, the ovule becomes a seed, and the fertilized egg within it divides and becomes an embryo. The other fertilized cell develops into the endosperm, which stores food for the embryo.

12 The ovary develops into a fruit. Fruits help protect and disperse seeds.

13 Seed germination continues the life cycle. A seed starts to germinate when it takes up water and expands. The embryo resumes growth and absorbs nutrients from the endosperm. An embryonic root emerges, and a shoot pushes upward and expands its leaves.

14 Asexual reproduction produces plant clones. Asexual reproduction can occur by fragmentation or outgrowths of root systems. Propagating plants asexually from cuttings or bits of tissue can increase agricultural productivity but reduces genetic diversity.

15 Evolutionary adaptations help some plants to live very long lives. The continued growth produced by meristems and the protection provided by dense wood can help some trees live thousands of years.

Connecting the Concepts

1. Create a diagram that shows the relationships between the following: root system, root hairs, shoot system, leaves, petioles, blades, stems, nodes, internodes, flowers.

Testing Your Knowledge

Multiple Choice

2. Which of the following is closest to the center of a woody stem? (*Explain your answer.*)
   a. vascular cambium
   b. primary phloem
   c. secondary phloem
   d. primary xylem
   e. secondary xylem
3. A pea pod is formed from ______. A pea inside the pod is formed from ______.
   a. an ovule . . . a carpel
   b. an ovary . . . an ovule
   c. an ovary . . . a pollen grain
   d. an anther . . . an ovule
   e. endosperm . . . an ovary
4. While walking in the woods, you encounter an unfamiliar nonwoody flowering plant. If you want to know whether it is a monocot or eudicot, it would *not* help to look at the
   a. number of seed leaves, or cotyledons, present in its seeds.
   b. shape of its root system.
   c. number of petals in its flowers.
   d. arrangement of vascular bundles in its stem.
   e. size of the plant.
5. In angiosperms, each pollen grain produces two sperm. What do these sperm do?
   a. Each one fertilizes a separate egg cell.
   b. One fertilizes an egg, and the other fertilizes the fruit.
   c. One fertilizes an egg, and the other is kept in reserve.
   d. Both fertilize a single egg cell.
   e. One fertilizes an egg, and the other fertilizes a cell that develops into stored food.

Matching

6. Attracts pollinator	a. pollen grain
7. Develops into seed	b. ovule
8. Protects flower before it opens	c. anther
9. Produces sperm	d. ovary
10. Produces pollen	e. sepal
11. Houses ovules	f. petal

Describing, Comparing, and Explaining

12. How does a fruit develop from a flower?
13. Name two kinds of asexual reproduction. Explain two advantages of asexual reproduction over sexual reproduction. What is the primary drawback?
14. What part of a plant are you eating when you consume each of the following: celery stalk, peanut, strawberry, lettuce, beet?

Applying the Concepts

15. Plant scientists are looking for the wild ancestors of potatoes, corn, and wheat. Why is this search important?

Review Answers

1. Here is one possible concept map:

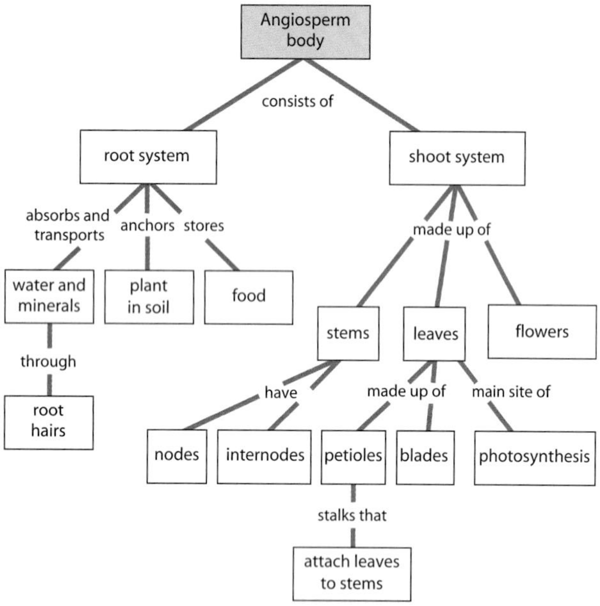

2. d (The vascular cambium forms to the outside of the primary xylem. The secondary xylem forms between primary xylem and the vascular cambium. The secondary phloem and primary phloem are outside the vascular cambium.) **3.** b **4.** e **5.** e **6.** f **7.** b **8.** e **9.** a **10.** c **11.** d

12. Pollen is deposited on the stigma of a carpel, and a pollen tube grows to the ovary at the base of the carpel. Sperm travel down the pollen tube and fertilize egg cells in ovules. The ovules grow into seeds, and the ovary grows into the flesh of the fruit. As the seeds mature, the fruit ripens and falls (or is picked).

13. Fragmentation of bulbs and sprouting from roots are examples of asexual reproduction. Asexual reproduction is less wasteful and costly than sexual reproduction and less hazardous for young plants. The primary disadvantage of asexual reproduction is that it produces genetically identical offspring, decreasing genetic variability that can help a species survive times of environmental change.

14. Celery stalk: leaf stalk (petiole); peanut: seed (ovule); strawberry: fruit (ripened ovary); lettuce: leaf blades; beet: root

15. Modern methods of plant breeding and propagation have increased crop yields but have decreased genetic variability, so plants have become more vulnerable to epidemics. Primitive varieties of crop plants could contribute to gene banks and be used for breeding new strains.

Plant Nutrition and Transport

From Chapter 32 of *Campbell Biology: Concepts & Connections*, Seventh Edition. Jane B. Reece, Martha R. Taylor, Eric J. Simon, Jean L. Dickey.

Plant Nutrition and Transport

BIG IDEAS

Brian Capon

The Uptake and Transport of Plant Nutrients
(1–5)

Plants absorb and transport substances—such as water, minerals, CO_2, and sugar—required for growth.

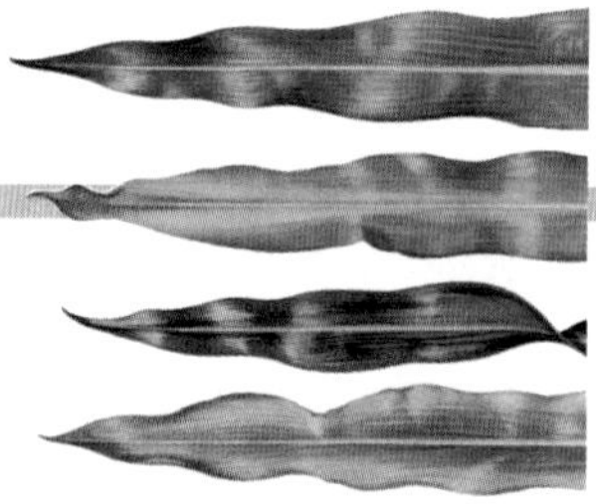

Plant Nutrients and the Soil
(6–11)

Plants require many essential minerals for proper health.

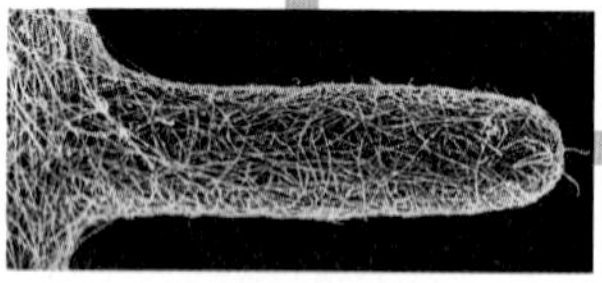

R. L. Peterson/ Biological Photo Service

Plant Nutrition and Symbiosis
(12–14)

Plants have evolved relationships with bacteria, fungi, animals, and other plants.

Igor Plotnikov/Shutterstock

On August 29, 2005, Hurricane Katrina—the costliest natural disaster in U.S. history—slammed into New Orleans. After levees failed during the storm surge, most of the city was flooded with noxious runoff containing raw sewage, heavy metals, toxic chemicals, and petroleum. The effect on the local environment was devastating, especially the contamination of the city's topsoil.

Years later, cleanup efforts continue. In addition to bulldozers and wrecking crews, help is coming from a natural source: plants. Plants can efficiently move minerals and other compounds from the soil into the roots and eventually through the body of the plant. Most often, the compounds absorbed are nutrients that help the plant grow. But some species of plants (such as the sunflowers shown above) can absorb lead, zinc, and other heavy metals in concentrations that would kill most plants. These plants are ideally suited for detoxifying soil because the heavy metals they remove become concentrated in the plant body, which can be easily discarded.

Volunteers from a grassroots relief organization called Common Ground have planted tens of thousands of sunflowers in New Orleans's hardest hit wards. Once fully grown, the metal-laden crops are harvested and hauled away to a hazardous-waste landfill. Besides easing cleanup efforts, sunflowers provide an uplifting message to the community, bringing beauty and life back to an area weary with grief and loss.

In using plants to clean up toxic wastes, we are benefiting from millions of years of plant evolution. Unable to move about in search of food, plants have evolved amazing abilities to pull water and nutrients out of the soil and air. In this chapter, we explore how plants obtain essential nutrients and how they transport them throughout their roots, stems, and leaves.

The Uptake and Transport of Plant Nutrients

1 Plants acquire nutrients from air, water, and soil

Watch a plant grow from a tiny seed, and you can't help wondering where all the mass comes from. If you had to take a guess, what would you think is the source of the raw materials that make up a plant's body? Soil? Water? Air?

Aristotle thought that soil provided all the substance for plant growth. The 17th-century physician Jan Baptista van Helmont performed an experiment to test this hypothesis. He planted a willow seedling in a pot containing 91 kg of soil. After five years, the willow had grown into a tree weighing 76.8 kg, but only 0.06 kg of soil had disappeared from the pot. Van Helmont concluded that the willow had grown mainly from added water. A century later, an English physiologist named Stephen Hales postulated that plants are nourished mostly by air.

As it turns out, there is truth in all these early hypotheses about plant nutrition; air, water, and soil all contribute to plant growth **(Figure 1A)**. A plant's leaves absorb carbon dioxide (CO_2) from the air; in fact, about 96% of a plant's dry weight is organic (carbon-containing) material built mainly from CO_2. Meanwhile, a plant gets water (H_2O), minerals, and some oxygen (O_2) from the soil.

What happens to the materials a plant takes up from the air and soil? The sugars that a plant makes by photosynthesis are composed of the elements carbon, oxygen, and hydrogen. The carbon and oxygen used in photosynthesis come from atmospheric CO_2 and that the hydrogen comes from water molecules. Plant cells use the sugars made by photosynthesis in constructing all the other organic materials they need, but primarily for carbohydrates. The trunk of the giant sequoia tree in **Figure 1B**, for instance, consists mainly of sugar derivatives, such as the cellulose of cell walls.

Plants use cellular respiration to break down some of the sugars they make, obtaining energy from them in a process that consumes O_2. A plant's leaves take up some O_2 from the air, but Figure 1A does not show this because plants are actually net producers of O_2, giving off more of this gas than they use. When water is split during photosynthesis, O_2 gas is produced and released through the leaves. The O_2 being taken up from the soil by the plant's roots in Figure 1A is actually atmospheric O_2 that has diffused into the soil; it is used in cellular respiration in the roots themselves.

A plant's ability to move water from its roots to its leaves and its ability to deliver sugars to specific areas of its body are staggering feats of evolutionary engineering. Figure 1B highlights an extreme example; the topmost leaves of a giant sequoia can be over 100 m (300 feet) above the roots. In the next four modules, we follow the movement of water, dissolved mineral nutrients, and sugar throughout the plant body.

? What inorganic substance is obtained in the greatest quantities from the soil?

● Water

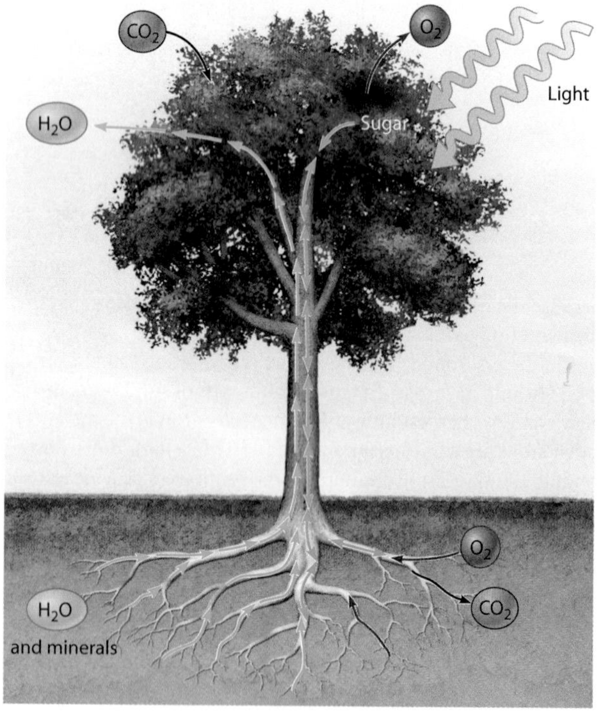

▲ **Figure 1A** The uptake of nutrients by a plant

▼ **Figure 1B** A giant sequoia, a product of photosynthesis

Mario Verin/Photolibrary

2 The plasma membranes of root cells control solute uptake

With its surface area enormously expanded by thousands of root hairs (**Figure 2A**), a plant root has a remarkable ability to extract water and minerals from soil. Recall from Module 31.3 that root hairs are extensions of epidermal cells that cover the root. Root hairs provide a huge surface area in contact with nutrient-containing soil; in fact, the root hairs of a single sunflower plant, if laid end to end, could stretch for several miles.

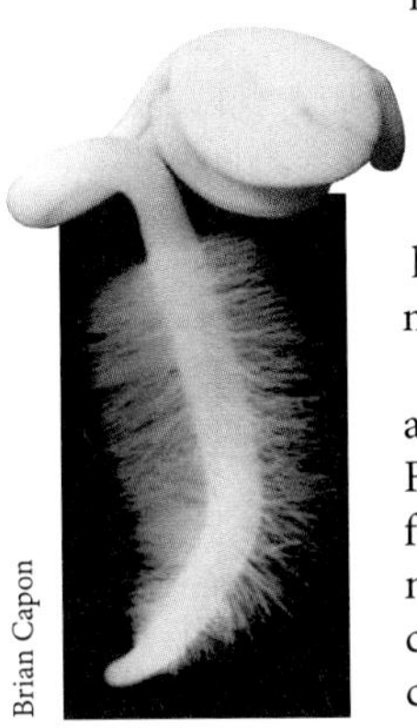

▲ **Figure 2A**
Root hairs of a radish seedling

All substances that enter a plant root are in solution (that is, dissolved in water). For water and solutes to be transported from the soil throughout the plant, they must move through the epidermis and cortex of the root and then into the water-conducting xylem tissue in the central cylinder of the root. Any route the water and solutes take from the soil to the xylem requires that they pass through some of the plasma membranes of the root cells. Because plasma membranes are selectively permeable, only certain solutes reach the xylem. This plasma membrane selectivity provides a checkpoint that transports needed minerals from the soil and keeps many unneeded or toxic substances out.

You can see two possible routes to the xylem in the bottom part of **Figure 2B**. The long blue arrow indicates an *intracellular* route. Water and selected solutes cross the cell wall and plasma membrane of an epidermal cell (usually at a root hair). The cells within the root are all interconnected by plasmodesmata; there is a continuum of living cytoplasm among the root cells. Therefore, once inside the epidermal cell, the solution can move inward from cell to cell without crossing any other plasma membranes, diffusing through the interconnected cytoplasm all the way into the root's endodermis. An endodermal cell then discharges the solution into the phloem, from which it enters the xylem (light purple).

The long red arrow in Figure 2B indicates an alternative route. This route is *extracellular*; the solution moves inward within the hydrophilic walls and extracellular spaces of the root cells but does not enter the cytoplasm of the epidermis or cortex cells. The solution does not cross any plasma membranes, and there is no selection of solutes until the solution reaches the endodermis. Here, a continuous waxy barrier called the **Casparian strip** stops water and solutes from entering the xylem through the cell walls, instead forcing them to cross a plasma membrane into an endodermal cell (short blue arrow). Ion selection occurs at this membrane instead of in the epidermis, and once the selected solutes and water are in the endodermal cell, they can be discharged into the xylem.

Actually, water and solutes rarely follow just the two kinds of routes in Figure 2B. They may take any combination of these routes, and they may pass through numerous plasma membranes and cell walls en route to the xylem. Because of the Casparian strip, however, there are no nonselective routes; the water and solutes must cross a selectively permeable plasma membrane at some point. In other words, no solutes enter the vascular tissue unchecked; all must pass through at least one checkpoint. Next, we see how water and minerals move upward within the xylem from roots to shoots.

? What is the function of the Casparian strip?

● It regulates the passage of minerals (inorganic ions) into the xylem by blocking access via cell walls and requiring all minerals to cross a selectively permeable plasma membrane.

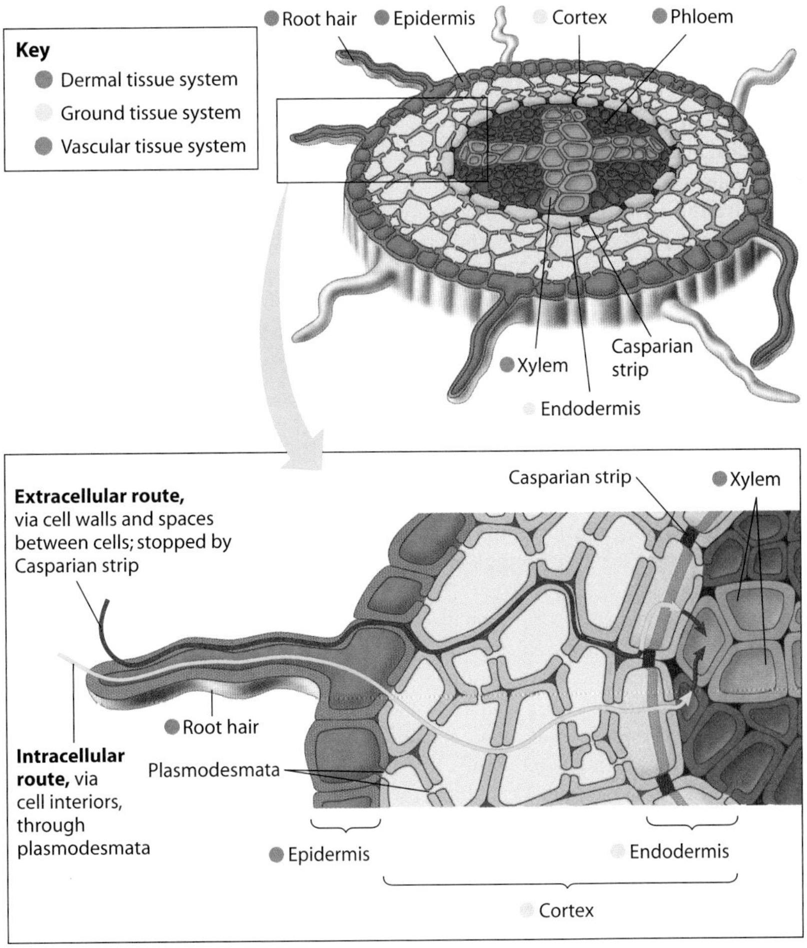

▲ **Figure 2B** Routes of water and solutes from soil to root xylem

3 Transpiration pulls water up xylem vessels

As a plant grows upward toward sunlight, it needs to extract an increasing supply of water and dissolved mineral ions from the soil. Imagine having to haul a 5-gallon water bucket weighing 40 pounds up five flights of stairs. Now imagine doing that every 20 minutes, all day long. To thrive, a typical tree must constantly transport that quantity of water from its roots to the rest of the plant.

Xylem tissue of angiosperms includes two types of conducting cells: tracheids and vessel elements. When mature, both types of cells are dead, consisting only of cell walls, and both are in the form of very thin tubes arranged end to end. Because the cells have openings in their ends, a solution of water and inorganic nutrients, called **xylem sap**, can flow through these tubes. Xylem sap flows all the way up from the plant's roots through the shoot system to the tips of the leaves.

What force moves xylem sap up against the downward pull of gravity? Is it pushed or pulled upward? Plant biologists have found that the roots of some plants do exert a slight upward push on xylem sap. The root cells actively pump inorganic ions into the xylem, and the root's endodermis holds the ions there. As ions accumulate in the xylem, water tends to enter by osmosis, pushing xylem sap upward ahead of it. This force, called **root pressure**, can push xylem sap up a few meters.

For the most part, however, xylem sap is not pushed from below by root pressure but pulled upward. Plant biologists have determined that the pulling force is **transpiration**, which is the loss of water from the leaves and other aerial parts of a plant by evaporation.

Figure 3 illustrates transpiration and its effect on water movement in a tree. ❶ Water molecules (blue arrows) leave the leaf through a stoma, a microscopic pore on the surface of a leaf. When the stoma is open, water diffuses out of the leaves because the surrounding air is usually drier than the inside of the leaf. You can see this yourself on chilly mornings when the cold air causes the evaporating water to condense as dew.

Transpiration can pull xylem sap up the tree because of two special properties of water: cohesion and adhesion. Both of these properties arise from the charge polarity of water molecules. **Cohesion** is the sticking together of molecules of the same kind. ❷ In the case of water, hydrogen bonds stick the H_2O molecules to one another. ❸ The cohering water molecules within the xylem tubes form continuous strings, extending all the way from the leaves down to the roots. In contrast, **adhesion** is the sticking together of molecules of different kinds. ❹ Water molecules tend to adhere to hydrophilic cellulose molecules in the walls of xylem cells via hydrogen bonds.

What effect does transpiration have on a string of water molecules that tend to adhere to the walls of the xylem tubes? Because the air outside the leaf is much drier than the moist interior of the leaf, the water molecule at the end of the string diffuses outward, exiting the leaf. Cohesion resists the pulling force of diffusion, but it is not strong enough to overcome it. The outer water molecule breaks off, and the opposing forces of cohesion and transpiration put tension on the remainder of the string of water molecules. As long as transpiration continues, the string is kept tense and is pulled upward as one molecule exits the leaf and the one right behind it is tugged up into its place. Adhesion of the water molecules to the walls of the xylem cells assists the upward movement of the xylem sap by counteracting the downward pull of gravity.

Plant biologists call this explanation for the ascent of xylem sap the **transpiration-cohesion-tension mechanism**. We can summarize it as follows: Transpiration exerts a pull

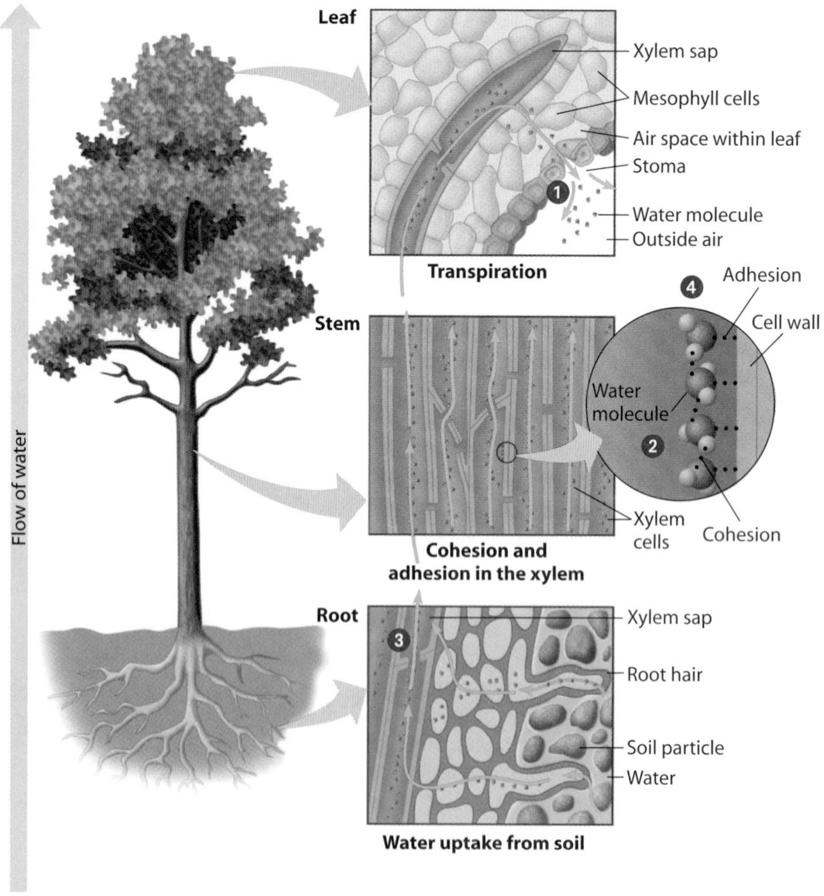

▲ **Figure 3** The flow of water up a tree

on a tense, unbroken chain of water molecules that is held together by cohesion and helped upward by adhesion. Transpiration is an efficient means of moving large volumes of water upward from roots to shoots. In fact, via transpiration, xylem sap can move up a tree at speeds approaching 50 m/hr. And yet, all this transport of xylem sap requires no energy expenditure by the plant. Physical properties—cohesion, adhesion, and evaporation—move water and dissolved minerals from a plant's roots to its shoots.

? Describe the roles of cohesion and adhesion in the ascent of xylem sap.

Cohesion causes water molecules to stick together in a continuous string; adhesion helps xylem sap to stick to the inside walls of the xylem cells.

4 Guard cells control transpiration

Adaptations that increase photosynthesis—such as large leaf surface areas—have the serious drawback of increasing water loss by transpiration. A plant must make a trade-off between its tremendous need for water and its need to make food by photosynthesis. As long as water moves up from the soil fast enough to replace the water that is lost, transpiration presents no problem. But if the soil dries out and transpiration exceeds the delivery of water to the leaves, the leaves will wilt. Unless the soil and leaves are rehydrated, the plant will eventually die.

The leaf stomata, which can open and close, are evolutionary adaptations that help plants adjust their transpiration rates to changing environmental conditions. As shown in **Figure 4**, a pair of guard cells flank each stoma. The guard cells control the opening of a stoma by changing shape, like an inflatable set of doors that widen or narrow the gap between the two cells.

What actually causes guard cells to change shape and thereby open or close stomata? Figure 4 illustrates the principle. A stoma opens (left) when its guard cells gain potassium ions (K^+, red dots) and water (blue arrows) from neighboring cells (shown in light gray). The cells actively take up K^+, and water then enters by osmosis. When the vacuoles in the guard cells gain water, the cells become more turgid and bowed. The cell wall of a guard cell is not uniformly thick, and the cellulose molecules are oriented in such a way that the cell buckles away from its companion guard cell when it is turgid. The result is an increase in the size of the gap (stoma) between the two cells. Conversely, when the guard cells lose K^+, they also lose water by osmosis and become flaccid and less bowed, closing the space between them (right).

In general, guard cells keep the stomata open during the day and closed at night. During the day, CO_2 can enter the leaf from the atmosphere and thus keep photosynthesis going when sunlight is available. At night, when there is no light for photosynthesis and therefore no need to take up CO_2, the stomata are closed, saving water.

At least three cues contribute to stomatal opening at dawn. One is light, which stimulates guard cells to accumulate K^+ and become turgid. A low level of CO_2 in the leaf can have the same effect. A third cue is an internal timing mechanism—a biological clock—found in the guard cells; even if you keep a plant in a dark closet, stomata will continue their daily rhythm of opening and closing.

Even during the day, the guard cells may close the stomata if the plant is losing water too fast. This response reduces further water loss and may prevent wilting, but it also slows down CO_2 uptake and photosynthesis—one reason that droughts reduce crop yields. In summary, guard cells arbitrate the photosynthesis-transpiration compromise on a moment-to-moment basis by integrating a variety of stimuli.

? Some leaf molds secrete a chemical that causes guard cells to accumulate K^+. How does this help the mold infect the plant?

Accumulation of K^+ by guard cells causes the stomata to stay open. The mold can then grow into the leaf interior via the stomata.

Jeremy Bur Photo Researchers

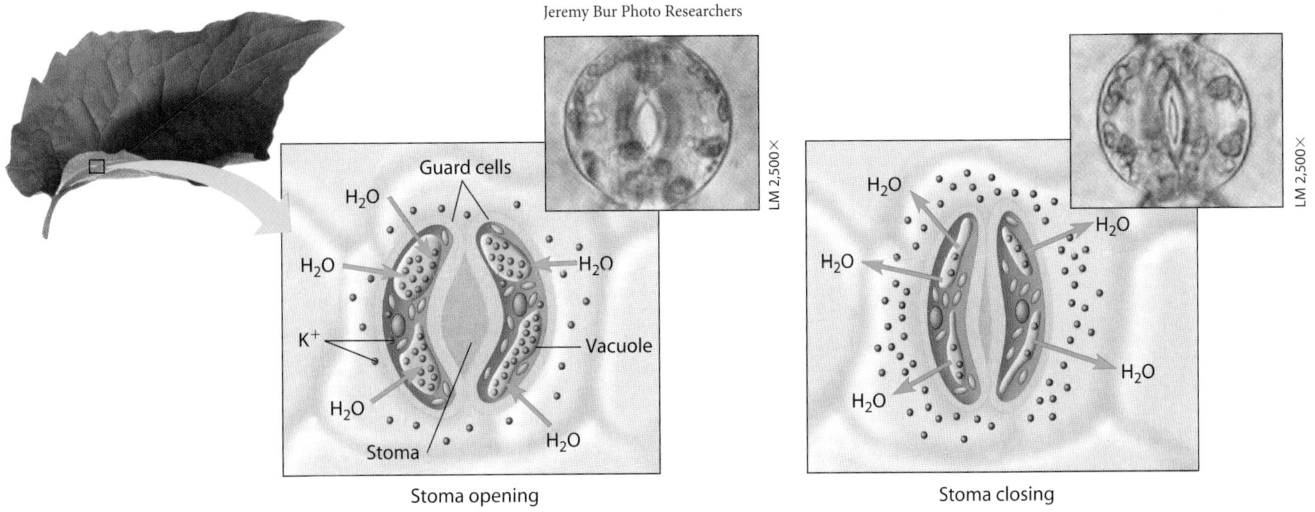

Frank Greenaway/Dorling Kindersley

▲ **Figure 4** How guard cells control stomata

5 Phloem transports sugars

A plant has two separate transport systems: xylem (the topic of Module 3) and phloem. Xylem transports xylem sap (water and dissolved minerals), while the main function of phloem is to transport the products of photosynthesis. In angiosperms, phloem contains food-conducting cells called sieve-tube elements arranged end to end to form long tubes. The art and micrograph in **Figure 5A** show two sieve-tube elements and the sieve plate between them. Perforations in sieve plates connect the cytoplasm of these living cells into one continuous solution. Sugary liquid called **phloem sap** can thus move freely from one cell to the next. Phloem sap may contain inorganic ions, amino acids, and hormones in transit from one part of the plant to another, but its main solute is usually the disaccharide sugar sucrose.

In contrast to xylem sap, which only flows upward from the roots, phloem sap moves throughout the plant in various directions. However, sieve tubes always carry sugars from a sugar source to a sugar sink. A **sugar source** is a plant organ that is a net producer of sugar, by photosynthesis or by breakdown of starch. Leaves are the primary sugar sources in most mature plants. A **sugar sink** is an organ that is a net consumer or storer of sugar. Growing roots, buds, stems, and fruits are sugar sinks. A storage organ, such as a tuber or a bulb, may be a source or sink, depending on the season. When stockpiling carbohydrates in the summer, it is a sugar sink. After breaking dormancy in the spring, it is a source as its starch is broken down to sugar, which is carried to the growing shoot tips of the plant. Thus, each food-conducting tube in phloem tissue has a source end and a sink end, but these may change with the season or the developmental stage of the plant.

What causes phloem sap to flow from a sugar source to a sugar sink? Flow rates may be as high as 1 m/hr, which is much too fast to be accounted for by diffusion. (By diffusion alone, phloem sap would travel less than 1 meter per year.) Biologists have tested a number of hypotheses for phloem sap movement. A hypothesis called the **pressure flow mechanism** is now widely accepted for angiosperms. **Figure 5B** illustrates how this mechanism works, using a beet plant as an example. The pink dots in the phloem tube represent sugar molecules; notice their concentration gradient from top to bottom. The blue color represents a parallel gradient of water pressure in the phloem sap.

At the sugar source (the beet leaves, in this example), ❶ sugar is loaded from a photosynthetic cell into a phloem tube by active transport. Sugar loading at the source end raises the solute (sugar) concentration inside the phloem tube. ❷ The high solute concentration draws water into the tube by osmosis, usually from the xylem. The inward flow of water raises the water pressure at the source end of the phloem tube.

At the sugar sink (the beet root, in this case), both sugar and water leave the phloem tube. ❸ As sugar departs the phloem, lowering the solute (sugar) concentration at the sink end, ❹ water follows by osmosis back into the xylem. The exit of

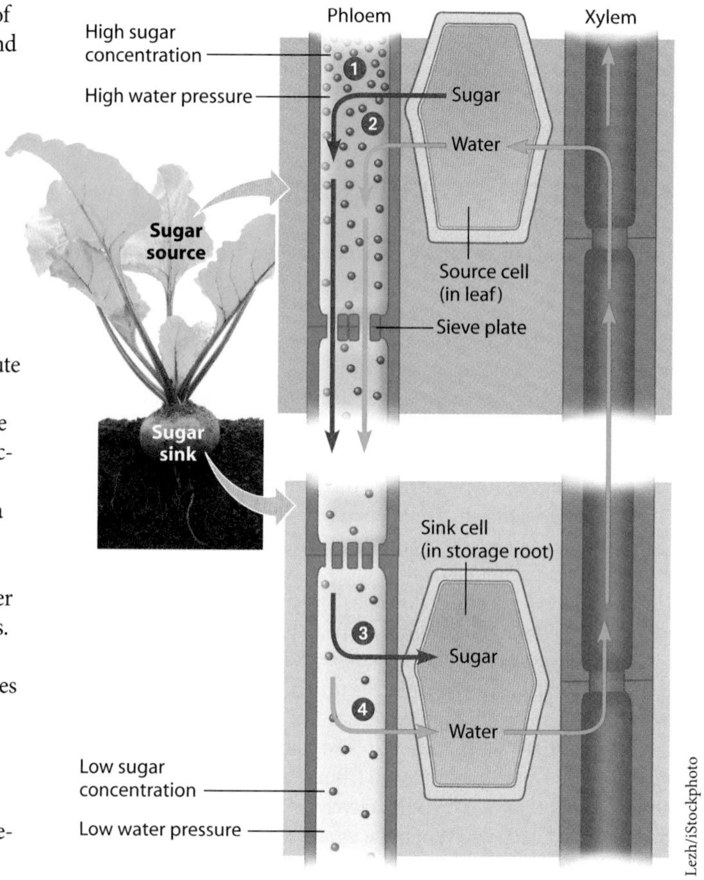

▲ **Figure 5B** Pressure flow in plant phloem from a sugar source to a sugar sink (and the return of water to the source via xylem)

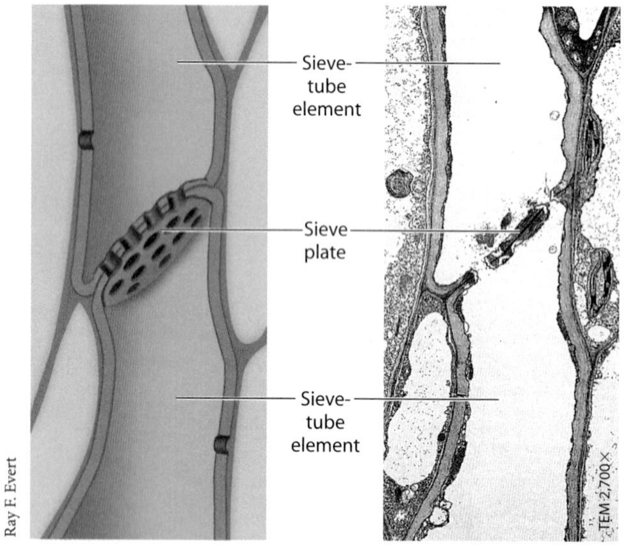

▲ **Figure 5A** Food-conducting cells of phloem

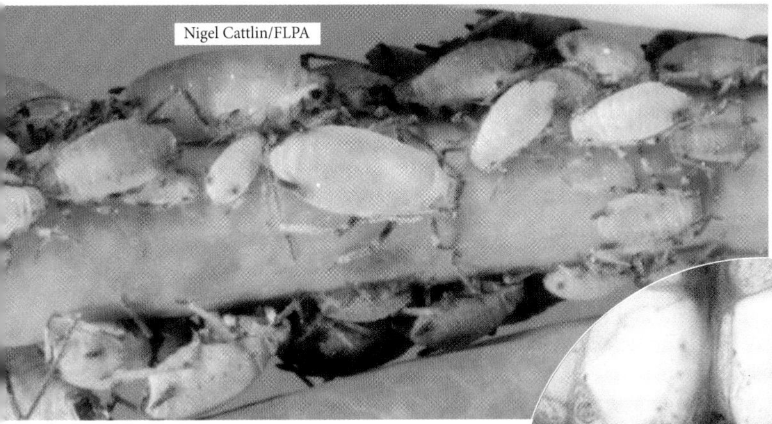

Aphids feeding on a branch

Aphid with phloem droplet

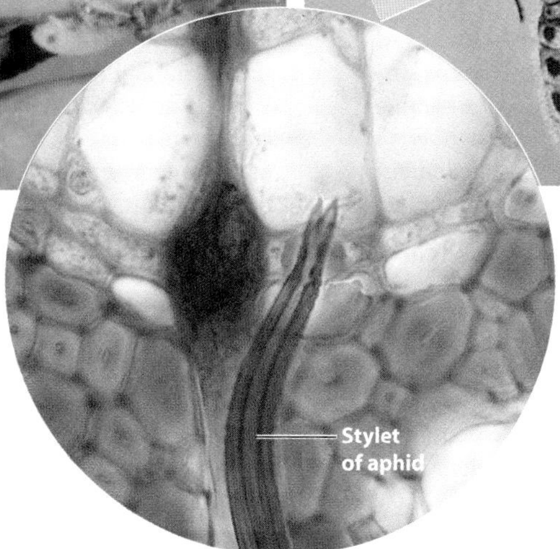

Aphid's stylet inserted into a phloem cell

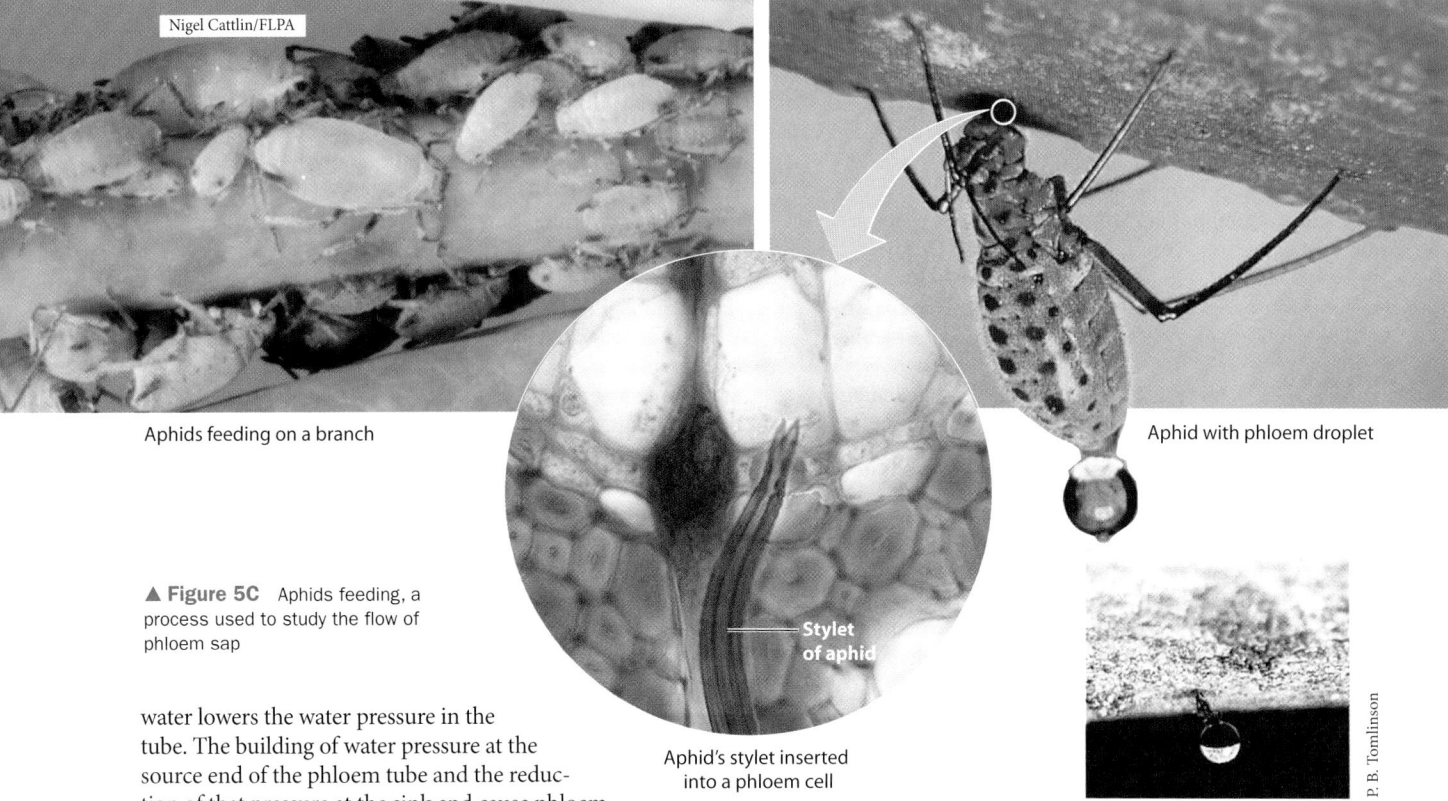

Severed stylet dripping phloem sap

▲ **Figure 5C** Aphids feeding, a process used to study the flow of phloem sap

water lowers the water pressure in the tube. The building of water pressure at the source end of the phloem tube and the reduction of that pressure at the sink end cause phloem sap to flow from source to sink—down a gradient of water pressure. Sieve plates allow free movement of solutes as well as water. Thus, sugar is carried along from source to sink at the same rate as the water. As indicated on the right side of Figure 5B, xylem tubes recycle the water back from sink to source.

The pressure flow mechanism explains why phloem sap always flows from a sugar source to a sugar sink, regardless of their locations in the plant. However, the mechanism is somewhat difficult to test because most experimental procedures disrupt the structure and function of the phloem tubes. Some of the most interesting studies have taken advantage of natural phloem probes: insects called aphids.

Figure 5C shows how an aphid feeds by inserting its needle-like mouthpart, called a stylet, into the phloem of a tree branch. The aphid is releasing from its anus a drop of phloem sap lacking some solutes that the insect's digestive tract has removed for food. The micrograph in the center shows an aphid's stylet inserted into one of the plant's food-conducting cells. The pressure within the phloem force-feeds the aphid, swelling it to several times its original size. Plant biologists have used this process to study the flow of phloem sap. While the aphid is feeding, it can be anesthetized and severed from its stylet. The stylet then serves the researcher as a miniature tap that drips phloem sap for hours (similar to the way a tap in a pressurized keg of beer allows the liquid to flow out for hours). The photograph at the lower right shows a droplet of phloem sap on the cut end of a stylet. Studies using this technique support the pressure flow model: The closer the stylet is to a sugar source, the faster the sap flows and the greater its sugar concentration. This is what we would expect if pressure is generated at the source end of the phloem tube by the active pumping of sugar into the tube.

? Contrast the forces that move phloem sap with the forces that move xylem sap.

Pressure is generated at the source end of a sieve tube by the loading of sugar and the resulting osmotic flow of water into the phloem. This pressure pushes phloem sap from the source end to the sink end of the tube. In contrast, transpiration generates a pulling force that drives the ascent of xylem sap.

• • •

We now have a broad picture of how a plant absorbs substances from the soil and transports materials from one part of its body to another: Water and inorganic ions enter from the soil and are distributed by xylem. The xylem sap is pulled upward by transpiration. Carbon dioxide enters leaves through stomata and is incorporated into sugars. A second transport system, phloem, distributes the sugars. Pressure flow drives the phloem sap from leaves and storage sites to other parts of the plant, where the sugars are used or stored.

Plants convert raw materials to organic molecules by photosynthesis. We have yet to say much about the kinds of inorganic nutrients a plant needs and what it does with them. This is the subject of plant nutrition, which we discuss in the next section.

Plant Nutrients and the Soil

6 Plant health depends on a complete diet of essential inorganic nutrients

In contrast to animals, which require a diet of complex organic (carbon-containing) foods, plants survive and grow solely on CO_2 and inorganic substances (that is, plants are autotrophs). The ability of plants to assimilate CO_2 from the air, extract water and inorganic ions from the soil, and synthesize organic compounds is essential not only to the survival of plants but also to the survival of humans and other animals.

A chemical element is considered an **essential element** if a plant must obtain it from its environment to complete its life cycle—that is, to grow from a seed and produce another generation of seeds. A method called hydroponic culture can be used to determine which chemical elements are essential nutrients. As shown in **Figure 6**, plants are grown without soil by bathing the roots in mineral solutions. Air is bubbled into the water to give the roots oxygen for cellular respiration. By omitting a particular element from the medium, a researcher can test whether that element is essential to the plant.

If the element left out of the solution is an essential nutrient, then the incomplete medium will make the plant abnormal in appearance compared with control plants grown on a complete nutrient medium. The most common symptoms of a nutrient deficiency are stunted growth and discolored leaves. Hydroponic culture studies have helped identify 17 essential elements needed by all plants. Most research has involved crop plants and houseplants; little is known about the nutritional needs of uncultivated plants.

Nine of the essential elements are called **macronutrients** because plants require relatively large amounts of them. Six of the nine macronutrients—carbon, oxygen, hydrogen, nitrogen, sulfur, and phosphorus—are the major ingredients of organic compounds forming the structure of a plant. These six elements make up almost 98% of a plant's dry weight. The other three macronutrients—calcium, potassium, and magnesium—make up another 1.5%.

How does a plant use calcium, potassium, and magnesium? Calcium has several functions. For example, it is important in the formation of cell walls, and it combines with certain proteins to form the glue that holds plant cells together in tissues. Calcium also helps maintain the structure of cell membranes and helps regulate their selective permeability. Potassium is crucial as a cofactor required for the activity of several enzymes. (A cofactor is an atom or molecule that cooperates with an enzyme in catalyzing a reaction.) Potassium is also the main solute for osmotic regulation in plants potassium ion movements regulate the opening and closing of stomata. Magnesium is a component of chlorophyll and thus essential for photosynthesis.

Elements that plants need in very small amounts are called **micronutrients**. The eight known micronutrients are chlorine, iron, manganese, boron, zinc, copper, nickel, and molybdenum. Micronutrients function in plants mainly as cofactors. Iron, for example, is a component of cytochromes, proteins in the electron transport chains of chloroplasts and mitochondria. Micronutrients can generally be used over and over, so plants need only minute quantities of these elements. The requirement for molybdenum, for example, is so modest that there is only one atom of this rare element for every 60 million atoms of hydrogen in dried plant material. Yet a deficiency of molybdenum or any other micronutrient can weaken or kill a plant.

▲ **Figure 6** A hydroponic culture experiment

? You conduct an experiment like the one in Figure 6 to test whether a certain plant species requires a particular chemical element as a micronutrient. Why is it important that the glassware be completely clean?

● Because micronutrients are required in only minuscule amounts, even the smallest amount of dirt in the experimental flask may contain enough of the element you are testing to allow normal growth and invalidate your results.

CONNECTION 7 Fertilizers can help prevent nutrient deficiencies

The quality of soil—especially the availability of nutrients—determines the health of a growing plant and, for plants we consume, the quality of our own nutrition ("You are what you eat"). **Figure 7A** shows a healthy corn leaf (top) as well as three leaves from plants suffering various nutrient deficiencies. Such mineral shortages can stunt plant growth, and if grain is produced at all, it will likely have low nutritional value. In this way, nutritional deficiencies in plants can be passed on to livestock or human consumers.

Fortunately, the symptoms of nutrient deficiency are often distinctive enough for a grower to diagnose its cause. Many growers make visual diagnoses and then check their conclusions by having soil and plant samples chemically analyzed at a state or local laboratory.

Nitrogen shortage is the most common nutritional problem for plants. Soils are usually not deficient in nitrogen, but they are often deficient in the nitrogen compounds that plants can use: dissolved nitrate ions (NO_3^-) and ammonium ions (NH_4^+). Stunted growth and yellow-green leaves, starting at the tips of older leaves, are signs of nitrogen deficiency (see Figure 7A, second leaf from top). Other common nutrient shortfalls in plants include phosphorus and potassium deficiencies.

Once a diagnosis of a nutrient deficiency is made, treating the problem is usually simple. **Fertilizers** are compounds given to plants to promote growth. Fertilizers come in two basic types: inorganic and organic. Inorganic fertilizers (also called mineral fertilizers) can contain naturally occurring inorganic compounds (such as mined limestone or phosphate rock) or synthetic inorganic compounds (such as ammonium nitrate). Inorganic fertilizers come in a wide variety of formulations, but most emphasize their "N-P-K ratio," the amounts of the three nutrients most often deficient in depleted soils: nitrogen (N), phosphorus (P), and potassium (K). For example, a 100-pound bag of 5–6–7 fertilizer contains 5 pounds of nitrogen (often as ammonium or nitrate), 6 pounds of phosphorus (as phosphoric acid), and 7 pounds of potassium (as the mineral potash), plus 82 pounds of filler. Many crops benefit from an all-purpose 5–5–5 formula, but some plants thrive only under special fertilizer formulations.

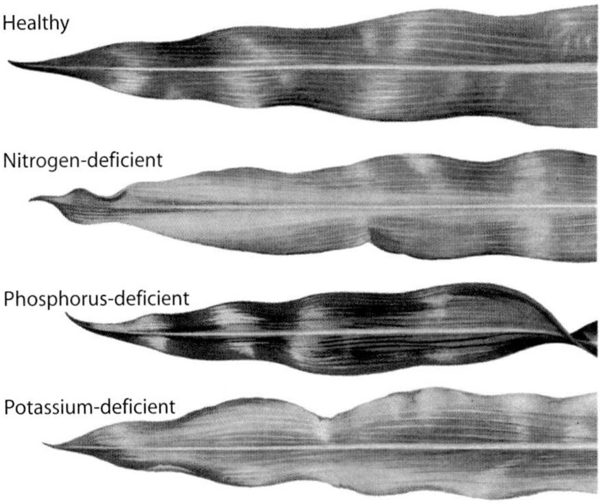

▲ **Figure 7A** The most common mineral deficiencies, as seen in corn leaves

Organic fertilizers are composed of chemically complex organic matter such as **compost**, which is a soil-like mixture of decomposed organic matter. (Organic fertilizers may or may not be *certified organic*, which means that the product meets a strict set of guidelines—see Module 10.) Many gardeners maintain a free-standing compost pile or an enclosed compost bin to which they add leaves, grass clippings, yard waste, and kitchen scraps (but not most animal products, such as meat, fat, bone, or animal droppings). Over time, the vegetable matter breaks down due to the action of naturally occurring microbes, fungi, and animals (**Figure 7B**). Occasional turning and watering will speed the composting process, producing homegrown fertilizer in several months. The compost can then be applied to outdoor gardens or indoor pots.

? **What is the most common nutrient deficiency in plants? What are the signs?**

Nitrogen deficiency; stunted growth and yellowing leaves

▼ **Figure 7B** Steam produced by the metabolic activity of organisms within a compost pile

Paul Rapson/Photo Researchers

8 Fertile soil supports plant growth

Along with climate, the major factor determining whether a plant can grow well in a particular location is the quality of the soil. Fertile soil can support abundant plant growth by providing conditions that enable plant roots to absorb water and dissolved nutrients.

Distinct layers of soil are visible in a road cut or deep hole, such as the cross section shown in **Figure 8A**. You can see three distinct soil layers, called horizons, in the cut. The A horizon, or **topsoil**, is subject to extensive weathering (freezing, drying, and erosion, for example). Topsoil is a mixture of rock particles of various sizes, living organisms, and **humus**, the remains of partially decayed organic material produced by the decomposition of dead organisms, feces, fallen leaves, and other organic matter by bacteria and fungi. The rock particles provide a large surface area that retains water and minerals while also forming air spaces containing oxygen that can diffuse into plant roots. Fertile topsoil is home to an astonishing number—about 5 billion per teaspoon—and variety of bacteria, algae and other protists, fungi, and small animals such as earthworms, roundworms, and burrowing insects. Along with plant roots, these organisms loosen and aerate the soil and contribute organic matter to the soil as they live and die. Nearly all plants depend on bacteria and fungi in the soil to break down organic matter into inorganic molecules that roots can absorb. Besides providing nutrients, humus also tends to retain water while keeping the topsoil porous enough for good aeration of the plant roots. Topsoil is rich in organic materials and is therefore most important for plant growth. Plant roots branch out in the A horizon and usually extend into the next layer, the B horizon.

The soil's B horizon contains many fewer organisms and much less organic matter than the topsoil and is less subject to weathering. Fine clay particles and nutrients dissolved in soil water drain down from the topsoil and often accumulate in the B horizon. The C horizon is composed mainly of partially broken-down rock that serves as the "parent" material for the upper layers of soil.

Figure 8B illustrates the intimate association between a plant's root hairs, soil water, and the tiny particles of topsoil. The root hairs are in direct contact with the water that surrounds the particles. The soil water is not pure but a solution containing dissolved inorganic ions. Oxygen diffuses into the water from small air spaces in the soil. Roots absorb this soil solution.

Cation exchange is a mechanism by which the root hairs take up certain positively charged ions (cations). Inorganic cations—such as calcium (Ca^{2+}), magnesium (Mg^{2+}), and potassium (K^{+})—adhere by electrical attraction to the negatively charged surfaces of clay particles. This adhesion helps prevent these positively charged nutrients from draining away during heavy rain or irrigation. In cation exchange **(Figure 8C)**, root hairs release hydrogen ions (H^{+}) into the soil solution. The hydrogen ions displace cations on the clay particle surfaces, and root hairs can then absorb them.

In contrast to cations, negative ions (anions)—such as nitrate (NO_3^-)—are usually not bound tightly by soil particles. Unbound ions are readily available to plants, but they tend to drain out of the soil quickly due to irrigation or rainfall. If they do, the soil may become deficient in nitrogen.

It may take centuries for a soil to become fertile through the breakdown of rock and the accumulation of organic material. The loss of soil fertility is one of our most pressing environmental problems, as we discuss next.

? How do roots actively increase the availability of mineral nutrients that are cations?

By secreting hydrogen ions, which displace cations from soil particles

Agricultural Research Service/USDA

▲ **Figure 8A** Three soil horizons visible beneath grass

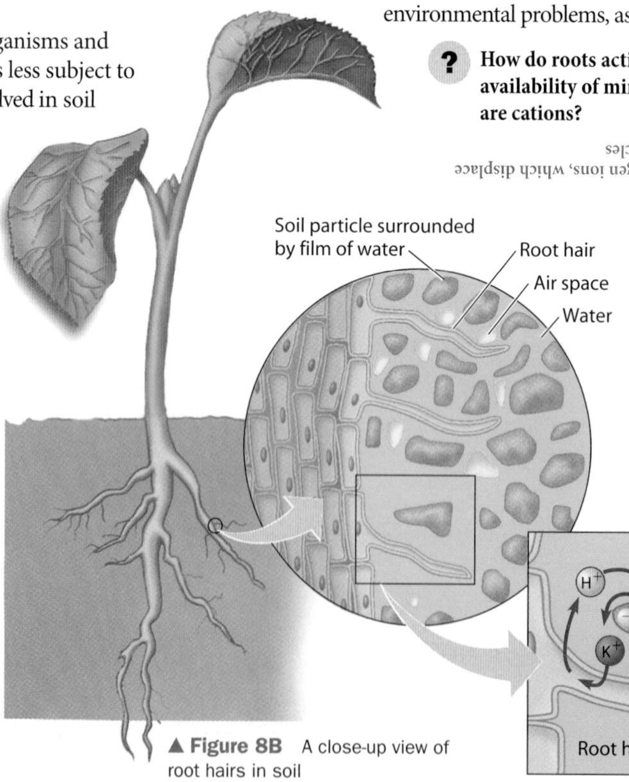

▲ **Figure 8B** A close-up view of root hairs in soil

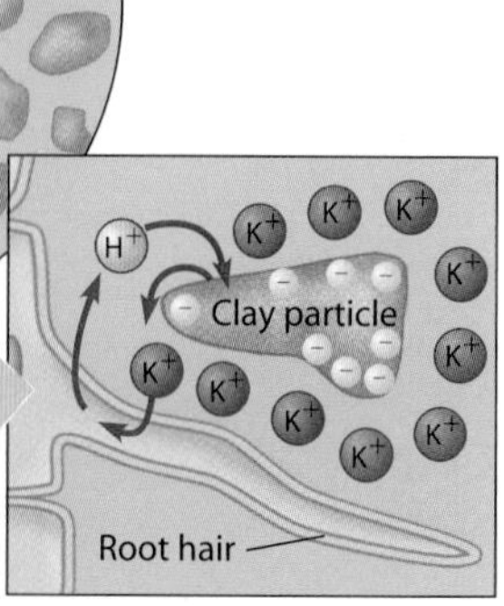

▲ **Figure 8C** Cation exchange

CONNECTION 9 Soil conservation is essential to human life

Our survival as a species depends on soil. But erosion and chemical pollution threaten this vital resource throughout the world. As the human population continues to grow and more land is cultivated, farming practices that conserve soil fertility will become essential to our survival. Three critical goals of soil conservation are proper irrigation, prevention of erosion, and prudent fertilization.

Irrigation can turn a desert into a garden, but farming in dry regions is a huge drain on water resources: Globally, about 75% of all freshwater use is dedicated to crop irrigation. Yet irrigation can be very wasteful, and overextraction of groundwater can cause various environmental problems, such as the sudden appearance of sinkholes **(Figure 9A)**. Instead of wastefully flooding fields, modern irrigation often employs perforated pipes that drip water slowly into the soil close to plant roots. This drip irrigation uses less water, allows the plants to absorb most of the water, and reduces water loss from evaporation and drainage.

Erosion—the blowing or washing away of soil—is a major cause of soil degradation because nutrients are carried away by wind and streams. In the 1930s, the Great Plains of the United States suffered devastating dust storms that swept away huge amounts of topsoil after decades of inappropriate farming techniques and years of drought **(Figure 9B)**. The resulting disaster took a huge toll—both economic and human—that reshaped our country, a plight immortalized in John Steinbeck's novel *The Grapes of Wrath*.

To limit erosion today, farmers plant rows of trees as windbreaks, terrace hillside crops, and cultivate crops in a contour pattern that helps slow runoff of water and topsoil **(Figure 9C)**. Crops such as alfalfa and wheat provide good ground cover and protect the soil better than corn and other crops that are usually planted in more widely spaced rows.

Prehistoric farmers may have started fertilizing their fields after noticing that grass grew faster and greener where animals had defecated. In developed nations today, most farmers use inorganic, commercially produced fertilizers containing minerals that are either mined or prepared by industrial processes. These fertilizers are usually enriched in nitrogen, phosphorus, and potassium, the macronutrients most commonly deficient in farm and garden soils. Manure, fish meal, and compost (decaying plant matter) are common fertilizers that contain decomposing organic material. Before the nutrients in these substances can be used by plants, the organic material must be broken down by bacteria and fungi to inorganic nutrients that roots can absorb.

Whether from natural sources or a chemical factory, the minerals a plant extracts from the soil are in the same form. The difference is that naturally derived fertilizers release nutrients gradually, whereas minerals in inorganic commercial fertilizers are available immediately. However, because minerals from inorganic fertilizers are soluble in water, they may not be retained in the soil for long. Problems arise when fields are overfertilized with inorganic products and excess nutrients are not taken up by plants. The excess minerals are often leached from the soil by rainwater or irrigation. This mineral runoff may pollute groundwater, streams, and lakes.

Agricultural researchers are developing ways to maintain crop yields while reducing the use of costly fertilizer. One approach is to genetically engineer "smart" plants that inform the grower when a nutrient deficiency is imminent but before damage has occurred. One such plant contains a gene that causes leaf cells to produce a blue pigment when the phosphorus content of plant tissues declines. When leaves of these smart plants develop a blue tinge, the farmer knows it is time to add phosphorus-containing fertilizer.

? Why do fertilizers containing organic materials generally contaminate water resources less than inorganic fertilizers?

● Fertilizers containing organic materials release mineral nutrients gradually as they decompose, so there is less likelihood of the minerals leaching into the groundwater or running off into streams and lakes.

U.S. Geological Survey, Denver

▲ **Figure 9A** A sinkhole caused by overuse of groundwater for irrigation

NOAA

▲ **Figure 9B** A dust storm in the American Great Plains during the 1930s

Kevin Horan/Stone/Getty Images

▲ **Figure 9C** Planting to prevent soil erosion in a hilly area

CONNECTION 10 Organic farmers follow principles of sustainable agriculture

If you find tomatoes at a local farmers' market labeled "organic" and also find tomatoes at a giant grocery store chain marked the same way, can you be sure they were both *grown* the same way? What does an "organic" label mean? In the United States, to use the term *organic* or to bear the "USDA Organic" seal, food must be grown and processed according to strict guidelines established and regulated by the U.S. Department of Agriculture (USDA).

Organic farming involves agricultural practices that promote biological diversity by maintaining soil quality through natural methods (rotating crops, planting cover crops, amending the soil with organic matter, and using few or no synthetic fertilizers), providing habitat for predators of pests rather than relying mainly on synthetic pesticides, and avoiding genetically modified organisms. Yearly inspections ensure proper organic farming practices, accurate record keeping, and a buffer of land between organic farms and neighboring conventional farms. The primary goal of organic farming is to achieve **sustainable agriculture**, a system embracing farming methods that are conservation-minded, environmentally safe, and profitable.

Farmers in the U.S. have dedicated over 4 million acres to organic farming (**Figure 10**). Some of this land is managed by small family operations, some by large businesses. The U.S. organic farming industry is growing at a rate of 20% per year, making it one of the fastest-growing segments of agriculture.

The ultimate aim of many organic farmers is to restore as much to the soil as is drawn from it, creating fields that are bountiful and self-sustaining. Many have chosen organic farming to both protect the environment and answer the growing demand for more naturally produced foods. The benefits of organic farming are clear: fewer synthetic chemicals in the environment and less risk of exposing farmworkers and wildlife to potential toxins. And since organic fruits and vegetables are often picked when ripe and sold locally, rather than treated with preservatives and shipped long distances, they can be fresher and better tasting than conventional produce.

But while organic farming is spreading, it hasn't replaced conventional agriculture: Only about 1% of U.S. cropland is certified organic, and only about 2% of the U.S. food supply is grown using organic methods. Furthermore, an organic label is no guarantee of safety or extra health benefits. Scientists disagree, for example, about the nutritional differences, if any, between organic and conventional produce.

▲ **Figure 10** An organic farmer harvesting sweet corn

Ralf-Finn Hestoft/Corbis

Still, organic farmers continue to improve their practices. Some are trying new growing methods that promote greater biological diversity among the plants and wildlife in their fields. Others are looking for better natural fertilizers that increase crop yields. The future of farming, they say, lies in working toward two goals simultaneously: feeding the world's people and promoting a healthy environment.

? If you buy some "organic" apples, does that tell you anything about how they were grown? About how healthy they are?

An organic label indicates that the grower has been certified to be following standards meant to promote long-term agricultural sustainability. The organic designation does not mean that the produce is healthier than conventionally grown produce.

CONNECTION 11 Agricultural research is improving the yields and nutritional values of crops

Eight hundred million people suffer from malnutrition. Every day, 40,000 people, including 16,000 children, die of hunger. This death rate is truly staggering. A person dies of hunger somewhere in the world *about every 2 seconds*. Advocates of plant biotechnology believe that the genetic engineering of crop plants is the key to overcoming the pressing problem of world hunger.

The most limited resource for food production is land. The size of the human population is steadily increasing while the amount of farmland is decreasing. Thus, improving crop yields is a major goal of plant biotechnology.

The commercial adoption by farmers of genetically modified (GM) crops has been one of the most rapid advances in the history of agriculture. These crops include transgenic varieties of cotton and corn that contain genes from the bacterium *Bacillus thuringiensis*. These transgenes encode a protein (*Bt* toxin) that effectively controls a number of serious insect pests. The use of such plant varieties greatly reduces the need for spraying crops with chemical insecticides. Although *Bt* toxin is harmless to humans, its use is controversial due, in part, to concerns about its effects on helpful insects.

▶ **Figure 11** Researchers with high-protein rice

Considerable progress also has been made in the development of transgenic varieties of cotton, corn, soybeans, sugar beets, and wheat that are tolerant to a number of herbicides. The cultivation of these plants may reduce production costs and enables farmers to "weed" crops with herbicides that do not damage the transgenic crop plants. This can reduce tillage, which can cause erosion of soil. Researchers are also engineering plants with enhanced resistance to disease. For example, a transgenic papaya resistant to a ring spot virus saved Hawaii's papaya industry.

The nutritional quality of plants is also being improved. Golden Rice, a transgenic variety with a few daffodil genes that increase synthesis of vitamin A, is being developed. Plant breeding has also resulted in new varieties of corn, wheat, and rice that are enriched in protein (**Figure 11**). Such modified crops may be particularly important because protein deficiency is the most common cause of malnutrition around the world. However, many of these "super" varieties have an extraordinary demand for nitrogen, usually supplied in the form of commercial fertilizer. Unfortunately, these fertilizers are expensive to produce. Thus, the countries that most need high-protein crops are usually the ones least able to afford to grow them.

There is debate about the environmental effects of GM crops. Decisions about developing GM crops should be based on sound science rather than on reflexive fear or blind optimism.

? Why is research on the protein content of crop plants so important to human health worldwide?

● Because the most common form of malnutrition is protein deficiency, and most people in the world get most of their protein from plants.

Plant Nutrition and Symbiosis

12 Most plants depend on bacteria to supply nitrogen

As discussed in Module 7, nitrogen deficiency is the most common nutritional problem in plants. This is puzzling when you consider that the atmosphere is nearly 80% nitrogen. But even though plants are bathed in gaseous nitrogen (N_2), they cannot absorb it directly from the air. To be used by plants, N_2 must be converted to ammonium (NH_4^+) or nitrate (NO_3^-).

Within soil, ammonium and nitrate are produced from atmospheric N_2 or from organic matter by bacteria. As shown in **Figure 12**, certain soil bacteria, called nitrogen-fixing bacteria, convert atmospheric N_2 to ammonia (NH_3), a metabolic process called **nitrogen fixation**. In soil, ammonia picks up an H^+ to form an ammonium ion (NH_4^+). A second group of bacteria, called ammonifying bacteria, adds to the soil's supply of ammonium by decomposing organic matter.

Plant roots can absorb nitrogen as ammonium. However, plants acquire their nitrogen mainly in the form of nitrate (NO_3^-), which is produced in the soil by a third group of soil bacteria called nitrifying bacteria. After nitrate is absorbed by roots, enzymes within plant cells convert the nitrate back to ammonium, which is then incorporated into amino acids. Thus, were it not for soil bacteria, most plants would starve for nitrogen even though they are surrounded by it in huge quantities.

? What is the danger in applying a compound that kills bacteria to the soil around plants?

● The compound might kill soil bacteria that make nitrogen available to plants, causing nitrogen deficiency.

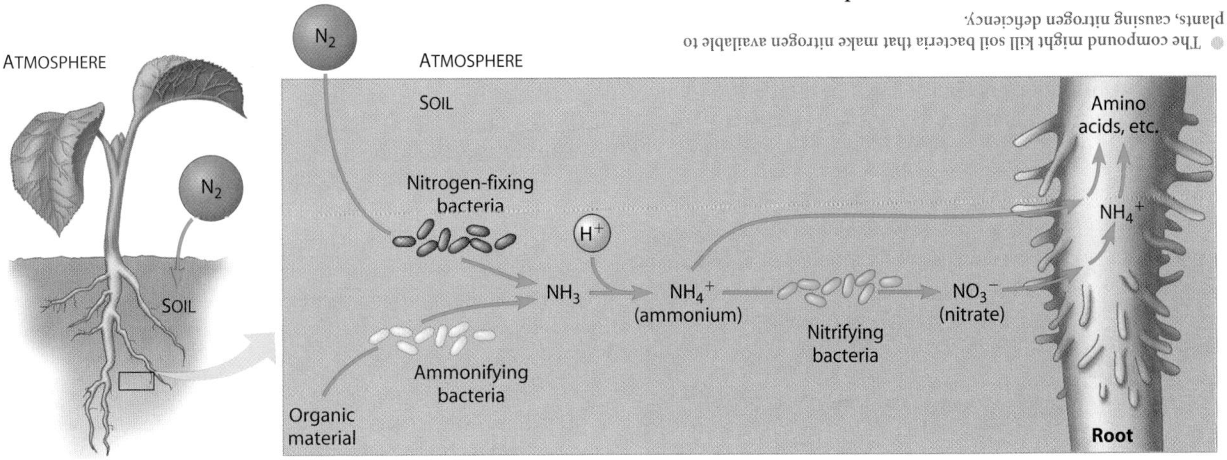

▲ **Figure 12** The roles of bacteria in supplying nitrogen to plants

EVOLUTION CONNECTION

13 Plants have evolved symbiotic relationships that are mutually beneficial

Reliance on the soil for nutrients that may be in short supply makes it imperative that the roots of plants have a large surface area for absorption. As we have seen, root hairs add a great deal of surface to plant roots. About 80% of living plant species gain even more absorptive surface by teaming up with fungi.

The micrograph in **Figure 13A** shows a root of a eucalyptus tree. The root is covered with a twisted mat of fungal filaments. Together, the roots and the fungus form a mutually beneficial relationship called a **mycorrhiza** ("fungus root"). The fungus benefits from a steady supply of sugar supplied by the host plant. In return, the fungus increases the surface area for water uptake and selectively absorbs phosphate and other minerals from the soil and supplies them to the plant. The fungi of mycorrhizae also secrete growth factors that stimulate roots to grow and branch, as well as antibiotics that may protect the plant from pathogens in the soil.

Evolution of mutually beneficial symbiotic associations between roots and fungi was a critical step in the successful colonization of land by plants. Evidence for this is found in the fact that fossilized roots from some of the earliest plants show that they already contained mycorrhizae. When terrestrial ecosystems were young, the soil was probably not very rich in nutrients. The fungi of mycorrhizae, which are more efficient at absorbing minerals than the roots, would have helped nourish the pioneering plants. Even today, the plants that first become established on nutrient-poor soils, such as abandoned farmland or eroded hillsides, are usually heavily colonized with mycorrhizal fungi.

However, roots can be transformed into mycorrhizae only if they are exposed to the appropriate species of fungus. For example, if seeds are collected in one environment and planted in foreign soil, the plants may show signs of malnutrition resulting from the absence of the plants' natural mycorrhizal partners. Farmers may avoid this problem by inoculating seeds with spores of appropriate mycorrhizal fungi.

Plants also form mutually beneficial symbiotic relationships with other organisms besides fungi. Some plant species maintain close association with nitrogen-fixing bacteria. For example, the roots of plants in the legume family—including peas, beans, peanuts, alfalfa, and many other plants that produce their seeds in pods—have swellings called nodules (**Figure 13B**). Within these nodules, plant cells have been "infected" by nitrogen-fixing bacteria of the genus *Rhizobium*, which means "root living." *Rhizobium* bacteria reside in cytoplasmic vesicles formed by the root cell (visible in the inset micrograph). Each legume is associated with a particular strain of *Rhizobium*. Other nitrogen-fixing bacteria are found in the root nodules of some plants that are not legumes, such as alders.

The relationship between a plant and its nitrogen-fixing bacteria is mutually beneficial ("You scratch my back, I'll scratch yours"). The plant provides the bacteria with carbohydrates and other organic compounds. The bacteria have enzymes that catalyze the conversion of atmospheric N_2 to ammonium ions (NH_4^+), a form readily used by the plant. When conditions are favorable, root nodule bacteria fix so much nitrogen that the nodules secrete excess NH_4^+, which increases the fertility of the soil. This is one reason farmers practice crop rotation, one year planting a nonlegume, such as corn, and the next year planting a legume, such as alfalfa. The legume crop may be plowed under so that it will decompose as "green manure," reducing the need for fertilizer.

E. H. Newcomb and S. R. Tandon/Biological Photo Service

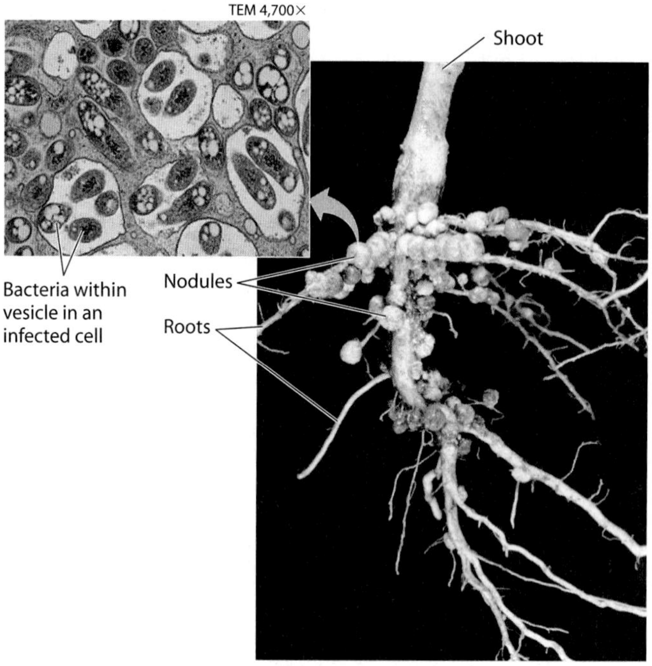

Scimat/Photo Researchers

▲ **Figure 13B** Root nodules on a soybean plant

R. L. Peterson/Biological Photo Service

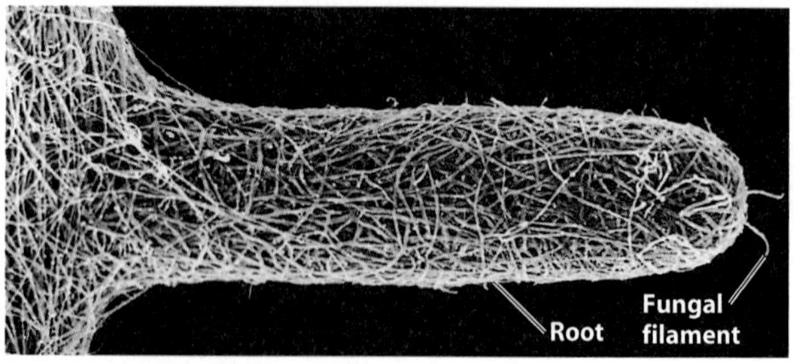

▲ **Figure 13A** Part of a eucalyptus mycorrhiza

? How do the nitrogen-fixing bacteria of root nodules benefit from their symbiotic relationship with plants?

The bacteria receive organic nutrients produced by the plant.

14 The plant kingdom includes epiphytes, parasites, and carnivores

Almost all plant species have mutually beneficial symbiotic associations with soil fungi, bacteria, or both. Though rarer, there are also plant species with nutritional adaptations that take advantage of other organisms. For example, an epiphyte is a plant that grows on another plant, usually anchored to branches or trunks of living trees. Examples of epiphytes include staghorn ferns and orchids, like the one shown in **Figure 14A**. Epiphytes absorb water and minerals from rain.

Clay Perry/Photolibrary

▲ **Figure 14A** An orchid plant, a type of epiphyte, growing on the trunk of a tree

Unlike epiphytes, parasitic plants absorb sugars and minerals from their living hosts. **Figure 14B** shows a parasitic plant called dodder (the yellow-orange threads that are wound around the green plant). Dodder cannot photosynthesize; it obtains organic molecules from other plant species, using roots that tap into the host's vascular tissue.

Kevin Schafer/CorbisS

▲ **Figure 14B** Dodder growing on a pickleweed

Figure 14C shows part of an oak tree parasitized by mistletoe, the plant traditionally tacked above doorways during the Christmas season. All the leaves you see here are mistletoe; the oak has lost its leaves for winter. Mistletoe is photosynthetic, but it supplements its diet by siphoning sap from the vascular tissue of the host tree.

A few plants are carnivores. They grow in acid bogs and other habitats where soils are poor in nitrogen and other minerals. In these acidic soils, organic matter decays so slowly that there is little inorganic nitrogen available for plant roots to take up. Carnivorous plants obtain most of their nitrogen and some minerals by killing and digesting insects and other small animals.

Few species illustrate the correlation of structure and function better than carnivorous plants. The sundew plant has modified leaves, each bearing many club-shaped hairs (**Figures 14D**). A sticky, sugary secretion at the tips of the hairs attracts insects and traps them. The presence of an insect triggers the hairs to bend and the leaf to enfold its prey. The hairs then secrete digestive enzymes, and the plant absorbs nutrients released as the insect is digested.

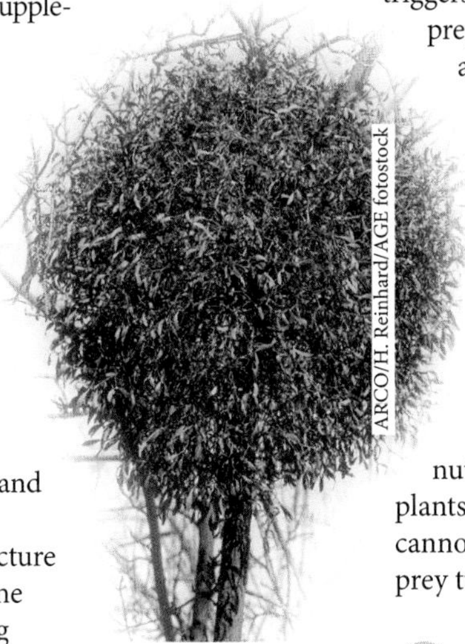
ARCO/H. Reinhard/AGE fotostock

▲ **Figure 14C** Mistletoe growing on an oak tree

Jean-Philippe Delobelle/Photolibrary

▲ **Figure 14D** A sundew plant trapping a winged ant

James H. Robinson/Photo Researchers

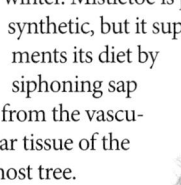

▲ **Figure 14E** A Venus flytrap capturing a fly

The Venus flytrap (**Figure 14E**) has hinged leaves that close around small insects, usually ants and grasshoppers. As insects walk on the insides of these leaves, they touch sensory hairs that trigger closure of the trap. The leaf then secretes digestive enzymes and absorbs nutrients from the prey.

Using insects as a source of nitrogen is a nutritional adaptation that enables carnivorous plants to thrive in soils where most other plants cannot. Fortunately for animals, such predator-prey turnabouts are rare!

? Carnivorous plants are most common in locales where the soil is deficient in _____ and _____.

● nitrogen . . . minerals

REVIEW

For Practice Quizzes, BioFlix, MP3 Tutors, and Activities, go to www.masteringbiology.com.

Reviewing the Concepts

The Uptake and Transport of Plant Nutrients (1–5)

1 Plants acquire nutrients from air, water, and soil. As a plant grows, its roots absorb water, minerals (inorganic ions), and some O_2 from the soil. Its leaves take in carbon dioxide from the air.

2 The plasma membranes of root cells control solute uptake. Root hairs greatly increase a root's absorptive surface. Water and solutes can move through the root's epidermis and cortex by going either through cells or between them. However, all water and solutes must pass through the selectively permeable plasma membranes of cells of the endodermis to enter the xylem (water-conducting tissue) for transport upward.

3 Transpiration pulls water up xylem vessels. Transpiration can move xylem sap, consisting of water and dissolved inorganic nutrients, to the top of the tallest tree.

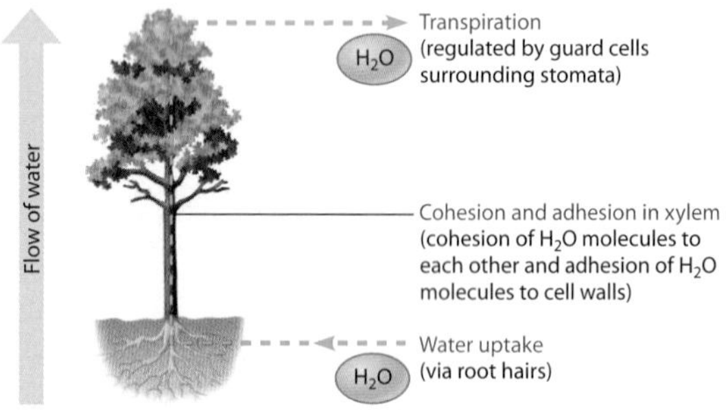

4 Guard cells control transpiration. By changing shape, guard cells generally keep stomata open during the day (allowing transpiration) but closed at night (preventing excess water loss).

5 Phloem transports sugars. By a pressure flow mechanism, phloem transports food molecules made by photosynthesis. At a sugar source, sugar is loaded into a phloem tube. The sugar raises the solute concentration in the tube, and water follows, raising the pressure in the tube. As sugar is removed at a sugar sink, water follows. The increase in pressure at the sugar source and decrease at the sugar sink cause phloem sap to flow from source to sink.

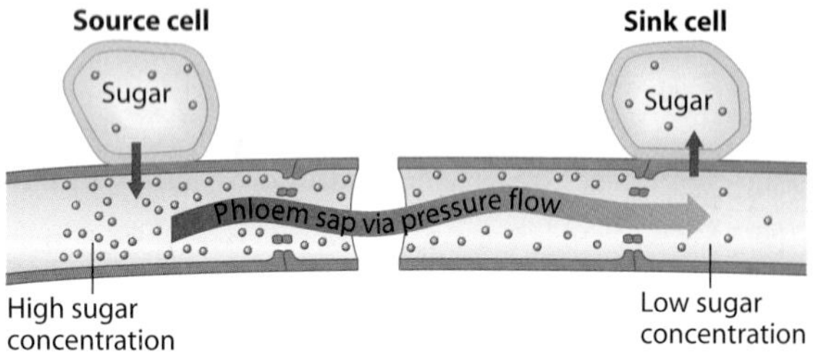

Plant Nutrients and the Soil (6–11)

6 Plant health depends on a complete diet of essential inorganic nutrients. A plant must obtain usable sources of the chemical elements—inorganic nutrients—it requires from its surroundings. Macronutrients, such as carbon and nitrogen, are needed in large amounts, mostly to build organic molecules. Micronutrients, including iron and zinc, act mainly as cofactors of enzymes.

7 Fertilizers can help prevent nutrient deficiencies. Nutrient deficiencies can often be recognized and then fixed by using appropriate fertilizers.

8 Fertile soil supports plant growth. Fertile soil contains a mixture of small rock and clay particles that hold water and ions and also allow O_2 to diffuse into plant roots. Humus (decaying organic material) provides nutrients and supports the growth of organisms that enhance soil fertility. Anions (negatively charged ions), such as nitrate (NO_3^-), are readily available to plants because they are not bound to soil particles. However, anions tend to drain out of soil rapidly. Cations (positively charged ions), such as K^+, adhere to soil particles. In cation exchange, root hairs release H^+, which displaces cations from soil particles; the root hairs then absorb the free cations.

9 Soil conservation is essential to human life. Water-conserving irrigation, erosion control, and the prudent use of herbicides and fertilizers are aspects of good soil management.

10 Organic farmers follow principles of sustainable agriculture. To earn the certified organic designation, food must be grown and processed following a strict set of guidelines.

11 Agricultural research is improving the yields and nutritional values of crops. Through traditional plant breeding and DNA technologies, researchers are developing new varieties of crop plants with improved yields and nutritional value.

Plant Nutrition and Symbiosis (12–14)

12 Most plants depend on bacteria to supply nitrogen. Relationships with other organisms help plants obtain nutrients. Bacteria in the soil convert atmospheric N_2 to forms that can be used by plants.

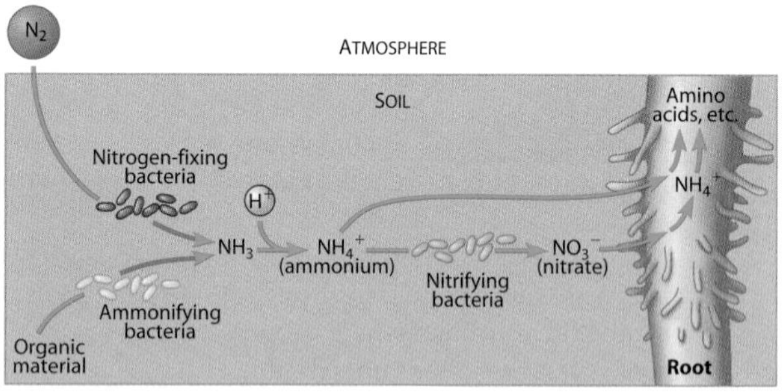

13 Plants have evolved symbiotic relationships that are mutually beneficial. Many plants form mycorrhizae, mutually beneficial associations between roots and fungi. A network of fungal threads increases a plant's absorption of nutrients and water, and the fungus receives some nutrients from the plant. Legumes and certain other plants have nodules in their roots that house nitrogen-fixing bacteria.

14 The plant kingdom includes epiphytes, parasites, and carnivores. Epiphytes are plants that grow on other plants. Parasitic plants siphon sap from host plants. Carnivorous plants can obtain nitrogen by digesting insects.

Connecting the Concepts

1. Fill in the blanks in this concept map to help you tie together key concepts concerning transport in plants.

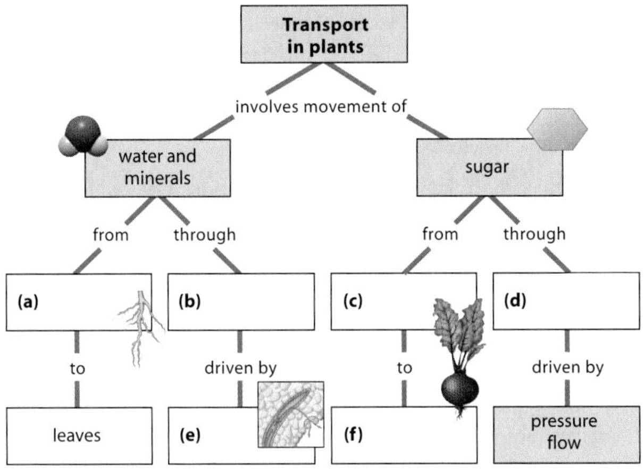

Testing Your Knowledge

Multiple Choice

2. Plants require the smallest amount of which of the following nutrients?
   a. oxygen
   b. phosphorus
   c. carbon
   d. iron
   e. hydrogen

3. Which of the following activities of soil bacteria does *not* contribute to creating usable nitrogen supplies for plant use?
   a. the fixation of atmospheric nitrogen
   b. the conversion of ammonium ions to nitrate ions
   c. the decomposition of dead animals
   d. the assembly of amino acids into proteins
   e. the generation of ammonium from proteins in dead leaves

4. By trapping insects, carnivorous plants obtain ______, which they need ______. (*Pick the best answer.*)
   a. water . . . because they live in dry soil
   b. nitrogen . . . to make sugar
   c. phosphorus . . . to make protein
   d. sugars . . . because they can't make enough by photosynthesis
   e. nitrogen . . . to make protein

5. A major long-term problem resulting from flood irrigation is the
   a. drowning of crop plants.
   b. accumulation of salts in the soil.
   c. erosion of fine soil particles.
   d. encroachment of water-consuming weeds.
   e. excessive cooling of the soil.

Describing, Comparing, and Explaining

6. Explain how guard cells limit water loss from a plant on a hot, dry day. How can this be harmful to the plant?

Applying the Concepts

7. Acid rain contains an excess of hydrogen ions (H^+). One effect of acid rain is to deplete the soil of plant nutrients such as calcium (Ca^{2+}), potassium (K^+), and magnesium (Mg^{2+}). Offer a hypothesis to explain why acid rain washes these nutrients from the soil. How might you test your hypothesis?

8. In some situations, the application of nitrogen fertilizer to crops has to be increased each year because the fertilizer decreases the rate of nitrogen fixation in the soil. Propose a hypothesis to explain this phenomenon. Describe a test for your hypothesis. What results would you expect from your test?

9. Transpiration is fastest when humidity is low and temperature is high, but in some plants it seems to increase in response to light as well. During one 12-hour period when cloud cover and light intensity varied frequently, a scientist studying a certain crop plant recorded the data in the table below. (The transpiration rates are grams of water per square meter of leaf area per hour.)

Time (hr)	Temperature (°C)	Humidity (%)	Light (% of full sun)	Transpiration Rate ($g/m^2 \cdot hr$)
8 AM	14	88	22	57
9	14	82	27	72
10	21	86	58	83
11	26	78	35	125
12 PM	27	78	88	161
1	33	65	75	199
2	31	61	50	186
3	30	70	24	107
4	29	69	50	137
5	22	75	45	87
6	18	80	24	78
7	13	91	8	45

Do these data support the hypothesis that the plants transpire more when the light is more intense? If so, is the effect independent of temperature and humidity? Explain your answer. (*Hint*: Look for overall trends in each column, and then compare pairs of data within each column and between columns.)

10. Agriculture is by far the biggest user of water in arid western states, including Colorado, Arizona, and California. The populations of these states are growing, and there is an ongoing conflict between cities and farm regions over water. To ensure water supplies for urban growth, cities are purchasing water rights from farmers. This is often the least expensive way for a city to obtain more water, and some farmers can make more money selling water than growing crops. Discuss the possible consequences of this trend. Is this the best way to allocate water for all concerned? Why or why not?

Review Answers

1. a. roots; b. xylem; c. sugar source; d. phloem; e. transpiration; f. sugar sink
2. d 3. d 4. e 5. b
6. If the plant starts to dry out, K^+ is pumped out of the guard cells. Water follows by osmosis, the guard cells become flaccid, and the stomata close. This prevents wilting, but it keeps leaves from taking in carbon dioxide, which is needed for photosynthesis.
7. Hypothesis: The hydrogen ions in acid precipitation displace positively charged nutrient ions from negatively charged clay particles. Test: In the laboratory, place equal amounts and types of soil in separate filters. The pore size of the filter must not allow any undissolved soil particles to pass through. Spray (to simulate rain) soil samples in the filters with solutions of different pH (for example, pH 5, 6, 7, 8, 9). Determine the concentration of nutrient ions in the solutions. (The only variable in the solutions should be the hydrogen ion concentration. Ideally, the solutions would contain no dissolved nutrient ions.) Collect fluid that drips through soil samples and filters. Determine the hydrogen ion concentration and the nutrient ion concentration in each sample of fluid. Prediction: If the hypothesis is correct, the fluid collected from the soil samples exposed to pH lower than 5.6 (acid rain) will contain the highest concentration of positively charged nutrient ions.
8. Hypothesis: When fixed nitrogen levels increase to a certain level in the soil, it slows the metabolism of (or otherwise harms or kills) the nitrogen-fixing bacteria that provide usable nitrogen to the crops. Test: Expose cultures of nitrogen-fixing bacteria (symbiotic and nonsymbiotic ones found in soil) to solutions of different concentrations of fixed nitrogen (that is, NO_3^- and NH_4^+). Determine the concentrations of nitrogen-fixing enzymes produced by the surviving bacteria in each sample. Prediction: If your hypothesis is correct and the fixed nitrogen concentration is high enough to cause harm in some of the samples, you would expect the enzyme concentration to be measurably lower in samples whose fixed nitrogen concentration is above the level that causes harm.
9. The hypothesis is supported if transpiration varies with light intensity when humidity and temperature are about the same. These conditions are seen at two places in the table; at hours 11 and 12, recordings for temperature and humidity are about the same, but light intensity increased markedly from 11 to 12, as did the transpiration rate. The recordings made at hours 3 and 4 show the same effects. Also, the recordings made at hours 1 and 2 generally support the hypothesis. Here, both temperature and humidity decreased, so you might expect the transpiration rate to stay about the same or perhaps increase because the temperature decrease is small; however, the transpiration rate dropped, as did the light intensity.
10. Some issues and questions to consider: How were the farmers assigned or sold "rights" to the water? How is the price established when a farmer buys or sells water rights? Is there enough water for everyone who "owns" it? What kinds of crops are these farmers growing? What will the water be used for in the city? Are there other users with no rights, such as wildlife? Is any effort being made to curb urban growth and conserve water? Should millions of people be living in what is essentially a desert? What are the reasons for farming desert land?

Control Systems in Plants

Control Systems in Plants

BIG IDEAS

Kristin Piljay

Plant Hormones
(1–8)

Chemical signals help regulate many important processes within a plant's body.

Martin Shields/Photo Researchers

Responses to Stimuli
(9–13)

Plants respond to environmental stimuli in a variety of ways.

Brian Leatart/Getty Images

Soy is everywhere. Walk the aisles of your local supermarket, and chances are you can find soy sauce (made from fermented soybeans), soy milk (ground soybeans mixed with water and flavorings), tofu (cooked pureed soybeans), and miso (fermented soybean paste used for seasoning). Soy offers a number of dietary benefits. Soy protein contains all the essential amino acids, and it reduces the risk of heart disease. Soy is rich in antioxidants and fiber, is low in fat, and reduces levels of LDL ("bad cholesterol") and triglycerides while maintaining HDL ("good cholesterol").

Soybeans also contain phytoestrogens, a class of plant hormones. Like humans, plants use hormones as chemical signals that control growth and development. Phytoestrogens are chemically similar to human estrogen. In menopausal women, phytoestrogens may help reduce the negative effects of lower estrogen production, such as the risk of osteoporosis. However, the benefits and risks of supplementary phytoestrogens are not fully known. For example, moderate levels of estrogen relieve menopausal symptoms, ward off heart disease, and sustain bone mass, but high levels appear to increase the risk of breast and ovarian cancers. It is hard to assess the risks of dietary supplements because they are not subject to federal regulation.

Whatever their health benefits to us, we do know that phytoestrogens and other plant hormones are crucial to the life of plants. In this chapter, we explore the diverse roles of plant hormones: how they affect plant growth, movement, flowering, fruit development, and even defense.

Plant Hormones

1 Experiments on how plants turn toward light led to the discovery of a plant hormone

▲ **Figure 1A** A houseplant growing toward light

The shoots of a houseplant sitting on a windowsill grow toward light (**Figure 1A**). If you rotate the plant, it soon reorients its growth until the leaves again face the window. Any growth response that results in plant organs curving toward or away from stimuli is called a **tropism** (from the Greek *tropos*, turn).

The growth of a shoot in response to light is called **phototropism**. Phototropism has an obvious evolutionary advantage, directing growing seedlings and the shoots of mature plants toward the sunlight that drives photosynthesis.

Microscopic observations of plants growing toward light reveal the cellular mechanism that underlies phototropism. **Figure 1B** shows a grass seedling curving toward light that comes from one side. As the enlargement shows, cells on the darker side of the seedling are larger—actually, they have elongated faster—than those on the brighter side. The different cellular growth rates made the shoot bend toward the light. If a seedling is illuminated uniformly from all sides or if it is kept in the dark, the cells all elongate at a similar rate and so the seedling grows straight upward.

Studies of plant responses to light led to the first evidence of a plant hormone. A **hormone** is a chemical signal produced in one part of the body and transported to other parts, where it acts on target cells to change their functioning, such as growth rate. Like animals, plants also have hormones that regulate body activities. Some key discoveries about plant hormones emerged from a series of classic experiments conducted by two scientists with a very famous name.

Showing That Light Is Detected by the Shoot Tip In the late 1800s, Charles Darwin and his son Francis conducted some of the earliest experiments on phototropism. They observed that grass seedlings could bend toward light only if the tips of their shoots were present. The first five grass plants in **Figure 1C** summarize the Darwins' findings. ❶ When they removed the tip of a grass shoot, the shoot did not curve toward the light. ❷ The shoot also remained straight when the Darwins placed an opaque cap on its tip. ❸ However, the shoot curved normally when they placed a transparent cap on its tip or ❹ an opaque shield around its base. The Darwins concluded that the tip of the shoot was responsible for sensing light. They also recognized that the growth response, the bending of the shoot, occurs in cells that are below the tip. Therefore, they speculated that some signal must be transmitted from the tip downward to the growth region of the shoot.

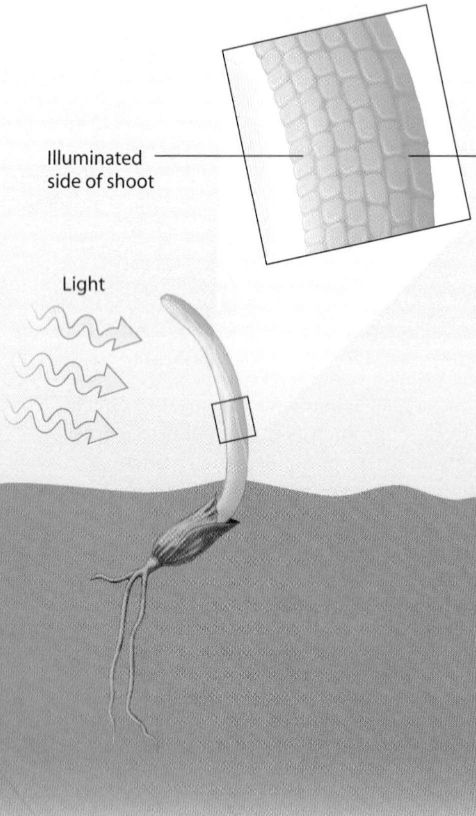

▲ **Figure 1B** Phototropism in a grass seedling

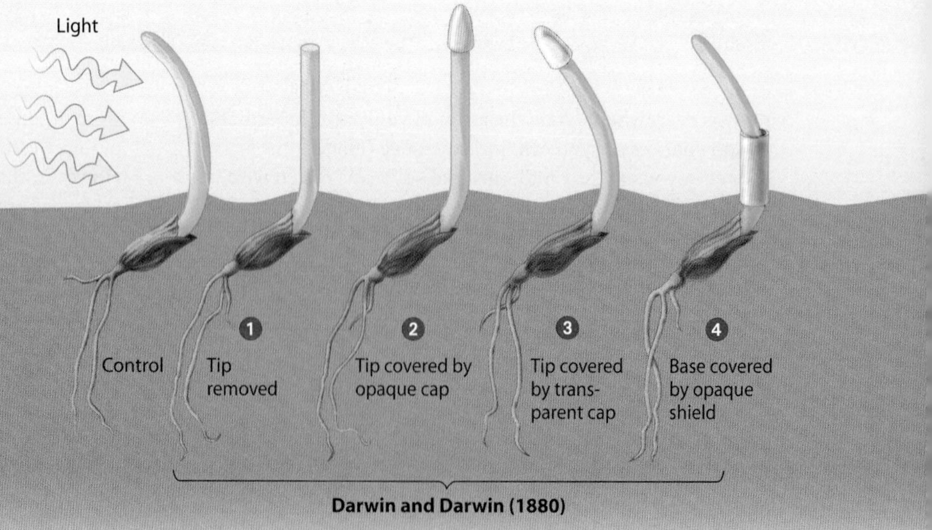

▲ **Figure 1C** Early experiments on phototropism: detection of light by shoot tips and evidence for a chemical signal

A few decades later, Danish plant biologist Peter Boysen-Jensen further tested the chemical signal idea of the Darwins. ❺ In one group of seedlings, Boysen-Jensen inserted a block of gelatin between the tip and the lower part of the shoot. The gelatin prevented cellular contact but allowed chemicals to diffuse through. The seedlings with gelatin blocks behaved normally, bending toward light. ❻ In a second set of seedlings, Boysen-Jensen inserted a thin piece of the mineral mica under the shoot tip. Mica is an impermeable barrier, and the seedlings with mica had no phototropic response. These experiments supported the hypothesis that the signal for phototropism is a mobile chemical.

Isolating the Chemical Signal In 1926, Frits Went, a Dutch graduate student, modified Boysen-Jensen's techniques and extracted the chemical messenger for phototropism in grasses. As shown in **Figure 1D**, Went first removed the tips of grass seedlings and placed them on blocks of agar, a gelatin-like material. He reasoned that the chemical messenger (pink in the figure) from the shoot tips should diffuse into the agar and that the blocks should then be able to substitute for the shoot tips. Went tested the effects of the agar blocks on tipless seedlings, which he kept in the dark to eliminate the effect of light. ❶ First, he centered the treated agar blocks on the cut tips of a batch of seedlings. These plants grew straight upward. They also grew faster than the decapitated control seedlings, which hardly grew at all. Went concluded that the agar had absorbed the chemical messenger produced in the shoot tip and that the chemical had passed into the shoot and stimulated it to grow. ❷ He then placed agar blocks off center on another batch of tipless seedlings. These plants bent away from the side with the chemical-laden agar block, as though growing toward light. ❸ Control seedlings with blank agar blocks (whether offset or not) grew no more than the control. Went concluded that the agar block contained a chemical produced in the shoot tip, that this chemical stimulated growth as it passed down the shoot, and that a shoot curved toward light because of a higher concentration of the growth-promoting chemical on the darker side. Went called this chemical messenger, or hormone, auxin. In the 1930s, biochemists determined the chemical structure of Went's auxin.

The classical hypothesis for what causes grass shoots to grow toward light, based on these early experiments, is that an uneven distribution of auxin moving down from the shoot tip causes cells on the darker side to elongate faster than cells on the brighter side. Studies with plants other than grass shoots, however, do not always support this hypothesis. For example, there is no evidence that light from one side causes an uneven distribution of auxin in the stems of sunflowers or other eudicots. There is, however, a greater concentration of substances that may act as growth inhibitors on the lighted side of a stem.

The discovery of auxin stimulated discoveries of a wide variety of plant hormones. The next six modules will introduce some of the major plant hormones and the roles they play in the lives of plants.

? How do the experiments illustrated in Figures 1C and 1D provide evidence that phototropism depends on a chemical signal?

Light is detected by the shoot tip, but the bending response occurs farther down the shoot. The fact that the signal can pass through a barrier that prevents cell contact but allows chemicals to pass suggests that the signal is a chemical.

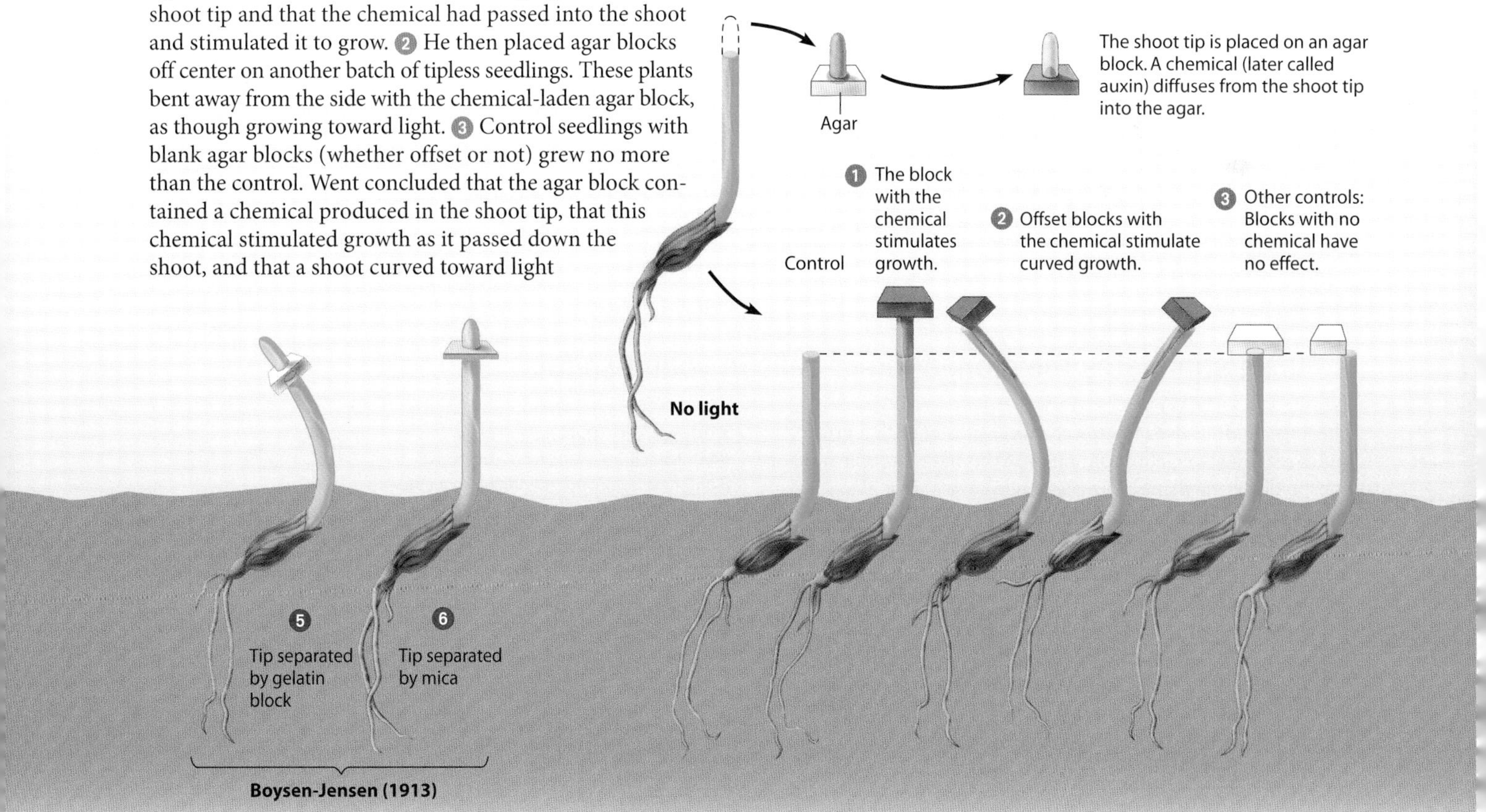
No light
Agar
The shoot tip is placed on an agar block. A chemical (later called auxin) diffuses from the shoot tip into the agar.
Control
1 The block with the chemical stimulates growth.
2 Offset blocks with the chemical stimulate curved growth.
3 Other controls: Blocks with no chemical have no effect.
5 Tip separated by gelatin block
6 Tip separated by mica
Boysen-Jensen (1913)

▲ **Figure 1D** Went's experiments: isolation of the chemical signal

2 Five major types of hormones regulate plant growth and development

Plant hormones are produced in very low concentrations, but a tiny amount of hormone can have a profound effect on growth and development. The binding of a hormone to a specific cell-surface receptor triggers a signal transduction pathway that amplifies the hormonal signal and leads to one or more responses within the cell. In general, plant hormones control whole-body activities such as growth and development by affecting the division, elongation, and differentiation of cells.

Plant biologists have identified five major types of plant hormones, previewed in **Table 2.** There are other hormones important to plants that we will not discuss in this chapter, and some of the "hormones" listed in the table actually represent a group of related hormones; such cases will be mentioned in the text. Notice that all five types of hormones stimulate or inhibit cell divison and elongation, thereby influencing growth.

As Table 2 indicates, each hormone has multiple effects, depending on its site of action, its concentration, and the developmental stage of the plant. In most situations, no single hormone acts alone. Instead, it is usually the balance of several hormones—their relative concentrations—that controls the growth and development of a plant. These interactions will become apparent as we survey the five major types of plant hormones and their functions.

? Hormones elicit cellular responses by binding to cell-surface receptors and triggering ______.

signal transduction pathways

TABLE 2 | MAJOR TYPES OF PLANT HORMONES

Hormone (Module)	Major Functions	Where Produced or Found in the Plant
Auxins (3)	Stimulate stem elongation; affect root growth, differentiation, branching, development of fruit, apical dominance, phototropism, and gravitropism (response to gravity)	Meristems of apical buds, young leaves, embryos within seeds
Cytokinins (4)	Affect root growth and differentiation; stimulate cell division and growth; stimulate germination; delay aging	Made in the roots and transported to other organs
Gibberellins (5)	Promote seed germination, bud development, stem elongation, and leaf growth; stimulate flowering and fruit development; affect root growth and differentiation	Meristems of apical buds and roots, young leaves, embryos
Abscisic acid (ABA) (6)	Inhibits growth; closes stomata during water stress; helps maintain dormancy	Leaves, stems, roots, green fruits
Ethylene (7)	Promotes fruit ripening; opposes some auxin effects; promotes or inhibits growth and development of roots, leaves, and flowers, depending on species	Ripening fruits, nodes of stems, aging leaves and flowers

3 Auxin stimulates the elongation of cells in young shoots

The term **auxin** is used for any chemical substance that promotes seedling elongation, although auxins have multiple functions in flowering plants. The major natural auxin in plants is indoleacetic acid, or IAA, although several other compounds, including some synthetic ones, have auxin activity. When we use the term *auxin* in this text, however, we are referring specifically to IAA.

Figure 3A shows the effect of auxin on a mustard plant called *Arabidopsis thaliana*. The plant on the right is a wild-type *Arabidopsis*. The plant on the left contains a mutation in the auxin gene. The result of this mutation is a lack of auxin protein, which in turn results in stunted growth.

The apical meristem at the tip of a shoot is a major site of auxin synthesis. As auxin moves downward, diffusing from cell to cell, it stimulates growth of the stem by making

▶ **Figure 3A** The effect of auxin (IAA): comparing a wild-type *Arabidopsis* plant (right) with one that underproduces auxin (left)

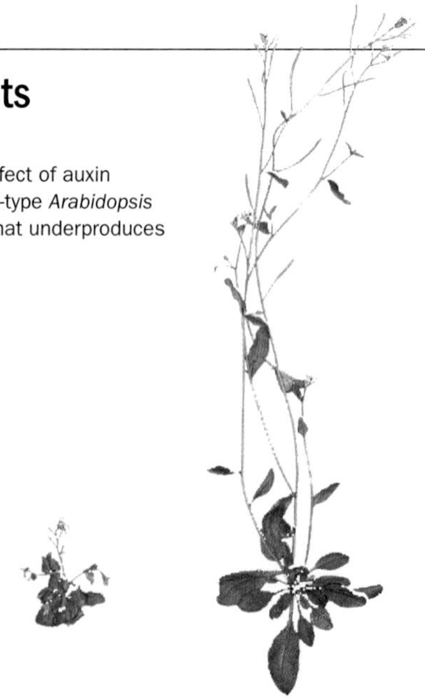

William M. Gray

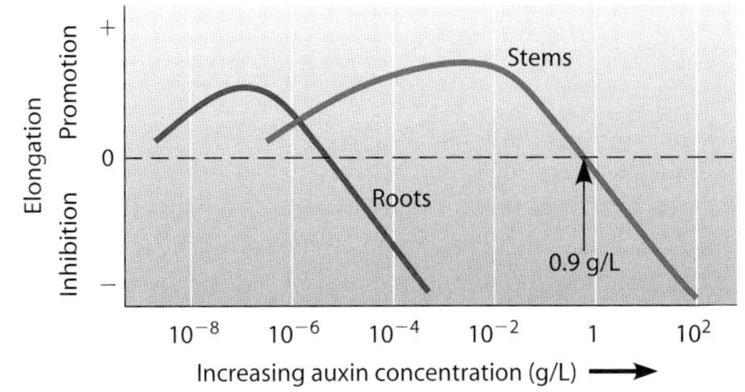

▲ **Figure 3B** The effect of auxin concentration on cell elongation

cells elongate. As the blue curve in **Figure 3B** shows, auxin promotes cell elongation in stems only within a certain concentration range. Above a certain level (0.9 g of auxin per liter of solution, in this case), it usually inhibits cell elongation in stems, probably by inducing the production of ethylene, a hormone that generally counters the effects of auxin (see Module 7).

The red curve on the graph shows the effect of auxin on root growth. An auxin concentration too low to stimulate shoot cells will cause root cells to elongate. On the other hand, an auxin concentration high enough to make stem cells elongate is in the concentration range that inhibits root cell elongation. These effects of auxin on cell elongation reinforce two points: (1) the same chemical messenger may have different effects at different concentrations in one target cell, and (2) a given concentration of the hormone may have different effects on different target cells. In fact, some herbicides (including the infamous Vietnam War defoliant Agent Orange) contain synthetic auxins at high concentrations that kill plants through hormonal overdose.

How does auxin make plant cells elongate? One hypothesis is that auxin initiates elongation by weakening cell walls. As shown in **Figure 3C**, ❶ auxin may stimulate certain proteins in a plant cell's plasma membrane to pump hydrogen ions into the cell wall, lowering the pH within the wall. The low pH ❷ activates enzymes that separate cross-linking molecules from cellulose microfibrils in the wall. ❸ The cross-linking molecules are now more exposed to enzymes that loosen the cell wall. ❹ The cell then swells with water and elongates because its weakened wall no longer resists the cell's tendency to take up water via osmosis. After this initial elongation caused by the uptake of water, the cell sustains the growth by synthesizing more cell wall material and cytoplasm. These processes are also stimulated by auxin.

Auxin produces a number of other effects in addition to stimulating cell elongation and causing stems and roots to lengthen. For example, it induces cell division in the vascular cambium, thus promoting growth in stem diameter. Also, auxin produced by developing seeds promotes the growth of fruit. Sometimes commercial fruit growers need to add synthetic auxins. For instance, greenhouse tomatoes typically have fewer seeds than tomatoes grown outdoors, resulting in poorly developed fruits. Therefore, growers spray synthetic auxins on tomato vines to induce normal fruit development, making it possible to grow commercial tomatoes in greenhouses.

? **Suppose you had a tiny pH electrode that could measure the pH of a plant cell's wall. How could you use it to test the hypothesis in Figure 3C for how auxin stimulates cell elongation?**

The hypothesis predicts that addition of auxin to the cell should lower the pH of the cell wall (make it more acidic). You could test this prediction with your pH electrode by measuring the cell wall pH in the presence of (experimental group) or absence of (control group) auxin.

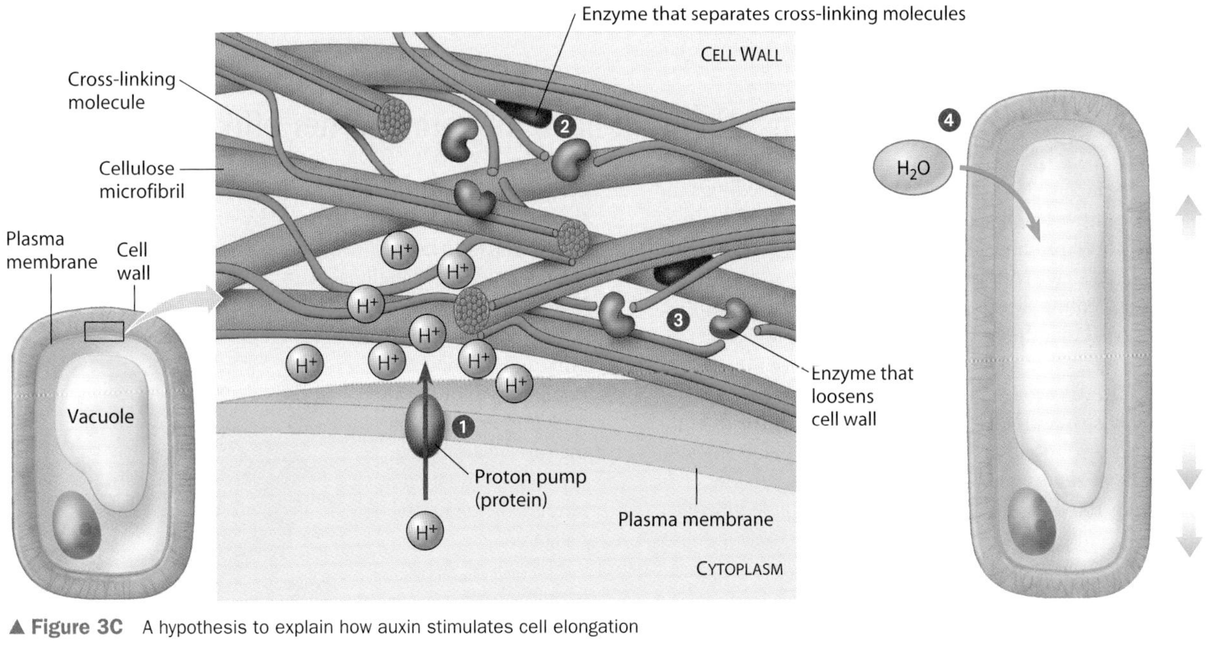

▲ **Figure 3C** A hypothesis to explain how auxin stimulates cell elongation

4 Cytokinins stimulate cell division

Cytokinins are a group of closely related hormones that promote cytokinesis, or cell division. A number of cytokinins have been extracted from plants; the most common is zeatin, so named because it was first discovered in corn (*Zea mays*). Also, several synthetic cytokinins have been manufactured. Natural cytokinins are produced in actively growing tissues, particularly in roots, embryos, and fruits. Cytokinins made in the roots reach target tissues in stems by moving upward in xylem sap.

Plant biologists have found that cytokinins enhance the division, growth, and development of plant cells grown in culture. Cytokinins also retard the aging of flowers and leaves. Thus, florists use cytokinin sprays to keep cut flowers fresh.

Cytokinins and auxin interact in the control of apical dominance, the ability of the terminal bud of a shoot to suppress the growth of the axillary buds. **Figure 4** shows the results of a simple experiment that demonstrates this effect. Both basil plants pictured are the same age. The one on the left has an intact terminal bud on the main shoot; the one on the right had the terminal bud of the main shoot removed several weeks earlier. In the plant on the left, apical dominance resulted in lengthwise growth but inhibited growth of the axillary buds (the buds that produce side branches). As a result, the shoot grew in height but did not branch out to the sides very much. In the plant on the right, the lack of a terminal bud on the main shoot resulted in the activation of the axillary buds, making the plant grow more branches and become bushy.

◀ **Figure 4** Apical dominance in a basil plant resulting from the action of auxin and cytokinins

The leading hypothesis to explain the hormonal regulation of apical dominance—known as the direct inhibition hypothesis—proposes that auxin and cytokinin act antagonistically (with the action of each one opposing the action of the other) in regulating axillary bud growth. According to this view, auxin transported down the shoot from the terminal bud directly inhibits axillary buds from growing, causing a shoot to lengthen at the expense of lateral branching. Meanwhile, cytokinins entering the shoot system from roots counter the action of auxin by signaling axillary buds to begin growing. Thus, the ratio of auxin to cytokinin is viewed as the critical factor in controlling axillary bud inhibition. Many observations and experiments are consistent with the direct inhibition hypothesis. It now appears, however, that the effects of auxin are partially indirect. Recent research suggests that the flow of auxin down the shoot triggers the synthesis of a newly-discovered plant hormone that represses bud growth. Thus, the direct inhibition hypothesis does not account for all experimental findings, and it is likely that plant biologists have not uncovered all the pieces of this puzzle. The role of plant hormones in apical dominance remains an area of active research.

? **According to the direct inhibition hypothesis, the status of axillary buds—dormant or growing—depends on the relative concentrations of _______, which inhibits axillary bud growth, and _______ moving up from the roots, which stimulate axillary bud growth.**

auxin . . . cytokinins

5 Gibberellins affect stem elongation and have numerous other effects

Farmers in Asia have long noticed that some rice seedlings in their paddies grew so tall and spindly that they toppled over before they could produce grain. In the 1920s, Japanese scientists discovered that a fungus of the genus *Gibberella* caused this "foolish seedling disease." By the 1930s, Japanese scientists had determined that the fungus produced hyperelongation of rice stems by secreting a chemical, which was given the name **gibberellin**. In the 1950s, researchers discovered that gibberellin exists naturally in plants, where it is a growth regulator. Foolish seedling disease occurs when rice plants infected with the *Gibberella* fungus get an overdose of gibberellin.

More than 100 different gibberellins have been identified in plants. Roots and young leaves are major sites of gibberellin production. One of the main effects of gibberellins is to stimulate cell elongation and cell division in stems and leaves. This action generally enhances that of auxin and can be demonstrated when certain dwarf plant varieties (such as the dwarf pea plant variety studied by Mendel grow to normal height after treatment with gibberellins (**Figure 5A**).

Gibberellin-induced stem elongation can also cause bolting, the rapid growth of a floral stalk (**Figure 5B**). Bolting can frustrate gardeners because it often renders the plant unpalatable.

▶ **Figure 5A** Reversing dwarfism in pea plants with gibberellins

In many plants, both auxin and gibberellins must be present for fruit to develop. The most important commercial application of gibberellins is in the production of the Thompson variety of seedless grapes. Gibberellins make the grapes grow farther apart in a cluster (which increases air flow and decreases rotting) and produces larger grapes, a trait desired by most shoppers (**Figure 5C**).

Gibberellins are also important in seed germination in many plants. Many seeds that require special environmental conditions to germinate, such as exposure to light or cold temperatures, will germinate when sprayed with gibberellins. In nature, gibberellins in seeds are probably the link between environmental cues and the metabolic processes that renew growth of the embryo. For example, when water becomes available to a grass seed, it causes the embryo in the seed to release gibberellins, which promote germination by mobilizing nutrients stored within the seed. In some plants, gibberellins seem to be interacting antagonistically with another hormone, abscisic acid, which we discuss next.

Kristin Piljay

◀ **Figure 5B** A parsley plant bolting, a result of too much gibberellin

Fred Jensen

▶ **Figure 5C** Gibberellin-treated grapes (left) and untreated grapes (right)

? A gibberellin deficiency probably caused the dwarf variety of pea plants studied by Gregor Mendel. Given the role of gibberellin as a growth hormone, why does it make sense that a heterozygote (with one copy of the normal gene and one copy of the mutant gene) would not be dwarf?

Because minute amounts of a hormone are usually enough to elicit a strong response, a heterozygous plant with one functioning copy of the gene would still produce enough gibberellin to result in normal growth.

6 Abscisic acid inhibits many plant processes

In the 1960s, one research group studying bud dormancy and another team investigating leaf abscission (the dropping of autumn leaves) isolated the same compound, **abscisic acid (ABA)**. Ironically, ABA is no longer thought to play a primary role in either bud dormancy or leaf abscission (for which it was named), but it is a plant hormone of great importance in other functions. In contrast to the growth-stimulating hormones we have studied so far—auxin, cytokinins, and gibberellins—ABA *slows* growth. ABA often counteracts the actions of growth hormones; it is the ratio of ABA to one or more growth hormones that often determines the final outcome.

Seed dormancy (a period of inactivity of the seed) is an evolutionary adaptation that ensures that a seed will germinate only when there are favorable conditions of light, temperature, and moisture. What prevents a seed dispersed in autumn from germinating immediately, only to be killed by winter? For that matter, what prevents seeds from germinating in the dark, moist interior of the fruit? ABA is the answer. Levels of ABA may increase 100-fold during seed maturation. The high levels of ABA in maturing seeds inhibit germination. Many types of dormant seeds therefore only germinate when ABA is removed or inactivated in some way. For example, some seeds require prolonged exposure to cold to trigger ABA inactivation. In these plants, the breakdown of ABA in the winter is required for seed germination in the spring. The seeds of some desert plants remain dormant in parched soil until a downpour washes ABA out of the seeds, allowing them to germinate. For example, the California poppies in **Figure 6** grew from seeds that germinated just after a hard rain.

As we saw in the previous module, gibberellins promote seed germination. For many plants, the ratio of ABA to gibberellins determines whether the seed will remain dormant or germinate.

In addition to its role in dormancy, ABA is the primary internal signal that enables plants to withstand drought. When a plant begins to wilt, ABA accumulates in its leaves and causes stomata to close rapidly. This closing of stomata reduces transpiration and prevents further water loss. In some cases, water shortage can stress the root system before the shoot system. ABA transported from roots to leaves may function as an "early warning system."

? Which two hormones regulate seed dormancy and germination? What are their opposing effects?

Abscisic acid maintains seed dormancy; Gibberellins promote germination.

Design Pics Inc./Alamy

▼ **Figure 6** The Mojave Desert in California blooming after a rain

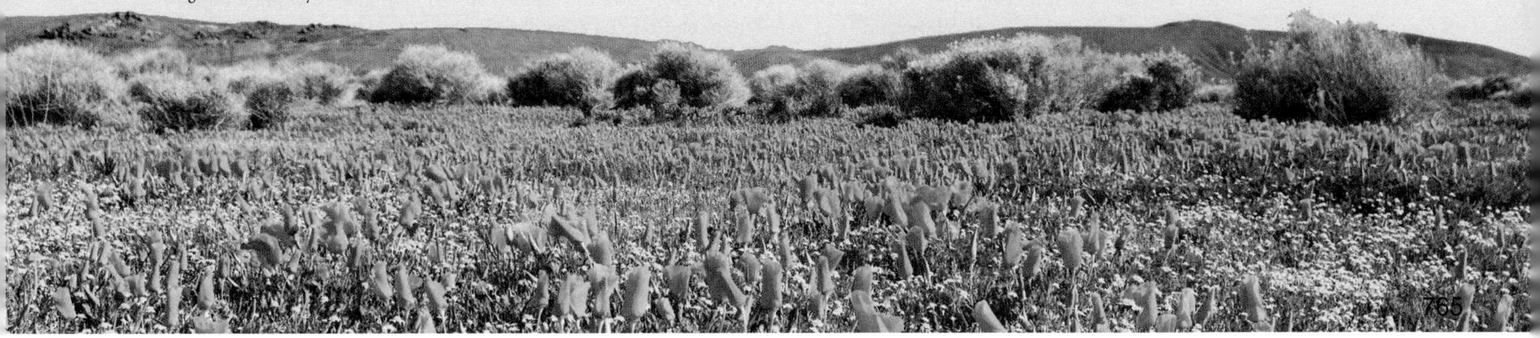

7 Ethylene triggers fruit ripening and other aging processes

During the 1800s, leakage from gas streetlights caused the leaves on nearby trees to drop prematurely. In 1901, scientists demonstrated that this effect was due to the gas **ethylene**, a by-product of coal combustion. We now know that plants produce their own ethylene, which functions as a hormone that triggers a variety of aging responses, including fruit ripening and programmed cell death. Ethylene is also produced in response to stresses such as drought, flooding, injury, and infection.

Fruit Ripening A burst of ethylene production in a fruit triggers its ripening. In an example of positive feedback, ethylene triggers ripening, which then triggers more ethylene production, resulting in a surge of released gas. And because ethylene is a gas, the signal to ripen can spread from fruit to fruit: One bad apple really can spoil the bunch! The ripening process includes the enzymatic breakdown of cell walls, which softens the fruit, and the conversion of starches and acids to sugars, which makes the fruit sweet. The production of new scents and colors attracts animals, which eat the fruits and disperse the seeds.

You can make some fruits ripen faster if you store them in a bag so that the ethylene gas accumulates. **Figure 7A** shows two unripe bananas that were stored for the same time period in bags under different conditions: alone and with an ethylene-releasing peach. As you can see, the ethylene released by the ripening peach resulted in a riper (darker) banana. On a commercial scale, many kinds of fruit—tomatoes, for instance—are often picked green and then ripened in huge storage bins into which ethylene gas is piped.

Kristin Piljay

▲ **Figure 7A** The effect of ethylene on the ripening of bananas

In other cases, growers take measures to slow down the ripening action of natural ethylene. Stored apples are often flushed with CO_2, which inhibits ethylene synthesis. Also, air is circulated around the apples to prevent ethylene from accumulating. In this way, apples picked in autumn can be stored for sale the following summer.

The Falling of Leaves Like fruit ripening, the changes that occur in deciduous trees each autumn—color changes, drying, and the loss of leaves—are also aging processes. Leaves lose their green color because chlorophyll is broken down during autumn. Fall colors result from a combination of new red pigments made in autumn and the exposure of yellow and orange pigments that were already present in the leaf but masked by dark green chlorophyll. Autumn leaf drop is an adaptation that helps keep the tree from drying out in winter. Without its leaves, a tree loses less water by evaporation when its roots cannot take up water from the frozen ground. Before leaves fall, many essential elements are salvaged from them and stored in the stem, where they can be recycled into new leaves the following spring.

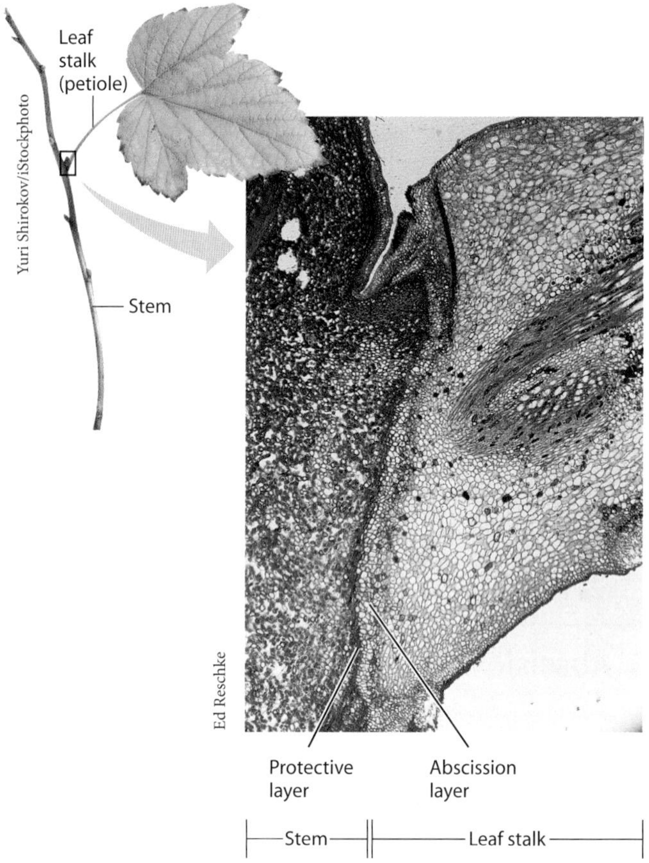

▲ **Figure 7B** The abscission layer at the base of a leaf

When an autumn leaf falls, the base of the leaf stalk separates from the stem. The separation region is called the abscission layer. As indicated in **Figure 7B**, the abscission layer consists of a narrow band of cells with thin walls that are further weakened when enzymes digest the cell walls. The leaf drops off when its weight, often helped by wind, splits the abscission layer apart. Notice the layer of cells next to the abscission layer. Even before the leaf falls, these cells form a leaf scar on the stem. Dead cells covering the scar help protect the plant from infectious organisms.

Leaf drop is triggered by environmental stimuli, including the shortening days of autumn and, to a lesser extent, cooler temperatures. These stimuli apparently cause a change in the balance of ethylene and auxin. The auxin prevents abscission and helps maintain the leaf's metabolism, but as a leaf ages, it produces less auxin. Meanwhile, cells begin producing ethylene, which stimulates formation of the abscission layer. The ethylene primes the abscission layer to split by promoting the synthesis of enzymes that digest cell walls in the layer.

We have now completed our survey of the five major types of plant hormones. Before moving on to the topic of plant behavior, let's look at some agricultural uses of these chemical regulators.

? Plant biologists sometimes refer to ethylene as the "aging hormone." What is the basis for this term?

Many of ethylene's functions, including fruit ripening and leaf abscission, are associated with aging-like changes in cells.

CONNECTION

8 Plant hormones have many agricultural uses

Rudolf Madár/Shutterstock

Although a lot remains to be learned about plant hormones, much of what we do know has a direct application to agriculture. As already mentioned, the control of fruit ripening and the production of seedless fruits are two of several major uses of these chemicals. Plant hormones also enable farmers to control when plants will drop their fruit. For instance, synthetic auxins are often used to prevent orange and grapefruit trees from dropping their fruit before they can be picked. **Figure 8** shows a fruit grower spraying a grove with auxins. The quantity of auxin must be carefully monitored because too much of the hormone may stimulate the plant to release more ethylene, making the fruit ripen and drop off sooner. Indeed, large doses of auxins are sometimes used to *promote* premature fruit drop. With apple and olive trees, for example, auxin may be sprayed on some fruit to cause them to drop prematurely; the remaining fruit will grow larger. Ethylene is used similarly on peaches and plums, and it is sometimes sprayed on berries, grapes, and cherries to loosen the fruit so it can be picked by machines.

In combination with auxin, gibberellins are used to produce seedless fruits, as mentioned in Module 5. Sprayed on other kinds of plants, at an earlier stage, gibberellins can have the opposite effect: the *promotion* of seed production. A large dose of gibberellins will induce many biennial plants, such as carrots, beets, and cabbage, to flower and produce seeds during their first year of growth, rather than in their second year as is normally the case.

Stockbyte

▲ **Figure 8** Using auxins to prevent early fruit drop

Research on plant hormones has also been used to develop herbicides. One of the most widely used weed killers is the synthetic auxin 2,4-D, which disrupts the normal balance of hormones that regulate plant growth. Monocots can rapidly inactivate this herbicide, but eudicots cannot, so 2,4-D can be used to selectively remove dandelions and other broadleaf eudicot weeds from a lawn or grainfield. By applying herbicides to cropland, a farmer can reduce the amount of tillage required to control weeds, thus reducing soil erosion, fuel consumption, and labor costs.

Modern agriculture relies heavily on the use of synthetic chemicals. Without chemically synthesized herbicides to control weeds and synthetic plant hormones to help grow and preserve fruits, less food would be produced, and food prices could increase. At the same time, there is growing concern that the heavy use of certain artificial chemicals in food production may pose environmental and health hazards. One of these chemicals is dioxin, a by-product of 2,4-D synthesis. Though 2,4-D itself does not appear to be toxic to mammals, dioxin is a serious hazard when it leaks into the environment, causing birth defects, liver disease, and leukemia in laboratory animals. Also, many consumers are concerned that foods produced using artificial hormones may not be as tasty or nutritious as those raised naturally. At present, however, organic foods are relatively expensive to produce. The issue of how our food should be grown involves both economics and ethics: Should we continue to produce cheap, plentiful food using artificial chemicals that may cause problems, or should we put more of our agricultural effort into farming without these potentially harmful substances, recognizing that foods may be less plentiful and more expensive as a result?

? What is the main commercial incentive to treat products with plant hormones? What behavior of consumers helps drive this use of hormone sprays in agriculture?

The need to compete in the market by increasing production and lowering cost; shopping for the lowest price for produce

Responses to Stimuli

9 Tropisms orient plant growth toward or away from environmental stimuli

Having surveyed the hormones that carry signals within a plant, we now shift our focus to the responses of plants to physical stimuli from the environment. A plant cannot migrate to water or a sunny spot, and a seed cannot maneuver itself into an upright position if it lands upside down in the soil. Because of their immobility, plants must respond to environmental stimuli through developmental and physiological mechanisms. Tropisms are directed growth responses that cause parts of a plant to grow toward a stimulus (a positive tropism) or away from a stimulus (a negative tropism). In Module 1, we discussed positive phototropism, the growth of a plant shoot toward light. Two other types of tropisms are gravitropism, a response to gravity, and thigmotropism, a response to touch.

Response to Light As we saw in Module 1, the mechanism for phototropism is a greater rate of cell elongation on the darker side of a stem. In grass seedlings, the signal linking the light stimulus to the cell elongation response is auxin. Researchers have shown that illuminating a grass shoot from one side causes auxin to migrate across the shoot tip from the bright side to the dark side. The shoot tip contains a protein pigment that detects the light and somehow passes the "message" to molecules that affect auxin transport. (We discuss protein light receptors in Module 12.)

Response to Gravity **Gravitropism,** the directional growth of a plant in response to gravity, is responsible for the fact that no matter how a seed lands on the ground, shoots grow upward (negative gravitropism) and roots grown downward (positive gravitropism). The corn seedlings in **Figure 9A** were both germinated in the dark. The one on the left was left untouched; notice that its shoot grew straight up and its root straight down. The seedling on the right was germinated in the same way, but two days later it was turned on its side so that the shoot and root were horizontal. By the time the photo was taken, the shoot had turned back upward, exhibiting a negative response to gravity, and the root had turned down, exhibiting positive gravitropism.

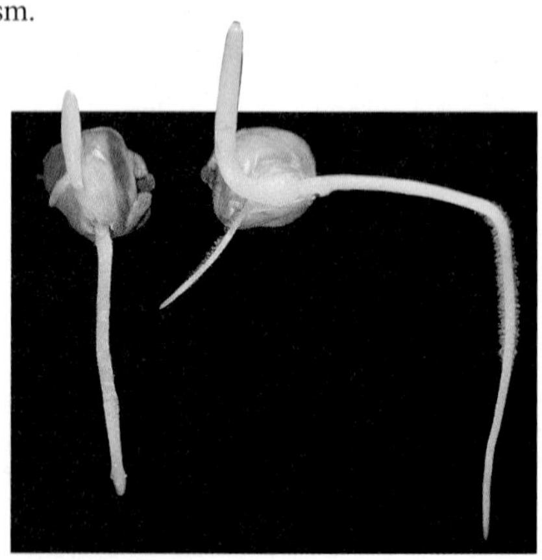
Michael Evans

▶ **Figure 9A** Gravitropism in a corn seedling

Martin Shields/Photo Researchers

▲ **Figure 9B** The "sensitive plant" *Mimosa pudica*

One hypothesis for how plants tell up from down is that gravity pulls special organelles containing dense starch grains to the low points of cells. The uneven distribution of organelles may in turn signal the cells to redistribute auxin. This effect has been documented in roots. A higher auxin concentration on the lower side of a root inhibits cell elongation (see the red line in Figure 3B). As cells on the upper side continue to elongate, the root curves downward. This tropism continues until the root is growing straight down.

Response to Touch **Thigmotropism**, directional growth in response to touch, is illustrated when the tendril of a pea plant (which is actually a modified leaf) contacts a string or wire and coils around it for support. Tendrils grow straight until they touch an object. Contact then stimulates the cells to grow at different rates on opposite sides of the tendril (slower in the contact area), making the tendril coil around the support. Most climbing plants have tendrils that respond by coiling and grasping when they touch rigid objects. Thigmotropism enables these plants to use such objects for support while growing toward sunlight.

Some plants show remarkable abilities to respond to touch. When stimulated by contact, the "sensitive plant" (*Mimosa pudica*) rapidly folds its leaflets together (**Figure 9B**). This response, which takes only a second or two, results from a rapid loss of turgor by cells within the joints of the leaf. It takes about 10 minutes for the cells to regain their turgor and restore the open form of the leaf.

Tropisms all have one function in common: They help plant growth stay in tune with the environment. In the next module, we see that plants also have a way of keeping time with their environment.

? **Why are tropisms called "growth responses"?**

Because the movement of a plant organ toward or away from an environmental stimulus generally takes place by growing. An organ bends when cells on one side grow faster than cells on the other side, and it elongates in one direction when cells grow evenly.

10 Plants have internal clocks

Your pulse rate, blood pressure, body temperature, rate of cell division, blood cell counts, alertness, urine composition, metabolic rate, sex drive, and responsiveness to medications all fluctuate rhythmically with the time of day. Plants also display rhythmic behavior; examples include the opening and closing of stomata and the "sleep movements" of many species that fold their leaves or flowers in the evening and unfold them in the morning. Some of these cyclic variations continue even under artificially constant conditions, implying that plants have a built-in ability to sense time.

Innate Biological Rhythms An innate biological cycle of about 24 hours is called a **circadian rhythm** (from the Latin *circa*, about, and *dies*, day). A circadian rhythm persists even when an organism is sheltered from environmental cues. A bean plant, for example, exhibits sleep movements at about the same intervals even if kept in constant light or darkness. Thus, circadian rhythms occur with or without external stimuli such as sunrise and sunset. Research on a variety of organisms indicates that circadian rhythms are controlled by internal timekeepers known as **biological clocks**.

Although a biological clock continues to mark time in the absence of environmental cues, it requires daily signals from the environment to remain tuned to a period of *exactly* 24 hours. This is because innate circadian rhythms generally differ somewhat from a 24-hour period. Consider bean plants, for instance. As shown in **Figure 10**, the leaves of a bean plant are normally horizontal at noon and folded downward at midnight. When the plant is held in darkness, however, its sleep movements change to a cycle of about 26 hours.

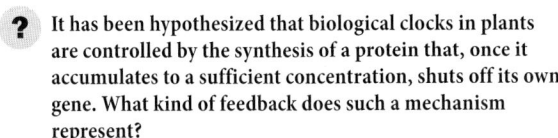

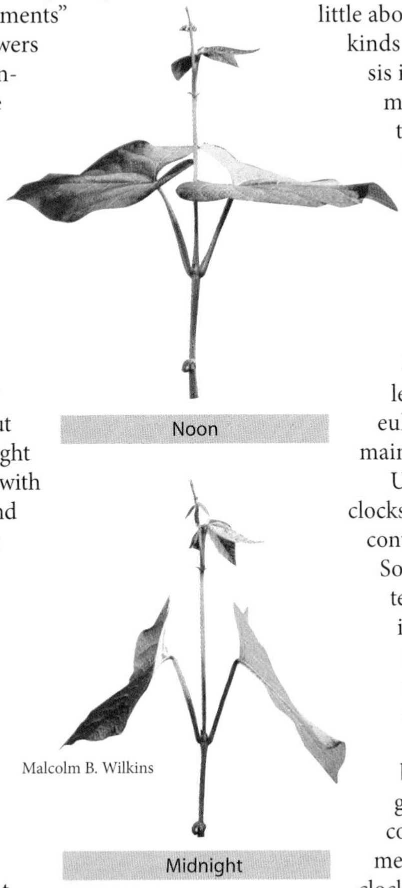

▲ **Figure 10** Sleep movements of a bean plant

The light/dark cycle of day and night provides the cues that usually keep biological clocks precisely synchronized with the outside world. But a biological clock cannot immediately adjust to a sudden major change in the light/dark cycle. We observe this problem ourselves when we cross several time zones in an airplane: When we reach our destination, we have "jet lag"; our internal clock is not synchronized with the clock on the wall. Moving a plant across several time zones produces a similar lag. The plant will, for example, display leaf movements that are synchronized to the clock in its original location. In the case of either the plant or the human traveler, resetting the clock usually takes several days.

The Nature of Biological Clocks Just what is a biological clock? Researchers are actively investigating this question. In humans and other mammals, the clock is located within a cluster of nerve cells in the hypothalamus of the brain. But for most other organisms, including plants, we know little about where the clocks are located or what kinds of cells are involved. A leading hypothesis is that biological timekeeping in plants may depend on the synthesis of a protein that regulates its own production through feedback control. After the protein accumulates to a sufficient concentration, it turns off its own gene. When the concentration of the protein falls, transcription restarts. The result would be a cycling of the protein's concentration over a roughly 24-hour period—a clock! Some research indicates that such a molecular mechanism may be common to all eukaryotes. However, much research remains to be done in this area.

Unlike most metabolic processes, biological clocks and the circadian rhythms they control are affected little by temperature. Somehow, a biological clock compensates for temperature shifts. This adjustment is essential, for a clock that speeds up or slows down with the rise and fall of outside temperature would be an unreliable timepiece.

In attempting to answer questions about biological clocks, it is essential to distinguish between the clock and the processes it controls. You could think of the sleep movements of leaves as the "hands" of a biological clock, but they are not the essence of the clockwork itself. You can restrain the leaves of a bean plant for several hours so that they cannot move. But on release, they will rush to the position appropriate for the time of day. Thus, we can interfere with an organism's rhythmic activity, but its biological clock goes right on ticking. In the next module, we'll learn more about the interface between a plant's internal biological clock and the external environment.

? It has been hypothesized that biological clocks in plants are controlled by the synthesis of a protein that, once it accumulates to a sufficient concentration, shuts off its own gene. What kind of feedback does such a mechanism represent?

Negative feedback, where the result of a process (in this case, a protein) shuts off that process (the transcription and translation of the gene)

11 Plants mark the seasons by measuring photoperiod

A biological clock not only times a plant's everyday activities, but may also influence seasonal events that are important in a plant's life cycle. Flowering, seed germination, and the onset and ending of dormancy are all stages in plant development that usually occur at specific times of the year. The environmental stimulus plants most often use to detect the time of year is called **photoperiod**, the relative lengths of day and night.

Plants whose flowering is triggered by photoperiod fall into two groups. One group, the **short-day plants**, generally flower in late summer, fall, or winter, when light periods shorten. Chrysanthemums and poinsettias are examples of short-day plants. In contrast, **long-day plants**, such as spinach, lettuce, iris, and many cereal grains, usually flower in late spring or early summer, when light periods lengthen. Spinach, for instance, flowers only when daylight lasts at least 14 hours. Some plants, such as dandelions, tomatoes, and rice, are day-neutral; they flower when they reach a certain stage of maturity, regardless of day length.

In the 1940s, researchers discovered that flowering and other responses to photoperiod are actually controlled by the length of continuous *darkness*, not the length of continuous daylight. That is, "short-day" plants will flower only if it stays dark long enough, and "long-day" plants will flower only if the dark period is short enough. Therefore, it would seem more appropriate to call the two types "long-night" plants and "short-night" plants. However, the day-length terms are embedded firmly in the literature of plant biology and so will be used here.

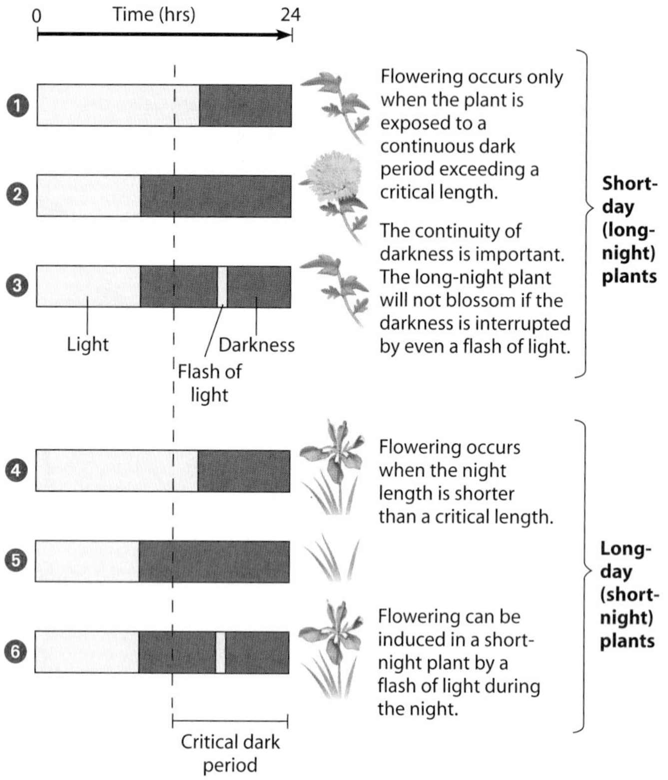

▲ **Figure 11** Photoperiodic control of flowering

Figure 11 illustrates the evidence for the night-length effect and shows the difference between the flowering response of a short-day plant and a long-day plant. The top part of the figure represents short-day plants. Notice that a short-day plant will not flower ❶ until it is exposed to a *continuous* dark period ❷ exceeding a critical length (about 14 hours, in this case). The continuity of darkness is important. The short-day plant will not blossom if the nighttime part of the photoperiod is interrupted ❸ by even a brief flash of light. There is no effect if the daytime portion of the photoperiod is broken by a brief exposure to darkness.

Florists apply this information about short-day plants to bring us flowers out of season. Chrysanthemums, for instance, are short-day plants that normally bloom in the autumn, but their blooming can be stalled until Mother's Day in May by punctuating each long night with a flash of light, thus turning one long night into two short nights. Easter lilies are also forced by growers to bloom at one particular time of year, which may vary by as much as five weeks from year to year, due to the relationship of Easter to the lunar calendar.

The bottom part of the figure demonstrates the effect of night length on a long-day plant. In this case, flowering occurs when the night length is *shorter* ❹ than a critical length (less than 10 hours, in this example). A dark interval that is too long ❺ will prevent flowering. Additionally, flowering can be induced in a long-day plant by a flash of light ❻ during the night.

Every species of plant has a critical night length, but how that critical night length affects flowering varies with the type of plant. In short-day plants, the critical night length is the *minimum* number of hours of darkness required for flowering; less darkness prevents flowering. In long-day plants, this critical night length is the *maximum* number of hours of darkness required for flowering; less darkness promotes flowering.

? A particular short-day plant won't flower in the spring. Suppose a short dark interruption splits the long-light period of spring into two short-light periods. What result do you predict?

The plants still won't flower because it is actually night length, not day length, that counts in the photoperiodic control of flowering.

12 Phytochromes are light detectors that may help set the biological clock

The discovery that photoperiod (specifically night length) determines the seasonal responses of plants leads to another question: How does a plant actually measure photoperiod? Much remains to be learned, but photoreceptive pigments called phytochromes are part of the answer. **Phytochromes** are proteins with a light-absorbing component. Because the light absorbed is at the red end of the spectrum, while light in the blue/green range is reflected, the molecules appear blue or bluish green.

Phytochromes were discovered during studies on how different wavelengths of light affect seed germination. Red light, with a wavelength of 660 nm, was found to be most effective at increasing germination. Light of a longer wavelength, called far-red light (730 nm, near the edge of human visibility), both inhibited germination and reversed the effect of red light.

How do phytochromes respond differently to different wavelengths of light? The key to this ability is that a phytochrome molecule changes back and forth between two forms that differ slightly in structure. One form, known as P_r, absorbs red light, and the other, known as P_{fr}, absorbs far-red light. As diagrammed in **Figure 12A**, when the P_r form absorbs red light (660 nm), it is quickly converted to P_{fr}, and when P_{fr} absorbs far-red light (730 nm), it is slowly converted back to P_r.

Each night, new phytochrome molecules are synthesized only in the P_r form. Thus, molecules of P_r slowly accumulate in the continuous darkness that follows sunset. After sunrise, the red wavelengths of sunlight cause much of the phytochrome to be rapidly converted from the P_r form to P_{fr}. Sunlight contains both red light and far-red light, but the conversion to P_{fr} is faster than the conversion to P_r. Therefore, the ratio of P_{fr} to P_r increases in the sunlight. It is this sudden increase in P_{fr} each day at dawn that resets a plant's biological clock. Interactions between phytochrome and the biological clock enable plants to measure the passage of night and day. In doing so, the clock monitors photoperiod and, when detecting seasonal changes in day and night length, cues responses such as seed germination, flowering, and the beginning and ending of bud dormancy.

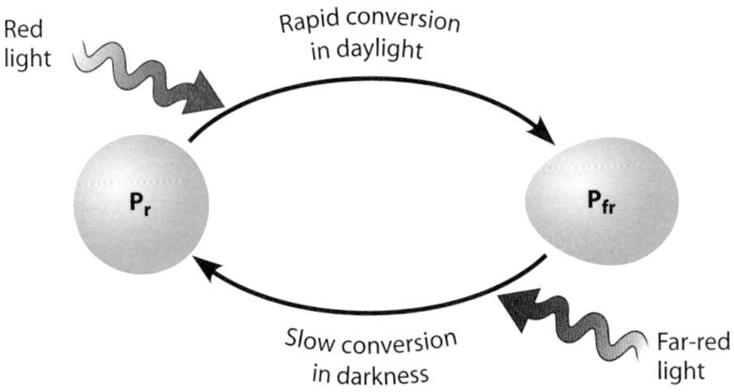

▲ **Figure 12A** Interconversion of the two forms of phytochrome

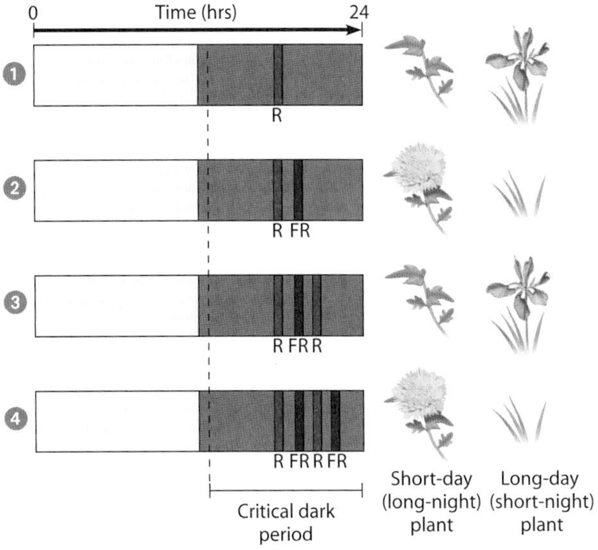

▲ **Figure 12B** The reversible effects of red and far-red light

The consequences of this phytochrome switch are shown in **Figure 12B**. Bar 1 shows the results we saw in the previous module for both short-day and long-day plants that receive a flash of light during their critical dark period. The letter R on the light flash stands for red light. The other three bars show how flashes of far-red (FR) light affect flowering. Bar 2 reveals that the effect of a flash of red light that interrupts a period of darkness can be reversed by a subsequent flash of far-red light: Both types of plants behave as though there is no interruption in the night length. Bars 3 and 4 indicate that no matter how many flashes of red or far-red light a plant receives, only the wavelength of the *last* flash of light affects the plant's measurement of night length.

Plants also have a group of blue-light photoreceptors that control such light-sensitive plant responses as phototropism and the opening of stomata at daybreak. Light is an especially important environmental factor in the lives of plants, and diverse receptors and signaling pathways have evolved that mediate a plant's responses to light.

? How do phytochrome molecules help the plant recognize dawn each day?

Phytochrome molecules are mainly in the P_r form during the night. Dawn is signaled by the sudden conversion of P_r to P_{fr} due to the absorption of the red wavelengths of sunlight.

EVOLUTION CONNECTION

13 Defenses against herbivores and infectious microbes have evolved in plants

Plants are at the base of most food webs and are subject to attack by a wide range of **herbivores** (animals that eat mainly plants). Also, some pathogens can damage or kill plants. However, a wide variety of defenses, many regulated by hormones, have evolved in plants.

Defenses Against Herbivores Plants counter herbivores with physical defenses, such as thorns, and chemical defenses, such as distasteful or toxic compounds. For example, some plants produce an amino acid called canavanine. Canavanine resembles arginine, one of the 20 amino acids normally used to make proteins. If an insect eats a plant containing canavanine, the molecule is incorporated into the insect's proteins in place of arginine. Because canavanine is different enough from arginine to change the shape and function of proteins, the insect is harmed. Some plants even recruit predatory animals that help defend the plants against certain herbivores. For example, insects called parasitoid wasps can kill caterpillars that feed on plants (**Figure 13A**). ❶ When a caterpillar bites into the plant, the combination of physical damage to the plant and a chemical in the caterpillar's saliva triggers ❷ a signal transduction pathway within the plant cells. The pathway leads to a specific cellular response: ❸ the synthesis and release of gases that ❹ attract the wasp. ❺ The wasp injects its eggs into the caterpillar. When the eggs hatch, the wasp larvae eat their way out of the caterpillar, killing it.

Defenses Against Pathogens A plant, like an animal, is subject to infection by pathogenic microbes: viruses, bacteria, and fungi. A plant's first line of defense against infection is the physical barrier of the plant's "skin," the epidermis. However, microbes can cross this barrier through wounds or through natural openings such as stomata. Once infected, the plant uses chemicals as a second line of defense. Plant cells damaged by

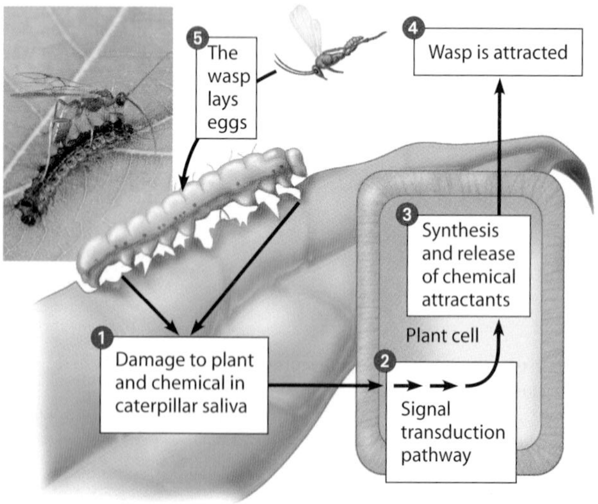

▲ **Figure 13A** Recruitment of a wasp in response to an herbivore

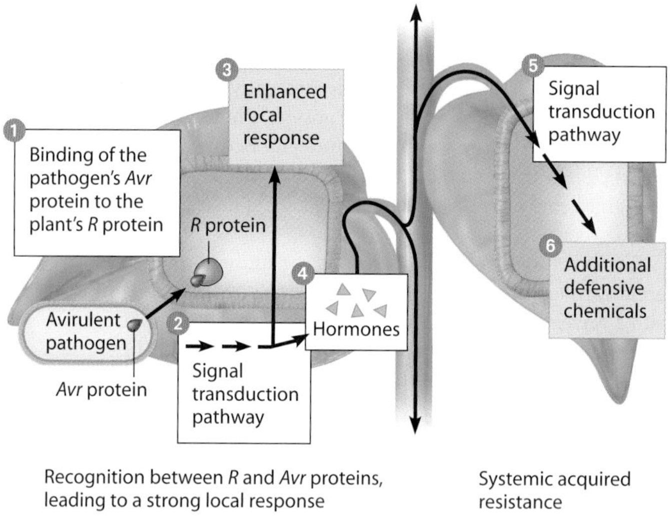

▲ **Figure 13B** Defense responses against an avirulent pathogen

the infection release microbe-killing molecules and chemicals that signal nearby cells to mount a similar chemical defense. In addition, infection stimulates chemical changes in the plant cell walls, which toughen the walls and thus slow the spread of the microbes within the plant.

This chemical defense system is enhanced by the plant's inherited ability to recognize certain pathogens. A kind of "compromise" has coevolved between plants and most of their pathogens: The pathogen gains enough access to its host to perpetuate itself without severely harming the plant. Otherwise, hosts and pathogens would soon perish together. The plant is said to be resistant to that pathogen, and the pathogen is said to be avirulent for the plant.

This resistance to destruction by a specific pathogen is based on the ability of the plant and the microbe to make a complementary pair of molecules. A plant has many *R* genes (for *resistance*), and each pathogen has a set of *Avr* genes (for *avirulence*). Researchers hypothesize that an *R* gene encodes an *R* protein, a receptor protein in the plant's cells, and that the complementary *Avr* gene encodes an *Avr* protein, a signal molecule of the pathogen that binds specifically to the *R* protein in the plant cell. Recognition of pathogen-derived molecules by *R* proteins triggers a signal transduction pathway that leads to both local and plant-wide defense responses.

Figure 13B shows this interaction and subsequent events in the plant. ❶ The binding of the pathogen's signal molecule (magenta) to the plant's receptor (purple) triggers ❷ a signal transduction pathway, which leads to ❸ a defense response that is much stronger than would occur without the matchup of the *R* and *Avr* proteins. The cells at the site of infection mount a vigorous chemical defense, tightly seal off the area, and then kill themselves.

The defense response at the site of infection helps protect the rest of the plant in yet another way. Among the signal molecules produced there are ❹ hormones that sound an alarm throughout the plant. ❺ At destinations distant from the original site, these hormones trigger signal transduction pathways leading to ❻ the production of additional defensive chemicals. This defense response, called **systemic acquired resistance**, provides protection against a diversity of pathogens for days.

Researchers suspect that one of the alarm hormones is salicylic acid, a compound whose pain-relieving effects led early cultures to use the salicylic-acid-rich bark of willows (*Salix*) as a medicine. Aspirin is a chemical derivative of this compound. With the discovery of systemic acquired resistance, biologists have learned one function of salicylic acid in plants.

? What is released at a site of infection that triggers the development of general resistance to pathogens?

Hormones

REVIEW

For Practice Quizzes, BioFlix, MP3 Tutors, and Activities, go to www.masteringbiology.com.

Reviewing the Concepts

Plant Hormones (1–8)

1 Experiments on how plants turn toward light led to the discovery of a plant hormone. Hormones coordinate the activities of plant cells and tissues. Experiments carried out by Darwin and others showed that the tip of a grass seedling detects light and transmits a signal down to the growing region of the shoot.

2 Five major types of hormones regulate plant growth and development. By triggering signal transduction pathways, small amounts of hormones regulate plant growth and development.

3 Auxin stimulates the elongation of cells in young shoots. Plants produce the auxin IAA in the apical meristems at the tips of shoots. At different concentrations, auxin stimulates or inhibits the elongation of shoots and roots. It may act by weakening cell walls, allowing them to stretch when cells take up water. Auxin also stimulates the development of vascular tissues and cell division in the vascular cambium, promoting growth in stem diameter.

4 Cytokinins stimulate cell division. Cytokinins, produced by roots, embryos, and fruits, promote cell division. Cytokinins from roots may balance the effects of auxin from apical meristems, causing lower buds to develop into branches.

5 Gibberellins affect stem elongation and have numerous other effects. Gibberellins stimulate the elongation of stems and leaves and the development of fruits. Gibberellins released from embryos function in some of the early events of seed germination.

6 Abscisic acid inhibits many plant processes. Abscisic acid (ABA) inhibits germination. The ratio of ABA to gibberellins often determines whether a seed remains dormant or germinates. Seeds of many plants remain dormant until their ABA is inactivated or washed away. ABA also acts as a "stress hormone," causing stomata to close when a plant is dehydrated.

7 Ethylene triggers fruit ripening and other aging processes. As fruit cells age, they give off ethylene gas, which hastens ripening. A changing ratio of auxin to ethylene, triggered mainly by shorter days, probably causes autumn color changes and the loss of leaves from deciduous trees.

8 Plant hormones have many agricultural uses. Auxin can delay or promote fruit drop. Auxin and gibberellins are used to produce seedless fruits. A synthetic auxin called 2,4-D kills weeds. There are questions about the safety of using such chemicals.

Responses to Stimuli (9–13)

9 Tropisms orient plant growth toward or away from environmental stimuli. Plants sense and respond to environmental changes. Tropisms are growth responses that change the shape of a plant or make it grow toward or away from a stimulus.

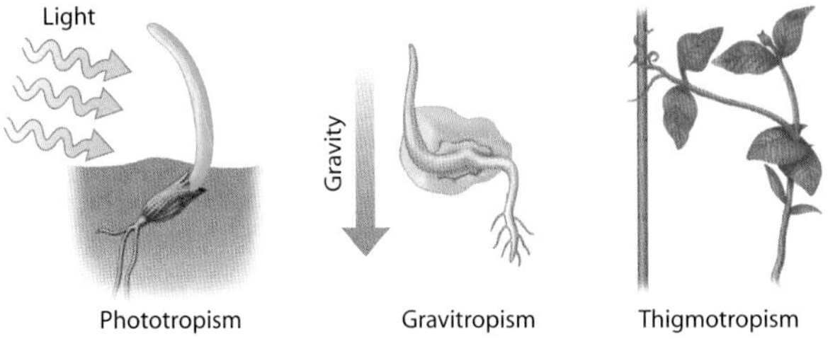

Phototropism, bending in response to light, may result from auxin moving from the light side to the dark side of a stem. A response to gravity, or gravitropism, may be caused by settling of organelles on the low sides of shoots and roots, which may trigger a change in hormone distribution. Thigmotropism, a response to touch, is responsible for coiling of tendrils and vines around objects.

10 Plants have internal clocks. An internal biological clock controls sleep movements and other daily cycles in plants. These cycles, called circadian rhythms, persist with periods of about 24 hours even in the absence of environmental cues, but such cues are needed to keep them synchronized with day and night.

11 Plants mark the seasons by measuring photoperiod. The timing of flowering is one of the seasonal responses to photoperiod, the relative lengths of night and day.

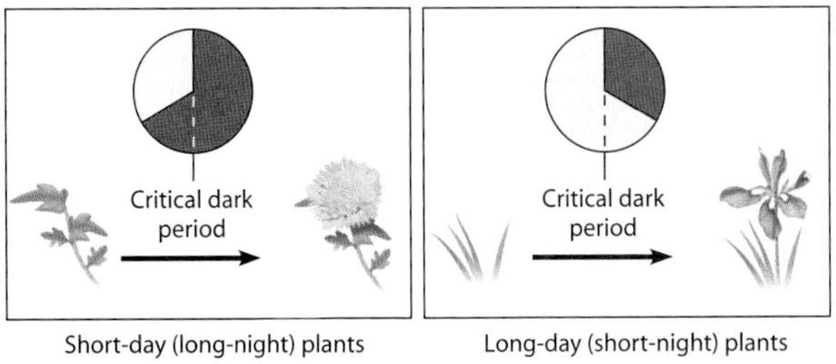

12 Phytochromes are light detectors that may help set the biological clock. Light-absorbing proteins called phytochromes may help plants set their biological clock and monitor photoperiod.

13 Defenses against herbivores and infectious microbes have evolved in plants. Plants use chemicals to defend themselves against both herbivores and pathogens. So-called avirulent plant pathogens interact with host plants in a specific way that stimulates both local and systemic defenses in the plant. Local defenses include microbe-killing chemicals and the sealing off of the infected area. Hormones trigger generalized defense responses in other organs (systemic acquired resistance).

Connecting the Concepts

1. Test your knowledge of the five major classes of plant hormones (auxins, cytokinins, gibberellins, abscisic acid, ethylene) by matching one hormone to each lettered box. (Note that some hormones will match up to more than one box.)

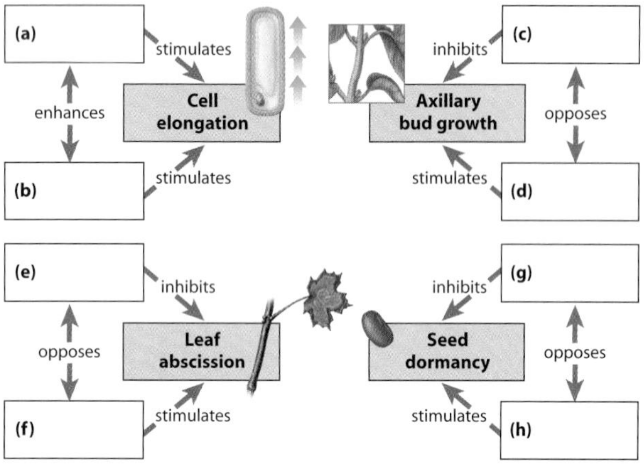

Testing Your Knowledge

Multiple Choice

2. During winter or periods of drought, which one of the following plant hormones inhibits growth and seed germination?
   a. ethylene
   b. abscisic acid
   c. gibberellin
   d. auxin
   e. cytokinin
3. A certain short-day plant flowers only when days are less than 12 hours long. Which of the following would cause it to flower?
   a. a 9-hour night and 15-hour day with 1 minute of darkness after 7 hours
   b. an 8-hour day and 16-hour night with a flash of white light after 8 hours
   c. a 13-hour night and 11-hour day with 1 minute of darkness after 6 hours
   d. a 12-hour day and 12-hour night with a flash of red light after 6 hours
   e. 24 hours of continuous light
4. Auxin causes a shoot to bend toward light by
   a. causing cells to shrink on the dark side of the shoot.
   b. stimulating growth on the dark side of the shoot.
   c. causing cells to shrink on the lighted side of the shoot.
   d. stimulating growth on the lighted side of the shoot.
   e. inhibiting growth on the dark side of the shoot.
5. In autumn, the amount of ______ increases and the amount of ______ decreases in fruits and leaf stalks, causing a plant to drop fruit and leaves.
   a. ethylene . . . auxin
   b. gibberellin . . . abscisic acid
   c. cytokinin . . . abscisic acid
   d. auxin . . . ethylene
   e. gibberellin . . . auxin
6. Plant hormones act by affecting the activities of
   a. genes.
   b. membranes.
   c. enzymes.
   d. genes, membranes, and enzymes.
   e. genes and enzymes.
7. Buds and sprouts often form on tree stumps. Which hormone would stimulate their formation?
   a. auxin
   b. cytokinin
   c. abscisic acid
   d. ethylene
   e. gibberellin
8. A plant's defense response at the site of initial infection by a pathogen will be especially strong if
   a. the pathogen is virulent.
   b. the plant makes a receptor protein that recognizes a signal molecule from the microbe.
   c. the pathogen is a fungus.
   d. the plant has an *Avr* gene that is the right match for one of the microbe's *R* genes.
   e. the right combination of hormones travel from the infection site to other parts of the plant.

Matching

9. Bending of a shoot toward light	a. phytochrome
10. Growth response to touch	b. photoperiod
11. Cycle with a period of about 24 hours	c. sleep movement
12. Pigment that helps control flowering	d. circadian rhythm
13. Relative lengths of night and day	e. thigmotropism
14. Growth response to gravity	f. phototropism
15. Folding of plant leaves at night	g. gravitropism

Describing, Comparing, and Explaining

16. If apples are to be stored for long periods, it is best to keep them in a place with good air circulation. Explain why.
17. Write a short paragraph explaining why a houseplant becomes bushier if you pinch off its terminal buds.

Applying the Concepts

18. A plant nursery manager tells the new night security guard to stay out of a room where chrysanthemums (which are short-day plants) are about to flower. Around midnight, the guard accidentally opens the door to the chrysanthemum room and turns on the lights for a moment. How might this affect the chrysanthemums? How could the guard correct the mistake?
19. A plant biologist observed a peculiar pattern when a tropical shrub was attacked by caterpillars. After a caterpillar ate a leaf, it would skip over nearby leaves and attack a leaf some distance away. Simply removing a leaf did not trigger the same change nearby. The biologist suspected that a damaged leaf sent out a chemical that signaled other leaves. How could this hypothesis be tested?
20. Imagine the following scenario: A plant biologist has developed a synthetic chemical that mimics the effects of a plant hormone. The chemical can be sprayed on apples before harvest to prevent flaking of the natural wax that is formed on the skin. This makes the apples shinier and gives them a deeper red color. What kinds of questions do you think should be answered before farmers start using this chemical on apples? How might the scientist go about finding answers to these questions?

Review Answers

1. a. auxin; b. gibberellin; c. auxin; d. cytokinin; e. auxin; f. ethylene; g. gibberellin; h. abscisic acid
2. b 3. c 4. b 5. a 6. d 7. b 8. b 9. f 10. e 11. d 12. a 13. b 14. g 15. c
16. Fruits produce ethylene gas, which triggers the ripening and aging of the fruit. Ventilation prevents a buildup of ethylene and delays its effects.
17. The terminal bud produces auxins, which counter the effects of cytokinins from the roots and inhibit the growth of axillary buds. If the terminal bud is removed, the cytokinins predominate, and lateral growth occurs at the axillary buds.
18. The red wavelengths in the room's lights quickly convert the phytochrome in the chrysanthemums to the P_{fr} form, which inhibits flowering in a long-night plant. The chrysanthemums will not flower unless the security guard can set up some far-red lights. Exposure to a burst of far-red light would convert the phytochrome to the P_r form, allowing flowering to occur.
19. The biologist could remove leaves at different stages of being eaten to see how long it takes for changes to occur in nearby leaves. The "hormone" could be captured in an agar block, as in the phototropism experiments in Module 1, and applied to an undamaged plant. Another experiment would be to block "hormone" movement out of a damaged leaf or into a nearby leaf.
20. Some issues and questions to consider: Is the hormone safe for human consumption? What are its effects in the environment? Could its production produce impurities or wastes that might be harmful? What kinds of tests need to be done to demonstrate its safety? How much does it cost to make and use? Are the benefits worth the costs and risks? Is it worth using an artificial chemical on food simply to improve its appearance?

    A scientist could seek answers by studying the stability of the hormone in a variety of laboratory simulations of natural conditions. The toxicity of the hormone, the materials used to produce it, and its breakdown products could be determined in laboratory tests.

The Biosphere: *An Introduction to Earth's Diverse Environments*

From Chapter 34 of *Campbell Biology: Concepts & Connections*, Seventh Edition. Jane B. Reece, Martha R. Taylor, Eric J. Simon, Jean L. Dickey.

The Biosphere: *An Introduction to Earth's Diverse Environments*

BIG IDEAS

franzfoto.com/Alamy

The Biosphere
(1–5)

The distribution and abundance of life in the biosphere is influenced by living and nonliving components of the environment.

James Randklev/Image Bank/Getty Images

Aquatic Biomes
(6–7)

In marine biomes, the salt concentration is generally around 3%. In freshwater biomes, the salt concentration is typically less than 1%.

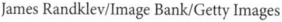

Dennis Frates/Workbook Stock/Getty Images

Terrestrial Biomes
(8–18)

The distribution of terrestrial biomes is primarily determined by temperature and rainfall.

Garry Weare/Lonely Planet Images

At the top of the world, the rocky slopes of the eastern Himalayas seem an unlikely place for living things to flourish. But in spite of the frigid air, thin soil, and brief growing season, spring fills the alpine meadows with greenery and the colorful blooms of dozens of plant species. And where there are plants to eat, animals thrive, too. Nimble-footed blue sheep and other hoofed animals graze on the hardy herbaceous plants of the meadow and in turn may be eaten by the reclusive snow leopard.

While the top of the world may seem like an improbable place to find life, the bottom of the world—the darkest depths of the ocean—seems like an impossible environment. For most life on Earth, the sun is the main source of energy. Yet life also thrives in the dark world of hydrothermal vents, sites near the adjoining edges of giant plates of Earth's crust where molten rock and hot gases surge upward from Earth's interior. Towering chimneys, some as tall as a nine-story building, emit scalding water and hot gases such as hydrogen sulfide. This environment is home to organisms capable of overcoming its challenges. Life at hydrothermal vents depends on chemoautotrophic sulfur bacteria, most of which obtain energy by oxidizing hydrogen sulfide. Many of the animals in these communities obtain food from chemoautotrophic sulfur bacteria harbored within their bodies. These animals are food for others in the hydrothermal vent community, which also includes sea anemones, giant clams, shrimps, crabs, and a few fishes.

From the roof of the world to the deepest ocean, Earth's diverse environments are bursting with life. Investigating the interactions of organisms with their environment offers a context for new discoveries, a means of understanding familiar forms of life, and insight into our own role as inhabitants of Earth.

The Biosphere

1 Ecologists study how organisms interact with their environment at several levels

Ecology (from the Greek *oikos*, home) is the scientific study of the interactions of organisms with their environment. Ecologists describe the distribution and abundance of organisms—where they live and how many live there. Because the environment is complex, organisms can potentially be affected by many different variables. Ecologists group these variables into two major types, biotic factors and abiotic factors. **Biotic factors**, which include all of the organisms in the area, are the living component of the environment. **Abiotic factors** are the environment's nonliving component, the physical and chemical factors such as temperature, forms of energy available, water, and nutrients. An organism's **habitat**, the specific environment it lives in, includes the biotic and abiotic factors present in its surroundings.

As you might expect, field research is fundamental to ecology. But ecologists also test hypotheses using laboratory experiments, where conditions can be simplified and controlled. Some ecologists take a theoretical approach, devising mathematical and computer models that enable them to simulate large-scale experiments that are impossible to conduct in the field.

Ecologists study environmental interactions at several levels. At the **organism** level, they may examine how one kind of organism meets the challenges and opportunities of its environment through its physiology or behavior. For example, an ecologist working at this level might study adaptations of the Himalayan blue poppy (*Meconopsis betonicifolia*, **Figure 1A**) to the freezing temperatures and short days of its abiotic environment.

Another level of study in ecology is the **population**, a group of individuals of the same species living in a particular geographic area. Blue poppies living in a particular Himalayan alpine meadow would constitute a population (**Figure 1B**). An ecologist studying blue poppies might investigate factors that affect the size of the population, such as the availability of chemical nutrients or seed dispersal.

A third level, the **community**, is an assemblage of all the populations of organisms living close enough together for potential interaction—all of the biotic factors in the environment. All the organisms in a particular alpine meadow would constitute a community (**Figure 1C**). An ecologist working at this level might focus on interspecies interactions, such as the competition between poppies and other plants for soil nutrients or the effect of plant-eaters on poppies.

The fourth level of ecological study, the **ecosystem**, includes both the biotic and abiotic components of the environment (**Figure 1D**). Some critical questions at the ecosystem level concern how chemicals cycle and how energy flows between organisms and their surroundings. For an alpine meadow, one ecosystem-level question would be, How rapidly does the decomposition of decaying plants release inorganic molecules?

Some ecologists take a wider perspective by studying **landscapes**, which are arrays of ecosystems. Landscapes are usually visible from the air as distinctive patches. For example, the Himalayan alpine meadows are part of a mountain landscape that also includes conifer and broadleaf forests.

▲ **Figure 1A** An organism

▲ **Figure 1B** A population

▲ **Figure 1C** A community

▲ **Figure 1D** An ecosystem

A landscape perspective emphasizes the absence of clearly defined ecosystem boundaries; energy, materials, and organisms may be exchanged by ecosystems within a landscape.

The **biosphere**, which extends from the atmosphere several kilometers above Earth to the depths of the oceans, is all of Earth that is inhabited by life. It is the home of us all, and as you will learn throughout this unit, our actions have consequences for the entire biosphere.

? List the biotic and abiotic factors in the hydrothermal vent ecosystems described in the chapter introduction.

The biotic factors include chemosynthetic bacteria, sea anemones, giant clams, shrimps, crabs, and fishes. Abiotic factors include hydrogen sulfide and boiling hot temperatures.

CONNECTION 2 The science of ecology provides insight into environmental problems

> The "control of nature" is a phrase conceived in arrogance, born of the Neanderthal age of biology and philosophy, when it was supposed that nature exists for the convenience of man.
>
> —Rachel Carson, *Silent Spring*

An airplane trip almost anywhere in the United States reveals the extent to which people have changed the landscape **(Figure 2A)**. But the cities, farms, and highways visible from the air are only the tip of the vast and largely unseen iceberg of human impact on the environment. Overgrazing, deforestation, and overcultivation of land have caused massive soil erosion. Industrial and agricultural practices have spread chemical pollutants throughout the biosphere. Human-made products have contaminated the groundwater, water that has penetrated the soil and may eventually resurface in the drinking supply. In some areas, groundwater is being depleted by the ever-expanding human population. Hundreds of species of organisms are in danger of joining the growing list of those that we have already driven to extinction. Most worrisome of all, our activities are producing potentially catastrophic changes in the global climate.

Although most of our environmental problems have been many decades in the making, few people were aware of them until the 1960s. Technology had been advancing rapidly, and humankind seemed poised on the brink of freedom from several age-old bonds. New chemical fertilizers and pesticides, for example, showed great promise for increasing agricultural productivity. Massive applications of fertilizer and extensive aerial spraying of pesticides such as DDT brought astonishing increases in crop yields. Pesticides were also used to combat disease-carrying insects, including mosquitoes (which transmit malaria), body lice (typhus), and fleas (plague).

Rich Reid/NGS Image Collection

▲ **Figure 2A** An aerial view of a landscape changed by humans

But two events raised questions about the long-term effects of widespread DDT use. The first event was the evolution of DDT resistance in insects, rendering this widely used pesticide less effective. The second event was the publication of a book called *Silent Spring* by Rachel Carson **(Figure 2B)**. As DDT lost its killing power and the chemical companies introduced ever more potent pesticides, Carson sought to alert the general public to the dangers of global pesticide use. Although Carson and her work were harshly criticized by the agriculture and pesticide industries, scientists had already begun to document the harmful effects of DDT on predatory birds. Moreover, traces of the pesticide were turning up worldwide, thousands of miles from where it had been applied. DDT even showed up in human milk.

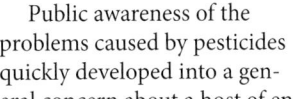

Alfred Eisenstaedt/Time Life Pictures/Getty Images

▲ **Figure 2B** Rachel Carson

Public awareness of the problems caused by pesticides quickly developed into a general concern about a host of environmental issues. The 1970s brought a series of legislative acts aimed at curbing pollution and cleaning up the environment. Despite these efforts, serious issues clearly remain.

The science of ecology can provide the understanding needed to resolve environmental problems. But these problems cannot be solved by ecologists alone, because they require making decisions based on values and ethics. On a personal level, each of us makes daily choices that affect our ecological impact. Legislators and corporations, motivated by environmentally aware voters and consumers, must address questions that have wider implications: How should land use be regulated? Should we try to save all species or just certain ones? What alternatives to environmentally destructive practices can be developed? How can we balance environmental impact with economic needs?

But analyzing environmental issues and planning for better practices begin with an understanding of the basic concepts of ecology, so let's start to explore them now.

? **Why can't ecologists alone solve environmental problems?**

● The science of ecology can inform the decision-making process, but solving environmental problems involves making ethical, economic, and political judgments that are outside the realm of science.

Text quotation from Rachel Carson, *Silent Spring* (New York: Houghton Mifflin, 1962).

3 Physical and chemical factors influence life in the biosphere

You have learned that life thrives in a wide variety of habitats, from the mountaintops to the seafloor. To be successful, the organisms that live in each place must be adapted to the abiotic factors present in those environments.

Energy Sources All organisms require a source of energy to live. Solar energy from sunlight, captured during the process of photosynthesis, powers most ecosystems. Lack of sunlight is seldom the most important factor limiting plant growth for terrestrial ecosystems, although shading by trees does create intense competition for light among plants growing on forest floors. In many aquatic environments, however, light is not uniformly available. Microorganisms and suspended particles, as well as the water itself, absorb light and prevent it from penetrating beyond certain depths. As a result, most photosynthesis occurs near the water's surface.

In dark environments such as caves or hydrothermal vents, bacteria that extract energy from inorganic chemicals power ecosystems. As we mentioned in the chapter introduction, sulfur bacteria perform this function in hydrothermal vent communities, where many of the animals either feed directly on the sulfur bacteria or derive nutrition from bacteria living inside their bodies. For example, tube worms **(Figure 3A)** have no mouth or digestive tract. The red tip extending from the white casing is a respiratory surface that acquires oxygen and sulfide from the water. Bacteria living in a specialized organ in the worm's body get energy from the sulfide—a lot of energy. These worms can grow to be over 2 m (6.5 feet) long.

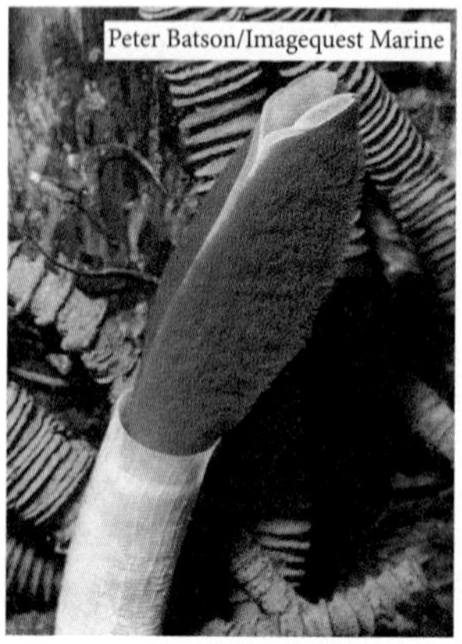
Peter Batson/Imagequest Marine

▲ **Figure 3A** The respiratory surface of a giant tube worm

Temperature Temperature is an important abiotic factor because of its effect on metabolism. Few organisms can maintain a sufficiently active metabolism at temperatures close to 0°C, and temperatures above 45°C (113°F) destroy the enzymes of most organisms. However, extraordinary adaptations enable some species to live outside this temperature range. For example, archaeans living in hot springs have enzymes that function optimally at extremely high temperatures. Mammals and birds, such as the snowy owl in **Figure 3B**, can remain considerably warmer than their surroundings and can be active at a fairly wide range of temperatures. Amphibians and reptiles, which gain most of their warmth by absorbing heat from their surroundings, have a more limited distribution.

Water Water is essential to all life. Thus, for terrestrial organisms, drying out in the air is a major danger. Watertight coverings were key adaptations enabling plants and vertebrates to be successful on land. Aquatic organisms are surrounded by water; their problem is solute concentration. Freshwater organisms live in a hypotonic medium, while the environment of marine organisms is hypertonic. As we saw in Module 25.4, animals maintain fluid balance by a variety of mechanisms.

▶ **Figure 3B** A snowy owl

Vincent Munier/naturepl.com

Inorganic Nutrients The distribution and abundance of photosynthetic organisms, including plants, algae, and photosynthetic bacteria, depend on the availability of inorganic nutrients such as nitrogen and phosphorus. Plants obtain these from the soil. Soil structure, pH, and nutrient content often play major roles in determining the distribution of plants. In many aquatic ecosystems, low levels of nitrogen and phosphorus limit the growth of algae and photosynthetic bacteria.

Other Aquatic Factors Several abiotic factors are important in aquatic, but not terrestrial, ecosystems. While terrestrial organisms have a plentiful supply of O_2 from the air, aquatic organisms must depend on oxygen dissolved in water. This is a critical factor for many species of fish. Trout, for example, require high levels of dissolved oxygen. Cold, fast-moving water has a higher oxygen content than warm or stagnant water. Salinity, current, and tides may also play a role in aquatic ecosystems.

Other Terrestrial Factors On land, wind is often an important abiotic factor. Wind increases an organism's rate of water loss by evaporation. The resulting increase in evaporative cooling can be advantageous on a hot summer day, but it can cause dangerous wind chill in the winter. In some ecosystems, fire occurs frequently enough that many plants have adapted to this disturbance.

Next we examine the interaction between one animal species and the abiotic and biotic factors of its environment.

? **Why are birds and mammals, but not amphibians and reptiles other than birds, found in the Himalayan alpine meadows?**

● As ectotherms (see Module 25.1), reptiles and amphibians do not have adaptations that enable them to withstand the cold temperatures of the alpine habitat.

EVOLUTION CONNECTION

4 Organisms are adapted to abiotic and biotic factors by natural selection

One of the fundamental goals of ecology is to explain the distribution of organisms. The presence of a species in a particular place has two possible explanations: The species may have evolved from ancestors living in that location, or it may have dispersed to that location and been able to survive once it arrived. The magnificent pronghorn "antelope" (*Antilocapra americana,* **Figure 4**) is the descendant of ancestors that roamed the open plains and shrub deserts of North America more than a million years ago. The animal is found nowhere else and is only distantly related to the many species of antelope in Africa. What selective factors in the abiotic and biotic environments of its ancestors produced the adaptations we see in the pronghorn that roams North America today?

▲ **Figure 4** A pronghorn (*Antilocapra americana*)

The pronghorn's present-day habitat, like that of its ancestors, is arid, windswept, and subject to extreme temperature fluctuations both daily and seasonally. Individuals able to survive and reproduce under these conditions left offspring that carried their alleles forward into subsequent generations. Thus, we can infer that many of the adaptations that contribute to the success of present-day pronghorns must also have contributed to the success of their ancestors. For example, the pronghorn has a thick coat made of hollow hairs that trap air, insulating the animal in cold weather. If you drive through Wyoming or parts of Colorado in the winter, you will see herds of these animals foraging in the open when temperatures are well below 0°C. In hot weather, the pronghorn can raise patches of this stiff hair to release body heat.

The biotic environment, which includes what the animal eats and any predators that threaten it, is also a factor in determining which members of a population survive and reproduce. The pronghorn's main foods are small broadleaf plants, grasses, and woody shrubs. Over time, characteristics that enabled the ancestors of the pronghorn to exploit these food sources more efficiently became established through natural selection. As a result, the teeth of a pronghorn are specialized for biting and chewing tough plant material. Like the stomach of a cow, the pronghorn's stomach contains cellulose-digesting bacteria. As the pronghorn eats plants, the bacteria digest the cellulose, and the animal obtains most of its nutrients from the bacteria.

While many factors in the pronghorn's environment have been fairly consistent throughout its evolutionary history, one aspect has changed significantly. Until around 12,000 years ago, one of the pronghorn's major predators was probably the American cheetah, a fleet-footed feline that bears some similarities to the more familiar African cheetah. The now-extinct American cheetah was one of many ferocious predators of Pleistocene North America, along with lions, jaguars, and saber-toothed cats with 7-inch canines. Ecologists hypothesize that the selection pressure of the cheetah's pursuit led to the pronghorn's blazing speed, which far exceeds that of its main present-day predator, the wolf. With a top speed of 97 km/h (60 mph), the pronghorn is easily the fastest mammal on the continent, and an adult pronghorn can keep up a pace of 64 km/h (40 mph) for at least 30 minutes. Unable to match the pronghorn's extravagant speed, wolves typically take adults that have been weakened by age or illness.

Like many large herbivores that live in open grasslands, the pronghorn also derives protection from living in herds. When one pronghorn starts to run, its white rump patch seems to alert other herd members to danger. Other adaptations that help the pronghorn foil predators include its tan and white coat, which provides camouflage, and its keen eyes, which can detect movement at great distances. Thus, the adaptations shaped by natural selection in the distant past still serve as protection from the predators of today's environment.

If the pronghorn's environment changed significantly, the adaptations that contribute to its current success might not be as advantageous. For example, if an increase in rainfall turned the open plains into woodlands, where predators would be more easily hidden by vegetation and could stalk their prey at close range, the pronghorn's adaptations for escaping predators might not be as effective. Thus, in adapting populations to local environmental conditions, natural selection may limit the distribution of organisms.

In the next module, we see how global climatic patterns determine temperature and precipitation, the major abiotic factors that influence the distribution of organisms. We examine the biotic components of the environment more closely in other chapters in this unit.

? **What is the role of the environment in adaptive evolution?**

● The individuals whose phenotypes are best suited to the environment (including both abiotic and biotic factors) will pass their alleles to the next generation. But individuals with other phenotypes may not. For example, if the biotic environment includes wolves, a pronghorn that is not able to run as long as the rest of the herd will probably not survive to reproduce.

5 Regional climate influences the distribution of terrestrial communities

When we ask what determines whether a particular organism or community of organisms lives in a certain area, the climate of the region—especially temperature and precipitation—is often a crucial part of the answer. Earth's global climate patterns are largely determined by the input of radiant energy from the sun and the planet's movement in space.

Figure 5A shows that because of its curvature, Earth receives an uneven distribution of solar energy. The sun's rays strike equatorial areas most directly (perpendicularly). Away from the equator, the rays strike Earth's surface at a slant. As a result, the same amount of solar energy is spread over a larger area. Thus, any particular area of land or ocean near the equator absorbs more heat than comparable areas in the more northern or southern latitudes.

The seasons of the year result from the permanent tilt of the planet on its axis as it orbits the sun. As **Figure 5B** shows, the globe's position relative to the sun changes through the year. The Northern Hemisphere, for instance, is tipped most toward the sun in June. The more direct angle and increased intensity of the sun result in the long days of summer in that hemisphere. But during this time, days are short and it is winter in the Southern Hemisphere. Conversely, the Southern Hemisphere is tipped farthest toward the sun in December, creating summer there and causing winter in the Northern Hemisphere. The **tropics**, the region surrounding the equator between latitudes 23.5° north (the Tropic of Cancer) and 23.5° south (the Tropic of Capricorn), experience the greatest annual input and least seasonal variation in solar radiation.

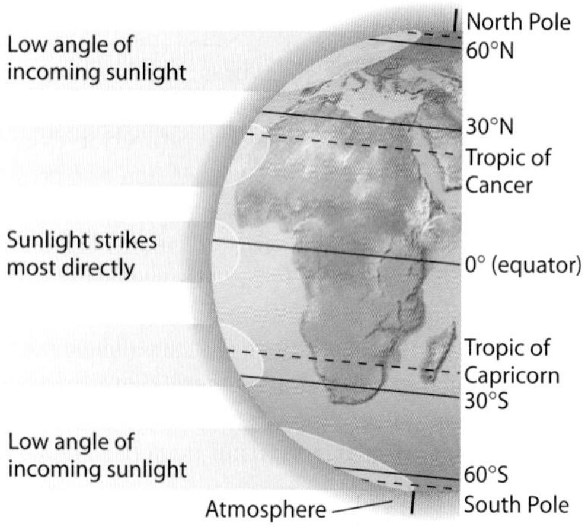

▲ **Figure 5A** How solar radiation varies with latitude

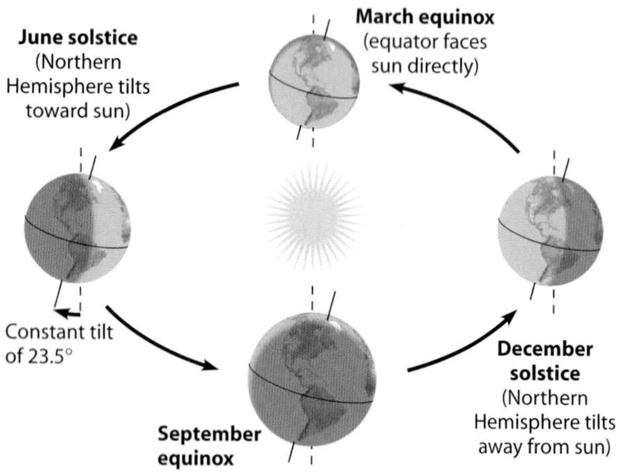

▲ **Figure 5B** How Earth's tilt causes the seasons

Figure 5C shows some of the effects of the intense solar radiation near the equator on global patterns of rainfall and winds. Arrows indicate air movements. High temperatures in the tropics evaporate water from Earth's surface. Heated by the direct rays of the sun, moist air at the equator rises, creating an area of calm or of very light winds known as the **doldrums**. As warm equatorial air rises, it cools and releases much of its water content, creating the abundant precipitation typical of most tropical regions. High temperatures throughout the year and ample rainfall largely explain why rain forests are concentrated near the equator.

After losing their moisture over equatorial zones, high-altitude air masses spread away from the equator until they cool and descend again at latitudes of about 30° north and south. This descending dry air absorbs moisture from the land. Thus, many of the world's great deserts—the Sahara in North Africa and the Arabian on the Arabian Peninsula, for example—are centered at these latitudes. As the dry air descends, some of it spreads back toward the equator. This movement creates the cooling **trade winds**, which dominate the tropics. As the air moves back toward the equator, it warms and picks up moisture until it ascends again.

The latitudes between the tropics and the Arctic Circle in the north and the Antarctic Circle in the south are called **temperate zones**. Generally, these regions have seasonal variations in climate and more moderate temperatures than the tropics or the polar zones. Notice in Figure 5C that some of the descending air heads into the latitudes above 30°. At first

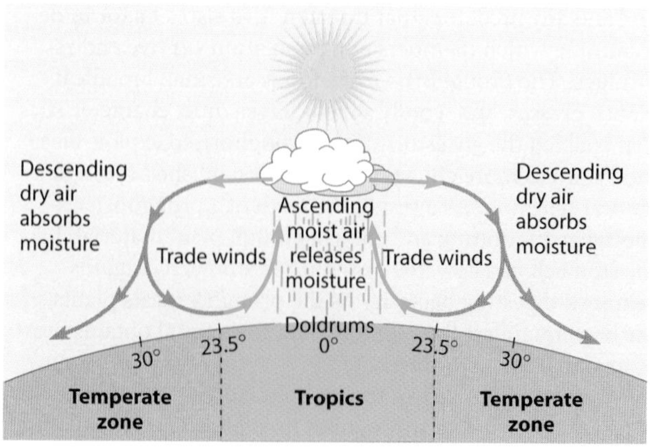

▲ **Figure 5C** How uneven heating causes rain and winds

these air masses pick up moisture, but they tend to drop it as they cool at higher latitudes. This is why the north and south temperate zones, especially latitudes around 60°, tend to be moist. Broad expanses of coniferous forest dominate the landscape at these fairly wet but cool latitudes.

Figure 5D shows the major global air movements, called the **prevailing winds**. Prevailing winds (pink arrows) result from the combined effects of the rising and falling of air masses (blue and brown arrows) and Earth's rotation (gray arrows). Because Earth is spherical, its surface moves faster at the equator (where its diameter is greatest) than at other latitudes. In the tropics, Earth's rapidly moving surface deflects vertically circulating air, making the trade winds blow from east to west. In temperate zones, the slower-moving surface produces the **westerlies**, winds that blow from west to east.

A combination of the prevailing winds, the planet's rotation, unequal heating of surface waters, and the locations and shapes of the continents creates **ocean currents**, river-like flow patterns in the oceans (**Figure 5E**). Ocean currents have a profound effect on regional climates. For instance, the Gulf Stream circulates warm water northward from the Gulf of Mexico and makes the climate on the west coast of Great Britain warmer during winter than the coast of New England, which is actually farther south but is cooled by a branch of the current flowing south from the coast of Greenland (not shown in figure).

Landforms can also affect local climate. Air temperature declines by about 6°C with every 1,000-m increase in elevation, an effect you've probably experienced if you've ever hiked up a mountain. **Figure 5F** illustrates the effect of mountains on rainfall. This drawing represents major landforms across the state of California, but mountain ranges cause similar effects elsewhere. California is a temperate area in which the prevailing winds are westerlies. As moist air moves in off the Pacific Ocean and encounters the westernmost mountains (the Coast Range), it flows upward, cools at higher altitudes, and drops a large amount of water. The world's tallest trees, the coastal redwoods, thrive here. Farther inland, precipitation increases again as the air moves up and over higher mountains (the Sierra Nevada). Some of the world's deepest snow packs occur here. On the eastern side of the Sierra, there is little precipitation, and the dry descending air also absorbs moisture. This effect, called a rain shadow, is responsible for the desert that covers much of central Nevada.

Climate and other abiotic factors of the environment control the global distribution of organisms. The influence of these abiotic factors results in **biomes**, major types of ecological associations that occupy broad geographic regions of land or water. Terrestrial biomes are determined primarily by temperature and precipitation—similar assemblages of plant and animal types are found in areas that have similar climates. Aquatic biomes are defined by different abiotic factors; the primary distinction is based on salinity. Marine biomes, which include oceans, intertidal zones, coral reefs, and estuaries, generally have salt concentrations around 3%, while freshwater biomes (lakes, streams and rivers, and wetlands) typically have a salt concentration of less than 1%. We describe several aquatic biomes in the next two modules.

? What causes summer in the Northern Hemisphere?

Because of the fixed angle of Earth's axis relative to the orbital plane around the sun, the Northern Hemisphere is tilted toward the sun during the portion of the annual orbit that corresponds to the summer months.

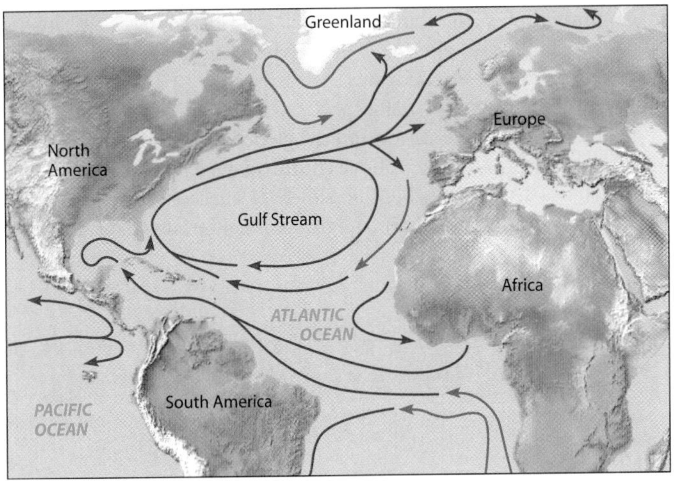

▲ **Figure 5E** Atlantic Ocean currents (red arrows indicate warming currents; blue arrows indicate cooling currents)

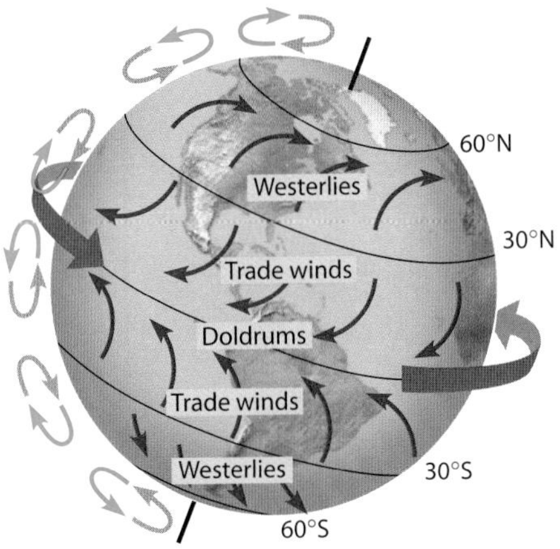

▲ **Figure 5D** Prevailing wind patterns

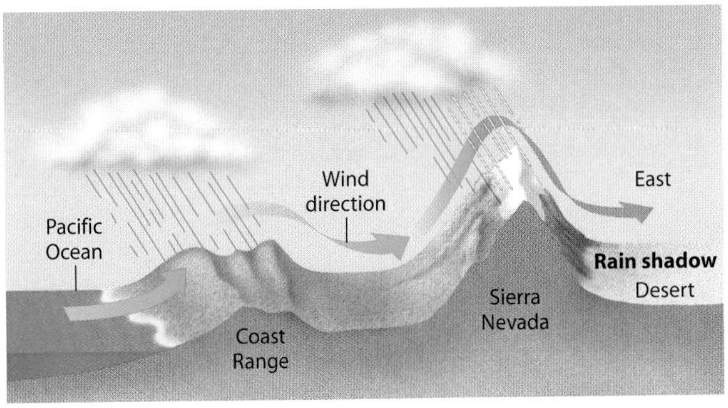

▲ **Figure 5F** How mountains affect precipitation (California)

Aquatic Biomes

6 Sunlight and substrate are key factors in the distribution of marine organisms

Gazing out over a vast ocean, you might think that it is the most uniform environment on Earth. But marine ecosystems can be as different as night and day. The deepest ocean, where hydrothermal vents are located, is perpetually dark. In contrast, the vivid coral reefs are utterly dependent on sunlight. Habitats near shore are different from those in mid-ocean, and the substrate, which varies with depth and distance from shore, hosts different communities from the open waters.

The **pelagic realm** of the oceans includes all open water, and the substrate—the seafloor—is known as the **benthic realm** (**Figure 6A**). The depth of light penetration, a maximum of 200 m (656 feet), marks the **photic zone**. In shallow areas such as the submerged parts of continents, called **continental shelves**, the photic zone includes both the pelagic and benthic realms. In these sunlit regions, photosynthesis by **phytoplankton** (microscopic algae and cyanobacteria) and multicellular algae provides energy and organic carbon for a diverse community of animals. Sponges, burrowing worms, clams, sea anemones, crabs, and echinoderms inhabit the benthic realm of the photic zone. **Zooplankton** (small, drifting animals), fish, marine mammals, and many other types of animals are abundant in the pelagic photic zone.

Coral reefs, a visually spectacular and biologically diverse biome, are scattered around the globe in the photic zone of warm tropical waters above continental shelves, as shown in **Figure 6B**. A reef is built up slowly by successive generations of coral animals—a diverse group of cnidarians that secrete a hard external skeleton—and by multicellular algae encrusted with limestone. Unicellular algae live within the corals, providing the coral with food. Coral reefs support a huge variety of invertebrates and fishes.

Digital Vision/Getty Images

▲ **Figure 6B** A coral reef with its immense variety of invertebrates and fishes

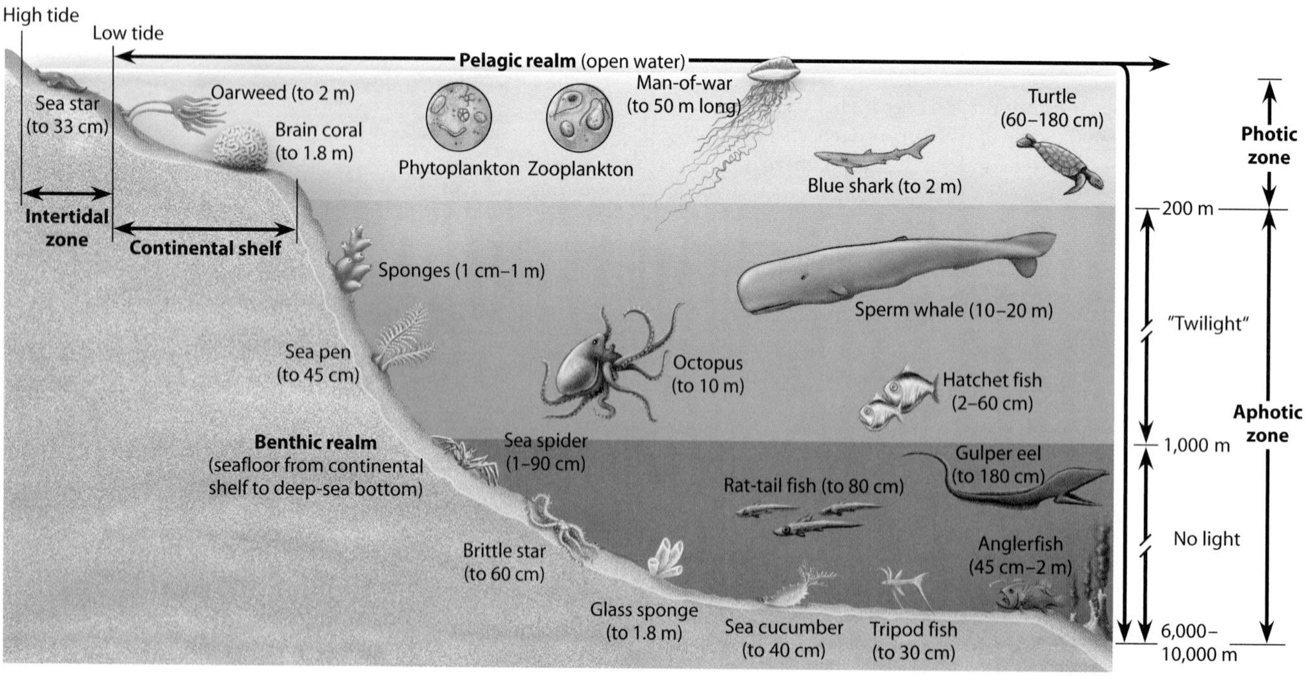

▲ **Figure 6A** Ocean life (zone depths and organisms not drawn to scale)

Below the photic zone of the ocean lies the **aphotic zone**. Although there is not enough light for photosynthesis between 200 and 1,000 m (0.6 mile), some light does reach these depths. This dimly lit world, sometimes called the twilight zone, is dominated by a fascinating variety of small fishes and crustaceans. Food sinking from the photic zone provides some sustenance for these animals. In addition, many of them migrate to the surface at night to feed. Some fishes in the twilight zone have enlarged eyes, enabling them to see in the very dim light, and luminescent organs that attract mates and prey.

Below 1,000 m, the ocean is completely and permanently dark. Adaptation to this environment has produced bizarre-looking creatures, such as the angler fish shown in **Figure 6C**. The scarcity of food probably explains the strangely outsized mouths of the angler and other fishes that inhabit this region of the ocean, a feature that allows them to grab any available prey, large or small. Inwardly angled teeth ensure that once caught, prey do not escape. The angler fish improves its chances of encountering prey by dangling a lure lit by bioluminescent bacteria. Most benthic organisms here are deposit feeders, animals that consume detritus on the substrate. Crustaceans, polychaete worms, sea anemones, and echinoderms such as sea cucumbers, sea stars, and sea urchins are common. Because of the scarcity of food, however, the density of animals is low—except at hydrothermal vents, where chemoautotrophic bacteria support an abundance of life.

SOC/Image Quest Marine

▲ **Figure 6C**
An angler fish

The marine environment also includes distinctive biomes where the ocean interfaces with land or with fresh water. In the **intertidal zone**, where the ocean meets land, the shore is pounded by waves during high tide and exposed to the sun and drying winds during low tide. The rocky intertidal zone is home to many sedentary organisms, such as algae, barnacles, and mussels, which attach to rocks and are thus prevented from being washed away when the tide comes in. On sandy beaches, suspension-feeding worms, clams, and predatory crustaceans bury themselves in the ground.

Figure 6D shows an **estuary**, a biome that occurs where a freshwater stream or river merges with the ocean. The saltiness of estuaries ranges from nearly that of fresh water to that of the ocean. With their waters enriched by nutrients from the river, estuaries are among the most productive biomes on Earth. Oysters, crabs, and many fishes live in estuaries or reproduce in them. Estuaries are also crucial nesting and feeding areas for waterfowl.

Wetlands constitute a biome that is transitional between an aquatic ecosystem—either marine or freshwater—and a terrestrial one. Covered with water either permanently or periodically, wetlands support the growth of aquatic plants (see Figure 7C). Mudflats and salt marshes are coastal wetlands that often border estuaries.

For centuries, people viewed the ocean as a limitless resource, harvesting its bounty and using it as a dumping ground for wastes. The impact of these practices is now being felt in many ways, large and small. From worldwide declines in commercial fish species to dying coral reefs to beaches closed by pollution, danger signs abound. Because of their proximity to land, estuaries and wetlands are especially vulnerable. Many have been completely replaced by development on landfill. Other threats include nutrient pollution, contamination by pathogens or toxic chemicals, alteration of freshwater inflow, and introduction of non-native species. Coral reefs have suffered from many of the same problems. In addition, overfishing has upset the species balance in some reef communities and greatly reduced diversity, and the widespread demise of reef-building corals in some regions has been attributed to global warming.

Freshwater biomes share many characteristics with marine biomes and experience some of the same threats. We introduce freshwater biomes in the next module.

? Oil from the 2010 Deepwater Horizon disaster in the Gulf of Mexico has polluted estuaries in Louisiana. Why does this pollution affect other animals in addition to those that live permanently in the estuaries?

Many species, including fishes and waterfowl, feed or reproduce in estuaries.

James Randklev/Image Bank/Getty Images

▲ **Figure 6D** An estuary in Georgia

7 Current, sunlight, and nutrients are important abiotic factors in freshwater biomes

Freshwater biomes cover less than 1% of Earth's surface and contain a mere 0.01% of its water. But they harbor a disproportionate share of biodiversity—an estimated 6% of all described species. Moreover, we depend on freshwater biomes for drinking water, crop irrigation, sanitation, and industry.

Freshwater biomes fall into two broad categories: standing water, which includes lakes and ponds, and flowing water, such as rivers and streams. Because these biomes are embedded in terrestrial landscapes, their characteristics are intimately connected with the soils and organisms of the ecosystems that surround them.

Lakes and Ponds In lakes and large ponds, as in the oceans, the communities of plants, algae, and animals are distributed according to the depth of the water and its distance from shore. Phytoplankton grow in the photic zone, and rooted plants often inhabit shallow waters near shore **(Figure 7A)**. If a lake or pond is deep enough or murky enough, it has an aphotic zone where light levels are too low to support photosynthesis. In the benthic realm, large populations of microorganisms decompose dead organisms that sink to the bottom. Respiration by microbes removes oxygen from water near the bottom, and in some lakes, benthic areas are unsuitable for any organisms except anaerobic microbes.

Temperature may also have a profound effect on standing water biomes. During the summer, deep lakes have a distinct upper layer of water that has been warmed by the sun and does not mix with underlying, cooler water.

The mineral nutrients nitrogen and phosphorus typically determine the amount of phytoplankton growth in a lake or pond. Many lakes and ponds receive large inputs of nitrogen and phosphorus from sewage and runoff from fertilized lawns and farms. These nutrients may produce a heavy growth ("bloom") of algae, which reduces light penetration. When the algae die and decompose, a pond or lake can suffer severe oxygen depletion, killing fish that are adapted to high-oxygen conditions.

Rivers and Streams Rivers and streams generally support communities of organisms quite different from those of lakes and ponds. A river or a stream changes greatly between its source (perhaps a spring or snowmelt) and the point at which it empties into a lake or the ocean. Near the source, the water is usually cold, low in nutrients, and clear **(Figure 7B)**. The channel is often narrow, with a swift current that does not allow much silt to accumulate on the bottom. The current also inhibits the growth of phytoplankton; most of the organisms found here are supported by the photosynthesis of algae attached to rocks or by organic material, such as leaves, carried into the stream from the surrounding land. The most abundant benthic animals are usually arthropods, such as small crustaceans and insect larvae, that have physical and behavioral adaptations that enable them to resist being swept away. Trout, which locate their insect prey mainly by sight in the clear water, are often the predominant fishes.

Don Fink/Shutterstock

▲ **Figure 7B** A stream in the Great Smoky Mountains, Tennessee

Downstream, a river or stream generally widens and slows. The water is usually warmer and may be murkier because of sediments and phytoplankton suspended in it. Worms and insects that burrow into mud are often abundant, as are waterfowl, frogs, and catfish and other fishes that find food more by scent and taste than by sight.

Wetlands Freshwater wetlands range from marshes, as shown in **Figure 7C**, to swamps and bogs. Like marine wetlands, freshwater wetlands are rich in species diversity. They provide water storage areas that reduce flooding and improve water quality by filtering pollutants. Recognition of their ecological and economic value has led to government and private efforts to protect and restore wetlands.

? **Why does sewage cause algal blooms in lakes?**

The sewage adds nutrients, such as nitrates and phosphates, that stimulate growth of algae.

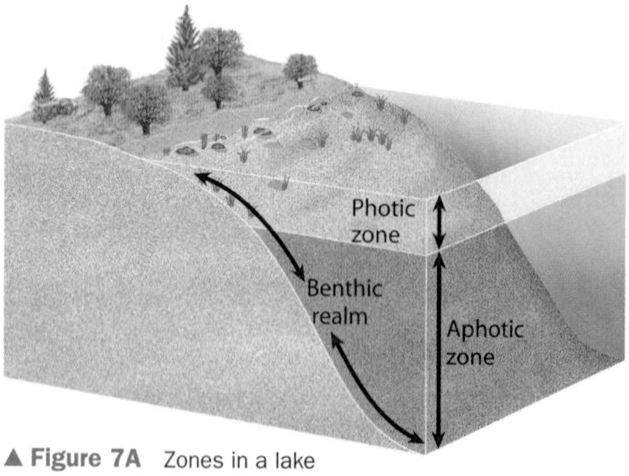

▲ **Figure 7A** Zones in a lake

Jean Dickey

▲ **Figure 7C** A marsh at Kent State University in Ohio

Terrestrial Biomes

8 Terrestrial biomes reflect regional variations in climate

Terrestrial ecosystems are grouped into nine major types of biomes, which are distinguished primarily by their predominant vegetation. By providing food, shelter, nesting sites, and much of the organic material for decomposers, plants build the foundation for the communities of animals, fungi, and microorganisms that are characteristic of each biome. The geographic distribution of plants, and thus of terrestrial biomes, largely depends on climate, with temperature and precipitation often the key factors determining the kind of biome that exists in a particular region.

Figure 8 shows the locations of the biomes. If the climate in two geographically separate areas is similar, the same type of biome may occur in both places; notice on the map that each kind of biome occurs on at least two continents. Each biome is characterized by a type of biological community, rather than an assemblage of particular species. For example, the species living in the deserts of the American Southwest and in the Sahara Desert of Africa are different, but all are adapted to desert conditions. Widely separated biomes may look alike because of convergent evolution, the appearance of similar traits in independently evolved species living in similar environments.

There is local variation within each biome that gives the vegetation a patchy, rather than uniform, appearance. For example, in northern coniferous forests, snowfall may break branches and small trees, causing openings where broadleaf trees such as aspen and birch can grow. Local disturbances such as storms and fires also create openings in many biomes.

The current concern about global warming is generating intense interest in the effect of climate on vegetation patterns. Using powerful new tools such as satellite imagery, scientists are documenting shifts in latitudes of biome borders, changes in snow and ice coverage, and changes in length of the growing season. At the same time, many natural biomes have been fragmented and altered by human activity. A high rate of biome alteration by humans is correlated with an unusually high rate of species loss throughout the globe.

Now let's begin our survey of the major terrestrial biomes. To help you locate the biomes, an orientation map color-coded to Figure 8 is included with each module. Icons indicate a relative temperate range (in red) and average annual precipitation (in blue) for each biome. Icons also identify biomes in which fire plays a significant role.

? Test your knowledge of world geography: Which biome is most closely associated with a "Mediterranean climate"?

Chaparral

▶ **Figure 8**
Major terrestrial biomes

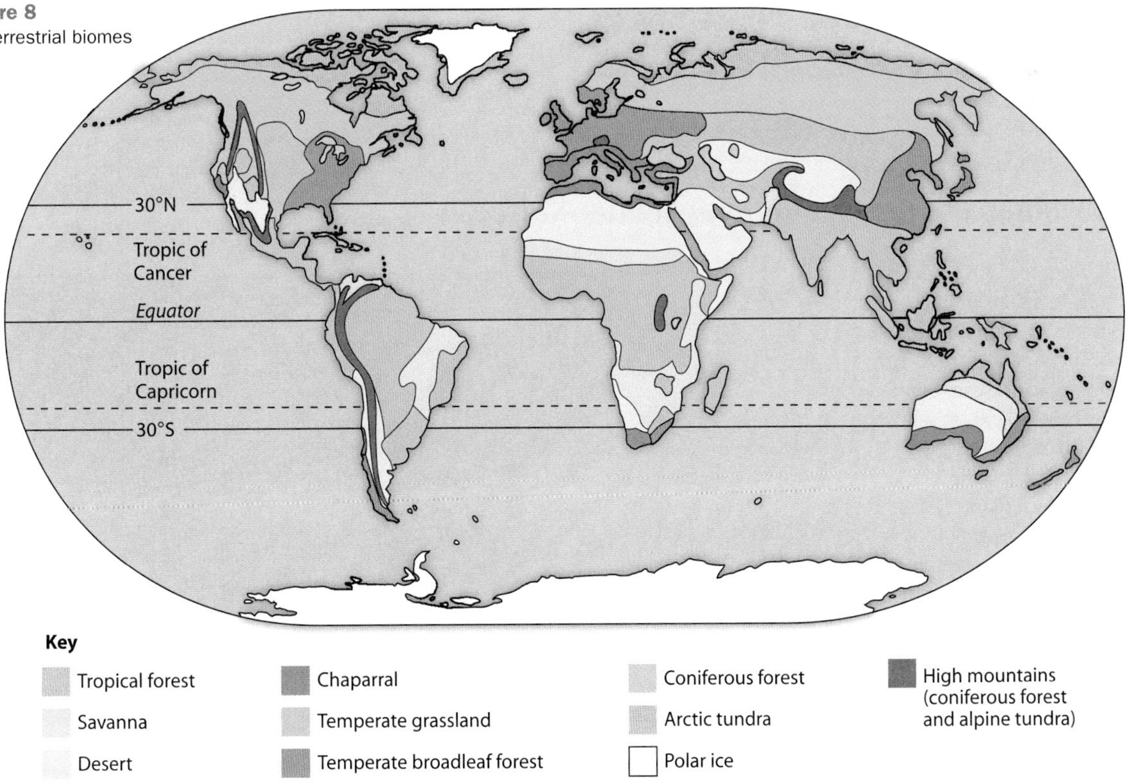

Adapted from Heinrich Walter and Siegmar-Walter Breckle. 2003. *Walter's Vegetation of the Earth*, fig. 16, p. 36. Springer-Verlag, © 2003.

9 Tropical forests cluster near the equator

Tropical forests occur in equatorial areas where the temperature is warm and days are 11–12 hours long year-round. Rainfall in these areas is quite variable, and this variability, rather than temperature or day length, generally determines the vegetation that grows in a particular tropical forest. In areas where rainfall is scarce or there is a prolonged dry season, tropical dry forests predominate. The plants found there are a mixture of thorny shrubs and deciduous trees and succulents. Tropical rain forests are found in very humid equatorial areas where rainfall is abundant (200–400 cm, or 79–157 inches, per year).

Tropical rain forest, such as the lush area on the island of Borneo shown in Figure 9, is among the most complex of all biomes, harboring enormous numbers of different species. Up to 300 species of trees can be found in a single hectare (2.5 acres). The forest structure consists of distinct layers that provide many different habitats: emergent trees growing above a closed upper canopy, one or two layers of lower trees, a shrub understory, and a sparse ground layer of herbaceous plants. Because of the closed canopy, little sunlight reaches the forest floor. Many trees are covered by woody vines growing toward the light. Other plants, including bromeliads and orchids, gain access to sunlight by growing on the branches or trunks of tall trees. Many of the animals also dwell in trees, where food is abundant. Monkeys, birds, insects, snakes, bats, and frogs find food and shelter many meters above the ground.

The soils of tropical rain forests are typically poor. High temperatures and rainfall lead to rapid decomposition. However, the nutrients released are quickly taken up by the luxuriant vegetation or washed away by the frequent rains.

Human impact on the world's tropical rain forests is an ongoing source of great concern. It is a common practice to clear the forest for lumber or simply burn it, farm the land for a few years, and then abandon it. Mining has also devastated large tracts of rain forest. Once stripped, the tropical rain forest recovers very slowly because the soil is so nutrient-poor.

▲ **Figure 9** Tropical rain forest

? Why are the soils in most tropical rain forests so poor in nutrients that they can only support farming for a few years after the forest is cleared?

The conditions favor rapid decomposition of organic litter in the soil and immediate uptake of the resulting nutrients by plants. Thus, most of the ecosystem's nutrients are tied up in the vegetation that is cleared away before farming.

10 Savannas are grasslands with scattered trees

Figure 10, a photograph taken in the Serengeti Plain in Tanzania, shows a typical **savanna**, a biome dominated by grasses and scattered trees. The temperature is warm year-round. Rainfall averages 30–50 cm (about 12–20 inches) per year, with dramatic seasonal variation. Poor soils and lack of moisture inhibit the establishment of most trees. Grazing animals and frequent fires, caused by lightning or human activity, further inhibit invasion by trees. Grasses survive burning because the growing points of their shoots are below ground. Savanna plants have also been selected for their ability to survive prolonged periods of drought. Many trees and shrubs are

Eric Isselée/Shutterstock

▲ **Figure 10** Savanna

deciduous, dropping their leaves during the dry season, an adaptation that helps conserve water.

Grasses and forbs (small broadleaf plants) grow rapidly during the rainy season, providing a good food source for many animal species. Large grazing mammals must migrate to greener pastures and scattered watering holes during seasonal drought. The dominant herbivores in savannas are actually insects, especially ants and termites. Also common are many burrowing animals, including mice, moles, gophers, snakes, ground squirrels, worms, and numerous arthropods.

Many of the world's large herbivores and their predators inhabit savannas. African savannas are home to giraffes, zebras, and many species of antelope, as well as to lions and cheetahs. Several species of kangaroo are the dominant mammalian herbivores of Australian savannas.

? How do fires help to maintain savannas as grassland ecosystems?

By repeatedly preventing the spread of trees and other woody plants; grasses survive because the growing points of their shoots are underground.

11 Deserts are defined by their dryness

Deserts are the driest of all terrestrial biomes, characterized by low and unpredictable rainfall (less than 30 cm—12 inches—per year). The large deserts in central Australia and northern Africa have average annual rainfalls of less than 2 cm, and in the Atacama Desert in Chile, the driest place on Earth, there is often no rain at all for decades at a time. But not all desert air is dry. Coastal sections of the Atacama and of the Namib Desert in Africa are often shrouded in fog, although the ground remains extremely dry.

As we discussed in Module 5, large tracts of desert occur in two regions of descending dry air centered around the 30° north and 30° south latitudes. At higher latitudes, large deserts may occur in the rain shadows of mountains (see Figure 5F); these encompass much of central Asia east of the Caucasus Mountains, and Washington and Oregon east of the Cascade Mountains. The Mojave Desert, shown in **Figure 11**, is in the rain shadow of the Sierra Nevada, along with much of the rest of Southern California and Nevada.

Some deserts, as represented by the temperature icon in Figure 11, are very hot, with daytime soil surface temperatures above 60°C (140°F) and large daily temperature fluctuations. Other deserts, such as those west of the Rocky Mountains, are relatively cold. Air temperatures in cold deserts may fall below −30°C (−22°F).

The cycles of growth and reproduction in the desert are keyed to rainfall. The driest deserts have no perennial vegetation at all, but less arid regions have scattered deep-rooted shrubs, often interspersed with water-storing succulents such as cacti. The leaves of some plants, including the Joshua tree shown in Figure 11, have a waxy coating that prevents water loss. Desert plants typically produce great numbers of seeds, which may remain dormant until a heavy rain triggers germination. After periods of rainfall (often in late winter), annual plants in deserts may display spectacular blooms.

Like desert plants, desert animals are adapted to drought and extreme temperatures. Many live in burrows and are active only during the cooler nights, and most have special adaptations that conserve water. Seed-eaters such as ants, many birds, and rodents are common in deserts. Lizards, snakes, and hawks eat the seed-eaters.

The process of **desertification**, the conversion of semi-arid regions to desert, is a significant environmental problem. In northern Africa, for example, a burgeoning human population, overgrazing, and dryland farming are converting large areas of savanna to desert.

? Why isn't "cold desert" an oxymoron?

Because deserts are defined by low precipitation and dry soil, not by temperature

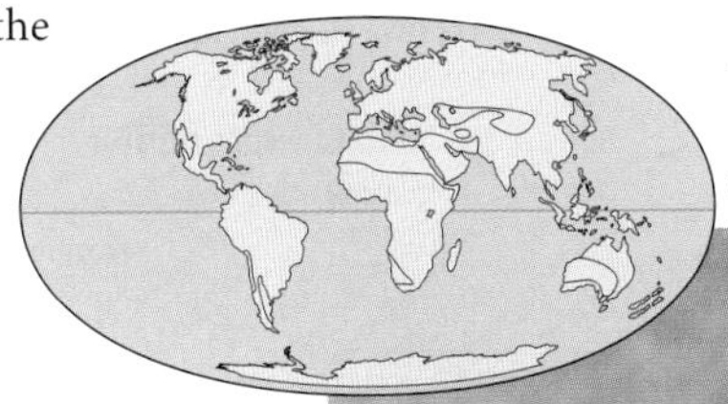

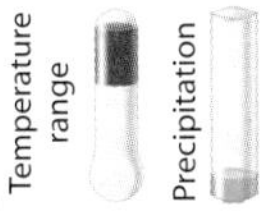

Dennis Frates/Workbook Stock/Getty Images

▲ **Figure 11** Desert

12 Spiny shrubs dominate the chaparral

Chaparral (a Spanish word meaning "place of evergreen scrub oaks") is characterized by dense, spiny shrubs with tough, evergreen leaves. The climate that supports chaparral vegetation results mainly from cool ocean currents circulating offshore, which produce mild, rainy winters and hot, dry summers. As a result, this biome is limited to small coastal areas, including California, where the photograph in **Figure 12** was taken. The largest region of chaparral surrounds the Mediterranean Sea; "Mediterranean" is another name for this biome. In addition to the perennial shrubs that dominate chaparral, annual plants are also commonly seen, especially during the wet winter and spring months. Animals characteristic of the chaparral include browsers such as deer, fruit-eating birds, and seed-eating rodents, as well as lizards and snakes.

Chaparral vegetation is adapted to periodic fires, most often caused by lightning. Many plants contain flammable chemicals and burn fiercely, especially where dead brush has accumulated. After a fire, shrubs use food reserves stored in the surviving roots to support rapid shoot regeneration. Some chaparral plant species produce seeds that will germinate only after a hot fire. The ashes of burned vegetation fertilize the soil with mineral nutrients, promoting regrowth of the plant community. Houses do not fare as well, and firestorms that race through the densely populated canyons of Southern California can be devastating.

? **What is one way that homeowners in chaparral areas can protect their neighborhoods from fire?**

They can keep the area clear of dead brush, which is highly flammable.

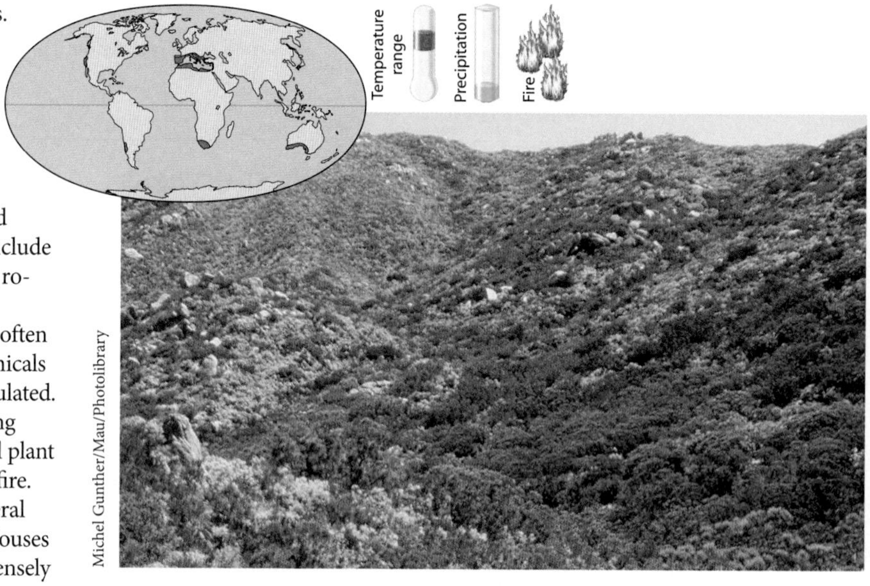

▲ **Figure 12** Chaparral

13 Temperate grasslands include the North American prairie

Temperate grasslands have some of the characteristics of tropical savannas, but they are mostly treeless, except along rivers or streams, and are found in regions of relatively cold winter temperatures. Precipitation, averaging between 25 and 75 cm (about 10–30 inches) per year, with periodic severe droughts, is too low to support forest growth. Fires and grazing by large mammals also inhibit growth of woody plants but do not harm the belowground grass shoots.

Large grazing mammals, such as the bison and pronghorn of North America and the wild horses and sheep of the Asian steppes, are characteristic of grasslands. Without trees, many birds nest on the ground, and some small mammals dig burrows to escape predators. Enriched by glacial deposits and mulch from decaying plant material, the soil of grasslands supports a great diversity of microorganisms and small animals, including annelids and arthropods.

The amount of annual precipitation influences the height of grassland vegetation. Shortgrass prairie is found in relatively dry regions; tallgrass prairie occurs in wetter areas. **Figure 13** shows a mixed-grass prairie in Alberta, Canada. Little remains of North American prairies today. Most of the region is intensively farmed, and it is one of the most productive agricultural regions in the world.

? **What factors inhibit woody plants from growing in temperate grasslands?**

Low rainfall, fires, and grazing by large mammals

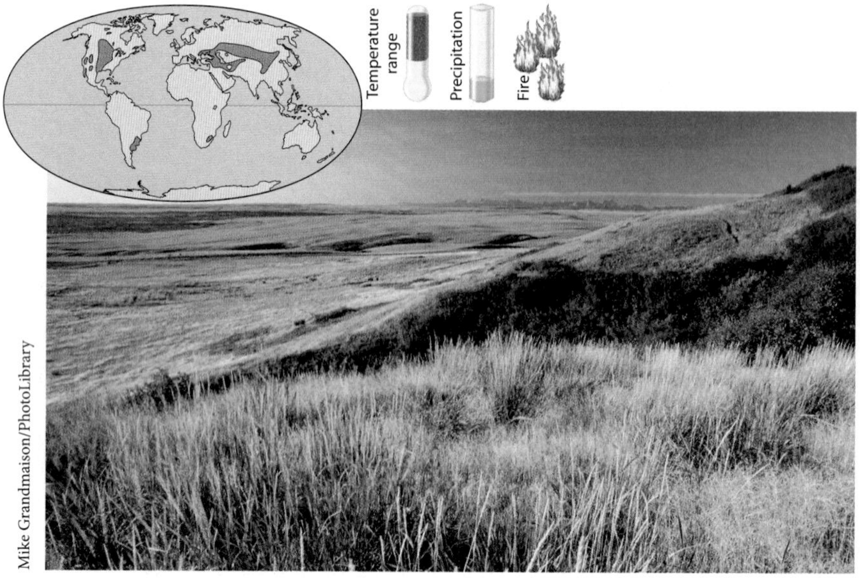

▲ **Figure 13** Temperate grassland

14 Broadleaf trees dominate temperate forests

Temperate broadleaf forests grow throughout midlatitude regions, where there is sufficient moisture to support the growth of large trees. In the Northern Hemisphere, deciduous trees (trees that drop their leaves seasonally) characterize temperate broadleaf forests. Some of the dominant trees include species of oak, hickory, birch, beech, and maple. The mix of tree species varies widely, depending on such factors as the climate at different latitudes, topography, and local soil conditions. **Figure 14** features a photograph taken during the spectacular display of autumn color in Great Smoky Mountains National Park in North Carolina.

Temperatures in temperate broadleaf forests range from very cold in the winter (−30°C) to hot in the summer (30°C). Annual precipitation is relatively high—75–150 cm (30–60 inches)—and usually evenly distributed throughout the year as either rain or snow. These forests typically have a growing season of 5 to 6 months and a distinct annual rhythm. Trees drop their leaves and become dormant in late autumn, preventing the loss of water from the tree at a time when frozen soil makes water less available. The trees produce new leaves in the spring.

The canopy of a temperate broadleaf forest is more open than that of a tropical rain forest, and the trees are not as tall or as diverse. However, their soils are richer in inorganic and organic nutrients. Rates of decomposition are lower in temperate forests than in the tropics, and a thick layer of leaf litter on forest floors conserves many of the biome's nutrients.

Numerous invertebrates live in the soil and leaf litter. Some vertebrates, such as mice, shrews, and ground squirrels, burrow for shelter and food, while others, including many species of birds, live in the trees. Predators include bobcats, foxes, black bears, and mountain lions.

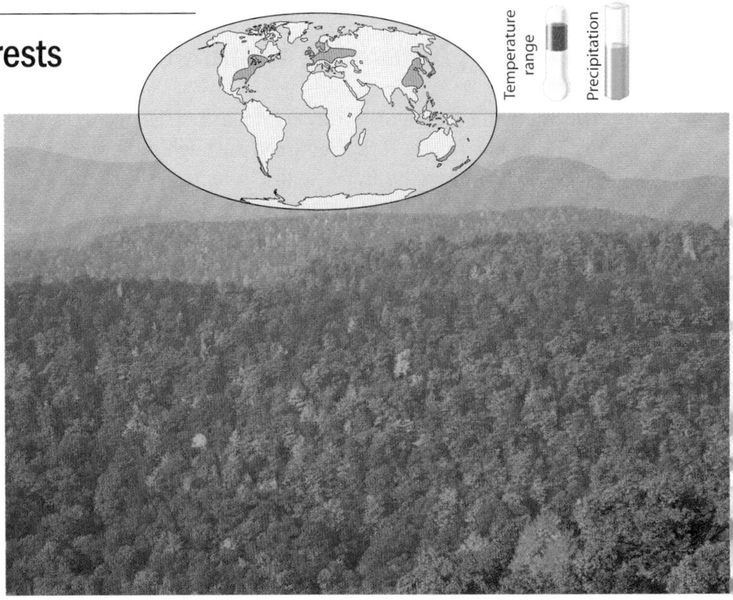

Kennan Ward/Corbis

▲ **Figure 14** Temperate broadleaf forest

? How does the soil of a temperate broadleaf forest differ from that of a tropical rain forest?

The soil in temperate broadleaf forests is rich in inorganic and organic nutrients, while the soil in tropical rain forests is low in nutrients.

15 Coniferous forests are often dominated by a few species of trees

Cone-bearing evergreen trees, such as spruce, pine, fir, and hemlock, dominate **coniferous forests**. The northern coniferous forest, or **taiga**, is the largest terrestrial biome on Earth, stretching in a broad band across North America and Asia south of the Arctic Circle. **Figure 15** shows taiga in Finland. Taiga is also found at cool, high elevations in more temperate latitudes, as in much of the mountainous region of western North America.

The taiga is characterized by long, cold winters and short, wet summers, which are sometimes warm. The soil is thin and acidic, and the slow decomposition of conifer needles makes few nutrients available for plant growth. Most of the precipitation is in the form of snow. The conical shape of many conifers prevents too much snow from accumulating on their branches and breaking them. Animals of the taiga include moose, elk, hares, bears, wolves, grouse, and migratory birds.

The **temperate rain forests** of coastal North America (from Alaska to Oregon) are also coniferous forests. Warm, moist air from the Pacific Ocean supports this unique biome, which, like most coniferous forests, is dominated by a few tree species, such as hemlock, Douglas fir, and redwood. These forests are heavily logged, and the old-growth stands of trees may soon disappear.

? How does the soil of the northern coniferous forests differ from that of a broadleaf forest?

The soil is thinner, nutrient-poor, and acidic because conifer needles decompose slowly in the low temperatures.

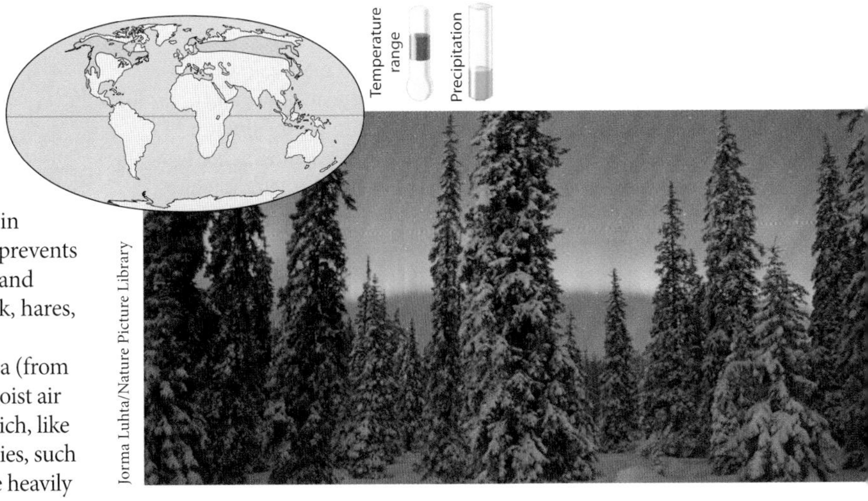

Jorma Luhta/Nature Picture Library

▲ **Figure 15** Coniferous forest

16 Long, bitter-cold winters characterize the tundra

Tundra (from the Russian word for "marshy plain") covers expansive areas of the Arctic between the taiga and polar ice.

Figure 16 shows the arctic tundra in central Alaska in the autumn. The climate here is often extremely cold, with little light for much of the autumn and winter. The arctic tundra is characterized by **permafrost**, continuously frozen subsoil—only the upper part of the soil thaws in summer. The arctic tundra may receive as little precipitation as some deserts. But poor drainage, due to the permafrost, and slow evaporation keep the soil continually saturated.

Permafrost prevents the roots of plants from penetrating very far into the soil, which is one factor that explains the absence of trees. Extremely cold winter air temperatures and high winds also contribute to the exclusion of trees. Vegetation in the tundra includes dwarf shrubs, grasses and other herbaceous plants, mosses, and lichens. During the brief, warm summers, when there is nearly constant daylight, plants grow quickly and flower in a rapid burst. High winds and cold temperatures create plant communities called alpine tundra on very high mountaintops at all latitudes, including the tropics. Although these communities are similar to arctic tundra, there is no permafrost beneath alpine tundra.

Animals of the tundra withstand the cold by having good insulation that retains heat. Large herbivores include musk oxen and caribou. The principal smaller animals are rodents called lemmings and a few predators, such as the arctic fox and snowy owl. Many animals are migratory, using the tundra as a summer breeding ground. During the brief warm season, the marshy ground supports the aquatic larvae of insects, providing food for migratory waterfowl, and clouds of mosquitoes often fill the tundra air.

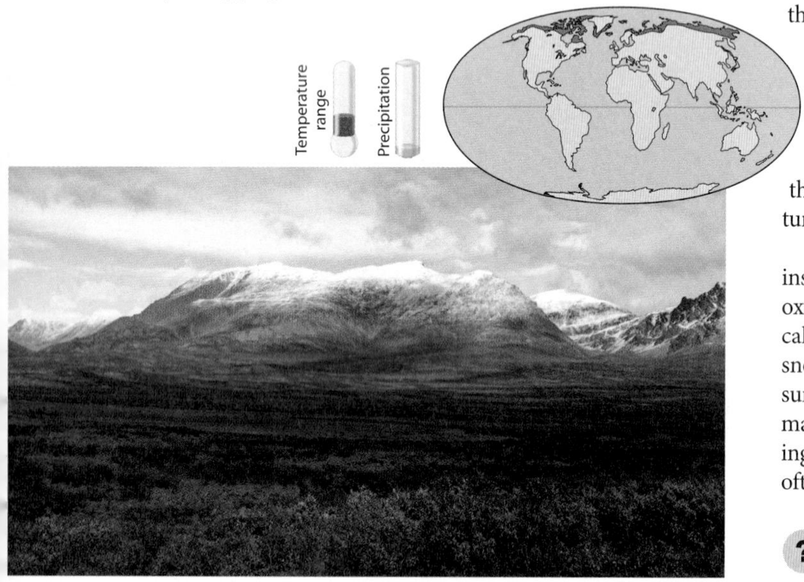

▲ **Figure 16** Tundra   Frontier Sights/Shutterstock

? What three abiotic factors account for the rarity of trees in arctic tundra?

Long, very cold winters (short growing season), high winds, and permafrost

17 Polar ice covers the land at high latitudes

In the Northern Hemisphere, **polar ice** covers land north of the tundra; much of the Arctic Ocean is continuously frozen as well. In the Southern Hemisphere, polar ice covers the continent of Antarctica (**Figure 17**), which is surrounded by a ring of sea ice, and numerous islands.

The temperature in these regions is extremely cold year-round, and precipitation is very low. Only a small portion of these landmasses is free of ice or snow, even during the summer. Nevertheless, small plants, such as mosses, and lichens manage to survive, and invertebrates such as nematodes, mites, and wingless insects called springtails inhabit the frigid soil.

The terrestrial polar biome is closely interconnected with the neighboring marine biome. Seals and marine birds, such as penguins, gulls, and skuas, feed in the ocean and visit the land or sea ice to rest and to breed. In the Northern Hemisphere, sea ice provides a feeding platform for polar bears.

? How does the vegetation found in polar ice regions compare with tundra vegetation?

Neither biome is hospitable to plants because of the cold temperatures. However, tundra supports the growth of small shrubs, while polar ice vegetation is limited to mosses and lichens.

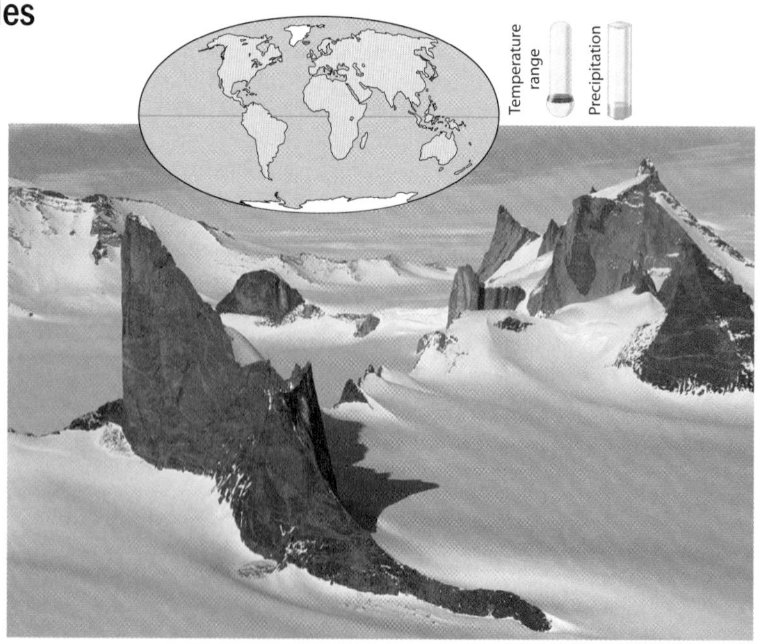

Gordon Wiltsie/National Geographic Image Collection

▲ **Figure 17** Polar ice

18 The global water cycle connects aquatic and terrestrial biomes

Ecological subdivisions such as biomes are not self-contained units. Rather, all parts of the biosphere are linked by the global water cycle, illustrated in **Figure 18**, and by nutrient cycles. Consequently, events in one biome may reverberate throughout the biosphere.

Recall from Module 5 that solar energy helps drive the movements of water and air in global patterns. In addition, precipitation and evaporation, as well as transpiration from plants, continuously move water between the land, oceans, and atmosphere. Over the oceans (left side of Figure 18), evaporation exceeds precipitation. The result is a net movement of water vapor to clouds that are carried by winds from the oceans across the land. On land (right side of the figure), precipitation exceeds evaporation and transpiration. Excess precipitation forms systems of surface water (such as lakes and rivers) and groundwater, all of which flow back to the sea, completing the water cycle.

Just as the water draining from your shower carries dead skin cells from your body along with the day's grime, the water washing over and through the ground carries traces of the land and its history. For example, water flowing from land to sea carries with it silt (fine soil particles) and chemicals such as fertilizers and pesticides. The accumulation of silt, aggravated by the development of coastal areas, has muddied the waters of some coral reefs, dimming the light available to the photosynthetic algae that power the reef community. Chemicals in surface water may travel hundreds of miles by stream and river to the ocean, where currents then carry them even farther from their point of origin. For instance, in the 1960s, researchers began finding traces of DDT in marine mammals in the Arctic, far from any places DDT had been used (see Module 2). Airborne pollutants such as nitrogen oxides and sulfur oxides, which combine with water to form acid precipitation, are also distributed by the water cycle.

Human activity also affects the global water cycle itself in a number of important ways. One of the main sources of atmospheric water is transpiration from the dense vegetation making up tropical rain forests. The destruction of these forests changes the amount of water vapor in the air. This, in turn, will likely alter local, and perhaps global, weather patterns. Pumping large amounts of groundwater to the surface for irrigation affects the water cycle, too. This practice can increase the rate of evaporation over land, resulting in higher humidity as well as depleting groundwater supplies.

? **What is the main way that living organisms contribute to the water cycle?**

Plants move water from the ground to the atmosphere via transpiration.

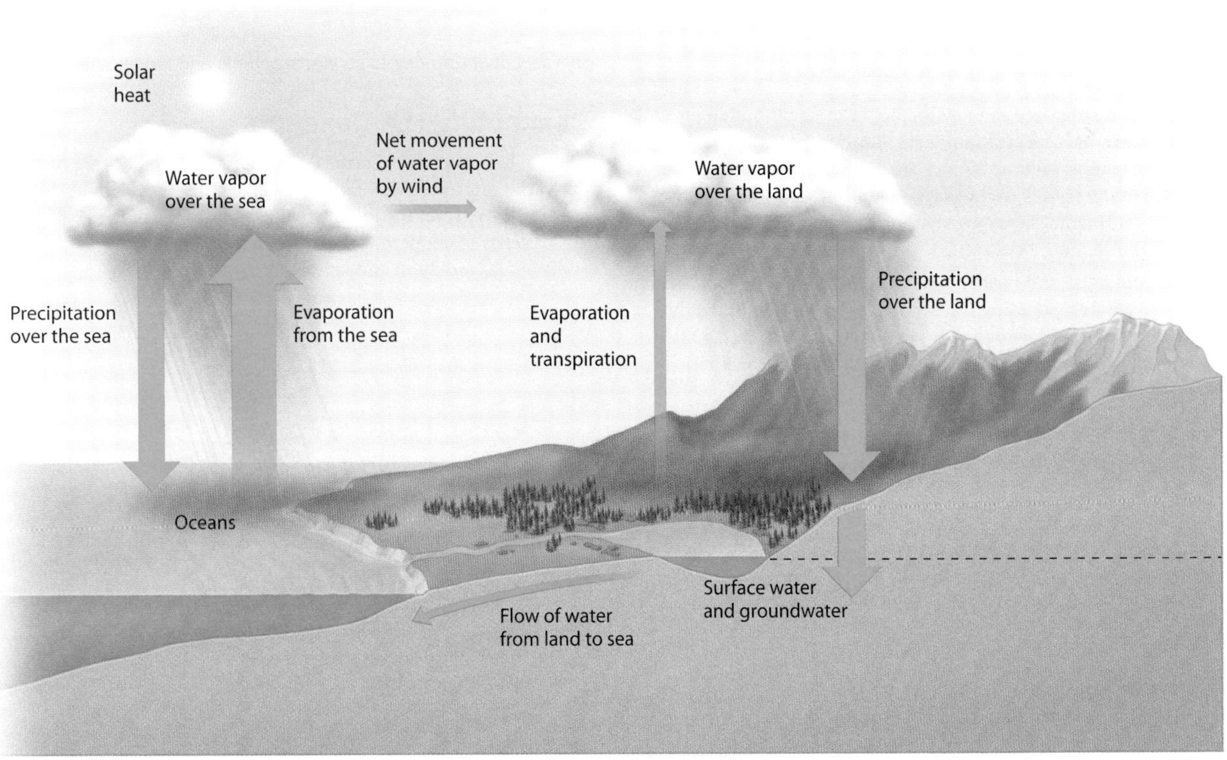

▲ **Figure 18** The global water cycle

REVIEW

For Practice Quizzes, BioFlix, MP3 Tutors, and Activities, go to www.masteringbiology.com.

Reviewing the Concepts

The Biosphere (1–5)

1 Ecologists study how organisms interact with their environment at several levels.

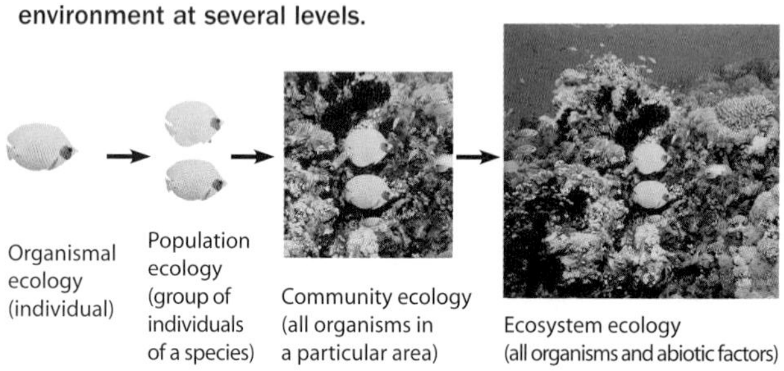

2 The science of ecology provides insight into environmental problems.

3 Physical and chemical factors influence life in the biosphere. Major factors include energy sources, temperature, the presence of water, and inorganic nutrients.

4 Organisms are adapted to abiotic and biotic factors by natural selection. The pronghorn's adaptations show the variety of factors that can affect an organism's fitness.

5 Regional climate influences the distribution of terrestrial communities. Most climatic variations are due to the uneven heating of Earth's surface as it orbits the sun, setting up patterns of precipitation and prevailing winds. Ocean currents influence coastal climate. Landforms such as mountains affect rainfall.

Aquatic Biomes (6–7)

6 Sunlight and substrate are key factors in the distribution of marine organisms. Marine biomes are found in both the pelagic and benthic realms, and the biomes are further distinguished by the availability of light. Coral reefs are found in warm, shallow waters above continental shelves. Other marine biomes are estuaries, wetlands, and the intertidal zone.

7 Current, sunlight, and nutrients are important abiotic factors in freshwater biomes. Standing water biomes (lakes and ponds) differ in structure from flowing water biomes (rivers and streams), and communities vary accordingly. Wetlands include marshes, swamps, and bogs.

Terrestrial Biomes (8–18)

8 Terrestrial biomes reflect regional variations in climate.

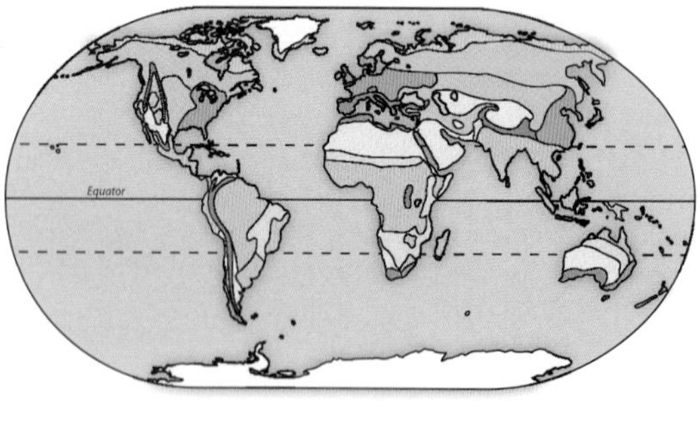

9 Tropical forests cluster near the equator.

10 Savannas are grasslands with scattered trees.

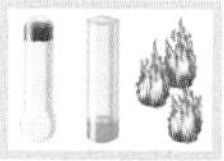

11 Deserts are defined by their dryness.

12 Spiny shrubs dominate the chaparral.

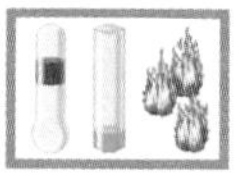

13 Temperate grasslands include the North American prairie.

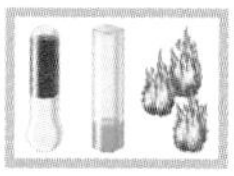

14 Broadleaf trees dominate temperate forests.

15 Coniferous forests are often dominated by a few species of trees.

16 Long, bitter-cold winters characterize the tundra.

17 Polar ice covers the land at high latitudes.

18 The global water cycle connects aquatic and terrestrial biomes.

Connecting the Concepts

1. You have seen that Earth's terrestrial biomes reflect regional variations in climate. But what determines these climatic variations? Interpret the following diagrams in reference to global patterns of temperature, rainfall, and winds.
   a. Solar radiation and latitude:

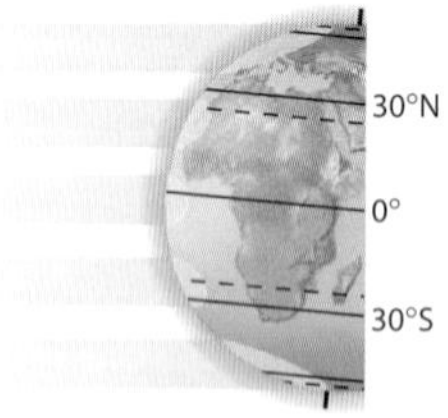

   b. Earth's orbit around the sun:

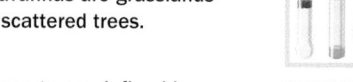

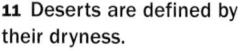

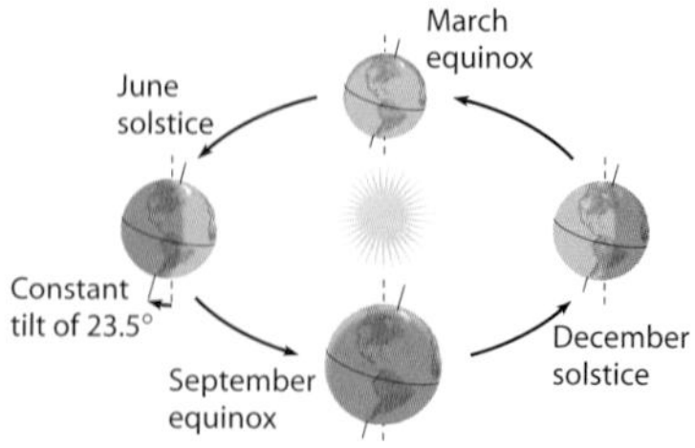

c. Global patterns of air circulation and rainfall:

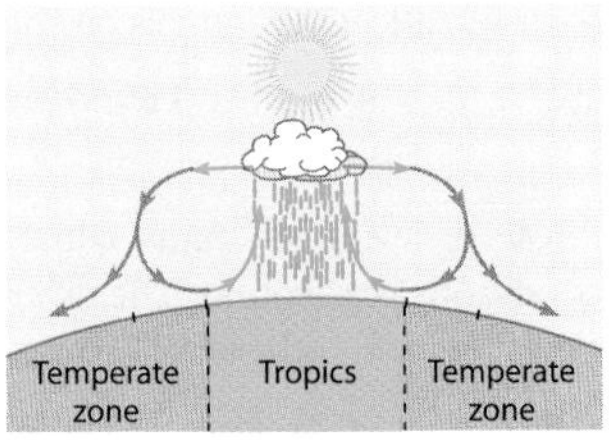

2. Aquatic biomes differ in levels of light, nutrients, oxygen, and water movement. These abiotic factors influence the productivity and diversity of freshwater ecosystems.
   a. Productivity, roughly defined as photosynthetic output, is high in estuaries, coral reefs, and shallow ponds. Describe the abiotic factors that contribute to high productivity in these ecosystems.
   b. How does extra input of nitrogen and phosphorus (for instance, by fertilizer runoff) affect the productivity of lakes and ponds? Is this nutrient input beneficial for the ecosystem? Explain.

Testing Your Knowledge

Matching

3. The most complex and diverse biome
4. Ground permanently frozen
5. Deciduous trees such as hickory and birch
6. Mediterranean climate
7. Spruce, fir, pine, and hemlock trees
8. Home of ants, antelopes, and lions
9. North American plains

a. chaparral
b. savanna
c. taiga
d. temperate broadleaf forest
e. temperate grassland
f. tropical rain forest
g. arctic tundra

Multiple Choice

10. Changes in the seasons are caused by
    a. the tilt of Earth's axis toward or away from the sun.
    b. annual cycles of temperature and rainfall.
    c. variation in the distance between Earth and the sun.
    d. an annual cycle in the sun's energy output.
    e. the periodic buildup of heat energy at the equator.
11. What makes the Gobi Desert of Asia a desert?
    a. The growing season there is very short.
    b. Its vegetation is sparse.
    c. It is hot.
    d. Temperatures vary little from summer to winter.
    e. It is dry.
12. Which of the following sea creatures might be described as a pelagic animal of the aphotic zone?
    a. a coral reef fish
    b. a giant clam near a deep-sea hydrothermal vent
    c. an intertidal snail
    d. a deep-sea squid
    e. a harbor seal
13. Why do the tropics and the windward side of mountains receive more rainfall than areas around latitudes 30° north and south and the leeward side of mountains?
    a. Rising warm, moist air cools and drops its moisture as rain.
    b. Descending air condenses, creating clouds and rain.
    c. The tropics and the windward side of mountains are closer to the ocean.
    d. There is more solar radiation in the tropics and on the windward side of mountains.
    e. Earth's rotation creates seasonal differences in rainfall.
14. Phytoplankton are the major photosynthesizers in
    a. the benthic realm of the ocean.
    b. streams.
    c. the ocean photic zone.
    d. the intertidal zone.
    e. the aphotic zone of a lake.
15. An ecologist monitoring the number of gorillas in a wildlife refuge over a 5-year period is studying ecology at which level?
    a. organism
    b. population
    c. community
    d. ecosystem
    e. biosphere
16. Many plant species have adaptations for dealing with periodic fires. Such fires are typical of a
    a. chaparral.
    b. savanna.
    c. temperate grassland.
    d. temperate broadleaf forest.
    e. a, b, or c

Describing, Comparing, and Explaining

17. Tropical rain forests are the most diverse biomes. What factors contribute to this diversity?
18. What biome do you live in? Describe your climate and the factors that have produced that climate. What plants and animals are typical of this biome? If you live in an urban or agricultural area, how have human interventions changed the natural biome?

Applying the Concepts

19. In the climograph below, biomes are plotted by their range of annual mean temperature and annual mean precipitation. Identify the following biomes: arctic tundra, coniferous forest, desert, grassland, temperate forest, and tropical forest. Explain why there are areas in which biomes overlap on this graph.

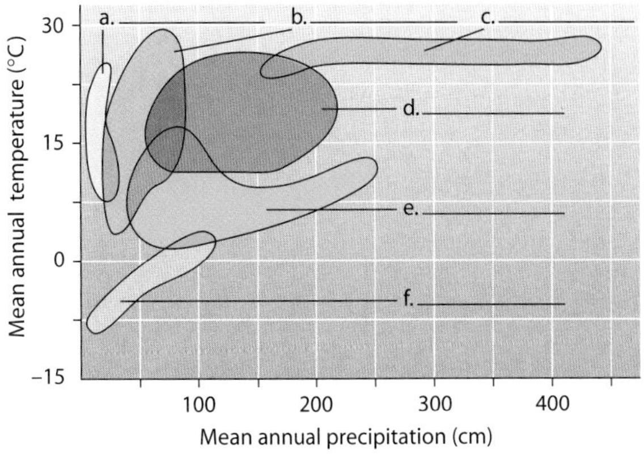

20. The North American pronghorn looks and acts like the antelopes of Africa. But the pronghorn is the only survivor of a family of mammals restricted to North America. Propose a hypothesis to explain how these widely separated animals came to be so much alike.

Review Answers

1. a. The shape of Earth results in uneven heating, such that the tropics are warm and polar regions are cold.
   b. The seasonal differences of winter and summer in temperate and polar regions are produced as Earth tips toward or away from the sun during its orbit around the sun.
   c. Intense solar radiation in the tropics evaporates moisture; warm air rises, cools, and drops its moisture as rain; air circulates, cools, and drops around 30°N and S, warming as it descends and evaporating moisture from land, which creates arid regions.
2. a. All three areas have plenty of sunlight and nutrients.
   b. The addition of nitrogen or phosphorus "fertilizes" the algae living in ponds and lakes, leading to explosive population growth. (These nutrients are typically in short supply in aquatic ecosystems.) The effects are harmful to the ecosystem. Algae cover the surface, reducing light penetration. When the algae die, bacterial decomposition of the large amount of biomass can deplete the oxygen available in the pond or lake, which may adversely affect the animal community.
3. f 4. g 5. d 6. a 7. c 8. b 9. e 10. a 11. e 12. d 13. a 14. c 15. b 16. e
17. Tropical rain forests have a warm, moist climate, with favorable growing conditions year-round. The diverse plant growth provides various habitats for other organisms.
18. After identifying your biome by looking at the map in Figure 8, review Module 5 on climate and the module that describes your biome (Modules 9–17).
19. a. desert; b. grassland; c. tropical forest; d. temperate forest; e. coniferous forest; f. arctic tundra. Areas of overlap have to do with seasonal variations in temperature and precipitation.
20. Through convergent evolution, these unrelated animals adapted in similar ways to similar environments—temperate grasslands and savanna.

Behavioral Adaptations to the Environment

Behavioral Adaptations to the Environment

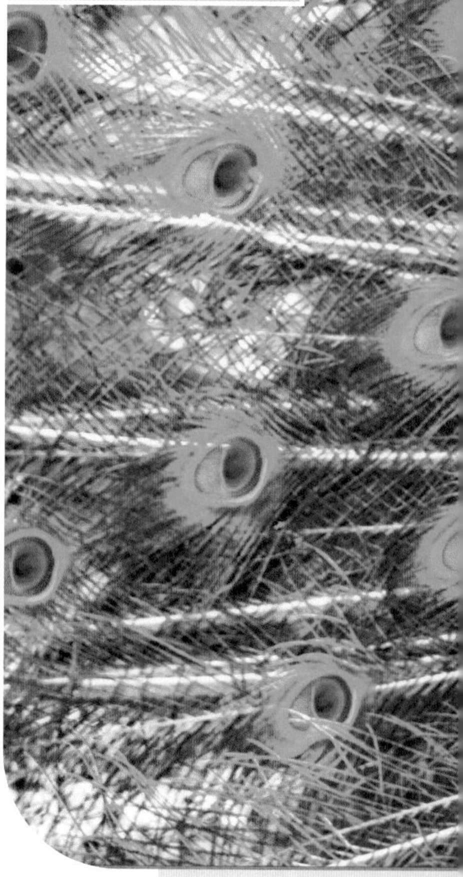

BIG IDEAS

Theo Allofs/Danita Delimont

The Scientific Study of Behavior
(1–3)

Behavioral ecologists study behavior in an evolutionary context, asking both *how* behaviors are triggered and *why* behaviors occur.

John Cancalosi/Photolibrary

Learning
(4–11)

Learning, which encompasses a broad range of behaviors, enables animals to change their behaviors in response to changing environmental conditions.

Carol Walker/Nature Picture Library

Survival and Reproductive Success
(12–16)

Many animal behaviors are adaptations that improve the ability to obtain food or increase reproductive success.

Rupert Barrington/Nature Picture Library

Social Behavior and Sociobiology
(17–23)

The behaviors of animals that live in groups may include territoriality, conflict resolution, dominance hierarchies, and altruism.

AlexeyVis/iStockphoto

When composer Cole Porter wrote the famous lines, "Birds do it, bees do it/Even educated fleas do it," he was referring to one of the most fundamental activities in the animal world—mating. Many animals, including humans, expend enormous amounts of energy persuading members of the opposite sex to mate with them. The peacock (*Pavo cristatus*) shown above is a case in point. During mating season, male peacocks assemble on a communal display ground and shake their flamboyant tail feathers for a shrewd audience of females. Only a fraction of the males are judged satisfactory. Few of those are popular enough to attract more than one female, but a particularly handsome specimen might score several copulations with different females. The sperm passed during these brief couplings are the male's entire contribution to his future offspring.

In contrast, consider the behavior of the small, mouse-like prairie vole (*Microtus ochrogaster*). A prairie vole relationship begins when a female sniffs the scent of a potential male partner. The smell causes her to become sexually receptive, which leads to repeated matings over the next 24 or more hours. During this honeymoon period, biochemical responses in the brains of both partners cement a lasting bond between them. Afterward, the pair associate closely and exclusively with each other, sharing a nest and care of their young. The relationship endures throughout life. If one partner dies, the other is unlikely to bond with a new mate.

In this chapter, you'll learn how evolutionary biologists explain such strikingly different adaptations for maximizing reproductive success. You'll also learn about several other categories of behavior (including learning!) and explore the roles of genetics and the environment in determining behavior.

The Scientific Study of Behavior

1 Behavioral ecologists ask both proximate and ultimate questions

Behavior encompasses a wide range of activities. At its most basic level, a behavior is an action carried out by muscles or glands under the control of the nervous system in response to an environmental cue. Collectively, behavior is the sum of an animal's responses to internal and external environmental cues. Although we commonly think of behavior in terms of observable actions—for instance, a courtship dance or an aggressive posture—other activities are also considered behaviors. Chemical communication, such as secreting a chemical that attracts mates or marks a territory, is a form of behavior. Learning is also considered a behavioral process.

The foundation for the scientific study of behavior was established in the mid-20th century with the work of Karl von Frisch and Konrad Lorenz, of Austria, and Niko Tinbergen, of the Netherlands. These researchers shared a Nobel Prize in 1973 for their discoveries in animal behavior. Von Frisch, who pioneered the use of experimental methods in behavior, studied honeybee behavior in detail. You'll learn about some of his work in Module 13. Konrad Lorenz, often regarded as the founder of behavioral biology, emphasized the importance of studying and comparing the behavior of various animals in response to different stimuli. Tinbergen worked closely with Lorenz, concentrating on experimental studies of inborn behavior and simple forms of learning.

Behavioral ecology is the study of behavior in an evolutionary context. Behavioral ecologists draw on the knowledge of a variety of disciplines to describe the details of animal behaviors and investigate how they develop, evolve, and contribute to the animal's survival and reproductive success.

The questions investigated by behavioral ecologists fall into two broad categories. **Proximate questions** concern the immediate reason for a behavior—how it is triggered by **stimuli** (environmental cues that cause a response), what physiological or anatomical mechanisms play a role, and what underlying genetic factors are at work. For example, researchers studying the mating pattern of prairie voles might ask, "How do voles choose their mates?" or "How does the act of mating cause voles to form lifelong bonds with their partners?" or "How are genetic factors involved in mating behavior?" Generally, proximate questions help us understand *how* a behavior occurs. **Proximate causes** are the answers to such questions about the immediate mechanism for a behavior.

Ultimate questions address *why* a particular behavior occurs. As a component of the animal's phenotype, behaviors are adaptations that have been shaped by natural selection. The answers to ultimate questions, or **ultimate causes**, are evolutionary explanations—they lie in the adaptive value of the behavior. For example, researchers think that at some point in their evolutionary history, the ancestors of prairie voles had numerous sexual partners, like the overwhelming majority of mammals. Why did natural selection favor the change in mating behavior? To explore this ultimate question, researchers test hypotheses on the adaptive value of prairie voles' bonding behavior. For example, perhaps male prairie voles that form lasting bonds have greater reproductive success because they prevent other males from getting close to their mates. An experiment to determine whether a female vole would have sexual intercourse with other males if she were not constantly accompanied by her mate would shed light on one possible adaptive value of bond formation by males.

In the next module, we look at a type of behavior that demonstrates the complementary nature of proximate and ultimate questions.

? When you touch a hot plate, your arm automatically recoils. What might be the proximate and ultimate causes of this behavior?

● The proximate cause is a simple reflex, a neural pathway linking stimulation of receptors in your finger to a motor response by muscles of your arm and hand; the ultimate cause is natural selection for a behavior that minimizes damage to the body, thereby contributing to survival and reproductive success.

2 Fixed action patterns are innate behaviors

One important proximate question is how a behavior develops during an animal's life span. Lorenz and Tinbergen were among the first to demonstrate the importance of **innate behavior**, behavior that is under strong genetic control and is performed in virtually the same way by all individuals of a species. Many of Lorenz's and Tinbergen's studies were concerned with behavioral sequences called **fixed action patterns (FAPs)**. A FAP is an unchangeable series of actions triggered by a specific stimulus. Once initiated, the sequence is performed in its entirety, regardless of any changes in circumstances. Consider a coffee vending machine as an analogy. The purchaser feeds money into the machine and presses a button. Having received this stimulus, the machine performs a series of actions: drops a cup into place; releases a specific volume of coffee; adds cream; adds sugar. Once the stimulus—in this case, the money—triggers the mechanism, it carries out its complete program. Likewise, FAPs are behavioral routines that are completed in full.

Figure 2A (top of facing page) illustrates one of the FAPs that Lorenz and Tinbergen studied in detail. The bird is the graylag goose, a common European species that nests in shallow depressions on the ground. If the goose happens to bump one of her eggs out of the nest, she always retrieves it in the same manner. As shown in the figure, she stands up, extends her neck, uses a side-to-side head motion to nudge the egg back with her beak, and then sits down on the nest again. If the

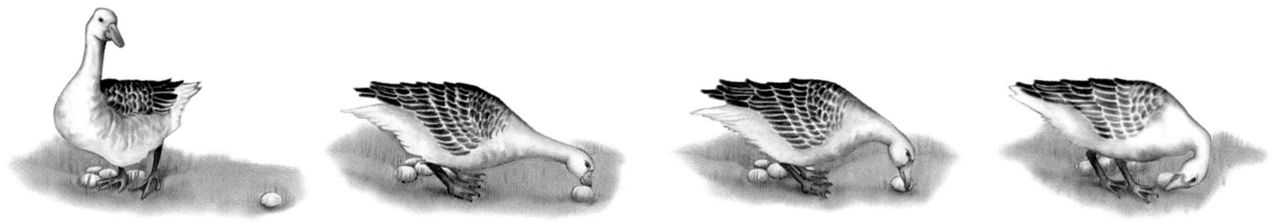

▲ **Figure 2A** A graylag goose retrieving an egg—a FAP

egg slips away (or is pulled away by an experimenter) while the goose is retrieving it, she continues as though the egg were still there. Only after she sits back down on her eggs does she seem to notice that the egg is still outside the nest. Then she begins another retrieval sequence. If the egg is again pulled away, the goose repeats the retrieval motion without the egg. A goose would even perform the sequence when Lorenz and Tinbergen placed a foreign object, such as a small toy or a ball, near her nest. The goose performs the series of actions regardless of the absence of an egg, just as a coffee vending machine does when the cup dispenser is empty—the coffee, cream, and sugar are poured anyway.

In its simplest form, a fixed action pattern is an innate response to a certain stimulus. For the graylag goose, the stimulus for egg retrieval is the presence of an egg (or other object) near the nest. Such relatively simple, innate behaviors seem to occur in all animals. When a baby bird senses that an adult bird is near, it responds with a FAP: It begs for food by raising its head, opening its mouth, and cheeping. In turn, the parent responds with another FAP: It stuffs food into the gaping mouth. Humans perform FAPs, too. Infants grasp strongly with their fingers in response to a touch stimulus on the palm of the hand. They smile in response to a face or even something that vaguely resembles a face, such as two dark spots in a circle.

Although a single FAP is typically a simple behavior, complex behaviors can result from several FAPs performed sequentially. Many vertebrates engage in courtship rituals that consist of chains of FAPs, as you'll learn in Module 14. The completion of a single FAP by one partner cues the other partner to begin its next FAP. There is some flexibility in these patterns. For example, a segment of the pattern might be repeated if the partner does not readily respond.

What might be the ultimate causes of FAPs? Automatically performing certain behaviors may maximize fitness to the point that genes that result in variants of that behavior do not persist in the population. For example, there are some things that a young animal has to get right on the first try if it is to stay alive. Consider kittiwakes, gulls that nest on cliff ledges. Unlike other gull species, kittiwakes show an innate aversion to cliff edges; they turn away from the edge. Chicks in earlier generations that did not show this edge-aversion response would not have lived to pass the genes for their risk-taking behavior on to the next generation.

Fixed action patterns for reproductive behaviors are also under strong selection pressure. One example is the behavior of mated king penguins. Each member of the pair takes a turn incubating their egg while its mate feeds (**Figure 2B**). Standing face-to-face, the pair must execute a delicate series of maneuvers to pass the egg from the tops of one penguin's feet to the tops of its partner's feet, where the egg will incubate in a snug fold of the abdominal skin. (You may have seen emperor penguins engage in this behavior in the film *The March of the Penguins.*) If either partner makes a mistake, the egg may roll onto the ice, where it can freeze in seconds, eliminating the pair's only chance of successful reproduction for the year.

Innate behaviors are under strong genetic control, but the animal's performance of most innate behaviors also improves with experience. And despite the genetic component, input from the environment—an object or sensory stimulus, for example—is required to trigger the behavior. In the next module, we look at the interaction of genes and environment in producing a behavior.

? How would you explain FAPs in the context of proximate and ultimate causes?

The proximate cause of a FAP is often a simple environmental cue. The ultimate cause is that natural selection favors behaviors that enable animals to perform tasks essential to survival without any previous experience.

Theo Allofs/Danita Delimont

▲ **Figure 2B** A pair of king penguins transferring their egg

3 Behavior is the result of both genetic and environmental factors

Evolutionary explanations for behavior assume that it has a genetic basis. Many scientific studies have corroborated that specific behaviors do indeed have a genetic component. Until recently, though, the heritability of a trait could only be estimated by traditional methods such as constructing pedigrees and performing breeding experiments. With the tools of molecular genetics, scientists have begun to investigate the roles of specific genes in behavior.

Groundbreaking experiments with fruit flies have led to the discovery of genes that govern learning, memory, internal clocks, and courtship and mating behaviors. For example, a male fruit fly courts a female with an elaborate series of actions that include tapping the female's abdomen with a foreleg and vibrating one of his wings in a courtship "song." Researchers identified a master gene known as *fruitless* (*fru* for short), which in males encodes a protein that switches on a suite of genes responsible for courtship behavior (**Table 3**, on facing page). Male fruit flies that possess a mutated version of *fru* attempt to court other male flies. In females, the protein encoded by *fru* is different from the male version of the protein. But when researchers used genetic engineering to produce female flies that made the male *fru* protein, the females behaved like normal males, vigorously courting other females.

Some of the genes governing fruit fly behavior have counterparts in mice and even in humans. The genetic underpinnings of courtship and mating behaviors in mammals are probably quite different from those of flies, but research on genes implicated in learning and memory has yielded promising results. By studying fruit fly mutants with colorful names like *dunce*, *amnesiac*, and *rutabaga*, scientists have identified key components of memory storage and have even used genetic engineering to produce fruit flies that have exceptionally good memories. Similar genes and their protein products have been identified in mice and humans, sparking a flood of research on memory-enhancing drugs. Such drugs could improve the quality of life for people suffering from Alzheimer's and other neurological diseases that impair memory.

Phenotype depends on the environment as well as genes. Many environmental factors, including diet and social interactions, can modify how genetic instructions are carried out. In some animals, even an individual's sex can be determined by the environment. For example, sex determination in some reptiles depends on the temperature of the egg during embryonic development.

Let's look at a study that illustrates the influence of environment on behavior (**Figure 3**). Some female Norway rats (*Rattus norvegicus*) spend a great deal of time licking and grooming their offspring (called pups), while others have little interaction with their pups. Pups raised by these "low-interaction" mothers tend to be more sensitive as adults to stimuli that trigger the "fight-or-flight" stress response and thus are more fearful and anxious in new situations. Female pups from these litters become low-interaction mothers themselves. On the other hand, pups with "high-interaction"

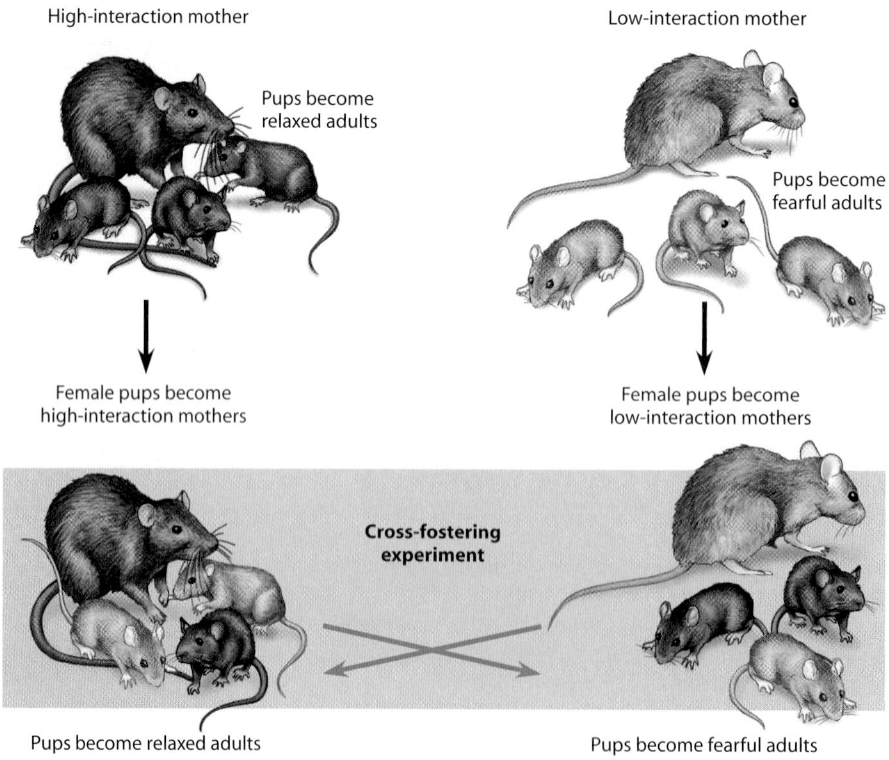

▶ **Figure 3** A cross-fostering experiment with rats

TABLE 3 | COURTSHIP BEHAVIOR OF FRUIT FLIES WITH NORMAL OR MUTATED *FRUITLESS (FRU)* GENE

	Male	Genetically Altered Female
Normal male *fru*	Courts females	Courts females
Mutated male *fru*	Courts males	——

mothers are more relaxed in stressful situations as adults, and the female pups from these litters become high-interaction mothers.

To investigate whether these responses to stress are entirely determined by genetics or are influenced by the interactions with their mothers, researchers performed "cross-fostering" experiments. They placed pups born to high-interaction mothers in nests with low-interaction mothers, and pups born to low-interaction mothers in nests with high-interaction mothers. As adults, the cross-fostered rats responded to stress more like their foster mothers than like their biological mothers. The results showed that the pups' environment—in this case maternal behavior—was the determining factor in their anxiety level. Remarkably, these experiments also demonstrated that behavioral changes can be passed to future generations, not through genes, but through the social environment. When the female rats that had been fostered gave birth, they showed the same degree of interaction with their pups as their foster mothers had shown with them. In further studies, researchers learned that interaction with the mother changes the pattern of gene expression in the pups, thus affecting the development of parts of the neuroendocrine system that regulate the fight-or-flight response.

These experiments and others provide evidence that behavior is the product of both genetic *and* environmental factors. Indeed, the interaction of genes and the environment appears to determine most animal behaviors. One of the most powerful ways that the environment can influence behavior is through learning, the topic we consider next.

? Without doing any experiments, how might you distinguish between a behavior that is mostly controlled by genes and one that is mostly determined by the environment?

In many cases, genetically controlled behavior would not differ much between populations, regardless of the environment, while the environmentally controlled behavior would differ widely across populations located in different environments.

Learning

4 Habituation is a simple type of learning

Learning is modification of behavior as a result of specific experiences. Learning enables animals to change their behaviors in response to changing environmental conditions. As Table 4 indicates, there are various forms of learning, ranging from a simple behavioral change in response to a single stimulus, to complex problem solving that uses entirely new behaviors.

One of the simplest forms of learning is **habituation**, in which an animal learns not to respond to a repeated stimulus that conveys little or no information. There are many examples of habituation in both invertebrate and vertebrate animals. The cnidarian *Hydra*, for example, contracts when disturbed by a slight touch; it stops responding, however, if disturbed repeatedly by such a stimulus. Similarly, a scarecrow stimulus will usually make birds avoid a tree with ripe fruit for a few days. But the birds soon become habituated to the scarecrow and may even land on it on their way to the fruit tree. Once habituated to a particular stimulus, an animal still senses the stimulus—its sensory organs detect it—but the animal has learned not to respond to it.

In terms of ultimate causation, habituation may increase fitness by allowing an animal's nervous system to focus on stimuli that signal food, mates, or real danger and not waste time or energy on a vast number of other stimuli that are irrelevant to survival and reproduction.

Proximate and ultimate causes of behavior are also evident in imprinting, the type of learning we introduce in the next module.

TABLE 4 | TYPES OF LEARNING

Learning Type	Defining Characteristic
Habituation	Loss of response to a stimulus after repeated exposure
Imprinting	Learning that is irreversible and limited to a sensitive time period in an animal's life
Spatial learning	Use of landmarks to learn the spatial structure of the environment
Cognitive mapping	An internal representation of the spatial relationships among objects in the environment
Associative learning	Behavioral change based on linking a stimulus or behavior with a reward or punishment; includes trial-and-error learning
Social learning	Learning by observing and mimicking others
Problem solving	Inventive behavior that arises in response to a new situation

? Your new roommate hums continuously while studying. You found this habit extremely annoying at first, but after a while you stopped noticing it. What kind of learning accounts for your tolerance of your roommate's humming?

Habituation

5 Imprinting requires both innate behavior and experience

Learning often interacts closely with innate behavior. Some of the most interesting examples of such an interaction involve the phenomenon known as imprinting. **Imprinting** is learning that is limited to a specific time period in an animal's life and that is generally irreversible. The limited phase in an animal's development when it can learn certain behaviors is called the **sensitive period**.

In classic experiments done in the 1930s, Konrad Lorenz used the graylag goose to demonstrate imprinting. When incubator-hatched goslings spent their first few hours with Lorenz, rather than with their mother, they steadfastly followed Lorenz (Figure 5A) and showed no recognition of their mother or other adults of their species. Even as adults, the birds continued to prefer the company of Lorenz and other humans to that of geese.

Lorenz demonstrated that the most important imprinting stimulus for graylag geese was movement of an object (normally the parent bird) away from the hatchlings. The effect of movement was increased if the moving object emitted some sound. The sound did not have to be that of a goose, however; Lorenz found that a box with a ticking clock in it was readily and permanently accepted as a "mother."

Just as a young bird requires imprinting to know its parents, the adults must also imprint to recognize their young. For a day or two after their own chicks hatch, adult herring gulls will accept and even defend a strange chick introduced into their nesting territory. However, once imprinted on their offspring, adults will kill any strange chicks that arrive later.

Not all examples of imprinting involve parent-offspring bonding. Newly hatched salmon, for instance, do not receive parental care but seem to imprint on the complex mixture of odors unique to their stream. This imprinting enables adult salmon to find their way back to their home stream to spawn after spending a year or more at sea.

For many kinds of birds, imprinting plays a role in song development. For example, researchers studying song development in white-crowned sparrows found that male birds memorize the song of their species during a sensitive period (the first 50 days of life). They do not sing during this phase, but several months later they begin to practice this song, eventually learning to reproduce it correctly. The birds do not need to hear the adult song during their practice phase; isolated males raised in soundproof chambers learned to sing normally as long as they had heard a recorded song of their species during the sensitive period (Figure 5B, top sonogram). In contrast, isolated males that did not hear their species' song until they were more than 50 days old sang an abnormal song (Figure 5B, bottom). Researchers also discovered a purely genetic component of white-crowned sparrow song development: Isolated males exposed to recorded songs of *other* species during the sensitive period did not adopt these foreign songs. When they later learned to sing, these birds sang an abnormal song similar to that of the isolated males of their own species that had heard no recorded bird songs.

The ability of parents and offspring to keep track of each other and the ability of male songbirds to attract mates are examples of behaviors that have direct and immediate effects on survival and reproduction. Imprinting provides a way for such behavior to become more or less fixed in an animal's nervous system.

Thomas McAvoy/Time & Life Pictures/Getty Images

▲ **Figure 5A** Konrad Lorenz with geese imprinted on him

? Explain why we say that imprinting has both innate and learned components.

● Its innate component is the tendency to imprint on a stimulus during a sensitive period. The imprinting itself is a form of learning.

Frank Lane Picture Agency/Corbis

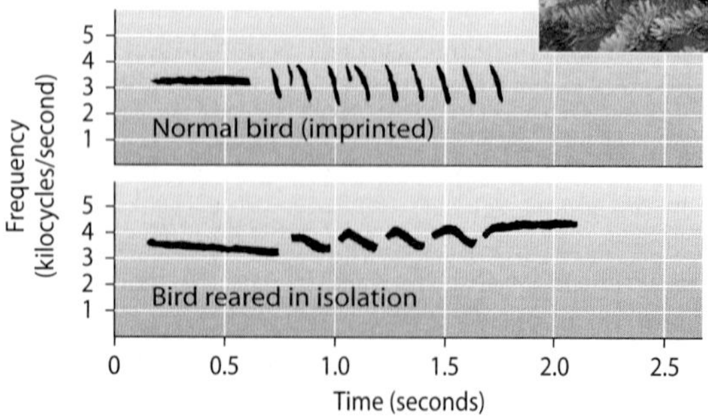

▲ **Figure 5B** The songs of male white-crowned sparrows that heard an adult sing during the sensitive period (top) and after the sensitive period (bottom)

CONNECTION

6 Imprinting poses problems and opportunities for conservation programs

In attempting to save species that are at the edge of extinction, biologists sometimes try to increase their numbers in captivity. Generally, the strategy of a captive breeding program is to provide a safe environment for infants and juveniles, the stages at which many animals are most vulnerable to predation and other risks. In some programs, adult animals are caught and kept in conditions that are conducive to breeding. Offspring are usually raised by the parents and may be kept for breeding or released back to the wild. In other programs, parents are absent, as when eggs are removed from a nest. Artificial incubation is often successful, but without parents available as models for imprinting, the offspring may not learn appropriate behaviors. The effort to save the whooping crane is one example of a program that has successfully used surrogate parents, though they are a bit unusual.

The whooping crane (*Grus americana*) is a migratory waterfowl that reaches a height of about 1.5 m and has a white body with black-tipped wings that spread out over 2 m. Its name comes from its distinctive call. Once common in North American skies during their north-south migrations, whooping cranes were almost killed off by habitat loss and hunting. By the 1940s, only 16 wild birds returned from their summer breeding ground in Canada to one of their wintering areas on the Texas coast. Because a whooping crane doesn't reach sexual maturity until it is 4 years old, new generations of birds are not produced quickly. And although whooping cranes often lay two eggs, parents usually successfully rear only one chick. Protections for whooping cranes were established, and in 1967, U.S. wildlife officials launched long-term recovery and captive breeding efforts.

At first, biologists used sandhill cranes as surrogate parents for whooping crane chicks. All went well until the whooping cranes reached maturity. Having imprinted on the sandhill cranes, they showed no interest in breeding with their own kind. Biologists realized that they needed another approach. With plans to set up a separate breeding colony of whooping cranes in Wisconsin, they turned to Operation Migration to help get the birds to a winter nesting site in Florida. This bird advocacy group had developed ways to hatch geese and sandhill cranes, teach the birds to recognize a small, lightweight plane as a parent figure, and train them to follow the plane along migratory routes.

In 2001, Operation Migration applied its techniques to whooping cranes. Incubating crane eggs were serenaded with recorded sounds of the plane's engine. When the chicks emerged from their shells, the first thing they saw was a hand puppet, shaped and painted in the form of an adult whooping crane. As the chicks grew, the same type of puppet guided them through exercise and training (**Figure 6A**). But now the puppet was attached to a plane that rolled along the ground, coaxing the chicks to follow it. To make sure the birds bonded with the puppet and plane and not humans, pilots and other members of Operation Migration were cloaked in hooded suits. Eventually, the birds started following the plane on short flights (**Figure 6B**).

▲ **Figure 6A** A whooping crane chick interacting with a puppet "parent" USGS

October 2001 brought the real test. Would the young whooping cranes follow the plane from their protected grounds in Wisconsin along a migratory route to Florida? The trip took 48 days but ultimately proved successful. Each flight day, the young cranes lined up eagerly behind the plane, wings raised, ready to follow their "parent" to the next stop. And the next spring, five of the eight young cranes retraced the route to Wisconsin on their own. Operation Migration has since taught several more generations of whooping cranes to migrate, boosting the species' chances of survival. At the end of 2009, there were about 384 whooping cranes in the wild.

? What features of whooping cranes have made their recovery difficult?

They do not breed until 4 years of age and usually only raise one chick each breeding season. As migratory birds, they require resources and protection in two habitats.

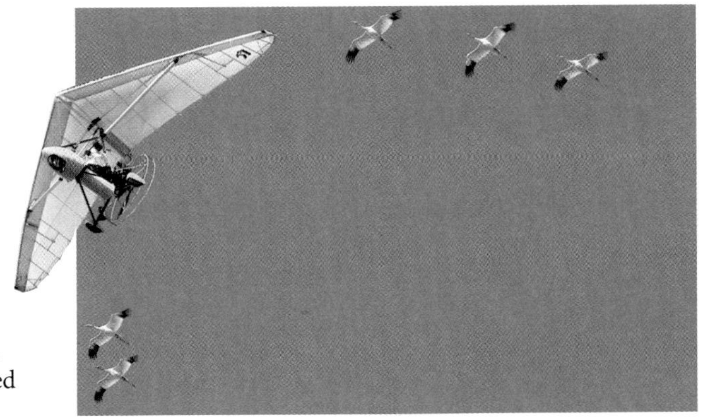

Star Banner, Doug Engle/AP Photo

▲ **Figure 6B** Whooping cranes following a surrogate parent

7 Animal movement may be a simple response to stimuli or require spatial learning

Moving in a directed way enables animals to avoid predators, migrate to a more favorable environment, obtain food, and find mates and nest sites. The simplest kinds of movement do not involve learning. A random movement in response to a stimulus is called a **kinesis** (plural, *kineses*; from the Greek word for "movement"). A kinesis may be merely starting or stopping, changing speed, or turning more or less frequently. Sow bugs (**Figure 7A**), which are the only terrestrial crustaceans, are not as well protected from drying out as their insect cousins. Consequently, they typically live in moist habitats, such as the underside of a rock or log. In a dry area, sow bugs exhibit kinesis, becoming more active and moving about randomly. The more they move, the greater their chance of finding a moist area. Once they are in a more favorable environment, their decreased activity tends to keep them there.

pzAxe/Shutterstock

▲ **Figure 7A** A sow bug (also known as a roly-poly or wood louse)

In contrast to kineses, a **taxis** (plural, *taxes*; from the Greek *tasso*, put in order) is a response directed toward (positive taxis) or away from (negative taxis) a stimulus. For example, many stream fish, such as salmon, exhibit positive taxis in the current; they automatically swim or orient in an upstream direction (**Figure 7B**). This orientation keeps them from being swept away by the current and keeps them facing in the direction food is likely to come from.

In **spatial learning**, animals establish memories of landmarks in their environment that indicate the locations of food, nest sites, prospective mates, and potential hazards. **Figure 7C** illustrates a classic experiment Tinbergen performed to investigate spatial learning in an insect called the digger wasp. The female digger wasp builds its nest in a small burrow in the ground. She will often excavate four or five separate nests and fly to each one daily, cleaning them and bringing food to the larvae in the nests. Before leaving each nest, she hides the entrance. To test his hypothesis that the digger wasp uses landmarks to keep track of her nests, 1 Tinbergen placed a circle of pinecones around a nest opening and waited for the mother wasp to return and tend the nest.

James Watt/Photolibrary

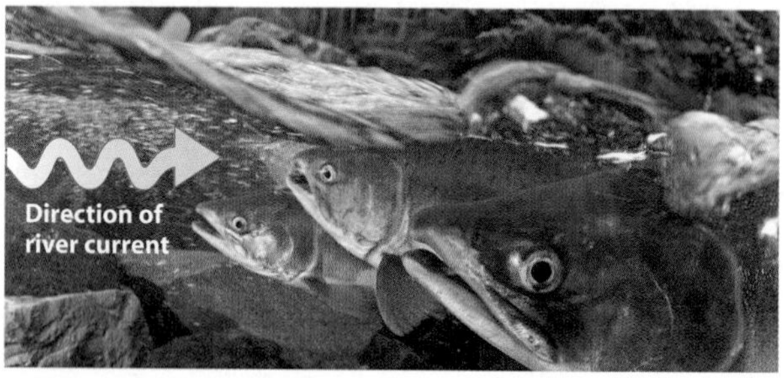

▲ **Figure 7B** Salmon exhibiting positive taxis to a current as they swim upstream to spawn

After the wasp flew away, 2 Tinbergen moved the pinecones a few feet to one side of the nest opening. The next time the wasp returned, she flew to the center of the pinecone circle instead of to the actual nest opening.

This experiment indicated that the wasp did use landmarks and that she could learn new ones to keep track of her nests. Next, to test whether the wasp was responding to the pinecones themselves or to their circular arrangement, 3 Tinbergen arranged the pinecones in a triangle around the nest and made a circle of small stones to one side of the nest opening. This time, the wasp flew to the stones, indicating that her cue was the *arrangement* of the landmarks rather than their appearance.

While many animals find their way by learning the particular set of landmarks in their area, others appear to use more sophisticated navigation mechanisms, as we see next.

? Planarians move directly away from light into dark places. What type of movement is this?

Negative taxis (directed movement away from light)

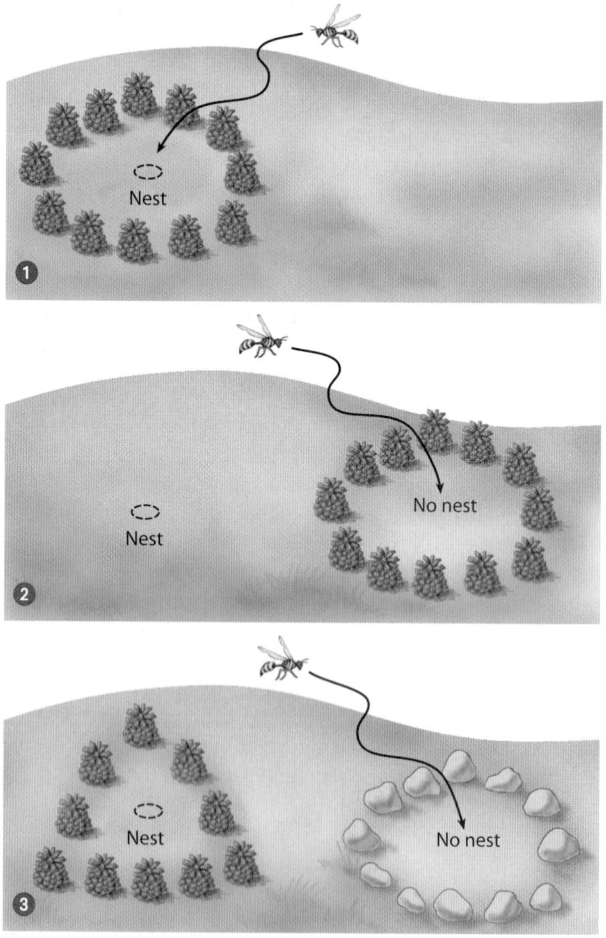

▲ **Figure 7C** The nest-locating behavior of the digger wasp

Adapted from Lawrence G. Mitchell, John A. Mutchmor, and Warren D. Dolphin, *Zoology*. Copyright © 1988 Pearson Education, Inc., publishing as Pearson Benjamin Cummings.

8 Movements of animals may depend on cognitive maps

An animal can move around its environment using landmarks alone. Honeybees, for instance, might learn 10 or so landmarks and locate their hive and flowers in relation to those features. An even more powerful mechanism is a **cognitive map**, an internal representation, or code, of the spatial relationships among objects in an animal's surroundings.

It is actually very difficult to distinguish experimentally between an animal that is simply using landmarks and one that is using a true cognitive map. The best evidence for cognitive maps comes from research on the family of birds that includes jays, crows, and nutcrackers. Many of these birds hide food in caches. A single bird may store nuts in thousands of caches that may be widely dispersed. The bird not only finds each cache, but also keeps track of food quality, bypassing caches in which the food was relatively perishable and would have decayed. It would seem that these birds use cognitive maps to memorize the locations of their food stores.

The most extensive studies of cognitive maps have involved animals that exhibit **migration**, the regular back-and-forth movement of animals between two geographic areas. Migration enables many species, such as the whooping cranes discussed in Module 6, to access food resources throughout the year and to breed or winter in areas that favor survival.

A notable long-distance traveler is the gray whale. During the summer, these giant mammals feast on small, bottom-dwelling invertebrates that abound in northern oceans. In the autumn, they leave their feeding grounds north of Alaska and begin the long trip south along the North American coastline to winter in the warm lagoons of Baja California (Mexico). Females give birth there before migrating back north with their young. The yearly round-trip, some 20,000 km, is the longest made by any mammal.

Researchers have found that migrating animals stay on course by using a variety of cues. Gray whales seem to use the coastline to pilot their way north and south. Whale watchers sometimes see gray whales stick their heads straight up out of the water, perhaps to obtain a visual reference point on land. Gray whales may also use the topography of the ocean floor, cues from the temperature and chemistry of the water, and magnetic sensing to guide their journey.

Many birds migrate at night, navigating by the stars the way early sailors did. Navigating by the sun or stars requires an internal timing device to compensate for the continuous daily movement of celestial objects. Consider what would happen if you started walking one day, orienting yourself by keeping the sun on your left. In the morning, you would be heading south, but by evening you would be heading back north, having made a circle and gotten nowhere. A calibration mechanism must also allow for the apparent change in position of celestial objects as the animal moves over its migration route.

At least one night-migrating bird, the indigo bunting, seems to avoid the need for a timing mechanism by fixing on the North Star, the one bright star in northern skies that appears almost stationary. **Figure 8** illustrates an experimental setup that was used to study the bunting's navigational mechanism. During the migratory season, wild and laboratory-reared buntings were placed in funnel-like cages in a planetarium. Each funnel had an ink pad at its base and was lined with blotting paper. When a bird stepped on the ink pad and then tried to fly in a certain direction, it tracked ink on the paper. The researchers found that wild buntings tracked ink in the direction of the North Star, as did those raised in the lab and introduced to the northern sky in a planetarium. Furthermore, researchers found that buntings raised under a planetarium sky with a different fixed-location star oriented to that star. Apparently, buntings learn to construct a star map and guide on a stationary star.

Some animals appear to migrate using only innate responses to environmental cues. For example, each fall, a new generation of monarch butterflies flies about 4,000 km over a route they've never flown before to specific wintering sites. Studies of other animals, including some songbirds, show the interaction of genes and experience in migration. Research efforts continue to reveal the complex mechanisms by which animals traverse Earth.

? Why is a timekeeping mechanism essential for navigating by the stars?

Because the positions of the stars change with time of night and season

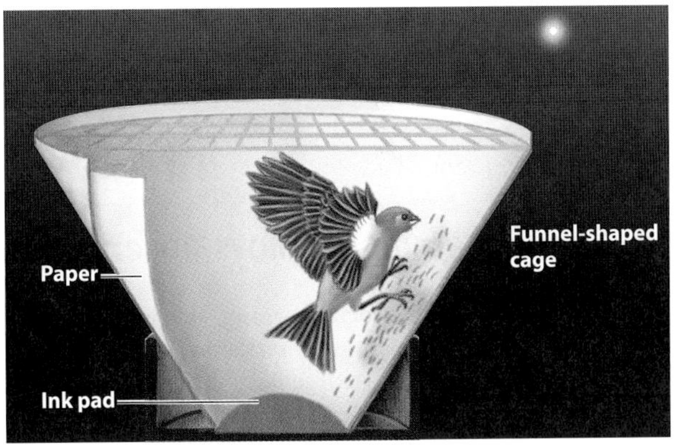

This asset is intentionally omitted from this text.

▲ **Figure 8** An indigo bunting in an experimental funnel (top); a researcher setting up funnels in the planetarium (bottom)

9 Animals may learn to associate a stimulus or behavior with a response

Associative learning is the ability to associate one environmental feature with another. In one type of associative learning, an animal learns to link a particular stimulus to a particular outcome. If you keep a pet, you have probably observed this type of associative learning firsthand. A dog or cat will learn to associate a particular sound, word, or gesture (stimulus) with a specific punishment or reward (outcome). For example, the sound of a can being opened may bring a cat running for food.

In the other type of associative learning, called **trial-and-error learning**, an animal learns to associate one of its own behaviors with a positive or negative effect. The animal then tends to repeat the response if it is rewarded or avoid the response if it is harmed. For example, predators quickly learn to associate certain kinds of prey with painful experiences. A porcupine's sharp quills and ability to roll into a quill-covered ball are strong deterrents against many predators. Coyotes, mountain lions, and domestic dogs often learn the hard way to avoid attacking porcupines nose-first (**Figure 9**).

Memory is the key to all associative learning. The prairie vole described in the chapter opening provides insight into the mechanisms involved. During mating, three neurochemical events occur simultaneously: Neural reward circuits are activated; olfactory signals identify the partner; and hormones are released in the brain. The brain of the monogamous prairie vole has dense clusters of hormone receptors in the area associated with social memory. As a result, the brain forms a memory that connects the reward to the current partner's scent.

John Cancalosi/Photolibrary

▲ **Figure 9** Trial-and-error learning by a dog

Other species of vole are promiscuous—they mate with multiple partners and form no lasting bonds. Promiscuous voles experience the same biochemical reward during the act of mating, but have very few hormone receptors in the area of the brain associated with social memory. Thus, the brain fails to forge a link between the partner and the reward.

? How might the fact that many bad-tasting or stinging insect species have similar color patterns benefit both the insects and the animals that may prey on them?

Potential predators associate insects displaying that coloration with a negative effect, and consequently, all these insects are less likely to be preyed on.

10 Social learning employs observation and imitation of others

Another form of learning is **social learning**—learning by observing the behavior of others. Many predators, including cats, coyotes, and wolves, seem to learn some of their basic hunting tactics by observing and imitating their mothers.

Studies of the alarm calls of vervet monkeys in Amboseli National Park, in Kenya, provide an interesting example of how performance of a behavior can improve through social learning. Vervet monkeys (*Cercopithecus aethiops*) are about the size of a domestic cat. They give distinct alarm calls when they see leopards, eagles, or snakes, all of which prey on vervets. When a vervet sees a leopard, it gives a loud barking sound; when it sees an eagle, it gives a short two-syllable cough; and the snake alarm call is a "chutter." Upon hearing a particular alarm call, other vervets in the group behave in an appropriate way: They run up a tree on hearing the alarm for a leopard (vervets are nimbler than leopards in trees); look up on hearing the alarm for an eagle; and look down on hearing the alarm for a snake (**Figure 10**).

Infant vervet monkeys give alarm calls, but in a relatively undiscriminating way. For example, they give the "eagle" alarm on seeing any bird, including harmless birds such as

Richard Wrangham

▲ **Figure 10** On seeing a python (foreground), a vervet monkey gives a distinct "snake" alarm call (inset).

Alissa Crandall/Corbis

bee-eaters. With age, the monkeys improve their accuracy. In fact, adult vervet monkeys give the eagle alarm only on seeing an eagle belonging to either of the two species that eat vervets. Infants probably learn how to give the right call by observing other members of the group and receiving social confirmation. For instance, if the infant gives the call on the right occasion—an eagle alarm when there is an eagle overhead—another member of the group will also give the eagle call. But if the infant gives the call when a bee-eater flies by, the adults in the group are silent. Thus, vervet monkeys have an initial, unlearned tendency to give calls on seeing potentially threatening objects in the environment. Learning fine-tunes the calls so that by adulthood, vervets give calls only in response to genuine danger and are prepared to fine-tune the alarm calls of the next generation. However, neither vervets nor any other species comes close to matching the social learning and cultural transmission that occurs among humans.

? What type of learning in humans is exemplified by identification with a role model?

Social learning (observation and imitation)

11 Problem-solving behavior relies on cognition

A broad definition of **cognition** is the process carried out by an animal's nervous system to perceive, store, integrate, and use information gathered by the senses. One area of research in the study of animal cognition is how an animal's brain represents physical objects in the environment. For instance, some researchers have discovered that many animals, including insects, are capable of categorizing objects in their environment according to concepts such as "same" and "different." One research team has trained honeybees to match colors and black-and-white patterns. Other researchers have developed innovative experiments, in the tradition of Lorenz and Tinbergen, for demonstrating pattern recognition in birds called nuthatches. These studies suggest that nuthatches apply simple geometric rules to locate their many seed caches.

Some animals have complex cognitive abilities that include **problem solving**—the process of applying past experience to overcome obstacles in novel situations. Problem-solving behavior is highly developed in some mammals, especially dolphins and primates. If a chimpanzee is placed in a room with a banana hung high above its head and several boxes on the floor, the chimp will "size up" the situation and then stack the boxes in order to reach the food. One way many animals learn to solve problems is by observing the behavior of other individuals. For example, young chimpanzees can learn from watching their elders how to crack oil palm nuts by using two stones as a hammer and anvil (**Figure 11A**).

Problem-solving behavior has also been observed in some bird species. For example, researchers placed ravens in situations in which they had to obtain food hanging from a string. Interestingly, the researchers observed a great deal of variation in the ravens' solutions. The raven in **Figure 11B** used one foot to pull up the string incrementally and the other foot to secure the string so the food didn't drop. An excellent test of human cognition and problem-solving behavior is in the construction of experiments that allow us to explore the cognition and problem-solving behavior of other animals!

? Besides problem solving, what other type of learning is illustrated in Figure 11A?

Social learning (observation)

Clive Bromhall/Photolibrary

▲ **Figure 11A** A chimpanzee solving a problem

Bernd Heinrich

▲ **Figure 11B** A raven solving a problem

Survival and Reproductive Success

12 Behavioral ecologists use cost-benefit analysis to study foraging

Michael Hutchinson/Nature Picture Library

◀ **Figure 12B** A dung fly

Because adequate nutrition is essential to an animal's survival and reproductive success, we should expect natural selection to refine behaviors that enhance the efficiency of feeding. Food-obtaining behavior, or **foraging**, includes not only eating, but also any mechanism an animal uses to search for, recognize, and capture food items.

Animals forage in a great many ways. Some animals are "generalists," whereas others are "specialists." Crows, for instance, are extreme generalists; they will eat just about anything that is readily available—plant or animal, alive or dead. In sharp contrast, the koala of Australia, an extreme feeding specialist, eats only the leaves of a few species of eucalyptus trees. As a result, it is restricted to certain areas and is extremely vulnerable to habitat loss. Most animals are somewhere in between crows and koalas in the range of their diet. The pronghorn, for example, eats a variety of plants, including forbs, grasses, and woody shrubs.

Often, even a generalist will concentrate on a particular item of food when it is readily available. The mechanism that enables an animal to find particular foods efficiently is called a **search image**. If the favored food item becomes scarce, the animal may develop a search image for a different food item. (People often use search images; for example, when you look for something on a kitchen shelf, you probably scan rapidly to find a package of a certain size and color rather than reading all the labels.)

Whenever an animal has food choices, there are trade-offs involved in the selection. The amount of energy required to locate, capture, subdue, and prepare prey for consumption may vary considerably among the items available. Some behavioral ecologists use an approach known as cost-benefit analysis, comparing the positive and negative aspects of the alternative choices, to evaluate the efficiency of foraging behaviors. According to the predictions of **optimal foraging theory**, an animal's feeding behavior should provide maximal energy gain with minimal energy expense and minimal risk of being eaten while foraging. A researcher tested part of this theory by studying insectivorous birds called wagtails (**Figure 12A**).

In England, wagtails are commonly seen in cow pastures foraging for dung flies (**Figure 12B**). The researcher collected data on the time required for a wagtail to catch and consume dung flies of different sizes (**Figure 12C**). He then calculated the number of calories the bird gained per second of "handling" time for the different sizes of flies. The smallest flies were easily handled but yielded few calories, while the caloric value of large flies was offset by the energy required to catch and consume them. An optimal forager would be expected to choose medium-sized flies most often. The researcher tested this prediction by observing wagtails as they foraged for dung flies in a cow pasture. As **Figure 12D** shows, he found that wagtails did select medium-sized flies most often, even though they were not the most abundant size class.

Predation is one of the most significant potential costs of foraging. Studies have shown that foraging in groups, as done by herds of antelopes, flocks of birds, or schools of fish, reduces the individual's risk of predation. And for some predators, such as wolves and spotted hyenas, hunting in groups improves their success. Thus, group behavior may increase foraging efficiency by both reducing the costs and increasing the benefits of foraging.

? Early humans were hunter-gatherers, but evidence suggests that they obtained more nutrition from gathering than hunting. How does this finding relate to optimal foraging theory?

● Meat is very nutritious, but hunting poses relatively high costs in effort and risk compared with the gathering of plant products and dead animals.

▶ **Figure 12A**
A wagtail

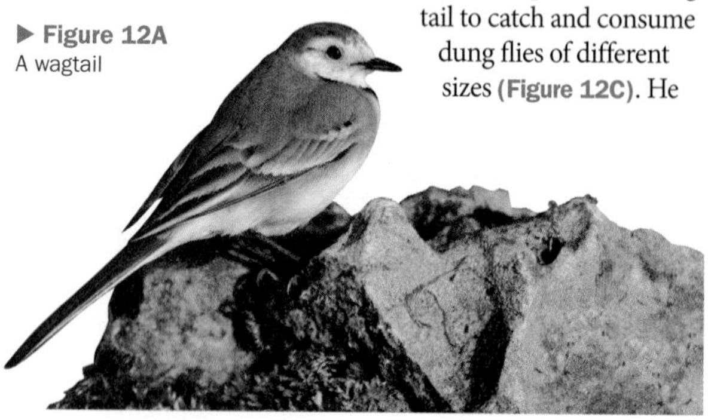

Jose B. Ruiz/Nature Picture Library

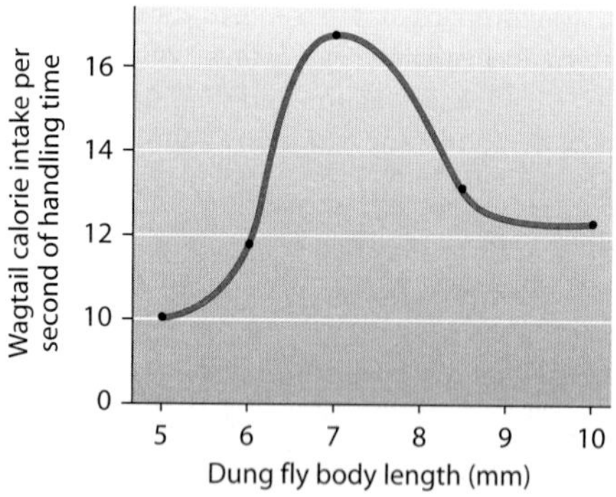

◀ **Figure 12C** The relationship between the calories gained by the wagtail and the size of its prey (the dung fly)

Graph from N. B. Davies. 1977. Prey selection and social behaviour in wagtails (Aves: Motacillidae). *Journal of Animal Ecology* 46: 37-57, fig. 9. Reproduced with permission of Blackwell Publishing Ltd

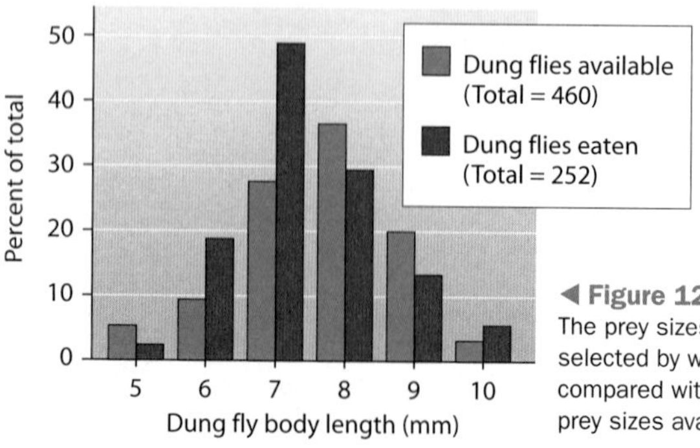

◀ **Figure 12D** The prey sizes selected by wagtails compared with the prey sizes available

Graph adapted from N. B. Davies. 1977. Prey selection and social behaviour in wagtails (Aves: Motacillidae). *Journal of Animal Ecology* 46: 37-57, fig. 8. Reproduced with permission of Blackwell Publishing Ltd.

13 Communication is an essential element of interactions between animals

Interactions between animals depend on some form of signaling between the participating individuals. In behavioral ecology, a **signal** is a stimulus transmitted by one animal to another animal. The sending of, reception of, and response to signals constitute animal **communication**, an essential element of interactions between individuals. In general, the more complex the social organization of a species, the more complex the signaling required to sustain it.

What determines the type of signal animals use to communicate? Most terrestrial mammals are nocturnal (active at night), which makes visual displays relatively ineffective. Many nocturnal mammals use odor and auditory (sound) signals, which work well in the dark. Birds, by contrast, are mostly diurnal (active in daytime) and use visual and auditory signals. Humans are also diurnal and likewise use mainly visual and auditory signals. Therefore, we can detect the bright colors and songs birds use to communicate. If we had the well-developed olfactory abilities of most mammals and could detect the rich world of odor cues, mammal-sniffing might be as popular with us as bird-watching.

What types of signals are effective in aquatic environments? A common visual signal used by territorial fishes is to erect their fins, which is generally enough to drive off intruders. Electrical signals produced by certain fishes communicate hierarchy or status. Fish may also use sound to communicate. For example, the males of some species make noises to attract mates. Marine mammals use sound for courtship, territoriality, and maintaining contact with their group.

Animals often use more than one type of signal simultaneously. **Figure 13A** shows a ring-tailed lemur, also featured on the book's cover. These tree-dwelling primates of Madagascar live in social groups averaging 15 individuals. Lemurs use visual displays, scent, and vocalizations to communicate with other members of the group. The animal shown here is communicating aggression with its prominent tail. Prior to this display, it smeared its tail with odorous secretions from glands on its forelegs. By waving its scented tail over its head, the lemur transmits both visual and chemical signals.

▶ **Figure 13A** A lemur communicating aggression

J.P. Varin/Jacana/Photo Researchers

One of the most amazing examples of animal communication is found in honeybees, which have a complex social organization characterized by division of labor. Adult worker bees leave the hive to forage, bringing back pollen and nectar for the nest. When a forager locates a patch of flowers, she regurgitates some nectar that the others taste and smell, then communicates the location of the food source by performing a "dance."

Beginning with studies by Karl von Frisch, researchers have conducted numerous experiments to decipher the meanings of different honeybee dances. A pattern of movements called the waggle dance communicates the distance and direction of food **(Figure 13B)**. The dancer runs a half circle, then turns and runs in a straight line back to her starting point, buzzing her wings and waggling her abdomen as she goes. She then runs a half circle in the other direction, followed by another waggling run to the starting point. The length of the straight run and the number of waggles indicate the distance to the food source. The angle of the straight run relative to the vertical surface of the hive is the same as the horizontal angle of the food in relation to the sun (30° in Figure 13B). Once the other workers have learned the location of the food source, they leave the hive to forage there.

? What types of signals do honeybees use?

Visual and chemical

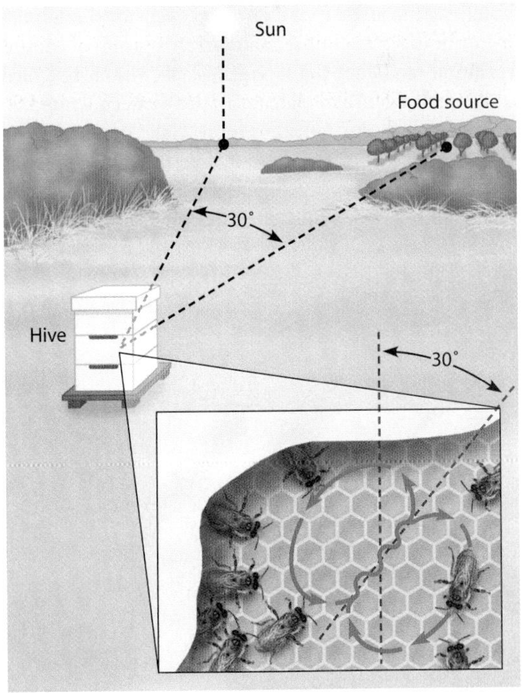

▲ **Figure 13B** A returning honeybee forager performing the waggle dance inside the hive

14 Mating behavior often includes elaborate courtship rituals

Animals of many species tend to view members of their own species as competitors to be driven away. Even animals that forage and travel in groups often maintain a distance from their companions. Thus, careful communication is an essential prerequisite for mating. In many species, prospective mates must perform an elaborate courtship ritual, which confirms that individuals are of the same species, of the opposite sex, physically primed for mating, and not threats to each other. Courtship rituals are common among vertebrates as well as some groups of invertebrates, such as insects and cephalopods. The male cuttlefish in **Figure 14A**, for instance, is using part of his enormous repertoire of color changes to signal a female.

Figure 14B shows the courtship behavior of the common loon, which breeds on secluded lakes in the northern United States and Canada. The courting male and female swim side by side, performing a series of displays, ❶ frequently turning their heads away from each other. (In contrast, a male loon defending his territory charges at an intruder with his beak pointed straight ahead.) ❷ The birds then dip their beaks in the water and submerge their heads. ❸ The male invites the female onto land by turning his head backward with his beak down. There, they copulate. Each movement in this complex behavior is a FAP (see Module 2). When a FAP is executed successfully by one partner, it triggers the next FAP in the other partner. Thus, the entire routine is a chain of FAPs that must be performed flawlessly if mating is to occur.

▲ **Figure 14B** Courtship of the common loon

In some species, including the peacocks mentioned in the opening essay, courtship is a group activity in which members of one or both sexes choose mates from a group of candidates. For example, consider the sage grouse, a chicken-like bird that inhabits high sagebrush plateaus in the western United States. Every day in early spring, 50 or more males congregate in an open area, where they strut about, erecting their tail feathers in a bright, fanlike display **(Figure 14C)**. A booming sound produced in the male's inflated air sac accompanies the show. Dominant males usually defend a prime territory near the center of the area. Females arrive several weeks after the males. After watching the males perform, a female selects one, and the pair copulates. Usually, all the females choose

Steve Bloom Images/Alamy

▲ **Figure 14A** A male cuttlefish displaying courtship colors

Carol Walker/Nature Picture Library

▲ **Figure 14C** A courtship display by a male sage grouse

From *The Common Loon: Spirit of Northern Lakes* by Judith W. McIntyre, drawings by Anne Olson, p. 97. (Minneapolis, MN: University Minnesota Press, 1988). Copyright © 1988 by University of Minnesota Press. Used by permission of the publisher.

dominant males, so only about 10% of the males actually mate. In choosing a dominant male, a female sage grouse may be giving her offspring, and thus her own genes, the best chance for survival. Research on several species of animals has shown a connection between a male's physical characteristics and the quality of his genes.

? What categories of signals do the cuttlefish, loon, and sage grouse use to communicate with potential mates?

All three use visual signals. The sage grouse also uses an auditory signal.

15 Mating systems and parental care enhance reproductive success

Courtship and mating are not the only elements of reproductive success. In order for genes to be passed on to successive generations, the offspring produced by the union must themselves survive and reproduce. Therefore, the needs of the young are an important factor in the evolution of mating systems. Animal mating systems fall into three major categories: **promiscuous** (no strong pair-bonds or lasting relationships between males and females), **monogamous** (a bond between one male and one female, with shared parental care), and **polygamous** (an individual of one sex mating with several of the other). Polygamous relationships most often involve a single male and many females, although in some species this is reversed, and a single female mates with several males.

Most newly hatched birds cannot care for themselves and require a large, continuous food supply that a single parent may not be able to provide. A male may leave more viable offspring by helping a single mate than by seeking multiple mates. This may explain why most birds are monogamous, though often with a new partner each breeding season. (And in many species of birds, one or both members of the pair also seek other sexual partners.) On the other hand, birds whose young can feed and care for themselves almost immediately after hatching, such as pheasants and quail, gain less benefit from monogamy—a polygamous or promiscuous mating system may maximize their reproductive success.

In the case of mammals, the lactating female is often the only food source for the young, and males usually play no role in caring for their offspring. The prairie voles described in the chapter introduction are highly unusual mammals in regard to both monogamy and parental care.

Parental care involves significant costs, including energy expenditure and the loss of other mating opportunities. For females, the investment almost always pays off in terms of reproductive success—young born or eggs laid definitely contain the mother's genes. But even in the case of a normally monogamous relationship, the young may have been fathered by a male other than the female's usual mate. As a result, certainty of paternity may be a factor in the evolution of male mating behavior and parental care.

The certainty of paternity is relatively low in most species with internal fertilization because the acts of mating and birth (or mating and egg laying) are separated over time. This may help explain why species in which males are the sole parental caregiver are rare in birds and mammals. However, the males of many species with internal fertilization do engage in behaviors that appear to increase their certainty of paternity, such as guarding females from other males. In species such as lions, where males protect the females and young, a male or small group of males typically guard many females at once in a harem. The aggression that the male prairie vole displays toward intruders is another example of this mate-guarding behavior.

Certainty of paternity is much higher when egg laying and mating occur together, as happens when fertilization is external. This connection may explain why parental care in aquatic invertebrates, fishes, and amphibians, when care occurs at all, is at least as likely to be by males as by females. **Figure 15** shows a male jawfish exhibiting paternal care of eggs. Jawfish, which are found in tropical marine habitats, hold the eggs they have fertilized in their mouths, keeping them aerated (by spitting them out and sucking them back in) and protected from predators until they hatch. Seahorses have the most extreme method of ensuring paternity. Females lay their eggs in a brood pouch in the male's abdomen, where they are fertilized by his sperm. When the eggs hatch a few weeks later, the pouch opens and the male pushes them out with pumping movements.

Keep in mind that when behavioral ecologists use the phrase *certainty of paternity*, they do not mean that animals are aware of paternity when they behave a certain way. Parental behaviors associated with certainty of paternity exist because they have been reinforced over generations by natural selection. Individuals with genes for such behaviors reproduced more successfully and passed those genes on to the next generation.

? Based on what you know of the peacock's mating behavior, what would you predict about the need to care for young?

Peacocks are promiscuous, and the male does not contribute to caring for the offspring. It is likely that their young can feed and care for themselves immediately after hatching, like those of pheasants and quail.

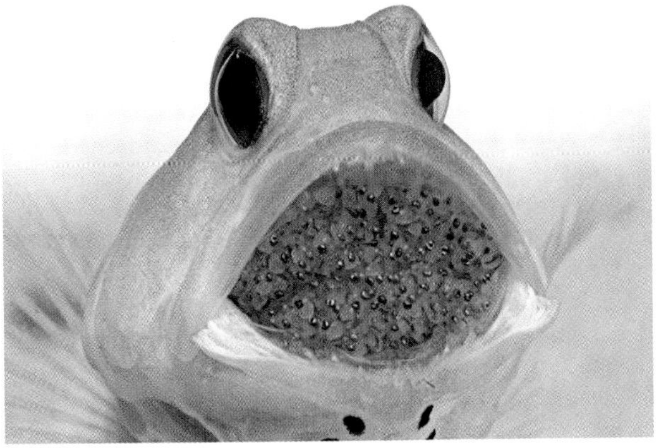

Image Quest Marine

▲ **Figure 15** A male jawfish with his mouth full of eggs

CONNECTION 16 Chemical pollutants can cause abnormal behavior

Appropriate behavior is the cornerstone of success in the animal world. So, something is amiss when fish are lackadaisical about territorial defense, salamanders ignore mating cues, birds exhibit sloppy nest-building techniques, and mice take inexplicable risks. Scientists have linked observations of these abnormal behaviors, as well as many others, to endocrine-disrupting chemicals. Endocrine disruptors are a diverse group of substances that affect the vertebrate endocrine system by mimicking a hormone or by enhancing or inhibiting hormone activity. Endocrine disruptors enter ecosystems from a variety of sources, including discharge from paper and lumber mills and factory wastes such as dioxin (a by-product of many industrial processes) and PCBs (organic compounds used in electrical equipment until 1977). Agriculture is another major source of pollutants—DDT and other pesticides are endocrine disruptors. Traces of birth control pills and other hormones are commonly found in wastewater from sewage treatment plants. Endocrine disruptors are especially worrisome pollutants because they persist in the environment for decades and become concentrated in the food chain.

Hundreds of studies have demonstrated the effects of endocrine disruptors on vertebrate reproduction and development. Like hormones, endocrine disruptors also affect behavior. For example, some male fish attract females during the breeding season by defending territories. Males have high levels of androgens (male hormones) during this time. Researchers showed that the intensity of nest-guarding behavior in male sticklebacks (a type of fish) dropped after they were exposed to pollutants that mimic the female hormone estrogen. Male sticklebacks' performance of courtship rituals was also impaired.

Another series of studies showed that the anatomy of female mosquitofish was masculinized by endocrine disruptors. Female mosquitofish that were exposed to pollutants discharged from a paper mill **(Figure 16A)** developed the fin modification that males use to transfer sperm to females **(Figure 16B)**. The masculinized females also behaved like males, waving the fin back and forth in front of females in the typical courtship behavior. Female fish living downstream from the paper mill were masculinized at contaminated sites, while female fish living in uncontaminated water near the same mill were normal.

Although the effects of endocrine disruptors on reproductive behavior have received the most attention, endocrine disruptors also affect other kinds of behavior by acting on thyroid hormones and neurological functions. For example, spatial learning ability was impaired in young monkeys exposed to PCBs.

Could endocrine disruptors in drinking water or food affect humans, too? Answers are not yet clear, but in late 2009, the Environmental Protection Agency (EPA) ordered the manufacturers of several dozen chemicals to begin testing their products' potential as endocrine disruptors.

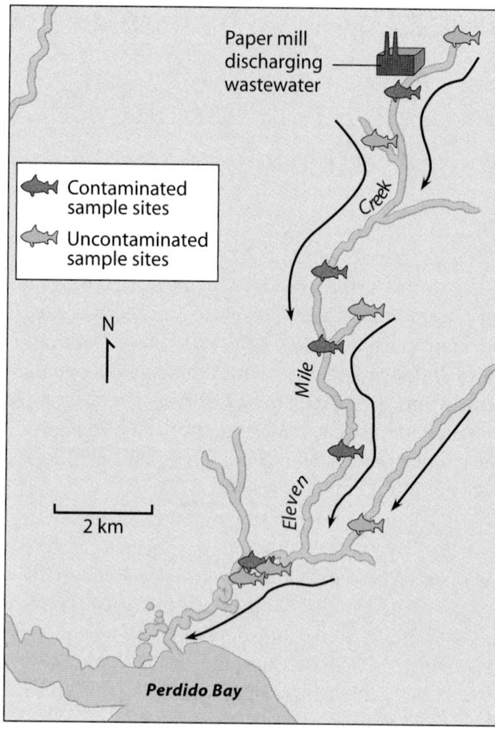

▲ **Figure 16A** Map of Eleven Mile Creek, in Escambia County, Florida, an area used to study the effect of endocrine disruptors on mosquitofish

Mike W. Howell

▲ **Figure 16B** A normal female mosquitofish and a male showing the modified fin used in courtship and sperm transfer

? How would ineffective courtship behavior affect the fitness of a male fish?

A male whose courtship display is perceived by females as inferior will not be successful in attracting mates and will thus be less likely to produce offspring. Therefore, the male's fitness would be reduced.

Map adapted from W. M. Howell et al. 1980. Abnormal expression of secondary sex characteristics in a population of mosquitofish, *Gambusia affinis holbrooki*: Evidence for environmentally-induced masculinization. *Copeia* 4, 676-681, fig. 5. Used with permission

Social Behavior and Sociobiology

17 Sociobiology places social behavior in an evolutionary context

Biologists define **social behavior** as any kind of interaction between two or more animals, usually of the same species. The courtship behaviors of loons and sage grouse are examples of social behavior. Other social behaviors observed in animals are aggression and cooperation.

Many animals migrate and feed in large groups (flocks, packs, herds, or schools). The pronghorns, for example, derive protection from feeding in herds. Many watchful eyes increase the chance that a predator will be spotted before it can strike. When alarmed, a pronghorn flares out the white hairs on its rump, sending a danger signal to other members of its herd. Predators, too, may benefit from traveling in a group. Wolves usually hunt in a pack consisting of a tightly knit group of family members. Hunting in packs enables them to kill large animals, such as moose or elk, that would be unattainable by an individual wolf.

The discipline of **sociobiology** applies evolutionary theory to the study and interpretation of social behavior—the study of how social behaviors are adaptive and how they could have evolved by natural selection. We discuss several aspects of social behavior in the next several modules.

? Why is communication essential to social behavior?

Group members must be able to transfer information—for example, to signal danger or to cooperate in obtaining food.

18 Territorial behavior parcels out space and resources

Many animals exhibit territorial behavior. A **territory** is an area, usually fixed in location, that individuals defend and from which other members of the same species are usually excluded. The size of the territory varies with the species, the function of the territory, and the resources available. Territories are typically used for feeding, mating, rearing young, or combinations of these activities.

Figure 18A shows a nesting colony of gannets in Quebec, Canada. Space is at a premium, and the birds defend territories just large enough for their nests by calling out and pecking at other birds. As you can see, each gannet is literally only a peck away from its closest neighbors. Such small nesting territories are characteristic of many colonial seabirds. In contrast, most cats, including jaguars, leopards, cheetahs, and even domestic cats, defend much larger territories, which they use for foraging as well as breeding.

Individuals that have established a territory usually proclaim their territorial rights continually; this is the function of most bird songs, the noisy bellowing of sea lions, and the chattering of squirrels. Scent markers are frequently used to signal a territory's boundaries. The male cheetah in **Figure 18B**, a resident of Africa's Serengeti National Park, is spraying urine on a tree. The odor will serve as a chemical "No Trespassing" sign. Other males that approach the area will sniff the marked tree and recognize that the urine is not their own. Usually, the intruder will avoid the marked territory and a potentially deadly confrontation with its proprietor.

Not all species are territorial. However, for those that are, the territory can provide exclusive access to food supplies, breeding areas, and places to raise young. Familiarity with a specific area may help individuals avoid predators or forage more efficiently. In a territorial species, such benefits increase fitness and outweigh the energy costs of defending a territory.

? Why is the territory of a gannet so much smaller than the territory of a cheetah?

The gannet uses its territory only for raising young, not for foraging. Cheetahs use their territories for foraging as well as for breeding.

CreativeNature.nl/Shutterstock

▲ **Figure 18A** Gannet territories

C & M Denis-Huot/Peter Arnold, Inc./PhotoLibrary

▲ **Figure 18B** A cheetah marking its territory

19 Agonistic behavior often resolves confrontations between competitors

In many species, conflicts that arise over limited resources, such as food, mates, or territories, are settled by **agonistic behavior** (from the Greek *agon*, struggle), including threats, rituals, and sometimes combat that determine which competitor gains access to the resource. An agonistic encounter may involve a test of strength, such as when male moose lower their heads, lock antlers, and push against each other. More commonly, animals engage in exaggerated posturing and other symbolic displays that make the individual look large or aggressive. For example, the colorful Siamese fighting fish (*Betta splendens*) spreads its fins dramatically when it encounters a rival, thus appearing much bigger than it actually is. Eventually, one individual stops threatening and becomes submissive, exhibiting some type of appeasement display—in effect, surrendering.

Because violent combat may injure the victor as well as the vanquished in a way that reduces reproductive fitness, we would predict that natural selection would favor ritualized contests. And, in fact, this is what usually happens in nature. The rattlesnakes pictured in Figure 19, for example, are rival males wrestling over access to a mate. If they bit each other, both would die from the toxin in their fangs, but they are engaged in a pushing, rather than a biting, match. One snake usually tires before the other, and the winner pins the loser's head to the ground. In a way, the snakes are like two people who settle an argument by arm wrestling instead of resorting to fists or guns. In a typical case, the agonistic ritual inhibits further aggressive activity. Once two individuals have settled a dispute by agonistic behavior, future encounters between them usually involve less conflict, with the original loser giving way to the original victor. Often the victor of an agonistic ritual gains first or exclusive access to mates, and so this form of social behavior can directly affect an individual's evolutionary fitness.

? Why is "fighting to the death" an unusual form of agonistic behavior among animals?

● Because ritualized posturing or nonlethal combat can usually produce a winner without injuries that would lower reproductive fitness for the winner and eliminate it altogether for the loser.

Rupert Barrington/Nature Picture Library

▲ **Figure 19** Ritual wrestling by rattlesnakes

20 Dominance hierarchies are maintained by agonistic behavior

Many animals live in social groups maintained by agonistic behavior. Chickens are an example. If several hens unfamiliar with one another are put together, they respond by chasing and pecking each other. Eventually, they establish a clear "pecking order." The alpha, or top-ranked, hen in the pecking order (the one on the right in Figure 20) is dominant; she is not pecked by any other hens and can usually drive off all the others by threats rather than actual pecking. The alpha hen also has first access to resources such as food, water, and roosting sites. The beta, or second-ranked, hen similarly subdues all others except the alpha, and so on down the line to the omega, or lowest, animal.

Pecking order in chickens is an example of a **dominance hierarchy**, a ranking of individuals based on social interactions. Once a hierarchy is established, each animal's status in the group is fixed, often for several months or even years. Dominance hierarchies are common, especially in vertebrate populations. In a wolf pack, for example, there is a dominance hierarchy among the females, and this hierarchy may control the pack's size. When food is abundant, the alpha female mates and also allows others to do so. When food is scarce, she usually monopolizes the males for herself and keeps other females from mating. In other species, such as savanna baboons and red deer, the dominant male monopolizes fertile females. As a result, his reproductive success is much greater than that of lower-ranking males.

As you will learn in the next module, however, not all social behaviors pit one individual against another.

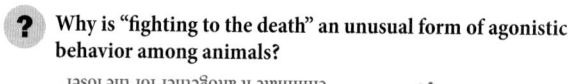

© Zoo Imaging Photography/Alamy

▲ **Figure 20** Chickens exhibiting pecking order

? Dog trainers like Cesar Millan, TV's "Dog Whisperer," advise dog owners to be sure the dog understands that the owner is the alpha. How might this facilitate obedience?

● Like wolves, dogs are pack animals. They are more likely to respond to commands from a higher-ranking individual.

EVOLUTION CONNECTION

21 Altruistic acts can often be explained by the concept of inclusive fitness

Many social behaviors are selfish. Behavior that maximizes an individual's survival and reproductive success is favored by selection, regardless of how much the behavior may harm others. For example, superior foraging ability by one individual may leave less food for others. **Altruism**, on the other hand, is defined as behavior that reduces an individual's fitness while increasing the fitness of others in the population.

Altruistic behavior is often evident in animals that live in cooperative colonies. For example, workers in a honeybee hive are sterile females who labor all their lives on behalf of the queen. When a worker stings an intruder in defense of the hive, the worker usually dies. The animals in **Figure 21A** are highly social rodents called naked mole rats. Almost hairless and nearly blind, they live in colonies in underground chambers and tunnels in southern and northeastern Africa. With a social structure resembling that of honeybees, each colony has only one reproducing female, called the queen. The queen mates with one to three males, called kings. The rest of the colony consists of nonreproductive females and males who forage for roots and care for the queen, her young, and the kings. While trying to protect the queen or kings from a snake that invades the colony, a nonreproductive naked mole rat may lose its own life.

If altruistic behavior reduces the reproductive success of self-sacrificing individuals, how can it be explained by natural selection? It is easy to see how selfless behavior might be selected for when it involves parents and offspring. When parents sacrifice their own well-being to ensure the survival of their young, they are maximizing the survival of their own genes. But reproducing is only one way to pass along genes; helping a close relative reproduce is another. Siblings, like parents and offspring, have half their genes in common, and an individual shares one-fourth of its genes with the offspring of a sibling.

The concept of **inclusive fitness** describes an individual's success at perpetuating its genes by producing its own offspring *and* by helping close relatives, who likely share many of those genes, to produce offspring. Altruism increases inclusive fitness when it maximizes the reproduction of close relatives. The natural selection favoring altruistic behavior that benefits relatives is called **kin selection**. Thus, the genes for altruism may be propagated if individuals that benefit from altruistic acts are themselves carrying those genes.

Stephen Kraseman/Peter Arnold, Inc./PhotoLibrary

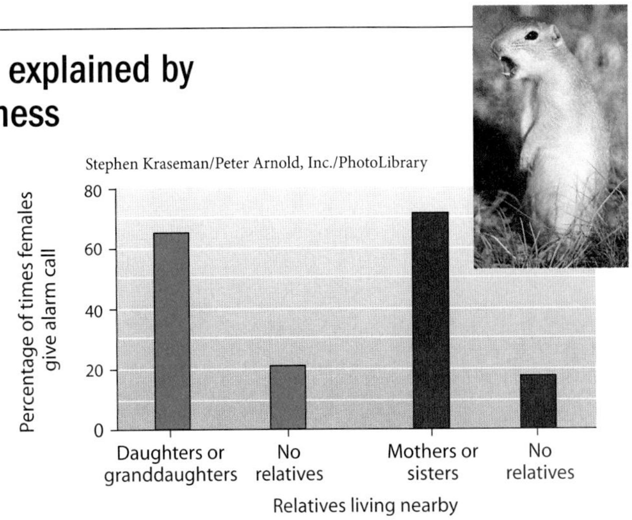

▲ **Figure 21B** The frequency of alarm calls by Belding's ground squirrels (inset photo) when close relatives are nearby or absent

A classic study of Belding's ground squirrels, which live in regions of the western United States, provided empirical support for kin selection. Female ground squirrels generally live near the burrow where they were born, while males move farther away. As a result, female ground squirrels tend to have kin nearby, but males are not related to their neighbors. Upon seeing a predator such as a coyote or weasel approach, a squirrel often gives a high-pitched alarm call (**Figure 21B**, inset) that alerts nearby squirrels, which then retreat to their burrows. Field observations have confirmed that the conspicuous alarm call identifies the caller's location and increases the risk of being killed. In the study mentioned, most alarm calls were given by female Belding's squirrels. Figure 21B shows which females issued these warnings. Each set of bars compares the percentage of times that squirrels who did and did not have close relatives nearby gave an alarm call when a predator approached. Squirrels whose close relatives lived nearby were much more likely to call, and they were as likely to warn mothers or sisters as they were to alert their descendants.

Kin selection also explains the altruistic behavior of bees in a hive, which all share genes with the queen. Their work (or even death) in support of the queen helps ensure that many of their genes will survive. In the case of mole rats, researchers have found that all the individuals in a naked mole rat colony are closely related. The nonreproductive members are the queen's descendants or siblings; by enhancing a queen's chances of reproducing, they increase the chances that some genes identical to their own will be propagated.

Jennifer Jarvis

▲ **Figure 21A** The queen of a naked mole rat colony nursing offspring while surrounded by other individuals of the colony

? **What is the ultimate cause of altruism between kin?**

Natural selection reinforces such altruistic behavior through the reproductive success of closely related individuals that have many genes in common with the altruist, including genes for altruism.

SCIENTIFIC DISCOVERY

22 Jane Goodall revolutionized our understanding of chimpanzee behavior

People are fascinated by the humanlike qualities of primates, especially chimpanzees. With their expressive faces and playful behavior, chimpanzees have long been stars in the entertainment industry. For example, a chimpanzee named Cheeta appeared in dozens of Tarzan movies from the 1930s through the 1960s. In the early days of television, NBC's *Today Show* attracted few viewers until a chimpanzee joined the show as the host's sidekick. Like their human counterparts, however, these chimpanzees were only playing a role. No one knew how they behaved in the wild, without the influence of trainers or zookeepers. A young Englishwoman named Jane Goodall changed that.

In 1960, paleoanthropologist Louis Leakey, who discovered *Homo habilis* and its stone tools, thought that studying the behavior of our closest relatives might provide insight into the behavior of early humans. Jane Goodall had no scientific background—she did not even have a college degree, though she later earned a Ph.D. from Cambridge—when Leakey hired her to observe chimpanzees in the wild.

Through countless hours of carefully recorded observations in an area that was later designated the Gombe Stream Research Center, near Lake Tanganyika, Goodall documented the social organization and behavior of chimpanzees (**Figure 22A**). Most astonishingly, she reported that chimpanzees make and use tools by fashioning plant stems into probes for extracting termites from their mounds. Until that time, scientists thought that toolmaking defined humans, setting them apart from other animals.

Goodall discovered other important aspects of chimpanzee life, too. She learned that chimpanzees eat meat as well as plants. She described the close bond between chimpanzee mothers and their offspring, who are constantly together for several years after birth. She also recorded the daily foraging activities undertaken by small groups consisting of a mother and her offspring, sometimes with one or two males (**Figure 22B**).

Michael Nichols/National Geographic Stock

▲ **Figure 22A** Jane Goodall observing interactions between two chimpanzees (*Pan troglodytes*)

Martin Harvey/Photolibrary

▲ **Figure 22B** A female chimpanzee foraging with an infant

In areas of abundant food, where these separate groups often came together, Goodall observed the establishment of a male dominance hierarchy through agonistic displays. The alpha periodically reasserted his status with an intimidating charging display, during which, in Goodall's words, he "races across the ground, hurls rocks, drags branches, leaps up, and shakes the vegetation." Goodall witnessed several dramatic episodes in which a subordinate male overthrew the reigning alpha, as well as numerous unsuccessful challenges. These clashes, which cause tension among the group members, were usually followed by some kind of reconciliation behavior. After a conflict, a chimpanzee would make peace through gestures such as embracing or grooming a defeated rival. Grooming—picking through the fur and removing debris or parasites—is the social glue of a chimpanzee community. As Goodall put it, "It improves bad relationships and maintains good ones."

Jane Goodall's pioneering research added immeasurably to our scientific knowledge of primate behavior. She has also spent half a century communicating another aspect of what she learned during her decades in Gombe: "The most important spin-off of the chimp research is probably the humbling effect it has on us who do the research. We are not, after all, the only aware, reasoning beings on this planet." Through her many books, lecture tours, and television appearances, Goodall has educated the public about our closest relatives. For more than 30 years, the Jane Goodall Institute has promoted issues related to the conservation of primates and their habitats, an increasing concern as deforestation, war, and the "bushmeat" trade (the practice of hunting wild animals, including primates, for food) take a rising toll on Africa's great apes. In addition, Goodall has been at the forefront of efforts to ban the use of invasive research methods on chimpanzees and to ensure the humane and ethical treatment of captive chimpanzees in research facilities and zoos.

? Why is the study of chimpanzee behavior relevant to understanding the origins of certain human behaviors?

Because chimpanzees and humans share a common ancestor

23 Human behavior is the result of both genetic and environmental factors

Variations in behavioral traits such as personality, temperament, talents, and intellectual abilities make each person a unique individual. What are the roles of nature (genes) and nurture (environment) in shaping these behaviors? Let's look at how scientists distinguish between genetic and environmental influences on behavioral variations in humans.

Twins provide a natural laboratory for investigating the origins of complex behavioral traits. In general, researchers attempt to estimate the heritability of a trait, or how much of the observed variation can be explained by inheritance. Twin studies compare identical twins (**Figure 23A**), who have the same DNA sequence and are raised in the same environment, with fraternal twins (**Figure 23B**), who share an environment but only half of their DNA sequence. Some twin studies compare identical twins who were raised in the same household with identical twins who were separated at birth, a design that allows researchers to study the interactions of different environments with the same genotype. However, separated twins are very rare.

Results from twin studies consistently show that for complex behavioral traits such as general intelligence and personality characteristics, genetic differences account for roughly half the variation among individuals. The remainder of the variation can be attributed mostly to each individual's unique environment. Thus, neither factor—genes nor environment—is more important than the other.

Determining that a trait is heritable does not mean that scientists have identified a gene "for" the trait. Genes do not dictate behavior. Instead, genes cause tendencies to react to the environment in a certain way. For example, the hormone receptor gene in prairie voles does not produce a protein that causes them to be monogamous. Rather, the receptor protein produced by the gene connects the dopamine reward experienced during mating with the scent of its partner, and the prairie vole responds to the presence of that scent with behaviors that keep it in close contact with its mate.

Let's look at genes related to social bonding as an example of how genetics could play a role in human behavioral variations. Like the brains of other mammals, human brains produce the same hormones and hormone receptors that are implicated in social recognition and bonding in voles. Humans also have considerable individual variation in the key segment of DNA that determines receptor density. As a result, human brains vary in their sensitivity to hormones involved in bonding. As a social species, our well-being depends on relationships with others. Research on voles is helping scientists understand how our brains process the information used to form these bonds. These studies also provide insight into autism, a disorder characterized by difficulty forming social attachments. Scientists hypothesize that variation in a hormone receptor gene may play a role in some types of autism.

Dominic Chan/WENN.com/Newscom.

▲ **Figure 23A** Identical twins, actors James and Oliver Phelps

Janet Mayer/Splash News/Newscom.

◀ **Figure 23B** Actress Scarlett Johansson and her fraternal twin, Hunter

The mechanisms and underlying genetics of behavior are proximate causes (see Module 1). Scientists are also exploring the ultimate causes of human behavior. Sociobiology, the area of research introduced in Module 17, centers on the idea that social behavior evolves, like anatomical traits, as an expression of genes that have been perpetuated by natural selection. When applied to humans, this idea might seem to imply that life is predetermined. But it's unlikely that most human behavior is directly programmed by our genome. Unlike other animals, human offspring have an extraordinarily long period of development after birth. Children interact with a rich social environment, consisting of parents and other family members, peers, teachers, and society in general. The abilities to learn, to innovate, to advance technologically, and to participate in complex social networks have been key elements in the phenomenal success of the human species. It is much more likely that natural selection favored mechanisms that enabled humans to operate on the fly, that is, to use experience and feedback from the environment to adjust their behavior according to the circumstances.

? A researcher conducted a study on "pseudo-twins," unrelated children of the same age who were raised in the same household. Results showed no correspondence between the IQs of pseudo-twins. What do these results indicate about the influence of genes and environment on IQ?

The results indicate a strong genetic component to IQ. The children shared an environment but none of their DNA sequences. If environment were the main influence, you would expect the IQ scores to be similar.

REVIEW

For Practice Quizzes, BioFlix, MP3 Tutors, and Activities, go to www.masteringbiology.com.

Reviewing the Concepts

The Scientific Study of Behavior (1–3)

1 Behavioral ecologists ask both proximate and ultimate questions. Behavioral ecology is the study of behavior in an evolutionary context, considering both proximate (immediate) and ultimate (evolutionary) causes of an animal's actions. Natural selection preserves behaviors that enhance fitness.

2 Fixed action patterns are innate behaviors. Innate behavior is performed the same way by all members of a species. A fixed action pattern (FAP) is an unchangeable series of actions triggered by a specific stimulus. FAPs ensure that activities essential to survival are performed correctly without practice.

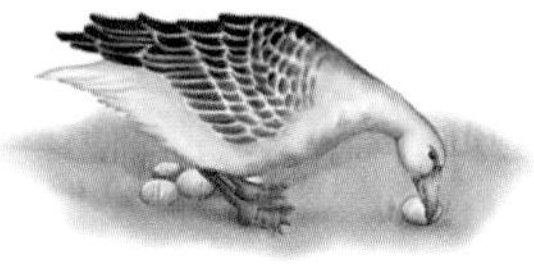

3 Behavior is the result of both genetic and environmental factors. Genetic engineering has been used to investigate genes that influence behavior. Cross-fostering experiments are useful for studying environmental factors that affect behavior.

Learning (4–11)

4 Habituation is a simple type of learning. Learning is a change in behavior resulting from experience. Habituation is learning to ignore a repeated, unimportant stimulus.

5 Imprinting requires both innate behavior and experience. Imprinting is irreversible learning limited to a sensitive period in the animal's life.

6 Imprinting poses problems and opportunities for conservation programs.

7 Animal movement may be a simple response to stimuli or require spatial learning. Kineses and taxes are simple movements in response to a stimulus. Spatial learning involves using landmarks to move through the environment.

8 Movements of animals may depend on cognitive maps. Cognitive maps are internal representations of spatial relationships of objects in the surroundings. Migratory animals may move between areas using the sun, stars, landmarks, or other cues.

9 Animals may learn to associate a stimulus or behavior with a response. In associative learning, animals learn by associating external stimuli or their own behavior with positive or negative effects.

10 Social learning employs observation and imitation of others.

11 Problem-solving behavior relies on cognition. Cognition is the process of perceiving, storing, integrating, and using information. Problem-solving behavior involves complex cognitive processes.

Survival and Reproductive Success (12–16)

12 Behavioral ecologists use cost-benefit analysis to study foraging. Foraging includes identifying, obtaining, and eating food. Optimal foraging theory predicts that feeding behavior will maximize energy gain and minimize energy expenditure and risk.

13 Communication is an essential element of interactions between animals. Signaling in the form of sounds, scents, displays, or touches provides means of communication.

14 Mating behavior often includes elaborate courtship rituals. Courtship rituals advertise the species, sex, and physical condition of potential mates.

15 Mating systems and parental care enhance reproductive success. Mating systems may be promiscuous, monogamous, or polygamous. The needs of offspring and certainty of paternity help explain differences in mating systems and parental care by males.

16 Chemical pollutants can cause abnormal behavior. Endocrine disruptors are chemicals in the environment that may cause abnormal behavior as well as reproductive abnormalities.

Social Behavior and Sociobiology (17–23)

17 Sociobiology places social behavior in an evolutionary context. Social behavior is any kind of interaction between two or more animals.

18 Territorial behavior parcels out space and resources. Animals exhibiting this behavior defend their territories.

19 Agonistic behavior often resolves confrontations between competitors. Agonistic behavior includes threats, rituals, and sometimes combat.

20 Dominance hierarchies are maintained by agonistic behavior.

21 Altruistic acts can often be explained by the concept of inclusive fitness. Kin selection is a form of natural selection favoring altruistic behavior that benefits relatives. Thus, an animal can propagate its own genes by helping relatives reproduce.

22 Jane Goodall revolutionized our understanding of chimpanzee behavior. In decades of fieldwork, she described many aspects of chimpanzee social behavior. She also discovered that chimpanzees make and use tools.

23 Human behavior is the result of both genetic and environmental factors.

Connecting the Concepts

1. Complete this map, which reviews the genetic and environmental components of animal behavior and their relationship to learning.

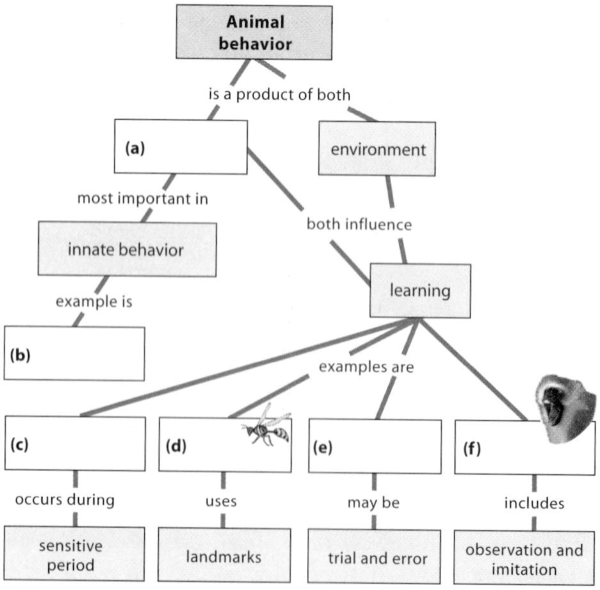

Testing Your Knowledge

Multiple Choice

2. Although many chimpanzee populations live in environments containing oil palm nuts, members of only a few populations use stones to crack open the nuts. The most likely explanation for this behavioral difference between populations is that
   a. members of different populations differ in manual dexterity.
   b. members of different populations have different nutritional requirements.
   c. members of different populations differ in learning ability.
   d. the use of stones to crack nuts has arisen and spread through social learning in only some populations.
   e. the behavioral difference is caused by genetic differences between populations.

3. Pheasants do not feed their chicks. Immediately after hatching, a pheasant chick starts pecking at seeds and insects on the ground. How might a behavioral ecologist explain the ultimate cause of this behavior?
   a. Pecking is a fixed action pattern.
   b. Pheasants learned to peck, and their offspring inherited this behavior.
   c. Pheasants that pecked survived and reproduced best.
   d. Pecking is a result of imprinting during a sensitive period.
   e. Pecking is an example of habituation.

4. A blue jay that aids its parents in raising its siblings is increasing its
   a. reproductive success.
   b. status in a dominance hierarchy.
   c. altruistic behavior.
   d. inclusive fitness.
   e. certainty of paternity.

5. Ants carry dead ants out of the anthill and dump them on a "trash pile." If a live ant is painted with a chemical from dead ants, other ants repeatedly carry it, kicking and struggling, to the trash pile, until the substance wears off. Which of the following best explains this behavior?
   a. The chemical triggers a fixed action pattern.
   b. The ants have become imprinted on the chemical.
   c. The ants continue the behavior until they become habituated.
   d. The ants can learn only by trial and error.
   e. The chemical triggers a negative taxis.

Describing, Comparing, and Explaining

6. Almost all the behaviors of a housefly are innate. What are some advantages and disadvantages to the fly of innate behaviors compared with behaviors that are mainly learned?

7. In Module 3, you learned that Norway rat offspring whose mothers don't interact with them much grow up to be fearful and anxious in new situations. Suggest a possible ultimate cause for this link between maternal behavior and stress response of offspring. (*Hint*: Under what circumstances might high reactivity to stress be more adaptive than being relaxed?)

8. A chorus of frogs fills the air on a spring evening. The frog calls are courtship signals. What are the functions of courtship behaviors? How might a behavioral ecologist explain the proximate cause of this behavior? The ultimate cause?

Applying the Concepts

9. Crows break the shells of certain molluscs before eating them by dropping them onto rocks. Hypothesizing that crows drop the molluscs from a height that gives the most food for the least effort (optimal foraging), a researcher dropped shells from different heights and counted the drops it took to break them.

Height of drop (m)	Average number of drops required to break shell	Total flight height (number of drops × height per drop)
2	55	110
3	13	39
5	6	30
7	5	35
15	4	60

   a. The researcher measured the average drop height for crows and found that it was 5.23 m. Does this support the researcher's hypothesis? Explain.
   b. Describe an experiment to determine whether this feeding behavior of crows is learned or innate.

10. Scientists studying scrub jays found that it is common for "helpers" to assist mated pairs of birds in raising their young. The helpers lack territories and mates of their own. Instead, they help the territory owners gather food for their offspring. Propose a hypothesis to explain what advantage there might be for the helpers to engage in this behavior instead of seeking their own territories and mates. How would you test your hypothesis? If your hypothesis is correct, what kind of results would you expect your tests to yield?

11. Researchers are very interested in studying identical twins who were raised apart. Among other things, they hope to answer questions about the roles of inheritance and upbringing in human behavior. Why do identical twins make such good subjects for this kind of research? What do the results suggest to you? What are the potential pitfalls of this research? What abuses might occur in the use of these data if the studies are not evaluated critically?

12. Do animals think and feel the same kinds of things we do? These questions bear on animal rights, a subject much in the news. Many important biological discoveries have come from experiments performed on animals, yet some animal rights activists believe that all animal experimentation is cruel and should be stopped. They have harassed researchers, vandalized laboratories, and set animals free. Why are animals used in experiments? Are there uses of animals that should be discontinued? What kinds of guidelines should researchers follow in using animals in experiments, and who should establish and enforce the guidelines?

Review Answers

1. a. genes; b. fixed action pattern (FAP); c. imprinting; d. spatial learning; e. associative learning; f. social learning
2. d 3. c 4. d 5. a
6. Main advantage: Flies do not live long. Innate behaviors can be performed the first time without learning, enabling flies to find food, mates, and so on without practice. Main disadvantage: Innate behaviors are rigid; flies cannot learn to adapt to specific situations.
7. In a stressful environment, for example, where predators are abundant, rats that behave cautiously are more likely to survive long enough to reproduce.
8. Courtship behaviors reduce aggression between potential mates and confirm their species, sex, and physical condition. Environmental changes such as rainfall, temperature, and day length probably lead frogs to start calling, so these would be the proximate causes. The ultimate cause relates to evolution. Fitness (reproductive success) is enhanced for frogs that engage in courtship behaviors.
9. a. Yes. The experimenter found that 5 m provided the most food for the least energy because the total flight height (number of drops × height per drop) was the lowest. Crows appear to be using an optimal foraging strategy.
   b. An experiment could measure the average drop height for juvenile and adult birds, or it could trace individual birds during a time span from juvenile to adult and see if their drop height changed.
10. One likely hypothesis is that the helper is closely related to one or both of the birds in the mated pair. Because closely related birds share relatively many genes, the helper bird is indirectly enhancing its own fitness by helping its relatives raise their young. (In other words, this behavior evolved by kin selection.) The easiest way to test the hypothesis would be to determine the relatedness of the birds by DNA analysis. If birds are closely related, their DNA should be more similar than those of more distantly related or unrelated birds.
11. Identical twins are genetically the same, so any differences between them are due to environment. Thus, the study of identical twins enables researchers to sort out the effects of "nature" and "nurture" on human behavior. The data suggest that many aspects of human behavior are inborn. Some people find these studies disturbing because they seem to leave less room for free will and self-improvement than we would like. Results of such studies may be carelessly cited in support of a particular social agenda.
12. Animals are used in experiments for various reasons. A particular species of animal may have features that make it well suited to answer an important biological question. Squids, for example, have a giant nerve fiber that made possible the discovery of how all nerve cells function. Animal experiments play a major role in medical research. Many vaccines that protect humans against deadly diseases, as well as drugs that can cure diseases, have been developed using animal experiments. Animals also benefit, as vaccines and drugs are developed for combating their own pathogens. Some researchers point out that the number of animals used in research is a small fraction of those killed as strays by animal shelters and a minuscule fraction of those killed for human food. They also maintain that modern research facilities are models of responsible and considerate treatment of animals. Whether or not this is true, the possibility that at least some kinds of animals used in research suffer physical pain as a result, and perhaps mental anguish as well, raises serious ethical issues. Some questions to consider: What are some medical treatments or products that have undergone testing in animals? Have you benefited from any of them? Are there alternatives to using animals in experiments? Would alternatives put humans at risk? Are all kinds of animal experiments equally valuable? In your opinion, what kinds of experiments are acceptable, and what kinds are unnecessary? What kinds of treatment are humane, and what kinds are inhumane?

Population Ecology

Population Ecology

BIG IDEAS

Population Structure and Dynamics (1–8)

Population ecology is concerned with characteristics that describe populations, changes in population size, and factors that regulate populations over time.

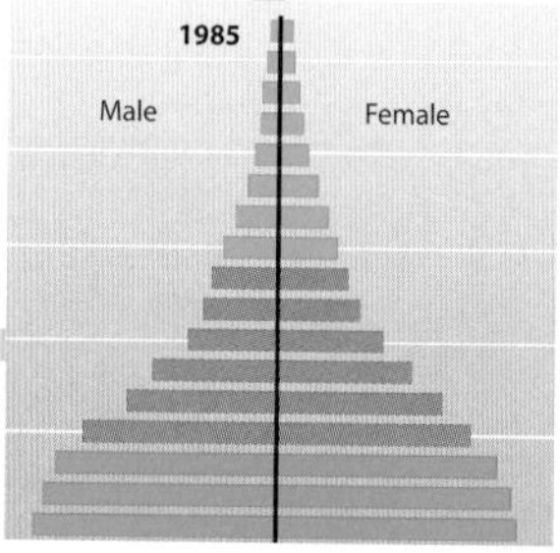

The Human Population (9–11)

The principles of population ecology can be used to describe the growth of the human population and its limits.

Antoine Dervaux/Photolibrary

Emperor penguins, shown above, are one of the few species adapted to life on the polar ice. Their outer feathers form a waterproof cover for foraging in the icy sea, while a downy underlayer provides insulation. Beneath their skin, their bodies are swaddled in a thick layer of fat that adds insulation as well as storing energy. Reproduction is especially challenging in this environment. After the female lays an egg, she passes it to the male for safekeeping while she trudges many miles to the sea to feed. Throughout the frigid Antarctic winter, the male penguins huddle together in a shifting mass of bodies, each protecting its precious egg from the bitter cold and howling winds. Weeks later, the females return, and the males trek to the sea to replenish their depleted energy stores. In the following months, the parents make multiple trips to the sea to fetch food, which they regurgitate for their hungry offspring. Even chicks that receive food regularly are not assured of survival—predatory birds known as giant petrels carry off many of the young.

As an individual, each penguin faces the rigors of the Antarctic climate, the threat of predators, and the struggle to reproduce. In terms of population ecology, however, the fates of individuals are merged into group characteristics—each chick that escapes being a petrel's prey feeds into the percentage of chicks that survive their first year; the breeding success of one pair of birds feeds into the growth rate of the population.

In this chapter, you'll learn about the structure and dynamics of populations and the factors that regulate populations over time. As ecologists gain greater insight into natural populations, we become better equipped to develop sustainable food sources, assess the impact of human activities, and balance human needs with the conservation of biodiversity and resources.

Population Structure and Dynamics

1 Population ecology is the study of how and why populations change

Ecologists usually define a **population** as a group of individuals of a single species that occupy the same general area. These individuals rely on the same resources, are influenced by the same environmental factors, and are likely to interact and breed with one another. For example, the emperor penguins living near Dumont d'Urville Station, where *March of the Penguins* was filmed, are a population. When a researcher chooses a population to study, he or she defines it by boundaries appropriate to the species being studied and to the purposes of the investigation.

Population ecology is concerned with changes in population size and the factors that regulate populations over time. A population ecologist might use statistics such as the number and distribution of individuals to describe a population. Population ecologists also examine population dynamics, the interactions between biotic and abiotic factors that cause variation in population sizes. One important aspect of population dynamics—and a major topic for this chapter—is population growth. The penguin population at Dumont d'Urville Station increases through births and the immigration of penguins from nearby colonies. Deaths and the emigration of individuals away from Dumont d'Urville Station decrease the population. Population ecologists might investigate how various environmental factors, such as availability of food, predation by killer whales, or the extent of sea ice, affect the size, distribution, or dynamics of the population.

Population ecology plays a key role in applied research. For example, population ecology provides critical information for identifying and saving endangered species. Population ecology is being used to manage wildlife populations and to develop sustainable fisheries throughout the world. The population ecology of pests and pathogens provides insight into controlling their spread. Population ecologists also study human population growth, one of the most critical environmental issues of our time.

? What is the relationship between a population and a species?

A population is a localized group of individuals of a single species.

2 Density and dispersion patterns are important population variables

Two important aspects of population structure are population density and dispersion pattern. **Population density** is the number of individuals of a species per unit area or volume—the number of oak trees per square kilometer (km^2) in a forest, for instance, or the number of earthworms per cubic meter (m^3) in forest soil. Because it is impractical or impossible to count all individuals in a population in most cases, ecologists use a variety of sampling techniques to estimate population densities. For example, they might base an estimate of the density of alligators in the Florida Everglades on a count of individuals in a few sample plots of 1 km^2 each. The larger the number and size of sample plots, the more accurate the estimates. In some cases, population densities are estimated not by counts of organisms but by indirect indicators, such as number of bird nests or rodent burrows.

Within a population's geographic range, local densities may vary greatly. The **dispersion pattern** of a population refers to the way individuals are spaced within their area. A **clumped dispersion pattern**, in which individuals are grouped in patches, is the most common in nature. Clumping often results from an unequal distribution of resources in the environment. For instance, plants or fungi may be clumped in areas where soil conditions and other factors favor germination and growth. Clumping of animals often results from uneven food distribution. For example, the sea stars shown in **Figure 2A** group together where food is abundant. Clumping may

Matthew Banks/Alamy

▲ **Figure 2A** Clumped dispersion of ochre sea stars at low tide

David Jay Zimmerman/Corbis

▲ **Figure 2B** Uniform dispersion of sunbathers at Coney Island

mashe/Shutterstock

▲ **Figure 2C** Random dispersion of dandelions

also reduce the risk of predation, or be associated with social behavior.

A **uniform dispersion pattern** (an even one) often results from interactions between the individuals of a population. For instance, some plants secrete chemicals that inhibit the germination and growth of nearby plants that could compete for resources. Animals may exhibit uniform dispersion as a result of territorial behavior. Figure 2B (on the previous page) shows the uniform dispersion of sunbathers at a popular New York beach.

In a **random dispersion pattern**, individuals in a population are spaced in an unpredictable way, without a pattern. Plants, such as dandelions (Figure 2C, previous page), that grow from windblown seeds might be randomly dispersed. However, varying habitat conditions and social interactions make random dispersion rare.

Estimates of population density and dispersion patterns enable researchers to monitor changes in a population and to compare and contrast the growth and stability of populations in different areas. The next module describes another tool that ecologists use to study populations.

? What dispersion pattern would you predict in a forest population of termites, which live in damp, rotting wood?

Clumped (in fallen logs or dead trees).

3 Life tables track survivorship in populations

Life tables track survivorship, the chance of an individual in a given population surviving to various ages. Starting with a population of 100,000 people, Table 3 shows the number who are expected to be alive at the beginning of each age interval, based on death rates in 2004. For example, 93,735 out of 100,000 people are expected to live to age 50. The chance of surviving to age 60, shown in the last column of the table, is 0.939. The chance of surviving to age 90, however, is only 0.412. The life insurance industry uses life tables to predict how long, on average, a person will live. Population ecologists have adopted this technique and constructed life tables for various other species. By identifying the most vulnerable stages of an organism's life, life table data help conservationists develop effective measures for maintaining a viable population.

Life tables can be used to construct **survivorship curves**, which plot survivorship as the proportion of individuals from an initial population that are alive at each age (Figure 3). By using a percentage scale instead of actual ages on the *x*-axis, we can compare species with widely varying life spans on the same graph. The curve for the human population shows that most

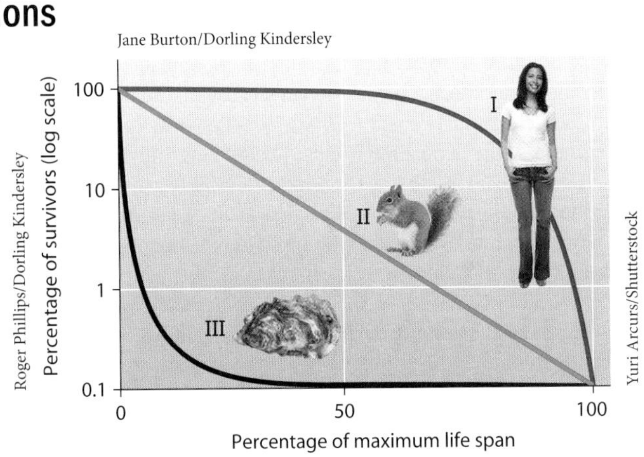

▲ **Figure 3** Three types of survivorship curves

people survive to the older age intervals, as we saw in the life table. Ecologists refer to the shape of this curve as Type I survivorship. Species that exhibit a Type I curve—humans and many other large mammals—usually produce few offspring but give them good care, increasing the likelihood that they will survive to maturity.

In contrast, a Type III curve indicates low survivorship for the very young, followed by a period when survivorship is high for those few individuals who live to a certain age. Species with this type of survivorship curve usually produce very large numbers of offspring but provide little or no care for them. Some fishes, for example, can produce millions of eggs at a time, but most offspring die as larvae from predation or other causes. Many invertebrates, such as oysters, also have Type III survivorship curves.

A Type II curve is intermediate, with survivorship constant over the life span. That is, individuals are no more vulnerable at one stage of the life cycle than at another. This type of survivorship has been observed in some invertebrates, lizards, and rodents.

? How does the chance of survival change with age in organisms with a Type III survivorship curve?

The chance of survival is initially low but increases after an individual reaches a certain age.

TABLE 3 | LIFE TABLE FOR THE U.S. POPULATION IN 2004

Age Interval	Number Living at Start of Age Interval (*N*)	Number Dying During Interval (*D*)	Chance of Surviving Interval 1 - (*D*/*N*)
0–10	100,000	871	0.991
10–20	99,129	419	0.996
20–30	98,709	933	0.991
30–40	97,776	1,259	0.987
40–50	96,517	2,781	0.971
50–60	93,735	5,697	0.939
60–70	88,038	11,847	0.865
70–80	76,191	22,267	0.708
80–90	53,925	31,706	0.412
90+	22,219	22,219	0.000

Data from Center for Health Statistics, 2004 CDC.

4 Idealized models predict patterns of population growth

Wild Dales Photography - Simon Phillpotts/Alamy

Population size fluctuates as new individuals are born or immigrate into an area and others die or emigrate. Some populations—for example, trees in a mature forest—are relatively constant over time. Other populations change rapidly, even explosively. Consider a single bacterium that divides every 20 minutes. There would be two bacteria after 20 minutes, four after 40 minutes, eight after 60 minutes, and so on. In just 12 hours, the population would approach 70 billion cells. If reproduction continued at this rate for a day and a half—a mere 36 hours—there would be enough bacteria to form a layer a foot deep over Earth's entire surface. Using idealized models, population ecologists can predict how the size of a particular population will change over time under different conditions.

The Exponential Growth Model The rate of population increase under ideal conditions, called exponential growth, can be calculated using the simple equation $G = rN$. The G stands for the growth rate of the population (the number of new individuals added per time interval); N is the population size (the number of individuals in the population at a particular time); and r stands for the **per capita rate of increase** (the average contribution of each individual to population growth; per capita means "per person").

How do we estimate the per capita rate of increase? Population growth reflects the number of births minus the number of deaths (the model assumes that immigration and emigration are equal). Suppose a population of rabbits has 100 individuals, and there are 50 births and 20 deaths in one month. The net increase is 30 rabbits. The per capita increase in the population, or r, is 30/100, or 0.3.

In a population growing in an ideal environment with unlimited space and resources, r is the maximum capacity of members of that population to reproduce. Thus, the value of r depends on the kind of organism. For example, rabbits have a higher r than elephants, and bacteria have a higher r than rabbits.

TABLE 4A EXPONENTIAL GROWTH OF RABBITS, $r = 0.3$

Time (months)	N	$G = rN$
0	20	6
1	26	8
2	34	10
3	44	13
4	57	17
5	74	22
6	96	29
7	125	38
8	163	49
9	212	64
10	276	83
11	359	108
12	467	140

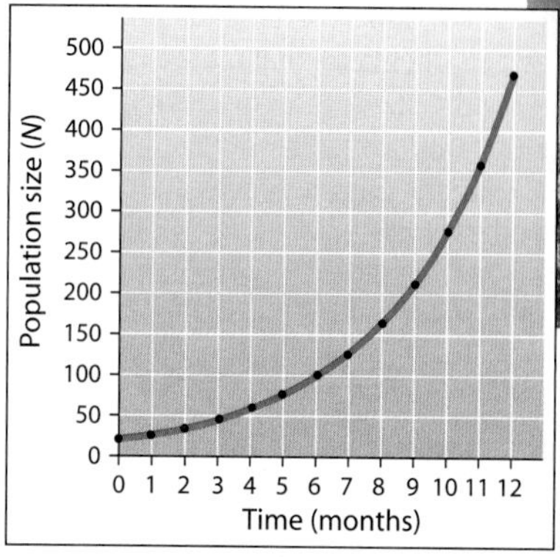

▲ **Figure 4A** Exponential growth of rabbits

When a population is expanding without limits, r remains constant and the rate of population growth depends on the number of individuals already in the population (N). In **Table 4A**, a population begins with 20 rabbits. The growth rate (G) for this population, using $r = 0.3$, is shown in the right-hand column. Notice that the larger the population size, the more new individuals are added during each time interval.

Graphing these data in **Figure 4A** produces a J-shaped curve, which is typical of exponential growth. The lower part of the J, where the slope of the line is almost flat, results from the relatively slow growth when N is small. As the population increases, the slope becomes steeper.

The **exponential growth model** gives an idealized picture of unlimited population growth. There is no restriction on the abilities of the organisms to live, grow, and reproduce. Even elephants, the slowest breeders on the planet, would increase exponentially if enough resources were available. Although elephants typically produce only six young in a 100-year life span, Charles Darwin estimated that it would take only 750 years for a single pair to give rise to a population of 19 million. But any population—bacteria, rabbits, or elephants—will eventually be limited by the resources available.

Limiting Factors and the Logistic Growth Model In nature, a population that is introduced to a new environment or is rebounding from a catastrophic decline in numbers may grow exponentially for a while. Eventually, however, one or more environmental factors will limit its growth rate as the population reaches its maximum sustainable size. Environmental factors that restrict population growth are called **limiting factors**.

You can see the effect of population-limiting factors in the graph in **Figure 4B** (see top of facing page), which illustrates the growth of a population of fur seals on St. Paul Island, off the coast of Alaska. (For simplicity, only the mated bulls were counted. Each has a harem of a number of females, as shown in the photograph.) Before 1925, the seal population on the island remained low because of uncontrolled hunting, although it changed from year to year. After hunting was controlled, the population increased rapidly until about 1935,

Roy Corral/Corbis

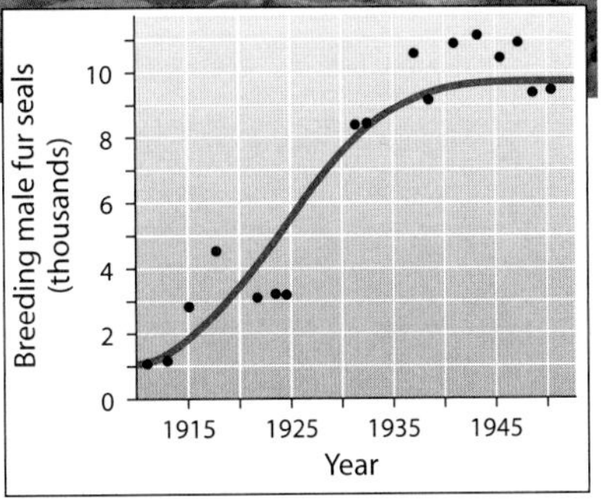

▶ **Figure 4B** Growth of a population of fur seals

when it began to level off and started fluctuating around a population size of about 10,000 bull seals. At this point, a number of limiting factors, including some hunting and the amount of space suitable for breeding, restricted population growth.

The fur seal growth curve fits the **logistic growth model**, a description of idealized population growth that is slowed by limiting factors as the population size increases. **Figure 4C** compares the logistic growth model (red) with the exponential growth model (blue). As you can see, the logistic curve is J-shaped at first, but gradually levels off to resemble an S.

To model logistic growth, the formula for exponential growth, rN, is multiplied by an expression that describes the effect of limiting factors on an increasing population size:

$$G = rN\frac{(K - N)}{K}$$

This equation is actually simpler than it looks. The only new symbol in the equation is K, which stands for carrying capacity. **Carrying capacity** is the maximum population size that a particular environment can sustain ("carry"). For the fur seal population on St. Paul Island, for instance, K is about 10,000 mated males. The value of K varies, depending on the species and the resources available in the habitat. K might be considerably less than 10,000 for a fur seal population on a smaller island with fewer breeding sites. Even in one location, K is not a fixed number. Organisms interact with other organisms in their communities, including predators, parasites, and food sources, that may affect K. Changes in abiotic factors may also increase or decrease carrying capacity. In any case, the concept of carrying capacity expresses an essential fact of nature: Resources are finite.

Table 4B demonstrates how the expression $(K - N)/K$ in the logistic growth model produces the S-shaped curve. At the outset, N (the population size) is very small compared to K (the carrying capacity). Thus, $(K - N)/K$ nearly equals K/K, or 1, and population growth (G) is close to rN—that is, exponential growth. As the population increases and N gets closer to carrying capacity, $(K - N)/K$ becomes an increasingly smaller fraction. The growth rate slows as rN is multiplied by that fraction. At carrying capacity, the population is as large as it can theoretically get in its environment; at this point, $N = K$, and $(K - N)/K = 0$. The population growth rate (G) becomes zero.

What does the logistic growth model suggest to us about real populations in nature? The model predicts that a population's growth rate will be small when the population size is *either* small or large, and highest when the population is at an intermediate level relative to the carrying capacity. At a low population level, resources are abundant, and the population is able to grow nearly exponentially. At this point, however, the increase is small because N is small. In contrast, at a high population level, limiting factors strongly oppose the population's potential to increase. There might be less food available per individual or fewer breeding territories, nest sites, or shelters. These limiting factors cause the birth rate to decrease, the death rate to increase, or both. Eventually, when the birth rate equals the death rate, the population stabilizes at the carrying capacity (K).

It is important to realize that the logistic growth model presents a mathematical ideal that is a useful starting point for studying population growth and for constructing more complex models. Like any good starting hypothesis, the logistic model has stimulated research, leading to a better understanding of the factors affecting population growth. We take a closer look at some of these factors next.

? In logistic growth, at what population size (in terms of K) is the population increasing most rapidly? Explain why.

When N is $\frac{1}{2}K$. At this population size, there are more reproducing individuals than at lower population sizes and still lots of space or other resources available for growth.

WizData/Shutterstock

Joshua Lewis/Shutterstock

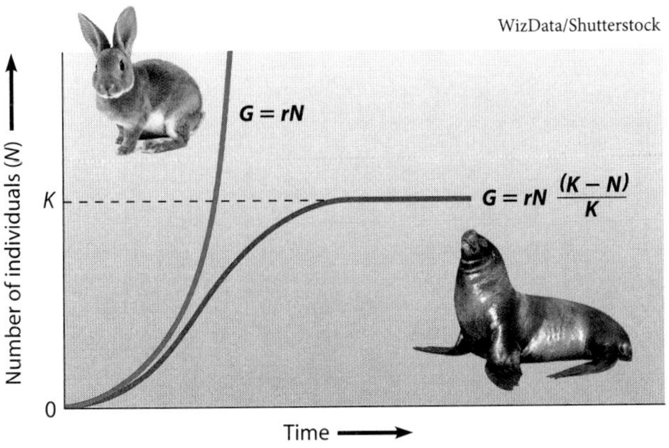

▲ **Figure 4C** Logistic growth and exponential growth compared

TABLE 4B EFFECT OF K ON GROWTH RATE AS N APPROACHES K, $K = 1{,}000$, $r = 0.1$

N	rN	$(K - N)/K$	$G = rN(K - N)/K$
10	1	0.99	0.99
100	10	0.9	9.00
400	40	0.6	24.00
500	50	0.5	25.00
600	60	0.4	24.00
700	70	0.3	21.00
900	95	0.05	0.25
1,000	100	0.00	0.00

5 Multiple factors may limit population growth

The logistic growth model predicts that population growth will slow and eventually stop as population density increases. That is, at higher population densities, the birth rate decreases, the death rate increases, or both. What are the possible causes of these density-dependent changes in birth and death rates?

Several **density-dependent factors**—limiting factors whose intensity is related to population density—appear to limit growth in natural populations. The most obvious is **intraspecific competition**—competition between individuals of the same species for limited resources. As a limited food supply is divided among more and more individuals, birth rates may decline as individuals have less energy available for reproduction. In **Figure 5A**, clutch size (the number of eggs a female bird lays in a "litter") declines as the population density, and therefore the number of competitors, increases.

Density-dependent factors often depress a population's growth by increasing the death rate. In a laboratory experiment with flour beetles, for example, survivorship declined with increasing population density **(Figure 5B)**. In a natural setting, plants that grow close together may experience increased mortality as competition for resources increases. And those that survive will likely produce fewer flowers, fruits, and seeds than uncrowded individuals. In an animal population, the death rate may climb as a result of increased disease transmission under crowded conditions or the accumulation of toxic waste products. Predation may also be an important cause of density-dependent mortality. A predator may concentrate on a particular kind of prey as that prey becomes abundant.

A limiting factor may be something other than food or nutrients. In many vertebrates that defend a territory, the availability of space may limit reproduction. For instance, the number of nesting sites on rocky islands may limit the population size of seabirds such as gannets. Or, like a game of musical chairs, the number of safe hiding places may limit a prey population by exposing some individuals to a greater risk of predation.

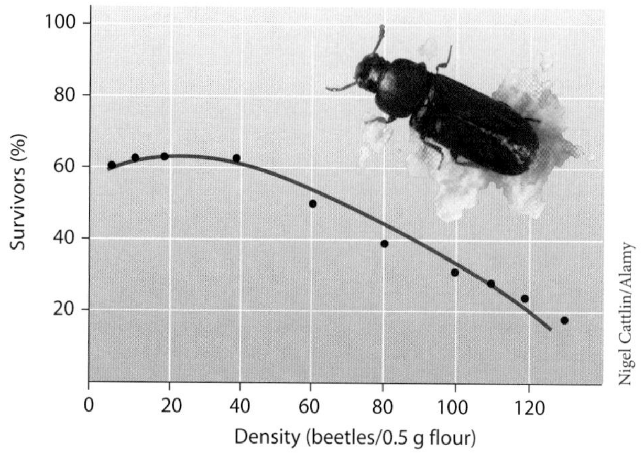

▲ **Figure 5B** Decreasing survival rates with increasing density in a population of flour beetles

For some animal species, physiological factors may regulate population size. White-footed mice in a small field enclosure will multiply from a few to a colony of 30 to 40 individuals, but reproduction then declines until the population ceases to grow. This drop in reproduction occurs even when additional food and shelter are provided. High population densities in mice appear to induce a stress syndrome in which hormonal changes can delay sexual maturation, cause reproductive organs to shrink, and depress the immune system. In this case, high densities cause both a decrease in birth rate and an increase in death rate. Similar effects of crowding have been observed in wild populations of other rodents.

In many natural populations, abiotic factors such as weather may affect population size well before density-dependent factors become important. A population-limiting factor whose intensity is unrelated to population density is called a **density-independent factor**. If we look at the growth curve of such a

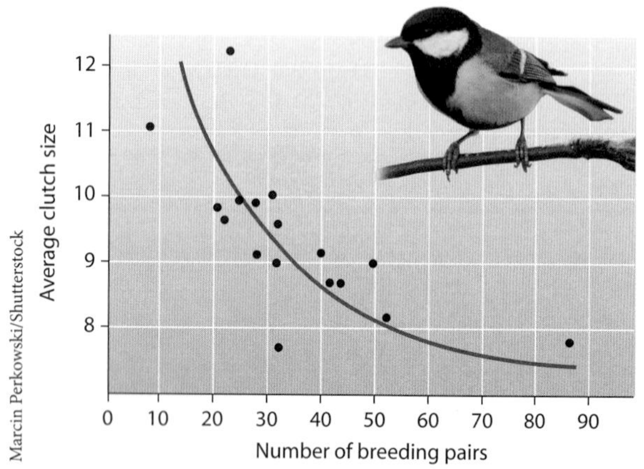

▲ **Figure 5A** Decreasing birth rate with increasing density in a population of great tits

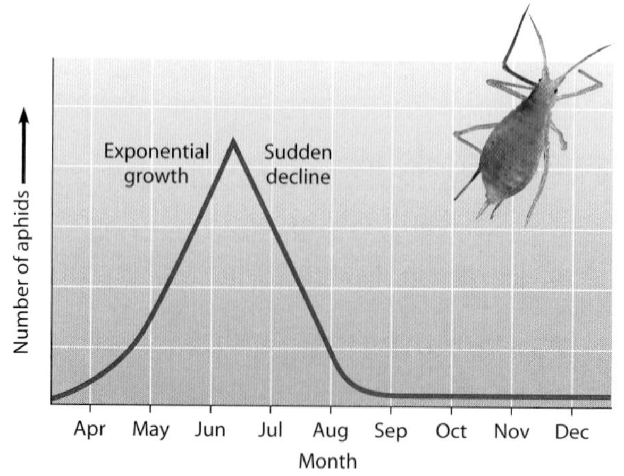

▲ **Figure 5C** Weather change as a density-independent factor limiting aphid population growth

population, we see something like exponential growth followed by a rapid decline, rather than a leveling off. **Figure 5C** (on the previous page) shows this effect for a population of aphids, insects that feed on the sugary phloem sap of plants. These and many other insects undergo virtually exponential growth in the spring and then rapidly die off when the weather turns hot and dry in the summer. A few individuals may survive, and these may allow population growth to resume if favorable conditions return. In some populations of insects—many mosquitoes and grasshoppers, for instance—adults die off entirely, leaving only eggs, which initiate population growth the following year. In addition to seasonal changes in the weather, disturbances—such as fire, storms, and habitat disruption by human activity—can affect a population's size regardless of its density.

Over the long term, most populations are probably regulated by a mixture of factors. Some populations remain fairly stable in size and are presumably close to a carrying capacity that is determined by biotic factors such as competition or predation. Most populations for which we have long-term data, however, show fluctuations in numbers. Thus, the dynamics of many populations result from a complex interaction of both density-dependent factors and density-independent abiotic factors such as climate and disturbances.

? List some of the factors that may reduce birth rate or increase death rate as population density increases.

Food and nutrient limitations, insufficient territories, increase in disease and predation, accumulation of toxins

6 Some populations have "boom-and-bust" cycles

Some populations of insects, birds, and mammals undergo dramatic fluctuations in density with remarkable regularity. "Booms" characterized by rapid exponential growth are followed by "busts," during which the population falls back to a minimal level. A striking example is the boom-and-bust growth cycles of lemming populations that occur every three to four years. (Lemmings are small rodents that live in the tundra.) Some researchers hypothesize that natural changes in the lemmings' food supply may be the underlying cause. Another hypothesis, as discussed in Module 5, is that stress from crowding during the "boom" may reduce reproduction, causing a "bust."

Figure 6 illustrates another well-known example—the cycles of snowshoe hare and lynx. The lynx is one of the main predators of the snowshoe hare in the far northern forests of Canada and Alaska. About every 10 years, both hare and lynx populations show a rapid increase followed by a sharp decline.

What causes these boom-and-bust cycles? Since ups and downs in the two populations seem to almost match each other on the graph, does this mean that changes in one directly affect the other? For the hare cycles, there are three main hypotheses. First, cycles may be caused by winter food shortages that result from overgrazing. Second, cycles may be due to predator-prey interactions. Many predators other than lynx, such as coyotes, foxes, and great-horned owls, eat hares, and the combination of predators might overexploit their prey. Third, cycles could be affected by a combination of limited food resources and excessive predation. Recent field experiments support the hypothesis that the 10-year cycles of the snowshoe hare are largely driven by excessive predation, but are also influenced by fluctuations in the hare's food supplies.

For the lynx and many other predators that depend heavily on a single species of prey, the availability of prey can have a strong influence on population size. Thus, the 10-year cycles in the lynx population probably do result at least in part from the 10-year cycles in the hare population. As the lynx population declines, the prey population—released from predator pressure—rebounds. Long-term studies are the key to unraveling the complex causes of such population cycles.

Now that we have looked at patterns of population growth, we turn our attention to the differences in reproductive patterns of populations and how they are shaped by natural selection.

? In one experiment, providing more food to hares increased their population density, but the population continued to show cyclic collapses. What might you conclude from these results?

Hare population cycles are not primarily caused by food shortages.

Alan Carey/Photo Researchers

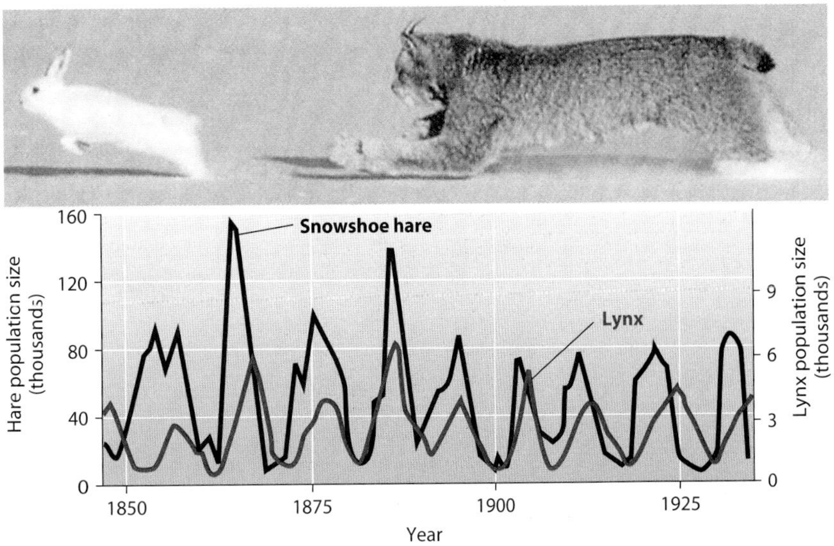

▲ **Figure 6** Population cycles of the snowshoe hare and the lynx

EVOLUTION CONNECTION

7 Evolution shapes life histories

The traits that affect an organism's schedule of reproduction and death make up its **life history**. Some key life history traits are the age of first reproduction, the frequency of reproduction, the number of offspring, and the amount of parental care given. Natural selection cannot optimize all of these traits simultaneously because an organism has limited time, energy, and nutrients. For example, an organism that gives birth to a large number of offspring will not be able to provide a great deal of parental care. Consequently, the combination of life history traits in a population represents trade-offs that balance the demands of reproduction and survival. Because selective pressures vary, life histories are very diverse. Nevertheless, ecologists have observed some patterns that are useful for understanding how life history characteristics have been shaped by natural selection.

One life history pattern is typified by small-bodied, short-lived animals (for example, insects and small rodents) that develop and reach sexual maturity rapidly, have a large number of offspring, and offer little or no parental care. A similar pattern is seen in small, nonwoody plants such as dandelions that produce thousands of tiny seeds. Ecologists hypothesize that selection for this set of life history traits occurs in environments where resources are abundant, permitting exponential growth. It is sometimes called ***r*-selection** because *r* (the per capita rate of increase) is maximized. Most *r*-selected species have an advantage in habitats that experience unpredictable disturbances, such as fire, floods, hurricanes, drought, or cold weather, which create new opportunities by suddenly reducing a population to low levels. Human activity is a major cause of disturbance, producing road cuts, freshly cleared fields and woodlots, and poorly maintained lawns that are commonly colonized by *r*-selected plants and animals.

In contrast, large-bodied, long-lived animals (such as bears and elephants) develop slowly and produce few, but well-cared-for, offspring. Plants with comparable life history traits include coconut palms, which produce relatively few seeds that are well stocked with nutrient-rich material—the plant's version of parental care. Ecologists hypothesize that selection for this set of life history traits occurs in environments where the population size is near carrying capacity (*K*), so it is sometimes called ***K*-selection**. Population growth in these situations is limited by density-dependent factors. Because competition for resources is keen, *K*-selected organisms gain an advantage by allocating energy to their own survival and to the survival of their descendants. Thus, *K*-selected organisms are adapted to environments that typically have a stable climate and little opportunity for rapid population growth.

The hypothesis of *r*- and *K*-selection has been criticized as an oversimplification, and most organisms fall somewhere between the extremes. However, this hypothesis has stimulated a vigorous subfield of ecological research on the evolution of life histories.

A long-term project in Trinidad has provided direct evidence that life history traits can be shaped by natural selection. For years, researchers have been studying guppy populations living in small, relatively isolated pools. As shown in **Figure 7**, some guppy populations live in pools with predators called killifish, which eat mainly small, immature guppies (Pool 1).

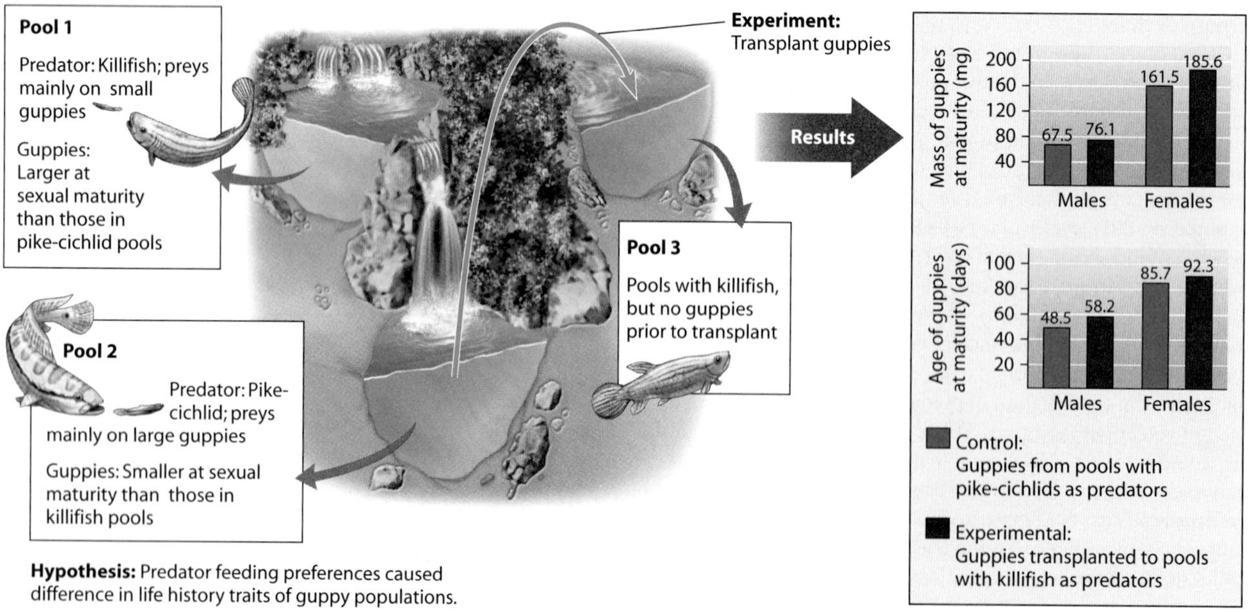

▲ **Figure 7** The effect of predation on the life history traits of guppies

Other guppy populations live where larger fish, called pike-cichlids, eat mostly mature, large-bodied guppies (Pool 2). Guppies in populations exposed to these pike-cichlids tend to be smaller, mature earlier, and produce more offspring at a time than those in areas with killifish. Thus, guppy populations differ in certain life history traits, depending on the kind of predator in their environment. For these differences to be the result of natural selection, the traits should be heritable. And indeed, guppies from both populations raised in the laboratory without predators retained their life history differences.

To test whether the feeding preferences of different predators caused these differences in life histories by natural selection, researchers introduced guppies from a pike-cichlid habitat into a guppy-free pool inhabited by killifish (Pool 3). The scientists tracked the weight and age at sexual maturity in the experimental guppy populations for 11 years, comparing these guppies with control guppies that remained in the pike-cichlid pools. The average weight and age at sexual maturity of the transplanted populations increased significantly as compared with the control populations. These studies demonstrate not only that life history traits are heritable and shaped by natural selection, but also that questions about evolution can be tested by field experiments.

As we have seen, population ecology involves theoretical model building as well as observations and experiments in the field. Next we look at how the principles of population ecology can be applied to conservation and management.

? Refer to Module 3. Which type of survivorship curve would you expect to find in a population experiencing *r*-selection? *K*-selection?

Type III for a population experiencing *r*-selection; Type I for *K*-selection

CONNECTION 8 Principles of population ecology have practical applications

Principles of population ecology can help guide us toward resource management goals, such as increasing populations we wish to harvest or save from extinction or decreasing populations we consider pests. Wildlife managers, fishery biologists, and foresters try to use **sustainable resource management**: harvesting crops without damaging the resource. In terms of population growth, this means maintaining a high population growth rate to replenish the population. According to the logistic growth model, the fastest growth rate occurs when the population size is at roughly half the carrying capacity of the habitat. Theoretically, a resource manager should achieve the best results by harvesting the populations down to this level. However, the logistic model assumes that growth rate and carrying capacity are stable over time. Calculations based on these assumptions, which are not realistic for some populations, may lead to unsustainably high harvest levels that ultimately deplete the resource. In addition, economic and political pressures often outweigh ecological concerns, and the amount of scientific information is frequently insufficient.

Fish, the only wild animals still hunted on a large scale, are particularly vulnerable to overharvesting. For example, in the northern Atlantic cod fishery, estimates of cod population sizes were too high, and the practice of discarding young cod (below legal size) at sea caused a higher mortality rate than predicted. The fishery collapsed in 1992 and has not recovered (**Figure 8**). Following the decline of many other fish and whale populations, resource managers are trying to minimize the risk of resource collapse by setting minimum population sizes or imposing protected, harvest-free areas. For species that are in decline or facing extinction, resource managers may try to provide additional habitat or improve the quality of existing habitat to raise the carrying capacity and thus increase population growth.

Reducing the size of a population may also be a challenging task. Simply killing many individuals will not usually decrease the size of a pest population. Many insect and weed species have life history traits that are *r*-selected and have adaptations

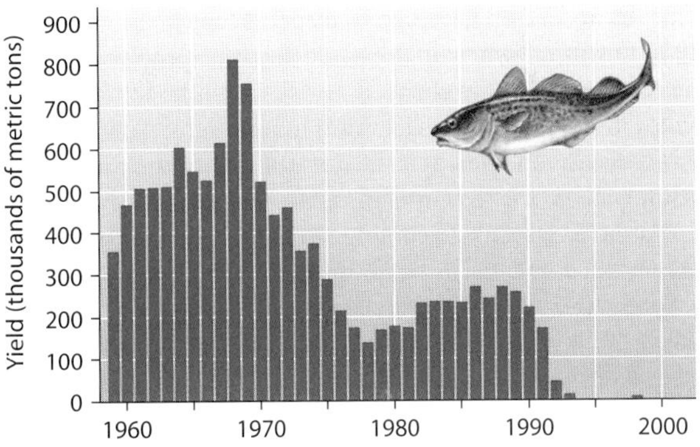

▲ **Figure 8** Collapse of northern cod fishery off Newfoundland

that promote rapid population growth. Also, most pesticides kill both the pest and their natural predators. Because prey species often have a higher reproductive rate than predators, pest populations rapidly rebound before their predators can.

Integrated pest management (IPM) uses a combination of biological, chemical, and cultivation methods to control agricultural pests. IPM relies on knowledge of the population ecology of the pest and its associated predators and parasites, as well as crop growth dynamics.

As you've learned, there are many factors that influence a population's size. To effectively manage any population, we must identify those variables, account for the unpredictability of the environment, consider interactions with other species, and weigh the economic, political, and conservation issues. These same issues apply to the growth of the human population, which we explore next.

? Explain why managers often try to maintain populations of fish and game species at about half their carrying capacity.

To protect wildlife from overharvest yet maintain lower population levels so that growth rate is high and mortality from resource limitation is reduced

Data from Fisheries and Oceans, Canada, 1999.

The Human Population

9 The human population continues to increase, but the growth rate is slowing

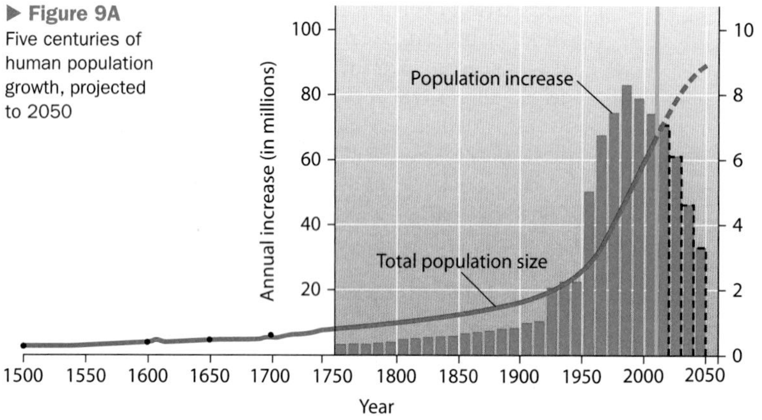

▶ **Figure 9A** Five centuries of human population growth, projected to 2050

Data from United Nations, the World at Six Billion, 2007.

In the few seconds it takes you to read this sentence, 21 babies will be born somewhere in the world and nine people will die. The statistics may have changed a bit since this book was printed, but births will still far outnumber deaths. An imbalance between births and deaths is the cause of population growth (or decline), and as the red curve in **Figure 9A** shows, the human population is expected to continue increasing for at least the next several decades. The bar graph in Figure 9A tells a different part of the story. The number of people added to the population each year has been declining since the 1980s. How do we explain these patterns of human population growth?

Let's begin with the rise in population from 480 million people in 1500 to the current population of more than 6.8 billion. In our simplest model (see Module 4), population growth depends on r (per capita rate of increase) and N (population size). Because the value of r was assumed to be constant in a given environment, the growth rate in the examples we used in Module 4 depended wholly on the population size. Throughout most of human history, the same was true of people. Although parents had many children, mortality was also high, so r (birth rate − death rate) was only slightly higher than 0. As a result, population growth was very slow. (If we extended the x-axis of Figure 9A back in time to year 1, when the population was roughly 300 million, the line would be almost flat for 1,500 years.) The 1 billion mark was not reached until the early 19th century.

As economic development in Europe and the United States led to advances in nutrition and sanitation and later, medical care, people took control of their population's rate of increase (r). At first, the death rate decreased, while the birth rate remained the same. The net rate of increase rose, and population growth began to pick up steam as the 20th century began. By mid-century, improvements in nutrition, sanitation, and health care had spread to the developing world, spurring growth at a breakneck pace as birth rates far outstripped death rates.

As the world population skyrocketed from 2 billion in 1927 to 3 billion just 33 years later, some scientists became alarmed. They feared that Earth's carrying capacity would be reached and that density-dependent factors (see Module 5) would maintain that population size through human suffering and death. But the overall growth rate peaked in 1962. In the more developed nations, advanced medical care continued to improve survivorship, but effective contraceptives held down the birth rate. As a result, the overall growth rate of the world's population began a downward trend.

Demographic Transition The world population is undergoing a change known as a **demographic transition**, a shift from zero population growth in which birth rates and death rates are high but roughly equal, to zero population growth characterized by low but roughly equal birth and death rates. **Figure 9B** shows the demographic transition of Mexico, which is projected to approach zero population growth with low birth and death rates in the next few decades. Notice that the death rate dropped sharply from 1925 to 1975 (the spike corresponds to the worldwide flu epidemic of 1918–1919), while the birth rate remained high until the 1960s. This is a typical pattern for demographic transitions.

Because economic development has occurred at different times in different regions, worldwide demographic transition is a mosaic of the changes occurring in different countries. The most developed nations have completed or are nearing completion of their demographic transitions. In these countries collectively, the rate of increase per 1,000 individuals was estimated at 0.4 in 2009 (**Table 9**, on the next page). In the developing world, death rates have dropped, but high birth rates persist. As a result, these populations are growing rapidly. Of the more than 74 million people added to the world in 2009, more than 71 million were in developing nations.

Reduced family size is the key to the demographic transition. As women's status and education increase, they delay

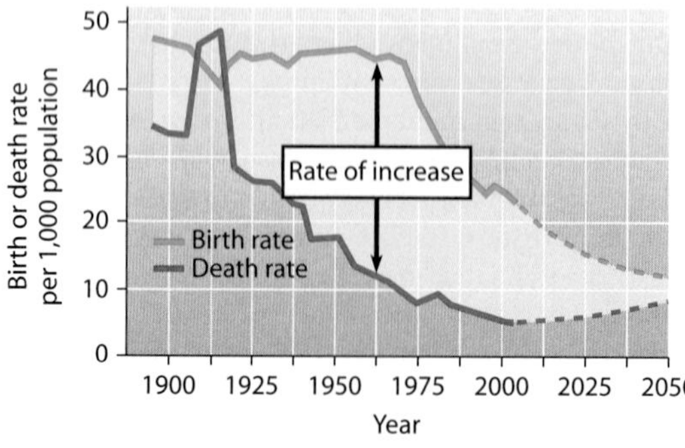

▲ **Figure 9B** Demographic transition in Mexico

Data from Population Reference Bureau, 2000 and U. S. Census Bureau International Data Base, 2003.

TABLE 9 | POPULATION CHANGES IN 2009 (Estimated)

Population	Birth Rate (per 1,000)	Death Rate (per 1,000)	Rate of Increase (per 1,000)
World	19.5	8.3	11.2
More developed nations	10.9	10.5	0.4
Less developed nations	21.4	7.8	13.6

Data from United Nations, the World at Six Billion, 2007.

reproduction and choose to have fewer children. This phenomenon has been observed in both developed and developing countries, wherever the lives of women have improved. Given access to affordable contraceptive methods, women generally practice birth control, and many countries now subsidize family planning services and have official population policies. In many other countries, however, issues of family planning remain socially and politically charged, with heated disagreement over how much support should be provided for family planning.

Age Structures A demographic tool called an age-structure diagram is helpful for predicting a population's future growth. The **age structure** of a population is the number of individuals in different age-groups. **Figure 9C** shows the age structure of Mexico's population in 1985, its estimated 2010 age structure, and its projected age structure in 2035. In these diagrams, green represents the portion of the population in their prereproductive years (0–14), pink indicates the part of the population in prime reproductive years (15–44), and blue is the proportion in postreproductive years (45 and older). Within each of these broader groups, each horizontal bar represents the population in a 5-year age-group. The area to the left of each vertical center line represents the number of males in each age-group; females are represented on the right side of the line.

An age structure with a broad base, such as Mexico's in 1985, reflects a population that has a high proportion of children and a high birth rate. On average, each woman is substantially exceeding the replacement rate of two children per couple. As Figure 9B shows, the birth rate and the rate of increase have dropped 25 years later, but the population continues to be affected by its earlier expansion. This situation, which results from the increased proportion of women of childbearing age in the population, is known as **population momentum**. Girls 0–14 in the 1985 age structure (outlined in yellow) are in their reproductive years in 2010, and girls who are 0–14 in 2010 (outlined in purple) will carry the legacy of rapid growth forward to 2035. Putting the brakes on a rapidly expanding population is like stopping a freight train—the end result takes place long after the decision to do it was made. Even when fertility is reduced to replacement rate, the total population will continue to increase for several decades. The percentage of individuals under the age of 15 gives a rough idea of future growth. In the developing countries, about 29% of the population is in this age-group. In contrast, roughly 16% of the population of developed nations is under the age of 15. Population momentum also explains why the population size in Figure 9A continues to increase even though fewer people are added to the population each year.

In the next module, we examine the age structure of the United States.

? During the demographic transition from high birth and death rates to low birth and death rates, countries usually undergo rapid population growth. Explain why.

The death rate declines before the birth rate declines, creating a period when births greatly outnumber deaths. This also sets up population momentum.

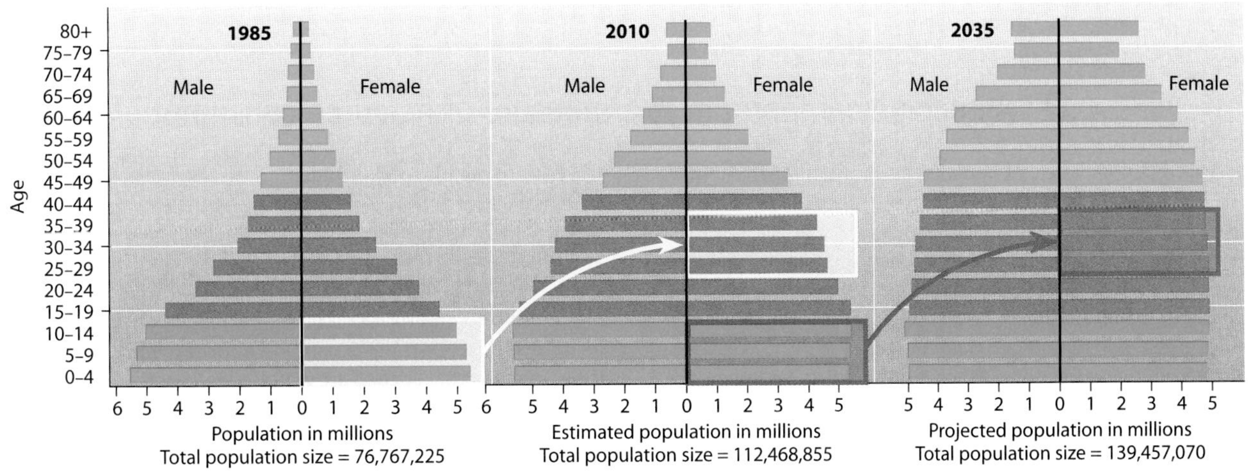

▲ **Figure 9C** Population momentum in Mexico

Data from U. S. Census Bureau, International Data Base.

CONNECTION 10 Age structures reveal social and economic trends

Age-structure diagrams not only reveal a population's growth trends, they also indicate social conditions. For instance, an expanding population has an increasing need for schools, employment, and infrastructure. A large elderly population requires that extensive resources be allotted to health care. Let's look at trends in the age structure of the United States from 1985 to 2035 (**Figure 10**).

The large bulge in the 1985 age structure (tan screen) corresponds to the "baby boom" that lasted for about two decades after World War II ended in 1945. The large number of children swelled school enrollments, prompting construction of new schools and creating a demand for teachers. On the other hand, graduates who were born near the end of the boom faced stiff competition for jobs. Because they make up such a large segment of the population, boomers have had an enormous influence on social and economic trends. They also produced a boomlet of their own, seen in the 0–4 age-group in 1985 and the bump (purple screen) in the 2010 age structure.

Where are the baby boomers now? The leading edge has reached retirement age, which will place pressure on programs such as Medicare and Social Security. In 2010, 60% of the population was between 20 and 64, the ages most likely to be in the workforce, and 13% of the population was over 65. In 2035, the percentages are projected to be 54 and 20. In part, the increase in the elderly population is because people are living longer. The percentage of the population over 80, which was 2.5% in 1985, is projected to rise to nearly 6%—more than 23 million people—in 2035.

? Point out an example of population momentum in Figure 10.

The 1981–1995 "boomlet" is a consequence of rapid reproduction in 1946–1965, as girls born during the baby boom entered their reproductive years.

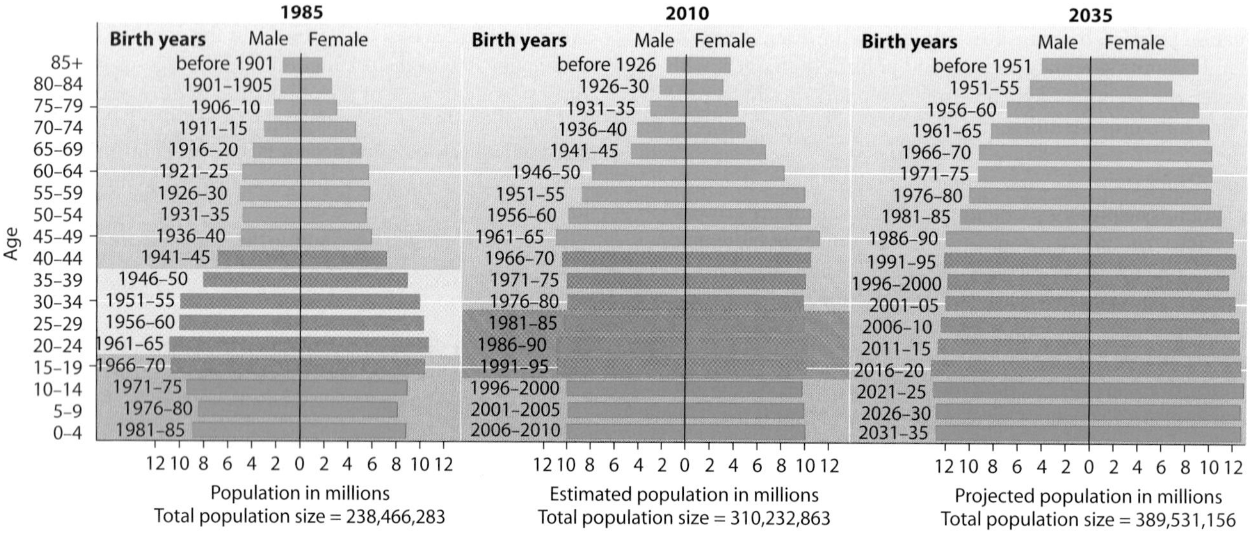

▲ **Figure 10** Age structures for the United States in 1985, 2010 (estimated), and 2035 (projected)

Data from U. S. Census Bureau, International Data Base.

CONNECTION 11 An ecological footprint is a measure of resource consumption

How large a population of humans can Earth hold? In Module 9, we saw that the world's population is growing exponentially, though at a slower rate than it did in the last century. The rate of increase, as well as population momentum, predict that the populations of most developing nations will continue to increase for the foreseeable future. The U.S. Census Bureau projects a global population of 8 billion within the next 20 years and 9.5 billion by the mid-21st century. But these numbers are only part of the story. Trillions of bacteria can live in a petri dish *if* they have sufficient resources. Do we have sufficient resources to sustain 8 or 9 billion people? To accommodate all the people expected to live on our planet by 2025, the world will have to double food production. Already, agricultural lands are under pressure. Overgrazing by the world's growing herds of livestock is turning vast areas of grassland into desert. Water use has risen sixfold over the past 70 years, causing rivers to run dry, water for irrigation to be depleted, and levels of groundwater to drop. And because so much open space will be needed to support the expanding human population, many thousands of other species are expected to become extinct.

The concept of an ecological footprint is one approach to understanding resource availability and usage. An **ecological footprint** is an estimate of the amount of land required to provide the raw materials an individual or a nation consumes,

This asset is intentionally omitted from this text.

Peter Menzel Photography

▲ **Figure 11A** Families in India (left) and the United States (right) display their possessions

including food, fuel, water, housing, and waste disposal. When the total area of ecologically productive land on Earth is divided by the global population, we each have a share of about 2.1 global hectares (1 hectare, or ha, = 2.47 acres; a *global hectare* is a hectare with world-average ability to produce resources and absorb wastes). According to the World Wildlife Fund, in 2005 (the most recent year for which data are available), the average ecological footprint for the world's population was 2.7 global hectares (gha)—we have already overshot the planet's capacity to sustain us.

The United States has a bigger ecological footprint (9.4 gha per person) than its own land and resources can support (5 gha per person)—it has a large ecological deficit. Looking at **Figure 11A**, it is not difficult to understand why. Compared with a family in rural India, Americans have an abundance of possessions. Americans also consume a disproportionate amount of food and fuel. By this measure, the ecological impact of affluent nations such as the United States is potentially as damaging as unrestrained population growth in the developing world. So the problem is not just overpopulation, but overconsumption. **Figure 11B** shows the ecological footprint of each country. The world's richest countries, with 15% of the global population, account for 36% of humanity's total footprint. Some researchers estimate that providing everyone with the same standard of living as the United States would require the resources of 4.5 planet Earths.

? What is your ecological footprint? Do a Web search to find a site that calculates personal resource consumption.

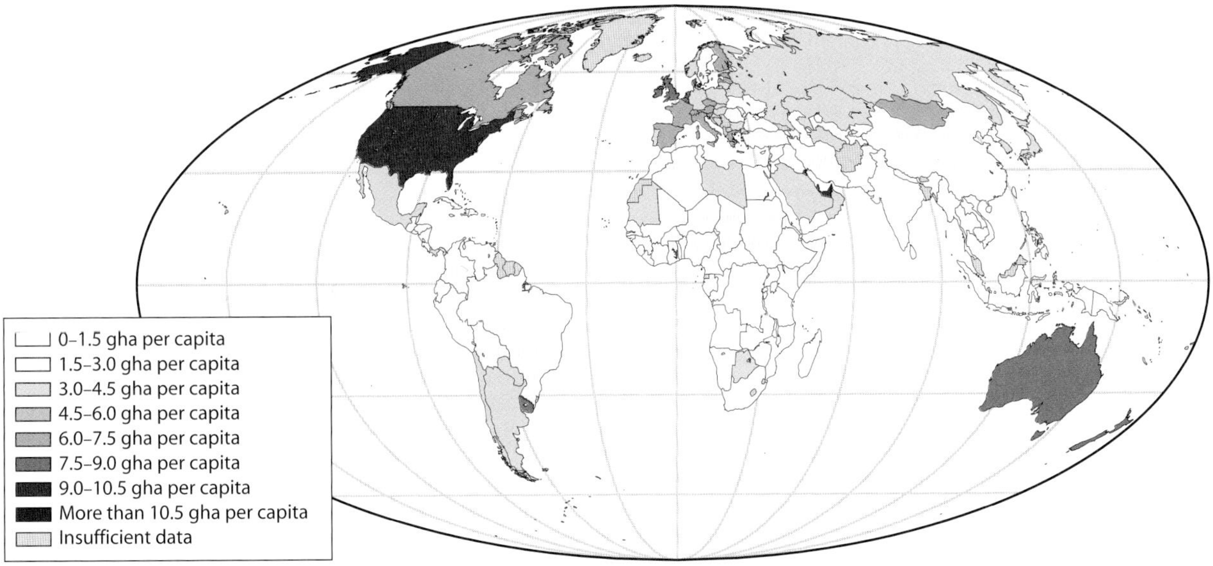

▲ **Figure 11B** Ecological footprints around the world

From *The Ecological Footprint Atlas* 2009 by B. Ewing et al., p. 31, Map 2. © 2009 Global Footprint Network. All rights reserved. Used by permission.

REVIEW

For Practice Quizzes, BioFlix, MP3 Tutors, and Activities, go to www.masteringbiology.com.

Reviewing the Concepts

Population Structure and Dynamics (1–8)

1 Population ecology is the study of how and why populations change.

2 Density and dispersion patterns are important population variables. Population density is the number of individuals in a given area or volume. Environmental and social factors influence the spacing of individuals in various dispersion patterns: clumped (most common), uniform, or random.

3 Life tables track survivorship in populations. Life tables and survivorship curves predict an individual's statistical chance of dying or surviving during each interval in its life. The three types of survivorship curves reflect species' differences in reproduction and mortality.

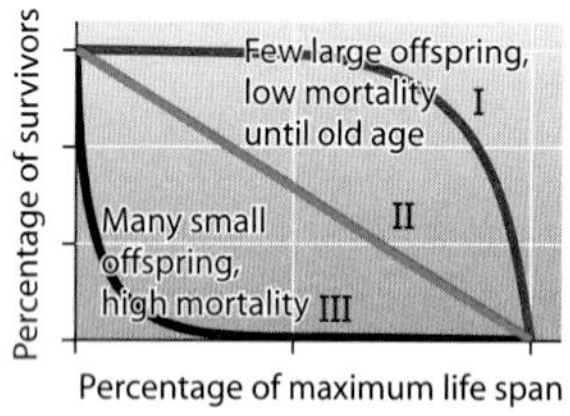

4 Idealized models predict patterns of population growth. Exponential growth is the accelerating increase that occurs when growth is unlimited. The equation $G = rN$ describes this J-shaped growth curve; G = the population growth rate, r = an organism's inherent capacity to reproduce, and N = the population size. Logistic growth is the model that represents the slowing of population growth as a result of limiting factors and the leveling off at carrying capacity, which is the number of individuals the environment can support. The equation $G = rN(K - N)/K$ describes a logistic growth curve, where K = carrying capacity and the term $(K - N)/K$ accounts for the leveling off of the curve.

5 Multiple factors may limit population growth. As a population's density increases, factors such as limited food supply and increased disease or predation may increase the death rate, decrease the birth rate, or both. Abiotic, density-independent factors such as severe weather may limit many natural populations. Most populations are probably regulated by a mixture of factors, and fluctuations in numbers are common.

6 Some populations have "boom-and-bust" cycles. Boom-and-bust cycles alternate population growth and decline at regular intervals.

7 Evolution shapes life histories. Natural selection shapes a species' life history, the series of events from birth through reproduction to death. Populations with so-called r-selected life history traits produce many offspring and grow rapidly in unpredictable environments. Populations with K-selected traits raise few offspring and maintain relatively stable populations. Most species fall between these extremes.

8 Principles of population ecology have practical applications. For example, resource managers use population ecology to determine sustainable yields.

The Human Population (9–11)

9 The human population continues to increase, but the growth rate is slowing. The human population grew rapidly during the 20th century and currently stands at more than 6.8 billion. Demographic transition, the shift from high birth and death rates to low birth and death rates, has lowered the rate of growth in developed countries. In the developing nations, death rates have dropped, but birth rates are still high. The age structure of a population—the proportion of individuals in different age-groups—affects its future growth. Population momentum is the continued growth that occurs despite reduced fertility and is a result of girls in the 0–14 age-group of a previously expanding population reaching their childbearing years.

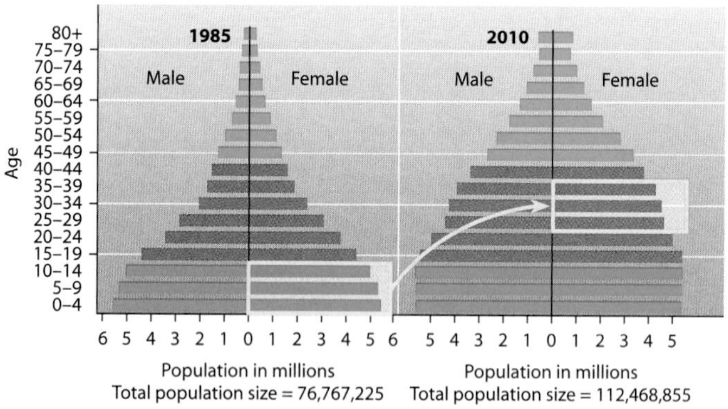

10 Age structures reveal social and economic trends.

11 An ecological footprint is a measure of resource consumption. An ecological footprint estimates the amount of land required by each person or country to produce all the resources it consumes and to absorb all its wastes. The global ecological footprint already exceeds a sustainable level. There is a huge disparity between resource consumption in more developed and less developed nations.

Connecting the Concepts

1. Use this graph of the idealized exponential and logistic growth curves to complete the following.
   a. Label the axes and curves on the graph.
   b. Give the formula that describes the blue curve.
   c. What does the dotted line represent?
   d. For each curve, indicate and explain where population growth is the most rapid.
   e. Which of these curves best represents global human population growth?

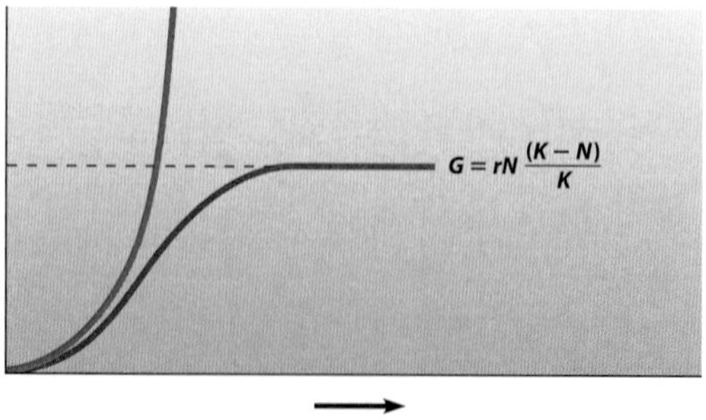

2. The graph at the top of the next page shows the demographic transition for a hypothetical country. Many developed countries that have achieved a stable population size have

undergone a transition similar to this. Answer the following questions concerning this graph.

a. What does the blue line represent? The red line?
b. This diagram has been divided into four sections. Describe what is happening in each section.
c. In which section(s) is the population size stable?
d. In which section is the population growth rate the highest?

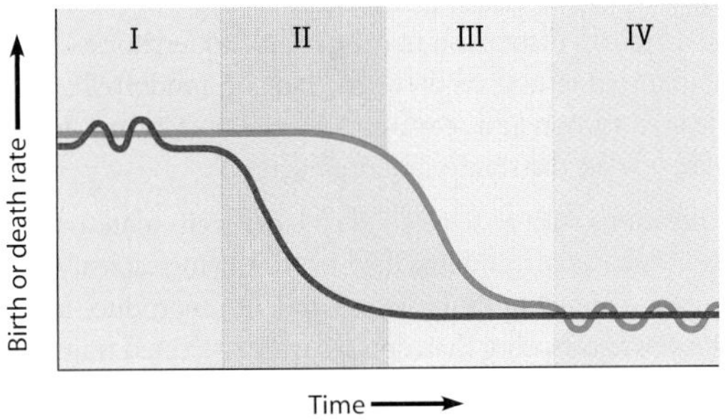

Testing Your Knowledge

Multiple Choice

3. After seeds have sprouted, gardeners often pull up some of the seedlings so that only a few grow to maturity. How does this practice help produce the best yield?
   a. by increasing *K*
   b. by decreasing *r*
   c. by changing the population's age structure
   d. by reducing intraspecific competition
   e. by adding a density-independent factor to the environment
4. To figure out the human population density of your community, you would need to know the number of people living there and
   a. the land area in which they live.
   b. the birth rate of the population.
   c. whether population growth is logistic or exponential.
   d. the dispersion pattern of the population.
   e. the carrying capacity.
5. The term $(K - N)/K$
   a. is the carrying capacity for a population.
   b. is greatest when *K* is very large.
   c. is zero when population size equals carrying capacity.
   d. increases in value as *N* approaches *K*.
   e. accounts for the overshoot of carrying capacity.
6. With regard to its rate of growth, a population that is growing logistically
   a. grows fastest when density is lowest.
   b. has a high intrinsic rate of increase.
   c. grows fastest at an intermediate population density.
   d. grows fastest as it approaches carrying capacity.
   e. is always slowed by abiotic factors.
7. Which of the following represents a demographic transition?
   a. A population switches from exponential to logistic growth.
   b. A population reaches zero population growth when the birth rate drops to zero.
   c. There are equal numbers of individuals in all age-groups.
   d. A population exhibits boom-and-bust cycles.
   e. A population switches from high birth and death rates to low birth and death rates.
8. Skyrocketing growth of the human population appears to be mainly a result of
   a. migration to thinly settled regions of the globe.
   b. a drop in death rate due to sanitation and health care.
   c. better nutrition boosting the birth rate.
   d. the concentration of humans in cities.
   e. social changes that make it desirable to have more children.
9. According to data on ecological footprints,
   a. the carrying capacity of the world is 10 billion.
   b. the carrying capacity of the world would increase if all people ate more meat.
   c. the current demand on global resources by industrialized countries is less than the resources available in those countries.
   d. the United States has a larger ecological footprint than its own resources can provide.
   e. nations with the largest ecological footprints have the fastest population growth rates.

Describing, Comparing, and Explaining

10. What are some factors that might have a density-dependent limiting effect on population growth?
11. What is survivorship? What does a survivorship curve show? Explain what the three survivorship curves tell us about humans, squirrels, and oysters.
12. Describe the factors that might produce the following three types of dispersion patterns in populations.

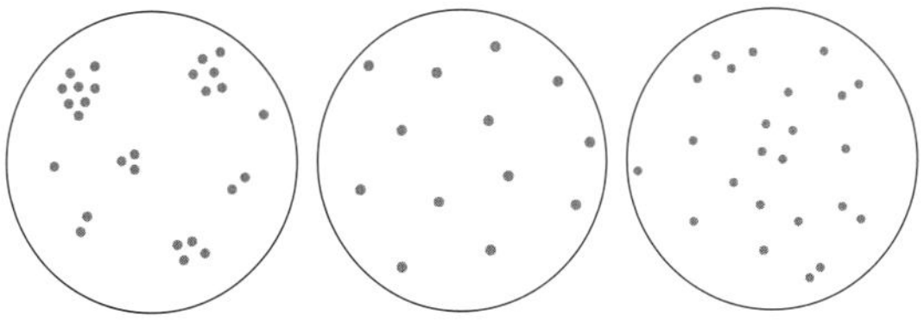

Applying the Concepts

13. The mountain gorilla, spotted owl, giant panda, snow leopard, and grizzly bear are all endangered by human encroachment on their environments. Another thing these animals have in common is that they all have *K*-selected life history traits. Why might they be more easily endangered than animals with *r*-selected life history traits? What general type of survivorship curve would you expect these species to exhibit? Explain your answer.
14. How does the age structure of the U.S. population explain the current surplus in the Social Security fund? If the system is not changed, why will the surplus give way to a deficit sometime in the next few decades?
15. Many people regard the rapid population growth of developing countries as our most serious environmental problem. Others think that the growth of developed countries, though slower, is actually a greater threat to the environment. What kinds of environmental problems result from population growth in (a) developing countries and (b) developed countries? Which do you think is the greater threat? Why?

Review Answers

1. a. The x-axis is time; the y-axis is the number of individuals (N). The blue curve represents exponential growth; red is logistic growth.
   b. $G = rN$
   c. The carrying capacity of the environment (K)
   d. In exponential growth, population growth continues to increase as the population size increases. In logistic growth, the population grows fastest when the population is about $\frac{1}{2}$ the carrying capacity—when N is large enough so that rN produces a large increase, but the expression $(K - N)/K$ has not yet slowed growth as much as it will as N gets closer to K.
   e. Exponential growth curve, although the worldwide growth rate is slowing
2. a. The blue line is birth rate; the red line is death rate.
   b. I. Both birth and death rates are high. II. Birth rate remains high; death rate decreases, perhaps as a result of increased sanitation and health care. III. Birth rate declines, often coupled with increased opportunities for women and access to birth control; death rates are low. IV. Both birth and death rates are low.
   c. I and IV
   d. II, when death rate has fallen but birth rate remains high
3. d 4. a 5. c 6. c 7. e 8. b 9. d
10. Food and resource limitation, such as food or nesting sites; accumulation of toxic wastes; disease; increase in predation; stress responses, such as seen in some rodents
11. Survivorship is the fraction of individuals in a given age interval that survive to the next interval. It is a measure of the probability of surviving at any given age. A survivorship curve shows the fraction of individuals in a population surviving at each age interval during the life span. Oysters produce large numbers of offspring, most of which die young, with a few living a full life span. Few humans die young; most live out a full life span and die of old age. Squirrels have approximately constant mortality and about an equal chance of surviving at all ages.
12. Clumped is the most common dispersion pattern, usually associated with unevenly distributed resources or social grouping. Uniform dispersion may be related to territories or inhibitory interactions between plants. A random dispersion is least common and may occur when other factors do not influence the distribution of organisms.
13. Populations with K-selected life history traits tend to live in fairly stable environments held near carrying capacity by density-dependent limiting factors. They reproduce later and have fewer offspring than species with r-selected traits. Their lower reproductive rate makes it hard for them to recover from human-caused disruption of their habitat. We would expect species with K-selected life histories to have a Type I survivorship curve (see Module 3).
14. The largest population segment, the baby boomers, is currently in the workforce in their peak earning years, paying into the Social Security system. However, they are approaching the end of their contributing years, and the smaller working numbers following them will provide less money, driving the fund into a deficit.
15. Some issues and questions to consider: How does population growth in developing countries relate to food supply, pollution, and the use of natural resources such as fossil fuels? How are these things affected by population growth in developed countries? Which of these factors are most critical to our survival? Are they affected more by the growth of developing or developed countries? What will happen as developing countries become more developed? Will it be possible for everyone to live at the level of the developed world?

Communities and Ecosystems

From Chapter 37 of *Campbell Biology: Concepts & Connections*, Seventh Edition. Jane B. Reece, Martha R. Taylor, Eric J. Simon, Jean L. Dickey.

Communities and Ecosystems

BIG IDEAS

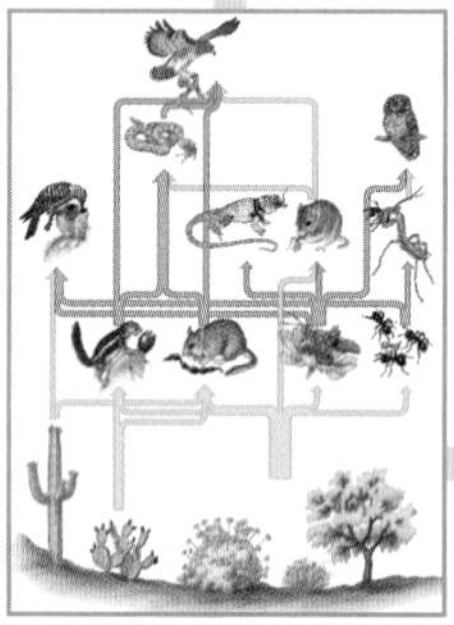

Community Structure and Dynamics (1–13)

Community ecologists examine factors that influence the species composition and distribution of communities and factors that affect community stability.

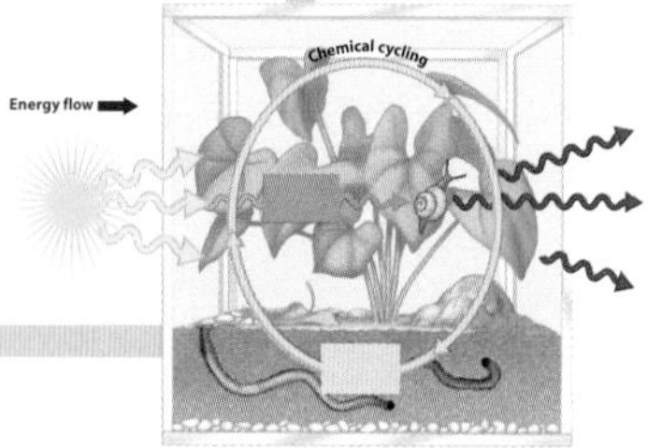

Ecosystem Structure and Dynamics (14–23)

Ecosystem ecology emphasizes energy flow and chemical cycling.

Chad Ehlers/Photolibrary

As the human population has expanded, we have changed natural ecosystems to serve our needs. Only about a quarter of Earth's land surfaces remain untouched by human alterations, and our activities have also had an enormous impact on aquatic biomes.

What is the value of natural ecosystems? Most people appreciate the direct benefits provided by certain ecosystems. For example, some of the resources we use, including water and food such as fish and shellfish, come from natural or near-natural ecosystems. Many people enjoy outdoor recreation such as hiking or whitewater rafting in pristine ecosystems, while others appreciate nature in less strenuous ways. As the 2010 oil spill in the Gulf of Mexico dramatically demonstrated, some ecosystems have obvious economic value. Billions of dollars were lost by fishing, recreation, and other industries as a result of the disaster.

But human well-being also depends on less obvious services that healthy ecosystems provide. The coastal wetlands affected by the Gulf oil spill normally act as a buffer against hurricanes, reduce the impact of flooding, and filter pollutants. The wetlands also furnish nesting sites for birds and marine turtles, and breeding areas and nurseries for a wide variety of fish and shellfish. Other services provided by natural ecosystems include recycling nutrients, preventing erosion and mudslides, and controlling agricultural pests and pollinating crops.

In this chapter, you'll examine the interactions among organisms and how those relationships determine the features of communities. On a larger scale, you'll explore the dynamics of ecosystems. And throughout the chapter, you'll learn how an understanding of these ecological relationships can help us manage Earth's resources wisely.

Community Structure and Dynamics

1 A community includes all the organisms inhabiting a particular area

A population is a group of interacting individuals of a particular species. We now move one step up the hierarchy of nature to the level of the community. A biological **community** is an assemblage of all the populations of organisms living close enough together for potential interaction. Ecologists define the boundaries of the community according to the research questions they want to investigate. For example, one ecologist interested in wetland communities might study the shoreline plants and animals of a particular marsh, while another might investigate only the benthic (bottom-dwelling) microbes.

A community can be described by its species composition. Community ecologists seek to understand how abiotic factors and interactions between populations affect the composition and distribution of communities. For example, a community ecologist might compare the benthic microbes of a marsh located in the temperate zone with those of a tundra marsh.

Community ecologists also investigate community dynamics, the variability or stability in the species composition of a community caused by biotic and abiotic factors. For example, a community ecologist might study changes in the species composition of a wetlands community in Louisiana after a hurricane.

Community ecology is necessary for the conservation of endangered species and the management of wildlife, game, and fisheries. It is vital for controlling diseases, such as malaria, bird flu, and Lyme disease, that are carried by animals. Community ecology also has applications in agriculture, where people attempt to control the species composition of communities they have established.

? What is the relationship between a community and a population?

A community is a group of populations that interact with each other.

2 Interspecific interactions are fundamental to community structure

Organisms also engage in **interspecific interactions**—relationships with individuals of other species in the community—that greatly affect population structure and dynamics. In Table 2, interspecific interactions are classified according to the effect on the populations concerned, which may be helpful (+) or harmful (–).

Members of a population may compete for limited resources such as food or space. **Interspecific competition** occurs when populations of two different species compete for the same limited resource. For example, desert plants compete for water, while plants in a tropical rain forest compete for light. Squirrels and black bears are among the animals that feed on acorns in a temperate broadleaf forest in autumn. When acorn production is low, the nut is a limited resource for which squirrels and bears compete. In general, the effect of interspecific competition is negative for both populations (–/–). However, it may be far more harmful for one population than the other. Interspecific competition is responsible for some of the disastrous effects of introducing non-native species into a community, a topic we will explore further in Module 13.

In **mutualism**, both populations benefit (+/+). Plants and mycorrhizae and herbivores and the cellulose-digesting microbes that inhabit their digestive tracts are examples of mutualism between symbiotic species. Mutualism can also occur between species that are not symbiotic. For example, flowers and their pollinators are mutualists.

There are three categories of interactions in which one species exploits another species (+/–). In **predation**, one species (the predator) kills and eats another species (the prey). **Herbivory** is consumption of plant parts or algae by an animal. Both plants and animals may be victimized by parasites or pathogens. Thus, parasite-host and pathogen-host interactions are also +/–.

In the next several modules, you will learn more about these interspecific interactions and how they affect communities. You will also discover how interspecific interactions can act as powerful agents of natural selection.

TABLE 2 | INTERSPECIFIC INTERACTIONS

Interspecific Interaction	Effect on Species 1	Effect on Species 2	Example
Competition	–	–	Squirrels/ black bears
Mutualism	+	+	Plants/ mycorrhizae
Predation	+	–	Crocodiles/fish
Herbivory	+	–	Caterpillars/ leaves
Parasites and pathogens	+	–	Heartworm/dogs; *Salmonella*/ humans

? Populations of eastern bluebirds declined after the introduction of non-native house sparrows and European starlings. All three species nest in tree cavities. Suggest how an interspecific interaction could explain the bluebird's decline.

Based on the information given, interspecific competition for nest sites is a plausible explanation.

3 Competition may occur when a shared resource is limited

Each species in a community has an **ecological niche**, defined as the sum of its use of the biotic and abiotic resources in its environment. For example, the ecological niche of a small bird called the Virginia's warbler (**Figure 3A**) includes its nest sites and nest-building materials, the insects it eats, and climatic conditions such as the amount of precipitation and the temperature and humidity that enable it to survive. In other words, the ecological niche encompasses everything the Virginia's warbler needs for its existence.

Doug Backlund

▲ **Figure 3A** A Virginia's warbler (*Vermivora virginiae*)

−/−

Interspecific competition occurs when the niches of two populations overlap and both populations need a resource that is in short supply. Ecologists can study the effects of competition by removing all the members of one species from a study site. For example, in central Arizona, the niche of the orange-crowned warbler (**Figure 3B**) overlaps in some respects with the niche of the Virginia's warbler. When researchers removed either species, the remaining species was significantly more successful in raising their offspring. Thus, interspecific competition has a direct effect on reproductive fitness in these birds.

In general, competition lowers the carrying capacity for competing populations because the resources used by one population are not available to the other population. In 1934, Russian ecologist G. F. Gause demonstrated the effects of interspecific competition using three closely related species of ciliates: *Paramecium caudatum*, *P. aurelia*, and *P. bursaria*. He first determined the carrying capacity for each species under laboratory conditions. Then he grew cultures of the two species together. In a mixed culture of *P. caudatum* and *P. bursaria*, population sizes stabilized at lower numbers than each achieved in the absence of a competing species—competition lowered the carrying capacity of the environment. On the other hand, in a mixed culture of *P. caudatum* and *P. aurelia*, only *P. aurelia* survived. Gause concluded that the requirements of these two species were so similar that they could not coexist under those conditions; *P. aurelia* outcompeted *P. caudatum* for essential resources.

Tim Zurowski /All Canada Photos/Alamy

▲ **Figure 3B** An orange-crowned warbler (*Vermivora celata*)

? **Which do you think has more severe effects, competition between members of the same species or competition between members of different species? Explain why.**

Competition between members of the same species is more severe because members of the same species have exactly the same niche. Thus, they compete for exactly the same resources.

4 Mutualism benefits both partners

+/+

Reef-building corals and photosynthetic dinoflagellates (unicellular algae) provide a good example of how mutualists benefit from their relationship. Coral reefs are constructed by successive generations of colonial coral animals that secrete an external calcium carbonate ($CaCO_3$) skeleton. Deposition of the skeleton must outpace erosion and competition for space from fast-growing seaweeds. Corals could not build and sustain the massive reefs that provide the food, living space, and shelter to support the splendid diversity of the reef community without the millions of dinoflagellates that live in the cells of each coral polyp (**Figure 4**). The sugars that the dinoflagellates produce by photosynthesis provide at least half of the energy used by the coral animals. In return, the dinoflagellates gain a secure shelter that provides access to light. They also use the coral's waste products, including CO_2 and ammonia (NH_3), a valuable source of nitrogen for making proteins. Unicellular algae have similar mutually beneficial relationships with a wide variety of other marine invertebrates, including sponges, flatworms, and molluscs.

Jurgen Freund/Nature Picture Library

▲ **Figure 4** Coral polyps

? **When corals are stressed by environmental conditions, they expel their dinoflagellates in a process called bleaching. How is widespread bleaching likely to affect coral reefs?**

Without their dinoflagellate mutualists, corals do not have enough energy to maintain the reef structure. Bleached reefs will die.

EVOLUTION CONNECTION

5 Predation leads to diverse adaptations in prey species

+/−

Predation benefits the predator but kills the prey. Because predation has such a negative impact on reproductive success, numerous adaptations for predator avoidance have evolved in prey populations through natural selection.

Insect color patterns, including camouflage, provide protection against predators. Camouflage is also common in other animals. As **Figure 5A** shows, the gray tree frog (*Hyla arenicolor*), an inhabitant of the southwestern United States, becomes almost invisible on a gray tree trunk.

Other protective devices include mechanical defenses, such as the sharp quills of a porcupine or the hard shells of clams and oysters. Chemical defenses are also widespread. Animals with effective chemical defenses usually have bright color patterns, often yellow, orange, or red in combination with black. Predators learn to associate these color patterns with undesirable consequences, such as noxious taste or a painful sting, and avoid potential prey with similar markings. The vivid orange and black pattern of monarch butterflies **(Figure 5B)** warns potential predators of a nasty taste. Monarchs acquire and store the unpalatable chemicals during the larval stage, when the caterpillars feed on milkweed plants.

? Explain why predation is a powerful factor in the adaptive evolution of prey species.

The prey that avoid being eaten will most likely survive and reproduce, passing alleles for antipredator adaptations on to their offspring.

S.J. Krasemann/Photolibrary

▲ **Figure 5A** Camouflage: a gray tree frog on bark

Jean Dickey

▲ **Figure 5B** Chemical defenses: the monarch butterfly

EVOLUTION CONNECTION

6 Herbivory leads to diverse adaptations in plants

+/−

Although herbivory is not usually fatal, a plant whose body parts have been eaten by an animal must expend energy to replace the loss. Consequently, numerous defenses against herbivores have evolved in plants. Thorns and spines are obvious anti-herbivore devices, as anyone who has plucked a rose from a thorny rosebush or brushed against a spiky cactus knows. Chemical toxins are also very common in plants. Like the chemical defenses of animals, toxins in plants tend to be distasteful, and herbivores learn to avoid them. Among such chemical weapons are the poison strychnine, produced by a tropical vine called *Strychnos toxifera*; morphine, from the opium poppy; nicotine, produced by the tobacco plant; mescaline, from peyote cactus; and tannins, from a variety of plant species. A variety of sulfur compounds, including those that give Brussels sprouts and cabbage their distinctive taste, are also toxic to herbivorous insects and mammals such as cattle. (The vegetables we eat are not toxic because the amount of chemicals in them has been reduced by crop breeders.) Some plants even produce chemicals that cause abnormal development in insects that eat them. Chemical companies have taken advantage of the poisonous properties of certain plants to produce the pesticides called pyrethrin and rotenone. Nicotine is also used as an insecticide.

Some herbivore-plant interactions illustrate the concept of **coevolution**, a series of reciprocal evolutionary adaptations in two species. Coevolution occurs when a change in one species acts as a new selective force on another species, and the resulting adaptations of the second species in turn affect the selection of individuals in the first species. **Figure 6** (top right of next page) illustrates an example of coevolution between an herbivorous insect (the caterpillar of the butterfly *Heliconius*, top left) and a plant (the passionflower, *Passiflora*, a tropical vine). *Passiflora* produces toxic chemicals that protect its leaves from most insects, but *Heliconius* caterpillars have digestive enzymes that break down the toxins. As a result, *Heliconius* gains access to a food source that few other insects can eat.

These poison-resistant caterpillars seem to be a strong selective force for *Passiflora* plants, and defenses have evolved in some species. For instance, the leaves of some *Passiflora* species produce yellow sugar deposits that look like *Heliconius* eggs. You can see two eggs in the top right photograph of Figure 6 and two egg-like sugar deposits in the bottom photo. Female butterflies avoid laying their eggs on leaves that already have eggs, presumably ensuring that only a few caterpillars will hatch and feed on any one leaf. Because the butterfly often mistakes the yellow sugar deposits for eggs, *Passiflora* species with the yellow deposits are less likely to be eaten.

The story of *Passiflora* is even more complicated, however. The egg-like sugar deposits, as well as smaller ones scattered over the leaf, attract ants and wasps that prey on *Heliconius* eggs and larvae. Thus, adaptations that appear to be coevolutionary responses between just two species may in fact involve interactions among many species in a community.

? People find most bitter-tasting foods objectionable. Why do you suppose we have taste receptors for bitter-tasting chemicals?

Individuals having bitter taste receptors presumably survived better because they could identify potentially toxic food when they foraged.

Lawrence E. Gilbert/Biological Photo Service

▲ **Figure 6** Coevolution: *Heliconius* and the passionflower vine (*Passiflora*)

7 Parasites and pathogens can affect community composition

+/−

A parasite lives on or in a host from which it obtains nourishment. Several invertebrate parasites, including flukes and tapeworms and a variety of nematodes live inside a host organism's body. External parasites include arthropods such as ticks, lice, mites, and mosquitoes, which attach to their victims temporarily to feed on blood or other body fluids. Plants are also attacked by parasites, including nematodes and aphids, tiny insects that tap into the phloem and suck plant sap (**Figure 7**). Pathogens are disease-causing bacteria, viruses, fungi, or protists that can be thought of as microscopic parasites.

The potentially devastating effects of parasites and pathogens on cultivated plants, livestock, and humans are well known, but ecologists know little about how these interactions affect natural communities. Non-native pathogens, whose impact is rapid and often dramatic, have provided some opportunities to study the effects of pathogens on communities. In one example, ecologists studied the consequences of an epidemic of chestnut blight that wiped out virtually all American chestnut trees during the first half of the 20th century; the disease is caused by a protist. Chestnuts were massive canopy trees that dominated many forest communities in North America. Their loss had a significant impact on species composition and community structure. Overall, the diversity of tree species increased as trees that had formerly competed with chestnuts, such as oaks and hickories, became more prominent. The dead chestnut trees furnished niches for other organisms, such as insects, cavity-nesting birds, and eventually decomposers. On the other hand, populations of organisms that depended heavily on living chestnut trees for their food and shelter declined.

A fungus-like protist that causes a disease called sudden oak death is currently spreading on the West Coast. More than a million oaks have been lost so far, causing the decline of bird populations. Despite its name, sudden oak death affects many other species as well, including the majestic redwood and Douglas fir trees and flowering shrubs such as rhododendron and camellia. Because the epidemic is in its early stages, its full effect on forest communities will not be known for some time.

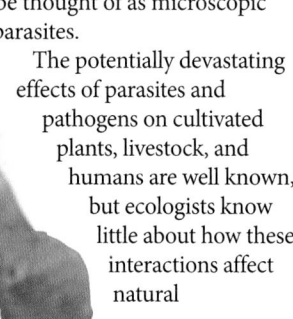

Joel Sartore/Photolibrary

◀ **Figure 7** Aphids parasitizing a plant

? Use your knowledge of interspecific interactions to explain why tree diversity increased after all the chestnuts died.

Chestnuts had many of the same niche characteristics as other trees, but apparently chestnuts were superior competitors. After they died, the remaining species may have had fewer niche similarities, or they may have been more equal as competitors, allowing more species to coexist.

8 Trophic structure is a key factor in community dynamics

Every community has a trophic structure, a pattern of feeding relationships consisting of several different levels. The sequence of food transfer up the trophic levels is known as a **food chain**. This transfer of food moves chemical nutrients and energy from organism to organism up through the trophic levels in a community.

Figure 8 compares a terrestrial food chain and an aquatic food chain. In this figure, the trophic levels are arranged vertically, and the names of the levels appear in colored boxes. The arrows connecting the organisms point from the food to the consumer. Starting at the bottom, the trophic level that supports all others consists of autotrophs ("self-feeders"), which ecologists call **producers**. Photosynthetic producers use light energy to power the synthesis of organic compounds. Plants are the main producers on land. In water, the producers are mainly photosynthetic unicellular protists and cyanobacteria, collectively called phytoplankton. Multicellular algae and aquatic plants are also important producers in shallow waters. In a few communities, such as deep-sea hydrothermal vents, the producers are chemosynthetic prokaryotes.

All organisms in trophic levels above the producers are heterotrophs ("other-feeders"), or consumers, and all consumers are directly or indirectly dependent on the output of producers. Herbivores, which eat plants, algae, or phytoplankton, are **primary consumers**. Primary consumers on land include grasshoppers and many other insects, snails, and certain vertebrates, such as grazing mammals and birds that eat seeds and fruits. In aquatic environments, primary consumers include a variety of zooplankton (mainly protists and microscopic animals such as small shrimps) that eat phytoplankton.

Above primary consumers, the trophic levels are made up of carnivores and insectivores, which eat the consumers from the level below. On land, **secondary consumers** include many small mammals, such as the mouse shown here eating an herbivorous insect, and a great variety of birds, frogs, and spiders, as well as lions and other large carnivores that eat grazers. In aquatic ecosystems, secondary consumers are mainly small fishes that eat zooplankton.

Higher trophic levels include **tertiary** (third-level) **consumers**, such as snakes that eat mice and other secondary consumers. Most ecosystems have secondary and tertiary consumers. As the figure indicates, some also have a higher level, **quaternary** (fourth-level) **consumers**. These include hawks in terrestrial ecosystems and killer whales in the marine environment.

Not shown in Figure 8 is another trophic level—consumers that derive their energy from **detritus**, the dead material produced at all the trophic levels. Detritus includes animal wastes, plant litter, and the bodies of dead organisms. Different organisms consume detritus in different stages of decay. **Scavengers**, which are large animals, such as crows and vultures, feast on carcasses left behind by predators or speeding cars. The diet of **detritivores** is made up primarily of decaying organic material. Examples of detritivores include earthworms and millipedes. **Decomposers**, mainly prokaryotes and fungi, secrete enzymes that digest molecules in organic material and convert them to inorganic forms. Enormous numbers of microscopic decomposers in the soil and in the mud at the bottom of lakes and oceans break down most of the community's organic materials to inorganic compounds that plants or phytoplankton can use. The breakdown of organic materials to inorganic ones is called **decomposition**. By breaking down detritus, decomposers link all trophic levels. Their role is essential for all communities and, indeed, for the continuation of life on Earth.

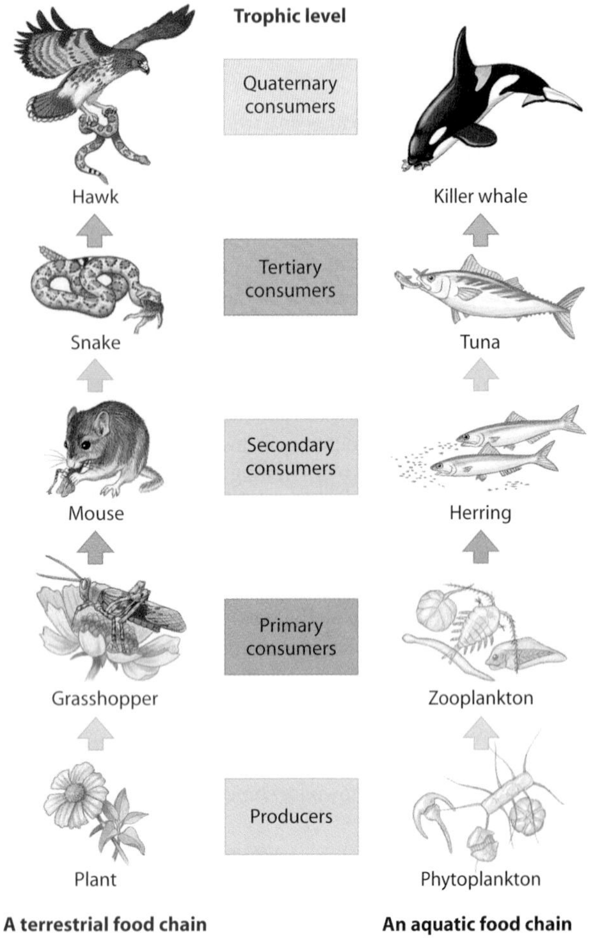

▲ **Figure 8** Two food chains

? I'm eating a cheese pizza. At which trophic level(s) am I feeding?

Primary consumer (flour and tomato sauce) and secondary consumer (cheese, a product from cows, which are primary consumers)

9 Food chains interconnect, forming food webs

A more realistic view of the trophic structure of a community is a **food web**, a network of interconnecting food chains. Figure 9 shows a simplified example of a food web in a Sonoran desert community. As in the food chains of Figure 8, the arrows indicate the direction of nutrient transfer ("who eats whom") and are color-coded by trophic level.

Notice that a consumer may eat more than one type of producer, and several species of primary consumers may feed on the same species of producer. Some animals weave into the web at more than one trophic level. The lizard and mantid are strictly secondary consumers, eating insects. The woodpecker on the left, however, is a primary consumer when it eats cactus seeds and a secondary consumer when it eats ants or grasshoppers. The hawk at the top of the web is a secondary, tertiary, or quaternary consumer, depending on its prey. Food webs, like food chains, do not typically show detritivores and decomposers, which obtain energy from dead organic material from all trophic levels.

We have now looked at how populations in a community interact with each other. In the next few modules, we consider factors that affect the community as a whole.

? Even consumers at the highest level of a food web eventually become food for _____.

detritivores and decomposers

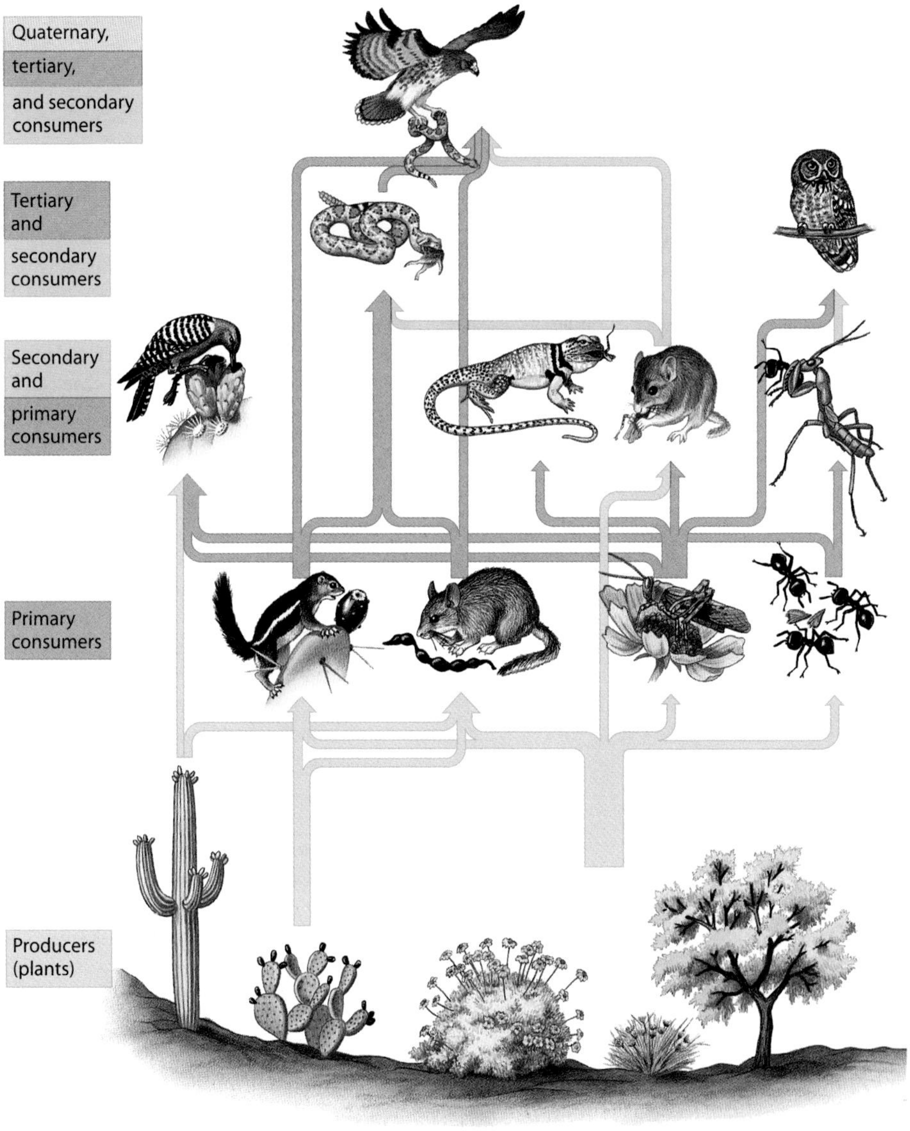

▲ **Figure 9** A food web

10 Species diversity includes relative abundance and species richness

A community's **species diversity** is defined by two components: species richness, or the number of different species in a community, and relative abundance, the proportional representation of a species in a community. To understand why both components are important for describing species diversity, imagine walking through woodlot A on the path shown in Figure 10A. You would pass by four different species of trees, but most of the trees you encounter would be the same species. Now imagine walking on the path through woodlot B in Figure 10B. You would see the same four species of trees that you saw in woodlot A—the species richness of the two woodlots is the same. However, woodlot B would probably seem more diverse to you, because no single species predominates. As Table 10 shows, the relative abundance of one species in woodlot A is much higher than the relative abundances of the other three species. In woodlot B, all four species are equally abundant. As a result, species diversity is greater in woodlot B.

Plant species diversity in a community often has consequences for the species diversity of animals in the community. For example, suppose a species of caterpillar only eats the leaves of a tree that makes up just 5% of woodlot A. If the caterpillar is present at all, its population may be small and scattered. Birds that depend on those caterpillars to feed their young may be absent. But the caterpillars would easily be able to locate their food source in woodlot B, and their abundance would attract birds as well. By providing a broader range of habitats, a diverse tree community promotes animal diversity.

Species diversity also has consequences for pathogens. Most pathogens infect a limited range of host species or may even be

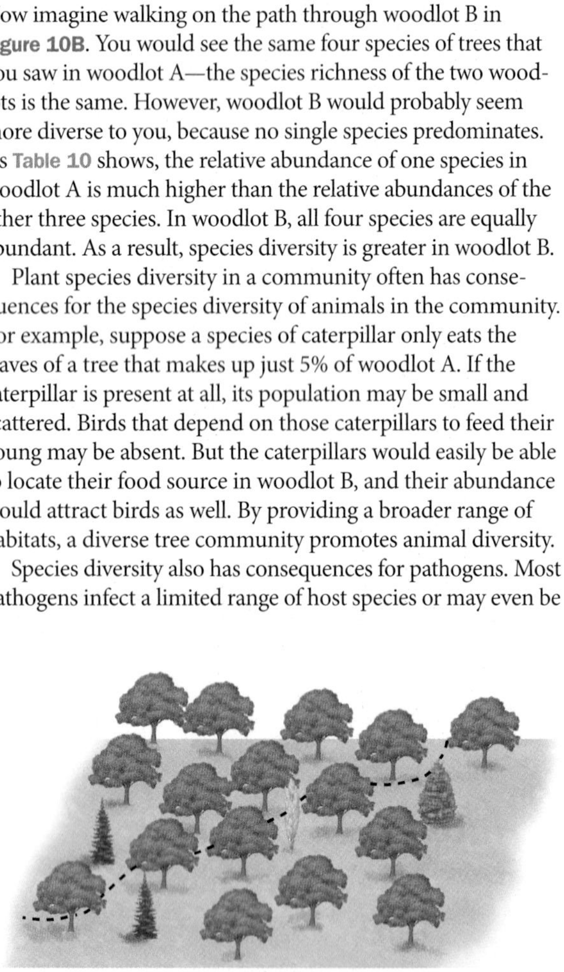

▲ **Figure 10A** Species composition of woodlot A

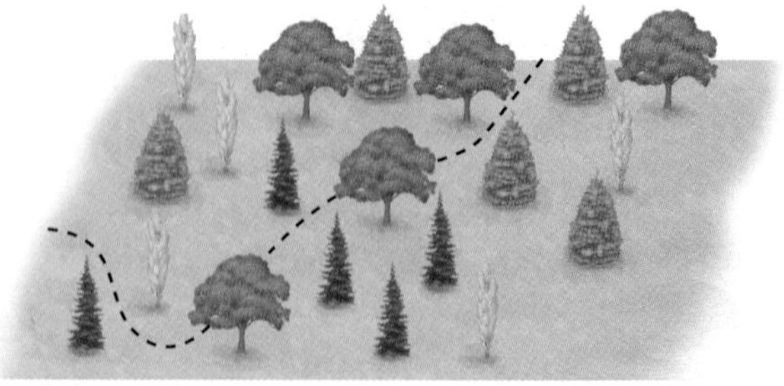

▲ **Figure 10B** Species composition of woodlot B

TABLE 10 | RELATIVE ABUNDANCE OF TREE SPECIES IN WOODLOTS A AND B

Species	Relative Abundance in Woodlot A (%)	Relative Abundance in Woodlot B (%)
	80	25
	10	25
	5	25
	5	25

restricted to a single host species. When many potential hosts are living close together, it is easy for a pathogen to spread from one to another. In woodlot A, for example, a pathogen that infects the most abundant tree would rapidly be transmitted through the entire forest. On the other hand, the more isolated trees in woodlot B are more likely to escape infection.

Low species diversity is characteristic of most modern agricultural ecosystems. For efficiency, crops and trees are often planted in monoculture—a single species grown over a wide area. Monocultures are especially vulnerable to attack by pathogens and herbivorous insects. Also, plants grown in monoculture have been bred for certain desirable characteristics, so their genetic variation is typically low, too. As a result, a pathogen can potentially devastate an entire field or more. Between 1845 and 1849, a pathogen wiped out a monoculture of genetically uniform potatoes throughout Ireland. As a result, thousands of people died of starvation, and thousands more left the country.

To combat potential losses, many farmers and forest managers rely heavily on chemical methods of controlling pests. Modern crop scientists have bred varieties of plants that are genetically resistant to common pathogens, but these varieties can suddenly become vulnerable, too. In 1970, pathogen evolution led to an epidemic of a disease called corn leaf blight that resulted in a billion dollars of crop damage in the United States. Some researchers are now investigating the use of more diverse agricultural ecosystems—polyculture—as an alternative to monoculture.

? Which would you expect to have higher species diversity, a well-maintained lawn or one that is poorly maintained? Explain.

A lawn that is poorly maintained would have higher species diversity. A well-maintained lawn should have low species diversity. While a lawn that is cared for may not be a perfect monoculture, any weeds that are present would have low relative abundance. The opposite is true if the lawn is not cared for.

11 Keystone species have a disproportionate impact on diversity

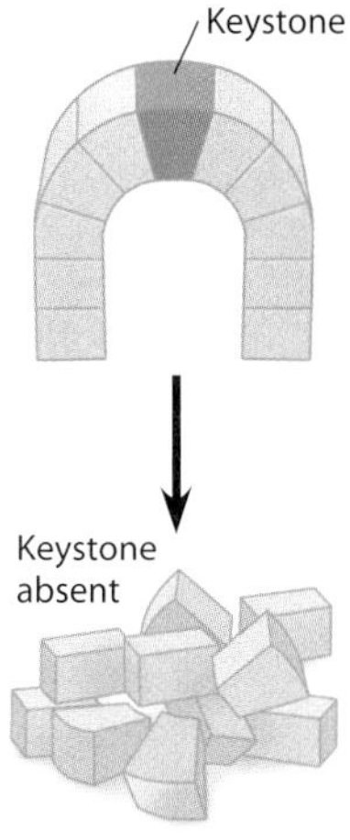

▲ **Figure 11A** Arch collapse with removal of keystone

In Module 10, you saw that the abundance of dominant species such as a forest tree can have an impact on the diversity of other species in the community. But less abundant species may also exert control over community composition. A **keystone species** is a species whose impact on its community is much larger than its biomass or abundance would indicate. The word "keystone" comes from the wedge-shaped stone at the top of an arch that locks the other pieces in place. If the keystone is removed, the arch collapses (**Figure 11A**). A keystone species occupies a niche that holds the rest of its community in place.

To investigate the role of a potential keystone species in a community, ecologists compare diversity when the species is present and when it is absent. Experiments by Robert Paine in the 1960s were among the first to provide evidence of the keystone species effect. Paine manually removed a predator, a sea star of the genus *Pisaster* (**Figure 11B**), from experimental areas within the intertidal zone of the Washington coast. The result was that *Pisaster*'s main prey, a mussel of the genus *Mytilus*, outcompeted many of the other shoreline organisms (algae, barnacles, and snails, for instance) for the important resource of space on the rocks. The number of different organisms present in experimental areas dropped from more than 15 species to fewer than 5 species.

The keystone concept has practical application in efforts to restore or rehabilitate damaged ecosystems. In 1983, a disease swept through the coral reefs of the Caribbean, killing huge numbers of the long-spined sea urchin, *Diadema antillarum* (**Figure 11C**). In the following decade, the area of reef covered by living coral animals plummeted, along with overall species diversity. Fleshy seaweeds replaced the low turf of encrusted red algae that is vital to reef building (**Figure 11D**). The thick growth of seaweed also prevented light from reaching the symbiotic dinoflagellates that corals depend on for food. These dramatic changes in the reef community revealed that *Diadema* is a keystone species whose herbivorous habits have two major effects. First, its grazing suppresses the seaweed populations. (Because *Diadema* normally shares this role with herbivorous fishes, it is an especially important species on the many Caribbean reefs that have been overfished.) Second, the urchins scrape patches of substrate clear of algae, providing platforms for coral larvae to settle. *Diadema* populations have been slow to rebound from the catastrophic die-off. However, recognition of this organism's key role in the community has prompted conservationists to artificially replenish urchin populations in some areas to help restore damaged reefs.

William E. Townsend/Photo Researchers

▶ **Figure 11B** A *Pisaster* sea star, a keystone species, eating a mussel

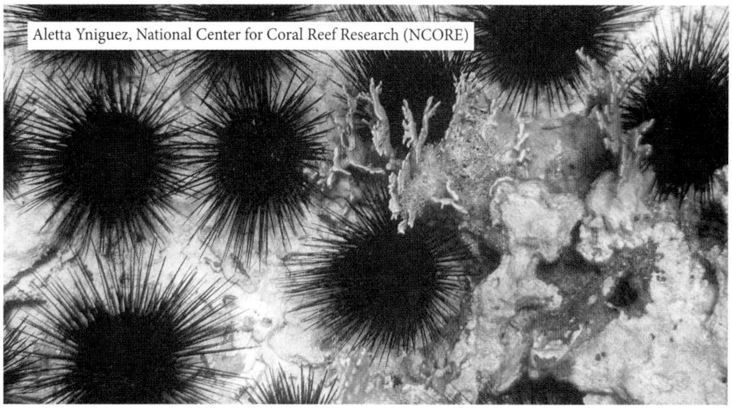

Aletta Yniguez, National Center for Coral Reef Research (NCORE)

▲ **Figure 11C** *Diadema* sea urchins grazing on a reef

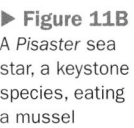

Reinhard Dirscherl/Alamy

▲ **Figure 11D** A reef overgrown by fleshy seaweeds

? Removing saguaro cacti from the Sonoran desert community (see Module 9) would have a drastic impact, and yet saguaro is not considered a keystone species. Why not?

Saguaro is abundant and makes up a large part of the community. Keystone species have a large effect relative to their representation in the community just as a keystone is a small but vital piece of the arch.

12 Disturbance is a prominent feature of most communities

Early ecologists viewed biological communities as more or less stable in structure and species composition. But like many college campuses, where some construction or renovation project is always under way, many communities are frequently disrupted by sudden change. **Disturbances** are events such as storms, fires, floods, droughts, or human activities that damage biological communities and alter the availability of resources. The types of disturbances and their frequency and severity vary from community to community.

Although we tend to think of disturbances in negative terms, small-scale disturbances often have positive effects. For example, when a large tree falls in a windstorm, it creates new habitats. For instance, more light may now reach the forest floor, giving small seedlings the opportunity to grow; or the depression left by its roots may fill with water and be used as egg-laying sites by frogs, salamanders, and numerous insects.

Communities change drastically following a severe disturbance that strips away vegetation and even soil. The disturbed area may be colonized by a variety of species, which are gradually replaced by a succession of other species, in a process called **ecological succession**. When ecological succession begins in a virtually lifeless area with no soil, it is called **primary succession**. Examples of such areas are the rubble left by a retreating glacier or fresh volcanic lava flows (**Figure 12A**). Often the only life-forms initially present are autotrophic bacteria. Lichens and mosses, which grow from windblown spores, are commonly the first large photosynthesizers to colonize the area. Soil develops gradually as rocks break down and organic matter accumulates from the decomposed remains of the early colonizers. Lichens and mosses are gradually overgrown by larger plants that sprout from seeds blown in from nearby areas or carried in by animals. Eventually, the area is colonized by plants that become the community's prevalent form of vegetation. Primary succession can take hundreds or thousands of years.

Jean Dickey

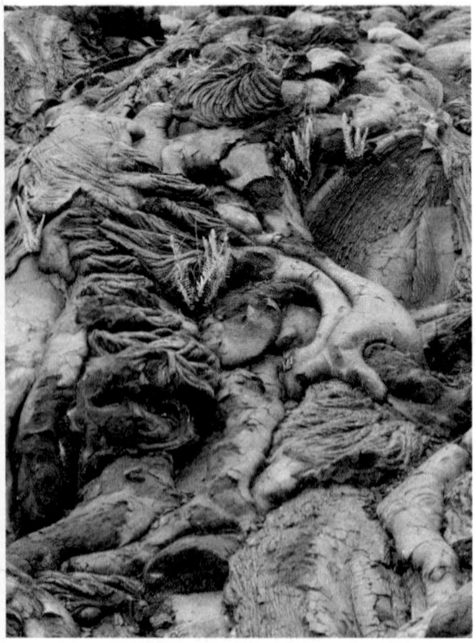

▲ **Figure 12A** Primary succession on a lava flow

Secondary succession occurs where a disturbance has cleared away an existing community but left the soil intact. For example, secondary succession occurs as areas recover from fires or floods. Some disturbances that lead to secondary succession are caused by human activities. Even before colonial times, people were clearing the forests of eastern North America for agriculture and settlements. Some of this land was later abandoned as the soil was depleted of its chemical nutrients or the residents moved west to new territories. Whenever human intervention stops, secondary succession begins.

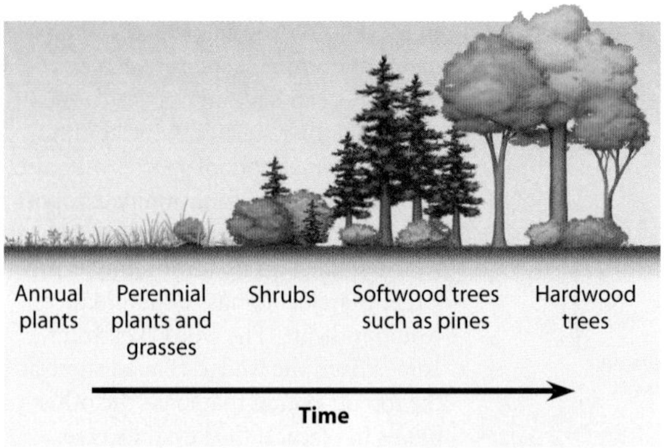

▲ **Figure 12B** Stages in the secondary succession of an abandoned farm field

Numerous studies have documented the stages by which an abandoned farm field returns to forest (**Figure 12B**). A recently disturbed site provides an environment that is favorable to *r*-selected species—plants and animals that reach reproductive age rapidly, produce huge numbers of offspring, and provide little or no parental care. Interspecific competition is not a major factor during the very early stages of succession, which are dominated by weedy annual species such as crabgrass and ragweed. Within a few years, perennial grasses and small broadleaf plants cover the field. (An annual plant completes its life cycle in a single year. Perennial plants live for many years.) Softwood trees, especially pines, begin to invade within 5 years, turning the area into a pine forest in roughly 10 to 15 years. But pine seedlings, which need high levels of light to grow, don't do well in the understory. The seedlings of many hardwood species are more shade tolerant, and thus trees such as oak and maple begin to replace pine as competition becomes a significant force in determining the composition of the community. The final mixture of species depends on local abiotic factors such as soil and topography. Because animals depend on plants for food and shelter, the animal community undergoes successional changes, too. The diversity of bird species, for example, increases dramatically as trees replace herbaceous plants.

Understanding the effects of disturbance in communities is especially important today; People are the most widespread and significant agents of disturbance. Disturbances may also create opportunities for undesirable plants and animals that people transport to new habitats, which is the topic of the next module.

? What is the main abiotic factor that distinguishes primary from secondary succession?

Absence (primary) versus presence (secondary) of soil at the onset of succession

CONNECTION 13 Invasive species can devastate communities

For as long as people have traveled from one region to another, they have carried organisms along, sometimes intentionally and sometimes by accident. Many of these non-native species have established themselves firmly in their new locations. Furthermore, many have become **invasive species**, spreading far beyond the original point of introduction and causing environmental or economic damage by colonizing and dominating wherever they find a suitable habitat. In the United States alone, there are hundreds of invasive species, including plants, mammals, birds, fishes, arthropods, and molluscs. Worldwide, there are thousands more. Invasive species are a leading cause of local extinctions. The economic costs of invasive species are enormous—an estimated $120 billion a year in the United States. Regardless of where you live, an invasive plant or animal is probably living nearby.

Not every organism that is introduced to a new habitat is successful, and not every species that is able to survive in its new habitat becomes invasive. There is no single explanation for why any non-native species turns into a destructive pest, but community ecology offers some insight. Interspecific interactions act as a system of checks and balances on the populations in a community. Every population is subject to multiple negative effects, whether from competitors, predators, herbivores, or pathogens, that curb its growth rate. Without biotic factors such as these to check population growth, a population will continue to expand until limited by abiotic factors.

In 1859, 12 pairs of European rabbits (*Oryctolagus cuniculus*) were released on a ranch in southern Australia by a European who wanted to hunt familiar game. The animals quickly became a nuisance. In 1865, 20,000 rabbits were killed on the ranch. By 1900, *several hundred million* rabbits were distributed over much of the continent (**Figure 13A**). The rabbit invasion was a catastrophe in several ways. Their activities destroyed farm and grazing land by eating vegetation down to, and sometimes including, the roots (**Figure 13B**). Especially in arid regions, the loss of plant cover led to soil erosion. In addition, rabbits dug extensive underground burrows that made grazing treacherous for cattle and sheep. They also competed directly with native herbivorous marsupials. After many fruitless attempts to control the rabbit population, in 1950 the Australian government turned to **biological control**, the intentional release of a natural enemy to attack a pest population. A virus lethal to rabbits was introduced into the environment. The rabbits and virus then underwent several coevolutionary cycles as the rabbits became more resistant to the disease and the virus became less lethal. The government managed to stave off a complete resurgence of the rabbit population by introducing new viral strains, but in 1995, they had to switch to a different pathogen to maintain control.

▲ **Figure 13B** A familiar sight in early 20th-century Australia

Embassy of Australia

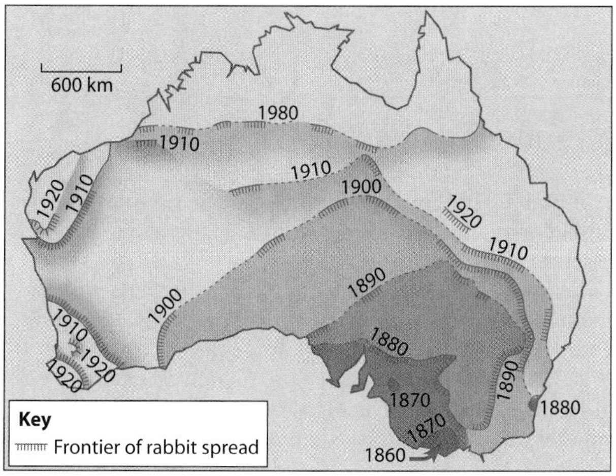

▲ **Figure 13A** The spread of rabbits in Australia

Source: www.dest.gov.au/archive/Science/pmsec/14meet/rcd1.html.

Coevolution is just one potential pitfall of biological control. The imported enemy may not be as successful in the new environment as the target species.
It may not disperse widely enough, or its population growth rate may not be high enough to overtake a rapidly expanding population. Caution is especially warranted because the control agent may turn out to be as invasive as its target. For example, cane toads (**Figure 13C**) imported to control an agricultural pest in Australia became a widespread threat to native wildlife.

▲ **Figure 13C** A cane toad (*Bufo marinus*)

Ace Stock Limited/Alamy

In the next modules, we broaden our scope to look at ecosystems, the highest level of ecological complexity.

? What is the ecological basis for biological control of pests?

By having a negative effect on population growth rate, a natural enemy keeps the target population in check.

Ecosystem Structure and Dynamics

14 Ecosystem ecology emphasizes energy flow and chemical cycling

An **ecosystem** consists of all the organisms in a community as well as the abiotic environment with which the organisms interact. Ecosystem ecologists are especially interested in **energy flow**, the passage of energy *through* the components of the ecosystem, and **chemical cycling**, the transfer of materials *within* the ecosystem.

The terrarium in **Figure 14** represents a familiar type of ecosystem and illustrates the fundamentals of energy flow. Energy enters the terrarium in the form of sunlight ( ). Plants (producers) convert light energy to chemical energy ( ) through the process of photosynthesis. Animals (consumers) take in some of this chemical energy in the form of organic compounds when they eat the plants. Decomposers, such as bacteria and fungi in the soil, obtain chemical energy when they decompose the dead remains of plants and animals. Every use of chemical energy by organisms involves a loss of some energy to the surroundings in the form of heat ( ). Because so much of the energy captured by photosynthesis is lost as heat, the ecosystem would run out of energy if it were not powered by a continuous inflow of energy from the sun. A few ecosystems—for example, hydrothermal vents—are powered by chemical energy obtained from inorganic compounds.

In contrast to energy flow, chemical cycling ( ) involves the transfer of materials *within* the ecosystem. While most ecosystems have a constant input of energy from sunlight, the supply of the chemical elements used to construct molecules is limited. Chemical elements such as carbon and nitrogen are cycled between the abiotic component of the ecosystem, including air, water, and soil, and the biotic component of the ecosystem (the community). Plants acquire these chemical elements in inorganic form from the air and soil and use them to build organic molecules. Animals, such as the snail in Figure 14, consume some of these organic molecules. When the plants and animals become detritus, decomposers return most of the elements to the soil and air in inorganic form. Some elements are also returned to the soil as the by-products of plant and animal metabolism.

In summary, both energy flow and chemical cycling involve the transfer of substances through the trophic levels of the ecosystem. However, energy flows through, and ultimately out of, ecosystems, whereas chemicals are recycled within ecosystems. We explore these fundamental ecosystem dynamics in the rest of the chapter.

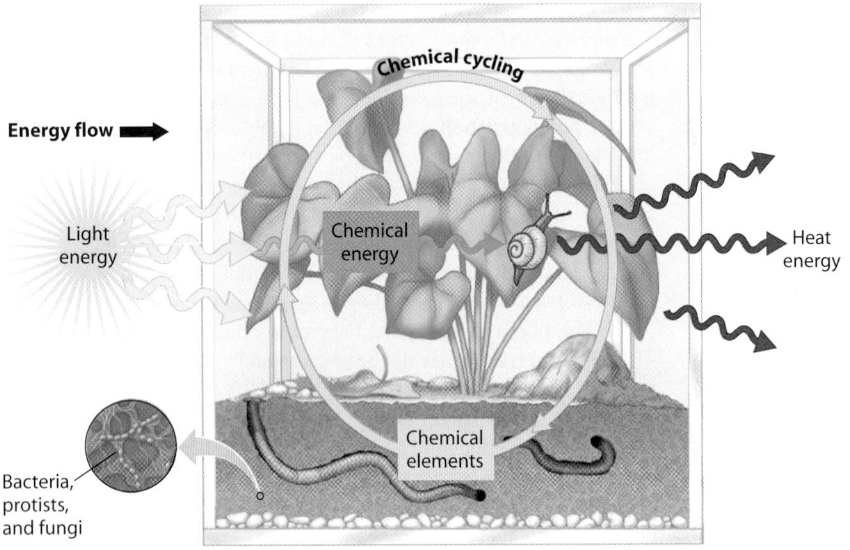

▲ **Figure 14** A terrarium ecosystem

? How do chemical cycles in an ecosystem differ from food chains in a community?

● Chemicals pass through one or more abiotic components of an ecosystem as well as passing through the biotic components (food chain).

15 Primary production sets the energy budget for ecosystems

Each day, Earth receives about 10^{19} kcal of solar energy, the energy equivalent of about 100 million atomic bombs. Most of this energy is absorbed, scattered, or reflected by the atmosphere or by Earth's surface. Of the visible light that reaches plants, algae, and cyanobacteria, only about 1% is converted to chemical energy by photosynthesis.

Ecologists call the amount, or mass, of living organic material in an ecosystem the **biomass**. The amount of solar energy converted to chemical energy (in organic compounds) by an ecosystem's producers for a given area and during a given time period is called **primary production**. It can be expressed in units of energy or units of mass. The primary production of the entire biosphere is roughly 165 billion tons of biomass per year.

Different ecosystems vary considerably in their primary production as well as in their contribution to the total production of the biosphere. **Figure 15**, at the top of the next page, contrasts the net primary production of a number of different ecosystems. (Net primary production refers to the amount of biomass produced minus the amount used by producers as fuel for their own cellular respiration.) Tropical rain forests are among the most productive terrestrial ecosystems and contribute a large portion of the planet's overall production of biomass. Coral reefs also have very high production, but their contribution to global production is small because they cover such a small area. Interestingly, even though the open ocean has very low production, it contributes the most to Earth's total net primary production because of its huge size—it covers 65% of Earth's surface area.

? **Deserts and semidesert scrub cover about the same amount of surface area as tropical forests but contribute less than 1% of Earth's net primary production, while rain forests contribute 22%. Explain this difference.**

The primary production of tropical rain forests is over 20 times greater than that of desert and semidesert scrub ecosystems.

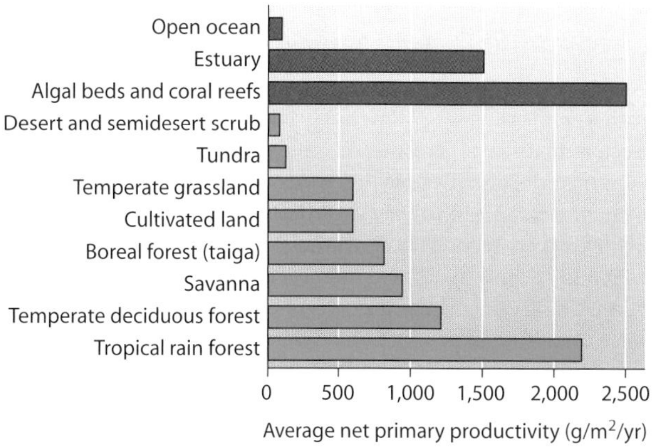

▲ **Figure 15** Net primary production of various ecosystems

16 Energy supply limits the length of food chains

When energy flows as organic matter through the trophic levels of an ecosystem, much of it is lost at each link in a food chain. Consider the transfer of organic matter from a producer to a primary consumer, such as the caterpillar shown in **Figure 16A**. The caterpillar might digest and absorb only about half the organic material it eats, passing the indigestible wastes as feces. Of the organic compounds it does absorb, the caterpillar typically uses two-thirds as fuel for cellular respiration. Only the chemical energy left over after respiration—15% of the organic material the caterpillar consumed—can be converted to caterpillar biomass. Thus, a secondary consumer that eats the caterpillar only gets 15% of the biomass (and the energy it contains) that was in the leaves the caterpillar ate.

Figure 16B, called a *pyramid of production*, illustrates the cumulative loss of energy with each transfer in a food chain. Each tier of the pyramid represents the chemical energy present in all of the organisms at one trophic level of a food chain. The width of each tier indicates how much of the chemical energy of the tier below is actually incorporated into the organic matter of that trophic level. Note that producers convert only about 1% of the energy in the sunlight available to them to primary production. In this idealized pyramid, 10% of the energy available at each trophic level becomes incorporated into the next higher level. The efficiencies of energy transfer usually range from 5 to 20%. In other words, 80–95% of the energy at one trophic level never transfers to the next.

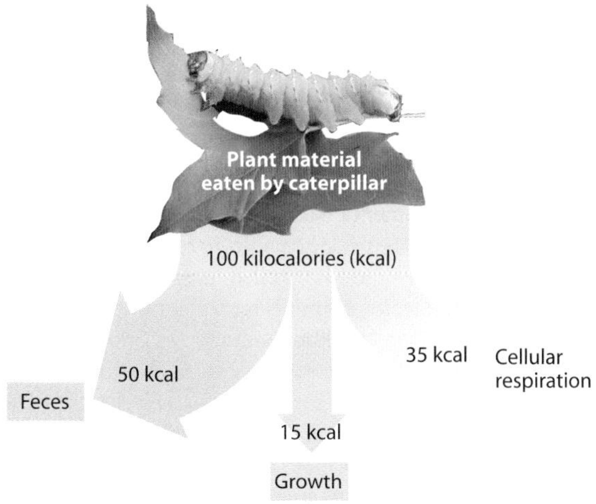

▲ **Figure 16A** The fate of leaf biomass consumed by a caterpillar

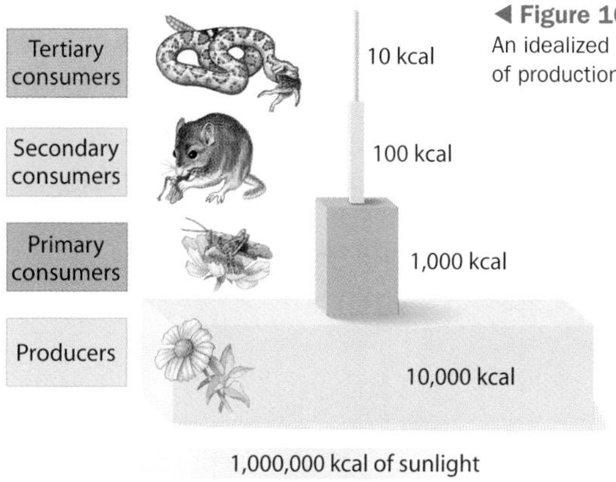

◀ **Figure 16B** An idealized pyramid of production

An important implication of this stepwise decline of energy in a trophic structure is that the amount of energy available to top-level consumers is small compared with that available to lower-level consumers. Only a tiny fraction of the energy stored by photosynthesis flows through a food chain all the way to a tertiary consumer. This explains why top-level consumers such as lions and hawks require so much geographic territory; it takes a lot of vegetation to support trophic levels so many steps removed from photosynthetic production. Pyramids of production help us understand why most food chains are limited to three to five levels; there is simply not enough energy at the very top of an ecological pyramid to support another trophic level.

? **Approximately what proportion of the energy produced by photosynthesis makes it to the snake in Figure 16B?**

1/1,000 of the 10,000 kcal produced by photosynthesis [(0.1 × 0.1 × 0.1)(10,000 kcal) = 10 kcal]

CONNECTION 17 A pyramid of production explains the ecological cost of meat

The dynamics of energy flow apply to the human population as much as to other organisms. As omnivores, people eat both plant material and meat. When we eat grain or fruit, we are primary consumers; when we eat beef or other meat from herbivores, we are secondary consumers. When we eat fish like trout and salmon (which eat insects and other small animals), we are tertiary or quaternary consumers.

The pyramid of production on the left in Figure 17 indicates energy flow from producers to vegetarians (primary consumers). The energy in the producer trophic level comes from a corn crop. The pyramid on the right illustrates energy flow from the same corn crop, with people as secondary consumers, eating beef. These two pyramids are generalized models, based on the rough estimate that about 10% of the chemical energy available in a trophic level appears at the next higher trophic level. Thus, the pyramids indicate that the human population has about 10 times more energy available to it when people eat corn than when they process the same amount of corn through another trophic level and eat corn-fed beef.

Eating meat of any kind is both economically and environmentally expensive. Compared with growing plants for direct human consumption, producing meat usually requires that more land be cultivated, more water be used for irrigation, more fossil fuels be burned, and more chemical fertilizers and pesticides be applied to croplands used for growing grain. In many countries, people cannot afford to buy much meat and are vegetarians by necessity. Sometimes religion also plays a role in the decision. In India, for example, about 80% of the population practice Hinduism, a religion that discourages meat-eating. India's meat consumption was roughly 5.2 kg (11.5 pounds) per person annually in 2002 (the most recent year for which statistics are available). In Mexico, where many people are too poor to eat meat daily, per capita consumption in 2002 was 58.6 kg (129 pounds) per year. That is a large amount compared with India, but less than half that of the United States, where the per capita rate was 124.8 kg (275 pounds) in 2002.

We turn next to the subject of chemical nutrients. Unlike energy, which is ultimately lost from an ecosystem, all chemical nutrients cycle within ecosystems.

? Why does demand for meat also tend to drive up prices of grains such as wheat and rice, fruits, and vegetables?

The potential supply of plants for direct consumption as food for humans is diminished by the use of agricultural land to grow feed for cattle, chickens, and other meat sources.

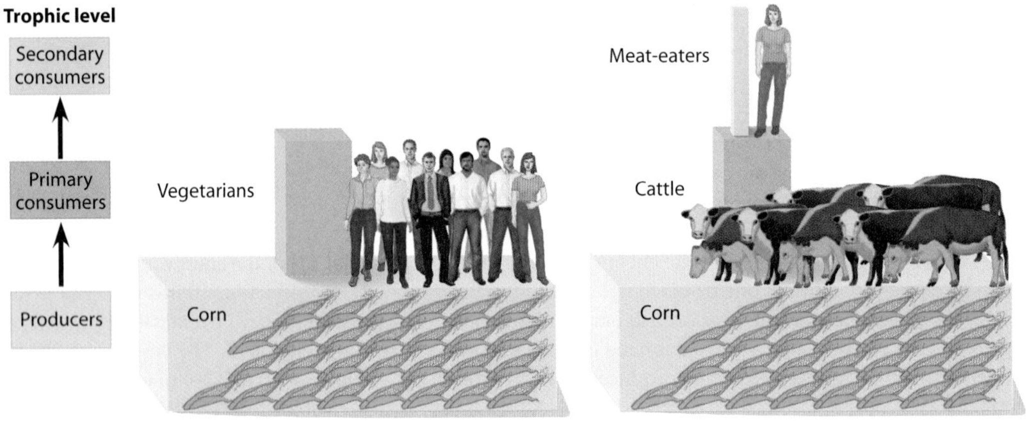

▶ Figure 17 Food energy available to people eating at different trophic levels

18 Chemicals are cycled between organic matter and abiotic reservoirs

The sun (or in some cases Earth's interior) supplies ecosystems with a continual influx of energy, but aside from an occasional meteorite, there are no extraterrestrial sources of chemical elements. Life, therefore, depends on the recycling of chemicals. While an organism is alive, much of its chemical stock changes continuously as nutrients are acquired and waste products are released. Atoms present in the complex molecules of an organism at the time of its death are returned to the environment by the action of decomposers, replenishing the pool of inorganic nutrients that plants and other producers use to build new organic matter.

Because chemical cycles in an ecosystem include both biotic and abiotic (geologic and atmospheric) components, they are called **biogeochemical cycles.** Figure 18, at the top of the next page, is a general scheme for the cycling of a nutrient within an ecosystem. Note that the cycle has **abiotic reservoirs**, where chemicals accumulate or are stockpiled outside of living organisms. The atmosphere, for example, is an abiotic reservoir for carbon. Phosphorus, on the other hand, is available only from the soil.

Let's trace the general biogeochemical cycle in Figure 18. ❶ Producers incorporate chemicals from the abiotic reservoirs into organic compounds. ❷ Consumers feed on the producers, incorporating some of the chemicals into their own bodies. ❸ Both producers and consumers release some chemicals

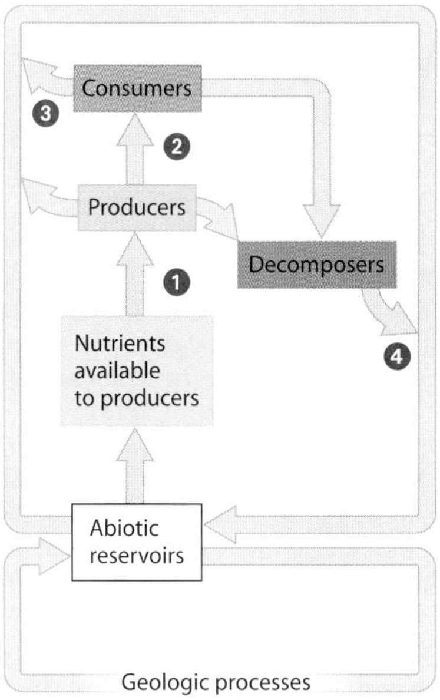

▶ **Figure 18** A general model of the biogeochemical cycling of nutrients

back to the environment in waste products (CO_2 and nitrogenous wastes of animals). ❹ Decomposers play a central role by breaking down the complex organic molecules in detritus such as plant litter, animal wastes, and dead organisms. The products of this metabolism are inorganic compounds such as nitrates (NO_3^-), phosphates (PO_4^{3-}), and CO_2, which replenish the abiotic reservoirs. Geologic processes such as erosion and the weathering of rock also contribute to the abiotic reservoirs. Producers use the inorganic molecules from abiotic reservoirs as raw materials for synthesizing new organic molecules (carbohydrates and proteins, for example), and the cycle continues.

Biogeochemical cycles can be local or global. Soil is the main reservoir for nutrients in a local cycle, such as phosphorus. In contrast, for those chemicals that exist primarily in gaseous form—carbon and nitrogen are examples—the cycling is essentially global. For instance, some of the carbon a plant acquires from the air may have been released into the atmosphere by the respiration of an organism on another continent.

In the next three modules, we look at the cyclic movements of carbon, phosphorus, and nitrogen. As you study the cycles, look for the four basic steps we have cited, as well as the geologic processes that may move chemicals around and between ecosystems. In the diagrams, the main abiotic reservoirs are highlighted in white boxes.

? Which boxes in Figure 18 represent biotic components of an ecosystem?

Consumers, producers, and decomposers

19 The carbon cycle depends on photosynthesis and respiration

Carbon, the major ingredient of all organic molecules, has an atmospheric reservoir and cycles globally. Carbon also resides in fossil fuels and sedimentary rocks, such as limestone ($CaCO_3$), and as dissolved carbon compounds in the oceans.

As shown in **Figure 19**, the reciprocal metabolic processes of photosynthesis and cellular respiration are mainly responsible for the cycling of carbon between the biotic and abiotic worlds. ❶ Photosynthesis removes CO_2 from the atmosphere and incorporates it into organic molecules, which are ❷ passed along the food chain by consumers. ❸ Cellular respiration by producers and consumers-returns CO_2 to the atmosphere. ❹ Decomposers break down the carbon compounds in detritus; that carbon, too, is eventually released as CO_2.

On a global scale, the return of CO_2 to the atmosphere by cellular respiration closely balances its removal by photosynthesis. However, ❺ the increased burning of wood and fossil fuels (coal and petroleum) is raising the level of CO_2 in the atmosphere. This increase in CO_2 is leading to significant global warming.

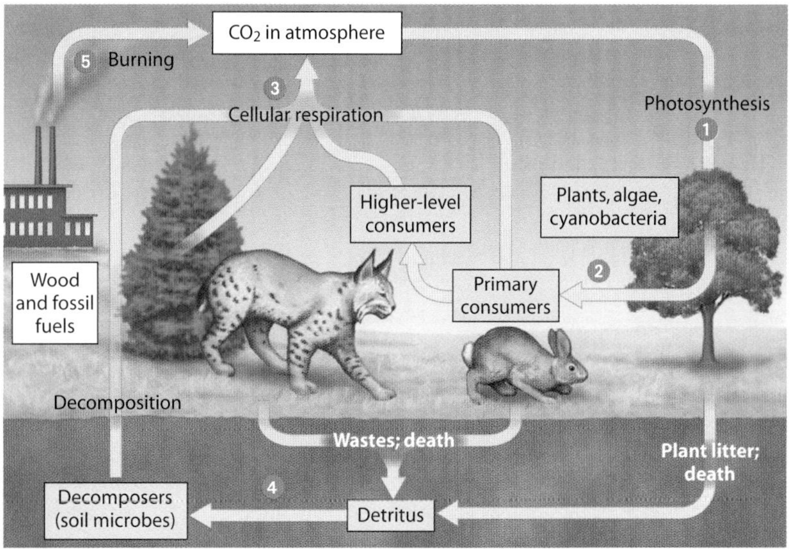

▲ **Figure 19** The carbon cycle

? What would happen to the carbon cycle if all the decomposers suddenly went on "strike" and stopped working?

Carbon would accumulate in organic mass, the atmospheric reservoir of carbon would decline, and plants would eventually be starved for CO_2.

20 The phosphorus cycle depends on the weathering of rock

Organisms require phosphorus—usually in the form of the phosphate ion (PO_4^{3-})—as an ingredient of nucleic acids, phospholipids, and ATP and (in vertebrates) as a mineral component of bones and teeth. In contrast to the carbon cycle and the other major biogeochemical cycles, the phosphorus cycle does not have an atmospheric component. Rocks are the only source of phosphorus for terrestrial ecosystems; in fact, rocks that have high phosphorus content are mined for agricultural fertilizer.

At the center of **Figure 20**, ❶ the weathering (breakdown) of rock gradually adds inorganic phosphate (PO_4^{3-}) to the soil. ❷ Plants assimilate the dissolved phosphate ions in the soil and build them into organic compounds. ❸ Consumers obtain phosphorus in organic form by eating plants. ❹ Phosphates are returned to the soil by the action of decomposers on animal waste and the remains of dead plants and animals. ❺ Some phosphate drains from terrestrial ecosystems into the sea, where it may settle and eventually become part of new rocks. This phosphorus will not cycle back into living organisms until ❻ geologic processes uplift the rocks and expose them to weathering, a process that takes millions of years.

Because phosphates are transferred from terrestrial to aquatic ecosystems much more rapidly than they are replaced, the amount in terrestrial ecosystems gradually diminishes over time. Furthermore, much of the soluble phosphate released by weathering quickly binds to soil particles, rendering it inaccessible to plants. As a result, the phosphate availability is often quite low and commonly a limiting factor. Mycorrhizal fungi that facilitate phosphorus uptake are essential to many plants, especially those living in older, highly weathered soils. Soil erosion from land cleared for agriculture or development accelerates the loss of phosphates.

Farmers and gardeners often use crushed phosphate rock, bone meal (finely ground bones from slaughtered livestock), or guano to add phosphorus to the soil. Guano, the droppings of seabirds and bats, is mined from densely populated colonies or caves, where meters-deep deposits have accumulated. As you'll learn in Module 22, however, runoff of large amounts of phosphate fertilizer pollutes aquatic ecosystems.

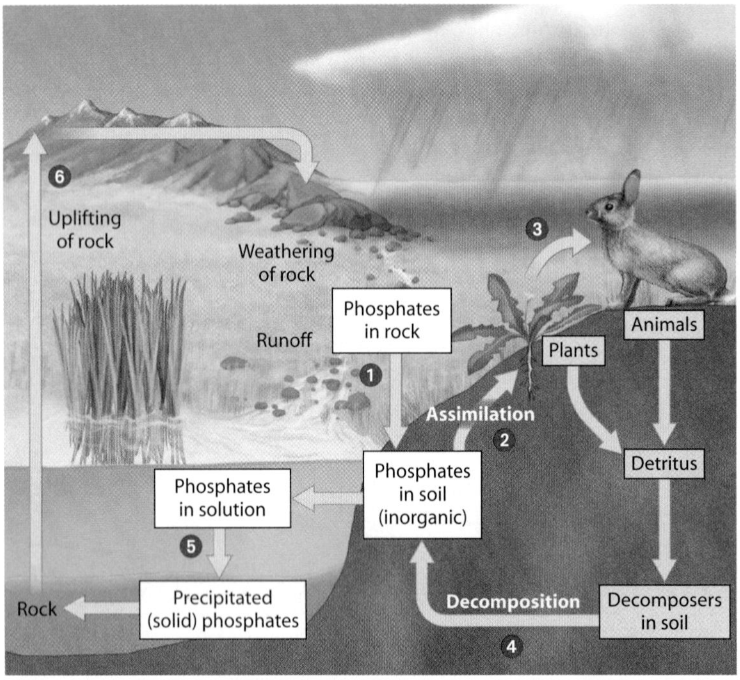

▲ **Figure 20** The phosphorus cycle

? Over the short term, why does phosphorus cycling tend to be more localized than either carbon or nitrogen cycling?

● Because phosphorus is cycled almost entirely within the soil rather than transferred over long distances via the atmosphere

21 The nitrogen cycle depends on bacteria

As an ingredient of proteins and nucleic acids, nitrogen is essential to the structure and functioning of all organisms. In particular, it is a crucial and often limiting plant nutrient. Nitrogen has two abiotic reservoirs, the atmosphere and the soil. The atmospheric reservoir is huge; almost 80% of the atmosphere is nitrogen gas (N_2). However, plants cannot absorb nitrogen in the form of N_2. The process of **nitrogen fixation**, which is performed by some bacteria, converts N_2 to compounds of nitrogen that can be used by plants. Without these organisms, the natural reservoir of usable soil nitrogen would be extremely limited.

Figure 21, on the facing page, illustrates the actions of two types of nitrogen-fixing bacteria. Starting at the far right in the figure, ❶ some bacteria live symbiotically in the roots of certain species of plants, supplying their hosts with a direct source of usable nitrogen. The largest group of plants with this mutualistic relationship is the legumes, a family that includes peanuts, soybeans, and alfalfa. A number of non-legume plants that live in nitrogen-poor soils have a similar relationship with bacteria. ❷ Free-living nitrogen-fixing bacteria in soil or water convert N_2 to ammonia (NH_3), which then picks up another H^+ to become ammonium (NH_4^+).

❸ After nitrogen is "fixed," some of the NH_4^+ is taken up and used by plants. ❹ Nitrifying bacteria in the soil also convert some of the NH_4^+ to nitrate (NO_3^-), which is more readily

5 absorbed by plants. Plants use the nitrogen they assimilate to synthesize molecules such as amino acids, which are then incorporated into proteins.

6 When an herbivore (represented by the rabbit in Figure 21) eats a plant, it digests the proteins into amino acids, then uses the amino acids to build the proteins it needs. Higher-order consumers gain nitrogen from their prey. Nitrogen-containing waste products are formed during protein metabolism; consumers excrete some nitrogen as well as incorporate some into their body tissues. Mammals, such as the rabbit, excrete nitrogen as urea, a substance that is widely used as an agricultural fertilizer.

Organisms that are not consumed eventually die and become detritus, which is decomposed by prokaryotes and fungi. 7 Decomposition releases NH_4^+ from organic compounds back into the soil, replenishing the soil reservoir of NH_4^+ and, with the help of nitrifying bacteria (step 4), NO_3^-. Under low-oxygen conditions, however, 8 soil bacteria known as denitrifiers strip the oxygens from NO_3^-, releasing N_2 back into the atmosphere and depleting the soil reservoir of usable nitrogen. Aerobic denitrification produces a different gas, N_2O.

Although not shown in the figure, some NH_4^+ and NO_3^- are made in the atmosphere by chemical reactions involving N_2 and ammonia gas (NH_3). The ions produced by these chemical reactions reach the soil in precipitation and dust, which are crucial sources of nitrogen for plants in some ecosystems.

Human activities are disrupting the nitrogen cycle by adding more nitrogen to the biosphere each year than natural processes. Combustion of fossil fuels in motor vehicles and coal-fired power plants produces nitrogen oxides (NO and NO_2). Nitrogen oxides react with other gases in the lower atmosphere to increase the production of ozone. Unlike the protective ozone layer in the upper atmosphere, ground-level ozone is a health hazard. Exposure to ozone, which irritates the respiratory system, can cause coughing and breathing difficulties. It is especially dangerous for people with respiratory problems such as asthma. In many regions, ozone alerts are common during hot, dry summer weather. Nitrogen oxides also combine with water in the atmosphere to become nitric acid. The Clean Air Act Amendments of 1990 diminished acid precipitation from sulfur emissions, but environmental damage from nitric acid precipitation is causing new concern.

Modern agricultural practices are another major source of nitrogen. Animal wastes from intensive livestock production release ammonia into the atmosphere. Farmers use enormous amounts of nitrogen fertilizer to supplement natural nitrogen fixation by bacteria. Worldwide, the application of synthetic nitrogen fertilizer has increased 100-fold since the late 1950s. However, less than half the fertilizer is taken up by the crop plants. Some nitrogen escapes to the atmosphere, where it forms NO_2 or nitrous oxide (N_2O), an inert gas that lingers in the atmosphere and contributes to global warming. As you'll learn in the next module, nitrogen fertilizers also pollute aquatic systems.

? What are the abiotic reservoirs of nitrogen? In what form does nitrogen occur in each reservoir?

Atmosphere: N_2; soil: NH_4^+ and NO_3^-

▶ **Figure 21**
The nitrogen cycle

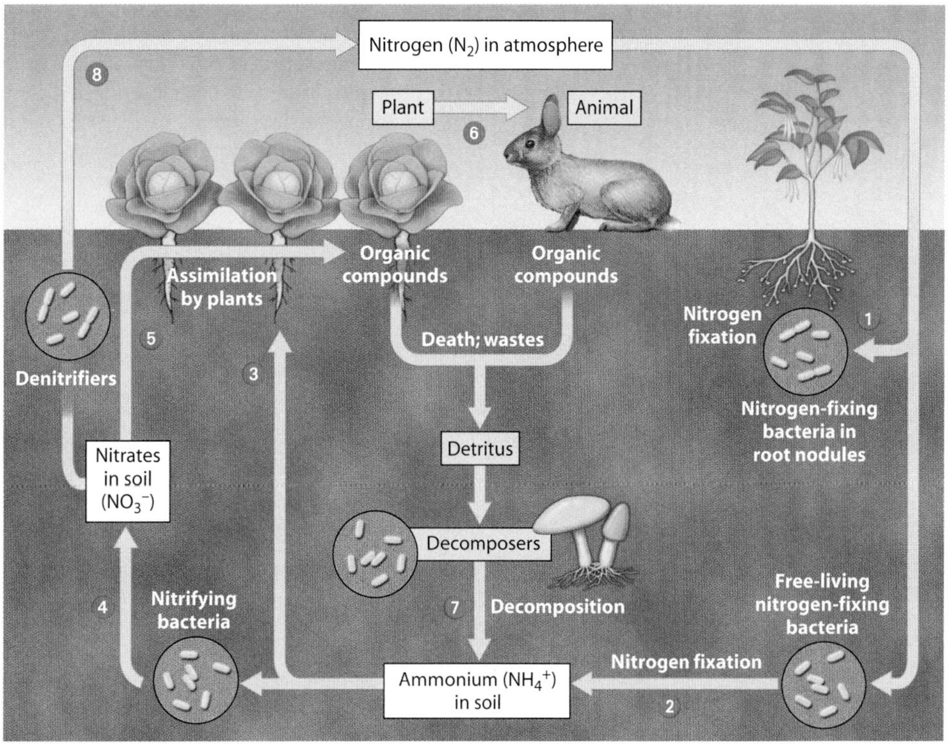

CONNECTION | 22 A rapid inflow of nutrients degrades aquatic ecosystems

Low levels of nutrients, especially phosphorus and nitrogen, often limit the growth of algae and cyanobacteria—and thus primary production—in aquatic ecosystems. Standing-water ecosystems (lakes and ponds) gradually accumulate nutrients from the decomposition of organic matter and fresh influx from the land. As a result, primary production increases naturally over time in a process known as eutrophication. Human activities that add nutrients to aquatic ecosystems accelerate this process and also cause eutrophication in rivers, estuaries, coastal waters, and coral reefs.

You might think that an increase in primary production would be beneficial to a biological community. After all, Figure 15 shows that coral reefs and tropical rain forests, ecosystems renowned for spectacular species diversity, have the greatest net primary production. But rapid eutrophication actually lowers species diversity. In some ecosystems, cyanobacteria replace green algae as primary producers. These prokaryotes, which are often encased in a slimy coating, form extensive mats on the surface of the water that prevent light from penetrating the water (**Figure 22A**). Some species of cyanobacteria can fix nitrogen, which gives them an additional advantage when phosphate is the pollutant and nitrogen is scarce. Other ecosystems are overrun by blooms of unicellular diatoms, toxin-producing dinoflagellates, or multicellular algae. These heavy growths, or "blooms," of cyanobacteria or algae greatly reduce oxygen levels at night, when the photosynthesizers respire. As the cyanobacteria and algae die, microbes consume a great deal of oxygen as they decompose the extra biomass. Thus, rapid nutrient enrichment results in oxygen depletion of the water. Fishes that have a high oxygen requirement cannot survive in such an environment.

In many areas, phosphate pollution comes from agricultural fertilizers. Phosphates are also a common ingredient in pesticides. Other major sources of phosphates include outflow from sewage treatment facilities and runoff of animal waste from livestock feedlots (where hundreds of animals are penned together). Sewage treatment facilities may discharge large amounts of dissolved inorganic nitrogen compounds into rivers or streams when extreme conditions (such as unusually high rainfall) overwhelm their capacity. Agricultural sources of nitrogen include feedlots and the large amounts of inorganic nitrogen fertilizers that are routinely applied to crops, lawns, and golf courses. Plants take up some of the nitrogen compounds in fertilizer, and denitrifiers convert some to atmospheric N_2 or N_2O, but nitrate is not bound tightly by soil particles and is easily washed out of the soil by rain or irrigation. As a result, chemical fertilizers often exceed the soil's natural recycling capacity.

In an example of how far-reaching this problem can be, nitrogen runoff from midwestern farm fields has been linked to a "dead zone" observed each summer in the Gulf of Mexico. Vast algal blooms, indicated in red and orange in **Figure 22B**, extend outward from where the Mississippi River deposits its nutrient-laden waters. As the algae die, decomposition of the huge quantities of biomass diminishes the supply of dissolved oxygen over an area that ranges from 13,000 to 22,000 km^2, or roughly 5,000 to 8,500 square miles. Oxygen depletion disrupts benthic communities, displacing fishes and invertebrates that can move and killing organisms that are attached to the substrate. More than 400 recurring and permanent coastal dead zones totaling approximately 245,000 km^2 have been documented in seas worldwide.

Michael Marten/SPL/Photo Researchers

▲ **Figure 22A** Algal growth on a pond resulting from nutrient pollution

? How would excessive addition of mineral nutrients to a lake eventually lead to the loss of many fish species?

The nutrients initially cause population explosions of algae and cyanobacteria. Their respiration and that of the decomposers of all the detritus as the algae and cyanobacteria die consume most of the lake's oxygen, which the fish require.

NASA/Goddard Space Flight Center

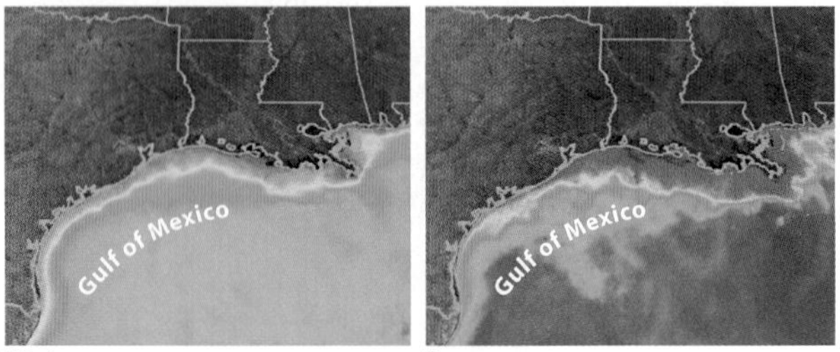

▲ **Figure 22B** Concentrations of phytoplankton in winter and summer. Red and orange indicate highest concentrations of phytoplankton.

CONNECTION 23 Ecosystem services are essential to human well-being

In the chapter introduction, we mentioned that people rely on numerous services that natural ecosystems provide. In addition to supplying fresh water and some foods, healthy ecosystems recycle nutrients, decompose wastes, and regulate climate and air quality. Wetlands buffer coastal populations against tidal waves and hurricanes, reduce the impact of flooding rivers, and filter pollutants. Natural vegetation helps retain fertile soil and prevent landslides and mudslides.

Ecosystems that we create are also essential to our well-being. For example, agricultural ecosystems supply most of our food and fibers. Although we manage these ecosystems, they are modifications of natural ecosystems and make use of ecosystem services, such as control of agricultural pests by natural predators and pollination of crops. Soil fertility, the foundation for crop growth, depends on nutrient cycling, another ecosystem service. But agricultural methods introduced over the past several decades have pushed croplands beyond their natural capacity to produce food. Large inputs of chemical fertilizers are needed to supplement soil nutrients. Synthetic pesticides are used to control the population growth of crop-eating insects and pathogens that take advantage of vast monocultures of crop species. Herbicides are applied to kill weeds that would compete with crop plants for water and nutrients. In many areas, crops require additional water supplied by irrigation.

These agricultural practices have resulted in enormous increases in food production, but at the expense of natural ecosystems and the services they provide. The detrimental effects of nutrient runoff, discussed in the previous module, are affecting both freshwater and marine ecosystems as fertilizer use increases. Pesticides may kill beneficial organisms as well as pests, and as you learned previously, chemicals that persist in the environment can be carried far from their point of origin. Perhaps most worrisome is the deterioration of fertile soil. Clearing and cultivation expose land to wind and water that erode the rich topsoil. Erosion and soil degradation are especially severe in grassland, savanna, and some forest ecosystems where low amounts of precipitation and high rates of evaporation result in low levels of soil moisture. In recent years, dust storms sweeping across overcultivated areas have removed millions of tons of topsoil from these stressed ecosystems. In China, for example, overgrazing and other poor agricultural practices are turning 900 square miles of land—an area the size of Rhode Island—into desert each year (**Figure 23A**). Irrigation of arid land enables farmers to grow crops but leaves a salty residue that eventually prevents plant growth. In addition, population growth in these regions places increasing demands on the already scarce water supply.

Human activities also threaten many forest ecosystems and the services they provide. Every year, more and more land is cleared for agriculture. Some of this land is needed to feed the growing human population, but replacing worn-out cropland accounts for much of the deforestation occurring today. Forests are also cut down to provide timber and fuel wood; many people in nonindustrialized countries use wood for heating and cooking (**Figure 23B**). The most immediate impact of deforestation is soil erosion. In Haiti, for example, where less than 2% of the original tree cover remains, heavy rains inevitably bring flooding and mudslides that damage crops. During recent hurricanes, floodwaters surging down stripped hillsides caused thousands of deaths. With much of the soil destabilized by the devastating earthquake in 2010, massive landslides are likely to cause further ecological and economic destruction during storms.

Fang xinwu/Newscom

▲ **Figure 23A** A dust storm in Changling, China

The growing demand of the human population for food, fibers, and water has largely been satisfied at the expense of other ecosystem services, but these practices cannot continue indefinitely. **Sustainability** is the goal of developing, managing, and conserving Earth's resources in ways that meet the needs of people today without compromising the ability of future generations to meet theirs.

? How can clear-cutting a forest (removing all trees) damage the water quality of nearby aquatic ecosystems?

Without the growing trees to assimilate minerals from the soil, more of the minerals run off and end up polluting water resources.

▶ **Figure 23B** A woman in Haiti gathering wood to process into charcoal

Thony Belizaire/AFP/Getty Images

REVIEW

For Practice Quizzes, BioFlix, MP3 Tutors, and Activities, go to www.masteringbiology.com.

Reviewing the Concepts

Community Structure and Dynamics (1–13)

1 A community includes all the organisms inhabiting a particular area. Community ecology is concerned with factors that influence the species composition and distribution of communities and with factors that affect community stability.

2 Interspecific interactions are fundamental to community structure. Interspecific interactions can be categorized according to their effect on the interacting populations.

3 Competition may occur when a shared resource is limited.

4 Mutualism benefits both partners.

5 Predation leads to diverse adaptations in prey species.

6 Herbivory leads to diverse adaptations in plants. Some herbivore-plant interactions illustrate coevolution or reciprocal evolutionary adaptations.

7 Parasites and pathogens can affect community composition.

8 Trophic structure is a key factor in community dynamics. Trophic structure can be represented by a food chain.

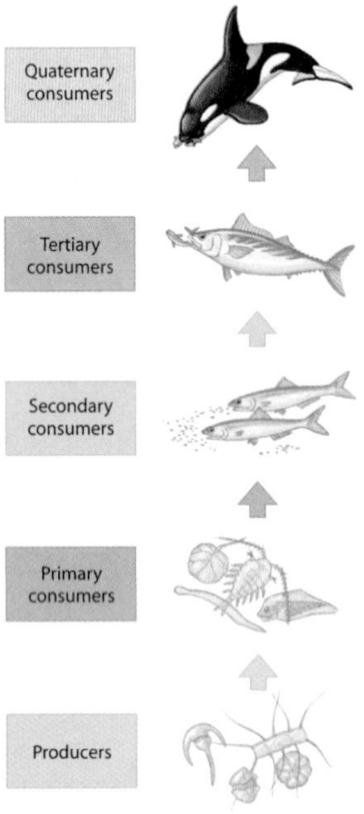

9 Food chains interconnect, forming food webs.

10 Species diversity includes relative abundance and species richness. Thus, diversity takes into account both the number of species in a community and the proportion of the community that each species represents.

11 Keystone species have a disproportionate impact on diversity. Although a keystone species has low biomass or relative abundance, its removal from a community results in lower species diversity.

12 Disturbance is a prominent feature of most communities. Ecological succession is a transition in species composition of a community. Primary succession is the gradual colonization of barren rocks. Secondary succession occurs after a disturbance has destroyed a community but left the soil intact.

13 Invasive species can devastate communities. Organisms that have been introduced to non-native habitats by human actions and have established themselves at the expense of native communities are considered invasive. The absence of natural enemies often allows rapid population growth of invasive species.

Ecosystem Structure and Dynamics (14–23)

14 Ecosystem ecology emphasizes energy flow and chemical cycling. An ecosystem includes a community and the abiotic factors with which it interacts.

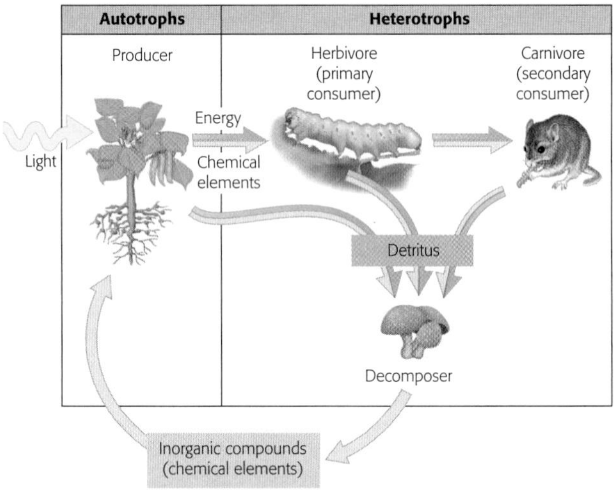

15 Primary production sets the energy budget for ecosystems.

16 Energy supply limits the length of food chains. A pyramid of production shows the flow of energy from producers to primary consumers and to higher trophic levels. Only about 10% of the energy stored at each trophic level is available to the next level.

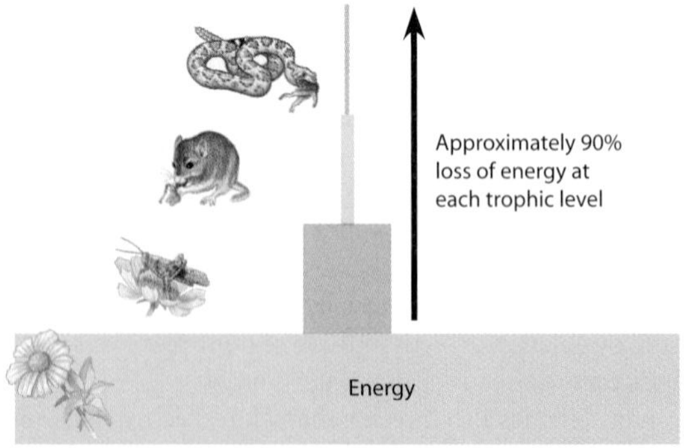

17 A pyramid of production explains the ecological cost of meat. A field of corn can support many more human vegetarians than meat-eaters.

18 Chemicals are cycled between organic matter and abiotic reservoirs.

19 The carbon cycle depends on photosynthesis and respiration.

20 The phosphorus cycle depends on the weathering of rock.

21 The nitrogen cycle depends on bacteria. Various bacteria in soil (and root nodules of some plants) convert gaseous N_2 to compounds that plants can use, such as ammonium (NH_4^+) and nitrate (NO_3^-).

22 A rapid inflow of nutrients degrades aquatic ecosystems. Nutrient input from fertilizer and other sources causes rapid eutrophication, resulting in decreased species diversity and oxygen depletion of lakes, rivers, and coastal waters.

23 Ecosystem services are essential to human well-being. Although agricultural and other managed ecosystems are necessary to supply our needs, we also depend on services provided by natural ecosystems.

Connecting the Concepts

1. Fill in the blanks in the table below summarizing the interspecific interactions in a community.

Interspecific Interaction	Effect on Species 1	Effect on Species 2	Example
	+	−	
	−	−	
	+	−	
	+	−	
	+	+	

2. Fill in the blanks in the table below summarizing terrestrial nutrient cycles.

	Carbon	Phosphorus	Nitrogen
Main abiotic reservoir(s)			
Form in abiotic reservoir			
Form used by producers			
Human activities that alter cycle			
Effects of altering cycle			

Testing Your Knowledge

Multiple Choice

3. Which of the following groups is absolutely essential to the functioning of an ecosystem?
   a. producers
   b. producers and herbivores
   c. producers, herbivores, and carnivores
   d. detritivores
   e. producers and decomposers

4. The open ocean and tropical rain forests contribute the most to Earth's net primary production because
   a. both have high rates of net primary production.
   b. both cover huge surface areas of Earth.
   c. nutrients cycle fastest in these two ecosystems.
   d. the ocean covers a huge surface area and the tropical rain forest has a high rate of production.
   e. Both a and b are correct.

5. Which of the following organisms is mismatched with its trophic level?
   a. algae—producer
   b. fungi—decomposer
   c. phytoplankton—primary consumer
   d. carnivorous fish larvae—secondary consumer
   e. eagle—tertiary or quaternary consumer

6. Which of the following best illustrates ecological succession?
   a. A mouse eats seeds, and an owl eats the mouse.
   b. Decomposition in soil releases nitrogen that plants can use.
   c. Grasses grow in a deserted field, followed by shrubs and then trees.
   d. Imported pheasants increase in numbers, while local quail disappear.
   e. Overgrazing causes a loss of nutrients from soil.

7. To ensure adequate nitrogen for a crop, a farmer would want to *decrease* ______ by soil bacteria.
   a. cellular respiration
   b. nitrification
   c. denitrification
   d. nitrogen fixation
   e. b and d

Describing, Comparing, and Explaining

8. Explain how seed dispersal by animals is an example of mutualism in some cases.
9. What is rapid eutrophication? What steps might be taken to slow this process?
10. Local conditions, such as heavy rainfall or the removal of plants, may limit the amount of nitrogen, phosphorus, or calcium available, but the amount of carbon available in an ecosystem is seldom a problem. Explain.
11. In Southeast Asia, there's an old saying: "There is only one tiger to a hill." In terms of energy flow in ecosystems, explain why big predatory animals such as tigers and sharks are relatively rare.
12. For which chemicals are biogeochemical cycles global? Explain.
13. What roles do bacteria play in the nitrogen cycle?

Applying the Concepts

14. An ecologist studying plants in the desert performed the following experiment. She staked out two identical plots, which included a few sagebrush plants and numerous small, annual wildflowers. She found the same five wildflower species in roughly equal numbers on both plots. She then enclosed one of the plots with a fence to keep out kangaroo rats, the most common grain-eaters of the area. After two years, to her surprise, four of the wildflower species were no longer present in the fenced plot, but one species had increased dramatically. The control plot had not changed. Using the principles of ecology, propose a hypothesis to explain her results. What additional evidence would support your hypothesis?
15. Sometime in 1986, near Detroit, a freighter pumped out water ballast containing larvae of European zebra mussels. The molluscs multiplied wildly, spreading through Lake Erie and entering Lake Ontario. In some places, they have become so numerous that they have blocked the intake pipes of power plants and water treatment plants, fouled boat hulls, and sunk buoys. What makes this kind of population explosion occur? What might happen to native organisms that suddenly must share the Great Lakes ecosystem with zebra mussels? How would you suggest trying to solve the mussel population problem?

Review Answers

1.

Interspecific Interaction	Effect on Species 1	Effect on Species 2	Example (many other answers possible)
Predation	+	−	Crocodile/fish
Competition	−	−	Squirrel/black bear
Herbivory	+	−	Caterpillar/leaves
Parasites and pathogens	+	−	Heartworn/dog; *Salmonella*/person
Mutualism	+	+	Plant/mycorrhizae

2.

	Carbon	Phosphorus	Nitrogen
Main abiotic reservoir(s)	Atmosphere	Rocks	Atmosphere, soil
Form in abiotic reservoir	Carbon dioxide (CO_2)	Phosphate (PO_4^{3-}) bound with other minerals in rock	N_2 in atmosphere; ammonium (NH_4^+) or nitrate (NO_3^-) in soil
Form used by producers	Carbon dioxide (CO_2)	Phosphate (PO_4^{3-})	NH_4^+ or NO_3^-
Human activities that alter cycle	Burning wood and fossil fuels	Agriculture (fertilizers, feedlots, pesticides, soil erosion)	Agriculture (fertilizers, feedlots, soil erosion); combustion of fossil fuels; manufacture of nitrogen fertilizer
Effects of altering cycle	Global warming	Eutrophication of aquatic ecosystem; nutrient-depleted soils	Eutrophication of aquatic ecosystems; nutrient-depleted soils; global warming; smog; depletion of ozone layer; acid precipitation

3. e 4. d 5. c 6. c 7. c

8. Plants benefit by having their seeds distributed away from the parent. Animals benefit when the seeds contains food, as in fleshy fruits.

9. Rapid eutrophication occurs when bodies of water receive nutrient pollution (for example, from agricultural runoff) that results in blooms of cyanobacteria and algae. Respiration from these organisms and their decomposers depletes oxygen levels, leading to fish kills. Reducing this type of pollution will require controlling the sources of excess inorganic nutrients, for example, runoff from feedlots and fertilizers.

10. The abiotic reservoir of the first three nutrients is the soil. Carbon is available as carbon dioxide in the atmosphere.

11. These animals are secondary or tertiary consumers, at the top of the production pyramid. Stepwise energy loss means not much energy is left for them; thus, they are rare and require large territories in which to hunt.

12. Chemicals with a gaseous form in the atmosphere, such as carbon and nitrogen, have a global biogeochemical cycle.

13. Nitrogen fixation of atmospheric N_2 into ammonium; decomposition of detritus into ammonium; nitrification of ammonium into nitrate; denitrification (by denitrifiers) of nitrates into N_2

14. Hypothesis: The kangaroo rat is a keystone species in the desert. (Apparently, herbivory by the rats kept the one plant from outcompeting the others; removing the rats reduced plant diversity.) Additional supporting evidence: Observations of the rats preferentially eating dominant plants; finding that the dominant plant recovers from herbivore damage faster.

15. Some issues and questions to consider: What relationships (predators, competitors, parasites) might exist in the mussels' native habitat that are altered in the Great Lakes? How might the mussels compete with Great Lakes organisms? Might the Great Lakes species adapt in some way? Might the mussels adapt? Could possible solutions present problems of their own?

Conservation Biology

From Chapter 38 of *Campbell Biology: Concepts & Connections*, Seventh Edition. Jane B. Reece, Martha R. Taylor, Eric J. Simon, Jean L. Dickey.

Conservation Biology

BIG IDEAS

Charlie Riedel/AP Photo

The Loss of Biodiversity
(1–6)

Biodiversity is declining rapidly worldwide as a result of human activities.

Rick & Nora Bowers/Alamy

Conservation Biology and Restoration Ecology
(7–13)

Biologists are applying their knowledge of ecology to slow the loss of biodiversity and help define a sustainable future.

Arco Images GmbH/Alamy

How does a species become extinct? Here's one scenario. Imagine a species whose habitat constantly shrinks, squeezing populations into ever-smaller areas with ever-diminishing resources. As a population dwindles, the loss of even a single member to hunting, disease, or natural disaster brings the population closer to the brink. And then one day, that population is simply gone. Imagine a relentless succession of such occurrences—one population after another passing quietly out of existence, until none remains. This is the series of events unfolding as tigers slide toward extinction.

A hundred years ago, scientists estimate, about 100,000 tigers (*Panthera tigris*) could be found in the wild. Now that number has plummeted to around 3,200. Three of the world's nine tiger subspecies have disappeared entirely, and one has not been seen for the past 25 years. Tigers now occupy just 7% of their original range, and even that remaining sliver is decreasing. In Indonesia and Malaysia, tropical forests that are home to two tiger subspecies are being replaced by plantations for palm oil, paper, and rubber. In Russia, logging in temperate forests is destroying the habitat of the Siberian tiger. People moving into rural areas of South Asia are encroaching on the habitat of the Bengal tiger. And throughout their range, tigers are at risk from poachers eager to sell the big cats' bones and internal organs, considered potent ingredients in some traditional Asian medicines.

In this chapter, you will learn about one of the major ecological challenges of our time—the rapid loss of biodiversity that is a result of our dominance over the environment. As you learn about the fight to save our biological heritage, you will see that conservation biology touches all levels of ecology, from a single tiger to the forest it roams.

The Loss of Biodiversity

1 Loss of biodiversity includes the loss of ecosystems, species, and genes

The decline of tiger populations is just one example of the worldwide loss of biodiversity. Why do we care about losing species, especially ones that are less charismatic than the magnificent tiger? One reason is what Harvard biologist E. O. Wilson calls biophilia, our sense of connection to nature and to other forms of life. And many people share a moral belief that other species have an inherent right to life. Our dependence on vital ecosystem services also gives us practical reasons for preserving biodiversity.

Biodiversity encompasses more than individual species—it includes ecosystem diversity, species diversity, and genetic diversity. Let's examine each level of diversity to see what we stand to lose if the decline is not stopped.

Ecosystem Diversity The world's natural ecosystems are rapidly disappearing. Nearly half of Earth's forests are gone, and thousands more square kilometers disappear every year. Grassland ecosystems in North America, where millions of bison roamed as recently as the 19th century, have overwhelmingly been lost to agriculture and development.

The temperate coniferous forest of the Klamath-Siskiyou Wilderness **(Figure 1A)** is located in a region spanning parts of California and Oregon that is extraordinarily rich in ecosystem diversity. In addition to the distinctive chaparral ecosystem, forests of sequoia, redwood, and Douglas fir, coastal dunes, salt marshes, and a wide variety of other ecosystems are found in this rapidly vanishing treasure trove of biodiversity. Only about a quarter of the original area remains in its natural state.

Aquatic ecosystems are also threatened. For example, an estimated 20% of the world's coral reefs, ecosystems known for their species richness and productivity, have been destroyed by human activities, and 15% are in danger of collapse within the next two decades. The deteriorating state of freshwater ecosystems is particularly worrisome. Tens of thousands of species live in lakes and rivers, and these ecosystems supply food and water for many terrestrial species, as well—including us.

As natural ecosystems are lost, so are essential services. Water purification is one of the services provided free of charge by healthy ecosystems. As water moves slowly through forests, streams, and wetlands, pollutants and sediments are filtered out. Whether taken from surface waters such as lakes or subsurface sources (groundwater), the drinking water supplied by public water systems typically has passed through this natural filtration process. In some places, including New York City, no further filtration is required, although the water is chlorinated to kill microorganisms. As farm fields and housing developments replaced the naturally diverse ecosystems in New York City's watershed, the land's ability to purify water deteriorated. The additional pollution from agricultural runoff and sewage reduced water quality to the point where the city had to take action. Officials considered spending $8 billion to build a filtration plant, which would cost a further $1 million per day to operate. They decided to invest in lower cost ecosystem services instead. Actions included more tightly restricting land use in the watershed, purchasing land to preserve natural ecosystems, and helping landowners better manage their land to protect the watershed. As a result of these measures, the quality of naturally filtered water supplied to New York City remains high.

Species Diversity When ecosystems are lost, populations of the species that make up their biological communities are also lost. A species may disappear from a local ecosystem but remain in others; for example, a population of tigers may be lost from one region of India while other populations survive elsewhere. Ecologists refer to the loss of a single population of a species as **extirpation**. Although extirpation and declining population sizes are strong signals that a species is in trouble, it may still be possible to save it. **Extinction** means that all populations of a species have disappeared, an irreversible situation.

▼ **Figure 1A** The Klamath-Siskiyou Wilderness, home to a wide variety of ecosystems

Mike Dobel/Alamy

Created from data from The IUCN Red List of Threatened Species™ at iucnredlist.org.

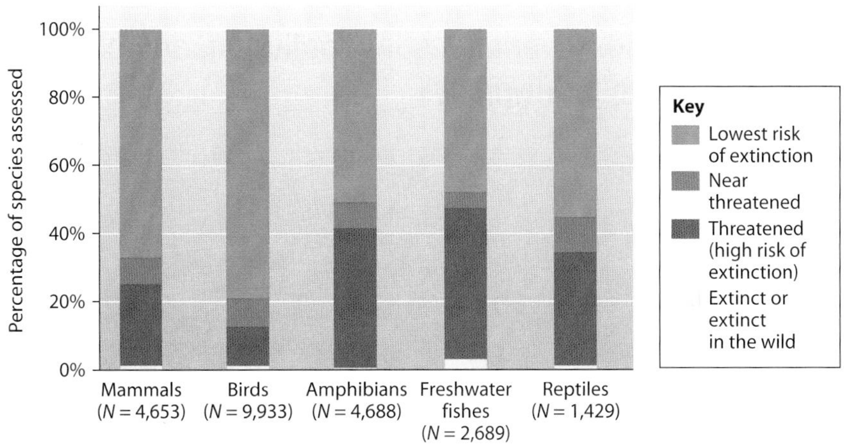

▲ **Figure 1B** Results of the 2009 IUCN assessment of species at risk for extinction (*N* = the number of species assessed)

How rapidly are species being lost? Because biologists are uncertain of the total number of species that exist, it is difficult to determine the actual rate of species loss. Some scientists estimate that current extinction rates are around 100 times greater than the natural rate of extinction. The International Union for Conservation of Nature (IUCN) is a global environmental network that keeps track of the status of species worldwide. **Figure 1B** shows the 2009 IUCN assessment of five major groups of animals. Notice the large proportions of amphibians and freshwater fishes that are considered threatened, further indications of the declining health of freshwater ecosystems.

Because of the network of community interactions among populations of different species within an ecosystem, the loss of one species can have a negative impact on the overall species richness of the ecosystem. Keystone species illustrate this effect. Other species modify their habitat in ways that encourage species diversity. In prairie ecosystems, for instance, plant and arthropod diversity is greatest near prairie dog burrows, where the soil has been altered by the animal's digging (**Figure 1C**). Abandoned burrows provide homes for cottontail rabbits, burrowing owls, and other animals. Thus, extirpation of prairie dogs results in lower species diversity in prairie communities.

In the United States, the Endangered Species Act protects species and the ecosystems on which they depend. Many other nations have also enacted laws to protect biodiversity, and an international agreement protects some 33,000 species of wild animals and plants from trade that would threaten their survival.

Species loss also has practical consequences for human well-being. Many drugs have been developed from substances found in the natural world, including penicillin, aspirin, antimalarial agents, and anticancer drugs. Dozens more potentially useful chemicals from a variety of organisms are currently being investigated. For example, researchers are testing possible new antibiotics produced by microbial symbionts of marine sponges; painkillers extracted from a species of poison dart frog; and anti-HIV and anticancer drugs derived from compounds in from rain forest plants.

▼ **Figure 1C** A group of young black-tailed prairie dogs (*Cynomys ludovicianus*) near their burrow

Jeff March/Alamy

Genetic Diversity The genetic diversity within and between populations of a species is the raw material that makes microevolution and adaptation to the environment possible—a hedge against future environmental changes. If local populations are lost and the total number of individuals of a species declines, so, too, do the genetic resources for that species. Severe reduction in genetic variation threatens the survival of a species.

The enormous genetic diversity of all the organisms on Earth has great potential benefit for people, too. Breeding programs have narrowed the genetic diversity of crop plants to a handful of varieties, leaving them vulnerable to pathogens. For example, researchers are currently scrambling to stop the spread of a deadly new strain of wheat stem rust, a fungal pathogen that has devastated harvests in Africa and central Asia. Resistance genes found in the wild relatives of wheat (**Figure 1D**) may hold the key to the world's future food supply. Many researchers and biotechnology leaders are enthusiastic about the possibilities that "bioprospecting" for potentially useful genes in other organisms holds for the development of new medicines, industrial chemicals, and other products.

Now that you have some insight into the nature and value of biodiversity, let's examine in more detail the causes for its decline.

James King-Holmes/Photo Researchers

▲ **Figure 1D** Einkorn wheat, a wild relative of modern cultivated varieties

? What are two reasons to be concerned about the impact of the biodiversity crisis on human welfare?

The environmental degradation threatening other species may also harm us. We are dependent on biodiversity, both directly through use of organisms and their products and indirectly through ecosystem services.

CONNECTION

2 Habitat loss, invasive species, overharvesting, pollution, and climate change are major threats to biodiversity

The human population has been growing exponentially for more than 100 years. We have supported this growth by using increasingly effective technologies to capture or produce food, to extract resources from the environment, and to build cities. In industrialized countries, we consume far more resources than are required to meet our basic requirements for food and shelter. Thus, it should not surprise you to learn that human activities are largely responsible for the current decline of biodiversity. In this section, we examine the major factors that threaten biodiversity.

Habitat Loss Human alteration of habitats poses the single greatest threat to biodiversity throughout the biosphere. Agriculture, urban development, forestry, mining, and environmental pollution have brought about massive destruction and fragmentation of habitats. Deforestation continues at a blistering pace in tropical and coniferous forests (**Figure 2A**).

The amount of human-altered land surface is approaching 50%, and we use over half of all accessible surface fresh water. The natural course of most of the world's major rivers has been changed. Worldwide, tens of thousands of dams constructed for flood control, hydroelectric power, drinking water, and irrigation have damaged river and wetland ecosystems. Some of the most productive aquatic habitats in estuaries and intertidal wetlands have been overrun by commercial and residential development. The loss of marine habitat is severe, especially in coastal areas and coral reefs.

Invasive Species Ranking second behind habitat loss as a threat to biodiversity are invasive species, which disrupt communities by competing with, preying on, or parasitizing native species. The lack of interspecific interactions that keep the newcomer populations in check is often a key factor in a non-native species becoming invasive. Meanwhile, a newly arrived species is an unfamiliar biotic factor in the environment of native species. Natives are especially vulnerable when a new species poses an unprecedented threat. In the absence of an evolutionary history with predators, for example, animals may lack defense mechanisms or even a fundamental recognition of danger. The Pacific island of Guam was home to 13 species of forest birds—but no native snakes—when brown tree snakes (**Figure 2B**) arrived as stowaways on a cargo plane. With no competitors, predators, or parasites to hinder them, the snakes proliferated rapidly on a diet of unwary birds. Four of the native species of birds were extirpated, although they survive on nearby islands. Three species of birds that lived nowhere else but Guam are now extinct. As the populations of two other species of birds became perilously low, officials took the remaining individuals into protective custody; they now exist only in zoos. The brown tree snake also eliminated species of seabirds and lizards.

Corbis/SuperStock

▲ **Figure 2A** Clear-cut areas in Mount Baker-Snoqualmie National Forest, Washington

David M Dennis/Photolibrary

▲ **Figure 2B** A brown tree snake (*Boiga irregularis*)

Overharvesting The third major threat to biodiversity is overexploitation of wildlife by harvesting at rates that exceed the ability of populations to rebound. Such overharvesting has threatened some rare trees that produce valuable wood, such as mahogany and rosewood. Animal species whose numbers have been drastically reduced by excessive commercial harvest, poaching, or sport hunting include tigers, whales, rhinoceroses, Galápagos tortoises, and numerous fishes. In parts of Africa, Asia, and South America, wild animals are heavily hunted for food, and the African term "bushmeat" is now used to refer generally to such meat. As once-impenetrable forests are opened to exploitation, the commercial bushmeat trade has become one of the greatest threats to primates, including gorillas, chimpanzees, and many species of monkeys, as well as other mammals and birds (**Figure 2C**). No longer hunted only for local use, large quantities of bushmeat are sold at urban markets or exported worldwide, including to the United States.

Aquatic species are suffering overexploitation, too. Many edible marine fish and seafood species are in a precarious state. Worldwide, fishing fleets are working farther offshore and harvesting fish from greater depths in order to obtain hauls comparable to those of previous decades.

Ho New/Reuters

▲ **Figure 2C** Lemurs killed by poachers for sale in a bushmeat market

Pollution Pollutants released by human activities can have local, regional, and global effects. Some pollutants, such as oil spills, contaminate local areas (**Figure 2D**). The global water cycle can transport pollutants—for instance, pesticides used on land—from terrestrial to aquatic ecosystems hundreds of miles away. Pollutants that are emitted into the atmosphere, such as nitrogen oxides from the burning of fossil fuels, may be carried aloft for many of miles before falling to earth in the form of acid precipitation.

Ozone depletion in the upper atmosphere is another example of the global impact of pollution. The **ozone layer** protects Earth from the harmful ultraviolet rays in sunlight. Beginning in the mid-1970s, scientists realized that the ozone layer was gradually thinning. The consequences of ozone depletion for life on Earth could be quite severe, not only increasing skin cancers, but harming crops and natural communities, especially the phytoplankton that are responsible for a large proportion of Earth's primary production. International agreements to phase out the production of chemicals implicated in ozone destruction have been effective in slowing the rate of ozone depletion. Even so, complete ozone recovery is probably decades away.

In addition to being transported to areas far from where they originate, many toxins produced by industrial wastes or applied as pesticides become concentrated as they pass through the food chain. This concentration, or **biological magnification**, occurs because the biomass at any given trophic level is produced from a much larger toxin-containing biomass ingested from the level below. Thus, top-level predators are usually the organisms most severely damaged by toxic compounds in the environment. In the Great Lakes food chain shown in **Figure 2E**, the concentration of industrial chemicals called PCBs increased at each successive trophic level. The PCB concentration measured in the eggs of herring gulls, top-level consumers, was almost 5,000 times higher than that measured in phytoplankton. Many other synthetic chemicals that cannot be degraded by micro-organisms also become concentrated through biological magnification, including DDT and mercury. Mercury, a by-product of plastic production and coal-fired power plants, enters the food chain after being converted to highly toxic methylmercury by benthic bacteria.

Charlie Riedel/ AP Photo

▲ **Figure 2D** A brown pelican on the Louisiana coast suffering the effects of the 2010 British Petroleum oil rig explosion

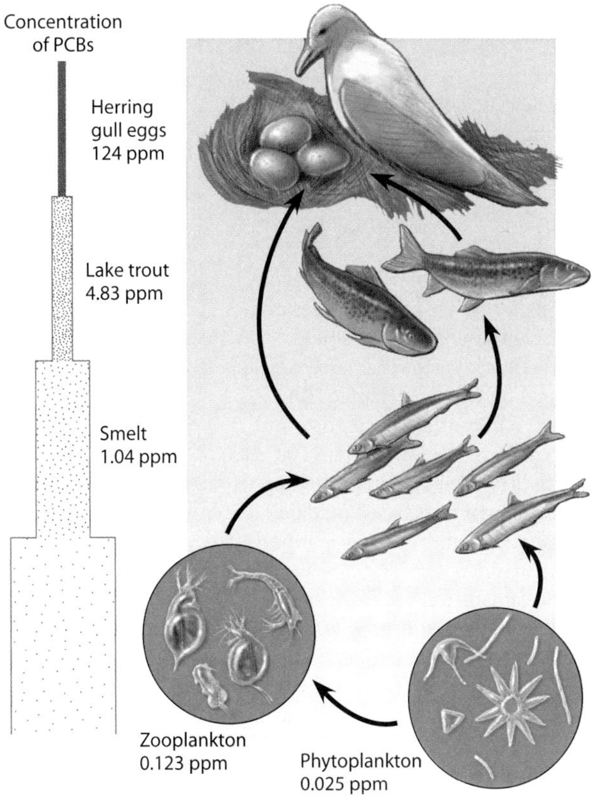

▲ **Figure 2E** Biological magnification of PCBs in a food web, measured in parts per million (ppm)

Since people are top-level predators, too, eating fish from contaminated waters can be dangerous.

Recently, scientists have recognized a new type of aquatic pollutant: plastic particles that are small enough to be eaten by zooplankton. Many body washes and facial cleansers include plastic "microbeads" to boost scrubbing power. (To see if your shower products contain plastic, check the list of ingredients for polyethylene.) Too small to be captured by wastewater treatment plants, these microparticles enter the watershed and eventually wash out to sea. Larger particles called preproduction pellets or "nurdles," used in making plastic products, are also common marine pollutants. Nurdles may be broken down to microbead size in the ocean. Toxins such as PCBs and DDT adhere to these plastic spheres. Thus, toxins may be concentrated first on microparticles and then again by biological magnification.

Global Climate Change According to many scientists, the changes in global climate that are occurring as a result of global warming are likely to become a leading cause of biodiversity loss. In the next four modules, you'll learn about some of the causes and consequences of global climate change.

? List four threats to biodiversity and give an example of each.

Habitat loss—deforestation; invasive species—brown tree snake; overharvesting—bushmeat; pollution—biological magnification of PCBs, DDT, and mercury. (Other examples could be used.)

CONNECTION 3 Rapid warming is changing the global climate

The scientific debate about global warming is over. The vast majority of scientists now agree that rising concentrations of greenhouse gases in the atmosphere, such as carbon dioxide (CO_2), methane (CH_4), and nitrous oxide (N_2O), are changing global climate patterns. This was the overarching conclusion of the assessment report released by the Intergovernmental Panel on Climate Change (IPCC) in 2007. Thousands of scientists and policymakers from more than 100 countries participated in producing the report, which is based on data published in hundreds of scientific papers.

The signature effect of increasing greenhouse gases is the steady increase in the average global temperature, which has risen 0.8°C (1.5°F) over the last 100 years, with 0.6°C of that increase occurring over the last three decades. Further increases of 2–4.5°C (3.6–8.1°F) are likely by the end of the 21st century, depending on the rate of future greenhouse gas emissions. Ocean temperatures are also rising, in deeper layers as well as at the surface. But the temperature increases are not distributed evenly around the globe. Warming is greater over land than sea, and the largest increases are in the northernmost regions of the Northern Hemisphere. In **Figure 3A**, red areas indicate the greatest temperature increases. In parts of Alaska and Canada, for example, the temperature has risen 1.4°C (2.5°F) just since 1961. Some of the consequences of the global warming trend are already clear from rising temperatures, unusual precipitation patterns, and melting ice.

National Parks Service U.S. Geological Survey, Denver U.S. Geological Survey, Denver

▲ **Figure 3B** Grinnell Glacier in Glacier National Park, 1938 (left), 1981 (center), and 2005 (right)

Many of the world's glaciers are receding rapidly, including mountain glaciers in the Himalayas, the Alps, the Andes, and the western United States. Glacier National Park in northwest Montana will need a new name by 2030, when its glaciers are projected to disappear entirely. For example, almost all of the Grinnell Glacier is now a meltwater lake **(Figure 3B)**. The permafrost that characterizes the tundra biome is also melting.

Permanent Arctic sea ice is shrinking; each summer brings increased melting and thinner ice. The massive ice sheets of Greenland and Antarctica are thinning and collapsing. If this melting trend accelerates, rising sea levels would cause catastrophic flooding of coastal areas worldwide.

Warm weather is beginning earlier each year. Cold days and nights and frosts have become less frequent; hot days and nights have become more frequent. Deadly heat waves are increasing in frequency and duration.

Precipitation patterns are changing, bringing longer and more intense drought to some areas. In other regions, a greater proportion of the total precipitation is falling in torrential downpours that cause flooding. Hurricane intensity is increasing, fueled by higher sea surface temperatures.

Many of these changes will have a profound impact on biodiversity, as we explore in Modules 5 and 6. In the next module, we examine the causes of rising greenhouse gas emissions.

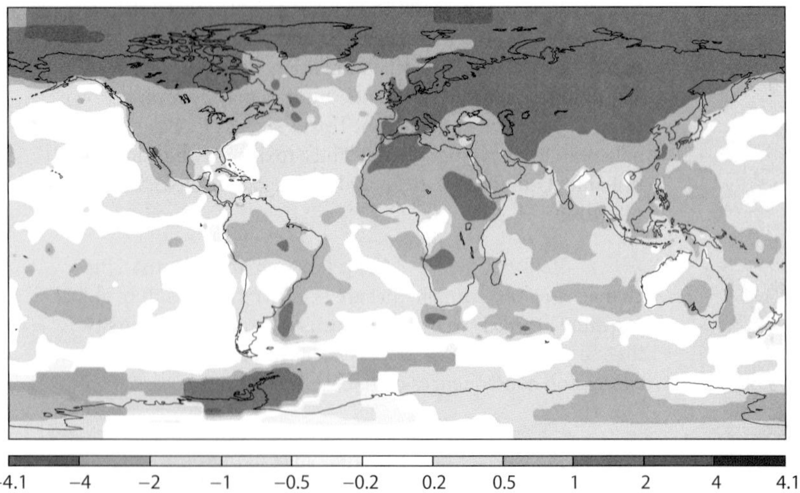

▲ **Figure 3A** Differences in temperature during 2000–2009 relative to temperatures during 1951–1980 (in °C)

? From the map in Figure 3A, which biomes are likely to be most affected by global warming, and why?

The high-latitude biomes of the Northern Hemisphere, tundra and taiga, and the polar ice biomes will be most affected. Those biomes are experiencing the greatest temperature change. Also, the organisms that live there are adapted to cold weather and a short growing season, so their survival is on the line.

CONNECTION

4 Human activities are responsible for rising concentrations of greenhouse gases

Without its blanket of natural greenhouse gases such as CO_2 and water vapor to trap heat, Earth would be too cold to support most life. However, increasing the insulation that the blanket provides is making the planet uncomfortably warm, and that increase is occurring rapidly. For 650,000 years, the atmospheric concentration of CO_2 did not exceed 300 parts per million (ppm); the preindustrial concentration was 280 ppm. Today, atmospheric CO_2 is approximately 385 ppm. The levels of nitrous oxide (N_2O) and methane (CH_4), which also trap heat in the atmosphere, have increased dramatically, too (**Figure 4A**). CO_2 and N_2O are released when fossil fuels—oil, coal, and natural gas—are burned. N_2O is also released when nitrogen fertilizers are used in agriculture. Livestock and landfills are among the factors responsible for increases of atmospheric CH_4. The consensus of scientists, as reported by the IPCC, is that rising concentrations of greenhouse gases—and thus, global warming—are the result of human activities.

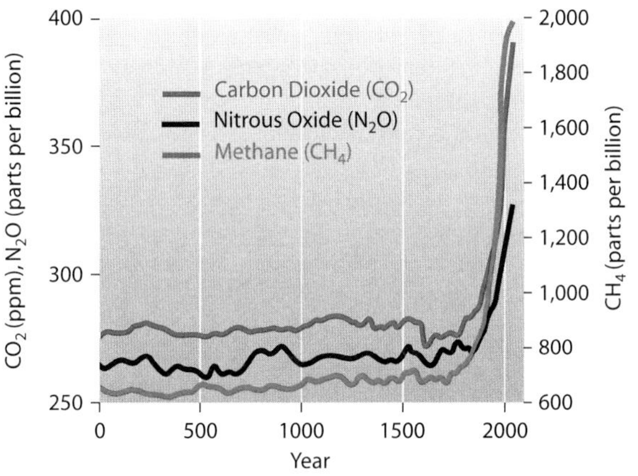

▲ **Figure 4A** Atmospheric concentrations of CO_2, N_2O (*y*-axis, left), and CH_4 (*y*-axis, right), as of 2009

Let's take a closer look at CO_2, the dominant greenhouse gas. Atmospheric CO_2 is a major reservoir for carbon. (CH_4 is also part of that reservoir.) CO_2 is removed from the atmosphere by the process of photosynthesis and stored in organic molecules such as carbohydrates (**Figure 4B**). Thus, biomass, the organic molecules in an ecosystem, is a biotic carbon reservoir. The carbon-containing molecules in living organisms may be used in the process of cellular respiration, which releases carbon in the form of CO_2. Nonliving biomass may be decomposed by microorganisms or fungi that also release CO_2. Overall, uptake of CO_2 by photosynthesis roughly equals the release of CO_2 by cellular respiration. CO_2 is also exchanged between the atmosphere and the surface waters of the oceans.

Fossil fuels consist of biomass that was buried under sediments without being completely decomposed. The burning of fossil fuels and wood, which is also an organic material, can be thought of as a rapid form of decomposition. While cellular respiration releases energy from organic mol-ecules slowly and harnesses it to make ATP, combustion liberates the energy rapidly as heat and light. In both processes, the carbon atoms that make up the organic fuel are released in CO_2.

The CO_2 flooding into the atmosphere from combustion of fossil fuels may be absorbed by photosynthetic organisms and incorporated into biomass. But deforestation has significantly decreased the number of CO_2 molecules that can be accommodated by this pathway. CO_2 may also be absorbed into the ocean. For decades, the oceans have been absorbing considerably more CO_2 than they have released, and they will continue to do so, but the excess CO_2 is beginning to affect ocean chemistry. When CO_2 dissolves in water, it becomes carbonic acid. Recently, measurable decreases in ocean pH have raised concern among biologists. Organisms that construct shells or exoskeletons out of calcium carbonate ($CaCO_3$), including corals and many plankton, are most likely to be affected as decreasing pH reduces the concentration of the carbonate ions.

Greenhouse gas emissions are accelerating. From 2000 to 2005, global CO_2 emissions increased four times faster than in the preceding 10-year span. At this rate, further climate change is inevitable.

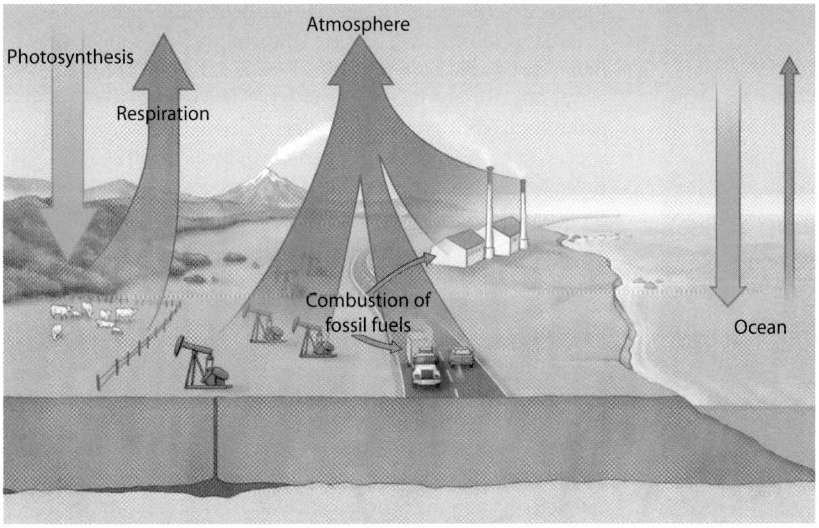

▲ **Figure 4B** Carbon cycling

? The amount of CO_2 you are responsible for releasing every year is called your *carbon footprint*. Search for an online calculator that estimates your carbon footprint. What are the primary sources of the CO_2 you generate?

Transportation and home energy use are the two major categories contributing to the footprint.

From Climate Change 2007: The Physical Science Basis. Working Group I Contribution to the Fourth Assessment Report of the Intergovernmental Panel on Climate Change, FAQ 2.1, Figure 1. Cambridge University Press. Reprinted by permission of IPCC, c/o World Meteorological Organization.

5 Global climate change affects biomes, ecosystems, communities, and populations

The distribution of terrestrial biomes, which is primarily determined by temperature and rainfall, is changing as a consequence of global warming. Melting permafrost is shifting the boundary of the tundra as shrubs and conifers are able to stretch their ranges into the previously frozen ground. Prolonged droughts will increasingly extend the boundaries of deserts. Great expanses of the Amazonian tropical rain forest will gradually become savanna as increased temperatures dry out the soil.

The combined effects of climate change on components of forest ecosystems in western North America have spawned catastrophic wildfire seasons **(Figure 5A)**. In these mountainous regions, spring snowmelt releases water into streams that sustain forest moisture levels over the summer dry season. With the earlier arrival of spring, snowmelt begins earlier and dwindles away before the dry season ends. As a result, the fire season has been getting longer since the 1980s. In addition, drought conditions have made trees more vulnerable to insect and pathogen attack; vast numbers of dead trees add fuel to the flames. Fires burn longer, and the number of acres burned has increased dramatically. As dry conditions persist and snowpacks diminish, the problem will worsen.

The earlier arrival of warm weather in the spring is disturbing ecological communities in other ways. In many animal and plant species, certain annual spring events are triggered by temperature increases. With temperatures rising earlier in the year, a variety of species, including some birds and frogs, have begun their breeding season earlier. Satellite images show earlier greening of the landscape, and flowering occurs sooner. For other species, day length is the environmental cue that spring has arrived. Because global climate change affects temperature but not day length, interactions between species may become out of sync. For example, plants may bloom before pollinators have emerged, or eggs may hatch before a dependable food source for the young is available. Because the magnitude of seasonal shifts increases from the tropics to the poles, migratory birds may also experience timing mismatches. For instance, birds arriving in the Arctic to breed may find that the period of peak food availability has already passed.

Greg Cortopassi/dailycamera.com

▲ **Figure 5A** A wildfire racing down a mountainside near Boulder, Colorado, in September 2010

Tom Murphy/WWI/Peter Arnold, Inc./PhotoLibrary

▲ **Figure 5B** A polar bear (*Ursus maritimus*) on melting pack ice in Canada

Warming oceans threaten tropical coral reef communities. When stressed by high temperatures, corals expel their symbiotic algae in a phenomenon called bleaching. Corals can recover if temperatures return to normal, but they cannot survive prolonged temperature increases. When corals die, the community is overrun by large algae, and species diversity plummets.

The distributions of populations and species are also changing in response to climate change. The distribution of a species may be determined by its adaptations to the abiotic conditions in its environment. With rising temperatures, the ranges of many species have already shifted toward the poles or to higher elevations. For example, researchers in Europe and the United States have reported that the ranges of more than two dozen species of butterflies have moved north by as much as 150 miles. Shifts in the ranges of many bird species have also been reported; the Inuit peoples living north of the Arctic Circle have sighted birds such as robins in the region for the first time.

However, species that live on mountaintops or in polar regions have nowhere to go. Researchers in Costa Rica have reported the disappearance of 20 species of frogs and toads as warmer Pacific Ocean temperatures reduce the dry-season mists in their mountain habitats. In the Arctic, polar bears **(Figure 5B)**, which stalk their prey on ice and need to store up body fat for the warmer months, are showing signs of starvation as their hunting grounds melt away. Similarly, in the Antarctic, the disappearance of sea ice is blamed for recent decreases in populations of Emperor and Adélie penguins.

Global climate change has been a boon to some organisms, but so far the beneficiaries have been species that have a negative impact on humans. For example, in mountainous regions of Africa, Southeast Asia, and Central and South America, the ranges of mosquitoes that carry diseases such as malaria, yellow fever, and dengue are restricted to lower elevations by frost. With rising temperatures and fewer days of frost, these mosquitoes—and the diseases they carry—are appearing at higher elevations. In another example, longer summers in western North America have enabled bark beetles to complete their life cycle in one year instead of two, promoting beetle outbreaks that have destroyed millions of acres of conifers. Undesirable plants such as poison ivy and kudzu have also benefited from rising temperatures.

Environmental change has always been a part of life; in fact, it is a key ingredient of evolutionary change. In the next module, we consider the evidence of evolutionary adaptation to global warming.

? How might timing mismatches caused by climate change affect an individual's reproductive fitness?

Any factor that reduces the number of offspring an organism produces may affect fitness. Examples: flowers emerging too soon or too late for pollinators; birds that arrive too late in the season to find food for offspring

EVOLUTION CONNECTION

6 Climate change is an agent of natural selection

Global climate change is already affecting habitats thoughout the world. Why do some species appear to be adapting to these changes while others, like the polar bear, are endangered by them?

In the previous module, we described several ways in which organisms have responded to global climate change. For the most part, those examples can be attributed to **phenotypic plasticity**, the ability to change phenotype in response to local environmental conditions. Differences resulting from phenotypic plasticity are within the normal range of expression for an individual's genotype. Phenotypic plasticity allows organisms to cope with short-term environmental changes. On the other hand, phenotypic plasticity is itself a trait that has a genetic basis and can evolve. Researchers studying the effects of climate change on populations have detected microevolutionary changes in phenotypic plasticity.

William Osborn/Nature Picture Library

▲ **Figure 6A** A great tit (*Parus major*)

A common bird in Europe, the great tit (**Figure 6A**) is the third link in a food chain that has been altered by climate change. As warm weather arrives earlier in the spring, tree leaves emerge earlier and caterpillars, which use the swelling buds and unfolding leaves as their food source, hatch sooner. The reproductive success of great tits depends on having an ample supply of these nutritious caterpillars to feed their offspring. Like many other birds, great tits have some phenotypic plasticity in the timing of their breeding, which helps them synchronize their reproduction with the availability of caterpillars. The range and degree of plasticity vary among great tits, and this variation has a genetic basis. Researchers have found evidence of directional selection (see Module 13.13) favoring individuals that have the greatest phenotypic plasticity and lay their eggs earlier, when the abundance of food gives their offspring a better chance of survival.

In another example, scientists studied reproduction in a population of red squirrels (**Figure 6B**) in the Yukon Territory of Canada, where spring temperatures have increased by approximately 2°C in the last three decades. These researchers also found earlier breeding times in the spring. Over a period of 10 years, the date on which female squirrels gave birth advanced by 18 days, a change of about 6 days per generation. Using statistical analysis, the scientists determined that phenotypic plasticity was responsible for most of the shift in breeding times. However, a small but significant portion of the change (roughly 15%) could be attributed to microevolution, directional selection for earlier breeding. The researchers hypothesize that red squirrels born earlier in the year are larger and more capable of gathering and storing food in the autumn and thus have a better chance of successful reproduction the following spring.

▶ **Figure 6B** A red squirrel (*Tamiasciurus hudsonicus*) eating the seeds from a spruce cone

Paul McCormick/Getty Images

From the scant evidence available at this time, it appears that some populations, especially those with high genetic variability and short life spans, may adapt quickly enough to avoid extinction. In addition to the studies on phenotypic plasticity in great tits and red squirrels, researchers have also documented microevolutionary changes in traits such as dispersal ability and timing of life cycle events in insect populations. However, evolutionary adaptation is unlikely to save long-lived species such as polar bears and penguins that are experiencing rapid habitat loss. The rate of climate change is incredibly fast compared with major climate shifts in evolutionary history, and if it continues on its present course, thousands of species—the IPCC estimates as many as 30% of plants and animals—will likely become extinct.

? How does a short generation time hasten the process of evolutionary adaptation?

Each generation has the potential for testing new phenotypes in the environment, allowing natural selection to proceed rapidly.

Conservation Biology and Restoration Ecology

7 Protecting endangered populations is one goal of conservation biology

As we have seen in this unit, many of the environmental problems facing us today are consequences of human enterprises. But the science of ecology is not just useful for telling us how things have gone wrong. Ecological research is the foundation for finding solutions to these problems and for reversing the negative consequences of ecosystem alteration. Thus, we end the ecology unit with a section that highlights some of these applications of ecological research.

Conservation biology is a goal-oriented science that seeks to understand and counter the loss of biodiversity. Some conservation biologists focus on protecting populations of threatened species. This approach requires an understanding of the behavior and ecological niche of the target species, including its key habitat requirements and interactions with other members of its community. Threats posed by human activities are also assessed. With this knowledge, scientists can design a plan to expand or protect the resources needed. For example, the territory size required to support a tiger varies with the abundance of prey. Consequently, preserves set aside for Siberian tigers in Russia, where prey are scarce, must be 10 times as large as those provided for Bengal tigers in India.

The case of the black-footed ferret (**Figure 7A**) provides an example of the population approach to conservation. Little was known about this elusive nocturnal predator until the mid-20th century, and by then it was almost too late—population decline was already under way. Black-footed ferrets, one of three ferret species worldwide and the only one found in North America, feed almost exclusively on prairie dogs (see Figure 1C). Over the past century, prairie dogs have been extirpated from most of their former range by land-use changes and by poisoning or shooting. Epidemics of sylvatic plague, the animal version of bubonic plague, have devastated populations of black-footed ferrets as well as their prey. When an outbreak threatened to wipe out the last known population of black-footed ferrets, conservation biologists captured 18 remaining individuals and began breeding them in captivity to rebuild population numbers. Genetic variation, a prerequisite for adaptive evolutionary responses to environmental change, is a concern, given the bottleneck effect of near-extinction (see Module 13.11). Matings in the captive breeding facilities are carefully arranged to maintain as much genetic diversity as possible in the ferret populations.

Rick & Nora Bowers/Alamy

▲ **Figure 7A** A black-footed ferret (*Mustela nigripes*)

In 1991, biologists began reintroducing captive-bred black-footed ferrets into the wild. Research carried out during these efforts has improved the success rate of reintroductions. For example, scientists found that the predatory behavior of ferrets has both innate and learned components, a discovery that led to more effective methods of preparing captive-bred animals to survive in the wild. Today, about 400 adult ferrets are living in the wild at sites scattered from Canada to Mexico. Despite the successes achieved thus far, however, the future of the black-footed ferret is far from secure. Biologists continue to monitor and manage the populations and their habitats.

DLILLC/Corbis

▲ **Figure 7B** A Mauna Kea silversword (*Argyroxiphium sandwicense*)

Captive breeding programs are being used for numerous other species whose population numbers are perilously low. For example, you learned about efforts to save the whooping crane in Module 35.6. In Hawaii, biologists have planted thousands of greenhouse-grown silverswords (*Argyroxiphium sandwicense*; **Figure 7B**) on the cinder cone of the volcano Mauna Kea in hopes of reestablishing wild populations. Once so abundant that observers mistook their silvery color for snow on the distant peak, silverswords were grazed to near-extinction by goats and sheep that people had brought to the island.

By using a variety of methods, biologists have improved the conservation status of some endangered species, reintroduced many species to areas where they had been extirpated, and reversed declining population trends for others. However, we will not be able to save every threatened species. One way to select worthwhile targets is to identify and protect keystone species that may help preserve entire communities. And in many situations, conservation biologists must look beyond individual species to ecosystems.

? What do you think is the first priority for conservation biologists when they select a site for ferret reintroduction?

The presence of a sufficiently large population of prairie dogs

8 Sustaining ecosystems and landscapes is a conservation priority

One of the most harmful effects of habitat loss is population fragmentation, the splitting and consequent isolation of portions of populations. As you saw in Figure 2A, for example, logging carves once-continuous forest into a patchwork of disconnected fragments. For many species, the world instantly shrinks to a fraction of its former size. Populations are reduced, and so are resources such as food and shelter. To counteract the effects of fragmentation, conservation biology often aims to sustain the biodiversity of entire ecosystems and landscapes. Ecologically, a **landscape** is a regional assemblage of interacting ecosystems, such as a forest, adjacent fields, wetlands, streams, and streamside habitats. **Landscape ecology** is the application of ecological principles to the study of the structure and dynamics of a collection of ecosystems.

Edges, or boundaries between ecosystems, are prominent features of landscapes. The photograph in **Figure 8A** shows a landscape area in Yellowstone National Park that includes grassland and forest. Human activities, such as logging and road building, often create edges that are more abrupt than those delineating natural landscapes. Such edges have their own sets of physical conditions and thus their own communities of organisms. Some organisms thrive in edges because they require resources from the two adjacent areas. For instance, whitetail deer browse on woody shrubs found in edge areas between woods and fields, and their populations often expand when forests are logged or interrupted with housing developments.

Calvin Larsen/Photo Researchers

▲ **Figure 8B** A male brown-headed cowbird (*Molothrus ater*)

Communities where human activities have generated many edges often have less diversity and are dominated by a few species that are adapted to edges. In one example, populations of the brown-headed cowbird (**Figure 8B**), an edge-adapted species that lays its eggs in the nests of other birds, are currently expanding in many areas of North America. Cowbirds forage in open fields on insects disturbed by or attracted to cattle and other large herbivores; the cowbirds also need forests, where they can parasitize the nests of other birds. Increasing cowbird parasitism and loss of habitats are correlated with declining populations of several songbird species.

Where habitats have been severely fragmented, a **movement corridor**, a narrow strip or series of small clumps of high-quality habitat connecting otherwise isolated patches, can be a deciding factor in conserving biodiversity. In areas of heavy human use, artificial corridors are sometimes constructed. In many areas, bridges or tunnels have reduced the number of animals killed as they try to cross highways (**Figure 8C**).

Corridors can also promote dispersal and reduce inbreeding in declining populations. Corridors are especially important to species that migrate between different habitats seasonally. In some European countries, amphibian tunnels have been constructed to help frogs, toads, and salamanders cross roads to access their breeding territories.

On the other hand, a corridor can be harmful—as, for example, in the spread of diseases, especially among small subpopulations in closely situated habitat patches. The effects of movement corridors between habitats in a landscape are not completely understood, and researchers continue to study them.

? How can "living on the edge" be a good thing for some species, such as whitetail deer and cowbirds?

● Such animals use a combination of resources from the two ecosystems on either side of the edge.

Yann Arthus-Bertrand/Corbis

▲ **Figure 8A** A landscape in Yellowstone National Park with distinct edges

Joel Sartore/National Geographic Image Collection

▲ **Figure 8C** A wildlife bridge in Banff National Park, Canada

9 Establishing protected areas slows the loss of biodiversity

Conservation biologists are applying their understanding of population, community, ecosystem, and landscape dynamics in establishing parks, wilderness areas, and other legally protected nature reserves. Choosing locations for protection often focuses on **biodiversity hot spots**. These relatively small areas have a large number of endangered and threatened species and an exceptional concentration of **endemic species**, those that are found nowhere else. Together, the "hottest" of Earth's biodiversity hot spots, shown in **Figure 9A**, total less than 1.5% of Earth's land but are home to a third of all species of plants and vertebrates. For example, all lemurs are endemic to Madagascar, which is home to more than 50 species. In fact, almost all of the mammals, reptiles, amphibians, and plants that inhabit Madagascar are endemic. There are also hot spots in aquatic ecosystems, such as certain river systems and coral reefs.

Because endemic species are limited to specific areas, they are highly sensitive to habitat degradation. Thus, biodiversity hot spots can also be hot spots of extinction. They rank high on the list of areas demanding strong global conservation efforts.

Concentrations of species provide an opportunity to protect many species in very limited areas. However, the "hot spot" designation tends to favor the most noticeable organisms, especially vertebrates and plants. Invertebrates and microorganisms are often overlooked. Furthermore, species endangerment is a truly global problem, and it is important that a focus on hot spots not detract from efforts to conserve habitats and species diversity in other areas.

Neo Edmund/Shutterstock

Migratory species pose a special problem for conservationists. For example, monarch butterflies occupy much of the United States and Canada during the summer months, but migrate in the autumn to specific sites in Mexico and California, where they congregate in huge numbers. Overwintering populations are particularly susceptible to habitat disturbances because they are concentrated in small areas. Thus, habitat preservation must extend across all of the sites that monarchs inhabit in order to protect them. The situation is similar for many species of migratory songbirds, waterfowl, marine mammals, and sea turtles.

Roger Kirkpatrick/V&W/Image Quest Marine

▲ **Figure 9B** An adult loggerhead turtle (*Caretta caretta*) swimming off the coast of Belize

Sea turtles, such as the loggerhead turtle **(Figure 9B)**, are threatened both in their ocean feeding grounds and on land. Loggerheads take about 20 years to reach sexual maturity, and great numbers of juveniles and adults are drowned at sea when caught in fishing nets. The adults mate at sea, and the females migrate to specific sites on sandy beaches to lay their eggs. Buried in shallow depressions, the eggs are susceptible to predators, especially raccoons. And many egg-laying sites have become housing developments and beachside resorts. An ongoing international effort to conserve sea turtles focuses on protecting egg-laying sites and minimizing the death rates of adults and juveniles at sea.

Currently, governments have set aside about 7% of the world's land in various forms of reserves. One major conservation question is whether it is better to create one large reserve or a group of smaller ones. Far-ranging animals with low-density populations, such as tigers, require extensive habitats. As conservation biologists learn more about the requirements for achieving minimum population sizes to sustain endangered species, it is becoming clear that most national parks and other reserves are far too small. Given political and economic realities, it is unlikely that many existing parks will be enlarged, and most new reserves will also be too small. In the next two modules, we look at two approaches to this problem.

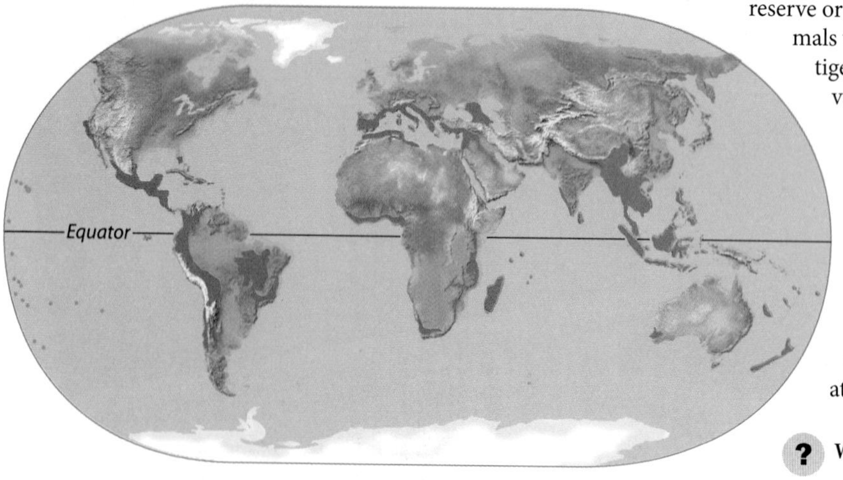

▲ **Figure 9A** Earth's terrestrial biodiversity hot spots (pink)

? What is a biodiversity hot spot?

A relatively small area with a disproportionate number of endangered and threatened species, many of which are endemic

10 Zoned reserves are an attempt to reverse ecosystem disruption

Conservation of Earth's natural resources is not purely a scientific issue. The causes of declining biodiversity are rooted in complex social and economic issues, and the solutions must take these factors into account. Let's look at how the small Central American nation of Costa Rica is managing its biodiversity.

Despite its small size (about 51,000 km^2, the size of New Hampshire and Vermont combined), Costa Rica is a treasure trove of biodiversity. Its varied ecosystems, which extend over mountains and two coasts, are home to at least half a million species. As **Figure 10A** shows, the entire country is a biodiversity hot spot. Since the 1970s, the Costa Rican government and international agencies have worked together to preserve these unique assets. Approximately 25% of Costa Rica's territory is currently protected in some way.

One type of protection is called a **zoned reserve**, an extensive region of land that includes one or more areas undisturbed by humans. The lands surrounding these areas continue to be used to support the human population, but they are protected from extensive alteration. As a result, they serve as a buffer zone, or shield, against further intrusion into the undisturbed areas. A primary goal of the zoned reserve approach is to develop a social and economic climate in the buffer zone that is compatible with the long-term viability of the protected area.

Costa Rica is making progress in managing its reserves so that the buffer zones provide a steady, lasting supply of forest products, water, and hydroelectric power and also support sustainable agriculture. An important goal is providing a stable economic base for people living there. Destructive practices that are not compatible with long-term ecosystem stability and from which there is often little local profit are gradually being discouraged. Such destructive practices include massive logging, large-scale single-crop agriculture, and extensive mining.

However, a recent analysis showed mixed results for Costa Rica's system of zoned reserves. The good news is that negligible deforestation has occurred within and just beyond protected parkland boundaries. However, some deforestation has occurred in the buffer zones, with plantations of cash crops such as banana and palm replacing the natural vegetation. Conservationists fear that continuing these practices will isolate protected areas, restricting gene flow and decreasing species and genetic diversity.

Costa Rica's commitment to conservation has resulted in a new source of income for the country—**ecotourism**, travel to natural areas for tourism and recreation **(Figure 10B)**. People from all over the world come to experience Costa Rica's spectacular range of biodiversity, generating thousands of jobs and a significant chunk of the country's revenue. Worldwide, ecotourism has grown into a multibillion-dollar industry as tourists flock to the world's remaining natural areas. Whether ecotourism dollars ultimately help conserve Earth's biodiversity, however, remains to be seen.

? Why is it important for zoned reserves to prevent large-scale alterations of habitat in the buffer zones? Why is it also important to support sustainable development for the people living there?

Large-scale disruptions could impact the nearby undisturbed areas. Preservation is a realistic goal only if it is compatible with an acceptable standard of living for the local people.

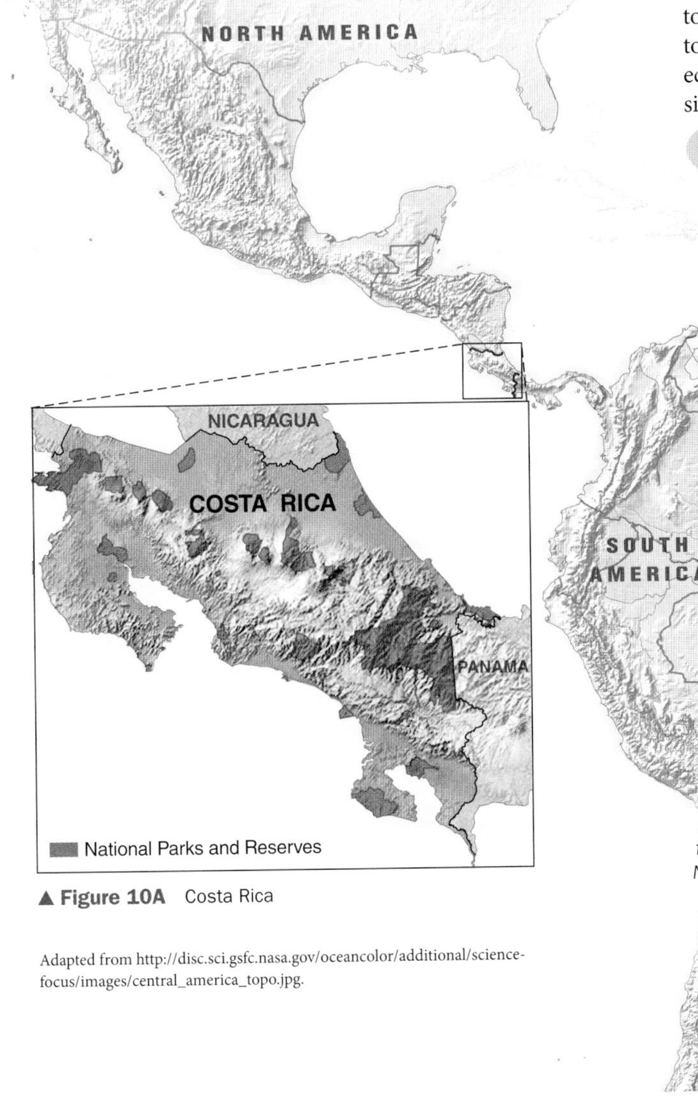

▲ **Figure 10A** Costa Rica

Adapted from http://disc.sci.gsfc.nasa.gov/oceancolor/additional/science-focus/images/central_america_topo.jpg.

Chris Fredriksson/Alamy

▲ **Figure 10B** Ecotourism: Seeing the tropical rain forest by boat in Costa Rica's Tortuguero National Park

CONNECTION

11 The Yellowstone to Yukon Conservation Initiative seeks to preserve biodiversity by connecting protected areas

If many existing reserves are too small to sustain a large number of threatened species, how can biologists include the land around reserves in conservation efforts? In North America, one ambitious biodiversity plan is creating innovative ways to give wild creatures more room. The plan builds on lessons from research on a howling predator that once roamed a vast stretch of the northern Rocky Mountains.

This predator was a single individual, a gray wolf dubbed Pluie that was captured by scientists in western Canada in 1991. The biologists fitted the 5-year-old female with a radio tracking collar and released her—routine work in studies of threatened animals. But the scientists were stunned by what they learned from Pluie. Over the next two years, this wolf roamed over an area of more than 100,000 km^2 (38,600 square miles), crossing between Canada and the United States and traveling between protected reserves and lands where she was fair game for killing. In 1995, her story came to a bloody end. While moving through lands outside the boundary of a nearby national park, Pluie, her mate, and one of her pups were shot (legally) by a hunter.

Biologists who had studied Pluie realized that the wolf's life captured all the promise—and all the pitfalls—of efforts to protect her. She had thrived for years within the sporadic shelter of parks and other protected territory. But such lands were never big enough to hold her. Like others of her species (*Canis lupus*), Pluie needed more room. Reserves could shield animals briefly, the scientists realized. True protection would have to include paths of safe passage between reserves.

This conclusion inspired the creation of the Yellowstone to Yukon Conservation Initiative (Y2Y), one of the world's most ambitious conservation biology efforts. The initiative aims to preserve the web of life that has long defined the Rocky Mountains of Canada and the northern United States. This area is dotted with famous parks—Canada's Banff National Park and Wyoming's Yellowstone National Park among them—but scientists behind Y2Y now say that those areas alone cannot protect native species from human threats.

Y2Y seeks to knit together a string of parks and reserves, creating a vast 3,200-km wildlife corridor stretching down from Alaska across Canada to northern Wyoming (**Figure 11A**). The idea is not to create one giant park, but rather to connect parks with protected corridors where wildlife can travel safely.

Many of the signature species that live in this vast region, such as grizzly bears (**Figure 11B**), lynx, moose, and elk, don't confine themselves to human boundaries. But few have as great a range as the wolf. If Y2Y can provide safe passage for gray wolves, it will have also created secure zones for other animals in the Rockies.

Gray wolves (**Figure 11C**) once roamed all of North America. These carnivorous hunters live in packs that protect pups and search cooperatively for food. A pack may have a territory of about 130 km^2 or range much farther to find prey. The wolf's hunting prowess kept it the top predator of North American ecosystems as long as the human population was small. Things changed when large numbers of people migrated from Europe and pushed into the continent.

Deeming wolves a dangerous predator and competitor that threatened people and livestock, settlers in the United States launched widespread campaigns to wipe out wolves. By the early 20th century, gray wolves were nearly extinct in the lower 48 states, with only a few hundred surviving in northern Minnesota. More managed to stay alive in the wilds of less populated western Canada and Alaska.

Scientists gradually realized that widespread damage rippled through habitats after wolves had been removed. Without a predator to control their numbers, populations of elk and deer grew unchecked. As these increasing numbers of herbivores foraged for food, vegetation that sheltered smaller animals was damaged. Other animals, such as ravens and foxes, had once fed

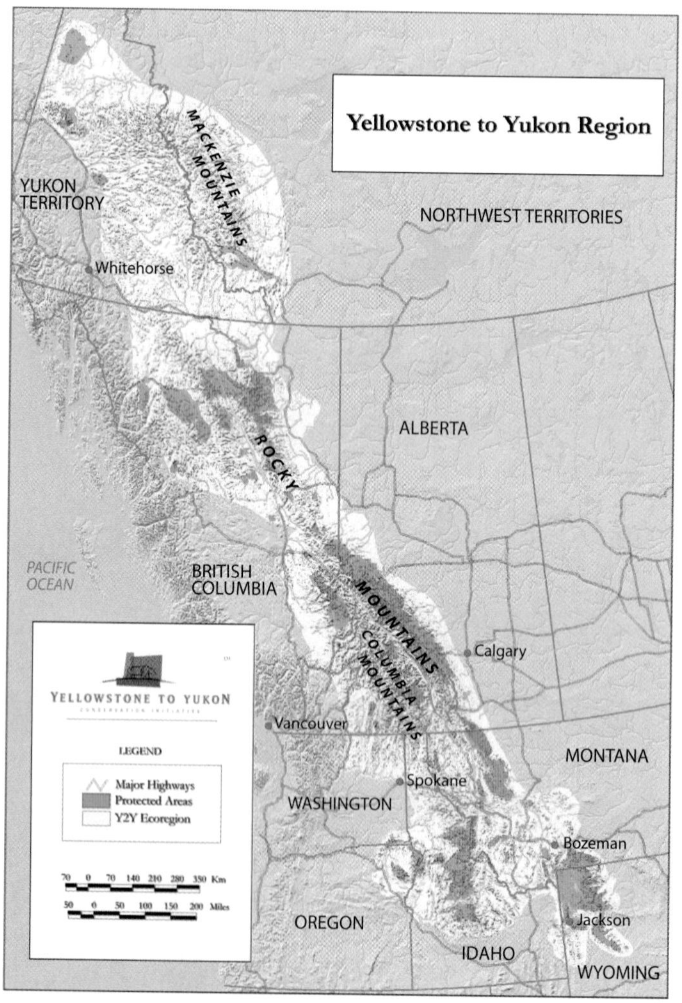

▲ **Figure 11A** A map of the Yellowstone to Yukon Conservation Initiative region, with protected areas shown in green

on the carcasses of wildlife killed by wolves and were now left without an important source of food. Gray wolves, biologists determined, were a keystone species—a species critical to the balance and maintenance of an ecosystem.

That understanding led to one of the most important and controversial conservation biology efforts in the Yellowstone to Yukon area. In 1991, the U.S. Fish and Wildlife Service launched a campaign to bring wolves back to Yellowstone National Park, a reserve that hadn't sheltered the animals in at least 50 years. After careful planning, which included compensation for ranchers who feared losing their cattle and sheep, about 60 wolves from Canada were released in the park in 1995 and 1996.

As wolf howls once again echoed through the Wyoming darkness, the wolf quickly became a hopeful symbol for Y2Y backers. Yellowstone's wolves formed new packs and raised pups. By 2004, scientists counted 12 wolf packs inside the park, totaling about 300 wolves. And the wolf's return has brought more than howling. Park officials noted significant environmental improvements as wolves once again roamed Yellowstone. As wolves killed elk, moose, and deer, streambeds and other lands near waterways started to shelter a greater variety of plants and animals. Fewer hoofed animals meant more grasses and taller trees. Those plants brought more birds, along with more water-dwelling beaver. In all, park biologists report that the wolf's return has affected at least 25 different species.

True to their nature, Yellowstone's wolves haven't followed human borders; six packs have been found just outside the park. Meanwhile, the migrations of Canadian wolves, along with smaller release programs, have brought the animals back to Idaho and Montana. In June 2004, a Yellowstone wolf was found hundreds of kilometers away in Colorado—a reminder that travels like Pluie's are common to wolves throughout the Y2Y region.

Such successes also bring risks—and reminders from scientists that wolves need safe corridors. The Colorado wolf was discovered dead by the side of a highway, most likely the victim of a car. Its appearance sparked angry protests from ranchers in the state, who said that any new wolves that appear should either be shot or shipped back to Yellowstone. Meanwhile, wildlife advocates maintain that wolves should be allowed to migrate naturally.

That argument reflects a broader debate about how to treat wolves as they return to their old ranges. As populations in the northern Rockies recovered, Federal officials removed gray wolves from the endangered species list in 2009. In August 2010, however, that decision was overturned when conservationists won a lawsuit charging that state management practices had failed to ensure sustainable population sizes.

The biologists involved in the Yellowstone to Yukon Conservation Initiative are studying wildlife population dynamics on a landscape scale to help support regional conservation planning. The initiative is backing a range of research projects to determine the requirements for maintaining terrestrial and aquatic ecosystems in the Y2Y region. Their efforts to connect habitats include wildlife bridges, such as the one in Banff National Park (see Figure 8C). This cross-border initiative is also providing broader lessons for global conservation. In Southeast Asia, for example, plans for a tiger reserve in Myanmar (Burma) include proposals to link the Myanmar reserve with others in neighboring countries.

In addition to creating reserves to protect species and their habitats from human disruptions, conservation efforts also attempt to restore ecosystems degraded by human activities. We look at the field of restoration ecology next.

? **For what reasons are gray wolves considered a keystone species?**

Wolves regulate the populations of their prey, preventing these herbivores from damaging vegetation and degrading habitat for other members of the community. Wolf kills also provide food for other animals.

▶ Figure 11C A gray wolf

Corbis/Fotolia

▼ Figure 11B A grizzly bear with cubs in Yellowstone National Park

NPS Photo

CONNECTION

12 The study of how to restore degraded habitats is a developing science

For centuries, humans have altered and degraded natural areas without considering the consequences. But as people have gradually come to realize the severity of some of the consequences of ecosystem alteration, they have sought ways to return degraded areas to their natural state. The expanding field of **restoration ecology** uses ecological principles to develop methods of achieving this goal.

One of the major strategies in restoration ecology is bioremediation, the use of living organisms to detoxify polluted ecosystems. For example, bacteria have been used to clean up oil spills and old mining sites. Bacteria are also employed to metabolize toxins in dump sites. Plants have successfully extracted potentially toxic metals such as zinc, nickel, lead, and cadmium from contaminated soil. Researchers are using trees and lichens to clean up soil polluted with uranium.

Some restoration projects have the broader goal of returning ecosystems to their natural state, which may involve replanting vegetation, fencing out non-native animals, or removing dams that restrict water flow. Hundreds of restoration projects are currently under way in the United States. One of the most ambitious endeavors is the Kissimmee River project in south central Florida.

The Kissimmee River was once a meandering shallow river that wound its way through diverse wetlands from Lake Kissimmee southward into Lake Okeechobee (**Figure 12A**). Periodic flooding of the river covered a wide floodplain during about half of the year, creating wetlands that provided critical habitat for vast numbers of birds, fishes, and invertebrates. As often happens, however, people saw the floodplain as wasted land that could be developed if the flooding were controlled. Between 1962 and 1971, the U.S. Army Corps of Engineers converted the 166-km wandering river into a straight canal 9 m deep, 100 m wide, and 90 km long. This project drained approximately 31,000 acres of wetlands, with significant negative impacts on fish and wetland bird populations. Spawning and foraging habitats for fishes were eliminated, and important sport fishes, such as largemouth bass, were replaced by nongame species more tolerant of the lower oxygen concentration in the deeper canal. The populations of waterfowl declined by 92%, and the number of bald eagle nesting territories decreased by 70%. Without the marshes to help filter and reduce agricultural runoff, phosphorus and other excess nutrients were transported through Lake Okeechobee into the Everglades ecosystem to the south.

As these negative ecological effects began to be recognized, public pressure to restore the river grew. In 1992, Congress authorized the Kissimmee River Restoration Project, one of the largest landscape restoration projects and ecological experiments in the world. As Figure 12A shows, the plan involves removing water control structures such as dams, reservoirs, and channel modifications and filling in about 35 km of the canal. The first phase of the project was completed in 2004, and the entire project is slated to be completed in 2015. In **Figure 12B**, the natural curves of the river are a

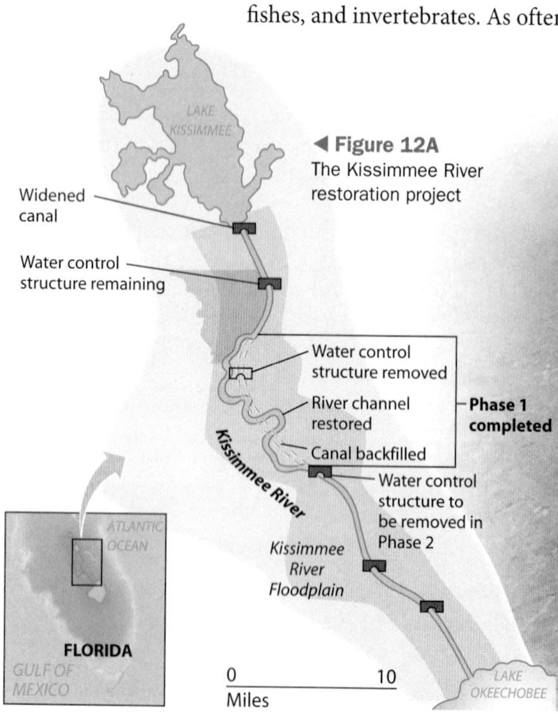

◀ **Figure 12A** The Kissimmee River restoration project

▼ **Figure 12B** Restoring the natural water flow patterns of the Kissimmee River

South Florida Water Management District (WPB)

pleasing contrast to the artificial linearity of the backfilled canal. Birds and other wildlife have returned in unexpected numbers to the 11,000 acres of wetlands that have been restored. The marshes are filled with native vegetation, and game fishes again swim in the river channels. However, 2006 was the driest season on record in south central Florida. As the drought continued, the southward flow of the Kissimmee River stopped. Lake Okeechobee, which depends on the Kissimmee basin for more than half its water supply, hit record lows. The rapidly worsening water shortage in southern Florida has renewed attention to the urgent need to complete an even more ambitious project, the restoration of the Everglades.

? How will the Kissimmee River Restoration Project improve water quality in the Everglades ecosystem?

The wetlands filter agricultural runoff and prevent excess nutrients from entering the Everglades.

13 Sustainable development is an ultimate goal

The demand for the "provisioning" services of ecosystems, such as food, fibers, and water, is increasing as the world population grows and becomes more affluent. Although these demands are currently being met, they are satisfied at the expense of other critical ecosystem services, such as climate regulation and protection against natural disasters. Clearly, we have set ourselves and the rest of the biosphere on a precarious path into the future. How can we best manage Earth's resources to ensure that all generations inherit an adequate supply of natural and economic resources and a relatively stable environment?

Many nations, scientific societies, and private foundations have embraced the concept of sustainable development. The Ecological Society of America, the world's largest organization of ecologists, endorses a research agenda called the Sustainable Biosphere Initiative. The goal of this initiative is to acquire the basic ecological information necessary for the intelligent and responsible development, management, and conservation of Earth's resources. The research agenda includes devising ways to sustain the productivity of natural and artificial ecosystems and studying the relationship between biological diversity, global climate change, and ecological processes.

Sustainable development doesn't only depend on continued research and application of ecological knowledge. It also requires us to connect the life sciences with the social sciences, economics, and humanities. Conservation and restoration of biodiversity is only one side of sustainable development; the other key facet is improving the human condition. Public education and the political commitment and cooperation of nations, especially the United States, are essential to the success of this endeavor.

The image of the ring-tailed lemur on this book's cover and in **Figure 13** serves as a reminder of what we stand to lose if we fail to recognize and solve the ecological crises at hand. All of Madagascar's 37 species of lemurs are threatened with extinction, as are dozens of other species endemic to the island. Less than 20% of the original vegetation of Madagascar remains, and deforestation for agriculture, logging, and fuel wood continue. The country's steadily growing human population is beset by political and economic woes. Consequently, the future of Madagascar's splendid biodiversity may be bleak. On the other hand, national and international initiatives in sustainable agriculture, ecotourism, conservation, and ecosystem restoration could turn the tide and save this showcase of biodiversity.

Inaki Relanzon/Nature Picture Library

▲ **Figure 13** The ring-tailed lemur (*Lemur catta*)

Biology is the scientific expression of the human desire to know nature. We are most likely to save what we appreciate, and we are most likely to appreciate what we understand. By learning about the processes and diversity of life, we also become more aware of our dependence on healthy ecosystems. An awareness of our unique ability to alter the biosphere and jeopardize the existence of other species, as well as our own, may help us choose a path toward a sustainable future.

The risk of a world without adequate natural resources for all its people is not a vision of the distant future. It is a prospect for your children's lifetime, or perhaps even your own. But although the current state of the biosphere is grim, the situation is far from hopeless. Now is the time to aggressively pursue more knowledge about life and to work toward long-term sustainability.

? Why is a concern for the well-being of future generations essential for progress toward sustainable development?

Sustainable development is a long-term goal—longer than a human lifetime. Preoccupation with the here and now is an obstacle to sustainable development because it discourages behavior that benefits future generations.

REVIEW

For Practice Quizzes, BioFlix, MP3 Tutors, and Activities, go to www.masteringbiology.com.

Reviewing the Concepts

The Loss of Biodiversity (1–6)

1 Loss of biodiversity includes the loss of ecosystems, species, and genes. While valuable for its own sake, biodiversity also provides food, fibers, medicines, and ecosystem services.

2 Habitat loss, invasive species, overharvesting, pollution, and climate change are major threats to biodiversity. Human alteration of habitats is the single greatest threat to biodiversity. Invasive species disrupt communities by competing with, preying on, or parasitizing native species. Harvesting at rates that exceed a population's ability to rebound is a threat to many species. Human activities produce diverse pollutants that may affect ecosystems far from their source. Biomagnification concentrates synthetic toxins that cannot be degraded by organisms.

3 Rapid warming is changing the global climate. Increased global temperature caused by rising concentrations of greenhouse gases is changing climatic patterns, with grave consequences.

4 Human activities are responsible for rising concentrations of greenhouse gases. Much of the increase is the result of burning fossil fuels.

5 Global climate change affects biomes, ecosystems, communities, and populations. Organisms that live at high latitudes and high elevations are experiencing the greatest impact.

6 Climate change is an agent of natural selection. Phenotypic plasticity has minimized the impact on some species, and a few cases of microevolutionary change have been observed. However, the rapidity of the environmental changes makes it unlikely that evolutionary processes will save many species from extinction.

Conservation Biology and Restoration Ecology (7–13)

7 Protecting endangered populations is one goal of conservation biology. Conservation biology is a goal-driven science that seeks to understand and counter the rapid loss of biodiversity. Some conservation biologists direct their efforts at increasing endangered populations.

8 Sustaining ecosystems and landscapes is a conservation priority. Conservation efforts are increasingly aimed at sustaining ecosystems and landscapes. Edges between ecosystems have distinct sets of features and species. The increased frequency and abruptness of edges caused by human activities can increase species loss. Movement corridors connecting isolated habitats may be helpful to fragmented populations.

9 Establishing protected areas slows the loss of biodiversity. Biodiversity hot spots have high concentrations of endemic species.

10 Zoned reserves are an attempt to reverse ecosystem disruption. Zoned reserves are undisturbed wildlands surrounded by buffer zones of compatible economic development. Costa Rica has established many zoned reserves. Ecotourism has become an important source of revenue for conservation efforts.

11 The Yellowstone to Yukon Conservation Initiative seeks to preserve biodiversity by connecting protected areas. The success of this innovative international research and conservation effort hinged on the reintroduction of gray wolves.

12 The study of how to restore degraded habitats is a developing science. Restoration ecology uses ecological principles to return degraded areas to their natural state, a process that may include detoxifying polluted ecosystems, replanting native vegetation, and returning waterways to their natural course. Large-scale restoration projects attempt to restore damaged landscapes. The Kissimmee River Restoration Project is restoring river flow and wetlands, thus improving wildlife habitat.

13 Sustainable development is an ultimate goal. Sustainable development seeks to improve the human condition while conserving biodiversity. It depends on increasing and applying ecological knowledge as well as valuing our linkages to the biosphere.

Connecting the Concepts

1. Complete the following map, which organizes some of the key concepts of conservation biology.

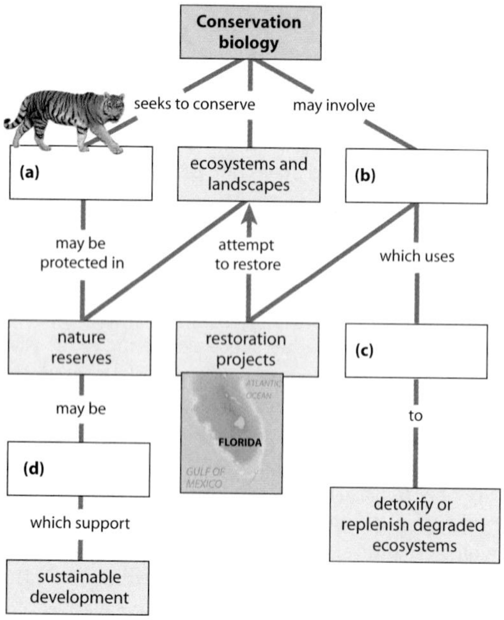

Testing Your Knowledge

Multiple Choice

2. Which of these statements best describes what conservation biologists mean by the "the rapid loss of biodiversity"?
   a. Introduced species, such as starlings and zebra mussels, have rapidly expanded their ranges.
   b. Harvests of marine fishes, such as cod and bluefin tuna, are declining.
   c. The current species extinction rate is as much as 1,000 times greater than at any time in the last 100,000 years.
   d. Many potential medicines are being lost as plant species become extinct.
   e. The number of hot spots worldwide is rapidly declining.
3. Which of the following poses the single greatest threat to biodiversity?
   a. invasive species
   b. overhunting
   c. movement corridors
   d. habitat loss
   e. pollution
4. Which of the following is characteristic of endemic species?
   a. They are often found in biodiversity hot spots.
   b. They are distributed widely in the biosphere.
   c. They require edges between ecosystems.
   d. Their trophic position makes them very susceptible to the effects of biological magnification.
   e. They are often keystone species whose presence helps to structure a community.
5. Ospreys and other top predators are most severely affected by pesticides such as PCBs because they
   a. are especially sensitive to chemicals.
   b. have rapid reproductive rates.
   c. have very long life spans.
   d. store the pesticides in their tissues.
   e. consume prey in which pesticides are concentrated.
6. Movement corridors are
   a. the routes taken by migratory animals.
   b. strips or clumps of habitat that connect isolated fragments of habitat.
   c. landscapes that include several different ecosystems.
   d. edges, or boundaries, between ecosystems.
   e. buffer zones that promote the long-term viability of protected areas.
7. With limited resources, conservation biologists need to prioritize their efforts. Of the following choices, which should receive the greatest attention for the goal of conserving biodiversity?
   a. the black-footed ferret
   b. a commercially important species
   c. all endangered vertebrate species
   d. a declining keystone species in a community
   e. all endangered species
8. Which of the following statements about protected areas is *not* correct?
   a. We now protect 25% of the land areas of the planet.
   b. National parks are only one type of protected area.
   c. Most reserves are smaller in size than the ranges of some of the species they are meant to protect.
   d. Management of protected areas must coordinate with the management of lands outside the protected zone.
   e. Biodiversity hot spots are important areas to protect.

Describing, Comparing, and Explaining

9. What are the three levels of biological diversity? Explain how human activities threaten each of these levels.
10. What are the so-called greenhouse gases? How are they important to life on Earth?
11. What are the causes and possible consequences of global warming? Why is international cooperation necessary if we are to solve this problem?

Applying the Concepts

12. Biologists in the United States are concerned that populations of many migratory songbirds are declining. Evidence suggests that some of these birds might be victims of pesticides. Most of the pesticides implicated in songbird mortality have not been used in the United States since the 1970s. Suggest a hypothesis to explain the current decline in songbird numbers.
13. You may have heard that human activities cause the extinction of one species every hour. Such estimates vary widely because we do not know how many species exist or how fast their habitats are being destroyed. You can make your own estimate of the rate of extinction. Start with the number of species that have been identified. To keep things simple, ignore extinction in the temperate latitudes and focus on the 80% of plants and animals that live in the tropical rain forest. Assume that destruction of the forest continues at a rate of 1% per year, so the forest will be gone in 100 years. Assume (optimistically) that half the rain forest species will survive in preserves, forest remnants, and zoos. How many species will disappear in the next century? How many species is that per year? Per day? Recent studies of the rain forest canopy have led some experts to predict that there may be as many as 30 million species on Earth. How does starting with this figure change your estimates?
14. The price of energy does not reflect its real costs. What kinds of hidden environmental costs are not reflected in the price of fossil fuels? How are these costs paid, and by whom? Do you think these costs could or should be figured into the price of oil? How might that be done?
15. Research your country's per capita carbon emissions. Compare your carbon footprint with the average for your country. (See the question at the end of Module 4). How can individuals reduce the carbon emissions for which they are directly responsible? Make a list of actions that you are willing to take to reduce your carbon footprint.
16. Until recently, response to environmental problems has been fragmented—an antipollution law here, incentives for recycling there. Meanwhile, the problems of the gap between the rich and poor nations, diminishing resources, and pollution continue to grow. Now people and governments are starting to envision a sustainable society. The Worldwatch Institute, a respected environmental monitoring organization, estimates that we must reach sustainability by the year 2030 to avoid economic and environmental disaster. To get there, we must begin shaping a sustainable society during this decade. In what ways is our present system not sustainable? What might a sustainable society be like? Do you think a sustainable society is an achievable goal? Why or why not? What is the alternative? What might we do to work toward sustainability? What are the major roadblocks to achieving sustainability? How would your life be different in a sustainable society?

Review Answers

1. a. species at risk of extinction; b. restoration ecology; c. bioremediation; d. zoned reserves
2. c 3. d 4. a 5. e 6. b 7. d 8. a
9. Genetic, species, and ecosystem diversity. As populations become smaller, genetic diversity is usually reduced. Genetic diversity is also threatened when local populations of a species are extirpated owing to habitat destruction or other assaults or when entire species are lost. Many human activities have led to the extinction of species. The greatest threats include habitat loss, invasive species, and overharvesting. Species extinction or population extirpation may alter the structure of whole communities. Pollution and other widespread disruptions may lead to the loss of entire ecosystems.
10. Greenhouse gases in the atmosphere, including carbon dioxide, methane, and nitrous oxide, absorb infrared radiation and thus slow the escape of heat from Earth. This is called the greenhouse effect (see Module 7.13). Without greenhouse gases in the atmosphere, the temperature at the surface of Earth would be much colder and less hospitable for life.
11. Fossil fuel consumption, industry, and agriculture are increasing the quantity of greenhouse gases—such as CO_2, methane, and nitrous oxide—in the atmosphere. These gases are trapping more heat and raising atmospheric temperatures. Increases of 2–5°C are projected over the next century. Logging and the clearing of forests for farming contribute to global warming by reducing the uptake of CO_2 by plants (and adding CO_2 to the air when trees are burned). Global warming is having numerous effects already, including melting polar ice, permafrost, and glaciers, shifting patterns of precipitation, causing spring temperatures to arrive earlier, and reducing the number of cold days and nights. Future consequences include rising sea levels and the extinction of many plants and animals. Global warming is an international problem; air and climate do not recognize international boundaries. Greenhouse gases are primarily produced by industrialized nations. Cooperation and commitment to reduce use of fossil fuels and to reduce deforestation will be necessary if the problem of global warming is to be solved.
12. These birds might be affected by pesticides while in their wintering grounds in Central and South America, where such chemicals may still be in use. The birds are also affected by deforestation throughout their range.
13. About 1.8 million species have been named and described. Assume that 80% of all living things (not just plants and animals) live in tropical rain forests. This means that there are 1.44 million species there. If half the species survive, this means that 0.72 million species will be extinct in 100 years, or 7,400 per year. This means that 19+ species will disappear per day, or almost one per hour. If there are 30 million species on Earth, 24 million live in the tropics, and 12 million will disappear in the next century. This is 120,000 per year, 329 per day, or 14 per hour.
14. Some issues and questions to consider: How does the use of fossil fuels affect the environment? What about oil spills? Disruption of wildlife habitat for construction of oil fields and pipelines? Burning of fossil fuels and possible climate change and flooding from global warming? Pollution of lakes and destruction of property by acid precipitation? Health effects of polluted air on humans? How are we paying for these "side effects" of fossil fuel use? In taxes? In health insurance premiums? Do we pay a nonfinancial price in terms of poorer health and quality of life? Could oil companies be required to pick up the tab for environmental effects of fossil fuel use? Could these costs be covered by an oil tax? How would this change the price of oil? How would a change in the price of oil change our pattern of energy use, our lifestyle, and our environment?
15. Data on carbon emissions can be found at a number of websites, including the United Nations Statistics Division (http://unstats.un.org/unsd/default.htm) and the World Resources Institute (http://www.wri.org/climate/). The per capita rankings of the United States and Canada are generally very high, along with other developed nations. Transportation and energy use are the major contributors to the carbon footprint. Any actions you can take to reduce these will help. For example, if you have a car, you can try to minimize the number of miles you drive by consolidating errands into fewer trips and using an alternative means of transportation (public transportation, walking, biking) whenever possible. To reduce energy consumption, be aware of the energy you use: Turn off lights, disconnect electronics that draw power when on standby, do laundry in cold water, for example. Websites such as http://www.ucsusa.org/ (Union of Concerned Scientists) and http://climatecrisis.net offer simple suggestions such as changing to energy-efficient lightbulbs. A Web search will turn up plenty of sites.
16. Some issues and questions to consider: How do population growth, resource consumption, pollution, and reduction in biodiversity relate to sustainability? How do poverty, economic growth and development, and political issues relate to sustainability? Why might developed and developing nations take different views of a sustainable society? What would life be like in a sustainable society? Have any steps toward sustainability been taken in your community? What are the obstacles to sustainability in your community? What steps have you taken toward a sustainable lifestyle? How old will you be in 2030? What do you think life will be like then?

Metric Conversion Table

Measurement	Unit and Abbreviation	Metric Equivalent	Approximate Metric-to-English Conversion Factor	Approximate English-to-Metric Conversion Factor
Length	1 kilometer (km)	= 1,000 (10^3) meters	1 km = 0.6 mile	1 mile = 1.6 km
	1 meter (m)	= 100 (10^2) centimeters	1 m = 1.1 yards	1 yard = 0.9 m
		= 1,000 millimeters	1 m = 3.3 feet	1 foot = 0.3 m
			1 m = 39.4 inches	
	1 centimeter (cm)	= 0.01 (10^{-2}) meter	1 cm = 0.4 inch	1 foot = 30.5 cm
				1 inch = 2.5 cm
	1 millimeter (mm)	= 0.001 (10^{-3}) meter	1 mm = 0.04 inch	
	1 micrometer (μm)	= 10^{-6} meter (10^{-3} μm)		
	1 nanometer (nm)	= 10^{-9} meter (10^{-3} μm)		
	1 angstrom (Å)	= 10^{-10} meter (10^{-4} μm)		
Area	1 hectare (ha)	= 10,000 square meters	1 ha = 2.5 acres	1 acre = 0.4 ha
	1 square meter (m^2)	= 10,000 square centimeters	1 m^2 = 1.2 square yards	1 square yard = 0.8 m^2
			1 m^2 = 10.8 square feet	1 square foot = 0.09 m^2
	1 square centimeter (cm^2)	= 100 square millimeters	1 cm^2 = 0.16 square inch	1 square inch = 6.5 cm^2
Mass	1 metric ton (t)	= 1,000 kilograms	1 t = 1.1 tons	1 ton = 0.91 t
	1 kilogram (kg)	= 1,000 grams	1 kg = 2.2 pounds	1 pound = 0.45 kg
	1 gram (g)	= 1,000 milligrams	1 g = 0.04 ounce	1 ounce = 28.35 g
			1 g = 15.4 grains	
	1 milligram (mg)	= 10^{-3} gram	1 mg = 0.02 grain	
	1 microgram (μg)	= 10^{-6} gram		
Volume (Solids)	1 cubic meter (m^3)	= 1,000,000 cubic centimeters	1 m^3 = 1.3 cubic yards	1 cubic yard = 0.8 m^3
			1 m^3 = 35.3 cubic feet	1 cubic foot = 0.03 m^3
	1 cubic centimeter (cm^3 or cc)	= 10^{-6} cubic meter	1 cm^3 = 0.06 cubic inch	1 cubic inch = 16.4 cm^3
	1 cubic millimeter (mm^3)	= 10^{-9} cubic meter (10^{-3} cubic centimeter)		
Volume (Liquids and Gases)	1 kiloliter (kL or kl)	= 1,000 liters	1 kL = 264.2 gallons	
	1 liter (L or l)	= 1,000 milliliters	1 L = 0.26 gallon	1 gallon = 3.79 L
			1 L = 1.06 quarts	1 quart = 0.95 L
	1 milliliter (mL or ml)	= 10^{-3} liter	1 mL = 0.03 fluid ounce	1 quart = 946 mL
		= 1 cubic centimeter	1 mL = $\frac{1}{4}$ teaspoon	1 pint = 473 mL
			1 mL = 15–16 drops	1 fluid ounce = 29.6 mL
				1 teaspoon = 5 mL
	1 microliter (μL or μl)	= 10^{-6} liter (10^{-3} milliliters)		
Time	1 second (s)	= $\frac{1}{60}$ minute		
	1 millisecond (ms)	= 10^{-3} second		
Temperature	Degrees Celsius (°C)		$°F = \frac{9}{5}°C - 32$	$°C = \frac{5}{9}(°F - 32)$

From Appendix 1 of *Campbell Biology: Concepts & Connections*, Seventh Edition. Jane B. Reece, Martha R. Taylor, Eric J. Simon, Jean L. Dickey.

APPENDIX: The Periodic Table

APPENDIX: The Periodic Table

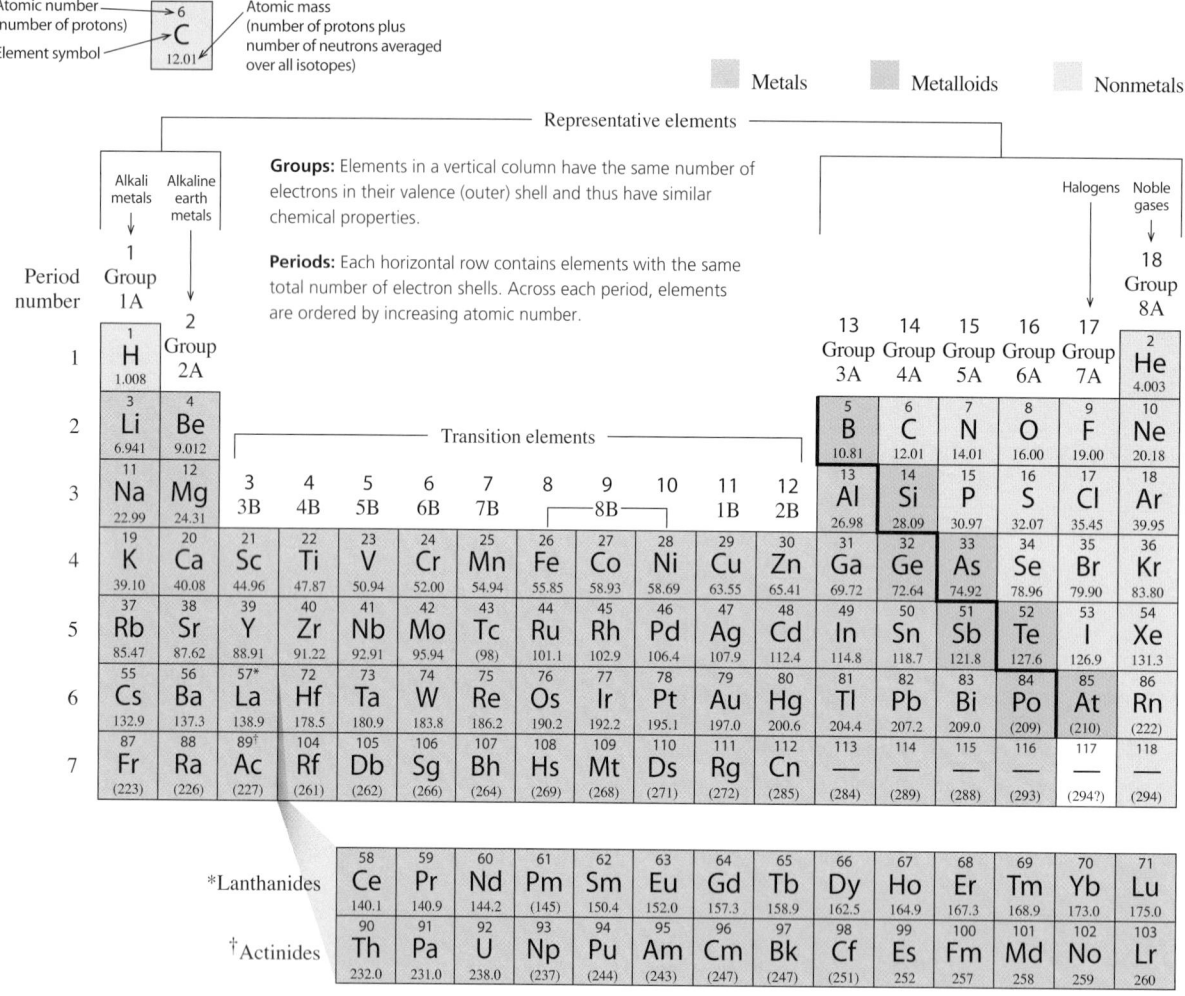

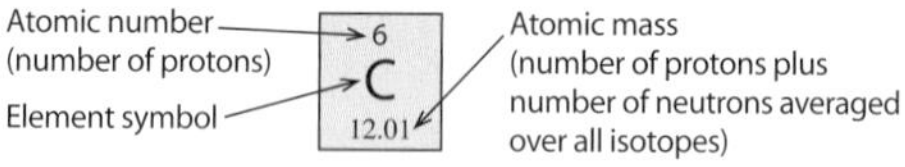

Metals | Metalloids | Nonmetals

Representative elements

Alkali metals | Alkaline earth metals | Halogens | Noble gases

Groups: Elements in a vertical column have the same number of electrons in their valence (outer) shell and thus have similar chemical properties.

Periods: Each horizontal row contains elements with the same total number of electron shells. Across each period, elements are ordered by increasing atomic number.

Transition elements

Period number	1 Group 1A	2 Group 2A	3 3B	4 4B	5 5B	6 6B	7 7B	8 8B	9 8B	10 8B	11 1B	12 2B	13 Group 3A	14 Group 4A	15 Group 5A	16 Group 6A	17 Group 7A	18 Group 8A
1	1 H 1.008																	2 He 4.003
2	3 Li 6.941	4 Be 9.012											5 B 10.81	6 C 12.01	7 N 14.01	8 O 16.00	9 F 19.00	10 Ne 20.18
3	11 Na 22.99	12 Mg 24.31											13 Al 26.98	14 Si 28.09	15 P 30.97	16 S 32.07	17 Cl 35.45	18 Ar 39.95
4	19 K 39.10	20 Ca 40.08	21 Sc 44.96	22 Ti 47.87	23 V 50.94	24 Cr 52.00	25 Mn 54.94	26 Fe 55.85	27 Co 58.93	28 Ni 58.69	29 Cu 63.55	30 Zn 65.41	31 Ga 69.72	32 Ge 72.64	33 As 74.92	34 Se 78.96	35 Br 79.90	36 Kr 83.80
5	37 Rb 85.47	38 Sr 87.62	39 Y 88.91	40 Zr 91.22	41 Nb 92.91	42 Mo 95.94	43 Tc (98)	44 Ru 101.1	45 Rh 102.9	46 Pd 106.4	47 Ag 107.9	48 Cd 112.4	49 In 114.8	50 Sn 118.7	51 Sb 121.8	52 Te 127.6	53 I 126.9	54 Xe 131.3
6	55 Cs 132.9	56 Ba 137.3	57* La 138.9	72 Hf 178.5	73 Ta 180.9	74 W 183.8	75 Re 186.2	76 Os 190.2	77 Ir 192.2	78 Pt 195.1	79 Au 197.0	80 Hg 200.6	81 Tl 204.4	82 Pb 207.2	83 Bi 209.0	84 Po (209)	85 At (210)	86 Rn (222)
7	87 Fr (223)	88 Ra (226)	89† Ac (227)	104 Rf (261)	105 Db (262)	106 Sg (266)	107 Bh (264)	108 Hs (269)	109 Mt (268)	110 Ds (271)	111 Rg (272)	112 Cn (285)	113 — (284)	114 — (289)	115 — (288)	116 — (293)	117 — (294?)	118 — (294)

*Lanthanides	58 Ce 140.1	59 Pr 140.9	60 Nd 144.2	61 Pm (145)	62 Sm 150.4	63 Eu 152.0	64 Gd 157.3	65 Tb 158.9	66 Dy 162.5	67 Ho 164.9	68 Er 167.3	69 Tm 168.9	70 Yb 173.0	71 Lu 175.0
†Actinides	90 Th 232.0	91 Pa 231.0	92 U 238.0	93 Np (237)	94 Pu (244)	95 Am (243)	96 Cm (247)	97 Bk (247)	98 Cf (251)	99 Es 252	100 Fm 257	101 Md 258	102 No 259	103 Lr 260

Name (Symbol)	Atomic Number	Name (Symbol)	Atomic Number	Name (Symbol)	Atomic Number	Name (Symbol)	Atomic Number	Name (Symbol)	Atomic Number
Actinium (Ac)	89	Copernicium (Cn)	112	Iridium (Ir)	77	Palladium (Pd)	46	Sodium (Na)	11
Aluminum (Al)	13	Copper (Cu)	29	Iron (Fe)	26	Phosphorus (P)	15	Strontium (Sr)	38
Americium (Am)	95	Curium (Cm)	96	Krypton (Kr)	36	Platinum (Pt)	78	Sulfur (S)	16
Antimony (Sb)	51	Darmstadtium (Ds)	110	Lanthanum (La)	57	Plutonium (Pu)	94	Tantalum (Ta)	73
Argon (Ar)	18	Dubnium (Db)	105	Lawrencium (Lr)	103	Polonium (Po)	84	Technetium (Tc)	43
Arsenic (As)	33	Dysprosium (Dy)	66	Lead (Pb)	82	Potassium (K)	19	Tellurium (Te)	52
Astatine (At)	85	Einsteinium (Es)	99	Lithium (Li)	3	Praseodymium (Pr)	59	Terbium (Tb)	65
Barium (Ba)	56	Erbium (Er)	68	Lutetium (Lu)	71	Promethium (Pm)	61	Thallium (Tl)	81
Berkelium (Bk)	97	Europium (Eu)	63	Magnesium (Mg)	12	Protactinium (Pa)	91	Thorium (Th)	90
Beryllium (Be)	4	Fermium (Fm)	100	Manganese (Mn)	25	Radium (Ra)	88	Thulium (Tm)	69
Bismuth (Bi)	83	Fluorine (F)	9	Meitnerium (Mt)	109	Radon (Rn)	86	Tin (Sn)	50
Bohrium (Bh)	107	Francium (Fr)	87	Mendelevium (Md)	101	Rhenium (Re)	75	Titanium (Ti)	22
Boron (B)	5	Gadolinium (Gd)	64	Mercury (Hg)	80	Rhodium (Rh)	45	Tungsten (W)	74
Bromine (Br)	35	Gallium (Ga)	31	Molybdenum (Mo)	42	Roentgenium (Rg)	111	Uranium (U)	92
Cadmium (Cd)	48	Germanium (Ge)	32	Neodymium (Nd)	60	Rubidium (Rb)	37	Vanadium (V)	23
Calcium (Ca)	20	Gold (Au)	79	Neon (Ne)	10	Ruthenium (Ru)	44	Xenon (Xe)	54
Californium (Cf)	98	Hafnium (Hf)	72	Neptunium (Np)	93	Rutherfordium (Rf)	104	Ytterbium (Yb)	70
Carbon (C)	6	Hassium (Hs)	108	Nickel (Ni)	28	Samarium (Sm)	62	Yttrium (Y)	39
Cerium (Ce)	58	Helium (He)	2	Niobium (Nb)	41	Scandium (Sc)	21	Zinc (Zn)	30
Cesium (Cs)	55	Holmium (Ho)	67	Nitrogen (N)	7	Seaborgium (Sg)	106	Zirconium (Zr)	40
Chlorine (Cl)	17	Hydrogen (H)	1	Nobelium (No)	102	Selenium (Se)	34		
Chromium (Cr)	24	Indium (In)	49	Osmium (Os)	76	Silicon (Si)	14		
Cobalt (Co)	27	Iodine (I)	53	Oxygen (O)	8	Silver (Ag)	47		

APPENDIX:
The Amino Acids of Protein

HYDROPHOBIC (Nonpolar)

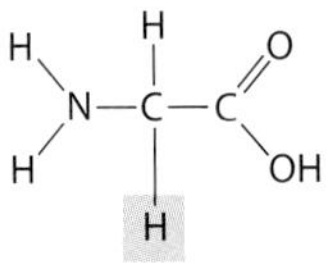
GLYCINE (Gly)

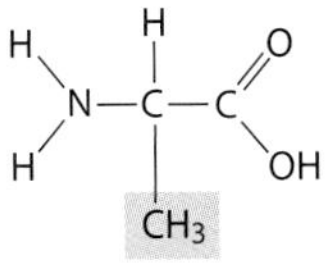
ALANINE (Ala)

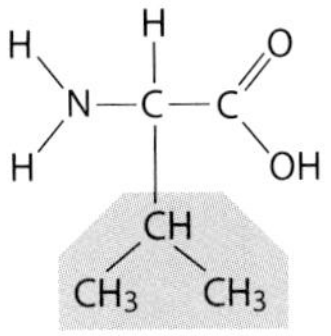
VALINE (Val)

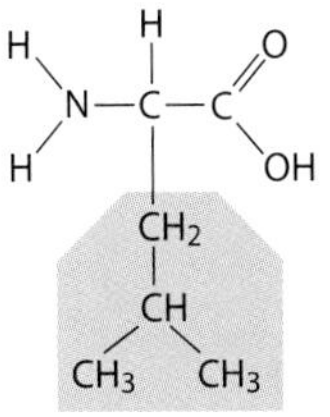
LEUCINE (Leu)

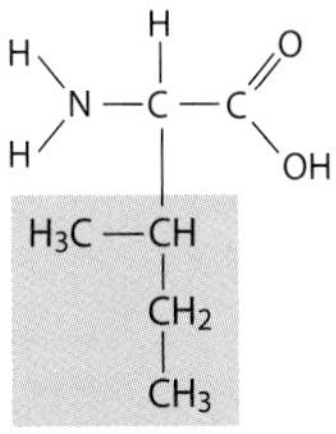
ISOLEUCINE (Ile)

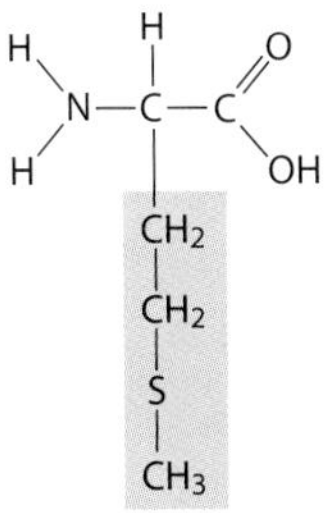
METHIONINE (Met)

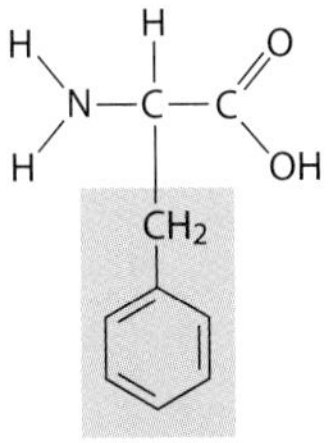
PHENYLALANINE (Phe)

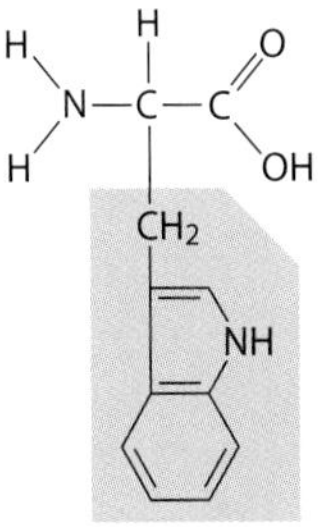
TRYPTOPHAN (Trp)

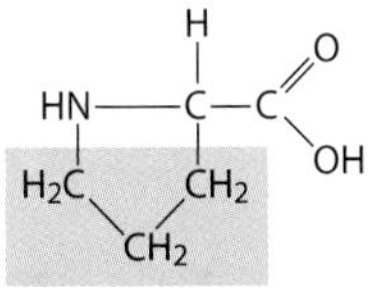
PROLINE (Pro)

HYDROPHILIC (polar or charged)

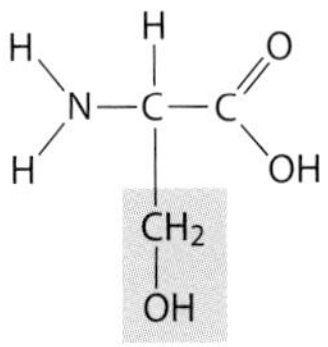
SERINE (Ser)

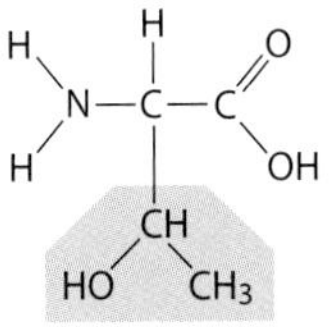
THREONINE (Thr)

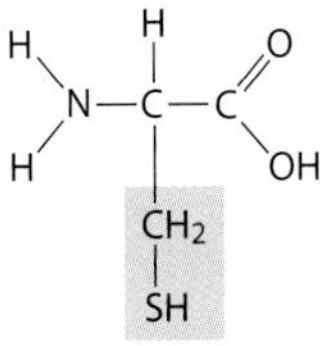
CYSTEINE (Cys)

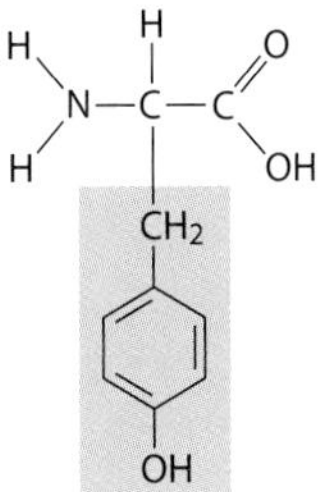
TYROSINE (Tyr)

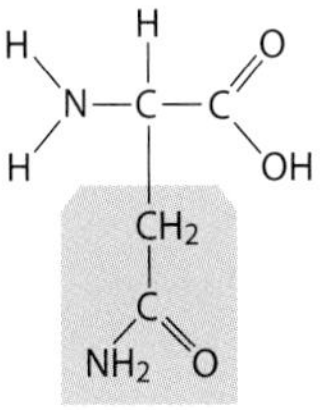
ASPARAGINE (Asn)

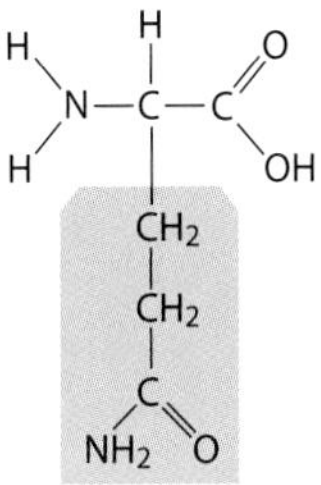
GLUTAMINE (Gln)

Acidic

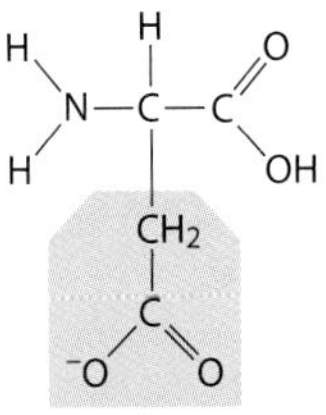
ASPARTIC ACID (Asp)

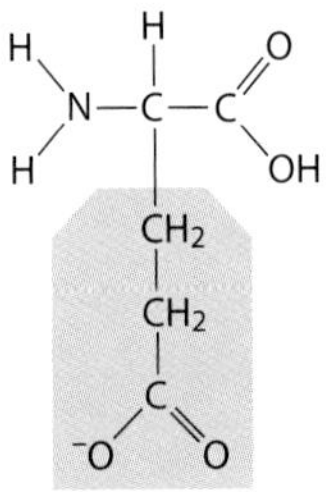
GLUTAMIC ACID (Glu)

Basic

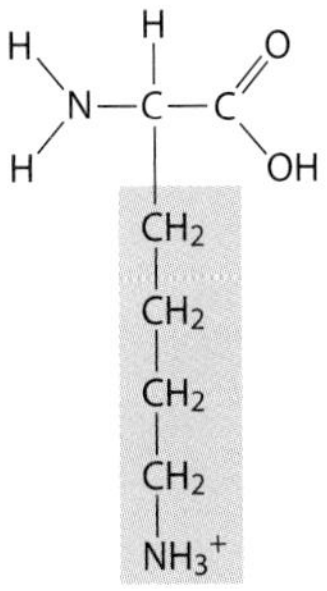
LYSINE (Lys)

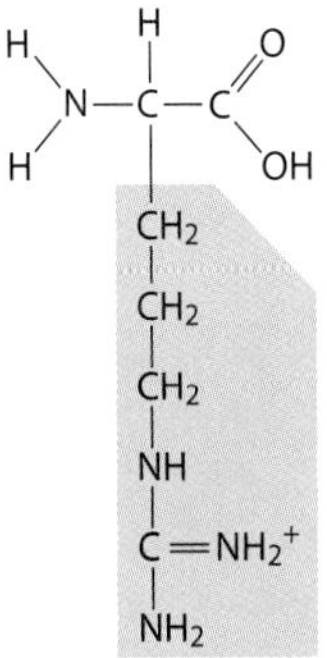
ARGININE (Arg)

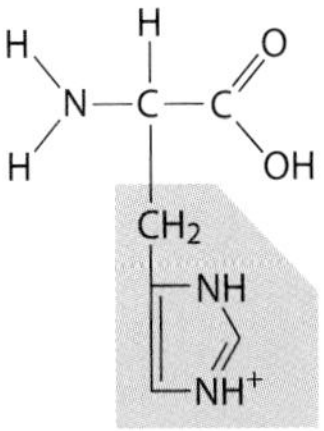
HISTIDINE (His)

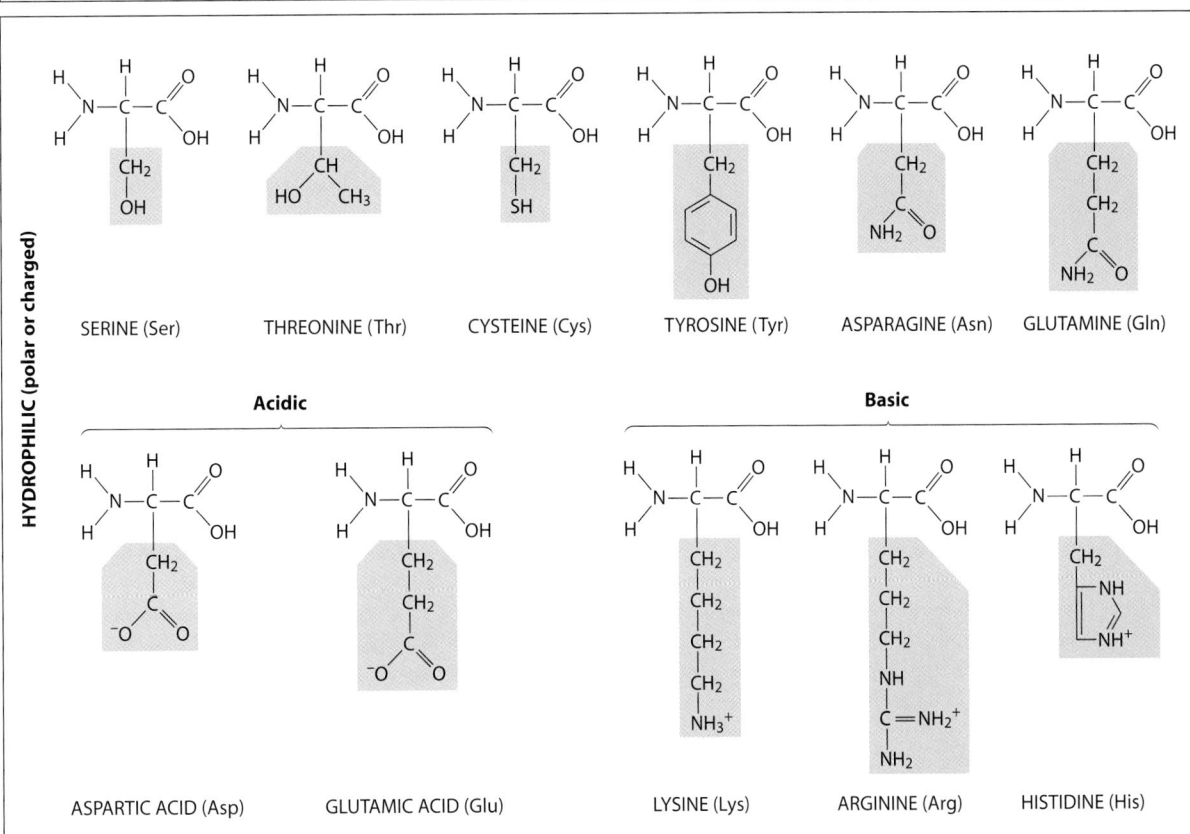

APPENDIX 3

Index

E

F

P

T

W

X

Y

Z